# DISCOVER BIOLOGY

**Fifth Edition**

Anu Singh-Cundy

Western Washington University

Michael L. Cain

Bowdoin College

Jennie Dusheck

Contributing Author

W. W. NORTON & COMPANY

NEW YORK • LONDON

W.W. Norton & Company has been independent since its founding in 1923, when William Warder Norton and Mary D. Herter Norton first published lectures delivered at the People's Institute, the adult education division of New York City's Cooper Union. The firm soon expanded its program beyond the Institute, publishing books by celebrated academics from America and abroad. By mid-century, the two major pillars of Norton's publishing program—trade books and college texts—were firmly established. In the 1950s, the Norton family transferred control of the company to its employees, and today—with a staff of four hundred and a comparable number of trade, college, and professional titles published each year—W.W. Norton & Company stands as the largest and oldest publishing house owned wholly by its employees.

Editor: Betsy Twitchell

Senior Production Manager: Christopher Granville

Cover Design: Leah Clark

Development Editor: Carol Pritchard-Martinez

Managing Editor, College: Marian Johnson

Associate Managing Editor, College: Kim Yi

Project Editor: Christine D'Antonio

Marketing Manager: John Kresse

Copy Editor: Stephanie Hiebert

Electronic Media Editors: Rob Bellinger and Patrick Shriner

Media Editorial Assistant: Carson Russell

Print Ancillary Editors: Matthew Freeman and Callinda Taylor

Editorial Assistant: Cait Callahan

Photo Researchers: Stephanie Romeo and Elyse Rieder

Illustrations for the Fifth Edition: Imagineering, Inc.

Composition: Preparé

Manufacturing: Courier

Library of Congress Cataloging-in-Publication Data
Singh-Cundy, Anu.
  Discover biology / Anu Singh-Cundy, Michael L. Cain ; Jennie Dusheck, contributing author.—5th ed.
    p. cm.
  Includes index.
  ISBN 978-0-393-93570-7 (pbk.)
  1. Biology—Textbooks. I. Cain, Michael L. (Michael Lee), 1956- II. Dusheck, Jennie. III. Title.
  QH308.2.D57 2012
  570—dc23
                        2011043361

W. W. Norton & Company, Inc., 500 Fifth Avenue, New York, N.Y. 10110
www.wwnorton.com

W. W. Norton & Company Ltd., Castle House, 75/76 Wells Street, London W1T 3QT
1  2  3  4  5  6  7  8  9  0

# From Science to Scientific Literacy

Taking into consideration the feedback of over a hundred reviewers, the authors of *Discover Biology* have extensively revised the book's content to address the needs of introductory biology students. Here's what reviewers and adopters are saying:

"This book is a great resource, and was written with the student in mind. The simple yet complete figures, many examples, and recent news and events are amazing in conveying the information to the students of all learning styles. I honestly enjoy reading the chapters and reviewing the figures." —Michael Wenzel, California State University–Sacramento

"The engaging writing style and level are perfect for a nonmajors biology course."
—Holly Ahern, Adirondack Community College

"*Discover Biology* is an excellent text for nonmajors. It is a very readable text that provides applications and relates material to everyday concepts students can understand. Students find it easy to read and not intimidating. The basics are covered well enough that students come to class ready to build on what they have read." —Francie Cuffney, Meredith College

"I chose it because it makes the material accessible to nonmajor students with stories, mysteries, and current news." —Jason Oyadomari, Finlandia University

"A well-written, issue-based text that promotes biological literacy. This text relates scientific concepts to current topics and encourages the development of critical thinking skills."
—Cynthia Littlejohn, University of Southern Mississippi

"I use *Discover Biology* because of its conciseness, currency, clarity, and content."
—Edison Fowlkes, Hampton University

"My students enjoy the text and find it helpful in delivering the central concepts in an informative and applicable way." —Keith Crandall, Brigham Young University

"I think the strengths of the text lie in its incorporation of current events and stories that link biological topics to relevant issues in the students' lives."
—Melinda Ostraff, Brigham Young University

# Our Goals: Engage, Learn, Apply

Using stem cells to repair or replace damaged organs. Discovering every last branch on the vast evolutionary tree of life. Seeking a cure for cancer. These are but a few of the many reasons why biology is a gripping subject for us and, we hope, for students using this book. These topics are simultaneously intensely interesting and critically important. Because the scientific understanding of fundamental biological principles is growing by leaps and bounds, this is an exciting time to write, teach, and learn about all areas of the biological sciences.

But the very things that make biology so interesting—the rapid pace of new discoveries and the many applications of these discoveries by human societies—can make it a difficult subject to teach and to learn. The problem is only exacerbated by the wide variation in backgrounds and interests of the students who take this course. When we set out to write the Fifth Edition of *Discover Biology*, we asked ourselves, "How can we convey the excitement, breadth, and relevance of biology to this varied group of students without burying them in an avalanche of information?"

We answered this question in several ways. In considering which topics and details to include from the vast group of possibilities, our goals were

- To pique students' interest with engaging and current "real-world" examples
- To take the feedback of the extensive reviews we received for every chapter to present core biological concepts at just the right level of detail for nonscience students
- To demonstrate the application of fundamental concepts in a way that helps build students' scientific literacy with tools that will serve them well regardless of the educational or career path they follow

How we achieve each of these goals is described in detail in the guided tour that follows this introduction.

## What's New in the Fifth Edition

NEW STORIES. Each chapter in *Discover Biology* has always been bookended by an engaging story relating to the chapter's topic. The chapter opens with an interesting contemporary story that helps students answer the all-too-common question, Why should I care about biology? The topic of that story is revisited at the end of the chapter, demonstrating to students how reading the chapter gives them a deeper understanding of biology in the world around them. For the Fifth Edition, science writer and veteran textbook author Jennie Dusheck developed new or substantially revised stories for every chapter that highlight topics relevant to today's students.

NEW APPLICATIONS. We believe it's important that introductory biology students learn to actively apply core concepts relevant to issues in our rapidly changing world. "Biology in the News" features—excerpts from actual news articles from 2009–11 paired with analysis and critical thinking questions—serve as a capstone for every chapter. We selected new article excerpts and wrote new analysis for the Fifth Edition in order to focus this feature on contemporary and relevant topics that students will have actually read or heard about in the news.

In this edition we also include two extended applications—one on cancer and stem cells (Chapter 11), the other on global change (Chapter 25)—and we moved the Fourth Edition's "Interlude" chapters online, where students can access them free of charge at DiscoverBiology.com.

NEW ART. For the Fifth Edition, every figure in the text was updated with two goals in mind. First, we increased the clarity and visual appeal of the figures by using a more vibrant color palette and by adding engaging photos that show students the connection between the concept being illustrated and its relevance in the real world. Second, to every figure we added a consistent and pedagogically useful system of banners, labels, and bubble captions that will help students identify the big picture before diving into the details.

NEW COVERAGE. You may notice that *Discover Biology*'s table of contents is unique, with the diversity of life on Earth covered in Unit 1, before chemistry and cells. We structured the book in this way to immediately grab students' attention with the enormous variety of life on this planet. We find this is an effective way to engage students with the other equally important material in the course. However, we wrote these chapters so that they could be taught at any point in the course, giving instructors maximum flexibility. For the Fifth Edition, we updated the diversity-of-life content and expanded it from one chapter into three to more adequately capture the amazing variety of life on Earth.

NEW ONLINE RESOURCES. As the Internet becomes an increasingly rich and powerful tool for teaching and learning, our interest in offering online review, assessment, and application resources has grown in turn. The Fifth Edition of *Discover Biology* is accompanied by the most extensive and diverse resource package for instructors and students yet. We are especially excited about the SmartWork online homework course, which presents high-quality questions specific to *Discover Biology*, paired with answer-specific feedback for students and easy authoring tools for instructors. A full list of the instructor and student ancillaries is included after the book's guided tour.

# GUIDED TOUR

## Fascinating Stories

A story on a contemporary topic, from the possibility of arsenic-based life-forms to HeLa cells and Henrietta Lacks, opens each chapter. The story is then revisited in more depth at the end of the chapter, showing students how a mastery of the chapter's content makes them more scientifically literate. All the chapter stories, developed by science writer Jennie Dusheck, are either brand-new or substantially revised for the Fifth Edition.

---

### 11 Stem Cells, Cancer, and Human Health

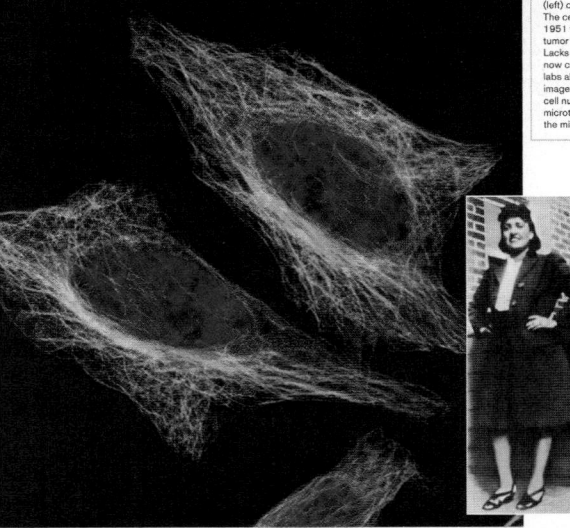

**HELA CELLS ARE A LINE OF CANCER CELLS.** HeLa cells (left) can grow in a lab dish. The cell line was established in 1951 from a slice of a cervical tumor taken from Henrietta Lacks (inset). HeLa cells are now cultivated in thousands of labs all over the world. In this image, the DNA in the HeLa cell nucleus is stained blue, the microtubules are green, and the mitochondria are red.

#### Henrietta Lacks's Immortal Cells

In 1951, a poor, stay-at-home mother named Henrietta Lacks arrived at Johns Hopkins medical school to ask about bleeding at the wrong time of the month. Her doctor took one look at the bright purple tumor on Lacks's cervix—the entrance to the uterus—and feared the worst. He soon determined that she had malignant cervical cancer and broke the news to her.

Just 31 and mother to five children, Henrietta Lacks had little idea that she would not live to see her children grow up. But she would have been even more surprised to learn that the cancer cells growing in her body would end up in thousands of biomedical laboratories and grow vigorously decades after her death.

When Lacks returned to the hospital for treatment, a surgeon sewed envelopes of radioactive radium into her cervix, the state-of-the-art treatment of the day. Most of Lacks's cells were dividing at the slow, highly regulated pace that is normal for human cells. But her cancer cells were different.

Also at Johns Hopkins were two researchers, named George and Margaret Gey, who had been trying for could study their behavior. In 1951, all of the Geys' attempts had so far ended in complete failure, and they were eagerly trying to grow any human cells they could get their hands on. To help out, Lacks's surgeon took a slice of Lacks's tumor and delivered it to the Geys.

The cells from the tumor, called "HeLa" cells after Henrietta Lacks, were a dramatic success, turning out to be among the fastest-dividing human cells ever known and, more important, the first human cells ever grown in the lab outside a human body. But besides being able to survive in a test tube and divide quickly, the cells had one more amazing attribute: they were immortal; they would keep dividing seemingly forever.

What made Lacks's cells immortal? How have those cells contributed to our understanding of cancer biology?

We will return to these questions in the closing segment

2

3

---

**APPLYING WHAT WE LEARNED**

#### How HeLa Cells Changed Biomedicine

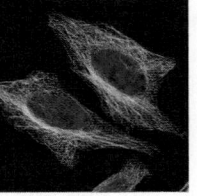

When Henrietta Lacks's tumor cells arrived in the Gey lab in early 1951, a young technician dropped them into test tubes on top of clots of chicken blood and put them in an incubator to stay warm. She didn't expect much, since all the cells she did this with seemed to be dooner or later, usually sooner.

But the Gey lab was in for a surprise. Lacks's cells had an astounding ability to divide and thrive in laboratory conditions. In just 24 hours, the cells had doubled. A day later, their numbers had doubled again. By day 4, the cells were outgrowing their new test tube homes and the lab tech had to rustle up some more test tubes for them.

It was the first time anyone had successfully cultured human cells. The Geys happily celebrated their success, and other researchers were soon demanding "HeLa cells" for their experiments. The cells have continued to grow and divide in test tubes and glass dishes for another 60 years—essentially forever, in biomedical terms.

What made HeLa cells immortal? In this chapter, we have seen that in most normal cells, the tips of chromosomes, called telomeres, shorten a little each time the cell divides. If the telomeres are short enough, the cell can no longer divide, and, after a certain number of divisions, most cells stop dividing and die. Shortened telomeres help prevent uncontrolled cell division, or cancer. The only normal cells in the human body that divide indefinitely are stem cells and germ line cells, the cells that make sperm or eggs.

But HeLa cells are anything but normal. Henrietta Lacks had been exposed to human papillomavirus (HPV),

a sexually transmitted virus that transfers its viral DNA into human skin cells. The result was a line of cells with DNA that was different from both the virus and the original human cells. For example, normal human cells have 46 chromosomes, while HeLa cells contain double that number. And unlike normal cells, HeLa cells make the enzyme telomerase, which repairs the chromosomes' telomeres, preventing them from shortening during cell division (see Figure 11.14). Because the telomeres of HeLa cells do not shorten, HeLa cells are immortal; they can theoretically grow and divide forever.

Since 1951, HeLa cells have been used in thousands of experiments, leading to more than 60,000 scientific papers by 2010. Henrietta Lacks's cells have contributed to the rise of biomedical companies worth billions of dollars. A single anticancer drug developed using HeLa cells had revenues of more than $300 million in 2007. What follows are a few highlights of HeLa cell research.

In 1960, Soviet researchers sent HeLa cells into space, long before any human astronauts ventured from Earth. HeLa cells were used to grow viruses, helping to spawn an entire field of biology called virology, the study of viruses. The mumps, measles, polio, and AIDS viruses were all grown in HeLa cells. When a private company first began mass-producing HeLa cells in a former Frito-Lay factory, Henrietta Lacks's cancer cells became a commodity, like corn or lumber. HeLa cells were used to develop ways to transport live cells in bulk, so researchers could mail cells from one lab to another. HeLa cells were used to show that HPV infection can cause cervical cancer and to study other forms of cancer, as well as in the development of anticancer drugs such as Herceptin. In fact, one way to treat cancer is to deactivate the telomerase enzyme so that cancer cells stop dividing and die off—like normal cells.

cancer cells divide

Gene mutations are the root cause of all cancers. The great majority of human cancers are caused by the accumulation of multiple somatic mutations over the course of life.

Environmental and lifestyle factors play a large role in human cancers.

22    **CHAPTER 11** STEM CELLS, CANCER, AND HUMAN HEALTH

---

# New Art Program

Every piece of art has been updated in the Fifth Edition for accessibility and visual appeal. A consistent and pedagogically useful system of banners, labels, and bubble captions has been added to every figure. First, black banners and gray part labels help students identify and understand the "big picture" of the concept being illustrated. Then, extensive bubble captions guide students through the figure's most important elements, helping them develop a more complete understanding of the concept being illustrated.

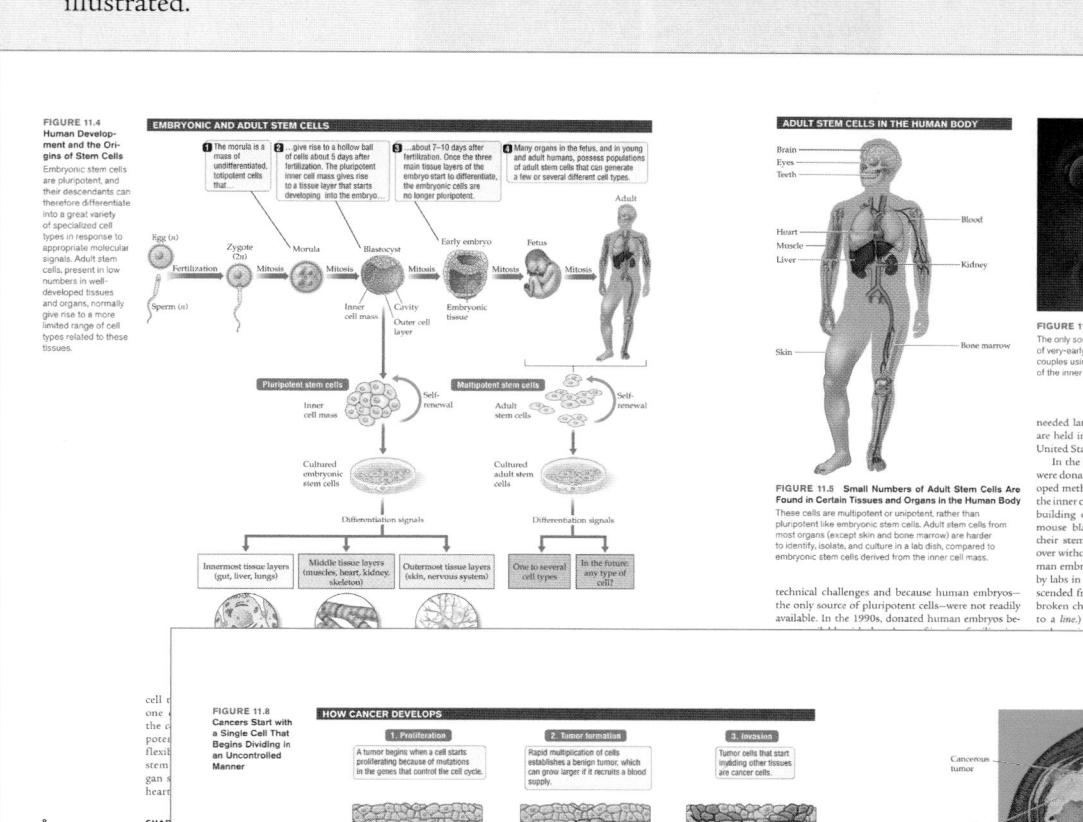

# Unique Pedagogy

Distinctive pedagogical features throughout each chapter promote long-term retention of key concepts and a deeper understanding of new terminology.

**PRONUNCIATION GUIDES** are included next to new and unfamiliar terms in the chapter narrative, with the expectation that if students can pronounce key terms, they will be more likely to speak up in lecture or lab.

**CONCEPT CHECKS** at the end of many chapter sections ask students to identify and think about important concepts. The answers are provided, but upside-down, striking the perfect balance between challenging students and making sure they get the answer.

**HELPFUL TO KNOW** boxes demystify new concepts and terms, making it more likely students will retain them for the test and beyond.

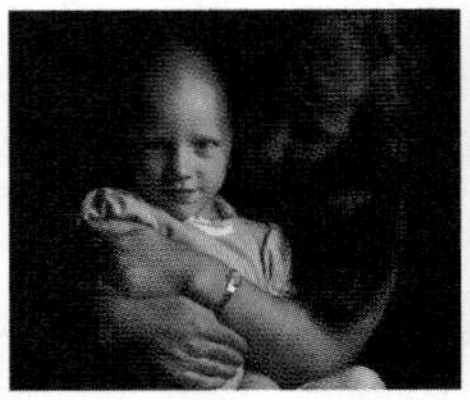

FIGURE 11.13 **Hair Loss Is a Common Side Effect of Chemotherapy**
Chemotherapy attacks any and all rapidly dividing cells and results in collateral damage, including loss of cells that grow and maintain hair follicles. Dormant stem cells survive the treatment, and they will regrow the lost hair in about 3 months.

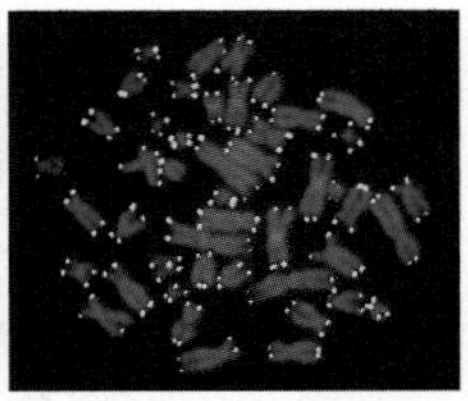

FIGURE 11.14 **Telomerases Confer Cellular Immortality by Preventing Chromosome Ends from Fraying**
The tips of chromosomes, here shown in yellow, are eroded each time a somatic cell divides, limiting the life span of a cell. An enzyme called telomerase overcomes this obstacle to perpetual cell division by lengthening the tips as needed. Embryonic stem cells and cancer cells produce telomerase and can therefore divide endlessly. Clinical trials are under way to see whether crippling telomerase will limit the runaway growth of cancer cells.

### The challenge in cancer treatment is to destroy malignant cells selectively

About 40 years ago, President Richard Nixon declared a war on cancer in the United States by making anticancer research a high priority. Since then, some major victories have been won, thanks to improvements in radiation and drug therapies. Whereas in the early twentieth century very few individuals survived cancer, today roughly 40 percent of patients are alive 5 years after treatment has begun. Nevertheless, the war against cancer is far from over, and the need for powerful new treatments to stop or kill the malignant cells is as urgent as ever. **Cryosurgery** makes use of extremely cold temperatures to kill abnormal cells, usually precancerous cells or cancers that are confined to a small region, such as the cervix. **Hormone therapy**, which manipulates the hormone environment in the body to stop or slow cancer cells, is used in some hormone-responsive cancers, which include some types of breast and prostate cancer.

The greatest challenge in battling cancer is the selective destruction of rogue cells while sparing healthy cells. The standard plan of attack today relies on high-energy radiation (**radiation therapy**), high doses of chemical poisons (**chemotherapy**), or both in sequence to kill any and all rapidly dividing cells. The side effects of radiation therapy and chemotherapy are terrible because this all-out assault also kills many innocent bystanders, cells necessary for the normal functions of the human body. Alopecia (hair loss) is the most visible of the many side effects of these therapies (FIGURE 11.13). Hair falls out because the cells that grow and maintain hair are destroyed; but *dormant* stem cells survive the treatment, and they regrow the hair in about 3 months. Cells that divide to produce red blood cells are killed, leading to the fatigue and weakness of anemia. The cells that line the entire digestive tract are regularly replaced by cell division, and the loss of that cell supply produces illness in the mouth, stomach, and intestines.

The good news is that discoveries in basic cell biology, along with the large investment in cancer research, have produced a variety of innovative strategies for destroying malignant cells selectively. One line of attack is to selectively disable proteins that give cancer cells their unusual immortality, the capacity to divide indefinitely as long as oxygen and nutrients are available.

An enzyme called *telomerase* **[teh-LOH-muh-rays]** is key to cell immortality. The tips of our chromosomes, known as **telomeres [TEH-luh-mirz]**, are whittled down each time a cell divides (FIGURE 11.14), *unless* telomerase steps in to repair the ends. The only cells



**Concept Check Answers**

**1.** A stem cell is undifferentiated, it can renew itself, and under the right conditions, its descendants can differentiate into specialized cell types.
**2.** An adult stem cell generates a relatively small number of specialized cell types; an embryonic stem cell can give rise to any of the cell types in the adult body.

agencies to support and conduct embryonic stem cell research. Advocates of stem cell research maintain that federal support for hES research will advance basic knowledge and spur therapeutic applications of stem cell technology.

Since adult stem cells exist and have been used for therapeutic purposes, what need is there for research on human embryonic stem cells? Advocates of hES research point to the limited developmental flexibility of adult stem cells and the fact that these cells are small, scarce, and usually hard to identify and difficult to grow in a lab dish. The controversies may be laid to rest, however, if the potential of a new type of stem cell—the induced pluripotent stem cell—comes to be realized.

### Induced pluripotent stem cells are derived from differentiated cells

The development of *induced pluripotent stem cells* is one of the most exciting discoveries in cell biology and biomedicine in recent years. An **induced pluripotent stem cell (iPSC)** is any cell, even a highly differentiated cell in the adult body, that has been genetically reprogrammed to mimic the pluripotent behavior of embryonic stem cells. Human embryos do not have to be destroyed to generate iPSC lines, yet the cells appear to have nearly the same developmental flexibility that embryonic stem cells have. The technology therefore appeals to all camps.

Scientists developed iPSCs by carefully studying the key genes that give an embryonic stem cell its unique properties. In 2006, researchers introduced just four critical genes into mouse skin cells (fibroblasts), using harmless viruses as delivery vehicles. The introduced genes forced the mature cell type to start behaving almost exactly like embryonic cells. In 2007, similar techniques were used to create an iPSC line from human skin cells. More recently, iPSC lines have been raised by the introduction of key proteins, not genes, into mature skin cells.

The developers of iPSCs are quick to point out that the technology is still in its infancy and many advances are needed before iPSC-based therapies can be tested in human subjects. For example, mice treated with iPSC therapy seem to be more prone to developing cancer than are mice treated with embryonic stem cell technology. In time, a better understanding of the control of cell division and differentiation in stem cells will probably resolve this problem. Stem cell researchers emphasize the vital importance of studying all types of stem cells because each avenue of study has added, and continues to add, to our understanding of how, when, and where cells divide or differentiate. Understanding how ce[...] critical in treating cancer, the subject we turn to next.

**Concept Check**

1. What are the unique properties of a stem cell?
2. Compare the developmental flexibility of adult stem cells to that of embryonic stem cells.

## 11.2 Cancer Cells: Good Cells Gone Bad

Cancer accounts for more than 500,000 deaths in the United States each year. Over the course of a lifetime, an American male has a nearly one in two chance of being diagnosed with cancer. American women fare slightly better, with a one in three chance of developing cancer. There are more than 200 different types of cancer, but the big four—lung, prostate, breast, and colon cancers—account for more than half of all cancers combined (Table 11.2). Although the past decade has seen improvements in treatment and prevention, more than a million Americans are diagnosed with some form of cancer every year. In the United States, one in four deaths is due to cancer, and more than 8 million Americans alive today have been diagnosed with cancer and are either cured or undergoing treatment. More than 1,500 Americans die from cancer each day. Only heart disease kills more people in the United States. The National Cancer Institute estimates that the collective price tag for the various forms of cancer is more than $100 billion per year.

In the following case study of cell division gone awry, we explore the biology of cancer by considering how a cancer is launched when normal restraints on cell division and cell movement are lost. We will see that the vast majority of cancers arise from the gradual accumulation of mutations in cells of the adult body, often as a result of DNA damage caused by environmental factors. We will examine some of the environmental factors known to contribute to cancer, and we will remind ourselves that, as with any disease, prevention is better than having to find cures.

### Cancer develops when cells lose normal restraints on division and migration

Cancer represents a breakdown in the cooperative functioning of the cells in a multicellular organism.



| TYPE OF CANCER | OBSERVATION | ESTIMATED NEW CASES IN 2010 | ESTIMATED DEATHS IN 2010 |
|---|---|---|---|
| Lung cancer | Accounts for 28 percent of all cancer deaths and kills more women than breast cancer does | 222,520 | 157,300 |
| Prostate cancer | The second leading cause of cancer deaths in men (after lung cancer) | 217,730 | 32,050 |
| Breast cancer | The second leading cause of cancer deaths in women (after lung cancer) | 207,090 | 39,840 |
| Colon and rectal cancer | The number of new cases is leveling off as a result of early detection and polyp removal | 142,570 | 51,370 |
| Malignant melanoma | The most serious and rapidly increasing form of skin cancer in the United States | 68,130 | 8,700 |
| Leukemia | Often thought of as a childhood disease, this cancer of white blood cells affects more than 10 times as many adults as children every year | 43,050 | 21,840 |
| Ovarian cancer | Accounts for 3 percent of all cancers in women | 22,200 | 16,200 |

Every cancer begins with a single rogue cell that starts dividing with wild abandon, giving rise to a mass of abnormal cells. The cell mass formed by the inappropriate proliferation of cells is known as a **tumor**. Tumors that remain confined to one site are **benign tumors**. Because they can be surgically removed in most cases, benign tumors are generally not a threat to survival. However, a benign tumor that is growing actively is like a cancer-in-training. With the passage of time, the descendants of these abnormal cells can become increasingly abnormal: they change shape, increase in size, and quit their normal job. These **precancerous cells** look abnormal enough that pathologists (disease experts) looking through a microscope can usually pick them out from normal cells by their size and shape (FIGURE 11.7). Tumor cells often produce proteins—known as *tumor markers*—that normal cells of that type do not make or make in much lower quantity. The presence of tumor markers in blood, urine, and other fluids and tissues is used to screen for the *possibility* of some types of cancer.

Most cells in the adult animal body are firmly anchored in one place. Tumor cells on the path to cancer start producing enzymes (matrix metalloproteinases) that break up the adhesion proteins that attach a cell to the extracellular matrix or to other cells in the tissue. Most human cells stop dividing if they are detached from their surroundings—a phenomenon known as **anchorage dependence**. But some tumor cells that have broken loose from their moorings may acquire anchorage *in*dependence, the ability to divide even when released from the normal attachment sites.

As tumor cells progress toward a cancerous state, they start secreting substances that cause new blood vessels to form in their vicinity in a process known as

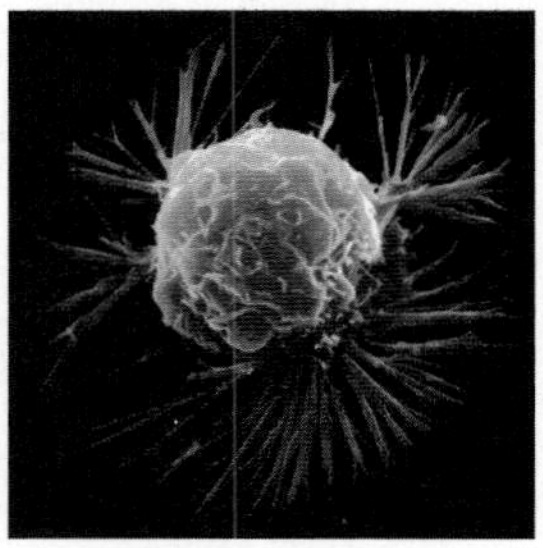

FIGURE 11.7 **A Home-Grown Monster**
This color-enhanced photograph, captured with a scanning electron microscope, shows a breast cancer cell. A lab technician can recognize it as a cancer cell because of its large size, abnormally rounded shape, and altered cell surface.

**Helpful to Know**

The word "cancer" comes from the Old English for "spreading sore," which in turn comes from the Latin for "crab," probably because the enlarged blood vessels on a cancerous tumor were seen as resembling the legs of a crab. The study of cancer is called "oncology," which comes from *onkos*, Greek for "lump" or "mass," as in a tumor mass. Genes that promote cancer formation are called "oncogenes."

# Applied Features

**BIOLOGY MATTERS** boxes in virtually every chapter connect biology to real-life relevant topics that students care about: their health, society, and the environment.

---

**BIOLOGY** MATTERS

## Avoiding Cancer by Avoiding Chemical Carcinogens

Although chemical pollutants are often the most feared of carcinogens in the public mind, experts estimate that only about 2 percent of human cancers can be blamed on environmental carcinogens, while what we eat and drink accounts for more than 30 percent of our cancer risk. Tobacco use is responsible for about 30 percent of the cancers diagnosed in the United States.

Many explanations have been offered for the observation that those who eat a lot of animal products have higher rates of some cancers than those who consume less or none. Some carcinogenic pollutants become concentrated along the food chain and are found in higher concentrations in meat, fish, and dairy than in plant foods. The saturated fat in meat and dairy is an independent risk factor for cancers of the breast, prostate, and colon. Animal products also lack the cancer-protective substances—such as fiber and antioxidants—that presumably help those who eat a mostly plant-based diet. Those who eat a lot of red meat—about 4 ounces, or the equivalent of one steak, a day—may experience iron overload, a known risk factor in some types of cancer.

When animal flesh is cooked at high temperatures, as in grilling, broiling, or deep frying, certain amino acids that are abundant in meat, poultry, and fish, are converted into a family of carcinogenic compounds known as heterocyclic amines (HCAs). High-temperature cooking of fatty foods creates yet another class of potent carcinogens, called polycyclic **[pah-lee-SYKE-lik]** aromatic hydrocarbons, or PAHs. If you like your meat well done, you might consider cutting back or at least resorting to safer cooking methods much of the time. Microwaving food lightly before high-temperature cooking reduces

HCA and PAH formation. So does marinating in antioxidant-rich sauces—think berries, red wine, and herbs like rosemary.

> Antioxidants in some fruits and herbs may reduce HCA formation.
> Amino acids
> High temperature
> Heterocyclic amines (HCAs)

Antioxidants in some fruits, and herbs such as rosemary, may reduce HCA formation in grilled meats.

When it comes to cancer risk, processed meats are even worse than charbroiled red meat. Nitrites are commonly used to preserve the color and fresh appearance of hot dogs, sausages, bologna, cold cuts, and other processed meats. In acidic environments, such as the stomach, the nitrites react with proteins to form carcinogenic compounds called nitrosamines. Vitamin C (ascorbic acid) is often added to processed meats because this antioxidant dampens the conversion of nitrites to nitrosamines; even so, it may not be a bad idea to swig some orange juice when chowing down on a hot dog.

Excessive alcohol consumption is strongly linked to cancers of the mouth and esophagus, and a smaller risk of liver, breast, and colorectal cancer. Doctors advise no more than one drink per day for women, and two for men, if we drink at all. A standard drink in the United States is defined as 0.6 ounce (13.7 grams) of pure eth-

anol, which amounts to one 12-ounce beer at 4 percent alcohol, or a 5-ounce glass of wine at 11 percent alcohol.

The link between cancer and smoking is the most dramatic illustration of how chemical exposure can transform healthy cells into dangerous ones. Lung cancer was a rare cancer just before the turn of the twentieth century, when few people smoked tobacco. Now, with nearly a third of the world's population lighting up, lung cancer is the most common and the deadliest cancer worldwide, killing more than 1.2 million people annually. Both tobacco and marijua[na]...a compound calle[d]...erful cancer-causi[ng]...marijuana cigarette con[tains]...more benzopyrene than...of tobacco. And marijua[na]...haled more deeply and [held]...than cigarette smoke. E[ven]...much more addictive th[an]...and that means that pe[ople]...cigarettes are far more [likely to continue] the habit for a long tim[e]...marijuana smoker will c[ontinue]...rettes per day than eve[n]...smoker does. The botto[m line is that] both types of smoke are [harmful].

The good news is [that cancer risk] can dramatically reduce[d]...risk. People who quit sm[oking before] 50 reduce their risk of d[ying in the next] 15 years by half. Rega[rdless, people] who quit smoking live [longer than those who] continue to smoke. Whil[e tobacco is a highly] addictive drug in tobacc[o, making quit-] ing difficult, all should fin[d comfort in the fact] that one in five America[ns]...

---

**BIOLOGY IN THE NEWS** features serve as a capstone to each chapter, reinforcing the chapter opening and closing story by providing students with an example of how they might encounter a related issue in their own lives—by reading about it in the news. The news article excerpts from 2009–11 are paired with author analysis and questions that stimulate critical thinking.

---

**BIOLOGY IN THE NEWS**

## Boys to Men: Unequal Treatment on HPV Vaccine

BY MICHELLE ANDREWS

The vaccine that prevents 70 percent of cervical cancers got the thumbs up under the health overhaul law as one of the preventive benefits that must be provided free to girls and young women between the ages 9 and 26.

Boys and young men, however, don't get the same free coverage under the law, even though the human papilloma virus (HPV) vaccine is also approved for the prevention of genital warts in males.

Why the difference? "Genital warts aren't life threatening," says Debbie Saslow, director of breast and gynecologic cancer for the American Cancer Society.

Merck, which manufactures one of the two FDA-approved HPV vaccines, is conducting research to see if the vaccine pre-

vents genital cancers in men, says Saslow. In the meantime, though, the vaccine has already been shown to prevent cervical, vaginal and vulvar cancers in women. "It's a matter of cost-effectiveness," says Saslow.

The vaccine is pricey, requiring three shots over a six-month period, at about $130 each.

The human papilloma virus, which is transmitted through sexual contact, is extremely common, accounting for an estimated 500,000 cases of genital warts every year. "It's the common cold of the genital tract," says Saslow.

There are several treatments to eliminate genital warts. And, the body can rid [itself] of the virus, although it's possible to get re-infected.

The health law requires that immunizations that are recommended by the Advisory Committee on Immunization Practices [ACIP] be provided without charge to patients in new health plans starting this fall. ACIP is a group of 15 experts appointed by the Secretary of Health and Human Services.

While the HPV vaccine is recommended for all girls and young women between 9 and 26, the committee made a "permissive" recommendation for boys and men in the same age range, says Lance Rodewald, director of the immunization services division for the Centers for Disease Control and Prevention. Basically, that means it's OK to vaccinate males for genital warts, but it's not essential.

Human papillomavirus (HPV) is the most common sexually transmitted disease in the United States. At any given time, about half of men and 15 percent of women are infected with HPV. Often people's immune systems fight off the infection. But at some time in their lives, most adults become infected with one or more of the 30–40 strains that infect the human genital tract.

Only a few of these strains can cause cancer—mostly cervical cancer, but sometimes cancer of the tonsils and anus, in both men and women. The strains of the virus that cause cancer are different from the ones that cause genital warts. Nearly all cases of cervical cancer result from HPV infections, and overwhelmingly, women contract those HPV infections from their male sexual partners.

The HPV vaccine was initially recommended for girls and young women only. This recommendation was unusual, since most vaccines are prescribed for everyone. Populations where nearly everyone is vaccinated are much more resistant than populations where large numbers of people remain unvaccinated. For example, smallpox has been completely eradicated throughout the world as a result of aggressive vaccination programs, and polio is close to being eradicated. Even in the rare cases where only part of the population is threatened by a disease, we normally vaccinate everyone. For example, rubella (German measles) is a harmless infection in most people. But if a fetus is infected, it may die or be born with severe birth defects. So, to protect developing fetuses, health agencies try to make sure everyone is vaccinated against rubella.

One reason for not vaccinating boys and men against HPV is cost. If the government recommends a vaccine, typically health insurers are required to cover it. An optional vaccine—like the one for HPV in boys—need not be covered. The rubella vaccine costs less than $10 per person; the HPV vaccine, more than $300. Another reason not to vaccinate boys is their lower level of risk.

Since the invention of the Pap smear—a cheap test for cervical cancer—the death rate from cervical cancer in the United States has dropped to about 4,000 deaths per year. If caught early, cervical cancer is easy to cure. In contrast, in developing countries where Pap smears are hard to get, some 370,000 women are diagnosed with cervical cancer each year. Half of those women die of the disease each year.

### Evaluating the News

1. If all 156 million women in the United States were vaccinated against HPV, how many deaths from cervical cancer could be prevented, and at what cost per life saved?

2. Do women who have been vaccinated need Pap smears? Explain.

3. Discuss the pros and cons of vaccinating boys and young men against HPV. What considerations do you think are the most important in making a decision?

**SOURCE:** NPR, December 7, 2010, http://www.npr.org/blogs/health/2010/12/07/131881133/boys-to-men-unequal-treatment-on-hpv-vaccine.

# Visually Rich Assignments in SmartWork

WWNORTON.COM/SMARTWORK

**SmartWork**, Norton's online homework system, offers high-quality questions in an easy-to-access and easy-to-use interface. With a heavy emphasis on visual learning, **SmartWork** questions include answer-specific feedback and are fully customizable. Powerful instructor tools, including the ability to easily author original questions, automatic grading, and item analysis, provide real-time assessment of student progress.

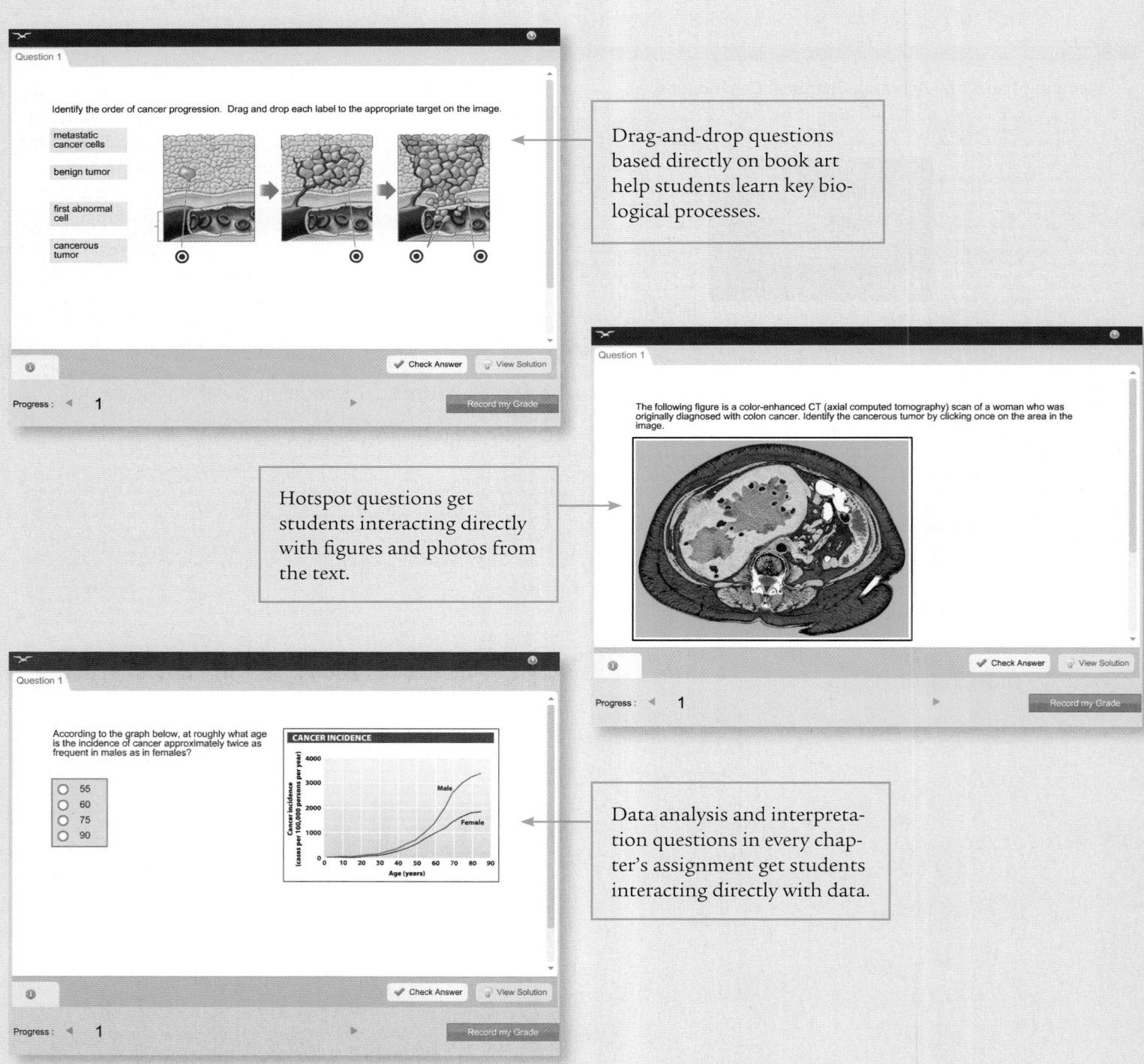

Drag-and-drop questions based directly on book art help students learn key biological processes.

Hotspot questions get students interacting directly with figures and photos from the text.

Data analysis and interpretation questions in every chapter's assignment get students interacting directly with data.

# The Most Robust FREE and OPEN Student Website

DISCOVERBIOLOGY.COM

**StudySpace** includes extensive review and application features for every chapter that will help students prepare for exams without having to register, log in, or pay extra.

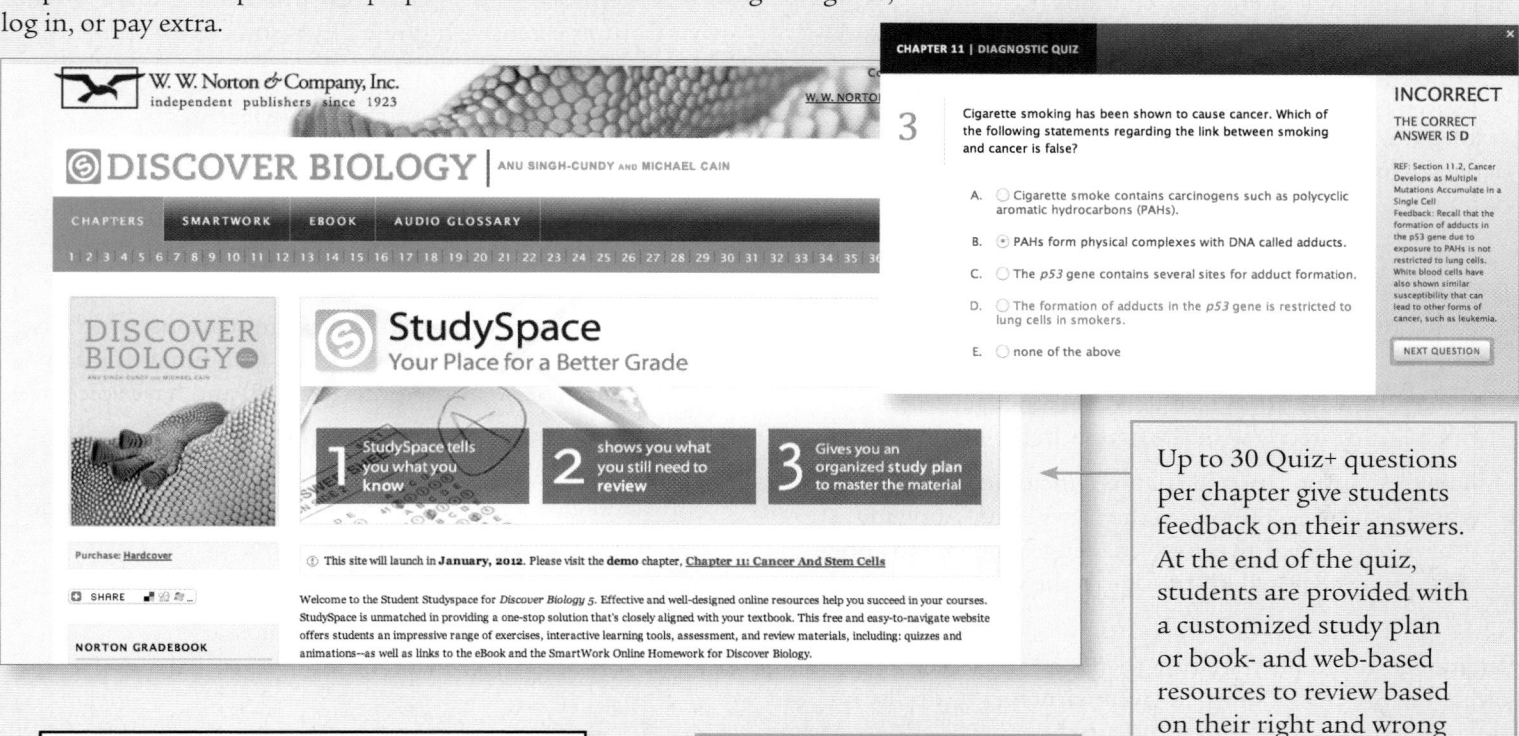

Up to 30 Quiz+ questions per chapter give students feedback on their answers. At the end of the quiz, students are provided with a customized study plan or book- and web-based resources to review based on their right and wrong answers.

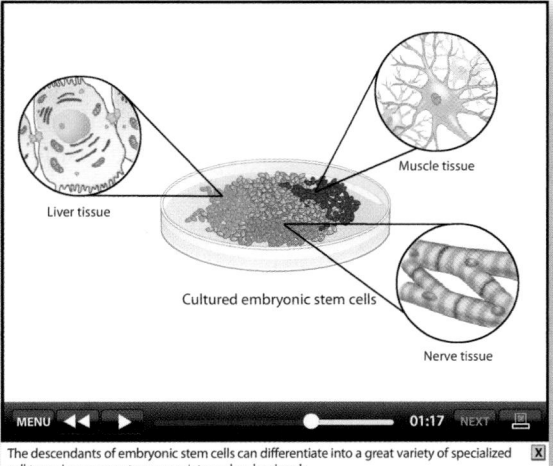

Totally revised animations based on the new Fifth Edition art are integrated with activities that help students assess their learning.

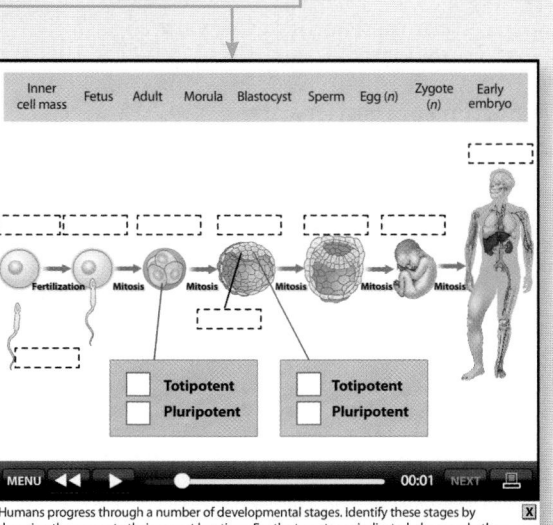

## NEW! smartwork for Biology

SmartWork online homework contains high-quality and easy-to-use, visually oriented questions that include answer-specific feedback that helps students right when they need it most. Drag-and-drop and hotspot questions based on text figures—as well as questions based on brief animations derived from key text figures—all reinforce the visual aspects of biology.

The SmartWork questions for every chapter follow the engage, learn, apply structure in the text:

- **Engage and Learn:** A large proportion of the SmartWork questions are drag-and-drop, hotspot, graph-interpretation, and animation-based questions that get students interacting directly with the figures in the book.

- **Apply:** Questions in each chapter—including those derived from the book's "Biology in the News" and "Applying What We Learned" features—challenge students to apply their newly acquired knowledge to the world in which they live.

**Additional Questions.** Two other types of questions are offered in SmartWork, which give instructors flexibility in teaching and assessing their students:

- **Reading Quizzes:** Ten brief questions (without feedback) covering the breadth of each chapter can be assigned for instructors who want to test if their students have studied the text **BEFORE** coming to class.

- **Test Bank:** Questions from the Test Bank (also without feedback) are offered for instructors who want to assign them outside of a high-stakes testing setting.

**Learn more at** wwnorton.com/smartwork

## DiscoverBiology.com

DISCOVERBIOLOGY.COM is a **FREE** and **OPEN** student website that features extensive high-quality resources for student study and review.

The free and open, feature-rich website features:

- Quiz+ questions with feedback and site references

- Updated What's News quizzes, which link to recent news articles and provide students with a few interpretation questions

- Comprehensively updated process animations, which use the stunning Fifth Edition new art. All animations now feature a consistent user interface (UI) and have voice-over.

- **FREE** and **OPEN** access to fully applied chapters on biodiversity, the human genome, human evolution, smoking, and global food distribution (formerly the Interlude chapters in the Fourth Edition)

- Study Plans for each chapter

- Vocabulary Flashcards and Audio Glossary

- Biology in the News feed

## Study Guide

The Study Guide, free when packaged with the textbook, provides a guided tour of each chapter of the text, using summaries, thought questions, and practice questions to help students achieve an even greater understanding of the material.

## 📖 Ebook and Chapter Select Ebook

Norton ebooks are an affordable and convenient alternative to the printed book. Ebook chapters are also available for $2 each through Norton's Chapter Select program at **nortonebooks.com**.

## NEW! Instructor Presentation Resources

Discover Biology comes with **FOUR** types of presentation tools:

1. **Lecture Slides.** These slides feature selected art from the text with detailed lecture outlines and links to the animations on **DiscoverBiology.com**.

2. **Art Slides.** These PowerPoint sets contain all art from the book.

3. **Unlabeled Art Slides.** The figure labels are removed to allow students to "fill in the blank."

4. NEW! **Active Art Slides.** These slides feature complex pieces of artwork broken down into moveable, editable components. This function allows instructors to customize figures within PowerPoint, choosing exactly what their students will see in lecture.

## Test Bank

Revised using the Norton Assessment Guidelines, the Test Bank features over 2,700 high quality problems. The Test Bank achieves a high degree of integration with the textbook by featuring selected art from the text as figure interpretation questions and includes questions about all of the "Biology in the News" features. Each problem in the Test Bank is classified according to Norton's taxonomy of knowledge types:

1. Factual questions test students' basic understanding of facts and concepts.

2. Applied questions require students to apply knowledge in the solution of a problem.

3. Conceptual questions require students to engage in qualitative reasoning and to explain why things are as they are.

Questions are further classified by section and difficulty, making it easy to construct tests and quizzes that are meaningful and diagnostic. The question types are multiple-choice, true-false, and completion. Available as PDF, Word, and in the Instructor Resource Folder.

## Instructor Resource Folder

This folder contains three discs:

1. **Instructor resource DVD**, which contains:
   - Lecture Slides, with links to offline versions of the animations
   - Art slides
   - Art slides without labels
   - Active art slides
   - All art in JPG format
   - Animations

2. **Full Edition Test Bank** in ExamView format

3. **Core Edition Test Bank** in ExamView format

## Downloadable Resources

WWNORTON.COM/INSTRUCTORS features instructional content for use in lecture and distance education, including coursepacks, test-item files, PowerPoint lecture slides, images, figures and more.

- **FREE**, customizable Blackboard, WebCT, Moodle, Angel, and D2L coursepacks
- All four types of Lecture Slides
- Art from book, in PowerPoint and JPG
- Glossary
- Test Bank
- Website quizzes

## Coursepacks

Available at no cost to professors or students, Norton coursepacks bring high-quality Norton digital media into a new or existing online course. Content includes chapter-based assignments from **DiscoverBiology.com**, reading quizzes (also found in SmartWork), the test bank, plus links to the process animations.

# Acknowledgments

We would like to thank the many people who provided critical commentary on our revisions. Their insights helped us make *Discover Biology* a better book.

## Reviewers of the Fifth Edition

Holly Ahern, Adirondack Community College
Mac Alford, University of Southern Mississippi
Marilyn Banta, Texas State University–San Marcos
Robert Bevins, Georgetown College
Randy Brewton, University of Tennessee–Knoxville
Christine Buckley, Rose-Hulman Institute of Technology
Aaron Cassill, University of Texas at San Antonio
Keith Crandall, Brigham Young University
Helen Cronenberger, University of Texas at Austin
Chad Cryer, Austin Community College
Francie Cuffney, Meredith College
Gregory Dahlem, Northern Kentucky University
Don Dailey, Austin Peay State University
Angela Davis, Danville Area Community College
Susan Epperson, University of Colorado at Colorado Springs
Susan Farmer, Abraham Baldwin Agricultural College
Linda Fergusson-Kolmes, Portland Community College
Paul Florence, Jefferson Community & Technical College
April Ann Fong, Portland Community College, Sylvania Campus
Edison Fowlks, Hampton University
Jennifer Fritz, University of Texas at Austin
Caitlin Gille, Pasco-Hernando Community College
Tamar Goulet, University of Mississippi
John Griffis, University of Southern Mississippi
Cindy Gustafson-Brown, University of California–San Diego
Ronald Gutberlet, Salisbury University
Jill Harp, Winston-Salem State University
Anne-Marie Hoskinson, Minnesota State University, Mankato
Tonya Huff, Riverside Community College
Meshagae Hunte-Brown, Drexel University
Brenda Hunzinger, Lake Land College
Karen Jackson, Jacksonville University
Sayna Jahangiri, Folsom Lake College
Jane Jefferies, Brigham Young University
Denim Jochimsen, University of Idaho
Mark Johnson, Georgetown College
Anthony Jones, Tallahassee Community College
Joshua King, Central Connecticut State University
Yolanda Kirkpatrick, Pellissippi State Community College
Jennifer Landin, North Carolina State University
Neva Laurie-Berry, Pacific Lutheran University
Paula Lemons, University of Georgia
Margaret Liberti, SUNY Cobleskill
Cynthia Littlejohn, University of Southern Mississippi
Suzanne Long, Monroe Community College
Monica Macklin, Northeastern State University
Lisa Maranto, Prince George's Community College
Boriana Marintcheva, Bridgewater State College
Catarina Mata, Borough of Manhattan Community College
Susan Meiers, Western Illinois University
James Mickle, North Carolina State University

James Mone, Milllersville University
Elizabeth Nash, Long Beach Community College
Jon Nickles, University of Alaska–Anchorage
John Niedzwiecki, Belmont University
Zia Nisani, Antelope Valley College
Ikemefuna Nwosu, Lake Land College
Brady Olson, Western Washington University
Melinda Ostraff, Brigham Young University
Jason Oyadomari, Finlandia University
Don Padgett, Bridgewater State College
Patricia Phelps, Austin Community College
Joel Piperberg, Millersville University of Pennsylvania
Todd Primm, Sam Houston State University
Ashley Rall McGee, Valdosta State University
Stuart Reichler, University of Texas at Austin
Mindy Reynolds-Walsh, Washington College
Michelle Rogers, Austin Peay State University
Lori Ann (Henderson) Rose, Sam Houston State University
Allison Roy, Kutztown University
Michael Rutledge, Middle Tennessee State University
Tara Scully, George Washington University
Brian Seymour, Sonoma State University
Erica Sharar, Irvine Valley College
Mary Lou Soczek, Fitchburg State University
Michael Sovic, Ohio State University
Ruth Sporer, Rutgers–Camden
Kirsten Swinstrom, State Rosa Junior College
Kristina Teagarden, West Virginia University
Holly Walters, Cape Fear Community College
Teresa Weglarz-Hall, University of Wisconsin–Fox Valley
Michael Wenzel, California State University–Sacramento
Jennifer Wiatrowski, Pasco-Hernando Community College
Antonia Wijte, Irvine Valley College
Daniel Williams, Winston-Salem State University
Edwin Wong, Western Connecticut State University
Donald Yee, University of Southern Mississippi
Calvin Young, Fullerton College

## Reviewers of Previous Editions

Michael Abruzzo, California State University–Chico
James Agee, University of Washington
Laura Ambrose, University of Regina
Marjay Anderson, Howard University
Angelika M. Antoni, Kutztown University of Pennsylvania
Idelisa Ayala, Broward College
Caryn Babaian, Bucks County College
Neil R. Baker, Ohio State University
Sarah Barlow, Middle Tennessee State University
Christine Barrow, Prince George's Community College
Gregory Beaulieu, University of Victoria

Craig Benkman, New Mexico State University
Elizabeth Bennett, Georgia College and State University
Stewart Berlocher, University of Illinois–Urbana
Robert Bernatzky, University of Massachusetts–Amherst
Nancy Berner, University of the South
Janice M. Bonner, College of Notre Dame of Maryland
Juan Bouzat, University of Illinois–Urbana
Bryan Brendley, Gannon University
Randy Brewton, University of Tennessee–Knoxville
Peggy Brickman, University of Georgia
Sarah Bruce, Towson University
Neil Buckley, SUNY Plattsburgh
Art Buikema, Virginia Tech University
John Burk, Smith College
Kathleen Burt-Utley, University of New Orleans
Wilbert Butler, Jr., Tallahassee Community College
David Byres, Florida Community College at Jacksonville–South Campus
Naomi Cappuccino, Carleton University
Kelly Cartwright, College of Lake County
Heather Vance Chalcraft, East Carolina University
Van Christman, Ricks College
Jerry Cook, Sam Houston State University
Francie Cuffney, Meredith College
Kathleen Curran, Wesley College
Judith D'Aleo, Plymouth State University
Vern Damsteegt, Montgomery College
Paul da Silva, College of Marin
Garry Davies, University of Alaska–Anchorage
Sandra Davis, University of Louisiana–Monroe
Kathleen DeCicco-Skinner, American University
Véronique Delesalle, Gettysburg College
Pablo Delis, Hillsborough Community College
Lisa J. Delissio, Salem State College
Alan de Queiroz, University of Colorado
Jean de Saix, University of North Carolina–Chapel Hill
Joseph Dickinson, University of Utah
Gregg Dieringer, Northwest Missouri State University
Deborah Donovan, Western Washington University
Christian d'Orgeix, Virginia State University
Harold Dowse, University of Maine
John Edwards, University of Washington
Jean Engohang-Ndong, Brigham Young University–Hawaii
Jonathon Evans, University of the South
William Ezell, University of North Carolina–Pemberton
Deborah Fahey, Wheaton College
Richard Farrar, Idaho State University
Marion Fass, Beloit College
Tracy M. Felton, Union County College
Richard Finnell, Texas A&M University
Ryan Fisher, Salem Sate College
Susan Fisher, Ohio State University
April Fong, Portland Community College
Kathy Gallucci, Elon University
Wendy Garrison, University of Mississippi
Gail Gasparich, Towson University
Aiah A. Gbakima, Morgan State University
Dennis Gemmell, Kingsborough Community College
Alexandros Georgakilas, East Carolina University
Kajal Ghoshroy, Museum of Natural History–Las Cruces
Beverly Glover, Western Oklahoma State College
Jack Goldberg, University of California–Davis
Andrew Goliszek, North Carolina Agricultural and Technological State University
Glenn Gorelick, Citrus College
Tamar Goulet, University of Mississippi
Bill Grant, North Carolina State University
Harry W. Greene, Cornell University
Laura Haas, New Mexico State University

Barbara Hager, Cazenovia College
Blanche Haning, University of North Carolina–Chapel Hill
Robert Harms, St. Louis Community College–Meramec
Chris Haynes, Shelton State Community College
Thomas Hemmerly, Middle Tennessee State University
Nancy Holcroft-Benson, Johnson County Community College
Tom Horvath, SUNY Oneonta
Daniel J. Howard, New Mexico State University
Laura F. Huenneke, New Mexico State University
James L. Hulbert, Rollins College
Karel Jacobs, Chicago State University
Robert M. Jonas, Texas Lutheran University
Arnold Karpoff, University of Louisville
Paul Kasello, Virginia State University
Laura Katz, Smith College
Andrew Keth, Clarion University of Pennsylvania
Tasneem Khaleel, Montana State University
John Knesel, University of Louisiana–Monroe
Will Kopachik, Michigan State University
Olga Kopp, Utah Valley University
Erica Kosal, North Carolina Wesleyan College
Hans Landel, North Seattle Community College
Allen Landwer, Hardin-Simmons University
Katherine C. Larson, University of Central Arkansas
Shawn Lester, Montgomery College
Harvey Liftin, Broward County Community College
Lee Likins, University of Missouri–Kansas City
Craig Longtine, North Hennepin Community College
Melanie Loo, California State University–Sacramento
Kenneth Lopez, New Mexico State University
David Loring, Johnson County Community College
Ann S. Lumsden, Florida State University
Blasé Maffia, University of Miami
Patricia Mancini, Bridgewater State College
Lisa Maranto, Prince George's Community College
Roy Mason, Mount San Jacinto College
Joyce Maxwell, California State University–Northridge
Phillip McClean, North Dakota State University
Quintece Miel McCrary, University of Maryland–Eastern Shore
Amy McCune, Cornell University
Bruce McKee, University of Tennessee
Bob McMaster, Holyoke Community College
Dorian McMillan, College of Charleston
Alexie McNerthney, Portland Community College
Susan Meacham, University of Nevada, Las Vegas
Gretchen Meyer, Williams College
Steven T. Mezik, Herkimer County Community College
Brook Milligan, New Mexico State University
Ali Mohamed, Virginia State University
Daniela Monk, Washington State University
Brenda Moore, Truman State University
Ruth S. Moseley, S. D. Bishop Community College
Jon Nickles, University of Alaska–Anchorage
Benjamin Normark, University of Massachusetts–Amherst
Douglas Oba, University of Wisconsin–Marshfield
Mary O'Connell, New Mexico State University
Jonas Okeagu, Fayetteville State University
Alexander E. Olvido, Longwood University
Marcy Osgood, University of Michigan
Donald Padgett, Bridgewater State College
Penelope Padgett, University of North Carolina–Chapel Hill
Kevin Padian, University of California–Berkeley
Brian Palestis, Wagner College
John Palka, University of Washington
Anthony Palombella, Longwood College
Snehlata Pandey, Hampton University
Murali T. Panen, Luzerne County Community College
Robert Patterson, North Carolina State University

Nancy Pelaez, California State University–Fullerton
Pat Pendarvis, Southeastern Louisiana University
Brian Perkins, Texas A&M University
Patrick Pfaffle, Carthage College
Massimo Pigliucci, University of Tennessee
Jeffrey Podos, University of Massachusetts–Amherst
Robert Pozos, San Diego State University
Ralph Preszler, New Mexico State University
Jim Price, Utah Valley University
Jerry Purcell, Alamo Community College
Richard Ring, University of Victoria
Barbara Rundell, College of DuPage
Ron Ruppert, Cuesta College
Lynette Rushton, South Puget Sound Community College
Shamili Sandiford, College of DuPage
Barbara Schaal, Washington University
Jennifer Schramm, Chemeketa Community College
John Richard Schrock, Emporia State University
Kurt Schwenk, University of Connecticut
Harlan Scott, Howard Payne University
Erik Scully, Towson University
Tara A. Scully, George Washington University
David Secord, University of Washington
Marieken Shaner, University of New Mexico
William Shear, Hampden-Sydney College
Cara Shillington, Eastern Michigan University
Barbara Shipes, Hampton University
Mark Shotwell, Slippery Rock University
Shaukat Siddiqi, Virginia State University
Jennie Skillen, College of Southern Nevada
Donald Slish, SUNY Plattsburgh
Julie Smit, University of Windsor
James Smith, Montgomery College
Philip Snider, University of Houston
Julie Snyder, Hudson High School
Ruth Sporer, Rutgers–Camden
Jim Stegge, Rochester Community and Technical College
Richard Stevens, Monroe Community College
Neal Stewart, University of North Carolina–Greensboro

Tim Stewart, Longwood College
Bethany Stone, University of Missouri
Nancy Stotz, New Mexico State University
Steven Strain, Slippery Rock University
Allan Strand, College of Charleston
Marshall Sundberg, Emporia State University
Alana Synhoff, Florida Community College
Joyce Tamashiro, University of Puget Sound
Steve Tanner, University of Missouri
Josephine Taylor, Stephen F. Austin State University
John Trimble, Saint Francis College
Mary Tyler, University of Maine
Doug Ure, Chemeketa Community College
Rani Vajravelu, University of Central Florida
Roy Van Driesche, University of Massachusetts–Amherst
Cheryl Vaughan, Harvard University
John Vaughan, St. Petersburg College
William Velhagen, Longwood College
Mary Vetter, Luther College
Alain Viel, Harvard Medical School
Carol Wake, South Dakota State University
Jerry Waldvogel, Clemson University
Elsbeth Walker, University of Massachusetts–Amherst
Daniel Wang, University of Miami
Stephen Warburton, New Mexico State University
Carol Weaver, Union University
Paul Webb, University of Michigan
Cindy White, University of Northern Colorado
Peter Wilkin, Purdue University North Central
Daniel Williams, Winston-Salem State University
Elizabeth Willott, University of Arizona
Peter Wimberger, University of Puget Sound
Allan Wolfe, Lebanon Valley College
David Woodruff, University of California–San Diego
Louise Wootton, Georgian Court University
Silvia Wozniak, Winthrop University
Robin Wright, University of Washington
Carolyn A. Zanta, Clarkson University

## Thanks to the *Discover Biology* Team

As always, revising this textbook was a monumental task, and we are thankful to the many editors, researchers, and assistants at W. W. Norton who helped shepherd this book through the significant revisions in text, photos, and artwork that you see here. In particular, we'd like to thank our editor, Betsy Twitchell, for helping us plan and execute this ambitious revision. Her enthusiasm for our book and keen market sense have been invaluable to us. Our developmental editor, Carol Pritchard-Martinez, contributed enormously to the vision and execution of the revision. Carol's thoughtfulness and enthusiasm are truly amazing. Thanks to our eagle-eyed copy editor, Stephanie Hiebert, a most superbly meticulous, perceptive, and skillful wordsmith. Thanks also to Kim Yi and Christine D'Antonio for seamlessly coordinating the movement and synthesis of the innumerable parts of this book. Our thanks also to Chris Granville for skillfully overseeing the final assembly into a tangible, beautiful book. Photo researchers Stephanie Romeo and Elyse Rieder also contributed enormously to the visual appeal of this beautiful revision. Media editors Rob Bellinger and Patrick Shriner's herculean efforts in developing the media have resulted in the highest-quality and most robust media package this book has ever had. With marketing manager John Kresse's tireless advocacy of this book in the marketplace, we're confident that it will reach as wide an audience as possible. Callinda Taylor progressed from assisting us in all aspects of this project to managing the improvement of the book's print ancillaries, most notably the Test Bank. Cait Callahan deserves thanks for making sure that the many parallel tracks of reviewing, revising, and correcting eventually converged at the right time and place. Finally, we would like to thank our families for support during the long process that is a textbook revision, especially Don, Ryan, and Erika Singh-Cundy.

# About the Authors

**Anu Singh-Cundy** received her PhD from Cornell University and did postdoctoral research in cell and molecular biology at Penn State. She is an associate professor at Western Washington University, where she teaches a variety of undergraduate and graduate courses, including organismal biology, cell biology, plant developmental biology, and plant biochemistry. She has taught introductory biology to nonmajors for over 15 years and is recognized for pedagogical innovations that communicate biological principles in a manner that engages the nonscience student and emphasizes the relevance of biology in everyday life. Her research focuses on cell–cell communication in plants, especially self-incompatibility and other pollen–pistil interactions. She has published over a dozen research articles and has received several awards and grants, including a grant from the National Science Foundation.

**Michael L. Cain** received his PhD in ecology and evolutionary biology from Cornell University and did postdoctoral research at Washington University in molecular genetics. He taught introductory biology and a broad range of other biology courses at New Mexico State University and the Rose-Hulman Institute of Technology for 13 years. He is now writing full-time and is affiliated with Bowdoin College in Maine. Dr. Cain has published dozens of scientific articles on such topics as genetic variation in plants, insect foraging behavior, long-distance seed dispersal, and factors that promote speciation in crickets. Dr. Cain is the recipient of numerous fellowships, grants, and awards, including the Pew Charitable Trust Teacher–Scholar Fellowship and research grants from the National Science Foundation.

**Contributing author Jennie Dusheck** has a BA in integrative biology from the University of California, Berkeley; a master's by thesis in ecology and evolutionary biology from the University of California, Davis; and a graduate degree in science communication from the University of California, Santa Cruz. After 5 years of independent field and laboratory work for UC Berkeley, UC Davis, NASA (a space shuttle experiment), and the National Park Service, and 5 years publishing a newsletter on science research for UC Santa Cruz, Dusheck embarked on a career as a freelance science writer. She is coauthor (with Allan Tobin) of the college biology textbook *Asking about Life*; contributing author to *Nature of Life*, by Postlethwait and Hopson, and to Holt's middle-school text *Life Science*; author of news stories for *Science, Nature, and Natural History*; and author of curriculum for low-income eighth-graders, open-source online texts for high school, and interactive digital learning for premed students.

# Contents

## 36  Plant Growth and Reproduction  776

# DISCOVER BIOLOGY

**Fifth Edition**

# The Nature of Science and the Characteristics of Life

**ARSENIC BACTERIA.** Geomicrobiologist Felisa Wolfe-Simon collects mud from the bottom of Mono Lake, California. Bacteria from the lake tolerate high concentrations of salt and arsenic.

# Earthbound Extraterrestrial? Or Just Another Microbe in the Mud?

At the end of 2010, NASA put the national science news media through a roller-coaster ride of emotions. On a Monday, the national space agency announced a Thursday press conference "to discuss an astrobiology finding that will impact the search for evidence of extraterrestrial life."

News reporters get press releases all day long, but this one sounded like major news about extraterrestrial life, if not on Mars or one of the moons of Jupiter, then maybe on one of the hundreds of planets outside our solar system. All week long, science bloggers and news reporters traded rumors. What had NASA found?

When Thursday finally rolled around, NASA finally made its big announcement, and the excitement turned to puzzlement. A researcher in California had discovered a bacterium that could grow in high concentrations of poisonous arsenic. Not only that, but the bacterium appeared to build toxic arsenic right into its DNA, which no other organism does. DNA in all organisms—from modern humans to ancient bacteria—is all the same, chemically speaking. But NASA's announcement suggested that this one bacterium was replacing the phosphorus in DNA with arsenic. It was comparable, they said, to something you'd find on another planet.

The news media ran with the story. The *Huffington Post* wrote, "In a bombshell that upends long-held assumptions about the basic building blocks of life, scientists have discovered a whole new type of creature: a microbe that lives on arsenic."

Felisa Wolfe-Simon, the NASA-funded researcher, told reporters, "What we think are fixed constants of life are not." And NASA's breathless press release—titled "*Get Your Biology Textbook… and an Eraser!*"—said that the discovery "begs a rewrite of biology textbooks by changing our understanding of how life is formed from its most basic elemental building blocks."

What exactly was NASA's claim? What are the implications of the discovery, both for life here on Earth and for life elsewhere in the universe?

But wait. Don't start erasing just yet. In this chapter we'll learn how biologists go about answering questions—that is, the basics of the scientific method—and about the characteristics of life.

**MAIN MESSAGE** The scientific method is an evidence-based system for generating knowledge about our world. All living organisms have some characteristics in common.

## KEY CONCEPTS

- Science is a body of knowledge about the natural world and an evidence-based process for generating that knowledge. A scientific fact is a direct and repeatable observation of a particular aspect of the natural world.

- Scientific inquiry begins with observations of nature. The scientific method involves generating and testing hypotheses about those observations. Biology is the scientific study of the living world.

- Hypotheses can be tested with observational studies, experiments, or both. A hypothesis cannot be proven with absolute certainty.

- A scientific theory is an explanation about the natural world that has been repeatedly confirmed in diverse ways and is provisionally accepted as part of scientific knowledge because it has stood the test of time.

- All living organisms are composed of one or more cells, reproduce using DNA, acquire energy from their environment, sense and respond to their environment, maintain their internal state, and evolve.

- Biological evolution is a change in the overall genetic characteristics of a group of organisms over successive generations. According to the theory of evolution by natural selection, characteristics that make a population better adapted to its environment will persist in that population.

- Life on Earth can be studied on many levels, from atom to biosphere.

**THIS BOOK IS ABOUT YOU,** the rest of the living world around you, and the intricate web that connects living beings to one another and to their surroundings. This exploration of the living world will contribute to scientific literacy: it will help you understand how science works and introduce you to some of the key principles of modern science. A multitude of controversies confronting society today require scientific deliberation, such as the definitions of when life begins and when it ends, the therapeutic use of embryonic stem cells, genetic testing and the confidentiality of personal genetic data, the taxing of "junk food" and the banning of trans fats in restaurants, the protection of endangered species, and the search for more sustainable sources of energy. Our opinions on these complex issues are often influenced by personal values and individual concerns, as well as by prevailing cultural values and commercial and political interests. A shared understanding of the underlying science offers the hope of rational debate and constructive social action on these complex issues.

**FIGURE 1.1 Curiosity Is at the Heart of Scientific Inquiry**

We begin this chapter with a look at science as a way of knowing and as a way of accumulating knowledge about the natural world. Next we turn our attention to **biology**, the scientific study of life, by asking what, exactly, is meant by that powerfully evocative word: "life." We will see that all living things, diverse though they are, are related and have certain characteristics in common, and that all living organisms are part of an interlinked pattern we call the hierarchy of life.

As we explore the grand story of life in the chapters to come, you may be reminded of things that held you spellbound as a child and that may intrigue you still. You will encounter questions you may have pondered while strolling through the woods, walking down a grocery aisle, waiting in a doctor's office, or setting the recycling bins by the curb. Are the dinosaurs still with us in the form of birds? Is your friend's fondness for hang gliding rooted in his DNA? Should you have your DNA tested to learn more about your ancestry or your risk for certain diseases? Why are big, fierce creatures like sharks and tigers less common than the sardines in the sea and spotted deer in the jungle? Should you go for organic veggies despite the higher sticker price, or is conventional produce OK? Are biofuels really ecofriendly? Will there be polar bears in the wild when your grandchildren are growing up? There is much to explore, to question, and to discover.

## 1.1 The Nature of Science

You probably asked a lot of questions when you were a child: What is that? How does it work? Why does it do that? We are driven by a deep-seated curiosity about the world around us, a tendency to ask questions that we seem to express most freely when we are children (**FIGURE 1.1**). Through the ages, that spirit of inquiry has been the main driving force behind science. Almost all of us lean toward science whether we realize it or not. The urge to observe and understand nature—from the tracks of heavenly bodies to the migration routes of game animals to the special qualities of native herbs and trees—clearly had survival value for our ancestors. For example, many ancient cultures based their annual calendar on the arc of the sun in the sky, which in turn dictated the agricultural cycle of sowing and reaping. The few surviving communities of hunter-gatherers have such a deep and practical familiarity with the animals and plants in their habitat that biologists are rushing to catalog and analyze the irreplaceable know-how of indigenous peoples before it disappears in the face of "modernization."

The many benefits of **technology**, which refers to the practical application of scientific techniques and principles, are obvious to every Internet-savvy teenager who tweets and texts as if her life depended on it. But more than a provider of technological wonders, creature comforts, lifesaving medicines, and food security, science is a way of understanding the world. The scientific way of looking at the world—let's call it scientific thinking—is logical, strives for objectivity, and values evidence over all other ways of discovering the truth. Scientific thinking is one of the most egalitarian of human endeavors because it is owned by no group, tribe, or nation, nor is it presided over by any human authority that is elevated above ordinary humans. And scientific thinking is by no means the domain of scientists alone. Most colleges and universities in North America require non–science majors to take at least some science classes with the expectation that graduates with a scientific way of looking at the world will have a well-rounded and socially productive outlook. For society at large, a scientifically literate citizenry translates into well-informed voters, wise consumers, and responsible stewards of the nation and our world.

## Science is a body of knowledge and a process for generating that knowledge

*Science* takes its name from *scientia*, one of the Latin words for "knowledge." Science is a particular kind of knowledge, one that deals with the natural world. By "natural world" we mean the observable universe around us—that which can be seen or measured or detected in some way by humans. Some aspects of the natural world cannot be observed directly but we can deduce their existence by observing their effects. For example, although electrons cannot be seen, and many events in the past history of life have not been directly observed by humans, the *effects* produced by atomic structures and evolutionary processes are readily detected, which is how we have learned so much about these phenomena. We can define **science** as a body of knowledge about the natural world and an evidence-based process for acquiring that knowledge. In describing the nature of science, we might say that scientific knowledge

- Deals with the natural world, which can be detected, observed, and measured

- Is based on evidence that can be demonstrated through observations and/or experiments
- Is subject to independent validation and peer review
- Is open to challenge by anyone at any time on the basis of evidence
- Is a self-correcting enterprise

As implied in the definition, science is much more than a mountain of knowledge about our world. Science is a particular system for *generating* knowledge. The processes that generate scientific knowledge have traditionally been called the **scientific method**, a label that originated with nineteenth-century philosophers of science. Despite the singular noun, the scientific method is not *one* single sequence of steps, or a set recipe, that all practitioners of the scientific method follow in a rigid manner. Instead, the term is meant to embody the core *logic* of how science works. Some people prefer to speak of the "process of science," rather than the scientific method. Whatever we call it, the procedures that generate scientific knowledge can be applied in a broad range of disciplines—such as social psychology and archaeology.

**FIGURE 1.2** shows a concept map of the scientific method. A *concept map* is a diagram illustrating how the components of a particular structure or organization or process relate to each other; concept maps help us visualize how parts fit together and flow from one another. With the concept map in hand, let's examine the elements of the scientific method more closely and consider some examples of the scientific method in action.

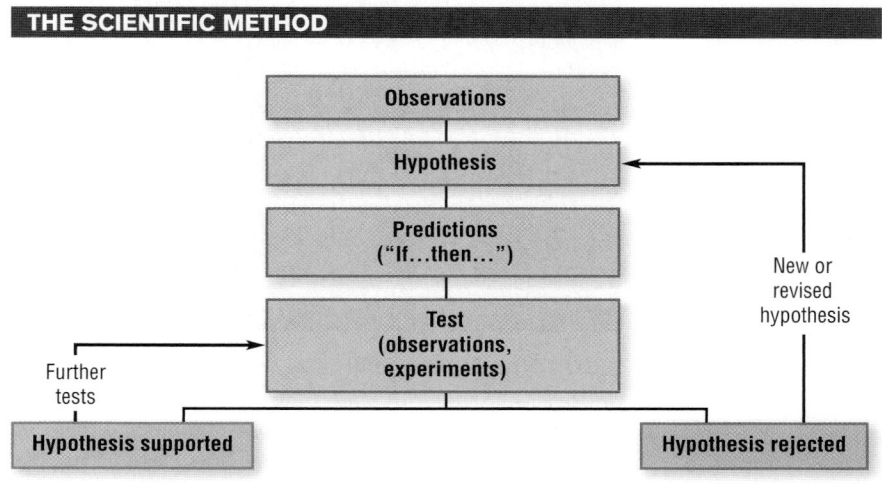

**THE SCIENTIFIC METHOD**

**FIGURE 1.2  Scientific Hypotheses Must be Testable**

## Observations are the wellspring of science

Scientific inquiry generally begins as an attempt to explain observations about the natural world. An **observation** is a description, measurement, or record of any object or phenomenon. A biologist can study nature in many different ways: by looking through a microscope, diving to the ocean floor, walking through a meadow, studying satellite images of forest cover, running chemical tests with sophisticated instruments, or using remote cameras to photograph a secretive animal. To be of any use in science, *an observation must be reproducible*; independent observers should be able to see or detect the object or phenomenon at least some of the time. Sightings of Sasquatch ("Big Foot"), lack credibility precisely because fans of "cryptobiology" have failed to produce samples or recorded images or sounds that stand up to scrutiny by independent observers.

Consider how an observation might lead to scientific inquiry. In the 1990s, fishermen, riverfront home owners, and state officials observed that huge numbers of fish were periodically being killed in mysterious die-offs in the rich and productive waterways of North Carolina's Albemarle-Pamlico estuary. (An estuary is the place where a river system drains into the sea.) Over successive summers, in different parts of the estuary system, millions of fish were found floating belly-up, their bodies covered with bleeding sores (**FIGURE 1.3**). What was causing these sporadic, unpredictable fish kills? The fish die-offs were an urgent economic concern for the hundreds of thousands of people employed in the fish and shellfish industry, and they distressed the many citizens who enjoy the beauty and recreational opportunities offered by the waterways of the Albemarle-Pamlico system, the second largest estuary in North America. As this chapter unfolds, we will describe how the scientific method led to a resolution of this whodunit.

## Scientific hypotheses must be testable and falsifiable

In science, just as in everyday life, observations generally lead to questions, and questions lead to potential

---

**THE SCIENTIFIC METHOD IN ACTION**

**The observations**
- Sporadic fish kills in estuarine rivers
- Water from fish kill site kills healthy aquarium fish.
- *Pfiesteria piscicada* multiplies in calm, warm water in the presence of live or dead fish or fish excreta.

**The hypothesis**
The toxic form of a single-celled organism, *Pfiesteria piscicada*, is the cause of the fish kills.

**The predictions**
*If* the hypothesis is correct, *then*
(1) toxic *Pfiesteria* is abundant at fish kill sites but not at other sites in the river system, and
(2) aquarium fish will be killed if exposed to purified toxic *Pfiesteria*.

(a)

(b)

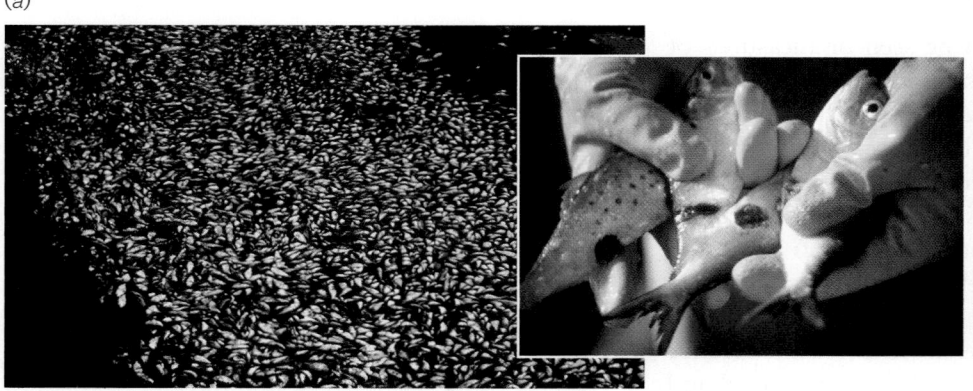

*Pfiesteria piscicada,*
a dinoflagellate

**FIGURE 1.3 Mystery of the Dying Fish: Observations, Hypothesis, and Predictions**
(*a*) Massive fish die-offs were seen in many North Carolina rivers in the 1990s. Approximately a million fish were affected, as this photograph taken in 1991 of Blount Bay, in the Pamlico Estuary of North Carolina, illustrates. Scientists observed bloody sores on the bodies of the dead fish (inset). (*b*) *Pfiesteria piscicada*, a single-celled dinoflagellate was determined to be the cause.

explanations. A **scientific hypothesis** is an informed, logical, and plausible explanation for observations of the natural world. We know what is plausible, or possible, if we have a good understanding of what is already known, which is why a new scientific hypothesis is often called an *educated* guess. Keeping established knowledge in mind keeps us from "reinventing the wheel" and helps us choose among alternative explanations to identify one that is not only possible but also, in our own judgment, the most probable.

A well-constructed hypothesis should be precise enough to make unambiguous predictions that can be expressed as "if-then" statements. Clear-cut predictions are absolutely critical for the next step in the scientific method: hypothesis testing. In science, a hypothesis is useless unless it is testable. In addition, a scientific hypothesis should be constructed in such a way that it is at least *potentially falsifiable*, or refutable. The core of the scientific method, scientific hypotheses

- Are educated guesses that seek to explain observed phenomena
- Make clear predictions that can be arranged in "if-then" statements
- Must be testable repeatedly and independently
- Must be potentially falsifiable
- Can never be proven, only supported

In other words, it should be possible to design tests able to demonstrate that a hypothesis is wrong. We shall see, however, that no amount of testing can *prove* a hypothesis with complete certainty.

Generating a hypothesis to explain an observation is not always easy. For some time, researchers were thoroughly stumped by the North Carolina fish kills, unable to come up with explanations of what might be causing them. Dr. JoAnn Burkholder, a professor at North Carolina State University, wondered if a tiny single-celled organism called *Pfiesteria* [**fih**-STEER-**ee-uh**] might be responsible. *Pfiesteria* is a dinoflagellate, an organism classified among the Protista, a large and diverse group of mostly single-celled organisms. As it happened, several years earlier some of Dr. Burkholder's colleagues had been dismayed to find that their laboratory fish were dying suddenly after exposure to local river water. When Dr. Burkholder looked into the problem, she found that a microscopic dinoflagellate, which she identified as *Pfiesteria*, greatly increased in numbers in the laboratory aquariums just before the fish started to die.

When the fish die-offs began happening in the wild, Dr. Burkholder came up with the hypothesis that the same tiny dinoflagellate, *Pfiesteria*, that appeared to have killed the laboratory fish was causing the fish die-offs in local rivers. Dr. Burkholder had several clues to go by. For example, she knew that this highly changeable organism—known to exist in 26 different forms—morphs into a dormant form and drops to the river bottom when the water is disturbed—for example, by a storm coming in from the sea. Interestingly, the fish kills invariably took place on warm days with calm waters.

No matter how creative and plausible the hypothesis, the *burden of testing a scientific hypothesis rests on the proponent* of that hypothesis. As a first step toward confirming her hypothesis, Dr. Burkholder made predictions that she could test. She asked herself, "*If* my hypothesis is correct and *Pfiesteria* is causing fish die-offs in local rivers, *then* what else can I expect to happen?" Her first prediction was that the fish-killing form of *Pfiesteria* would be found in abundance in the river water during times when fish die-offs were happening and would be scarce there when fish die-offs were not happening. Her second prediction was that the same *Pfiesteria* would be capable of killing healthy fish if introduced into the aquariums that housed fish in her laboratory.

Note that Dr. Burkholder's hypothesis and predictions are potentially falsifiable. If the population of *Pfiesteria* was found to *decline* before every fish kill, or exposure to *Pfiesteria* improved the health of various native fish, we would not only reject Dr. Burkholder's hypothesis as an inadequate explanation of the fish kills; we would know it to be a false statement. *Pfiesteria* would be exonerated.

## Hypotheses can be tested with observations or experiments or both

Testing the predictions of a hypothesis is a crucial step in the scientific method. We can test a hypothesis through observational studies or experimental studies. In studying nature, whether through observations or experiments or both, scientists focus on **variables**, which are characteristics of any object or individual organism that can change. Observations help us identify variables that are potentially interesting or that could potentially explain the patterns we might observe in nature. We can then design experiments to help us tease out the relationships between any two variables, or a small handful of variables, that we find interesting or that we believe may further our understanding of natural processes.

Observational studies can be purely *descriptive*, reporting information (**data**; singular "datum") about what is found in nature: where, when, how much. Mapping the types of sea creatures found in different zones on a rocky shore, listing the flowers in bloom through the growing season in an alpine meadow, counting how many birds of prey and how many perching birds are found on an island—all these are examples of descriptive observational studies. Observational studies can also be *analytical*—looking for patterns in nature and addressing how or why those patterns came to exist. A species of small barnacles is always found above the high-tide mark, but a related species is seen only low on the shoreline. Yellow-flowered plants bloom earlier than red-flowered ones in a particular alpine meadow. Perching birds always outnumber birds of prey. These are examples of patterns that may emerge when we analyze descriptive data. Observational studies usually rely on both descriptive and analytical data to test predictions made by a hypothesis.

The first prediction that emerged from Dr. Burkholder's hypothesis about *Pfiesteria* and fish kills could be tested through observational studies. Dr. Burkholder and her colleagues observed that *Pfiesteria* were indeed found swarming in river regions where fish were dying, but not in those same regions when fish were not dying—upholding her first prediction. But the question arises: Might it just be coincidence that *Pfiesteria* happened to be where the dead fish were? That is, might the *Pfiesteria* have proliferated just by chance? The researchers used the tools of **statistics**—a mathematical science that uses probability theory to estimate the reliability of data—to show it was very unlikely that the link between the fish and *Pfiesteria* was just a random event, an accidental occurrence. Nevertheless, it would be invalid to conclude from the data that *Pfiesteria* was *causing* the fish kills. After all, maybe something else—pollutants, for example—could be causing a population boom in *Pfiesteria* and that same thing (pollutants) could also be poisoning the fish. The most the researchers could conclude was that high levels of *Pfiesteria* were *correlated* with fish kills.

**Correlation** means that two or more aspects of the natural world behave in an interrelated manner: if one shows a particular value (large numbers of dead fish with open sores in a particular stretch of the river under certain weather conditions), we can predict a particular value for the other aspect (large numbers of fish-killing *Pfiesteria* in the same stretch of the river under the same weather conditions). But *correlation does not establish causation*. We cannot assume from the correlation that any one aspect of nature we are observing is the *cause* of a change in another aspect of nature. Observational studies suggest possible causes for a phenomenon (such as fish kills), but they do not *establish* a cause-effect relationship. Dr. Burkholder's observational studies bolstered her hypothesis but did not establish a causal link between an increase in *Pfiesteria* populations and mass killing of fish in the estuary.

## Experiments are the gold standard for establishing causality

One way that scientists test their predictions is by devising and conducting *experiments*. An **experiment** is a repeatable manipulation of one or more aspects of the natural world by experimenters. As noted earlier, any characteristic of any thing or any individual that is capable of changing is called a variable. The attributes of a corn plant that change through the growing season— plant height, number of leaves, number of cobs—are examples of variables. The changeable aspects of the plant's environment—amount of rainfall, hours of sunshine, number of cobs lost to raccoons—are also variables. Variables are often referred to as factors or conditions.

In a scientific experiment, the researcher typically manipulates a single variable, known as the **independent variable** or manipulated variable. Any variable that responds, or could potentially respond, to the changes in the independent variable is called the **dependent variable** or responding variable. If we think of the independent variable as the cause, then the dependent variable is the effect. In the simplest experimental design, the researcher manipulates a single independent variable (such as fertilizer applied to a cornfield) and tracks how that manipulation changes the value of one other variable (for example, total yield, measured by weighing all the corncobs at the end of the season). For the sake of efficiency, the experimenter may track some additional dependent variables at the same time (number of leaves, root mass, plant height), but technically, each of those responses constitutes a separate experiment (**FIGURE 1.4**).

Usually we manipulate a single independent variable in one experiment because if we manipulate two or more such variables (fertilizer application *and* water supply, for example), it is difficult to tell whether the dependent variable (plant height) is responding to the one variable (fertilizer) or the other variable (water supply) or to a complicated mix of the two (interaction between fertilizer application and water supply).

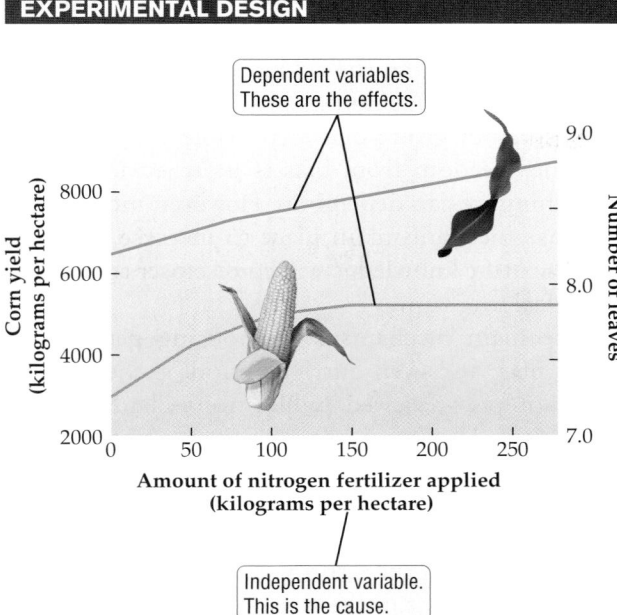

FIGURE 1.4 **Testing How Fertilizer Affects Corn Yield**
The graph shows the effect of applying nitrogen fertilizer on average corn yield over three seasons (left-hand *y*-axis). The right-hand *y*-axis shows the average number of leaves on 10 randomly selected plants from each season's crop 5 months after sowing. One hectare = 2.5 acres of land.

However, certain advanced statistical methods can be used to tease out the simultaneous effects that may be produced when there is more than one independent variable in a single experiment.

A controlled experiment is a common and particularly useful experimental design. In a **controlled experiment**, the researcher measures the value of the dependent variable for two groups of study subjects that are comparable in all respects except that one group is exposed to a systematic change in the independent variable and the other group is not. Typically, the researcher obtains a sufficiently large sample of study subjects—corn plants or aquarium fish, for example—and assigns them randomly to two groups. Randomization helps ensure that the two groups are comparable to start with. One group, the **control group**, is maintained under a standard set of conditions with *no change in the independent variable*. The other group, known as the experimental or **treatment group**, is maintained under the same standard set of conditions as the control group but is manipulated in a way that *changes the independent variable*. In a well-designed experiment, the control and treatment groups are as similar to each other as possible and all variables, other than the independent variable, are held constant.

To test the second prediction of her hypothesis—that aquarium fish would die if exposed to purified toxic *Pfiesteria*—Dr. Burkholder and her colleagues conducted controlled experiments. They used different species of fish maintained in aquariums under the same environmental conditions—same salinity (salt levels), same temperature, and same degree of disturbance, for example. Water obtained from fish kill sites caused rapid death in these aquarium fishes, but water from unaffected sites in the same river had no harmful effects. In one of many related experiments, researchers chose tilapia fish of similar size, age, and gender and assigned equal numbers to either a control group or a treatment group exposed to one of three different levels of the treatment. Three different concentrations of toxic *Pfiesteria*, purified from water collected at the fish kill sites, were added to tanks containing the three treatment groups, while the control group was not exposed to the dinoflagellate. The researchers counted the number of fish that had died after 16 hours—the dependent variable—in the treatment group and the control groups. The fish exposed to high levels of *Pfiesteria* developed open sores and died in as few as 5 hours, and the fish exposed to lower levels of *Pfiesteria* died at lower rates, but all the fish in the control group remained healthy for the duration of the experiment (**FIGURE 1.5**).

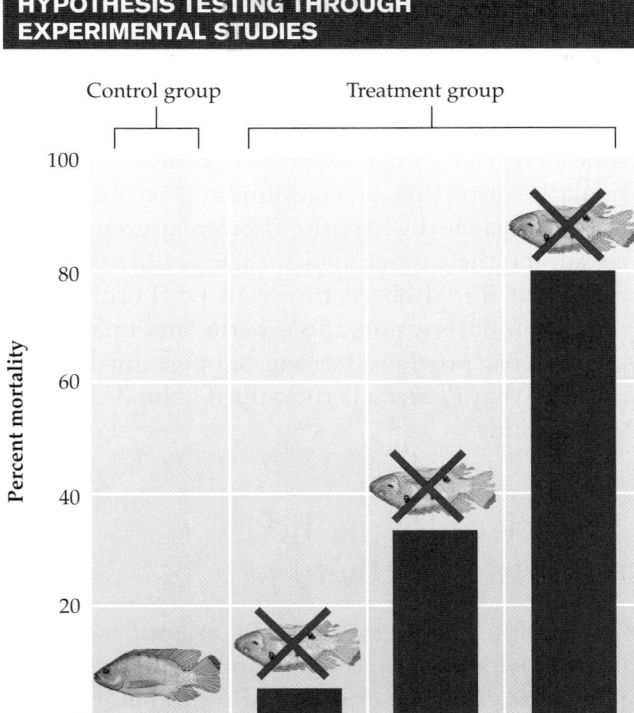

FIGURE 1.5 **Effect of Purified *Pfiesteria* on Survival of Aquarium Fish**
Researchers exposed a fish called tilapia to three different levels of the toxic dinoflagellate for 16 hours and counted the number of fish that died in that time span. The controls in this experiment were fish that were not exposed to *Pfiesteria* at all. All of the controls were alive at the end of the study period.

The observational studies produced results consistent with Dr. Burkholder's hypothesis, but the experimental studies furnished stronger evidence of causality. Not only was toxic *Pfiesteria* abundant at fish kills and uncommon at other sites, as the observational studies showed, but exposing healthy fish to toxic *Pfiesteria* from the fish kill sites caused the same symptoms and lethality as were observed in the wild fish during the fish die-offs.

## Testing does not prove a hypothesis with absolute certainty

When scientists test a prediction of a hypothesis and find it upheld, the hypothesis is said to be *supported*. A supported hypothesis is one in which scientists can be relatively confident, but they cannot say that the hypothesis has been proven true. Because a hypothesis could fail when subjected to a different test, no hypothesis can be proven true. For example, a cherished hypothesis might fail a retest because the outcome in the retest happens to be influenced by a variable that the experimenter was unaware of and therefore failed to hold constant. A person would have to know everything about everything to be completely certain that the outcome of her experiment would always be the same! Such certainty, of course, is not possible. Albert Einstein famously said, "No amount of experimentation can ever prove me right; a single experiment can prove me wrong."

When a prediction is not upheld, the hypothesis is reexamined and changed, or discarded. Sometimes, especially when experimental results are particularly surprising or confounding, scientists not only reconsider the hypothesis but may go back and reexamine their experiments, their predictions, and even their initial observations. In Dr. Burkholder's case, both observation and experiments upheld the predictions, providing strong support for the hypothesis that *Pfiesteria* is the culprit behind the massive fish kills.

## The scientific method requires objectivity

An absolute requirement of the scientific method is that evidence be based on observations or experiments or both. Furthermore, the observations and experiments that furnish the evidence must be subject to testing by others: independent researchers should be able to make the same observations, or obtain the same results if they use the same conditions. In addition, the evidence must be collected in as objective a fashion as possible—that is, as free of personal or group bias as possible. As you might imagine, freedom from bias is more an ideal than something we can depend on. However, modern science has mechanisms in place to increase the odds that scientific knowledge will come closer to meeting that ideal over time.

The main mechanism for policing personal or group bias, and even outright fraud, is the requirement for peer-reviewed publication. Claims of evidence that are confined to a scientist's notebook or the "blogosphere" do not meet the criterion of peer-reviewed publication. Peer-reviewed publications are scientific journals that publish original research only after it has passed the scrutiny of experts who have no direct involvement in the research under review. The experts who serve as referees are usually anonymous, and they could be competitors with a vested interested in finding fault with the work. Science is a social enterprise, and both the collaborative and competitive tendencies of scientists strengthen the everyday practice of science.

## The acceptance of scientific hypotheses is provisional

Like all other scientific studies, Dr. Burkholder's work raises as many questions as it answers, and her studies have generated much subsequent research. Scientists still are not sure exactly how *Pfiesteria* kills fish—by means of a toxin (a harmful or poisonous substance), by attacking the fish physically, or by another method. Dr. Burkholder hypothesized a toxin as the agent of destruction. Recent studies from other laboratories have found no evidence of a toxin coming from fish-killing strains of *Pfiesteria*. However, Dr. Burkholder maintains that these other labs tested the wrong strains of *Pfiesteria* and, in addition, handled them improperly. In other words, she questions whether their experiments were really a proper test. This kind of disagreement is not unusual at all and is a good illustration of how science advances through discussion and debate and further research stimulated by the exchanges.

One of the greatest strengths of science is that scientific knowledge is tentative and therefore open to challenge at any time by anybody. Far from being written in stone, or proven beyond the shadow of a doubt, even well-established scientific ideas can be overturned if new evidence against the prevailing view comes to light.

## A scientific theory is a body of knowledge that has stood the test of time

Outside of science, people often use the word "theory" to mean an unproven explanation. If something unusual occurs, someone might say, "I have a theory about how that happened." The theory could be anything from a wild guess to a well-considered explanation, but either way, it is "just a theory."

Scientists use the term "theory" to mean something very different. If an idea is merely one of many plausible explanations, it is a hypothesis. A *scientific theory*, however, is a hypothesis, or a group of related hypotheses, that have received substantial confirmation through diverse lines of investigation by independent researchers. Further, competing hypotheses are ruled out by common consensus. When experts in the field accept the surviving hypothesis, or set of related hypotheses, the theory becomes an accepted part of scientific knowledge because it has a very high probability of being correct. We can define a **scientific theo-**ry as a major component of scientific knowledge that has been repeatedly confirmed in diverse ways and is provisionally accepted by those who have knowledge of the discipline as the best available description of the truth about the phenomenon in question.

Scientific theories are not "iffy" ideas; they have such a high level of certainty that we base our everyday actions on them. For example, the germ theory of disease, formally verified by Robert Koch in 1890, is the basis for treating infections and maintaining hygiene in the modern world (**FIGURE 1.6**). Anthropogenic climate change is another example of a scientific theory. According to this theory, Earth's climate has warmed over the past century and that warming is caused mainly by human activities. The vast majority of scientists agree that this theory is well supported by many observations and scientific experiments from fields as diverse as climatology, geology, and ecology. The anthropogenic climate change theory is especially compelling because some of its predictions have been borne out. Climate models developed by proponents of the theory predicted the melting of glaciers in many parts of the world, as well as a reduction in the summer sea ice in the Arctic Ocean, well before these predicted outcomes could be reliably measured.

What about the term "facts"? In casual conversation, we typically use the term to mean things that are

---

**APPLYING THE GERM THEORY**

(a)

(b)

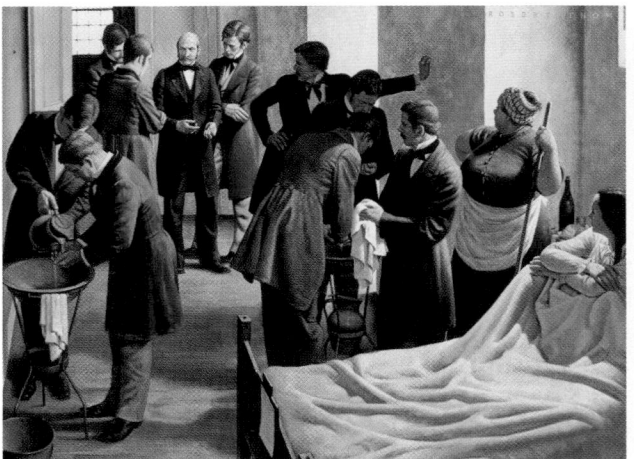 

**FIGURE 1.6  The Germ Theory of Disease Is a Scientific Theory**

The germ theory of disease holds that some diseases are caused by microorganisms, minute organisms visible only with a microscope. Application of the germ theory, through such measures as scrupulous hand washing (a), cut death rates in hospital wards by half in the late nineteenth century. The germ theory remains a cornerstone of modern medicine and hygiene, and people put this scientific theory into practice in everyday life (b). [Photo credit for (a) Robert Thom (Grand Rapids, MI, 1915–1979, Michigan). Semmelweis-Defender of Motherhood. Oil on Canvas. Collection of the University of Michigan Health System. Gift of Pfizer, Inc. UMHS.26.]

known to be true, for example, as opposed to things that are "just a theory." A **scientific fact** is a direct and repeatable observation of any aspect of the natural world. An example of such an observation might be that an apple, when dropped, falls to the ground—not up into the sky. Another fact is that fish were dying in huge numbers in North Carolina estuaries in the 1990s. A fact that some people find difficult to accept, even though it can be easily and reproducibly demonstrated, is that biological evolution happens: the inherited characteristics of groups of organisms on Earth change over time. The factual nature of biological evolution may become easier to appreciate after our discussion of the topic later in this chapter and especially in Unit 3.

How do facts in science relate to theories in science? In science, a theory, such as the climate change theory, is not mere speculation or an educated guess. Instead, a theory is based on decades of accumulated facts—confirmed observations. The climate change theory, for example, is based on direct measurements that show that Earth's climate has warmed significantly over the last century and continues to do so now. Likewise, although biological evolution is a fact—it can be observed over the course of a few days in lab dishes containing bacteria—the *mechanisms* that drive biological evolution are regarded as theories. Charles Darwin's theory that living organisms evolve *through the means of natural selection* is a scientific theory built on many facts, including the fact that groups of living beings change with respect to their inherited characteristics over successive generations. Massive amounts of observational and experimental data in the more than 150 years since Darwin have confirmed that natural selection can and does drive the evolution of life on Earth, but modern science has revealed other mechanisms that can also contribute to evolutionary change (more on this in Unit 3).

**Helpful to Know**

There are many academic disciplines within biology. Physiology is the study of how organisms function. Biochemistry focuses on the chemistry of life. Ecology studies interactions among organisms and with their surroundings.

## 1.2 The Characteristics of Living Organisms

The science that will occupy our attention in the rest of this book is biology, the science of life. But what is life? Though many have tried, no one has produced a simple, single-sentence, airtight definition of life that encompasses the great diversity of living forms—from massive redwoods to microscopic bacteria and every-

thing in between. But all living organisms are thought to be the descendants of a single common ancestor that arose billions of years ago. So, not surprisingly, living things share certain features that characterize life. All living organisms

1. Are composed of one or more cells
2. Reproduce using DNA
3. Obtain energy from their environment to support metabolism
4. Sense their environment and respond to it
5. Maintain a constant internal environment (homeostasis)
6. Can evolve as groups

Biologists recognize these shared features, detailed in the discussion that follows, as the key hallmarks of life.

## Living organisms are composed of cells

The first organisms were single cells that existed billions of years ago. The **cell** is the smallest and most basic unit of life, the fundamental building block of all living things. Cells are tiny, self-contained units enclosed by a water-repelling layer called the **plasma membrane**. All organisms are made of one or more cells. A bacterium consists of a single cell. In contrast, the human body is composed of approximately 10 trillion cells.

Large organisms, such as monkeys and oak trees, are made up of many different kinds of specialized cells and are known as **multicellular organisms**. The human body, for example, is composed of trillions of cells with specialized functions, including skin cells, muscle cells, cells that fight disease, and brain cells. Another way to visualize the cell as the fundamental unit of life is to think about how all living organisms are organized. In the human body, for example, cells are grouped into tissues that are organized into internal organs. For the body to function properly, those organs must be not only present but also arranged in a particular way spatially, with the stomach in its proper place, the brain positioned just so, and so on. Likewise, close examination of a flower reveals that the parts are far from randomly organized. A living organism must maintain its characteristic spatial organization to function properly. And the cell is the fundamental and most basic level of that organization.

# Science and the Citizen

For scientists and nonscientists alike, knowing how nature works can be exciting and fulfilling. In addition, applications of scientific knowledge influence all aspects of modern life. Every time we take medicine, text a friend, or run on a treadmill, we are enjoying the benefits of science.

The public is not simply a consumer of science and its spin-offs. The citizen can shape the course of science and influence what technology is used where and how. Before we examine the many ways in which the relationship between science and the citizen promotes social well-being, let's first consider what science *cannot* do.

## The Scientific Method Has Limits

As powerful as the scientific method is, it is restricted to seeking natural causes to explain the workings of our world. So there are areas of inquiry that science cannot address. The scientific method cannot tell us, for example, what is morally right or wrong. Science can inform us about how men and women differ physically, but it cannot identify the morally correct way to act on that information. Science cannot speak to the existence of God or any other supernatural being. Nor can science tell us what is beautiful or ugly, which poems are lyrical, or which paintings most inspiring. So although science can exist comfortably alongside different belief systems—religious, political, and personal—it cannot answer all their questions.

According to a 2010 poll by the Pew Research Center, 61 percent of the American public sees no conflict between science and their own beliefs. The same poll shows that 85 percent of the American public views science as having a mostly positive effect on society.

## Public-funded Research Contributes to the Advancement of Science

In North America, the vast majority of *basic research* in science is funded by the federal government—that is, by the taxpayers. Basic research is intended to expand the fundamental knowledge base of science. Many industries and businesses spend a great deal of money on *applied research*, which seeks to commercialize the knowledge gained from basic research. The new drugs, diagnostic tests, and medical technology that biomedical companies introduce each year are, ultimately, the fruit of the public investment in basic research.

In the United States, Capitol Hill appropriates more than $20 billion each year for basic research in the life sciences, including biomedicine and agriculture. These funds are disbursed mainly to four federal agencies: the National Institutes of Health (NIH), the National Science Foundation (NSF), the U.S. Department of Energy (DOE), and the U.S. Department of Agriculture (USDA). Some of these agencies have their own research institutes and laboratories, but each of them also awards funds to university researchers, who conduct the bulk of the basic research. Researchers must compete vigorously for the limited funds, and this competition helps ensure that the public money goes toward supporting high-quality science. How much money is allocated, as well as how the funding priorities are set, is strongly influenced by public opinion and even by social activism (as has been the case with HIV-AIDS research, breast cancer research, and, with more limited success, embryonic stem cell research).

## Scientific Literacy Strengthens Democracy

We are often called upon to vote on issues that have a scientific underpinning. The table lists some of the science-related ballot measures that have been put to the vote during state and local elections in the United States in recent years. Although our personal values and political leanings are likely to influence how we vote, most would agree that the underlying science should be taken into consideration.

## Some Statewide Ballot Measures[a] on Science-Related Issues

| INITIATIVE/REFERENDUM | STATE | YEAR INTRODUCED | INTENT OF PROPOSED INITIATIVE OR REFERENDUM |
|---|---|---|---|
| Proposition C | Missouri | 2008 | To require public utilities to generate 10% of their energy from renewable sources |
| Medical Marijuana Initiative | Maine | 2009 | To legalize sale of limited amounts of marijuana to patients with doctor's prescription |
| Proposition 23 | California | 2010 | To suspend the state's Global Warming Act (AB2), which puts limits on greenhouse emissions |
| Initiative 1107 | Washington | 2010 | To repeal the 2-cent sales tax on candy, soda pop, and bottled water |

[a]A ballot measure is a referendum or initiative that is put to the vote in state or local elections. A referendum originates with the state legislature, whereas an initiative is brought forward by a petition from citizens (who could be backed by special interests). Citizen initiatives are given different names in different states ("proposition," "issue," or "measure," for example), and not all states have a system of citizen initiatives.

## Living organisms reproduce themselves via DNA

One of the key characteristics of living organisms is **reproduction**, the ability to generate new individuals like themselves. The new individuals generated through reproduction are called **offspring**. Single-celled organisms, such as bacteria, can reproduce by dividing into two cells that are virtually identical copies of themselves. In contrast, multicellular organisms can reproduce in a variety of ways. Humans and other mammals, for example, can reproduce by having sex, in which a specialized reproductive cell in the male, called a sperm, joins with a female's specialized reproductive cell, an egg, in the process known as **fertilization**. A certain amount of time later—about 9 months in humans—the female gives birth to the offspring.

Many plants also use **sexual reproduction** to produce offspring: sperm cells (contained in pollen) are delivered to the female organ of the flower, where they proceed to fertilize the egg cells. The female organs mature into fruit containing seeds, which in turn contain miniature offspring produced by the fusion of egg and sperm. The miniature plant emerges as a seedling when the seed sprouts (germinates). Some multicellular organisms can multiply themselves through **asexual reproduction**—that is, without the involvement of specialized reproductive cells such as sperm and egg. Sponges can bud off new individuals, and many plants can produce side shoots that can break off and develop into new individuals.

**DNA**, or deoxyribonucleic [**dee**-*OX*-ee-*RYE*-**boh-noo**-*CLAY*-**ic**] acid, is a large and complex chemical that is stored in just about every cell of every living organism. DNA can be thought of as a set of instructions for building an organism. DNA is the hereditary, or genetic, material in that it transfers information from parents to offspring, which is why it is essential for reproduction. DNA is a complex molecule made up of many atoms held together in a ladderlike pattern and twisted into a spiral along its length—a structure known as the double helix (**FIGURE 1.7**). In the cells of plants and animals, DNA is housed inside a special structure called the **nucleus** (plural "nuclei"), which is bounded by its own membranes.

A **gene** is a segment of DNA that codes for a distinct genetic characteristic, such as having an O blood type or a dimpled chin. Human cells have 46 separate DNA molecules organized as **chromosomes**. Scattered along those 46 pieces of DNA are approximately 25,000 genes. Life, no matter how simple or how complex, uses this inherited genetic code to direct the

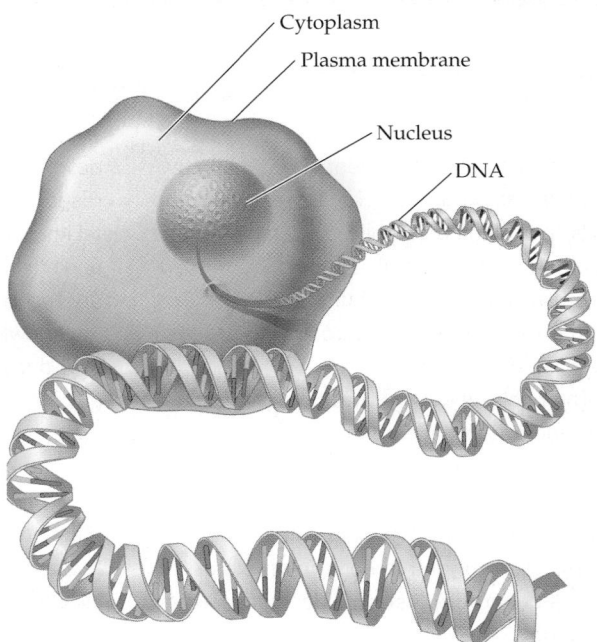

Cytoplasm

Plasma membrane

Nucleus

DNA

**FIGURE 1.7  DNA Is an Information-Storing Molecule**
Organisms pass genetic information to their young by transferring copies of DNA to their offspring.

structure, function, and behavior of every cell, as well as to transmit the same information from parent to offspring. If we think of a genetic characteristic as a computer application, then the gene is the particular software that runs that "app."

## Living organisms obtain energy from their environment

All organisms need energy to persist. Organisms use a wide variety of methods to capture this energy from their environment. The capture, storage, and use of energy by living organisms is known as **metabolism**. If living organisms exist anywhere else in our universe, they may be very different in chemistry, appearance, and behavior from earthly life-forms. But certain universal laws (theories of thermodynamics) enable us to predict that even extraterrestrial life-forms must possess metabolism.

Plants are among the organisms that can absorb the energy of sunlight, convert it into chemical energy, and then use that chemical energy to manufacture food, using a metabolic process known as **photosynthesis**. (We will discuss photosynthesis in detail in Chapter 9.) Some bacteria can also capture energy from the sun through photosynthesis. In addition, certain bacteria can harness energy from chemical sources such as iron or ammonia through an entirely different chemical reaction. And many organisms, including animals like us, fungi (mushroom-producing organisms and their kin), and certain single-celled organisms, gather energy by consuming other organisms (**FIGURE 1.8**).

Organisms that obtain metabolic energy from the *nonliving* part of their environment are called **producers** or **autotrophs**. Photosynthetic organisms, such as plants, are examples of producers that use light energy to drive the manufacture of food inside their cells. The food that producers make is acquired by consumers. **Consumers** (also called **heterotrophs**) are organisms that acquire food either directly from producers or from other consumers, whose energy ultimately derives from producers. Animals are a familiar example of consumers. A **food web** is a description of the nutritional relationship between all the producers and consumers in a particular natural environment, or habitat. A **food chain** is any simplified portion of a food web that portrays a sequence of who eats whom in a particular habitat.

## Living organisms sense their environment and respond to it

Living organisms sense many aspects of their external environment, from the direction of sunlight (as the sunflowers in Figure 1.8 illustrate) to the presence of food (**FIGURE 1.9**) and mates. Like humans, many animals can smell, hear, taste, touch, and see the environment around them. Some organisms can sense things that humans are not good at perceiving, such as ultraviolet light, electrical fields, and ultrasonic sounds. Some bacteria can even act like a living compass, sensing which direction is north and which direction is up or down by means of magnetic particles within them. All organisms gather information about their internal and external environments by sensing it, and then they respond appropriately for their continued well-being.

## Living organisms actively maintain their internal conditions

Living organisms sense and respond not only to external conditions, but also to their internal conditions. Most cells, and many multicellular organisms as well, maintain remarkably constant internal conditions—a process known as **homeostasis**. An example of homeostasis in humans is the maintenance of a constant internal body temperature of about 98.6°F. When

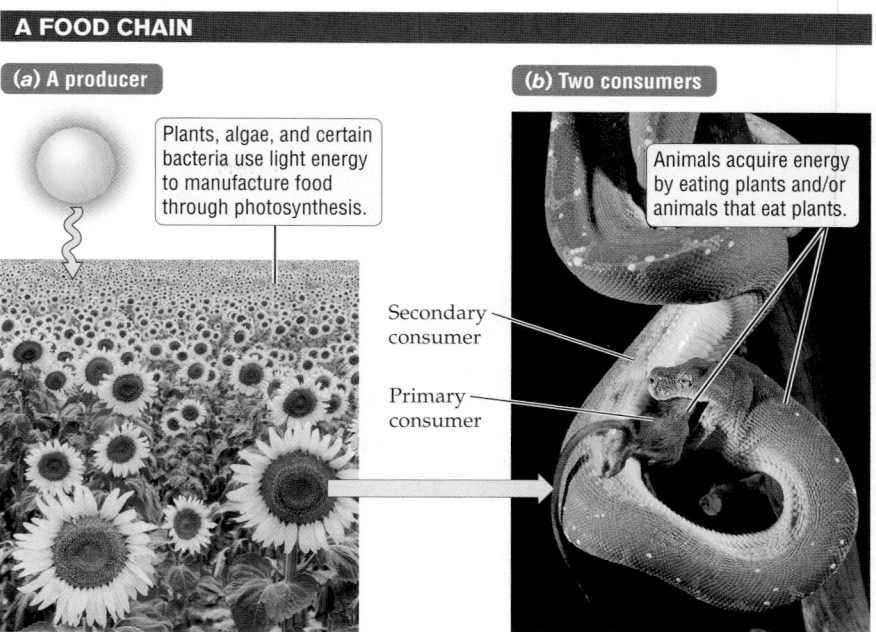

**A FOOD CHAIN**

**(a) A producer**

Plants, algae, and certain bacteria use light energy to manufacture food through photosynthesis.

**(b) Two consumers**

Animals acquire energy by eating plants and/or animals that eat plants.

Secondary consumer

Primary consumer

**FIGURE 1.8  Energy Capture by Organisms**
Whereas plants can capture energy from sunlight through photosynthesis (*a*), animals must get their energy by eating other organisms (*b*). The green tree python pictured in (*b*) is ingesting a source of energy (a rat) that itself derived energy from eating plants.

heat or cold threatens to alter our inner temperature, our bodies quickly respond—for example, by sweating to cool us or by shivering to warm us.

## Groups of living organisms can evolve

The fact of biological evolution has very likely been apparent to people since ancient times. As hunters know, for example, if they take down bucks with large antlers before the rut begins, many more males with smaller racks will get to mate than would have been the case without the hunters' intervention. The

**FIGURE 1.9
Organisms Sense and Respond to Their Environment**
This monarch butterfly detects suitable flowers by "smelling" the plant chemicals with its feet.

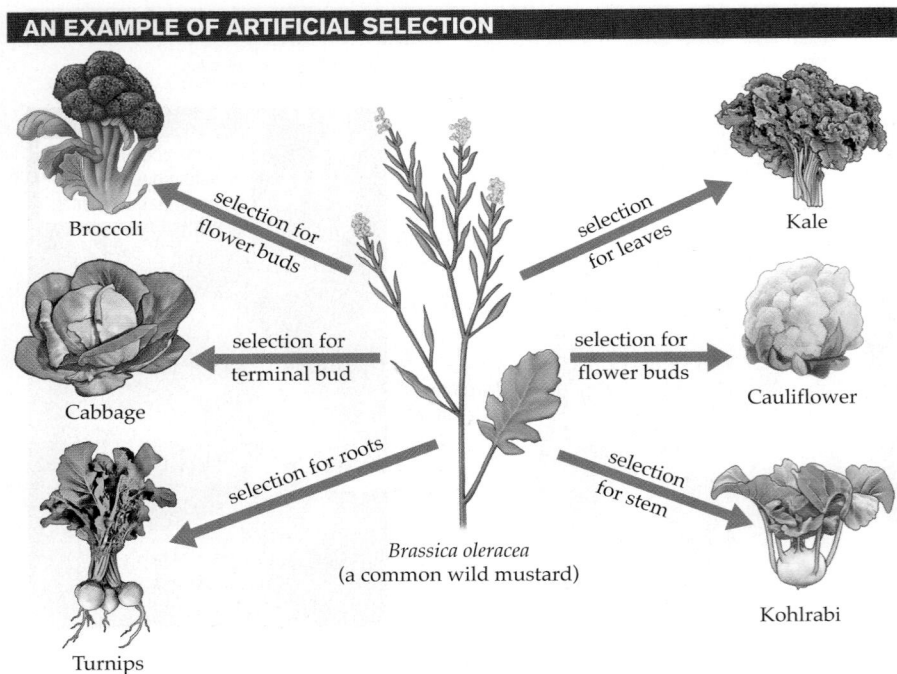

**FIGURE 1.10  Evolution of Wild Mustard through Domestication**
In artificial selection, humans are the selective agents. The many different varieties of cruciferous plants were created by humans through selective breeding of genetically varied wild mustard plants. Individuals with desired genetic characteristics were bred with each other until their descendants became distinctly different from the ancestor, the wild mustard.

consequence of this selective culling of large-antlered males is that in the next generation fewer fawns will grow up to sport large racks. When humans are responsible for a generational change in the overall inherited characteristics of a group of organisms, we call it **artificial selection**. Artificial selection is just one mechanism by which biological evolution can take place. Humans have not only recognized the fact of biological evolution; we have exploited the principles of biological evolution for thousands of years to shape the evolution of domesticated plants such as wheat, corn, and the mustard plant (**FIGURE 1.10**), as well as pets and livestock such as dogs, cats, and horses.

**Biological evolution** is a change in the overall genetic characteristics of a group of organisms over multiple generations of parents and offspring. By "overall genetic characteristics," we mean the sum total of all the inherited characteristics that are observable in the different individuals that make up a group of organisms. Antler size, coat color, running speed, and maternal care are examples of genetic characteristics that are readily observed. If we see a change in the commonness, or frequency, of one or more of these characteristics from one generation to the next, we say that the group as a whole has evolved. Notice that it is *groups* of organ-

isms that evolve, not an individual organism such as a certain white-tailed deer or a particular red oak tree.

A **species** consists of all organisms that interbreed in their natural surroundings to produce fertile offspring (that is, offspring that can themselves reproduce), but that do not, or cannot, breed with other organisms. For example, all pronghorn belong to a single species, as do all monarch butterflies, and all sugar maple trees. All species known to science are given two-part Latin names, such as *Antilocapra americana* for the pronghorn. A **population** is a group of organisms within a species that shares a common habitat, like all the pronghorn in the high plains of western Wyoming.

We can consider evolution at the level of a population (all the pronghorn in western Wyoming), or a whole species (the pronghorn, *Antilocapra americana*), or a larger group, such as ruminants (animals with four-chambered stomachs, such as cows and pronghorn), or the yet larger grouping of two-hoofed animals (which includes all ruminants, and deer and pigs too), and the still larger grouping of mammals (all animals that suckle their young, including all hoofed animals).

What would *cause* the overall genetic characteristics of a group of organisms to change over the generations? **Natural selection** is an evolutionary mechanism that changes the genetic composition of a population from one generation to the next by favoring the survival and reproduction of individuals best suited to their natural environment. A well-camouflaged pronghorn is better suited to its sagebrush habitat than is a poorly camouflaged animal, which is more likely to be picked out by predators. In a process known as **adaptation**, individuals with advantageous genetic characteristics, which match them to their habitat, survive and reproduce better than competing individuals that lack such characteristics. A genetic characteristic that leads to adaptation is an **adaptive trait**. Another commonly used term is "fitness"; an adaptive trait gives the individual possessing that trait a *fitness advantage* over individuals lacking the trait, meaning that the fitter individual enjoys greater success in survival and reproduction.

Natural selection is the mechanism that causes an adaptive trait (for example, camouflage coloring) to become more common in a population over the generations. Notice that although an adaptive trait is a characteristic of an individual, natural selection causes the *population as a whole* to become better adapted to its environment over the generations. By favoring individuals with adaptive traits over those with maladaptive ones, natural selection drives the evolution of the whole population by making adaptive traits more common in succeeding generations.

(a) Genetic variability

Original population

(b) Differential reproduction

(c) Adaptation

Next generation

Time

Some animals run faster than others.

Faster animals are more likely to survive and reproduce.

Fast runners are more common than in the original population. Therefore, this population has evolved from one generation to the next.

**FIGURE 1.11  Groups of Living Organisms Can Evolve**

Natural selection, exerted through predators looking for a meal, favors the survival and reproduction of fast pronghorn over the slower animals in the population. The next generation of pronghorn has evolved, and is better adapted to the presence of predators, because it contains many more fast runners than the previous generation had.

Pronghorn, which live on the dry plains and grasslands of western North America, are exquisitely adapted to their environment (Table 1.1). Being fleet of foot is one of their many adaptive traits for living in open country. Pronghorn are the fastest runners in the New World, capable of sprinting at 50 miles per hour (mph) and sustained running for many miles at 30 mph. By contrast, the African cheetah is exhausted after a 10-second dash at 70 mph. The speediness of pronghorn as a species is the result of natural selection. The original population of pronghorn must have been genetically varied with respect to running speed. The presence of predators led to differential survival and reproduction: swifter individuals were more likely to outrace their predators, whereas slower individuals were more likely to be eaten. Fast runners produced more offspring (because more of the swift pronghorn survived to have offspring), and the trait consequently became more common in the descendant population (**FIGURE 1.11**).

The evolution of fast-running pronghorn is an example of natural selection in which predators exert the selection, or act as the *selective agent*. Interestingly, there are no predators in the Americas now that can run at anywhere near the top speed of the average pronghorn. Scientists hypothesize that the American cheetah, now extinct but known from the fossil record of North America, acted as the agent of natural selection to drive the evolution of high-speed running in pronghorn.

The genetic variation in populations, and the genetic changeability of these populations, was noted by many naturalists in the eighteenth and nineteenth centuries. Charles Darwin, and his English compatriot Alfred Wallace, proposed a mechanism—natural selection—to explain how the overall inherited

**TABLE 1.1  Adaptive Traits in Pronghorn**

- Camouflage coloration
- High running speed
- Large eyes set high on the head for 320-degree field of vision
- Very large lungs and heart for sustained running
- Hollow hair for insulation against winter cold
- Strong grinding teeth (molars) for breaking up tough plant food
- Four-chambered stomach for digesting plant food with the help of bacteria
- Long intestines for digesting and absorbing nutrient-poor plant material

Charles Darwin
(1809–1882)

Alfred Wallace
(1823–1913)

characteristics of a population can change over long periods of time. Darwin recognized that, over time, two populations descended from a single ancestral species can become so different from each other that they lose the ability to interbreed and produce fertile offspring. In other words, the divergence in overall genetic characteristics causes two descendant populations to become distinctly different species. This insight led Darwin to the conclusion that all present-day species are descended "with modification" from ancestral species: humans and all other mammals are descended from a common ancestor that lived many millions of years ago, all animals are descended from a common ancestor that lived even further back in time, and all animals and plants have a common ancestor yet further back in the history of life.

Charles Darwin presented detailed and convincing evidence for natural selection in his 1859 classic, *On the Origin of Species*. Darwin proposed that in the struggle to survive and the contest to reproduce, individuals possessing certain inherited characteristics are favored by nature in that they produce more offspring than do individuals lacking the advantageous characteristics. Darwin offered natural selection as the mechanism by which nature exerts the favoritism that causes a population as a whole to become better adapted to its surroundings. Although we now know of other mechanisms that can cause a change in the overall genetic characteristics of a group of organisms, Darwin's theory of evolution through natural selection has been confirmed by thousands of observations and experiments over more than 150 years.

## 1.3 The Biological Hierarchy

One way of grappling with the vast complexity of the living world is to organize it according to the many levels at which it can be studied, from the minutest detail to the broadest and most all-encompassing level. The **biological hierarchy** is essentially a linear concept map for visualizing the breadth and scope of life, from the smallest structures that are meaningful in biology to the broadest interactions between living and nonliving systems that we can comprehend (**FIGURE 1.12**). The biological hierarchy has many levels of organization, ranging from *atoms* at the lowest level up to the entire *biosphere* at the highest level. In scale, the hierarchy ranges from less than one ten-billionth of a meter (the approximate size of an atom) to 12 million meters (the diameter of Earth).

At its lowest level, the biological hierarchy begins with **atoms**, which are the building blocks of matter, the material of which the universe is composed. Two or more atoms held together by strong chemical bonds become a **molecule**, the next level in the hierarchy. We use the term **biomolecules** to refer to molecules that are found in living cells. Carbon atoms are prominent in biomolecules, which is why we say that life on Earth is carbon based. DNA, the genetic material that carries the code for building an organism, is an example of a biomolecule. **Proteins**, which contain nitrogen atoms, are a large and very important category of biomolecules.

As noted earlier, the **cell** is the basic unit of life; and some organisms, such as bacteria, consist of only a single cell. Multicellular organisms also form tissues, the next level in the biological hierarchy. A **tissue** is a group of cells that performs a unique but fairly narrow set of tasks in the body. Plants and animals have many different types of tissues, each with unique functions. Nervous tissue, for example, performs the important function of transmitting electrical signals in the animal body. Muscle tissue can contract, enabling animals to move the body.

Plants and animals have **organs**, which are body parts composed of different types of tissues functioning in a coordinated manner to perform a broader range of functions than any one tissue can carry out on its own. An organ has a distinct boundary, a particular shape, and a specific location in the body. The heart and brain are examples of organs in the animal body. In animals, groups of organs are networked into **organ systems**, which extend through large regions of the body instead of being confined to a particular region of the body. An organ system performs a greater range of functions than a single organ does and, moreover, delivers services through much of the animal body. The stomach, liver, and intestines are organs that are integrated within the organ system known as the digestive system. All the organ systems in the animal body come together to work as a well-knit whole that we recognize as a single **individual**.

Each individual is a member of a **population**, a group of organisms of the same species living and interacting in a shared environment. As noted earlier, different populations that can interbreed are considered to be members of the same species; for example, all the cougars on Earth make up a single species, just as all the pronghorn are one species. The populations of different species that live and interact with one another in a particular environment are a biological **community**. The grasses, sagebrush, insects, pronghorn, coyotes, cougars, and eagles on the high plains in the Wind River Range form a community.

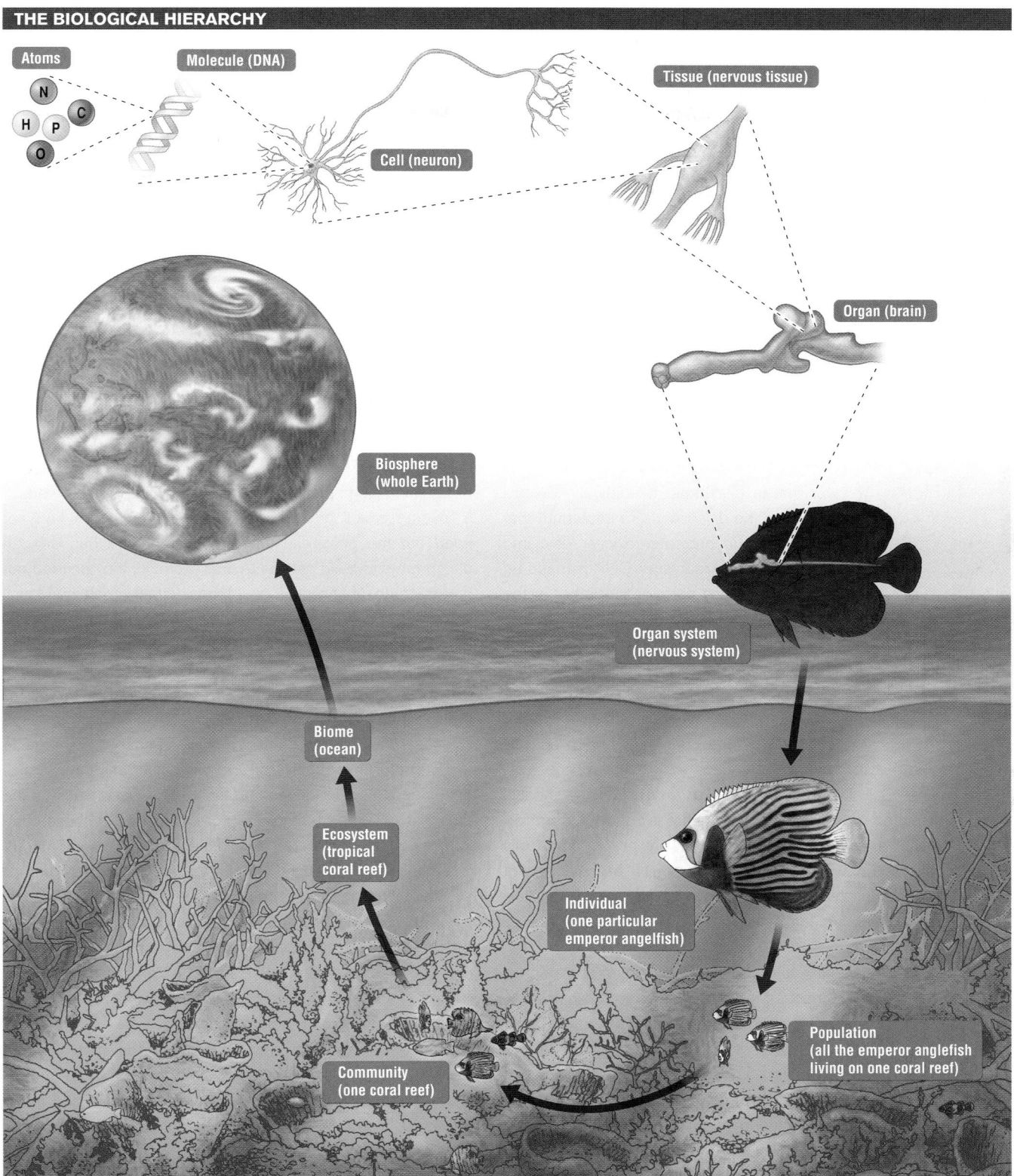

**FIGURE 1.12  The Biological Hierarchy Extends from the Atom to the Biosphere**
Levels of biological organization can be traced from atoms and molecules found in organisms all the way up through levels of increasing organization to the biosphere, which includes all living organisms and their nonliving environment. Aquarium enthusiasts may recognize the emperor angelfish pictured in the hierarchy.

A particular physical environment and all the communities in it together make up an **ecosystem**. For example, a river ecosystem includes the river itself, as well as the communities of organisms living in it and along its banks. At the next level are **biomes**, which are large regions of the world defined by shared physical characteristics, especially climate, and a distinctive community of organisms. The Arctic tundra is an example of a land-based (*terrestrial*) biome, and coral reefs are an example of a water-based (*aquatic*) biome. Finally, at the highest level of the biological hierarchy, all biomes become part of one **biosphere**, which is defined as all the world's living organisms and the places where they live.

> **Concept Check**

1. Explain what is wrong with the following statement: Over time, a male white-tailed deer evolves larger antlers.

2. Unscramble this scrambled biological hierarchy: community, organ system, ecosystem, atom, tissue, individual, biosphere, organ, cell, biome, population, molecule.

---

## APPLYING WHAT WE LEARNED

# Researchers Wrangle over Bacteria

Recall from the beginning of this chapter that in December of 2010, NASA announced the discovery of a bacterium that seemed to incorporate toxic arsenic into its DNA. In all organisms, DNA is made up of the same few parts put together in the same way, and DNA does not contain arsenic. But NASA's announcement suggested that this one bacterium was replacing the phosphorus in DNA with arsenic. It was a radical idea, but could it be true?

Felisa Wolfe-Simon and her colleagues had found a bacterium that could survive what for most organisms would be deadly amounts of toxic arsenic. In the lab, Wolfe-Simon grew the bacteria in increasing amounts of arsenic with almost no phosphorus. Without phosphorus, cells cannot make new DNA; and without new DNA, cells cannot divide and multiply. Yet despite being deprived of phosphorus, these particular bacterial cells continued to multiply. How were they doing it?

Arsenic has chemical properties like those of phosphorus, and Wolfe-Simon proposed that the bacteria were simply substituting arsenic for phosphorus. And if bacteria on Earth could pull off such a switch, there was no telling what life on other planets might do. Biologists say that all life on Earth requires six major elements: carbon, hydrogen, nitrogen, oxygen, sulfur, and phosphorus. If a bacterium could do without phosphorus, maybe none of the six were absolutely required for life. Such a conclusion meant that life might flourish on many more planets than previously thought.

It seemed like an eye-opening discovery, and the media played it up (despite the lack of actual extraterrestrials). But a number of scientists challenged Wolfe-Simon's conclusions, which were published at the same time in a peer-reviewed scientific journal. They wanted more evidence and wrote letters, tweets, and blogs that openly criticized Wolfe-Simon's work.

Some critics argued that Wolfe-Simon's team had not ruled out the possibility that the bacteria still had enough phosphorus to build new DNA. For example, they pointed out that the "phosphorus-free" test tubes actually contained small amounts of phosphorus from dying bacteria and other sources. Wolfe-Simon's team needed to determine if that was enough phosphorus for the bacteria to multiply, they said. Mono Lake, the lake where the bacteria live, contains large amounts of phosphorus. With such a rich supply of phosphorus, some scientists said, there could be no evolutionary advantage to evolving the unique ability to live without it. Still other critics demanded to see evidence that the arsenic was actually in the bacterial DNA and functioning normally. None of these challenges could be answered immediately.

"We cannot indiscriminately wade into a media forum for debate at this time," said one of Wolfe-Simon's coauthors on the paper, Ronald Oremland. "If we are wrong, then other scientists should be motivated to reproduce our findings. If we are right (and I am strongly convinced that we are) our competitors will agree and help to advance our understanding of this phenomenon. I am eager for them to do so."

NASA's press office may have hyped the story, but ultimately Wolf-Simon's hypothesis will be evaluated on its own merits. Science is a process of asking questions and trying to answer those questions. Wolfe-Simon and her colleagues took a chance on an interesting question: What do organisms actually need to live? Even if their research doesn't end up telling us that bacteria can live on arsenic instead of phosphorus, it will tell us something—perhaps about how bacteria deal so well with minimal phosphorus and large amounts of toxins like arsenic. The criticism by other researchers is a healthy part of the way science works.

# Poisoned Debate Encircles a Microbe Study's Result

BY DENNIS OVERBYE, *New York Times*

The announcement that NASA experimenters had found a bacterium that seems to be able to subsist on arsenic in place of phosphorus—an element until now deemed essential for life—set off a cascading storm of criticism on the Internet, first about alleged errors and sloppiness in the paper published in *Science* by Felisa Wolfe-Simon and her colleagues, and then about their and NASA's refusal to address the criticisms.

The result has been a stormy brew of debate about the role of peer review, bloggers and the reliability of NASA, at least as it pertains to microbiological issues, almost as toxic as the salty and arsenic waters of Mono Lake in California, from which Dr. Wolfe-Simon of the U.S. Geological Survey scooped up some bacteria last year.

Seeking evidence that life could follow a different biochemical path than what is normally assumed, Dr. Wolfe-Simon grew them in an arsenic-rich and phosphorus-free environment, reporting in the paper and a NASA news conference on Dec. 2 that the bacterium, strain GFAJ-1 of the Halomonadaceae family of Gammaproteobacteria, had substituted arsenic for phosphorus in many important molecules in its body, including DNA.

But the ink had hardly dried on headlines around the world when microbiologists, who have been suspicious of NASA ever since the agency announced that it had found fossils of microbes in a meteorite from Mars in 1996, began shooting back, saying the experimenters had failed to provide any solid evidence that arsenic had actually been incorporated into the bacterium's DNA.

In a scathing commentary on her blog RRResearch, Rosie Redfield, a microbiologist at the University of British Columbia, in Vancouver, ran through a long list of what she said were errors and omissions in the paper, which she summarized in the end as "lots of flim-flam, but very little reliable information."

Among other mistakes, she and others say, the experimenters failed to wash the bug's DNA before testing it for arsenic, thus leaving the possibility that the arsenic detected there was just stuck to the outside of the giant molecule, like mud on the bottom of a shoe, a process she described as "Microbiology 101." Over the course of a week her blog, which normally has a few hundred visitors a day, recorded almost 90,000 hits before the furor died down. She has also sent a letter to Science.

Things got nastier when NASA and Dr. Wolfe-Simon refused to respond to such criticisms, which quickly leapt from Dr. Redfield and others' blogs to Wired and Slate and The Observatory, a blog covering the science press for the Columbia Journalism Review. According to CBC News, Dwayne Brown, a NASA spokesman, said that the agency wouldn't debate science with bloggers and would stick to peer-reviewed literature . . .

In the interest of stimulating healthy debate, she [Dr. Wolfe-Simon] said, Science was making the paper available free of charge (although registration is required) for a couple of weeks. The experimenters are compiling a list of answers for frequently asked questions, which can be sent to gfajquestions@gmail.com and will eventually be posted online.

Dr. Redfield said Mr. Brown's reaction was silly. "We are the peers," she said. Conversation and arguing have always been an important part of how science has been done, she said. Once upon a time it was by mail, and was private and slow. "Now," she said, "the conversation is carried out in public in ways everyone can see."

In the nineteenth and twentieth centuries, scientists worked by themselves or in small groups. They corresponded through written letters that took days or even weeks to arrive, personal phone calls, and meetings that took place just a few times a year. Even as recently as 10 years ago, most biologists who disagreed with a paper's conclusions would have thought the matter over, talked to colleagues, written a considered critique, and submitted the letter to the editor of the scientific journal. If the letter were published at all, the original author would have been given time to respond so that the attack and the rebuttal appeared together.

Today, communications among scientists occur far more rapidly. Scientists are in near-constant communication with one another. A scientific paper goes online, and fellow researchers and casual readers alike may immediately respond with tweets, blogs, and comments. And, as on the rest of the Internet, these responses may be enthusiastic or riddled with cutting remarks.

## Evaluating the News

**1.** If Wolfe-Simon's conclusions are validated, will we need to change the list of the basic characteristics of life?

**2.** Wolfe-Simon initially refused to respond to online criticisms. Do you consider her reasons valid? Why or why not? Would your thinking and answer change if the topic were a financial or social issue? Explain.

**SOURCE:** *New York Times*, December 14, 2010.

# Summary

## 1.1 The Nature of Science

- Science is both a body of knowledge about the natural world and an evidence-based process for generating that knowledge.
- The scientific method represents the core logic of the process by which scientific knowledge is generated. The scientific method requires that we (1) make observations, (2) devise a hypothesis to explain the observations, (3) generate predictions from that hypothesis, and (4) test those predictions.
- We can test hypotheses by making further observations or by performing experiments (controlled, repeated manipulations of nature) that will either uphold the predictions or show them to be incorrect.
- In a scientific experiment, the independent variable is the one that is manipulated by the investigator. Any variable that can potentially respond to the changes in the independent variable is called the dependent variable.
- Correlation means that two or more aspects of the natural world behave in an interrelated manner: if one shows a particular value, we can predict a particular value for the other aspect. However, correlation betweeen two variables does not necessarily mean that one is the cause of the other.
- A hypothesis cannot be proven true; it can only be upheld or not upheld. If the predictions of a hypothesis are not upheld, the hypothesis is rejected or modified. If the predictions are upheld, the hypothesis is supported.
- A scientific theory is a major idea that has been supported by many different observations and experiments.
- A scientific fact is a direct and repeatable observation of any aspect of the natural world.

## 1.2 The Characteristics of Living Organisms

- All living organisms have these characteristics in common: (1) They are built of cells; some are single-celled and some are multicellular. (2) They reproduce, using DNA to pass genetic information from parent to offspring. (3) They take in energy from their environment. (4) They sense and respond to their environment. (5) They maintain constant internal conditions. (6) They can evolve as groups.
- Natural selection is a key evolutionary mechanism: it changes the overall genetic characteristics of a population over successive generations by favoring the survival and reproduction of individuals that are best suited to their environment. Natural selection makes a population or species better adapted to its environment.

## 1.3 The Biological Hierarchy

- The biological hierarchy refers to the many levels at which life can be studied: atom, molecule, cell, tissue, organ, organ system, individual, population, community, ecosystem, biome, biosphere.
- The individuals of a given species in a particular area constitute a population. Populations of different species in an area make up a community. Communities, along with the physical habitat they live in, constitute ecosystems.
- Ecosystems make up biomes, large regions of the world that are defined by the climate and the distinctive communities found there. All the biomes on Earth make up our one single biosphere.

# Key Terms

adaptation (p. 16)
adaptive trait (p. 16)
artificial selection (p. 16)
asexual reproduction (p. 14)
atom (p. 18)
autotroph (p. 15)
biological evolution (p. 16)
biological hierarchy (p. 18)
biology (p. 4)
biome (p. 20)
biomolecule (p. 18)
biosphere (p. 20)
cell (p. 12)
chromosome (p. 14)
community (p. 18)

consumer (p. 15)
control group (p. 9)
controlled experiment (p. 9)
correlation (p. 8)
data (p. 8)
dependent variable (p. 8)
DNA (p. 14)
ecosystem (p. 20)
experiment (p. 8)
fertilization (p. 14)
food chain (p. 15)
food web (p. 15)
gene (p. 14)
heterotroph (p. 15)
homeostasis (p. 15)

independent variable (p. 8)
individual (p. 18)
metabolism (p. 14)
molecule (p. 18)
multicellular
  organism (p. 12)
natural selection (p. 16)
nucleus (p. 14)
observation (p. 6)
offspring (p. 14)
organ (p. 18)
organ system (p. 18)
photosynthesis (p. 14)
plasma membrane (p. 12)
population (p. 16)

producer (p. 15)
protein (p. 18)
reproduction (p. 14)
science (p. 5)
scientific fact (p. 12)
scientific hypothesis (p. 7)
scientific method (p. 5)
scientific theory (p. 11)
sexual reproduction (p. 14)
species (p. 16)
statistics (p. 8)
technology (p. 5)
tissue (p. 18)
treatment group (p. 9)
variable (p. 7)

# Self-Quiz

1. A scientific hypothesis is
   a. an educated guess explaining an observation.
   b. a prediction based on an observation.
   c. a scientific theory.
   d. an idea that can be proven beyond all doubt through experimentation.

2. A scientific theory
   a. is so well established that it is not open to challenge on the basis of new evidence.
   b. is a major explanation that has been supported by many and diverse lines of evidence.
   c. is considered a scientific fact in that it is a direct and repeatable observation of nature.
   d. is a scientific hypothesis that can be tested through experimental but not observational studies.

3. Of the steps listed here, which one would come immediately before an investigator who is following the scientific method rejects her hypothesis?
   a. Making a prediction.
   b. Developing a new or revised hypothesis.
   c. Making observations of nature.
   d. Conducting controlled, repeated manipulations of nature.

4. Which of the following questions could *not* be used to develop a testable hypothesis?
   a. Can arsenic replace phosphorus in the DNA of a bacterium?
   b. Is there a relationship between high consumption of soda pop and obesity?
   c. Should smokers pay the same premium for medical insurance as nonsmokers?
   d. Do imported grapes have higher levels of pesticides than grapes grown in the United States?

5. Which of the following is manipulated by the experimenter in a controlled experiment?
   a. confounding variable
   b. control group
   c. dependent variable
   d. independent variable

6. Which of the following is a property shared by all life-forms on Earth?
   a. use of DNA for reproduction
   b. presence of a cell wall composed of long chains of sugars
   c. storing of genetic information in the nucleus
   d. ability to capture energy from the nonliving environment for all metabolic needs

7. Natural selection
   a. is a random process in which some individuals survive and others die off over time.
   b. is readily observed within a single generation of individuals.
   c. tends to make a whole population better adapted to its surroundings.
   d. occurs at the level of populations but not at the level of species.

8. An organ in the human body
   a. is the basic unit of life.
   b. has a particular shape and unique location in the body.
   c. consists of a single tissue type.
   d. is composed of two or more organ systems.

9. The biome
   a. consists of two or more tissues conducting specialized functions in an integrated manner.
   b. encompasses all organisms on Earth plus their environment.
   c. extends over large regions of Earth that share similar climate and plant communities.
   d. consists of members of one species that share the same habitat.

# Analysis and Application

1. Describe three characteristics of science as a way of knowing. Can science answer all types of questions that humans might raise? Explain.

2. What is meant by the statement: correlation is not causation? Give an example of two variables that you know are correlated but where neither one is likely to be the cause of the other.

3. Describe one observation, one hypothesis, and one experiment from JoAnn Burkholder's work.

4. What is a scientific fact? How does it differ from a scientific theory?

5. Consider a biological community that includes grasses, lions, sunshine, antelope, and ticks (small blood-sucking animals related to spiders). Identify the producers and consumers and arrange them in a food chain.

6. What are the levels of the biological hierarchy? Arrange them in their proper relationship with respect to one another, from smallest to largest. Give an example for each level that you know from your own experience.

7. The manufacturers of toning shoes make a variety of fitness claims for their products, most promising that you will burn more calories and improve muscle function if you wear the shoes regularly. Form small groups with your classmates and have each group pick a different line of toning shoes. Study the advertising for your toning-shoe brand, and then design an experiment to test the claims made the manufacturer. You can probably come up with some creative ways for measuring calorie consumption, and you are allowed any imaginary gadgetry (a "muscle tonometer," for example) to measure the function of any tissue or organ system. As a group, critique each other's experimental design. Is the hypothesis stated clearly? Does the prediction follow from it? Does the experiment test the prediction, with the least possibility of ambiguity? Can tests falsify the hypothesis? Are the test results likely to be reliable enough that average American consumers can base their buying decisions on them? Now compare your experimental design with one conducted by professional researchers sponsored by the American Council on Exercise (available free at www.acefitness.org) or by *Consumer Reports*.

# 2 Biological Diversity, Bacteria, and Archaea

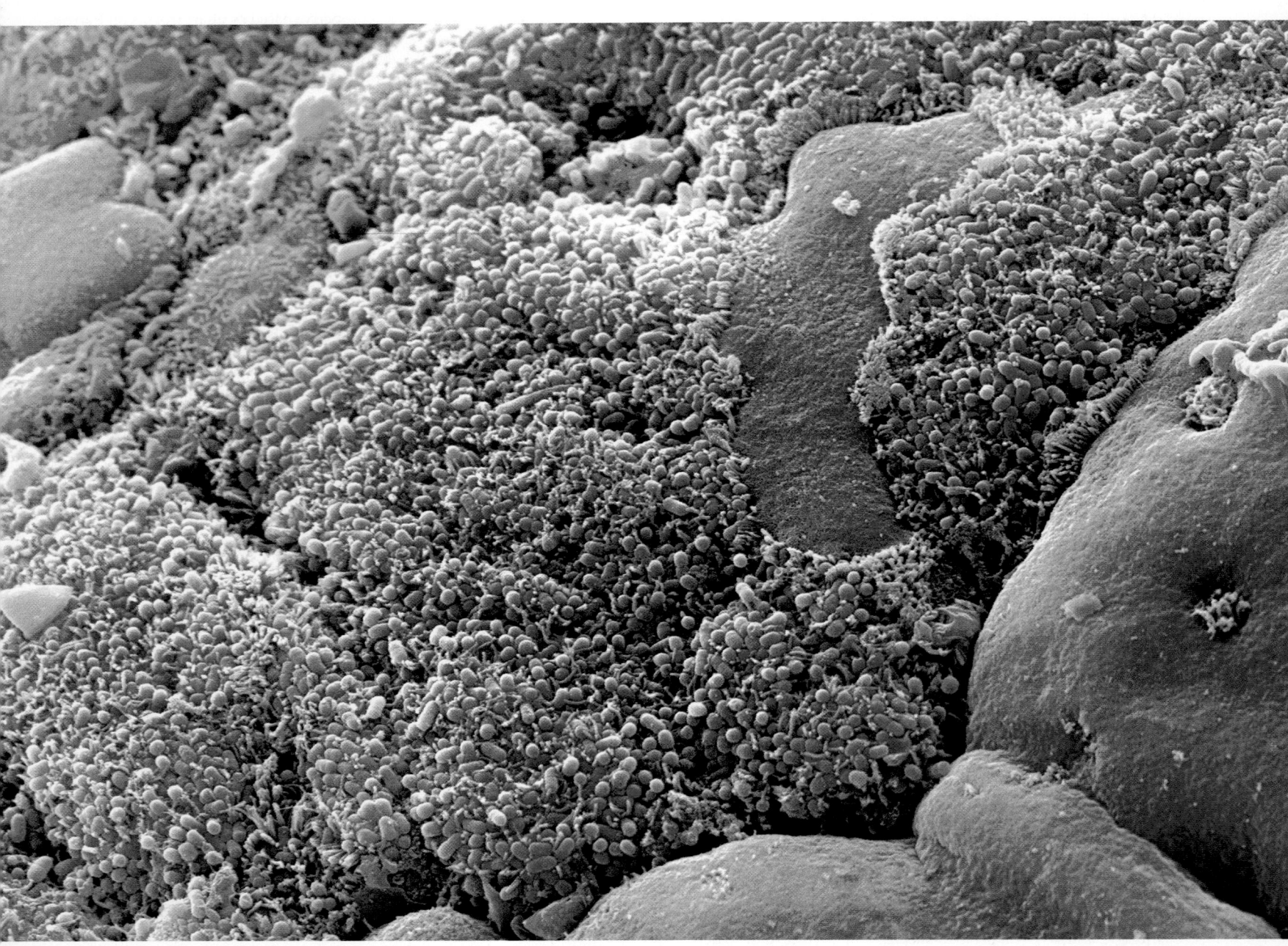

**GUT BACTERIA.** Microscopic bacteria called *Escherichia coli* (*E. coli*), shown here in blue and purple, grow in the human small intestine (green) as part of a community of hundreds of kinds of microorganisms.

# A Hitchhiker's Guide to the Human Body

There's more to you than meets the eye. The next time you're feeling lonely, think about the fact that of the trillions of cells in your body, about 90 percent belong to other individuals, living everywhere on your skin and deep inside your intestines. In the furthest reaches of your gut live some 10 trillion individual bacteria, archaeans, and fungi.

Even more live higher in the gut, as well as in your mouth and in every crevice of your skin. Some of these organisms inhabit moist tropical regions such as your underarms, while others specialize in oily habitats like the sides of your nose. The greatest biodiversity of skin bacteria and fungi lie not in any of the places you might think, but on your forearms. The least diverse region is the ecological desert that is the back of your ears.

Taken together, about a thousand different species of microbes live in you and on you—several hundred species in your mouth alone. Most of these tiny companions are bacteria or archaeans. But some are yeasts (fungi), not to mention a smattering of protozoans and even arthropods like the tiny mites that live, mate, and die in the roots of your eyelashes.

Virtually all organisms share their bodies with other organisms, in various kinds of symbiosis—the shared existence of two or more species. Most of the bacteria and other organisms that live with us are harmless; many are essential to good health. For example, many of those in the gut help digest food and make vitamins. And many bacteria act as bouncers, shouldering aside harmful organisms.

Where does this huge diversity of organisms hide in our bodies? How do we know which are good and which are bad?

All these organisms constitute the human flora. To better understand the difference between germs that we need and germs that may harm us, let's take a look at the huge diversity of bacteria. Before we begin, we'll consider how biologists organize and classify living organisms.

**MAIN MESSAGE** Of the three domains of life, Bacteria and Archaea are the most ancient, most diverse, and most abundant.

## KEY CONCEPTS

- Biologists look for shared inherited similarities to decipher evolutionary relationships among groups of organisms.

- The three-domain system organizes all life on Earth into three large categories, or domains: Bacteria, Archaea, and Eukarya.

- The Linnaean taxonomy classifies life into a hierarchy of groups, placing each species into a series of ever-larger groups— families, phyla, kingdoms, and so on. At least six kingdoms of life are commonly recognized: Bacteria, Archaea, Protista, Fungi, Plantae, and Animalia.

- Bacteria and Archaea are single-celled organisms informally known as prokaryotes. Prokaryotes are the most numerous in terms of number of individuals and diversity of types. They are the most widespread of all living groups and are essential to the web of life as both producers and consumers. Bacteria are especially important as decomposers.

- Viruses are not classified into any kingdom or domain. Viruses are infectious agents containing genetic material, often DNA, wrapped in protein layers. Viruses lack cellular organization and independent metabolism.

**FIGURE 2.1 How Many Organisms Do You See Here?**

Can you distinguish the nonliving from the living in these photos? Can you classify any of these organisms as plants or animals? Some of the organisms pictured here are single-celled, others are multicellular. Some are microscopic (not readily seen without a microscope), but most are visible to the naked eye. For the answers, turn to page 28.

**YOU MAY HAVE ENCOUNTERED SOME LIVING THING** somewhere that made you wonder, what in the world is that? If you cannot recall such a moment, we invite you to look at **FIGURE 2.1** and play a game of "Animal, Vegetable, Mineral, or Thing." Can you identify the living organisms in the portrait gallery? Which ones would you classify as plants? Which are animals? What criteria did you apply to decide that something was a plant, not an animal? Do you think some of these organisms should be placed in yet other categories, apart from plants or animals? What might those categories be? How would you set the boundaries of those categories; in other words, what characteristics must an organism display to be placed in the groupings you have in mind? And in lumping organisms into a particular category, are you going by their appearance, or by the way they obtain energy from their environment, or by their evolutionary descent from a recent common ancestor?

Biologists are faced with much the same puzzle when they observe the bewildering variety of life. In this chapter we begin with some thoughts about the common heritage and dazzling diversity of life on Earth. We follow that introduction with a description of the classification scheme known as the *Linnaean hierarchy*, which was begun in the 1700s and still finds use today. We will see that the goal of modern classification schemes is to map all known organisms on the **tree of life**, a branched, treelike diagram that depicts the evolutionary history of life from its origin to the present time. Finally, we take a closer look at two of the earliest and most awe-inspiring branches on the tree of life: Bacteria and Archaea. We will note that not only are these single-celled groups astonishingly diverse and remarkably successful, but some of them are champions of extreme living.

## 2.1 The Unity and Diversity of Life

The enormous variety of life on Earth is still far from being completely known, counted, or named. No chapter, including this one, could provide a comprehensive examination of all the world's species. Only a

**ANIMAL, VEGETABLE, MINERAL . . . ?**

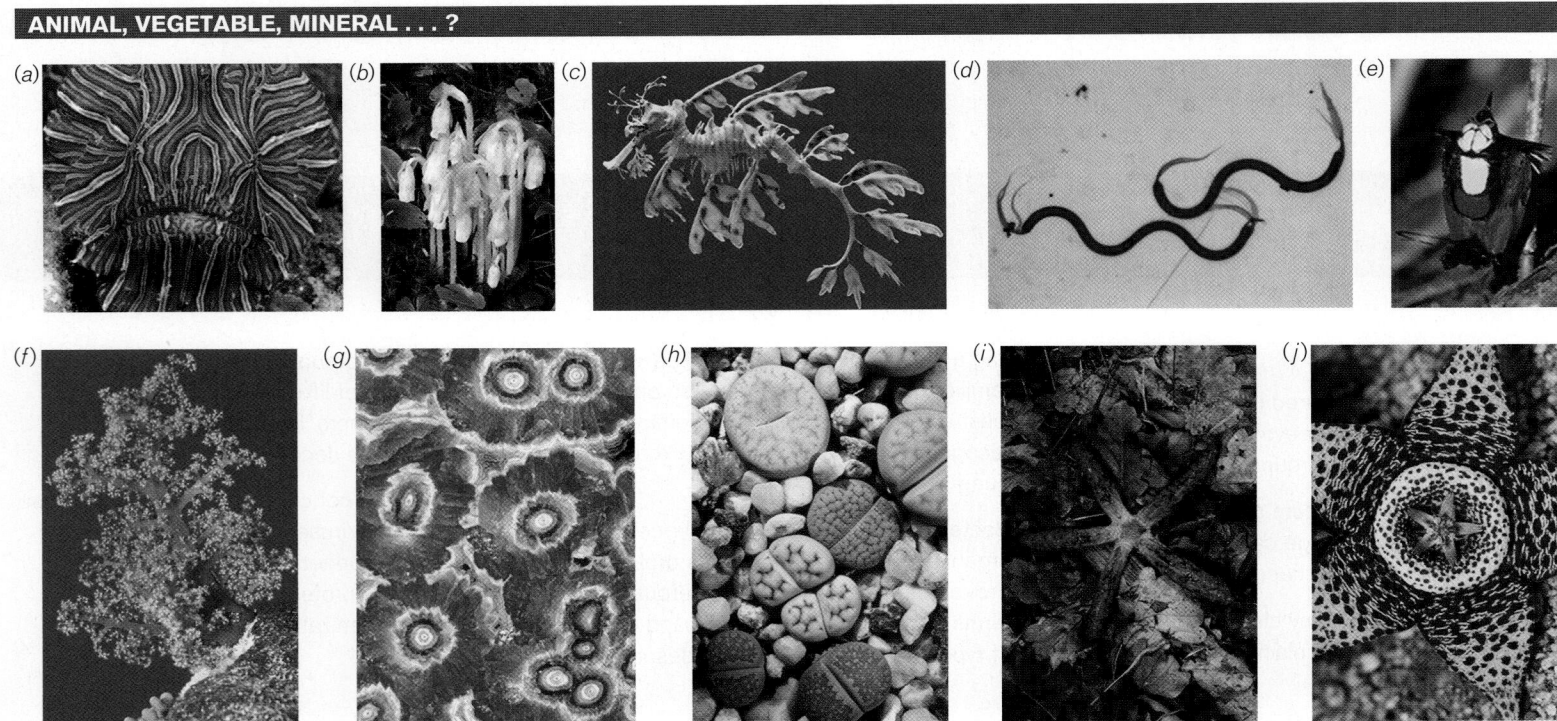

tiny fraction of the astounding array of living things can be introduced in our survey of the major groups of organisms in this and the next two chapters. Before we begin, we will orient ourselves by considering how biologists organize the major groups and the rationale behind the classification schemes in use today.

## A common origin explains the unity of life on Earth

At the time Earth formed, some 4.5 billion years ago, the planet was lifeless. Going by chemical clues, life may have evolved by 3.5 billion years ago. However, the first sure signs of cellular life are fossilized bacteria from the Gunflint Formation of Ontario, Canada. This rock formation is about 2 billion years old. The question of how life arose from nonlife is one of the greatest riddles in biology, but scientists have little doubt that all life on Earth is descended, with evolutionary modification, from a common ancestor similar to the fossil bacteria from the ancient Canadian rocks.

As noted in Chapter 1, all living organisms share a basic set of characteristics. Life shares this set of common properties because all living organisms descended from a common ancestor, also known as the universal ancestor. This hypothetical ancestral cell is placed at the base of the tree of life. Bacteria and another single-celled group, Archaea, are among the earliest branches on the tree of life. All other organisms are placed in a third category, known as the Eukarya, which includes us and all animals and plants. Archaea, Bacteria, and Eukarya constitute the broadest groupings of life, known as **domains**. **FIGURE 2.2** depicts an *evolutionary tree* showing the descent of these three basic groups from a universal ancestor. An **evolutionary tree** is a branching diagram that models evolutionary relationships among groups of organisms according to similarities and differences in their DNA, in their physical features, in their biochemical characteristics, or in some combination of these.

## Evolutionary divergence explains the diversity of life on Earth

Life on Earth evolves. Since life on Earth began, the universal ancestor has given rise to many lines of descent, or **lineages**, that have evolved into millions of different types of organisms, many of them now extinct (see the box on page 29 for a discussion of how quickly species can be decimated).

An evolutionary tree maps the relationship between ancestral groups and their descendants, the way your family tree describes your relationship to your mother, grandmother, and great grandmother. An evolutionary tree clusters the groups that are most closely related by common descent on neighboring branches, the way you and your siblings would be depicted on one branch of your family tree and your cousins on another branch.

In an evolutionary tree, the organisms under consideration (Bacteria, Archaea, and Eukarya in Figure 2.2) are depicted as if they were leaves at the tips of the tree branches. A **node** marks the moment in time when an ancestral group split, or diverged, into two separate lineages (such as Archaea and Eukarya). The node represents the **most recent common ancestor** of the two lineages in question—that is, the most *immediate* ancestor that *both* lineages share. A given ancestor and *all* its descendants make up a **clade**, or branch, on the evolutionary tree. Archaea and Eukarya and their most recent common ancestor make up a clade, but Bacteria and Archaea and their most recent ancestor do not constitute a clade, because such a grouping would leave out some of the descendants—namely, Eukarya. To continue the analogy to a family tree, you and your siblings and your mother would constitute a clade on your family tree, but any cousins on your mother's side would not be part of the same clade. If we go further back in history, however, and define your maternal grandmother as the ancestor of interest, then she, your

Bacteria-like shapes have been found in stromatolites 3.4 billion years old. However, some researchers say the preservation is too poor for reliable identification. Stromatolites, such as these modern-day forms in Shark Bay, Western Australia, are pillowlike deposits of sediments that can trap bacterial biofilms.

**EVOLUTIONARY TREE**

Present   Bacteria          Archaea                    Eukarya

This is the most recent common ancestor of Archaea and Eukarya.

Each node marks the divergence of an ancestor into two different lineages.

The horizontal lines have no particular meaning. They serve simply to facilitate reading of the diagram.

Past          Universal
              ancestor

**FIGURE 2.2**
**Evolutionary Tree of Domains**

This tree shows one model of the relationships of the three domains. At the root of the tree is the universal ancestor, from which all living things descended. Of the three surviving lineages, the first split came between the Bacteria and the lineage that would give rise to the Archaea and Eukarya. The next split was between the Archaea and the Eukarya, making Archaea and Eukarya more closely related to each other than either group is to Bacteria.

mother, your maternal aunt, and all her kids would be in the same clade as you and your siblings.

How do we know which group of organisms belong where on an evolutionary tree—which "leaves" belong next to each other in the same clade? The most useful characteristics for discerning evolutionary relationships are unique features that originated in a group's most recent common ancestor. Those features are then shared by the descendant groups, having been passed down, or *derived*, from the most recent common ancestor. **Shared derived traits** are evolutionary novelties shared by an ancestor and its descendants but not seen in groups that are not direct descendants of that ancestor. One way biologists pick out close relatives is to look for shared derived traits that are unique to groups of organisms. For example, fur and mammary glands are shared derived traits that are unique to mammals and the most recent common ancestor of mammals; no nonmammals on the tree of life display these traits.

Evolutionary trees imply a time sequence going from ancestors to the descendants at the tips of the branches. The time sequence may have a scale—in millions of years, for example—but most trees depicting large-scale evolution do not have a precise timescale, because the information simply is not available. The evolutionary trees shown in this book imply time on the vertical axis (arrow in Figure 2.2), but usually no scale is shown, because we do not know, for example, how many billions of years ago Archaea diverged from Eukarya. In evolutionary trees of this type, however, the base of the tree represents the past and the "leaves" of the tree are in the present.

## The extent of Earth's biodiversity is unknown

The term **biodiversity** embraces all the world's living things, as well as all the variety in their interactions with the living and nonliving components of the ecosystems they inhabit. The variety of life can be described at the level of genes, or species, or the way ecosystems function. For example, you could inventory biodiversity at the level of DNA variation by estimating the full range of genetic information that is held collectively in the DNA of all the grasshoppers in a patch of prairie. You could measure the biodiversity of a place by totaling up all species of all the organisms found there—all the insect species in a prairie ecosystem, for example. Biodiversity also embraces the variety in the interactions of organisms with each other

and with the nonliving part of their environment. To measure biodiversity on the prairie, you could determine the variety in food chains and the interactions among the food chains by making a comprehensive record of who eats whom in that ecosystem.

In spite of intense worldwide interest, scientists do not know the exact number of species alive today. Estimates range from 3 million to 100 million species. Most estimates, however, fall in the range of 3–30 million. So far, a total of about 1.5 million species have been collected, identified, named, and placed in the most basic system for biological classification, the Linnaean hierarchy. Despite this massive cataloging effort, which has taken more than two centuries, many researchers believe that we have barely scratched the surface. Some estimates suggest that 90 percent or more of all living organisms remain to be identified and named by biologists. But, as described in the box on page 29, Earth's biodiversity is disappearing even as scientists race to catalog it.

## All of life on Earth can be sorted into three distinct domains

To organize the mind-numbing diversity of life in our biosphere, biologists start by sorting all life-forms into the broadest of all groupings, domains, which form the highest hierarchical level in the organization of life (see Figure 2.2). The domains describe the most basic and ancient divisions among living organisms. There are three domains of life: **Bacteria**, which includes familiar disease-causing bacteria; **Archaea [ahr-***KEE***-uh]**, which consists of single-celled organisms best known for living in extremely harsh environments; and **Eukarya [yoo-***KAYR***-ee-uh]**, which includes all the rest of the living organisms, from amoebas to plants to fungi to animals.

The three-domain scheme was highly controversial when first proposed, but it is now widely accepted. Although the Archaea were once thought to be just another type of Bacteria, DNA studies have revealed that they are distinct from both Bacteria and Eukarya, and therefore a domain in their own right. Although Bacteria and Archaea belong to two different domains, and despite the fact that Archaea are in some ways more like Eukarya than like Bacteria, the two non-Eukarya domains have traditionally been lumped under a common label: *prokaryotes*. **Prokaryotes** is an informal label for Bacteria and Archaea; that is, the term is not part of the modern system of biological classification. "Prokaryotes" is a tag of convenience,

# The Many Threats to Biodiversity

The history of life on Earth includes a handful of drastic events, known as *mass extinctions*, during which huge numbers of species went extinct. Today, even as researchers struggle to get a total species count, many biologists assert that we are on our way toward a new mass extinction. In fact, many biologists say that the ongoing extinction—if it continues unabated—will lead to the most rapid mass extinction in the history of Earth. As with the total number of species on the planet, extinction rates are estimates. But even using conservative calculations, species are being lost at a staggering pace. The cause of the contemporary extinction is clear: the activities of the ever-increasing number of people living on Earth.

### Habitat Loss and Deterioration are the Biggest Threats to Biodiversity

Foremost among the direct threats to biodiversity is the destruction or deterioration of habitats. As human homes, farms, and industries spring up where natural areas once existed, habitats suited to nonhuman species continue to disappear or become radically altered. For many people the term "habitat loss" conjures up images of burning rainforest in the Amazon, but the problem is much more widespread and much closer to home.

Every time a suburban development of houses goes up where once there was a forest or field, habitat is destroyed. So widespread is the impact of growing human populations in urban and suburban areas that species are disappearing even from parks and reserves in heavily populated areas. For example, ecologists studied a large preserve in the midst of increasing suburban development outside Boston; there they found that 150 of the park's native plant species had disappeared. The immediate cause of the loss of species was most likely trampling and other disturbances, as more and more people—likely including many nature lovers—used the park. But the increasing number of homes in the area and the decreasing number of nearby natural areas—from which seeds could have come to repopulate the park—also played an important role in the loss of species.

### Introduced Foreign Species Can Wipe Out Native Species

Also threatening existing species is the introduction of nonnative species. Researchers estimate that 50,000 such *introduced species* have entered the United States since Europeans arrived. Some of these are *invasive species*, so called because they sweep through a landscape, competing with and wiping out native species.

In Hawaii, introduced pigs that have escaped into and are living in the wild are devouring native plant species. Domesticated cats and mongooses, also introduced species, have killed many of Hawaii's native birds, especially the ground-dwelling species whose nests are easy targets. Purple loosestrife, eucalyptus trees, and Scotch broom are invasive plant species that are choking out native plants in various parts of the United States.

### Climate Changes Also Threaten Species

Recent changes in climate, which the vast majority of scientists now agree are caused largely by human activities, constitute another threat affecting many species. In Austria, for example, biologists have found whole communities of plants moving slowly up the Alps; apparently this movement is a response to global warming, as these plants are able to survive only at ever-higher, ever-cooler elevations. These plant communities are moving at an average rate of about a meter per decade. If the climate continues to warm, these alpine plants—which exist nowhere else in the world—will eventually run out of mountaintop and go extinct. And in areas where organisms do not have cool mountains to climb, many species have begun moving to and living in increasingly northern latitudes to escape the heat.

### Human Population Growth Underlies Many, If Not All, of the Major Threats to Biodiversity

The biggest threat overall to nonhuman species is the growth of human populations. Our growth is what spurs continuing habitat deterioration as natural areas are converted to the farms, roads, and factories needed to support human life. The effects of our growing population are further magnified by the fact that more resources are being used *per person* now than in the past. In the box in Chapter 3 (page 65), we discuss what we stand to lose when we lose biodiversity.

Asian carp often leap from the water when disturbed. These natives of Asia can grow to 100 pounds and more than 4 feet long. They are voracious eaters and they reproduce rapidly. They have invaded the Mississippi River and now threaten the Great Lakes ecosystem.

rather than one with much evolutionary relevance. As used in this textbook, "prokaryotes" means simply "organisms that are not Eukarya."

## The Eukarya are sorted into four different kingdoms

In addition to recognizing three broad domains of life, biologists group life into six distinct **kingdoms**, the second-highest level in the hierarchical classification of life. Bacteria and Archaea are each kingdoms unto their own, in addition to constituting two distinct domains of life (**FIGURE 2.3**). The six-kingdom system of classification is especially useful for sorting the diverse groups of single-celled and multicellular organisms that make up the Eukarya. The unifying feature of all the Eukarya, which includes all plants and animals, is that the DNA in their cells is enclosed in a structure called the **nucleus**.

Members of the Eukarya are commonly referred to as **eukaryotes**, to distinguish them from the prokaryotes (Bacteria and Archaea). The four kingdoms of the Eukarya are Protista, a diverse group that includes amoebas and algae; Plantae [*PLAN*-tee], which encompasses all plants; Fungi [*FUN*-jye], which includes mushrooms, molds, and yeasts; and Animalia, which encompasses all animals. The members of each of these kingdoms share a set of characteristics, or evolutionary innovations, that adapted the organisms to their environment and thereby enabled them to live and reproduce successfully. The adaptations are produced by particular genes, and therefore DNA comparisons offer one way of demarcating distinct groups and identifying lineages with a common most recent ancestor. The tree of life in **FIGURE 2.4** organizes all six kingdoms of life into an evolutionary tree based on relationships deduced from shared evolutionary innovations and similarities in DNA.

## 2.2 The Linnaean System of Biological Classification

Modern biologists have devised the three-domain and six-kingdom classification systems for pigeonholing all of life. But when it comes to sorting life below the level of the kingdom, many biologists still use a hierarchical classification system that was first introduced in the 1700s. The **Linnaean [lih-*NEE*-un] hierarchy** is a system of biological classification devised by a Swedish naturalist named Carolus Linnaeus [lih-*NEE*-us]. The species is the smallest unit (lowest level) of classification in the Linnaean hierarchy (**FIGURE 2.5**). Closely related species are grouped together to form a **genus** (plural "genera"). Using these two categories in the hierarchy, every species is given a unique, two-word Latin name, called its **scientific name**. The first word of the name identifies the genus to which the organism belongs; the second word defines the species. For example, humans are called *Homo sapiens*: *Homo* ("man") is the genus to which we belong, and *sapiens* [*SAY*-pee-enz] ("wise") is our species name. We are the only living species in our genus. Other species in the genus include *Homo erectus* ("upright man") and *Homo habilis* ("handy man"), both of which are extinct.

In the Linnaean hierarchy, each species is placed in successively larger and more inclusive categories

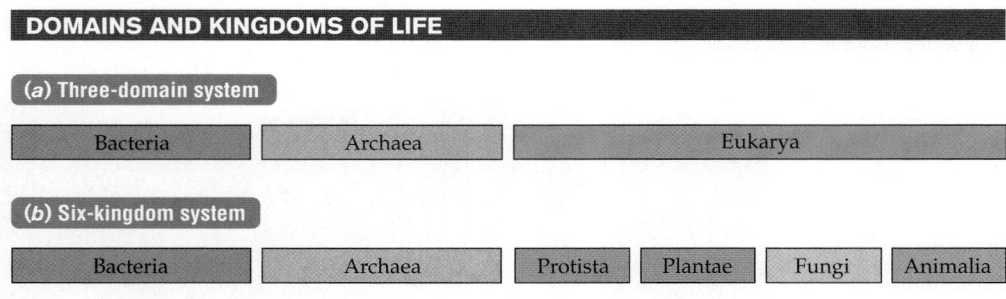

**DOMAINS AND KINGDOMS OF LIFE**

**(a) Three-domain system**

| Bacteria | Archaea | Eukarya |

**(b) Six-kingdom system**

| Bacteria | Archaea | Protista | Plantae | Fungi | Animalia |

**FIGURE 2.3 The Three Domains of Life Are Divided into Six Kingdoms**

This book employs both the three-domain system (*a*) and the widely used six-kingdom system (*b*) for classifying life. The domain Bacteria is equivalent to the kingdom Bacteria, and the domain Archaea is equivalent to the kingdom Archaea. The domain Eukarya encompasses four kingdoms in the six-kingdom scheme: Protista (protists, which include organisms such as amoebas and algae), Plantae (plants), Fungi (including yeasts and mushroom-producing species), and Animalia (animals).

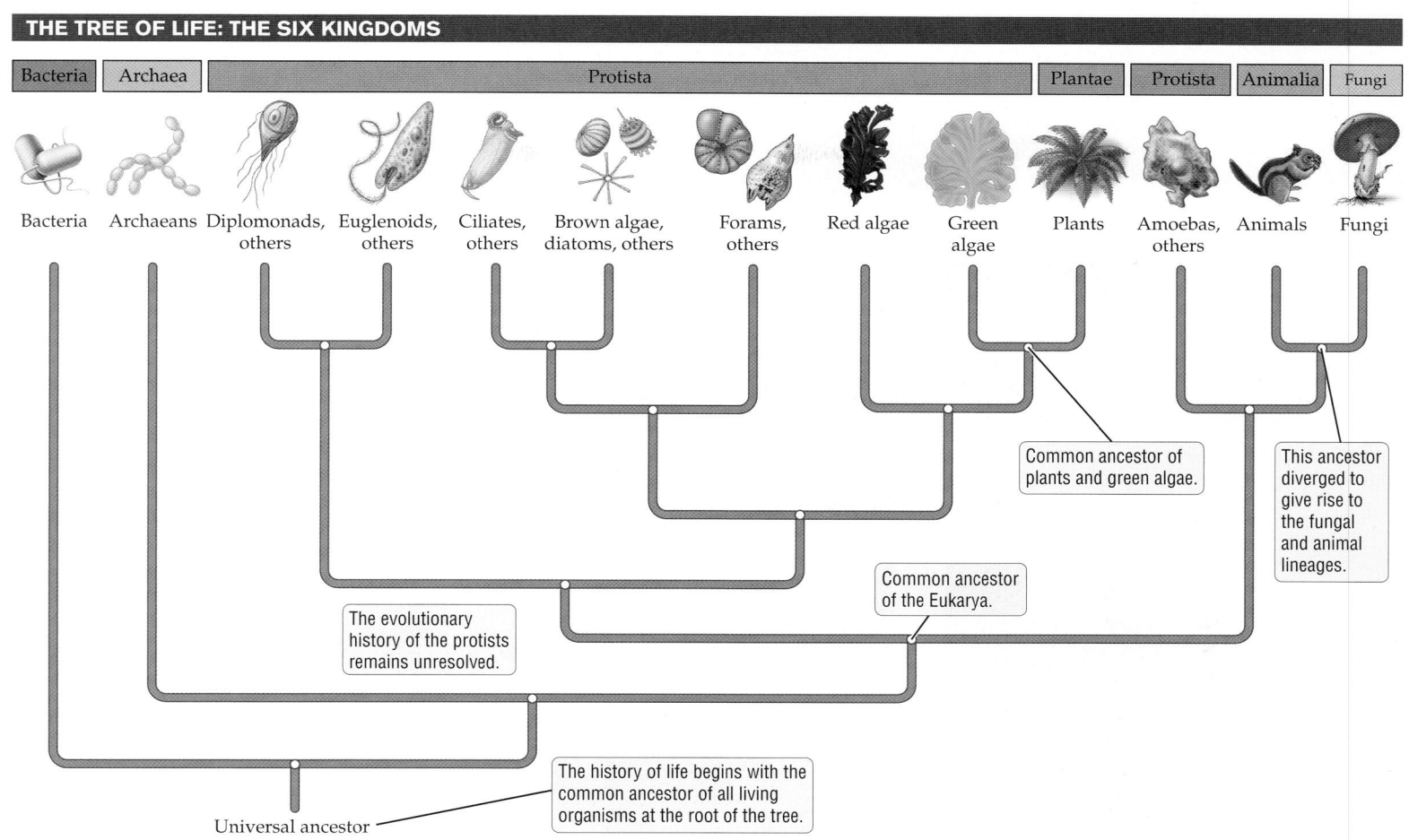

| Bacteria | Archaea | Protista | | | | | | | Plantae | Protista | Animalia | Fungi |

Bacteria | Archaeans | Diplomonads, others | Euglenoids, others | Ciliates, others | Brown algae, diatoms, others | Forams, others | Red algae | Green algae | Plants | Amoebas, others | Animals | Fungi

Common ancestor of plants and green algae.

This ancestor diverged to give rise to the fungal and animal lineages.

Common ancestor of the Eukarya.

The evolutionary history of the protists remains unresolved.

Universal ancestor

The history of life begins with the common ancestor of all living organisms at the root of the tree.

**FIGURE 2.4  The Three Domains and Six Kingdoms of Life**
This evolutionary tree shows the hypothesized relationships among the six kingdoms, as well as the three domains. Each group branching off the tree can be thought of as a cluster of close relatives, or clade.

beyond the genus. Closely related genera are grouped together into a **family**. Closely related families are grouped into an **order**. Closely related orders are grouped into a **class**. Closely related classes are grouped into a **phylum** [*FYE*-**lum**] (plural "phyla"). Finally, closely related phyla are grouped together into a **kingdom**. Two of the kingdoms that Linnaeus designated are still recognized in modern biology: Plantae (or simply, "Plants") and Animalia (or "Animals" in plain English).

**Taxonomy** is the branch of biology that deals with the naming of organisms and with their classification in the Linnaean hierarchy. Biologists refer to a group of organisms at any of these various levels of classification as a taxonomic group or, more simply, a **taxon** (plural "taxa"). Using Figure 2.5 as an example, we see

that the species *Rosa chinensis* is a taxon, but so are the higher levels of classification to which it belongs: Rosaceae, Rosales, Dicotyledones, and so on, up to Plantae. Each of these is a taxon, or taxonomic group, to which the China rose belongs, along with all the other organisms grouped with it in that taxon.

### Concept Check

1. What is an evolutionary tree?

2. What is the Linnaean hierarchy, and how does it relate to the domain system for organizing life?

3. Which of the six kingdoms in the Linnaean hierarchy is *not* represented by just a single branch in Figure 2.4, and why?

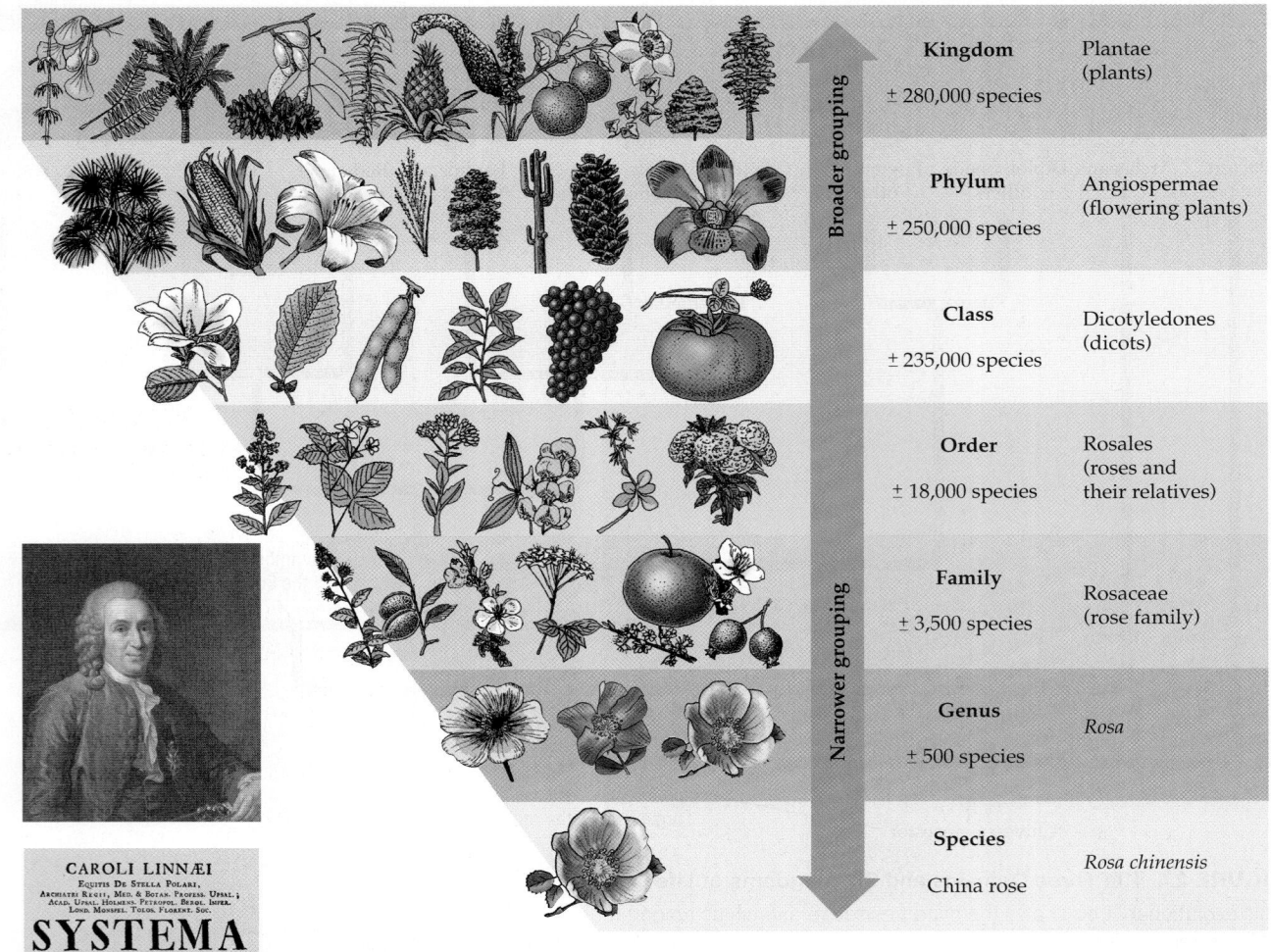

**FIGURE 2.5 The Linnaean Hierarchy Places Organisms In Successively Largely Categories**

The smallest unit of classification is the species (here, the China rose, whose scientific name is *Rosa chinensis*). This species belongs to the genus *Rosa*, which includes other roses. The genus *Rosa* lies within the family Rosaceae [**roh-ZAY-see-ee**], which lies within the order Rosales [**roh-ZAH-leez**], within the class Dicotyledones [**dye-kah-tuh-LEE-dun-eez**], within the phylum Angiospermae and the kingdom Plantae. We can use the same categories—from species to kingdom—to classify all organisms. This classification system was first devised by the Swedish naturalist Carolus Linnaeus (inset).

## 2.3 Bacteria and Archaea: Tiny, Successful, and Abundant

Two groups that have been extremely successful in colonizing Earth are the Bacteria and the Archaea. Bacteria are probably most familiar as single-celled disease-causing organisms such as *Streptococcus pneumoniae* [**noo-MOH-nee-eye**], which can cause pneumonia in humans. However, most bacteria are not harmful to humans. Archaea are single-celled organisms as well, but they have a different evolutionary history, and none of them are known to cause disease in humans or any other organism.

### Archaeans constitute a distinct domain of life

Discovered in the 1970s, archaeans are single-celled organisms that superficially resemble bacteria. Some archaeans are **extremophiles** (literally, "lovers of the extreme"): they thrive in boiling-hot geysers, highly acidic

| TABLE 2.1 | Lifestyles of the Extreme | |
| --- | --- | --- |
| **TYPE OF ENVIRONMENT** | **TYPE OF EXTREME ORGANISM** | **EXAMPLE** |
| High salt (10× saltiness of ocean) | Halophiles | *Halobacterium salinarum* |
| High acidity (pH 0–3) | Acidophiles | *Acidianus infernus* |
| High temperature (80°C–121°C) | Thermophiles | *Pyrodictium abyssi* |
| Low temperature (0°C–15°C) | Psychrophiles | *Methanogenium frigidum* |
| High pressure (100–380 atmospheres) | Piezophiles | *Pyrolobus fumarii* |
| Low moisture (less than 5% water) | Xerophiles | *Pyrococcus furiosus* |

**NOTE:** All examples in this table are members of the domain Archaea; however, some bacterial species can also live and grow in some of these extreme environments.

waters, highly salted environments, and the freezing-cold seas off Antarctica (see Table 2.1 for examples of archaeal extremism). When life first appeared, Earth was still hot and the atmosphere laced with toxic gases—like methane and ammonia—produced by intense volcanic activity. Many archaeans would have been right at home under those conditions. Modern-day archaeans certainly dominate in the pockets of Earth where hostile conditions persist today. Despite their reputation for extreme living, archaeans can also be found in locations closer to home, such as the hot-water heater in most houses or a fistful of average garden soil or any human's intestines.

In some aspects of their cellular metabolism, archaeans are more similar to organisms in the domain Eukarya than in the domain Bacteria. Archaeans possess some exceptional chemistry, however, including a unique type of fat in their plasma membranes that makes them stand apart from organisms in the other two domains. DNA comparisons have convinced most biologists that the archaeans belong in a domain of their own. Chemical clues in fossil remains suggest that Archaea had evolved by about 2.7 billion years ago, but the most ancient prokaryotic fossils are difficult to interpret and assign conclusively to one or the other domain. Therefore, biologists are still uncertain about the placement of Archaea in the tree of life. The evolutionary tree in Figure 2.2 presents one hypothesis; it shows descendants of the universal ancestor splitting into Bacteria and another lineage, and the other lineage itself diverging into Archaea and Eukarya later in the history of life.

Although they are distinct domains, Bacteria and Archaea are similar in that the organisms contained in both domains are generally microscopic (cannot be seen without the aid of a microscope), although we will tell you about some giant bacteria shortly. Both bacteria and archaeans are regarded as single-celled organisms, although some can aggregate to form dense mats on the ocean floor or a cottony film on unbrushed teeth.

Further, both are widespread and extremely abundant, and they display an astonishing diversity in metabolism. Because of these general similarities, and because of their status as the "non-Eukarya," we begin our introduction to the major groups of life by describing the prokaryotes collectively, pointing out significant differences between the two domains as they come up.

## Prokaryotes represent biological success

Most people tend to think of the living world in terms of butterflies, tigers, and orchids, giving little thought to microscopic organisms, even though the vast majority of life on Earth is in fact single-celled and prokaryotic. Scientists estimate that the number of prokaryotes on Earth is about 5,000,000,000,000,000,000,000,000,000,000 (5 nonillion = $5 \times 10^{30}$). The success of prokaryotes is due, in part, to how quickly they reproduce. Prokaryotes typically reproduce by splitting in two—a process called binary fission. Overnight, a single bacterium of the common species *Escherichia coli* (usually referred to by the abbreviated form *E. coli*), which normally lives harmlessly in the human gut, can divide to produce a population of 16 million bacteria (**FIGURE 2.6**).

Prokaryotes are also the most widespread of organisms, able to live in many places where few other forms of life can exist, such as the lightless ocean depths, the insides of boiling-hot geysers, and 2 miles deep in a mine shaft. Although the archaeans are exceptionally suited to extreme environments, some bacteria, too, live in harsh environments. Prokaryotes are also abundant in milder habitats, from doorknobs to kitchen sinks. In one teaspoon of garden soil, there are many millions of several thousand different types of bacteria. Scientists estimate that 1 square centimeter of healthy human skin is

## Helpful to Know

Microscopic organisms, or *microbes*, are invisible to the naked eye because they are less than 0.2 millimeter (200 microns, or μm) in diameter. The great majority of prokaryotes are microbes, but so are many members of the kingdom Protista and some of the kingdom Fungi. The science of microbes is *microbiology*.

## Helpful to Know

Scientific names are printed in italics. The first letter of the genus name is always capitalized, but the species name is entirely lowercase. Within a given discussion, once a scientific name has been introduced in full (as in *Escherichia coli*), the genus may be symbolized by just the first letter in subsequent references to the species (for example, *E. coli*).

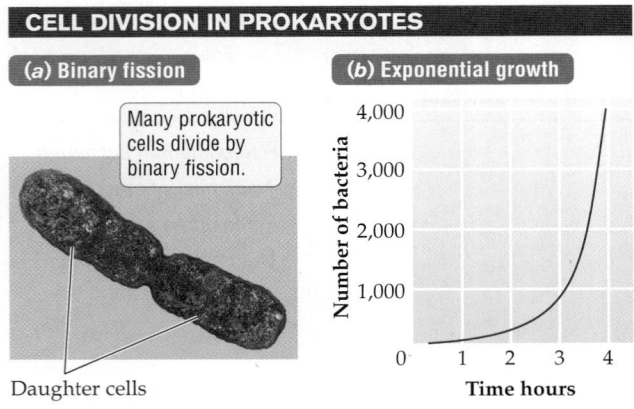

**CELL DIVISION IN PROKARYOTES**

**(a) Binary fission**

Many prokaryotic cells divide by binary fission.

Daughter cells

**(b) Exponential growth**

Number of bacteria

4,000

3,000

2,000

1,000

0   1   2   3   4

**Time hours**

**FIGURE 2.6** Through Binary Fission, *E. coli* Multiplies at an Astounding Rate

(*a*) *E. coli* divides into two every 20 minutes, given the right conditions, especially the ready availability of food and the temperature of the human body (37°C). (*b*) This graph shows how the population of *E. coli* increases, starting from a single cell that turns into two cells in 20 minutes, which together become four cells at 40 minutes. With doubling of cell numbers every 20 minutes, there are 1,000 cells after 3 hours and 20 minutes have gone by. This type of growth, in which the number of cells added in each unit of time increases at an accelerating pace, is known as exponential or geometric growth.

home to between 1,000 and 10,000 different types of bacteria.

Some bacteria, such as *E. coli*, are readily grown in the laboratory as a pure culture. A **bacterial culture** is a lab dish that contains bacteria living off a specially formulated food supply. However, the great majority of the world's prokaryotes are not culturable; that is, we do not know how to grow them in a lab dish with special food. Further, many bacteria and archaeans are so small that they are difficult to see except with exceptionally powerful microscopes. Scientists therefore resort to DNA analysis not only to detect but also to study and classify the many millions of unseen and unculturable species of bacteria and archaeans that we now know populate every imaginable habitat on land and in the ocean.

## Prokaryotes occupy a great diversity of habitats

Prokaryotes can live in many places where no other organisms can live. Bacteria and archaeans are the most abundant organisms in the open ocean, where they play a crucial role in the ecology of our biosphere. Although some of the Bacteria thrive in

unusual environments, Archaea is the group best known for the extreme lifestyles of some of its members. Some are extreme thermophiles (*thermo*, "heat"; *phile*, "lover") that live in geysers, hot springs, and hydrothermal vents, which are cracks in the seafloor that spew boiling water. The cells of most organisms cannot function at such high temperatures, but thermophiles have come up with evolutionary innovations—for example, proteins that are not destroyed at high heat—that enable them to succeed where others cannot. Others are extreme halophiles (*halo*, "salt"), thriving in very salty, high-sodium environments where nothing else can live—for example, in the Dead Sea and on fish and meat that have been heavily salted to keep most bacteria away.

Many prokaryotes need oxygen gas to survive; that is, they are **aerobes** (*aero*, "air"; *bios*, "life"). Many other prokaryotes are **anaerobes** (*an*, "without"), which not only survive without oxygen but may actually be poisoned by the gas. *Clostridium botulinum*, which is responsible for the deadly food poisoning called botulism, is an example of an anaerobic bacterium. The ability to exist in both oxygen-rich and oxygen-free environments is another reason why prokaryotes inhabit such a diversity of habitats. Prokaryotes live and work in the oxygen-poor muck of swamps and sewage tanks, on the seafloor, and in natural gas and petroleum deposits buried deep in Earth's crust.

Among the anaerobic archaeans are several species of **methanogens** (*methano*, "methane"; *gen*, "producer"), archaeans that produce methane gas as a by-product of their metabolism. Some methanogens inhabit animal guts and produce the methane gas that causes human flatulence (intestinal gas) and cow burps. The methanogens that inhabit wetlands are the single largest natural source of atmospheric methane, also known as swamp gas. Landfills are another potent source of methane.

## Prokaryotes are mainly single-celled, but some show social behaviors

Bacteria and Archaea are quite variable in shape, with some having shapes like spheres, called cocci [KAHK-eye] (singular "coccus"); rods, called bacilli [buh-SIL-eye] (singular "bacillus"); or corkscrews, called spirilla [spye-RIL-uh] (singular "spirillum"). But they all share a basic structural plan (**FIGURE 2.7**). Most bacteria and many archaeans have a protective cell wall that surrounds the plasma membrane. Some have an ad-

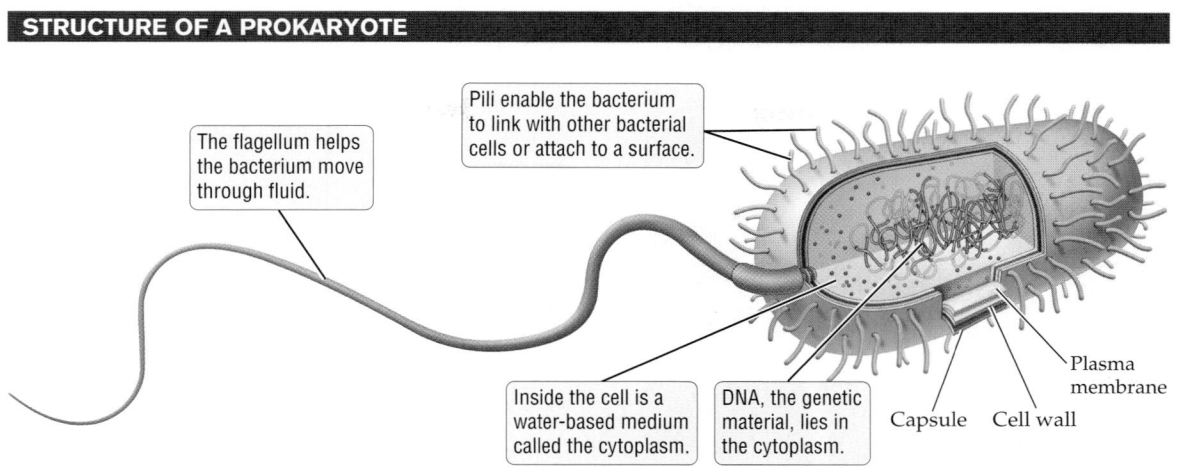

The flagellum helps the bacterium move through fluid.

Pili enable the bacterium to link with other bacterial cells or attach to a surface.

Inside the cell is a water-based medium called the cytoplasm.

DNA, the genetic material, lies in the cytoplasm.

Capsule   Cell wall

Plasma membrane

**FIGURE 2.7**
**Prokaryotic Cells Lack a Nucleus**
Prokaryotic cells tend to be about 10 times smaller than eukaryotic cells, and generally have much less DNA.

ditional wrapping, called a capsule. The capsule, made of slippery biomolecules, helps disease-causing bacteria evade the defense system (known as the immune system) that attempts to protect the animal body from foreign invaders.

The surface of some bacteria is covered in many short hairlike projections called **pili [PYE-lye]** (singular "pilus"). Bacteria use their pili to link together to form bacterial mats, or to attach to surfaces in their environment, including the cells inside a host animal they have infected. Some bacteria have one or more long, whiplike structures called **flagella** (singular "flagellum") that spin like a boat's propeller to push the bacterium through liquid.

As noted earlier, prokaryotes do not have a true nucleus. They typically have much less DNA than do cells of eukaryotic organisms. Eukaryotic genetic material is often full of what appears to be extra DNA, popularly known as junk DNA, that serves no clear function. In contrast, prokaryotic genetic material is composed mainly of DNA that is actively used for the survival and reproduction of the bacterial cell.

Although prokaryotes are regarded as single-celled organisms, some form colonies, produced by the splitting of one original cell. Many bacteria and archaeans form long chains of cells, called filaments, produced by the splitting of an original cell. For example, the cyanobacterium *Nostoc* forms long filaments of photosynthetic cells (see **FIGURE 2.8**).

Some prokaryotes have a system of cell-to-cell communication, called **quorum sensing**, that enables them to sense and respond to bacterial density (the number of other bacteria in the neighborhood). Quorum-sensing bacteria can coordinate their behavior, for example, by forming tough aggregates called *biofilms* that are made up of the same or different species. Bacteria form biofilms on everything from shower curtains to the tissues

of the organisms they infect. Some disease-causing bacteria begin to multiply rapidly only upon sensing that their numbers are high enough to overwhelm the host organism's immune system with a sudden burst of cell division. The ability of some bacteria to swarm together and display coordinated behavior has led some researchers to suggest that these forms possess a basic type of multicellular organization.

## Prokaryotes reproduce asexually

Prokaryotes typically reproduce by splitting in two in the process called *binary fission*. **Binary fission** is a form of asexual reproduction, since one single individual produces the offspring. The DNA in the parent cell is copied before fission, and one copy is transferred to each of the resulting offspring, known as daughter cells. The genetic information in the daughter cells is virtually identical to that of the parent cell, as is invariably the case in asexual reproduction. As Figure 2.6 illustrates, bacteria can reproduce quickly. A single *E. coli* cell, which can divide in two in just 20 minutes, can become a billion (1,000,000,000) cells in about 10 hours, as long as the cells continue to receive all the nutrients they need. The main reason the world is not completely buried in bacteria is that under natural conditions the growth of bacteria is rapidly curtailed by the lack of adequate resources.

Some types of bacteria, but not any archaeans, can undergo *sporulation*. Sporulation is the formation of spores, thick-walled dormant structures that can survive boiling and freezing. *Clostridium tetani* (which causes a life-threatening infection known as tetanus) and *Bacillus anthracis* (which causes an often fatal disease called anthrax) both form spores.

Although sexual reproduction has not been seen in prokaryotes, these enormously diverse organisms are

## Structure

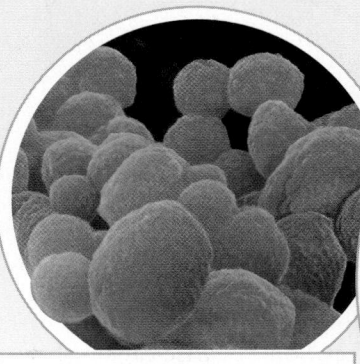

Mycoplasmas are among **the smallest organisms**; many are less than 0.5 μm in diameter. For comparison, the head of a pin is 2 mm, or 2,000 μm, across. Most mycoplasmas are parasites of plants and animals.

**Did you know?**

In 2010, a team led by scientist-entrepreneur Craig Venter created the first artificial life-form, nicknamed "Synthia." The scientists used machines to manufacture the 485 genes known to exist in *Mycoplasma mycoides* and transferred them into a DNA-free *Mycoplasma capricolum* cell.

*Epulopiscium fishelsoni* is **a giant among prokaryotes**, measuring about 600 μm in length. The bacterium lives inside the gut of the surgeonfish.

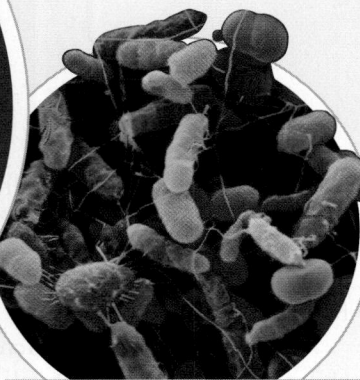

Bacteria and Archaea range from balloonlike forms to flattened triangles. Shown here are **rod and cocci bacteria** found on a dollar bill. A wide variety of bacteria are found on paper money, and they may spread infections.

## Habitat

**Did you know?**

"Red herring" was the name given to salted fish spoiled by *Halobacterium*, which is red in color. Stench from the rotted fish worked well to divert foxhounds, giving rise to a common expression that means "a diversionary tactic, intended to mislead someone."

**Psychrophiles** live at temperatures **close to freezing**. *Psychrobacter* is just one of many bacteria found in this subglacial stream in the Swiss Alps.

**FIGURE 2.8**
**Prokaryotes Are Extremely Diverse in Structure, Habitat, and Modes of Energy Acquisition**

**Hyperthermophiles** inhabit some of the **hottest places** on Earth. *Pyrodictium abyssi* is an archaean that lives at about 110°C at extremely high pressure around deep-sea hydrothermal vents, such as the one shown in this photo.

**Halophiles** can live in water 10 times **saltier than the sea**. Seawater is evaporated to make salt in these lagoons by San Francisco Bay. The pink and purple tints come from an abundance of *Halobacterium*, an archaean that uses colored pigments to turn sunshine into chemical energy.

# Energy Acquisition

Producers

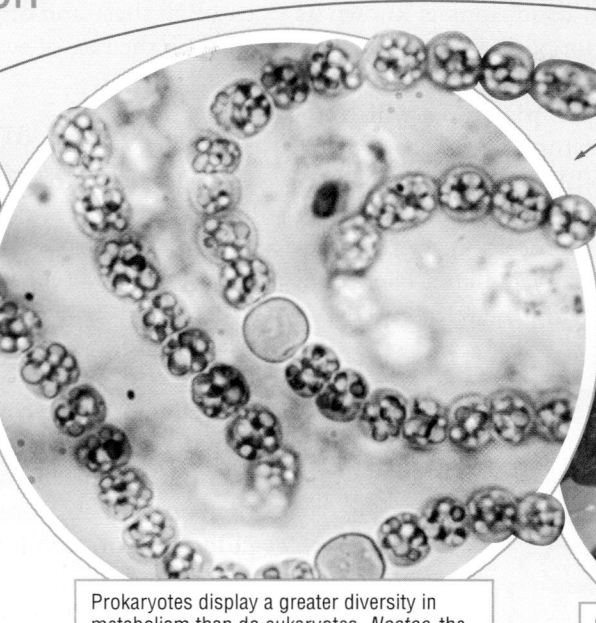

**Chemolithotrophs** tap the energy in minerals to make food. The "**rock eaters**" in these highly acidic hot springs in Yellowstone National Park obtain energy from hydrogen sulfide and arsenite.

Prokaryotes display a greater diversity in metabolism than do eukaryotes. *Nostoc*, the cyanobacterium in this photo, **uses light energy** to make food through photosynthesis. It can also turn nitrogen gas from the atmosphere into organic molecules—something no eukaryote can do.

Some prokaryotes **can live completely independently from sunshine**. *Desulfotomaculum*, discovered about 2 miles underground in a South African gold mine, is a bacterium that taps sulfur minerals, as well as hydrogen gas released from water by radioactivity, as sources of energy for converting carbon dioxide into food.

Did you know?

Dentists use a red vegetable dye to disclose the presence of bacterial plaque. The child featured in the photo below is not brushing to her full dental hygiene potential.

**Decomposers** are consumers that break down the remains of **dead organisms** and absorb the chemicals that are released. Millions of species of Bacteria and Archaea are hard at work in these holding tanks at a sewage treatment plant in Houston, Texas.

Some prokaryotes are **extreme collaborators**. The root nodules on these bean plants are colonized by *Rhizobium* bacteria that obtain sugar from the host plant. The bacteria return the favor by converting atmospheric nitrogen gas into a form that the plant can use to build proteins and other vital molecules.

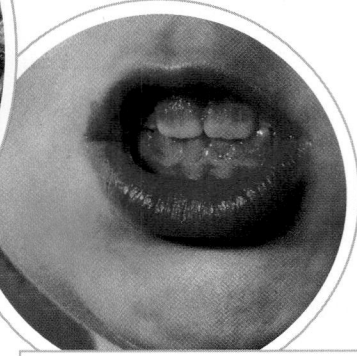

More than 500 species of bacteria live **in the human mouth**. Some, such as *Streptococcus mutans*, are decidedly unfriendly, causing tooth decay and gum disease.

Consumers

capable of capturing bits of DNA from their environment and incorporating them into their own genetic material. The transfer of genetic material between different species under natural conditions is known as **lateral gene transfer** (or horizontal gene transfer). One mechanism through which Bacteria and Archaea trade genetic material involves **plasmids**, which are loops of extra DNA in the cytoplasm of prokaryotes (and some eukaryotes as well). A bacterium may actively trade DNA with another bacterium, usually of the same species, through a process known as **bacterial conjugation** (FIGURE 2.9). When a bacterium dies, the cell may burst open and the released DNA—

plasmid or the main genetic material (chromosomal DNA)—may be taken up by another bacterium of the same, or even a different, species. Lateral gene transfer through these and other mechanisms appears to have sped up the rate of evolution in prokaryotes.

## Prokaryotes are unrivaled in metabolic diversity

All organisms require nutrients and energy to survive, grow, and reproduce. The atoms in nutrients go into building the organism. Life on Earth is carbon based: carbon atoms create the basic framework of all the critical biomolecules of life, such as proteins and DNA. Carbon-containing molecules of biological origin are known as **organic molecules**. In contrast, chemicals that lack carbon atoms, or have no more than one carbon atom, are known as **inorganic molecules**. Iron ore, sand, and carbon dioxide are examples of inorganic molecules.

All organisms need a source of energy to create the complex order within their cells and to run all their activities, including the building of biomolecules. While some organisms (like us) get both energy and nutrients directly from organic molecules (food molecules, in our case), others obtain energy and nutrients from their *nonliving* environment and then use these inputs to manufacture organic molecules "from scratch." As noted in Chapter 1, plants, protists, and some prokaryotes are producers, or **autotrophs**, meaning that they make organic molecules (food) on their own, rather than taking them from other organisms. Plants, for example, absorb the energy of sunlight and take in carbon dioxide (the gas in the air that humans and other animals exhale) to conduct photosynthesis. In the process of photosynthesis, plants combine carbon dioxide and water, using light energy as the fuel to power this energy-hungry chemical reaction. The reaction yields sugars and releases oxygen gas that we animals must inhale in order to survive.

Some bacteria photosynthesize. Cyanobacteria [*sye*-an-oh-. . .] are commonly found in the green slime that people commonly call pond scum, but they are known to inhabit just about every environment on Earth. Cyanobacteria are the only prokaryotes that produce oxygen gas as a by-product of photosynthesis (FIGURE 2.10). Organisms that make food using light energy and inorganic carbon (carbon atoms from the nonliving part of their environment) are known as **photoautotrophs**. The first part of the term—*photo* (meaning "light," as in the energy of sunlight)—indicates

**BACTERIAL CONJUGATION**

Donor bacterium    Recipient bacterium

Cell wall
Plasma membrane
Cytoplasm
Chromosomal DNA

Plasmid DNA    Conjugation tube

The donor bacterium attaches to a recipient.

The membranes of the two cells fuse to form a conjugation tube.

DNA is transferred to the recipient through the tube.

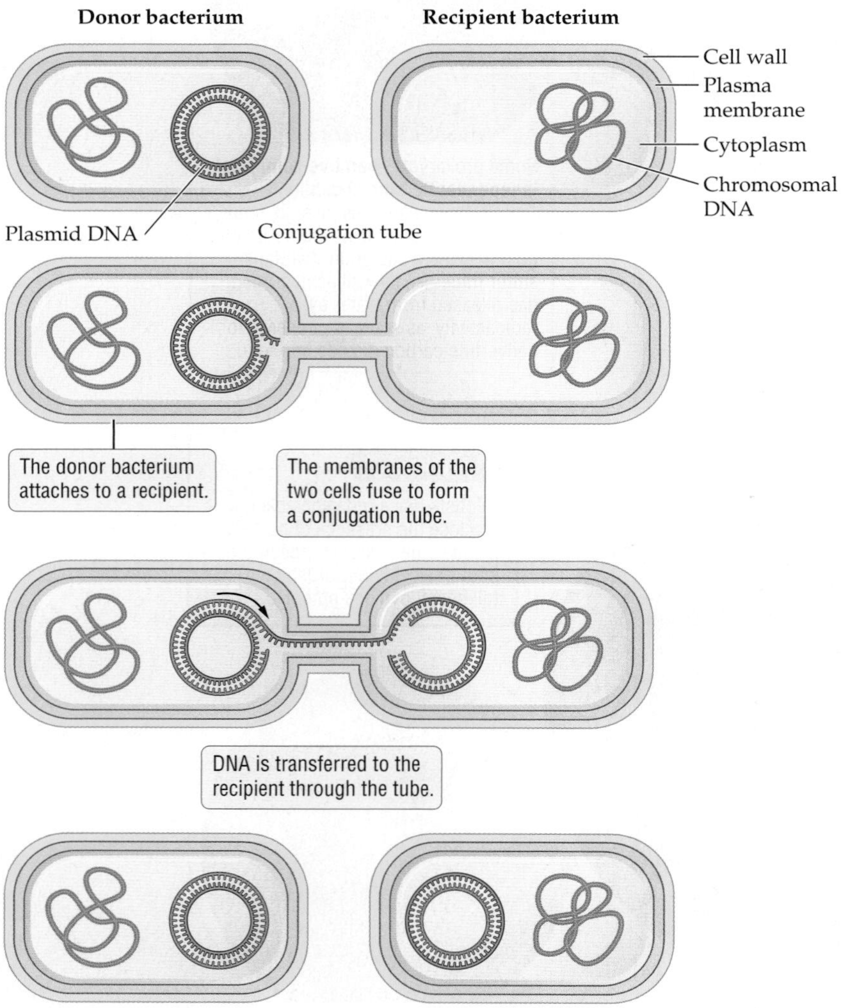

**FIGURE 2.9 Lateral Gene Transfer Accelerates the Rate of Evolution in Prokaryotes**
Bacterial conjugation is one mechanism by which DNA is transferred from one bacterium to another. This diagram depicts the transfer of plasmid DNA, but chromosomal DNA can also be transferred through conjugation.

**FIGURE 2.10 Pond Scum Contains Bacteria That Photosynthesize**

Photosynthetic bacteria, called cyanobacteria or "blue-green algae," can be found growing as slimy mats on freshwater ponds. The green mats may also include true algae, which are photosynthetic protists.

how the organism obtains energy. The second part of the term—*auto* ("self")—indicates that the organism does not depend on other organisms for its carbon but uses carbon dioxide. The term *troph* means "to eat."

Curiously, some autotrophic prokaryotes get their energy not from light but from *inorganic chemicals* in their environment, including such unlikely materials as iron ore, hydrogen sulfide, and ammonia (**FIGURE 2.11**). These prokaryotes use carbon dioxide in the air as a carbon source. Given the rules for naming that have been illustrated so far, can you guess what they are called? They are known as **chemoautotrophs**, organisms that make food from carbon dioxide and energy extracted from chemicals in their environment. Some of the bacteria and archaeans that inhabit hydrothermal vents in the lightless depths of the ocean are chemoautotrophs. The food they produce supports an extraordinary ocean floor ecosystem that includes animals such as vent clams, tube worms, and even octopi.

When humans and other animals need nutrients, we consume other species. Our bodies then break down the tissues of the other species, from which we get both energy (in the form of chemical bonds, which we will learn more about in Unit 2) and carbon (in the form of carbon-containing molecules). In fact, many organisms, including all animals, all fungi, and some protists, get their nutrients by consuming other organisms—or, in humans, by what we call eating. Many prokaryotes also get their carbon from other organisms—that is, from organic sources—and are therefore classified as consumers or **heterotrophs**.

**FIGURE 2.11 This Chemoautotroph Has an Appetite for Metal**

The crusty orange and yellow puddle is a colony of the organism known as *Sulfolobus* [**sul-*FAH*-luh-bus**], an archaean that gets its carbon from carbon dioxide, as plants do. This archaean, however, gets its energy in an unusual way—not by harnessing sunlight (as plants do), or by eating other organisms (as animals do), but by chemically processing inorganic chemicals such as iron ore. This chemoautotroph is living in a volcanic vent in Japan.

**Chemoheterotrophs** are organisms that obtain energy *and* carbon from organic molecules. The first part of the term—*chemo*—denotes where the organism gets its energy, in this case from organic molecules, derived ultimately from living or dead organisms. The second part of the term—*hetero* ("other")—describes where the organism gets its carbon atoms, in this case ultimately from other organisms. All animals and fungi, and many protists, are chemoheterotrophs (Table 2.2).

| TABLE 2.2 | Modes of Nutrition among Prokaryotes | | |
|---|---|---|---|
| **TYPE OF NUTRITION** | **SOURCE OF ENERGY** | **SOURCE OF CARBON** | **EXAMPLE**[a] |
| Photoautotroph | Light | Carbon dioxide | Cyanobacteria |
| Chemoautotroph | Inorganic chemicals (such as iron ore) | Carbon dioxide | *Thiobacillus ferrooxidans* |
| Chemoheterotroph | Organic molecules | Organic molecules | *Escherichia coli* |
| Photoheterotroph | Light | Organic molecules | *Heliobacterium chlorum* |

**NOTE:** The real key to the success of prokaryotes is the great diversity of ways in which they obtain and use nutrients.

[a]All the examples in this table are members of Bacteria. The domain Archaea has all the listed modes of nutrition *except* photoautotrophy.

Many bacteria and archaeans are chemoheterotrophs. Compared to eukaryotic chemoheterotrophs, prokaryotes can use a greater variety of carbon sources. Some prokaryotes can live off the carbon-rich molecules in petroleum. (Petroleum, you may be aware, represents the fossilized remains of ancient organisms that have been transformed into liquid by intense heat and pressure generated by geologic processes.)

Some prokaryotes, both bacteria and archaeans, are **photoheterotrophs**, meaning that they use light as an *energy* source (as do plants) but derive their carbon from organic molecules (as opposed to deriving it from carbon dioxide as plants do). For example, the salt-tolerant halobacteria (which are archaeans, despite their name) absorb sunlight using a pigment called bacteriorhodopsin, the way plants absorb light energy with their green-colored chlorophyll pigment. The absorbed light energy is used to make an energy-rich chemical (called ATP) that then drives the manufacture of food molecules using carbon atoms derived from organic molecules they take up from their environment. Bacteriorhodopsin is red to purple in color, and large *Halobacterium* populations often lend a beautiful tint to salt lakes and to the evaporation ponds used in salt manufacturing (see Figure 2.8).

## Prokaryotes changed the world with oxygen-producing photosynthesis

There was very little oxygen gas on the early Earth, which was a hothouse thick with gases like carbon dioxide, methane, and ammonia. The evolution of oxygen-generating photosynthesis, probably about 2.5 billion years ago, changed the chemistry of Earth and, in so doing, changed the living world forever.

The first photosynthetic organisms on the planet were prokaryotes. Cyanobacteria, in particular, changed Earth's chemistry by evolving a form of photosynthesis that produces oxygen gas as a by-product (oxygenic photosynthesis). Oxygen gas accumulated in the air and water, and the levels rose from next to nothing to almost 10 percent about 2.0 billion years ago. Eukaryotes appear at about that time in the fossil record, suggesting that the oxygen generated by cyanobacteria may have facilitated the evolution of eukaryotes, especially multicellular forms. Eukaryotic cells are generally larger than prokaryotic cells and therefore need more energy. The evolution of a new, oxygen-dependent type of metabolism (known as *cellular respiration*) may have made the critical difference by delivering enough energy to large-celled chemoheterotrophic eukaryotes.

Green algae, and later plants, are believed to have descended from an evolutionary lineage that goes back to cyanobacteria. These eukaryotic photoautotrophs conduct photosynthesis through much the same oxygen-producing mechanism that cyanobacteria use. The oxygen released by eukaryotic photosynthesizers raised the levels of the gas almost to their present-day levels, about 21 percent, some 600 million years ago.

## Prokaryotes play important roles in the biosphere and in human society

Because of the wide range of evolutionary innovations they possess—particularly those that enable them to obtain nutrients in a variety of ways—prokaryotes play numerous and important roles in ecosystems and in human society. Bacteria that are producers, like cyanobacteria, are at the base of the food chain in many aquatic ecosystems, such as the open ocean.

Many heterotrophic bacteria and archaeans are **decomposers**, which are consumers that extract nutrients from the remains of dead organisms and from waste products such as urine and feces. Decomposers play a crucial role in **nutrient recycling**: by breaking down dead organisms or waste products, decomposers release the chemical elements locked in the biological material and return them to the environment; the released elements, such as carbon dioxide or nitrogen or phosphorus, are used by autotrophs and eventually heterotrophs as well. Decomposer prokaryotes include oil-eating bacteria used to clean up ocean oil spills and bacteria that live on sewage, breaking down waste so that it can be released into the environment in a safer form.

Bacteria can directly aid plants as well. Plants need nitrogen in the form of ammonia or nitrate, which they cannot make themselves. Plants benefit from bacteria that can take nitrogen, a gas in the air, and convert it to ammonia, in a process known as **nitrogen fixation**. Most nitrogen-fixing bacteria live free in the soil or water. But some, such as species of *Rhizobium*, form elaborate and intimate associations with certain plants, which house them inside special outgrowths of the roots (root nodules, seen in Figure 2.8).

Because prokaryotic metabolism is so enormously diverse, prokaryotes produce an astonishing diversity of metabolic by-products, ranging from antibiotics used as medicine to the acetone in nail polish remover. Some oxygen-utilizing heterotrophic bacteria resort to a special type of metabolism, known as **fermentation**, usually in low-oxygen environments. Whether conducted by prokaryotes or by eukaryotes such as yeasts, fermentation results in the accumulation of a variety of end products, depending on the species and the food molecules being broken down. The breakdown products range from acetic acid (the main acid in vinegar) to butyric acid (which is responsible for the distinctive flavor of Swiss cheese). Yogurt, buttermilk, soy sauce, some cheeses, and pickled vegetables (such as kimchi) are some of the foods that are made with the help of bacterial fermentation.

The extraordinary metabolic diversity of prokaryotes explains their ability to live in an astonishing diversity of habitats, and that metabolic versatility often comes to our aid in resolving mistakes made by humans. **Bioremediation** is the use of organisms to clean up environmental pollution. Prokaryotes have been especially useful in cleaning up oil spills. Some chemoheterotrophic bacteria and archaeans can use the organic molecules in petroleum as a source of both energy and organic molecules, turning the contaminant molecules into harmless carbon dioxide and water in the process. Oil spill sites are often sprayed with fertilizer (usually containing nitrogen and sulfur) to encourage the explosive growth of bioremediator prokaryotes (**FIGURE 2.12**). The fertilizer components are nutrients that are usually not available in sufficient quantity naturally to support rapid proliferation of prokaryotic cells.

**FIGURE 2.12 Bioremediation by Prokaryotes**
Workers spray fertilizer on an oil-contaminated shore to stimulate the growth of oil-degrading bacteria.

## Some bacteria cause disease

Although the great majority of bacteria are harmless, and many are actually beneficial to humans, some cause mild to deadly disease. Organisms that cause disease in other organisms are called **pathogens**. Interestingly, Archaea are not known to be pathogens of *any* organism.

With their ability to use almost anything as food, bacteria infiltrate crops, stored foods, and domesticated livestock, in addition to sickening or killing humans. Like most other pathogens, pathogenic bacteria tend to be quite host-specific, meaning they infect a specific type of organism. Bacteria that infect plants, for example, do not affect humans. Some bacterial infections, such as anthrax, can be communicated to humans from another animal species (cattle, in the case of anthrax). However, most bacterial species infect just one or a few closely related species.

Some bacteria, such as those that cause "flesh-eating disease" (necrotizing fasciitis) are the stuff of nightmares. The invading bacteria, which are often mixtures of different species but usually include *Staphylococcus pyogenes*, produce an **exotoxin**, which is a poison that an organism releases into its surroundings. The exotoxin kills tissues, and because the bacteria can spread rapidly, the tissue death can be extensive enough to kill a person in a day or two after the symptoms appear.

Fortunately, most people appear to be able to fend off these deadly bacteria under normal conditions, and necrotizing fasciitis is therefore rare.

Across the world, bacterial infections kill more than 2 million people each year, mainly from tuberculosis, typhoid, and cholera. In the developed world, better sanitation makes bacterial infections less common. Bacterial food poisoning, bacterial pneumonia, and "strep throat" (caused by *Streptococcus* species) are the most common bacterial infections in the United States. Federal agencies estimate that 5 million people per year are sickened by food-borne bacteria, mainly species of *Campylobacter* and *Salmonella*. Certain bacteria, including some that cause food poisoning, such as *Salmonella* and *Vibrio cholerae*, produce endotoxins. An **endotoxin** is a component of bacterial cell walls that triggers illness. Endotoxins can produce fever, blood clots, and toxic shock (a sudden, sometimes fatal, drop in blood pressure).

Antibiotics are commonly used to combat bacterial infections. Antibiotics are molecules secreted by one microorganism to kill or slow the growth of another microorganism. **Antibiotics** are naturally produced by a variety of fungi. As decomposers and pathogens, bacteria and fungi often occupy the same ecological "niche," a term that refers to the habitat and food requirements of an organism. The two are therefore in direct competition for resources, and antibiotics are the fungal weapons of choice against the bacterial world. However, some species of bacteria also deploy antibiotics in warfare against other bacteria, and a number of commercial antibiotics (such as streptomycin) come from bacteria rather than fungi.

In 1928, Alexander Fleming described antibiotics from a mold called *Penicillium*. In 1939, Ernst Chain and Howard Florey purified the antibiotic, penicillin, and launched the antibiotic era. Since then, many natural and synthetic (human-made) antibiotics have been put to use in fighting bacterial infections. To appreciate the social transformation wrought by antibiotics, visit an old cemetery and observe the large numbers of young children that families used to lose before the 1940s.

Using antibiotics inappropriately—not completing a prescribed course, or overusing them in farm animals—can lead to selection for antibiotic resistance, reducing the options available for fighting bacterial infections. It is important to realize that antibiotics are ineffective against viruses (discussed next), which can cause diseases similar to bacterial infections (some types of pneumonia and some types of food poisoning, for example).

## 2.4 Viruses: Nonliving Infectious Agents

You may have noticed the absence of *viruses* from our tree of life and from the three-domain and six-kingdom system of biological classification. What *is* a virus? A **virus** is a microscopic, noncellular infectious particle. Most viruses are little more than genetic material wrapped in proteins, yet they can attack and devastate organisms in every kingdom of life—bacteria, archaeans, protists, fungi, plants, and animals.

## Viruses lack cellular organization

Like living organisms, viruses can have DNA, they can reproduce, and they evolve. Yet viruses lack some of the key characteristics of life, which is why, after many years of head scratching, most scientists today regard viruses as nonliving infectious particles. For one thing, viruses are not made up of cells. A virus is much simpler than a cell, usually consisting of small piece of genetic material, like DNA, that is wrapped in a coat of large biomolecules called proteins (**FIGURE 2.13**). Some viruses also have an envelope, an oily layer rather like the plasma membrane, enclosing the central core of genetic material and protein.

Another difference, compared to living organisms, is that viruses lack the many structures within cells

**STRUCTURE OF THE INFLUENZA VIRUS**

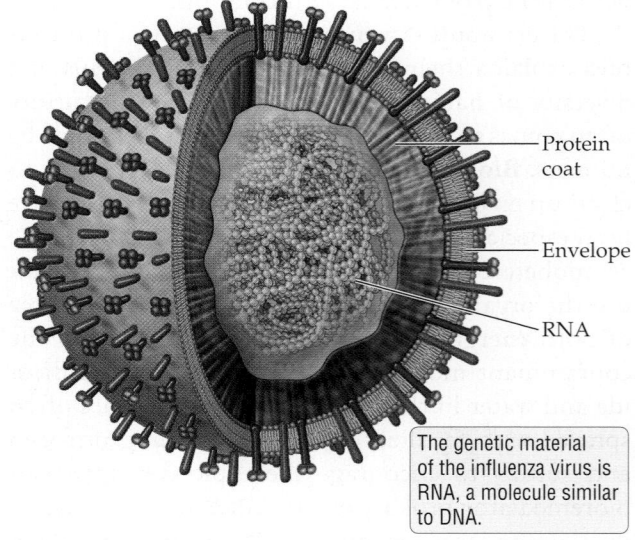

Protein coat

Envelope

RNA

The genetic material of the influenza virus is RNA, a molecule similar to DNA.

**FIGURE 2.13  Viral Structure**

that are necessary for critical cellular functions such as homeostasis, reproduction, and metabolism. To gain these functions, viruses make the cells of other organisms do the work for them. They accomplish this feat by invading those cells, releasing their genetic material into the cytoplasm, and essentially "hijacking" the host cell's metabolism. A third unusual feature of viruses, unlike living organisms, is that the genetic material they pass from one generation to the next is not always DNA. Some viruses use a related molecule, known as **RNA** (ribonucleic acid).

## Viruses are classified by structure and type of infection

A special classification system, similar to the Linnaean hierarchy, is used to classify viruses. Viruses are generally classified by the type of genetic material they possess (type of DNA or RNA molecule), their shape and structure, the type of organism (host) they infect, and the disease they produce. **FIGURE 2.14** shows some of the variations in viral shape and structure. The variant forms of a particular type of virus are called **viral strains**, or serotypes. The common cold, for example, is caused most often by one of the many strains of the rhinovirus, a member of the picornavirus family. Viral strains are sometimes named after the place where they were identified. For example, the Ebola virus is named after the Ebola Valley in the Democratic Republic of the Congo, in Africa. More often, newly discovered strains of otherwise well-known viruses are named simply with letters and numbers (such as the H1N1 strain of the influenza virus). Like bacterial pathogens, viruses tend to be highly host-specific. However, some viruses can "jump" from one host species to another, evolving into new strains as they go. The avian influenza virus ("bird flu virus") infects birds, but it is occasionally passed on to humans who handle infected birds.

Some viruses, known as *retroviruses*, can insert their genetic material into a host cell's DNA and then lie dormant for long periods of time, even for the life span of the infected organism. Pathogenic retroviruses usually become active after a while (what triggers the change is largely unknown), multiply to huge numbers, and then escape the host cell. Viral offspring escape from a host cell either by causing it to burst open or by budding off from the cell wrapped in a layer of the host cell's plasma membrane.

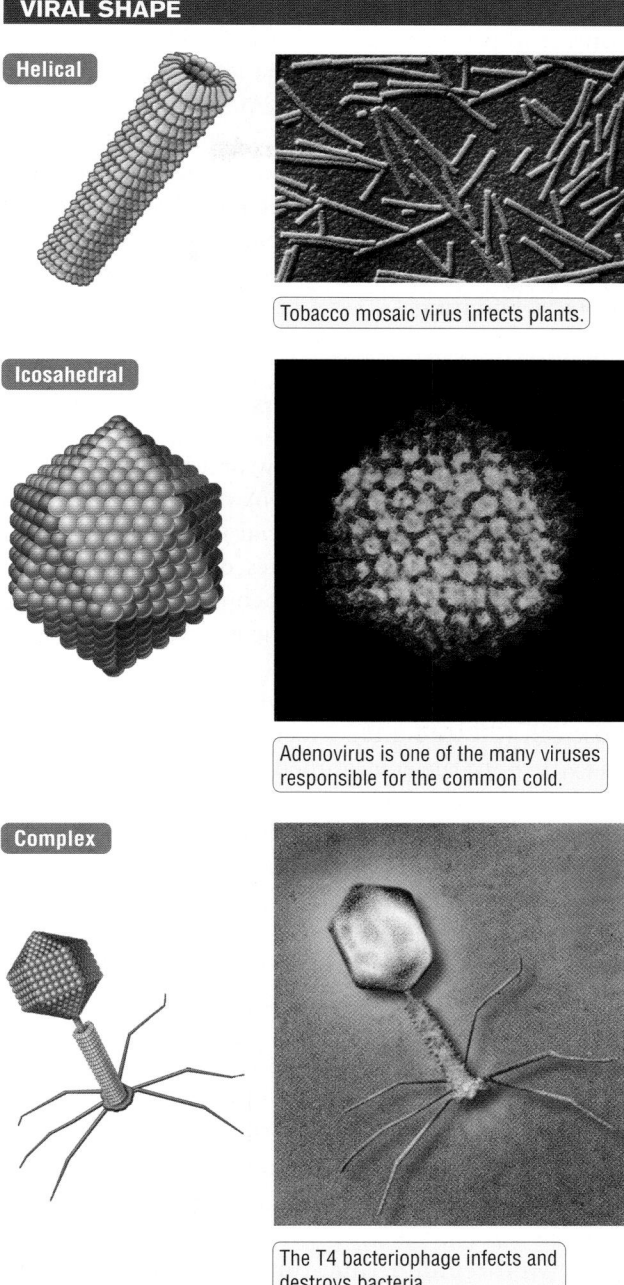

**VIRAL SHAPE**

Helical

Tobacco mosaic virus infects plants.

Icosahedral

Adenovirus is one of the many viruses responsible for the common cold.

Complex

The T4 bacteriophage infects and destroys bacteria.

**FIGURE 2.14  Viruses Can Be Classified by Their Shape**

The HIV-1 virus, which causes HIV-AIDS in humans, is a retrovirus that invades immune cells by fusing with the host cell's membrane and releasing the genetic material (which, in the case of HIV, is RNA) into the cytoplasm. The viral genetic material is converted into DNA and then integrated into the host cell's DNA. In most infected adults, the immune system beats back the virus enough that no disease symptoms are seen for the next 2–20 years. However, the continued proliferation of the virus during this time means that in most people the virus gets the upper hand eventually, as the critical immune cells begin dying faster than they can be replaced. Once the number of these immune cells falls below a certain threshold, the body's defense capacity declines sharply, making the patient extremely vulnerable to opportunistic infections by other pathogens, including other viruses, bacteria, and fungi.

## Flu viruses evolve rapidly

Many people are laid up each winter by the influenza virus. Influenza viruses are RNA viruses, but they do not insert their genetic material into the host cell's DNA. However, the virus takes over the host cell's protein and RNA-making machinery, directing the cell to make many copies of virus particles. The particles then exit the cell enclosed in a "bubble" made by the plasma membrane, without destroying the host cell (**FIGURE 2.15**). Most people start shedding the virus into the environment 2–3 days after becoming infected, which is usually a day before symptoms appear, and they remain infective for about 7 days after first catching the virus. The virus can survive on a doorknob for a few days, and on a moist surface for about 2 weeks.

Influenza viruses attach to cells in the nose, throat, and lungs of humans and in the intestines of birds. Flu sufferers cough and sneeze, develop a fever, and ache all over. Runny noses suit the virus very well because they help disseminate the virus from one victim to the next. Some flu symptoms are the result of the actions of immune cells on a mission to attack and destroy the virus. Fever, for example, is a defensive strategy: turning up body temperature slows viral reproduction. But very high fever is often an overreaction that can damage the body; doctors therefore have to exercise judgment in deciding whether to fight a fever with medication or let nature take its course.

Why are people susceptible to the influenza virus year after year? The reason is that as this microscopic virus proliferates throughout the body's cells—whether for days or for weeks—it is evolving rapidly. Viruses evolve into new strains so quickly that sometimes an antiviral drug developed to fight an older strain becomes useless against a new strain. Remember, anti-

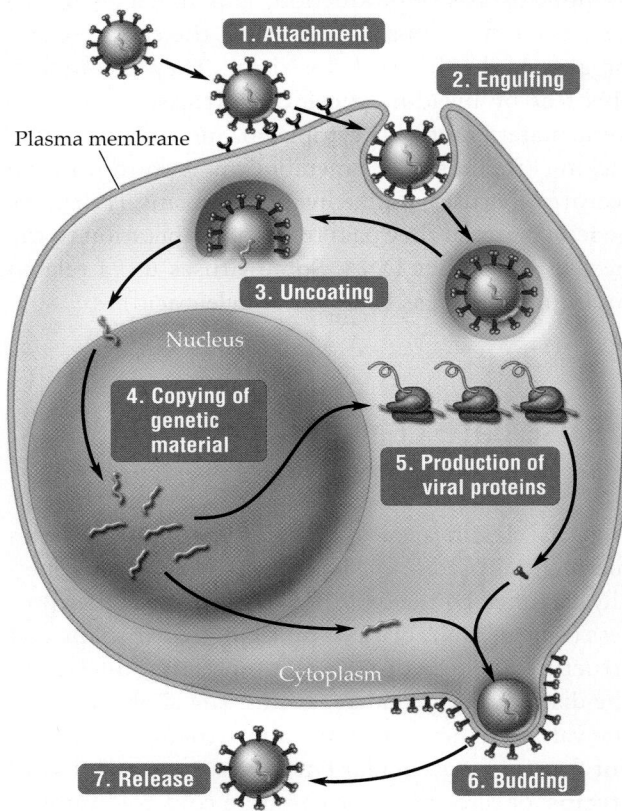

**LIFE CYCLE OF A FLU VIRUS**

1. Attachment
2. Engulfing
Plasma membrane
3. Uncoating
Nucleus
4. Copying of genetic material
5. Production of viral proteins
Cytoplasm
7. Release
6. Budding

**FIGURE 2.15  Viruses Reproduce inside Their Host Cells.**

biotics are useless against viral infections. Instead, life-threatening viral infections, such as HIV-AIDS, are treated with drug cocktails that disable viral reproduction by interfering with the copying of their genetic material. The antiviral drug cocktails can have serious side effects, which is why they are reserved for serious illness.

> **Concept Check**

1. What characteristics of prokaryotes make them so numerous and so successful?

2. What are some differences between archaeans and bacteria?

3. Why are viruses not included in the tree of life?

# All of Us Together

At the beginning of this chapter we learned that a typical human body plays host to about a thousand different species of organisms besides ourselves. Recent studies suggest the number may be several thousand. Virtually all are too small to be seen without a microscope. All are symbionts, organisms that live with us—some essential to good health, some harmless, and others harmful parasites that hurt us.

Bacteria and other microbes can definitely be good. Mice raised in germ-free environments have to eat 30 percent more calories than regular mice to make up for the absence of germs that help digest food. Babies born by C-section (surgically) have different microorganisms than those born vaginally, and the C-section babies are more susceptible to dangerous infections. Some bacteria found in human milk relax the muscles of the infant gut (perhaps preventing colic), and other bacteria in human milk increase weight gain. Children treated with antibiotics early in life seem more likely to develop asthma. Overweight adults have different microorganisms in their guts than do slimmer people, and when people lose weight, the biodiversity of their gut changes.

Which microbes are essential to our health, and why, is a rapidly advancing area of modern biology. Researchers funded by the federal Human Microbiome Project are in the process of cataloging the hundreds of microorganisms that are part of the human body.

These organisms belong to every major taxonomic group discussed in this chapter except Plantae. Representing the Animalia, for example, are eight-legged mites that live in hair follicles and in the oil-secreting sebaceous glands of the skin. At night, these one-third-millimeter-long arthropods come out and walk about on the surface of our skin, speedily covering a few inches each hour. Other animals known to live on or in the human body include a variety of parasitic lice and worms.

Representing the Protista are the parasites *Giardia lamblia*, which gives diarrhea to hikers who drink from contaminated streams, and *Toxoplasma gondii*, often contracted from raw meat or cat feces. About 10 percent of Americans and one-third of the world's population carry *T. gondii*. Some evidence suggests that infection with *T. gondii* can alter the personality and behavior of the animal playing host to this single-called eukaryote.

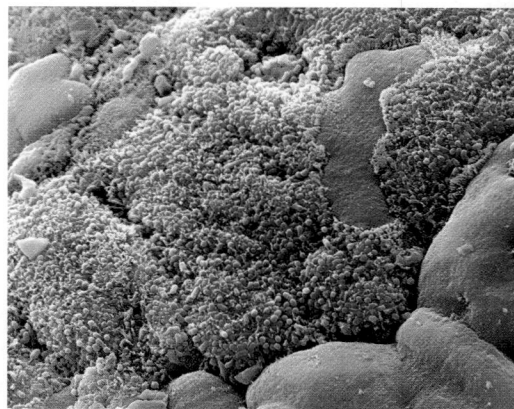

Representing the Fungi are a dozen or more genera of yeasts, molds, and other fungi that inhabit the skin between the toes and other nutritious habitats. Species of the genus *Malassezia* consume oils secreted by the skin, so they live in oily habitats such as the scalp, face, and shoulders.

The overwhelming majority of organisms that live in and on humans, however, are prokaryotes—mostly Bacteria, but a few Archaea as well. Archaean methanogens—extremophiles that generate methane—live in our guts, generating methane, and also in our gums. And then there are the bacteria, hundreds of them—including *Escherichia coli* (*E. coli*), which lives in the colon; numerous kinds of *Bacteroides*, a genus that makes up about a third of all gut bacteria; and others that live on our teeth and make slimy plaque.

Yet even though microbes are clearly essential to good health, some of those that live with us every day sometimes turn on us. Some 40 kinds of staphylococci [sta-**fih-loh-**kah-**keye**] harmlessly inhabit our skin and mucous membranes. But one strain of *Staphylococcus aureus* stands ever-ready to invade a cut and initiate an infection—anything from a pimple on an earlobe to a fatal infection deep in the heart or lungs. Yet this exact strain of staph is a normal resident of our bodies, occurring naturally in 25 percent of humans.

# Exploring the Bacterial Zoo

BY LAURAN NEERGAARD, *Buffalo News*

Antibiotics can temporarily upset your stomach, but now it turns out that repeatedly taking them can trigger long-lasting changes in all those good germs that live in your gut, raising questions about lingering ill effects.

Nobody yet knows if that leads to later health problems. But the finding is the latest in a flurry of research raising questions about how the customized bacterial zoo that thrives in our intestines forms—and whether the wrong type or amount plays a role in ailments from obesity to inflammatory bowel disease to asthma.

Don't be grossed out: This is a story in part about, well, poop. Three healthy adults collected weeks of stool samples so that scientists could count exactly how two separate rounds of a fairly mild antibiotic caused a surprising population shift in their microbial netherworld—as some original families of germs plummeted and other types moved in to fill the gap.

It's also a story of how we coexist with trillions of bacteria, fungi and other microbes in the skin, the nose, the diges-tive tract, what scientists call the human microbiome. Many are beneficial, even indispensable, especially the gut bacteria that play an underappreciated role in overall health.

"Gut communities are fundamentally important in the development of our immune system," explains Dr. David Relman of Stanford University, who led the antibiotic study published earlier this month in Proceedings of the National Academy of Sciences. "Let's not take them for granted." . . .

Antibiotics aren't choosy and can kill off good germs as well as bad ones. But Relman and fellow research scientist Les Dethlesfsen wondered how hardy gut bacteria are, how well they bounce back. So they recruited healthy volunteers who hadn't used antibiotics in at least the past year to take two five-day courses of the antibiotic Cipro, six months apart.

The volunteers reported no diarrhea or upset stomach, yet their fecal samples showed a lot going on beneath the surface. Bacterial diversity plummeted as a third to half of the volunteers' original germ species were nearly wiped out, although some other species moved in. Yet about a week after stopping the drug, two of the three volunteers had their bacterial levels largely return to normal. The third still had altered gut bacteria six months later.

The surprise: Another die-off and shift happened with the second round, but this time no one's gut bacteria had returned to the pre-antibiotic state by the time the study ended two months later.

Of course, antibiotics aren't the only means of disrupting our natural flora. Other research recently found that babies born by Caesarean section harbor quite different first bacteria than babies born vaginally, offering a possible explanation for why C-section babies are at higher risk for some infections. Likewise, the gut bacteria of premature infants contains more hospital-style germs than a full-term baby's.

The big issue is when such differences will matter, something so far, "we're not really smart enough to know," Relman says.

Human beings naturally harbor hundreds of prokaryotes, fungi, protists, and even other animals. The vast majority of these organisms are bacteria that live in our mouths and intestines, including the colon. This article suggests that when we take antibiotics to kill off a bacterium that is making us sick, the antibiotic also changes the flora (bacteria and fungi) that live with us. Biologists who study the human "microbiome" have begun to talk about these communities of bacteria in ecological terms.

There's no evidence that there's one perfect set of bacteria. As far as anyone knows, two people with different bacteria could both be equally healthy. But it's clear that antibiotics have a huge effect on the flora of the gut. For example, recent research has shown that people who take antibiotics have a harder time fighting off a flu virus than people who have not taken an antibiotic. If the lungs are sterile, as doctors have long thought, how could bacteria in the gut help fight flu viruses in the respiratory tract? One possible answer is that the lungs have a natural com-munity of bacteria that repels viruses, just like the bacterial communities of the skin and the gut.

## Evaluating the News

**1.** After the first round of treatment with the antibiotic Cipro, two of the three volunteers in this study recovered their natural flora within a week. What happened after the volunteers took 5 days of Cipro again 6 months later?

**2.** Suggest some reasons why the volunteers reacted differently in the two rounds of antibiotic treatment. If the bacteria were back after a week, where could they have come from? Were they ever really gone? Why would the result be so much different after 6 months?

**3.** Design an experiment that would answer one of the questions posed in question 2.

**SOURCE:** *Buffalo News* (New York), September 28, 2010.

# Summary

## 2.1 The Unity and Diversity of Life

- Biologists use evolutionary trees to model ancestor-descendant relationships among different organisms. The tips of branches represent existing groups of organisms, and each node represents the moment when an ancestor split into two descendant groups.
- The most basic and ancient branches of the tree of life define three domains: Bacteria, Archaea, and Eukarya. All life-forms fall into one of these three domains. The domains are further divided into six kingdoms: the prokaryotic Bacteria and Archaea; and the eukaryotic Protista, Fungi, Plantae, and Animalia
- Closely related groups of organisms share distinctive features that originated in their most recent common ancestor. These shared derived traits are used to identify lineages of closely related organisms.

## 2.2 The Linnaean System of Biological Classification

- The Linnaean hierarchy is a classification system for organizing life-forms. In this scheme, every species of organism has a two-part scientific name indicating its genus and species.
- The lowest level of the Linnaean hierarchy is the species. Each species falls into ever-more inclusive groups: genera, families, orders, classes, phyla, and kingdoms.

## 2.3 Bacteria and Archaea: Tiny, Successful, and Abundant

- The non-Eukarya, commonly called prokaryotes, fall into two domains: Bacteria and Archaea. All prokaryotes are microscopic, single-celled organisms, but the Bacteria and Archaea differ in significant ways, such as in their DNA, plasma membrane structure, and metabolism.
- Prokaryotes can reproduce extremely rapidly and are the most numerous life-forms on Earth. They also have the most widespread distribution. Some prokaryotes, including many archaeans, thrive in extreme environments. Thermophiles, for example, live in extremely hot places, and halophiles in very salty places.
- Prokaryotes exhibit unmatched diversity in methods of getting and using energy and nutrients. Prokaryotes can be chemoheterotrophs, photoautotrophs, chemoautotrophs, or photoheterotrophs.
- Prokaryotes perform key tasks in ecosystems, including photosynthesizing, providing nitrate to plants, and decomposing dead organisms. Prokaryotes are useful to humanity in many ways (for example, in cleaning up oil spills and helping with our digestion), but some of them also cause deadly diseases.

## 2.4 Viruses: Nonliving Infectious Agents

- A virus is a microscopic, noncellular infectious particle.
- Viruses lack some of the characteristics of living organisms: they are not made of cells, and they lack the structures necessary to perform certain activities essential to life. Because viruses exhibit only some of the characteristics of living organisms, most biologists consider them to be nonliving.
- Viruses do have genetic material, and they do evolve. Some viruses use DNA as genetic material, others use a related molecule called RNA.

# Key Terms

aerobe (p. 34)
anaerobe (p. 34)
antibiotic (p. 42)
Archaea (p. 28)
autotroph (p. 38)
Bacteria (p. 28)
bacterial conjugation (p. 38)
bacterial culture (p. 34)
binary fission (p. 35)
biodiversity (p. 28)
bioremediation (p. 41)
chemoautotroph (p. 39)
chemoheterotroph (p. 39)
clade (p. 27)
class (p. 31)

decomposer (p. 40)
domain (p. 27)
endotoxin (p. 42)
Eukarya (p. 28)
eukaryote (p. 30)
evolutionary tree (p. 27)
exotoxin (p. 41)
extremophile (p. 32)
family (p. 31)
fermentation (p. 41)
flagellum (p. 35)
genus (p. 30)
heterotroph (p. 39)
inorganic molecule (p. 38)
kingdom (p. 30)

lateral gene transfer (p. 38)
lineage (p. 27)
Linnaean hierarchy (p. 30)
methanogen (p. 34)
most recent common
  ancestor (p. 27)
nitrogen fixation (p. 41)
node (p. 27)
nucleus (p. 30)
nutrient recycling (p. 40)
order (p. 31)
organic molecule (p. 38)
pathogen (p. 41)
photoautotroph (p. 38)
photoheterotroph (p. 40)

phylum (p. 31)
pilus (p. 35)
plasmid (p. 38)
prokaryote (p. 28)
quorum sensing (p. 35)
RNA (p. 43)
scientific name (p. 30)
shared derived trait (p. 28)
taxon (p. 31)
taxonomy (p. 31)
tree of life (p. 26)
viral strain (p. 43)
virus (p. 42)

# Self-Quiz

1. Which of the following can be concluded from Figure 2.4?
   a. Archaeans and protists had a common ancestor more recently than plants and animals did.
   b. Plants and animals diverged more recently than fungi and animals did.
   c. Animals and fungi are more closely related to each other than either is to plants.
   d. Archaeans are the most ancient group known.

2. A node (represented in Figure 2.4 by a circle) represents
   a. species that are extinct.
   b. the descendant lineage.
   c. the most recent common ancestor of two or more descendant lineages.
   d. the shared derived feature.

3. Which of the following groupings list only domains?
   a. Eukarya, Bacteria, and Animalia
   b. Plantae, Protista, and Archaea
   c. Archaea, Bacteria, and Eukarya
   d. Bacteria, Archaea, and Ciliates

4. Most bacteria
   a. have a nucleus.
   b. can be cultured in a lab dish with special nutrients.
   c. are smaller than the average eukaryotic cell.
   d. are multicellular.

5. Archaeans
   a. can be autotrophs but not heterotrophs.
   b. are responsible for a number of human diseases.
   c. lack cellular organization.
   d. can be found in the human body.

6. Quorum sensing
   a. is the transfer of plasmid DNA from one bacterium to another.
   b. enables bacteria to form biofilms.
   c. is the formation of thick-walled dormant structures, called spores, under conditions unfavorable for growth.
   d. enables bacteria to switch from cellular respiration to fermentation when they sense that oxygen levels are low.

7. Viruses
   a. divide by a form of cell division known as binary fission.
   b. are considered nonliving because they contain no hereditary material, such as DNA.
   c. infect humans but not bacteria.
   d. lack the ability to acquire energy independently.

8. According to the Linnaean system of classification,
   a. all life-forms can be divided into two domains: prokaryotes and eukaryotes.
   b. viruses are classified in the same domain as bacteria.
   c. phylum is a broader, more inclusive category than order.
   d. the scientific name of each unique organism consists of two parts: the genus and the family.

# Analysis and Application

1. Name one of your favorite wild animals and one of your favorite wild plants. Find the scientific names of each, and research their taxonomic classification (that is, determine the phylum, class, order, and family they belong to). Name two other species that are classified in the same order, but *not* the same family, as your favorites. Are there any species that belong to the same family as your favorites but *not* to the same order? Explain.

2. Describe the ecological importance of bacteria and archaeans.

3. Choose one of these two hypotheses:
   1. Viruses are living organisms.
   2. Viruses are not living organisms.

   Given what you know about the characteristics of life, (a) state some testable predictions about viruses based on the hypothesis you chose; and (b) propose an experiment to test one of your predictions. (For example, if you chose hypothesis 1, then one prediction might be, "Viruses require energy from their environment." An experiment to test this prediction might be described like this: "I will provide one group of viruses with a lot of energy in the form of nutrients, heat, and light, and another group of viruses with no energy. If the first group of viruses multiplies and the second group does not, these observations will support my prediction and hypothesis.") In dreaming up your experiment, don't hold back.

4. The tree on the next page depicts evolutionary relationships, based on DNA comparisons, among some archaeal genera. The growth conditions they prefer are color-coded. The icon next to each genus indicates cell shape. Judging by the evolutionary relationships implied in the tree, would you say that cell shape is a suitable criterion for classifying members of the Archaea? Are the habitat preferences a good taxonomic criterion? Explain your answer. The archeans whose name begins with "Methano . . . " are methanogens. Do all the methanogens in this tree form a clade of their own? Can we assume that *Desulfurococcus* is the most complex and recently evolved genus because it is farthest to the right in this evolutionary tree?

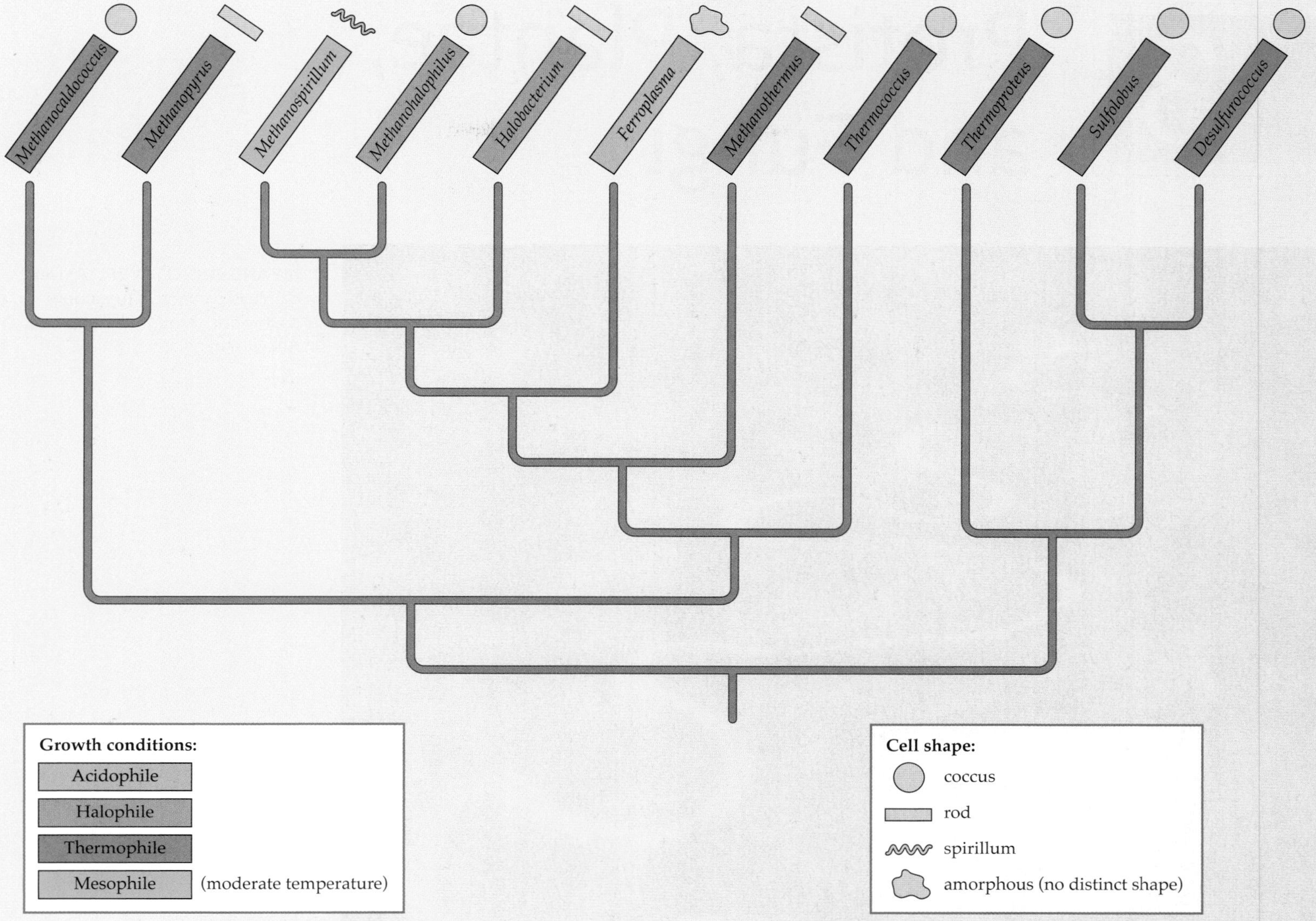

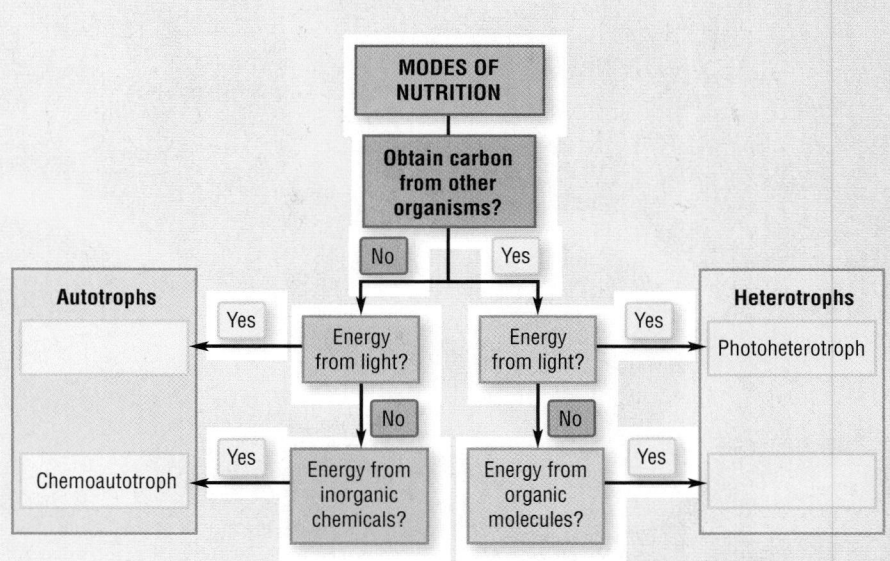

Methanocaldococcus
Methanopyrus
Methanospirillum
Methanohalophilus
Halobacterium
Ferroplasma
Methanothermus
Thermococcus
Thermoproteus
Sulfolobus
Desulfurococcus
Growth conditions:
Acidophile
Halophile
Thermophile
Mesophile
(moderate temperature)
Cell shape:
coccus
rod
spirillum
amorphous (no distinct shape)

5. Examine the flowchart on the right depicting the four main types of nutrition found among bacteria. Complete the diagram by labeling the two empty boxes.

MODES OF NUTRITION
Obtain carbon from other organisms?
No
Yes
Autotrophs
Yes
Energy from light?
Energy from light?
Yes
Heterotrophs
Photoheterotroph
No
No
Yes
Chemoautotroph
Energy from inorganic chemicals?
Energy from organic molecules?
Yes

# 3 Protista, Plantae, and Fungi

MEANDERING RIVER. A river displays distinctive meanders as it winds through the upper Amazon basin in Peru.

# Did Plants Teach Rivers to Wander?

Look at a river flowing across a gentle landscape and one of the first things you'll notice about it is that it meanders from side to side, winding in ever-wider loops. At every loop, one side is steeply cut and the other side has a low beach of sand or rocks.

Over geologic time, the meanders themselves sweep from side to side, touching first one side of a valley, then the other. Gradually the meanders move downstream, like the coils of a snake moving from head to tail. River meanders wander across floodplains, creating new deposits of sand and gravel, tree-lined sloughs, and lakes that teem with wildlife. Rivers gradually create rich, flat bottomland where marshes, meadows, and farms appear. As rivers sculpt the landscape, they promote the diversity of plants, fungi, animals, and other organisms. Meandering rivers support more bird species than straight rivers do. Rivers are a valuable part of the ecology of Earth.

Rivers that have been damaged by development, deforestation, or floods sometimes lose their curves. Instead of helping to create fertile soils, they wash them away. As a result, re-creating a river's meandering shape has been a major goal of river restoration projects that bring back natural vegetation and habitat. Yet amazingly, only recently have scientists discovered what makes rivers meander.

Until now, scientists didn't know whether rivers' meanders were caused by things like the volume of water or by qualities of the rocky sediments that make up the soil. Some scientists tried to explain meandering with mathematical equations. More hands-on researchers built artificial indoor streams using different kinds of gravel, sand, and mud—all of which refused to meander. Out in the field, restoration ecologists tried digging winding channels that looked like meandering streams through fields, only to have them form wide, shallow streams in the next big storm.

What were scientists missing? Why do natural rivers meander, and what will you learn in this chapter that relates to how rivers meander?

Why do rivers meander? The answer has to do with how plants with roots came to dominate on land 425 million years ago—a topic we'll explore in this chapter. The greening of Earth's continents through the evolution of plants is one of the most significant events in the history of Earth.

**MAIN MESSAGE**  Subcellular compartmentalization and sexual reproduction are among the evolutionary innovations of the Eukarya

## KEY CONCEPTS

- The organisms in the domain Eukarya are distinguished by having a nucleus and complex subcellular organization. Sexual reproduction and multicellularity are among the key evolutionary innovations of eukaryotes.

- The domain Eukarya encompasses several large groups of organisms— plants, fungi, and animals among them. Eukaryotes that are neither plants nor fungi, and not animals either, are placed in the catchall category Protista (protists).

- Plants are descended from green algae and have evolved numerous evolutionary innovations to adapt to life on land. Vascular tissue enabled ferns, gymnosperms, and flowering plants to grow tall. Seeds first evolved among the gymnosperms. Angiosperms evolved flowers, and they enclose their seeds in the fruit.

- Fungi include yeasts, molds, and mushrooms. The fungi are distinguished by their mode of nutrition: they acquire their nutrients by absorption, digesting their food outside of their bodies. As decomposers, fungi play a critical role in recycling nutrients in ecosystems.

- A lichen is a mutually beneficial association between a fungus and a photosynthetic microbe, usually a green alga or a cyanobacterium. Most plant roots in natural habitats form close associations, called mycorrhizae, with beneficial fungi.

THE MIND-BOGGLING DIVERSITY OF LIFE began with the original prokaryotic ancestor that probably arose about 3.5 billion years ago. We noted in Chapter 2 the amazing array of bacteria and archaeans that have descended from that universal ancestor and that now colonize nearly every imaginable habitat, from ocean depths to mountaintops. The evolution of the Eukarya, the third domain of life, was the next momentous event in the history of life.

In this chapter we explore three of the four kingdoms of Eukarya: Protista, Plantae, and Fungi. (We will review the fourth eukaryotic kingdom—Animalia—in the next chapter). The catchall kingdom of Protista is the first stop in our journey through the world of the Eukarya. Then we survey the kingdoms Plantae

**FIGURE 3.1 Fossil Eukaryote**

*Grypania spiralis*, the oldest fossil eukaryote currently known, dates to 2.1 billion years ago. It resembles certain modern-day multicelluar red algae.

and Fungi, which appear at nearly the same time in the fossil record, working sometimes as friends and sometimes as foes to transform the land. We begin by describing how the eukaryotes arose and what sets them apart from prokaryotes.

# 3.1 The Dawn of Eukarya

If you study the chart (Milestones in the History of Life on Earth) printed inside the back cover of this book, you will notice that prokaryotes dominated the scene for about a billion years after life evolved on Earth. On the basis of chemical clues from ancient rocks, some scientists suggest that eukaryotes—organisms with a nucleus—had evolved by 2.7 billion years ago. But the oldest fossil that is indisputably a eukaryote comes from iron-rich mud shale that formed 2.1 billion years ago. The fossil *Grypania spiralis* (**FIGURE 3.1**) is similar to modern-day red algae that form long ribbons of relatively large cells.

The evolution of the Eukarya was a major milestone in the history of life, not only because this domain spawned organisms like us, but also because eukaryotes represent new ways of organizing cell structure and novel strategies for propagating life. These are the key evolutionary innovations of eukaryotes:

- The presence of a nucleus and many other membrane-enclosed internal compartments
- Larger average cell size compared to prokaryotes
- Sexual reproduction
- Multicellularity (in some groups)

## Eukaryotes have subcellular compartmentalization and larger cells

The defining feature of the Eukarya is that they possess a true nucleus: instead of lying free in the cytoplasm, eukaryotic DNA is enclosed in two concentric layers of cell membranes that together make up the nuclear envelope. Further, eukaryotes have a greater variety of other membrane-enclosed cytoplasmic compartments, or subcellular compartments, compared to the prokaryotes. Some prokaryotes do have cytoplasmic compartments, including spherical structures, called vesicles, that store nutrients

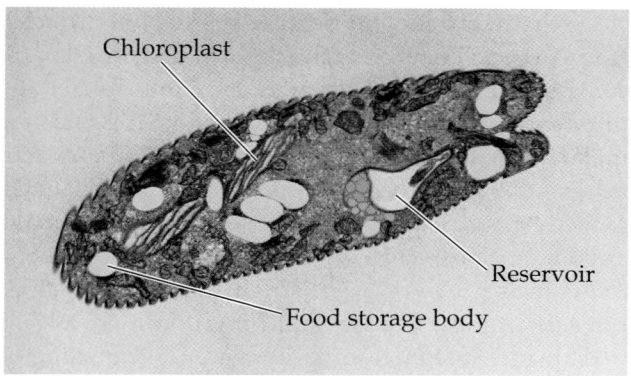

Chloroplast

Reservoir

Food storage body

**FIGURE 3.2 Internal Organization in *Euglena***

The compartments seen in this green alga, *Euglena gracilis*, include structures specialized for conducting photosynthesis (chloroplasts) and storing food. The protist uses a long, whiplike structure (the flagellum) to swim about. The flagellum is not visible in this color-enhanced electron microscope photograph. The reservoir is a pocket in which flagella are anchored. The following additional membrane-enclosed compartments are also visible in this photograph: mitochondria (purple), lipid bodies (dark orange), Golgi apparatus (blue). The functions of these organelles are described in Chapter 6.

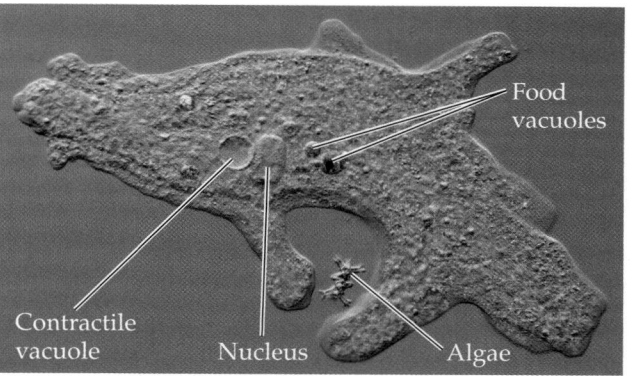

Food vacuoles

Contractile vacuole

Nucleus

Algae

**FIGURE 3.3 An Amoeba Digesting Its Prey**

The prey being ingested by the amoeba are a type of single-celled algae known as desmids.

and other substances. However, as illustrated by the *Euglena* in **FIGURE 3.2**, the average eukaryotic cell displays much greater complexity in subcellular compartmentalization. Each type of internal compartment in *Euglena* specializes in conducting a unique set of functions. Through specialization and division of labor, these subcellular compartments can function with greater efficiency.

Compartmentalization of the cell interior also enabled novel cell functions not seen in prokaryotes. Consider, for example, that unlike prokaryotes, many heterotrophic eukaryotes can engulf their prey and digest them internally, as the amoeba in **FIGURE 3.3** is doing. These eukaryotes possess an elaborate system of internal membrane compartments for digesting engulfed prey, ridding the cell of waste, and storing surplus food. Prokaryotes, in contrast, digest their prey externally—a more wasteful way of obtaining nutrients.

On average, eukaryotic cells are 10 times wider than prokaryotic cells, and their cell volume is a thousandfold greater. The great majority of prokaryotes measure about 1 micrometer or less in width, and generally speaking, prokaryotes cannot afford to get much larger than that. As cell diameter increases, cell volume increases more dramatically than surface area: surface area increases as a squared function, but volume increases as a cubic function. It takes much longer for vital nutrients and crucial molecules to move from one point to the next in a large volume of cytoplasm. In the highly competitive environment in which most of them live, prokaryotes must get nutrients in and waste out as quickly as possible. A larger cell volume would hamper that exchange. Consequently, unless there are significant payoffs in being larger, cells tend to be small.

Most human cells measure about 10 micrometers across, and the fossil eukaryote in Figure 3.1 is not unlike its many present-day algal relatives in being about half a millimeter wide. Like any other adaptive trait, large size is likely to be useful only in particular contexts—that is, within the framework of a given habitat and in a particular set of circumstances. Within limits, a larger cell volume translates into higher metabolic capacity for a cell, meaning that the cell can acquire and store more food. Being larger is often advantageous for predatory organisms, those heterotrophs that live off other organisms. The amoeba in Figure 3.3 can readily engulf whole bacterial cells and even other eukaryotes, such as single-celled algae.

Getting big can also benefit potential prey species. The *Euglena* in Figure 3.2 is too big a meal for the amoeba, for example. Incidentally, the researchers who discovered *Epulopiscium*, the giant fish gut bacterium in Figure 2.8, hypothesize that large size protects this unusual prokaryote from being eaten by other prokaryotic and eukaryotic residents that share its home in the intestines of the surgeonfish.

## Sexual reproduction increases genetic diversity

Sexual reproduction is perhaps the most outstanding evolutionary innovation of the Eukarya. By combining genetic information from two individuals, the parents, **sexual reproduction** produces offspring that are genetically different from each other and from both parents. Sexual reproduction is one means by which natural populations become genetically diverse. As we noted in Chapter 1, genetic variation is the raw material for evolution by natural selection. If the environment changes, a genetically diverse population is more likely to evolve adaptively than is a genetically uniform population. For example, if a new virus strain sweeps through a genetically diverse deer population, chances are that some individuals will be resistant to that virus and these animals will survive and reproduce. Resistant individuals will therefore become more common in the next generation, which, by definition, will have evolved.

## SEXUAL REPRODUCTION IN EUKARYOTES

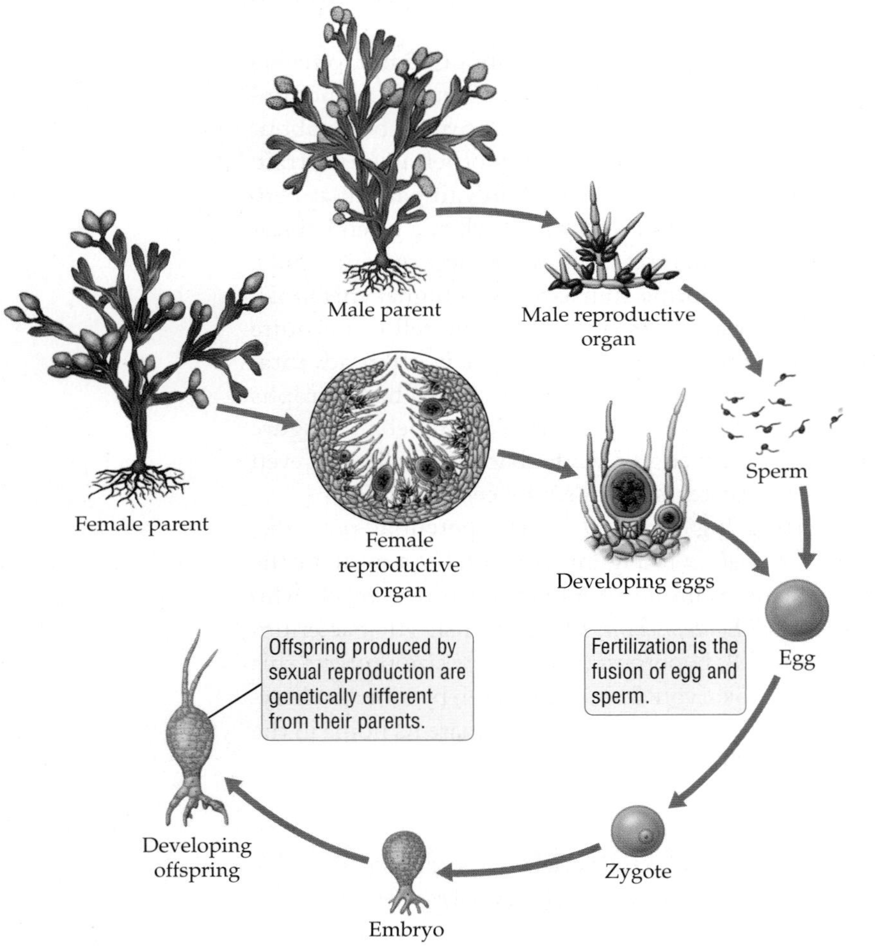

**FIGURE 3.4 An Example of Sexual Reproduction**
The rockweed, a brown alga, produces eggs and sperm that unite to create offspring, in a life cycle that resembles that of animals. Other seaweeds have different, and more complex, life cycles.

Sexual reproduction requires the fusion of nuclei from two different sex cells, known as **gametes**. In some species, the gamete-producing individuals are different enough from each other to be classified as male and female (which is the case in humans and other animals). The gametes produced by female animals are called *eggs*, and those produced by male animals are *sperm*. However, the gamete-producing individuals need not be strikingly different (which is the case in most protists and fungi). In many algae, most plants, and even some animals, one individual—known as a *hermaphrodite*—can produce both male and female gametes. Most hermaphroditic species have mechanisms that prevent self-breeding, so that a given individual's male gametes fertilize another individual's female gametes, not its own. **FIGURE 3.4** illustrates the role of sexual reproduction in the life cycle of a brown alga known as rockweed. The rockweed life cycle resembles the way sexual reproduction works in animals, but protists actually display many variations on the basic theme of sexual reproduction that is depicted in this figure.

**Asexual reproduction**, which generates genetically identical offspring, is also common among the Eukarya. As with sexual reproduction, there is variety in asexual reproduction among protists. *Euglena* and its relatives can produce **clones**, genetically identical offspring, by partitioning a single cell into many minicells, and then releasing the minicells as free-swimming offspring called *zoospores*. Protists like *Amoeba* split into two in a process similar to binary fission in prokaryotes (see Chapter 2). Many large multicellular algae ("seaweeds") can fragment into pieces, each piece developing into a new individual the way cuttings from some plants can grow into whole new plants.

## Multicellularity evolved independently in several eukaryotic lineages

Most protists are single-celled, but multicellular forms evolved several times among different lineages of the eukaryotes, including some groups that are currently lumped under Protista. The fungi include some single-celled species and many multicellular forms. Plants and animals are both exclusively multicellular.

A multicellular organism is a well-integrated assemblage of genetically identical cells in which different groups of cells perform distinctly specialized functions. Multicellular organisms therefore enjoy the benefits of cell specialization, which produces more efficient functioning through division of labor. Multicellularity also enables the individual organism to grow large, which can be advantageous for evading potential predators. A bigger individual can often gather resources from its environment more effectively than can a smaller individual. Having more resources, such as light or food, usually translates into producing more surviving offspring, the ultimate measure of biological success. (But remember, no single strategy is adaptive in all habitats under all circumstances; being small and nimble can be more adaptive than being larger for a given species living in its special niche in a particular habitat.)

The giant kelp, which can grow to 60 meters, illustrates the adaptive benefits of multicellular organization (**FIGURE 3.5**). Cells at the base form a special tissue, the holdfast, that anchors the giant seaweed and keeps it from being washed out to sea. Seaweed blades are broad and flat, the better to capture light for photosynthesis. Loosely arranged cells form swollen bladders (technically, *pneumatocysts*) that trap air and give buoyancy to the kelp body; without these flotation devices, the kelp's large body would collapse to the seafloor, where it would perish for lack of sunlight. Being large presents a challenge: how to quickly deliver food from one part of the body to another. Food-conducting tubes in the interior of this brown alga enable transport of photosynthetic sugars from the blades to all other parts of the kelp.

## 3.2 Protista: The First Eukaryotes

The kingdom **Protista** is an artificial grouping, defined only by what members of this group are *not*: protists are neither plants nor animals nor fungi, nor are they bacteria or archaeans. The kingdom was first proposed by nineteenth-century biologists attempting to deal with puzzling and poorly understood forms of life: species as disparate as amoebas that make people sick, dinoflagellates that spin like a top and cause deadly red tides, mustard-yellow masses of slime that creep along on tree trunks, and monstrous

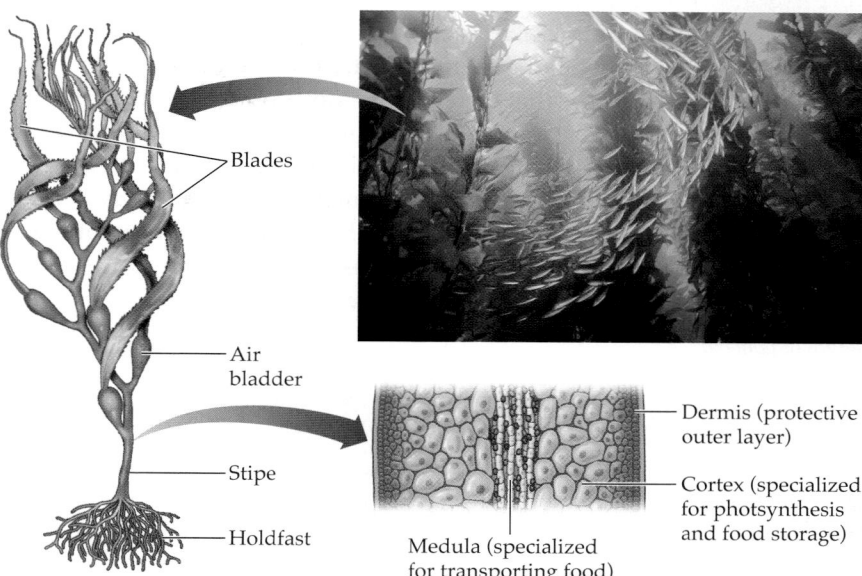

**MULTICELLULAR ORGANIZATION IN EUKARYOTES**

Blades

Air bladder

Stipe

Holdfast

Medula (specialized for transporting food)

Dermis (protective outer layer)

Cortex (specialized for photsynthesis and food storage)

**FIGURE 3.5 Specialized Cells Perform Different Functions in the Giant Kelp**

seaweeds that some said could trap sailing ships in a faraway tideless sea.

Today, nearly all biologists agree that Protista is an artificial kingdom, composed of groups with very different evolutionary histories. A number of classification schemes have been proposed that split the protists into separate kingdoms, with some members sometimes assigned to the Plantae, Fungi, or Animalia. However, there is no consensus yet on how many new kingdoms should be carved out, which groups should be placed in them, and what the new kingdoms should be called. We therefore retain "Protista" as a label of convenience. For our purposes, Protista is a catchall category of eukaryotic organisms that have not been formally assigned to other kingdoms or to separate kingdoms of their own.

## Protists are not a natural grouping

Much remains unclear about the evolutionary relationships of the protists to one another and to other living organisms. As a result, there are a number of competing hypotheses, and therefore different evolutionary trees, postulating the relationships of the protists. The evolutionary tree in **FIGURE 3.6** depicts a recent interpretation of the main lineages that are deduced to have evolved from the ancestor of all eukaryotes. The tree is based on many different lines of evidence, including comparisons of cell structure, metabolic chemistry, and the DNA code; and it makes clear that protists do not constitute just one distinct branch (clade) on the tree of life. The presence of plants, animals, and fungi on various branches of this tree shows that Protista is not a natural grouping; that is, it is not a cluster of organisms that are all more closely related to each other than to any organism that falls *outside* the category of protists.

Some groups traditionally placed in Protista are actually more closely related to plants or animals than they are to other protists. For example, red and green algae share a most recent common ancestor with land plants, and these three groups therefore form a distinct branch on the evolutionary tree. The lineages that gave rise to animals and fungi diverged from each other more recently than did the lineages that gave rise to animals, fungi, *and* amoebozoans (which include amoebas and slime molds). Because they shared a common ancestor more recently (as Figure 3.6 shows), animals and fungi are closer to each other

evolutionarily than either of them is to amoebas and slime molds.

The major groups of protists have traditionally been imagined as falling into two broad categories: the **protozoans**, which are nonphotosynthetic and motile (capable of moving); and the **algae** (singular "alga"), which are photosynthetic and may or may not be motile. The evolutionary tree in Figure 3.6 reveals that the categorization into protozoans and algae is also artificial in that it is evolutionarily meaningless. For example, species of *Euglena* are often green and photosynthetic, but this group is about as distantly related to green algae and plants as it is to amoebas and slime molds. Instead, *Euglena* is more closely related to diplomonads such as *Giardia*, a colorless, single-celled parasite that lives in animal guts and causes painful diarrhea, as campers who drink untreated water from infested streams know all too well. In other words, *Euglena* and *Giardia* are in a distinct clade that is neither plantlike nor animal-like; they belong in a group of their own and are likely to gain their own kingdom when the dust finally settles on the reorganization of the protists.

## THE KINGDOM PROTISTA

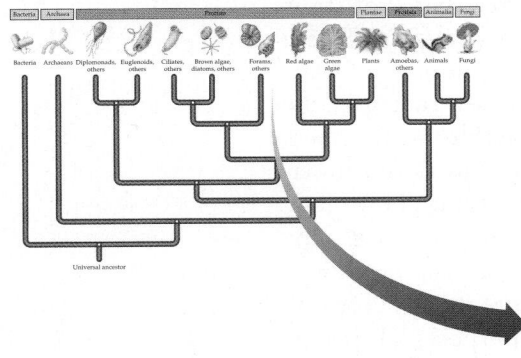

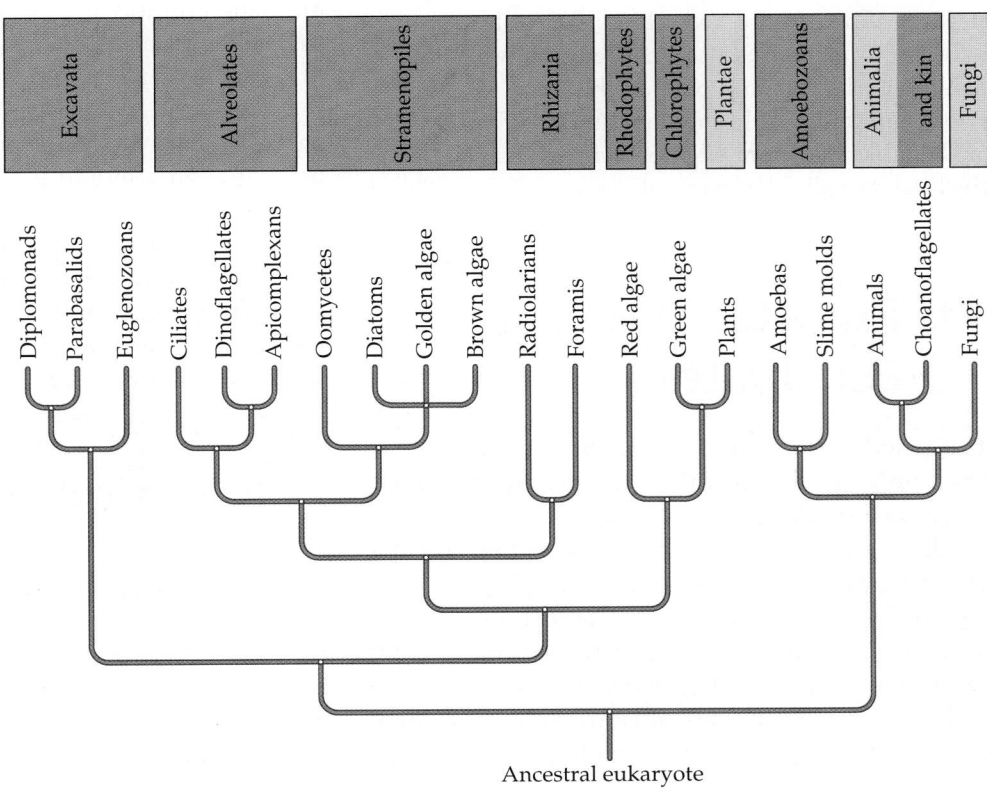

**FIGURE 3.6**

**Evolutionary Tree of the Protista**

The tree shows evolutionary relationships hypothesized for some of the major groups of protists. The colored boxes show labels commonly used for some of the "supergroups" among the protista.

## Most protists are single-celled and microscopic

In part because they are not a single evolutionary lineage, protists are diverse in size, shape, cellular organization, modes of nutrition, and life cycle (see Figure 3.8, pp. 58–59). Most protists are single-celled and microscopic, and these forms are often pigeon-holed with prokaryotes under the tag of "microbes." Most microbial protists are motile. They can swim with the help of one or more *flagella*, or by waving a carpet of tiny hairs called *cilia*. Some microbial protists, like the amoeba in Figure 3.3, crawl on a solid surface with the help of cellular projections called *pseudopodia* (literally, "false feet").

Some protists, especially the ones that live as parasites within the bodies of multicellular organisms, are single cells bounded by nothing more than a flexible plasma membrane. Others are covered in protective sheets, heavy coats, chalky plates, or other types of armor. Diatoms, for example, are renowned for the exquisite beauty of their glassy coverings.

How do the armored protists stay afloat, despite their heavy casing? Many produce oily chemicals that make them buoyant because oil is lighter than water. When the protists die, the heavy casings shower down to the ocean floor, creating deposits thick and pure enough to be mined. Diatomaceous earth, used in swimming pool filters, is mined from such deposits. If you examine scrapings from diatomaceous earth under a microscope, you will see a kaleidoscope of glassy shapes, the remains of ancient diatoms.

The white color of the famed white cliffs of Dover, on the southern shores of England, comes from the casings of fossil coccolithophores. The intricately patterned armor plates of coccolithophores are made of calcium carbonate, or chalk.

Red, green, and brown seaweed are protists with multicellular bodies. Some groups have evolved from free-living single cells into multicellular associations that function to varying degrees like more complex multicellular individuals. Among the more interesting of these multicellular-like associations of protists are the slime molds, protists that were originally mistaken for fungi. Commonly found on rotting vegetation, slime molds are protists that eat bacteria and live their lives in two phases: as independent, single-celled creatures and as members of a multicellular association (**FIGURE 3.7**).

## Protists are autotrophs, heterotrophs, or mixotrophs

The protists known as algae play a vital role as producers, especially in oceans, lakes, rivers, and streams (**FIGURE 3.8**). Algae are autotrophic protists that carry out oxygen-generating photosynthesis: they use energy from sunlight to combine carbon dioxide and water, making sugar molecules in the process and releasing oxygen gas. Roughly half of the photosynthesis that occurs on Earth takes place in the oceans; and algae, together with photosynthetic bacteria, are responsible for nearly all of that prodigious activity. In other words, algae and photosynthetic bacteria carry out as much photosynthesis in the world's oceans as all the crops, forests, and other types of vegetation do on land. Seaweeds create food-rich habitats in coastal areas, but much of the photosynthesis in open water—both fresh water and salt water—is carried out by free-floating, single-celled algae called **phytoplankton**. **Plankton** [from the Greek *planktos*, "drifting"] is a general term for microbes that drift at or near the surface of water bodies. Diatoms and coccolithophores are some of the most abundant marine phytoplankton.

Variety in diatom shapes

Coccolithophores and the white cliffs of Dover, England

### LIFE CYCLE OF A SLIME MOLD

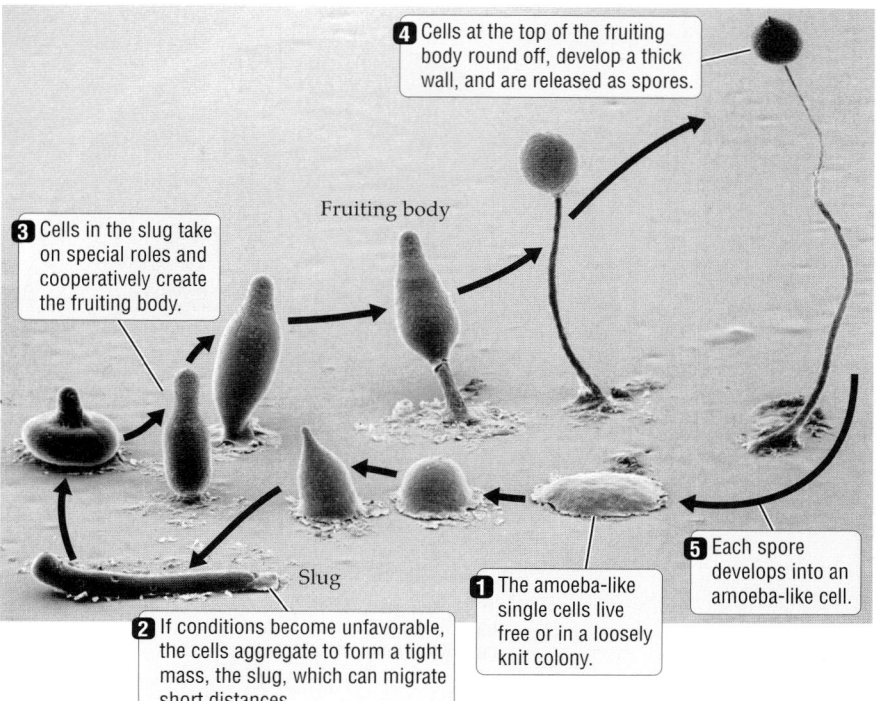

**4** Cells at the top of the fruiting body round off, develop a thick wall, and are released as spores.

**3** Cells in the slug take on special roles and cooperatively create the fruiting body.

Fruiting body

**5** Each spore develops into an amoeba-like cell.

**1** The amoeba-like single cells live free or in a loosely knit colony.

**2** If conditions become unfavorable, the cells aggregate to form a tight mass, the slug, which can migrate short distances.

Slug

FIGURE 3.7 **Social Behavior in a Slime Mold**

## Life Strategies

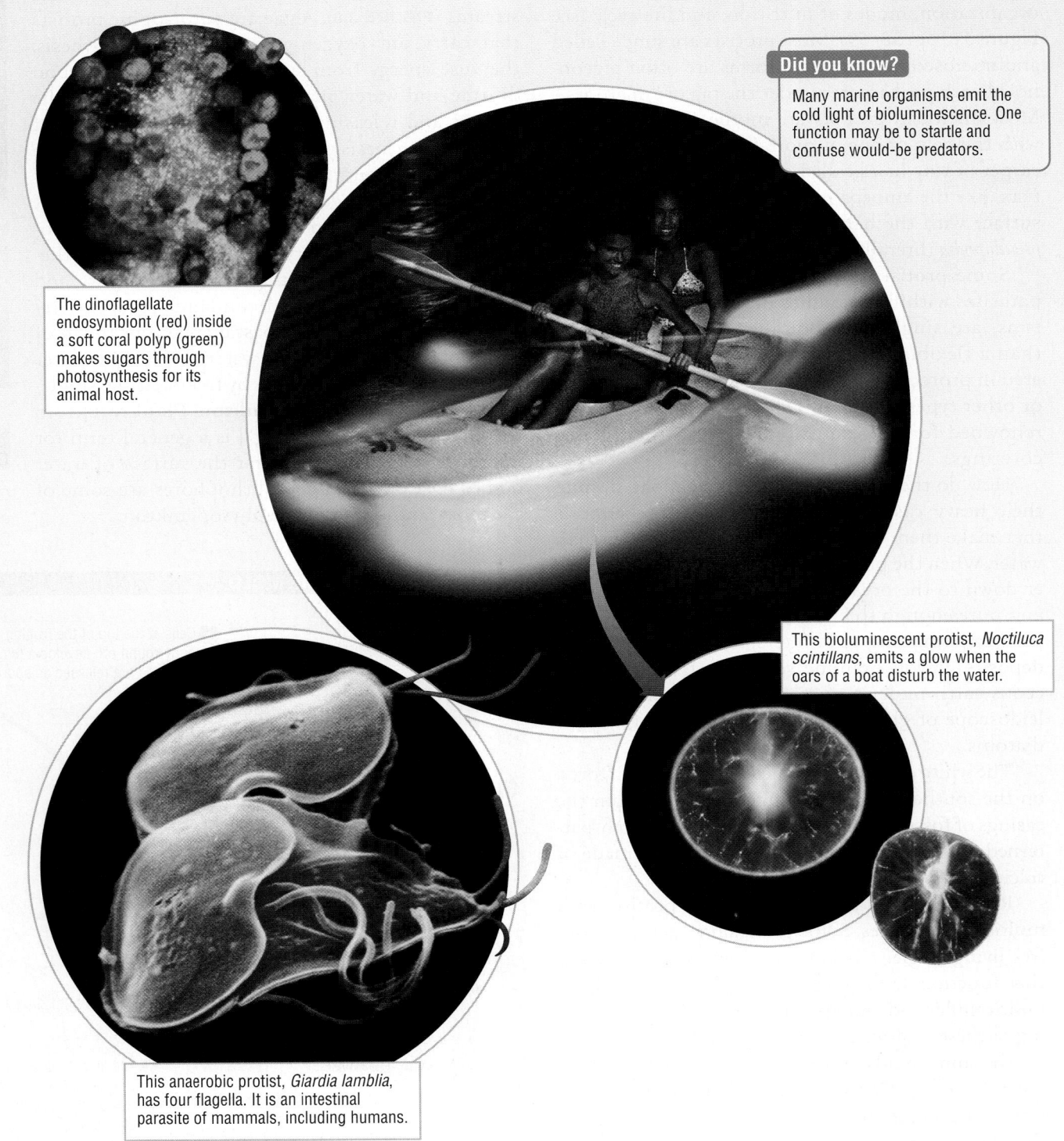

The dinoflagellate endosymbiont (red) inside a soft coral polyp (green) makes sugars through photosynthesis for its animal host.

**Did you know?**
Many marine organisms emit the cold light of bioluminescence. One function may be to startle and confuse would-be predators.

This bioluminescent protist, *Noctiluca scintillans*, emits a glow when the oars of a boat disturb the water.

This anaerobic protist, *Giardia lamblia*, has four flagella. It is an intestinal parasite of mammals, including humans.

**FIGURE 3.8  Extreme Diversity: The Protists**

# Structure

The three starlike shapes with red centers are foraminiferans, seen here with other plankton.

This multicellular red alga, *Antithamnion plumula*, has special cells that attach it to coastal rocks. The bulbous structures are reproductive organs.

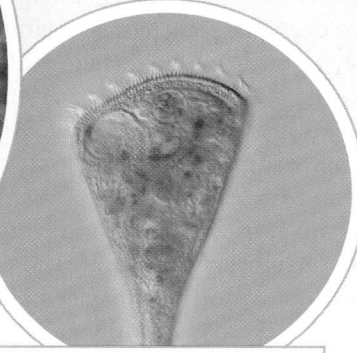

*Stentor coeruleus* is a ciliated protozoan common in freshwater worldwide. Up to 2 mm in length, it is one of the largest single-celled organisms.

# Energy Acquisition

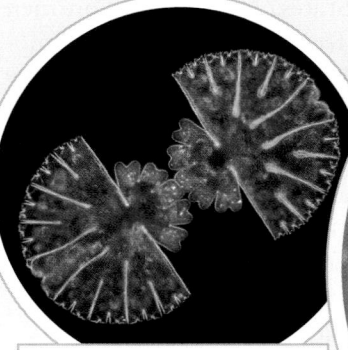

Photosynthesis in *Micrasterias thomasiana*, a single-celled desmid (a type of alga), is enabled by its many chloroplasts (green). One cell has divided to produce the two cells seen here.

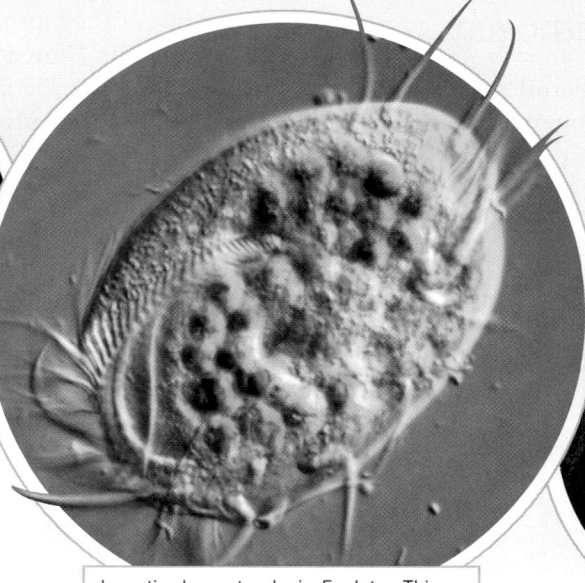

Ingestive hererotrophy in *Euplotes*. This ciliated protist has engulfed more than a dozen algal cells (green) that are at various stages of degradation. This single-celled protozoan can "walk" on its cilia.

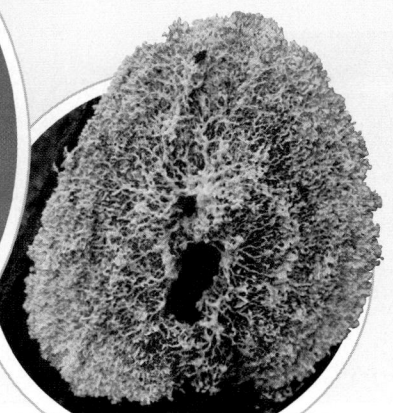

Absorptive heterotrophy is exemplified by the scrambled-egg slime mold (*Fuligo septica*), growing on a rotting log.

**Concept Check Answers**

**1.** Among the novel features of eukaryotes are complex subcellular compartments, sexual reproduction, and multicellularity in some forms.

**2.** Greater efficiency through specialization and division of labor among subcellular compartments, and novel cellular functions such as internal digestion of food.

**3.** Rather than forming one distinct clade on the tree of life, Protista includes groups that are descended from many different lineages and that are more closely related to organisms in other kingdoms of life, such as plants, animals, or fungi, than to other protists.

Many protists are consumers, or heterotrophs, in that they live off other organisms. Heterotrophic prokaryotes and protists, together with microscopic animals, form **zooplankton**. Foraminiferans (or, more commonly, "forams") and radiolarians are among the important planktonic protists in the ocean, occurring in such vast numbers that their remains are a major component of seafloor sediments.

Many heterotrophic protists function as decomposers in terrestrial (land-based) or aquatic (water-based) ecosystems. **Decomposers** are heterotrophs that play a crucial role in an ecosystem: they break down waste or dead material, releasing vital nutrients into the nonliving environment, where those nutrients can be taken up by producers and fed back into the food chain.

Some protists are nutritional opportunists, or **mixotrophs**, organisms that can use energy and carbon from a variety of sources to fuel their growth and reproduction. Mixotrophic algae function as photoautotrophs *or* as heterotrophs, depending on environmental conditions. *Euglena* and some of its relatives, as well as many dinoflagellates (such as the *Pfiesteria* we met in Chapter 1), are mixotrophs. When light and mineral nutrients such as nitrogen are abundant, they may live photoautotrophically; but if the resources become scarce, these versatile protists switch to engulfing prey organisms such as bacteria or absorbing organic molecules from their surroundings.

## Some protists are pathogens

Although most protists are harmless, many of the best-known protists are **pathogens** (disease-causing agents). Among them are toxic dinoflagellates that sometimes experience huge population explosions, known as blooms. Dinoflagellate blooms sometimes cause the water to turn red from a reddish pigment concentrated inside the cells of these protists (**FIGURE 3.9**). Dinoflagellates can produce a variety of harmful chemicals (toxins), including some that cause nerve and muscle paralysis in humans and other mammals. Some of these toxins accumulate in shellfish and cause *paralytic shellfish poisoning* in humans and wild animals who eat the shellfish. The toxin, which is not destroyed by cooking, can be stored in the shellfish for weeks and even for a year or more in some species of clams. Scientists do not have a full understanding of what causes the sudden blooms, but the frequency of such blooms has been rising around the world, and pollution from fertilizer runoff and sewage are thought to be among the culprits.

Protists left their mark on human history forever when a water mold (an oomycete, mistakenly referred to as a fungus sometimes) attacked potato crops in Ireland in the 1800s, causing the disease known as potato blight. The resulting widespread loss of potato crops caused a devastating famine and a major emigration of Irish people to the United States in the 1840s.

*Plasmodium*, an apicomplexan **[ay-pee-kum-*PLEX*-un]**, causes malaria, which kills millions of people around the world each year—more than does any other infectious disease except AIDS. *Trichomonas vaginalis* **[*TRIH*-kuh-*MOH*-nus *VAJ*-ih-*NAL*-is]**, which belongs to a protist group most closely related to diplomonads, causes one of the most common sexually transmitted diseases in the United States. About 7 million men and women have the infection, known as trichomoniasis **[*TRIH*-kuh-muh-*NYE*-uh-sis]**, which is readily cured with medications.

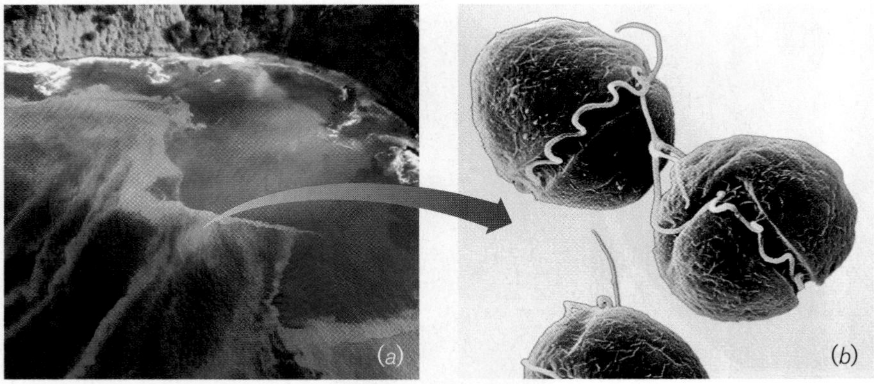


**FIGURE 3.9 Red Tides Can Close Beaches**

Red tides (*a*) are caused by a large increase in the population of pigmented protists such as dinoflagellates. Some dinoflagellates, such as the *Gymnodinium* seen here (*b*), produce nerve toxins that can poison mammals.

**Concept Check**

1. Name three evolutionary innovations of Eukarya.
2. What is the adaptive value of subcellular compartmentalization?
3. Why is Protista an artificial grouping of organisms?

# 3.3 Plantae: The Green Mantle of Our World

Life on Earth began in the water, where it stayed for nearly 3 billion years. It was only when the kingdom **Plantae**—the plants—evolved that life took to land in a big way. The first plants evolved about 470 million years ago, most likely from a lineage of multicellular green algae that made the move to land and became successfully established in freshwater habitats. In colonizing the land, plants turned barren ground into a green paradise in which a whole new world of land-dwelling organisms, including humans, could then evolve.

Plants are multicellular autotrophs that are mostly terrestrial (land-dwelling). Like photosynthetic protists (algae), plants use chloroplasts in order to photosynthesize: they make sugars from carbon dioxide and water molecules, using energy from light to drive the chemical reaction. Oxygen gas is a by-product of photosynthesis that is vital to animals like us. Most photosynthesis in plants takes place in their leaves, which typically have a broad, flat surface—a design that maximizes light interception. Because plants are producers, they form the basis of essentially all food webs on land.

Plants reproduce both asexually and sexually. The distinctive life cycle of plants is explained in detail in Unit 7. Although their life cycle is distinctly different from that of animals, plants are like animals in that they produce embryos: the fusion of egg and sperm produces a single cell, called a *zygote*, which then divides to produce a multicellular structure called an *embryo*.

Today the diversity of the Plantae, a kingdom within the domain Eukarya, ranges from the most ancient lineages—liverworts and mosses—to ferns, which evolved next; to gymnosperms; and finally, to the most recently evolved plant lineage, the angiosperms, or flowering plants (**FIGURE 3.10**). Mosses, liverworts, and hornworts—informally known as **bryophytes**—were among the earliest land plants. These "amphibians of the plant world" still thrive in moist habitats throughout the world, and some can even withstand drying out or freezing during the non–growing season. *Gymnosperms* are familiar to us as the conifers, or cone-bearing trees, such as pines and firs, that dominate in the colder regions of the world. Tropical gymnosperms, such as the palmlike cycads in Florida, are probably less familiar to people from nontropical regions. The *angiosperms*, or flowering plants, are familiar to everyone because we depend on them for food, clothing, building material, paper, medicines, and many other products.

## Plants had to adapt to life on land

Organisms on land had to adapt to challenges not faced by organisms living in water. The biggest challenge was how to obtain and conserve water. Plants have a waxy covering, known as the **cuticle**, that covers their aboveground parts (**FIGURE 3.11**). A waxy cuticle holds in moisture and thereby keeps plant tissues from drying out, even when exposed to sun and air all through the day. The cuticle is so effective as weatherproofing that grocers like to prolong the shelf life of vegetables such as cucumbers and tomatoes by dipping them in waxy material, supplementing the natural waxes already present on their surface. Plants that live in relatively dry climates tend to have exceptionally thick cuticles.

If a leaf has a waxy coating, how is the carbon dioxide that is so crucial for photosynthesis brought into leaf cells? Air enters leaf cells through many minute openings, or air pores, in the cuticle. In liverworts, the air pores are simple gaps in the surface layer of cells. Many mosses, and all gymnosperms and angiosperms, have more elaborate pores called **stomata [stoh-*MAH*-tuh]** (singular "stoma," Greek for "mouth") that open and close to regulate the flow of gases into and out of the leaf. Each stoma is bordered by a pair of **guard cells**, which can inflate or deflate like water balloons. When the guard cells are inflated with water, they buckle out and the opening between them is revealed; but when the guard cells lose water, they flop against each other and thereby close off the opening. Although they are essential for photosynthesis, stomata also put a plant at risk of drying out because moisture-laden air escapes from a leaf even as external air containing carbon dioxide enters an open stoma. As you might expect, plants regulate their stomata with

*Marchantia polymorpha,* a liverwort

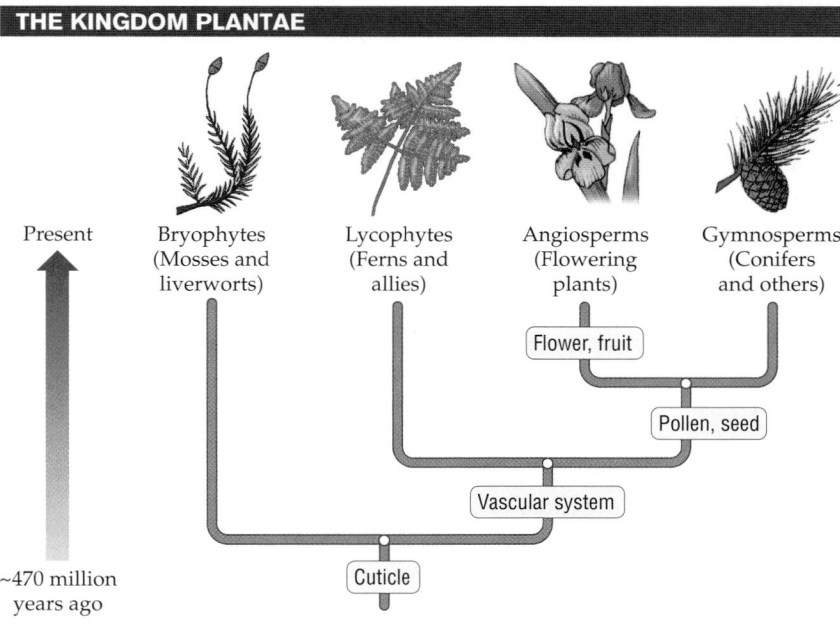

**THE KINGDOM PLANTAE**

Present

Bryophytes (Mosses and liverworts)    Lycophytes (Ferns and allies)    Angiosperms (Flowering plants)    Gymnosperms (Conifers and others)

Flower, fruit

Pollen, seed

Vascular system

~470 million years ago

Cuticle

**FIGURE 3.10** **Evolutionary Tree of the Plantae**

**FIGURE 3.11 The Plant Body Consists of the Shoot and Root Systems**

Shown here is a bell pepper plant (*Capsicum annuum*). Because it is a member of the angiosperms, the last of the major plant groups to evolve, this one plant illustrates all the evolutionary innovations that distinguish plants.

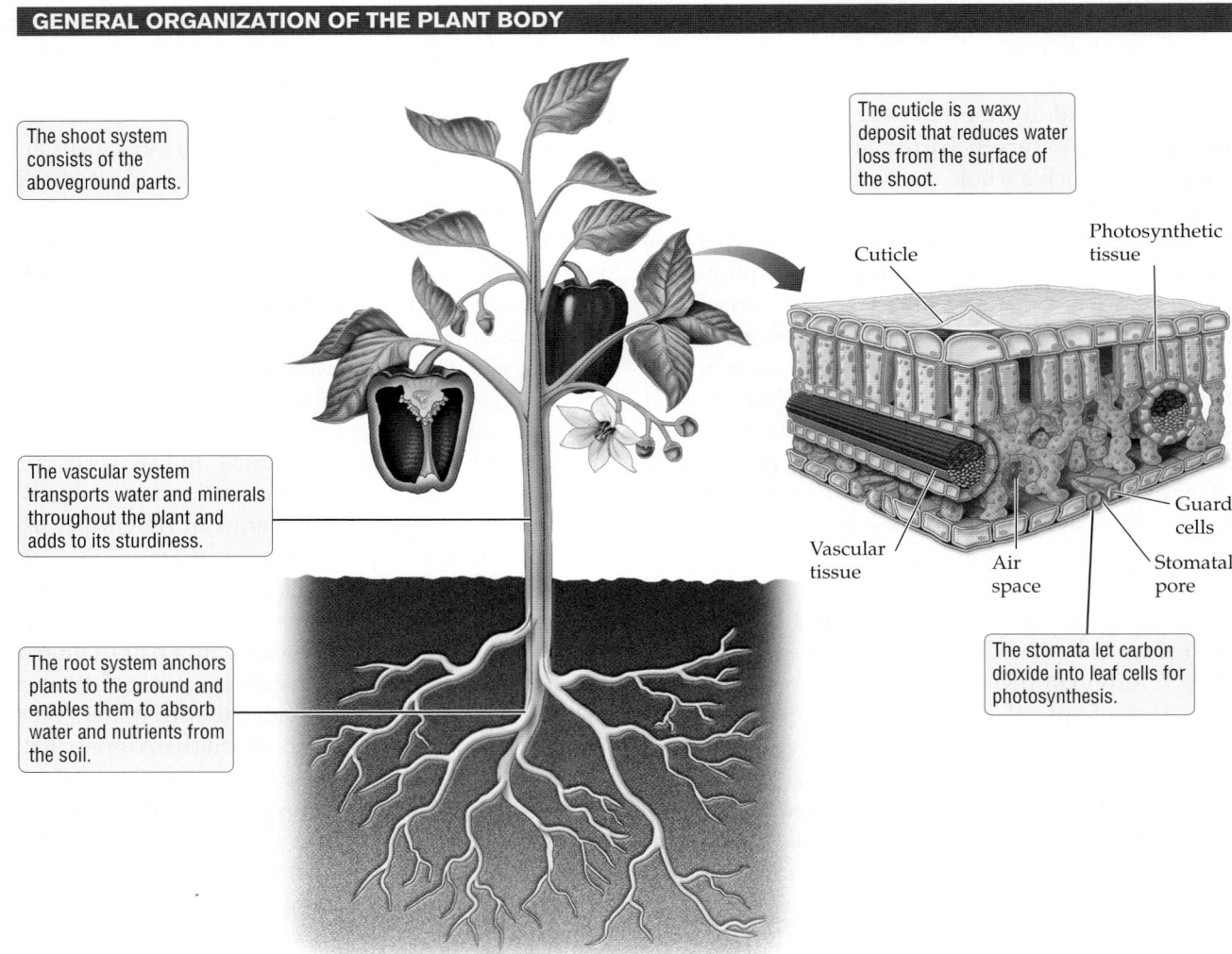

The shoot system consists of the aboveground parts.

The cuticle is a waxy deposit that reduces water loss from the surface of the shoot.

The vascular system transports water and minerals throughout the plant and adds to its sturdiness.

The root system anchors plants to the ground and enables them to absorb water and nutrients from the soil.

Cuticle

Photosynthetic tissue

Vascular tissue

Air space

Guard cells

Stomatal pore

The stomata let carbon dioxide into leaf cells for photosynthesis.

utmost care: in most plants, stomata close at night or if the plant experiences water stress (inadequate water supply) at any time during the day.

## Lignin enabled plants to grow tall

The first plants, the ancestors of present-day liverworts and mosses, grew as ground-hugging carpets of greenery. To this day, reproductive structures—such as the capsules raised on stiff stalks—are the only vertical structures that these groups produce. Like green algae, plant cells have strong but flexible cell walls composed of a substance known as **cellulose**. Cellulose cell walls give structural strength to all types of plant cells, including those of the low-growing mosses and liverworts.

To grow taller—the way ferns, rosebushes, and pine trees do—plants had to evolve yet another type of strengthening material, called lignin. One of the strongest materials in nature, **lignin** links cellulose

fibers in the cell wall to create a rigid network that resembles an ancient knight's armor of chain mail. Wood is strong because it is made of cells whose cell walls are reinforced with lignin. Lignified tissues enabled plants to reach for the sky, like the 300-foot redwoods that soar into the ocean mists along the coast of northern California. Growing tall raises its own set of challenges: how to raise fluids, such as water, from ground level to the crown of a bush or tree. As we shall see, the evolution of lignin—in fernlike plants that lived about 420 million years ago—went hand in hand with the evolution of good plumbing.

## The vascular system enables plants to move fluids efficiently

Bryophytes have relatively thin bodies, often just a few cells thick, and sprawling on wet surfaces enables these plants to absorb water through a wicking action known as capillarity. Many bryophytes have tufts of threadlike cells (*rhizoids*) on their lower surface that grow into the soil and soak up water a few centimeters deeper in the ground. These simple strategies are effective in delivering water, and the mineral nutrients dissolved in it, to a relatively thin plant body that is close to the ground. However, absorption by direct contact or capillary action cannot transport fluids effectively in a plant that rises a foot (0.3 meter) or more aboveground.

About 425 million years ago, plants evolved a network of tissues, called the **vascular system**, that includes tubelike structures specialized for transporting fluids. Vascular tissues that specialize in transporting food molecules, such as sugars, are called **phloem** [*FLOH*-em]. Vascular tissues that specialize in transporting water and dissolved nutrients are known as **xylem** [*ZYE*-lem]. Xylem and phloem are usually bundled together in branching strands that snake throughout the plant body, pervading every organ the way our blood vessels pass within close reach of every cell in our body. If you turn over a leaf, you will see the bundles of xylem and phloem—commonly called "leaf veins"—branching into an ever-finer pattern throughout the leaf tissues. The inside of a tree trunk is taken up mostly by bundles

of xylem—which you recognize as wood—with phloem in a ring under the corky layers that make up the bark. The water-conducting tubes of xylem are reinforced with lignin, which is why wood is strong enough to build with. Roots—the water-absorbing organ system found in all plants except bryophytes—also have an extensive vascular system. Root xylem brings water from the soil to the aboveground parts of the plant, and root phloem delivers sugars made in the leaves to the nonphotosynthetic tissues below ground.

## The evolution of seeds contributed to the success of gymnosperms

**Gymnosperms** [*JIM*-noh-spermz] evolved about 365 million years ago. Conifers (cone-bearing plants) are the most diverse and abundant gymnosperms today: spruce, fir, pine, and larch are the dominant vegetation in large swaths of the northern lands, including much of Canada, northern Europe, and Siberia.

Gymnosperms were the first plants to evolve **pollen**, a microscopic structure that contains sperm cells (**FIGURE 3.12**). All plant lineages that appeared

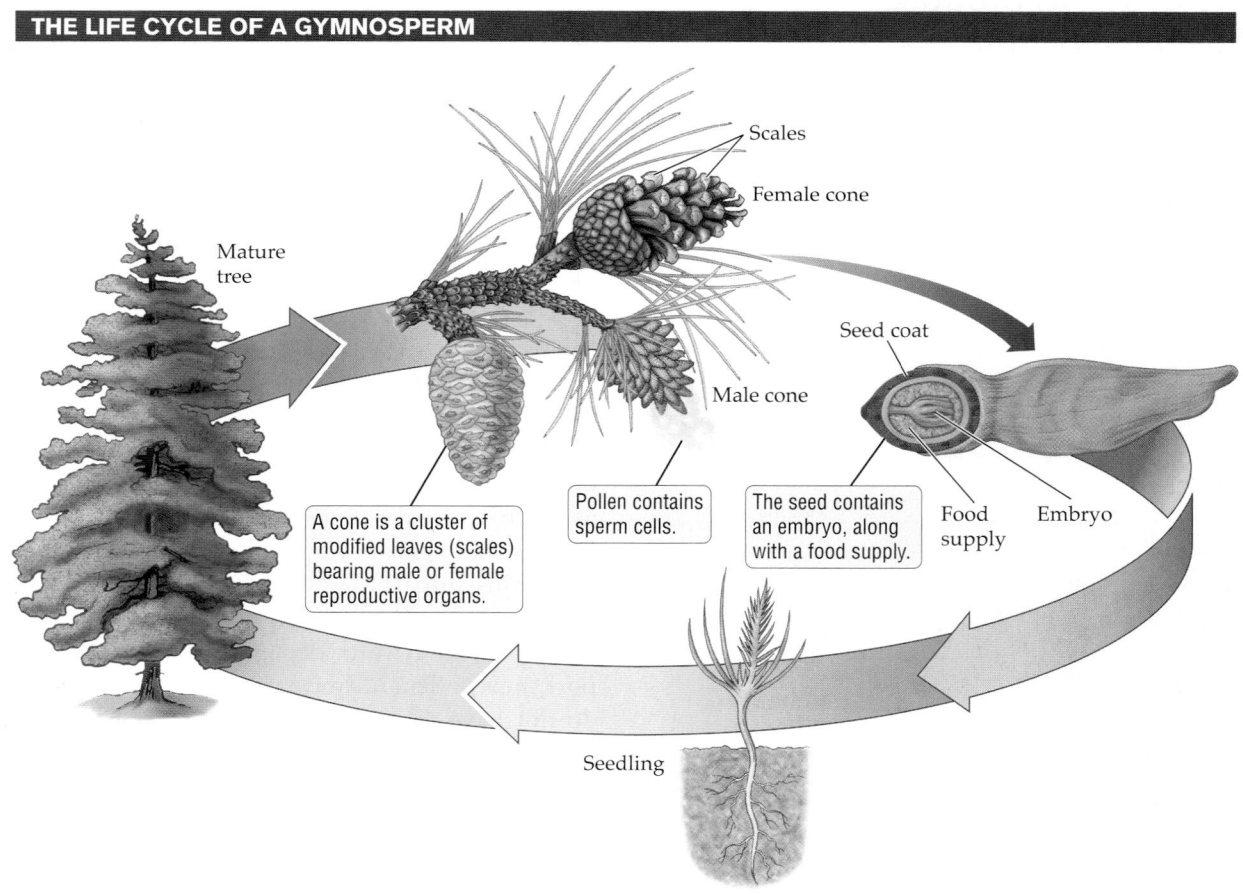

**THE LIFE CYCLE OF A GYMNOSPERM**

Scales

Female cone

Mature tree

Seed coat

Male cone

A cone is a cluster of modified leaves (scales) bearing male or female reproductive organs.

Pollen contains sperm cells.

The seed contains an embryo, along with a food supply.

Food supply

Embryo

Seedling

**FIGURE 3.12**
**Gymnosperms Were the First Plants to Produce Pollen and Seeds**

before the gymnosperms produce flagellated sperm, which need at least a film of water in order to swim to a nearby egg cell. The sperm package that is pollen is dry and powdery and can be lofted into the air in massive quantities. The evolution of pollen freed the gymnosperms and their close cousins, the angiosperms, from a dependence on water for fertilization.

Gymnosperms were the first plants to evolve the **seed**, which consists of the plant embryo and a supply of stored food, encased in a protective covering called the seed coat (see Figure 3.12). Gymnosperms (*gymno*, "naked"; *sperm*, "seed") were the dominant plants 250 million years ago, and the evolution of seeds was an important part of their success. Seeds could be disseminated farther away, so they would not be in competition with the mother plant for sunlight or for water and mineral nutrients in the soil. Pine, fir, spruce, and hemlock produce winged seeds that ride wind currents to drift away from the mother tree. Seeds contain stored food that plant embryos use to grow before they are able to make their own via photosynthesis. Seeds also provide embryos with protection from drying or rotting, and from attack by predators.

## Angiosperms produce flowers and fruit

Flowering plants, or **angiosperms** [AN-jee-oh-spermz], are a relatively recent development in the history of life. The first flowering plants evolved about 145 mil-

lion years ago, shortly after another upstart group, the mammals, appeared on the Mesozoic scene. Today, with about 250,000 species, angiosperms are the most dominant and diverse group of plants on our planet. Angiosperms include orchids, grasses, corn plants, and apple and maple trees. Highly diverse in size and shape, angiosperms live in a wide range of habitats—from mountaintops to deserts to salt marshes and fresh water.

The key evolutionary innovation of angiosperms and the key to their success is the **flower**, a structure that evolved through modification of the conelike reproductive organs of gymnosperms. Flowers are structures that enhance sexual reproduction in angiosperms by bringing male gametes (sperm cells) to the female gametes (egg cells) in highly efficient ways. In the most dramatic example, animals are enlisted to carry pollen from flower to flower.

Most flowers are bisexual in that the male and female structures, known as the stamen and carpel, respectively, are present in the same flower. Stamens consist of sacs, often borne on long stalks, that produce pollen (**FIGURE 3.13**). Pollen grains are most familiar as the dustlike particles that some plants release into the air in spring and summer and that can cause allergies in some people. The receipt of pollen by the carpel is known as *pollination*. Pollen grains germinate on the receptive surface of the carpel (*stigma*), producing a long extension called the *pollen tube*. Sperm cells travel in the growing pollen tube to the egg cells inside the base of the carpel (*ovary*). The fusion of a sperm with an egg, in the process known as **fertilization**, leads to a multicellular embryo. As in gymnosperms, the embryo of angiosperms is enclosed in protective layers, which collectively form the seed.

Some angiosperms are like the overwhelming majority of gymnosperms in that they rely on wind currents for pollination. Angiosperms that depend on wind pollination—all grasses, and trees such as birch and oak—tend to have small, dull-colored flowers that spew prodigious quantities of tiny pollen grains into the air. Angiosperms that produce brightly colored or strongly scented flowers, usually with a nectar reward, depend on animals to transport their pollen. Although animal-pollinated plants do not have to invest energy in producing massive amounts of tiny pollen, they must devote a good deal of energy to making special products—bright petals, odors, and the sugary liquid known as nectar—to lure their pollinators. Pollen delivery by animals is

**FIGURE 3.13**
**The Flower**

A flower is really a meeting place for plant gametes. Pollen contains sperm cells. The egg cells are held inside roughly oval structures, called ovules, that turn into seeds after fertilization.

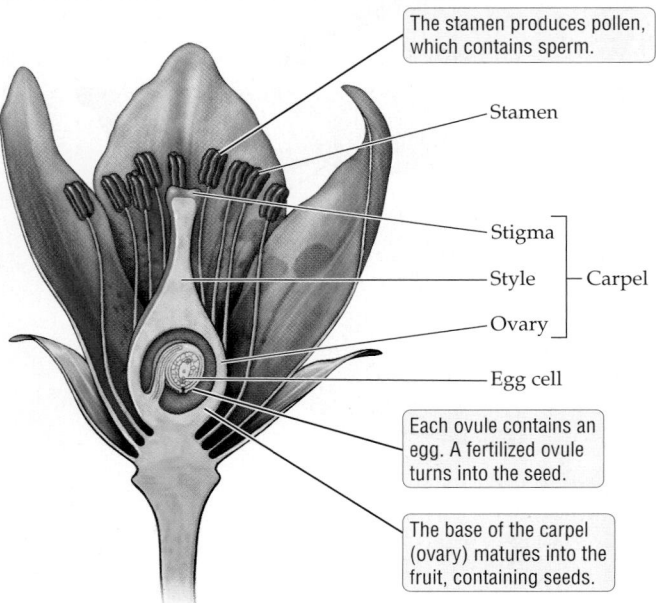

**THE REPRODUCTIVE STRUCTURES OF AN ANGIOSPERM**

The stamen produces pollen, which contains sperm.

Stamen

Stigma

Style ⎤
Ovary ⎦ Carpel

Egg cell

Each ovule contains an egg. A fertilized ovule turns into the seed.

The base of the carpel (ovary) matures into the fruit, containing seeds.

# The Importance of Biodiversity

Many people wonder whether the loss of one beetle here or a buttercup there really makes a difference to humanity. To answer these questions, it helps to look at the situation from the perspective of biologists who have long wrestled with how to assess the value of diverse species within particular habitats. One question that biologists are particularly interested in answering is how, if at all, biodiversity affects the forests, wetlands, oceans, rivers, and other wild ecosystems of the world. Since the 1990s, researchers have been studying how biodiversity can contribute to the health and stability of ecosystems.

### Biodiversity Can Improve the Function of Ecosystems

From tiny experimental ecosystems in an English laboratory to experimental prairies in the midwestern United States, researchers have found that the more species an ecosystem has, the healthier it appears to be. For one thing, species diversity maximizes the use of available resources because different species are good at using differing resources; for example, some thrive in the sun-drenched portions of a habitat, whereas others can make the best use of the areas that lie mostly in the shade.

Evidence also suggests that the more species there are in an ecosystem, the more resilient that ecosystem is. For example, the greater the number of species present in a patch of prairie, the more easily that area can return to a healthy state following a drought. Scientists have also found that an increased diversity of species in an area leads to a lower the incidence of disease and lower rates of invasion by introduced species.

Furthermore, diversity can even lead to more diversity: researchers have found that greater plant diversity in a plot of ground leads to greater insect diversity as well.

### Biodiversity Provides People with Goods and Services

Why should we care if various ecosystems are healthy? The biosphere provides us with many goods and services. Even the most basic requirements of human life are provided by other organisms. Plants produce the oxygen we breathe; they also provide us with food and many other necessities. One-fourth of all prescription drugs dispensed by pharmacies are extracted from plants. Quinine, used as an antimalarial drug, comes from a plant called yellow cinchona [sing-*KOH*-nuh]. Taxol, an important drug for treating cancer, comes from the Pacific yew tree. Bromelain [*BROH*-muh-lun], a substance that controls tissue inflammation, comes from pineapples.

The Madagascar rosy periwinkle (*Catharanthus roseus*) is a source of anticancer drugs.

Whole ecosystems full of species can provide what are known as ecosystem services. For example, the coast redwood trees in northern California intercept fog, mist, and rain, and channel the water onto and into the ground. Where these forests have been cut down, moisture-laden air masses are much less likely to linger, and more likely to evaporate in the sun, so that much less water ends up entering the ground and reservoirs. Without trees on the ground, hillsides erode more easily and rivers become filled with sediments.

The value of seemingly useless marshes full of reed and grass species became all too apparent when Hurricane Katrina barreled through the Gulf Coast in 2005. A healthy marsh acts much like a barrier island, blocking storm surges while also creating areas for fish to spawn and birds to nest. But the wetlands outside New Orleans had been so reduced by humans—for flood control, navigation, agriculture, oil drilling, and other human activities—that storm surges easily destroyed the rest. Hurricane Katrina caused some marshy areas east of the Mississippi River to lose 25 percent of their land areas, thereby interfering not only with natural fisheries and birdlife, but also reducing the power of remaining marshes to act as natural filters for the water moving through them, providing water-cleaning services to growing human populations.

Nature's bounty of life is great, but with so many species yet undiscovered, the vast majority of the living world's wealth remains untapped. If the sheer numbers of species yet to be discovered are any indication, much more awaits—beauty, food, shelter, medicine—if we can just find it before it disappears.

**FIGURE 3.14 Plants Get Around in a Variety of Ways**
Plants have evolved many ways of spreading to new areas. The seed inside this coconut fruit can float for hundreds of miles until it reaches a new beach where it can take root and grow.

## Helpful to Know

What is the difference between a fruit and a veggie? A fruit is the mature *ovary* of an angiosperm and normally contains seeds inside it. "Vegetable" is not a scientific term. In everyday life, people tend to use the word "vegetables" to refer to plant structures that do not taste sweet. This popular usage ropes in storage roots such as carrots and radishes, plant leaves such as lettuce and spinach leaves, but also some true fruits—such as tomatoes and cucumbers—that happen to be nonsweet.

highly effective because many pollinators specialize in the plant species they visit and therefore tend to carry pollen to the carpel of the same species, instead of wasting it on nontarget species. Wind pollination works best in species that grow in thick, near-pure stands, but animal-pollinated species can afford to be scattered across the landscape.

Gymnosperms produce "naked" seeds that sit bare, unwrapped in any additional layers, on the modified leaf (the scale in a pine cone). In angiosperms, the modified leaf evolved into the ovary wall, which consists of tissue layers that enclose the egg-bearing structures, or *ovules*, and protect them as they develop. *Angio* means "vessel," referring to the ovary. After fertilization, the ovules develop into seeds, and the ovary wall that enclosed them becomes the fruit wall. A **fruit**, therefore, is a mature ovary with seeds inside it.

What is the adaptive value of the fruit, which encloses the seeds of angiosperms? In addition to protecting immature seeds from would-be seed predators, the angiosperm fruit wall is a seed dispersal device. In some species, the fruit wall dries out and flings the seeds from the spent carpel with considerable force. For example, the fruit of the squirting cucumber discharges its seeds a distance of 3–6 meters (10–20 feet) in a stream of mucilage. The fruit wall

that surrounds the coconut has a leathery jacket that resists salt water, as well as fibrous, air-filled layers that make the heavy seed as buoyant as a flotation device (**FIGURE 3.14**), permitting seeds to travel in the ocean to colonize new habitats.

The most effective seed dispersal strategies are those that employ animals as a delivery service. Some fruit walls, such as those of burweeds, have sharp hooks or Velcro-like barbs that enable the fruits to hitch a ride on animal fur. Fleshy fruits become palatable when they ripen, after the seeds inside have matured and developed a tough, protective seed coat. Animals that eat fruit often excrete the seeds in their feces well away from the mother plant. The nutrient-rich wastes provide a good place for excreted seeds to begin their new life.

## Plants are the basis of land ecosystems and provide many valuable products

It is difficult to overstate the significance of plants, which play the role of producers. Nearly all organisms on land ultimately depend on plants for food, either directly by eating plants or indirectly by eating other organisms (such as animals) that eat plants. Many organisms live on or in plants, or in soils largely made up of decomposed plants. Aquatic plants, such as the water lilies and duckweed in **FIGURE 3.15**, provide food and shelter for organisms that range from bacteria to adult animals and their larvae.

Flowering plants provide humans with materials such as cotton for clothing and with pharmaceuticals such as morphine. Nearly all agricultural crops are flowering plants, and the entire floral industry rests on the reproductive structures of angiosperms. Gymnosperms such as pine, spruce, and fir are the basis of forestry industries, providing wood and paper. As valuable as plants are when harvested, they are also valuable when left in nature. By soaking up rainwater in their roots and other tissues, for example, plants prevent runoff and erosion that can contaminate streams. Plants also recycle carbon dioxide and produce the gas we breathe, oxygen. The importance of plants not only to humans but to all of Earth's ecosystems mandates efforts to conserve plant biodiversity. However, humans have drastically and rapidly modified Earth's plant landscapes with cities, agriculture, and industries (see the box on page 65).

## Size and Structure

**Did you know?**

A baobab tree can hold more than 10,000 gallons of water in its hollow trunk, which local people use as water storage tanks.

The leaf of the Amazon lily, *Victoria regia*, is the largest in the world.

The duckweed, *Wolffia globosa*, is the smallest plant. Its entire body is just under 1 mm in length.

The baobab tree (*Adansonia rubrostipa* from Madagascar is seen here) can measure more than 150 feet in circumference. The trees drop their leaves just before the dry season.

*Sequoiadendron giganteum*, the giant sequoia, can reach 300 feet tall, with a circumference of 56 feet.

## Reproduction

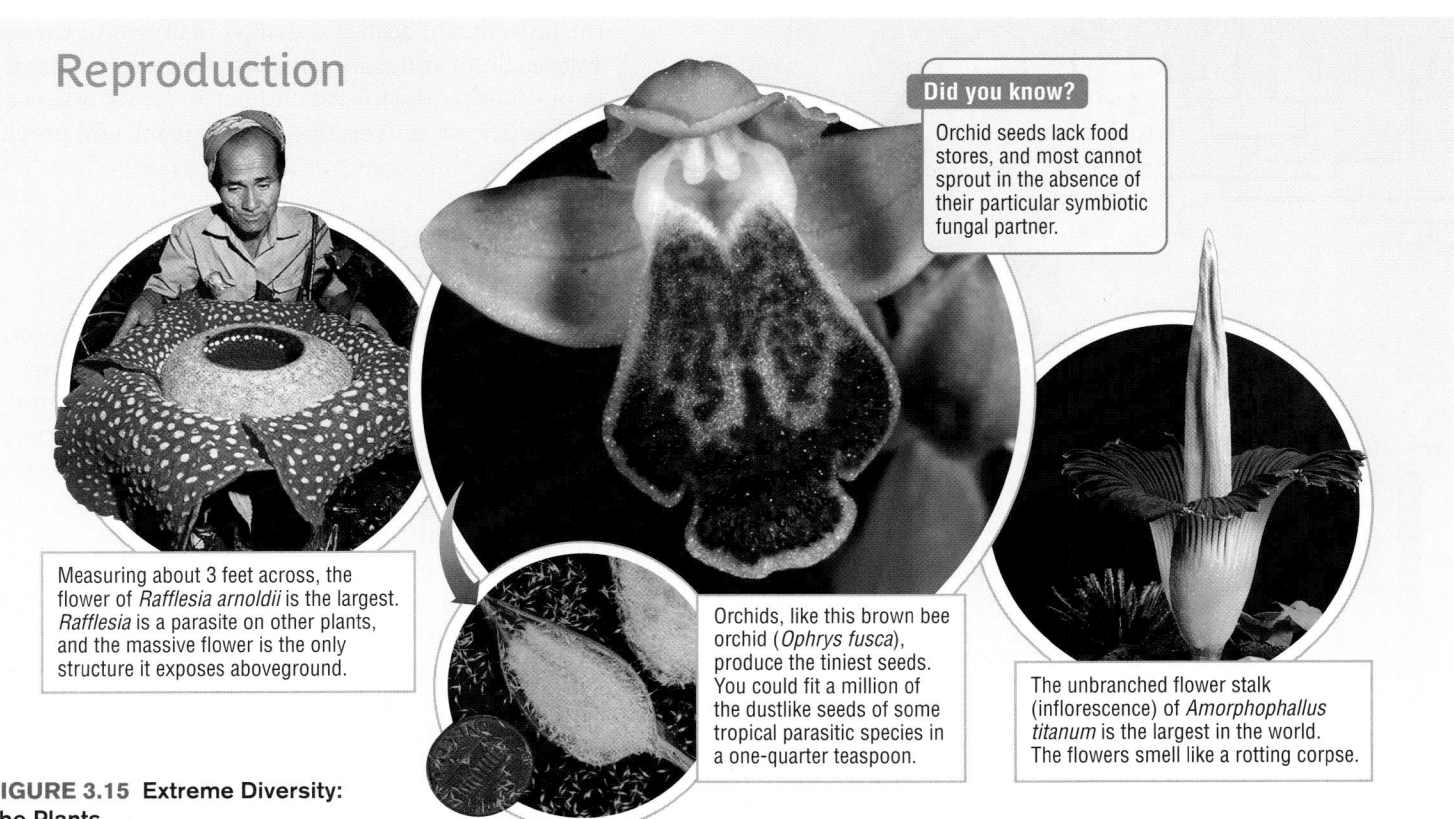

**Did you know?**

Orchid seeds lack food stores, and most cannot sprout in the absence of their particular symbiotic fungal partner.

Measuring about 3 feet across, the flower of *Rafflesia arnoldii* is the largest. *Rafflesia* is a parasite on other plants, and the massive flower is the only structure it exposes aboveground.

Orchids, like this brown bee orchid (*Ophrys fusca*), produce the tiniest seeds. You could fit a million of the dustlike seeds of some tropical parasitic species in a one-quarter teaspoon.

The unbranched flower stalk (inflorescence) of *Amorphophallus titanum* is the largest in the world. The flowers smell like a rotting corpse.

**FIGURE 3.15 Extreme Diversity: The Plants**

## 3.4 Fungi: A World of Decomposers

Most people are familiar with fungi as the mushrooms on their pizza or lawns. However, fungi also include single-celled yeasts that ferment beer and make bread rise, and the threadlike mold that grows on cheese and bread. However, because much of the fungal body remains hidden from view, most people fail to realize the extent to which fungi permeate our world, and fungi remain poorly understood organisms.

The kingdom **Fungi**, in the domain Eukarya, consists of *absorptive* heterotrophs: fungi digest organic material outside the body and absorb the molecules released as breakdown products. In contrast, all animals and most protist consumers are *ingestive* heterotrophs: eukaryotes that bring prey organisms or organic material into the body, or into the cell, and break it down internally. Unlike animal cells, fungal cells have a protective cell wall that wraps around the plasma membrane and encases the cell. But like *some* animals, such as insects and lobsters, fungi produce a tough material called **chitin** that strengthens and protects the body. Fungi are similar to animals in that they store surplus food energy in the form of a biomolecule called **glyco-**

**gen**, the same molecule that serves as an energy reserve in our skeletal muscle cells. Before the tools of DNA analysis became widely available, many biologists believed that fungi were more closely related to plants than to animals. However, DNA comparisons show that fungi diverged from animals more recently than both groups diverged from plants.

Because fungi do not fossilize well, their early evolutionary history is shrouded in mystery. Reconstructing the evolutionary history of eukaryotes from DNA data, scientists estimate that the common ancestor of fungi and animals diverged from all other eukaryotes about 1.5 billion years ago, and fungi diverged from their closest cousins, the animals, about 10 million years after that. Fungi were aquatic for much of their evolutionary history and made the journey from water to land about 500 million years ago, shortly before the colonization of land by plants. The majority of fungal species fall into three main groups (**FIGURE 3.16**): **zygomycetes**, which were the first fungal group on land; **ascomycetes**, informally known as sac fungi; and the perhaps more familiar **basidiomycetes**, or club fungi, which evolved about 400 million years ago. Each of these groups differs in—and is named for—its unique reproductive structures.

Fungi play several roles in terrestrial ecosystems. Many are decomposers, consumers that live off nonliving organic material. Playing the role of garbage processor and recycler, these fungi speed the return of the nutrients in dead and dying organisms to the ecosystem. Some fungi are **parasites** (organisms that live in or on other organisms and harm them); others are mutualists (organisms that benefit from, and provide benefits to, the organisms they associate with).

## Fungi are adapted for absorptive heterotrophy

Most fungi are multicellular, but there are some single-celled species, which are collectively known as "yeasts." A key evolutionary innovation of the multicellular fungi is their body form, which is exceptionally suited for absorptive heterotrophy. The main body of a multicellular fungus is called a **mycelium [mye-SEE-lee-um]** (plural "mycelia"), which consists of an extensive mat of highly branched strands (**FIGURE 3.17**). Each mycelial strand, known as a **hypha [HYE-fuh]** (plural "hyphae" [HYE-fee]) is a slender threadlike filament consisting of multiple cells arranged in a row. In some species, the hyphal cells are incompletely separated. Instead of a complete partition, adjacent cells in the hyphae of basidiomycete fungi are separated by

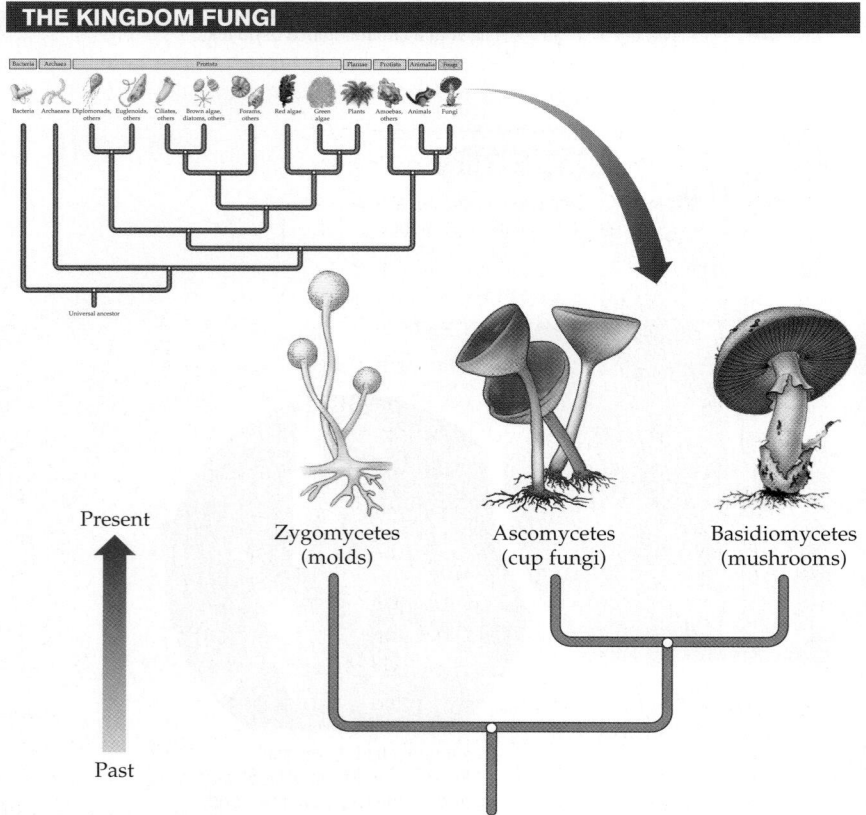

**THE KINGDOM FUNGI**

Present

Past

Zygomycetes (molds)  Ascomycetes (cup fungi)  Basidiomycetes (mushrooms)

**FIGURE 3.16 Evolutionary Tree of the Fungi**

a structure called a *septum*. Openings in the septum allow organelles—even the nucleus sometimes—to move from one cell compartment to another.

Mycelia extend deep into the medium the fungus is growing in, weaving into the soil, a rotting tree stump, or the compost that commercial mushroom growers commonly use. The mycelia of parasitic fungi send fine branches through the living tissues of their host. The hyphae, which are typically 10 micrometers (0.01 millimeter) wide, are in intimate and extensive contact with the medium they permeate. Mycelia therefore have an enormous surface area for taking up food and nutrients from their environment. A single mycelium can have a surface area equivalent to three tennis courts.

Like animals and all chemoheterotrophs, fungi rely entirely on other organisms for both energy and carbon. Fungal hyphae release special digestive proteins to break down organic material, and even living tissues, through which they grow. The hyphae then absorb the nutrients for the fungus to use. Some fungi have evolved to become predators, actively trapping small animals, such as the tiny worms known as nematodes, with sticky secretions or a noose made from three cells arranged in a ring.

## Fungi have unique ways of reproducing

Fungi can reproduce both asexually and sexually. Some species, such as the yeasts, appear to multiply only asexually. Baker's yeast, *Saccharomyces cerevisiae* [SAK-uh-roh-MYE-seez sair-uh-VEE-see-eye], are single-celled fungi that divide in a lopsided manner, known as budding, to produce new daughter cells that are genetically identical to the parent cell. Most multicellular fungi can reproduce asexually through fragmentation—that is, by simply breaking off from the mother colony. You can start your own culture of shiitake mushrooms if you obtain some spawn, plugs of mycelium cut out from an established mycelial mat.

Many fungi, like the greenish blue *Penicillium* that is so common on blocks of cheese, produce asexual *spores* (**FIGURE 3.18**). A **spore** is a reproductive structure, usually thick walled, that can survive for long periods of time in a dormant state and will sprout under favorable conditions to produce the body of the organism (mycelium, in the case of fungi). Fungal spores may be single-celled or multicellular and are usually encased in a thick wall that is resistant to decay and keeps the cytoplasm inside from drying out. The powdery material that speckles moldy bread

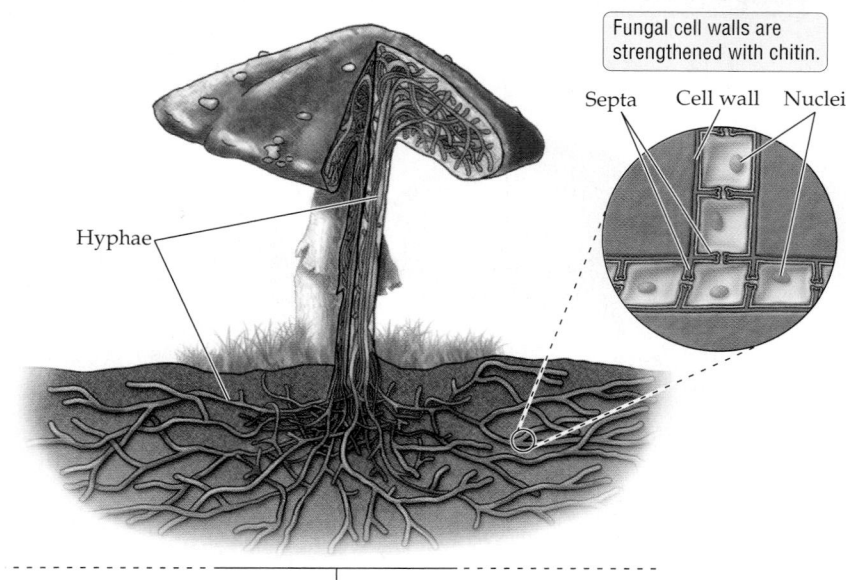

**THE FRUITING BODY OF A BASIDIOMYCETE FUNGUS**

Fungal cell walls are strengthened with chitin.

Septa  Cell wall  Nuclei

Hyphae

Mycelium

**FIGURE 3.17  The Basic Structure of a Multicellular Fungus**
Mats of hyphae, known collectively as a mycelium, form the main body of a fungus. Each hypha is a row of cells separated by septa. Openings in the septa allow organelles to move from one compartment to another. The fungal cell walls encasing the hyphae contain chitin, the same material that makes up the hard outer skeleton of insects.

or the walls of a damp basement consists of fungal spores. Fungal spores can travel great distances and have been found to survive in the stratosphere at altitudes greater than 30,000 feet.

Sexual reproduction in fungi is varied and complex. Fungi do not have distinct male and female individuals. Instead, a sexually reproducing mycelium belongs to one of two mating types, usually designated as plus (+) or minus (−) mating types. The two mating types are not visually different, but differences in their DNA translate into differences in their chemistry, which in turn governs mating behavior. Each mating type can mate successfully only with a different mating type. Opposite mating types come together to form a *fruiting body* that may be large enough to be readily observed. A mushroom or cup fungus or puffball is a fruiting body in which opposite mating types have come together to fuse their DNA and produce offspring.

Fungal fruiting bodies release the offspring as sexual spores that are scattered into the world by wind, water, and animals, the way asexual spores are. Spores released from a fruiting body raised above the growing medium are better able to catch a ride on wind currents or attract animals (as with flies drawn to stinkhorn mushrooms) that can carry them far and wide (**FIGURE 3.19**). Once carried to new locales,

**FIGURE 3.18  Fungi Spread via Spores**
This puffball fungus is expelling a cloud of spores into the air.

## Life Strategies

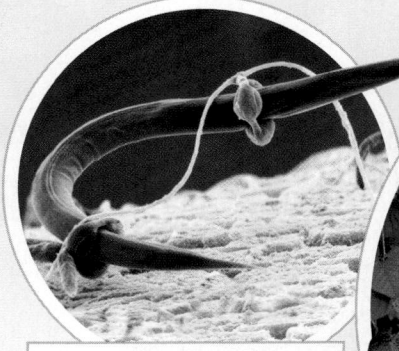

This predatory fungus, *Arthrobotrys anchonia*, is a killer with a lasso. It has trapped a nematode worm in a lasso made of three inflatable cells.

This bioluminescent fungus, *Mycena lampadis*, grows in Australia.

Like the more famous *Psilocybe*, this showy flame cap (*Gymnopilus spectabilis*) produces hallucinogenic chemicals to deter animals from nibbling it.

The death cap fungus, *Amanita phalloides*, contains a deadly toxin. Ingesting just half of one cap can kill a person.

## Reproduction

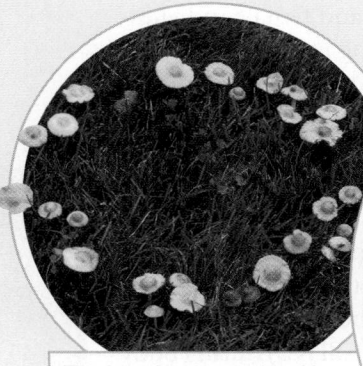

This fairy ring was produced by *Marasmius oreades*, a basidiomycete. Fairy rings start from a single spore at the center that sprouts mycelia that continue to grow outward until they reach a barrier. The largest fairy ring, located in France, measures nearly half a mile in diameter.

The best shot in the fungal world is *Pilobolus*, a zygomycete that lives in dung. Certain worm larvae that complete their development in the guts of herbivores are known to climb up the 1-centimeter-long stalk of the fruiting body. They hitch a ride on the spore mass as it rockets out of the dung heap and lands on adjacent vegetation, where it can be ingested by a cow or a horse.

**Did you know?**

The long-stalked fruiting body of *Pilobolus* works like a light-sensing rocket launcher. The stalk grows toward light, bearing the black spore mass (the rocket) at the tip. As water pressure builds in the stalk, the tip explodes, shooting the spores as far as 8 feet.

The aptly named stinkhorn mushroom, *Phallus impudicus*, is a basidiomycete that attracts flies. The insects get covered with the sticky spores and scatter them as they fly to other locations.

**FIGURE 3.19 Extreme Diversity: The Fungi**

these sexual spores can begin growing as new, separate individuals that are genetically diverse because they contain DNA shuffled and blended from two different parental mycelia.

## Fungi play a key role as decomposers

Fungi are among the most important decomposers in terrestrial ecosystems, recycling a large proportion of the dead and dying organisms on land. They play a crucial role in nutrient recycling because they break down organic matter, such as leaf litter or the dead bodies of other organisms, releasing the nutrients as inorganic chemicals. Fungi are particularly effective in breaking down plant remains, including lignin, the tough material that strengthens shrubs and trees and that no other organism except certain bacteria can decompose. The inorganic chemicals released by such decomposition become available to producers such as autotrophic bacteria, algae, and plants, which absorb the inorganic chemicals and use them to manufacture food "from scratch." All other organisms in the food web depend, directly or indirectly, on the food made by these producers. You could say that whereas producers inject food into a food chain, decomposers break it apart and return the ingredients to the nonliving environment, making them potentially available for another run through the interlinked food chains that constitute the food web.

What would the world look like if all the fungi disappeared? Great piles of leaf litter and brush and whole tree trunks would accumulate on the forest floor, and even animal carcasses would decay more slowly with mainly bacteria and some protists available to decompose them. What would the world look like if fungi were around but there was a sudden, colossal die-off among *other* organisms, especially plants? Paleobiologists, who study ancient life-forms, believe that this scenario has been realized in the past history of Earth. The mass extinctions that mark the end of certain geologic periods—such as the Permian, which ended about 250 million years ago—were often accompanied by a fungal spike. A *fungal spike* is a sudden and massive increase in the abundance of fossil spores belonging to decomposing fungi. The explanation is that the death of large numbers of organisms, triggered by volcanic activity or meteor impact or some other cause, set up a veritable feast for decomposing fungi, which multiplied explosively as their food base expanded suddenly.

## Fungi can be dangerous parasites

Some fungi are parasites. Parasitic fungi grow their hyphae through the tissues of living organisms, causing diseases in animals (including humans) and plants (including crops). Mammals have a complex immune system that appears to defend them well against fungi. In humans with a healthy immune system, fungi cause no more than mild disease, such as athlete's foot or ringworm. But in humans with a compromised immune system, fungal diseases can be deadly, like the pneumonia caused by the fungus *Pneumocystis carinii* [*NOO*-moh-*SIS*-tis kuh-*REE*-nee], the leading killer of people suffering from AIDS.

Fungi are the most significant parasite of plants: they are responsible for two-thirds of all plant diseases, causing more crop damage than do bacteria, viruses, and insect pests combined. *Ceratocystis ulmi* [*SAIR*-uh-toh-*SIS*-tis *OOL*-mee] causes Dutch elm disease, and this fungal import from Asia has nearly eliminated the American elm, which once formed arching canopies over streets all across the United States. Likewise, *Cryphonectria parasitica*, another Asian import, has all but wiped out the American chestnut, a handsome native hardwood that can grow to 200 feet. Rusts and smuts are serious fungal pests of crop plants, especially grain crops such as wheat and rice. Insects are another group of organisms that appear to be especially susceptible to fungal pathogens. Agricultural scientists are trying to use fungi that specialize in killing insects to protect crop plants from insect pests (**FIGURE 3.20**).

**FIGURE 3.20 Fungal Parasites**
Some fungi are parasites and make their living by attacking the tissues of other living organisms. This beetle, a weevil in Ecuador, has been killed by a *Cordyceps* [*KOHR*-duh-seps] fungus, the stalks of which are growing out of its back.

## Fungi can benefit human society

Although some fungi can be costly to human society, other fungi are beneficial, providing us with pharmaceuticals, including antibiotics such as penicillin. Yeasts such as *Saccharomyces cerevisiae* can feed on sugars and produce two important products: alcohol and the gas carbon dioxide—crucial to the rising of bread, the brewing of beer, and the fermenting of wine. Species of *Aspergillus* are used to ferment soybean extracts in the manufacture of soy sauce and tamari. Fungi also provide highly sought-after delicacies such as truffles. Truffles grow underground, and when the closed, saclike fruiting body is fully mature, it releases a distinctive aroma to attract animals that ingest the reproductive structures whole and disseminate the spores in their scat. Dogs or pigs are often trained to sniff out the subterranean fruiting bodies, whose pungent aroma is also appreciated by many food connoisseurs.

## 3.5 Lichens and Mycorrhizae: Collaborations between Kingdoms

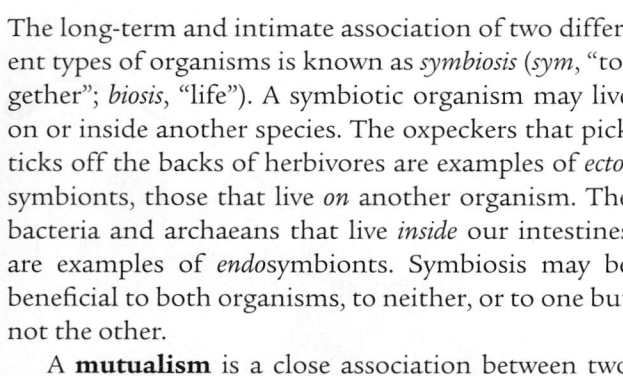

The long-term and intimate association of two different types of organisms is known as *symbiosis* (*sym*, "together"; *biosis*, "life"). A symbiotic organism may live on or inside another species. The oxpeckers that pick ticks off the backs of herbivores are examples of *ecto*symbionts, those that live *on* another organism. The bacteria and archaeans that live *inside* our intestines are examples of *endo*symbionts. Symbiosis may be beneficial to both organisms, to neither, or to one but not the other.

A **mutualism** is a close association between two species that benefit both symbiotic partners. Despite their fearsome image as destroyers of the world, fungi have evolved beneficial partnerships with nearly every kingdom of life: with photosynthetic bacteria, with photosynthetic protists (algae), with plants, and also with some animals.

Crust-like lichens on the bark of a maple tree

## Lichens contain a fungus and a photosynthetic microbe

A **lichen** [*LYE*-kun] is a mutualistic association between a photosynthetic microbe and a fungus. The lichen-forming microbe can be a unicellular green alga, a cyanobacterium, or both together. Ascomycetes, with their distinctive cup-shaped fruiting bodies (see Figure 3.16) are the most common in lichens, although basidiomycetes (club fungi) can also form lichens. Much of the body of the lichen is created by packed mycelial strands, with algal or cyanobacterial cells embedded in the mycelial mat. The fungus receives sugars and other carbon compounds from its photosynthetic partner. In return, the fungus produces lichen acids, a mixture of chemicals that scientists believe may function to protect both the fungus and the alga from being eaten by predators.

Lichens grow very slowly, typically increasing in size by less than a centimeter per year. They can multiply by fragmentation, or by disseminating dry powdery packages, called *soredia*, that consist of a few photosynthetic cells wrapped in fungal mycelia. The fungal partner can also reproduce sexually, producing fruiting bodies that launch sexual spores into the air or water (see Figure 3.18). Some fungal spores can sprout to produce mycelia that continue to grow independently, but other species must be colonized by a photosynthetic alga or cyanobacterium, or else they perish after a while.

The lichen body is thin and has no protective sheath like the cuticle of a plant or the "armor plating" of some protists. Nor does it have any mechanism for ridding the body of wastes or toxic substances. Therefore, lichens readily absorb and accumulate air- or water-borne pollutants. Lichens are destroyed by acid rain, heavy-metal pollutants, and organic toxins, which is why lichens tend to disappear in heavily industrialized areas with poor pollution controls.

Each unique association between a certain fungus and a particular photosynthetic symbiont is given a "species" name, usually based on the name of the fungus involved in the partnership. More than 30,000 "species" of lichens have been described from a variety of habitats, including cold and dry places and warm and moist ecosystems. Lichens are especially abundant in the Arctic tundra, where they form an important food source for herbivores such as reindeer. Lichens are often pioneers in barren en-

**FIGURE 3.21** Mycorrhizae Are Mutualisms between Fungi and Plant Roots

vironments. Lichen acids produced by the fungal symbiont wear down a rocky surface, facilitating soil formation. Soil particles build up from the slow weathering of rock, and over time other life-forms, including pioneering plant species, gain a toehold in the newly made soil.

## Mycorrhizae are beneficial associations between a fungus and plant roots

Plants probably would not have been as successful on land if they had not entered into a mutualistic relationship with fungi almost immediately on arrival. Some of the earliest fossil plants—liverworts that lived about 460 million years ago—appear to have had fungal endosymbionts in their roots. Today, the vast majority of plants in the wild have mutualistic fungi, known as mycorrhizal (*mykos*, "fungus"; *rhiza*, "root") fungi, associated with their root systems. Truffles, morels, and chanterelles—highly prized as food by some—are the reproductive structures of mycorrhizal fungi.

**Mycorrhizae** [*MYE*-koh-*RYE*-zee] (singular "mycorrhiza") are mutualistic associations between fungal mycelia and the root system of a plant. Mycorrhizal fungi form thick, spongy mats of mycelium on and in the roots of their plant hosts (**FIGURE 3.21**) and also extend into the surroundings, sometimes permeating several acres of the soil around the root. Mycelia are thinner, more extensively branched, and in closer contact with the soil, than even the thinnest branches on a plant root. As a result, a mycelial mat plumbs far more water and mineral nutrients, such as phosphorus and nitrogen, than the plant's root system could absorb on its own. In return for sharing absorbed water and mineral nutrients, the fungus obtains sugars that the plant manufactures through photosynthesis. **FIGURE 3.22** illustrates the impact of mycorrhizal associations on the growth of tomato plants.

Mycorrhizal fungi also improve soil quality by changing its chemical and physical properties. The mycorrhizal mantle protects roots from potentially harmful soil pathogens, including other types of fungi, bacteria, and root-nibbling animals. A predatory mycorrhizal fungus called *Laccaria* attracts and kills small arthropods (insects and their relatives) called springtails, and it shares with its plant host the nitrogen that the mycelia absorb from the digested animal. Mycorrhizal fungi are known to network plants in natural stands, often connecting the root systems of unrelated species, such as birches and pines, and transferring food from one to the other. Mycorrhizal fungi assist in providing nutrients to orchid seeds, which are tiny and lack stored food. The embryo within a newly sprouted orchid seed could not survive without the mycorrhizal network linking the seedling to mature photosynthesizing plants, from which the seedling draws nourishment until it can photosynthesize on its own.

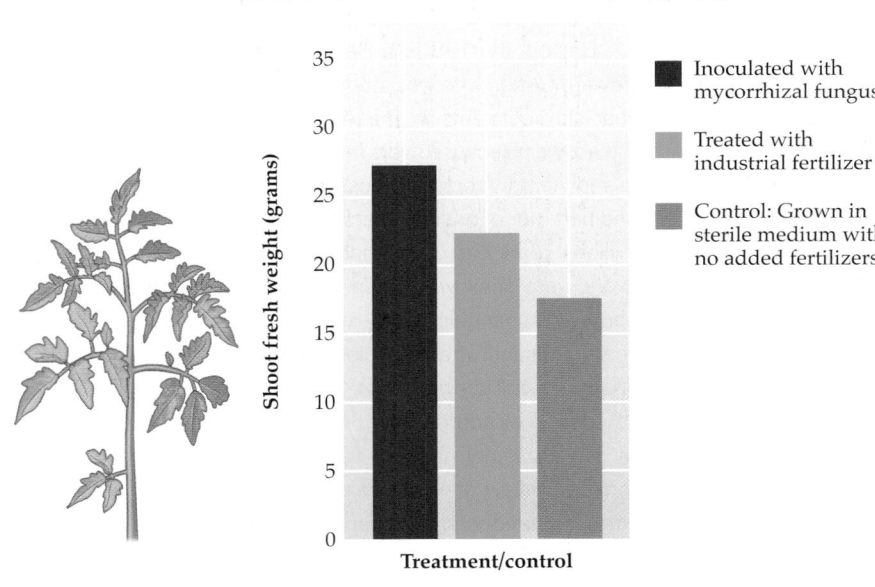

**EFFECT OF MYCORRHIZAL ASSOCIATIONS ON PLANT GROWTH**

Shoot fresh weight (grams)

- Inoculated with mycorrhizal fungus
- Treated with industrial fertilizer
- Control: Grown in sterile medium with no added fertilizers

Treatment/control

**FIGURE 3.22** Mycorrhizal Associations Benefit Growth of Tomato Plants

Certain practices in intensive agriculture, such as soil fumigation, destroy native mycorrhizal fungi. Plants grow very poorly in fumigated fields unless large amounts of industrial fertilizer are added. Mycorrhizal fungi, dependent as they are on plant roots, cannot be grown and multiplied in a lab dish, but biologists and organic farmers have cultured fungal isolates in soil that are sold commercially for use by home gardeners and organic farmers.

## APPLYING WHAT WE LEARNED

# The Root of the Problem: Why Rivers Meander

At the beginning of this chapter we learned that the meanders of rivers are important to the formation of flat valleys, rich sediments, and other features of Earth's surface that promote the diversity of plants and other organisms. When a habitat is destroyed by development or flooding, one thing that happens is that rivers and streams stop flowing through narrow, meandering channels and instead spread out into shallow fingers called "braided streams." Ideally, restoration ecologists would like to be able to rebuild rivers to restore their valuable meanders. But until recently, it wasn't clear what caused rivers to meander.

Back in the 1980s, a Bucknell University geologist name Edward Cotter, who was studying ancient 450- to 250-million-year-old sediments in the Appalachian Mountains, guessed the answer. Because rivers leave sediments that are preserved in sedimentary rocks, it's possible to tell a lot about the shape and behavior of ancient rivers, even those that are hundreds of millions years old. Cotter had noticed that early Appalachian rivers were shallow, braided streams running through gravel bars. River channels and meanders didn't appear for millions of years—not until after the colonization of land by *vascular* plants about 425 million years ago.

Cotter wondered whether, for the first few billion years of Earth's history, all rivers flowing over gently sloping terrain formed shallow, braided streams and were incapable of meandering. Meandering, he suspected, depended entirely on the presence of vascular plants, which have roots that can hold soil in place. But although Cotter had an intriguing idea, he didn't have any strong evidence.

In recent decades, researchers have begun looking a lot harder at river meandering. In 2009, Christian Braudrick, a graduate student at the University of California, Berkeley, built a miniature river in a 20-by-50-foot box. To his delight, it meandered not just for a few hours or days, but for an entire year. By using just the right slope, just the right sized sand, and one exciting new ingredient—alfalfa sprouts—he succeeded where countless others had failed. The sprouts' roots stabilized the sand enough to keep the water from cutting through it. Over the course of a year, the tiny river formed five new meanders, each of which migrated downstream as it grew and disappeared. It was a dramatic breakthrough.

Meanwhile, at Dalhousie University, geologists Martin Gibling and Neil Davies decided to evaluate the evidence for and against the theory that vascular plants sculpt modern rivers. Davies and Gibling read and analyzed 144 scientific papers that discussed the shapes of ancient rivers. The two researchers concluded that until vascular plants evolved and came to dominate on land, ancient rivers left thin, wide blankets of gravel that were consistent with shallow, braided streams, not meandering rivers. Some of these ancient rivers were as much as 1,000 times wider than they were deep. (For comparison, the Mississippi River is only about 50 times as wide as it is deep.) To further test their theory, the two researchers are now looking to see whether rivers changed back to braided streams after a mass extinction that wiped out many terrestrial plants 250 million years ago.

Mounting evidence suggests that the living and nonliving features of our planet dynamically shape one another. When plants colonized the land, their roots stabilized soils perhaps for the first time, creating tough, erosion-resistant banks capable of channeling wide, shallow streams into beautiful, meandering rivers that help create rich soil and diverse vegetation. Besides promoting the diversity of terrestrial life, these deep rivers later gave rise to major human civilizations by allowing humans to travel along waterways, carrying goods and information.

# Plants That Earn Their Keep

BY KIRK JOHNSON, *New York Times*

Could airport security gardens be the wave of the future? ("Please have photo ID and boarding pass ready and walk past the rhododendrons.") How about a defensive line of bomb-sniffing tulips in Central Park in New York, or at the local shopping mall's indoor waterfall, or lining the streets of Baghdad?

Researchers at Colorado State University said Wednesday that they had created the platform for just such a plant-kingdom early warning system: plants that subtly change color when exposed to minute amounts of TNT in the air.

They are redesigned to drain off chlorophyll—the stuff that makes them green—from leaves, blanching to white when bomb materials are detected.

"It had to be simple, something [anyone] could recognize," said June Medford, a professor of biology at Colorado State, referring to the idea of linking a plant's chemical response to its color, visible to the naked eye.

The research [paper], published in the peer-reviewed online science journal PLoS One, and financed mostly by the Departments of Defense and Homeland Security, said that plants are uniquely suited by evolution to chemical analysis of their environment, in detecting pests, for example.

Plants in the lab, when modified to sense TNT, the most commonly used explosive, reacted to levels one one-hundredth of anything a bomb-sniffing dog could muster, the paper said.

The trick, still in refinement, is how to make sure the plant's signal is clear enough and fast enough to be of use.

"Right now, response time is in the order of hours," said Linda Chrisey, a program manager at the Office of Naval Research, which hopes to use the technology to help protect troops from improvised explosive devices.

Practical application, she said, requires a signal within minutes, and a natural

reset system back to healthy green in fairly short order.

Professor Medford said she thought both goals were attainable, perhaps within three years—the goal that military backers are pushing for, she said—but more likely in five to seven years.

One scientist who read the scientific paper on Wednesday and was not involved in the project said he was concerned that the difference between all-clear green and TNT-detected white might be too subtle or subject to false inputs.

"What you want is something that is extreme on-and-off and reliable, and I don't think they're there yet," said Sean R. Cutler, an associate professor of plant cell biology at the University of California, Riverside. "It's a very interesting work-in-progress."

Everyone knows plants have no nervous system, no brain, no ability to register pain or sense the world in the way we animals do. Or do they? Research in the last few years suggests that plants communicate information by means of subtle electrical signals. Plants definitely don't "think," but they do monitor and respond to their environment. For example, if insects chew on a leaf, the leaf releases chemicals that send a signal that other leaves soon "hear." Leaves on the same plant and even leaves on other plants may respond by releasing chemicals that toughen leaves or make them taste bad to insects.

For the Department of Defense, then, it made sense to fund research on plants that could monitor human environments for explosives and other chemicals. One element of such a system has to be a signal from plants to people, a signal that people can understand. Plants naturally have a system for pulling the green pigment chlorophyll—used for photosynthesis—from their leaves. This "degreening" is easy for us to see and happens in as little as 2 hours.

Another necessary element is a way for the plant to detect a chemical—whether TNT, a whiff of nerve gas, or an air pollutant. The

response process is triggered by switchlike "receptors" on the surfaces of cells, which all organisms have. June Medford found a way to make plant receptors respond to TNT. It took some of the plants 2 days to turn white—much too long to warn anyone about a person walking by with a suitcase of explosives. But Medford hopes to make her plants respond in minutes instead of hours or days. The receptors can be made to respond to a variety of chemicals and can be added to virtually any plant, from grass to trees.

## Evaluating the News

**1.** Since the pigment chlorophyll is essential for photosynthesis in plants, what would eventually happen to a genetically modified plant whose leaves turned white and remained white?

**2.** Other researchers have been using rats to sniff out land mines and other explosives. List some advantages that plants might have over rats for this purpose, and some disadvantages.

**SOURCE:** *New York Times,* January 26, 2011.

# Summary

## 3.1 The Dawn of Eukarya

- The Eukarya have traditionally been divided into four major kingdoms: Protista, Plantae, Fungi, and Animalia.
- Eukaryotes possess a true nucleus; they have complex subcellular compartments, which enable larger cell size.
- Sexual reproduction is another evolutionary innovation of the Eukarya; however, many eukaryotes reproduce asexually as well.
- The Eukarya evolved multicellularity separately in several lineages, which endowed these groups with the benefits of functional specialization among different cell types.

## 3.2 Protista: The First Eukaryotes

- The Protista (protists), the most ancient category of eukaryotes, is a highly diverse group.
- Protista is a catchall category that lumps together many evolutionarily distinct lineages, some of which are only distantly related to others in the category.
- Most protists are single-celled and microscopic.
- Protists can be photosynthetic autotrophs, heterotrophs, or mixotrophs.
- Algae are photosynthetic protists. Nonphotosynthetic motile protists are commonly called protozoans.
- Sexual reproduction, a key evolutionary innovation of the protists, combines genetic information from two parents to produce offspring that are genetically different from each other and from both parents. Many protists can reproduce asexually as well.
- Some protists are poisonous and some are pathogens of other organisms, including humans.

## 3.3 Plantae: The Green Mantle of Our World

- The Plantae (plants) are multicellular photosynthetic eukaryotes that are adapted for life on land.
- Plants evolved a waxy covering, the cuticle, that reduces water loss.
- Plant cells have cell walls stiffened with a strong but flexible biomolecule called cellulose. The evolution of another cell wall-strengthening biomolecule, lignin, enabled plants to resist the pull of gravity.
- The vascular system, containing bundles of water-conducting xylem tubes and food-conducting phloem tubes, was also critically important in enabling plants to grow tall. All plants except bryophytes have a vascular system.

- Pollen and seeds first evolved among the gymnosperms. Pollen deliver sperm cells to the female reproductive structures. Each seed contains an immature plant, or embryo, along with a food supply enclosed in a protective seed coat.
- Angiosperms evolved flowers, reproductive structures with numerous innovations to aid in pollen dispersal. The fruit wall protects the seeds inside it and often develops special elaborations that aid in seed dispersal. Many angiosperms recruit animals to deliver pollen and also to disperse their seeds.
- As producers, plants are the ultimate food source for nearly all terrestrial organisms.

## 3.4 Fungi: A World of Decomposers

- Fungi are eukaryotic absorptive heterotrophs with chitin-containing cell walls. DNA comparisons show that fungi are more closely related to animals than to any other kingdom of life.
- Although some fungi are single-celled, most are multicellular.
- Multicellular fungi have a unique body plan: threadlike branched hyphae that penetrate organic material. The organic material is digested externally, and the breakdown products are absorbed as food. Fungi are important as decomposers in terrestrial ecosystems.
- Fungi reproduce both asexually and sexually. The products of fertilization become encased in thick walls to form spores, which are remarkably resistant to harsh conditions and able to disperse long distances via animals, wind, and water.
- There are at least three main groups of fungi—zygomycetes, ascomycetes, and basidiomycetes—and each group is characterized by distinctive reproductive structures, or fruiting bodies.
- Some fungi are parasites and cause serious disease, especially in plants and insects. Others are beneficial.

## 3.5 Lichens and Mycorrhizae: Collaborations between Kingdoms

- Mutualisms are close associations between two species that benefit both organisms.
- In lichens, algae or cyanobacteria live in a mutually beneficial association with a fungal partner.
- Mycorrhizal fungi live in and on the roots of most wild plants, and they assist their plant hosts in absorbing water and mineral nutrients from the soil. In return, the plants supply sugars to the fungi.

# Key Terms

alga (p. 56)
angiosperms (p. 64)
ascomycetes (p. 68)
asexual reproduction (p. 54)

basidiomycetes (p. 68)
bryophytes (p. 61)
cellulose (p. 62)
chitin (p. 68)

clone (p. 54)
cuticle (p. 61)
decomposer (p. 60)
fertilization (p. 64)

flower (p. 64)
fruit (p. 66)
Fungi (p. 68)
gamete (p. 54)

glycogen (p. 68)
guard cell (p. 61)
gymnosperms (p. 63)
hypha (p. 68)
lichen (p. 72)
lignin (p. 62)
mixotroph (p. 60)

mutualism (p. 72)
mycelium (p. 68)
mycorrhiza (p. 73)
parasite (p. 68)
pathogen (p. 60)
phloem (p. 63)
phytoplankton (p. 57)

plankton (p. 57)
Plantae (p. 60)
pollen (p. 63)
Protista (p. 55)
protozoan (p. 56)
seed (p. 64)
sexual reproduction (p. 53)

spore (p. 69)
stomata (p. 61)
vascular system (p. 63)
xylem (p. 63)
zooplankton (p. 60)
zygomycetes (p. 68)

# Self-Quiz

1. Eukaryotes differ from prokaryotes in which of the following ways?
   a. Eukaryotes do not have membrane-enclosed compartments in their cells.
   b. Eukaryotes exhibit a much greater diversity in modes of nutrition.
   c. Eukaryotes have a nucleus.
   d. Eukaryotes are more widespread.

2. Which of these evolutionary innovations enabled larger cell size?
   a. autotrophic mode of nutrition
   b. multicellularity
   c. sexual reproduction
   d. subcellular compartmentalization

3. Which of these groups contains *only* multicellular species?
   a. algae
   b. protists
   c. bryophytes
   d. yeasts

4. Which of these groups consists *entirely* of autotrophic species?
   a. fungi
   b. protists
   c. gymnosperms
   d. animals

5. Protists
   a. are all descended from a single evolutionary lineage and are more closely related to each other than to any other kingdom of life.
   b. are mostly multicellular, with a few single-celled species.
   c. are more diverse in modes of nutrition and life cycle characteristics than fungi are.
   d. were the first pioneers on land and dominate the terrestrial environment to this day.

6. Fungi grow by extending their
   a. hyphae.
   b. septa.
   c. basidiomycetes.
   d. coelomic cavities.

7. Which of the following constitutes an evolutionary innovation that enabled plants to become taller?
   a. a cuticle
   b. a mycelial network
   c. the presence of cell walls reinforced with chitin
   d. the presence of mycorrhizal associations

8. Mycorrhizal fungi are
   a. beneficial to plants because they help plants stay dry.
   b. harmful to plants because they secrete acids.
   c. beneficial to plants because they help in absorbing minerals.
   d. harmful to plants because they degrade lignin.

# Analysis and Application

1. Is sexual reproduction seen among the prokaryotes? Is it known among each of the four kingdoms of Eukarya described in this chapter? What might be the adaptive benefits of sexual reproduction compared to asexual reproduction?

2. Name a multicellular protist. How does it benefit this protist to be multicellular instead of single-celled?

3. What are red tides? Could humans be contributing to red tides? Explain.

4. Fungi are like us in that they are heterotrophs. How is their heterotrophic mode of nutrition different from ours? In terms of evolutionary relatedness, are fungi closer to plants or to animals? What evidence can you cite in support of your answer?

5. Imagine that an especially potent strain of a broad-spectrum fungal virus (mycovirus) destroys almost all the major fungi in your part of the world. What consequences would you expect? Describe the scene as if you were a TV news reporter touring your county a year after the massive fungal die-off.

6. Two of the major challenges facing plants when they colonized land were (a) obtaining and retaining water and (b) fighting gravity to grow taller. What evolutionary innovations enabled plants to deal with these challenges?

7. Lichens are so sensitive to pollution that they can be used as indicators of air quality. What is a lichen, and why are lichens like canaries in a gold mine when it comes to environmental pollution?

# Animalia

**SPONGES ARE THE SIMPLEST ANIMALS.**
Like other animals, sponges develop from
an embryo and feed off other organisms. But
sponges' lack of symmetry or complex tissues
and organs make them look like blobs unless,
like this sponge, they appear to have a face.

# Who We Are

Take a swim at any beach and you will likely swallow a gulp or two of water that contains a clue to the origin of animals. It may sound far-fetched, but most water contains tiny microorganisms that are distant cousins to all animals.

A single gallon of seawater can contain millions of single-celled choanoflagellates, whose whiplike flagella both push them through the water and beat bacteria and food particles into a collar that traps food. Lakes and oceans are filled with tiny microorganisms, called plankton, that drift with currents and serve as food for larger organisms. But choanoflagellates have some special traits that set them apart from other plankton. Not only can choanoflagellates propel themselves about with their tiny flagella; many of them also have an ability to form colonies—cooperative groups that live and work together. Cooperation, it turns out, is a central theme in the evolution of life.

For the first few billion years of life on Earth, all living organisms were single cells. But sometime between about a billion years ago and 700 million years ago, cells began forming cooperative groups, or colonies. From these early colonies evolved multicellular organisms such as plants and animals. In fact, one of the ancestors of modern colonial choanoflagellates was also the ancestor of all the animals that have ever lived. The earliest animals were soft, multicellular eukaryotes similar to sponges and jellyfishes. Later came worms, slugs, squids, sharks, and bony fishes, followed on land by amphibians, reptiles, and mammals. Even humans are the descendants of tiny colonial microorganisms.

Your own body consists of trillions of individual cells that somehow manage to communicate and cooperate second by second for an entire lifetime. Your nerve and muscle cells coordinate with your brain to help you swallow water or to allow you to breathe at the right time while swimming laps. Amazingly, the communication systems that human cells use to cooperate are similar to those that choanoflagellates use.

How do biologists know animals are related to choanoflagellates? What can humans and other animals possibly have in common with plankton?

At the end of this chapter we'll find out what answers choanoflagellates can provide.

---

**MAIN MESSAGE**
Animals are multicellular ingestive heterotrophs that display a remarkable diversity in form and behavior

## KEY CONCEPTS

- Animals are multicellular, ingestive heterotrophs that evolved by about 700 million years ago.

- Animal cells lack cell walls, but most are bound to an extracellular matrix and, through cell junctions, to each other.

- Key evolutionary innovations of animals include specialized tissues, organs, and organ systems; complete body cavities; body segmentation; and an astounding range of behaviors.

- Insects are the most species-rich group of all organisms. Molluscs are the most diverse animals in the sea.

- Fish were the first vertebrates, and amphibians were the first tetrapods.

- Reptiles and birds produce an amniotic egg, a major adaptation for terrestrial living.

- The success of mammals is attributed to an exceptional investment in nourishment and care of the young.

## Helpful to Know

*Zoion* is Greek for "animal" and the root word in "zoology," the science of animals (*logia*, "study of"). Animals are also referred to as "fauna," just as plants are generically known as "flora." "Metazoans" is another general label for animals.

**WHAT DO YOU THINK ARE THE QUINTESSENTIAL QUALITIES** of an animal? As food for thought, examine the photos in **FIGURE 4.1**. Would you say that all animals are capable of moving? Can all animals see?

(a) Yeti crab, *Kiwa hirsuta*

(b) Sea squirt, a tunicate

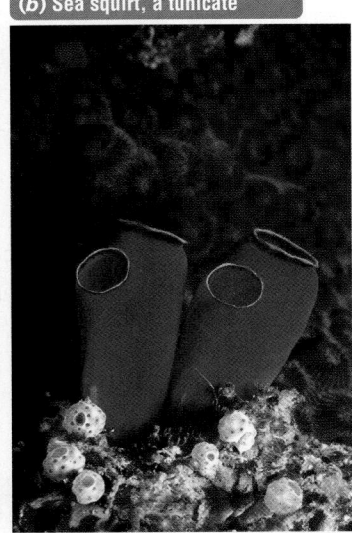

(c) Sea slugs or nudibranchs

(d) Christmas tree worm, *Spirobranchus giganteus*

**FIGURE 4.1  A Gallery of Animalia**

Can you guess which of these animals is most closely related to us?

Do they all have a brain? Which of the animals in Figure 4.1 is our closest relative? Why?

In this chapter we explore the captivating diversity of the world of animals. We will see that animal bodies are organized in many different ways and will contemplate the pros and cons of these varied body plans. We will note that the evolution of organ systems has given rise to complex behaviors, including our own ability to reflect on the nature of all life.

# 4.1 The Evolutionary Origins of Animalia

The kingdom **Animalia** comprises a diverse variety of organisms, including sponges, corals, worms, snails, insects, sea stars, and flashy beasts like Komodo dragons, Bengal tigers, and us. Animals are multicellular ingestive heterotrophs, obtaining energy and carbon by *ingesting* food—that is, by bringing food inside their multicellular bodies and digesting it internally. All animals are consumers, and some are important decomposers in the ecosystems they inhabit.

Some animals, such as sponges, have a just a few cell types and no distinct tissues, but most animals have two or three main tissue layers that appear during embryonic development and give rise to a structurally complex body. Most animals exhibit *locomotion*, the ability to move from place to place on their own. Although the adult forms of some aquatic animals—such as sponges and sea squirts—are fixed in place, or *sessile*, these species produce motile forms at some point in their life cycle. For example, the sessile sea squirts shown in Figure 4.1*b* have an immature form (larva) that swims about like a tadpole.

Animals very likely evolved by about 700 million years ago, toward the close of the Precambrian (review the geologic timescale printed on the inside back cover of this book). DNA-based estimates suggest that the lineage of sponges goes back at least that far. The first animals are thought to have descended from flagellated protists similar to the present-day protist group known as choanoflagellates (see Figure 3.6). DNA comparisons show that choanoflagellates are the closest living relatives of animals.

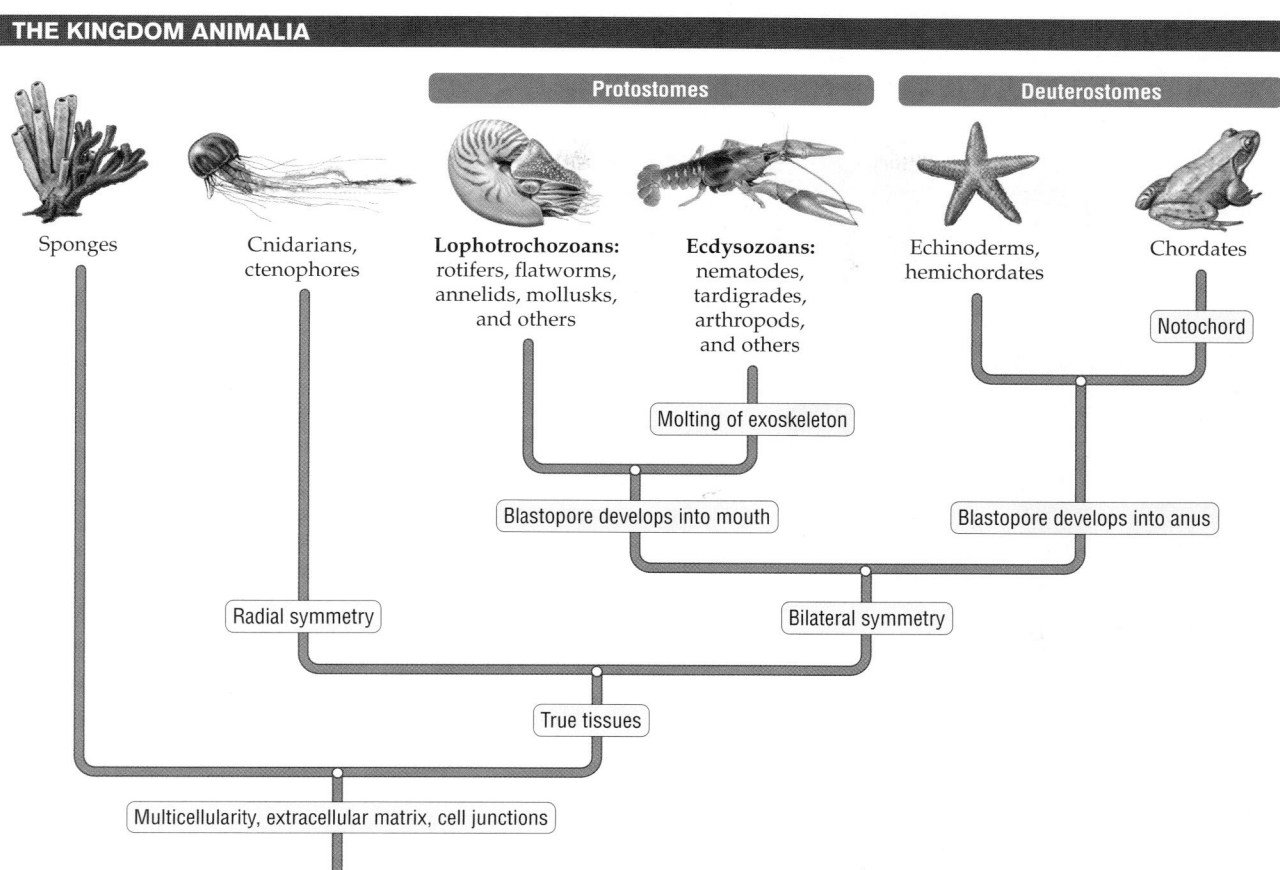

FIGURE 4.2
**Evolutionary Tree
of the Animalia**
The unique
evolutionary
innovations of each
lineage are shown with
boxed labels.

The sponges, the most ancient of animal lineages, were the first to branch off the evolutionary tree (**FIGURE 4.2**). Cnidarians [**nye**-*DAIR*-**ee-unz**], a group that includes jellyfish, sea anemones, and corals, evolved next. The remaining animal phyla fall into two broad groups: the *protostomes* and *deuterostomes*, which are distinguished by different patterns of embryonic development (to be described shortly).

Protostomes comprise more than 20 separate subgroups, including mollusks (such as snails and clams), annelids (segmented worms), and arthropods (including crustaceans, spiders, and insects). Deuterostomes [*DOO*-**ter-oh**-*STOHMZ*] include echinoderms [**ee**-*KYE*-**noh-dermz**], which are sea stars and their relatives; and the chordates, which include us. The chordates constitute a large phylum encompassing all animals with backbones, such as fishes, birds, and humans. The phylum also includes a few subgroups of less familiar animals, such as sea squirts and lancelets, that have a nerve cord along the back of the body but no backbone. Chordates that possess a backbone are known as **vertebrates**, and all other phyla of animals are informally lumped together under **invertebrates**.

However, as Figure 4.2 illustrates, "invertebrates" is not an evolutionarily meaningful label, because it merges lineages with divergent evolutionary histories and different sets of evolutionary innovations.

## 4.2 Characteristics of Animals

As you will learn in Chapter 5, animal cells differ from those of plants and fungi in that they lack cell walls. Instead, many cells in the animal body are enveloped in, or attached to, a feltlike layer known as the *extracellular matrix*. Most cells in the animal body are also firmly attached to one another by Velcro-like patches of proteins located at special sites called *cell junctions*. The proteins of the extracellular matrix and the cell junction bind the multicellular body of an animal. Many of these binding proteins are unique to animals, but some are also found in our closest relatives, the choanoflagellates.

FIGURE 4.3
**Sponges Have Specialized Cells but Lack Tissues**

Sponges are loose associations of cells. Although some of these cells are specialized, none are organized as tissues as they are in most other animals.

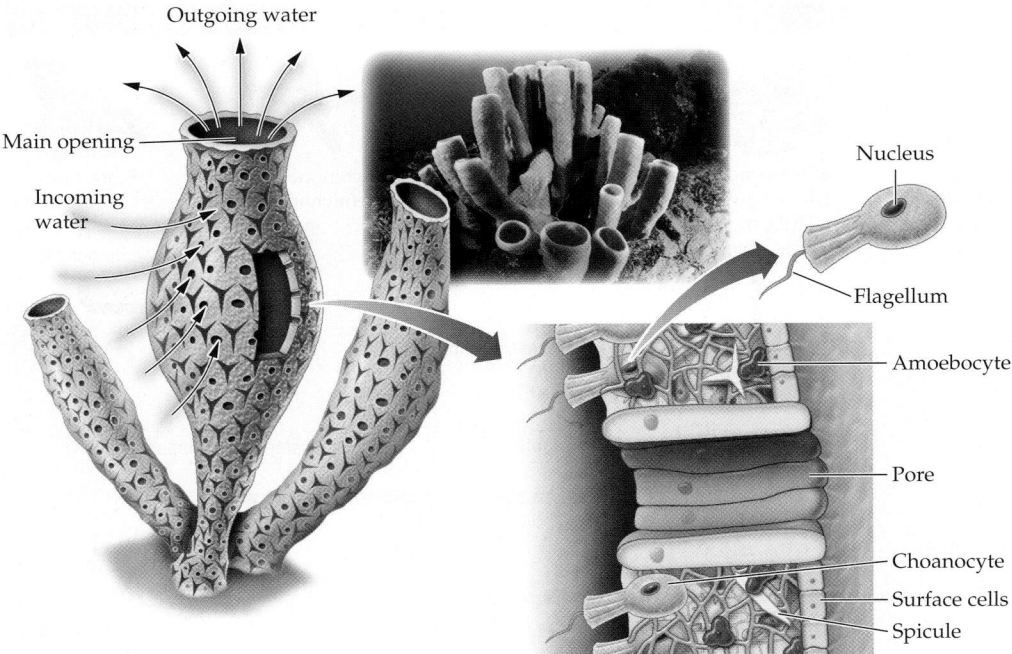

**SPECIALIZED CELLS IN SPONGES**

Outgoing water

Main opening

Incoming water

Nucleus

Flagellum

Amoebocyte

Pore

Choanocyte

Surface cells

Spicule

## Most animals have true tissues

Sponges are among the simplest of animals. Although they do have some specialized cell types, sponges lack true tissues. Recall from Chapter 1 that a tissue is a group of cells that works in a coordinated fashion to

**TRUE TISSUES IN CNIDARIANS**

Gut cavity

The body wall tissues, derived from ectoderm, are protective.

A network of nerve cells extends through the jellylike mesoglea and into the adjacent tissue layers.

The cells lining the gut cavity are derived from the endoderm and facilitate digestion by releasing proteins that break down food.

Tentacles

**FIGURE 4.4 Jellyfish Have True Tissue Layers**
Cnidarians (including jellyfish) were one of the earliest groups to evolve true tissues. These tissues include the ectoderm (*ecto*, "outer"; *derm*, "skin") and the endoderm (*endo*, "inner"). For clarity, these two layers are color-coded blue and yellow, respectively. Sandwiched between them is an inner (red) layer of secreted material known as the mesoglea (*meso*, "middle"; *glea*, "jelly").

perform a set of unique functions. A sponge is a loose collection of cells, with each cell functioning largely independently from other cells (**FIGURE 4.3**).

The body of the simplest sponges consists of an outer layer of flattened cells and a large pouchlike central chamber. Sponges with more complex organization have a thicker body wall permeated by a network of canals. The cells that line the canals and interior chambers are called *choanocytes*, or collar cells, and they strongly resemble the single-celled protists known as choanoflagellates. Water enters the body from pores in the surface layer and leaves through the large opening in the central chamber. The choanocytes use sticky secretions to snare prey, such as bacteria, drifting by in the water current. Most sponges have spiny structures, called *spicules*, made of glassy or chalky material; these sharp structures, which can resemble shards of glass, are a deterrent for would-be grazers and also shape and strengthen the animal's body. Certain crawling cells (*amoebocytes*) inside the sponge body are exceptionally versatile in the range of functions they perform and can even regenerate all the cell types in the animal's body. Because they have no tissues or organs, sponges lack specialized sensory structures, such as eyes. Sponges lack muscle cells and nerve cells and have no centralized information-processing organ like the brain.

An important evolutionary innovation of animals is the development of true tissues. The cnidarians are among the first animal groups to have evolved true tissues. Jellyfish (**FIGURE 4.4**) exhibit a two-layered pattern of tissue organization: an outer layer of tissues makes up the body wall, and an inner layer lines the gut cavity. These cnidarians have a network of interconnected nerve cells that enable rapid communication from one part of the body to another. Although they do not have eyes like ours that can form images (camera eyes), cnidarians do have light-sensitive regions, known as *eyespots*, that can sense light and shade.

Muscle tissue resides between the inner and outer layers in most species of cnidarians. Muscle tissue is unique to animals and found in all phyla of living animals except the sponges and placozoans (which constitute a puzzling phylum with a single species). The coordinated action of the specialized tissues—the nerve net and muscle tissue, for example—facilitates complex behaviors. For example, some jellyfish impale small animals with special stinging cells located on their armlike tentacles, and then use the tentacles to bring the captured prey to the mouth.

## Animals exhibit unique patterns of embryo development

**Development** refers to a sequence of predictable changes that occur over the life cycle of an organism as it grows and matures to the reproductive stage. Animal development varies from the relatively simple sequence of events that mark the life cycle of a sponge to the complex sculpting of tissues and organ systems in mammalian embryo development. Embryo development in animals involves *cell migration*, the wholesale movement of cells from one region to another. The movement of cells within the developing embryo gives rise to distinct *embryonic cell layers* (also known as germ layers) that in turn generate all the different tissue types in the adult body.

Most animals generate offspring through **sexual reproduction** involving the fusion of egg and sperm (an event known as *fertilization*), although some species also reproduce asexually. The single-celled *zygote* that results from fertilization turns into a *blastula*, a hollow sphere with a single layer of cells on the periphery. The movement of cells into the interior of the hollow sphere creates a dent, called the *blastopore*, and with this inward migration of cells the blastula is transformed into a *gastrula* (**FIGURE 4.5**).

In most protostomes (*proto*, "first"), tissues at or near the blastopore give rise to the mouth of the adult animal. In protostomes with a complete gut, a second opening that forms on the opposite side from the blastopore develops into the anus. In deuterostomes, by contrast, the blastopore generally gives rise to the anus, not the mouth (see Figure 4.5). A secondary opening, often but not always at the opposite end from the blastopore, gives rise to the mouth in deuterostomes (*deutero*, "secondary").

The cells on the outer surface of the gastrula become the *ectoderm*, the embryonic cell layer that generates the outer tissues of the animal body, such as the body wall and the nerve network of the jellyfish in Figure 4.4. The innermost embryonic cell layer, created by the inward migration of cells at the blastopore in most animal embryos, forms the *endoderm* and gives rise to the digestive system (shown in yellow in Figures 4.4 and 4.5). Most protostomes and all deuterostomes have a third embryonic cell layer, called the *mesoderm*, that arises from clumps of cells near the blastopore in most protostomes or, in many deuterostomes, from pouches that form at the far end of the pocket created by ingrowth of the blastopore. The mesoderm gives rise to tissues such as muscle and reproductive structures; in chordates, the mesoderm gives rise to the skeleton as well.

## Most animals have symmetrical bodies

The organization of the body—or body plan—often distinguishes one group of animals from another. All animals except the sponges have a distinct body symmetry. Animals other than sponges can be divided into two main groups: those with *radial symmetry*, and those with *bilateral symmetry*.

**EMBRYO DEVELOPMENT IN ANIMALS**

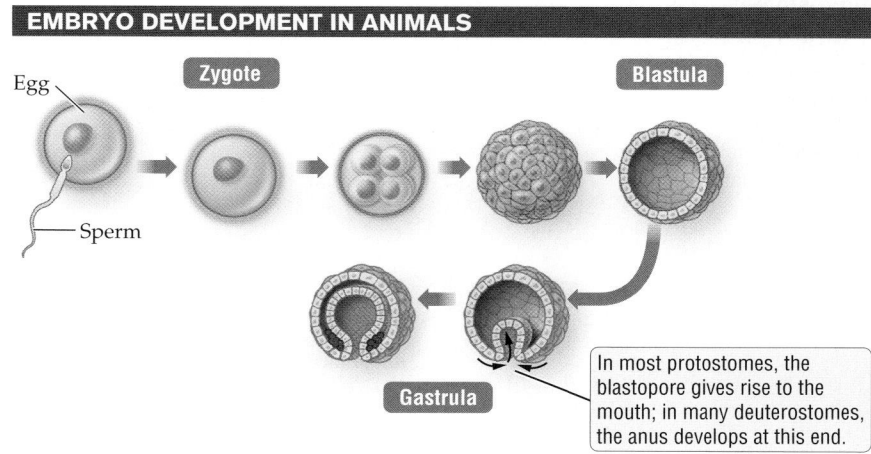

In most protostomes, the blastopore gives rise to the mouth; in many deuterostomes, the anus develops at this end.

**FIGURE 4.5** Early Embryo Development Distinguishes Protostomes and Deuterostomes

FIGURE 4.6 Most Animals Display Either Radial or Bilateral Symmetry

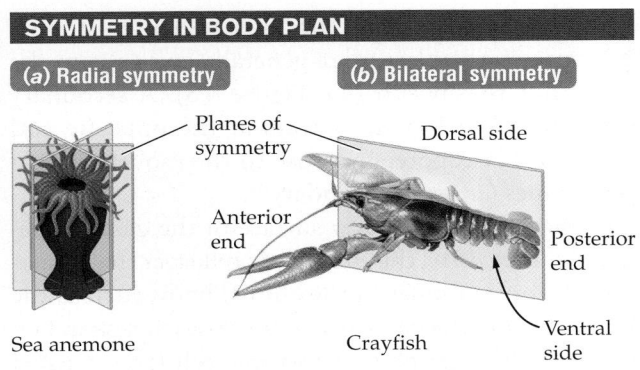

**SYMMETRY IN BODY PLAN**

**(a) Radial symmetry**

Planes of symmetry

Sea anemone

**(b) Bilateral symmetry**

Dorsal side

Anterior end

Posterior end

Ventral side

Crayfish

The body of an animal with **radial symmetry** can be sliced symmetrically along any number of vertical planes that pass through the center of the animal (**FIGURE 4.6a**). A sea anemone has radial symmetry because it can be cut, like a pie, to produce body parts that are nearly identical. Cnidarians (sea anemones, jellyfish, and corals, among others) and ctenophores (comb jellies) are among the animals that display radial symmetry. Radial symmetry gives an animal sweeping, 360-degree access to its environment. The animal can snare food drifting in from any direction of the compass, and it can also sense and respond to danger from any of these directions. Radial symmetry is adaptive for sessile animals and those that drift in currents without being able to propel themselves in a preferred direction; because these animals cannot track down food or flee from a predator, they must be prepared to deal with whatever comes their way from any direction.

In animals with **bilateral symmetry**, only one plane passes vertically from the top to the bottom of the animal, dividing the body into two halves that mirror each other (**FIGURE 4.6b**). Bilateral animals have distinct right and left sides, with near-identical body parts on each side. The top of a bilateral animal is the *dorsal* side, and its bottom surface is the *ventral* side. Bilateral animals almost always have a clear-cut front end, the *anterior* end, and a distinct back end, the *posterior* end.

Bilateral symmetry is seen in virtually all protostomes and deuterostomes, at least at some developmental stage in the life cycle. The symmetrical arrangement of body parts on either side of a central body facilitates movement in bilateral animals. The paired arrangement of limbs or fins, for example, enables quick and efficient movement on land or in water. Locomotion is a key evolutionary innovation of animals and one that has sparked a wide range of behaviors, including varied ways of capturing prey, eating prey, avoiding being captured, attracting mates, caring for young, and migrating to new habitats.

Structures used for eating, and those involved in sensing and responding to the environment, tend to be concentrated at the anterior end in bilateral animals—an evolutionary trend known as **cephalization** (*kephale*, "head"). A cephalized animal moves efficiently and displays a wide variety of adaptive behaviors because it can receive and process information rapidly from the most important direction: the direction in which it is traveling. In other words, cephalization enables animals to look where they are going!

## Most animals have organs and organ systems

Most protostomes and all deuterostomes evolved organs and organ systems, which enabled animals to function yet more efficiently. Recall that organs are body parts composed of two or more tissues that work in a concerted fashion to carry out a set of unique body functions. Usually organs have a defined boundary and a characteristic size, shape, and location in the body (see Chapter 1). **FIGURE 4.7** shows some of the organs found in the body of a flatworm, a protostome with one of the simplest body plans. The flatworm is a hermaphrodite, meaning that it contains both sperm-producing organs, known as *testes*, and egg-producing organs, known as *ovaries*. Clusters of nerve cells in the anterior are organized into a simple brain, which is a centralized information-processing organ found only in cephalized animals.

Recall from Chapter 1 that an organ system is composed of two or more organs that work in an in-

FIGURE 4.7
Protostomes and Deuterostomes Have Organ Systems

The flatworm is a hermaphrodite, meaning that it contains both male and female structures.

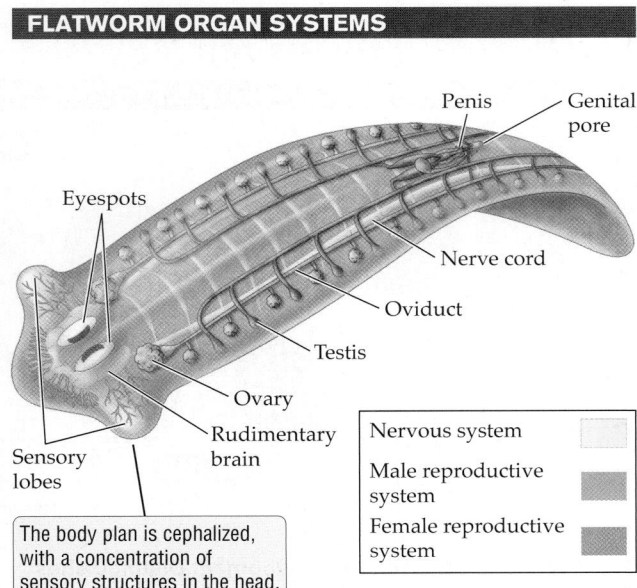

**FLATWORM ORGAN SYSTEMS**

Penis

Genital pore

Eyespots

Nerve cord

Oviduct

Testis

Ovary

Rudimentary brain

Sensory lobes

The body plan is cephalized, with a concentration of sensory structures in the head.

| Nervous system | |
| Male reproductive system | |
| Female reproductive system | |

tegrated manner to perform a unique set of functions in the body. The flatworm brain is networked with nerves that extend in a ladderlike pattern through the length of the body. The simple brain and the nerves together constitute the *nervous system*. As minimal as it seems, the nervous system of flatworms is more complex than the nerve net of cnidarians. Certain nerves deliver sensory input to the brain, which processes the information. Nerves that specialize in relaying output signals rapidly carry the response generated in the brain to the rest of the body for action.

Like the nervous system, the *reproductive system* of the flatworm is composed of organs linked to accessory structures. Eggs produced in the ovaries move through tubes called oviducts and exit the animal through the genital pore, for example. Flatworms also have a simple *excretory system* (not shown in Figure 4.7), specialized for removing excess water. The water is collected by specialized cells, called flame cells, routed through a system of collecting tubes, and then discharged to the outside.

Flatworms have a saclike gut (not shown in Figure 4.7) that is rather like the simple pouch of cnidarians and ctenophores; food and waste must enter and leave through the same opening in such dead-end guts. In contrast, most other protostomes and all deuterostomes have a complete *digestive system*, the organ system specialized for breaking down food, absorbing usable nutrients, and expelling undigested material. For example, the digestive system of vertebrates, such as fishes, consists of a one-way tube that includes a stomach and two functionally different intestinal regions, as well as accessory organs such as the pancreas and liver.

The *respiratory system*, which facilitates uptake of oxygen and removal of carbon dioxide gas, is structured in varied ways among the different groups of protostomes and deuterostomes. Many protostomes and all deuterostomes have a *circulatory system*, which delivers essential gases and nutrients to, and whisks away waste from, all the cells in the body. The *muscular system* works in conjunction with an external skeleton in some protostomes, such as the arthropods, or with the bony internal *skeletal system* of vertebrates, to facilitate movement. The defensive system, or *immune system*, varies from simple to complex across the animal phyla, as does the *endocrine system*, which consists of organs that produce signaling molecules called hormones.

## Some animals evolved complex body cavities

The evolution of a body cavity—an interior space of the body that holds many organs—was another major step in the evolution of the animal body. Flatworms are among the few protostomes that are *acoelomate* (lacking a body cavity), so their internal organs lie crowded between the body wall (derived from the ectoderm) and the gut (derived from the endoderm) in tightly packed layers of mesoderm-derived tissues. In *pseudocoelomate* animals, which include roundworms, many of the organs are suspended in a *pseudocoelom*, a fluid-filled space that lies between the gut and the mesoderm-derived tissues, such as the muscles of the body wall (**FIGURE 4.8**). Only *coelomate* animals have a true **coelom**, a body cavity that lies *within the*

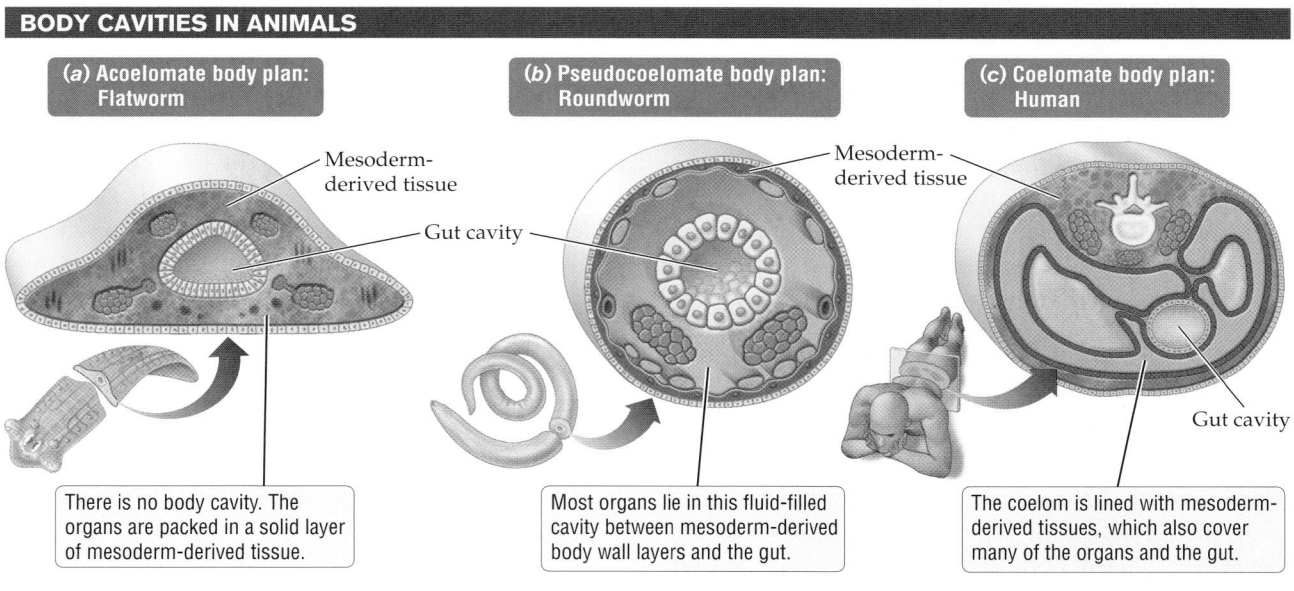

**BODY CAVITIES IN ANIMALS**

**(a) Acoelomate body plan: Flatworm**

Mesoderm-derived tissue

Gut cavity

There is no body cavity. The organs are packed in a solid layer of mesoderm-derived tissue.

**(b) Pseudocoelomate body plan: Roundworm**

Mesoderm-derived tissue

Most organs lie in this fluid-filled cavity between mesoderm-derived body wall layers and the gut.

**(c) Coelomate body plan: Human**

Gut cavity

The coelom is lined with mesoderm-derived tissues, which also cover many of the organs and the gut.

**FIGURE 4.8 Some Protostomes and All Deuterostomes Have a True Coelom**

## FIGURE 4.9 Body Segments Evolved Diverse Functions

Segmentation, a body plan in which segments repeat, enables the evolution of diverse uses of appendages, as shown here in the body of a lobster.

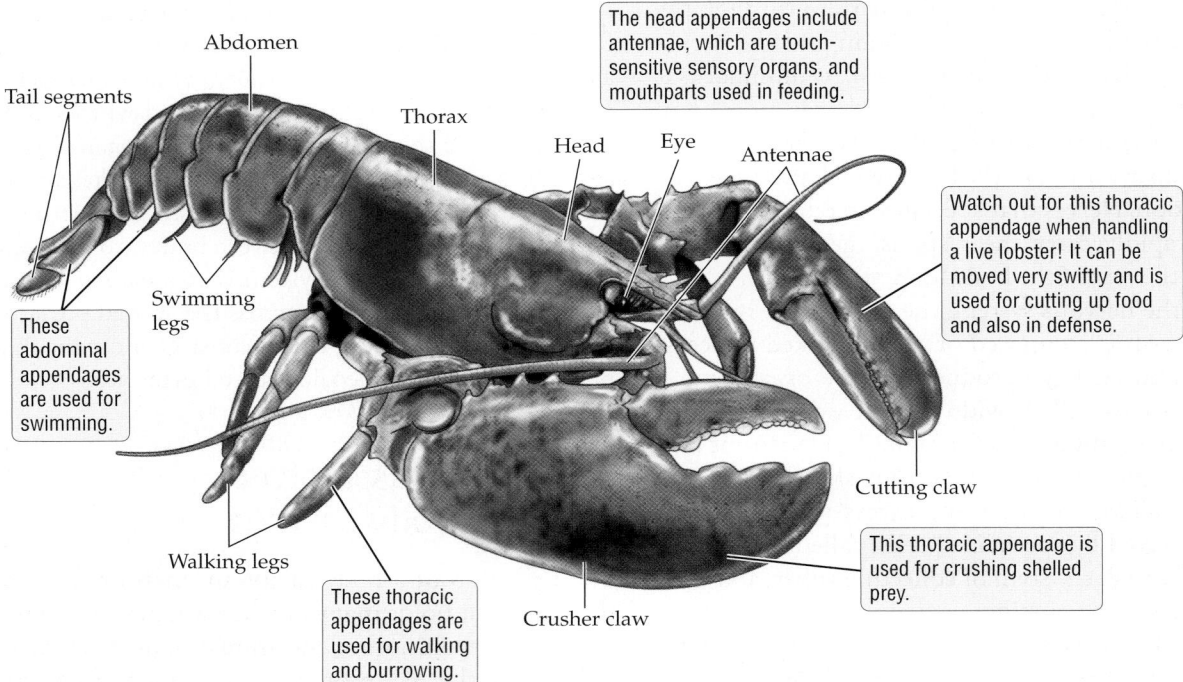

**ADAPTIVE VALUE OF SEGMENTATION**

Abdomen

Tail segments

Thorax

Head

Eye

Antennae

The head appendages include antennae, which are touch-sensitive sensory organs, and mouthparts used in feeding.

Watch out for this thoracic appendage when handling a live lobster! It can be moved very swiftly and is used for cutting up food and also in defense.

Swimming legs

These abdominal appendages are used for swimming.

Walking legs

These thoracic appendages are used for walking and burrowing.

Crusher claw

Cutting claw

This thoracic appendage is used for crushing shelled prey.

---

mesoderm. Tissues derived from the mesoderm therefore line the internal surface of the coelom and also wrap around all the organs and the hollow tube of the gut that is suspended in the cavity. Many protostomes and all deuterostomes have a true coelom.

The evolution of body cavities made it possible for an animal's internal organs to grow more freely and function independently, as the organs became liberated from the body wall on the outside and the gut on the inside. Body cavities also provide padding and protection for the organs, as well as turgidity and mechanical support for the entire body. Pseudocoelomate animals, such as nematode worms, use their fluid-filled body cavity as a *hydrostatic skeleton*. The constriction of the surrounding muscles squeezes the fluid in the pseudo-coelom, which flexes the body to one side or the other, the way a squeezed balloon bulges out in one direction or the other depending on where it is being compressed. The sequential flexion of the body produces the wriggling movements that the animal uses for locomotion.

## Segmentation enabled division of labor among body parts

Many animals have segmented bodies; that is, their body plan consists of repeated units known as *segments*. Specialized body parts, known as *appendages*, originate, often in pairs, from specific segments of the body. Over evolutionary time, the segments and the appendages that spring from them have acquired diverse form and function, enabling the animal body to adapt to new habitats or acquire new modes of life. The varied uses of segments and their appendages are beautifully exemplified by the lobster in **FIGURE 4.9**.

The evolution of just the posterior segments of arthropods illustrates how evolution can take a basic structure, such as the segmented body plan, and modify it to produce many variations over time. The last segment in the body of arthropods has evolved into the delicate abdomen of the butterfly, the piercing abdomen of the wasp (which has a needlelike structure for inserting and laying eggs deep in another animal's body), and the delicious tail of the lobster. The front appendage of vertebrates is another good example of variation on a basic structure. This appendage has evolved as an arm in humans, a wing in birds, a flipper in whales, an almost nonexistent nub in snakes, and a front leg in salamanders and lizards.

### Concept Check

1. Name four evolutionary innovations of animals.

2. Are humans and jellyfish bilateral? What are some advantages of bilaterality?

---

## 4.3 The First Invertebrates: Sponges, Jellyfish, and Relatives

**THE KINGDOM ANIMALIA**

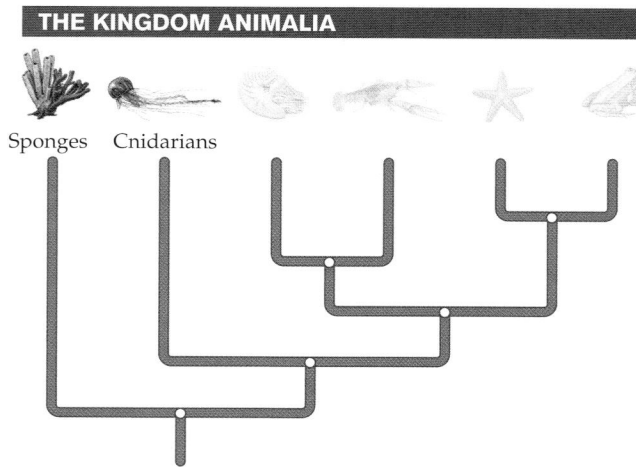

Sponges    Cnidarians

Examine the evolutionary tree of animals in Figure 4.2 and you will see that we share our world with some ancient lineages of animals that diverged very early in animal evolution, before the most recent common ancestor of protostomes and deuterostomes put in an appearance. These ancient animals include the sponges (phylum Porifera); jellyfish, corals, hydras, and their relatives (phylum Cnidaria); and the comb jellies (phylum Ctenophora). In this section we will take a closer look at these ancient animal lineages.

### Sponges have a simple body plan

Sponges are found in shallow tidal zones and at great depths in the open ocean, and from the poles to the tropics. Some sponges are absorbent and yellow and porous, to be sure, but real sponges are much more varied than that. Nearly every color of the rainbow is seen in living sponges. There are three main groups of sponges—desmosponges, glass sponges, and calcareous sponges—which differ in the type of extracellular strengthening material they produce. Calcareous sponges encase themselves in heavy deposits of calcium carbonate, the main ingredient in limestone; many species also have glassy spicules for strengthening and defense. Desmosponge species that lack the sharp spicules but have an abundance of the tough but flexible *spongin* (an extracellular matrix protein) were once widely used as bath sponges. Today, most commercial sponges are made from synthetic (human-made) material to protect natural sponges from being overharvested.

Most members of the phylum Porifera (literally, "pore-bearers") are filter feeders: the choanocytes lining the canals and interior chambers beat their flagella, creating a current that draws water through the pores on the surface (see Figure 4.3). Sponges feed on bacteria, amoebas, and other tiny organisms in their aquatic environment, filtering a ton of water just to get enough food to grow one ounce of tissue.

### Cnidarians and ctenophores have radial symmetry

There are about 10,000 species of cnidarians and about 150 species of ctenophores (comb jellies), most of them marine. *Cnidarians* include corals and sea anemones (anthozoans), hydrozoans, and jellyfish (scyphozoans). *Ctenophores* look superficially like jellyfish, but they differ from cnidarians in having a complete gut, a tube with a mouth at one end and anal pores at the other end from which undigested material exits the gut. Cnidarians and ctenophores have these characteristics in common:

- Radial symmetry
- Two distinct tissue layers separated by a jellylike extracellular matrix (mesoglea)
- A nerve network consisting of interlinked nerve cells
- Lack of organs and organ systems

The adult form of some cnidarians is a sessile cylindrical *polyp*; the adult anemone is a polyp, as is the adult coral animal. Other adult cnidarian species have a motile, bell-shaped *medusa*, which is essentially an upside-down, free-floating polyp. The medusae drift in currents but may move actively by contracting the muscle cells in the "bell." Some species have both a polyp and a medusa stage in their life cycle.

Most cnidarians have one or two rows of tentacles around an opening that leads to a blind gastrovascular cavity with a single opening at the top that serves as both the mouth and the anus. Cells lining the cavity secrete digestive juices to break down food entering the cavity. The cavity also circulates food and oxygen and flushes out waste produced by the body.

The name "Cnidaria" comes from the Greek word for "nettle," a stinging plant. Cnidarians are characterized by stinging cells (*cnidocytes*) that they use to immobilize prey and to protect themselves from predators.

A desmosponge. About 90 percent of living sponges are desmosponges.

## 4.4 The Protostomes

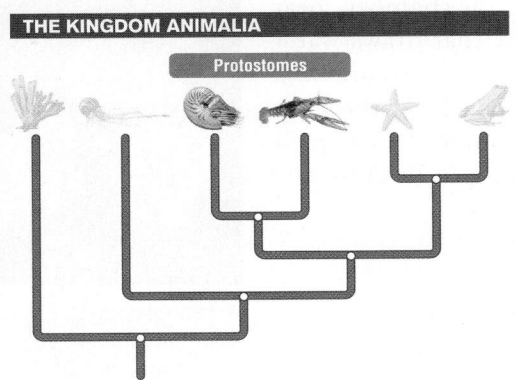

**THE KINGDOM ANIMALIA**

Protostomes

Protostomes form the largest branch on the evolutionary tree of animals, encompassing a staggering diversity that ranges from the feathery Christmas tree worms to the colorful sea slugs (nudibranchs) seen in Figure 4.1. Protostomes range in size from microscopic rotifers, which are smaller than many single-celled protists, to mollusks like the colossal squid, which can grow to 14 meters (46 feet). **Protostomes** share the following characteristics:

- Bilaterality at some stage in the life cycle
- Three embryonic cell layers
- Development of the mouth from the embryonic blastopore
- Cephalization, with an anterior brain and a ventral nervous system

Next we will consider just a small sampling of the more widespread and species-rich groups of protostomes.

## Rotifers and flatworms lack a true coelom

If you can obtain some pond water and have the use of a microscope, you could have a lot of fun observing *rotifers* (literally, "wheel-bearers"). Most species in this phylum propel themselves using a ring of fine projections (*cilia*) that sprout from the head like a crown (**FIGURE 4.10**). Their near-transparent bodies are about 0.1 millimeter long on average, and they are sometimes mistaken for ciliated protists. Rotifers are important decomposers, using their crown of cilia to whisk organic particles and fish waste into the mouth. They also graze on algae and are very effective in keeping fish tanks clean. Rotifers are preyed upon by many aquatic animals, including bryozoans, jellyfish, and sea stars.

Flatworms (phylum Platyhelminthes) are acoelomate animals with a dead-end gut (one that lacks an anus). They look nothing like rotifers, but DNA comparisons show that the two phyla are closely related. Flatworms are either free-living (such as the several thousand species of freshwater planarians) or parasitic (such as flukes and tapeworms). Planarians are distinctly cephalized, with a triangular head bearing sensory cells and a pair of eyespots (pigments in the eyespots give the animal its cross-eyed appearance, which you can see in Figure 4.7). They lack respiratory and circulatory systems, and gas exchange takes place across the skin. The thin, flattened shape of the animal facilitates gas exchange because the cells are close to the skin surface, where they can take up oxygen and release waste carbon dioxide.

A parasitic blood fluke that migrates from snails to humans is responsible for schistosomiasis, a devastating illness found in parts of Asia, Africa, and South America. Tapeworms, another serious parasite of humans, attach to the intestinal walls of their hosts—mainly pigs and cattle—with hooks and suckers on their anterior end, and then migrate through the bloodstream to lodge in muscle tissue. Humans eating underdone meat can acquire the parasite, which proceeds to reproduce in the intestines, producing fertilized eggs that pass out in feces to continue the infection cycle.

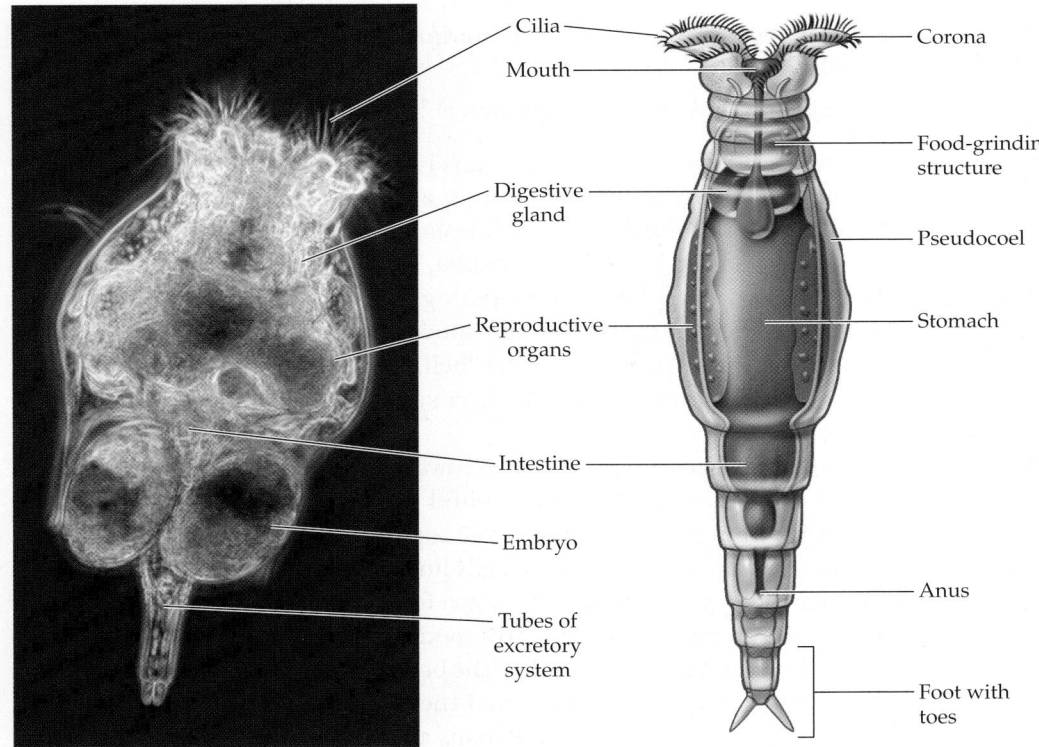

Cilia
Mouth
Digestive gland
Reproductive organs
Intestine
Embryo
Tubes of excretory system

Corona
Food-grinding structure
Pseudocoel
Stomach
Anus
Foot with toes

**FIGURE 4.10 Rotifers Are Smaller Than Some Protists**

## Annelids are coelomate worms with segmented bodies

All worms with true coeloms and segmented bodies are placed in the phylum **Annelida** (literally, "ringed ones"). The earthworms that emerge from their flooded underground burrows after a hard rain are the most familiar members of this phylum. Like most annelids, earthworms have a simple brain connected to two *nerve cords* (bundles of nerves) that run along the ventral side of the body (**FIGURE 4.11**). The skin of earthworms is thin and moist and used for gas exchange. Earthworms have a closed circulatory system, which means the blood moves in closed tubes (blood vessels) at all times, instead of draining into a body cavity at any point.

Earthworms have a well-developed digestive system, with a *crop* for storing food and a *gizzard* in which ingested leaf litter is ground with swallowed grit for better nutrient extraction. Worms are essential decomposers, especially of plant matter, in most terrestrial ecosystems besides those in cold northern latitudes and desert environments. Worm castings, the undigested remains that exit from the anus, are rich in mineral nutrients and humic acid and increase the fertility of the soil. Earthworms turn the soil through their burrowing activity, bringing plant matter from the surface to deeper layers. Worm activity also improves soil aeration, the mixing of air into the soil, which benefits plant roots.

There are several thousand species of marine annelids, most of which live partially buried in marine sediments. The showy Christmas tree worm in Figure 4.1*d* belongs to the *polychaete* group of marine annelids. Like many other members of the tube worm group, the Christmas tree worm secretes tubes of chitin that it can withdraw into when threatened. Some tube worms live at great depths on the ocean floor, often at deep-sea thermal vents, which are cracks in the ground that spew hot water.

## Mollusks constitute the largest marine phylum

One of the largest and most diverse group of protostomes consists of the mollusks, which include familiar shellfish, snails, slugs, squid, and octopi. **Mollusca**

**THE ANNELID BODY PLAN**

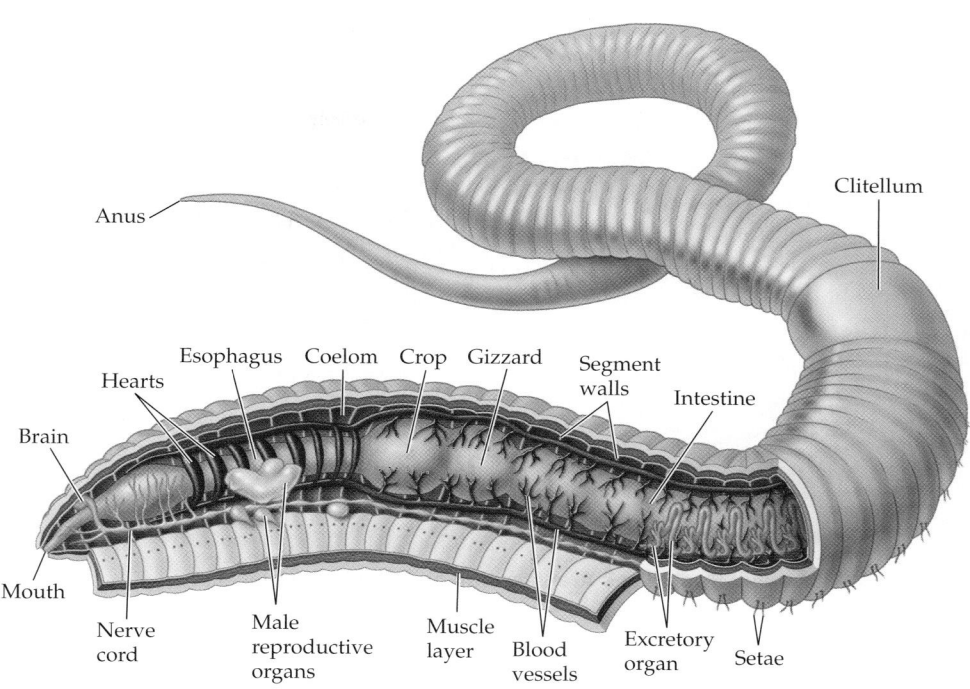

**FIGURE 4.11** **Segmentation Is a Key Feature of the Annelids**
Annelids have a true coelom and well-developed organ systems.

is the most diverse phylum of animals, next to the arthropods (spiders, insects, and their relatives). Most mollusks are saltwater animals, although many thrive in fresh water and some live in moist terrestrial environments. From chitons to conchs to colossal squid, mollusks are enormously varied in size, shape, and details of the life cycle. The mollusk body plan, illustrated in **FIGURE 4.12**, includes the following basic characteristics:

- A muscular *foot* at the base of body
- A compact grouping of internal organs called the *visceral mass*
- A *mantle* enclosing the body cavity and the visceral mass

The thick, muscular foot is used for locomotion in many mollusks. The foot is also used for burrowing in mollusks like clams, and it is modified into several tentacles in mollusks like squid and octopi. Most of the internal organs of a typical mollusk are bunched together in a visceral mass. The outer layers of the body wall form a flap of tissue called the mantle that encloses a body cavity (*mantle cavity*) and also forms a protective cover over the visceral mass. In some

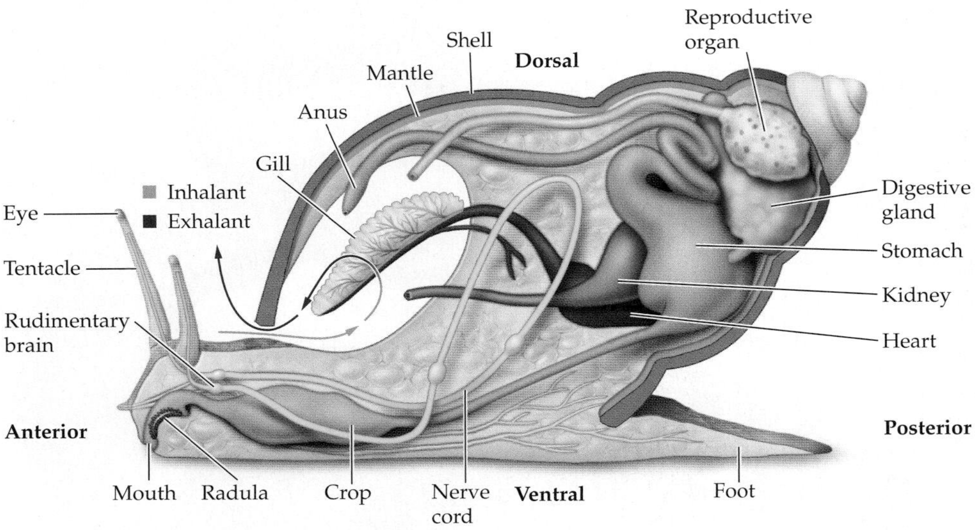

**FIGURE 4.12 Mollusk Body Plans Are Variation on a Basic Theme**
Although they show astounding diversity, mollusks have certain commonalities in their body plans. For example, all mollusks are bound by a tense sheet of tissue called the mantle, which wraps around the visceral mass, the compact lump formed by the internal organs. Most mollusks have a muscular foot used in locomotion.

species the mantle secretes a protective shell. The *gills*, which are specialized for gas exchange, protrude into the mantle cavity; the gills absorb oxygen from and release carbon dioxide into the water that is pumped into and out of the cavity. Mollusks have a circulatory system made up of a simple heart and blood vessels, as well as a fluid-filled cavity. Mollusks have a complete digestive system and an excretory system with structures called nephridia that filter waste substances from blood, combine them with excess water, and expel both from the body in a fluid called urine. The nervous system varies from relatively simple in bivalves to highly complex in octopi.

Some marine experts estimate that over 23 percent of marine species are mollusks. We introduce here just three large—and probably familiar—groups: the *bivalves* (shellfish), the *gastropods* (which include snails), and the *cephalopods* (which include squid and octopi).

BIVALVES. *Bivalves* are mollusks that shelter their soft bodies inside a hard, hinged shell (**FIGURE 4.13a**). Bivalves include familiar "shellfish" such as clams, oysters, scallops, cockles, and mussels. A bivalve animal opens its shell just enough to draw in water, often through a specialized intake tube, the *incurrent siphon*. The incoming water passes over the gills, enabling the animal to absorb oxygen and rid itself of waste carbon dioxide. In filter-feeding species, which are the major-

ity among bivalves, fine cilia on the surface of gills secrete mucus that traps incoming plankton and send the food-laden mucus into the mouth. The water is routed through the mantle cavity and then exits the shell, sometimes through a specialized *excurrent siphon*. Some bivalves, such as scallops, can move by jet propulsion: they shoot water out the rear by clapping their valves forcefully, thereby moving the animal forward like a jet.

Strong muscles hold valves shut when the animal is threatened by a predator or by the risk of drying out. Steamed clams and mussels gape because cooking destroys the muscles that contract to shut the two halves of the shell. The secretion of the valve material by the mantle is sensitive to growing conditions, and in cold seas growth often occurs in a start-and-stop cycle that reflects the warming and cooling of the environment with the change of seasons. The periodic growth of these species creates growth bands that can be used to estimate the age of the animals, the way growth rings in a pine tree can tell us how many growing seasons the tree has lived through.

GASTROPODS. Most *gastropods* have a spiral calcareous shell on the dorsal side of the animal (**FIGURE 4.13b**). Gastropods (literally, "belly-foot" animals) include slugs, snails, periwinkles, limpets, whelks, abalones, and the sea slugs (nudibranchs) pictured in Figure 4.1c. Some gastropods, such as garden slugs and sea slugs, lack a shell. In land snails and slugs, the gill is modified into a lung with an extensive network of fine blood vessels (capillaries). In some cases the bright coloration of nudibranchs (literally, "naked gills") is camouflage; in other cases it is a warning to would-be predators that they are poisonous.

Some gastropods use the thick foot for gliding on a slime trail. In others, the foot does double duty as a digging tool, and in some species it assists in swimming as well. Many gastropods scrape food, such as algae, with the rasping action of a grooved, tonguelike structure called the *radula* (see Figure 4.12). The radula has evolved additional functions in some species. In cone shells, for example, the radula is modified into a venom-injecting tube. Stings from these tropical gastropods can be fatal.

CEPHALOPODS. *Cephalopods* (literally, "head-foot" animals) are exclusively marine and include shelled species such as the nautilus; species with reduced shells, such as squid and cuttlefish; and species that have no shell

(a) A bivalve

(b) A gastropod

(c) A cephalopod

**FIGURE 4.13 Mollusks Dominate the Sea**

(a) Bivalve: The rows of simple eyes (at shell edges) of the Atlantic bay scallop can detect motion by sensing changes in light levels. (b) Gastropod: At 40 pounds and 30 inches long, the Australian trumpet snail is the largest gastropod in the world. (c) Cephalopod: The octopus Paul is credited with correctly predicting the outcomes of all international games played by the German national soccer team. Paul died in 2010 of natural causes.

at all, such as the octopus. In most species, the foot is modified into several clasping arms, or tentacles.

Cephalopods are thought to be the brainiest invertebrates. They have a well-developed brain and an extensive system of branched nerves. Octopi can learn to navigate a maze and have been known to use tools (such as carrying away a coconut shell and later using it as a hideout). An octopus at a German zoo learned how to open screw-top jars with its tentacles after watching zoo staff unscrewing the lids on jars containing shrimp, a favorite food. A famous octopus named Paul is shown in **FIGURE 4.13c.**

The octopus has a pair of large, image-forming eyes (camera eyes) that rival ours in complexity, although vertebrate eyes evolved independently from those of any mollusk. The paired eyes have lenses to focus light and an inner surface (*retina*) capable of forming detailed images. In contrast to vertebrate eyes, which focus by changing the shape of the lens, the typical cephalopod eye focuses by moving the lens away from or toward the retina.

The sophisticated nervous system of cephalopods is an adaptation to their predatory lifestyle, particularly the need to swim fast and track down prey, which range from crabs to schools of fish. Many cephalopods can change color, either for camouflage or to communicate with other individuals, such as potential mates.

Most cephalopods rely on jet propulsion for speedy locomotion, using strong contractions of the mantle cavity to shoot out water. Many cephalopods expel large clouds of ink, which they store in a sac below the gills, to help them escape from predators; in addition, some species blanch as they shoot out the ink—a disappearing act that must further confuse the predator.

## Some protostomes shed their outer covering to enable growth

One group of protostomes, the ecdysozoans (*ecdysis*, "getting out of"), have the habit of shedding their outer covering on a regular basis. Ecdysozoans have a protective noncellular layer, the cuticle, composed of organic material secreted by the outermost layer of skin cells. The cuticle is relatively thin in some phyla but forms a thick, platelike exoskeleton in other phyla. Juveniles encased in the cuticle cannot grow unless they shed the covering in the process known as **molting** (ecdysis), grow rapidly, and then secrete a new cuticle to protect themselves.

The phylum **Arthropoda** is the largest phylum grouped under the "ecdysozoan" label, and it is also the largest phylum among all eukaryotes. Arthropods (*arthro*, "jointed"; *pod*, "foot") are animals with jointed body parts, including crustaceans, chelicerates (horseshoe crabs and spiders), millipedes and centipedes (myriapods), and insects and their relatives (hexapods). We consider some interesting nonarthropod ecdysozoans—the tardigrades and nematodes—before turning our attention to the dramatically diverse arthropods.

Tardigrades, commonly known as water bears, are soft-bodied, usually microscopic, segmented animals with three to four pairs of fleshy, unjointed legs. Found in deep oceans, beach sand, freshwater sediments, garden soil, and the film of water that covers lichens and mosses, tardigrades are survival specialists. Two million or more of these animals may occur in one square meter of a mossy bank. Tardigrades can survive high and low extremes in temperature, extremely high levels of radiation, and even desiccation.

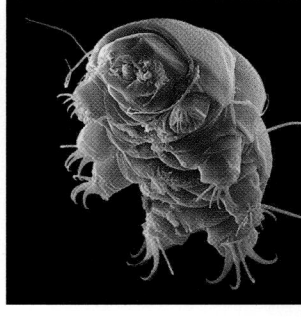

A tardigrade

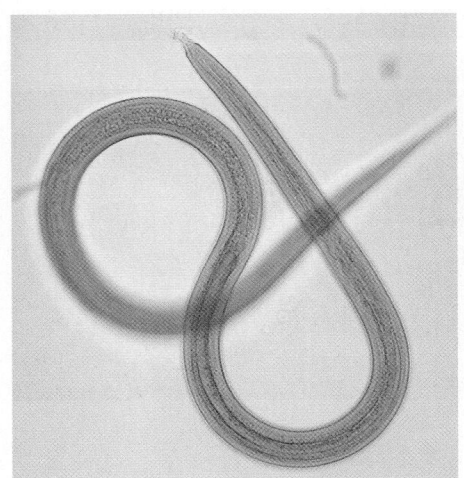

**FIGURE 4.14**

**Nematodes**

Nematodes are often hermaphrodites, commonly reproduce about 4 days after hatching, and live for 3–4 weeks. Much of the body is filled with reproductive structures, which include two ovaries and two sperm-producing structures.

The phylum Nematoda (roundworms) consists of at least 25,000 species, widely distributed in freshwater, marine, and terrestrial environments. Most of the species are free-living soil dwellers, but many are parasitic. Soil nematodes often live at extremely high density: billions of them may be found in a single acre of fertile farmland. Nematodes range in size from microscopic to gigantic: the largest nematode known measures 9 meters and lives inside the placenta of sperm whales!

A soil nematode called *Caenorhabditis elegans* (**FIGURE 4.14**) has become a star in the modern study of animal development. The entire DNA of the species has been decoded, and it is very easy to manipulate the animal's genes using the tools of genetic engineering. The transparent body of the adult worm, about 1 millimeter long, contains exactly 959 cells, including 302 nerve cells, and researchers have been able to track the development of every one of these cells from the zygote stage. Studying the genes that control development in this animal has taught us a great deal about how human cells become specialized for unique functions and how nerve cells communicate with one another.

## Arthropoda is the most species-rich phylum

More than a million species of arthropods, mainly crustaceans and insects, have been described. Members of the group are also extremely abundant: about $10^{18}$ (a billion billion) arthropods are estimated to be alive at any given time! Arthropods have a hard outer cuticle, the **exoskeleton** (*exo*, "outer"), that is made of *chitin* [*KYE*-**tin**], the same biomolecule found in the cell walls of fungi. Arthropods shed their exoskeleton in order to grow. Enzymes secreted by the underlying skin cells break down the base of the cuticle, which splits, allowing the animal to crawl out (see Figure 4.15*b*). The new cuticle that forms is soft and pliable initially, enabling the animal to grow. Until the growth stops and the cuticle hardens, the animal is vulnerable to predation and desiccation. Between molts, most arthropods stockpile food stores to fuel the growth that occurs immediately after the exoskeleton is shed.

Why are arthropods so successful? One characteristic that has contributed to the evolutionary diversity of this phylum is the segmented body plan. Over time, individual body segments have evolved different combinations of legs, antennae, and other specialized appendages, resulting in a huge number of different types of animals, many of them exceptionally well adapted to their habitat (see Figure 4.9). The tough exoskeleton provides waterproofing on land and protection against many potential predators in all types of habitats. The rigid exoskeleton also anchors muscles, which in turn enable rapid and precise movements of the different types of appendages, including the jointed legs that facilitate efficient locomotion. The arthropods share these general features:

- Jointed appendages that facilitate quick and precise movements of body parts
- A cuticle that forms a hard exoskeleton
- A segmented body plan at some stage of the life cycle
- A three-part body plan, consisting of an anterior head and thorax and a posterior abdomen

Four major arthropod groups are illustrated in **FIGURE 4.15**. We will discuss each of them next.

(*a*) Crustacean  (*b*) Insect  (*c*) Arachnid  (*d*) Myriapod

**FIGURE 4.15  Arthropods**

(*a*) Crustaceans (lobsters, shrimp, and crabs) are primarily water dwellers with 10 or more legs. (*b*) Insects have 6 legs. (*c*) Arachnids (mites, ticks, spiders, and scorpions) have 8 legs. (*d*) Myriapods (millipedes and centipedes) live on land and have many more than 10 legs.

CRUSTACEANS. **Crustaceans** are aquatic arthropods that are especially diversified in the marine environment (**FIGURE 4.15a**), although freshwater species are also common. Shrimp, lobster, and crabs are among the most familiar crustaceans. Copepods, and the shrimplike krill, are microscopic crustaceans of enormous importance in aquatic food chains because a great variety of animals, including whales, feed on these abundant and prolific animals. Barnacles are among the few crustaceans that are sessile as adults.

As noted earlier, the many different appendages of crustaceans such as the lobster perform a great variety of functions (see Figure 4.9), from sensing their world to capturing food and eating it. Many crustaceans have a flap of exoskeleton—the carapace—that covers and protects the head and thorax. Most crustaceans produce free-swimming planktonic larvae that are important food for larger animals.

INSECTS. No animal group is more species-rich than the insects (**FIGURE 4.15b**). More than half of the nearly 1.7 million known species of eukaryotes are insects. All the remaining animals make up only about 300,000 species (**FIGURE 4.16**). Insects are mainly terrestrial and are distinguished by the ability of many species to fly. Including grasshoppers, beetles, butterflies, and ants, among others, the insects are probably the best-known arthropod group. They have a three-part body plan with six legs attached to the thorax, but none to the abdominal segments in contrast with many crustaceans (**FIGURE 4.17a**).

Insects were among the earliest animals on land, and they exhibit many adaptations for a landlubber existence. The exoskeleton, which prevents the soft body tissues from drying out, is one such adaptation.

## INSECTS: THE MOST SPECIES-RICH GROUP

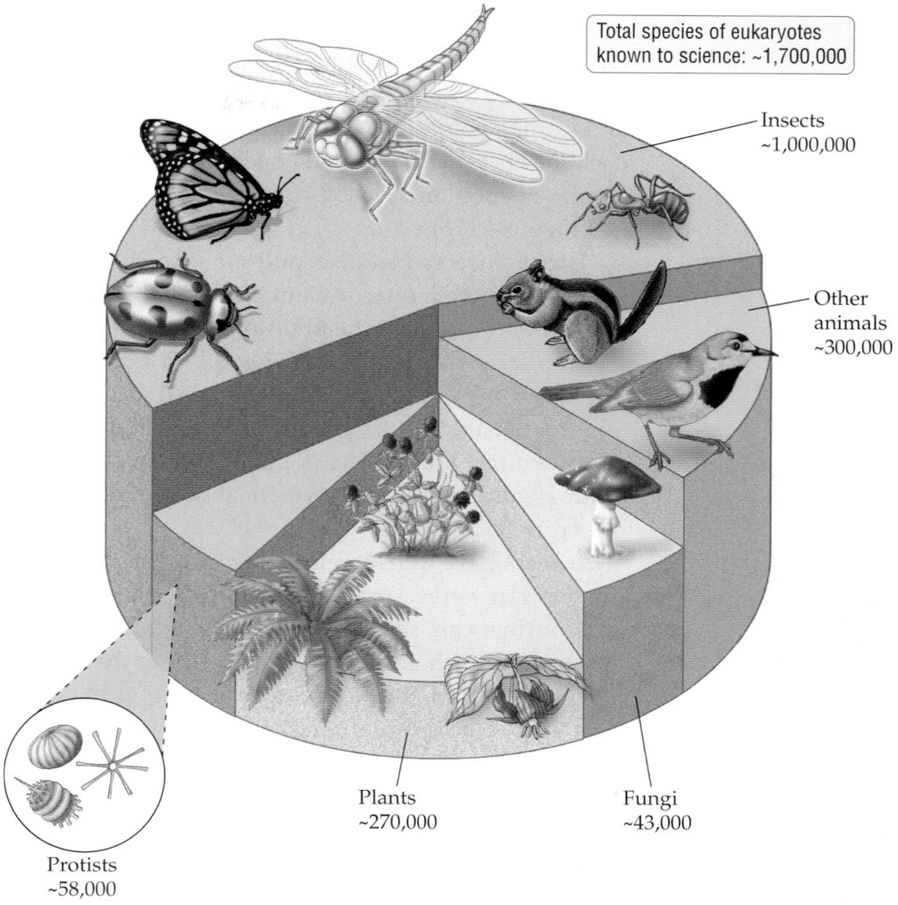

Total species of eukaryotes known to science: ~1,700,000

Insects ~1,000,000

Other animals ~300,000

Plants ~270,000

Fungi ~43,000

Protists ~58,000

**FIGURE 4.16 Of the Eukaryotes Known to Science, Insect Are the Most Diversified**

In contrast to the external gills of crustaceans, the gas exchange surfaces for insects are internal, so the moist surfaces are protected from desiccation (**FIGURE 4.17b**).

## BODY PLAN AND ORGAN SYSTEMS: INSECTS

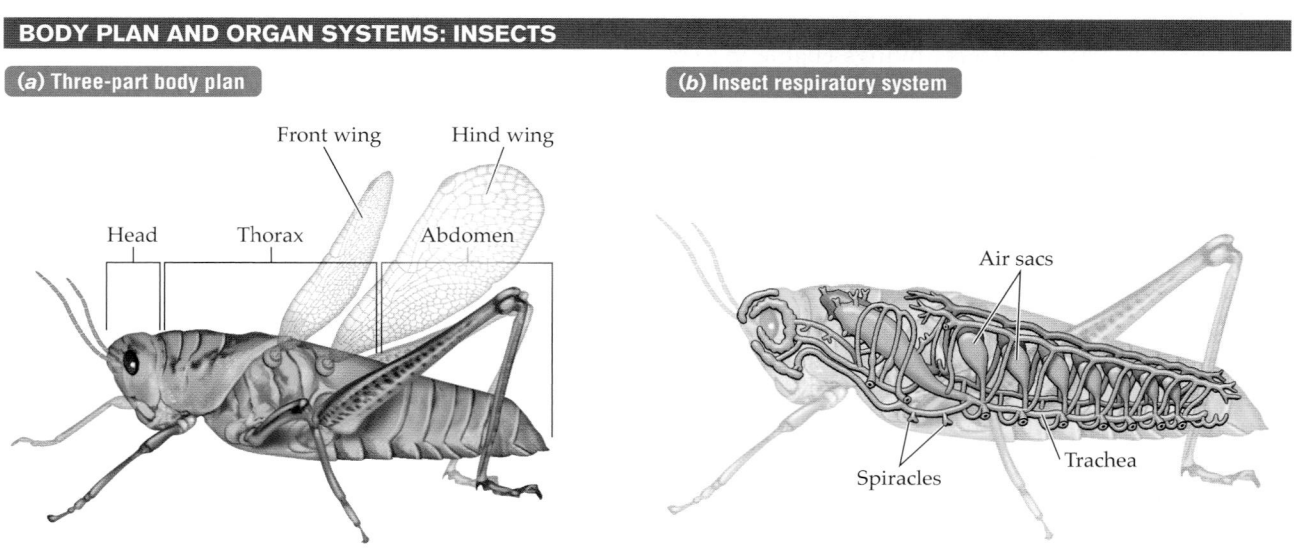

(a) Three-part body plan

Front wing    Hind wing

Head    Thorax    Abdomen

(b) Insect respiratory system

Air sacs

Spiracles    Trachea

**FIGURE 4.17
Insects Have a Three-Part Body Plan and an Enclosed Respiratory System**

As an adaptation to life on dry land, insects evolved a series of breathing tubes that branch inward from openings (*spiracles*) on the surface of the body. The largest tubes, called *tracheae*, branch repeatedly; the smallest of these, called *tracheoles*, allow gases to move directly to and from the bloodlike fluid (hemolymph) that fills the coelom. The valvelike outside openings of the tracheae can be opened and closed to regulate oxygen uptake (see Figure 4.17*b*).

Larger insects can also pull air into the tracheal system, or push it out, by using muscles to alternately expand and contract the abdomen. If an insect happens to land nearby, you may want to take the opportunity to watch its respiratory system in action. You may be able to observe the pumping action of the abdomen (the body section farthest from the head), as air is admitted into the tracheal system or pushed out from it.

Insects have well-developed visual systems. Some insects can detect motion far better than we can. Dragonflies can see and snatch mosquitoes (their prey) when both are flying in midair—no mean feat, if you recall from experience how difficult it is to swat a fly. One simple visual system found in many arthropods consists of **simple eyes** that can distinguish light and dark, and sometimes distance and color. Insects have, in addition to simple eyes, a pair of **compound eyes** that can form images. Honeybees, for example, use their compound eyes to see patterns of color on flowers. A compound eye consists of many individual light-receiving units, each with its own lens and small cluster of photoreceptors. In dragonflies, the light-receiving units are densely clustered on certain portions of the eye, which, like the human retina, can form especially sharp images.

Flight has very likely been a big factor in the extraordinary success of insects. Wings enable insects to escape predators and unfavorable conditions, to locate food and mates far and wide, and to disperse offspring over a broad landscape. Most insects have two pairs of stiff, gauzy wings, which may have evolved from the anterior appendages of a crustacean-like ancestor. Flies have only one pair of wings, with the second pair modified into stubby balancing organs (haltares). In beetles, the thin and gauzy second pair of wings is tucked under thick and sturdy forewings. When at rest, butterflies hold their four wings folded vertically, but moths hold them flat. Some insects, such as lice, lost their wings over evolutionary time.

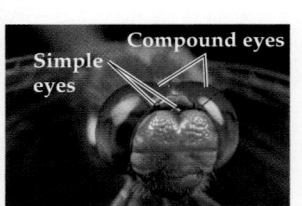

Simple eyes and compound eyes in a dragonfly

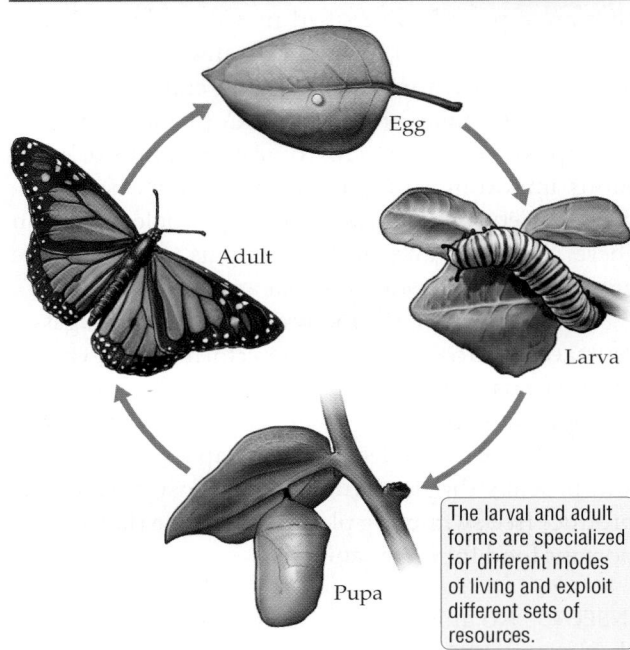

The larval and adult forms are specialized for different modes of living and exploit different sets of resources.

**FIGURE 4.18 Butterflies Are among the Insects That Undergo Complete Metamorphosis**

Many insects have complex life cycles in which the immature insect looks nothing like the adult. Caterpillars look very different from their adult forms, for example. The multistep developmental changes through which immature forms of animals are transformed into adults are collectively called **metamorphosis** (**FIGURE 4.18**). In arthropods, the transition from one developmental stage to the next is generally accompanied by molting. In some species, including grasshoppers and cockroaches, the developmental changes are gradual; such species are said to have *incomplete metamorphosis*. In other species, such as butterflies, the transition from one developmental stage to the next is dramatic, with the immature forms bearing little resemblance to the adults; these insects are said to have *complete metamorphosis*.

Why do some animals produce markedly different forms at various stages of their life cycle? Why do butterflies live much of their lives as the stub-legged leaf-munching machine that is the caterpillar, and then transform into a nectar-sipping, gossamer-winged adult? By having two very different body forms in their life cycle, metamorphic insects can pack two very different but highly successful and highly specialized modes of living into the life cycle of one animal. The body plan of the larval stage (the caterpillar) is well suited for voracious eating in a limited area, whereas the winged adult (the butterfly) is best suited for seeking mates and the best locations for depositing eggs. Together, these very different modes of living acquire a greater variety, and therefore quantity, of available resources than the one body form could on its own.

ARACHNIDS. *Arachnids* [uh-*RAK*-nidz] are mostly terrestrial (**FIGURE 4.15c**), although freshwater and marine species are known. Mites are small arachnids that eat tiny particles of organic matter, and although some are noxious parasites, especially of plants, most are beneficial as decomposers. The dust mites that inhabit most homes live mainly on the skin cells we shed regularly (some people are allergic to their droppings). Ticks are tiny, spiderlike arachnids that are carriers for a number of infectious organisms, such as the bacterium that causes Lyme disease and the one that causes Rocky Mountain spotted fever.

The sensory hairs that cover the body of most spiders (see Figure 4.15c), and that some people find loathsome, simply help the animals feel their way through the world. Spiders partially digest their prey by injecting digestive juices into them and then sucking up the liquefied remains with tubelike mouthparts. Many spiders inject venom to subdue prey or deter predators. Spiders do not produce enough venom to kill a healthy person, although the bite of some spiders is painful and can cause significant tissue damage. Most spiders produce strong, flexible strands of protein—spider silk—that they use to spin webs, to climb from place to place, to spin cocoons for their young, and to wrap food leftovers. Many spiders have several (usually three or four) pairs of simple eyes on their heads. Jumping spiders, which hunt visually, can form sharp color images with the largest of their eight eyes.

Scorpions have a many-segmented trunk that usually ends in a venom-injecting stinger. Scorpions are unusual among arthropods in that they give birth to live young.

MYRIAPODS. The *myriapods* include the centipedes and millipedes (**FIGURE 4.15d**). Centipedes are long-bodied arthropods with one pair of legs per segment. Most centipedes are predators of small animals such as insects, and some are venomous. Millipedes, in contrast, have two legs per segment and eat vegetable matter.

## 4.5 The Deuterostomes— I: Echinoderms, Chordates, and Relatives

What do sea stars, sea squirts, and people have in common? Although we look very different, live in very different places, and behave in diverse ways, we are all deuterostomes. The deuterostome heritage of these groups is not readily apparent at the level of the whole animal, but **deuterostomes** share the following characteristics:

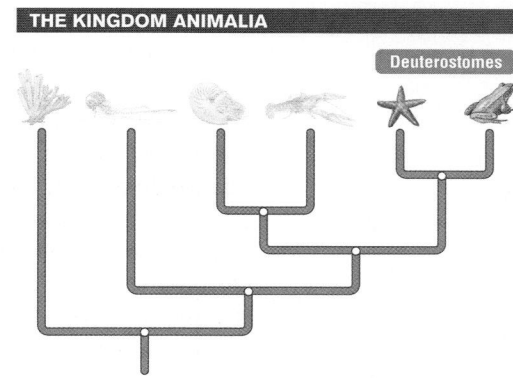

THE KINGDOM ANIMALIA

Deuterostomes

- The embryonic blastopore gives rise to the anus, and the mouth forms secondarily.
- There is a hollow nerve cord located on the dorsal side (rather than ventrally).

DNA comparisons offer the strongest support for the idea that deuterostomes are a distinct clade, a branch of the evolutionary tree that contains an ancestor and all its descendants but none from any other clade. All deuterostomes have a true coelom and the adult individual is generated by three distinct embryonic cell layers (germ layers) that arise during gastrulation. If it has a skeletal support system, a deuterostome wears it on the inside, in contrast to the external support structures (exoskeleton) of some protostomes. The deuterostomes include a number of species-poor groups—for example, the phylum Hemichordata (literally, "half chordates"). Hemichordates, such as acorn worms, are filter-feeding animals that anchor themselves in marine sediments.

Sea stars, sea urchins, and sea cucumbers are examples of animals that belong to phylum Echinodermata (*echino*, "spiny"; *derm*, "skin"). In the remainder of this section we will take a closer look at the echinoderms and also at the phylum Chordata, the group that encompasses sea squirts, tigers, and humans.

## Echinoderms use a water vascular system for locomotion and gas exchange

Echinoderms such as sea stars, sea urchins, and sand dollars are radially symmetrical as adults, but their larvae are bilaterally symmetrical. The loss of bilateral

Echinoderms, like this necklace seastar from the Red Sea, are closely related to the vertebrates.

symmetry may have been an adaptation to a slow-moving lifestyle in adulthood, when access to the environment from all sides would be more adaptive than a head-tail body plan. Some echinoderms, such as the sea lilies and sea feathers, are sessile as adults. Others, like sea urchins and their nonspiky and flattened relatives, the sand dollars, are able to move slowly.

Echinoderms such as sea stars and brittle stars have a mouth on the ventral side, an anus on the dorsal side, and five or more spokelike arms. Like some other echinoderms, sea stars have thick plates of calcified tissue that function like an internal skeleton just under the skin.

Below the branched sacs of the digestive system is a system of water-filled canals; it consists of a central ring, with branches radiating out into the arms and into the stumpy protrusions known as *tube feet*. Seawater entering the water vascular system is squeezed by the contraction of the surrounding muscles, and that compression, in turn, flexes the body, enabling the animal to move or to attach to its surroundings. The tube feet can latch onto a choice prey item through suction, and the body muscles can generate enough force to pry open a bivalve. The circulation of water through the canals also facilitates gas exchange, ushering in oxygen-rich water and spewing out the carbon dioxide–laden wastewater.

## Chordates possess a dorsal notochord at some stage of their life cycle

You might find it hard to believe that the brightly colored sessile sea squirt in Figure 4.1*b* is our closest relative in that photo gallery, but you might be more convinced if you could peer inside the juvenile sea squirt (**FIGURE 4.19*b***). Like all other **chordates**, the tadpolelike larva of sea squirts (more formally known as tunicates) has these characteristics:

- A dorsal rod of strengthening tissue, the notochord
- Pharyngeal pouches, which develop on either side of the throat in the embryo
- A post-anal tail

The **notochord** is composed of large cells that collectively form a strong but flexible bar running dorsally along the length of the animal. The notochord provides support for the rest of the body, the way a ridge beam supports the structure of a frame house. So, where is the human notochord? In some chordates, and all vertebrates, the notochord is lost during early development and its function transferred to stronger skeletal structures, such as the backbone. In humans, traces of the notochord survive in the flattened pads of tissue—the intervertebral discs—that act as cushions between the vertebrae, the bones that make up our backbone.

Like the notochord, *pharyngeal pouches* are found in the early embryo of all chordates. The **pharyngeal pouches** first appear as pockets of tissue on either side of the embryonic pharynx. The pharynx, commonly known as the throat, is the passageway posterior to the mouth and leading into the windpipe (trachea) and food tube (esophagus). In fish and larval amphibians, the pharyngeal pouches deepen into the mesoderm and ectoderm layers until they merge with the body surface to create openings, called gill slits, in which the gills develop. In other vertebrates, the pharyngeal pouches fail to extend all the way to the body surface, and instead cells derived from these pouches give rise to a diversity of structures in the developing embryo. In mammalian embryos, for example, the pharyngeal pouches give rise to parts of the larynx (voice box) and trachea, as well as certain glands in the neck and chest (the thyroid and thymus glands, to name some).

The phylum Chordata includes several small subgroups of marine organisms, but we will focus on the vertebrates, animals with an internal *vertebral column* ("backbone") composed of a series of strong, hollow, cylindrical sections, known as *vertebrae*. The vertebral column encloses the nerve cord that all deuterostomes have on the dorsal side of the body.

The first vertebrates, which had evolved by the Cambrian period, were jawless fishes. A few groups of jawless fishes have survived to the pres-

**THE CHORDATE BODY PLAN**

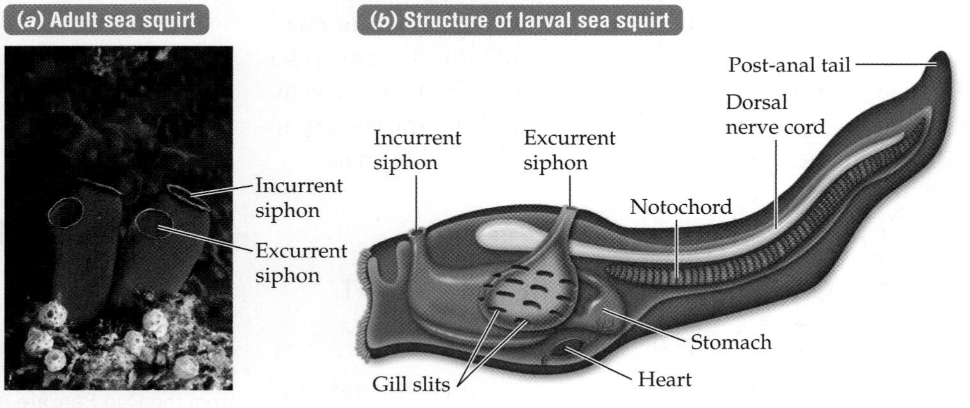

(*a*) Adult sea squirt

(*b*) Structure of larval sea squirt

Incurrent siphon

Excurrent siphon

Incurrent siphon

Excurrent siphon

Post-anal tail

Dorsal nerve cord

Notochord

Stomach

Gill slits

Heart

**FIGURE 4.19 The Notochord Is Unique to the Chordates**
Although the adult form of the sea squirt *Ciona* (*a*) looks very different from that of other chordates, such as people, the chordate hallmarks are evident in the larval form of the animal (*b*).

FIGURE 4.20
Evolutionary Tree of
the Chordates

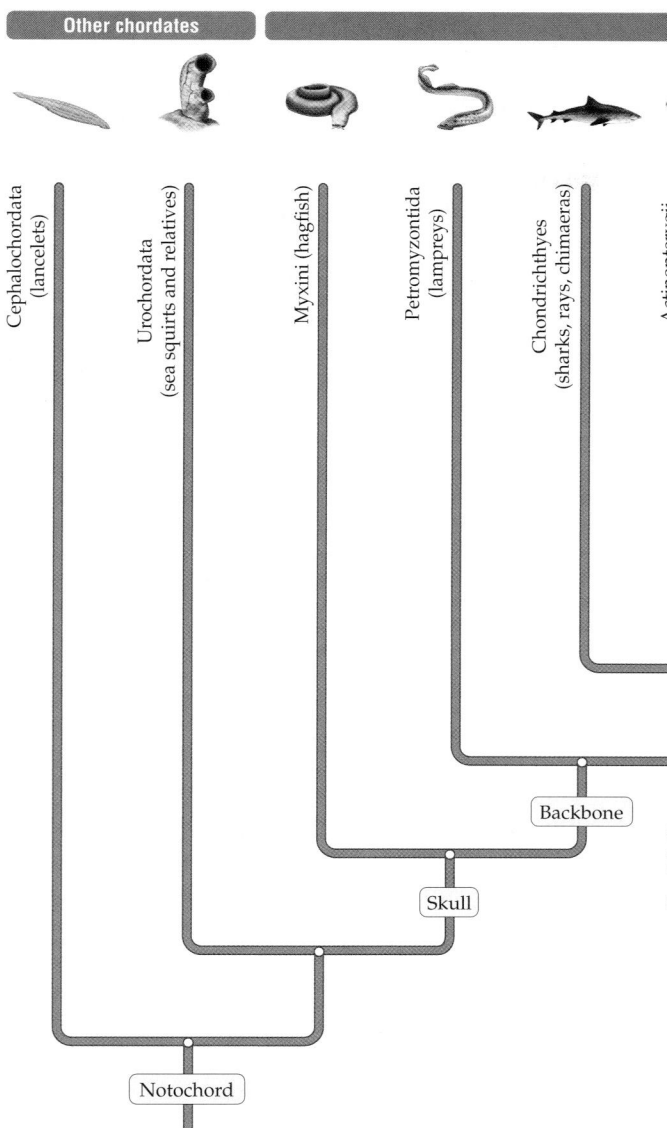

**Other chordates** | **Vertebrates**

Cephalochordata (lancelets)

Urochordata (sea squirts and relatives)

Myxini (hagfish)

Petromyzontida (lampreys)

Chondrichthyes (sharks, rays, chimaeras)

Actinopterygii (ray-finned fishes)

Actinistia (coelacanths)

Amphibia (frogs, salamanders)

Reptilia (turtles, snakes, birds)

Mammalia (mammals)

Scales

Fur, mammary glands

Amniotic egg

Four limbs

Lobed fins

Bony skeleton

Jaws

Backbone

Skull

Notochord

ent, most notably the lampreys. Some lampreys are parasites that attach themselves to other fish and scrape off flesh with the rasping structures inside the mouth. Vertebrates include fishes, amphibians (frogs and salamanders), reptiles (snakes, lizards, turtles, and crocodiles), birds, and mammals (including humans and kangaroos). We consider these groups of vertebrates next.

## 4.6 The Deuterostomes— II: Vertebrates

Fish first appeared in the Cambrian, and they diversified rapidly as the rest of Paleozoic era unfolded. When the Silurian began, 443 million years ago, jawless fishes dominated the warm oceans (refer to the time line in the inside cover of this book). The vertebrate characteristics (**FIGURE 4.20**) that first evolved among the jawless fishes include the following:

- A strong internal skeleton held together by a vertebral column
- An anterior braincase, or skull
- A closed circulatory system, with a pumping organ, the heart

The skeleton of the first vertebrates was made from a strong but flexible tissue, called *cartilage*, that also extended over the brain in the form of a protective braincase, or skull. A muscular heart pumped blood through a network of blood vessels. Fine extensions of the blood vessels (capillaries) permeated the gills, which in these early fish were supported on an archlike framework of cartilage. The lamprey is one representative of jawless fishes found in today's oceans (**FIGURE 4.21a**).

## Jaws and a bony skeleton represent major steps in vertebrate evolution

The next great leap in vertebrate evolution came when two or three of the anterior gill arches became modified into hinged jaws. Jaws were a major evolutionary advance because jawed predators could grab, overpower, and swallow prey more efficiently than could jawless animals. The evolution of teeth

(*a*) A lamprey

(*b*) A cartilaginous fish

(*c*) A ray-finned fish

(*d*) A lobe-finned fish

**FIGURE 4.21 Fishes May Be Jawless or Jawed**

(*a*) Lampreys and hagfishes are jawless fishes, but the three other categories of fish shown here have a hinged jaw attached to the skull. (*b*) Sharks have thousands of teeth. The large main teeth are continually shed and replaced by smaller teeth in the inner rows. Shark teeth retract into the gums when the mouth closes. (*c*) Ray-finned fish have a lateral line, a sensory organ that enables fish to detect movement by sensing pressure. (*d*) The muscular lobed fins of the coelacanth have leglike bones inside. An ancestor of a fish like this one is thought to have given rise to four-legged vertebrates (tetrapods).

made jaws yet more effective because they enabled animals to seize, tear, and cut up food. With their fearsome rows of serrated teeth (**FIGURE 4.21b**), sharks had appeared on the world stage by the mid Paleozoic. The evolution of jawed fishes ushered the demise of the once-mighty jawless fishes, which are represented today by just a few members, such as hagfishes and lampreys.

Another major step in the evolution of fish, and vertebrates in general, was the replacement of the cartilage-based skeleton with a denser tissue strengthened by calcium salts: bone. Although the descendants of cartilaginous fishes—sharks, skates, and rays—are still with us today, *bony fishes* are far more diversified and widespread in both marine and freshwater environments. With more than 40,000 species, a majority of them marine, fish are the most diversified vertebrates today.

Most fish have two sets of paired fins, and typically several unpaired fins as well. The fins, which are modified appendages, are used for stability and swimming. The fins of cartilaginous fishes are extensions of their tough, sandpaper-textured skin. *Ray-finned fishes* evolved slender, spikelike extensions of their bony skeleton to serve as a support framework for their fins (**FIGURE 4.21c**). *Lobe-finned fishes* are a third group of jawed fishes distinguished by paired fins having a thick muscular lobe supported by joint-like bones (**FIGURE 4.21d**). Muscles in each lobe move the fins independently. Only eight species of lobe-

finned fishes survive today, including coelacanths and lungfishes.

The body surface of bony fishes is typically covered by thin, flattened *scales*, which, along with the slime secreted by skin glands, reduces resistance to water flow over their streamlined bodies. The gills of ray-finned fishes are covered by a movable flap, the *operculum*. Fanning the operculum improves the flow of water over the gills, which enhances gas exchange. Sharks and other cartilaginous fishes do not have an operculum, and many of them must gulp regularly or swim constantly to pump enough oxygen-rich water over their gills.

Bone is heavier than cartilage, and to stay afloat with a minimum expenditure of energy, bony fishes have evolved gas-filled swim bladders (**FIGURE 4.22**). By controlling the amount of gas in the swim bladders, bony fishes can stay at a preferred depth without having to swim actively to maintain that position. Swim bladders are probably modified air storage sacs that first evolved in fish inhabiting oxygen-poor swamps or coastal mud. Lungfishes and coelacanths (both lobe-finned fishes) are living species that have air storage sacs that function like a scuba diver's oxygen tank: the fish gulp mouthfuls of air and store it in their lunglike sacs to supplement the oxygen supply from the gills. Scientists believe that a pair of air storage sacs in a bony vertebrate ancestor evolved into a pair of dorsal swim bladders in the bony-fish lineage, and the same two

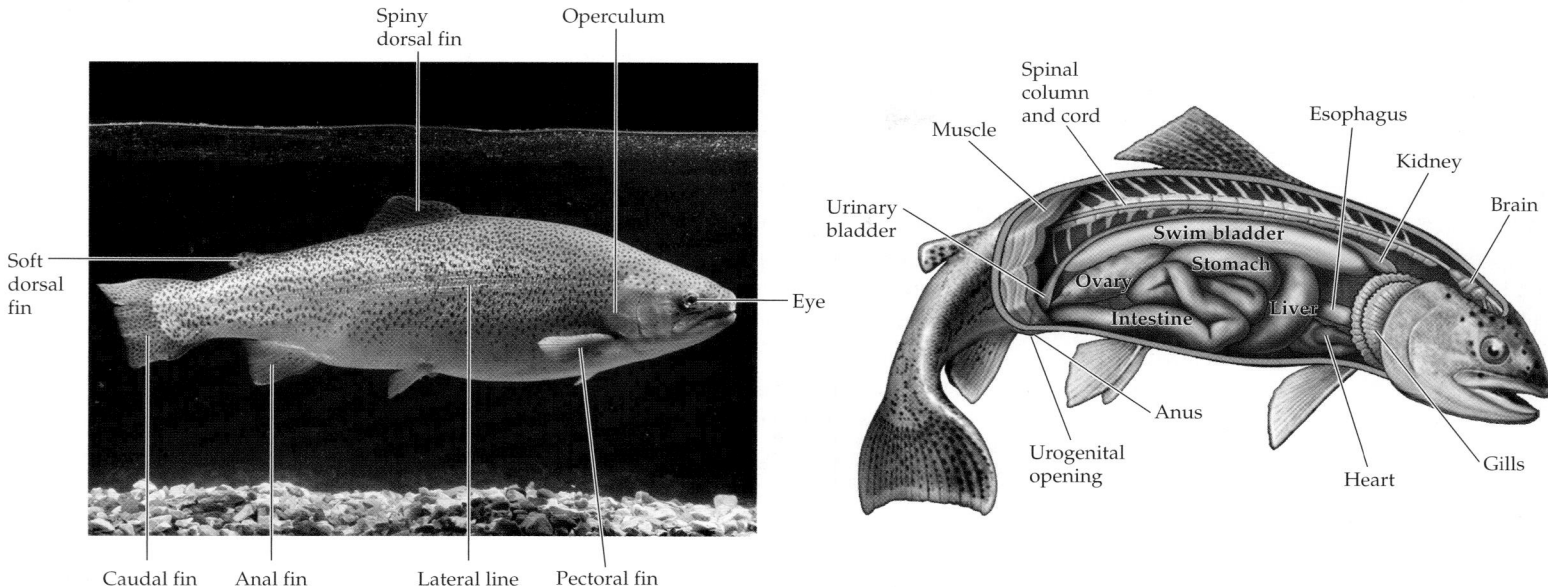

Spiny dorsal fin

Operculum

Spinal column and cord

Muscle

Esophagus

Kidney

Brain

Urinary bladder

Swim bladder

Soft dorsal fin

Stomach

Ovary

Eye

Intestine

Liver

Caudal fin    Anal fin    Lateral line    Pectoral fin

Anus

Urogenital opening

Heart

Gills

**FIGURE 4.22  Bony Fishes Have a Strong Internal Skeleton**
The swim bladder provides buoyancy, reducing the energy required to stay afloat.

sacs were modified into a pair of lungs in the lineage that gave rise to *tetrapods*, the four-legged vertebrates that colonized land.

## Amphibians breathe through both the lungs and the skin surface

Terrestrial vertebrates with four limbs are known as **tetrapods** (*tetra*, "four"; *pod*, "foot"). This Devonian lineage is thought to have given rise to all the mainly terrestrial vertebrates, beginning with the amphibians. The coelacanth, a "living fossil" that was thought to have become extinct in the Devonian, has just such jointed, lobelike fins, attached to the skeleton by one large bone. A fossil named *Tiktaalik*, discovered recently in the Canadian north, appears to be a transitional fish with a weight-bearing limb design and forelimb bones similar to the wrist and finger bones seen in amphibians.

The several thousand species of amphibians include frogs and salamanders, and the less familiar wormlike, legless caecilians, which are confined to the humid tropics. Most of these amphibians (*amphi*, "double"; *bios*, "mode of life") are semiaquatic, and even the most terrestrial of them must return to water to lay eggs. Many amphibians, such as frogs, have a life cycle characterized by complete metamorphosis: fertilized eggs (zygotes) hatch into tailed tadpoles that are entirely aquatic until they develop limbs, lose the tail, and metamorphose into the more terrestrial adult. Whereas tadpoles use gills for gas exchange, adult frogs breathe through a pair of lungs and also exchange gases through the moist surface of the skin and mouth. The advent of lungs was, of course, a crucial milestone in vertebrates' transition onto land.

## Reptiles exhibit adaptations to a drier environment

The amphibious life limits vertebrates to habitats in which bodies of water are found at least seasonally. Reptiles were the first vertebrates to head into drier environments, and they evolved a number of adaptive traits to deal with the risk of dehydration. The major evolutionary innovations that first appeared among the reptiles include the following:

- Skin covered in waterproof scales
- A water-conserving excretory system
- The amniotic egg, with stored food and a waterproof shell
- Internal fertilization

Tiktaalik fossil

# Goodbye, Catch of the Day?

Biologists have amassed a wealth of data on species already gone or on their way out, and even conservative analyses suggest that huge numbers of species have been, and continue to be, lost. According to Edward O. Wilson, a biologist at Harvard University, at the current rate of rainforest destruction 27,000 additional species will be doomed to extinction each year—an average of 74 per day, or 3 every hour. And although rainforests are particularly rich in biodiversity, they are just one of many different habitats—characteristic places or types of environments in which different species can live.

Although estimating extinction rates for the world as a whole can be difficult, scientists have definitively documented the extinction, caused by humans, of many hundreds of particular species in the last few thousand years. Twenty percent of the freshwater fish species known to be alive in recent history either have gone extinct already or are nearly extinct. One large-scale study showed that 20 percent of the world's species of birds that existed 2,000 years ago are no longer alive. Of the remaining bird species, 10 percent are estimated to be endangered—that is, in danger of extinction.

Although it may be tempting to assume that little of this extinction is happening close to home, evidence suggests otherwise. In the United States, frogs are disappearing in Yosemite National Park. In North America overall, 29 percent of freshwater fishes and 20 percent of freshwater mussels are endangered or extinct.

Humans eat many different kinds of species, but not all meals have the same impact on the planet's biodiversity. In particular, a number of fish and other marine species that humans consume are being overharvested and are in rapid decline. Eating these species only pushes them into steeper declines and could drive them into extinction. In addition, some species, like salmon, are raised on aquatic farms. But while that might seem like the perfect solution to overfishing wild species, these farms sometimes create high levels of ocean pollution, threatening other species.

So, when we are shopping at the grocery store or scanning a menu at the restaurant, how can we decide what might be both delicious and harmless to eat? Various conservation organizations have put together lists of seafood that are best to eat and best to avoid, often printed on handy wallet-sized cards. Shown here are some of the recommendations from the wallet card of the marine conservation organization known as Blue Ocean Institute. Just by choosing wisely, the next time you crave seafood, you can help preserve the planet's biodiversity.

| ENJOY | BE CAREFUL | AVOID |
|---|---|---|
| Arctic Char | Crabs (Blue, Snow, and Tanner) | Chilean Sea Bass |
| Clams, Mussels, and Oysters (farmed) | Monkfish | Cod (Atlantic) |
| Mackerel | Rainbow Trout (farmed) | Halibut (Atlantic) |
| Mahi-Mahi (pole- and troll-caught) | Sea Scallops | Salmon (farmed) |
| Salmon (wild Alaskan) | Swordfish | Sharks |
| Striped Bass | Tuna: Albacore, Bigeye, Yellowfin, | Shrimp (imported) |
| Tilapia (U.S. farmed) | and Skipjack (canned or | Tuna: Bluefin (Atlantic) |
| | longline-caught) | |

**SOURCE:** The Blue Ocean Institute, "Guide to Ocean Friendly Seafood," September 2007.

A lineage of tetrapods with these characteristics gave rise to the reptiles in the Carboniferous period, about 354 million years ago. The line of descent that gave rise to mammals separated from the reptile lineage roughly 225 million years ago. Reptiles dominated on land, in salt water and fresh water, and even in the skies, for much of the Mesozoic era (**FIGURE 4.23**), which is therefore known as the Age of Reptiles.

One line of reptiles, the dinosaurs, became spectacularly successful in the middle of the Mesozoic, although their fortunes were to decline toward the end of the era, 65 million years ago. Dinosaurs ranged greatly in size: the horned-faced *Microceratops* stood about 0.5 meter tall (about a foot and a half), while another plant eater, *Argentinosaurus*, was a 100-ton behemoth that measured 37 meters (about 120 feet) from nose to tail. Although *Tyrannosaurus rex*, at 40 feet in length and 40 tons in weight, tends to steal the limelight, some of the smaller predators that hunted in packs, like the 20-foot-long *Utahraptor* and the chicken-sized *Velociraptor*, were every bit as ferocious.

Turtles (which are aquatic) and tortoises and box turtles (which are terrestrial) are among the oldest reptiles. The dorsal side of the rib cage is modified into a tough, leathery protective shell in these reptiles. Tuataras resemble lizards but represent a distinct lineage whose members lived in the Mesozoic and survive today as "living fossils" in New Zealand. Snakes are a lineage of legless reptiles.

The **amniotic egg** is the most magnificent evolutionary innovation to appear among the reptiles (**FIGURE 4.24**). The developing embryo is surrounded and protected by layers of extraembryonic membranes that also promote gas exchange and store waste. The calcium-rich protective shell retards moisture loss but is porous enough to allow the entry of life-giving oxygen and the release of waste carbon dioxide. The amniotic egg hoards food in the form of a large *yolk* mass, which enables the young to achieve a relatively advanced level of development before they emerge (hatch) from the shell.

The body of reptiles is covered in overlapping layers of scales (**FIGURE 4.25**), made chiefly of a protein called *keratin*, that reduce water loss from the skin surface. Gases are exchanged exclusively through lungs, which have a larger surface area than the lungs of amphibians. Like amphibians, reptiles have a three-chambered heart in which oxygen-poor blood that is pumped to the lungs is kept partially separated from oxygen-rich blood that is pumped to the rest of the body. Reptiles have a well-developed excretory system, with a pair of kidneys and an extensive system of tubes that help concentrate nitrogen-containing waste (urea or uric acid or both) and get rid of them in a relatively small quantity of urine. Living reptiles do not generate extra metabolic heat to warm their bodies. Instead, they are **ectotherms**, meaning that their body temperature matches that of their environment. In cool temperatures, reptiles like lizards bask in the sun or sprawl on sun-warmed rocks; when the temperature rises, they seek shade.

## Birds are adapted for flight

In the mid Mesozoic, about 175 million years ago, a lineage of feathered theropod dinosaurs, which include raptors and ornithomimids, evolved into birds. Like other theropods, this lineage had an upright, or *bipedal*, stance, with large powerful legs and smaller forelimbs. Their scales were modified into feathers made of keratin. The feathers evolved not so much for flight initially, but as a layer of insulation that kept the animals warm. In time, the feathers attached to the forelimbs became modified for stronger, sustained flight, with the tail feathers acting as stabilizers.

**FIGURE 4.23 Jurassic Park**
Dinosaurs reached the height of their dominance during the Jurassic, the middle period of the Mesozoic era.

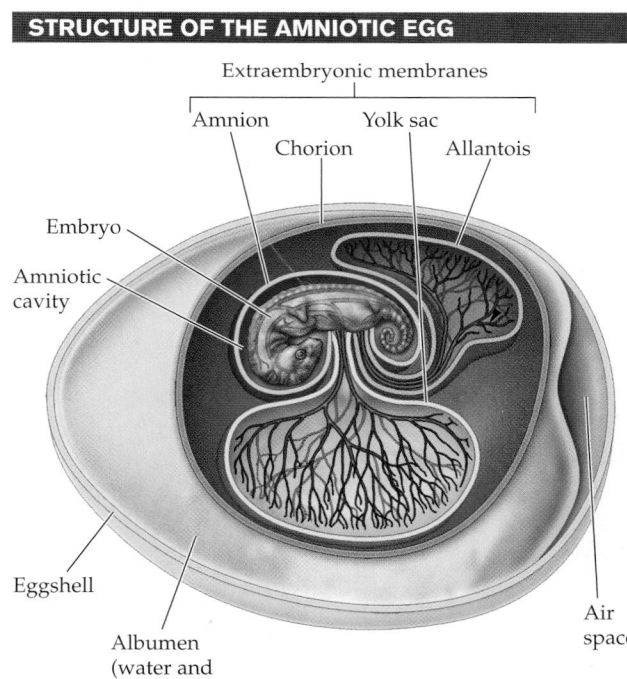

**STRUCTURE OF THE AMNIOTIC EGG**

Extraembryonic membranes
Amnion
Chorion
Yolk sac
Allantois
Embryo
Amniotic cavity
Eggshell
Albumen (water and stored protein)
Air space

**FIGURE 4.24**
**The Amniotic Egg Was a Crucial Evolutionary Innovation for Life on Land**

Transitional fossils, such as the 150-million-year-old *Archaeopteryx* discovered in a Bavarian quarry, offer a rare and wonderful glimpse of the evolutionary transformation of feathered dinosaurs into birds. Like its theropod relatives, *Archaeopteryx* had teeth in its beak, although modern birds have lost them

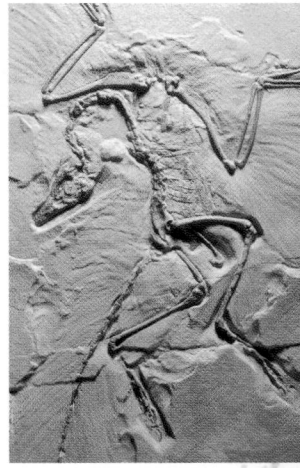

*Archaeopteryx*

## Size

Pygmy marmosets are the smallest primates.

This tiny carp from Indonesia is the world's smallest vertebrate. It is about 8 millimeters (0.3 inch) long.

The blue whale is the largest animal that has ever lived. It can grow to 110 feet and weigh 190 tons.

The collosal squid is the largest invertebrate. This rare cephalopod is believed to grow to 14 meters (46 feet) and can weigh 500 kilograms (1,100 pounds).

## Locomotion

The three-toed sloth sleeps for 10 hours a day, stays in one tree for life, and if chased by a predator, can manage a top speed of 2 meters per minute (6.5 feet per minute).

**FIGURE 4.25 Extreme Diversity: The Animals**

The Arctic tern flies 22,000 miles round-trip every year from its Arctic breeding grounds to summer and feed in Antarctica.

The black marlin has been clocked at 130 kilometers per hour, or 87 miles per hour.

# Reproduction

Clown fish are born male; dominant individuals change gender to become female if a resident female dies.

Asian elephant mothers have the longest gestation (pregnancy): about 22 months. The long duration enables the newborn to become tall enough to reach with its trunk and nurse. Mothers nurse their calves for 2–3 years.

Komodo dragons are among the reptiles that can produce young without mating and fertilization. This phenomenon, known as parthenogenesis, has been reported in sharks, as well as in other fishes and in many invertebrates.

**Did you know?**

The 17-year cicadas live underground as larvae feeding on plant roots. After exactly 17 years, the larvae emerge in large numbers, molt, and metamorphose into adults. The male sings its famously loud and raspy song to attract females, neither sex eats, and both die shortly after mating and egg laying.

# Life Span

The mayfly has no gut and lives for about 1 day.

Galápagos turtles are known to live to 150 years. They can go for 18 months without food or water.

The giant clam is believed to live for upwards of 150 years. Algae live inside its mantle, and the clam obliges its mutualistic partners by keeping the valves open during the day for maximum photosynthesis.

**Did you know?**

The Arctic frog freezes solid in the winter and slowly thaws out in the spring. It contains special chemicals that keep ice crystals from growing inside the cells and destroying them. Like many cold-climate ectotherms, Arctic flounder contain antifreeeze proteins.

# Eating

Koalas live on the leaves of just a handful of *Eucalyptus* species. There are more than 600 species of *Eucalyptus* in Australia.

This new species of catfish from Peru eats wood from trees that fall into the rainforest rivers where it makes its home.

This deep-sea anglerfish dangles a bioluminescent lure that grows out from its head. The light, produced by bacteria that colonize the fleshy lure, attracts other predatory fish, which become a meal for the angler.

as a weight-minimizing adaptation. The three claws on the forelimbs have also been lost in most modern birds, although hoatzins, a puzzling species from the Amazon, have two claws that they use for clambering in trees when they are chicks. *Archaeopteryx* perched on tree branches with clawed hind limbs covered in scales that resemble those of modern birds, and its feathers seem virtually identical to those of modern birds in their system of branching and interlocking barbs.

Beyond hollow bones and light, toothless jaws (beaks), modern birds exhibit other adaptations for flight, such as a reduction in internal organs; birds, for example, have only one ovary, instead of the two that are characteristic of all other tetrapods. Unlike the living amphibians and reptiles, birds have a four-chambered heart that resembles ours: oxygenated and nonoxygenated blood is kept in separate chambers, and the blood that is pumped to the rest of the body therefore contains high levels of oxygen, supporting a high metabolic rate. The respiratory system of birds is considered even more efficient than ours. Birds have a one-way flow of air through their respiratory passages, so that incoming air never mixes with outgoing air. In contrast, the air we inhale and exhale takes the same route (although in opposite directions), so there is some mixing of the incoming and outgoing airstreams. Like the mammals we discuss next, birds generate extra metabolic heat to warm their bodies (that is, they are **endotherms**) and they maintain a near-constant internal temperature (they are **homeotherms**).

## Mammals came into prominence when dinosaur populations declined

Through much of the Mesozoic, when dinosaurs roamed all the continents, there were small, hairy creatures that kept a low profile as they scurried about in the vegetation living mostly on insects. The mammalian lineage split off from the reptilian line about 225 million years ago, in the early Mesozoic. With the decline of the dinosaurs toward the end of Mesozoic, mammals came into their own and diversified into a spectacular diversity of forms—from koalas to giant sloths—in the last 70 million years of Earth history. The following characteristics of mammals have contributed to their success:

- Hair on the body and endothermy
- Sweat glands that cool the body through evaporation
- Young nourished with milk from mammary glands
- Internal fertilization and parental care

There are about 5,000 species of mammals, divided into three broad categories (**FIGURE 4.26**). The nonplacental egg-laying mammals are classified as **monotremes** (*mono*, "one"; *treme*, "hole," referring to a common opening for the anus and urinary system). Monotremes (**FIGURE 4.26a**) today consists of just one platypus species and several echidna species, all

**MAMMALIAN DIVERSITY**

(a) A monotreme  (b) A marsupial  (c) A eutherian

**FIGURE 4.26 Three Categories of Mammals**
(a) Monotremes like the platypus are characterized by milk-producing mammary glands in females. Unlike all other mammals, however, monotremes lay eggs instead of giving birth to live young. (b) Kangaroos are marsupial mammals that give birth to relatively immature young that finish developing in a pouch (marsupium). (c) Eutherians give birth to well-developed young and suckle from the mammary glands.

confined to Australia and New Guinea. The **marsupials** are animals that protect and feed their newborns with milk in an external pocket or pouch (*marsupium*, "pouch"). Marsupials (**FIGURE 4.26b**) are found mainly in Australia and New Zealand, with a few species in the Americas. The North American, or Virginia, opossum is the only marsupial in North America. More than 95 percent of mammals alive today are **eutherians** (*eu*, "true"; *therion*, "beast") a category that includes us. A unifying characteristic of eutherians (**FIGURE 4.26c**) is that the young are nourished inside the mother's body through a special organ called the placenta and are therefore born in a relatively well-developed state.

## Mammals can live in diverse habitats because they regulate body temperature

All birds and mammals are endotherms and homeotherms: they use metabolic energy to generate heat, and they maintain a near-constant body temperature. The reptilian scales of the mammalian ancestor evolved into long, keratin-containing strands that we recognize as hair (we call it fur when it is especially thick and luxuriant on the bodies of nonhuman animals such as foxes and mink). Muscles in the skin can raise the hair to trap a thicker layer of air next to the skin, thereby increasing the insulating properties of body hair.

Endothermy combined with hair has enabled mammals to colonize cold regions of the world and to remain active at temperatures too low for most other animal groups except some birds. Hair is reduced on some aquatic mammals, such as whales, but the fetuses of even these species sport a good head of hair. Many adult whales have hair on the chin or snout (in right whales, for example), and the bumps (tubercles) on the snout of a humpback whale are large hair follicles, the living hair roots embedded in skin. Aquatic mammals, and cold-climate mammals in general, often have thick layers of fat, or blubber, under the skin, which provides energy as well as insulation. Hair is often modified into long whiskers that have a sensory function: the follicles are attached to nerve cells that enable the animals to pick up physical sensations.

Only mammals have sweat glands. Evaporation of sweat cools terrestrial mammals and enables them to maintain a moderate body temperature even in extremely hot and dry environments such as the desert. Aquatic animals and many fur-bearing mammals lack sweat glands.

## Parental care contributed to the success of mammals

Like all reptiles and birds, and in contrast to most amphibians, mammals have internal fertilization, meaning that the male deposits the sperm inside the female's body and fertilization occurs within. In monotremes, the embryo develops within a shelled egg that is deposited outside the body. In marsupials, gestation (pregnancy) lasts for a short period—about 4 weeks in kangaroos—and the offspring are born in a relatively immature stage. The joeys, as the infant marsupials are called, have strong limbs that they use to clamber into a pouch on the ventral side of the mother's body.

All eutherians, and some marsupials, are placental animals. Embryonic tissues and maternal tissues combine to form a very special structure, the **placenta**, with an extensive blood supply. As the offspring develops in the special chamber known as the *uterus* (womb), it receives nutrients and oxygen across the placenta, which also removes waste chemicals and carbon dioxide. Nourished and protected inside the mother's body, eutherian offspring are born at a much more advanced stage of development than are marsupial young. The newborns of many herbivores, such as deer and caribou, can run within hours of birth—a necessary adaptation that enables the young to keep up with the herd and potentially stay out of reach of predators. Like some reptiles and most birds, mammals guard their newborns until they become capable of living independently.

**Mammary glands**, which are modified sweat glands, are the most distinctive feature of mammals. They produce a liquid rich in fat, proteins, salts, and other nutritive substances that nourish the newborn. Monotremes secrete the fluid from the glands directly on the fur, where it is lapped up by the newborns after they hatch from their shells. Marsupial females have a nipple in each pouch that the newborn attaches to and feeds from. Eutherians nurse their newborns from two or more nipples on the ventral side of the body. The fat and sugar composition of eutherian milk varies: whale and seal milk has 10 times as much fat as cow's milk and almost no milk sugar (lactose).

Eutherians are highly successful in terms of the variety of habitats they occupy and the range of sizes they display, from shrews that weigh less than a gram to elephants that tip the scales at more than 5 tons. They have replaced dinosaurs as the top predators in most terrestrial habitats, and they thrive in both marine and freshwater habitats. Bats are the only mammals that can fly, although a number of eutherians and marsupials can glide.

Learning is developed to the highest degree among mammals, although many birds also transmit learned behaviors (such as song). Some eutherians are social animals, living in groups or herds in which guard duty and even rearing of the young are shared. Social eutherians tend to have comparatively large brains, which enable complex behaviors, which in turn pave the way for exploiting a greater range of resources and habitats.

Primates evolved about 56 million years ago from ancestors that were probably small, tree-dwelling (arboreal), insect-eating mammals with grasping feet and some opposable digits (fingers and toes that can be moved to meet at the tips). About 35 million years ago, the primate lineage split into the mostly arboreal and nocturnal *prosimians* (lemurs, lorises, and tarsiers) and the *anthropoids*, which include Old and New World

monkeys and all the apes. About 6 million years ago, the hominid lineage, which includes us, split from the lineage that gave rise to chimpanzees and bonobos. We consider primate and hominid evolution in greater detail in Chapter 20.

> ### Concept Check
>
> 1. (a) Which animal group lacks tissues? (b) Which animal group has tissues but lacks organs? (c) Which major animal groupings have bilateral symmetry?
> 2. Which invertebrates are considered the "brainiest"?
> 3. Which single adaptation is believed to have contributed tremendously to the biological success of insects, and why?
> 4. What do sea squirts have in common with us?
> 5. What is the adaptive value of the amniotic egg?

---

APPLYING WHAT WE LEARNED

# Clues to the Evolution of Multicellularity

At the beginning of this chapter we met our tiny cousins the choanoflagellates. Just as we share a grandparent with a cousin (and a great grandparent with a second cousin), humans and all other animals share a distant ancestor with the choanoflagellates. This ancestor, which lived between 700 million and 1 billion years ago, evolved the ability to form colonies of communicating cells. But while animals are multicellular all of the time, with specialized cells (and usually tissues and organs), choanoflagellates stayed single-celled and kept their colony-forming abilities as just an option—like a biology study group.

One way we know that choanoflagellates are our kin is that their individual cells look and work just like the special feeding cells of the simplest of all animals, the sponges (see Figure 4.3). Sponges are marine animals that attach themselves to a rock or other solid base and filter tiny organisms out of the water. Sponges lack a nervous system, blood circulation, or even a one-way digestive tract like ours, but they do have an interior space lined with cells called *choanocytes*, also known as "collar cells." (*Choano* comes from the Greek word for "collar," and *cyte* just means "cell.") The choanocytes lining the interior cavity of a sponge look amazingly like choanoflagellates and function in a similar way, waving their whiplike flagella to pump water and food where it's needed.

Another clue to our kinship with choanoflagellates is that the choanoflagellates that form colonies do so using the same process that animals use to form embryos. A fertilized egg—whether it's from a human, a hamster, a fish, or a fly—divides into two cells, which stay in contact with each other. Those 2 cells divide again into 4 cells (and then 8, 16, and 32). But even though the cells divide, they stick together in a clump. They don't normally break away and go their separate ways. In the same way, colonial choanoflagellates repeatedly divide and stick together to create a cooperative group. There are other ways to form primitive colonies. For example, in slime molds, separate cells come together to form a sluglike mass. The divide-and-stick-together development of animals and colonial choanoflagellates is a shared trait that supports the theory that they are related.

Still another clue that choanoflagellates are related to animals is that animals as different as flies and mice share dozens of nearly identical proteins with choanoflagellates. Intriguingly, nearly all of those proteins are involved in helping animal cells stick to one another or talk to one another. It's a talent that is very useful in a multicellular organism.

Surprisingly, some choanoflagellates that have these sticky proteins are not colonial. What do they use the proteins for? One possibility is that they use the sticky proteins to capture food particles. In some species, a long time ago, the food-snaring proteins may have been re-purposed to become the glue that binds a colony of individual cells. At least one such colonial lineage evolved into a close-knit community of interdependent cells, with division of labor: a multicellular animal. All animals, from sponges to people, are likely descended from such an ancestor.

# Canadian Scientists Wipe Away Misconceptions about Sponges

BY RANDY BOSWELL, *Postmedia News*

Canadian researchers who probed the traits of a freshwater sponge from Vancouver Island say their findings about the species' "skin" could rewrite the history of animal life and illuminate a primordial family connection between humans and the porous organisms best known for mopping up kitchen spills.

A study by three University of Alberta biologists, which appears in a journal published by the U.S.-based Public Library of Science, shows how the outer tissue of the B.C. specimen acts much like the protective layer of skin that distinguishes almost all other animals, including humans, from the seemingly flow-through sponges.

The discovery, the research team concludes, could eventually force scientists to reclassify sponges closer to our own "eumetazoan" clade [lineage] of animals, and to rethink humanity's evolutionary roots among these absorbent creatures of the deep.

"It doesn't quite make them into Sponge Bob," study coauthor Sally Leys said on Monday. "But it very much does put sponges into the fold with the rest of us."

The U of A team, including Emily Adams and Greg Goss, gathered samples of the common species spongilla lacustris [sic] from Sarita and Rosseau lakes near Bamfield, B.C., about 120 kilometres northwest of Victoria.

Leys said the advantage of collecting sponges from Vancouver Island is that their habitats typically don't ice over in winter—allowing access year-round—and that colder weather triggers a degree of shrinkage and dormancy that makes the specimens easier to handle in experiments.

The researchers tested the sponge's "epithelial" membrane to determine whether it can effectively block certain molecules from penetrating the organism's interior—the way a mammal's skin or an insect's outer layer does. They found the sponge's membrane provided a "good, tight seal" akin to how a chimpanzee's skin protects against unwanted microbes and chemical invaders.

"It shows that sponges share a physiology with other animals and are not just some odd offshoot," said Leys.

Sponges, fossils of which have been found from about 550 million years ago, are among the earliest known complex creatures to appear following the evolution of life from unicellular to multi-celled organisms.

For decades, researchers have been trying to decide how sponges are related to all other animals—which biologists call "eumetazoans." As we saw in this chapter, eumetazoans have symmetry—a left and a right, for example, or perhaps a radial symmetry like that of a jellyfish. They also have specialized cells, tissues, and organs and include every kind of familiar animal, from worms, slugs, and insects to elephants and humans. But unlike eumetazoans, sponges lack symmetry, and they lack organs such as a heart, liver, brain, or stomach.

By eumetazoan standards, sponges are no more than blobs. Yet they belong in the kingdom Animalia. They feed off other living organisms, develop from embryos, have a motile stage, and have cells that lack a cell wall. Their embryos even have an extracellular matrix. For this reason, they are animals and, therefore, our relatives.

The question has been what *kind* of relatives. Scientists have disagreed about whether all sponges are descended from a lineage that is separate from the one that gave rise to us eumetazoans—making them a sort of "sister group"—or if instead the eumetazoans actually descended from some kind of ancient sponge.

If we eumetazoans are descended from one of several lineages of sponges, then we must be more related to some sponges than to others. What's more, it means we are actually descended from sponges.

Increasingly, it's beginning to look as if that is the case. More and more evidence suggests that eumetazoans are descended from a sponge, and this news article presents just one more line of evidence.

In general, biologists are reaching agreement on this topic, but in one area some have found reason to feud. Some taxonomists say that if eumetazoans are descended from sponges, then eumetazoans are sponges. This is the same kind of argument that makes people say that birds are dinosaurs, since birds are descended from one lineage of dinosaurs. But other biologists insist that this is nonsense, that it is silly to say that everything descended from a sponge is a sponge.

## Evaluating the News

**1.** What traits of sponges indicate that they are animals? How do they differ from eumetazoan animals?

**2.** Sponges have a certain kind of cell that resembles a choanoflagellate. What do those cells do in sponges, and how does that function compare with what choanoflagellates do?

**3.** Do you think birds are dinosaurs? Defend your position.

**SOURCE**: *Postmedia News*, Canada.com, December 14, 2010.

# CHAPTER REVIEW

## Summary

### 4.1 The Evolutionary Origins of Animalia

- The Animalia are multicellular, ingestive heterotrophs. Most animals are capable of locomotion at some stage in the life cycle.
- Sponges are the most ancient animal lineage. Cnidarians evolved next. Sponges lack distinct symmetry; cnidarians have radial symmetry.
- Animals with bilateral symmetry are divided into two main groups—protostomes and deuterostomes—on the basis of their pattern of embryo development. Protostomes include mollusks, annelids, and arthropods. Deuterostomes include echinoderms and the chordates, which include us.
- The chordates include all animals with a dorsal nerve cord. Chordates with a backbone are vertebrates, and all other phyla of animals are informally designated as invertebrates.

### 4.2 Characteristics of Animals

- All animals except sponges have true tissues. All animals other than sponges and cnidarians have organs and organ systems.
- Animal development is characterized by cell migration and the formation of embryonic layers that generate all the different tissue types in the adult body.
- After fertilization, the single-celled zygote turns into a hollow ball, which forms a blastopore as cell migration begins. In most protostomes, the blastopore end gives rise to the mouth. In most deuterostomes, the anus develops from the blastopore end.
- Although some animals lack body cavities (are acoelomate), most have either a pseudocoelom or a true coelom.
- Many animals have segmented bodies; segments and their various appendages have become adapted to perform diverse functions in different species.

### 4.3 The First Invertebrates: Sponges, Jellyfish, and Relatives

- Sponges have specialized cells, but no tissues or distinct body symmetry.
- Cnidarians and ctenophores have true tissues and are radially symmetrical.

### 4.4 The Protostomes

- The evolutionary innovation of complete body cavities first evolved in the protostomes.

- Rotifers and flatworms lack a true coelom. Annelids are coelomate worms with segmented bodies.
- Most mollusks have a hard shell. Bivalves have hinged shells; gastropods have a dorsally located spiral shell. Cephalopods have a well-developed nervous system and are capable of complex learning tasks.
- Arthropods have jointed body parts, and must shed their protective layer, the cuticle, in order to grow.
- Arthropods include crustaceans, insects, arachnids, and myriapods. Crustaceans are the most diversified aquatic animals, whereas insects are the most species-rich phylum on Earth. Crustaceans and insects have a tough exoskeleton that protects them from desiccation and possibly from predation.

### 4.5 The Deuterostomes—I: Echinoderms, Chordates, and Relatives

- Deuterostomes have a dorsal nerve cord, and in most the blastopore gives rise to the anus.
- Echinoderms are deuterostomes that use a water vascular system for locomotion and gas exchange.
- Chordates have a dorsal notochord at some stage of their life cycle.

### 4.6 The Deuterostomes—II: Vertebrates

- Vertebrates (which include fishes, amphibians, reptiles, birds, and mammals) are distinguished by having an internal backbone and an anterior braincase.
- The evolution of jaws and a bony skeleton were major steps in vertebrate evolution. These innovations first occurred in fish.
- Fish have paired fins that enable rapid swimming maneuvers.
- Amphibians breathe through both lungs and the skin surface. Their life cycles are often characterized by complete metamorphosis.
- Reptiles have evolved adaptive traits that reduce the risk of desiccation: waterproof scales, a water-conserving excretory system, and an amniotic egg.
- Birds evolved from theropod dinosaurs. They are adapted for flight, with a bipedal stance, feathers modified from scales, and hollow bones.
- Mammals are subdivided into three categories: monotremes, marsupials, and eutherians. Mammals have hair on the body, endothermy and homeothermy, mammary glands, and sweat glands.
- The success of mammals can be attributed to long gestation periods and enhanced investment in parental care.

## Key Terms

amniotic egg (p. 101)
Animalia (p. 80)
Annelida (p. 89)
Arthropoda (p. 91)
bilateral symmetry (p. 84)
cephalization (p. 84)
chordate (p. 96)
coelom (p. 85)
compound eye (p. 94)

crustacean (p. 93)
deuterostome (p. 95)
development (p. 83)
ectotherm (p. 101)
endotherm (p. 104)
eutherian (p. 105)
exoskeleton (p. 92)
homeotherm (p. 104)
invertebrate (p. 81)

mammary gland (p. 105)
marsupial (p. 105)
metamorphosis (p. 94)
Mollusca (p. 89)
molting (p. 91)
monotreme (p. 104)
notochord (p. 96)
pharyngeal pouch (p. 96)
placenta (p. 105)

protostome (p. 88)
radial symmetry (p. 84)
sexual reproduction (p. 83)
simple eye (p. 94)
tetrapod (p. 99)
vertebrate (p. 81)

# Self-Quiz

1. Which animal group is the most abundant in number of individuals and number of species?
   - a. insects
   - b. birds
   - c. protists
   - d. mammals

2. Which of these statements about animals is *not* true?
   - a. Animals are ingestive heterotrophs.
   - b. Animal cells are enclosed in a cell wall made of polysaccharides.
   - c. All animals have at least some specialized cell types.
   - d. Most animal cells are attached to an extracellular matrix.

3. Which of the following groups was the first to take to the air?
   - a. bats
   - b. birds
   - c. protostomes
   - d. certain reptiles

4. In deuterostomes,
   - a. radial symmetry is absent.
   - b. the mouth does not develop from the blastopore.
   - c. the notochord is absent.
   - d. the body is acoelomate.

5. Segmentation is beneficial
   - a. to insects because it helps them stay dry.
   - b. to arthropods because it facilitates specialization among body parts.
   - c. to annelids because it is necessary for coelomate organization.
   - d. to sponges because it led to cephalization.

6. True tissues
   - a. are found in all animals.
   - b. are thought to be absent in sponges.
   - c. consist of two or more organs that work together in an integrated manner to carry out specific functions.
   - d. consist of loose collection of cells that function independently from each other.

7. Chordates are distinguished from all other animals in that all of them
   - a. have a dorsal nerve chord and a post-anal tail.
   - b. possess mammary glands.
   - c. have an anterior skull and well-developed jaws.
   - d. have a backbone.

8. An amniotic egg
   - a. is a characteristic of all tetrapods.
   - b. is believed to have first evolved in jawless fishes.
   - c. is found in birds, but not reptiles such as snakes and crocodiles.
   - d. contains membranes that facilitate gas exchange.

# Analysis and Application

1. What were the major challenges facing insects when they colonized land? What evolutionary innovations did insects use to deal with these challenges?

2. The popular television show *SpongeBob SquarePants* features a sponge and his pet snail, Gary. Compare and contrast a sponge and a snail with respect to tissue and organ system organization, body symmetry, coelomate organization, and cephalization. Can you name any cartoon character from the TV series that would be classified as a protostome if it were a real animal? Name a chordate character from the TV show (hint: this cartoon character wears a diving helmet).

3. What is the evolutionary significance of the segmentation of the animal body? Name an animal that illustrates the adaptive value of segmentation.

4. Animals are typically mobile. What tissue types and organ systems enabled locomotion in animals? Compare locomotion in fish and tetrapods. What is the adaptive value of locomotion?

5. What evolutionary innovations adapted birds for flight? List the characteristics shared by birds and reptiles such as crocodiles and dinosaurs. Which of these characteristics are also found in mammals?

6. List the main structures in a hen's egg. What is the function of each of these structures?

7. Did feathers first evolve as an adaption for flight? Cite evidence that supports your answer.

8. How is complete metamorphosis different from incomplete metamorphosis? What is the adaptive rationale for complete metamorphosis (in other words, why do some animals go through such elaborate and dramatic changes in their life cycle)?

9. What *unique* adaptations do mammals possess for coping with cold and hot environments?

10. Compare how female monotremes, marsupials, and eutherians nurture their young.

# The Chemistry of Life

**AN ANTI-TRANS FAT RALLY.** A young protestor makes her views known, as the New York City Board of Health holds public hearings on a plan to ban trans fats from city restaurants.

# How the "Cookie Monster" Tackled Trans Fats

On January 1, 2011, California became the first state to ban trans fats in restaurants, bakeries, hospitals, and other places that prepare food. Although for years the restaurant industry had argued that such laws would keep them from making the foods that customers wanted and even put some restaurants out of business, the date when the law went into effect passed with barely a ripple. Restaurants didn't go out of business and nobody went without hot, salty fries.

Five years earlier, New York City's Board of Health had stunned and horrified restaurant owners by banning trans fats from the city's famous restaurants, giving restaurateurs just 6 months to stop frying in trans fat–laden cooking oils, and 18 months to eliminate trans fats from all food. An outraged spokesman for the National Restaurant Association argued that the city had "no business banning a product the Food and Drug Administration has already approved."

The restaurant association had a point: the momentum for banning trans fats hadn't come from the federal government or even from the prestigious American Heart Association, but from a California public interest attorney named Stephen Joseph. Joseph's father had died of heart disease, and as Joseph learned how important trans fats are in the development of heart disease he decided to take action. In 2003, he filed a lawsuit against Nabisco, asking the company to stop selling Oreo cookies containing trans fats to children. Nabisco's parent

> What is a trans fat and how is it different from other fats? What are fats? Are all fats bad for our health?

company, Kraft Foods, told CNN, "We stand behind Oreo, a wholesome snack people have known and loved for more than 90 years."

Joseph—dubbed "the Cookie Monster" by the *San Francisco Chronicle*—argued that Oreo cookies made with trans fats were literally dangerous. Talk show hosts called the suit "ridiculous" and suggested banning lawyers instead. Within days, Joseph withdrew his lawsuit. But, as we'll see, the Cookie Monster attorney had the last laugh.

In this chapter we learn about the building-block molecules that make up all living things. One group of building blocks is the lipids, including the trans fats. Let's take a look at how living organisms are put together.

**MAIN MESSAGE** Life on Earth is based on carbon-containing molecules, and four types of these—carbohydrates, proteins, lipids, and nucleic acids—are common to all life-forms on our planet.

## KEY CONCEPTS

- Living organisms are composed of atoms held together by chemical bonds. Four elements—oxygen, carbon, hydrogen, and nitrogen—account for about 96 percent of the weight of a living cell.

- A molecule contains two or more atoms linked through covalent bonds. An ionic bond is the attraction between atoms that have opposite electrical charge.

- Water molecules associate with each other through weak attractions known as hydrogen bonds. The unique properties of water, the primary medium for life-supporting chemical reactions, have a profound influence on the chemistry of life.

- A chemical reaction occurs when chemical bonds between atoms are formed or broken. To sustain life, thousands of different types of chemical reactions must occur inside even the simplest cell.

- An acid releases hydrogen ions, and a base accepts them. Many chemical reactions within a cell are sensitive to the levels of acids and bases.

- Four main classes of molecules are common to all living organisms: carbohydrates, proteins, lipids, and nucleic acids. The functions of these critical molecules range from providing energy to storing hereditary information.

## Helpful to Know

The *mass* of an object is a measure of all the material in it. The more mass an object has, the more difficult it is to move it. *Weight* measures how strongly the mass is pulled on by gravity. On Earth, the mass of an object is the same as its weight.

**FOR ALL ITS REMARKABLE DIVERSITY,** life as we know it is built from a rather limited variety of atoms. The fact that all cells share this limited range of atomic ingredients reminds us of the common evolutionary heritage of all life on Earth.

In this chapter we begin our exploration of cellular life by identifying the chemical components shared by all cells. We shall see that atoms can associate with each other to form assemblages called molecules. We use the term **biomolecules** to refer to molecules found in living cells. The cell is built from many different types of biomolecules, small and large. Sugars and amino acids are small biomolecules that may be familiar to you. DNA and protein are examples of larger molecules made by linking smaller molecules. All biomolecules have a backbone of carbon atoms, which is why we say that life on our planet is carbon based.

Our first step toward understanding the structure and function of the major biomolecules will be to examine the structure of atoms and how they join to form molecules. We will consider why life is crucially dependent on water, and why most of the chemical reactions that are vital for life occur in a watery environment. We will discuss why carbon is an exceptional element and why it accounts for much of the dry mass of living things. In short, this chapter is about how living organisms function at the chemical level, the focus of a branch of science called *biochemistry*. Many of the topics introduced in this chapter serve as a foundation for the deeper investigation of life, at every level of the biological hierarchy, that follows in later chapters.

# 5.1 Matter, Elements, and Atomic Structure

What is the world made of? The answer is: *matter*. **Matter** is defined as anything that has mass and occupies space. Think of it as the "stuff" that the universe is composed of. At least 92 different types of matter occur naturally in the universe; each unique type is known as an element. An **element** is a pure substance with distinctive physical and chemical properties, and it cannot be broken down to other substances by ordinary chemical methods. Each element is identified by a one- or two-letter symbol; for example, oxygen is identified as O, calcium as Ca.

Hydrogen (H) is the most abundant element in the universe. Silicon (Si) makes up 28 percent of Earth's crust, but less than 0.001 percent of the human body (**FIGURE 5.1**). Much of the silicon in rock and sand is combined with oxygen atoms (as silicates, $SiO_3$), which is why the oxygen content of Earth's crust is also high. Just four elements—oxygen (O), carbon (C), hydrogen (H), and nitrogen (N)—account for more than 96 percent of the mass of the average cell. Water, which contains hydrogen and oxygen atoms, makes up 70 percent of the mass of such a cell. Water is also abundant on the surface of our planet, 71 percent of which is covered by oceans. In contrast, carbon is scarce in Earth's crust, but it is the third most abundant element in a cell. The overall chemical composition of living beings is therefore distinctly different from that of the nonliving part of Earth.

An **atom** is defined as the smallest unit of an element that still has the distinctive chemical properties of that element. Atoms are so small that more than a trillion of them could easily fit on the head of a pin. Because there are 92 naturally occurring elements, there are also 92 different types of atoms. The uniqueness of an element comes from the special characteristics of the atoms that compose it.

What makes the atoms of one element different from those of another? The answer lies in the specific combination of three atomic components. The first two components are electrically charged: **protons** have a positive charge (+), and **electrons** have a negative charge (−). As the name of the third component implies, **neutrons** lack an electrical charge; they are electrically neutral. These three components—especially electrons—determine the physical and chemical properties of an element and how its atoms interact with other atoms.

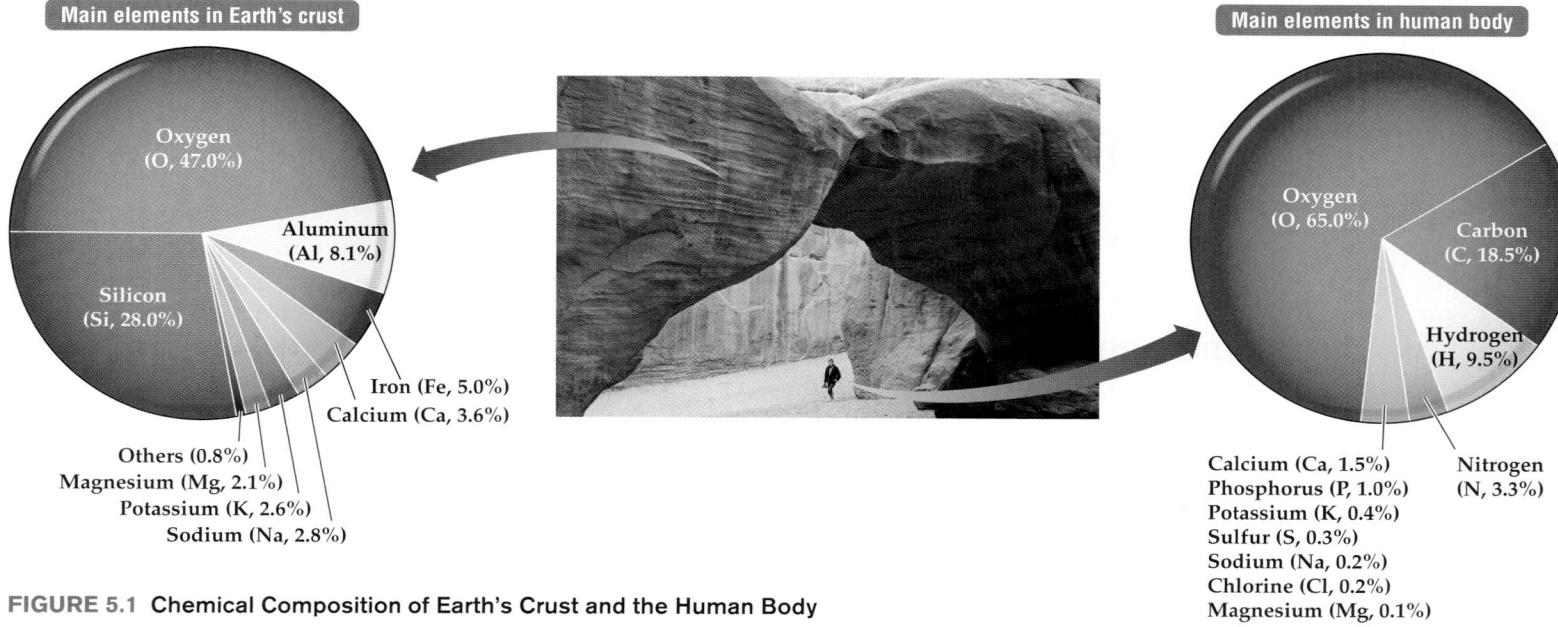

Main elements in Earth's crust

Oxygen
(O, 47.0%)

Aluminum
(Al, 8.1%)

Silicon
(Si, 28.0%)

Iron (Fe, 5.0%)
Calcium (Ca, 3.6%)

Others (0.8%)
Magnesium (Mg, 2.1%)
Potassium (K, 2.6%)
Sodium (Na, 2.8%)

Main elements in human body

Oxygen
(O, 65.0%)

Carbon
(C, 18.5%)

Hydrogen
(H, 9.5%)

Calcium (Ca, 1.5%)      Nitrogen
Phosphorus (P, 1.0%)    (N, 3.3%)
Potassium (K, 0.4%)
Sulfur (S, 0.3%)
Sodium (Na, 0.2%)
Chlorine (Cl, 0.2%)
Magnesium (Mg, 0.1%)

**FIGURE 5.1  Chemical Composition of Earth's Crust and the Human Body**

A single atom has a dense central core, called the **nucleus [*NOO*-klee-us]** (plural "nuclei"), that contains one or more protons and is therefore positively charged. Except in the case of the most common form of hydrogen, the nucleus also contains one or more neutrons. One or more negatively charged electrons move around the nucleus in defined volumes of space known as **electron shells (FIGURE 5.2)**. If a hydrogen nucleus were the size of a marble, the electron would move around it in a space as big as the Houston Astrodome. As a whole, the positive charge on the nucleus balances the negative charges of the electrons, making atoms electrically neutral.

## The size and structure of atoms can be described with numbers

The distinctive aspects of the atom of each element can be summarized by numbers that describe the atom's structure and mass. The number of protons in an atom's nucleus is the **atomic number** of that particular element. Hydrogen, having a single proton, has an atomic number of 1; carbon, which has six protons, has an atomic number of 6. The sum of an atom's protons and neutrons, another distinguishing feature of each element, is its **atomic mass number**. The mass of an electron is negligible—only about 1/2,000 that of a proton or neutron. Protons and neutrons have about the same mass, so the atomic mass number of an element is based on the total number of protons and neutrons contained in the nucleus of the atom.

Hydrogen has a single proton and no neutrons, so the atomic mass number for hydrogen is the same as its atomic number: 1, written $^1$H. In contrast, the nucleus of a carbon atom contains six protons and six neutrons, giving carbon an atomic mass number of 12 ($^{12}$C). A carbon atom has about 12 times as much mass as a hydrogen atom. Here are the atomic numbers (in black) and atomic mass numbers (in

### THE STRUCTURE OF ATOMS

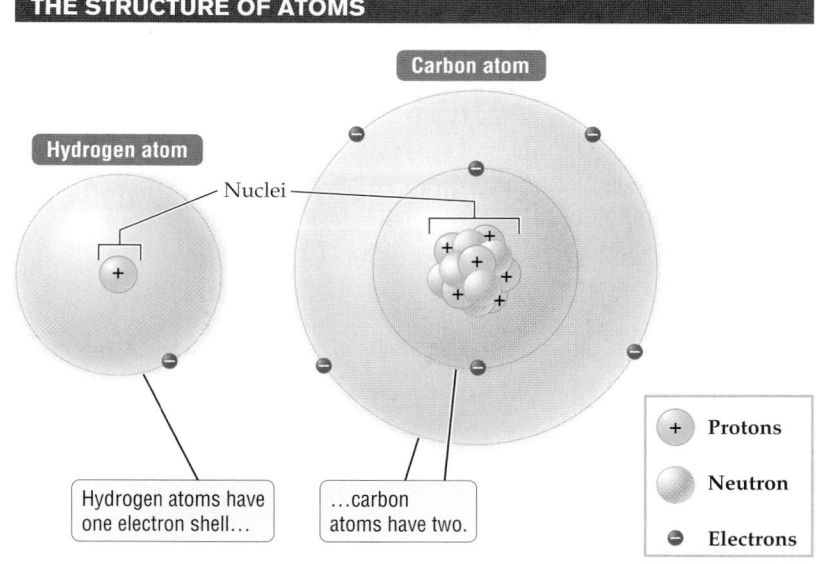

Carbon atom

Hydrogen atom

Nuclei

Hydrogen atoms have one electron shell…

…carbon atoms have two.

+ Protons

Neutron

– Electrons

**FIGURE 5.2  Atomic Structure**

The electrons, protons, neutrons, and nuclei of these hydrogen and carbon atoms are shown greatly enlarged in relation to the size of the whole atom. An electron shell is a simplified way of representing the space that electrons move in as they orbit the nucleus.

green) of the four atoms that are most abundant in biomolecules:

Atomic number

| 1 | 6 | 7 | 8 |
|---|---|---|---|
| H | C | N | O |
| 1 | 12 | 14 | 16 |
| Hydrogen | Carbon | Nitrogen | Oxygen |

Atomic mass number

Atomic number

| 6 | 6 | 6 |
|---|---|---|
| C | C | C |
| 12 | 13 | 14 |
| Carbon-12 | Carbon-13 | Carbon-14 |

Atomic mass number

## Some elements exist in variant forms called isotopes

**Helpful to Know**

The atoms commonly found in living cells are shown in the Periodic Table of the Elements, p. T2.

Each element is uniquely distinguished by the number of protons in its nucleus. However, the atoms of some elements come in different forms, known as *isotopes*, that vary only in the number of neutrons in their nuclei. **Isotopes** of an element have the same atomic number, but different atomic mass numbers. In other words, the isotopes of an element have the same number of protons and electrons but differ in the number of neutrons, and therefore in mass. For example, over 99 percent of the carbon atoms found in atmospheric carbon dioxide gas ($CO_2$) have an atomic mass number of 12 ($^{12}C$). However, a tiny fraction—slightly less than 1 percent—of those carbon atoms exist as an isotope containing seven neutrons instead of six. These seven neutrons, together with the six protons, give an atomic mass number of 13 ($^{13}C$). This isotope is therefore referred to as carbon-13. Carbon-14 ($^{14}C$) is an even rarer form of carbon; it makes up about 0.01 percent of the carbon in the atmosphere. How many neutrons does carbon-14 have?

The answer is: 8 (subtract the atomic number from the atomic mass number). Notice that all three isotopes have the same atomic number. All isotopes have the same number of electrons too. It is only with respect to neutrons that isotopes differ.

Certain isotopes, called **radioisotopes** [RAY-dee-oh-EYE-soh-topes], have unstable nuclei that change ("decay") into simpler forms, releasing high-energy radiation in the process. For example, carbon-14 is a radioisotope. Only a fraction of known isotopes are radioactive; some of these, such as carbon-14, phosphorus-32, and the two radioisotopes of hydrogen (deuterium, $^2H$; and tritium, $^3H$), have important uses in both research and medicine.

The radiation given off by radioisotopes can be detected in various ways, ranging from simple film exposure to the use of sophisticated scanning machines. Because radioisotopes can be detected, their location and quantity can be tracked fairly easily—a characteristic that makes them useful in medical diagnostics. For example, the thyroid gland takes up iodine for producing a special type of hormone required by the body. When a low dose of an iodine radioisotope (iodine-131) is administered to patients with thyroid disease, physicians can use an imaging device to see how the radioisotope is taken up by the thyroid (**FIGURE 5.3**). If a patient is found to be suffering from cancer of the thyroid,

**FIGURE 5.3 Radioisotopes Are Useful in Medical Imaging**

(*a*) The thyroid gland, located in the neck, helps regulate metabolism. The element iodine accumulates in the gland and is necessary for normal thyroid function. (*b*) This image is a visualization of the thyroid gland in a patient with goiter, an enlargement of the thyroid gland, often caused by iodine deficiency. Small amounts of radioactive iodine were given to the patient, and the accumulated radioisotope was visualized with a gamma-ray scan.

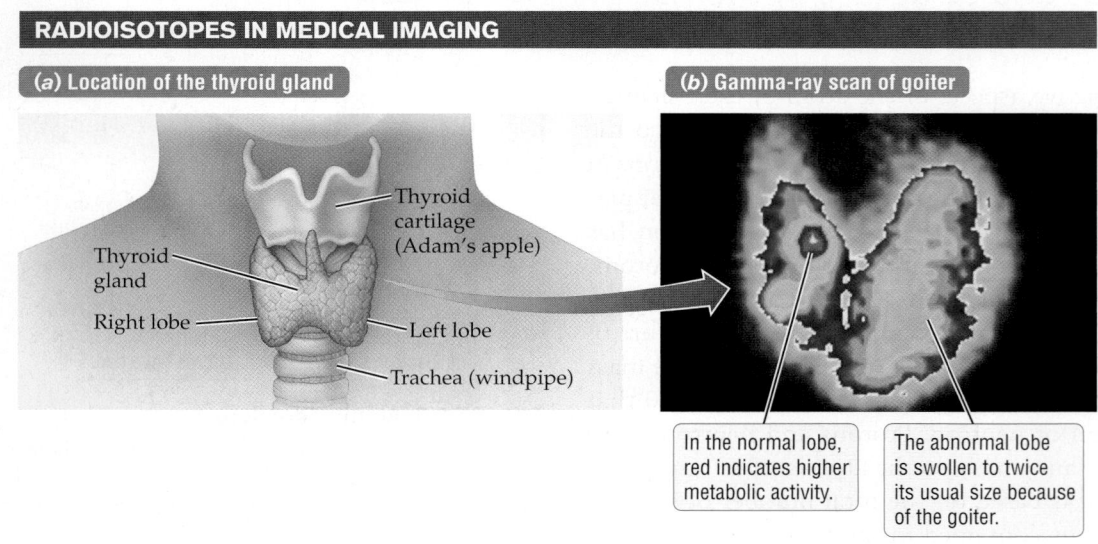

RADIOISOTOPES IN MEDICAL IMAGING

(*a*) Location of the thyroid gland

Thyroid cartilage (Adam's apple)
Thyroid gland
Right lobe
Left lobe
Trachea (windpipe)

(*b*) Gamma-ray scan of goiter

In the normal lobe, red indicates higher metabolic activity.

The abnormal lobe is swollen to twice its usual size because of the goiter.

repeated doses of iodine-131 can be administered as a therapy because the accumulation of radioactivity in the thyroid tends to kill the cancer cells.

## 5.2 Covalent and Ionic Bonds

The number of electrons in an atom and how they are packed around the nucleus are the main determinants of an atom's chemical behavior. Some elements are said to be chemically inert because their atoms tend not to interact with other atoms by losing, gaining, or sharing electrons. Most elements, however—and all elements that are biologically important—have atoms that are far more social. Such atoms have a tendency to donate electrons, accept electrons, and even share electrons, if the right type of atom becomes available under the right conditions. The interaction that causes two atoms to associate with each other is known as a **chemical bond**.

A **molecule** is an assemblage of atoms in which at least two of the atoms are linked through electron sharing. Electron sharing creates an exceptionally strong chemical bond known as a *covalent bond*. A molecule contains a minimum of two atoms, but larger molecules can be composed of millions of atoms. A molecule may consist of atoms of the same element (say, oxygen atoms only), or it may contain atoms from two or more elements (two hydrogen atoms and one oxygen atom in the water molecule, for example).

When an atom loses one or more of its negatively charged electrons, it acquires an overall (net) positive charge. Likewise, when an atom gains one or more electrons, it becomes negatively charged. Atoms that have acquired electrical charge because they have lost or gained electrons are called **ions**. Ions with opposite charge experience an electrical attraction, and such ionic interactions have an important role in the chemistry of living cells. The chemical attraction between negatively charged and positively charged ions is a type of chemical bond called an *ionic bond*. A *salt* is made up of ions from at least two different elements; the ions are held together exclusively by the electrical attraction between them.

A **chemical compound** is a substance that contains atoms from two or more *different* elements, each in a precise ratio. A molecule of water, for example, is a compound of hydrogen and oxygen atoms. Chemists have developed a simple shorthand, known as a **chemical formula**, to represent the atomic composition of molecules and salts. The formulas use the letter symbol of each element and a subscript number to the right of the symbol to show how many atoms of that element are contained in the molecules or salts. For example, the chemical formula for a water molecule is $H_2O$. The chemical formula for table sugar (sucrose), which has 12 carbons, 22 hydrogens, and 11 oxygens per molecule, is $C_{12}H_{22}O_{11}$. Chemical formulas describe ionic compounds (such as table salt) as well. Table salt has equal numbers of sodium ions ($Na^+$) and chloride ions ($Cl^-$). The molecular formula for salt is therefore NaCl.

### Covalent bonds form by electron sharing between atoms

A molecule contains at least two atoms held together by **covalent bonds**, which are chemical bonds formed by the sharing of electrons between atoms. A single covalent bond represents the sharing of *one pair* of electrons between two atoms (**FIGURE 5.4a**). In a structural formula, which displays how atoms are connected within a molecule, a single covalent bond is indicated by a single straight line (**FIGURE 5.4b**). The atoms joined in a covalent bond can be of the same type (Figure 5.4a), or they can be atoms of two different elements (which is the case for all the molecules in Figure 5.4b except hydrogen gas and oxygen gas).

What drives atoms to share electrons with one another? To answer that question, consider how electrons are distributed in the space around the nucleus. The electrons of every atom move in volumes of space that can be visualized as concentric layers, called shells, around the nucleus (see Figure 5.2). The maximum number of electrons in any shell is fixed. When all its shells are filled to capacity, an atom is in its most stable state. Atomic shells are filled starting from the innermost shell, which can hold two electrons at most. The next shell outside it can hold a maximum of eight electrons (see Figure 5.4a).

Atoms that have unfilled outer shells can achieve a more stable state by interacting in ways that will achieve maximum occupancy of the outermost shell (called the *valence shell*). One way for an atom to achieve greater stability is to fill its outermost shell by sharing one or more of its outer-shell electrons with a neighboring atom. Each atom in this arrangement contributes one electron to every pair of shared electrons. Each pair of electrons that is shared between the two atoms constitutes one covalent bond.

The number of covalent bonds an atom can form is equal to the number of electrons needed to fill its outermost shell. Consider the electron sharing that occurs between hydrogen and oxygen in a water molecule. Hydrogen has one electron in its single shell, so

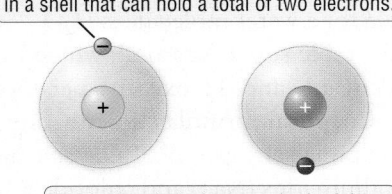

Each hydrogen atom has a single electron contained in a shell that can hold a total of two electrons.

By sharing their electrons, both of these hydrogen atoms have filled their shells.

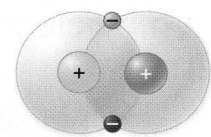

Hydrogen gas (H₂)

The outermost shell of an oxygen atom has six electrons but can hold a maximum of eight.

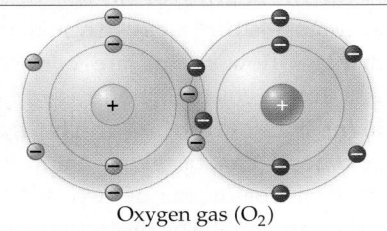

Oxygen gas (O₂)

By sharing *two* pairs of electrons, each oxygen atom fills its outermost electron shell to capacity.

| ATOM | SYMBOL | NUMBER OF POSSIBLE BONDS | SAMPLE MOLECULES | |
|---|---|---|---|---|
| Hydrogen | H | 1 | H—H | Hydrogen gas ($H_2$) |
| Oxygen | O | 2 | O=O | Oxygen gas ($O_2$) |
| Sulfur | S | 2 | H—S—H | Hydrogen sulfide ($H_2S$) |
| Nitrogen | N | 3 | H—N—H with H above | Ammonia ($NH_3$) |
| Carbon | C | 4 | H—C—H with H above and below | Methane ($CH_4$) |

**FIGURE 5.4 Covalent Bonds and Electron Shells**

Atoms in biologically important molecules have as many as four electron shells. The innermost shell holds a maximum of two electrons, and the next shell outside it holds a maximum of eight. (*a*) A single covalent bond is formed by the sharing of one pair of electrons between two atoms. Two hydrogen atoms can share one pair of electrons, forming one covalent bond. An oxygen atom requires two additional electrons to fill its outermost shell and can therefore form two covalent bonds. (*b*) The number of covalent bonds an atom can form depends on the number of electrons needed to fill its outermost shell to capacity. A carbon atom can form four bonds because it has only four electrons in its outermost shell and requires four more electrons to fill that shell.

that shell is one electron short of maximum occupancy and therefore maximum stability. The inner shell of oxygen is filled, but its outer shell is not: it has six electrons, when it can hold as many as eight. This situation can be resolved by mutual borrowing, on a "time sharing" basis, between two hydrogen atoms and an oxygen atom: each hydrogen atom shares its one electron with oxygen, and the oxygen atom shares two of its electrons, one with each of the hydrogen atoms:

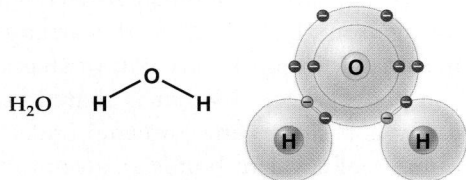

So the atoms contribute electrons in a way that makes the outer shells of all three atoms complete, at least on a shared basis. This kind of sharing requires an in-

timate association between the atoms, which is why covalent bonds are so strong.

A **double bond**, representing two covalent bonds, exists when *two pairs* of electrons are shared between two atoms. Some atoms may even share *three pairs* of electrons, in which case they are said to have **triple bonds**. Nitrogen gas, also known as molecular nitrogen ($N_2$) to distinguish it from elemental nitrogen (N), consists of two nitrogen atoms bound by triple covalent bonds (three pairs of shared electrons):

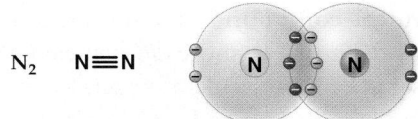

The physical and chemical properties of a molecule are often profoundly influenced by the shape of that molecule in three-dimensional space. A *ball-*

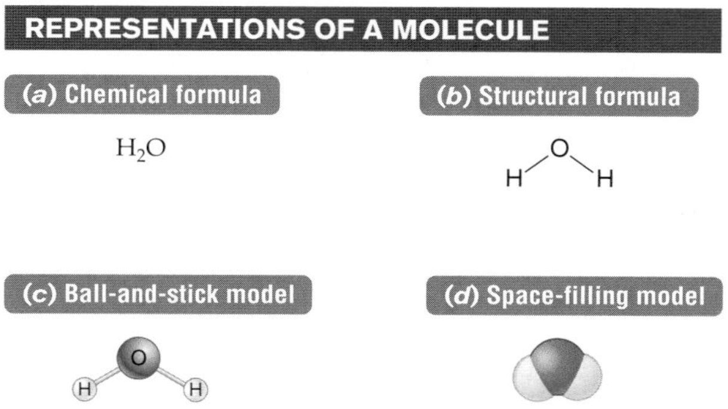

## REPRESENTATIONS OF A MOLECULE

**(a) Chemical formula**

$H_2O$

**(b) Structural formula**

H—O—H

**(c) Ball-and-stick model**

H—O—H

**(d) Space-filling model**

**FIGURE 5.5 Covalent Bonds between Atoms Can Be Represented in Various Ways**

By convention, hydrogen is shown in white and oxygen in red.

*and-stick model* shows the angles made by adjacent covalent bonds in a molecule (**FIGURE 5.5c**). Atoms are represented by spheres ("balls"), and a single covalent bond is represented by a line or rod ("stick"). A *space-filling model* is yet another way to depict the shape of a molecule in three-dimensional space (**FIGURE 5.5d**). Unlike a ball-and-stick representation of a molecule, a space-filling model shows the width of atoms (atomic radii) and the distance between nuclei (bond length) in accurate proportions.

## Ionic bonds form between atoms of opposite charge

Atoms that have acquired a net electrical charge can associate through another type of chemical bond, the **ionic bond**, in which ions with opposite electrical charge are mutually attracted. In the process of ionization (ion formation), one or more electrons are transferred from one neutral atom to another. For example, an electron in the outer shell of a neutral sodium atom can be transferred to the outer shell of a neutral chlorine atom, as shown in **FIGURE 5.6**. The loss of an electron from the sodium atom and the gain of an electron by chlorine converts both neutral atoms into ions with a *net* charge that is equal and opposite. For maximum stability, the ions created through such a charge transfer must remain closely associated. Compounds consisting of charged atoms that are held exclusively through ionic bonds are known as **salts**.

In ionic solids such as crystals of table salt (NaCl), ions are closely packed in an orderly pattern known as a crystal lattice (**FIGURE 5.7**). When salt is added to water, the ionic bonds between the ions are disrupted. For reasons we explore in the next section, water molecules surround both types of charged ions in

## HOW IONS ARE FORMED

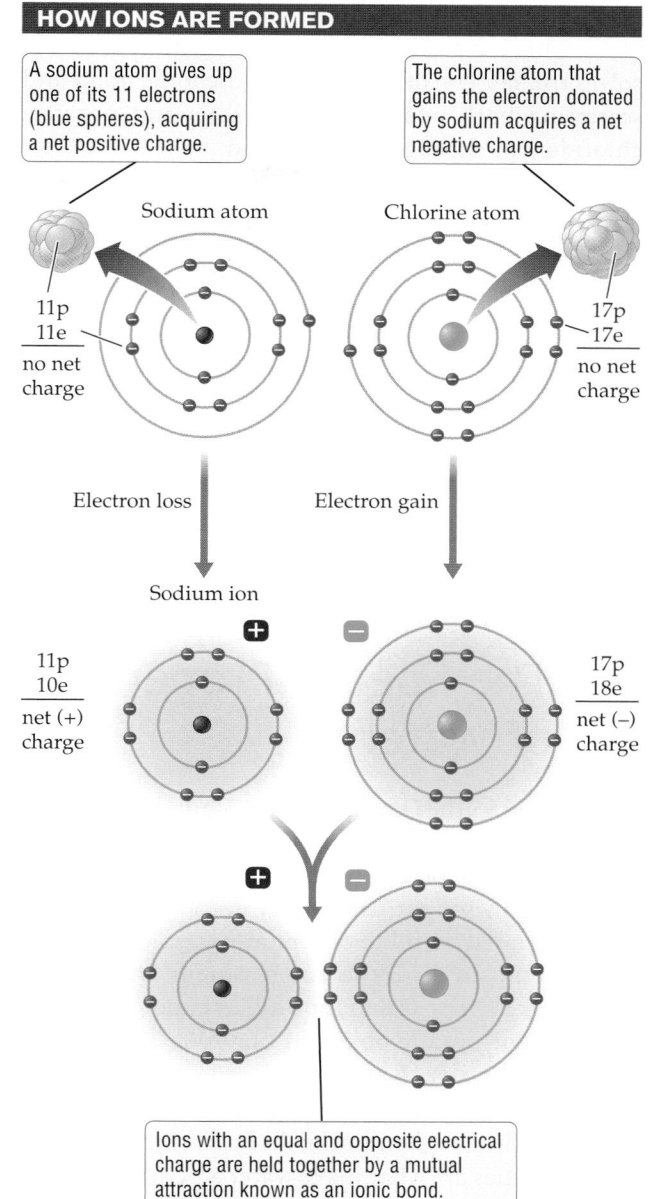

A sodium atom gives up one of its 11 electrons (blue spheres), acquiring a net positive charge.

The chlorine atom that gains the electron donated by sodium acquires a net negative charge.

Sodium atom

Chlorine atom

11p
11e
no net charge

17p
17e
no net charge

Electron loss

Electron gain

Sodium ion

11p
10e
net (+) charge

17p
18e
net (−) charge

Ions with an equal and opposite electrical charge are held together by a mutual attraction known as an ionic bond.

**FIGURE 5.6 Ions Are Created through the Loss or Gain of Electrons**

Negatively charged ions are often given special names, such as "chloride" for the chlorine ion (Cl⁻) and "fluoride" for the fluorine ion (F⁻).

## STRUCTURE OF SALT CRYSTALS

Table salt crystals are held together by the mutual attraction between positively charged sodium ions (Na⁺) and negatively charged chlorine ions (Cl⁻).

Crystals of NaCl, table salt

Positively charged ion

Negatively charged ion

**FIGURE 5.7 Salts Are Ionic Compounds**

NaCl. This interaction with water breaks up and dissolves the salt crystals, weakening the ionic bond and scattering both positive sodium ions and negative chloride ions throughout the liquid.

> ### Concept Check
>
> 1. An atom of iron (Fe) contains 26 electrons and 30 neutrons. (a) How many protons does it have? (b) How many uncharged particles? (c) What is its atomic mass number?
>
> 2. What is a molecule? How many atoms of oxygen (symbol O) are present in each molecule of table sugar (sucrose), whose chemical formula is $C_{12}H_{22}O_{11}$?
>
> 3. Explain how ions are formed.

# 5.3 | The Special Properties of Water

Life evolved in the oceans about 3.5 billion years ago, and that aquatic ancestry is reflected in the makeup of all living beings today. The average cell is about 70 percent water by weight, and nearly every chemical process associated with life occurs in water. Metabolism ceases in most cells if the water content drops below 50 percent. Life is abundant on Earth, but it appears to be absent on any object in our solar system that lacks liquid water. Water has chemical and physical properties unmatched by any other known substance and, as we will explain in the discussion that follows, these qualities are what make water a perfect medium for sustaining life.

## Water is a polar molecule

As we have seen, each molecule of water is made up of two hydrogen atoms and one oxygen atom held together by covalent bonds. We described the covalent bond as one in which electrons are shared. Sometimes, however, electrons are not *equally* shared by the two atoms joined by a covalent bond. In some molecules, one atom has greater electron-grabbing power (technically known as *electronegativity*). The more electronegative atom exerts a stronger pull on the shared electrons, resulting in a polar covalent bond. **Polar molecules** contain polar covalent bonds created by lopsided electron sharing. The uneven distribution of

electrical charge in polar molecules makes one end of the molecule slightly negative and the opposite end (or *pole*) slightly positive.

Water is a polar molecule because the electrical charge is distributed unevenly over its boomerang shape. The nucleus of the oxygen atom pulls on the shared electrons more powerfully than do the nuclei of the two hydrogen atoms. The oxygen end of the molecule therefore develops a partial negative charge, while a partial positive charge arises in the pole where the hydrogen atoms are located. In the following schematic representation of the water molecule, the two poles are shown in green and blue:

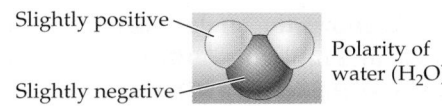

Slightly positive

Slightly negative

Polarity of water ($H_2O$)

Because opposite electrical charges attract, the slightly positive hydrogen atoms of one water molecule are drawn toward the slightly negative oxygen atom of a neighboring water molecule. A **hydrogen bond** is the weak electrical attraction between a hydrogen atom with a partial positive charge and a partially negative neighboring atom. Intermolecular hydrogen bonds develop readily among adjacent water molecules, with the partially positive pole of one water molecule tugging at, and being tugged upon, by the partially negative pole of a neighboring water molecule:

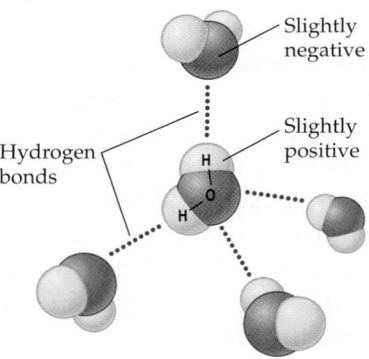

Slightly negative

Slightly positive

Hydrogen bonds

Although a single hydrogen bond is about 20 times weaker than a covalent bond, collectively the cross-linking of many water molecules through many hydrogen bonds amounts to a potent force. The polarity of water and the resultant hydrogen bonding explain nearly all of the special properties of water that we explore next.

## Water is a solvent for charged or polar substances

Water molecules can form hydrogen bonds with other polar molecules, which is why polar compounds dissolve in water. Such compounds are said to be **soluble** in water because they dissolve—that is, mix completely with the water. Ions also dissolve readily in water because water molecules can form a network, or *hydration shell*, around the ions. For example, when table salt is added to water, the solid crystals dissolve. The crystals break apart as the ions in the salt crystal are surrounded by water molecules and held in solution throughout the liquid (**FIGURE 5.8**).

Since dissolved compounds abound in and around living cells, chemists and biologists use specific terms to describe these mixtures: a **solution** is any combination of a **solute** (a dissolved substance) and a **solvent** (the fluid into which the solute has dissolved). A cup of black coffee has hundreds of different types of solutes in it, caffeine being one of them. Stirring a spoonful of table sugar (sucrose) into the cup adds yet another solute. Water, which makes up more than 90 percent of the volume of that cup, is the solvent in which the many solutes are dissolved. Water is also the solvent of life: the cytoplasm contains thousands of different types of solutes, and they are dissolved in a thick solution that is about 70 percent water in most cells. Because so many biologically important ions and molecules function as solutes dissolved in a watery (aqueous) environment, water is commonly called the universal solvent.

Because of their polar nature, water molecules will not interact with uncharged or nonpolar substances. Electron sharing is largely symmetrical in **nonpolar molecules**; therefore, polar covalent bonds are absent. When added to water, nonpolar molecules fail to go into solution and tend to cluster among their own kind instead. This is exactly what happens when olive oil (composed of nonpolar molecules) is added to vinegar, an aqueous solution containing ions and polar molecules. To make a salad dressing, the two substances have to be shaken vigorously; otherwise the oil separates from the vinegar. The distribution of electrons among carbon atoms and hydrogen atoms within oil molecules is nearly equal, making these molecules nonpolar and therefore insoluble. Waxes are also nonpolar; automobile enthusiasts wax their cars not just to make them look shiny, but also to repel water and reduce the risk that their hot rods will be marred by rust.

### HOW SALTS DISSOLVE IN WATER

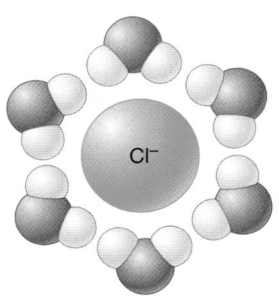

**Solutes:** Na⁺ (sodium ions)
Cl⁻ (chloride ions)
**Solvent:** $H_2O$ (water)
**Solution:** solutes + solvent

The negative pole of the water molecule orients to the positive ion.

The positive pole of the water molecule orients to the negative ion.

**FIGURE 5.8**
**Charged Substances Dissolve in Water to Form Solutions**

Molecules that associate with water (such as sugar and salt) are called **hydrophilic** (*hydro*, "water"; *philic*, "loving"); molecules that are excluded from water (such as oil and wax) are called **hydrophobic** (*phobic*, "fearing"). The reason that olive oil does not dissolve easily in vinegar is that the ingredients in the oil are hydrophobic and water molecules therefore do not interact with them. Being lighter than water, the "shunned" oil molecules clump together and float on top of the vinegar.

## Hydrogen bonding accounts for the physical properties of liquid water and ice

Water can exist in all three states of matter: liquid, solid, and gas. The interplay of hydrogen bonds explains the physical properties of the liquid and solid phases of water. At moderate temperatures, hydrogen bonds among water molecules are constantly forming and breaking. This nonstop jostling is the reason that water is a liquid at room temperature: the molecules cannot be packed together tightly enough to form a solid.

At 0°C, water molecules have less energy and cannot move about as vigorously. A more stable network of hydrogen bonds emerges as water turns into ice. A given mass of liquid water occupies more space when it turns into ice because the molecules are spaced farther apart in ice, locked into an orderly pattern known

Oil molecules are hydrophobic. They are excluded from water and tend to clump together.

**Olive oil**

**Vinegar**

Vinegar molecules are hydrophilic. They are held in solution by water molecules.

FIGURE 5.9
**Hydrogen Bonding between Water Molecules**

A hydrogen bond is the weak attraction between a hydrogen atom with partial positive charge and the partially negative region of any polar molecule. In vapor form, water molecules move too rapidly, and are too far apart, to form hydrogen bonds.

**HYDROGEN BONDING**

Hydrogen bonding in liquid water

Hydrogen bonds are constantly forming…

…and breaking.

Hydrogen bonding in ice

When water freezes, the hydrogen bonds become more rigid as water molecules become stacked in a three-dimensional network of hexagons, forming an ice crystal.

Water molecules are placed farther apart in an ice crystal, making ice less dense and able to float on liquid water.

## Helpful to Know

*Heat* refers to the *total energy* in a body due to the movement of the atoms within it. *Temperature* is a measure of the *average speed* with which the atoms are moving. Size affects the total heat the object possesses, but its temperature does not. A bathtub with 30°C water has more heat than a teacup with 30°C water, but water molecules move at the same average speed in both.

as a crystal lattice (**FIGURE 5.9**). As a consequence, ice is 9 percent less dense than liquid water (density is mass divided by volume), and that is why ice floats on water. If ice were denser than liquid water, it would sink in a northern lake in winter and the lake would then freeze from the bottom up. Instead, the ice that forms on the lake surface acts like an insulating blanket, enabling aquatic organisms to survive the winter in the liquid water below.

## Water moderates temperature swings

The heat capacity of water is exceptionally large, which means that a given mass of water can absorb or release a much larger amount of heat than can the same mass of any other common substance. **Heat capacity** is a measure of the heat energy required to raise the temperature of a specific volume of water by a fixed amount. It takes a lot of heat energy to increase the temperature of water because a large proportion of the energy input goes toward breaking hydrogen bonds, before the temperature (average speed) of water molecules can rise. For cells, the practical consequence is that it takes a relatively large amount of heat to increase the temperature of their watery contents.

Because of its large heat capacity, water is also a very effective heat reservoir. If air temperatures start

to drop and water cools, some of the stored heat is released, warming up the environment in the process. Seaside locales experience milder winters for this very reason. Cells, too, are buffered from a sharp decrease in internal temperature when the mercury drops. The large heat capacity of water therefore protects cells from drastic changes in internal temperature in response to fluctuations in the external temperatures.

## The evaporation of water has a cooling effect

When heat energy is supplied to a liquid, some molecules become energetic enough to make the transition from the liquid to the gaseous state—a phenomenon known as **evaporation**. In the case of water, a substantial amount of heat energy must first be invested in snapping the network of hydrogen bonds before water molecules in the liquid state can move fast enough to escape as water vapor (steam). Water has an exceptionally high *heat of vaporization*, which is another way of saying that evaporating a given mass of water requires a substantial input of heat energy.

The evaporation of water molecules removes heat from a surface, cooling it in the process. *Evaporative cooling* refers to the lowering of temperature associated with the evaporation of liquids. The fastest molecules evaporate first because they alone have enough energy to break loose from the surface of the liquid. The molecules left behind are the most sluggish of the group, which is why they did not escape in the first place. The departure of the high-speed molecules causes the average temperature of the remaining liquid to drop, because the molecules that stay behind are slower and therefore cooler. Sweaty humans and panting dogs are cooled when body heat goes toward disrupting hydrogen bonds and increasing the speed of water molecules in human sweat or canine slobber.

## Hydrogen bonding accounts for the cohesion of water molecules

**Cohesion** is the attractive force that binds the same types of atoms or molecules. Substances with high cohesion resist breaking. The tendency of water molecules to cling to each other through hydrogen bonding generates strong cohesion. As a result, a column of water in a pipe or tube is strong enough to withstand a substantial pulling force. This cohesive strength is critical for the ascent of water into the canopy of tall trees. Water is lifted to a treetop in the xylem, which consists of narrow, tubelike cells that occupy most of the interior in a tree trunk. The lifting force is generated by the sun-driven evaporation of water from the surface of a plant, a process known as *transpiration* (discussed in Chapter 36). However, it is the cohesive property of water molecules that keeps the water column in the xylem tubes intact and continuous. Without their cohesive strength, water molecules would be ripped apart by the transpirational pull and the xylem tubes would become clogged with short stretches of water separated by large air bubbles.

*Surface tension* is an important property of water resulting from the cohesiveness of water molecules. Water has the highest surface tension of any liquid except mercury. **Surface tension** is a force that tends to minimize the surface area of water at an air-water boundary. Think of water in a cup and visualize the water molecules on the surface forming an elastic sheet, like the fabric in a trampoline. This sheet of water does not interact with the molecules in the air at all, but it experiences a substantial inward tug because of hydrogen bonding with the water molecules below. The pull of hydrogen bonds generates a force—surface tension—that holds the water surface taut, resisting any stretching or breaking of that sur-

**SURFACE TENSION**

The weight of a water spider, with its long, water-repellent legs, dimples the water surface.

Surface tension opposes the spider's weight, preventing the spider from breaking through the sheet of hydrogen-bonded water molecules at the air-water boundary.

**FIGURE 5.10**

**Hydrogen Bonding between Water Molecules Contributes to the High Surface Tension of Water**

The cohesion of water molecules gives rise to surface tension, a force that resists the stretching of the water surface at any air-water boundary.

face. Surface tension is strong enough to support very light objects: aquatic insects like water spiders, or a paper clip in a cup filled to the brim. In the case of the insect, its long legs distribute its weight, and the hydrophobic material on the legs repels water; the legs dimple the water surface without breaking through the taut sheet of water molecules at the air-water boundary (**FIGURE 5.10**).

You do not feel the surface tension of water in a swimming pool if you dive in neatly, with your fingertips leading the way. But if you expose more of your body surface, as you do when you belly flop into the pool, you might feel the collective smack of the water molecules under surface tension at the air-water boundary.

## 5.4 Chemical Reactions

Many biological processes require atoms to break existing connections or form new ones. The process of breaking existing chemical bonds and/or creating new chemical bonds is known as a **chemical reaction**. A **reactant** is a substance that undergoes a chemical reaction, either alone or in conjunction with other reactants. The alteration of electron-sharing patterns through a chemical reaction yields at least one chemical substance that is different from the reactants, and any newly formed substances are called the **products** of the chemical reaction.

The standard notation for chemical reactions, the *chemical equation*, displays reactants to the left of an arrow and the products to the right of that arrow. Hydrogen and oxygen, for example, can combine to

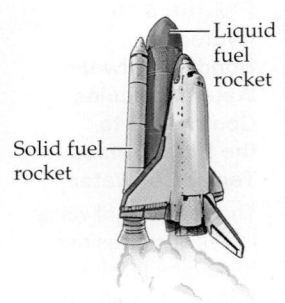

Liquid fuel rocket

Solid fuel rocket

produce water molecules in an explosive reaction. The reaction releases so much energy that it is used in the liquid fuel rocket that helps propel the space shuttle into orbit:

$$2\,H_2 + O_2 \longrightarrow 2\,H_2O + Energy$$
(Reactants)        (Product)

The arrow in the equation indicates that the molecules on the left side of the equation are converted to the product, water. The numbers in front of the molecules define how many molecules participate in the reaction. In this case, two molecules of molecular hydrogen ($H_2$) combine with one molecule of molecular oxygen ($O_2$) to produce two molecules of water (note that when a single molecule of a substance is intended, the numeral 1 is generally omitted).

Some chemical reactions release energy, but others will not occur without an input of energy. The manufacture of ammonia, a key ingredient in many synthetic (human-made) fertilizers, requires large amounts of energy, usually provided by fossil fuels such as natural gas or petroleum. Nitrogen and hydrogen gases are combined to produce ammonia ($NH_3$). The chemical equation for this reaction is

$$Energy$$
$$\downarrow$$
$$3\,H_2 + N_2 \longrightarrow 2\,NH_3$$
(Reactants)        (Product)

N-P-K refers to the percentage of nitrogen (N), phosphorus (P), and potassium (K) in the fertilizer.

Chemical bonds in atoms are rearranged during a chemical reaction, but the process can neither create nor destroy atoms. Therefore, the reaction must begin and end with the same number of atoms of each element. In the reaction depicting ammonia manufacture above, there are six hydrogen atoms for each pair of nitrogen atoms among the reactants, and all six hydrogen atoms and both the nitrogen atoms are accounted for in the two molecules of product (ammonia).

## 5.5   The pH Scale

Almost all chemical reactions that support life occur in water. Some of the most important are those that involve two classes of compounds: *acids* and *bases*. An **acid** is a hydrophilic compound that releases hydrogen ions ($H^+$) when it dissolves in water. The following equation depicts how an acid, represented by AH, releases hydrogen ions in an aqueous solution:

$$AH \rightleftharpoons A^- + H^+$$

**Bases** are hydrophilic compounds as well, but unlike acids they *accept* hydrogen ions from aqueous surroundings. ("Alkali" is an older, less precise, term that is sometimes used interchangeably with "base.") The equation below depicts how a base, represented by the letter B, absorbs $H^+$ ions in an aqueous solution:

$$B + H_2O \rightleftharpoons BH^+ + OH^-$$

Because a base removes $H^+$ ions from solution, it has the overall effect of *reducing* the concentration of free $H^+$ ions in an aqueous solution (and increasing the concentration of $OH^-$ ions). In contrast, acids donate $H^+$ ions to water and therefore *increase* the concentration of free $H^+$ ions in an aqueous solution.

Hydrogen ion concentration is commonly expressed on a scale from 0 to 14, where 0 represents an extremely high concentration of free hydrogen ions and 14 represents the lowest concentration. On this scale, called the **pH scale**, each number represents a 10-fold increase or decrease in the concentration of hydrogen ions (**FIGURE 5.11**). In pure water, the concentrations of free hydrogen ions and hydroxide ions are equal, and the pH is said to be neutral, or in the middle of the scale, at pH 7. The addition of acids to pure water raises the concentration of free hydrogen ions, making the solution more acidic and pushing the pH below the neutral value of 7. Adding a base lowers the concentration of free hydrogen ions in the solution, making the resulting solution more basic and raising the pH above 7.

We have all encountered acidic and basic substances. Mildly acidic solutions have a tangy taste, and moderately basic solutions often feel soapy. Our stomach juices are able to break down food because they are very acidic (about pH 2). At this low pH, noncovalent bonds within and between molecules are disrupted by the high concentration of free hydrogen ions, and even the covalent bonds of some biomolecules are broken. At the other extreme, a very basic substance, such as oven cleaner (pH 13.5), can also break or disrupt biomolecules. This is why extremes of pH can be caustic, causing chemical burns on the skin.

Most living systems function best at an internal pH close to neutral. Any change in pH to a value significantly below or above 7 adversely affects many biological processes. Because hydrogen ions can move so freely from one molecule to another during normal life processes, organisms must have ways of preventing dramatic changes in the pH levels of their internal environments. Substances called **buffers** meet this need by maintaining the concentration of hydrogen ions within narrow limits. They do so by releasing hydrogen ions when the surroundings become too basic (excessive $OH^-$ ions, high pH) and accepting hydrogen ions when the surroundings become too acidic (excessive $H^+$ ions, low pH).

Our blood pH is maintained between pH 7.34 and 7.45 in part by a buffer system made up of carbonic acid ($H_2CO_3$) and bicarbonate ions ($HCO_3^-$). This buffer system will resist a change in pH caused by the addition of small amounts of acid or base. However, the effectiveness of a buffer is overcome if *large* amounts of an acid or base are added. The blood buffer system can be overwhelmed, for example, by an excess of hydrogen ions (acidosis) or an excess of hydroxide ions (alkalosis). Our lungs and kidneys assist in regulating the blood concentration of hydrogen ions, carbonic acid, and bicarbonate, which is why acidosis or alkalosis can develop when the function of these organs is impaired.

> ### Concept Check

1. New York City and Pittsburgh are at the same latitude (about 40° north), but Pittsburgh is farther inland. Which one has a more moderate climate, with less difference between average midwinter and midsummer temperatures? Why?

2. Why don't oil and water mix?

3. The following reaction describes the burning of methane ($CH_4$), or "swamp gas": $CH_4 + 2\ O_2 \rightarrow CO_2 + 2\ H_2O$. (a) How many molecules of molecular oxygen ($O_2$) are needed to produce one molecule of carbon dioxide ($CO_2$) through this reaction? (b) Write the structural formula for carbon dioxide, given that the atomic numbers for carbon and oxygen are 6 and 8, respectively.

4. Which has a higher concentration of free hydrogen ions: vinegar, pH 2.8; or coffee, pH 5.5?

---

## 5.6 The Chemical Building Blocks of Life

If all the water in any living organism were removed, four major classes of molecules would remain, all of them critical for living cells: carbohydrates, nucleic acids, proteins, and lipids (informally known as "fats" and "oils"). Each of these biologically important molecules is built on a framework of covalently linked carbon atoms associated with hydrogen. Oxygen, nitrogen, phosphorus, and sulfur atoms are also found in some of these molecules.

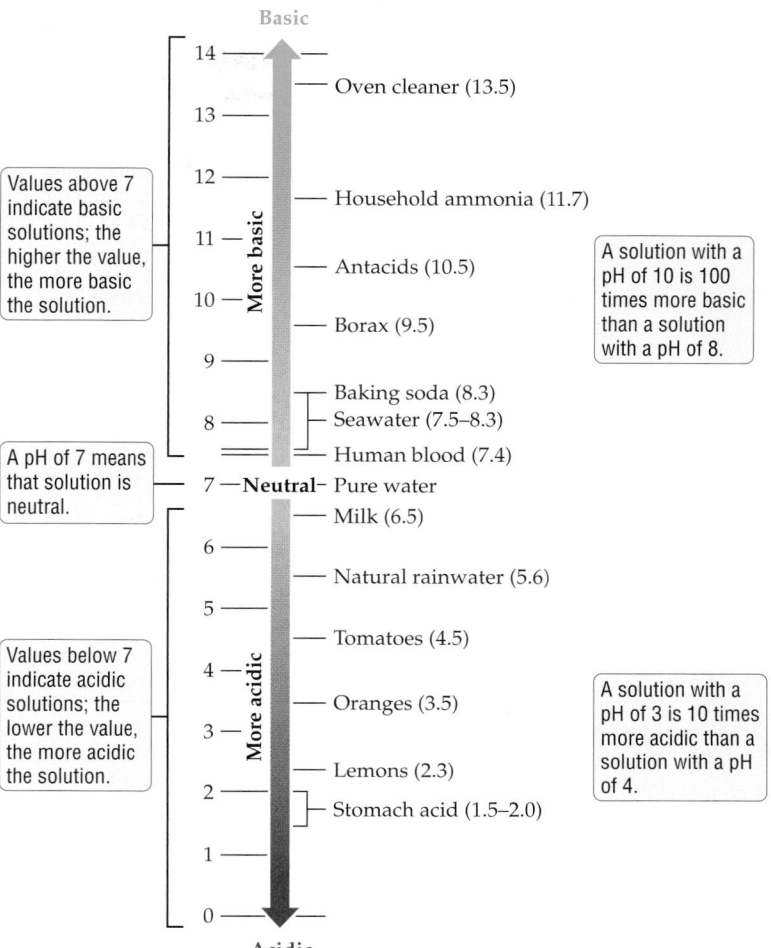

**THE pH SCALE**

Values above 7 indicate basic solutions; the higher the value, the more basic the solution.

A solution with a pH of 10 is 100 times more basic than a solution with a pH of 8.

A pH of 7 means that solution is neutral.

Values below 7 indicate acidic solutions; the lower the value, the more acidic the solution.

A solution with a pH of 3 is 10 times more acidic than a solution with a pH of 4.

14 — Oven cleaner (13.5)
13
12 — Household ammonia (11.7)
11 — Antacids (10.5)
10 — Borax (9.5)
9
— Baking soda (8.3)
8 — Seawater (7.5–8.3)
— Human blood (7.4)
7 — **Neutral** — Pure water
— Milk (6.5)
6
— Natural rainwater (5.6)
5
— Tomatoes (4.5)
4 — Oranges (3.5)
3
— Lemons (2.3)
2 — Stomach acid (1.5–2.0)
1
0

Basic / More basic / More acidic / Acidic

**FIGURE 5.11 The pH Scale Indicates Hydrogen Ion Concentration, a Measure of Acidity versus Basicity**

Carbon is the predominant element in living systems partly because it can form large molecules that contain thousands of atoms. A single carbon atom can form strong covalent bonds with up to four other atoms. Even more important, carbon can bond to carbon, forming long chains, branched molecules, or even rings (Table 5.1). The diversity of biological processes depends on the wide variety of small and large molecular structures that can be built from a carbon-carbon framework and a small handful of other types of atoms. No other element is as versatile as carbon in the sheer diversity of complex molecules that can be assembled from it.

Biomolecules that include at least one carbon-hydrogen bond are referred to as **organic molecules**. The cell contains many types of small organic molecules (where "small" is defined as containing up to 20 atoms or so). Sugars and amino acids are

| TABLE 5.1 | The Versatility of Carbon Atoms | | |
|---|---|---|---|
| **NAME** | **STRUCTURAL FORMULA** | **NOTES** | |

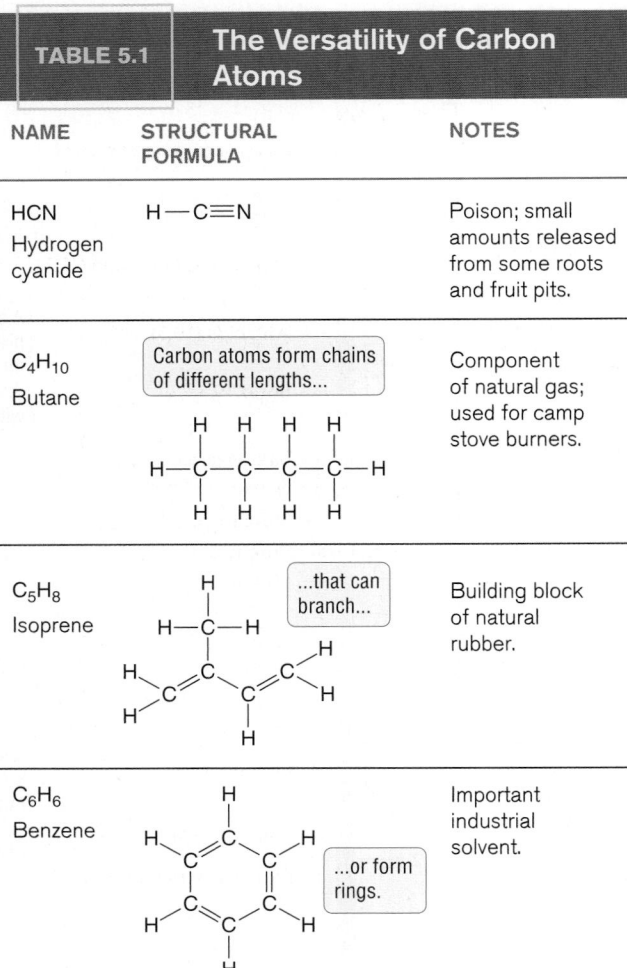

| NAME | STRUCTURAL FORMULA | NOTES |
|---|---|---|
| HCN Hydrogen cyanide | H—C≡N | Poison; small amounts released from some roots and fruit pits. |
| $C_4H_{10}$ Butane | Carbon atoms form chains of different lengths... | Component of natural gas; used for camp stove burners. |
| $C_5H_8$ Isoprene | ...that can branch... | Building block of natural rubber. |
| $C_6H_6$ Benzene | ...or form rings. | Important industrial solvent. |

## BUILDING POLYMERS FROM MONOMERS

**Monomers**

If five different monomers are available...

**Polymers**

Covalent bonds

...they can be linked to make 15,625 *different* kinds of polymers that are each six units long.

**FIGURE 5.12 Polymers Are Long Chains Made from Repeating Units Known as Monomers**
A handful of monomers can be assembled into a great variety of polymers. The formula is: number of available monomers$^n$, where $n$ is the chain length (total number of monomers in each polymer). In this example, we have $5^6$, which is 15,625.

examples of small organic molecules. Small organic molecules can link up via covalent bonds to create larger assemblies of atoms, called **macromolecules** (*macro*, "large"). Starch and proteins are examples of macromolecules. Small molecules that serve as repeating units in a macromolecule are called **monomers** (*mono*, "one"; *mer*, "part"). Many macromolecules in a cell are made up of hundreds of monomers covalently bonded to one another. Macromolecules that contain monomers as building blocks are called **polymers** (*poly*, "many") (**FIGURE 5.12**). Polymers account for most of an organism's dry weight (its weight after all water is removed) and are essential for every structure and chemical process that we associate with life.

In living organisms, fewer than 70 different biological monomers combine in an endless variety of ways to produce polymers with many different properties. Polymers are therefore a step up from monomers in organizational complexity, and they

have chemical properties that are not manifested in a monomer. Furthermore, the properties of organic polymers depend on the properties of attached clusters of atoms, called *functional groups* (Table 5.2). As the name implies, **functional groups** are clusters of covalently bonded atoms that have the same distinctive chemical properties no matter what molecule they are in. Some functional groups help establish covalent linkages between monomers; others have more general effects on the chemical characteristics of a polymer. The properties of each of the four general classes of small organic molecules critical for life in every organism—sugars, nucleotides, amino acids, and fatty acids—are strongly influenced by the specific functional groups within them.

## 5.7 Carbohydrates

*Sugars* are familiar as compounds that make foods taste sweet. Although not all sugars are perceived as sweet by human taste buds, most sugars are important food sources and also serve as a means of storing energy. **Sugars** and their polymers are referred to as

| TABLE 5.2 | Important Functional Groups | | |
|---|---|---|---|
| **FUNCTIONAL GROUP** | **STRUCTURAL FORMULA** | **BALL-AND-STICK MODEL** | |
| Amino group | —NH₂<br><br>—N⟨H / H | Bond to carbon atom | |
| Carboxyl group | —COOH<br><br>—C(=O)OH | | |
| Hydroxyl group | —OH | | |
| Phosphate group | —PO₄<br><br>—O—P(=O)(O⁻)O⁻ | | |

**carbohydrates**. The name comes from the ratio of C, H, and O in the compounds, in which for each carbon atom (*carbo*) there are two hydrogens and one oxygen (corresponding to a molecule of water, a *hydrate*).

The simplest sugar molecules are called **monosaccharides** [*MAH*-noh-*SAK*-uh-ridez] (*mono*, "one"; *sacchar*, "sugar"). Like most carbohydrates, monosaccharides are made of units containing carbon, hydrogen, and oxygen atoms in the ratio of 1:2:1—that is, one carbon atom to two hydrogen atoms to one oxygen atom. This ratio can also be expressed as the molecular formula $(CH_2O)_n$, with *n* ranging from 3 to 7. This means that the many different types of monosaccharides found in nature have anywhere from three to seven carbon atoms.

Monosaccharides are often referred to by the number of carbon atoms they contain; for example, a sugar with the molecular formula $(CH_2O)_5$ is known as a five-carbon sugar. Note that the parentheses work the same way in this notation as in multiplication. A more common way to express the molecular formula for this sugar is $C_5H_{10}O_5$. When monosaccharides with five or more carbon atoms are dissolved in water, the sugar molecules may exist in chain form or ring form. Here are the chain and ring forms of a five-carbon sugar called ribose:

**Ribose: chain form**  **Ribose: ring form**

The one monosaccharide that is found in almost all cells is **glucose** ($C_6H_{12}O_6$). Glucose has a key role as an energy source within the cell, and nearly all the chemical reactions that produce energy for living organisms involve the manufacture or breakdown of this sugar. Fructose, fruit sugar, has the same molecular formula as glucose ($C_6H_{12}O_6$), but the atoms are connected in a different pattern, resulting in very different physical and chemical properties. Fructose is nearly twice as sweet as glucose, and it is widely used as a sweetener in processed foods because corn is a cheap source of the sugar.

Scientists often use a shorthand notation to represent the ring form of carbon-containing molecules: the symbols for the carbon atoms, and most of the hydrogen and oxygen atoms, are left out and their presence at the appropriate positions is simply implied. Here are the structural formula and shorthand notation for glucose:

**Glucose**

Monosaccharides can combine to form larger, more complex molecules. Two covalently joined monosaccharides form a **disaccharide** (*di*, "two"). Our familiar table sugar, sucrose, is a disaccharide built by linking a molecule of glucose and a molecule of fructose, with the removal of a water molecule (**FIGURE 5.13**). Chemical reactions in which a water molecule is *released* as two monomers become covalently linked are known as dehydration reactions. In the reverse reaction, called a *hydrolytic reaction*, a water molecule is *added* to break the covalent bond linking two monomers. Sucrose is broken down in our intestines through a hydrolytic reaction, and the released monomers are absorbed by the intestinal wall and eventually delivered to the bloodstream.

  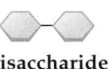

**Monosaccharide**
(e.g., glucose, fructose)

**Disaccharide**
(e.g., lactose, sucrose)

**Polysaccharide**
(e.g., glycogen, starch, cellulose)

The prefixes *macro* ("large") and *micro* ("small")—as in "macromolecules" and "microorganisms"—are common in scientific terminology. Often they show up in pairs of opposing words; for example, "macroscopic" and "microscopic."

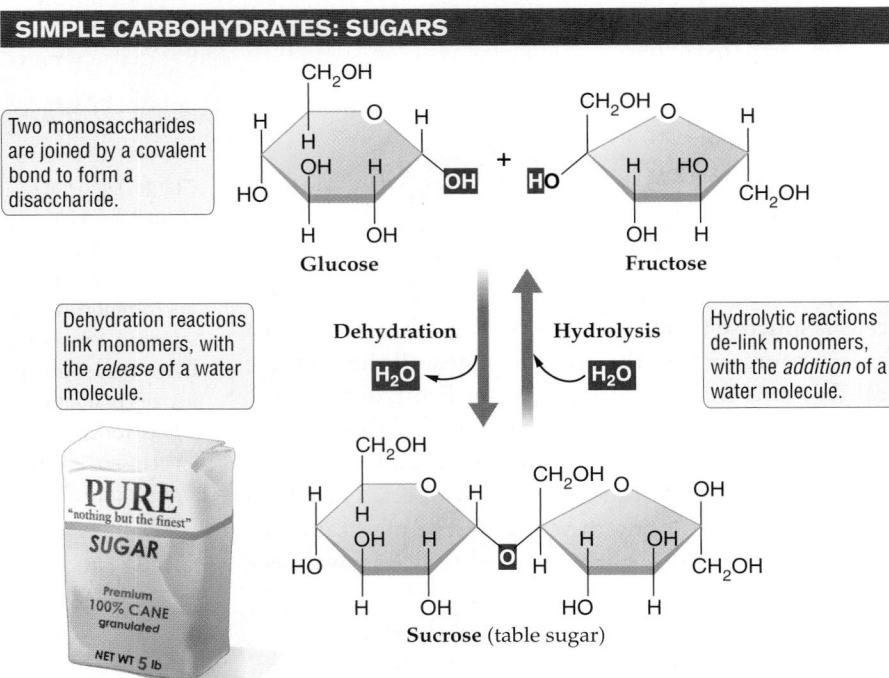

Two monosaccharides are joined by a covalent bond to form a disaccharide.

Dehydration reactions link monomers, with the *release* of a water molecule.

Hydrolytic reactions de-link monomers, with the *addition* of a water molecule.

**Dehydration**    **Hydrolysis**

Glucose    Fructose

Sucrose (table sugar)

**FIGURE 5.13 Monosaccharides Can Bond Together to Form Disaccharides**
Glucose and fructose are sugar monomers that, when linked by a covalent bond, form the disaccharide sucrose, or table sugar.

**Polysaccharides** are large polymers built by linking many monosaccharides. Like most polymers, they are put together via dehydration reactions and taken apart by hydrolytic reactions. Polysaccharides perform a variety of functions in living organisms (**FIGURE 5.14**). **Cellulose**, for example, is a polysaccharide that is bundled into strong parallel fibers that help support the plant body (**FIGURE 5.14a**). Cotton fabric, made from special cells on the surface of cotton seeds, is mostly cellulose. Carbohydrates are polysaccharides that provide metabolic energy, as we have already seen in the case of glucose. Starch—abundant in a dish of mashed potatoes or steamed rice—is a polysaccharide that serves as an energy store inside plant cells (**FIGURE 5.14b**).

Cellulose and starch are both built from glucose, but they differ in how the monosaccharides are linked. Take a closer look at the linkages (known as *glycosidic bonds*) between the monomers in Figure 5.14, especially the orientation of the linking oxygen atom, O. Starch is water-soluble and easily broken down in our digestive system to release glucose for our energy needs. Cellulose, on the other hand, is not water-soluble, which is fortunate for owners of 100 percent cotton clothes who would literally lose their shirts in the wash otherwise. Unlike starch, cellulose cannot be digested in the human gut, and only

some bacteria and fungi make the biomolecules needed to break it down. Some herbivores, known as ruminants, have special stomachs in which multitudes of microorganisms thrive by breaking down the nutrients in the cellulose that their grass-eating hosts consume.

Glycogen is the main storage polysaccharide in animal cells (**FIGURE 5.14c**), although, as we shall see later, most of the surplus energy ingested by animals is stockpiled in the form of storage lipids ("fat") rather than carbohydrate. The majority of the glycogen reserve in our bodies is stored inside liver cells and skeletal muscle cells. About 14 hours without food, or a couple of hours of intense exercise, will use up all the glycogen stored in the typical adult. The fatigue that follows glycogen depletion is called "hitting the wall" in the world of long-distance runners and bikers.

As Figure 5.14c shows, glycogen is a polymer of glucose and very similar to starch in its chemical structure, but with many more side branches. Highly branched polysaccharides yield their energy more rapidly than do unbranched ones. Storage polysaccharides are dismantled from one end of each sugar chain, and because each branched polysaccharide molecule has many more of these free ends, it is whittled down all at once from many different ends. It therefore releases more energy in a shorter time span than does the same mass of an unbranched polymer.

## 5.8 Proteins

Of the many different kinds of chemical compounds found in living organisms, **proteins** are among the most familiar, since we often hear and read about proteins in connection with diet and nutrition. These polymers make up more than half the dry weight of animal cells. We can categorize proteins by some of the main functions they perform:

- *Storage.* Bird eggs and plant seeds contain storage proteins, whose function is to supply the building blocks needed for growth, in this case the growth of the emergent chicks or seedlings.

- *Structure.* Our own bodies contain thousands of different types of proteins with diverse functions. Some of these proteins form physical structures and are therefore classified as structural proteins. Collagen is a strong but flexible structural protein that sheaths skin cells and is also found in bone and cartilage. Human hair, and the fur

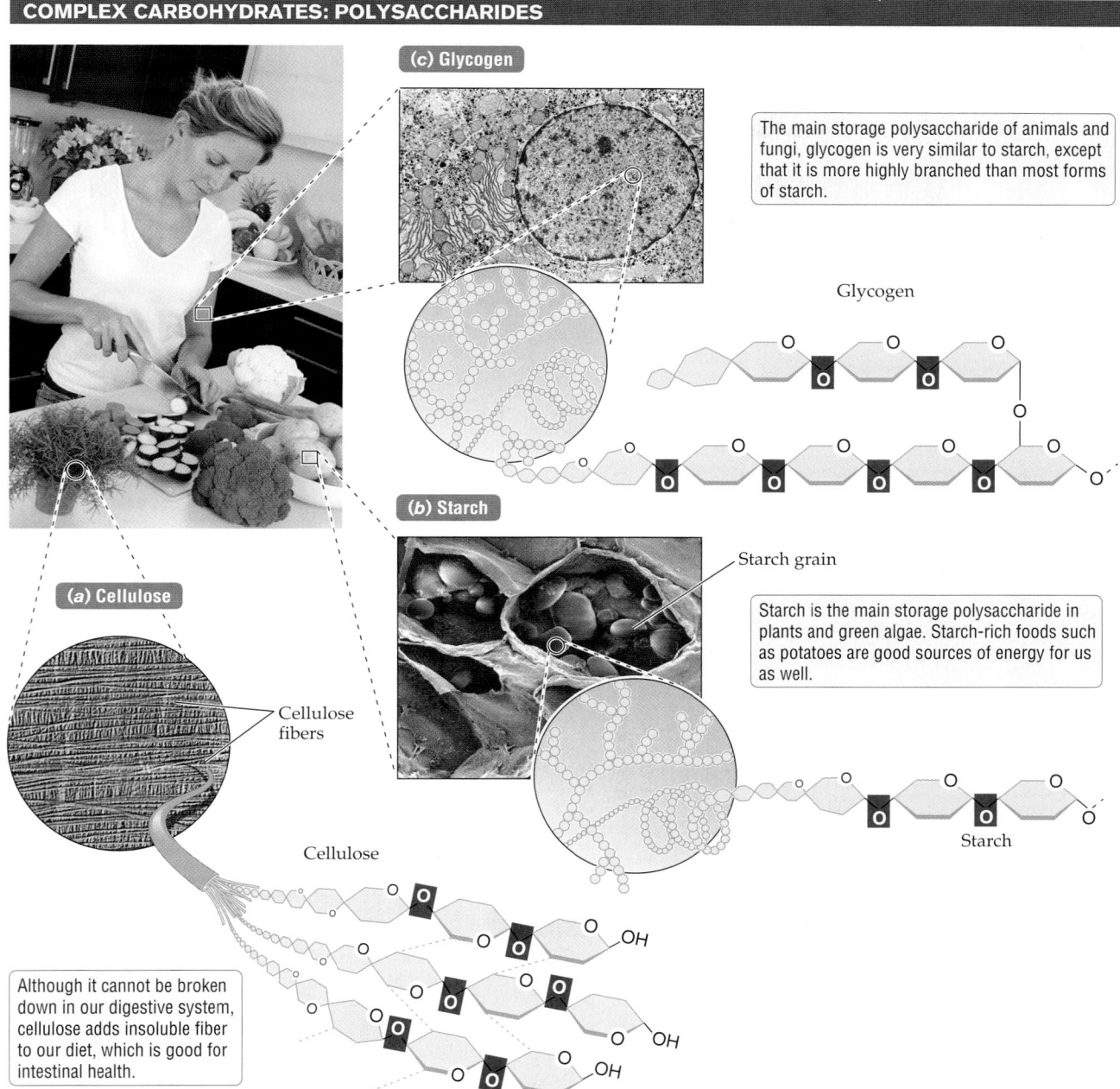

**FIGURE 5.14**

**Monosaccharides Can Bond Together to Form Polysaccharides**

Cellulose, starch, and glycogen are all polymers built from glucose subunits.

**(c) Glycogen**

The main storage polysaccharide of animals and fungi, glycogen is very similar to starch, except that it is more highly branched than most forms of starch.

Glycogen

**(b) Starch**

Starch grain

Starch is the main storage polysaccharide in plants and green algae. Starch-rich foods such as potatoes are good sources of energy for us as well.

Starch

**(a) Cellulose**

Cellulose fibers

Cellulose

Although it cannot be broken down in our digestive system, cellulose adds insoluble fiber to our diet, which is good for intestinal health.

of mammals, contains keratin, another family of structural proteins.

- *Transport.* Some proteins ferry substances such as nutrients within the body. Hemoglobin, packed inside disc-shaped red blood cells, is a protein that binds oxygen and helps move it throughout the body.

- *Catalysis.* Substances that speed up chemical reactions are called catalysts. Almost all chemical reactions in living organisms are catalyzed by proteins known as **enzymes**.

## Proteins are built from amino acids

**Amino acids** are the monomers from which proteins are built. Twenty different amino acid monomers can be arranged in a multitude of ways to construct a huge variety of proteins. The "anatomy" of these 20 amino acids is similar. As shown in **FIGURE 5.15a**, all amino acids consist of an "alpha" carbon attached to a hydrogen atom, a chemical side chain called the *R group*, and two functional groups: an *amino group*

**FIGURE 5.15**

## The Structure and Diversity of Amino Acids

(*a*) Amino acids are the building blocks of proteins. Twenty different amino acids can be found in proteins, each differing from the other only in the nature of its R group. (*b*) Each of the 20 amino acids has a different R group bonded to the alpha carbon. The R group is responsible for the distinctive properties of each amino acid.

**AMINO ACIDS**

**(*a*) The general structure of an amino acid monomer**

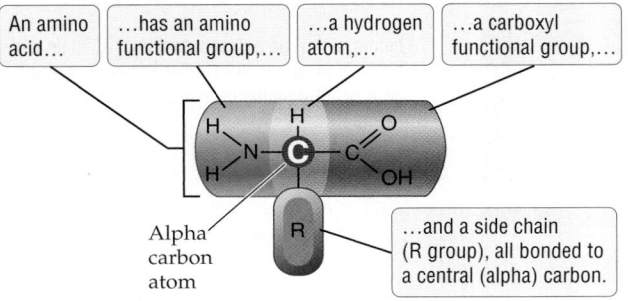

An amino acid…   …has an amino functional group,…   …a hydrogen atom,…   …a carboxyl functional group,…

Alpha carbon atom

…and a side chain (R group), all bonded to a central (alpha) carbon.

**(*b*) The 20 amino acids found in proteins**

**Hydrophobic amino acids** are nonpolar or uncharged and are repelled by water.

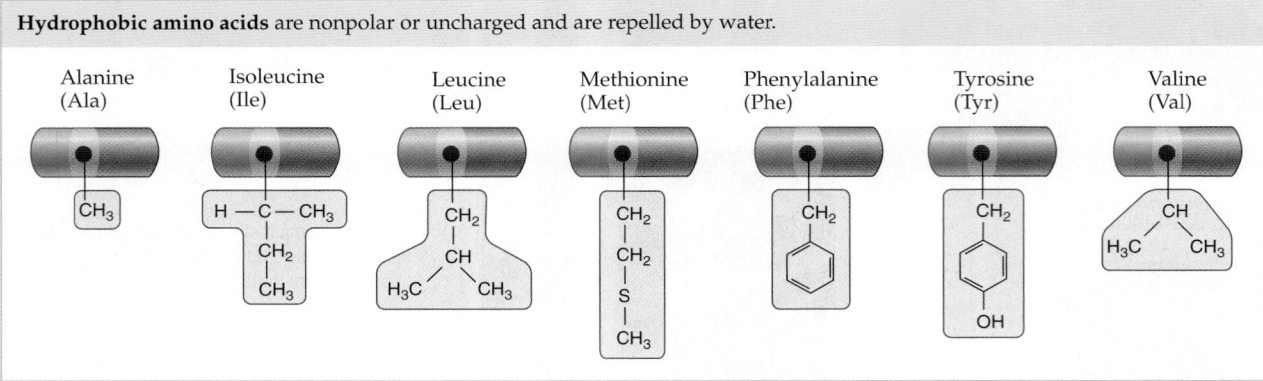

Alanine (Ala)   Isoleucine (Ile)   Leucine (Leu)   Methionine (Met)   Phenylalanine (Phe)   Tyrosine (Tyr)   Valine (Val)

**Hydrophilic amino acids** are either charged or polar and can interact with water molecules.

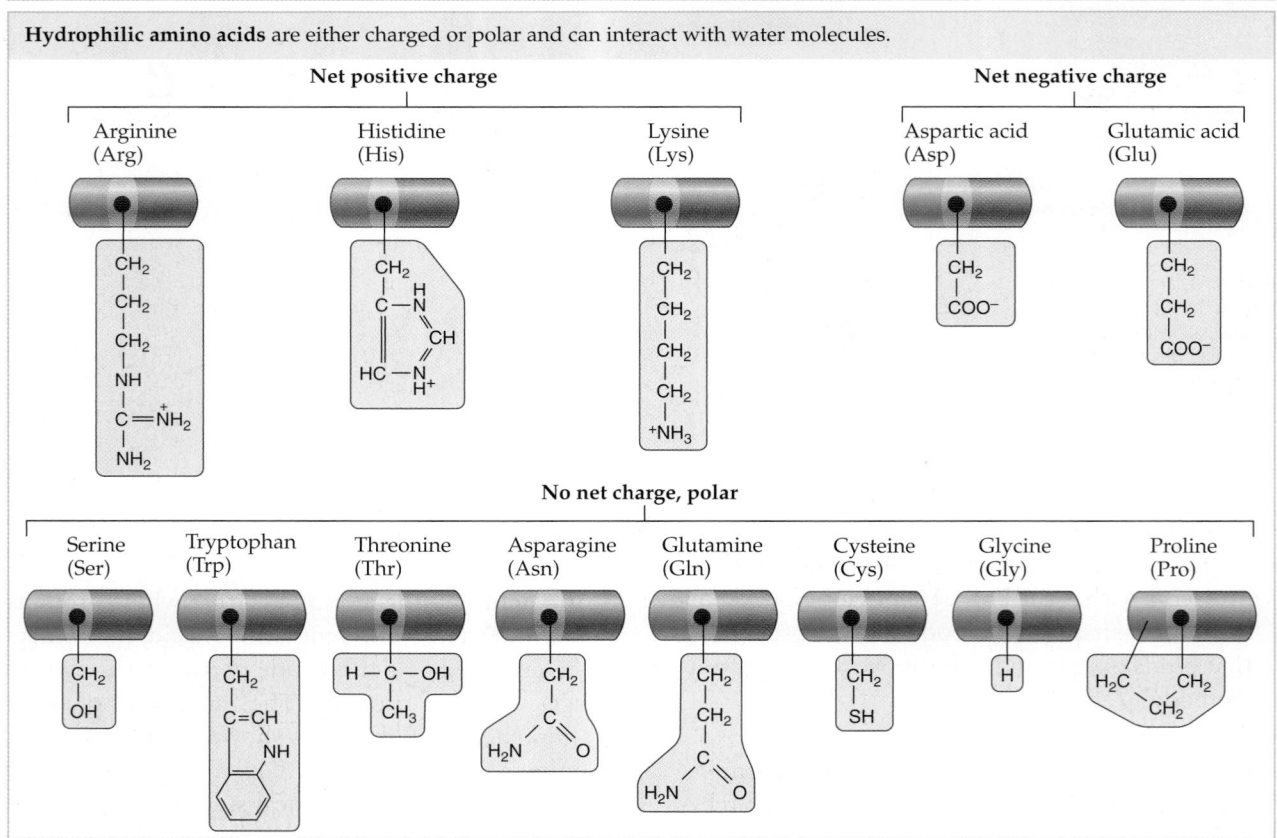

Net positive charge

Arginine (Arg)   Histidine (His)   Lysine (Lys)

Net negative charge

Aspartic acid (Asp)   Glutamic acid (Glu)

No net charge, polar

Serine (Ser)   Tryptophan (Trp)   Threonine (Thr)   Asparagine (Asn)   Glutamine (Gln)   Cysteine (Cys)   Glycine (Gly)   Proline (Pro)

(—NH$_2$) and a *carboxyl group* (—COOH). In this group of 20 amino acids, one amino acid differs from another only with respect to the type of R group present. Each of the 20 possible R groups is unique to one amino acid. R groups vary in terms of size, acidic or basic properties, and whether they are hydrophobic or hydrophilic. The R groups found in amino acids range from just one atom (in glycine), to complex arrangements of carbon chains (in arginine, for example) and ring structures (in tryptophan, for example), as depicted in **FIGURE 5.15b**.

Linear chains of amino acids are covalently linked to create a polymer known as a **polypeptide**. Every protein consists of one or more polypeptides. In a polypeptide chain, the amino group of one amino acid is covalently linked to the carboxyl group of another via a covalent linkage called a **peptide bond** (**FIGURE 5.16**). A polypeptide may contain hundreds to thousands of amino acids held together by peptide bonds. These polypeptides are built from the same pool of 20 possible amino acids, so the crucial difference between one polypeptide and another is the *sequence* in which the amino acids are linked. Two polypeptides may differ in the amounts of various amino acids found in the chain; for instance, lysine may be absent in one polypeptide but extremely abundant in another. The thousands of different polypeptides found in the average cell also vary enormously in overall length—that is, in the total number of amino acids in each.

How can just 20 amino acids generate the millions of different proteins found in nature? Well, consider this: the number of different sentences that can be written using the 26 letters of the English alphabet seems to have no bounds. If we think of the protein alphabet as having 20 amino acid letters (only six fewer than in the English alphabet), we can see that an enormous number of different protein sentences is possible. The complexity and diversity of life is fundamentally dependent on this variety in protein structure and function.

## A protein must be correctly folded to be functional

The sequence of amino acids in a polypeptide is known as the **primary structure** (**FIGURE 5.17a**). A polypeptide must acquire a higher level of organization, beyond its primary structure, before it can function as a protein or part of a protein. In that sense, the primary structure is to a protein what a sheet of paper is to an origami crane. The secondary structure is the next level in the organization of a protein. The **secondary structure** of a protein is created by the *regional* folding of the amino acid chain into specific three-dimensional patterns (**FIGURE 5.17b**). Alpha (α) helices and beta (β) sheets are two of the most common types of secondary structure. An *alpha helix* is a spiral pattern, like curled ribbon on a gift package. The spirals in an alpha helix are created and maintained by hydrogen bonds. A *beta sheet* is created when the polypeptide backbone is bent into "ridges" and "valleys," as in a paper fan.

In addition to their secondary structure, most polypeptides must have another level of folding, to create a tertiary structure, before they can function as a protein (**FIGURE 5.17c**). The **tertiary structure** of a polypeptide is a very specific three-dimensional shape that is attained not merely through local patterns of folding, as in the secondary structure, but through interactions between *distantly placed* segments of the polypeptide chain. Tertiary structure is stabilized by noncovalent associations such as ionic bonds, and often also by covalent links between distant amino acids (for example, through a *disulfide bond* between the sulfur atoms of distantly spaced cysteines).

Some proteins are composed of more than one polypeptide, in which case they have yet another level of organization, called the **quaternary structure** [KWAH-ter-nair-ee], that must be achieved before the protein can become biologically active.

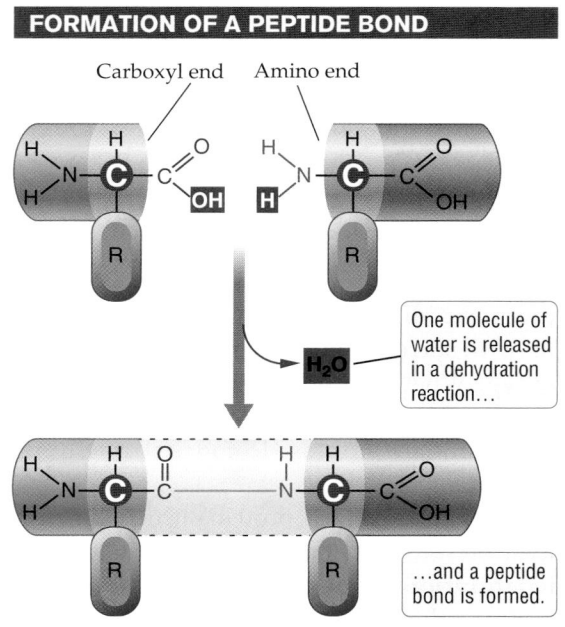

**FORMATION OF A PEPTIDE BOND**

Carboxyl end    Amino end

One molecule of water is released in a dehydration reaction…

…and a peptide bond is formed.

**FIGURE 5.16  Peptide Bonds Link Amino Acids**

**FIGURE 5.17**

**The Four Levels of Protein Structure**

All proteins have a primary structure composed of a linear chain of amino acids known as a polypeptide (*a*). Regional coiling and folding of the polypeptide chain gives a protein its secondary structure (*b*). Further folding, stabilized by long-distance interactions, gives a protein its tertiary structure (*c*). Proteins composed of more than one polypeptide have a quaternary structure stabilized by bonds between the chains (*d*).

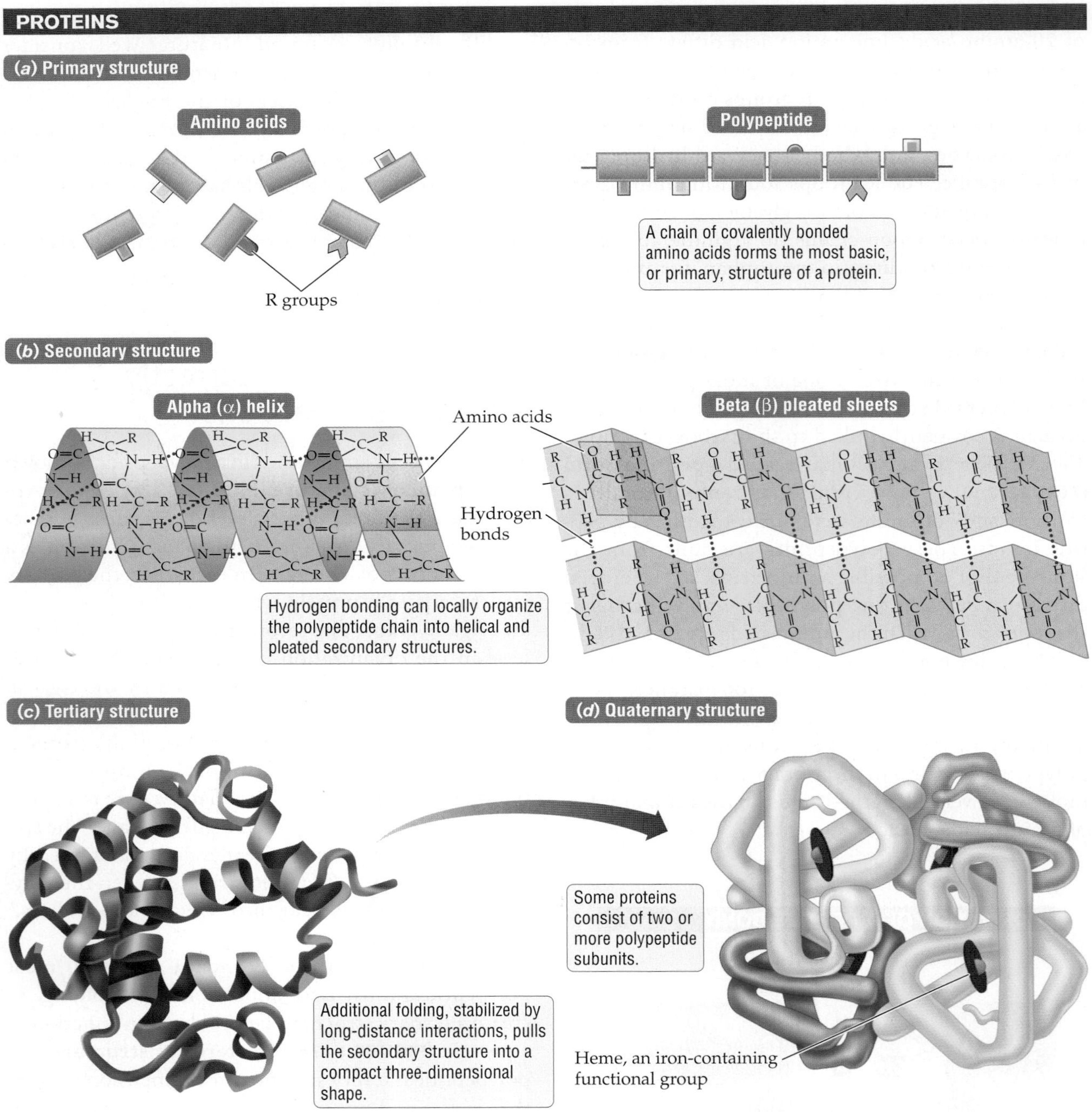

**(a) Primary structure**

Amino acids

R groups

Polypeptide

A chain of covalently bonded amino acids forms the most basic, or primary, structure of a protein.

**(b) Secondary structure**

Alpha (α) helix

Amino acids

Hydrogen bonds

Beta (β) pleated sheets

Hydrogen bonding can locally organize the polypeptide chain into helical and pleated secondary structures.

**(c) Tertiary structure**

Additional folding, stabilized by long-distance interactions, pulls the secondary structure into a compact three-dimensional shape.

**(d) Quaternary structure**

Some proteins consist of two or more polypeptide subunits.

Heme, an iron-containing functional group

Protein denaturation

Hemoglobin, which transports oxygen in our blood, is an example of a protein with quaternary structure (**FIGURE 5.17***d*).

The activity of most proteins is critically dependent on their three-dimensional structure. Extreme temperature, pH, and salt concentration can change or destroy the structure of a protein, and consequently change or destroy its activity. How? Most properly folded proteins tend to have hydrophilic R groups exposed on the surface and hydrophobic R groups buried deep inside the folded structure. This precise arrangement allows the protein to form hydrogen bonds with surrounding water molecules and remain dissolved. When a protein is heated beyond a certain temperature, its weak noncovalent bonds break and the protein unfolds, losing its orderly three-dimensional shape.

The destruction of a protein's three-dimensional structure, resulting in loss of protein activity, is known as **denaturation**. Some proteins are more easily denatured than others. Conditions that can denature proteins include low pH, high pH, high salt concentration, or high temperature. We witness the denaturation of egg proteins when we cook eggs. At room temperature, albumin, the predominant protein in eggs, is dissolved in the aqueous egg white. Albumin is denatured as the egg is cooked, causing the polypeptides to clump together into the solid mass we recognize as the whites in a fried egg. Most denatured proteins cannot regain their original three-dimensional structure, which is why you cannot "unscramble" an egg!

## 5.9 Lipids

**Lipids** are hydrophobic molecules made by living cells. They are built from chains or rings of *hydrocarbon*, which consist of carbon and hydrogen atoms. Fatty acids, glycerides, sterols, and waxes are examples of lipids that we will discuss in greater detail shortly. Most lipids are built from one or more **fatty acids**. A fatty acid has a long hydrocarbon chain that is strongly hydrophobic. At the other end of the hydrocarbon chain is a carboxyl group, a functional group that is polar and therefore hydrophilic.

The hydrocarbon chains found in fatty acids contain many carbon atoms—from 16 to 22 in the types common in our foods—that are linked in various ways. Fatty acids in which *all* the carbon atoms in the hydrocarbon chain are linked by *single* covalent bonds are called **saturated fatty acids** because each carbon in them is bonded to the maximum number of hydrogen atoms (**FIGURE 5.18a**). In an **unsaturated fatty acid** one or more of these carbon atoms are linked by double bonds; because some of the carbon atoms are not bonded to a full complement of hydrogen atoms, the hydrocarbon chain is unsaturated (**FIGURE 5.18b**).

The significance of the double bonds in unsaturated fatty acids goes beyond a mere difference in the number of hydrogen atoms in the hydrocarbon chain. Hydrocarbon chains linked exclusively by single bonds tend to be straight, but the presence of double bonds can introduce kinks into the hydrocarbon chain. The consequence of these differences in shape is that the straight-chain saturated fatty acids can pack together very tightly, forming solids or semisolids at room temperature. Unsaturated fatty acids with kinks cannot pack tightly, so these lipids tend to be liquid at room temperature.

## FATTY ACIDS

**(a) A saturated fatty acid**

It contains no double bonds in its fatty acid chain.

**(b) An unsaturated fatty acid**

It has one or more double bonds in its fatty acid chain.

**Stearic acid** (straight chain)

**Oleic acid** (bent chain)

Unsaturated lipids, such as those in olive oil, are liquid at room temperature because bends in the fatty acid chains prevent them from being closely packed together.

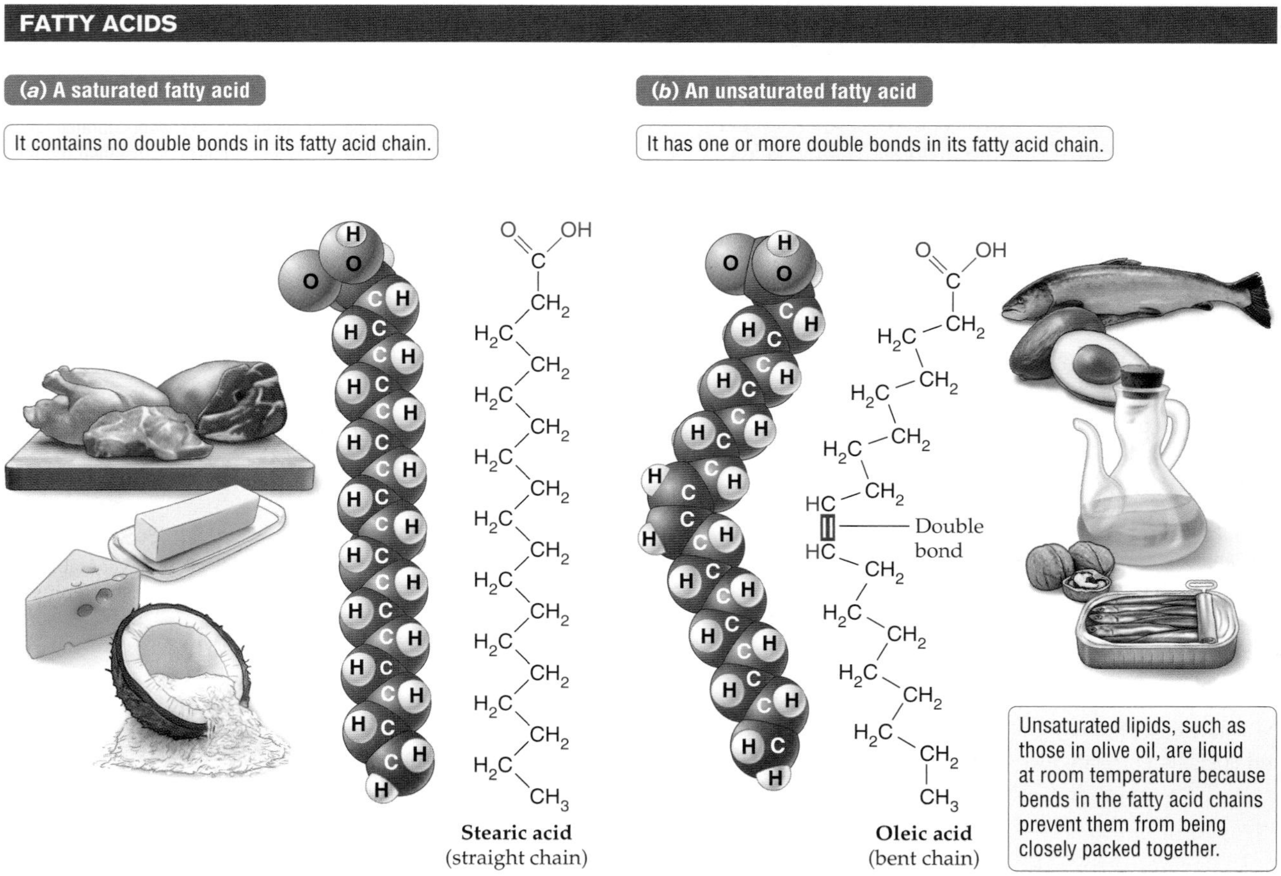

**FIGURE 5.18**
**Saturated and Unsaturated Fatty Acids Are the Two Main Types of Fatty Acids**

The space-filling models of (a) stearic acid and (b) oleic acid show that a saturated fatty acid is a straight molecule, whereas an unsaturated fatty acid molecule has a bend in it. Saturated fatty acids can pack tightly to form a solid at room temperature, whereas unsaturated lipids cannot.

# Dietery Lipids: The Good, the Bad, and the Truly Ugly

Our bodies can make nearly all the lipids we need from the organic molecules we consume as food. However, a moderate intake of lipids, especially certain types of lipids, is an important part of a healthy diet. Nutritionists recommend that we consume moderate amounts of unsaturated fatty acids, such as those found in olive oil and canola oil. One class of unsaturated fatty acids, known as *omega-3 fatty acids*, is especially renowned for its health benefits. Flaxseed and walnuts, certain algae, and cold-water fish are good sources of omega-3 fatty acids. There is ample evidence to show that EPA (eicosapentaenoic acid) and DHA (docosahexaenoic acid), the main omega-3 fatty acids in fish oil, have anti-inflammatory effects in the human body and protect against heart disease.

How saturated lipids affect human health is a complex and contentious subject. Harmful effects are seen in laboratory rats fed a diet rich in saturated lipids; the effects include increased risk of heart disease and some types of cancer. Studies of human populations, however, present a confusing picture, in part because human subjects are difficult to study and the experimental approaches are often less than ideal because of limits imposed by ethical and cost considerations. Some populations (such as Amish farmers and South Pacific island communities) have low rates of heart disease despite a very high intake of saturated lipids. It is possible that these populations are protected from potential negative effects because their total calorie consumption is lower or because they exercise more.

A recent metanalysis (an analysis that pools results from many different studies) suggests that the *ratio* of saturated to unsaturated lipids in our diet has a significant influence on disease risk. According to the metanalysis, consuming large amounts of saturated lipids, and relatively low amounts of unsaturated lipids, is correlated with greater risk of heart disease. Despite the uncertainties, most nutrition experts say no more than 7% of our total calories should come from saturated fat.

Better understood are the harmful effects of trans unsaturated lipids, known as trans fats. Trans fats contain unsaturated fatty acids whose hydrophobic tails are relatively straight compared to the bent shape of the more common (cis) type of unsaturated fatty acids. Because their straighter chains are readily compacted, trans fats are semisolid at room temperature.

The overwhelming majority of trans fat in the American diet comes from partial *hydrogenation* of vegetable oils. The goal of this industrial process is to convert the unsaturated lipids in liquid vegetable oils to semisolid products such as margarine. As the oil is treated with hydrogen gas ($H_2$), some of the cis fatty acid molecules are turned into saturated fatty acids through complete hydrogenation (and therefore loss) of the carbon-carbon double bonds. In some of the cis fatty acid molecules, however, the double bonds remain but swivel in a way that straightens the fatty acid tail; these straight-chain unsaturated fatty acids are trans fats.

Trans fats have been popular in the processed-food industry because they are cheaper than the alternatives and less prone to becoming rancid. Foods prepared with trans fats last well on the shelf and do not need expensive refrigeration. Cis unsaturated fatty acids, in contrast, are very susceptible to attack by oxygen.

By the 1990s, alarm bells were being sounded about the health risks associated with the consumption of trans fats. Many studies have confirmed that eating a significant amount of trans fat increases one's risk of heart disease. Since January 2007, all Nutrition Facts labels on packaged foods must disclose the amount of trans fat in each serving. Health experts say that there is no safe dose for artificial trans fats, and that consumers should minimize their intake and preferably eliminate these lipids from their diet.

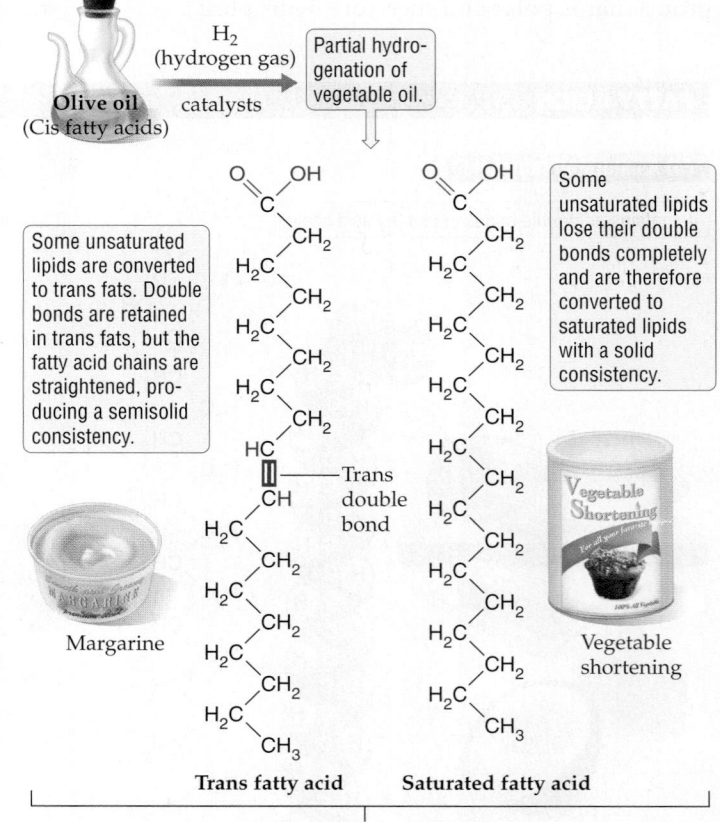

Olive oil (Cis fatty acids)

$H_2$ (hydrogen gas)
catalysts

Partial hydrogenation of vegetable oil.

Some unsaturated lipids are converted to trans fats. Double bonds are retained in trans fats, but the fatty acid chains are straightened, producing a semisolid consistency.

Some unsaturated lipids lose their double bonds completely and are therefore converted to saturated lipids with a solid consistency.

Trans double bond

Margarine

Trans fatty acid

Vegetable shortening

Saturated fatty acid

Straight chain molecules

## Animals store surplus energy as triglycerides

Familiar foods such as butter and olive oil are actually complex mixtures of different types of lipids, with small amounts of other substances, such as milk protein in butter and vitamin E in most vegetable oils. The lipids in butter and olive oil include fatty acids of different types and a class of lipids called glycerides. A glyceride contains one to three fatty acids covalently bonded to a three-carbon molecule called glycerol. **Triglycerides**, which consist of three fatty acids bonded to a glycerol (**FIGURE 5.19**), are the most common glyceride in our diet. Triglycerides built largely from saturated fatty acids tend to be solid at room temperature and are informally known as *fats*. Because butter and lard are rich in triglycerides containing saturated fatty acids, these fatty foods are solid at room temperature. Triglycerides rich in unsaturated fatty acids tend to be liquid at room temperature and are informally known as *oils*. Triglycerides containing unsaturated fatty acids are abundant in canola oil, olive oil, and flaxseed oil, which is why all of these lipids are liquid at room temperature. Some tropical "oils," such as those derived from coconut and palm kernels, are solid at room temperature because they have as much or more saturated lipids as butter and lard. These examples illustrate the inconsistency in how we use everyday words such as "fat" and "oil" and why special terminology, with a more precise meaning, is valuable in science.

A wide variety of organisms store surplus energy in the form of triglycerides, usually deposited in the cytoplasm as lipid droplets (Figure 5.19). Lipids are efficient as storage reserves because they pack slightly more than twice the energy found in an equal weight of carbohydrate or protein, while occupying only one-sixth the volume. Carbohydrates and proteins take up more space inside a cell because they are hydrophilic; these macromolecules are extensively associated with water molecules, and all these extra molecules add bulk.

## Phospholipids are important components of cell membranes

Glycerides of another group, the phospholipids, are major components of the **plasma membrane**, which is the outermost boundary of a cell. Most cells also have internal membranes, which make up the boundaries of internal compartments. Cell membranes are built mainly from **phospholipids**, molecules consisting of two fatty acids joined to a glycerol bearing a phosphate group. All phospholipids have a hydrophilic "head"

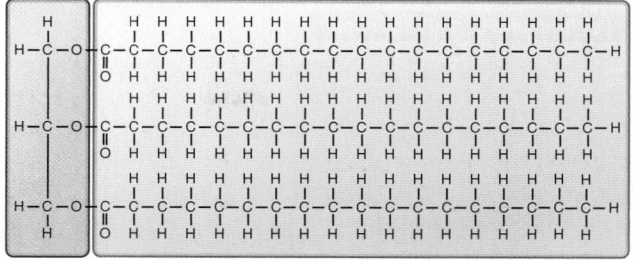

**(a) Triglyceride**

Glycerol is a three-carbon molecule.

The fatty acid chains in a triglyceride may be saturated or unsaturated.

**(b) Human fat cells**

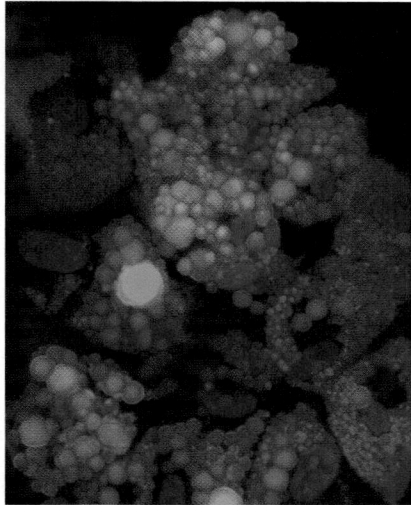

**FIGURE 5.19 Triglycerides Contain Three Fatty Acids Bound to a Glycerol**
(a) Glycerides consist of a three-carbon sugar alcohol called glycerol, bound to one, two, or three fatty acids. Triglycerides have three fatty acids, one linked to each of the three carbons of glycerol. The triglyceride depicted here is glyceryl tristearate, the most common storage lipid in animal cells. (b) Color-enhanced photograph of human adipocytes, which are cells specialized for storing triglycerides (green deposits). The blue structure is the nucleus.

containing a negatively charged phosphate group, and a hydrophobic "tail" consisting of two long fatty acid chains (**FIGURE 5.20a**). The head group also includes other functional groups (choline, in Figure 5.20a) that differ from one type of phospholipid to another.

Because of their dual character, phospholipids exposed to water spontaneously arrange themselves in a double-layer sheet known as a **phospholipid bilayer** (**FIGURE 5.20b**). The double layers are arranged so that the hydrophilic head groups of the phospholipids are exposed to the watery world on either side while the hydrophobic tails are tucked inward, away from the water. Nearly all cell membranes are organized as lipid bilayers. One phospholipid bilayer constitutes a single biological membrane, or unit membrane. Membranes establish the boundaries of living cells and of compartments within cells. They control the exchange of ions and molecules between the cells and their external environment, and also between various compartments within a cell.

FIGURE 5.20
**Membranes Contain Double Sheets of Phospholipids**

Phospholipids spontaneously orient themselves into a double-layer sheet in which their hydrophilic head groups are oriented toward the watery environments of the cell interior and exterior, while the hydrophobic fatty acid chains convene in the middle of the "sandwich."

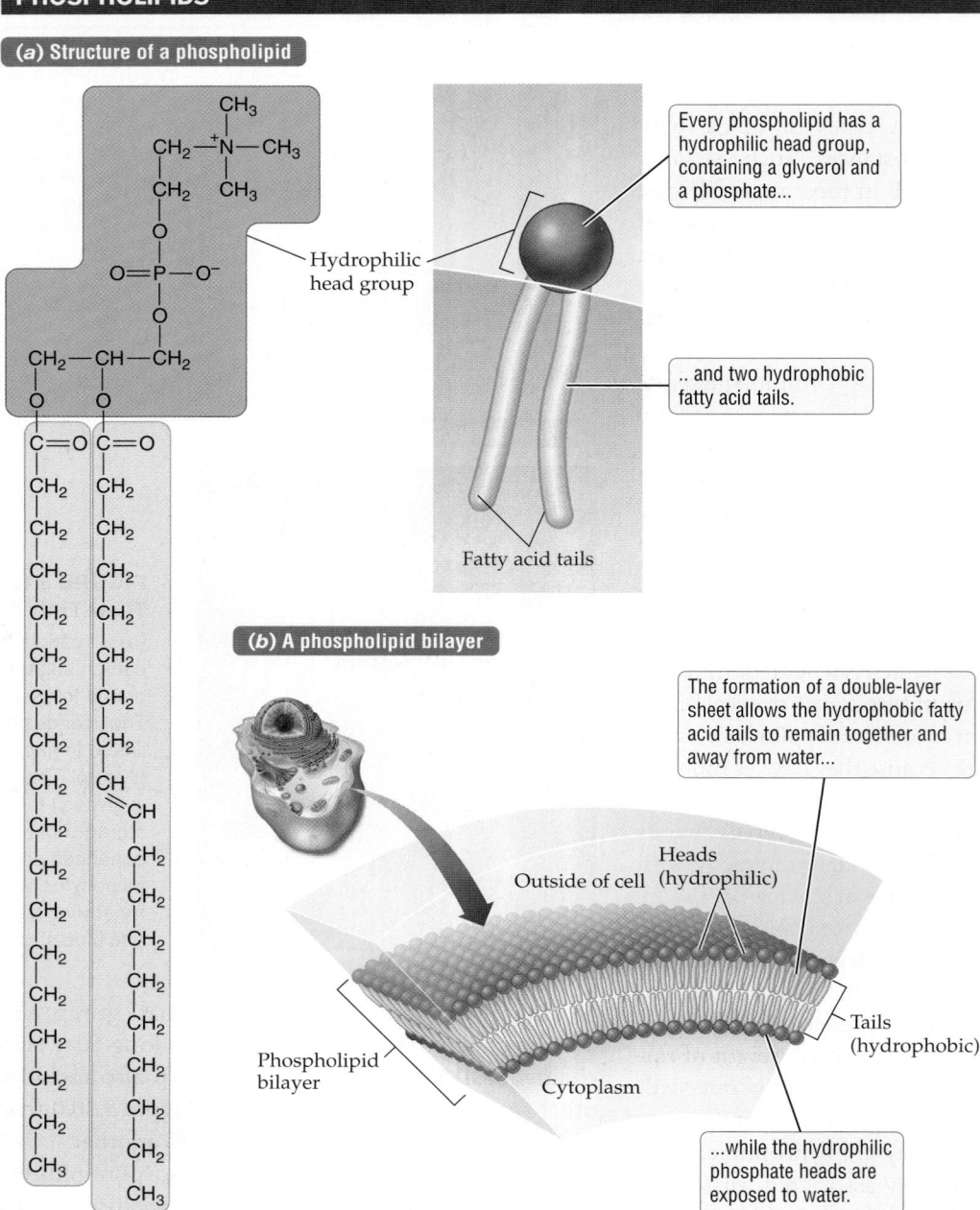

**PHOSPHOLIPIDS**

**(a) Structure of a phospholipid**

Hydrophilic head group

Every phospholipid has a hydrophilic head group, containing a glycerol and a phosphate...

.. and two hydrophobic fatty acid tails.

Fatty acid tails

**(b) A phospholipid bilayer**

The formation of a double-layer sheet allows the hydrophobic fatty acid tails to remain together and away from water...

Outside of cell

Heads (hydrophilic)

Tails (hydrophobic)

Phospholipid bilayer

Cytoplasm

...while the hydrophilic phosphate heads are exposed to water.

## Sterols play vital roles in a variety of life processes

Cholesterol, testosterone, estrogen, vitamin D—all of these are lipids with enough star quality, or medical notoriety, that they turn up on the evening news on a regular basis. Although they are widely divergent in the functions they perform, all four molecules are classified in a group of lipids known as **sterols** (also known as *steroids*). All sterols have the same fundamental structure: four hydrocarbon rings fused to each other. They differ in the number, type, and position of functional groups,

and in the carbon side chains linked to the four hydrocarbon rings (**FIGURE 5.21**).

Cholesterol is a necessary component in the cell membranes of many animals. Cholesterol strengthens membranes and helps maintain their fluidity when temperature changes. The human liver can manufacture all the cholesterol we need. Ingesting a large excess from dietary sources can be harmful because surplus cholesterol tends to accumulate in the lining of our blood vessels, which can lead to cardiovascular disease.

Cholesterol is the "starting" molecule for the manufacture of many other sterols, including vitamin D

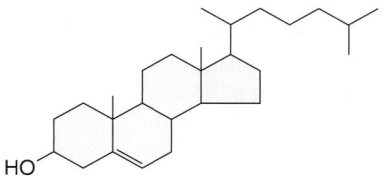

Four fused hydrocarbon rings, basic structure of all sterols

Testosterone

Cholesterol

**FIGURE 5.21  Sterols Are Lipids Built from Four Fused Hydrocarbon Rings**
All sterols share the same basic four-ring structure but have different groups of atoms (functional groups) attached to these rings. Testosterone is a hormone that controls male sexual characteristics in many animals, including the male wood duck (photo, left). Cholesterol is an important constituent of the cell membranes of all birds and mammals, and of many other animals as well.

and bile salts. Vitamin D is important in the growth and maintenance of many tissues in the body, especially bone and muscle. It is partially manufactured by skin cells in response to ultraviolet radiation, but the process is completed in the liver and kidneys. Bile salts are green, bitter-tasting lipids made by the liver and stored in the gall bladder. Bile salts aid in the digestion of fats. The bile is secreted into the small intestine when food arrives there.

Cholesterol is also the "starting" molecule in the production of steroid *hormones*, including sex hormones such as estrogen and testosterone. **Hormones** are signaling molecules that are active at very low concentrations and control a great variety of processes in plants and animals. The sex hormones, such as estrogen and testosterone, promote the development and maintenance of the reproductive system in animals. Testosterone, in its several natural forms and numerous synthetic forms, is an anabolic steroid (*anabolic*, "putting together"). Among its many effects is the promotion of muscle growth. The use of anabolic steroids by competitive athletes is seen as unfair advantage, and the drugs are banned by all major sports organizations. The regular use of anabolic steroids is associated with significant health risks, including higher odds of heart attack, stroke, liver damage, and liver and kidney cancer.

## 5.10 Nucleotides and Nucleic Acids

**Nucleotides** are important monomers in all organisms because they are the building blocks of the hereditary material. Some types of nucleotides also serve as energy-delivering molecules. Nucleotides have three components: a **nitrogenous base** (nitrogen-containing base) that is covalently bonded to a five-carbon sugar, which in turn is covalently bonded to a **phosphate group**, a functional group consisting of a phosphate atom and four oxygen atoms (**FIGURE 5.22**).

Five different nucleotides serve as the components for a class of polymers called nucleic acids. **Nucleic acids** in living cells are of two kinds: deoxyribonucleic acid (**DNA**) and ribonucleic acid (**RNA**). DNA is distinguished from RNA both by the type of sugar in its nucleotides and by two of the nitrogenous bases that bond with that sugar. Ribose, the sugar in RNA, differs from deoxyribose, the sugar in DNA, in that it has one more oxygen atom (see Figure 5.22). Five different kinds of nitrogenous bases are found in nucleic acids: adenine, cytosine, guanine, thymine, and uracil. Thymine is found only in DNA, and uracil is found only in RNA. The nucleotides in RNA and DNA are bonded through covalent linkages (known as *phosphodiester bonds*) between the sugar and phosphate groups of each successive nucleotide, creating a chain of nucleotides, or *polynucleotide*. RNA consists of a single polynucleotide chain (or "strand"); DNA is composed of two polynucleotide chains (is "double-stranded"), which are twisted in a spiral pattern to form the *DNA double helix*.

Nucleotides play two essential functions in the cell: information storage and energy transfer. By "information storage" we mean storage of hereditary information. Every organism has nucleic acid "software" dictating how that organism will live, grow, reproduce, and respond to the external world around it. The information is coded by the precise order in which nucleotides are joined together in the nucleic acid polymer. DNA

**FIGURE 5.22**

## Nucleotides Are the Building Blocks of Nucleic Acids

Each nucleotide consists of a five-carbon sugar linked to a nitrogenous base and one or more phosphate groups. The bases adenine, guanine, cytosine, and thymine, when linked to the sugar deoxyribose, form the building blocks of DNA. The bases adenine, guanine, cytosine, and uracil, when linked to the sugar ribose, form the building blocks of RNA.

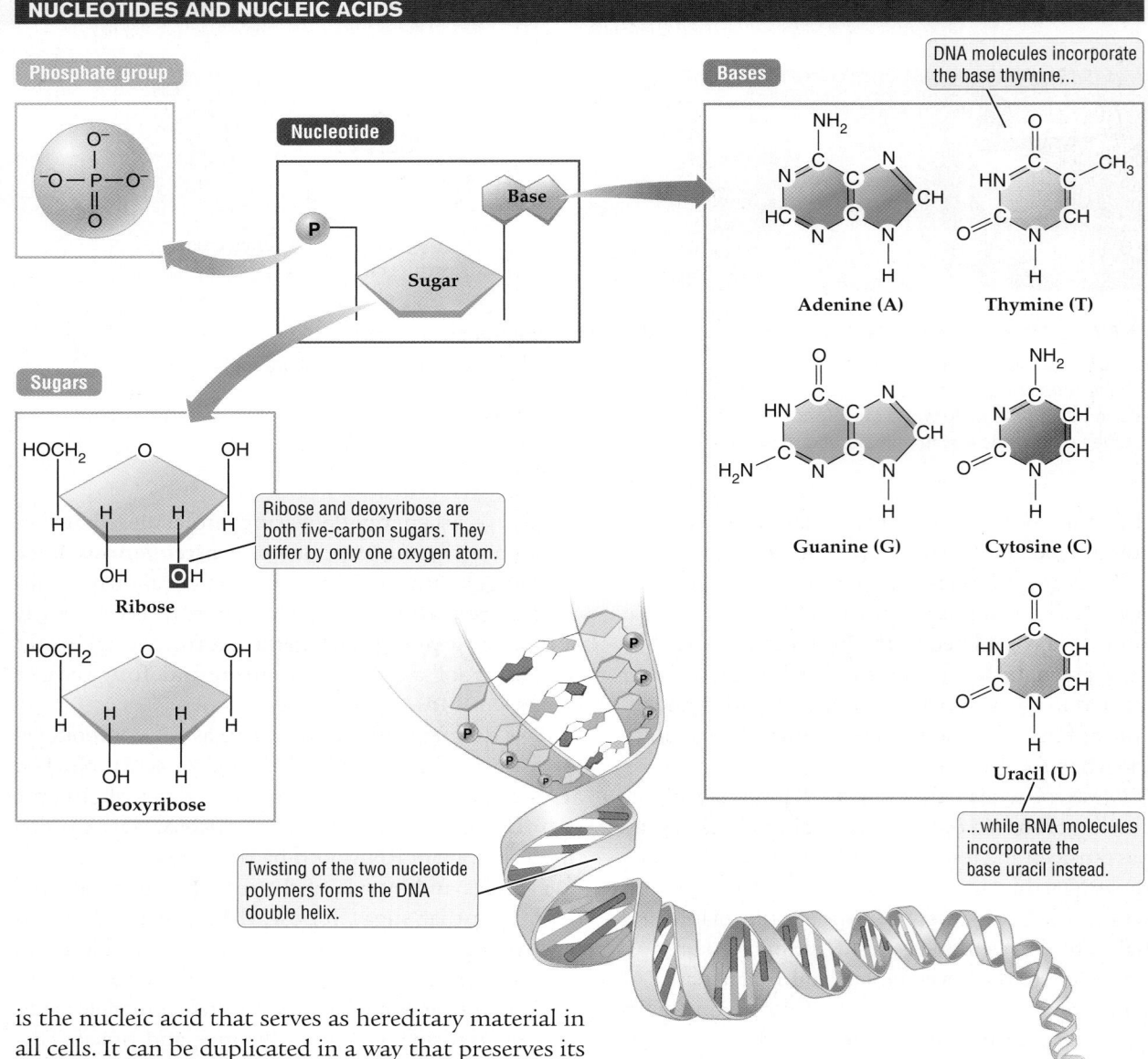

**NUCLEOTIDES AND NUCLEIC ACIDS**

Phosphate group

Nucleotide

P — Sugar — Base

Bases

DNA molecules incorporate the base thymine...

Adenine (A)    Thymine (T)

Guanine (G)    Cytosine (C)

Uracil (U)

...while RNA molecules incorporate the base uracil instead.

Sugars

Ribose and deoxyribose are both five-carbon sugars. They differ by only one oxygen atom.

Ribose

Deoxyribose

Twisting of the two nucleotide polymers forms the DNA double helix.

is the nucleic acid that serves as hereditary material in all cells. It can be duplicated in a way that preserves its sequence of nucleotides and therefore the information coded in it. The transmission of the DNA code from one generation to the next is the basis of heredity, also known as *genetics*.

Some types of nucleotides function as energy delivery molecules, or energy carriers. The most universal of these energy carriers is the nucleotide known as **adenosine triphosphate**, or **ATP** (**FIGURE 5.23**). The ATP molecule is similar to the adenine-containing nucleotide that is one of the building blocks of RNA. ATP is the universal energy carrier for living organisms, and there are many chemical reactions in every cell that could not proceed without energy delivered by ATP. The energy of ATP is stored in the covalent bonds that link the three phosphate groups. The breaking of the bond between two phosphate groups releases energy that is used to power other chemical

reactions. Animal cells must extract energy from food molecules and use that energy to generate ATP from a precursor called ADP (adenosine monophosphate) and phosphate groups. Plants, algae, and certain bacteria, have the additional ability to convert ADP and phosphate into ATP using light energy.

> ### Concept Check
>
> 1. Which of the following is a polysaccharide and a key structural component of plant cell walls: glucose, sucrose, monosaccharide, cellulose, glycogen?
>
> 2. What is protein denaturation?
>
> 3. How is a saturated fatty acid different from an unsaturated one in terms of chemical structure?

ATP stores energy in the covalent bonds that link the phosphate groups.

Adenine

Ribose

**ATP**

This symbol will be used throughout this book for this important molecule.

Most cellular processes in your body are fueled by ATP. ATP provides the energy for thinking, moving, and growing.

**FIGURE 5.23** The Nucleotide ATP Serves as an Energy Carrier in Every Living Cell

The phosphate groups in ATP are held together by energy-rich covalent bonds. Energy is released when these bonds are broken, and the released energy powers a great variety of biological processes, including the motion of the skimboarder pictured here.

## APPLYING WHAT WE LEARNED

# How Bad Are Trans Fats?

Back in 2003, attorney Stephen Joseph sued Kraft Foods, the makers of Oreo cookies, demanding that the food conglomerate stop selling Oreo cookies containing trans fats. At the time, most people had never heard of trans fats. But they'd heard of Oreo cookies, and Joseph's suit was an instant sensation. Joseph didn't care if everyone laughed at him. What mattered was that people wanted to know what trans fats were; and when they found out, they took a second, doubtful look at their Oreos.

Just days after filing the lawsuit, however, Joseph withdrew it. Media coverage of his lawsuit and of trans fats was so widespread, he said, that he could no longer argue in court that consumers didn't know the cookies contained dangerous trans fats. Just as important, within a day of the first major news stories, Kraft reversed itself and announced it would phase out the dangerous fats. Since the "Cookie Monster" lawyer had won his point through publicity alone, he had no reason to pursue an expensive lawsuit. In fact, by the end of 2005, just 31 months later, Kraft said it had voluntarily eliminated or reduced trans fats in both Oreo cookies and all of its other products.

In this chapter we learned that fats and other lipids are an essential part of all living organisms. Every cell membrane contains lipids, and we use the energy in fat. Even a lean person is at least 15 percent fat. Yet for decades, scientists have suspected a link between the fats we eat and heart disease.

The nature of that association has been surprisingly hard to pin down. Because saturated animal fats were thought to contribute to cardiovascular disease, most people's grandparents and great grandparents were told that margarine and other artificial "spreads" were healthier for the heart than butter. But until recently, nearly all margarines were made from partially hydrogenated vegetable oils, which virtually always contain high levels of trans fats.

Trans fats change the ratio of two forms of cholesterol in our blood, increasing the risk of heart disease. As a result, trans fats are much worse for us than the saturated fats in butter. Experts estimate that trans fats contribute to 30,000–100,000 deaths from heart disease per year. In fact, a tiny increase in the consumption of industrial trans fats—2 percent, as measured in calories—results in a 20–30 percent increase in the risk of death from heart disease.

The U.S. Food and Drug Administration (FDA) recommends that Americans consume minimal amounts of trans fats—as little as possible. Animal fats—including those in butter, ice cream, milk, cheese, and yogurt—contain small amounts of naturally occurring trans fats, but researchers are not yet sure if these are as dangerous as the industrial trans fats in partially hydrogenated oils.

Today, nutrition labels in the United States must list how many grams of trans fats a "serving" contains. But don't assume that "zero" means zero. If a food contains less than 0.5 gram of trans fats per serving, the manufacturer may list that as "0 g"—even if the food contains up to 0.49 gram of trans fats per serving. If a "serving" is small enough, almost any food can be listed as having "0 grams of trans fats." As a result, it's easy to consume enough trans fats to affect your health without knowing it. A person who consumed five servings of something that contained 0.49 gram of trans fat would exceed the 2-gram daily maximum recommended by the American Heart Association. In general, it's safe to assume that any food containing partially hydrogenated oils contains trans fats—even if the package says "Zero Trans Fats."

# Do Trans Fats Cause Depression?

BY ELIZABETH LOPATTO, *Bloomberg News*

People who eat more trans fats from cheese, milk or processed foods may have a 48 percent increased risk of depression, compared with those who consume almost no trans fats, a study from Spain showed.

Olive oil, by contrast, appeared to have a slight protective effect against the mental illness among 12,059 Spanish participants in the research, according to a report in the journal *PLoS One*.

The study is the first to analyze the effects of dietary fat on depression, the authors wrote. Research has already linked trans fats, which are created through a process that adds hydrogen to oil, to increased heart risk.

"Our findings suggest that trans fat intake, a well known risk factor for cardiovascular disease, might also have a detrimental effect on depression,"

wrote the authors, who were led by Almudena Sanchez-Villegas, a researcher at the University of Las Palmas de Gran Canaria in Spain.

The effects of trans fats on mood may be more amplified in a study of Americans, who eat more processed foods, a major source of trans fats, the authors wrote. In the Spanish study, most of the trans fats came from milk and cheese, and contributed only 0.4 percent of total energy. In an American diet, about 2.5 percent of energy intake is trans fat, mostly from artificial foods, the authors wrote.

## Depression Risk

The 48 percent increase in depression risk was found among those whose diet consisted of more than 0.6 percent of

calories from trans fats, compared with almost no trans fats.

Trans-fat and overall dietary intake were measured by a 136-item food questionnaire filled out by study participants that the authors said showed validity for assessing fat content. Nutrient amounts were calculated using the latest available information from food composition tables for Spain, they said. The questionnaires were given at the beginning of the study and again a second time, then analyzed.

The authors suggest that depression may be related to low-grade inflammation, which is commonly observed in patients with depression. Trans fats increase inflammation, which is also a risk factor for cardiovascular disease, as well as raising bad cholesterol and lowering good cholesterol, the authors wrote.

Researchers have known for years that trans fats substantially increase the risk of heart attack and even strokes. Could trans fats also affect our mental health? This study seemed to show a large increase in the rate of depression in people who consumed trans fats.

How could a single class of molecules—trans fats—affect parts of the body as different as the heart and the nervous system? The answer lies in the way the body uses molecules. The chemical building blocks introduced in this chapter each play hundreds of different roles. For example, we've seen that cells use cholesterol to build cell membranes, as well as a starting material for making vitamins, hormones, and bile (which we use to digest food).

Think of the way a high school student plays different roles depending on the situation—second violin in the school orchestra, babysitter for a younger sibling at home, and reliable member of a church choir. It's the same person, just doing different things. Similarly, molecules perform widely different roles in different parts of the body. Since lipids play roles as hormones and nerve cell membranes that help transmit nerve signals, trans fats could conceivably affect mood. How that might work is unknown. The researchers speculate that the mechanism is related to inflammation.

Why do they suggest inflammation? One reason may be that inflammation—the aching, redness, and swelling associated with injuries and insect bites—is initiated in the body by a set of molecules called prostaglandins. Cells build prostaglandins from fatty acids. In fact, aspirin and drugs such as ibuprofen stop pain and inflammation by blocking the synthesis of prostaglandins. So it's at least possible that trans fats alter the way prostaglandins are synthesized, influencing inflammation.

## Evaluating the News

**1.** This article compares people who consumed 0.6 gram or more of their calories in the form of trans fats to those who consumed almost none. What was the increased risk of depression in the trans fat–eating group?

**2.** This article reports that Americans consume 2.5 percent of their calories in the form of trans fats, well above the 0.6 percent cut off for "high" trans-fat consumption in this study. If you ate 2,000 calories a day, 2.5 percent would be how many calories? Now look up the number of grams in a calorie of fat and translate that into grams of trans fats.

**SOURCE:** *Bloomberg News*, January 26, 2011.

# Summary

## 5.1 Matter, Elements, and Atomic Structure

- The physical world is composed of matter, which is anything that has mass and occupies space. Matter consists of 92 different types of chemical elements, each with unique properties.
- An atom is the smallest unit of an element that has the chemical properties of that element. Atoms contain positively charged protons, uncharged neutrons, and negatively charged electrons. The number and arrangement of electrons in the atom of an element determines the chemical properties of that element.
- The atomic number of an element is the number of protons in its nucleus, and its atomic mass number is the sum of the number of protons and neutrons.
- Isotopes of an element have different numbers of neutrons but the same number of protons. Radioisotopes are isotopes that give off radiation.

## 5.2 Covalent and Ionic Bonds

- The chemical interactions that cause atoms to associate with each other are known as chemical bonds.
- When an atom loses or gains electrons, it becomes a positively or negatively charged ion, respectively. Ions of opposite charge are held together by ionic bonds, and atoms that are bound exclusively through such bonds are known as salts.
- Covalent bonds are formed by the sharing of electrons between atoms. Atoms share electrons with other atoms to fill their outermost electron shells to capacity. The bonding properties of an atom are determined by the number of electrons in its outermost shell. A molecule contains at least two atoms that are held together by covalent bonds.
- Chemical compounds contain atoms from at least two different elements. All salts are compounds, but a molecule is a compound only if it contains atoms from at least two different elements.

## 5.3 The Special Properties of Water

- Hydrogen bonds are weak associations between two molecules such that a partially positive hydrogen atom within one molecule is attracted to a partially negative region of the other molecule. Partial electrical charges result from the unequal sharing of electrons between atoms, giving rise to polar molecules.
- Water is a polar molecule. Hydrogen bonding between water molecules accounts for the special properties of water, including its high heat capacity and high heat of vaporization, properties that enable water to moderate temperature swings.
- Water is a universal solvent for ions and polar molecules, which are hydrophilic and therefore readily dissolve in water. Nonpolar molecules cannot associate with water and are therefore hydrophobic. Nonpolar molecules are excluded by water, causing them to clump together.
- Water has the highest surface tension of any liquid except mercury. Surface tension is a force that tends to minimize the surface area of water at an air-water boundary, and it influences many biological phenomena.

## 5.4 Chemical Reactions

- In chemical reactions, bonds between atoms are formed or broken. Although the participants in a chemical reaction (reactants) are modified to give rise to new ions or molecules (products), atoms are neither created nor destroyed in the process.
- Some chemical reactions release energy; others cannot proceed without an input of energy.

## 5.5 The pH Scale

- The life-supporting chemical reactions of a cell are conducted in a watery medium. Acids donate hydrogen ions in a solution; bases accept hydrogen ions.
- The concentration of free hydrogen ions in water is expressed by the pH scale.
- Buffers help maintain a constant pH in an aqueous solution.

## 5.6 The Chemical Building Blocks of Life

- Carbon atoms can link with each other and with other atoms to generate a great diversity of compounds.
- The four main biomolecules are carbohydrates, proteins, lipids, and nucleic acids.

## 5.7 Carbohydrates

- Carbohydrates include simple sugars (monosaccharides), as well as disaccharides and more complex polymers (polysaccharides).
- Carbohydrates provide energy and physical support for living organisms.

## 5.8 Proteins

- Amino acids are the building blocks of proteins. A chain of amino acids linked together makes a polypeptide, which constitutes the primary structure of a protein.
- The 3-D shape of a protein is critical for its biological function.

## 5.9 Lipids

- Lipids are hydrophobic substances containing one or more rings or chains of hydrocarbons. Fatty acids, the building blocks of most lipids, are saturated or unsaturated, depending on the absence or presence, respectively, of double covalent bonds in their hydrocarbon chains.
- Triglycerides are important for energy storage.
- Phospholipids are the basic components of biological membranes.
- Sterols include cholesterol and sex hormones.

## 5.10 Nucleotides and Nucleic Acids

- Each nucleotide consists of a five-carbon sugar, a nitrogenous base, and a phosphate group.
- Nucleotides are the building blocks of the nucleic acids DNA and RNA. DNA polymers, made up of four types of nucleotides, form the blueprint for life and govern the physical features and chemical reactions of a living organism.
- ATP is an energy-rich molecule that delivers energy for a great variety of cellular processes.

# Key Terms

acid (p. 122)
amino acid (p. 127)
atom (p. 112)
atomic mass number (p. 113)
atomic number (p. 113)
ATP (adenosine triphosphate) (p. 136)
base (p. 122)
biomolecule (p. 112)
buffer (p. 122)
carbohydrate (p. 125)
cellulose (p. 126)
chemical bond (p. 115)
chemical compound (p. 115)
chemical formula (p. 115)
chemical reaction (p. 121)
cohesion (p. 121)
covalent bond (p. 115)
denaturation (p. 130)
disaccharide (p. 125)
DNA (p. 135)

double bond (p. 116)
electron (p. 112)
electron shell (p. 113)
element (p. 112)
enzyme (p. 127)
evaporation (p. 120)
fatty acid (p. 131)
functional group (p. 124)
glucose (p. 125)
heat capacity (p. 120)
hormone (p. 135)
hydrogen bond (p. 118)
hydrophilic (p. 119)
hydrophobic (p. 119)
ion (p. 115)
ionic bond (p. 117)
isotope (p. 114)
lipid (p. 131)
macromolecule (p. 124)
matter (p. 112)
molecule (p. 115)

monomer (p. 124)
monosaccharide (p. 125)
neutron (p. 112)
nitrogenous base (p. 135)
nonpolar molecule (p. 119)
nucleic acid (p. 135)
nucleotide (p. 135)
nucleus (p. 113)
organic molecule (p. 123)
peptide bond (p. 129)
pH scale (p. 122)
phosphate group (p. 135)
phospholipid (p. 133)
phospholipid bilayer (p. 133)
plasma membrane (p. 133)
polar molecule (p. 118)
polymer (p. 124)
polypeptide (p. 129)
polysaccharide (p. 126)
primary structure (p. 129)
product (p. 121)

protein (p. 126)
proton (p. 112)
quaternary structure (p. 129)
radioisotope (p. 114)
reactant (p. 121)
RNA (p. 135)
salt (p. 117)
saturated fatty acid (p. 131)
secondary structure (p. 129)
soluble (p. 119)
solute (p. 119)
solution (p. 119)
solvent (p. 119)
sterol (p. 134)
sugar (p. 124)
surface tension (p. 121)
tertiary structure (p. 129)
triglyceride (p. 133)
triple bond (p. 116)
unsaturated fatty acid (p. 131)

# Self-Quiz

1. The neutral atoms of a single element
   a. all have the same number of electrons.
   b. can form linkages only with other atoms of the same element.
   c. can have different numbers of electrons.
   d. can never be part of a chemical compound.

2. Two atoms can form a covalent bond
   a. by sharing protons.
   b. by swapping nuclei.
   c. by sharing electrons.
   d. by sticking together because they have opposite electrical charges.

3. Which of the following statements about molecules is true?
   a. A single molecule cannot have atoms from two different elements.
   b. Atoms in a molecule are linked only via ionic bonds.
   c. Molecules are found only in living organisms.
   d. Molecules can contain as few as two atoms.

4. Which of the following statements about ionic bonds is *not* true?
   a. They cannot exist without water molecules.
   b. They are not the same as hydrogen bonds.
   c. They involve electrical attraction between atoms with opposite charge.
   d. They are known to exist in crystals of table salt, NaCl.

5. Hydrogen bonds are especially important for living organisms because
   a. they occur only inside of organisms.
   b. they are stronger than covalent bonds and maintain the physical stability of molecules.

   c. they enable polar molecules to dissolve in water, which is the universal medium for life processes.
   d. once formed, they never break.

6. Glucose is an important example of a
   a. protein.                c. lipid.
   b. carbohydrate.           d. nucleic acid.

7. Peptide bonds in proteins
   a. connect amino acids to sugar monomers.
   b. bind phosphate groups to adenine.
   c. connect amino acids together.
   d. connect nitrogenous bases to ribose monomers.

8. An alpha helix is an example of
   a. primary protein structure.
   b. secondary protein structure.
   c. tertiary protein structure.
   d. quaternary protein structure.

9. Sterols are classified as
   a. sugars.                 c. nucleotides.
   b. amino acids.            d. lipids.

10. Unlike saturated fatty acids, unsaturated fatty acids
   a. are solid at room temperature.
   b. pack more tightly because they have straight chains.
   c. have one or more double bonds in their hydrocarbon chain.
   d. have the full complement (maximum number) of hydrogen atoms covalently bonded to each carbon atom in the hydrocarbon chain.

11. Which of the following statements about the nature of matter is *false*?
    a. Anything that is matter must occupy space.
    b. Anything that is matter must have mass.
    c. There about 92 different elements and 20 types of matter that occur naturally in our universe.
    d. The smallest unit of matter that has all the properties of a particular element is the atom.

12. Lactase is a protein whose function is to break apart a milk sugar called lactose. RNA polymerase is a protein whose function is to join nucleotides together to create RNA molecules. The structure and function of these two proteins is different most likely because
    a. there are different amounts of monosaccharides in the backbone of the two proteins.
    b. the two proteins differ in their sensitivity to pH.
    c. one protein is more abundant in the cytoplasm than the other.
    d. the two proteins differ in their amino acid sequence.

13. The structure of hydrogen cyanide (HCN) is shown below. Hydrogen cyanide
    a. is an organic molecule.
    b. is a salt.
    c. contains a carbon atom that shares all the electrons in its outermost electron shell with one atom of nitrogen.
    d. contains a carbon atom that shares two pairs of electrons with a hydrogen atom.
    $$H-C\equiv N$$

14. The water molecule ($H_2O$)
    a. is nonpolar.
    b. can form a network around both negatively charged ions and positively charged ions.
    c. can form hydrogen bonds with hydrophobic as well as hydrophilic chemicals.
    d. has symmetrical sharing of electrons between the oxygen atom and each of the two hydrogen atoms.

# Analysis and Application

1. What is a monomer, and what is its relationship to a polymer? Should lipids be regarded as polymers? Why or why not?

2. A sample of pure water contains no added acids or bases. Predict the pH of the water and explain your reasoning.

3. What are hydrogen bonds? Explain how the polarity of water molecules contributes to their tendency to form hydrogen bonds.

4. Describe the chemical properties of carbon atoms that make them especially suitable for forming so many different molecules of life.

5. Describe one function that is relevant to biological processes for each of the following compounds: carbohydrates, nucleic acids, proteins, lipids.

6. Identify one macromolecule that is especially abundant in each of the main ingredients of the hamburger sandwich in the figure below. Name the building blocks that each macromolecule is composed of, and name at least one important function it performs in the human body.

7. Figure 5.1 on p. 113 shows a man approaching Sand Dune Arch in Arches National Park near Moab, Utah. Name two elements that are abundant in the man's body but scarce in the Sand Dune Arch. Name a small organic molecule that contains both these elements. What type of polymer, if any, can be formed by this organic molecule?

8. Name the solvent, and some of the solutes, we might find in a cup of sweetened black coffee topped with whipped cream. Which of these substances are hydrophilic? Which are hydrophobic?

9. The photo shows an Inuit fisherman with Artic char he has pulled from his ice fishing hole. Explain how fish can thrive under the ice covering this lake. Why has the lake not frozen solid from the bottom up? Compare the arrangement of water molecules in the lake ice and the liquid water below it.

# Cell Structure and Internal Compartments

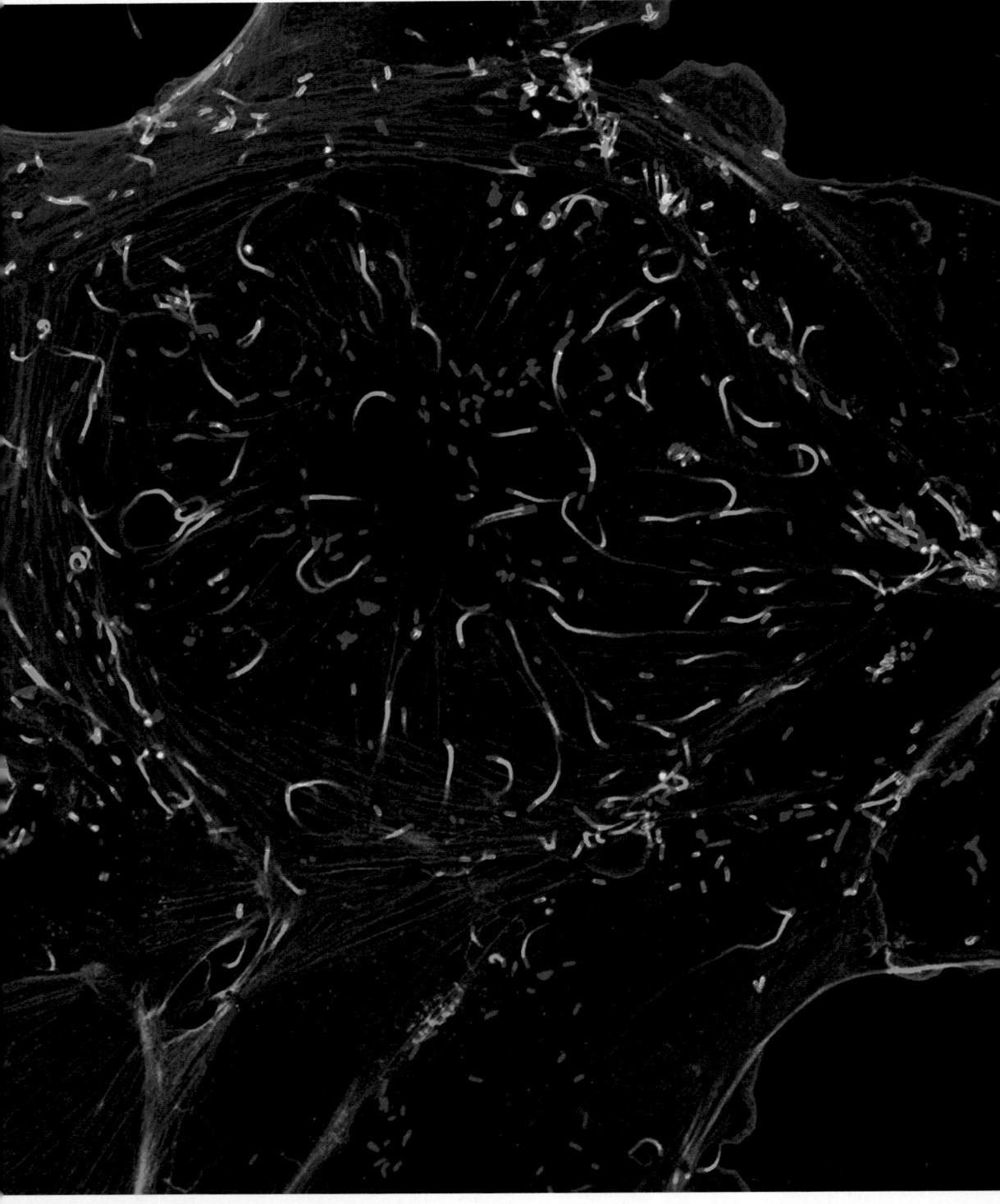

# Wanted: Long-Term Roommate; Must Help Keep House and Have Own DNA

Of the hundreds of trillions of cells in your body, only a fraction are actually you. The rest are mostly bacteria inhabiting different parts of your body, such as your skin, mouth, stomach, and intestines. Although most of these fellow travelers are an important part of who we are, not all of them are benign or helpful. Parasites, which grow and multiply at our expense, include not only animals such as worms, but also disease-causing microbes. We have defensive immune cells that attack these invaders, but some parasitic microbes actually hide inside our cells and take over their internal machinery.

Consider *Listeria monocytogenes* [lis-*TEER*-ee-uh *MAH*-noh-sye-*TAH*-juh-neez], a bacterium that can lurk in raw milk, meat, and fish. In the United States, this bacterium causes nearly 2,000 cases of severe food poisoning each year, killing about 400 people. Inside the body, giant immune cells called macrophages attack and engulf *L. monocytogenes*. But the bacterium narrowly escapes being digested by the macrophage and then handily turns the tables on its attacker. While still in the macrophage, *L. monocytogenes* uses the hapless macrophage's own cellular skeleton to invade other human cells.

Surprisingly, this business of one cell diving inside another is not only common but also ancient. Many kinds of bacteria, both nasty and nice, infect or inhabit the cells of animals, plants, insects, amoebas, and even other bacteria. Aphids—insects that commonly infest houseplants—harbor a bacterial parasite that is itself infected by a bacterium. And the malaria parasite *Plasmodium falciparum* (a eukaryote) has been filmed burglarizing human red blood cells.

The hijacking of one organism's cellular machinery and energy resources by another organism is a form of *parasitism*. And parasitism is just one form of symbiosis, in which organisms of different species live together in close association. The ability of cells to survive inside other cells is a clue to an amazing story of how eukaryotic cells came

What is the evolutionary link between prokaryotes and eukaryotes? Are our cells constructed from ancient prokaryotic invaders?

to be. At the end of this chapter we'll look more closely at these primal relationships. But first, let's examine the general structures of prokaryotic and eukaryotic cells.

**MAIN MESSAGE** Prokaryotic and eukaryotic cells are both bounded by a plasma membrane that encloses an aqueous cytoplasm, but eukaryotic cells have a greater variety of membrane-enclosed compartments.

## KEY CONCEPTS

- All living organisms are made up of one or more basic units called cells.

- Most cells are small. Small size optimizes surface area relative to volume.

- Multicellularity enables larger body size, with efficient functioning through division of labor among specialized cell types.

- The plasma membrane forms the boundary of a cell. It controls the movement of

substances into and out of the cell and determines how the cell communicates with the external world.

- Prokaryotes are single-celled organisms that lack a nucleus. Eukaryotes are single-celled or multicellular organisms whose cells have a nucleus and several other internal compartments specialized for unique functions.

- The membrane-enclosed compartments of the eukaryotic cell have diverse functions,

including the manufacture of lipids and proteins, the sorting and targeting of membrane proteins and secreted proteins, the digestion and recycling of macromolecules, and the generation of energy to fuel cellular activities.

- The cytoskeleton is a network of protein cables and cylinders. Three types of cytoskeletal structures play an important role in giving shape and mechanical strength to a cell and enabling it to move.

## Helpful to Know

*Organelle* is Latin for "little organs," as in the tiny organs of a cell. Some biologists reserve the term for membrane-enclosed compartments alone. In recent years, most cell biologists have gravitated toward the usage you see in this textbook: an organelle is any cytoplasmic structure that carries out a unique cellular function. A cytoplasmic *structure* should be understood as a discrete functional unit composed of several to many macromolecules.

**CERTAIN LARGE BIOMOLECULES** are common to all life-forms. By themselves, these building blocks of life, which were the focus of Chapter 5, are inanimate. Only when they come together to constitute the highly organized, energy-dependent, self-replicating structure that is the **cell** can we recognize a living entity. From bacteria to blue whales, the cell is the smallest and simplest unit of life.

This chapter explores life at the level of the cell. After a broad overview of the unity and diversity of living cells, followed by a comparison of cellular organization in prokaryotes and eukaryotes, we examine the internal structures and compartments that enable a cell to function as an efficient and well-coordinated whole. We will begin the tour at the outer boundary of the cell and then work our way inward.

## 6.1 Cells: The Smallest Units of Life

Every living organism is composed of one or more cells. All cells living today came from a preexisting cell. These two concepts are the main tenets of the **cell theory**, proposed in the middle of the nineteenth century. Today the cell theory is one of the unifying principles of biology, and we regard the cell as the smallest unit of life on Earth. If we disassemble a cell, we find that the constituent parts do not retain the distinctive characteristics of life, such as the capacity to reproduce.

A cell is a highly organized unit: it has an aqueous interior containing high concentrations of many different types of biomolecules, and a lipid-based boundary known as the **plasma membrane**. The contents of a cell internal to the plasma membrane are col-

lectively the **cytoplasm**. The cytoplasm contains a thick fluid called the **cytosol** that is composed of a multitude of ions and biomolecules mixed in water. Embedded in the cytosol, or adrift in it, are structures called *organelles* that are vital for cell function. An **organelle** is a cytoplasmic structure that performs a unique function in the cell. Some organelles are wrapped in one or more lipid-based membranes. The **nucleus** (plural "nuclei"), the largest organelle in our cells, contains DNA enveloped in double membranes. The **mitochondrion** (plural "mitochondria"), often dubbed the powerhouse of the cell because it supplies energy, is another example of an organelle bounded by two membranes. **Ribosomes**, which are key components of the cell's protein-making machinery, are minute organelles that lack membranes. Many thousands of ribosomes are found in the cytosol of both prokaryotic and eukaryotic cells.

The cell or cells that constitute the individual organism come in a wonderful diversity of shapes, sizes, life strategies, and behaviors (**FIGURE 6.1**). Prokaryotes—bacteria and archaeans—are generally regarded as single-celled organisms (**FIGURE 6.1a**). Protists, grab bag grouping that they are, may be single-celled like *Paramecium* (**FIGURE 6.1b**), or multicellular like red seaweed (**FIGURE 6.1c**). Fungi include single-celled species, like baker's yeast (*Saccharomyces cerevisiae*), and many multicellular forms such as *Penicillium* (**FIGURE 6.1d**) and the familiar mushrooms. All members of the plant and animal kingdoms are multicellular (**FIGURE 6.1e and f**).

## The microscope is a window into the life of a cell

Our awareness of cells as the basic units of life is based largely on our ability to see them. The instrument that opened the eyes of the scientific world to the existence of cells—the *light microscope*—was invented in the last quarter of the sixteenth century. The key components of early light microscopes were ground-glass lenses that bent incoming rays of light to produce magnified images of tiny specimens (**FIGURE 6.2**).

The study of cells began in the seventeenth century when Robert Hooke examined a piece of cork under a microscope and noticed that it was made up of little compartments. Hooke called the compartments "cells," after the Latin word for a monk's tiny room.

Although the light microscope has a place in the early history of biology, similar instruments are just as important in ongoing research today. The quality of

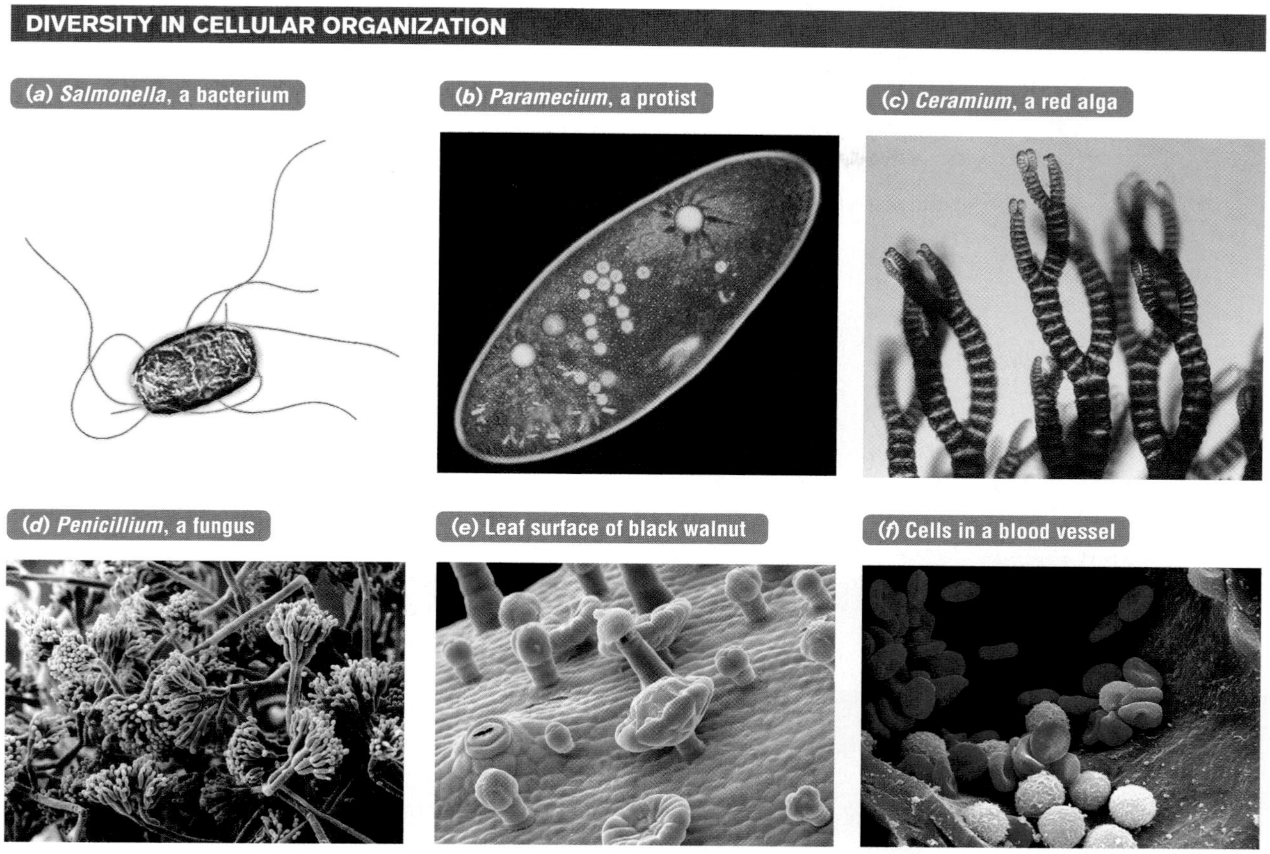

**(a) *Salmonella*, a bacterium**

**(b) *Paramecium*, a protist**

**(c) *Ceramium*, a red alga**

**(d) *Penicillium*, a fungus**

**(e) Leaf surface of black walnut**

**(f) Cells in a blood vessel**

**FIGURE 6.1**

**An Individual Organism May Consist of a Single Cell or Very Many Cells**

(*a*) *Salmonella typhimurium*, a single-celled prokaryote that is a common cause of food poisoning. (*b*) *Paramecium caudatum*, a single-celled eukaryote that lives in freshwater. (*c*) *Ceramium pacificum*, a multicellular red alga. (*d*) *Penicillium camembertii*, a multicellular fungus, with spores (green) used for asexual reproduction. (*e*) Surface view of a black walnut (*Juglans nigra*) leaf, with stomata and protective hair. (*f*) Red blood cells and white blood cells inside an arteriole, one of the smaller blood vessels in the human body.

lenses, however, has improved significantly since that time: the 200- to 300-fold magnification achieved in the seventeenth century has been improved to well over the 1,000-fold magnification achieved by today's standard light microscopes. This degree of magnification enables us to distinguish structures as small as 1/2,000,000 of a meter, or 0.5 micrometer (μm). Modern light microscopes reveal not just animal and plant cells (5–100 μm), but also organelles such as mitochondria, which are about 1 μm wide. Nowadays, cell biologists can mark membranes, organelles, even an individual biomolecule such as a particular protein, with an assortment of fluorescent dyes. The dyes report on the position of the tagged structure or biomolecule by glowing in brilliant colors when illuminated with the appropriate wavelength of light.

Since the 1930s, an even more dramatic increase in magnification has been achieved by replacing visible light with streams of electrons that are focused by powerful magnets instead of by glass lenses. *Electron microscopes*, as these instruments are called, can magnify a specimen more than 100,000 times (**FIGURE 6.3**). One version of the instrument—the *transmission electron microscope* (*TEM*)—can produce detailed views of the smallest organelles. For example,

a ribosome, which is little more than a bundle of proteins enmeshed with strands of RNA, is easily visualized with a TEM. Biological material must be sliced into extremely thin sections to prepare it for TEM

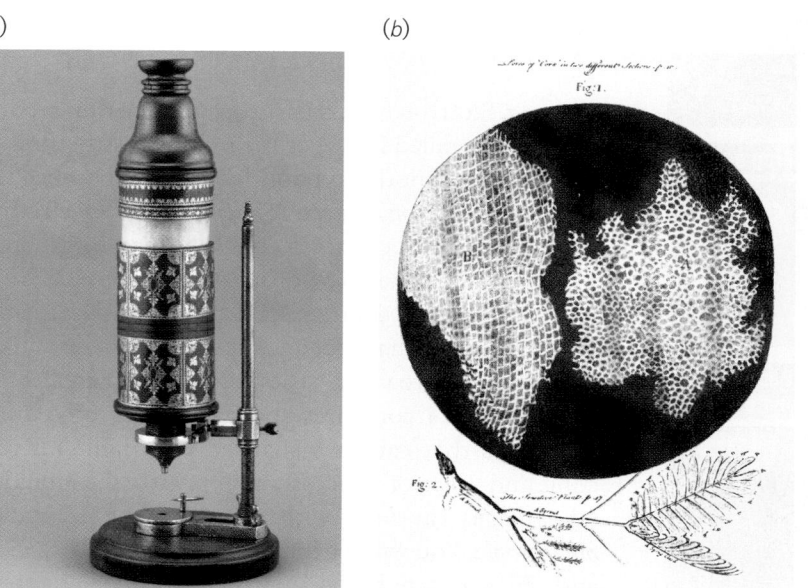

(*a*)  (*b*)

**FIGURE 6.2** **Light Microscope Used by Robert Hooke (1635–1703)**

(*a*) Hooke's microscope. (*b*) A piece of cork examined under Hooke's microscope.

**Light microscope**

**Electron microscope**

Heated filament (source of electrons)

Beam of electrons

Specimen

Lenses

Specimen

Beam of light

(a)

(b)

(c)

**FIGURE 6.3 Microscopy Enables Us to Visualize Cells and Cell Structures**

The photos show human mast cells imaged through light microscopy (a), transmission electron microscopy (b), and scanning electron microscopy (c). Mast cells are immune cells, part of the body's defense against invaders.

**(a) An ostrich egg**

**(b) Frog egg cells**

**(c) Bacteria on a needle**

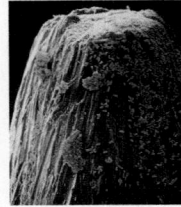

Cells vary in size.

imaging. Electron beams pass through the thin section to form a detailed image of the minutest structures in the specimen. Another type of electron microscope—the *scanning electron microscope* (*SEM*)—moves an electron beam back and forth across the specimen to generate a three-dimensional view of its surface. Because SEM images have great depth of field, all visible parts of the specimen are in sharp focus. A specimen does not have to be sectioned to visualize its surfaces with an SEM.

**FIGURE 6.4** compares the size ranges of cellular structures that can be visualized with light microscopy and electron microscopy or with the unaided eye. Images captured with a microscope are called *micrographs*. You will see many examples of light, SEM, and TEM micrographs in this and other chapters. Computer software is often used to add color to structures seen in a micrograph. Without colorization,

SEM and TEM produce black-and-white images that nonexperts may find difficult to interpret.

## The ratio of surface area to volume limits cell size

Some cells are giants. There are nerve cells in your body that extend all the way from the base of the spine to your big toe, a distance of about 1 meter (3 feet) in the average adult. The motor neurons, as these exceptionally long cells are known, are only about 10 μm at the widest point, so they cannot be seen without a microscope.

Egg cells, engorged with food for the young that might develop from them in the future, are often quite large. The ostrich egg—about 6 inches (15 centimeters) long and weighing in at 1.5 kilograms (3 pounds)—is the largest single cell on the planet today. (The shell of a bird's egg, and the membranes immediately inside the shell, are not part of the egg cell but are instead deposited on top of the egg cell's plasma membrane during development.) Frog eggs, about 1 millimeter in diameter, are easily seen with the unaided eye.

Big cells are not the norm, however. The great majority of cells are too small to be seen with the naked eye; they are microscopic. On average, prokaryotic cells are smaller than eukaryotic cells. Most bacteria average about 1 μm in width, and most animal cells are about 10 μm in diameter. Plant cells tend to be larger, most of them falling in the 20- to 100-μm range.

Why are most cells small? Every cell must exchange materials with its environment, and that exchange becomes more of a challenge as cell size increases. Our cells, for example, must pick up nutrients such as glucose and oxygen from the fluid that surrounds them; they must also get rid of waste products, such as carbon dioxide and urea, by releasing them into the surrounding fluid. The exchange of materials takes place across the plasma membrane. A big cell has more cytoplasm and more metabolic activity, and therefore a greater need for nutrients and a larger waste output, than does a smaller cell.

Simple geometry dictates, however, that as the width of a typical cell increases, the volume increases by a greater proportion than the surface area does, as illustrated in **FIGURE 6.5**. Notice that when the length of the side of a cube is doubled, volume increases eightfold but the surface area increases only fourfold. The relationship applies to spheres as well: as the diameter increases, the *ratio* between the surface area and volume decreases. The surface area–to–volume ratio puts an upper limit of about 100 μm on the width of most cells.

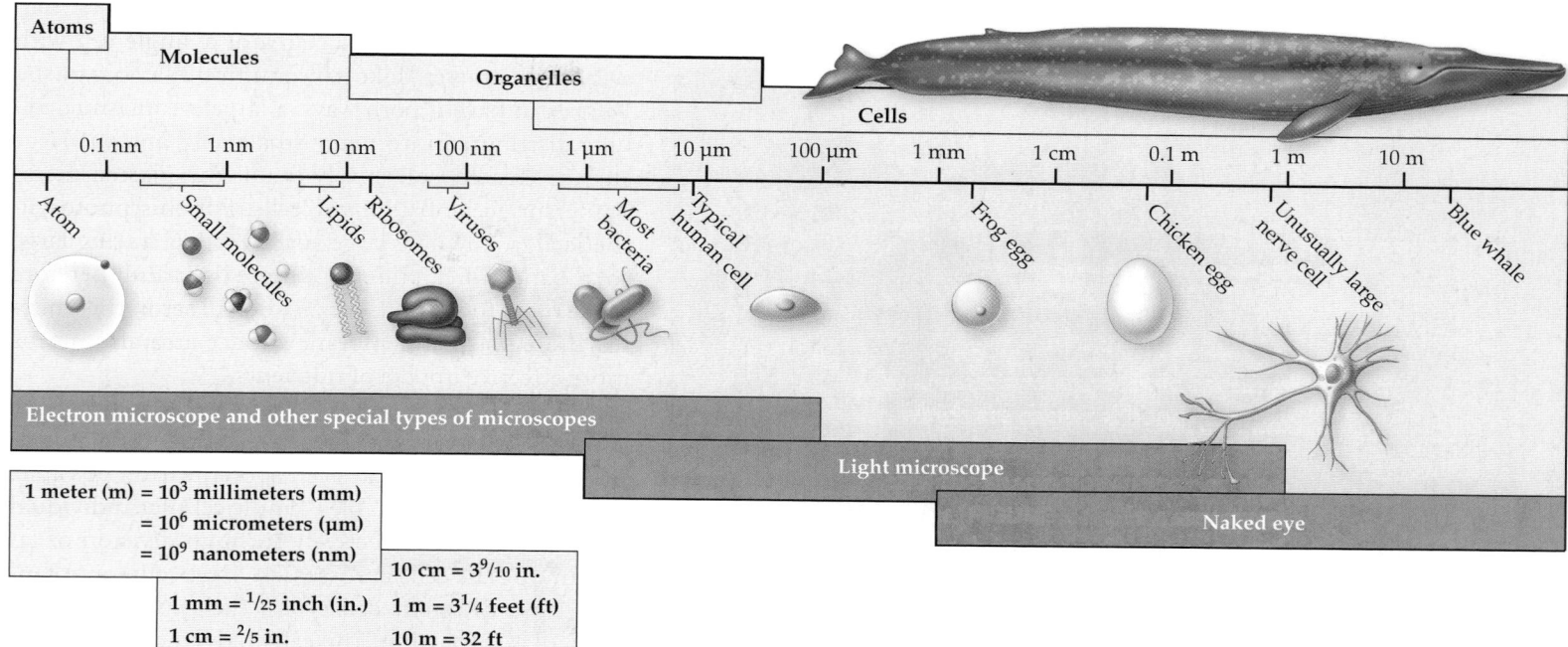

1 meter (m) = $10^3$ millimeters (mm)
= $10^6$ micrometers (μm)
= $10^9$ nanometers (nm)

| | |
|---|---|
| 1 mm = $^1/_{25}$ inch (in.) | 10 cm = $3^9/_{10}$ in. |
| 1 cm = $^2/_5$ in. | 1 m = $3^1/_4$ feet (ft) |
| | 10 m = 32 ft |

**FIGURE 6.4** Most Cells Are Microscopic

Some cells are freed from the limit by their extraordinary geometry: Because of their long and narrow shape, motor neurons have about the same cytoplasmic volume as the average animal cell but a much larger surface area. Giant eggs, too, are not constrained by this limit; the center of a large egg is crammed with metabolically inactive food stores, with the cytoplasm confined to a narrow peripheral band, where it presents a large surface for taking up oxygen and releasing carbon dioxide.

## Multicellularity enables larger body size and efficiency through division of labor

Does size matter? Bacteria and other microscopic organisms demonstrate beyond a shadow of a doubt that being tiny can be an extremely successful life strategy. However, being bigger than the competition can deliver benefits, depending on how and where an organism lives. A large individual may be better positioned to recruit resources from its environment than would be a smaller individual. For example, a predatory single-celled organism can capture and engulf smaller cells with relative ease. A single *Paramecium* (see Figure 6.1*b*) can ingest several thousand bacteria in a day. But with a diameter of 2 millimeters, *Volvox carteri* (**FIGURE 6.6**) is too big a meal for *Paramecium*. The trick that saves *V. carteri* from the likes of *Paramecium* is larger size made possible by multicellularity. Because it is bigger, *V. carteri* is also better able to store nutrients, such as phosphorus, which translates into reproductive success.

### HOW SURFACE AREA CHANGES RELATIVE TO VOLUME

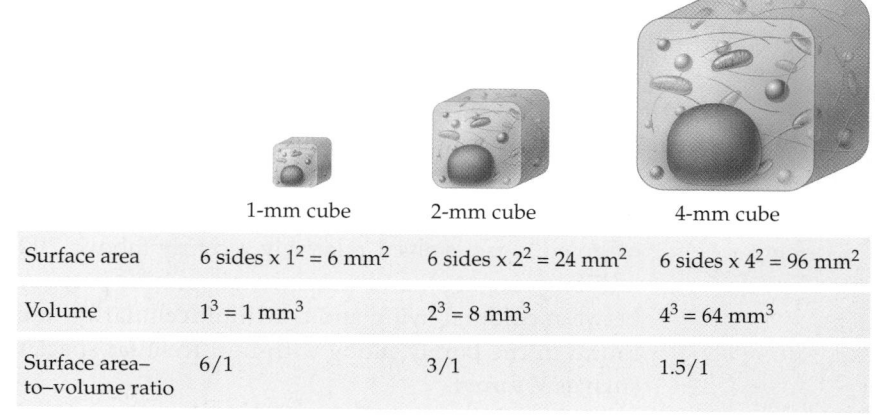

| | 1-mm cube | 2-mm cube | 4-mm cube |
|---|---|---|---|
| Surface area | 6 sides x $1^2$ = 6 mm$^2$ | 6 sides x $2^2$ = 24 mm$^2$ | 6 sides x $4^2$ = 96 mm$^2$ |
| Volume | $1^3$ = 1 mm$^3$ | $2^3$ = 8 mm$^3$ | $4^3$ = 64 mm$^3$ |
| Surface area–to–volume ratio | 6/1 | 3/1 | 1.5/1 |

**FIGURE 6.5** Limits to Cell Size
As the width of a cell increases, the volume increases more steeply than the surface area. Cells exchange nutrients and wastes across the cell surface and must have a large enough surface area for that exchange to take place rapidly and efficiently.

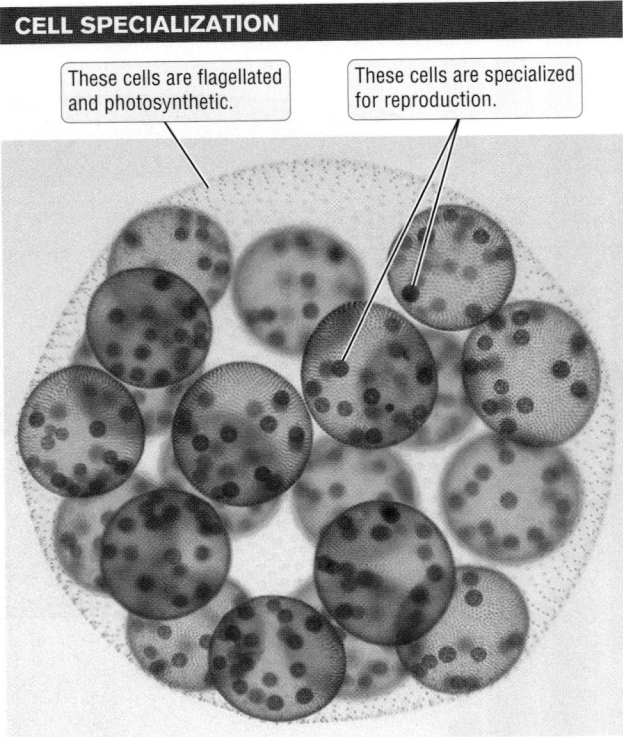

These cells are flagellated and photosynthetic.

These cells are specialized for reproduction.

**FIGURE 6.6 Cell Specialization Is One Benefit of Multicellularity**
Recent studies show that *Volvox carteri*, a green alga, is a multicellular organism with specialized cell types that function in an integrated manner.

A **multicellular organism** consists of an interdependent group of cells that are genetically identical because of their developmental origin from a single cell and whose constituent cells are incapable of living independently. If we separated the cells making up a multicellular organism, the individual cells would not survive on their own in nature. Multicellularity evolved several times among different groups of Protista, probably from colonial species. A colonial organism is a loose federation of cells that cooperate for mutual benefit, but whose cells are also capable of independent existence. According to researchers at the Salk Institute, multicellularity among the volvocine family of green algae evolved relatively recently (about 200 million years ago), and transitional species—partway between colonial living and true multicellularity—are found in the family, along with multicellular species such as *V. carteri*.

Although a multicellular organism can be as large as a giant redwood or a blue whale, the cells in it are themselves quite small, so they do not run afoul of the limits imposed by the ratio of surface area to volume. Instead of a *single cell* that is 2 mm across, *V. carteri* is a closely integrated assemblage of cells, each cell no more than 10 μm across. The surface area of each of the individual cells, added together, is substantially greater than the surface area of a single cell with a 2-mm diameter. Like any multicellular organism, *Volvox* can have it both ways: a large cytoplasmic volume distributed into many small cells, and the large surface created collectively by all the plasma membrane surfaces of its many cells. For this photosynthetic alga, having a large surface is like having large solar panels: it can intercept more light and therefore make more food for itself. As we shall see in this chapter, more plasma membrane surface area also means more effective uptake of nutrients.

The adaptive benefits of multicellularity go beyond getting food or avoiding becoming food; multicellularity is what makes cell specialization possible. Cell specialization enables a multicellular individual to function more effectively through division of labor among the cells it possesses. Multicellular organisms have different cell types, each specialized for particular jobs. All of the cells have the same DNA, but different subsets of the DNA-based information are read out (*expressed*) in different cell types, causing them to acquire "skill sets" unique to the cell type. Because its structure and function is completely dedicated to its unique tasks, a specialized cell does its job more efficiently than does a cell that has to be a jack-of-all-trades and master of none. The outer ring of cells in *Volvox* are small, photosynthesize, and have whiplike structures (flagella) that lash about to move the organism closer to the light at the pond surface. The cells in *Volvox*'s interior, on the other hand, are dedicated to reproduction. They can afford to be large, to dispense with photosynthesis and flagella, and to be specialized only for the production of egg or sperm.

The evolution of multicellularity, about 1.5 billion years ago, is one of the big events in the story of life (see Milestones in the History of Life on Earth, printed on the inside back cover). Multicellularity evolved independently among various eukaryotic groups. Multicellular fungi, plants, and animals evolved from very different groups of protists, and their bodies are therefore built from very different cell types organized in their own unique ways. The greatest diversity of cell types is seen in mammals, the most complex animals. The human body has 220 different types of cells. Each *differentiated* (specialized) cell performs unique functions, in addition to conducting some basic activities that all cells must possess to be alive. The diversity of specialized cell functions is illustrated by the many differentiated cells in a human blood vessel (see Figure 6.1f). The disclike red blood cells, with their large surface area, are specialized for delivering oxygen to all the cells in the body. The bloblike white blood cells defend us by engulfing invading microbes. The blood vessel walls contain an even greater diversity of cell types, including endothelial cells that line the blood vessel and muscle cells that contract or relax to reduce or increase blood flow.

## 6.2 The Plasma Membrane

A key characteristic of every cell is the existence of a plasma membrane separating that cell from its surrounding environment. Most of the chemical reactions required for sustaining life take place within the cytoplasm, the main compartment formed by this boundary. The lipid boundary created by the plasma membrane has the effect of enclosing and concentrating necessary raw materials in a limited space, thereby facilitating chemical processes.

Recall from Section 5.9 that biological membranes, such as the plasma membrane, consist of a bilayer of phospholipids, in which all the hydrophilic head groups are exposed to the aqueous environments outside the cell or toward the cytoplasm, while the hydrophobic tails of the phospholipids congregate in the interior of the membrane (see Figure 5.20). If the plasma membrane had no function other than to define the boundary of the cell and to confine its

contents, a simple phospholipid bilayer would suffice. However, the plasma membrane must also allow the cell to capture essential molecules while shutting out unwanted ones, it must release waste products but prevent needed molecules from leaving the cell, and it must interact with the outside world by receiving and sending signals as necessary. In other words, the plasma membrane must function as a *selectively permeable barrier* and also as a communication center. Most cells in the animal body are held firmly in place, and this anchoring function, too, is mediated by the plasma membrane. The diverse functions of the plasma membrane are enabled chiefly by the many different types of *membrane proteins* associated with, or embedded in, the phospholipid bilayer (**FIGURE 6.7**).

The selective permeability of the plasma membrane comes from the different types of transport proteins embedded in the phospholipid bilayer. *Transport proteins* are membrane-spanning proteins whose function is to facilitate the import or export of substances. Some transport proteins form tunnels that allow the passage of selected ions and molecules. The activity of membrane transport proteins is usually tightly regulated, so the protein moves its cargo of ions or molecules only when directed to do so by signals originating from the cytoplasm or from the external environment.

*Receptor proteins* act as sites for signal perception and as such they are key components of a cell's communication system. Some receptor proteins are found in all or nearly all cell types in the human body; others are unique to a particular cell type. Each receptor protein generally binds a specific type of signaling

**MEMBRANE PROTEINS**

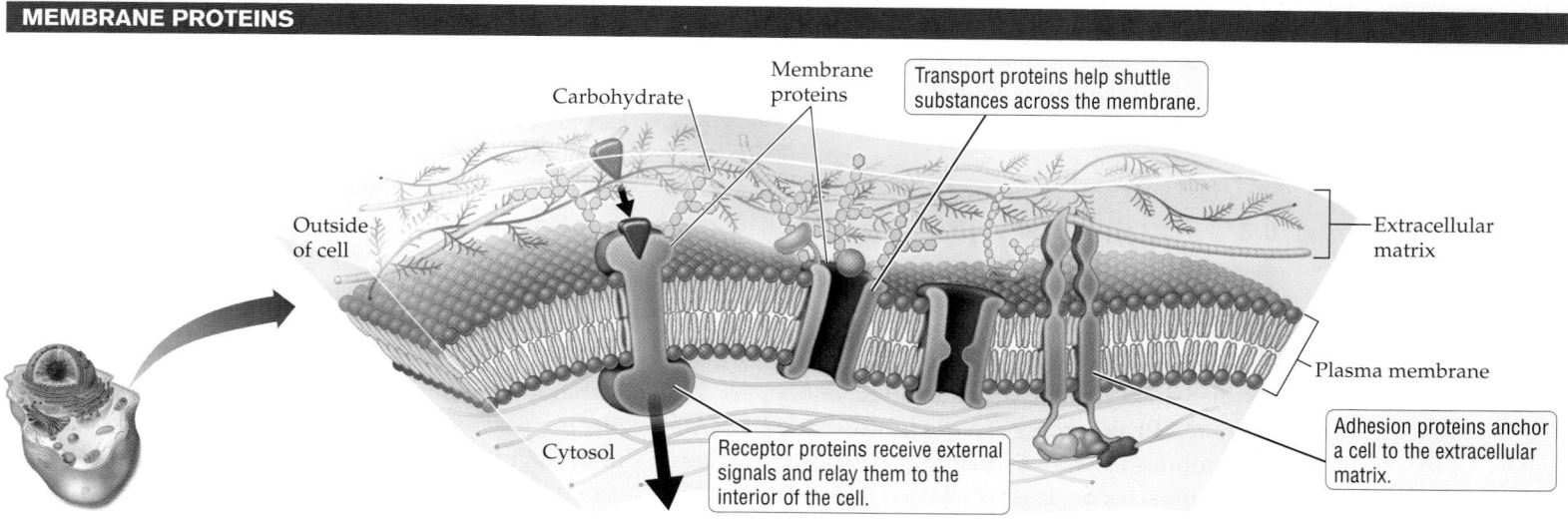

**FIGURE 6.7 The Many Functions of Membrane Proteins**

molecule. When a signal docks on the cell surface side of its target receptor, it triggers a change in cellular activity. Most of our cells have insulin receptors in their plasma membrane, and when the hormone binds to its receptor, a message is relayed into the cytoplasm that induces the cell to step up its import of glucose.

Most cells in the animal body are attached to other cells, or to a dense mat of biomolecules called the **extracellular matrix** (**ECM**) that is deposited on the outside surface of cells. Cell attachment is highly specific, with a given cell adhering only to particular cell types or to an extracellular matrix with a particular chemical composition. In many cases, *adhesion proteins* embedded in the plasma membrane are the main link between a cell and its extracellular neighborhood (see Figure 6.7). Chains of sugars are covalently linked to the cell surface side of adhesion proteins, and these carbohydrate groups help in both the recognition and the interlinking that are necessary for cell attachment. Our skin cells have transmembrane adhesion proteins, called integrins, that tether those cells to an ECM made mostly of collagen. Collagen is the most abundant protein in the human body, and destruction of collagen over time leads to wrinkling and sagging as we age.

Unless they are anchored to structures inside or outside the cell, most plasma membrane proteins are free to drift within the plane of the phospholipid bilayer. The **fluid mosaic model** describes the plasma membrane as a highly mobile mixture of phospholipids, other types of lipids, and many different types of membrane proteins. The flexibility and free-flowing nature of the plasma membrane is critical for many cellular functions. The flexibility of the plasma membrane facilitates whole-cell movement, for example. Cells that move from place to place—whether white blood cells in pursuit of invaders or *Amoeba proteus* in search of a juicy alga—could not crawl about if their cell membranes were rigid and unchangeable.

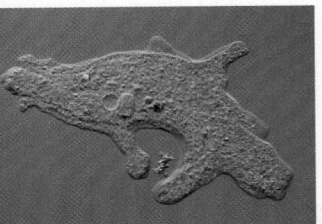

*Amoeba proteus*, a single-celled protist, engulfing an algal cell

## 6.3 Prokaryotic and Eukaryotic Cells

As noted in Chapter 2, organisms can be informally classified into two broad categories: *prokaryotes* and *eukaryotes*. **Prokaryotes** are organisms whose DNA is not confined within a membrane-enclosed nucleus. **Eukaryotes** are organisms whose DNA is enclosed in a nucleus. Prokaryotes are single-celled; eukaryotes can be single-celled or multicellular (see Figure 6.1).

**FIGURE 6.8** shows generalized drawings of prokaryotic and eukaryotic cells. Prokaryotic cells, on average, are smaller than eukaryotic cells. For example, the well-studied bacterium *Escherichia coli*, a common resident of the human intestine, is only two-millionths of a meter, or 2 micrometers, long. About 125 *E. coli* would fit end to end across the period at the end of this sentence.

Most prokaryotes have a tough cell wall outside the plasma membrane (**FIGURE 6.8a**) that helps maintain the shape and structural integrity of the organism. Some bacteria have additional protective layers, the *capsule*, made of slippery polysaccharides.

Plants, fungi, and some protists (mostly algae) have cell walls made of polysaccharide, but the precise chemistry of the wall polysaccharides differs among these groups of eukaryotes and differs again from that of prokaryotic cell walls. Animal cells lack a polysaccharide cell wall, but many of them are encased in, or attached to, the meshwork of protein and carbohydrate polymers that makes up the extracellular matrix (shown in Figure 6.7).

A striking difference between prokaryotic and eukaryotic cells is the diversity and extent of internal compartments within the cytoplasm of eukaryotes. In addition to the nucleus, several other types of membrane-enclosed organelles, each with a distinct function, are seen in eukaryotic cells (**FIGURE 6.8b**). In contrast, membrane-enclosed organelles are lacking in most prokaryotes.

The average eukaryotic cell has roughly a thousand times the volume of the average prokaryotic cell. The partitioning of the cytoplasm into a variety of highly specialized membrane-enclosed compartments confers speed and efficiency through an intracellular division of labor. The different types of membrane-enclosed organelles serve very specific and unique functions. Because its structure is finely tuned for its particular functions, each organelle conducts its special tasks with great effectiveness. For example, each membrane-enclosed organelle stockpiles high concentrations of just the raw materials it needs for the special functions it performs. The chemical reactions can proceed more rapidly if all necessary chemical ingredients are concentrated, instead of being scattered over a large volume of cytoplasm.

Unique chemical environments can be maintained within a membrane-enclosed compartment. For example, the reactions that break apart polymers work best under highly acidic conditions, so the organelles that specialize in that task maintain a very low pH, even though the pH of the cytosol outside the organelles is close to neutral. Some chemical reactions produce by-products

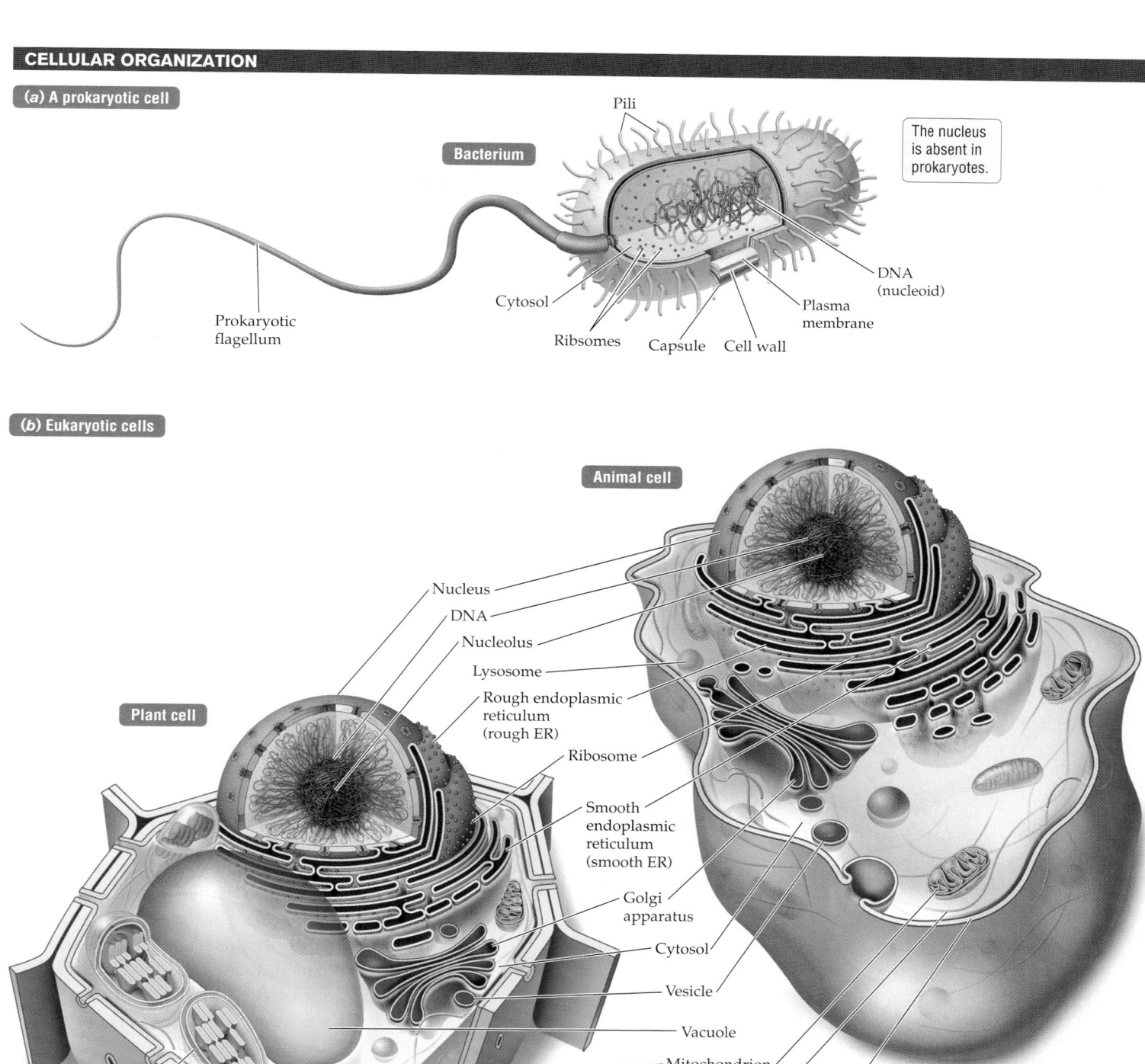

(a) A prokaryotic cell

Pili

Bacterium

The nucleus is absent in prokaryotes.

Prokaryotic flagellum

Cytosol

Ribsomes

Capsule

Cell wall

Plasma membrane

DNA (nucleoid)

(b) Eukaryotic cells

Animal cell

Plant cell

Nucleus

DNA

Nucleolus

Lysosome

Rough endoplasmic reticulum (rough ER)

Ribosome

Smooth endoplasmic reticulum (smooth ER)

Golgi apparatus

Cytosol

Vesicle

Vacuole

Mitochondrion

Cytoskeleton

Plasma membrane

Chloroplast

Plasmodesmata

Cell wall

Plant cells are distinguished from animal cells by the presence of chloroplasts, one or more vacuoles, and an extracellular cell wall.

FIGURE 6.8 Prokaryotic and Eukaryotic Cells Compared

that could interfere with other vital reactions or even poison the cell. Locking interfering or toxic substances into special compartments avoids such "collateral damage." We will explore the structure and function of some of the internal compartments of eukaryotes in greater detail in the next section.

> ### Concept Check

1. Which of these cell structures is found in eukaryotes but not in prokaryotes: plasma membrane, cytoplasm, ribosome, nucleus?

2. Name two functions performed by plasma membrane proteins.

3. What is meant by the fluid mosaic nature of the plasma membrane?

4. Eukaryotes have a variety of membrane-enclosed organelles. How is this type of internal organization beneficial?

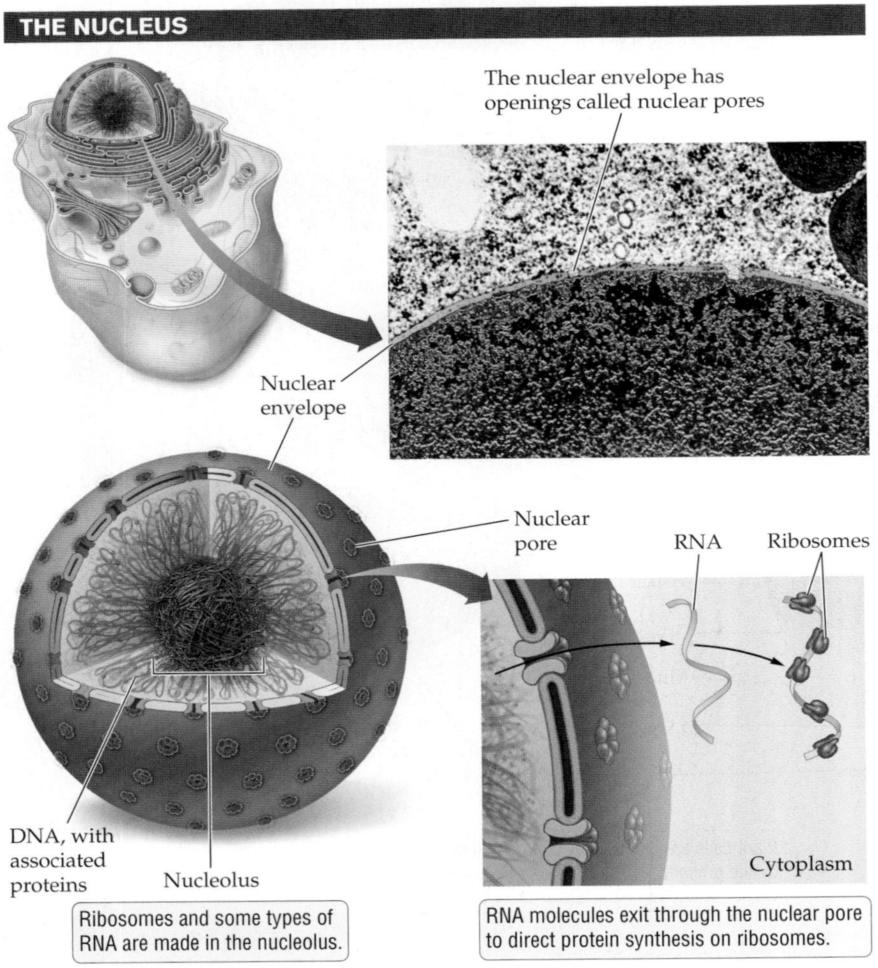

## THE NUCLEUS

The nuclear envelope has openings called nuclear pores

Nuclear envelope

Nuclear pore

RNA    Ribosomes

DNA, with associated proteins

Nucleolus

Cytoplasm

Ribosomes and some types of RNA are made in the nucleolus.

RNA molecules exit through the nuclear pore to direct protein synthesis on ribosomes.

**FIGURE 6.9  The Nucleus Contains DNA, the Genetic Material of the Cell**
The nucleus is enclosed within a double-membrane nuclear envelope. Nuclear pores provide a regulated passageway for molecules entering and exiting the nucleus.

## 6.4 Internal Compartments of Eukaryotic Cells

Imagine a large factory with many rooms housing different departments. Each department has a specific function and an internal organization that contributes to the overall mission of manufacturing certain products. Workers assembling a particular item are arranged in a specific order in a centralized assembly department, with packers and shipping agents taking over from them in the next department. A warehouse stores raw materials and finished products, a power station supplies energy, there is a site for waste disposal and recycling, and all the operations are overseen by an administrative office. A eukaryotic cell is just such a highly structured, highly efficient, and energy-dependent "factory." However, even the simplest amoeba is vastly more complex than any factory built by humans, and it has the capacity to reproduce itself—a quality unique to living systems. We will begin our tour of the eukaryotic cell at the nucleus, the cellular factory's "administrative office."

### The nucleus houses genetic material

No living cell can function without DNA, the code-bearing molecules that contain information necessary for building the cell, managing its day-to-day activities, and controlling its growth and reproduction. In eukaryotic cells, the clearly delineated membrane-enclosed nucleus houses most of the cell's DNA. (Toward the close of this chapter we will discuss certain eukaryotic organelles that have their very own DNA.) The nucleus functions like a highly responsive head office, well tuned to the talk on the shop floor. Hence, the readout of the DNA code can be modulated by signals received from other parts of the cell and even the world outside the cell.

The boundary of the nucleus, called the **nuclear envelope**, is made up of two concentric lipid bilayers (**FIGURE 6.9**). Inside the nuclear envelope, long strands of DNA are packaged with proteins into a remarkably small space. If all the 46 separate DNA double helices in one of your cells were laid end to end, they would have a combined length of 1.8 meters, or almost 6 feet. Fitting that much DNA into a space only about 5 micrometers (0.005 millimeter) in diameter is accomplished by winding it around spools of special proteins called *histones*. Each DNA double helix, wrapped around spools of the compacting proteins, constitutes one **chromosome**.

With the exception of certain reproductive cells, every cell in the average human contains 46 chromosomes.

The nuclear envelope contains thousands of small openings, called **nuclear pores** (see Figure 6.9). Nuclear pores allow free passage to ions and small molecules, but they regulate the entry of larger molecules such as proteins, admitting some and shutting out others. Information stored in DNA is conveyed to ribosomes, the protein-manufacturing units in the cytoplasm, by RNA molecules. RNA molecules carry the "recipe" for making a particular protein from the DNA to ribosomes, which actually assemble the protein. RNA molecules must pass through nuclear pores to reach the cytoplasm, where the ribosomes are located and proteins are built.

The nuclei of most cells contain one or more distinct regions, known as nucleoli (singular "nucleolus"). A *nucleolus* is a region of the nucleus that specializes in churning out large quantities of a special type of RNA, called rRNA (ribosomal RNA). In the nucleolus, rRNA is bundled with special proteins into partially assembled ribosomes, and these structures exit the nucleus through the nuclear pores (see Figure 6.9). The final, fully functional version of the ribosome is assembled in the cytoplasm. Nucleoli are most prominent in cells that are making large amounts of proteins and therefore have a high demand for ribosomes.

## The endoplasmic reticulum manufactures certain lipids and proteins

If the nucleus functions as the administrative office of the cell, the cytoplasm is the main factory floor manufacturing the great majority of proteins and other chemical components of the cell. However, certain types of lipids and proteins are made in the endoplasmic reticulum, which functions like a specialized department preparing items for being exported outside the cell or shipped to other parts of the cell, such as the plasma membrane. The **endoplasmic reticulum [reh-TIK-yoo-lum] (ER)** is an extensive and interconnected network of tubes and flattened sacs (**FIGURE 6.10**). The boundary of the endoplasmic reticulum is formed by a single membrane that is usually joined with the outer membrane of the nuclear envelope. The space inside the ER membrane is a **lumen**, a general term for the cavity inside any closed structure.

The membranes of the ER are classified into two types based on their appearance: smooth and rough (see Figure 6.10). Enzymes associated with the surface

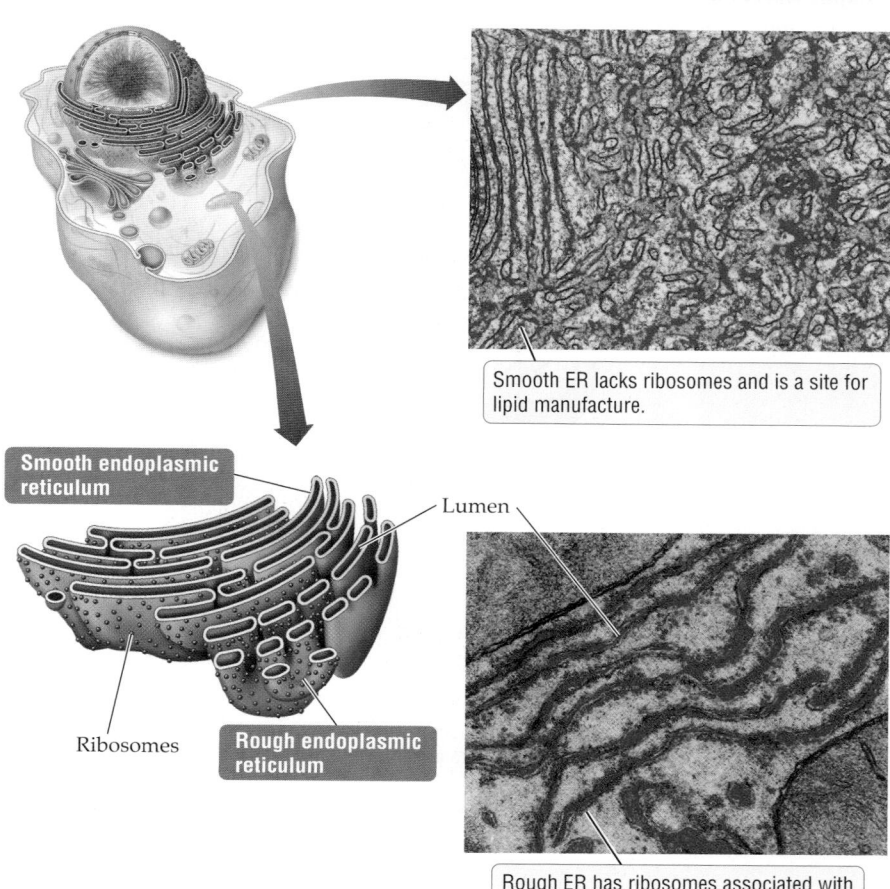

THE ENDOPLASMIC RETICULUM

Smooth ER lacks ribosomes and is a site for lipid manufacture.

Smooth endoplasmic reticulum

Lumen

Ribosomes

Rough endoplasmic reticulum

Rough ER has ribosomes associated with it and is a site for protein production.

**FIGURE 6.10 Some Types of Lipids and Proteins Are Made in the Endoplasmic Reticulum**

of the **smooth ER** manufacture various types of lipids destined for other cellular compartments, including the plasma membrane. In some cell types, smooth-ER membranes also break down organic compounds that could be toxic if they accumulated in the body. The potentially toxic compounds include certain plant chemicals and medicines. For example, the smooth ER in human liver cells has enzymes that break down caffeine, alcohol, and ibuprofen (the active ingredient in some painkillers). Animals that are mainly carnivorous tend to lack many of the ER enzymes that herbivores and omnivores use to degrade plant toxins, and some plant chemicals (such as those in chocolate) can therefore be harmful to them.

**Rough ER** is so named because of the bumpy appearance given to it by the many ribosomes attached on its cytosolic side. Some proteins manufactured on the rough ER go to other compartments within the cytoplasm; others are shipped to the cell surface,

No chocolate for Fido

**FIGURE 6.11**
**Cellular Materials Are Dispatched to a Wide Variety of Destinations via Vesicles**
Here, molecules are being shipped from the ER to the Golgi apparatus.

**TRANSPORT VESICLES**

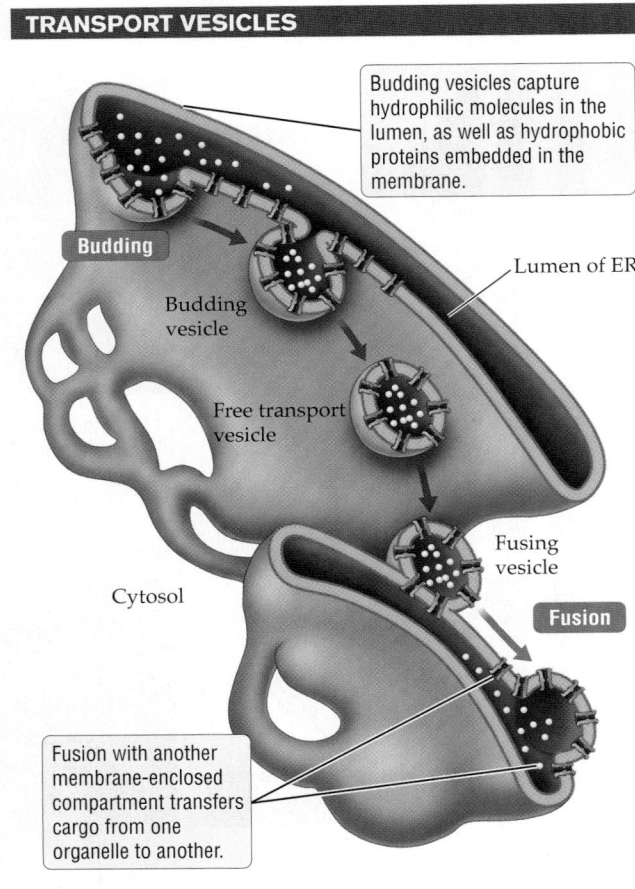

Budding vesicles capture hydrophilic molecules in the lumen, as well as hydrophobic proteins embedded in the membrane.

Lumen of ER

**Budding**

Budding vesicle

Free transport vesicle

Fusing vesicle

Cytosol

**Fusion**

Fusion with another membrane-enclosed compartment transfers cargo from one organelle to another.

either for incorporation into the plasma membrane or for release into the world outside the cell (the extracellular space).

## Transport vesicles move materials

How do macromolecules such as lipids, proteins, and carbohydrates move from one internal compartment to another, and even to the extracellular space? First the macromolecule in question is packaged into *transport vesicles*, which are like the carts used to move goods between different departments in a factory. A **transport vesicle** is a small, spherical, membrane-enclosed sac that moves material between cellular compartments (**FIGURE 6.11**). Hydrophobic cargo (such as lipids and hydrophobic proteins) is incorporated into the single membrane that forms the boundary of each transport vesicle. Hydrophilic material (mainly carbohydrates and hydrophilic proteins) is carried in the lumen of the transport vesicle.

A transport vesicle buds off from a membrane, such as the ER membrane, like a soap bubble emerging from a bubble blower. The vesicle delivers the cargo to its destination simply by fusing with the membrane of the target compartment, like a small soap bubble merging with a bigger one. The contents of a transport vesicle are determined by the compartment where it originated. For example, transport vesicles that pinch off from the ER enclose a small portion of the ER lumen and are enclosed by a patch of ER membrane.

**THE GOLGI APPARATUS**

Golgi stack

Golgi stack

Vesicle being formed

Free vesicle

Vesicle being received

Proteins and lipids are chemically modified as they transit from one Golgi compartment to the next.

**FIGURE 6.12 The Golgi Apparatus Routes Proteins and Lipids to Their Final Destinations**
Proteins and lipids are chemically modified, sorted, and shipped to their final destinations, inside or outside the cell, by the Golgi apparatus.

## The Golgi apparatus sorts and ships macromolecules

Another membranous organelle, the **Golgi apparatus**, directs proteins and lipids produced by the ER to their final destinations, either inside or outside the cell. The Golgi apparatus functions as a sorting station, much like the shipping department in a factory. In a shipping department, goods destined for different locations get address tags that indicate where they should be sent. Similarly, in the Golgi apparatus the addition of specific chemical groups to proteins and lipids helps target them to other destinations. The chemical tags include carbohydrate molecules and phosphate groups.

Under the electron microscope, the Golgi apparatus looks like a series of flattened membrane sacs stacked together and surrounded by many small transport vesicles (**FIGURE 6.12**). The vesicles move lipids and proteins from the ER to the Golgi apparatus and carry them between the various sacs of the Golgi.

## Lysosomes and vacuoles disassemble macromolecules

In animal cells, large molecules destined to be broken down are addressed to organelles called *lysosomes*. **Lysosomes** are membranous organelles that degrade macromolecules and release the subunits into the cytoplasm (*lyso*, "to break"; *soma*, "body"). In other words, lysosomes are the junkyard and recycling center of the cell. The macromolecules destined for destruction are delivered to lysosomes by transport vesicles. Even whole organelles, such as damaged mitochondria, can fuse with lysosomes and be taken apart in the lysosomal lumen. A single membrane forms the boundary of each lysosome (**FIGURE 6.13**). Inside it are a variety of enzymes, each specializing in degrading specific macromolecules, such as a particular class of lipid or protein. Many of the breakdown products—among them fatty acids, amino acids, and sugars—are transported across the lysosomal membrane and released into the cytoplasm for reuse. Lysosomes can adopt a variety of irregular shapes, but all are characterized by an acidic interior with a pH of about 5. (For comparison, the pH of the cytosol is close to 7.) Lysosomal enzymes work best at the low pH.

The plant organelles called **vacuoles** perform many of the same functions as the lysosomes of animal cells, as well as some others. Most mature plant cells have a central vacuole that can occupy more than a third of a plant cell's total volume (**FIGURE 6.14**).

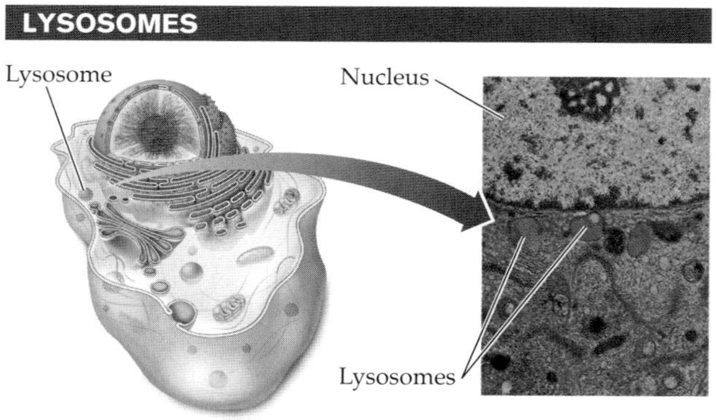

Lysosome    Nucleus

Lysosomes

**FIGURE 6.13 Lysosomes Degrade Macromolecules**
Lysosomes are found in animal cells. Lysosomes help to digest molecules taken up from outside and to break down cell components whose molecules can be repurposed.

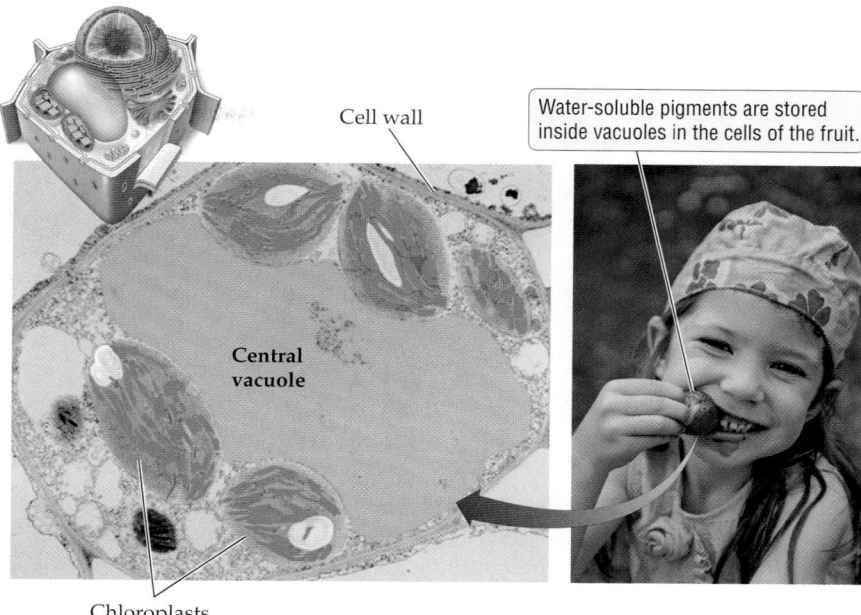

Cell wall

Water-soluble pigments are stored inside vacuoles in the cells of the fruit.

Central vacuole

Chloroplasts

**FIGURE 6.14 Plant Vacuoles Store, Recycle, and Provide Turgor**
Plant vacuoles contain enzymes for degrading large macromolecules. They also store water, ions, sugars, and other nutrients, and they may contain pigments that attract pollinators and/or toxins that deter herbivores. The fluid pressure that develops inside the vacuole gives turgidity to plant cells.

Besides containing enzymes that break down macromolecules, plant vacuoles store a variety of ions and water-soluble molecules: calcium ions, sugars, and colorful pigments, for example. Like lysosomes, most plant vacuoles have an acidic pH because they accumulate hydrogen ions. The fruit of the lemon is "impossible to eat" (as an old song proclaims) because it has cells ("juice sacs") with very large vacuoles containing mouth-puckering acids (pH about 2).

Some plant vacuoles stockpile noxious compounds that could deter feeding by herbivores. For example, the vacuoles of tobacco leaves accumulate nicotine, a nervous system toxin that is released when the leaf cells are damaged. Large vacuoles filled with water also contribute to the overall rigidity of the non-woody parts of a plant. The vacuolar contents exert a physical pressure, known as *turgor pressure*, against the cytoplasm, the plasma membrane, and the cell walls. Turgor pressure keeps a plant cell plumped up the way air pressure in an inner tube keeps a car tire inflated. Loss of turgor pressure leads to the droopy appearance of houseplants that have gone too long without water. Compartments similar to plant vacuoles are also found in fungi and some protists, though the different types of vacuoles differ in how they work and the functions they perform.

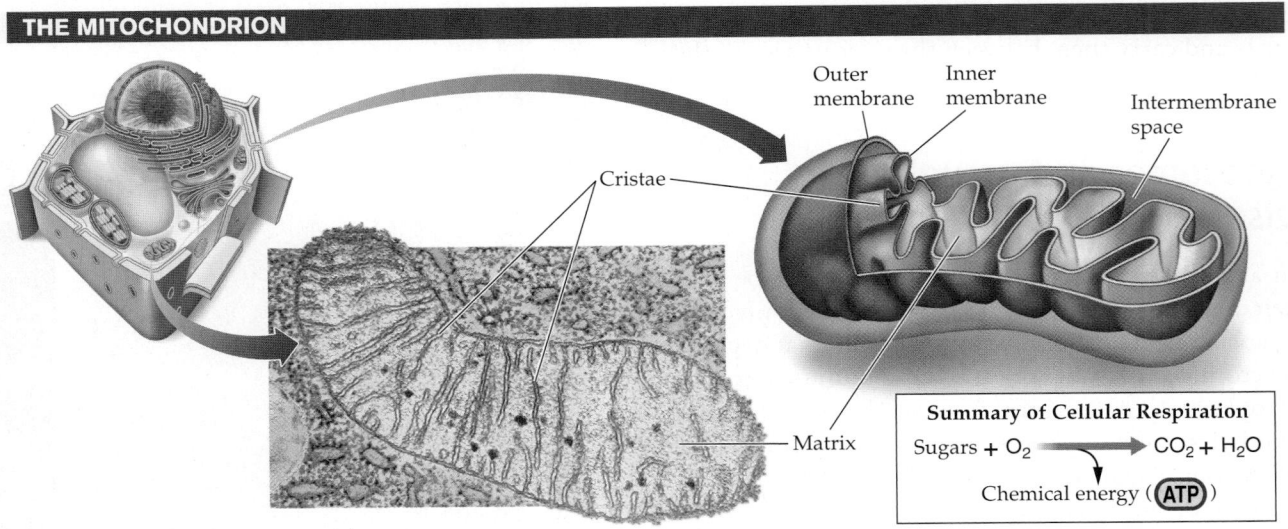

Outer membrane
Inner membrane
Intermembrane space
Cristae
Matrix

**Summary of Cellular Respiration**

$$Sugars + O_2 \longrightarrow CO_2 + H_2O$$

Chemical energy ( **ATP** )

**FIGURE 6.15 Mitochondria Generate Energy in the Form of ATP**

Each mitochondrion has a double membrane. The infoldings of the inner membrane (cristae) create a large surface area which enables many units of ATP-generating enzymes to be located there.

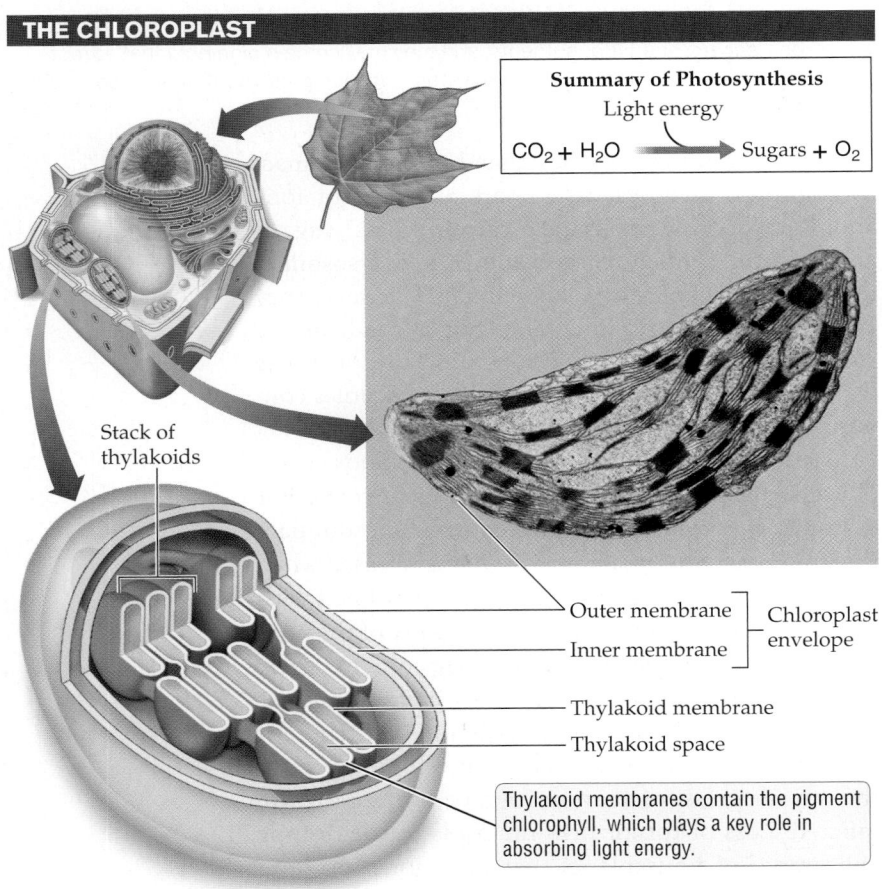

**Summary of Photosynthesis**

Light energy

$$CO_2 + H_2O \longrightarrow Sugars + O_2$$

Stack of thylakoids

Outer membrane ⎤ Chloroplast
Inner membrane ⎦ envelope

Thylakoid membrane
Thylakoid space

Thylakoid membranes contain the pigment chlorophyll, which plays a key role in absorbing light energy.

**FIGURE 6.16 Chloroplasts Capture Energy from Sunlight and Use It to Make Sugars**

Chloroplasts are found in green plant parts and in the protists known as algae.

## Mitochondria power the cell

So far, we have explored the administrative offices, factory floors, and shipping department within the cellular factory. However, none of the offices or specialized departments could function without a source of energy to run the machines that produce the goods and services. In most eukaryotic cells, the main source of this energy is the **mitochondrion**. This double-membrane organelle fuels cellular activities by extracting energy from food molecules. We shall see shortly that photosynthetic organisms, such as plants, have an additional unit, called the chloroplast, which uses sunlight to make energy-storing molecules ("food molecules"). However, all kinds of eukaryotes, photosynthetic or not, need mitochondria to convert the chemical energy in food molecules into a form useful for powering cellular activities.

Mitochondria are pod shaped and bound by double membranes—that is, by two distinctly different lipid bilayers. The space between the two membranes is called the **intermembrane space**. The inner mitochondrial membrane is thrown into large folds, called **cristae** [*KRIS*-tee] (singular "crista") that present a large surface area. The space interior to the cristae is called the **matrix** (**FIGURE 6.15**). Mitochondria use chemical reactions to transform the energy of food molecules into *ATP*, the universal cellular fuel. The energy stored in the covalent bonds of ATP is used, in turn, to power the many chemical reactions of the cell.

The production of ATP by mitochondria is critically dependent on the activities of proteins embedded

# Organelles and Human Disease

The normal functioning of cellular structures is crucial to our survival. The *complete* failure of any one of the organelle types in *all* cells of the body would be incompatible with life. Small errors in the workings of an organelle can make us ill, sometimes very ill. For example, mitochondrial malfunctions are implicated in a variety of *neurodegenerative diseases*, those that result in the death of brain or nerve cells. Defects in ER enzymes produce a host of disorders broadly classified as ER dysfunctions. Similarly, defects in Golgi apparatus enzymes lead to a variety of disorders, such as the carbohydrate-deficient glycoprotein (CDG) syndromes. Many of these conditions are genetic disorders, caused by a corruption of the DNA code that normally directs production of a functional protein. However, environmental substances can also damage organelles. *Asbestosis* is the destruction of lung tissue from exposure to asbestos, a mineral substance. Asbestos fibers are engulfed by immune cells, whose job is to defend the body from foreign substances. Once inside a cell, the fibers accumulate in lysosomes and, over time, damage the lysosomal membrane; release of lysosomal contents into the cytosol will kill any cell.

The vital importance of healthy lysosomes is underscored by the existence of more than

The movie *Lorenzo's Oil* tells the true story of a couple's battle to prolong the life of their son Lorenzo, who was born with a peroxisome disorder called adrenoleukodystrophy (ALD). Peroxisomes are membranous organelles that degrade lipids, and patients with ALD have a defective transporter in their peroxisome membrane that fails to import very long chain fatty acids (VLFAs) that are common in normal diets. The fatty acids pile up in cytosol; the accumulation is especially destructive in brain and nerve cells and in the hormone-producing organs known as adrenal glands. Untrained in scientific research, the couple developed a vegetable oil formula that slows the buildup of the fatty acids. Lorenzo died of pneumonia at age 30, more than two decades later than his doctors had predicted.

40 different types of lysosomal storage disorders in humans. These inherited conditions are caused by the malfunction of one or more of the many lysosomal enzymes whose job it is to degrade specific macromolecules. When a lysosomal enzyme is absent or fails to work properly, the macromolecule it would normally degrade piles up inside the lysosomes. The consequences are devastating, and most of these disorders are fatal in childhood.

Tay-Sachs disease is one such metabolic disorder. The lysosomal enzyme responsible for taking apart a membrane lipid found in brain tissue fails to do its job, with the result that large amounts of this lipid accumulate in nerve cells, compromising the function of these cells and eventually destroying them. Tay-Sachs disease is rare in the population as a whole, but it was once unusually common in some populations that tend to marry among themselves, such as certain French-Canadian communities and Orthodox Jews from eastern Europe. Through a combination of genetic testing and genetic counseling for individuals contemplating marriage, Tay-Sachs disease has been virtually eradicated among Jewish communities in North America and Israel.

---

in the cristae. The availability of the intermembrane space and the matrix as two separate and distinct compartments is also crucial for the process. With this unique setup, mitochondria are able to trap some of the chemical energy released when food molecules are broken down and to use some of that energy to synthesize energy-rich ATP, through a process that requires oxygen. In other words, oxygen gas ($O_2$) and molecules derived from food are the raw materials that fuel the mitochondrial power station. As in most human-made power stations, carbon dioxide ($CO_2$) and water ($H_2O$) are released as by-products of this energy-producing process, which is called **cellular respiration**. We will delve into the details of cellular respiration in Chapter 9.

## Chloroplasts capture energy from sunlight

Mitochondria provide life-sustaining ATP to eukaryotic cells (both plant and animal), but the cells of plants and the protists known as algae have additional organelles, called **chloroplasts** (**FIGURE 6.16**), that capture energy from sunlight and use it to manufacture food molecules. The light energy is first trapped in the form of energy carriers such as ATP, which are then used to assemble sugar molecules from carbon dioxide ($CO_2$) and water ($H_2O$) in a process called **photosynthesis**. The energy in these sugars is used directly by plant cells and indirectly by all organisms that eat plants. At this very moment, as you read this page, your brain and the muscles that move your eyes are using energy from food molecules that were originally produced in chloroplasts through photosynthesis.

During photosynthesis, water molecules are broken down, releasing oxygen gas. The oxygen produced in photosynthesis sustains life for us and many other life-forms. Mitochondria depend on a continual supply of that oxygen to produce ATP in a process that is essentially the reverse of photosynthesis. Photosynthesis *generates* oxygen as a by-product when *making* food molecules (sugars) from carbon dioxide and water; cellular respiration in the mitochondrion *consumes* oxygen while *breaking down* food molecules and releasing carbon dioxide and water as by-products.

Chloroplasts are enclosed by two concentric membranes that form the chloroplast envelope. Inside the envelope lies an internal network of membranes, some of which are arranged like stacked pancakes (see Figure 6.16). Each "pancake" in the stack is called a **thylakoid** [*THYE*-luh-koyd]. Embedded in the thylakoid membranes are special light-absorbing pigments, notably **chlorophyll**, that enable chloroplasts to capture energy from sunlight. Chlorophyll is green because it does *not* absorb green wavelengths, which bounce off and reach our eyes. (Our perception of an object as having a particular color is based on the wavelengths of light the object *reflects*, not the wavelengths it absorbs.) The light absorbed by chlorophyll (mainly blue and red wavelengths) is used to generate energy carriers such as ATP. Enzymes present in the space surrounding the thylakoids use the energy carriers to synthesize (manufacture) carbohydrates from water and from carbon dioxide that the plant absorbs from the air around it.

> **Concept Check**
>
> 1. Which macromolecules are made in (a) the smooth ER and (b) the rough ER?
> 2. What is the role of the Golgi apparatus?
> 3. List one important function performed by the lysosomes. What organelle performs a similar function in plant cells?
> 4. Chloroplasts and mitochondria both make ATP. What is the crucial difference in the function of these organelles?

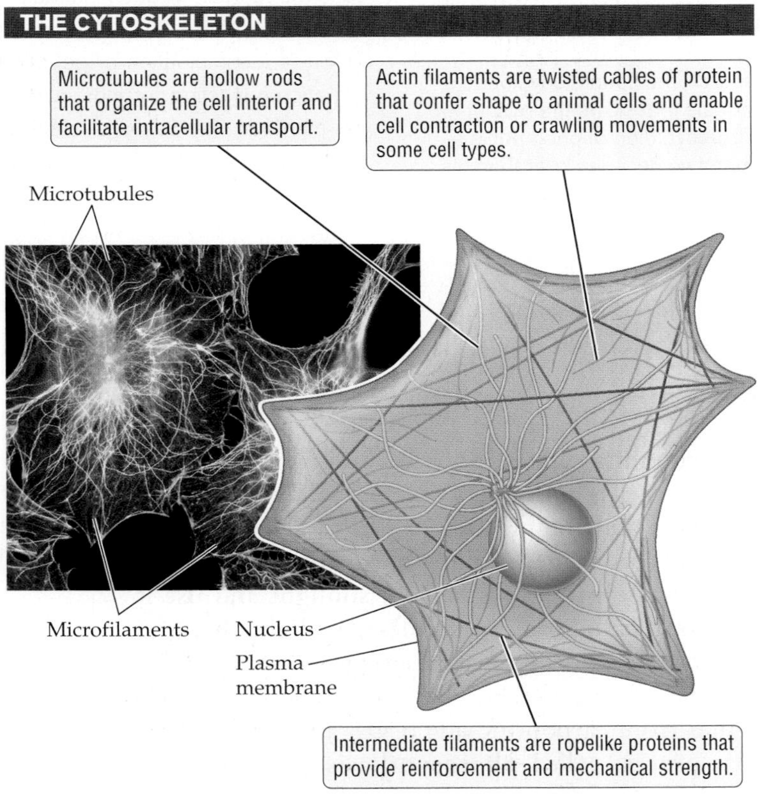

## 6.5 The Cytoskeleton

The eukaryotic cell is not simply a formless bag of membrane with cytosol and organelles sloshing around inside. A network of protein cylinders and filaments, collectively known as the **cytoskeleton**, organizes the interior of the cell, supports the intracellular movement of organelles such as transport vesicles, gives shape to wall-less cells, and even enables whole-cell movement in some cell types (**FIGURE 6.17**). Tubelike cytoskeletal structures (called *microtubules*) pin organelles such as the ER and Golgi in specific positions and create tracks on which transport vesicles and other cellular particles can move. Many animal cells are reinforced by ropelike bands of proteins (*intermediate filaments*) that strengthen the cell and keep it from distorting or even rupturing in response to mechanical forces. A meshwork of cytoskeletal structures (called *microfilaments*) creates a scaffold that supports the plasma membrane and gives shape and mechanical strength to cells that lack cell walls, such as those of animals and many protists. Wall-less cells can change shape rapidly if the cytoskeletal scaffolding is dismantled or rearranged.

Cells like *Amoeba* that creep along on a solid surface are also crucially dependent on the changeable nature of cytoskeletal components (mainly microfilaments), because cell crawling involves dramatic changes in cell shape, particularly the extension and retraction of the

**THE CYTOSKELETON**

Microtubules are hollow rods that organize the cell interior and facilitate intracellular transport.

Actin filaments are twisted cables of protein that confer shape to animal cells and enable cell contraction or crawling movements in some cell types.

Microtubules

Microfilaments    Nucleus

Plasma membrane

Intermediate filaments are ropelike proteins that provide reinforcement and mechanical strength.

**FIGURE 6.17 An Overview of the Cytoskeletal System**

plasma membrane at the leading and trailing edges of the cell, respectively. Organisms that can swim in a liquid, like the *Paramecium* and *Volvox* we met earlier, depend on cytoskeletal elements (microtubules) for the movement.

## The cytoskeleton consists of three basic components

Microtubules, intermediate filaments, and microfilaments are the key components of the cytoskeleton (**FIGURE 6.18**). **Microtubules** are relatively rigid, hollow cylinders of protein that help position organelles, move transport vesicles and other organelles, and generate force in cell projections such as the cilia or flagella found in some eukaryotic cells. **Intermediate filaments** are ropelike cables of protein, the nature of which can vary from one cell type to another, that provide mechanical reinforcement for the cell. **Microfilaments** are the thinnest and most flexible of the three types of cytoskeletal structures. Microfilaments consist of strands of protein and are involved in creating cell shape and generating the crawling movements displayed by some eukaryotic cells.

## Microtubules support movement inside the cell

Microtubules are the thickest of the cytoskeleton filaments, with a diameter of about 25 nanometers (nm). Each microtubule is a cylindrical structure made from protein subunits called **tubulin** [TOO-**byoo-lin**] (**FIGURE 6.18a**). A microtubule can grow or shrink in length by adding or losing tubulin monomers. This capacity allows microtubules to form dynamic structures capable of rapidly altering the internal organization of the cell, or of capturing organelles and hauling them through the cytosol. Most animal cells have a system of microtubules that radiate from the center of the cell and terminate at the cytosolic face of the plasma membrane (see Figure 6.17). The radial pattern of microtubules serves as an internal scaffold that helps position organelles such as the ER and the Golgi apparatus.

Microtubules also define the paths along which vesicles are guided in their travels from one organelle to another, or from an organelle to the cell surface. The ability of microtubules to act as "railroad tracks" for vesicles depends on the action of *motor proteins*, which attach to the vesicle by their "tail" end and associate with microtubules through their "head" end.

**THE CYTOSKELETON: THREE MAIN COMPONENTS**

**(a) Microtubules**

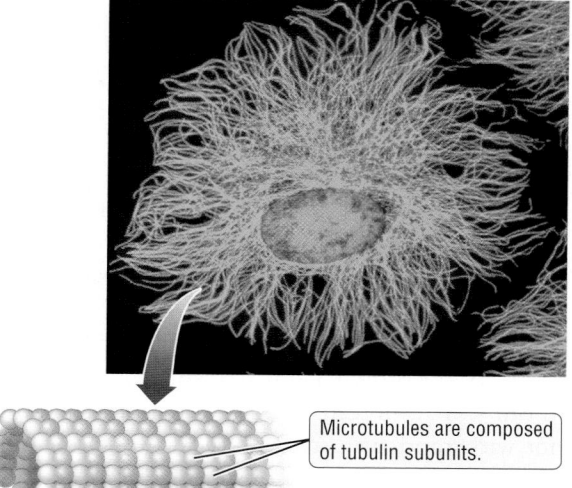

Microtubules are composed of tubulin subunits.

**(b) Intermediate filaments**

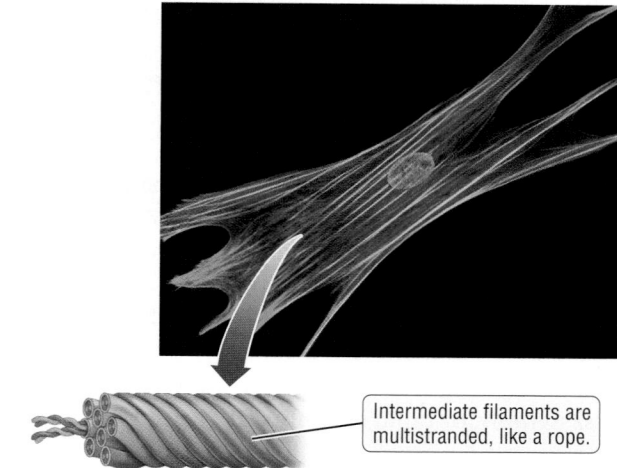

Intermediate filaments are multistranded, like a rope.

**(c) Microfilaments**

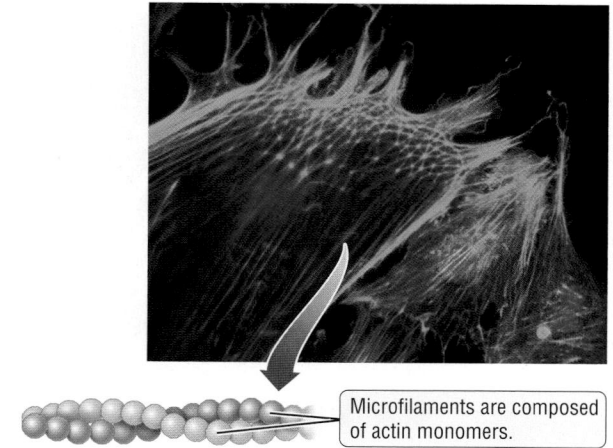

Microfilaments are composed of actin monomers.

**FIGURE 6.18**
**The Structure of Microtubules, Intermediate Filaments, and Microfilaments**
The cytoskeleton is composed of three basic units: (*a*) microtubules, (*b*) intermediate filaments, and (*c*) microfilaments.

### Helpful to Know

"Skeleton" may seem an odd word to apply to cell structures, since cells do not have bones as we do. However, biologists give the word a larger meaning: any structure or framework that supports and/ or protects a cell or an organism. "Cytoskeleton" refers to the structures within cells (*cyto*, "cell") that give them shape and structure. "Endoskeletons" (*endo*, "inner") are found in many animals, such as fish and mammals; and "exoskeletons" (*exo*, "outer") are found in insects and crustaceans.

Motor proteins convert the energy of ATP into mechanical force, which enables them to move along a microtubule in a specific direction, carrying attached cargo such as a vesicle.

## Intermediate filaments provide mechanical reinforcement

Intermediate filaments are a diverse class of ropelike filaments about 8–12 nm in diameter. They are thinner than microtubules but thicker than microfilaments (**FIGURE 6.18b**). Intermediate filaments serve as structural supports, in the way beams and girders support a building. For example, intermediate filaments consisting of the protein keratin strengthen the living cells in our skin. Skin cells lacking functional keratin cannot withstand even mild physical pressure, and their rupture results in severe blistering and other types of skin lesions. Intermediate filaments also provide mechanical reinforcement for internal cell membranes. The nuclear envelope, for example, is supported by an underlying meshwork of intermediate filaments.

## Microfilaments are involved in cell movement

Of the three filament types, microfilaments have the smallest diameter, about 7 nm (**FIGURE 6.18c**). Each microfilament is a cable formed by two polymeric strands twisted around each other. Each strand is built from monomers of the protein **actin**. Like microtubules, microfilaments are dynamic structures that can shorten or lengthen in either direction through rapid disassembly, or rapid assembly, at one or both ends.

Perhaps the best example of the rapid changes in microfilaments is illustrated by a cell creeping along on a solid surface. White blood cells, for example, begin crawling by throwing forward a bump of plasma membrane known as a *pseudopodium* (*pseudo*, "false"; *podium*, "foot") (plural "pseudopodia"). In quick succession, the opposite end of the cell retracts by lifting the trailing portion of the plasma membrane off the substratum and pulling it forward (**FIGURE 6.19**). The movements are made possible by the rapid rearrangements of microfilaments at the leading edge of the cell and at the trailing edge. At the leading edge, microfilaments lengthen in parallel arrays, pushing the plasma membrane out to form the pseudopodia. At this time, microfilaments at the trailing end of the cell display random organization, angling in all directions. Next, the randomly oriented microfilaments disassemble altogether, enabling the cell to pick up its trailing end in preparation for scooting forward. The plasma membrane at the trailing end detaches from the solid surface, while motor proteins generate forces that pull the entire rear of the cell forward.

Cell crawling enables protists such as amoebas and slime molds to find food and mating partners. Fibroblasts, which play an important role in the healing of skin wounds, migrate to the site of injury using the microfilament-based system of cell crawling. Cell migration is also crucial in the embryonic development of many animals. However, cell migration is a devastating step in the development of cancer because it enables cancer cells to invade other tissues and even spread through the body.

## Cilia and flagella enable whole cell movement

Many protists and animals have cells covered in a large number of hairlike projections, called **cilia** (singular "cilium"), that can be moved back and forth, like the oars of a rowboat, to move the whole cell through a liquid or to move a liquid over the cell surface. Inside each cilium is a flexible cytoskeletal apparatus created by bundles of microtubules. This apparatus consists of nine pairs ("doublets") of microtubules arranged in a ring around a central pair. Motor proteins interlinking the microtubules use the energy of ATP to flex the microtubules against each other, causing the whole cilium to bend. Many aquatic protists, like the *Paramecium* shown in Figure 6.1b, use cilia to move about in the waters they inhabit. Some cells use cilia not to move themselves around, but to move an overlying fluid layer. This is true of the cells that line portions of our

**HOW MICROFILAMENTS ENABLE CELL CRAWLING**

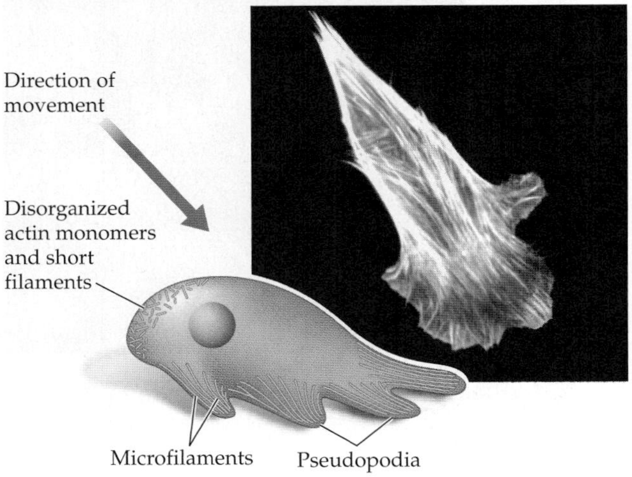

Direction of movement

Disorganized actin monomers and short filaments

Microfilaments    Pseudopodia

**FIGURE 6.19**
**Microfilaments Drive Some Types of Whole Cell Movement**

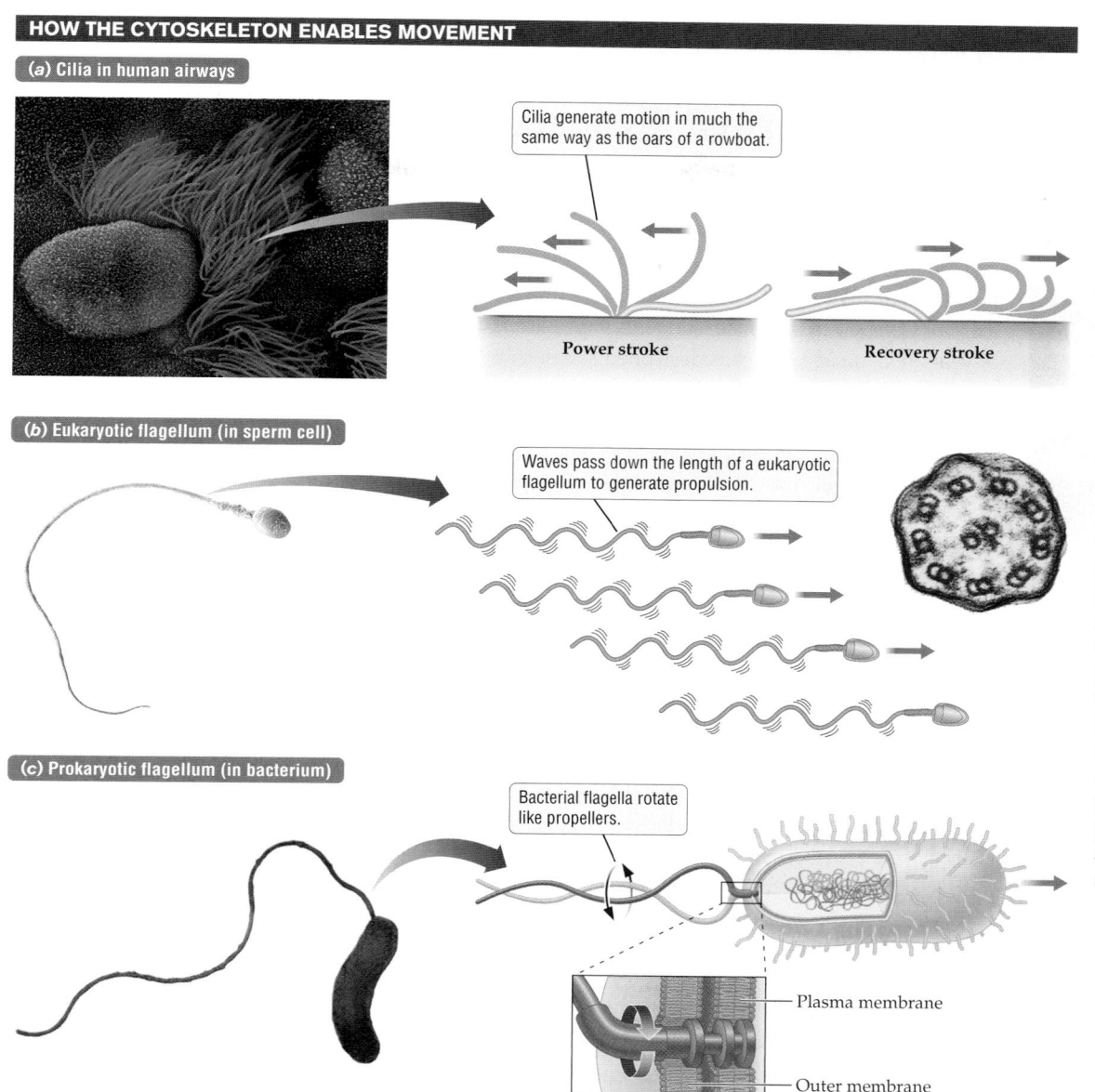

(a) Cilia in human airways

Cilia generate motion in much the same way as the oars of a rowboat.

Power stroke

Recovery stroke

(b) Eukaryotic flagellum (in sperm cell)

Waves pass down the length of a eukaryotic flagellum to generate propulsion.

(c) Prokaryotic flagellum (in bacterium)

Bacterial flagella rotate like propellers.

Plasma membrane

Outer membrane

**FIGURE 6.20**

**Cilia and Flagella Generate Movement**

Many organisms use cilia or flagella to generate movement. (a) Tufts of cilia are present on the cells that line our breathing tubes (bronchi). (b) Eukaryotic flagella, such those in sperm cells, are much longer than cilia. Eukaryotic cilia and flagella contain bundles of microtubules arranged in a 9+2 pattern (inset) and are covered by a plasma membrane. Prokaryotic flagella have a very different structure. (c) A prokaryotic flagellum, such as the one on this bacterium (*Bdellovibrio bacteriovorus*), consists of ropelike proteins attached to protein complexes anchored in the cell membranes.

respiratory passages (**FIGURE 6.20a**). The cells use the cilia on their surface to propel unwanted material, caught in a layer of mucus, out of the lungs and into the throat for elimination by coughing or swallowing.

Many bacteria, archaeans, and protists, as well as the sperm cells of some plants and all animals, can propel themselves through a fluid using one or more whiplike structures called **flagella [fluh-*JEL*-uh]** (singular "flagellum"). **Eukaryotic flagella** are lashed in a pattern that resembles the movement of a circus ringmaster's whip (**FIGURE 6.20b**). They are much longer than cilia, but the internal structure of the two is very similar, and both are covered by a lipid bilayer that is

an extension of the plasma membrane. **Prokaryotic flagella**, however, lack a membrane covering, have a very different internal structure, and are believed to have evolved separately from eukaryotic flagella (**FIGURE 6.20c**). Instead of the whiplike motion displayed by eukaryotic flagella, prokaryotic flagella spin in a rotary motion, rather like a boat's propellers.

> ## Concept Check
>
> 1. List three important functions performed by the cytoskeleton.
> 2. Compare eukaryotic cilia and flagella.

# The Evolution of Eukaryotes

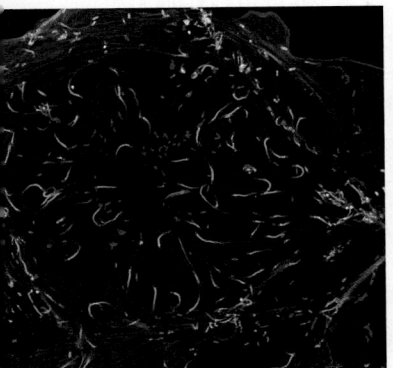

The entry of one cell into another is not always a hostile takeover. Sometimes when two cells merge, they enter into a long-term stable relationship that benefits both of them—a form of symbiosis known as a *mutualism*. Compelling evidence suggests that billions of years ago, prokaryotic cells developed just such a mutualistic arrangement with other cells, giving rise to the ancestors of modern eukaryotic cells.

Unlike most organelles, chloroplasts and mitochondria have their own DNA and divide—like prokaryotes—through fission, reproducing independently from the cells in which they live. This amazing and important fact was not fully understood or appreciated until 1967, when biologist Lynn Margulis proposed that chloroplasts and mitochondria might be former prokaryotes that had come to live inside a bigger cell.

In 1970 Margulis published a book elaborating on the idea that cells evolve through symbiosis, bolstering her arguments with detailed observations. A few years later, some researchers decided to test Margulis's idea. They predicted that if Margulis were right, a chloroplast's genes would resemble those of a photosynthetic bacterium more than those of the nucleus of the eukaryotic cell in which the chloroplast lived. The same would be true for mitochondrial DNA. Sure enough, the genes of both chloroplasts and mitochondria resembled those of certain prokaryotes.

It took years of tenacious argument for Margulis to fully persuade most other biologists. Today, modern biologists accept the "**endosymbiont theory**" that mitochondria, chloroplasts, and certain other organelles coevolved inside larger cells, which gave rise to the eukaryotic lineage.

Besides dividing by fission as prokaryotes do and having genes like those of certain prokaryotes, mitochondria and chloroplasts also resemble certain prokaryotes in other ways. First, both have an arrangement of membranes that looks exactly as if a bacterium became enveloped in a fold of eukaryotic cell membrane: in both organelles the outer membrane resembles the plasma membrane of eukaryotes, and the innermost membrane resembles the plasma membrane of prokaryotes. Second, both organelles have their own DNA that is separate from the cell's nuclear DNA. Each organelle's DNA is arranged in a small, closed loop—just like prokaryotic DNA. Finally, mitochondria and chloroplasts have their own ribosomes, which resemble those of prokaryotes.

These amazing resemblances support the theory that mitochondria and chloroplasts were once free-living prokaryotes that came to live inside ancient eukaryotes. According to the theory, a relatively large predatory cell engulfed a prokaryotic neighbor in a digestive vesicle (**FIGURE 6.21**). Instead of being digested, however, the smaller prokaryote survived and the two organisms coevolved, forming a symbiotic partnership. The first primitive eukaryotes probably arose between 2.7 billion and 2.1 billion years ago. Biologists hypothesize that large cells preyed on smaller ones by letting their outer cell membranes fold inward to form internal membranes and vacuole-like structures in which their prokaryotic prey could be engulfed.

But a prokaryote that survived and moved into a eukaryote host got a great deal: shelter from predators and a steady food supply. Prokaryotes with an outstanding capacity to convert food molecules into ATP evolved into mitochondria, sharing a cornucopia of ATP with their hosts. Meanwhile, photosynthetic cyanobacteria that moved in with a eukaryotic landlord paid "rent" in the form of sunlight-generated sugar.

**FIGURE 6.21 How Ancestral Eukaryotes Acquired Membrane-Enclosed Organelles**

Some organelles, such as mitochondria and chloroplasts, are likely descendants of engulfed prokaryotes. Other membrane-enclosed organelles, such as the endoplasmic reticulum, probably arose through an infolding of the plasma membrane.

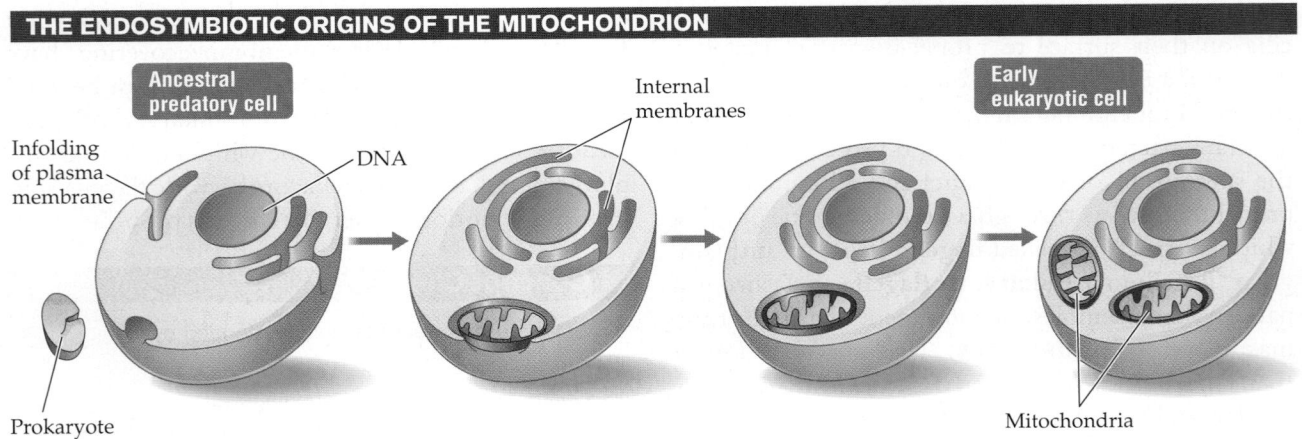

THE ENDOSYMBIOTIC ORIGINS OF THE MITOCHONDRION

Ancestral predatory cell

Internal membranes

Early eukaryotic cell

Infolding of plasma membrane

DNA

Prokaryote

Mitochondria

# New Mosquito Presents a New Challenge

BY AMINA KHAN, *Los Angeles Times*

Researchers have discovered a previously unknown subspecies of mosquito in West Africa that is highly susceptible to the malaria parasite and whose existence may stymie efforts to eradicate the deadly disease.

Unlike the indoor-dwelling mosquitoes that are the usual targets of malaria eradication efforts, members of the newly described subgroup of the species *Anopheles gambiae* live outdoors, which means they're more difficult to kill, according to the study published online Thursday in the journal *Science*.

"We've got egg on our face," said William Black, a medical entomologist at Colorado State University who was not involved in the study. "We've been working with this mosquito for so long … and right under our noses, here's this other form of mosquito," he said—one that could force researchers "to start thinking about what's going on outside of those huts."

Malaria is an incurable blood disease caused by parasites in the genus *Plasmodium*. Mosquitoes spread the parasite by drawing blood from infected humans, becoming carriers themselves and then transmitting the parasite to other humans they bite.

Malaria victims suffer headaches, high fever, chills, vomiting, anemia and sometimes death. There are about 250 million cases and nearly 1 million deaths each year, according to the World Health Organization.

*Anopheles gambiae* is thought to be the most significant malaria transmitter in Africa, and it has resisted repeated efforts to wipe it out, said Kenneth Vernick, a vector geneticist at the Pasteur Institute in Paris who co-wrote the study. Malaria control projects going back to the 1970s have apparently failed because they used tactics such as indoor sprays and chemically treated bed netting that only targeted mosquitoes inside human homes.

One of the reasons for focusing on indoor insects "is that people imagine that mosquitoes that are in proximity to people might be the most epidemiologically important to transmit malaria," Vernick said. But there's another, less scientific reason as well, he added: "Because it's easy."

Outdoor mosquitoes, on the other hand, are notoriously difficult to kill, let alone capture to study in experiments. Without the benefit of walls and a roof, Vernick said, "you don't have a defined space to spray-bomb and catch them and have them fall on a sheet."

Instead of chasing after the adult insects, the researchers decided to search for them in the larval stage, when they're easier to capture. Mosquitoes breed close to human habitations in puddles of standing water. In West Africa, standing bodies of water are hard to come by, so all types of mosquitoes would likely be forced to use common puddles to hatch their young.

The researchers traveled across Burkina Faso, collecting larvae from puddles around villages and growing thousands of mosquitoes over a four-year period. They genetically analyzed the young and found that only 43% of the larvae in the puddles were the indoor variety.

When the researchers fed the mosquitoes blood from humans who had malaria, 58% of the outdoor mosquitoes became infected with the parasite, compared with 35% of the indoor-resting ones.

Malaria is a tropical disease that sickens 200–300 million people each year and kills as many as a million. Almost 90 percent of malaria deaths occur in Africa, mainly in pregnant women and in children under five. DNA analysis suggests that *Plasmodium falciparum* has been invading human cells and coevolving with humans for at least 50,000 years—as along as modern humans have been around.

*Plasmodium falciparum* is a one-celled protist that causes malaria by cycling back and forth between humans and mosquitoes. A mosquito becomes infected when it sucks blood from a human infected with *P. falciparum*. The next time the mosquito bites another human, the parasite enters the human bloodstream and heads for the liver for 6 days of rest and reproduction. After that, 30,000–40,000 daughter cells of the protist slip out of the liver and into the bloodstream to invade human red blood cells. Each *P. falciparum* cell has a doughnut-shaped pore that it attaches to the outside of a human red blood cell. The parasite then squeezes through the pore and begins multiplying inside the red blood cell. The next time a mosquito takes a blood meal, the parasites move into the mosquito and another round of infection begins.

## Evaluating the News

1. How does the malaria parasite enter a human red blood cell?

2. Plasmodial parasites are relatives of *Amoeba*. Describe the type of cellular structures it uses to move about in the human body. Why do you think red blood cells and liver cells are susceptible to invasion, while other cells are left alone?

3. How do you think that knowing how these one-celled parasites invade a cell could help biomedical researchers come up with drugs and other treatments that might cure or treat malaria?

**SOURCE:** *Los Angeles Times*, February 4, 2011.

# Summary

## 6.1 Cells: The Smallest Units of Life

- Cells are the basic units of all living organisms.
- Most cells are small because the ratio of surface area to volume limits cell size. As a cell's width increases, its volume increases vastly more than its surface area, so a larger cell has proportionately less plasma membrane area to import and export substances but must support a much larger cytoplasmic volume.
- Broadly speaking, a larger organism is a more effective predator, less susceptible prey, and better able to obtain and store nutrients.
- A multicellular organism is a closely integrated group of cells, with a common developmental origin, whose constituent cells are incapable of living independently.
- Multicellularity enabled organisms to attain larger size, and it conferred the added advantage of greater efficiency through division of labor among the multiple cell types.

## 6.2 The Plasma Membrane

- Every cell is surrounded by a plasma membrane that separates the chemical reactions of life from the surrounding environment.
- According to the fluid mosaic model, the plasma membrane is a highly mobile assemblage of lipids and proteins, many of which can move within the plane of the membrane.
- Proteins in the plasma membrane perform a variety of functions. Receptor proteins facilitate communication, transport proteins mediate the movement of substances across the membrane, and adhesion proteins help cells attach to one another.

## 6.3 Prokaryotic and Eukaryotic Cells

- Living organisms are classified as either prokaryotes or eukaryotes.
- Prokaryotes are single-celled organisms lacking a nucleus and complex internal compartments. Eukaryotes may be single-celled or multicellular, and their cells typically possess many membrane-enclosed compartments, such as the nucleus.
- The cytoplasm is all the cell contents enclosed by the plasma membrane. It consists of an aqueous cytosol (a thick fluid that contains many ions and molecules) and organelles (internal structures with unique functions).
- By volume, eukaryotic cells can be 1,000 times larger than prokaryotic cells. They require internal compartments that concentrate and organize cellular chemical reactions for optimal function.

## 6.4 Internal Compartments of Eukaryotic Cells

- The nucleus contains DNA. It is bounded by the nuclear envelope, which has many pores. Information stored in DNA is conveyed by RNA molecules to the cytoplasm.
- Lipids are made in the smooth endoplasmic reticulum (ER). Some proteins are manufactured in the rough ER.
- Molecules move among organelles via vesicles that bud off one compartment to fuse with a target membrane.
- The Golgi apparatus receives proteins and lipids, sorts them, and directs them to their final destinations.
- Lysosomes break down large organic molecules such as proteins into simpler compounds that can be used by the cell. Vacuoles are similar to lysosomes but also store ions and molecules and lend physical support to plant cells.
- Mitochondria produce chemical energy for eukaryotic cells in the form of ATP.
- Chloroplasts harness the energy of sunlight to make sugars through photosynthesis.

## 6.5 The Cytoskeleton

- Eukaryotic cells depend on the cytoskeleton for structural support, and for the ability to move and change shape.
- The cytoskeleton consists of three types of filaments: microtubules, intermediate filaments, and microfilaments. Microtubules position organelles and can move them inside the cell. Microfilaments give shape to the cell and enable cell crawling. Intermediate filaments provide mechanical strength to cells.
- Some protists, sperm cells, archaeans, and bacteria move using cilia or flagella. Eukaryotic flagella are different in structure and action from prokaryotic flagella.

# Key Terms

actin (p. 160)
cell (p. 144)
cell theory (p. 144)
cellular respiration (p. 157)
chlorophyll (p. 158)
chloroplast (p. 157)
chromosome (p. 152)
cilium (p. 160)
crista (p. 156)
cytoplasm (p. 144)
cytoskeleton (p. 158)
cytosol (p. 144)

endoplasmic reticulum (ER) (p. 153)
endosymbiont theory (p. 162)
eukaryote (p. 150)
eukaryotic flagellum (p. 161)
extracellular matrix (ECM) (p. 150)
flagellum (p. 161)
fluid mosaic model (p. 150)
Golgi apparatus (p. 154)
intermediate filament (p. 159)
intermembrane space (p. 156)

lumen (p. 153)
lysosome (p. 155)
matrix (p. 156)
microfilament (p. 159)
microtubule (p. 159)
mitochondrion (p. 144)
multicellular organism (p. 148)
nuclear envelope (p. 152)
nuclear pore (p. 153)
nucleus (p. 144)
organelle (p. 144)

photosynthesis (p. 157)
plasma membrane (p. 144)
prokaryote (p. 150)
prokaryotic flagellum (p. 161)
ribosome (p. 144)
rough ER (p. 153)
smooth ER (p. 153)
thylakoid (p. 158)
transport vesicle (p. 154)
tubulin (p. 159)
vacuole (p. 155)

# Self-Quiz

1. In contrast to the average prokaryotic cells, eukaryotic cells
   a. have no nucleus.
   b. have many different types of internal compartments.
   c. have ribosomes in their plasma membranes.
   d. lack a plasma membrane.

2. Which of the following would be found in a plasma membrane?
   a. proteins
   b. DNA
   c. mitochondria
   d. endoplasmic reticulum

3. Which of the following organelles has ribosomes attached to it?
   a. Golgi apparatus
   b. smooth endoplasmic reticulum
   c. rough endoplasmic reticulum
   d. microtubule

4. Which organelle captures energy from sunlight?
   a. mitochondrion
   b. cell nucleus
   c. Golgi apparatus
   d. chloroplast

5. Which organelle uses oxygen to extract energy from sugars?
   a. chloroplast
   b. mitochondrion
   c. nucleus
   d. plasma membrane

6. Which organelle contains both thylakoids and cristae?
   a. chloroplast
   b. mitochondrion
   c. nucleus
   d. none of the above

7. The internal system of protein cables and cylinders that makes whole-cell movement possible is called the
   a. endoplasmic reticulum.
   b. cytoskeleton.
   c. lysosomal system.
   d. mitochondrial matrix.

8. Which of the following is *not* part of the cytoskeleton?
   a. pseudopodium
   b. intermediate filament
   c. microtubule
   d. microfilament

9. How is a prokaryotic flagellum different from a eukaryotic flagellum?
   a. It moves in a whiplike manner.
   b. It is not covered by plasma membrane.
   c. It evolved from eukaryotic flagella.
   d. It is composed of many cilia.

10. Which of the following organelles are thought to have arisen from primitive prokaryotes?
    a. endoplasmic reticulum and nucleus
    b. Golgi apparatus and lysosomes
    c. chloroplasts and mitochondria
    d. vacuoles and transport vesicles

# Analysis and Application

1. What features are common to all cells, and what is the function of each?
2. Describe the major components of the plasma membrane and explain why we say that the membrane has a fluid mosaic nature.
3. Compare mitochondria and chloroplasts in terms of their occurrence (in what type of cells, in what type of organisms), structure, and function.
4. Why are most cells small?.
5. What are the possible adaptive benefits of multicellularity?.
6. List the advantages of membrane-enclosed internal compartments.
7. Complete the table by listing the main functions of the cellular structures specified. In the column on the right, state whether the organelle is found in *most* prokaryotic, eukaryotic, plant, or animal cells.

| CELLULAR STRUCTURE | MAIN FUNCTION(S) | FOUND IN: PROKARYOTES? EUKARYOTES? PLANTS? ANIMALS? |
|---|---|---|
| Plasma membrane | | |
| Cytoplasm | | |
| Nucleus | | |
| Ribosome | | |
| Endoplasmic reticulum | | |
| Golgi apparatus | | |
| Lysosome | | |
| Mitochondrion | | |
| Chloroplast | | |
| Cytoskeleton | | |

# Cell Membranes, Transport, and Communication

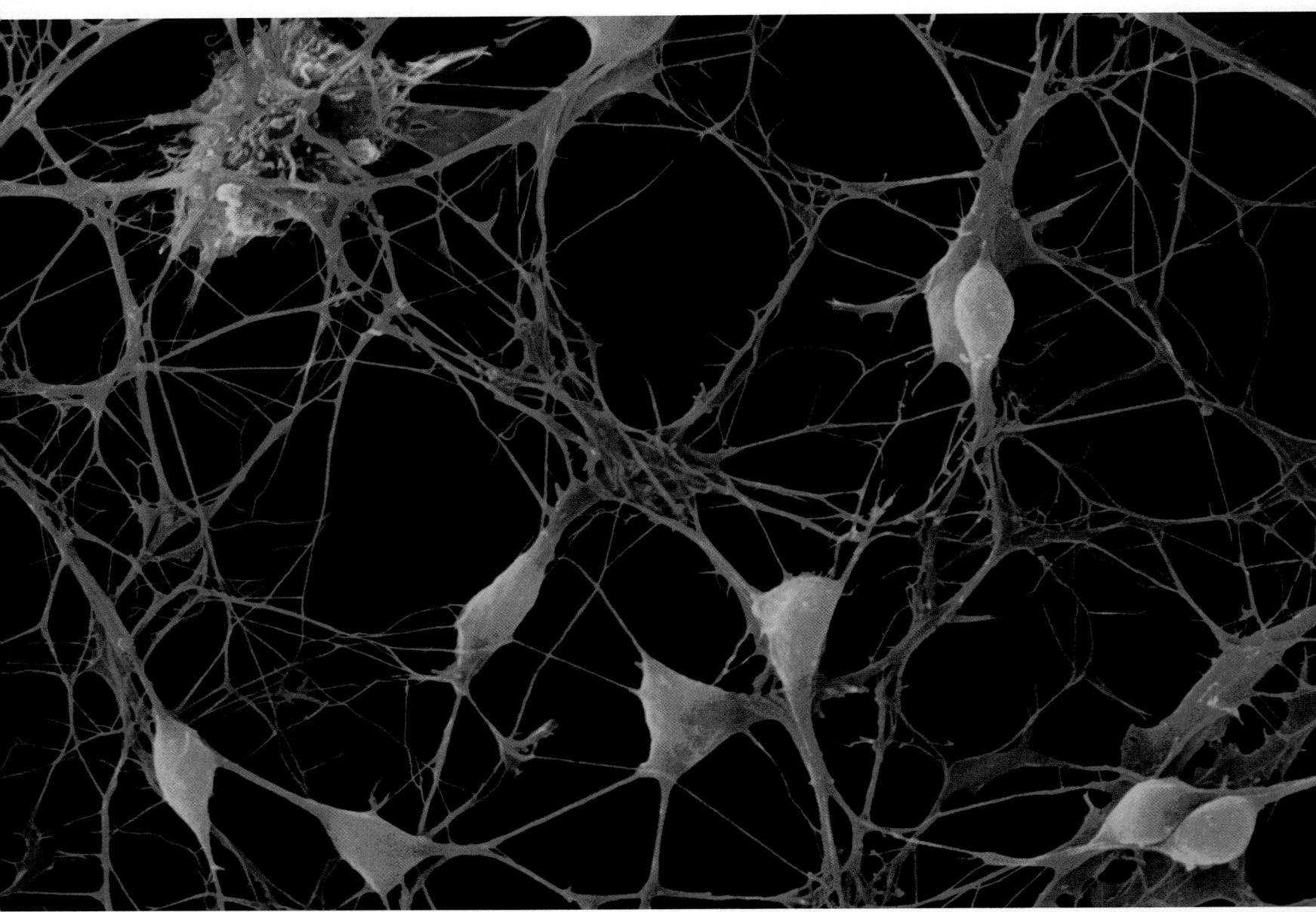

**BRAIN CELLS.** These cells from the brain cortex are growing in a lab dish. Neurons (orange) have large cell bodies and many projections used in communication. Glial cells (yellow) support and protect the neurons.

# Mysterious Memory Loss

One day in 1999, a retired astronaut and army flight surgeon named Duane Graveline returned home from his regular morning walk but then stopped in his driveway, feeling lost. His wife soon found him wandering in front of their house, but he seemed to have no idea who she was and, suspicious of her, he refused to come inside their house or get in his own car for a trip to the doctor. Eventually his wife got him to a doctor, but it was 6 hours after she found him before he recognized her (and his doctor) and seemed back to normal. Yet Graveline had no memory of what had happened to him. The doctor diagnosed "transient global amnesia"—meaning he temporarily forgot everything. While accurate, it wasn't a very helpful diagnosis. What had caused his amnesia?

Graveline suspected it might have been a drug he was taking. A few weeks earlier, he had reported to NASA's Johnson Space Center for his annual physical. The doctor there had told him that his cholesterol levels were on the high side and prescribed a "statin" drug. High cholesterol levels in the blood are associated with an increased risk of heart disease, the most common cause of death in older people. Lowering cholesterol with statins is standard medical practice, and Graveline, 68 years old and himself a doctor and medical researcher, accepted the statin prescription without question.

What could have caused this memory loss? Can lowering cholesterol affect the brain? What role does cholesterol play in cell membranes and cellular communications?

But after his bout of amnesia, he became suspicious and stopped taking the drug. A year went by with no more amnesia. When he returned for his next physical, the doctor dismissed Graveline's suspicion that the statin might have caused his amnesia and persuaded him to start taking it, although at a lower dose. Six weeks later, Graveline had another bout of amnesia in which he completely forgot that he was a doctor, a former astronaut, an author, and the father of four children. For 12 hours he couldn't remember anything that had happened to him after high school.

## MAIN MESSAGE

The plasma membrane controls how a cell exchanges materials with its surroundings and how it communicates with other cells.

## KEY CONCEPTS

- The movement of materials into and out of a cell across the plasma membrane is highly selective.

- Passive transport mechanisms move substances across the plasma membrane without energy input; active transport requires energy. Diffusion is the passive transport of a substance in response to a concentration gradient.

- In osmosis, water molecules diffuse across a selectively permeable membrane.

- Some materials can move within a cell in transport vesicles, and into and out of a cell through endocytosis and exocytosis.

- A multicellular organism is an assemblage of specialized cells.

- Neighboring cells in a multicellular organism are often connected through cell junctions, which can be specialized for attaching cells to their surroundings or for communicating with neighboring cells.

- Cells communicate with one another through signaling molecules and membrane-localized signal receptors. Signaling molecules enable cells to communicate over both short and long distances.

- The cellular response to a signaling molecule can be rapid (less than a second) or slow (over an hour).

- Hydrophilic signaling molecules bind to receptor proteins localized to the plasma membrane. Hydrophobic signaling molecules can cross cell membranes and bind to an intracellular receptor.

**MOST LIFE-SUSTAINING CHEMICAL REACTIONS** cannot take place outside of cells. To maintain an internal chemistry suitable for life, cells must carefully manage the traffic of substances across their membranes. All cells must have a way of moving materials into and out of themselves, as well as a way of controlling which materials can enter or leave at any given time.

In this chapter we consider how cells manage their relationship with the surroundings. We begin by examining the role of the plasma membrane as gate and gatekeeper for substances entering and leaving the cell. Then we consider how and why water enters or leaves a cell, and why managing its relationship with water is a matter of life or death for any cell. We show how cell membranes serve as manufacturing and packaging centers and also as "luggage" transporters. We discuss the ways cells are physically connected to one another, and how these connections facilitate communication between cells. We conclude with a look at the role of signaling molecules in cell communication.

## 7.1 The Plasma Membrane as Gate and Gatekeeper

We noted in Chapter 6 that the plasma membrane separates the inside of a cell from the environment outside. The plasma membrane is as universal a feature of life on Earth as the DNA-based genetic code. A double layer of lipids, the *phospholipid bilayer*, provides the structural framework for the plasma membrane (recall Figure 5.20). Although some biologically important materials can pass directly through the phospholipid bilayer, most cannot. Embedded in the phospholipid bilayer are many different types of proteins, which collectively make up more than half the weight of the typical plasma membrane. Some of the

membrane-spanning proteins, known as **transport proteins**, provide pathways by which materials can enter or leave cells.

The phospholipid bilayer and the transport proteins together act as a sophisticated filter, a **selectively permeable membrane** that controls which materials enter and leave the cell. Selective permeability means that some substances can cross the membrane at any and all times, others are excluded at all times, and yet others pass through the membrane aided by transport proteins when needed by the cell. Small molecules like oxygen gas and water can move across a biological membrane at any time. Larger substances, and those with a net or partial electrical charge, can traverse a membrane only with the aid of their special transport protein. Membrane proteins are usually regulated: a particular transport protein may be open for business or shut down, depending on the needs of the cell.

Because of the selectivity exercised by the plasma membrane, the environment of the cell interior is chemically very different from that of the cell exterior. Even in a multicellular organism, whose cells are continually bathed in some sort of fluid, the chemical composition of the cytoplasm is distinctly different from that of the extracellular environment (**FIGURE 7.1**). Our cells, for example, are washed by a fluid, such as blood, that has high concentrations of sodium and calcium ions. In contrast, the cytosol (the fluid part of cytoplasm) has very low concentrations of these ions and higher concentrations of potassium ions than are found in the extracellular fluid. A large portion of the metabolic energy spent by a cell goes toward maintaining the very special chemistry of the cell interior. The loss of selective membrane permeability is one of the surest signs of cell death.

## In diffusion, substances move passively down a concentration gradient

What drives the transport of any substance from one point to another? Two general rules can help us understand the movement of substances anywhere in our universe.

1. **Passive transport** is the spontaneous movement of a substance and can take place without any input of energy.

2. **Active transport** is the movement of a substance in response to an input of energy.

## SELECTIVE PERMEABILITY OF BIOLOGICAL MEMBRANES

**(a) The special chemistry of the cell interior**

Some substances (such as Na⁺) are abundant in the extracellular environment but scarce in the cytosol...

...others (such as K⁺) are present in high concentration in the cytosol but in low concentration in the extracellular environment.

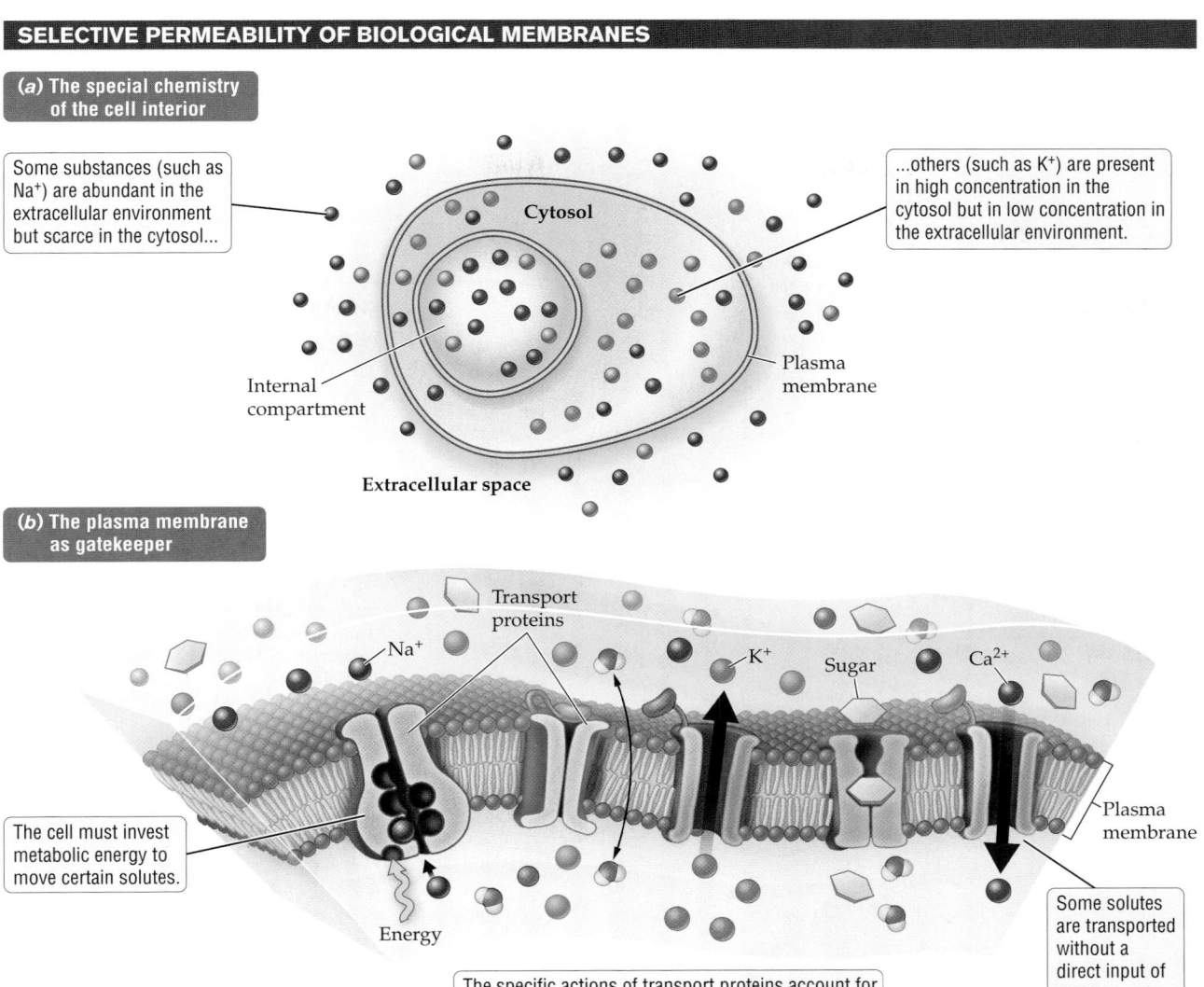

Cytosol

Internal compartment

Plasma membrane

Extracellular space

**(b) The plasma membrane as gatekeeper**

Transport proteins

Na⁺

K⁺

Sugar

Ca²⁺

Plasma membrane

The cell must invest metabolic energy to move certain solutes.

Energy

Some solutes are transported without a direct input of energy.

The specific actions of transport proteins account for the selective permeability of the plasma membrane.

**FIGURE 7.1 The Plasma Membrane Is a Barrier and a Gatekeeper**

(a) The chemistry of the cytosol is distinctly different from that of the extracellular environment, in part because the plasma membrane moves substances in a highly selective fashion. Some substances are shut out altogether, while others are allowed to enter or leave in a controlled fashion. (b) The selectivity of biological membranes is determined in large part by the types of membrane proteins in their phospholipid bilayer.

Passive and active transport are often described using the physical example of a ball moving down or up a hill, in which the ball represents a chemical substance (**FIGURE 7.2**). The ball rolls downhill on its own (passively), but it cannot roll uphill unless it is actively pushed, which requires energy.

**Diffusion** is the passive transport of a substance from a region where it is more concentrated to a region where it is less concentrated. Imagine emptying a packet of drink mix, with food coloring, into a pitcher of water. The atoms and molecules in any fluid are in constant, random motion, and as a result the ingredients in the drink mix immediately begin to blend with the water. You can watch the food coloring move *down* its concentration gradient: it spreads from the area of high concentration—the spot where the mix was

poured—into the surrounding water, where the drink mix ingredients are not abundant. The drink mix ingredients will spread, or diffuse, through the water even if you do not expend any energy to stir the mixture (**FIGURE 7.3**). Once the food coloring is distributed evenly, we say that *equilibrium* has been reached; the concentration differences driving the diffusion have disappeared and overall (net) movement of the colored chemicals has ceased. Although the chemicals in the drink mix will continue to move about in the water, the averages of these movements are equal in all directions and will therefore cancel themselves out.

Small substances—whether atoms or molecules or tiny particles of food dye—diffuse faster than larger ones. The rate, or speed, of diffusion increases with increasing temperature because any substance has

**(a) Passive transport**

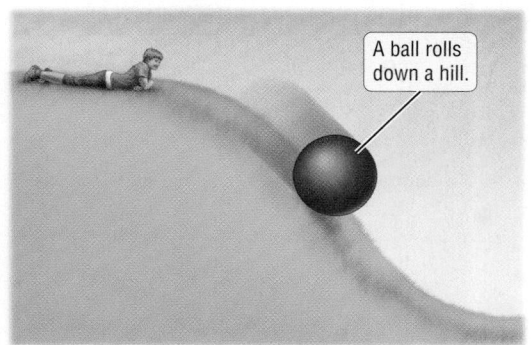

A ball rolls down a hill.

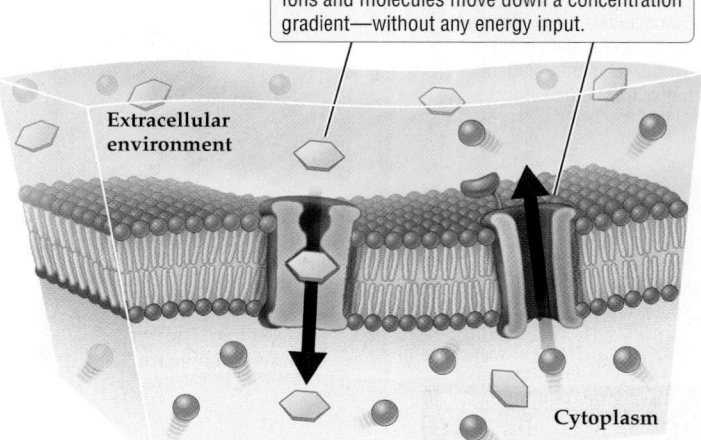

Ions and molecules move down a concentration gradient—without any energy input.

Extracellular environment

Cytoplasm

**(b) Active transport**

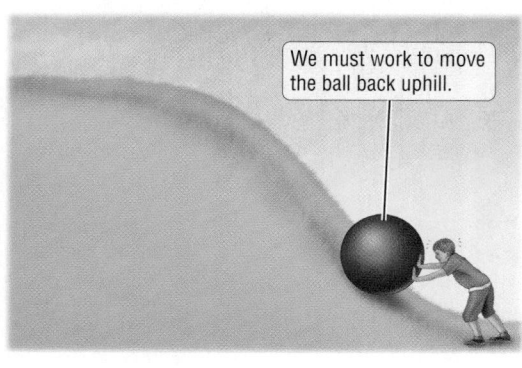

We must work to move the ball back uphill.

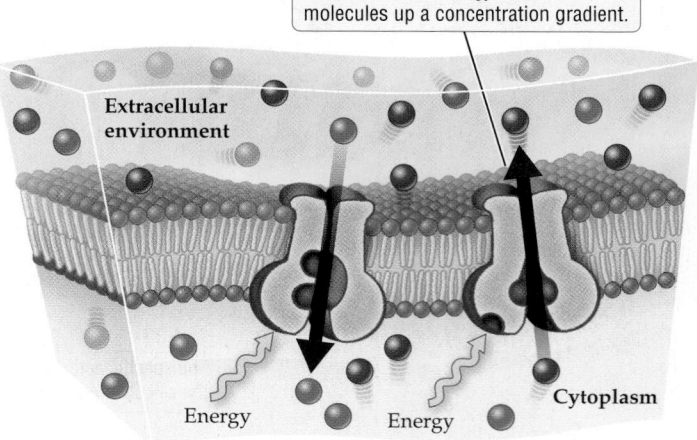

A cell must use energy to move molecules up a concentration gradient.

Extracellular environment

Energy

Energy

Cytoplasm

**FIGURE 7.2 Active versus Passive Movement of Substances**

Materials can move into and out of organisms either passively (without an input of energy) or actively (with an input of energy). (a) Substances can move passively through a membrane from a region where they are present at high concentration to a region of low concentration. (b) Energy is required to move substances from regions of low concentration to regions of high concentration.

more energy, and therefore moves faster, at warm temperatures than at cooler temperatures. The food coloring in the drink mix will diffuse into warm water faster than it will diffuse into cold water, for example. The steeper the concentration gradient between two points—that is, the greater the difference in concentration between those two points—the faster the substance diffuses.

Cells rely heavily on both passive and active transport to take up or get rid of nutrients, gases, and wastes. We shall see throughout this unit that the same passive process, diffusion, that causes the concentrated drink mix to spread through a pitcher of water plays a key role in the transfer of some molecules, such as water, oxygen, and carbon dioxide, into and out of cells. However, many of the ions and larger molecules the cell needs for its life-sustaining chemical reactions are present in fairly low concentrations in a cell's surroundings, but at relatively high concentrations within the cell, where they are accumulated for ready availability. The continued uptake of these substances requires the expenditure of energy because they must be brought in *against* a concentration gradient. It is safe to say that without active transport, or the energy to fuel it, no organism would survive very long.

## Some small molecules can diffuse through the phospholipid bilayer

Materials that readily cross a phospholipid bilayer on their own do so through simple diffusion. **Simple diffusion** is the passive transmembrane movement of a substance *without* the assistance of any membrane components. Water, oxygen, and carbon dioxide usually enter and leave cells by simple diffusion (see Figure 7.1*a*); these small, uncharged molecules slip by the large molecules in the phospholipid bilayer without much hindrance. Most hydrophobic molecules, even fairly large ones, can pass through cell membranes because they mix readily in the hydrophobic core of the phospholipid bilayer. Many early pesticides, such as DDT, were effective in killing insect pests precisely because they could easily get into their cells this way. These pesticides tend to accumulate in animal fat (unfortunately, not only in the target species but also in the predators who eat them) because they are hydrophobic substances.

Although some small molecules can slip by, the phospholipid bilayer is for the most part a barrier to large molecules and to substances with partial electrical charge (polar molecules) or a net electrical charge (ions). Being hydrophilic, and relatively large, these substances are repelled by the hydrophobic interior of the phospholipid bilayer. As we shall see in Section 7.3, special membrane proteins are needed to facilitate the transmembrane movement of all ions and all but the tiniest polar molecules.

## 7.2 | Osmosis

Water is the medium of life: most cells are at least 70 percent water, and nearly all cellular processes take place in an aqueous environment. So it will come as no surprise to learn that maintaining a proper water balance is vital for every cell. How do cells take up water, and how do they keep from shipping too much or too little water? Water molecules are so small,

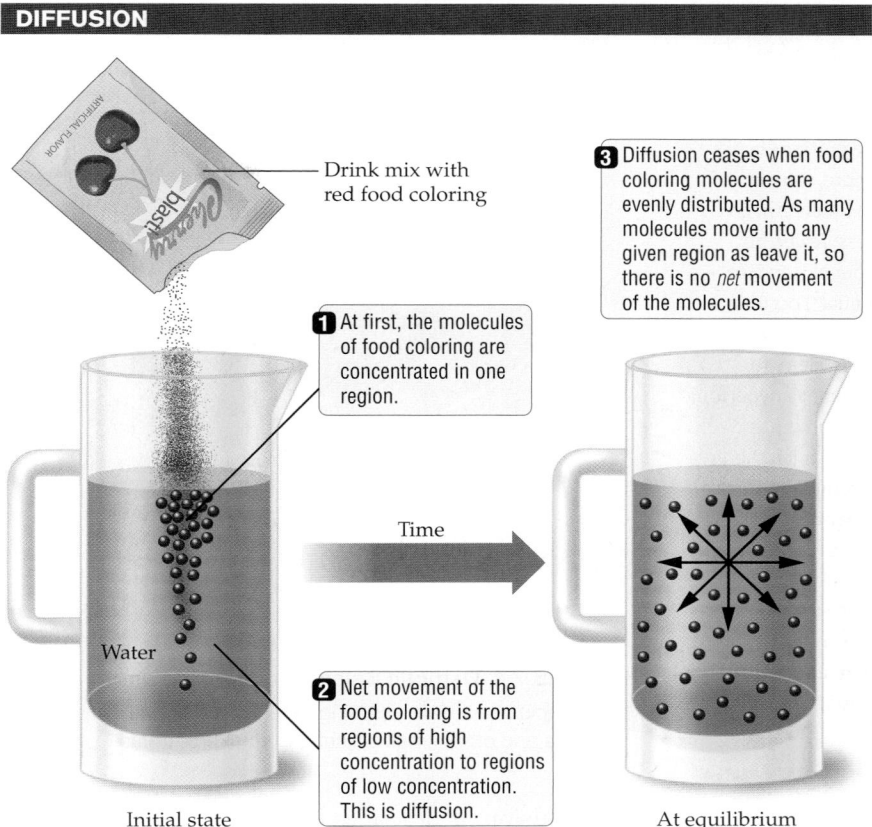

**DIFFUSION**

Drink mix with red food coloring

**3** Diffusion ceases when food coloring molecules are evenly distributed. As many molecules move into any given region as leave it, so there is no *net* movement of the molecules.

**1** At first, the molecules of food coloring are concentrated in one region.

Water

Time

**2** Net movement of the food coloring is from regions of high concentration to regions of low concentration. This is diffusion.

Initial state

At equilibrium

**FIGURE 7.3 Diffusion Is a Passive Process**
Diffusion is the spontaneous movement of a substance (such as the food coloring in the drink mix) from a region of high concentration to a region of low concentration without an input of energy. Net movement ceases, and equilibrium is reached, when the substance becomes uniformly distributed.

compared to the phospholipids and other hydrophobic components of the plasma membrane, that they can move across a plasma membrane readily, like a small child weaving among the adults in a crowded room. When a large inflow of water is needed, many cells can activate special tunnel-like protein complexes, called *aquaporins*, that facilitate rapid uptake of water across the plasma membrane. Usually, however, water molecules enter and leave a cell by wiggling through the phospholipid bilayer, driven by a passive process called *osmosis*.

**Osmosis** is the diffusion of water across a selectively permeable membrane. Because it is a type of diffusion—a passive process—no energy is expended when water enters or leaves a cell by osmosis. As in any type of diffusion, a concentration difference drives the net movement: water molecules move from a region where they are more abundant to a region where they are less abundant. But osmosis is a special case

## Helpful to Know

The prefixes *hyper* ("more, too many, too much"), *hypo* ("less, fewer, not enough"), and *iso* ("equal") occur in a number of scientific terms. Here, for example, a hypertonic environment is one in which the surroundings have a higher concentration of solutes than the cytosol of the cell. Notice that these are all terms describing the relationship of one solution to another.

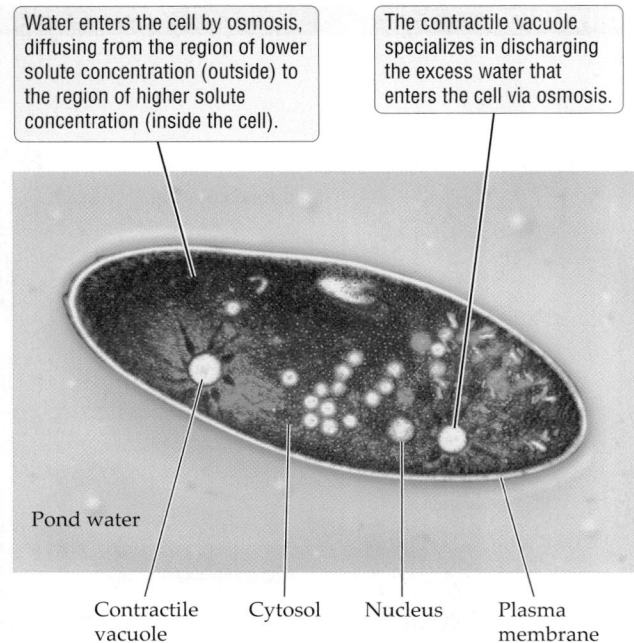

Water enters the cell by osmosis, diffusing from the region of lower solute concentration (outside) to the region of higher solute concentration (inside the cell).

The contractile vacuole specializes in discharging the excess water that enters the cell via osmosis.

Pond water

Contractile vacuole    Cytosol    Nucleus    Plasma membrane

**FIGURE 7.4  Osmotic Balance in *Paramecium***

*Paramecium* is a wall-less, single-celled protist that ingests bacteria and other small organisms at lake bottoms.

Pond-dwelling protists such as *Paramecium* take up water by osmosis because the concentration of water molecules is higher in pond water than it is inside the cell. The cell interior (cytosol) is about 70 percent water, and teeming with solute particles: there are many, many millions of ions and small organic molecules, such as sugars and amino acids, and many millions of larger macromolecules, such as proteins and nucleic acids. The cytoplasm, therefore, has a much higher concentration of solute particles than pond water has, and a proportionately lower concentration of water molecules. The difference in concentration of water molecules across the plasma membrane is what drives the net flow of water molecules from the pond water into the *Paramecium* (**FIGURE 7.4**). The convention is to describe osmosis with reference to the concentration of *solutes*, rather than the concentration of water molecules, which constitute the *solvent* in biological systems. So in practice, osmosis is defined as the net movement of water molecules across a selectively permeable membrane from a region of lower total solute concentration to a region of higher total solute concentration.

The water content of cells is continually affected by osmosis, and too much or too little water in a cell can be disastrous. Cells can find themselves in external environments that are too watery, not watery enough, or just right (**FIGURE 7.5**).

of diffusion: it describes the net movement of water across a membrane that is permeable to water molecules but not to most solutes.

**FIGURE 7.5**

**Water Moves into and out of Cells by Osmosis**

Osmosis is the diffusion of water across a selectively permeable membrane. Cells lose water in a hypertonic solution and gain water in a hypotonic solution. Our cells are bathed in an isotonic fluid to prevent osmotic swelling and bursting.

## OSMOSIS

|  | Isotonic solution: Cells neither gain nor lose water | Hypertonic solution: Cells lose water | Hypotonic solution: Cells gain water |
|---|---|---|---|
|  | Total solute concentration outside **equals** total solute concentration inside. | Total solute concentration outside **exceeds** total solute concentration inside. | Total solute concentration outside is **lower than** total solute concentration inside. |
| Animal cells | $H_2O$ | $H_2O$ | $H_2O$ |
| Plant cells | $H_2O$ | $H_2O$ | $H_2O$ |

Wall-less cells can gain so much water in a hypotonic solution that they burst.

Walled cells, such as those of plants, are protected from bursting in a hypotonic solution because the cell wall restricts continual water uptake.

# Osmosis in the Kitchen and Garden

A hypertonic environment spells doom for any metabolically active cell. Water is lost and the cell shrinks if the concentration of all the solutes on the outside exceeds the total solute concentration inside the cell. In walled cells, such as those of plants and fungi, the osmotic loss of water is known as *plasmolysis*. The cytoplasmic volume is drastically lowered as water leaves the cell and the plasma membrane pulls away from the cell wall of a plasmolyzed cell (see the plant cell in Figure 7.5). The crowding of the cytosolic components into a concentrated mass is often fatal. Proteins lose their vital three-dimensional shapes, macromolecular components clump together, and organelles are destroyed as the cytoplasm dehydrates.

For many millennia, in diverse cultures, cooks have intentionally created hypertonic environments to preserve food. Meat can be preserved by drying and salting. The salt hastens the drying by drawing out the water through osmosis. Salt also discourages the growth of bacteria and fungi: most bacterial cells or fungal strands that begin to grow on the salty surface will quickly wither from plasmolysis. Plasmolysis is the culinary weapon in many pickles, preserves, chutneys, jams, and jellies, as well. The sugar concentra-

tion is so high in most jams and jellies that fungal and bacterial cells suffer osmotic water loss and death from dehydration in the hypertonic environment. The acidity of the vinegar used in many pickles adds to the inhospitable environment for microorganisms (microbes).

Many bacteria and fungi produce thick-walled dormant structures called *spores*. The sprouting and growth of spores is prevented in a hypertonic environment, but the spores themselves can be extremely resistant and may survive if sterilization by boiling and steaming is inadequate. You can store honey in the pantry at room temperature because its high sugar content prevents spoilage. But spores of the deadly bacterium *Clostridium botulinum* can survive the hypertonic conditions and start growing if the external solute concentration drops. Infants are especially susceptible to the bacterium, which is why experts say untreated honey should not be given to children younger than 1 year. (Most commercial honey is treated to kill bacterial spores.)

High concentrations of fertilizer can kill a plant through irreversible plasmolysis. Plant food—or *fertilizer*, to use the more accurate term—is usually sold as a concentrated powder

or liquid that must be sufficiently diluted before it is applied to plant roots or foliage. Plants suffering from "fertilizer burn" have a wilted look because the hypertonic environment removes water from the plant by osmosis. Plants that grow in brackish water—sea grasses and mangrove trees, for example—cope with the saltiness by increasing the solute concentration of their cytosol. The road salt used for deicing pavement in winter is usually not concentrated enough to plasmolyze plant roots, but upon washing into lakes and ponds it can damage the delicately balanced osmotic equilibrium of aquatic species, especially wall-less protists and the eggs and larvae of animals.

---

- A **hypotonic solution** is an external medium that is more watery (has lower solute concentration) than the cytosol of the cell, so more water flows into the cell than out of it; unchecked, this movement can cause wall-less cells to burst.

- A **hypertonic solution** is an external medium that is less watery (has higher solute concentration ) than the cytosol, so more water flows out of the cell than into it. This movement causes the cell to shrink. For example, if a dog were to drink a lot of seawater, which is hypertonic for this land mammal, the cells lining its digestive tract could lose enough water that they would shrink dangerously, possibly resulting in a life-threatening

illness. Drinking an isotonic solution would present no such threat, however.

- An **isotonic solution** is "just right" in that its solute concentration is the same as that inside the cell under consideration. In this situation, the concentration of solutes is the same on both sides of the plasma membrane, so just as much water leaves the cell as enters it, with the result that there is no *net* transport of water across the membrane.

Most of our cells are bathed in an isotonic solution. Human blood, for example, is isotonic with the cells in the body. The fluid part of blood—known as *plasma*—contains many ions (such as sodium and calcium), many small organic molecules (such as glucose), and a variety of plasma proteins. Wall-less protists that live in a hypotonic environment, as the pond-dwelling *Paramecium* does, must "bail out" the excess water they take in using a special organelle, the contractile vacuole (see Figure 7.4). Walled cells, such as plant or fungal cells, do not burst in hypotonic solutions. Like the tire around an inner tube, the cell wall resists the fluid pressure (turgor pressure) that builds as water enters the cell. When the wall pressure equals the turgor pressure, the net uptake of water ceases.

Because organisms inhabit such a diverse range of habitats, cells do not necessarily find themselves in a perfect world of isotonic solutions at all times. Some organisms that live in a hypertonic world, such as ocean-dwelling fish, have special adaptations to help them counteract the tendency to lose water in a hypertonic environment. They actively transport salt out of their gills, while their kidneys help them retain more of the water they drink. Freshwater fish face the opposite challenge: a tendency to absorb too much water from their hypotonic surroundings. These fish have special transport proteins in the gills that enable them to absorb salts against a concentration gradient, while their kidneys help them excrete excess water. This constant balancing act to maintain an appropriate amount of salt and water inside each cell is known as **osmoregulation**. As illustrated by saltwater and freshwater fish, osmoregulation involves active transport and therefore requires energy.

## 7.3 Facilitated Membrane Transport

Hydrophilic substances such as ions, and larger molecules such as sugars and amino acids, cannot cross the plasma membrane without assistance. Even nutrients such as the simplest sugars and amino acids, which consist of no more than 30 atoms or so, are too large or too hydrophilic ("water-loving") to diffuse through the hydrophobic ("water-hating") core of the phospholipid bilayer. Despite being small, ions such as $H^+$ (hydrogen ions) or $Na^+$ (sodium ions) cannot get through, because they are repelled by the hydrophobic tails in the middle of the phospholipid bilayer. As a result, larger molecules, and all those that bear a partial or net electrical charge, need assistance to cross the plasma membrane. That assistance is rendered by membrane transport proteins. **Facilitated**

**diffusion** is the passive transmembrane movement of a substance with the assistance of membrane transport proteins. Two types of transport proteins help these substances move across the plasma membrane: *channel proteins* and *carrier proteins* (**FIGURE 7.6**).

**Channel proteins**, or *membrane channels*, as they are also known, enable substances of the right size and charge to move passively through the plasma membrane. In other words, channel proteins facilitate diffusion of substances down a concentration gradient (**FIGURE 7.6b**). Channel proteins form tunnels that span the thickness of the phospholipid bilayer. A channel protein has just the right width, and just the right chemistry, to enable the transit of its special cargo selectively. A calcium channel facilitates the transmembrane flow of calcium ions, for example, but it excludes potassium ions. Because channel proteins facilitate *diffusion*, a passive process, they operate without any direct input of energy.

Aquaporins, which facilitate the rapid transmembrane flow of water, are an example of channel proteins. Most channel proteins, however, specialize in moving some type of ion in or out of the cell. Most channel proteins are tightly regulated by the cell to maintain the unique chemical environment of the cytosol. Although the flow of ions through a channel is passive, and therefore requires no direct expenditure of energy, cells commonly use complex and energy-dependent processes to control the activity of channels and other membrane transport proteins. A channel protein called CFTR (for <u>c</u>ystic <u>f</u>ibrosis <u>t</u>rans-membrane conductance <u>r</u>egulator) specializes in the facilitated diffusion of chloride ions ($Cl^-$) across the plasma membrane of cells lining the lungs, small intestines, sweat glands, and other organs. The CFTR channel protein is defective in people with cystic fibrosis, a genetic disorder that affects one in 2,500 people of northern European descent. When the chloride channel fails to function properly, salt accumulates on the skin surface and on the lining of the lungs.

**Carrier proteins** function more like a revolving door than an open tunnel (**FIGURE 7.6c**). A carrier protein recognizes, binds, and transports a specific cargo molecule, such as a particular ion or a certain type of sugar. The selectivity comes from the fact that only this specific ion or molecule can fit into the folds on the surface of a given carrier protein. Once the appropriate cargo is bound to the carrier protein, the protein changes shape in such a way that the cargo now becomes exposed on the other side of the membrane. The shape change also decreases the carrier protein's affinity ("clinginess") for the cargo, causing the ion

| Type of transport: | Passive Transport | | Active Transport |
|---|---|---|---|
| **Mechanism:** | **(a) Simple diffusion**<br><br>Molecules slip between phospholipids. | **(b) Facilitated diffusion**<br><br>Diffusion is facilitated by channel proteins or carrier proteins. | **(c) Energy input required**<br><br>Active transport is facilitated by carrier proteins. |
| Outside hydrophilic heads / Hydrophobic tails / Cytoplasm | Water  Oxygen  Carbon dioxide | | Energy |
| Types of molecules that typically cross the membrane: | Molecules such as water, oxygen, and carbon dioxide; small hydrophobic molecules | **Channel proteins:** ions, water<br>**Carrier proteins:** ions, various charged or uncharged molecules | **Active carrier proteins:** ions, various charged or uncharged molecules |

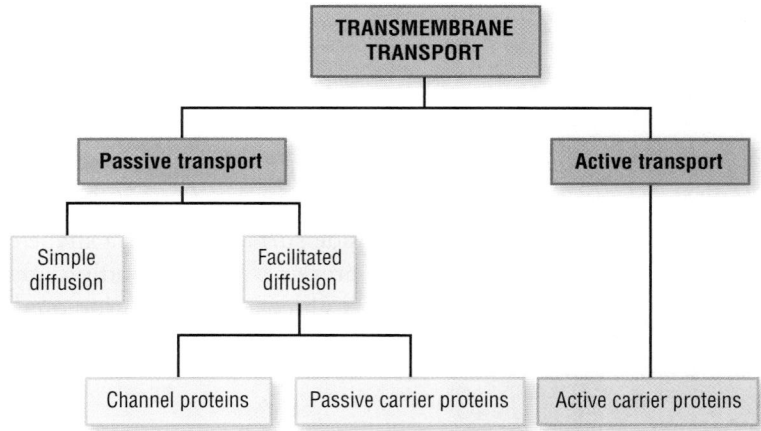

**TRANSMEMBRANE TRANSPORT**

- Passive transport
  - Simple diffusion
  - Facilitated diffusion
    - Channel proteins
    - Passive carrier proteins
- Active transport
  - Active carrier proteins

**FIGURE 7.6  The Plasma Membrane Controls What Enters and Leaves the Cell**

Proteins that span the plasma membrane (*b, c*) play an important role in moving materials into and out of cells. The flowchart summarizes the different types of processes that move materials across biological membranes.

or molecule to be let go. In this way, the molecule is picked up on one side of the membrane and discharged on the other side. Carrier proteins transport a great variety of substances across biological membranes: ions, amino acids, sugars, and nucleotides, for example. Carrier proteins are of two types:

- *Passive carrier proteins* move substances down a concentration gradient and therefore do not require energy.
- *Active carrier proteins* mobilize substances against a concentration gradient and cannot function without an input of energy.

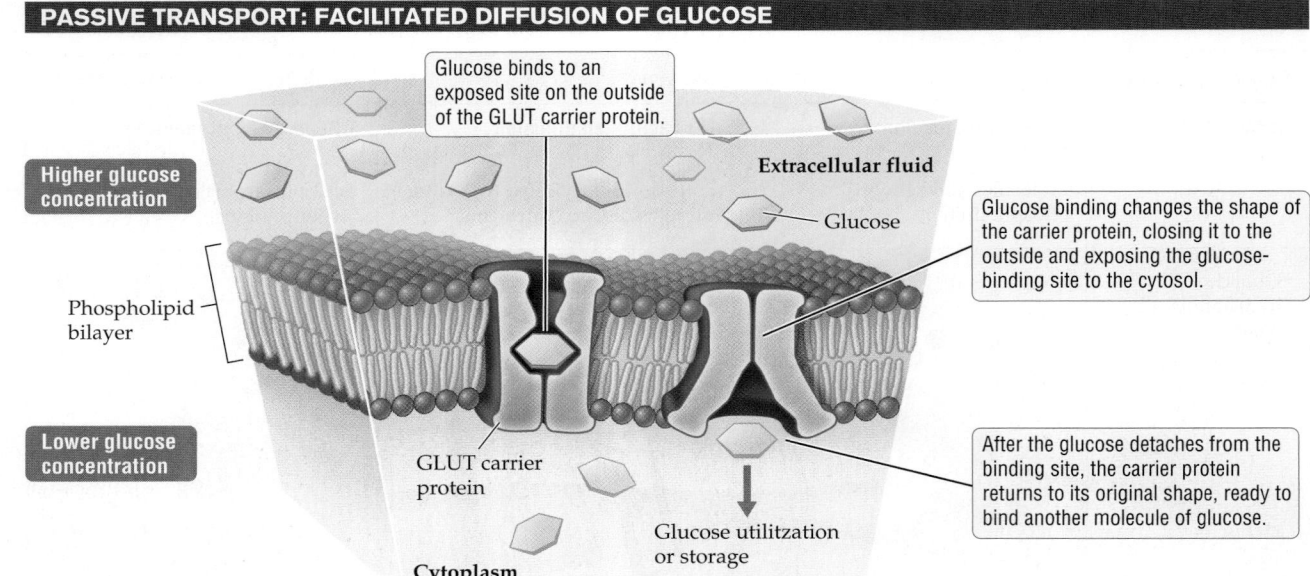

**FIGURE 7.7**
**Passive Carrier Proteins**

The facilitated diffusion of glucose into our cells is mediated by a class of passive carriers called GLUT proteins. No energy is required for the transport of glucose by carrier proteins. The sugar moves from the side of the membrane where it is at high concentration to the side where it is at lower concentration.

Glucose binds to an exposed site on the outside of the GLUT carrier protein.

Higher glucose concentration

Lower glucose concentration

Phospholipid bilayer

Extracellular fluid

Glucose

Glucose binding changes the shape of the carrier protein, closing it to the outside and exposing the glucose-binding site to the cytosol.

GLUT carrier protein

After the glucose detaches from the binding site, the carrier protein returns to its original shape, ready to bind another molecule of glucose.

Glucose utilitzation or storage

Cytoplasm

## Passive carrier proteins mediate facilitated diffusion

**Passive carrier proteins** assist in the diffusion of ions and molecules that are distributed unevenly between the two sides of a biological membrane. Glucose carriers known as GLUT proteins are an

Higher H⁺ concentration

Phospholipid bilayer

Extracellular fluid

Lower H⁺ concentration

ATP      ADP + P

Cytoplasm

The breakdown of ATP provides energy for the proton pump, an active carrier protein that...

...moves protons (H⁺) against a concentration gradient.

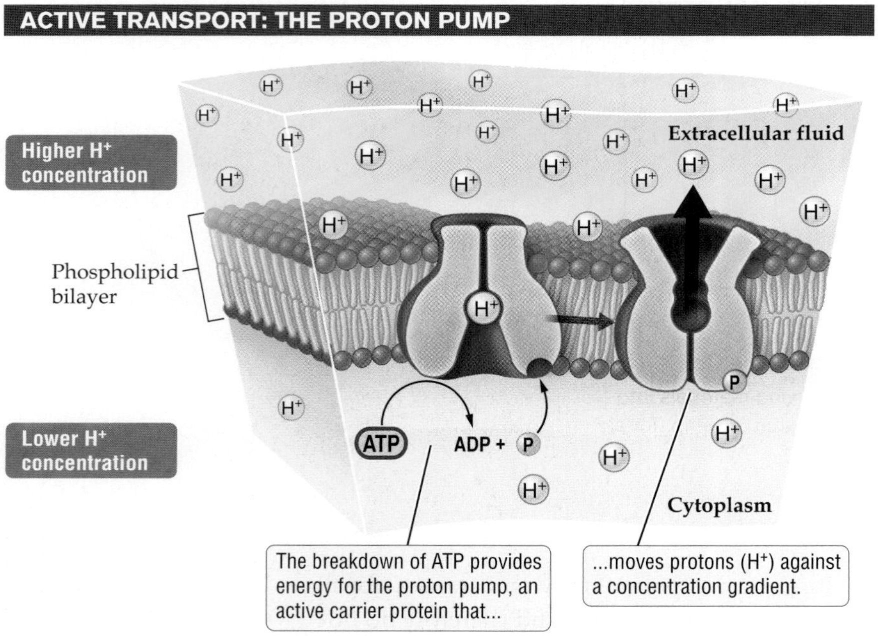

**FIGURE 7.8  Active Carrier Proteins**

Active carrier proteins use energy to move materials from regions of low concentration to regions of high concentration. When food arrives, proton pumps secrete hydrogen ions (H⁺) into your stomach.

example of passive carrier proteins (**FIGURE 7.7**). All cells in the human body need glucose for energy. The plasma membrane of every cell has glucose carrier proteins to absorb the sugar from the bloodstream. The majority of our cells can pick up glucose from the blood in a passive manner because blood normally contains about 10 times as much glucose as the cytoplasm of the average cell has. Glucose can simply "roll" down its concentration gradient, aided by GLUT proteins.

Like all other carrier proteins, GLUT proteins are shape-shifters: the docking of glucose on their outside surface triggers a shape change that exposes the protein's glucose-binding site to the cytosol. The lower glucose concentration on the cytosolic side triggers the release of the bound glucose.

## Active carrier proteins move materials against a concentration gradient

Molecules can cross a plasma membrane against a concentration gradient only by active transport. **Active carrier proteins**, also known as *membrane pumps*, can move molecules across the plasma membrane with the aid of an energy-rich molecule such as ATP. Like passive carrier proteins, active carrier proteins bind only certain ions or molecules: those that can fit into specific folds in the protein (**FIGURE 7.8**). In this case, however, the addition of energy induces a shape change in the active carrier protein that forcibly releases the molecule being transferred, regardless of the concentration of that molecule near the site of release. An energy-driven mechanism enables active carrier proteins to move ions or molecules from regions of low concentration to regions of high concentration.

Many ions and molecules are distributed across cell membranes in a lopsided manner, with a lot of the substance on one side and much less on the other side. These unevenly distributed substances include sodium, potassium, calcium, and hydrogen ions, and a variety of sugars and amino acids. It takes a specialized active carrier protein to move any such substance from the low-concentration side to the high-concentration side of a cell membrane. Because the cell must spend energy anytime these substances have to be pumped "uphill," the amount of energy a cell spends on active transport is quite substantial. It may surprise you to learn that 30–40 percent of the energy used by a resting human body goes into fueling active transport across plasma membranes!

The sodium-potassium pump is one of the most important active carrier proteins in our cells. It is present in the plasma membrane of virtually all the cells in our bodies, and it is so vital that most animal cells would die quickly if the sodium-potassium pump stopped working. The sodium-potassium pump creates and maintains the large, but opposite, concentration gradients of sodium and potassium ions across the plasma membrane in most animal cells. Blood and other body fluids have high concentrations of sodium ions ($Na^+$) but low concentrations of potassium ions ($K^+$). Within our cells, the situation is reversed: $Na^+$ is scarce in the cytoplasm but $K^+$ is plentiful. The sodium-potassium pump maintains these concentration differences by exporting sodium ions from the cell while importing potassium ions. It picks up $Na^+$ ions from the cytoplasm and moves them "uphill" to the outside of the cell, using energy from the breakdown of ATP.

## 7.4 Endocytosis and Exocytosis

In Chapter 6 we saw that many molecules are transported from place to place within a cell wrapped in small membrane packages called transport vesicles. Chemical substances are also packaged into such molecular "ferries" in preparation for being exported or imported to and from the plasma membrane.

In **exocytosis** [EX-oh-sye-TOH-sis] (*exo*, "outside"; *cyt*, "cell"; *osis*, "process"), cells release substances into their surroundings by fusing membrane-enclosed vesicles with the plasma membrane (**FIGURE 7.9**). The substance to be exported is packaged into transport vesicles by the ER-Golgi network of membranes inside the cell. As the transport vesicle approaches the plasma membrane, a portion of the vesicular membrane makes contact with the plasma membrane and fuses with it. In the process, the inside of the vesicle (the lumen) is opened to the exterior of the cell, discharging the contents. Many of the chemical messages released into the bloodstream in humans, and many other animals, are discharged via exocytosis by the cells that produce them. For example, after we eat a sugary snack, specialized cells in the pancreas exocytose [EX-oh-sye-TOHZ] the hormone insulin, which moves through the bloodstream to other cells and signals them to take up the glucose released from the snack.

The reverse of exocytosis is **endocytosis** (*endo*, "inside"). In this process, a section of plasma membrane bulges inward to form a pocket around extracellular fluid, selected molecules, or whole particles.

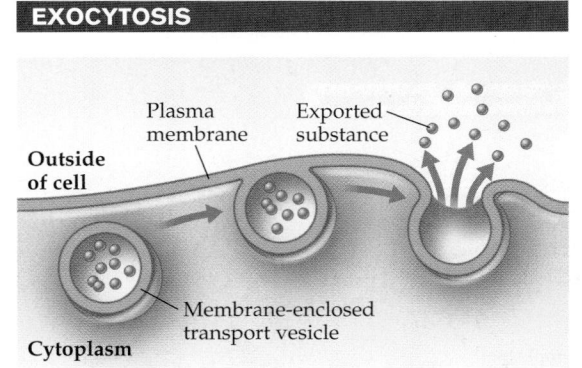

**EXOCYTOSIS**

**FIGURE 7.9 Cell Contents Are Exported through Exocytosis**

These transport vesicles fuse with the plasma membrane.

### Concept Check

1. Explain why ions, such as $Na^+$, cannot move across a phospholipid bilayer unassisted.

2. What is the difference between active and passive transport across a biological membrane?

The pocket deepens until the membrane it is made of breaks free and becomes a closed vesicle, now wholly inside the cytoplasm and enclosing extracellular contents (**FIGURE 7.10a**). Endocytosis can be nonspecific or specific. In the nonspecific case, all of the material in the immediate area is surrounded and taken in. One form of nonspecific endocytosis is **pinocytosis** [*PIN*-**oh**-...], often described as "cell drinking" because cells take in fluid in this way (**FIGURE 7.10e**). The cell does not attempt to collect particular solutions; the vesicle budding into the cell contains whatever solutes were dissolved in the fluid when the cell "drank."

Endocytosis can be so specific that only one type of molecule is enveloped and imported. How does a particular section of plasma membrane "know" what to endocytose? The answer lies in the presence in the membrane of specific **receptors**, specialized proteins that interact with specific substances in the exterior environment. In **receptor-mediated endocytosis**, specialized receptor proteins embedded in the plasma membrane determine which substances will be selected for incorporation into the vesicles arising from that membrane region (**FIGURE 7.10c**). The receptors select the cargo by recognizing specific surface characteristics of the material they bring in.

Our cells use receptor-mediated endocytosis to take up cholesterol-containing packages called low-density lipoprotein (LDL) particles. The liver

**FIGURE 7.10**
**Extracellular Substances Are Imported through Endocytosis**

(*a*) Endocytosis brings material from the outside of the cell to the inside, wrapped in membrane vesicles. (*b*) Receptor-mediated endocytosis is a highly selective process in which only certain extracellular molecules are recognized by, and bound to, special plasma membrane receptors. (*c, d*) Phagocytosis is endocytosis on a large scale. A macrophage (blue) engulfing an invading yeast cell (yellow) is seen in (*d*). Macrophages are part of the body's defense system. (*e*) Pinocytosis is nonspecific endocytosis of external fluid.

**ENDOCYTOSIS**

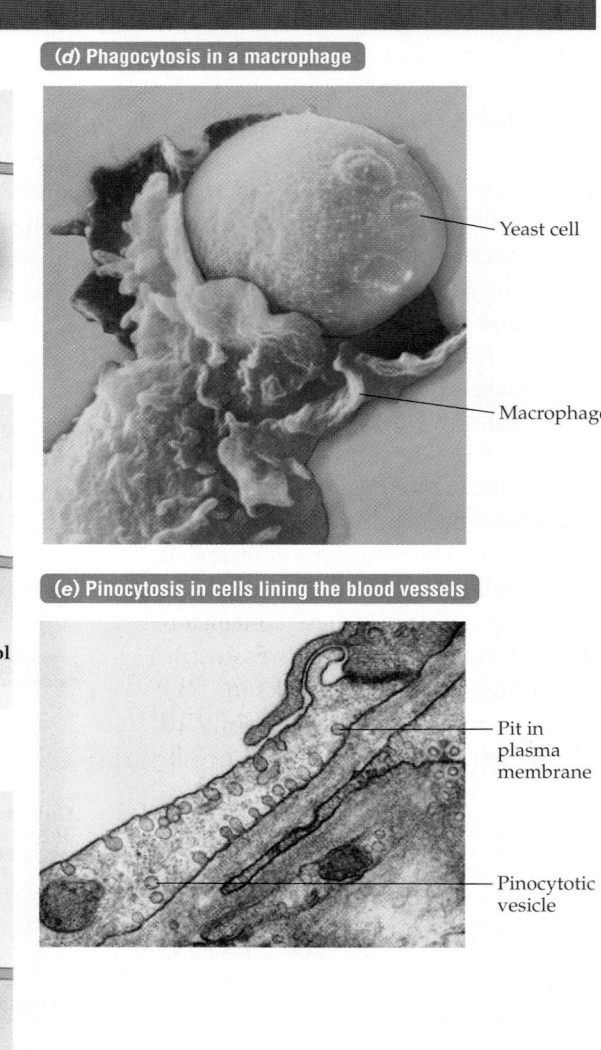

(*a*) **Mechanism of endocytosis**

Cytoplasm    Transport vesicle

(*b*) **Mechanism of receptor-mediated endocytosis**

LDL particle
Cholesterol
LDL receptor
Cholesterol processed
Cytoplasm

(*c*) **Mechanism of phagocytosis**

Bacterium (or food particles)
Pseudopodium
Cytoplasm
Vesicle
Macrophage

(*d*) **Phagocytosis in a macrophage**

Yeast cell
Macrophage

(*e*) **Pinocytosis in cells lining the blood vessels**

Pit in plasma membrane
Pinocytotic vesicle

produces cholesterol, and because it is hydrophobic, this lipid must be packaged with proteins that help it mix into an aqueous environment. Only as a part of the LDL particle can cholesterol be released into the bloodstream, bound for any cell in the body that needs this vital lipid.

LDL receptors in the plasma membrane recognize specific proteins (lipoproteins) that are exposed on the surface of LDL particles. The docking of an LDL particle with an LDL receptor triggers endocytosis of the entire complex. Endocytotic compartments eventually deliver the LDL particle–LDL receptor complex to the lysosome, an organelle that specializes in taking apart biomolecules and releasing the building blocks into the cytosol for reuse.

Patients who have familial hypercholesterolemia lack LDL receptors. Because the cells cannot take delivery of the LDL particles, cholesterol and other lipids are deposited in the lining of the blood vessels as fatty plaques. The fatty plaques undergo chemical changes (lipid oxidation) that aggravate the immune system, launching yet other changes (inflammation) that eventually cause hardening and narrowing of the blood vessels. Plaques can break loose from the blood vessel wall to form a clot that can block the flow of blood. A heart attack is precipitated when blood supply to the heart is blocked; a stroke results when blood flow in the brain is restricted.

**Phagocytosis**, or "cell eating," is a large-scale version of endocytosis in that it involves the ingestion of particles considerably larger than macromolecules, such as an entire bacterium or virus (**FIGURE 7.10c and d**). This remarkable process occurs in specialized cells, such as the white blood cells that defend us from infection. A single white blood cell can engulf a whole bacterium or yeast cell; this would be roughly equivalent to a person swallowing a big Thanksgiving Day turkey whole! As is the case with cholesterol uptake through internalization of LDL particles, receptors in the membrane of the white blood cell enable it to recognize and phagocytose harmful microorganisms.

## 7.5 Cellular Connections

Multicellular organisms benefit from having a variety of cells and tissues, each of which performs a narrow range of specialized tasks. Specialized cells, also known as *differentiated cells*, have specific chemical properties, and often a distinctive shape and structure as well, that enable them to conduct their functions more efficiently than any multitasking generalist cell could. By maximizing the efficiency of critical activities, an organism becomes better adapted to the challenges presented by its surroundings. But all the many cells in a multicellular body must be woven together properly, with appropriate means of communication among them, for the body to function as a well-knit, well-coordinated community of cells. In this section, we examine the physical connections that are commonly found between cells in a multicellular organism.

Multicellular organisms, from seaweed to walruses, have at least some cells that are interconnected in special ways. Plasma membrane structures that interconnect adjacent cells are known as **cell junctions**. Vertebrate animals possess three main types of cell junctions: *anchoring junctions*, *tight junctions*, and *gap junctions* (**FIGURE 7.11a**).

**CELLULAR CONNECTIONS**

**(a) Animal cells: cell junctions**

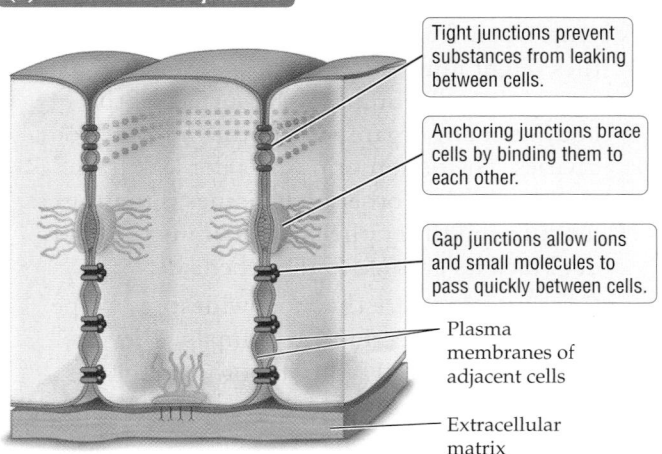

Tight junctions prevent substances from leaking between cells.

Anchoring junctions brace cells by binding them to each other.

Gap junctions allow ions and small molecules to pass quickly between cells.

Plasma membranes of adjacent cells

Extracellular matrix

**(b) Plant cells: plasmodesmata**

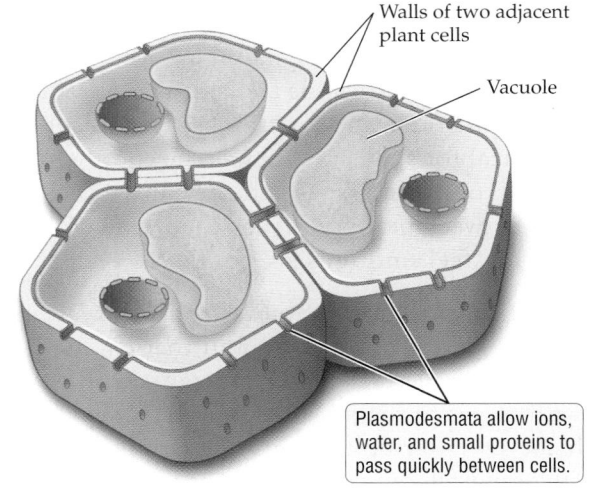

Walls of two adjacent plant cells

Vacuole

Plasmodesmata allow ions, water, and small proteins to pass quickly between cells.

**FIGURE 7.11 Cells in Multicellular Organisms Are Interconnected in Various Ways**
(a) Many animal cells are joined by different types of junctions.
(b) Plant cells are interconnected by plasmodesmata.

- **Anchoring junctions** (also known as *desmosomes*) are structures formed by patches of protein located for the most part on the cytoplasmic face of the plasma membrane. Extensions from each protein patch pass through the plasma membrane to "hook" with similar extensions protruding from an adjacent cell. The main function of anchoring junctions is to link cells together to brace them collectively against forces that would rupture an isolated cell. Anchoring junctions are especially abundant in tissues that experience heavy structural stress, such as heart muscle.
- **Tight junctions** are structures formed by belts of proteins that run along the plasma membrane. Adjacent cells are bound together by their belts of protein, collectively forming a leak-proof sheet of cells. Most molecules cannot pass from one side of the sheet to the other side because tight junctions keep these substances from creeping *between* cells, the way grout prevents water from creeping between bathroom tiles. Tight junctions are especially common in *epithelial cells*, found on the surface of the body, in many organs, and in the lining of body cavities. The reason urine does not leak from the bladder into other body tissues is that tight junctions in the epithelial cells lining the bladder block its passage to the other side.
- **Gap junctions** are the most widespread type of cellular connections in animals. They are found in most cell types in both vertebrates and invertebrates. Gap junctions are direct cytoplasmic connections between two cells. They consist of protein-lined tunnels that span the small intercellular space separating adjacent cells (see Figure 7.11*a*). Gap junctions allow the rapid and direct passage of ions and small molecules, including signaling molecules. Electrical signals can be transmitted extremely quickly through gap junctions, and this speed is critical for such activities as the coordinated activity of heart tissues and the communication between brain cells that enables us to think or feel emotions.

Recall that plant cells, *unlike animal cells*, are enclosed in a polysaccharide cell wall that covers their plasma membrane, like the plastic wrap you must battle with to get to the jewel case (plasma membrane) with the DVD (cytoplasm) inside. Wrapped as they are in a cell wall, can plant cells chat as intimately as animal cells connected through gap junctions? Yes, because plants use communication channels, called *plasmodesmata*, that are functionally similar to the gap junctions of animals. **Plasmodesmata [plaz-moh-*DEZ*-muh-tuh]** (singular "plasmodesma") are tunnels that breach the cell walls between two cells and connect their cytoplasm (**FIGURE 7.11***b*). They are lined by the merged plasma membranes of the two cells and provide a pathway for the direct and rapid flow of ions, water, and molecules such as small proteins.

## 7.6 Cell Signaling

The use of molecules to transmit signals between cells is widespread among multicellular organisms. In general, communication between cells is based on the release and perception of **signaling molecules**, which could be ions, small molecules the size of amino acids, or larger molecules such as proteins. The signaling molecule is sensed by another cell, the **target cell**, usually through the means of **receptor proteins** (or simply receptors) localized in the plasma membrane or somewhere in the cytoplasm. Signaling molecules and target cells are therefore the key components of any signaling system in the living world. Most signaling molecules are short-lived, being destroyed or removed from the vicinity of the target cell within seconds. Some, however, are long-lived, lasting in the body for many days.

A signaling molecule can activate several different types of target cells, and a single cell can be the target of a variety of signaling molecules. The specificity in signal perception and response comes from the receptor protein. Receptor proteins are located either on the plasma membrane of a target cell or somewhere in the cytoplasm. Within the cytoplasm, the receptor proteins may lie in the cytosol or inside an organelle such as the nucleus. Plasma membrane receptors bind their signaling molecules at the cell surface and must relay receipt of the signal to the cytoplasm through a series of cellular events collectively known as **signal transduction pathways**.

Signaling molecules that bind to receptor proteins inside the cell—that is, in the cytoplasm—must cross the plasma membrane and may enter membrane-enclosed compartments such as the nucleus (**FIGURE 7.12**). Because they must diffuse across the hydrophobic phospholipid bilayer of cell membranes, signaling molecules that bind to intracellular (within the cell) receptors tend to be hydrophobic lipids themselves.

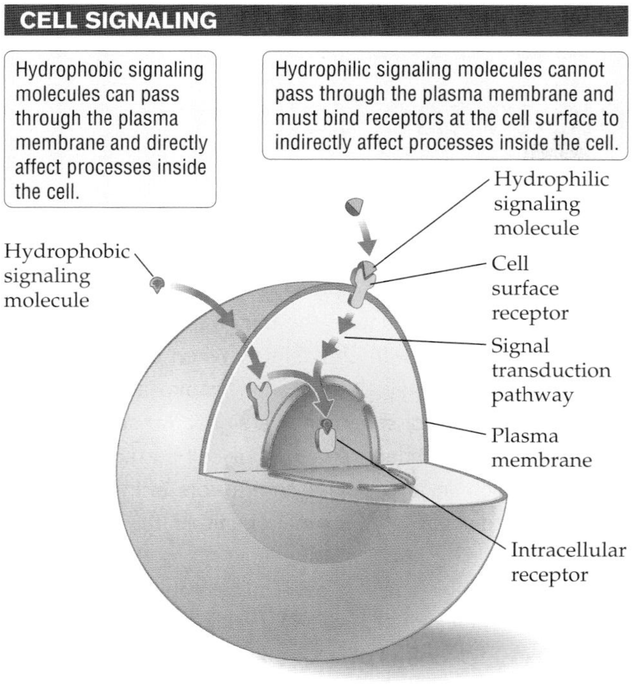

Hydrophobic signaling molecules can pass through the plasma membrane and directly affect processes inside the cell.

Hydrophilic signaling molecules cannot pass through the plasma membrane and must bind receptors at the cell surface to indirectly affect processes inside the cell.

Hydrophilic signaling molecule

Hydrophobic signaling molecule

Cell surface receptor

Signal transduction pathway

Plasma membrane

Intracellular receptor

**FIGURE 7.12** **Receptors for Signaling Molecules**
Intracellular receptors reside in the cytosol or the nucleus and bind to hydrophobic signaling molecules that can cross the plasma membrane. Cell surface receptors are embedded in the plasma membrane and bind to hydrophilic signaling molecules that cannot cross the membrane.

Multicellular organisms use a wide array of signaling systems. Some signaling systems act rapidly, so that the signal molecule elicits a response from the target cell within seconds. Other signaling systems involve signal molecules and specific target cells that together produce a slower response, on the order of hours in most cases. Some signals are narrowly targeted, released only near their target tissues, the way a cable TV show is sent only to subscribers. Other signals are widely dispersed throughout the body, like shows broadcast by satellite TV. Some signaling systems op-

erate over long distances; other signaling molecules do not have to travel far, because they are generated and released in the vicinity of their target cells. If you have ever jumped in response to a sudden noise, you have experienced the almost instantaneous work of the fast-acting signaling molecules, called *neurotransmitters*, released by nerve cells. If these signaling molecules acted slowly, a "jump" would take hours or even days to occur. Nerve signals are narrowly targeted, with a certain nerve releasing its neurotransmitters inside specific tissues, such as the heart and skeletal muscle cells.

All multicellular organisms use hormones to coordinate the activities of different cells and tissues. **Hormones** are long-lasting signaling molecules that can act over long distances. In contrast to nerve cell signals, most hormones are broadly disseminated through the body. Typically they act more slowly than neurotransmitters, but some, such as the hormone adrenaline, can act quite rapidly, triggering a response within seconds. Human growth hormone (hGH) is an example of a slower-acting signaling molecule. This hormone is at work, stimulating the growth of bones and other tissues, as we grow in size throughout childhood. If hGH acted as fast as a neurotransmitter, the term "growth spurt" would have a whole new meaning! In vertebrate animals, most hormones are produced by cells in one part of the body and transported through the bloodstream to target cells in another part of the body, ensuring rapid and widespread distribution.

> ## Concept Check
>
> 1. Which process is more selective in terms of the cargo transported: pinocytosis or receptor-mediated endocytosis?
>
> 2. What is the main function of tight junctions?
>
> 3. Would you expect the receptor for a small lipid signal, such as a hormone, to be located within the cytoplasm or at the cell surface? Would it require signal transduction? Explain.

**Concept Check Answers**

1. Receptor-mediated endocytosis.

2. To create a leak-proof sheet of cells.

3. Within the cytoplasm. Because lipid signals are hydrophobic and can cross the plasma membrane, they do not dock with cell surface receptors and, consequently, signal transduction is not required.

# Does Cholesterol Play a Role in the Cell Membranes of the Brain?

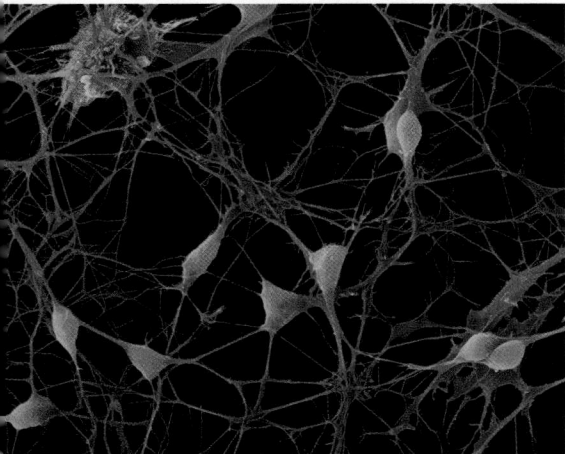

At the beginning of this chapter we learned about a retired flight surgeon and medical researcher named Duane Graveline who became convinced that a drug he was taking to lower his cholesterol had caused him to experience a temporary but dramatic case of amnesia—not once but twice. Was it really reasonable for him to think that cholesterol-lowering drugs might cause memory problems? Isn't cholesterol just some bad molecular actor that clogs arteries? Could reducing cholesterol cause amnesia?

Graveline himself didn't know. He wasn't in a position to begin a study in which thousands of people would be given the drug and an equal number of matched controls would take a placebo, and then all of them would be followed for several years. Still, like a lot of scientists, Graveline decided to follow his intuition that there was a connection. He stopped taking the drug and began reading everything he could find on cholesterol and the brain.

Over the years, it has become clear, both to Graveline and to increasing numbers of researchers, that cholesterol is deeply involved in brain function. Like other parts of the body, the brain is made of cells, each of which is encased in a cell membrane packed with cholesterol. In fact, cell membranes in the brain contain so much cholesterol that a quarter of all the cholesterol in the body is in the brain—all of it made in the brain. What the cholesterol is doing there is a subject of intense interest to biologists.

Cholesterol appears to be essential to forming connections between one nerve cell and the next—which is how we make and store memories. Our brains contain two kinds of nerve cells: neurons, which connect to one another to transmit information and form memories; and glial cells, which care for the neurons. One of the things that glial cells do is supply the neurons with cholesterol. When glial cells can't make cholesterol, their neurons don't make as many connections. Cholesterol speeds nerve signal transmission and plays a role in the shape changes of cell surface receptor proteins, which, in turn, play a central role in inflammation, Alzheimer's disease, and even addictions to alcohol and cigarettes.

Recall from this chapter that a cell membrane is a double layer of phospholipids that can smoothly envelop and transport material through endocytosis and exocytosis. The lipids and proteins that make up a cell membrane are not glued rigidly together like the walls of a house. Instead, a cell membrane is fluid.

Floating within this fluid are "lipid rafts"—stiffer aggregates of cholesterol and protein. These cholesterol rafts carry receptor proteins and other components involved in signal transduction. Cell biologists speculate that the positions of cholesterol rafts influence their interaction with the signal transduction pathway. For example, once a receptor protein on a lipid raft has been activated, it may become isolated, like a ship on the high seas. Because it is no longer accessible, the enzymes in the vicinity cannot deactivate the receptor protein.

But do statins—which interfere with the synthesis of cholesterol—have the capacity to cause memory problems like the ones that Duane Graveline experienced? For now, no one knows. Statin manufacturers argue that no large studies support the idea. A few small studies suggest that some statins may affect memory in some people. Dr. Graveline, for one, is more convinced than ever. Since 2000, he has not only lowered his blood cholesterol with exercise and changes to his diet; he has written four books about statin drugs, and he doesn't plan to let anyone forget about it.

# "Good" Cholesterol May Lower Alzheimer's Risk

BY RONI CARYN RABIN, *Well* blog, *New York Times*

A new study reports that older New York City residents who had very high blood levels of high-density lipoprotein, or HDL, the so-called "good" cholesterol, were at less than half the risk of developing dementia over time than those with the lowest levels.

The people who reaped the benefit had very high HDL blood levels that exceeded 56 milligrams per deciliter of blood, the study reported. They developed 60 percent fewer cases of Alzheimer's disease than people with the lowest HDL levels, of 38 milligrams or below. The differences between the two groups held even after the researchers adjusted the figures to account for other causal factors that influence the development of dementia, like vascular disease, as well as age, sex, education level, and genes that predispose to Alzheimer's.

"We think it's a causal relationship," said Dr. Christiane Reitz, the study's lead author and an assistant professor of neurology at Columbia University's Taub Institute for Research on Alzheimer's Disease and the Aging Brain. "At the baseline, when we recruited these people, they didn't have cognitive problems. We followed them, and they developed dementia during the follow-up period."

But the HDL was only protective at extremely high levels, Dr. Reitz said. "It really only makes a difference if you're higher than 56 [milligrams]," she said.

The report, published in the Archives of Neurology, is not the first to find that what's good for the body may also [be] good for the brain. Numerous studies have found that older people who walked the most were at lowest risk of developing vascular dementia, possibly because the regular exercise improves cerebral blood flow and lowers the risk of vascular disease.

Regular physical activity, in general, is believed to improve brain function, both by increasing blood flow to the brain and by stimulating the production of hormones and nerve growth factors involved in new nerve cell growth. Exercise also raises levels of "good" HDL cholesterol. Studies have found that animals that are kept physically active have better memories and more cells in their hippocampus, a part of the brain critical for memory. And exercise can help stave off or keep in check diseases like Type 2 diabetes, which increase the risk of developing dementia.

In Section 7.4 we learned that LDL particles are taken up by cells through receptor-mediated endocytosis and then digested—into cholesterol, fatty acids, glycerol, and amino acids. If LDL levels in the bloodstream are high—either because too much cholesterol is absorbed from the digestive tract, too much is produced by the liver, or not enough is removed by cells in the body—the LDL particles deposit cholesterol and other lipids in the arteries, increasing the risk of a heart attack. For this reason, LDL is called "bad" cholesterol. So what exactly are "good cholesterol" and "bad cholesterol"? And what makes them good or bad?

HDL is high-density lipoprotein, and LDL is low-density lipoprotein. Lipoproteins are a way for the body to transport fats—including cholesterol—in the water-based blood. Think of the lipoproteins as trucks. You can have light delivery vans that keep your local convenience store stocked with milk and corn chips. Or you can have big, heavy recycling trucks that pick up used bottles and newspapers from every house in the neighborhood.

Just like these trucks, cholesterol-transporting HDL and LDL perform two different jobs. HDL particles are the big, heavy recycling trucks; they collect cholesterol from body tissues and return it to the liver, ovaries, testes, and adrenal glands for recycling into other materials. In the liver, cholesterol is converted to bile, which helps us digest food. In the reproductive and adrenal glands, cholesterol is the raw material for making hormones.

In contrast, LDL particles are the light delivery vans. They carry cholesterol made in the liver to cells all over the body. When present in excess, LDL particles deliver cholesterol into the walls of arteries, which causes inflammation. Over time, a tough "plaque"—consisting of cholesterol and other lipids, fibrous material, and calcium—forms. The plaque gradually enlarges until the covering layers crack a little, stimulating the formation of blood clots. A blood clot can block the flow of blood to the heart or another critical organ. Anything that encourages the formation of plaques in the arteries is "bad."

The "good" HDL recycling trucks go to the arterial walls and remove cholesterol deposits that contribute to plaque formation. If you have enough HDL particles in your blood, they also reduce the risk of Alzheimer's disease.

## Evaluating the News

**1.** Describe what led scientists to believe there is a relationship between HDLs and dementia.

**2.** What can people do to increase their HDL levels?

**SOURCE:** *New York Times*, *Well* blog, December 16, 2010, well.blogs.nytimes.com.

# CHAPTER REVIEW

## Summary

### 7.1 The Plasma Membrane as Gate and Gatekeeper

- The plasma membrane is a selectively permeable membrane phospholipid bilayer with embedded proteins.
- In passive transport, cells carry substances across the plasma membrane without the direct expenditure of energy. Active transport by cells requires an energy input.
- Diffusion is the passive transport of a substance from a region where it is at a higher concentration to a region where it is at a lower concentration.

### 7.2 Osmosis

- Osmosis is the diffusion of water across a selectively permeable membrane. When placed in a hypotonic solution, a cell gains water. In a hypertonic solution, water moves out of cells. In an isotonic solution, there is no net uptake of water by the cell.
- Cells can actively balance their water content by osmoregulation.

### 7.3 Facilitated Membrane Transport

- Hydrophilic substances and larger molecules cannot cross the plasma membrane without the assistance of membrane-spanning transport proteins.
- Passive carrier proteins facilitate the passive transport of molecules and ions down a concentration gradient. Active carrier proteins move substances into or out of the cell against a concentration gradient and require an input of energy (from an energy source such as ATP) to do so.

### 7.4 Endocytosis and Exocytosis

- Cells export materials by exocytosis and import materials by endocytosis.
- In receptor-mediated endocytosis, receptor proteins in the plasma membrane recognize and bind the substance to be brought into the cell.

### 7.5 Cellular Connections

- Cell junctions hold cell communities together.
- Anchoring junctions attach adjacent cells and make them resistant to breaking forces. Tight junctions bind cells together to form leak-proof sheets. Gap junctions are cytoplasmic tunnels that allow the passage of small molecules.
- Plasmodesmata are cytoplasmic tunnels that connect neighboring plant cells.

### 7.6 Cell Signaling

- Receptor proteins respond to signaling molecules dispatched by other cells.
- Signal transduction pathways relay the signal within the cytoplasm.
- Hormones are long-distance signaling molecules that are broadly distributed in the body.

## Key Terms

active carrier
  protein (p. 176)
active transport (p. 168)
anchoring junction (p. 180)
carrier protein (p. 174)
cell junction (p. 179)
channel protein (p. 174)
diffusion (p. 169)
endocytosis (p. 177)

exocytosis (p. 177)
facilitated diffusion (p. 174)
gap junction (p. 180)
hormone (p. 181)
hypertonic solution (p. 173)
hypotonic solution (p. 173)
isotonic solution (p. 173)
osmoregulation (p. 174)
osmosis (p. 171)

passive carrier protein (p. 176)
passive transport (p. 168)
phagocytosis (p. 179)
pinocytosis (p. 178)
plasmodesma (p. 180)
receptor (p. 178)
receptor-mediated
  endocytosis (p. 178)
receptor protein (p. 180)

selectively permeable
  membrane (p. 168)
signal transduction
  pathway (p. 180)
signaling molecule (p. 180)
simple diffusion (p. 171)
target cell (p. 180)
tight junction (p. 180)
transport protein (p. 168)

## Self-Quiz

1. Which of the following are *not* part of the plasma membrane?
   a. proteins
   b. phospholipids
   c. receptors
   d. genes

2. A direct input of energy is needed for
   a. diffusion.
   b. active transport.
   c. osmosis.
   d. passive transport.

3. Which of the following can move across a plasma membrane through simple diffusion?
   a. oxygen gas ($O_2$)
   b. hydrogen ions ($H^+$)
   c. aspartic acid, a charged amino acid
   d. human growth hormone, a hydrophilic protein

4. Water would move out of a cell in
   a. a hypotonic solution.
   b. an isotonic solution.
   c. a hypertonic solution.
   d. none of the above

5. Channel proteins are different from carrier proteins in that they
   a. are needed for simple diffusion but not for facilitated diffusion.
   b. help in the transmembrane transport of water, but not of ions.
   c. cannot transport substances actively.
   d. cannot function without the direct input of energy in the form of ATP.

6. Which of the following describes movement of material out of a cell?
   a. pinocytosis
   b. phagocytosis
   c. endocytosis
   d. exocytosis

7. Which of these cellular connections creates a leak-proof sheet of cells, such as is found in the cells lining the urinary bladder?
   a. anchoring junction
   b. tight junction
   c. plasmodesmata
   d. gap junction

8. Animal cells can directly exchange water and other small molecules through
   a. gap junctions.
   b. microfilaments.
   c. anchoring junctions.
   d. tight junctions.

9. Cell signaling involves
   a. receptor proteins.
   b. signaling molecules.
   c. target cells.
   d. all of the above

10. A nerve signal (neurotransmitter)
    a. must travel through the bloodstream to reach target cells.
    b. acts on a target cell that is nearby.
    c. must be long-lived.
    d. must be hydrophobic in nature.

# Analysis and Application

1. Imagine you release some scented air freshener in one corner of a room. Is the spread of the scent molecules through the room an example of diffusion? When equilibrium is reached, probably many hours later, has diffusion ceased? Have the scent molecules stopped moving about? Explain.

2. *Paramecium* is a wall-less, single-celled protist that lives in freshwater ponds. Are its natural surroundings hypertonic, hypotonic, or isotonic with respect to the cell? What osmoregulatory problem does this organism face in pond water, and how does it cope with that problem?

3. A classmate has come down with strep throat caused by group A *Streptococcus* bacteria. She has a fever and sore throat, and the pus on her tonsils is a sign that her white blood cells are doing battle with the invading bacteria (the pus contains the remains of white blood cells that have died after doing their share to destroy the invaders). Explain the important role of cell membranes in the mechanism by which your white blood cells destroy invading bacteria.

4. The epithelial cells lining your intestines encounter a great variety of substances, many of which could produce ill effects if they were to cross the epithelial layer to enter the bloodstream. How do cell junctions in intestinal epithelial cells help keep potential toxins out of your bloodstream?

5. The figure shows the structure of an LDL particle. An LDL particle has a highly hydrophobic core that consists of cholesterol covalently linked with fatty acids ("esterified"). The shell-like surface is made up of phospholipids, unesterified cholesterol, and a single large protein (apolipoprotein B-100). What is the role of LDL particles? Which chemical component of the LDL particle is recognized by the LDL receptor? Describe how LDL particles are internalized by the many cells in the human body that import cholesterol.

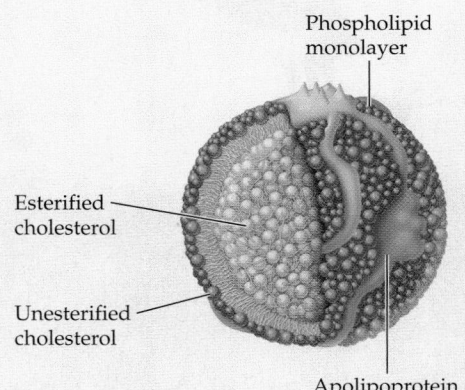

**Structure of LDL particle**

# Energy, Metabolism, and Enzymes

**METABOLIC KICK.** The kinetic energy of these spectacular moves comes from the chemical energy in the food the soccer players ate. It takes enzymes to unleash that energy in a usable form.

# Kick-Start Your Metabolic Engine!

"Fourteen ways to revive and boost your metabolism!" "Speed up your metabolism and lose weight fast!" You hear the sales pitch in advertisements for diet foods, energy drinks, and herbal supplements. At the checkout stand, every other magazine seems to promise weight loss through higher metabolism.

But what *is* metabolism? It's all the chemical reactions organisms use to capture, store, and use energy. All living organisms need energy to build and sustain every part of the body—from bones to DNA. Organisms use energy for everything from growing and reproducing to fighting off pathogens and predators. Energy fuels every process in the body, and all these energy transactions taken together are *metabolism*.

How fast your metabolism runs is a measure of how much energy you use. Physical activity takes energy, so the more active you are, the higher your metabolic rate is. If you're sitting still and you are a young woman of about 140 pounds, you're using as much energy per hour as a 75-watt lightbulb. The amount of energy you use while resting is your resting metabolic rate (RMR). RMR depends on factors like height and weight, muscle mass, age, and sex. Children have the highest RMR, and RMR tends to decline with age. Large people with more muscle have a higher RMR than those who are smaller or have less muscle. Because men are usually bigger than women and, on average, have more muscle, the average man has a higher RMR than the average woman. But there's no firm rule about this; an athletic woman likely has a higher RMR than a male couch potato.

What else increases metabolism? Will eating jalapeño peppers boost your metabolism? Can metabolism be too high? Does a lower RMR lead to a longer life?

Before we address these questions, let's explore why cells need energy and how they use it to create organized structures. In this chapter we'll see that a living cell is a highly organized, energy-dependent chemical factory whose thousands of reactions proceed with the help of enzymes.

## MAIN MESSAGE

It takes energy to create the complex order found inside a living cell; enzymes speed up the many chemical reactions that are necessary for generating and maintaining that order.

## KEY CONCEPTS

- Living organisms obey the universal laws of energy conversion and chemical change (the first and second laws of thermodynamics).

- The sun is the ultimate source of energy for most living organisms. Photosynthetic organisms capture energy from the sun and use it to synthesize sugars from carbon dioxide and water. Most organisms can break down sugars to release energy.

- Metabolism is all the chemical reactions that occur inside living cells. During catabolism, biomolecules are taken apart and energy is released. During anabolism, biomolecules are put together and an energy input is necessary.

- Enzymes greatly increase the rate of chemical reactions in cells.

- Metabolic pathways are sequences of enzyme-controlled chemical reactions.

- Enzymes position bound reactants so they collide more frequently to make products. The shape of an enzyme is critical for its biological activity.

- Enzymes display substrate specificity.

- The activity of enzymes is sensitive to temperature, pH, and salt concentration.

ALL LIVING CELLS REQUIRE ENERGY, which they must obtain from the living or nonliving components of their environment. Organisms use energy to manufacture the many chemical compounds that make up living cells, and for growth, reproduction, and defense. Thousands of different types of chemical reactions are required to sustain life in even the simplest cell. Most chemical reactions in a cell occur in chains of linked events known as **metabolic pathways**. The metabolic pathways that assemble or disassemble the key macromolecules of life, and their building blocks, are similar in all organisms—another sign of our common evolutionary heritage. Nearly all of those metabolic reactions are facilitated by *enzymes*. **Enzymes** are biological catalysts, biomolecules that speed up chemical reactions. Without the action of enzymes—most of which are proteins—metabolic reactions would be extremely slow and life as we know it could not exist.

In this chapter we examine the role played by energy in the chemical reactions that maintain living systems. We discuss the special properties of enzymes and explain how these remarkable biomolecules speed up chemical reactions that would otherwise be too slow to sustain life. We conclude with a look at metabolism in men and women, young and old, exercisers and couch potatoes, and ask how the pace of metabolism might be related to life span.

## 8.1 The Role of Energy in Living Systems

Any discussion about chemical processes in cells is at heart a discussion about the capture and use of *energy*. Every atom, molecule, particle, or object in the physical world possesses energy. We can define **energy** as the capacity of any object to do work. Work in turn can be defined as the capacity to bring about a change in a defined system. In the context of energy, the word "system" refers to any arbitrary portion of the universe we choose to demarcate. In speaking of energy in the living world, a system can be a biomolecule, an organelle, a single bacterial cell, a mat of algae in a pond, a community of organisms in an oak-maple woodland … and on up to the biosphere.

The energy of any system is an attribute of that system—a physical quantity associated with that system. The energy in a system can be recognized and expressed in many different ways, depending on which aspects of the system we wish to describe. An astronomer interested in the pull between a planet and its moon might focus on the *gravitational energy* of those bodies. Physicists study the *nuclear energy* released in the splitting or fusion of atomic nuclei. Chemists are interested in the energy stored in the bonds between clusters of atoms—such as molecules—and how that energy changes during a chemical reaction. Energy transformations are of great interest in biology too. We might want to know, for example, how much of the energy in our breakfast goes into beating an opponent to a soccer ball, how much is released as body heat, and how much is tucked away for future use.

The many different forms of energy can be organized into two broad categories: *potential energy* and *kinetic energy*. **Potential energy** is the energy stored in any system as a consequence of its position. A rock on a hilltop, water in a dam, and Lady Gaga's headgear—all have potential energy, a capacity to do work because of the position of these objects in relation to the rest of the universe around them. All three objects possess a type of potential energy we call gravitational energy. **Chemical energy** is another form of potential energy; it is the energy stored in atoms because of their position in relation to other atoms in the system under consideration. The covalent bonds that hold the atoms in a molecule, for example, store substantial amounts of chemical energy. A spoonful of table sugar (sucrose) harbors the chemical energy of many millions of carbon, hydrogen, and oxygen atoms linked via covalent bonds. A pinch of table salt contains the chemical energy of many millions of sodium and chloride ions chained together through ionic bonds.

**Kinetic energy** is the energy a system possesses as a consequence of its state of motion. Consider what happens when we use an electric blender to whip strawberries and ice cream into a smoothie. Some of the *electrical energy*, which is the energy associated with the flow of electrons, is turned into the *mechanical energy* of whirling blades in the blender. *Light energy*, which is one type of radiant energy, is the energy associated with the wavelike movement of packets of energy called *photons*. Electrical energy, mechanical energy, and light energy are all examples of kinetic energy.

**Heat energy**, also known as thermal energy, can be considered a type of kinetic energy. All atoms and molecules move to some degree, either vibrating in place or careening randomly from one point to another. As they move, these particles of matter can collide with other particles and transfer some of their energy to their target. The energy inherent in the random motion of particles in a system that can be *transferred* to other particles in the system is known as heat energy. In other words, heat energy is not the total energy that is stored in a particle of matter; it is that portion of the total energy of a particle of matter that *can flow* from one particle of matter to another.

One form of energy can often be converted into another form. Potential energy is released as kinetic energy when a rock rolls down a hill, water rushes down the spillway of a dam, or headgear comes crashing to the floor. A falling object may release *sound energy*, another type of kinetic energy. A hovering hummingbird beats its wings 30–80 times per second, fueled by the conversion of chemical energy stored in nectar into the kinetic energy of contracting muscles (**FIGURE 8.1**). If you lift a book and place it on a high shelf, some of the chemical energy from your breakfast is converted into the kinetic energy of moving muscles, and some of that kinetic energy comes to be stored as the potential energy of the book perched high on the bookshelf. Certain universal principles—known as the *laws of thermodynamics*—specify that our universe contains a fixed amount of energy, and that although energy can be converted from one form to another, it can neither be created nor destroyed. As we shall see in the next section, these powerful concepts apply to living organisms just as much as they describe energy transactions in the engine of a gas-electric hybrid car or in the center of a distant galaxy.

## The laws of thermodynamics apply to living systems

The universal laws of energy—the laws of thermodynamics—have shaped how life works. The laws dictate which chemical reactions can occur and under what circumstances. When applied to living systems, the laws of thermodynamics explain why the highly organized state of a living cell cannot exist without an input of energy. They set the ground rules for energy conversions inside a cell so that a rosebush can convert light energy into the chemical energy of sugars, and then use the chemical energy of sugars for all its life processes, from growing bigger to manufacturing the fragrances and brightly colored pigments of its blossoms. As we will see shortly, the laws of thermodynamics predict which chemical reactions in a cell will yield energy, and they stipulate that a certain portion of that released energy will be "wasted"—in that it will be unavailable for any cellular work.

The **first law of thermodynamics** states that energy can neither be created nor destroyed, but only converted from one form to another. According to the first law, also known as the *law of conservation of energy*, the total energy of any closed system remains the same over time. All the energy in our universe is associated with matter, and energy and matter are equivalent; since matter cannot arise from nothing or

**Helpful to Know**

The term "thermodynamics" comes from physics and is derived from Greek words meaning the movement (*dynamics*) of heat (*thermo*). Heat always moves from warmer to cooler areas, a basic principle governing all biological processes.

**ENERGY TRANSFORMATIONS IN LIVING SYSTEMS**

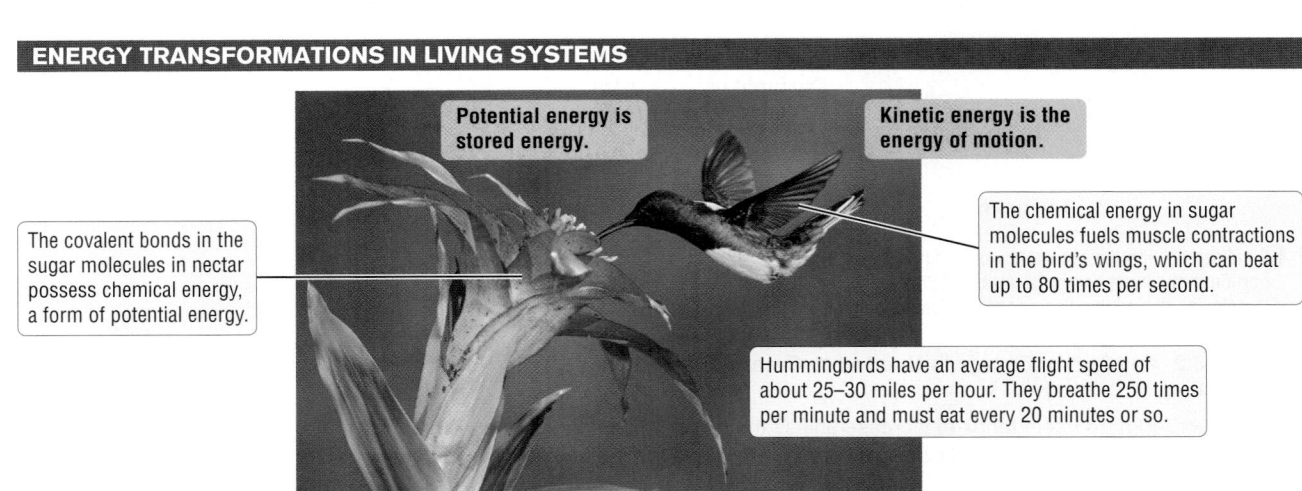

Potential energy is stored energy.

Kinetic energy is the energy of motion.

The covalent bonds in the sugar molecules in nectar possess chemical energy, a form of potential energy.

The chemical energy in sugar molecules fuels muscle contractions in the bird's wings, which can beat up to 80 times per second.

Hummingbirds have an average flight speed of about 25–30 miles per hour. They breathe 250 times per minute and must eat every 20 minutes or so.

**FIGURE 8.1**
**Potential Energy in Food Is Converted to Kinetic Energy in a Hummingbird's Body**

be turned into nothing, energy, too, cannot disappear without a trace or be created brand-new out of nothing. Energy can, however, be converted from one form into another. Some of the light energy that a plant absorbs can be turned into the chemical energy of sugar molecules in nectar. The chemical energy of a flower's nectar can be converted into electrical energy in a hummingbird's brain cells, into mechanical energy in its muscle cells, and into heat energy in any of its cells whose metabolism is stoked by the nectar.

The **second law of thermodynamics** describes how energy use or transformation in any system affects the rest of the universe. This law states that the natural tendency of the universe is to become less organized, more disorderly. Any system will exhibit this tendency unless energy from elsewhere in the universe is used to organize that system. If any part of the universe, whether a cell or a toolshed, displays a high level of order, we can be certain that energy has been captured from elsewhere to create and maintain that order (**FIGURE 8.2a**). In using energy to create internal order, a system reduces the order of its surroundings, so on the whole the universe becomes more disordered.

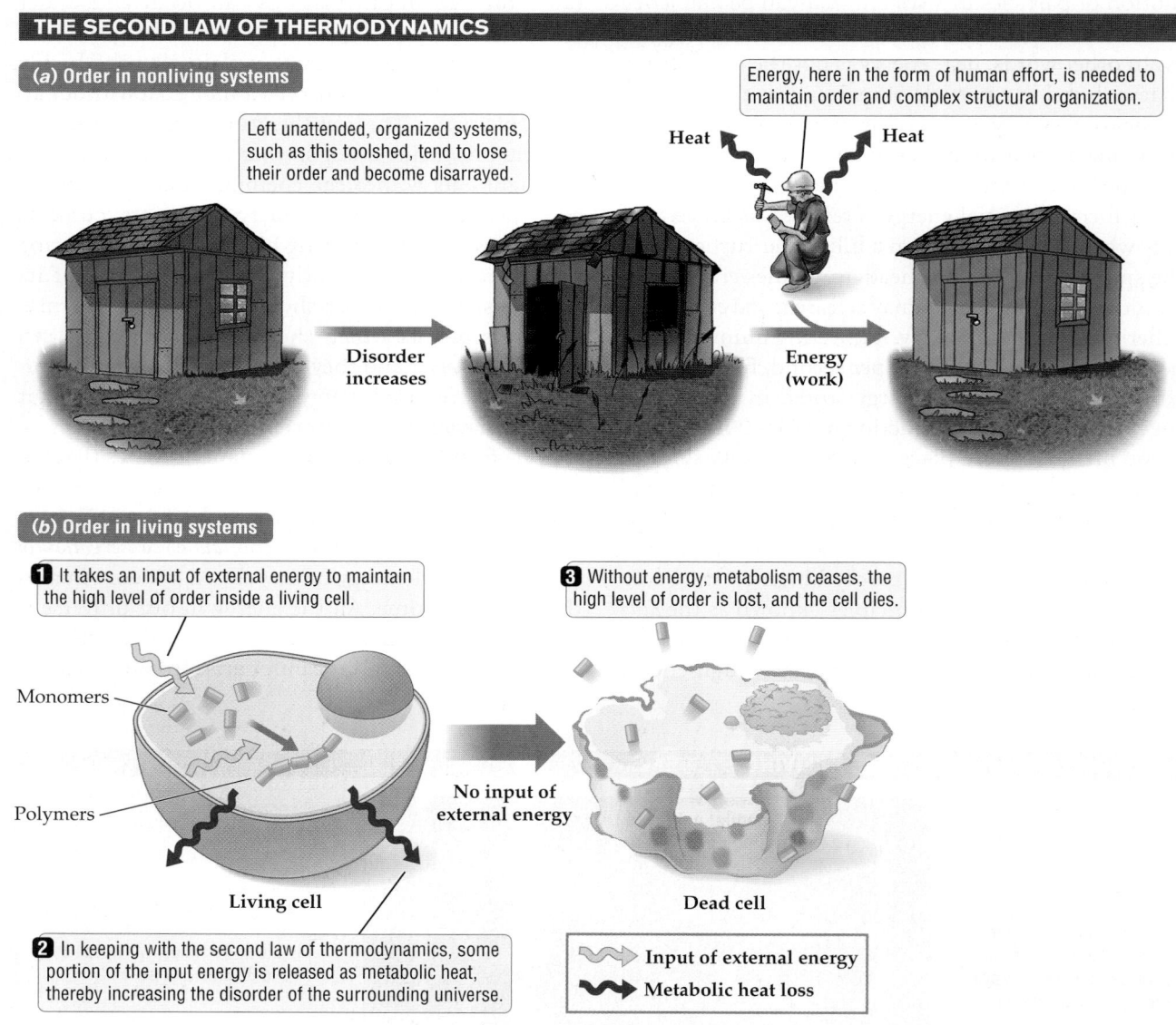

## THE SECOND LAW OF THERMODYNAMICS

(*a*) Order in nonliving systems

Left unattended, organized systems, such as this toolshed, tend to lose their order and become disarrayed.

Energy, here in the form of human effort, is needed to maintain order and complex structural organization.

Heat    Heat

Disorder increases

Energy (work)

(*b*) Order in living systems

**1** It takes an input of external energy to maintain the high level of order inside a living cell.

**3** Without energy, metabolism ceases, the high level of order is lost, and the cell dies.

Monomers

Polymers

No input of external energy

Living cell

Dead cell

**2** In keeping with the second law of thermodynamics, some portion of the input energy is released as metabolic heat, thereby increasing the disorder of the surrounding universe.

〰️ Input of external energy
〰️ Metabolic heat loss

**FIGURE 8.2  An Input of Energy Is Needed to Create or Maintain a High Level of Organization**

(*a*) The disorder of a system tends to increase unless that tendency is countered by an input of energy. (*b*) Living cells maintain their complex organization through continual input of energy from the environment.

As we saw in Chapters 5 and 6, a cell is made up of many chemical compounds assembled into highly ordered structures, such as proteins built from amino acids. The tremendous structural and functional complexity of the cell exists in the midst of a general tendency toward chaos. To counteract the natural tendency toward disorganization, the cell must capture, store, and use energy. One of the many implications of the second law of thermodynamics is that the capture, storage, or use of energy by living cells is never 100 percent efficient, so at least some of the invested energy is dissipated as a disordered and unusable form of energy called *metabolic heat*. In other words, through the very act of creating order within, living systems add to the disorder of the universe by releasing metabolic heat into the environment (**FIGURE 8.2b**). Consequently, only a portion of available energy, usually a relatively small portion, is available to fuel cellular processes, since a significant portion goes toward generating metabolic heat.

No muscle cell can convert *all* the chemical energy in a sugar molecule into the kinetic energy of a hummingbird's wing beat. Only a portion of the chemical energy in food molecules actually moves the special contractile proteins inside a muscle cell, because a good deal of that energy is turned into heat. Some of the released heat may warm the hummingbird, or overheat the home improvement enthusiast working on a crumbling toolshed. Eventually, all of the energy released from food molecules is radiated into the rest of the universe, rendering it a tiny bit more chaotic.

## The flow of energy connects living things with the environment

Where does the energy used to generate order in the cell come from? We know from the first law of thermodynamics that the cell cannot create energy from nothing; the necessary energy must come from outside the cell. In other words, energy must be transferred into the cell in some fashion. In the metabolic pathway known as **photosynthesis**, the energy for making sugar molecules comes from sunlight. Photosynthetic organisms trap light energy and use it to make sugars from carbon dioxide and water, thereby transforming light energy into chemical energy stored in the covalent bonds of sugar molecules. For most organisms that do not photosynthesize, energy comes from the chemical energy of food molecules, such as sugars and fats, gained from eating other organisms.

In the biosphere there is a tight relationship between energy and matter, and between organisms that produce energy derived from the sun (*producers*) and those that consume it (*consumers*). Producers are also known as *autotrophs*, literally "self-feeders" (*auto*, "self"). Consumers, including decomposers, are also known as *heterotrophs*, literally "eaters of other things" (*hetero*, "other"). As we saw in Chapter 1, producers make their own food, which we can define as biomolecules rich in energy and nutrients. Photosynthetic organisms—plants, algae, and certain bacteria—are producers that use sunlight to manufacture food; consumers obtain energy, as well as the building blocks of macromolecules, by eating other organisms or absorbing their dead remains. Photosynthetic organisms make up the bulk of producers found on land and in the sea. This means that, thanks to photosynthesis, the sun is the primary energy source in most ecosystems.

Energy streams through an ecosystem in a single direction, passing from producers to consumers, with some escaping as metabolic heat during every biological process and at every step in the food chain, as dictated by the second law of thermodynamics. In contrast to the one-way flow of energy, much of the matter in an ecosystem is recycled within it. Carbon atoms and other essential elements of living things pass from producers to consumers, and then back to producers after cycling through nonliving parts of the environment. For example, carbon-containing molecules from living cells are returned to the inanimate part of the ecosystem as organisms break down food molecules through an energy-releasing process known as **cellular respiration**. The carbon-carbon bonds in food molecules are broken during cellular respiration, and each carbon atom is released into the environment as a molecule of carbon dioxide ($CO_2$).

Consumers, including decomposers such as bacteria and fungi, are not the only organisms that degrade food molecules and release carbon dioxide through cellular respiration. Producers such as plants also rely on cellular respiration. Photosynthetic cells manufacture energy-rich sugar molecules, but they rely on cellular respiration to extract energy from these molecules to meet their daily energy needs.

Photosynthetic cells play a crucial role in extracting carbon from the inanimate world and returning it to living systems. What photosynthetic cells can do that most other cells cannot is absorb carbon dioxide from the environment and link its carbon atoms into the backbone of biological molecules. In this way, carbon atoms are continually cycled from carbon dioxide

FIGURE 8.3
**Photosynthesis and Cellular Respiration Are Complementary Processes**

Matter, in the form of carbon atoms, cycles among producers, consumers, and the environment.

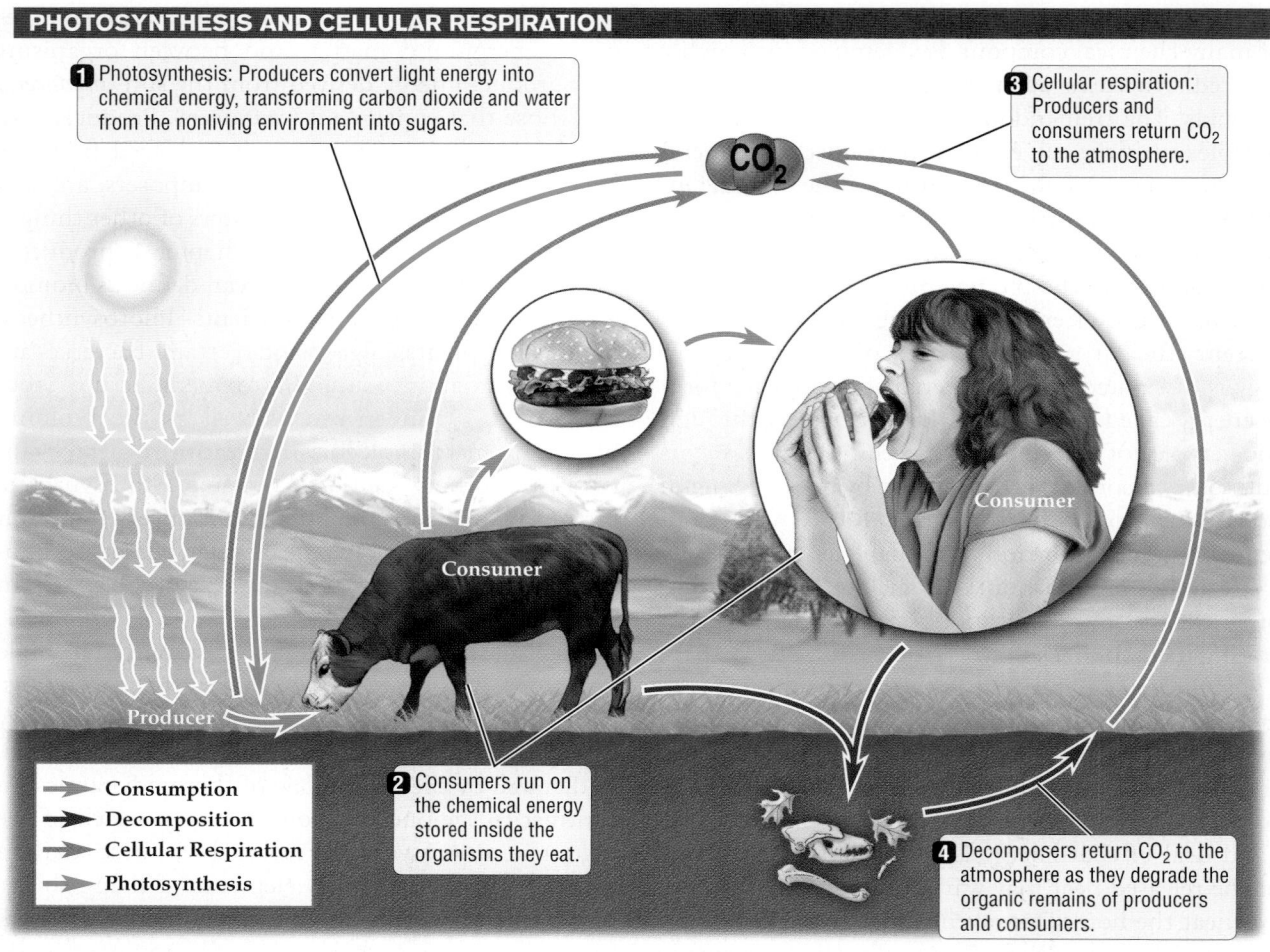

**PHOTOSYNTHESIS AND CELLULAR RESPIRATION**

**1** Photosynthesis: Producers convert light energy into chemical energy, transforming carbon dioxide and water from the nonliving environment into sugars.

**3** Cellular respiration: Producers and consumers return $CO_2$ to the atmosphere.

$CO_2$

Consumer

Consumer

Producer

→ Consumption
→ Decomposition
→ Cellular Respiration
→ Photosynthesis

**2** Consumers run on the chemical energy stored inside the organisms they eat.

**4** Decomposers return $CO_2$ to the atmosphere as they degrade the organic remains of producers and consumers.

in the atmosphere to sugars and other molecules made by producers, and then back to the atmosphere as carbon dioxide released by respiring producers and consumers. **FIGURE 8.3** illustrates the relationship between photosynthesis and cellular respiration. If you pay close attention to which molecules are used and which are released by photosynthesis and cellular respiration, respectively, you will see that the two are complementary processes.

## 8.2 Metabolism

As noted earlier, **metabolism** refers to all the chemical reactions within a living cell that capture, store, or use energy. All living cells require two main types of metabolism: *catabolism* and *anabolism* (**FIGURE 8.4**). **Catabolism** refers to the linked chain of reactions that release chemical energy in the process of breaking down complex biomolecules. Carbohydrates, and

lipids such as triglycerides, are the complex biomolecules most commonly degraded by these energy-releasing pathways. **Anabolism** refers to the linked chain of energy-requiring reactions that create complex biomolecules from smaller organic compounds. Anabolic reactions are also known as *biosynthetic pathways* (*bios*, "life"; *synthetikos*, "of composition") because complex biomolecules such as glycogen (a polymer of glucose) and triglycerides (a lipid containing three fatty acids) are put together from simpler building blocks during the course of these reactions.

## ATP delivers energy to anabolic pathways and is regenerated via catabolic pathways

Every living cell deploys the small, energy-rich organic molecule **ATP** (adenosine triphosphate) as an energy delivery service. Energy from ATP powers a variety of activities in the cell, such as moving molecules and ions in or out of the cell, moving organelles through the cytosol on tracks formed by cytoskeletal elements, and generating mechanical force during the contraction of a muscle cell. ATP also fuels the biosynthetic reactions of anabolism. Much of the usable energy in ATP is stored in its energy-rich phosphate bonds (**FIGURE 8.5**). Energy is released when a molecule of ATP loses its terminal phosphate group, breaking into a molecule of **ADP** (adenosine diphosphate) and a free phosphate. (Note that the adenosine part of the molecule is composed of the nitrogenous base adenine, coupled with the sugar ribose.)

Where does ATP come from? Loading a high-energy phosphate group on ADP transforms the otherwise sedate molecule into the live wire that is ATP. But turning ADP and phosphate groups into ATP—the universal energy currency of cells—takes metabolic energy. Every cell must have special energy-releasing catabolic pathways that can transform ADP and phosphate into ATP. Producers, such as plants, can use light energy to turn ADP and phosphate into ATP, during photosynthesis. In animals, cellular respiration is the most important ATP-generating pathway. Continual ATP production is an urgent priority for the human body; if it were halted, each cell would consume its entire supply of ATP in about a minute.

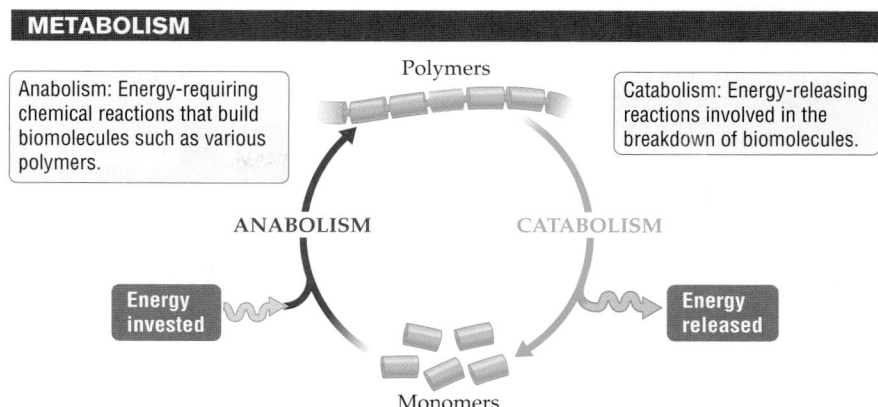

**METABOLISM**

Anabolism: Energy-requiring chemical reactions that build biomolecules such as various polymers.

Catabolism: Energy-releasing reactions involved in the breakdown of biomolecules.

Polymers

ANABOLISM            CATABOLISM

Energy invested

Energy released

Monomers

**FIGURE 8.4  Anabolic Reactions Build Macromolecules, Catabolic Reactions Degrade Them**

## Energy is extracted from food through a series of oxidation-reduction reactions

Many metabolic pathways, such as photosynthesis and cellular respiration, consist of a series of chemical reactions in which electrons are transferred from one molecule or atom to another. **Oxidation** is the loss of

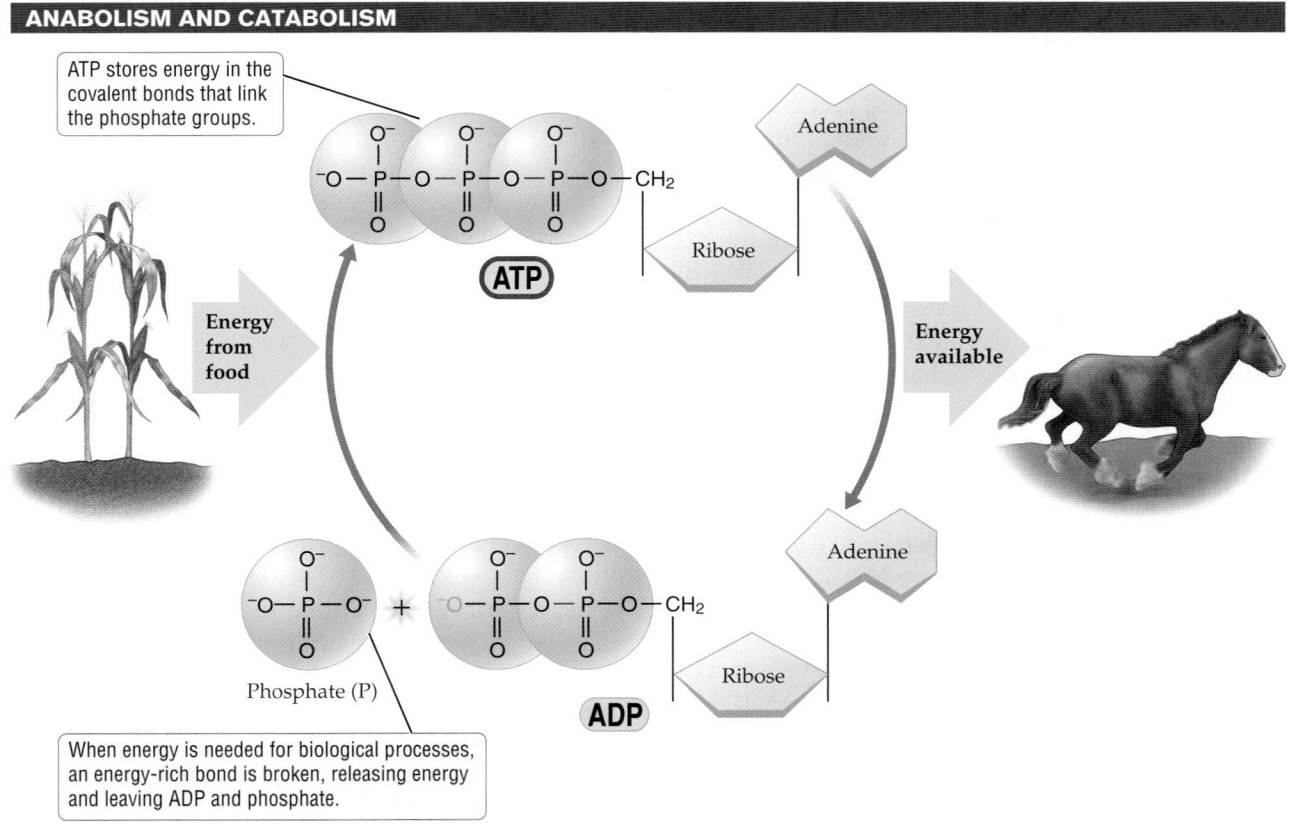

**ANABOLISM AND CATABOLISM**

ATP stores energy in the covalent bonds that link the phosphate groups.

Energy from food

ATP

Adenine

Ribose

Energy available

Phosphate (P)

ADP

Adenine

Ribose

When energy is needed for biological processes, an energy-rich bond is broken, releasing energy and leaving ADP and phosphate.

**FIGURE 8.5**

**ATP Functions as an Energy-Storing Molecule in All Cells**

When the terminal high-energy bond of ATP breaks, energy is released, and ATP is turned into ADP (adenosine diphosphate) with the release of a phosphate group.

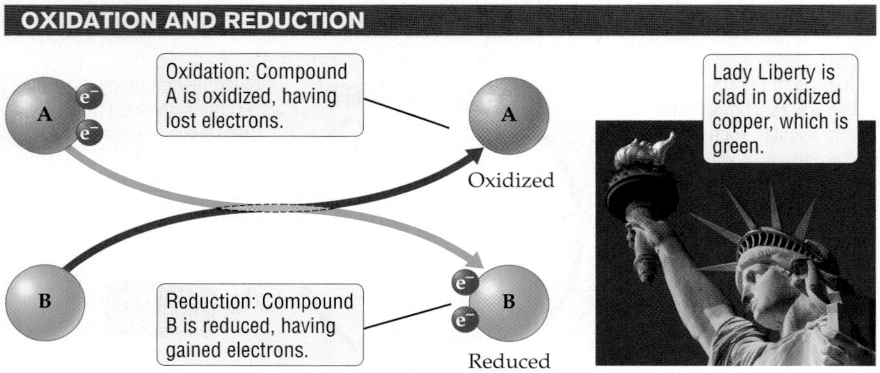

Oxidation: Compound A is oxidized, having lost electrons.

Oxidized

Reduction: Compound B is reduced, having gained electrons.

Reduced

Lady Liberty is clad in oxidized copper, which is green.

**FIGURE 8.6** Oxidation and Reduction Reactions Go Together

## Helpful to Know

Mnemonic aid for remembering oxidation and reduction:

- Electron Loss Means Oxidation: ELMO

- Gaining Electrons Reduction: GER!

electrons from a molecule, atom, or ion. **Reduction** is just the opposite: the gain of electrons by a molecule, atom, or ion (**FIGURE 8.6**). Because the two processes are complementary, oxidation and reduction go hand in hand, and the paired processes are called oxidation-reduction reactions, or simply **redox reactions**.

Although the modern definition of oxidation focuses on the loss of electrons, the term is derived from the nineteenth-century description of certain chemical reactions in which oxygen is added to an atom or molecule. Oxygen is a powerful oxidizer in that it can pull electrons from many other atoms. Iron combines with $O_2$ to form the crumbly red material we know as rust ($Fe_2O_3$). Because of the electron-drawing power of oxygen atoms, electron pairs shared between the two types of atoms are closer to the oxygen atoms than to the iron atoms, so the iron atoms can be regarded as existing in an oxidized state. In contrast, a hydrogen atom covalently bonded to a different atom, such as carbon or oxygen, will become "electron-poor" if the shared electrons are pulled more strongly by the partner atom. An atom or molecule that gains one or more hydrogen atoms becomes "richer" in electrons and is therefore considered reduced (remember, reduction is the *gain* of electrons).

Redox reactions take place, for example, in the oxygen-dependent catabolic pathways that extract energy from glucose molecules to make ATP in your cells. The summary equation for these pathways, collectively known as cellular respiration, can be written like this:

$$C_6H_{12}O_6 \ + \ 6\,O_2 \longrightarrow 6\,CO_2 \ + \ 6\,H_2O \ + \ energy$$

(glucose)　　(oxygen gas)　　(carbon dioxide)　　(water)　　(ATP; metabolic heat)

In this catabolic pathway, each of the six carbon atoms in glucose is *oxidized* as it gains oxygen atoms and

loses hydrogen atoms to become carbon dioxide. Each oxygen atom in $O_2$ strips two hydrogen atoms from glucose to become a molecule of water. The gain of hydrogen atoms by the oxygen atom signifies that it has been *reduced*.

In dramatic contrast, consider the summary equation for photosynthesis:

$$6\,CO_2 \ + \ 6\,H_2O \ + \ Energy \longrightarrow C_6H_{12}O_6 \ + \ 6\,O_2$$

(carbon dioxide)　　(water)　　(light)　　(glucose)　　(oxygen gas)

Photosynthesis is essentially the reverse of cellular respiration. In this anabolic pathway, carbon dioxide is *reduced* as it gains electrons and hydrogen atoms to be transformed into glucose. Water molecules lose electrons and hydrogen atoms and are therefore *oxidized* to molecular oxygen ($O_2$).

## Chemical reactions are governed by the laws of thermodynamics

How does the cell control such a powerful event as the oxidation of a glucose molecule and break it down into smaller, more manageable and useful steps? The answer lies in the stepwise nature of these metabolic reactions. Let's review some of the fundamental principles that govern chemical reactions, as represented by the following general example:

$$A + B \longrightarrow C + D$$

A and B are the starting materials, called **reactants**; C and D are the **products** formed by the reaction.

Recall that a chemical reaction involves the creation or rearrangement of chemical bonds in one or more substances. The second law of thermodynamics dictates that chemical reactions *can* occur spontaneously—that is, without an investment of energy—provided they generate products that are less organized, and therefore at a lower energy state, than the reactants. Put another way, a reaction can occur spontaneously (on its own) provided that some energy is lost. However, reactions that create products with higher total energy than the reactants simply will not occur without an input of energy to "push" the reactants in this nonspontaneous direction. In summary, chemical reactions that are "downhill" (because they go hand in hand with a loss of organization) can occur without any "outside help," but those that are "uphill" (because they generate a higher level of organization) will not occur unless energy is expended to make them happen.

Assuming that the reactants (A and B) have more total energy than the products (C and D), the second law says that A and B *can* be converted to C and D spontaneously—that is, without any outside input of energy. But that does not mean A and B *will* react with each other to any noticeable extent. The second law makes the reaction possible but says nothing about whether it is *probable*; it says nothing about the speed with which a reaction will take place or whether it will take place at all.

What does it take for atoms or molecules (such as A and B in our example) to react with each other? They must bump into each other often enough, fast enough, and in the correct orientation to allow their chemical bonds to be rearranged. This set of conditions constitutes an energy "hump," or energy barrier, that atoms and molecules must overcome *before* they can react with each other. One way to nudge reactants over the energy barrier is to invest a small amount of energy to get more of them moving faster and colliding more frequently. The minimum energy input that will enable atoms or molecules to overcome the energy barrier, thereby allowing them to react, is called the **activation energy**. The larger the fraction of atoms or molecules that overcome the energy barrier (because they possess enough activation energy), the faster a reaction will proceed.

Heat is one type of activation energy that can push atoms or molecules over the energy barrier so that they can react. Atoms and molecules move faster at higher temperatures, so raising the temperature will cause more of them to collide forcefully enough to react. When the energy of these atoms or molecules equals the activation energy, the reaction will proceed. The reason a safety match does not ignite spontaneously at room temperature is that its chemical ingredients react very slowly with molecular oxygen in the air. But a small rise in temperature—from friction against a rough surface, for example—can provide the activation energy necessary for the immediate combustion of the chemicals in the match head.

> **Concept Check**

1. What is metabolism?
2. What is activation energy?

## 8.3 Enzymes

Thousands of different kinds of chemical reactions take place in a living cell, and nearly all of them are mediated by enzymes. The great majority of enzymes are proteins, although certain RNA molecules are also known to function as enzymes. An enzyme is a **catalyst**, a chemical that speeds up chemical reactions without itself being changed in the course of the reaction. Specifically, an enzyme is a *biological* catalyst. In many ways, the action of an enzyme is similar to that of nonbiological catalysts such as platinum, an element used in automobile catalytic converters to detoxify waste gases generated by the combustion of gasoline. But enzymes are biomolecules, and as we will see shortly, these catalysts have special properties, such as sensitivity to extreme temperatures.

## Enzymes remain unaltered and are reused in the course of a reaction

An important characteristic of catalysts is that, unlike reactants, they remain chemically unaltered after the reaction is over (Table 8.1). Because enzyme molecules are used over and over, relatively small amounts of an enzyme are needed to catalyze a reaction. Most enzymes are highly specific in their action, catalyzing only one type of chemical reaction or, at most, a small number of very similar reactions.

The specific reactants that bind to a particular enzyme are called the **substrates** of that enzyme. Substrates bind to an enzyme in an orientation that favors the making and breaking of chemical bonds. Most enzymes have docking points—known as *active sites*—for one or more substrates. The size, geometry, and chemistry of an active site determine which substrate

| TABLE 8.1 | Properties of Enzymes |
| --- | --- |

**ENZYMES**

- Are usually proteins
- Increase the rate of chemical reactions, often by a million-fold or more
- Generally act on one or a few specific substrates
- Remain unchanged by the reaction
- Are reused over and over, catalyzing the transformation of many substrate molecules
- Are sensitive to temperature
- Are sensitive to their chemical environment, generally working best within a narrow range of pH and salt concentration
- May need the assistance of special cofactors (specific ions or molecules)
- May be inhibited by specific ions or molecules (inhibitors)
- Are usually tightly regulated within the cell or inside the body of a multicellular organism

**Concept Check Answers**

1. The many chemical reactions that yield, store, or consume energy in a living cell.
2. The minimum input of energy that is required to overcome the activation barrier, enabling a reaction to proceed.

**FIGURE 8.7**
**Enzymes as Molecular Matchmakers**

(*a*) An enzyme brings together two reactants (A and B) to form the product AB.
(*b*) Carbonic anhydrase catalyzes the reaction of carbon dioxide and water (substrates) to form bicarbonate (product).

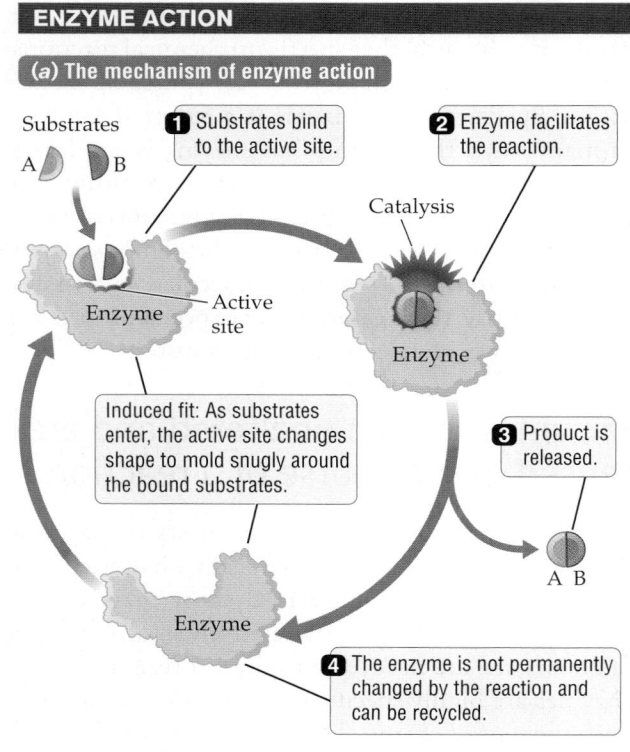

**ENZYME ACTION**

**(*a*) The mechanism of enzyme action**

Substrates
A     B

**1** Substrates bind to the active site.

**2** Enzyme facilitates the reaction.

Catalysis

Enzyme — Active site

Enzyme

Induced fit: As substrates enter, the active site changes shape to mold snugly around the bound substrates.

**3** Product is released.

A  B

Enzyme

**4** The enzyme is not permanently changed by the reaction and can be recycled.

**(*b*) The action of carbonic anhydrase**

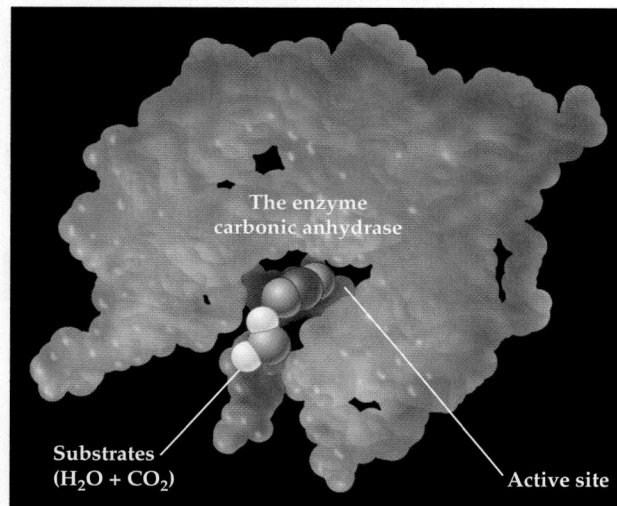

The enzyme carbonic anhydrase

Substrates ($H_2O + CO_2$)

Active site

Carbonic anhydrase eliminates carbon dioxide produced by cellular respiration by speeding up the reaction of carbon dioxide with water to form an ion, called bicarbonate ($HCO_3^-$), which dissolves readily in blood:

$$H_2O + CO_2 \xrightarrow{\text{Carbonic anhydrase}} HCO_3^- + H^+$$

catalyze. Enzymes are often named by the addition of "ase" to the name of the substrate they act on. Lactase, for example, is the digestive enzyme that breaks down the major milk sugar, lactose. You will find many more examples in the box on page 197.

The function of enzymes, like the function of most proteins, is crucially dependent on their three-dimensional shape. High temperatures denature (destroy) the three-dimensional configuration of most enzymes (see page 130, Chapter 5). Most human enzymes, for example, work best at the average core body temperature, 37°C (98.6°F), and many will lose activity at temperatures even 5°C above or below that optimum temperature. Extremes of pH—high acidity or high alkalinity—also disrupt the function of most enzymes (see Table 8.1), often by altering the chemistry of the active site. Some enzymes need particular ions or molecules—known as *cofactors*—to be maximally active. For example, carbonic anhydrase, an important blood enzyme, needs zinc ions as cofactors. Some enzymes need a particular salt concentration (ionic strength) for optimal activity, and work poorly at higher or lower salt concentrations. Enzyme function is seriously impaired at very high salt concentrations because large amounts of salt destroy the three-dimensional structure of proteins. The denaturing effect of salt is put to work in the manufacture of some cheese and tofu products: sea salt or calcium salts are used to curdle soy-milk proteins in making tofu, for example.

## The shape of an enzyme determines its function

The binding of an enzyme to its particular substrate depends on the three-dimensional shapes of both the substrate and the enzyme molecules. In the same way that a particular lock accepts only a key with just the right shape, each enzyme has an **active site** that fits only substrates with the correct three-dimensional shape and chemical characteristics (**FIGURE 8.7*a***). The shape of an active site is relatively flexible, and a substrate can tweak it to create an even better match between it and the enzyme. According to the **induced fit model** of substrate-enzyme interaction, as a substrate enters the active site the parts of the enzyme shift about slightly to enable the active site to mold itself around the substrate. This is similar to the way a limp glove (enzyme) takes on the shape of your hand (substrate) as you put it on. The ability of a substrate to induce a tighter and more accurate fit for itself in the active site of an enzyme stabilizes the interaction between the two and enables catalysis to proceed.

can bind to it, and this selectivity is the source of the substrate specificity of enzymes. One enzyme can bind only a particular substrate in the precise alignment necessary for product formation, which is why enzymes are specialists in terms of the reactions they

# Enzymes in Action

Enzymes are the workhorses of the cell. There are thousands of different types of enzymes in the human body, and if any one of them fails to function properly, some sort of illness is likely to follow. All 50 states have regulations that require newborns to be screened for several *inborn errors of metabolism* including *phenylketonuria*, a condition caused by the failure of an enzyme called phenylalanine hydroxylase (PAH). The enzyme catalyzes degradation of the amino acid phenylalanine. Without PAH activity, large amounts of brain-damaging chemicals (phenylketones) accumulate in the blood and urine. About one in 15,000 Americans is born with phenylketonuria; and if the condition is identified early, managing the patient's diet in the first 16 years of life can avert brain impairment. Affected individuals must avoid foods high in phenylalanine (meat, cheese, and legumes, as well as food containing the artificial sweetener aspartame) and take special amino acid formulations to prevent protein deficiency.

Enzymes help us digest the foods we eat, starting with the starch-degrading enzymes (*amylases*) in the mouth. The lining of the stomach produces protein-degrading enzymes (broadly known as *proteases*) that dismantle proteins under acidic conditions. The pancreas

and small intestine produce a variety of enzymes that degrade proteins (proteases) and lipids (lipases). In the small intestines, an enzyme called *lactase* breaks apart a milk disaccharide—*lactose*—to release the two constituent monosaccharides: glucose and galactose. About 90 percent of east Asians and Native Americans are *lactose intolerant* as adults because their small intestines stop making lactase once they are past childhood (see box on page 604). Undigested lactose is fermented by microbes in the large intestine, leading to intestinal discomfort. People who are lactose intolerant can enjoy dairy products supplemented with the lactase enzyme.

Some fruits contain proteases—papain in papayas and bromelain in pineapples—that are useful as meat tenderizers. The enzymes break

up some of the large fibrous proteins, such as collagen, that are abundant in tougher cuts of meat. The bromelain in fresh pineapples will also destroy any dessert made from gelatin, which is rich in an animal protein called collagen. Canned pineapples are no threat to gelatin desserts because canned fruit is treated with heat to prevent microbial growth, which inactivates the bromelain also.

Enzymes are widely used in household products, pharmaceuticals, food manufacturing, and industrial processes. Many clothes and dishwashing detergents contain amylases, proteases, and lipases, to help remove stains and residues from organic material. *Rennin*, an enzyme from the gut of cows and sheep, has been used for hundreds of years to make cheese. The enzyme denatures milk proteins, causing them to separate from the whey in clumps (curds) that are further processed to make the finished product. A number of carbohydrate-degrading enzymes are used at the mash stage of beer brewing, and proteases are used to remove cloudiness. Cellulases and pectinases—enzymes that break down polysaccharides—are used to clarify fruit juices. Enzymes are also important in the paper industry and are expected to assume greater importance in the emerging biofuel industry.

---

Carbonic anhydrase is a vital blood enzyme that speeds up the removal of carbon dioxide from our tissues. It accelerates the reaction of water and carbon dioxide by a factor of nearly 10 million. In fact, a single carbonic anhydrase molecule can process more than 10,000 molecules of carbon dioxide in just one second. Without it, carbon dioxide would react with water so slowly that little of it would dissolve in the blood and we would not be able to rid our bodies of carbon dioxide fast enough to survive. Because of its shape, carbonic anhydrase is able to bind both carbon dioxide and water in its active site. By bringing these two substrates together in exactly the right positions, the active site of carbonic anhydrase promotes the reaction (**FIGURE 8.7b**). If no enzyme were present, the two substrates would need to collide in just the right way before the reaction could take place. These sorts of molecular collisions do occur, but not nearly as frequently as would be required for the continual and rapid transfer of carbon dioxide from cells into the blood.

## Enzymes increase reaction rates by lowering the activation energy barrier

How do chemical reactions inside a cell overcome the activation energy barrier? Relying on heat as a source

## FIGURE 8.8
### Enzymes Reduce the Activation Energy Needed to Initiate a Reaction

(*a*) In the combustion (burning) of glucose, the reactants (glucose and oxygen) are at a higher energy level than the products (carbon dioxide and water), as represented by the solid red line. (*b*) In the analogy shown here, the reactants are represented by water in the reservoir of a dam. On the left, the dam (representing the energy barrier) is so high that most of the waves (reactants) cannot spill over it. But if the dam is lowered (as shown on the right), a large proportion of the waves (reactants) will overcome the barrier and run downhill (be turned into products).

**ENZYME CATALYSIS**

**(*a*) Activation energy**

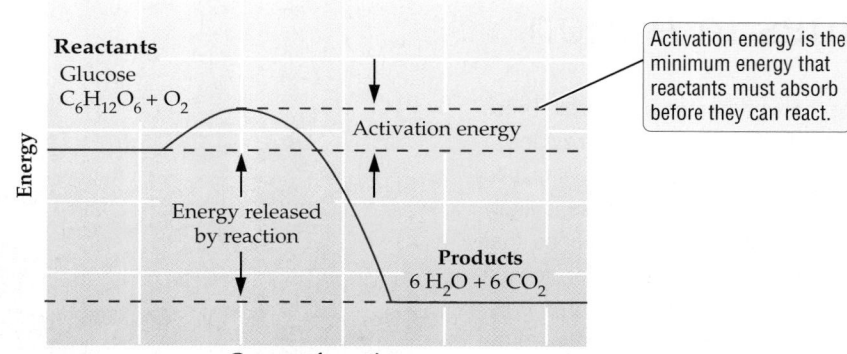

Many reactions would proceed very slowly without a catalyst. A slow reaction is one in which very few reactant molecules are converted into products per unit time.

Activation energy is the minimum energy that reactants must absorb before they can react.

**(*b*) Enzymes lower the activation energy barrier**

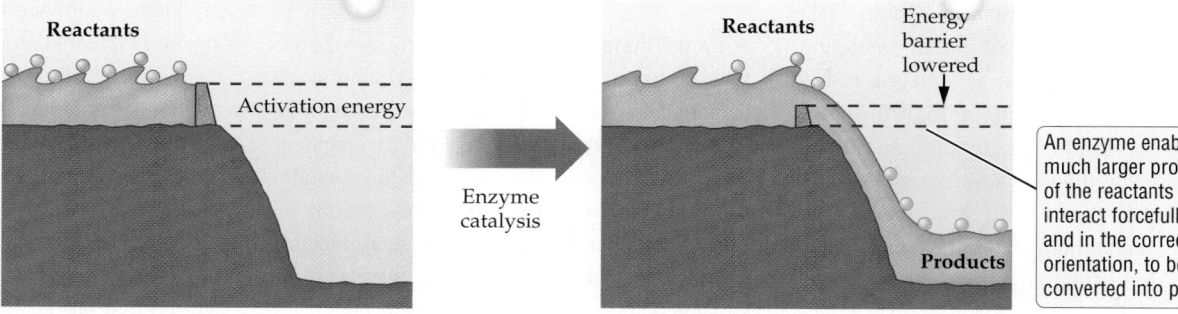

An enzyme enables a much larger proportion of the reactants to interact forcefully enough, and in the correct orientation, to be converted into products.

of activation energy is not a workable solution for most cellular processes, because heat acts indiscriminately. Although high temperatures will speed up almost all chemical reactions, the cell must exercise great selectivity in which chemical reactions are allowed to proceed at any given time. A catalyst lowers the activation barrier so that more of the reactants can cross it. As we have seen, the larger the proportion of reactants that make it over the energy barrier, the faster a reaction proceeds.

The great majority of chemical reactions in the cell take place when the activation barrier for the reaction is lowered by an enzyme (**FIGURE 8.8**). An enzyme lowers the activation barrier of a chemical reaction by binding tightly to the reactants and straining their chemical bonds in a way that promotes product formation. It is important to understand that an enzyme simply increases the rate, or speed, of a reaction that could occur on its own. An enzyme does not provide energy for a reaction that is thermodynamically uphill—that is, one that would create products at a higher energy state than the reactants. Because the enzyme is not altered by the chemical reaction in any way, an enzyme molecule can be used over and over again, which is why a small amount of enzyme can turn thousands of substrate molecules into product in less than a second.

## 8.4 Metabolic Pathways

So far, we have discussed the activity of a single enzyme acting alone to promote a single chemical reaction, but this state of affairs is not the most common in the cell. Typically, enzymes are involved in catalyzing steps in sequences of chemical reactions known as metabolic pathways. Metabolic pathways are widely used in the production of key biological molecules, including those that generate important chemical building blocks of the cell, such as amino acids and nucleotides. The pathways that harness energy from sunlight or food are also organized in a multistep sequence.

The most noteworthy advantage of multistep metabolic pathways is that they can proceed rapidly and efficiently because enzymes in the pathway are close together and the products of one enzyme-catalyzed step serve as substrates for the next reaction in the series. In other words, the products of the first reaction are instantly available in large amounts, and in close proximity, to act as substrate for the next enzyme and are therefore rapidly processed by the second enzyme-catalyzed reaction. The net result is that a multistep pathway of enzyme-catalyzed steps is "pushed" toward one specific outcome.

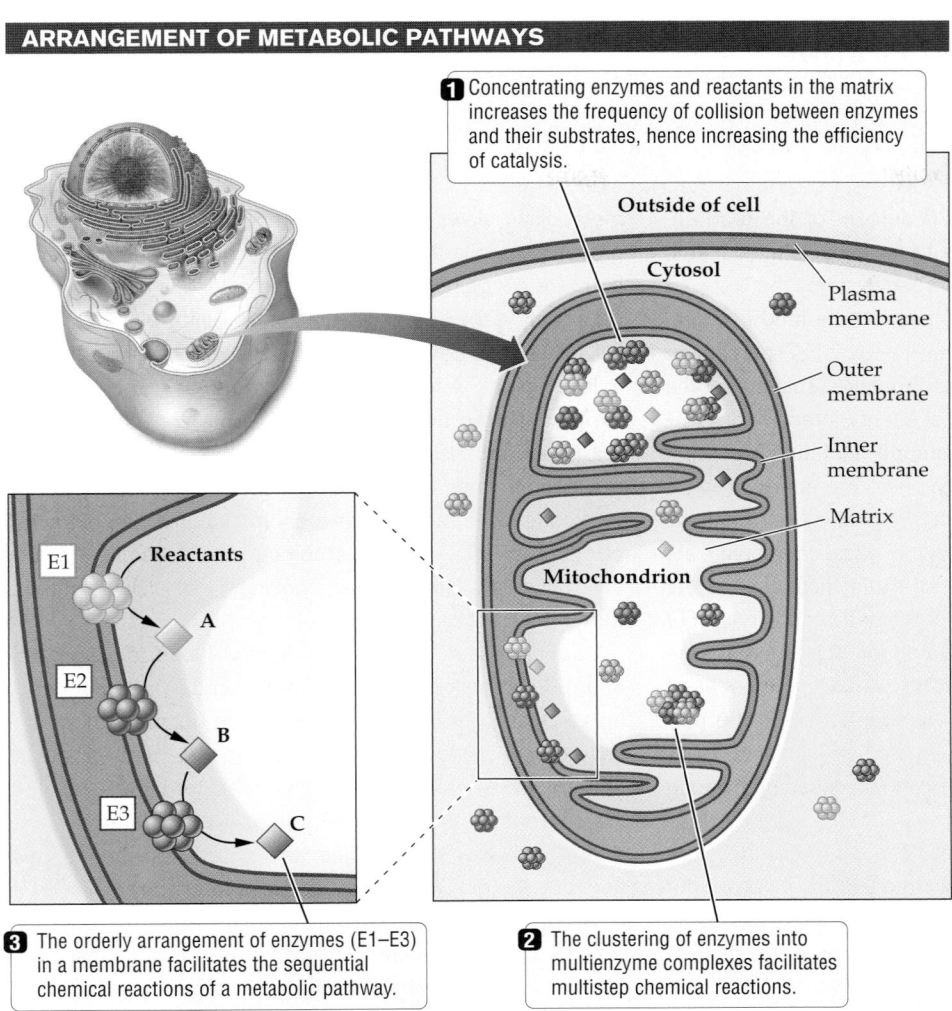

**1** Concentrating enzymes and reactants in the matrix increases the frequency of collision between enzymes and their substrates, hence increasing the efficiency of catalysis.

Outside of cell

Cytosol

Plasma membrane

Outer membrane

Inner membrane

Matrix

Mitochondrion

Reactants

E1
A

E2
B

E3
C

**3** The orderly arrangement of enzymes (E1–E3) in a membrane facilitates the sequential chemical reactions of a metabolic pathway.

**2** The clustering of enzymes into multienzyme complexes facilitates multistep chemical reactions.

**FIGURE 8.9 Metabolic Pathways Are Organized in Ways That Increase Their Efficiency**

Enzymes are often arranged in the cell in ways that facilitate the orderly series of chemical reactions that make up a metabolic pathway. These arrangements include the concentration of enzymes in organelles (in this case, a mitochondrion), their localization in membranes, and their clustering in multienzyme complexes.

Such a sequential arrangement of chemical reactions can be represented as follows:

$$A \xrightarrow{E1} B \xrightarrow{E2} C \xrightarrow{E3} D$$

The enzyme E1 catalyzes the conversion of A to B, enzyme E2 catalyzes the conversion of B to C, and so on, ensuring that D will be produced in the end. Metabolic pathways of this type are widely used in the production of key biological molecules, including those that generate important chemical building blocks of the cell, such as amino acids and nucleotides. The pathways that harness energy from food or sunlight are also organized in a multistep sequence.

The challenge faced by all enzyme-catalyzed reactions is the need for the enzyme and its specific substrates to encounter each other often enough. Because enzymes in multistep pathways are located close to one another in cells, the products of one reaction are in effect "aimed" at the next enzyme in the pathway. Another way of increasing the odds of encounters between an enzyme and its substrate is to contain and concentrate both of

them inside a membrane-enclosed compartment, such as a mitochondrion (**FIGURE 8.9**). As we saw in Chapter 6, organelles concentrate the proteins and chemical compounds required for specific biological processes. Mitochondria, for example, are the sites where food molecules are oxidized, generating most of the cell's ATP in the process. Enzymes present in the mitochondrial matrix participate in a sequence of reactions that produce large amounts of ATP. Other enzymes involved in the production of ATP are embedded in the inner mitochondrial membrane in a precise order.

> ## Concept Check
>
> 1. What is an enzyme? Describe two characteristics that are important to its function as a catalyst.
>
> 2. How does an enzyme affect the activation energy of a reaction?
>
> 3. How is the three-dimensional shape of an enzyme related to its function?

**Concept Check Answers**

1. An enzyme is a biological catalyst that speeds up chemical reactions. It is substrate specific and is reused because it remains unaltered in the course of the reaction.

2. An enzyme lowers the activation energy barrier, making it possible for more reactant molecules to engage in the reaction and thereby effectively speeding up the reaction.

3. The shape of the enzyme's active site must match the approximate shape of its substrate(s), the way a lock matches a key: the particular shape of the active site positions the substrates in ways that increase the likelihood of product formation, thereby speeding up the reaction.

# Metabolic Rates, Health, and Longevity

About 70 percent of the food energy we typically eat is used just to run the organ systems that keep us alive. The liver—a hotbed of metabolic activity—uses 27 percent of our energy intake. The brain—the next most energy-intensive organ—burns through another 20 percent of our calories. The energy it takes to sustain these basic functions is called the basal metabolic rate (BMR). BMR is measured under highly controlled conditions. For example, a person must not have eaten for 12 hours (because digestion consumes energy), must be awake, and must be at rest. The room must be a certain temperature, and the person must be calm. Once all these conditions are met, BMR is calculated from the amount of oxygen consumed during one hour.

Practically, it's difficult to meet all these conditions, so a looser measure is resting metabolic rate (RMR), which is your metabolic rate when you are sitting doing nothing. Metabolic rate goes up with physical activity, during the digestion of food, with ambient temperature, and even with nervous system stimulation such as watching an exciting movie.

A person with a large surface area–to–volume ratio has a higher metabolic rate than does a person with a small one. Muscle uses energy even at rest, so a person with more muscle expends more calories than a person the same height and weight who has less muscle (for more on exercise and BMR, see the "Biology in the News" feature on page 201). In fact, every pound of muscle uses about 16 Calories per day. So exercising burns calories directly and also builds muscle that burns more energy while you're sitting at your computer. Because most men have more muscle than women, men tend to have a higher BMR (see Table 8.2). As we get older, our metabolism slows—by 1–2 percent per decade of life. It's probably obvious that if we eat more calories than our bodies burn, we will gain weight. Balancing calorie intake and physical activity is the key to maintaining a healthy weight.

Can eating certain foods raise the BMR, making it easier to lose weight? A few studies involving small numbers of people have shown a temporary boost from consuming oolong tea, green tea, caffeine, or capsaicin—the chemical that gives chili peppers their fiery bite. For example, in one study, people who consumed capsaicin at every meal burned an extra 100 calories per day—about the number of calories in a golden delicious apple. Still, if you are looking for a quick weight-loss trick, tea and hot sauce isn't it.

Eating less and exercising more will lead to weight loss and reduced risk of diabetes and heart disease. But eating less has another benefit as well. Large numbers of studies have demonstrated that a super-low-calorie diet—about 25 percent fewer calories than is normally required—leads to life spans that are about 30 percent longer, in animals as different as worms, fruit flies, and mice. Even rhesus monkeys live longer when they eat less, and they have lower rates of cancer and other disease as well. There's every reason to think the same might be true in humans. Small numbers of people have undertaken a permanent calorie-restricted diet. Not surprisingly, it's a lifelong diet that hasn't caught on yet.

| TABLE 8.2 | Basal Metabolic Rates for Men and Women | | | |
| --- | --- | --- | --- | --- |
| | WEIGHT | | BMR | |
| | (KG) | (LB) | (WATTS/H) | (CALORIES/H) |
| Woman | 60 | 132 | 68 | 58 |
| Man | 70 | 154 | 87 | 75 |

# Phys Ed: A Workout for Your Bloodstream

BY GRETCHEN REYNOLDS, *Well* blog, *New York Times*

What does exercise do to your body? It may seem as if science, medicine and common sense answered that question long ago. But in fact, the precise mechanisms by which exercise alters your body—at a deep, molecular level—remain poorly understood. A number of analyses of the effect that exercise has on heart disease, for instance, have concluded that working out lessens a person's chances of developing heart problems far more than scientists can account for. They understand the physiological reasons for about 60 percent of the reduced risk. The rest is a mysterious if welcome bonus.

But a new study that gauged the metabolic effects of exercise may significantly advance our understanding of what's going on inside a body in motion. During the experiment, scientists actually saw how much being fit changes your ability to incinerate fat, moderate blood sugar and otherwise function well. They also uncovered proof, at once inspiring and cautionary, of just how complicated and pervasive exercise's consequences are.

In the experiment, published late last month in *Science Translational Medicine*, researchers from Harvard University and other institutions relied on a mass spectrometer to enumerate specific molecules

in the bloodstreams of people who'd been exercising. The molecules were metabolites, which drive or are the byproduct of metabolic changes in the body. Metabolism is, of course, the chemical process of keeping yourself alive. All of the biochemical processes that feed and nurture cells constitute your metabolism. What the researchers wanted to know is, how does your metabolism change during and after exercise?

For the work, the scientists drew blood from a group of normal, healthy adults, as well as from a separate group who'd been referred for exercise testing because of shortness of breath or suspected coronary-artery disease. This group was relatively unfit. Each of the groups was told to exercise for about 10 minutes on a treadmill or a stationary bicycle, then had more blood drawn. Finally, the scientists also examined blood samples from a group of runners who had finished the 2006 Boston Marathon.

What they found was that after 10 minutes of treadmill jogging or stationary-bicycle riding, the healthy adults showed enormous changes in the metabolites within their bloodstream, as did the less-fit group, although to a lesser degree. In particular, certain metabolites associated

with fat burning were elevated. The fit adults showed increases of almost 100 percent in many of these molecules. The less-fit group had increases in those same metabolites of about 50 percent. As for the marathoners, their blood contained up to 10 times more of the fat-burning markers.

These findings suggest that exercise has both "acute and cumulative" effects on your body's ability to use and burn fat, says Gregory Lewis, a cardiologist at Massachusetts General Hospital in Boston and an author of the study. After only 10 minutes of exercise, even the least fit showed evidence that their bodies were burning fat; the more fit, the more metabolic evidence of fat burning.

The researchers then took a number of the metabolites that had been elevated by exercise and infused them into mouse muscle cells in a laboratory dish. Almost immediately, the metabolites, in combination (but not individually) ignited a reaction that resulted in increased expression of a gene involved in cholesterol and blood-sugar regulation. In other words, the metabolites weren't just marking activity that was happening elsewhere in the body; they also may have been sparking some of that activity directly.

In this experiment, researchers measured the amounts of 200 different molecules in three groups of people. After 10 minutes of exercise, unhealthy people increased blood levels of certain exercise "metabolites" by about 50 percent, while fit adults increased these same molecules by 100 percent. A third group, marathon runners who had just completed a 3-to-5-hour, 26-mile marathon, increased metabolites 10 times. They were exercising strenuously a lot longer.

This article says that the metabolites caused an increase in fat burning—at least in mouse muscle cells. Recall that exercise boosts metabolism by increasing muscle, so that you burn more calories even at rest. This study suggests that exercise also helps burn fat not only when you're doing the exercise, but perhaps for some time afterward as well.

## Evaluating the News

**1.** Ten minutes of exercise in people who exercise regularly increased the blood "metabolites" how much? How about in unhealthy people?

**2.** If researchers could recognize healthy people just by the metabolites in their blood, how could they use that information to learn more about the best way to exercise?

**3.** What's wrong with comparing the marathon runners' metabolite levels to those of the other two groups? Design an experiment that would let you compare metabolites in people who ran for different amounts of time.

**SOURCE:** Excerpt from *New York Times*, *Well* blog, June 16, 2010, well.blogs.nytimes.com.

# Summary

## 8.1 The Role of Energy in Living Systems

- Energy is the capacity to do work. Work is the capacity to bring about a change in a system.
- Potential energy is the energy stored in any system as a consequence of its position. Chemical energy is one type of potential energy.
- Kinetic energy is the energy a system possesses as a consequence of its state of motion. Heat energy, a form of kinetic energy, is that portion of the total energy of a particle of matter that can flow to another particle.
- The first law of thermodynamics states that energy can be converted from one form to another but is never created or destroyed.
- The second law of thermodynamics states that the natural tendency of the universe is to become less organized. The creation of biological order therefore requires energy. The creation of internal order in living organisms is always accompanied by the transfer of disorder to the environment, generally in the form of metabolic heat.
- The sun is the source of energy fueling most living organisms. Producers capture the sun's energy through photosynthesis. Plants, algae, and some bacteria gain energy from their environment through photosynthesis. Many producers and consumers use cellular respiration to extract usable energy from food molecules.
- Matter, in the form of chemical elements such as carbon, cycles between living organisms and their environment.

## 8.2 Metabolism

- All the many chemical reactions involved in the capture, storage, and use of energy by living organisms are collectively known as metabolism.
- Energy-releasing breakdown reactions are catabolism; energy-requiring synthesis reactions are anabolism.

- The energy-rich molecule ATP supplies much of the energy needed to fuel cellular activities.
- In oxidation, electrons are lost from a molecule, atom, or ion; in reduction, electrons are gained.
- The minimum energy required to initiate a chemical reaction is called the activation energy. Most chemical reactions must overcome an activation energy barrier to proceed at an appreciable rate.

## 8.3 Enzymes

- Enzymes are biological catalysts that speed up chemical reactions. Enzymes position bound reactant molecules in such a way that they collide more often in the orientation that favors product formation. Like all catalysts, enzymes lower the activation energy barrier of a reaction.
- The activity of enzymes is highly specific. Each enzyme binds to a specific substrate or substrates and catalyzes a specific chemical reaction. The specificity of an enzyme is based on the three-dimensional shape and chemical characteristics of its active site.
- The three-dimensional shape of an enzyme, and therefore its activity, can be affected by temperature, pH, and salt concentration. Some enzymes must work with other chemicals, called cofactors, to be maximally effective.

## 8.4 Metabolic Pathways

- A metabolic pathway is a multistep sequence of chemical reactions, with each step catalyzed by a different enzyme.
- Metabolic pathways proceed rapidly and efficiently because all necessary components are placed close together, at high concentrations, and in the correct order. The products of one enzyme-catalyzed step serve as substrates for the next reaction in the series.

# Key Terms

activation energy (p. 195)
active site (p. 196)
ADP (p. 193)
anabolism (p. 192)
ATP (p. 192)
catabolism (p. 192)
catalyst (p. 195)

cellular respiration (p. 191)
chemical energy (p. 188)
energy (p. 188)
enzyme (p. 188)
first law of
  thermodynamics (p. 189)
heat energy (p. 189)

induced fit model (p. 196)
kinetic energy (p. 188)
metabolic pathway (p. 188)
metabolism (p. 192)
oxidation (p. 193)
photosynthesis (p. 191)
potential energy (p. 188)

product (p. 194)
reactant (p. 194)
redox reaction (p. 194)
reduction (p. 194)
second law of
  thermodynamics (p. 190)
substrate (p. 195)

# Self-Quiz

1. Which of the following statements is true?
   a. Cells can produce their own energy from nothing.
   b. Cells use energy only to generate heat and move molecules around.
   c. Cells obey the same physical laws of energy as the nonliving environment.
   d. Most animals obtain energy from minerals to fuel their metabolic needs.

2. Living organisms need energy to
   a. organize chemical compounds into complex biological structures.
   b. decrease the disorder of the surrounding environment.
   c. transform metabolic heat into kinetic energy.
   d. keep themselves separate from the nonliving environment.

3. The carbon atoms contained in organic compounds such as proteins
   a. are manufactured by cells for use in the organism.
   b. are recycled from the nonliving environment.
   c. differ from those found in $CO_2$ gas.
   d. cannot be oxidized under any circumstances.

4. Oxidation is the
   a. removal of oxygen atoms from a molecule.
   b. gain of electrons by an atom.
   c. loss of electrons by an atom.
   d. synthesis of complex molecules.

5. Which of these molecules is in a reduced state?
   a. $CO_2$
   b. $N_2$
   c. $O_2$
   d. $CH_4$

6. The minimum input of energy that initiates a chemical reaction
   a. is called activation energy.
   b. is independent of the laws of thermodynamics.
   c. is known as the activation barrier.
   d. always takes the form of heat.

7. Activation energy is most like
   a. the energy released by a ball rolling down a hill.
   b. the energy required to push a ball from the bottom of a hill to the top.
   c. the energy required to get a ball over a hump and onto a downward slope.
   d. the energy that keeps a ball from moving.

8. Enzymes
   a. provide energy for anabolic but not catabolic pathways.
   b. are consumed during the reactions that they speed up.
   c. catalyze reactions that would otherwise never occur.
   d. catalyze reactions that would otherwise occur much more slowly.

9. The active site of an enzyme
   a. has the same shape for all known enzymes.
   b. binds the products of reaction, not the substrate.
   c. does not play a direct role in catalyzing the reaction.
   d. can bring molecules together in a way that promotes a reaction between them.

10. Metabolic pathways
    a. always break down large molecules into smaller units.
    b. only link smaller molecules together to create polymers.
    c. are often organized as a multistep sequence of reactions.
    d. occur only in mitochondria.

# Analysis and Application

1. Think back to what you ate for breakfast. What has become of all the chemical energy locked in those food molecules? List the different types of energy transformations that have occurred in your body as you have gone about your day, starting with the chemical energy in your breakfast.

2. Describe the role of the second law of thermodynamics in living systems.

3. Compare anabolism and catabolism. Is photosynthesis an anabolic or catabolic process?

4. Explain what is wrong with this statement: An enzyme provides energy for reactions that cannot proceed without an investment of energy.

5. Explain the induced fit model of interaction between an enzyme and its substrate.

6. A Chinese herbal medicine called *ma huang* (*Ephedra sinica*) increases BMR significantly and was once widely used in weight loss pills, usually in combination with caffeine. Its use in supplements was banned in 2006, after reports of deaths among users of the herb. Research the metabolic effects of ephedrine, the active ingredient in the herb, to explain why a very high metabolism can kill.

# Photosynthesis and Cellular Respiration

**BIG-WAVE SURFER JAY MORIARITY.** In 2001, Jay Moriarity—renowned for a spectacular wipeout that ran on the cover of *Surfer* magazine—died while on a photo shoot in the Maldive Islands, in the Indian Ocean. But Moriarity didn't die surfing; he died instead meditating 45 feet below the surface of the water without oxygen—a common practice among surfers and divers called "static apnea."

# Every Breath You Take

How long can you hold your breath? A minute? Two minutes? Doctors fear brain death in anyone who has gone without oxygen for more than 5 minutes. But German engineering student Tom Siestas held his breath under water for 11 minutes 35 seconds and claimed the men's record for the rather intimidating sport of *static apnea* (literally, "motionless, without air"). Natalia Molchanova claimed the women's record with a time of 8 minutes 23 seconds.

Athletes like Siestas and Molchanova are extremely unusual—certainly in their training and possibly in their genetic makeup—which is why static apnea should be left to trained professionals. In fact, health professionals suspect that even experts may be putting their long-term health at risk with their arduous oxygen deprivation training regimens, not to mention the even more dangerous competitions. In 2001, famed big-wave surfer Jay Moriarity drowned while practicing static apnea as part of his regular training. He was last seen alive while meditating 45 feet below the surface of the Indian Ocean.

Static-apnea training includes endurance exercise (which depends heavily on high oxygen uptake) and training at high altitude (which also improves oxygen delivery).

Endurance athletes such as bicyclist Lance Armstrong have a phenomenal ability to deliver oxygen to their tissues. Before a competition, Armstrong doubles his caloric intake to 6,000 Calories per day, most of it from carbohydrates. In the digestive tract, carbohydrates break down into glucose molecules, a key molecule in this chapter.

Whether elite athletes or couch potatoes, we all need oxygen and glucose. But what do our cells do with them?

Is efficient use of oxygen and glucose the main difference between elite endurance athletes and the rest of us? What species is the static-apnea champion?

To tackle these questions, we begin this chapter by considering how organisms capture and use energy. Plants and other photosynthetic organisms use energy from sunlight to build high-energy sugar molecules and then break down those sugars when they need energy. Animals (including humans) acquire energy by eating plants (or by eating animals that eat plants). The chemical reactions that enable plants to build sugar molecules and all organisms to extract energy from sugar are the focus of this chapter.

> **MAIN MESSAGE** Photosynthesis and cellular respiration are complementary processes; these chemical pathways furnish chemical energy for cells.

## KEY CONCEPTS

- All cells need energy carriers, such as ATP, to store and deliver usable energy.

- In plants and protists, photosynthesis takes place in special organelles called chloroplasts.

- Photosynthesis uses sunlight and water to produce energy carriers during the light reactions, releasing oxygen gas in the

process. In the Calvin cycle reactions of photosynthesis, the energy carriers are used to manufacture sugars from carbon dioxide.

- Most eukaryotes rely on cellular respiration, which requires oxygen ($O_2$), to extract energy from sugars and other food molecules. Cellular respiration has three main stages: glycolysis occurs

in the cytoplasm; the Krebs cycle and oxidative phosphorylation take place in the mitochondrion.

- Fermentation enables certain organisms and certain cell types to generate ATP anaerobically—through glycolysis alone—when aerobic ATP production is hamstrung by an inadequate oxygen supply.

**ENERGY IS NECESSARY** for all types of cellular activity. **Metabolism** encompasses all the chemical reactions involved in the capture, storage, and utilization of energy in a cell, and it is a fundamental necessity for every living thing. In most ecosystems on Earth, the sun is the ultimate source of energy for living cells. Plants and other photosynthetic organisms, which trap energy from sunlight to manufacture their own food, are the **producers** in most ecosystems. Producers, in turn, support **consumers**, which acquire energy by eating producers or other consumers.

As discussed in Chapter 8, a **metabolic pathway** consists of a series of chemical reactions by which energy is transformed as biomolecules are changed, step-by-step, into either simpler or more complex forms. Many of these reactions are critically dependent on small molecules known as energy carriers, which function as an energy delivery service inside the cell. We begin this chapter by exploring the nature of energy carriers in order to appreciate two of the most important and widespread metabolic pathways on Earth: *photosynthesis*, which captures light energy to make sugars from carbon dioxide and water; and *cellular respiration*, which releases energy from food molecules to fuel cellular activities. We will see that photosynthesis gives off oxygen ($O_2$), a by-product that is crucial for the survival of consumers like us because it is necessary for cellular respiration. While photosynthesizers *use up* carbon dioxide ($CO_2$) and water ($H_2O$) to make sugars, consumers *release* these molecules as they extract energy from sugars during cellular respiration (**FIGURE 9.1**).

**FIGURE 9.1**
**The Relationship Between Photosynthesis and Cellular Respiration**

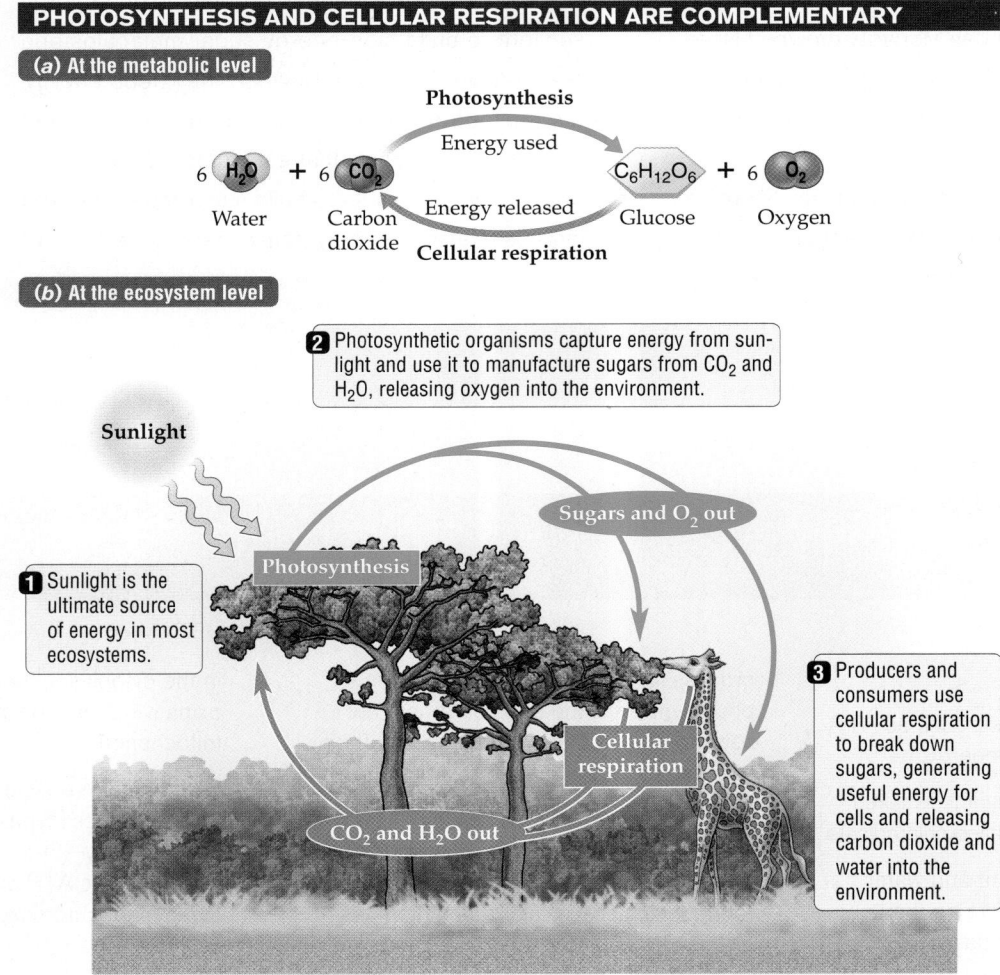

# 9.1 Molecular Energy Carriers

You may think you could run very nicely on bacon double cheeseburgers and triple fudge sundaes, but you should know that no food is of any *direct* use to the trillions of cells in your body. Cellular activities are powered not by bacon fat, beef protein, or carbohydrate from a sesame bun, but by tiny molecules that are generically known as *energy carriers*. A molecular energy carrier is rather like a rechargeable battery and the cell is like a miniature machine that runs on energy from many millions of these batteries. **Energy carriers** are small organic molecules specialized for receiving, storing, and delivering energy *within* the cell. They become "fully charged" when they receive energy from metabolic pathways that release energy, and they "lose the charge" as they deliver energy to the many thousands of chemical reactions and cellular activities that could not proceed without an infusion of energy

(**FIGURE 9.2**). Energy carriers are what make a cell tick, and any cell that runs out of fully charged energy carriers is a dead cell.

Of the common energy carriers in a cell, **ATP** (<u>a</u>denosine <u>tri</u>phosphate) is the most versatile: it delivers energy to the largest number and greatest diversity of cellular processes. ATP stores energy in the form of covalent bonds connecting its three phosphate groups (*triphosphate* = "3 phosphates"). ATP releases stored energy when it loses a terminal phosphate (P) to become ADP (<u>a</u>denosine <u>di</u>phosphate). The released energy is put to good use in the cell: it fuels myriad cellular activities ranging from the manufacture of biomolecules to cell division (see Figure 9.2).

What becomes of the low-energy ADP molecules that are spun off every time an ATP molecule discharges its energy? ADP can combine with phosphate and be converted back to ATP, but this is not a trivial reaction. The recharging requires very special energy-releasing pathways, such as photosynthesis and cellular respiration. In photosynthesis, the energy for

**CELLULAR ACTIVITIES ARE FUELED BY ENERGY CARRIERS**

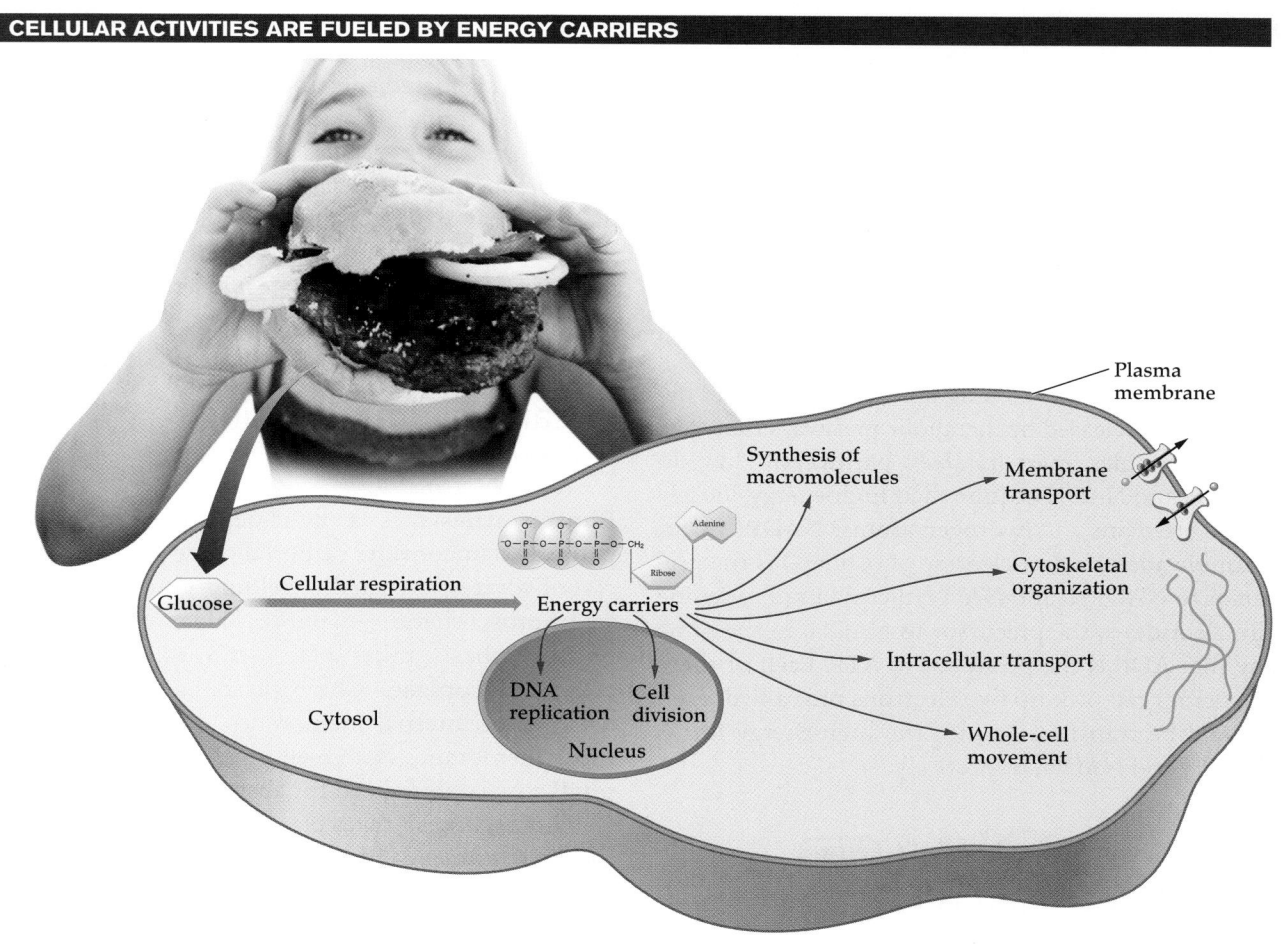

**FIGURE 9.2 You, Your Lunch, and Energy Carriers in the Cell**

Just about all cellular processes run on molecular energy carriers. Your body needs a mechanism to turn lunch into energy carriers.

turning ADP into ATP comes from sunlight. In cellular respiration, this energy comes from the breakdown of food molecules. Therefore, photosynthesis and cellular respiration are rather like battery chargers: they turn low-energy ADP molecules into high-energy ATP molecules:

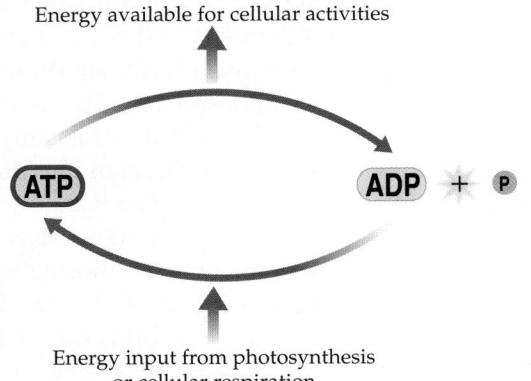

ATP is not the only energy carrier that cells rely on. NADPH and NADH are two of the other energy carriers we discuss in this chapter. A striking feature of energy carriers is that they are nearly universal in the living world. Each of these energy carriers is a specialist in terms of the amount of energy it carries and the types of chemical reactions to which it supplies energy and from which it receives energy.

**NADPH** and **NADH** hold energy in the form of certain loosely bound electrons and hydrogen atoms. (Recall that one hydrogen atom is the same as one electron plus one hydrogen ion, H$^+$.) NADPH delivers electrons and hydrogen ions (H$^+$) to metabolic pathways that build macromolecules (*anabolic* pathways); NADH specializes in picking up the electrons and hydrogen ions released by metabolic pathways that take macromolecules apart (*catabolic* pathways). How do NADPH and NADH acquire their high-energy, loosely bound electrons and hydrogen atoms? NADP$^+$ (<u>n</u>icotinamide <u>a</u>denine <u>d</u>inucleotide <u>p</u>hosphate) is the precursor to NADPH, and NAD$^+$ (<u>n</u>icotinamide <u>a</u>denine <u>d</u>inucleotide) is the precursor to NADH, in the same way that ADP is the precursor to ATP. Each of these precursors can pick up two electrons plus a hydrogen ion and be transformed into the high-energy forms NADPH and NADH, respectively:

So, we can think of ATP, NADPH, and NADH as different types of rechargeable batteries. ATP releases energy when its phosphate groups break off. NADPH and NADH deliver energy by donating electrons and hydrogen ions to chemical reactions that need them.

## 9.2 An Overview of Photosynthesis and Cellular Respiration

Photosynthesis and cellular respiration are two of the most important metabolic pathways in living organisms. Only producers can carry out photosynthesis, but whether they are producers or consumers, all eukaryotes and many prokaryotes need cellular respiration. Cellular respiration enables organisms to extract the chemical energy that is locked in the covalent bonds of food molecules and turn it into a directly usable form: the chemical energy of ATP.

The only photosynthetic producers on Earth are certain groups of bacteria, the protists known as algae, and plants. Photosynthetic organisms do not need to eat other organisms, because they make their food from scratch, using ingredients from their nonliving environment. Their recipe? Take six molecules of carbon dioxide (CO$_2$) and six molecules of water (H$_2$O), bake them with a few beams of light, and voilà, a molecule of sugar! As simple as we have made it sound, photosynthesis is so staggeringly complex that modern chemists, for all their wizardry, cannot duplicate it in a test tube.

### Light powers the manufacture of carbohydrates during photosynthesis

**Photosynthesis** is a light-dependent metabolic pathway that converts carbon dioxide and water into carbohydrates, eventually yielding glucose, a sugar. In photosynthetic eukaryotes—algae and plants—photosynthesis takes place in special organelles called **chloroplasts**. Chloroplasts have an extensive network of internal membranes, and embedded in those membranes is a green pigment called **chlorophyll** that is specialized for absorbing light energy.

Photosynthesis takes place in two principal stages: the *light reactions* and the *Calvin cycle* (**FIGURE 9.3**). During the **light reactions**, water molecules are split using light energy absorbed by chlorophyll molecules.

The splitting of water releases gaseous oxygen ($O_2$) as a by-product. Electrons and protons ($H^+$) extracted from water molecules are handed over to other molecules in an elaborate chain of chemical events that ultimately generates ATP and NADPH. The following illustration summarizes the main outcomes of the light reactions:

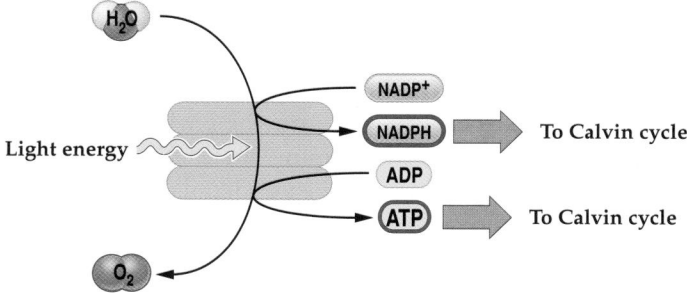

In the **Calvin cycle**, the second stage of photosynthesis, a series of enzyme-catalyzed chemical reactions converts carbon dioxide ($CO_2$) into sugar, using energy delivered by ATP and electrons and hydrogen ions donated by NADPH. In summary, the light reactions of photosynthesis generate energy carriers, which fuel the manufacture of sugars via the Calvin cycle. If we imagine the Calvin cycle as a candy factory, then the light reactions would be the power supply for the factory.

## Energy from sugars is used to make ATP during cellular respiration

**Cellular respiration** is an oxygen-dependent metabolic pathway through which food molecules, such as sugars, are broken down and the released energy is used to generate ATP. Oxygen ($O_2$) is necessary for the complete breakdown of sugars, and carbon dioxide ($CO_2$) and water ($H_2O$) are released as by-products. We refer to this process as *cellular* respiration to distinguish it from *whole body* respiration, which, in the case of animals with lungs, is the inhaling and exhaling of air (breathing). Breathing in animals, including humans, is directly related to cellular respiration. The air we breathe out is rich in carbon dioxide and water vapor, both by-products of cellular respiration. We must inhale air rich in oxygen because the gas is essential for turning all the available energy in sugar molecules into the energy stored in ATP.

Cellular respiration begins in the cytosol and is completed in the **mitochondrion**, an organelle enclosed in double membranes. Mitochondria are espe-

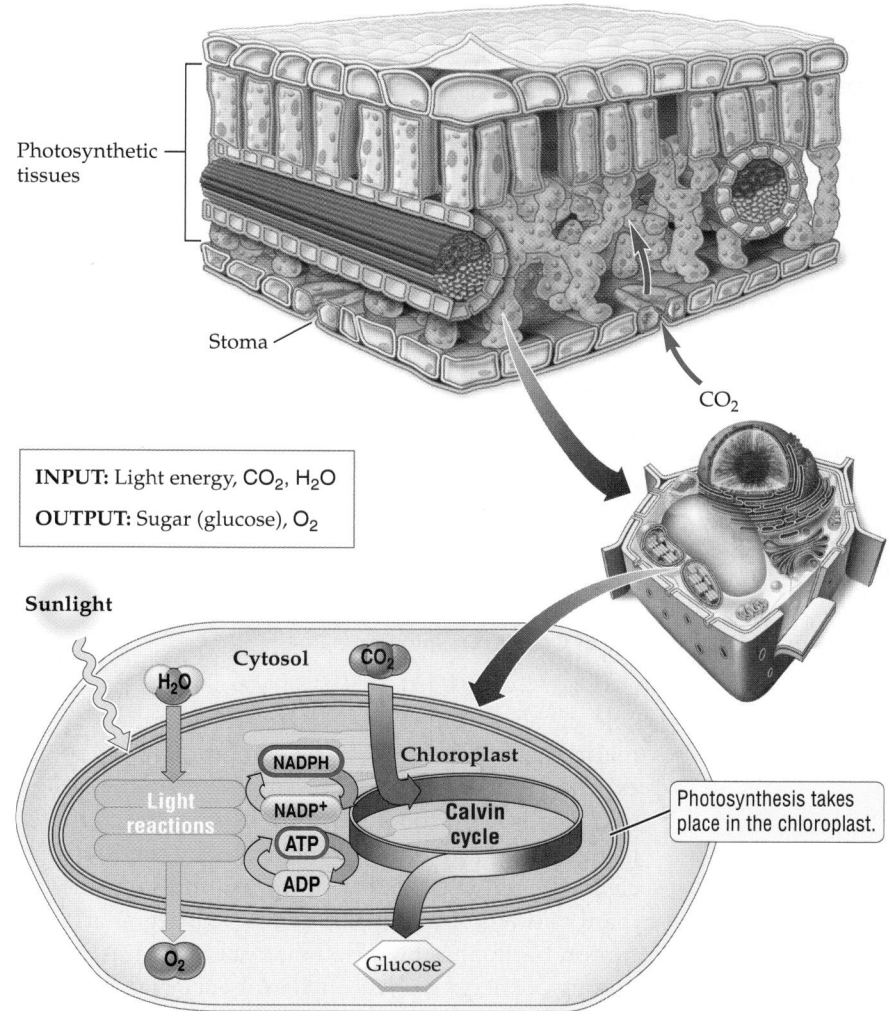

THE TWO STAGES OF PHOTOSYNTHESIS

Photosynthetic tissues

Stoma

$CO_2$

**INPUT:** Light energy, $CO_2$, $H_2O$

**OUTPUT:** Sugar (glucose), $O_2$

Sunlight

Cytosol

$H_2O$

$CO_2$

NADPH

Chloroplast

Light reactions

NADP$^+$

ATP

Calvin cycle

ADP

Photosynthesis takes place in the chloroplast.

$O_2$

Glucose

**FIGURE 9.3 An Overview of Photosynthesis**

cially abundant in cell types that have large energy demands, such as skeletal muscles, brain cells, and liver cells. Liver tissue, for example, contains over a thousand mitochondria per cell.

Cellular respiration takes place in three stages: *glycolysis*, the *Krebs cycle*, and *oxidative phosphorylation* (**FIGURE 9.4**). The first stage, glycolysis, takes place in the cytosol, the fluid portion of the cytoplasm. During **glycolysis [glye-*KAH*-luh-sis]**, sugars (mainly glucose) are split to make a three-carbon compound (*pyruvate*), releasing two molecules of ATP and two molecules of NADH for each glucose molecule that is split. Pyruvate enters the mitochondrion and is completely degraded through a sequence of enzyme-driven reactions known as the **Krebs cycle**.

INPUT: Glucose, $O_2$

OUTPUT: Chemical energy
(as ATP), $CO_2$, $H_2O$

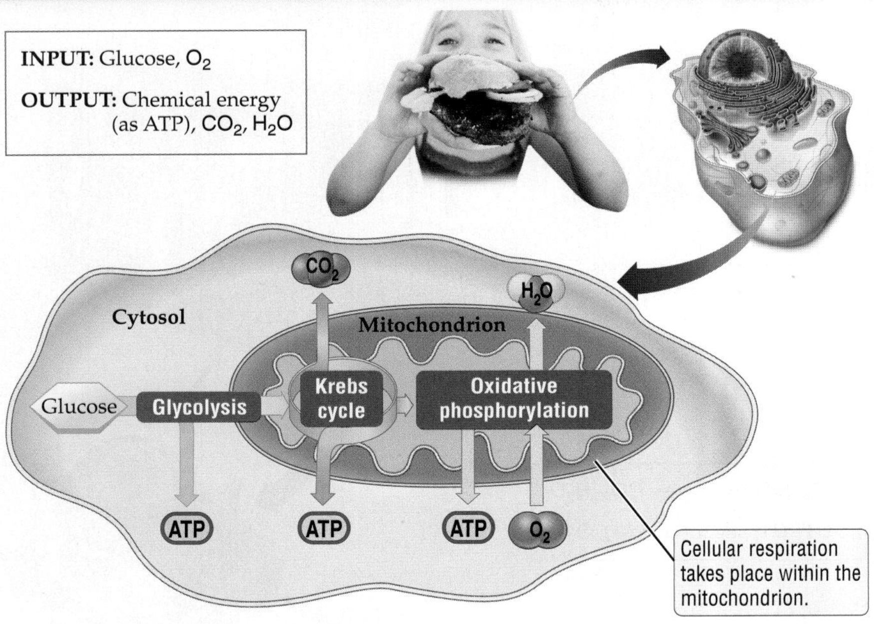

FIGURE 9.4  An Overview of Cellular Respiration

The carbon backbone of pyruvate is taken apart, releasing carbon dioxide. The carbon dioxide you are exhaling into the air right now originated in the carbon backbones of the food you ate, following the complete dismantling of those food molecules by the Krebs cycle in the many hundreds of mitochondria inside the many trillions of cells in your body. The degradation of carbon backbones by the Krebs cycle produces a large bounty of energy carriers, including ATP and NADH.

In the last and final step of cellular respiration, the chemical energy of NADH is converted into the chemical energy of ATP through a membrane-dependent process known as **oxidative phosphorylation**. Electrons and hydrogen atoms removed from NADH are handed over to molecular oxygen ($O_2$), creating water ($H_2O$). In the process, a large amount of ATP is generated. Oxidative phosphorylation generates at least 15 times more ATP than does glycolysis alone. We cannot survive more than a few minutes without oxygen because, in the absence of this gas, our cells cannot make enough ATP to run the many activities that rely on energy delivered by this extraordinary energy carrier.

> ### Concept Check

1. What is the role of energy carriers in the cell?

2. Explain what is wrong with this statement: Cellular respiration occurs in consumers like us, but not in plants.

## 9.3 Photosynthesis: Energy from Sunlight

The next time you walk outside, look at the plants around you and consider the critical role they play in supporting the web of life. Through photosynthesis, plants support humans and a great variety of other organisms, which depend on them for both food and oxygen. As we have seen, photosynthesis uses light energy to generate two types of energy carriers, ATP and NADPH, which are then used to synthesize sugars, releasing oxygen gas ($O_2$) into the environment in the process. Before we examine the mechanisms underlying these processes in greater detail, let's consider the nature of light.

## The color of an object is determined by the wavelengths of visible light it reflects

Light is a stream of massless particles, called photons, that have wavelike characteristics. Every photon contains a fixed amount of energy. The energy of a photon is related to its wavelength, the distance between one wave crest and the next. The energy, and therefore the wavelength, of photons covers a broad span, known as the electromagnetic spectrum (**FIGURE 9.5**). A photon with a *short wavelength* has *more energy* than a photon with a longer wavelength (compare gamma rays and radio waves in Figure 9.5). Visible light is the portion of the electromagnetic spectrum that our eyes can perceive. It includes all photons with wavelengths between 300 and 780 nanometers (nm); and blended together, this range of the spectrum looks white to us. Light travels at a speed of 186,282 miles per second in a vacuum, but it slows down when passing through matter, and the different wavelengths are slowed to different extents. Isaac Newton famously split the wavelengths within white light, using a glass prism, to reveal the "seven rainbow colors," with violet at a wavelength of 300 nm and red at 780 nm.

The color of an object is determined by the wavelengths of light that bounce off it and reach our eyes. A white object *reflects* all wavelengths in the visible part of the electromagnetic spectrum. A black object *absorbs* all of these wavelengths, so no visible light bounces off it to enter our eyes. Chlorophyll, the light-absorbing pigment in leaves, absorbs almost all of the blue and red wavelengths but reflects much of the green. A red apple reflects the red wavelengths, and blue wavelengths of light bounce off a blueberry to reach our eyes. Other colors we are familiar with are produced by a mixture of different wavelengths, to create an endless palette. Violet and blue wavelengths *both* bounce off an eggplant, and together they are interpreted by our eyes and brain as a rich purple.

## Chloroplasts are photosynthetic organelles

Photosynthesis in plants and protists takes place inside chloroplasts. While chloroplasts are located throughout the green parts of a plant, including young stems, leaves are especially well structured to assist in photosynthesis. The cells inside a leaf are packed with many chloroplasts that look like flattened green footballs. The outer layers of a leaf are pockmarked with many microscopic pores, called **stomata** (singular "stoma") that facilitate gas exchange. Carbon dioxide from the outside enters a leaf through open stomata to be used in the Calvin cycle reactions within chloroplasts. A plant absorbs water and dissolved minerals through its roots but gains *no energy* whatsoever from the soil.

Like mitochondria, chloroplasts are bounded by two concentric membranes. The inner of the two membranes encloses a gel-like fluid that makes up the **stroma** (**FIGURE 9.6**). Embedded in the stroma is a network of interconnected membrane-enclosed sacs. The sacs, known as **thylakoids**, lie one on top of another in stacks. Each thylakoid sac consists of a thylakoid membrane that encloses a thylakoid space. The distinctive arrangement of the thylakoid membranes is crucial for light capture and for generating ATP and NADPH, the main outcomes of the light reactions. The Calvin cycle reactions take place in the stroma, which contains the many enzymes, ions, and molecules needed for turning carbon dioxide into sugar with the help of energy carriers.

## The light reactions generate energy carriers

The thylakoid membrane is densely packed with many disclike clusters of pigments complexed with

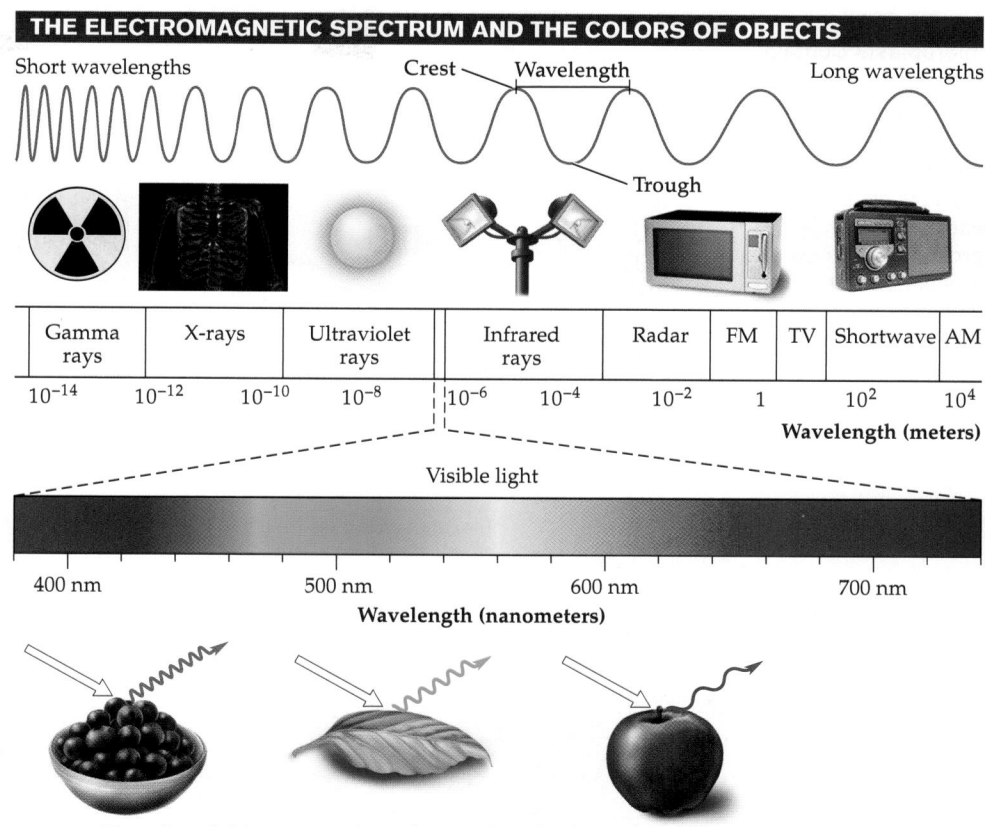

**THE ELECTROMAGNETIC SPECTRUM AND THE COLORS OF OBJECTS**

Short wavelengths    Crest    Wavelength    Long wavelengths

Trough

| Gamma rays | X-rays | Ultraviolet rays | Infrared rays | Radar | FM | TV | Shortwave | AM |

$10^{-14}$    $10^{-12}$    $10^{-10}$    $10^{-8}$    $10^{-6}$    $10^{-4}$    $10^{-2}$    $1$    $10^{2}$    $10^{4}$

**Wavelength (meters)**

Visible light

400 nm          500 nm          600 nm          700 nm

**Wavelength (nanometers)**

The color of objects comes from the wavelengths they reflect

**FIGURE 9.5 The Color of an Object Is Determined by the Wavelengths of Light It Reflects**

proteins. Each disclike grouping is known as an **antenna complex** (**FIGURE 9.7**) and contains different types of pigments, including chlorophylls *a* and *b* and carotenoids (see box on page 216). The antenna complex captures solar energy, especially blue and red wavelengths, and funnels it to an enzyme-chlorophyll complex known as the **reaction center**, where light reactions are initiated.

Electrons associated with certain chemical bonds in a chlorophyll molecule become more energized when they absorb light. The high-energy electron is picked up by an **electron transport chain** (**ETC**), a series of electron-accepting molecules embedded next to each other in the thylakoid membrane. As electrons pass from one component of the ETC to the next, small amounts of energy are released and used to generate ATP. What is the fate of the electrons that ride down the thylakoid ETCs? Eventually they are picked up by $NADP^+$, which, with the addition of protons ($H^+$) from the stroma, is transformed

FIGURE 9.6
**Chloroplasts Contain Membranes Studded with Chlorophyll**

Chlorophyll absorbs light energy, which is used to drive the synthesis of energy carriers. The energy carriers fuel the synthesis of sugar in the stroma of the chloroplast.

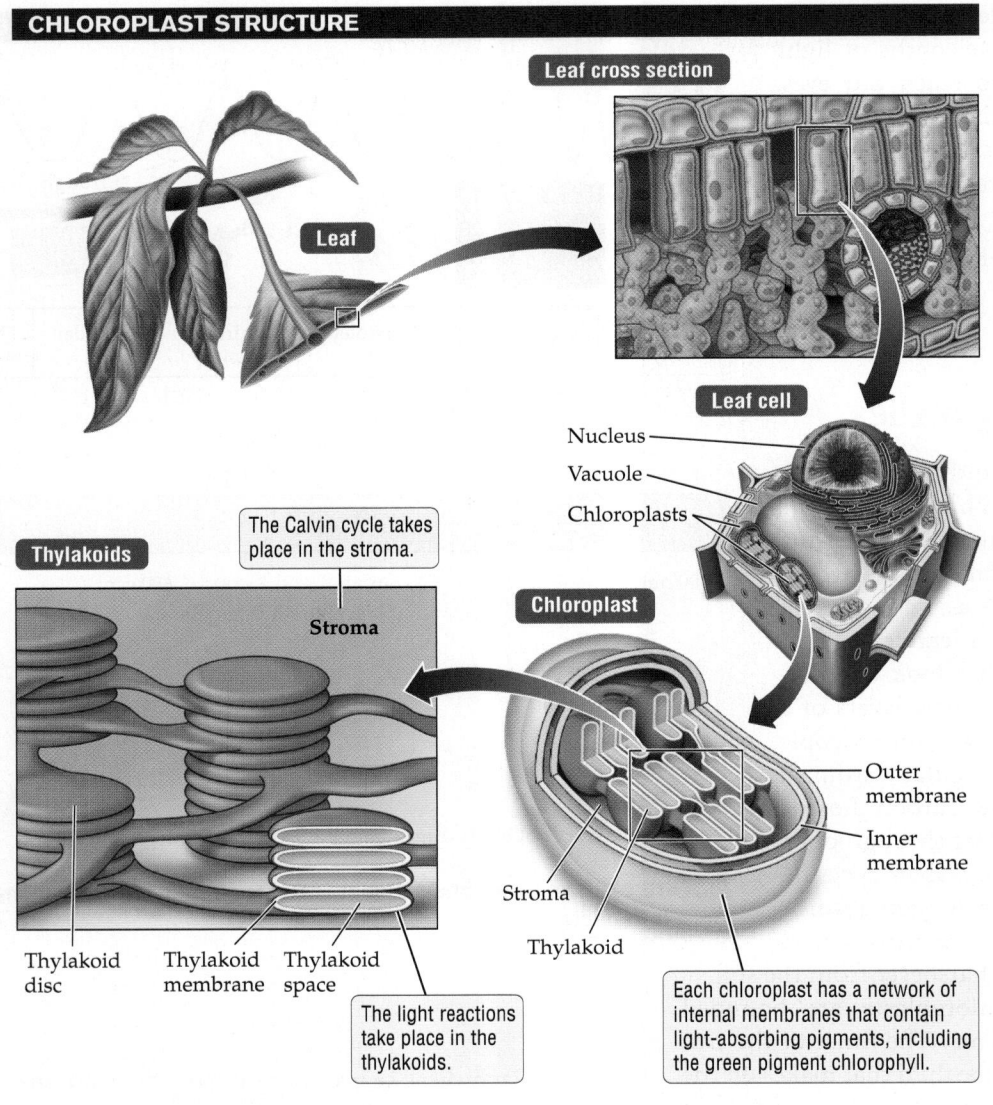

**CHLOROPLAST STRUCTURE**

Leaf cross section

Leaf

Leaf cell

Nucleus
Vacuole
Chloroplasts

Thylakoids

The Calvin cycle takes place in the stroma.

**Stroma**

Chloroplast

Outer membrane

Inner membrane

Stroma

Thylakoid

Thylakoid disc

Thylakoid membrane

Thylakoid space

The light reactions take place in the thylakoids.

Each chloroplast has a network of internal membranes that contain light-absorbing pigments, including the green pigment chlorophyll.

into NADPH, the second of the two energy carriers furnished by the light reactions.

The combination of an antenna complex and the associated reaction center is called a **photosystem**. Plant chloroplasts have two interlinked photosystems—photosystems I and II—that work in tandem (see Figure 9.7). (The numbering of the two photosystems reflects the order in which each system was discovered by researchers, not the order of steps during photosynthesis.) **Photosystem II** is associated with the splitting of water (*photolysis*) and generates electrons, $O_2$, and hydrogen ions (**FIGURE 9.8a**). **Photosystem I** receives electrons from photosystem II, and

after traveling down a relatively short electron transport chain, these electrons are donated to $NADP^+$ to generate NADPH (**FIGURE 9.8b**).

The transfer of high-energy electrons is at the heart of the light reactions. The journey begins at the reaction center of photosystem II. Absorbed light energy ejects high-energy electrons from a photosystem II reaction center chlorophyll (see Figure 9.8a). The high-energy electrons are picked up by the first component of the ETC, which transfers it to the next component in the chain, and so on. As the electrons travel down the ETC they lose energy, and some of that energy is used to drive the transport of protons (hydrogen ions, $H^+$)

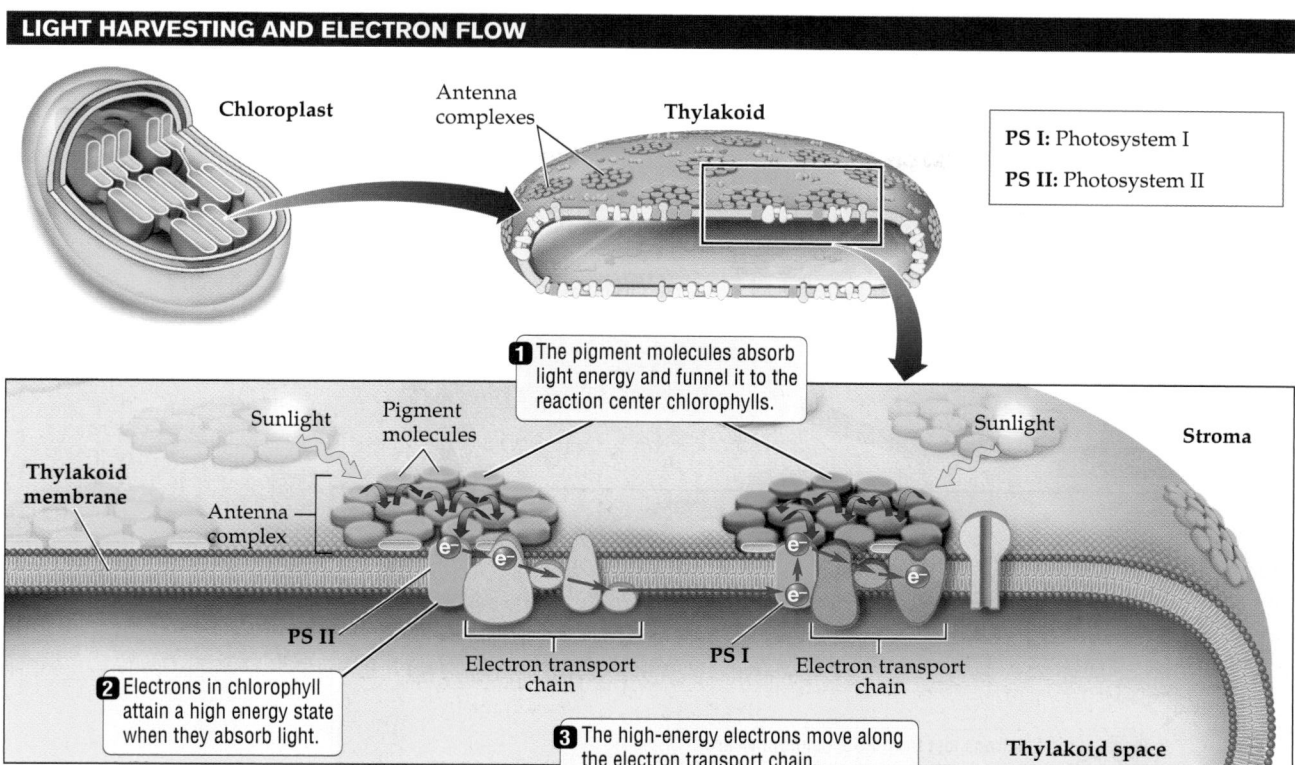

PS I: Photosystem I

PS II: Photosystem II

**1** The pigment molecules absorb light energy and funnel it to the reaction center chlorophylls.

Sunlight

Pigment molecules

Sunlight

Stroma

Thylakoid membrane

Antenna complex

PS II

Electron transport chain

PS I

Electron transport chain

**2** Electrons in chlorophyll attain a high energy state when they absorb light.

**3** The high-energy electrons move along the electron transport chain.

Thylakoid space

**FIGURE 9.7**
**The Light Reactions Are Conducted by Two Linked Photosystems**

across the thylakoid membrane. A thylakoid transport protein uses the energy released during electron transfer to move protons from the stroma into the thylakoid space (**FIGURE 9.8c**). As protons accumulate inside the thylakoid space, their concentration builds up relative to the hydrogen ion concentration in the stroma, creating a **proton gradient** (an imbalance in the proton concentration) across the thylakoid membrane.

Pumping of ions to create a concentration gradient is a common means of harnessing energy for cellular processes. In this case, the gradient is used to manufacture ATP. As explained in Chapter 7, all dissolved substances, including protons, tend to move from a region of higher concentration to one of lower concentration. So the protons in the thylakoid space have a spontaneous tendency to move back down the proton gradient to the stroma. Since the thylakoid membrane will not allow protons to pass through it, the only way for them to cross the membrane is through a large channel-containing protein complex called **ATP synthase**, which spans the thylakoid membrane. As protons rush through the ATP synthase channel, the potential energy stored in their

concentration gradient is converted into chemical energy: enzymes associated with the lollipop-like head of ATP synthase catalyze the addition of a phosphate group on ADP, converting it to ATP (see Figure 9.8c).

Electrons lose energy as they travel down the electron transport chain from photosystem II until they reach the reaction center of photosystem I. There, light gathered by the photosystem I antenna complexes boosts each arriving electron to a very high energy level. Next, these energized electrons flow through the short ETC and are then donated to $NADP^+$.

In addition to the two electrons received from the ETC, each $NADP^+$ takes up one proton ($H^+$) from the stroma, which converts it to NADPH. Realize that these electrons and protons picked up by $NADP^+$ come ultimately from water molecules: the electrons traveling down the ETC were initially ejected from photosystem II, which replaced them by extracting electrons from water molecules to form hydrogen ions and $O_2$ (as pictured in Figure 9.8a). In sum, the two photosystems and the two ETCs are arranged so that they synchronize perfectly to generate ATP and NADPH, making oxygen as a by-product.

FIGURE 9.8
How the Light
Reactions Generate
Energy Carriers

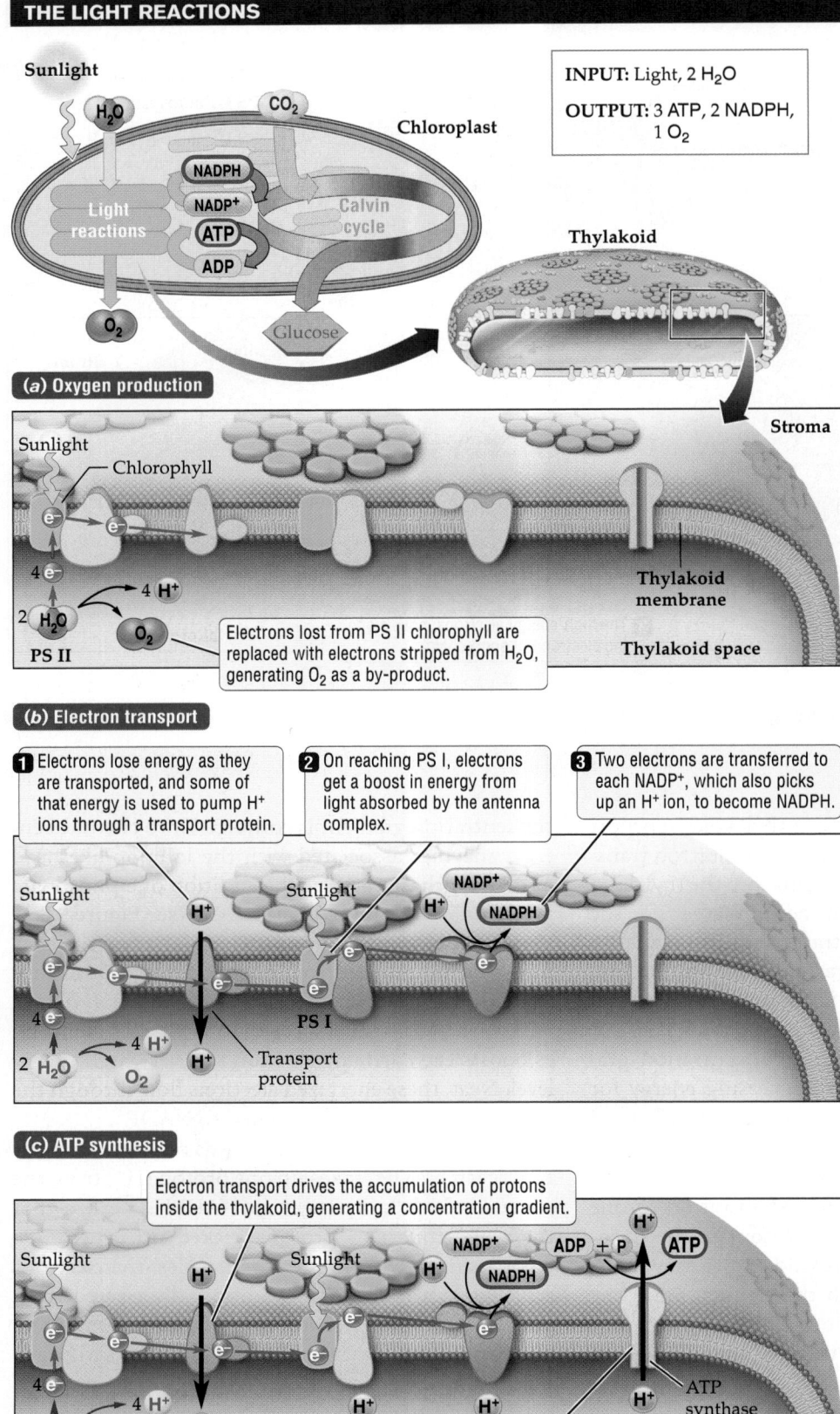

**THE LIGHT REACTIONS**

Sunlight

Chloroplast

INPUT: Light, 2 $H_2O$

OUTPUT: 3 ATP, 2 NADPH,
1 $O_2$

$H_2O$   $CO_2$

NADPH

NADP+

Light
reactions

Calvin
cycle

ATP

ADP

$O_2$

Glucose

Thylakoid

**(a) Oxygen production**

Stroma

Sunlight

Chlorophyll

4 e−

2 $H_2O$

4 H+

$O_2$

**PS II**

Thylakoid
membrane

Thylakoid space

Electrons lost from PS II chlorophyll are
replaced with electrons stripped from $H_2O$,
generating $O_2$ as a by-product.

**(b) Electron transport**

**1** Electrons lose energy as they
are transported, and some of
that energy is used to pump H+
ions through a transport protein.

**2** On reaching PS I, electrons
get a boost in energy from
light absorbed by the antenna
complex.

**3** Two electrons are transferred to
each NADP+, which also picks
up an H+ ion, to become NADPH.

Sunlight

Sunlight

H+

NADP+

H+

NADPH

4 e−

2 $H_2O$

4 H+

$O_2$

**PS I**

H+

Transport
protein

**(c) ATP synthesis**

Electron transport drives the accumulation of protons
inside the thylakoid, generating a concentration gradient.

Sunlight

Sunlight

H+

H+

NADP+

ADP + P

H+

ATP

NADPH

4 e−

2 $H_2O$

4 H+

$O_2$

H+

H+

H+

H+

H+

ATP
synthase

ATP synthase permits accumulated protons to rush back into
the stroma, and the energy released drives ATP formation.

## The Calvin cycle reactions manufacture sugars

The energy carriers produced by the light reactions—ATP and NADPH—are used in the Calvin cycle reactions. The Calvin cycle is a series of enzymatic reactions that take place in the stroma of the chloroplast and synthesize sugars from carbon dioxide and water (**FIGURE 9.9**). This process, also known as **carbon fixation**, illustrates the interconnectedness of life-forms and their environment. By capturing inorganic carbon atoms from $CO_2$ gas and fixing (incorporating) them into organic compounds such as sugars, the Calvin cycle reactions make carbon from the nonliving world available to the photosynthetic producer and eventually to other living organisms as well.

The Calvin cycle reactions are catalyzed by enzymes present in the stroma. The most abundant of these enzymes is **rubisco**. Rubisco catalyzes the first reaction of the Calvin cycle, in which a molecule of the one-carbon compound $CO_2$ combines with a five-carbon compound called ribulose 1,5-bisphosphate [*RYE-byoo-lohss ... biss-FAHS-fayt*], or RuBP for short, to eventually produce two three-carbon compounds.

This reaction can be expressed as an equation displaying just the carbon atoms in the compounds involved: $1C + 5C = 2 \times 3C$. This first reaction in carbon fixation is followed by a multistep cycle catalyzed by many different enzymes. These reactions manufacture sugars for use by the cell and also regenerate RuBP. RuBP is absolutely necessary to keep the Calvin cycle running because it is the acceptor molecule for $CO_2$. Rubisco links RuBP to $CO_2$ to produce a three-carbon organic acid. The conversion of the organic acid into a sugar (G3P) in subsequent steps requires the input of energy from ATP, together with electrons and hydrogen ions delivered by NADPH (see Figure 9.9).

Three turns of the Calvin cycle bring in the three carbon atoms needed to make one molecule of a three-carbon sugar called glyceraldehyde 3-phosphate [**gliss-er-AL-duh-hide...**], or G3P for short. We can follow this process by tracking the number of carbon atoms as they get rearranged into different compounds at each step of the cycle (see Figure 9.9). For every 3 molecules of $CO_2$ ($3 \times 1C = 3C$) that combine with 3 molecules of ribulose 1,5-bisphosphate ($3 \times 5C = 15C$), 6 molecules of the three-carbon compound are produced ($6 \times 3C = 18C$). These molecules even-

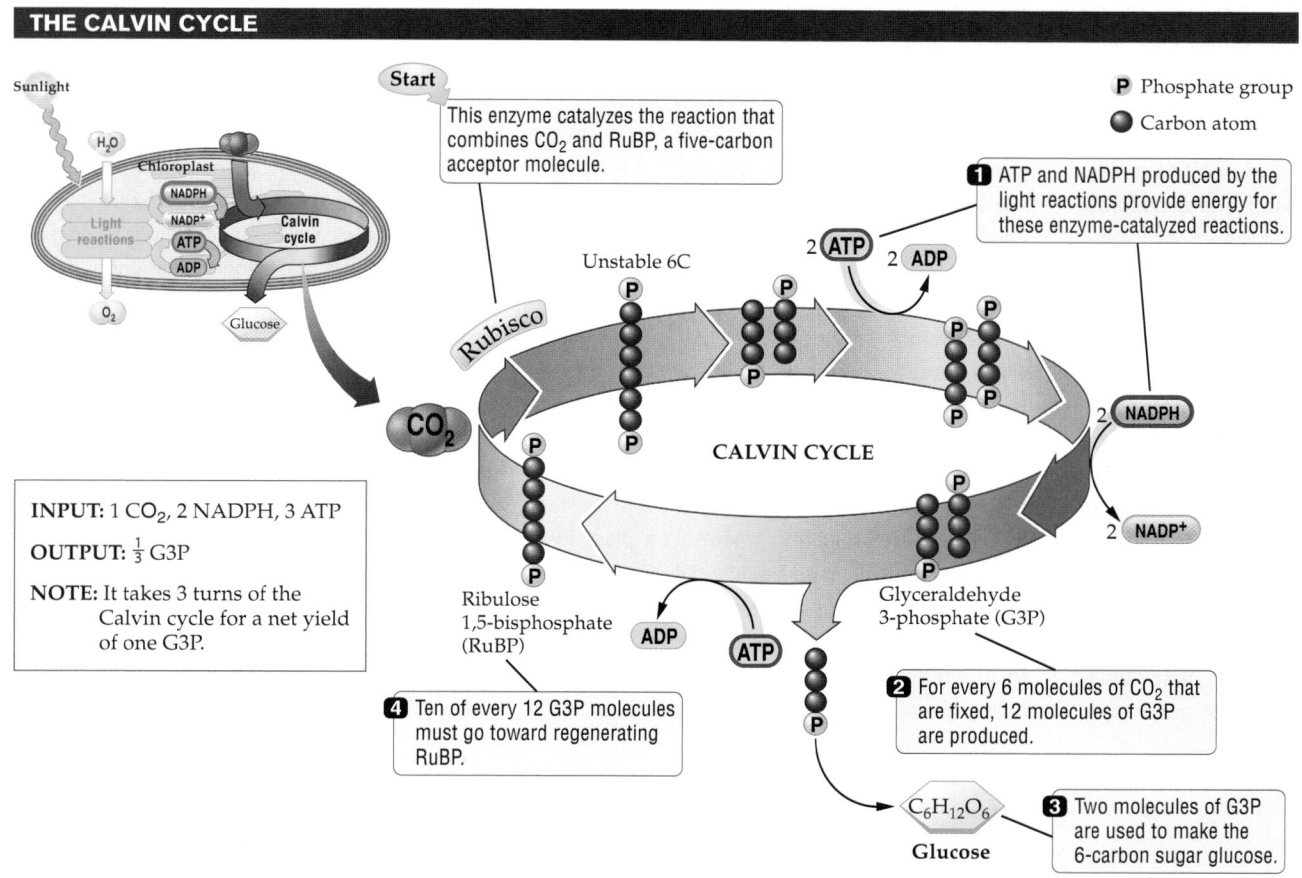

**THE CALVIN CYCLE**

Sunlight

$H_2O$

Chloroplast
NADPH
NADP+
ATP
ADP
Light reactions
Calvin cycle

$O_2$

Glucose

**Start**

This enzyme catalyzes the reaction that combines $CO_2$ and RuBP, a five-carbon acceptor molecule.

Rubisco

$CO_2$

Unstable 6C

2 ATP    2 ADP

❶ ATP and NADPH produced by the light reactions provide energy for these enzyme-catalyzed reactions.

**P** Phosphate group

● Carbon atom

**CALVIN CYCLE**

2 NADPH

2 NADP+

**INPUT:** 1 $CO_2$, 2 NADPH, 3 ATP

**OUTPUT:** $\frac{1}{3}$ G3P

**NOTE:** It takes 3 turns of the Calvin cycle for a net yield of one G3P.

Ribulose 1,5-bisphosphate (RuBP)

ADP

ATP

Glyceraldehyde 3-phosphate (G3P)

❹ Ten of every 12 G3P molecules must go toward regenerating RuBP.

❷ For every 6 molecules of $CO_2$ that are fixed, 12 molecules of G3P are produced.

$C_6H_{12}O_6$
**Glucose**

❸ Two molecules of G3P are used to make the 6-carbon sugar glucose.

**FIGURE 9.9**
**The Calvin Cycle Converts Carbon Dioxide into Sugar**

The Calvin cycle reactions fix carbon by turning $CO_2$ into sugar molecules using energy delivered by ATP, and electrons and protons delivered by NADPH.

# The Rainbow Colors of Plant Pigments

Plant pigments color our world. We like to be surrounded by greenery and take delight in the bright displays in a flower garden. What do these brightly colored pigments do for the plant, and what can they do for us?

A biological pigment is a carbon-containing molecule with a distinctive color. Chlorophyll and carotenoids are the two main families of sun-catcher pigments in the plant chloroplast. The two types of chlorophyll found in plants—chlorophyll *a* and chlorophyll *b*—are green to yellowish green; green light

Rainbow colors and plant pigments.

bounces off these molecules, while blue and red light are strongly absorbed. Because the two chlorophylls absorb slightly different wavelengths of blue and red, collectively they harvest light over a broader range than just the one pigment could.

Carotenoids, which are yellow-orange pigments, also help harvest light, but they have the additional function of protecting photosynthetic cells from too much sun and from highly reactive chemicals that are often spun off as a side effect of the light reactions. If a leaf absorbs more sunshine than it can utilize in photosynthesis, the excess energy can actually damage cellular structures—a phenomenon called photodamage. Carotenoids act as a safety valve for absorbed light energy: they pick up the excess and release it gently as heat. Carotenoids are found in all oxygen-producing photosynthetic organisms, including bacteria and algae. Shrimp get their pink flush from the carotenoid-containing algae they eat, and salmon and flamingos get their pink hues from the algae or shrimp they eat.

Metabolic processes that involve oxygen gas ($O_2$)—photosynthesis and cellular respiration, for example—have the potential for generating

by-products called free radicals. *Free radicals* are highly reactive chemicals that can damage cellular macromolecules—lipids and proteins, in particular. Many carotenoids act as *antioxidants*, molecules that disarm free radicals by reacting with them. The antioxidant activity of carotenoids has been shown to boost the immune system and protect against cancer in lab animals, but its impact on human health is poorly understood. Some carotenoids (such as lutein from leafy greens) accumulate in the retina of the eye and are believed to be good for eye health. Some carotenoids, such as the beta-carotene in carrots, are converted into vitamin A, which plays a variety of important roles at all stages of human development but especially in maintaining vision, bone, and skin, and the immune system. Like chlorophyll, carotenoids are fat-soluble molecules, and you will absorb more of them from a green salad or a pumpkin pie if your recipe calls for some fat.

Most water-soluble pigments in a plant reside in the vacuole (see Section 6.4). Flavonoids are a large family of mostly vacuolar pigments whose color ranges from red, blue, and purple, to … white! They are responsible for the bright colors of many flowers, fruits, and vegetables.

Their diverse functions include attracting pollinators to flowers, luring seed-dispersing animals to fruit, and acting as sunscreens and antioxidants. Flavonoids are found in brightly colored fruits and vegetables, in coffee and tea, in wine grapes, and even in whole grains such as wheat berries and popcorn. Flavonoids have antioxidant, anticancer, and antidiabetic activity in the animal body, which is why health experts recommend at least five daily servings of vegetables and fruits and about six servings of whole grains.

In cold parts of the world, broad-leaved plants drop their leaves in autumn to protect against excessive water loss from their large leaf surfaces. Most of the large macromolecules in the leaf are broken down and sent for recycling and storage in the tree trunk or root. The degradation of chlorophyll reveals the carotenoids, which linger to protect the cells as the programmed dismantling of the foliage continues. The unmasked carotenoids are responsible for the flaming yellows of nut trees, cottonwoods, and birches. Some species also step up synthesis of protective flavonoids called anthocyanins, and these pigments help create the show-stopping fall displays of sugar maples, scarlet oaks, and sumac, among many others.

tually produce 3 RuBP molecules (3 × 5C = 15C) and 1 molecule of glyceraldehyde 3-phosphate (3C). As the arithmetic indicates, it takes three turns of the cycle to make one molecule of G3P, with the other carbon atoms constantly recycled to maintain the pool of RuBP, without which the cycle would come to a halt. The formation of one molecule of G3P is fueled by the input of 9 molecules of ATP and 6 molecules of NADPH.

Glyceraldehyde 3-phosphate is the building block of glucose and all the other carbohydrates that a cell might need to manufacture. Most of the G3P made in the chloroplasts is exported from these organelles and eventually consumed in various chemical reactions in the same or other cells. Some of the exported molecules of G3P are used to make glucose and fructose, which in turn are used to manufacture sucrose (table sugar). Sucrose is an important food source for all the cells in a plant and is transported from the leaves, where photosynthesis takes place, to other parts of the plant. Significant amounts of sucrose are stored in the vacuoles of sugarcane stems and sugar beet roots, which is why these two crops are the mainstay of the sugar industry worldwide.

Not all the G3P made in the chloroplasts is shipped out. Some of it is converted into starch by enzymes in the stroma. Starch, a polymer of glucose, is an important form of stored energy in plants (see Section 5.7). It accumulates in chloroplasts during the day and is then broken down to simple sugars at night. The sugars are broken down by cellular respiration to generate ATP for the cell's nighttime energy needs. Fruits, seeds, roots, and tubers such as potatoes are rich in stored starch, which provides the energy needs of these nonphotosynthetic tissues. The energy-rich nature of these plant parts explains why they are such an important food source for animals.

## 9.4 Cellular Respiration: Energy from Food

Cellular respiration extracts energy from food molecules to generate ATP—a vital process in producers and consumers alike. In this section we take a closer look at each of the three major stages of cellular respiration: glycolysis, the Krebs cycle, and oxidative phosphorylation.

## Glycolysis is the first stage in the cellular breakdown of sugars

"Glycolysis" means literally "sugar splitting." From an evolutionary standpoint, it was probably the earliest means of producing ATP from food molecules, and it is still the primary means of energy production in many prokaryotes. However, the energy yield from glycolysis is small because sugar is only partially degraded through this process. In most eukaryotes, glycolysis is just the first step in energy extraction from sugars, and additional reactions in the mitochondrion help achieve the complete degradation of carbon backbones, typically yielding at least 15 times as much ATP as does glycolysis.

Glycolysis takes place in the cytosol (**FIGURE 9.10**). Through a series of enzyme-catalyzed reactions, glucose is converted to a six-carbon intermediate, which is then split into two molecules of a three-carbon sugar (G3P). In successive steps, each G3P molecule is converted into a three-carbon organic acid called **pyruvate [pye-ROO-vayt]**.

The net energy yield for glycolysis is calculated as follows: For each molecule of glucose consumed during glycolysis, 4 molecules of ADP are turned into ATP, and electrons and hydrogen atoms are donated to two molecules of $NAD^+$, generating 2 molecules of NADH. Since the early steps of glycolysis consume 2 molecules of ATP per glucose molecule, a single glucose molecule produces a net yield of 2 ATP molecules and 2 NADH molecules (see Figure 9.10). Note that glycolysis does not require oxygen ($O_2$). Much energy remains in the pyruvate molecules furnished by glycolysis. For maximum energy extraction, pyruvate must enter the mitochondrion to be broken down through the extremely efficient, oxygen-dependent ATP-generating pathways of the Krebs cycle and oxidative phosphorylation. Before we track the complete dismantling of pyruvate for maximum energy yield, let's consider some variations on the theme of glycolysis.

## Fermentation facilitates ATP production through glycolysis when oxygen is absent

Glycolysis does not require $O_2$, which means it is an **anaerobic [AN-ayr-OH-bik]** process. Glycolysis was probably the main source of energy for early life-forms in the oxygen-poor atmosphere of primitive Earth. It is still the only means of generating ATP for

Roughly half the table sugar used in the United States comes from sugar beets (pictured here); the rest, from sugarcane.

FIGURE 9.10
**Glycolysis Converts Glucose into Pyruvate**

In glycolysis, each six-carbon glucose is converted into two molecules of pyruvate, a three-carbon organic acid.

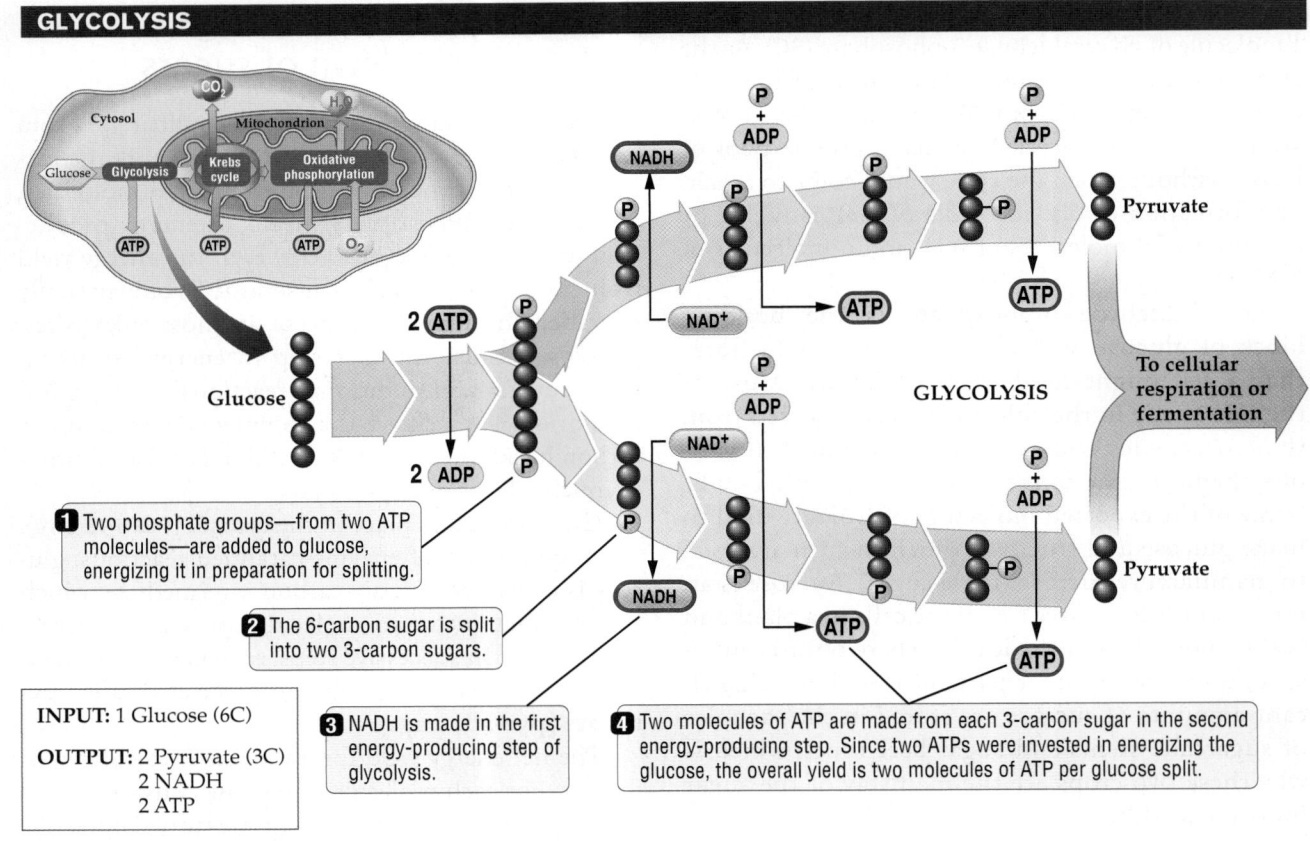

**GLYCOLYSIS**

**1** Two phosphate groups—from two ATP molecules—are added to glucose, energizing it in preparation for splitting.

**2** The 6-carbon sugar is split into two 3-carbon sugars.

**3** NADH is made in the first energy-producing step of glycolysis.

**4** Two molecules of ATP are made from each 3-carbon sugar in the second energy-producing step. Since two ATPs were invested in energizing the glucose, the overall yield is two molecules of ATP per glucose split.

**INPUT:** 1 Glucose (6C)

**OUTPUT:** 2 Pyruvate (3C)
2 NADH
2 ATP

---

some anaerobic organisms. Many anaerobic bacteria that live in oxygen-deficient swamps, in sewage, or in deep layers of soil are actually poisoned by oxygen. Most anaerobic organisms extract energy from organic molecules using a *fermentation* pathway. **Fermentation** begins with glycolysis, followed by a special set of reactions (postglycolytic reactions) whose only role is to help perpetuate glycolysis.

During fermentation, the pyruvate and NADH produced by glycolysis remain in the cytosol, instead of being imported by mitochondria. The postglycolytic reactions convert pyruvate into other molecules, such as alcohol or lactic acid, depending on the types of fermentation pathways used by the cell type. In the process, NADH is converted back to $NAD^+$, an adequate supply of which is necessary to keep glycolysis running (Figure 9.10 shows the point at which $NAD^+$ is used in glycolysis). The cell has a finite pool of $NAD^+$, and if all of it were converted into NADH, glycolysis would cease for lack of enough $NAD^+$. The postglycolytic reactions are a clever way of averting this problem: these reactions remove electrons and hydrogen atoms from NADH, resulting in $NAD^+$ formation, which restores the cell's limited pool of

this vital metabolic precursor. In regenerating $NAD^+$ in the postglycolytic reactions, the electrons and hydrogen ions from NADH create two- or three-carbon compounds, with alcohol and lactic acid being two of the most common by-products of fermentation.

Yeasts are single-celled fungi that resort to alcoholic fermentation when oxygen is absent or low. Yeast strains are used in the production of beer, wine, and other alcoholic products. Fermentation by anaerobic yeasts converts pyruvate into an alcohol called ethanol, releasing $CO_2$ gas, which gives beer its foamy effervescence (**FIGURE 9.11a**). The gas plays an important role in bread making with baker's yeast: the $CO_2$ released by fermentation expands the dough and creates small holes that contribute a light texture to the baked product.

Fermentation is not limited to single-celled anaerobic organisms. It can also occur in the human body in certain cell types, such as muscle cells. A burst of strenuous exercise places a huge ATP demand on muscle cells. The ATP demand may be too large to be met by cellular respiration alone, which is limited by the amount of oxygen that can be delivered by blood vessels in a short span of time.

If the oxygen supply is inadequate to support high enough rates of cellular respiration, severely taxed muscle cells resort to anaerobic ATP production. The rate of glycolysis is boosted to make higher-than-normal amounts of glycolytic ATP, but this extra ATP production can be sustained only through post-glycolytic fermentation of pyruvate (**FIGURE 9.11b**). The postglycolytic pathway converts pyruvate to lactate, regenerating NADH to sustain the high rates of glycolysis (which, remember, is the only source of extra ATP in oxygen-starved tissues).

The short-duration, high-intensity effort required of sprinters and weight lifters gets a very significant assist from anaerobic ATP production sustained through lactic acid fermentation in muscle cells. The burning pain felt in overtaxed muscles stems from the acidity of lactic acid, which irritates nerve endings. Lactic acid is swept to the liver by the bloodstream, where it is reconverted to pyruvate and fed into mitochondria for oxygen-dependent energy extraction. We huff and puff after strenuous exercise because we incur an "oxygen debt" when our muscles resort to lactic acid fermentation. Very soon, we must bring in extra oxygen to make up for the oxygen deficit as the chemical energy of lactic acid is salvaged through cellular respiration. Lactic acid fermentation is of little use in prolonged exercise. "Aerobic exercise," such as long-distance running or biking, is dependent on the moderate but sustained ATP supply that cellular respiration provides.

## Cellular respiration in the mitochondrion furnishes much of the ATP needed by most eukaryotes

As long as oxygen is available, most eukaryotes use cellular respiration to satisfy their relatively large ATP needs. Mitochondria break up pyruvate through a series of reactions and package the released energy into many molecules of ATP. ATP production in the mitochondrion is crucially dependent on oxygen—that is, the mitochondrial portion of cellular respiration is a strictly **aerobic** (oxygen-dependent) process.

Highly aerobic tissues tend to have high concentrations of mitochondria and a rich blood supply to deliver the large amounts of oxygen needed to support their activity. Muscle cells in the human heart, for example, have an exceptionally large number of mitochondria to produce the enormous amounts of ATP needed to keep the heart beating every second

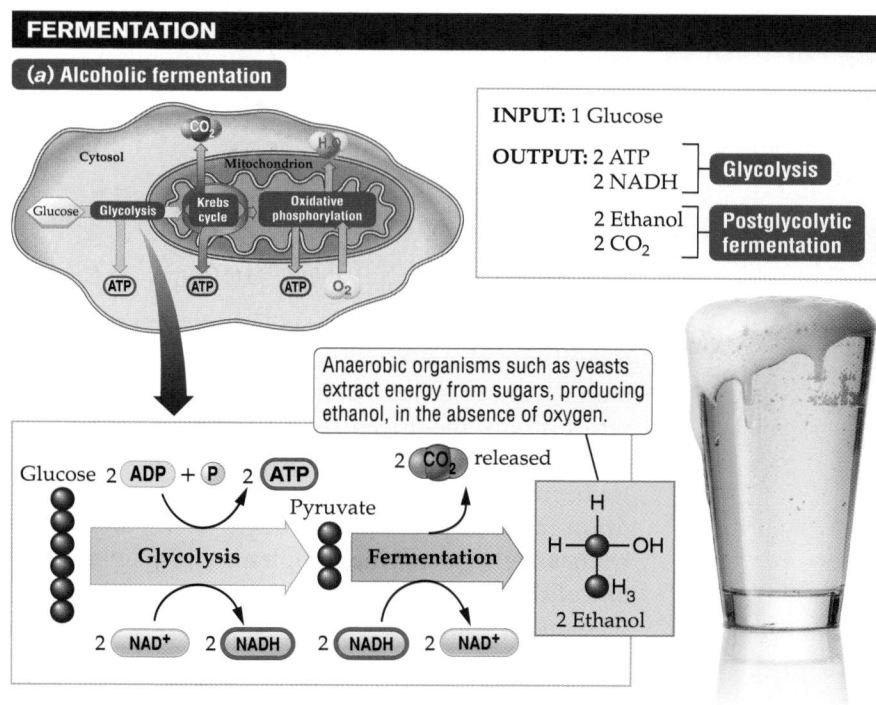

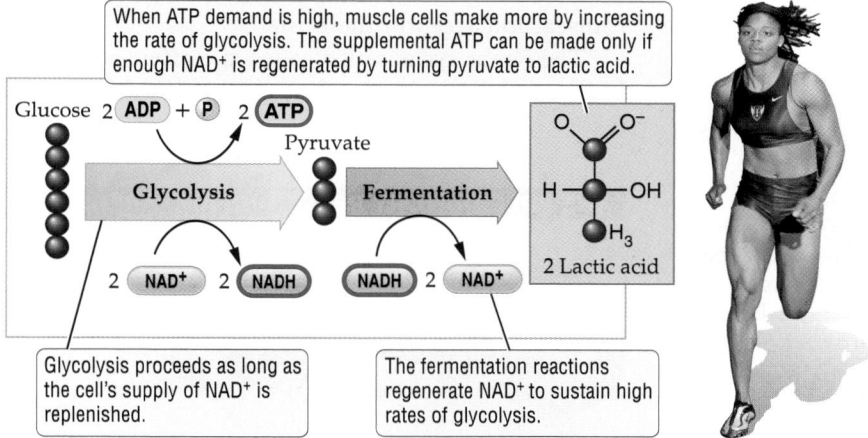

**FIGURE 9.11 Ethanol and Lactic Acid Are By-products of Fermentation**
When the oxygen supply is inadequate for supporting ATP production through cellular respiration, fermentation supports ATP production through glycolysis alone. (a) Strains of yeasts, which are single-celled fungi, are used in the brewing of alcoholic beverages such as beer. When oxygen is excluded from the fermentation tanks, the yeasts resort to fermentation of sugars, producing ethanol and $CO_2$ as by-products of the postglycolytic steps. (b) A similar process occurs in our muscles during short bursts of strenuous exercise, except that the postglycolytic reactions turn pyruvate into a three-carbon organic acid known as lactic acid, and no $CO_2$ is made.

of every day of our lives. Naked mole rats, which live underground in oxygen-poor burrows and must dig a lot every day, have cellular and physiological adaptations that enhance the supply of oxygen to support ATP production by muscle cells too. Compared with the muscles of laboratory rats, the muscles of naked mole rats have 50 percent more mitochondria and 30 percent more blood capillaries.

Naked mole rat

## The Krebs cycle releases carbon dioxide and generates energy carriers

The end product of glycolysis—pyruvate—is transported into the mitochondria and enters the second major stage of cellular respiration, the Krebs cycle, a series of enzyme-driven reactions that take place in the mitochondrial matrix (**FIGURE 9.12**). Before the cycle begins, however, pyruvate entering the mitochondrion must be processed through several *preparatory reactions*. A large enzyme complex in the matrix breaks one of the carbon-carbon covalent bonds in pyruvate, releasing a molecule of $CO_2$ and leaving behind a two-carbon unit known as an acetyl group. The same enzyme complex attaches this acetyl group to a "carbon carrier" known as *coenzyme A (CoA)*, producing a molecule called *acetyl CoA*. The Krebs cycle begins with acetyl CoA donating the two-carbon acetyl group to a four-carbon acceptor molecule.

The Krebs cycle is also called the **citric acid cycle** because citric acid is the first product in this looped, enzyme-driven pathway. *Citric acid*, a six-carbon compound, is produced when the two-carbon acetyl group from acetyl CoA is added to a four-carbon acceptor. CoA is liberated in this process to recruit yet more acetyl groups for the cycle. Citric acid is converted into a six-carbon intermediate that releases a carbon and two oxygen atoms ($CO_2$) to become a five-carbon compound. This five-carbon molecule is then degraded to a four-carbon molecule, releasing one $CO_2$. As the covalent bonds are broken, the energy stored in them is used to drive the formation of energy carriers: ATP, NADH, and a chemical cousin of NADH called $FADH_2$. As we will see shortly, the energy locked in the NADH and $FADH_2$ will be used to make many molecules of ATP during the third and final stage of cellular respiration.

The Krebs cycle is like the Grand Central Station of metabolism because it is the meeting point of several different types of degradative (catabolic) and biosynthetic (anabolic) pathways (see Section 8.2). For example, carbon skeletons derived from other types of food molecules, such as lipids, can be degraded through the Krebs cycle. Lipids such as triglycerides (see Section 5.9) are broken into fatty acids and glycerol in the cytosol; upon entering the mitochondrion, these molecules are converted to acetyl CoA by special enzymes. The lipid-derived acetyl CoA is indistinguishable from

**FIGURE 9.12**
**The Krebs Cycle Releases Carbon Dioxide and Generates Energy Carriers**

The Krebs cycle, also called the citric acid cycle, occurs in the mitochondrial matrix.

---

**Krebs Cycle**

**INPUT:** Acetyl CoA
**OUTPUT:** 3 NADH
1 $FADH_2$
1 ATP
2 $CO_2$

---

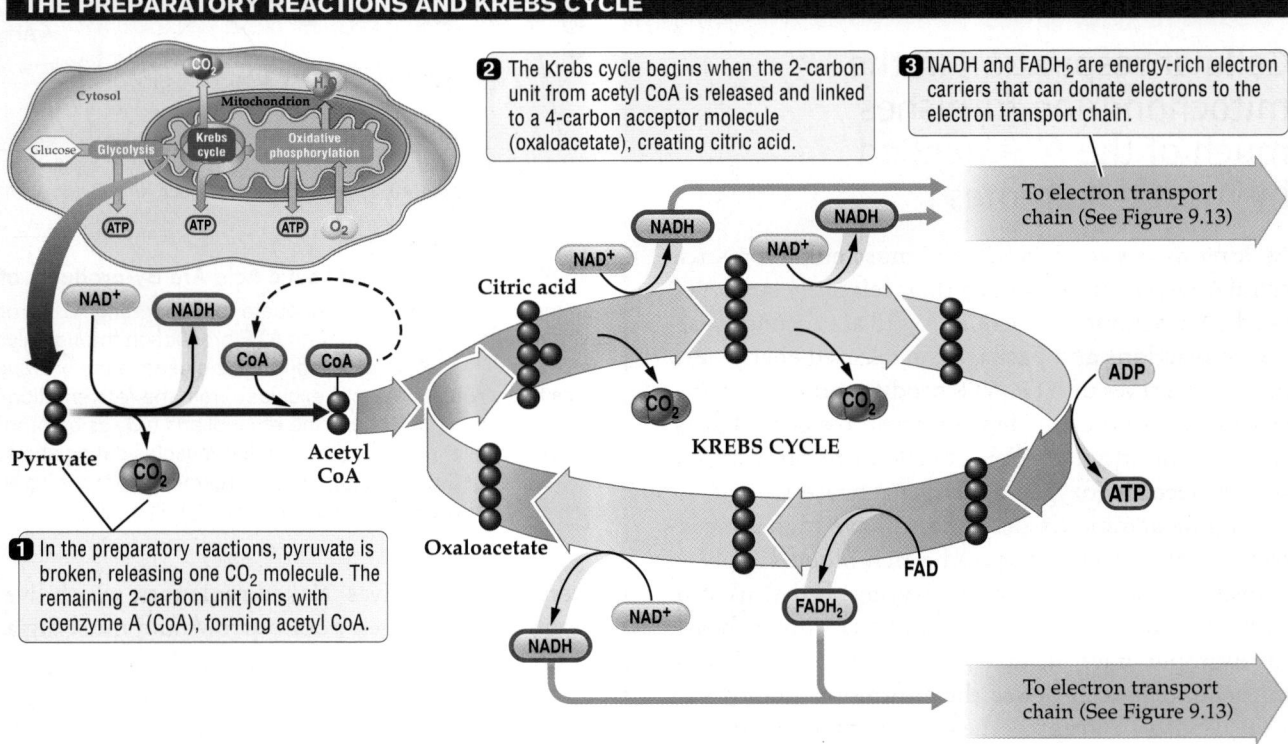

THE PREPARATORY REACTIONS AND KREBS CYCLE

2 The Krebs cycle begins when the 2-carbon unit from acetyl CoA is released and linked to a 4-carbon acceptor molecule (oxaloacetate), creating citric acid.

3 NADH and $FADH_2$ are energy-rich electron carriers that can donate electrons to the electron transport chain.

To electron transport chain (See Figure 9.13)

KREBS CYCLE

1 In the preparatory reactions, pyruvate is broken, releasing one $CO_2$ molecule. The remaining 2-carbon unit joins with coenzyme A (CoA), forming acetyl CoA.

To electron transport chain (See Figure 9.13)

that made from pyruvate during the preparatory reactions, and its carbon atoms are broken up in the same manner by the Krebs cycle.

## Oxidative phosphorylation uses oxygen to produce ATP in quantity

The largest output of ATP is generated in mitochondria during the third and last stage of cellular respiration: oxidative phosphorylation. The enzymatic reactions associated with Krebs cycle take place in the mitochondrial matrix. Oxidative phosphorylation, however, takes place in the many folds (*cristae*) of the inner mitochondrial membrane. Folds in the mitochondrial membrane create a large surface area on which are embedded many electron transport chains (ETCs) and many units of the elaborate ATP-manufacturing machine called ATP synthase. NADH and FADH$_2$ made by the Krebs cycle, and also the NADH generated during glycolysis (see Figure 9.10), diffuse to the inner membrane and donate their high-energy electrons to the ETC. Energy is released as electrons travel the ETC, and some of that energy drives the addition of a phosphate group on ADP (a *phosphorylation* reaction), thereby generating ATP. The phosphorylation of ADP is an oxygen-dependent process—hence the name *oxidative* phosphorylation (**FIGURE 9.13** on page 222). Let's take a closer look at the link between electron transfer through the ETC and phosphorylation of ATP.

The electron transport chains found in the inner mitochondrial membrane are similar in function to those found in the thylakoid membranes of chloroplasts (you may find it helpful to revisit Figure 9.8c). In mitochondria, electrons donated by NADH and FADH$_2$ are passed along a series of ETC components (**FIGURE 9.13a**), and the energy released is used to pump protons (H$^+$) through channel proteins from the matrix into the intermembrane space, creating a proton gradient across the inner membrane (**FIGURE 9.13b**). As in chloroplasts, the proton gradient is depleted periodically as protons gush through the ATP synthases (**FIGURE 9.13c**). Each ATP synthase is a multiprotein complex consisting of a membrane channel and associated motor proteins and enzymes. The movement of protons through the ATP synthase channel activates

enzymes that catalyze the phosphorylation of ADP to from ATP.

The striking similarities in the way ATP is produced in chloroplasts and mitochondria illustrate that diverse metabolic pathways can evolve through modifications of the same basic machinery. Both of these organelles use ETCs built from at least some similar components, both use energy released during electron transfer to move protons across an inner membrane, and the two use very similar ATP synthases to harness the energy of the proton gradient to make ATP.

Note the fate of the electrons, off-loaded by NADH and FADH$_2$, that travel down the ETCs of the inner mitochondrial membrane. They are accepted by oxygen (O$_2$), which combines them with H$^+$ picked up from the mitochondrial matrix to form water (H$_2$O). In other words, electron transfer along the ETC terminates with O$_2$, which serves as the final electron acceptor (see Figure 9.13b). Without oxygen to whisk the electrons away, the electron traffic on the ETC "highway" would come to a standstill. Without electron transfer along the ETC, protons cannot be pumped into the intermembrane space; and without the energy of the proton gradient, ADP cannot be phosphorylated to make ATP. Like all other aerobic organisms, we need oxygen because we cannot make enough ATP without it, so anything that interferes with that process is a danger to life. Hydrogen cyanide is a deadly poison because it binds to the last protein component in the electron transport chain, preventing it from handing electrons to oxygen, and thereby arresting energy production in mitochondria.

Cellular respiration (glycolysis, the Krebs cycle, and oxidative phosphorylation) has a net yield of about 30-32 ATP molecules per molecule of glucose. Mitochondrial respiration is therefore more productive than glycolysis alone, which yields only 2 ATP molecules per molecule of glucose consumed.

> ### Concept Check

1. Compare the functions of photosystems I and II.

2. What is rubisco, and what role does it play in photosynthesis?

3. Why do we need oxygen (O$_2$) to live?

FIGURE 9.13

**The Mitochondrial Electron Transport Chain and ATP Synthase Generate ATP through Oxidative Phosphorylation**

Oxidative phosphorylation is the last stage in cellular respiration, and it produces the most ATP of any metabolic pathway. (*a*) Electrons donated by NADH and FADH$_2$ enter the ETC. (*b*) A proton gradient is generated. (*c*) Proton flow through ATP synthase catalyzes the production of ATP.

## OXIDATIVE PHOSPHORYLATION

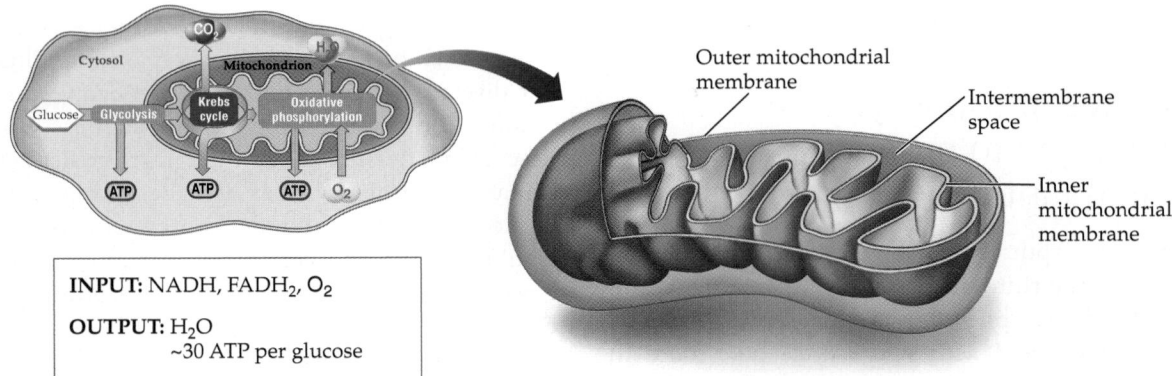

**INPUT:** NADH, FADH$_2$, O$_2$

**OUTPUT:** H$_2$O
~30 ATP per glucose

### (*a*) Electron transport

High-energy electrons are donated to the ETC embedded in the inner mitochondrial membrane.

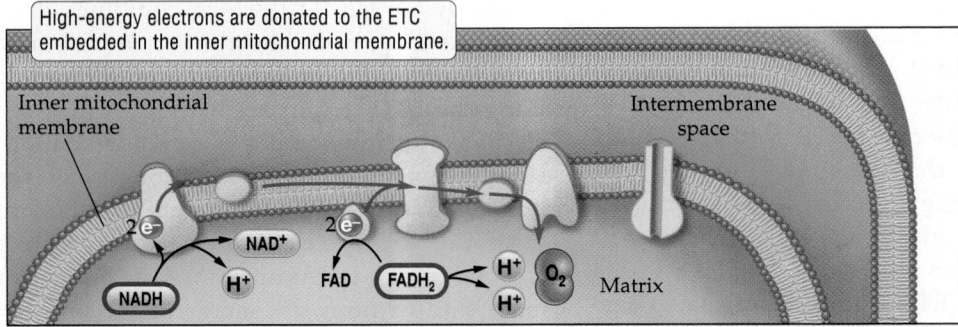

### (*b*) Proton gradient

As electrons move down the ETC, pumping of protons from the matrix to the intermembrane space creates a proton gradient across the inner mitochondrial membrane.

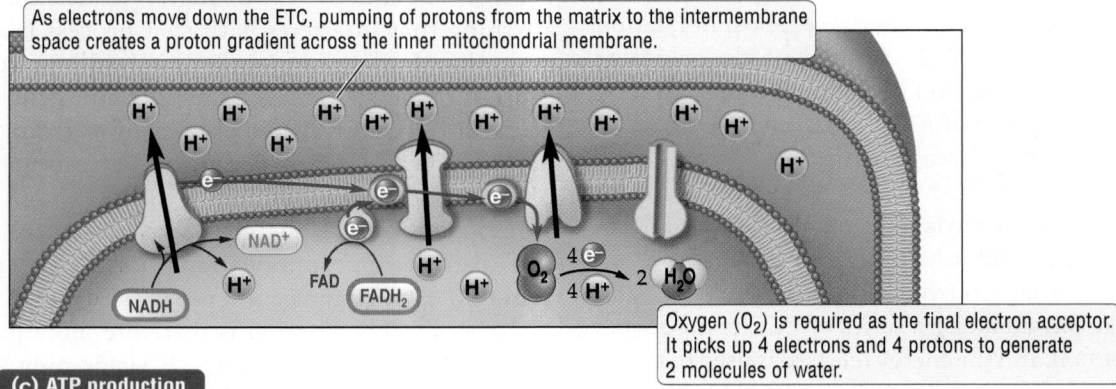

Oxygen (O$_2$) is required as the final electron acceptor. It picks up 4 electrons and 4 protons to generate 2 molecules of water.

### (*c*) ATP production

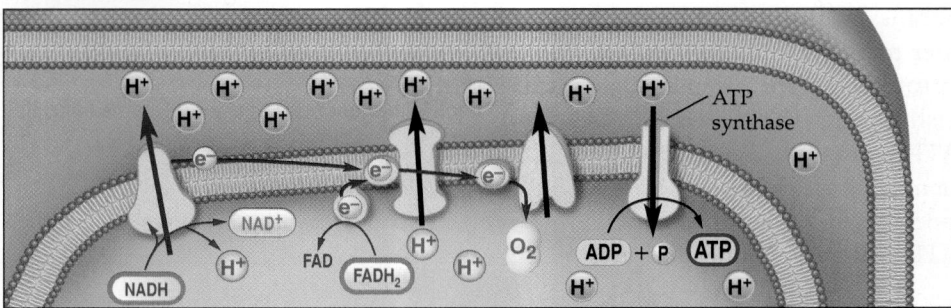

The passage of accumulated protons from the intermembrane space to the matrix through ATP synthase drives the phosphorylation of ADP to produce ATP.

# Waiting to Exhale

Breathing is such a vital activity that if we stop breathing for even a few minutes, disaster will strike. Waste carbon dioxide from cellular respiration begins to accumulate in the body if it is not exhaled from the lungs. The buildup of carbon dioxide in turn makes the blood acidic. The brain detects the blood's increase in pH and sends signals to the body to take action. Blood pressure soars, the heart beats faster, and we gasp for air. People prone to anxiety attacks are especially sensitive and tend to panic at lower levels of $CO_2$ exposure than the average person does.

Competitive free divers must train their bodies not only to become accustomed to less oxygen but also to ignore the panic and tightness in their chests as they fight the urge to breathe. Stressful thoughts or emotions can increase heart rate and therefore oxygen consumption, so they often meditate to clear their minds. Recall that surfer Jay Moriarity died while meditating far below the surface of the ocean.

Diving into cool water activates the powerful diving reflex, which diverts blood from the body to the heart and brain, preserving the function of these vital organs while reducing overall ATP use. Before a competition, breath holders reduce resting metabolic rate—and therefore oxygen consumption—by losing weight and then fasting for several hours beforehand. During the competition, they minimize ATP use by remaining as calm and still as possible.

Competitive free divers also increase their ability to withstand low oxygen levels by training at high altitudes or by sleeping in low-oxygen tents. Air at 8,000 feet has only 74 percent as much oxygen as air at sea level. The human body

responds to lower oxygen levels in several ways: It increases the number and size of our red blood cells, whose job it is to ferry oxygen throughout the body. It increases the amount of hemoglobin— the oxygen-binding protein inside the red blood cells, as well as the number of tiny blood vessels that supply the tissues. This acclimation takes a week or two to complete and disappears within 2 weeks of returning to low altitudes.

Enormous lung capacity is another asset in both free diving and endurance sports such as long-distance running, swimming, or biking. Lung capacity varies with size, sex, age, and aerobic fitness. Taller people have larger lungs than shorter people. Women's lungs are, on average, 20–25 percent smaller than men's, and we all tend to lose lung capacity as we age. The average human lung holds about 5 liters of air. Elite Australian swimmer Grant Hacket has a lung capacity of 13 liters.

Engaging in aerobic exercise, living at high altitude, doing controlled breathing exercises, and playing a wind instrument such as a trumpet or tuba can all increase the efficiency of the respiratory system (mainly the lungs) and the circulatory system (the heart and blood vessels) and reduce age-related declines in lung capacity. Aerobic exercise such as running or biking and high altitudes increase the number of mitochondria in skeletal muscle cells and also the blood supply to these muscles. The enhanced circulation delivers more oxygen per second.

Cuvier's beaked whales are the static-apnea champions of the animal world. They are known to stay underwater for up to 85 minutes, diving to depths as great as 1,900 meters (6,230 feet). These breath-holding divers hold four times as much oxygen in their blood and muscles as humans do. Their body temperature drops as they dive, cooling the brain by 3°C to reduce its energy demands.

# Scientists Cite Fastest Case of Human Evolution

BY NICHOLAS WADE, *New York Times*

Tibetans live at altitudes of 13,000 feet, breathing air that has 40 percent less oxygen than is available at sea level, yet suffer very little mountain sickness. The reason, according to a team of biologists in China, is human evolution, in what may be the most recent and fastest instance detected so far.

Comparing the genomes of Tibetans and Han Chinese, the majority ethnic group in China, the biologists found that at least 30 genes had undergone evolutionary change in the Tibetans as they adapted to life on the high plateau. Tibetans and Han Chinese split apart as recently as 3,000 years ago, say the biologists, a group at the Beijing Genomics Institute led by Xin Yi and Jian Wang. The report appears in Friday's issue of *Science*.

If confirmed, this would be the most recent known example of human evolutionary change. Until now, the most recent such change was the spread of lactose tolerance—the ability to digest milk in adulthood—among northern Europeans about 7,500 years ago. But archaeologists say that the Tibetan plateau was inhabited much earlier than 3,000 years ago and that the geneticists' date is incorrect.

When lowlanders try to live at high altitudes, their blood thickens as the body tries to counteract the low oxygen levels by churning out more red blood cells. This overproduction of red blood cells leads to chronic mountain sickness and to lesser fertility—Han Chinese living in Tibet have three times the infant mortality of Tibetans.

The Beijing team analyzed the 3 percent of the human genome in which known genes lie in 50 Tibetans from two villages at an altitude of 14,000 feet and in 40 Han Chinese from Beijing, which is 160 feet above sea level. Many genes exist in a population in alternative versions. The biologists found about 30 genes in which a version rare among the Han had become common among the Tibetans. The most striking instance was a version of a gene possessed by 9 percent of Han but 87 percent of Tibetans.

Such an enormous difference indicates that the version typical among Tibetans is being strongly favored by natural selection. In other words, its owners are evidently leaving more children than those with different versions of the gene.

The gene in question is known as hypoxia-inducible factor 2-alpha, or HIF2a, and the Tibetans with the favored version have fewer red blood cells and hence less hemoglobin in their blood.

The finding explains why Tibetans do not get mountain sickness but raises the question of how they compensate for the lack of oxygen if not by making extra red blood cells.

According to this article, the same physiological changes that allow people to acclimate to low oxygen conditions can also cause problems. The researchers compared the genes and physiology of people who live at very high altitudes (Tibetans living at 13,000 feet) with those of people originally from sea level (Han Chinese from Beijing). For example, increasing the size and number of red blood cells leads to altitude sickness, also called mountain sickness. Populations of lowlanders who live in Tibet have an infant mortality rate three times that of native Tibetans.

Among 30 genes that differed in how common they were, one—*HIF2a*—stood out: 87 percent of Tibetans had the gene but only 9 percent of Han Chinese did. HIF2a is one of several kinds of molecules that respond to falling oxygen levels in the body, or hypoxia. HIFs can increase the density of red blood cells, actually making the blood thicker; increase blood pressure in the lungs; increase the growth of tiny blood vessels called capillaries, which carry oxygenated blood to cells; and, finally, increase cellular metabolism.

Large increases in the density of red blood cells have a downside. At very high altitudes, such increases can lead to blood so thick that a blood clot can occur and cause a stroke. Tibetans have a version of HIF2a that does not cause the blood to become dangerously thick.

## Evaluating the News

**1.** How much higher is the infant mortality rate of ethnic Han living at high altitudes in Tibet than that of ethnic Tibetans living in the same region?

**2.** During hypoxia, HIFs have a variety of effects on the body. According to this article, one effect HIF2a does *not* have on native Tibetans is to increase the number of red cells and the total amount of hemoglobin in the blood. In what other ways could Tibetan physiology compensate for the low-oxygen conditions to ensure that cells have enough oxygen to make ATP?

**SOURCE:** *New York Times*, July 1, 2010.

# Summary

## 9.1 Molecular Energy Carriers

- Energy carriers store energy and deliver it for cellular activities.
- ATP is the most commonly used energy carrier. Photosynthesis and cellular respiration transform ADP and phosphate into ATP.
- The energy carriers NADPH and NADH donate electrons and hydrogen ions to metabolic pathways.
- NADPH is used in biosynthetic (anabolic) pathways such as photosynthesis. NADH participates in degradative (catabolic) pathways such as cellular respiration.

## 9.2 An Overview of Photosynthesis and Cellular Respiration

- In chemical terms, photosynthesis is the reverse of cellular respiration.
- Photosynthesis occurs in producers only. The light reactions make ATP and NADPH, splitting water molecules and releasing oxygen gas. The energy carriers are used to convert carbon dioxide into sugar molecules during the Calvin cycle reactions.
- Cellular respiration occurs in producers and consumers. It begins in the cytoplasm and is completed in the mitochondrion. Small amounts of ATP and NADH are made during the first stage, glycolysis, during which sugar molecules are broken to make pyruvate, a three-carbon compound.
- The next two stages of cellular respiration take place inside the mitochondrion. Carbon dioxide is released during the degradation of pyruvate via the preparatory reactions and the Krebs cycle, which yields NADH, $FADH_2$, and ATP.
- The final stage of cellular respiration is oxidative phosphorylation, during which many molecules of ATP are made using a membrane-dependent, oxygen-utilizing process.

## 9.3 Photosynthesis: Energy from Sunlight

- Photosynthesis takes place in chloroplasts—light reactions in the thylakoid membrane, and Calvin cycle reactions in the stroma.

- In the light-reactions, energy is absorbed using pigment molecules that include chlorophyll.
- Electrons are stripped from chlorophyll and replaced with electrons from water molecules, releasing oxygen ($O_2$). ATP is generated as electrons flow along the electron transport chain (ETC) that links photosystem II to photosystem I. In the last step, electrons are accepted by $NADP^+$, which picks up protons ($H^+$ ions) to become NADPH.
- Calvin cycle reactions use the ATP and NADPH produced by the light reactions to turn $CO_2$ into G3P, which is converted to six-carbon sugars in the cytosol. Rubisco catalyzes the fixation of $CO_2$ in the stroma.

## 9.4 Cellular Respiration: Energy from Food

- Cellular respiration requires oxygen and has three stages: glycolysis, the Krebs cycle, and oxidative phosphorylation.
- Glycolysis occurs in the cytosol and splits each glucose molecule into two molecules of pyruvate. It yields 2 ATP and 2 NADH.
- In fermentation, pyruvate is converted into carbon compounds such as $CO_2$ and alcohol (as in fermentation by yeasts) or lactic acid (as in skeletal muscles). The postglycolytic fermentation reactions regenerate $NAD^+$, essential for continued glycolysis when the oxygen supply is inadequate.
- In the presence of oxygen, the pyruvate from glycolysis enters the mitochondria, where it is degraded while energy carriers are generated and $CO_2$ is released.
- The Krebs cycle is a series of enzyme-catalyzed reactions that produces: 2 $CO_2$, 3 NADH, 1 $FADH_2$, and 1 ATP.
- Oxidative phosphorylation generates about 30–32molecules of ATP from each glucose. Electrons unloaded by NADH and $FADH_2$ travel an ETC, creating a proton gradient. ATP synthase phosphorylates ADP to make ATP, as protons rush through it.

# Key Terms

aerobic (p. 219)
anaerobic (p. 217)
antenna complex (p. 211)
ATP (p. 207)
ATP synthase (p. 213)
Calvin cycle (p. 209)
carbon fixation (p. 215)
cellular respiration (p. 209)
chlorophyll (p. 208)

chloroplast (p. 208)
citric acid cycle (p. 220)
consumer (p. 206)
electron transport chain (ETC) (p. 211)
energy carrier (p. 207)
fermentation (p. 218)
glycolysis (p. 209)
Krebs cycle (p. 209)
light reactions (p. 208)

metabolic pathway (p. 206)
metabolism (p. 206)
mitochondrion (p. 209)
NADH (p. 208)
NADPH (p. 208)
oxidative phosphorylation (p. 210)
photosynthesis (p. 208)
photosystem (p. 212)
photosystem I (p. 212)

photosystem II (p. 212)
producer (p. 206)
proton gradient (p. 213)
pyruvate (p. 217)
reaction center (p. 211)
rubisco (p. 215)
stoma (p. 211)
stroma (p. 211)
thylakoid (p. 211)

# Self-Quiz

1. The chemical that functions as an energy-carrying molecule in all organisms is
   a. carbon dioxide.
   b. water.
   c. RuBP.
   d. ATP.

2. The main function of the Calvin cycle reactions is to produce
   a. carbon dioxide.
   b. sugars.
   c. NADPH.
   d. ATP.

3. The oxygen produced in photosynthesis comes from
   a. carbon dioxide.
   b. sugars.
   c. pyruvate.
   d. water.

4. The light reactions in photosynthesis require
   a. oxygen.
   b. chlorophyll.
   c. rubisco.
   d. carbon fixation.

5. Glycolysis occurs in
   a. mitochondria.
   b. the cytosol.
   c. chloroplasts.
   d. thylakoids.

6. The electrons needed to replace those lost from chlorophyll in the light reactions of photosynthesis ultimately come from
   a. sugars.
   b. channel proteins.
   c. water.
   d. electron transport chains.

7. Which of the following statements is *not* true?
   a. Glycolysis produces most of the ATP required by aerobic organisms like us.
   b. Glycolysis produces pyruvate, which is consumed by the Krebs cycle.
   c. Glycolysis occurs in the cytosol of the cell.
   d. Glycolysis is the first stage of cellular respiration.

8. The Krebs cycle reactions
   a. take place in the cytoplasm.
   b. convert glucose to pyruvate.
   c. generate ATP with the help of an enzyme complex called ATP synthase.
   d. yield ATP, NADH, and $FADH_2$.

9. Which of the following is essential for oxidative phosphorylation?
   a. rubisco
   b. NADH
   c. carbon dioxide
   d. chlorophyll

10. Oxidative phosphorylation
    a. produces less ATP than glycolysis.
    b. produces simple sugars.
    c. depends on the activity of ATP synthase.
    d. is part of the photosystem I electron transport chain.

# Analysis and Application

1. The Calvin cycle reactions are sometimes called the "dark reactions" to contrast them with the light reactions. Can the Calvin cycle be sustained in a plant that is kept in total darkness for several days? Why or why not?

2. In both chloroplasts and mitochondria, the transfer of electrons down an ETC involves hydrogen ions and leads to a similar outcome. Describe that outcome, and explain how it contributes to the production of ATP in each of these organelles.

3. Explain what is wrong with this statement: The postglycolytic fermentation reactions are a significant source of energy because they generate ATP and NADH through the degradation of pyruvate.

4. Dinitrophenol (DNP) belongs to a class of metabolic poisons known as uncoupling agents. DNP shuttles protons ($H^+$ ions) freely across biological membranes, thereby destroying any existing proton gradient. How would exposure to DNP affect ATP synthesis by mitochondria? Extra challenge: DNP raises body temperature and leads to rapid weight loss. Doctors prescribed DNP as a weight loss drug in the 1930s, until the death of several patients led to a ban. Explain why DNP causes weight loss in humans.

5. Compare the energy yield of glycolysis and cellular respiration. Which pathway releases more usable energy and why?

6. Explain the role of mitochondrial membranes in cellular respiration. Why do you suppose that the inner mitochondrial membrane is extensively folded, but the outer membrane is not? What structure in the chloroplast performs the same function that the intermembrane space does in a mitochondrion?

7. Compare photosynthesis and cellular respiration. Name the organelles in which these processes occur. Use the diagram below to show which atmospheric gases are released and which are consumed by each of these processes. What is the role in photosynthesis of the energy carriers generated by the light reactions? Which of the three main stages of cellular respiration also generate both of these energy carriers?

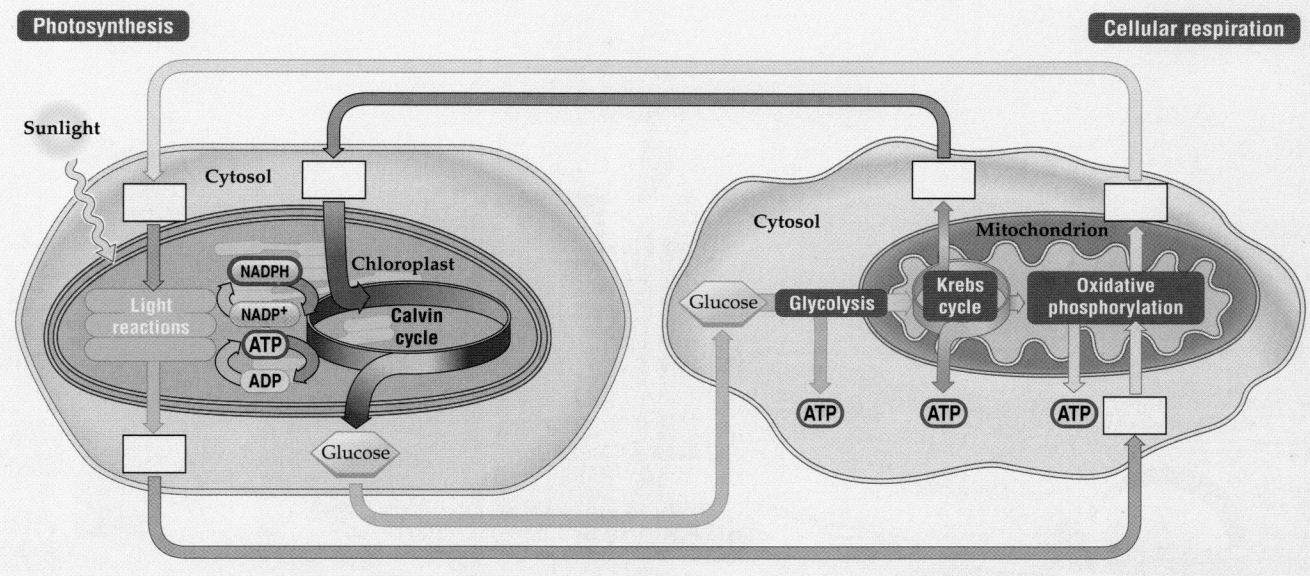

# Cell Division

**IT ISN'T GRASS.** A giant bloom of sticky green algae, thick and vast enough to halt sailboats and windsurfers, threatened to stall sailing events at the Olympic Games in China in 2008.

# Olympic-Class Algal Bloom

In June 2008, as the People's Republic of China put the finishing touches on elaborate preparations for the Summer Olympics, a minor natural disaster struck. It was China's first chance to host the 200 countries attending the Olympics, and China had already spent nearly $2 billion to build 37 venues for 28 summer sports. Most of the events would take place in or near the inland capital city of Beijing. But one sport, sailing, would be held at Qingdao (or Tsingtao), a major port city on the Yellow Sea, 450 miles southeast of Beijing.

Olympic sailing events traditionally appear in the media as beautiful blue-water events, with sails bent to the wind. But starting in May, small, green mats of algae appeared here and there near the shore in Qingdao Bay. At first it was just a minor annoyance. But by the end of June, world-class sailors from all over the world had arrived to practice for the August Olympic Games, and they were beginning to complain. The problems started with dense fog and nearly windless days. But then football field–sized rafts of algae began drifting across the racecourse; mats of the sticky, green stuff wrapped around the keels of expensive racing boats, bringing them to a halt, and even entangled Olympic windsurfers. The sailing blogosphere was buzzing with complaints.

How could a handful of tiny cells multiply fast enough to cover 1,500 square miles in a few weeks? Would it be possible to get rid of the bloom in time for the Olympics?

Algae are photosynthetic organisms—eukaryotes, like us—that float near the surface of water, their cells dividing rapidly in the presence of nutrients and sunshine. In the summer of 2008, conditions in the Yellow Sea must have been ideal for these algae, because despite the best efforts of Olympic organizers, the algal cells continued to divide and multiply at a crushing pace, eventually covering a third of the planned Olympic sailing racecourse. Altogether, the algae covered 1,500 square miles of ocean. It was the world's largest drifting algal bloom.

---

**MAIN MESSAGE** Cell division is the means by which organisms grow, maintain their tissues, and pass genetic information from one generation to the next.

## KEY CONCEPTS

- A dividing cell splits to produce daughter cells that inherit genetic information, in the form of DNA, from the original cell.

- The cell cycle is a set sequence of events that take place in the life span of a cell.

- A chromosome consists of one DNA molecule compacted by packaging proteins.

- DNA is replicated during interphase.

- Cell division is necessary for reproduction, and for growth and repair in a multicellular body.

- Mitotic divisions generate genetically identical daughter cells.

- Meiosis, necessary for sexual reproduction, produces four daughter cells, each with half of the chromosome set found in the parent cell.

- Meiosis generates gametes that are genetically diverse.

- Crossing over and independent assortment of homologous chromosomes contribute to the genetic diversity of gametes.

**CELL DIVISION IS A DISTINCTIVE PROPERTY** of all life-forms. Without cell division, there would be no eggs, no babies, no grown-ups, no circle of life. Your life began as a single fertilized egg cell, and it took many billions of cell divisions to grow you to a baby and to take you from babyhood to adulthood. Even now, mil-

**ROLE OF CELL DIVISION**

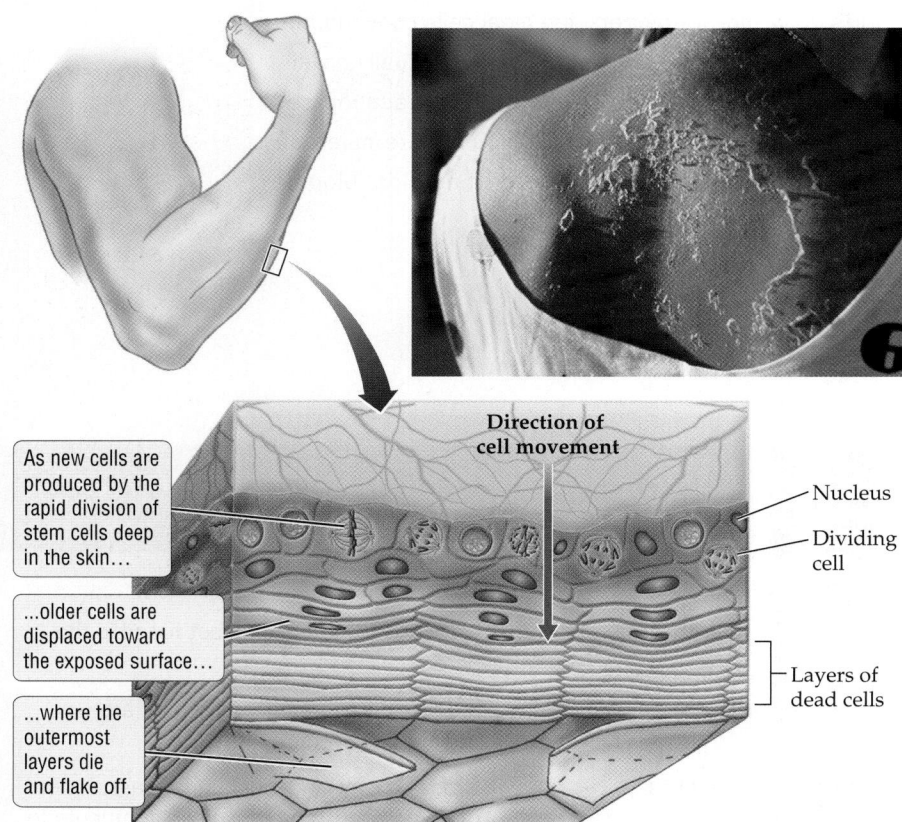

As new cells are produced by the rapid division of stem cells deep in the skin...

...older cells are displaced toward the exposed surface...

...where the outermost layers die and flake off.

Direction of cell movement

Nucleus

Dividing cell

Layers of dead cells

**FIGURE 10.1 Cell Division Replenishes the Skin**
Rapid cell division in the deepest layer of the skin is necessary to replace skin cells (called keratinocytes) lost at the surface of the skin. The loss is due to the normal programmed death of the outer layers of mature keratinocytes, but it can also be the consequence of severe DNA damage, as in a peeling sunburn (inset).

lions of cell divisions take place in your body every day. The purpose of most of these cell divisions is to replace cells that have died in the line of duty—cells that have outlived their useful lives or that have been damaged for some reason. Cell divisions are needed to increase the ranks of immune cells—those that do battle with invading organisms such as bacteria and viruses in response to an infection. A very special type of cell division—called *meiosis*—produces the egg cells in a female and the sperm cells in a male, which join together to produce plant or animal offspring.

This chapter discusses how cells replace themselves through cell division and highlights the essential role of cell divisions in asexual and sexual reproduction, the mechanisms through which whole organisms propagate themselves. After examining the *cell cycle*, which describes what a cell is up to through the course of its life span, we turn the spotlight on the two main types of cell division: *mitotic division* and *meiosis*. We will see that mitotic divisions replace a single cell with two daughter cells that are just like the original; and that meiosis, by contrast, is all about generating daughter cells that are genetically diverse.

# 10.1 Why Cells Divide

Cells divide for two basic reasons: (1) to reproduce organisms and (2) to grow and repair a multicellular organism. For example, multicellular organisms like us need cell divisions to add new cells to the body (**FIGURE 10.1**). **Cell division** generates daughter cells from a parent cell, with the transfer of genetic information—in the form of DNA—from the parent to the daughter cells. All types of organisms need cell division to create the next generation of individuals. New individuals, which receive DNA from the parents, are called *offspring*.

Some organisms make offspring through **asexual reproduction**, which generates *clones*, offspring that are genetically identical to the parent. Bacteria and archaeans reproduce only through asexual means. Some multicellular organisms, including many plants and even some animals, reproduce asexually as well. The majority of eukaryotes, however, produce offspring through sexual reproduction. In **sexual reproduction**, genetic information from two individuals of opposite mating types is combined to produce offspring. Offspring resulting from sexual reproduction are similar, but not identical, to the parents.

Asexual reproduction and sexual reproduction both propagate the species (Table 10.1), but sexual re-

| TABLE 10.1 | Biological Relevance of Cell Division | |
| --- | --- | --- |
| **TYPE OF CELL DIVISION** | **OCCURS IN** | **FUNCTION** |
| Binary fission | Prokaryotes (Bacteria and Archaea) | Asexual reproduction |
| Mitotic division | Eukaryotes: single-celled or multicellular | Asexual reproduction |
| | Eukaryotes: multiceullular | Growth of individual; repair and replacement of cells and tissues |
| Meiosis | Eukaryotes: single-celled or multicellular | Sexual reproduction |

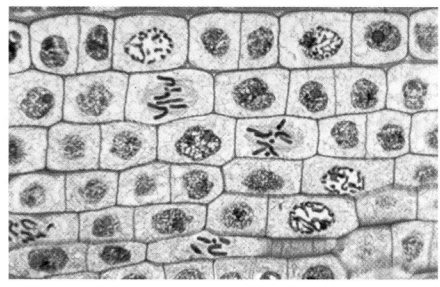
Mitosis in onion roots

production adds a great deal of genetic diversity to the population. Genetic diversity arises because DNA from two different individuals is shuffled and combined in offspring, giving each offspring a unique mix of the genetic characteristics of its two parents. A genetically diverse population is more likely to have among its members some that can adapt to changing conditions. It is more likely, for example, that resistance to a new virus strain will emerge in a genetically diverse population than within a population of individuals that are essentially clones. On the other hand, asexual reproduction is often a highly cost-effective strategy, especially in a relatively stable environment. The go-it-alone reproducer has no need to spend precious energy, and possibly risk life and limb, in the search for mates.

## Many bacteria use binary fission for asexual reproduction

It is likely that the first organisms to appear on Earth reproduced asexually, producing offspring genetically identical to the parent cell. The strategies they used to propagate themselves probably resembled the mechanisms seen in modern-day prokaryotes, the single-celled organisms that make up the domains Bacteria and Archaea. Many prokaryotes reproduce asexually through a mechanism called **binary fission** (literally "splitting in two").

The genetic material of most bacteria and archaeans takes the form of one single loop of DNA. (We will see later that this DNA molecule is complexed with proteins to form a *chromosome*, but for now let's focus on the DNA alone.) The first step in binary fission is the duplication of this DNA, giving rise to two DNA molecules (**FIGURE 10.2**). The cell now expands and a partition appears roughly at the center of the cell. The partition, consisting of plasma membrane and usually

cell wall material as well, segregates the two DNA molecules into separate cytoplasmic compartments. Each compartment expands until it breaks loose of the other, so that two daughter cells now replace the parent cell. Binary fission is an asexual form of reproduction because it creates daughter cells that are genetically identical to each other and the parent cell. The daughter cells are identical because each inherits an exact copy of the DNA code that was present in the parent cell.

**BINARY FISSION**

Parent cell

Cell wall
Plasma membrane
Circular DNA molecule

DNA replication and segregation

DNA is replicated giving rise to two circular DNA molecules.

Cytoplasmic division

New cell wall

The cell expands and a partition is created that isolates the two DNA molecules into separate cytoplasmic compartments.

Cell separation

Two daughter cells

**FIGURE 10.2 Cell Division in a Prokaryote**
Many prokaryotes, including many types of bacteria, propagate themselves asexually through a type of cell division known as binary fission.

FIGURE 10.3
Cell Division in
a Eukaryote

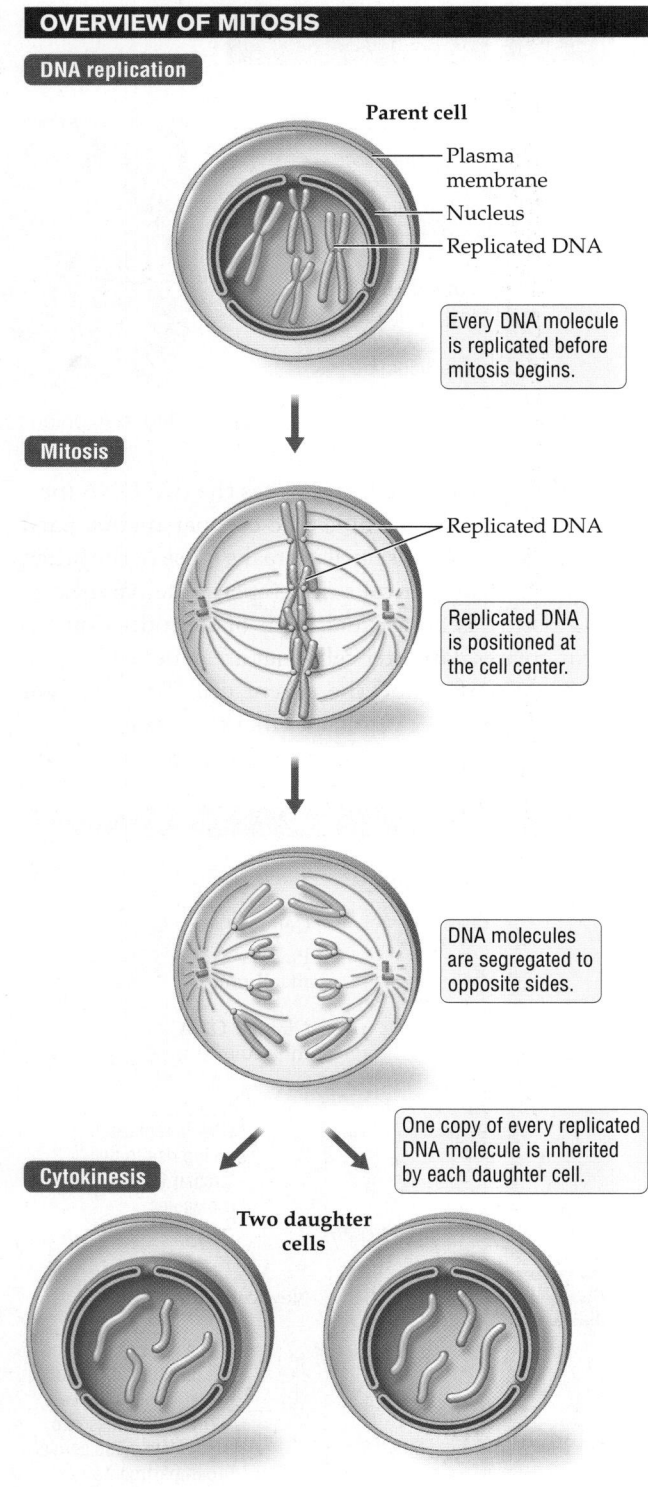

**OVERVIEW OF MITOSIS**

**DNA replication**

Parent cell

Plasma membrane
Nucleus
Replicated DNA

Every DNA molecule is replicated before mitosis begins.

**Mitosis**

Replicated DNA

Replicated DNA is positioned at the cell center.

DNA molecules are segregated to opposite sides.

One copy of every replicated DNA molecule is inherited by each daughter cell.

**Cytokinesis**

Two daughter cells

## Eukaryotes use mitosis to generate identical daughter cells

Cell division in eukaryotes is more complicated than binary fission because eukaryotic cells have many molecules of DNA that have to be replicated and then distributed evenly between the two daughter cells. Eukaryotic DNA lies in the nucleus, wrapped in the double layer of membranes that make up the nuclear envelope. In most eukaryotes, the nuclear envelope is disassembled in the dividing cell and then reassembled in each of the daughter cells toward the end of cell division. **Mitotic division** is the process that generates two genetically identical daughter cells from a single parent cell in eukaryotes. A mitotic division begins with **mitosis [mye-TOH-sis]** which refers to the division of the nucleus. Mitosis is followed by **cytokinesis [SYE-toh-kih-NEE-sis]**, the splitting of the original cytoplasm into two new daughter cells. A parent cell sets up for an upcoming mitotic division by duplicating its DNA well before mitosis gets under way. During mitosis, an elaborate cytoskeletal machinery works in a precise sequence to deliver one complete copy of the replicated DNA to opposite sides of the original cell. The cytoplasm between the two copies of DNA is then divided in the process of cytokinesis, giving rise to two daughter cells (**FIGURE 10.3**).

Mitotic divisions can serve both the eukaryotic organism's need to replace itself (to reproduce) and its need to add new cells to the body. Single-celled eukaryotes use mitotic divisions for asexual reproduction, in much the same way that prokaryotes reproduce asexually through binary fission. Many multicellular eukaryotes use mitotic divisions to reproduce asexually as well, including most seaweed, fungi, and plants, and some animals, such as sponges and flatworms. While most multicellular organisms reproduce by sexual means—dividing their nuclei by *meiosis* rather than by mitosis—all multicellular organisms rely on mitotic divisions for the growth of tissues and organs and the body as a whole, and for repairing injured tissue and replacing worn-out cells.

## Meiosis is necessary for sexual reproduction

**Meiosis [mye-OH-sis]** is a specialized type of cell division that makes sexual reproduction possible. In animals, meiosis in the female body generates daughter cells that mature into eggs, while meiosis in the male animal produces sperm cells. Egg and sperm are examples of sex cells, or **gametes**.

The non–sex cells in a multicellular organism are called **somatic cells** (from *soma*, Greek for "body"). The somatic cells of plants and animals have twice as much genetic information as is found in a sex cell. This double set of genetic information is known as the diploid set, represented by 2*n*. Meiosis (*meio*, "less") *reduces* the amount of genetic information transmitted by the parent cell to the daughter cells by half, so that only one set of the genetic information is inherited by each daughter cell. This single set of genetic information is called the haploid set, represented by the letter *n*.

Because they are products of meiosis—a reduction division—egg and sperm each have a haploid set (*n*) of genetic information. Mating between male and female individuals can result in **fertilization**, the merging of egg and sperm to create a single cell, the **zygote**. The zygote inherits one haploid set of genetic material from the egg and another from the sperm, and that is how fertilization restores the complete diploid set of genetic information to the offspring. In animals, the zygote divides mitotically to create a mass of developing cells known as the **embryo**. The embryo develops organs to become a fetus and eventually matures into a juvenile and then into an adult individual.

## Cell divisions grow, maintain, and reproduce the human body

Your body arose from a single-celled zygote, formed by the fusion of a particular egg and a certain sperm cell. Mitotic divisions in the zygote give rise to a ball of cells, the embryo (**FIGURE 10.4**). Cells in the very young embryo are not noticeably different from each other; but as the embryo develops, many of the cells in it acquire unique properties and highly specialized functions. The process through which a daughter cell becomes different from the parent cell is known as **cell differentiation**. Heart muscle cells and neurons (a type of nerve cell) are two examples of the 220 differentiated cell types in the adult human. All the genetic material that was in the zygote is present in all cells of the embryo and all the somatic cells of the adult body, no matter how differentiated the cell type is.

A small group of gamete-producing cells, called **germ line cells**, is set aside very early in embryonic development. A subset of the germ line cells—called gametocytes—eventually undergo meiosis to produce gametes: egg cells in females and sperm in males. Germ line cells are set aside as a special lineage of cells within a week after conception, and they undergo relatively few mitotic divisions and cell movements, compared to the upheaval of dividing and migrating cells that surrounds them in the developing embryo. In about the fifteenth week after conception, the germ line cells migrate into the reproductive organs: a pair of ovaries in a developing female and a pair of testes in a developing male. In a developing female, about a million gametocytes launch into meiosis, but then the process is put on hold until puberty! So, girls are born

FIGURE 10.4

**MITOSIS AND MEIOSIS IN HUMANS**

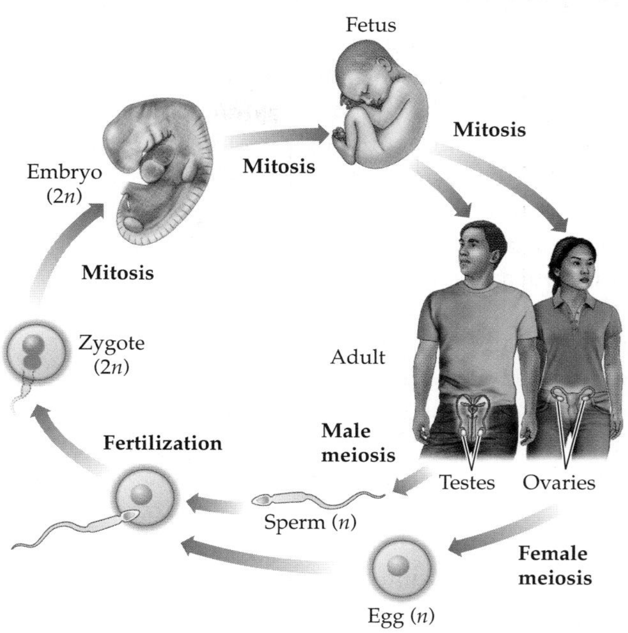

FIGURE 10.4
Mitosis and Meiosis
Play Vital Roles
in the Human Life
Cycle

with about a million gametocytes that are arrested in an early stage of meiosis. Typically, one such arrested gametocyte *resumes* meiosis each month in one of the ovaries of a fertile female. In males, meiosis does not begin until hormonal changes produced by puberty signal millions of gametocytes, in both testes, to begin producing sperm. Male meiosis occurs daily and continues well into old age, but in women the supply of functioning gametocytes dwindles and usually disappears by age 50.

As organ systems develop, about 15 weeks after conception, an embryo matures into a **fetus**. At birth a baby has most of the specialized cell types that an adult has. Although most cells in the newborn have differentiated to perform a narrow set of tasks, *adult stem cells* in the various organs remain unspecialized. **Adult stem cells** are not committed to specialized functions, but some of their descendants can differentiate into certain types of specialized cells. Mitosis in adult stem cells contributes to the growth of the body, and to the regeneration and repair of tissues throughout our lifetime.

## 10.2 The Cell Cycle

The **cell cycle** is a set sequence of events that make up the life of a typical eukaryotic cell that is capable of dividing. The cell cycle extends over the life span of a cell, from the moment of its origin to the time it divides to produce two daughter cells. The time it takes to complete the cycle depends on the organism, the type of cell, and the life stage of the organism. Dividing cells in tissues that require frequent replenishing, such as the skin or the lining of the intestine, take about 12 hours to complete the cell cycle. Cells in most other actively dividing tissues in the human body require about 24 hours to complete the cycle. By contrast, a single-celled eukaryote such as yeast can complete the cell cycle in just 90 minutes.

There are two main stages in the cell cycle—*interphase* and *cell division*—each marked by distinctive cell activities. **Interphase** is the longest stage of the cell cycle. Most cells spend 90 percent or more of their life span in interphase. During this phase, the cell takes in nutrients, manufactures proteins and other substances, expands in size, and conducts its special functions. In cells that are destined to divide, preparations for cell division also begin during interphase. A critical event in these preparations is the copying of all the DNA molecules, which contain the organism's genetic information in the form of genes.

Cell division is the last stage in the life of an individual cell. Cell division is not only the most rapid stage of the cell cycle; it is also the most dramatic in visual terms. In tissues with many rapidly dividing cells, such as onion root tips or fish embryos, the events of cell division can be readily seen with an ordinary light microscope. The discussion that follows offers a closer look at the major events (*phases*) of the cell cycle.

## DNA is replicated in S phase

In cells capable of dividing, interphase can be divided into three main intervals: $G_1$, S, and $G_2$ (**FIGURE 10.5**). These intervals are defined by distinctive cellular events. The *replication* (copying) of DNA, requiring the manufacture (*synthesis*) of new DNA, occurs in **S phase** ("S" stands for "synthesis"). The **$G_1$ phase** (for "Gap 1") is the first phase in the life of a "newborn" cell. The **$G_2$ phase** (for "Gap 2") begins after S phase and before the start of division. Early cell biologists bestowed the term "gap" on the $G_1$ and $G_2$ phases because they believed those phases to be less significant periods in the life of a cell compared to S phase and cell division. We now know that many crucial events occur during the two "gap" phases and that some of these events have a profound effect on the precision and stability of the cell division stage.

$G_1$ and $G_2$ are important phases for two reasons. They are often periods of growth, during which both the size of the cell and its protein content increase. Furthermore, each phase prepares the cell for the phase immediately following it, serving as a checkpoint ensuring that the cell cycle will not progress to the next phase unless all conditions are suitable.

**FIGURE 10.5 The Cell Cycle Consists of Two Major Stages: Interphase and Cell Division**

The cell prepares itself for division by increasing in size and producing proteins needed for division during $G_1$ and $G_2$ phases, and by replicating its DNA during S phase. Mitotic cell division consists of mitosis and cytokinesis, which result in two daughter cells that are genetically identical to the parent cell.

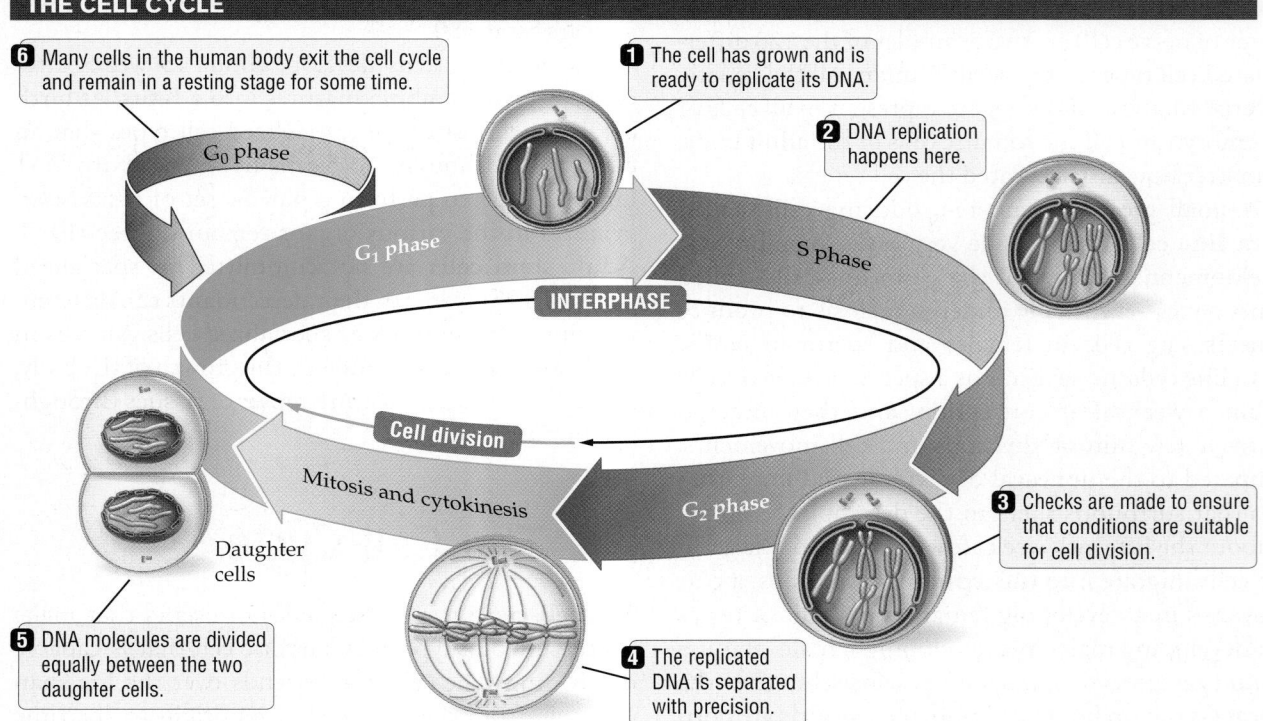

THE CELL CYCLE

6 Many cells in the human body exit the cell cycle and remain in a resting stage for some time.

1 The cell has grown and is ready to replicate its DNA.

2 DNA replication happens here.

$G_0$ phase

$G_1$ phase

S phase

INTERPHASE

Cell division

Mitosis and cytokinesis

$G_2$ phase

3 Checks are made to ensure that conditions are suitable for cell division.

Daughter cells

5 DNA molecules are divided equally between the two daughter cells.

4 The replicated DNA is separated with precision.

## Most cell types in the adult body do not divide

Not all of our cells complete the cell cycle. Many of our 220 different cell types start differentiating (becoming specialized) shortly after entering $G_1$, and they then exit the cell cycle to enter a nondividing state that cell biologists named the **$G_0$ phase** (see Figure 10.5). The $G_0$ phase can last for periods ranging from a few days to the lifetime of the organism. Most liver cells stay in the $G_0$ phase much of the time but reenter the cell cycle about once a year, on average, to make up for cells that have died because of normal wear and tear. The liver is highly active metabolically and is also on the front line in dealing with toxins, from antibiotics to alcohol, that we commonly ingest. The exceptional regenerative capacity of the liver compensates for the cell damage the organ suffers in conducting its routine detoxification activities. The liver's regenerative capacity rests in part on a large pool of $G_0$ cells that can reenter the cell cycle when needed.

Some cells, such as those that form the lens of the eye, remain in $G_0$ for life, as part of a nondividing tissue. Many of the cell types in the brain have also exited the cell cycle, which is why neurons lost as a result of physical trauma or chemical damage are not readily replaced. Some highly specialized cell types not only exit the cell cycle, but intentionally self-destruct in a process called *programmed cell death* (described in the box on page 236).

## The cell cycle is tightly regulated

The decision to divide is a momentous one, even for single-celled organisms. Cell division is metabolically expensive. It would be a reckless move for a yeast cell to launch into mitosis at a time of food scarcity, and natural selection tends to weed out such maladaptive behaviors. In vertebrates like us, there is the additional threat of runaway cell division, which can turn into a cancer. A cancer begins with a single cell that breaks loose of normal restraints on cell division and starts dividing rapidly to establish a colony of rogue cells. A clump of such cells is called a **tumor**, and tumor cells turn into **cancer cells** if they begin to invade neighboring tissues. As cancer cells spread through the body, they disrupt the normal function of tissues and organs; unchecked, cancer cells can cause death through failure of multiple organ systems. (Chapter 11 discusses cancer in more detail.)

It is little wonder, then, that the cell cycle is carefully controlled in the healthy individual. The decision to divide the cell is made during the $G_1$ phase of the cell cycle, in response to internal and external signals. In humans, external signals that influence the commitment to divide include hormones and proteins called growth factors. As we explain in Chapter 11, some hormones and growth factors act like the gas pedal in a car and push a cell toward cell division, while others act like a brake and hinder cell division. Special *cell cycle regulatory proteins* are activated after a cell receives and interprets a signal to divide. The proteins "throw the switch" that enables the cell to pass the critical checkpoints and progress from one phase of the cell cycle to the next (**FIGURE 10.6**). For example, upon receiving the appropriate signals, cell cycle regulatory proteins advance the cell from $G_1$ to S phase by triggering DNA replication and other processes associated with the synthesis phase.

Cell cycle regulatory proteins also respond to negative internal or external control signals. Internal signals will pause a cell in $G_1$, barring entry to S phase, under any of the following conditions: the cell is too small, the nutrient supply is inadequate, or DNA is damaged. $G_2$ arrests in the same circumstances, as well as when the DNA duplication that begins in S phase is incomplete for any reason. So, it is as if the cell cycle comes with start buttons and pause buttons, but no reverse buttons. The cell cycle can progress in only one direction—toward mitosis and the completion of cytokinesis; otherwise it stalls indefinitely

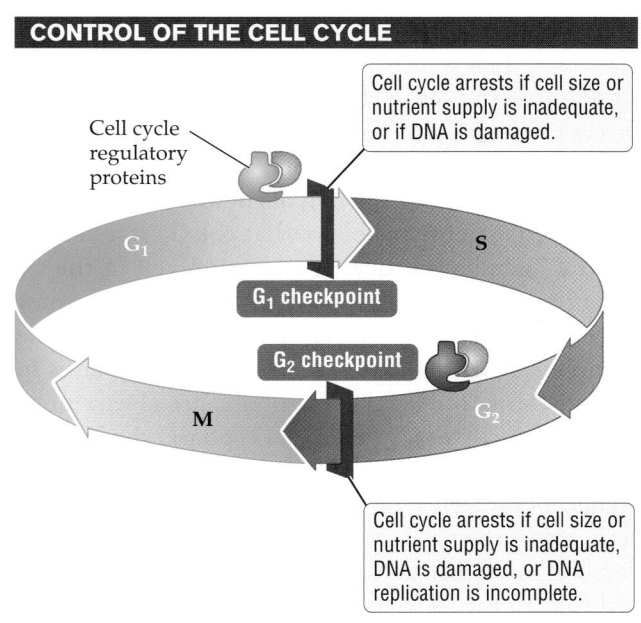

**CONTROL OF THE CELL CYCLE**

Cell cycle regulatory proteins

Cell cycle arrests if cell size or nutrient supply is inadequate, or if DNA is damaged.

$G_1$

S

$G_1$ checkpoint

$G_2$ checkpoint

M

$G_2$

Cell cycle arrests if cell size or nutrient supply is inadequate, DNA is damaged, or DNA replication is incomplete.

**FIGURE 10.6 Cell Cycle Regulatory Proteins Help Control the Cell Cycle**

Only two of the known cell cycle checkpoints are depicted in this diagram. Checkpoints are known to operate in S phase and partway through mitosis as well.

# Programmed Cell Death: Going Out in Style

Some cells die young. Neutrophils, the most abundant of the white blood cells, live for a day or two before they destroy themselves from the inside out. The keratinocytes in your skin (see Figure 10.1) kill themselves about 27 days after the mitotic division that produced them. Your red blood cells are demolished about 120 days after being made in the bone marrow, the destruction occurring mostly in a small organ called the spleen. And cellular suicide begins early in development; it is especially common in the fetal stage. About 50 percent of the neurons that develop in the brain of a human fetus are lost before birth.

It may seem like mayhem, but this cellular suicide is an elegantly controlled process that performs a valuable function in the body. The stepwise dismantling of a cell, controlled and conducted by the cell itself, is known as *programmed cell death* (*PCD*). In contrast to PCD, *necrosis* is the messy, disorganized death of a cell because of injury or infection that the cell is unable to resist. A paper cut kills some skin cells by rupturing them, and if bacteria enter the cut, their toxins can cause yet more cell death. Necrotic cells spill their cytoplasmic contents,

provoking a reaction, called *inflammation*, by the body's immune system. An infected cut appears red and swollen and warm to the touch, the three telltale signs of inflammation.

Programmed cell death in animal cells follows an orderly sequence, the details of which vary depending on the cell type. *Apoptosis* (plural "apoptoses"), a particular form of PCD in animals, often begins with mitochondrial damage, followed by the activation of protein-destroying enzymes (called caspases) that digest the cell from the inside. The cell shrinks and the DNA is broken into fragments as the cell begins to die. Cell remnants are typically engulfed by *phagocytes*, the immune system's cleanup crew.

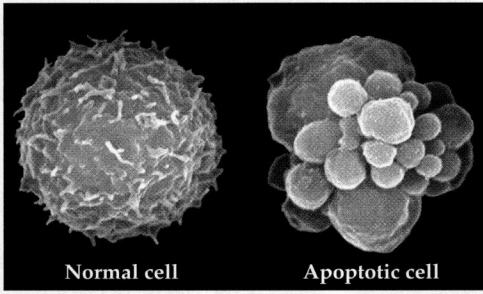

Normal cell    Apoptotic cell

Programmed cell death does away with cells that are no longer needed. In addition, PCD sculpts tissues and organs during development. For example, the fingers and toes of a human fetus are webbed initially, and were it not for the death of the webbing tissue between the digits, we would have paddlelike hands and feet.

Mitosis generates more than 200 billion nerve cells in the fetal brain, which extend fine projections (dendrites) that must link with target cells in order to become functional. The hit-and-miss nature of these projections means that many nerve cells fail to make an appropriate connection; superfluous nerve cells then proceed to self-destruct through PCD. At this point, we are very happy to bust the common myth that people lose thousands of neurons daily throughout their adult lives. Although certain support cells, known as astrocytes (glial cells), are constantly lost through PCD and regenerated through mitosis in the brain, most of us keep the same population of roughly 100 billion neurons well into old age, barring disease or injury.

---

Because the cell cycle is controlled through many layers of checks and balances—cell division promoters and cell division inhibitors, and multiple checkpoints, for example—the risk of runaway cell division is reduced. Scientists believe that the $G_0$ state represents a "safe haven" against cancer. A cell in the $G_0$ phase and a nondividing cell in the $G_1$ phase may appear to behave similarly; the key difference is the complete absence of cell cycle regulatory proteins in $G_0$ cells. In contrast, these proteins are always present inside cells that are in the $G_1$ phase, though the proteins may be lying dormant for lack of a go-ahead signal. Because it *lacks* the cell cycle regulatory proteins altogether, a $G_0$ cell is further removed from the capacity to divide and therefore less likely than a $G_1$ cell to turn into a rogue.

> **Concept Check**
>
> 1. In what way is binary fission similar to mitotic cell division? Give one difference.
> 2. What is the function of mitosis?

## 10.3 The Chromosomal Organization of Genetic Material

DNA in the nucleus is not a disorganized tangle of naked nucleotide polymers. Instead, each long, double-stranded DNA molecule is attached to proteins that help pack it into a more compact physical structure called a **chromosome**. The packing is necessary because each DNA molecule is enormously long, even in the simplest cells. If all of the 46 different DNA molecules in one of your skin cells were lined up end to end, they would make a double helix nearly 2 meters (about 6 feet) long. How can that much DNA be stuffed into a nucleus, which, in a human cell, has a diameter slightly under 5 micrometers (0.005 millimeter)? Extreme packaging is the answer.

Each DNA double helix winds around special DNA packaging proteins to create a DNA-protein complex known as **chromatin**. Chromatin is further looped and compressed to form an even more compact structure called a chromosome (**FIGURE 10.7**). We will learn much more about genes, chromosomes, and DNA packaging and replication in Unit 3. For now, it is enough to know that each chromosome is a compacted DNA-protein complex, and within it is a single long molecule of DNA that bears many genes.

Before cell division can proceed, the DNA of the parent cell must be replicated (copied) so that each daughter cell can receive a complete set of chromosomes. DNA is replicated during S phase, resulting in two identical double helices, known as **sister chromatids**, that remain linked to each other until the later stages of mitosis. Therefore, as mitosis begins, the nucleus of a human cell contains twice the usual amount of DNA, since each of the 46 chromosomes now consists of two identical sister chromatids, held together especially firmly in a region called the **centromere** (**FIGURE 10.8**). At the beginning of mitosis, the chromatin becomes packed and condensed even more tightly than during interphase, which is why chromosomes are most easily seen at the cell division stage.

## DNA PACKAGING

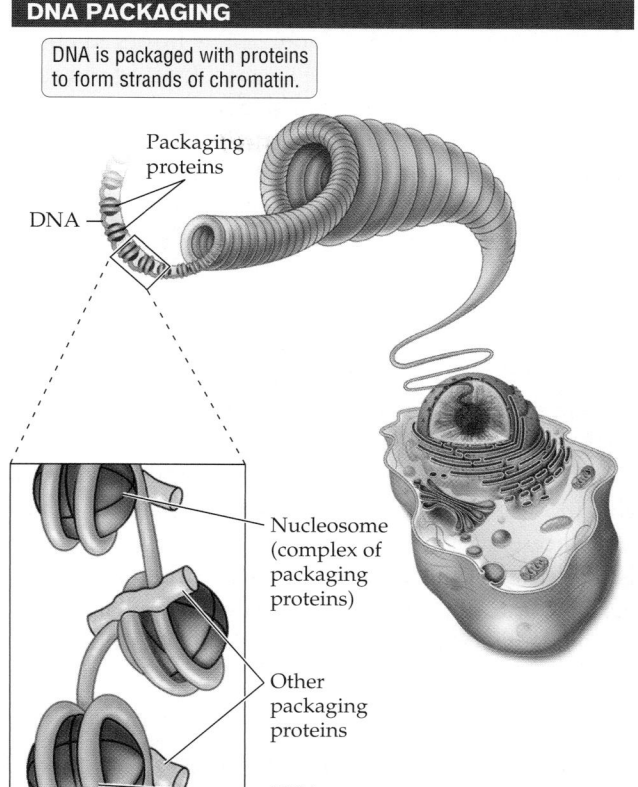

DNA is packaged with proteins to form strands of chromatin.

Packaging proteins

DNA

Nucleosome (complex of packaging proteins)

Other packaging proteins

DNA

FIGURE 10.7 The Packing of DNA into a Chromosome

### Helpful to Know

"Chromosome" means literally "colored body" (*chromo*, "color"; *some*, "body"). Although they did not know the function of chromosomes, nineteenth-century microscopists could see them in dividing cells by staining them with certain dyes. Nowadays, cell biologists use a variety of fluorescent dyes to visualize specific molecules (including DNA) and particular organelles. You will see many fluorescent photographs in this and other chapters.

## The karyotype describes all the chromosomes in a nucleus

Every species has its own characteristic number of chromosomes in the nucleus of each of its cells. As noted earlier, somatic cells are all those cells in a multicellular organism that are not gametes (egg or sperm, in animals) or the direct precursors of gametes (germ line cells). We can think of somatic cells as the "generic" cells in the body, those not involved in sex-

## STRUCTURE OF A REPLICATED CHROMOSOME

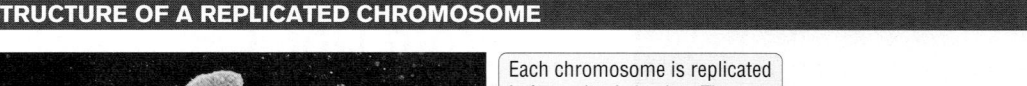

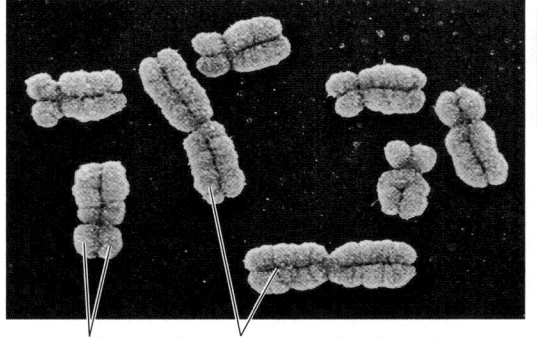

Sister chromatids in one replicated homologue

One pair of replicated homologous chromosomes

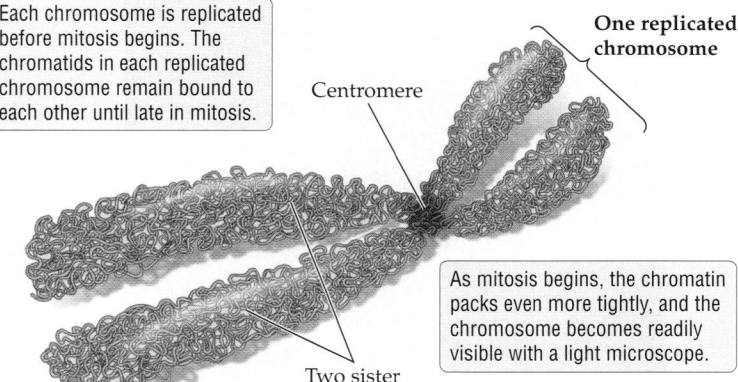

Each chromosome is replicated before mitosis begins. The chromatids in each replicated chromosome remain bound to each other until late in mitosis.

One replicated chromosome

Centromere

As mitosis begins, the chromatin packs even more tightly, and the chromosome becomes readily visible with a light microscope.

Two sister chromatids

FIGURE 10.8 Each Replicated Chromosome Consists of Two Identical Sister Chromatids

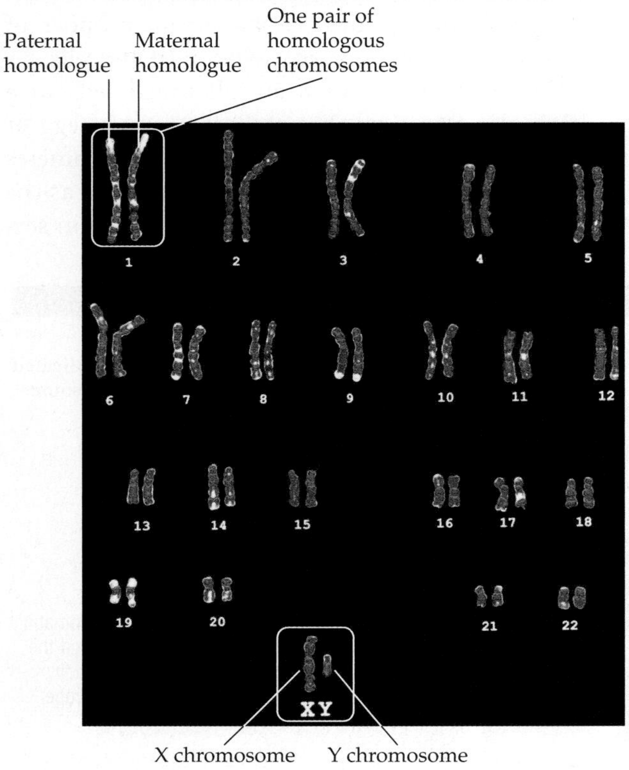

**FIGURE 10.9 The Karyotype Identifies All the Chromosomes of a Species**

The 46 chromosomes in this micrograph represent the karyotype of a human male. With the help of computer graphics, photos of the chromosomes have been aligned so that the two members of each homologous pair are placed next to each other. The non–sex chromosomes (known as autosomes) are numbered. The sex chromosomes are represented by letters (XY, in the case of this individual, a male).

ual reproduction. Somatic cells of plants and animals may have anywhere from two to a few hundred chromosomes, depending on the species. During mitosis, when chromosomes are compacted to the maximum extent, the different types of chromosomes in a somatic cell can often be identified under a microscope by their size and distinctive shape.

A display of all the chromosomes in a somatic cell is the **karyotype** (**FIGURE 10.9**). Karyotypes are generally made from microscopic observations of mitotic cells, because chromosomes are more easily seen in such cells than in interphase cells. The karyotype of a human somatic cell shows a total of 46 chromosomes. Each somatic cell in a horse has 64 chromosomes; in a corn plant, 20 chromosomes. The total number of chromosomes per somatic cell has no particular significance other than being an identifying characteristic of the species. It does not reflect the number of genes found in that species, nor does it say anything about the species' structural or behavioral complexity.

## Most human cells have two copies of each type of chromosome

A distinctive feature of most eukaryotes, compared to prokaryotes, is that their somatic cells contain

**THE HUMAN KARYOTYPE**

Paternal homologue — Maternal homologue — One pair of homologous chromosomes

X chromosome    Y chromosome

two matched copies of every type of chromosome. Two matched chromosomes make up a pair of **homologous [hoh-*MAH*-luh-gus] chromosomes**. Returning to the example of the human karyotype, our 46 chromosomes are actually a double set consisting of 23 *pairs* of homologous chromosomes. You inherited one set (23 chromosomes) from your mother and the other set (the remaining 23) from your father, to create a double set of 46, or 23 pairs. In 22 of these homologous pairs (numbered 1–22), the two *homologues* (individual members of the pair) are alike in length, shape, and the location and type of genes they carry. But the twenty-third pair can be an odd couple, consisting of two very dissimilar homologues: an X chromosome and a Y chromosome (see Figure 10.9).

The X and Y chromosomes are called **sex chromosomes** because, in mammals and some other vertebrates, these chromosomes determine the sex of the individual animal. In humans and other mammals, individuals with two X chromosomes in their cells are female, and those with one X and one Y chromosome are male. The X chromosome is considerably longer than the Y chromosome and carries many more genes. Most, if not all, of the few genes found on the Y chromosome appear to be involved in controlling the development of male characteristics. All of the many genes that are unique to the X chromosome are also present in normal males, because their cells contain one X chromosome, but males (with their XY combination of sex chromosomes) have just one copy of the X-specific genes. Normal females, with their two copies of the X chromosome (XX), have two copies of all the X-specific genes.

## 10.4 Mitosis and Cytokinesis: From One Cell to Two Identical Cells

The climax of the cell cycle is cell division, which, in the case of mitotic divisions, consists of two steps: mitosis and cytokinesis. These steps are not discrete in time; cytokinesis overlaps with the last phase of mitosis. The central event of mitosis is the equal distribution of the parent cell's replicated DNA into two daughter nuclei. This process, called DNA segregation, requires the coordinated actions of many different types of proteins, including those that make up the cytoskeleton.

Mitosis is divided into four main phases, each of them defined by easily identifiable events that are visible under the light microscope (**FIGURE 10.10** on pages 240–41):

1. *Prophase*
2. *Metaphase*
3. *Anaphase*
4. *Telophase*

Recall that the cell sets up for mitosis well beforehand, and all the chromosomes in the nucleus have been replicated in S phase before mitosis begins. Each replicated chromosome consists of two identical DNA molecules—the sister chromatids—held together along their length, and especially tightly at the constriction known as the centromere.

You could say that the objective of mitosis is to separate all the sister chromatids and deliver *one of each* to the opposite ends of the parent cell. The elaborate chromosomal choreography of mitosis has evolved to minimize the risk of mistakes in the equal and symmetrical partitioning of the replicated genetic material. Normally, no daughter cell winds up short a chromosome, nor does it acquire duplicates. Each daughter cell inherits exactly the same information that the parent cell possessed in the $G_1$ phase of its life—no more, no less.

## Chromosomes are compacted during early prophase

DNA molecules undergo a high level of compaction during the first stage of mitosis, **prophase** (*pro*, "before"; *phase*, "appearance"), and by the end of this phase the DNA is 10 times more tightly wound than during interphase. In an interphase cell, each DNA molecule is dispersed through the nucleus in the form of a less compact string of chromatin that is too fine to be visualized by an ordinary light microscope. As the cell moves from $G_2$ phase into prophase, the chromatin is furled more tightly, so that each chromosome gets shorter and stouter, becoming readily visible in the nucleus. The functional value of this extra level of compaction is that the chromosomes can be lined up and sorted to the opposite poles of the cell without excessive tangling and breakage (yes, chromosomes can break). Imagine if you had to untangle a heap of cooked spaghetti and move exactly the same number of strands to opposite sides of a dinner plate. Would the task be easier if the strands were short and stumpy, like penne pasta, instead of long and floppy like spaghetti?

During prophase, important changes occur both in the cytoplasm and in the nucleus. Two cytoskeletal structures, called **centrosomes** (*centro*, "center"; *some*, "body"), begin to move through the cytosol, finally halting at opposite sides in the cell. As we shall see, this arrangement of centrosomes defines the opposite ends, or poles, of the cell. Most cells split along an imaginary line roughly halfway between the two centrosomes, so that each of the two daughter cells resulting from cytokinesis inherits one of the centrosomes.

At the same time that the centrosomes are moving toward the poles of the cell, cytoskeletal structures called microtubules are growing outward from each centrosome. *Microtubules* are long cylinders of special proteins (see Section 6.5), and during prophase some microtubules assemble themselves into an elaborate apparatus called the *mitotic spindle*. The microtubules of the mitotic spindle will later attach to the chromosomes and help move them through the cytosol to sort them into the two daughter cells. The **mitotic spindle**, with its spokelike array of microtubules, functions as a moving crew that hauls chromosomes through the cytoplasm and eventually deposits them at opposite ends of the parent cell.

## Chromosomes are attached to the spindle in late prophase

The nuclear envelope breaks down late in prophase (see Figure 10.10), during a step that cell biologists call *prometaphase*. With the nuclear envelope out of the way, the microtubules of the mitotic spindle, radiating out from the centrosome at each pole, seek out and attach to the now highly condensed chromosomes. The overall result is that each replicated chromosome is "captured" by the mitotic spindle and becomes linked to the two centrosomes through microtubules.

The physical structure of the centromere dictates how each chromosome will be attached to the spindle microtubules. Each centromere has two patches of protein, called *kinetochores* [kih-*NET*-oh-korz], that are oriented on opposite sides of the centromere. Each kinetochore forms a site of attachment for at least one microtubule, so that the two chromatids that make up a replicated chromosome end up being linked to the centrosomes at the opposite poles of the cell. The successful "capture" of each pair of sister chromatids by the mitotic spindle sets the stage for the proper positioning of these replicated chromosomes in the next phase of mitosis.

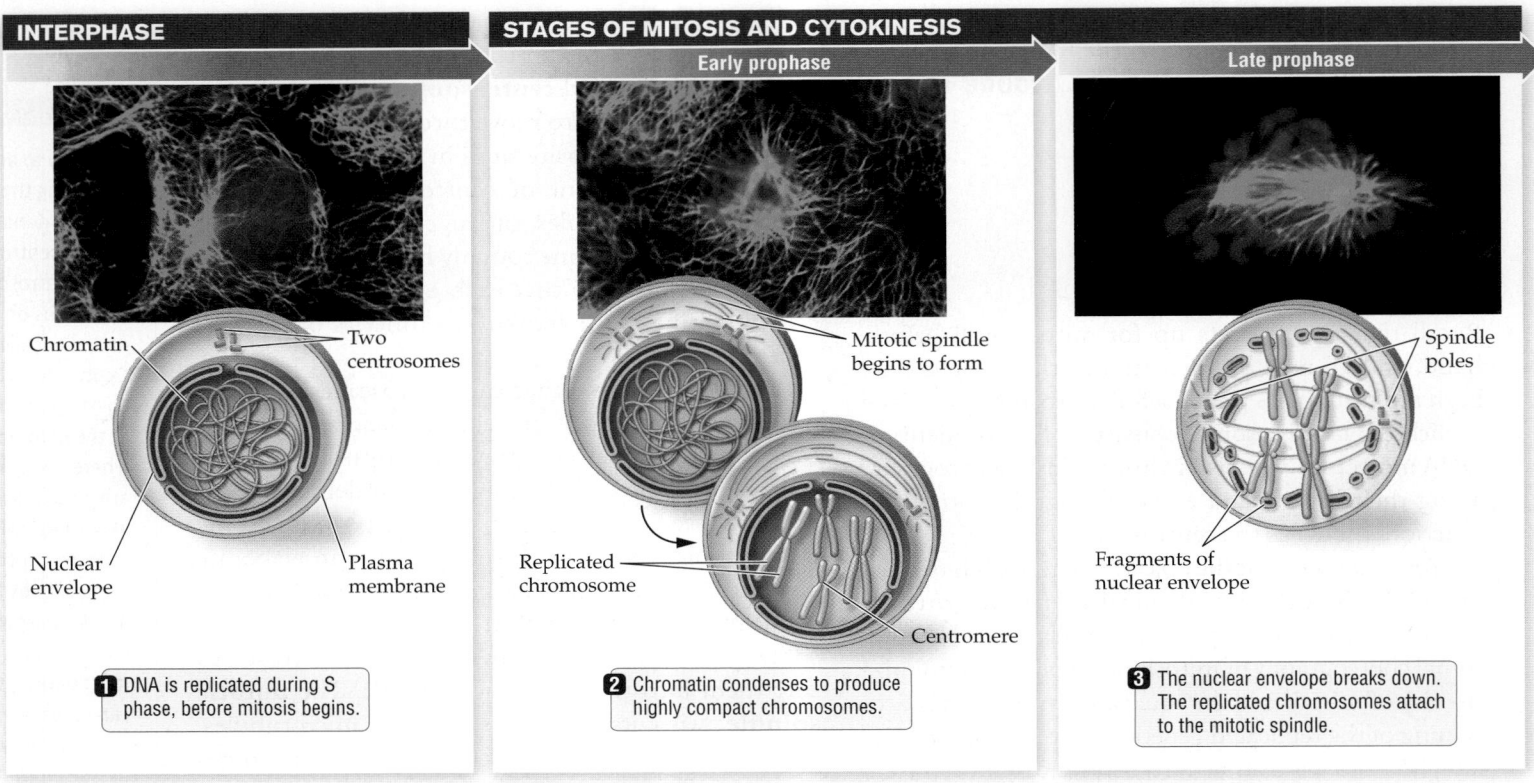

Chromatin

Two centrosomes

Nuclear envelope

Plasma membrane

**1** DNA is replicated during S phase, before mitosis begins.

Mitotic spindle begins to form

Replicated chromosome

Centromere

**2** Chromatin condenses to produce highly compact chromosomes.

Spindle poles

Fragments of nuclear envelope

**3** The nuclear envelope breaks down. The replicated chromosomes attach to the mitotic spindle.

**FIGURE 10.10** Mitosis and Cytokinesis Are the Two Main Stages of Mitotic Cell Division

## Chromosomes line up in the middle of the cell during metaphase

Once each replicated chromosome has become linked to both poles of the spindle, the length of the microtubule attachments is adjusted to move the chromosomes until they are arranged in a row. This stage of mitosis, when all the replicated chromosomes are arranged in a single plane, is called **metaphase** (*meta*, "after"). The plane in which the chromosomes are arranged is called the metaphase plate, which in most cells lies at the center (see Figure 10.10). The function of the elaborate mitotic machinery—mitotic spindle, chromosomes, and centrosomes—is to align each replicated chromosome in the proper position and to facilitate the equal and balanced segregation of chromatids to opposite poles of the cell.

## Chromatids separate during anaphase

During the next phase of mitosis, called **anaphase** (*ana*, "up"), the two chromatids in each pair of sister chromatids break free from each other. They are then dragged to opposite sides of the parent cell, resulting in the equal and orderly partitioning of the replicated genetic information. At the beginning of anaphase, the sister chromatids separate as the special proteins holding them together are broken down. Once separated, each chromatid is considered a new chromosome. The progressive shortening of the microtubules pulls the new chromosomes to opposite poles of the cell. This remarkable event results in equal segregation of the replicated chromosomes into the two daughter cells, paving the way for telophase and cytokinesis.

## New nuclei form during telophase

The next phase of mitosis, **telophase** (*telo*, "end"), begins when a complete set of daughter chromosomes arrives at a spindle pole. Major changes also occur in the cytoplasm—changes that set the stage for division of the cytoplasm, and the cell as a whole, to create two new cells. The spindle microtubules break down, and the nuclear envelope begins to form around the chromosomes that have arrived at each pole (see Figure 10.10). As the two new nuclei become increasingly distinct in the cell, the chromosomes within them start to unfold, becoming less distinct under the micro-

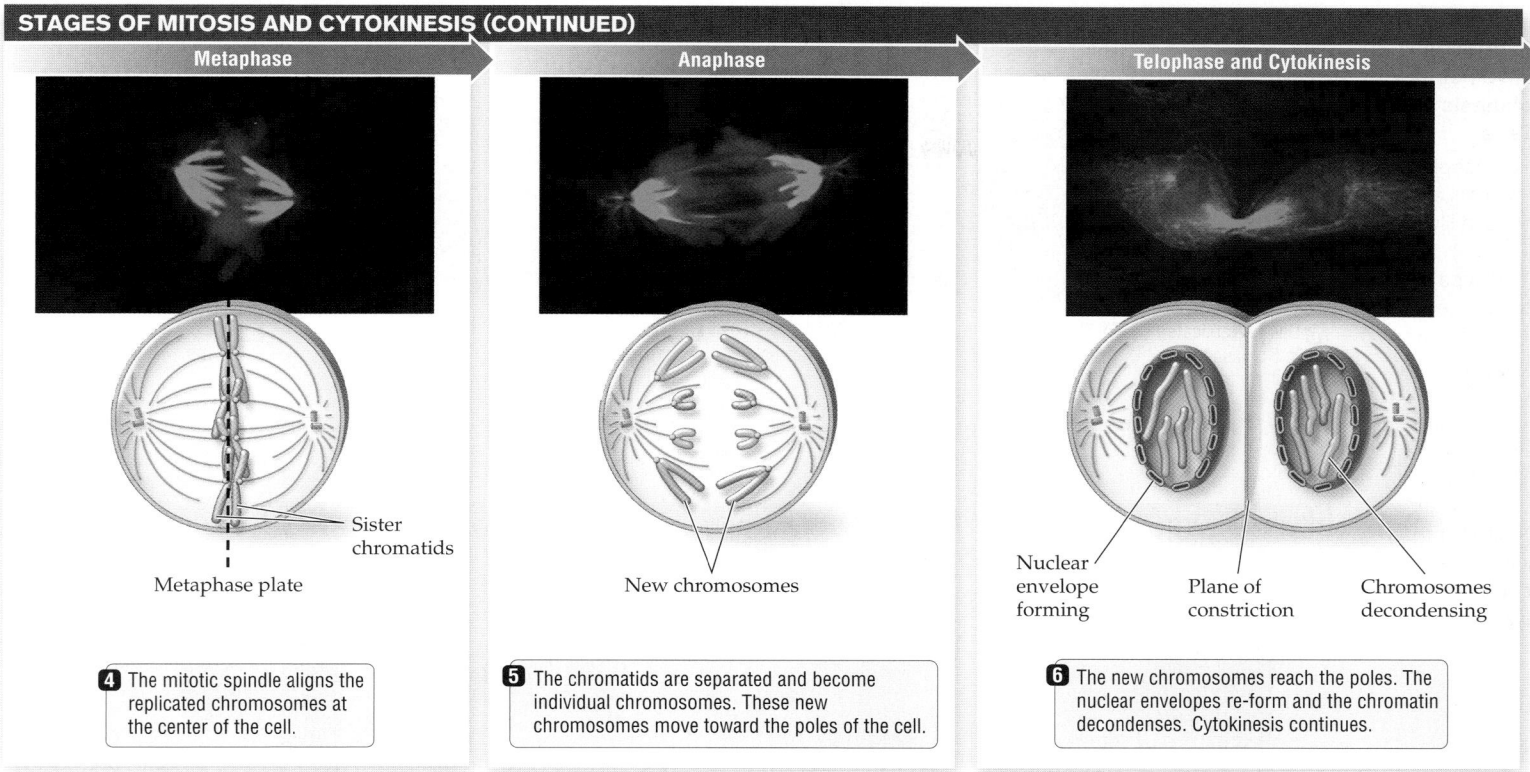

**Metaphase**

Sister chromatids

Metaphase plate

**4** The mitotic spindle aligns the replicated chromosomes at the center of the cell.

**Anaphase**

New chromosomes

**5** The chromatids are separated and become individual chromosomes. These new chromosomes move toward the poles of the cell.

**Telophase and Cytokinesis**

Nuclear envelope forming

Plane of constriction

Chromosomes decondensing

**6** The new chromosomes reach the poles. The nuclear envelopes re-form and the chromatin decondenses. Cytokinesis continues.

scope. Telophase is the last stage of mitosis, and cytokinesis begins even before telophase is quite complete.

## The cytoplasm is divided during cytokinesis

Cytokinesis (*cyto*, "cell"; *kinesis*, "movement") is the process of dividing the parent cell cytoplasm into two daughter cells. Animal cells divide by drawing the plasma membrane inward until it meets in the center of the cell, separating the cytoplasm into two compartments (**FIGURE 10.11**). The physical act of separation is performed by a ring of protein cables made of actin microfilaments. These form against the inner face of the plasma membrane like a belt at the equator of the cell. When the actin ring contracts, it draws in the plasma membrane, eventually pinching the cytoplasm and dividing it in two. Since the plane of constriction lies in the plane of the metaphase plate between the two newly formed nuclei, successful cytokinesis results in two daughter cells, each with its own nucleus.

A plant cell, however, has a relatively stiff cell wall around it that cannot be pulled in like the mouth of a drawstring bag. Plant cells achieve cytokinesis by erecting two new plasma membranes, separated by cell wall material, between the two newly segregated nuclei. Guided by cytoskeletal structures, a partition known

as a **cell plate** appears where the metaphase plate had been. The cell plate, consisting mostly of membrane vesicles, starts forming in telophase (**FIGURE 10.12**). Vesicles filled with cell wall components start to accumulate in the region that was previously the metaphase plate. These vesicles join together, fusing their membranes and mingling their cell wall components (mostly polysaccharides and some protein) to create two new plasma membranes separated by a newly formed cell wall.

Cytokinesis marks the end of the cell cycle. Once it is completed, the resulting daughter cells are free to enter the $G_1$ phase and start the process anew, or to differentiate into a specialized cell type and perhaps take a rest from cell division by entering the $G_0$ phase.

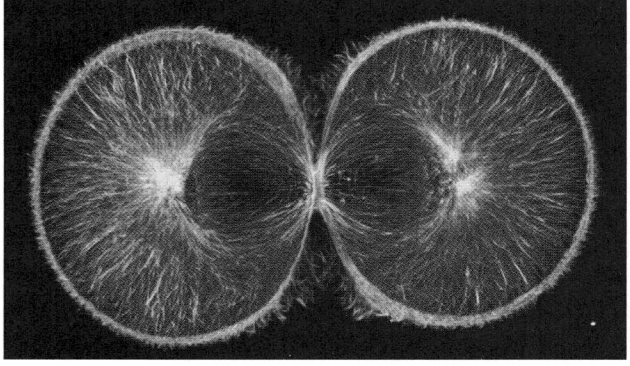

**FIGURE 10.11
Cytokinesis in an Animal Cell**

This fluorescence image shows cytokinesis in a sea urchin zygote that is dividing into two cells. Microtubules are orange, actin filaments blue.

(*a*) A microscopic
view of mitosis and
cytokinesis in lily
pollen. The cell plate
appears as a pale
line in the center of
the cell, in the last
photograph in the
series (telophase).
(*b*) A diagrammatic
view of the main
events in mitosis
and cytokinesis in
a plant cell. Plant
cells lack prominent
centrosomes, but have
structures that perform
the same function.

**CELL DIVISION IN PLANTS**

(*a*)

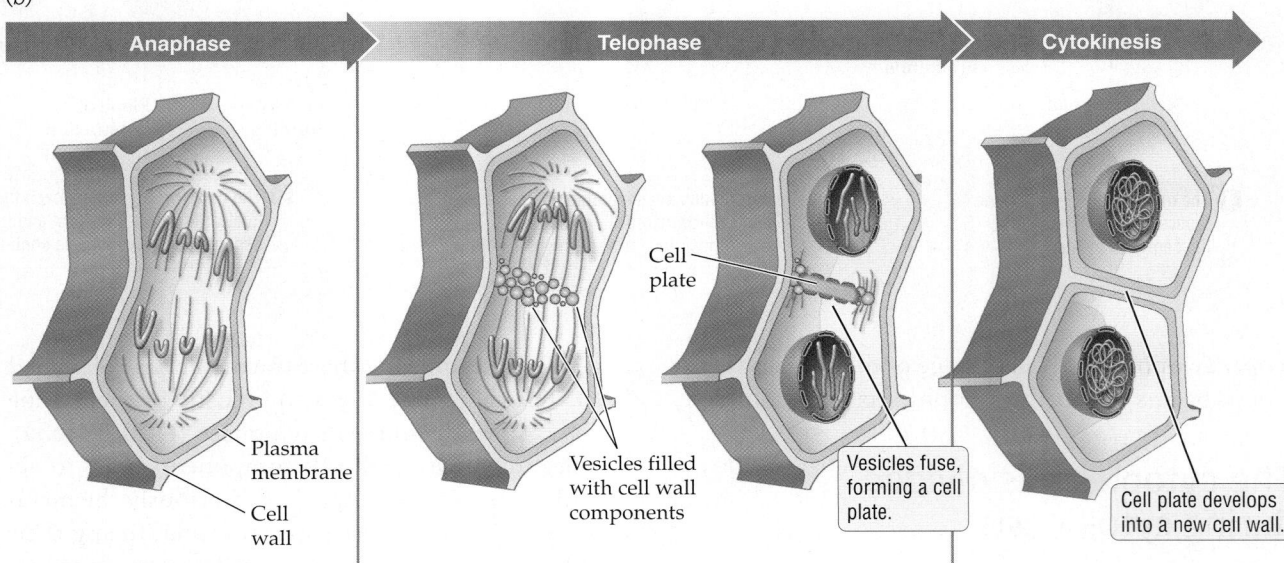

| Interphase | Early prophase | Late prophase | Metaphase | Anaphase | Telophase |
|---|---|---|---|---|---|
| **1** DNA is replicated in S phase. | **2** The chromosomes condense. The spindle assembles. | **3** The nuclear envelope breaks down. | **4** The spindle arranges chromosomes at the cell center. | **5** The sister chromatids separate. | **6** The cell plate forms. |

(*b*)

Anaphase → Telophase → Cytokinesis

Plasma membrane

Cell wall

Vesicles filled with cell wall components

Cell plate

Vesicles fuse, forming a cell plate.

Cell plate develops into a new cell wall.

Concept Check
Answers

slleɔ ɹǝʇɥbnɐp oʍʇ oʇui
'lləɔ əloɥʍ əɥʇ
ʎlǝʇɐɯıʇln pɐ 'ɯsɐldoʇʎɔ
əɥʇ ɟo buıuoıʇıʇɹɐd əɥʇ
sı sısǝuıʞoʇʎɔ ;uoısıʌıp
ɹɐǝlɔnu sı sısoʇıW **.3**

'lləɔ ɹǝʇɥbnɐp
ʇuǝɹǝɟɟıp ɐ oʇ pɐpɐǝɥ
ɥɔɐǝ 'sǝɯosoɯoɹɥɔ ʇuǝp
-uǝdǝpuı oʍʇ əɯoɔǝq
oʇ ǝʇɐɹɐdǝs ʎǝɥʇ uǝɥʍ
'ǝsɐɥdɐuɐ lıʇun sısoʇıɯ
buıɹnp pǝɥɔɐʇʇɐ uıɐɯǝɹ
ʇɐɥʇ 'ǝɯosoɯoɹɥɔ
lɐʇuǝɹɐd ǝlbuıs ɐ ɯoɹɟ
pǝʇɐɔıldǝɹ 'sǝlnɔǝloɯ
∀NG lɐɔıʇuǝpı oʍʇ ǝɹɐ
spıʇɐɯoɹɥɔ ɹǝʇsıS **.2**

'97 (p)
:97 (ɔ) :8ε (q) :8ε (ɐ) **.1**

> **Concept Check**
>
> 1. The house cat karyotype displays a total of **38** chromosomes. How many separate DNA molecules are present in a cat skin cell toward the end of (a) $G_0$ phase; (b) $G_1$ phase; (c) S phase; (d) $G_2$ phase?
>
> 2. What are sister chromatids? When during mitosis do they separate to become chromosomes in their own right?
>
> 3. What is the difference between mitosis and cytokinesis?

## 10.5 Meiosis: Halving the Chromosome Set to Make Gametes

As noted already, meiosis is a special type of cell division that produces daughter cells with half the chromosome count of the parent cell. Gametes are the

only cells in the animal body that are created through meiosis.

## Gametes contain half the chromosomes found in somatic cells

Sexual reproduction requires the fusion of two gametes in the process of fertilization. The successful union of an egg and a sperm creates a single-celled zygote, which develops into the multicellular embryo (**FIGURE 10.13**). Sexual reproduction produces offspring that are genetically different from their parents and siblings.

If the sperm and the egg both contained a complete set of chromosomes (46 for humans), the resulting zygote would have twice that chromosome number (92 for humans), and this karyotype would be passed down to all the cells of the embryo through mitotic divisions. Such an embryo would have twice as many chromosomes as either of its parents had, and therefore twice the number of genes characteristic of the species. The outcome of this genetic excess would be developmental chaos, generally resulting in death of the embryo. Therefore, for offspring to have the same karyotype as their parents, fertilization must yield the normal number of chromosomes in the zygote (46 chromosomes in the case of humans).

The simple solution to this problem is for the gametes to contain half of the full set of chromosomes found in somatic cells. Recall that somatic cells in eukaryotes possess *two copies*, or one homologous pair, for each type of chromosome found in the organism. If *only one copy* from every homologous pair is inherited by a gamete, the chromosome set is halved, and the gamete now possesses only *one copy* of all the genetic information. In humans, for example, all somatic cells contain 23 homologous pairs, for a total of 46 chromosomes. Each gamete a person produces, however, contains only one chromosome from each homologous pair, for a total of 23 chromosomes per gamete. Where the sex chromosomes are concerned, all eggs produced by a woman normally contain a single X chromosome, while 50 percent of the sperm produced by a man contain an X chromosome and the rest carry a Y chromosome. Because gametes contain only one copy of each type of chromosome, instead of having double copies, gametes are said to be **haploid** (*haploos*, "single"). The symbol *n* is traditionally used to indicate the number of chromosomes in a haploid cell (see Figure 10.13). In a human gamete, $n = 23$. Somatic cells, which have twice the number of chromosomes as gametes have, are said to be **diploid** (*di*, "double") because they have $2n$ chromosomes—that is, two of every kind of chromosome.

Because each gamete contains the haploid (*n*) number of chromosomes, the zygote formed by fertilization will contain $2n$ chromosomes—that is, the diploid set of chromosomes. In humans, that means a complete set consisting of 23 homologous chromosome pairs (46 chromosomes in all), including one pair of sex chromosomes. Furthermore, each pair of homologous chromosomes in the zygote will consist of one chromosome received from the father (**paternal homologue**) and one from the mother (**maternal homologue**), as shown in Figure 10.13. The equal contribution of chromosomes by each parent is the basis for genetic inheritance. We will investigate the details of inheritance in Unit 3.

Meiosis occurs in two stages—*meiosis I* and *meiosis II*— each involving one round of nuclear division followed by cytokinesis (**FIGURE 10.14**). **Meiosis I** reduces the chromosome set by *separating each homologous pair* into two different daughter cells. **Meiosis II** *separates sister*

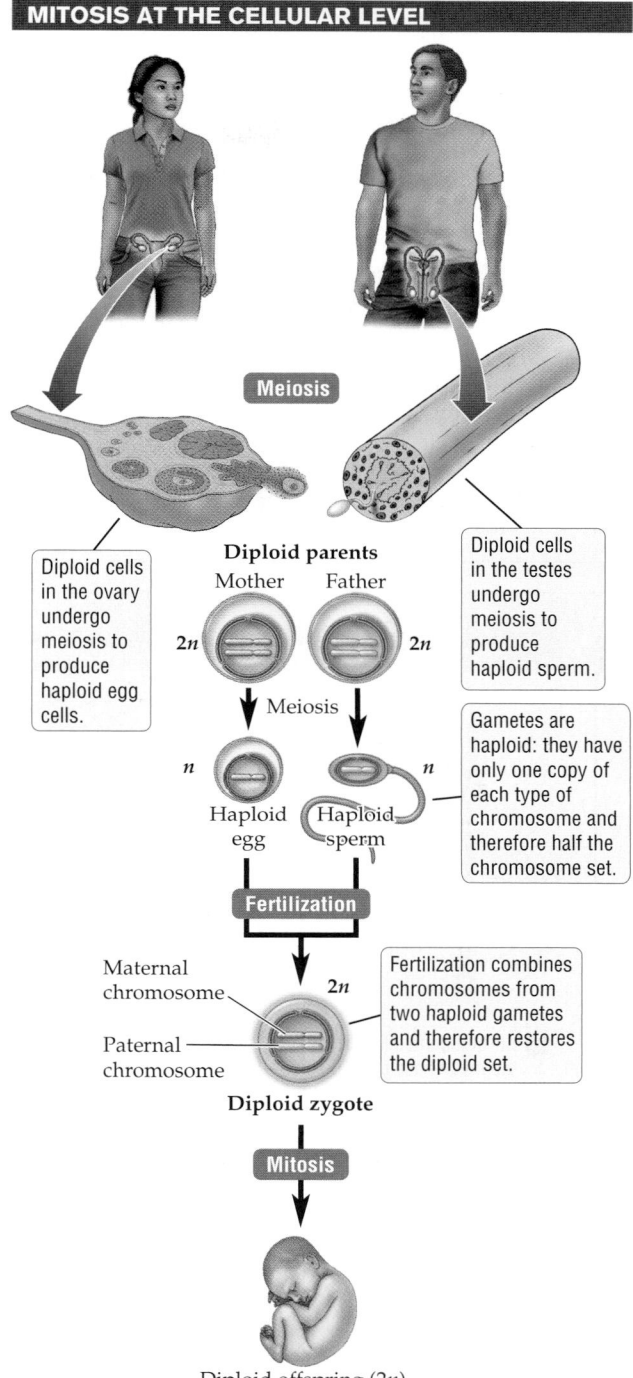

**MITOSIS AT THE CELLULAR LEVEL**

Meiosis

Diploid cells in the ovary undergo meiosis to produce haploid egg cells.

**Diploid parents**
Mother    Father

$2n$          $2n$

Diploid cells in the testes undergo meiosis to produce haploid sperm.

Meiosis

$n$          $n$

Gametes are haploid: they have only one copy of each type of chromosome and therefore half the chromosome set.

Haploid egg    Haploid sperm

**Fertilization**

Maternal chromosome

$2n$

Paternal chromosome

Fertilization combines chromosomes from two haploid gametes and therefore restores the diploid set.

**Diploid zygote**

**Mitosis**

Diploid offspring ($2n$)

**FIGURE 10.13**
**Sexual Reproduction Requires a Reduction in the Chromosome Set in Gametes**
The fusion of haploid sperm and egg at fertilization produces a zygote with the complete diploid ($2n$) chromosome set. Human somatic cells have 23 pairs of chromosomes. For clarity, only one homologous pair (consisting of a maternal and a paternal homologue) is shown here.

*chromatids* into two different daughter cells. The phases of meiosis are broadly similar to those of mitosis.

## Meiosis I reduces the chromosome number

Our tour of meiosis begins with the diploid cells in reproductive tissues that are responsible for the production of gametes. Well before meiosis begins,

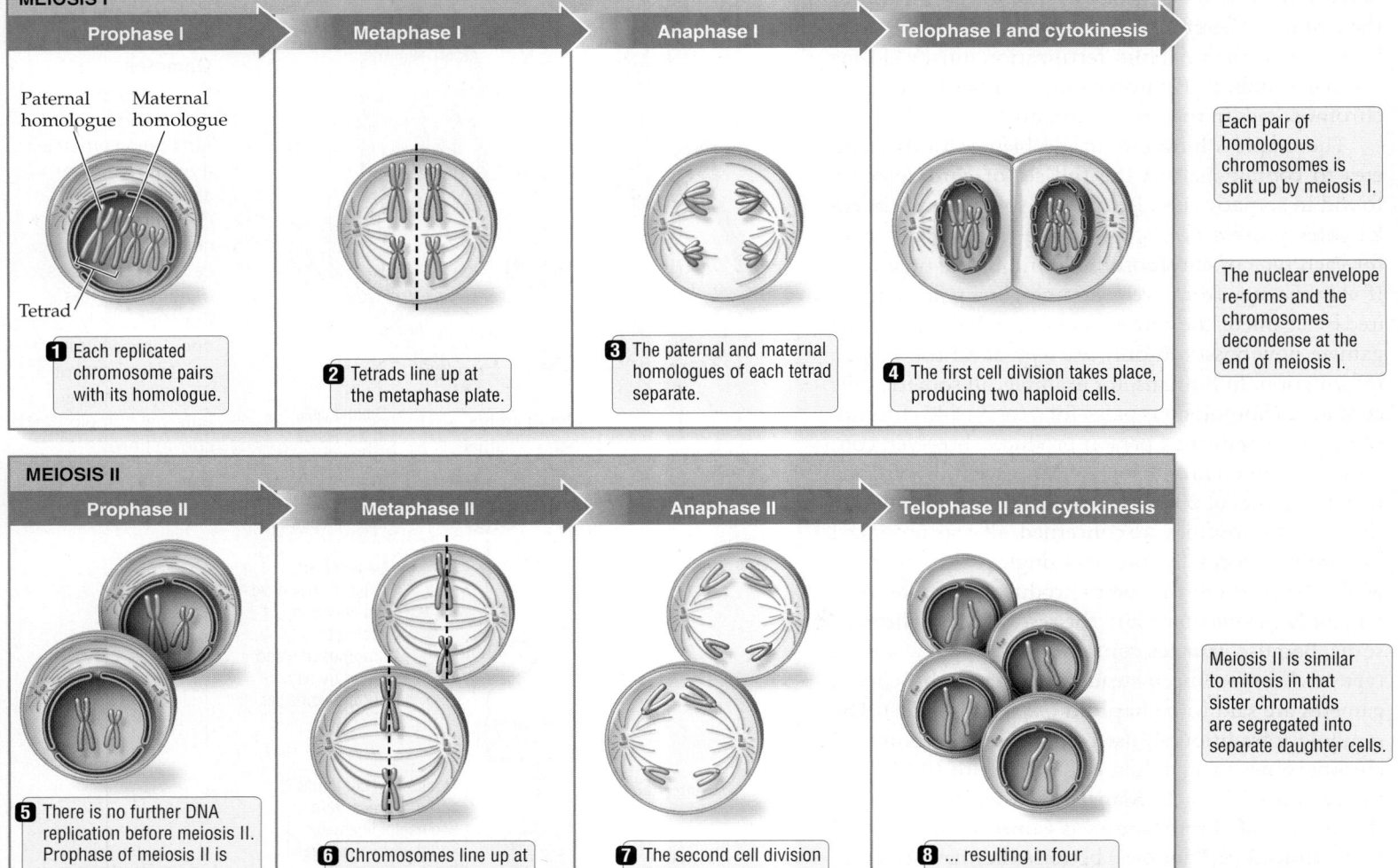

**MEIOSIS I**

**Prophase I**

Paternal homologue    Maternal homologue

Tetrad

**1** Each replicated chromosome pairs with its homologue.

**Metaphase I**

**2** Tetrads line up at the metaphase plate.

**Anaphase I**

**3** The paternal and maternal homologues of each tetrad separate.

**Telophase I and cytokinesis**

**4** The first cell division takes place, producing two haploid cells.

Each pair of homologous chromosomes is split up by meiosis I.

The nuclear envelope re-forms and the chromosomes decondense at the end of meiosis I.

**MEIOSIS II**

**Prophase II**

**5** There is no further DNA replication before meiosis II. Prophase of meiosis II is just like mitotic prophase.

**Metaphase II**

**6** Chromosomes line up at the metaphase plate.

**Anaphase II**

**7** The second cell division follows...

**Telophase II and cytokinesis**

**8** ... resulting in four haploid daughter cells.

Meiosis II is similar to mitosis in that sister chromatids are segregated into separate daughter cells.

**FIGURE 10.14  In Meiosis, Each Daughter Cell Receives Half the Chromosome Set**
The maternal and paternal homologues are paired during prophase I through metaphase I, and separated from each other during anaphase. Meiosis II is similar to mitosis in that the sister chromatids that compose each replicated chromosome are pulled apart. For simplicity and clarity, not all the steps described in the text are illustrated here.

all the chromosomes in the diploid precursor cell are replicated during S phase of the cell cycle. Therefore, as meiosis I begins, every replicated chromosome exists as two identical DNA molecules, with each DNA molecule constituting one chromatid. Two identical sister chromatids, bound to each other like Siamese twins, make up a *dyad*.

The first unique aspect of meiosis I, not seen at any stage of mitosis, is the coming together of each replicated homologous pair. In other words, early in meiosis I, each maternal homologue pairs off with its matching paternal homologue (see Figure 10.14). Furthermore, the sister chromatids in each replicated

chromosome (in each dyad) remain attached throughout meiosis I, instead of coming apart and going to opposite poles as they do in mitosis. It is the paternal and maternal homologues that are sorted into two separate daughter cells at the end of meiosis I. Put another way, mitosis brings about the separation and sorting of sister chromatids into different daughter cells, but meiosis I leads to the separation and sorting of each homologous chromosome pair so that daughter cells receive only one member of the pair.

The pairing off and orderly sorting of homologous chromosomes during meiosis I is what makes it possible for the resulting daughter cells to inherit exactly half the chromosome set of the parent cell. The paternal and maternal partners of each homologous chromosome pair align themselves next to each other during prophase of meiosis I, known as prophase I (see Figure 10.14). Each closely aligned pair of paternal and maternal homologues is collectively a **tetrad** (also known as a *bivalent*). This means that each tetrad, consisting of one replicated maternal chromosome and one replicated paternal chromosome, contains a total of *four* chromatids (four individual DNA molecules). At this point, an extraordinary process unfolds: the maternal and paternal members of the tetrad swap pieces of themselves! The exchange of genetic material between the non–sister chromatids in each pair of homologous chromosomes is brought about by a process called *crossing-over*, a subject we will return to after completing the tour of meiosis I and II.

Late in prophase I, a meiotic spindle develops and captures each tetrad. The microtubules from one centrosome attach themselves to only one member of the tetrad—either the maternal homologue or the paternal homologue, as shown in Figure 10.14. Next, in metaphase I, each tetrad is positioned at the metaphase plate. Anaphase I begins after all tetrads have been captured and positioned in one plane (the metaphase plate), generally at the equator of the cell. As the spindle microtubules begin to shorten during anaphase I, the paternal and maternal partners in each tetrad are pulled to opposite poles of the cell. Although this process looks superficially similar to anaphase of mitosis, during anaphase I it is homologous chromosome pairs—*not* sister chromatids—that are pulled apart and deposited at opposite poles of the cell.

After anaphase I of meiosis, the events of telophase I follow the same patterns seen in mitosis, with re-formation of the nuclear envelope and decompaction of the chromosomes. Cytokinesis of meiosis I results in two daughter cells, each with half the chromosome count of the parent cell. As we have seen, the chromosome set becomes halved because each daughter cell receives only one member of each homologous pair—either the maternal homologue (shown in pink in Figure 10.14) or the paternal homologue (shown in blue). Meiosis I is a *reduction division* because it halves the chromosome set, as one diploid parent cell ($2n$) becomes two haploid daughter cells ($n$).

## Meiosis II segregates sister chromatids into separate daughter cells

The two daughter cells formed at the end of meiosis I are not ready for prime time just yet: they cannot mature into gametes, because each of their chromosomes is still in a replicated state; each consists of two identical DNA molecules bound together as sister chromatids. Separating these sister chromatids into two different daughter cells is the sole purpose of meiosis II.

Each of the two haploid cells formed at the end of meiosis I go through a second round of nuclear and cytoplasmic divisions that makes up meiosis II. This time the phases of the division cycle are *almost exactly like those of mitosis*. In particular, sister chromatids separate at anaphase II, leading to an equal segregation of sister chromatids into two new daughter cells. In this manner, the two haploid cells produced by meiosis I give rise to a total of four haploid cells. The haploid cells differentiate into gametes containing half of the chromosome set found in the original diploid cell that underwent meiosis. As noted earlier, the reduction in chromosome number achieved through meiosis I offsets the combining of chromosomes when gametes fuse during fertilization. It is nature's way of maintaining the constant chromosome number of a species during sexual reproduction.

## Meiosis and fertilization contribute to genetic variation in a population

Meiosis and fertilization are the means of sexual reproduction in eukaryotes. Individuals in a population tend to be genetically different from each other because sexual reproduction leads to offspring that are genetically different not only from their parents but also from their siblings. You may resemble one or the other, or both, of your parents, but you cannot be genetically identical to either of them. Similarly, you may resemble a brother or sister, but you are not a clone of anyone, unless you have an identical twin. Genetic diversity in a population is important because genetic variation is the raw material on which evolution acts.

Where does the genetic variation in a population come from in the first place? *Mutations*, which are accidental changes in the DNA code, are the ultimate source of genetic variation in all types of organisms. The mutation of a given gene creates a different "flavor," or genetic variant, of that gene. Different versions of a particular gene, created ultimately through DNA mutations, are known as *alleles*.

Meiosis is exceptionally effective at shuffling the alleles that mutations create. We will see shortly that meiosis in a single individual can generate a staggering diversity of gametes, by shuffling alleles between homologous pairs and then sorting these scrambled

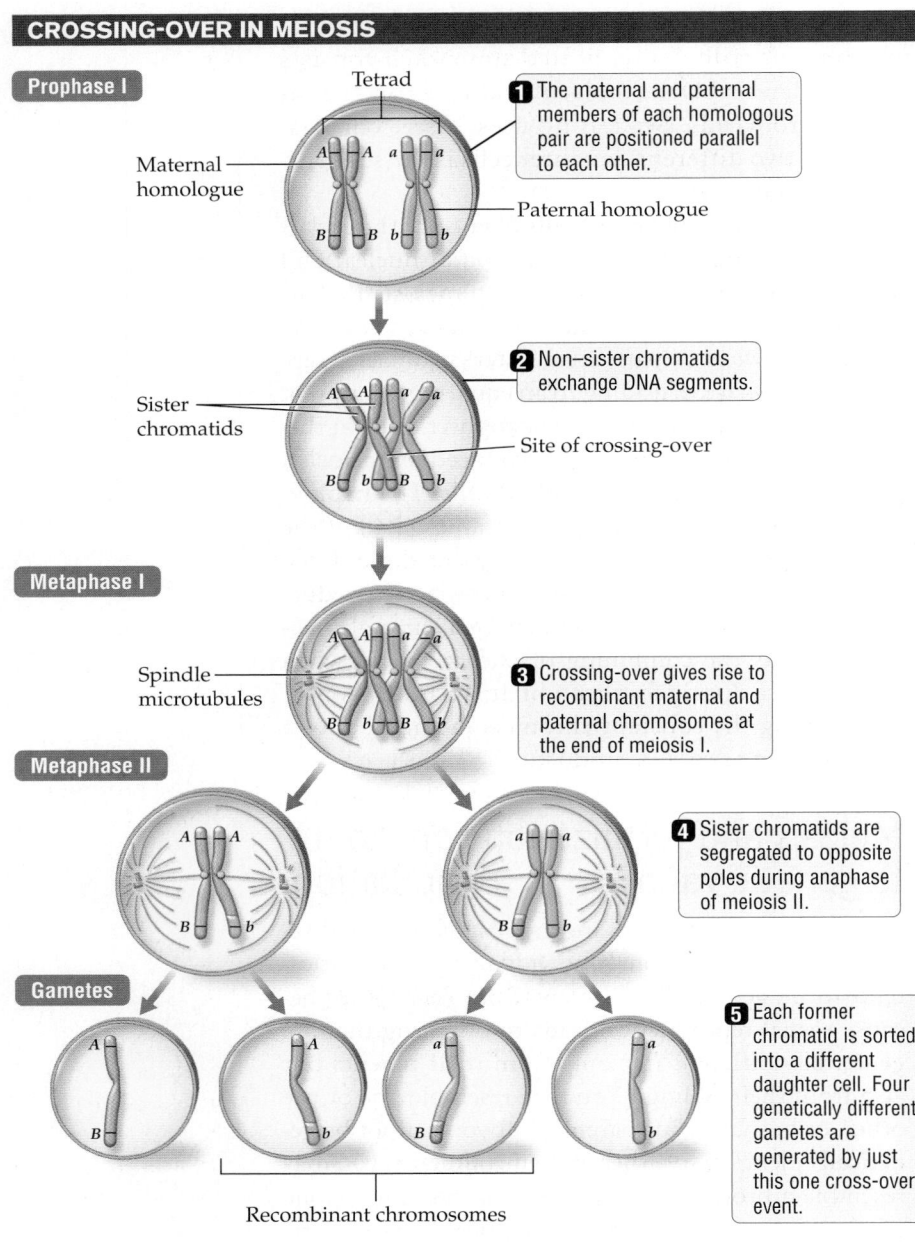

**Prophase I**

Tetrad

Maternal homologue

Paternal homologue

**1** The maternal and paternal members of each homologous pair are positioned parallel to each other.

Sister chromatids

**2** Non–sister chromatids exchange DNA segments.

Site of crossing-over

**Metaphase I**

Spindle microtubules

**3** Crossing-over gives rise to recombinant maternal and paternal chromosomes at the end of meiosis I.

**Metaphase II**

**4** Sister chromatids are segregated to opposite poles during anaphase of meiosis II.

**Gametes**

**5** Each former chromatid is sorted into a different daughter cell. Four genetically different gametes are generated by just this one cross-over event.

Recombinant chromosomes

**FIGURE 10.15 Crossing-over Produces Recombinant Chromosomes**

Crossing-over is the physical exchange of corresponding segments between the non–sister chromatids in a pair of homologous chromosomes that are aligned parallel to each other during prophase I. For clarity, only one maternal and one paternal chromatid are depicted here as exchanging segments. In human cells undergoing meiosis, most tetrads have one to three crossover sites, with longer chromosomes more likely to have multiple crossovers. A crossover site can be located at any point along the length of the paired homologues (tetrad), not just at the tips of the chromatids. The letters (A/a and B/b) represent alternative alleles of two genes, A and B. Note that the parental combinations of these alleles have been shuffled in the recombinant chromosomes.

homologues *randomly* into gametes. The randomness of fertilization adds to genetic diversity in sexually reproducing populations. Entirely new combinations of genetic information are created when an egg with a unique genetic makeup fuses with one of many genetically diverse sperm cells.

## Crossing-over shuffles alleles

Meiosis generates genetic diversity in two ways: *crossing-over* between the paternal and maternal members of each homologous pair, and *independent assortment* of the paternal and maternal homologues during meiosis I. We will consider crossing-over first.

**Crossing-over** is the name given to the physical exchange of chromosomal segments between non–sister chromatids in paired-off paternal and maternal homologues. Early in prophase I, every paternal and maternal homologue, now replicated and therefore consisting of two chromatids, finds its homologous partner and the two line up parallel to each other (see Figure 10.14, prophase panel). Crossing-over is initiated when a chromatid belonging to one homologue (say, the paternal one) makes contact with the chromatid across from it (belonging to the maternal homologue in this case). These *non*–sister chromatids contact each other at one or more random sites along their length. Special proteins at the crossover site act in such a way that they, in effect, exchange segments of the non–sister chromatids (**FIGURE 10.15**).

The swapped segments contain the same genes positioned in the same order. But, as we have seen, genes can exist in different versions, called alleles. Crossing-over exchanges alleles between the paternal and maternal chromatids. Therefore, chromatids produced by crossing-over are genetic mosaics, bearing new combinations of alleles compared to those originally carried by the paternal and maternal homologues in the diploid parent cell. The mosaic chromatid is said to be *recombined*, and the creation of new groupings of alleles through the exchange of DNA segments is known as **genetic recombination**. Without crossing-over, every chromosome inherited by a gamete would be just the way it was in the parent cell. Crossing-over between *just one* pair of homologous chromosomes enables meiosis to produce at least four genetically distinct gametes, as shown in Figure 10.15.

## The independent assortment of homologous pairs generates diverse gametes

The possibilities for creating genetically diverse gametes are not restricted to crossing-over. The **independent**

**assortment of chromosomes**—that is, the random distribution of the different homologous chromosome pairs into daughter cells during meiosis I—also contributes to the genetic variety of the gametes produced. It comes about because each homologous chromosome pair in a given meiotic cell orients itself independently—without regard to the alignment of any other homologous pair—when it lines up at the metaphase plate during meiosis I.

To understand why the independent alignment of homologous pairs produces random patterns in the subsequent sorting of paternal and maternal homologues, and therefore genetically varied gametes, consider a cell with just two homologous chromosome pairs ($n = 2$, $2n = 4$). During metaphase I, there are two ways of arranging each homologous pair at the metaphase plate: Option A would place the maternal homologue "to the left" and the paternal homologue "to the right" for *both* chromosome pairs. Option B would keep the first homologous pair in the same orientation as in option A but reverse the orientation of the second homologous pair, so this second couple would be oriented with the paternal homologue "on the left" and the maternal homologue "on the right." As **FIGURE 10.16** shows, option A would sort the maternal and paternal homologues of the two pairs in one particular pattern, creating two different types of gametes. Orienting the two homologous pairs accord-

ing to option B would produce a *different* pattern for the assortment of paternal and maternal homologues, leading to two types of gametes, both with a different combination of homologues than is seen in gametes created by option A.

What all this means is that four types of gametes, each with a different combination of paternal and maternal homologues, can potentially be generated through meiosis in a diploid cell containing just two pairs of homologous chromosomes ($n = 2$). How many types of gametes can be made by a meiotic parent containing three pairs of homologous chromosomes ($n = 3$)? Since there are two ways to arrange each of the three pairs of homologues, $2^3$ (or 8) patterns are possible; therefore, eight different types of gametes could be created.

What about meiosis in human cells, with $n = 23$? Since each of the 23 pairs of homologous chromosomes can be oriented one of two ways, there are $2^{23}$, or 8,388,608, different ways of combining homologues in the gametes. Of these 8,388,608 ways of mixing and matching homologues, only one is the combination found in the original diploid cell that underwent meiosis. As we saw in the case of crossing-over, the independent assortment of chromosomes creates gametes that are likely to be different from the parent, and also from each other, with respect to the mix of chromosomes they contain.

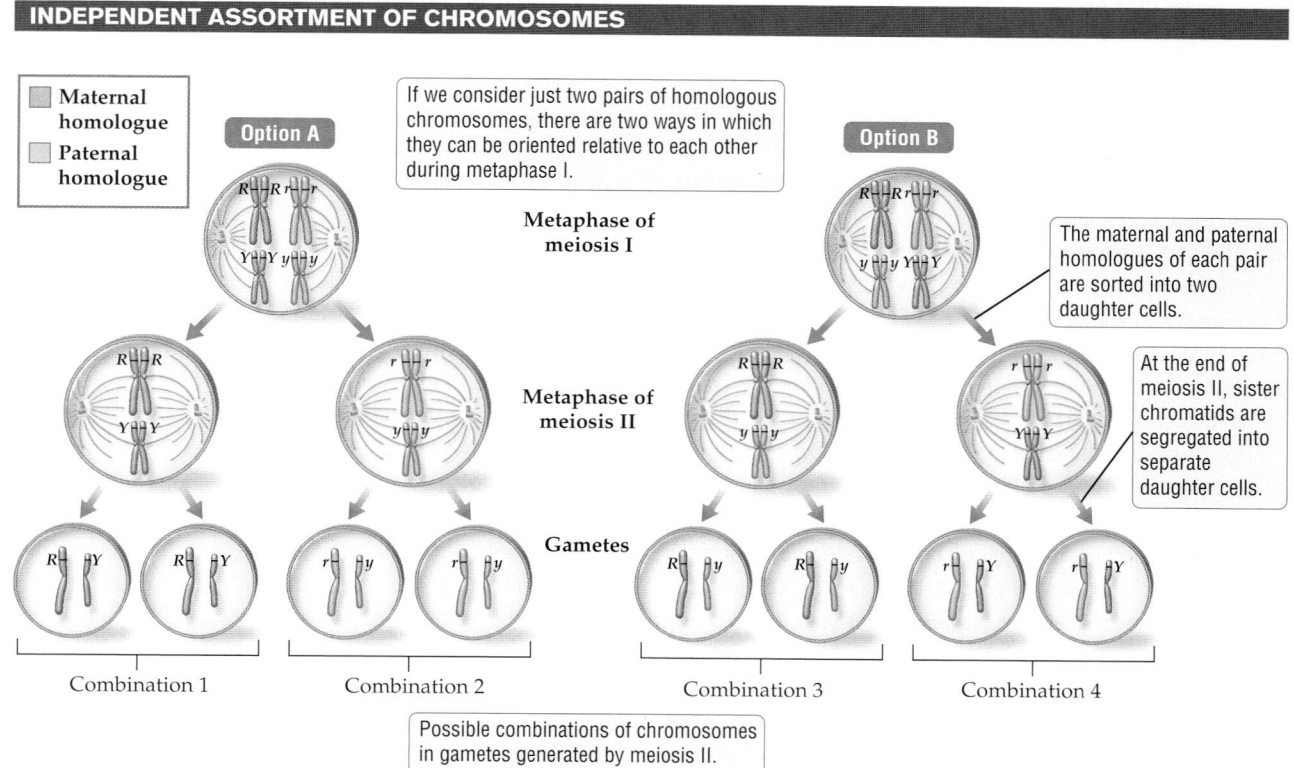

**INDEPENDENT ASSORTMENT OF CHROMOSOMES**

Maternal homologue
Paternal homologue

**Option A**

If we consider just two pairs of homologous chromosomes, there are two ways in which they can be oriented relative to each other during metaphase I.

**Option B**

Metaphase of meiosis I

The maternal and paternal homologues of each pair are sorted into two daughter cells.

Metaphase of meiosis II

At the end of meiosis II, sister chromatids are segregated into separate daughter cells.

Gametes

Combination 1    Combination 2    Combination 3    Combination 4

Possible combinations of chromosomes in gametes generated by meiosis II.

**FIGURE 10.16**
**The Random Assortment of Homologous Chromosomes Generates Chromosomal Diversity among Gametes**

## Concept Check Answers

1. Egg cells have a haploid number of chromosomes, which would be 19 (38 ÷ 2) for house cats (*n* = 19).

2. At the end of meiosis I, each daughter cell has 23 replicated homologues, and therefore 46 separate DNA molecules.

3. Meiosis II separates and symmetrically segregates the two sister chromatids of each duplicated chromosome.

4. Meiosis generates genetically diverse gametes through two different mechanisms: recombination of alleles via crossing-over, and random assortment of homologous chromosomes.

Finally, fertilization has the potential to add a tremendous amount of genetic variation to the variation already produced by crossing-over and the independent assortment of chromosomes. In the previous paragraph we noted that each gamete represents one of over 8 million possible outcomes of independent assortment during meiosis in humans. That is, every sperm cell and every egg represents one out of more than 8 million types with respect to chromosomal combinations. If we set aside the variation caused by crossing-over, this means there are more than 64 trillion genetically different possibilities for offspring (8 million possible sperm × 8 million possible eggs) each time a sperm fertilizes an egg. The total number of humans that have ever lived is estimated to be no more than 100 billion. Therefore, the chance that two siblings will be genetically identical is less than one in 64 trillion. (The exception is identical twins; because they arise from a single zygote, the chance that they will be genetically identical is 100 percent.) The random events in meiosis and the random fertilization of a certain egg by a particular sperm give each of us our genetic uniqueness.

## Concept Check

1. If the diploid number for house cats is 38, how many chromosomes are present in the egg cells they produce?

2. How many molecules of DNA are present in each daughter cell at the end of meiosis I in humans?

3. Meiosis I reduces the diploid chromosome set. What role does meiosis II play in gamete formation?

4. How does meiosis contribute to genetic variation in a population?

---

APPLYING WHAT WE LEARNED

# The Great Divide

At the beginning of this chapter, we learned about giant mats of green algae that covered vast areas of the Yellow Sea just weeks before the 2008 Summer Olympics sailing races. It was the world's largest drifting algal bloom. With only weeks left, Chinese Olympic organizers frantically mobilized heavy equipment, fishing boats, and soldiers to help clear the embarrassing mats of algae that clogged beaches and, most important, the Olympic racecourse. Even volunteer swimmers went out into the water to drag armloads of algae back to shore. In all, 130,000 people and 1,000 boats worked to collect and bury an estimated 1.5 million tons of algae less than a month before the Olympic Games. It was a dramatic accomplishment.

But where did all that algae come from? How could a few small, green mats at the beginning of June transform themselves into 1,500 square miles of living cells in just a few weeks? Even before the algal bloom was cleared, biologists identified the alga as a species of *Ulva*, a kind of "sea lettuce" that thrives in oceans all over the world. Most sea lettuces are edible, and many people eat them regularly in soups and salads. This particular species, *Ulva prolifera*, had grown from a few green mats in late May to 22 tons of biomass by early July. Most of the growth, researchers said, occurred in less than 2 weeks.

How was it that despite a massive effort, the slimy growth at first continued to outgrow all efforts to eliminate it? How did these tiny marine organisms do it? The answer is cell division. Through a combination of mitosis and meiosis, the algal cells doubled again and again, multiplying geometrically.

One thing that helps cells divide rapidly is nutrients. As mentioned in Section 10.2, nutrients are essential for cell growth during interphase. *Ulva prolifera* grows best in warm seas with lots of sunlight and nutrients. In China's agricultural regions, nitrogen and phosphorus from rice, fish, and pig farms flow down major rivers to the sea, flooding coastal seas with nutrients. Scientists have known for decades that such dense concentrations of nutrients are a major cause of algal blooms—a process called "eutrophication." But after the disastrous 2008 Olympic algal bloom, Chinese biologists took a closer look at what makes *U. prolifera* grow.

The researchers found that sea lettuce seems to like being shredded. In quiet, sheltered areas, individual leaves can grow as long as 15 inches, but in open waters, where waves tumble the leaves, they are typically much smaller. Tearing the leaves into smaller pieces seems to stimulate the cells along the torn edge to divide through meiosis to make haploid spores, which then grow into new individuals. In fact, *U. prolifera* shed spores most efficiently when they are torn up into bits and pieces no bigger than the star on a keyboard—whether by pounding waves, the propellers of boats, or sea lettuce–munching animals.

# Puzzle Solved: How a Fatherless Lizard Species Maintains Its Genetic Diversity

BY SINDYA N. BHANOO, *New York Times*

More than 40 years ago, Bill Neaves, then a young Ph.D. student, discovered how an all-female, asexual species of the whiptail lizard came to be. He found that the lizard was a cross between the female species of one type of lizard and the male species of another.

But what has puzzled him for years is how this all-female species maintains its high level of genetic variation, a contribution to evolutionary fitness that typically comes from sexual reproduction.

Despite reproducing without a male partner, this lizard species has a strong presence in the wild. Now, Aracely Lutes, a graduate student at the Stowers Institute for Medical Research in Kansas City, Mo., where Dr. Neaves works, has figured out the missing piece of the puzzle. The findings were reported Sunday in the journal *Nature*.

In a sexually reproducing lizard species, each lizard has 23 chromosomes from its mother, and 23 from its father.

During reproduction in the all-female species, Ms. Lutes found, all 46 of the mother's chromosomes are duplicated, resulting in 92 chromosomes in each egg cell.

Those chromosomes then pair with their identical duplicates, and after two cell divisions, a mature egg with 46 chromosomes is produced. Since crossing-over during the cell divisions occurs only between pairs of identical chromosomes, the lizard that develops from the unfertilized egg is identical to its mother.

Importantly, each new lizard is also a replication of the original. Since the original was genetically diverse, pulling its chromosomes from two different species, its carbon copy descendants are, too.

It is because of this that the species has thrived over the generations. But unlike sexually reproducing species, these lizards will probably never evolve into a stronger species, Dr. Neaves said.

They "may be really well suited for the desert Southwest," he said, "but when the next ice age comes, it may be that all of them get wiped out."

In Section 10.5, we saw that egg and sperm cells are created through a process called meiosis, in which a cell duplicates its chromosomes and then divides twice. Meiosis halves the usual number of chromosomes to produce four haploid (*n*) daughter cells, which can develop into eggs or sperm. Normal human cells have 46 chromosomes (23 pairs), but because of meiosis, eggs and sperm have only 23 chromosomes. When the haploid egg and sperm fuse at fertilization, each contributes 23 chromosomes to form a one-celled embryo with 46 chromosomes.

In this article we learn about a species of whiptail lizard that is parthenogenetic; it reproduces without sex, males, or sperm. All the members of this species are female. Scientists have known about parthenogenetic lizards for many years, but until now it wasn't clear how these lizards made a one-celled embryo with the right number of chromosomes.

Just as in humans, the diploid (2*n*) cells of these lizards have 46 chromosomes. But the cells that divide to form eggs begin meiosis with twice that many chromosomes: 92. That is, the cells are 4*n*. After meiosis, the resulting eggs are diploid, with 46 chromosomes, instead of haploid, with 23. Because these diploid egg cells already have the normal number of chromosomes, they do not need to fuse with a sperm before developing into young lizards.

How these whiptail lizards double the number of chromosomes still isn't clear. A couple of possibilities are that the germ cells go through two rounds of DNA replication just before meiosis; or, alternatively, that two germ cells fuse to form a 4*n* cell.

If sexual reproduction increases genetic diversity, how do parthenogenetic whiptails maintain genetic diversity? Researchers aren't completely sure. In this chapter we learned that new mutations can create lots of different versions of genes, called alleles. Meiosis reshuffles alleles, and fertilization combines the shuffled alleles into virtually unlimited combinations. The result is genetically unique offspring.

Whiptails have lots of alleles, and their genes can mutate even more, but neither new alleles nor preexisting ones get reshuffled. Normally, meiosis reshuffles alleles by crossing-over between the paternal and maternal homologues during meiosis I. In whiptail lizards, crossing-over occurs instead between genetically identical sister chromosomes.

## Evaluating the News

**1.** In the whiptail lizards described in this article, how many chromosomes are present in normal somatic cells, in germ cells about to begin meiosis, and in the egg cell?

**2.** Because of their genetic history and the way they reproduce, these whiptail lizards have lots of different versions of their genes, or alleles. Would you expect the daughters of a mother lizard to be very similar to one another or very different? Explain your reasoning.

**SOURCE:** *New York Times*, February 23, 2010.

# Summary

## 10.1 Why Cells Divide

- Cell division is necessary for growth and repair in multicellular organisms, and for asexual and sexual reproduction in all types of organisms.
- Many prokaryotes divide through binary fission, a form of asexual reproduction.
- Mitotic divisions produce daughter cells that are genetically identical to each other and to the parent cell.
- Meiosis is critical for sexual reproduction. In animals, the products of meiosis are sex cells, called gametes, that fuse during fertilization to give rise to offspring.

## 10.2 The Cell Cycle

- The cell cycle is the set sequence of events over the life span of a eukaryotic cell.
- Interphase and cell division are the two main stages of the cell cycle. Interphase is the longest, and consists of $G_1$, S, and $G_2$. DNA is replicated in S phase.
- The cell cycle is carefully regulated. Checkpoints ensure that the cycle does not proceed if conditions are not right.

## 10.3 The Chromosomal Organization of Genetic Material

- Each chromosome contains a single DNA molecule, bearing many genes, and compacted by packaging proteins.
- The somatic cells of eukaryotes have two of each type of chromosome, forming matched pairs called homologous chromosomes.
- One homologue in each homologous pair is inherited from the maternal parent, the other from the paternal parent. In mammals, the sex chromosomes (X and Y) determine gender. Females have two copies of the X chromosome; males have one X and one Y chromosome.

## 10.4 Mitosis and Cytokinesis: From One Cell to Two Identical Cells

- During mitotic cell divisions, the parent cell's replicated DNA is distributed in such a way that each daughter cell receives all the genetic information that was present in the parent cell.
- DNA replication produces two identical sister chromatids that are held together firmly at the centromere.
- The four main phases of mitosis are prophase, metaphase, anaphase, and telophase. Through these phases, the chromosomes of a parent cell are condensed, positioned appropriately, and separated to opposite ends.
- During cytokinesis, the cytoplasm of the parent cell is physically divided to create two daughter cells.

## 10.5 Meiosis: Halving the Chromosome Set to Make Gametes

- Meiosis—consisting of two rounds of nuclear and cytoplasmic divisions—produces haploid ($n$) gametes containing only one chromosome from each homologous pair.
- During meiosis I, the maternal and paternal members of each homologous pair are sorted into two different daughter cells.
- Meiosis II is similar to mitosis in that sister chromatids are segregated into separate daughter cells at the end of cytokinesis.
- Meiosis produces genetically diverse gametes through two means: crossing-over and the independent assortment of homologous chromosomes.
- Meiosis and fertilization introduce genetic variation in a population.

# Key Terms

adult stem cell (p. 233)
anaphase (p. 240)
asexual reproduction (p. 230)
binary fission (p. 231)
cancer cell (p. 235)
cell cycle (p. 233)
cell differentiation (p. 233)
cell division (p. 230)
cell plate (p. 241)
centromere (p. 237)
centrosome (p. 239)
chromatin (p. 237)
chromosome (p. 237)

crossing-over (p. 246)
cytokinesis (p. 232)
diploid (p. 243)
embryo (p. 233)
fertilization (p. 233)
fetus (p. 233)
$G_0$ phase (p. 235)
$G_1$ phase (p. 234)
$G_2$ phase (p. 234)
gamete (p. 232)
genetic recombination (p. 246)
germ line cell (p. 233)
haploid (p. 243)

homologous chromosome (p. 238)
independent assortment
  of chromosomes (p. 246)
interphase (p. 234)
karyotype (p. 238)
maternal homologue (p. 243)
meiosis (p. 232)
meiosis I (p. 243)
meiosis II (p. 243)
metaphase (p. 240)
mitosis (p. 232)
mitotic division (p. 232)
mitotic spindle (p. 239)

paternal homologue (p. 243)
prophase (p. 239)
S phase (p. 234)
sex chromosome (p. 238)
sexual reproduction (p. 230)
sister chromatids (p. 237)
somatic cell (p. 232)
telophase (p. 240)
tetrad (p. 245)
tumor (p. 235)
zygote (p. 233)

# Self-Quiz

1. In the cell cycle, DNA is duplicated in the
   a. $G_1$ phase.
   b. S phase.
   c. $G_2$ phase.
   d. division stage.

2. Which of the following statements is true?
   a. Chromatin is more compacted in prophase than during the $G_2$ phase.
   b. The key event of S phase is the segregation of sister chromatids.
   c. The mitotic spindle first appears during anaphase.
   d. The cell increases in size during metaphase.

3. Which of the following statements is *not* true?
   a. DNA is packaged into chromatin with the help of proteins.
   b. All chromosomes in the somatic cell of a particular species have the same shape and size.
   c. Each chromosome contains a single DNA molecule.
   d. Somatic cells in the animal body are diploid.

4. Which of the following correctly represents the order of the phases in the cell cycle?
   a. mitosis, S phase, $G_1$ phase, $G_2$ phase
   b. $G_0$ phase, $G_1$ phase, mitosis, S phase
   c. S phase, mitosis, $G_2$ phase, $G_1$ phase
   d. $G_1$ phase, S phase, $G_2$ phase, mitosis

5. In fertilization, gametes fuse to form
   a. a tetrad zygote.
   b. a haploid zygote.
   c. a diploid zygote.
   d. a triploid zygote.

6. Human gametes contain
   a. twice the number of chromosomes than our skin cells have.
   b. only sex chromosomes.
   c. half the number of chromosomes than our skin cells have.
   d. only X chromosomes.

7. The reduction division is
   a. prophase of mitosis.
   b. anaphase II of meiosis.
   c. metaphase II of mitosis.
   d. meiosis I.

8. Meiosis results in
   a. four haploid cells.
   b. two diploid cells.
   c. four diploid cells.
   d. two haploid cells

# Analysis and Application

1. What is the functional value of the regular checkpoints that are a feature of the cell cycle?

2. Horses have a karyotype of 64 chromosomes. How many separate DNA molecules are present in a horse cell that is in the $G_2$ phase just before mitosis? How many separate DNA molecules are present in each cell produced at the end of meiosis I in a horse?

3. Can anaphase take place before metaphase during mitosis? Why or why not?

4. Compare mitotic cell division to meiosis by completing the table on the right. Write "True" or "False" in each column as appropriate, following the example in the first row.

5. For sexually reproducing organisms, how would the chromosome numbers of offspring be affected if gametes were produced by mitosis instead of meiosis?

6. Compare and contrast meiosis I and meiosis II. Which of the two is more similar to mitosis and why?

7. Explain how crossing-over gives rise to recombinant chromosomes. At what stage of meiosis does crossing-over occur? Explain the biological significance of crossing-over.

| | MITOTIC CELL DIVISION | MEIOSIS |
|---|---|---|
| 1. In humans, the cell undergoing this type of division is diploid. | True | True |
| 2. A total of four daughter cells are produced when one parent cell undergoes this type of division. | | |
| 3. Our skin makes more skin cells using this type of cell division. | | |
| 4. The daughter cells are genetically identical to the parent cell in this type of division. | | |
| 5. This type of division involves two nuclear divisions. | | |
| 6. This type of division involves two rounds of cytokinesis. | | |
| 7. Maternal and paternal homologues pair up to form tetrads at some point during this type of division. | | |
| 8. Sister chromatids separate from each other at some point during this type of cell division. | | |

# Stem Cells, Cancer, and Human Health

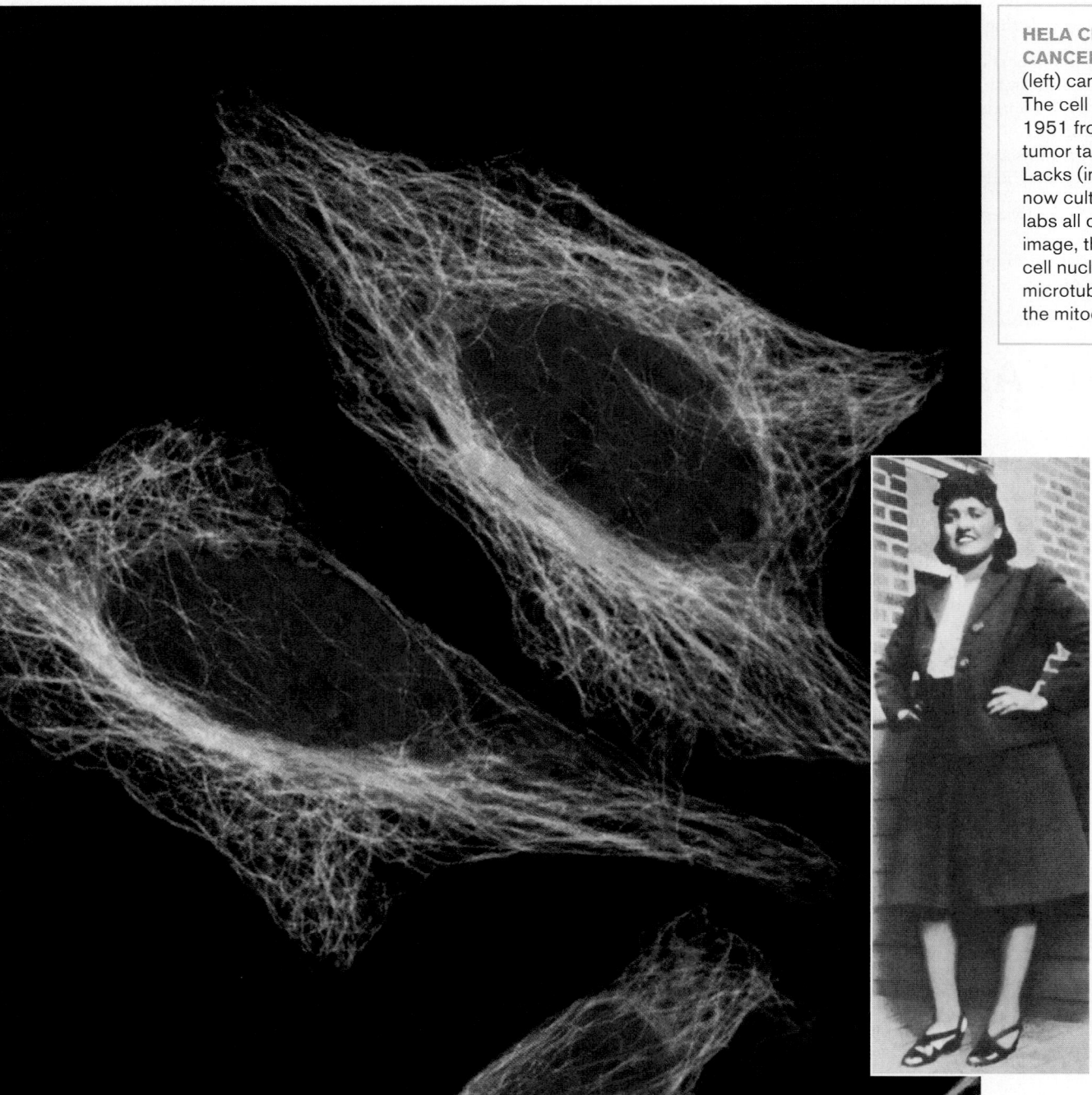

**HELA CELLS ARE A LINE OF CANCER CELLS.** HeLa cells (left) can grow in a lab dish. The cell line was established in 1951 from a slice of a cervical tumor taken from Henrietta Lacks (inset). HeLa cells are now cultivated in thousands of labs all over the world. In this image, the DNA in the HeLa cell nucleus is stained blue, the microtubules are green, and the mitochondria are red.

# Henrietta Lacks's Immortal Cells

In 1951, a poor, stay-at-home mother named Henrietta Lacks arrived at Johns Hopkins medical school to ask about bleeding at the wrong time of the month. Her doctor took one look at the bright purple tumor on Lacks's cervix— the entrance to the uterus—and feared the worst. He soon determined that she had malignant cervical cancer and broke the news to her.

Just 31 and mother to five children, Henrietta Lacks had little idea that she would not live to see her children grow up. But she would have been even more surprised to learn that the cancer cells growing in her body would end up in thousands of biomedical laboratories and grow vigorously decades after her death.

When Lacks returned to the hospital for treatment, a surgeon sewed envelopes of radioactive radium into her cervix, the state-of-the-art treatment of the day. Most of Lacks's cells were dividing at the slow, highly regulated pace that is normal for human cells. But her cancer cells were different.

Also at Johns Hopkins were two researchers, named George and Margaret Gey, who had been trying for years to "culture," or grow, human cells in test tubes so they could study their behavior. In 1951, all of the Geys' attempts had so far ended in complete failure, and they were eagerly trying to grow any human cells they could get their hands on. To help out, Lacks's surgeon took a slice of Lacks's tumor and delivered it to the Geys.

The cells from the tumor, called "HeLa" cells after Henrietta Lacks, were a dramatic success, turning out to be among the fastest-dividing human cells ever known and, more important, the first human cells ever grown in the lab outside a human body. But besides being able to survive in a test tube and divide quickly, the cells had one more amazing attribute: they were immortal; they would keep dividing seemingly forever.

What made Lacks's cells immortal? How have those cells contributed to our understanding of cancer biology?

We will return to these questions in the closing segment of this chapter.

---

**MAIN MESSAGE**  Stem cells are a source of new cells over the life of an individual; cancer cells divide without restraint and invade other tissues

## KEY CONCEPTS

- Stem cells are a self-renewing pool of undifferentiated cells.

- Embryonic stem cells have the greatest developmental flexibility because their descendants can differentiate into any of the cell types in the adult body. Cells derived from adult stem cells have a narrower developmental fate.

- Cell division is normally tightly controlled by a system of external and internal checks and balances. A cell that breaks loose of the normal restraints on cell division and starts dividing rapidly produces a mass called a benign tumor.

- Tumor cells that gain the ability to invade surrounding tissue are cancer cells. Cancer cells that metastasize—spread to other organs—are especially dangerous.

- Gene mutations are the root cause of all cancers. The great majority of human cancers are caused by the accumulation of multiple somatic mutations over the course of life.

- Environmental and lifestyle factors play a large role in human cancers.

THE CELL IS THE BASIC UNIT OF ALL LIFE on Earth. In the preceding chapters, we explored how cells are organized and how they obtain energy from their surroundings. We noted that life propagates itself through cell divisions, and cell divisions are also crucial for maintaining and growing the multicellular body. In this chapter we illustrate how an appreciation of cellular principles helps us understand our own bodies. We will see how cell division contributes to our health and well-being and how dysfunction in these processes can lead to serious illness. We begin with a brief review of how genes work and how eukaryotic cells divide and differentiate.

**DNA** is a molecule that stores genetic information as a precise sequence of four nucleotides, which are the building blocks of the genetic code. We could say DNA is the software that runs all the hardware of the cell. A **gene** is a segment of DNA that codes for a distinct inherited characteristic. Imagine the roughly 25,000 genes in human cells as many different applications built into one software package.

A human adult has 220 different cell types, all of them tracing back to a single fertilized egg cell. **Cell differentiation** is the developmental process by which a cell acquires a specialized function. During development, a cell is prompted by its internal genetic program and by the external cues it receives to differentiate into a particular cell type.

All cells in the human body have the same DNA sequence, but because **gene expression** varies among different cell types, a different set of genes is read out in every unique type. For example, a nerve cell (neuron), expresses a very different set of genes than does a white blood cell. Extending the computer analogy, the same DNA software is installed in all cells, but different cell types run a different mix of the available applications.

Every cell alive today arose from the division of a preexisting cell. In **mitotic division**, a parent cell divides to become two genetically identical daughter cells (see Section 10.4). Chromosomes (each one a highly packaged DNA molecule) have been replicated before mitosis begins; the two identical copies of each chromosome are methodically sorted to opposite ends of the parent cell during mitosis so that when the cell divides, one copy of every chromosome is inherited by each of the two daughter cells.

In this chapter we wrap up our discussion of cellular principles with a look at *stem cells* and *cancer cells*, both of which are profoundly important in human health and disease. Stem cells do nothing but divide, and in the process they shape the lives of embryos, fetuses, children, and adults. We explore some of the biomedical applications of stem cell research and examine why stem cell technology is sometimes controversial. Cancer is a story about good cells gone bad. Cancer cells not only divide with wild abandon, but they overrun their neighbors and commandeer resources such as oxygen and nutrients to the point of starving the hardworking normal cells in the body. We will examine the stages through which a tumor cell morphs into a cancer cell. We explore why cancer becomes more common as we get older, and what we can do to reduce our risk of cancer.

## 11.1 Stem Cells: Dedicated to Division

The job of stem cells is to divide. Some of the daughter cells go into the family business to become stem cells themselves—a process that is crucial for maintaining an ever-ready pool of stem cells. Some daughter cells, however, set out on a career path that turns them into highly specialized cell types.

Stem cells are crucial for human development and also for maintaining the body in adulthood. Stem cells helped grow you from a tiny ball of cells to an adult with about 10 trillion cells. Stem cells are found in a wide variety of tissues and organs in the human body, including bone marrow (**FIGURE 11.1**). Stem cells in the bone marrow divide regularly, and some of their descendants become red blood cells, or blood-clotting cells called platelets, or one of eight different types of immune cells whose job is to fight infection. Every day, skin stem cells deliver a fresh batch of descendants that differentiate into brand-new skin cells, replacing the cells that continually slough off the body surface. Surgeons can remove the diseased portion of a person's liver, and previously dormant stem cells in the remaining tissue will jump into action to regenerate most of the missing tissue.

The loss of stem cells contributes to such signs of aging as graying hair and a weakening of the immune system. The depletion of crucial stem cell populations may even put an upper limit on the maximum life span of humans, which is probably about 123 years. The

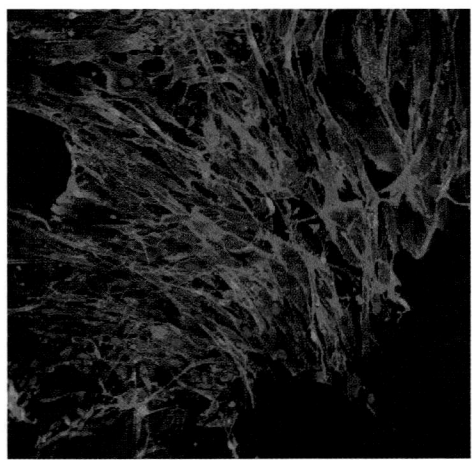

**FIGURE 11.1 Bone Marrow Stem Cells**

These bone marrow mesenchymal stem cells are among the most versatile of adult stem cells. The descendants of mesenchymal stem cells can become bone or cartilage cells. Other stem cells in the bone marrow give rise to red blood cells, white blood cells, and small cell fragments called platelets that are important in blood clotting.

manipulation of human stem cells holds enormous potential for advancing human health, but stem cell research has also stirred impassioned debate and political controversy. In this section we take a closer look at the special characteristics of stem cells, and the promise and controversy surrounding stem cell technology.

## Stem cells are a source of new cells

**Stem cells** are undifferentiated cells with the extraordinary attribute of self-renewal, which means that a stem cell can undergo mitotic divisions to make more of itself. Some of the daughter cells generated by mitotic division become stem cells themselves to maintain or grow the stem cell population. However, other daughter cells can graduate to the status of a highly specialized cell type, such as a skin cell or a fat cell (**FIGURE 11.2**). There are many different types of stem cells, and how a daughter cell turns out depends on the gene expression repertoire of its parent stem cell and the additional cues the cell receives from its physical and chemical environment.

There are two main classes of stem cells: *embryonic stem cells* and *adult stem cells*. **Embryonic stem cells**, found only in embryos, can potentially give rise to *all* the 220 cell types known to exist in the human body. **Adult stem cells**, by contrast, are undifferentiated cells that can yield some or all of the specialized cell types of the tissue or organ in which they are found. Brain stem cells give rise to neurons and two other types of brain cells (oligodendrocytes and astrocytes), but they cannot produce other cell types, such as red blood cells or kidney cells. You might say that embryonic stem cells are like a high school, graduating people in all walks of life, while adult stem cells are like a law school that produces different kinds of attorneys but does not award business or engineering degrees.

Human embryonic stem cells, and some types of human adult stem cells, can be grown in a laboratory procedure known as **cell culture**. When scientists spread the stem cells in a special nutrient broth in a plastic petri dish, the cells undergo mitotic divisions to make more stem cells. Exposing these cells to particular physical and chemical signals induces them to differentiate into specific cell types, such as skin cells or neurons.

## DIVERSE CELL TYPES FROM STEM CELLS

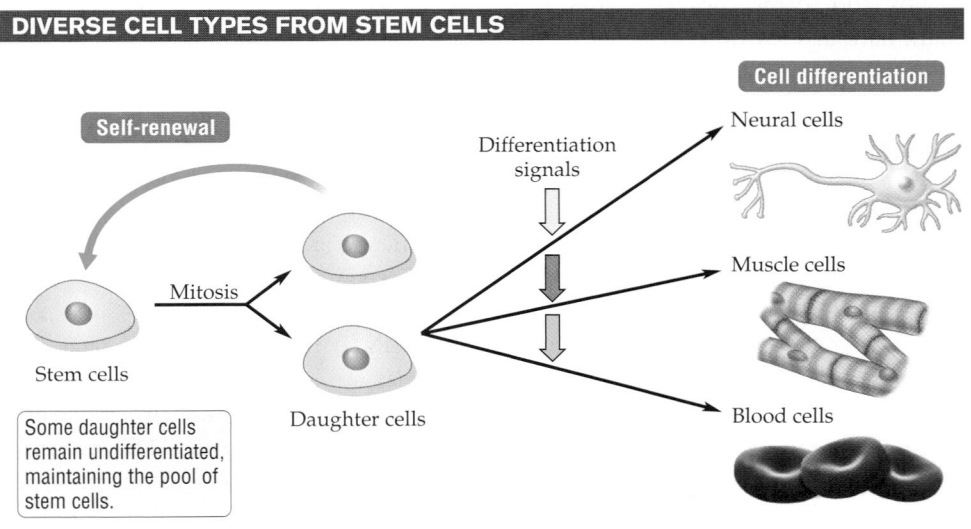

**FIGURE 11.2 Stem Cells Can Renew Themselves**

Stem cells are undifferentiated cells that have the unique capacity of self-renewal. Given the right mix of physical and chemical signals (differentiation signals), their descendants can differentiate into highly specialized cell types.

## Stem cell technology offers much hope and some successes

Stem cells have taught us much about cell biology and human development, although a great deal remains to be learned. The vast majority of the 220 different human cell types cannot be cultured in a lab dish and are therefore difficult to study in isolation. An important goal of stem cell research is to culture the appropriate stem cells in a lab dish and experiment with what it takes to get the cells to differentiate into a particular cell type. Such studies would help us understand the basic processes controlling cell division and differentiation and how alterations in those processes lead to birth defects or diseases like cancer.

Beyond basic research, human tissues made from stem cells can speed up drug discovery and lower the cost of developing new medications (Table 11.1). Human tissues raised from stem cells can be used to screen candidate drugs for effectiveness or toxicity. For example, a drug company can screen potential heart medications by trying them out on human heart tissue grown from stem cells in a lab dish. Researchers can test for potential side effects by looking for toxicity against other, nontarget, cell types generated from stem cells. Conventional drug testing in lab animals is time-consuming and expensive; the cost of testing a new drug in mice often runs to $3 million or so. Because mice are not people, drugs must be tested in humans as well before a pharmacy near you can dispense them; the price tag for large-scale human studies often amounts to $1 billion or more.

Stem cells offer a strategy for repairing or replacing tissues and organs damaged by injury or disease—a field known as **regenerative medicine**. Stem cell technology gives a ray of hope to patients with Parkinson's disease, like actor Michael J. Fox and former heavyweight boxing champion Muhammad Ali (**FIGURE 11.3**).

Parkinson's disease is a progressively disabling condition marked by uncontrolled body movements and speech impairments. The symptoms are caused by the death of dopamine-producing cells in a region of the brain that regulates movement. Dopamine is a neurotransmitter, a signaling molecule that enables brain cells to communicate with each other. Researchers cured Parkinson's disease in rats by implanting embryonic stem cells into the brains of affected animals; the stem cells differentiated into dopamine-producing cells and led to substantial recovery of motor function. We cannot hope to replicate this success in humans anytime soon, because there is a great deal more we need to learn about manipulating human stem cells. But in some instances the knowledge gap has narrowed and the successes have been spectacular.

Adult stem cells have been used for more than 50 years to treat people with some types of blood disorders, including blood cancers such as leukemia. Adult stem cells are also used in the treatment of severe burns. Skin stem cells can be isolated from any patch of healthy skin that survives on a burn victim, and under the right conditions in the lab the cells quickly generate sheets of skin tissue that can be grafted onto the patient's body. Because it comes from the patient's own stem cells, the lab-grown skin tissue is not attacked by the immune system.

Hundreds of clinical experiments are currently under way in the United States and other countries to test various stem cell therapies, mostly using adult stem cells, although some trials using human embryonic stem cells are also in progress. Stem cell therapy for patients newly diagnosed with type 1 diabetes has been particularly encouraging. The disease damages insulin-producing cells in the pancreas. In a recent trial, children with the disease were treated with adult stem cells that had been multiplied through cell culture. Most of the children were then able to produce enough insulin on their own and no longer needed injections of this vital hormone.

A number of recent trials report substantially improved heart function in heart attack patients treated with adult stem cells isolated from muscle tissue or blood vessels. Clinical trials began in July 2010 to see whether injecting certain cultured tissues (oligodendrocyte progenitor cells), raised in a dish from human embryonic stem cells, will help patients with spinal cord injuries. Hopes are high that stem cells can be used to treat Parkinson's disease, Alzheimer's disease, stroke, amyotrophic lateral sclerosis (ALS), muscular dystrophy, arthritis, Crohn disease, and various vision and hearing defects.

| TABLE 11.1 | Applications of Stem Cell Technology |
|---|---|

**Basic research**

Understanding cell division and differentiation

Understanding the biology of stem cells

Understanding the biology of cancer

Understanding human development

**Biomedical applications**

Drug development: testing new drugs in cultured human tissues

Regenerative medicine: repairing or replacing damaged or diseased tissues

Much more stem cell research is needed before the hoped-for treatments can be turned into reality. Human adult stem cells from organs other than skin, fat tissue, and bone marrow are usually difficult to obtain in sufficient quantity, since living humans are the only source; even when a fairly large tissue sample is available, adult stem cells are few in number, hard to recognize, and difficult to isolate from the surrounding tissue. Most adult stem cells are tricky to manipulate in a lab dish, mainly because we just do not know enough about cell differentiation to be able to control it at will in the many different types of adult stem cells. Pure masses of human embryonic stem cells can be obtained from donated embryos and endlessly multiplied in a dish; and given their broad developmental potential, they can teach us a great deal about differentiation in human stem cells. However, the use of these tissues in research or therapy is highly controversial. To understand why the debate over human embryonic stem cells is so bitterly polarized, we must examine where embryonic and adult stem cells come from.

## Embryonic stem cells are found only in very early stages of development

Each of us is a product of fertilization, the union of an egg and sperm that creates a single cell, the **zygote**. It takes several mitotic divisions to convert the zygote into a ball of cells, called a *morula* (Latin for "mulberry"). A subset of cells in that tiny ball turns into the **embryo**, the mass of cells that gives rise to the adult body. The cells in the embryo divide and differentiate swiftly, generating the main tissue types of all the organ systems of the body. The human embryo is regarded as a **fetus** once all the major organs have formed, about 3 months after fertilization. From that point until birth, the organs grow and mature through billions of cell divisions, followed by the differentiation of many of the resulting daughter cells.

In the first 3–4 days after fertilization, the morula is composed of **totipotent [toh-*TIP*-uh-tent]** cells, so called because they can give rise to *any* cell type in the organism, including the protective birth sac that surrounds the developing embryo in the case of mammals like us. As cell divisions continue, the morula is transformed into a hollow sphere, known as a **blastocyst**, consisting of about 150 cells. Inside this hollow ball is a group of about 30 cells, known as the **inner cell mass**, which gives rise to the embryo proper (**FIGURE 11.4**). The inner cell mass is **pluripotent [pluh-*RIP*-uh-tent]**,

**FIGURE 11.3 Still Fighting for a Cure**
About half a million Americans suffer from Parkinson's disease, including boxer Muhammad Ali and actor Michael J. Fox. The genes we inherit, and environmental influences such as exposure to pesticides or physical trauma, are both known to affect our risk of developing the disease. Testifying before a U.S. Senate subcommittee in 2002, the two men joined forces to plead for federal support of stem cell research.

which means these cells are capable of producing all of the cell types in the adult body but not, in the case of mammals, the tissues that make up the birth sac. The inner cell mass is composed entirely of embryonic stem cells, pluripotent stem cells that give rise to the embryo and can potentially generate any cell type in the adult body.

About 5–7 days after fertilization, the blastocyst attaches to the lining of the uterus (womb) in a process known as implantation. Mitosis continues at a rapid pace, but many of the daughter cells begin differentiating into specialized cell types as the three main tissue layers of the embryo emerge. About 10 weeks after fertilization, most of the organ systems are established and the embryo is transformed into a fetus. The developmental flexibility of most fetal cells declines as development proceeds, meaning that the cells become fated to differentiate into one of an increasingly narrow range of cell types. Small populations of cells in various tissues and organs are retained as adult stem cells.

Adult stem cells, also known as somatic stem cells, are said be **multipotent [mul-*TIP*-uh-tent]** or **unipotent [yoo-*NIP*-uh-tent]**: they can differentiate into a smaller repertoire of differentiated

**FIGURE 11.4**

Embryonic stem cells are pluripotent, and their descendants can therefore differentiate into a great variety of specialized cell types in response to appropriate molecular signals. Adult stem cells, present in low numbers in well-developed tissues and organs, normally give rise to a more limited range of cell types related to these tissues.

**EMBRYONIC AND ADULT STEM CELLS**

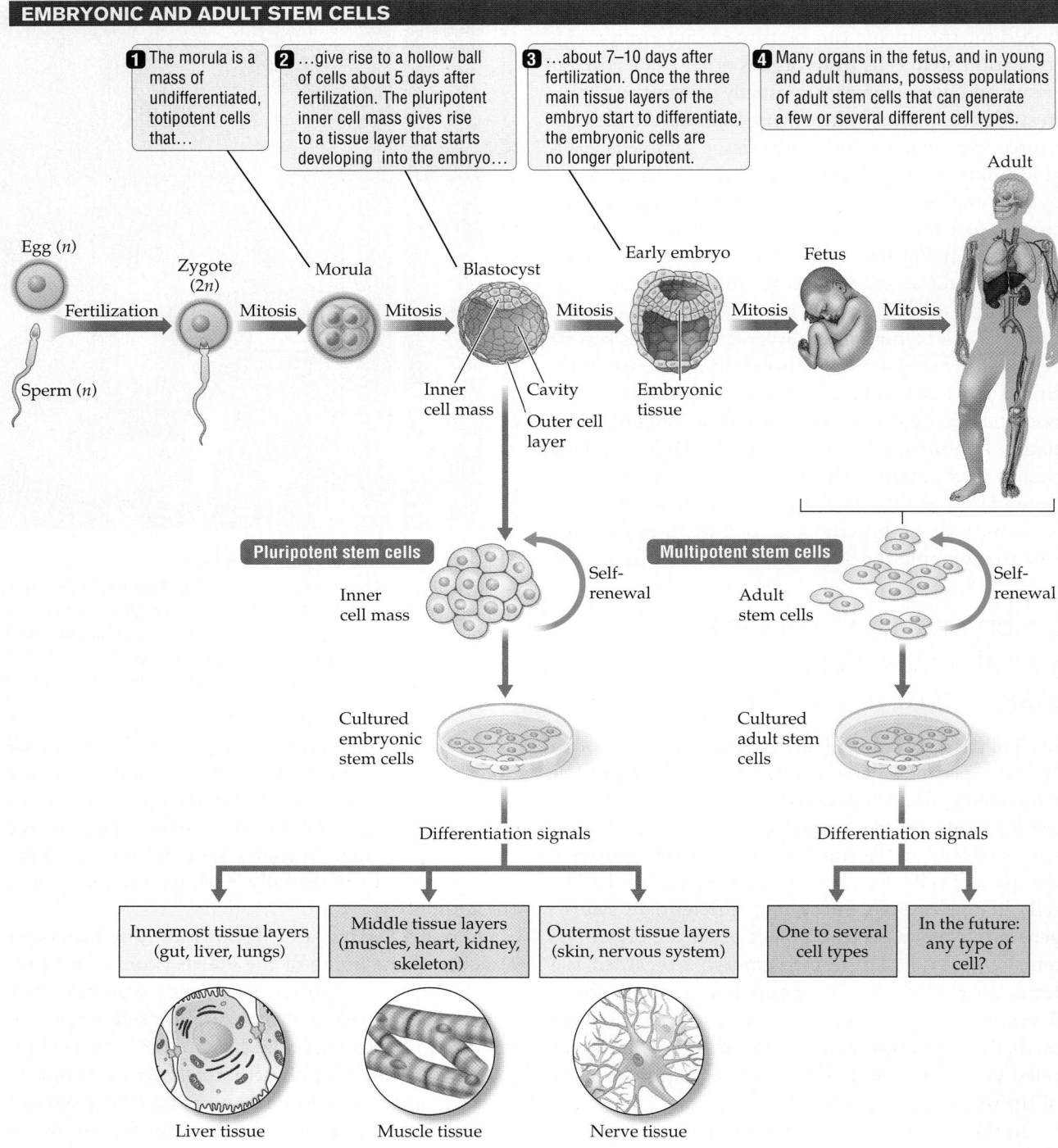

**1** The morula is a mass of undifferentiated, totipotent cells that…

**2** …give rise to a hollow ball of cells about 5 days after fertilization. The pluripotent inner cell mass gives rise to a tissue layer that starts developing into the embryo…

**3** …about 7–10 days after fertilization. Once the three main tissue layers of the embryo start to differentiate, the embryonic cells are no longer pluripotent.

**4** Many organs in the fetus, and in young and adult humans, possess populations of adult stem cells that can generate a few or several different cell types.

Egg (*n*)
Sperm (*n*)
Fertilization
Zygote (2*n*)
Mitosis
Morula
Mitosis
Blastocyst
Inner cell mass
Cavity
Outer cell layer
Mitosis
Early embryo
Embryonic tissue
Mitosis
Fetus
Mitosis
Adult

Pluripotent stem cells
Inner cell mass
Self-renewal
Cultured embryonic stem cells
Differentiation signals

Multipotent stem cells
Adult stem cells
Self-renewal
Cultured adult stem cells
Differentiation signals

Innermost tissue layers (gut, liver, lungs)
Middle tissue layers (muscles, heart, kidney, skeleton)
Outermost tissue layers (skin, nervous system)

One to several cell types
In the future: any type of cell?

Liver tissue
Muscle tissue
Nerve tissue

cell types (multipotent) and, in some cases, only one cell type (unipotent). Adult stem cells lack the *complete* developmental flexibility of the totipotent cells in the morula, and the *nearly complete* flexibility of the pluripotent inner cell mass. Adult stem cells are known to exist in many human organ systems, including skin, gut, liver, brain, and heart (**FIGURE 11.5**).

## The use of embryonic stem cells is controversial

The inner cell mass of a blastocyst is the only source of embryonic stem cells. Scientists succeeded in culturing embryonic stem cells from mouse blastocysts as far back as 1981. But human embryonic stem cells would not be cultured until 1998, both because of the

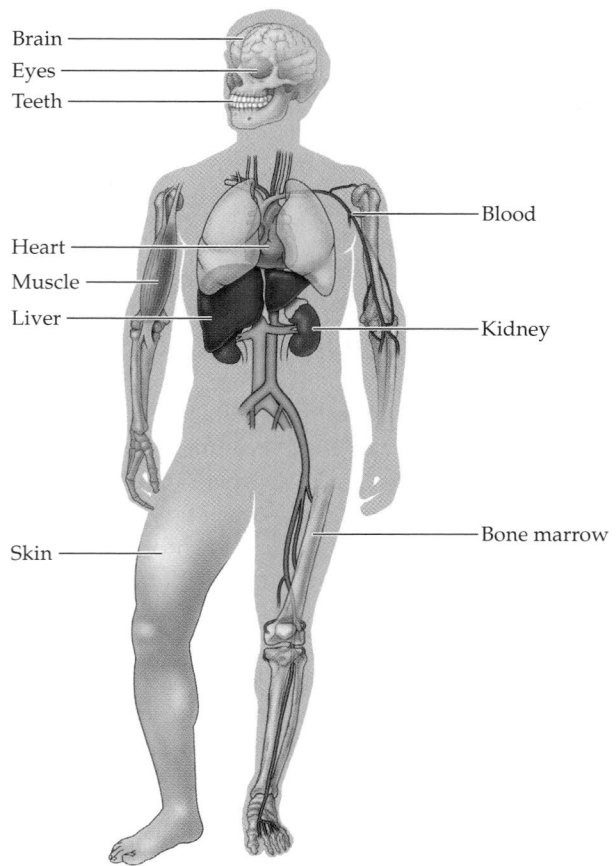

Brain
Eyes
Teeth

Blood

Heart
Muscle
Liver

Kidney

Skin

Bone marrow

**FIGURE 11.5 Small Numbers of Adult Stem Cells Are Found in Certain Tissues and Organs in the Human Body**
These cells are multipotent or unipotent, rather than pluripotent like embryonic stem cells. Adult stem cells from most organs (except skin and bone marrow) are harder to identify, isolate, and culture in a lab dish, compared to embryonic stem cells derived from the inner cell mass.

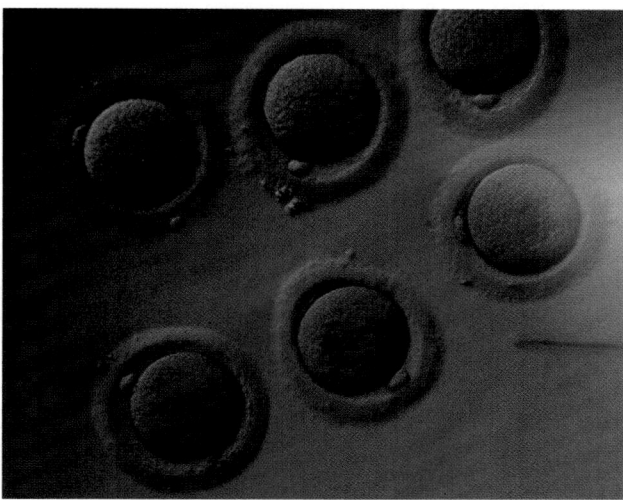

**FIGURE 11.6 Human Embryonic Stem Cells**
The only source of human embryonic stem cells is the pool of very-early-stage human embryos (blastocysts) donated by couples using in vitro fertilization technology. The harvesting of the inner cell mass destroys the embryo.

technical challenges and because human embryos—the only source of pluripotent cells—were not readily available. In the 1990s, donated human embryos became available with the advent of in vitro fertilization (IVF), the technology behind "test tube babies."

In IVF, egg and sperm are combined in a lab dish and development is allowed to proceed to the blastocyst stage. A cultured blastocyst must be implanted in the woman seeking fertility treatment, or else frozen for later use, because it cannot be kept alive in a dish at later stages of development. Because the process is arduous for the mother-to-be and the failure rate can be high, many eggs are removed from the woman and fertilized in vitro (literally "in glass"), although only one or a few blastocysts can be implanted in her womb. The "extra" embryos are frozen in case they are needed later. More than 400,000 human blastocysts are held in deep freezers at fertility clinics across the United States.

In the late 1990s, a number of "extra" blastocysts were donated by couples to researchers who had developed methods for growing human cells derived from the inner cell mass in a nutrient medium (**FIGURE 11.6**), building on knowledge gained from manipulating mouse blastocysts. The pluripotent cells displayed their stem cell characteristics by dividing over and over without differentiating. Several lines of such human embryonic stem cells (hES cells) were developed by labs in the United States and elsewhere. (Cells descended from a common parent cell through an unbroken chain of mitotic divisions are said to belong to a *line*.) The announcement in 1998 that human embryonic stem cells had been multiplied in a dish, and had been made to differentiate into certain cell types by the addition of certain signaling molecules, was met with both elation and condemnation.

Opponents of human embryonic stem cell research believe that the life of a human being begins at conception. They contend that an embryo has moral status and that it is unethical to use its cells for someone else's benefit. In 2001, the U.S. government cut off funding for any research involving human embryos and banned the use of public funds for creating new hES lines. In 2009, an executive order issued by a new administration reversed the ban and allowed federal agencies to support and conduct embryonic stem cell research. Advocates of stem cell research maintain that federal support for hES research will advance basic knowledge and spur therapeutic applications of stem cell technology.

Since adult stem cells exist and have been used for therapeutic purposes, what need is there for research on human embryonic stem cells? Advocates of hES research point to the limited developmental flexibility of adult stem cells and the fact that these cells are small, scarce, and usually hard to identify and difficult to grow in a lab dish. The controversies may be laid to rest, however, if the potential of a new type of stem cell—the induced pluripotent stem cell—comes to be realized.

## Induced pluripotent stem cells are derived from differentiated cells

The development of *induced pluripotent stem cells* is one of the most exciting discoveries in cell biology and biomedicine in recent years. An **induced pluripotent stem cell (iPSC)** is any cell, even a highly differentiated cell in the adult body, that has been genetically reprogrammed to mimic the pluripotent behavior of embryonic stem cells. Human embryos do not have to be destroyed to generate iPSC lines, yet the cells appear to have nearly the same developmental flexibility that embryonic stem cells have. The technology therefore appeals to all camps.

Scientists developed iPSCs by carefully studying the key genes that give an embryonic stem cell its unique properties. In 2006, researchers introduced just four critical genes into mouse skin cells (fibroblasts), using harmless viruses as delivery vehicles. The introduced genes forced the mature cell type to start behaving almost exactly like embryonic cells. In 2007, similar techniques were used to create an iPSC line from human skin cells. More recently, iPSC lines have been raised by the introduction of key proteins, not genes, into mature skin cells.

The developers of iPSCs are quick to point out that the technology is still in its infancy and many advances are needed before iPSC-based therapies can be tested in human subjects. For example, mice treated with iPSC therapy seem to be more prone to developing cancer than are mice treated with embryonic stem cell technology. In time, a better understanding of the control of cell division and differentiation in stem cells will probably resolve this problem. Stem cell researchers emphasize the vital importance of studying all types of stem cells because each avenue of study has added, and continues to add, to our understanding of how, when, and where cells divide or differentiate. Understanding how cells divide is especially critical in treating cancer, the subject we turn to next.

> ### Concept Check
>
> 1. What are the unique properties of a stem cell?
> 2. Compare the developmental flexibility of adult stem cells to that of embryonic stem cells.

## 11.2 Cancer Cells: Good Cells Gone Bad

Cancer accounts for more than 500,000 deaths in the United States each year. Over the course of a lifetime, an American male has a nearly one in two chance of being diagnosed with cancer. American women fare slightly better, with a one in three chance of developing cancer. There are more than 200 different types of cancer, but the big four—lung, prostate, breast, and colon cancers—account for more than half of all cancers combined (Table 11.2). Although the past decade has seen improvements in treatment and prevention, more than a million Americans are diagnosed with some form of cancer every year. In the United States, one in four deaths is due to cancer, and more than 8 million Americans alive today have been diagnosed with cancer and are either cured or undergoing treatment. More than 1,500 Americans die from cancer each day. Only heart disease kills more people in the United States. The National Cancer Institute estimates that the collective price tag for the various forms of cancer is more than $100 billion per year.

In the following case study of cell division gone awry, we explore the biology of cancer by considering how a cancer is launched when normal restraints on cell division and cell movement are lost. We will see that the vast majority of cancers arise from the gradual accumulation of mutations in cells of the adult body, often as a result of DNA damage caused by environmental factors. We will examine some of the environmental factors known to contribute to cancer, and we will remind ourselves that, as with any disease, prevention is better than having to find cures.

## Cancer develops when cells lose normal restraints on division and migration

Cancer represents a breakdown in the cooperative functioning of the cells in a multicellular organism.

TABLE 11.2    Selected Human Cancers in the United States

| TYPE OF CANCER | OBSERVATION | ESTIMATED NEW CASES IN 2010 | ESTIMATED DEATHS IN 2010 |
| --- | --- | --- | --- |
| Lung cancer | Accounts for 28 percent of all cancer deaths and kills more women than breast cancer does | 222,520 | 157,300 |
| Prostate cancer | The second leading cause of cancer deaths in men (after lung cancer) | 217,730 | 32,050 |
| Breast cancer | The second leading cause of cancer deaths in women (after lung cancer) | 207,090 | 39,840 |
| Colon and rectal cancer | The number of new cases is leveling off as a result of early detection and polyp removal | 142,570 | 51,370 |
| Malignant melanoma | The most serious and rapidly increasing form of skin cancer in the United States | 68,130 | 8,700 |
| Leukemia | Often thought of as a childhood disease, this cancer of white blood cells affects more than 10 times as many adults as children every year | 43,050 | 21,840 |
| Ovarian cancer | Accounts for 3 percent of all cancers in women | 22,200 | 16,200 |

Every cancer begins with a single rogue cell that starts dividing with wild abandon, giving rise to a mass of abnormal cells. The cell mass formed by the inappropriate proliferation of cells is known as a **tumor**. Tumors that remain confined to one site are **benign tumors**. Because they can be surgically removed in most cases, benign tumors are generally not a threat to survival. However, a benign tumor that is growing actively is like a cancer-in-training. With the passage of time, the descendants of these abnormal cells can become increasingly abnormal: they change shape, increase in size, and quit their normal job. These **precancerous cells** look abnormal enough that pathologists (disease experts) looking through a microscope can usually pick them out from normal cells by their size and shape (**FIGURE 11.7**). Tumor cells often produce proteins—known as *tumor markers*—that normal cells of that type do not make or make in much lower quantity. The presence of tumor markers in blood, urine, and other fluids and tissues is used to screen for the *possibility* of some types of cancer.

Most cells in the adult animal body are firmly anchored in one place. Tumor cells on the path to cancer start producing enzymes (matrix metalloproteinases) that break up the adhesion proteins that attach a cell to the extracellular matrix or to other cells in the tissue. Most human cells stop dividing if they are detached from their surroundings—a phenomenon known as **anchorage dependence**. But some tumor cells that have broken loose from their moorings may acquire anchorage *in*dependence, the ability to divide even when released from the normal attachment sites.

As tumor cells progress toward a cancerous state, they start secreting substances that cause new blood vessels to form in their vicinity in a process known as

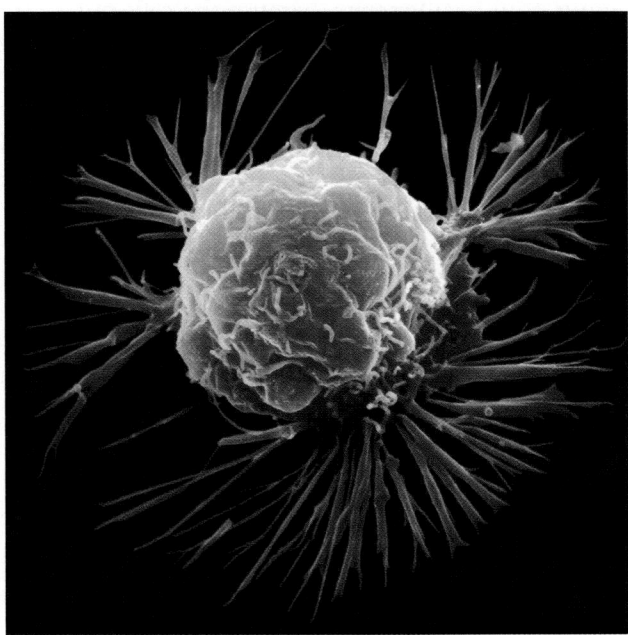

**FIGURE 11.7 A Home-Grown Monster**
This color-enhanced photograph, captured with a scanning electron microscope, shows a breast cancer cell. A lab technician can recognize it as a cancer cell because of its large size, abnormally rounded shape, and altered cell surface.

## Helpful to Know

The word "cancer" comes from the Old English for "spreading sore," which in turn comes from the Latin for "crab," probably because the enlarged blood vessels on a cancerous tumor were seen as resembling the legs of a crab. The study of cancer is called "oncology," which comes from *onkos*, Greek for "lump" or "mass," as in a tumor mass. Genes that promote cancer formation are called "oncogenes."

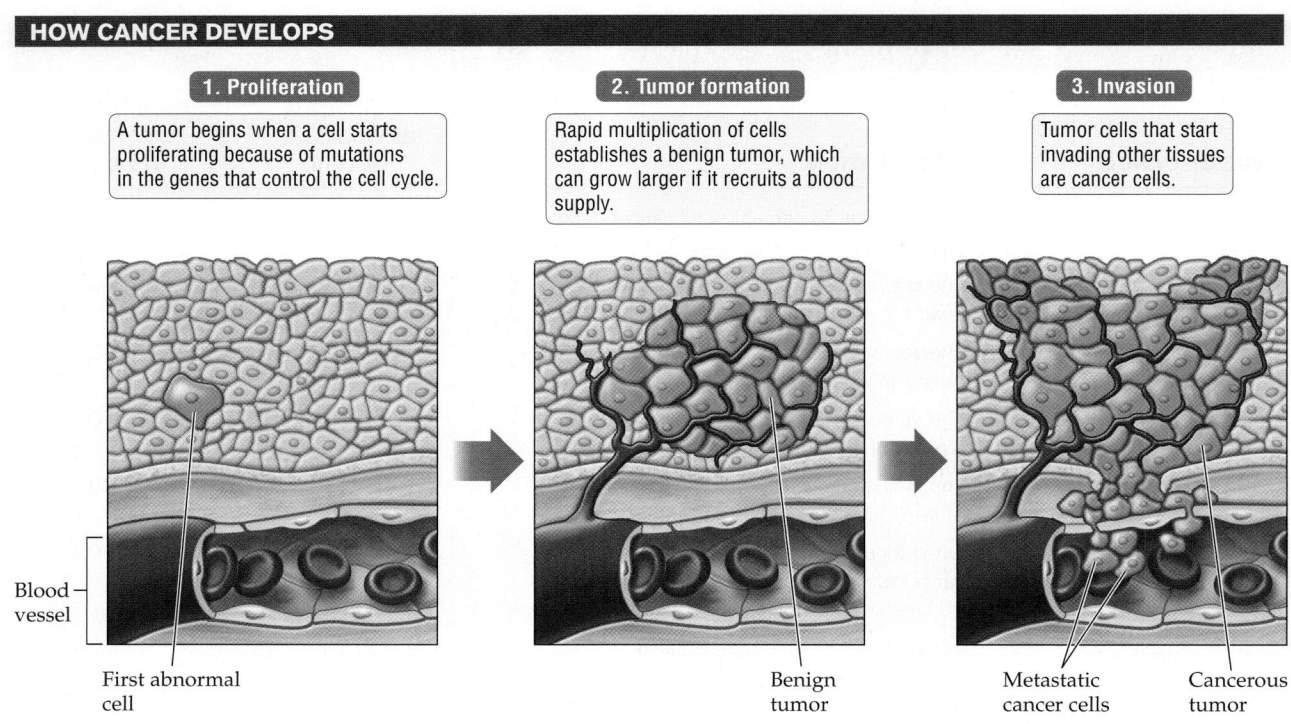

## HOW CANCER DEVELOPS

**1. Proliferation**

A tumor begins when a cell starts proliferating because of mutations in the genes that control the cell cycle.

**2. Tumor formation**

Rapid multiplication of cells establishes a benign tumor, which can grow larger if it recruits a blood supply.

**3. Invasion**

Tumor cells that start invading other tissues are cancer cells.

Blood vessel

First abnormal cell

Benign tumor

Metastatic cancer cells

Cancerous tumor

**angiogenesis** [AN-jee-oh-JEN-uh-sis] (*angio*, "vessel"; *genesis*, "creation of"). Angiogenesis recruits a blood supply for the tumor, important for delivering nutrients and whisking away waste. Without angiogenesis, solid tumors would not be able to grow larger than about 1 or 2 millimeters, because the cells buried inside would be starved for nutrients and poisoned by waste products if the tumor were to increase. Angiogenesis, however, enables malignant tumors to grow bigger. As we explain next, the network of fine blood vessels in and around the tumor is also an easy exit route for tumor cells escaping the home port.

When tumor cells gain anchorage independence and start invading other tissues, they are transformed into **cancer cells**, also known as malignant cells. At their very worst, cancer cells break loose from their neighborhood and head to other organs to set up satellite colonies there. The rogue cells enter the bloodstream or invade the lymphatic system, the network of tubes in which immune cells are made, stored, and move about. They squeeze between the cells that make up the walls of blood vessels or lymphatic vessels to emerge in distant locations throughout the body, where they set up new tumors. The tumor at the initial location is called a **primary tumor**, and the new tumors it spawns at distant sites are known as **secondary tumors** (**FIGURE 11.8**).

The spread of a disease from one organ to another is known as **metastasis** [meh-TAS-tuh-sis] (*meta*, "to

change"; *stasis*, "a set condition") (**FIGURE 11.9**). Metastasis typically occurs at later stages in cancer development (Table 11.3); and because there are tumors in multiple organs, each of them capable of further metastasis, a cancer that has metastasized is difficult to fight. Some types of cancers, such as basal cell or squamous cell skin cancers, rarely metastasize and are therefore relatively easy to treat. Malignant melanoma is the most dangerous form of skin cancer because it arises in pigment-bearing skin cells, called melanocytes, that migrate locally as a normal part of their function and are therefore that much more prone to metastasis when they become malignant.

Some cancers typically spread to fairly predictable locations because certain routes offer less resistance

| TABLE 11.3 | Typical Steps in Cancer Progression |
| --- | --- |

1. Cell proliferation

2. Loss of cell adhesion

3. Loss of anchorage dependence

4. Tissue invasion

5. Angiogenesis

6. Metastasis

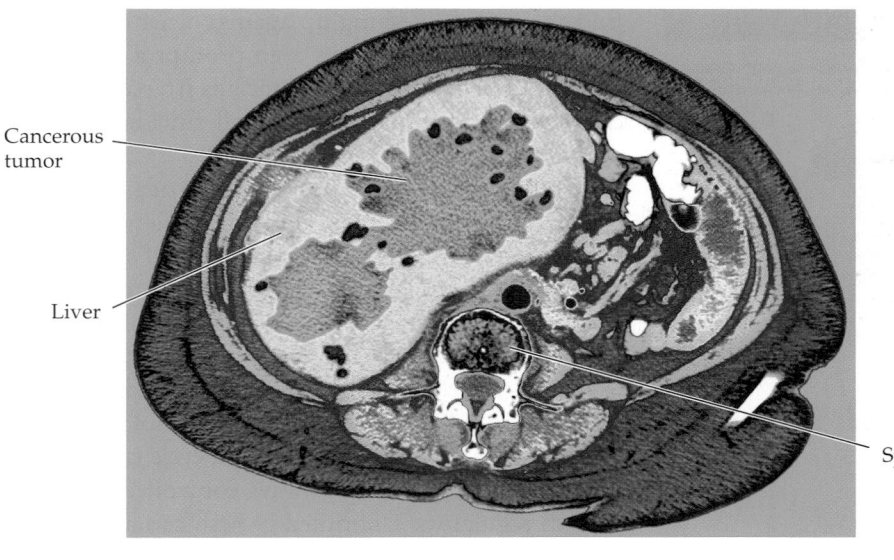

Cancerous tumor

Liver

Spine

**FIGURE 11.9 Cancer Metastasis**

This color-enhanced CT (axial computed tomography) scan shows a large cancerous tumor (dark green) in the liver (light green) of a woman who was originally diagnosed with colon cancer. The cancerous cells metastasized from the large intestine to the liver to create the secondary tumors seen here. The patient's back and spine are at the bottom center of the CT scan, a computer-generated image that represents a slice through the body.

to the movement of malignant cells than do others. A cancer that arises in the colon (the large intestine) spreads most easily to the liver, for example (see Figure 11.9). Breast cancer cells commonly metastasize to the bones, lungs, liver, or brain. Because the cells that make up an organ have certain distinctive properties, pathologists can examine the malignant cells in a secondary tumor and tell which organ they came from. Knowing the origin of metastatic tumors is often valuable in choosing among treatment options.

Wherever they establish themselves, cancer cells multiply furiously, overrunning their neighbors and monopolizing oxygen and nutrients to the extent that they starve normal cells in the vicinity. Without restraints on their growth and migration, cancer cells take over, steadily destroying tissues, organs, and organ systems. The normal function of these organs is seriously impaired, and cancer deaths are ultimately caused by the failure of vital organs.

## Cell division is controlled by positive and negative growth regulators

Rampant cell division, or proliferation, is the first abnormal behavior a cell displays on its way to becoming cancerous. Normally, all aspects of cell behavior, especially the frequency with which a cell divides, are closely regulated in a healthy body. Controlling which cell divides, where and when, is quite literally a matter of life and death in something as complex as the human body.

As noted in Chapter 10, the cell cycle is controlled by a variety of external and internal signals. **Positive growth regulators** is a blanket term for all those factors that stimulate cell division. **Negative growth regulators** are any and all factors that put the brakes on cell division. Signaling molecules called **hormones**, and proteins called **growth factors**, are common regulators of cell division in the human body. Some hormones act as positive growth regulators in certain tissues. Human growth hormone induces cell division at the ends of the long bones in children, and estrogen stimulates cell division in breast tissue, for example. Some hormones act as negative growth regulators of specific cell types. Cortisone, for example, inhibits cell division in immune cells and certain other cell types.

Some growth factors, such as epidermal growth factor (EGF), are positive growth regulators: they promote cell division in their target cells by pushing the cell cycle from $G_1$ to S phase (see Figure 10.5). Transforming growth factor beta (TGF-β) is an example of a negative growth regulator (**FIGURE 11.10**), a protein that arrests the cell cycle when cell division would be inappropriate—for example, if the DNA were damaged. Many of the external signals that regulate cell division commonly do so by binding to **receptor proteins** (see Section 6.2) that in turn trigger a stepwise sequence of protein activities, called a **signal transduction pathway**, which relays the external signal through the cytoplasm (see Figure 11.10). The protein components of signaling pathways that promote cell proliferation are examples of positive growth regulators. In contrast, protein components

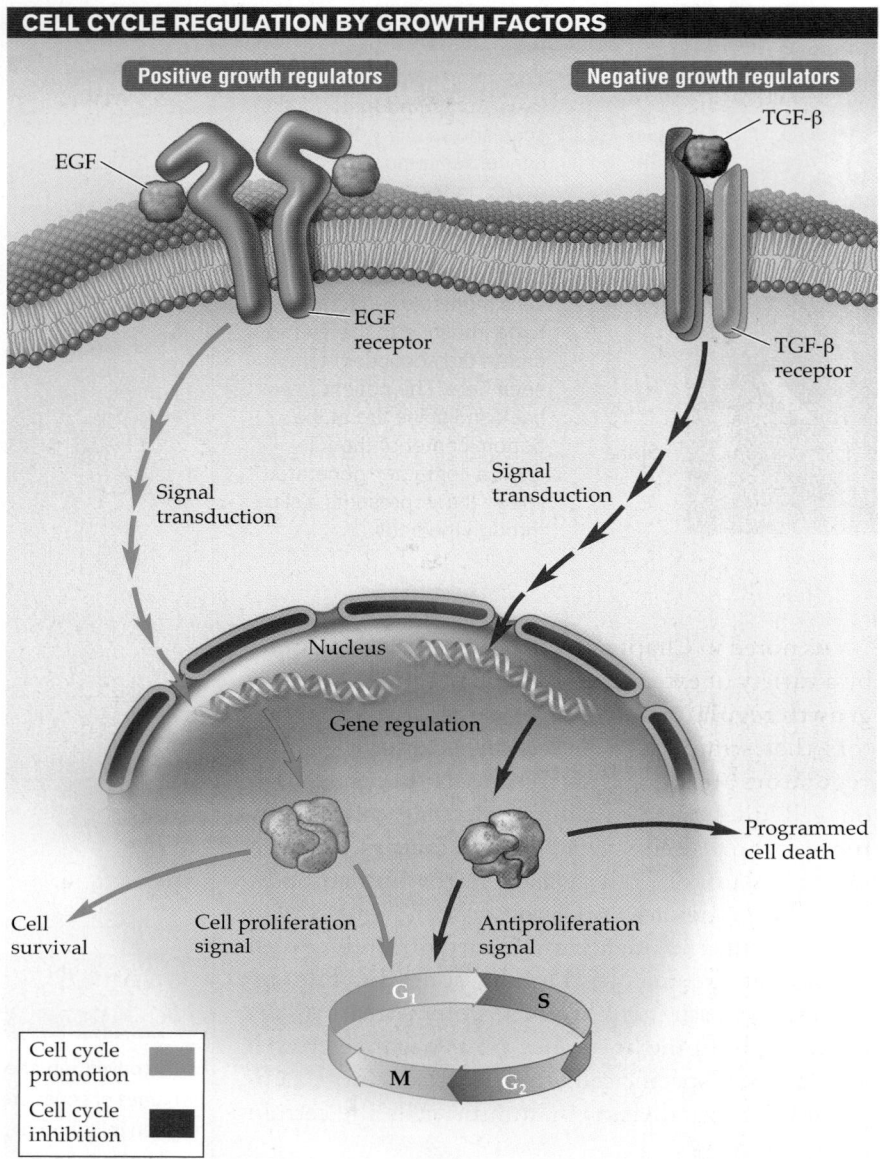

## CELL CYCLE REGULATION BY GROWTH FACTORS

Positive growth regulators

Negative growth regulators

EGF

TGF-β

EGF receptor

TGF-β receptor

Signal transduction

Signal transduction

Nucleus

Gene regulation

Programmed cell death

Cell survival

Cell proliferation signal

Antiproliferation signal

$G_1$  S

M  $G_2$

Cell cycle promotion

Cell cycle inhibition

**FIGURE 11.10 Growth Factors Can Be Positive or Negative Regulators of the Cell Cycle**

that transmit an antiproliferation message, such as that set in motion by TGF-β, are examples of negative growth regulators.

Under normal circumstances, positive and negative growth regulators are released only when and where they are needed. A paper cut on your finger will trigger the activity of positive growth regulators that causes skin cells to divide until the wound is closed. How do cells at the wound edge "know" when they have divided enough and it is time to quit? Negative growth regulators, such as TGF-β and its signal cascade components, halt the cell cycle when the newly made cells

find themselves pushing up against other cells. Negative growth regulators can even prompt a potentially dangerous cell, such as one that has irreparable DNA damage, to commit a form of cell suicide known as *programmed cell death* (see Chapter 10).

As these examples illustrate, the life of a cell is managed by a delicate interplay between a variety of positive and negative growth regulators. What if this system fails? A multicellular individual, such as a human, is a cooperative community of cells, so the failure of just one cell to respond appropriately to the balance of positive and negative growth regulators can have serious consequences for the organism as a whole. Runaway growth is the consequence if the cell cycle is excessively *stimulated* through the pathways controlled by *positive* growth regulators. Unbridled growth can also be triggered in a cell that *ignores* the antiproliferation message of *negative* growth regulators. In either case, excessive cell proliferation sets the stage for the development of cancer. But what would cause a cell to behave in such a reckless way, endangering the whole organism? We shall see in the next section that malfunctioning genes are responsible for the wayward behavior of tumor cells: A benign tumor cell harbors some damaged genes, leading to some bad behavior. A cancerous cell has yet more gene malfunctions, producing the most malicious cell behaviors.

## Gene mutations are the root cause of all cancers

A **mutation** is a change in DNA sequence (Table 11.4). We discuss how mutations arise in Chapter 14. For now, it should suffice to know that genes typically code for proteins and that a gene with a mutation is likely to produce an altered protein. If the mutation is severe enough, protein production might fail altogether. Most commonly, gene mutations reduce or eliminate the activity of the protein encoded by that gene. Such mutations are like a corrupted computer application: the code is garbled, so the application cannot run properly. In some instances, however, a gene mutation can alter the encoded protein in such a way that its activity is actually *increased*. DNA mutations can cause certain proteins to be made in larger-than-normal amounts, to be manufactured in cells that do not normally make them, or to be made at the wrong time. Mutations that put gene expression in "hyperdrive" are more like stealthware, unwanted applications that do things the original software was not designed to do.

| TABLE 11.4 | Origins of Mutations in Cancer-Critical Genes |
| --- | --- |

**Inheritance**

Mutated genes inherited from one or both parents; such hereditary cancers account for less than 10% of human cancers

**External agents (carcinogens) that can act as mutagens**

Chemical mutagens (agents that cause mutations in DNA)

Ionizing radiation (UV, X-rays, other high-energy radiation)

Viruses (involved in ~15% of human cancers) and other pathogens

Chronic injury (physical, chemical)

Lifestyle factors (e.g., diet and exercise)

**Internal processes that generate mutations**

Genetic instability due to errors in DNA replication or repair; accidental chromosome breaks and rearrangements

**Internal processes that allow uncontrolled cell regulation**

Errors in regulation of cell division and cell behavior

---

Genes that have been implicated in the development of cancer fall into two main classes: *proto-oncogenes* and *tumor suppressor genes*. Genes whose action results in cell proliferation are broadly classified as **proto-oncogenes**. All genes that code for positive growth regulators are proto-oncogenes. These genes can mutate in such a way that they become hyperactive, in which case they tend to trigger runaway cell proliferation. Proto-oncogenes are perfectly normal genes with essential roles in the body; it is only when they become *inappropriately active*, as a consequence of mutations, that they lead to excessive cell proliferation and tumor development. Proto-oncogenes that have become hyperactive as a result of DNA mutations are known as **oncogenes** (*onkos*, "tumor"). An oncogene is analogous to a stuck gas pedal: one hyperactive copy of the gene is all it takes to push the cell cycle forward excessively or in the wrong cell or at the wrong time. For example, the proto-oncogene that codes for the EGF receptor depicted in Figure 11.10 can mutate such that the receptor is turned on even in the absence of EGF. The rogue receptor, now classified as the product of an oncogene, will push the cell cycle forward even when it is not being "told" by EGF to stimulate cell division.

Genes that restrain cell division or cell migration are called **tumor suppressor genes**. All genes that code for negative growth regulators are tumor suppressor genes. The normal activity of such genes is to inhibit the cell cycle, stimulate repair of damaged DNA, promote cell adhesion, enforce anchorage dependence, prevent angiogenesis, or trigger cell suicide. Mutations in tumor suppressor genes can reduce or eliminate the normal activity of the encoded protein, in which case the cell bearing the mutations becomes capable of uncontrolled proliferation and possibly invasiveness as well. As with most genes, a cell has two copies of each tumor suppressor gene. The failure of one copy can reduce the effectiveness of tumor suppression, but the loss of both copies is usually much more serious because it means a complete inability to restrain cell proliferation. A tumor suppressor is analogous to the brakes in a car, and the loss of *both* copies of a tumor suppressor gene is like losing the main brake pedal and also the emergency brakes.

## Most human cancers are not hereditary

Hereditary cancers are those that are linked to inherited gene mutations. A child receiving an oncogene or a mutated tumor suppressor gene from either parent has a hereditary risk of cancer. Inheriting a gene linked to cancer does *not* mean that cancer is inevitable in that child's future. It simply means that she has an elevated risk of cancer compared to people who have not inherited the faulty gene.

The great majority of human cancers cannot be blamed on bad genes passed down from parent to child. In fact, only 1–5 percent of all human cancers can be traced to inherited gene defects. Some forms of breast cancer and colorectal cancer are among the handful of cancers that are based in heredity. Mutations in two tumor suppressor genes, *BRCA1* and *BRCA2*, are associated with a big increase in the risk of breast and ovarian cancer. A woman who inherits one faulty copy of the *BRCA1* gene has 60 percent odds of developing breast cancer, compared to a 12 percent risk faced by the average woman who does not have a faulty copy of *BRCA1*. However, only 5–10 percent of the women who are diagnosed with breast cancer have a mutated version of *BRCA1* or *BRCA2*, and most cases of breast cancer cannot be tied to any inherited gene mutation.

Hereditary mutations are suspected in people who have a family history of a particular type of cancer. A woman who has a harmful mutation in the *BRCA1* gene, for example, is likely to have multiple close family members—grandmother, mother, aunts, sisters—who have suffered from the disease. Cancers caused by hereditary mutations tend to appear earlier in life than nonhereditary cancers. Hereditary breast cancer often develops in women of reproductive age, whereas nonhereditary breast and ovarian cancers are rare in this age group and most common in postmenopausal women (those who are past the reproductive years, typically age 50 or older). Genetic tests are available to screen individuals at risk for some types of hereditary cancer, including hereditary breast cancer and familial adenomatous polyposis, a hereditary condition that almost always leads to colon cancer.

About 95 percent of the people diagnosed with cancer seem to have no inherited risk of the disease. That means their cancer was caused by a series of un-fortunate changes in their DNA, resulting from environmental agents such as viruses or toxic chemicals, from unavoidable cellular accidents such as cell division errors, or from a combination of these two sources of nonhereditary mutations, or **somatic mutations**. The requirement for multiple mutations within a single cell explains why cancer is rare among the young, becomes more common as we get older, and steeply increases in frequency past middle age (**FIGURE 11.11**). The longer we live, the more opportunity there is for well-behaved proto-oncogenes and tumor suppressor genes inside one cell to accumulate mutations. As we shall see in the next section, with every such mutation that accumulates, a good cell becomes steadily worse, until a full-fledged monster emerges in the form of a metastatic malignancy.

## Cancer develops as multiple mutations accumulate in a single cell

More than one in three Americans will be diagnosed with cancer at some point in their lives. Given such a high incidence, you might think that the human body is exceptionally prone to cancer. Actually, we have robust defenses against cancer. A number of safeguards—cell adhesion and anchorage dependence, for example—reduce the likelihood of runaway cell proliferation and tumor development, at least during the reproductive years. As we age, however, mutations start to accumulate in the genes that orchestrate our anticancer defenses, bringing us closer to the unlucky string of failures that result in a cancerous tumor.

Consider cancer of the colon, the large intestine. In many cases of colon cancer, the tumor cells contain at least one oncogene (which, remember, is a hyperactive version of the normal proto-oncogene) and several completely inactive tumor suppressor genes. The mutations in different genes that eventually lead to colon cancer usually occur over a period of years, and the gradual accumulation of these mutations often goes hand in hand with stepwise changes in cell behavior that mark the progression toward cancer.

We can illustrate the step-by-step sequence of chance mutations, and the accompanying changes in cell activities, by following the disease progression that is characteristic of colon cancer. In most cases, the first cancer-promoting mutation results in a

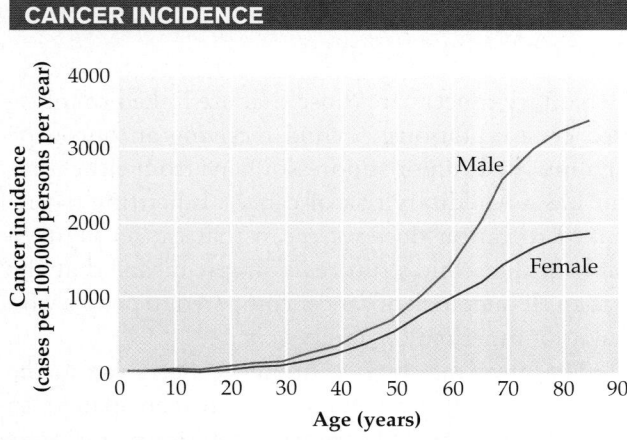

**CANCER INCIDENCE**

**FIGURE 11.11  The Incidence of Cancer Rises Sharply as We Grow Older**

The graph shows incidence rates for all cancers combined, by age group and gender. The incidence of a disease refers to the number of newly diagnosed cases per year for a defined unit of the population—in this case, per 100,000 men or women in each age group.

relatively harmless, or benign, growth described as a *polyp* (**FIGURE 11.12**). The cells that make up the polyp divide at an inappropriately rapid rate. These cells are the descendants of a single cell in the lining of the colon that has suffered one or more mutations, usually in a proto-oncogene that codes for proteins that stimulate cell proliferation. If a tumor suppressor gene fails in addition, cell division speeds up like a runaway car whose gas pedal is on the floorboard and whose brakes have failed. The polyps grow faster than ever before, but even so, most such polyps do not spread to other tissues and can be safely removed surgically at this stage. Colonoscopy, an examination of the lining of the colon, is advised for individuals over 50, earlier for those at high risk. (The procedure is not terribly pleasant, and offspring who make insensitive remarks when a middle-aged parent is about to undergo the procedure could be putting their inheritance at risk!)

A common mutation in many cancers is the complete inactivation of an especially critical tumor suppressor gene, named *p53*, which has multiple roles in guarding the integrity of cellular processes. Nicknamed the "guardian angel of the cell," the p53 protein not only prevents the cell from dividing at inappropriate times, it also halts cell division when there is DNA damage in the cell. By halting the cell cycle, p53 gives the cell an opportunity to repair the damage to its DNA, keeping mutated DNA from being passed on to daughter cells. If the repair process fails—for instance, because the damage is so extensive that it is beyond repair—p53 goes so far as to induce programmed cell death. Given the important guardian functions of the p53 protein, it is not surprising that more than half of all human cancers show a complete loss of p53 activity in tumor cells. The number goes as high as 80 percent in some types of cancer, such as colon cancer.

As a polyp grows larger, the odds increase that one or more cells within this larger population of abnormal cells will acquire additional mutations. Mutations that knock out tumor suppressor genes that code for cell adhesion proteins, and the emergence of oncogenes that produce hyperactive matrix metalloproteinases, will enable a cell to detach itself from its surroundings. Additional mutations that produce loss of anchorage dependence enable the detached cells to continue multiplying, paving the way for invasion of other tissues and the emergence of a frank cancer. The metastasis of colon cancer cells, typically to the liver, is the last and most destructive step in cancer progression.

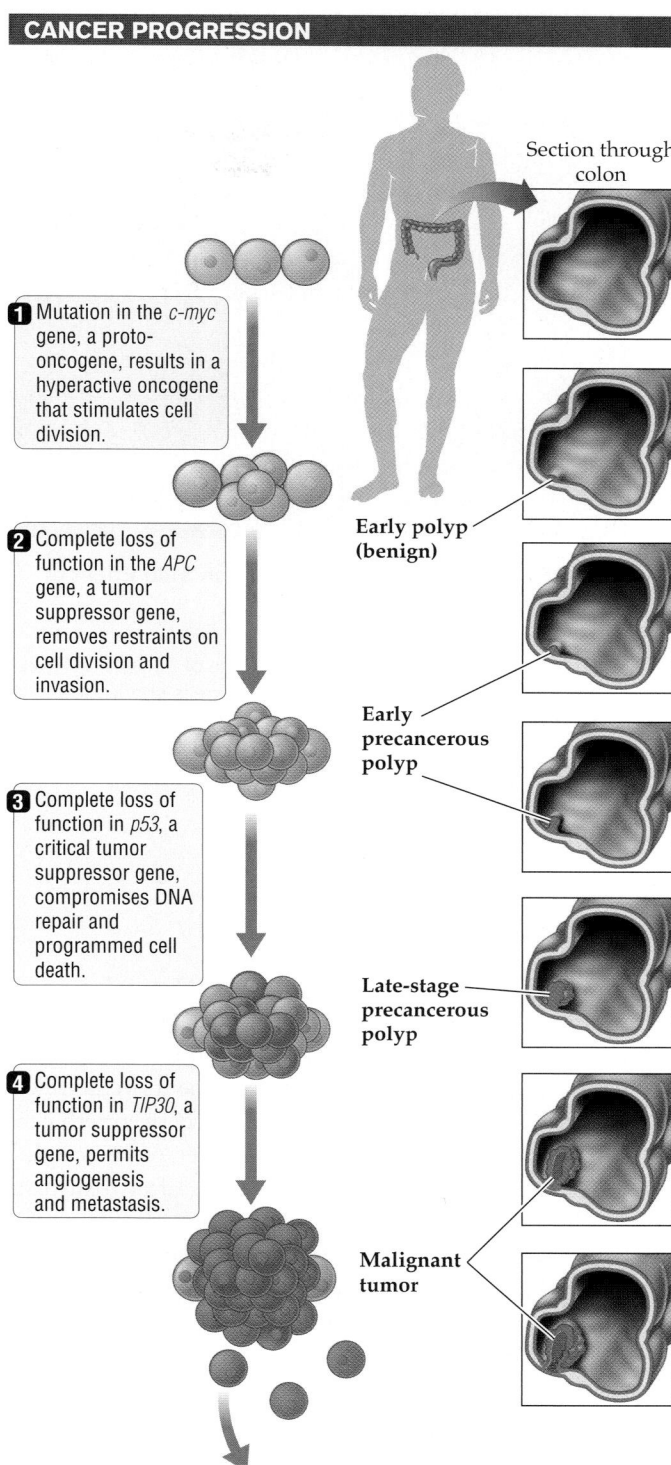

**CANCER PROGRESSION**

Section through colon

**1** Mutation in the *c-myc* gene, a proto-oncogene, results in a hyperactive oncogene that stimulates cell division.

**2** Complete loss of function in the *APC* gene, a tumor suppressor gene, removes restraints on cell division and invasion.

**3** Complete loss of function in *p53*, a critical tumor suppressor gene, compromises DNA repair and programmed cell death.

**4** Complete loss of function in *TIP30*, a tumor suppressor gene, permits angiogenesis and metastasis.

Early polyp (benign)

Early precancerous polyp

Late-stage precancerous polyp

Malignant tumor

To bloodstream

**FIGURE 11.12 Development of Colon Cancer Is a Multistep Process**

The sequential mutation of several genes that code for positive and negative growth regulators coincides with the progression from a benign polyp in the colon to a malignant tumor. The order in which the various proto-oncogenes and tumor suppressor genes mutate is not set, and it can vary from one patient to another, which is why every cancer is unique.

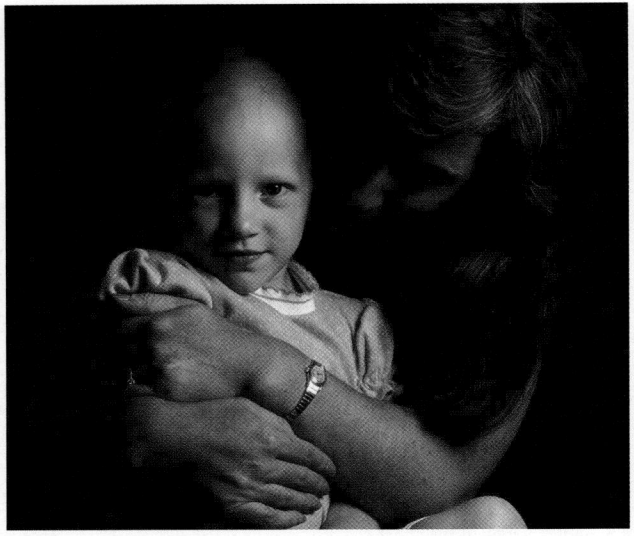

**FIGURE 11.13 Hair Loss Is a Common Side Effect of Chemotherapy**

Chemotherapy attacks any and all rapidly dividing cells and results in collateral damage, including loss of cells that grow and maintain hair follicles. Dormant stem cells survive the treatment, and they will regrow the lost hair in about 3 months.

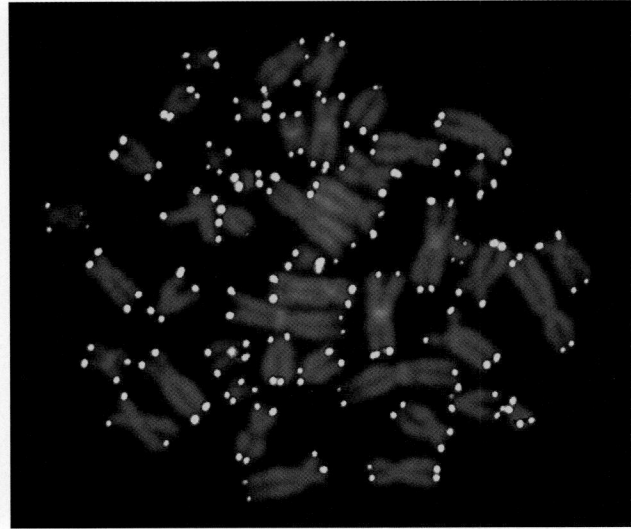

**FIGURE 11.14 Telomerases Confer Cellular Immortality by Preventing Chromosome Ends from Fraying**

The tips of chromosomes, here shown in yellow, are eroded each time a somatic cell divides, limiting the life span of a cell. An enzyme called telomerase overcomes this obstacle to perpetual cell division by lengthening the tips as needed. Embryonic stem cells and cancer cells produce telomerase and can therefore divide endlessly. Clinical trials are under way to see whether crippling telomerase will limit the runaway growth of cancer cells.

## The challenge in cancer treatment is to destroy malignant cells selectively

About 40 years ago, President Richard Nixon declared a war on cancer in the United States by making anticancer research a high priority. Since then, some major victories have been won, thanks to improvements in radiation and drug therapies. Whereas in the early twentieth century very few individuals survived cancer, today roughly 40 percent of patients are alive 5 years after treatment has begun. Nevertheless, the war against cancer is far from over, and the need for powerful new treatments to stop or kill the malignant cells is as urgent as ever. **Cryosurgery** makes use of extremely cold temperatures to kill abnormal cells, usually precancerous cells or cancers that are confined to a small region, such as the cervix. **Hormone therapy**, which manipulates the hormone environment in the body to stop or slow cancer cells, is used in some hormone-responsive cancers, which include some types of breast and prostate cancer.

The greatest challenge in battling cancer is the selective destruction of rogue cells while sparing healthy cells. The standard plan of attack today relies on high-energy radiation (**radiation therapy**), high doses of chemical poisons (**chemotherapy**), or both in sequence to kill any and all rapidly dividing cells. The side effects of radiation therapy and chemotherapy are terrible because this all-out assault also kills many innocent bystanders, cells necessary for the normal functions of the human body. Alopecia (hair loss) is the most visible of the many side effects of these therapies (**FIGURE 11.13**). Hair falls out because the cells that grow and maintain hair are destroyed; but *dormant* stem cells survive the treatment, and they regrow the hair in about 3 months. Cells that divide to produce red blood cells are killed, leading to the fatigue and weakness of anemia. The cells that line the entire digestive tract are regularly replaced by cell division, and the loss of that cell supply produces illness in the mouth, stomach, and intestines.

The good news is that discoveries in basic cell biology, along with the large investment in cancer research, have produced a variety of innovative strategies for destroying malignant cells selectively. One line of attack is to selectively disable proteins that give cancer cells their unusual immortality, the capacity to divide indefinitely as long as oxygen and nutrients are available.

An enzyme called *telomerase* [teh-*LOH*-muh-rays] is key to cell immortality. The tips of our chromosomes, known as **telomeres** [*TEH*-luh-mirz], are whittled down each time a cell divides (**FIGURE 11.14**), *unless* telomerase steps in to repair the ends. The only cells in our body that produce telomerase are stem cells, germ line cells (those that produce gametes through meiosis), and cancer cells. Researchers have identified chemicals that knock out telomerase activity, and these are being tested in clinical trials in the hopes that wearing out the chromosome ends in such cells will put a stop to their runaway proliferation.

As noted earlier, cancer cells are masters at recruiting nutrient supplies, and they commonly secrete chemicals that trigger the growth of new blood vessels in their vicinity—escape routes for metastasis. One experimental approach against cancer attempts to block angiogenesis, the recruitment of new blood vessels by a malignant tumor. For example, an antibody-based pharmaceutical called Avastin mops up the angiogenesis-recruiting chemicals that cancer cells release; the drug has been approved for treatment of colorectal cancer. Another inventive method uses genetically engineered viruses that infect and destroy specific types of cancer cells while leaving normal cells alone. These and several other novel anticancer strategies are now in clinical trials. The early results are encouraging, although a lot of groundwork will be necessary before any of them becomes a routine treatment for any type of cancer.

The biggest lesson learned from the past three decades of cancer research is a surprisingly simple one: instead of dealing with cancers only when they become monsters, we should try to reduce the odds that our cells will progress toward cancer in the first place. It is now abundantly clear that environmental factors, including lifestyle choices, have a very large impact on a person's risk of developing cancer. The overall incidence of cancer began to decline in the United States during the last 10 years not because of a billion-dollar technological innovation, but mostly because fewer people are smoking and therefore fewer people are developing tobacco-related cancers. Cancer prevention, through increased public awareness of the risk factors and the adoption of cancer-protective behaviors, is now a high priority in the war against cancer.

## Avoiding risk factors is the key to cancer prevention

The relative contributions of inherited and environmental factors to an individual's cancer risk have been debated for decades. In recent years, large-scale studies settled this issue by tracking cancer incidence in thousands of pairs of identical twins, who share the same genetic makeup. If inherited genetic defects are more important than environmental factors in causing cancer, then one would expect to see a very similar incidence of cancer in both twins. On the other hand, if environmental factors play a greater role, one would expect to see significant differences in cancer incidence due to differences in the twins' adult environments or habits. In one Scandinavian study that tracked more than 44,000 pairs of identical twins, environmental influences were by far the most important determinant of individual cancer risk.

Table 11.5 lists some of the factors that are known to increase our risk of cancer. Only two of the items

---

| **TABLE 11.5** | **Common Risk Factors for Cancer** |
| --- | --- |

**Unavoidable risks**

Growing older

Family history of cancer

**Avoidable risks**

Tobacco use

Excessive exposure to ultraviolet radiation

Poor diet quality, especially high consumption of red meat and processed meats

Obesity

Lack of physical activity

Exposure to certain hormones

Some viruses and bacteria

Excessive alcohol consumption

Chemical carcinogens

Ionizing radiation (such as X-rays)

Skin cancers are the most common cancers in the United States. The use of tanning booths is strongly associated with elevated risk of all types of skin cancer, including the most dangerous form, malignant melanoma.

# Avoiding Cancer by Avoiding Chemical Carcinogens

Although chemical pollutants are often the most feared of carcinogens in the public mind, experts estimate that only about 2 percent of human cancers can be blamed on environmental carcinogens, while what we eat and drink accounts for more than 30 percent of our cancer risk. Tobacco use is responsible for about 30 percent of the cancers diagnosed in the United States.

Many explanations have been offered for the observation that those who eat a lot of animal products have higher rates of some cancers than those who consume less or none. Some carcinogenic pollutants become concentrated along the food chain and are found in higher concentrations in meat, fish, and dairy than in plant foods. The saturated fat in meat and dairy is an independent risk factor for cancers of the breast, prostate, and colon. Animal products also lack the cancer-protective substances—such as fiber and antioxidants—that presumably help those who eat a mostly plant-based diet. Those who eat a lot of red meat—about 4 ounces, or the equivalent of one steak, a day—may experience iron overload, a known risk factor in some types of cancer.

When animal flesh is cooked at high temperatures, as in grilling, broiling, or deep frying, certain amino acids that are abundant in meat, poultry, and fish, are converted into a family of carcinogenic compounds known as heterocyclic amines (HCAs). High-temperature cooking of fatty foods creates yet another class of potent carcinogens, called polycyclic [pah-lee-SYKE-lik] aromatic hydrocarbons, or PAHs. If you like your meat well done, you might consider cutting back or at least resorting to safer cooking methods much of the time. Microwaving food lightly before high-temperature cooking reduces

HCA and PAH formation. So does marinating in antioxidant-rich sauces—think berries, red wine, and herbs like rosemary.

Antioxidants in some fruits and herbs may reduce HCA formation.

Amino acids

High temperature

Heterocyclic amines (HCAs)

Antioxidants in some fruits, and herbs such as rosemary, may reduce HCA formation in grilled meats.

When it comes to cancer risk, processed meats are even worse than charbroiled red meat. Nitrites are commonly used to preserve the color and fresh appearance of hot dogs, sausages, bologna, cold cuts, and other processed meats. In acidic environments, such as the stomach, the nitrites react with proteins to form carcinogenic compounds called nitrosamines. Vitamin C (ascorbic acid) is often added to processed meats because this antioxidant dampens the conversion of nitrites to nitrosamines; even so, it may not be a bad idea to swig some orange juice when chowing down on a hot dog.

Excessive alcohol consumption is strongly linked to cancers of the mouth and esophagus, and a smaller risk of liver, breast, and colorectal cancer. Doctors advise no more than one drink per day for women, and two for men, if we drink at all. A standard drink in the United States is defined as 0.6 ounce (13.7 grams) of pure eth-

anol, which amounts to one 12-ounce beer at 4 percent alcohol, or a 5-ounce glass of wine at 11 percent alcohol.

The link between cancer and smoking is the most dramatic illustration of how chemical exposure can transform healthy cells into dangerous ones. Lung cancer was a rare cancer just before the turn of the twentieth century, when few people smoked tobacco. Now, with nearly a third of the world's population lighting up, lung cancer is the most common and the deadliest cancer worldwide, killing more than 1.2 million people annually. Both tobacco and marijuana cigarettes contain a compound called benzopyrene, a powerful cancer-causing agent. An average marijuana cigarette contains about 50 percent more benzopyrene than does a cigarette made of tobacco. And marijuana smoke is typically inhaled more deeply and held in the lungs longer than cigarette smoke. But smoking tobacco is much more addictive than smoking marijuana, and that means that people who start smoking cigarettes are far more likely to be trapped in the habit for a long time. In addition, a regular marijuana smoker will consume far fewer cigarettes per day than even a moderate cigarette smoker does. The bottom line, however, is that both types of smoke are dangerous.

The good news is that stopping smoking can dramatically reduce an individual's cancer risk. People who quit smoking before the age of 50 reduce their risk of dying in the subsequent 15 years by half. Regardless of age, people who quit smoking live longer than those who continue to smoke. While nicotine, which is the addictive drug in tobacco, makes quitting smoking difficult, all should find inspiration in the fact that one in five Americans is a *former* smoker.

listed—the genes we inherit and the inevitability of getting older—are factors that nobody can do anything about. Everything else on the list, which is by no means exhaustive, is an environmental influence that we should try to avoid if we want to lower our risk.

A **carcinogen** [kar-*SIN*-uh-jen] is any physical, chemical, or biological agent that elevates the risk of cancer. Asbestos, implicated in an otherwise rare cancer called mesothelioma, is an example of a physical carcinogen. The long, fine crystals of the mineral, once widely used for fireproofing and in electrical insulation, enter the cells that line the lungs and damage the organelles. Hundreds of chemicals are known or suspected carcinogens in humans (see the box on page 270).

Cancer experts believe that bacteria and viruses contribute to about 10 percent of all cancer cases in humans. Cancers caused by infectious agents are more common in developing nations than in developed countries. Infection with *Helicobacter pylori* [*HELL*-ih-koh-*BAK*-ter pye-*LOH*-ree], a bacterium that thrives in the acidic environment of the stomach, is linked with increased odds of some types of stomach and intestinal cancers. Long-term infection with the hepatitis B or hepatitis C virus increases the risk of liver cancer, which, along with stomach cancer, is common in poor parts of the world but relatively rare in the United States.

Some of the more than 100 different types of human papillomavirus (HPV) produce proteins that disable tumor suppressor genes in the cells that the virus infects. HPV infection is the main cause of cervical cancer, and HPV is still detectable in the HeLa cell line established from Henrietta Lacks's cervical cancer cells. The Pap smear is a screening test that looks for abnormal precancerous or cancerous cells on the cervix, the lower end of the uterus that is exposed in the vagina. Public education and widespread screening in the United States have led to a large decline in death from cervical cancer, the malignancy that took the life of Henrietta Lacks.

A vaccine is now available that prevents infection by all the common strains of HPV. Best known by its commercial name, Gardasil, this vaccine reduces the risk of cervical, penile, and anal cancers, as well as genital warts. The U.S. Food and Drug Administration (FDA) recommends Gardasil, given as three injections over 6 months, for all persons 9–26 years old. For maximum effectiveness, the vaccination must be administered before a person becomes infected with the virus, which is why the FDA recommends vaccination even for children who are not likely to be sexually active.

In addition to carcinogens, the physiological state of the body can influence our susceptibility to cancer. Exposure to hormones and growth factors affects cancer risk because these signaling molecules have such a profound influence on cell proliferation. Taller people have a slightly heightened (excuse the pun) risk of some types of cancer, probably because they have higher levels of certain growth factors (in particular, insulin-like growth factor), compared to people of medium or short stature. We hasten to add that the increased risk is small and not something that tall people should worry about! More seriously, studies show that the more estrogen a woman is exposed to, the higher are her odds of developing breast cancer, and the risk is especially high for certain individuals and ethnic groups whose genetic profile makes them particularly vulnerable. High testosterone levels are similarly associated with prostate cancer risk in men.

Obesity is strongly linked with some types of cancer, including breast, prostate, and colorectal cancer. Fat cells are known to be a potent source of hormones and growth factors, which may explain the association. Lack of physical activity is independently associated with an increase in cancer risk, meaning that even normal-weight people who are couch potatoes are at greater risk for some types of cancer, including cancers of the breast, uterus, prostate, and lung, than are people who exercise regularly. How physical activity sways tumor development and progression is not understood, but the effect of exercise on metabolism, hormones such as insulin, and the immune system may all play a role. Diet is believed to have a substantial impact on cancer risk. People who eat a lot of animal fat and red meat have increased risk of colorectal, lung, esophageal, and liver cancers (see the box on page 270). Populations that eat a mostly plant-based diet—rich in whole grains, vegetables, and fruits—have lower rates of some types of cancer, including cancer of the colon, esophagus, pancreas, and kidney, than does the average American. Within a single generation, migrants from these populations who adopt a typical American diet acquire risks similar to that of the average American.

> ### Concept Check
>
> 1. How are cancer cells different from the cells in a benign tumor?
>
> 2. What is an oncogene, and how does it differ from a tumor suppressor gene?

# How HeLa Cells Changed Biomedicine

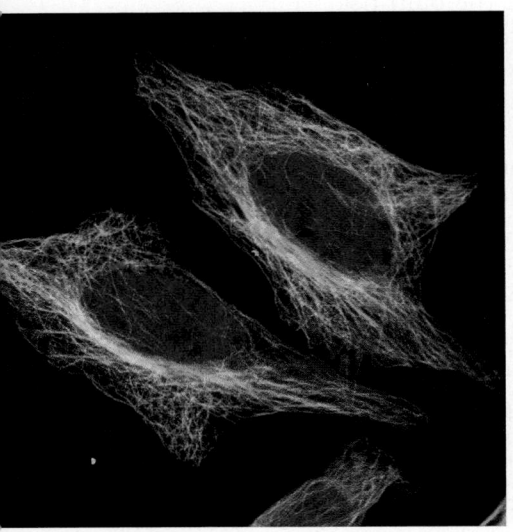

When Henrietta Lacks's tumor cells arrived in the Gey lab in early 1951, a young technician dropped them into test tubes on top of clots of chicken blood and put them in an incubator to stay warm. She didn't expect much, since all the cells she did this with seemed to die sooner or later, usually sooner.

But the Gey lab was in for a surprise. Lacks's cells had an astounding ability to divide and thrive in laboratory conditions. In just 24 hours, the cells had doubled. A day later, their numbers had doubled again. By day 4, the cells were outgrowing their new test tube homes and the lab tech had to rustle up some more test tubes for them.

It was the first time anyone had successfully cultured human cells. The Geys happily celebrated their success, and other researchers were soon demanding "HeLa cells" for their experiments. The cells have continued to grow and divide in test tubes and glass dishes for another 60 years—essentially forever, in biomedical terms.

What made HeLa cells immortal? In this chapter, we have seen that in most normal cells, the tips of chromosomes, called telomeres, shorten a little each time the cell divides. If the telomeres are short enough, the cell can no longer divide, and, after a certain number of divisions, most cells stop dividing and die. Shortened telomeres help prevent uncontrolled cell division, or cancer. The only normal cells in the human body that divide indefinitely are stem cells and germ line cells, the cells that make sperm or eggs.

But HeLa cells are anything but normal. Henrietta Lacks had been exposed to human papillomavirus (HPV),

a sexually transmitted virus that transfers its viral DNA into human skin cells. The result was a line of cells with DNA that was different from both the virus and the original human cells. For example, normal human cells have 46 chromosomes, while HeLa cells contain double that number. And unlike normal cells, HeLa cells make the enzyme telomerase, which repairs the chromosomes' telomeres, preventing them from shortening during cell division (see Figure 11.14). Because the telomeres of HeLa cells do not shorten, HeLa cells are immortal; they can theoretically grow and divide forever.

Since 1951, HeLa cells have been used in thousands of experiments, leading to more than 60,000 scientific papers by 2010. Henrietta Lacks's cells have contributed to the rise of biomedical companies worth billions of dollars. A single anticancer drug developed using HeLa cells had revenues of more than $300 million in 2007. What follows are a few highlights of HeLa cell research.

In 1960, Soviet researchers sent HeLa cells into space, long before any human astronauts ventured from Earth. HeLa cells were used to grow viruses, helping to spawn an entire field of biology called virology, the study of viruses. The mumps, measles, polio, and AIDS viruses were all grown in HeLa cells. When a private company first began mass-producing HeLa cells in a former Frito-Lay factory, Henrietta Lacks's cancer cells became a commodity, like corn or lumber. HeLa cells were used to develop ways to transport live cells in bulk, so researchers could mail cells from one lab to another. HeLa cells were used to show that HPV infection can cause cervical cancer and to study other forms of cancer, as well as in the development of anticancer drugs such as Herceptin. In fact, one way to treat cancer is to deactivate the telomerase enzyme so that cancer cells stop dividing and die off—like normal cells.

# Boys to Men: Unequal Treatment on HPV Vaccine

BY MICHELLE ANDREWS, NPR

The vaccine that prevents 70 percent of cervical cancers got the thumbs up under the health overhaul law as one of the preventive benefits that must be provided free to girls and young women between the ages 9 and 26.

Boys and young men, however, don't get the same free coverage under the law, even though the human papilloma virus (HPV) vaccine is also approved for the prevention of genital warts in males.

Why the difference? "Genital warts aren't life threatening," says Debbie Saslow, director of breast and gynecologic cancer for the American Cancer Society.

Merck, which manufactures one of the two FDA-approved HPV vaccines, is conducting research to see if the vaccine prevents genital cancers in men, says Saslow. In the meantime, though, the vaccine has already been shown to prevent cervical, vaginal and vulvar cancers in women. "It's a matter of cost-effectiveness," says Saslow.

The vaccine is pricey, requiring three shots over a six-month period, at about $130 each.

The human papilloma virus, which is transmitted through sexual contact, is extremely common, accounting for an estimated 500,000 cases of genital warts every year. "It's the common cold of the genital tract," says Saslow.

There are several treatments to eliminate genital warts. And, the body can rid [itself] of the virus, although it's possible to get re-infected.

The health law requires that immunizations that are recommended by the Advisory Committee on Immunization Practices [ACIP] be provided without charge to patients in new health plans starting this fall. ACIP is a group of 15 experts appointed by the Secretary of Health and Human Services.

While the HPV vaccine is recommended for all girls and young women between 9 and 26, the committee made a "permissive" recommendation for boys and men in the same age range, says Lance Rodewald, director of the immunization services division for the Centers for Disease Control and Prevention. Basically, that means it's OK to vaccinate males for genital warts, but it's not essential.

Human papillomavirus (HPV) is the most common sexually transmitted disease in the United States. At any given time, about half of men and 15 percent of women are infected with HPV. Often people's immune systems fight off the infection. But at some time in their lives, most adults become infected with one or more of the 30–40 strains that infect the human genital tract.

Only a few of these strains can cause cancer—mostly cervical cancer, but sometimes cancer of the tonsils and anus, in both men and women. The strains of the virus that cause cancer are different from the ones that cause genital warts. Nearly all cases of cervical cancer result from HPV infections, and overwhelmingly, women contract those HPV infections from their male sexual partners.

The HPV vaccine was initially recommended for girls and young women only. This recommendation was unusual, since most vaccines are prescribed for everyone. Populations where nearly everyone is vaccinated are much more resistant than populations where large numbers of people remain unvaccinated. For example, smallpox has been completely eradicated throughout the world as a result of aggressive vaccination programs, and polio is close to being eradicated. Even in the rare cases where only part of the population is threatened by a disease, we normally vaccinate everyone. For example, rubella (German measles) is a harmless infection in most people. But if a fetus is infected, it may die or be born with severe birth defects. So, to protect developing fetuses, health agencies try to make sure everyone is vaccinated against rubella.

One reason for not vaccinating boys and men against HPV is cost. If the government recommends a vaccine, typically health insurers are required to cover it. An optional vaccine—like the one for HPV in boys—need not be covered. The rubella vaccine costs less than $10 per person; the HPV vaccine, more than $300. Another reason not to vaccinate boys is their lower level of risk.

Since the invention of the Pap smear—a cheap test for cervical cancer—the death rate from cervical cancer in the United States has dropped to about 4,000 deaths per year. If caught early, cervical cancer is easy to cure. In contrast, in developing countries where Pap smears are hard to get, some 370,000 women are diagnosed with cervical cancer each year. Half of those women die of the disease each year.

## Evaluating the News

**1.** If all 156 million women in the United States were vaccinated against HPV, how many deaths from cervical cancer could be prevented, and at what cost per life saved?

**2.** Do women who have been vaccinated need Pap smears? Explain.

**3.** Discuss the pros and cons of vaccinating boys and young men against HPV. What considerations do you think are the most important in making a decision?

**SOURCE:** NPR, December 7, 2010, http://www.npr.org/blogs/health/2010/12/07/131881133/boys-to-men-unequal-treatment-on-hpv-vaccine.

# Summary

## 11.1 Stem Cells: Dedicated to Division

- Stem cells are undifferentiated cells that renew themselves and can generate descendants that differentiate into specialized cell types.
- Stem cell research advances our basic understanding of cell division and differentiation, as well as drug discovery and screening. Stem cell therapies include tissue engineering to repair and replace damaged tissues and organs.
- Embryonic stem cells, derived from the inner cell mass in the blastocyst, are pluripotent, giving rise to all cell types in the body. Adult stem cells, which arise in the fetus and persist in small numbers in various tissues and organs in children and adults, are multipotent or unipotent.
- Human embryonic stem cells, and some types of human adult stem cells, can be grown in a cell culture. Exposing culture cells to particular physical and chemical signals induces them to differentiate into specific cell types.
- The use of human embryonic stem cells is controversial. Opponents believe that the life of a human being begins at conception and it is unethical to use its cells for someone else's benefit.
- Induced pluripotent stem cells have the developmental flexibility of embryonic stem cells but are made by reprogramming fully differentiated cells.

## 11.2 Cancer Cells: Good Cells Gone Bad

- Over a lifetime, an American male has a nearly one in two chance, and an American woman a one in three chance, of developing cancer. There are more than 200 different types of cancer, but four of them—lung, prostate, breast, and colon cancers—account for more than half of all cancers combined.
- Cancer develops when cells lose normal restraints on cell division and migration.

- The cell mass formed by the inappropriate proliferation of cells is known as a tumor. Tumors grow larger when they recruit blood vessels through angiogenesis.
- Precancerous cells become cancerous (malignant) when they invade other tissues. Invasive cells detach themselves from their surroundings and lose anchorage dependence. Cancer cells can spread from a primary tumor to set up secondary tumors in other organs, in the process known as metastasis.
- Positive growth regulators stimulate cell division; negative growth regulators restrain it. Runaway cell proliferation is the consequence if the cell cycle is excessively stimulated through the pathways controlled by positive growth regulators while the antiproliferation message of negative growth regulators is ignored.
- Gene mutations are the root cause of all cancers. Most human cancers are not hereditary. Most cancers develop as multiple mutations accumulate in a single cell.
- Genes whose action results in cell proliferation are proto-oncogenes. Proto-oncogenes that have become hyperactive as a result of DNA mutations are known as oncogenes. Genes that restrain cell division or cell migration are called tumor suppressor genes.
- The challenge in cancer treatment is to destroy malignant cells selectively. Inhibiting tumor growth by preventing angiogenesis, and slowing runaway cell division by blocking telomerase activity, are some of the new approaches in clinical trials.
- Cancer prevention, through increased public awareness of the risk factors, avoidance of infectious agents such as viruses and carcinogens, and the adoption of cancer-protective behaviors, is now a high priority in the war against cancer.
- Diet is believed to have a substantial impact on cancer risk.

# Key Terms

adult stem cell (p. 255)
anchorage dependence (p. 261)
angiogenesis (p. 262)
benign tumor (p. 261)
blastocyst (p. 257)
cancer cell (p. 262)
carcinogen (p. 271)
cell culture (p. 255)
cell differentiation (p. 254)
chemotherapy (p. 268)
cryosurgery (p. 268)
DNA (p. 254)

embryo (p. 257)
embryonic stem cell (p. 255)
fetus (p. 257)
gene (p. 254)
gene expression (p. 254)
growth factor (p. 263)
hormone (p. 263)
hormone therapy (p. 268)
induced pluripotent stem cell
  (iPSC) (p. 260)
inner cell mass (p. 257)
metastasis (p. 262)

mitotic division (p. 254)
multipotent (p. 257)
mutation (p. 264)
negative growth regulator (p. 263)
oncogene (p. 265)
pluripotent (p. 257)
positive growth regulator (p. 263)
precancerous cell (p. 261)
primary tumor (p. 262)
proto-oncogene (p. 265)
radiation therapy (p. 268)
receptor protein (p. 263)

regenerative medicine (p. 256)
secondary tumor (p. 267)
signal transduction pathway (p. 263)
somatic mutation (p. 266)
stem cell (p. 255)
telomere (p. 268)
totipotent (p. 257)
tumor (p. 261)
tumor suppressor
  gene (p. 265)
unipotent (p. 257)
zygote (p. 257)

# Self-Quiz

1. Which of the following cell types displays the *least* developmental flexibility?
   a. cell in a morula
   b. inner cell mass in a blastocyst
   c. bone marrow stem cell
   d. neuron

2. Growth factors
   a. can only stimulate cell division, not restrain it.
   b. are signaling molecules that affect cell division.
   c. function as positive growth regulators, but not as negative growth regulators.
   d. function as negative growth regulators, but not as positive growth regulators.

3. An example of an oncogene is a gene that codes for a protein that
   a. inhibits angiogenesis.
   b. promotes anchorage dependence.
   c. pushes the cell cycle from $G_1$ to S phase.
   d. triggers programmed cell death.

4. Why is cancer more common in older people than in younger people?
   a. It takes many cells to produce a cancer and older people have more cells.
   b. The DNA of older people is more susceptible to chemical carcinogens.
   c. Our immune system becomes stronger as we age.
   d. Multiple genes must pick up harmful mutations for cancer to develop, and that takes time.

5. Which of the following is the last step in the development of a cancer?
   a. anchorage independence
   b. loss of cell adhesion
   c. metastasis
   d. cell proliferation

6. The majority of human cancers
   a. are hereditary, that is, caused by the inheritance of mutated genes.
   b. are the result of infection with oncogenic viruses.
   c. arise in cells that have to be in $G_1$ as a normal part of their function.
   d. are caused by the activity of proto-oncogenes.

7. Tumor suppressor genes
   a. normally increase the rate of cell proliferation.
   b. normally promote cell migration.
   c. include those that code for enzymes that degrade cell adhesion proteins.
   d. include those that code for proteins that recognize and repair DNA damage.

8. Of the scenarios listed below, which is *most* likely to increase the odds that a cancer will develop?
   a. One copy of proto-oncogene is mutated such that it has becomes non-functional.
   b. One copy of a tumor-suppressor gene is mutated such that it has becomes non-functional.
   c. One copy of an oncogene is hyperactive.
   d. Two copies of a tumor suppressor gene are hyperactive.

9. The severe side effects of radiation therapy against cancer cells come about because
   a. there is indiscriminate acceleration of the cell cycle in cells that would normally not divide because they are fully differentiated.
   b. the therapy is narrowly targeted against the biomolecule, such as a growth factor receptor, found to be responsible for the cancerous change.
   c. all rapidly dividing cells in the body are killed by the radiation.
   d. all stem cells in the body are killed by the radiation.

# Analysis and Application

1. Compare iPSCs with embryonic stem cells and adult stem cells, in terms of both the developmental flexibility of these cells and their origins.

2. In 2004, Californians passed Proposition 71, which generated $3 billion in funding for embryonic stem cell research and turned the state into a magnet for stem cell researchers. Do you think controversial issues in science should be resolved by panels of experts, decided by elected government officials, or put to a vote by the citizens, as was done in California? Explain your rationale.

3. It is possible to remove a cell from an early-stage morula without harming the embryo. Should it be permissible to remove a cell in such a nondestructive manner and use it to make new hES lines? Explain your viewpoint.

4. Colon cancer develops in a series of stages. Outline the stages and what happens in each stage.

5. In light of the clear link between tobacco use and cancer, many have questioned the right of tobacco companies to continue selling such a deadly substance. Consider the issues of personal freedom versus public health policy and explain what restraints, if any, you think should be placed on the sale of tobacco.

6. As environmental causes of cancer receive increasing attention, the warning labels on food have become longer and more ominous-sounding. Since many factors contribute to cancer, do you think that expanding food warning labels is an effective approach to reducing cancer risk? If so, how might one combat the public's tendency to ignore long and complex warning labels?

# Patterns of Inheritance

**A ROYAL MYSTERY.** Czar Nicholas II, seen here with the czarina and their five children, was the last czar of Russia. His youngest daughter, Princess Anastasia, is on the right in this 1910 photo.

# The Lost Princess

In the early hours of July 17, 1918, Nicholas II (the last czar of Russia), his family, and his servants were awakened and directed to the ground floor of a mansion where they were being held by Bolshevik secret police. The czar had abdicated his throne the previous year, and the family had been captured and moved to a succession of hideaways. Thinking the family was about to be moved again and expecting a wait, the czar called for chairs for Czarina Alexandra, his son Alexei, and himself. Then, with his daughters—Olga, Tatiana, Maria, and Anastasia—and servants assembled, the czar turned his attention to Bolshevik officer Yakov Yurovsky. But Yurovsky abruptly announced that they were all to be executed, and seconds later soldiers opened fire, bringing a brutal end to the 300-year-old Romanov dynasty.

Or had it ended? Two years later, a young woman with scars on her head and body was admitted to a Berlin mental institution. She seemed not to know her own name and became known as Anna Anderson. A fellow patient with social connections insisted that Anna was the czar's youngest daughter, Anastasia, who had miraculously escaped execution. Hopeful friends and relatives of the czar visited Anna but sadly concluded she was not Anastasia; she could not even speak Russian. In 1927, a private detective hired by Anastasia's uncle reported that Anna was Franziska Schanzkowska, a Polish factory worker injured when a grenade exploded in a weapons factory.

But the captivating legend of the Russian princess who escaped death had a life of its own, celebrated in books, magazines, and movies for more than 60 years. Minor European royalty insisted that they recognized Anna, and for the rest of her life, a succession of supporters claimed her as the lost princess Anastasia. Only a combination of detective work and genetics finally answered the question of Anna's background. To answer the questions raised here and to explore other aspects of inheritance, let's take a look at the principles of genetics, the focus of this chapter.

> What rules govern the inheritance of traits? How was genetics used to show whether Anna Anderson was Anastasia Romanov?

**MAIN MESSAGE** Inherited characteristics of organisms are governed by genes and may be influenced by environmental factors as well.

## KEY CONCEPTS

- Genetics is the study of inherited characteristics (genetic traits) and the genes that affect those traits.

- A gene is a stretch of DNA that affects one or more genetic traits. It is the basic unit of inheritance.

- A phenotype is the specific version of a genetic trait that is displayed by a particular individual.

- Alternative versions of a gene (alleles) arise by mutation. The genotype is the allelic makeup of an individual.

- Diploid cells have two copies of every gene—one copy inherited from the male parent, the other from the female parent. Homozygotes have the same two alleles for a particular gene; heterozygotes have two different alleles.

- A dominant allele controls the phenotype in a heterozygote. The masked allele is recessive.

- Mendel's laws of inheritance help us predict the phenotypes of offspring from the genotypes of the parents.

- Mendel's laws apply broadly to most sexually reproducing organisms, but geneticists have extended those basic laws to describe more complex patterns of inheritance.

- Many aspects of an organism's phenotype are determined by multiple genes that interact with one another and with the environment, so offspring with identical genotypes can have very different phenotypes.

**HUMANS HAVE USED THE PRINCIPLES OF IN-HERITANCE** for thousands of years. Noticing that offspring tend to be similar to their parents, people raised animals and plants by mating individuals with desirable characteristics and selecting the "pick of the litter" for further breeding. Our ancestors used such methods to domesticate wild animals and to develop agricultural crops from wild plants (see Figure 1.10). As a field of science, however, *genetics* did not begin until 1866, the year that an Augustinian monk named Gregor Mendel (**FIGURE 12.1**) published a landmark paper on inheritance in pea plants. Prior to Mendel's work, some aspects of inheritance were understood, but no one had conducted extensive and systematic experiments to explain the patterns in which inherited characteristics, or *genetic traits*, are passed from parent to offspring.

Mendel's extraordinary insights were made possible by his exceptional training. As a young monk, Mendel attended the University of Vienna, where he took courses ranging from mathematics to botany. Upon assuming his duties at the monastery of Saint Thomas, Mendel put his training in probability statistics and plant breeding to good use. He used mathematics to analyze the inheritance of seven different genetic traits in garden peas. From the patterns he observed, Mendel was able to deduce the fundamental principles that govern how genetic information is passed from one generation to the next.

Mendel's experiments with peas led him to some basic generalizations, now known as the laws of inheritance, that have stood the test of time. Even though Mendel's laws have been modified considerably by modern genetics, his predictions about the outcomes of certain types of mating experiments still hold true, and they apply broadly to most organisms that multiply themselves through sexual reproduction. Mendel proposed that the inherited characteristics of organ-

**FIGURE 12.1**
**Gregor Mendel**
**(1822–1884)**

isms are controlled by hereditary factors—now known as *genes*—and that one factor for each trait is inherited from each parent. Although he did not use the word "gene," Mendel was the first to propose the concept of the gene as the basic unit of inheritance.

In the more than 100 years since Mendel's work, we have learned a great deal about genes, especially about their physical and chemical properties. We now know that genes are located on chromosomes, which consist of DNA molecules complexed with packaging proteins (as described in Chapter 10). A **gene** is a stretch of DNA that governs one or more genetic traits. A single chromosome typically contains many hundreds of genes. Genes commonly contain instructions for the manufacture of proteins, and to a very large extent it is these proteins that shape the chemical, structural, and behavioral characteristics of the individual organism. In other words, the genetic traits we display are brought about mostly by proteins, which in turn are encoded by genes.

Before we describe Mendel's remarkable insights into the principles of inheritance, and the further elaborations of those principles in modern times, we will introduce some basic terminology used in genetics. Some of this vocabulary, and its associated concepts, did not exist in Mendel's time, which probably made it difficult for his contemporaries to appreciate the significance of his largely mathematical analysis. Perhaps the biggest obstacle for Mendel's contemporaries (who included Charles Darwin) was the lack of understanding of meiosis, especially the random assortment of chromosomes during meiosis I (see Section 10.5). Mendel's work was largely ignored for about 30 years after it was published. Upon its "rediscovery" in the early 1900s, his principles became the foundation for the modern discipline of **genetics**, the study of genes. Today, Mendel's principles have been extended to reveal, in much greater complexity, how genes, especially as influenced by the environment, shape the observable characteristics of organisms.

## 12.1 Essential Terms in Genetics

A **genetic trait** is any inherited characteristic of an organism that can be observed or detected in some manner. Genetic traits are controlled at least in part by genes, although, as we shall see later, the outward manifestation of a genetic trait can also be influenced by the environment. Sizes of pumpkins, stripe patterns of zebras, and the song of the meadowlark are examples of genetic traits. How many different genetic traits can you spot among the humans pictured in **FIGURE 12.2**? Physical traits such as height or the shape of the face are among the easiest to observe. Some biochemical traits, such as hair color, which is produced by a pigment called melanin, are also readily observed. It takes special tests to detect other biochemical traits, such as a person's blood type or susceptibility to certain diseases. Certain behavioral traits—shyness, extroversion, risk taking—are now known to be influenced by genes.

At the molecular level, a gene consists of a stretch of DNA. Geneticists often use italicized letters, symbols, and numbers to represent a particular gene, although other conventions also exist. For example, the *Orange* gene, which leads to orange hair in cats, is designated by an italicized *O*. More than a dozen different genes affect skin color in humans, and the shorthand for one key gene is *MC1R*. A gene called *IGF2* is crucial for human development and is among the many genes that affect height and weight in humans.

Some genetic traits are invariant or nearly invariant, meaning that they are about the same in all individuals in the population. For example, all the people in Figure 12.2 have the same number of eyes (two); that is, the number of eyes in humans is essentially invariant. Other traits you see in Figure 12.2 are quite variable. The display of a *particular version* of a genetic trait in a specific individual is the **phenotype** of that individual. For example, black, bay, and chestnut are different types of coat color in horses, and each of them is therefore a specific phenotype of the coat color trait. How many different phenotypes do you know of for the genetic traits you spotted in Figure 12.2?

## Diploid cells have two copies of every gene

As we described in Section 10.5, somatic cells in plants and animals are **diploid** in that they possess two copies of each type of chromosome. The two copies make up a **homologous pair** for each chromosome type. One member of the homologous pair is inherited from the male parent and is called the **paternal homologue**; the other member of the pair is inherited from the female parent and is called the **maternal homologue**. Human somatic cells, for example, have a diploid (double) set of 23 different chromosomes;

**FIGURE 12.2 Genetic Traits in Humans**
How many genetic traits can you identify in this crowd? Are the traits discrete, falling into clear-cut categories, or do they vary by degree in different people (the way height does)? Do you suppose that the traits you have identified are controlled by genes exclusively, or might they be influenced by environmental factors also?

that is, our somatic cells have 23 pairs of homologous chromosomes, for a total of 46 chromosomes.

Because there are two copies of each type of chromosome (with one exception), a diploid cell has two copies of every gene located on these chromosomes. One copy of the gene is located on the paternal homologue, the other copy on the maternal homologue, of each homologous pair (**FIGURE 12.3**). The one exception concerns the so-called sex chromosomes: the X and Y chromosomes. As discussed in Section 10.3, while female mammals have matched XX homologous chromosomes, male mammals have only one X chromosome, which is paired up with a dissimilar Y chromosome. This means that in all of their diploid cells, male mammals have *only one* copy of the genes located on the X chromosome, while females of the species have two copies, one on each of their two X chromosomes.

We also mentioned in Chapter 10 that gametes—sperm and egg cells—are **haploid** because these sex cells have only one set of chromosomes (half the total number of chromosomes found in a diploid cell). Because a gamete has only one copy of each pair of homologous chromosomes—either the paternal or the maternal homologue—it possesses *only one copy* of

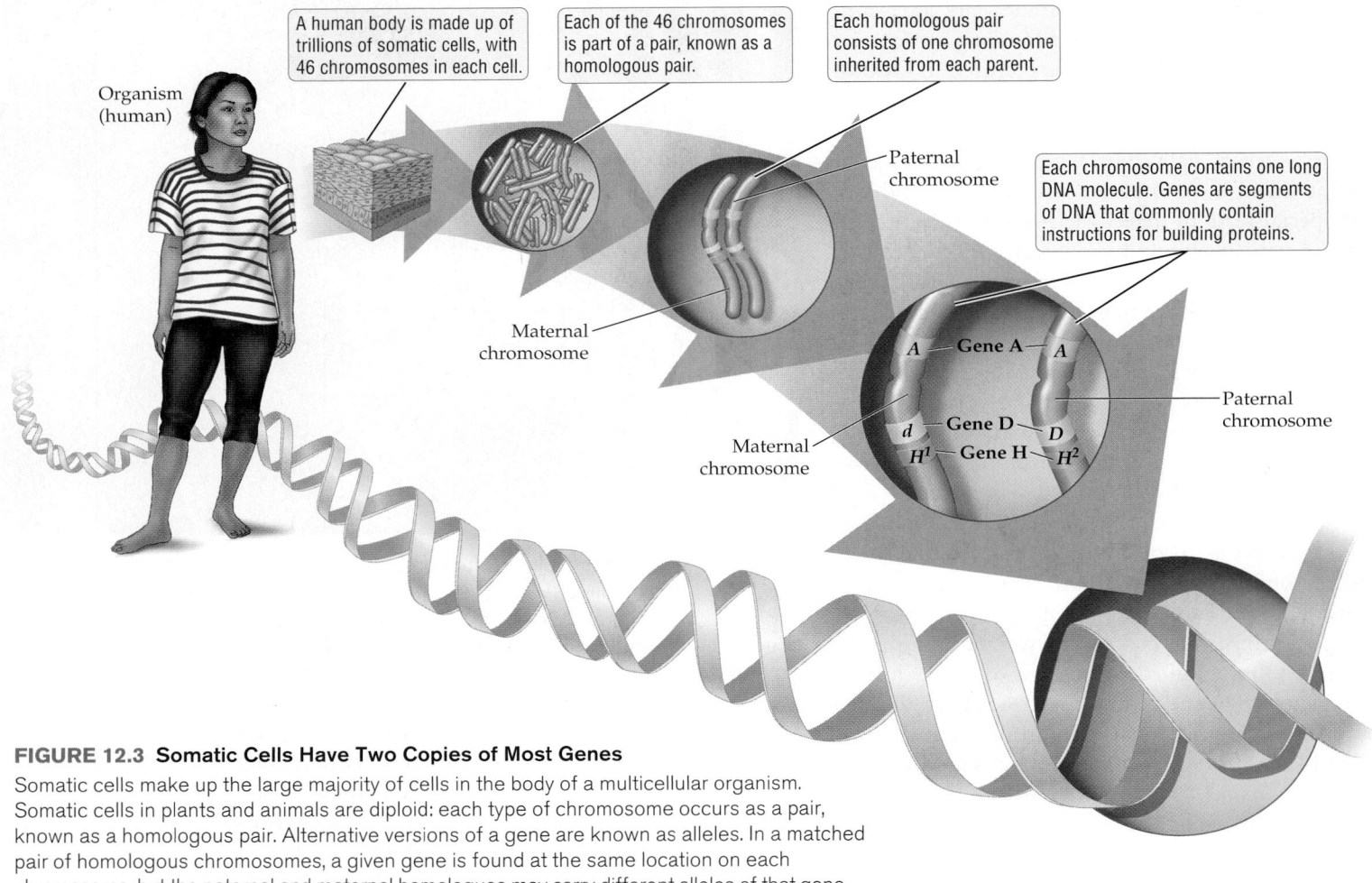

**FIGURE 12.3 Somatic Cells Have Two Copies of Most Genes**
Somatic cells make up the large majority of cells in the body of a multicellular organism. Somatic cells in plants and animals are diploid: each type of chromosome occurs as a pair, known as a homologous pair. Alternative versions of a gene are known as alleles. In a matched pair of homologous chromosomes, a given gene is found at the same location on each chromosome, but the paternal and maternal homologues may carry different alleles of that gene.

every gene. Gametes are created through meiosis. Meiosis generates haploid daughter cells by halving the diploid chromosome number of the parent cell (see Figure 10.13). A human gamete, for example, has only 23 chromosomes, half the diploid set of 46.

## Genotype directs phenotype

Different versions of a given gene are known as **alleles** of that gene. For example, the ABO blood groups in humans are controlled by at least three alleles of the $I$ gene: $I^A$, $I^B$, and $i$. Note that while a diploid cell contains at most two different alleles for a given gene, and a haploid cell can have only one of all the possible alleles of a gene, a population of individuals may collectively harbor many different alleles. For example, every one of the three alleles of the $I$ gene is probably

represented among the students in your classroom. The genetic diversity we see in natural populations comes about because each population contains many different alleles of its many genes. The main reason just about everyone in your classroom looks recognizably different from the next person is that each of you carries different alleles for many of the roughly 25,000 genes that all humans possess.

The **genotype** of an individual is the allelic makeup of that individual with respect to a specific genetic trait. In other words, a genotype is the pair of alleles an individual has that code for a given phenotype. It is the genotype that partly, or wholly, creates the phenotype. For example, the genotype $ii$ is responsible for the blood group phenotype known as type O.

An individual who carries two copies of the *same* allele (such as $I^AI^A$, $I^BI^B$, or $ii$) is said to be **homozygous** for that gene. An individual whose genotype consists of two *different* alleles for a given phenotype (as in $I^AI^B$, $I^Ai$, or $I^Bi$) is said to be **heterozygous** for that gene.

## Some phenotypes are controlled by dominant alleles

Some genes have alleles that show *dominance* when paired with another allele; that is, one allele can prevent a second allele from affecting the phenotype when the two alleles are paired together in a heterozygote. The allele that exerts a controlling influence on the phenotype, to the point of masking the effect of the allele with which it is paired, is said to be **dominant**. An allele that has no effect on the phenotype when paired with a dominant allele is said to be **recessive**. For example, in some breeds of cats the *L* allele controls hair length. The *L* allele is dominant over the *l* allele. A cat with one or two copies of the *L* allele (*LL* or *Ll*) has short hair; only if a cat is homozygous for the recessive allele (*ll*) will it have long hair (**FIGURE 12.4**). In this example, the *L* allele is dominant over the *l* allele, and it therefore masks the presence of the recessive allele in a heterozygote (*Ll*).

Of the thousands of human genetic traits, governed by an estimated 25,000 genes, fewer than 4,000 are known or suspected to be controlled by a single gene with a dominant and a recessive allele. Cleft chin, freckles, tongue-rolling, unattached earlobes,

"widow's peak," and "hitchhiker's thumb," are commonly offered as examples of human phenotypes controlled by a single, dominant allele; however, the genetics of these traits is more complex. Some of these traits (freckles, for example) are controlled by more than one gene. Others in this list are not discrete traits, meaning that their phenotypes cannot be reliably sorted into just two clear-cut categories (there are all kinds of hairline shapes and degrees of ear lobe attachment, for example). Nor should you infer from the term that a dominant allele is somehow better for the individual or more common in a population. For example, although the *I* allele of the gene that controls a person's ABO blood group is dominant over *i*, there is no evidence that a person with the dominant allele is better off than someone who has the recessive version of the gene. And across the globe, the *i* allele is far more common than either of the *I* alleles.

## Gene mutations are the source of new alleles

Different alleles of a gene arise by **mutation**, which we can define briefly here as any change in the DNA that makes up a gene (see Section 14.4 for a more detailed discussion). When a mutation occurs, the new allele may contain instructions for a protein whose form differs from the version specified by the original

**DOMINANT AND RECESSIVE ALLELES**

| Phenotype | Short hair | Short hair | Long hair |
|---|---|---|---|
| Genotype | *LL* homozygote | *Ll* heterozygote | *ll* homozygote |

**FIGURE 12.4 In Cats, Short Hair Is Dominant over Long Hair**
The *L* allele is dominant over the *l* allele and therefore confers short hair in both the homozygous and heterozygous states (*LL* and *Ll*). The long-hair phenotype, controlled by the *l* allele, is manifested only when unmasked in the *ll* homozygote.

allele. By specifying different versions of a protein, the different alleles of a gene produce genetic differences among the individuals in a population.

Mutations are sometimes harmful. For example, a mutation may lead to the production of a protein that performs a vital function poorly or not at all. Harmful or nonfunctional alleles tend to be recessive. Recessive alleles can be fairly common in a population because heterozygotes can harbor such alleles and pass them on to future generations without suffering any ill effects themselves. The one "good copy" of the normal allele masks the effect of the harmful recessive allele, so the heterozygote does not display the negative phenotype.

The most common mutations are those that are neither harmful nor beneficial to the individual. So-called neutral mutations arise, for instance, when a new allele specifies a protein that is nearly identical to the protein specified by the original allele. In some cases a cell can tolerate variation in the activity of a protein with no harm to the organism, as is the case for the *I* gene in humans. Occasionally, mutations produce alleles that improve on the original protein or carry out new, useful functions. Such beneficial mutations are the rarest of the three mutation types.

The nature of mutations is often misunderstood. Two misconceptions are particularly common: that mutations happen because the individual "needs" them and that all mutations are inherited. First, mutations occur at random with respect to their usefulness. There is no evidence, for example, that beneficial mutations occur because they are "needed" by the organism to cope with environmental challenges. Second, mutations can happen at any time and in any cell of the body. In multicellular organisms, however, only mutations that are present in gametes, or in the cells that ultimately produce gametes, can be passed on to offspring.

## Controlled crosses help us understand patterns of inheritance

A **genetic cross**, or just "cross" for short, is a controlled mating experiment performed to examine how a particular trait may be inherited. "Cross" can also be used as a verb, as in "individuals of genotype *AA* were crossed with individuals of genotype *aa*." The parent generation in a genetic cross is called the **P generation**. The first generation of offspring is called the **$F_1$ generation** ("F" is for "filial," a word that refers to a son or daughter). When the individuals of the $F_1$ gen-

| TABLE 12.1 | Essential Terms in Genetics |
|---|---|
| **TERM** | **DEFINITION** |
| Allele | One of two or more alternative versions of a gene that exist in a population of organisms. |
| Dominant allele | The allele that controls the phenotype when paired with a different allele in a heterozygote individual. |
| $F_1$ generation | The first generation of offspring in a breeding trial involving a series of genetic crosses. |
| $F_2$ generation | The second generation of offspring in a series of genetic crosses. |
| Gene | The basic unit of genetic information. Each gene consists of a stretch of DNA that is part of a chromosome, and it affects at least one genetic trait. |
| Genetic cross | A mating experiment, often performed to analyze the inheritance of a particular trait. |
| Genotype | The genetic makeup of an individual; more specifically, the two alleles of a given gene that affect a specific genetic trait in a given individual. |
| Heterozygote | An individual that carries one copy of each of two different alleles (for example, an *Aa* individual or a $C^W C^R$ individual). |
| Homozygote | An individual that carries two copies of the same allele (for example, an *AA*, *aa*, or $C^W C^W$ individual). |
| P generation | The parent generation in a breeding trial involving a series of genetic crosses. |
| Phenotype | The specific version of a genetic trait that is displayed by a given individual; for example, black, brown, red, and blond are phenotypes of the hair color trait in humans. |
| Recessive allele | An allele that does not affect the phenotype when paired with a dominant allele in a heterozygote. |
| Trait (genetic) | Any inherited feature of an organism that can be observed or detected; size, color and length of fur, and aggressive behavior are known examples of genetic traits. |

eration are crossed with each other, the resulting off-spring are said to belong to the **F₂ generation**.

Definitions of these important genetics terms are collected in Table 12.1. Study these terms carefully, and refer to the table as needed as you study the rest of this chapter.

## 12.2 Basic Patterns of Inheritance

Having defined some key genetics concepts and discussed how mutations produce new alleles, we are ready now to explore how genes are transmitted from parents to offspring. Prior to Mendel, many people argued that the traits of both parents were blended in their offspring, much as paint colors blend when they are mixed together. According to this notion, which was known as the theory of blending inheritance, offspring should be intermediate in phenotype to their two parents, and it should *not* be possible to see "lost" traits reappear in later generations.

Many observations, however—including Mendel's—do not match these predictions. The features of offspring often are not intermediate between those of their parents, and it is common for traits to skip a generation. How can such observations be explained? During 8 years of investigation, Gregor Mendel conducted many experiments to analyze inheritance in pea plants, and his results led him to reject the idea of blending inheritance. Mendel proposed instead that for each trait, offspring inherit two separate units of genetic information (now known as genes), one from each parent.

### Mendel's genetic experiments began with true-breeding pea plants

Peas are excellent organisms for studying inheritance. Ordinarily, peas self-fertilize; that is, an individual pea plant contains both male and female reproductive organs, and it fertilizes itself. But because peas can also be mated experimentally, Mendel was able to perform carefully controlled genetic crosses. In addition, peas have *true-breeding varieties*, which means that when these plants self-fertilize, *all* of their offspring have the same phenotype as the parent. For example, one variety has yellow seed and it produces only offspring with yellow seeds when it is bred with itself. True-breeding pets and farmyard animals are commonly called *purebreds*. A mating of two purebred Labrador retrievers can be expected to produce pups that *all* have the distinctive qualities of this breed.

According to what we know today about how genotypes affect phenotypes, individuals that breed true for a given phenotype must have a homozygous genotype, such as *PP* or *pp* for the flower color trait shown in **FIGURE 12.5**. Mendel based all his experiments on varieties that were homozygous—and therefore true-breeding—for traits such as plant height, flower position, flower color, and the color and shape of the seeds or of the pea pods. He raised these varieties by continually self-fertilizing a plant with a particular phenotype (such as purple flowers) over many generations, until he was sure he had a line of plants that would reliably produce offspring with the same phenotype (purple flowers) as the parent.

Mendel then crossed true-breeding lines with contrasting phenotypes (purple flowers and white flowers, for example) and meticulously recorded the phenotypes of all the offspring over two generations. He began with a set of original, true-breeding parents (P generation) and tracked the phenotypes through two generations of **hybrid** (non-true-breeding) offspring. For example, he crossed plants that bred true for purple flowers with plants that bred true for white flowers

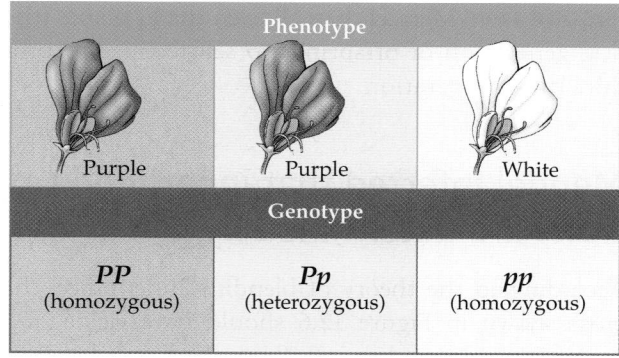

**FIGURE 12.5 True-Breeding Traits Have a Homozygous Genotype**
Flower color in peas is controlled by a gene with two alleles (*P* and *p*). Although there are three genotypes (*PP*, *Pp*, and *pp*), there are only two phenotypes (purple flowers and white flowers). There are fewer phenotypes because genotypes *PP* and *Pp* both produce purple flowers, while *pp* produces white flowers.

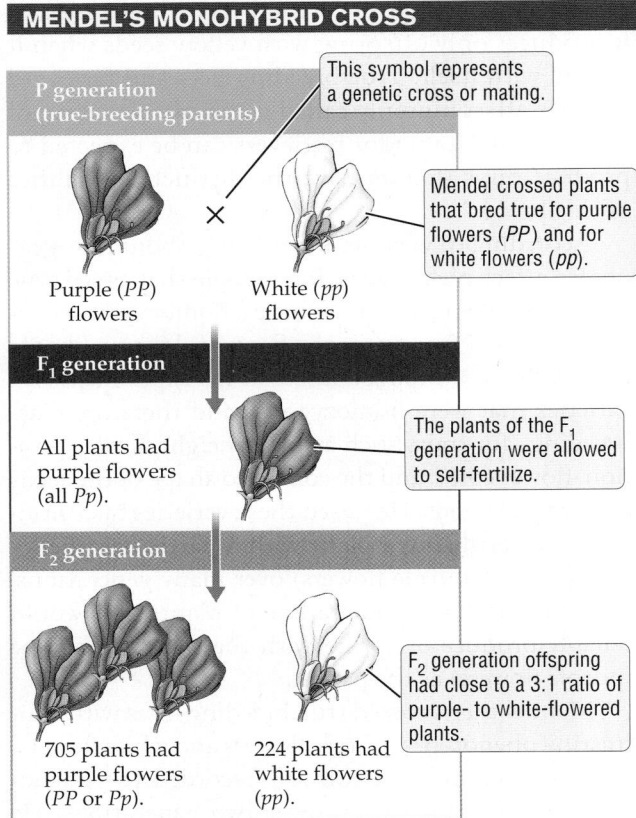

P generation (true-breeding parents)

This symbol represents a genetic cross or mating.

Purple (*PP*) flowers

White (*pp*) flowers

Mendel crossed plants that bred true for purple flowers (*PP*) and for white flowers (*pp*).

**F₁ generation**

All plants had purple flowers (all *Pp*).

The plants of the F₁ generation were allowed to self-fertilize.

**F₂ generation**

705 plants had purple flowers (*PP* or *Pp*).

224 plants had white flowers (*pp*).

F₂ generation offspring had close to a 3:1 ratio of purple- to white-flowered plants.

**FIGURE 12.6 Inheritance of a Single Trait over Three Generations**

Mendel crossed parent plants that were true-breeding (homozygous) for two discrete phenotypes (purple or white) of a particular trait (flower color). Such breeding trials are described as monohybrid crosses because the F₁ plants are hybrid (heterozygous) for a single trait (flower color).

(**FIGURE 12.6**). Mendel then allowed the F₁ plants (the first generation of offspring) to self-fertilize to produce the F₂ generation.

## Mendel inferred that inherited traits are determined by genes

According to the theory of blending inheritance, the cross shown in Figure 12.6 should have yielded F₁-generation plants bearing flowers of intermediate color. Instead, all the F₁ plants had purple flowers. Furthermore, Mendel noted that when the F₁ plants self-fertilized, about 25 percent of the F₂ offspring had white flowers. That is, white flowers appeared in a 1:3 ratio in the F₂ generation. Mendel realized that the reappearance of white flowers in the second generation was not consistent with the theory of blending inheritance.

Mendel studied seven true-breeding traits in peas, repeating the experiment illustrated in Figure 12.6 for each trait. In the F₂ generation, he repeatedly observed a 3:1 ratio of the dominant to the recessive phenotype for each of the traits under study. These results led him to propose a new theory of inheritance, in which the units of inheritance (which we now call genes) exist as discrete factors that do not lose their unique characteristics when crossed, as would colors of paints blended together. Using modern terminology, we can summarize Mendel's concepts as follows:

1. *Alternative versions of genes cause variation in inherited traits.* For example, peas have one version (allele) of a certain gene that causes flowers to be purple, and another version (a different allele) of the same gene that causes flowers to be white. One individual carries at most two different alleles for a particular gene.

2. *Offspring inherit one copy of a gene from each parent.* In his analysis of crosses like the one in Figure 12.6, Mendel reasoned that white flowers could not reappear in the F₂ generation unless the white-flower allele was present in F₁ plants to pass on to F₂ plants. He deduced that the pea plant must carry two copies of the flower color gene (one copy that caused white flowers and one copy that caused purple flowers). Mendel was right: with the exception of our gametes, somatic cells in the adult organism typically have one maternal and one paternal copy of each of their many genes (see Figure 12.3).

3. *An allele is dominant if it has exclusive control over the phenotype of an individual when paired with a different allele.* For example, plants that breed true for purple flowers must have two copies of the *P* allele (that is, they are of genotype *PP*), since otherwise they would occasionally produce white flowers. Similarly, plants that breed true for white flowers have two copies of the *p* allele (genotype *pp*). Working back from the phenotypes of both the F₁ and F₂ generations, Mendel correctly deduced that the F₁ plants in Figure 12.6 must have genotype *Pp*. He realized that the genotype of the F₁ plants was best explained if it was assumed that each such plant received a *P* allele from the *PP* parent with purple flowers and a *p* allele from the *pp* parent with white flowers. Since all the F₁ plants had purple flowers, one must also assume that the *P* allele is dominant over the *p* allele. The recessive allele, *p*, has no effect on the phenotype, because its effects are masked by the *P* allele's dominant impact on flower color.

4. *The two copies (alleles) of a gene segregate during meiosis and end up in different gametes.* Recall that each gamete receives only one copy of each gene in the process of meiosis. If an organism has two copies of the same allele for a particular trait, as in the homozygous varieties used by Mendel, all of its gametes will contain that allele. However, if the organism has two different alleles, as an individual of genotype *Pp* has, then 50 percent of the gametes will receive one allele and 50 percent will receive the other allele.

5. *Gametes fuse without regard to the alleles they carry.* When gametes fuse to form a zygote, they do so *randomly* with respect to the alleles they carry for a particular gene.

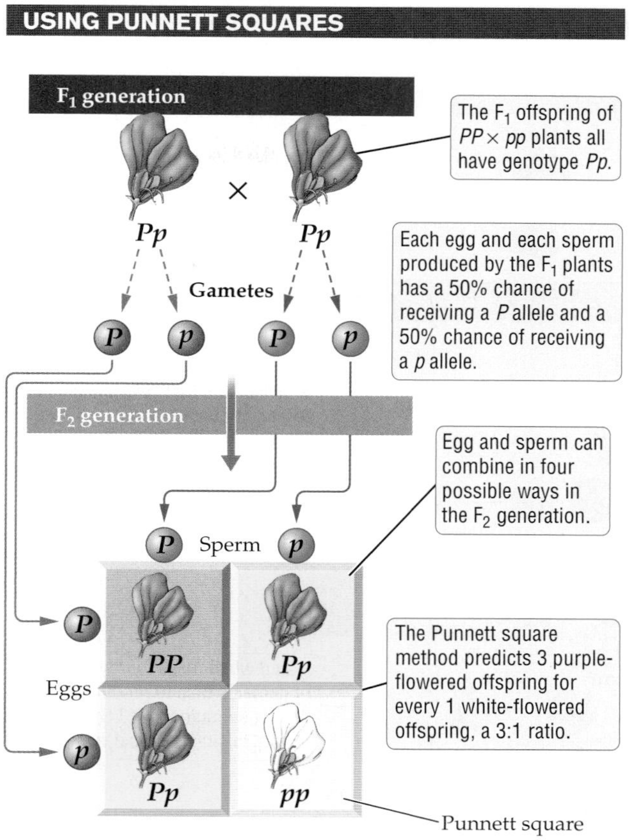

# 12.3 Mendel's Laws of Inheritance

Mendel summarized the results of his experiments in two laws: the *law of segregation* and the *law of independent assortment*. He was able to deduce the law of segregation from breeding experiments in which he tracked a *single trait* (such as flower color or plant height). The law of independent assortment was based on *two-trait* breeding experiments: crosses in which he tracked two completely different traits at the same time.

## Mendel's single-trait crosses revealed the law of segregation

The **law of segregation** states that the two copies of a gene are separated during meiosis and end up in different gametes. This law can be used to predict how a single trait will be inherited. To see how, revisit the experiment shown in Figure 12.6. In that experiment, Mendel crossed plants that bred true for purple flowers (genotype *PP*) with individuals that bred true for white flowers (genotype *pp*). This cross produced an $F_1$ generation composed entirely of **monohybrids**: they were all hybrids—heterozygotes (*Pp*)—with respect to the one trait, flower color. According to the law of segregation, when the $F_1$ plants reproduced, 50 percent of the pollen (sperm) should have contained the *P* allele, and the other 50 percent the *p* allele. The same is true for the eggs.

We can represent the separation of the two copies of a gene during meiosis, and their random recombining through fertilization, using a gridlike diagram called a **Punnett square** (**FIGURE 12.7**). The Punnett

**USING PUNNETT SQUARES**

The $F_1$ offspring of *PP* × *pp* plants all have genotype *Pp*.

Each egg and each sperm produced by the $F_1$ plants has a 50% chance of receiving a *P* allele and a 50% chance of receiving a *p* allele.

Egg and sperm can combine in four possible ways in the $F_2$ generation.

The Punnett square method predicts 3 purple-flowered offspring for every 1 white-flowered offspring, a 3:1 ratio.

**FIGURE 12.7 The Punnett Square Method Is Used to Predict All Possible Outcomes of a Genetic Cross**

Punnett squares chart the segregation (separation) of alleles into gametes and all the possible ways in which the alleles borne by these gametes can be combined to produce offspring.

square diagram shows all possible ways that two alleles can be brought together through fertilization. To create a Punnett square, list all possible genotypes of the male gametes across the top of the grid, writing each unique genotype just once. List all possible genotypes of the female gametes along the left edge of the grid, again writing each unique genotype only once. Next, fill in the grid by combining in each box (or "cell") the male genotype at the top of each column with the female genotype listed at the beginning of each row (follow the blue and pink arrows in Figure 12.7). In Figure 12.7, regardless of whether it has a *P* or a *p* allele, each sperm has an equal chance of fusing with an egg that has a *P* allele or an egg that has a *p* allele. The Punnett square shows all four ways in which the two different alleles in the sperm can combine with the two alleles found in the egg. The four genotypes shown within the Punnett square are all equally likely outcomes of this cross.

Using the Punnett square method, we can predict that ¼ of the $F_2$ generation is likely to have genotype *PP*, ½ to have genotype *Pp*, and ¼ to have genotype *pp*. Because the allele for purple flowers (*P*) is dominant, plants with *PP* or *Pp* genotypes have purple flowers, while *pp* genotypes have white flowers. Therefore, we

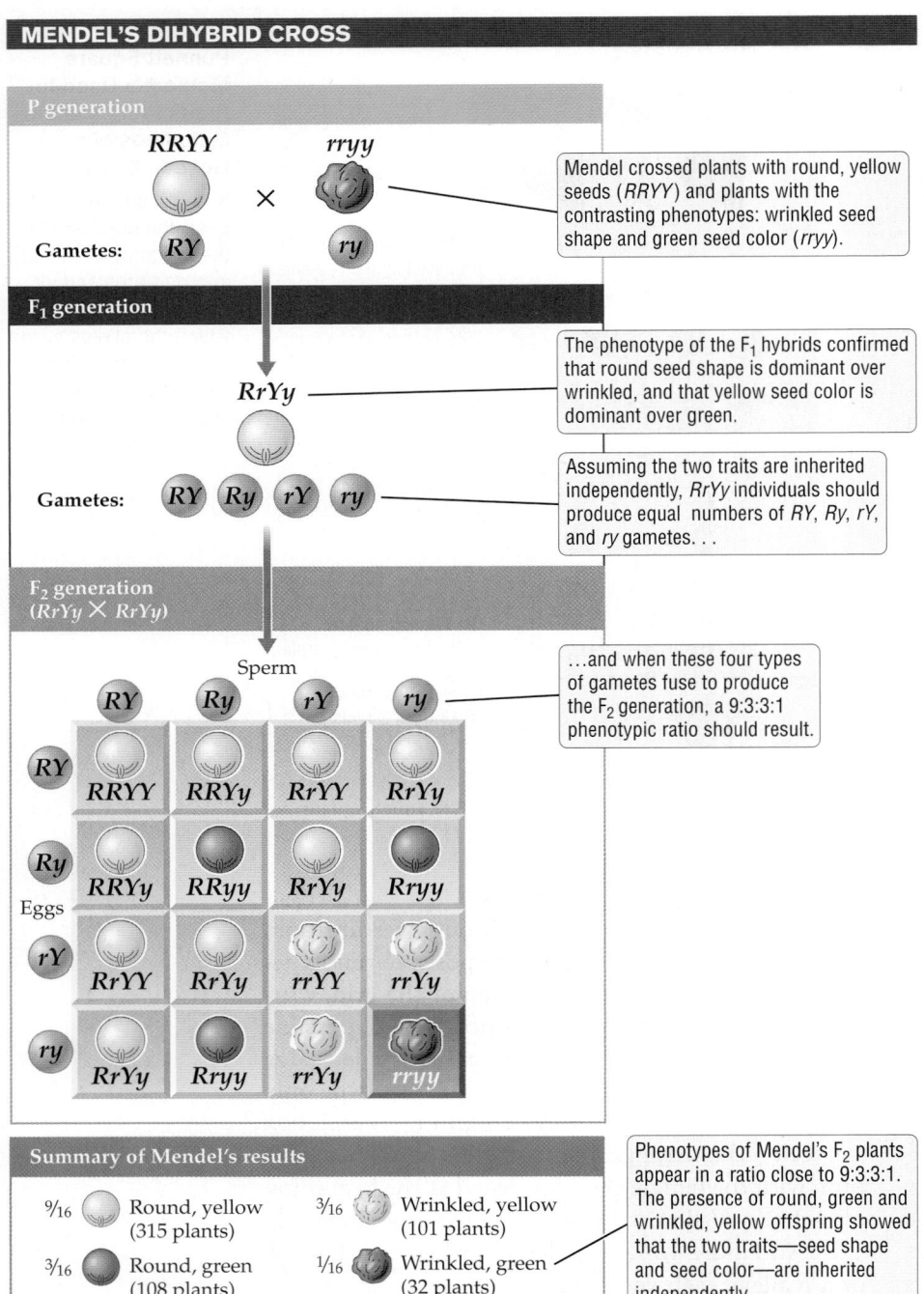

**P generation**

*RRYY* × *rryy*

Gametes: *RY* *ry*

Mendel crossed plants with round, yellow seeds (*RRYY*) and plants with the contrasting phenotypes: wrinkled seed shape and green seed color (*rryy*).

**F₁ generation**

*RrYy*

The phenotype of the F₁ hybrids confirmed that round seed shape is dominant over wrinkled, and that yellow seed color is dominant over green.

Gametes: *RY* *Ry* *rY* *ry*

Assuming the two traits are inherited independently, *RrYy* individuals should produce equal numbers of *RY*, *Ry*, *rY*, and *ry* gametes. . .

**F₂ generation**
**(*RrYy* × *RrYy*)**

Sperm

*RY* *Ry* *rY* *ry*

...and when these four types of gametes fuse to produce the F₂ generation, a 9:3:3:1 phenotypic ratio should result.

Eggs

| | *RY* | *Ry* | *rY* | *ry* |
|---|---|---|---|---|
| *RY* | *RRYY* | *RRYy* | *RrYY* | *RrYy* |
| *Ry* | *RRYy* | *RRyy* | *RrYy* | *Rryy* |
| *rY* | *RrYY* | *RrYy* | *rrYY* | *rrYy* |
| *ry* | *RrYy* | *Rryy* | *rrYy* | *rryy* |

**Summary of Mendel's results**

⁹⁄₁₆ Round, yellow (315 plants)

³⁄₁₆ Round, green (108 plants)

³⁄₁₆ Wrinkled, yellow (101 plants)

¹⁄₁₆ Wrinkled, green (32 plants)

Phenotypes of Mendel's F₂ plants appear in a ratio close to 9:3:3:1. The presence of round, green and wrinkled, yellow offspring showed that the two traits—seed shape and seed color—are inherited independently.

**FIGURE 12.8 Inheritance of Two Traits over Three Generations**

Mendel used two-trait breeding experiments to test the hypothesis that the alleles of *two different genes* are inherited independently from each other. In one set of experiments, illustrated here, Mendel tracked the seed shape trait controlled by the *R/r* alleles and the seed color trait controlled by the *Y/y* alleles. The real test of the hypothesis came when Mendel examined the phenotypes of the offspring produced by crossing the heterozygous F₁ plants (*RrYy*). As predicted by the hypothesis, two new phenotypic combinations were found among the F₂ offspring: plants that made round, green seeds (*R-yy*) and plants that made wrinkled, yellow seeds (*rrY-*). The bottom panel summarizes the ratio of the two parental phenotypes and the two novel, nonparental phenotypes. A two-trait breeding experiment in which the F₁ plants are double heterozygotes (heterozygous for both traits) is called a dihybrid cross.

predict that ¾ (75 percent) of the F₂ generation will have purple flowers and ¼ (25 percent) will have white flowers—a 3:1 ratio. This prediction is very close to the actual results that Mendel obtained. Of a total of 929 F₂ plants that Mendel raised, 705 (76 percent) had purple flowers and 224 (24 percent) had white flowers.

## Mendel's two-trait experiments led to the law of independent assortment

Mendel also performed experiments in which he simultaneously tracked the inheritance of *two* genetic traits by performing crosses with **dihybrids** (individuals heterozygous for two traits). For example, pea seeds can have a round or wrinkled shape, and they can be yellow or green. Two different genes control the two different traits: the *R* gene controls seed shape; the *Y* gene controls the color of the seed. Mendel wanted to know what would happen if true-breeding round, yellow-seeded individuals (genotype *RRYY*) were crossed with true-breeding wrinkled, green-seeded individuals (genotype *rryy*).

When Mendel performed the two-trait cross (**FIGURE 12.8**), all of the resulting F₁ plants had round, yellow seeds. From the phenotypes of this F₁ generation, Mendel could see that the allele for round seeds (denoted *R*) was dominant over the allele for wrinkled seeds (denoted *r*). Similarly, with respect to seed color, Mendel deduced that the allele for yellow seeds (*Y*) was dominant over the allele for green seeds (*y*).

Next, Mendel crossed large numbers of the F₁ plants (genotype *RrYy*) to raise a generation of F₂ plants. He knew that the phenotypes of the F₂ generation would answer the question that most interested him: *Is the inheritance of seed shape independent of the inheritance of seed color?* If his hunch was right, the distribution of the alleles of one gene into gametes would be independent of the distribution of the alleles of the other gene, so all possible combinations of the alleles would be found in the gametes (**FIGURE 12.9**). If this were not the case, and a particular seed color always went with a particular seed shape, we would *not* find novel combinations of phenotypes among the F₂ offspring. Instead, the F₂ offspring would show one of the two parental phenotypes: round, yellow seeds or wrinkled, green seeds.

**P-generation plants**

RRYY × rryy

Parents with the *RRYY*, or *rryy*, genotype produce only one type of gamete: with genotype *RY*, or genotype *ry*.

The dotted arrows trace all the ways in which alleles of the two genes can be combined in the gametes during meiosis.

Gametes: RY RY      ry ry

**F₁ hybrids**

RrYy × RrYy

These hybrid (non-true-breeding) F₁ generation plants, with genotype *RrYy*, produce four different types of gametes, each with a unique combination of *R/r* and *Y/y* alleles.

Gametes with these genotypes are expected if Mendel's law of independent assortment is correct: the *R* gene (*R/r*) and the *Y* gene (*Y/y*) segregate into gametes independently of each other.

Gametes: RY Ry rY ry      RY Ry rY ry

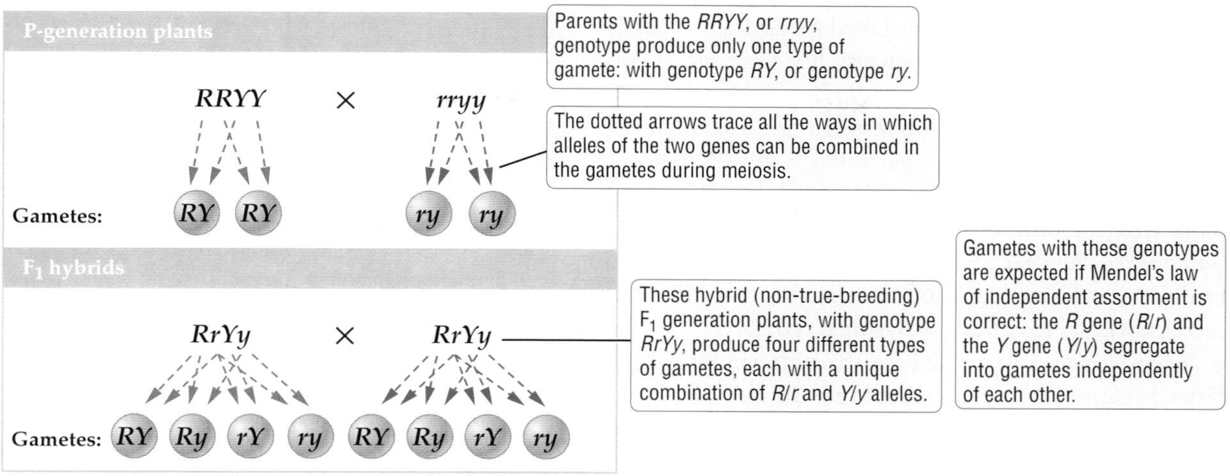

**FIGURE 12.9 The Alleles of Two Genes Are Sorted into Gametes Independently**

The 9:3:3:1 phenotypic ratios in the $F_2$ generation of Mendel's dihybrid cross are best explained by the law of independent assortment, according to which the alleles of the *R* gene segregate independently of the alleles of the *Y* gene during meiosis.

Mendel tested the two possibilities by crossing the *RrYy* $F_1$ plants with each other (see Figure 12.8). He obtained the following results in the $F_2$ generation: approximately 9/16 of the seeds were round and yellow, 3/16 were *round and green*, 3/16 were *wrinkled and yellow*, and 1/16 were wrinkled and green (a 9:3:3:1 ratio). Mendel's results were what we would expect if the genes for these two traits are inherited independently of each other. If the alleles of the *R* gene segregated independently from the alleles of the *Y* gene during gamete formation (as depicted in Figure 12.9), we would predict four phenotypes among the $F_2$ offspring: 9/16 displaying the dominant phenotypes for both traits, 1/16 with the recessive phenotypes of both traits, and two novel combinations of phenotypes not seen in either parent. The two nonparental combinations of phenotypes Mendel was looking for did turn up among the $F_2$ plants: 3/16 of the plants had round, green seeds; and another 3/16 had wrinkled, yellow seeds.

Mendel made similar crosses for various combinations of the seven traits he studied. His results led him to propose the **law of independent assortment**, which states that when gametes form, the two copies of any given gene (alleles) segregate during meiosis independently of any two alleles of other genes.

Mendel's observations are consistent with what we now know about the chromosomal basis of inheritance: homologous maternal and paternal chromosomes line up randomly at the metaphase plate during meiosis I (see Figure 10.16). Most of the seven genes Mendel was tracking in his breeding experiments happened to be on different pairs of homologous chromosomes; for example, the seed shape gene is located on chromosome pair 7, while the seed color gene sits on chromosome pair 1. As explained in Section 10.5, the maternal and paternal members of a pair of homologous chromosomes are randomly distributed into the daughter cells during gamete formation through meiosis. This means for any two pairs of homologous chromosomes—say, pair 7 and pair 1—there is no telling whether a particular gamete will receive the paternal copies of both 7 and 1, or whether the paternal copy of 7 will be combined with the maternal copy of chromosome 1 in that gamete.

Because the two members of each homologous pair are randomly sorted into gametes during meiosis and can be mixed in all possible combinations during fertilization, the alleles on these chromosomes can also be mixed and matched in all possible allelic combinations. That is how offspring can turn out with genotypes and phenotypes (such as *RRyy* and *rrYY* in Figure 12.8) that were not present in either parent. To this day, Mendel's law of independent assortment applies to the inheritance of two genes that are physically separated because they lie on different chromosomes. For reasons we will explore in Chapter 13, this law may not apply to a pair of genes located relatively close to each other on the *same* chromosome.

## Mendel's insights rested on a sound understanding of probability

The probability of an event is the chance that the event will occur. For example, there is a probability of 0.5 that a fair coin will turn up "heads" when it is tossed. A probability of 0.5 is the same thing as a 50 percent chance, or ½ odds, or a ratio of one heads to one tails (1:1). If you toss a penny only a few times, the observed percentage of heads may differ greatly from 50 percent. For example, if you tossed a coin only 10 times, you might get 7 heads, or 70 percent heads. However, if you tossed a coin 10,000 times, it would be very unusual to get 7,000 heads (70 percent heads). Each toss of a coin is an independent event, in the sense that the outcome of one toss does not affect the outcome of the next toss. The probability of getting two heads in a row is a product of the separate probabilities of each individual toss: $0.5 \times 0.5$, which is 0.25.

Mendel was able to deduce the patterns of inheritance because he conducted a large number of genetic crosses, which gave him data for a large number of offspring. He knew that he could not predict with certainty the phenotype of F₂ offspring from a single or even a few crosses. When performing genetic crosses, the experimenter has no control over which sperm and which egg fuse to produce an offspring. Mendel knew, however, that the eggs produced by a *Pp* plant have a 0.5 probability of carrying the *P* allele and a 0.5 probability of carrying the *p* allele; the same probabilities hold for the alleles carried by the sperm produced by such a plant. When two such plants are mated, the odds that an egg with a *p* allele will fuse with a sperm carrying a *p* allele are given by the rule of multiplication outlined in the previous paragraph: $0.5 \times 0.5$, yielding a 0.25 probability of generating a *pp* offspring.

It is important to understand that the ratios predicted by Mendel (for example, the 3:1 ratio illustrated in Figure 12.7) simply give the probability that a particular offspring will have a certain phenotype or genotype. We cannot know with certainty what the actual phenotype or genotype of a *particular* offspring is going to be (except when true-breeding individuals are crossed). Moreover, the probability that a particular offspring will display a specific phenotype is completely unaffected by how many offspring there are. The likelihood that we will see a 3:1 outcome, however, increases when we analyze a larger number of offspring. For example, if there is a ¼ chance that each offspring will be a homozygous recessive (*pp*) individual, we are more likely to observe that ratio in a large population of offspring than in a small sample of offspring. But for any four offspring produced by a particular mating, there is no guarantee that one will definitely have genotype *pp*. It is possible for none of the offspring to have genotype *pp*, or for more than one to have genotype *pp*. Such outcomes can occur because the 25 percent probability that an offspring will have genotype *pp* applies not only to the first off-

spring, but to the second, third, and fourth offspring as well. Hence, it is even possible (but not likely, since the chance is $0.25 \times 0.25 \times 0.25 \times 0.25$, or less than half of 1 percent) that *all* four offspring will have genotype *pp*. When many offspring are examined, the chance of obtaining unusual results (such as all of them having genotype *pp*) becomes very small.

### Concept Check

1. What is the ultimate source of genetic variation in a population of organisms?

2. For the offspring of a cross between an *Rr* plant and an *rr* plant, where *R* is dominant, predict the number and ratio of genotypes and phenotypes.

3. What central concept of genetics is outlined by Mendel's law of segregation?

4. Mendel found that wrinkled shape (controlled by the dominant allele *R*) is always inherited with green seed color (controlled by the dominant allele *Y*). Explain what is wrong with this statement.

## 12.4 Extensions of Mendel's Laws

Mendel's laws describe how genes are passed from parents to offspring. In some cases, such as the seven traits of pea plants that Mendel studied, these laws allow offspring phenotypes to be predicted accurately from parental genotypes. In particular, Mendel's laws enable us to make accurate predictions whenever a genetic trait is controlled by a single gene with two alleles—one dominant, the other recessive. But many traits are not under such simple genetic control. Geneticists refined and extended Mendel's laws through much of the twentieth century to explain more complex patterns of inheritance. They discovered, for example, that sometimes a single allele can produce a number of different phenotypes. The most complex patterns of inheritance are created when a phenotype is controlled by more than one gene, each with multiple alleles, and especially if the phenotype is also affected by environmental factors rather than by the genotype alone.

### Many alleles display incomplete dominance

For dominance to be complete, a single copy of the dominant allele must be enough to produce its phenotypic effect in a heterozygote; for example, one

$P$ allele ensures that even a $Pp$ pea plant has purple flowers, since $P$ is dominant over $p$. But in some allele combinations, no single allele completely dominates over the other when the two are paired together in a heterozygote. When neither allele is able to exert its full effect, the heterozygote displays an "in-between" phenotype, and the two alleles are said to display **incomplete dominance**. For example, two of the aleles that control flower color in snapdragons are incompletely dominant. When a homozygote with red flowers ($C^RC^R$) and a homozygote with white flowers ($C^WC^W$) are crossed, the heterozygous offspring ($C^RC^W$) produce pink flowers.

**INCOMPLETE DOMINANCE IN SNAPDRAGONS**

$C^RC^R$
Red flowers

$C^WC^W$
White flowers

$C^RC^W$
Pink flowers

Some of the genes controlling coat color in animals also display incomplete dominance. The coat color of the palomino horse is a case in point. **FIGURE 12.10** shows how two incompletely dominant alleles, $H^C$ and $H^{Cr}$, produce an intermediate phenotype in a heterozygote. The chestnut horse (which has a reddish brown coat) is homozygous for $H^C$ ($H^CH^C$). A cream-colored horse (known as "cremello") is produced when the horse is homozygous for $H^{Cr}$ ($H^{Cr}H^{Cr}$). The palomino horse, beloved in parades and shows for its golden coat and flaxen mane, is heterozygous for coat color ($H^CH^{Cr}$). In the palomino, the effect of the $H^C$ allele is "diluted" by the presence of the cream-color allele, $H^{Cr}$.

When incomplete dominance of alleles is involved, we can predict the genotypes and phenotypes of $F_1$ and $F_2$ offspring using Mendelian laws of inheritance and the Punnett square method, if we assign an intermediate phenotype to the heterozygotes. For example, if two heterozygous snapdragons ($C^RC^W$) are crossed, the odds are that ¼ of the offspring will have red flowers (genotype $C^RC^R$), ½ are likely to have pink flowers (genotype $C^RC^W$), and ¼ are likely to have white flowers (genotype $C^WC^W$). Work this out for yourself using the Punnett square method. You will see that Mendel's laws apply to alleles that show incomplete dominance, just as they apply to alleles that display a dominant-recessive relationship; the main difference is that the heterozygote ($C^RC^W$) looks different from a $C^RC^R$ individual (whereas a heterozygote would be indistinguishable from a $C^R$ homozygote if $C^R$ were dominant over $C^W$). We can also predict that when pink-flowered plants from the $F_1$ generation are bred with one another, we are likely to see among the $F_2$ offspring some plants with red

Phenotype: Chestnut
Genotype: $H^C H^C$

Phenotype: Palomino
Genotype: $H^C H^{Cr}$

Phenotype: Cremello
Genotype: $H^{Cr} H^{Cr}$

**FIGURE 12.10 Incomplete Dominance in Horses**
Palominos (heterozygous genotype $H^CH^{Cr}$) are intermediate in color to chestnuts ($H^CH^C$) and cremellos ($H^{Cr}H^{Cr}$) because in the heterozygote the $H^{Cr}$ allele "dilutes" the effect of the $H^C$ allele to produce the intermediate phenotype.

# Know Your Type

One of the most important inherited characteristics to know about yourself and your family members is blood type, since the majority of Americans will receive donated blood at some point in their lives. Blood transfusion is the transfer of whole blood or blood products to a patient. Whole blood or its components are used for many surgical procedures, as well as for ongoing treatment of chronic diseases. For example, a person who has been in a car accident and suffered massive blood loss may need as many as 100 pints of blood. Some patients with complications from severe sickle-cell disease, which affects 80,000 people in the United States, need as much as 4 pints of blood every month.

Whole-blood transfusions are uncommon nowadays; a patient is more likely to receive specific cell types (such as red blood cells or platelets) or plasma, the fluid portion of blood. The immune system of the patient may recognize molecules present in transfused blood as foreign and launch an attack against the blood. The molecules that the immune system attacks are known as antigens. The cell surface sugars responsible for the A, B, and AB blood groups are potential antigens. If red blood cells from a person with blood type A ($I^A I^A$ or $I^A i$) is transfused into a patient with blood type B ($I^B I^B$ or $I^B i$) or blood type O ($ii$), the patient's immune system will produce specific antigen-fighting proteins (called antibodies) that will launch an attack against their target antigen (the A type of cell surface sugars, in this example). The transfused cells clump together when attacked, leading to life-threatening clots. To prevent transfusion incompatibility, all donated blood is extensively tested, and patients receive blood products that are matched to avoid introducing antigens. The table summarizes how a recipient's blood will respond to a blood transfusion from donors with O, A, B, or AB blood type.

Experts estimate that 4.5 million Americans would die annually without lifesaving blood transfusions. Approximately 32,000 pints of blood are used each day in the United States, with someone receiving blood about every 3 seconds. Blood centers often run short of type O and type B blood, but shortages of all types of blood occur during the summer and winter holidays. We all expect blood to be there for us, but only a small fraction of those who can give actually do. Yet sooner or later, virtually all of us will face a situation in which we will need blood. And that time is all too often unexpected. To find out where you can donate, visit www.givelife.org.

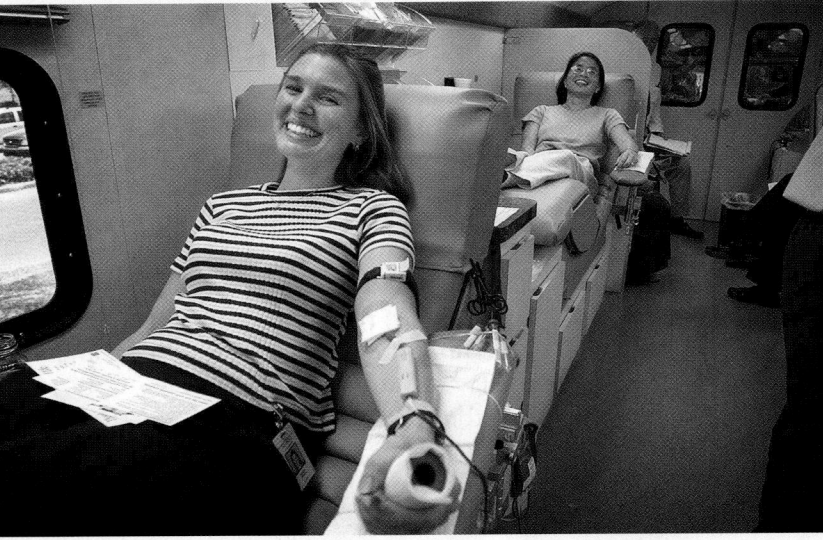

| Recipient's blood | | | Reactions with donor's red blood cells | | | |
|---|---|---|---|---|---|---|
| ABO blood type | Antigens produced | Antibodies produced | Donor: O type | Donor: A type | Donor: B type | Donor: AB type |
| O | None | Anti-A Anti-B | ✓ | ✗ | ✗ | ✗ |
| A | A | Anti-B | ✓ | ✓ | ✗ | ✗ |
| B | B | Anti-A | ✓ | ✗ | ✓ | ✗ |
| AB | A and B | None | ✓ | ✓ | ✓ | ✓ |

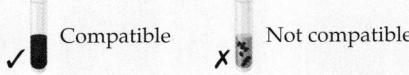

✓ Compatible    ✗ Not compatible

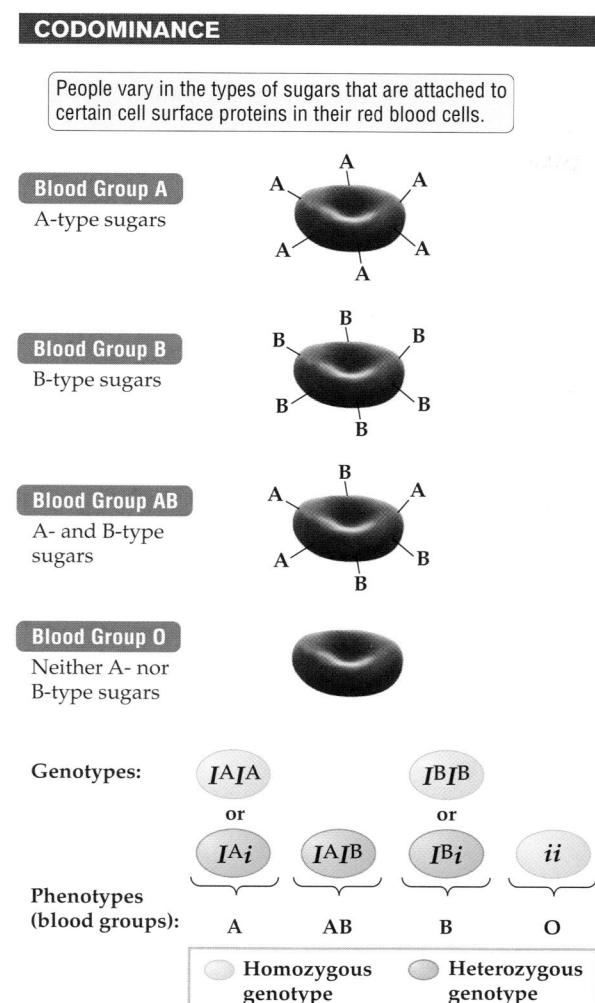

People vary in the types of sugars that are attached to certain cell surface proteins in their red blood cells.

**Blood Group A**
A-type sugars

**Blood Group B**
B-type sugars

**Blood Group AB**
A- and B-type sugars

**Blood Group O**
Neither A- nor B-type sugars

Genotypes: $I^A I^A$ or $I^A i$    $I^A I^B$    $I^B I^B$ or $I^B i$    $ii$

Phenotypes (blood groups):   A    AB    B    O

Homozygous genotype    Heterozygous genotype

**FIGURE 12.11 Genetic Basis of the ABO Blood Types in Humans**

The ABO blood types are determined by chains of sugars covalently attached to certain cell surface proteins on red blood cells. Type A red blood cells have a distinctive "A type" of sugar (chemically, *N*-acetylgalactosamine), whereas type B blood cells have a "B type" of sugar (galactose) at the end of the chain. The AB blood type reflects codominance: the red blood cells in type AB blood have about equal amounts of A-type sugars and B-type sugars on their cell surface. The *I* gene encodes an enzyme that comes in at least two allelic forms: the form encoded by the A allele adds A-type sugars to red blood cell surfaces, and the form encoded by the B allele adds B-type sugars. The *i* allele codes for a nonfunctional version of the enzyme that cannot attach *any* sugar to the cell surface protein. People who have two copies of the *i* allele (homozygotes) are said to have blood type O.

flowers and some with white flowers, as the alleles from the original true-breeding parents are redistributed in the F$_2$ generation.

## The alleles of some genes are codominant

A pair of alleles can also show **codominance**, in which the effect of both alleles is equally visible in the phenotype of the heterozygote. In other words, the influence of each codominant allele is fully displayed in the heterozygote, without being diminished or diluted by the presence of the other allele (as in incomplete dominance) or being suppressed by a dominant allele (as in the case of dominant-recessive alleles).

The ABO blood typing system we mentioned earlier provides an example of codominant alleles (**FIGURE 12.11**). Three alleles can determine a person's ABO blood type: the $I^A$, $I^B$, and $i$ alleles. Recall that genes commonly affect a genetic trait by coding for the manufacture of a specific protein. The $I^A$ allele codes for an enzyme that puts an "A type" of sugar on the surface of red blood cells (on certain proteins found in the plasma membrane of these cells). The $I^B$ allele makes a different version of the enzyme, one that inserts the "B type" of sugar on the blood cells. The $i$ allele simply fails to make a sugar-inserting enzyme that works, so both the A and the B types of sugars are absent from the red blood cells of a person with the $ii$ genotype; such a blood type is represented by the letter "O."

The first two of these alleles, $I^A$ and $I^B$, are codominant, producing an AB blood type in the heterozygote. The AB blood type is not halfway between A and B blood types, as would be the case if these alleles were incompletely dominant. Instead, people with the AB blood type ($I^A I^B$ genotype) have *both* A and B types of sugars on the surface of their red blood cells. In contrast, the $I^A I^A$ homozygote has only A-type sugars, and therefore an A blood type. Similarly, an $I^B I^B$ individual has only B-type sugars, and therefore blood type B. (The $i$ allele is recessive to the other two alleles, so the $I^A i$ individual has blood type A, the $I^B i$ individual has blood type B, and the $ii$ individual has blood type O.)

## A pleiotropic gene affects multiple traits

Mendel studied seven discrete traits—each a single, clear-cut genetic characteristic (such as flower color, seed color, seed shape) specified by a single gene. We now know, however, that some genes control functions of such central importance that abnormal functioning of the gene in question affects multiple biological processes. The situation in which a single

**Helpful to Know**

It is easy to confuse the terms "incomplete dominance" and "codominance." To help keep these two types of inheritance straight in your mind, remember that <u>in</u>complete dominance produces an <u>in</u>termediate phenotype. Codominant alleles are like cochairs of a committee: both are active in their own right, and neither is the boss of the other.

**FIGURE 12.12  A Child with Albinism**

A mother and her son, who has albinism, in their living room in Douala, Cameroon, Africa. Because it affects eyesight as well as pigmentation in skin, eyes, and hair, albinism is an example of a pleiotropic condition.

Marfan syndrome is a pleiotropic disorder in which many organ systems are affected because a single gene, coding for a protein called fibrillin-1, does not work properly. The protein is crucial for the normal function of connective tissues, which act as a gluing and scaffolding system for all types of organs, from bones to the walls of blood vessels. The one in 5,000 Americans diagnosed with Marfan syndrome show a wide range of phenotypes, depending on which allele of the fibrillin-1 gene they possess and depending on their other genetic characteristics. Most have problems with both vision and the skeleton. Many are tall and gangly, with long arms, legs, fingers, and toes. The nervous system, lungs, and skin are commonly affected. Weakening of the largest blood vessel carrying blood away from the heart is the most serious phenotype associated with the disorder. The rupture of this blood vessel is believed to have killed volleyball star Flo Hyman, who won the Olympic silver medal with her team in 1984 (**FIGURE 12.13**). Flo was nearly 6½ feet tall. Her condition was diagnosed only after her death in 1986, at the age of 31, during an exhibition game in Japan.

**FIGURE 12.13
Flo Hyman**

U.S. Olympic volleyball silver medalist Flora Jean ("Flo") Hyman, shown during practice in 1984. Complications from Marfan syndrome, a pleiotropic genetic disorder, contributed to her death in 1986.

gene influences a variety of different traits is called **pleiotropy** (*pleio*, "many"; *tropy*, "change"). A pleiotropic gene is one that can influence two or more distinctly different traits.

Pleiotropy is at work in the genetic condition known as albinism (**FIGURE 12.12**), in which traits as different as skin color and vision are affected by the action of a single gene. There are various forms of albinism, all of them marked by the absence, or reduced production, of a brown-black pigment called melanin. About one in 17,000 Americans has albinism. Most affected individuals produce very little melanin in the skin, hair, and eyes; contrary to popular belief, most of them have blue eyes, although a minority have red or violet eyes. Everyone diagnosed with albinism has eye problems, ranging from being "cross-eyed" to being legally blind. The gene involved in the most common type of albinism controls a step in the pathway for melanin production. It is easy to imagine how a malfunction in this gene would affect pigment deposition and therefore skin color, but what does that have to do with vision? Certain cells in the retina of the eye produce melanin, which is necessary for proper development of the eye, including the nerves that help the eye communicate with the brain.

## Alleles for one gene can alter the effects of another gene

Another twist in the inheritance of genetic traits comes from the interplay of different genes. Genes can interact in such a way that the phenotypic effect of gene A depends not only on the genotype of an individual with respect to gene A, but also the particular alleles of another gene, gene B. Such interactions among the alleles of different genes are common in all types of organisms. For example, in yeast (a single-celled fungus used in making bread and beer), each gene tested was found to interact with at least 34 other genes.

The term **epistasis [eh-pee-***STAY***-sis]** applies when the phenotypic effect of the alleles of one gene depends on the presence of certain alleles for another, independently inherited gene. Epistasis is seen among the many genes that control coat color in mice and other animals. Many of these genes code for enzymes involved in the multistep pathway that converts the amino acid tyrosine into melanin. One such gene, acting farther down the pathway, has a dominant allele (*B*) that leads to black fur and a recessive allele (*b*) that produces brown fur. But the effects of these alleles (*B* and *b*) can be eliminated completely, depending on which alleles of the *C* gene are present. The *C* gene codes for an enzyme (called tyrosinase) that acts at the very first step in this pathway. In mice with the *cc* genotype, the enzyme fails to do its work and the entire "assembly line" grinds to a halt. No melanin is made in the fur or eyes, resulting in an albino phenotype (**FIGURE 12.14**).

For a mouse with the *cc* genotype, it makes no difference whether the genotype with respect to the *B* gene is *BB* or *Bb* or *bb*; as long as the genotype is *BBcc*, *Bbcc*, or *bbcc*, the albino phenotype prevails. The *C* gene is therefore said to be epistatic to the *B* gene, meaning that alleles of the *C* gene can completely obscure any potential effect of any allele of the *B* gene. This does *not* mean that the alleles of the *C* gene are dominant over those of the *B* gene. The protein encoded by the *C* gene simply acts earlier in the pathway, so the way the alleles of the *C* gene function affects whether the phenotype controlled by any allele of *B* is displayed or not.

## The environment can alter the effects of a gene

The effects of many genes depend on internal and external environmental conditions, such as body temperature, carbon dioxide levels in the blood, external temperature, and amount of sunlight. For example, in Siamese cats the $C^t$ allele of the tyrosinase gene is sensitive to temperature, which means that the production of melanin depends on the temperature of the surroundings (**FIGURE 12.15**). The $C^t$ allele codes for a tyrosinase that works well at colder temperatures (35°C) but does not work at all at warmer temperatures (37°C). Because a cat's extremities tend to be colder than the rest of its body, melanin can be produced there, and hence the paws, nose, ears, and tail of a Siamese cat tend to be dark. If a patch of light hair is shaved from the body of a Siamese cat and the skin is covered with an ice pack, when the hair grows back it will be dark. Similarly, if dark hair is shaved from the tail and allowed to grow back under warm conditions, it will be light-colored.

Chemicals, nutrition, sunlight, and many other environmental factors can also alter the effects of genes. In plants, genetically identical individuals (clones) grown in different environments often exhibit phenotypic differences in a variety of traits, including height and the number of flowers they produce. For

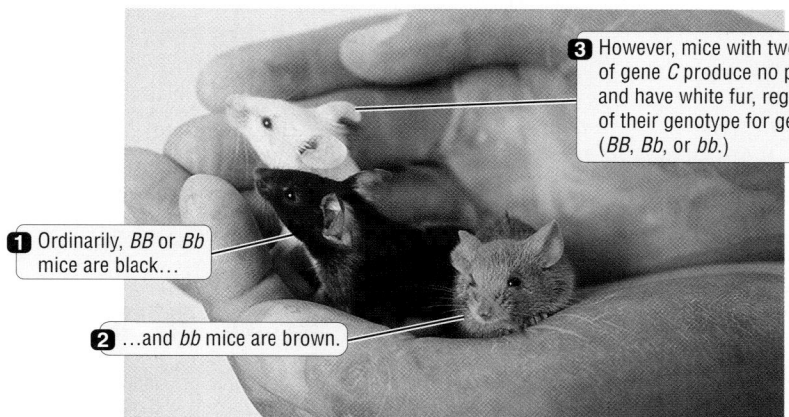

❶ Ordinarily, *BB* or *Bb* mice are black…

❷ …and *bb* mice are brown.

❸ However, mice with two *c* alleles of gene *C* produce no pigment and have white fur, regardless of their genotype for gene *B* (*BB*, *Bb*, or *bb*.)

**FIGURE 12.14  Alleles of One Gene May Affect the Phenotype Produced by Alleles of Another Gene**

In this example the *c* allele of gene *C* masks the alleles of another gene, *B*. In mice with the *CC* or *Cc* genotype, the dominant *B* allele directs the production of melanin, resulting in black fur, while the recessive *bb* genotype "dilutes" melanin accumulation so that brown fur results. Mice with the *cc* genotype are albinos because melanin production gets blocked at an early point in the pathway, before the *B* gene exerts its influence. The *cc* genotype always results in albinism, regardless of whether the genotype with respect to the *B* gene is *BB*, *Bb*, or *bb*.

example, plants on a windswept mountainside may be short and have few flowers, while clones of the same plants grown in a warm, protected valley are tall with many flowers. Similar effects are found in people. For example, a person who was malnourished as a child is likely to be shorter as an adult than if he or she had received good nutrition.

## Most traits are determined by two or more genes

Mendel analyzed traits that displayed straightforward genetic control, with two contrasting alleles of a single gene exerting complete control over the phenotype. Most traits, however, are **polygenic**; that is, they are governed by the action of more than one gene. Examples of polygenic traits include skin color, running speed, blood pressure, and body size in humans; and flowering time and seed number in plants. Let's look in more detail at one of these examples, the inheritance of skin color in humans.

**FIGURE 12.15
The Environment Can Alter the Effects of Genes**

Coat color in Siamese cats is controlled by a temperature-sensitive allele. The $C^t$ allele of the *C* gene directs melanin production only at lower temperatures (below 37°C). Melanin therefore accumulates only in the colder extremities—the snout, tail, lower legs, and edges of the ears—and not in the main trunk, which is at the core body temperature (37°C).

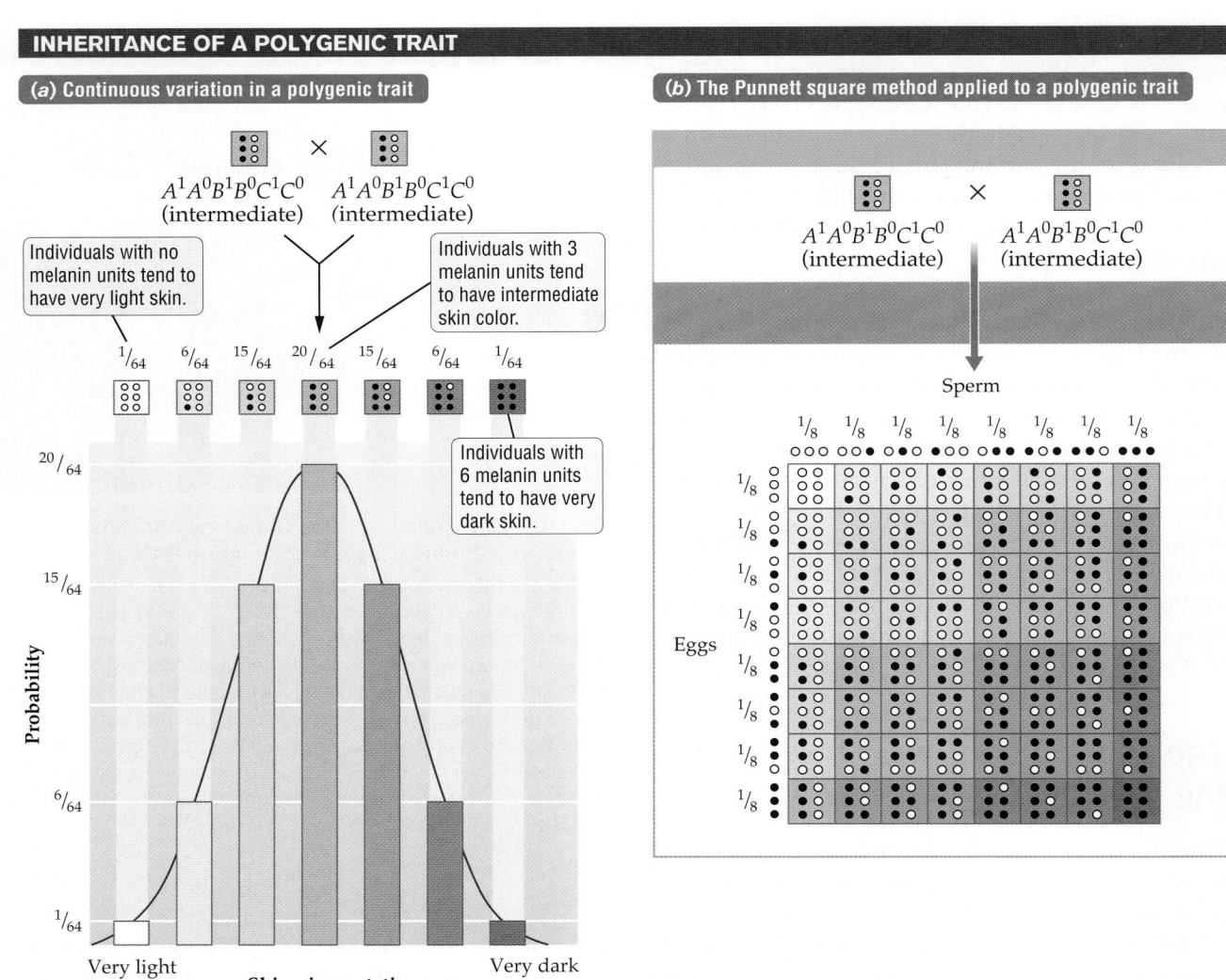

**(a) Continuous variation in a polygenic trait**

**(b) The Punnett square method applied to a polygenic trait**

**FIGURE 12.16 Three Genes Can Produce a Range of Skin Color in Humans**
In this model, skin color is influenced by the total number of melanin "units" specified by the person's genotype. (a) For each of the three hypothetical genes shown here (A, B, and C), the superscript 1 (for example, $A^1$) identifies an allele that leads to the production of one "unit" of melanin, while the superscript 0 (for example, $A^0$) identifies an allele that produces no melanin. The two alleles of each of these genes are incompletely dominant. Alleles that do not contribute to melanin production ($A^0$, $B^0$, and $C^0$) are represented by open circles, while alleles that do contribute to melanin production ($A^1$, $B^1$, and $C^1$) are represented by solid circles. The bar graph depicts the seven phenotypic outcomes arranged from lightest skin to darkest skin. Bar heights indicate the relative proportions of children of each phenotype. (b) The Punnett square shows the proportions of gametes with specific genotypes that are produced by two heterozygote ($A^1A^0B^1B^0C^1C^0$) parents and all the possible ways in which the eight different genotypes can come together during fertilization. Each of the seven different phenotypes (that is, children with 0, 1, 2, 3, 4, 5, or 6 melanin units) can be specified by one to several different genotypes. For example, children producing two melanin units can have any one of six different genotypes. Additional variation in skin color would result from different levels of sun exposure.

As we mentioned earlier, the pigment melanin is the main source of skin color in humans and most other mammals. Geneticists estimate that more than a dozen genes are involved in controlling how much melanin is deposited in the specialized pigment-bearing cells in human skin. For simplicity, we will assume that there are only three such genes (A, B, and C) and that each affects skin color equally. We will also assume that there are only two incompletely dominant versions of each gene: one version ($A^1$, $B^1$, $C^1$) that causes melanin production, and another ($A^0$, $B^0$, $C^0$) that prevents melanin production. In this model, the phenotype of skin color is controlled by how many "units" of the melanin-producing allele the person has; the more such units there are, the darker the skin is. If two individuals heterozygous for each of the three genes were to marry, a child of theirs could have any one of seven potential skin color phenotypes, as shown in **FIGURE 12.16**.

A suntan is another example of an environmental influence on the phenotype. In most people, melanin production increases in response to the ultraviolet radiation in sunlight. If we add different degrees of suntanning to the seven shades of skin color predicted on the basis of genotype alone, we get an even greater variety of skin colors—enough to fit a smooth bell curve to the bar graph shown in Figure 12.16*a*. In summary, human skin comes in just about every shade, from very light to very dark, because human skin color is a polygenic trait, and such a trait yields a greater range of phenotypes than does a trait controlled by a single gene. In addition, skin color is affected by environmental factors, adding enough variety to the palette of skin color phenotypes that there is almost continuous variation in this trait.

## 12.5 Complex Traits

Some traits are controlled by a single gene and are unaffected by environmental conditions. For such traits—flower color in pea plants, and the *I* gene alleles that control ABO blood types in humans, for example—we can predict the phenotypes of offspring just by knowing the genotype of the parents. Such traits are said to display Mendelian inheritance.

Chances are that just about all the traits you identified in Figure 12.2 are **complex traits**—that is, genetic traits whose pattern of inheritance cannot be predicted by Mendel's laws of inheritance. It is not that the inheritance of these traits violates Mendelian principles. Rather, there are additional complications, such as epistasis and environmental influences on the phenotype, that make it difficult to predict offspring phenotypes from the genotypes of the parents (**FIGURE 12.17**). For example, adult height in humans is affected by more than a hundred different genes, in addition to being influenced by environmental factors such as nutrition and stress. And before you were born, no one could have used the height of your parents to lay precise odds that you would be the height you are.

Most traits that are crucial for survival tend to be complex traits, generally displaying polygenic inheritance and phenotypes influenced by the environment. Why? One explanation is that polygenic control combined with environmental influence produces a great range of phenotypic classes that often grade smoothly into the next—a pattern known as *continuous variation*—and a population that shows substantial phenotypic diversity (continuous variation in size, for example) may stand to reap an evolutionary benefit if the environment changes in a way that threatens survival. The greater the phenotypic diversity, the higher the odds that a previously rare and undistinguished phenotypic class will present a survival advantage under the changed circumstances. According to this model, complex traits produce a greater range of phenotypes, which means more options for natural selection to work on.

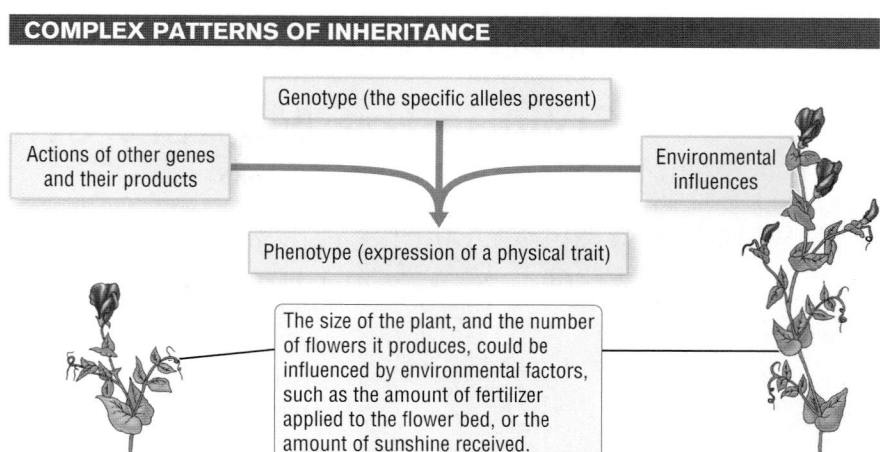

**COMPLEX PATTERNS OF INHERITANCE**

Genotype (the specific alleles present)

Actions of other genes and their products

Environmental influences

Phenotype (expression of a physical trait)

The size of the plant, and the number of flowers it produces, could be influenced by environmental factors, such as the amount of fertilizer applied to the flower bed, or the amount of sunshine received.

**FIGURE 12.17 Most Phenotypes Are Shaped by Interactions among Multiple Genes and the Environment**

The effect of a gene on an organism's phenotype can depend on a combination of the gene's own function, the function of other genes with which it interacts, and the impact of environmental factors. As a result, two individuals with the same genotype for a gene may show very different phenotypes, as illustrated by the differences in growth between the two pea plants depicted here.

### Concept Check

1. Allele *H* produces straight hair; allele *H'* produces curly hair. Individuals with the *HH'* genotype have wavy hair, somewhere between straight and curly. Are alleles *H* and *H'* codominant?

2. What is pleiotropy?

3. The ABO blood groups are determined by several alleles of the *I* gene. Is blood type a polygenic trait?

4. Explain why this statement is incorrect: Polygenic traits are affected by environmental influences, but single-gene traits are not.

# Solving the Mystery of the Lost Princess

During the 1917 Russian Revolution, Czar Nicholas II abdicated his throne, leaving his country in the hands of a government that was immediately overthrown by Bolshevik revolutionaries. As long as the czar and czarina were alive, they served as a rallying point for those who opposed the Bolsheviks, but their deaths would also trigger an outcry. To prevent anyone from being certain about the fate of the family, the Bolsheviks secretly buried the czar and his family and told people that Czarina Alexandra and her daughters were in hiding. The resulting uncertainty over the fate of Alexandra and her daughters set the stage for the legend of Anastasia (**FIGURE 12.18**).

In 1991, the grave of Czar Nicholas and his family was finally discovered. To verify their identities, investigators electronically superimposed photographs of the family's skulls onto old photographs and also compared the sizes of the skeletons with their clothes, which had been saved. Such tests strongly supported the hypothesis that the skeletons in the grave were those of the Russian royal family. Still, more definitive evidence was needed.

By the late twentieth century, biologists had figured out that genes are made of DNA and that DNA can be extracted from all sorts of living and once-living tissues, including bone. Since different families have different combinations of alleles, it is possible to tell if two people are related by comparing their alleles. When DNA from the skeletons was compared with DNA from the living descendants of the Romanovs, the results showed conclusively that the skeletal remains were those of the Russian royal family.

But missing from the Romanov grave were two sets of bones: those of Prince Alexei and those of one of two princesses (actually grand duchesses)—either Maria or Anastasia. It was hard to tell which because the two sisters were about the same height. Was it possible that Anastasia had escaped? To find out whether Anna Anderson could have been Anastasia, investigators obtained Anna Anderson's DNA from tissue samples saved by her doctors and then compared her alleles with those of the czar and czarina (**FIGURE 12.19**).

In human DNA, five alleles of one gene—called $A^1$, $A^2$, $A^3$, $A^4$, and $A^5$—are common in people of European descent. The czar had genotype $A^1A^2$, and the czarina had genotype $A^2A^3$. According to Mendel's laws, for Anna to have been the daughter of the czar and czarina, she could have had any one of the following four genotypes: $A^1A^2$, $A^1A^3$, $A^2A^2$, or $A^2A^3$. If this seems confusing, draw a Punnett square with the czar's contributions ($A^1$ and $A^2$) on the top and the czarina's ($A^2$ and $A^3$) on the left side.

Anna Anderson's genotype was $A^4A^5$, which meant the czar and czarina could not have been her parents. Three other sets of alleles yielded the same result, showing that Anna Anderson could not be Grand Duchess Anastasia. Furthermore, in 1995, comparisons between Anna Anderson's DNA and that of the descendants of Franziska Schanzkowska's family showed that Anna was almost certainly a member of the Polish family.

The clincher came in August 2007, when an archaeologist located the last two members of the Romanov family, who had been buried some distance away from their relatives. In 2009, DNA analysis confirmed that they were the skeletons of the Romanovs' son Alexei and one of his sisters. Since there is currently no way to tell Maria and Anastasia apart from their skeletons, we may never know for certain which grave Anastasia was buried in, but we do know that she died in July 1918 with her family.

**FIGURE 12.18 Anastasia and One of Her Three Sisters**
Grand Duchesses Maria (left) and Anastasia visiting a hospital for soldiers.

**FIGURE 12.19 Anna Anderson**
Franziska Schanzkowska, also known as Anna Anderson, in about 1931. Anna Anderson was a Polish factory work who was 5 years older than Grand Duchess Anastasia and spoke no Russian. Yet many people believed the two women were the same person.

# A Family Feud over Mendel's Manuscript on the Laws of Heredity

BY NICHOLAS WADE, *New York Times*

A long lost manuscript, one of the most important in the history of modern biology, has resurfaced as part of a dispute over its ownership. The manuscript is the account by Gregor Mendel of the pea-breeding experiments from which he deduced the laws of heredity and laid the foundations of modern genetics.

Mendel read his paper in 1865 at two meetings of the Natural History Society of Brünn. He was then an Augustinian monk, later the abbot, in the Abbey of St. Thomas in Brünn, now Brno in the Czech Republic.

The paper was published the next year in the Brünn Natural History Society's journal, but Mendel's work was largely ignored during his lifetime. It was only in 1900, 16 years after his death, that other researchers rediscovered Mendel's laws.

The original manuscript of Mendel's great work, called "Experiments on Plant Hybridization" in English, has suffered a longer obscurity, despite its historical significance. "From a conceptual view, it is the most remarkable scientific document in the history of the 19th century," Robert C. Olby, a historian of science at the University of Pittsburgh, said in an interview. "For the design and interpretation of an experiment, there is nothing to get near it. So it is priceless."

The priceless manuscript was discarded in 1911 by the Brünn Natural History Society and, luckily, rescued by a local high school teacher who retrieved it from a wastepaper basket in the society's library... [and] restored [it] to the society's files. During [World War II]... Mendel's manuscript disappeared for almost half a century.

The first people to hear of it again were... the descendants of [Mendel's] two sisters, Veronica and Theresia. At some point after 1988, Erich Richter, a Mendel descendant who is also an Augustinian monk known as Father Clemens, told other family members that he possessed Mendel's manuscript... and he wanted to place it legally in the family's possession.

... Dr. Maria Schmidt, a great-great-great-granddaughter of Veronica Mendel..., said that Father Clemens had recently changed his mind about the ownership of the manuscript after the head of the Augustinians in southern Germany and Austria, Father Dominic of Vienna, demanded that the manuscript be given to the order.

"There was tremendous pressure on Clemens," Dr. Schmidt said. "They said they would kick him out of the cloister. My father's cousin is 77 and has no property. He would have lost his car and apartment."

Father Clemens began to change his story about the ownership of the manuscript, suggesting it really belonged to the Augustinians, said William Taeusch, Dr. Schmidt's husband. "He started to say to the family, 'Aren't they the rightful owners?'... In the history of modern biology, Mendel's article is probably second in importance only to Darwin's "On the Origin of Species." Mendel made 40 reprints of his article, which he sent to scientific luminaries of the time. There are reports that one copy went to Darwin... [but that he did not read it]. Modern historians say that this account is wrong. [But] the history of biology could have been quite different had Darwin read Mendel's article.

---

Gregor Mendel originally sent his groundbreaking paper to 40 scientists, hoping they would read it and understand its importance. But virtually no one responded. One reason was that nobody knew who Mendel was. He had grown up on a farm, and only because one of his sisters gave up her dowry for him was he able to finish school at all. The one scientist willing to correspond with Mendel gave him consistently bad advice. Despite his brilliance and success, Mendel eventually gave up on science.

It's hard to imagine how different Mendel's life would have been if he had had a network of other researchers with whom to discuss his ideas. In science, as in so many other fields of endeavor, communication and contacts are critical. If Charles Darwin had read Mendel's paper, he may have recognized its value for his own ideas about evolution.

Darwin longed to know how traits were passed from parents to offspring. Mendel's theory of genes would have explained how the natural selection could work, and Darwin and his friends would have helped Mendel continue his work. As it was, Mendel died in obscurity, and genetics and evolutionary theory were not fully combined for another 70 years.

## Evaluating the News

**1.** The article says Mendel's paper was the second most important publication in nineteenth-century biology. What have you learned in this chapter that would support that statement? Why were Mendel's laws an important contribution to biology?

**2.** If you were Gregor Mendel today, what would you do to find a community of scientists who would understand and take an interest in your ideas?

**SOURCE:** *New York Times*, May 31, 2010

# Summary

## 12.1 Essential Terms in Genetics

- Genes—the basic units of inheritance—are segments of DNA that help determine an organism's inherited characteristics, or genetic traits.
- Diploid individuals generally have two copies of each gene: one inherited from the male parent, the other from the female parent.
- Alternative versions of a gene are called alleles. In a population of many individuals, a particular gene may have one, a few, or many alleles.
- The genotype is an individual's allelic makeup; the phenotype is the specific version of an observable trait that the individual displays.
- A dominant allele controls the phenotype of an individual even when paired with a different allele (heterozygote genotype). A recessive allele has no phenotypic effect when paired with a dominant allele.
- The different alleles of a gene arise by mutation; and by specifying different versions of the same protein, they generate hereditary differences among organisms.
- Some mutations are harmful, many have little effect, and a few are beneficial.

## 12.2 Basic Patterns of Inheritance

- Mendel's experiments with pea plants led him to reject the old notion of blending inheritance. Instead, they suggested that the inherited characteristics of organisms are controlled by specific units of inheritance, which we now know as genes.
- In modern terminology, Mendel's discoveries can be summarized as follows: (1) Alleles of genes account for the variation in genetic traits. (2) Offspring inherit one allele of a gene from each parent. (3) Alleles can be dominant or recessive. (4) The two copies of a gene separate into different (haploid) gametes. (5) Fertilization combines gametes in a random manner to give rise to diploid offspring.

## 12.3 Mendel's Laws of Inheritance

- Mendel's law of segregation states that the two copies of a gene (the two alleles) end up in different gametes during meiosis.
- According to the law of independent assortment, when gametes form during meiosis the alleles of one gene are distributed independently of the alleles of some other gene located on a different chromosome.

## 12.4 Extensions of Mendel's Laws

- For some traits, Mendel's laws may not predict the phenotype of the offspring because (1) many alleles show incomplete dominance or codominance; (2) one gene may affect more than one genetic trait (pleiotropy); (3) alleles for one gene can suppress the alleles of another gene (epistasis); (4) the environment can alter the phenotype of a gene; and (5) most traits are polygenic—that is, governed by two or more genes.

## 12.5 Complex Traits

- Most traits that are crucial for survival tend to be complex traits. Complex traits are influenced by multiple genes that interact with one another and with the environment. The inheritance of complex traits cannot be predicted by Mendel's laws, because Mendelian traits are controlled by a single gene that is little affected by other genes or by environmental conditions.
- Complex traits often generate a great range of phenotypic classes, or continuous variation in phenotypes.
- A population that shows substantial phenotypic diversity may stand to reap an evolutionary benefit if the environment changes in a way that threatens survival.

# Key Terms

allele (p. 280)
codominance (p. 291)
complex trait (p. 295)
dihybrid (p. 286)
diploid (p. 279)
dominant (p. 281)
epistasis (p. 292)
$F_1$ generation (p. 282)

$F_2$ generation (p. 283)
gene (p. 278)
genetic cross (p. 282)
genetic trait (p. 278)
genetics (p. 278)
genotype (p. 280)
haploid (p. 279)
heterozygous (p. 281)

homologous pair (p. 279)
homozygous (p. 280)
hybrid (p. 283)
incomplete dominance (p. 289)
law of independent assortment (p. 287)
law of segregation (p. 285)
maternal homologue (p. 279)
monohybrid (p. 285)

mutation (p. 281)
P generation (p. 282)
paternal homologue (p. 279)
phenotype (p. 279)
pleiotropy (p. 292)
polygenic (p. 293)
Punnett square (p. 285)
recessive (p. 281)

# Self-Quiz

1. Alternative versions of a gene for a given trait are called
   a. alleles.
   b. heterozygotes.
   c. genotypes.
   d. copies of a gene.

2. If *A* and *a* are two alleles of the same gene, then individuals of genotype *Aa* are
   a. homozygous.
   b. heterozygous.
   c. dominant.
   d. recessive.

3. The illustration on the right shows the seven traits of garden peas that Gregor Mendel analyzed by conducting large numbers of crosses and recording the phenotypes of all their offspring over two generations. Which of the following statements is true?
   a. When Mendel crossed plants that were true-breeding for the seed shape trait, the F$_2$ plants had round seeds and wrinkled seeds in a ratio of 9:3.
   b. When Mendel crossed true-breeding tall plants with true-breeding dwarf plants, all the F$_1$ plants displayed the dwarf phenotype.
   c. When Mendel crossed true-breeding plants with green pods and true-breeding plants with yellow pods, the F$_1$ plants had pods of an intermediate color, somewhere between green and yellow.
   d. When Mendel crossed homozygous tall plants and homozygous dwarf plants, the F$_2$ progeny included one dwarf plant for every three tall plants.

4. Coat color in horses shows incomplete dominance (see Figure 12.10). What is the predicted phenotypic ratio of chestnut to palomino to cremello if $H^C H^{Cr}$ individuals are crossed with other $H^C H^{Cr}$ individuals?
   a. 3:1                   c. 9:3:1
   b. 2:1:1                 d. 1:2:1

5. When the phenotypes controlled by two alleles are equally displayed in the heterozygote, the two alleles are said to show
   a. codominance.          c. incomplete dominance.
   b. complete dominance.   d. epistasis.

6. Traits that are determined by the action of more than one gene are
   a. recessive.            c. epistatic.
   b. not common.           d. polygenic.

7. Which term indicates that the phenotypic effects of alleles for one gene can be suppressed by the alleles of another, independently inherited gene?
   a. phenotypic variation  c. polygenic
   b. codominance           d. epistasis

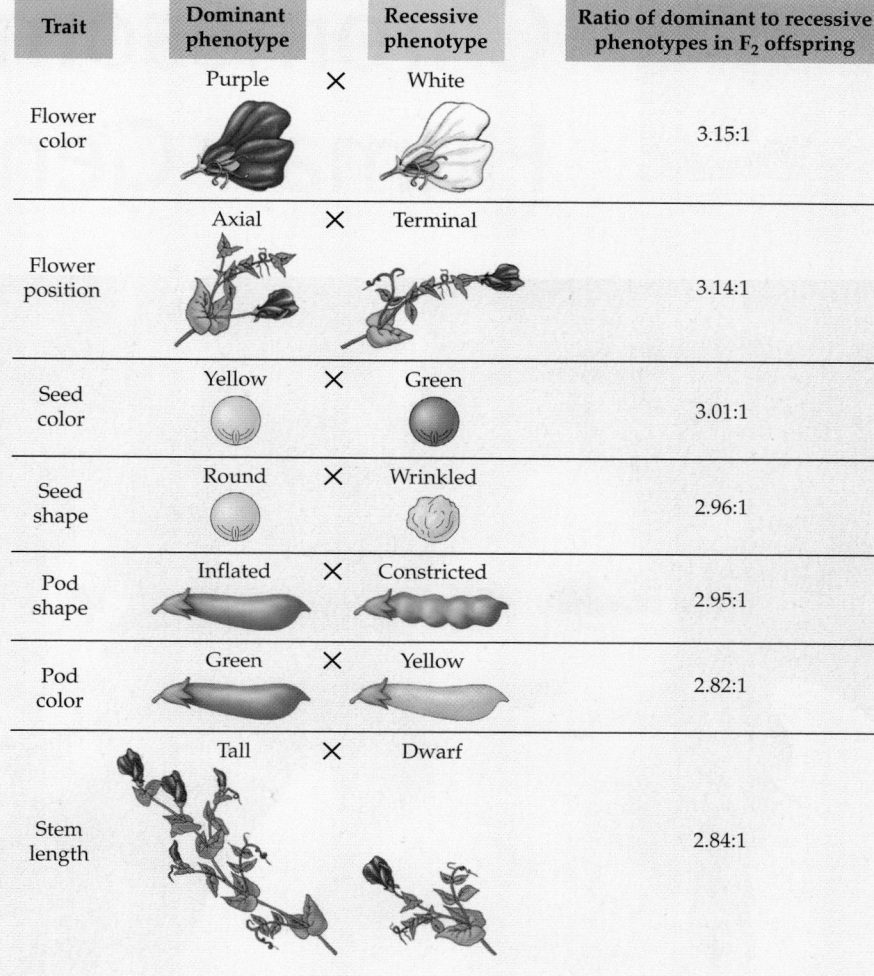

| Trait | Dominant phenotype | Recessive phenotype | Ratio of dominant to recessive phenotypes in F$_2$ offspring |
|---|---|---|---|
| Flower color | Purple × | White | 3.15:1 |
| Flower position | Axial × | Terminal | 3.14:1 |
| Seed color | Yellow × | Green | 3.01:1 |
| Seed shape | Round × | Wrinkled | 2.96:1 |
| Pod shape | Inflated × | Constricted | 2.95:1 |
| Pod color | Green × | Yellow | 2.82:1 |
| Stem length | Tall × | Dwarf | 2.84:1 |

# Analysis and Application

1. Describe what genes are and how they work. Include a summary of what we now know about (a) the chemical and physical structure of genes and (b) the information they encode. How many copies of each gene are found in the diploid cells in a woman's body? Explain.

2. Explain how new alleles arise, and how different alleles cause hereditary differences among organisms.

3. A purple-flowered pea plant could be genotype *PP* or *Pp*. What genetic cross could you make to determine the genotype of this plant? Explain how such a *test cross* enables you to deduce whether the purple-flowered plant is a homozygote or a heterozygote.

4. Can "identical" twins have different phenotypes? Why or why not?

5. Do the research to identify the four diseases responsible for more deaths than any other in North America. Is our risk for each of these diseases a genetic trait? If so, is it a complex trait? How would you explain a complex trait to your next-door neighbor? What is the practical relevance of the concept of complex traits in our everyday life?

Note: You will find additional problem sets on page 320.

# 13 Chromosomes and Human Genetics

**MISTY OTO WITH HER MOTHER AND DAUGHTER.** Misty Oto's mother, Rosie Shaw, died of Huntington's disease, a genetic disease passed through families on a single dominant allele. Oto, who works to educate people about Huntington's disease, has lost several family members to the disease. When her daughter was born, Oto did not know whether she or her daughter carried the disease gene.

# Kith and Kin

One of the first things Misty Oto remembers is her grandmother's death. Lots of kids have sad memories of losing a grandparent, but the disease that killed Misty's grandmother would profoundly shape Misty's life. Misty's grandmother died of Huntington's disease, an unusual genetic disease caused by a dominant allele. The only one of 10 children to inherit the disease, her grandmother passed the disease to all three of her daughters, including Misty's mother, Rosie. By the time Misty was 11, Rosie had begun to change from a vibrant, loving mother to an empty shell.

As Misty grew up, Huntington's decimated her family, taking her mother, both her aunts, and her brother. Yet Misty had not taken the genetic test that would tell her if she and her children had the Huntington's gene. None of Misty's four other siblings and the nine kids in the next generation had been tested. The gene for Huntington's disease was located in 1993, and a genetic test was developed soon after. But there is still no cure. As Misty told a radio interviewer, "Why would I even put myself through that process? And to know that I have this black cloud over my head and that there's nothing I could do about it."

But in 2010, Misty's older sister began thinking about the future. If she became ill with Huntington's, she would need long-term care. Who would care for her? She asked Misty to be her caregiver, and without thinking, Misty said yes. Only then, Misty recalled, did she realize that she herself might be too ill to help. For the first time, she considered being tested.

**What is the genetic basis of Huntington's disease? And why do 90 percent of at-risk people avoid the test?**

To begin to understand the genetic basis of Huntington's disease, we must explore the chromosomal basis of inheritance. Along the way, we'll find out why some genes tend to be inherited together; we'll learn why some genetic conditions are more common in males; and we'll examine the patterns of inheritance that explain some common and rare human genetic disorders, including Huntington's disease.

**MAIN MESSAGE** The inheritance of genes is affected by their precise location on specific chromosomes.

## KEY CONCEPTS

- A gene is a stretch of DNA that specifies a genetic trait. Each gene has a particular location on a chromosome.

- Human males have one X and one Y chromosome, and human females have two X chromosomes. A specific gene on the Y chromosome is required for human embryos to develop as males.

- Genes located near one another on the same chromosome tend to be inherited together, and they are said to be linked. Genes located far from one another on the same chromosome are often not linked.

- Many inherited genetic disorders in humans are caused by mutations of single genes.

- Dominant disorders whose harmful effects prevent a person from bearing children are generally rare, compared to recessive

disorders that similarly keep a person from reproducing.

- Recessive alleles located on the X chromosome often produce gender-specific phenotypes known as X-linked traits. Recessive X-linked conditions tend to be more common among males than females.

- Some human genetic disorders result from abnormalities in chromosome number or structure.

## Genes are located on chromosomes

Early in the twentieth century, geneticists gathered a great deal of experimental evidence in support of Weisman's hypothesis. The concept that genes are located on chromosomes came to be known as the **chromosome theory of inheritance**. Modern genetic techniques enable us to pinpoint which chromosome contains a particular gene and where on the chromosome that gene is located (**FIGURE 13.1**).

How are chromosomes, DNA, and genes related? As described in Chapters 10 and 12, chromosomes that pair during meiosis are called **homologous [hoh-*MAH*-luh-gus] chromosomes**. One member of each pair of homologous chromosomes is inherited from the female parent (the maternal chromosome), the other from the male parent (the paternal chromosome). Each chromosome consists of a single long DNA molecule attached to many bundles of packaging proteins (see Figure 10.7). Each gene is a small region of the DNA molecule, and there are many genes on each chromosome. For example, we humans are estimated to have about 25,000 genes located on one set of our chromosomes, which consists of 23 different types of chromosomes. On average, then, we have 25,000/23, or 1,086, genes per chromosome.

The physical location of a gene on a chromosome is called a **locus** (plural "loci"). With some exceptions (to be discussed shortly), every diploid cell has two copies of every gene, one on each of the chromosomes in a homologous pair, as **FIGURE 13.2** shows. Because a gene can

**OVER A HUNDRED YEARS AGO,** Gregor Mendel deduced that inherited traits are governed by discrete hereditary units. As we learned in Chapter 12, today these hereditary units are known as genes. We begin this chapter with a second cornerstone of modern genetics: the *chromosome theory of inheritance*. We then explain how an individual's sex is determined in humans and other mammals, and how new combinations of alleles, different from those of either parent, can occur in offspring. This information about chromosomes, sex determination, and new allele combinations sets the stage for the discussion of human genetic disorders in the remainder of the chapter.

## 13.1 The Role of Chromosomes in Inheritance

When Mendel published his theory of inheritance in 1866, he had no idea what genes were made of or where they were located within a cell. By 1882, studies using microscopes had revealed that threadlike structures—the *chromosomes*—exist inside dividing cells. The German biologist August Weismann hypothesized that the number of chromosomes was first reduced by half during the formation of sperm and egg cells, and then restored to its full number during fertilization. The discovery of meiosis in 1887 supported Weismann's hypothesis. Weismann also suggested that the hereditary material was located on chromosomes, but at that time there was no experimental evidence for or against that idea.

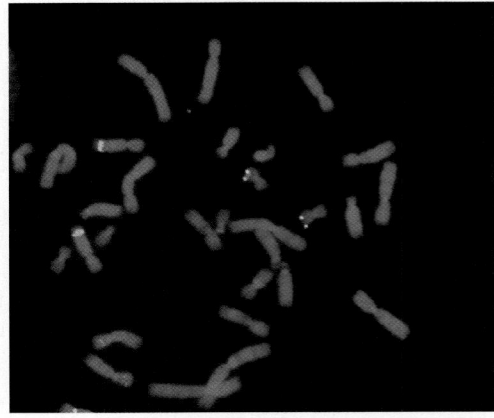

**FIGURE 13.1  Genes Are Located on Chromosomes**
A technique called fluorescence in situ hybridization (FISH) was used to show the location of three different genes on these chromosomes prepared from tumor cells in mitosis. Each of the three genes (*HER2*, green; *p16*, pink; *ZNK217*, gold) is known to be involved in cancer.

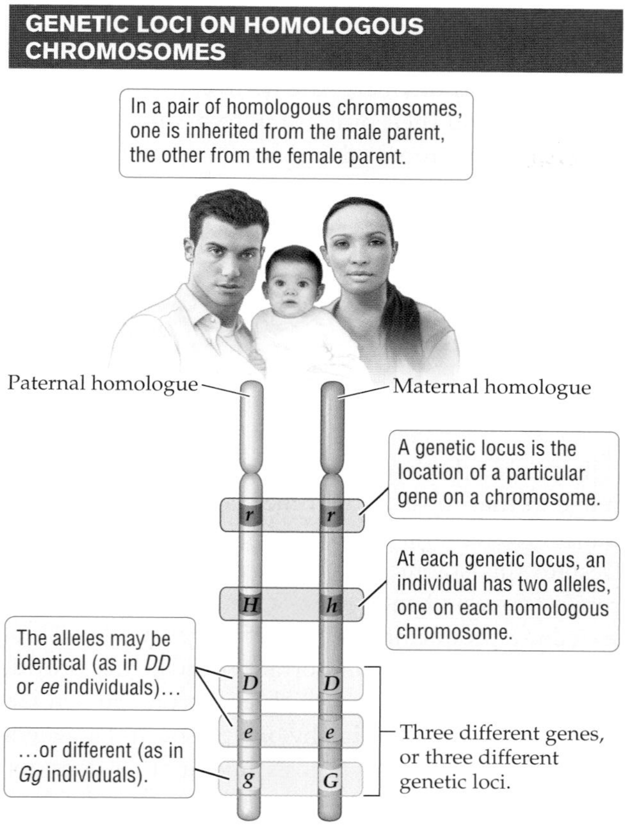

In a pair of homologous chromosomes, one is inherited from the male parent, the other from the female parent.

Paternal homologue — — Maternal homologue

A genetic locus is the location of a particular gene on a chromosome.

At each genetic locus, an individual has two alleles, one on each homologous chromosome.

The alleles may be identical (as in *DD* or *ee* individuals)…

…or different (as in *Gg* individuals).

Three different genes, or three different genetic loci.

**FIGURE 13.2  Genes Are Located on Chromosomes**
The genes shown here take up a larger portion of the chromosome than they would if they were drawn to scale. The average human chromosome has more than a thousand different genes interspersed with large stretches of noncoding DNA.

come in different versions, or alleles, a diploid cell can have two *different* alleles at a given genetic locus, in which case it has a heterozygous genotype for the gene at that locus. But if the two alleles at a genetic locus are identical, then the diploid cell has a homozygous genotype for the gene at that locus. As discussed in Chapter 12, the allelic makeup, or genotype, at a particular genetic locus influences the phenotype, or outward display of a genetic trait. We will see shortly that the inheritance of various genes can be affected by how close or far apart they are on a chromosome and whether they are located on a sex chromosome or an autosome (non–sex chromosome).

## Autosomes differ from sex chromosomes

As noted in Chapter 10, the two chromosomes that make up a homologous pair are exactly alike in terms of length, shape, and the genetic loci they carry. But in humans and many other organisms, the chromosomes that determine the sex of the organism are dissimilar.

These *sex chromosomes* are assigned different letter names. In humans, for example, males have one X chromosome and one Y chromosome (**FIGURE 13.3**), whereas females have two X chromosomes. The Y chromosome in humans is much smaller than the X chromosome, and few of its genes have a counterpart on the X chromosome. Since human males have one X and one Y chromosome, they have only one copy (instead of the usual two) of each gene that is unique to either the X or the Y chromosome.

Chromosomes that determine sex are called **sex chromosomes**; all other chromosomes are called **autosomes**. Human autosomes are labeled not with letters, but with the numbers 1 through 22 (for example, chromosome 4).

## In humans, maleness is specified by the Y chromosome

Because human females have two copies of the X chromosome, all the gametes (eggs) they produce contain one X chromosome. Males, however, have one X chromosome and one Y chromosome, so the odds are that half of their gametes (sperm) will contain an X chromosome and half will contain a Y chromosome (**FIGURE 13.4**). The sex chromosome carried by the sperm therefore determines the sex of the offspring. If a sperm carrying an X chromosome fertilizes an egg, the resulting child will be a girl; if a sperm carrying a Y chromosome fertilizes the egg, the child will be a boy.

Compared with the X chromosome, the Y chromosome has few genes. It does, however, carry one very

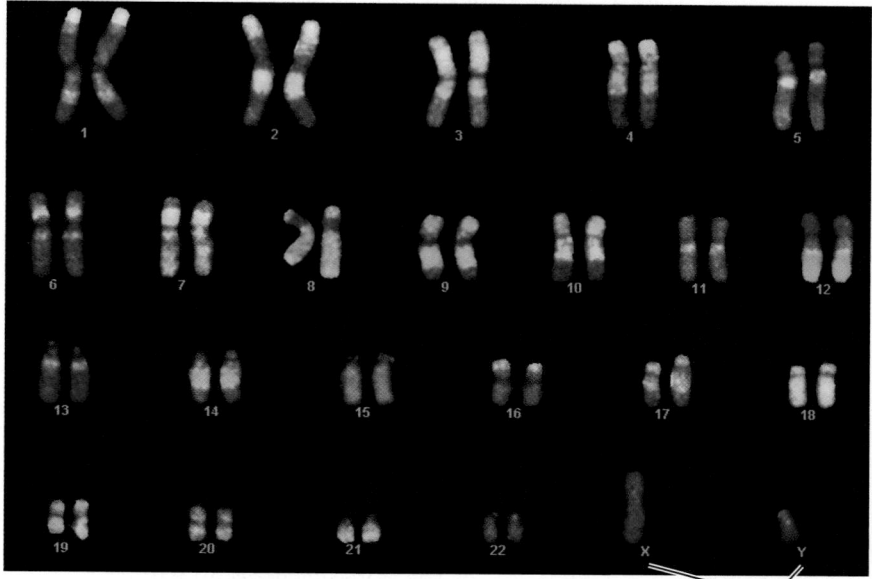

**FIGURE 13.3  Autosomes and Sex Chromosomes**
These chromosomes have been stained using a technique called FISH (fluorescence in situ hybridization).

While each of the paired autosomes are similar in shape and size, the X and Y chromosomes are not.

FIGURE 13.4 Dad's
Chromosomes
Determine Baby's
Gender

Human females have two X chromosomes, while human males have one X and one Y chromosome. If a baby receives an X chromosome from the father, it's a girl! If the baby receives a Y chromosome from Dad, it's a boy! Because a man's X and Y chromosomes segregate into different gametes during meiosis, the odds are 50 percent that a given sperm cell will contain an X or Y chromosome.

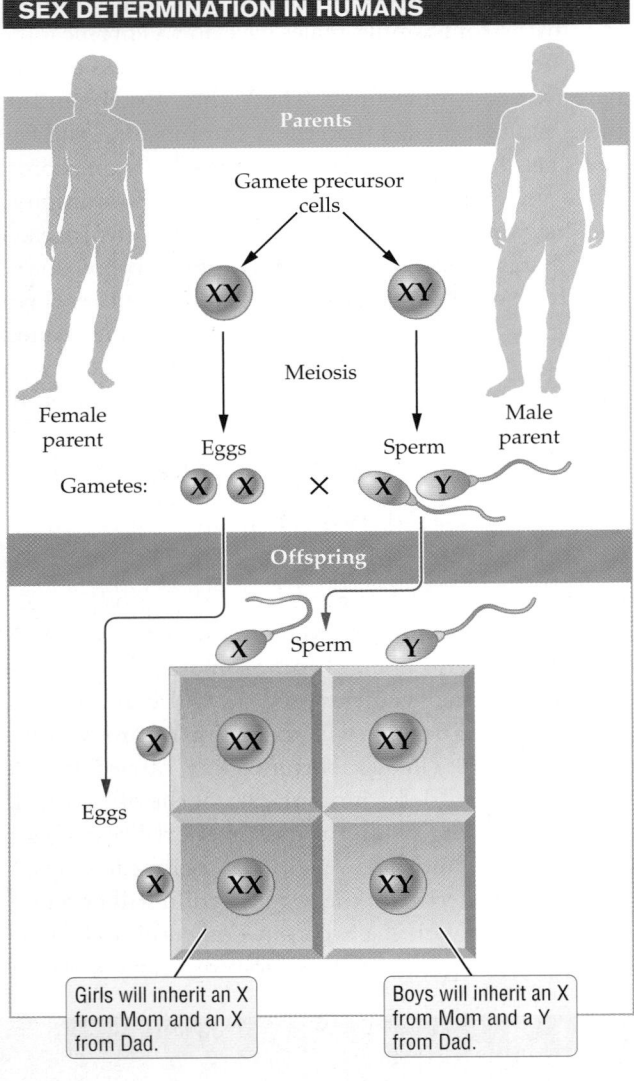

**SEX DETERMINATION IN HUMANS**

Parents

Gamete precursor cells

XX          XY

Meiosis

Female parent          Male parent

Gametes:    Eggs          Sperm

X  X    ×    X  Y

Offspring

X  Sperm  Y

X

Eggs

X    XX          XY

X    XX          XY

Girls will inherit an X from Mom and an X from Dad.

Boys will inherit an X from Mom and a Y from Dad.

important gene: the **SRY gene** (short for "<u>s</u>ex-determin<u>ing r</u>egion of <u>Y</u>"). The *SRY* gene functions as a master switch, committing the sex of the developing embryo to "male." In the absence of this gene, a human embryo develops as a female. The *SRY* gene does not act alone: in both males and females, other genes on the autosomes and sex chromosomes directly influence the development of the many sexual characteristics that distinguish men and women. The *SRY* gene plays a crucial role because when it is present, it causes the other genes to produce male sexual characteristics; whereas when it is absent, the other genes produce female sexual characteristics. For example, the *SRY* gene directs the development of testes, the male reproductive organs, in place of ovaries.

The role of the non-*SRY* genes in gender determination becomes evident in disorders of sexual development, a broad range of conditions in which a person appears to be, or identifies with, a gender different from the one predicted by his or her sex chromosomes.

For example, individuals with *androgen insensitivity syndrome* (*AIS*) are female in appearance and say they "feel female," even though they are XY in their chromosomal makeup, and their ovaries fail to develop normally. AIS is most commonly due to a genetic mutation that makes these individuals unable to respond normally to male hormones such as testosterone.

## 13.2 Origins of Genetic Differences between Individuals

Inheritance is both a stable and a variable process. It is stable in that most of the time genetic information is transmitted accurately from one generation to the next. Despite this stability, offspring are not exact genetic copies (clones) of their parents, so the individuals of a sexually reproducing species differ genetically. These genetic differences are important because they provide the genetic variation on which evolution can act. Genetic differences explain why one person has a genetic disorder, such as cystic fibrosis, and another does not. Genetic differences also explain why some individuals suffer severely from a disease, such as asthma, while others have a milder form. How do genetic differences among individuals arise? As we saw in Chapter 12, new alleles arise by mutation. Once formed, those alleles are shuffled or arranged in new ways by *crossing-over, independent assortment of chromosomes*, and *fertilization*.

**Crossing-over** is a reciprocal exchange of segments of nonsister chromatids in prophase I of meiosis (see Figure 10.15). Crossing-over is part of the reason why one set of parents typically produces offspring with a range of different genotypes—in other words, why siblings do not always look exactly alike. Every time meiosis occurs, crossing-over produces chromosomes with new combinations of alleles. These "recombined" chromosomes contain some alleles inherited from one parent and other alleles inherited from the other parent. By exchanging alleles between chromosomes, crossing-over causes offspring to have a genotype that differs from the genotype of either parent (**FIGURE 13.5**).

New combinations of alleles can also be produced by the **independent assortment of chromosomes**, the random distribution of maternal and paternal chromosomes into gametes during meiosis. Put another way, the paternal and maternal members of any given homologous pair are sorted into gametes independently of any other homologous pair. The independent assortment of homologous pairs comes about because the maternal and paternal homologues of each homologous pair are randomly oriented at the center of the cell (metaphase plate) just before they are sorted into separate daughter cells during meiosis. As a result the daughter cells, which mature into gametes, acquire a random mix of paternal and maternal homologues. Because the members of a homologous pair usually carry different alleles for at least some of the genes, a gamete may acquire a combination of alleles that was not present in the parental cell undergoing meiosis (see Figure 10.16).

Maternal and paternal homologues pair up with each other very early in meiosis (during prophase I). During metaphase I, the members of each homologous pair are randomly positioned by the spindle microtubules. By "random" we mean there is no pattern with respect to how the maternal and paternal homologues are arranged with respect to the metaphase plate; either could be to the "right" or to the "left" of the metaphase plate. To picture this, imagine a cell with a total of four chromosomes (haploid number = 2), so it has just two pairs of homologous chromosomes: pair 1 and pair 2. The illustration that follows depicts the two ways in which the maternal and paternal members of pair 1 and pair 2 can be arranged at the metaphase plate:

In pattern A, the maternal chromosomes of *both* pairs happen to be on the same side of the metaphase plate. In pattern B, the maternal homologues of the two pairs are on opposite sides. Whether the maternal and

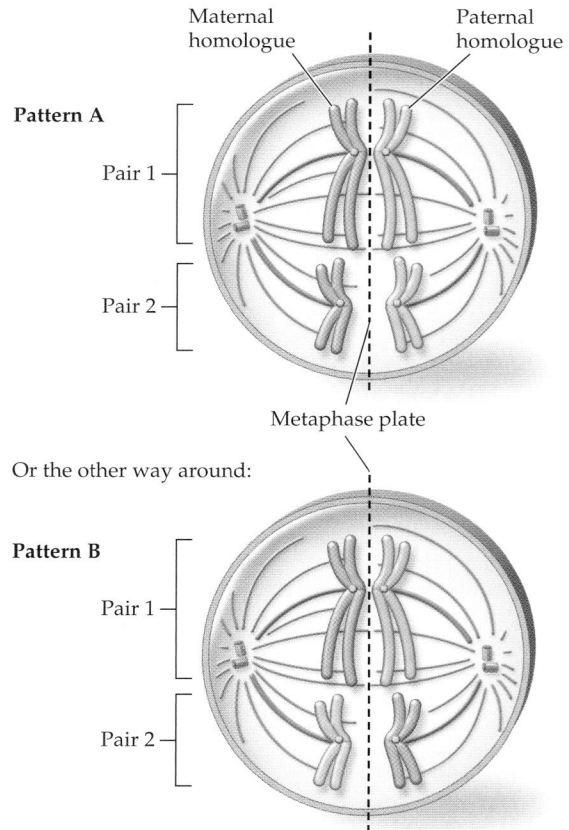

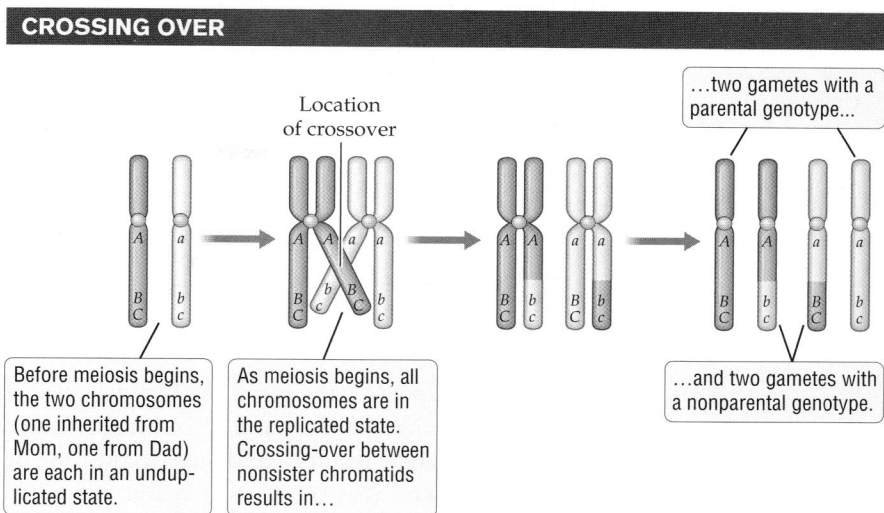

Location of crossover

...two gametes with a parental genotype...

Before meiosis begins, the two chromosomes (one inherited from Mom, one from Dad) are each in an unduplicated state.

As meiosis begins, all chromosomes are in the replicated state. Crossing-over between nonsister chromatids results in...

...and two gametes with a nonparental genotype.

**FIGURE 13.5  Segments of Chromosomes Are Exchanged in Crossing-Over**
Crossing-over takes place during prophase I. As a result of crossover between *A/a* and *B/b*, half of the gametes have a parental genotype (*ABC* or *abc*), while the other half have a nonparental genotype (*Abc* or *aBC*). In this example, there is no crossing-over between *B/b* and *C/c*.

paternal members of the two homologous pairs are positioned according to pattern A or to pattern B during meiosis in a particular cell is purely a matter of chance.

Note that patterns A and B will generate gametes containing a very different mix of homologues. Pattern A produces two types of gametes: one containing the maternal members of pair 1 and pair 2, the other containing the paternal members of pair 1 and pair 2. What combination of homologues is found in gametes generated by pattern B? Half the gametes will have the maternal member of pair 1 and the paternal member of pair 2; the other half will have the paternal member of pair 1 and the maternal member of pair 2.

The larger the number of homologous pairs in an organism, the greater the variety of patterns in which the homologues can be arranged during metaphase, and therefore the greater the variety of gametes that can be generated through meiosis. In the example just illustrated, we saw that just two homologous pairs can be arranged in two different ways, and can therefore generate $2^2$ (4) different types of gametes.

The diploid cells that undergo meiosis in the human body have 23 pairs of homologous chromosomes (diploid number = 46). Since each of the 23 pairs of homologous chromosomes lines up at random in one of two ways, there are $2^{23}$, or 8,388,608, different ways that our chromosomes can be arranged during meiosis I. This means a person can produce at least 8,388,608 different types of gametes—either egg or sperm—in terms of the mix of homologues found in each type of gamete. Like crossing-over, the independent assortment of chromosomes

Maternal homologue

Paternal homologue

**Pattern A**

Pair 1

Pair 2

Metaphase plate

Or the other way around:

**Pattern B**

Pair 1

Pair 2

generates gametes with unique combinations of the genetic information found in the parental chromosomes.

In Chapter 10 we explained how fertilization adds to the genetic variation produced by crossing-over and the independent assortment of chromosomes.

## 13.3 Genetic Linkage and Crossing-Over

As we saw in Chapter 12, the results of Mendel's experiments indicated that genes are inherited independently of one another. These results led Mendel to propose his law of independent assortment, which states that the two copies of one gene are separated into gametes independently of the two copies of other genes. Early in the twentieth century, however, results from several laboratories indicated that certain genes were often inherited together, contradicting the law of independent assort-

ment. Much of this work was done on fruit flies, which have a short reproductive cycle. The insects develop from egg to reproductive adult in about 2 weeks, which means that the inheritance of a trait can be tracked over many generations in a relatively short time span.

### Linked genes are located on the same chromosome

In his research on fruit flies, which began in 1909 at Columbia University in New York City, Thomas Hunt Morgan discovered genes that are inherited together. In one experiment, Morgan crossed a fruit fly that was homozygous for both a gray body (*G*) and wings of normal length (*W*) with another that was homozygous for both a black body (*g*) and wings that were greatly reduced in length (*w*). That is, he crossed *GGWW* flies with *ggww* flies to obtain flies of genotype *GgWw* in the F$_1$ generation. He then mated those *GgWw* flies with *ggww* flies. As **FIGURE 13.6** shows, Morgan's results

**FIGURE 13.6**
**Genes on the Same Chromosome May Not Assort Independently**

By crossing flies of genotype *GgWw* with flies of genotype *ggww*, Morgan found that the gene for body color (dominant allele *G* for gray, recessive allele *g* for black) is linked to the gene for wing length (dominant allele *W* for normal length, recessive allele *w* for reduced length). The parental genotypes are overrepresented because the *G* and *W* genes are linked to each other: the two genes are located relatively close to each other on the same chromosome.

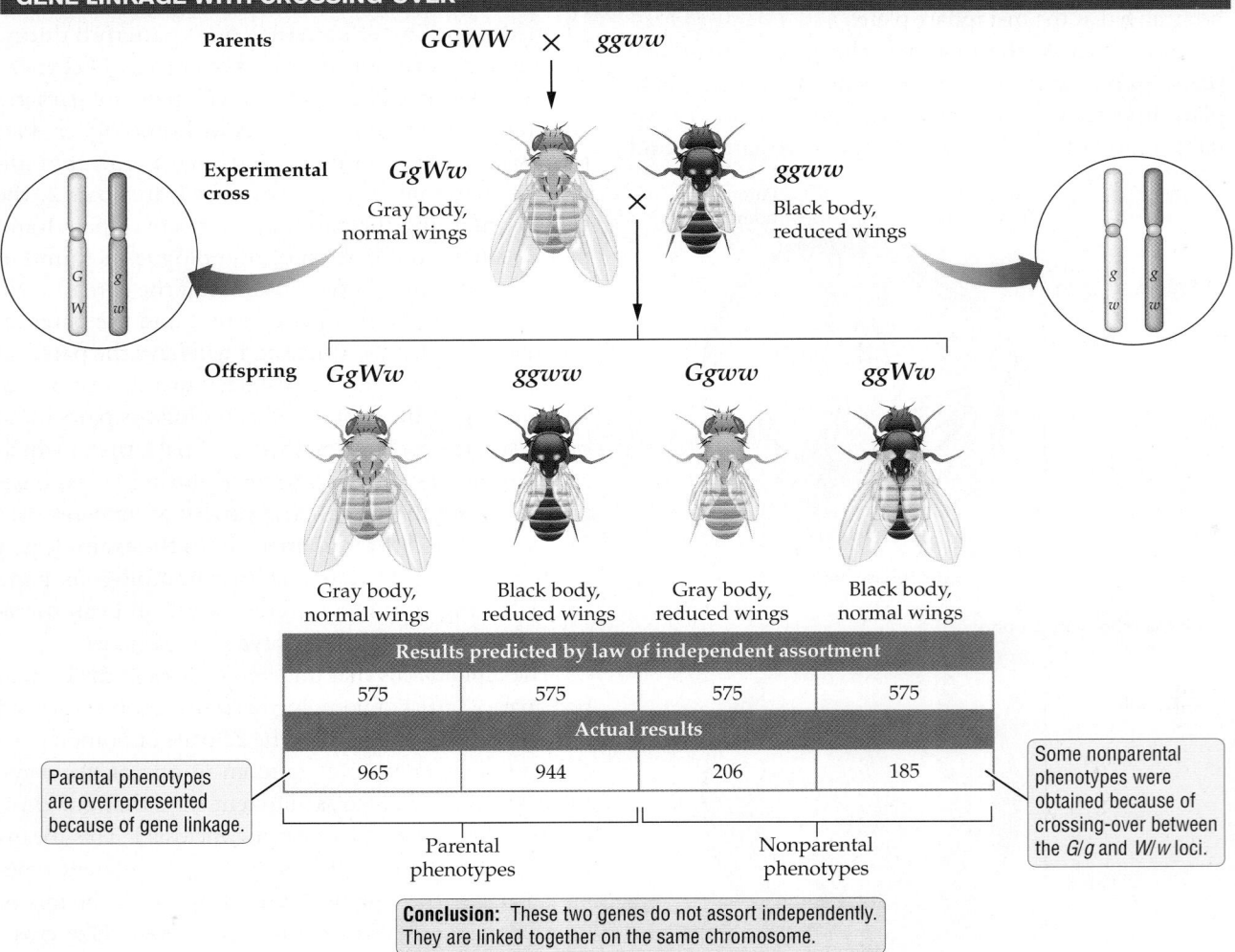

**GENE LINKAGE WITH CROSSING-OVER**

Parents   *GGWW* × *ggww*

Experimental cross   *GgWw* Gray body, normal wings × *ggww* Black body, reduced wings

Offspring   *GgWw*   *ggww*   *Ggww*   *ggWw*

Gray body, normal wings | Black body, reduced wings | Gray body, reduced wings | Black body, normal wings

| Results predicted by law of independent assortment | | | |
|---|---|---|---|
| 575 | 575 | 575 | 575 |
| **Actual results** | | | |
| 965 | 944 | 206 | 185 |

Parental phenotypes are overrepresented because of gene linkage.

Some nonparental phenotypes were obtained because of crossing-over between the *G/g* and *W/w* loci.

Parental phenotypes | Nonparental phenotypes

**Conclusion:** These two genes do not assort independently. They are linked together on the same chromosome.

were very different from the results that the law of independent assortment had led him to expect. What had happened?

Morgan concluded that the genes for body color and wing length are physically tied because they are located on the same chromosome. He inferred that genes close to one another are more likely to be inherited "in one lump" than are genes that are on completely separate chromosomes (such as those for seed color and seed shape that Mendel had studied). So, for the traits of body color and wing length that Morgan studied in fruit flies, the law of independent assortment does not hold. Genetic loci that are neighbors or positioned close to each other on the same chromosome tend to be inherited together and are said to be **genetically linked**. As we shall see shortly, some genes that are located far apart on the same chromosome are not linked. Genes located on different chromosomes also are not genetically linked.

## Crossing-over reduces genetic linkage

If the linkage between two genes on a chromosome could never be broken, the chromosomes inherited by a gamete would never differ from those of the parent that produced that gamete. Consider, for example, the offspring of the *GGWW* × *ggww* cross shown in Figure 13.6. Recall that the two genes (one with alleles *G* and *g*, the other with alleles *W* and *w*) are on the same chromosome. As a result, the *GgWw* flies would have inherited a *GW* chromosome from the *GGWW* parent and a *gw* chromosome from the *ggww* parent. If linkage were absolute, however, the *GgWw* flies would have been able to make gametes only with chromosomes like those in one of their parents; namely, *GW* gametes or *gw* gametes (**FIGURE 13.7**). In that case, half of the offspring from the *GgWw* × *ggww* cross shown in Figure 13.6 would have had genotype *GgWw*, and the

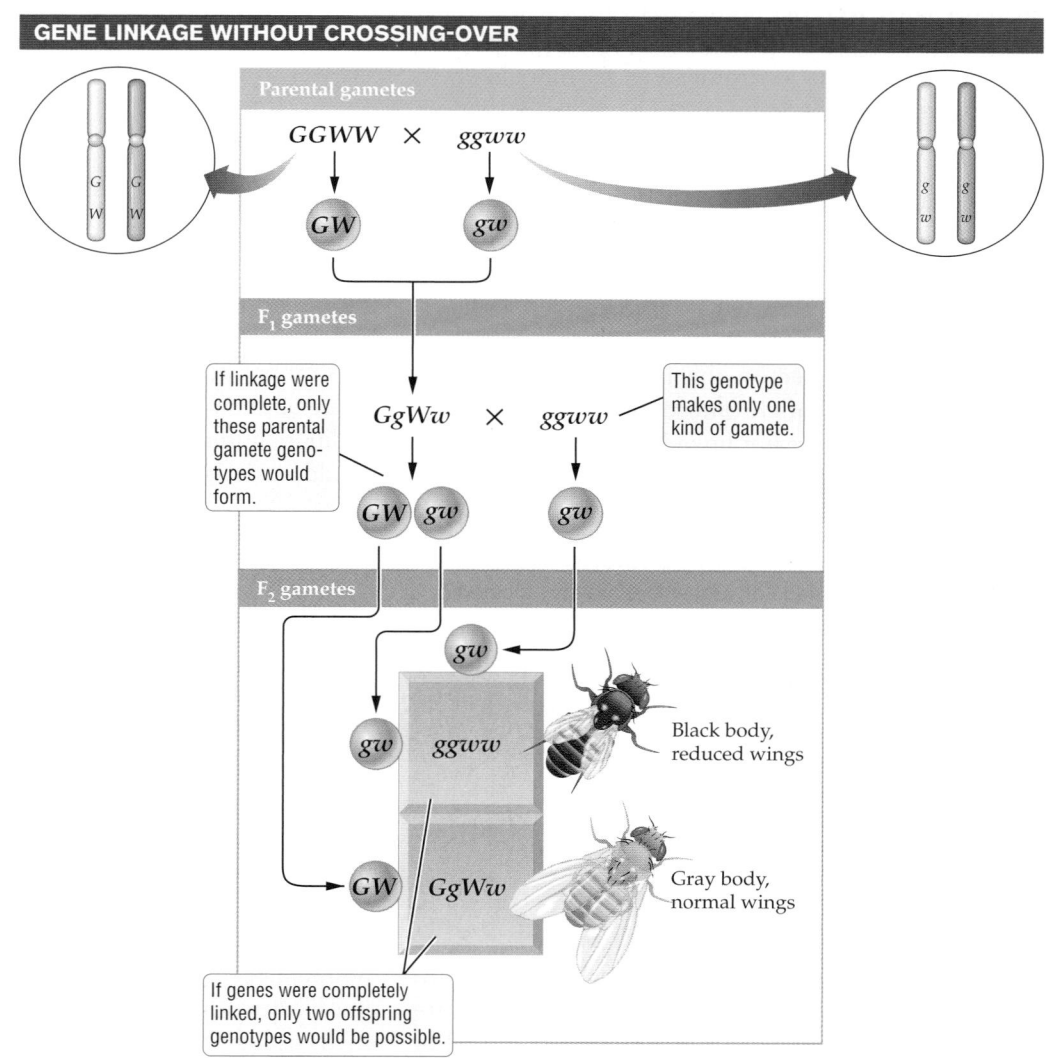

**GENE LINKAGE WITHOUT CROSSING-OVER**

Parental gametes

*GGWW* × *ggww*

*GW*          *gw*

F₁ gametes

If linkage were complete, only these parental gamete genotypes would form.

*GgWw* × *ggww*

This genotype makes only one kind of gamete.

*GW*  *gw*          *gw*

F₂ gametes

*gw*

*gw*  *ggww*          Black body, reduced wings

*GW*  *GgWw*          Gray body, normal wings

If genes were completely linked, only two offspring genotypes would be possible.

**FIGURE 13.7**
**Without Crossing-Over, Genes on the Same Chromosome Would Be Completely Linked**
If genes were completely linked, the F₂ progeny in Morgan's experiment would *all* have had exactly the same genotype as the parents: *GgWw* or *ggww*. But Morgan's experiment (see Figure 13.6) did not produce this result, indicating that there was some recombination between the two genetic loci.

other half would have had genotype *ggww*. Since the majority of the offspring did have those two genotypes, Morgan realized that the two genes are linked. But how can we explain the appearance of the *Ggww* and *ggWw* offspring, which have chromosomes (such as a *Gw* chromosome or a *gW* chromosome) that differ from those found in either parent?

To explain why—in spite of linkage—four genotypes (rather than two) were seen among the offspring of the *GGWW* × *ggww* cross, Morgan suggested that crossing-over had occurred; that is, some genes had been physically exchanged between homologous chromosomes during meiosis. This exchange generates gametes with combinations of alleles that differ from those found in either parent, such as the gametes that resulted in the *Ggww* and *ggWw* offspring shown in Figure 13.6.

Crossing-over can be compared to the cutting of a string at a few random locations stretching from one end of the string to the other. Two points that are far apart on the string will be separated from each other in most cuts, whereas points that are close to each other will rarely be separated. Similarly, genetic loci that are far apart on a chromosome are more likely to be separated by crossing-over than are genes that are close together. In fact, two genes on the same chromosome that are very far apart may be separated by crossing-over so often that they are not genetically linked. Such genes are inherited independently even though they are located on the same chromosome. Among the traits that Mendel studied in pea plants, we now know that the genes for flower color and seed color are both located on chromosome 1 (of the pea plant's seven pairs of chromosomes), but they are so far apart that they are not linked, which is why the law of independent assortment holds for these genes.

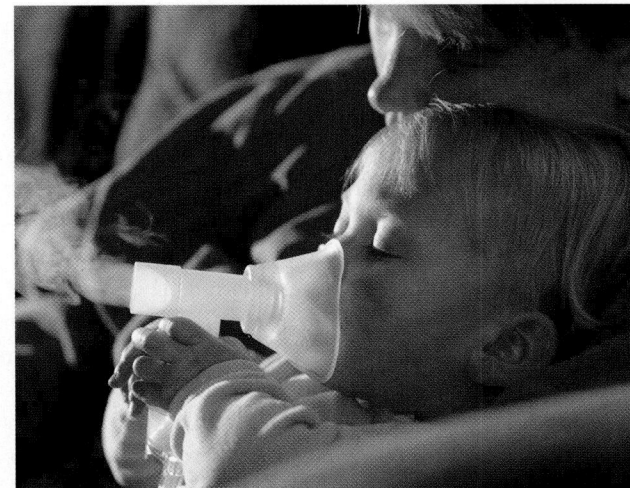

**FIGURE 13.8 Living with a Genetic Disorder**
A toddler with cystic fibrosis inhales medication into her lungs with the help of a nebulizer, which turns the medicine into a fine spray. Cystic fibrosis is a recessive genetic disorder in which mucus builds up in the lungs, digestive tract, and pancreas, causing chronic bronchitis, poor absorption of nutrients, and recurrent bacterial infections, often leading to death before the age of 35.

inherited gene mutations (**FIGURE 13.8**). It is important to study genetic disorders, since such studies could lead to the prevention or cure of much human suffering. But the study of human genetic disorders is beset by daunting problems: We humans have a long generation time, we select our own mates, and we decide whether to have children and how many children we want to have. Geneticists cannot perform experiments directly in humans to figure out how human genetic disorders are inherited. In addition, human families tend to be much smaller than would be ideal in a scientific study.

## Pedigrees are a useful way to study human genetic disorders

Geneticists often analyze *pedigrees* to study patterns of inheritance in humans. A **pedigree** is a chart similar to a family tree that shows genetic relationships among family members over two or more generations of a family's history. Pedigrees provide geneticists with a way to analyze information from many families in order to learn about the inheritance of a particular disorder. The pedigree shown in **FIGURE 13.9**, for example, shows the inheritance of brachydactyly. The condition involves a dominant gene, so it will be seen in all heterozygotes, and it is likely to appear in every generation when we analyze many pedigrees.

> **Concept Check**
>
> 1. What are genes made of, and where are they located?
>
> 2. How are sex chromosomes different from autosomes?
>
> 3. Will two genes that are close together on a chromosome show stronger genetic linkage than a pair of genes that are far apart? Explain.

## 13.4 Human Genetic Disorders

Many of us know someone with a genetic disorder, such as cystic fibrosis, sickle-cell anemia, a hereditary form of cancer, or one of the many other disorders caused by

FIGURE 13.9
**Patterns of Inheritance Can Be Analyzed in Family Pedigrees**

Generation

I 1 2

II 1 2 3 4 5 6 7 8

III 1 2 3 4 5 6 7 8 9 10 11

IV 1 2 3 4 5 6 7 8

○ Female   □ Male   ○□ Unaffected individual   ●■ Affected individual

Actress Megan Fox has one form of brachydactyly.

The pedigree shown here illustrates symbols commonly used by geneticists. The Roman numerals at left identify different generations. Numbers listed below the symbols identify individuals of a given generation. This pedigree charts the inheritance of brachydactyly, a dominant condition marked by short or clubbed fingers and toes. Brachydactyly was the first genetic condition to be analyzed, in 1903, by following pedigrees.

## Some genetic disorders are inherited

Humans can be afflicted by a variety of genetic disorders. Some of these disorders—including most cancers (see Chapter 11)—result from new mutations that occur in a person's cells sometime during that individual's life. **Somatic mutations** occur in cells other than the sex cells (gametes and their germ line precursors) and hence are not passed down to offspring. However, any mutation that is present in the gametes can be passed down from parent to child. Inherited genetic disorders can be caused by mutations in individual genes (**FIGURE 13.10**) or by abnormalities in chromosome number or structure. Clinical genetic tests can be used to determine whether a prospective parent carries an allele for certain genetic disorders, and variations on these methods can be used to test for genetic disorders long before a baby is born (see the box in Chapter 14, page 331).

The remainder of this chapter focuses on inherited genetic disorders that have relatively simple causes: those caused by mutations of a single gene and those caused by chromosomal abnormalities. As you read this material, however, it is important to bear in mind that the tendency to develop some diseases, such as heart disease, diabetes, and some inherited forms of cancer, is caused by interactions among multiple genes and the environment. For most diseases caused by multiple genes, the identity of the genes involved and how they lead to disease is poorly understood.

## 13.5 Autosomal Inheritance of Single-Gene Mutations

We organize our discussion of single-gene genetic disorders by whether the gene is located on an autosome or a sex chromosome. We further subdivide the discussion of autosomal disorders by whether the disease-causing allele is recessive or dominant. As we shall see, recessive genetic disorders with serious ill effects are much more common than dominant ones.

## Autosomal recessive genetic disorders are common

Several thousand human genetic disorders are inherited as recessive traits. Most of these disorders, such as cystic fibrosis, sickle-cell anemia, and Tay-Sachs disease, are caused by recessive mutations of genes located on autosomes.

Recessive genetic disorders vary in severity: some are lethal; others have relatively mild effects. Tay-Sachs disease (see chromosome 15 in Figure 13.10) is a lethal recessive genetic disorder in which the disease-causing allele encodes a defective version of a crucial enzyme. Because the enzyme does not work properly, lipids accumulate in brain cells, the brain begins to

### Helpful to Know

A *genetic condition* is simply a genetic trait that is out of the ordinary. A *genetic disorder* is a disease caused by gene malfunction; the illness or disability it produces can range from mild to life-threatening.

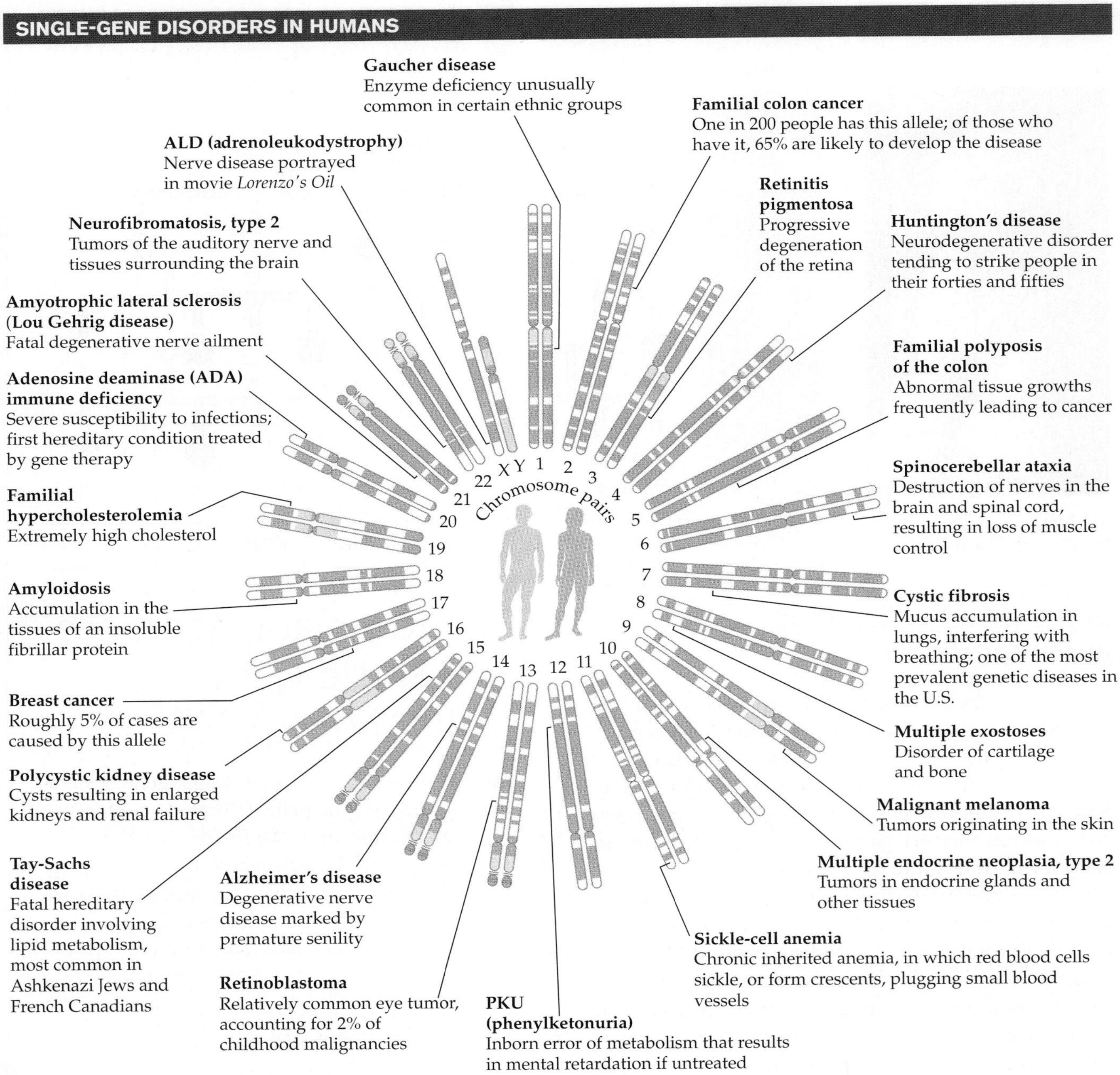

**Gaucher disease**
Enzyme deficiency unusually common in certain ethnic groups

**ALD (adrenoleukodystrophy)**
Nerve disease portrayed in movie *Lorenzo's Oil*

**Neurofibromatosis, type 2**
Tumors of the auditory nerve and tissues surrounding the brain

**Amyotrophic lateral sclerosis (Lou Gehrig disease)**
Fatal degenerative nerve ailment

**Adenosine deaminase (ADA) immune deficiency**
Severe susceptibility to infections; first hereditary condition treated by gene therapy

**Familial hypercholesterolemia**
Extremely high cholesterol

**Amyloidosis**
Accumulation in the tissues of an insoluble fibrillar protein

**Breast cancer**
Roughly 5% of cases are caused by this allele

**Polycystic kidney disease**
Cysts resulting in enlarged kidneys and renal failure

**Tay-Sachs disease**
Fatal hereditary disorder involving lipid metabolism, most common in Ashkenazi Jews and French Canadians

**Alzheimer's disease**
Degenerative nerve disease marked by premature senility

**Retinoblastoma**
Relatively common eye tumor, accounting for 2% of childhood malignancies

**Familial colon cancer**
One in 200 people has this allele; of those who have it, 65% are likely to develop the disease

**Retinitis pigmentosa**
Progressive degeneration of the retina

**Huntington's disease**
Neurodegenerative disorder tending to strike people in their forties and fifties

**Familial polyposis of the colon**
Abnormal tissue growths frequently leading to cancer

**Spinocerebellar ataxia**
Destruction of nerves in the brain and spinal cord, resulting in loss of muscle control

**Cystic fibrosis**
Mucus accumulation in lungs, interfering with breathing; one of the most prevalent genetic diseases in the U.S.

**Multiple exostoses**
Disorder of cartilage and bone

**Malignant melanoma**
Tumors originating in the skin

**Multiple endocrine neoplasia, type 2**
Tumors in endocrine glands and other tissues

**Sickle-cell anemia**
Chronic inherited anemia, in which red blood cells sickle, or form crescents, plugging small blood vessels

**PKU (phenylketonuria)**
Inborn error of metabolism that results in mental retardation if untreated

X Y 1 2 3 4 5 6 7 8 9 10 11 12 13 14 15 16 17 18 19 20 21 22
Chromosome pairs

**FIGURE 13.10  Genes Known to be Associated with Inherited Genetic Disorders in Humans**
About 2,000 Mendelian traits are known, and each has been mapped to one of the 23 pairs of human chromosomes. Mutations of single genes that lead to genetic disorders are found on the X chromosome and on each of the 22 autosomes in humans. For clarity, we show only one such genetic disorder per chromosome.

deteriorate during a child's first year of life, and death occurs within a few years. Toward the other end of the severity spectrum, adult-onset lactose intolerance is caused by a single recessive allele that leads to a shutdown in the production of lactase, the enzyme that digests milk sugar, in adolescence.

The only individuals who actually get a disorder caused by an autosomal recessive allele (say, *a*) are those who have two copies of that allele (*aa*). Usually, when a child inherits a recessive genetic disorder, both parents are heterozygous; that is, they both have genotype *Aa*. (It is also possible for one or both parents to have genotype *aa*.) Because the *A* allele is dominant and does not cause the disorder, heterozygous individuals are said to be **genetic carriers** of the disorder: they carry the disorder allele (*a*), but do not get the disorder.

If two carriers of a recessive genetic disorder have children, the patterns of inheritance are the same as for any recessive trait: ¼ of the children are likely to have genotype *AA*, ½ to have genotype *Aa*, and ¼ to have genotype *aa*. As **FIGURE 13.11** shows, each child has a 25 percent chance of not carrying the disorder allele (genotype *AA*), a 50 percent chance of being a carrier (genotype *Aa*), and a 25 percent chance of actually getting the disorder (genotype *aa*). Because the children of two carriers have a 75 percent chance of *not* displaying any symptoms, recessive disorders often "skip a generation" in a family pedigree.

These percentages reveal one way in which lethal recessive disorders such as Tay-Sachs disease can persist in the human population: although homozygous recessive individuals (with genotype *aa*) die long before they are old enough to have children, carriers (with genotype *Aa*) are not harmed by the disorder. In a sense, the *a* alleles can "hide" in heterozygous carriers, and those carriers are likely to pass the disorder allele to half of their children. Recessive genetic disorders also arise in the human population because new mutations produce new copies of the disorder alleles.

## Serious dominant genetic disorders are less common

A dominant allele (*A*) that causes a genetic disorder cannot "hide" in the same way that a recessive allele can. In this case, *AA* and *Aa* individuals get the disorder; only *aa* individuals are symptom-free. If one parent has a dominant genetic disorder, the children have 50 percent chance of inheriting it. In a pedigree, every individual who is affected has at least one parent who is affected.

When a dominant genetic disorder produces serious negative effects, the individuals that have the *A* allele may not live long enough to reproduce; hence, few of them pass the allele on to their children. As a result, dominant alleles that prevent a sufferer from reproducing are uncommon in the population. Such lethal dominant alleles are rarely handed down through the generations in a family line; instead, most such alleles appear in the population because of new mutations that arise during gamete formation.

If a dominant allele expresses its lethal effects later in life, however, after the person carrying

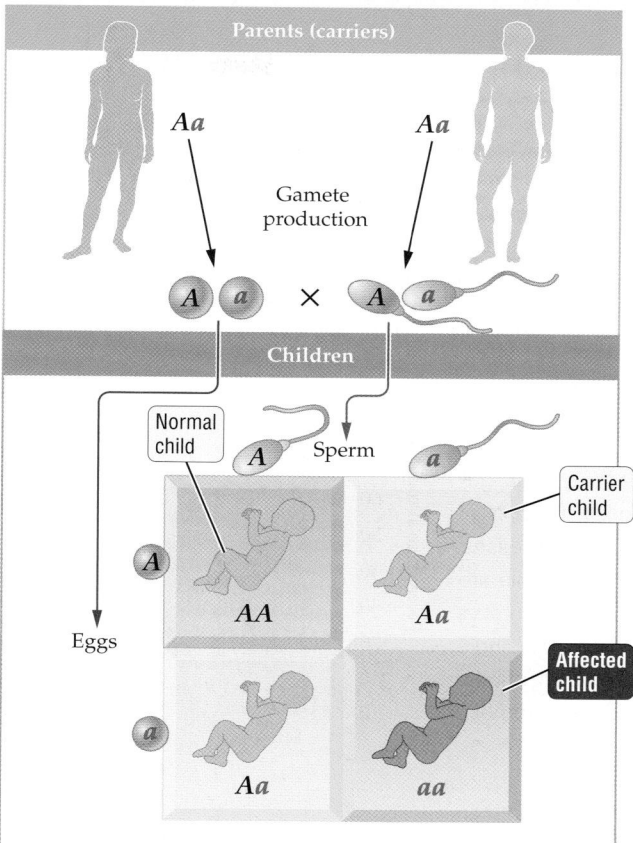

**FIGURE 13.11 Genetic Carriers Transmit an Autosomal Recessive Condition without Displaying It Themselves**
The patterns of inheritance for a human autosomal recessive genetic disorder are the same as for any other recessive trait (compare this figure with the pattern shown by Mendel's pea plants in Figure 12.7). Recessive disorder alleles are colored red and denoted *a*. Dominant, normal alleles are black and denoted *A*. Here, the parents are a carrier male (genotype *Aa*) and a carrier female (genotype *Aa*).

that allele has had an opportunity to reproduce, that allele is readily passed from one generation to the next. Huntington's disease (described at the beginning of this chapter) illustrates how such a dominant lethal allele can remain in the population. The symptoms of Huntington's disease begin relatively late in life, often after victims of the disorder have had children. Because the allele that causes the disorder can be passed on to the next generation before the victim dies, the disorder is more common than it would be if the symptoms began before childbearing age or if the disorder were to arise from new mutations alone.

# Most Chronic Diseases Are Complex Traits

A *disease* is a condition that impairs health. It may be caused by external factors, such as infection by viruses, bacteria, and other parasites, or injury produced by harmful chemicals or high-energy radiation. Nutrient deficiency can also lead to disease, the way inadequate vitamin C consumption produces scurvy. Disease may be caused also by internal factors, controlled by one or more genes. Diseases that are caused exclusively by gene malfunction are described as *genetic disorders*, to distinguish them from infections and other types of diseases. Cystic fibrosis and sickle-cell anemia are examples of inherited genetic disorders caused by errors in the activity of a single gene. Myotonic dystrophy, which affects muscle cells, is an example of a genetic disorder created by malfunctions in more than one gene.

The diseases that are most common in industrialized countries—heart disease, cancer, stroke, diabetes, asthma, and arthritis, for example—are caused by multiple genes interacting in complex ways with each other and with various external factors. Malfunctions in key genes make a person prone to developing these diseases, but environmental factors affect whether the disease will actually appear or how severe the symptoms will be. As shown in the graph, a large part of the estimated risk of developing colon cancer, stroke, coronary heart disease, and type 2 diabetes is avoidable. Lifestyle choices such as maintaining good nutrition, exercising regularly, and avoiding tobacco have

a significant impact on our risk of developing chronic diseases like these. (The word "chronic" means "unceasing"—a reference to the fact that once we develop one of these diseases, we have it for the rest of our lives.)

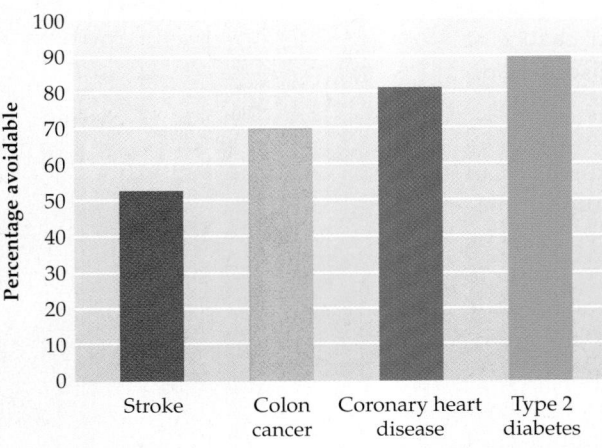

To reduce our risk of these diseases, we should maintain a healthy weight, engage in physical activity equivalent to at least 30 minutes of brisk walking per day, avoid smoking, and consume fewer than three alcoholic beverages per day. It helps to limit our intake of saturated fat and trans fat (see Chapter 5), as well as sugar and refined carbohydrates. Consuming enough folic acid (at least 400 milligrams a day) and dietary fiber (21–38 grams a day, depending on size and age) is linked to a lower risk of several chronic diseases, including heart disease and diabetes. People who eat fewer than three servings of red meat per week

lower their odds of getting colon cancer. Many independent studies have demonstrated that those who eat more vegetables, whole grains, and fruits have better health outcomes than do those who consume very little of these plant-based foods.

A major goal of modern genetics is to identify genes that contribute to human disease when they fail to function normally. Researchers have identified alleles associated with increased risk of a number of common ailments, including high blood pressure, heart disease, diabetes, Alzheimer's disease, several types of cancer, and schizophrenia. For example, a particular allele on chromosome 9 raises the risk of heart disease. Those who are homozygous for the risk allele are about 30–40 percent more likely to have coronary heart disease than are individuals who are homozygous for the harmless allele. Heterozygotes are about 15–20 percent more likely to have the disease than are individuals who lack the risk allele.

The hope is that genetic tests will tell us if we are predisposed to a disease before we become ill with it. A person carrying a risk allele could take preventive measures to reduce the risk of actually developing the condition, and treatment could be made more effective by customization to match the particular allele involved. This customized approach to treatment, or *personalized medicine*, is already being used to treat breast cancer and some other chronic diseases.

# 13.6 Sex-Linked Inheritance of Single-Gene Mutations

Roughly 1,200 of the estimated 25,000 human genes are found only on the X chromosome or only on the Y chromosome; such genes are said to be **sex-linked**. Because sex-linked genes are found on either the X chromosome or the Y chromosome—but not both—males receive only one copy of each sex-linked gene (whether it is on the X chromosome or the Y chromosome), while females receive two copies of genes on the X chromosome and no copies of genes on the Y chromosome. About 15 genes, however, are shared by the X and Y chromosomes. In each of these cases, males and females receive two copies of the gene, just as they do for all autosomal genes; as a result, these 15 genes are not sex-linked, even though they are found on the sex chromosomes.

The overwhelming majority of the approximately 1,100 human sex-linked genes are located on the X chromosome, while about 50 are located on the much smaller Y chromosome. Sex-linked genes on the X chromosome are said to be **X-linked**; similarly, all sex-linked genes on the Y chromosome are said to be **Y-linked**. Although there are no well-documented cases of disease-causing Y-linked genes, X chromosomes do contain genes known to be associated with genetic conditions that range from mild traits, such as red-green color blindness, to devastating genetic disorders like Duchenne muscular dystrophy. Sex-linked genes have different patterns of inheritance than do genes on autosomes.

Consider how an X-linked recessive allele for a human genetic disorder is inherited (**FIGURE 13.12**). We label the recessive disorder allele $a$, and in the Punnett square we write this allele as $X^a$ to emphasize the fact that it is on the X chromosome. Similarly, the dominant allele is labeled $A$ and is written as $X^A$ in the Punnett square. If a carrier female (genotype $X^AX^a$) has children with a normal male (genotype $X^AY$), each of their sons will have a 50 percent chance of getting the disorder (see Figure 13.12). This result differs greatly from what would happen if the same disorder allele ($a$) were on an autosome: in that case none of the children, male or female, would get the disorder, because then males could be heterozygous ($Aa$) as well, in which case they would be shielded from the harmful effects of the $a$ allele by the presence of the normal $A$ allele. Instead, what happens in X-linked disorders is

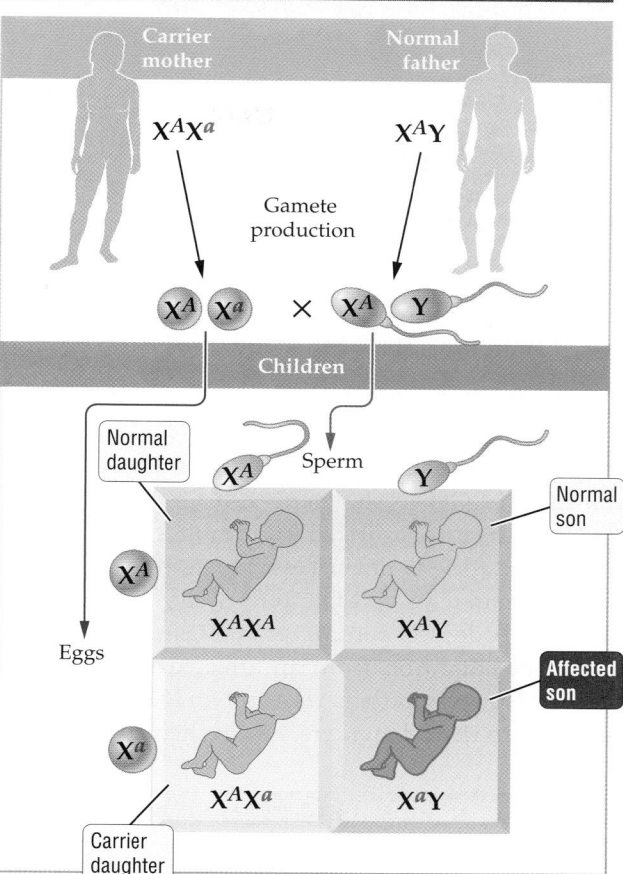

**INHERITANCE OF AN X-LINKED DISORDER**

Carrier mother — $X^AX^a$
Normal father — $X^AY$

Gamete production

$X^A$  $X^a$  ×  $X^A$  $Y$

Children

Sperm

Normal daughter — $X^A$

Normal son

Eggs

$X^A$ — $X^AX^A$  |  $X^AY$

$X^a$ — $X^AX^a$  |  $X^aY$ — Affected son

Carrier daughter

**FIGURE 13.12**
**X-Linked Recessive Conditions Tend to Be More Common in Males Than in Females**

The recessive disorder allele ($a$) is located on the X chromosome and is denoted by $X^a$. The dominant normal allele ($A$) on the X chromosome is denoted by $X^A$.

that males of genotype $X^aY$ suffer from the condition because the Y chromosome does not have a copy of that gene at all; in other words, because males cannot be heterozygous for any X-linked genes, the effects of an $a$ allele cannot be masked. In general, males are more likely than females to get recessive X-linked disorders, because they have to inherit only a single copy of the disorder allele to exhibit the disorder, whereas females must inherit two copies to be affected. In contrast, both sexes are equally likely to be affected by autosomal recessive disorders, since both have two copies of autosomal chromosomes and identical odds of being homozygous or heterozygous for a disorder allele.

X-linked genetic disorders in humans include hemophilia, a serious bleeding disorder in which minor cuts and bruises can cause a person to bleed to death, and Duchenne muscular dystrophy, a lethal disorder that causes the muscles to waste away, often leading to death at a young age. Both of these X-linked disorders are caused by recessive alleles. An example of a dominant X-linked disorder with a daunting name—

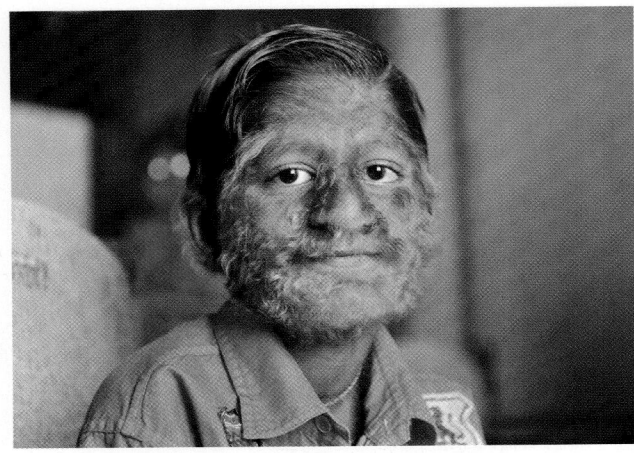

**FIGURE 13.13**
**Congenital Generalized Hypertrichosis**

An 11-year-old boy with CGH, a rare genetic disorder characterized by extreme hairiness of the face and upper body. CGH is caused by a dominant allele of a single gene on the X chromosome.

congenital generalized hypertrichosis [*HYE*-per-trih-*KOH*-sis], or CGH—is shown in **FIGURE 13.13**.

Note that an X-linked condition controlled by a *dominant* allele is not necessarily more common in males than in females. Male and female children both have equal odds (50 percent) of inheriting an X-linked dominant condition from an affected heterozygous mother. However, fathers transmit an X-linked condition only to their daughters, not to their sons. All the

daughters of a man with an X-linked dominant trait will display that trait.

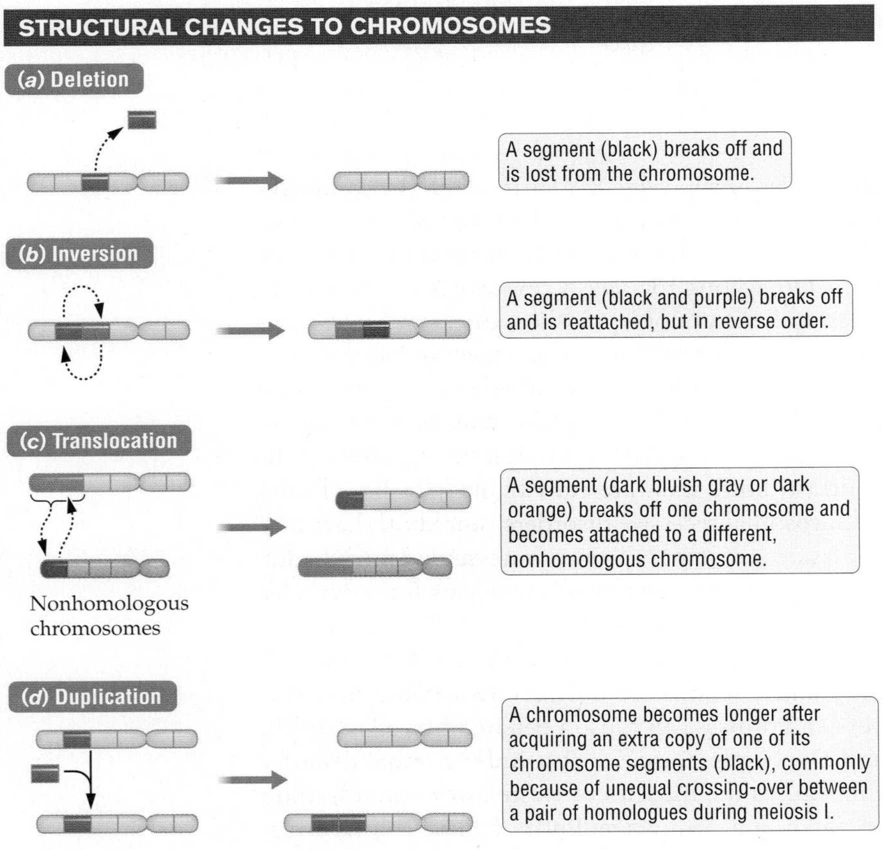

## 13.7 Inherited Chromosomal Abnormalities

Every species has a characteristic number of chromosomes, and each chromosome has a particular structure, with specific genetic loci arranged on it in a precise sequence. Any change in the chromosome number or structure, compared to what is typical for the species, is considered a chromosomal abnormality. Two main types of chromosomal changes are seen in humans and other organisms: changes in the overall number of chromosomes and changes in chromosome structure, such as a change in the length of an individual chromosome. A cell is especially vulnerable to developing chromosomal abnormalities when it is dividing: chromosomes can be misaligned, misdirected, or even ripped into pieces during the delicate business of lining them up in the center of the cell and then segregating them into daughter cells. For chromosomal abnormalities to be passed on from a parent to offspring, the chromosomal changes have to occur in the gametes or gamete-producing cells. However, relatively few human genetic disorders are caused by inherited chromosomal abnormalities, probably because most large changes in the chromosomes kill the developing embryo.

### The structure of chromosomes can change in several ways

When chromosomes are being aligned or separated during cell division, breaks can occur that alter the length of one or more chromosomes. In **deletion**, a piece breaks off and is lost from the chromosome (**FIGURE 13.14a**). Other times the broken piece is reattached, but incorrectly. For example, in an **inversion**, the fragment returns to the correct place on the original chromosome, but with the genetic loci in reverse order (**FIGURE 13.14b**). This would be similar to snapping a pencil in two and then reattaching the eraser end to the broken end of the other piece. In a **translocation**, a broken piece from one chromosome becomes attached to a different, nonhomologous chromosome (**FIGURE 13.14c**). Translocations are frequently reciprocal, meaning that two nonhomologous chromosomes are involved in a mutual exchange of fragments. This type of translocation is shown in Figure 13.14c; it is rather like breaking pieces off a blue pencil and an

**STRUCTURAL CHANGES TO CHROMOSOMES**

**(a) Deletion**

A segment (black) breaks off and is lost from the chromosome.

**(b) Inversion**

A segment (black and purple) breaks off and is reattached, but in reverse order.

**(c) Translocation**

Nonhomologous chromosomes

A segment (dark bluish gray or dark orange) breaks off one chromosome and becomes attached to a different, nonhomologous chromosome.

**(d) Duplication**

A chromosome becomes longer after acquiring an extra copy of one of its chromosome segments (black), commonly because of unequal crossing-over between a pair of homologues during meiosis I.

**FIGURE 13.14 Chromosome Rearrangements Can Cause Serious Genetic Disorders**

orange one, and then swapping the broken pieces between the two pencils, so that the blue one acquires a piece of the orange one and vice versa.

**Duplication** is a type of chromosomal abnormality in which a chromosome becomes longer because it ends up with two copies of a particular chromosome fragment (**FIGURE 13.14d**). There are several ways in which a chromosome can acquire duplicates of a particular stretch of the chromosome. Errors in crossing-over are one source of duplications. When the two members of a homologous pair line up to exchange chromosome segments during meiosis I (see Chapter 10), sometimes the trade is uneven in that one homologue receives a chromosomal fragment from the other homologue but fails to reciprocate by "giving back" an equivalent length of itself to its partner. The first homologue therefore ends up with duplicates of that stretch of the chromosome (one copy of its own, plus one acquired from its partner homologue), while the other homologue suffers a deletion.

Changes in chromosome structure can have dramatic effects. A break in the sex chromosomes can cause a change in the expected sex of the developing fetus. A deletion in which the portion of the Y chromosome that contains the *SRY* gene is lost produces an XY individual who develops as a female. A translocation that results in an X chromosome to which the Y chromosome's sex-determining region is attached produces an XX individual who develops as a male. XY females and XX males are always sterile (unable to produce offspring).

Changes to the structure of autosomes can have even more dramatic effects. Cri du chat [*KREE*-**doo**-*SHAH*] syndrome occurs when a child inherits a chromosome 5 that is missing a particular region (**FIGURE 13.15**). *Cri du chat* is French for "cry of the cat," which describes the characteristic mewing sound made by infants with this condition. Other characteristics of this condition are slow growth and a tendency toward severe intellectual disability, a small head, and low-set ears.

## Changes in chromosome number are often fatal

Unusual numbers of chromosomes—such as one or three copies instead of the normal two—can result when chromosomes fail to separate properly during meiosis. In humans, such changes in chromosome number often result in fetal death. At least 20 percent of human pregnancies abort spontaneously, largely as a result of such changes in chromosome number.

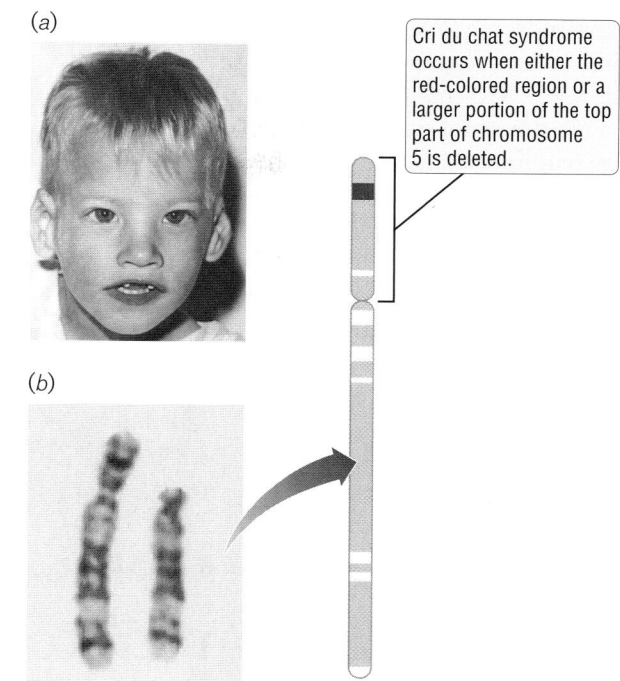

(a)

Cri du chat syndrome occurs when either the red-colored region or a larger portion of the top part of chromosome 5 is deleted.

(b)

**FIGURE 13.15  Cri du Chat Syndrome**
Cri du chat syndrome is caused by the deletion of a portion of the top part of chromosome 5.

There is only one genetic anomaly in which a person who inherits the wrong number of autosomes commonly reaches adulthood: Down syndrome. Individuals with this condition usually have three copies of chromosome 21, one of the smallest autosomes in humans. Down syndrome is also known as trisomy 21, where **trisomy** [*TRYE*-soh-mee] refers to the condition of having three copies of a chromosome (instead of the usual two). A small minority (3–4 percent) of Down syndrome cases occur when an extra piece of chromosome 21 breaks off during cell division and attaches to another chromosome. These individuals have two copies of chromosome 21 plus part of another chromosome 21. People with Down syndrome tend to be short and intellectually disabled, and they may have defects of the heart, kidneys, and digestive tract. With appropriate medical care, most people with this condition lead healthy lives, and many live to their sixties or seventies (their average life expectancy is 55). Live births can also result when an infant has three copies of chromosome 13, 15, or 18. However, such children have severe birth defects, and they rarely live beyond their first year.

Compared with having too few or too many autosomes, changes in the number of sex chromosomes have more minor effects. Klinefelter syndrome, for example, is a condition found in males that have an extra

**Helpful to Know**

Earlier we noted the use of the suffix *some* in the names of cellular particles, such as chromosomes. Biologists use the related suffix *somy* to name conditions involving unusual numbers of chromosomes; two examples are *trisomy* ("three bodies") and *monosomy* ("one body").

X chromosome (XXY males). Such men have a normal life span and normal intelligence, and they tend to be tall. Many XXY males also have small testicles (about one-third normal size) and reduced fertility, and some have feminine characteristics, such as enlarged breasts. Females with a single X chromosome (Turner syndrome) have normal intelligence and tend to be short (with adult heights of less than 150 centimeters, about 4 feet 11 inches), to be sterile, and to have a broad, webbed neck. Other changes in the number of sex chromosomes, as in XYY males and XXX females, also produce relatively mild effects. However, when there are two or more extra sex chromosomes, as in XXXY males or XXXX females, a wide range of problems can result, including profound intellectual disability.

> ## Concept Check

1. What is a genetic pedigree?
2. Why are dominant autosomal disorders that produce serious negative effects less common than serious recessive autosomal disorders?
3. In North America, the X-linked form of red-green color blindness affects about 8 percent of men, but only 0.5 percent of women. Why is this condition more common in men?

| APPLYING WHAT WE LEARNED |

# Testing for Huntington's Disease

Most genetic diseases are caused by a recessive allele, which means that for a person to become sick, they have to inherit two "bad" versions of the gene (alleles)—one from their mother and one from their father. But Huntington's disease (HD) is different. Because it's caused by a dominant allele, even people who inherit only one copy of the HD allele will get the disease (unless they die from an accident or some other disease). Each child of a Huntington's parent has a 50 percent chance of developing the disease.

In 1983, researchers mapped the HD gene by showing that it was linked to another gene on chromosome 4. By 1993, the HD gene had been located exactly, making it the first autosomal disease gene to be located with genetic linkage analysis. The gene was named *huntingtin* and the protein it coded for, huntingtin. Researchers soon developed a genetic test for the allele. And within a handful of years, they also developed strains of laboratory mice carrying a mutated *huntingtin* allele that developed symptoms similar to those of human Huntington's disease. The race was soon on to find a way to cure these mice and, after that, human patients.

Genetic counselors began to talk to the families of Huntington's patients about being tested for the disease. Genetic testing can help reduce the number of people with genetic diseases. For example, since genetic screening for Tay-Sachs disease became available in the 1970s, the disease has nearly disappeared from families that once had it. Unlike Huntington's, Tay-Sachs is caused by a recessive allele that affects young children, who typically die before the age of 4. In principle, both Huntington's disease and the HD allele could be eliminated in a single generation if no one carrying the HD allele had children.

But few people want to know if they have the allele. As Misty Oto explained to an interviewer in 2007, finding out that she definitely had HD was "not something I would want to know. I would feel guilty, not necessarily for my kids, but for [my husband]—because I would know what his future would hold, and I wouldn't want him to be trapped." Many people also worry that a positive result will affect their ability to get health insurance or keep a job. Sometimes a positive result leads to severe depression or even suicide. So, for a range of reasons, only about 10 percent of people at risk for Huntington's take the test.

Still, when Misty's older sister began to suspect she might have the disease, the two sisters decided to confront their fears and take the test. To Misty's relief, the test showed that she did not have an HD allele. That meant her husband and kids would not have to cope with the prospect of her developing this terrible and lingering disease. And it meant her kids would not grow up wondering if they would develop the disease. But Misty's happiness was short-lived. Her sister, who got her results later the same week, learned that she had Huntington's—a result Misty was unprepared for.

So far, there's little prospect of a cure for Huntington's, but researchers have learned a lot about the disease, including how the faulty allele is constructed, which environmental factors speed up or slow progression of the disease, and how different proteins and enzymes are involved. Eventually, researchers expect to find a way to at least delay the onset of symptoms, giving patients and their caregivers something to hope for.

# Stanford Students See What's in Their Genes

BY KATHRYN ROETHEL, *San Francisco Chronicle*

Konrad Karczewski, a bioinformatics student at Stanford University, learned quite a bit about himself over the summer. He found out that his risk of prostate cancer is 24 percent—seven percentage points higher than the average man's—and confirmed that his chance of having hypertension is higher than that of the general population, something he already suspected from his family history. The doctorate student also learned his responsiveness to certain medications.

Not that any of that necessarily bothered him. He knew what he was getting into when he signed up for a controversial Stanford Medical School summer course that allowed students to test their own DNA.

Potentially troubling genetic results like Karczewski's were at the center of a university task force's yearlong debate over whether to allow med students to study their own genotypes. Some faculty members worried students might learn results that would be too upsetting and felt personal genetic testing wasn't a necessary element for student learning.

But Karczewski disagrees with the critics.

"More knowledge is always better," he said. "I fully understood the implications of this test because I have a genetics background. It's important to remember that these predictions are not a diagnosis. It doesn't mean you are or aren't getting the disease. But [if] you're aware that there's a risk early on, you can get screening and read up on prevention."

The course was designed to teach future doctors and scientists to understand the commercially available genetic tests. All students had the choice of using their own genetic data or that of 12 anonymous patients. Class activities ranged from mapping patients' ancestry to certain geographic locations to calculating the prescription drug dosages patients are genetically predisposed to respond to.

Thirty-three students in a class of 60 chose to participate in their own genotyping. Ten others mapped their DNA before the class using commercial services like 23andMe.

Krystal St. Julien, a doctoral student in biochemistry, took the course but opted not to participate in genotyping. "I didn't know at the time if this was something I could handle," St. Julien said. "I kept hearing the words, 'This may do harm,' so I went back and forth a lot before I made my decision."

Yet, St. Julien is adamant that the course was valuable and should be offered again. And her opinion of genotyping changed during her eight weeks of classes. "I realized it probably wouldn't have been such a traumatic experience if I had studied my own DNA," she said.

Keyan Salari—who is working toward his doctoral degrees in medicine and genetics, and designed and co-taught the course—said he was pleased with how the class turned out. "The students were fantastic, and the classroom was really dynamic and interactive," he said. "I think students will walk away with a better understanding of genetic tests— their benefits and limitations—than most of the doctors practicing today."

---

Increasing numbers of private companies are marketing genetic information directly to individuals instead of through a doctor. Such companies take a few cells, often scraped from the inside of the cheek, and deliver a report listing a selection of alleles from among the 25,000 human genes. In a few cases, researchers have a pretty good idea of how different alleles affect our health, but the effects often depend on what other alleles are present and on environmental influences. Sometimes nothing is known at all. But a gene can have one or two alleles or hundreds, and we don't know what most of them do.

Diseases such as Huntington's, Tay-Sachs, and cystic fibrosis are strongly genetic, so knowing if you carry an allele for one of these can have a huge impact on your life. But the great majority of people in the United States— about 60 percent—will die of one of just three conditions: cardiovascular disease, cancer, or stroke. All of these are "lifestyle" diseases, diseases that have a significant environmental component. Genetic analysis may tell you that you are more likely than the average person to have a heart attack and less likely to get cancer. But to act on that information, you'd just do the things you already know you should do—minimize stress, don't smoke, exercise regularly, and eat a diet high in whole grains, fruits, and vegetables.

## Evaluating the News

**1.** Why did instructors at Stanford Medical School think this course might be a good idea?

**2.** Some experts think it's a mistake for individuals to know their genetic information without a genetic counselor or doctor to discuss what the information actually means. Discuss the advantages and disadvantages of such direct-to-consumer genetic testing.

**3.** Would you be tested? Would you choose not to have children if you knew you had the HD allele?

**SOURCE:** *San Francisco Chronicle*, September 8, 2010

# Summary

## 13.1 The Role of Chromosomes in Inheritance

- The chromosome theory of inheritance states that genes are located on chromosomes. The physical location of a gene on a chromosome is known as its genetic locus.
- Each chromosome is composed of a single DNA molecule and many associated proteins.
- Chromosomes that pair during meiosis are homologous chromosomes. A pair of homologous chromosomes has the same genetic loci.
- Chromosomes that determine the sex of the organism are called sex chromosomes; all other chromosomes are called autosomes. Humans have two types of sex chromosomes: X and Y. Males have one X and one Y chromosome. Females have two X chromosomes.
- A specific gene on the Y chromosome (the *SRY* gene) is required for human embryos to develop as males.

## 13.2 Origins of Genetic Differences between Individuals

- Genetic differences among individuals provide the genetic variation on which evolution can act. Mutations are the source of new alleles.
- Offspring differ genetically from one another and from their parents because (1) each pair of homologues exchanges random stretches of the chromosome during crossing-over, (2) the different homologous pairs are sorted into gametes independently of each other during meiosis, and (3) fertilization combines male and female genotypes randomly.

## 13.3 Genetic Linkage and Crossing-Over

- Genes that tend to be inherited together are said to be genetically linked.
- Two genes that are far apart on a chromosome are more likely to be shuffled by crossing-over than are genes located near one another, and they are therefore less likely to show genetic linkage.
- Mendel's law of independent assortment holds for genes located far apart on a chromosome, just as it does for a pair of genes located on two different chromosomes.

## 13.4 Human Genetic Disorders

- Pedigrees are useful for studying the inheritance of human genetic disorders.

- Humans suffer from a variety of genetic disorders, including some caused by mutations of a single gene and some caused by abnormalities in chromosome number or structure.

## 13.5 Autosomal Inheritance of Single-Gene Mutations

- Most genetic disorders are caused by a recessive allele (*a*) of a gene on an autosome. Only homozygous (*aa*) individuals get these disorders; heterozygous (*Aa*) individuals are merely carriers of the disorders.
- In dominant autosomal genetic disorders, both *AA* and *Aa* individuals are affected. If defects of the phenotype are so severe that the *AA* or *Aa* individual cannot reproduce, such a dominant genetic disorder is likely to be rare.
- Alleles that cause lethal dominant genetic disorders can remain in the population if symptoms begin late in life, as in Huntington's disease, or if new disorder alleles are produced by mutation during each generation.

## 13.6 Sex-Linked Inheritance of Single-Gene Mutations

- Because males inherit only one X chromosome, genes on sex chromosomes have different patterns of inheritance than do genes on autosomes.
- Genes found on one of the sex chromosomes but not on both may show sex-linked patterns of inheritance. Genes found only on the X chromosome are said to be X-linked; those found only on the Y chromosome are Y-linked.
- Males are more likely than females to have recessive X-linked genetic disorders, because males need to inherit only one copy of the disorder allele to be affected, whereas females must inherit two copies to be affected. In contrast, males and females are equally likely to be affected by autosomal genetic disorders or dominant X-linked disorders.

## 13.7 Inherited Chromosomal Abnormalities

- The structure of an individual chromosome can change through breakage during cell division, resulting in deletion, inversion, translocation, or duplication of a chromosome fragment.
- Changes in the number of autosomes in humans are usually lethal. Down syndrome, a form of trisomy in which individuals receive three copies of chromosome 21, is an exception to this rule.
- People who have one too many or one too few sex chromosomes may experience relatively minor effects, but if there are four or more sex chromosomes (instead of the usual two), serious problems can result.

# Key Terms

autosome (p. 303)
chromosome theory
   of inheritance (p. 302)
crossing-over (p. 304)
deletion (p. 314)
duplication (p. 315)

genetic carrier (p. 311)
genetic linkage (p. 307)
homologous
   chromosomes (p. 302)
independent assortment
   of chromosomes (p. 304)

inversion (p. 314)
locus (p. 302)
pedigree (p. 308)
sex chromosome (p. 303)
sex-linked (p. 313)
somatic mutation (p. 309)

*SRY* gene (p. 304)
translocation (p. 314)
trisomy (p. 315)
X-linked (p. 313)
Y-linked (p. 313)

# Self-Quiz

1. Genes commonly exert their effect on the phenotype by
   a. promoting DNA mutations.
   b. creating chromosomal structures such as centromeres.
   c. coding for a protein.
   d. all of the above

2. Which of the following is an autosomal dominant disorder in which the symptoms begin late in life and the nervous system is destroyed, resulting in death?
   a. Tay-Sachs disease
   b. Huntington's disease
   c. Down syndrome
   d. cri du chat syndrome

3. Crossing-over is more likely to occur between genes that are
   a. close together on a chromosome.
   b. on different chromosomes.
   c. far apart on a chromosome.
   d. on the Y chromosome.

4. Comparatively few human genetic disorders are caused by chromosomal abnormalities. One reason is that
   a. most chromosomal abnormalities have little effect.
   b. it is difficult to detect changes in the number or length of chromosomes.
   c. most chromosomal abnormalities result in spontaneous abortion of the embryo.
   d. it is not possible to change the length or number of chromosomes.

5. Nonparental genotypes can be produced by
   a. crossing-over and the independent assortment of chromosomes.
   b. linkage.
   c. autosomes.
   d. sex chromosomes.

6. Sometimes a segment of DNA breaks off from a chromosome and then returns to the correct place on the original chromosome, but in reverse order. This type of chromosomal structural change is called
   a. crossing-over.
   b. translocation.
   c. inversion.
   d. deletion.

7. Which of the following can most precisely be described as a master switch that commits the sex of the developing embryo to "male"?
   a. an X chromosome
   b. a Y chromosome
   c. an XY chromosome pair
   d. the *SRY* gene

# Analysis and Application

1. In terms of its physical structure, describe what a gene is and where it is located.

2. Consider the XY chromosome system by which sex is determined in humans. Do patterns of inheritance for genes located on the X chromosome differ between males and females? Why or why not?

3. In 1917, Shinoba Ishihara designed a series of colored plates for testing individuals for red-green color perception. In the images shown below, a person with red-green color blindness will not see any number in the Ishihara plate on the left, and will not be able to read the number in the plate on the right correctly. Can a woman have X-linked red-green color blindness? If so, how? Draw a three-generation pedigree of a woman with this condition. Include the woman's parents and her five children (three girls and two boys), and assume that her husband is not color-blind. Indicate all carriers in your pedigree by placing a dot in the center of the symbol representing them—a standard way of symbolizing carriers. What are the odds that this woman's sons will be color-blind? What are the odds that any of her children will be carriers for X-linked red-green color blindness?

4. Look carefully at Figure 13.6. Explain in your own words why the results shown there convinced Morgan that genes located near one another on a chromosome tend to be inherited together, or linked. What results would be expected if these genes were located on different chromosomes?

5. Explain how crossing-over occurs. Assume that genes *A*, *B*, and *C* are arranged in that order along a chromosome. From your understanding of crossing-over, do you think it will occur more often between genes *A* and *B* or between genes *A* and *C*?

6. Describe two mechanisms that give rise to nonparental genotypes in the next generation.

7. Are genetic disorders that are caused by single-gene mutations more common or less common in human populations than those caused by abnormalities in chromosome number or structure? Explain your answer.

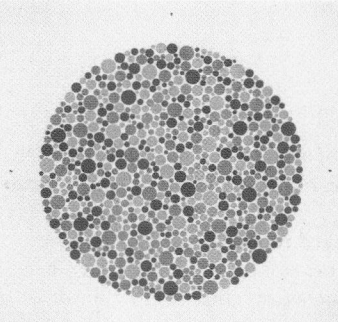
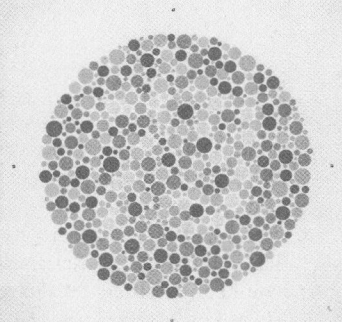

# Questions from Chapter 12

1. If you were to repeat one of Mendel's dihybrid crosses, which of the following statements about the $F_2$ generation would be consistent with your results (would hold true)?
   a. A total of four different genotypes is seen.
   b. A total of six different phenotypes is seen.
   c. All offspring with the same phenotype will possess the same genotype.
   d. It is not possible to tell the genotype of some offspring by looking at their phenotypes.

2. If a child has an AB blood type, the parents
   a. must both be heterozygotes.
   b. could be A or B, but both of them could not AB.
   c. would both have to be AB.
   d. could not have blood type O.

3. One gene has alleles $A$ and $a$, a second gene has alleles $B$ and $b$, and a third gene has alleles $C$ and $c$. List the possible gametes that can be formed from the following genotypes:
   a. *Aa*
   b. *BbCc*
   c. *AAcc*
   d. *AaBbCc*
   e. *aaBBCc*

4. For the same three genes described in problem 3, what are the predicted genotype and phenotype ratios of the following genetic crosses? (Following our standard notation, alleles written in uppercase letters are dominant over alleles written in lowercase letters; the phenotype produced by allele $A$ is therefore dominant over the phenotype controlled by allele $a$.)
   a. *Aa* × *aa*
   b. *BB* × *bb*
   c. *AABb* × *aabb*
   d. *BbCc* × *BbCC*
   e. *AaBbCc* × *AAbbCc*

5. Sickle-cell anemia is inherited as a recessive genetic disorder in humans. That means that, in terms of disease onset, the normal hemoglobin allele ($S$) is dominant over the sickle-cell allele ($s$). For two parents of genotype $Ss$, construct a Punnett square to predict the possible genotypes and phenotypes (does or does not have the disease) of their children. Also list the genotype and phenotype ratios. Each time two $Ss$ individuals have a child together, what is the chance that the child will have sickle-cell anemia?

6. Alleles for a gene ($C$) that determines the color of Labrador retrievers show incomplete dominance. Black labs have genotype $C^BC^B$, chocolate labs have genotype $C^BC^Y$, and yellow labs have genotype $C^YC^Y$. If a black lab and a yellow lab mated, what proportions of black, chocolate, and yellow coat colors would you expect to find in a litter of their puppies?

7. For any human genetic disorder caused by a recessive allele, let $n$ be the allele that causes the disease and $N$ be the normal allele.
   a. What are the phenotypes of $NN$, $Nn$, and $nn$ individuals?
   b. Predict the outcome of a genetic cross between two $Nn$ individuals. List the genotype and phenotype ratios that would result from such a cross.
   c. Predict the outcome of a genetic cross between an $Nn$ and an $NN$ individual. List the genotype and phenotype ratios that would result from such a cross.

8. For any human genetic disorder caused by a dominant allele, let $D$ be the allele that causes the disorder and $d$ be the normal allele (where the capital "$D$" stands for "disorder").
   a. What are the phenotypes of $DD$, $Dd$, and $dd$ individuals?
   b. Predict the outcome of a genetic cross between two $Dd$ individuals. List the genotype and phenotype ratios that would result from such a cross.
   c. Predict the outcome of a genetic cross between a $Dd$ and a $DD$ individual. List the genotype and phenotype ratios that would result from such a cross.

9. If blue flower color ($B$) is dominant over white flower color ($b$), what are the genotypes of the parents in the following genetic cross: blue flower × white flower, yielding only blue-flowered offspring?

10. The fruit pods of peas can be yellow or green. In one of his experiments, Mendel crossed plants homozygous for the allele for yellow fruit pods with plants homozygous for the allele for green fruit pods. All fruit pods in the $F_1$ generation were green. Which allele is dominant: the one for yellow or the one for green? Explain why.

# Questions from Chapter 13

11. Recall that human females have two X chromosomes and human males have one X chromosome and one Y chromosome.
    a. Do males inherit their X chromosome from their mother or from their father?
    b. If a female has one copy of an X-linked recessive allele for a genetic disorder, does she have the disorder?
    c. If a male has one copy of an X-linked recessive allele for a genetic disorder, does he have the disorder?
    d. Assume that a female is a carrier of an X-linked recessive disorder. With respect to the disorder allele, how many types of gametes can she produce?
    e. Assume that a male with an X-linked recessive genetic disorder has children with a female who does not carry the disorder allele. Could any of their sons have the genetic disorder? How about their daughters? Could any of their children be carriers for the disorder? If so, which sex(es) could they be?

12. Cystic fibrosis is a recessive genetic disorder. The disorder allele, which we will call *a*, is located on an autosome. What are the chances that parents with the following genotypes will have a child with the disorder?
    a. *aa × Aa*
    b. *Aa × AA*
    c. *Aa × Aa*
    d. *aa × AA*

13. Huntington's disease (HD) is a genetic disorder caused by a dominant allele—call it *A*—that is located on an autosome. What are the chances that parents with the following genotypes will have a child with HD?
    a. *aa × Aa*
    b. *Aa × AA*
    c. *Aa × Aa*
    d. *aa × AA*

14. Hemophilia is a recessive genetic disorder whose disorder allele—call it *a*—is located on the X chromosome. What are the chances that parents with the following genotypes will have a child with hemophilia?
    a. $X^A X^A × X^a Y$
    b. $X^A X^a × X^a Y$
    c. $X^A X^a × X^A Y$
    d. $X^a X^a × X^A Y$
    e. Do male and female children have the same chance of getting the disorder?

15. Explain why the terms "homozygous" and "heterozygous" do not apply to X-linked traits in males.

16. The chart below shows a representative family pedigree for the inheritance of phenylketonuria. Is this disorder a dominant trait or a recessive trait? Is the disorder allele located on an autosome or on the X chromosome? What are the genotypes of individuals 1 and 2 in generation I?

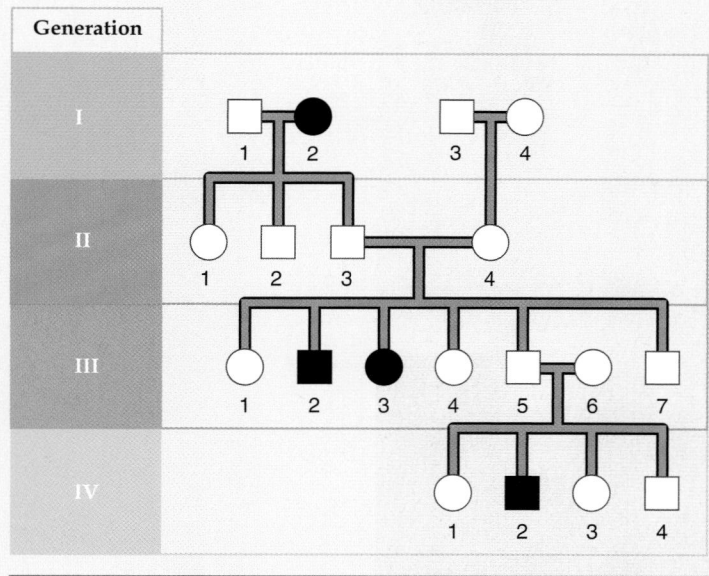

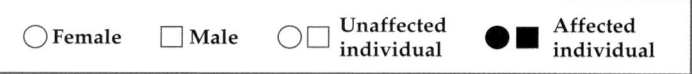

17. In the text we state that males are more likely than females to inherit recessive X-linked genetic disorders. Are males also more likely than females to inherit dominant X-linked genetic disorders? Illustrate your answer by constructing Punnett squares in which
    a. an affected female has children with a normal male.
    b. an affected male has children with a normal female.

18. Study the pedigree shown below. Is the disorder allele dominant or recessive? Is the disorder allele located on an autosome or on the X chromosome? To answer this question, assume that individual 1 in generation I, and individuals 1 and 6 in generation II, do not carry the disorder allele.

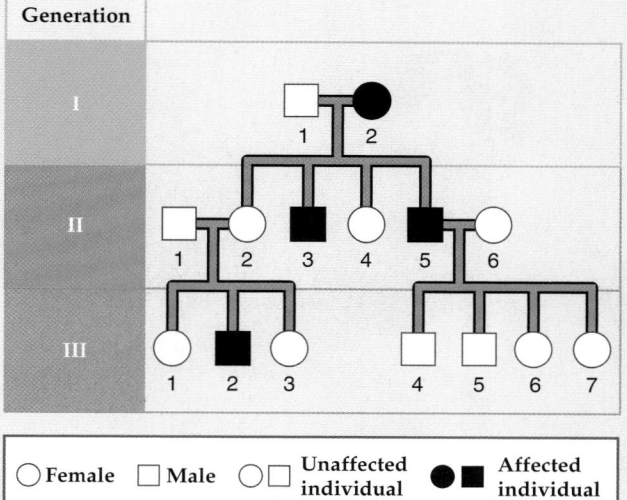

19. Imagine you are conducting an experiment on fruit flies and you are tracking the inheritance of two genes: one with alleles *A* or *a*, the other with alleles *B* or *b*. *AABB* individuals are crossed with *aabb* individuals to produce $F_1$ offspring, all of which have genotype *AaBb*. These *AaBb* $F_1$ offspring are then crossed with *aaBB* individuals. Construct Punnett squares and list the possible offspring genotypes that you would expect in the $F_2$ generation
    a. if the two genes were completely linked.
    b. if the two genes were on different chromosomes.

20. How many different types of gametes can be generated by a pea plant that has genotype *Aa Dd Ee gg*, where *A/a*, *D/d*, *E/e*, and *G/g* are four separate unlinked genetic loci? List all the genotypes that are possible among the gametes produced by this individual. How many different types of gametes are possible, and what would be their genotypes, if the *A* and *D* alleles were located very close together on the same chromosome and therefore completely linked?

# 14

# DNA and Genes

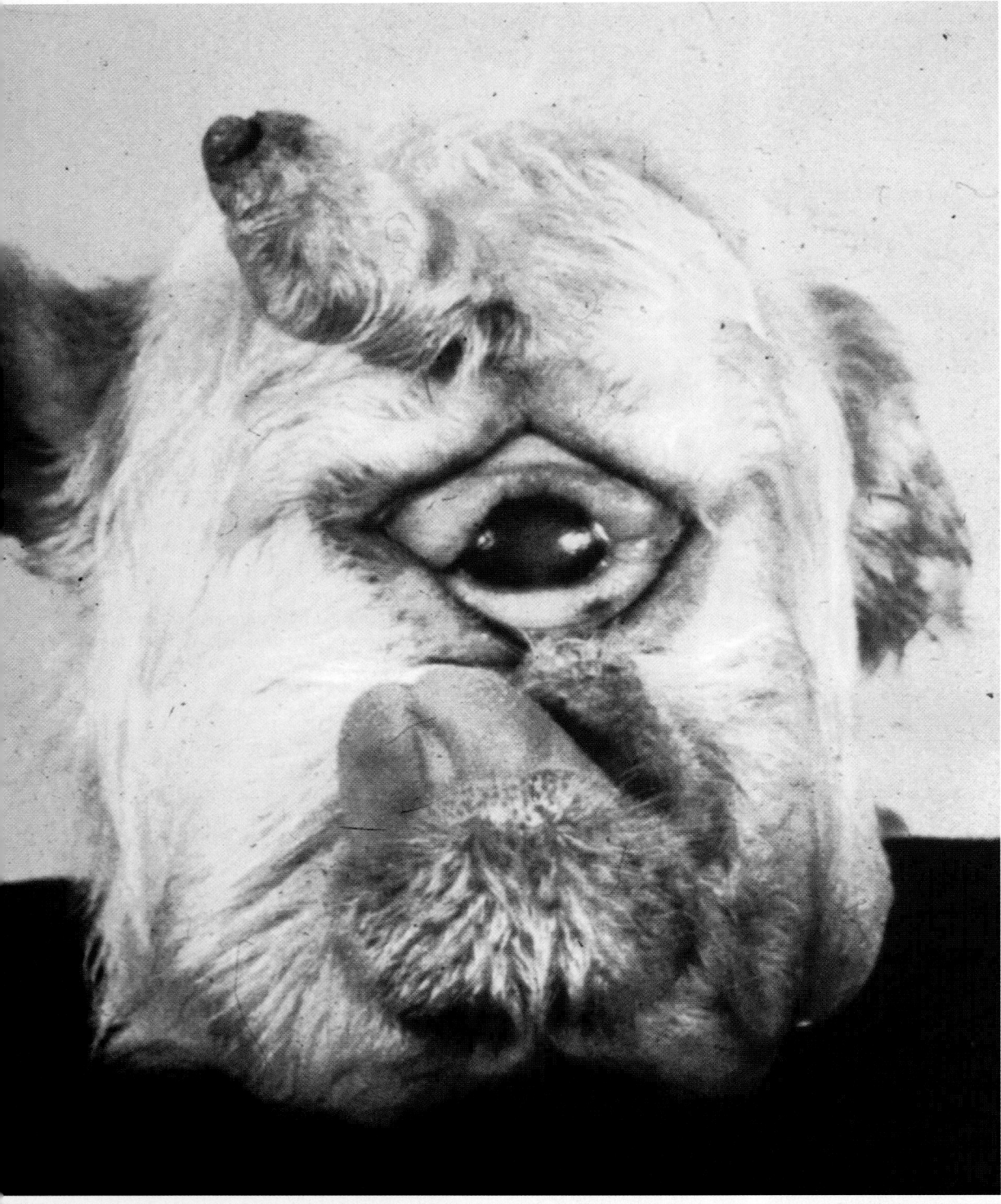

**HEAD OF A LAMB WITH CYCLOPIA.** This lamb suffered severe birth defects, including a single large eye and a fused, malformed brain, because its mother ate a wild plant called corn lily early in pregnancy. Corn lilies make a compound that shuts down the expression of certain genes.

# Greek Myths and One-Eyed Sheep

After the Trojan War, the Greek hero Odysseus encountered and outwitted the Cyclops Polyphemus, a giant with a single eye in the middle of his forehead. In the myth, Polyphemus locks Odysseus and his men in a cave and commences to eat them, two at a time. The men make a daring escape by getting the Cyclops drunk on wine, putting out his eye, and hiding among his sheep.

The Cyclops is a mythical creature, but its single eye resembles a birth defect, called cyclopia, that is rare in humans but sometimes common in sheep and cattle. Cyclopia also occurs in other vertebrates, including sharks, fishes, amphibians, and birds, as well as mammals such as cows, sheep, goats, cats, mice, monkeys, and even humans.

Why would an animal be born with one eye? Untangling the causes of this bizarre developmental defect brings us to one of the most exciting and fruitful areas of modern biological research: the control of gene expression. To develop normally from a single-celled zygote, an organism must express the right genes at the right time and place, producing just the right amounts of thousands of different proteins in exactly the right cells. It is a task of monumental and bewildering complexity. By and large, organisms accomplish this feat flawlessly throughout embryonic development, and each of us continues to perform the same feat every day of our lives.

However, even tiny missteps in the sequence of gene expression can result in disaster: an organism may not

What changes in gene expression would cause an embryo to develop into a cyclops? How do organisms regulate the expression of their genes? How do mutations and chemicals in the environment influence the expression of genes?

develop properly (resulting in cyclopia or other birth defects), or a group of cells may begin multiplying out of control (becoming cancerous). We'll revisit these issues at the end of this chapter. But first we consider how prokaryotic and eukaryotic cells regulate gene expression. Despite the risk of dangerous mistakes, our cells usually control the expression of our 25,000 genes with astonishing accuracy.

---

**MAIN MESSAGE**   DNA is the genetic material in all living cells. The expression of most genes is tightly controlled.

---

## KEY CONCEPTS

- Genes are composed of DNA. DNA consists of two polynucleotide strands held together by hydrogen bonds and twisted into a spiral shape.

- All cells in a multicellular body have essentially the same nucleotide sequence in their DNA. DNA sequences differ between individuals of one species and among different species. Sequence differences are the basis of inherited variation.

- During DNA replication, each strand of DNA serves as a template from which a new strand is copied.

- DNA in cells is subject to damage by various physical, chemical, and biological agents. Up to a point, such damage can be repaired.

- Prokaryotes have relatively little DNA compared to eukaryotes. By comparison to prokaryotes, eukaryotes tend to have not only more genes, but also a large amount of DNA that does not encode proteins.

- Gene expression begins with the activation of a gene and ends with a detectable phenotype influenced by that gene. Transcription and translation are steps in the gene expression pathway.

- In multicellular organisms, different sets of genes are activated in different cell types, and patterns of gene expression change dramatically at different stages of development. Aspects of the environment, such as short-term changes in food availability, can alter gene expression.

**GENES CONTROL THE INHERITANCE OF TRAITS** and are located on chromosomes, as we learned in the previous two chapters. However, this knowledge leaves several fundamental questions unanswered: How exactly does DNA store information? How is that information turned into a particular observable trait, or phenotype? How does a parent cell create enough DNA for two new cells, before it divides to form two daughter cells? What if mistakes are made in the copying of DNA? If all cells in a multicellular body have the same genes, why are heart muscle cells so very different from the nerve cells in the brain? Can a cell sense what is going on around it and tweak the activity of genes accordingly?

In this chapter we describe the physical structure of DNA and how this genetic material is copied. We consider how mistakes in DNA copying that are not "fixed" by DNA repair mechanisms can result in mutations and how mutations can give rise to genetic disorders. We tour the nucleus to inspect the orderly way in which DNA is packed inside it. We close by considering the frugal nature of cells: why not all the genes are working away in every cell all of the time, and how environmental cues and internal signals can turn genes on or off according to the needs of the organism.

# 14.1 An Overview of DNA and Genes

By the early 1900s, geneticists knew that genes control the inheritance of traits, that genes are located on chromosomes, and that chromosomes contain DNA and protein. The first step in the quest to understand the physical structure of genes was to determine whether the genetic material was DNA or protein. Initially, most geneticists thought that protein was the more likely candidate. Proteins are large, complex biomolecules, and it was not hard to imagine that they could store the tremendous amount of information needed to govern the lives of cells. DNA, on the other hand, was initially judged a poor candidate for the genetic material, mainly because DNA was thought to be a simple molecule whose composition varied little among species. Over time, through a number of key experiments, these ideas about DNA were shown to be wrong.

In 1928, a British medical officer named Frederick Griffith transformed a harmless strain of bacterium (the R strain) into a deadly strain (the S strain) by exposing the R strain to heat-killed strain-S bacteria. In 1944, Oswald Avery and his colleagues published a landmark paper demonstrating that only the DNA from heat-killed strain-S bacteria was able to transform strain-R bacteria into strain-S bacteria. Additional evidence came in 1952, when Alfred Hershey and Martha Chase demonstrated that the DNA of a virus, not its proteins, was responsible for taking over a bacterial cell and producing the next generation of viruses. With that, nearly all biologists were persuaded that the genetic material was DNA, not protein.

## DNA stores genetic information as a sequence of nucleotides

As noted in Chapter 5, **DNA** is a nucleic acid composed of two strands of polynucleotides that are twisted around an imaginary axis to form a spiral structure known as the *double helix* (see Figure 5.22). Recall that a polynucleotide is a long chain made up of building blocks called *nucleotides*. Four types of nucleotides are found in the DNA of all cells, from bacteria to blue whales. We will describe the structure of DNA in greater detail later in this chapter, but for now we represent the four nucleotides that occur in DNA by their one-letter abbreviations: A, T, C, and G. These four nucleotides form the four-letter code of DNA, the way 0 and 1 form the binary code of the machine language used in computers.

The **genome** of an organism or a particular individual is all the DNA-based information in the nucleoid of a prokaryote or the nucleus of a eukaryote. Before cell division, in S phase of the cell cycle (see

Section 10.2), the entire genome of a cell is copied in a process called DNA replication.

The DNA sequence information in the genome of one species is different from that of another species, which is what *makes* them different species. Human and chimpanzee genomes are about 95 percent identical if we match their nucleotide sequences (however, the estimates vary depending on the methods used). That sequence variation produces the substantial difference in appearance and behavior of these two primates. The genomes of two different individuals of the *same* species are much more similar than are the genomes of two different species. No matter their race or ethnicity, the genomes of two randomly selected, unrelated humans is likely to be about 99.6 percent identical. Identical twins have essentially the same DNA sequence information in their genomes.

## Most genes code for proteins, which generate phenotypes

If the genome is the software package that runs all the hardware of the cell, then a *gene* is like a specific computer application. A **gene** is a segment of DNA that codes for at least one distinct genetic trait. You could imagine the roughly 25,000 genes in human cells as many different applications built into one software package, the human genome.

How do we go from DNA-based information held in a gene to the manifestation of a particular genetic characteristic, such as the ability to digest the milk sugar lactose? In the living cell, information flows from the DNA that composes a gene to another information-carrying molecule: *RNA*. **RNA** is a single-stranded nucleic acid that is similar to DNA in some ways and different from it in other ways. One difference between RNA and DNA is that a different nucleotide—uracil, symbolized by U—replaces T in RNA.

Most of the RNA in a cell is of a type known as **messenger RNA** (**mRNA**). The molecule is called *messenger* RNA because its function is to deliver the genetic information from DNA to the ribosomes, the protein-making machinery in the cytosol. The conversion of the DNA-based information in a gene to a complementary RNA-based sequence is known as **transcription** (**FIGURE 14.1**). The lactase gene (*LCT*) is transcribed into a lactase mRNA, for example.

You may remember from Chapter 5 that both prokaryotic and eukaryotic cells contain millions of *ribosomes* in the cytoplasm. **Ribosomes** manufacture specific proteins, as directed by the code delivered by a particular mRNA molecule. The lactase mRNA molecule, for example, carries the code that instructs the ribosomes to put together a specific protein, the lactase enzyme. Recall that proteins are constructed from amino acids. The ribosome connects the 20 different amino acids that can be used to build a protein in a precise sequence dictated by the particular mRNA. The process by which ribosomes convert the genetic information in mRNA into proteins in known as **translation**. Translation by ribosomes is similar to speech translation by a bilingual human: ribosomes translate the language of nucleotide sequences (in mRNA) into the language of amino acid sequences (in protein). The mechanism of translation will be covered in more detail in Chapter 15.

Each type of protein has a unique amino acid sequence leading to a unique function that is often detectable as a genetic trait. In the case of lactase, the unique function is the breakdown of lactose (a disaccharide) into two simpler sugars (monosaccharides) that can be absorbed by the intestines and used as food. The manifestation of the information encoded in a gene as a specific genetic trait is known as **gene expression**. Expression of the *LCT* gene is evident when certain cells that line the small intestine secrete

**THE FLOW OF GENETIC INFORMATION**

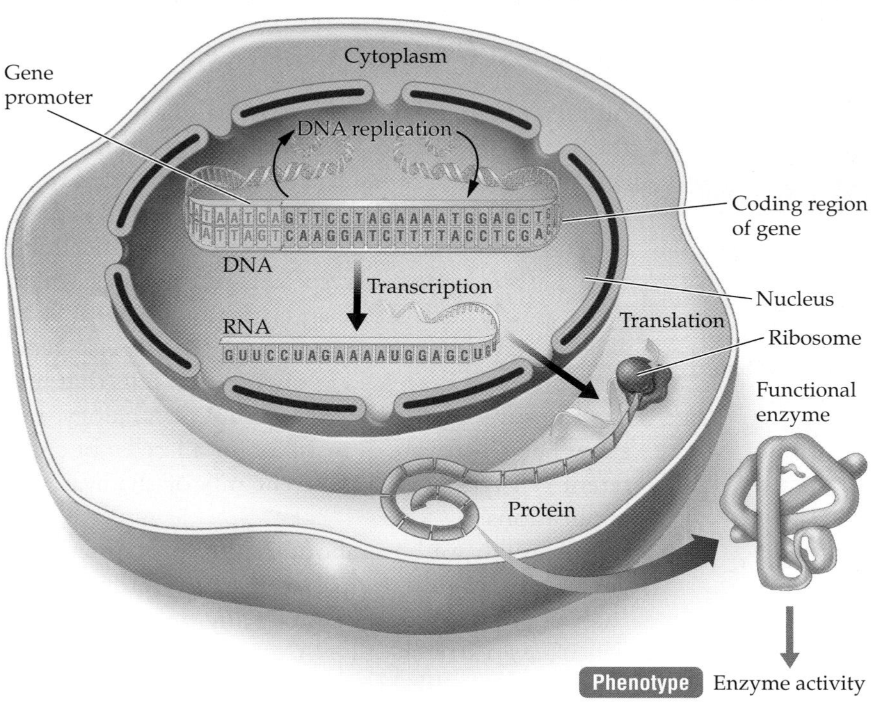

**FIGURE 14.1 From Gene to Phenotype at the Molecular Level**
The readout of information coded in DNA produces the phenotype, such as the activity of lactase enzyme in the small intestine of humans.

FIGURE 14.2
**Different Cell Types Have the Same Genes**

Although all cells within a multicellular organism have the same genes, these cells can differ greatly in structure and function because different genes are active in different types of cells.

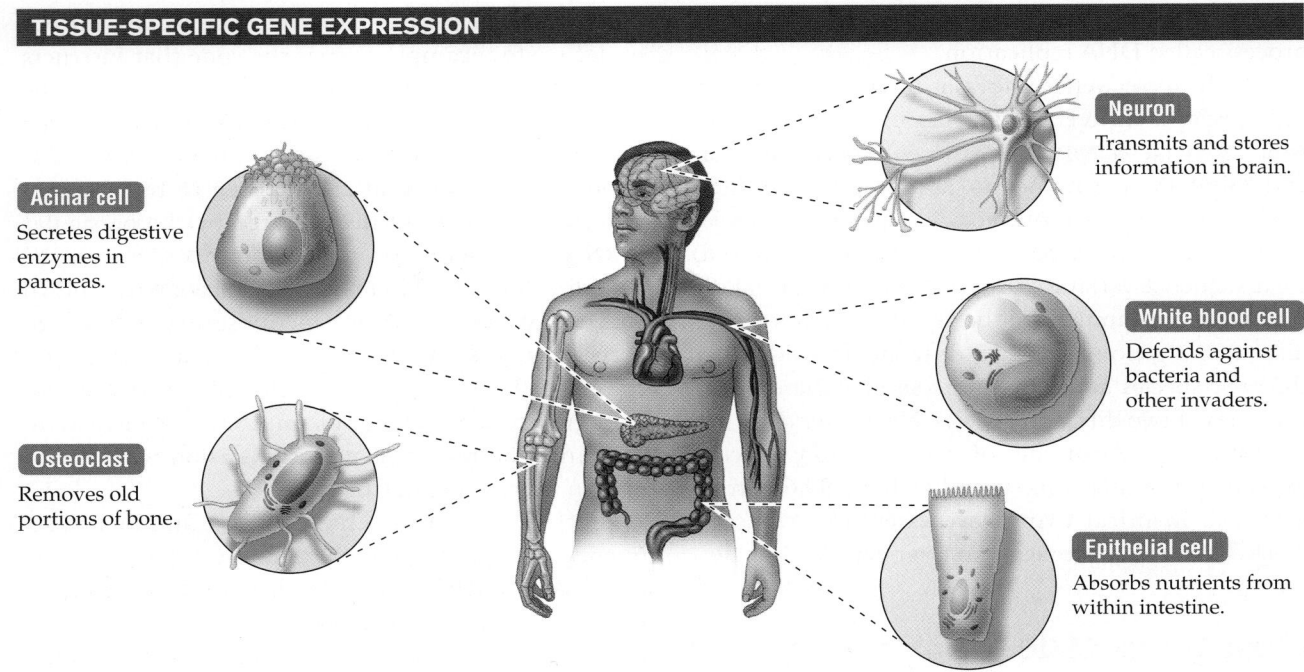

**TISSUE-SPECIFIC GENE EXPRESSION**

**Acinar cell**
Secretes digestive enzymes in pancreas.

**Osteoclast**
Removes old portions of bone.

**Neuron**
Transmits and stores information in brain.

**White blood cell**
Defends against bacteria and other invaders.

**Epithelial cell**
Absorbs nutrients from within intestine.

the lactase enzyme into the gut cavity. Gene expression requires transcription, translation, and functional activity of the protein (in this example, functional activity of the lactase enzyme).

## Different sets of genes are expressed in different cell types

All cells in a multicellular individual have essentially the same DNA-based information. For example, all cells in your body have the *LCT* gene that codes for lactase. But not all cells in your body need to express the *LCT* gene. Even though every cell in a multicellular body has the same genes, specialized cell types, which differ greatly in structure and function (**FIGURE 14.2**), express very different sets of genes. The lactase gene, for example, is expressed in only certain cells that line the intestine. Extending the computer analogy, the same DNA software is installed in all cells, but different cell types run a different mix of the available applications.

Although we have about 21,000 protein-coding genes, the average human cell expresses no more than about 10,000 of them at any one time. Most genes have a stretch of DNA, called a gene promoter, that functions like an on/off switch for transcription (recall, transcription is the process by which the information in DNA is conveyed to messenger RNA, mRNA). Many genes are regulated in a tissue-

specific manner, meaning their promoter is on in some cell types but off in other cell types. The expression of a gene can also be regulated quantitatively, instead of simply being on or off: gene activity can be tweaked upward (up-regulated) or downward (down-regulated), usually in response to signals that the cell receives.

Many genes are *developmentally regulated*, meaning that their expression can change, sometimes dramatically, as an organism grows and develops. At birth, the *LCT* gene is turned on in just about all of us, but in a majority of the world's population, the expression of the gene is steadily down-regulated between 5 and 10 years of age, when children were traditionally weaned from mother's milk in most cultures. The down-regulation of the *LCT* gene in adolescence and adulthood produces the condition most people call lactose intolerance, although the technical term is adult-type hypolactasia.

## 14.2 The Three-Dimensional Structure of DNA

Working at Cambridge University in England, the American James Watson and the Englishman Francis Crick deciphered the physical structure of DNA. In a two-page paper published in 1953, they proposed

### Helpful to Know

Biologists use several words to describe a gene whose product is made: the gene is "activated," is "turned on," or is being "expressed."

that DNA was a double helix, a structure that can be thought of as a ladder twisted into a spiral coil (**FIGURE 14.3**). Watson was 25 at the time; Crick, 37. In 1962, Watson and Crick were awarded the Nobel Prize in Physiology or Medicine for their discovery. They shared the prize with Maurice Wilkins, who had also worked to discover the structure of DNA. Missing from the Nobel ceremony was Rosalind Franklin, a gifted young scientist whose research had provided Watson and Crick with critical data. Rosalind Franklin died of cancer in 1958 at age 37. Nobel Prizes cannot be awarded to a deceased person, so we will never know whether she might have received a Nobel Prize for her contributions.

## DNA is built from two helically wound polynucleotides

As Watson and Crick described it, DNA is built from two parallel strands of repeating units called nucleotides. Each nucleotide is composed of the sugar deoxyribose, a phosphate group, and one of four **nitrogenous bases**: adenine (A), cytosine (C), guanine (G), or thymine (T). The way the two strands are connected is reminiscent of the rungs that connect the sides of a ladder (see Figure 14.3). The nucleotides of each strand are connected by covalent bonds between the phosphate of one nucleotide and the sugar of the next nucleotide. The "rungs" are created by hydrogen

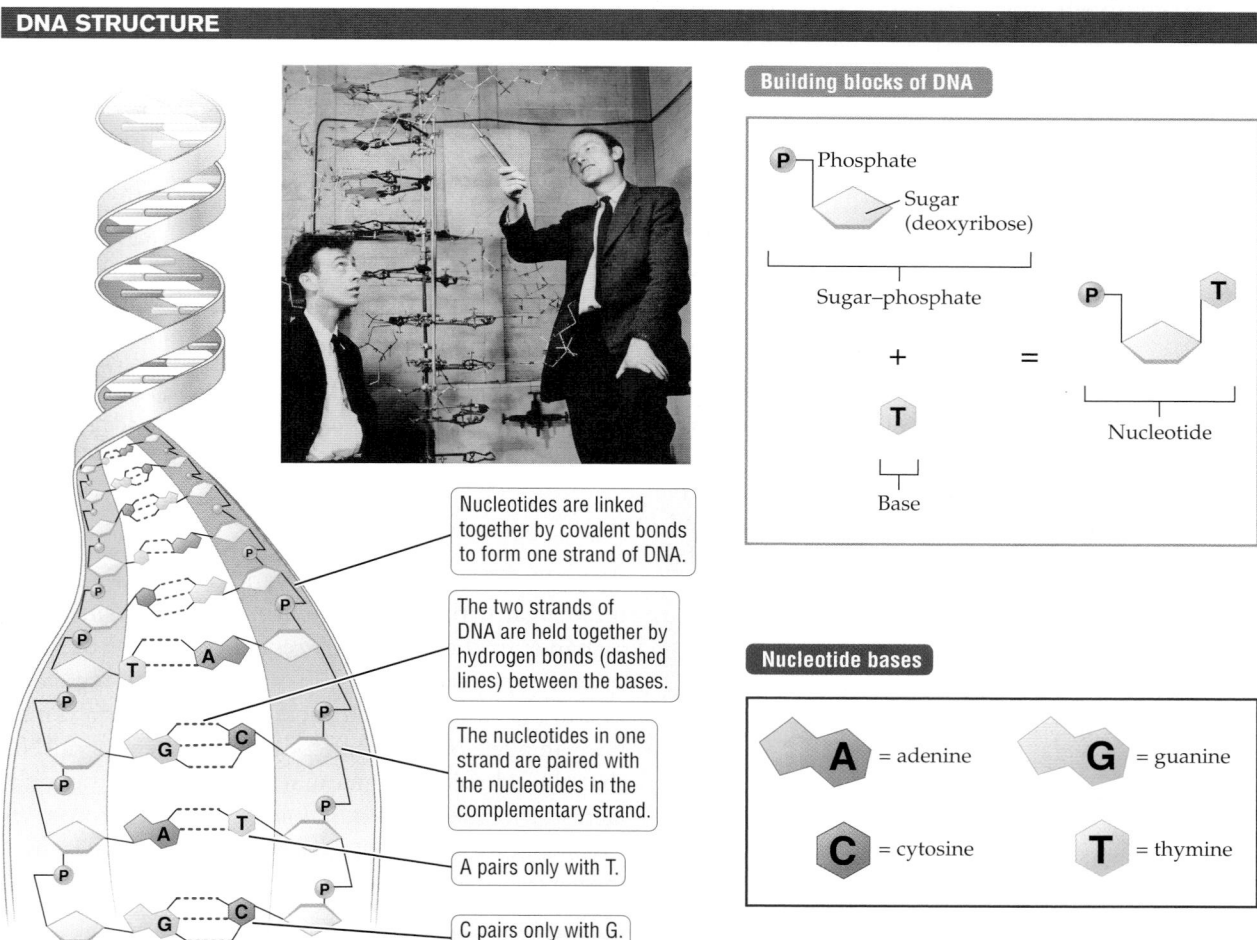

**DNA STRUCTURE**

**Building blocks of DNA**

Nucleotides are linked together by covalent bonds to form one strand of DNA.

The two strands of DNA are held together by hydrogen bonds (dashed lines) between the bases.

The nucleotides in one strand are paired with the nucleotides in the complementary strand.

A pairs only with T.

C pairs only with G.

**Nucleotide bases**

A = adenine  G = guanine

C = cytosine  T = thymine

**FIGURE 14.3  The DNA Double Helix and Its Building Blocks**

A nucleotide consists of three types of chemical groups: a phosphate, a sugar, and a nitrogen-containing base. DNA contains four types of nucleotides, varying only in the type of base found in them. DNA consists of two complementary strands of nucleotides that are twisted into a spiral around an imaginary axis, rather like the winding of a spiral staircase. The two strands are held together by hydrogen bonds between their complementary bases. (*Inset*) James Watson (left) and Francis Crick, with a model of the DNA double helix.

bonds that link the bases on one strand to the bases on the other strand. The term **base pair** (or nucleotide pair) refers to two bases held together by hydrogen bonds in a DNA molecule.

Watson and Crick proposed a set of base-pairing rules, stating that adenine on one strand could pair only with thymine on the other strand; similarly, cytosine on one strand could pair only with guanine on the other strand. These base-pairing rules have an important consequence: when the sequence of bases on one strand of the DNA molecule is known, the sequence of bases on the other, *complementary strand* of the molecule is automatically known. For example, if one strand consists of the sequence

ACCTAGGG,

then the complementary strand has to have the sequence

TGGATCCC.

Any other sequence in the complementary strand would violate the base-pairing rules.

## DNA's structure explains its function

We now know that the physical structure of DNA proposed by Watson and Crick is correct in all its essential elements. This structure has great explanatory power. For example, as we shall see in Section 14.3, the fact that adenine can pair only with thymine and that cytosine can pair only with guanine suggested a simple way in which the DNA molecule could be copied: each original strand could serve as template on which a new strand could be built.

Knowledge of the three-dimensional structure of DNA also suggested that the information stored in DNA could be represented as a long string of the four bases: A, C, G, and T. The four bases can be arranged in any order along a single strand of DNA. The fact that each DNA strand is composed of millions of these bases means that a tremendous amount of information can be stored in the order of the bases along the DNA molecule, or *DNA sequence*.

The sequence of bases in DNA differs between species and between individuals within a species (**FIGURE 14.4**). We now know that different alleles of a gene have different DNA sequences, and that differences in DNA sequence are the basis of inherited variation. For example, in people with adult lactase persistence (those who

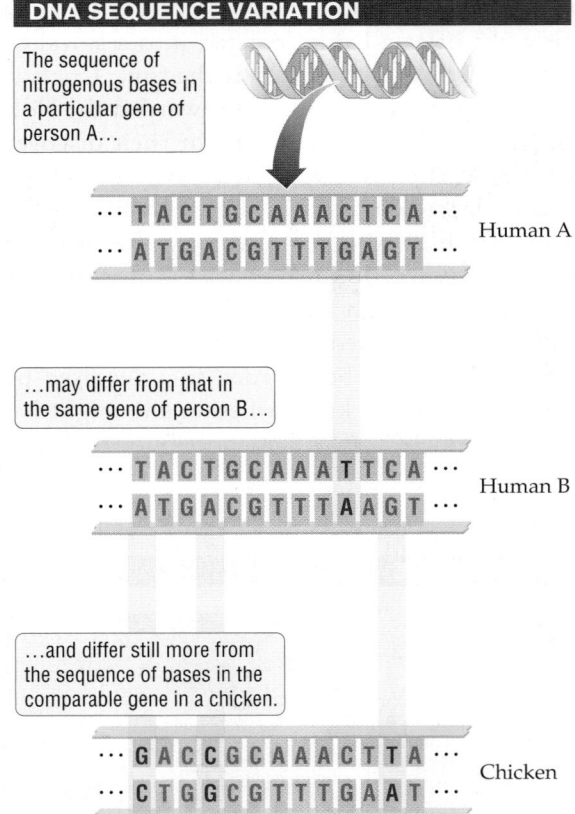

**DNA SEQUENCE VARIATION**

The sequence of nitrogenous bases in a particular gene of person A...

··· T A C T G C A A A C T C A ···
··· A T G A C G T T T G A G T ···   Human A

...may differ from that in the same gene of person B...

··· T A C T G C A A A T T C A ···
··· A T G A C G T T T A A G T ···   Human B

...and differ still more from the sequence of bases in the comparable gene in a chicken.

··· G A C C G C A A A C T T A ···
··· C T G G C G T T T G A A T ···   Chicken

**FIGURE 14.4  The Sequence of Bases in DNA Differs among Species and among Individuals within a Species**
Here the sequence of bases in a hypothetical gene is compared for two humans (A and B) and a chicken. Base pairs highlighted in yellow are variant; that is, they differ between the genes of persons A and B, and between the comparable human and chicken genes.

do *not* have lactose intolerance, and can therefore digest milk sugar without difficulty throughout adulthood), a change in the nucleotide sequence of the *LCT* gene promoter *prevents* that gene from being down-regulated in adolescence and adulthood. Rarely, a newborn may have an allele of *LCT* that codes for a defective mRNA that then directs the manufacture of a nonfunctional lactase; the resulting *lactase deficiency* is a much more serious condition than the discomfort caused by lactose intolerance in adulthood.

### Concept Check

1. What is the percentage of thymine (T) in a DNA double helix in which 20 percent of the nitrogenous bases are guanine (G)?

2. If all genes are composed of just four nucleotides, how can different genes carry different types of information?

# 14.3 How DNA Is Replicated

As Watson and Crick noted in their historic 1953 paper, the structure of the DNA molecule suggested a simple way that the genetic material could be copied. They elaborated on this suggestion in a second paper, also published in 1953. Because A pairs only with T, and C pairs only with G, each strand of DNA contains the information needed to duplicate the complementary strand. For this reason, Watson and Crick suggested that **DNA replication**—the copying of DNA—might work in the following way (**FIGURE 14.5**):

1. The hydrogen bonds connecting the two strands of the DNA molecule are broken as the two strands are unwound by special proteins.

2. Each strand is then used as a template for the construction of a new strand of DNA.

3. When this process is completed, there are two identical copies of the original DNA molecule, each with the same sequence of bases. Each copy is composed of a template strand of DNA (from the original DNA molecule) and one newly synthesized strand of DNA. This mode of replication is known as **semiconservative replication** because one "old" strand (the template strand) is retained, or conserved, in each new double helix.

Five years later, other researchers confirmed that DNA replication produces DNA molecules composed of one old strand and one new strand, as predicted by Watson and Crick. The key enzyme involved in the replication of DNA has now been identified as well: **DNA polymerase [puh-*LIM*-uh-rays]**.

The Watson-Crick model of DNA replication is elegant and simple, but the mechanics of actually copying DNA are far from simple. More than a dozen enzymes and proteins are needed to unwind the DNA, to stabilize the separated strands, to start the replication process, to attach nucleotides to the correct positions on the template strand, to "proofread" the results, and to join partly replicated fragments of DNA to one another.

Although DNA replication is a very complex task, cells can copy DNA molecules containing billions of nucleotides in a matter of hours—about 8 hours in people (over 100,000 nucleotides per second). This speed is achieved in part by starting the replication of the DNA molecule at thousands of different places at once. Despite the speed, cells make remarkably few mistakes when they copy their DNA.

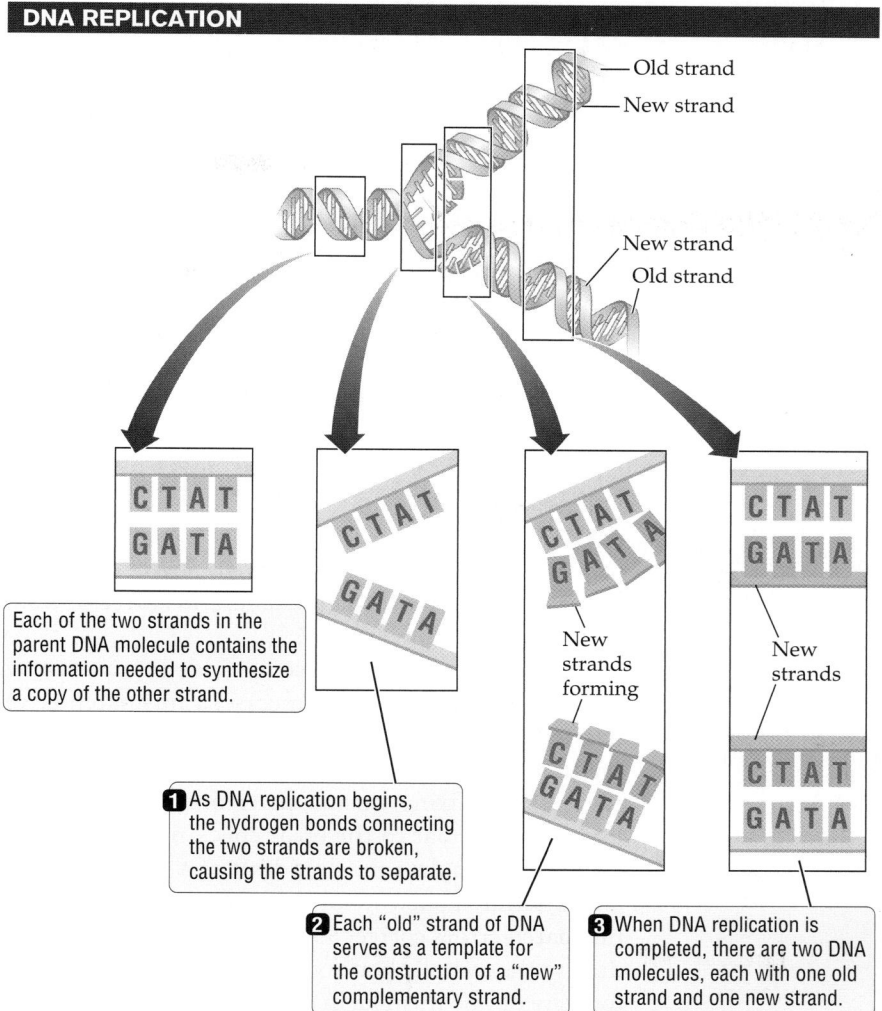

**DNA REPLICATION**

Old strand
New strand

New strand
Old strand

Each of the two strands in the parent DNA molecule contains the information needed to synthesize a copy of the other strand.

New strands forming

New strands

**1** As DNA replication begins, the hydrogen bonds connecting the two strands are broken, causing the strands to separate.

**2** Each "old" strand of DNA serves as a template for the construction of a "new" complementary strand.

**3** When DNA replication is completed, there are two DNA molecules, each with one old strand and one new strand.

**FIGURE 14.5 The Replication of DNA Is Semiconservative**
In this overview of DNA replication, the template DNA strands are blue, and the newly synthesized strands are orange. One strand (blue) from the parent double helix is conserved in each newly made daughter double helix (blue and orange).

# 14.4 Repairing Replication Errors and Damaged DNA

When DNA is copied, there are many opportunities for mistakes to be made. In humans, for example, more than 6 billion base pairs of template DNA must be copied each time a diploid cell divides. Two new strands must be made from each of the two template strands, so there are over 12 billion opportunities for mistakes. In addition, the DNA in cells is constantly being damaged by various sources. Replication errors and damage to the DNA—especially to essential genes—disrupt normal cell functions. If not repaired, this damage would lead to the death of many cells and, potentially, to the death of the organism.

## Few mistakes are made in DNA replication

The enzymes that copy DNA sometimes insert an incorrect base in the newly synthesized strand. For example, if DNA polymerase were to insert a cytosine (C) across from an adenine (A) located on the template strand, an incorrect C-A pair bond would form instead of the correct T-A pair bond (FIGURE 14.6). However, nearly all of these mistakes are corrected immediately by DNA polymerase itself, which "proofreads" the pair bonds as they form. This form of error correction is similar to what happens as you type a paper, realize you made a mistake, and correct it right away with the delete key.

When an incorrect base is added but escapes proofreading by DNA polymerase, a mismatch error has occurred. Mismatch errors occur about once in every 10 million bases. Cells contain repair proteins that specialize in fixing mismatch errors; these proteins play a role similar to the error checking you perform when you complete the first draft of a paper, print it, and carefully review it for mistakes. Proteins that fix mismatch errors correct 99 percent of those errors, reducing the overall chance of an error to the incredibly low rate of one mistake in every billion bases.

On the rare occasions when a mismatch error is not corrected, the DNA sequence is changed, and the altered sequence is reproduced the next time the DNA is replicated. A change to the sequence of bases in an organism's DNA is called a **mutation**. Mutations can also occur when cells are exposed to **mutagens** (substances or energy sources that alter DNA).

Mutations result in the formation of new alleles. Some of the new alleles that result from mutation are beneficial, but most are either neutral or harmful. Among the harmful alleles are those that cause cancer and other human genetic disorders, such as sickle-cell anemia and Huntington's disease. Note that our definition of mutation includes not only changes in the DNA sequence of a single gene, but also the larger-scale changes in DNA sequence that are created by chromosomal abnormalities (see Chapter 13). Changes in chromosome number or chromosome organization generally affect large numbers of genes, and such changes also amount to DNA mutations because they have the effect of adding, deleting, or rearranging nucleotide sequences.

## Normal gene function depends on DNA repair

Every day, the DNA in each of our cells is damaged thousands of times by chemical, physical, and biological agents. These agents include energy from radiation or heat, collisions with other molecules in the cell, attacks by viruses, and random chemical accidents (some of which are caused by environmental pollutants, but most of which result from normal metabolic processes). Our cells contain a complex set of repair proteins that fix the vast majority of this damage. Single-celled organisms such as yeasts have more than 50 different repair proteins, and humans probably have even more.

Although our cells are very good at repairing damaged DNA, some damage far exceeds the cell's ability to repair it. Humans exposed to 1,000 rads (radiation units) of energy die in a few weeks, in part because their DNA is damaged beyond repair. (A rad is a unit for measuring the amount of absorbed radiation; to give you a sense of scale, overall our tissues absorb less than 0.1 rad of energy during a dental X-ray.) Some of the people who initially survived the atomic blasts at Hiroshima and Nagasaki were exposed to about 1,000 rads. Over the next few weeks they died from acute radiation poisoning as cells in the bone marrow and digestive system died from the severe DNA damage.

**FIGURE 14.6**
**Mistakes in DNA Replication May Lead to Mutations**
DNA repair enzymes usually detect and fix mismatch errors such as the one shown here before the cell's DNA is replicated again. If the mismatch is not repaired before the next round of DNA replication, half the daughter helices made by this DNA will have a C-G base pair in place of the original A-T base pair. Such a change in the DNA sequence constitutes a mutation.

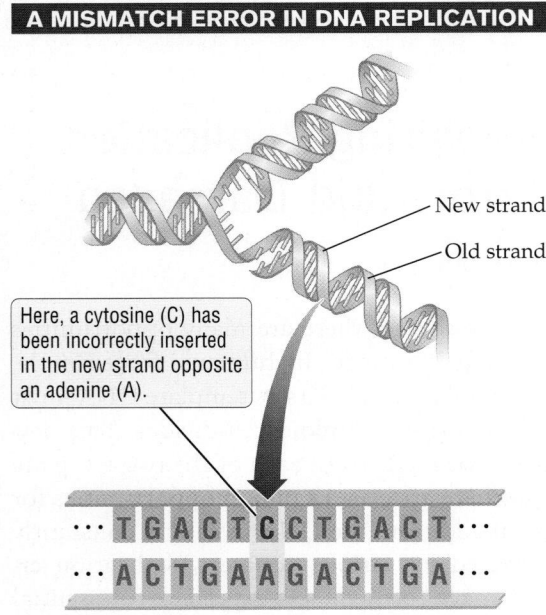

**A MISMATCH ERROR IN DNA REPLICATION**

New strand
Old strand

Here, a cytosine (C) has been incorrectly inserted in the new strand opposite an adenine (A).

··· T G A C T C C T G A C T ···
··· A C T G A A G A C T G A ···

# Prenatal Genetic Screening

How is the baby? This is one of the first questions we ask after a child is born. Usually everything is fine, but sometimes the answer can be devastating. Today, some parents choose to make use of one of several prenatal genetic screening methods to check their baby's health long before it is born.

The practice of prenatal screening has been around a surprisingly long time. In the 1870s, doctors occasionally withdrew some of the fluid in which the fetus is suspended to obtain information about its health. Modern versions of that practice have been standard medical procedure since the early 1960s. In *amniocentesis*, a needle is inserted through the abdomen into the uterus to extract a small amount of amniotic fluid from the pregnancy sac that surrounds the fetus. This fluid contains fetal cells (often sloughed-off skin cells) that can be tested for genetic disorders. Another method is *chorionic* [kohr-ee-AH-nik] *villus sampling* (*CVS*), in which a physician uses ultrasound to guide a narrow, flexible tube through a woman's vagina and into her uterus, where the tip of the tube is placed next to the villi, a cluster of cells that attaches the pregnancy sac to the wall of the uterus. Cells are removed from the villi by gentle suction and then tested for genetic disorders.

Risks associated with amniocentesis and CVS, including vaginal cramping, miscarriage, and premature birth, have declined quite dramatically in the past 10 years because of technological advances and more extensive training. Recent studies suggest that the risk of miscarriage after CVS and amniocentesis is essentially the same, about 0.06 percent. The tests are

widely used by parents who know they face an increased chance of giving birth to a baby with a genetic disorder. Older parents, for example,

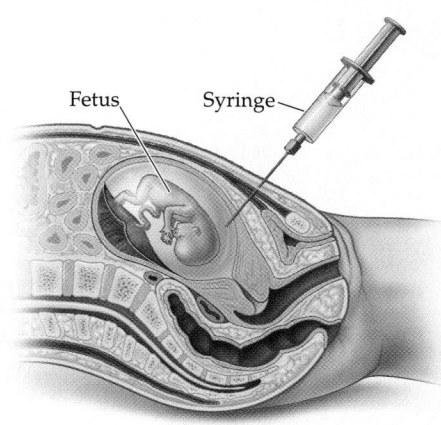

Fetus    Syringe

In amniocentesis, amniotic fluid, which contains fetal cells, is extracted from the uterus.

might want to test for Down syndrome, since the risk of that condition increases with the age of the mother. A couple in which one parent carries a dominant allele for a specific genetic disorder (such as Huntington's disease), or both parents are carriers for a recessive genetic disorder (such as cystic fibrosis), might also choose prenatal genetic screening.

Until recently, couples who elected to have such tests performed had only two choices if their fears were confirmed: they could abort the baby, or they could give birth to a child who would have a genetic disorder. Since 1989, however, a third option has been available to couples who are willing, and can afford, to have a child by in vitro fertilization (in which fertilization oc-

curs in a petri dish, after which one or more embryos are implanted into the mother's uterus). In *preimplantation genetic diagnosis* (*PGD*), one or two cells are removed from the developing embryo, usually 3 days after fertilization occurs. (It is important to perform PGD at this time because then the embryo typically has 4–12 loosely connected cells; in another day or two, the cells will begin to fuse more tightly, making PGD more difficult.) Next, the cell or cells removed from the embryo are tested for genetic disorders. Finally, one or more embryos that are free of disorders are implanted into the mother's uterus, and the rest of the embryos, including those with genetic disorders, are discarded.

PGD is typically used by parents who either have a serious genetic disorder or carry alleles for one—for example, cystic fibrosis or Huntington's disease. Like all other genetic screening methods, the use of PGD raises ethical issues. People who support the use of PGD think that amniocentesis and CVS provide parents with a bleak set of moral choices: if the fetus has a serious genetic disorder, the parents can either abort the baby or allow it to live a life that may be short and full of suffering. In their view, it is morally preferable to discard an embryo at the 4- to 12-cell stage than it is to abort a well-developed fetus, or to give birth to a child that will suffer the devastating effects of a serious genetic disorder. Those opposed to the use of PGD argue that the moral choices are the same, but in their view, once fertilization has occurred a new life has formed, and it is immoral to end that life, even at the 4- to 12-cell stage. What do you think?

Although 1,000 rads kills a person, such a dose would barely faze the bacterium *Deinococcus radiodurans* [DYE-noh-KAH-kus ray-dee-oh-DUR-unz]. This species is so efficient at repairing damage to DNA that a dose of 1,000,000 rads does not kill it but merely slows its growth. Even when the dose is raised to 3,000,000 rads—3,000 times more than a lethal dose for a person—a small percentage of the bacteria survive.

In humans, *Deinococcus*, and all other organisms, there are three steps in **DNA repair**: the damaged DNA must be (1) recognized, (2) removed, and (3) replaced. Different sets of repair proteins specialize in recognizing

**DNA REPAIR: RECOGNIZE, REMOVE, REPLACE**

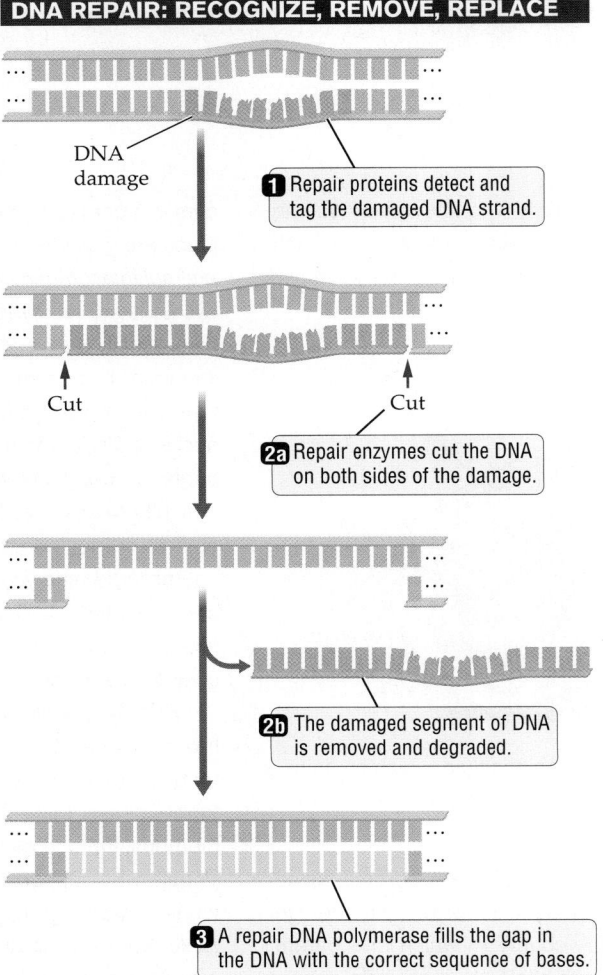

DNA damage

**1** Repair proteins detect and tag the damaged DNA strand.

Cut          Cut

**2a** Repair enzymes cut the DNA on both sides of the damage.

**2b** The damaged segment of DNA is removed and degraded.

**3** A repair DNA polymerase fills the gap in the DNA with the correct sequence of bases.

**FIGURE 14.7 Repair Proteins Fix DNA Damage**
Large complexes of DNA repair proteins work together to fix damaged DNA. They (1) recognize DNA damage, (2) remove the segment of DNA strand that is damaged, and (3) replace the missing segment through new DNA synthesis.

**3.** Proofreading by DNA polymerase and the DNA repair system reduce the chance of mutation. Neither mechanism is 100 percent effective, and mutations results when these systems fail.

**2.** In DNA replication, each new double helix is composed of one old strand (from the template DNA molecule) and one newly polymerized strand.

**1.** DNA polymerase catalyzes the formation of a polynucleotide strand complementary to a template DNA strand. It is localized in the nucleus in eukaryotic cells.

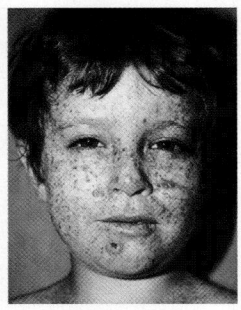

Child with XP. The growth on his chin is a skin cancer.

defects in DNA structure and removing the damaged segment of DNA by cutting it out with special enzymes (**FIGURE 14.7**). Once these first two steps have been accomplished, the final step is to add the correct sequence of nucleotides to fill the gap created by removal of the damaged DNA strand. This third step of the repair process uses the intact strand of DNA as template for re-creating the missing segment in the complementary strand.

Mutations are the consequence of failure in DNA repair. When an animal cell that is capable of multiplying through cell divisions acquires many mutations in a variety of genes, it becomes dangerous because such a cell is more likely to have malfunctions in genes that normally keep cell division under tight control. Consequently, it is more likely to become cancerous—a condition marked by runaway cell proliferation (see Chapter 11).

The importance of DNA repair mechanisms is also highlighted by genetic disorders in which the repair system itself is inactive or seriously inefficient. Xeroderma pigmentosum (XP) is a recessive genetic disorder in which even brief exposure to sunlight causes painful blisters. The allele that causes XP produces a nonfunctional version of one of the many human DNA repair proteins. The job of the normal form of this protein is to repair the kind of damage to DNA caused by UV light. The lack of this DNA repair protein makes individuals with XP highly susceptible to skin cancer. Several inherited tendencies to develop cancer, including some types of breast and colon cancer, also stem from less effective versions of genes that participate in DNA repair.

> **Concept Check**
>
> **1.** What key function is performed by DNA polymerase? Where is this enzyme found in eukaryotic cells?
>
> **2.** Describe the semiconservative nature of DNA replication.
>
> **3.** What types of mechanisms reduce the chance that the DNA in a cell will mutate? Are these mechanisms 100 percent effective?

## 14.5 Genome Organization

How big is the genome of a prokaryote? How big is the genome of a eukaryote? Does all of the DNA in an organism consist of genes? These questions call for a comparison of prokaryotes (Bacteria and Archaea) and eukaryotes (all other organisms) because the genomes of these two major groups are organized differently:

- First, a typical bacterium has several million base pairs of DNA, all in a single chromosome, whereas most eukaryotic cells have hundreds of millions to billions of base pairs distributed among several chromosomes.

- Second, most of the DNA in prokaryotes encodes proteins, and prokaryotic genes rarely contain noncoding segments of DNA. In contrast, eukaryotes have noncoding DNA *inside* most genes, as well as large amounts of noncoding DNA interspersed *between* genes.

- Finally, prokaryotic genes tend to be organized by function: the different genes needed for a given metabolic pathway are usually clustered together

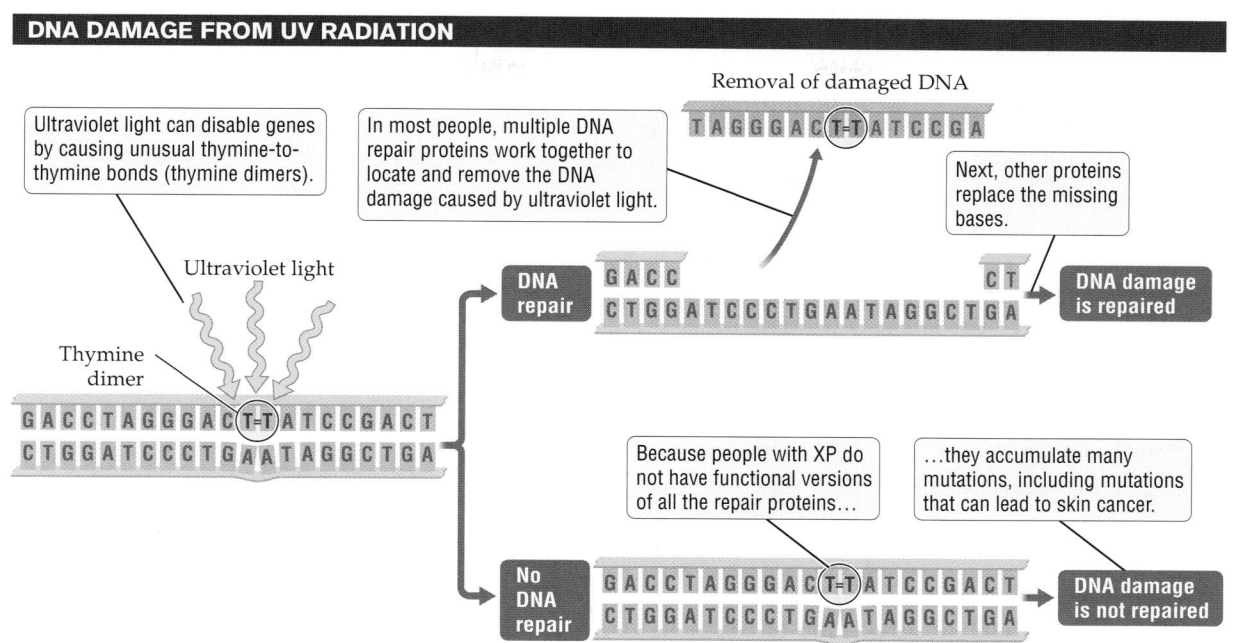

**FIGURE 14.8 The Importance of DNA Repair Mechanisms**

When DNA repair mechanisms fail to work properly, the consequences can be severe, as illustrated by the high frequency of skin cancers in people who have xeroderma pigmentosum (XP), a recessive genetic disorder in which cells are unable to make a protein used to repair DNA damage caused by ultraviolet light.

Ultraviolet light can disable genes by causing unusual thymine-to-thymine bonds (thymine dimers).

In most people, multiple DNA repair proteins work together to locate and remove the DNA damage caused by ultraviolet light.

Removal of damaged DNA

Next, other proteins replace the missing bases.

DNA repair

DNA damage is repaired

Ultraviolet light

Thymine dimer

Because people with XP do not have functional versions of all the repair proteins...

...they accumulate many mutations, including mutations that can lead to skin cancer.

No DNA repair

DNA damage is not repaired

and are turned on or turned off as one unit. In contrast, although some eukaryotic genes with related functions are grouped near one another on a chromosome, most are not organized in this way; genes with related functions may even be found on entirely different chromosomes.

Overall, the organization of eukaryotic DNA is more complex. Let's consider some of these differences in greater detail.

## Eukaryotes have more DNA per cell than most prokaryotes have

The genome of a prokaryote is typically contained in a single chromosome. The genome of a eukaryote is equivalent to all the genetic information contained in a haploid set of chromosomes, such as that found in a sperm or egg. Genome size is measured in base pairs. The genomes of prokaryotes vary in size from 0.6 million to 30 million base pairs. The genome sizes of single-celled eukaryotes show much greater variation, ranging from 12 million base pairs in yeast (a single-celled fungus) to over a trillion base pairs in a certain species of *Amoeba* (a single-celled protist). Most vertebrates have genomes that contain hundreds of millions to billions of base pairs. For example, puffer fish have only 400 million base pairs of DNA in their genome, while mammalian genomes range from 1.5 billion to 6.3 billion base pairs (with humans at 3.3 billion). Some salamanders have 90 billion base pairs.

As these examples illustrate, eukaryotes usually have far more DNA than prokaryotes have. Why? In part, the reason is that eukaryotes in general are structurally and behaviorally more complex than prokaryotes and hence need more genes to direct that complexity. A typical prokaryote has about 2,000 genes, although certain tiny bacteria have no more than about 500. Among the eukaryotes studied to date, the single-celled yeast *Saccharomyces cerevisiae* has 6,000 genes, the nematode worm *Caenorhabditis elegans* has 19,100 genes, several plant species have an estimated 20,000 genes each, and humans have about 25,000 genes.

## Genes constitute only a small percentage of the DNA in most eukaryotes

Although eukaryotes have roughly 3–15 times as many genes as a typical prokaryote has, this difference in gene number does not fully explain why eukaryotes often have hundreds to thousands of times more DNA than prokaryotes have. Only a small percentage of the DNA in eukaryotic genomes consists of genes, defined as DNA sequences that code for RNA and affect a phenotype. Some of the remaining DNA has regulatory functions—for example, controlling gene expression. Some of it has architectural functions, such as giving structure to chromosomes or positioning them at precise locations within the interphase nucleus. Much of the genome, however—in fact, a majority of the genome in most eukaryotes—has

This puffer fish has only 400 million base pairs of DNA in one set of its chromosomes, compared to the diver's 3.3 billion.

| TABLE 14.1 | Types of Eukaryotic DNA |
| --- | --- |
| **TYPE** | **DESCRIPTION** |
| Exons (in a gene) | Transcribed portions of a gene that code for the amino acid sequence of a protein |
| Noncoding DNA: | |
| Introns (in a gene) | Transcribed portions of a gene that do not code for the amino acid sequence of a protein, and whose corresponding sequence in RNA is removed before the RNA leaves the nucleus |
| Spacer DNA | DNA sequences that separate genes |
| Regulatory DNA | DNA sequences that control the expression of genes |
| Structural DNA | DNA sequences that have architectural function, creating structures such as the centromere of a chromosome |
| DNA of unknown function | DNA sequences that have no known function in the cell |
| Transposons ("jumping genes") | DNA sequences that can move from one position on a chromosome to another, or from one chromosome to another |

no apparent function. At least some of this *noncoding DNA* appears to be nonessential and is popularly called "junk DNA" because researchers have found no impact on the phenotype if this DNA is lost from the cell. Although various hypotheses have been proposed, we do not yet understand why so many eukaryotes have large amounts of DNA with no clear-cut function.

Scientists estimate that genes that encode proteins make up less than 1.5 percent of the human genome. Other genes in our cells encode different types of non-protein RNA molecules. The rest of our genome consists of various types of **noncoding DNA**, defined as DNA that does not code for any kind of functional RNA.

Noncoding DNA includes *introns* and *spacer DNA* (**FIGURE 14.9**; Table 14.1). Most eukaryotic genes are interrupted by stretches of DNA, called **introns**. Introns do not code for the amino acid sequence of a protein and must be spliced out of a newly made mRNA molecule before translation. The segments of a gene that actually code for amino acids are called **exons**. Exons are preserved in the mature, translation-ready mRNA, and ribosomes manufacture a "made-to-order" protein according to the code carried by the exons in the mRNA. **Spacer DNA** is noncoding DNA that separates one gene from another; spacer sequences found in eukaryotic genomes are exceptionally long compared to those found in the much more compact genomes of prokaryotes.

Prokaryotic and eukaryotic genomes also contain "jumping genes," or **transposons**: sequences that can move from one position on a chromosome to another, or even from one chromosome to another. The activity of a gene can be disrupted if a transposon inserts itself somewhere within that gene. Although transposons can

encode proteins, and these proteins are necessary for mobilizing a transposon from one location to another, the great majority of transposon sequences in the human genome are "DNA fossils" in that they do not make functional proteins and they are usually incapable of moving. Transposons may constitute a large proportion of a eukaryote's DNA. For example, transposons make up an estimated 36 percent of the human genome and more than 50 percent of the 5.4 billion base pairs in the corn genome. Most transposons appear to be viral in origin, consisting of genetic material from ancient viruses that have become noninfective (not transmitted between genetically unrelated individuals). Other transposons appear to be fragments of "selfish DNA," segments of

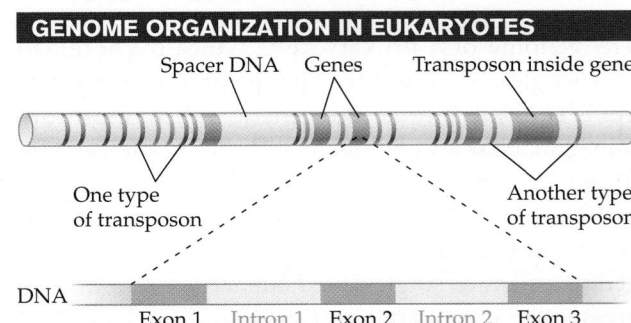

**GENOME ORGANIZATION IN EUKARYOTES**

Spacer DNA    Genes    Transposon inside gene

One type of transposon    Another type of transposon

DNA    Exon 1    Intron 1    Exon 2    Intron 2    Exon 3

**FIGURE 14.9 Eukaryotic Genomes Contain Both Coding and Noncoding DNA**
Eukaryotic genomes contain a small proportion of genes (purple) interspersed with a large amount of spacer DNA (light blue) and transposons (red and dark blue). Each of the two different types of transposons shown here is found in many copies throughout the human genome. Note that one transposon (dark blue) has inserted itself into one of the five genes. Most eukaryotic genes consist of coding regions (exons) interspersed with noncoding regions called introns.

the cell's own DNA that have acquired the ability to make copies of themselves and insert them at random locations in the genome.

## 14.6 DNA Packing in Eukaryotes

To be expressed, the information in a gene must first be transcribed into an RNA molecule. A gene cannot be transcribed unless the enzymes that guide transcription are able to reach that gene. The task may sound simple, but it is complicated by what may be the ultimate storage problem: how to store an enormous amount of genetic information (the organism's DNA) in a small space (the nucleus), and still be able to retrieve each piece of that information precisely when it is needed.

In humans and other eukaryotes, every chromosome in each cell contains one DNA molecule. These chromosomes hold a vast amount of genetic information. As we learned in Chapter 10, the haploid number of chromosomes in humans is 23; these chromosomes together contain about 3.3 billion base pairs of DNA. Stretched to its full length, the DNA from all 46 chromosomes in a single human cell would be more than 2 meters long (taller than most of us). That is a huge amount of DNA, especially considering that it is packed into a nucleus only 0.000006 meter (0.0002 inch) in diameter. The combined length of DNA in our bodies is staggering: the human body has about $10^{13}$ cells, each of which contains roughly 2 meters of DNA. Therefore, each of us has about $2 \times 10^{13}$ meters of DNA in the body—a length more than 130 times the distance between Earth and the sun.

How can our cells stuff such an enormous amount of DNA into such a small space? Cells use a variety of packaging proteins to wind, fold, and compress the DNA double helix, going through several levels of packing to create the DNA-protein complex we call a chromosome. Let's examine the packing of DNA in a metaphase chromosome, beginning with the DNA double helix, which is about 2 nanometers wide (depicted at the bottom of **FIGURE 14.10**). Short lengths of this double helix are wound around "spools" of proteins, known as *histone proteins*, to create a "bead-on-a-string" structure that is about 10 nanometers wide and consists of many histone "beads" linked by a DNA molecule that wraps around each bead. The bead-on-a-string structure is compressed into a more compact form, known as the 30-nanometer fiber, by yet other types of packaging proteins. This fiber is then looped back and forth to further condense

the chromosome in the interior of the interphase nucleus. During interphase, much of the genome exists as a looped 30-nanometer fiber, but any region of the chromosome that is being transcribed is "unpacked" into a bead-on-a-string state. Some portions of an interphase chromosome remain tightly condensed at all times because they contain no genes, although they may have other types of functional DNA, such as structural DNA.

At the beginning of cell division, whether by mitosis or meiosis, all chromosomes undergo yet another level of condensation. During prophase, each looped 30-nanometer fiber is further compressed to create a

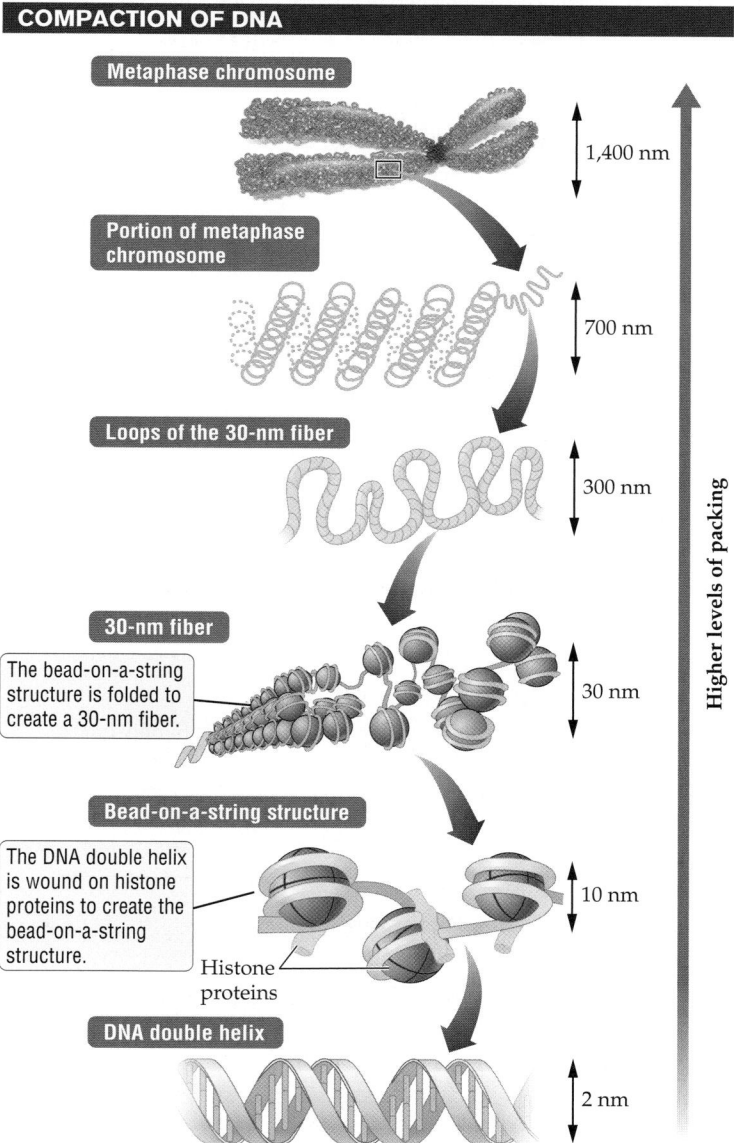

**COMPACTION OF DNA**

Metaphase chromosome — 1,400 nm

Portion of metaphase chromosome — 700 nm

Loops of the 30-nm fiber — 300 nm

30-nm fiber — 30 nm

The bead-on-a-string structure is folded to create a 30-nm fiber.

Bead-on-a-string structure — 10 nm

The DNA double helix is wound on histone proteins to create the bead-on-a-string structure.

Histone proteins

DNA double helix — 2 nm

Higher levels of packing

**FIGURE 14.10 DNA Packing in Eukaryotes**
The DNA of eukaryotes is highly organized by a complex packing system. Increasingly higher levels of DNA packing are illustrated, with the chromosome at metaphase representing the stage of the cell cycle at which DNA is most tightly packed.

cable that is considerably shorter and more than twice as thick. This mitotic (or meiotic) chromosome represents the highest level of packing that a chromosome can acquire. In this state of maximum condensation, these shorter, stouter chromosomes are less liable to become entangled and are therefore less likely to be ripped apart when they are aligned at the cell center during metaphase or segregated to opposite sides of the cell during anaphase and telophase.

> ### Concept Check
>
> 1. Each human cell contains over 1,000 times as much DNA as *E. coli* bacteria have. Do we have over 1,000 times as many genes as *E. coli* has? Explain.
> 2. What is meant by "gene expression"?
> 3. What is the functional value of the roughly twofold greater condensation that chromosomes undergo during prophase of mitosis or meiosis?

## 14.7 Patterns of Gene Expression

Gene expression is the means by which genes influence the structure and function of a cell or organism. In other words, gene expression enables a gene to in-

**GENE EXPRESSION IN RESPONSE TO THE ENVIRONMENT**

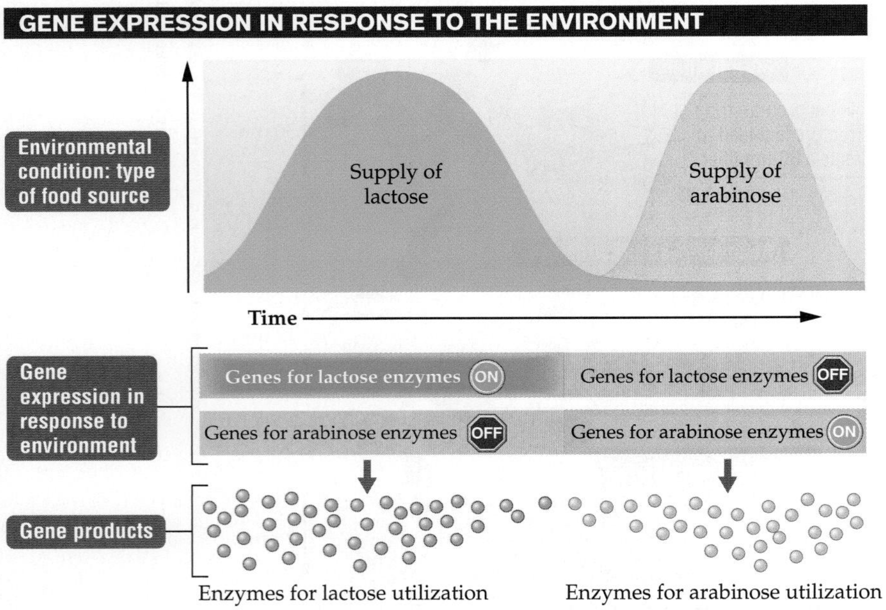

**FIGURE 14.11 Bacteria Express Different Genes as Food Availability Changes**
The sugars lactose and arabinose are not always available, but both can serve as food for the bacterium *E. coli*. The presence of the sugar triggers expression of the gene that codes for enzymes needed to metabolize that sugar.

fluence a particular phenotype, the actual display of a genetic trait. Different cells express different sets of genes, and within a given cell the pattern of gene expression can change over time. But what determines which genes an organism expresses at a particular time in a particular cell?

## Organisms can turn genes on or off in response to environmental cues

Single-celled organisms such as bacteria face a big challenge: they are directly exposed to their environment, and they have no specialized cells to help them deal with changes in that environment. One way they meet this challenge is to express different genes as conditions change. Bacteria respond to changes in nutrient availability, for example, by turning genes on or off. If the *E. coli* bacteria in a petri dish are given lactose (a sugar found in milk) as the only source of energy, within a matter of minutes they turn on genes that code for lactose-digesting enzymes (**FIGURE 14.11**). When the lactose is used up, the bacteria stop producing those enzymes. In effect, these bacteria reorganize gene expression to match the food resources available to them. When a resource runs out, they switch on genes that will enable them to exploit the next resource that becomes available (in Figure 14.11, the sugar arabinose). By producing the enzymes to process a particular food only when that food is available, bacteria avoid wasting energy and cellular resources making enzymes that are not needed.

Like single-celled organisms, multicellular organisms change which genes they express in response to internal signals (arising inside the body) or external cues present in the environment. For example, we humans change the genes we express when our blood sugar or blood pH levels change, enabling us to keep those levels from becoming too high or too low. Similarly, humans, plants, and many other organisms, when exposed to high temperatures, turn on genes encoding proteins that protect cells against heat damage.

## Different cells in eukaryotes express different genes

Throughout the life of a multicellular organism, different types of cells express different sets of genes. Whether a cell expresses a particular gene depends on whether that gene function is needed by that cell at that time under the prevailing conditions. Not surprisingly, a gene that encodes a specialized protein is expressed only in cells that need either to use that protein within the cell or to produce it for transport to other cells that will use it. For example, among the 220 cell types in the human body, red blood cells are the only cells that use the oxygen transport protein hemoglobin; therefore, as they develop to maturity, red blood cells are the only cells that express the gene for this protein (**FIGURE 14.12**). Similarly, the gene for crystallin, a protein that is the main component of the lens in our eyes, is expressed only in certain cells in the developing eye.

Some genes, known as **housekeeping genes**, play an essential role in the maintenance of cellular activities in all kinds of cells. These genes are therefore expressed most of the time by most cells in the body. For example, genes for rRNA, a key component of ribosomes, are expressed by almost all cells (see Figure 14.12). This is not surprising, since virtually all cells need to make proteins. Housekeeping genes tend to be highly conserved in evolutionary history, meaning that their base sequence and general function are similar in diverse

Expression of the *Noggin* gene in a mouse embryo is shown by the green stain. The gene controls brain and skeleton development in mammals.

groups of organisms. Genes that are critical for survival—because they help conduct a function as elemental as protein synthesis, for example—tend not to change much as new groups evolve from the ancestral forms.

## DIFFERENTIAL GENE EXPRESSION

| | Developing red blood cell | Eye lens cell (in embryo) | Pancreatic cell |
|---|---|---|---|
| **Hemoglobin gene** | ON | OFF | OFF |
| **Crystallin gene** | OFF | ON | OFF |
| **Insulin gene** | OFF | OFF | ON |
| **rRNA gene** | ON | ON | ON |

**FIGURE 14.12 Different Types of Cells Express Different Genes**
Some genes, such as those encoding hemoglobin, crystallin, and insulin, are active ("on") only in the types of cells that use or produce the protein encoded by the gene. Housekeeping genes, such as the rRNA gene, are active in most types of cells.

## 14.8 How Cells Control Gene Expression

Cells receive signals that determine which of their genes are turned on or off. Some of these signals are sent from one cell to another, as when one cell releases a signaling molecule that alters gene expression in another cell. Cells also receive signals from the organism's internal environment (for example, blood sugar level in humans) and external environment (for example, sunlight in plants). Overall, cells process information from a variety of signals, and that information affects which genes are expressed.

## Most genes are controlled at the transcriptional level

The most common way for cells to control gene expression is to regulate the transcription of the gene. In the absence of gene transcription, the RNA encoded by the gene is not made, the protein encoded by the RNA cannot be made, and any phenotype directly produced by the protein is therefore absent.

In general, transcription is controlled through two essential elements: **regulatory DNA** that can activate

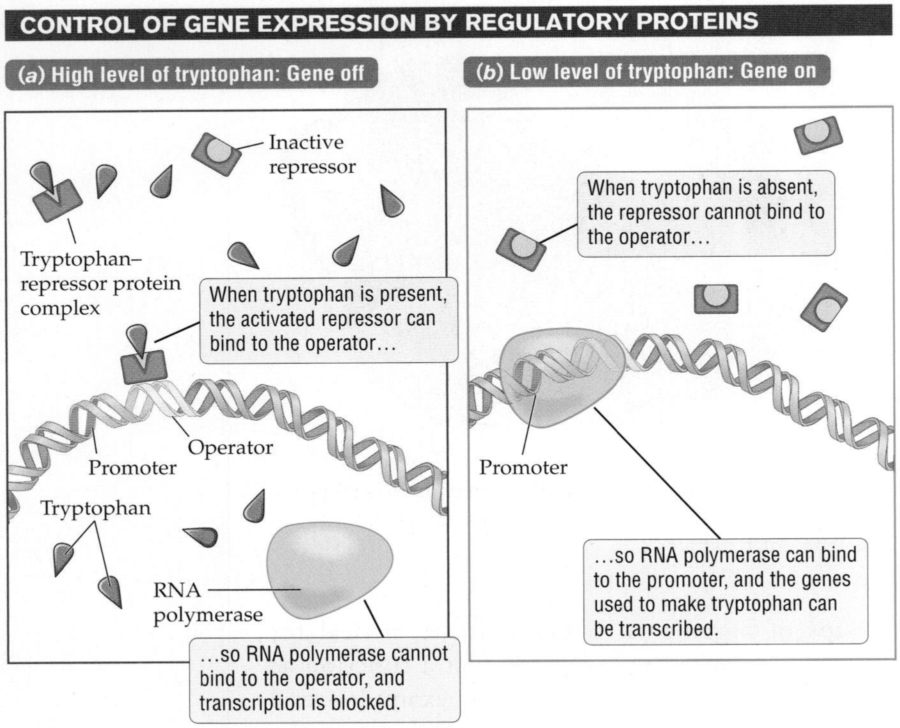

**(a) High level of tryptophan: Gene off**

Inactive repressor

Tryptophan–repressor protein complex

When tryptophan is present, the activated repressor can bind to the operator…

Operator

Promoter

Tryptophan

RNA polymerase

…so RNA polymerase cannot bind to the operator, and transcription is blocked.

**(b) Low level of tryptophan: Gene on**

When tryptophan is absent, the repressor cannot bind to the operator…

Promoter

…so RNA polymerase can bind to the promoter, and the genes used to make tryptophan can be transcribed.

**FIGURE 14.13 Repressor Proteins Turn Genes Off**

In the bacterium *E. coli*, a repressor protein interacts with an operator to control the transcription of a group of genes that encode the enzymes needed to make tryptophan. (*a*) When tryptophan is present, it binds to a repressor protein; this tryptophan–repressor protein complex then turns the genes off by binding to the operator. (*b*) On its own (that is, when tryptophan is absent), the repressor protein cannot bind to the operator, but RNA polymerase *can* bind to the promoter, which turns on the genes for tryptophan synthesis.

or inactivate gene transcription, and gene **regulatory proteins** that interact with signals from the environment and also with the regulatory DNA to promote or repress gene transcription. Gene regulatory proteins are also known as *transcription factors*.

The regulation of tryptophan synthesis in the bacterium *E. coli* illustrates how gene regulatory DNA and gene regulatory proteins alter gene transcription in response to internal or external cues. *E. coli* requires a supply of the amino acid tryptophan. If tryptophan is available in the external environment, the bacterium absorbs it rather than wasting cellular resources to make it. But if tryptophan is not readily available, the bacterium expresses a series of five genes that together encode the enzymes needed for making tryptophan "from scratch" inside the cytoplasm.

*E. coli* controls the five genes in the tryptophan-making pathway in the following way: When tryptophan is present in the environment, it binds to a *repressor protein* in the bacterial cell (**repressor proteins** are so named because they stop the expression of one or more genes). This tryptophan–repressor protein complex can then bind to the tryptophan *operator*. An

operator is a DNA sequence that controls the transcription of a gene cluster—in this case, the five genes needed to make tryptophan. When bound to the operator, the tryptophan–repressor protein complex blocks access to the *promoter* of the tryptophan genes (**FIGURE 14.13a**). As mentioned earlier, the **promoter** is the segment of a gene that recruits the RNA polymerase and guides it to the transcription start site. The operator and promoter are both examples of gene regulatory DNA. The repressor protein is an example of a gene regulatory protein.

In the presence of tryptophan, the tryptophan–repressor protein complex prevents RNA polymerase from binding to the promoter. But in the absence of tryptophan, the repressor protein cannot bind to the operator, so RNA polymerase is free to bind to the promoter, and the genes are transcribed (**FIGURE 14.13b**). As a result, the bacterial cells do not make tryptophan when it is present in the surrounding medium, but they do manufacture it when tryptophan levels are low. This control of gene expression ensures two things: the cell always has an adequate supply of tryptophan, and the cell does not waste resources producing tryptophan when it is readily available in the environment.

## Gene expression can be regulated at several levels

In eukaryotes, gene expression can be controlled at several points along the pathway from gene to protein to phenotype (**FIGURE 14.14**). Let's consider some of the ways gene expression is regulated along the gene expression pathway.

1. *Tight packing of DNA prevents it from being expressed* (Figure 14.14, control point 1). During interphase, parts of the chromosome are "unpacked" down to the bead-on-a-string state. The unpacking gives gene regulatory proteins and RNA polymerase access to gene promoters and other regulatory DNA sequences, making it possible for these regions of the genome to be transcribed. In contrast, the tightly packed regions of the chromosome are transcriptionally inactive because their gene regulatory DNA is not accessible. DNA packing is itself a regulated process, and most eukaryotic cells have special protein complexes whose function is to selectively condense or decondense portions of a chromosome, depending on the cell type or specific signals received by that cell.

2. *Regulation of transcription conserves resources* (Figure 14.14, control point 2). Regulation of transcrip-

## CONTROLLING GENE EXPRESSION AT DIFFERENT LEVELS

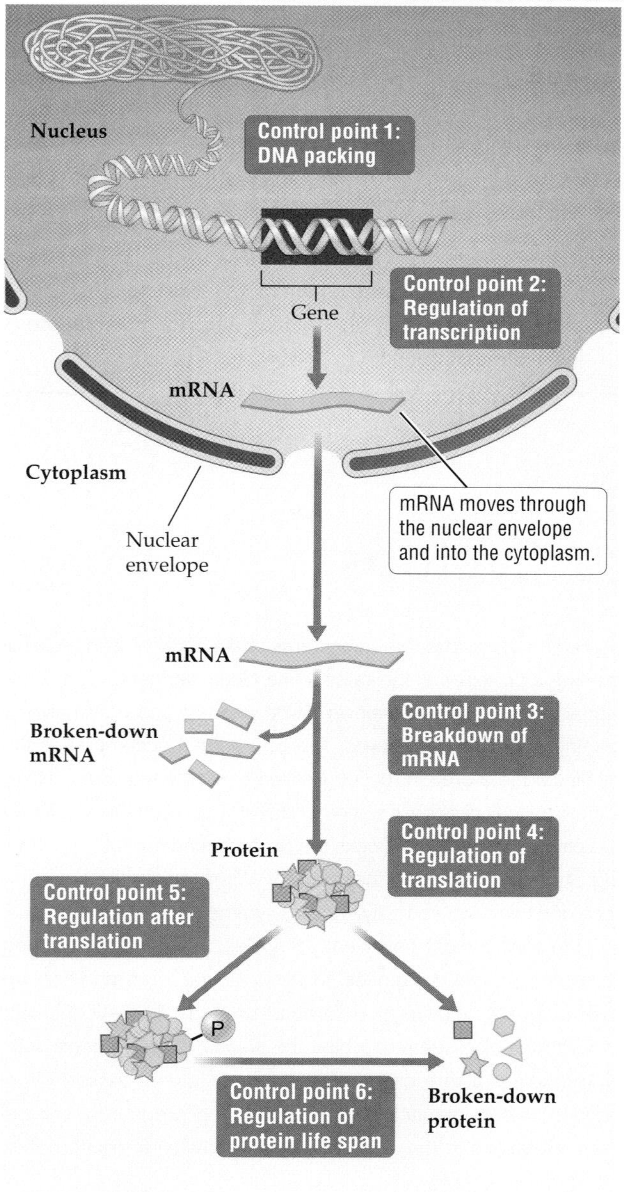

**Nucleus**

**Control point 1:
DNA packing**

Gene

**Control point 2:
Regulation of
transcription**

mRNA

**Cytoplasm**

Nuclear
envelope

mRNA moves through
the nuclear envelope
and into the cytoplasm.

mRNA

Broken-down
mRNA

**Control point 3:
Breakdown of
mRNA**

**Control point 4:
Regulation of
translation**

Protein

**Control point 5:
Regulation after
translation**

**Control point 6:
Regulation of
protein life span**

Broken-down
protein

**FIGURE 14.14 Steps at Which Gene Expression Can Be Regulated in Eukaryotes**

Each point in the pathway from gene to protein provides an opportunity for cells to regulate the production or activity of proteins.

tion is the most common means of controlling gene expression. Regulation of transcription is an efficient control mechanism because it enables the cell to conserve resources when it does not need the gene product. Transcriptional activation of gene expression is, however, relatively slow in eukaryotes: even in the best-case scenarios it takes 15–30 minutes for a protein to be made after a gene is transcriptionally activated.

3. *Regulation of mRNA breakdown prevents wasteful synthesis of proteins* (Figure 14.14, control point 3). Most mRNA molecules are broken down within a few minutes to hours after they are made; a few persist for days or weeks in the cytoplasm. The longer the life of an mRNA molecule, the more protein molecules can be made from it. By limiting the life span of many types of mRNA, a cell prevents the wasteful synthesis of proteins that are needed only temporarily. If circumstances change and the protein is needed, its mRNA can be quickly stabilized, and translation and accumulation of the protein may proceed rapidly. Time is saved because transcription does not have to be activated, so protein levels often start rising just a few minutes after the need for the protein is sensed.

4. *Regulation of translation keeps mRNA ready to direct rapid protein synthesis when needed* (Figure 14.14, control point 4). Specific RNA-binding proteins can attach to their target mRNA molecules and block translation. RNA-binding proteins enable the cell to deactivate mRNA whose protein product cannot be used immediately by the cell, but to retain the mRNA in a state of readiness should circumstances change and rapid protein synthesis be required. For example, some immune cells in the body make large amounts of mRNA for certain signaling proteins, called cytokines, but do not allow them to be translated. If these cells detect an invading bacterium, the translation block is immediately lifted. Cytokines are translated within minutes and poured into the bloodstream, where they act like an early warning system that prepares other components of the immune system to defend the body.

5. *Proteins can be directly regulated by modification following translation* (Figure 14.14, control point 5). Many proteins must be chemically modified before they can exert their effect on a phenotype. Some of the blood-clotting proteins, for example, are synthesized as inactive precursors; a segment of the protein must be cleaved off from such a protein before it can help plug a wound. A translated protein can also be rendered inactive by chemical modification or through the action of other molecules that bind to it. For example, liver cells respond to a hormone called glucagon by attaching a phosphate group on glycogen synthase, an enzyme that makes an energy-storing carbohydrate called glycogen. Glycogen synthase becomes inactive when it has a phosphate group on it. With the enzyme blocked, sugars cannot be stockpiled as glycogen and are instead released into the blood.

6. *Regulation of protein breakdown conserves resources and prevents damage* (Figure 14.14, control point 6). The activity of a protein provides one final opportunity to control the pathway from gene to protein. Most proteins in the cell have a limited life span, but a few, such as collagen and crystallin, last us through our lifetime. Proteins that are no longer needed, or are damaged, are taken apart and their amino acids recycled into new proteins. The recycling of unnecessary proteins enables the cell to invest resources where they are most needed. The excessive accumulation of proteins can even kill a cell. Brain cells, perhaps because of the complexity of their function, appear to be especially vulnerable to damage from protein deposits. Alzheimer's, Parkinson's, and Huntington's diseases are marked by the gradual death of brain cells after large protein clumps appear inside the cells, probably because of a failure to dispose of damaged proteins.

> ### Concept Check
>
> 1. Why are most genes controlled at the level of transcription?
> 2. If transcriptional control is the most favored method of gene regulation, why are not all genes controlled at the level of transcription?

---

## APPLYING WHAT WE LEARNED

# From Gene Expression to Cyclops

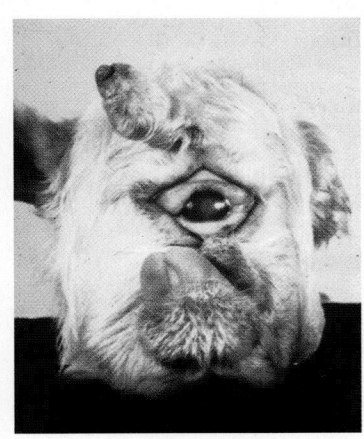

At the beginning of this chapter we described a rare birth defect called cyclopia, in which an animal is born with a single, large eye. Cyclopia may occur in as many as one in 250 embryos. Because most die before birth, the condition is rare in live births—about one in every 5,000–10,000 births. Mammals born with cyclopia usually die within a few hours of birth, but biologists have recorded some amazing exceptions—including a cyclopean skate (skates are related to sharks) that lived for several years and a trout that lived for 22 months. Even human babies with cyclopia have been known to live for several days.

Individuals with cyclopia have more problems than a missing eye; their whole face is malformed. Often they have either no nose or one that is only vaguely noselike, located above the eye or elsewhere on the head. In extreme cases, there is neither nose nor mouth. Inside the skull, the two halves of the forebrain have not divided normally into separate hemispheres. This malformed brain makes cyclopia a severe and fatal birth defect.

Cyclopia is sometimes common in animals that graze on wild plants. During the 1950s, 5–7 percent of all Utah lambs were born with cyclopia. Biologists eventually discovered that the lambs' mothers had eaten corn lilies on about the fourteenth day of pregnancy. Corn lilies—tall wildflowers that grow in the high mountain meadows of western North America—contain a molecule called cyclopamine.

Cyclopamine causes cyclopia by binding to a protein receptor that is part of a critical gene pathway in embryonic development. When cyclopamine binds to the receptor, it shuts down the expression of a series of genes in this pathway. The first gene is one that geneticists named *sonic hedgehog* (or *shh*). (Yes, geneticists named the gene after the video game character.) The *shh* gene expresses a protein with the same name, Sonic hedgehog (Shh), that signals the embryonic brain to divide into left and right halves—leading to left and right eyes. The Shh protein accomplishes this by binding to a protein receptor called Smoothened (Smo), which stimulates cell division. If a mother sheep eats corn lilies, however, the plant's cyclopamine blocks the Smoothened receptor, preventing Shh from binding to the receptor. Without Shh, the brain and face don't develop normally into two symmetrical halves.

Humans almost never eat corn lilies. In rare cases, though, women have been exposed to cyclopamine from drinking the milk of goats or cows that foraged on corn lilies. Alcohol and other drugs also seem to inhibit the *sonic hedgehog* gene pathway, leading to rare cases of cyclopia in human babies. Other causes of cyclopia include diabetes during pregnancy and genetic mutations in the *hedgehog* gene pathway (including other genes besides *shh*).

Amazingly, cyclopamine is turning out to be a valuable treatment for cancer. During development, the Shh protein promotes the rapid multiplication of certain cells, helping to shape the embryo. In adults, Shh also promotes rapid cell division. But because cells should multiply slowly or not at all in most adult tissues, the *sonic hedgehog* pathway is usually quiet in adults.

Sometimes, however, a gene in the *hedgehog* pathway can mutate as a result of damage from chemicals or ultraviolet radiation, causing abnormally rapid cell division, or cancer. Cancers influenced by the *hedgehog* pathway include certain cancers of the brain and the most common form of skin cancer: basal cell carcinoma. To slow cell division in such cancers, pharmaceutical companies have developed commercial drugs from cyclopamine that suppress the *hedgehog* pathway and slow cell division.

# Sluggish Cell Division May Help Explain Genital Defects

STAFF REPORT, *Sun-Sentinel*

Scientists have learned how a gene widely known for precisely positioning and sculpting various organs also controls the speed of cell division, a finding that could be useful for understanding the explosive growth of cancer cells or why increasing numbers of children are being born with genital and urinary tract malformations.

Writing in *Nature Communications*, researchers at the University of Florida say a gene memorably named Sonic hedgehog controls genital development by regulating a process known as the cell cycle—a biological event that regulates when, and how fast, cells divide to form hearts, brains, limbs and all the other complex structures needed to build an individual.

The findings in mice provide insight into the molecular mechanisms that underlie growth of urinary and reproductive organs in both sexes. Abnormalities of the genitalia and urinary tract are among the most common birth defects, according to the March of Dimes...

"The role of Sonic hedgehog during embryonic development is to set up...a process known as patterning," said senior author Martin Cohn..."The surprise is to find out how much patterning and growth are intertwined... Once we discovered that inactivation of Sonic hedgehog slowed down the cell cycle, it explained the big differences in growth and the structural defects we were finding in genitalia."

The knowledge may help scientists understand why an increasing number of boys are being born with birth defects called hypospadias, which involve incomplete formation of the urethral tube, resulting in an abnormally placed urethral opening on the underside of the penis. About one in 250 children has a urethral tube defect, more than double the frequency of 30 years ago.

The cell cycle controls whether a cell continues to give rise to more cells or stop[s] dividing and become[s] specialized with a specific function to carry out. Humans begin life as a single fertilized egg cell that eventually gives rise to countless cells in an adult. As each cell divides it must proceed through a growth phase, replicate its DNA and divide again, or it can be instructed to stop dividing and perform a specific function.

When scientists deleted the Sonic hedgehog gene in specific [mouse] tissues..., they discovered the cell cycle takes... about 14.4 hours instead of the usual 8.5 hours... As a result, fewer cells are produced and genital growth is reduced by about 75 percent. The shape of the genitals is also altered...

The *sonic hedgehog* pathway helps guide the development of not only the brain and the face, but also other parts of the body, including the genitals. Just as for the face, *sonic hedgehog* functions by altering the pace of cell division. If cells in one area are dividing faster than those in nearby areas, overall shape will change. During normal development, it is a good thing for the different parts of the body to grow at different rates, so that the body can take shape. But it's easy to see how small changes in how fast a few cells divide could dramatically alter that shape from normal to abnormal.

This article reports that birth defects of the genitals of boys have doubled in recent decades. One suspected cause is a group of common chemicals called phthalates. In lab studies, male rodents exposed to phthalates develop the same problem, hypospadias, at the same levels of exposure that humans experience every day.

Phthalates are in hundreds of household products, from hair spray and nail polish to plastic shower curtains. Rats exposed in the womb to phthalates have less testosterone, fewer testosterone receptors in the penis, more estrogen receptors, and less Shh (Sonic hedgehog) protein. Testosterone plays a critical role in development by increasing or decreasing the transcription of specific genes—including those in the *sonic hedgehog* pathway. When phthalates lower testosterone, they lower the activity of the *sonic hedgehog* gene, slowing cell division when it should be rapid. Women exposed to high levels of phthalates and hair spray have babies with 2–3 times the usual rate of hypospadias.

## Evaluating the News

**1.** How does expression of *Shh*, the *sonic hedgehog* gene, affect the cell cycle? Explain how slowing down the cell cycle can change the shapes of parts of the body such as the brain, face, eyes, and genitals.

**2.** If testosterone normally increases the activity of the *sonic hedgehog* pathway during development, what would happen to the cell cycle in the presence of phthalates?

**3.** Phthalates wash out of the body within a few days. In humans, the genitals develop between the eighth and fifteenth weeks of pregnancy. Assuming that you were pregnant and that phthalates play a role in the development of hypospadias, what steps might you take to reduce the risk of having a baby with hypospadias?

**SOURCE:** *Sun-Sentinel* (Fort Lauderdale, Florida), June 1, 2010.

# CHAPTER REVIEW

## Summary

### 14.1 An Overview of DNA and Genes

- DNA stores genetic information as a sequence of four nucleotides.
- The genome is all the DNA-based information in the nucleoid of a prokaryote or the nucleus of a eukaryote.
- A gene is segment of DNA that encodes at least one distinct genetic trait. Most genes code for proteins, which generate phenotypes.
- Genetic information flows from DNA to RNA in the process of gene transcription. In translation, ribosomes convert the genetic information in an mRNA molecule into a specific protein.
- Gene expression is the manifestation of a phenotype based on information encoded in the gene.
- Different sets of genes are expressed in different cell types. The gene promoter functions like an on/off switch for transcription.

### 14.2 The Three-Dimensional Structure of DNA

- Watson and Crick determined that DNA is a double helix formed by two polynucleotide strands containing nitrogenous bases: adenine (A), cytosine (C), guanine (G), and thymine (T).
- The two polynucleotide strands are held together by hydrogen bonds between each A and T, and each G and C.
- The sequence of bases in DNA, which differs among species and among individuals within a species, is the basis of inherited variation.

### 14.3 How DNA Is Replicated

- A complex of proteins guides the replication of DNA; the primary enzyme involved is DNA polymerase.
- The DNA helix is unwound, and DNA polymerase uses each strand as a template from which to build a new strand of DNA.
- DNA replication is semiconservative: it produces two copies of the original DNA molecule, each composed of one old strand and one newly synthesized daughter strand.

### 14.4 Repairing Replication Errors and Damaged DNA

- On rare occasions, mistakes occur during DNA replication. Most mistakes in the copying process are corrected, either immediately by "proofreading" or later by the mismatch repair system.
- Uncorrected mistakes in DNA replication are one source of mutations.
- The DNA in our cells is altered thousands of times every day by accidental chemical changes, and possibly by radiation and mutagens as well.
- Replication errors and damage to DNA are fixed by a complex set of DNA repair proteins.

### 14.5 Genome Organization

- Compared with eukaryotes, prokaryotes have a smaller genome, and it usually takes the form of a single chromosome. Most prokaryotic DNA encodes proteins, and functionally related genes in prokaryotes are grouped together in the genome.
- With regard to genome organization, eukaryotes differ from prokaryotes in several ways: (1) In eukaryotic cells, the DNA is distributed among several chromosomes. (2) Eukaryotes have more DNA per cell than prokaryotes have, in part because they have more genes than prokaryotes have. In addition, genes constitute only a small portion of the genome in many eukaryotes; the rest consists of noncoding DNA (including introns and spacer DNA) and transposons. (3) Eukaryotic genes with related functions often are not located near one another.

### 14.6 DNA Packing in Eukaryotes

- Cells can pack an enormous amount of DNA into a very small space because their DNA is highly organized by a complex packing system. In eukaryotes, segments of DNA are wound around histone spools, packed together tightly into a narrow fiber, and further folded into loops.
- Chromosomes are most tightly packed during mitosis and meiosis. The packing is looser during interphase, when most gene expression occurs.
- In DNA regions that are tightly packed, genes cannot be expressed, because the proteins necessary for transcription cannot reach the genes.

### 14.7 Patterns of Gene Expression

- In both prokaryotes and eukaryotes, genes are turned on and off selectively in response to short-term changes in environmental conditions.
- The different cell types in a multicellular organism express different sets of genes.
- Housekeeping genes, which have an essential role in the maintenance of cellular activities, are expressed in most cells of the body.

### 14.8 How Cells Control Gene Expression

- Most genes are regulated at the level of transcription. Transcription is controlled by regulatory DNA sequences that interact with regulatory proteins to switch genes on and off.
- Gene regulatory proteins link gene expression to internal and external signals. Some regulatory proteins (repressor proteins) inhibit transcription; others (activator proteins) promote it.

## Key Terms

base pair (p. 328)
DNA (p. 324)
DNA polymerase (p. 329)
DNA repair (p. 331)
DNA replication (p. 329)
exon (p. 334)
gene (p. 325)
gene expression (p. 325)
genome (p. 324)
housekeeping gene (p. 337)
intron (p. 334)
messenger RNA (mRNA) (p. 325)
mutagen (p. 330)
mutation (p. 330)
nitrogenous base (p. 327)
noncoding DNA (p. 334)

# Self-Quiz

1. The base-pairing rules for DNA state that
   a. any combination of bases is allowed.
   b. T pairs with C, A pairs with G.
   c. A pairs with T, C pairs with G.
   d. C pairs with A, T pairs with G.

2. DNA replication results in
   a. two DNA molecules, one with two old strands and one with two new strands.
   b. two DNA molecules, each of which has two new strands.
   c. two DNA molecules, each of which has one old strand and one new strand.
   d. none of the above

3. The DNA of cells is damaged
   a. thousands of times per day.
   b. by collisions with other molecules, chemical accidents, and radiation.
   c. not very often and only by radiation.
   d. both a and b

4. The DNA of different species differs in the
   a. sequence of bases.
   b. base-pairing rules.
   c. number of nucleotide strands.
   d. location of the sugar-phosphate portion of the DNA molecule.

5. If a strand of DNA has the sequence CGGTATATC, then the complementary strand of DNA has the sequence
   a. ATTCGCGCA.
   c. GCCATATAG.
   b. GCCCGCGCTT.
   d. TAACGCGCT.

6. Mutation
   a. can produce new alleles.
   b. can be harmful, beneficial, or neutral.
   c. is a change in an organism's DNA sequence.
   d. all of the above

7. In prokaryotes and eukaryotes, gene expression is most often controlled by regulation of
   a. the destruction of a gene's protein product.
   b. the length of time mRNA remains intact.
   c. transcription.
   d. translation.

8. The DNA of eukaryotes is packed most loosely at which time?
   a. during mitosis
   c. during interphase
   b. during meiosis
   d. both b and c

# Analysis and Application

1. A gene has two codominant alleles: $A^1$ and $A^2$. Each allele produces a different but related version of a protein found on the surface of a type of white blood cell. In physical terms, each allele is a segment of DNA. Explain how the DNA of one allele might differ from the DNA of the other allele. Describe the possible effects of these differences in the two DNA segments.

2. Explain why the structure of DNA proposed by Watson and Crick suggested a way DNA could be replicated.

3. Describe how the bases in DNA, mutations, and alleles that cause human genetic disorders are related.

4. Summarize how DNA repair works and why the repair mechanisms are essential for cells and whole organisms to function normally.

5. The total length of the DNA in a eukaryotic cell is hundreds of thousands of times greater than the diameter of the nucleus. Explain how so much DNA can be packed inside the nucleus.

6. Summarize the major differences between the genomes of prokaryotes and eukaryotes, emphasizing differences in the amount and function of DNA and the organization of genes.

7. Imagine you transferred a bacterium from an environment in which glucose was available as food to an environment in which the only source of food was the sugar arabinose. A specific enzyme is required to digest arabinose but not glucose. How do you think gene expression in the bacterium would change as a result of your action?

8. Explain how cells with the same genes can be so different (a) in structure and (b) in the metabolic tasks they perform.

9. Summarize how gene expression can be controlled at various steps in the process of generating a protein that is fully active.

# From Gene to Protein

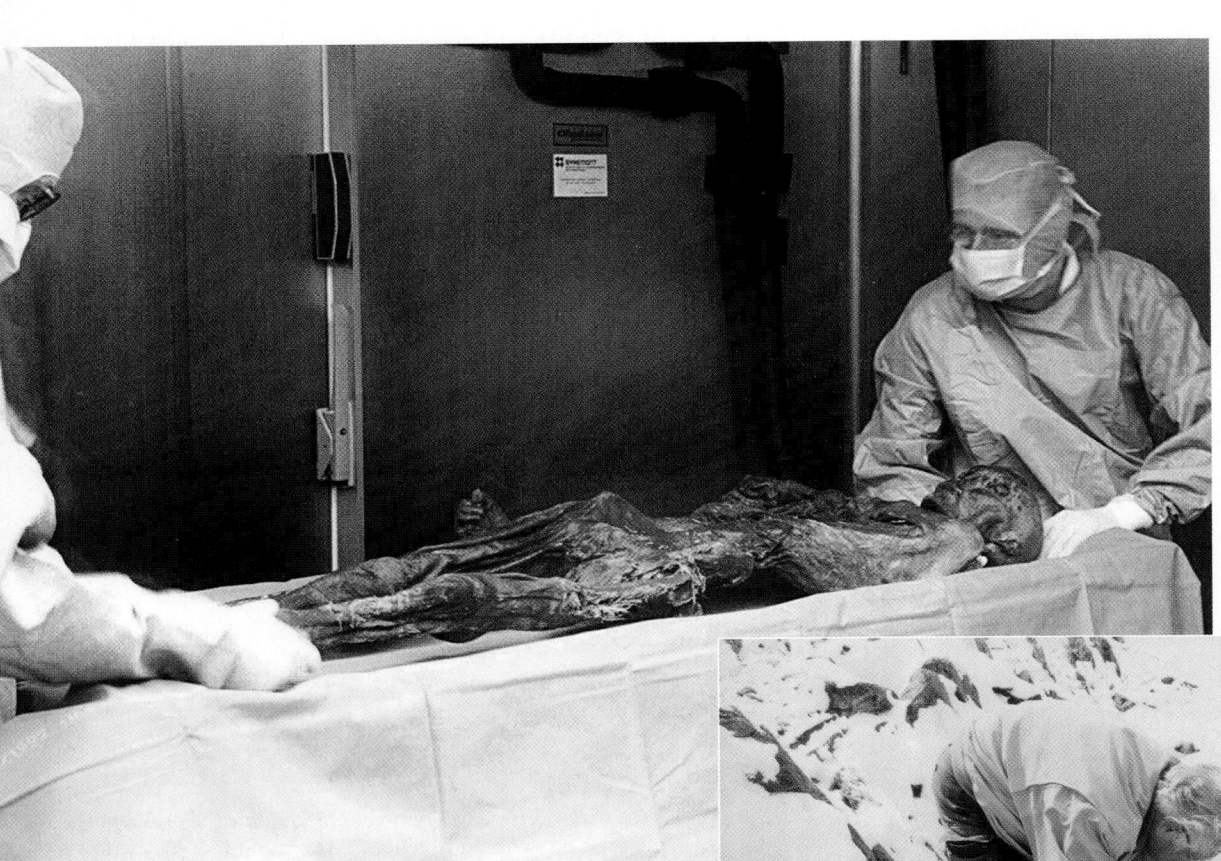

**ÖTZI, THE ICEMAN.** In 1991, hikers discovered the body of a man who lived thousands of years ago, lying at the base of a melting glacier. Long-frozen DNA sequences revealed how he was related to modern Europeans and how he may have died.

# The Man from the Copper Age

On a September day in 1991, a German couple hiking in a high mountain pass in the Italian Alps stumbled on an astonishing find: at their feet, in a pool of melting glacier water, lay a mummified corpse. Far from being even a few decades old, the unknown man turned out to be 5,200 years old—a visitor from the Copper Age promising a tantalizing peek into the past.

Extremely well preserved, the Iceman—as researchers dubbed him—was about 45 years old, 5 feet 5 inches tall, and 110 pounds. Near his naked, tattooed body were a beautifully made axe with a solid-copper blade, a flint dagger and woven sheath, a half-constructed longbow, a quiver full of arrows, and a full set of clothes that included a leather belt and pouch, sheepskin leggings and a jacket made of animal skin, a cape of woven grass, and calfskin shoes lined with grass. Among other items, he also carried pieces of fungus, one possibly a kind of prehistoric penicillin for treating infections, and another useful for starting a fire. He had brought berries and grains from lower elevations and was fully equipped for a mountain trek;

his backpack and other gear were beautifully crafted from 18 different kinds of wood.

Frozen high in the Alps for 5,200 years, the ancient Iceman's body was perfectly suited to scientific examination

How could biologists use his DNA to determine the true identity of this long-frozen man? Were his closest relatives modern Europeans? Or was he from some other part of the world?

and fully equipped with Copper Age tools and clothes. His discovery seemed almost too good to be true, and some people speculated that the body was an elaborate hoax; perhaps the Iceman was a transplanted Egyptian or pre-Columbian American mummy. To solve the mystery, scientists would have to study his DNA and place him in the human family tree. As unusual as the Iceman was, the way in which scientists used his DNA to learn more about him was a standard approach in biology.

**MAIN MESSAGE** Genes carry information that affects the production of proteins, and proteins play a key role in determining the inherited characteristics of an individual.

## KEY CONCEPTS

- Some genes encode RNA molecules as their final product, but most genes contain instructions for building proteins. In the latter case, the gene's DNA sequence encodes the amino acid sequence of its protein product.

- The flow of information from gene to protein requires two steps: transcription and translation.

- In eukaryotic cells, transcription occurs in the nucleus and produces a messenger RNA version of the sequence information stored in the gene. The mRNA moves from the nucleus to the cytoplasm, where it guides the manufacture of a protein by ribosomes, with assistance from transfer RNA.

- Translation occurs in the cytoplasm and converts the sequence of bases in an

mRNA molecule to the sequence of amino acids in a protein.

- Gene mutations can alter the sequence of amino acids in the protein encoded by that gene. Such changes, in turn, can alter the protein's function. Although changes in protein function are often harmful, occasionally they benefit the organism.

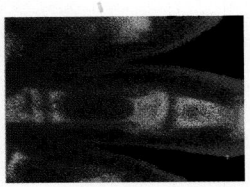

**RNA Molecules Carry Genetic Information**

The image shows the localization of certain mRNA that are abundant in joint tissue, here in the developing limbs of a fetal mouse. The pink stain marks mRNA that encode collagen, the green stain reveals mRNA that encode growth differentiation factor 5. These proteins are critical for development of bone and cartilage. DNA is stained blue in this image, prepared with in situ hybridization techniques.

**HOW DO GENES STORE THE INFORMATION** needed to build their final products: RNA and proteins? How do RNA molecules direct the manufacture of proteins? Knowing how genes work can help us understand how mutations produce new phenotypes, including disease phenotypes. We begin this chapter by describing how genetic information is encoded in genes and how the cell uses that information to build proteins. We then describe how a change in a gene can change an organism's phenotype. Our discussion of how cells use the information stored in genes to build proteins focuses mostly on eukaryotes; except where noted, events are similar in prokaryotes.

## 15.1 How Genes Work

Proteins are essential to life. They are used by cells and organisms in many ways: some provide structural support; others transport materials through the body; still others defend against disease-causing organisms. In addition, the many chemical reactions that life depends on are controlled by a crucial group of proteins: the enzymes. Enzymes and other proteins influence so many features of the organism that they, along with the organism's internal and external environment, are the key determinants of an organism's phenotype.

How do genes affect the phenotype of an organism? Early clues came at the beginning of the twentieth century from the work of British physician Archibald Garrod, who studied several inherited disorders in which metabolism was disrupted. In 1902 he argued that these metabolic disorders were caused by an inability of the body to produce specific enzymes. Garrod was particularly interested in alkaptonuria, a condition in which the urine of otherwise healthy in-fants turns black when exposed to air. He proposed that infants with alkaptonuria have a defective version of an enzyme that, in its normal form, breaks down the substance that causes the discoloration. But Garrod did not stop there; he and his collaborator, William Bateson, went on to suggest that in general, genes are responsible for the production of enzymes.

## Genes contain information for building RNA molecules

Garrod and Bateson were on the right track, but they were not entirely correct: genes control the production of all types of proteins, not just enzymes. Furthermore, not all genes contain code that specifies the order in which amino acids will be linked to make a protein. The so-called *noncoding genes* store instructions for building any one of a variety of ribonucleic acid (RNA) molecules with varied functions that do not include carrying code for the manufacture of a specific protein. In contrast, all *protein-coding genes* direct the production of messenger RNA (mRNA), which in turn directs the production of a particular protein in the cytoplasm. Recall that gene transcription is the process by which DNA sequence information is used as a template to drive the synthesis of a complementary RNA molecule, and that transcription is the first step in gene expression (see Chapter 14). At the molecular level, we can say that a **gene** is any DNA sequence that is transcribed into RNA.

RNA and DNA share a number of structural similarities, as well as several important differences. Both are nucleic acids consisting of nucleotides covalently bonded to one another. But whereas DNA molecules are double-stranded, the various types of RNA molecules are all single-stranded. While the two strands of a DNA molecule form a double helix, a single-stranded RNA molecule may fold back on itself to assume a variety of three-dimensional shapes. As in DNA, each nucleotide in RNA is composed of a sugar, a phosphate group, and one of four nitrogen-containing bases (**FIGURE 15.1**). However, the nucleotides in RNA and DNA differ in two respects: First, RNA uses the sugar ribose, whereas DNA contains the sugar deoxyribose. Second, in RNA the base uracil (U) replaces the base thymine (T), which is found only in DNA. The other three bases—adenine (A), cytosine (C), and guanine (G)—are the same in RNA and DNA. In general, RNA is chemically less stable than DNA, and most RNA molecules in the cell have a limited life in the cytoplasm (see Section 14.8). As the permanent store of

FIGURE 15.1  RNA Is a Single-Stranded Chain of Nucleotides

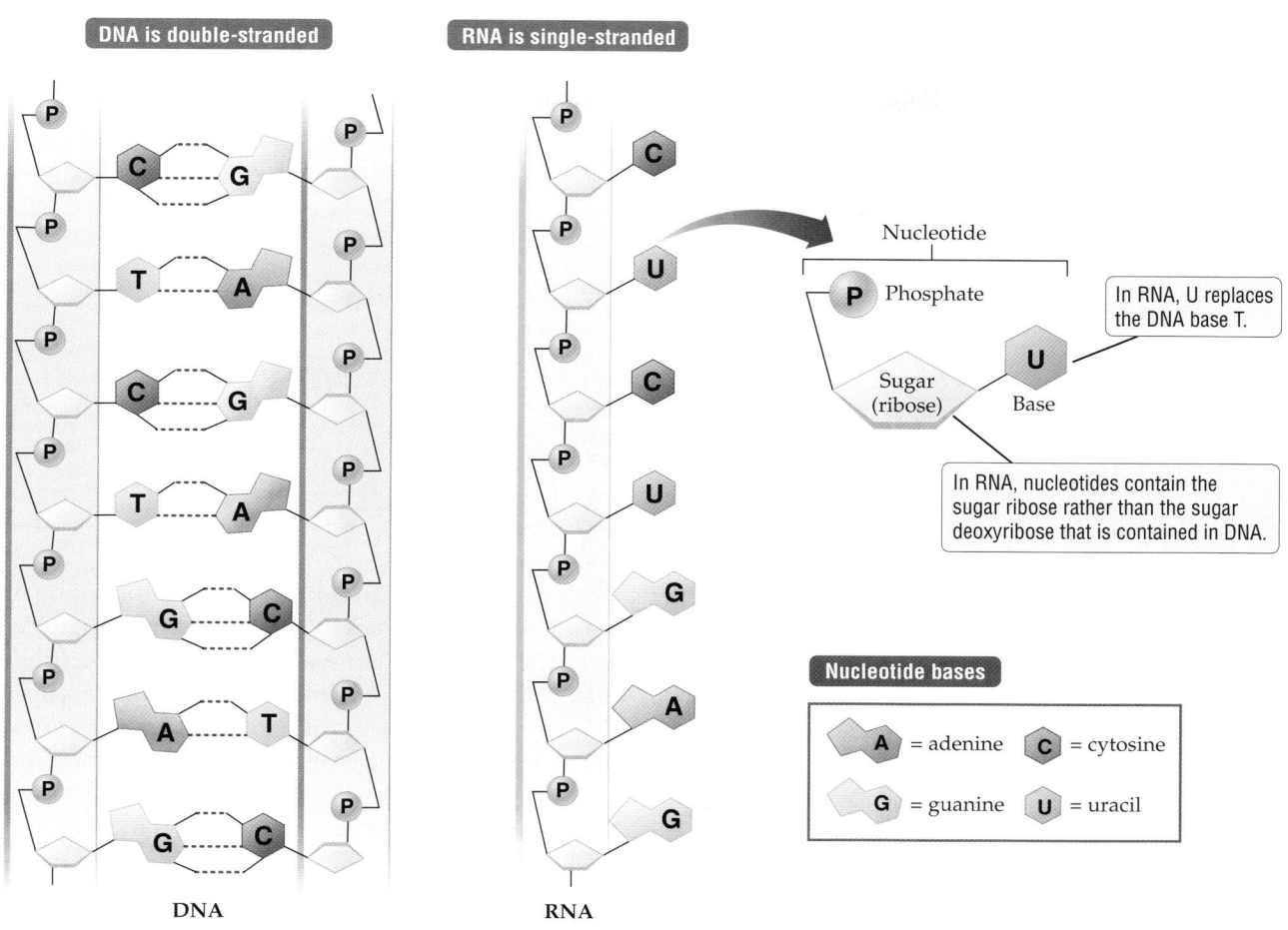

genetic information, the DNA in the nucleus of most cells is much more stable, being destroyed only if the cell itself is destined to die soon.

### Three types of RNA assist in the manufacture of proteins

The nucleic acids DNA and RNA play key roles in the construction of proteins. As already described, DNA directs the production of all these essential molecules, so the information for manufacturing a specific protein comes ultimately from DNA. Cells use three main types of RNA molecules to construct proteins: **messenger RNA (mRNA)**, **ribosomal RNA (rRNA)**, and **transfer RNA (tRNA)**. The function of each type is defined in Table 15.1 and discussed in more detail in the sections that follow. Cells also produce several other types of RNA that affect the production of proteins, but these more complicated functions are beyond the scope of this textbook.

## 15.2  How Genes Guide the Manufacture of Proteins

In both prokaryotes and eukaryotes, proteins are produced in two steps: *transcription* and *translation*. During **transcription** of a protein-coding gene, mRNA is made using the information in the DNA sequence of that gene. The base sequence of an mRNA is complementary to the DNA template it was transcribed from, and it carries the code for the amino acid sequence of a specific protein. During **translation**, amino acids are covalently linked in the precise sequence dictated by the base sequence of the mRNA. Translation requires not only mRNA, but also at least two other types of RNA: (1) rRNA, which is an important component of ribosomes, the "workbenches" on which proteins are assembled; and (2) more than 20 different types of tRNA molecules, each of which delivers a specific

TABLE 15.1          RNA Molecules and Their Functions

| TYPE OF RNA | FUNCTION | SHAPE |
|---|---|---|
| Messenger RNA (mRNA) | Specifies the order of amino acids in a protein using a series of three-base codons, where different amino acids are specified by particular codons. |  |
| Ribosomal RNA (rRNA) | As a major component of ribosomes, assists in making the covalent bonds that link amino acids together to make a protein. | 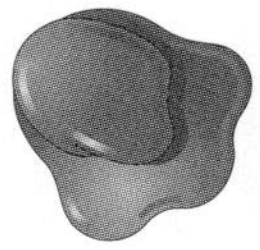 |
| Transfer RNA (tRNA) | Transports the correct amino acid to the ribosome, using the information encoded in the mRNA; contains a three-base anticodon that pairs with a complementary codon revealed in the mRNA. |  |

amino acid to the ribosomes for assembly into the protein in accordance with the information encoded in the base sequence of the mRNA.

## MESSENGER RNA DIRECTS PROTEIN SYNTHESIS

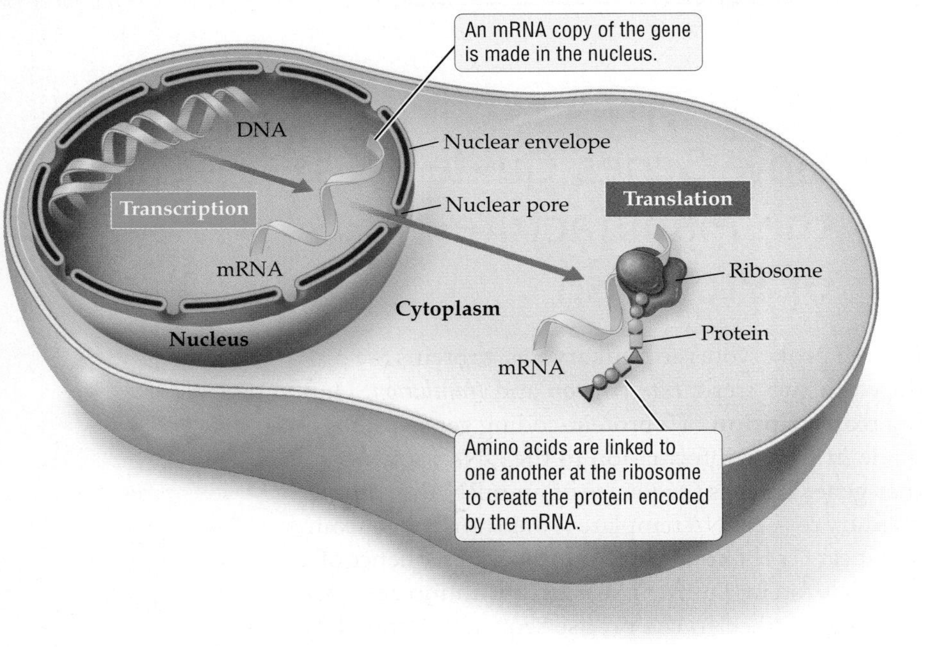

An mRNA copy of the gene is made in the nucleus.

DNA

Nuclear envelope

Transcription

Nuclear pore

Translation

mRNA

Ribosome

Cytoplasm

Protein

Nucleus

mRNA

Amino acids are linked to one another at the ribosome to create the protein encoded by the mRNA.

**FIGURE 15.2 Genetic Information Flows from DNA to RNA to Protein during Transcription and Translation**
The transcription of a protein-coding gene produces an mRNA molecule, which is then transported to the cytoplasm, where translation occurs and the protein is made with the help of ribosomes. Different amino acids in the protein being constructed at the ribosome are represented here by different colors and shapes.

To make a protein in eukaryotes, the information in a gene must be sent from the gene, located in the nucleus, to the site of protein synthesis: the ribosomes that drift freely in the cytoplasm (**FIGURE 15.2**). Messenger RNA, exiting the nucleus through a nuclear pore, is the intermediary molecule that transfers the information from the gene in the nucleus to the ribosomes. During gene transcription, the sequence of bases in one DNA strand (the template strand) of a gene is copied (transcribed) into mRNA. The information stored in the gene is not only transmitted to the cytoplasm, but it is substantially amplified through gene transcription, as each template DNA typically generates many hundreds of mRNA molecules when it is transcribed.

Once the mRNA molecule arrives in the cytoplasm, the information it contains must be translated, with the help of ribosomes, from the language of mRNA (nitrogenous bases) to the language of proteins (amino acids). The information is translated at the ribosomes by tRNA molecules. To do this, a three-base sequence on each tRNA molecule binds to its complementary sequence on the mRNA by forming hydrogen bonds with it; one end of the tRNA molecule carries the particular amino acid specified by the three-base sequence in the mRNA. We will examine the binding rules involved in this process shortly. For now, it is enough to know that the base sequence of the mRNA dictates which tRNAs are bound, and therefore which amino acids are delivered at the ribosomes. The ribosomes hold the mRNA, position the incoming tRNAs, and also join the amino acids delivered by the tRNAs to synthesize a protein with the exact sequence of amino acids called for by the base sequence of the mRNA.

## 15.3 Transcription: Information Flow from DNA to RNA

Gene transcription is similar to DNA replication in that one strand of DNA is used as a template from which a new strand—in this case, a strand of mRNA—is formed. However, transcription differs from DNA replication in three important ways:

1. A different enzyme guides the process. The key enzyme in DNA replication is DNA polymerase; the key enzyme in transcription is **RNA polymerase**.

2. In DNA replication, the entire DNA double helix is copied. In the transcription of a particular gene, only a small portion of the chromosome is copied into RNA.

3. Whereas DNA replication produces double-stranded DNA molecules, transcription produces a single-stranded RNA molecule that is complementary to one strand (the template strand) of the DNA double helix.

As discussed in Section 14.8, transcription of a gene begins when the enzyme RNA polymerase binds to a segment of DNA near the beginning of the gene, called a **promoter**. Although the promoters of different genes vary in size and sequence, all contain several specific sequences of 6–10 bases that enable the RNA polymerase to recognize and bind to them. Once bound to the promoter, the RNA polymerase unwinds the DNA double helix at the beginning of the gene, separating a short portion of the two strands. Then the enzyme begins to construct an mRNA molecule (**FIGURE 15.3**). Only one of the two DNA strands is used as a template, and this strand is called the **template strand** (lower strand in the magnified portion of Figure 15.3). If the opposite strand (upper strand in Figure 15.3) were used as template, it would result in a completely different sequence of amino acids, and hence a different protein. How does the RNA polymerase "choose" which strand to use as template? The RNA polymerase binds to the promoter in a specific orientation, thereby determining which strand it will be able to "read." The location and orientation of the promoter sequence guide the binding of the RNA polymerase, so ultimately the positioning of the promoter is what specifies which DNA strand will serve as the template.

The four kinds of bases in RNA pair with the four kinds of bases in DNA according to the rules described in Chapter 14: A in RNA pairs with T in DNA, C in RNA pairs with G in DNA, G in RNA pairs with C in DNA, and U in RNA pairs with A in DNA. As an RNA polymerase moves away from the gene promoter and travels down the template strand, another RNA polymerase can bind at the promoter and start synthesizing an mRNA "fast on the heels" of the previous RNA polymerase. At any given time, therefore, many RNA polymerases can be traveling down a DNA template, each synthesizing its own mRNA (at the rate of 60 bases per second in human cells).

The importance of rapid transcription is demonstrated by how quickly a death cap mushroom can kill the individual who eats it. The death cap mushroom produces an extremely toxic substance, called alpha-amanitin, that binds to RNA polymerase and reduces its travel speed to just 2 or 3 bases per second. Eating just one death cap mushroom can kill a person in about 10 days, primarily because the liver and pancreas fail. The cells in these organs are highly active in gene transcription, pouring out a great variety of enzymes and other proteins that are vital for life. A slowdown of transcriptional activity in these organs has devastating consequences.

Transcription terminates in prokaryotes when the RNA polymerase reaches a special sequence of bases called a **terminator** (see Figure 15.3). When the terminator sequence is copied into mRNA, it generates a three-dimensional shape, known as a hairpin, in the mRNA sequence. Hairpins in the mRNA destabilize the RNA polymerase and cause it to drop off the template. Transcription ends at this point, and the newly formed mRNA molecule separates from its DNA template.

Special terminator sequences signal the end of transcription in eukaryotes also, but the exact mechanism is more complicated than in prokaryotes. Eukaryotic mRNA also undergoes elaborate processing, a sequence of steps that modifies RNA and prepares it for export from the nucleus. Posttranscriptional RNA processing in eukaryotes includes chemical modification of both ends of the mRNA: a chemically unique nucleotide is added to create a cap structure at the end of the mRNA that is transcribed first, and a string of adenines is added at the opposite end to create what is known as the *poly-A tail*.

*RNA splicing* is another key event in RNA processing. Recall from Chapter 14 that eukaryotic cells contain a lot of noncoding DNA. Most eukaryotic genes have stretches of noncoding sequences, called **introns**, embedded within the nucleotide sequences that actually code for protein. The stretches of nucleotide

*Amanita phalloides*, the death cap mushroom

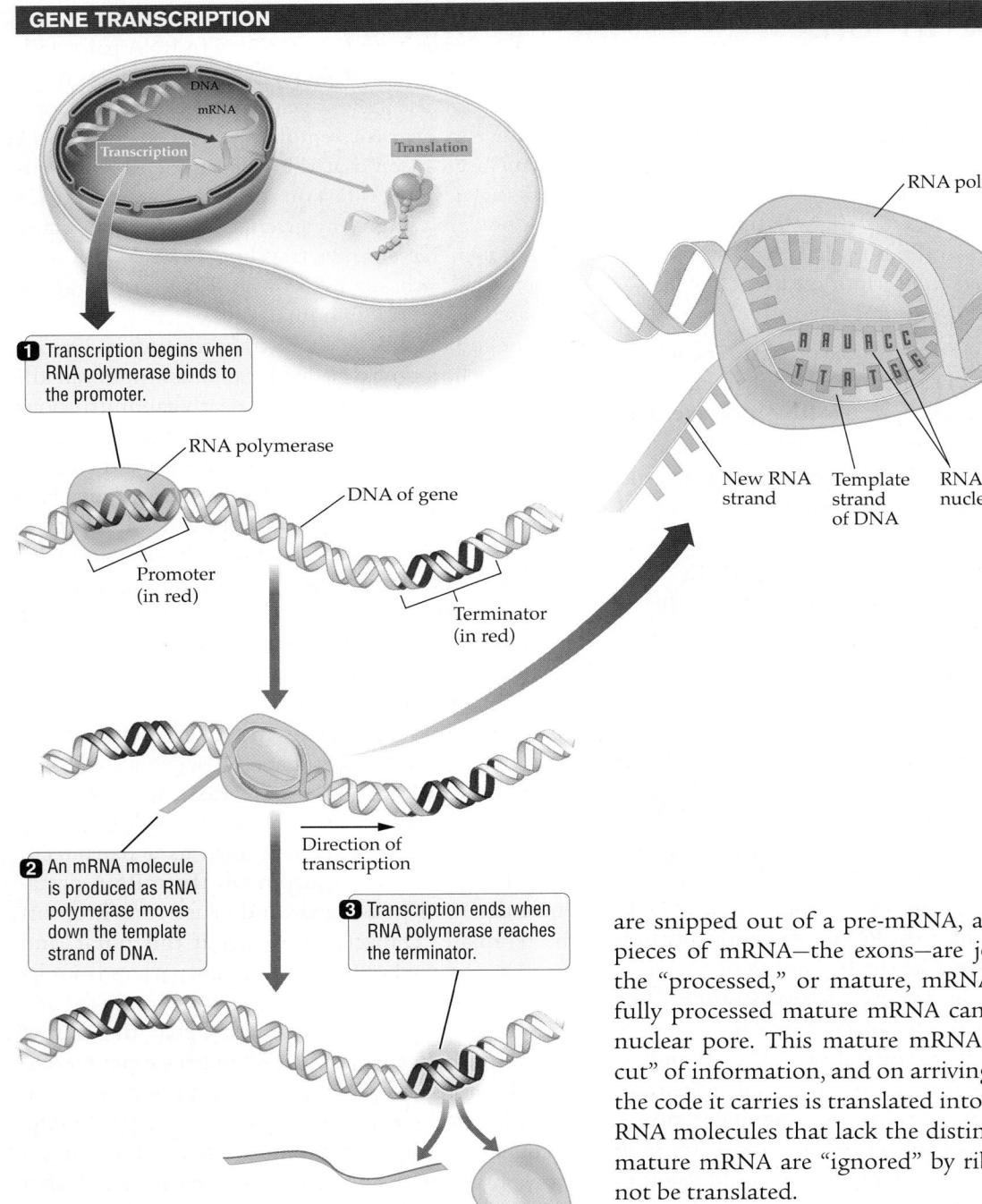

**GENE TRANSCRIPTION**

❶ Transcription begins when RNA polymerase binds to the promoter.

RNA polymerase

DNA of gene

Promoter (in red)

Terminator (in red)

RNA polymerase

New RNA strand

Template strand of DNA

RNA nucleotides

Direction of transcription

Direction of transcription

❷ An mRNA molecule is produced as RNA polymerase moves down the template strand of DNA.

❸ Transcription ends when RNA polymerase reaches the terminator.

are snipped out of a pre-mRNA, and the remaining pieces of mRNA—the exons—are joined to generate the "processed," or mature, mRNA. Normally, only fully processed mature mRNA can exit through the nuclear pore. This mature mRNA carries the "final cut" of information, and on arriving in the cytoplasm the code it carries is translated into a specific protein. RNA molecules that lack the distinctive features of a mature mRNA are "ignored" by ribosomes and cannot be translated.

## Concept Check

1. What is gene transcription?

2. Compare the chemical structures of RNA and DNA. Which is more stable chemically, and how is that stability consistent with its function?

3. The template strand of a gene has the base sequence TGAGAAGACCAGGGTTGT. What is the sequence of RNA transcribed from this DNA, assuming RNA polymerase travels from left to right on this strand?

sequences in a gene that carry instructions for building the protein are called **exons** (**FIGURE 15.4**). Because of this patchwork construction of eukaryotic genes, most newly transcribed mRNA (pre-mRNA) is also a patchwork of coding sequences intermixed within noncoding sequences. These pre-mRNA sequences are therefore like an uncut video recording with "extra footage." In **RNA splicing**, the introns

**Concept Check Answers**

3. ACUCUUCUGGU-CCAACA.

2. RNA is single-stranded; it contains ribose and the bases A, G, C, and U. DNA is double-stranded; it contains deoxyribose and A, G, C, and T. DNA is more stable—a property it must have to serve as the storehouse of genetic information.

1. The copying of genetic information from DNA to RNA by RNA polymerase, using one of the two strands of the DNA as template.

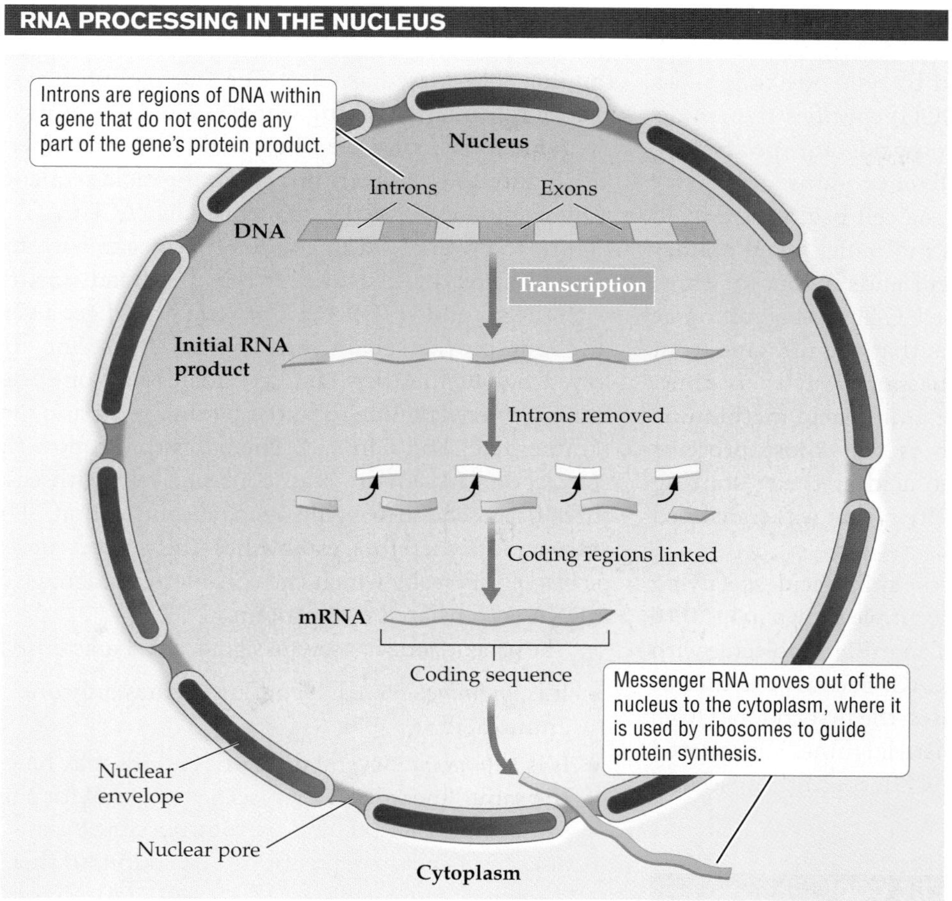

Introns are regions of DNA within a gene that do not encode any part of the gene's protein product.

Nucleus

Introns    Exons

DNA

Transcription

Initial RNA product

Introns removed

Coding regions linked

mRNA

Coding sequence

Messenger RNA moves out of the nucleus to the cytoplasm, where it is used by ribosomes to guide protein synthesis.

Nuclear envelope

Nuclear pore

Cytoplasm

**FIGURE 15.4 In Eukaryotes, Introns Must Be Removed Before an mRNA Leaves the Nucleus**

Most eukaryotic genes contain both coding sequences (exons) and noncoding sequences (introns). Before the mRNA transcribed from such genes can be exported to the cytoplasm, enzymes in the nucleus must remove the introns and link the remaining exons.

## 15.4 The Genetic Code

The information in a gene is encoded in its sequence of bases. As we learned in the previous section, the gene's DNA sequence is used as a template to produce an mRNA molecule. Recall that the final products of most genes are proteins, and that proteins consist of one or more folded strings of amino acids. How does mRNA encode (specify) the sequence of amino acids in a protein?

The information in an mRNA molecule is "read" by ribosomes in sets of three bases, and each unique sequence of three bases is called a **codon**. Since there are 64 different ways in which we can arrange four bases to create a three-base sequence, there are 64 possible codons. A language with only four letters (A, U, C, and G) would have 64 different words, if only three-letter words were allowed. The **genetic code** refers to the information specified by each of the 64 possible codons (the "meaning" of every word in the language). The entire genetic code is shown in **FIGURE 15.5**. Most of the 64 codons specify a particular amino acid, but some

**FIGURE 15.5 The 64 Possible Codons Specify Amino Acids or Signals That Start or Stop Translation**

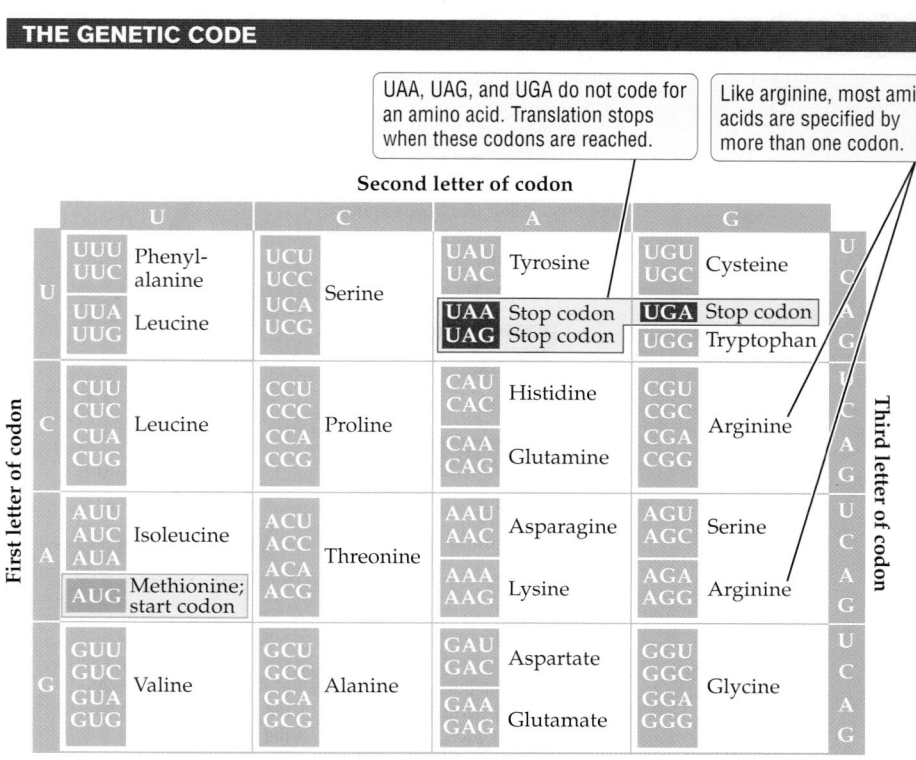

UAA, UAG, and UGA do not code for an amino acid. Translation stops when these codons are reached.

Like arginine, most amino acids are specified by more than one codon.

Second letter of codon

| First letter of codon | U | C | A | G | Third letter of codon |
|---|---|---|---|---|---|
| **U** | UUU UUC Phenyl-alanine / UUA UUG Leucine | UCU UCC UCA UCG Serine | UAU UAC Tyrosine / UAA Stop codon UAG Stop codon | UGU UGC Cysteine / UGA Stop codon UGG Tryptophan | U C A G |
| **C** | CUU CUC CUA CUG Leucine | CCU CCC CCA CCG Proline | CAU CAC Histidine / CAA CAG Glutamine | CGU CGC CGA CGG Arginine | U C A G |
| **A** | AUU AUC Isoleucine / AUA / AUG Methionine; start codon | ACU ACC ACA ACG Threonine | AAU AAC Asparagine / AAA AAG Lysine | AGU AGC Serine / AGA AGG Arginine | U C A G |
| **G** | GUU GUC GUA GUG Valine | GCU GCC GCA GCG Alanine | GAU GAC Aspartate / GAA GAG Glutamate | GGU GGC GGA GGG Glycine | U C A G |

act as signposts that communicate to the ribosomes where they should start or stop reading the mRNA. A few amino acids are specified by only one codon; for example, only one codon (UGG) specifies the amino acid tryptophan. Other amino acids are specified by anywhere from two to six different codons.

When reading the code, the cell begins at a fixed starting point on an mRNA molecule, called a **start codon** (the codon AUG), and ends at one of three **stop codons** (UAA, UAG, or UGA). By beginning at a fixed point, the cell ensures that the message from the gene is read precisely the same way every time. The start codon specifies the amino acid methionine [meh-*THYE*-oh-neen], which is why most proteins inside a cell have this amino acid at their "starting point," the portion of the protein that was translated first.

To examine how the rest of the amino acid–specifying codons are read, consider the example shown in **FIGURE 15.6**, which shows a portion of an mRNA molecule with the base sequence UUCACUCAG. Because the mRNA code is read in sets of three bases, the first codon (UUC) specifies one amino acid (phenylalanine, abbreviated as Phe), the next codon (ACU) specifies a second

amino acid (threonine, abbreviated as Thr), and the third codon (CAG) specifies a third amino acid (glutamine, abbreviated as Gln). The start codon plays a crucial role in establishing which trio of bases is interpreted as a codon by the ribosomal-tRNA machinery. Use Figure 15.5 to determine the amino acid sequence that would result if the sequence UUCACUCAG in Figure 15.6 were read in codons that began with the *second* U, not the first. If the code were read starting with the second U (UCACUC...), we would get a very different protein chain: one containing serine followed by leucine (Ser, Leu, ...), instead of one containing phenylalanine followed by threonine and glutamine (Phe, Thr, Gln, ...). The bases that follow the start codon (AUG) are read consecutively, with each three-base sequence being read as one codon. The start codon therefore establishes the correct order, or *reading frame*, by which the three-letter language of mRNA is translated into protein.

The genetic code has several significant characteristics

- It is *unambiguous*. Each codon specifies only one amino acid.

- It is *redundant*. Several different codons may have the same "meaning"; that is, they may call for the same amino acid. There are four possible bases at each of the three positions of a codon, so there are a total of 64 codons ($4 \times 4 \times 4 = 64$). However, there are only 20 amino acids, so some of these codons specify the same amino acid. For example, six different codons specify the amino acid serine (see Figure 15.5).

- It is virtually *universal*. Nearly all organisms on Earth use the same genetic code, which underscores the common descent of all organisms. The discovery of the genetic code and its near universality revolutionized our understanding of how genes work and helped pave the way for what is now a thriving biotech industry. There are a few minor exceptions to the universality of the genetic code: in certain species some of the 64 codons are read differently than they are in most other species.

**FIGURE 15.6** The Genetic Code Is Read in Sets of Three Bases, Each Set Constituting a Codon

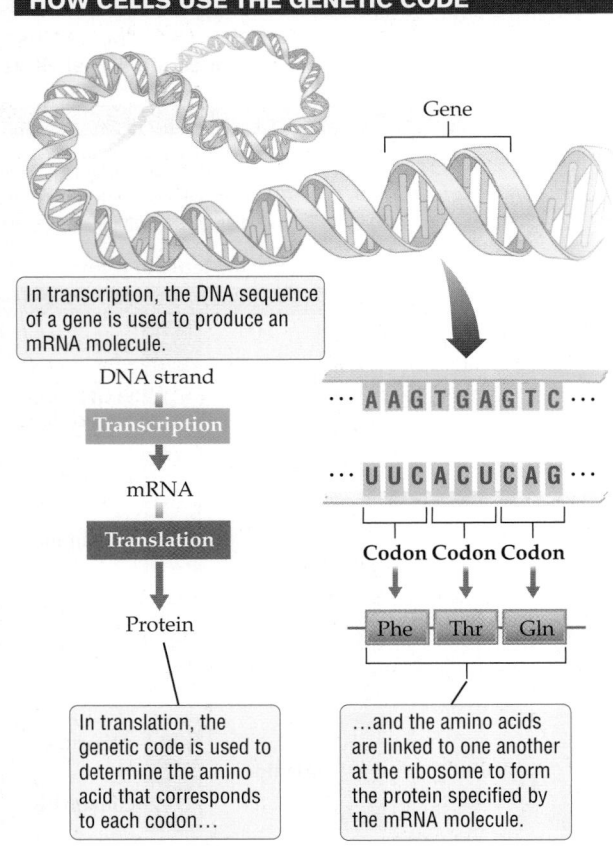

**HOW CELLS USE THE GENETIC CODE**

Gene

In transcription, the DNA sequence of a gene is used to produce an mRNA molecule.

DNA strand

Transcription

mRNA

Translation

Protein

··· A A G T G A G T C ···

··· U U C A C U C A G ···

Codon Codon Codon

Phe | Thr | Gln

In translation, the genetic code is used to determine the amino acid that corresponds to each codon...

...and the amino acids are linked to one another at the ribosome to form the protein specified by the mRNA molecule.

## 15.5 Translation: Information Flow from mRNA to Protein

The genetic code provides the cell with the equivalent of a dictionary for transforming the language of genes into the language of proteins. The conversion of a sequence of bases in mRNA to a sequence of amino acids in a protein is called translation. Translation is the second major step in the process by which genes specify the manufacture of specific proteins (see Figure 15.2). It occurs at ribosomes, which are composed of more than 50 different proteins and several different strands of rRNA (ribosomal RNA). Ribosomes link amino acids in a precise order to make a particular protein.

Another type of RNA, known as tRNA (transfer RNA), also plays a crucial role in the manufacture of proteins (protein synthesis). There are many types of tRNA, but they all have a similar three-dimensional structure, as shown in **FIGURE 15.7**. Each type of tRNA specializes in binding to one specific amino acid, and it recognizes and pairs with specific codons in the mRNA. At one end of a tRNA molecule is a site at which its specific amino acid attaches. In another region, each tRNA molecule has a special sequence of three nitrogenous bases that make up the **anticodon**. The three-base sequence of the anticodon determines which codons on the mRNA this tRNA can recognize through complementary base pairing. For example, the tRNA that binds the amino acid serine will recognize and pair with any AGC codon in an mRNA (see Figure 15.7).

Some tRNAs can recognize more than one codon because the base at the third position of their anticodon can "wobble," meaning it can pair with any one of two or three different bases in the codon. For example, one serine-bearing tRNA can pair with either UCU or UCC in the mRNA, while another serine-bearing tRNA pairs with either UCA or UCG. A third tRNA recognizes the two other serine-specifying codons: AGU and AGC (consult the genetic code in Figure 15.5). This flexibility in the pairing between some anticodons and codons means that a cell does not need 61 different tRNAs, one for each of the 61 amino acid–specifying codons. In fact, most organisms have only about 40 different tRNAs, because many tRNA anticodons can recognize and pair with more than one codon in the mRNA.

For translation to occur, as **FIGURE 15.8** illustrates, an mRNA molecule must first bind to a ribosome. The ribosomal machinery "scans" the mRNA until it finds a start codon, which is the first AUG codon in the mRNA sequence. Then the ribosomes recruit the appropriate tRNAs, as determined by the codons they encounter as they proceed to read the message in the mRNA. With all the necessary components held together in the required three-dimensional orientation, a special site on the ribosome facilitates the linking of one amino acid to another.

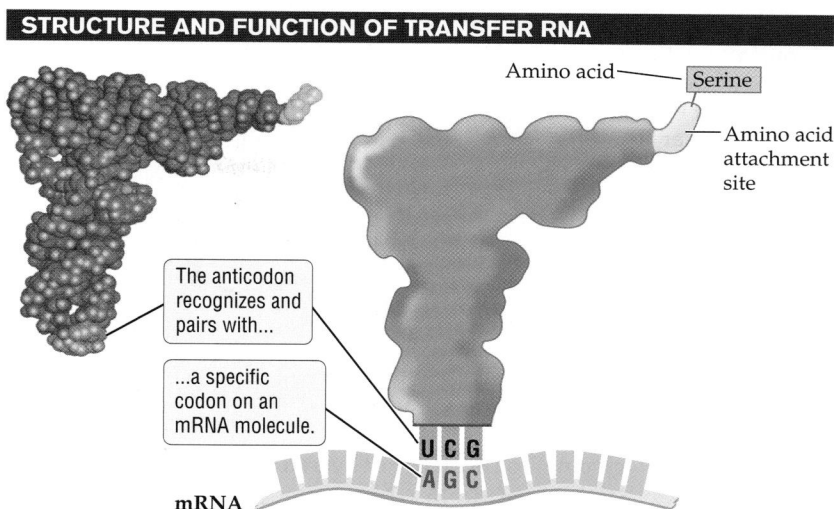

**STRUCTURE AND FUNCTION OF TRANSFER RNA**

Amino acid — Serine

Amino acid attachment site

The anticodon recognizes and pairs with...

...a specific codon on an mRNA molecule.

U C G
A G C

mRNA

**FIGURE 15.7 Transfer RNA Delivers Amino Acids Specified by mRNA Codons**
A space-filling model (left) and a diagrammatic version (right) illustrate the general structure of all tRNA molecules. Similar regions in the space-filling model and on the diagram are shown in matching colors. Each tRNA carries a specific amino acid (serine in this example) and has a specific anticodon sequence (UCG in this example) that binds to a complementary three-base sequence (the codon) in the mRNA.

Translation begins when the tRNA molecule that specializes in carrying the amino acid methionine recognizes the AUG of the start codon and pairs with it. Then other tRNA molecules come into play, one by one bringing to the ribosome the amino acids specified by each codon on the mRNA. In the example in Figure 15.8, after the tRNA carrying methionine binds to the start codon, the next codon to be recognized is GGG, which corresponds to the amino acid glycine. A tRNA molecule that specializes in carrying glycine recognizes the GGG codon and pairs with it through its anticodon. The ribosome now forms a covalent bond between the first amino acid (methionine) and the second amino acid (glycine). When the bond between the first two amino acids is formed, the first tRNA (the one bound to AUG) releases its amino acid (methionine). This tRNA, now freed from the amino acid it had been carrying, is ejected from the mRNA-ribosome complex, and its place is taken by the next tRNA (the one carrying glycine). The ribosome is now ready to read the third codon in the mRNA, which is UCC in our example. The tRNA that carries serine bears the complementary anticodon (AGG) and specifically pairs with the UCC codon. This serine-specific tRNA now occupies the site in the ribosome previously occupied by the glycine codon.

The ribosome links the amino acid chain it has built so far (consisting of just methionine and

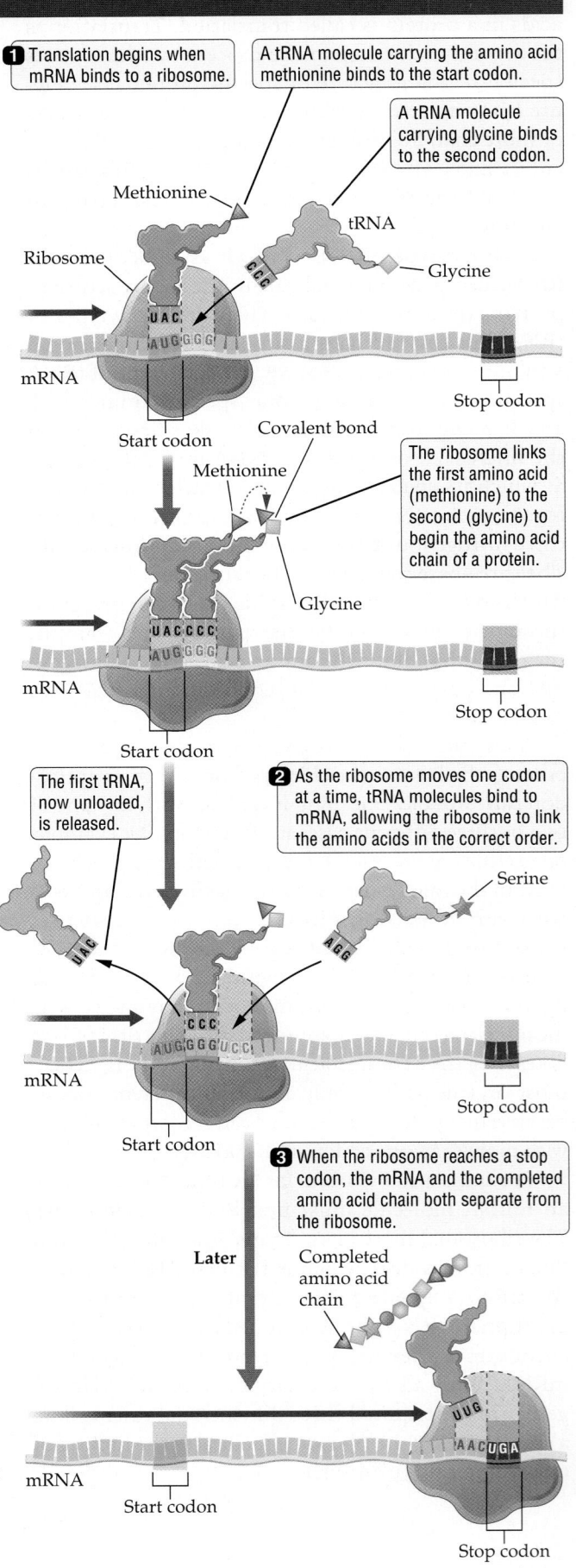

**①** Translation begins when mRNA binds to a ribosome.

A tRNA molecule carrying the amino acid methionine binds to the start codon.

A tRNA molecule carrying glycine binds to the second codon.

The ribosome links the first amino acid (methionine) to the second (glycine) to begin the amino acid chain of a protein.

The first tRNA, now unloaded, is released.

**②** As the ribosome moves one codon at a time, tRNA molecules bind to mRNA, allowing the ribosome to link the amino acids in the correct order.

**③** When the ribosome reaches a stop codon, the mRNA and the completed amino acid chain both separate from the ribosome.

glycine) to the newly delivered amino acid (serine), and the second tRNA (the glycine specialist) is now freed of its cargo and ejected from the ribosome-mRNA complex. This cycle continues: each codon encountered by the ribosome is recognized by a specific tRNA (the one whose anticodon can pair with that codon), and the ribosome adds the amino acid delivered by this tRNA to the growing amino acid chain. Finally, a stop codon is reached. The amino acid chain cannot be extended further, because none of the tRNAs will recognize and pair with any of the three stop codons. At this point the mRNA molecule and the completed amino acid chain both separate from the ribosome. The new protein folds into its compact, specific three-dimensional shape (discussed in Chapter 5).

## 15.6 The Effect of Mutations on Protein Synthesis

A mutation is a change in the base sequence of an organism's DNA. As noted in previous chapters of this unit, mutations range in extent from a change in the identity of a single base pair to the addition or deletion of one or more whole chromosomes. How do mutations affect protein synthesis? In answering this question, we will focus on mutations that occur within exons—the protein-encoding portions of a gene—rather than on mutations that occur in introns or on the large-scale mutations that disrupt entire chromosomes.

### Mutations can alter one or many bases in a gene's DNA sequence

There are three major types of mutations that can alter a gene's DNA sequence: *substitutions*, *insertions*, and *deletions*. For simplicity, we will first describe **point**

**mutations**, those in which a single base is altered. We will then discuss mutations that involve changes in multiple bases.

In a **substitution** mutation, one base is substituted for another in the DNA sequence of the gene. In the example shown in **FIGURE 15.9**, the sequence of the gene is changed when a thymine (T) is replaced by a cytosine (C). As the figure shows, this particular change causes the substitution of one amino acid for another: When the TAA sequence in the DNA is changed to CAA, the mRNA codon is changed from AUU to GUU; GUU is recognized by the valine-carrying tRNA, so a valine is inserted at this position in the protein, instead of an isoleucine (Ile).

**Insertion** or **deletion** mutations occur when a base is inserted into or deleted from a DNA sequence. Single-base insertions and deletions cause a genetic **frameshift**. Consider what happens in a multiple-choice test if you accidentally record the answer to a question twice on a machine-gradable answer sheet: all your answers from that point forward are likely to be wrong, since each is an answer to the previous question. This situation is equivalent to a frameshift caused by the *insertion* of a single base. If you forget to answer a question but record the next question's answer in the overlooked question's space, all your entries from that point on will be scrambled. This situation is equivalent to a frameshift caused by the *deletion* of a base. The insertion or deletion of one or two bases shifts all "downstream" codons by one or two bases. The entire downstream message is scrambled because the ribosomes assemble a very different sequence of amino acids from that point onward, compared to the protein encoded by the original DNA sequence and its corresponding mRNA (see Figure 15.9).

Insertions or deletions involving three bases do not shift the reading frame of an mRNA and therefore do not change the resulting protein as extensively as frameshift mutations do. Insertions and deletions can involve more than a few bases: sometimes thousands of bases may be added or deleted as a result of mutations. Large insertions or deletions almost always result in the synthesis of a protein that cannot function properly. Frameshift mutations, whether caused by point mutations or by large insertions or deletions, also alter the resulting protein so severely that it fails to function in most cases.

## Mutations can cause a change in protein function

Mutations alter the DNA sequence of a gene, which in turn alters the sequence of bases in any mRNA mol-

ecule made from that gene. Such changes can have a wide range of effects on the resulting protein.

A mutation that produces a frameshift usually prevents the protein from functioning properly because it alters the identity of *many* of the amino acids in the protein. Frameshift mutations can also stop protein synthesis before it is complete: if a frameshift converts a codon specifying an amino acid into a stop codon, a full-length version of the protein will not be made. Regardless of whether it causes a frameshift, a mutation that alters an active site in a protein (such as the substrate-binding site of an enzyme) is usually harmful. Such mutations change the way the protein interacts with other molecules, decreasing or destroying its function. Finally, a mutation that inserts or deletes a series of bases causes the protein to have extra or missing amino acids, which can change the protein's shape and hence its function.

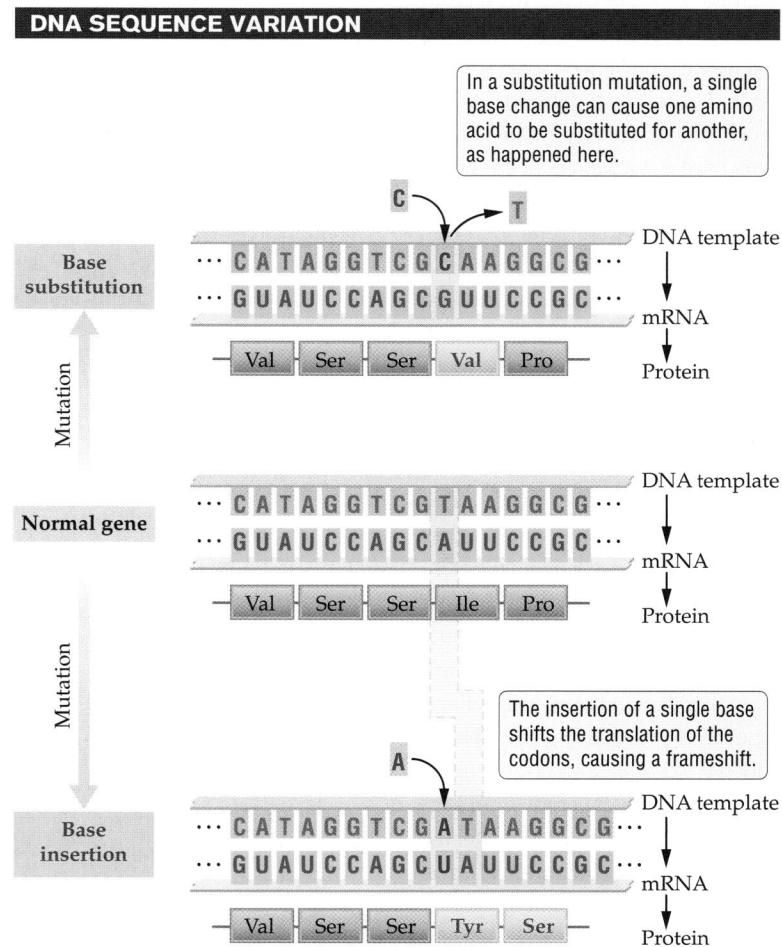

**DNA SEQUENCE VARIATION**

In a substitution mutation, a single base change can cause one amino acid to be substituted for another, as happened here.

**FIGURE 15.9 A Change in the DNA Sequence Translates into a Change in the Amino Acid Sequence of the Protein**
Two kinds of mutations are shown here: a substitution and an insertion. In each case, the mutation and its effects on transcription and translation are shown in red.

# One Allele Makes You Strong, Another Helps You Endure

If you have several thousand dollars to spare, you could have your entire genome sequenced. For a few hundred dollars, many genome science companies will tell you all about your genetic ancestry and also your genotype with respect to more than 50 different genes. Some of these genes may be associated with disease risk; others may be linked to less worrisome genetic traits, such as your attitude toward broccoli and other strong flavors. Welcome to the science of *personal genomics*, whose goal is to inspect and catalog an individual's total genetic makeup, or genome. The hope is that personal genomics will lead to personalized medicine, the practice of tailoring health care and disease prevention to a person's genotype. The field of personal genomics is not, however, without ethical issues. For example, some opponents have expressed concern about the marketing of genomics services directly to the consumer, asserting that personal genomic data that has not been filtered through or delivered by an expert such as a physician may be misunderstood or misconstrued by the consumer, with potentially harmful consequences.

Consider the difficulties in evaluating information you might acquire from a personal-genomics company concerning whether you or your child has a particular gene associated with athletic ability. An Australian company has developed a commercial test for athletic potential, based on the *R* and *X* alleles of the *ACTN3* gene. Skeletal muscles contain bundles of muscle cells, or muscle fibers. There are two main types of fibers: fast-twitch and slow-twitch. Fast-twitch fibers specialize in producing large bursts of power, but they tire quickly. Slow-twitch fibers are more efficient in extracting energy from sugars, and their power output can be sustained much longer. Most of us have roughly equal numbers of the two types of fibers in our skeletal muscles. In contrast, the muscles of elite athletes in strength-based sports, such as sprinting or weight lifting, may be 80 percent fast-twitch fibers. Those excelling in endurance sports, such

Florence Griffith Joyner is a sprinter, a strength athlete.

Paula Radcliffe runs marathons, an endurance sport.

as long-distance running or cycling, can have muscles that are 80 percent slow-twitch fibers.

Alpha-actinin-3, the protein encoded by the *ACTN3* gene, is made only in skeletal muscles. It anchors the contractile proteins so that muscle fibers can generate power. Australian scientists found two alleles of the *ACTN3* gene seemingly linked to athletic ability. The *R* allele codes for a functional alpha-actinin-3 protein; the *X* allele leads to the production of a shortened, nonfunctional version of the protein. The *X* allele of *ACTN3* is the result of a base substitution that changes an amino acid codon into a stop codon—a mutation that halts protein synthesis prematurely.

A study of Australian athletes showed that the *XX* genotype is rare in athletes in strength sports, but is found in about 24 percent of endurance athletes (see the table). The advantage of

the *X* allele in endurance sports is supported by experiments in which genetic engineering technology (see Chapter 16) was used to "knock out" the activity of both copies of the *ACTN3* gene in lab mice, creating "marathon mice."

Success in sports depends on many things, including psychological attributes such as personal drive, and top-level performance always requires extensive training and conditioning. The physical component of athletic success is likely influenced by many genes, not just one or two genes such as *ACTN3*. In one study, in fact, as many as 92 different genes were potentially associated with athletic ability and health-related fitness. This means that the predictive power of the *ACTN3* alleles is limited. Studies found that as many as 31 percent of the elite distance runners lacked the *X* allele, and 45 percent had only one copy.

Given the complexities in interpreting the genetic tests, would you want to know your allelic makeup? Would you want your children tested for the *ACTN3* gene, and if so, how would you use that information? Researchers justify studies of genes influencing physical performance by pointing out their relevance in conditions such as muscular dystrophy and other muscle diseases. Opponents say genetic tests of this sort will lead to abuse by overambitious athletes and pushy parents of would-be sports stars. Does the potential for medical benefit outweigh the risk for abuse?

## Association between Athletic Ability and Prevalence of *ACTN3* Alleles in Strength and Endurance Athletes

| GENOTYPE | PERCENTAGE OF THE GROUP THAT HAS THE GENOTYPE | | |
| --- | --- | --- | --- |
| | CONTROL (NONATHLETES) | STRENGTH-SPORT ELITE ATHLETES (SPRINTERS) | ENDURANCE-SPORT ELITE ATHLETES (DISTANCE RUNNERS) |
| *RR* | 30 | 50 | 31 |
| *RX* | 52 | 45 | 45 |
| *XX* | 18 | 6 | 24 |

Sometimes changing a few bases in a gene's DNA sequence has little or no effect. For example, if a single-base substitution mutation does not alter the amino acids specified by the gene, then the structure and function of the protein will not be changed. Although a change in the DNA sequence from GGG to GGA would alter the mRNA sequence from CCC to CCU, both CCC and CCU code for the same amino acid, proline (see Figure 15.5). In such cases, the substitution mutation is said to be "silent" because it produces no change in the structure (and hence the function) of the protein, and therefore no change in the phenotype of the organism.

In rare instances, a mutation may even be beneficial. Changes to the binding region of a protein, for example, might improve its efficiency or enable it to take on a new and useful function, such as reacting with a new substrate.

## 15.7 Putting It All Together: From Gene to Protein to Phenotype

According to current estimates, humans have approximately 25,000 genes, arranged linearly on 23 different chromosomes. About 21,000 of these genes code for proteins; the rest code for RNA molecules such as tRNA and rRNA. Here we review the major steps in how a cell proceeds from gene to protein to phenotype, focusing on genes that encode proteins. Remember, however, that transcription—the first step in the process leading from gene to protein—is similar in all genes, including the genes that specify tRNA and rRNA molecules. The difference is that the RNA is the final product of these noncoding genes, since tRNA and rRNA are not translated into protein.

Each gene is composed of a segment of DNA on a chromosome and consists of a sequence of the four bases adenine (A), cytosine (C), guanine (G), and thymine (T). The particular sequence of bases in the DNA of a protein-coding gene specifies the amino acid sequence of the gene's protein product.

Transcription and translation are the two major steps in the synthesis of a protein from the information in its corresponding gene (see Figure 15.2). In transcription, the sequence of bases in a gene is used as a template to produce mRNA molecules. The cell then transports the mRNA out of the nucleus to a ribosome in the cytoplasm, where translation occurs. In translation, the sequence of bases in each mRNA molecule is used as a template to synthesize the gene's protein product by stringing together the correct sequence of amino acids.

The proteins encoded by genes are essential to life. A mutation in a gene can alter the sequence of amino acids in the gene's protein product, and this change can disable or otherwise alter the function of the protein. When a critical protein is disabled, the entire organism may be harmed. In people with the genetic disorder sickle-cell anemia, for example, a single base in the gene that encodes hemoglobin is altered (**FIGURE 15.10**). Hemoglobin is a protein involved in the transport of oxygen by red blood cells. The red blood cells of people with sickle-cell anemia become curved and distorted under low-oxygen conditions, such as those found in narrow blood vessels like our capillaries. The distorted red blood cells clog these narrow blood vessels, thereby

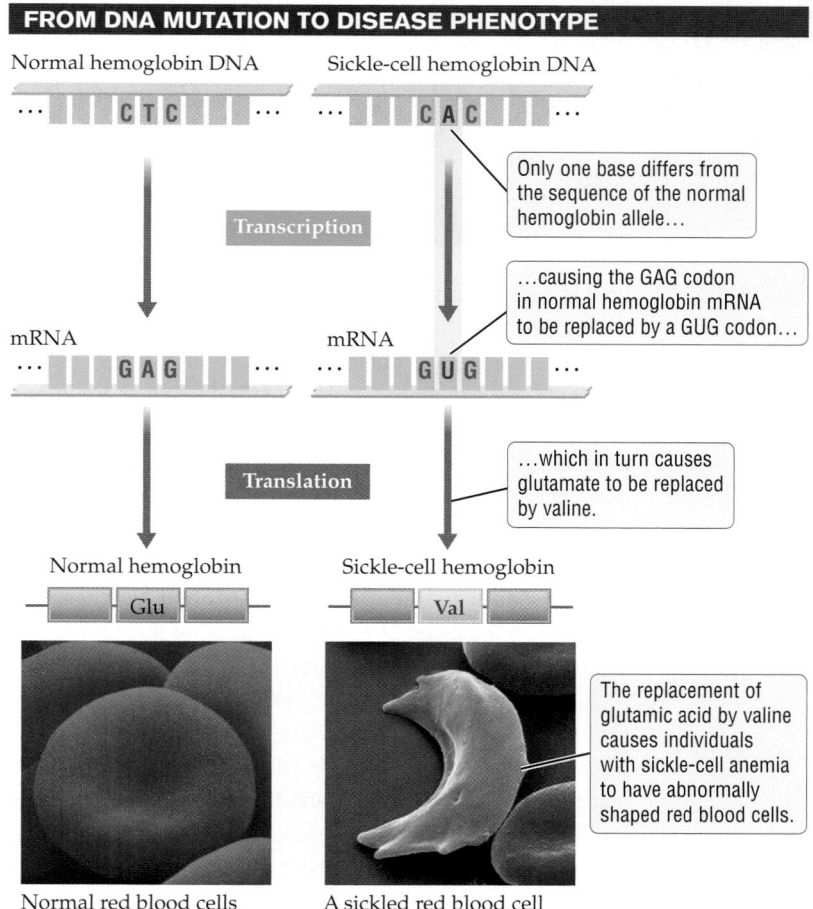

**FROM DNA MUTATION TO DISEASE PHENOTYPE**

Normal hemoglobin DNA

Sickle-cell hemoglobin DNA

··· C T C ···

··· C A C ···

Only one base differs from the sequence of the normal hemoglobin allele…

Transcription

mRNA

mRNA

…causing the GAG codon in normal hemoglobin mRNA to be replaced by a GUG codon…

··· G A G ···

··· G U G ···

Translation

…which in turn causes glutamate to be replaced by valine.

Normal hemoglobin

Sickle-cell hemoglobin

Glu

Val

Normal red blood cells

A sickled red blood cell

The replacement of glutamic acid by valine causes individuals with sickle-cell anemia to have abnormally shaped red blood cells.

**FIGURE 15.10  Mutations in a Single Base in the Hemoglobin Gene Can Lead to Sickle-Cell Anemia**

leading to a wide range of serious effects, including heart and kidney failure. If they receive little or no medical care, people with sickle-cell anemia usually die before they reach childbearing age. With intensive medical care, however, they can now live well into middle age and beyond.

## APPLYING WHAT WE LEARNED

# CSI: Copper Age

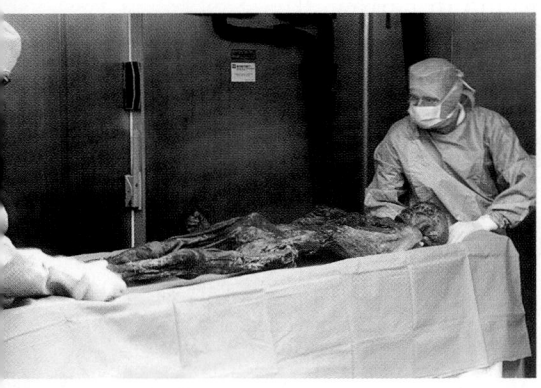

Soon after the Iceman was discovered, skeptics suggested that the frozen man was a hoax, a mummy transported from Egypt or Central America. A wealth of forensic information from his body and clothes offered compelling clues about his life and the manner of his death. But in the end, it was his DNA that confirmed his identity—a genuine Copper Age European. Ötzi—as he was called, because he was found in the Ötzal Alps—shared distinctive sequences of DNA with modern Europeans.

Blood stained an arrow in Ötzi's quiver, his knife, and his cloak. Researchers extracted DNA from long-frozen cells in the blood and sequenced regions of the DNA that are known to vary from person to person. By comparing these "highly variable" sequences of bases, biologists can distinguish the blood of one person from another—a technique known as DNA fingerprinting (see Chapter 16). An unpublished report claimed that one of Ötzi's arrows had blood from two different people, his knife had blood from a third person, and his cloak was stained with the blood of a fourth, suggesting that Ötzi died in a violent skirmish.

Ötzi's pure-copper axe, high levels of copper and arsenic in his hair, and blackened lungs were all clues that he may have been a copper smelter, an early adopter of Copper Age technology. Isotopes in his bones and teeth showed that he grew up 60 miles from the 10,000-foot-high pass where he died. Frozen in his gut was a last meal containing an ancient kind of wheat. On his left hand was a deep, days-old gash. He had a recent head injury, deep bruises on his side, a slash on his right forearm, and an arrowhead embedded in his left shoulder. It

was the arrow that likely killed him, slicing an artery, which led to massive blood loss and almost instant death. Within hours, a spring snowstorm buried him in snow, leaving him frozen in a glacier for 5,200 years.

To find out more, researchers turned once again to Ötzi's DNA. Most protein-coding genes are "conserved," meaning that they are roughly the same in everyone. Conserved genes don't accumulate many changes over time, because changes in nucleotide sequence alter the protein that is expressed, which, in turn, may drastically affect health. On average, the recipients of such mutated protein-coding genes are less likely to have children, so the altered DNA sequences gradually die out. This is the essence of natural selection, a topic we will explore in later chapters.

But not all sequences of DNA are conserved. Some vary enormously, and the variations in sequence are passed on to children. Over thousands of years, highly variable regions of DNA accumulate many mutations. In 2008, geneticists working on Ötzi's DNA discovered that he had a variable sequence, called K, that is about 12,000 years old. K is found in peoples of North Africa, South Asia, and Europe.

Imagine that all your relatives learned a special code word when they were little. Even if you were meeting a fourth or fifth cousin for the first time, you'd be able to recognize that person as a relative because you would both know the code word. In the same way, highly variable sequences of nucleotides mark people as descendants of a particular family of humans. Generation after generation, they pass these special sequences of nucleotides on to their children. K is carried by about 6 percent of ethnic Europeans. But K has continued to mutate, so the European K group includes three subgroups whose nucleotides mutated more recently. Ötzi belongs to a fourth genetic group, which is either extinct or extremely rare.

# Atrazine Hurts Animals

BY LINDSAY PETERSON, *Tampa Tribune*

The weed killer atrazine, widely used on Florida lawns and golf courses, interferes with the normal growth and development of amphibians and freshwater fish, says an analysis by researchers at the University of South Florida.

USF assistant professor Jason Rorh and postdoctoral fellow Krista McCoy analyzed more than 100 studies of atrazine, an herbicide banned in Europe in 2004 but still used in the United States and 80 other countries.

The studies showed consistent trends, Rohr said. "Perhaps the most striking were the highly consistent reductions in animal growth and changes in animal behavior associated with atrazine," Rohr said.

In a paper published last week in the peer-reviewed journal Environmental Health Perspectives, Rohr and McCoy wrote that atrazine didn't kill fish or amphibians but had significant "indirect and sub-lethal effects."

In study after study, they found evidence that the weed killer:

- Reduced the size of amphibians at or near metamorphosis.
- Impaired the ability of fish to smell and the ability of fish and amphibians to evade predators.
- Impaired the amphibians' ability to fight infection.
- Altered the reproductive development of male frogs.

Rohr said that for years, researchers have been looking at how atrazine affects the endocrine system of frogs, but he was surprised at the number of studies that showed other effects.

"The thing that was most alarming were the effects on disease risk and growth," he said.

There is more to learn about atrazine, Rohr and McCoy wrote in their paper, but the research shows that the herbicide's effects should be weighed against its benefits and the costs of using alternatives.

Atrazine prohibits photosynthesis in weeds after being absorbed. It is applied by ground and aerial spraying. One of the major manufacturers of atrazine is Switzerland-based Syngenta. Spokeswoman Sherry Ford said that Syngenta stands behind the product's safety.

The U.S. Environmental Protection Agency monitors studies on atrazine, some involving the risk atrazine poses to life in stream systems. A recent EPA review concluded atrazine does not inhibit the male reproductive system in amphibians.

Atrazine is an ingredient in products that have been marketed under the brand names Marksman, Coyote, Atrazina, Atrazol and Vectal.

---

Atrazine is one of the most widely used herbicides in the world. How does this weed killer alter development in frogs? The answer is related to transcription, the topic of this chapter. Long thought harmless to animals, atrazine is so widespread that it falls in rain on fields and cities, reaching levels as high as 40 parts per billion (ppb) in some corn-growing regions. After a heavy rain, creeks and ditches that receive runoff from atrazine-treated agricultural fields may have concentrations of atrazine that are 100 times normal.

Atrazine can profoundly affect development by interfering with the transcription of genes. Amphibians are exquisitely sensitive to atrazine. For example, among male tadpoles exposed to minute amounts of atrazine, a small percentage develop into female frogs that produce eggs. Of the exposed males, even those that don't become female have reduced testosterone levels, and impaired fertility and sex drive. These problems interfere with the frogs' ability to mate and reproduce. Federal drinking-water standards, which specify the cleanest water, set maximum atrazine concentrations at 3 ppb. Yet some male tadpoles in water with only 2.5 ppb atrazine become feminized frogs. At higher concentrations of atrazine, *all* the male frogs may be feminized.

How does this happen? Recall that for a gene to be expressed, it must be transcribed into messenger RNA. Regulatory proteins called *transcription factors* determine the number of copies of a gene to be transcribed, increasing or decreasing the amount of protein a cell makes.

In tadpoles, a transcription factor called DM-W stimulates the expression of the gene for an enzyme called aromatase, which helps cells make estrogen. Estrogen is a "female" hormone; if male tadpoles are exposed to enough estrogen, they develop into female frogs. When tadpoles are exposed to atrazine, expression of the DM-W transcription factor increases, causing cells to make more aromatase, which makes more estrogen.

It isn't only frogs that are affected by atrazine. In human cells grown in a laboratory, atrazine increases the expression of a transcription factor called SF-1, which activates a region of DNA called a promoter. This particular promoter is near the gene for the enzyme aromatase; the promoter helps RNA polymerase bind to the aromatase gene and transcribe it. In short, atrazine causes human cells to make more aromatase and, therefore, more estrogen.

## Evaluating the News

**1.** Explain what a gene promoter does. Specifically, what does the aromatase promoter do in human cells?

**2.** If atrazine impairs the ability of fish and amphibians to breed, fight infections, and escape predators, what overall effect could atrazine have on populations of these animals?

**SOURCE:** *Tampa Tribune*, June 30, 2010.

# Summary

## 15.1 How Genes Work

- Genes code for RNA molecules. Protein-coding genes code for mRNA, which contains instructions for building a protein.
- An RNA molecule consists of a single strand of nucleotides. Each nucleotide is composed of the sugar ribose, a phosphate group, and one of four nitrogen-containing bases. The bases found in RNA are the same as those in DNA, except that uracil (U) replaces thymine (T).
- At least three types of RNA (mRNA, rRNA, and tRNA) and many enzymes and other proteins participate in the manufacture of proteins.

## 15.2 How Genes Guide the Manufacture of Proteins

- In both prokaryotes and eukaryotes, protein synthesis takes place in two steps: transcription and translation. In transcription, an RNA molecule is made using the DNA sequence of the gene. In translation, ribosomes, mRNA, and tRNA molecules together direct protein synthesis. Ribosomes consist of rRNA and certain proteins.
- In eukaryotes, information for the synthesis of a protein is transmitted from the gene (located in the nucleus) to the site of protein synthesis, the ribosomes (located in the cytoplasm).

## 15.3 Transcription: Information Flow from DNA to RNA

- During transcription, one strand of the gene's DNA serves as a template for synthesizing many copies of mRNA.
- The key enzyme in transcription is RNA polymerase.
- Each gene has a sequence (a promoter) at which RNA polymerase begins transcription. Transcription stops after RNA polymerase encounters special transcription termination sequences.
- The mRNA molecule is constructed according to specific base-pairing rules: A, U, C, and G in mRNA pair with T, A, G, and C, respectively, in the template strand of DNA.
- Newly made, or "preliminary," eukaryotic mRNA must be processed while it is still in the nucleus to (among other things) splice out noncoding sequences of DNA (introns) found in many genes. The remaining, protein-encoding segments of mRNA (exons) are then joined in a "mature" mRNA molecule that is exported to the ribosomes for translation into protein.

## 15.4 The Genetic Code

- The information encoded by an mRNA is read in sets of three bases; each three-base sequence is a codon. Of the 64 possible codons, most specify a particular amino acid, but certain codons signal the start or stop of translation. The information specified by each codon is collectively called the genetic code.
- When reading the genetic code, ribosomes begin at a fixed starting point on the mRNA (the start codon) and stop reading the code when they encounter any one of the three stop codons.
- The genetic code is unambiguous (each codon specifies no more than one amino acid), redundant (several codons specify the same amino acid), and nearly universal (used in almost all organisms on Earth).

## 15.5 Translation: Information Flow from mRNA to Protein

- The codon sequence in each mRNA molecule determines the amino acid sequence of the protein it encodes.
- Translation occurs at ribosomes, which are composed of rRNA and more than 50 different proteins.
- Translation calls into play transfer RNA molecules that each specialize in carrying a particular amino acid. Each tRNA has a three-base sequence, called the anticodon, that recognizes and pairs with a specific codon in the mRNA by complementary base pairing. In this way, a specific tRNA molecule delivers a specific amino acid based on the codon message present in the mRNA.
- The ribosome holds the mRNA and tRNA in a manner that allows the amino acid carried by the tRNA to be covalently bonded to the growing amino acid chain. When translation is complete, the amino acid chain folds into the three-dimensional shape of the protein.

## 15.6 The Effect of Mutations on Protein Synthesis

- Many mutations are caused by the substitution, insertion, or deletion of a single base in a gene's DNA sequence.
- Insertion or deletion of a single base causes a genetic frameshift, resulting in a different sequence of amino acids in the gene's protein product. Mutations causing frameshifts usually destroy the protein's function.
- Some mutations have neutral effects, and a few even have beneficial effects.

## 15.7 Putting It All Together: From Gene to Protein to Phenotype

- Some genes code for RNA only, but a large number encode proteins.
- Proteins are essential to life. In conjunction with the environment, they determine an organism's phenotype.

# Key Terms

anticodon (p. 353)
codon (p. 351)
deletion (p. 355)
exon (p. 350)
frameshift (p. 355)
gene (p. 346)

genetic code (p. 351)
insertion (p. 355)
intron (p. 349)
messenger RNA (mRNA) (p. 347)
point mutation (p. 354)
promoter (p. 349)

ribosomal RNA (rRNA) (p. 347)
RNA polymerase (p. 349)
RNA splicing (p. 350)
start codon (p. 352)
stop codon (p. 352)
substitution (p. 355)

template strand (p. 349)
terminator (p. 349)
transcription (p. 347)
transfer RNA (tRNA) (p. 347)
translation (p. 347)

# Self-Quiz

1. For a protein-coding gene, what molecule carries information from the gene to the ribosome?
   a. DNA
   b. mRNA
   c. tRNA
   d. rRNA

2. During translation, each amino acid in the growing protein chain is specified by how many nitrogenous bases in mRNA?
   a. one
   b. two
   c. three
   d. four

3. Which molecule(s) deliver the amino acid specified by a codon to the ribosome?
   a. rRNA
   b. tRNA
   c. anticodons
   d. DNA

4. During transcription, which molecule(s) are produced?
   a. mRNA
   b. rRNA
   c. tRNA
   d. all of the above

5. A portion of the template strand of a gene has the base sequence CGGATAGGGTAT. What is the sequence of amino acids specified by this DNA sequence? (Use the information in Figure 15.5 and assume that the corresponding mRNA sequence will be read from left to right.)
   a. alanine, tyrosine, proline, isoleucine
   b. arginine, tyrosine, tryptophan, isoleucine
   c. arginine, isoleucine, glycine, tyrosine
   d. none of the above

6. Which of the following is responsible for creating the covalent bonds that link the amino acids of a protein in the order specified by the gene that encodes that protein?
   a. tRNA
   b. mRNA
   c. rRNA
   d. ribosome

7. In a single mutation, the fourth, fifth, and sixth bases are deleted from a gene that encodes a protein with 57 amino acids. Which of the following would be expected to happen?
   a. The resulting frameshift would prevent protein synthesis.
   b. A protein with 56 amino acids would be constructed.
   c. A protein differing from the original one—but still with 57 amino acids—would be constructed.
   d. A protein with 54 amino acids would be constructed.

8. Most eukaryotic genes contain one or more segments that are transcribed but not translated. Each such segment is known as
   a. a start codon.
   b. a promoter.
   c. an intron.
   d. an exon.

# Analysis and Application

1. What is a gene? In general terms, how does a gene store the information it contains?

2. Discuss the different products specified by genes. What are the function(s) of each of these products?

3. Describe the flow of genetic information from gene to phenotype.

4. How is the information contained in a gene transferred to another molecule? Why must the molecule that "carries" the information stored in the gene be transported out of the nucleus?

5. What is RNA splicing? Does it occur in both eukaryotes and prokaryotes? Explain your answer.

6. Describe the roles played by rRNA, tRNA, and mRNA in translation.

7. In a gene encoding a tRNA molecule, scientists have discovered a mutation that appears to be responsible for a series of human metabolic disorders. The mutation occurred at a base located immediately next to the anticodon of the tRNA—a change that destabilized the ability of the tRNA anticodon to bind to the correct mRNA codon. Why might a single-base mutation of this nature result in a series of metabolic disorders?

8. Write a paragraph explaining to someone with little background in biology what a mutation is and why mutations can affect protein function.

# 16

# DNA Technology

THE EDUNIA. This transgenic petunia, part plant and part human, carries a gene for a human antibody extracted from the cells of artist Eduardo Kac. Kac created the "Edunia" to express some of his ideas about biotechnology.

# Eduardo Kac's "Plantimal"

Brazilian artist Eduardo Kac is fascinated by biotechnology. Kac is well known for his role publicizing a white rabbit named Alba whose cells expressed a jellyfish gene that made the rabbit glow a fluorescent green. In 2008, Kac introduced a new work of biotechnological art that dramatically expresses his thoughts about combining genes from different organisms. Kac planned and created a garden petunia that carries a gene from his own body.

The "Edunia" is a brilliant-pink petunia whose petals express a gene from Kac's own genome. The petunia's ability to express a human gene illustrates the shared heritage of plants and animals. Because plants and animals evolved from a common ancestor that lived 1.6 billion years ago, all of them use the same cellular machinery to transcribe and translate genes into proteins. Not only can the Edunia express Kac's gene, but the plant also passes that human gene to its offspring in the form of seeds. In other words, the human gene is a permanent addition to the petunia's genome.

Kac chose a gene that would specifically illustrate another idea. The gene he selected was for an antibody from his immune system. Antibodies protect us from infections by recognizing and disarming bacteria, viruses, and other foreign entities. As Kac writes on his website, "The gene I selected is responsible for the identification of foreign bodies. In this work, it is precisely that which identifies and rejects the other that I integrate into the other, thus creating a new kind of self that is partially flower and partially human."

How can biologists move DNA from one organism to another? What are the implications of this kind of transgenic technology for medicine?

In this chapter we will study how DNA can be manipulated using some of the tools and techniques of DNA technology. We will explore how biotechnology has advanced our understanding of genetics and generated medical treatments and commercial products, not to mention artistic ones. At the end of the chapter we'll also learn how Eduardo Kac created his "plantimal" Edunia.

## MAIN MESSAGE

DNA technology makes it possible to isolate a gene and produce many copies of it, and then to introduce it into whole organisms.

## KEY CONCEPTS

- "DNA technology" refers to the process of modifying and manipulating DNA and the use of such techniques for research, health, forensic, and commercial purposes.

- We clone genes by isolating them, joining them with other pieces of DNA, and introducing such recombinant DNA into a host cell, such as a bacterial cell, that can generate many copies of the genes.

- DNA fingerprinting produces an individual's genetic profile and is widely used to establish the identity of victims, criminals, and others.

- The goal of reproductive cloning is to generate offspring that are a genetic copy of a selected individual.

- Genetic engineering is the introduction of foreign DNA into an organism such that the introduced DNA becomes incorporated into the genetic material of the recipient. Expression of the introduced gene usually changes one or more phenotypes in the recipient, which is considered a genetically modified organism (GMO).

- Automated DNA sequencing machines can read the DNA code with great speed.

- The polymerase chain reaction (PCR) is a revolutionary technique that can make billions of copies from a few molecules of DNA.

- DNA technology provides many benefits, but its use also raises ethical concerns and poses potential risks to human society and natural ecosystems.

**INDIRECTLY, PEOPLE HAVE BEEN MANIPULAT-ING THE GENETIC MATERIAL** of other organisms for thousands of years. This fact is well illustrated by the many differences between domesticated species and their wild ancestors. For example, because of genetic changes brought about through selective breeding, dog breeds often look very different from one another and from their wild ancestor, the wolf. Similarly, food plants such as corn bear little resemblance to the wild species from which they arose.

Although we have a long history of manipulating the genetic characteristics of other organisms through selective breeding (**FIGURE 16.1a**), the past 40 years have witnessed a huge increase in the power, precision, and speed with which we can alter the genes of organisms ranging from bacteria to mammals. We can now select a particular gene, produce many copies of it, and then transfer it into a living organism of our choice. In doing so, we can alter DNA directly and rapidly in ways that do not happen naturally (**FIGURE 16.1b**).

We begin this chapter with a broad overview of DNA technology. We then turn to some of the many practical applications of this technology, including DNA fingerprinting and genetic engineering. We also consider some of the ethical issues and social and environmental risks associated with DNA technology. We close with a peek into the DNA technologist's toolkit to help you better understand the techniques and procedures used to manipulate DNA and to create genetically modified organisms.

**FIGURE 16.1**
**Traditional Breeding versus Direct DNA Manipulation**

When it comes to strange appearances, conventional breeding methods and DNA technology can both generate some extraordinary phenotypes. (*a*) This British Belgian Blue bull was developed through conventional selective breeding practices for heavily muscled, low-fat meat. (*b*) These "naked" chickens were genetically altered using the tools of DNA technology. The goal of the research project was to develop low-fat poultry that would be cheap and environmentally friendly, in part because there would be no feathers to remove and dispose of.

(*a*)

(*b*)

## 16.1 The Brave New World of DNA Technology

DNA technology has produced dramatic advances in biology, and the wide-ranging applications of this technology touch many aspects of our daily life today, from the foods we eat to the diagnosis we might hear in a doctor's office. DNA technology began in the 1970s, as chemists and biologists began to decipher the way DNA is modified and replicated within cells, especially by bacteria and viruses infecting their host cells. Scientists could then use nature's own toolkit—a host of DNA-modifying enzymes and other proteins—to read the DNA code, to alter it at will, and even to create it from scratch in a test tube.

### Revolutionary techniques are the foundation of DNA technology

The strategies and techniques that scientists use to analyze and manipulate DNA are broadly known as **DNA technology**. DNA is a polymer of nucleotides and it serves as the genetic material in all organisms. Although the nucleotide (base) sequence of DNA can vary greatly between species, as well as among individuals of a species, the general chemical structure of the

DNA molecule (see Figure 14.3) is the same in all species. This consistency means that similar laboratory techniques can be used to isolate and analyze DNA from organisms as different as bacteria and people.

DNA can be readily extracted from most cells and tissues. Extracted DNA can be cut into smaller fragments with certain enzymes (called restriction enzymes), separated in a gel-like material, stained with certain dyes, and viewed under special lights. A revolutionary technique known as the polymerase chain reaction (PCR) makes it possible to amplify a target DNA so that billions of copies can be generated from a single molecule of the desired DNA. The coding information in a piece of DNA—its nucleotide sequence—can be determined by robotlike machines called DNA sequencers. Such machines have made it possible to decipher *all* the information held in the chromosomes (**genomes**) of many species of bacteria, fungi, plants, and animals. In the year 2000, teams of scientists from across the globe published the nucleotide sequence of almost the entire human genome, as part of the Human Genome Project. Since then, the genomes of more than a thousand prokaryotes and a few hundred eukaryotes have been completely sequenced. This nucleotide sequence information is available in public databases that anyone may inspect and analyze through the Internet.

Fragments of DNA can be joined together with the help of special enzymes, creating an artificial assembly of genetic material known as **recombinant DNA** (**FIGURE 16.2**). Many bacteria, for example, have closed

## GENETIC ENGINEERING: RECOMBINANT DNA AND DNA CLONING

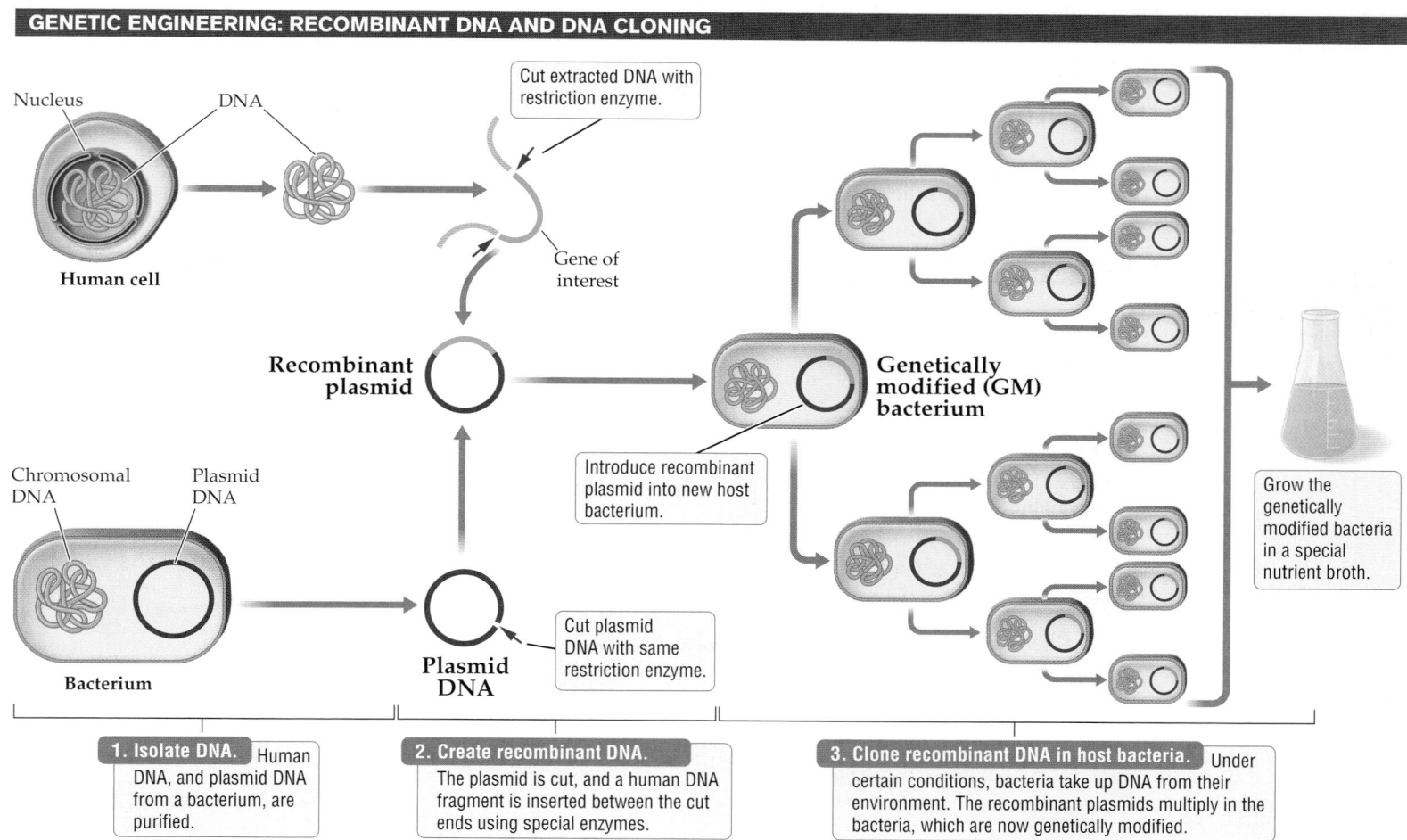

**1. Isolate DNA.** Human DNA, and plasmid DNA from a bacterium, are purified.

**2. Create recombinant DNA.** The plasmid is cut, and a human DNA fragment is inserted between the cut ends using special enzymes.

**3. Clone recombinant DNA in host bacteria.** Under certain conditions, bacteria take up DNA from their environment. The recombinant plasmids multiply in the bacteria, which are now genetically modified.

**FIGURE 16.2 Recombinant DNA Can Be Propagated in Bacteria**
This diagram illustrates the general scheme for genetic engineering, using the introduction of a human gene into a bacterium as an example. Once a recombinant plasmid has been created, it can be introduced into a desired strain of bacterium in a number of different ways. The bacterium makes many copies of the recombinant plasmid, which are inherited by the daughter cells every time the bacterium divides. DNA cloning is the propagation of foreign DNA in a suitable host, such as a bacterium.

loops of DNA, called **plasmids**, in addition to the single large chromosome typical of prokaryotic cells; it is a relatively simple matter to extract these plasmids from the bacterium and insert a foreign gene into them to create recombinant DNA molecules. **DNA cloning**, or gene cloning, is the introduction of recombinant DNA into a host cell, followed by the copying and propagation of the introduced DNA in the host cells and all its offspring. Gene cloning has many commercial and medical uses because it can be used to generate proteins of value on a large scale (**FIGURE 16.3**). DNA cloning is also used in research to multiply a particular type of recombinant DNA so that a large amount of this DNA is made available for further analysis and manipulation. Bacteria are the most common host cells in DNA cloning. Introduced recombinant plasmids can multiply rapidly, creating hundreds of copies of the desired DNA within the cytoplasm of a host bacterium.

## DNA technology has transformed our world

In the past three decades, DNA technology has helped convict hundreds of killers and has exonerated over 200 people who were wrongfully imprisoned before DNA-based forensic technology became widely available. Those who like to explore genealogies as a hobby can buy DNA kits that will let them compare their DNA with that of an obliging stranger in another city who happens to have the same last name. For a few hundred dollars, some companies will create a DNA profile of you: they will list your genotype for hundreds of genetic loci, including alleles associated with such traits as having a good verbal memory, or perfect pitch, or a tendency to take risks.

DNA technology has enabled us to identify many critical genes, and it has enhanced our understanding of gene function in diverse cell types, in a great range of organisms. Hundreds of genetic disorders and infectious diseases in humans can now be diagnosed through genetic tests (see the box in Chapter 14, page 331). *DNA microarrays*, informally known as gene chips, are a relatively recent innovation in DNA technology that have made it possible to understand which sets of genes are expressed ("turned on") in specific cell types under certain conditions, and how the normal pattern might be altered when a person is sick. Rather than relying on a one-size-fits-

**FIGURE 16.3**
**Cloned DNA Serves Many Purposes**

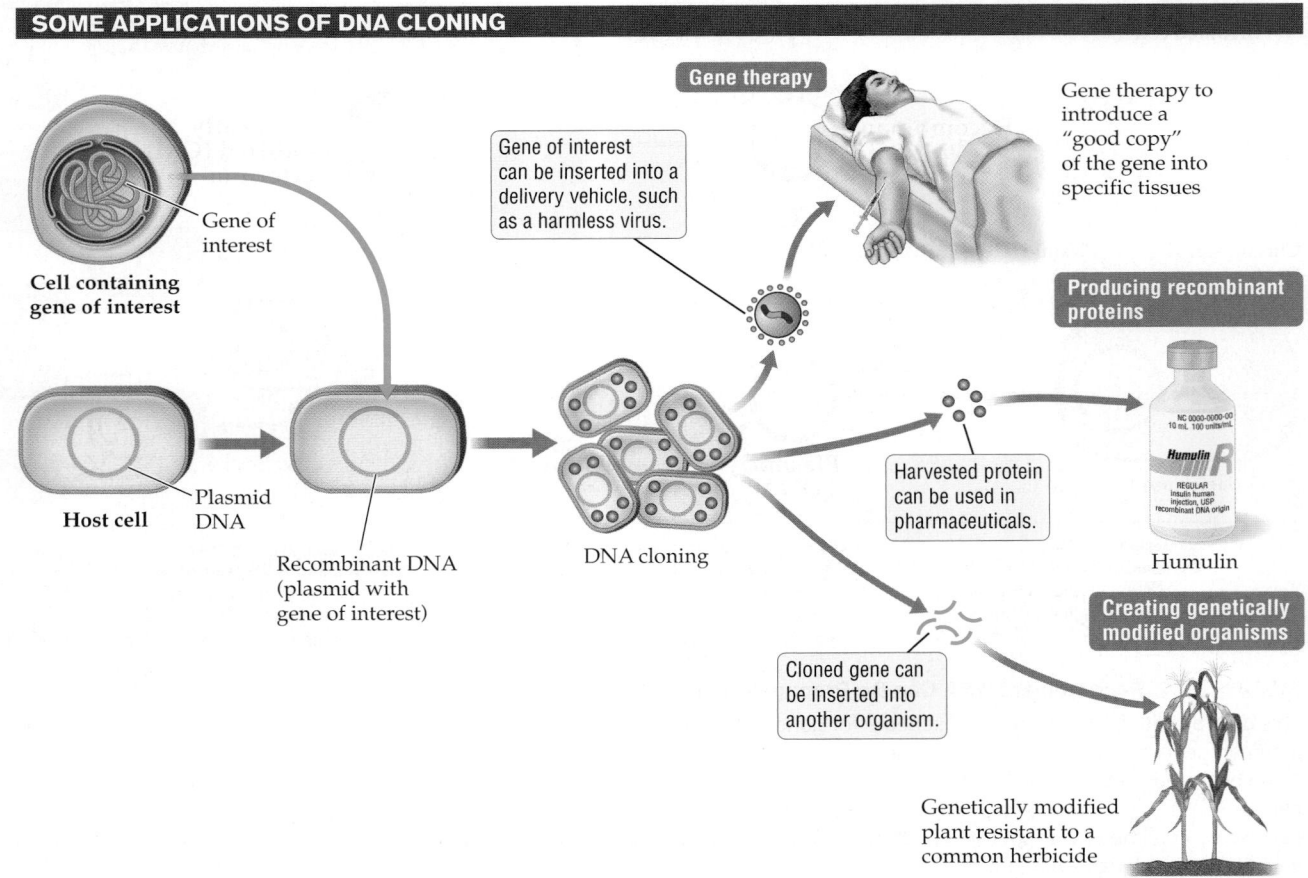

SOME APPLICATIONS OF DNA CLONING

Cell containing gene of interest

Gene of interest

Host cell

Plasmid DNA

Recombinant DNA (plasmid with gene of interest)

DNA cloning

Gene therapy

Gene of interest can be inserted into a delivery vehicle, such as a harmless virus.

Gene therapy to introduce a "good copy" of the gene into specific tissues

Producing recombinant proteins

Harvested protein can be used in pharmaceuticals.

Humulin

Cloned gene can be inserted into another organism.

Creating genetically modified organisms

Genetically modified plant resistant to a common herbicide

all approach to treatment, such techniques make it possible to determine exactly which genes are behaving in destructive ways in a certain patient and which drug will be most effective in correcting the problem. *Personalized medicine*—the practice of tailoring treatments and therapies to a patient's DNA profile—has been launched in a limited way and is likely to become quite common in the years ahead.

**Genetic engineering** is the permanent introduction of one or more genes into a cell, a certain tissue, or a whole organism, leading to a change in at least one genetic characteristic in the recipient. The organism receiving the DNA is said to be a **genetically modified organism** (**GMO**), sometimes known as a genetically engineered organism (GEO). Many lifesaving pharmaceuticals, such as human insulin, human growth hormone, human blood-clotting proteins, and anticancer drugs, are manufactured by genetically modified (GM) bacteria or GM mammalian cells grown in laboratory dishes. Most of the corn, soybean, canola, and cottonseed products consumed in North America are likely to be products of genetic engineering. At the very forefront of DNA technology, and still in its infancy, is the strategy of curing human genetic disorders through gene therapy. **Gene therapy** is the use of genetic engineering techniques to alter the characteristics of specific tissues and organs in the human body, with the goal of treating serious genetic disorders or diseases.

DNA technology goes beyond genetic testing and genetic engineering. The ability to analyze DNA from different species, and from individuals in populations of the same species, has helped us understand biological processes at all levels in the biological hierarchy. DNA analysis has played a large role in helping us reconstruct the history of life on Earth (see Chapter 2), and it has revealed, perhaps more clearly than any other approach, the common ancestry and the genetic diversity of life. Researchers have used DNA from fossil material and from diverse groups of modern-day humans to conclude that anatomically modern humans evolved in southeastern Africa and spread out from this one area of Africa some 50,000 years ago, displacing other early humans (such as Neandertals) as they proceeded to colonize the world. Anthropologists suggest that some Neandertals may have been redheads, because DNA from the fossil bones of these human relatives contains an allele that is responsible for red hair and pale skin in modern humans. A geneticist in Spain extracted DNA from the bones of Christopher Columbus and is attempting to compare it to DNA volunteered by thousands of people, in-

cluding many named Colom or Colombos, who hope that genetic analysis will show them to be related to the Great Navigator.

Today the classification of organisms into related groups (such as genus and family) relies heavily on a comparison of DNA sequences. Wildlife biologists use DNA analysis to measure genetic diversity in populations of endangered species, and they use that information to design conservation strategies. As another example of the use of DNA technology in understanding wild animals and how they live, a team of researchers concluded that beaked whales do not live on only squid as was widely believed. These elusive marine mammals eat certain species of bony fishes as well, since DNA from the fish could be identified in the well-digested meals recovered from the whales' guts. The uses of DNA technology go beyond biology, medicine, and law enforcement, extending to history, art, and even sport. To certify the authenticity of sports memorabilia, all of the balls used in Super Bowl XXXIV (January 2000) were marked with a unique DNA fragment that can be "read" with a special laser-based detector.

> **Concept Check**
>
> 1. What is recombinant DNA?
> 2. What is genetic engineering? Describe a pharmaceutical application of genetic engineering.

## 16.2 DNA Fingerprinting

The process of identifying DNA unique to a species, or a specific individual within a species, is called **DNA fingerprinting**. DNA fingerprinting is a method for generating a DNA-based identity, or DNA profile, of an individual or species. DNA fingerprinting can be used to detect the contamination of food or water by certain harmful microorganisms, and to identify remains of endangered species found in the possession of alleged poachers. It is used to match organ donors with patients seeking organ transplants, and to identify victims of mass attacks such as the one on the World Trade Center on September 11, 2001.

DNA fingerprinting is such a powerful tool in forensics today that it has traveled beyond the real world of law enforcement to be dramatized in crime novels, TV dramas, and movie thrillers. A laboratory technician can take a biological sample such as blood, tissue,

**Concept Check Answers**

1. A single DNA molecule created by linking DNA from two different sources.
2. Genetic engineering is the permanent introduction of one or more genes into a cell or whole organism, producing a change in at least one phenotype in the recipient. It has been used to move human genes into bacterial cells for the production of valuable pharmaceutical products, such as insulin and growth hormone.

or semen from a crime scene and develop a DNA fingerprint, or profile, of the person from whom the sample came. That profile can then be compared with another profile—for example, that of a crime victim or suspect—to see whether they match.

It is theoretically possible for two people to have the same DNA profile. Therefore, a match such as the one in **FIGURE 16.4**, between a victim's DNA profile and blood stains found on a suspect's clothes, does not provide definitive proof that the two samples are from the same person. In most cases, however, the probability that two people will have the same DNA profile by sheer accident is between one in 100,000 and one in a billion. (The actual odds, within this range, depend on the methods used to create the DNA profile, as we will see shortly.)

DNA fingerprinting takes advantage of the fact that all individuals (except identical twins and other multiples) are genetically unique. To distinguish between different individuals, scientists examine certain regions of the human genome that are known to vary greatly from one person to the next. Such highly variable regions include noncoding portions of our DNA, such as introns and spacer DNA, which tend to be quite different in size and nucleotide sequence among different individuals.

DNA fingerprinting can be done by various methods, including PCR amplification. (PCR and other tools of DNA technology are described in Section 16.7.) We can use PCR to amplify certain highly variable regions of the human genome. This type of PCR generates a pattern of amplified DNA fragments that is likely to vary from one person to another. The more regions we compare by PCR analysis, the less likely it is that there will be a coincidental match between all of the PCR fragments generated from the DNA of two individuals who are not twins. The Federal Bureau of Investigation (FBI) and most state law enforcement agencies use PCR-based amplification from 13 different regions of the human genome to produce a near-unique DNA profile of an individual. DNA fingerprints of missing persons, victims of unsolved crimes, and people convicted of serious crimes are maintained in a DNA database named CODIS (Combined DNA Index System). The CODIS database contains more than 5 million DNA profiles, including PCR-based fingerprints of biological material collected at the site of unsolved crimes.

## 16.3 Reproductive Cloning of Animals

The goal of **reproductive cloning** is to produce an offspring (baby) that is a genetic copy of a selected individual. In 1996, Dolly the sheep, born on a Scottish farm, became the first mammal to be created through reproductive cloning (**FIGURE 16.5**).

There are three key steps in reproductive cloning. First, an egg cell is obtained from a cytoplasmic donor (a Scottish Blackface ewe, in Dolly's case) and its nucleus is removed (enucleation). Next, an electrical current is used to fuse the enucleated egg with a somatic cell from a nuclear donor (for example, a Finn Dorset ewe, Dolly's genetic parent). Chemicals are used to activate this product of cell fusion—that is, to trick it into dividing so that it begins to form an embryo. Finally, the embryo is transferred to the uterus of a surrogate mother, where, if all goes well, it continues to develop, ultimately resulting in the birth of a healthy baby animal. Such a baby—referred to as a clone—is genetically identical to the ewe that provided the donor nucleus, not to the egg cell donor or the surrogate mother. To date, reproductive clones have been developed in a variety of mammals, including sheep, pigs, mice, cows, horses, dogs, and cats.

**FIGURE 16.4  DNA Fingerprinting Can Be Used to Identify Criminals**

The DNA profile on the far left is that of the defendant (D) in a murder trial. The profile on the far right is that of the victim (V). The defendant's jeans and shirt were splattered with blood. The fingerprint of the DNA extracted from the blood splatters is an exact match with the victim's DNA profile.

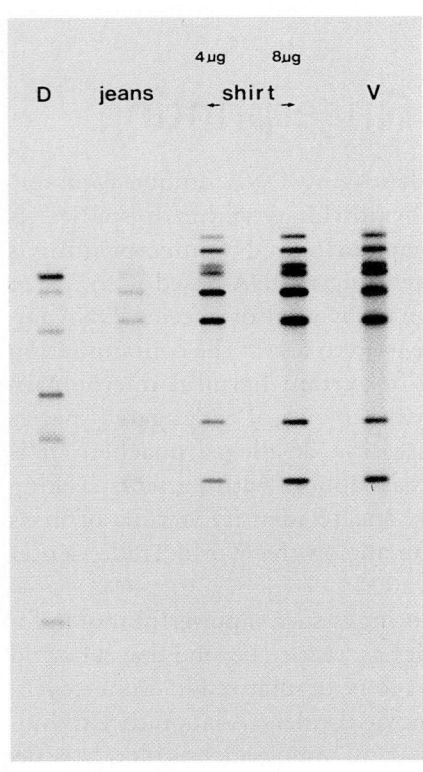

**FIGURE 16.5**
**Cloning Sheep: How Dolly Came to Be**

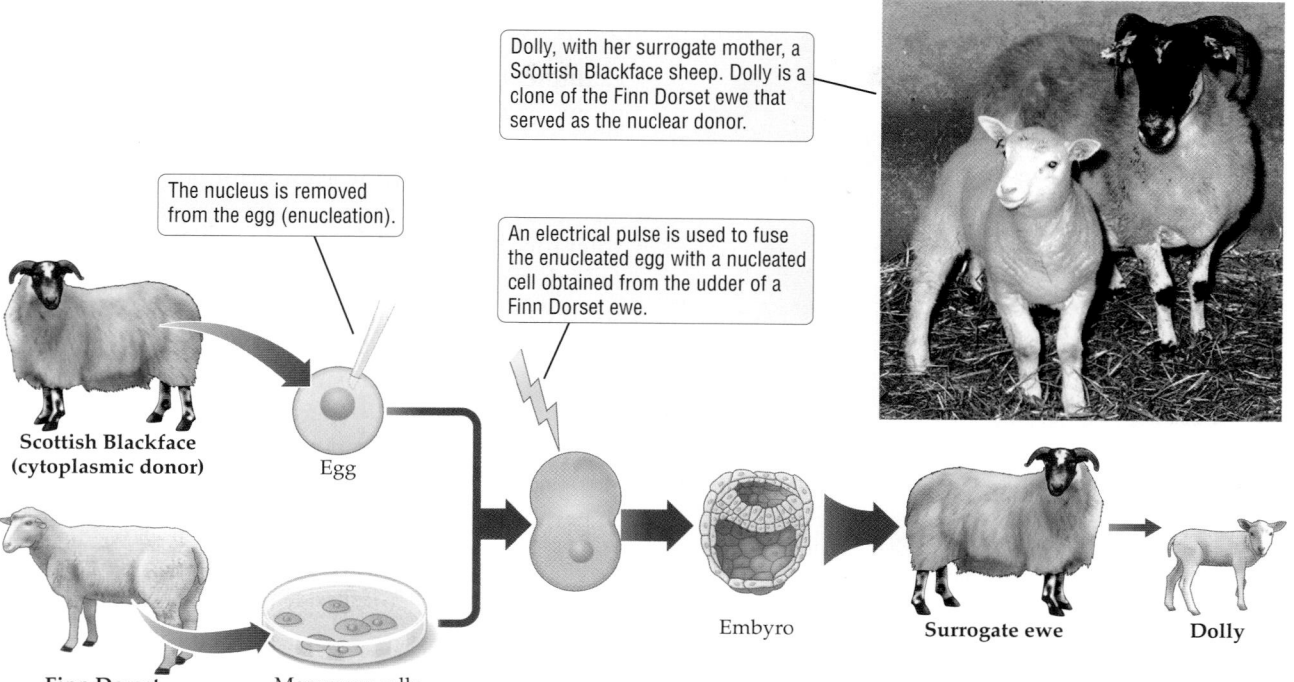

Dolly, with her surrogate mother, a Scottish Blackface sheep. Dolly is a clone of the Finn Dorset ewe that served as the nuclear donor.

The nucleus is removed from the egg (enucleation).

An electrical pulse is used to fuse the enucleated egg with a nucleated cell obtained from the udder of a Finn Dorset ewe.

**Scottish Blackface (cytoplasmic donor)**

Egg

**Finn Dorset (nuclear donor)**

Mammary cells

Embyro

**Surrogate ewe**

**Dolly**

Why would anyone want to clone a sheep, a pig, or a cow? Reproductive cloning can be used to produce multiple copies of an organism that has useful characteristics. For example, a company in South Dakota has cloned calves that are genetically engineered to produce human disease-fighting proteins called immunoglobulins. The ultimate aim is to create herds of genetically identical cows, each of which would serve as a "biological factory," producing large quantities of commercially valuable immunoglobulin proteins.

Reproductive cloning is also being used to produce pigs that could save the lives of people in need of organ transplants. Each year, thousands of people die while waiting for an organ transplant. Pig organs are roughly the same size as human organs and could work well in people, except for one major problem: the human immune system rejects them as foreign. In recent studies, scientists have used reproductive cloning to produce pigs whose organs lack a key protein that stimulates the human immune system to attack. This work represents an important step toward the production of pig clones whose organs could be used to save the lives of people who would otherwise die for lack of a suitable transplant organ.

## 16.4 Genetic Engineering

As noted in Section 16.1, genetic engineering is the permanent introduction of one or more genes into a cell, a certain tissue, or a whole organism, leading to a change in at least one genetic characteristic in the recipient. The organism receiving the DNA is a genetically modified organism (GMO), and the gene introduced into a GMO is called a *transgene* (*trans*, "across from"), so GMO individuals are also known as **transgenic organisms**.

What are the objectives of genetic engineering? For one, if a gene can be transferred between species, it can often make a functional protein product in the new species. For example, jellyfish have a gene enabling them to produce flashes of fluorescent light that may ward off attackers. This gene, which codes for a small light-producing protein known as green fluorescent protein (GFP), has been transferred to and expressed in organisms as different as bacteria, plants, and rabbits (**FIGURE 16.6**). By attaching the GFP-producing gene to a gene of interest, researchers can track the product of the gene of interest by the glow it creates.

**FIGURE 16.6 Alba, a White Rabbit Genetically Modified to Glow**

A gene coding for a jellyfish protein called green fluorescent protein (GFP) was introduced into the fertilized egg from which this rabbit developed. Attaching this gene to genes that encode small components of the cell that would otherwise be very difficult to visualize has led to tremendous advances in our understanding of basic cell biology. The three scientists who contributed to the discovery and use of GFP in biology and biomedicine shared the 2008 Nobel Prize in Chemistry.

In other words, GFP has become invaluable as a fluorescent marker for gene expression in both basic and applied research.

The deliberate transfer of a gene from one species to another is one example of genetic engineering. As illustrated in Figure 16.2, genetic engineering involves isolating a DNA sequence (usually a gene) from one source of DNA and inserting it into a DNA molecule from another source. The inserted fragment of DNA can be introduced into the DNA of the same species or that of a different species.

Recombinant DNA or foreign genes can be introduced into a cell or whole organism in many different ways. We have already seen how plasmids can be used to transfer a gene from humans or other organisms to bacteria. Plasmids can also be used to transfer genes to plant cells or animal cells. In some species, including many plants and some mammals, genetically modified (GM) adults can be generated, or cloned, from these altered cells (see the box on page 374). Other means of gene transfer include viruses, which can be used to "infect" cells with genes from other species, and gene guns that fire microscopic pellets coated with the gene of interest into target cells.

Genetic engineering is commonly used to alter the genetic characteristics of the recipient organism, often focusing on a particular aspect of performance or productivity. Atlantic salmon, for example, have been given genes that cause them to grow up to six times faster than normal (**FIGURE 16.7**). Crop plants have been genetically engineered for a wide variety of traits, including increased yield, insect resistance, disease resistance, frost tolerance, drought tolerance, herbicide resistance (which enables crops to survive the application of weed-killing chemicals), increased shelf life, and improved nutritional value. GM crops now occupy more than 200 million acres worldwide—a coverage that includes over 20 percent of the world's soybeans, corn, cotton, and canola.

**GENETICALLY MODIFIED SALMON**

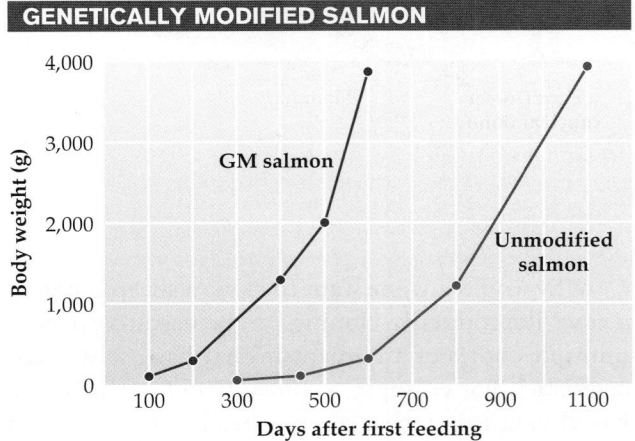

**FIGURE 16.7 Genetically Modified Organisms— Good or Bad?**

Genetically modified salmon eat more food and grow more rapidly compared to unmodified fish.

| TABLE 16.1 | Methods of Production and Uses for Some Products of Genetic Engineering | |
|---|---|---|
| **PRODUCT** | **METHOD OF PRODUCTION** | **USE** |
| **PROTEINS** | | |
| Human insulin | *E. coli* | Treatment of diabetes |
| Human growth hormone | *E. coli* | Treatment of growth disorders |
| Taxol biosynthesis enzyme | *E. coli* | Treatment of ovarian cancer |
| Luciferase (from firefly) | Bacterial cells | Testing for antibiotic resistance |
| Human clotting factor VIII | Mammalian cells | Treatment of hemophilia |
| Adenosine deaminase (ADA) | Human cells | Treatment of ADA deficiency |
| **DNA SEQUENCES** | | |
| Sickle-cell probe | DNA synthesis machine | Testing for sickle-cell anemia |
| *BRCA1* probe | DNA synthesis machine | Testing for breast cancer mutations |
| HD probe | *E. coli* | Testing for Huntington's disease |
| Probe *M13*, among many others | *E. coli*, PCR | DNA fingerprinting in plants and animals |
| Probe 33.6, among many others | *E. coli*, PCR | DNA fingerprinting in humans |

**NOTE:** To make each product, either the DNA sequence that codes for the product is inserted into host cells, such as *E. coli* or mammalian cells, or the product is made using one of several automated procedures, such as DNA synthesis or PCR (described in Section 16.7).

Another common use of genetic engineering is to churn out large amounts of a gene product, usually one with therapeutic or commercial value (Table 16.1). In 1978 the gene for human insulin was transferred into *E. coli* bacteria, and this hormone became the first genetically engineered product to be mass-produced. Before GM insulin, diabetics had to use insulin extracted from pigs and cows to control their blood sugar levels. The animal-derived hormone was often in short supply and could cause allergic reactions in some people. GM insulin is safer and less expensive, and every year more than 300,000 Americans who suffer from insulin-dependent diabetes use the GM product to control their disease.

Another application of genetic engineering is the manufacture of certain types of vaccines. A vaccine is any substance that stimulates the immune system in such a way that it shields the body from future attack by a specific invading organism. Bacteria have been genetically engineered to produce large amounts of the distinctive proteins that are present on the surface of many disease-causing organisms, including a number of viruses, bacteria, and infectious protists. When injected into the human body, the GM versions of these cell surface proteins stimulate the immune system to recognize the disease organisms that normally carry such proteins. The GM proteins act like a vaccine because they prepare the immune system to quickly fend off a real invasion by the disease organism. Such a GM approach has been used to develop effective vaccines for flu and malaria, among other diseases. There are ongoing attempts to develop an AIDS vaccine using genetically engineered surface proteins from the AIDS virus.

## 16.5 Human Gene Therapy

On September 14, 1990, 4-year-old Ashanthi DeSilva made medical history when she received intravenous fluid that contained genetically modified versions of her own white blood cells. She suffered from adenosine deaminase (ADA) deficiency, a genetic disorder that severely limits the ability of the body to fight disease and can make ordinary infections, such as colds or flu, lethal. The disorder is caused by a mutation in a single gene that affects the ability of white blood cells to fight off infections. Earlier, doctors had removed some of Ashanthi's white blood cells and added the normal ADA gene to them, in an attempt to fix the lethal genetic defect through genetic engineering. Ashanthi responded very well to the

Ashanthi DeSilva has ADA. Gene therapy for ADA may have contributed to her remarkable good health.

**1** A blood sample is taken.

**5** White blood cells containing the corrected ADA gene are returned to the patient.

White blood cells containing corrected ADA gene

**4** The patient's white blood cells are infected with GM viruses.

**2** White blood cells are isolated from the sample.

**3** A corrected ADA gene is inserted into a harmless virus.

Corrected ADA gene

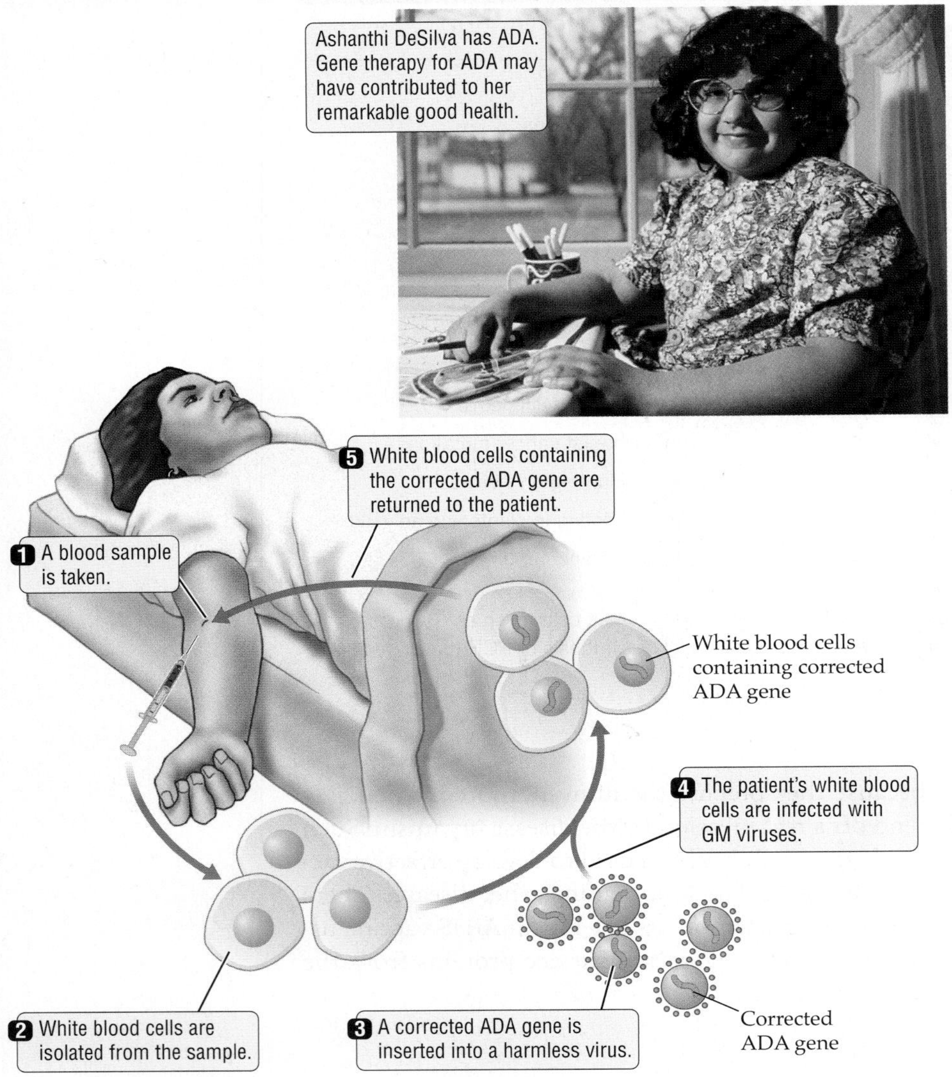

**FIGURE 16.8 Gene Therapy Has Been Used to Treat ADA Deficiency**

treatment and now leads an essentially normal life (**FIGURE 16.8**).

The treatment that Ashanthi received was the first gene therapy experiment on a person. Human gene therapy seeks to correct genetic disorders through genetic engineering and other methods that can alter gene function. The possibility of curing even the worst genetic disorders by reaching into our cells and restoring gene function, or turning off a troublesome gene, is a bold and captivating prospect. As such, gene therapy has attracted much media attention—some of it, unfortunately, bordering on excess. Take Ashanthi's case: In addition to gene therapy, she received other treatments for ADA deficiency. Hence, contrary to some reports in the media, her remarkable good health cannot be attributed to gene therapy alone.

Overall, more than 600 gene therapy experiments have been conducted worldwide, but there are relatively few success stories so far. Why? Has gene therapy been oversold? Until recently, proponents of gene therapy could point to a pioneering French study and answer with a confident no. That study, published in 2000, described how researchers cured X-SCID—a disease similar to ADA deficiency that cripples the immune system—by inserting a healthy version of the gene into bone marrow cells. Of 11 children with X-SCID who were treated by gene therapy alone, 9 were cured. For the first time, scientists had achieved the holy grail of gene therapy: they had cured a human genetic disorder solely by fixing the gene that caused it. Unfortunately, three of the children who were cured of X-SCID went on to develop leukemia, a form of cancer that strikes white blood cells. One of these children has died. What went wrong? By accident, the "good copy" of the gene was inserted near and promoted the expression of a gene that causes cells to divide rapidly, increasing the risk of cancers such as leukemia.

The three cases of leukemia followed close on the heels of another terrible incident: in 1999, a young man participating in a gene therapy experiment died from an allergic reaction to the virus used to deliver the engineered gene. The combined effect of these tragedies sent shock waves through the gene therapy field. Worldwide, many gene therapy trials were placed on hold, and there were calls to abandon gene therapy efforts altogether.

Scientists have pressed on, however, focusing their attention on the biggest hurdle: finding a way to deliver the engineered gene safely and effectively to just the right cells. Harmless viruses are often deployed as delivery vehicles (technically, vectors), but the use of viruses for this purpose presents challenges. The human body defends itself so well against viruses that the recombinant viruses are often destroyed before they can deliver a "good copy" of the gene to enough of the target cells. Researchers at Cedars-Sinai Medical Center have developed a particularly stealthy version of a harmless virus that is almost invisible to the body's defense system. The new vector is being tested in clinical trials aimed at treating Parkinson's disease, a devastating condition marked by progressive brain degeneration. Worrying about unforeseen consequences, the researchers applied the gene therapy to only one side of each patient's brain. One year later, brain scans revealed improved brain activity in the treated side of the brain, while the untreated side showed a decline, compared to the pretrial status of brain function.

Scientists are also working on strategies to prevent the introduced gene from being inserted into the wrong tissues or in a wrong location within the DNA of the target cells (one that could increase the risk of cancer, for example). With respect to X-SCID, for instance, researchers are currently modifying the delivery virus so that it targets fewer types of cells and will therefore be less likely to cause cancer. Other scientists have focused on trying to understand how introduced genes are integrated into the DNA in human cells. This new understanding provides hope that the modified strategies now being proposed will make it possible to insert foreign genes into human cells in a safer, more controlled manner.

In just the past few years, *RNA interference* has emerged as a tremendously promising tool for conducting gene therapy. **RNA interference** (**RNAi**) is a mechanism that selectively blocks the expression of a given gene. In RNAi, small chunks of RNA are able to silence genes that share nucleotide sequence similarity with them. The mechanism is used by many plants and animals to regulate their own genes, and it can be manipulated by DNA technologists to turn genes off in a target cell or organism. RNAi offers a potential cure for genetic disorders caused by the inappropriate activity, or overactivity, of a gene. Clinical trials are currently under way to test the effectiveness of RNAi for shutting off viral genes in people who are infected with hepatitis B or have RSV pneumonia, a lethal viral infection. RNAi-based gene therapy is also being tested against macular degeneration, which is caused by overgrowth of blood vessels in the eye and is the leading cause of blindness in the elderly. Although viral vectors were employed in most of these trials, therapeutic RNA could also be delivered by direct injection into the cytoplasm or by packaging of the RNA into special lipids or polymers. These alternative methods for delivering gene-silencing RNA are under intense investigation because if the delivery system works, there will be no need to insert any foreign DNA into the recipient's DNA.

# 16.6 Ethical and Social Dimensions of DNA Technology

DNA technology provides many benefits to human society. At the same time, the immense power and scope of genetic engineering raises ethical concerns in the minds of some and poses potential risks, especially to the genetic integrity of wild populations. At the most basic level, some people ask how we can assume we have the right to alter the DNA of other species. Some see no ethical conflict in altering the DNA of a bacterium or a virus but object to changing the genome of a food plant or of a sentient animal such as a dog or a chimpanzee.

Few people find fault with the use of GM bacteria to produce lifesaving pharmaceuticals such as insulin for diabetics, blood-clotting proteins for hemophiliacs, and clot-dissolving enzymes for stroke victims. GM plants, on the other hand, are bitterly opposed by some groups, especially in Europe. In some European countries, foods containing GM products must be labeled as such, kept separate from non-GM products, and monitored through the entire food production chain. Critics of GM foods worry that the presence of a GM protein in a common food might cause a severe allergic reaction in an unsuspecting consumer. Proponents counter that no adverse reactions have been authenticated in the United States, where millions of people have been eating GM foods for more than a decade.

Some environmentalists worry that engineering crops to be resistant to herbicides might promote increased use of herbicides, some of which could be harmful to the environment. Supporters of GMO technology say that herbicide-resistant plants would be good for soil health because farmers would not have to use soil-damaging tilling methods to uproot weeds if the weeds in a standing crop could be controlled by an herbicide spray instead. Many crops, including corn and cotton, have been engineered to produce Bt toxin, named after the bacterium (*Bacillus thuringiensis*) in which it was discovered. There were fears that the presence of this protein insecticide in corn pollen would be harmful to monarch butterflies, but recent studies have found no evidence of ill effects on this or other insect species.

Critics of genetic engineering have long argued that genes from genetically modified plants or animals could spread to wild species, potentially wreaking environmental havoc. Some have expressed concern that GMOs will escape from the bounds of farm fields, barnyards, and fish pens to contaminate natural ecosystems with genomes that have been altered by humans. Escaped GM salmon could threaten wild fish stocks not only by interbreeding with them and thereby reducing their natural diversity, but also by outcompeting them for resources and thereby driving them toward extinction.

Of the world's 13 most important crop plants, 12 (all except corn) can mate and produce offspring

# Have You Had Your GMO Today?

The use of genetically modified organisms (GMOs) in agriculture and food production has expanded dramatically since biotech crops were first commercially grown in 1996. So, too, has controversy over the risks and benefits of genetically modified (GM) foods. The United States grows more than 50 percent of the GM crops that are raised worldwide; Canada grows about 6 percent.

Because the United States has no labeling requirements for GM foods, many consumers are unaware that about 75 percent of all processed foods available in U.S. grocery stores may contain ingredients from genetically engineered plants. Breads, cereal, frozen pizzas, hot dogs, and soda are just a few of them. Corn syrup,

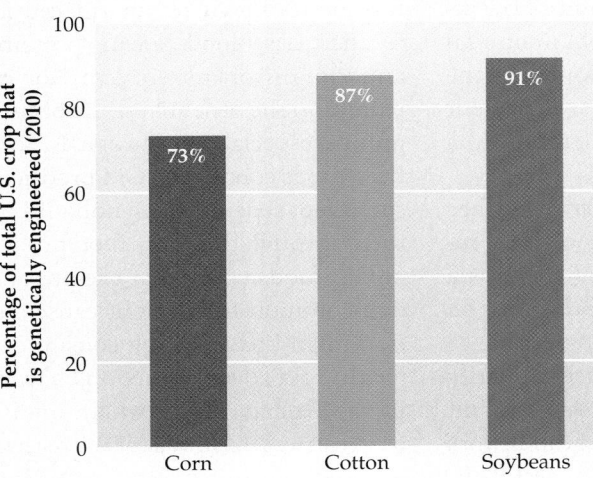

derived from corn, is a common ingredient in many juices and sodas, and GM corn syrup is used in most brand-name sodas. Soybean oil, cottonseed oil, and corn syrup are ingredients used extensively in processed foods. Soybeans, cotton, and corn dominated the 135 million acres of genetically engineered crops planted in the United States in 2010, according to the U.S. Department of Agriculture (USDA). Through genetic engineering, these plants have been made to ward off pests and to tolerate herbicides used to kill weeds. Other crops, such as squash, potatoes, and papaya, have been engineered to resist plant diseases.

with a wild plant species in some region where they are grown. If a crop plant is genetically engineered to be resistant to an herbicide, the potential exists for the resistance gene to be transferred (by mating) from the crop to the wild species. There is a risk that by engineering our crops to resist herbicides, we will unintentionally create "superweeds" resistant to the same herbicides.

In a recent study, scientists used a commercial-scale (400-acre) plot to examine the spread of genes for resistance to the herbicide Roundup. These resistance genes had been inserted into a variety of GM bent grass developed by the Monsanto and Scotts companies, which hope to use the engineered bent grass as turf in golf courses. Grasses are wind-pollinated, raising the concern that resistance genes could spread long distances if pollen blown by wind from GM bent grass were to fertilize any one of a dozen other grass species that can breed with bent grass. That fear appears to have been realized: DNA tests discovered the modified genes in non-GM bent grass located up to 13 miles away from a test plot; the genes were also found in other grass species growing wild up to 9 miles from the test plot. Representatives from Monsanto and Scotts suggested that these results are not

all that worrisome, since any plants that contain the Roundup resistance genes can be controlled by other herbicides.

The U.S. Department of Agriculture (USDA) has ordered that an environmental impact study be conducted on GM bent grass before the agency decides whether Scotts can sell it. This is the first time that the USDA has demanded such a review for a GM crop. Experts say a way around the problem is to create GM grasses with sterile pollen, making it much less likely that genes from such grasses could escape to wild species.

In some circles, the heated debate over GMOs tends to center on political and socioeconomic issues. The use of "terminator genes" in a GM plant can theoretically prevent that plant from making viable seed. Proponents say that "terminator technology" would be an effective barrier against the runaway spread of GMOs. Opponents of GMOs see this technology as a veiled attempt by seed companies to control the supply of seed, since farmers would have to buy seed from the company each year instead of saving their own.

Critics of genetic engineering point to the social costs of using such technology. They cite bovine growth hormone (BGH), which is mass-produced by GM bacteria, as a case in point. Among its other effects, BGH increases milk production in cattle. Before the introduction of genetically engineered BGH in the 1980s, milk surpluses already were common. The use of BGH by large milk producers has created even larger milk surpluses, driving down the price of milk and forcing small producers of milk—the traditional family farms—out of business. As some see it, the lower milk prices for consumers are not worth the social cost of driving small dairy farms into bankruptcy. Others believe that all types of commercial enterprises, including family farms, should sink or swim on their own and unsuccessful businesses should not be rescued because of social sentiment.

As these examples illustrate, the debate over the social dimensions of GMOs often gets caught up in a much larger discussion about the politics of food and the economics of modern agriculture. With respect to altering human DNA, nearly everyone accepts that the use of genetic engineering in humans for the purpose of curing a horrible disease is ethical. What about the use of genetic engineering to make less critical changes, such as enhancing physical or mental traits? For example, if it were possible to do so, would it be ethical to alter the future intelligence, personality, looks, or sexual orientation of our children before birth? According to a 1990s March of Dimes survey, more than 40 percent of Americans would make such modifications if given the chance. Is it fair for parents to make such decisions on behalf of their children? The genetic engineering of humans still faces substantial challenges, so there may be a few more years to resolve the dilemmas raised by such questions before the technology arrives at a clinic near you.

## 16.7 A Closer Look at Some Tools of DNA Technology

Extracting DNA is a fairly straightforward process. Scientists first break open the cell membranes to release the cell contents. Then they use specific chemicals to strip away other macromolecules, such as proteins and lipids, until just the DNA remains. The many enzymes that are used to cut and join DNA, and to amplify it through such methods as PCR, come from natural sources. Cells, and even viruses, deploy a diverse and versatile collection of enzymes in order to replicate

DNA (see Chapter 14), swap chromosome segments during crossing-over (see Chapter 13), invade a cell, or fight off an invading virus or other parasite.

As researchers have come to understand these natural processes better, they have been able to "borrow" these same enzymes to manipulate DNA in a test tube. Some of these enzymes can function under extreme conditions, such as at near-boiling temperatures, because the organisms from which they are derived are adapted to living in some of the most hostile habitats on the planet. As you read the details of some of the procedures that are widely used in DNA technology, keep in mind that most of these "tricks" are inspired by nature. Just about everything in the DNA technologist's tool chest comes from living organisms or viruses and has therefore been honed by evolutionary processes.

## Enzymes are used to cut and join DNA

Each of us has 3.3 billion base pairs of DNA on 23 unique chromosomes. The DNA molecule in each chromosome is so large (140 million base pairs, on average) that after DNA has been extracted from a cell, it must be broken into smaller pieces before it can be analyzed further. DNA can be split into more manageable pieces by **restriction enzymes**, which cut the DNA at highly specific sites. A restriction enzyme called *Alu*I, for example, cuts DNA everywhere the sequence AGCT occurs, but nowhere else (**FIGURE 16.9**). There are hundreds of restriction enzymes, each of which recognizes and cuts DNA if its unique target sequence is present in that DNA, and it cuts the DNA

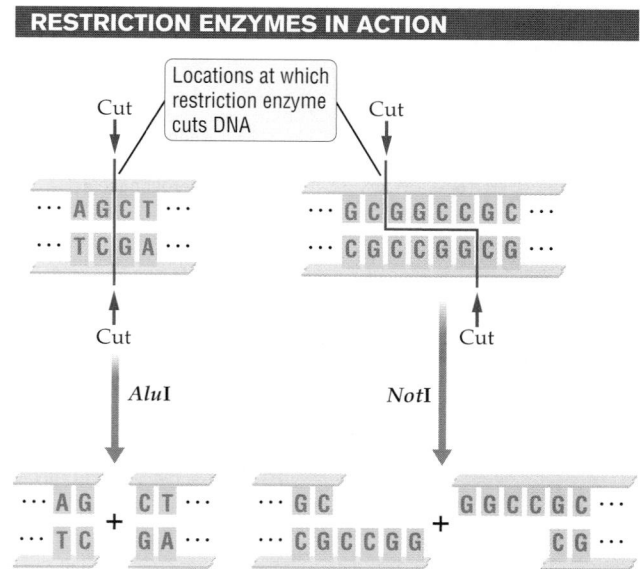

RESTRICTION ENZYMES IN ACTION

Locations at which restriction enzyme cuts DNA

**FIGURE 16.9**
**Restriction Enzymes Cut DNA at Specific Places**

The restriction enzyme *Alu*I specifically binds to and then cuts the DNA molecule wherever the sequence AGCT occurs. Another restriction enzyme, *Not*I, specifically binds to and cuts the DNA sequence GCGGCCGC. Each enzyme binds to and cuts its own special target sequence, and no other.

only at those sites. Restriction enzymes were discovered in bacteria in the late 1960s. Bacteria appear to have evolved restriction enzymes to do battle against foreign DNA, such as viral DNA. Infecting viruses begin by injecting their DNA into a bacterium. The bacterium then deploys its restriction enzymes in an attempt to chop up the viral DNA, thereby "restricting" viral growth.

DNA **ligase** is another important enzyme for making recombinant DNA. DNA ligases join two DNA fragments. DNA ligases are commonly used to insert one piece of DNA, such as a human gene, into another DNA molecule, such as a plasmid extracted from bacterial cells (as illustrated in Figure 16.2), creating a recombinant DNA molecule. For instance, the gene coding for green fluorescent protein (GFP) can be cut out of jellyfish DNA with restriction enzymes and "pasted" into a bacterial plasmid with DNA ligase, creating a recombinant plasmid. The recombinant plasmid can then be inserted into bacterial cells by means of genetic engineering, and if the recombinant plasmid is constructed in the right manner, the resulting GM bacteria will glow when exposed to blue light.

## Gel electrophoresis sorts DNA fragments by size

Once a DNA sample has been cut into fragments by one or more restriction enzymes, researchers often use a process called *gel electrophoresis* to help them see and analyze the fragments. In **gel electrophoresis**, fragments of DNA are placed into "wells" in a gelatin-like slab called a "gel." When an electrical current passes through the gel, it causes the DNA (which has a negative electrical charge) to move toward the positive end of the gel (**FIGURE 16.10**). (Recall that opposite charges attract each other.) Long fragments of DNA pass through the gel with more difficulty and therefore move more slowly than short pieces. The distance a fragment travels through the gel is related to its speed of movement. After a fixed time period, the shorter, more rapidly moving fragments are found toward the positive end of the gel, and the longer, more slowly moving fragments are located closer to the negative end. Because DNA is invisible to the human eye, the fragments must be stained or labeled in special ways before they can be seen.

By using restriction enzymes and gel electrophoresis together, researchers can examine differences in DNA sequences. For example, the restriction enzyme *Dde*I cuts the normal hemoglobin allele into two pieces, but it cannot cut the sickle-cell allele. This difference provides a simple test for the disease allele (**FIGURE 16.11**).

## DNA sequencing and DNA synthesis are key tools in biotechnology

DNA sequencing enables researchers to identify the sequence of nucleotides in a DNA fragment, a gene, or even the entire genome of an organism. Sequences can be determined by several methods, the most efficient of which rely on automated sequencing machines (**FIGURE 16.12**). One of these machines can identify over a million bases per day, making it possible to determine the sequence of a single gene quickly. DNA can also be sequenced manually by slower but still highly effective methods.

**FIGURE 16.10 DNA Fragments Can Be Separated by Gel Electrophoresis**

Gel electrophoresis is used to separate and visualize biomolecules. When subjected to an electrical current, DNA fragments move through a gel at different rates, depending on their size. Larger molecules move slowly, while smaller molecules move quickly. To enable visualization of the DNA, the gel is commonly stained with a DNA-binding dye ethidium bromide that glows pink when exposed to ultraviolet light.

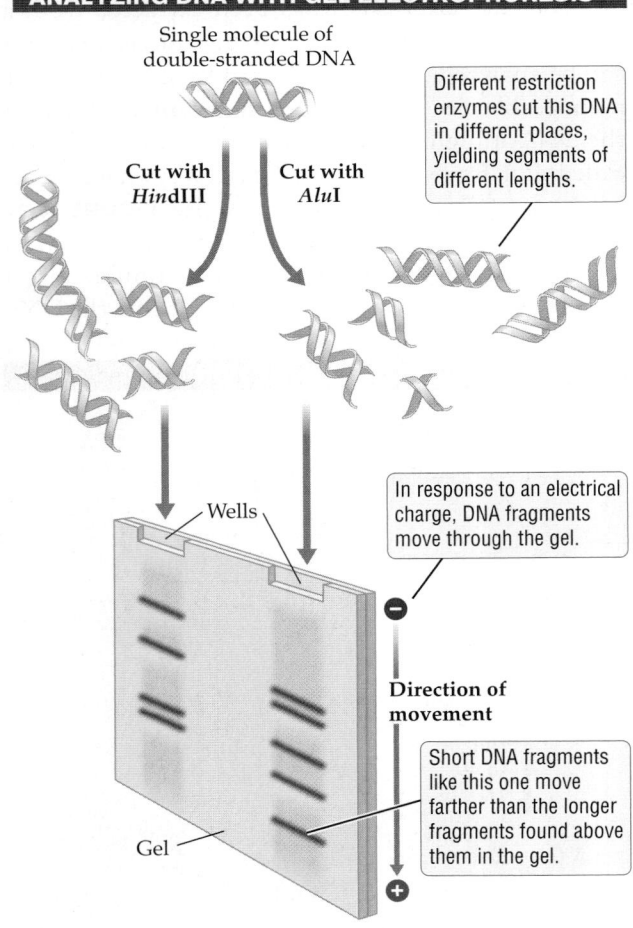

ANALYZING DNA WITH GEL ELECTROPHORESIS

Single molecule of double-stranded DNA

Cut with *Hind*III

Cut with *Alu*I

Different restriction enzymes cut this DNA in different places, yielding segments of different lengths.

Wells

In response to an electrical charge, DNA fragments move through the gel.

Direction of movement

Short DNA fragments like this one move farther than the longer fragments found above them in the gel.

Gel

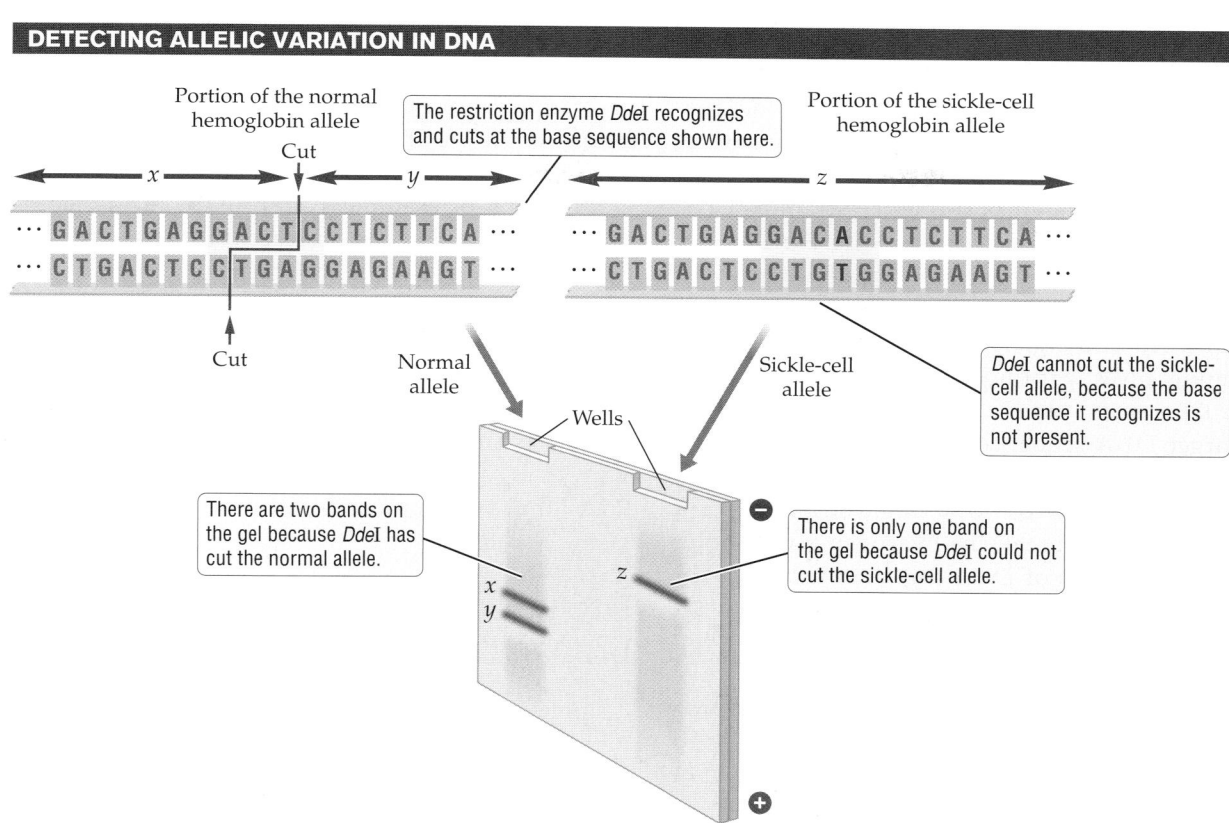

The restriction enzyme *Dde*I recognizes and cuts at the base sequence shown here.

Portion of the normal hemoglobin allele

Cut

*x*  *y*  *z*

Portion of the sickle-cell hemoglobin allele

··· GACTGAGGACTCCTCTTCA ···
··· CTGACTCCTGAGGAGAAGT ···

··· GACTGAGGACACCTCTTCA ···
··· CTGACTCCTGTGGAGAAGT ···

Cut

Normal allele

Sickle-cell allele

*Dde*I cannot cut the sickle-cell allele, because the base sequence it recognizes is not present.

Wells

There are two bands on the gel because *Dde*I has cut the normal allele.

*x*
*y*

*z*

There is only one band on the gel because *Dde*I could not cut the sickle-cell allele.

Electrophoresis of DNA fragments

**FIGURE 16.11**

**Restriction Enzymes and Gel Electrophoresis Can Be Used to Identify the Sickle-Cell Allele**

The restriction enzyme *Dde*I cuts DNA wherever it encounters the sequence GACTC (shown in yellow in the top panels). Only one such sequence occurs in the normal allele of the hemoglobin gene. A single base pair mutation (in which T-A becomes A-T) causes sickle-cell anemia; this mutation occurs in the GACTC sequence, and changes the sequence to GACAC. Since *Dde*I cannot recognize the mutant sequence GACAC, it does not cut the DNA at this location. As a result, the sickle-cell allele produces only one band on the gel, whereas the normal hemoglobin allele shows up as two bands on the gel, representing two fragments of different sizes.

Sequencing technology was greatly advanced by the Human Genome Project, which determined the sequence of the entire human genome. The genome sequences of many other organisms have now been determined as well, making it possible to detect genes that different organisms have in common. Gene maps are constructed showing where genes reside on chromosomes. Isolating sequences that are critical to human health—that is, genes that affect health and disease—is now an area of intense research. The relatively new, related field of **proteomics**—the study of the full set of proteins encoded by genes—promises to

**FIGURE 16.12 Machines Can Sequence DNA**

Automated DNA sequencing machines can rapidly determine the nucleotide (base) sequence of a DNA fragment. Here a scientist examines a computer display showing part of an automated DNA sequencing gel. Each of the four chemical bases in DNA is represented by a different color (red, green blue, or yellow).

bring even greater insight into the organization and functioning of complex organisms like us.

Machines can also be used to create, or synthesize, a DNA fragment with a specific, made-to-order nucleotide sequence. In less than an hour, a DNA synthesis machine can produce single-stranded DNA segments hundreds of nucleotides long. Synthesized single-stranded DNA segments, usually fewer than 30 nucleotides long, can be used as **DNA probes** that can help find a gene of interest in a sample of DNA under investigation. Probes may be synthesized in the laboratory, with a sequence complementary to a target DNA sequence. In **DNA hybridization** the target DNA is cut into fragments by restriction enzymes and converted to a single-stranded form by treatments that break the hydrogen bonds holding the two strands of DNA together. If the single-stranded probe encounters a complementary sequence of single-stranded DNA in the target, it will bind to it (a process called *hybridization*). The probe can be labeled with a radioactive or fluorescent tag to make it easier to identify the DNA fragments to which it binds. Probes are commonly used in DNA profiling (discussed in Section 16.2).

Nucleic acid hybridization of a short segment of artificially synthesized single-stranded DNA called a *primer* is an essential step in initiating an important laboratory technique called the *polymerase chain reaction* (*PCR*). PCR essentially automates replication of a DNA target in a test tube, resulting in billions of copies of the DNA starting from just a few molecules of the target.

## PCR is used to amplify small quantities of target DNA

The **polymerase chain reaction** (**PCR**) uses a special type of DNA polymerase to make billions of copies of a targeted sequence of DNA in just a few hours. To amplify a piece of DNA by PCR, researchers must use two short segments of synthetic DNA, called **DNA primers**. Each primer is designed to bind to (hybridize with) one of the two ends of each strand of the target DNA by complementary base pairing. By the series of steps shown in **FIGURE 16.13**, DNA polymerase then produces many copies of the DNA that is flanked by the two primers. To amplify a target DNA, scientists must have at least some information about its nucleotide sequence. Without

this knowledge, they cannot synthesize the specific primers required in every PCR reaction.

The power of PCR technology lies in the fact that it can amplify extremely small amounts of DNA—amounts extracted from just a few cells or a single blood stain, for example. As a result, PCR has come to be widely used in basic research and in fields as diverse as medical diagnostics, forensics, paternity testing, paleoanthropology, and the authentication of delicacies such as caviar and expensive vintage wine. The technique became so successful so quickly that in 1991, only 6 years after the first paper on PCR was published, the PCR patent was sold for $300 million. Two years later, in 1993, Kary Mullis won a Nobel Prize for the development of PCR.

## DNA cloning is a means of propagating recombinant DNA

A single copy of a gene is difficult to analyze and manipulate for such purposes as gene sequencing or genetic engineering. Biologists generally clone such a DNA fragment by isolating it, linking it (with the help of DNA ligase) with other DNA fragments to create a recombinant molecule, and then introducing the recombinant molecule into a host cell that will make many identical copies of it. **Cloning** is making a copy that is genetically identical to the original, so copying a whole organism is also a form of cloning. Whole-organism cloning includes such traditional methods as propagating houseplants by making cuttings of them or, as we saw in the example of Dolly (see Figure 16.5), using DNA technology to make genetically identical copies of animals.

DNA that is cloned can be sequenced, transferred to other cells or organisms, or used as a probe in DNA hybridization experiments. In addition, with cloning and gene sequencing as tools, researchers can use the genetic code to determine the amino acid sequence of the gene's protein product. Knowledge of a gene's product can provide vital clues to the gene's function, as it did in the case of the HD gene involved in Huntington's disease (see Chapter 13). For this reason, DNA cloning is a key step in the study of genes that cause inherited genetic disorders and cancers.

To clone a gene or other piece of DNA, scientists use restriction enzymes to cut the DNA from the source cells (as in, for example, the case of the gene for GFP from jellyfish cells). Next, they randomly in-

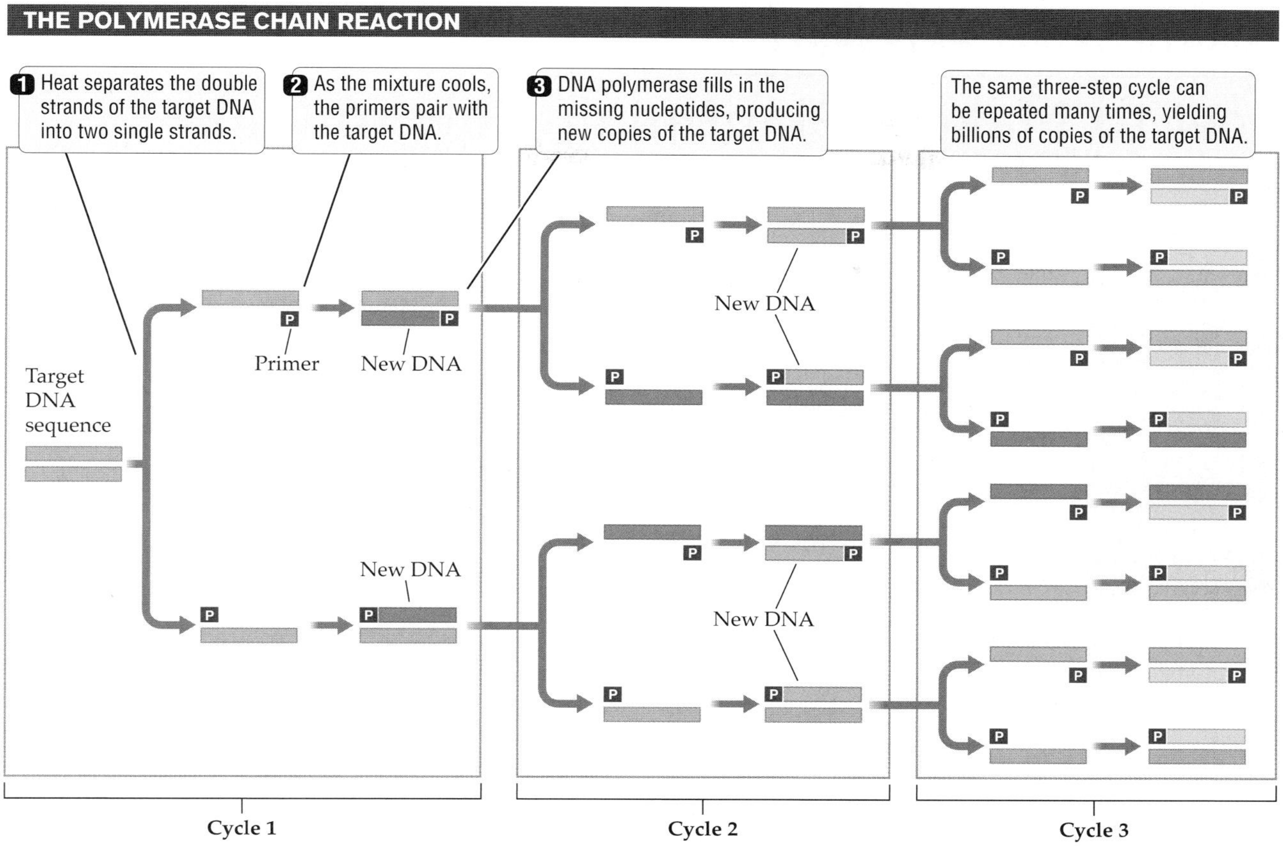

**1** Heat separates the double strands of the target DNA into two single strands.

**2** As the mixture cools, the primers pair with the target DNA.

**3** DNA polymerase fills in the missing nucleotides, producing new copies of the target DNA.

The same three-step cycle can be repeated many times, yielding billions of copies of the target DNA.

Target DNA sequence

Primer   New DNA

New DNA

New DNA

New DNA

Cycle 1          Cycle 2          Cycle 3

**FIGURE 16.13  Small Amounts of DNA Can Be Amplified More than a Millionfold through PCR**

Short primers (red) that can pair with the two ends of a gene of interest (bluish gray) are mixed in a test tube with a sample of the target DNA, the enzyme DNA polymerase, and all four nucleotides (A, C, G, and T). A machine then processes the mixture through the three steps shown, in which the temperature is first raised and then lowered, to double the number of double-stranded versions of the template sequence. The doubling process can be repeated many times (only three cycles are shown here). For clarity, color coding is used to identify the template DNA sequence in the first cycle (bluish gray), the DNA that is newly made in each cycle (shades of orange), and the primers (red).

sert each of the many resulting DNA fragments into at least one copy of a DNA vector. Recall from Section 16.1 that a plasmid vector is a loop of bacterial DNA designed to serve as a "DNA vehicle" for transferring cloned DNA fragments from one cell to another. The DNA of certain viruses, or the chromosomes of bacterial or yeast cells, can also be used as a DNA vector, especially when large DNA fragments need to be cloned. (Recall in the case of Ashanthi DeSilva, illustrated in Figure 16.8, that gene therapy was administered with the help of a viral vector.)

The few examples explored in this chapter only hint at the many ways in which biotechnology is changing medicine, agriculture, and the environment.

DNA manipulation is changing our world, often for the better—but not always. The well-informed citizen will want to understand both the potential positive and negative impacts of DNA technology on our world.

### Concept Check

1. What roles do restriction enzymes and DNA ligases play in DNA technology?

2. What is DNA fingerprinting?

3. Explain why the release of GMOs could pose potential risks to natural ecosystems.

# How to Make a Plantimal, How to Make a Little Girl Well

At the beginning of this chapter we met Eduardo Kac's transgenic petunia "Edunia," a work of art inspired by biotechnology. After Kac conceived his idea, he had technicians take a sample of his blood, isolate DNA from his blood cells, and locate a particular sequence of DNA, called immunoglobulin kappa, or IGκ, a sequence that encodes antibodies. In all, four DNA sequences were inserted into the petunia: (1) the IGκ DNA; (2) a gene for bacterial resistance to an antibiotic; (3) a gene for plant resistance to the same antibiotic; and (4) a plant promoter that instructed the petunia to express the gene only in its veins.

Why were so many genes required to make this project work? Let's look at the details. The carrier for the genes was a plasmid, a small circle of DNA into which foreign genes can be inserted to make recombinant DNA (see Figure 16.2). The recombinant plasmids infected the bacteria, which infected the plant cells and carried the genetically modified plasmid into the plant cells. To eliminate any bacteria that the plasmids hadn't infected, the bacteria were treated with the antibiotic. All the bacteria carrying the plasmid (with the antibiotic resistance gene) survived; all the bacteria without the plasmid died.

Next, the plasmid-infected bacteria infected the petunia cells. To eliminate plant cells that the bacteria/plasmids hadn't infected, the plant cells were treated with the antibiotic too. Only the plant cells carrying the plasmid survived. Tiny clumps of recombinant petunia cells grew into young plants and, after a few months, flowering petunias. The veins in their pink petals express Kac's immunoglobulin gene, uniting plant and human genes in a single organism.

The steps that Kac used to create Edunia are the same ones that biotechnologists use every day to create other transgenic organisms. Modern genetic engineering applies hundreds of techniques and procedures. One major area of research is gene therapy.

In 1990, Ashanthi DeSilva became the first person ever to be treated with gene therapy. DeSilva was a 4-year-old girl with a genetic disease in which a mutation disrupts the expression of a single enzyme needed by the white blood cells of the immune system. To treat DeSilva, her doctor removed some of her white blood cells, added a normal version of the enzyme gene to them, and then injected her with the modified cells. Today, DeSilva leads a normal life (**FIGURE 16.14**), but since she also receives doses of the enzyme itself, it's hard to be sure how effective the gene therapy has been.

The possibility of curing genetic disorders by reaching into our cells and restoring gene function is a captivating idea. Early on, researchers claimed that gene therapy would cure everything from cystic fibrosis to heart disease. But after more than 600 gene therapy experiments, researchers still had logged few successes.

Most forms of gene therapy can only treat genetic diseases, not cure them. To make the Edunia, a few genetically modified cells were grown into whole plants. But gene therapy seeks to repair gene function in whole humans made up of billions of cells. Rather than trying to modify every cell in the body, researchers alter only the specific cells that are malfunctioning—DeSilva's white blood cells, for example. But the genetically modified cells eventually die off, leaving the body's own unmodified cells. As a result, gene therapy needs to be repeated; it's not a permanent cure. And it's not necessarily safe. Genes are typically transferred using harmless viruses that infect human cells. But in a few cases, volunteers have died as a result of fatal immune reactions to these viruses.

**FIGURE 16.14 Ashanthi DeSilva**
In 1990, when Ashanthi DeSilva was 4 years old, researchers tried an experimental treatment, injecting her with cells that had been genetically modified to make a critical enzyme that her own cells were unable to make. Today, DeSilva is a healthy young woman with a career in international development.

# Drugmakers' Fever for the Power of RNA Interference Has Cooled

BY ANDREW POLLACK, *New York Times*

When RNA interference first electrified biologists several years ago, pharmaceutical companies rushed to harness what looked like a swift and surefire way to develop new drugs.

Billions of dollars later, however, some of those same companies are now losing their enthusiasm for $RNA_i$, as it is called. And that is raising doubts about how quickly, if at all, the Nobel Prize–winning technique for turning off specific genes will yield the promised bounty of innovative medicines.

The biggest bombshell was dropped in November, when the Swiss pharmaceutical giant Roche said it would end its efforts to develop drugs using $RNA_i$, after it had invested half a billion dollars in the field over four years.

Just last week, as part of a broader research cutback, Pfizer decided to shut down its 100-person unit working on $RNA_i$ and related technologies. Abbott Laboratories has also quietly shelved its $RNA_i$ drug development work.

"In 2005 and 2006, there was a very sudden buildup of expectation that $RNA_i$ was going to cure many diseases in a very short time frame," said Dr. Johannes Fruehauf, vice president for research at Aura Biosciences, a small company pursuing the field. "Some of the hype, I believe, is going away and a more realistic view is setting in."

The issue is that while drugs working through the $RNA_i$ mechanism can indeed shut off genes, it has been difficult to deliver such drugs to the cells where they are needed. At a time when hard-pressed pharmaceutical companies are already scaling back research expenditures, $RNA_i$ is losing out to alternatives that seem closer to producing marketable drugs…

Around 2005, RNA interference emerged as a promising tool for treating potentially any disease associated with gene expression. Interfering RNA ($RNA_i$) is an exotic kind of RNA that shuts off ("silences") genes that cells suspect of being viruses. In previous chapters we saw that messenger RNA (mRNA) carries genetic information from the DNA so that it can be translated into proteins. But RNA can serve another purpose as well.

You might recall that messenger RNA is single-stranded, while our DNA is double-stranded. Some viruses, which infect cells by injecting their genes into the cells, use RNA as their genetic material instead of DNA. But viral RNA is double-stranded, not single-stranded. So if a cell detects double-stranded RNA, the cell acts to break up the suspicious RNA. Cells also use $RNA_i$ to regulate gene expression, deactivating or modifying RNA that has been transcribed but not yet translated into a protein.

Pharmaceutical companies were initially excited about using $RNA_i$ to shut off malfunctioning genes, investing billions of dollars in RNA interference on the assumption that the technology was advanced enough to quickly deliver treatments. The idea was to make a double-stranded version of a disease-related gene, triggering the natural gene-silencing process inside cells.

As with other forms of gene therapy, however, there are huge technical challenges still to be overcome. For example, it's not always easy to deliver the drugs to the right cells. Sometimes the drugs simply don't work well. And, perhaps because they look like viruses, $RNA_i$ drugs can trigger immune reactions. Sometimes the immune reactions are merely an unwanted side effect of treatment. Other times the immune reaction may work better than the gene silencing itself. So it's unclear how the $RNA_i$ is working—through gene silencing or by stimulating an immune response against a misbehaving target. In a review of 35 studies of $RNA_i$ drugs, only two studies included controls that could rule out the possibility that the drugs worked by stimulating the immune system rather than by silencing particular genes.

Like most businesses, pharmaceutical companies want products that can be fully developed and sold right away. As a result, many are abandoning research on interfering RNA. That doesn't mean, however, that $RNA_i$ won't become a useful and important technology in the future. Because basic research can take a long time to develop into marketable products, governments often fund research that doesn't have an immediate use but might someday. For example, the federal government funds 36 percent of all medical research in the United States.

## Evaluating the News

**1.** In many $RNA_i$ studies, patients get better, but it's not clear whether $RNA_i$ is the reason. Explain what else could be causing patients to get better and why researchers can't tell whether $RNA_i$ has had an effect.

**2.** RNA interference remains an important new technology. If the private sector (the pharmaceutical industry) is reducing its investment in $RNA_i$ technology, would you support government grants to university research labs for $RNA_i$ research? Briefly discuss your reasons.

**SOURCE:** *New York Times*, February 7, 2011.

# CHAPTER REVIEW

## Summary

### 16.1 The Brave New World of DNA Technology

- Scientists can manipulate DNA using a variety of laboratory techniques. Because the structure of DNA is the same in all organisms, these techniques work in much the same way on the DNA of all species.
- A gene is said to be cloned if it has been isolated and many copies of it have been made in a suitable host cell, such as a bacterium.
- A recombinant DNA molecule is produced when DNA from one source is spliced into a DNA molecule from another source.
- Genetic engineering introduces cloned DNA into cells, tissues, or whole organisms to create genetically engineered organisms (GMOs).

### 16.2 DNA Fingerprinting

- DNA fingerprinting, which creates a unique genetic profile for each individual, is widely used in criminal cases to establish the identities of victims and criminals.
- DNA fingerprinting of humans makes use of highly variable regions of the genome. PCR can be used to generate a pattern of amplified DNA fragments that is likely to vary from person to person.

### 16.3 Reproductive Cloning of Animals

- The goal of reproductive cloning is to produce an offspring (baby) that is a genetic copy of a selected individual. Dolly the sheep was the first mammal to be created through reproductive cloning.
- In reproductive cloning, the nucleus is removed from an egg cell, and an electrical current is used to fuse the enucleated egg with a somatic cell from a nuclear donor. The product of cell fusion forms an embryo that is then transferred to the uterus of a surrogate mother.
- Reproductive cloning can be used to produce multiple copies of an organism that has useful characteristics, or to conserve rare and endangered animals.

### 16.4 Genetic Engineering

- In genetic engineering, a DNA sequence (often a gene) is isolated, modified, and inserted back into the same species or into a different species.
- Genetic engineering is used to alter the phenotype (especially the performance or productivity) of the GMO or to produce many copies of a DNA sequence, a gene, or a gene product.

### 16.5 Human Gene Therapy

- The goal of human gene therapy is to correct genetic disorders through genetic engineering and other methods that can alter gene function.
- Gene therapy in humans continues to present challenges, but many new advances have been made recently, especially in developing safe procedures for introducing genes into the patient; hundreds of gene therapy trials are currently under way.
- Few people object to the use of gene therapy to cure severe genetic disorders. The technology, however, faces many challenges, especially in finding safe and effective means for introducing a cloned gene into a patient.

### 16.6 Ethical and Social Dimensions of DNA Technology

- DNA technology offers potential benefits but also raises ethical questions and poses potential environmental risks. Opponents of genetic engineering are concerned that genes from GM plants or animals could spread to wild species, potentially wreaking environmental havoc.
- Much controversy surrounds the development of genetically modified plants, particularly in Europe. GM foods do not have to be labeled in the United States and are common in the American diet.

### 16.7 A Closer Look at Some Tools of DNA Technology

- Restriction enzymes are used to break DNA into small pieces. Gel electrophoresis separates the resulting DNA fragments by size.
- Ligases are enzymes that join pieces of DNA.
- Labeled DNA probes are used in DNA hybridization experiments to detect the presence of a particular gene in a sample of DNA.
- Automated sequencers greatly speed up the processes of sequencing and synthesizing DNA.
- To clone a gene, a vector such as a plasmid is used to transfer DNA fragments from the donor organism to a host organism, such as a bacterium.
- To amplify a gene by PCR, primers (short segments of DNA that are complementary to the beginning and end of the target gene) are synthesized and used to produce billions of copies of the target gene in a few hours.

# Key Terms

cloning (p. 378)
DNA cloning (p. 366)
DNA fingerprinting (p. 367)
DNA hybridization (p. 378)
DNA primer (p. 378)
DNA probe (p. 378)

DNA technology (p. 364)
gel electrophoresis (p. 376)
gene therapy (p. 367)
genetic engineering (p. 367)
genetically modified organism
   (GMO) (p. 367)

genome (p. 365)
ligase (p. 376)
plasmid (p. 366)
polymerase chain reaction
   (PCR) (p. 378)
proteomics (p. 377)

recombinant DNA (p. 365)
reproductive cloning (p. 368)
restriction enzyme (p. 375)
RNA interference
   (RNAi) (p. 373)
transgenic organism (p. 369)

# Self-Quiz

1. Which of the following cut(s) DNA at highly specific target sequences?
   a. DNA ligase
   b. DNA polymerase
   c. restriction enzymes
   d. RNA polymerase

2. The propagation of recombinant DNA in a culture of *E.coli* is known as
   a. PCR.
   b. DNA hybridization.
   c. DNA ligation.
   d. DNA cloning.

3. Genetic engineering
   a. can be used to make many copies of recombinant DNA introduced into a host cell.
   b. can be used to alter the inherited characteristics of an organism.
   c. raises ethical questions in the minds of some people.
   d. all of the above

4. When DNA fragments are placed on an electrophoresis gel and subjected to an electrical current, which fragments move the farthest in a given time?
   a. the smallest
   b. the largest

   c. PCR fragments
   d. mRNA fragments

5. A short, single-stranded sequence of DNA whose bases are complementary to a portion of the DNA on another DNA strand is called
   a. a DNA hybrid.
   b. a clone.
   c. a DNA probe.
   d. an mRNA.

6. Small loops of nonchromosomal DNA that are found naturally in bacteria are called
   a. plasmids.
   b. primers.
   c. amplimers.
   d. clones.

7. If the DNA sequences at the beginning and end of a gene are known, which of the following methods can be used to produce billions of copies of the gene in a few hours?
   a. $RNA_i$
   b. reproductive cloning
   c. therapeutic cloning
   d. PCR

# Analysis and Application

1. Discuss the extent to which our current ability to manipulate DNA differs from what people have done for thousands of years to produce a wide range of domesticated species, such as dogs, corn, and cows.

2. Judy and David are a couple considering whether to have children. Judy knows that her mother's sister died of sickle-cell anemia, which means Judy's grandparents passed the sickle-cell alleles on to her aunt, and Judy herself may have inherited the sickle-cell allele from them. The story on David's side is similar: two of his aunts have sickle-cell anemia. Describe in detail the DNA technology procedures that would enable Judy and David to know with certainty whether they carry the sickle-cell allele.

3. What is DNA cloning? Describe how you might clone a jellyfish GFP gene into a bacterial cell.

4. Discuss some practical benefits of DNA cloning.

5. What is genetic engineering? How is it accomplished? Select one example of genetic engineering and describe its potential advantages and disadvantages.

6. Is it ethical to modify the DNA of a bacterium? A single-celled yeast? A worm? A plant? A cat? A human? Give reasons for your answers.

7. Supporters of genetic engineering claim that GM plants are likely to be safer, in many ways, than crops bred through conventional methods. They claim that GM crops undergo extensive scrutiny and environmental impact studies, while non-GM crops do not. Are you persuaded by that argument? Should new varieties of conventional crops be overseen more strictly, despite the economic cost of doing so?

8. Are some modifications to the DNA of humans not acceptable? Assuming you think so, what criteria would you use to draw the line between acceptable and unacceptable changes?

**VAMPIRE FINCH.** This subspecies of the sharp-billed ground finch lives on two of the Galápagos Islands. In addition to eating insects, the birds drink nectar from flowers, the contents of eggs, and sometimes, the blood of other birds. Scientists speculate that the strange diet may be driven by the scarcity of fresh water on the islands inhabited by this subspecies.

# Finches Feasting on Blood

A cute little finch that drinks blood? Creepily enough, the sharp-beaked ground finch, also known as the vampire finch, does exactly that when it gets thirsty. Six hundred miles off the coast of South America lie the Galápagos Islands [guh-*LAH*-puh-gohs], an isolated group of volcanic islands that are home to hundreds of plants and animals found nowhere else on Earth. Unthinkably large tortoises that weigh up to 900 pounds sway across a hot, dry landscape; 4-foot-long crested iguanas dive into the surf to feed, and then climb ashore to lie in the sun and digest the masses of seaweed in their bellies; and if the weather turns especially dry and there's little to drink, the vampire finches hop onto the tails of large sea birds and sip their blood.

It's no wonder that Charles Darwin was fascinated by the animals of the Galápagos. Four years earlier and just out of college, Darwin had landed a job as ship's naturalist for the British survey ship HMS *Beagle*. By the time he arrived at the islands in the fall of 1835, he was 26 and a seasoned naturalist who'd already spent years in South America. "Nothing could be less inviting," he wrote of the dry, volcanic islands. But the Galápagos plants and animals fascinated him and he collected them avidly—both for study on the ship and for when he would return home, by way of Tahiti, Australia, and South Africa.

Despite Darwin's interest, it was only on his return to England that he understood the significance of what he had collected in the Galápagos. Many of the species were found on only a single island and nowhere else in the world. And a scattering of odd finches and blackbirds he had collected turned out to hold the key to a mystery that had begun to intrigue Darwin.

How do populations change? How do new species arise? Why are there so many unique species in the Galápagos Islands?

In this chapter we will see that biological evolution explains a great deal about life on Earth, and we will explore the answers to Darwin's questions.

**MAIN MESSAGE** Strong evidence for evolution is provided by fossils, features of existing organisms, continental drift, direct observations of genetic change, and species formation.

## KEY CONCEPTS

- Evolution is change in the overall genetic characteristics of groups of organisms over successive generations. Populations evolve; individuals do not.

- For evolution to occur, there must be genetic variation among the individuals in a population. DNA mutations introduce genetic variation into a population.

- Individuals with advantageous genetic characteristics may survive and reproduce at a higher rate than other individuals do—a process known as natural selection. The characteristics of individuals that produce more offspring become more common in succeeding generations.

- Adaptive traits are characteristics that improve an organism's performance in its particular environment. Natural selection causes a population to become better suited to its environment—a process known as adaptation.

- When one species splits into two, the two descendant species share many features because they evolved from a common ancestor.

- The great diversity of life on Earth has resulted from the repeated splitting of a single species into two or more species.

- An enormous amount of evidence shows that evolution has occurred. The fossil record, the features of existing organisms, patterns of continental drift, direct observations of genetic change, and the present-day formation of new species— all of these provide strong evidence for evolution.

EARTH TEEMS WITH LIVING THINGS. One of the most striking aspects of the planet's many organisms is the beautiful fit that so many exhibit for life in their particular environments. A hawk with its broad wings and powerful muscles can soar easily through the sky, a flower's colorful petals and sweet scent quickly attract pollinators, and an insect's body may look so much like the leaves around it that it is almost perfectly hidden from the view of hungry predators. How did organisms come to be so well matched to their surroundings? And why are there so many kinds of animals, plants, fungi, and other organisms? That is, why is there such a great diversity of life? And within that diversity, why do organisms share so many characteristics? The answer to all these questions is the same: biological evolution.

This chapter presents a broad overview of biological evolution, starting with ideas about the origins of living organism that had been percolating even before Charles Darwin and another Englishman, Alfred Wallace, began their globe-trotting explorations of the diversity of life. We then provide an overview of the mechanisms that cause evolution, the power of evolution to explain the characteristics of living things, the evidence that shows evolution is happening, and the impact of evolutionary thought on human society. We close the chapter with an example of evolution in action that also explains why there are so many odd finches in the Galápagos.

## 17.1 Descent with Modification

Through the ages, most cultures have viewed our planet, and all life on it, as fixed and unchanging. The Greek philosopher Aristotle (384–322 BC) saw the living world as unalterable and classified life-forms in 11 grades according to their level of perfection, with plants toward the bottom and humans at the pinnacle of perfection. The Greek philosophers had a profound influence on Western civilization, and the Aristotelian view of nature dominated for hundreds of years. The literal interpretation of scripture, especially the book of Genesis, shaped Judeo-Christian views about the origins of life, and these were embellished by a succession of biblical scholars. James Ussher, a seventeenth-century archbishop of Armagh in Northern Ireland, claimed to know the exact date that all life was created: October 23, 4004 BC.

## With the industrial revolution came doubts about the constancy of the world

The industrial revolution, launched in the middle to late eighteenth century, brought with it a new understanding of landforms as geologists explored Earth's crust in search of coal and minerals. Fossils, the remains or imprints of past life-forms, turned up regularly as rocks were mined and quarries were dug. A number of scholars began to advance the idea that Earth was much older than the 6,000 years claimed by some, that not all present-day life-forms had existed in the early history of Earth, and that species have changed and new species have appeared over time. Sharing his ideas with just a small readership, Comte George-Louis Leclerc de Buffon, wrote in his 1759 book, *Histoire Naturelle*, that species change over time either because of chance or because the changes are imposed by the environment.

Buffon's protégé, Jean-Baptiste Lamarck, was more vocal in his belief that life-forms change over time. Like other thinkers before him, Lamarck noted that many features of a particular organism enable it to function well in its particular habitat. He observed that the long necks of giraffes enable those animals to browse the tops of trees. Today we use the term "adaptation" to describe the evolutionary process that matches the inherited characteristics of an organism to its environment.

Lamarck missed the mark, however, when he tried to explain *why* life-forms change and *how* a group of organisms becomes better adapted to its surroundings. Writing in 1809, the year Charles Darwin was born, Lamarck claimed that the use or disuse of body parts is an inherited trait, and that the inheritance of such *acquired*

(a)

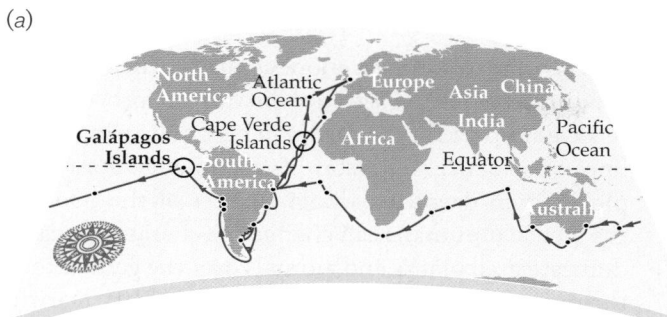

(b)

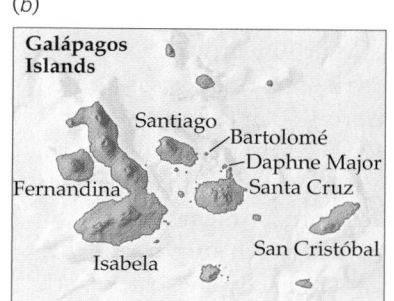

(c)

(d)

**FIGURE 17.1 Charles Darwin Was Ship's Naturalist on the HMS *Beagle***
(*a*) The course sailed by the *Beagle*. (*b*) The Galápagos Islands, located 1,000 kilometers to the west of Ecuador. (*c*) A giant Galápagos tortoise. (*d*) Charles Darwin.

*characteristics* by offspring causes the next generation to change—that is, to evolve. Giraffes have long necks, Lamarck argued, because the ancestral animals stretched their necks to browse on high branches; and because the stretched-neck trait was passed on to offspring, necks got longer with each generation. Lamarck's compatriot Georges Cuvier (1769–1832) was among many who pointed out that Lamarck's explanation for adaptation was demonstrably wrong. Workmen with bulging muscles do not necessarily father children with large muscles, and training a border collie to run fast would not ensure that its descendants would also be fleet of foot.

Fossils had been known to Greek and Chinese scholars from antiquity, but Cuvier's study of animal fossils was the most extensive up to that point. The fact that some species become extinct was firmly established by Cuvier. As a scholar of the comparative anatomy of vertebrates, Cuvier could see that some of the extinct forms were closely related to present-day animals, but he rejected the idea that one species can give rise to others.

In 1830 the Scottish geologist Charles Lyell published his acclaimed *Principles of Geology*, in which he detailed how the surface of Earth had been shaped by the slow action of rivers, glaciers, earthquakes, and

volcanoes. Darwin received the second volume of Lyell's treatise when the *Beagle* docked near Buenos Aires in 1832. Alfred Wallace had read Lyell's works before he set off for the tropics, and, like Darwin, he was strongly influenced by Lyell's account of change and transformation in the history of Earth.

## Darwin offers a unifying explanation: descent with modification

After pondering the oddities of the Galápagos, and the many other biological observations he made during his 5-year voyage on the HMS *Beagle* (**FIGURE 17.1**), Darwin concluded that species were not, as was generally thought at that time, the unchanging result of separate acts of creation. Instead, he would come to a bold new conclusion—one that would explain his odd little finches, as well as both the unity and diversity of all life on Earth. Species, he proposed, had descended with modification from ancestor species; that is, an ancestral lineage had changed over time to give rise to entirely new species. Darwin's compatriot and contemporary Alfred Wallace had studied the diversity of related species on the many different islands of the Malaysian archipelago, and he, too, concluded that new species arise from ancestral forms.

Charles Darwin and Alfred Wallace went further than any scholar before them in proposing a *mechanism* for the evolution of new species: **natural selection**,

In everyday conversation, the term "fittest" means the person who is strongest or in the best physical shape. When biologists talk about the survival of the fittest, however, they mean something quite different. The fittest are not necessarily the strongest individuals, but those individuals who are the bearers of advantageous inherited traits of any kind that enable them to survive and reproduce more than others. Hence, natural selection is the survival of the fittest.

which we can define as a nonrandom evolutionary process that adapts a population to its environment. Both men were influenced by an essay written by the Reverend Thomas Malthus, in which the clergyman stated that the growth of populations is limited by the availability of resources such as food and by adversities such as disease. According to Malthus, there is a struggle for survival when individuals produce more offspring than their environment can support, and there are winners and losers in such a struggle.

Both Darwin and Wallace were struck by the same idea: if there is a struggle for survival, with winners and losers, then the characteristics of the winners should be better represented in the descendant generation because the winners will live to produce more offspring than will the losers. As a consequence of such *differential reproduction*, the descendant population would come to have a different set of inherited characteristics than the original population had. Over the generations, environmental pressures could so modify the descendants that they would emerge as a distinctly different form—that is, a new species. Mutual friends, which included Charles Lyell, arranged to have Darwin's and Wallace's ideas presented jointly to the Linnaean Society of London in 1858. The presentation received scant attention, and it was not un-

til Darwin published his detailed and monumental work, *On the Origin of Species*, in 1859, that the world took notice.

Darwin tends to get more credit for the theory of evolution by natural selection because he provided extensive support for the theory in his book, using many lines of evidence to bolster his arguments. For example, Darwin devotes a whole chapter of the book to the fact that humans can change the characteristics of domesticated plants and animals over the generations through artificial selection (see Figure 1.10). Darwin reasoned that natural forces, acting on individuals competing for existence, could also produce such dramatic changes in a descendant population.

Groups of organisms evolve, Darwin said, when natural selection favors individuals with advantageous inherited characteristics (**FIGURE 17.2**). Characteristics that enable the individual to function well in its particular environment would enable that individual to survive and reproduce better than those in the population that lack the advantageous characteristics. Over the generations, the proportion of individuals possessing the beneficial characteristics grows, meaning that the population has evolved as a result of natural selection. When natural selection acts differently on two populations—favoring different sets of characteristics because each population is exposed to a different set of environmental conditions, for example—they evolve along separate paths and may emerge as two new species descended from one ancestral population.

It is hard to overstate the importance of Darwin's work. Evolution is biology's most powerful explanation for why living things are as they are, why they look, act, sound, breathe, grow—why they do everything—as they do. The theory of evolution through natural selection has stood the test of time, although, as we recount in this chapter, we now also know of other means by which organisms can evolve.

Evolution can be examined at the population level (**microevolution**), focusing on evolutionary changes that occur over relatively short periods of time. We will highlight this small-scale perspective of evolution in Chapter 18. Population-level changes produce large-scale changes, such as the splitting of a single population into two different species, a process known as *speciation*. **Macroevolution** is the study of the history of life from the perspective of speciation and other large-scale consequences of population-level changes. Macroevolution is essentially the history of the formation and extinction of species and higher taxonomic groups over time, and it will be the focus of

**DESCENT WITH MODIFICATION**

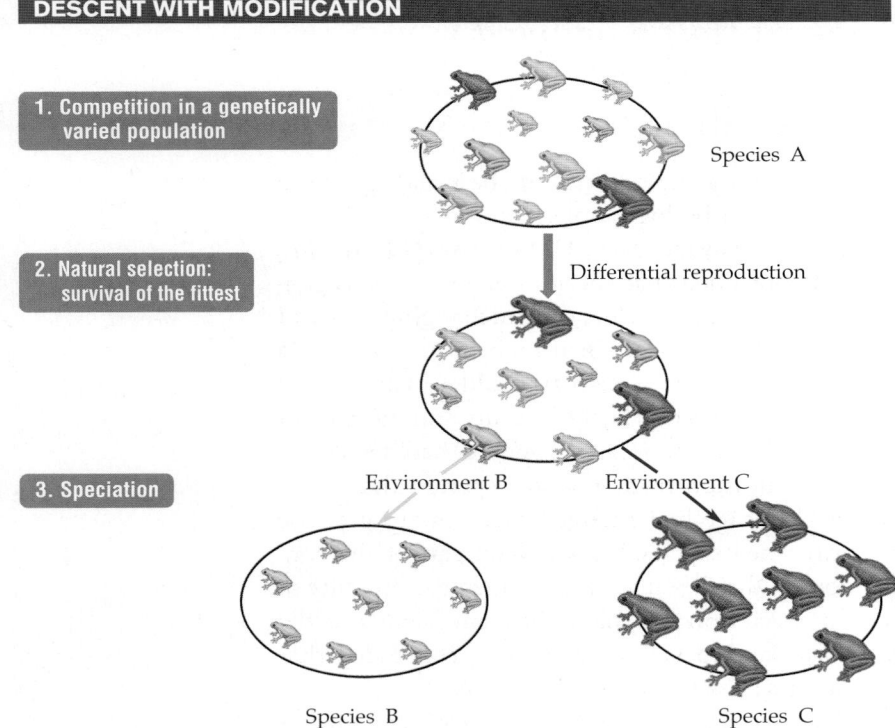

1. Competition in a genetically varied population

Species A

Differential reproduction

2. Natural selection: survival of the fittest

3. Speciation

Environment B    Environment C

Species B    Species C

**FIGURE 17.2 Darwin Proposed Natural Selection as the Mechanism behind Species Formation**

Chapter 19. The rise and fall of different groups of organisms over time provides us with a grand view of the history of life on Earth, and we will tell that story in Chapter 20.

## 17.2 Mechanisms of Evolution

There are many different ways to define evolution. In Chapter 1 we defined **biological evolution** broadly as a change in the overall inherited characteristics of a group of organisms over multiple generations of parents and offspring. We noted that evolution occurs at the level of the population, and not at the level of the individual.

With the knowledge of genetics gained in Unit 3, we can define evolution as a change in the *gene pool*. The **gene pool** is the sum of all the genetic information carried by all the individuals in a population. As we will learn in this section, four mechanisms can change the gene pool of a population: (1) mutation, (2) gene flow, (3) genetic drift, and (4) natural selection.

### Mutations introduce genetic variation in a population

In a natural population, individuals vary in their structural, biochemical, and behavioral traits. Much of that variation is under genetic control, and such DNA-based differences in observable traits constitute **genetic variation** in a population. **FIGURE 17.3** shows genetic variation in the color patterns of happy-face spiders, which are found only on four of the islands of the Hawaiian archipelago: Oahu, Molokai, Maui, and Hawaii. Dr. Rosemary Gillespie and her colleagues discovered that the *yellow* morph is the most common, making up two-thirds of all the color patterns on each island. Although the *yellow* morph is common to the four islands, the non-*yellow* morphs vary in distribution, with each island harboring some unique color patterns in its happy-face spider population.

Where does genetic variation come from? **Mutations**, which are random changes in the DNA sequence, are the original source of all genetic variation. The DNA variants produced by mutation are known as **alleles** (see Chapter 12). The allelic makeup of an individual is the **genotype** of that individual. The var-

ied color patterns on the backs of happy-face spiders are produced by many different alleles of at least one, but probably more than one, gene. New mutations that are inherited cause a population to evolve because the addition of new alleles changes the gene pool—the overall genetic composition of the population.

Mutations that occur in an organism's gametes (the spider's egg or sperm cells, as opposed to mutations in the animal's muscle cells, for example) can be passed on to the next generation. Gene mutations are caused by various accidents, such as mistakes in DNA replication, collisions of the DNA molecule with other molecules, or damage from heat or chemical agents. Despite the efficiency with which repair proteins fix damage to DNA and correct errors in DNA replication (see Chapter 14), mutations occur regularly in all organisms. Humans, for example, have two copies each (one copy from each parent) of approximately 25,000 genes. On average, between two and three of these 50,000 gene copies have mutations that make them different from those of either parent. Mutations and the genetic variation they produce do not appear because an organism "needs" them; instead, mutations occur at random and are not directed toward any goal.

Sexual reproduction adds to genetic variety in a population because the alleles of the two parents are shuffled and combined in new ways in each offspring. In sexually reproducing organisms, as we saw in Chapters 10 and 12, the alleles generated by mutation are grouped in new arrangements by crossing-

**GENETIC VARIATION**

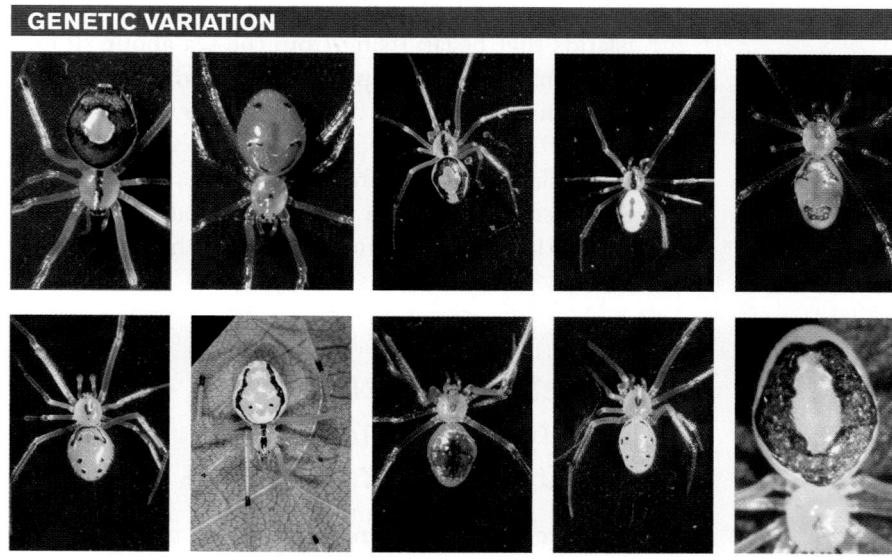

**FIGURE 17.3  Genetic Variation in Happy-Face Spiders**
Markings on the backs of happy-face spiders, which inhabit four of the Hawaiian Islands.

over, independent assortment of chromosomes, and fertilization; collectively, these three processes are known as **sexual recombination**. Because of sexual recombination, offspring have novel groupings of alleles that differ from the allelic combinations in either parent. Sexual recombination in a population does not by itself cause a population to evolve, because it does not change the gene pool, the *sum* of all the genetic variation that exists in a population. However, by rearranging the genetic variation produced originally by mutation, sexual recombination creates new combinations of traits that serve as raw material for other evolutionary processes.

## Gene flow moves genes between populations

**Gene flow** is the movement of genes from one population to another. When migrants from a different population breed with a resident population, they may introduce new alleles to the resident's gene pool. If its gene pool is altered by the introduction of novel alleles, the resident population has evolved. Emigration, or out-migration, can also change the gene pool of a population if it brings with it a decline in the frequency (commonness) of alleles harbored by the emigrants.

The *red blob* morph of the happy-face spider is found only on Oahu. If the *red blob* morph migrates from its homeland to live and reproduce in nearby Molokai, the gene pool of the Molokai spiders will change and that population will have evolved. The Hawaiian archipelago consists of a chain of volcanic islands that began rising from the sea about 70 million years ago. The happy-face spider probably evolved on Oahu, the oldest of the four islands where the species is found, and it may have dispersed to the other three islands shortly after it emerged as a new species. However, Dr. Gillespie has determined that there is very little, if any, gene flow among the happy-face spiders on the four islands today.

## Genetic drift generates differential reproduction through accidental events

**Genetic drift** is a random process that, by sheer accident, causes individuals with a unique set of characteristics to die off, while others with a differ-

ent set of characteristics survive and reproduce. The characteristics of the successful reproducers will be well represented, while the characteristics of the lost individuals become less common or will disappear in succeeding generations. Genetic drift is a random process, and it can cause the gene pool of a population to fluctuate randomly over time, rather than being pushed in a particular direction. Genetic drift is more likely to alter the gene pool of a small population than that of a large population. Because it *can* change the gene pool, genetic drift *can* cause a population to evolve.

Imagine that a hurricane destroys much of the happy-face spider population on Molokai, the smallest of the four islands that harbor the species. Almost all the *yellow* morphs are wiped out, but by sheer chance most of the previously uncommon *red smile* morphs survive. The *red smile* morphs are likely to produce more offspring overall than the almost-vanquished *yellow* morphs, in which case the gene pool will change substantially, and the *red smile* morphs will outnumber the *yellow* morphs in the next generation. As **FIGURE 17.4** shows, such chance differences in survival can alter the genetic makeup of the descendant population.

## Natural selection generates adaptation in a population

Evolution in populations is by no means restricted to chance events. The fourth evolutionary mechanism, natural selection, is not a random process. Unlike the other three mechanisms of evolution, natural selection is a directional process because it shifts the genetic characteristics of a population along a specific path—one that leads to adaptation—over successive generations.

Individuals in a natural population must compete for food, mates, living space, and other resources that help an organism survive long enough to reproduce. As pointed out by Malthus, and noted by Darwin and Wallace, organisms typically produce more offspring than can survive to themselves reproduce. In this competitive environment, any individual with an advantageous inherited characteristic—one that enables the individual to function better than others in that habitat—is more likely to survive and reproduce and to pass those characteristics on. Meanwhile, an individual that lacks these advantageous characteristics, or has disadvantageous characteristics, is less likely

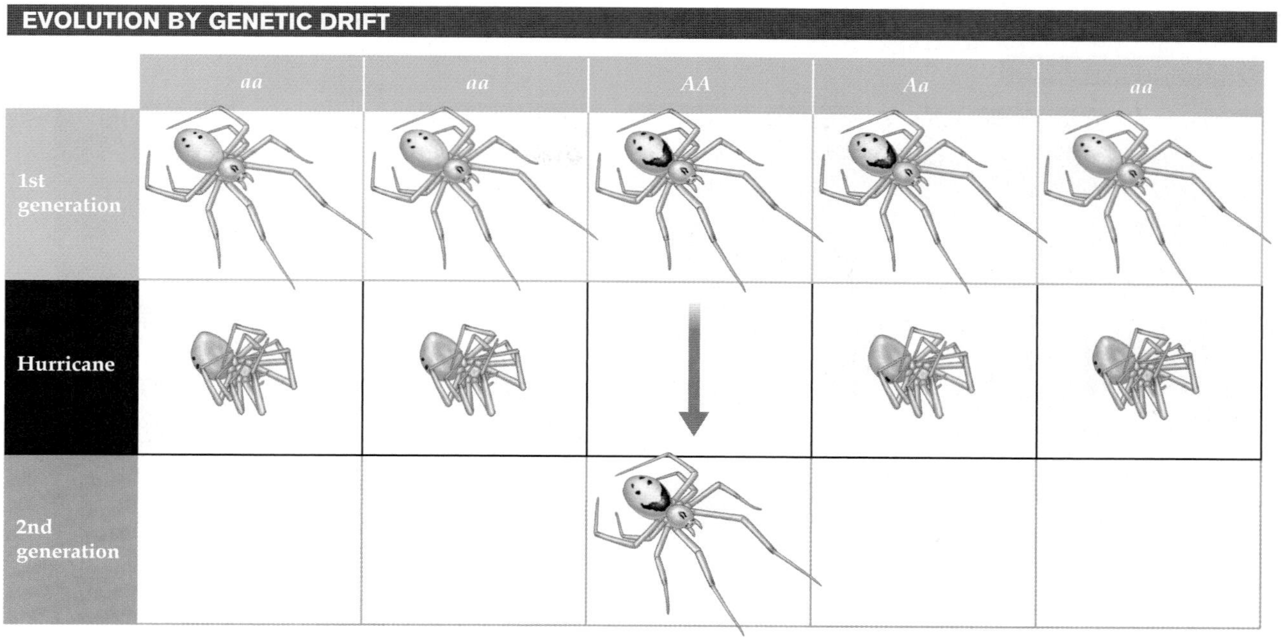

**FIGURE 17.4 Genetic Drift Can Drastically Alter the Genetic Makeup of a Population**

Chance events can determine which individuals survive and reproduce. Here, a hurricane destroys most of the spiders on a small island. By chance alone, nearly all the surviving spiders have two *A* alleles (genotype *AA*), causing the proportion of *A* alleles in the population to change from 30 percent to 100 percent in a single generation.

to survive and reproduce and to pass on those characteristics. As a result, advantageous characteristics become more common among the offspring, and disadvantageous characteristics disappear or become less common over successive generations.

Natural selection causes the gene pool of descendants to become different from that of the original population, and the descendant population therefore evolves. Individuals who possess **adaptive traits**—genetic characteristics that enable the individuals to function well in a competitive environment—become more common among the descendants. The evolutionary process by which a population as a whole becomes better matched to its habitat is known as **adaptation**. Over time, natural selection can cause a population to evolve so that more and more individuals have adaptive traits, and fewer and fewer have the disadvantageous, or maladaptive, traits.

There are many examples of adaptation as a result of natural evolution. **FIGURE 17.5a** depicts the evolution of pale fur in populations of beach mice on barrier islands off the coast of Florida. Dr. Hopi Hoekstra and her colleagues studied beach mice sub-species unique to each of five barrier islands off the Gulf coast and three off the Atlantic coast of the state. All eight subspecies have light coloration, which is an adaptive trait on the white-sand beaches that make up their habitat. Using DNA analysis, the scientists deduced that the eight subspecies descended from oldfield mice (*Peromyscus polionotus*) from the mainland over a span of about 6,000 years, when the barrier islands were formed. Darker coat patterns are common among oldfield mice, most likely because they afford better camouflage in the darker soils and denser vegetation of the coast. However, dark-colored mice are at a disadvantage on the sandhills of the barrier islands because predators such as hawks and herons can see them more easily.

The researchers arranged clay models of mice with pale coats and dark coats on the beach and found that the ones with dark coloration were 50 percent more likely to have missing parts or be missing altogether (**FIGURE 17.5b**). The pattern of attack was reversed when the models were set out on dark soils, with lighter models being attacked 50 percent more often than the paler versions. Tracks in the

**Helpful to know**

Adaptation is the evolutionary *process* that makes a population better suited to its habitat. However, the word "adaptation" is also used for an advantageous *characteristic*, an inherited structural, biochemical, or behavioral feature that enables an organism to function well and therefore to survive and reproduce better than competitors lacking the characteristic; in this latter sense, "adaptation" is synonymous with "adaptive trait."

## FIGURE 17.5

**Natural Selection Adapts a Population to Its Environment**

Researchers studied subspecies of a deer mouse (*Peromyscus polionotus*) that inhabit each of eight barrier islands off the coast of Florida. Darker and more extensive areas of pigmentation are most common among the mainland subspecies (called the oldfield mouse). (*a*) The barrier island subspecies, known as beach mice, tend to have paler coats and reduced areas of pigmentation on the back. (*b*) Researchers used clay models of oldfield mice and beach mice to test the hypothesis that predators will see and attack coat patterns that contrast with their background.

**ADAPTATION THROUGH NATURAL SELECTION**

**(a) Adaptation in coat color**

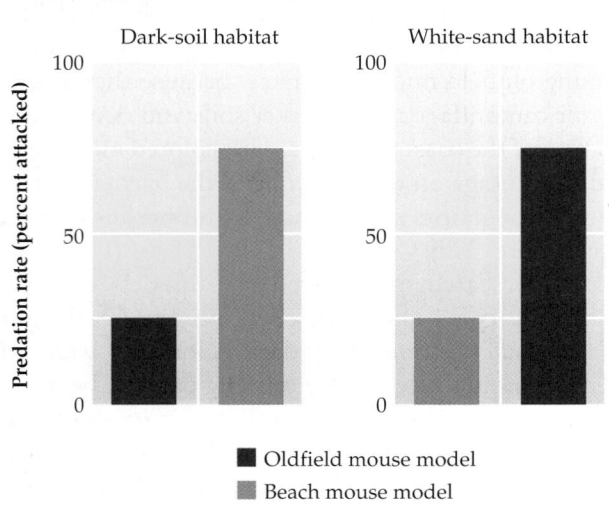

Mainland

Salt marsh

**Migration**

Barrier island

**(b) Predation on clay models of oldfield and beach mice**

Dark-soil habitat | White-sand habitat

Predation rate (percent attacked)

■ Oldfield mouse model
■ Beach mouse model

sand revealed that birds and carnivores such as coyotes were about equally represented among the attackers. The scientists demonstrated that beach mice populations on the barrier islands have mutations in one or more of three different genes, including the melanocortin receptor gene (*MC1R*), all of which depress pigment formation in different parts of the skin. Note that natural selection does not craft perfection for all time and all circumstances. Mutations in the *MC1R* gene may be advantageous on the barrier islands, but the same mutations are maladaptive on the mainland.

Like genetic drift, natural selection acts through differential reproductive success: some individuals produce more offspring than others; therefore, their characteristics become more common among the descendants. But in natural selection, differential reproduction is governed not by accidental events, but by whether or not individuals possess adaptive traits. Note, however, that the process of natural selection does not *create* advantageous traits. That is, natural selection does not cause the mutations (such as a change in the *MC1R* gene) that give individuals advantageous traits. Mutations that cause advantageous traits happen randomly. If a mutation produces an advantageous trait, then natural selection can act upon it. Through the process of natural selection, individuals that have such advantageous traits will survive and reproduce more than others and so, over successive generations, individuals with such advantageous traits will come to dominate a population.

> ### Concept Check

1. What is the ultimate source of genetic diversity in a population?

2. In what way are genetic drift and natural selection similar, and in what way are they different?

## 17.3 Evolution Can Explain the Unity and Diversity of Life

Life on Earth is distinguished by exquisite matches between organisms and their environments, by a great diversity of species, and by many puzzling examples of organisms that differ greatly in numerous respects yet share certain key characteristics. The fact that life evolves can explain all these features of life on Earth. In the following discussion we consider two of life's most striking features: (1) shared characteristics and (2) diversity.

## Organisms share characteristics as a result of common descent

The natural world is filled with many seemingly puzzling examples of very different organisms sharing certain characteristics that we would not imagine they should necessarily share. Consider the appendages we see in the wing of a bat, the arm of a human, and the flipper of a whale. They all have five digits and contain the same set of bones (**FIGURE 17.6**). But why should limbs that look so different and have such different functions share the same set of bones? Many living organisms also show the puzzling characteris-

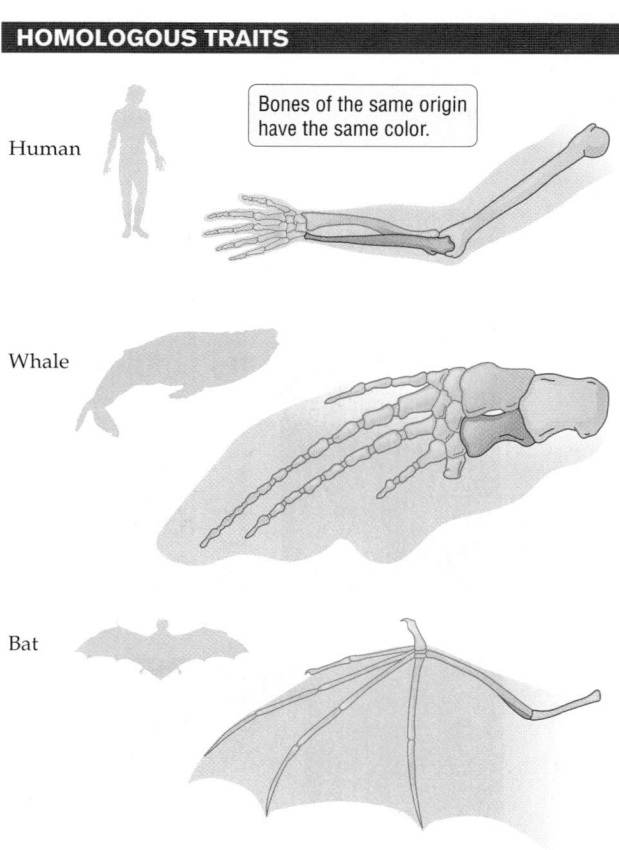

**HOMOLOGOUS TRAITS**

Bones of the same origin have the same color.

Human

Whale

Bat

**FIGURE 17.6 Shared Characteristics**
The human arm, the whale flipper, and the bat wing are homologous structures, all of which have what are essentially a matching set of five digits and a matching set of arm bones that have been altered by evolution for different functions.

FIGURE 17.7
A Python Has
Rudimentary
Hind Legs Visible
Externally and
Internally

**VESTIGIAL ORGANS**

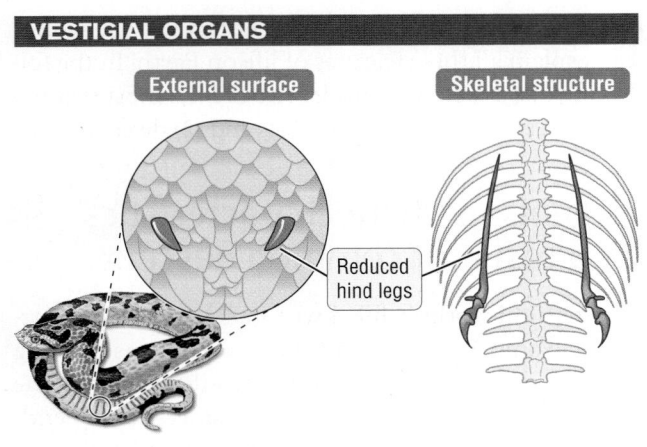

External surface

Skeletal structure

Reduced hind legs

tic of having features that appear to be of no use to them, such as **vestigial organs**, which are reduced or degenerate parts whose function is hard to discern. For example, why do we humans have a reduced tailbone and the remnants of muscles for moving a tail? Why do whales have remnants of hip bones when they have no leg bones requiring support? And why do some snakes have rudimentary leg bones but no legs (**FIGURE 17.7**)? The answer to all these questions is the same: evolution.

Many similarities among organisms are due to the fact that the organisms evolved from a common ancestor. When one species splits into two, the two species that result share many features because they evolved from a common ancestor. Features of organisms related to one another through common descent are said to be **homologous [hoh-*MAH*-luh-gus]**. For example, the wing of a bat, the arm of a human, and the flipper of a whale share the same set of homologous bones (see Figure 17.6). Similarly, some snakes have rudimentary leg bones because they evolved from reptiles with legs (see Figure 17.7), and humans have rudimentary bones and muscles for a tail because our (distant) ancestors had tails.

Organisms can also share features as a result of **convergent evolution**, which occurs when natural selection causes distantly related organisms to evolve similar structures in response to similar environmental challenges. For example, the cacti found in North American deserts share many convergent features with distantly related plants found in African and Asian deserts (**FIGURE 17.8a–c**). Similarly, although both sharks (a kind of cartilaginous fish) and dolphins (mammals) have bodies streamlined for aquatic life,

**FIGURE 17.8  The Power of Natural Selection**

(*a–c*) These three plants evolved from very different groups of leafy plants. They resemble one another because of convergent evolution: each was similarly driven by natural selection to adapt to life in a desert. Their shared structures (fleshy stems, spines, reduced leaves) are therefore analogous, not homologous. (*a*) *Euphorbia* **[yoo-*FOHR*-bee-uh]** belongs to the spurge family and can be found in Africa. (*b*) *Echinocereus* **[ee-*KYE*-noh-*SEER*-ee-us]** is a cactus found in North America. (*c*) *Hoodia*, a fleshy milkweed, can be found in Africa. (*d,e*) Convergent evolution can be a powerful force shaping animals as well. Here we see how natural selection has caused two distantly related animals—sharks (*d*) and dolphins (*e*)—to look very similar. Sharks are a kind of fish; dolphins are mammals.

**CONVERGENT EVOLUTION**

(*a*)  (*b*)  (*c*)

(*d*)  (*e*)

these species are only very distantly related, and their overall similarities result from convergent evolution, not common descent (**FIGURE 17.8d–e**). When species share characteristics because of convergent evolution, not common descent, those characteristics are said to be **analogous [uh-*NAL*-uh-gus]**.

## The diversity of life results from the splitting of one species into two or more species

Earth is home to millions of species. Why are there so many different kinds of living things? Here, too, evolution provides a simple, clear explanation: the diversity of life is a result of the repeated splitting of one species into two or more species—a process called **speciation**.

How do new species evolve? Speciation can happen in several different ways. One of the most important is the geographic separation, or geographic isolation, of populations. Consider two populations of a species that are isolated from each other, as by a mountain range or other barrier that prevents individuals from moving between the populations—or, in the case of the separate Galápagos islands, by the sea. The habitats in which the two populations of the original species reside will be different. Over time, natural selection may cause each population to become better adapted to its own particular environment on its own side of the barrier, leading to changes in the genetic makeup of both populations. Chance events may also cause genetic drift. Such chance events may cause the genetic makeup of the populations to begin to diverge. Eventually, so many genetic changes may accumulate between the two populations that individuals from the two populations are no longer able to reproduce with each other.

As we learned in Chapter 1, species are often defined in terms of the ability to interbreed: a species is a group of populations whose members can reproduce with each other but not with members of other such groups. Geographical isolation can lead to speciation because two populations that are physically separated may accumulate so many different types of genetic changes that they no longer interbreed.

## 17.4 The Evidence for Biological Evolution

Surveys taken during the past 10 years reveal that almost half the adults in the United States do not believe that humans evolved from earlier species of animals. The results of these surveys are startling because evolution has been a settled issue in science for nearly 150 years. The vast majority of scientists of all nations, races, and creeds think that the evidence for evolution is very strong. In his landmark book, *On the Origin of Species*, Charles Darwin argued convincingly that organisms are descended with modification from common ancestors. The scientific issues and questions being studied and debated by biologists today are not about *whether* evolution occurs but concern the details of *how* it occurs.

For example, biologists are still debating the relative importance of natural selection and other mechanisms of evolution (such as genetic drift), but they do not dispute whether evolution occurs. Today's scientific debate about the details of the workings of evolution can be compared to a dispute over what causes wars. Although we might argue about the causes, we all know that wars happen.

Why do scientists find the case for evolution so convincing? As we saw in Chapter 1, a scientific hypothesis must lead to predictions that can be tested, and hypotheses about evolution are no exception. Scientists have tested many predictions about evolution and have found them to be strongly supported by the evidence. Six lines of evidence provide compelling support for biological evolution: (1) fossils, (2) traces of evolutionary history in existing organisms, (3) similarities and divergences in DNA, (4) direct observations of genetic change in populations, (5) continental drift, and (6) the present-day formation of new species.

## Evolution is strongly supported by the fossil record

**Fossils** are the preserved remains (or their impressions) of formerly living organisms. The fossil record enables biologists to reconstruct the history of life on Earth, and it provides some of the strongest evidence that species have evolved over time. For example, the fossil record shows that structurally simpler organisms preceded the more complex forms in life's history. Multicellular organisms are absent in the oldest rocks that contain single-celled forms. The rocks that entombed the oldest fossil amphibian contain fish fossils, but none of reptiles or mammals.

As we saw in Chapter 2, the evolutionary relationships among organisms—their pattern of descent from a common ancestor—can often be determined by comparison of their anatomical characteristics.

# Can't Live with 'Em, Can't Live without 'Em

Evolution is a natural process, but it also occurs in response to human interventions like the increasing development of antibacterial (and, to a lesser extent, antiviral) consumer products. Once largely limited to hospitals and other places at high risk for infections, antibacterial agents are now routinely added to soaps, lotions, dishwashing liquids, and other cleaning products. A study published in *Annals of Internal Medicine* estimated that 75 percent of liquid soaps and 29 percent of bar soaps available in the United States contain antibacterial ingredients. Furthermore, many additional products are now coated or impregnated with antibacterial agents, including facial tissues, cutting boards, toothbrushes, bedding, and children's toys.

So what's the problem? Wouldn't anyone prefer to avoid disease-causing bacteria? If it were that simple, the answer would be yes, but these antibacterial products also kill the beneficial bacteria that surround us in our environment and, according to the World Health Organization, contribute directly to the spread of antibiotic-resistant bacteria. Note that as always, natural selection does not create advantageous traits. Natural selection does not cause antibiotic-resistant bacteria to arise out of nonresistant bacteria. Instead, once antibiotics are used, bacteria that are already resistant gain a huge advantage over nonresistant bacteria because the resistant bacteria are not killed off by the antibiotics. Resistant bacteria outreproduce the other bacteria, and so, through the process of natural selection, these bacteria quickly come to dominate bacterial populations.

In 2000, the American Medical Association (AMA) advised consumers to avoid extensive use of "antibacterial soaps, lotions, and other household products" and also called for greater regulation of antibacterial products. In 2005, the Food and Drug Administration (FDA) further announced that these soaps have no benefit over ordinary soap and water for ridding hands of bacteria. In fact, numerous studies have begun to show that the use of antibacterial consumer products does not decrease the frequency of illness in people that use them. The following guidelines (based on information from the American Society for Microbiology) will help you stay healthy and keep your home clean, while avoiding the risks associated with antibacterial products:

- Plain old soap and hot water remain the best for washing your hands, body, and dishes.
- Wash your hands thoroughly and often before you:
  Prepare or eat food
  Treat a cut or wound or tend to someone who is sick
  Insert or remove contact lenses
- and after you:
  Use the bathroom
  Handle uncooked foods, particularly raw meat, poultry, or fish
  Change a diaper
  Blow your nose, cough, or sneeze
  Touch a pet, especially reptiles and exotic animals
  Handle garbage
  Tend to someone who is sick or injured
- Limit your use of antibacterial products to situations when you are most at risk, such as when you are unable to wash your hands.
- Use bleach to clean your bathroom.
- Use separate cutting boards for raw meat and foods that may not be cooked before eating (for example, fruits and vegetables).
- Wash all fruits and vegetables either with soapy water (rinse thoroughly, of course) or with one of the new fruit and vegetable washes.
- Wash all kitchen surfaces, dishes, and utensils in hot, soapy water. Make sure to rinse thoroughly. If possible, put everything (including cutting boards) in the dishwasher.
- Every time you run your dishwasher, throw in the kitchen sponge.
- Do not wipe counters with a sponge that has been sitting on your sink. This can deposit even more bacteria on countertops. Use paper towels or replace your dish rag every day with a clean one.

The use of antibiotics for raising farm animals has contributed to the rise of "superbugs," bacterial strains resistant to multiple antibiotics.

Applying this technique to fossils, we find that the fossil record contains excellent examples of how major new groups of organisms arose from previously existing organisms. We will discuss one of these examples, the evolution of mammals from reptiles, in Chapter 20. Such fossils, which provide evidence for descent with modification by showing how new organisms evolved from ancestral organisms, exist for many other groups as well, including fishes, amphibians, reptiles, birds, and humans.

Finally, the evolutionary and ecological history of some groups is mirrored in an almost seamless fossil record that shows many intermediate forms—that is, species with some similarities to the ancestral group and some similarities to the descendant species. For example, scientists can track the evolutionary history of the horse in North America and match the changes in the anatomy of descendant species with changes in the continent's climate over the past 60 million years. *Hyracotherium*, the earliest known member of the horse family (the Equidae, or equids), lived at a time when much of North America was covered in lush rainforest. *Hyracotherium* was a dog-sized browser that walked on four small toes on its front feet and three on the hind feet. Many of the equid species that appeared later had teeth adapted for grazing on tough grasses and legs adapted for fast running (**FIGURE 17.9**), consistent with a change in the climate.

Geologic and climatological evidence shows that about 32 million years ago, the continental interior became drier and prairies replaced forests in much of the central plains of North America. Compared to *Hyracotherium*, the legs are longer in many of the fossil equids that first appear at this time in the central plains. In addition to long legs for running speed, the eye sockets were placed farther back and higher on the head, so these equids had a wide field of view for spotting predators in the open grasslands. *Merychippus*, which appeared about 17 million years ago, ran on a single large toe, just as present-day horses, asses, and zebras do. Its cheek teeth (molars) had large, flat grinding surfaces for breaking up tough grasses—another similarity

Fossils like this *Triceratops* skeleton show evidence of organisms now no longer alive.

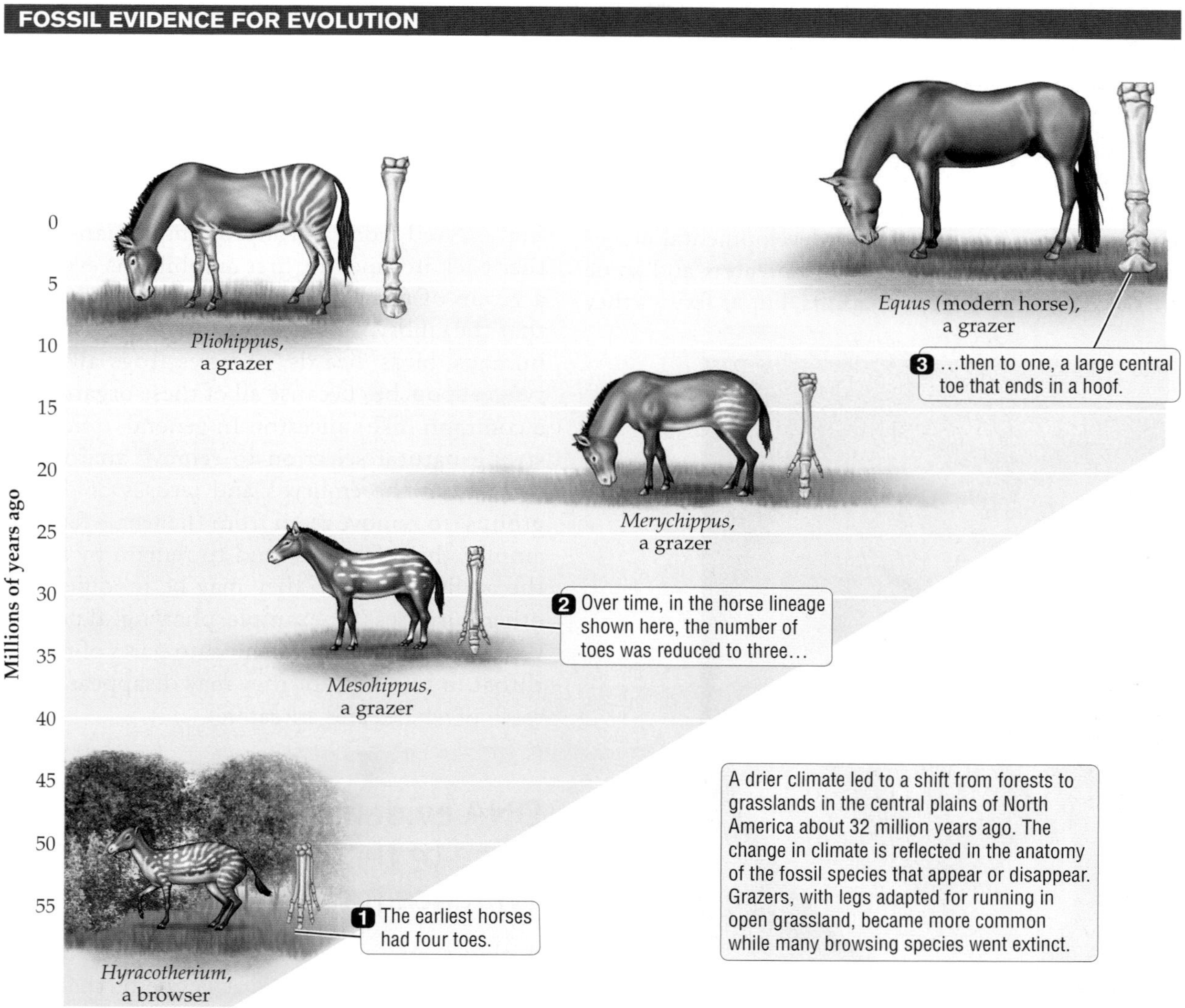

## FOSSIL EVIDENCE FOR EVOLUTION

Millions of years ago

0
5
10
15
20
25
30
35
40
45
50
55

*Pliohippus*, a grazer

*Equus* (modern horse), a grazer

**3** ...then to one, a large central toe that ends in a hoof.

*Merychippus*, a grazer

**2** Over time, in the horse lineage shown here, the number of toes was reduced to three...

*Mesohippus*, a grazer

**1** The earliest horses had four toes.

A drier climate led to a shift from forests to grasslands in the central plains of North America about 32 million years ago. The change in climate is reflected in the anatomy of the fossil species that appear or disappear. Grazers, with legs adapted for running in open grassland, became more common while many browsing species went extinct.

*Hyracotherium*, a browser

**FIGURE 17.9**
**Fossils Reveal Descent with Modification in Horse Lineages**
The evolutionary tree of the horse family (Equidae) is highly branched, as deduced from the fossil remains of more than 200 different species. Progressive changes in the anatomy, especially of the leg bones and teeth, are seen in some lineages.

with modern equids. A lineage related to *Merychippus* evolved into the modern horse (genus *Equus*), which came to inhabit temperate grasslands across northern Eurasia and North America about 5 million years ago. The last of the North American equids became extinct toward the end of the Pliocene epoch, about 12,000 years ago. Whether the demise of the North American horse was hastened by human arrivals from Asia at this time or by another climate event, or some combination of the two, is an issue that has yet to be settled.

## Organisms contain evidence of their evolutionary history

A major prediction of evolution is that organisms should carry within themselves evidence of their evolutionary past—and they do. Evidence of evolution is seen not only in homologous structures and vestigial organs, but in common patterns of embryonic development.

Patterns of growth in the very earliest stages of life can provide evidence of an organism's evolutionary past. Upon fusion of sperm and egg, an animal embryo begins to grow and develop (see Section 4.2). The manner in which an embryo develops, especially at the early stages, may mirror early developmental stages of ancestral forms. For example, anteaters and some whales do not have teeth as adults, but as fetuses they

do. Why should these organisms develop teeth during fetal development, only to reabsorb them? Or consider the observation that the embryos of fishes, amphibians, reptiles, birds, and mammals (including humans), all develop pharyngeal pouches. In fishes, the pouches develop into gills that the adults use to "breathe" underwater. But why should the embryos of organisms that breathe air develop pharyngeal pouches?

Our understanding of evolution provides an answer to these puzzles: similarities in patterns of development are caused by descent from a common ancestor. Complex structures in descendant species are generally elaborations of structures that existed in the ancestor: rather than evolving new organs "from scratch," new species inherit structures that may have been modified not only in form but sometimes even in function. The evolutionary repurposing of organs is sometimes evident as the embryo passes from the lockstep ancestral modes of development to the newly evolved patterns of development at later stages.

Fossil evidence suggests that anteater and whale fetuses have teeth because anteaters and whales evolved from organisms with teeth. Similarly, fossil evidence indicates that the first mammals and the first birds each evolved from reptiles, although from different groups of reptiles. In addition, the first reptiles evolved from a group of amphibians. Even farther back in time, the first amphibians evolved from a group of fishes. So this line of evidence indicates that the embryos of air-breathing organisms such as humans, birds, lizards, and tree frogs all have pharyngeal pouches because all of these organisms share a common (fish) ancestor. In general, unless there is strong natural selection to remove anatomical features from the embryos and fetuses of descendant groups (to remove teeth from the whale fetus, for example), these features tend to remain by default. In the adults, however, they may be modified to serve other purposes (for example, pharyngeal pouches develop into gills in fishes, but into parts of the ear and throat in humans), or they may disappear (as do the teeth of whales and anteaters).

## DNA evidence provides some of the most compelling evidence for evolution

Within every organism is another piece of evidence for evolution—DNA—and it is one of the strongest. DNA is universally used by living things as the he-

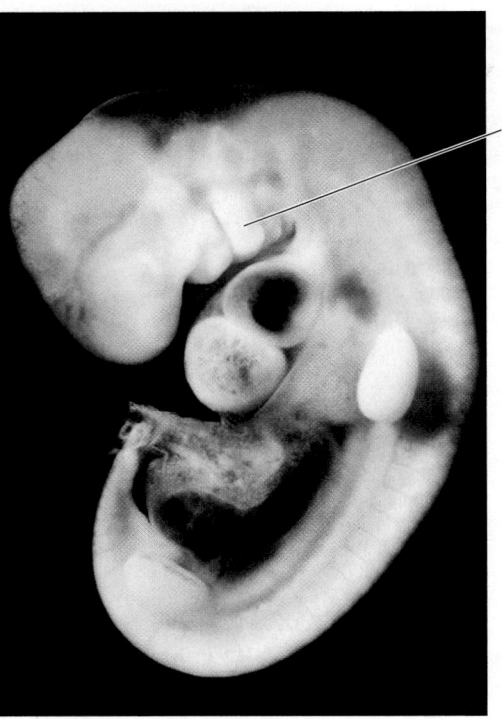

Human embryo, about 30 days after fertilization

Gill pouch

reditary or genetic material. In addition, with minor exceptions, all organisms use the same genetic code that we learned about in Chapter 15. That is, organisms translate DNA sequence information into a specific sequence of amino acids in much the same way. The fact that organisms as different as bacteria, redwood trees, and humans use DNA and the same genetic code is further evidence that the great diversity of living things descended or evolved from a common ancestor.

As we saw in Chapter 2, the evolutionary relationships among organisms—their patterns of descent from a common ancestor—can often be determined from anatomical features. These patterns of descent can be used to make predictions about the similarity among organisms of molecules such as DNA and proteins. Biologists have predicted that the DNA sequences and protein sequences of organisms that share a more recent common ancestor should be more similar—and they consistently are (**FIGURE 17.10**). If organisms were not related to one another by common descent, there would be no reason to expect

that DNA comparisons would both predict the same evolutionary relationships. The fact that we see these separate lines of evidence—anatomical features, and DNA and proteins—yielding the same result over and over again for diverse groups of organisms is strong evidence for evolution.

## Direct observation reveals genetic changes within species

In thousands of studies, naturalists and biologists from the Victorian era to the present day have observed populations changing in their overall genetic characteristics over successive generations—in the wild, in agricultural settings, and in the laboratory. As Darwin recognized, the breeding experiments conducted by humans provide direct, concrete evidence for evolution. Consider how humans have altered the gray wolf, *Canis lupus*, to produce the few hundred breeds of dogs that are recognized by various kennel societies today (**FIGURE 17.11**).

**DNA EVIDENCE FOR EVOLUTION**

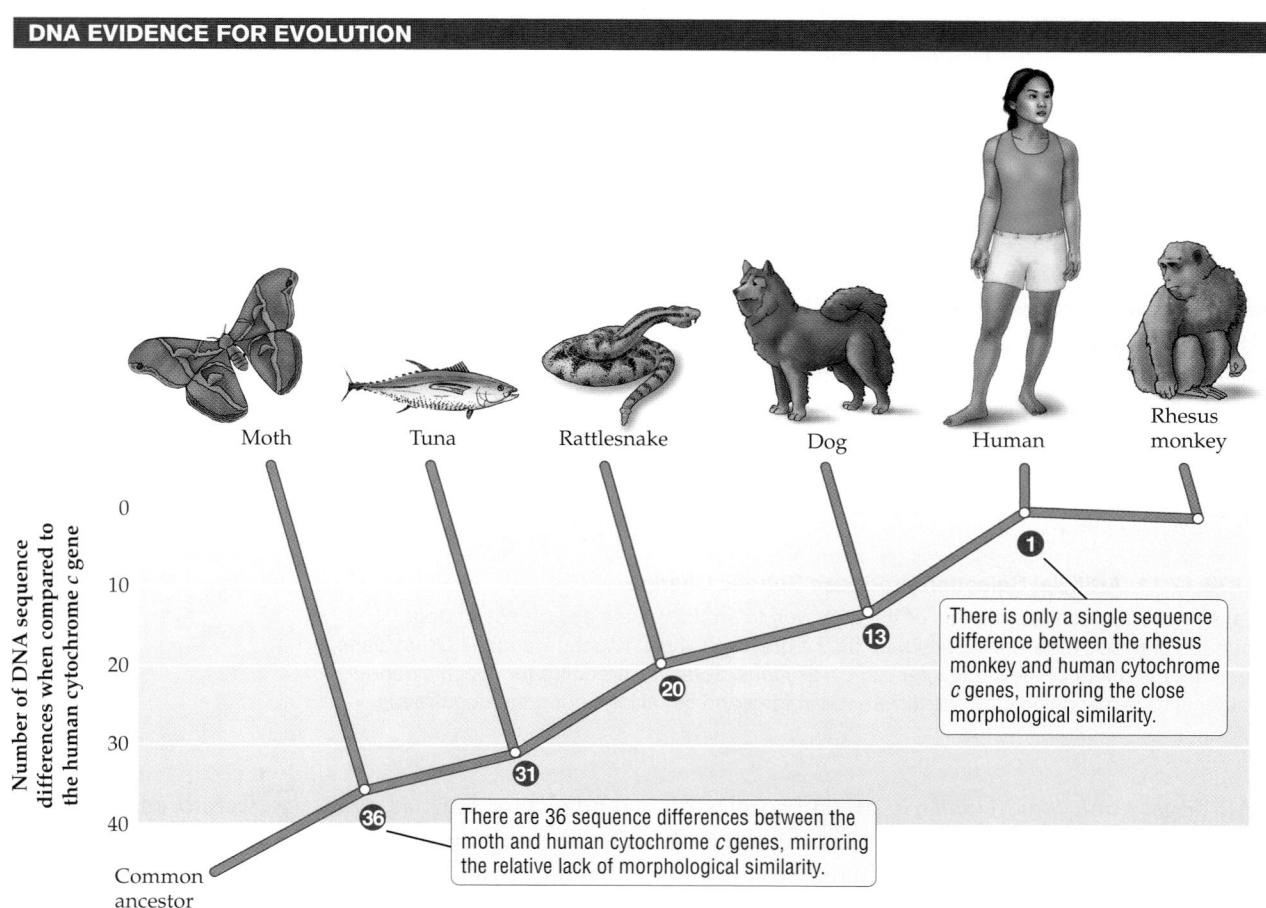

Moth    Tuna    Rattlesnake    Dog    Human    Rhesus monkey

Number of DNA sequence differences when compared to the human cytochrome *c* gene

0

10

20

30

40

Common ancestor

There is only a single sequence difference between the rhesus monkey and human cytochrome *c* genes, mirroring the close morphological similarity.

There are 36 sequence differences between the moth and human cytochrome *c* genes, mirroring the relative lack of morphological similarity.

**FIGURE 17.10**
**Independent Lines of Evidence Yield the Same Result**

Cytochrome *c* is an enzyme that functions in aerobic respiration and is found in all eukaryotes. A pattern of evolutionary relationships among animals is based on the number of DNA sequence differences between the cytochrome *c* gene found in humans and other organisms. The pattern of evolutionary relationships shown here, in which humans are most closely related to rhesus monkeys and least closely related to moths, matches the pattern derived independently from anatomical features.

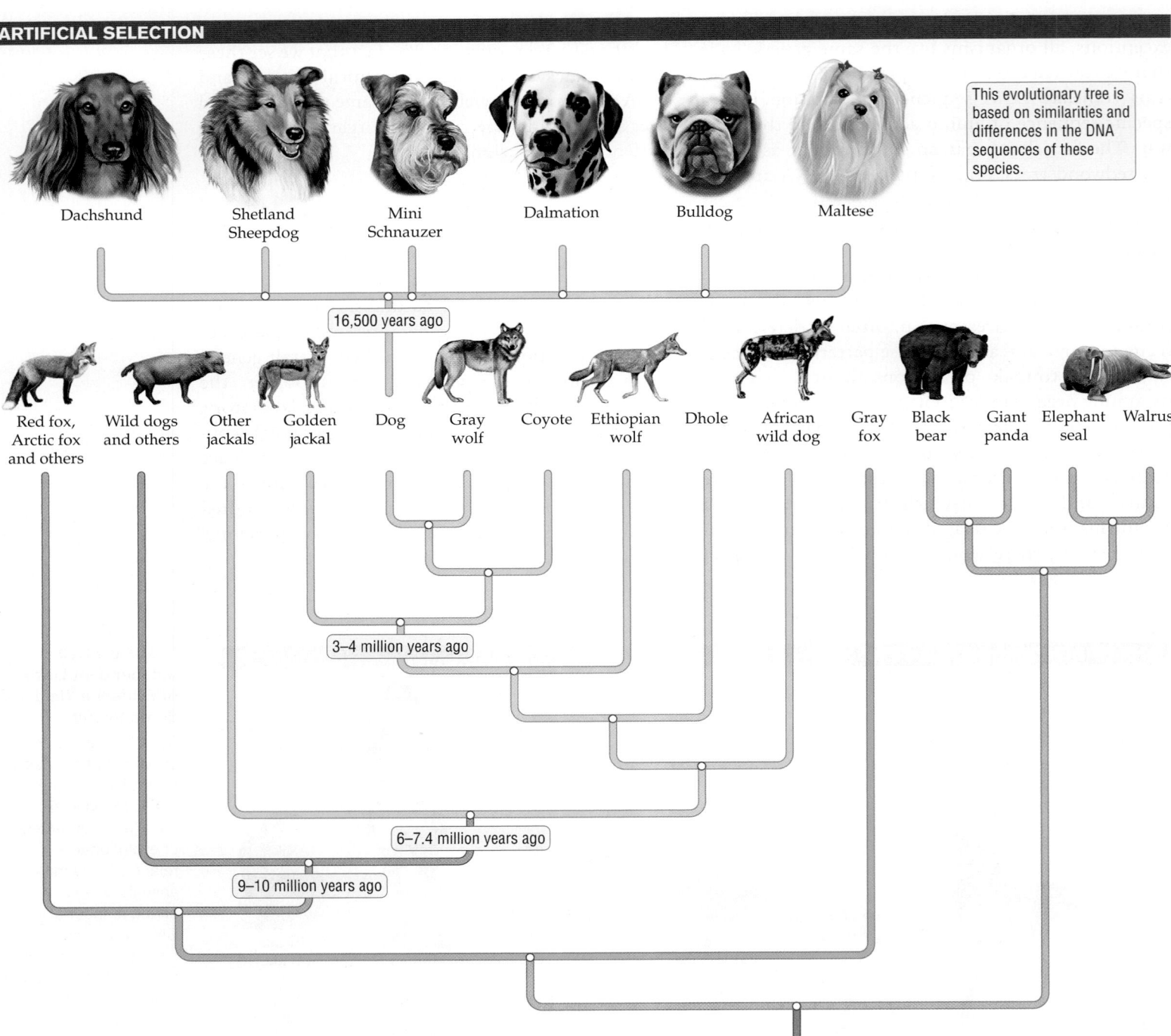

This evolutionary tree is based on similarities and differences in the DNA sequences of these species.

Dachshund  Shetland Sheepdog  Mini Schnauzer  Dalmation  Bulldog  Maltese

16,500 years ago

Red fox, Arctic fox and others  Wild dogs and others  Other jackals  Golden jackal  Dog  Gray wolf  Coyote  Ethiopian wolf  Dhole  African wild dog  Gray fox  Black bear  Giant panda  Elephant seal  Walrus

3–4 million years ago

6–7.4 million years ago

9–10 million years ago

**FIGURE 17.11 Artificial Selection Produces Genetic Change**

Humans have directed the evolution of the gray wolf to produce the many breeds of dogs. Despite the dramatic differences exhibited by the gray wolf, the Chihuahua, and the Great Dane, all three are members of the same species: *Canis lupus*. Selective breeding for specific genetic characteristics is responsible for the difference in form and behavior among the dog breeds.

All dogs, *Canis lupus familiaris*, are subspecies of the gray wolf. Humans began domesticating the gray wolf about 16,000 years ago. Individuals with desired qualities, such as friendliness toward owners and a tendency to bark at strangers, were bred selectively. By allowing only individuals with certain inherited characteristics to breed—a process called **artificial selection**—humans have crafted enormous evolutionary changes within this species. The tremendous variation that we have produced within dogs, ornamental flowers, and many other species illustrates the power of artificial selection to bring about evolutionary change. Natural selection can produce similar evolutionary changes, as shown by the often striking match between organisms and their environment (see Figure 17.5) and by case studies such as that of the medium ground finch, which we will examine toward the end of this chapter.

## Continental drift and evolution explain the geographic locations of fossils

Earth's continents move over time, by a process called **continental drift**, or plate tectonics. Each year, for example, the distance between South America and Africa increases by about 3 centimeters, a little more than an inch. Although they are separating from one another now, about 250 million years ago South America, Africa, and all of the other landmasses of Earth had drifted together to form one giant continent, called **Pangaea [pan-*JEE*-uh]**. Beginning about

200 million years ago, Pangaea slowly split up to form the continents we know today (see Figure 20.7).

We can use knowledge about evolution and continental drift to make predictions about the geographic locations where fossils will be found. For example, organisms that evolved when Pangaea was intact could have moved relatively easily between what later became widely separated regions, such as Antarctica and India. For that reason, we can predict that their fossils should be found on most or all continents. In contrast, the fossils of species that evolved after the breakup of Pangaea should be found on only one or a few continents, such as the continent on which they originated and on any connected or nearby landmasses.

Predictions about the geographic distributions of fossils have proven correct, and they provide another important line of evidence for evolution. For example, today the lungfish *Neoceratodus fosteri* [**nee-oh-suh-*RAD*-uh-dus *FAW*-ster-ee**] is found only in northeastern Australia, but its ancestors lived during the time of Pangaea, and fossils of those ancestors are found on all continents except Antarctica (**FIGURE 17.12**). At the other extreme, fossil evidence shows that modern horses in the genus *Equus* first evolved in North America about 5 million years ago (see Figure 17.9),

**BIOGEOGRAPHIC EVIDENCE FOR EVOLUTION**

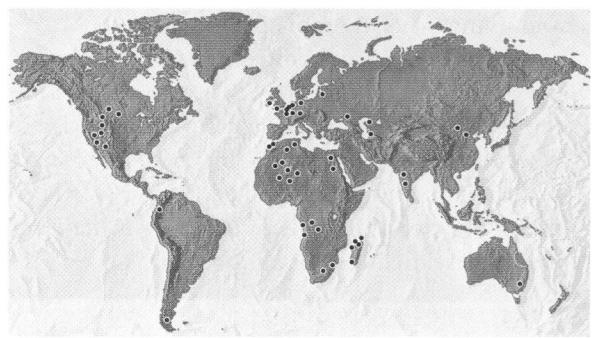

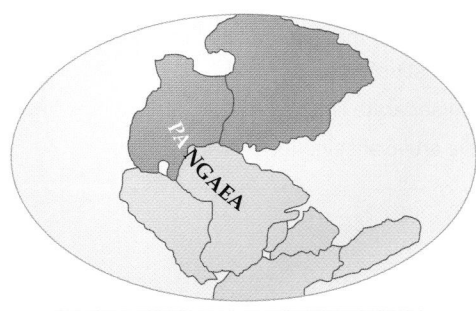

Portions of the supercontinent Pangaea began to drift apart about 200 million years ago.

**FIGURE 17.12  The Biogeography of a Lungfish Reflects Its Evolutionary Past**

Ancestors of the freshwater lungfish *Neoceratodus fosteri* (*inset*) lived during the time of Pangaea. *N. fosteri* fossils have been found on all continents except Antarctica. Currently found only in the orange-shaded region of northeastern Australia, this species is the only surviving member of its family. Red dots indicate places where fossils of *N. fosteri*'s ancestors have been found.

long after the breakup of Pangaea. The modern horse migrated to Eurasia through the land bridge that spanned the Bering Strait between eastern Siberia and Alaska during each of the recent ice ages. However, the land bridge that today connects North and South America was formed roughly 3 million years ago. We would predict that *Equus* fossils found in South America should be less than 3 million years old; and to date, all such discoveries have been less than 3 million years old. *Equus* fossils have not been found in Africa, Australia, or Antarctica, just as we would predict on the basis of the fossil history of *Equus* and changes in Earth's crust over that evolutionary time span.

## Formation of new species can be observed in nature and produced experimentally

Biologists have directly observed the formation of new species from previously existing species. The first experiment in which a new species was formed took place in the early 1900s, when the primrose *Primula kewensis* [PRIM-yoo-luh kee-WEN-sis] was produced. Scientists have also observed the formation of new species in nature. For example, two new species of salsify [SAL-suh-fee] plants were discovered in Idaho and eastern Washington in 1950. Neither of the new species had been found in those regions or anywhere else in the world as recently as 1920. Genetic data reveal that both of the new species evolved from previously existing species, and field surveys indicate that this event occurred between 1920 and 1950. The two new species continue to thrive, and one of them has become common since its discovery in 1950.

## 17.5 The Impact of Evolutionary Thought

The evolution of species was a radical idea in the mid-nineteenth century, and the argument that the form and function of organisms could be explained by natural selection was even more radical. These ideas not only revolutionized biology but also had profound effects on other fields, ranging from literature to philosophy to economics.

The idea of Darwinian evolution had a profound effect on religion as well. Evolution was viewed initially as a direct attack on Judeo-Christian religion, and this presumed attack prompted a spirited counterattack by many prominent members of the clergy. Today, however, most religious leaders and most scientists view evolution and religion as compatible but distinct fields of inquiry. The Catholic Church, for example, accepts that evolution explains the physical characteristics of humans but maintains that religion is required to explain our spiritual characteristics. Similarly, although the vast majority of scientists accept the scientific evidence for evolution, many of those same scientists have religious beliefs. Overall, most scientists recognize that religious beliefs are up to the individual and that science cannot answer questions regarding the existence of God or other matters of religious import.

The emergence of evolutionary thought has also had an effect on human technology and industry. For example, an understanding of evolution has proven essential as farmers and researchers have sought to prevent or slow the evolution of resistance to pesticides by insects. Information about the evolutionary relationships among organisms can also be used to increase the efficiency of the search for new antibiotics and other pharmaceuticals, food additives, pigments, and many other valuable products.

Dr. Eugenie Scott, who directs the National Center for Science Education (NCSE), has pointed out that evolution is central to the teaching of biology because it serves as common organizing principle for all of the life sciences and related fields. In her words, "Evolution is the periodic table of biology—it makes biology make sense, makes it hang together."

> **Concept Check**
>
> 1. What is another reason—besides descent from a common ancestor—that two species might display similar characteristics?
>
> 2. List five separate lines of evidence that life has evolved.
>
> 3. What is artificial selection?

# Darwin's Finches: Evolution in Action

During Charles Darwin's 5-year voyage around the world, he began, tentatively, to question his assumption that species could not change. There had been suggestions that species evolve, hints that simmered in the back of his mind. For example, Galápagos Islands locals told him they could tell which island a tortoise came from by its size and shape. Every island had its own tortoise. And Darwin had noticed the same thing with the island's mockingbirds, about which he wrote, "Each variety is constant on its own island. This is a parallel fact to the one mentioned about the Tortoises." Months later, he concluded a similar discussion in his notes by saying that "such facts would undermine the stability of species." Darwin was thinking that species might change over time.

Yet the strongest evidence for evolution came after Darwin returned to England and turned over his collection of birds to the renowned ornithologist John Gould. Darwin thought that each island had a different subspecies of mockingbird; Gould said no, the birds were separate species. Darwin thought that the 12 birds he had brought back were a mix of blackbirds, grosbeaks, a wren, and some finches; Gould said no, the 12 birds were all ground finches. Darwin was stunned to learn that the birds were so closely related. As he later wrote, "One might really fancy that . . . one species had been taken and modified for different ends." He guessed that all the finches in the Galápagos Islands were descended from a single species that had split into a dozen new species from the mainland.

Today we know that Darwin's conclusion was correct. The first finch species colonized the islands 3 million years ago, and over time, groups of finches on different islands evolved the ability to handle different types of foods—whether big seeds or little ones, insects, or fruits—and began to differentiate from one another. In short, the finches evolved to fill different ecological roles that on the mainland were already filled by other birds.

The Galápagos Islands serve as a natural laboratory in which scientists can study how evolution works. The islands lie squarely on the equator in a cold ocean current. From January to May, the weather is warm and rainy. The rest of the year the islands are generally dry. In 1977, researchers Peter and Rosemary Grant were on one of the islands studying medium ground finches when drought struck. Plants withered (**FIGURE 17.13**), seeds became scarce, and the seed-eating medium ground finches starved, their population plummeting from 1,200 to 180.

The plants that were able to reproduce during the dry spell were mainly drought-tolerant species that produce large seeds with hard seed coats: seeds that the finches would generally ignore in better times. Now, the large seeds of these species, especially of a plant named *Tribulus cistoides*, became the main food source for the surviving birds. Was there anything special about these survivors living mainly off *T. cistoides*? The Grants compared the average beak size of the pre-drought (1976) population of medium ground finches and the descendants of the drought-survivors a year after the worst of the drought had passed (in 1978). Their analysis revealed that in just one generation, average beak size in the post-drought population had increased by about half a millimeter (**FIGURE 17.14**).

A thick, more powerful beak was adaptive under drought conditions because it enabled its owners to make good use of the available food. Natural selection rapidly culled the less fit individuals—as most of the individuals with small beaks died of starvation—favoring the survival of individuals with stout beaks. Because beak size is an inherited trait, the greater reproductive success of large-beaked individuals meant that large-beaked finches became more common in the next generation; therefore, the post-drought population of finches had

Before drought

After drought

**FIGURE 17.13**
**A Drought Results in Rapid Evolutionary Change**

The 1977 drought on Daphne Major in the Galápagos Islands had a dramatic effect on the plant life there, setting the stage for natural selection to cause rapid evolutionary change in birds that depended on the plants for food.

## FIGURE 17.14
### Evolutionary Change within a Species

Scientists have been studying the medium ground finch (*a*) on the Galápagos island Daphne Major for more than 30 years. Average beak size increased in the population after a severe drought in 1977. When rainfall resumed, natural selection drove the evolution of smaller beaks (*b*). The orange-colored band indicates the range of values that would be expected without evolutionary change.

**ADAPTATION TO DROUGHT: CHANGE IN BEAK SIZE**

(*a*) **Medium ground finch**

(*b*) **Beak size over study period**

evolved. Exceptionally heavy rains in the early 1980s brought luxuriant plant growth, an abundance of small seeds with softer seed coats, and a decline in the large-seeded drought-adapted plants. In response to this change in the food supply, the finch population evolved yet again: the average beak size dropped to the pre-drought size over the rest of the 30-year study period.

Comparison of DNA from the 13 finch species on Galápagos, and one species from the Cocos Islands to the north, have confirmed Darwin's hunch about the origins of these

birds. All 14 species are descended from a single ancestral form, very likely an insect-eating warbler finch from nearby Equador that migrated to the islands between 2 and 3 million years ago. As the members of the ancestral population colonized diverse habitats, with varied vegetation, they adapted to local conditions, especially in their eating habits (**FIGURE 17.15**). A lineage of ground finches evolved that specializes in eating seeds. Tree finches use their blunt-ended beaks like pruning shears to snip off vegetation. The short stubby beak of the vegetarian finch is suited to a diet of buds, flowers, and fruits. The cactus finch deploys a cactus spine in its tweezers-like beak to pry insects from nooks and crannies.

A subspecies of the sharp-billed ground finch that is found on Wolf Island and on Darwin Island has acquired a taste for blood. The behavior may have evolved from this finch's habit of picking parasites off basking marine iguanas and large sea birds such as blue-footed boobies. Sometimes, usually when food is scarce, the finches will peck at the thin skin on the back of the seabirds and lap up the blood that oozes out.

**EVOLUTIONARY HISTORY OF THE GALÁPAGOS FINCHES**

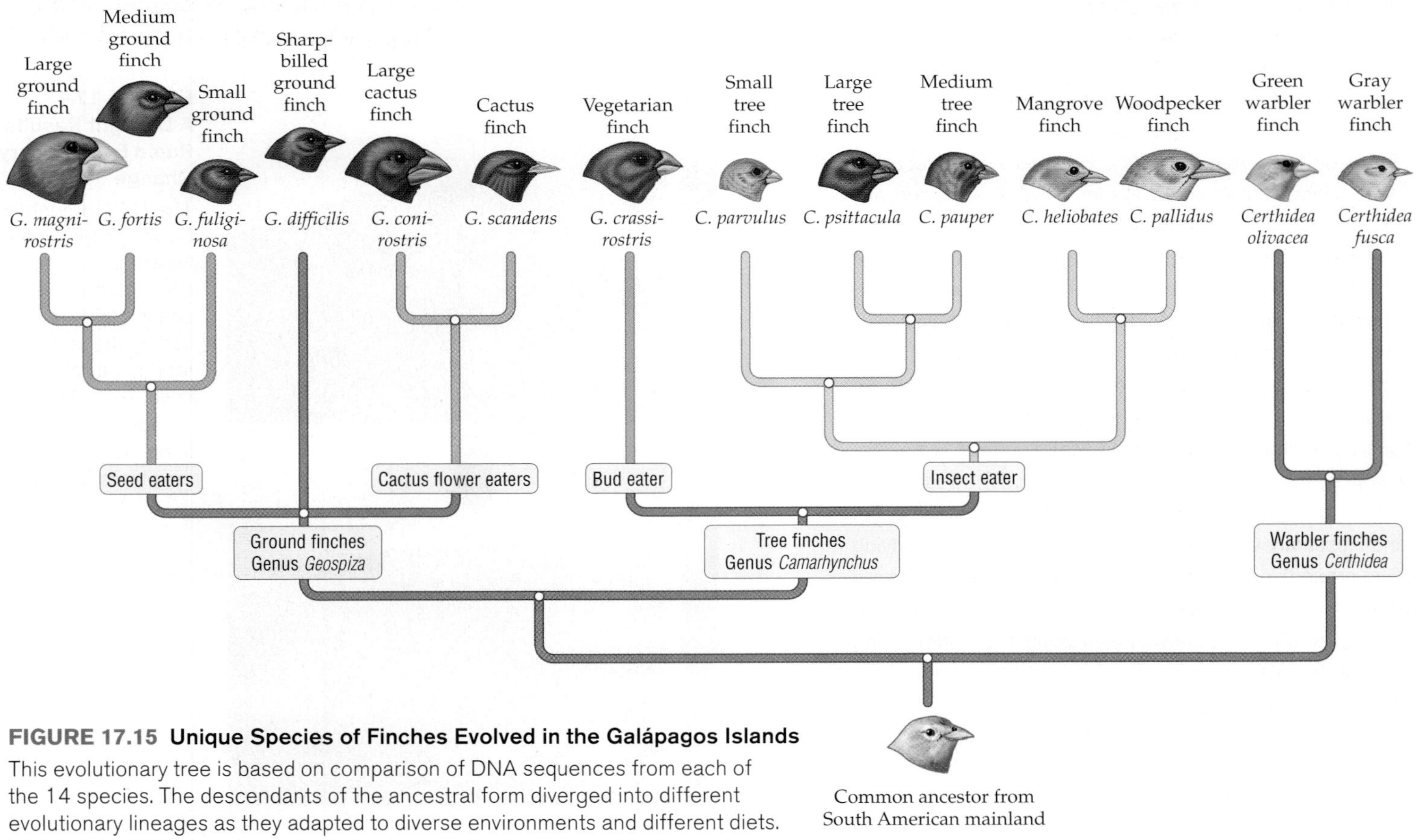

Large ground finch — *G. magnirostris*

Medium ground finch — *G. fortis*

Small ground finch — *G. fuliginosa*

Sharp-billed ground finch — *G. difficilis*

Large cactus finch — *G. conirostris*

Cactus finch — *G. scandens*

Vegetarian finch — *G. crassirostris*

Small tree finch — *C. parvulus*

Large tree finch — *C. psittacula*

Medium tree finch — *C. pauper*

Mangrove finch — *C. heliobates*

Woodpecker finch — *C. pallidus*

Green warbler finch — *Certhidea olivacea*

Gray warbler finch — *Certhidea fusca*

Seed eaters

Cactus flower eaters

Bud eater

Insect eater

Ground finches Genus *Geospiza*

Tree finches Genus *Camarhynchus*

Warbler finches Genus *Certhidea*

Common ancestor from South American mainland

**FIGURE 17.15  Unique Species of Finches Evolved in the Galápagos Islands**

This evolutionary tree is based on comparison of DNA sequences from each of the 14 species. The descendants of the ancestral form diverged into different evolutionary lineages as they adapted to diverse environments and different diets.

# As Mammals Supplanted Dinosaurs, Lice Kept Pace

BY NICHOLAS WADE, *New York Times*

Biologists have found a new way to peer back... millions of years in time, illuminating the catastrophic period in which the dinosaurs perished and birds and mammals arose.

The new approach rests on reconstructing the family tree of lice. Vincent S. Smith, a louse taxonomist at the Natural History Museum in London, has found that the tree stretches so far back in time that the host of the first louse would have been a dinosaur, probably one of the theropod dinosaurs that were the ancestors of birds.

Dr. Smith and his colleagues reconstructed the louse family tree by analyzing DNA from present-day louse species that parasitize birds and mammals. Most lice are specialists, feeding on a single species to whose fur or feathers their claws are adapted. The adaptation is so precise that when a louse's host species evolves into a new one, the louse will diversify into different species, too.

The human head louse, for instance, evolved from the chimpanzee louse when the ancestors of humans and chimps split apart some five million years ago. The human pubic louse, on the other hand, is related to the gorilla louse, from which it parted company some 13 million years ago. Species of human lice thus mirror the splits in the tree of ape and human evolution.

In the same way, species of lice on living animals [mammals] and birds reflect the splits in animal [mammal] and bird ancestry back to the time that lice first arose...

The assembled family tree shows that lice started to radiate into new species well before the end of the Cretaceous period, Dr. Smith and his colleagues report in the current issue of Biology Letters. The finding implies that their hosts, both mammals and birds, had also begun to flourish and speciate [radiate] before the reign of the dinosaurs was over.

The new tree bears on a longstanding dispute about the rise of birds and mammals. One school holds that both groups proliferated early in the Cretaceous period, which began 145 million years ago, and that many lineages survived the cataclysm that brought the Cretaceous and the dinosaurs to a sudden end: the strike of a large asteroid 65 million years ago. The opposing view is that mammals and birds did not become successful and radiate into many different species until after the demise of the dinosaurs...

---

Lice are tiny parasitic insects that suck blood from the skin of birds and mammals. As we learned in Section 17.3, species can split into two or more new species; that is, one species can diverge into many. In the Galápagos Islands, for example, a finch species that spreads to another island becomes isolated from its relatives. Over time, the two populations can evolve independently until they are quite different—one species with a heavy beak and a large body, for example, and the other species with a small, sharp beak and a lighter body. Where there was one species of finch there are now two.

In the same way, a single species of louse can split into multiple species. Each louse species is already adapted to a different species of bird or mammal. Imagine a finch species separating into two populations—say, on two separate islands. The two populations can evolve independently, eventually becoming two species. When the finch populations become separated, the single louse species living on the finches also becomes separated into two populations. These two populations of lice can evolve into two separate species, diverging in parallel with the finches. Studying the evolution of lice is a way of double-checking what we know about the evolution of the other animals with which they evolved.

This article discusses an effort by a researcher to shed light on the question of whether birds (and mammals) diversified into many different species after the dinosaurs went extinct 65 million years ago or earlier, while the dinosaurs were still around. The idea is to build a family tree of lice and work backward to the animals on which they lived.

## Evaluating the News

**1.** Birds evolved from a dinosaur ancestor and diversified into many species. Mammals evolved from an older reptile and later evolved into many species. This article describes a disagreement about when each of these groups diverged. What are the two main hypotheses?

**2.** What evidence in this article supports the hypothesis that dinosaurs had lice?

**3.** Explain why human head lice are more closely related to chimpanzee lice than to gorilla lice.

**SOURCE:** *New York Times*, April 6, 2011.

# CHAPTER REVIEW

## Summary

### 17.1 Descent with Modification

- Evolution can be broadly defined as change in the genetic characteristics of populations of organisms over successive generations.
- A number of scholars had advanced the idea that present-day species are descended from more ancient forms. Charles Darwin and Alfred Wallace offered a mechanism for the evolution of new species through modification of preexisting forms: natural selection.
- According to the theory of natural selection, individuals with adaptive traits (characteristics that enable them to function well in a competitive environment) produce more offspring, and therefore their characteristics become more common in descendant generations.
- Microevolution is the study of evolutionary processes at the level of populations. Macroevolution focuses on the larger-scale patterns—such as the evolution and extinction of species and major groups of organisms—that are a consequence of population-level changes.

### 17.2 Mechanisms of Evolution

- Individuals in populations differ genetically in their morphological, biochemical, and behavioral characteristics. The gene pool is the sum of all the genetic information carried by all the individuals in a population. Evolution can be defined as a change in the gene pool of a population over successive generations.
- Mutations in DNA are the ultimate source of all genetic variation. Mutations are random events, and as such they are not goal directed. Sexual recombination creates new combinations of the genetic variation (alleles) produced by DNA mutations.
- Genetic drift is a random process that may cause individuals with one set of inherited characteristics to survive or reproduce better than other individuals. Therefore, evolutionary changes caused by genetic drift are due to chance alone.
- Gene flow is the movement of genes from one population to another. A resident population can evolve if its gene pool is altered by the introduction or loss of novel alleles as individuals immigrate or emigrate.
- Natural selection is the greater reproductive success of individuals with advantageous genetic characteristics (adaptive traits), compared to competing individuals with other characteristics. The characteristics of individuals that produce more offspring become more common in the succeeding generations. Natural selection is a directional (nonrandom) process, and it generates adaptation in descendant populations. Adaptation is the process by which a population become better matched to its environment.

### 17.3 Evolution Can Explain the Unity and Diversity of Life

- Shared characteristics suggest either descent from a common ancestor or convergent evolution. Shared characteristics that result from common descent are said to be homologous; those that result from convergent evolution are said to be analogous.
- The diversity of life is a result of speciation, the repeated splitting of one species into two or more species.

### 17.4 The Evidence for Biological Evolution

- The fossil record provides clear evidence for the evolution of species over time.
- Within organisms, evidence for evolution includes remnant anatomical structures (vestigial organs), patterns of embryonic development, the universal use of DNA and the genetic code, and molecular (DNA and protein) similarities among organisms.
- As predicted by our understanding of evolution and continental drift, fossils of organisms that evolved when the present-day continents were all part of the supercontinent Pangaea have a wider geographic distribution than do fossils of more recently evolved organisms.

### 17.5 The Impact of Evolutionary Thought

- Darwin's ideas on evolution and natural selection revolutionized biology, overturning the views that species do not change with time.
- Evolutionary biology has many practical applications in agriculture, industry, and medicine.

## Key Terms

adaptation (p. 391)
adaptive trait (p. 391)
allele (p. 389)
analogous (p. 395)
artificial selection (p. 400)
biological evolution (p. 389)

continental drift (p. 401)
convergent evolution (p. 394)
fossil (p. 395)
gene flow (p. 390)
gene pool (p. 389)
genetic drift (p. 390)

genetic variation (p. 389)
genotype (p. 389)
homologous (p. 394)
macroevolution (p. 388)
microevolution (p. 388)
mutation (p. 389)

natural selection (p. 387)
Pangaea (p. 401)
sexual recombination (p. 390)
speciation (p. 395)
vestigial organ (p. 394)

# Self-Quiz

1. Which of the following provides evidence for evolution?
   a. direct observation of genetic changes in populations
   b. sharing of characteristics between organisms
   c. the fossil record
   d. all of the above

2. In natural selection,
   a. the genetic composition of the population changes randomly over time.
   b. new mutations are generated over time.
   c. all individuals in a population are equally likely to contribute offspring to the next generation.
   d. individuals that possess particular inherited characteristics consistently survive and reproduce at a higher rate than other individuals.

3. Adaptive traits
   a. are features of an organism that hinder its performance in its environment.
   b. are rare in most natural populations.
   c. are favored by natural selection.
   d. result from genetic drift.

4. The fossil record shows that the first mammals evolved 220 million years ago. The supercontinent Pangaea began to break apart 200 million years ago. Therefore, fossils of the first mammals should be found
   a. on most or all of the current continents.
   b. only in Antarctica.
   c. on only one or a few continents.
   d. only in Africa.

5. The fact that the flipper of a whale and the arm of a human both have five digits and the same set of bones can be used to illustrate that
   a. genetic drift can cause the evolution of populations.
   b. organisms can share characteristics simply because they share a common ancestor.
   c. whales evolved from humans.
   d. humans evolved from whales.

6. The Galápagos Islands provide examples of
   a. microevolution only.
   b. macroevolution only.
   c. both micro- and macroevolutionary change.
   d. none of the above

7. Differences in survival and reproduction caused by chance events can cause the genetic makeup of a population to change randomly over time. This process is called
   a. mutation.
   b. natural selection.
   c. macroevolution.
   d. genetic drift.

8. The splitting of one species into two or more species is called
   a. speciation.
   b. macroevolution.
   c. common descent.
   d. adaptation.

9. Features of organisms that are related to one another through common descent are
   a. convergent.
   b. homologous.
   c. divergent.
   d. analogous.

10. Artificial selection is the process by which
   a. natural selection fails to act in wild populations.
   b. humans prevent natural selection.
   c. humans allow only organisms with specific characteristics to breed.
   d. humans cause genetic drift in domesticated populations.

# Analysis and Application

1. Explain what evolution is and why we state that populations evolve but individuals do not.

2. A population of lizards lives on an island and eats insects found in shrubs. Because the shrubs have narrow branches, the lizards tend to be small (so they can move effectively among the branches). A group of lizards from this population migrates to a nearby island where the vegetation consists mostly of trees; the branches of those trees are thicker than the shrubs on which the lizards formerly fed. A few of the lizards that migrate to the new island are slightly larger than the others; large size is not a disadvantage for moving in the trees (because the branches are thicker) and is advantageous when males compete with other males to mate with females. Assuming that large size is an inherited characteristic, explain what is likely to happen to the average size of the lizards in their new home, and why.

3. How does evolution explain (a) adaptive traits, (b) the great diversity of species, and (c) the many examples in which otherwise dissimilar organisms share certain characteristics?

4. Why are scientists throughout the world convinced that evolution happens? Consider at least three of the lines of evidence discussed in this chapter.

5. Although biologists agree that evolution occurs, they debate which mechanisms are most important in causing evolutionary change. Does this mean that the theory of evolution is wrong?

6. Genetic drift occurs when chance events cause some individuals in a population to contribute more offspring to the next generation than other individuals contribute. In which populations—small ones or large ones—is the effect of such chance events likely to be greater? (*Hint:* Examine Figure 17.4. Consider whether the proportion of the *A* allele in the population would be likely to change from 30 percent [3 of the 10 alleles originally present were *A* alleles] to 100 percent if there were 1,000 spiders instead of 5 spiders in the population.)

# Evolution of Populations

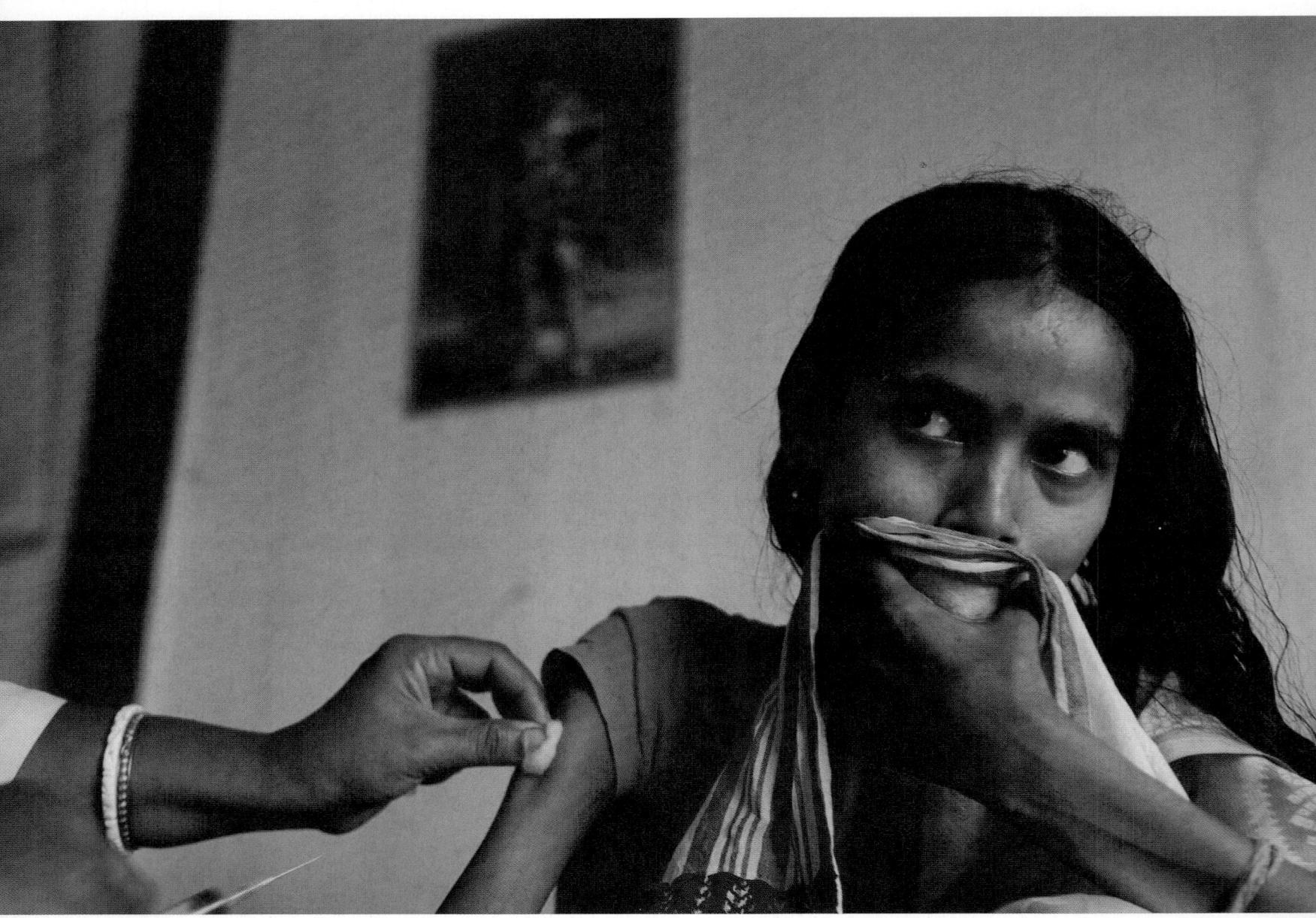

**A WOMAN WITH TUBERCULOSIS.** One-third of the world's population may be infected with tuberculosis (TB), a "mycobacterium." As the rate of TB infection has increased, so has the number of cases of antibiotic-resistant TB, a result of natural selection.

# Evolution of Resistance

As recently as the 1940s, Americans died of diseases that are nearly unheard of today. A healthy young father might cut his knee, develop an infection, and die of blood poisoning. A young mother might contract meningitis and be gone in 2 weeks. Today, such patients would be treated within days and sent home from the hospital perfectly healthy. The discovery of antibiotics—drugs that, like magic bullets, destroy bacteria but mostly leave our own cells alone—has made us take such dramatic recoveries for granted. Since the 1940s, antibiotics have cured millions of people of deadly bacterial infections—including gangrene, pneumonia, syphilis, tuberculosis, and blood infections (septicemia). When antibiotics were first introduced in the United States in 1944, they cut deaths from pneumonia by 75 percent and from rheumatic fever by 90 percent.

Antibiotics seemed so miraculous that some doctors prescribed them for everything, even viral infections and cancer, which are unaffected by these drugs. And yet, from the moment the drugs came into use, bacteria began to evolve resistance to them. In 1940, before penicillin was even available to the public, researchers observed that gut bacteria, *E. coli*, had already begun to evolve the ability to survive in the presence of penicillin—a phenomenon called *resistance*. When Alexander Fleming received a Nobel Prize for the discovery of penicillin in 1945, he warned that bacteria could evolve resistance to antibiotics.

Today we can no longer take these wonder drugs for granted. Many strains of bacteria—including those that cause strep throat, rheumatic fever, staph infections, and TB—are resistant to penicillin, tetracycline, or other antibiotics. All these bacteria have adapted to the presence of antibiotics in their environment. Antibiotic resistance is so widespread that health officials worry that bacterial infections may once again ravage human populations, as they did before the 1940s.

How does bacterial resistance to antibiotics evolve? What does it mean to adapt?

In this chapter we'll learn how small mutations in a gene can become more common in populations of organisms by making some individuals more likely to survive and reproduce. We'll see how mutations can make bacteria and other organisms resistant to antibiotics.

**MAIN MESSAGE** Allele frequencies in populations can change over time as a result of mutation, gene flow, genetic drift, and natural selection.

## KEY CONCEPTS

- Biological evolution is a change in allele frequencies in a population of organisms over successive generations. Individual organisms do not evolve; populations do.

- Four mechanisms can cause populations to evolve: mutation, gene flow, genetic drift, and natural selection.

- DNA mutations create genetic variation in a population's gene pool, which in turn provides the raw material for other evolutionary processes.

- Gene flow is the exchange of alleles between populations. By introducing or removing alleles from a population, immigrants and emigrants can change allele frequencies in a population.

- Genetic drift is a change in allele frequencies produced by random differences in survival and reproduction among the individuals in a population. Genetic drift is more likely to cause evolution in a small population than in a large one.

- Some populations can evolve very rapidly (in months to a few years). Rapid evolution has resulted in insect resistance to pesticides and microbial resistance to antibiotics and antiviral medications.

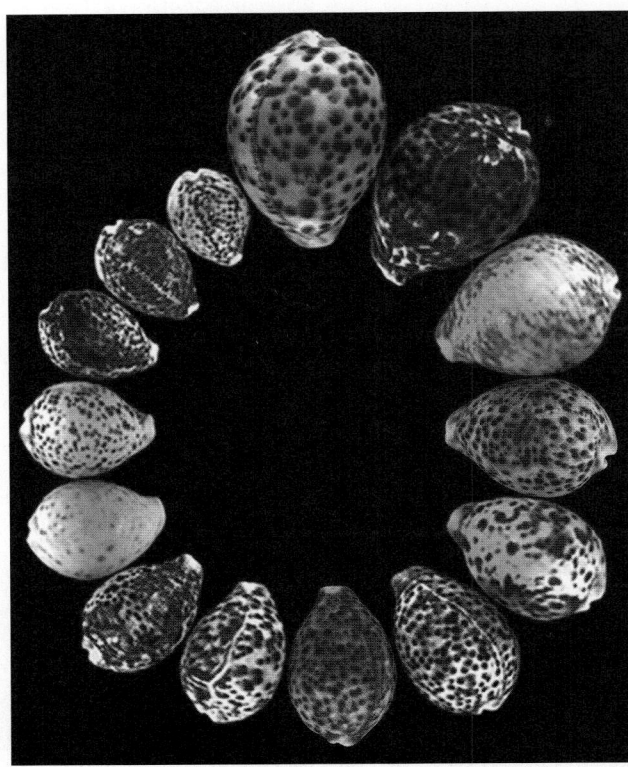

**FIGURE 18.1 Morphological Variation in Tiger Cowries,** *Cypraea tigris*

The individuals within a population often vary greatly in many morphological traits, just as these tiger cowries do in their size and color patterns. These snails are widespread throughout the Pacific and Indian oceans, from Hawaii to the eastern shores of Africa.

**IS EVOLUTION TOO SLOW TO BE OBSERVED?** That was once the prevalent thinking, but over the last 80 years, in thousands of studies, biologists have observed and documented the evolution of populations. These studies show that populations evolve slowly in some cases and very rapidly in others (within a few generations, covering a time span of months to a few years). Similarly, new species form slowly in some cases but rapidly in others (in a year to a few thousand years).

In this chapter we describe evolutionary changes that take place in populations, sometimes over short time spans. In particular, we emphasize that evolution is the manifestation of changes in the frequencies (proportions) of alleles in a population from generation to generation. You might think of "allele frequency" as a measure of the relative abundance of that allele in a population, or its "commonness" in that population. The allele that is more abundant than another will be inherited more often than an uncommon allele. If there is no change in overall allele frequencies in a population, no changes will be seen in the attributes of a population from generation to generation. But when allele frequencies in a population do change, the attributes of the population change over time; that is, the population evolves. A population is the smallest scale at which evolution can occur, and a change in allele frequencies in a population over successive generations is called **microevolution**. Microevolution occurs when there is a significant increase or decrease in the frequency of a particular allele in a particular population over several generations.

We noted previously that genetic variation is the raw material of evolution (**FIGURE 18.1**). Our discussion of microevolution begins with definitions of two essential terms: *allele frequency* and *genotype frequency*. We will briefly revisit how mutation generates new alleles, and how this process affects both genotype frequencies and allele frequencies. With that information in hand, we will then discuss four mechanisms that can cause allele frequencies to change over time: mutation, gene flow, genetic drift, and natural selection. We end the chapter by considering how organisms that cause disease evolve resistance to drugs, along with ways to combat infectious disease by slowing the evolution of such resistance.

## 18.1 Alleles and Genotypes

Evolution can be defined as a change in allele frequencies. **Allele frequency** refers to the proportion, or percentage, in a population of a particular allele, such as an *A* or *a* allele. Similarly, **genotype frequency** refers to the proportion, or percentage, in a population of a particular genotype, such as the genotypes *AA*, *Aa*, or *aa*.

For traits that are inherited in a simple Mendelian manner (see Chapter 12), we can calculate genotype frequencies in a population quite easily. Imagine that flower color in a population of 1,000 plants is determined by a single gene that has two alleles: a dominant allele, R, for red flower color; and a recessive allele, r, for white flower color. If the population contains 160 RR individuals, 480 Rr individuals, and 360 rr individuals, we can calculate the frequencies for the three genotypes (RR, Rr, and rr) by dividing their numbers by the total number of individuals in the population (1,000). The genotype frequency for RR, for example, would be 160/1,000, or 0.16. Rr and rr would have frequencies of 0.48, and 0.36, respectively. These three genotype frequencies add up to 1.0—as they should, because RR, Rr, and rr are the only three genotypes possible and a frequency of 1.0 is equivalent to 100 percent. Hence, one way to check our genotype calculations is to make sure they add up to 1.0.

Allele frequencies can be computed by the following method, which we illustrate for the R allele (though we could have chosen the r allele instead). Our plant population has 1,000 individuals, each of which has two alleles of the flower color gene. Therefore, the total number of alleles in the population equals 2,000. Each of the 160 RR individuals carries two R alleles, for a total of 320 R alleles among the RR individuals. The Rr individuals have one R allele each, for a total of 480 R alleles; and the rr individuals have no R alleles. Therefore, there are 800 (320 + 480 + 0) R alleles in the population. After calculating the total number of R alleles in the population, we can calculate the frequency of the R allele by dividing the number of R alleles by the total number of alleles in the population: (800 R alleles)/(2,000 total alleles) = 0.4. Since the gene for flower color has only two alleles, R and r, the sum of their frequencies must equal 1.0. Hence, the frequency of the r allele is 1.0 − 0.4 = 0.6. We can check this value by performing a calculation for the r allele similar to the one we did for the R allele.

flow, (3) genetic drift, and (4) natural selection (see Chapter 17). Here we provide a brief review of how these mechanisms work, and in the sections that follow we offer a more detailed discussion of how each mechanism contributes to microevolution.

A *mutation* is a change in the sequence of any segment of DNA in an organism, and it is the only means by which new alleles are generated. *Gene flow* is the movement or exchange of alleles between populations. Mutation and gene flow both introduce new alleles into a population and thereby change allele frequencies in that population. Microevolution occurs as the individuals bearing the new alleles reproduce to contribute to the next generation, which, as a consequence, harbors a different proportion of all the alleles compared to the original population.

*Genetic drift* is a random change in allele frequencies produced by differences in survival and reproduction that are a matter of chance alone. Genetic drift can result in alleles being lost from a population or becoming widespread in a population *without* respect to the adaptive value of the alleles involved. Although genetic drift is driven by chance alone, the fact that it can affect allele frequencies in succeeding generations means that it can cause evolution to occur. Finally, individuals that have certain beneficial alleles may have a survival and reproductive advantage over individuals that do not have those alleles. In this case, *natural selection* will cause the frequency of beneficial alleles to increase, thereby driving evolutionary change.

If we know the genotype frequencies in a population, there is a way to test whether they are changing—that is, to determine whether evolution is occurring. The test is based on a formula called the Hardy-Weinberg equation, which is described in the box on page 412. The Hardy-Weinberg equation can test whether gene frequencies are changing, even if we do not know which mechanisms are causing allele frequencies to change.

## 18.2 Four Mechanisms That Cause Populations to Evolve

Evolution occurs when allele frequencies in a population change over time. Allele frequencies change primarily by four mechanisms: (1) mutation, (2) gene

# Testing Whether Evolution Is Occurring in Natural Populations

When biologists study populations in nature, it can be difficult to immediately determine whether or not evolution is occurring in a given population. In 1908, Godfrey Hardy and Wilhelm Weinberg independently developed a quick calculation based on Mendelian genetics that can reveal whether a population is evolving.

Suppose we are tracking evolution of a single gene with just two alleles: *A* and *a*. We will need to keep track of the allele frequencies of both alleles. The letter *p* traditionally is used to denote the allele frequency of *A*, and the letter *q* is used to denote the frequency of *a*. Because this gene has exactly two alleles, we know that the sum of the two allele frequencies is $p + q = 1$.

The genetic calculation to determine whether evolution is occurring in a population relies on the *Hardy-Weinberg equation*, which has the general form

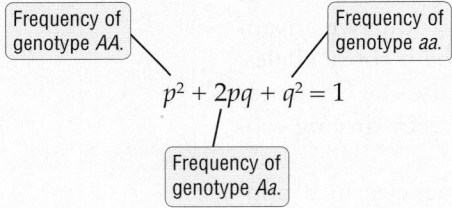

Frequency of genotype *AA*.

Frequency of genotype *aa*.

$$p^2 + 2pq + q^2 = 1$$

Frequency of genotype *Aa*.

The equation is an expression of the principle known as *Hardy-Weinberg equilibrium*, which states that the amount of genetic variation in a population will remain constant from one generation to the next in the absence of disturbing factors. That is, the Hardy-Weinberg equation predicts the genotype frequencies of a population that is *not* evolving (i.e., "at equilibrium"). The assumption inherent in the Hardy-Weinberg equation is that none of the factors that cause allele frequencies to change (mutation, gene flow, genetic drift, and natural selection) affect the proportions of alleles in the gene pool (the sum of all the alleles in the population at any one time). We use the equation to test this assumption at one genetic locus having two alleles (*a* and *A*). There are exactly three genotypes (*AA*, *aa*, and *Aa*) for this locus, whose frequencies must sum to 1 (equivalent to 100 percent). As indicated in the equation above, $p^2$ represents the frequency of the homozygous genotype *AA*, $q^2$ represents the frequency of the homozygous genotype *aa*, and $2pq$ represents the frequency of the heterozygous genotype *Aa*.

The Hardy-Weinberg approach enables us to test whether actual genotype frequencies in a real population match those predicted by the equation. If the actual frequencies differ considerably from the frequencies predicted by the Hardy-Weinberg equation for a nonevolving population, we conclude that the actual population *is* evolving and that one or more of the four evolutionary mechanisms (mutation, gene flow, genetic drift, or natural selection) are at work. If the population is evolving, we may expect that allele frequencies will change over the generations until equilibrium is reached.

To find out whether genotype frequencies in a real population differ from those predicted by the Hardy-Weinberg equation, we must first determine the genotypes of individuals in the population. Suppose we have a population of 1,000 individuals, 460 with genotype *AA*, 280 with genotype *Aa*, and 260 with genotype *aa*. Bearing in mind that each individual has 2 alleles for each gene, we are examining a gene pool of 2,000 alleles. The frequency of the *A* allele (denoted by the letter *p*) in this gene pool is calculated as a percentage of all the alleles in the gene pool:

$$p = [(2 \times 460) + 280]/2{,}000 = 0.6$$

Once we know the value of *p*, we can easily calculate the value of *q*, because we know that $p + q = 1$. Since $1.0 - 0.6 = 0.4$, we also know that *q*, the observed frequency of the *a* allele, is 0.4.

Next we plug allele frequencies into the Hardy-Weinberg equation to derive genotype frequencies. If the population is not evolving, the Hardy-Weinberg equation will hold and the observed genotype frequencies will match its predictions. With $p = 0.6$ and $q = 0.4$, the Hardy-Weinberg equation projects the following genotype frequencies:

$$p^2 = 0.6 \times 0.6 = 0.36$$
$$= \text{frequency of the } AA \text{ genotype}$$

$$2pq = 2 \times 0.4 \times 0.6 = 0.48$$
$$= \text{frequency of the } Aa \text{ genotype}$$

$$q^2 = 0.4 \times 0.4 = 0.16$$
$$= \text{frequency of the } aa \text{ genotype}$$

Given 1,000 individuals in the population, the Hardy-Weinberg equation predicts that if the population is not evolving, we should find 360 ($0.36 \times 1{,}000$) *AA* individuals, 480 ($0.48 \times 1{,}000$) *Aa* individuals, and 160 ($0.16 \times 1{,}000$) *aa* individuals. In fact, however, the actual population in our case has 460 *AA* individuals, 280 *Aa* individuals, and 260 *aa* individuals.

The differences between the actual and expected genotype frequencies just described are large. A biologist who obtained real data like those in this example would conclude that the actual genotype frequencies differed significantly from those in the Hardy-Weinberg equation, and that the population was evolving. Next, a researcher would begin to wonder what evolutionary mechanisms might be driving the population away from the predictions of the Hardy-Weinberg equation. After you finish reading this chapter, look again at the differences between the observed and the expected genotype frequencies in this example. Can you suggest one or more evolutionary mechanisms that might explain these differences?

## 18.3 Mutation: The Source of Genetic Variation

In prior chapters we discussed how **mutation** (a change in DNA sequence) creates new alleles at random, thereby providing the raw material for evolution. In this sense, all evolutionary change depends ultimately on mutation. Only mutations that occur in the germ line cells—the cell lineage that produces gametes such as eggs and sperm—can contribute to evolution.

We can also view mutation in another way: as a mechanism that drives microevolution by changing allele frequencies in a population. Because mutations occur so infrequently in any particular gene, however, they generally do not have a large effect on allele frequencies on their own. But in some cases, new mutations, combined with other evolutionary mechanisms, play a critical role in the evolution of populations. For example, HIV (the human immunodeficiency virus), which causes AIDS, has a high mutation rate, causing populations of the virus to evolve even within a single patient's body. Some of these mutations may enable the virus to resist new clinical treatments. Combating the virus is difficult because it is a "moving target."

Another example of the role mutation plays in evolution is illustrated by the mosquito *Culex pipiens* [KYOO-lex PIP-yenz]. Genetic evidence indicates that resistance to organophosphate pesticides in *C. pipiens* was caused by a single mutation that occurred in the 1960s. Since that time, mosquitoes blown by storms or moved accidentally by people have carried the initially rare mutant allele from its place of origin in Africa or Asia to both North America and Europe. This mutant allele is highly advantageous to the mosquito: individuals that have the nonmutant allele die when exposed to organophosphate pesticides. When the mutant allele is introduced by gene flow into a population exposed to organophosphate pesticides, natural selection favors the mutant, resulting in a rapid increase in the frequency of the mutant allele and leading to the evolution of resistance within the new population.

Mutations like those that enable disease agents or pests to resist our best efforts to kill them are obviously beneficial to the organisms in which they occur. Most mutations, however, are either harmful to their bearers or have little effect. In general, the effect of a mutation often depends on the environment in which the organism lives. For example, certain mutations that provide houseflies with resistance to the pesti-

**FIGURE 18.2 DDT Was Initially Described as a "Benefactor of All Humanity"**
Developed as the first of the modern insecticides early in World War II, DDT was initially used with great effect to combat malaria, typhus, and the other insect-borne human diseases. Often termed the "miracle" pesticide, DDT came into wide agricultural and commercial use. The photo shows DDT fogging machines spraying DDT at Jones Beach in New York in the 1940s. By 1972, however, the Environmental Protection Agency had banned DDT because of the devastating side effects of extensive DDT use, including insect resistance and adverse environmental impacts.

cide DDT also reduce their rate of growth. Flies that grow more slowly take longer to mature and do not produce as many offspring in their lifetimes as do flies that grow at the normal rate. In the absence of DDT, such mutations are harmful. When DDT is sprayed, however, these mutations provide an advantage great enough to offset the disadvantage of slow growth. As a result of DDT spraying, the mutant alleles have spread throughout the housefly populations and can now be found globally (**FIGURE 18.2**).

## 18.4 Gene Flow: Exchanging Alleles between Populations

When individuals move from one population to another, alleles are exchanged between these populations, and this transfer of alleles is known as **gene flow**

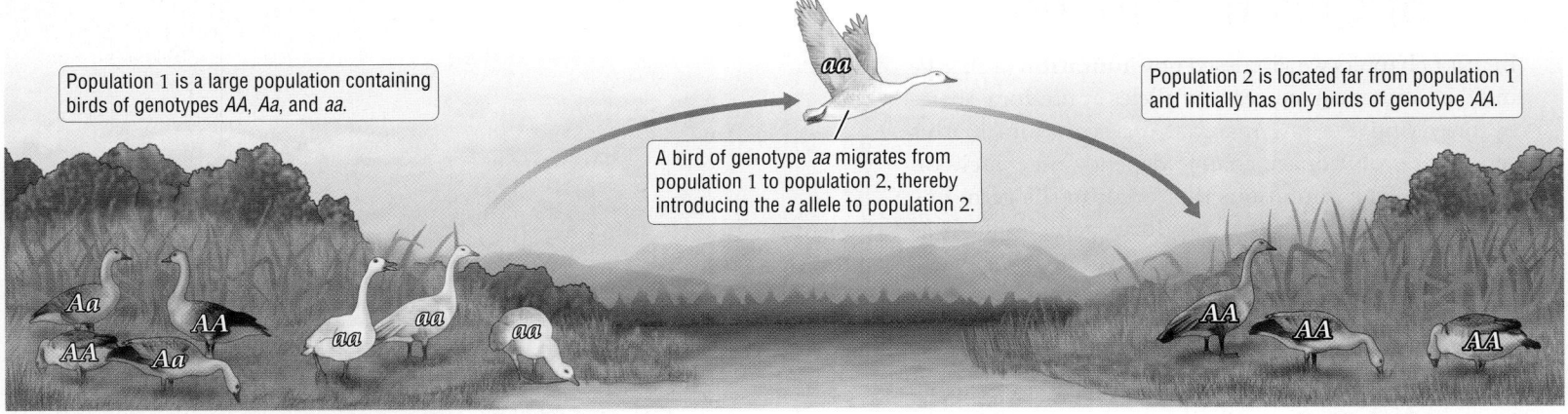

Population 1 is a large population containing birds of genotypes *AA*, *Aa*, and *aa*.

A bird of genotype *aa* migrates from population 1 to population 2, thereby introducing the *a* allele to population 2.

Population 2 is located far from population 1 and initially has only birds of genotype *AA*.

Population 1

Population 2

**FIGURE 18.3** **Migrants Can Move Alleles from One Population to Another**

(**FIGURE 18.3**). Gene flow can also occur when only gametes move from one population to another, as happens when wind or pollinators like insects transport pollen from one population of plants to another.

Gene flow can play a role similar to that of mutation by introducing new alleles into a population. Introductions of new alleles can have dramatic effects. For example, in the case of the mosquito *Culex pipiens*, discussed in the previous section, a new allele that made the mosquito resistant to organophosphate pesticides spread by gene flow across three continents. This spread of a new mutant allele enabled billions of mosquitoes to survive the application of pesticides that otherwise would have killed them.

Because it consists of an exchange of alleles between one population and another, two-way gene flow tends to make the genetic composition of different populations more similar. A mutual exchange of alleles through gene flow can *counteract* the effects of mutation, genetic drift, and natural selection, which otherwise tend to make populations more different from one another. In some plant species, for example, neighboring populations live in very different environments yet remain genetically similar because pollinating insects maintain gene flow between them. Without gene flow, genetic drift could cause the gene pools of the two populations to diverge randomly over evolutionary time. If the habitats of the two populations are distinctly different, natural selection might drive the

two populations along different evolutionary paths by favoring different alleles in each habitat. However, mutual gene flow between these populations could lessen or completely counteract the divergence in gene pools driven by natural selection.

## 18.5 Genetic Drift: The Effects of Chance

As noted in Chapter 17, chance events may determine which individuals contribute offspring to the next generation (see Figure 17.4). As a result, chance events can cause alleles from the parent generation to be selected at random for inclusion in the next generation. **Genetic drift** is a change in allele frequencies because of random differences in survival and reproduction from one generation to the next.

### Genetic drift affects small populations

The chance events that cause genetic drift are much more important in small populations than in large populations. To understand why, consider what happens when a coin is tossed. It would not be all that unusual to get four heads in five tosses; that result has about a 15 percent chance of occurring. But it would be astonishing to get 4,000 heads in 5,000 tosses. Even though the frequency of heads is the same in both cases (80 percent), the chance of getting many more, or many less, than the expected 50 percent heads is much greater if the coin is tossed a few times than if it is tossed thousands of times.

In natural populations, the number of individuals in a population has an effect similar to the number of times a coin is tossed. Consider the small population of wildflowers shown in **FIGURE 18.4**. By chance alone, some individuals leave offspring and others do not. In this example, such chance events alter the allele frequencies of a gene with two alleles (*R* and *r*). The changes are rapid. One of the alleles (*r*) is lost from the population in just two generations. The other allele (*R*) reaches **fixation**, a frequency of 100 percent. When a population has many individuals, the likelihood that each allele will be passed on to the next generation greatly increases. If the population in Figure 18.4 had had many more individuals, it is unlikely that chance events could have caused such dramatic changes in so short a time.

Genetic drift also occurs in large populations, but in these cases its effects are more easily overcome by natural selection and other evolutionary mechanisms. In large populations, genetic drift causes relatively little change in allele frequencies over time.

What types of chance events cause genetic drift? One important source of genetic drift is the random alignment of alleles during gamete formation, which causes (by chance alone) some alleles, but not others, to be passed on to offspring. Another source of genetic drift is chance events associated with the survival and reproduction of individuals. In this case, even though a particular genotype may increase in frequency from one generation to the next, it is important to remember that the increase is due to chance events (as in Figure 18.4), not to that genotype's having an advantage over other genotypes because of the alleles it carries (as would occur in natural selection).

Overall, genetic drift can affect the evolution of small populations in two ways:

1. It can reduce genetic variation within small populations because chance alone eventually causes one of the alleles to reach fixation. The fixation of alleles can happen rapidly in small populations, but in large populations it takes a long time.

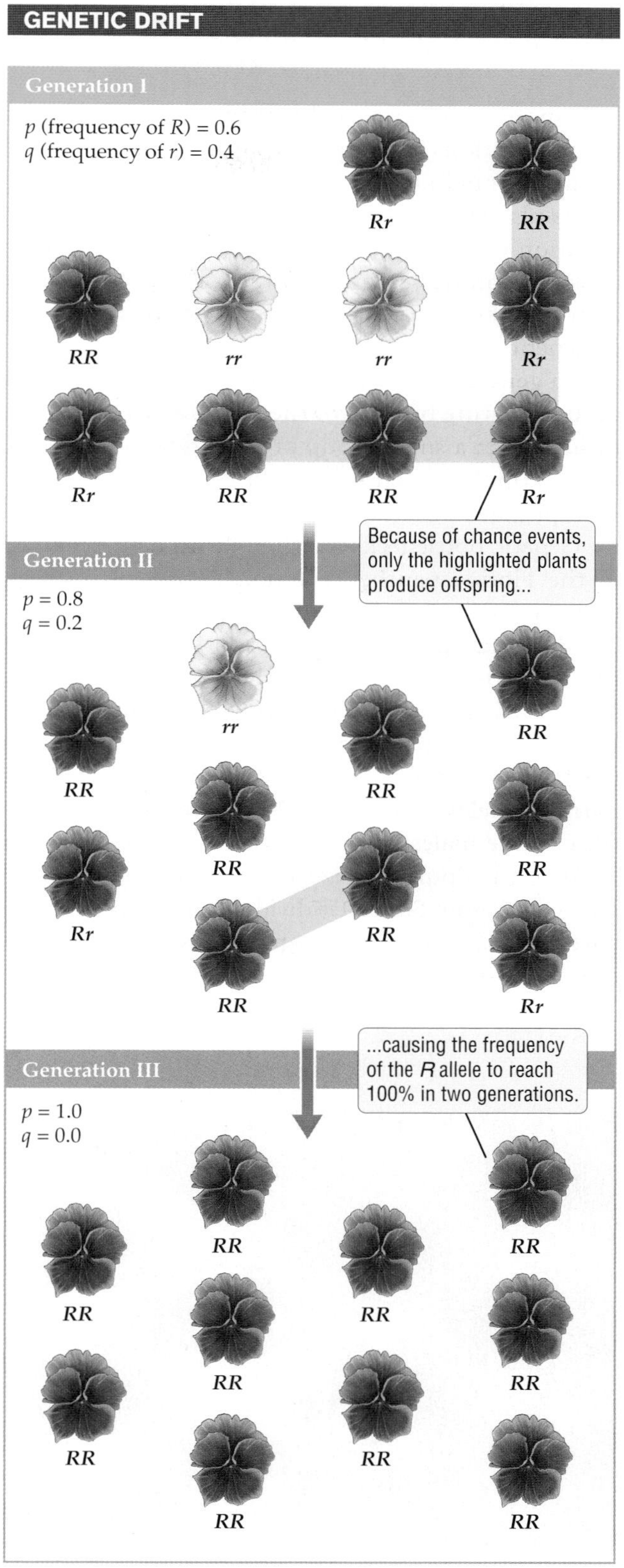

**GENETIC DRIFT**

**Generation I**

*p* (frequency of *R*) = 0.6
*q* (frequency of *r*) = 0.4

*Rr*   *RR*   *RR*   *rr*   *rr*   *Rr*   *Rr*   *RR*   *RR*   *Rr*

Because of chance events, only the highlighted plants produce offspring...

**Generation II**

*p* = 0.8
*q* = 0.2

*rr*   *RR*   *RR*   *RR*   *RR*   *RR*   *Rr*   *RR*   *RR*   *Rr*

...causing the frequency of the *R* allele to reach 100% in two generations.

**Generation III**

*p* = 1.0
*q* = 0.0

*RR*   *RR*   *RR*   *RR*   *RR*   *RR*   *RR*   *RR*   *RR*   *RR*

**FIGURE 18.4**
**Genetic Drift Occurs by Chance**

In this small population of wildflowers, chance events determine which plants leave offspring, without regard to which individuals are better equipped for survival or reproduction. Here, chance events cause the frequency of the *R* allele to increase from 60 percent to 100 percent in two generations. Note that this particular outcome is just one of many ways that genetic drift could have caused allele frequencies to change at random over time.

2. It can lead to the fixation of alleles that are neutral, harmful, or beneficial. As emphasized in Chapter 17, only natural selection consistently leads to adaptive evolution.

## Genetic bottlenecks can threaten the survival of populations

The importance of genetic drift in small populations has implications for the preservation of rare species. If the number of individuals in a population falls to very low levels, genetic drift may lead to a loss of genetic variation or to the fixation of harmful alleles, either of which can hasten extinction of the species. When a drop in the size of a population leads to such a loss of genetic variation, the population is said to experience a **genetic bottleneck**. Genetic bottlenecks often occur in nature because of the **founder effect**, which results when a small group of individuals establishes a new population far from existing populations (for example, on an island).

Genetic bottlenecks are thought to have occurred in the Florida panther, the northern elephant seal, and the African cheetah. In the case of the endangered Florida panther, population sizes plummeted to about 30–50 individuals in the 1980s. At that time, biologists discovered that male Florida panthers had low sperm counts and abnormally shaped sperm (**FIGURE 18.5**), probably as a result of the fixation of harmful alleles by genetic drift. The resulting low fertility in the males is thought to have contributed to the drop in population size. Panther numbers have increased to about 80–100 individuals in recent years, in part because of breeding programs designed to reduce the effects of genetic drift.

Although Florida panthers, northern elephant seals, and African cheetahs all show signs of having experienced genetic bottlenecks, each of these examples poses a problem: We do not know how much genetic variation was present before the population decreased in size. Hence, there is no way to be sure whether the observed low levels of genetic variation in these animals really were caused by a decrease in population size or were just a natural feature of the organism.

Recent studies on greater prairie chickens in Illinois avoided this problem by comparing the DNA of modern birds with the DNA of their pre-bottleneck ancestors, obtained from (nonliving) museum specimens. There were millions of greater prairie chickens in Illinois in the nineteenth century, but by 1993 the conversion of their prairie habitat to farmland had caused their numbers to drop to only 50 birds in two isolated populations (**FIGURE 18.6**). This drop in number caused a genetic bottleneck: the modern birds lacked 30 percent of the alleles found in the museum specimens, and they suffered poor reproductive success compared with other prairie chicken populations that had not experienced a genetic bottleneck (only 56 percent of their eggs hatched, versus 85–99 percent in other populations). From 1992 to 1996, 271 birds were introduced to Illinois from large populations in Minnesota, Kansas, and Nebraska, in order to increase both the size and the genetic variation of the Illinois populations. By 1997, the number of males in one of the two remaining Illinois popula-

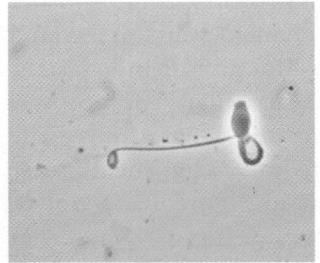
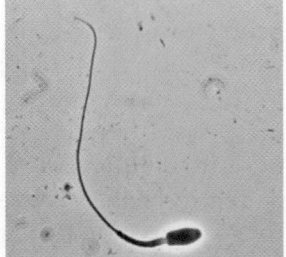

Abnormal panther sperm    Normal panther sperm

**FIGURE 18.5 Abnormally Shaped Sperm in the Rare Florida Panther**
Florida panthers have more abnormal sperm than do cats from other cougar populations—a possible effect of the fixation of harmful alleles. (*Insets*) Abnormal sperm and normal sperm, for comparison.

By 1993, only 50 greater prairie chickens remained in Illinois, causing both the number of alleles and the percentage of eggs that hatched to decrease.

| | ILLINOIS | | KANSAS | MINNESOTA | NEBRASKA |
|---|---|---|---|---|---|
| | PRE-BOTTLENECK (1933) | POST-BOTTLENECK (1993) | NO BOTTLENECK | | |
| Population size | 25,000 | 50 | 750,000 | 4,000 | 75,000 – 200,000 |
| Number of alleles at six genetic loci | 31 | 22 | 35 | 32 | 35 |
| Percentage of eggs that hatch | 93 | 56 | 99 | 85 | 96 |

Pre-bottleneck Illinois (1820)

Post-bottleneck Illinois (1993)

In 1820, the grasslands in which greater prairie chickens live covered most of Illinois.

In 1993, less than 1% of the grassland remained, and the birds could be found only in these two locations.

**FIGURE 18.6  A Genetic Bottleneck Can Cause a Population to Crash**
The Illinois population of greater prairie chickens dropped from 25,000 birds in 1933 to only 50 birds in 1993. This drop in population size caused a loss of genetic variation and a drop in the percentage of eggs that hatched. Here, the modern, post-bottleneck Illinois population is compared with the 1933 pre-bottleneck Illinois population, as well as with populations in Kansas, Minnesota, and Nebraska that never experienced a bottleneck.

tions had increased from a low of 7 to more than 60 birds, and the hatching success of eggs had risen to 94 percent.

## 18.6  Natural Selection: The Effects of Advantageous Alleles

In **natural selection**, individuals with particular inherited characteristics survive and reproduce at a higher rate than other individuals in a population. Natural selection acts by favoring some phenotypes over others, as when mosquitoes that are resistant to a pesticide survive at a higher rate and produce more offspring than do other mosquitoes. Although natural selection acts directly on the phenotype, not on the genotype, the alleles that code for forms of a trait favored by natural selection tend to become more common in the offspring generation than in

the parent generation. For example, if large size is an inherited characteristic, and if natural selection consistently favors large individuals, then alleles that cause large size will tend to become more common in the population over time.

### Even natural selection does not always lead to evolutionary change

Of the four mechanisms of evolution, natural selection is the only one that consistently improves the reproductive success of the organism in its environment. As shown by the study of the medium ground finch described in Chapter 17, natural selection can sometimes cause traits to evolve rapidly in response to changes in the environment.

However, even when natural selection favors one allele over another, it does not necessarily lead to evolutionary change. The other evolutionary mechanisms—genetic drift, gene flow, and mutation—may oppose its effects and prevent allele frequencies from changing. It also bears repeating that unless individuals within a population differ genetically, and unless some of them have beneficial mutations on which selection can act, natural selection is powerless. For example, if none of the mosquitoes in a population carry alleles for resistance to a pesticide, then natural selection cannot promote the evolution of resistance to that pesticide.

## There are three types of natural selection

Three patterns of natural selection are commonly observed: *directional selection*, *stabilizing selection*, and *disruptive selection*. Whatever the pattern, all types of natural selection operate by the same principle: individuals with certain forms of an inherited phenotypic trait tend to have better survival rates and to produce more offspring than do individuals with other forms of that trait.

In **directional selection**, individuals at one extreme of an inherited phenotypic trait have an advantage over other individuals in the population. For example, if large individuals produce more offspring than do small individuals, there will be directional selection for large body size (**FIGURE 18.7a**).

Directional selection is seen in the rise and fall of populations of dark-colored versus light-colored variants in some moth species. For example, dark-colored forms of the peppered moth were favored by natural selection when industrial pollution blackened the bark of trees in Europe and North America. The color of these moths is a genetically determined trait, and the allele for dark color is dominant over alleles for light color. Peppered moths are active at night and rest on trees during the day. The proportion of the dark-colored moths increased after industrialization, apparently because they were harder for predators such as birds to find against the blackened bark of trees. The rise in the proportion of dark-colored forms took less than 50 years. The first dark-colored moth was found near Manchester, England, in 1848. By 1895, about 98 percent of the moths near Manchester were dark-colored, and proportions of over 90 percent were common in other heavily industrialized areas of England—for example, nearby Liverpool (**FIGURE 18.8**).

Today, however, light-colored moths outnumber dark-colored moths. The reason, once again, is directional selection; but today it is the light-colored moths that are favored. This turnaround coincides with passage of the Clean Air Act in England in 1956. As the air quality improved, the bark of trees became lighter, and light-

**FIGURE 18.7**
**Directional, Stabilizing, and Disruptive Selection Affect Phenotypic Traits Differently**

The graphs in the top row show the relative numbers of individuals with different body sizes in a population before selection. The phenotypes favored by selection are shown in yellow. The graphs in the bottom row show how each type of natural selection affects the distribution of body size in the population.

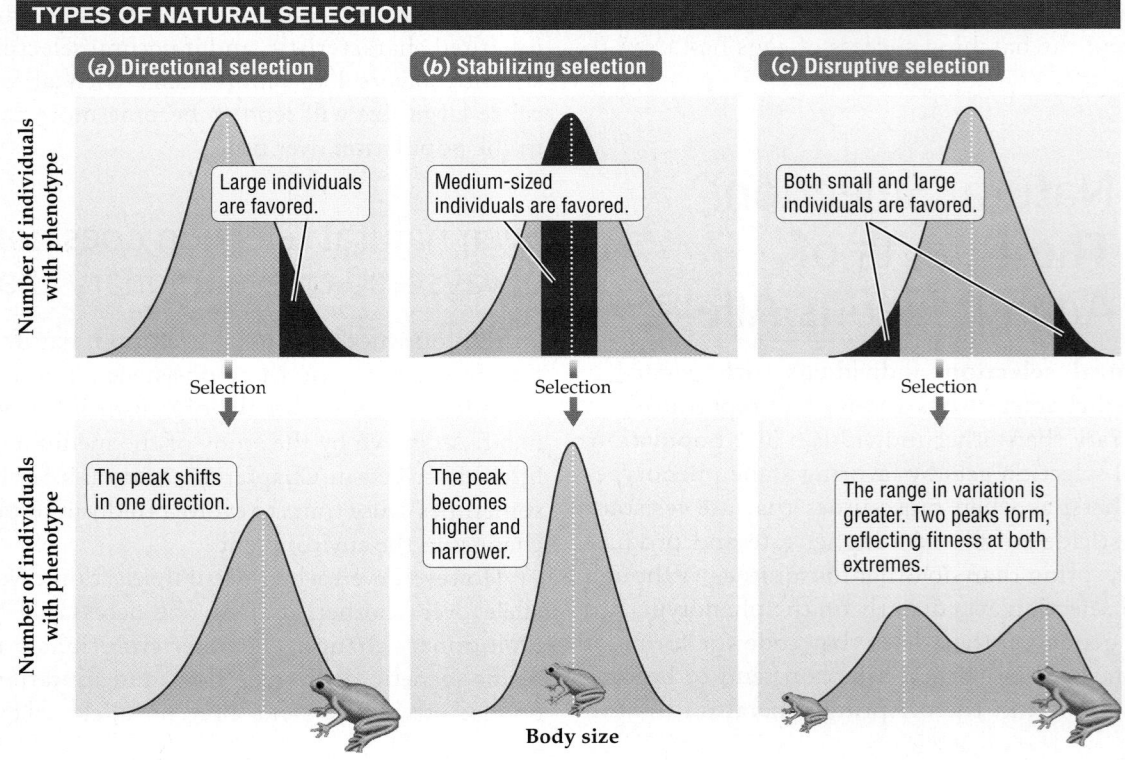

TYPES OF NATURAL SELECTION

(a) Directional selection — Large individuals are favored. — Selection — The peak shifts in one direction.

(b) Stabilizing selection — Medium-sized individuals are favored. — Selection — The peak becomes higher and narrower.

(c) Disruptive selection — Both small and large individuals are favored. — Selection — The range in variation is greater. Two peaks form, reflecting fitness at both extremes.

Number of individuals with phenotype

Body size

Dark-colored moth    Light-colored moth

**FIGURE 18.8 Directional Selection in the Peppered Moth**

The frequency of dark-colored peppered moths declined dramatically from 1959 to 1995 in regions near Liverpool, England, and Detroit, Michigan. Before 1959, dark-colored moths had risen in frequency in both England and the United States after industrial pollution had blackened the bark of trees, causing dark-colored moths to be harder for bird predators to find than light-colored moths. A reduction in air pollution following clean-air legislation enacted in 1956 in England and in 1963 in the United States apparently removed this advantage, leading the dark-colored moths to decline in a similar way in the two regions. (No data were collected in Detroit for the 30-year period 1963–1993.)

colored moths became harder for predators to find than dark-colored moths. As a result, the proportion of dark-colored moths plummeted (see Figure 18.8).

In **stabilizing selection**, individuals with intermediate values of an inherited phenotypic trait have an advantage over other individuals in the population (**FIGURE 18.7b**). Birth weight in humans provides a classic example. Historically, light or heavy babies did not survive as well as did babies of average weight, and as a result there was stabilizing selection for intermediate birth weights (**FIGURE 18.9**). By the late 1980s, however, selection against small and large babies had decreased considerably in some countries with advanced medical care, such as Italy, Japan, and the United States. This reduction in the strength of sta-

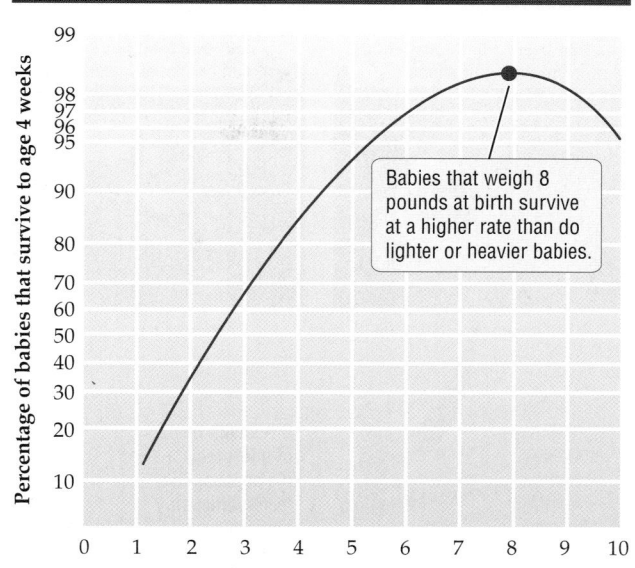

Babies that weigh 8 pounds at birth survive at a higher rate than do lighter or heavier babies.

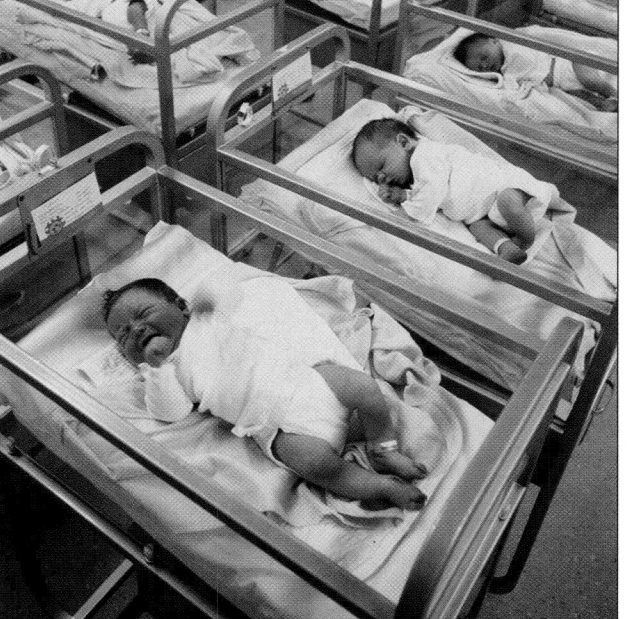

bilizing selection was caused by advances in the care of very light premature babies and by increases in the use of cesarean deliveries for babies that are large relative to their mothers (and hence pose a risk of injury to mother and child at birth).

In **disruptive selection**, individuals with either extreme of an inherited phenotypic trait have an advantage over individuals with an intermediate phenotype (**FIGURE 18.7c**). This type of selection is probably not common, but it appears to affect beak size within a population of African seed crackers, in which birds

**FIGURE 18.9**
**Stabilizing Selection for Human Birth Weight**

This graph is based on data for 13,700 babies born from 1935 to 1946 in a hospital in London. In countries that can afford intensive medical care for newborns, the strength of stabilizing selection has been reduced in recent years: because of recent improvements in the care of premature babies and increases in the number of cesarean deliveries of large babies, a graph of such data collected today would be flatter (less rounded) at its peak than the graph shown here.

## FIGURE 18.10
**Disruptive Selection for Beak Size**

In African seed crackers, differences in feeding efficiencies may cause differences in survival. Among a group of young birds hatched in one year, only those with a small or large beak size survived the dry season, when seeds were scarce; all the birds with intermediate beak sizes died. Therefore, natural selection favored both large-beaked and small-beaked birds over birds with intermediate beak sizes. Red bars indicate the beak sizes of young birds that survived the dry season; blue bars indicate the beak sizes of young birds that died.

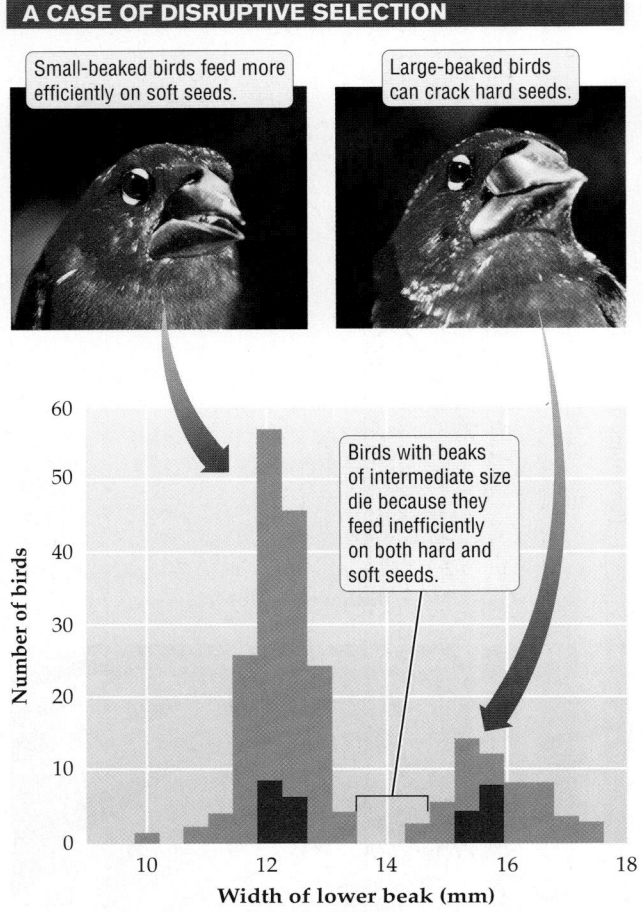

**A CASE OF DISRUPTIVE SELECTION**

Small-beaked birds feed more efficiently on soft seeds.

Large-beaked birds can crack hard seeds.

Birds with beaks of intermediate size die because they feed inefficiently on both hard and soft seeds.

*Number of birds* (y-axis), **Width of lower beak (mm)** (x-axis)

with large or small beaks survive better than birds with intermediate-sized beaks (**FIGURE 18.10**).

## 18.7 Sexual Selection: Where Sex and Natural Selection Meet

The males and females of many species differ greatly in size, appearance, or behavior. Many of these differences seem to be related to the ability of individuals to obtain mates. In lions, for example, males are considerably larger than females; and males fight, sometimes violently, for the privilege of mating with females. Since larger males are stronger and tend to be more successful in fights for females, natural selection may have favored large size in males, but not in females, leading to the substantial size difference between male and female lions.

When individuals differ in inherited characteristics that cause them to differ in their ability to get mates, a special form of natural selection, called **sexual selection**, is at work. Sexual selection favors individuals

**FIGURE 18.11 Elaborate Male Courtship Displays Influence Mate Choice**

Male peacocks with larger eyespots on their tails are thought to produce offspring with better survival rates.

that are good at getting mates and, as with the lions we just described, it often helps explain differences between males and females in size, courtship behavior, and other aspects of morphology and behavior. We say a species exhibits **sexual dimorphism** when males and females are distinctly different in appearance.

Ironically, some characteristics that increase an individual's chance of mating can *decrease* its chance of survival. For example, male túngara frogs perform a complex mating call that may or may not end in one or more "chucks." Females prefer to mate with males that emit chucks, but frog-eating bats use that same sound to help them locate their prey. As a result, a frog's attempt to locate a mate can end in disaster.

In many species, the members of one sex—often females—are choosy about whether to mate. The choosy partner acts as the brake, while the suitor, who tries to convince its reluctant partner to mate, acts as the accelerator. In birds, for example, brightly colored males may perform elaborate displays in their attempts to woo a mate (**FIGURE 18.11**). In other species, males may attract attention by other means, such as calling vigorously; females then select as their mates the males with the loudest calls.

For the process just described to make sense, we would expect that if the choosy partner bases her (or, occasionally, his) choice of a mate on a trait such as color or calling vigor, then that trait should serve as a good indicator of the quality of the mate. A high-quality mate, for example, might be especially healthy and perhaps more likely to produce healthy offspring or to be better at guarding a nest of young or at gathering food. A number of studies have provided support for this hypothesis. In blackbirds, females choose males with orange beaks more often than males with yellow beaks. It turns out that orange-beaked males have had fewer infections than yellow-beaked males; by selecting males with orange beaks, the females are selecting males in good health. Similarly, in mice, females can tell from the odor of a male's urine how many parasites he is harboring, and they use this information to select their mates.

In the two examples just mentioned, females use a trait such as beak color or urine odor to tell how healthy males currently are. In other cases, females seem to go beyond simply looking at a male's current bill of health, to assess whether he has "good genes." Female gray tree frogs, for example, prefer to mate with males that give long mating calls (**FIGURE 18.12a**). In an elegant experiment, scientists collected unfertilized eggs from wild females, sperm from males with long calls, and sperm from males with short calls. The eggs of each female were then divided into two batches, one of which was fertilized with sperm from long-calling males, the other with sperm from short-calling males. The offspring of long-calling male frogs grew faster, were bigger, and survived better than the

(a)                    (b)

**FIGURE 18.12**
**Competing for a Mate Can Be Hazardous**
(a) The call that this male gray tree frog issues to attack a mate can be emulated by predators to attract their prey. (b) Only the male staghorn beetle sports the large horn, which it uses to joust with other males. In most cases, the beetle with the longer horn wins more jousts, fertilizes more females, and therefore leaves more descendants. However, jousting carries the risk of serious injury or death.

offspring of short-calling male frogs. Apparently the long-calling males were both genetically and reproductively superior to the short-calling males.

### Concept Check

1. How is sexual selection a form of natural selection?
2. What mechanism of evolution is the only one that *consistently* leads a population toward adaptive evolution?

**Concept Check Answers**

1. Sexual selection favors individuals with genetic traits that are attractive to potential mates, which often include the ability to outcompete rivals.
2. Natural selection.

---

APPLYING WHAT WE LEARNED

# Flesh-Eating Bacteria and Antibiotic Resistance

Our bodies are naturally swarming with bacteria. Each of us is a community of 1,000 or more species, including trillions of individual microbes. Most are harmless, and the majority contribute to good health. But a few are pathogenic and can begin dividing and multiplying rapidly—infecting and destroying skin, muscles, or organs.

One of the worst pathogenic bacteria is MRSA, a "flesh-eating" strain of staphylococcus [*STAF*-ih-loh-*KAH*-kus], or "staph." MRSA (short for "methicillin-resistant *Staphylococcus aureus*") carries genes that make it resistant to most antibiotics (including methicillin). Some 40 other strains of staph harmlessly inhabit our skin and mucous membranes. Even *S. aureus* lives naturally on the skin of 32 percent of Americans, and nearly 1 percent carry the MRSA strain without knowing it. Yet when MRSA infects muscles or organs, it is so virulent that it kills 20 percent of its victims.

In the United States, about 94,000 people develop MRSA infections each year. Of these, 86 percent are the result of health care—for example, surgery in a hospital or clinic, where antibiotic-resistant pathogens are rampant. Why is MRSA so common in medical settings? Because antibiotics are in constant use in hospitals, bacteria that are sensitive to antibiotics quickly die off, leaving only those with adaptations for growing in the presence of the antibiotics.

In other words, antibiotics select for resistance. People with antibiotic-resistant MRSA may be perfectly healthy until

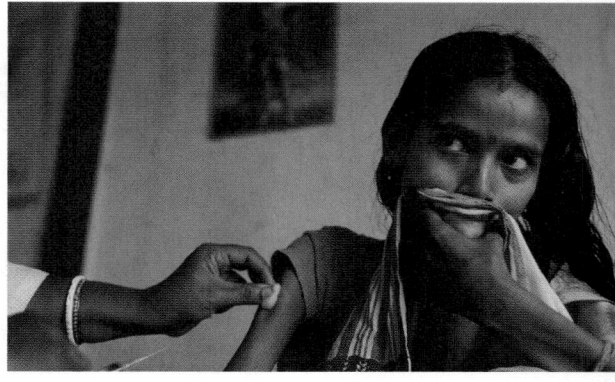

## Helpful to Know

The word "antibiotic" has roots in the Greek words meaning "against life." Because antibiotics work only on bacteria—and not against a wide range of disease agents, as many people think—we might do better to refer to them by the more accurate term "antibacterials," to avoid confusing them with substances used to kill fungi ("antifungals") and viruses ("antivirals").

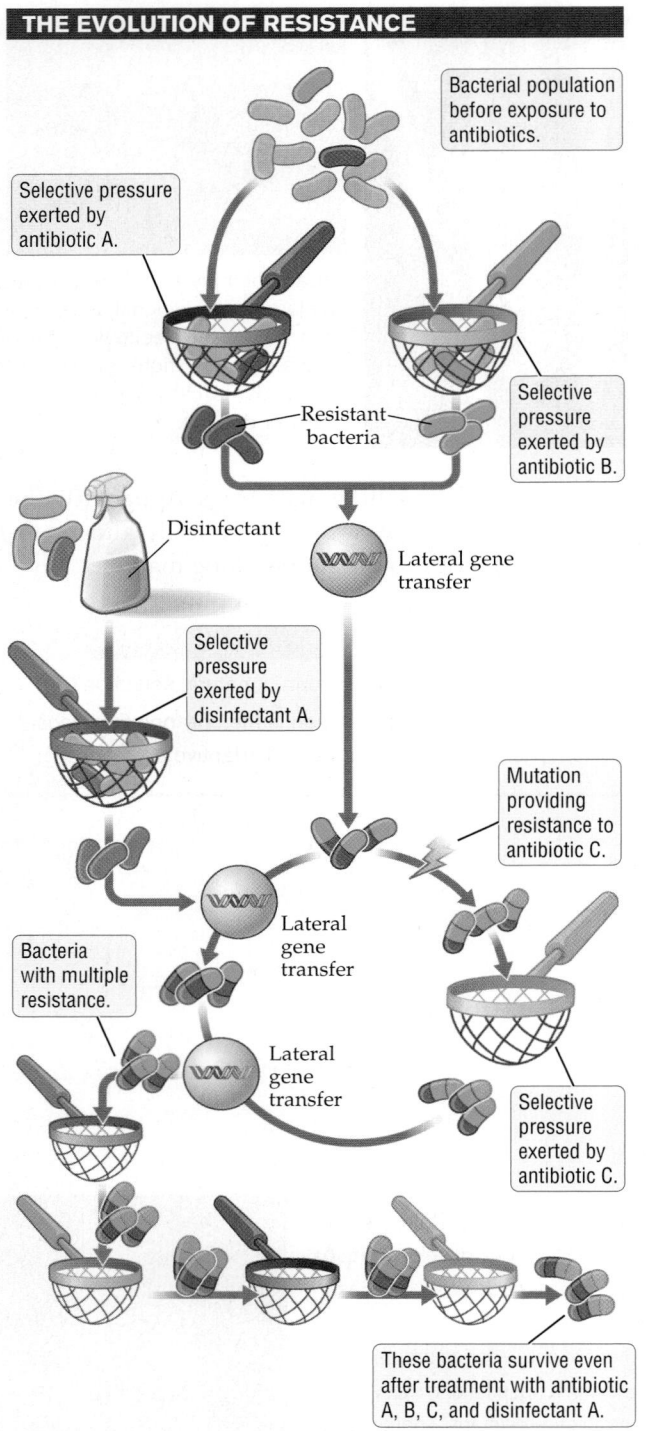

Bacterial population before exposure to antibiotics.

Selective pressure exerted by antibiotic A.

Resistant bacteria

Selective pressure exerted by antibiotic B.

Disinfectant

Lateral gene transfer

Selective pressure exerted by disinfectant A.

Mutation providing resistance to antibiotic C.

Bacteria with multiple resistance.

Lateral gene transfer

Lateral gene transfer

Selective pressure exerted by antibiotic C.

These bacteria survive even after treatment with antibiotic A, B, C, and disinfectant A.

**FIGURE 18.13 Natural Selection Results in Resistance to Antibiotics**

Before antibiotic use was common, most bacteria could be killed by antibiotics. Early use of antibiotics selected those few individuals that were resistant. Over time, new mutations and repeated episodes of selection—by the use of either antibiotics (represented here by sieves) or disinfectants—caused the frequency of resistant individuals to increase. In addition, transfer of genes within and among bacterial species allowed some bacteria to develop resistance to multiple antibiotics.

they take an antibiotic. When the antibiotic kills off their other, nonresistant bacteria, the MRSA bacteria suddenly have no competitors and begin multiplying faster (see **FIGURE 18.13** for an illustration of how antibiotic resistance develops in response to antibiotic use). If they happen upon a break in the skin, they can begin literally eating the flesh. They can even gain access to the body through hair follicles, the tiny holes from which hairs emerge. Indeed, any medical procedure that breaks the skin can potentially lead to infection if health care workers have not adequately washed their hands and sterilized equipment that comes anywhere in contact with a patient.

Out in the world and in our homes, antibiotics are usually far less common. In an environment where antibiotics are rare, resistant bacteria lose their selective advantage. If a bacterial strain's resistance genes are energy costly, nonresistant bacteria may gradually displace it. Biologists have begun exploring the idea of designing antibiotics for which resistance would be as expensive to the bacterium as possible. That way, even if resistance evolved, the resistant strain would quickly disappear as soon as the antibiotic was withdrawn.

Where do bacteria get resistance genes? Ordinary soil is full of both bacteria and fungi that compete with one another. Many of them naturally secrete antibiotics that kill their rivals. In fact, the first commercial antibiotics and fungicides were discovered by screening soil samples. It's natural for bacteria to encounter antibiotics secreted by fungi—for example, penicillin. At some point in the past, a random mutation in a bacterial genome turned out to confer resistance to penicillin. The mutation quickly took hold among bacteria growing near a penicillin-producing mold. Just as some dogs have yellow fur and others have black fur, some bacteria carry the resistance gene and some don't.

Bacteria share resistance genes with other strains of bacteria by means of plasmids. As we saw in Section 16.1, plasmids can carry such genes from one bacterial species to another—a process called *lateral gene transfer*. For example, if you were taking an antibiotic that killed most of the bacteria in your intestines, you could have a harmless bacterium with resistance that survived the antibiotic. It would not hurt you. But if a dangerous pathogen showed up, the harmless bacterium could pass its resistance genes to the pathogen, making it resistant and almost unstoppable.

# U.S. Meat Farmers Brace for Limits on Antibiotics

BY ERIK ECKHOLM, *New York Times*

Piglets hop, scurry and squeal their way to the far corner of the pen, eyeing an approaching human. "It shows that they're healthy animals," Craig Rowles, the owner of a large pork farm here, said with pride.

Mr. Rowles says he keeps his pigs fit by feeding them antibiotics for weeks after weaning, to ward off possible illness in that vulnerable period. And for months after that, he administers an antibiotic that promotes faster growth with less feed.

Dispensing antibiotics to healthy animals is routine on the large, concentrated farms that now dominate American agriculture. But the practice is increasingly condemned by medical experts who say it contributes to a growing scourge of modern medicine: the emergence of antibiotic-resistant bacteria…

Now, after decades of debate, the Food and Drug Administration appears poised to issue its strongest guidelines on animal antibiotics yet, intended to reduce what it calls a clear risk to human health. The guidelines, which are voluntary and will not have the binding force of regulations, [recommend an]… end [to] farm uses of the drugs simply to promote faster animal growth and call for tighter oversight by veterinarians…

… [But] the federal proposal has struck a nerve among major livestock producers, who argue that a direct link between farms and human illness has not been proved. The producers are vigorously opposing … [the guidelines] even as many medical and health experts call… [them] too timid.

Scores of scientific groups, including the American Medical Association and the Infectious Diseases Society of America, are calling for even stronger action that would bar most uses of key antibiotics in healthy animals, including use for disease prevention, as with Mr. Rowles's piglets. Such a bill is gaining traction in Congress.

… Proponents of strong controls note that the European Union barred most nontreatment uses of antibiotics in 2006 and that farmers there have adapted without major costs. Following a similar path in the United States, they argue, would have barely perceptible effects on consumer prices…

Genetic studies of drug-resistant E. coli strains found on poultry and beef in grocery stores and strains in sick patients have found them to be virtually identical, and further evidence also indicated that the resistant microbes evolved on farms and were transferred to consumers, said Dr. James R. Johnson, an infectious-disease expert at the University of Minnesota… "For those of us in the public health community, the evidence is unambiguously clear," Dr. Johnson said. "Most of the E. coli resistance in humans can be traced to food-animal sources."

All commercial raw meat is usually covered with bacteria. That's why we take precautions to clean cutting boards between cutting up raw meat and chopping vegetables for a salad. Some of these bacteria, including salmonella, are hazardous to us all by themselves. But other bacteria are harmless unless they happen to be antibiotic resistant. For example, most people have staph bacteria on their skin. So if you are exposed to a few staph bacteria while making hamburgers, that's not a big deal. But a 2011 study of grocery store meats revealed that 96 percent of staph bacteria were resistant to at least one antibiotic and 52 percent were resistant to three or more antibiotics.

In recent decades, in environments saturated with antibiotics—whether hospitals or pig farms—bacteria have accumulated strings of eight or more resistance genes against multiple antibiotics. These strings of different resistance genes are often inherited together. Bacteria that are resistant to five antibiotics have a selective advantage if they are exposed to *any one* of the five antibiotics. This makes them much more likely to thrive and persist in an antibiotic-rich environment.

If you take home a package of meat containing such bacteria, you normally cook the meat, killing the bacteria. You probably wash your cutting board and your hands too, but chances are high that a few bacteria will escape and colonize your skin, perhaps in the crook of your elbow, perhaps finding a way into your nostrils. And then they can share their resistance genes with your own bacteria. Now you have bacteria that are resistant to at least one antibiotic and possibly several different ones.

## Evaluating the News

**1.** Describe evidence presented in the article that supports the idea that meat animals treated with antibiotics are a major source of antibiotic resistance pathogens in people.

**2.** MRSA kills more Americans each year than AIDS. Since commonly used antibiotics no longer work against MRSA, doctors are now using less common antibiotics usually reserved for extreme cases of antibiotic resistance. Over time, how will selection alter the effectiveness of rarely used antibiotics? What can doctors do to slow the evolution of additional resistance in MRSA?

**SOURCE:** *New York Times*, September 14, 2010.

# Summary

## 18.1 Alleles and Genotypes

- An allele's frequency is the proportion of that allele in a population. Calculating allele frequencies is an important part of determining whether a population is evolving.
- A genotype's frequency is the proportion of that genotype in a population.

## 18.2 Four Mechanisms That Cause Populations to Evolve

- Allele frequencies in populations can change over time as a result of mutation, gene flow, genetic drift, or natural selection.
- The Hardy-Weinberg equation can be used to test whether one or more of these four mechanisms is causing a population to evolve.

## 18.3 Mutation: The Source of Genetic Variation

- All evolutionary change depends ultimately on the production of new alleles by mutation.
- Mutations cause little direct change in allele frequencies over time.
- Mutations can, however, stimulate the rapid evolution of populations by providing new genetic variation on which natural selection, genetic drift, or gene flow can act.

## 18.4 Gene Flow: Exchanging Alleles between Populations

- Gene flow can introduce new alleles into a population, providing new genetic variation on which evolution can work.
- Gene flow makes the genetic composition of different populations more similar.

## 18.5 Genetic Drift: The Effects of Chance

- Genetic drift causes random changes in allele frequencies over time.
- Genetic drift can cause small populations to lose genetic variation.

- Genetic drift can cause the fixation of harmful, neutral, or beneficial alleles.
- In a genetic bottleneck, a drop in population size causes genetic drift to have pronounced effects, including reduced genetic variation or the fixation of alleles. A genetic bottleneck can threaten the survival of a population.

## 18.6 Natural Selection: The Effects of Advantageous Alleles

- In natural selection, individuals that possess certain forms of an inherited phenotypic trait tend to survive better and produce more offspring than do individuals that possess other forms of that trait.
- Natural selection is the only evolutionary mechanism that consistently favors alleles that improve the reproductive success of the organism in its environment.
- There are three types of natural selection: directional, stabilizing, and disruptive.

## 18.7 Sexual Selection: Where Sex and Natural Selection Meet

- Sexual selection occurs when individuals with different inherited characteristics differ in their ability to get mates.
- Sexual selection underlies many differences between males and females, such as differences in size and courtship behavior.
- Some forms of a trait favored by sexual selection can lead to decreased survival.
- Sexual selection can occur when members of one sex—often females—are choosy about which individuals they will or will not mate with.

# Key Terms

allele frequency (p. 410)
directional selection (p. 418)
disruptive selection (p. 419)
fixation (p. 415)

founder effect (p. 416)
gene flow (p. 413)
genetic bottleneck (p. 416)
genetic drift (p. 414)

genotype frequency (p. 410)
microevolution (p. 410)
mutation (p. 413)
natural selection (p. 417)

sexual dimorphism (p. 420)
sexual selection (p. 420)
stabilizing selection (p. 419)

# Self-Quiz

1. A population of 1,500 individuals has 375 individuals of genotype *AA*, 750 individuals of genotype *Aa*, and 375 individuals of genotype *aa*. The genotype frequencies for genotypes *AA*, *Aa*, and *aa* are
   a. 0.33, 0.33, 0.33
   b. 0.25, 0.50, 0.25
   c. 0.375, 0.75, 0.375
   d. 0.125, 0.25, 0.125

2. A population of toads has 280 individuals of genotype *AA*, 80 individuals of genotype *Aa*, and 60 individuals of genotype *aa*. What is the frequency of the *a* allele?
   a. 0.24
   b. 0.33
   c. 0.14
   d. 0.07

3. A study of a population of the goldenrod *Solidago altissima* [sah-luh-*DAY*-goh al-*TIS*-ih-muh] finds that large individuals consistently survive at a higher rate than small individuals. Assuming size is an inherited trait, the most likely evolutionary mechanism at work here is
   a. disruptive selection.
   b. directional selection.
   c. stabilizing selection.
   d. natural selection, but it is not possible to tell whether it is disruptive, directional, or stabilizing.

4. Use the Hardy-Weinberg equation (see the box on page 412) to solve the following problem: If the frequency of the *A* allele is 0.7 and the frequency of the *a* allele is 0.3, what is the expected frequency of individuals of genotype *Aa* in a population that is not evolving?
   a. 0.21
   b. 0.09
   c. 0.49
   d. 0.42

5. Over time, a population of birds ranges in size from 10 to 20 individuals. If allele frequencies were observed to change in a random way from year to year, which of the following would be the most likely cause of the observed changes in gene frequency?
   a. stabilizing selection
   b. disruptive selection
   c. genetic drift
   d. mutation

6. Two large populations of a species found in neighboring locations with different environments are observed to become genetically more similar over time. Which evolutionary mechanism is the most likely cause of this trend?
   a. gene flow
   b. mutation
   c. natural selection
   d. genetic drift

7. Assume that individuals of genotype *Aa* are intermediate in size and that they leave more offspring than either *AA* or *aa* individuals do. This situation is an example of
   a. directional selection.
   b. disruptive selection.
   c. stabilizing selection.
   d. sexual selection.

8. The process by which differences in the inherited characteristics of individuals cause them to differ in their ability to get mates is most accurately called
   a. natural selection.
   b. reproductive success.
   c. mate choice.
   d. sexual selection.

# Analysis and Application

1. Select one of the four evolutionary mechanisms discussed in this chapter (mutation, gene flow, genetic drift, or natural selection), and describe how it can cause allele frequencies to change from generation to generation.

2. In your own words, define and explain the following: gene flow, genetic drift, natural selection, sexual selection.

3. One way to prevent a small population of a plant or animal species from going extinct is to deliberately introduce some individuals from a large population of the same species into the smaller population. In terms of the evolutionary mechanisms discussed in this chapter, what are the potential benefits and drawbacks of transferring individuals from one population to another? Do you think biologists and concerned citizens should take such actions?

4. Consider the toads in question 2 of the Self-Quiz. How do the numbers of toads with genotypes *AA*, *Aa*, and *aa* compare with the numbers you would expect on the basis of the Hardy-Weinberg equation (see the box on page 412)? Discuss the factors that could cause any differences you find.

5. Explain the reasoning behind the following statement: "Our attempts to kill bacteria are having the unintended effect of actually promoting the evolution of resistance." If we shifted our efforts away from killing bacteria and toward reducing our exposure to them or slowing their growth, why might such a change in our approach slow the evolution of antibiotic resistance?

6. Genetic drift is more likely to produce evolutionary change in a small population than in a large population. Explain why that is the case.

# Speciation and the Origins of Biological Diversity

**ELECTRIC-YELLOW CICHLIDS.** This particular cichlid species is native to Lake Malawi in East Africa. Africa alone is believed to have at least 1,600 species of cichlids. Hundreds more have evolved in other parts of the world, from Madagascar, India, and Syria to Central America, Mexico, and the southern United States.

# Cichlid Mysteries

The surface of Earth changes slowly but dramatically over time. Chains of islands rise from the sea; new lakes form and old ones disappear; mountains thrust upward, miles above sea level; and rivers cut massive canyons that divide continents. Such changes separate populations of organisms, alter the environments in which they live, and set the stage for evolution.

One of the most remarkable examples of rapid evolutionary change is provided by the cichlid fishes of Lake Victoria, the largest of the Great Lakes of East Africa and the second-largest freshwater lake in the world. Since it first formed 400,000 years ago, Victoria has dried up and filled with water again and again. It last filled about 15,000 years ago, and until the 1970s it was home to about 500 species of cichlid fishes—more kinds of fish than in all the lakes and rivers of Europe. In the 1970s, researchers reported so many fish that—in just 10 minutes—they were able to catch a thousand fish of 100 different species. All of these cichlid fishes descended from just two ancestor species from nearby Lake Kivu, and all had evolved in 15,000 years.

Beginning in the 1960s, humans began degrading Lake Victoria. Rapidly growing populations logged for-ests, causing heavy erosion of sediments into the lake; drained sewage and industrial chemicals into the lake, causing eutrophication (the overgrowth of algae); over-fished the lake; and introduced a giant predatory fish, the gigantic Nile perch, which ate cichlids faster than they could reproduce.

The Nile perch ate so well that the giant fish helped drive 200 species of cichlids to extinction in a single decade, the 1980s. Biologists wrung their hands in

How could so many species have evolved in just 15,000 years? What hope was there that the cichlids might adapt to the presence of perch and polluted lake water?

anguish over the destruction of so much biodiversity. Yet the lake's evolutionary history might have given them reason to hope. In this chapter we'll learn how new species arise and how mass extinctions can sometimes accelerate evolutionary change.

---

**MAIN MESSAGE**   Two important results of evolutionary change are adaptations and new species, both of which increase the diversity of living things.

---

## KEY CONCEPTS

- An adaptive trait is an inherited characteristic that matches an individual to its particular environment and thereby improves that individual's chances of surviving and reproducing.

- Natural selection leads to adaptation, the process that improves the match between a population of organisms and its environment over successive generations. Adaptive evolution can take many thousands of years or occur in as short a time as a few years or even

months. Adaptation does not produce "'perfection."

- The biological species concept defines a species as a group of natural populations that can potentially interbreed in nature to produce fertile offspring and that is reproductively isolated from other such groups.

- Speciation, the process by which one species splits to form two or more new species, is usually a by-product of genetic divergence between populations. The genetic divergence is promoted by mutation, genetic drift, and natural selection.

- Speciation often occurs when populations of a species become geographically isolated. Such isolation limits gene flow between the populations, making genetic divergence more likely.

- Speciation can also occur in the absence of geographic isolation—for example, when gene flow is blocked because of differences in mating behavior.

- Species formation generally takes thousands of years, but it has been observed in the span of a single generation also.

**KNOWING THAT POPULATIONS AND SPECIES CAN EVOLVE** explains adaptation, the diversity of life, and the shared characteristics of life. We discussed this great explanatory power of evolutionary biology in Chapter 17. In this chapter we return to two of these themes: adaptation and biodiversity. We examine the nature of adaptive traits and discuss how natural selection causes such traits to accumulate in a population. Then we reconsider the concept of species and how to define species effectively. We focus in the remainder of the chapter on the mechanisms that cause populations to diverge to the point that they become different species and how those mechanisms have led to the spectacular diversity of life on Earth.

## 19.1 Adaptation: Adjusting to Environmental Challenges

**FIGURE 19.1**
**Behavioral Adaptations in Weaver Ants Benefit All Members of the Colony**
Weaver ants collaborate to pull nest leaves together.

An **adaptive trait** is an inherited characteristic that enables an individual to function well in its particular environment and therefore to survive and reproduce better than competitors lacking the characteristic. Adaptive traits—which could be structural features, part of the biochemical makeup of the individual, or particular behaviors—improve how an individual performs in its environment. How quickly an individual grows compared to other competing individuals, how well it is camouflaged to escape predators, how successful it is in attracting mates—all these are examples of inherited traits that could be adaptive in certain populations in particular environments.

Individuals with adaptive traits are more successful in reproduction (have greater *reproductive fitness*) compared to individuals that lack those traits, and as a result individuals with adaptive traits become more common in descendant populations—a process known as **natural selection** (see Chapter 17). Natural selection causes a descendant population to become better matched, as a whole, to its environment. Because the descendant population evolves through the means of natural selection, the process is also known as *adaptive evolution*.

The term **adaptation** is commonly applied to adaptive traits *or* the process of evolution through natural selection. So, an adaptation can mean an adaptive trait, one that is advantageous to an individual or a species. Or we can apply the term more broadly to the *process* of natural selection: the evolutionary process that produces the good match we might observe between a population of organisms and their environment.

## Adaptations can take many different forms

The natural world offers many striking examples of adaptations. Consider the fascinating weaver ants (**FIGURE 19.1**). These ants construct nests out of living leaves. Building the nest requires the coordinated actions of many individual ants. Some draw the edges of leaves together, while others weave them in place by moving silk-spinning larvae (immature ants) back and forth over the seam of the two leaves. These complex behaviors are innate, or "hardwired," rather than learned. These gene-based behaviors illustrate how a simple evolutionary mechanism—natural selection—can produce a complex *behavioral* adaptation (cooperative nest building, which benefits the ants by providing them with shelter).

In another example of adaptation, the caterpillars (larvae) of a certain moth species develop bodies with different shapes depending on which part of their food plant (an oak tree) they feed on—the flowers or the leaves. Those that feed on flowers grow to resemble oak flowers; those that feed on leaves grow to resemble oak twigs (**FIGURE 19.2**). In this way, the larvae develop so that they match the background they are feeding and living on, making them more difficult for predators to see.

Natural selection has also shaped some astonishing adaptations that facilitate reproduction. Instead of using nectar to lure pollinators, the flowers of some orchids release chemical attractants (called *pheromones*) and mimic the shape, texture, and color of female

(a)

Caterpillars that hatch in the spring and eat oak flowers resemble oak flowers.

(b)

Caterpillars that hatch in the summer and eat oak leaves resemble oak twigs.

**FIGURE 19.2  An Adaptation That Offers Protection from Predators**

Although they do not look alike, the two caterpillars shown here are the same species of moth: *Nemoria arizonaria* [nee-*MOHR*-ee-uh air-ih-zoh-*NAIR*-ee-uh]. *N. arizonaria* caterpillars differ in appearance depending on what they feed on—the flowers or leaves of oak trees. Experiments have demonstrated that chemicals in the leaves control the switch that determines whether the caterpillars will mimic flowers (*a*) or twigs (*b*).

wasps. Male wasps attracted to the orchid attempt to mate with the flowers (**FIGURE 19.3**). In the course of these attempts, the insects become coated with pollen, which they then transfer from one plant to another.

Species interactions such as that between plant and pollinator can affect the balance and distribution of species in a community; that is, they have ecological as well as evolutionary consequences. Sometimes an interaction between two species so strongly influences their survival that the two species evolve in tandem—a phenomenon known as **coevolution**. The term "coevolution" encompasses a wide variety of ways in which the evolution of an adaptation in one species causes a reciprocal adaptation in another species.

**FIGURE 19.3  An Adaptation That Facilitates Reproduction**

The flowers of this orchid have evolved to look strikingly like the females of a wasp species that can be found flying in the area. The match is so good that the orchids are able to achieve pollination by being "mated" by a fooled male wasp.

## All adaptations share certain key characteristics

Look carefully at the eye of the four-eyed fish, *Anableps anableps*, in **FIGURE 19.4**. Although this fish really has only two eyes, they function as four, enabling the fish to see clearly through both air and water. The four-eyed fish is a surface feeder, so the ability to see above water helps it locate prey such as insects. Its unique eyes also enable it to scan simultaneously for predators attacking from above (such as birds) or below (such as other fish). The four-eyed fish can also walk on land, and it often escapes trouble by jumping out of the water, so although it would be interesting to watch, this fish would make a poor choice for a home aquarium.

Although there are literally millions of examples of adaptations crafted by natural selection, the four-eyed fish and the other examples discussed so

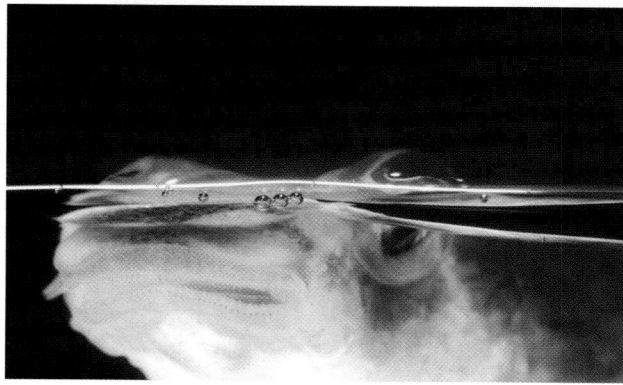

**FIGURE 19.4  A Fish with Four Eyes?**

The four-eyed fish (*Anableps anableps*) really has only two eyes, but each eye has a special design that lets it see clearly both in air and in water.

far illustrate the most important characteristics of an evolutionary adaptation:

- *Adaptations show a close match between organism and environment* (consider the caterpillars in Figure 19.2).
- *Adaptations are often complex* (consider the eye of *Anableps anableps* [Figure 19.4] and the nest-building behavior of weaver ants [Figure 19.1]).
- *Adaptations help the organism accomplish important functions*, such as feeding, defense against predators, and reproduction. (This point is illustrated by all of the examples we have discussed.)

## Populations can adjust rapidly to environmental change

Adaptive evolution can occur, and often does, over long periods of time. For example, as climates have slowly changed and ice ages have come and gone over the millennia, populations of organisms have continued to evolve and adapt to their changing environments during those thousands of years. But biologists have found that populations can undergo dramatic adaptive evolution surprisingly rapidly as well. For example, male guppies in the mountain streams of Trinidad and Venezuela have bright and variable colors that attract females. But the bright colors that help the males succeed in attracting mates also make them easier for predators to find. How do guppy populations evolve in response to such conflicting pressures?

Field observations show that guppies from streams where few predators lurk are brightly colored, but guppies from streams with more predators are drab in comparison. Predators can influence guppy coloration because, with more predators present, a higher proportion of the most brightly colored guppies are likely to be eaten each generation, so they do not pass their genes on to the next generation. The color of populations of guppies can evolve very rapidly in response to predators. When guppies that are brightly colored because they live in areas with few predators are experimentally transferred to an area with many predators, or vice versa, the color patterns in each population can evolve to match the new conditions within 10–15 generations (14–23 months).

The ability to evolve rapidly in response to changing environmental conditions is not limited to guppies. For example, consider the soapberry bug, which feeds on the seeds contained in the fruit of the soapberry plant, as illustrated in **FIGURE 19.5a**. In soapberry bug populations in Florida, beak length has evolved rapidly to match the size of the fruits that contain the seeds on which these insects feed (**FIGURE 19.5b**). Similarly, as we noted Chapter 17, beak sizes in the medium ground finch evolved extremely rapidly in response to drought in the Galápagos Islands and the subsequent change in the size of available seeds. In addition, as described in Chapter 18, it takes only a few months or years for viruses, bacteria, and insects to evolve resistance to our best efforts to kill them. These examples illustrate an important point: while evolution by natural selection can improve the adaptations of organisms over very long periods of time, it can do so over surprisingly short periods of time as well.

**AN EXAMPLE OF RAPID EVOLUTION**

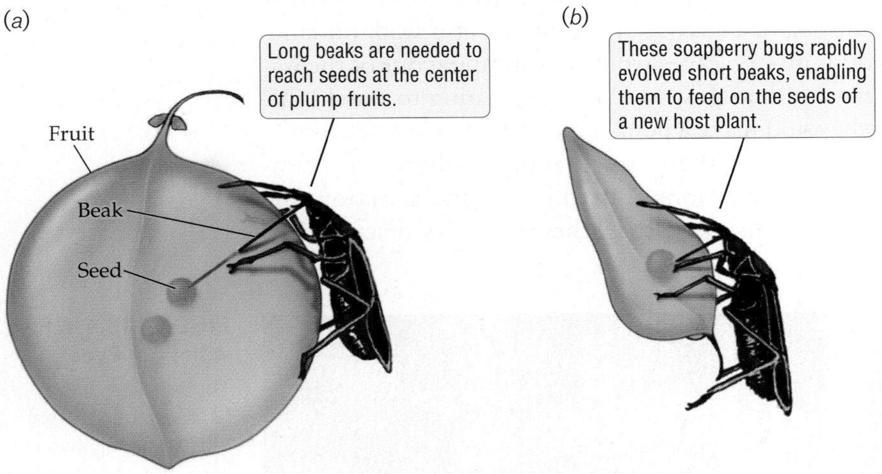

(a)

Long beaks are needed to reach seeds at the center of plump fruits.

Fruit

Beak

Seed

(b)

These soapberry bugs rapidly evolved short beaks, enabling them to feed on the seeds of a new host plant.

**FIGURE 19.5 Soapberry Bugs Have Rapidly Adapted to Changes in the Fruits They Eat**

(a) In Florida, soapberry bugs traditionally fed on seeds of a native plant species, the balloon vine. The bugs had to pierce to the center of the balloon vine's fruit to reach the seeds at the center. (b) Over the past 30–50 years, some populations of soapberry bugs have evolved short beaks, enabling them to feed on seeds within the narrower fruit of an introduced species, the golden rain tree.

## 19.2 Adaptation Does Not Craft Perfect Organisms

As impressive as the adaptations we see in nature may be, we do not mean to suggest that natural selection results in a perfect match between an organism and its environment. In many cases, genetic constraints, developmental constraints, or ecological trade-offs prevent further improvements in an organism's adaptation. Scientists estimate that 99 percent of all

# How We Affect the Evolution of Others

Human impact on the world is so large that Nobel Laureate Paul Crutzen has coined a term for it: "Anthropocene," the epoch in Earth's history marked by readily detected global changes caused by human activity. Because we have such a large effect on our environment, we also have a large effect on the evolutionary trajectory of other species.

Humans greatly affect how allele frequencies in populations of other species change over time. In many cases, we exert strong natural selection on other species, and they change genetically in response to our actions. Fishermen and hunters often target the largest individuals in a population, and such harvest selection can alter the evolutionary course of the surviving population. In an analysis of 29 species ranging from Atlantic cod to wild ginseng plants, Canadian researchers showed that the rate of evolutionary change is three times higher in species that are subject to harvest selection than in other species. Because older and larger individuals are likely to be harvested before they have a chance to produce offspring, individuals that reproduce at a younger age and smaller size gain a selective advantage. In this manner, adaptive evolution leads to early maturity in these species—on average, individuals reproduce at a smaller size and younger age than they did before they were subjected to hunting pressure.

When we modify or destroy the habitat in which a species lives, the number of individuals in a population is often reduced, sometimes dramatically. With such population reductions, genetic drift can reduce genetic variation and cause the fixation of harmful alleles, as in the greater prairie chicken (see Figure 18.6). Similarly, when we destroy portions of what once were large regions of continuous habitat, as when we convert native grasslands to farmland, we create geographic barriers that affect gene flow between natural populations.

The average body mass and horn length of bighorn sheep has declined in Alberta and other provinces because of selective culling of large males by hunters.

Human effects on macroevolution are most easily seen in human-caused extinctions of species. Large numbers of species are in danger of extinction as a result of habitat destruction, introductions of invasive species, and overharvesting. For many different types of organisms, threats such as these are causing a dramatic increase in extinction rates. For example, the extinctions of birds and mammals known to have occurred in the last 400 years suggest that human actions have increased extinction rates in these groups by 100–1,000 times the usual (or "background") rates found in the fossil record.

Could humans induce mass extinctions such as the one that wiped out the dinosaurs? A comparison of current extinction rates with those of prior mass extinctions indicates that humans have not yet induced extinctions to the degree seen in mass extinctions of the past (we have currently driven far fewer than 50 percent of the known species on Earth extinct, whereas more than 50 percent of species perished in each previous mass extinction). Scientists estimate, however, that if the current trends continue, extinction rates over the next 100 years will be at least 10,000 times the background rate for mammals, birds, plants, and many other organisms. From current rates of deforestation, for example, scientists estimate that virtually all of the world's tropical forests will be cut down in the next 50 years. If this happens, many of the species that live in tropical forests will go extinct. Since over 50 percent of the world's species live in tropical forests, the single action of removing tropical forests may cause extinctions of species on a scale that does rival mass extinctions of the past.

species that have ever lived are now extinct. Every extinct species is a silent testament to a failure to adapt in the face of adversity. In this section we look at some of the many reasons why an organism might fail to become perfectly adapted to all environments for all time.

## Lack of genetic variation can limit adaptation

For a population to become better adapted over successive generations, there must be genetic variation for traits that can enhance the match between the organism and its environment. In some cases, the absence of such genetic variation places a direct limit on the ability of natural selection to generate adaptation in descendant populations. For example, the mosquito *Culex pipiens* is now resistant to organophosphate pesticides, but this resistance is based on a single mutation that occurred in the 1960s, as we saw in Chapter 18. Before this mutation occurred, adaptation to these pesticides could not take place, and billions of mosquitoes died because their populations lacked the particular alleles that could confer resistance to the pesticides.

## The varied effects of developmental genes can limit adaptation

Changes in genes that control development can have dramatic effects on the particular manifestation of a trait, or phenotype. A change in the developmental program of an organism often influences more than one genetic trait, potentially generating many different types of novel phenotypes. Changes in genes that control development can therefore have many effects,

some of which may be advantageous while others may harm or kill the organism.

The multiple effects of developmental genes can limit the ability of the organism to evolve in certain directions, which in turn may limit what adaptive evolution can achieve. For example, the larval stages of some insects, such as beetles and moths, lack wings or well-developed eyes—two important adaptations that the adult forms of these insects have (**FIGURE 19.6**). Beetle and moth larvae have a wide range of lifestyles, so wings or well-developed eyes probably would benefit many of these larvae. Given that the adult and larval forms of these insects carry exactly the same genetic instructions, why are the genes for wings and eyes not "turned on" in the larvae? Recall that many genes express proteins that have not just one but sometimes many functions in the body—a phenomenon known as *pleiotropy* (see Chapter 12). If the expression of a gene has a deleterious effect at a certain stage of the organism's life, its expression is likely to be repressed. Expressing the genes that control wing or eye production in adults may have harmful—perhaps even fatal—effects in larvae. The lack of wings and eyes in beetle and moth larvae are therefore a result of developmental limitations.

## Ecological trade-offs can limit adaptation

To survive and reproduce, organisms must perform many functions, such as finding food and mates, avoiding predators, and surviving the challenges posed by the physical environment. Within the realm of what is genetically and developmentally possible, natural selection increases the overall ability of the organism to survive and reproduce. However, the many and often conflicting demands that organisms face

**FIGURE 19.6**
**A Developmental Limitation**

The larval form (*a*) of this moth does not have wings or functional eyes, two important adaptations that are found in the adult form (*b*).

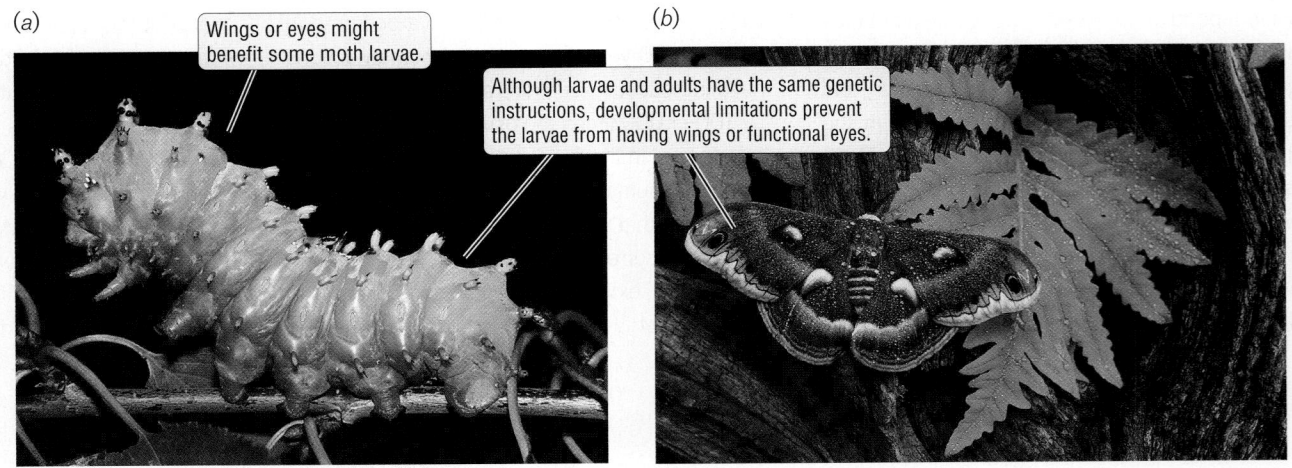

(*a*) Wings or eyes might benefit some moth larvae.

(*b*) Although larvae and adults have the same genetic instructions, developmental limitations prevent the larvae from having wings or functional eyes.

result in trade-offs, or compromises, in their ability to perform important functions.

High levels of reproduction, for example, are often associated with decreased longevity. In some cases, such a trade-off is due to relatively subtle costs of reproduction: resources directed toward reproduction are not available for other uses, such as storing energy to help the organism survive a cold winter. In red deer, for example, females that reproduced the previous spring have a higher rate of death during winter than do females that did not reproduce.

Costs associated with reproduction can sometimes be immediate and dramatic, as illustrated by the mating calls of the túngara frog. You may recall from Chapter 18 that it is not just females frogs that are lured by the mating call of the male túngara frog. Bats who prey on these frogs are attracted by the call as well (**FIGURE 19.7**). In general, the widespread existence of trade-offs between reproduction and other important functions ensures that organisms are not perfect, for the simple reason that it is not possible to be the best at all things at once.

**FIGURE 19.7  Does Love or Death Await?**
Male túngara frogs face an ecological trade-off: the same type of call that is most successful at attracting females also makes it easier for predatory bats to locate calling males.

> ### Concept Check
>
> 1. Why do organisms show so many adaptations to life in their environment?
>
> 2. What is the range of time over which adaptive evolution has been shown to take place?

## 19.3  What Are Species?

Before we discuss the processes that have generated the great diversity of species on Earth, we must define what a species is. It may surprise you to learn that there are several ways to answer that question. While the term "species" is commonly applied to members of a group that can mate with one another to produce fertile offspring, not all species can be defined by their ability to interbreed. For example, the ability to interbreed does not apply to the many species, including all prokaryotes, that reproduce asexually. Various "species concepts" have been advanced by biologists that collectively help us understand what defines a species.

### Species are often morphologically distinct

Species are commonly recognized by their appearance, the way we can readily distinguish polar bears (*Ursus maritimus*) from all brown bears (*Ursus arctos*).

Long before we understood that species might be distinguished by their genomes (total information in the DNA), species were defined in terms of morphology, or external form. That is, two organisms were classified as members of different species if they looked sufficiently different. The **morphological species concept** is based on the notion that most species can be identified as a separate and distinct group of organisms by the unique set of morphological characteristics they possess. Indeed, morphology is often the only way we can identify and distinguish fossil species. However, the morphological species concept does not always work well.

Sometimes distinct and separate species have members that look very much alike. For example, researchers from the Smithsonian Institution were startled to discover that the three species of *Starksia* blennies they had been studying in the Caribbean islands were really *10* different species, as DNA comparisons showed. Each of the 10 species of these reef fish inhabits a quite different habitat in the western Atlantic, and the species cannot interbreed; however, even experts cannot tell all 10 species apart by appearance alone.

Conversely, different populations can vary in appearance (sometimes dramatically) yet be assigned to the same species. All the brown bears of the world, for example—including the grizzly bear of inland Alaska and the much larger coastal brown bear—belong to one species: *Ursus arctos*. Similarly, all of the many color morphs of the happy-face spider (see Figure 17.3) are members of the same species, *Theridion grallator*, and some co-occur on the same island in the Hawaiian archipelago.

## Species are reproductively isolated from one another

Eastern meadowlark

Western meadowlark

In most cases, members of different species cannot reproduce with each other under natural conditions. When two species are prevented from interbreeding, we say that "barriers to reproduction" exist between those species, and that the species are **reproductively isolated** from one another. Barriers to reproduction are often divided into two categories (Table 19.1):

- Barriers that prevent a male gamete (like a human sperm cell) and a female gamete (like a human egg cell) from fusing to form a zygote are **prezygotic barriers**.
- Barriers that prevent zygotes from developing into healthy and fertile offspring are called **postzygotic barriers**.

A wide variety of cellular, anatomical, physiological, or behavioral mechanisms generate barriers to reproduction, but they all have the same overall effect: few or no alleles are exchanged between species. Barriers to reproduction ensure that the members of a species share a unique genetic heritage: a particular set of genes and alleles that is typical of the species but different from that of all other species. Because members of a species exchange alleles with one another but not with members of other species, they usually remain phenotypically similar to one another and different from members of other species.

The concept of reproductive isolation is the key to the most commonly used definition of a species: the *biological species concept*. The **biological species concept** defines a **species** as a group of natural populations that can interbreed to produce fertile offspring but that are reproductively isolated from other such groups. Note that reproductive isolation is not necessarily the same as geographic isolation. The biological species concept defines species as populations that could interbreed if they were in contact with one another but do not interbreed when they have no opportunity to do so. For example, the eastern and western meadowlark are different species, although they are very similar in morphology. The two species do not interbreed, even though their range overlaps in parts of the upper Midwest. These grassland species are reproductively isolated because each sings a distinctly different song and a female will mate only with the male that sings the melody unique to her species.

The biological species concept has important limitations. For example, it is of no use when we are defining fossil species, since no information can be obtained about whether two fossil forms were reproductively isolated from each other. So, as mentioned earlier, fossil species are defined on the basis of morphology. Nor does our definition apply to organisms that reproduce mainly by asexual means—for example, bacteria and dandelions.

The biological species concept also fails to work well for the many plant and animal species that are obviously distinct and separate species—sometimes even looking quite different from one another—but that remain able to mate in nature to produce fertile offspring. Distinct species that are able to interbreed in nature are said to **hybridize**, and their offspring are called **hybrids**. Although they can reproduce with each other, species that hybridize often look very different from one another (**FIGURE 19.8**) or are distinct ecologi-

| TABLE 19.1 | Barriers That Can Reproductively Isolate Two Species in the Same Geographic Region | | |
|---|---|---|---|
| **TYPE OF BARRIER** | **DESCRIPTION** | | **EFFECT** |
| **PREZYGOTIC BARRIERS** | | | |
| Ecological isolation | The two species breed in different portions of their habitat, at different seasons, or at different times of the day. | | Mating is prevented. |
| Behavioral isolation | The two species respond poorly to each other's courtship displays or other mating behaviors. | | Mating is prevented. |
| Mechanical isolation | The two species are physically unable to mate. | | Mating is prevented. |
| Gametic isolation | The gametes of the two species cannot fuse, or they survive poorly in the reproductive tract of the other species. | | Fertilization is prevented. |
| **POSTZYGOTIC BARRIERS** | | | |
| Zygote death | Zygotes fail to develop properly, and they die before birth. | | No offspring are produced. |
| Hybrid performance | Hybrids survive poorly or reproduce poorly. | | Hybrids are not successful. |

**Gray oak tree**

Gray oak

Hybrids

Gambel
oak

**Gambel oak tree**

**FIGURE 19.8 Some Species Interbreed yet Remain Distinct**
The gray oak and the Gambel oak can mate to produce fertile hybrids in regions where they co-occur. However, the gene flow in nature is limited enough that overall the two species remain phenotypically distinct.

cally. For example, they are usually found in different environments, or they differ in how they perform important biological functions, such as obtaining food.

The various species concepts are not mutually exclusive; they just focus on different attributes of species. The biological species concept, despite its limitations, remains the most useful definition for most biologists; and as a result, most biologists define species this way, so this is the definition we will use in this book. Remember, however, that while reproductive isolation (the cornerstone of the biological species concept) is very useful in establishing a species definition, the concept of a species in nature can be considerably more complicated.

## 19.4 Speciation: Generating Biodiversity

The tremendous diversity of life on Earth is caused by **speciation**, the process in which one species splits to form two or more species that are reproductively isolated from one another. The study of speciation is fundamental to understanding the diversity of life on Earth.

How do new species form? The crucial event in the formation of new species is the evolution of reproductive isolation, which requires that populations that once could interbreed diverge enough that they are no longer able to do so. As we saw in Chapter 18, however, populations within a species can be linked through gene flow, which tends to keep them genetically similar to one another. How does reproductive isolation develop within a species, whose members interbreed and therefore share a common set of genes and alleles?

## Speciation can be explained by the same mechanisms that cause the evolution of populations

Speciation is commonly thought of as a secondary, incidental consequence of the evolution of populations. In essence, over time populations evolve genetic differences from one another—because of mutation, genetic drift, or natural selection—and these genetic differences sometimes result in reproductive isolation.

This idea can be illustrated by the results of an experiment with fruit flies. A population of flies was separated into smaller populations, all of which were placed in similar environments, except that some flies were fed maltose (a simple sugar) while others were fed starch (starch is a polymer composed of many glucose molecules "stitched" together). Over time, the flies raised on these two different foods started to become reproductively isolated from each other (**FIGURE 19.9**). This reproductive isolation occurred simply because the flies adapted to living on two different kinds of food. That is, the fly populations changed genetically over time as the populations

**FIGURE 19.9 When What You Eat Affects Who You Love**

Fruit flies in an initial sample were separated into four populations and raised on two different kinds of food (starch or maltose) for several generations. In the experimental group, flies from the populations that had become adapted to feed on starch were then given the opportunity to mate with other flies that had adapted to feed on starch or with flies adapted to feed on maltose. As the mating frequencies show, scientists found that flies adapted to feed on starch preferred mating with other flies adapted to starch. Flies adapted to maltose likewise preferred other flies adapted to maltose. These preferences are the early stages of reproductive isolation that can eventually lead to speciation.

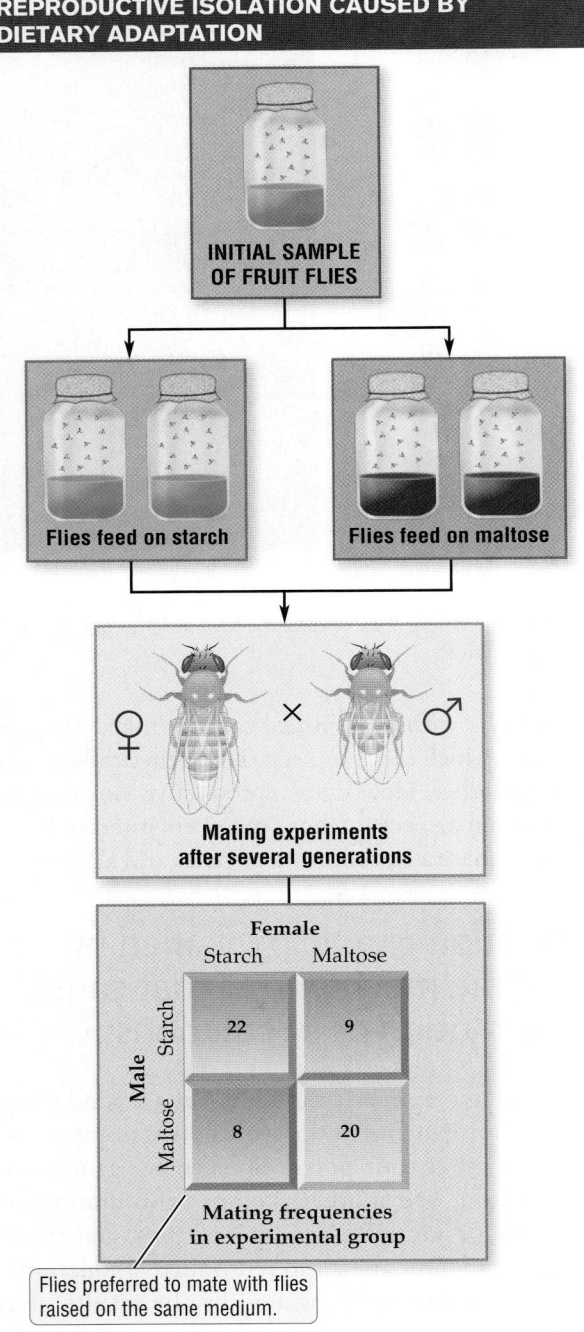

**REPRODUCTIVE ISOLATION CAUSED BY DIETARY ADAPTATION**

INITIAL SAMPLE OF FRUIT FLIES

Flies feed on starch

Flies feed on maltose

Mating experiments after several generations

**Female**

| | Starch | Maltose |
|---|---|---|
| **Starch** | 22 | 9 |
| **Maltose** | 8 | 20 |

Male

**Mating frequencies in experimental group**

Flies preferred to mate with flies raised on the same medium.

adapted to living on maltose or starch, respectively. The fly populations most likely changed genetically—that is, they evolved—in response to natural selection for the ability to grow and develop quickly on maltose versus starch. Those genetic changes had the side effect of causing reproductive isolation. That is, flies that were adapted to feed on maltose preferred to mate with flies that were also adapted to feed on maltose. The flies adapted to feed on starch likewise preferred to mate with flies adapted to feed on starch.

The experiment was not continued long enough for this budding reproductive isolation to become complete—that is, for speciation to occur. The flies from the maltose population and the starch population were still the same species and could still mate and reproduce with one another. But the experiment illustrates how reproductive isolation can begin to evolve as a by-product of other evolutionary changes. In this case, the other change was adaptation to living on maltose or starch. In the wild, there is always the potential for this kind of evolution among populations of a single species as populations adapt to living, for example, in different climates or to hunting different foods or avoiding different predators.

As suggested by Figure 19.9, natural selection can cause populations to diverge genetically when populations located in different environments face different selection pressures—in the case of our fruit flies, the selection to grow quickly on maltose versus starch. Populations can also diverge from one another as a result of mutation and genetic drift. In contrast, gene flow always operates to limit the genetic divergence of populations. For populations to accumulate enough genetic differences to cause speciation, the factors that promote divergence must have a greater effect than does the amount of ongoing gene flow.

## Speciation can result from geographic isolation

A new species can form when populations of a single species become separated, or geographically isolated, from one another. This process can begin when a newly formed geographic barrier, such as a river or a mountain chain, isolates two populations of a single species. Such **geographic isolation** can also occur when a few members of a species colonize a region that is difficult to reach, such as an island located far outside the usual geographic range of the species. For example, as we saw in Chapter 17, Darwin's finches on the separate Galápagos islands were geographically isolated from one another and from finches on the South American mainland by ocean waters.

The distance required for geographic isolation to occur varies tremendously from species to species, depending on how easily the species can travel across any given barrier. Populations of squirrels and other rodents that live on opposite sides of the Grand Canyon—a formidably deep and large barrier for a rodent—have diverged considerably (**FIGURE 19.10**). Meanwhile, populations of birds—which can easily fly across the canyon—have not. In general, geographic isolation is

## AN EXAMPLE OF GEOGRAPHIC ISOLATION

Kaibab squirrel (*Sciurus aberti kaibabensis*)

Abert's squirrel (*Sciurus aberti aberti*)

North rim

South rim

**FIGURE 19.10** The Grand Canyon Is a Geographic Barrier for Squirrels

The Kaibab squirrel is confined to the ponderosa pine forests on the north rim of the Grand Canyon. Abert's squirrel lives on the south rim and also farther south on the Colorado Plateau and in the southern Rocky Mountains to Mexico. The Kaibab population became isolated from the Abert's squirrels when the Colorado River cut a canyon—as deep as 6,000 feet in some places. With gene flow between them blocked, probably beginning about 5 million years ago, the Kaibab population diverged from the ancestral Abert's squirrel. The two squirrels used to be considered separate species, but now most experts classify them as subspecies of the Abert's squirrel, *Sciurus aberti*.

said to occur whenever populations are separated by a distance that is great enough to limit gene flow.

However they arise, geographically isolated populations are essentially disconnected genetically; there is little or no gene flow between them. For this reason, mutation, genetic drift, and natural selection can more easily cause isolated populations to diverge genetically from one another. If the populations remain isolated long enough, they can evolve into new species. The formation of new species from geographically isolated populations is called **allopatric speciation** (*allo*, "other"; *patric*, "country") (**FIGURE 19.11**).

Much evidence indicates that geographic isolation can lead to speciation. One case is the observation that in many groups of organisms, the number of species is greatest in regions where strong geographic barriers increase the potential for geographic isolation. Examples include species that live in mountainous regions (such as birds in New Guinea) or on island chains (such as plants in the Hawaiian Islands).

One particularly interesting example of a possible case of speciation by geographic isolation is the recent discovery of the so-called hobbit people of the island of Flores in Indonesia. These very tiny but very human-like creatures, also known as the "little people" of Indonesia, were discovered in 2004. Measuring 3 feet tall as adults, *Homo floresiensis* lived as recently as 17,000 years ago on their isolated island along with giant Komodo dragon lizards and pygmy elephants. The size of a typical human toddler, these people appeared to use stone

tools and fire. Anatomical studies have persuaded most anthropologists that these little islanders are a new and separate species, not just tiny members of our species, *Homo sapiens* (**FIGURE 19.12**).

A second line of evidence for the importance of geographic isolation in speciation comes from cases in which individuals of a population found at one extreme end of a species' geographic range reproduce poorly with individuals of a population at the other end, even though individuals from both ends reproduce well with individuals from intermediate portions of the species' range. A special case of this phenomenon occurs when the populations loop around a geographic barrier; in this case, **ring species** form, in which populations at the two ends of the loop are in contact with one another, yet individuals from these populations cannot interbreed. Ring species have been found in salamander populations that loop around the mountains of the Sierra Nevada in California.

## Speciation can occur without geographic isolation

Most speciation events are thought to occur by allopatric speciation, because speciation requires the cessation of gene flow between populations and the potential for gene flow is much greater between populations whose geographic ranges overlap or are adjacent to one another than between populations that

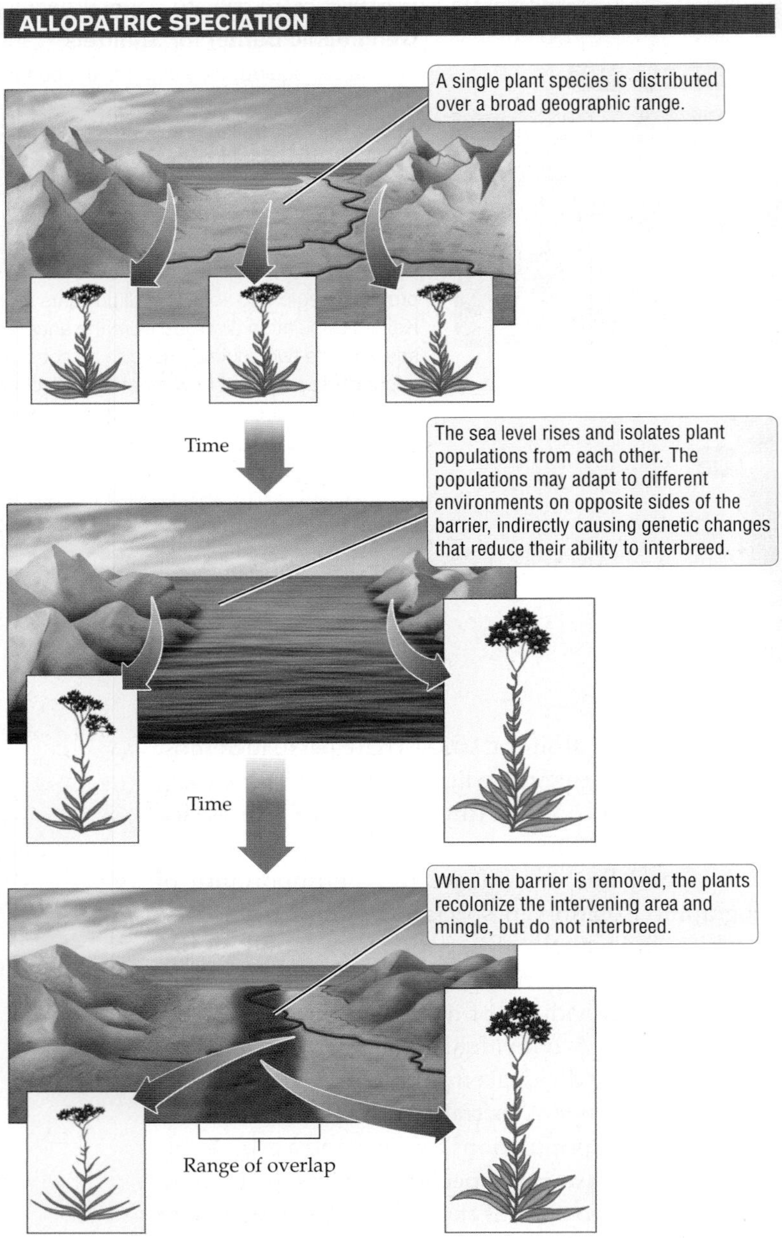

A single plant species is distributed over a broad geographic range.

Time

The sea level rises and isolates plant populations from each other. The populations may adapt to different environments on opposite sides of the barrier, indirectly causing genetic changes that reduce their ability to interbreed.

Time

When the barrier is removed, the plants recolonize the intervening area and mingle, but do not interbreed.

Range of overlap

**FIGURE 19.11 Physical Barriers Can Produce Speciation by Blocking Gene Flow**

New species can form when populations are separated by a geographic barrier, such as a rising sea.

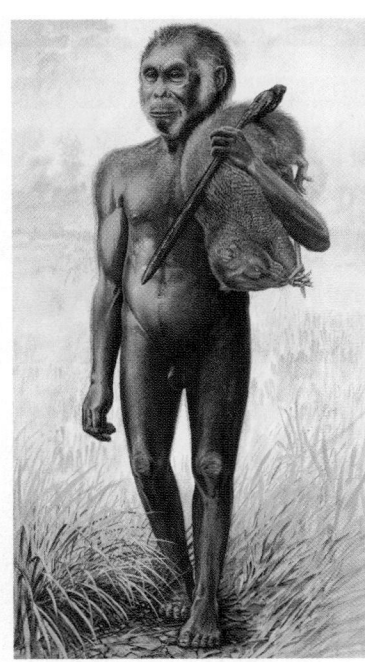

**FIGURE 19.12**
**Island Isolation May Have Driven the Evolution of *Homo floresiensis***

The foot structure and other anatomical features suggest that *H. floresiensis* was a distinctly different species, not a scaled-down version of *H. erectus* or *H. sapiens*.

In plants, rapid changes in chromosome numbers can cause sympatric speciation. New plant species can form in a single generation as a result of **polyploidy**, a condition in which an individual has more than two sets of chromosomes, usually because of the failure of chromosomes to separate during meiosis (discussed in Section 13.7). Polyploidy can also occur when a hybrid spontaneously doubles its chromosome number. Polyploidy is invariably fatal in people, but it is not lethal in many plant species. Although it does not kill the plant, a doubling of the chromosomes can lead to reproductive isolation because the chromosome number in the gametes of the new polyploid no longer matches the number in the gametes of either of its parents. Although relatively few plant species originate directly in this way, polyploidy has had a large effect on life on Earth: more than half of all plant species alive today are descended from species that originated by polyploidy. A few animal species also appear to have originated by polyploidy, including several species of lizards and fishes and one mammal (an Argentine rat).

Evidence is mounting that sympatric speciation can occur in animals by means other than polyploidy as well. For example, there is compelling evidence that new cichlid species have formed in the absence of geographic isolation in the lakes of Africa. The formation of new species within these small, environmentally uniform lakes provides strong evidence for sympatric speciation.

Strong evidence in support of sympatric speciation has also been found in other animals. Researchers believe that North American populations of the apple

are geographically isolated from one another. Therefore, at one time it was thought highly unlikely that new species could arise in any way other than by being geographically isolated—that is, via allopatric speciation. Today, however, it is well established that plants can form new species in the absence of geographic isolation, and recent work has provided convincing evidence that animals can as well. The formation of new species in the absence of geographic isolation is called **sympatric speciation** (*sym*, "together").

maggot fly, *Rhagoletis pomonella* [**rag-oh-**LEE-**tis poh-muh-**NELL-**uh**], are in the process of diverging into new species, even though their geographic ranges overlap. Historically, *Rhagoletis* usually ate native hawthorn fruits, but in the mid-nineteenth century these flies were first recorded as pests on apples, an introduced nonnative species. *Rhagoletis* populations that feed on apples are now genetically distinct from populations that feed on hawthorns. Members of the apple and hawthorn populations mate at different times of the year and usually lay their eggs only on the fruit of their particular food plant. As a result, there is little gene flow between the apple and hawthorn populations. In addition, researchers have identified alleles that benefit flies that feed on one host plant but are detrimental to flies that feed on the other host plant. Natural selection operating on these alleles acts to limit whatever gene flow does occur. Over time, the ongoing research on *Rhagoletis* may well provide a dramatic case history of sympatric speciation.

## 19.5 Rates of Speciation

How quickly can a new species evolve? When speciation is caused by polyploidy or other types of rapid chromosomal change, new species can form in a single generation. New species also appear to have formed relatively rapidly in the case of some African cichlid fishes. As we saw at the beginning of this chapter, scientists have described some 500 species of cichlids from Lake Victoria, and genetic analyses indicate that all of them descended from just two ancestor species over the past 100,000 years.

Many cichlids in Lake Victoria use color as a basis for mate choice: females prefer to mate with males of a particular color (an example of sexual selection; see Section 18.7). Further, cichlids have unusual jaws that can be modified relatively easily over the course of evolution to specialize on new food items. This feature of their biology causes the cichlids in Lake Victoria to vary greatly in form and feeding behavior (**FIGURE 19.13**). If female mate choice caused two populations to begin to be reproductively isolated from each other, the resulting lack of gene flow could allow the populations to specialize on different sources of food, which would make it increasingly likely that they would continue to diverge and form new species.

The cichlid rate of speciation is unusual. DNA evidence suggests that in most plants and animals speciation occurs more slowly. Freshwater fishes can speciate in a timespan ranging from 3,000 years (in pupfish)

to over 9 million years in characins [KAIR-**uh-sinz**], a group that includes carp, piranha, and many aquarium fish. As we saw in the cichlids, rates of speciation and rates of extinction are affected by various factors, including how fast reproductive isolation occurs and whether there is a sudden or gradual change in the physical environment, for example.

Once gene flow between populations of a species has been interrupted, speciation rates can be affected by factors other than the reproductive barrier itself, such as the frequency and severity of genetic events (e.g., polyploidy affects speciation much more quickly than random mutation), doubling time (the time it takes for a population to double in size) and behavior (such as the processes by which mates are chosen).

**SYMPATRIC SPECIATION**

*Haplochromis chilotes*
(feeds on insects)

*Haplochromis macrognathus*
(feeds on other fishes)

*Macropleurodus bicolor*
(feeds on snails and other mollusks)

*Astatotilapia elegans*
(generalized bottom feeder)

**FIGURE 19.13**
**Food Preferences and Female Mate Preferences May Have Driven Speciation Among Lake Victoria Cichlids**

The four species shown here illustrate some of the differences in feeding behavior and morphology among Lake Victoria cichlids.

Some populations can be geographically iso-lated for a long time without evolving reproductive isolation. North American and European sycamore trees, for example, have been separated for more than 20 million years, yet the two populations remain mor-phologically similar and can interbreed. Fossil evi-dence and DNA analysis shows that polar bears (*Ursus maritum*) and the brown bear (*Ursus arctos*) diverged into two distinctly different species about 200,000 years ago. Reports of hybrid bears have been increas-ing in recent years and DNA tests on odd-looking bears shot by hunters confirm that they are hybrids be-tween polar bears and grizzly bears. The ongoing loss of summer sea ice in the Arctic, one of many effects of climate change, may be driving the polar bear farther south in search of food, while grizzlies appear to be expanding their range northward. A gene flow that all but ceased 200,000 years has resumed, and the mighty hunter of the frozen north may come to be replaced by pizzly bears, prizzly bears, and grolar bears.

> ## Concept Check

1. What key characteristic defines a separate species, according to the biological species concept?
2. What role does geographic isolation play in speciation?
3. Why was sympatric speciation once thought to be unlikely?

---

### APPLYING WHAT WE LEARNED

# Lake Victoria: Center of Speciation

In the 1980s a population explosion of in-troduced Nile perch led to the extinction or near extinction of two-thirds of cichlid fish species in Lake Victoria, in East Africa. This catastrophic mass extinction drastically re-organized the lake's ecology. Cichlids, which had made up 80 percent of all the fish in the lake, now made up just 1 percent.

This was not the first time the lake had experienced such a disaster. Despite its vast size (nearly 27,000 square miles), Lake Victoria has dried up three times since it first formed 400,000 years ago. Each time the lake dried up, its fish species either went extinct or survived in only a few small ponds. Yet every time the lake refilled, the cichlids recolonized the lake and diversified into hundreds of new spe-cies in a classic example of adaptive radiation.

Like other fish, cichlids are masters of speciation. Half of all species of vertebrates are fish. Of the roughly 30,000 spe-cies of fish, nearly 10 percent are cichlids. How could 500 species of cichlids evolve in one lake in just 15,000 years? If the fish had evolved in a set of separate lakes, as in allopatric speciation, it would be easier to understand, but cichlids seem to speciate while living side by side, in sympatry.

One key aspect of cichlid biology partly explains the fish-es' rapid diversification. All animals tend to evolve the ability to see best in the color of light that bathes their environment. At the surface of a lake, red light dominates, while deeper down, blue light dominates. The eyes of fish species that feed deeper in the water are most sensitive to blue light. These blue-seeing fish usually stay deep, where they can see best. In addition, deep-water, blue-seeing females tend to choose bluish males as mates because bluer males are easier to see than reddish ones (which look black). In contrast, fish that feed near the surface see reds better, and likewise the females choose redder mates. The combination of cichlids' specialized color vision and the range of light color in the water helps to reproductively isolate each cichlid species, setting the stage for speciation.

But human development around Lake Victoria has changed all this. The water has become so filled with pollut-ants, sediment, and algae that the fish can hardly see. Instead of separating into distinct mating groups according to depth and light color, the fish swim wherever they like. Unable to see one another's colors, the fish crossbreed with other species—a process called hybridization.

Evolutionary biologists think that this species blending may be a first step in the process of adaptive radiation. Because about two-thirds of Lake Victoria's cichlid species are extinct, the lake is awash in empty niches that an evolving fish can ex-ploit. One species, *Yssichromis pyrrhocephalus*, seems to have jumped at this opportunity. Before the lakewide extinctions, this cichlid fed on plankton floating on the surface of the lake. It had nearly gone extinct when, in 2008, biologists discovered a surviving population that fed by rooting around in the mud at the bottom of the lake. The new *Yssichromis pyrrhocephalus* had gills for absorbing oxygen that were two-thirds larger than before—possibly because eutrophication had reduced oxygen levels in the water. *Y. pyrrhocephalus* 2.0 also had a strikingly different head shape, which apparently helped it prey on tough invertebrates in the muddy lake bottom. The fish had also changed its behavior: instead of avoiding the voracious Nile perch, it somehow managed to live in close proximity to this dangerous predator; how, biologists didn't know.

# Did Neanderthals Mate with Humans?

BY KATE BECKER, PBS NOVA

Modern humans and Neanderthals: Did they or didn't they? The sordid truth is out, and it's not what scientists expected. The closest-ever look at the Neanderthal genome reveals that yes, we did interbreed. But scientists are still fuzzy on the where, the when, and the why ...

So, our ancestors made babies with Neanderthals. But that's not the whole story: Only some modern populations have Neanderthal parentage. Africans don't seem to have any distinctively Neanderthal DNA. So what does that tell us about where and when modern humans and Neanderthals decided to commingle?

It must have been after some populations left Africa. Archeologists think that humans and Neanderthals lived side by side in Europe for about 15,000 years, until the Neanderthals died out about 30,000 years ago, so it seems reasonable to think that over the course of fifteen millennia these two populations would have done some canoodling.

But here's the weird part: The genetic clues point much farther back, between 60,000 and 100,000 years ago, when modern humans were settled in the Middle East and European and East Asian populations hadn't split yet. There is still no scientific consensus on whether Neanderthals and modern humans shared territory during this time. But dating DNA divergence is a complicated statistical game, so some archeologists and paleontologists, who are used to working with the kind of evidence you can hold in your hand—not the kind that comes chugging out of a computer algorithm—are skeptical.

But this research isn't just a high-tech paternity test. By comparing modern DNA with Neanderthal DNA, scientists can also uncover genes that are distinctively human. So far, the team has found about 100 genes that appear in modern humans but not in Neanderthals. They are involved in everything from skin to cognition to metabolism.

As Ian Tattersall, a paleontologist at the American Museum of Natural History, told the *New York Times*, "This is probably not the authors' last word" on the ties that bind us to Neanderthals.

---

Humans (*Homo sapiens*) and Neandertals (*Homo neanderthalensis*) are the tips of two lineages that evolved in Africa and separated half a million years ago. The ancestors of Neandertals migrated north to Europe and Asia, where they lived and evolved into Neandertals over hundreds of thousands of years. Our own ancestors stayed in Africa until about 100,000 years ago, when they, too, began migrating north out of Africa.

Neandertals didn't die out until about 30,000 years ago, which means both they and humans lived in Europe and Asia at the same time for perhaps thousands of years. But exactly when and where remains controversial. Some researchers argue that the two species lived side by side in the Middle East about 60,000 years ago, and perhaps in Europe as well, as recently as 24,000 years ago. Other researchers doubt that there was much overlap at all.

In this study, researchers extracted DNA from the 38,000-year-old bones of three Neandertal women from a cave in Croatia and roughed out a draft map of the Neandertal genome. Comparing the DNA of *H. neanderthalensis* and *H. sapiens*, the researchers identified 100 genes that had changed since the two lineages separated back in Africa.

The researchers also found that humans today still have some of these Neandertal genes. Any human whose ancestral group evolved outside of Africa carries between 1 and 4 percent genes that are recognizably Neandertal. What this means is that non-African humans were mating with Neandertals at some time in our past. Why only non-Africans? Although Neandertals were descended from a lineage that came out of Africa, Neandertals themselves never lived in Africa, so it would have been unlikely for the two groups to interbreed. And in fact, Africans carry no Neandertal genes.

We know that two separate species of humans—*Homo sapiens* and *Homo neanderthalensis*—hybridized, just like the cichlid fishes discussed in this chapter. But when did they interbreed? One clue depends on logic. All groups of non-Africans carry about the same amount of Neandertal DNA, whether they are European, Asian, or Papua New Guinean. Because these lineages split about 45,000 years ago, it's safe to conclude that the human-Neandertal cross-mating took place *before* that split—in other words, more than 45,000 years ago. Information in the genes themselves suggests that the interbreeding may have taken place even earlier, between 100,000 and 60,000 years ago, most likely in the Middle East.

## Evaluating the News

**1.** Do you think Neandertals and anatomically modern humans should be classified as separate species? Offer evidence to support your case.

**2.** Thinking about what you have learned in the last three chapters, including about the cichlid fishes in this chapter, what kind of population-wide situations might lead to two species hybridizing?

**3.** If you were a researcher, what question about Neandertals and humans would you want to ask next?

**SOURCE:** PBS NOVA, May 10, 2010.

# Summary

## 19.1 Adaptation: Adjusting to Environmental Challenges

- An adaptive trait is an inherited characteristic—structural, biochemical, or behavioral—that enables an organism to function well in its environment and that therefore increases survival and reproductive success.
- Natural selection results in adaptation, the fit between organisms and their environment. Because it causes the proportion of individuals with adaptive traits to increase over the generations, natural selection is also known as adaptive evolution.
- Adaptations can evolve over very long and very short periods of time.

## 19.2 Adaptation Does Not Craft Perfect Organisms

- Adaptive evolution can be limited by genetic constraints: lack of genetic variation gives natural selection little or nothing to act on.
- Adaptive evolution can be limited by developmental constraints: the multiple effects of developmental genes can prevent the organism from evolving in certain directions.
- Adaptive evolution can be limited by ecological trade-offs: conflicting demands faced by organisms can compromise their ability to perform important functions.

## 19.3 What Are Species?

- The morphological species concept recognizes that species can often be distinguished by their morphological features alone.
- The biological species concept defines species as a group of interbreeding natural populations that is reproductively isolated from other such groups.

- The biological species concept has important limitations: it does not apply to fossil species (which must be identified by morphology), to organisms that reproduce mainly by asexual means, or to organisms that hybridize extensively in nature.

## 19.4 Speciation: Generating Biodiversity

- The crucial event in the formation of a new species is the evolution of reproductive isolation.
- Speciation usually occurs as a by-product of the genetic divergence of populations from one another caused by natural selection, genetic drift, or mutation.
- Most new species are thought to arise through allopatric speciation, which occurs when populations are geographically isolated from one another long enough for reproductive isolation to evolve. Extended periods of reproductive isolation do not guarantee allopatric speciation, however.
- Speciation that occurs in the absence of geographic isolation is called sympatric speciation. Sympatric speciation can occur when part of a population diverges genetically from the rest of the population.
- Polyploidy is one way that many plants and some animals evolve new species during a single generation.

## 19.5 Rates of Speciation

- Rates of speciation vary. Speciation occurs rapidly in some cases but requires hundreds of thousands to millions of years in other cases.
- Rates of speciation depend both on how fast reproductive isolation is established (how long it takes to prevent gene flow between populations) and on factors that influence how quickly the species branches into a new lineage.

# Key Terms

adaptation (p. 428)
adaptive trait (p. 428)
allopatric speciation (p. 437)
biological species concept (p. 434)
coevolution (p. 429)

geographic isolation (p. 436)
hybrid (p. 434)
hybridize (p. 434)
morphological species concept (p. 433)

natural selection (p. 428)
polyploidy (p. 438)
postzygotic barrier (p. 434)
prezygotic barrier (p. 434)
reproductive isolation (p. 434)

ring species (p. 437)
speciation (p. 435)
species (p. 434)
sympatric speciation (p. 438)

# Self-Quiz

1. Species that have overlapping geographic ranges but do not interbreed in nature are said to be
   a. geographically isolated.
   b. reproductively isolated.
   c. influenced by genetic drift.
   d. hybrids.

2. Which of the following evolutionary mechanisms acts to slow down or prevent the evolution of reproductive isolation?
   a. natural selection
   b. gene flow
   c. mutation
   d. genetic drift

3. The splitting of one species to form two or more species most commonly occurs
   a. by sympatric speciation.
   b. by genetic drift.
   c. by allopatric speciation.
   d. suddenly.

4. The time required for populations to diverge to form new species
   a. varies from a single generation to millions of years.
   b. is always greater in plants than in animals.
   c. is never less than 100,000 years.
   d. is never more than 1,000 years.

5. Adaptations
   a. match organisms closely to their environment.
   b. are often complex.
   c. help the organism accomplish important functions.
   d. all of the above

6. Prezygotic and postzygotic barriers to reproduction have the effect of
   a. reducing genetic differences between populations.
   b. increasing the chance of hybridization.
   c. preventing speciation.
   d. reducing or preventing gene flow between species.

7. Evidence suggests that sympatric speciation may have occurred or may be in progress in all of the following cases *except*
   a. the apple maggot fly
   b. squirrels on opposite sides of the Grand Canyon

   c. cichlid fishes
   d. polyploid plants (or their ancestors)

8. The diploid number of chromosomes in plant species A is 8; the diploid number in plant species B is 16. If plant species C originated when a hybrid between A and B spontaneously doubled its chromosome number, what is the most likely number of diploid chromosomes in C?
   a. 8
   b. 12
   c. 24
   d. 48

9. The biological species concept
   a. can be applied to organisms that reproduce asexually.
   b. can be applied to fossil life-forms.
   c. would classify two natural populations, A and B, as separate species if a A and B are separated by a geographical barrier.
   d. would classify two natural populations, A and B, as separate species if A and B are unable to exchange genes even if they co-occur.

10. Lake Victoria cichlids
    a. exhibit low rates of speciation compared to most other fishes.
    b. exhibit high rates of sepeciation because of polyploidy.
    c. have evolved into many species in part because females choose mates by the specific color patterns the males display.
    d. have diverged into many species because females have different feeding behaviors than males.

# Analysis and Application

1. Select an organism (other than humans) that is familiar to you. List two adaptations of that organism. Explain carefully why each of these features is an adaptation.

2. What is adaptive evolution? Apply your understanding of adaptive evolution to organisms that cause infectious human diseases, such as bacterial species that cause plague or tuberculosis. How do our efforts to kill such organisms affect their evolution? Are the evolutionary changes we promote usually beneficial or harmful for us? Explain your answer.

3. Imagine that a species legally classified as rare and endangered is discovered to hybridize with a more common species. Since the two species interbreed in nature, should they be considered a single species? Since one of the two species is common, should the rare species no longer be legally classified as rare and endangered?

4. Should species that look different and are ecologically distinct, such as the oaks in Figure 19.8, be classified as one species or two? These

oak species hybridize in nature. Should species that hybridize in nature be considered one species or two?

5. High winds during a tropical storm blow a small group of birds to an island previously uninhabited by that species. Assume that the island is located far from other populations of this species, and that environmental conditions on the island differ from those experienced by the birds' parent population. Is natural selection or genetic drift (or both) likely to influence whether the birds on the island form a new species? Explain your answer.

6. Hundreds of new species of cichlids evolved within the confines of a large lake in Africa known as Lake Victoria. Some of these species live in different habitats within the lake and rarely encounter one another. Would you consider such species to have evolved with or without geographic isolation?

7. How can new species form by sympatric speciation? Why is it harder for speciation to occur in sympatry than in allopatry?

# The Evolutionary History of Life

**THE ROSS ICE SHELF, ANTARCTICA.** The Antarctic continent is so cold and dry that few organisms can survive here on land year-round. And yet, millions of years ago the continent supported an ecosystem with giant ferns, amphibians, dinosaurs, and flightless birds.

# Puzzling Fossils in a Frozen Wasteland

Antarctica is a crystal desert, an ice-covered land in which warmth and liquid water are nearly absent. Few organisms can survive in this cold and arid landscape, and those that do are small or live along the coast or are able to survive months of being frozen solid. Antarctica is almost entirely covered by an ice sheet that is up to a mile thick. In the few places that are not permanently covered with ice, just two species of flowering plants eke out a living—a modest grass and a mosslike pearlwort. Mosses, lichens, and tiny invertebrates also survive in the ice and cutting winds. The frigid coasts of the continent support a thriving community of plankton, fish, whales, seals, and penguins. But most are visitors that eat fish or plankton and fly or swim away when winter comes. On land, the largest year-round terrestrial animal is a flightless 5-millimeter fly.

In the interior of the continent, organisms are even tinier: in most places the only living things are microscopic bacteria and protists. A few interior valleys are nearly ice-free and seem less forbidding. But dry, freezing winds of up to 200 miles per hour blast away any remaining traces of water. In these Dry Valleys, even bacteria survive by living inside of rocks where a bit of water remains or within ice beneath the ground surface.

Despite the near lifelessness of modern Antarctica, paleontologists have uncovered fossils of trees as tall as 22 meters; of dinosaurs and other reptiles; of mammals and terror birds, a fast-running flightless bird that stood 3.5 meters tall—not to mention ferns, freshwater fishes, large amphibians, and aquatic beetles. In this chapter we'll

How could animals and plants adapted to warm, wet conditions have survived on a continent that is so barren today? What happened in Antarctica?

take a look back at the spectacular history of life on Earth. We'll find out how whole continents have moved, how the global climate has changed from warm and dry to frozen to tropical and back again, and how those dramatic changes have stimulated the evolution of life on Earth.

## MAIN MESSAGE

The geology and climate of our planet have changed over long time spans, with dramatic changes in life-forms, including mass extinctions and rapid bursts of evolution.

## KEY CONCEPTS

- The fossil record documents the history of life on Earth and provides clear evidence of evolution.

- Early photosynthetic organisms released oxygen to the atmosphere, setting the stage for evolution of the first eukaryotes and the first multicellular organisms.

- The Cambrian explosion was an astonishing increase in animal diversity that occurred about 530 million years ago, when most of the major living animal phyla appeared in the fossil record in a relatively short time span.

- The colonization of land by plants, fungi, and animals marked the beginning of another major increase in the diversity of life.

- The history of life can be summarized by the rise and fall of major groups of protists, plants, and animals. This history has been greatly influenced by continental drift, mass extinctions, and adaptive radiations.

- The first mammals evolved about 220 million years ago, at roughly the same time as dinosaurs, but their adaptive radiation began only after the demise of the dinosaurs about 65 million years ago.

- Primates, with their opposable thumbs and relatively large brains, had evolved by 65 million years ago.

- Bipedalism was a major step in the evolutionary history of the human family. Brain size increased in the lineage that led to *Homo erectus* and *Homo sapiens*.

We begin with a look at how the history of life on Earth is documented in the fossil record, followed by a summary of major events in the history of life. We then consider some of the forces that increase and decrease biodiversity over the long term—plate tectonics, mass extinctions, and adaptive radiations—and examine their effects with regard to the rise of our own group, the mammals.

## 20.1 The Fossil Record: A Guide to the Past

**EARTH ABOUNDS WITH LIFE.** About 1.7 million species have been described, and millions more await discovery. Though these numbers are large, the species alive today are thought to represent far less than 1 percent of all the species that have ever lived. How did they come to be—these "endless forms, most beautiful and most wonderful," as Darwin described them? We track that story of life in this chapter.

Fossils are the preserved remains or impressions of individual organisms that lived in the past (**FIGURE 20.1**). In many fossils, portions of the bodies of dead organisms are replaced with minerals. Petrified wood is a good example. In petrified fossils, the original body form is preserved but the fossil contains material not found in the living organism. Fossils are often found in sedimentary rock (rock built from layers of sediments that have hardened),

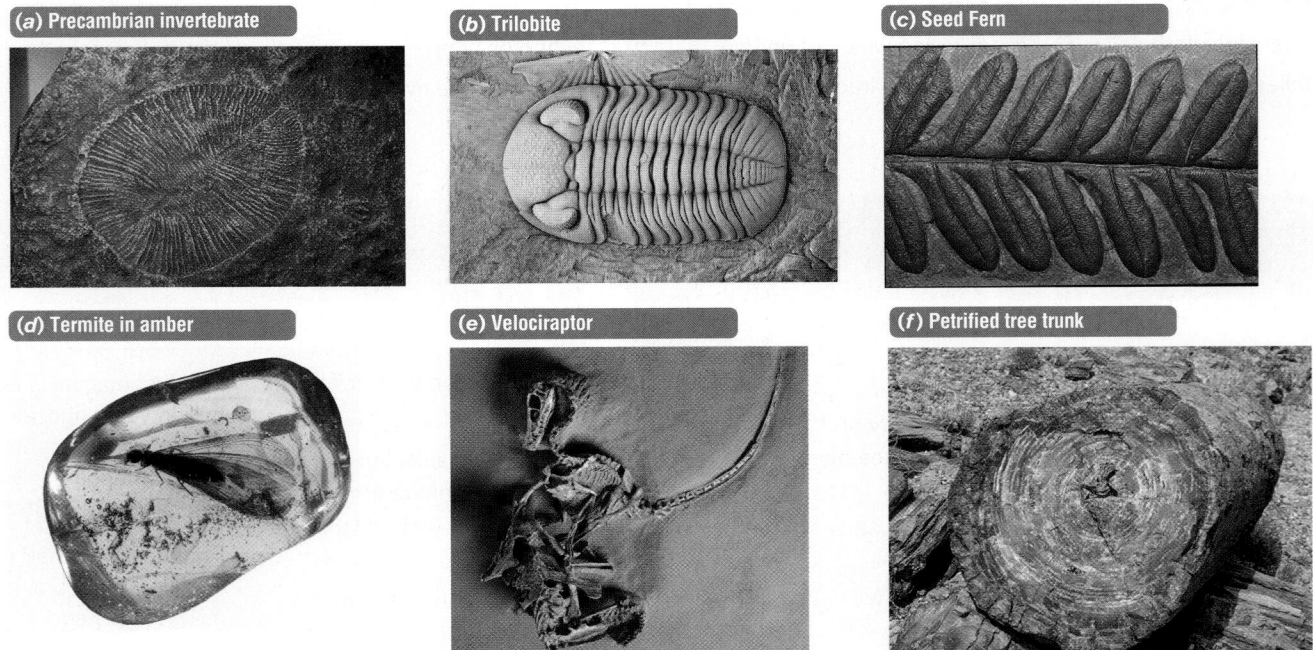

(a) Precambrian invertebrate

(b) Trilobite

(c) Seed Fern

(d) Termite in amber

(e) Velociraptor

(f) Petrified tree trunk

**FIGURE 20.1 Fossils through the Ages**
(a) Soft-bodied animals such as this one dominated life on Earth 600 million years ago (mya). (b) A fossil of a trilobite [*TRYE*-**loh-byte**] that lived in the Devonian period (410–355 mya). Note the furrowlike rows of lenses on each of the two large eyes. (c) Leaf of a 300-million-year-old seed fern found near Washington, DC. The fossil formed during the Carboniferous [*KAHR*-**buh-***NIF*-**er-us**] period (355–290 mya). The great forests of this period led to formation of the fossil fuels (oil, coal, and natural gas) that we use today as sources of energy. (d) This 20-million-year-old termite is preserved in amber, the fossilized resin of a tree. (e) A fossil of a *Velociraptor* entangled with a *Protoceratops*, which bit down on the predator's claw, locking both in a death grip. (f) Petrified wood. Here we see how what was once solid wood has fossilized into solid rock.

but they can also form in several other ways. Insects, for example, have been found in amber, the fossilized resin or pitch that comes from plants such as the pine tree (see Figure 20.1d). Animals, including mammoths and a 5,000-year-old man (see the photograph on page 344), have also been recovered from permafrost or found in melting glaciers.

As noted in Chapter 17, the fossil record documents the history of life and is central to the study of evolution. Fossils provided the first compelling evidence that past organisms were unlike organisms alive today, that many forms have disappeared from Earth completely, and that life has evolved through time.

The relative depth or distance from the surface of Earth at which a fossil is found is referred to as its order in the fossil record. The ages of fossils correspond to their order: older fossils are found in deeper, older rock layers. The order in which organisms appear in the fossil record agrees with our understanding of evolution based on other evidence, providing strong support for evolution. For example, analyses of the morphology (external form and internal structure), DNA sequences, and other characteristics of living organisms indicate that bony fishes gave rise to amphibians, which later gave rise to reptiles, which still later gave rise to mammals. That is exactly the order in which fossils from these groups appear in the fossil record. The fossil record also provides excellent examples of the evolution of major new groups of organisms, such as the evolution of mammals from reptiles (look ahead to Figure 20.12).

Although knowing the order of various organisms and groups of organisms in the fossil record is very helpful, it can provide only *relative* ages of fossils. That is, it can reveal which fossils are older than others. In some cases, we can approximate a fossil's age better by using **radioisotopes**, which are unstable, isotopic forms of elements that decay to more stable forms at a constant rate over time (see Section 5.1). For example, for a given amount of the radioisotope carbon-14 ($^{14}$C), half of the total decays to the stable element carbon-12 every 5,730 years. By measuring the amount of $^{14}$C that remains in a fossil, scientists can estimate the age of the fossil. Carbon-14 can be used to date only relatively recent fossils: too little $^{14}$C remains to date fossils formed more than 70,000 years ago. But elements such as uranium-235, which has a half-life of 700 million years, can be used to date much older materials. If, as commonly occurs, a fossil does not contain any radioisotopes, methods like carbon or uranium dating enable us to determine an approximate date for the fossil by dating rocks found above and below the fossil.

## The fossil record is not complete

The fossil record shows clearly that there have been great changes in the groups of organisms that have dominated life on Earth over time. As we will see throughout this chapter, these changes have been caused by the extinction of some groups and the expansion of other groups. Although many fossils have been found, however, the fossil record still has many gaps. Because most organisms decompose rapidly after death, very few of them form fossils. Even if an organism is preserved initially as a fossil, a variety of common geologic processes (such as erosion and extreme heat or pressure) can destroy the rock in which it is embedded. Furthermore, fossils can be difficult to find. Given the unusual circumstances that must occur for a fossil to form, remain intact, and be discovered by scientists, a species could evolve, thrive for millions of years, and become extinct without our ever finding evidence of its existence in the fossil record.

Still, although gaps in the fossil record remain, each year new discoveries fill in some of those gaps. One evolutionary event that has long been of interest is the evolution of sea-dwelling whales from land-dwelling mammals. This event is one gap in the fossil record that has recently begun to be filled in by newly discovered fossils.

## Fossils reveal that whales are closely related to a group of hoofed mammals

Let's take a closer look at what recent fossil discoveries revealed about one group of mammals: the whales. The origin of whales has long puzzled biologists. Most mammals live on land, and it is hard to imagine how a land mammal could be transformed into something as seemingly different as a whale. But recently discovered fossils provide a glimpse of how that adaptive transformation occurred (**FIGURE 20.2a**).

The bone structure of an early whale ancestor, *Pakicetus* [PAK-**uh**-SEE-**tus**], suggests that it probably spent most of its time on land. However, *Pakicetus* shared features (such as unusual bones in its inner ear) with modern whales and with the more whalelike creatures shown in Figure 20.2a. Over many generations, the ancestors of whales became increasingly similar to modern whales: their legs became smaller, and their overall shape took on the streamlined form of a fully aquatic mammal.

## EVOLUTION OF THE WHALES

### (a) Changes in overall anatomy

**Pakicetus**
1.8 meters long, 53 mya

**Ambulocetus**
4.2 meters long, 49 mya

**Rodhocetus**
3 meters long, 48 mya

**Dorudon**
4.5 meters long, 40 mya

**Orcinus orca (killer whale)**
4.5–9.1 meters long, 00 mya

### (b) Changes in anklebone structure

Artiodactyls    Primitive
                whales

This anklebone, from an extinct, hyena-like animal, is typical of non-artiodactyls.

**FIGURE 20.2  Shape-Shifters**

(*a*) It took roughly 15 million years for whale ancestors to make the transition from life on land to life in water. The oldest whale ancestor shown here, *Pakicetus*, lived on land 53 million years ago. *Ambulocetus* [**am-byoo-loh-SEE-tus**] had strong, well-developed legs and probably was semiaquatic, living at the water's edge and hunting in much the same way that a crocodile does today. In *Rodhocetus*, the body was more streamlined and the front legs were shaped more like flippers. By 40 mya, *Dorudon* was fully aquatic. These drawings are based on reconstructed fossil skeletons, which are superimposed on two of the whale ancestors. Compare the whale ancestors in this sequence with *Orcinus*, a modern toothed whale. (*b*) The anklebones of two whale ancestors—*Pakicetus* (shown here) and *Rodhocetus* (not shown)—are similar in shape to those of artiodactyls (hoofed mammals), but very different from those found in most other mammals. Each scale bar represents 1 centimeter.

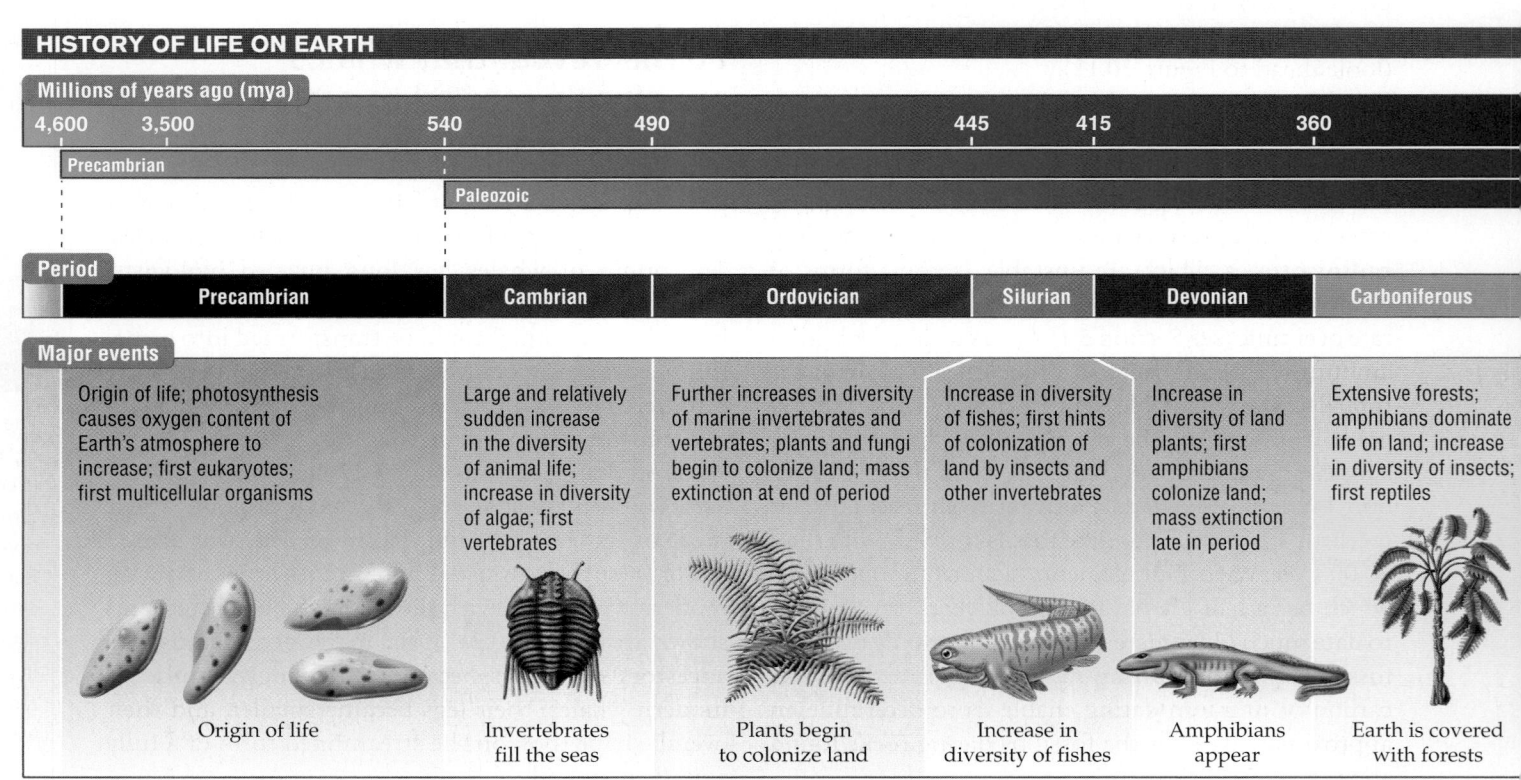

## HISTORY OF LIFE ON EARTH

**Millions of years ago (mya)**

| 4,600 | 3,500 | | 540 | 490 | | 445 | 415 | | 360 |
|---|---|---|---|---|---|---|---|---|---|

Precambrian

Paleozoic

**Period**

| Precambrian | Cambrian | Ordovician | Silurian | Devonian | Carboniferous |
|---|---|---|---|---|---|

**Major events**

| Origin of life; photosynthesis causes oxygen content of Earth's atmosphere to increase; first eukaryotes; first multicellular organisms | Large and relatively sudden increase in the diversity of animal life; increase in diversity of algae; first vertebrates | Further increases in diversity of marine invertebrates and vertebrates; plants and fungi begin to colonize land; mass extinction at end of period | Increase in diversity of fishes; first hints of colonization of land by insects and other invertebrates | Increase in diversity of land plants; first amphibians colonize land; mass extinction late in period | Extensive forests; amphibians dominate life on land; increase in diversity of insects; first reptiles |
|---|---|---|---|---|---|
| Origin of life | Invertebrates fill the seas | Plants begin to colonize land | Increase in diversity of fishes | Amphibians appear | Earth is covered with forests |

The recently discovered fossils also confirm the results of genetic analyses—namely, that whales are most closely related to the artiodactyls [AHR-tee-oh-DAK-tulz], which are even-toed, hoofed mammals such as camels, cows, pigs, deer, and hippopotamuses. In all artiodactyls, the anklebone has an unusual shape, in which both the top and bottom surfaces of the bone resemble those of a pulley (**FIGURE 20.2b**). In 2001, the anklebones of several whale ancestors, including *Pakicetus* and *Rodhocetus* [ROH-doh-SEE-tus], were discovered. These early whales had anklebones with the same unusual shape that the artiodactyl anklebones had. Since this shape is an adaptation for running on land, it is highly unlikely that whale ancestors developed such bones as a result of convergent evolution. Instead, these new fossils strongly suggest that whale ancestors had such bones because they shared a (recent) common ancestor with the artiodactyls.

## 20.2 The History of Life on Earth

In this section we focus on three of the main events in the history of life on Earth: the origin of cellular organisms, the beginning of multicellular life, and the colonization of land. **FIGURE 20.3** provides a sweeping overview of this history.

## The first single-celled organisms arose at least 3.5 billion years ago

Our solar system and Earth formed 4.6 billion years ago. The oldest known rocks on Earth (3.8 billion years old) contain carbon deposits that hint at life. Cell-like structures have been found in layered mounds called stromatolites [stroh-MAT-uh-lytes] that formed 3.5 billion years ago. Projections based on DNA analysis also support the idea that life had appeared on Earth by 3.5 billion years ago.

Eukaryotes first appear in the fossil record at about 2.1 billion years ago (see Chapter 3). After the origin of prokaryotes about 3.5 billion years ago, it took well over a billion years for the first eukaryotes to evolve. During this long period, the evolution of eukaryotes may have been limited in part by low levels of oxygen in the atmosphere. Chemical analyses of very old rocks indicate that Earth's atmosphere initially contained almost no oxygen. Roughly 2.8 billion years ago, however, a group of bacteria evolved a type of photosynthesis that releases oxygen as a by-product. As a result, the oxygen concentration in the atmosphere increased over time (**FIGURE 20.4**).

**FIGURE 20.3 The Geologic Timescale and the Major Events in the History of Life** The history of life can be divided into 12 major geologic time periods, beginning with the Precambrian (4,600–540 mya) and extending to the Quaternary (2.6 mya to the present). This timescale is not drawn to scale; to do so while including the Precambrian would require extending the diagram off the book page to the left by more than 5 *feet*.

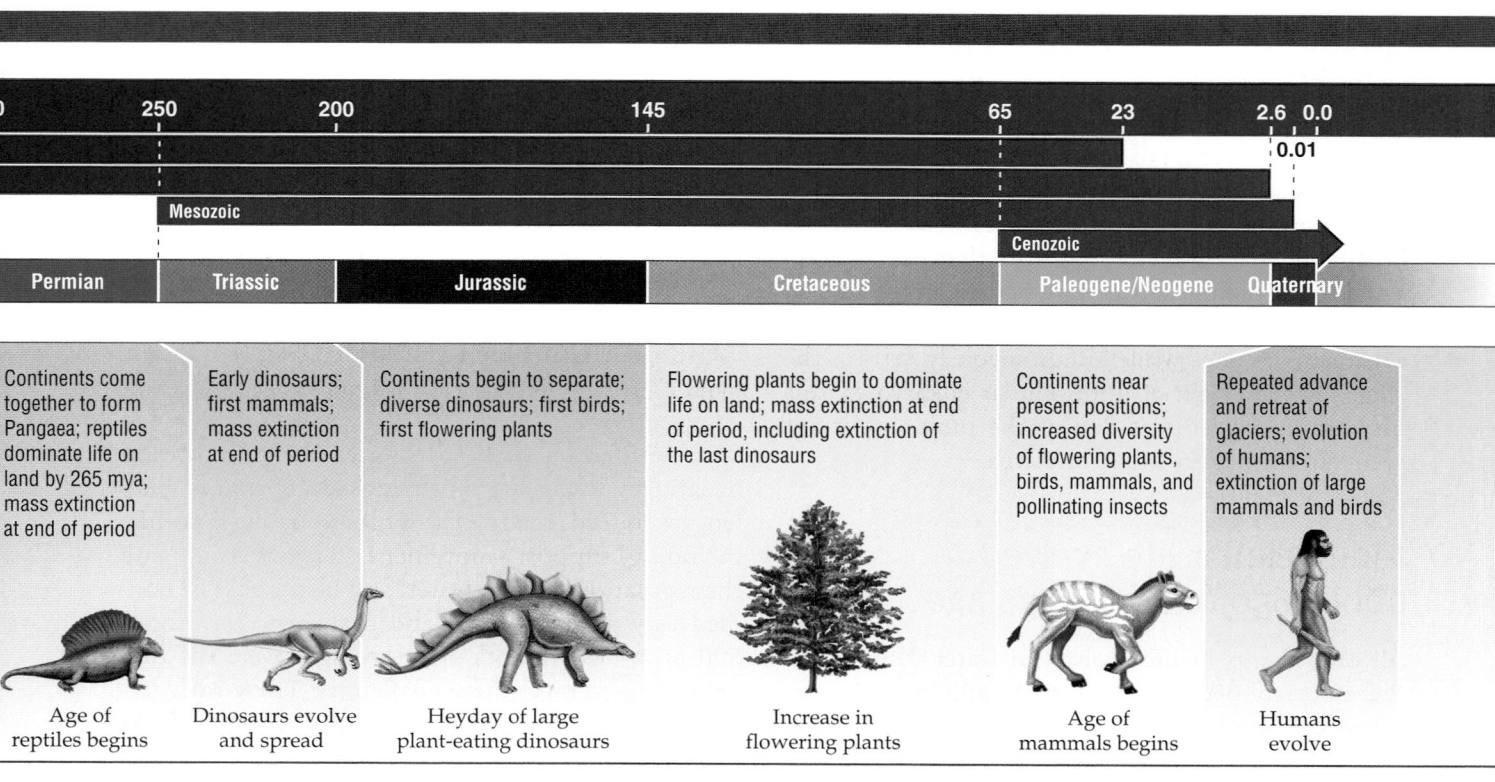

FIGURE 20.4
**Oxygen on the Rise**

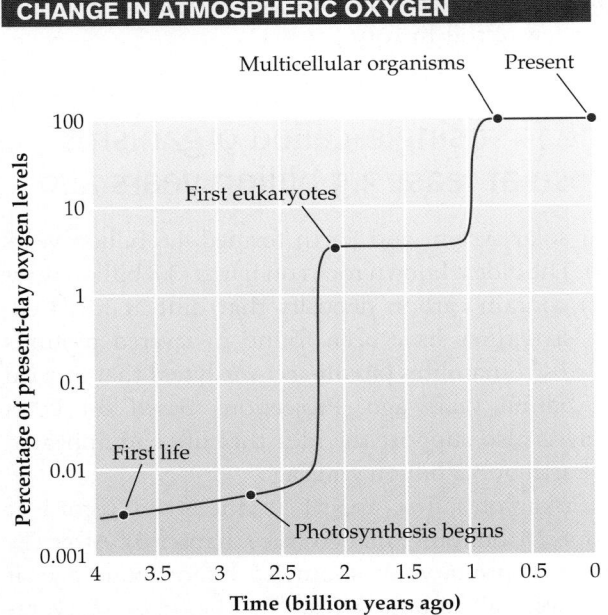

**CHANGE IN ATMOSPHERIC OXYGEN**

Eukaryotic cells are larger than most prokaryotic cells. Oxygen and other materials spread more slowly through a large cell than through a small cell. Overall, because of their relatively large size, eukaryotic cells would not have been able to get enough oxygen to meet their needs until the atmospheric concentration of oxygen reached at least 2–3 percent of present-day levels. Once those levels were reached, about 2.1 billion years ago, the first single-celled eukaryotes—organisms that resembled some modern algae—evolved (see Figure 20.4). When oxygen levels reached their current levels, the evolution of larger and more complex multicellular organisms also became possible.

Oxygen was toxic to many early forms of life, and as the oxygen concentration in the atmosphere increased, numerous early prokaryotes went extinct or became restricted to environments that lack oxygen. Because the biologically driven increase in the oxygen concentration of the atmosphere drove many early organisms extinct while simultaneously setting the stage for the origin of multicellular eukaryotes, this increase in oxygen was one of the most important events in the history of life on Earth.

## Multicellular life evolved about 650 million years ago

All early forms of life evolved in water. During the Precambrian period, about 650 million years ago (mya), the number of organisms appearing in the fossil record increased. At that time, much of Earth was covered by shallow seas, which were filled with plankton (protists, small multicellular animals, and single-celled and multicellular algae that float freely in the water). Later in the Precambrian, by about 600 mya, larger, soft-bodied multicellular animals had evolved (see Figure 20.1a). These animals were flat and appear to have crawled or stood upright on the seafloor, probably feeding on living plankton or their remains. No evidence indicates that any of these animals preyed on the others. Many of these early multicellular animals may have belonged to groups of organisms that are no longer found on Earth.

Later, during the Cambrian period, the world experienced an astonishing burst of evolutionary activity. During the early to middle Cambrian (530 mya), there was a dramatic increase in the diversity of life, known as the **Cambrian explosion**, in which large forms of most of the major living animal phyla, as well as other phyla that have since become extinct, appear in the fossil record. The word "explosion" here refers to a rapid increase in the number and diversity of species, not to a physical explosion. And that is because, in geologic terms, the Cambrian explosion was extremely rapid, lasting only 5–10 million years. It was a blink of an eye in geologic terms, when compared, for example, with the roughly 1.4 billion years it took for eukaryotes to evolve from prokaryotes.

The Cambrian explosion was one of the most spectacular events in the evolutionary history of life. It changed the face of life on Earth: from a world of relatively simple, slow-moving, soft-bodied scavengers and herbivores, suddenly there emerged a world filled with large, mobile predators. The presence of predators sped up the evolution of Cambrian herbivores, judging by the variety of scales and shells and other protective body coverings typical of many Cambrian, but not pre-Cambrian, fossils (**FIGURE 20.5**).

## Colonization of land followed the Cambrian explosion

Because life first evolved in water, the colonization of land by living organisms posed enormous challenges. Indeed, many of the functions basic to life, including support, movement, reproduction, and the regulation of ions, water, and heat, must be handled very differently on land than in water. About 500 mya, descendants of green algae were the first organisms to meet these challenges. These early terrestrial colonists had few cells and a simple body plan, but from them land plants evolved and diver-

## DIVERSIFICATION OF ANIMALS

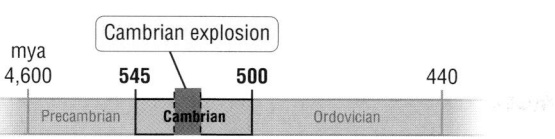

Cambrian explosion

mya
4,600    **545**    **500**    440

Precambrian | **Cambrian** | Ordovician

Before the Cambrian explosion, Earth was dominated by soft-bodied scavengers and herbivores.

After the Cambrian explosion, life was dominated by more complex animals, including predators and well-defended herbivores.

sified greatly. By the end of the Devonian [dih-*VOH-nee-un*] period (360 mya), Earth was covered with plants. Like plants today, the plants of the Devonian included low-lying spreading species, short upright species, shrubs, and trees.

As new groups of land plants arose, they evolved some key innovations, including a waterproof cuticle, vascular systems, structural support tissues (wood), leaves and roots of various kinds, seeds, the tree growth form, and specialized reproductive structures. These and other important changes enabled plants to cope with life on land. Waterproofing, stems with efficient transport mechanisms, and roots, for example, were important features that helped plants acquire and conserve water while living on land (see Section 3.3).

Fungi are thought to have made their way onto land soon afterward, according to new studies. For example, scientists have found fossils of terrestrial fungi—both fossil hyphae and spores—that are 455–460 million years old.

The first definite fossils of terrestrial animals are of spiders and millipedes that date from about 410 mya, although there are hints of land animals as early as 490 mya. Many of the early animal colonists on land were predators; others, such as millipedes, fed on living plants or decaying plant material. Insects, which are currently the most diverse group of terrestrial animals, first appeared roughly 400 mya, and they played a major role on land by 350 mya.

The first vertebrates to colonize land were amphibians, the earliest fossils of which date to about 365 mya. Early amphibians resembled, and probably descended from, lobe-finned fishes (**FIGURE 20.6**). Amphibians were the most abundant large organ-

## EVOLUTION OF TETRAPODS

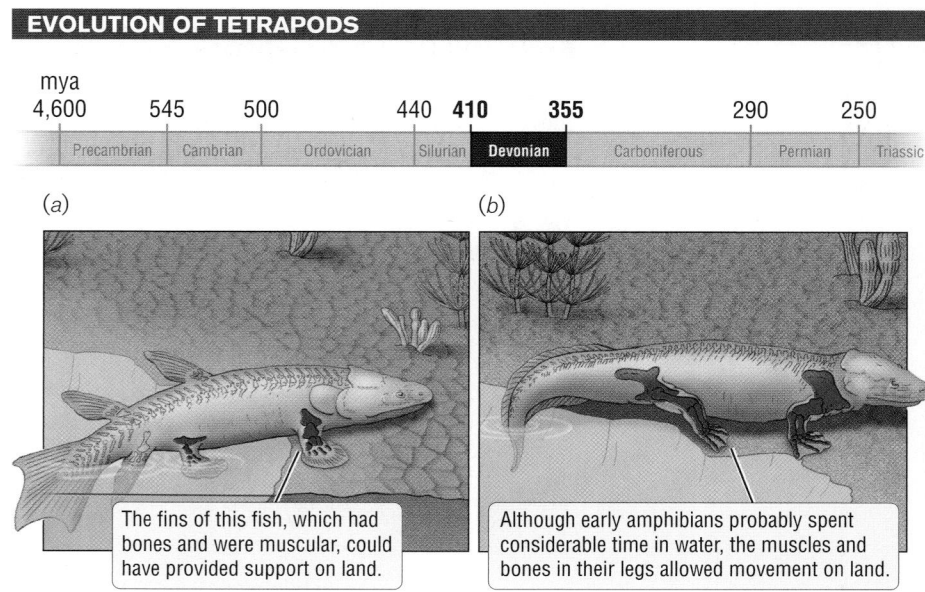

mya
4,600    545    500    440  **410**    **355**    290    250

Precambrian | Cambrian | Ordovician | Silurian | **Devonian** | Carboniferous | Permian | Triassic

(a)                (b)

The fins of this fish, which had bones and were muscular, could have provided support on land.

Although early amphibians probably spent considerable time in water, the muscles and bones in their legs allowed movement on land.

**FIGURE 20.6 The First Amphibians**

(*a*) Amphibians probably evolved over long periods of evolutionary time from a lobe-finned fish ancestor like the one shown here. (*b*) This early amphibian was reconstructed from a 365-million-year-old (late Devonian) fossil.

isms on land for about 100 million years. In the late Permian period, the reptiles, which had evolved from a group of reptile-like amphibians, rose to become the most common vertebrate group. Reptiles were the first group of vertebrates that could reproduce without returning to open water (for example, to lay eggs). As discussed in Chapter 4, the evolution of the amniotic egg was a major event in the history of life because it established a new evolutionary branch, the amniotes, which includes all reptiles, birds, and mammals. Reptiles, including the dinosaurs, dominated vertebrate life on land for 200 million years (265–65 mya), and they remain important today. Dinosaurs arose about 230 mya and dominated the planet from about 200 mya to about 65 mya. Mammals, the vertebrate group that currently dominates life on land, evolved from reptiles roughly 220 mya (see Figure 20.3).

> ### Concept Check
>
> 1. Why was it significant that early bacteria evolved the ability to carry out photosynthesis?
>
> 2. How did the Cambrian explosion change life on Earth?

## 20.3 The Effects of Plate Tectonics

The enormous size of the continents may lead us to think of them as immovable. But this notion is not correct. The continents move slowly relative to one another, and over hundreds of millions of years they travel considerable distances (**FIGURE 20.7**). This movement of the continents over time is called **plate tectonics** or continental drift. The continents can be thought of as plates of solid matter that "float" on the surface of Earth's mantle, a hot layer of semisolid rock.

How can something as big as a continent move from place to place? Two forces cause the continental plates to move. First, hot plumes of liquid rock rise from Earth's mantle to the surface and push the continents away from one another. This process can cause the seafloor to spread, as it is doing between North America and Europe, which are separating at a rate of 2.5 centimeters per year. This process can also cause bodies of land to break apart, as is currently happening in Iceland and East Africa. Second, where two plates collide, one plate can sometimes slip underneath the other and begin to sink into the mantle below. As the now hidden end of the continental plate sinks down and slowly melts, it gradually pulls the rest of the plate down along with it, causing the rest of the plate to continue to move or drift.

**FIGURE 20.7**
**Movement of the Continents over Time**

The continents move over time, as shown by these "snapshots" illustrating the breakup of the supercontinent Pangaea. Earlier movements of the continents led to the gradual formation of Pangaea, a process that was complete by 250 mya.

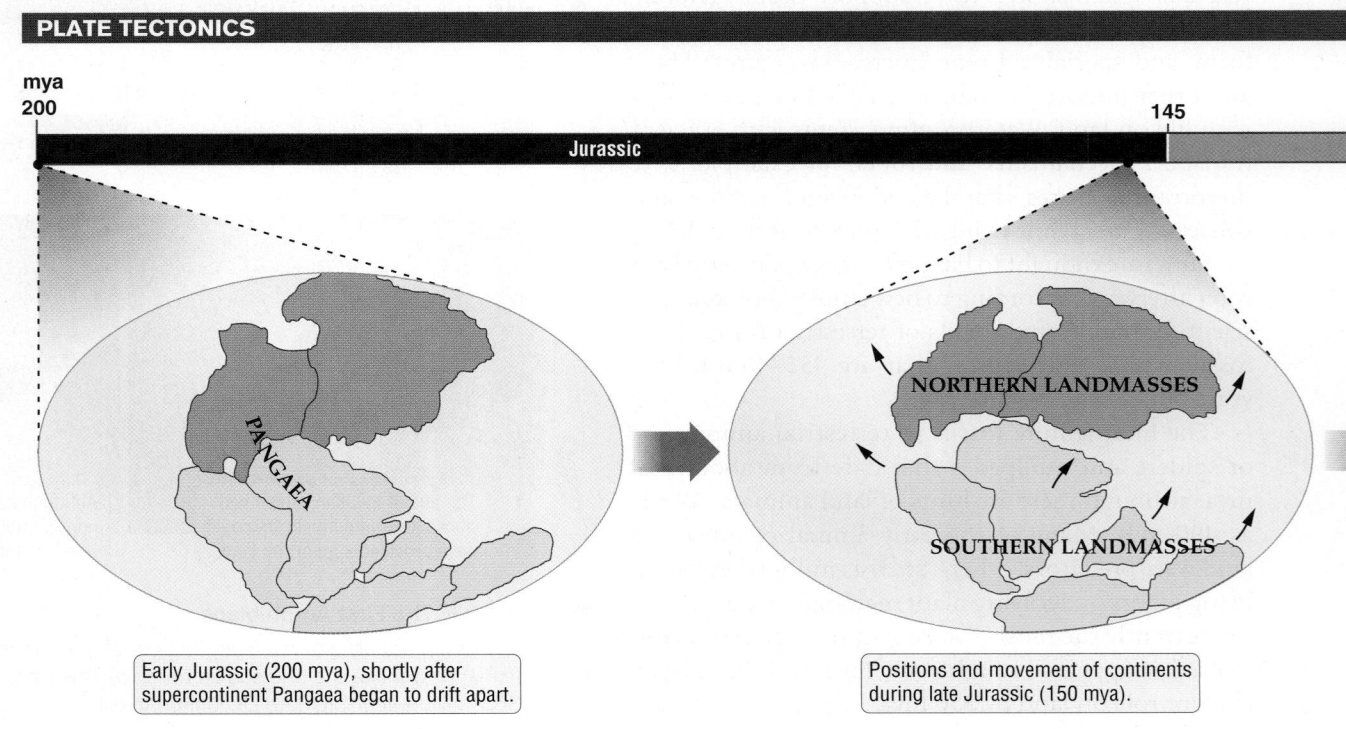

**PLATE TECTONICS**

mya
200 — Jurassic — 145

Early Jurassic (200 mya), shortly after supercontinent Pangaea began to drift apart.

Positions and movement of continents during late Jurassic (150 mya).

Patterns of plate tectonics—most notably the breakup of the ancient supercontinent Pangaea—have had dramatic effects on the history of life. Pangaea began to break apart early in the Jurassic period (about 200 mya), ultimately separating into the continents we know today (see Figure 20.7). As the continents drifted apart, populations of organisms that once were connected by land became isolated from one another.

As we noted in Chapter 19, geographic isolation reduces or eliminates gene flow, thereby promoting speciation. The separation of the continents was geographic isolation on a grand scale, and it led to the formation of many new species. Among mammals, for example, kangaroos, koalas, and other marsupials unique to Australia evolved in geographic isolation on that continent, which broke apart from Antarctica and South America about 40 mya.

Plate tectonics also affects climate, which has a profound effect on the evolution of life by altering what kinds of organisms natural selection will favor. For example, consider animals adapted to life in a warm tropical climate. If the shifting of continents moves the land those animals live on to a much colder climate, it will drastically alter which animals survive and thrive—that is, which are more likely to have an advantage in survival and reproduction. In addition, shifts in the positions of the continents alter ocean currents, and these currents have a major influence on the global climate. At various times, changes in the global climate caused by the movements of the continents have led to the extinctions of many species.

## 20.4 Mass Extinctions: Worldwide Losses of Species

As the fossil record shows, species have gone extinct throughout the long history of life. The rate at which this has happened—that is, the number of species that have gone extinct during a given period—has varied over time, from low to very high. At the upper end of this scale, the fossil record shows that there have been five **mass extinctions**, periods of time during which great numbers of species went extinct throughout most of Earth. Each of these upheavals left a permanent mark on the history of life, driving more than 50 percent of Earth's species to extinction. **FIGURE 20.8** shows the effects of the five mass extinctions on animal life alone. Though difficult to determine, the causes of the five mass extinctions are thought to include such factors as climate change, massive volcanic eruptions,

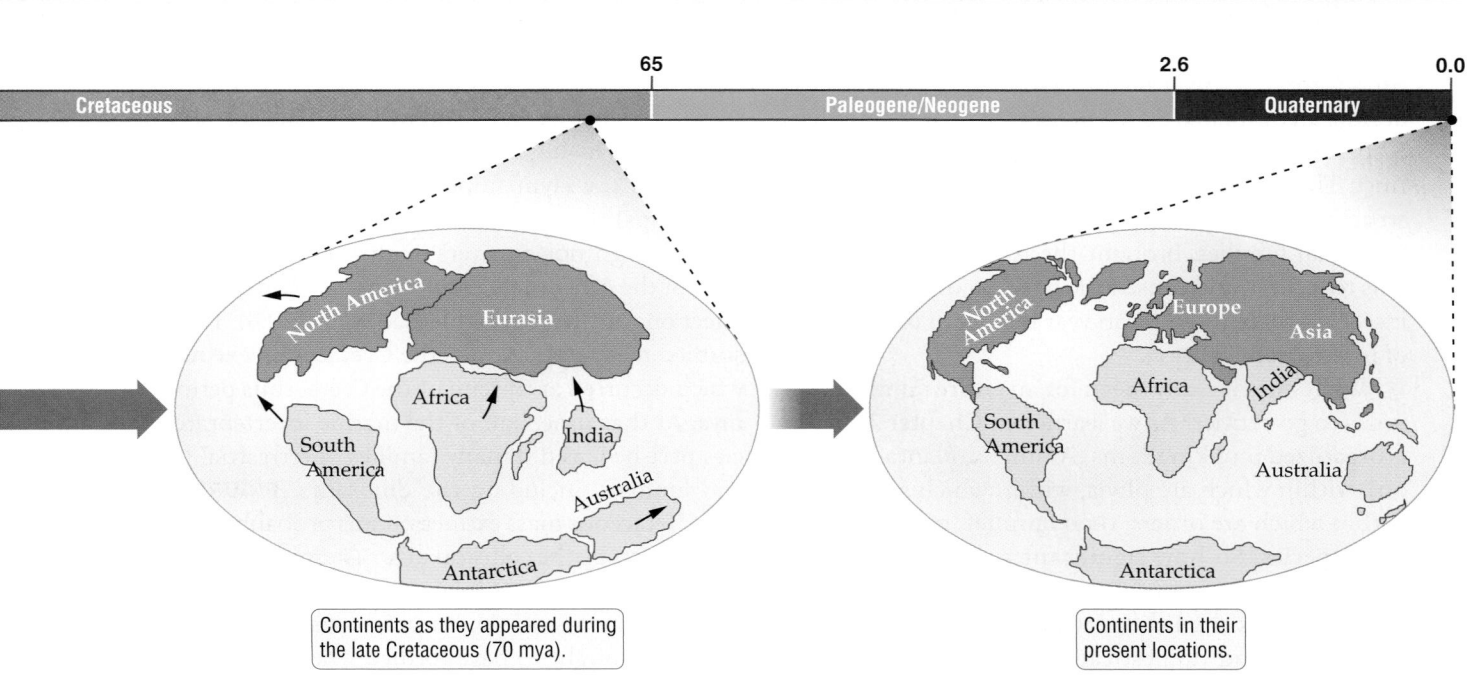

Continents as they appeared during the late Cretaceous (70 mya).

Continents in their present locations.

**FIGURE 20.8**
**The Five Mass Extinctions Drastically Reduced the Diversity of Animals**

In addition to both marine and terrestrial animals, plant groups (not shown) were severely affected by the five mass extinctions that have occurred in Earth's history. After each extinction, life again diversified.

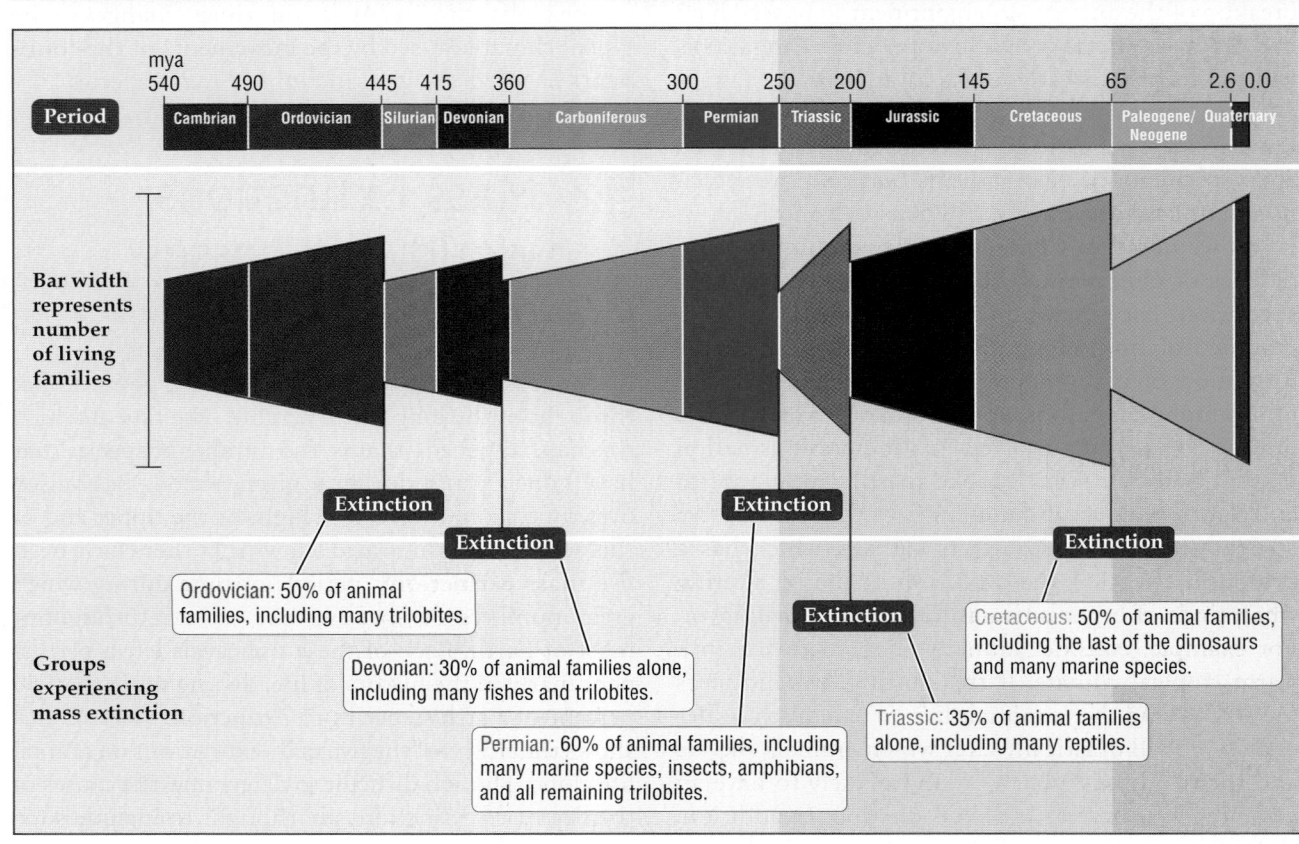

**MASS EXTINCTIONS**

mya

| Period | Cambrian | Ordovician | Silurian | Devonian | Carboniferous | Permian | Triassic | Jurassic | Cretaceous | Paleogene/ Neogene | Quaternary |

Bar width represents number of living families

Groups experiencing mass extinction

Extinction — Ordovician: 50% of animal families, including many trilobites.

Extinction — Devonian: 30% of animal families alone, including many fishes and trilobites.

Permian: 60% of animal families, including many marine species, insects, amphibians, and all remaining trilobites.

Extinction — Triassic: 35% of animal families alone, including many reptiles.

Extinction — Cretaceous: 50% of animal families, including the last of the dinosaurs and many marine species.

---

asteroid impacts, changes in the composition of marine and atmospheric gases, and changes in sea levels.

Of the five major mass extinctions, the largest occurred at the end of the Permian period, 250 mya. The **Permian extinction** radically altered life in the oceans. Among marine invertebrates (animals without backbones), an estimated 50–63 percent of existing families, 82 percent of genera, and 95 percent of species went extinct. The Permian mass extinction was also highly destructive on land. It removed 62 percent of the existing terrestrial families, brought the reign of the amphibians to a close, and caused the only major extinction of insects in their 400-million-year history (8 of 27 orders of insects went extinct).

What does it really mean for an entire family of animals to go extinct? As we learned in Chapter 2, all life is organized into kingdoms (Animalia, Plantae, and so on), within which are phyla, within which are classes, within which are orders, then families, genera, and finally species. So how significant a part of the living world is a single family? Today, extinction of an entire family such as the Felidae would mean extinction of all the wild and domesticated cats, from lions to ti-

gers to leopards to cougars to regular old kitty cats. Other familiar families include the Canidae (all wild and domesticated dogs) and the Ursidae (bears). And those are just single families. As described already, in some groups, such as the marine invertebrates, at least half the families disappeared altogether. Orders of insects were lost as well. Today, losing an entire order could mean losing the Lepidoptera (all butterflies and moths) or the Hymenoptera (including all bees, ants, and wasps).

Although not as severe as the Permian extinction, each of the other mass extinctions also had a profound effect on the diversity of life (see Figure 20.8). The best-studied mass extinction is the **Cretaceous extinction**, which occurred at the end of the Cretaceous period, 65 mya. At that time, half of the marine invertebrate species perished, as did many families of terrestrial plants and animals, including the dinosaurs (**FIGURE 20.9**). The Cretaceous mass extinction was probably caused at least in part by the collision of an asteroid with Earth. A 65-million-year-old, 180-kilometer-wide crater lies buried in sediments off the Yucatán coast of Mexico; this crater is thought to have formed when an asteroid 10

# Is a Mass Extinction Under Way?

The International Union for Conservation of Nature (IUCN), known also as the World Conservation Union, maintains what it calls its Red List, which identifies the world's threatened species. To be defined as such, a species must face a high to extremely high risk of extinction in the wild. The 2010 Red List contains 18,351 species threatened with extinction, of a total of 55,926 species assessed. That means about 33 percent of the species evaluated by the IUCN are threatened. Because this assessment accounts for only about 1 percent of the world's 1.7 million described species, the total number of species threatened with extinction worldwide would actually be much larger. In 2000, the IUCN list showed 11,046 species threatened with extinction. The number of threatened species has therefore increased nearly 66 percent in ten years.

Among the major groups assessed for the Red List, the number of threatened species ranges between 12 and 72 percent. For example, 12 percent of birds, 21 percent of mammals, 30 percent of amphibians, and 72 percent of the plants in the magnolia family are listed as threatened.

The Iberian lynx (*Lynx pardinus*) is the most critically endangered feline in the world. About 100 individuals survive in small populations in Spain.

The Red List is based on an easy-to-understand system for categorizing extinction risk; it is also objective, yielding consistent results when used by different people. These two attributes have earned the Red List international recognition as an effective method to assess extinction risk.

For many taxonomic groups, the status of relatively few of the described species has been evaluated. For example, only 771 out of 950,000 described insect species have been assessed in terms of their survival risk. If the threatened species listed by the IUCN do go extinct and the percentage of species under threat in other taxonomic groups turns out to be similar to those listed, then the percentages of species that will go extinct will approach the proportions lost in some of the previous mass extinctions.

For more information about the Red List, visit www.iucnredlist.org.

## Number of Species at Risk of Extinction in Some Major Groups of Organisms

|  | MAMMALS | BIRDS | REPTILES | AMPHIBIANS | FISHES | MOLLUSKS | OTHER | PLANTS | TOTAL |
|---|---|---|---|---|---|---|---|---|---|
| Canada | 12 | 15 | 3 | 1 | 32 | 2 | 10 | 2 | 77 |
| United States | 37 | 74 | 32 | 56 | 177 | 273 | 258 | 245 | 1,152 |

**NOTE:** Based on the IUCN Red List, 2010.

kilometers wide struck Earth. An asteroid of this size would have caused great clouds of dust to hurtle into the atmosphere; this dust would have blocked sunlight around the globe for months to years, causing temperatures to drop drastically and thereby driving many species extinct.

The effects of mass extinctions on the diversity of life are twofold. First, as noted earlier, entire groups of organisms perish, changing the history of life forever. Second, the extinction of one or more dominant groups of organisms can provide new ecological and evolutionary opportunities for groups of organisms that previously were of relatively minor importance, dramatically altering the course of evolution.

(*a*)

(*b*)

**FIGURE 20.9 Gone for Good**

(*a*) A dog sits next to a reconstruction of the head of a *Mapusaurus* dinosaur—what may have been the largest carnivore ever to walk the Earth. (*b*) This *Allosaurus* skeleton reveals how sharp, pointed, large, and numerous a dinosaur's teeth could be.

Helpful to
Know

Here, "radiation"
is not related to
radioactivity or
radioisotopes. In this
case, the prefix *radi*
conveys the idea of
expanding outward
(analogous to rays
of light)—not just
geographically but
ecologically, into
new roles.

## 20.5 Adaptive Radiations: Increases in the Diversity of Life

After each of the five mass extinctions, some of the surviving groups of organisms diversified to replace those that had become extinct. These bursts of evolution were just as important to the future course of evolution as were the extinctions themselves. For example, when the dinosaurs went extinct 65 mya, the mammals diversified greatly in size and in ecological role (**FIGURE 20.10**). If mammals had not diversified to replace the dinosaurs, humans probably would not exist and the history of life over the past 65 million years would have been very different.

When a group of organisms expands to take on new ecological roles and to form new species and higher taxonomic groups, that group is said to have undergone an **adaptive radiation**. Some of the great adaptive radiations in the history of life occurred after mass extinctions, as when the mammals diversified to replace the dinosaurs. In such cases, the adaptive radiations may have been caused by the reduction in competition that occurs after a dominant group of organisms (such as the dinosaurs) goes extinct.

In other cases, adaptive radiations have occurred after a group of organisms has acquired a new adaptation that enables it to use its environment in new ways. The first terrestrial plants, for example, possessed adaptations that helped them thrive on land, a new and highly challenging environment. The descendants of those early colonists radiated greatly, forming many new species and higher taxonomic groups that were able to live in a broad range of new environments (from desert to Arctic to tropical regions).

**FIGURE 20.10**
**They Became Giants**

Early mammals (such as *Morganucodon* [**mohr-***GUH*-**noo-***KUH*-**dahn**]) were small and are thought to have been nocturnal. Following the extinction of the dinosaurs, the diversity and size of mammals increased to include large forms such as those shown here. Because of the huge range in size between the smallest (*Morganucodon*, about the size of a shrew) and the largest (the blue whale, which can reach 25–30 meters long), none of these animals are drawn to scale.

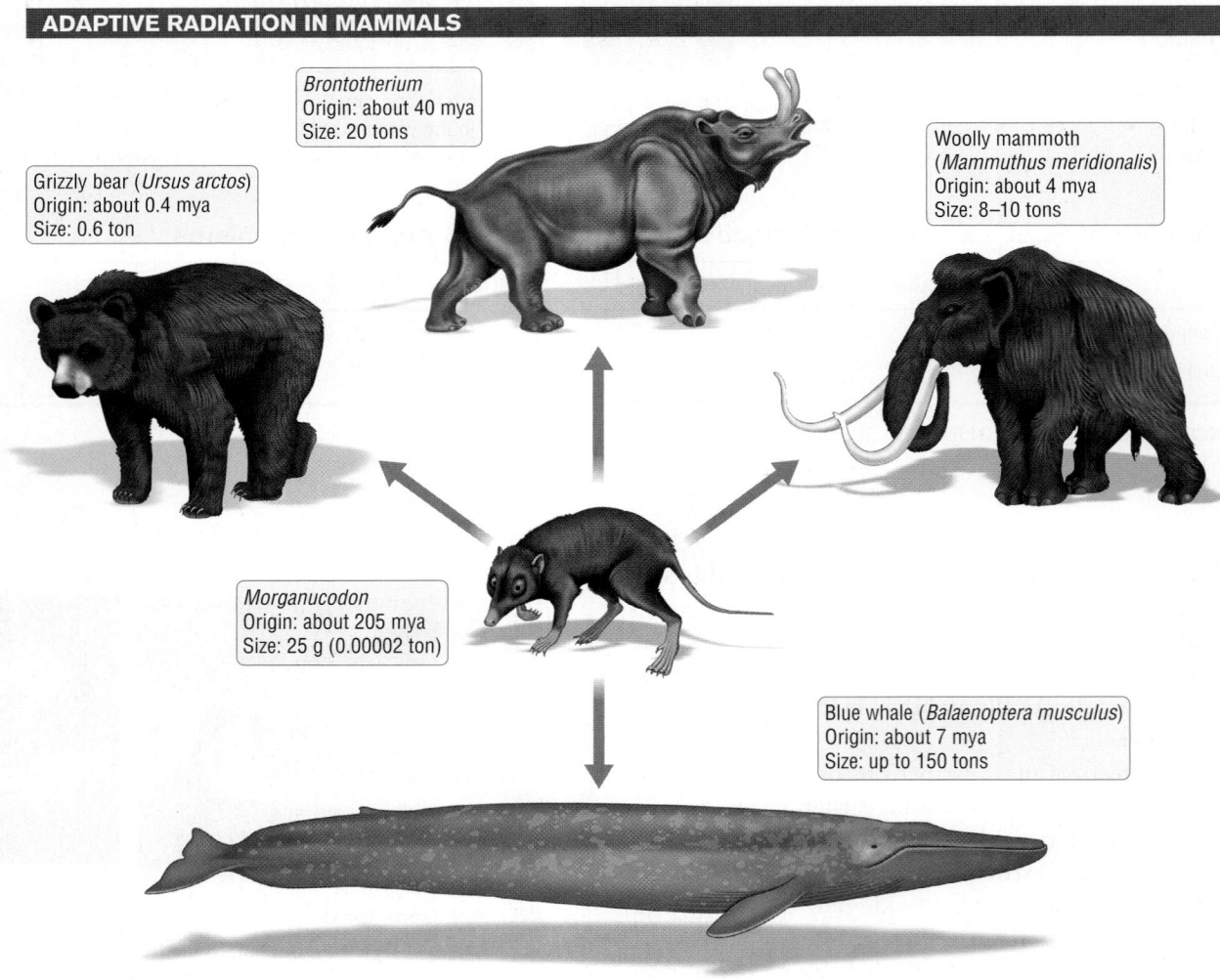

ADAPTIVE RADIATION IN MAMMALS

*Brontotherium*
Origin: about 40 mya
Size: 20 tons

Woolly mammoth
(*Mammuthus meridionalis*)
Origin: about 4 mya
Size: 8–10 tons

Grizzly bear (*Ursus arctos*)
Origin: about 0.4 mya
Size: 0.6 ton

*Morganucodon*
Origin: about 205 mya
Size: 25 g (0.00002 ton)

Blue whale (*Balaenoptera musculus*)
Origin: about 7 mya
Size: up to 150 tons

The term "adaptive radiation" can refer to relatively small evolutionary expansions, as seen in the radiation of finches in the Galápagos Islands (see Figure 17.15). It can also refer to much larger expansions, such as those that followed the movement of vertebrates onto land or the origin and rapid diversification of flowering plants.

## 20.6 The Origin and Adaptive Radiation of Mammals

The fossil record shows that mammals evolved from reptiles. Living mammals differ from living reptiles in many respects, including the way they move (**FIGURE 20.11**), the nature of their teeth, and the structure of their jaws. However, it is difficult to draw the line between mammals and reptiles in the fossil record. Some fossil species have features intermediate between the two groups; such fossils beautifully illustrate an evolutionary shift from one major group of organisms to another.

### The mammalian jaw and teeth evolved from reptilian forms in three stages

Let's examine the evolution of mammals from reptiles by focusing on two traits that are easily observed in fossils: the nature of their teeth and the structure of their jaws. Compared with reptiles, mammals have complex teeth and jaws. For example, mammalian teeth differ considerably from one portion of the jaw to another: some teeth are specialized for tearing (incisors), others for hunting or defense (canines), still others for grinding (molars). In contrast, reptilian teeth change little in form or function from one position along the jaw to another (see Figure 20.12a).

The reptilian jaw has a hinge at the back for the attachment of muscles that simply snap the top and bottom of the jaw together. In mammals, the hinge has moved forward and is controlled by strong cheek muscles, including some positioned in front of the hinge. As a result, mammalian jaws can be both more

**EVOLUTION OF GAIT**

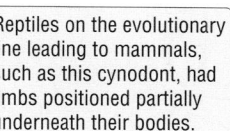

The limbs of reptiles extend from the sides of their bodies.

Reptiles on the evolutionary line leading to mammals, such as this cynodont, had limbs positioned partially underneath their bodies.

The limbs of mammals are even more vertically oriented.

**FIGURE 20.11  A Gradual Change in the Way Animals Move**
The legs of most living reptiles stick out to the sides of their bodies, giving them a sprawling gait. Over time, the legs of mammal-like reptiles became positioned under the body, leading eventually to the vertical orientation of the legs and the upright gait of living mammals.

powerful and more precisely controlled than the simple snap-shut jaws of reptiles. To see why, think of how easily you could close a door by pulling it shut with a handle located near its hinges (analogous to the reptilian method) versus pulling it shut with a handle located away from the hinges, closer to the actual position of a doorknob (analogous to the mammalian method).

How did these differences in the teeth and jaws of mammals and reptiles arise? The fossil record shows that these changes arose gradually, over the course of about 80 million years (from about 300 to 220 mya). During this time, there were three key steps in the transition from reptile to mammal, as shown in **FIGURE 20.12**. First, a group of reptiles evolved to have an opening in the bones behind the eye called the temporal fenestra (**FIGURE 20.12a**). In the living animal, a muscle passed through this opening and increased the power with which the jaw could be closed. Second, a group of reptiles known as the therapsids [thuh-*RAP*-sidz] evolved a larger temporal fenestra (**FIGURE 20.12b**), and hence more powerful jaw muscles, and their teeth showed the first signs of specialization.

The third step occurred when jaws very similar to mammalian jaws arose in one subgroup of the therapsids, the early **cynodonts** [*SYE*-noh-dahnts] (**FIGURE 20.12c**). In these animals—the last in a long

**Helpful to Know**

"Cynodont" comes from Greek roots meaning "dog-toothed." In cynodonts, the canine teeth (from *canis*, Latin for "dog") were prominent, though not as large as in some of their predecessors.

**FIGURE 20.12 From Reptile to Mammal**

Over an 80-million-year period, the jaws and teeth of the reptilian ancestors of mammals gradually changed to resemble those of living mammals. In addition to those shown here, there are dozens of other fossil species of mammal-like reptiles—species with features that are intermediate to those of reptiles and mammals. When fossils from all of these species are lined up next to one another, the transition from reptile to mammal appears very smooth. Here, red shows the size and position of the dentary bone, which ultimately formed the entire lower jaw in mammals. The muscles that close the jaw pass through an opening called the temporal fenestra (tf); a larger temporal fenestra allows for larger and more powerful jaw muscles. In reptiles (a), the hinge of the jaw is formed by the articular/quadrate (art/q) bones; in mammals (g), the hinge is formed by two entirely different bones: the dentary/squamosal (d/sq) bones. Advanced cynodonts and early mammals had two hinges: the reptilian (art/q) hinge and the mammalian (d/sq) hinge.

tf - temporal fenstra structure
art - articular bones
q - quadrate bones
d - dentary bones
sq - squamosal bones

## EVOLUTION OF JAW STRUCTURE

**(a) Ancestral reptile (*Haptodus*)**

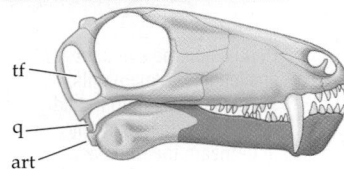

Eye socket
tf
sq
q
art

These reptiles had a temporal fenestra (tf), large jaw muscles, multiple bones in the lower jaw, and single-point teeth. They also had a single hinge (art/q) at the back of the jaw.

**(b) Therapsid (*Biarmosuchus*)**

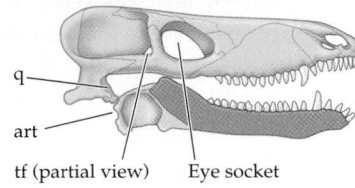

tf
q
art

Therapsids had a larger temporal fenestra (tf), large canine teeth, long faces, and a single hinge (art/q) at the back of the jaw.

**(c) Early cynodont (*Procynosuchus*)**

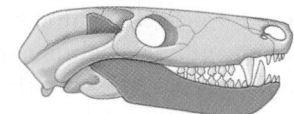

q
art
tf (partial view)    Eye socket

In early cynodonts, the temporal fenestra (here only partial view is shown) was further enlarged, allowing for very powerful jaw muscles.

**(d) Cynodont (*Thrinaxodon*)**

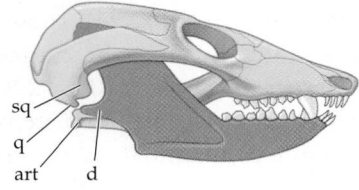

The dentary (the major jaw bone, colored in red) became enlarged, and the back teeth had multiple cusps.

**(e) Advanced cynodont (*Probainognathus*)**

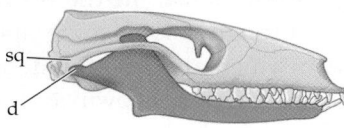

sq
q
art    d

In advanced cynodonts, complex teeth with multiple cusps enhanced chewing. The jaw had two hinges: art/q and d/sq.

**(f) Early mammal (*Morganucodon*)**

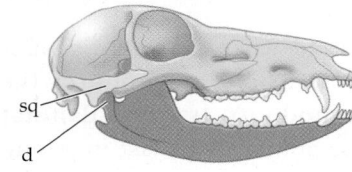

sq
d

*Morganucodon* had typical mammalian teeth. The jaw had two hinges, but the reptilian art/q hinge was reduced (not visible in this diagram).

**(g) Tree shrew (*Tupaia*)**

sq
d

In tree shrews and other mammals, the teeth are highly specialized. The lower jaw is composed of a single bone, and the jaw has one hinge (d/sq).

---

line of mammal-like reptiles—the teeth became still more specialized, and the hinge of the jaws moved forward, using a different set of bones from those used in other reptiles (**FIGURE 20.12c–e**). Changes in the jaw hinge are particularly clear in some cynodont species (**FIGURE 20.12f**) that had jaws with both a reptilian hinge (in reduced form) and a mammal-like hinge. Over time, mammals lost the reptilian hinge (**FIGURE 20.12g**), the bones of which evolved to become bones in the inner ear.

## Mammals increased in size after the extinction of the dinosaurs

There were many species of mammal-like reptiles in the early Triassic period, 245 mya. By 200 mya, however, the mammal-like reptiles had declined as other reptiles, most notably the dinosaurs, came to dominate Earth. Although the mammal-like reptiles became extinct, they left behind the first mammals as their descendants. The earliest mammals, which were small, rodent-sized organisms, evolved about 220 mya—in geologic terms not long after the first dinosaurs, which arose 230 mya.

Throughout the long reign of the dinosaurs, most mammals remained small. Many appear to have been nocturnal (active at night), because they had large eye sockets, as do many living nocturnal organisms. By being nocturnal and small, early mammals may have been to dinosaurs what a mouse is to a lion: hard to notice and too small to eat.

Fossil and genetic evidence suggests that several of the orders of living mammals diverged from one another between 100 and 85 mya, well before the extinction of the dinosaurs. But most of the major radiations within these and other groups of mammals did not occur until after the dinosaurs went extinct (65 mya). After the dinosaurs were gone, the mammals radiated greatly to include many new forms that were large and active by day (see Figure 20.10). Some land mammals reached enormous sizes. An example is the extinct Beast of Baluchistan, which was over three times as large as an elephant. Other mammals (such as whales; see Figure 20.2) became specialized for life in water, while still others (bats) became specialized for flight and hunting at night. One group of mammals, the primates, became specialized for life in trees and evolved especially large brains.

## 20.7 Human Evolution

Within the Linnaean hierarchy of all living things, we human beings—*Homo sapiens*—are members of the animal kingdom. Among the many phyla of the animal kingdom, we are members of the phylum consisting of the chordates, which includes all animals with a backbone. Within the chordates, we are members of the class consisting of the mammals. We share with all other mammals certain unique features, including body hair (which provides insulation in many mammals) and milk produced by mammary glands. Within the mammals, we are part of the order consisting of the *primates* (**FIGURE 20.13**). Like all other **primates**, we have flexible shoulder and elbow joints, five functional fingers and toes, thumbs that are **opposable** (that is, they can be placed opposite other fingers), flat nails (instead of claws), and brains that are large in relation to our body size. Within the primates we are members of the ape family (the **hominids**), and within that family we belong to the genus *Homo*. The genus *Homo* comprises a number of human species, including our own, *Homo sapiens*, the only species to have survived to modern times.

Genetic analyses and a series of spectacular fossil discoveries have led scientists to believe that the human lineage diverged from that of chimpanzees about 5–7 million years ago (see Figure 20.13). Similarly, a combination of genetic analyses and fossil discoveries suggests that the evolutionary lineage leading to humans diverged from the lineage leading to gorillas about 7–8 mya, and from the lineage leading to orangutans about 12–16 mya.

The broad conclusion that emerges from studies published over the last 40 years is that we are not just closely related to apes, we *are* apes. We share many characteristics with apes, especially chimpanzees, including the use of tools, a capacity for symbolic language, and the performance of deliberate acts of deception. These similarities—as well as our many differences—make the story of how humans evolved all the more interesting. We begin that story by describing the origin of the first **hominins**, the branch of the ape family that includes humans and our now extinct relatives. The members of this lineage have one or more humanlike features—such as thick tooth enamel or upright posture—that set them apart from the other apes.

## Walking upright was a big step in hominin evolution

Primates are thought to have originated by 65 million years ago from small nocturnal mammals, similar to tree shrews, that ate insects and lived in trees. The fossil evidence of primate origins is sketchy, however, and the first definite primate fossils are 56 million years old. These early primates resembled modern lemurs. Over time, the primates diversified greatly, eventually giving rise to the first hominids, roughly 12–16 million years ago. Hominins diverged from the chimpanzee lineages 5–7 million years ago. The earliest fossil hominins display a mix of anatomical features, including some chimpanzee-like and some humanlike characteristics.

A major step in hominin evolution was the shift from being quadrupedal (moving on four legs) to being **bipedal** (walking upright on two legs)—a change that occurred long before hominins evolved large brains. Many skeletal changes accompanied the switch to walking upright, including the loss of

The ancestor of all mammals may have resembled a tree shrew like this one.

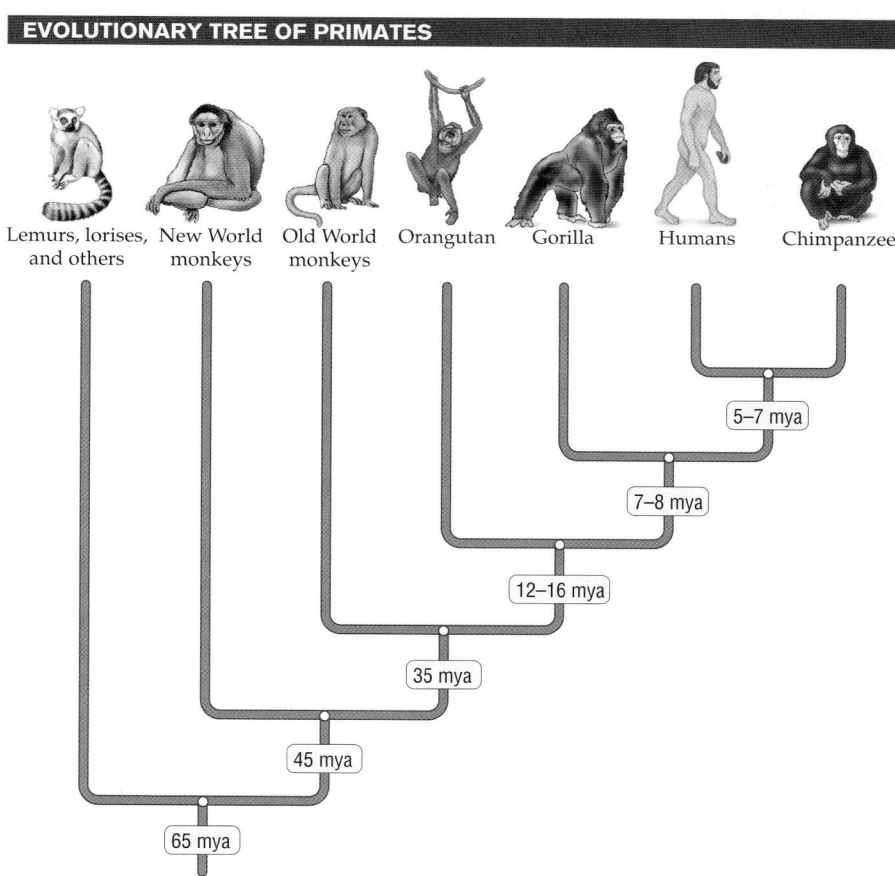

**EVOLUTIONARY TREE OF PRIMATES**

Lemurs, lorises, and others | New World monkeys | Old World monkeys | Orangutan | Gorilla | Humans | Chimpanzee

5–7 mya
7–8 mya
12–16 mya
35 mya
45 mya
65 mya

**FIGURE 20.13 The Primates Include Lemurs, Monkeys, and Apes**

**(a) Foot bones of early hominin**

**(b) Fossil footprints of early hominin**

Partially opposable big toe

These hominin foot bones, 3–3.5 million years old, were discovered in 1995.

These bones (shown in tan) are based on fossils of a similar age.

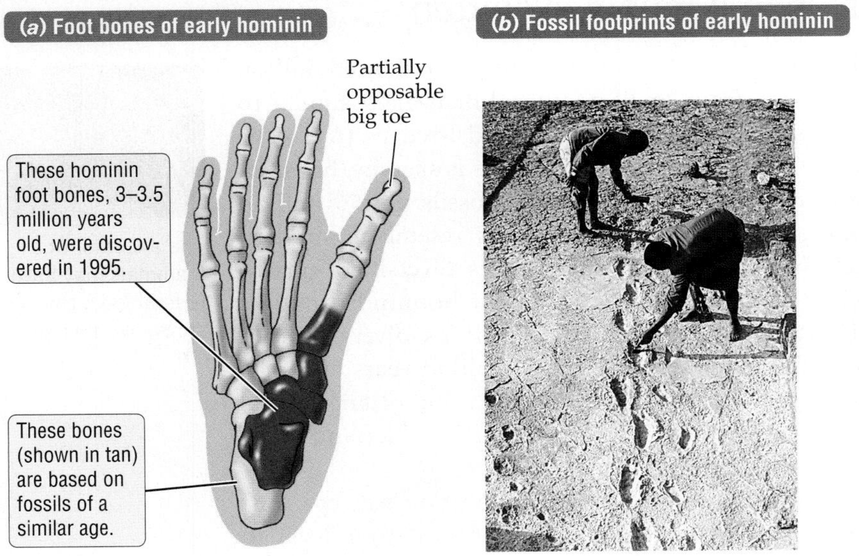

**FIGURE 20.14   Early Hominins Had an Upright Stance and Partially Opposable Big Toes**

(a) Fossilized foot bones show that some hominins living 3–3.5 million years ago walked upright but had partially opposable big toes. That is, the big toe could be placed opposite some toes (as is done when we use our thumb and pointer finger to pick up a pencil) but not all of the other toes. (b) These footprints of two early hominins walking upright, side by side, were found in Africa.

opposable toes (the big toe is opposable in all the other living hominids—all except humans, that is, as you will immediately realize if you try touching your little toe with the big toe on the same foot).

It is not necessary—or particularly adaptive—for an organism living in the trees to try to walk upright. Moreover, for organisms that live primarily in trees, the loss of opposable toes that accompanied bipedalism would be a handicap, since such toes can helpfully grasp branches during climbing. On the ground, however, walking upright provides several advantages, including freeing the hands for carrying food, tools, and weapons. In addition, walking upright elevates the head instantly, enabling the walker to see farther and over more things. It is likely that the evolution of an upright posture was linked to a switch from life in the trees to life on the ground. Exactly when this change took place has been a matter of some debate. The discovery of a foot bone in Ethiopia suggests that the first full-time walker with the first modern foot was *Australopithecus afarensis* (see Figure 20.14), of "Lucy" fame. You may have heard of this famous member of *A. afarensis*, a species of apelike hominins that lived a couple of million years before humans

evolved. The foot bone fossil is a metatarsal, one of five long bones that connect the large bones in the back of the foot to those of the toes (see Figure 20.14). The bone has been determined to be 3.2 million years old, suggesting that apes left the trees much sooner than was previously thought.

The shift to life on the ground was probably not sudden or complete. The skeletal structure of some of the oldest fossil hominins (dating from 4.4 million years ago) indicates that they walked upright. However, foot bones and fossilized footprints that are 3–3.5 million years old show that the hominins living at that time still had partially opposable big toes (**FIGURE 20.14**), perhaps because they continued to use trees some of the time.

The earliest known hominin is *Sahelanthropus tchadensis* [**sah-hel-**AN**-throh-pus chuh-**DEN**-sis**], known from a 6- to 7-million-year-old skull discovered in 2002. Other early hominins include *Ardipithecus ramidus* [**ahr-duh-**PITH**-uh-kus** RAM**-uh-dus**] who lived 4.4 million years ago, and several *Australopithecus* [AW**-stray-loh-**PITH**-uh-kus**] species that are 3.0–4.2 million years old. All of these fossil hominins are thought to have walked upright. Their brains were still relatively small (less than 400 cubic centimeters in volume), and their skulls and teeth were more similar to those of other apes than to those of humans. **FIGURE 20.15** shows just how radically brain size evolved in the hominins, since a typical modern human has a brain volume of about 1,400 cubic centimeters.

## A larger brain is the hallmark of the human genus, *Homo*

The oldest *Homo* fossil fragments were found in Africa and date from 2.4 million years ago (mya), suggesting that the earliest members of the genus *Homo* originated in Africa 2–3 mya. More complete early *Homo* fossils exist from the period 1.9–1.6 mya; these fossils have been given the species name *Homo habilis* [HAB**-uh-lis**]. The oldest *H. habilis* fossils resemble those of *Australopithecus africanus*, a slightly more recent hominin than *A. afarensis* (see Figure 20.15). In more recent *H. habilis* fossils, the face is not pulled forward as much, and the skull is more rounded. In these and other ways, more recent *H. habilis* specimens have features that are intermediate between those of *A. africanus* and *Homo erectus*, a species that evolved after *H. habilis*. *H. habilis* fossils therefore provide an excellent record of the evolutionary shift from ancestral hominin characteristics (in *Australopithecus*) to more recent ones (in *H. erectus*).

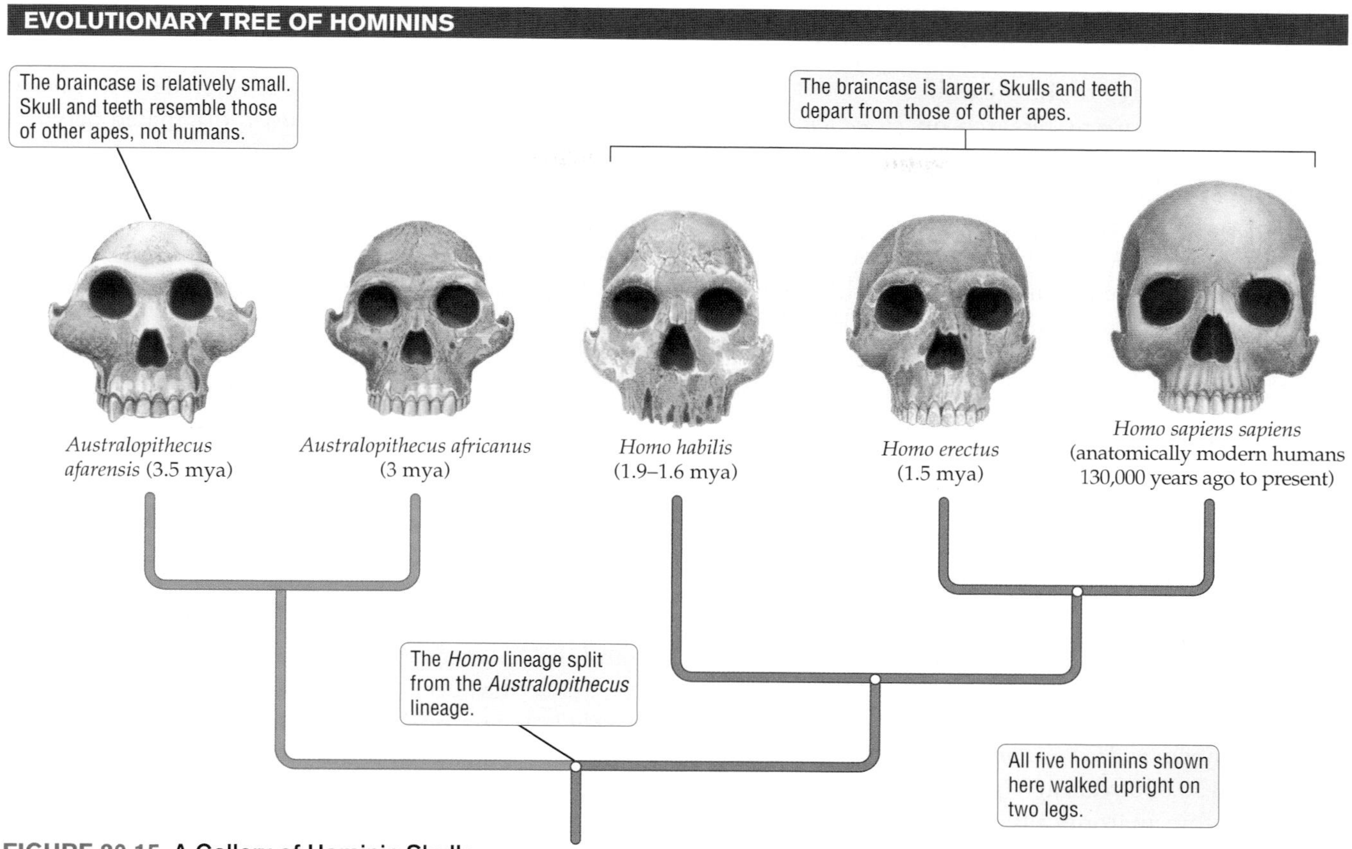

The braincase is relatively small. Skull and teeth resemble those of other apes, not humans.

The braincase is larger. Skulls and teeth depart from those of other apes.

*Australopithecus afarensis* (3.5 mya)

*Australopithecus africanus* (3 mya)

*Homo habilis* (1.9–1.6 mya)

*Homo erectus* (1.5 mya)

*Homo sapiens sapiens* (anatomically modern humans 130,000 years ago to present)

The *Homo* lineage split from the *Australopithecus* lineage.

All five hominins shown here walked upright on two legs.

**FIGURE 20.15 A Gallery of Hominin Skulls**

This tree shows the evolutionary relationships and the skulls of five of the many hominin species. A more complete evolutionary tree of hominins would be "bushier," with multiple side branches emerging at different times.

Taller and more robust than *H. habilis*, *H. erectus* also had a larger brain and a skull more like that of modern humans. It is likely that by 500,000 years ago *H. erectus* could use, but not necessarily make, fire. In addition, *H. erectus* probably hunted large species of game animals (**FIGURE 20.16**). The evidence to support the latter conclusion includes the remarkable 2010 discovery in Germany of three 400,000-year-old spears, each about 2 meters long and designed for throwing with a forward center of gravity (like a modern javelin).

*H. erectus* or an earlier form of *Homo* migrated from Africa about 2 million years ago. *Homo* fossils dating from the period 1.9–1.7 mya have been found in Java, in the central Asian republic of Georgia, and in China. As mentioned in Chapter 19, what may be a new miniature species of *Homo* was discovered on an Indonesian island in 2004. Described by its discoverers as *Homo floresiensis*, these "little people" appear to have lived on that island from 95,000 to 12,000 years ago. Other fossil evidence indicates that *H. erectus* lived on

**FIGURE 20.16**

**Tool Use and Social Organization in Hominins**

Both *Homo habilis* and *Homo erectus* likely worked in groups using stone tools when dealing with large kills.

nearby islands from 1 million to 25,000 years ago, while *H. sapiens* lived in the same general region from 60,000 years ago to the present.

Overall, current research on *H. habilis*, *H. erectus*, and other early *Homo* species indicates that there were more species of *Homo* than was once thought, and that several of these species existed in the same places and times. A complete hominin evolutionary tree would therefore be much "bushier" than the version shown in Figure 20.15. More research and evidence will be necessary before general agreement is reached regarding the number of early *Homo* species and their evolutionary relationships.

## Modern humans spread out of Africa to populate the rest of the world

All humans alive today are *Homo sapiens*. More specifically, we are all "anatomically modern humans," also known as *Homo sapiens sapiens*, a group that arose some 130,000 years ago (see the skull in Figure 20.16). But before anatomically modern humans arose, there existed earlier humans, known as "archaic" *Homo sapiens*. Who were these early humans?

The fossil record indicates that archaic *H. sapiens* bore features intermediate between those of *H. erectus* and those of the anatomically modern *H. sapiens sapiens*. The first archaic *H. sapiens* originated between 400,000 and 300,000 years ago. These ancestors of anatomically modern humans developed new tools and new ways of making tools, used new foods, built complex shelters, and controlled the use of fire.

What became of archaic *H. sapiens*? Early populations eventually gave rise to both the Neandertals (*H. sapiens neanderthalensis*, an advanced type of archaic *H. sapiens* that lived from 230,000 to 30,000 years ago) and us—that is, anatomically modern humans (*H. sapiens sapiens*). Although the oldest fossils of anatomically modern humans date from 130,000 years ago and have been found in Africa (**FIGURE 20.17**), more recent fossils of anatomically modern humans have been found in such places as Israel (115,000 years old), China (60,000 years old), Australia (56,000 years old), and the Americas (13,000–18,000 years old).

There has been considerable controversy over exactly how anatomically modern humans arose from archaic *H. sapiens* and how we came to be found all across the globe. Two conflicting hypotheses have been proposed. According to the **out-of-Africa hypothesis** (**FIGURE 20.18a**), anatomically modern humans first evolved in Africa from unknown populations of archaic *H. sapiens*, which evolved, also in Africa, from *H. erectus*. They then spread from Africa to the rest of the world, completely replacing all other *Homo* populations, including *H. erectus*, the Neandertals, and *H. floresiensis*. In contrast, the **multiregional hypothesis** (**FIGURE 20.18b**) proposes that modern humans evolved over time from *H. erectus* populations located throughout the world. According to this hypothesis, regional differences among human populations developed early, but worldwide gene flow caused these different populations to evolve modern characteristics simultaneously and to remain a single species.

Which of these hypotheses is correct? Let's consider some of the evidence. According to the multiregional hypothesis, when different populations of early humans came into contact, extensive gene flow should have caused them to become more similar to one another. In this case we would not expect different types of early humans to coexist in the same geographic region yet remain distinct for long periods of time. In fact, however, Neandertals and more modern humans coexisted in western Asia for about 80,000 years. Even as recently as 12,000–25,000

### HUMAN MIGRATIONS FROM AFRICA

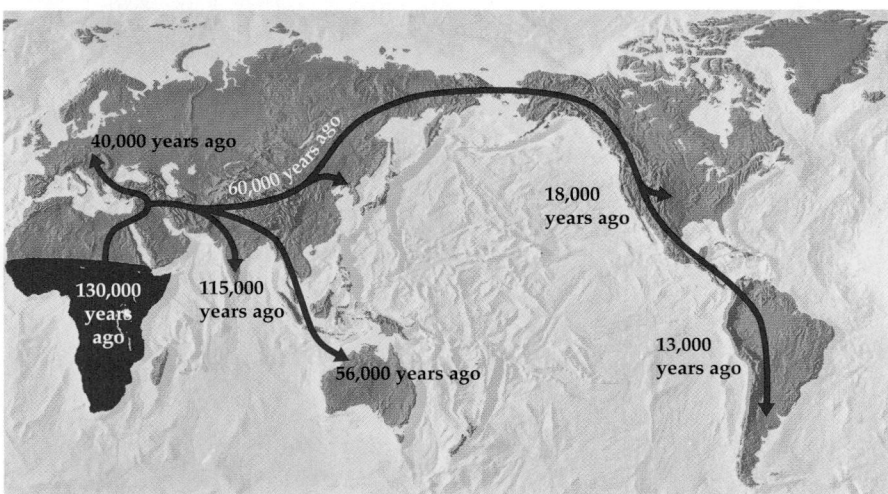

40,000 years ago

60,000 years ago

18,000 years ago

130,000 years ago

115,000 years ago

56,000 years ago

13,000 years ago

**FIGURE 20.17 Anatomically Modern Humans Evolved in Africa**
The earliest known fossil and archaeological specimens of *Homo sapiens sapiens*, the first anatomically modern humans, come from Africa. The dates provided give the age of the earliest evidence that anatomically modern humans lived in different regions of the world. These dates are continually challenged by new evidence that scientists must then work to confirm.

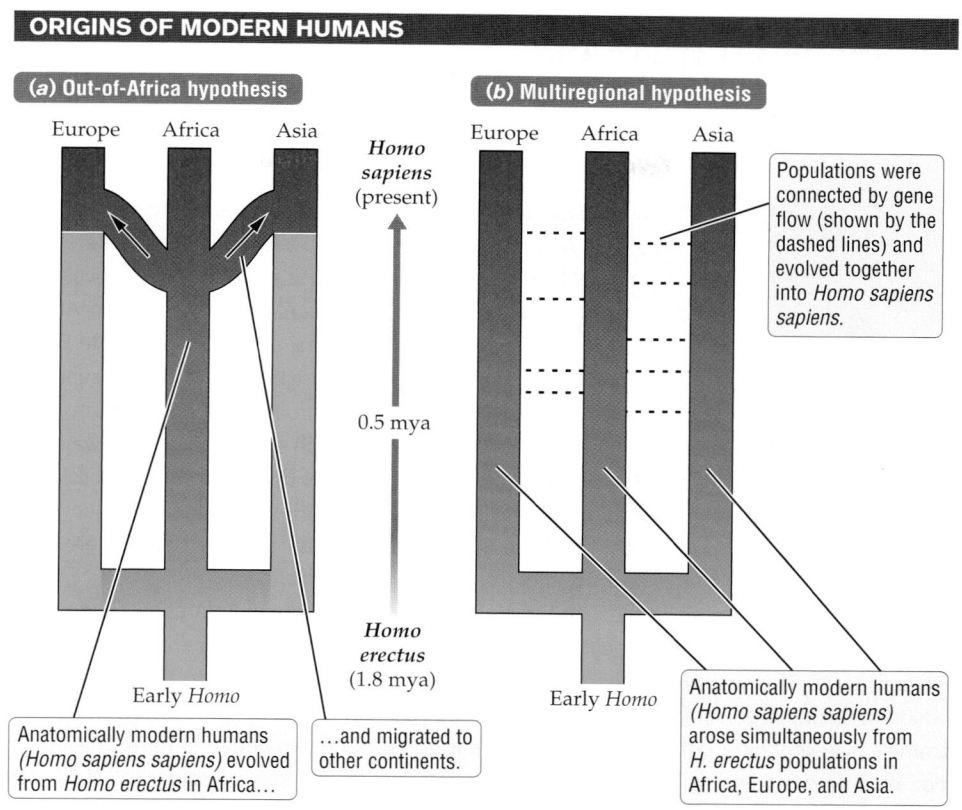

**(a) Out-of-Africa hypothesis**

Europe    Africa    Asia

Anatomically modern humans (*Homo sapiens sapiens*) evolved from *Homo erectus* in Africa...

...and migrated to other continents.

Early *Homo*

**(b) Multiregional hypothesis**

Europe    Africa    Asia

Populations were connected by gene flow (shown by the dashed lines) and evolved together into *Homo sapiens sapiens*.

*Homo sapiens* (present)

0.5 mya

*Homo erectus* (1.8 mya)

Early *Homo*

Anatomically modern humans (*Homo sapiens sapiens*) arose simultaneously from *H. erectus* populations in Africa, Europe, and Asia.

**FIGURE 20.18**

**Two Competing Hypotheses about the Origins of Anatomically Modern Humans**

(*a*) The out-of-Africa hypothesis: *Homo sapiens sapiens*, or anatomically modern humans (red), originated in Africa within the past 200,000 years and then migrated to Europe and Asia, replacing *Homo erectus* (blue) and archaic *H. sapiens* populations. (*b*) The multiregional hypothesis: Populations of *H. erectus* in Africa, Asia, and Europe evolved simultaneously into anatomically modern humans (*H. sapiens sapiens*).

years ago, *H. sapiens* may have shared some parts of its range with *H. floresiensis* and *H. erectus*. This evidence from the fossil record indicates that modern humans overlapped in time yet remained distinct from *H. erectus*, *H. floresiensis*, and Neandertal populations. This finding calls into question the extensive gene flow assumed by the multiregional hypothesis, because if there were indeed such gene flow, it should have been impossible to maintain such distinct lineages side by side.

The best fossil evidence for the shift from archaic to modern *H. sapiens* comes from Africa, providing some support for the out-of-Africa hypothesis. Recent analyses of human DNA sequences are also consistent with the out-of-Africa hypothesis. However, the "complete replacement" part of the hypothesis may not be correct. Fossils have been found that some scientists interpret as showing a mix of Neandertal and modern human characteristics. Similarly, DNA studies indicate that 1–4% of the genetic makeup of modern non-Africans may have come from the Neandertals.

In summary, many scientists think that anatomically modern humans arose in Africa and spread from there to other parts of the world. However, the details of the origin of modern humans continue to be a controversial issue. This debate is especially active concerning the extent to which early *Homo sapiens* interbred with, and hence did not completely replace, more ancient *Homo* populations.

## Concept Check

1. Name two ways in which mass extinctions can affect the history of life on Earth.

2. Did mammals co-occur with dinosaurs? What were the probable characteristics of the first mammals?

3. Did larger brain size precede bipedalism in the evolution of hominids? What might have been the adaptive benefits of bipedalism?

**Concept Check Answers**

1. First, the extinction itself causes the loss of many species, genera, and families. Second, evolutionary radiation of new groups (such as mammals after dinosaurs) often follows mass extinctions.

2. Yes. Judging by the oldest mammalian fossils, the first mammals were small, nocturnal, insectivorous animals that probably resembled modern-day shrews.

3. No. Bipedalism evolved before a marked increase in brain size in the hominid lineage. On ground, an upright stance offered views over long distances and freed up the hands to carry food, tools, and weapons.

# When Antarctica Was Green

At the beginning of the chapter we saw that Antarctica, a barren, frozen continent, with some of the lowest temperatures on Earth, was once home to plants and animals that might have lived in subtropical climates like that of modern Florida or Southeast Asia. What happened in Antarctica? How did such lush, subtropical life thrive in a place that is so icy and barren today?

Antarctic fossils of dinosaurs, forests, and tropical marine organisms are a vivid testimony to our dynamic world. These fossils reveal great changes over time, ranging as they do from Cambrian marine organisms to early land plants to birds and mammals. The very different organisms that have lived in Antarctica at different times illustrate the broad changes in the history of life described in this chapter, including the Cambrian explosion 530 million years ago, the colonization of land 400 million years ago, and the different periods of domination by amphibians, reptiles, and mammals.

The Antarctic fossil record also reveals the striking contrast between the diverse life-forms that once lived in Antarctica and the few that live there today. The small number of present-day organisms in Antarctica is due in part to continental drift. During the late Paleozoic, great forests thrived in the mild climate of the supercontinent Gondwana—modern Australia, Antarctica, South America, and Africa combined—which extended all the way from the equator to the south pole. But as Gondwana broke into several continents during the Mesozoic, the plants and animals on each new continent began to evolve separately—a classic example of allopatric speciation.

When Antarctica became fully separated from Australia and South America, a cold circumpolar current developed around Antarctica, isolating it from the rest of the world. The new continent became dramatically colder, and an immense ice cap began to form. The ice cap, in turn, caused Earth's climate to grow colder still. As Antarctica moved toward its present position over the South Pole, it grew ever colder. Once it became separated from Australia and South America, about 40 million years ago, the warm-adapted ferns, dinosaurs, and mammals of Antarctica were trapped on a continent drifting south. As the continent continued to cool, reaching its present position over the South Pole about 25 million years ago, most of these forms of life died out.

Recent research has revealed that as recently as 14 million years ago, alpine meadows and mosses filled the Dry Valleys of Antarctica. Then, in a short time, the temperature dropped precipitously—by 8°C—freezing virtually every living thing. In 14 million years, none have ever thawed again, and paleontologists marvel at the detail preserved in these frozen and fossilized Dry Valley organisms. Today, the Dry Valleys can barely support a few bacteria and are considered a good model for what life on Mars might be like.

The same continental movements that brought destruction to terrestrial life in Antarctica sowed the seeds of evolutionary diversity elsewhere. The rerouting of ocean currents, which contributed to the formation of the Antarctic ice cap, produced the largest difference in temperature between the poles and the tropics that Earth has ever known. The wide range of new habitats that resulted from this temperature difference set the stage for adaptive radiations in many organisms, including humans.

# Maybe Those Frog Teeth Weren't Useless After All

BY SINDYA N. BHANOO, *New York Times*

Dollo's Law, . . . proposed by the scientist Louis Dollo in the 1800s, states that when a particular trait is lost in a species, it never comes back.

It's one explanation for why humans no longer have tails, birds and turtles are toothless and snakes have stayed limbless.

But a new analysis, done by a researcher at Stony Brook University, found that while frogs lost teeth in the lower jaw at least 200 million years ago, a particular type of marsupial tree frog regained those lower teeth about 20 million years ago.

"It's a very clear-cut case of re-evolution because of the large time span," said John Wiens, the Stony Brook biologist who authored the paper in the journal *Evolution*.

Dr. Wiens analyzed DNA samples of 170 modern and fossilized frogs to approximate the dates of loss and re-evolution of the teeth.

Most frogs have teeth on their upper jaws, which may have made the re-evolution in the tree frog, known as *Gastrotheca guentheri*, easier, Dr. Wiens said.

"They already had teeth in the upper jaw, so they had the enamel, dentine and other necessities," he said. "There was a way to facilitate new teeth after 200 million years."

The species is the only known modern frog species with lower teeth, though certain other species with upper teeth do have toothlike structures on the lower jaw.

"That's a big question now: What's keeping the other frogs from developing real teeth on the lower jaw?" Dr. Wiens said.

---

Biologists know that traits can evolve more than once, but can a complex trait that has been lost return? The nineteenth-century French paleontologist Louis Dollo thought not, and recently some modern biologists have been arguing that Dollo was right. Biologists have seen examples of evolution reversing itself in many kinds of animals. For example, lizards that lost digits (fingers) regained them later in their evolution; snakes that lost their limbs later evolved short limbs with digits (and then went extinct); and some snails have lost and then regained the coil in their shells. Most snakes lay eggs, but snake lineages that evolved to give birth to live young later evolved to lay eggs again.

As this article hints, the genes that help an organism to make a particular structure—whether it's teeth, a coiled shell, fingers, or eggshells—are frequently still in the genome, just waiting to be used again. A dramatic example is the induction of tooth buds in developing chicks. Birds have not had teeth since they diverged from the dinosaurs about 150 million years ago. Unlike frogs, no birds have any teeth—hence the phrase "as rare as hen's teeth." And yet biologists can experimentally induce chicks developing inside eggs to grow primitive teeth.

Frog lost the teeth in their lower jaws even longer ago. In this research, biologist John Wiens compared the DNA sequences of 170 living amphibians (but not fossils as the article states) to show that the teeth in the lower jaw were lost at least 230 million years ago. One species of frog—a tree frog that lives in the Andes Mountains of South America—regained those lost teeth in the last 5–17 million years.

Given that evolution involves the interplay between a complex genome and an environment that selects for different traits at different times and places, it should be no surprise that traits lost can sometimes be regained. Still, that a trait can have been lost for more than 200 million years and then return is an amazing testament to the library of information that resides in the cells of every living organism.

## Evaluating the News

**1.** According to Wiens, why might frogs have kept the ability to form teeth in the lower jaw?

**2.** Humans have lost various traits that other apes still have, such as powerful jaws, large teeth, and long toes for grasping branches. Drawing on what you have learned in the previous four chapters on evolution, when did humans diverge from other apes (chimpanzees, gorillas, and orangutans), and how likely do you think it is that any of these traits could reappear in a human lineage? What circumstances might trigger their evolution?

**3.** The article uses the phrase "re-evolution" to describe a lineage reacquiring a lost trait through evolution. Is there a difference between evolution and re-evolution? Explain your reasoning.

**Source:** *New York Times*, February 8, 2011, http://www.nytimes.com.

# Summary

## 20.1 The Fossil Record: A Guide to the Past

- The fossil record documents the history of life on Earth. Fossils reveal that past organisms were unlike living organisms, that many species have gone extinct, and that the dominant groups of organisms have changed significantly over time.
- The order in which organisms appear in the fossil record is consistent with our understanding of evolution gained from other kinds of evidence, including morphology and DNA sequences. Sometimes the approximate age of a fossil can be determined through analysis of radioisotopes.
- Although the fossil record is not complete, it provides excellent examples of the evolution of major new groups of organisms.

## 20.2 The History of Life on Earth

- The first single-celled organisms resembled bacteria and probably evolved about 3.5 billion years ago.
- The release of oxygen by photosynthetic bacteria caused oxygen concentrations in the atmosphere to increase. Rising oxygen concentrations made possible the evolution of single-celled eukaryotes about 2.1 billion years ago. Multicellular eukaryotes followed about 650 mya.
- Life on Earth changed dramatically during the Cambrian explosion (530 mya), when large predators and well-defended herbivores suddenly appear in the fossil record.
- The land was first colonized by plants (about 500 mya), fungi (about 460 mya), and invertebrates (about 410 mya), which were followed later by vertebrates (about 365 mya).

## 20.3 The Effects of Plate Tectonics

- The shifting of Earth's crust, or plate tectonics, has had profound effects on the history of life on Earth.
- The separation of the continents over the past 200 million years has led to geographic isolation on a grand scale, promoting the evolution of many new species.
- At different times, climate changes caused by the movements of the continents have led to the extinctions of many species.

## 20.4 Mass Extinctions: Worldwide Losses of Species

- There have been five mass extinctions during the history of life on Earth.
- The extinction of a dominant group of organisms can provide new opportunities for other groups.

## 20.5 Adaptive Radiations: Increases in the Diversity of Life

- In adaptive radiations, a group of organisms diversifies greatly and takes on new ecological roles.
- Adaptive radiations can be caused by the reduction in competition that follows a mass extinction. They can also occur when a group of organisms evolves to fill new ecological roles.

## 20.6 The Origin and Adaptive Radiation of Mammals

- Mammals evolved from reptiles about 220 mya, evolving features such as vertically oriented legs, specialized teeth, and powerful jaws.
- Throughout the long reign of the dinosaurs (200–65 mya), most mammals remained small and nocturnal. Following the extinction of the dinosaurs, the mammals radiated to include many species that were large and active by day.

## 20.7 Human Evolution

- Like all other primates, we have flexible shoulder and elbow joints, five functional fingers and toes, opposable thumbs, flat nails (instead of claws), and brains that are large in relation to our body size. Within the primates we belong to the hominin subfamily, which is characterized by bipedalism.
- According to the out-of-Africa hypothesis, anatomically modern humans first evolved in Africa and then spread from Africa to the rest of the world, replacing other *Homo* populations.
- According to the multiregional hypothesis, modern humans evolved from *Homo erectus* populations located throughout the world. Regional differences among human populations developed early, but worldwide gene flow caused these different populations to remain a single species.

# Key Terms

adaptive radiation (p. 456)
bipedal (p. 459)
Cambrian
  explosion (p. 450)

Cretaceous extinction (p. 454)
cynodont (p. 457)
hominid (p. 459)
hominin (p. 459)

mass extinction (p. 453)
multiregional hypothesis (p. 462)
opposable (p. 459)
out-of-Africa hypothesis (p. 462)

Permian extinction (p. 454)
plate tectonics (p. 452)
primate (p. 459)
radioisotope (p. 447)

# Self-Quiz

1. Continental drift
   a. can occur when liquid rock rises to the surface and pushes continents away from one another.
   b. no longer occurs today.
   c. has led to the geographic isolation of many populations, thereby promoting speciation.
   d. both a and c

2. The fossil record
   a. documents the history of life.
   b. provides examples of the evolution of major new groups of organisms.
   c. is not complete.
   d. all of the above

3. Mass extinctions
   a. are always caused by asteroid impacts.
   b. are periods of time in which many species go extinct worldwide.
   c. have little lasting effect on the history of life.
   d. affect only terrestrial organisms.

4. The Cambrian explosion
   a. caused a spectacular increase in the size and complexity of animal life.
   b. caused a mass extinction.
   c. was the time during which all living animal phyla suddenly appeared.
   d. had few consequences for the later evolution of life.

5. The history of life shows that
   a. biodiversity has remained constant for about 400 million years.
   b. extinctions have little effect on biodiversity.
   c. macroevolution is greatly influenced by mass extinctions and adaptive radiations.
   d. macroevolution can be understand solely in terms of the evolution of populations.

6. Which of the following are radioactive forms of elements that decay to more stable forms over time?
   a. X-rays
   b. carbon-12 and carbon-14
   c. radioisotopes
   d. adaptive radiations

7. Large-scale evolution characterized by the rise and fall of major groups of organisms is called
   a. macroevolution.
   b. microevolution.
   c. mass extinction.
   d. adaptive radiation.

8. Which of the following terms most specifically describes what occurs when a group of organisms expands to take on new ecological roles, forming new species and higher taxonomic groups in the process?
   a. speciation
   b. evolution
   c. mass extinction
   d. adaptive radiation

9. Early bacteria evolved the ability to carry out photosynthesis, changing the levels of atmospheric oxygen. This development was significant for the evolutionary history of life because
   a. the decreased atmospheric oxygen set the stage for the evolution of the eukaryotes.
   b. the increased atmospheric oxygen set the stage for the evolution of eukaryotes.
   c. the increased atmospheric oxygen led to life's first mass extinction.
   d. the decreased atmospheric oxygen led to life's first mass extinction.

10. Which of the statements that follow is true?
   a. Humans and dinosaurs coexisted for part of Earth's history.
   b. Dinosaurs and mammals lived at the same time during part of Earth's history.
   c. Birds evolved from dinosaurs, and mammals evolved from birds.
   d. Humans evolved from chimpanzees.

# Analysis and Application

1. The fossil record provides clear examples of the evolution of new groups of organisms from previously existing organisms. Describe the major steps of one such example.

2. How did the evolution of photosynthesis affect the history of life on Earth?

3. What is the Cambrian explosion, and why was it important?

4. Life arose in water. Explain why the colonization of land represented a major evolutionary step in the history of life. What challenges—and opportunities—awaited early colonists of land?

5. Mass extinctions can remove entire groups of organisms, seemingly at random—even groups that possess highly advantageous adaptations. How can this be?

6. Evidence from the fossil record indicates that it usually takes 10 million years for adaptive radiation to replace the species lost during a mass extinction. Discuss this observation in light of your understanding of the speciation process (see Chapter 19). What does it suggest about the consequences of species losses that are occurring today?

7. Fossil evidence indicates that in the relatively recent past (about 30,000 years ago), anatomically modern humans, or *Homo sapiens sapiens*, may have shared the planet with at least three other distinct hominids: *H. erectus*, archaic *H. sapiens* like the Neandertals, and *H. floresiensis*. If one or more of these species were alive today, how would their existence affect the world as we know it?

# 21 The Biosphere

**A CARGO SHIP PUMPS BALLAST WATER.** This ship in Queensland, Australia, pumps ballast water while being loaded with coal bound for Japan. Dumping ballast water into local waters introduces aquatic species from other parts of the world. Zebra mussels are one of the most economically destructive invasive species in North America. Today, most ships replace ballast water while still at sea, where the saltiness of the ocean seawater kills most freshwater stowaways.

# Invasion of the Zebra Mussels

Walk outside on a spring day, and the wild riot of life is plain to see in even the biggest city: weeds crack and slowly lift chunks of sidewalk, dandelion seeds float by, and insects buzz through the air, while in every yard worms squirm through the soil beneath our feet. The rich abundance of life that surrounds us is a tiny fraction of the biosphere, a global system of life including all of Earth's organisms and their physical environment. From the depths of the oceans to the tops of the highest peaks, all of the trillions of individual organisms on Earth are part of the biosphere.

Human actions can have enormous impacts on the biosphere. Consider the seemingly harmless practice in which ships take up ballast water in one port and discharge it in another. If you live near a shipping port, you may have seen ships come into port and discharge ballast water by the ton, water gushing from the sides of the ship in giant streams. Whether such ships carry new laptops or furniture, they need ballast to keep from tipping over and to maintain the right depth in the water. Water is great ballast because it's easy for ships to load up with more or discard excess as needed.

But each time a ship dumps ballast water, it may also dump hundreds of aquatic species—everything from bacteria and viruses to small schools of fish. The result has been the transfer of aquatic species all over the world. One especially destructive species is the zebra mussel, a freshwater shellfish from Eurasia that arrived in the Great Lakes in the 1980s. Zebra mussels have flourished so well in their North American home away from home that they have displaced native species and caused billions of dollars in damage by clogging pipes and covering the hulls of boats.

Why does the zebra mussel cause more trouble in North America than in its original home? And how can we prevent this and other similar problems in the future?

In this chapter, we will discover how organisms interact with one another and with their environment on broad geographical scales, including the broadest of them all, our biosphere.

> **MAIN MESSAGE**    Organisms interact with one another and their physical environment, forming a web of interconnected relationships that is heavily influenced by climate.

## KEY CONCEPTS

- Ecology is the study of interactions between organisms and their environment. All ecological interactions occur in the biosphere, which consists of all living organisms on Earth, together with the environments they inhabit.

- Climate has a major effect on the biosphere. Climate is determined by incoming solar radiation, global movements of air and water, and major features of Earth's surface.

- The biosphere can be divided into large terrestrial and aquatic areas, called biomes, based on their climatic and ecological characteristics.

- Terrestrial biomes are usually named for the dominant plants that live there. The locations of terrestrial biomes are determined mainly by climate.

- Aquatic biomes are usually characterized by the environmental conditions that prevail there, especially salt content. They are heavily influenced by the surrounding terrestrial biomes and by climate.

- Because components of the biosphere interact in complex ways, human actions that affect the biosphere can have unexpected side effects.

A VIEW OF EARTH FROM SPACE highlights the beauty and fragility of the **biosphere**, which consists of all organisms on Earth, together with the physical environments in which they live. The biosphere is crucial to our survival and well-being, because we humans depend on it for food and raw materials, and many aspects of human society. In this unit we discuss ecology, the branch of science devoted to understanding how the biosphere works.

**Ecology** can be defined as the scientific study of interactions between organisms and their environment, where the environment of an organism includes both *biotic* factors (other organisms) and *abiotic* (nonliving) factors. Ecologists are interested in how the two parts of the biosphere—organisms and the environments in which they live—interact with and affect each other. In the chapters of this unit, our study of ecology covers several levels of the biological hierarchy: individual organisms, populations, communities, ecosystems, biomes, and the biosphere (see Figure 1.12).

All ecological interactions, at whatever level they occur, take place in the biosphere, the focus of this chapter. We begin by discussing why ecology is important and what types of information ecologists must have to understand how the biosphere works. We go on to discuss climate and other factors that shape the biosphere. Then we take a brief look at the variety of terrestrial and aquatic biomes that arise from those factors.

## Helpful to Know

Ecology and economy are intertwined in word derivation as well as in practice. *Eco* is Latin for "household," so "ecology" literally means "study of the household," and "economy" literally means "management of the household."

## 21.1 Ecology: Understanding the Interconnected Web

The science of ecology helps us understand the natural world in which we live. Beyond enriching our lives intellectually, such an understanding is becoming increasingly important because we are changing our world in ways that can be expensive—and in some cases, difficult or even impossible—to fix. Consider species such as zebra mussels that are accidentally or deliberately brought by people to new geographic regions. In the United States, people have introduced thousands of nonnative species, some of which have become pests (*invasive species*) that collectively cause an estimated $120 billion in economic losses each year—a huge cost, similar to the economic impact of smoking ($150 billion per year). By studying the ecology of invasive species, we can understand how people help them spread to new regions, why they increase dramatically in abundance, how they affect natural communities, and how they cause economic disruption—all of which can be helpful in limiting the damage that these species cause.

The impact of humans extends to the biosphere. Consider chlorofluorocarbons (CFCs), which are synthetic chemicals used as refrigerants, in aerosol sprays, and in foam manufacture. CFCs rise to the upper reaches of Earth's atmosphere, where their chlorine atoms react with and destroy a gas called ozone ($O_3$). Our atmosphere possesses an ozone-containing layer that absorbs more than 90 percent of the most powerful ultraviolet radiation (UV-B), so relatively few of these DNA-damaging rays from the sun normally make it through. However, chlorine atoms released from CFCs destroy the ozone, and therefore the UV-screening properties, of this atmospheric layer (see the box on page 472).

All organisms interact with their environment. These interactions go both ways: organisms affect their environment (as when a beaver builds a dam that blocks the flow of a stream and creates a pond), and the environment affects organisms (as when an extended drought limits the growth of plant species that the beaver depends on for food). Because the interactions go both ways, the organisms and physical environments of the biosphere can be thought of as forming a web of interconnected relationships.

Thinking of organisms and their environment as an interconnected web helps us understand the consequences of human actions. Consider what happened when people used fencing, poison, and hunting to remove dingoes from a large region of rangeland in Australia. Dingoes, a type of wild dog, were removed to prevent them from eating sheep. As **FIGURE 21.1** shows, in areas where dingoes were removed the population of their preferred prey, red kangaroos, increased dramatically (by a factor of 166). Kangaroos decrease the food available to sheep because kangaroos and sheep like to eat the same plants. In addition, in times of drought, kangaroos resort to a behavior not found in sheep: they dig up and eat belowground plant parts. This behavior

(a)

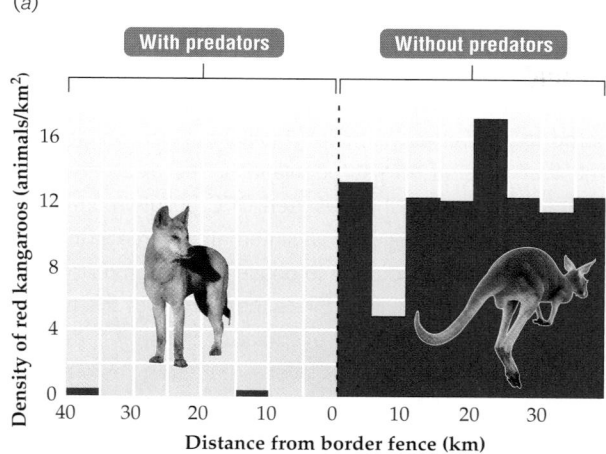

(b)

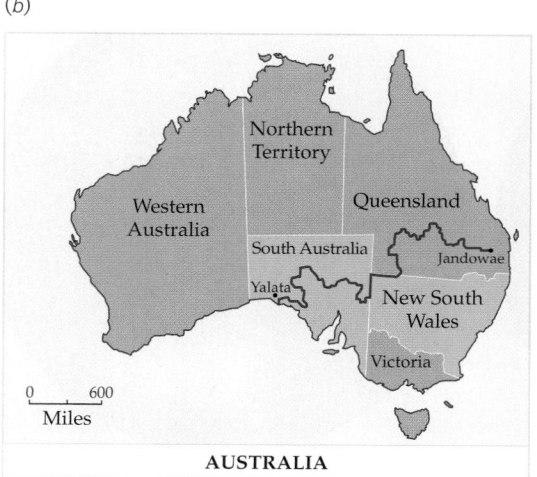

**AUSTRALIA**

**FIGURE 21.1 An Explosion in Numbers**

(a) Red kangaroo numbers increased by a factor of 166 when their dingo predators were removed from the Australian rangelands located south of (b) the world's longest fence (red line).

has the potential to change the number and types of plant species found in rangeland, thereby further increasing the effect that kangaroos have on sheep.

When people removed dingoes, the subsequent effects were not what they expected or desired, because red kangaroos outcompeted their sheep for food. With the advantage of hindsight, the negative side effects of removing dingoes seem predictable, because changes that affect one part of the biosphere (such as removal of dingoes) can have a ripple effect to produce changes elsewhere (such as increases in red kangaroos and decreases in the food available to sheep).

The examples given in this section illustrate how natural systems can be viewed as an interconnected web. To understand such connections further, it is useful to learn more about the physical factors that affect the distribution, abundance, and diversity of life in our biosphere.

## 21.2 Climate's Large Effect on the Biosphere

The terms "weather" and "climate" should not be used interchangeably. **Weather** refers to the temperature, precipitation (rainfall and snowfall), wind speed and direction, humidity, cloud cover, and other physical conditions of Earth's lower atmosphere at a specific place over a short period of time. **Climate** is more inclusive in area and time frame: it refers to the prevailing weather conditions experienced in a region over relatively long periods of time (30 years or more). Weather, as we all know, changes quickly and is hard to predict. Because it describes the state of the atmosphere over longer time periods—over decades or centuries, for example—the climate of a place is more predictable than the weather there.

Climate has major effects on ecological communities because organisms are more strongly influenced by climate than by any other feature of their environment. On land, for example, whether a particular region is desert, grassland, or tropical forest depends primarily on such features of climate as temperature and precipitation. In the next section we look at the factors that determine the climate of a given geographic region.

## Incoming solar radiation shapes climate

Sunlight strikes Earth directly at the equator, but at a slanted angle near the North and South poles. This difference causes more solar energy to reach the equator, as well as the regions on either side of it (the tropics), than the poles. Tropical regions receive 2.5 as much solar radiation as reaches polar regions, making them much warmer than the poles. Tropical regions show small seasonal fluctuations in temperature, so organisms that live there experience a relatively warm, stable climate throughout the year. Generally speaking, sunlight and warmth promote photosynthesis and thereby increase the productivity of plants and other light-dependent producers. Productivity is a measure of the energy that producers are able to store in the form of biological material, or biomass. Consumers, including all animals and all types of decomposers, depend on the productivity of producers.

# Wearing Thin: The Attack on Earth's Ozone Shield

The ozone layer is the portion of the atmosphere that encircles our planet from an elevation of about 10–50 kilometers (33,000–160,000 feet). This atmospheric layer is relatively rich in a gas called ozone ($O_3$), which blocks out almost 99 percent of the DNA-damaging ultraviolet radiation from the sun. Ozone molecules are created when ultraviolet light strikes molecular oxygen ($O_2$) in the air, releasing free oxygen atoms (O) that then combine with oxygen molecules ($O_2$) to produce ozone ($O_3$).

In 1974, Mario Molina and F. Sherwood Rowland wrote an influential paper in which they warned that the chlorofluorocarbons (CFCs) widely used in aerosol spray cans were damaging Earth's ozone layer. Chlorine atoms from these organic pollutants interact with ozone ($O_3$) molecules, triggering a chain reaction that results in the destruction of thousands of additional ozone molecules, as one of the by-products (chlorine monoxide, ClO) reacts with yet more ozone molecules. A single CFC molecule can lead to the destruction of about 100,000 molecules of ozone. As ozone is depleted through these chain reactions, the thickness of the protective layer decreases.

The complex airflow patterns of our upper atmosphere tend to funnel CFCs, as well as other atmospheric pollutants, over the polar regions. The resulting ozone depletion creates a "hole" in the ozone column above the poles. Low temperatures in the upper atmosphere promote the chemical processes that generate the highly reactive chlorine atoms. CFC accumulation and ozone destruction tend to be especially severe over Antarctica compared to the Arctic because of the wind patterns that prevail in the Southern Hemisphere, especially in the southern spring (September).

Thinning of the ozone layer may have many consequences, including a higher incidence of skin cancer, reduced yields for some crops, and reduced populations of phytoplankton, the small photosynthetic aquatic organisms on which all other aquatic organisms depend for food.

Decreases in phytoplankton can result in decreased fish populations. The discovery of the relationship between CFC pollution and ozone depletion led to an international treaty (the Montreal Protocol) to phase out the production and use of CFCs. Since the signing of the treaty in 1987, the rate of ozone depletion has declined, but the ozone layer is far from restored. The member nations of the United Nations have pledged to phase out CFCs completely by 2030.

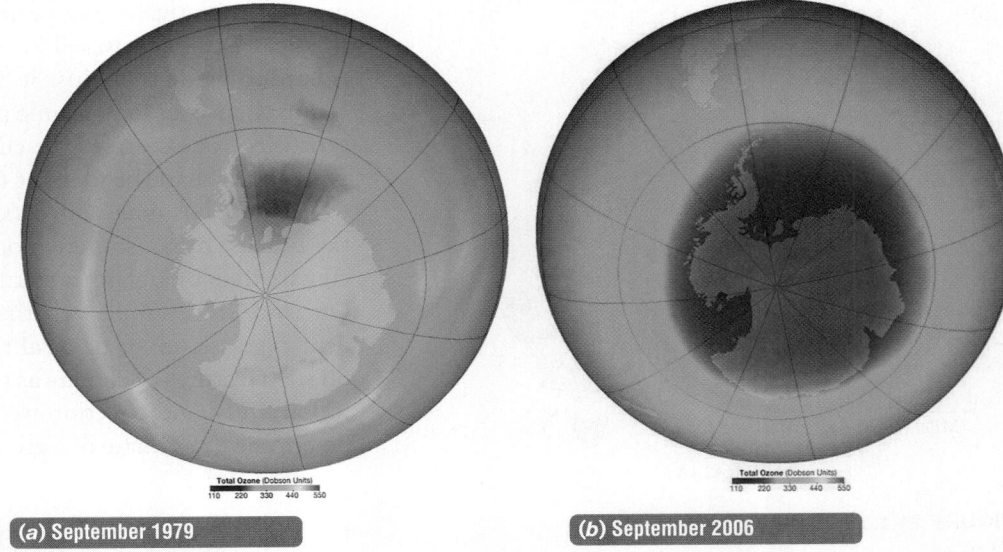

**(a) September 1979**  **(b) September 2006**

**The Antarctic Ozone Hole**

These computer-generated maps of the southern end of Earth depict the total concentration of ozone in the air column above Antarctica for the months of (a) September 1979 and (b) September 2006. Ozone concentrations were measured (in Dobson units) by instruments aboard NASA's *Aura* satellite. The ozone "hole" is the extreme thinning of the ozone layer to values below 150 Dobson units (deep purple color). At 29 million square miles, the 2006 ozone hole above Antarctica was a record setter. The September 1979 image represents near-normal conditions, prior to significant loss of ozone in the upper atmosphere. The ozone hole waxes and wanes from year to year because the concentration of CFCs over the polar regions is influenced in complex ways by weather conditions.

## Wind and water currents affect climate

Near the equator, intense sunlight causes water to evaporate from Earth's surface. The warm, moist air rises because heat causes it to expand and therefore to be less dense, or lighter, than air that has not been heated. The warm, moist air cools as it rises. Because cool air cannot hold as much water as warm air can, much of the moisture is "wrung out" from a cooling air mass and falls as rain (see Figure 21.2).

Usually, cool air sinks. The cool air above the equator cannot sink immediately, however, because of the warm air rising beneath it. Instead, the cool air is

Cool, dry air sinks.

Cell 2 north

North Pole

The clash of moist, warm air and polar cold fronts produces rain.

Some of the cool, dry air sinks, and is then deflected both north and south.

Arctic tundra

Coniferous forest

60°

Forest and grassland

30°

Desert

Forest

Cell 1 north

0° — Equator

Tropical rainforest

Cell 1 south

Forest

30°

Desert

Warm, moist air cools as it rises, causing rainfall.

Forest and grassland

60°

Antarctic tundra

South Pole

Cell 2 south

Moist air rises; rain falls.

Cold, dry air sinks.

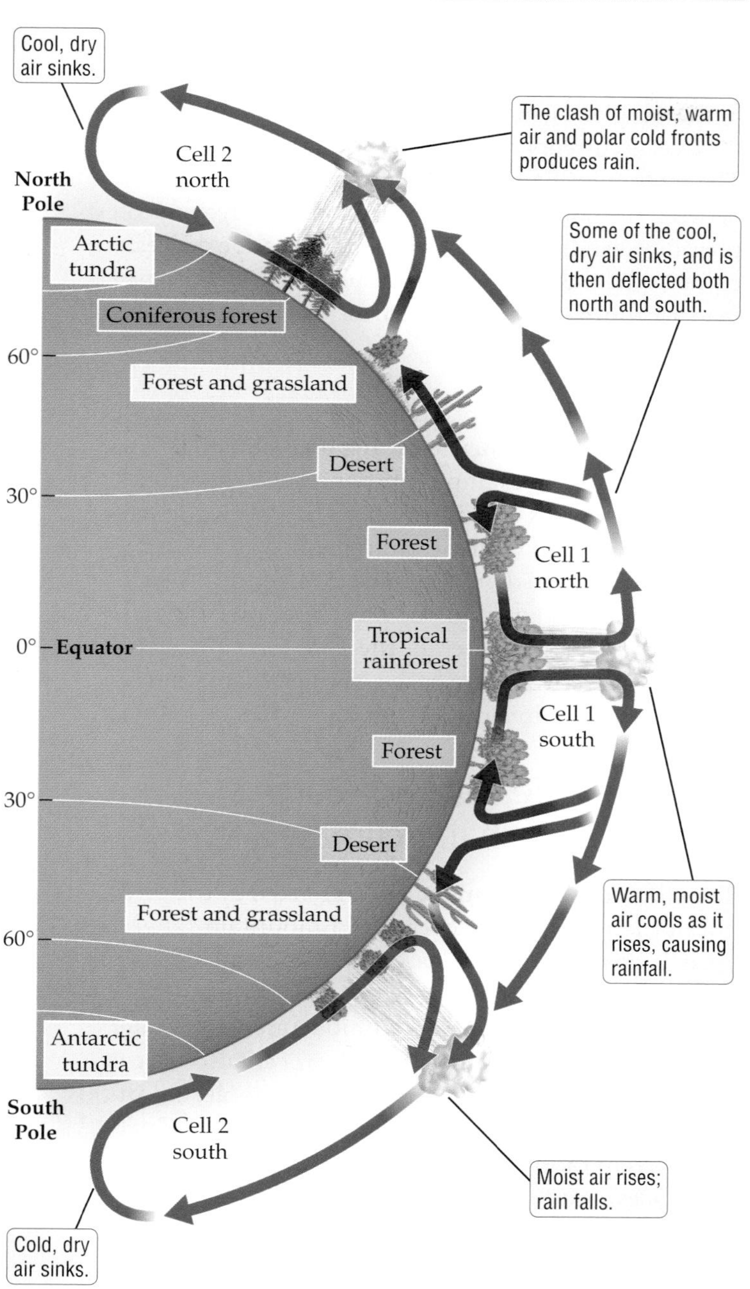

**FIGURE 21.2 Earth Has Four Giant Convection Cells**

Two giant convection cells are located in the Northern Hemisphere and two in the Southern Hemisphere. In each convection cell, relatively warm, moist air rises, cools, and then releases moisture as rain or snow.

drawn to the north and south, tending to sink back to Earth at about 30° latitude. Part of the air mass flows back toward the equator, and as it does so it absorbs moisture from Earth's surface. By the time it reaches the equator, the air is once more warm and moist, so it rises, repeating the cycle.

Earth has four giant **convection cells** in which warm, moist air rises and cools, dry air sinks (**FIGURE 21.2**). Two of the four convection cells are located in tropical regions and two in polar regions, where they generate relatively consistent wind patterns. In temperate regions (roughly 30°–60° latitude), winds are more variable, and there are no stable convection cells. Precipitation occurs when cool, dry air from polar regions collides with warm, moist air moving north, so most temperate regions receive ample moisture.

The winds produced by the four giant convection cells do not move straight north or straight south relative to the land spinning below. Instead, because of Earth's rotation, these winds appear to curve as they travel near Earth's surface (**FIGURE 21.3**). Winds traveling toward the equator appear to curve to the right. To an Earth-bound observer, these winds seem

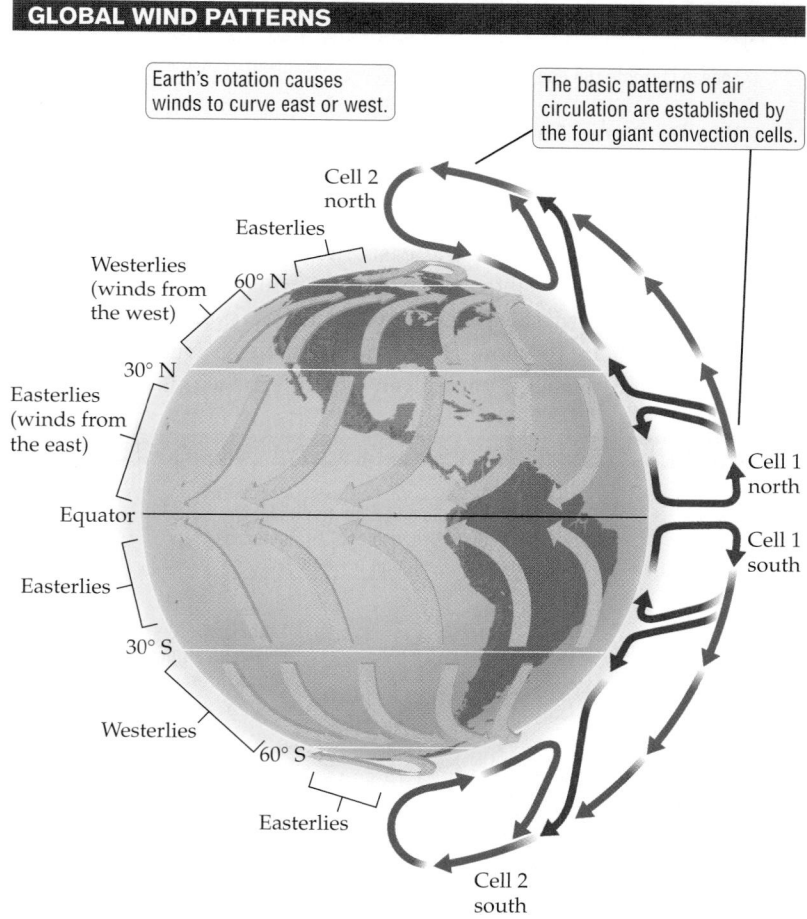

Earth's rotation causes winds to curve east or west.

The basic patterns of air circulation are established by the four giant convection cells.

Cell 2 north

Easterlies

Westerlies (winds from the west)

60° N

30° N

Easterlies (winds from the east)

Cell 1 north

Cell 1 south

Equator

Easterlies

30° S

Westerlies

60° S

Easterlies

Cell 2 south

**FIGURE 21.3 Prevailing Winds Are Determined by Global Patterns of Air Circulation**

Earth's rotation causes winds to curve to the east or west. The direction in which they curve depends on their latitude, but for most regions on Earth, the seasonal winds blow from a consistent direction.

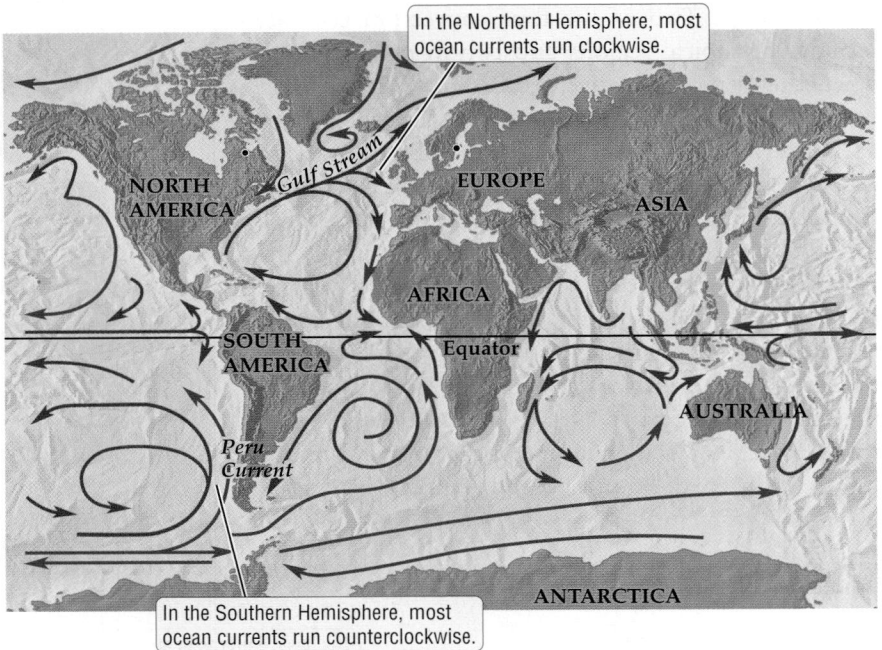

In the Northern Hemisphere, most ocean currents run clockwise.

In the Southern Hemisphere, most ocean currents run counterclockwise.

**FIGURE 21.4 Why Do Ocean Temperatures Vary?**
Ocean currents can be cold (blue) or warm (red), depending on a combination of factors, including water depth and latitude.

**THE RAIN SHADOW EFFECT**

On the windward side of the mountain, air rises and cools. Because cool air holds less water than warm air, rain or snow falls.

Prevailing winds pick up moisture from bodies of water.

On the leeward side of the mountain, air descends and warms, producing little rain or snow.

Ocean          Mountain range     Rain shadow area

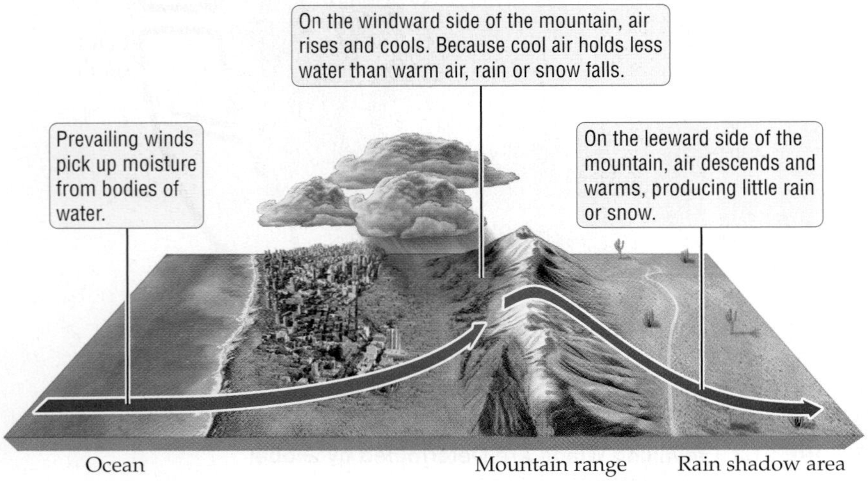

**FIGURE 21.5 The Leeward Side of a Mountain Is Usually Dry**
The side of a high mountain that faces the prevailing winds (the windward side) receives more precipitation than the side of the mountain that faces away from the prevailing winds (the leeward side). The leeward side is therefore said to be in a rain shadow.

from a consistent direction, and these predictable patterns of air movement are known as prevailing winds. In southern Canada and much of the United States, for example, winds blow mostly from the west, so storms in these regions usually move from west to east.

Ocean currents also have major effects on climate. The rotation of Earth, differences in water temperature between the poles and the tropics, and the directions of prevailing winds all contribute to the formation of ocean currents. In the Northern Hemisphere, ocean currents tend to run clockwise between the continents; in the Southern Hemisphere, they tend to run counterclockwise (**FIGURE 21.4**).

Ocean currents carry a huge amount of water and can have a great influence on regional climates. The Gulf Stream, for example, moves twenty-five times as much water as is carried by all the world's rivers combined. Without the warming effect of the water carried by this current, the climate in countries such as Great Britain and Norway would be sub-Arctic instead of temperate. Overall, the Gulf Stream causes cities in western Europe to be warmer than cities at similar latitudes in North America.

## The major features of Earth's surface also shape climate

The climate of a place may also be affected by the presence of large lakes, the ocean, and mountain ranges. Heat is absorbed and released more slowly by water than by land. Because they retain heat comparatively well, large lakes and the ocean moderate the climate of the surrounding lands. Mountains can also have a large effect on a region's climate. For example, mountains often produce a **rain shadow** effect, in which little precipitation falls on the side of the mountain that faces away from the prevailing winds (**FIGURE 21.5**). In the Sierra Nevada of North America, five times as much precipitation falls on the western side of the mountains (which faces toward winds that blow in from the ocean) as on the eastern side, where the lack of precipitation contributes to the formation of deserts. Mountain ranges in northern Mexico, South America, Asia, and Europe also create rain shadows.

to blow from the east; hence, such winds are called easterlies ("from the east"). Similarly, winds that travel toward the poles curve to the left; and because they seem to blow from the west, these winds are called westerlies. At any given location the winds usually blow

> **Concept Check**
>
> 1. Can knowledge gained from ecological research impact economic decisions? Explain, using an example.
> 2. What causes convection cells to form in Earth's atmosphere, and what effects do they have on climate?

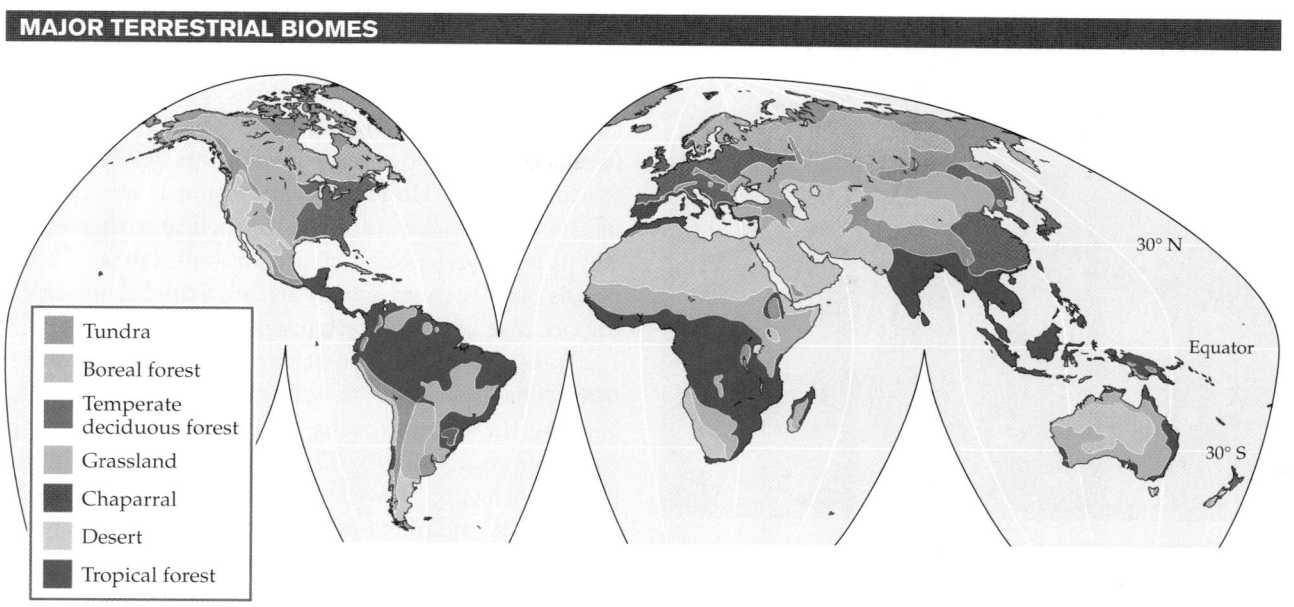

**FIGURE 21.6** **The Distribution of Biomes Is Affected by Climate, Latitude, and Disturbance**
Biomes do not begin and end abruptly. Instead, they generally transition into one another. Disturbances such as storms, fires, and human activities can alter biomes.

Legend:
- Tundra
- Boreal forest
- Temperate deciduous forest
- Grassland
- Chaparral
- Desert
- Tropical forest

## 21.3 Terrestrial Biomes

Ecologists categorize large areas of the biosphere into distinct regions, called **biomes**, based on the unique climatic and ecological features of each such region. Because climate strongly influences the life-forms that can live in a particular place, each biome is associated with a characteristic set of plant and animal species. A biome may encompass more than one ecosystem and typically stretches across large swaths of the globe.

Biomes on land—*terrestrial biomes*—are usually named after the dominant vegetation in the area. *Aquatic biomes* are classified on the basis of physical and chemical features, such as salt content. In this section we explore some of Earth's major terrestrial biomes: *tundra, boreal forest, temperate deciduous forest, grassland, chaparral, desert,* and *tropical forest* (**FIGURE 21.6**).

### The location of terrestrial biomes is determined by climate

Climate is the single most important factor controlling the location of natural terrestrial biomes. The climate of the area—most important, the temperature and the amount and timing of precipitation—allows some species to thrive and prevents other species from living there. Overall, the effects of temperature and moisture on different species cause particular

biomes to be found under a consistent set of conditions (**FIGURE 21.7**). Terrestrial biomes change from the equator to the poles, and from the bottom to the top of mountains.

**TYPICAL DISTRIBUTION OF BIOMES**

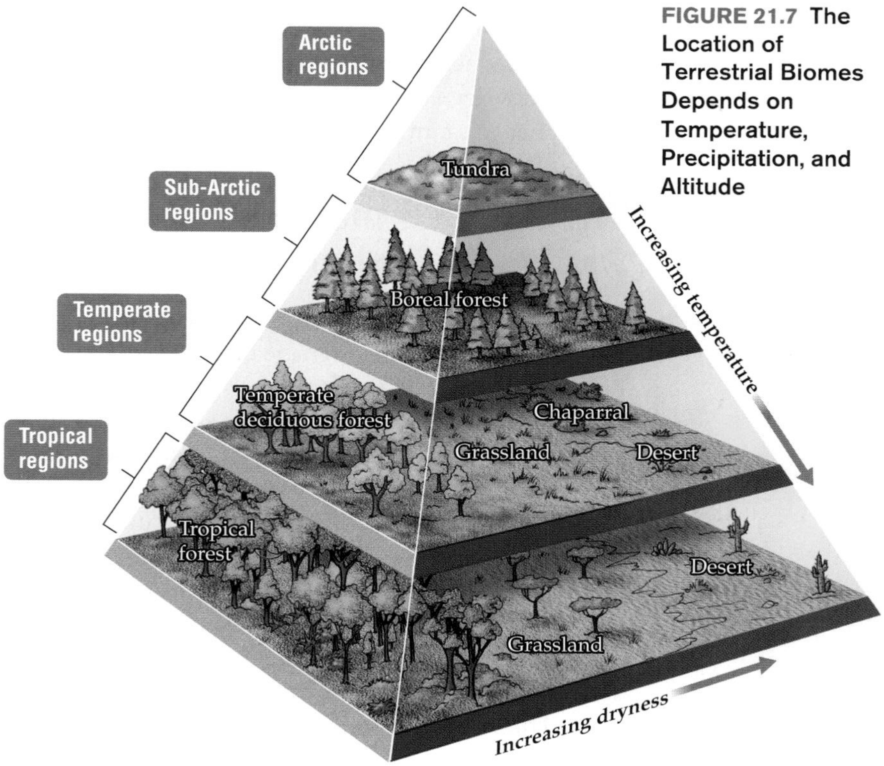

**FIGURE 21.7** **The Location of Terrestrial Biomes Depends on Temperature, Precipitation, and Altitude**

Arctic regions

Sub-Arctic regions

Temperate regions

Tropical regions

Tundra

Boreal forest

Temperate deciduous forest

Chaparral

Grassland

Desert

Tropical forest

Desert

Grassland

Increasing temperature

Increasing dryness

**FIGURE 21.8 Tundra: Denali National Park, Alaska**
Tundra is found at high latitudes and high elevations, and it is dominated by low-growing shrubs and nonwoody plants that can cope with a short growing season.

Climate can exclude species from a region directly or indirectly. Species that cannot tolerate the climate of a region are directly excluded from that region. Species that can tolerate the climate but are outperformed by other organisms that are better adapted to that climate are excluded indirectly. Although climate places limits on where biomes can be found, the actual extent and distribution of biomes in the world today are very strongly influenced by human activities. We will return to the effects of humans on natural biomes when we discuss global change in Chapter 25. But now let's tour seven great terrestrial biomes of our planet to examine the distinctive features of each.

## The tundra is marked by cold winters and a short growing season

The word **tundra** comes from *tunturi*, which means "treeless plain" in the language of the Sami people of Finland. The Arctic tundra covers nearly one-fourth of Earth's land surface, encircling the North Pole in a vast sweep that includes half of Canada and Alaska and sizable portions of northern Europe and Russia. A similar habitat, known as the alpine tundra, is encountered above the tree line in high mountains.

Winter temperatures in the Arctic tundra can dip to –50°C (–58°F), although they average about –34°C (–29°F). Even in summer, temperatures seldom climb above 12°C (54°F) and freezing weather is not uncommon. The land is frozen for 10 months of the year, thawing to a depth of no more than a meter (about

3 feet) during the short summer. Below these surface layers of soil is **permafrost**, permanently frozen soil that may be a quarter of a mile deep. Precipitation in the Arctic tundra ranges from 15 to 25 centimeters (up to 10 inches) per year—lower than in many of the world's deserts. However, evaporation is low because of the cold. In the summer, ice melt and thawing of the upper layers of soil create an abundance of bogs, ponds, and streams, which are prevented from draining because of the underlying permafrost.

Trees are absent or scarce in the tundra (**FIGURE 21.8**) because of the short growing season and because the permafrost is a barrier to the deep taproots of woody plants. The vegetation is dominated by low-growing flowering plants, such as grasses, sedges, and members of the heath family. The rocky landscape is covered in mosses and lichens, which are an important food source for herbivores such as reindeer (known as caribou in North America). Rodents such as lemmings, voles, and Arctic hare provide food for carnivores like Arctic foxes and wolves. Bears and musk oxen are among the few large mammals. Insects abound in the summer, along with the migratory birds that feed on them. There are few amphibians and even fewer reptiles.

## A few coniferous species dominate in the boreal forest

The **boreal forest** (from *borealis*, "northern") is the largest terrestrial biome. It is also known as taiga [*TYE*-**guh**], based on the Mongolian word for coniferous forests. It includes the sub-Arctic landmass immediately south of the tundra, covering a broad belt of Alaska, Canada, northern Europe, and Russia, approximately between 60° and 50° north latitude. Winters in the boreal forest are nearly as cold as in the tundra and last about half the year. Summers are longer and warmer than in the tundra, with temperatures reaching as high as 30°C (86°F) in some boreal forests of the world. Typically, the soil is thin and nutrient-poor. Rainfall is low in most boreal forests, but it may be high in some areas, most notably the Pacific Coast rainforests of North America. Because evaporation rates are low in these cold, northern latitudes, plants in the boreal forest generally receive adequate moisture during the growing season.

Conifers, which are cone-bearing trees with needlelike leaves, dominate the boreal forest vegetation (**FIGURE 21.9**). Spruce and fir are the most common species in the North American boreal forest; pine

**FIGURE 21.9 Boreal Forest: Banff National Park, Alberta**
Boreal forests are dominated by coniferous (cone-bearing) trees that grow in northern or high-altitude regions with cold, dry winters and mild summers.

and larch are abundant in Scandinavia and Russia. Broad-leaved species such as birch, alder, willow, and aspen are also found, especially in the southern ranges of this biome. Plant diversity is relatively low, except in the rainforests of the Pacific Coast, where the seaside climate is milder and soils are richer. The large herbivores of the boreal forest include elk and moose. Small carnivores, such as weasels, wolverines, and martens, are common. Larger carnivores include lynx and wolves. Bears, which are omnivores, are found throughout the world's boreal forests.

## Temperate deciduous forests have fertile soils and relatively mild winters

**Temperate deciduous forests** are a familiar forest type for most people who live in North America. They also constitute the dominant biome in large parts of Europe and Russia and parts of China and Japan. Temperate deciduous forests occur in regions with a distinct winter, with subfreezing conditions that may last 4 or 5 months. However, winter temperatures are not as harsh as in the Arctic and sub-Arctic biomes, and summers are typically much warmer. Temperatures commonly range from lows of $-30°C$ ($-22°F$) in winter to summertime highs above $30°C$ ($86°F$). Annual rainfall varies from 60 centimeters (nearly 24 inches) to more than 150 centimeters (59 inches), and precipitation is distributed evenly through much of the year.

Deciduous trees—that is, trees that drop their leaves in the cold season—are the dominant vegetation in this biome (**FIGURE 21.10**). These forests display greater species diversity than do the tundra and boreal biomes. Oak, maple, hickory, beech, and elm are common in these woodlands, which also harbor an understory of shade-tolerant shrubs and herbaceous plants forming a ground cover. Coniferous species, such as pine and hemlock, also occur, but they are not the dominant trees. The fauna includes squirrels, rabbits, deer, raccoons, beavers, bobcats, mountain lions, and bears. There are many different fishes, and amphibians and reptiles are common.

## Grasslands appear in regions with good soil but relatively little moisture

The **grassland** biome is characteristic of regions that receive about 25–100 centimeters (about 10–40 inches) of precipitation annually. The moisture levels are insufficient for vigorous tree growth, but they are not as low as in a desert. Grasslands are found in both temperate and tropical latitudes. The prairies of North America, the steppe in Russia and central Asia, and the pampas of South America are examples of temperate grasslands. Tropical and subtropical grasslands are known as savanna. Grasses dominate in this biome, although the landscape may be dotted with shrubs and small trees, as in

**FIGURE 21.10**
**Temperate Deciduous Forest: Pocono Mountains, Pennsylvania**
Temperate deciduous forests are dominated by trees and shrubs adapted to relatively rich soil, snowy winters, and moist, warm summers.

the African savanna. Soils in some grasslands, especially the prairies of North America and the pampas of Argentina, are exceptionally deep and fertile. As a result, most of these areas have been converted to agriculture, and today they are some of the most productive grain-growing regions of the world.

Before they succumbed to the plow, the grasslands of central North America formed the largest stretch of grassland biome in the world. The northern and central parts of the Great Plains receive moderate rainfall—about 100 centimeters annually—enough to support the growth of "hat-high" grasses like big bluestem, which averages about 6 feet in height. The tallgrass prairies, as these grasslands are called, once stretched from Manitoba in Canada, through the Dakotas and Nebraska, south to Kansas and Oklahoma. They also extended into neighboring states to the east, as far as the western edge of Indiana.

Only about 1 percent of the original prairie remains, most of it in protected areas. Grasses, and herbaceous plants such as coneflower and shooting star, predominate in these grasslands. That predominance is maintained by prairie fires, which destroy shrubs and trees but not the underground roots and stems of prairie plants. Burrowing rodents like voles and prairie dogs (a type of ground squirrel) aerate the soil, thereby improving growing conditions for the root systems of prairie plants. There are many butterflies and other insects, and the greatest diversity of mammals in North America is found here.

Rainfall declines farther west, and the prairie is replaced by mixed grassland and then short grassland

along the eastern foothills of the Rocky Mountains from Montana, Wyoming, and Colorado south to New Mexico. Drought-resistant grasses, such as buffalo grass, are the dominant vegetation (**FIGURE 21.11**), and bison and pronghorn are the larger herbivores of the short grasslands.

## Chaparral is characteristic of regions with wet winters and hot, dry summers

The **chaparral** [SHAP-uh-RAL] is a shrubland biome dominated by dense growths of scrub oak and other drought-resistant plants (**FIGURE 21.12**). It is found in regions with a "Mediterranean climate," characterized by cool, rainy winters and hot, dry summers. Annual rainfall ranges from 25 to 100 centimeters (about 10–40 inches). Chaparral occurs in regions of southern Europe and North Africa bordering the Mediterranean Sea, and also in coastal California, southwestern Australia, and along the west coast of Chile and South Africa.

The soil is relatively poor in these habitats, and most species are adapted to hot, dry conditions. Many chaparral plants have thick, leathery leaves that reduce water loss. Low moisture and high temperatures in summer make the chaparral exceptionally susceptible to wildfires. Common vegetation in the California chaparral includes scrub oak, pines, mountain mahogany, manzanita, and the chemise bush. The California quail is an iconic bird of the chaparral. Rodents such as jackrabbits and gophers are common, and there are many species of lizards and snakes. Mammals include deer, the peccary (a piglike animal), lynx, and mountain lions.

## The scarcity of moisture shapes life in the desert

The **desert** biome makes up one-third of Earth's land surface. It is characterized by less than 25 centimeters (10 inches) of precipitation annually. The defining feature of a desert is the scarcity of moisture, not high temperatures. Antarctica, which receives less than 2 centimeters of precipitation per year, is the largest cold desert in the world. The Sahara desert in northern Africa is the largest hot desert. Because desert air lacks moisture, it cannot retain heat and therefore cannot moderate daily temperature fluctu-

**FIGURE 21.11**
**Grassland: Eastern Colorado**

Grasslands are common throughout the world and are dominated by grasses, although scattered trees are found in some, such as the tropical grasslands known as the savanna. Pictured here is buffalo grass (*Buchloe dactyloides*) in a short grassland.

**FIGURE 21.12 Chaparral: North of San Francisco, California**

Chaparral is characterized by shrubs and small, nonwoody plants that grow in regions with cool, rainy winters and hot, dry summers.

**FIGURE 21.13 Desert: Near Phoenix, Arizona**

Deserts form in regions with low precipitation, usually 25 centimeters (10 inches) per year or less. The photo shows saguaro cacti (tall green columns) and other plants in the Sonoran Desert.

ations. As a result, temperatures can be above 45°C (113°F) in the daytime and then plunge to near freezing at night.

Desert plants have small leaves; the reduced surface area minimizes water loss. Succulents, such as cacti, store water in their fleshy stems or leaves (**FIGURE 21.13**). Some desert plants produce enormously long taproots that are able to reach subsurface water. Most animals in the desert are nocturnal, hiding in burrows during the heat of the day and emerging at night to feed. Jackrabbits have large ears that act like a radiator to help dissipate heat. Desert mammals, such as the desert fox, have light-colored fur, which deflects some of the radiant energy from the sun. The kangaroo rat loses very little water in urine because its kidneys enable it to recover water with exceptional efficiency. The rodent's respiratory passages wring moisture from exhaled air by cooling it. A similar mechanism enables the camel's nose to recover moisture from the air it breathes out.

## Tropical forests have high species diversity

The **tropical forest** biome is characterized by warm temperatures and about 12 hours of daylight all through the year. Rainfall may be heavy throughout the year or occur only during a pronounced wet season. Tropical rainforests, which are tropical forests that remain wet year-round, may receive in excess of 200 centimeters (80 inches) of rain annually. Because organic matter decomposes rapidly at warm temperatures, there is little leaf litter in tropical forests. Soils in this

biome tend to be nutrient-poor for two main reasons: First, a large percentage of nutrients are locked up in the living tissues (biomass) of organisms, especially large trees. Second, heavy rains tend to leach out soil nutrients, depleting certain minerals in particular.

The abundance of sunshine and moisture makes tropical rainforests the most productive terrestrial habitat (**FIGURE 21.14**). They are also hot spots for di-

**FIGURE 21.14 Tropical Forest: El Yunque National Forest, Puerto Rico**

Tropical forests form in warm regions with either seasonally heavy rains or year-round rain. Tropical rainforests, which receive abundant moisture throughout the year, are some of the most productive ecosystems on Earth. They have a rich diversity of trees, vines, and shrubs.

versity of life-forms (biodiversity). Tropical rainforests currently occupy about 6 percent of Earth's land area but harbor nearly 50 percent of its plant and animal species. South America has the largest tracts of tropical rainforest, especially in Brazil, Peru, and Bolivia. Large areas of Southeast Asia and equatorial Africa are also covered in tropical rainforests.

Tropical rainforests are under severe threat from logging and clearing for livestock grazing and other types of agricultural activity. More than half of the original tropical rainforest has been lost. About 2 acres of tropical rainforest are lost every second. Loss of rainforest biomes does not just mean loss of habitat for much of the world's biodiversity, but also loss of a significant carbon dioxide sink. A carbon dioxide ($CO_2$) sink is a means of sequestering carbon dioxide. The main natural sinks are the oceans and plants and other organisms that use photosynthesis to remove carbon dioxide from the atmosphere by incorporating it into biomass. The loss of the rainforests is likely to worsen global warming, because global warming is associated with increasing $CO_2$ in Earth's atmosphere.

## 21.4 Aquatic Biomes

Life evolved in water billions of years ago, and aquatic ecosystems cover about 75 percent of Earth's surface. Aquatic biomes are shaped primarily by the physical characteristics of their environment, such as salt content, water temperature, water depth, and the speed of water flow. Within an aquatic biome, we can recognize various habitats, or ecological zones—defined by their nearness to the shore, water depth, and the depth to which light penetrates. The bottom surface of any body of water, for example, is classified as the **benthic zone**.

Two main types of aquatic biomes can be distinguished on the basis of salt content: freshwater and marine. Lakes, rivers, and wetlands are examples of ecosystems within the *freshwater biome*. Estuaries, coral reefs, the coastal region, and the open ocean are examples of ecosystems within the *marine biome*.

## Aquatic biomes are influenced by terrestrial biomes and climate

Aquatic biomes, especially lakes, rivers, wetlands, and the coastal portions of marine biomes, are heavily influenced by the terrestrial biomes that they border or through which their water flows. High and low points of the land, for example, determine the locations of lakes and the speed and direction of water flow. In addition, when water drains from a terrestrial biome into an aquatic biome, it brings with it dissolved nutrients (such as nitrogen, phosphorus, and salts) that were part of the terrestrial biome. Rivers and streams carry nutrients from terrestrial environments to the ocean, where they may stimulate large increases in the abundance of phytoplankton.

Aquatic biomes also are strongly influenced by climate. Climate helps determine the temperature, depth, and salt content of the world's oceans, for example. Such physical conditions of the ocean have dramatic effects on the organisms that live there, and hence climate has a powerful effect on marine life. Consider the El Niño [ell-*NEEN*-yoh] events that are often reported in the news. These events begin when warm waters from the west deflect the cold Peru Current along the Pacific coast of South America (**FIGURE 21.15**). The results of this change are spectacular, including dramatic decreases in numbers of fish,

**FIGURE 21.15**
**Ocean Currents Change during El Niño Events**

During an El Niño event, winds from the west push warm surface water from the western Pacific to the eastern Pacific. The resulting changes in sea surface temperatures cause changes in ocean currents (shown here in blue for cold, red for warm). El Niño events cause many additional changes (not shown here), altering wind patterns, sea levels, and patterns of precipitation throughout the world.

**EL NIÑO EVENTS**

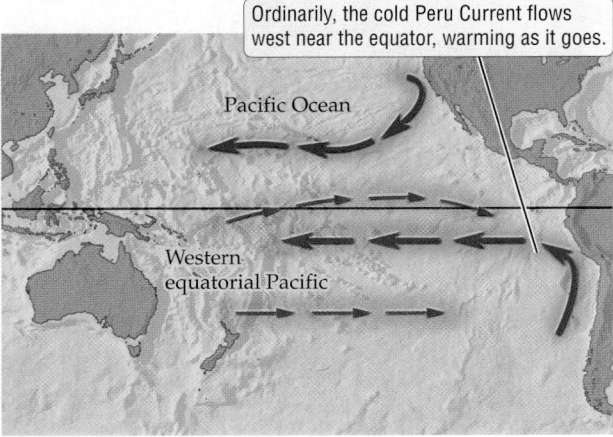

Ordinarily, the cold Peru Current flows west near the equator, warming as it goes.

Pacific Ocean

Western equatorial Pacific

During an El Niño event, warm water flows from west to east and turns aside the Peru Current.

Weakened Peru Current

die-offs of seabirds, storms along the Pacific coast of North America that destroy underwater "forests" of a brown alga called kelp, crop failures in Africa and Australia, and drops in sea level in the western Pacific that kill huge numbers of coral reef animals.

## Aquatic biomes are also influenced by human activity

Like terrestrial biomes, aquatic biomes are strongly influenced by the actions of humans. Portions of some aquatic biomes, such as wetlands and estuaries, are often destroyed to allow for development projects. Rivers, wetlands, lakes, and coastal marine biomes are negatively affected by pollution in most parts of the world. Aquatic biomes also suffer when humans destroy or modify the terrestrial biomes they occupy. For example, when forests are cleared for timber or to make room for agriculture, the rate of soil erosion increases dramatically because trees are no longer there to hold the soil in place. Increased erosion can cause streams and rivers to become clogged with silt, which harms or kills invertebrates, fishes, and many other species.

Next we take a closer look at six ecosystems that are commonly found within the freshwater or marine biome: *lake, river, wetland, estuary, coastal region,* and *oceanic region*.

## Lakes, rivers, and wetlands are part of the freshwater biome

**Lakes** are standing bodies of water that are surrounded by land and, according to some authorities, at least 2 hectares (5 acres) in size (**FIGURE 21.16**). The productivity of a lake, and the abundance and distribution of its life-forms, is strongly influenced by nutrient concentrations, water depth, and the extent to which the lake water is mixed. Northern lakes tend to be clear because they usually have low nutrient concentrations and therefore do not support vigorous growth of photosynthetic plankton (floating microscopic organisms). Lakes with higher nutrient concentrations appear more turbid because plankton thrive there. In temperate regions, seasonal changes in temperature cause the oxygen-rich water near the top of a lake to sink in the fall and the spring, bringing oxygen to the bottom of the lake. This seasonal turnover also delivers nutrients from the bottom sediments to the surface layers where they enhance the growth of photosynthetic organisms. In tropical regions, seasonal differences in temperature are not great enough to cause a similar mixing of water. This lack of

**FIGURE 21.16  Lakes, Rivers, and Wetlands**

Lakes are land-locked bodies of standing water. They vary in size from one-fiftieth of a square kilometer (5 acres) to thousands of square kilometers. In contrast, rivers are bodies of fresh water that move continuously in a single direction, and wetlands are characterized by shallow waters that flow slowly over lands that border rivers, lakes, or ocean waters.

mixing causes the deep waters of tropical lakes to have low oxygen levels and relatively few forms of life.

**Rivers** are bodies of fresh water that move continuously in a single direction. The physical characteristics of a river tend to change along its length. At its source—whether glacier, lake, or underground spring—the current is stronger and the water colder. The waters in these upper reaches are highly oxygenated because $O_2$, like most other gases, dissolves more readily in cold water and the turbulence created by rapids and riffles causes more of the gas to dissolve in water. As they approach their emptying point, rivers become wider, slower, warmer, and less oxygenated.

**Wetlands** are characterized by standing water shallow enough that rooted plants emerge above the water surface. A bog is a freshwater wetland with stagnant, acidic, oxygen-poor water. Because bogs are nutrient-poor, their productivity and species diversity are low. In contrast, marshes and swamps are highly productive wetlands. A marsh is a grassy wetland, but swamps are dominated by trees and shrubs.

## Estuaries and coastal regions are highly productive parts of the marine biome

The marine biome is the largest biome on our planet, since it includes the open oceans, which cover about three-fourths of Earth's surface. This biome is critically

**FIGURE 21.17 Estuaries**
Estuaries are tidal ecosystems where rivers flow into the ocean. They are usually classified as part of the marine biome. This photo shows a salt marsh at sunrise in Rachel Carson National Wildlife Refuge, Maine.

the rain that falls on terrestrial biomes comes from water evaporated off the surface of the world's oceans.

**Estuaries** are the shallowest of the marine ecosystems. An estuary is a region where a river empties into the sea. It is marked by the constant ebb and flow of fresh water and salt water, and all organisms that thrive on its bounty have to be able to tolerate daily and seasonal fluctuations in salinity (salt levels). The water depth fluctuates with ocean tides and river floods, but most of the time it is shallow enough that light can reach the bottom. The plentiful light, the abundant supply of nutrients delivered by the river system, and the regular stirring of nutrient-rich sediments by water flow create a rich and diverse community of photosynthesizers. Grasses and sedges are the dominant vegetation in most estuaries (**FIGURE 21.17**). The producers provide food and shelter for a varied and prolific community of invertebrates and vertebrates, including crustaceans, shellfish, and fishes. The abundance and diversity of life make estuaries one of the most productive ecosystems on our planet.

The **coastal region** stretches from the shoreline to the edge of the continental shelf, which is the undersea extension of a continent (**FIGURE 21.18**). The coastal region is among the most productive marine ecosystems because of the ready availability of nutrients and oxygen. Nutrients delivered by rivers and washed off the surrounding land accumulate in the coastal region.

important for us, for all other biomes, and for our planet in general. Simply because of the sheer size of this biome, photosynthetic plankton in the oceans release more oxygen (and absorb more $CO_2$) through photosynthesis than do all the terrestrial producers combined. Much of

**FIGURE 21.18 Ecological Zones in the Marine Biome**
This diagram depicts a cross-sectional view of the land and water, progressing from the shoreline toward the open ocean (oceanic region). The coastal region stretches out to sea as far as the continental shelf, which is the underwater extension of a continent's rim. Productivity, and the abundance of life-forms, declines with increasing water depth because sunlight, which producers need for photosynthesis, diminishes with depth. The sunlit zone, or photic zone, extends to a depth of 200 meters (656 feet). The waters are dimly lit up to 1,000 meters but lie in complete darkness at greater depths. Productivity also decreases with distance away from the shore because nutrient levels typically decline farther out to sea. The well-lit waters of the open ocean are less productive than the well-lit regions of coastal areas, and the deepest layers of the ocean (the abyssal zone) are typically the least productive of all aquatic habitats.

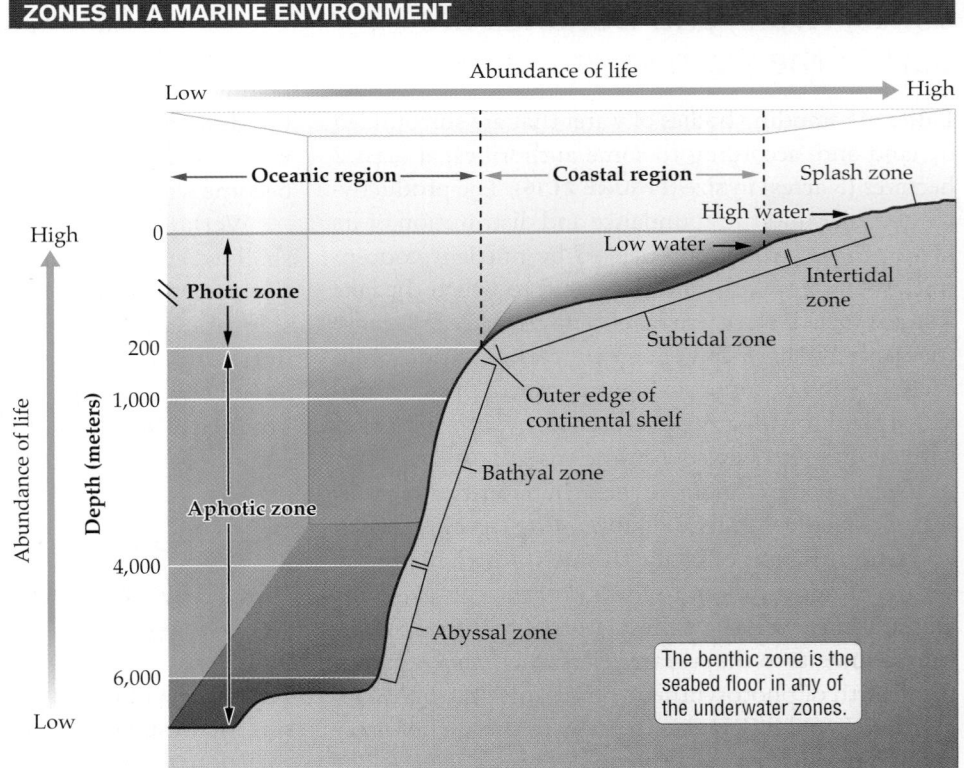

ZONES IN A MARINE ENVIRONMENT

Nutrients that settle to the bottom are stirred up by wave action, tidal movement, and the turbulence produced by storms. The nutrients support the growth of photosynthetic producers, which inhabit the well-lit upper layers of coastal waters to a depth of about 80 meters (260 feet). The vigorous mixing by wind and waves also adds atmospheric oxygen to the water. A majority of Earth's marine species live in the coastal region. Not surprisingly, most of the world's highly productive fisheries are also located along coasts.

The **intertidal zone** is the part of the coast that is closest to the shore, where the ocean meets the land. This ecological zone spans the uppermost reaches of ocean waves during the highest tide and the shoreline that remains submerged at the lowest of low tides. In other words, the intertidal zone extends from the highest tide mark to the lowest tide mark.

The intertidal zone is a challenging environment for plants and animals because they must cope with being submerged and exposed to dry air on a twice-daily basis, in addition to being pounded by surf and sand (**FIGURE 21.19a**). The organisms that inhabit the upper regions of the intertidal zone are subject to predation by shorebirds and other animals when the tide is out. Despite these challenges, a diverse community of seaweeds, worms, crabs, sea stars, sea anemones, mussels, and other species is adapted to living in this zone.

The benthic zone of coastal regions may lie as deep as 200 meters (656 feet) below the water surface. The coastal benthic zone is a relatively stable habitat, rich in sediments containing the dead and decaying remains of organisms (detritus). The detritus forms the basis of a food web that supports a wide variety of consumers, including sponges, worms, sea stars, sea fans, sea cucumbers, and many fishes (**FIGURE 21.19b**).

## Productivity in the oceanic region is limited by nutrient availability

The open ocean, or **oceanic region**, begins about 40 miles offshore, where the continental shelf, and therefore the coastal region, ends. The open oceans form a vast, complex, interconnected ecosystem that we know relatively little about. Although they are well lit and have sufficient oxygen, the surface layers of the open ocean are much less productive than estuarine and coastal waters, because they are relatively nutrient-poor. Nutrients are lacking because detritus tends to settle on the seafloor and the nutrients locked in it are not readily stirred up from the great depths of the open ocean.

(a)

(b)

Giant tube worms (*Riftia pachyptila*) on a hydrothermal vent in the abyssal zone of the deep oceans.

**FIGURE 21.19 The Marine Biome**

(a) Intertidal zones are found in coastal regions where the tides rise and fall on a daily basis, periodically submerging a portion of the shore. (b) Benthic zones, located on the bottom surfaces of lakes, rivers, wetlands, estuaries, coastal regions, and oceanic regions, are home to a variety of consumers, many of which feed on dead organisms drifting down from the upper, better-lit zones.

Where the continental shelf ends, the seafloor plunges steeply to a depth of approximately 6,000 meters (almost 20,000 feet). The cold, dark waters at these great depths constitute the **abyssal zone**. Few organisms can survive at the great pressures and low temperatures (about 3°C, or 37°F) of the abyssal zone. However, complex communities of archaeans and invertebrates are known to be associated with hydrothermal vents in geologically active regions of the deep oceans. A submarine hydrothermal vent is a crack in the seafloor that releases hot water containing dissolved minerals. Certain archaeans can extract energy from these minerals (see Figure 2.11), and these single-celled prokaryotes form the basis of a food web that supports invertebrates such as giant tube worms (see Figure 21.19b), clams, and shrimp.

> ### Concept Check
>
> 1. Which factors help shape the climate in a region? How does climate affect life-forms?
>
> 2. What are some of the physical conditions that influence life in aquatic biomes?

# How Invasive Mussels Can Harm Whole Ecosystems

International cargo ships that discharge ballast water in American ports introduced the Eurasian zebra mussels to the waterways of North America. Millions of the little one-inch-long mussels colonized lakes and rivers beginning in the Great Lakes and riding all the way to California on recreational fishing boats. Zebra mussels compete with native species and block or damage the cooling pipes of factories and power plants, causing both ecological and economic damage.

Mussels are filter feeders that live by filtering algae and other organic matter out of the water and consuming it. Zebra mussels have made Great Lakes waters so clear that there's little for other aquatic species to eat. They also attach themselves by the thousands to the backs of native mussels, preventing the natives from filtering feeding. In Lake St. Clair and Lake Erie, native mussels have all but vanished.

Any introduced species that becomes a major pest in a new environment is called an *invasive species*. Invasive species typically cause no harm in their native environment—the environment in which they originally evolved. Zebra mussels, for example, are native to the Caspian and Black seas and the freshwater rivers of southeastern Russia. There, a wide range of predators and parasites are adapted to feed on zebra mussels and keep their numbers in check. In Europe and North America, however, zebra mussels have few predators, and populations have exploded.

Ecologists have cataloged hundreds of examples of destructive invasive species all over the world. But not all introduced species are as destructive as zebra mussels. For example, California is home to 4,200 native plants and 1,800 introduced plants. Of the 1,800 introduced plant species, only 200 are considered invasive pests. Researchers concluded that the other 1,600 species (almost 90 percent) were not invasive, basing their conclusions on 13 criteria—including, for example, how easily the plants spread and what impact they have on native plant communities. The problem is that the 200 invasive ones are nearly impossible to control.

The effects of humans on the biosphere are not limited to introductions of invasive species. When we dam rivers, break up the landscape with roads and subdivisions, or eliminate top predators such as the gray wolf, we create situations we rarely anticipate until the problem is upon us. Dams can prevent fish from swimming upstream to spawn and stop nutrient-rich silt from passing into estuaries. Roads and housing developments prevent animals from migrating freely to forage for food, breed, or carry valuable genes to other populations. In the oceans off the North American coast, eliminating sea otters led to the overgrowth of sea urchins, which then destroyed kelp forest ecosystems.

The biosphere is an interconnected web of relations among different species and ecosystems. When we alter one component of the biosphere, we inevitably alter other parts. When the United States and China burn coal, for example, we raise carbon dioxide levels in the atmosphere throughout the world, contributing to global warming and the acidification of oceans all over the planet. Currents of air and water can carry pollutants around the world, causing damage to distant ecosystems. Agricultural runoff enters rivers, which carry pollutants into rich coastal ecosystems.

Humans, with our large numbers and heavy exploitation of natural resources, inevitably damage the natural world. Still, we are beginning to understand the biosphere. How can we use what we know about how the biosphere works to solve current environmental problems and prevent future ones? We have no easy answers to such questions, but in the next four chapters we'll examine how a sound understanding of ecological principles, combined with public awareness, citizen pressure and political will can help correct environmental mistakes of the past and prevent such blunders in the future.

# Our Ocean Backyard: Lost Cargo Tracks Ocean Currents

BY GARY GRIGGS, *Santa Cruz Sentinel*

The currents along our shoreline move things around in many different directions, sometimes north, sometimes south and sometimes onshore or offshore, depending on the season and the winds.

Whether nutrients or plankton, pollutants or lost people, oil spills or flotsam, the ocean currents can transport stuff thousands of miles. Surface currents are driven by the wind, and overall, the current patterns in the Pacific, the Atlantic and the Indian oceans are pretty similar and reasonably well understood. In the months and years ahead, we may well begin to see some of the debris washed offshore by the Japanese tsunami in March along our shoreline, just as beachcombers periodically find Japanese fishing floats along the beaches of California and Oregon.

On Jan. 29, 1992, a large container ship left Hong Kong, destined for Tacoma, Wash. Several thousand miles into the voyage, near the International Date Line, the ship hit severe storm conditions, not too uncommon in the Pacific in winter, and lost 12 shipping containers.

One of these was stuffed full of 29,000 Chinese-made plastic bath toys, yellow ducks, blue turtles, green frogs and red beavers. When the container broke open, the floating toys were carried off by the currents, and in the following years, traveled in different directions and reached some surprising places.

Their first landfall came 10 months later when more than 100 of the toys washed onto the beaches of southeast Alaska. In the fall of 1995, nearly four years later, after apparently circling counterclockwise around the entire north

Pacific and passing over the original drop site, several thousand of the floaters passed through the Bering Straits between Alaska and the Soviet Union.

This group continued, passing eastward into the Arctic Ocean, where they were trapped in the ice.

Simultaneously, another large group of the plastic bath toys floated south, ending up on the coastline of South America and also on beaches of Indonesia and Australia.

Moving with the ice, some of the Arctic travelers began reaching the Atlantic Ocean in 2000, where they began to thaw out and move southward. Soon, plastic ducks were seen bobbing in the waves from Maine to Massachusetts. In 2001, nearly 10 years after they were released, some of them were tracked in the area where the Titanic sank.

The story of the plastic bath toys traveling all over the world shows that things we drop in the ocean don't necessarily just go away, especially if they're made of plastic. At any moment, the world's shipping industry is moving 5–6 million shipping containers, and on average, one of them falls off a ship every hour. Tons of human-manufactured waste and trash are also deliberately dumped from ships and boats and washed into oceans from freshwater rivers.

If we are in any doubt about the persistence of human plastics, we have only to look at the Great Pacific Garbage Patch, a Texas-sized region of the Pacific Ocean where ocean currents have been carrying trash for years. Most of the plastic has broken down into countless tiny particles; in some areas, the density of plastic is greater than the density of plankton.

The bath toy story demonstrates the power of ocean currents to move material from one side of the globe to the other—whether warm or cold water, nutrients, or natural detritus such as logs and mats of algae. Sooner or later, ocean currents carry everything somewhere else. The plastic toys toured Alaska, the Arctic Ocean, South America, Australia, Indonesia, and Europe. It's fair to assume that everything from fish and bacteria to pollutants have a similar chance of traveling all over the world.

## Evaluating the News

**1.** Bacteria and other organisms break down organic matter into individual molecules that go to building new plants and animals. In contrast, most plastics undergo photodegradation, a process in which ultraviolet light from the sun breaks the plastic into tiny pieces, which may be consumed by aquatic animals. What kind of aquatic animals would consume plastic particles? What animals farther up the food chain could be affected?

**2.** The author of this article suggests that debris from the giant Japanese tsunami might eventually wash ashore in some of the same places that the bath toys appeared. Using the timeline described for the bath toys, estimate when debris from the March 2011 tsunami could show up in different regions.

**3.** The Great Pacific Garbage Patch was not actually observed until after scientists predicted its existence in a 1988 paper. Despite its enormous size and density, it is invisible in satellite photos because it consists of millions of tiny particles of plastic suspended in miles of seawater. Discuss how researchers could determine the size and depth of the Great Pacific Garbage Patch.

**SOURCE:** *Santa Cruz Sentinel,* May 7, 2011, http://www.santacruzsentinel.com.

# Summary

## 21.1 Ecology: Understanding the Interconnected Web

- Ecology is the scientific study of interactions between organisms and their environment. It helps us understand the natural world around us and our relationship with it.
- Organisms affect—and are affected by—their environment, which includes not only their physical surroundings but also other organisms. As a result, a change that affects one organism can affect other organisms and the physical environment.
- It is helpful to think of organisms and their environment as forming a web of interconnected relationships.
- Because components of the biosphere depend on one another in complex ways, human actions that affect the biosphere can have surprising side effects.

## 21.2 Climate's Large Effect on the Biosphere

- Weather describes the physical conditions of Earth's lower atmosphere in a specific place over a short period of time. Climate describes a region's long-term weather conditions.
- Climate depends on incoming solar radiation. Tropical regions are much warmer than polar regions because sunlight strikes Earth directly at the equator, but at a slanted angle near the poles.
- Climate is strongly influenced by four giant convection cells that generate relatively consistent wind patterns over much of the Earth.
- Human actions can affect climate, which impacts almost all aspects of the biosphere.

- Ocean currents carry an enormous amount of water and can have a large effect on regional climates. Regional climates are also affected by major features of Earth's surface, as when mountains create rain shadows.

## 21.3 Terrestrial Biomes

- Climate is the most important factor controlling the potential (natural) location of terrestrial biomes. Climate can exclude a species from a region directly (if it finds the climate intolerable) or indirectly (if other species in the region outcompete it for resources).
- Human activities heavily influence the actual location and extent of terrestrial biomes.
- Some of the major terrestrial biomes are tundra, boreal forest, temperate deciduous forest, grassland, chaparral, desert, and tropical forest.

## 21.4 Aquatic Biomes

- Aquatic biomes are usually characterized by physical conditions of the environment, such as temperature, salt content, and water movement.
- Aquatic biomes are strongly influenced by the surrounding terrestrial biomes, by climate, and by human actions.
- Two major aquatic biomes are recognized on the basis of salt content: the freshwater and marine biomes. The freshwater biome includes lakes, rivers, and wetlands; estuaries, the coastal region, and the open ocean are ecosystems that lie within the marine biome.
- Human activities, such as dams on rivers and the application of fertilizer and pesticides on lawns and farm fields, can affect aquatic biomes.

# Key Terms

abyssal zone (p. 483)
benthic zone (p. 480)
biome (p. 475)
biosphere (p. 470)
boreal forest (p. 476)
chaparral (p. 478)

climate (p. 471)
coastal region (p. 482)
convection cell (p. 473)
desert (p. 478)
ecology (p. 470)
estuary (p. 482)

grassland (p. 477)
intertidal zone (p. 483)
lake (p. 481)
oceanic region (p. 483)
permafrost (p. 476)
rain shadow (p. 474)

river (p. 481)
temperate deciduous forest (p. 477)
tropical forest (p. 479)
tundra (p. 476)
weather (p. 471)
wetland (p. 481)

# Self-Quiz

1. Which of the following is *not* a level of the biological hierarchy commonly studied by ecologists?
   a. ecosystem
   b. community
   c. organelle
   d. population

2. Earth has four stable regions ("cells") in which warm, moist air rises, and cool, dry air sinks back to the surface. Such cells are known as
   a. temperate cells.
   b. latitudinal cells.
   c. rain shadow cells.
   d. giant convection cells.

3. The biosphere consists of
   a. all organisms on Earth only.
   b. only the environments in which organisms live.
   c. all organisms on Earth and the environments in which they live.
   d. none of the above

4. What aspect(s) of climate most strongly influence the locations of terrestrial biomes?
   a. rain shadows
   b. temperature and precipitation
   c. only temperature
   d. only precipitation

5. Winds that blow from the west across warm waters in the Pacific Ocean become warm and moist. By analogy to what happens in a rain shadow, what do you think would happen if such warm, moist winds blew across the cold Peru Current (see Figure 21.15)?
   a. An El Niño event would occur.
   b. The winds would continue to pick up moisture from the ocean currents.
   c. The warm, moist winds would cool, causing rain to fall.
   d. The warm, moist winds would cool, but rain would not fall.

6. Which of the following represents a large area of the globe with unique climatic and ecological features?
   a. a population
   b. a community
   c. the biosphere
   d. a biome

7. The ozone layer in the upper atmosphere
   a. is depleted when it interacts with chlorine atoms.
   b. is created by the accumulation of human-made pollutants.
   c. has been increasing in thickness since the 1960s.
   d. is currently thickest over the Antarctic and thinnest over the equator.

8. Wetlands, ponds, and streams are common in summertime in the Arctic tundra because
   a. the permafrost thaws completely, releasing large amounts of water.
   b. water from melted snow and ice is prevented from draining by the underlying permafrost.
   c. precipitation in the Arctic is high, exceeding 80 inches per year.
   d. tundra trees grow best in wet, boggy ground.

9. Temperate deciduous forests
   a. experience cool, rainy winters and hot, dry summers.
   b. are maintained by frequent fires.
   c. are not dominated by coniferous trees.
   d. were the predominant biome in Great Plains states such as Nebraska before they were settled by Europeans.

10. Estuaries are highly productive because
    a. they are dominated by trees and shrubs, which have a large biomass.
    b. they receive nutrients delivered by rivers and stirred up by tide action.
    c. they lie between the high-tide mark and low-tide mark in the coastal region.
    d. archaeans that extract energy from minerals are especially abundant in this type of ecosystem.

# Analysis and Application

1. This chapter suggests that the organisms and environments of the biosphere can be thought of as forming an "interconnected web." Why do ecologists think of the biosphere this way? Give an example illustrating the interconnections between organisms and their environment.

2. Explain in your own words how global patterns of air and water movement can cause local events to have far-reaching ecological consequences. Give an example that shows how local ecological interactions can be altered by distant events.

3. Name seven of the major terrestrial biomes. How many of those biomes are located within 100 kilometers (about 60 miles) of your home? Describe the chief climatic and ecological characteristics of one terrestrial biome, and explain how those characteristics affect which life-forms are found in that biome.

4. Using examples, explain the following statement: The defining characteristic of a desert is low moisture, not high temperature.

5. Explain why desert plants generally have smaller leaves than plants that are native to tropical rainforests. Describe some adaptations of desert animals.

6. How does spring and fall turnover contribute to the productivity of temperate lakes?

7. What accounts for the high productivity of estuaries and coastal regions? Explain why the open ocean has lower productivity than these two regions of the marine biome.

# 22 Growth of Populations

**EASTER ISLAND WAS ONCE COVERED BY A LUSH FOREST.** Long ago the Rapanui people cut down all the trees on Easter Island—to build houses, to fuel cooking fires, and to move giant stone statues, visible in this photograph. The unsustainable use of resources led to an ecosystem collapse, and the island could no longer support the thousands of Rapanui who once lived there.

# The Tragedy of Easter Island

Imagine living on a beautiful subtropical island in the South Pacific, 2,000 miles from the nearest continent and 1,300 miles from the next island. The temperature and weather are idyllic, with soft ocean breezes. Thirty species of seabirds nest on the island, and a forest of palms and flowering trees is home to parrots, herons, and other birds. Frigate birds wheel overhead, stealing fish from other seabirds. The ocean is rich with sea life, the soil is fertile, and human families raise vegetables and chickens.

It may sound romantic, but now imagine that the island is only 15 miles long and the population of humans is growing out of control. There's not enough food for everyone; the trees have nearly all been cut down; and without wood, islanders cannot even build boats to escape. This place was Easter Island.

Today, Easter Island is a small, barren grassland, just 166 square kilometers in area, about the size of St. Louis, Missouri. The island has lost its forest, lost its fertile soil, and even lost its generous rainfall. Scattered around the island are nearly a thousand ancient stone statues representing ancestors, some of them 30 feet tall and 75 tons. Some of the statues, called Moai, stand upright but unfinished, as if the sculptors dropped their tools midstroke. Other Moai are complete but lie toppled in the grass, often broken. Hundreds more lie scattered along the coast.

In this chapter we examine the characteristics of populations, including how and why populations increase or decrease in size—topics that are the main focus of population ecology. Toward the end of the chapter we return to the story of Easter Island and think about what happens when humans use more resources than the biosphere can support. Will the biosphere continue to support all of us?

Who carved these statues? What happened to the people who made them? What happened to the forest and the birds? And what lessons can we apply to the human population of Earth?

**MAIN MESSAGE**    No population can continue to increase in size for an indefinite period of time.

## KEY CONCEPTS

- A population is a group of interacting individuals of a single species located within a particular area.

- Populations increase in size when birth and immigration rates exceed death and emigration rates, and they decrease when the reverse is true.

- A population that increases by a constant proportion from one generation to the next exhibits exponential growth.

- Eventually, the growth of all populations is limited by environmental factors such as lack of space, food shortages, predators, disease, and habitat deterioration.

- Different populations may exhibit different patterns of growth over time, including exponential growth (J-shaped curves), logistic growth (S-shaped curves), population cycles, and irregular fluctuations.

- The world's human population is increasing exponentially. Rapid human population growth cannot continue indefinitely; either we will limit our own growth, or the environment will do it for us.

HOW MANY ORGANISMS LIVE in a particular environment, and why? While ecology studies the interactions between organisms and with their environment (as we learned in Chapter 21), questions about the number of organisms in a particular place are the concern of **population ecology**. The answers to such questions not only provide insight into the natural world, but also are essential for solving real-world problems, such as protecting rare species or controlling pest species. To set the stage for our study of the factors that influence how many individuals are in a population, we begin by defining what populations are. We then describe how populations grow over time, and we consider the limits to growth that all populations face, including the human population.

## 22.1 What Is a Population?

A **population** is a group of interacting individuals of a single species located within a particular area. The human population of Easter Island, for example, consists of all the people who live on the island.

Ecologists usually specify the number of individuals in a population by **population size** (the total number of individuals in the population) or by **population density** (the number of individuals per unit of area). To calculate population density, we divide the population size by the total area. To illustrate, let's return to the Easter Island example: in the year 1500, the population size was 7,000 people. If we divide that number by 166 square kilometers (the size of the island), we get a population density of 42 people per square kilometer (7,000/166 = 42). The density of people that lived on Easter Island in 1500 was therefore higher than the 33 people per square

kilometer (87 per square mile) who lived in the United States in 2010.

Easter Island is an easy example for determining what constitutes a population: islands have well-defined boundaries, and human individuals are relatively easy to count. But often it is more difficult to determine the size or density of a population. Suppose a farmer wants to know whether the aphid population damaging a crop is increasing or decreasing (**FIGURE 22.1**). Aphids are small and hard to count. More important, it is not obvious how the aphid population should be defined. What do we mean by "a particular area" in this case? Aphids can produce winged forms that can fly considerable distances, so how do we know which aphids to count? Should we count only the aphids in the farmer's field? What about the aphids in the next field over?

In general, what constitutes a population is often not as clear-cut as in the Easter Island example. Overall, the area appropriate for defining a particular population depends on the questions being asked and on aspects of the biology of the organism of interest, such as how far and how rapidly it moves.

Aphids insert their mouthparts into a plant and withdraw nutrients, thus damaging the plant.

Aphids have infested this rose in large numbers.

**FIGURE 22.1 Populations of Aphids Can Cause Extensive Crop Damage**

Aphids are small insects with sucking mouthparts. They are pests on many plant species, which they infest in such large numbers that they can be difficult to count.

## 22.2 Changes in Population Size

Populations tend to change in size over time—sometimes increasing, sometimes decreasing. For example, in one year abundant rainfall and plant growth may cause mouse populations to increase; in the next, drought and food shortages may cause mouse populations to decrease dramatically. Such changes in the population sizes of organisms can have important consequences for people. For example, an increase in the number of deer mice, carriers of hantavirus, is thought to have been responsible for the 1993 outbreak of this deadly disease in the southwestern United States.

Whether a population increases or decreases in size depends on the number of births and deaths in the population, as well as on the number of individuals that immigrate to (enter) or emigrate from (leave) the population. A population increases in size whenever the number of individuals entering (by birth and immigration) is greater than the number of individuals leaving (by death and emigration). We can express this relationship in equation form:

$$\text{Birth} + \text{immigration} > \text{death} + \text{emigration}$$

The environment plays a key role in the increase or decrease of a population because birth, death, im-

migration, and emigration rates are affected by environmental factors. Consider how features of the environment caused monarch butterfly populations to fluctuate wildly in 2002. Each spring, monarchs make a spectacular migration that begins in the mountains west of Mexico City (where the butterflies overwinter) and ends in eastern North America (**FIGURE 22.2**). On January 13, 2002, the monarchs' overwintering sites were hit with an unusual storm that first drenched the butterflies with rain and then subjected them to freezing cold. The combination of wet and cold proved lethal: an estimated 70–80 percent of the butterflies—roughly 500 million of them—died overnight, the worst die-off in 25 years.

Fortunately for the butterflies, this huge increase in winter death rates was followed by an equally spectacular rise in birth rates during the summer of 2002. Monarch birth rates shot up that year because it turned out to be a great summer for the monarch's primary food plant, milkweed. Because of this chance good fortune, butterflies that had survived the winter produced so many young in the summer of 2002 that monarch numbers quickly rebounded to almost the historic average.

### Concept Check

1. What is population density? Explain why it can be difficult to measure.

2. What aspects of a population determine whether it increases or decreases in size?

### Helpful to Know

Remembering the meanings of two common prefixes can help you distinguish three similar-sounding terms. "Migration" comes from the Latin meaning "to move." The prefix *in* or *im* can mean "into," while *ex* or *e* means "out of, away from." So "immigration" is movement into an area, and "emigration" is movement out of an area.

### Concept Check Answers

1. Population density is the total number of interacting individuals of a single species, divided by the particular area they inhabit. It may be difficult to measure because individuals may be hard to detect, may move between populations, and may inhabit a complex, hard-to-define area.

2. Birth and immigration increase population size; death and emigration reduce it. Environmental factors influence these characteristics, so they have a strong impact on population size.

Monarch Butterfly
Fall Migration Patterns

**FIGURE 22.2**
**Before the Crash**
Huge numbers of monarch butterflies overwinter in mountains west of Mexico City and then migrate each spring to eastern North America. In 2002, 70–80 percent of the overwintering butterflies died in an unusual winter storm.

## 22.3 Exponential Growth

Like monarch butterflies, many organisms produce vast numbers of young. If even a small fraction of those young survive to reproduce, a population can grow extremely rapidly.

An important type of rapid population growth is **exponential growth**, which occurs when a population increases by a constant proportion ($\lambda$) over a constant time interval, such as 1 year (**FIGURE 22.3**). We can represent exponential growth from one year to the next by the equation

$$N_{\text{Next Year}} = \lambda \times N_{\text{This Year}}$$

where $N$ is the number of individuals in the population and $\lambda$ (lambda) is the proportional increase in population size, a constant multiplier that determines the population size from one year to the next. For example, if $\lambda = 1.5$ and the current population size is 40, then the population size in the following year will be 60 (40 × 1.5), and in successive years it will be 90 (60 × 1.5), and then 135 (90 × 1.5), and so on.

In exponential growth, the proportional increase ($\lambda$) is constant, but the numerical increase—the number of individuals added to the population—becomes larger with each generation. For example, the population in Figure 22.3 doubles every generation (that is, $\lambda = 2$). With respect to its *numerical* increase, however, the population increases vary: the population increases by only 1 individual between generations 1 and 2, but by 16 individuals between generations 5 and 6. When plotted on a graph, exponential growth forms a **J-shaped curve**, as seen in Figure 22.3.

The time it takes a population to double in size—the **doubling time**—can be used as a measure of how fast the population is growing. We like it when our bank accounts double rapidly, but when populations grow exponentially in nature, problems eventually result, as we shall see in the following discussion.

Endless exponential growth of a population is not seen under natural conditions, because sooner or later populations run out of vital resources, such as food or living space, that are needed to sustain them. However, a population can increase exponentially over a shorter window of time. Exponential growth is sometimes seen when individuals migrate to, or are introduced into, a new area. Consider the following tale of woe: In 1839, a rancher in Australia imported from South America a species of *Opuntia* (prickly pear cactus) and used it as a "living fence" (a thick wall of this cactus is nearly impossible for human or beast to cross). Unlike a real fence, however, the *Opuntia* cactus did not stay in one place; it spread rapidly throughout the landscape. As the cactus spread, whole fields were turned into "fence," crowding out cattle and destroying good rangeland.

In about 90 years, *Opuntia* cacti spread across eastern Australia, covering more than 243,000 square kilometers (over 60 million acres) and causing great economic damage. All attempts at control failed until 1925, when scientists introduced a moth species, appropriately named *Cactoblastis cactorum*, whose caterpillars feed on the growing tips of the cactus. This moth killed billions of cacti, successfully bringing the cactus population under control (**FIGURE 22.4**). Introducing nonnative species to control a problematic invader (biological control) is fraught with risk because the control agent (the cactus moth, in

**FIGURE 22.3 The Size of a Population That Is Growing Exponentially Increases at a Constant Rate**

In this hypothetical population, each individual produces two offspring, so the population increases by a constant rate: it doubles with each generation. The number of individuals added to the population increases each generation, resulting in the J-shaped curve that is characteristic of exponential growth. Exponential-growth curves are always J shaped, but the curve's steepness varies depending on the rate of increase.

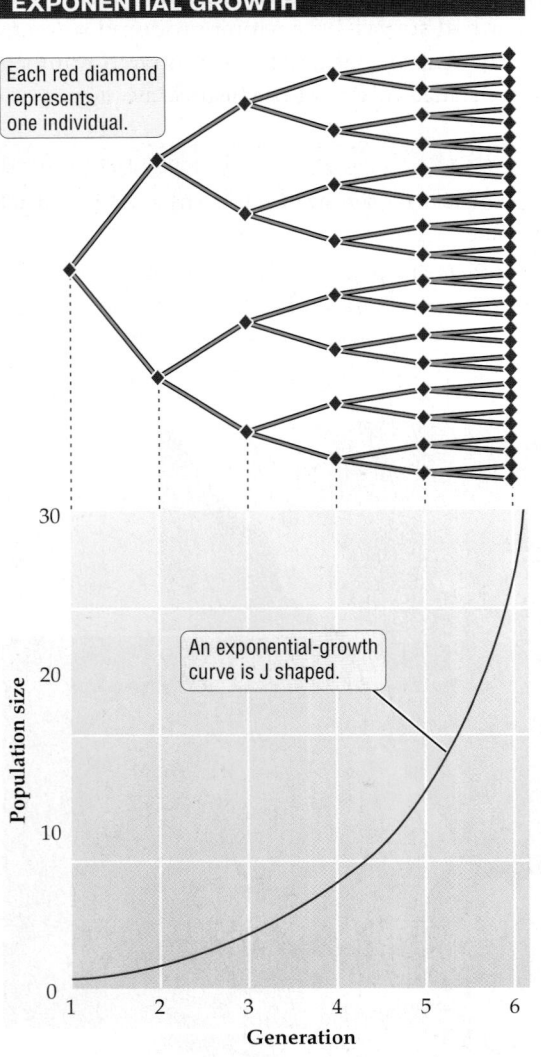

**EXPONENTIAL GROWTH**

Each red diamond represents one individual.

An exponential-growth curve is J shaped.

Population size

Generation

(a)

(b)

this case) could multiply exponentially and become a problem itself. Fortunately, the cactus moth is a specialist feeder, preferring cacti, which are native to the Americas, over any other plant the Australian outback had to offer. The moth's success in eradicating the cactus led to its own demise, as, after the prickly pear cactus population had been decimated, moth numbers plummeted because of lack of food. Today, both the cactus and the moth exist in eastern Australia in low numbers.

Overall, the *Opuntia* population in Australia increased exponentially at first; then it declined even more rapidly after introduction of the moth. Exponential growth has also been observed in other species introduced by people to new areas, as well as in species that have expanded naturally to new areas.

## Growth is limited by essential resources and other environmental factors

The most obvious reason that populations cannot continue to increase indefinitely is simple: food and other resources become diminished. Imagine that a few bacteria are placed in a closed jar containing a source of food. The bacteria absorb the food and then divide, and their offspring do the same. The population of bacteria grows exponentially, and in short order the jar contains billions of bacteria. Eventually, however, the food runs out and metabolic wastes build up. All the bacteria die.

This example may seem extreme because it involves a closed system: no new food is added, and the bacte-

## 22.4 Logistic Growth and the Limits on Population Size

A giant puffball mushroom can produce up to 7 trillion offspring (**FIGURE 22.5**). If all of these offspring survived and reproduced at this same (maximal) rate, the descendants of a giant puffball would weigh more than Earth in just two generations. Humans and *Opuntia* cacti have much longer doubling times than giant puffballs, but given enough time, they, too, can produce an astonishing number of descendants. Obviously, however, Earth is not covered with giant puffballs, *Opuntia* cacti, or even humans. These examples illustrate an important general point: no population can increase in size indefinitely. Limits exist.

**FIGURE 22.5 Will They Overrun Earth?**
Given the number of spores it produces, a giant puffball mushroom has the potential to produce 7 trillion offspring in a single generation. However, relatively few of those spores land in a habitat suited for their growth. Large giant puffballs weigh 40–50 kilograms each; a medium-sized example is shown here.

ria and the metabolic wastes cannot go anywhere. In many respects, however, the real world is similar to a closed system. Space and nutrients, for example, exist in limited amounts. In the *Opuntia* example of Section 22.3, even if humans had not introduced the *Cactoblastis* moth, the cactus population could not have sustained exponential growth indefinitely. Eventually, growth of the cactus population would have been limited by an environmental factor, such as a lack of suitable **habitat** (the type of environment in which an organism lives).

## Logistic population growth is the norm in the real world

The exponential-growth model assumes that resources are unlimited, which is unlikely over the long run. The logistic-growth model takes into consideration changes in growth rates that occur as resources become limited. **Logistic growth** is represented by an **S-shaped curve**, in which a population grows nearly exponentially at first but then stabilizes at the maximum population size that can be supported indefinitely by the environment. This maximum population size that can be sustained in a given environment is known as the **carrying capacity** (**FIGURE 22.6**). The growth rate of the population decreases as the population size nears the carrying capacity because resources such as food and water begin to be in short supply. At the carrying capacity, the population growth rate is zero.

In the 1930s, the Russian ecologist G. F. Gause carried out experiments on *Paramecium caudatum*, a common protist. He found that laboratory populations of paramecia increased to a certain size and then remained there (see Figure 22.6). In these experiments, Gause added new nutrients to the protists' liquid medium at a steady rate and removed the old solution at a steady rate. At first, the population increased rapidly in size. But as the population continued to increase, the paramecia used nutrients so rapidly that food began to be in short supply, slowing the growth of the population. Eventually the birth and death rates of the protists equaled each other and the population size stabilized. In contrast to natural systems, there was no immigration or emigration in Gause's experiments. In natural systems, populations reach and remain at a constant population size when birth plus immigration equals death plus emigration for extended periods of time.

Like laboratory populations of bacteria and paramecia, natural populations also experience limits (**FIGURE 22.7**). Their growth can be held in check by a number of environmental factors, including food shortages, lack of space, disease, predators, habitat de-

**LOGISTIC GROWTH**

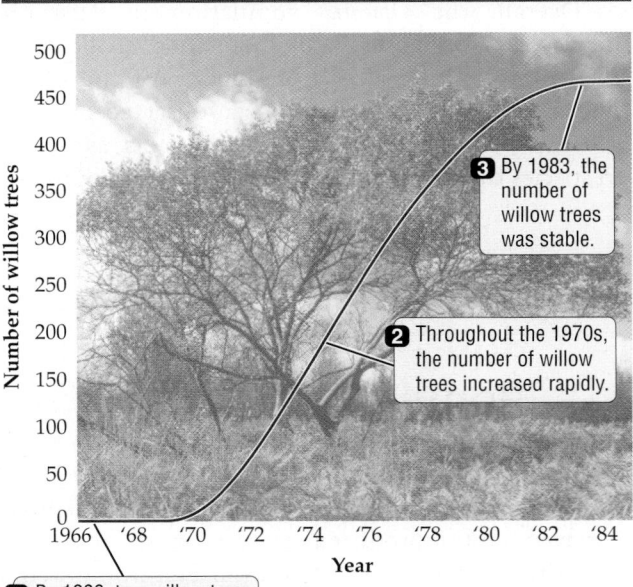

**❸** By 1983, the number of willow trees was stable.

**❷** Throughout the 1970s, the number of willow trees increased rapidly.

**❶** By 1966, two willow trees were growing in the area.

**FIGURE 22.7  A Logistic Curve in a Natural Population**
At a site in Australia, rabbits heavily grazed young willow trees, preventing willows from growing in the area. The rabbits were removed in 1954. By 1966, two willows had taken root in the area, presumably from seed blown in or carried in by animals. They increased rapidly in number, and the population then leveled off at about 475 trees.

**CARRYING CAPACITY**

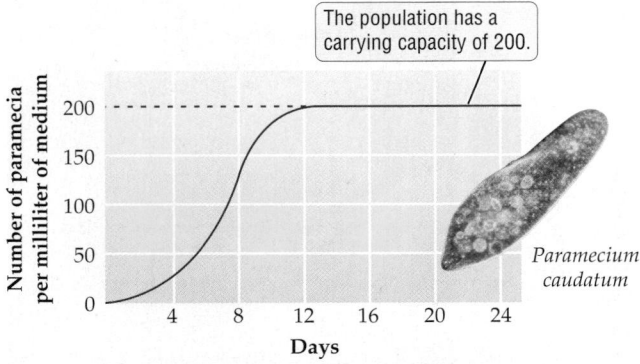

The population has a carrying capacity of 200.

*Paramecium caudatum*

**FIGURE 22.6  Carrying Capacity Is the Maximum Population Size a Particular Environment Can Sustain**
A laboratory population of the single-celled protist *Paramecium caudatum* increases rapidly at first and then stabilizes at the maximum population size that can be supported indefinitely by its environment—that is, at its carrying capacity. This growth pattern can be graphed as an S-shaped curve.

terioration, weather, and natural disturbances. When a population has many individuals, birth rates may drop or death rates may increase; either effect may limit the growth of the population, and sometimes both effects occur. Let's take a brief look at how this works.

Any area contains a limited amount of food and other essential resources. Therefore, as the number of individuals in a population increases, fewer resources are available to each individual. As resources diminish, each individual, on average, produces fewer offspring than when resources are plentiful, causing the birth rate of the population to decrease.

In addition, when a population has many individuals, disease spreads more rapidly (because individuals tend to encounter one another more often), and predators may pose a greater risk (because many predators prefer to hunt abundant sources of food). Disease and predators obviously increase the death rate.

Large populations can also damage or deplete their resources. If a population exceeds the carrying capacity of its environment, it may damage that environment so badly that the carrying capacity is lowered for a long time. A drop in the carrying capacity means that the habitat cannot support as many individuals as it once could. Such habitat deterioration may cause the population to decrease rapidly (**FIGURE 22.8**).

## Some growth-limiting factors depend on population density; others do not

Food shortages, lack of space, disease, predators, and habitat deterioration—all these factors influence a population more strongly as it grows and therefore increases in density. The birth rate, for example, may decrease or the death rate may increase when the population has many individuals. When birth and death rates change as the density of the population changes, such rates are said to be **density-dependent**. In natural populations, the number of offspring produced (**FIGURE 22.9**) and the death rate are often density-dependent.

In other cases, populations are held in check by factors that are not related to the density of the population; such factors cause the population to change in a **density-independent** manner. Density-independent factors can prevent populations from reaching high densities in the first place. Year-to-year variation in weather, for example, may cause con-

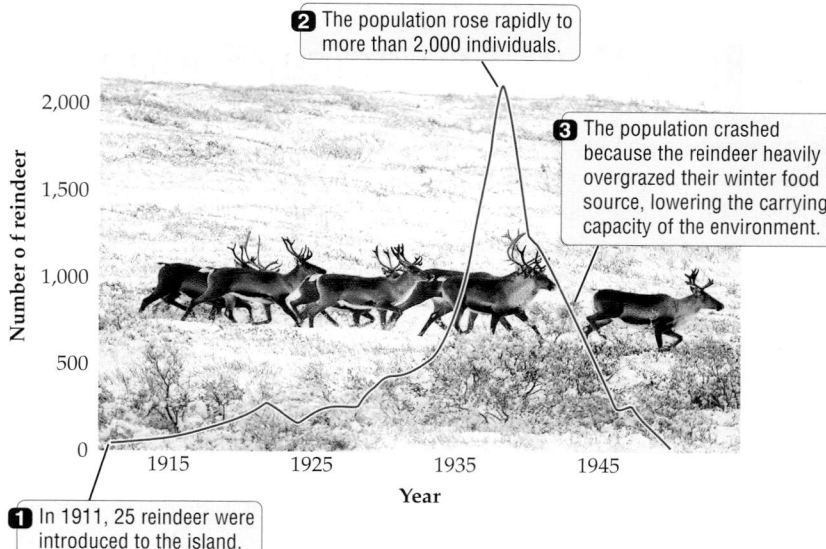

**HABITAT DETERIORATION LIMITS GROWTH**

**2** The population rose rapidly to more than 2,000 individuals.

**3** The population crashed because the reindeer heavily overgrazed their winter food source, lowering the carrying capacity of the environment.

**1** In 1911, 25 reindeer were introduced to the island.

**FIGURE 22.8 Boom and Bust**

When reindeer were introduced to Saint Paul Island, off the coast of Alaska, in 1911, their population increased rapidly at first and then crashed. By 1950, only eight reindeer remained.

ditions to be suitable for rapid population growth only occasionally. Poor weather conditions may reduce the growth of a population directly (by freezing the eggs of an insect, for example) or indirectly (by decreasing the number of plants available as food to that insect). Natural disturbances such as fires and floods also limit the growth of populations in a density-independent way. Finally, the effects of environmental pollutants such as DDT are density-independent; such pollutants can threaten natural populations with extinction.

**DENSITY-DEPENDENT GROWTH**

**FIGURE 22.9**
**The Effects of Overcrowding in Plantain Weed,** *Plantago major*

The number of seeds produced per plant drops dramatically under increasingly crowded conditions in plantain, a small herbaceous plant.

# How Big Is Your Ecological Footprint?

Our use of resources and our overall impact on the planet have increased even faster than our population size. For example, from 1860 to 1991 the human population increased 4-fold, but our energy consumption increased 93-fold. Many people think that our society, like that of Easter Island, is not based on the *sustainable* use of resources. The term "sustainable" describes an action or process that can continue indefinitely without using up resources or causing serious damage to the environment.

One measure of sustainability is the *Ecological Footprint*, which is the area of biologically productive land and water that an individual or a population requires to produce the resources it consumes and to absorb the waste it produces. Scientists compute the Ecological Footprint using standardized mathematical procedures and express it in global hectares (gha). According to recent estimates, the Ecological Footprint of the average person in the world is 2.7 gha, which is about 30 percent higher than the 2.1 hectares that would be needed for each of the world's 7 billion people if they were supported in a sustainable manner. Overall, such estimates suggest that, since the late 1970s, people have been using resources faster than they can be replenished—a pattern of resource use that, by definition, is not sustainable. As the world population grows, the amount of biologically productive land available per person continues to decline, increasing the speed at which Earth's resources are consumed.

The Ecological Footprint of individuals in some countries, such as the United States (8.0 gha per person) and the United Kingdom (5.45 gha per person), is three to five times what is sustainable. As the chart indicates, the per capita consumption of Earth's resources by different countries is most directly related to energy demand, affluence, and a technology-driven lifestyle. However, population size also has a large overall impact on sustainability. For example, in 2007 the *total* Ecological Footprint of the Chinese population (1.3 billion) was 2,959 gha, compared to 2,468 for the U.S. population of 308 million people. Like overconsumption, overpopulation has a severe negative impact on ecosystems, and pollution and habitat degradation is often most severe in densely populated countries that otherwise have a low per capita Ecological Footprint. As people in populous countries such as China and India become wealthier, their overall and per capita footprint is growing rapidly.

What is *your* Ecological Footprint? If you are a typical college student, your footprint is probably close to the U.S. average of 8.0 gha (about 20 acres). It would take about 4.5 planet Earths to support the human population if everyone on Earth enjoyed the same lifestyle that you do. Your Ecological Footprint depends on four main types of resource use:

1. Carbon footprint, or energy use
2. Food footprint, or the land and energy and water it takes to grow what you eat and drink
3. Built-up land footprint, which includes the building infrastructure (from schools to malls) that support your lifestyle
4. Goods and services footprint, which includes your use of everything from home appliances to paper products

If you drive a gas guzzler, live in a large suburban house, routinely eat higher up on the food chain (more beef than chicken), and do not recycle much, your footprint is likely to be higher than that of a person who uses public transport, shares an apartment, eats mostly plant-based foods, and sends relatively little to the local landfill. You can estimate your impact on the planet by using one of the many "footprint calculators" on the Internet, such as one offered by the Global Footprint Network. Most of us can significantly reduce our Ecological Footprint with little reduction in the quality of life, while bestowing an outsized benefit on our planet.

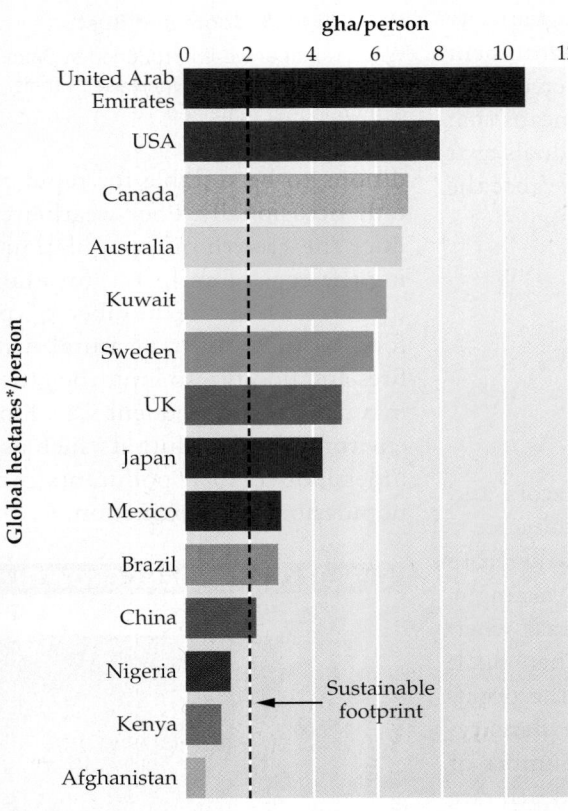

*One hectare is 2.47 acres, about the size of a football field.

Ecological Footprints across the World

# 22.5 Patterns of Population Growth

Different populations may exhibit a number of different growth patterns over time. Here we discuss four such patterns: exponential J-shaped curves, logistic S-shaped curves, population cycles, and irregular fluctuations.

Under favorable conditions, the population size of any species increases rapidly. An initial period of rapid exponential population growth can be seen in growth patterns that are J-shaped (see Figure 22.3) or S-shaped (see Figure 22.7). With the J-shaped growth pattern, rapid population growth may continue until resources are depleted, causing the population size to drop dramatically (see Figure 22.8). In contrast, with the logistic S-shaped growth pattern, the rate of population growth slows as the population size nears the carrying capacity. Predators, disease, and other factors may then keep the population near the carrying capacity for a long time.

As we have seen, populations change in size over time, increasing at some times and decreasing at others. Even populations with an S-shaped growth pattern do not remain indefinitely at a single, stable population size; instead, they fluctuate slightly over time, yet remain close to the carrying capacity.

In some cases, the population sizes of two species change together in a tightly linked cycle. Such **population cycles** can occur when at least one of the two species involved is very strongly influenced by the other. The Canadian lynx, for example, depends on the snowshoe hare for food, so lynx populations increase when hare populations increase, and they decrease when hare populations drop (**FIGURE 22.10**).

In relatively few examples from nature do populations of two species show regular cycles like those of the hare and the lynx. As illustrated dramatically by monarch butterflies, however, the populations of most species do rise and fall over time—just not as regularly as in Figure 22.10. **Irregular fluctuations** are far more common in nature than is the smooth rise to a stable population size shown in Figure 22.7.

Finally, different populations of the same species may experience different patterns of growth. Understanding the reasons for these differences can provide critical information on how best to manage endangered or economically important species. The first step

**POPULATION CYCLES**

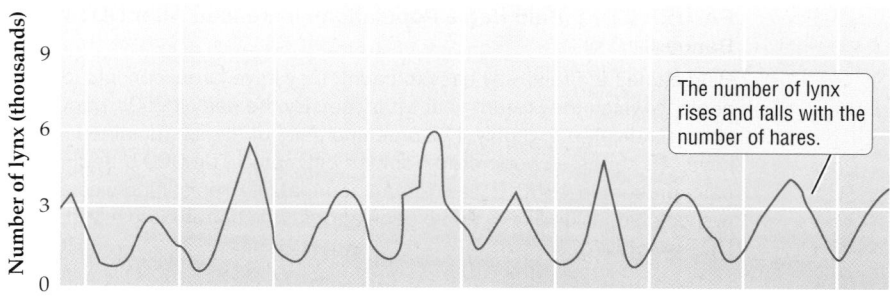

The number of lynx rises and falls with the number of hares.

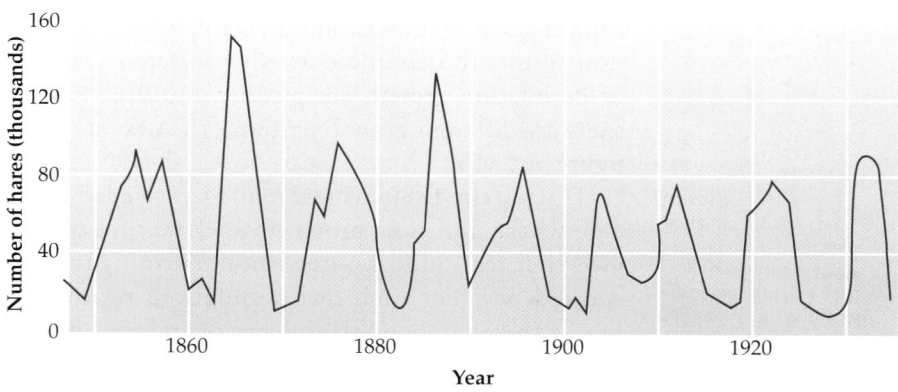

**FIGURE 22.10 Populations of Two Species Occasionally Increase and Decrease Together**

The Canadian lynx depends on the snowshoe hare for food, so the number of lynx is strongly influenced by the number of hares. Experiments conducted in the early twentieth century indicate that hare populations are limited by their food supply and by their lynx predators. (The researchers supplemented their observations with population estimates based on the numbers of hare and lynx pelts sold by trappers to the Hudson's Bay Company, Canada.)

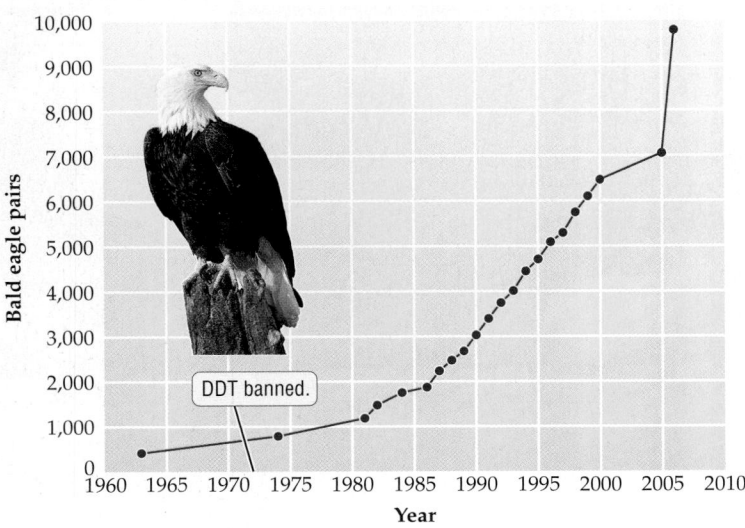

**FIGURE 22.11 Bald Eagle Populations Increased After DDT Was Banned**

Bald eagles are relatively easy to count: they have large, conspicuous nests to which they return year after year. By the early 1960s, population counts indicated that only 417 breeding pairs of eagles remained in the lower 48 states—a huge drop from the estimated 100,000 breeding pairs present in 1800. It was determined that DDT poisoning was directly responsible for declining eagle populations, which prompted a ban on DDT in 1972.

toward such an understanding is to perform population counts to determine whether different patterns of population growth are present (**FIGURE 22.11**). If there are different growth patterns, the next step is to figure out why.

During the 1980s, forest managers needed to decide where and how much (if any) mature or old-growth forest could be cut without harming the rare spotted owl. For each owl population, researchers first gathered data on the birth rate and the amount of habitat used by each individual. The researchers then used these data to predict how the growth of spotted owl populations would be affected by the number, size, and location of patches of the bird's preferred habitat: old-growth forest. (Patches are portions of a particular habitat that are surrounded by a different habitat or habitats.) The researchers discovered that the total area of these forest patches, and how they are arranged in the landscape, has a big impact on the growth rates of owl populations. (**FIGURE 22.12**).

**Concept Check Answers**

**1.** No. Exponential growth is almost always limited by space and/or resource availability in the habitat.

**2.** The population with the J-shaped growth curve increased in size exponentially. The population with the S-shaped growth pattern grew rapidly at first, and then its growth rate slowed as the population size approached the carrying capacity.

**3.** In a density-independent manner. Population size was limited not by the density of the population but by a natural disturbance that prevented the population from reaching the carrying capacity.

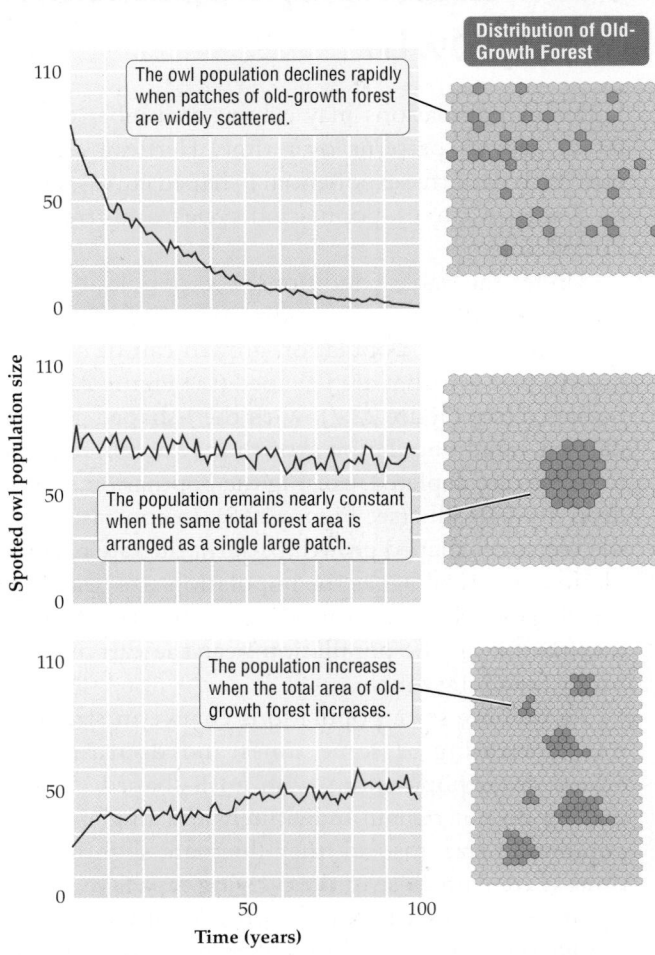

Distribution of Old-Growth Forest

The owl population declines rapidly when patches of old-growth forest are widely scattered.

The population remains nearly constant when the same total forest area is arranged as a single large patch.

The population increases when the total area of old-growth forest increases.

Spotted owl population size

Time (years)

**FIGURE 22.12 Same Species, Different Outcomes**

Different populations of the endangered spotted owl are predicted to show different patterns of growth over time, depending on the arrangement and area of their preferred habitat: old-growth forest. Patches of old-growth forest are shown in blue in the diagrams on the right.

## Concept Check

**1.** Can a population grow in an exponential manner indefinitely? Explain.

**2.** How is a population that shows an S-shaped growth curve different from one that exhibits a J-shaped growth curve?

**3.** A few seeds of plantain arrive in a vacant lot and start growing exponentially over the next three seasons. A late frost in the fourth season destroys most of the individuals in the population. Was the size of this population limited in a density-dependent manner or a density-independent manner? Explain.

# What Does the Future Hold?

At the beginning of this chapter we visited Easter Island, or Rapa Nui, which today is a nearly barren volcanic island. Before human colonization, Easter Island was a rich—but fragile—forest ecosystem. What happened and what role humans played in this dramatic change is still being debated. But a rich mine of archaeological evidence suggests that a combination of rapid population growth and habitat destruction led to the extinction of major groups of plants and animals and a human population crash. According to archaeological evidence, humans most likely first colonized Easter Island between AD 1000 and AD 1200, when a group of Polynesians arrived in canoes carrying domesticated chickens and crop plants such as yams, bananas, and taro root.

Almost as soon as the islanders arrived, they began cutting down the native palm forest. The trees were useful for building boats and houses and for megaprojects such as rolling 75-ton statues around the island. By about AD 1200, nearly half the trees were gone and the population of the island had reached about 2,000 people. By about 1300, the population is thought to have peaked at between 7,000 and 20,000. Archaeologists have calculated that such numbers must have existed to build and transport the hundreds of giant statues that still populate the island. By 1450, the palm trees were extinct. And by 1650, 20 other species of trees and shrubs had vanished as well.

Without wood, the islanders could not build wood houses; to cook, they burned grass or other plants. Before they cut down the forest, the islanders had relied on the palms and other trees to make baskets, sails, mats, roofing, rope, cloth, and wood carvings. The palm forests had also protected the soil and shaded crops, as well as providing nuts to eat and sap to drink. With the gradual removal of the forest, annual rainfall declined, limiting what they could grow. But despite the decline in rainfall, soil erosion increased. The island's soil, exposed by the loss of forest, washed away during winter storms at a rate of 3 meters per year starting along the shoreline and gradually working up the mountain slopes. Landslides buried houses and gardens.

The thousands of Rapanui people on the tiny island were in dire straits. Without wood to build new boats, they could not leave the island; thousands were trapped on an island with dwindling resources. They struggled to deal with the drastic changes by moving inland to areas that still had soil. By AD 1400, they were building mountainside terraces for growing taro root; they struggled to protect the remaining soil by paving it with stones—laying approximately a billion stones averaging 2 kilograms each.

By some estimates, the Rapanui cut 6 million trees in just 300 years, severely reducing the island's carrying capacity. As food production dropped, families probably starved; the death rate increased and the birth rate declined. By 1600, the population had dropped back to about 2,000 people. The island no longer had the resources to support the huge populations required to carve and move the gigantic statues. Indeed, according to oral tradition, the Rapanui stopped carving statues by 1680. During fierce intertribal wars over the next 200 years, the people toppled most of the hundreds of statues and abandoned their former homes to hide in caves.

As on Easter Island, many of the problems facing humans today relate to excess population growth and destruction of the environment. The human population is still growing at a spectacular rate (**FIGURE 22.13**). The global human population passed the 7 billion mark in 2011 and continues growing by about 200,000 people a day.

By the year 2025, the global human population is projected to surpass 8 billion people. Even if the global birth rate dropped tomorrow morning from the current 2.5 children per woman to a replacement level of 2.1 children per woman—that is, each person on Earth exactly replaced—the human population would continue to grow for decades because of *population momentum*: Because growing populations have more children than stable populations have, the "excess" children have to grow up and have their own children before the population stops growing.

As the global human population grows, we use more resources—including burning more fossil fuels, cutting forests faster, emptying the oceans of marine life more quickly—and generate more pollution and waste. Fresh water is becoming scarcer throughout the world, global fisheries are collapsing from too much fishing, and if current rates of logging continue, scientists estimate that, just as on Easter Island, all of Earth's tropical forests will be gone in less than 150 years.

All the evidence suggests that the current human impact on Earth is *un*sustainable. Since the 1970s, the human population has been using resources faster than they can be replenished—a pattern of resource use that, by definition, is unsustainable (**FIGURE 22.14**). If people spend more than they earn, they have unsustainable spending habits. Humans, and especially Americans, are using global resources unsustainably.

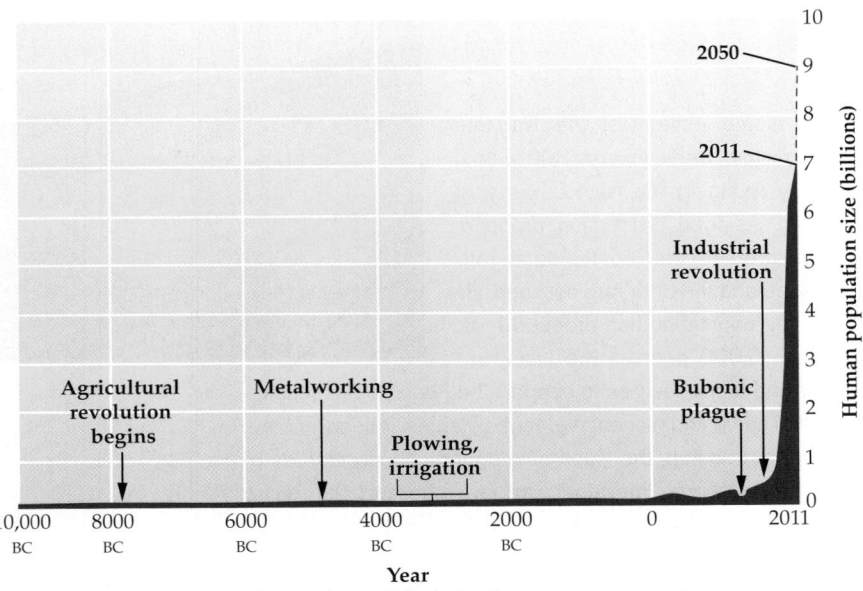

People in wealthy countries leave a larger footprint on the planet than do those in poor countries. The United States is the third largest population in the world, and because its per capita footprint is so immense, it uses more than twice the total resources that India uses. Specifically, India has a footprint of 0.9 hectare per person, multiplied by 1.2 billion people—or 1.1 billion hectares for the whole country. In contrast, the United States has a footprint of 8 hectares per person, multiplied by 309 million—or 2.5 billion hectares.

Will humans limit their own growth? Or will we continue to use too many resources, undermine the biosphere's ability to support us, and experience a population crash? There are hopeful signs: human population growth has slowed from a peak rate of 2.2 percent per year in 1963 to only 1.1 percent per year in 2009. But even a 1 percent growth rate translates into 70 million more people each year. To limit the impact of the human population, we must address the interrelated issues of population growth, overuse of resources, environmental deterioration, and sustainable development.

Hope for the future of our species (and the biosphere) lies in realistically evaluating the problems we face, and then addressing those problems. In the end, it is up to all of us to help ensure that humankind does not repeat on a grand scale the tragic lessons of Easter Island.

**FIGURE 22.13  Rapid Growth of the Human Population**

An indication of our growing numbers on Earth, this image from space shows the planet brightly lit by its most populous cities' lights. The brightness correlates well with population densities in industrialized countries but underrepresents population densities in countries such as China and India, which have sizable populations with poor access to modern amenities. The image was created by researchers at NASA using satellite image data.

**FIGURE 22.14  Our Rising Ecological Impact**

The global ecological impact of people has increased steadily. This graph compares human demands on the biosphere in each year between 1961 and 1999 with the capacity of the biosphere to regenerate itself. Human demand has exceeded the biosphere's entire regenerative capacity since the late 1970s.

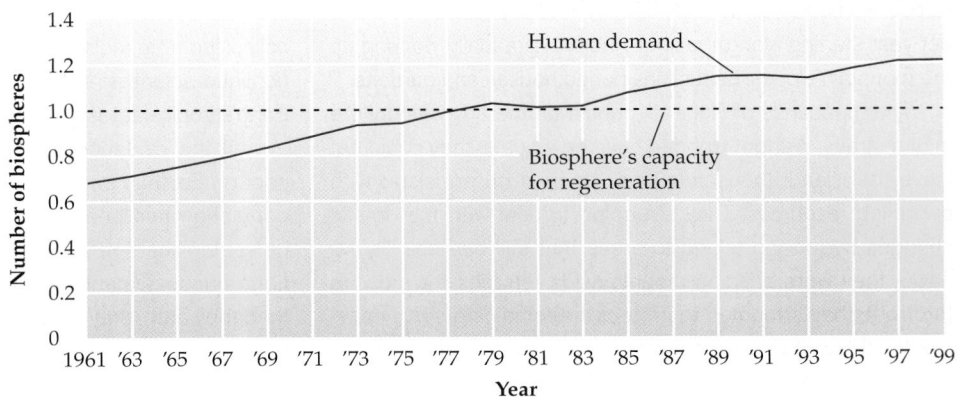

# China Population: 1.3 Bn

BY RESHMA PATIL, *Hindustan Times*

Almost half the people in China, the world's most populous nation, now live in cities and more than twice the numbers of Chinese students go to college compared to a decade ago. But new data reveals the second-largest economy is fast ageing as its population growth slows, sparking debate at home on competition from a younger Indian workforce.

Census results released on Thursday put China's population at 1.34 billion in 2010 with 73.9 million more people since 2000. India, which added 181 million in the last decade to hit 1.21 billion, will surpass the Chinese population size by 2025.

"China's ageing population is increasing quickly while the birth rate remains low," said chief statistician Ma Jiantang, referring to "tensions and challenges" ahead.

China's population growth slowed to an annual 0.57 per cent during the last decade, down from 1.07 per cent in the nineties . . .

The lower birth rate . . . shrank average household size from 3.4 persons to 3.1. China's family planning policy limiting urban families to one child and rural families to two children prevented about 400 million births since 1980. This sixth census has churned internal debate on allowing two children per family.

President Hu Jintao said this week that China will continue to aim for a low birth rate. He did not rule out changes to the one-child policy. "The opponents (of the policy) say if the depleting work force is not replenished, China will have difficulty to compete with countries like India," said a People's Daily report.

China has long had the largest population on Earth. In the 1980s, as experts predicted a global population explosion and China struggled to feed an ever-expanding population, its government decided to slow population growth by limiting the number of children that families could have. According to China's one-child policy, most families were allowed to have only one child. A few exceptions included certain farm families, ethnic minorities, and some families where both parents were themselves only children. In a few parts of China, however, families were allowed to have two children but were required to wait until the mother was 25 and also to space their two children several years apart. Delaying the births of both children had about the same effect on overall population growth rate as having just one child sooner.

China's population has continued to increase in size, but not nearly as fast as it would have without these restrictions. The total population of China is expected to continue to grow slightly until 2030, when it will probably peak and begin to decline somewhat. In contrast, India is growing three times as fast (1.4 percent per year) and will likely surpass China in total population by 2035. The United States is the third largest country by population, and has an annual growth rate of about 1 percent. Like India, the United States is still growing rapidly, mainly because of immigration, and the U.S. population is expected to reach about 420 million people by 2050.

Some critics object to China's one-child policy, arguing that individuals should be allowed to have as many children as they want. More recent objections have focused on the change in age structure. As in other countries with a stable population, including Germany and Japan, the ratio of older people to young people in China has increased. Economists worry that an excess of grandparents and a shortage of younger workers could slow economic growth.

Another issue is the sex ratio. Many Chinese families prefer boys to girls and use ultrasound during pregnancy to selectively abort girls. As a result, the ratio of boys to girls at birth is about 118 to 100, far higher than the usual rate of about 105 to 100 in most parts of the world. In China, one successful solution to this problem has been to allow parents of girls to have another baby. Giving families a second chance to have a boy means that fewer girls are aborted.

## Evaluating the News

**1.** How will a skewed sex ratio, with more males than females, affect overall population growth?

**2.** The analysis mentions that having two children after age 25 and spaced several years apart slowed overall growth rate in China about as much as having one child very early. How might you use this information to recommend policy changes that would slow population growth in India?

**3.** In this chapter we talked about the carrying capacity of environments for populations of organisms. How would you estimate the carrying capacity of Earth for humans? What factors would have to be considered?

**SOURCE:** *Hindustan Times*, April 28, 2011, http://www.hindustantimes .com.

# Summary

## 22.1 What Is a Population?

- A population is a group of interacting individuals of a single species located within a particular area.
- Two basic concepts used in studying populations are population size (the total number of individuals in the population) and population density (the number of individuals per unit of area).
- What constitutes an appropriate area for determining a population depends on the questions of interest and the biology of the organism under study.

## 22.2 Changes in Population Size

- All populations change in size over time. Populations increase when birth and immigration rates are greater than death and emigration rates, and they decrease when the reverse is true.
- Because birth, death, immigration, and emigration rates are all affected by environmental factors, the environment plays a key role in changing the size of a population.

## 22.3 Exponential Growth

- A population grows exponentially when it increases by a constant proportion from one generation to the next. Exponential growth produces a J-shaped curve.
- The doubling time is one measure of how fast a population is growing.
- Populations may grow at an exponential rate when organisms are introduced into or migrate to a new area.

## 22.4 Logistic Growth and Limits on Population Size

- Because the environment contains a limited amount of space and resources, no population can continue to increase in size indefinitely.
- Some populations increase rapidly at first and then level off and stabilize at the carrying capacity, the maximum population size that their environment can support. This logistic growth pattern is represented by an S-shaped curve.
- Density-dependent environmental factors limit the growth of a population more strongly when the density of the population is high. Such factors include food shortages, diminishing space, disease, predators, and habitat deterioration.
- Density-independent factors, such as weather and natural disturbances, limit the growth of populations without regard to their density.

## 22.5 Patterns of Population Growth

- Different populations (including those of the same species) can exhibit different patterns of growth over time, including exponential growth, logistic growth, population cycles, and irregular fluctuations.
- Two populations of different species can change together in tightly linked cycles when one or both species are strongly influenced by the other.
- In natural systems, a growth pattern of irregular fluctuations is much more common than a logistic growth pattern or tightly linked cycles.
- Understanding why different populations have different patterns of growth can provide critical information on how best to manage endangered species.

# Key Terms

carrying capacity (p. 494)
density-dependent (p. 495)
density-independent (p. 495)
doubling time (p. 492)

exponential growth (p. 492)
habitat (p. 494)
irregular fluctuation (p. 497)
J-shaped curve (p. 492)

logistic growth (p. 494)
population (p. 490)
population cycle (p. 497)
population density (p. 490)

population ecology (p. 490)
population size (p. 490)
S-shaped curve (p. 494)

# Self-Quiz

1. A group of interacting individuals of a single species located within a particular area is
   a. a biosphere.
   b. an ecosystem.
   c. a community.
   d. a population.

2. A population of plants has a density of 12 plants per square meter and covers an area of 100 square meters. What is the population size?
   a. 120
   b. 1,200
   c. 12
   d. 0.12

3. A population that is growing exponentially increases
   a. by the same number of individuals each generation.
   b. by a constant proportion each generation.
   c. in some years and decreases in other years.
   d. none of the above

4. In a population with an S-shaped growth curve, after an initial period of rapid increase the number of individuals
   a. continues to increase exponentially.
   b. drops rapidly.
   c. remains near the carrying capacity.
   d. cycles regularly.

5. The growth of populations can be limited by
   a. natural disturbances.
   b. weather.
   c. food shortages.
   d. all of the above

6. Factors that limit the growth of populations more strongly at high densities are said to be
   a. density-dependent.
   b. density-independent.
   c. exponential factors.
   d. sustainable.

7. The maximum number of individuals in a population that can be supported indefinitely by the population's environment is called the
   a. exponential size.
   b. J-shaped curve.
   c. sustainable size.
   d. carrying capacity.

8. A population that initially has 40 individuals grows exponentially with an (annual) proportional increase ($\lambda$) of 1.6. What is the size of the population after 3 years? (*Note*: Round down to the nearest individual.)
   a. 16
   b. 163
   c. 192
   d. 102,400

# Analysis and Application

1. Explain why it can be difficult to determine what constitutes a population.

2. Populations increase in size when birth and immigration rates are greater than death and emigration rates. Keeping this basic principle in mind, what actions do you think a scientist or policy maker might take to protect a population threatened by extinction?

3. Assume that a population grows exponentially, increasing by a constant proportion of 1.5 per year. If the population initially contains 100 individuals, it will contain 150 individuals in the next year. Graph the number of individuals in the population versus time for the next 5 years, starting with 150 individuals in the population.

4. Population growth cannot increase indefinitely. (a) What environmental factors prevent unlimited growth? (b) Why is it common for populations of species that enter a new region to grow exponentially for a period of time?

5. Describe the difference between density-dependent and density-independent factors that limit population growth. Give two examples of each.

6. Different populations of a species can have different patterns of population growth. Explain how an understanding of the causes of these different patterns can help managers protect rare species or control pest species.

7. List five specific actions that you can take to limit the growth or impact of the human population.

# 23 Ecological Communities

**A BABY RAT SITS ATOP A KITTEN'S HEAD.** When the parasite *Toxoplasma gondii* infects a rat, the rat loses its natural fear of cats and will readily approach a cat. The cat eats the rat and itself becomes infected by the parasite. Very young animals like the ones in this photo tend to be less fearful than adults; these two are probably not infected.

# Fatal Feline Attraction

Imagine an alien from outer space that enters people's brains and changes their behavior, sometimes even driving them crazy. Millions of people secretly harbor the infection. It kills babies in the womb and infects so easily that a person can get it just by making a hamburger. Does this sound like a bad plot for a science fiction story? Amazingly enough, it's all true except that the "alien" is a parasite that lives right here on Earth.

*Toxoplasma gondii* is a single-celled parasite, a relative of the parasite that causes malaria. "Toxo," as it's informally called, can infect nearly any warm-blooded animal, including birds and mammals. Most often contracted from raw meat, Toxo causes more deaths and illness than any other food-borne pathogen except *Salmonella*. And 60 million Americans are infected with the parasite.

Toxo's primary host, cats, get the parasite by eating rodents or, like us, raw meat. When rats and mice accidentally contact cat feces, they become infected. But then *Toxoplasma gondii* does something unexpected; it enters the rats' brains and makes them unafraid of cats. In addition, male rats (but not female rats) actually become sexually excited by the smell of cat urine and go out of their way to find it. It's a bad move for the rats. But for the parasite—which ends up in a new cat where it can reproduce—it all ends happily.

So, does the parasite affect our behavior too? You won't like the answer. Most people infected with the parasite either don't notice any symptoms or have mild flulike symptoms for a few weeks, from which they recover completely. But just as in rats, the parasite can take up residence in the brain and alter our behavior.

How can *Toxoplasma gondii* hijack the chemistry of the brain and force humans and other animals to do things they wouldn't normally do? How does the parasite benefit from altering the behavior of its hosts? Is this the way different species normally interact?

As we'll see in the course of this chapter, the species that make up a community interact in all kinds of ways. But the species interactions involving *T. gondii* are a bit more bizarre than is usual.

**MAIN MESSAGE** Communities change naturally as a result of interactions between and among species and as a result of interactions between species and their physical environment.

## KEY CONCEPTS

- A community is an association of populations of different species that live in the same area.

- The diversity of a community has two components: the number of different species that live in the community, and the relative abundances of those species.

- Community structure is influenced by a wide range of factors, such as climate, disturbance, and species interactions.

- Species interactions are classified by whether they help, harm, or have no effect on one or both of the species involved.

- The four primary types of species interactions are mutualism, commensalism, exploitation (predation and parasitism), and competition.

- Species interactions help determine where organisms live and how abundant they are, and therefore strongly influence community structure.

- Species interactions drive natural selection and evolution. Competition can cause greater differences to evolve between species. Two species that interact may trigger evolutionary change in each other

(coevolution) as a consequence of their interactions.

- All communities change over time. As species colonize new or disturbed habitat, they tend to replace one another in a directional and fairly predictable process called succession.

- Humans transform natural habitats in many ways. Communities can reassemble after some forms of human-caused disturbance, but the process may take from a few years to many centuries.

species in the community. **FIGURE 23.2** compares the diversity of two communities that have the same number of species; because community A is dominated by a single, highly abundant species, it is considered less diverse than community B, in which all species are equally common.

Most communities contain many species, and the interactions among those species can be complex. Ecologists seek to understand how interactions among organisms influence natural communities. Interactions among organisms have huge effects on natural communities. For example, as we saw in Chapter 22, the moth *Cactoblastis cactorum*, by feeding on the cactus *Opuntia*, caused *Opuntia* populations to crash throughout a large region of Australia. Overall, interactions among organisms have an influence at every level of the biological hierarchy at which ecology is studied.

Ecologists also study how human actions affect communities. People are having profound effects on many different kinds of ecological communities. When we cut down tropical forests we destroy entire communities of organisms, and when we give antibiotics to a cow we alter the community of microorganisms that live in its digestive tract. To prevent such actions from having effects that we do not anticipate or want, we must understand how ecological commu-

**AN ASSOCIATION OF DIFFERENT SPECIES** that live in the same area is known as an ecological **community**. Communities vary greatly in size and complexity, from the community of microorganisms that inhabits a small temporary pool of water, to the community of plants that lives on the floor of a forest, to a forest community that stretches for hundreds of kilometers (**FIGURE 23.1**). Whatever its size or type, an ecological community can be characterized by its species composition, or *diversity*. The **diversity** of a community has two components: **species richness**, which refers to the total number of different species that live in the community; and **relative species abundance**, which describes how common individuals of a species are compared to individuals of other

**ECOLOGICAL COMMUNITIES**

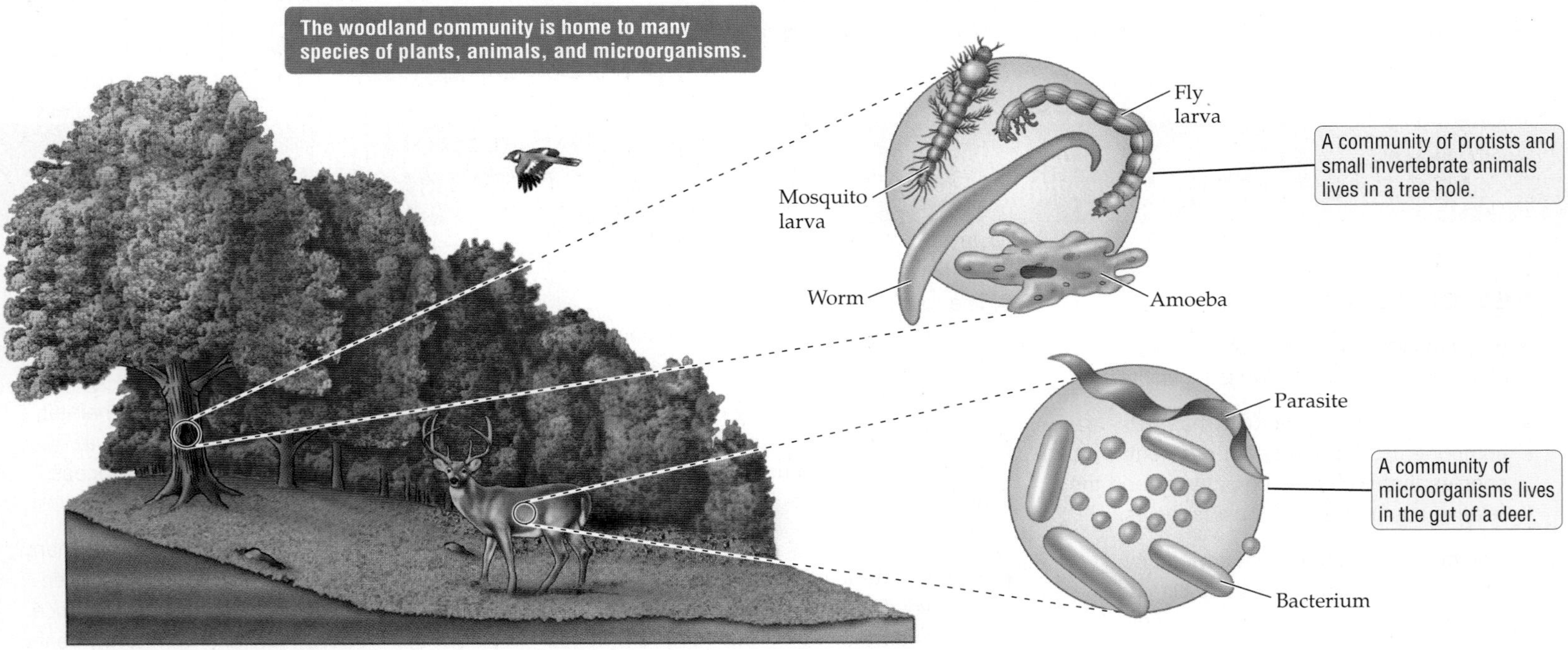

**FIGURE 23.1  Temperate Forests in North America Contain Many Types of Woodland Communities**
Smaller communities can be nested within a larger community. This beech-maple woodland community contains the community of a temporary pool of water in a tree hole and the community of a deer's gut, among others.

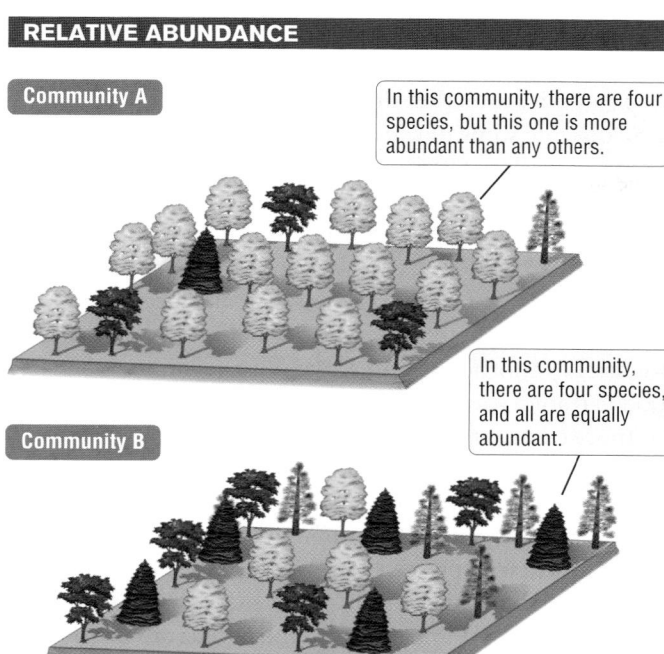

**Community A**

In this community, there are four species, but this one is more abundant than any others.

In this community, there are four species, and all are equally abundant.

**Community B**

**FIGURE 23.2 Which Community Has Greater Diversity?**
Community A is dominated by a single species; in community B, all four species are equally represented. Therefore, ecologists consider community A less diverse than community B.

nities work. In this chapter we describe how species interact with each other, and the factors that influence which species are found in a community. We pay particular attention to how communities change over time and how they respond to disturbance, including disturbance caused by people.

## 23.1 Species Interactions

The millions of species on Earth can interact in many different ways. In this chapter we classify interactions among organisms into four broad categories, based on whether the interaction is beneficial (+), harmful (−), or without benefit or harm (0) to each of the interacting species:

1. *Mutualism* (+/+), interactions in which both species benefit

2. *Commensalism* (+/0), interactions in which one species benefits at no cost to the other

3. *Exploitation* (+/−), interactions in which one species benefits and the other is harmed

4. *Competition* (−/−), interactions in which both species may be harmed

Each type of interaction plays a key role in determining where organisms live and how abundant they are. Species interactions can drive natural selection and evolution, thereby altering the composition of communities over both short and long spans of time. The study of an ecological community is a study of the relationships among organisms in a defined area. We will see how community stability is a function of species interactions and how changes in interactions among organisms can alter ecological communities.

We consider first the fascinating realm of cooperative relationships among species: mutualistic interactions that benefit both species. We then examine several types of antagonistic relationships, in which one species benefits at the expense of another.

## There are many types of mutualism

**Mutualism** is an association between two species in which both species benefit more than it costs them to interact with each other. Mutualism is common and important to life on Earth: many species receive benefits from, and provide benefits to, other species. These benefits increase the survival and reproduction of both of the interacting species.

Mutualism can occur when two or more organisms of different species live together—an association known as **symbiosis**. Insects such as aphids and mealybugs, both of which feed on the nutrient-poor sap of plants, often have a mutualistic, symbiotic association with bacteria that live within their cells. The bacteria receive food and a home from the insects, and the insects receive nutrients that the bacteria (but not the insects) can synthesize from sugars in the plant sap.

Nature abounds with varieties of mutualism; here we describe only some of the most common types. In *gut inhabitant mutualism*, organisms that live in an animal's digestive tract receive food from their host and benefit the host by digesting foods, such as wood or cellulose, that the host otherwise could not use. The interaction between a mealybug and bacterial species living in its gut is an example of this type of mutualism. So, too, is the interaction between termites and their gut bacteria, which enable the termites to digest wood. We humans, too, benefit from gut inhabitant mutualism because some of the hundreds of species of bacteria that colonize our large intestines manufacture beneficial nutrients, such as vitamin K.

Mutualism in which each partner has evolved to alter its behavior to benefit the other species is called *behavioral mutualism*. The relationship between certain

### Helpful to Know

"Symbiosis" (from *sym*, "with"; *biosis*, "life") describes some (but not all) mutualistic and parasitic interactions. Individuals of one species, for at least part of their life cycle, must live close to, in, or on individuals of another species for the interaction to be considered symbiotic.

Outside its burrow, the shrimp keeps one antenna on the goby. Sudden movements by the fish alert the shrimp to danger.

**FIGURE 23.3 Friends in Need**
Each *Alpheus* shrimp builds a burrow for shelter, which it shares with a goby fish. The fish provides an early-warning system to the nearly blind shrimp when the shrimp leaves the burrow to feed.

shrimp and fishes is a good example of behavioral mutualism (**FIGURE 23.3**). Shrimp of the genus *Alpheus* live in an environment with plenty of food but little shelter. They dig burrows to hide in, but they see poorly, so they are vulnerable to predators when they leave their burrows to feed. These shrimp have formed a fascinating relationship with some goby fishes in the genera *Cryptocentrus* and *Vanderhorstia*. When a shrimp ventures out of its burrow to eat, it keeps an antenna in contact with an individual goby with which it has formed a special relationship. If a predator or other disturbance causes the fish to make a sudden movement, the shrimp darts back into the burrow. The goby acts as a "Seeing Eye" fish for the shrimp, warning it of danger.

**FIGURE 23.4 Pollinator Mutualism**

In the inset photo, a yucca moth (*Tegeticula yuccasella*) is moving pollen from the anthers (orange) to the pistil of the yucca plant (*Yucca torreyi*) seen in the larger photo.

The yucca plant and yucca moth are obligate mutualists. Without the yucca moth, the plant is not pollinated. Without yucca seeds as food for its young, the yucca moth cannot complete its life cycle.

In return, the shrimp shares its burrow with the goby, thereby providing the fish with a safe haven.

Aggressive ants that defend their hosts from predators provide some of the best-known examples of *protection mutualism*. Certain species of *Acacia* trees play host to fierce ants that fend off herbivores. The trees secrete a nectarlike food from the leaf tips and produce swollen structures to shelter the insects. Larvae of the *Plebejus* butterfly produced a sugary "honeydew" that serves as food for the ants that defend them from predators—another example of a mutualism in which food is traded for protection.

In *seed dispersal mutualism*, an animal such as a bird or mammal eats a fruit that contains plant seeds and then later defecates the seeds far from the parent plant. Such dispersal by animals is the primary way that many plant species reach new areas of favorable habitat. For example, most of the plant species that live on isolated oceanic islands (those that are farther than 1,000 kilometers from land) are thought to have arrived there by bird dispersal of their seeds.

In *pollinator mutualism*, an animal such as a honeybee transfers pollen (which contains sperm) from one flower to the female reproductive organs (carpels) of another flower of the same species. Without such **pollinators**, many plants could not reproduce. To ensure that pollinators come to their flowers, plants offer a food reward, such as pollen or nectar, and both species benefit from the interaction. Pollinator mutualism is critical in both natural and agricultural ecosystems. For example, the apples we buy at the supermarket are available only because honeybees pollinate the flowers of apple trees, enabling the trees to produce their fruit.

MUTUALISTS ARE IN IT FOR THEMSELVES. Although both species in a mutualism benefit from the relationship, what is good for one species may come at a cost to the other. For example, a species may use energy or increase its exposure to predators when it acts to benefit its mutualistic partner. From an evolutionary perspective, mutualism evolves when the benefits of the interaction outweigh the costs for both species.

Consider the pollinator mutualism between the yucca plant and the yucca moth. A female yucca moth collects pollen from yucca flowers, flies to another group of flowers, and lays her eggs at the base of the carpel in a newly opened flower. After she has laid her eggs, the female moth climbs up the carpel and deliberately places the pollen she collected earlier onto the stigma of the flower. By this act she fertilizes the eggs of that second yucca plant (**FIGURE 23.4**). (For a review of the reproductive parts of a flower, see Figure 3.13.) When the moth larvae hatch, they feed on the seeds of the yucca plant.

In this mutualism, the plant gets pollinated (a reproductive benefit provided to the plant by the moth) and the moth eats some of its seeds (a food benefit provided to the moth by the plant). In fact, plant and pollinator each depend absolutely on the other—the yucca is the moth's only source of food, and this moth is the only species that pollinates the yucca—so this association is mutualistic, not parasitic. But there are costs for both species. Let's examine these costs more closely.

In a cost-free situation for the plant, the moth would transport pollen but would not destroy any of the plant's seeds. In a cost-free situation for the moth, the moth would produce as many larvae as possible, and they would consume many of the plant's seeds. In actuality, an evolutionary compromise has been reached: the moth usually lays only a few eggs per flower, and the plant tolerates the loss of a few of its seeds. Yucca plants have a defense mechanism that helps keep this compromise working: if a moth lays too many eggs in one of the plant's flowers, the plant can selectively abort that flower, thereby killing the moth's eggs or larvae.

MUTUALISM CAN DETERMINE THE DISTRIBUTION AND ABUNDANCE OF SPECIES. Mutualism can influence species **distribution**, the geographic area over which a species is found. Mutualism can also affect species **abundance**, the number of individuals of a species in a defined habitat. This can happen in two ways. First, because each species in a mutualism survives and reproduces better where its partner is found, the two species strongly influence each other's distribution and abundance. For example, because certain yucca plants and yucca moths depend absolutely on each other, each species is found only where the other is present.

Second, a mutualism can have indirect effects on the distribution and abundance of species that are not part of the mutualism. It can affect how a community is organized and what types of species are found in it, where those species are found, and in what numbers. Coral reefs, for example, are unique habitats that are home to many different organisms, including certain species of fishes, mollusks, crustaceans, and echinoderms such as sea stars. Corals are soft-bodied animals, most of which house photosynthetic algae—their mutualistic partners—inside their bodies (**FIGURE 23.5**). The corals provide the algae with a home and several essential nutrients, such as phosphorus, and the algae provide the corals with carbohydrates produced by photosynthesis. Because the corals that build the reefs depend on their mutualisms with algae, the many other species that live in coral reefs depend on those mutualisms indirectly.

**FIGURE 23.5  The Home a Mutualism Built**
The great diversity of life in tropical reefs depends on corals, many of which benefit from mutualistic associations with algae. The clownfish sheltering within their host sea anemone (photo on right) gain protection from predators repelled by the stinging cells on the sea anemone's tentacles. The clownfish are unharmed by the stinging cells because their bodies are coated with a thick mucus. The anemone uses nutrients from waste excreted by the fish. Also, the clownfish eat or chase away grazers that would otherwise nibble on the anemone.

**FIGURE 23.6 Commensalism**
This gray whale's rostrum (snout) is covered in barnacles.

## Only one partner benefits in commensalism

Among the residents of the coral reef are some that engage in another kind of beneficial partnership, called **commensalism**, in which one partner benefits while the other is neither helped nor harmed. Man-of-war fishes (Nomeidae)—which have acquired immunity from the deadly tentacles of jellyfish—evade predators by congregating among jellyfish. The man-of-war fishes clearly depend on this interaction with jellyfish, but the jellyfish get nothing out of it. Barnacles, as another example, attach themselves to whales (**FIGURE 23.6**). The whales are not harmed, but the filter-feeding hitchhikers get ferried around the ocean and may find more food than if they were stuck in one place.

### Concept Check

1. The yucca moth pollinates the yucca plant and depends on the plant for food. How would you classify this type of species interaction? Is this a cost-free interaction for both moth and plant? Explain.

2. Cattle egrets trail livestock, sometimes perching on their backs for a better view, to pick up insects stirred up by the grazing animals. How would you classify this type of species interactions? Who benefits in this type of interaction?

## In exploitation, one member benefits while another is harmed

**Exploitation** encompasses a variety of interactions in which one species (the exploiter) benefits and the other (the species that is exploited, usually for food) is harmed. Exploiters are generally consumers falling into three main groups:

1. **Herbivores** are consumers that eat plants or plant parts.
2. **Predators** are animals (or, in rare cases, plants) that kill other animals for food; the animals that are eaten are called **prey**.
3. **Parasites** are consumers that live in or on the organisms they eat (which are called **hosts**). An important group of parasites are **pathogens**, which cause disease in their hosts.

The three major types of exploitation are very different from one another. Whereas predators (such as wolves) kill their food organisms immediately, parasites (such as fleas) usually do not. Although the three types of exploitation have obvious and important differences, we focus here on some general principles applying to all three.

### CONSUMERS AND THEIR FOOD ORGANISMS CAN EXERT STRONG SELECTION PRESSURE ON EACH OTHER

The presence of consumers in the environment has caused many species to evolve elaborate strategies to avoid being consumed—yet another example of species interactions affecting evolutionary outcomes. Many plants, for example, produce spines and toxic chemicals as defenses against herbivores. Some plants rely on **induced defenses**, responses that are directly stimulated by an attack from herbivores. Spine production is an induced defense in some cactus species: an individual cactus that has been partially eaten, or grazed, is much more likely to produce spines than is an individual that has not been grazed (**FIGURE 23.7**)

Many prey organisms have evolved bright colors or striking patterns (**warning coloration**, also known as aposematic coloration) to warn potential predators that they are heavily defended (**FIGURE 23.8a**). Such warning coloration can be highly effective. Blue jays, for example, quickly learn not to eat brightly colored monarch butterflies, which contain chemicals that, in birds (and people), cause nausea and, at high doses,

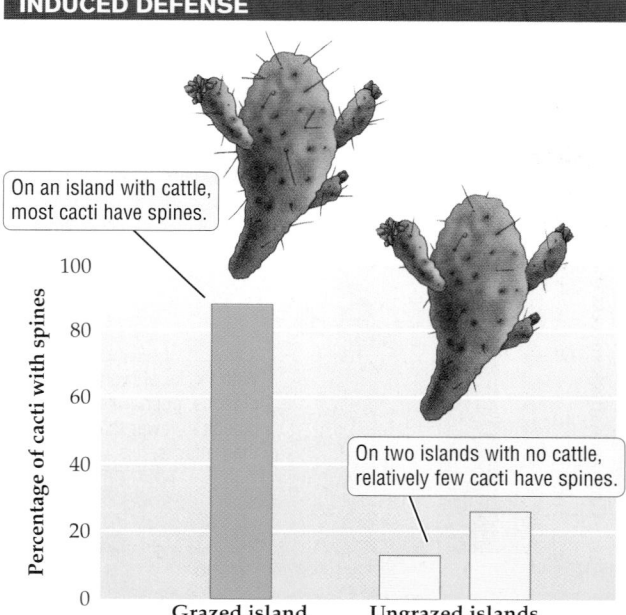

On an island with cattle, most cacti have spines.

On two islands with no cattle, relatively few cacti have spines.

Percentage of cacti with spines

100
80
60
40
20
0

Grazed island    Ungrazed islands

**FIGURE 23.7 Spines on Some Cacti Are an Induced Defense**

On three islands off the coast of Australia, the percentage of cacti with spines is higher on the island that has cattle than on the two islands that do not. Field and laboratory experiments show that grazing by cattle directly stimulates the production of spines in this species of cactus.

sudden death from heart failure. Other prey have evolved to avoid predators by being hard to find or hard to catch (**FIGURE 23.8b**). **Mimicry** is a type of adaptation arising from predator-prey interactions in which a species evolves to imitate the appearance of something unappealing to its would-be predator (**FIGURE 23.8c**). In addition, the potential hosts for parasites have evolved molecular defenses (immune systems) to help them fight off the effects of microbial diseases and parasitic infections.

Interactions among species can drive evolutionary change in the interacting species—a concept known as **coevolution**. Put another way, two species that interact may trigger evolutionary change in each other as a consequence of their interactions. For example, the many ways in which producers have evolved to protect themselves against consumers indicate that consumers often apply strong selection pressure on their food organisms. Selection occurs in the other direction as well. If a plant or prey species evolves a particularly powerful defense against attack, its consumers, in

**(a) Warning coloration**

The poison dart frog is among the most toxic animals on Earth.

**(b) Camouflage**

Can you find the insect in this photo?

**(c) Mimicry**

**FIGURE 23.8 Adaptive Responses to Predation**

(a) The bright colors of this poison dart frog warn potential predators of the deadly chemicals contained in its tissues. (b) With its long legs outstretched, this lichen-mimic katydid lies motionless on lichen-covered branches to escape detection in daytime by lizards, birds, and other predators. A relative of the cricket, the katydid forages at night. (c) The viceroy butterfly (left) mimics the color and pattern of the monarch butterfly (right), which contains toxic compounds. Confusing it with the monarch, predators tend to leave the viceroy alone.

**FIGURE 23.9 Come and Get Us**
Although a single musk ox may be vulnerable to predators such as wolves, a group that forms a circle makes a difficult target.

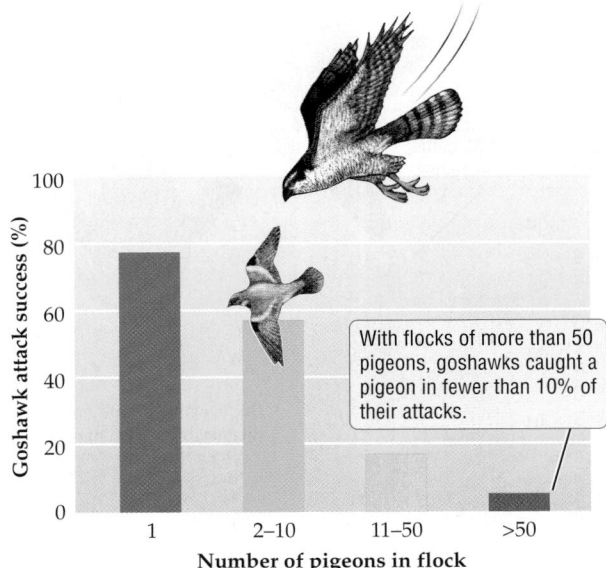

With flocks of more than 50 pigeons, goshawks caught a pigeon in fewer than 10% of their attacks.

**FIGURE 23.10 Safety in Numbers**
The success of goshawk attacks on wood pigeons decreases greatly when there are many pigeons in a flock.

turn, experience strong selection pressure to overcome that defense, an evolving situation sometimes called an *evolutionary arms race*. Many defenses work against all consumers except a few species that have evolved the ability to overcome them. Consider the rough-skinned newt, whose skin contains unusually large amounts of the potent neurotoxin TTX (tetrodotoxin)—enough to kill 25,000 mice. The newt is so toxic that only one predator, the garter snake, can tolerate its poison well enough to eat the newt and survive.

CONSUMERS CAN ALTER THE BEHAVIOR OF EXPLOITED ORGANISMS. The behaviors of animals who live or feed in groups likely evolved as a response to predation. In some cases, several prey individuals acting together may be able to repel attacks from predators (**FIGURE 23.9**). Large groups of prey may also be able to provide better warning of a predator's attack. Because more individuals can watch for predators, a large flock of wood pigeons detects the approach of a goshawk [GAHS-**hawk**] (a predatory bird) much sooner than a single pigeon does. The success rate of goshawk attacks drops from nearly 80 percent when they are attacking single pigeons to less than 10 percent when they are attacking flocks of more than 50 birds (**FIGURE 23.10**).

Many parasites cause behaviors in host organisms that benefit the parasite. For example, pinworms, once ingested, migrate to the colon and lay eggs around the anus, which causes itching. Scratching the infected area causes the unsuspecting host to transfer the eggs to common surfaces, where they can be picked up

to initiate a new round of infection. The rabies virus spreads by causing aggression in host animals, which transfer the virus via their saliva to animals that they bite. The protist *Toxoplasma gondii* causes its rat host to become more curious and less fearful. Such changes make infected rats easier prey for cats, the other host of the protist.

CONSUMERS CAN RESTRICT THE DISTRIBUTION AND ABUNDANCE OF THEIR FOOD ORGANISMS. The American chestnut used to be a dominant tree species across much of eastern North America. Within its range, anywhere from one-quarter to one-half of all trees were chestnuts. They were capable of growing to large size: trunks up to 10 feet in diameter were noted by settlers in colonial times. In 1900, however, a fungus that causes a disease called chestnut blight was introduced into the New York City area. This fungus spread rapidly, killing most of the chestnut trees in eastern North America. Today the American chestnut survives throughout its former range only in isolated patches, primarily as sprouts that arise from the base of otherwise dead trunks. With few exceptions, the new sprouts die back from reinfection with the fungus before they can grow large enough to generate new seeds.

The effect of chestnut blight on the American chestnut shows how a consumer (the fungus) can limit the distribution and abundance of its food organism (the chestnut): in this case, a formerly dominant tree species was virtually eliminated from its entire range. The effects of consumers on the distribution and abundance of the species they eat are also shown by what can happen when a food organism is separated from its consumers. Nonnative species introduced by people to new regions sometimes disrupt the ecological communities there. Some introduced species increase rapidly in number in their new areas in part because they have many fewer parasites there than in their original homes (**FIGURE 23.11**).

CONSUMERS CAN DRIVE THEIR FOOD ORGANISMS TO EXTINCTION Exploitation can drive food organisms to extinction. The effect of chestnut blight on the American chestnut provides one clear example: although the chestnut tree is not extinct through its entire range, many local populations have been driven to extinction. Similarly, *Cactoblastis* moths drove many populations of the *Opuntia* cactus in Australia

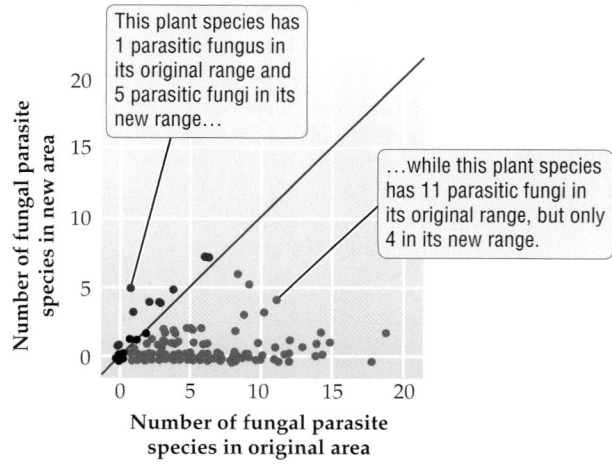

**POPULATIONS GROW WHEN PREDATION IS LOW**

This plant species has 1 parasitic fungus in its original range and 5 parasitic fungi in its new range…

…while this plant species has 11 parasitic fungi in its original range, but only 4 in its new range.

Number of fungal parasite species in new area

Number of fungal parasite species in original area

**FIGURE 23.11 Nonnative Species Leave Their Predators at Home**

Generally, nonnative species have fewer parasites and other predators in their new homes than in their original homes. Each point on the graph represents a different plant species that has been introduced to a new area. Points below the diagonal line represent plants with fewer fungal parasites in their new home than in their old home; points above the diagonal line represent the opposite. Points falling on the diagonal line indicate plants showing no difference in the number of fungal parasite species between their new and old ranges.

extinct (see Figure 22.4). If a consumer eats only one species and then drives a population of the species it eats to extinction, the consumer must either locate a new population of food organisms or go extinct itself. This is exactly what happened to *Cactoblastis* in eastern Australia: the moth drove most populations of the cactus it eats extinct, and now both species are found in low numbers.

## In competition, both species are negatively affected

In interspecific **competition**, each of two interacting species has a negative effect on the other. Competition is most likely when two species share an important resource, such as food or space, that is limited. An ecological **niche** is the sum total of the conditions and resources a species or population needs in order to survive and reproduce successfully in its particular habitat. Competition occurs when the niches of two species overlap. When two species compete, each has a negative effect on the other because each uses resources (such as food or living space) that otherwise could have been used by its competitor. This is true even when one species is so superior as a competitor that it ultimately drives the other species extinct. In such **competitive exclusion**, the inferior competitor continues to use some resources that could have been used by the superior competitor, until it loses so much ground to the superior rival that it becomes extinct.

There are two main types of competition:

1. In **interference competition**, one organism directly excludes another from the use of a resource. For example, individuals from two species of birds may fight over the tree holes that they both use as nest sites.

2. In **exploitative competition**, species compete indirectly for a shared resource, each reducing the amount of the resource available to the other. For example, two plant species may compete for a resource that is in short supply, such as nitrogen in the soil.

COMPETITION CAN LIMIT THE DISTRIBUTION AND ABUNDANCE OF SPECIES. Competition between species often has important effects on natural populations. These effects, as shown by a great deal of field evidence, include limiting the distributions and abundances of species. Let's explore two examples.

# Introduced Species: Taking Island Communities by Stealth

The Hawaiian Islands are the most isolated chain of islands on Earth. Because the islands are so remote, entire groups of organisms that live in most communities never reached them. For example, there are no native ants or snakes in Hawaii, and there is only one native mammal (a bat, which was able to fly to the islands).

The few species that did reach the Hawaiian Islands found themselves in an environment that lacked most of the species from their previous communities. The sparsely occupied habitat and the lack of competitor species resulted in the evolution of many new species and many unique natural communities.

Island communities are particularly vulnerable to the effects of introduced species. Relatively few species colonize newly formed islands, and those species then evolve in isolation. For this reason, species on islands may be ill equipped to cope with new predators or competitors that are brought by people from the mainland. In addition, introduced species often arrive without the predator and competitor species that held their populations in check on the mainland. On islands the potential exists for populations of introduced species to increase dramatically and become invasive.

In some cases, invasive species can destroy entire communities. The introduction of beard grass to Hawaii (as forage for cattle) is a case in point. By the late 1960s, beard grass had invaded the seasonally dry woodlands of Hawaii Volcanoes National Park. Before that time, fires had occurred there every 5.3 years on average, and each fire had burned an average of 0.25 hectare (about five-eighths of an acre). Since the introduction of beard grass, fires have occurred at a rate of more than one per year, and the average burn area of each fire has increased to more than 240 hectares (about 600 acres). Beard grass recovers well from large, hot fires, but the native trees and shrubs of the seasonally dry woodland do not. The fires are now so frequent and intense that the seasonally dry woodlands that once thrived in the park have disappeared.

There is no hope of restoring the native community, but ecologists are now trying to construct a new community that is tolerant of fire yet contains native trees and shrubs. This is a difficult challenge, and it is uncertain whether the effort will succeed. If not, what was once woodland is likely to remain indefinitely as open meadows filled with introduced grasses.

**Great Diversity from a Single Ancestor**
Hawaiian silverswords are found only in the Hawaiian Islands. Genetic evidence indicates that this diverse group of plant species evolved from a single ancestor (a tarweed from California). Although the three silversword species shown here are closely related, they live in very different habitats and differ greatly in form.

---

Along the coast of Scotland, the larvae of two species of barnacles—*Semibalanus balanoides* [...**buh**-LAY-**nus** BAH-luh-NOY-**deez**] and *Chthamalus stellatus* [**thuh**-MAY-**lus steh**-LAY-**tus**]—both settle on rocks on high and low portions of the shoreline. However, as adults *Semibalanus* individuals appear only on the lower portion of the shoreline, which is more frequently covered by water; and *Chthamalus* individuals are found only on the higher portion of the shoreline, which is more frequently exposed to air (**FIGURE 23.12**). In principle, the distributions of these two barnacles could have been caused either by competition or by environmental factors. In an experimental study, however, ecologists discovered that *Chthamalus* could thrive on low portions of the shoreline, but only when *Semibalanus* was removed. Hence, competition with *Semibalanus* ordinarily prevents *Chthamalus* from living low on the shoreline. This interaction is an example of interference competition because *Semibalanus* individuals often crush the smaller and more delicate *Chthamalus* individuals. The distribution of *Semibalanus*, on the other hand, depends mainly on environmental factors: the increased heat and dryness found at higher levels of the shoreline prevent *Semibalanus* from surviving there.

A second case of competition affecting distribution and abundance concerns wasps of the genus *Aphytis* [**ay**-FYE-**tus**]. These wasps attack scale insects, which can cause serious damage to citrus trees. Female wasps lay eggs on a scale insect, and when the wasp larvae hatch, they pierce the scale insect's outer skeleton and then consume its body parts.

In 1948, the wasp *Aphytis lingnanensis* was released in southern California to curb the destruction of citrus trees caused by scale insects. A closely related wasp, *A. chrysomphali*, was already living in that region at the time. *A. lingnanensis* was released in the hope that it would provide better control of scale insects than *A. chrysomphali* did. *A. lingnanensis* proved to be a superior competitor (**FIGURE 23.13**), in most locations driving *A. chrysomphali* to extinction by exploitative competition. As hoped for, *A. lingnanensis* also provided better control of scale insects.

Although competition between species is very common, note that it does not necessarily follow when two species share resources or space. This is especially true when the resources are abundant. Competition among leaf-feeding insects, for example, is relatively uncommon for this reason. A huge amount of leaf material is available for the insects to eat, and usually there are too few insects to cause their food to be in short supply. As long as their food remains abundant, little competition occurs.

Natural selection can also drive competing species to use their common niche in different ways—a phenomenon known as **niche partitioning**. The differential use of space or resources minimizes competition, enabling species to coexist despite their potential for competition. For example, many prairie plants that grow together in dense stands can coexist through niche partitioning. A shallow-rooted prairie grass mines the soil close to the surface to obtain water and mineral nutrients, while a neighboring coneflower plant taps deeper layers of soil with its deep root. Niche partitioning can be achieved through resource use at different times. For example, one species of spiny mouse forages in the daytime, while another forages at night, reducing direct competition between the two.

COMPETITION CAN INCREASE THE DIFFERENCES BE-TWEEN SPECIES As Charles Darwin realized when he formulated the theory of evolution by natural selection, competition between species can be intense when the two species are very similar in form. For example, birds whose beaks are similar in size eat seeds of similar sizes and therefore compete intensely, whereas birds whose beaks differ in size eat seeds of different sizes and compete less intensely. Intense competition between similar species may result in **character displacement**, in which the forms of the competing species evolve to become more different over time. By reducing the similarity in form between species, character displacement reduces the intensity of competi-

## INTERFERENCE COMPETITION

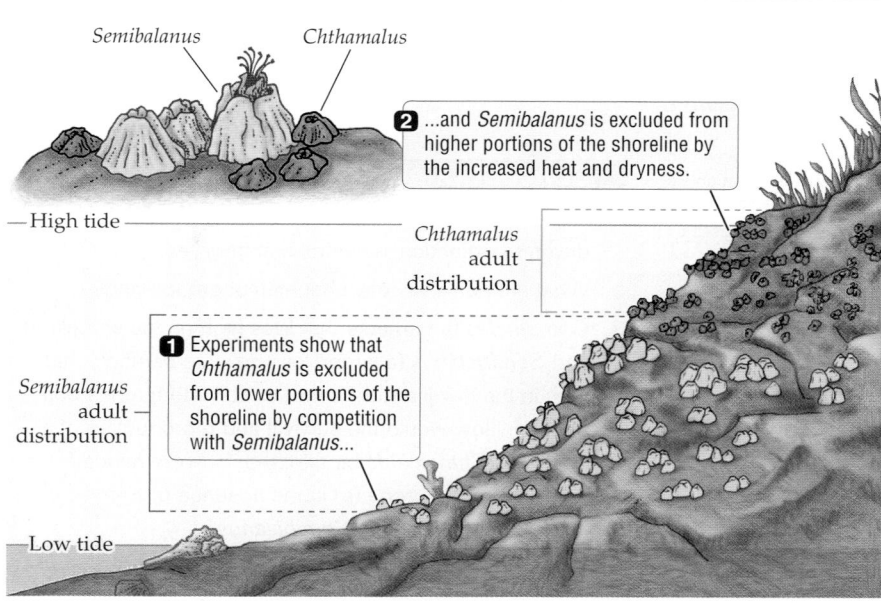

**FIGURE 23.12** **What Keeps Them Apart?**
On the rocky coast of Scotland, the larvae of *Semibalanus* and *Chthamalus* barnacles settle on rocks on both high and low portions of the shoreline. However, adult *Semibalanus* barnacles are not found on high portions of the shoreline, and adult *Chthamalus* individuals are not found on low portions.

## EXPLOITATIVE COMPETITION

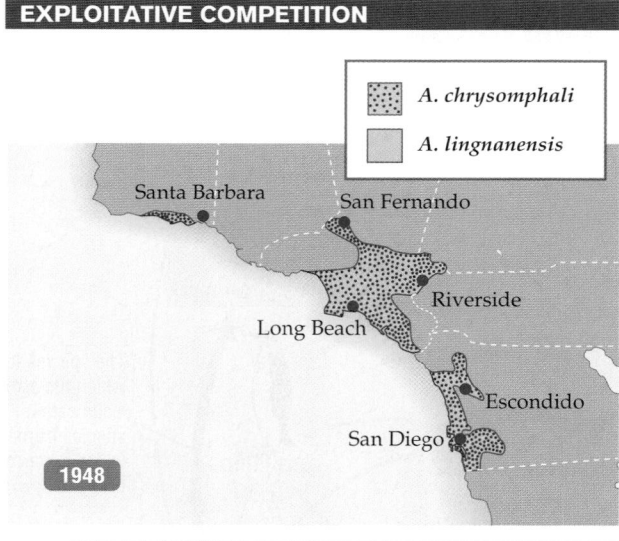

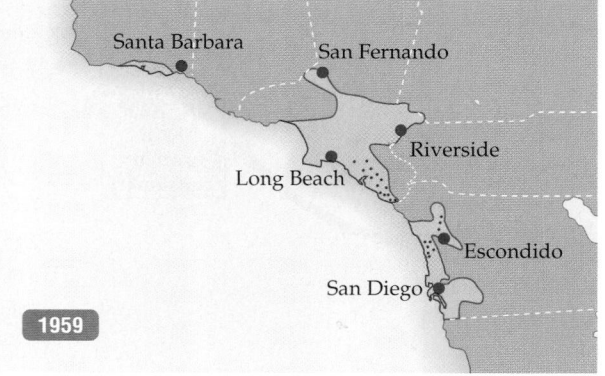

**FIGURE 23.13** **A Superior Competitor Moves In**

After being introduced to southern California in 1948, the wasp *Aphytis lingnanensis* rapidly drove its competitor, *A. chrysomphali*, extinct in most locations. Both species of wasps prey on scale insects that damage citrus crops (such as lemons and oranges).

tion. As we saw in Chapter 17, however, species can evolve in this way only if their populations vary genetically for traits (in this case, beak size) on which natural selection can act.

Concept Check
Answers
absent.
3. *Chthamalus* would thrive low on the shoreline as long as *Semibalanus* was success.
living reduces predation species show that group defense; studies of some presence and for collective warnings about a predator's opportunities for timely
2. Group living provides herbivory.
of which may be induced by defensive strategies, some chemicals, spines, and other themselves with toxic communities protect
1. Most plants in natural

> ### Concept Check
>
> 1. Explain why the plants that herbivores graze are not driven to extinction in natural communities.
>
> 2. What are some adaptive benefits of group living?
>
> 3. *Chthamalus* (a barnacle that lives high on the shoreline) and *Semibalanus* (a larger, less delicate barnacle that lives in the low intertidal zone) exhibit interference competition. How would the survival and reproduction of a colony of *Chthamalus* be affected if it were relocated to the low intertidal zone (a) in the absence of *Semibalanus* and (b) intermixed with *Semibalanus*?

## A MARINE FOOD WEB

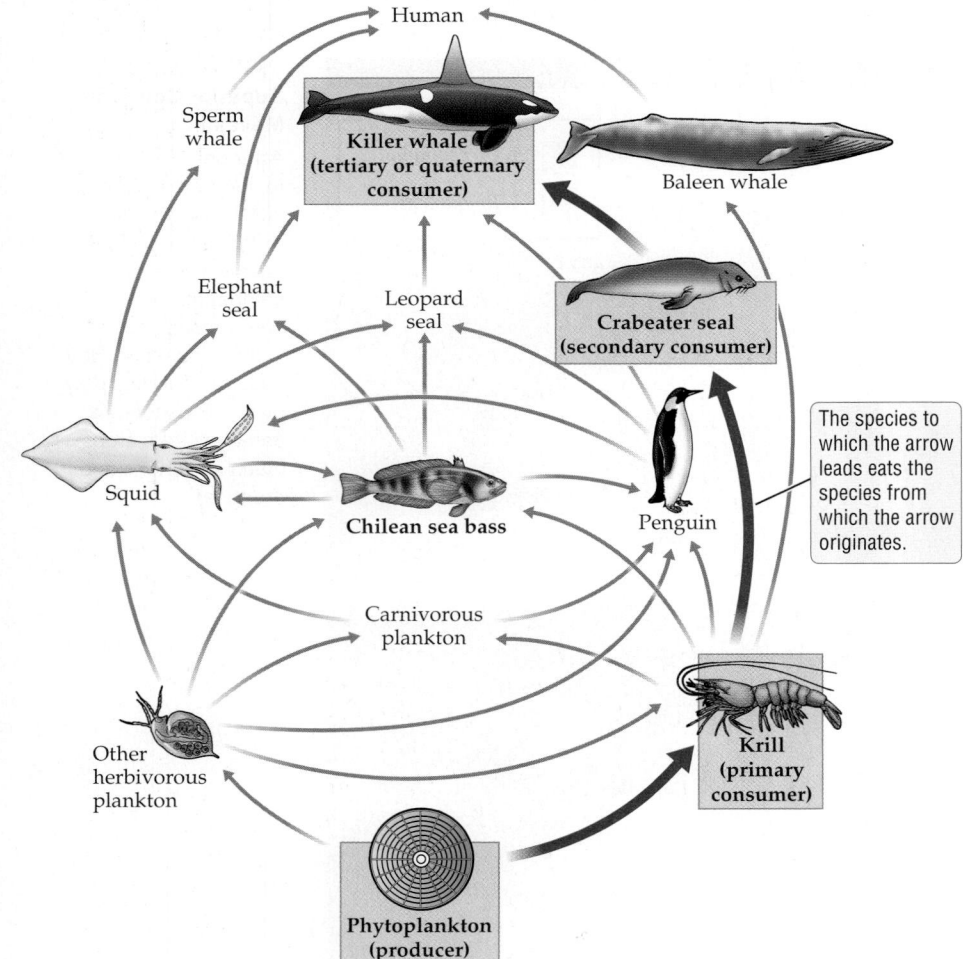

The species to which the arrow leads eats the species from which the arrow originates.

# 23.2 How Species Interactions Shape Communities

We have seen how interactions among organisms help determine their distribution and abundance. Interactions among organisms also have large effects on the communities and ecosystems in which those organisms live.

When dry grasslands are overgrazed by cattle, for example, grasses may become less abundant and desert shrubs may become more abundant. These changes in the abundances of grasses and shrubs can change the physical environment. The rate of soil erosion may increase because shrubs do not stabilize soil as well as grasses do. Ultimately, if overgrazing is severe, the ecosystem can change from a dry grassland to a desert.

A variety of factors other than species interactions can affect species diversity in a community. A fire can wipe out trees and the species they harbor and allow grasses and other sun-loving plants to flourish for the first time. Urbanization reduces habitat, diverts water, and introduces pollution that affects species composition and density in a community. Any event—whether natural or human caused—that changes the habitat in which species live will affect their chances of survival. Any change in species diversity in a community will have a ripple effect throughout the community.

## Food chains transfer nutrients through the community

How much food is available and who eats whom are important aspects of any community. The feeding relationship in a community can be illustrated by a **food chain**, which describes a linear sequence of who eats whom in a community. The movement of energy and nutrients throughout a community is shown in a **food web**, in which the various and often overlapping food chains of the community are connected (**FIGURE 23.14**).

**FIGURE 23.14 Food Webs Summarize the Movement of Food through a Community**

Food webs are composed of many specific sequences, known as food chains, that show one species eating another. To make it easier to follow a single sequence of feeding relationships, one of the food chains in this food web is highlighted with red arrows and orange boxes.

Food webs and the ecological communities they describe are based on a foundation of *producers*. **Producers** are organisms that use energy from an external source, such as the sun, to produce their own food without having to eat other organisms or their remains. On land, photosynthetic plants, which harvest energy from the sun, are the major producers. In aquatic biomes, a wide range of organisms serve as producers, including phytoplankton in the oceans, algae in intertidal zones and lakes, and bacteria in deep-sea hydrothermal vents.

**Consumers** are organisms that obtain energy by eating all or parts of other organisms or their remains. Important groups of consumers include decomposers (discussed in Chapter 24) and the herbivores, predators, and parasites (including pathogens) described earlier in this chapter. **Primary consumers** are organisms such as cows or grasshoppers that eat producers. **Secondary consumers** are organisms such as humans or birds that feed on primary consumers as part or all of their diet. This sequence of organisms eating organisms that eat other organisms can continue: a bird that eats a spider that ate a beetle that ate a plant is an example of a **tertiary consumer**. In the food chain highlighted in Figure 23.14, the killer whale can be a tertiary consumer or the next level up, a **quaternary consumer**.

## Keystone species have profound effects on communities

Certain species have a disproportionately large effect, relative to their own abundance, on the types and abundances of the other species in a community; these influential species are called **keystone species**. Keystone species are usually noticed when they are removed or they disappear from an ecosystem, resulting in dramatic changes to the rest of the community.

In an experiment conducted along the rocky Pacific coast of Washington State, ecologist Robert Paine demonstrated that the sea star *Pisaster ochraceus* [pih-ZAS-ter oh-KRAY-see-us] is a keystone species in its intertidal-zone community. He removed sea stars from one site and left an adjacent, undisturbed site as a control. In the absence of sea stars, all of the original 18 species in the community, except mussels, disappeared (**FIGURE 23.15**). When the sea stars were present, they ate the mussels, thereby keeping the number of mussels low enough that the mussels did not crowd out the other species.

How a predator maintains diversity

A *Pisaster* sea star feeding on a mussel.

Loss of a keystone species reduces diversity

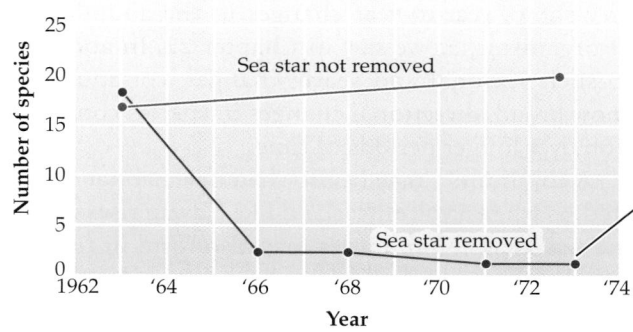

Sea stars completely eliminated mussels in submerged areas of this marine community, enabling other intertidal species to thrive there.

When the sea star *Pisaster* was removed from a community experimentally, the number of species dropped from 18 to 1, a mussel.

**FIGURE 23.15  The Star of the Community**
The sea star *Pisaster ochraceus* is a predator that feeds on mussels and thereby prevents the mussels from crowding out other species in their community.

In general, the term "keystone species" can include any producer or consumer of relatively low abundance that has a large influence on its community. The most abundant or dominant species in a community (such as the corals in a coral reef or the mussels in Paine's intertidal zone) also have large effects on their communities, but because their abundance is not low, they are not considered keystone species.

We usually do not know in advance which species are keystone species. It is often not until after people remove a species from a community, and then observe

### Helpful to Know

In genetics, specific traits of organisms, such as height, beak size, or the chemical structure of proteins, are sometimes referred to as "characters." Therefore, in character displacement, a specific trait (such as beak size) becomes more different over time in populations of two species that compete for resources.

large changes in that community, that a species is discovered to have been a keystone species. When people removed rabbits from a region in England, for example, grasslands with a variety of plant species, including many nongrasses, were converted (unintentionally) into grasslands consisting of just a few species of grasses for the most part. This change occurred because the rabbits were no longer present to hold the grasses in check. In the absence of rabbits, the grasses crowded out other plant species.

## 23.3 How Communities Change over Time

All communities change over time. The number of individuals of different species in a community often changes as the seasons change. For example, although butterflies might be abundant in summer, we would not find any flying in a North Dakota field in the middle of winter. Similarly, every community shows year-to-year changes in the abundances of organisms, as we saw in Chapter 22. In addition to such seasonal and yearly changes, communities show broad, directional changes in species composition over longer periods of time.

A community may begin when new habitat is created, as when a volcanic island like Hawaii rises out of the sea. New communities may also form in regions that have been disturbed, as by a fire or hurricane.

Some species arrive early in such new or disturbed habitat. These early colonists tend to be replaced later by other species, which in turn may be replaced by still other species. These later arrivals replace other species because they are better able to grow and reproduce under the changing conditions of the habitat.

## Succession establishes new communities and replaces disturbed communities

The process by which species in a community are replaced over time is called **succession**. In a given location, the order in which species will replace one another is fairly predictable (**FIGURE 23.16**). Such a sequence of species replacements sometimes ends in a *mature community*, which, for a particular climate and soil type, is a community whose species composition remains stable over long periods of time. But in many—perhaps most—ecological communities, disturbances such as fires or windstorms occur so frequently that the community is constantly changing in response to a disturbance event, and a mature community never forms.

**Primary succession** occurs in newly created habitat, as when an island rises from the sea or when rock and soil are deposited by a retreating glacier. In such a situation, the process begins with a habitat that contains no species. The first species to colonize the new habitat usually have one of two advantages over other species: either they can disperse more rapidly (and

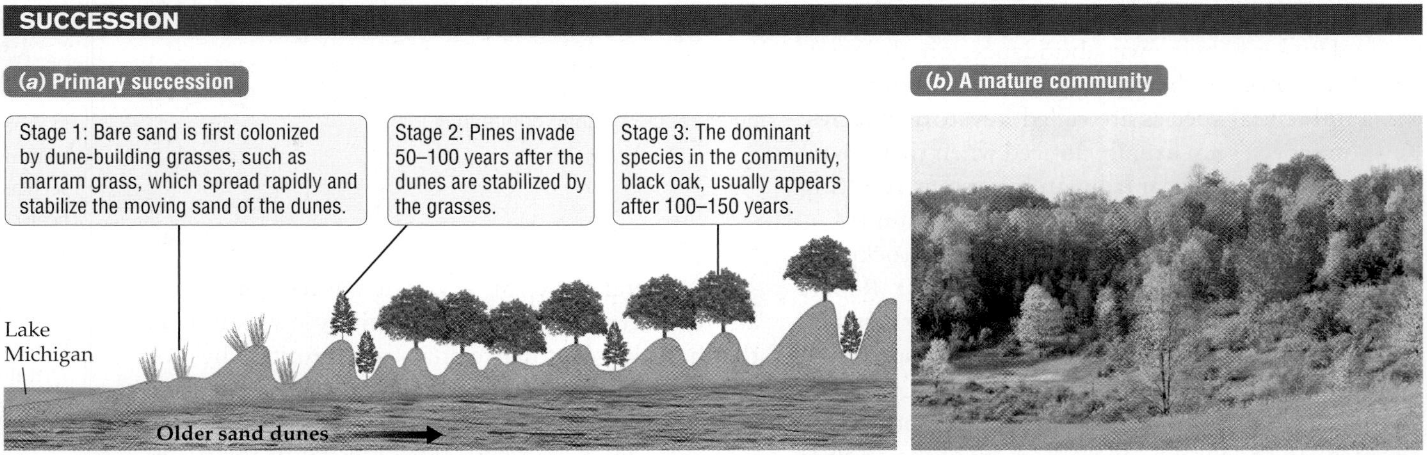

## SUCCESSION

**(a) Primary succession**

Stage 1: Bare sand is first colonized by dune-building grasses, such as marram grass, which spread rapidly and stabilize the moving sand of the dunes.

Stage 2: Pines invade 50–100 years after the dunes are stabilized by the grasses.

Stage 3: The dominant species in the community, black oak, usually appears after 100–150 years.

Lake Michigan

Older sand dunes

**(b) A mature community**

### FIGURE 23.16 Sand Becomes Woodland

When strong winds cause sand dunes to form at the southern end of Lake Michigan, succession often leads to a community dominated by black oak. Succession on such dunes occurs in three stages and forms communities of black oak that have lasted up to 12,000 years. Under different local environmental conditions, succession on Michigan sand dunes (a) can lead to the establishment of stable communities as different as grasslands, swamps, and sugar maple forests (b).

**(a) 1988**

**(b) 1992**

**(c) A mature lodgepole-pine forest**

**FIGURE 23.17** **The Slow but Steady Regrowth of Lodgepole-Pine Forest in Yellowstone National Park**
These photographs, taken in different locations, show the regrowth of lodgepole-pine forest following the large fire that struck Yellowstone National Park in 1988 (*a*, *b*), compared to unburned, mature forest (*c*).

hence reach the new habitat first), or they are better able to grow and reproduce under the challenging conditions of the newly formed habitat. In some cases of primary succession, the first species to colonize the area alter the habitat in ways that enable later-arriving species to thrive. In other cases, the early colonists hinder the establishment of other species.

**Secondary succession** is the process by which communities regain the successional state that existed before a disturbance, as when natural vegetation recolonizes a field that has been taken out of agriculture, or when a forest grows back after a fire (**FIGURE 23.17**). In contrast to primary succession, habitats undergoing secondary succession often have well-developed soil containing seeds from species that usually predominate late in the successional process. The presence of such seeds in the soil can considerably shorten the time required for the later stages of succession to be reached.

## Communities change as climate changes

Some groups of species stay together for long periods of time. For example, an extensive plant community once stretched across the northern parts of Asia, Europe, and North America. As the climate grew colder during the past 60 million years, plants in these com-

munities migrated south, forming communities in Southeast Asia and southeastern North America that are similar to one another—and similar in composition to the community from which they originated. The iconic magnolias of the southeastern United States and very similar species in southeastern Asia are remnants of that ancient intercontinental community.

Although groups of plants can remain together for millions of years, the community located in a particular place changes as the climate of that place changes. The climate at a given location can change over time for two reasons: global climate change and continental drift.

Consider first the climate of Earth as a whole, which changes over time. What we experience today as a "normal" climate is warmer than what was typical during the previous 400,000 years. Over even longer periods of time, the climate of North America has changed greatly (**FIGURE 23.18**), causing dramatic changes in the plant and animal species that live there. For example, fossil evidence indicates that 35 million years ago, the areas of southwestern North America that are now deserts were covered with tropical forests. Historically, changes in the global climate have been due to relatively slow natural processes, such as the advance and retreat of glaciers. However, evidence is mounting that human activities are now causing rapid changes in the global climate (see Chapter 25).

Second, as the continents move slowly over time (see Figure 20.7), their climates change. To give a dra-

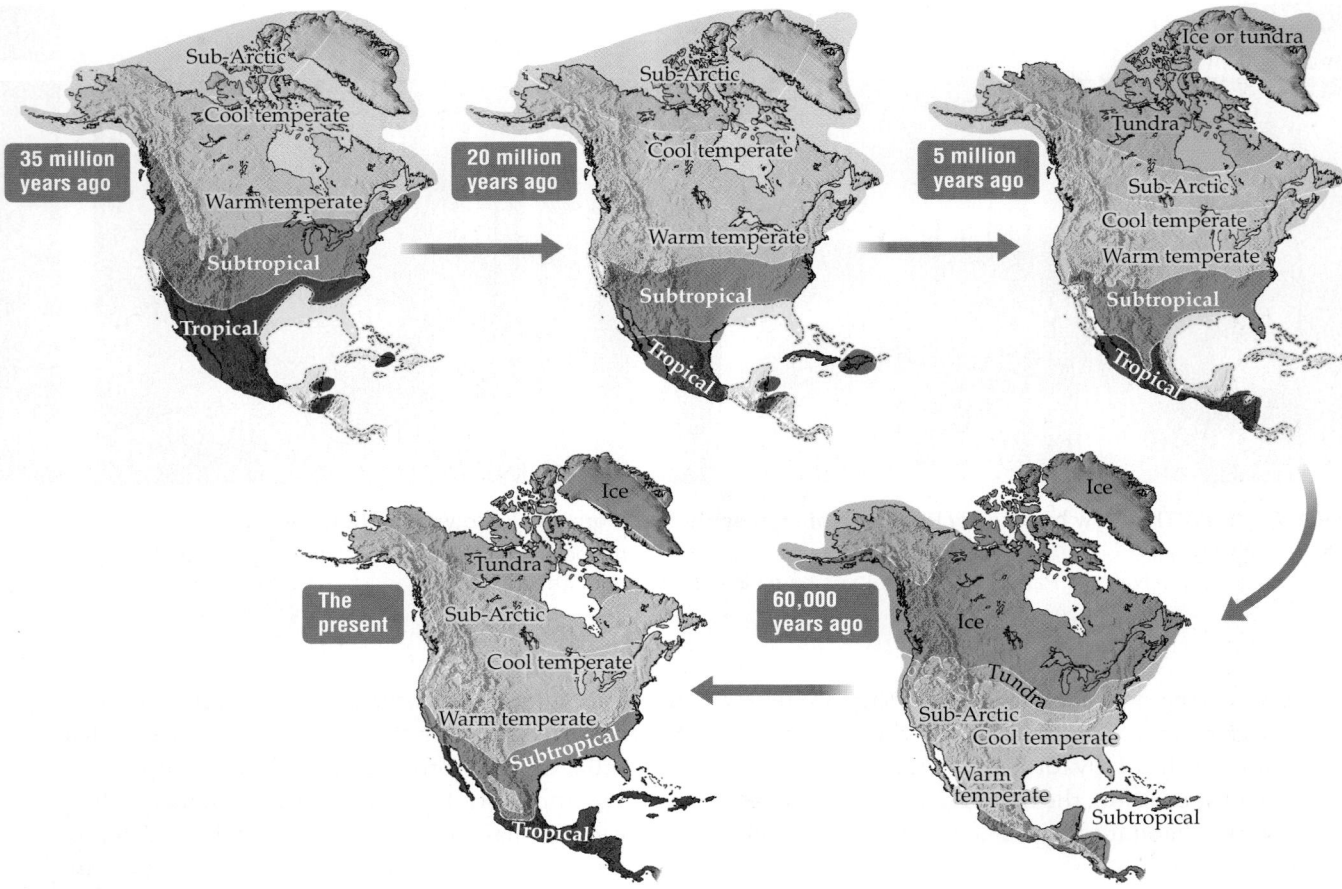

**FIGURE 23.18 Climates Change, Communities Change**

The climate of North America has changed greatly during the past 35 million years. As the climate has changed, the communities found in particular places have changed as well. The white regions surrounded by dotted lines were below sea level at the time indicated.

matic example of continental drift, 1 billion years ago Queensland, Australia, which is now located at 12° south latitude, was located near the North Pole. Roughly 400 million years ago, Queensland was at the equator. Since the species that thrive at the equator and in the Arctic are very different, continental drift has resulted in large changes in the communities of Queensland over time.

> ### Concept Check
>
> 1. What is a keystone species?
> 2. Describe the distinctive characteristics of species that tend to be the first to colonize a new habitat.
> 3. Compare primary succession with secondary succession.

## 23.4 Human Impacts on Community Structure

Ecological communities are subject to many natural forms of disturbance, such as fires, floods, and windstorms. As we saw with the example of Yellowstone Park (see Figure 23.17), following a disturbance secondary succession can reestablish the previously existing community. In this way communities can and do bounce back from some forms of disturbance. Depending on the community, the time required to regain a prior state varies from years to decades or centuries.

Communities have been exposed over long periods of time to natural forms of disturbance, such as windstorms. In contrast, people may introduce entirely new forms of disturbance, such as the dumping of hot wastewater into a river by a nuclear power plant. Human actions may also alter the frequency of an otherwise natural form of disturbance—for example, causing a dramatic increase or decrease in the frequency of fires or floods.

## Communities can reassemble after some human-caused disturbances

Can communities regain their previous state after disturbances caused by people? For some forms of human-caused disturbance, the answer is: yes. Throughout the eastern United States, for example, there are many places where forests were cut down and the land used for farmland; years later, the farmland was abandoned. Second-growth forests have grown on these abandoned farms, often within 40–60 years after farming stopped. Second-growth forests are not identical to the forests that were originally present. The sizes and abundances of the tree species are different, and fewer plant species grow beneath the trees of a second-growth forest than beneath a virgin forest (one that has never been cut down). However, the second-growth forests of the eastern United States already have recovered partially from cutting. If current trends continue, over the next several hundred years there will be fewer and fewer differences between such forests and the original forests.

## People can cause long-term damage to communities

It is encouraging that complex ecological communities northeastern forests can recover rapidly from disturbances caused by people. However, communities do not always recover so quickly from human-caused disturbances, as a few examples will show.

- Northern Michigan once was covered with a vast stretch of white-pine and red-pine forest. Between 1875 and 1900, nearly all of these trees were cut down, leaving only a few scattered patches of virgin forest. The loggers left behind large quantities of branches and sticks, which provided fuel for fires of great intensity. In some locations, the pine forests of northern Michigan have never recovered from the combination of logging and fire.
- Throughout South America and Southeast Asia, large tracts of tropical forest have been converted

into grasslands by a combination of logging and fire. Scientists estimate that it will take tropical forest communities hundreds to thousands of years to recover from such changes.

- In some areas of the American Southwest, grazing by cattle has transformed dry grasslands into desert shrublands (**FIGURE 23.19**). How do cattle cause such large changes? Overgrazing and trampling by cattle decrease the abundance of grasses in the community. With less grass to cover the soil and hold it in place, the soil dries out and erodes more rapidly. Desert shrubs thrive under these new soil conditions, but grasses do not. These changes in soil characteristics can make it very difficult to reestablish grasses, even when the cattle are removed.

(a) This dry grassland is in southern New Mexico.

(b) This nearby, former dry grassland is now a desert.

**FIGURE 23.19 Overgrazing Can Convert Grasslands into Deserts**
(a) More than 200 years ago, large regions of the American Southwest were covered with dry grasslands. (b) Most of these grasslands have been converted into desert shrublands, in large part because of overgrazing by cattle.

Grazing can have a dramatic effect on dry grasslands, but how do you think the effects of grazing would compare to the effects of an atomic bomb? The first aboveground explosion of an atomic bomb occurred on July 16, 1945, at the Trinity site in New Mexico. Fifty years later, dry grasslands destroyed by the bomb blast (but never grazed) had recovered. In contrast, nearby dry grasslands that had been heavily grazed (but not destroyed by the bomb blast) had not recovered, even though they had not been grazed since the time of the blast. The plant community recovered more rapidly from the effects of a nuclear explosion than from the effects of grazing—a dramatic example of how strongly ecological interactions can affect natural communities.

## APPLYING WHAT WE LEARNED

# How a Parasite Can Hijack Your Brain

At the beginning of this chapter we learned about a parasite that changes the behavior of its host. Specifically, male rats infected with *Toxoplasma gondii* find the smell of cat urine irresistible. When the unfortunate rat locates a cat, the cat eats the rat and—voilà—the cat is infected by the parasite. Cats (including lions and leopards) are the only animals in which *T. gondii* can sexually reproduce, making cats the parasite's primary host. When an infected cat passes feces containing the parasite, the parasite remains infectious for up to 2 years.

*T. gondii* also infects humans—causing two forms of toxoplasmosis. In the first form, an initial ("acute") infection can cause (1) no symptoms; (2) mild, flulike symptoms such as fever and enlarged lymph nodes; or (3) serious illness such as fever, seizures, and eye inflammation in those with compromised immune systems. Pregnant women are at special risk from acute Toxo infections, since the parasite can cause birth defects.

Most people do not even notice the initial infection. Once an untreated infection is over, the parasite enters the brain and muscle and forms cysts, creating a permanent "latent" infection. Despite the lack of medical symptoms, people with latent toxoplasmosis are measurably different. Early studies of people carrying antibodies to the parasite show that they have, on average, slower reaction times, higher rates of automobile accidents, increased rates of anxiety, and resistance to new situations. One study has even shown that women with a latent infection are more likely to give birth to sons.

And just as in male and female rats, men and women react differently to *T. gondii* infection. Some studies suggest that infected men are more likely than uninfected men to ignore rules and to be suspicious, jealous, or strongly opinionated. Infected men have three to four times the rate of automobile accidents as do normal men. In contrast, infected women are said to be more warmhearted, outgoing, and conscientious than are uninfected women. *T. gondii* infection has also been tied to higher rates of the mental disorder schizophrenia.

How does Toxo alter the brain? The verdict is still out, but very recently, researchers may have discovered some answers. The brains of many animals, including mammals, secrete a chemical called dopamine that is involved in the sense of pleasure. In response to a "rewarding" experience such as food, sex, or cocaine, the brain releases dopamine, which makes an animal feel good. (Dopamine is also involved in other important activities, such as thought, movement, and sleep.)

Single-celled organisms do not need dopamine and do not normally carry genes for making dopamine. Yet the single-celled *T. gondii* carries a gene for making an enzyme that makes dopamine. Once *T. gondii* has infected an animal's brain, the parasite floods the brain with dopamine, apparently causing poor choices in both male humans and male rats. But rats at least don't lose their fear of all dangerous situations—just of cats. In fact, *T. gondii* goes a step further and makes infected rats actually crave the smell of cat urine.

It's obvious how changing the behavior of rats improves *T. gondii*'s chances of spreading to more cats. It's less obvious if the parasite's effect on humans is just a side effect that doesn't help the parasite or if in our evolutionary past, when our ancestors were prey for leopards, *T. gondii* made human brains take risks, such as approaching leopards.

Today, nearly a quarter of the U.S. population is infected. Although most media accounts say that the parasite is contracted from cleaning cat litter boxes, the Department of Agriculture reports that half of all *T. gondii* cases are contracted from handling raw meat or from consuming undercooked meat, unpasteurized milk, and unwashed fruits and vegetables.

# How California Almonds May Be Hurting Bees

BY NICOLE MONTESANO, *Yamhill Valley News Register*

Perhaps the most significant of all the single crops affecting honeybees in the United States is the California almond.

California is the world's largest producer of almonds, and it takes 1.3 million beehives—each populated by tens of thousands of bees—to pollinate them, according to the U.S. Department of Agriculture.

Migratory beekeepers, who move truckloads of hives from state to state to pollinate seasonal crops, converge in California every year in early spring.

Oregon State University honeybee researcher Ramesh Sagili believes that's a problem.

He is studying bee nutrition, and preliminary results show drastic effects when bees are restricted to just one type of pollen at a time—specifically, almond pollen.

There's no question that almonds need bees.

"Without bees, there are literally no almonds," Sagili said.

But the enormous almond crop may be harming bees.

Honeybees, Sagili said, ordinarily will gather some six or eight different types of pollen in any one season. Bees require 10 different amino acids, he said, and if one is lacking from one particular type of pollen, they rely on getting it from another.

The hive workers that forage for pollen don't eat it. They take it back to the hive, where it is consumed by nurse bees, who process it into food for the larvae. Rich in protein, pollen plays an essential role in larvae development.

In California today, the almond crop covers hundreds of thousands of acres. There is nothing else for either the honeybees or native pollinators to gather.

"There's nothing else in bloom. Maybe a few weeds," Sagili said. "So we are restricting the diets of honeybees, and several other types of bees as well."

To test that theory, Sagili took a couple of hives to the California almond fields,

fitted out with a mesh the bees had to climb through that knocked the pollen off their legs and into a drawer below.

"We collected 20 pounds," he said. He brought it back and fed it to a hive enclosed in a flight cage, to ensure the bees had no access to any other pollen source. In another cage, he fed a control hive a more typical honeybee diet. He then took samples from each hive, over several weeks, and dissected them. The results were dramatic. Bees from the diet-restricted hive produced a less protein-rich food for the larvae and had substantially reduced immune systems. The colony grew more slowly than the control hive. "It was a huge difference we saw. These bees which are feeding on a single pollen source are not doing very well," Sagili said.

Further studies are planned, he said, and he plans to look at other pollen sources as well, including pollen from Oregon's blueberry crop.

---

Honeybees are critical to American agriculture, yet the fuzzy pollinators have been in sharp decline in recent years. It's likely that a long list of factors is causing the decline, sometimes called "colony collapse disorder" by beekeepers. The specific factor discussed in this news article is the limited diet forced on domestic honeybees.

Flowering plants depend heavily on bees, flies, and other pollinators to help carry pollen from one flower to another and to ensure a new generation of plants. Bees, in turn, count on nectar for energy and on pollen to feed their larvae. But if the research described here is correct, declines in the diversity of both wild flowering plants and crop plants could mean that honeybee larvae are malnourished. Honeybees exposed to "monocultures" of a single crop such as almonds have no chance to collect pollen from different kinds of plants, so critical nutrients are missing from their diets.

Domestic honeybees are not the only pollinators in decline. Native pollinators—such as other bees, flies, and moths—are also becoming more rare. The declines in bees and native pollinators means that there are fewer insects to pollinate the remaining flowers. Without these pollinators, some plant species cannot reproduce and could be in danger of going extinct.

Honeybees and other bees visit and pollinate a wide variety of plants. But many plants have coevolved tightly with a single pollinator. If a plant species becomes rare, then the pollinator becomes rare, which means the plant will likely go extinct. Such relationships form the framework of ecological communities everywhere.

## Evaluating the News

**1.** The research covered in this article indicates that beekeepers transport bee colonies to various agricultural regions when crops begin to flower. How might moving honeybees over wide geographic regions affect bee colony health?

**2.** The article suggests that honeybees must collect pollen from diverse plants to avoid malnutrition. What remedies can you suggest to address this problem?

**SOURCE:** *Yamhill Valley News Register*, April 23, 2011, http://www.newsregister.com.

# Summary

## 23.1 Species Interactions

- An ecological community is an association of populations of different species that live in the same area. Communities vary greatly in size and complexity and are characterized by both species richness and species abundance.
- Species interactions are classified by whether they hurt, harm, or have no effect on the species involved. In mutualism, both parties benefit. In an exploitative interaction, one party benefits while the other party is harmed. In competition, each of two interacting species has a negative effect on the other.
- Mutualism evolves when the benefits of the interaction to both partners are greater than its costs for both partners. In commensalism, one species benefits while the other is neither harmed nor benefited.
- In exploitation, one species benefits while the other is harmed. Exploiters are consumers. Consumers include herbivores, predators, and parasites.
- Consumers can be a strong selective force, leading their food organisms to evolve various ways to avoid being eaten. Many plants have evolved induced defenses, such as the growth of spines, that are directly stimulated by attacking herbivores. Food organisms, in turn, exert selection pressure on their consumers, which evolve ways to overcome the defenses of the species they eat.
- Consumers can restrict the distribution and abundance of the species they eat, thereby strongly affecting community structure.
- Competition can have a strong effect on the distribution and abundance of species. In interference competition, one species directly excludes another species from the use of a resource. In exploitative competition, species compete indirectly, each reducing the amount of a resource available to the other species.

## 23.2 How Species Interactions Shape Communities

- Communities can be described by food webs, which summarize the interconnected food chains describing who eats whom in a community.

- Producers obtain their energy from an external source, such as the sun. Consumers get their energy by eating all or parts of other organisms. Primary consumers eat producers, and secondary consumers feed on primary consumers.
- A keystone species has a large effect on the composition of a community relative to its abundance. Keystone species alter the interactions among organisms in a community, and they change the types or abundances of species in the community.

## 23.3 How Communities Change over Time

- All communities change over time. Directional changes that occur over relatively long periods of time have two main causes: succession and climate change.
- Primary succession occurs in newly created habitat. Secondary succession occurs as a community regains its previous state after a disturbance.

## 23.4 Human Impacts on Community Structure

- Humans transform natural habitats in many ways, from turning forests into farmland, to diverting water, polluting air and water, introducing new species, and altering the global climate.
- Communities can bounce back from some forms of natural and human-caused disturbance. The time required for restoration varies from years to decades or centuries.
- Degradation of water bodies resulting from eutrophication can be reversed if the sources of nutrient enrichment are removed.
- Understanding the consequences of community change can help people make decisions that take community stability into account.

# Key Terms

abundance (p. 509)
character displacement (p. 515)
coevolution (p. 511)
commensalism (p. 510)
community (p. 506)
competition (p. 513)
competitive exclusion (p. 513)
consumer (p. 517)
distribution (p. 509)
diversity (p. 506)

exploitation (p. 510)
exploitative competition (p. 513)
food chain (p. 516)
food web (p. 516)
herbivore (p. 510)
host (p. 510)
induced defense (p. 510)
interference competition (p. 513)
keystone species (p. 517)
mimicry (p. 511)

mutualism (p. 507)
niche (p. 513)
niche partitioning (p. 515)
parasite (p. 510)
pathogen (p. 510)
pollinator (p. 508)
predator (p. 510)
prey (p. 510)
primary consumer (p. 517)
primary succession (p. 518)

producer (p. 517)
quaternary consumer (p. 517)
relative species abundance (p. 506)
secondary consumer (p. 517)
secondary succession (p. 519)
species richness (p. 506)
succession (p. 518)
symbiosis (p. 507)
tertiary consumer (p. 517)
warning coloration (p. 510)

# Self-Quiz

1. The advantages received by a partner in a mutualism can include
   a. food.
   b. protection.
   c. increased reproduction.
   d. all of the above

2. The shape of a fish's jaw influences what the fish can eat. Researchers found that the jaws of two fish species were more similar when they lived in separate lakes than when they lived together in the same lake. The increased difference in jaw structure when the fishes live in the same lake may be an example of
   a. warning coloration.
   b. character displacement.
   c. mutualism.
   d. exploitation.

3. Experiments with the barnacle *Semibalanus balanoides* showed that
   a. where this species was found on the shoreline was not influenced by physical factors.
   b. competition with *Chthamalus* restricted *Semibalanus* to high portions of the shoreline.
   c. competition with *Chthamalus* restricted *Semibalanus* to low portions of the shoreline.
   d. this species restricted *Chthamalus* to high portions of the shoreline.

4. A low-abundance species that has a large effect on the composition of an ecological community is called a
   a. predator.
   b. herbivore.
   c. keystone species.
   d. dominant species.

5. Organisms that can produce their own food from an external source of energy without having to eat other organisms are called
   a. suppliers.
   b. consumers.
   c. producers.
   d. keystone species.

6. A directional process of species replacement over time in a community is called
   a. global climate change.
   b. succession.
   c. competition.
   d. community change.

7. A single sequence of feeding relationships describing who eats whom in a community is a
   a. life history.
   b. keystone relationship.
   c. food web.
   d. food chain.

8. The process in which the enrichment of water by nutrients causes bacterial populations to increase and oxygen concentrations to decrease is called
   a. eutrophication.
   b. disturbance.
   c. fertilization.
   d. nutrient loading.

# Analysis and Application

1. Mutualism typically has costs for both of the species involved. Why, then, is mutualism so common?

2. Consumers affect the evolution of the organisms they eat, and vice versa. Explain how this interaction occurs, and illustrate your reasoning with an example described in the text.

3. Rabbits can eat many plants, but they prefer some plants over others. Assume that the rabbits in a grassland containing many plant species prefer to eat a species of grass that happens to be a superior competitor. If the rabbits were removed from the region, which of the following do you think would be most likely to happen?
   a. The plant community would have fewer species.
   b. The plant community would have more species.
   c. The plant community would remain largely unchanged.
   Explain and justify your answer.

4. Describe how each of the following factors influences ecological communities: (a) species interactions, (b) disturbance, (c) climate change, (d) continental drift.

5. Provide an example of how the presence or absence of a species in a community can alter a feature of the environment, such as the frequency of fire.

6. Consider two forms of human disturbance to a forest:
   a. All trees are removed, but the soils and low-lying vegetation are left intact.
   b. The trees are not removed, but a pollutant in rainfall alters the soil chemistry to such an extent that the existing trees can no longer thrive.
   Which form of disturbance do you think would require the longest recovery time before a healthy forest community could once again be found at the site? Explain your assumptions in answering this question, and justify the conclusion you reach.

7. Do you think it is ethically acceptable for people to change natural communities so greatly that it takes thousands of years for the communities to recover? Why or why not?

# 24 Ecosystems

**A MASSIVE OIL SLICK.** In 2010, an explosion at the BP *Deepwater Horizon* deep-sea oil-drilling rig in the Gulf of Mexico resulted in a gigantic oil spill that coated thousands of square miles of ocean and destroyed marine life throughout the northern Gulf of Mexico and along 125 miles of coastline.

# *Deepwater Horizon*: Death of an Ecosystem?

On April 20, 2010, an oil-drilling rig in the Gulf of Mexico exploded, killing 11 people and initiating the largest oil spill in U.S. history. A fire on the *Deepwater Horizon* raged for 36 hours before the state-of-the-art oil rig sank, extinguishing the flames. From April to July, when the well leased by BP was finally capped, the gushing well a mile below the surface and 40 miles off the Louisiana coast released nearly 5 million barrels of crude oil. Within days, the oil that floated to the surface spread over thousands of square miles. The well was a rich source of natural gas (which had caused the explosion), and a 22-mile-long underwater plume of oil, gas, and organic solvents such as benzene and xylene spread a half mile beneath the surface of the ocean. Fishery experts predicted major damage to the Gulf shrimp and fishing industries, and ecologists predicted unprecedented damage to the Gulf ecosystem food chains.

The 1989 crash of the *Exxon Valdez* oil tanker in Alaska had spilled only 5 percent as much oil, yet it killed 250,000 seabirds, thousands of marine mammals, and billions of salmon and herring eggs—leading to long-term declines in fish populations that supported humans, as well as sea otters, killer whales, and other marine mammals. Twenty years later, oil still coats Alaskan rocks and sand, and it continues to leach slowly into the ocean, contaminating filter-feeding animals such as mussels, as well as animals farther up the food chain.

Knowing all this, Gulf Coast residents in 2010 feared the worst. Coast Guard personnel wept at the destruction, and fishermen said they'd never be able to fish the Gulf again.

Would the damage from the *Deepwater Horizon* oil spill be as bad as that from the *Exxon Valdez* oil spill in Alaska? How are ecosystems affected by such catastrophes?

Before we tackle the long-term impact of catastrophes such as the *Deepwater Horizon* spill, we will look at how healthy ecosystems function. In particular, we'll see how energy flows through ecosystems, driving the cycling of carbon, nitrogen, water, and other materials. And we'll see how much we depend on the services provided by ecosystems.

**MAIN MESSAGE** Nutrients are recycled in ecosystems, but energy flows through them in one direction. Ecosystems provide essential services to humans, but these services are frequently altered by human activities.

## KEY CONCEPTS

- An ecosystem consists of a community of organisms together with the physical environment in which those organisms live. Energy, materials, and organisms can move from one ecosystem to another.

- Energy enters an ecosystem when producers capture it from an external source, such as the sun. A portion of the energy captured by producers is lost as heat at each step in a food chain. As a result, energy flows in only one direction through ecosystems.

- Earth has a fixed amount of nutrients. If nutrients were not recycled between organisms and the physical environment, life on Earth would cease. Human activities affect the cycling of some nutrients.

- Ecosystems provide humans with essential services, such as nutrient recycling, at no cost. Our civilization depends on these and many other ecosystem services.

- Human activities frequently damage ecosystem processes, reducing the value of these services and thereby incurring both environmental and economic costs.

## 24.1 How Ecosystems Function: An Overview

An **ecosystem** consists of communities of organisms together with the physical environment they share. The assemblage of interacting organisms—prokaryotes, protists, animals, fungi, and plants—constitute the **biotic** world within an ecosystem ("biotic" means "having to do with life"). The physical environment that surrounds the biotic community—the atmosphere, water, and Earth's crust—is the **abiotic** part of an ecosystem. An ecosystem is therefore the sum of its biotic and abiotic components. Ecosystems do not always have sharply defined physical boundaries. Instead, ecologists recognize an ecosystem by the distinctive ways in which it functions, especially the means by which energy is acquired by the biotic community (**FIGURE 24.1**).

Ecosystems may be small or very large: a puddle teeming with protists is an ecosystem, as is the Atlantic Ocean. Smaller ecosystems can be nested inside larger, more complex ecosystems. In fact, global patterns of air and water circulation (discussed in Chapter 21) may be viewed as linking all the world's organisms into one giant ecosystem, the biosphere. Whether they are large or small, ecosystems can be challenging to study because organisms, energy, and nutrients often move from one ecosystem to another.

**TO SURVIVE, ALL ORGANISMS NEED ENERGY**, as well as materials to construct and maintain their bodies. For their energy needs, most organisms depend directly or indirectly on solar energy, an abundant supply of which reaches Earth each day. By contrast, materials such as the carbon, hydrogen, oxygen, and other elements of which we are made are added to our planet in relatively small amounts (in the form of meteoric matter from outer space). Earth, therefore, has an essentially fixed amount of materials for organisms to use. This simple fact means that for life to persist, natural systems must recycle materials. In this chapter we focus on these two essential aspects of life—energy and materials—as we discuss *ecosystem ecology*, the study of how energy and materials are used in natural systems.

**FIGURE 24.1**
**Ecosystems in the Snake River valley, near Jackson Hole, Wyoming**

How many ecosystems can you discern in this photo? What is the energy base in each of these ecosystems? Name some abiotic factors that are likely to affect the organisms living in the ecosystems you identified.

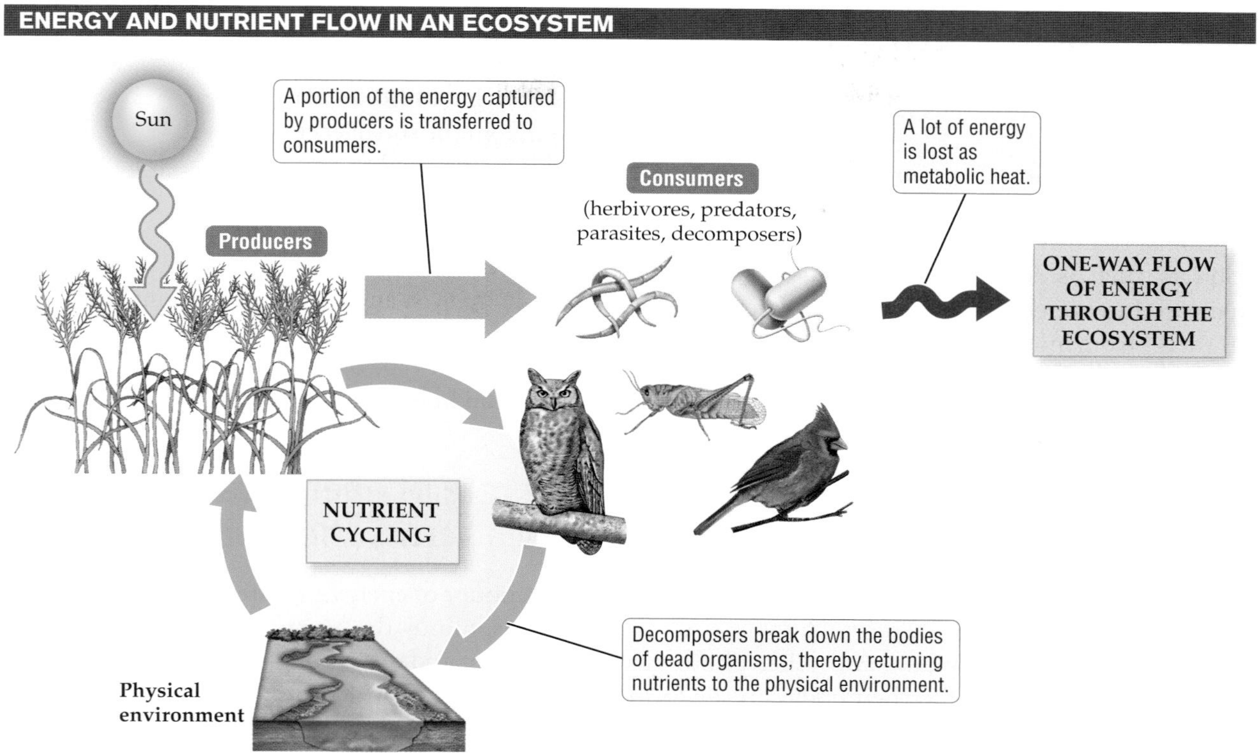

**FIGURE 24.2 How Ecosystems Work**

At each step in a food chain, a portion of the energy captured by producers is lost as metabolic heat (indicated by the red arrows). Therefore, energy flows through the ecosystem in a single direction and is not recycled (yellow, orange, and red arrows). In contrast, nutrients such as carbon and nitrogen cycle between organisms and the physical environment (blue arrows).

Labels within figure:

Sun

A portion of the energy captured by producers is transferred to consumers.

Producers

Consumers
(herbivores, predators, parasites, decomposers)

A lot of energy is lost as metabolic heat.

ONE-WAY FLOW OF ENERGY THROUGH THE ECOSYSTEM

NUTRIENT CYCLING

Physical environment

Decomposers break down the bodies of dead organisms, thereby returning nutrients to the physical environment.

**FIGURE 24.2** gives a broad overview of how ecosystems work—of how energy and nutrients move through natural systems. First, examine the orange and red arrows in the figure, which show the movement of energy through the ecosystem. At each step in a food web (see Section 23.2), a portion of the energy captured by producers is lost as metabolic heat (red arrows). Metabolic heat is a by-product of chemical reactions within a cell. Cellular respiration, the chemical process that most cells use to extract energy from food molecules (see Chapter 9), is responsible for much of the metabolic heat released by any oxygen-utilizing organism. Organisms lose a lot of energy as metabolic heat, as revealed by the fact that a small room crowded with people rapidly becomes hot. Because of this steady loss of heat, *energy flows in only one direction through ecosystems*: it enters Earth's ecosystems from the sun (in most cases) and leaves them as metabolic heat.

In contrast to energy, **nutrients**—the chemical elements required by living organisms—are largely recycled between living organisms and the physical environment. While Earth receives a constant stream of light energy from the sun, Earth does not acquire more nutrients on a daily basis, the way it receives light energy. A constant and finite pool of chemical elements cycles through the land, water, and air, within and between ecosystems. These elements may pass from rocks and mineral deposits to the soil and water and then to producers, various

consumers, and decomposers—and back again to the abiotic world. As seen in Figure 24.2 (blue arrows), nutrients are absorbed from the environment by producers, cycled among different types of consumers for varying lengths of time, and eventually returned to the environment when the ultimate consumers—decomposers—break down the dead bodies of organisms. Ecologists and earth scientists use the term "nutrient cycle" or "biogeochemical cycle" to describe the passage of a chemical element through the abiotic and biotic worlds.

The physical, chemical, and biological processes that link the biotic and abiotic worlds in an ecosystem are known as *ecosystem processes*. Energy capture through photosynthesis, release of metabolic heat, decay of biomass through the action of decomposers, and movement of nutrients from living organisms to the abiotic world around them are all examples of ecosystem processes. The activity of producers, in particular, profoundly influences ecosystem processes and therefore the characteristics of an ecosystem (see Figure 24.3). Ecologists often demarcate an ecosystem according to the types of producers it contains and the community of consumers that these producers support. A duckweed-covered pond, a saltwater marsh, a tallgrass prairie, and a beech-maple woodland are all examples of ecosystems that can be defined by the specific types of producers that capture energy and supply it to a characteristic assemblage of consumers.

## Helpful to Know

In discussing the biology of humans and other animals, "nutrients" refers to vitamins, minerals, essential amino acids, and essential fatty acids (see Chapter 27). In the ecosystem context of this chapter, however, "nutrients" has a more restricted meaning, referring only to the essential elements required by producers.

## 24.2 Energy Capture in Ecosystems

Life cannot exist without a source of energy to support it. Life on Earth depends directly or indirectly on the capture of solar energy by producers. In some unusual habitats, such as deep-sea vents and hot springs, the main producers are prokaryotes that can extract chemical energy from inorganic substances such as iron. Our discussion of energy capture will focus on ecosystems that depend mainly on solar energy, because these are common in both terrestrial (land-based) and aquatic (saltwater and freshwater) regions of the world.

### Ecosystems depend on energy captured by producers

The energy captured by plants and other photosynthetic organisms is stored in their bodies in the form of chemical compounds, such as carbohydrates. Herbivores (which eat plants and other producers), predators (which eat herbivores and other predators), and decomposers (which consume the remains of dead organisms) all depend indirectly on the solar energy originally captured by plants and other producers (such as photosynthetic bacteria and algae).

To better understand these points, imagine that all the plants in a terrestrial ecosystem suddenly vanished. Although bathed in sunlight, herbivores and predators alike would starve because they could not use that energy to produce food. This line of thinking also helps us understand why some environments can support more animals than other environments can. In a tropical forest, for example, many plants can capture energy from the sun (**FIGURE 24.3a**). As a result, a large amount of energy from the sun is stored in chemical forms that can be used as food by animals. In contrast, in environments with few plants (such as Arctic or desert regions, as shown in **FIGURE 24.3b**), relatively little energy is captured from the sun. Hence, less food is available in such environments, and fewer animals can live there.

Assessing the overall amount of energy captured by plants is an important first step in determining how a terrestrial ecosystem works: it influences the amount of plant growth and hence the amount of food available to other organisms. Each of these factors, in turn, influences the type of terrestrial ecosystem found in a region and how that ecosystem functions.

### The rate of energy capture varies across the globe

The amount of energy captured by photosynthetic organisms, minus the amount they expend on cellular respiration and other maintenance processes, is called **net primary productivity** (**NPP**). You can think of NPP as the energy acquired through photosynthesis that is available for growth and reproduction in a plant or other producer. Although NPP is defined in terms of energy, it is usually easier to estimate it as the amount of new **biomass** (the mass of organisms) produced by the photosynthetic organisms in a given area during a specified period of time. In a grassland ecosystem, for example, ecologists would estimate NPP by measuring the average amount of new grass and other plant matter produced in a square meter each year. Such NPP estimates based on biomass can be converted to units based on energy.

NPP is not distributed evenly across the globe. On land, NPP tends to decrease from the equator toward the poles (**FIGURE 24.4a**). This decrease occurs because the amount of solar radiation available to plants also decreases from the equator toward the poles (as we saw in

**FIGURE 24.3**
**Producers Are the Energy Base in an Ecosystem**

(a) In tropical rainforests, the abundant producers (plants) store a lot of chemical energy, which in turn is available to consumers. (b) In deserts, because of the sparse plant life, relatively little chemical energy is available for consumers.

(a)

(b)

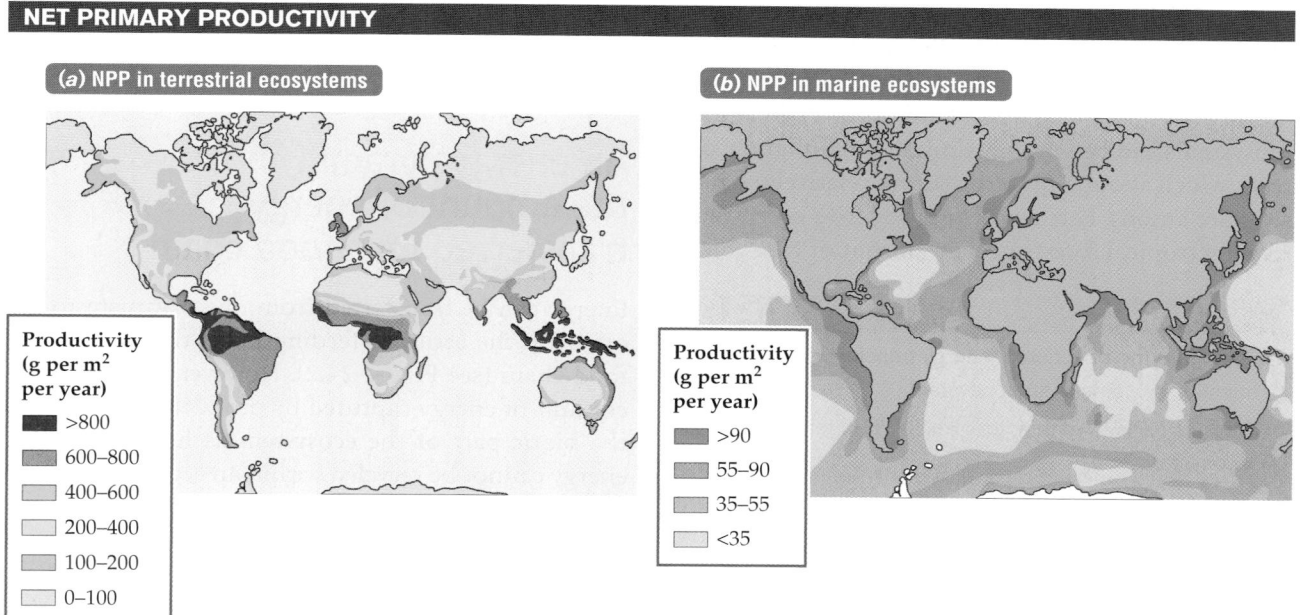

(a) NPP in terrestrial ecosystems

(b) NPP in marine ecosystems

Productivity
(g per m² per year)

- >800
- 600–800
- 400–600
- 200–400
- 100–200
- 0–100

Productivity
(g per m² per year)

- >90
- 55–90
- 35–55
- <35

**FIGURE 24.4**
**Net Primary Productivity Varies Greatly in Both Terrestrial and Marine Ecosystems**

NPP is commonly measured as grams of new biomass made by plants or other producers each year in a square meter of area.

Chapter 21). But there are many exceptions to this general pattern. For example, there are large regions of very low NPP in northern Africa, central Asia, central Australia, and the southwestern portion of North America. Each of these regions is a site of one of the world's major deserts.

The low NPP in deserts underscores the fact that sunlight alone is not sufficient to produce high NPP; water is also required. In addition to water and sunlight, productivity on land can be limited by temperature and the availability of nutrients in the soil. The most productive terrestrial ecosystems are tropical rainforests; the least productive are deserts and tundra (including some mountaintop communities).

The global pattern of NPP in marine ecosystems (**FIGURE 24.4b**) is very different from that on land. There is little tendency for NPP to decrease from the equator toward the poles. Instead, the general pattern relates to distance from shore: the productivity of marine ecosystems is often high in ocean regions close to land but relatively low in the open ocean, which is, in essence, a marine "desert." Calling the open ocean a desert makes sense largely because nutrients needed by aquatic photosynthetic organisms are in short supply there. Nutrients released by the death and decay of organisms tend to settle on the deep-ocean floor, in which case they are not immediately available to photosynthetic producers, which live near the surface. In some coastal areas, wind and ocean currents drive cold, nutrient-rich layers of water to the surface to replace warm, nutrient-depleted water—a process known as **upwelling**. Regions that experience upwelling have high NPP because producers that live there are less nu-

trient limited than are producers in similar ecosystems that lack upwelling. Temperature gradients drive a similar turnover of nutrients every fall and spring in many temperate lakes, making these lakes more productive than lakes that remain unmixed throughout the year.

Nutrients delivered by streams and rivers account for the high productivity of many coastal areas relative to the low productivity of the open oceans. Nutrients drained off the land stimulate the growth and reproduction of phytoplankton, the small photosynthetic producers that form the foundation of aquatic food webs. Estuaries—regions where rivers empty into the sea—are some of the most productive habitats on the planet precisely because the rich nutrient supply supports large populations of producers, which in turn nourish large populations of consumers (**FIGURE 24.5**). Wetlands such as swamps and marshes can also match the productivity levels of tropical forests and farmland. They trap soil sedi-

**FIGURE 24.5**
**The Most Productive Ecosystems**

Wetlands like this estuary in Virginia can harness energy from sunlight and oxygen and translate it into high net primary productivity on a par with terrestrial systems like tropical forests and agricultural fields.

ments rich in nutrients and organic matter, thereby promoting the growth of flooding-tolerant plants and phytoplankton, which in turn feed a complex community of consumers. As on land, the NPP in aquatic ecosystems can be strongly limited by sunlight. Coral reefs, which abound in warm, sunny, relatively shallow seas, are among Earth's most productive ecosystems, rivaling tropical forests in their output of NPP.

> ### Concept Check
>
> 1. Compare the direction of energy transfer with the movement of nutrients in an ecosystem.
>
> 2. What is NPP? What factors limit NPP in terrestrial ecosystems?
>
> 3. Explain why the open ocean has lower NPP than do coastal areas.

## 24.3 Energy Flow through Ecosystems

As described in Chapter 23, organisms capture energy in two major ways, depending on whether they are producers or consumers. Producers get their energy from abiotic sources, such as the sun. Photosynthetic producers—plants, algae, and photosynthetic bacteria—use light energy to turn carbon dioxide and water into biomass. Biomass is a storehouse of chemical energy. Consumers extract energy from the biomass of other organisms.

## An energy pyramid shows the amount of energy transferred up a food chain

Energy can be transferred from one organism to another up the series of feeding levels that make up a food chain (see Figure 24.2). However, eventually every unit of energy captured by producers is lost from the biotic part of the ecosystem as heat. Therefore, energy cannot be recycled within an ecosystem.

As biomass is transferred from one feeding level to the next, the chemical energy locked in that biomass also passes from one feeding level to the next. However, the available biomass, and therefore the available energy, declines with each successive feeding level in a food chain. To illustrate, let's follow the fate of energy from the sun after it strikes the surface of a grassland. A portion of the energy captured by grasses is transferred to the herbivores that eat the grasses, and then to the predators that eat the herbivores. The transfer of energy from grasses to herbivores to predators is not perfect, however. When a unit of energy is used by an organism to fuel its metabolism, some of that energy is lost as unrecoverable heat, as we have seen. Also, the chemical energy locked in the biomass of some producers—woody plants, for example—is inaccessible as an energy source if the primary consumer is unable to digest it. Therefore, as energy is transferred up a food chain, portions of the energy originally captured by photosynthesis are steadily lost. Because of this steady loss of energy, more energy is available at lower levels than at higher levels in a food chain.

The amounts of energy available to organisms in an ecosystem are often represented in what is called an **energy pyramid**. Each level of the pyramid corresponds to a step in a food chain and is called a **trophic level**. **FIGURE 24.6** shows four trophic levels: grass is on the first level, the grasshopper is on the second, the insect-eating bird is on the third, and the bird-eating bird is on the fourth. On average, roughly 10 percent of the energy at one trophic level is transferred to the next trophic level. The energy that is not transferred to the next trophic level is the energy that is not consumed (for example, when we eat an apple we eat only a small part of the apple tree), is not taken up by the body (for example, we cannot digest the cellulose that is contained in the apple), or has been lost as metabolic heat.

**FIGURE 24.6**
**Trophic Levels**
Of each 10,000 kilocalories (kcal; 1 kilocalorie = 1,000 calories) of energy from the sun captured by producers, primary consumers store only about 10 percent. Roughly 10 percent of the energy at each trophic level is then transferred to the next trophic level.

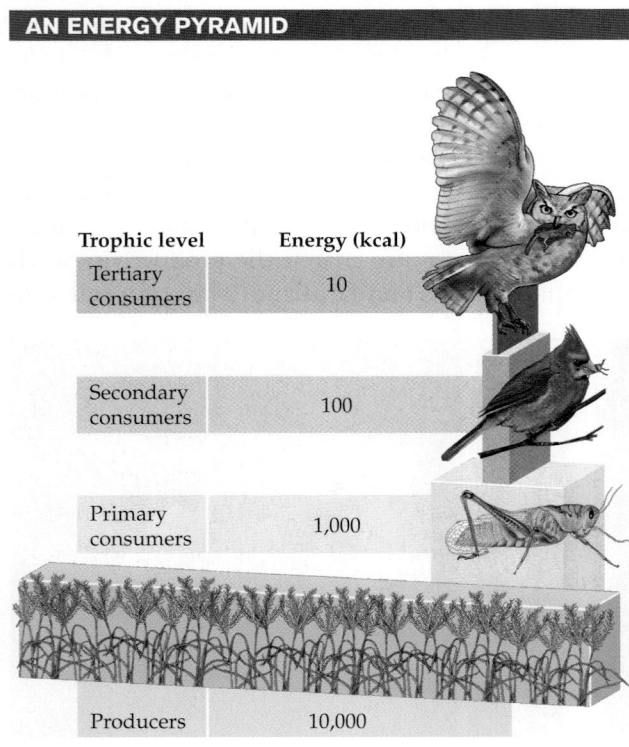

**AN ENERGY PYRAMID**

| Trophic level | Energy (kcal) |
|---|---|
| Tertiary consumers | 10 |
| Secondary consumers | 100 |
| Primary consumers | 1,000 |
| Producers | 10,000 |

## Secondary productivity is highest in areas of high NPP

The rate of new biomass production by *consumers* is called **secondary productivity**. Because consumers depend on producers for both energy and materials, secondary productivity is highest in ecosystems with high net primary productivity. A tropical forest, for example, has a much higher NPP than tundra has. For this reason, per unit of area there are many more herbivores and other consumers in tropical forests than in tundra, and hence secondary productivity is much higher in tropical forests than in tundra.

In natural ecosystems, new biomass made by plants and other producers is consumed by herbivores and by decomposers. In some ecosystems, 80 percent of the biomass produced by plants is used directly by decomposers. Eventually, since all organisms die, all biomass made by producers, herbivores, predators, and parasites is consumed by decomposers (**FIGURE 24.7**). In some instances, people bypass the decomposers, as when we use crops or agricultural refuse to produce fuels.

## 24.4 Biogeochemical Cycles

Nutrients—chemical elements such as carbon, hydrogen, oxygen, and nitrogen—are used by producers and other organisms to construct their bodies. Producers obtain these and other essential nutrients from the soil, water, or air in the form of ions such as nitrate ($NO_3^-$) or inorganic molecules such as carbon dioxide ($CO_2$). Consumers obtain them by eating producers or other consumers. Because these nutrients are essential for life, their availability and movement through ecosystems influence many aspects of ecosystem function.

Nutrients are transferred between organisms (the biotic community) and the physical environment (the abiotic world) in cyclical patterns called **nutrient cycles** or **biogeochemical cycles**. **FIGURE 24.8** provides a general description of how nutrient cycles work. Nutrients can be stored for long periods of time in abiotic reservoirs such as rocks, ocean sediments, or fossilized remains of organisms. The nutrients stored in such abiotic reservoirs are not readily accessible to producers. Weathering of rocks, geologic uplift, human actions, and other forces can move nutrients back and forth between such reservoirs and **exchange pools**—abiotic sources such as soil, water, and air where nutrients *are* available to producers.

**THE ROLE OF DECOMPOSERS**

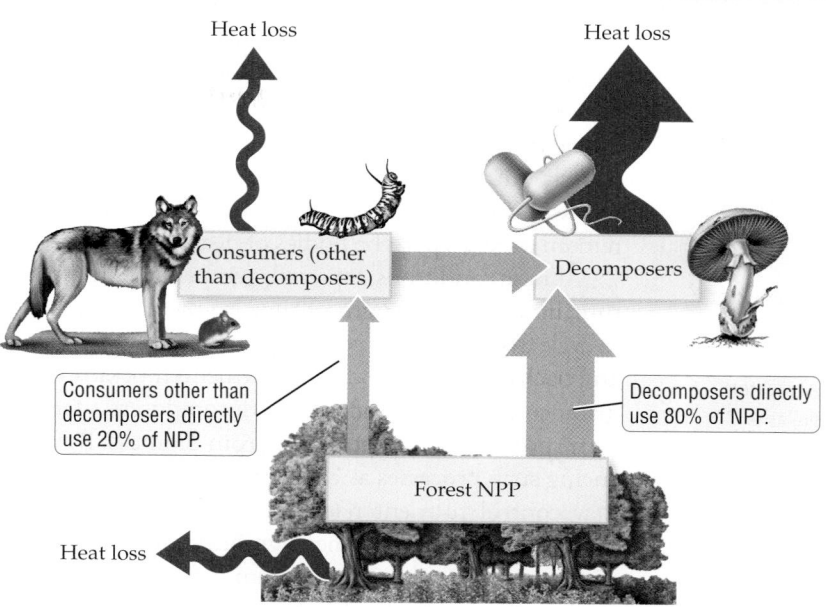

**FIGURE 24.7  Give Thanks to the Decomposers**

Without decomposers, not only would nature's waste pile up, but the soil would become devoid of nutrients. Decomposers such as bacteria and fungi recycle more than 50 percent of net primary productivity in ecosystems of all types. In the forest represented here, 80 percent of NPP is used directly by decomposers, and the remaining 20 percent is used by other consumers (such as herbivores and predators).

**NUTRIENT CYCLING**

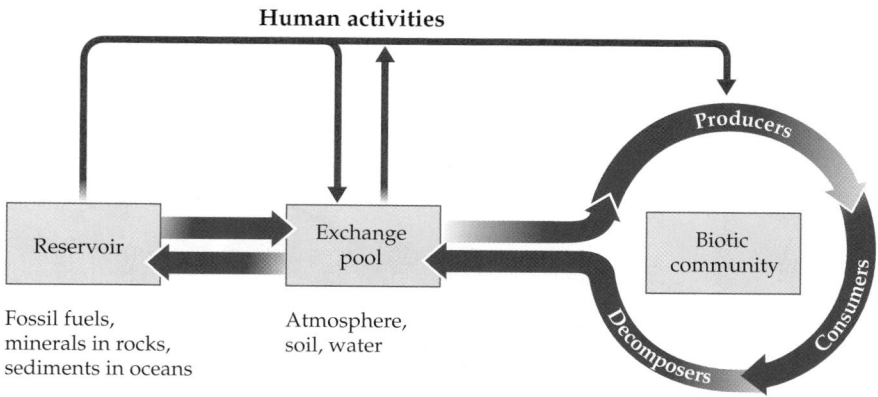

**FIGURE 24.8  Nutrient Cycling**

Nutrients cycle among reservoirs inaccessible to producers, exchange pools available to producers, and living organisms. Human activities, such as fossil fuel use and fertilizer synthesis, can move nutrients between reservoirs and exchange pools, altering their availability to producers and changing nutrient cycles.

Helpful to
Know

"Reservoir" has
two meanings: It
can refer to a body
of water held aside
for use by people.
More generally, a
reservoir is any place
or substance that
holds something (not
necessarily water)
in storage, as is
meant with regard to
nutrient cycling.

Once captured by a producer, a nutrient can be passed from the producer to an herbivore, then to one or more predators or parasites, and eventually to a decomposer. Decomposers break down once-living tissues into simple chemical components, thereby returning nutrients to the physical environment. Without decomposers, nutrients could not be repeatedly reused, and life would cease because all essential nutrients would remain locked up in the bodies of dead organisms.

Abiotic conditions, especially temperature and moisture, influence the length of time it takes for a nutrient to cycle from the biotic community to the exchange pool and back again to a producer. Warmer temperatures, for instance, increase the activity of decomposers, which in turn speeds up nutrient release from biomass. By influencing such processes as weathering and runoff, rainfall may control nutrient release from nutrient reservoirs and also determine whether significant amounts of nutrients are lost from one ecosystem (a forest community, for example) to another (such as a stream).

How long it takes for a nutrient to cycle from a producer to the physical environment and back to another producer varies from one type of nutrient to another. Some chemical elements are transported more quickly and over longer distances than others, depending on whether or not they have an *atmospheric cycle*. Carbon, hydrogen, oxygen, nitrogen, and sulfur are all essential nutrients that have atmospheric cycles. Nutrients that have an **atmospheric cycle** can exist as a gas under natural conditions, which means they can be released into the atmosphere, or be absorbed from it, by the biotic and abiotic components of an ecosystem. Once in the atmosphere, these nutrients can be transported by wind from one region of Earth to another. When nutrients are transported long distances in this way, they can affect nutrient cycles in distant ecosystems.

Phosphorus is one of the few major nutrients that does not have an atmospheric cycle and displays an exclusively *sedimentary cycle* instead. A nutrient is said to have a **sedimentary cycle** if it is not commonly found as a gas under natural conditions and moves mostly through land and water, rather than the atmosphere, in its passage through an ecosystem. Nutrients with sedimentary cycles tend to move slowly and are not dispersed as widely as those with atmospheric cycles.

## Carbon cycling between the biotic and abiotic worlds is driven by photosynthesis and respiration

Living cells are built mostly from organic molecules— molecules that contain carbons bonded to hydrogen atoms. Next to oxygen, carbon is the most plentiful element in cells by weight. Every one of the main macromolecules in an organism has a backbone of carbon atoms (see Chapter 5). The transfer of carbon within biotic communities, between living organisms and their physical surroundings, and within the abiotic world is known as the global **carbon cycle** (**FIGURE 24.9**).

Although carbon makes up a large part of biomass, the element is not abundant in the atmosphere. Carbon, in the form of $CO_2$ gas, makes up only about 0.04 percent of Earth's atmosphere, although that

**FIGURE 24.9**

**THE CARBON CYCLE**

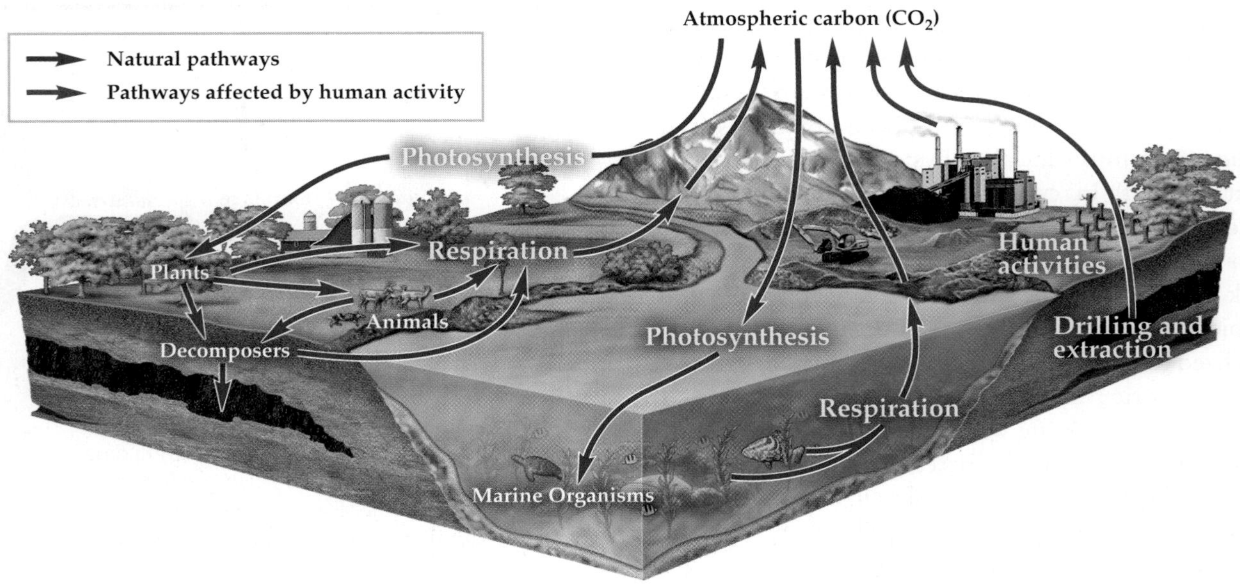

Natural pathways
Pathways affected by human activity

Atmospheric carbon ($CO_2$)

Photosynthesis

Respiration

Plants

Animals

Decomposers

Photosynthesis

Human activities

Drilling and extraction

Respiration

Marine Organisms

percentage has been creeping upward every year for the last 100 years or so (the effect of increasing $CO_2$ on global warming is covered in Chapter 25). The oceans represent the largest store of carbon on our planet. Most of this carbon is dissolved inorganic carbon (such as bicarbonate ions, $HCO_3^-$), and a minor portion is held in the biomass of marine organisms. Earth's crust generally contains only about 0.038 percent carbon by weight, but it also has regions with carbon-rich sediments and rocks formed from the remains of ancient marine and terrestrial organisms (see Figure 24.9). The deep layers of carbon-containing rocks and sediments constitute an inaccessible reservoir of carbon in natural ecosystems.

Some of the organic matter from ancient organisms has been transformed by geologic processes into deposits of fossil fuel, such as petroleum, coal, and natural gas. When we extract these fossil fuels and burn them to meet our energy needs, the carbon locked in these deposits for several hundred million years is released into the atmosphere as carbon dioxide (shown by a red arrow in Figure 24.9). Humans today have the power to change the global carbon cycle but, as we discuss in Chapter 25, altering an ecosystem process that is so vital, large-scale, and still poorly understood might prove dangerous for natural ecosystems and for human welfare.

Living organisms, in both aquatic and terrestrial ecosystems, acquire carbon mostly through photosynthesis. Aquatic producers, such as photosynthetic bacteria and algae, can absorb dissolved carbon dioxide (in the form of bicarbonate or carbonate ions) and convert it into organic molecules using sunlight as a source of energy. Plants, the most important producers in terrestrial ecosystems, absorb $CO_2$ from the atmosphere and transform it into food with the help of sunlight and water. The extraction of energy from food molecules, by producers and consumers alike, releases $CO_2$ and returns it to the abiotic world. Cellular respiration is the main pathway by which most organisms extract energy from organic molecules, although some organisms, including many decomposers, use other pathways that do not depend on molecular oxygen.

Decomposers release a great deal of the carbon contained in the dead organisms they live on, but some of it typically remains in the ecosystem as partially decayed organic matter. Leaf litter, humus, and peat are examples of partially decayed organic matter that form an important store of soil carbon. The top layers of soil in northern coniferous forests have large amounts of organic matter because decomposition is slow in these cold, wet regions. The vast peat deposits found in the Arctic tundra and northern boreal forests represent a large carbon store in the biosphere. Because decomposition is faster at warmer temperatures, in terrestrial ecosystems typical of the tropics the soil carbon levels are lower and carbon cycles more rapidly.

## Biological nitrogen fixation is the most important source of nitrogen for biotic communities

Nitrogen is a key component of amino acids, proteins, and nucleic acids such as DNA. It is therefore an essential element for all living organisms. The fact that this nutrient is not naturally abundant in soil and water limits the growth of producers in most ecosystems that are unaltered by human actions.

Earth's atmosphere is the largest store of nitrogen in the biosphere: $N_2$ gas (molecular nitrogen) makes up 78 percent of the air we breathe. Energy from lightning converts some of the atmospheric $N_2$ into nitrogen compounds that mix with rainwater to form nitrate ions ($NO_3^-$), as shown in **FIGURE 24.10**. Nitrate ions are the most common form of nitrogen in soil and water, although smaller amounts of ammonium ions ($NH_4^+$) also occur naturally. Nitrate is highly soluble and can therefore be easily lost from an ecosystem through runoff.

Lightning contributes only a small amount of the nitrogen found in soil and water, with much of it originating in a remarkable metabolic process, known as biological *nitrogen fixation*, that can be carried out only by some prokaryotes. Waste produced by animals, and the death and decay of organisms, adds ammonium ions to soil and water (see Figure 24.10). Bacteria common in most ecosystems rapidly convert ammonium ions to nitrate, which is why ammonium ions do not accumulate at high levels in most habitats. Denitrifying bacteria, common in oxygen-poor environments, convert nitrate into molecular nitrogen ($N_2$) or into $N_2O$ (nitrous oxide), returning nitrogen to the vast atmospheric pool.

The conversion of molecular nitrogen ($N_2$) into ammonium ions ($NH_4^+$) by bacteria is known as biological **nitrogen fixation**. Some nitrogen-fixing bacteria live free in the soil, and some fix nitrogen in a mutualistic symbiosis with certain plants, including legumes (members of the bean family), alder trees, and an aquatic fern called *Azolla*. The plant host obtains ammonium from its nitrogen-fixing partner and supplies to the prokaryote food energy derived from photosynthesis. Nitrogen fixed by bacteria is added to soil or water when these organisms, and any host species they might associate with, die and decompose. Consumers acquire nitrogen when they eat plants that have acquired nitrogen from the soil or through their association with nitrogen-fixing bacteria. The action of

**FIGURE 24.10**

**THE NITROGEN CYCLE**

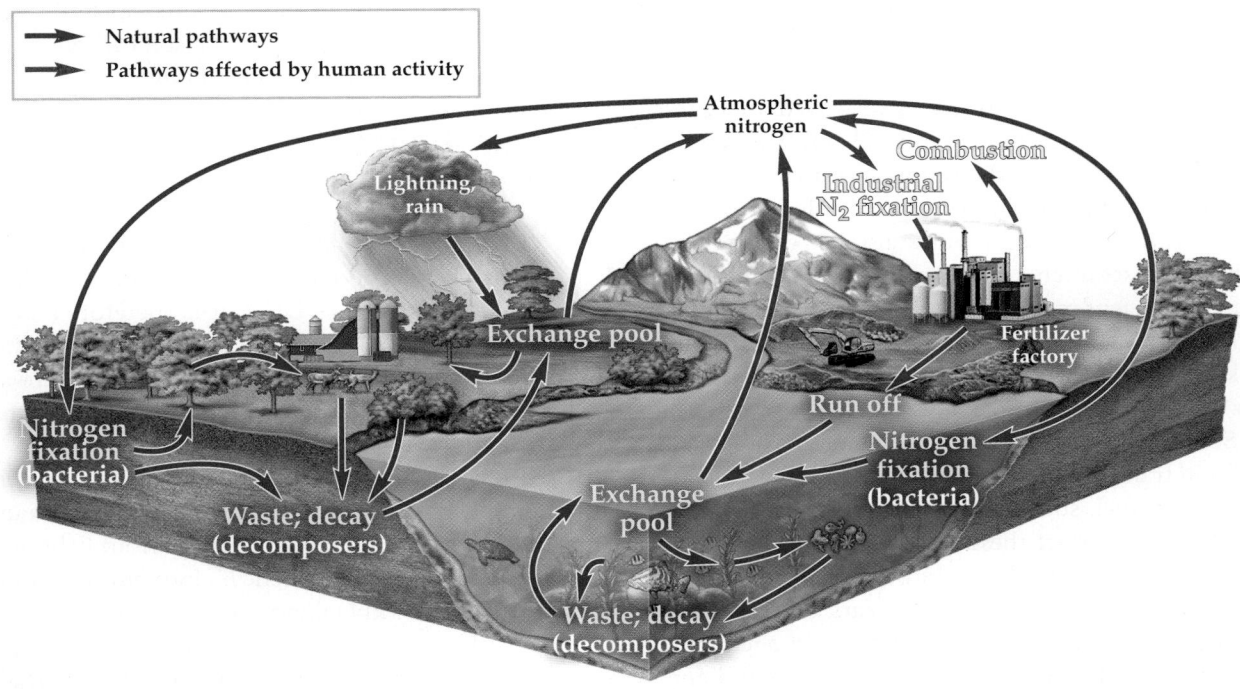

Natural pathways
Pathways affected by human activity

Nitrogen-fixing
nodules on a pea plant

decomposers transfers nitrogen from the biotic community to the abiotic component of an ecosystem.

Like carbon, nitrogen is said to have an atmospheric cycle because the element passes through a gaseous phase during its cycling through an ecosystem. Like all other nutrients with an atmospheric cycle, nitrogen can be transported long distances in the biosphere. Humans manufacture fertilizer through industrial nitrogen fixation, a process that combines nitrogen gas ($N_2$) with hydrogen gas ($H_2$) to make ammonium compounds. The manufacturing process relies on high temperature and pressure generated by the burning of fossil fuels. Applying nitrogen fertilizer to crop plants boosts their productivity, but the runoff can stimulate NPP in aquatic ecosystems in ways that are harmful to the biotic community. Excess nitrogen can disturb terrestrial communities as well, sometimes decreasing diversity (species number and abundance). This shift in diversity has been noted in some grassland communities, in which a few species that responded exceptionally vigorously to supplemental nitrogen were able to outcompete the other species.

## Sulfur is one of several important nutrients with an atmospheric cycle

Sulfur is a component of certain amino acids, many proteins, and some polysaccharides and lipids. It is also found in other organic compounds that have important roles in metabolism. Sulfur displays an atmospheric cycle because it moves easily among terrestrial ecosystems, aquatic ecosystems, and the atmosphere. Sulfur enters the atmosphere from terrestrial and aquatic ecosystems in three natural ways (**FIGURE 24.11**): as sulfur-containing compounds in sea spray; as a metabolic by-product (the gas hydrogen sulfide, $H_2S$) released by some types of bacteria; and, least important in terms of overall amount, as a result of volcanic activity.

About 95 percent of the sulfur that enters the atmosphere from the world's oceans does so in the form of strong-smelling sulfur compounds (such as dimethyl sulfide) that are breakdown products of organic molecules made by phytoplankton. The odorous compounds are lofted into the air by wave action and contribute to the smell we associate with seaside air. Hydrogen sulfide is another smelly gas. It is generated by metabolic reactions in bacteria that inhabit oxygen-poor environments such as swamps and sewage.

Sulfur enters terrestrial ecosystems through the weathering of rocks and as atmospheric sulfate ($SO_4^{2-}$) that mixes with water and is deposited on land as rain. Sulfur enters the ocean in stream runoff from land and, again, as sulfate lost from the atmosphere and falling as rain. Once in the ocean, sulfur cycles within marine ecosystems before being lost in sea spray or deposited in sediments on the ocean bottom. Like most other nutrients with atmospheric cycles, sulfur cycles through

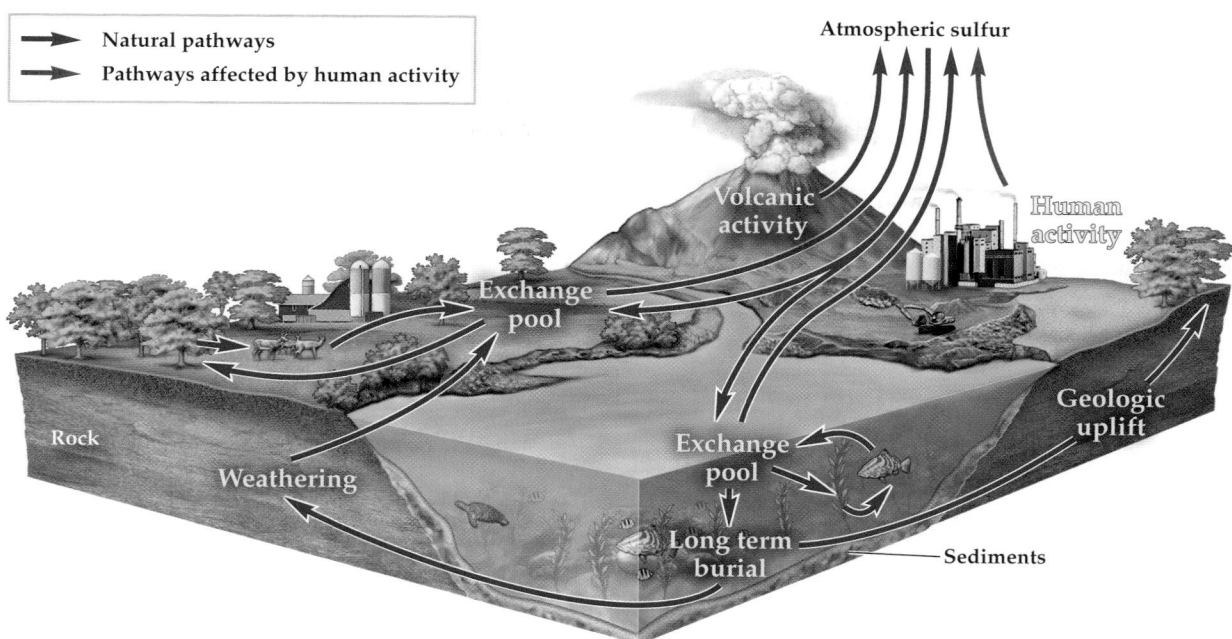

terrestrial and aquatic ecosystems relatively quickly. As we will see shortly, human activities can also add sulfur to the atmosphere, often with adverse consequences for biotic communities and human economic interests.

## Phosphorus is the only major nutrient with a sedimentary cycle

Phosphorus is a vital nutrient because it is a component of crucial macromolecules, such as DNA. Phosphorus is important to ecosystems because it strongly affects net primary productivity, especially in aquatic ecosystems. NPP usually increases, for example, when phosphorus is added to lakes. As we will see in Section 24.5, such an increase in productivity can have undesirable effects, including eutrophication, which leads to the death of aquatic plants, fishes, and invertebrates (see Figure 24.9 and look ahead to Figure 24.13).

Among the major nutrients that cycle within ecosystems, phosphorus is the only one with a sedimentary cycle (**FIGURE 24.12**). Soil conditions usually do not allow bacteria to carry out the chemical reactions required for

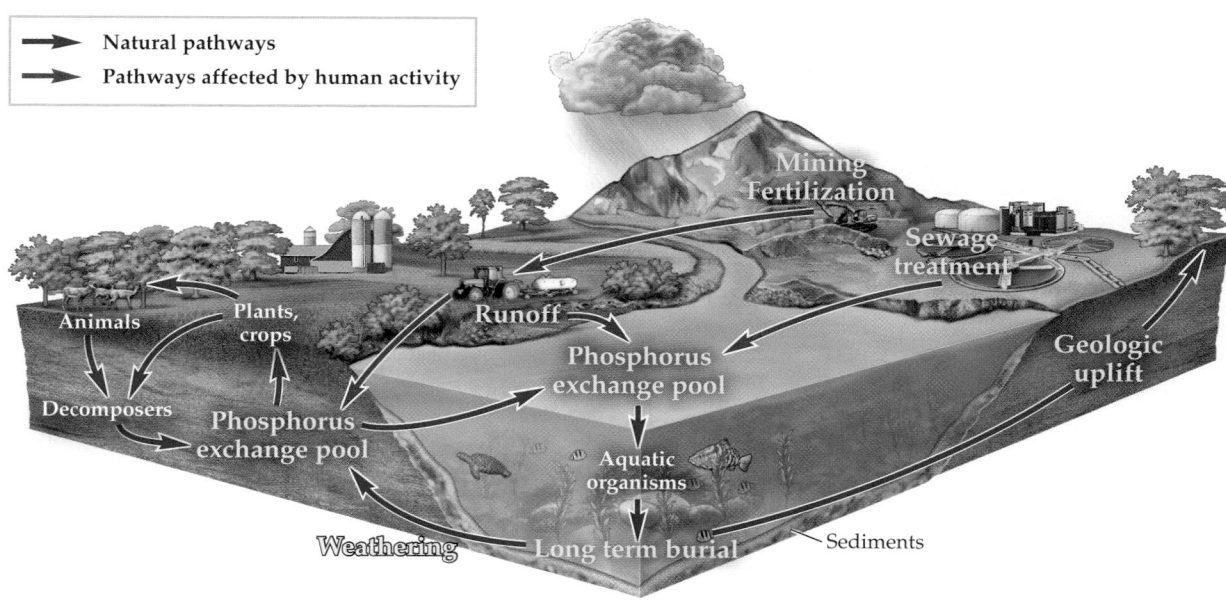

the production of a gaseous form of phosphorus (phosphine, $PH_3$). Nutrients like phosphorus first cycle within terrestrial and aquatic ecosystems for variable periods of time (from a few years to many thousands of years); then they are deposited on the ocean bottom as sediments. Nutrients may remain in sediments, unavailable to most organisms, for hundreds of millions of years. Eventually, however, the bottom of the ocean is thrust up by geologic forces to become dry land, and once again the nutrients in the sediments may be available to organisms. Sedimentary nutrients usually cycle very slowly, so they are not replaced easily once they are lost from an ecosystem.

## 24.5 Human Actions Can Alter Ecosystem Processes

Humans have been disrupting ecological communities for many hundreds, perhaps thousands, of years. The history of the Moai-building Polynesian settlers of Easter Island (see Chapter 22) demonstrates that such disruptions can produce tragic consequences for humans. The disruptions of preindustrial times are dwarfed, however, by the scale of the ecological changes that humans have brought about in the past 200 years. In this section we examine some of the ecosystem processes that are readily altered by human activities, reserving a broader discussion of the human impact on the biosphere for the next chapter.

## Human activities can increase or decrease NPP

Human activities can change the amount of energy captured by ecosystems on local, regional, and global scales. For example, rain can cause fertilizers to wash from a farm field into a stream, which then flows into a lake. The addition of extra nutrients to lakes or streams or offshore waters leads to **eutrophication** (literally, "overfeeding"). In a eutrophic lake, NPP increases because the added nutrients cause photosynthetic algae to become more abundant, so they absorb more energy from the sun.

It is not necessarily a good thing when human actions increase NPP. In some instances, nutrient enrich-

### DEAD ZONE IN THE GULF OF MEXICO

(a)

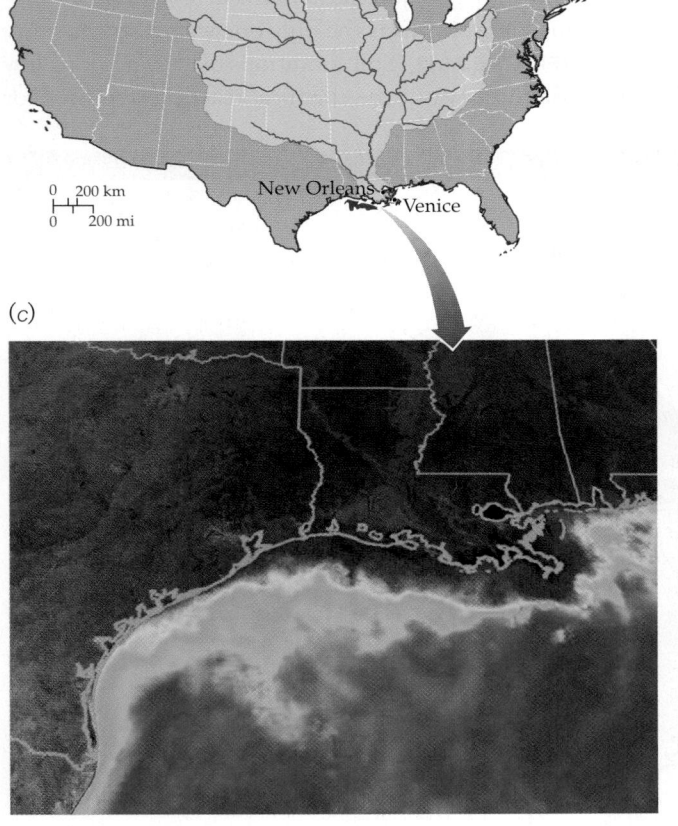

(b)

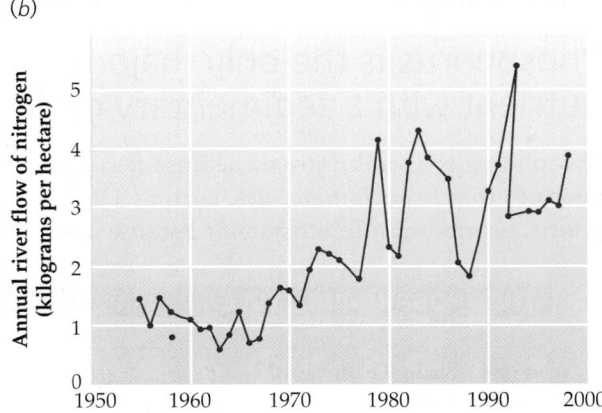

(c)

**FIGURE 24.13 Nutrient Loading Can Deplete Oxygen in Aquatic Ecosystems**

The Mississippi and Atchafalaya river basins drain nearly 40 percent of U.S. land area (a). Since the 1970s, large amounts of nitrogen (N) from fertilizer, sewage, and industrial by-products have entered waters within these watersheds. The resulting nutrient addition can be measured as the amount of nitrogen flowing past particular points on the Mississippi per year (b). This nitrogen then drains into the Gulf of Mexico, where each summer the extra nutrients create a large dead zone (c) in which virtually all animals are killed. The image, based on satellite photos from NASA, shows ocean turbidity and chlorophyll concentrations. Red and orange represent high concentrations of phytoplankton and river sediment.

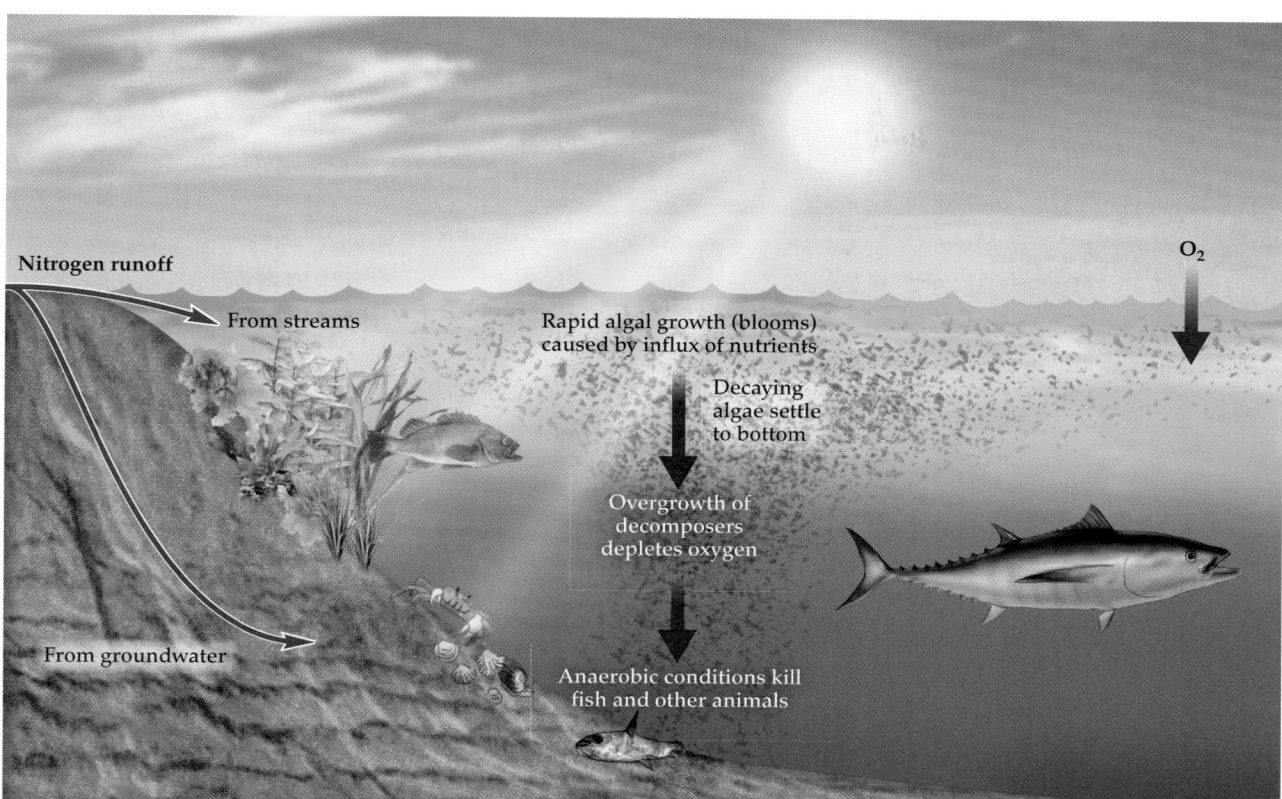

FIGURE 24.14
**Eutrophication Can Be Too Much of a Good Thing**
Large inputs of nutrients such as nitrogen and phosphorus can cause a population boom in photosynthetic plankton (algae). Dying algae in turn trigger explosive growth of oxygen-consuming decomposers, choking off the oxygen supply for animals.

*(Figure labels: Nitrogen runoff; From streams; Rapid algal growth (blooms) caused by influx of nutrients; Decaying algae settle to bottom; Overgrowth of decomposers depletes oxygen; From groundwater; Anaerobic conditions kill fish and other animals; $O_2$)*

ment has increased NPP to such an extent that large bodies of water have become nearly devoid of animal life. For example, each summer large amounts of nitrogen and phosphorus are carried by the Mississippi and other rivers to the Gulf of Mexico (**FIGURE 24.13**). Algal populations (and hence NPP) increase in these nutrient-rich waters, which float on top of the colder, saltier waters of the gulf. As the algae die, they drift into deeper waters, where bacteria decompose their bodies, using oxygen in the process (**FIGURE 24.14**). A larger-than-normal "rain" of dead algae triggers a population boom among decomposers, and dissolved oxygen declines as a result. Oxygen levels may drop so low that virtually all animals die or flee, creating a large "dead zone" (see Figure 24.13c). This dead zone threatens to diminish the fish and shellfish industry in the gulf, which produces an annual catch worth about $500 million. In the summer of 2002, the dead zone reached its largest size ever—about 22,000 square kilometers (8,500 square miles)—covering an area greater than the state of Massachusetts.

Human activities can also change NPP on land. For example, NPP decreases when logging and fire convert tropical forest to grassland. Globally, scientists estimate that human activities leading to such land conversions have decreased NPP in some regions while increasing it in other regions, but the net effect is a 5 percent decrease in NPP worldwide.

## Human activities can alter nutrient cycles

Human activities can have major effects on nutrient cycles. Ecologists have shown, for example, that the clear-cutting of a forest, followed by spraying with herbicides to prevent regrowth, causes the forest to lose large amounts of nitrate, an important source of nitrogen for plants (**FIGURE 24.15**).

On a larger geographic scale, nutrients such as nitrogen and phosphorus used to fertilize crops can be carried by streams to a lake or an ocean hundreds of kilometers away, where they can increase NPP and cause the body of water to become eutrophic. Finally, on a still larger geographic scale, many human activities, such as shipping crops and wood to distant locations, transport nutrients around the globe. Many human activities also release chemicals into the air, which global wind patterns then move over long distances.

When people alter atmospheric nutrient cycles, the effects are often felt across international borders. Consider sulfur dioxide ($SO_2$), which is released into the atmosphere when we burn fossil fuels such as oil and coal. The burning of fossil fuels has altered the sulfur cycle greatly: annual human inputs of sulfur into the atmosphere are more than one and a half times the inputs from all natural sources combined.

(a)

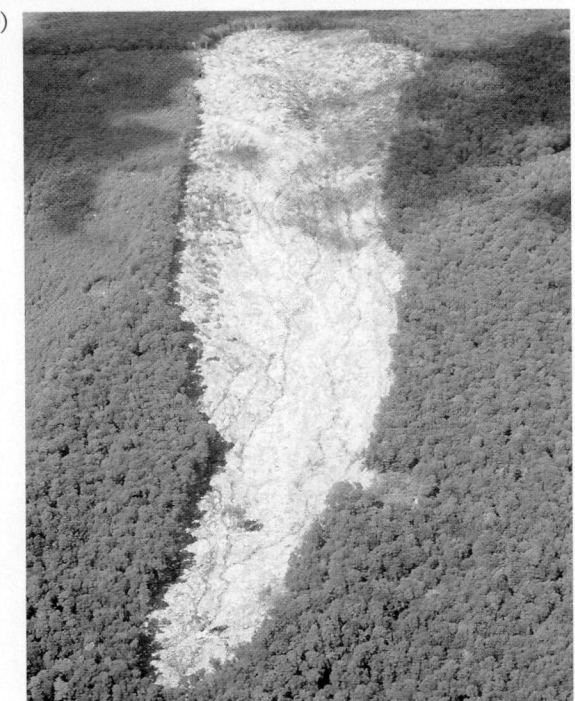

(b)

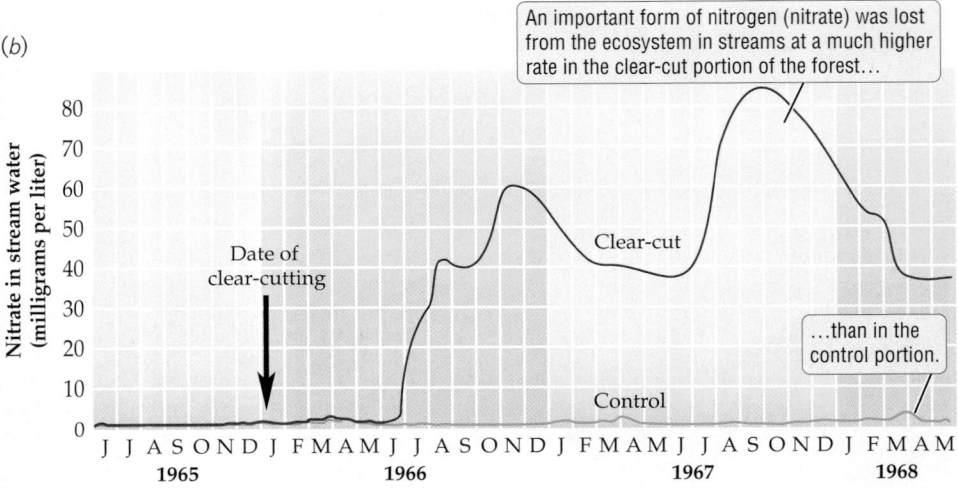

An important form of nitrogen (nitrate) was lost from the ecosystem in streams at a much higher rate in the clear-cut portion of the forest...

...than in the control portion.

**FIGURE 24.15 Altering Nutrient Cycles in a Forest Ecosystem**

(a) This portion of the Hubbard Brook Experimental Forest in New Hampshire has been clear-cut. (b) A portion of the forest was first clear-cut and then sprayed with herbicides for 3 years to prevent regrowth. A second portion of the forest was not clear-cut or sprayed and served as a control plot. Nitrate, a form of nitrogen that is important to plants, was lost from the ecosystem in streams at a much higher rate in the clear-cut portion of the forest than in the control portion. (The loss of nitrate was measured in milligrams of nitrate in the stream per liter of water.)

Most human inputs of sulfur into the atmosphere come from heavily industrialized areas such as central Europe and eastern North America. Once in the atmosphere, $SO_2$ combines with oxygen and water to become sulfuric acid ($H_2SO_4$), which then returns to the land in rainfall. Rainfall normally has a pH of 5.6, but sulfuric acid (as well as nitric acid, $HNO_3$, caused by nitrogen-containing pollutants) has caused the pH of rain to drop to values as low as 2 or 3 in the United States, Canada, Great Britain, and Scandinavia (see Chapter 5 for a review of pH). Rainfall with a low pH is called **acid rain**.

Acid rain can have devastating effects on human-made structures (such as statues) and on natural ecosystems. Acid rain has drastically reduced fish populations in thousands of Scandinavian and Canadian lakes. Much of the acid rain that falls in these lakes is caused by sulfur dioxide pollution that originates in other countries (such as Great Britain, Germany, and the United States). Acid rain has also caused extensive damage to forests in North America and Europe (**FIGURE 24.16**).

**FIGURE 24.16 Acid Rain Can Damage Many Ecosystems**

(a) This spruce forest in the Jizerské Mountains of the Czech Republic was killed by acid rain. (b) Thanks to regulations imposed by the 1990 Clean Air Act, the amount of sulfur dioxide, a major contributor to acid rain, emitted into the atmosphere each year in the United States has fallen by more than 50 percent since 1980.

ACID RAIN

(a) Effect of acid rain

(b) Effect of Clean Air Act

# Is There a Free Lunch? Ecosystems at Your Service

Next time you drink a glass of water, think for a moment about where your water comes from. If you are like many of us, you may not know. Does it come from surface waters, such as rivers, lakes, or reservoirs? Or does it come from deep underground? Consider New York City. The 8 million people who live there get about 90 percent of their water from the Catskill Watershed, with the remainder coming from the Croton Watershed. Together, these watersheds store 580 billion gallons of water in 19 reservoirs and 3 controlled lakes. Over 1.3 billion gallons of this water is delivered to New York each day. The water flows by gravity to the city in a vast set of pipes, some of them large enough to drive a bus through.

For years, New Yorkers drank high-quality water, essentially for free: their water was kept pure by the root systems, soil microorganisms,

and natural filtration processes of forests in the Catskill and Croton watersheds. By the late 1980s, however, pollutants such as sewage, fertilizers, pesticides, and oil had begun to overwhelm these purification processes, causing the quality of the water to decline. In the early 1990s the city embarked on an ambitious but simple plan: protect the watershed's environment so that natural systems could resume supplying the city with clean water, for about a tenth of the cost that a conventional water treatment project would have cost. The city is buying land that borders rivers and streams in the Catskills, and protecting the land from development to minimize the flow of fertilizers, pesticides, and other pollutants into the water.

New York City's decision to invest in the long-term health of the watershed shows that what is good for the environment can also be sound economic policy.

Ecological communities serve human communities in varied and valuable ways. The ecosystem processes and resources that benefit humankind are called *ecosystem services*. For example, floodplains provide us with a free service: they act as safety valves for major floods, preventing even larger floods. Floodplains normally function as huge sponges: when streams and rivers overflow, floodplains absorb the excess water, thereby preventing even more severe floods from occurring farther downstream. Dikes and levees are often constructed, and rivers are often diverted in an attempt

Flood devastation in the Midwest

to protect homes or industrial areas located in what were once floodplains. But in preventing rivers from overflowing into floodplains, we reduce the ability of the ecosystem to handle periods of heavy rainfall.

Although they are "free," ecosystem services should never be taken for granted. Critical ecosystem services include removal of pollutants from the air by plants, pollination of plants by insects (essential for about 30 percent of U.S. food crops), maintenance of breeding grounds for shellfish and fish in marine ecosystems, prevention of soil erosion by plants, screening of dangerous ultraviolet light by the atmospheric ozone layer, and moderation of the climate by vegetation. Intact and healthy natural ecosystems hold cultural, recreational, aesthetic, and spiritual value for many.

Ecosystem services are essential for maintaining healthy ecological communities, and they also provide people with enormous economic benefits. Globally, the value of ecosystem services provided by lakes and rivers alone has been estimated by some researchers to be a staggering $1.7 trillion each year. Other examples of the value of ecosystem services include the world's fish catch (valued at between $50 billion and $100 billion per year) and the billions of dollars' worth of crops that could not be produced if plants were not pollinated by insects.

New York City's water supply system

Albany

Oneonta

Catskill Watershed

Kingston

Liberty

Poughkeepsie

Ellenville

Catskill Aqueduct

Croton Watershed

PENNSYLVANIA

NEW YORK

NEW JERSEY

Croton Aqueduct

New York City

MASSACHUSETTS

CONNECTICUT

Hudson River

- Catskill Watershed area
- Croton Watershed area
- Rivers and reservoirs
- Aqueducts and tunnels
- State borders

The international nature of the acid rain problem has led nations to agree to reduce sulfur emissions. In the United States, annual sulfur emissions were cut by more than 50 percent between 1980 and 2008 (see Figure 24.16). Such reductions are a very positive first step, but the problems resulting from acid rain will be with us for a long time: acid rain alters soil chemistry and therefore has effects on ecosystems that will last for many decades after the pH of rainfall returns to normal levels.

**Concept Check**

1. Does the amount of energy transferred up a food chain increase, decrease, or stay the same? Explain.

2. How does runoff from farm fields and outflow from sewage treatment facilities contribute to the "dead zone" in the Gulf of Mexico?

## APPLYING WHAT WE LEARNED

# What Happens When the Worst Happens?

The Gulf of Mexico is one of the most productive fisheries in the world, a maritime treasure that provides major ecosystem services to humans. The Gulf supports a multibillion-dollar fishing industry that produces shrimp, crabs, oysters, and fish. Gulf marshlands protect coastal cities from flooding, clean waste from water, and nurture marine life. Scientists estimate that 75–90 percent of animal species in Gulf Coast waters—including fish, oysters, anchovies, and shrimp—spend at least part of their lives in estuaries and marshes. The oil released by the spill could potentially kill thousands of young marine animals and marsh plants.

A year after the disastrous oil spill, scientists took stock. One government official optimistically reported that the Gulf would be fully recovered by 2012. Indeed, the methane from the well seemed to have vanished altogether, and researchers reported that microbes had consumed it. And just days after the gushing well was finally capped, the surface oil had all but vanished from satellite photos.

Where had the oil gone? Some of it apparently dispersed naturally through the churning action of waves and ocean currents. By some estimates, the bright Gulf sunshine had evaporated a quarter of the oil. Some went up in flames after BP set fire to it. And BP poured about 2 million gallons of a chemical "dispersant" into the ocean; the dispersant broke the thick globs of oil into droplets that spread through the water column instead of congealing at the surface. Large amounts of oil fell to the seafloor, where, biologists reported, thick mucus-like layers of oil covered ocean beds miles from the spill site.

Damage to beaches and marshes could have been far worse. Ocean currents had carried the bulk of the oil away from the coast. And Louisiana had released Mississippi River water into coastal marshes, possibly pushing the oil offshore. Crab and shrimp fisheries, some fishery biologists announced, would be fine within a year or two, as would most fish populations.

Yet, other biologists remained pessimistic. To be sure, a months-long fishing moratorium had taken the pressure off marine populations. But laboratory work demonstrated that the mixture of oil and dispersant could kill, stunt, or sicken juvenile and adult fish and crabs. With fewer of these animals near the bottom of the food chain, there would be less food later for red snapper and bluefin tuna to eat. The food chain appeared to have some damaged links.

Other scientists argued that the oil was still around, buried under beaches or coating the seafloor, where it would be suffocating tube worms, corals, sea urchins, crustaceans, and other marine creatures that play vital roles in the Gulf ecosystem. One biologist described offshore reefs, where fish breed, as "dead"; and another compared parts of the Gulf seafloor to a "graveyard." Half of Louisiana's oyster beds were destroyed—partly by oil and partly by the surge of fresh Mississippi River water that had flooded the oyster beds for months—and could take a decade to recover.

Ecosystem damage from the *Exxon Valdez* spill continued for years after, and oil damage may continue in the Gulf as well . After both spills, researchers recorded unusual numbers of fish with skin sores, parasites, strange pigmentation patterns, damaged organs, and oil compounds in their organs. After the *Exxon Valdez* oil spill, a herring fishery near the spill collapsed, and 20 years later it still had not recovered. Meanwhile, oil sitting on the Gulf seafloor is not degrading as quickly as expected, and ecologists are still studying the aftermath of the spill.

# Lawrence Berkeley Lab Scientists Tinker with Microbes to Battle Climate Change

BY SUZANNE BOHAN, *Contra Costa Times*

Microbes will take center stage next week as the unusual protagonists in a free talk with local scientists tackling climate change.

The one-celled creatures are now part of a major research initiative at Lawrence Berkeley Laboratory to explore ways to recapture more of the 6 billion tons of heat-trapping carbon dioxide human activities generate annually.

The researchers . . . are working with the Joint Genome Institute in Walnut Creek, a federal facility for genetic sequencing of microbes and other life-forms.

The terrestrial ecosystem, meaning everything on land—from plants to soil and other geological features—absorbs about one-third of atmospheric carbon. Much of the credit goes to microbes, said Donald DePaolo, an associate lab direc-

tor who heads the Berkeley Lab's Carbon Cycle 2.0 initiative.

"Microbes are doing a huge job of taking carbon out of the atmosphere," DePaolo said. But how microbes, usually bacteria, do their work is still largely a mystery, he said.

The Carbon Cycle 2.0 initiative pursues scientific advances that would allow humanity to produce enough energy for a comfortable existence for the planet's population, without unleashing vast quantities of carbon dioxide.

At Monday's talk, two scientists from the genome institute and one Berkeley Lab scientist will share new research on microbes and their role in greenhouse gas emissions . . .

Terry Hazen, a Berkeley Lab scientist, will describe the capacity of bacteria to consume oil during the Gulf of Mexico

spill last year. Surprised scientists also found that the bacteria—once they finished off the oil—turned their appetite to the huge quantities of methane released by the oil spill. Methane is 20 times more powerful a heat-trapping gas than $CO_2$.

"The methane started to deplete very, very fast," he said. That knowledge eases some concerns that ocean warming might release methane stored deep below the ocean floor. Instead, it's possible—although not certain—that bacteria might consume at least some of it.

Hazen will also describe work in the Arctic, where melting tundra could lead to huge releases of methane by bacteria consuming thawed plant material long stored in the once-frozen ground. They're experimenting with ideas for stopping the bacteria from releasing methane.

Microbes play a central role in ecosystems. For most of the history of Earth, bacteria, archaeans, and other microbes have dominated biogeochemical cycles—including carbon and nitrogen cycles. Phytoplankton, for example, form the basis of marine food chains, fixing carbon and generating half of all new oxygen in the atmosphere. Microbes provide a huge array of other ecosystem services, including fixing atmospheric nitrogen that plants need to build proteins; recycling nitrogen-rich compounds from organic materials; and, as mentioned in this article, removing carbon dioxide from the atmosphere.

This last ecosystem service is one we need more than ever. In the last 100 years, humans have joined microbes in playing a major role in the carbon, nitrogen, and water cycles. Most noticeably, our activity has caused a gigantic increase in greenhouse gases that cause global climate change. Two of these gases are methane and carbon dioxide.

Some of the bacteria discussed in this article broke down oil and methane that was released in the 2010 Gulf of Mexico oil spill, giving researchers hope that as Earth continues to warm, such microbes could help slow the release of methane from the oceans and Arctic tundra. Since methane is a potent greenhouse gas, microbes that break down methane could help slow global warming. But just how much methane and oil was gobbled up by bacteria in the Gulf—and how the microbes did it—is still not well

understood. Future research will be required before we can feel confident that microbes can help us counter climate change and clean up the planet.

## Evaluating the News

**1.** As the Arctic warms, vast areas of frozen peat—partly decayed vegetation that serves as a fuel in some areas—thaws and rots. The microbes that break down peat release methane. Why is it important to stop bacteria from turning peat into methane?

**2.** The studies of oil-eating microbes mentioned in this article were conducted in a water column extending more than 1,000 meters (>3,000 feet) below the surface. The oil-degrading microbe species found at depths of 1,000–1,500 meters were completely different from those near the surface (within 150 meters). List some of the conditions to which deep-ocean microbes would have to be adapted in order to consume oil at the surface. What factors might play a role in the rate at which deep-ocean microbes consume oil relative to microbes at the surface?

**3.** Because the oil-eating microbes stuck around even after the oil was gone, researchers suggested that the presence of the microbes could be used to detect a recent oil spill. Discuss how that might be useful.

**SOURCE:** *Contra Costa Times*, May 6, 2011, www.contracostatimes.com.

# Summary

## 24.1 How Ecosystems Function: An Overview

- Energy and materials can move from one ecosystem to another.
- A substantial portion of the energy captured by producers is lost as metabolic heat at each step in a food chain. Therefore, energy moves through ecosystems in just one direction.
- Nutrients are recycled in ecosystems. They pass from the environment to producers to various consumers, and then back to the environment when the ultimate consumers—decomposers—break down the bodies of dead organisms.

## 24.2 Energy Capture in Ecosystems

- Energy capture in an ecosystem is measured as net primary productivity; assessing the amount of NPP in an area is an important first step in determining the type of ecosystem found there and how it functions.
- On land, NPP tends to decrease from the equator toward the poles. Temperature, moisture, and nutrient availability all influence NPP.
- In marine ecosystems, NPP tends to decrease from relatively high values where the ocean borders land to low values in the open ocean (except where upwelling provides scarce nutrients to marine organisms). Aquatic ecosystems on land (such as wetlands) can also show high NPP.

## 24.3 Energy Flow through Ecosystems

- An energy pyramid depicts the amounts of energy available to organisms at different trophic levels in an ecosystem, with each successive level harvesting only about 10 percent of the energy from the level below.
- Secondary productivity is highest in areas of high net primary productivity.

## 24.4 Biogeochemical Cycles

- Nutrients are transferred between organisms and the physical environment in what are called nutrient cycles or biogeochemical cycles.

- Decomposers return nutrients from the bodies of dead organisms to the physical environment.
- Nutrients that enter the atmosphere easily (in gaseous form) have atmospheric cycles, which occur relatively rapidly and can transfer nutrients between distant parts of the world.
- Nutrients that do not enter the atmosphere easily (such as phosphorus) have sedimentary cycles, which usually take a long time to complete.
- Decomposers play an important role in biogeochemical cycles. For example, in the carbon cycle they unleash much of the carbon locked inside the remains of dead organisms and release it as $CO_2$.
- Bacteria play an important role in the nitrogen cycle. In biological nitrogen fixation, $N_2$ is converted to $NH_4^+$ by bacteria.
- Sulfur has an atmospheric cycle, entering land and water through weathering of rocks. Sulfur-containing compounds produced by phytoplankton are lofted into the atmosphere in sea spray.
- Phosphorus is a component of vital macromolecules such as DNA. Phosphorus inputs usually boost NPP in an ecosystem. Phosphorus has a sedimentary cycle because it does not generally enter the atmosphere.

## 24.5 Human Actions Can Alter Ecosystem Processes

- Human activities can alter nutrient cycles, as well as net primary productivity, on local, regional, and global scales.
- The addition of extra nutrients (especially nitrogen and phosphorus) to lakes, streams, and coastal waters leads to eutrophication, marked by a population boom among producers.
- Human inputs to the sulfur cycle exceed those from all natural sources combined, creating problems of international scope, such as acid rain.

# Key Terms

abiotic (p. 528)
acid rain (p. 540)
atmospheric cycle (p. 534)
biogeochemical cycle (p. 533)
biomass (p. 530)

biotic (p. 528)
carbon cycle (p. 534)
ecosystem (p. 528)
energy pyramid (p. 532)
eutrophication (p. 538)

exchange pool (p. 533)
net primary productivity (NPP) (p. 530)
nitrogen fixation (p. 535)
nutrient (p. 529)

nutrient cycle (p. 533)
secondary productivity (p. 533)
sedimentary cycle (p. 534)
trophic level (p. 532)
upwelling (p. 531)

# Self-Quiz

1. The amount of energy captured by photosynthesis, minus the amount lost as metabolic heat, is
   a. secondary productivity.
   b. consumer efficiency.
   c. net primary productivity.
   d. photosynthetic efficiency.

2. The movement of nutrients between organisms and the physical environment is called
   a. nutrient cycling.
   b. ecosystem services.
   c. net primary productivity.
   d. a nutrient pyramid.

3. Free services provided to humans by ecosystems include
   a. prevention of severe floods.
   b. prevention of soil erosion.
   c. filtering of pollutants from water and air.
   d. all of the above

4. Each step in a food chain is called
   a. a trophic level.
   b. an exchange pool.
   c. a food web.
   d. a producer.

5. What type of organism consumes 50 percent or more of the NPP in all ecosystems?
   a. herbivore
   b. decomposer
   c. producer
   d. predator

6. Sources of nutrients that are available to producers, such as soil, water, or air, are
   a. called essential nutrients.
   b. called exchange pools.
   c. considered eutrophic.
   d. called limiting nutrients.

7. Which of the following is the most representative term for an organism that gets its energy by eating all or parts of other organisms or their remains?
   a. decomposer
   b. predator
   c. consumer
   d. producer

8. Nutrients that cycle between terrestrial and aquatic ecosystems and are then deposited on the ocean bottom
   a. have a short cycling time.
   b. have an atmospheric cycle.
   c. are more common than nutrients with a gaseous phase.
   d. have a sedimentary cycle.

9. Eutrophication refers to
   a. reduced NPP because of low nutrient levels.
   b. increased NPP because of low concentration of nitrogen.
   c. increase in the numbers of secondary and tertiary consumers.
   d. a population boom in producers because of increased nutrient levels.

# Analysis and Application

1. Some people think the current U.S. Endangered Species Act should be replaced with a law designed to protect ecosystems, not species. The intent of such a law would be to focus conservation efforts on what its advocates think really matters in nature: whole ecosystems. Given how ecosystems are defined, do you think it would be easy or hard to determine the boundaries of what should and should not be protected if such a law were enacted? Give reasons for your answer.

2. What prevents energy from being recycled in ecosystems?

3. Why do coastal areas have higher NPP than the open ocean? The commercial fisheries in the Gulf of Alaska are among the most productive in the world. Alaskan fishing boats haul in more than a billion dollars worth of salmon, pollock, herring, halibut, cod, crabs, and shrimp each season. How can such cold, northern waters be so productive?

4. What essential role do decomposers play in ecosystems?

5. Explain why human alteration of nutrient cycles can have international effects.

6. Describe some key ecosystem services and discuss the extent to which human economic activity depends on such services.

7. The table shows the land area needed to raise 1 kilogram (35 ounces) of edible portions of the foods listed. Explain why it takes more land to produce a kilogram of chicken meat than a kilogram of wheat.

| FOOD | LAND AREA NEEDED (m²) TO PRODUCE 1.0 kg EDIBLE PORTION | CALORIES PROVIDED BY 1.0 kg EDIBLE PORTION |
| --- | --- | --- |
| Milk | 9.8 | 610 |
| Beef (grain-fed) | 7.9 | 2,470 |
| Eggs | 6.7 | 1,430 |
| Chicken (broiler) | 6.4 | 1,650 |
| Wheat | 1.5 | 3,400 |

# Global Change

**TOP OF THE FOOD CHAIN, TOP OF THE WORLD.** As global warming melts the Arctic ice cap, receding summer ice is increasingly leaving polar bears stranded on land or ice floes. Some bears are moving south into Canada and Alaska, but global warming may also affect the bears in a surprising way—by damaging marine food chains from the bottom up.

# Is the Cupboard Bare?

Polar bears are in trouble. The big, white bears live by hunting seals on vast stretches of sea ice covering the Arctic Ocean. Yet each summer, a larger area of the ice cap melts, leaving the bears stranded on land, often with hungry cubs. Until the ice returns in winter, the giant bears fast or eat garbage at Canadian and Russian garbage dumps. As the planet heats up, the ice melts earlier each spring and freezes later each fall, extending the bears' months of fasting by weeks. Unless they abandon the Arctic and begin migrating south into Canada and Siberia to compete with grizzly bears each summer, polar bears may go extinct.

Surprisingly, global warming could also wallop the bears from a completely different direction. In the summer of 2010, ecologists reported that global populations of tiny but critically important organisms called phytoplankton had declined by 40 percent since the 1950s. Like plants, phytoplankton live by photosynthesizing—that is, making sugar molecules from carbon dioxide and water using energy from the sun. On land, the big photosynthesizers are trees and other plants. But in the world's oceans, photosynthesis is carried out by phytoplankton—a mix of bacteria, algae, diatoms, and other protists.

Floating greenly at the surface of millions of square miles of ocean, phytoplankton perform half of all the photosynthesis on the planet and supply half of all the new oxygen in Earth's atmosphere. Photosynthetic phytoplankton are the foundation for all the great marine food webs. Every marine animal—from tiny shrimplike krill to fish, seals, and polar bears—depends on the energy and building blocks stored in phytoplankton. A 40 percent decline

*What could cause such a decline in phytoplankton? Could a shortage of phytoplankton lead to a collapse of marine food chains?*

in phytoplankton would lead to a drastic decline in krill and marine fish. Without fish, there would be no seals, to say nothing of polar bears, poised precariously at the top of the energy pyramid.

Yet, in 2010 researchers reported that phytoplankton had declined in 8 out of 10 regions of the ocean since 1950. In this chapter we'll learn how the expanding human population is dramatically affecting the biosphere.

**MAIN MESSAGE** Global change caused by human actions is occurring at a rapid pace. At the current pace, there will be severe consequences for other species and eventually for humans as well.

## KEY CONCEPTS

- The effects of human actions on the world's lands and waters are thought to be the main causes of the current high rate of species extinctions.

- Human activities have added natural and synthetic chemicals to the environment, and these additions in turn have altered how natural chemicals cycle through ecosystems.

- Human inputs to the global nitrogen cycle now exceed those of all natural sources

combined. If unchecked, these changes to the nitrogen cycle are expected to have negative effects on many ecosystems.

- The concentration of carbon dioxide ($CO_2$) gas in the atmosphere is increasing at a dramatic rate, largely because of the burning of fossil fuels.

- There is strong evidence that human activity has increased the concentrations of $CO_2$ and certain other gases in the atmosphere and that in turn has led to the

observed rise in average global temperature over the past century.

- The warming of Earth's climate has led to melting of polar and glacial ice, rising of sea levels, acidification of the oceans, and range shifts for many species.

- Climate models predict an increase in the frequency of severe weather and changes in rainfall patterns. Many species are likely to become extinct over the next 100 years.

## Helpful to Know

"Global warming" and "climate change" are related but not synonymous. *Global warming* is a significant increase in the average surface temperature of Earth over decades or more. *Climate change* is a long-term alteration in Earth's climate, and it includes such phenomena as global warming, change in rainfall patterns, and increased frequency of violent storms. *Global change* is an even broader concept, encompassing all types of worldwide environmental change, including large-scale pollution and loss of biodiversity.

**ON LAND AND SEA AND SKY,** our world is changing, and the change is occurring at a pace unprecedented in human history. Much of this change has been caused by humans, and most of it has taken place in just the last 100 years. Statements by politicians, talk show hosts, and others can give the impression that worldwide change in the environment—**global change**—is a controversial topic. Such statements cause many in the general public to think that global change may not really be occurring, or cause them to wonder whether anything really needs to be done about it.

This impression of controversy is unfortunate because we know with certainty that global change is occurring. We can readily demonstrate that invasions of non-native species have increased worldwide, large losses of biodiversity have occurred, and pollution has altered ecosystems throughout the world (see examples in Chapters 22–24). The data show that increased accumulation of greenhouse gases has led to global warming, which in turn has changed the very climate of Earth.

In this chapter we describe how people influence global change. We first discuss changes in land and water use, and changes in the cycling of nutrients through ecosystems. Next we focus on one of the most serious ecological issues of our time: **climate change**, which is a large-scale and long-term alteration in Earth's climate. Although Earth has gone through many changes in its average climate over its 4.5-billion-year history, the speed of the change that has taken place in the past 100 years is without precedent in the climate record. Moreover, there is compelling evidence that the climate change in recent history has been caused to a large extent by human actions, and its consequences are likely to be negative for many people and ecosystems around the world.

People make many physical and biotic changes to the land surface of Earth, which collectively are referred to as **land transformation** or land-use change. Such changes include the destruction of natural habitat to allow for resource use (as when a forest is clear-cut for lumber), agriculture, or urban growth. Land transformation also includes many human activities that alter natural habitat to a lesser degree, as when we graze cattle on grasslands.

Similarly, **water transformation** refers to physical and biotic changes that people make to the waters of our planet. For example, we have drastically altered the way water cycles through ecosystems. People now use more than half of the world's accessible fresh water, and we have altered the flow of nearly 70 percent of the world's rivers. Since water is essential to all life, our heavy use of the world's waters has many and far-reaching effects, including changing where water is found and altering which species can survive at a given location.

Many of our effects on the lands and waters of Earth are local in scale, as when we cut down a single forest or pollute the waters of one river. However, such local effects can add up to have a global impact.

## There is ample evidence of land and water transformation

Aerial photos, satellite data, changing urban boundaries, and local instances of the destruction of natural habitats show how humanity is changing the face of Earth (**FIGURE 25.1**). Together, many such lines of evidence show that land transformation and water transformation are occurring, are caused by human actions, and are global in scope.

In modifying land and water for our own use, we have dramatically affected many ecosystems. Examples of human effects on ecosystems include the ongoing destruction of tropical rainforests (**FIGURE 25.2**) and the conversion of once vast grasslands in the American Midwest to cropland. Half of the world's wetlands, from mangrove swamps to northern peat bogs, have been lost in the last 100 years. During the

200-year period beginning in the 1780s, wetlands declined in every state in the United States. Thanks to protective legislation, widespread public outreach, and initiatives that encourage landowners and public groups to conserve and restore wetlands, the loss of wetland was sharply curtailed in the last decade of the twentieth century.

About 50 percent of the world's population lives within 3 miles of the coast, making coastal ecosystems extremely vulnerable to human activity. Estuaries, saltwater marshes, mangrove swamps, and coastal shelf waters are among the most productive ecosystems on Earth. Most of the world's coastline is under siege today from urban development, sewage, nutrient runoff from farm fields, chemical pollution, and unsustainable harvesting of shellfish and fish.

## Land and water transformation have important consequences

As we alter land and water to produce goods and services for an increasing number of people, we use a very large share of the world's resources. Estimates suggest that humans now control (directly and indirectly) roughly 30–35 percent of the world's total net primary productivity (NPP) on land. As described in Chapter 24, NPP is the new growth that producers generate in a unit area per year. By controlling such a large portion of the world's land area and resources, we have reduced the amount of land and resources available to other species, causing some to go extinct. Water transformation has similar effects. When people overfish or pollute Earth's waters, we may cause dramatic changes in the abundances and types of species found in the world's aquatic ecosystems.

The transformation of land and water has other effects as well. One of these effects is change in local climate. For example, when a forest is cut down, the local temperature may increase and the humidity may decrease. Such climatic changes can make it less likely that the forest will regrow if the logging stops. In addition, as we will see shortly, the cutting and burning of forests increases the amount of carbon dioxide in the atmosphere—an aspect of global change that can alter the climate worldwide.

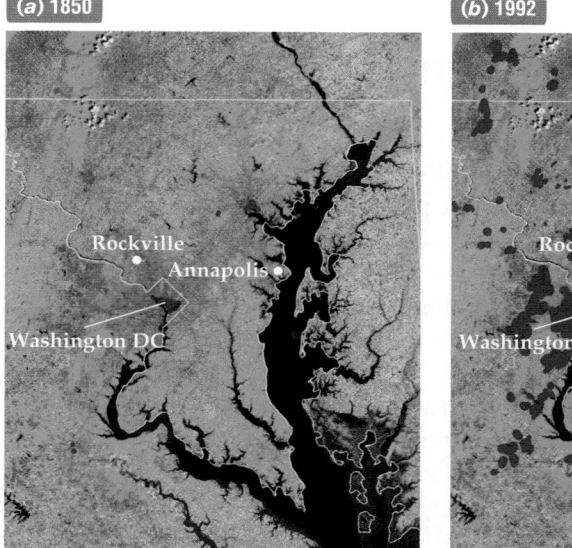

**FIGURE 25.1 Land Transformation**
Change in the boundaries (in red) of urban regions near Washington, DC, between 1850 (*a*) and 1992 (*b*).

**FIGURE 25.2 Disappearing Forests**
This graph, based on data from the United Nations Food and Agriculture Organization (FAO), shows that forest cover has shrunk in most regions of the world, except Europe. Asia's relatively good standing is due largely to extensive reforestation efforts in China over the past few years. The inset shows cattle grazing on land previously covered by Amazonian rainforest, in Para, Brazil.

## 25.2 Changes in the Chemistry of Earth

As described in Chapter 24, life on Earth depends on, and is heavily influenced by, the cycling of nutrients in ecosystems. Net primary productivity often depends on the amount of nitrogen and phosphorus available to producers, and the amount of sulfuric acid in rainfall has many effects on ecological communities. The nitrogen and phosphorus that stimulate eutrophication, and the sulfur in acid rain, are just two of many examples of naturally occurring chemicals that cycle through ecosystems.

### Bioaccumulation concentrates pollutants up the food chain

Synthetic chemicals (artificial chemicals made by humans) and other substances that humans release into the air, water, and soil may also cycle through ecosystems. Like nutrients that organisms gain from food or by direct absorption from their environment, chemicals released by humans can *bioaccumulate* in an individual. **Bioaccumulation** of a substance is the deposition of the substance within an organism at concentrations higher than in the surrounding abiotic environment. Substances that bioaccumulate tend to be stable chemicals that are deposited within cells and tissues faster than they can be eliminated (in urine, for instance).

Many synthetic chemicals—especially organic molecules found in pesticides, plastics, paints, solvents, and other industrial products—tend to bioaccumulate in cells and tissues. Long-lived organic molecules of synthetic origin that bioaccumulate in organisms, and that can have harmful effects, are broadly classified as **persistent organic pollutants** (**POPs**). Some of the most damaging POPs that are widespread in our biosphere include different types of PCBs (polychlorinated biphenyls) and dioxins. Because many of these pollutants have an atmospheric cycle, they can be transported over vast distances across the globe to contaminate food chains in remote places where the chemicals have never been used.

Heavy metals such as mercury, cadmium, and lead can also bioaccumulate in a wide variety of organisms, both producers and consumers. Mercury is a toxic chemical that occurs naturally in small amounts. However, the concentration of this heavy metal in air, water, and soil has increased threefold since the industrial revolution, mainly from the burning of coal and the incineration of mercury-containing waste. Mercury enters the food chain when bacteria absorb it from soil or water and convert it to an organic form known as methylmercury. Methylmercury is much more toxic than inorganic forms of mercury, in part because the organic form bioaccumulates more readily, being stored in muscle tissues of shellfish, fish, and humans. Methylmercury bioaccumulated by bacteria is passed on to consumers, such as zooplankton (microscopic aquatic animals), that feed on mercury-accumulating bacteria. In this way, the methylmercury is progressively transferred to other consumers throughout the food web. The FDA has issued an advisory suggesting that pregnant women in particular abstain from eating mackerel, shark, swordfish, and tilefish, as these large predatory fishes tend to accumulate higher levels of mercury.

The increase in the tissue concentrations of a bioaccumulated chemical at successively higher trophic levels in a food chain is known as **biomagnification**. Bioaccumulation and biomagnification might seem similar at first glance, but whereas bioaccumulation is the accumulation of a substance in individual organisms within a trophic level, biomagnification is the increase in tissue concentrations of a chemical as organic matter is passed from one trophic level to the next in a food chain.

Chemicals that show biomagnification (and therefore bioaccumulation as well) are resistant to degradation in the body or in the environment and are not easily excreted by animals, usually because they bind to macromolecules such as proteins or fats. PCBs, for example, are hydrophobic molecules that combine with fat and become locked within fatty tissues. Predators in the next trophic level acquire the chemical when they eat the fatty tissues of the prey species. Because predators consume large quantities of prey, and lose little if any of the chemical, its concentration builds up in their tissues over time. That is why top predators—those that feed at the end of a food chain—usually have the highest tissue concentration of biomagnified chemicals. **FIGURE 25.3** illustrates the 25 million–fold biomagnification of PCBs that has been recorded in some northern lakes. An important aspect of biomagnification is that pollutants that may be present in minuscule amounts in the abiotic environment, such as the water in a lake, can build up to damaging, even lethal, concentrations in the top predators of the food chain.

The pesticide DDT is an example of a POP that is bioaccumulated and biomagnified along a food chain. Until its use was banned in 1972, DDT was extensively sprayed in the United States to control mosquitoes and protect crops from insect pests. The pesticide ended up in lakes and streams, where it was taken up by plankton, such as algae, which were in turn ingested by zooplankton. As the pesticide moved up the food chain, from zooplankton to shell-fish to birds of prey such as ospreys and bald eagles, its tissue concentrations increased by hundreds of thousands of times. DDT disrupts reproduction in a variety of animals, but predatory birds were especially hard-hit. The chemical interferes with calcium deposition in the developing egg, resulting in thin, fragile eggshells that break easily.

DDT is an example of an **endocrine disrupter**, a chemical that interferes with hormone function to produce negative effects, such as reduced fertility, developmental abnormalities, immune system dysfunction, and increased risk of cancer. Bisphenol A (found in plastics that many water bottles are made of) and phthalates (found in everything from soft toys to cosmetics) are examples of endocrine disrupters that can be readily detected in the tissues of most Americans. In laboratory animals, bisphenol increases the risk of diabetes, obesity, reproductive problems, and various cancers. Phthalate exposure is associated with lowered sperm counts and defects in development of the male reproductive system. There are many reports of endocrine disrupters reducing fertility and causing developmental abnormalities in wild populations, especially among amphibians and reptiles. What is less clear at present is whether endocrine disrupters are harming human health to a significant degree. At a minimum, there is no assurance that long-term exposure to multiple endocrine disrupters, even at low doses of each, is necessarily safe for us.

## Many pollutants cause changes in the biosphere

Some of the POPs that we have poured into our environment have toxic effects on organisms, possibly including humans. Other POPs damage ecosystem processes on such a broad scale that there is little doubt about their negative impact on humans, and on a whole host of organisms from producers to top predators. The addition of **CFCs**, **chlorofluorocarbons**

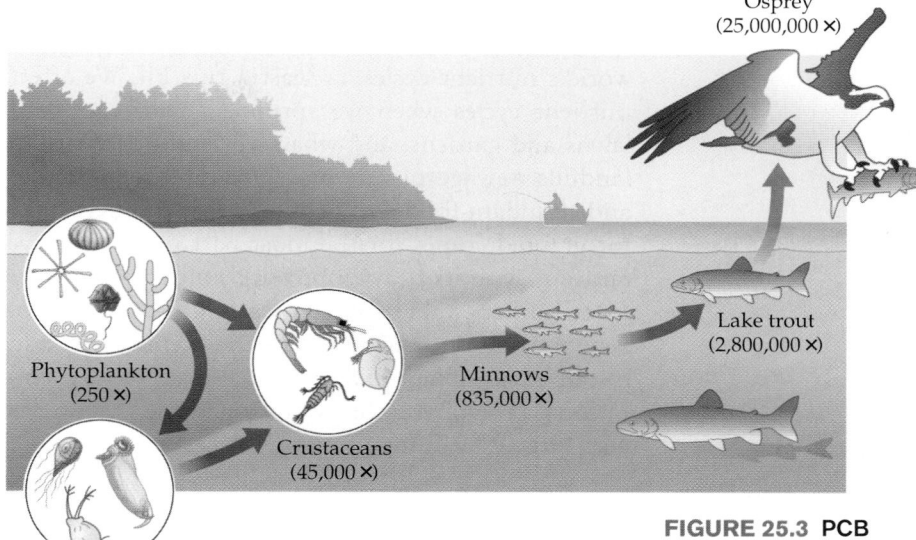

**BIOMAGNIFICATION**

Osprey (25,000,000 ×)

Lake trout (2,800,000 ×)

Minnows (835,000 ×)

Crustaceans (45,000 ×)

Phytoplankton (250 ×)

Zooplankton (500 ×)

**FIGURE 25.3 PCB Levels Become More Concentrated in Consumers Higher in the Food Chain**

[*KLOHR*-oh-*FLOHR*-oh-...], to the atmosphere is one of the most wide-ranging changes that humans have made to the chemistry of Earth. CFCs have caused a decrease in the thickness of the atmospheric ozone layer across the globe, and contributed to the ozone hole above Antarctica (see Section 21.1). Because the ozone layer shields the planet from harmful ultraviolet light (which can cause mutations in DNA), damage to the ozone layer poses a serious threat to all life.

Fortunately, the international community responded quickly to this threat, and the ozone layer has recently begun to show signs of a recovery. Clearly, in some cases we have succeeded in slowing down or undoing the harm caused by chemical pollution or the alteration of nutrient cycles (the mitigation of acid rain, discussed in Chapter 24, is another example). But in other cases, such as those of the global nitrogen cycle and the global carbon cycle, great challenges lie ahead.

> ### Concept Check
>
> 1. Describe some of the causes of the degradation of coastal ecosystems.
>
> 2. Compare and contrast bioaccumulation and biomagnification. What are some distinctive characteristics of chemicals that tend to bioaccumulate?

**Concept Check Answers**

1. Coastal ecosystems are affected by urban development, sewage dumping, excessive nutrient runoff, chemical pollution, and overfishing.

2. In bioaccumulation, a chemical accumulates within an organism's tissues at concentrations higher than in the surroundings; it takes place within a trophic level. In biomagnification, accumulated chemicals are passed on from one consumer to the next; that is, from one trophic level to the next. Chemicals that bioaccumulate are long-lived and not easily secreted, often because they bind to proteins or fats.

## 25.3 Changes in Global Nutrient Cycles

Nearly all of us have had a hand in changing the world's nutrient cycles, at least a tiny bit. We affect nutrient cycles when we sprinkle fertilizer on our lawns and gardens, and when we send our waste to landfills, sewage plants, or septic tanks. The cheap and abundant food that people in rich countries take for granted comes for the most part from intensive farming, with its heavy input of fertilizer and energy from nonrenewable sources. Carbon dioxide, nitrogen, phosphorus, and sulfur are the chemicals we add to our environment in largest quantities. Let's examine how the global cycling of nitrogen and carbon has been altered by human actions.

### Humans use technology to fix nitrogen

As described in Chapter 24, nitrogen is a crucial element for all organisms, but only certain bacteria can convert atmospheric nitrogen gas ($N_2$) into an organic form (ammonia) through the process of nitrogen fixation. Producers such as plants are crucially dependent on this biological nitrogen fixation, and all consumers are too, because they get their nitrogen from consuming plants and other producers.

In recent years, the amount of nitrogen fixed by human activities has exceeded the amount fixed by all natural processes combined (**FIGURE 25.4**). Much of this nitrogen fixation by humans is the result of the industrial production of fertilizers. A significant portion of the fertilizer that is applied on farm fields is broken down by bacterial action and released to the atmosphere as $N_2$ (nitrogen gas) and $N_2O$ (nitrous oxide). Other major sources of *anthropogenic* (human-generated) nitrogen fixation include car engines, in which heat from combustion converts some of the $N_2$ found in the air to nitric oxide (NO) and nitrogen dioxide ($NO_2$). These gases enter the atmosphere as engine exhaust, combine with oxygen and water in the air, and then fall to the ground as nitrate ($NO_3^-$) dissolved in rain.

The potential effects of changing the nitrogen cycle are far-reaching. When nitrogen is added to terrestrial communities, NPP usually increases, but the number of species often decreases (**FIGURE 25.5**). Species richness is affected because species best able to use the extra nitrogen outcompete other species. For example, the addition of nitrogen to grasslands in the Netherlands that historically were poor in nitrogen resulted in the loss of more than 50 percent of the species from some of those communities. Similarly, when

(a)

(b)

**FIGURE 25.5  A Nitrogen Addition Experiment**
Native grasslands in Minnesota often have 20–30 plant species per square meter. (a) No nitrogen was added to this control plot, and it lost no species between 1984 and 1994. (b) Researchers added nitrogen to this nearby experimental plot during the same time period. Most of the native species disappeared from this plot, and an introduced species, European quack grass, became dominant.

**FIGURE 25.4**
**Human Impact on the Global Nitrogen Cycle**

Nitrogen is fixed naturally by bacteria and by lightning at a rate of about 130 teragrams (Tg) per year (1 Tg = 1,012 grams, or 1.1 million tons). Human activities such as the production of fertilizers now fix more nitrogen than do all natural sources combined.

**GLOBAL NITROGEN FIXATION**

Nitrogen fixation by human activities now exceeds that by all natural sources combined.

N fixation by natural processes

The amount of nitrogen fixed by human activities increased greatly from 1940 to 1990.

N fixation by human processes

*(y-axis)* Global nitrogen fixation (Tg per year) — 0, 50, 100, 150, 200
*(x-axis)* Year — 1920, 1940, 1960, 1980, 1990, 2000

nitrogen is added to nitrogen-poor aquatic ecosystems, such as many ocean communities, productivity increases but many species are lost if eutrophication follows (see Chapter 24). In general, an increase in productivity caused by the addition of nitrogen is not necessarily a good thing for the ecosystem.

## Atmospheric carbon dioxide levels have risen dramatically

Although $CO_2$ makes up less than 0.04 percent of Earth's atmosphere, it is far more important than its low concentration might suggest. As we saw in earlier chapters, $CO_2$ is an essential raw material for photosynthesis, on which most life depends. $CO_2$ is also the most important of the atmospheric gases that contribute to global warming. Therefore, scientists took notice in the early 1960s when new measurements showed that the concentration of $CO_2$ in the atmosphere was rising rapidly.

Scientists have been directly measuring the concentration of $CO_2$ in the atmosphere since 1958. By measuring $CO_2$ concentrations in air bubbles trapped in ice for hundreds to hundreds of thousands of years, scientists have also estimated the concentration of $CO_2$ in both the recent and the relatively distant past (**FIGURE 25.6**). For ice formed recently, direct measurements of $CO_2$ in the air match estimates from ice bubbles, giving us confidence that the ice bubble

measurements for the past are accurate. Both types of measurements show that $CO_2$ levels have risen greatly during the past 200 years. Overall, of the current yearly increase in atmospheric $CO_2$ levels, about 75 percent is due to the burning of fossil fuels. Logging and burning of forests are responsible for most of the remaining 25 percent, but industrial processes such as cement manufacturing also make a significant contribution.

The recent increase in $CO_2$ levels is striking for two reasons. First, the increase happened quickly: the concentration of $CO_2$ increased from 280 to 380 parts per million (ppm) in roughly 200 years. Measurements from ice bubbles show that this rate of increase is greater than even the most sudden increase that occurred naturally during the past 420,000 years. Second, although the concentration of $CO_2$ in the atmosphere has ranged from about 200 to 300 ppm during the past 420,000 years, $CO_2$ levels are now higher than those estimated for any time during this period. Global $CO_2$ levels have changed very rapidly in recent years, reaching concentrations unmatched in the last 420,000 years. In the middle of 2011, global carbon dioxide concentrations stood at 394 ppm, with the levels increasing at the rate of about 3 ppm per year.

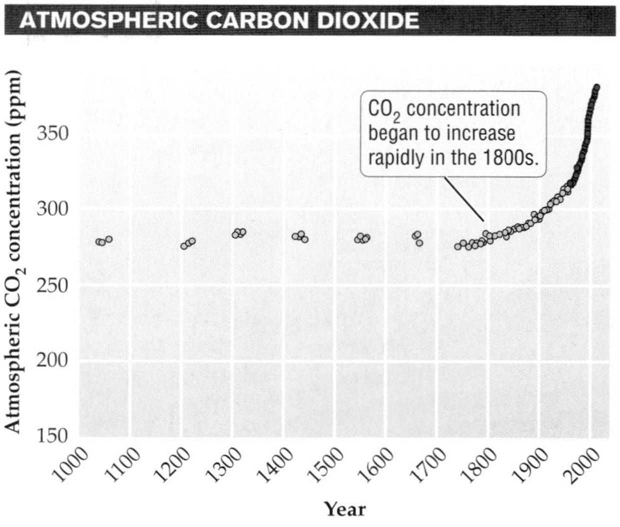

## Increased carbon dioxide concentrations have many biological effects

An increase in the concentration of $CO_2$ in the air can have large effects on plants (**FIGURE 25.7**). Many plants increase their rate of photosynthesis and use

### ATMOSPHERIC CARBON DIOXIDE

CO₂ concentration began to increase rapidly in the 1800s.

**FIGURE 25.6 Atmospheric $CO_2$ Levels Are Rising Rapidly**
Atmospheric $CO_2$ levels (measured in parts per million, or ppm) have increased greatly in the past 200 years. The red circles are direct measurements at the Mauna Loa Observatory in Hawaii, at 11,135 feet above sea level. The green circles indicate $CO_2$ levels measured from bubbles of air trapped in ice that formed many hundreds of years ago.

(a)   (b)   (c)

**FIGURE 25.7 High $CO_2$ Levels Can Increase Plant Size**
These three *Arabidopsis thaliana* plants all have the same genotype but were grown under different $CO_2$ concentrations: (a) 200 ppm, a level similar to that found roughly 20,000 years ago; (b) 350 ppm, the level found in 1988; and (c) 700 ppm, a predicted future level. Notice that plants grew larger at higher $CO_2$ concentrations.

water more efficiently, and therefore grow more rapidly, when more $CO_2$ is available. When $CO_2$ levels remain high, some plant species keep growing at higher rates, but others drop their growth rates over time. As $CO_2$ concentrations in the atmosphere rise, species that maintain rapid growth at high $CO_2$ levels might outcompete other species in their current ecological communities or invade new communities.

Differences in how individual species respond to higher $CO_2$ levels may cause changes to entire communities. However, it is difficult (at best) to predict exactly how communities will change under higher $CO_2$ levels. Increased $CO_2$ levels in the atmosphere have contributed to the warming of Earth's climate, as we shall see in Section 25.4. As both temperatures and $CO_2$ levels change, many different competitive and exploitative interactions may also change, but usually in ways that will not be known in advance. As we learned in Chapter 23, when interactions among species change, entire communities can change dramatically.

## 25.4 Climate Change

Some gases in Earth's atmosphere, such as carbon dioxide ($CO_2$), water vapor ($H_2O$), methane ($CH_4$), and nitrous oxide ($N_2O$), absorb heat that radiates from Earth's surface to space. These gases are called **greenhouse gases** because they function much as the walls of a greenhouse or the windows of a car do: they let in sunlight but trap heat. **FIGURE 25.8** illustrates how these gases contribute to the **greenhouse effect** that warms the surface of Earth.

About one-third of the solar radiation received by Earth is bounced back by the upper layers of the atmosphere. The rest is absorbed by the land and oceans, and to a lesser degree by the air. The warmed Earth emits some of its heat as long wavelengths of energy, known as *infrared radiation*. Some infrared radiation escapes Earth's atmosphere, but a good deal is absorbed by the greenhouse gases. Much of the heat absorbed by greenhouse gases is effectively trapped on Earth because when it is reemitted, it does not have sufficient energy

**FIGURE 25.8 How Greenhouse Gases Warm the Surface of Earth**

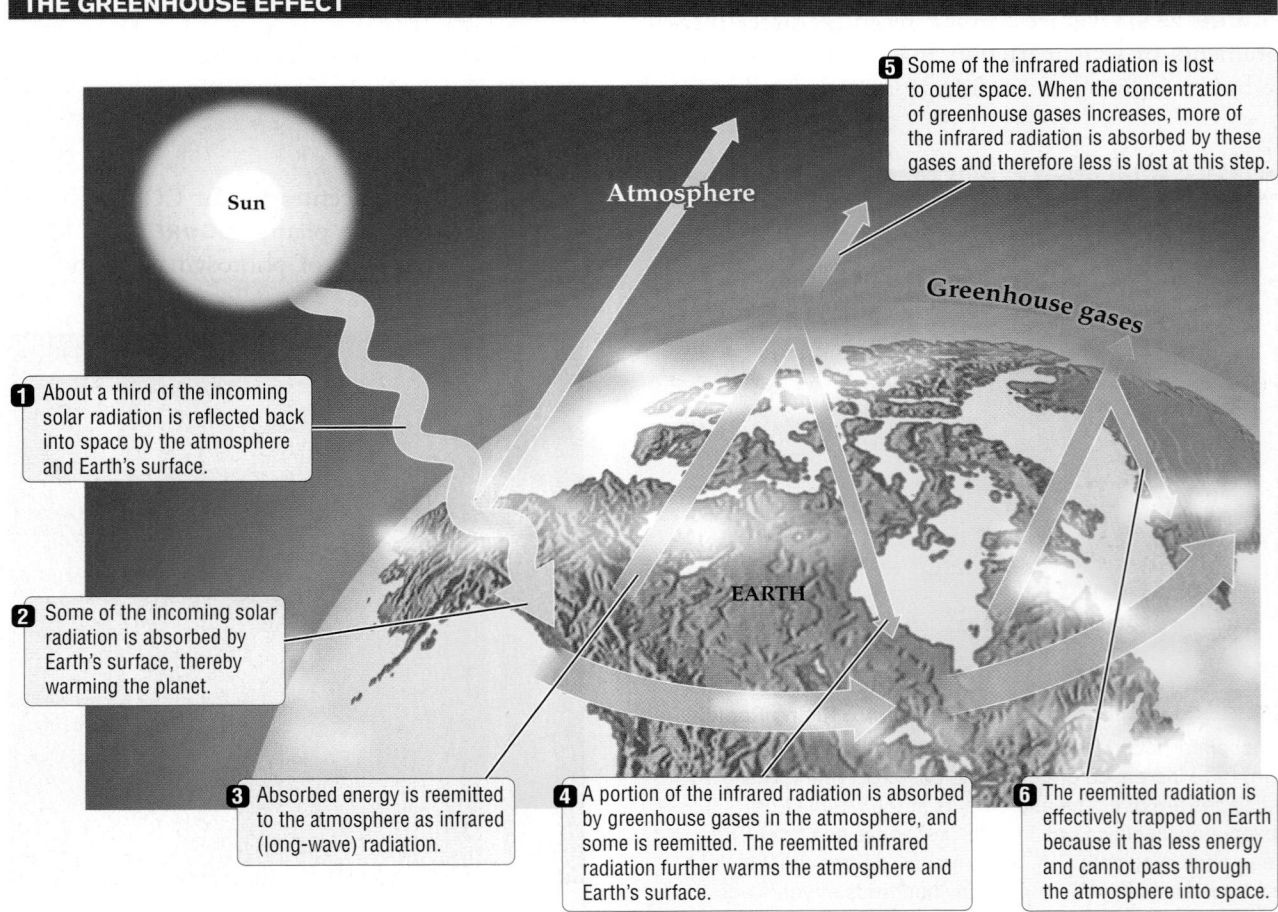

THE GREENHOUSE EFFECT

Sun

Atmosphere

Greenhouse gases

EARTH

**5** Some of the infrared radiation is lost to outer space. When the concentration of greenhouse gases increases, more of the infrared radiation is absorbed by these gases and therefore less is lost at this step.

**1** About a third of the incoming solar radiation is reflected back into space by the atmosphere and Earth's surface.

**2** Some of the incoming solar radiation is absorbed by Earth's surface, thereby warming the planet.

**3** Absorbed energy is reemitted to the atmosphere as infrared (long-wave) radiation.

**4** A portion of the infrared radiation is absorbed by greenhouse gases in the atmosphere, and some is reemitted. The reemitted infrared radiation further warms the atmosphere and Earth's surface.

**6** The reemitted radiation is effectively trapped on Earth because it has less energy and cannot pass through the atmosphere into space.

to pass through the atmosphere and escape into outer space. Greenhouse gases have existed in Earth's atmosphere for more than 4 billion years and have played an important part in maintaining temperatures that are warm enough for life to thrive over most of Earth's surface. Scientists reconstructing the past climate of our planet have noted a near-perfect correlation between the levels of carbon dioxide (the most potent of the greenhouse gases) and the surface temperature on Earth. These data show that as the concentration of greenhouse gases in the atmosphere increases, more heat is trapped, raising temperatures on Earth.

## Global temperatures are rising

Carbon dioxide is the most important of the greenhouse gases because so much of it enters the atmosphere. As far back as the 1960s, scientists predicted that the ongoing increases in atmospheric $CO_2$ concentrations would cause temperatures on Earth to rise. This aspect of global change, known as **global warming**, has provoked controversy in both the media and the political arena. Although year-to-year variation in the weather can make it hard to persuade everyone that the climate really is getting warmer, the overall trend in the data (**FIGURE 25.9**) has convinced the great majority of the world's climatologists and other scientists. A 2007 report from the United Nations–sponsored Intergovernmental Panel on Climate Change (IPCC) concluded that global surface temperatures rose by an average of 0.75°C between 1906 and 2005, with land warming more than the oceans, and higher rates of warming in the northern latitudes compared to the more tropical and equatorial regions of the planet. The IPCC also concluded that the increase in global temperatures since the mid-twentieth century is very likely a result of human-caused (an-

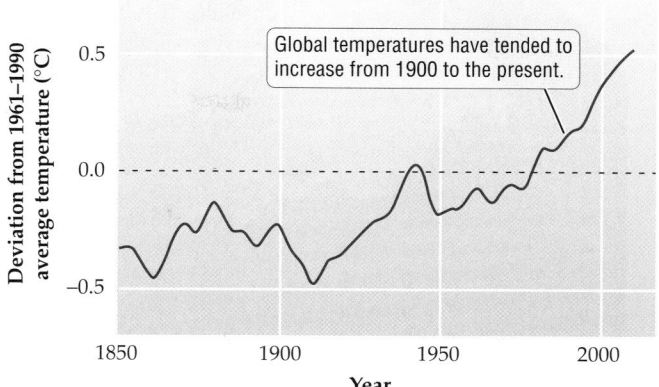

**GLOBAL TEMPERATURE CHANGE**

Global temperatures have tended to increase from 1900 to the present.

**FIGURE 25.9 Global Temperatures Are on the Rise**
Global air temperatures are plotted relative to the average temperature between 1961 and 1990 (dashed line). Portions of the curve below and above the dashed line represent lower-than-average and higher-than-average temperatures, respectively.

thropogenic) increases in the concentration of $CO_2$ and other greenhouse gases in the atmosphere.

The IPCC's conclusion is based on statistical analyses showing that the recent rise in global temperatures represents a significant trend, not just ordinary variation in the weather. Hundreds of studies published since 1995 indicate that the warming has been caused largely by human activities. For example, computer simulations based on data for the second half of the twentieth century were able to predict the observed 0.1°C rise in temperature of the top 2,000-meter layer of the world's oceans only when human activities (such as greenhouse gas emissions) were included in the calculations. The robustness of modern climate models is underscored by the fact that some of the predictions made by the computer simulations (Table 25.1) are now apparent at a statistically significant level.

| TABLE 25.1 | Some Consequences of Climate Change |
|---|---|
| **ABIOTIC CHANGES** | **SOME BIOTIC CONSEQUENCES** |
| Increase in near-surface and ocean temperatures | Ecosystem disruption, loss of ecosystem services; species extinction |
| Melting of glaciers | Spring floods, summer drought in glacier-fed regions |
| Loss of summer sea ice | Species extinction, loss of cultural and economic resources |
| Rise in sea levels (from melting ice, thermal expansion) | Loss of habitat, human habitation, and livelihood |
| Ocean acidification | Loss of marine organisms with calcified structures, coral bleaching; damage to fisheries |
| Increased frequency of severe weather | Habitat destruction; loss of human life, economic damage |
| Change in rainfall pattern, drought in some regions | Ecosystem degradation; severe agricultural and other economic loss |

(a) 1979

(b) 2005

**FIGURE 25.10**
**The Extent of Polar Sea Ice Has Declined Sharply**

Summer sea ice in the Arctic has declined by almost 25 percent compared to preindustrial levels. Climate change has affected wind and ocean currents in different ways across the globe, which explains why the Antarctic ice sheet is relatively stable. The satellite-based illustrations show the extent of the polar ice cap in the Arctic and the ice sheet on Greenland in the summer of 1979 (a) and the summer of 2005 (b). Courtesy NASA.

## Some predicted consequences of climate change are now being seen

Long-term and large-scale changes in the state of Earth's climate are broadly known as **climate change**. Global warming is one component of climate change, and some of its effects on the biosphere are now evident. Consistent with the warming trend, satellite images show that Arctic sea ice has been declining by 2.7 percent per decade since 1978 (**FIGURE 25.10**). Sea levels rose by an average of 1.8 millimeters per year between 1961 and 1993, and they have been rising by an average of 3.1 millimeters per year since then. Thermal expansion—the increase in volume as water warms up—has contributed to sea level rise, as has the melting of glaciers (**FIGURE 25.11**) and polar ice. As atmospheric carbon dioxide lev-

els rise, more of the gas is absorbed by the oceans, leading to ocean acidification. Since the industrial revolution, the pH of the world's oceans has declined from an average value of about 8.25 to 8.14 currently.

The additional heat energy that warmer temperatures generate, especially over the tropical oceans, is expected to increase the frequency of severe weather and to lengthen the storm season. Since the middle of the twentieth century, the number of tropical storms sweeping into North America has not changed significantly, but the number of class 3 and class 4 hurricanes has nearly tripled. Rainfall patterns have changed in some regions of the world—there is more rain in the eastern United States and northern Europe, and less in parts of the Mediterranean, northeastern and southern Africa, and parts of South Asia. Some recent climate simulations predict that global warming will worsen ozone depletion—with highest increase in UV radiation in tropical rather than polar regions—because of alterations in wind flow patterns in the upper atmosphere.

**FIGURE 25.11**
**Many Glaciers Are in Retreat**

The photos compare the extent of Shepard Glacier over time in Glacier National Park, Montana. Most of the world's glaciers are in retreat, although some, especially in parts of South America and Central Asia, are either stable or growing slightly.

1913

2005

## Climate change has brought many species to the brink

Studies show that recent temperature increases have also changed the biotic (living) component of ecosystems. Many northern ecosystems are shifting poleward at a rate of about 0.42 kilometer (a quarter of a mile) per year, as species migrate north in an attempt to find their "comfort zone." For example, as temperatures increased in Europe during the twentieth century, dozens of bird and butterfly species shifted their geographic ranges to the north. Similarly, the length of the growing season has increased for plants in northern latitudes as temperatures have warmed since 1980. However, some species—Arctic and alpine plants and animals, for example—have nowhere else to go (**FIGURE 25.12**). Canadian researchers have recorded a 60 percent decline in caribou and reindeer populations worldwide. There is higher calf mortality among the herds, and the animals suffer more from attacks by biting insects, whose populations have climbed.

High temperatures combined with lower pH result in *coral bleaching*, caused by a loss of the algal symbiotic partner and often resulting in the death of the coral animal as well. About a third of the tropical coral reefs have been destroyed in the last few decades, succumbing to the collective onslaught of coral bleaching, pollution, and physical damage from an increase in severe storms.

Although the magnitude of warming is much larger in the northern latitudes, scientists expect a more severe impact on tropical ecosystems, despite the smaller rise in temperature in the southern latitudes. Plants and animals in the moist tropics are adapted to a stable habitat and therefore live very close to the limits of their tolerance. Any change in that previously stable environment—increased temperature and reduced moisture, for example—puts them in jeopardy. In general, species with specialized habitat requirements are most vulnerable. Experts studying vulnerabilities of species in particular habitats warn that only 18–45 percent of the plants and animals native to the moist tropics are likely to survive beyond 2100. According to the International Union for Conservation of Nature (IUCN), 35 percent of the world's birds, 52 percent of amphibians, and 71 percent of warm-water reef-building corals have characteristics that make them especially vulnerable to the impacts of climate changes.

Marine organisms that protect themselves with plates or shells made of calcium carbonate ($CaCO_3$) are at risk from ocean acidification. The drop in pH changes ocean chemistry in such a way that these species cannot build their chalky protection, so they are at greater risk of death from damage or predation.

Will some species emerge as winners? Organisms that have broader tolerances and can live in a variety of habitats are most likely to emerge unscathed and even to expand their range. Versatility and wide tolerances are the calling cards of weedy plants and animal pests. Duke University researchers found that supplementing a forest ecosystem with extra carbon dioxide not only led to a boom in the growth of poison ivy, but the plants also became "itchier" because they produced a more potent form of the rash-inducing chemical urushiol. Another study predicts that the venomous brown recluse spider will continue to expand its range northward. Mosquito species that carry dengue fever virus and West Nile virus, and that were previously confined to the tropics, are also spreading north.

(a)

(b)

**FIGURE 25.12**
**Running Out of Altitude**
Alpine species such as the mountain androsace (a) and pika (b) are adapted to cold, high-elevation habitats. These species have been increasing their range upslope in the past few decades, but scientists fear that they will soon run out of options as they reach the tops of mountain peaks. Pikas live in boulder-strewn talus slopes in the highest reaches of the alpine zone. With their dense fur, these rodents suffer heat stress at temperatures that we would consider cool.

## Climate change will likely have severe consequences

Because there is no end in sight to the rise in $CO_2$ levels, the current trend of increasing global temperatures seems likely to continue. How will increased temperatures affect life on Earth in the future? Not surprisingly, the effects will depend on how much, and how fast, global warming occurs.

Computer models predict that by the end of the twenty-first century, average temperatures on Earth will have risen by anywhere from 1.1°C to 6.4°C (2°F–11.5°F) above the average global temperatures that prevailed between 1980 and 1990. The projections are based on a "business as usual" scenario, with no checks on the current trends in greenhouse gas emissions. The broad range reflects best estimates based on differing assumptions with regard to some aspects of climate change that scientists are still uncertain about. The most optimistic climate models project a *minimum* increase of 1.1°C by the end of the century. According to the most pessimistic models, the increase *could* be as high as 6.4°C, although a 4°C (7.2°F) increase is more probable.

What are the implications of such temperature increases for ecosystems and for human well-being?

Even an optimistic 1.8°C (3.2°F) increase in surface temperatures is likely to raise sea levels by as much as 0.38 meter (about 1.2 feet) and reduce ocean pH by at least 0.14 pH unit in about a hundred years from now. Summer sea ice in the Arctic is likely to disappear entirely by the end of the century. Extreme weather, including hurricanes, floods, and severe drought, is expected to become more common. Many species are expected to become extinct. Agricultural productivity is expected to increase in the northern latitudes but decrease in most other parts of the world.

A global temperature increase of 4°C or so will intensify the severity of these effects. Sea levels, for instance, are likely to rise as much as 0.59 meter (almost 2 feet). That may not sound like much, but combined with storm surges, it is enough to wipe out some island nations and destroy many coastal communities across the globe (**FIGURE 25.13**). A 4°C rise in global surface temperatures will wreak large-scale alterations in Earth's biomes. Some species will migrate, others will adapt, but a very large number will probably become extinct.

## Timely action can avert the worst-case scenarios

The world's agricultural systems will be severely strained, and it is unlikely that there will be enough food to nourish the extra 4–5 billion who are expected to join the human population between now and the end of this century. A recent analysis projects that by 2050, climate change will result in $125 billion in economic losses, will displace 26 million people, and will contribute to 300,000 deaths each year.

Although climate change is already under way, experts say the worst-case scenarios can be averted by timely action using technology that is currently available. Battling climate change will require reduced use of fossil fuels, increased energy efficiency, and increased reliance on renewable energy such as cellulose-based ethanol and solar power. Innovative carbon capture methods have been developed, and more are under development, to reduce atmospheric $CO_2$ levels. In one such strategy, carbon dioxide from factories and power plants is turned into oil by algae, and the oil is converted to biodiesel. Improvements in waste management—reducing the release of greenhouse gases by landfills, for example—will be necessary. Agricultural practices will have to change, placing greater emphasis on sustainability, improved manure management to minimize the release of

**RISE IN SEA LEVELS**

(a)

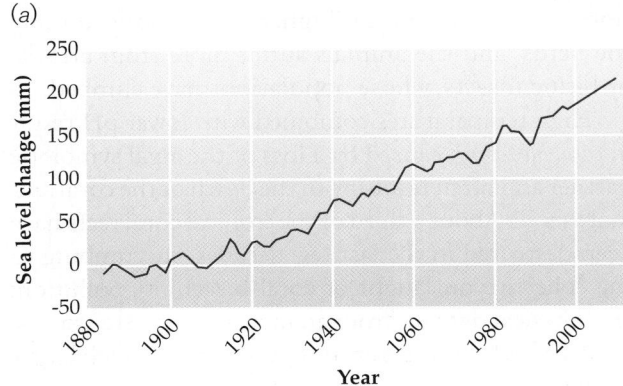

(b)

**FIGURE 25.13 Thermal Expansion Combined with Ice Melt Has Raised Sea Levels**

(a) Average sea levels have risen by about 200 millimeters since the start of the industrial revolution in 1880. By 2100, sea levels are expected to be between 500 and 1,400 millimeters higher than preindustrial levels. (b) The prime minister of the Maldives held an underwater cabinet meeting in 2009 to draw international attention to the threat that climate change presents to his island nation. The Maldive Islands, an archipelago of atolls and islets in the Indian Ocean, are so low-lying that they are likely to be submerged by 2100 if sea levels continue to rise at their current rate.

methane, and fertilizer application techniques that reduce the emission of $N_2O$ (nitrous oxide, a greenhouse gas that is released in the breakdown of synthetic fertilizers). Halting deforestation in tropical countries and increasing forest cover worldwide will be crucial.

Some renewable-energy technologies may need government support (such as tax credits) to enable them to compete with conventional energy sources. New regulations, such as higher fuel economy standards for vehicles and stringent energy efficiency codes for appliances, are needed; and these will require political will, resting on public support. Efforts to curb global

# Toward a Sustainable Society

Many different lines of evidence suggest that the current human impact on the biosphere is not sustainable. An action or process is *sustainable* if it can be continued indefinitely without serious damage being caused to the environment. Consider our use of fossil fuels. Although fossil fuels provide abundant energy now, our use of these fuels is not sustainable: they are not renewable, and hence supplies will run out, perhaps sooner rather than later (see Figure 1). Already, the volume of new sources of oil discovered worldwide has dropped steadily from over 200 billion barrels during the period from 1960 to 1965, to less than 30 billion barrels during 1995–2000. In 2007, the world used about 31 billion barrels of oil, but only 5 billion barrels of new oil was discovered in that year.

Actions that cause serious damage to the environment are also considered unsustainable, in part because our economies depend on clean air, clean water, and healthy soils. People currently use over 50 percent of the world's annual supply of available fresh water, and demand is expected to rise as populations increase. Many regions of the world already experience problems with either the amount of water available or its quality and safety. Declining water resources are a serious issue today, and experts are worried that matters may get much worse.

To illustrate the problem, let's look at water pumped from underground sources, or *groundwater*. How does the rate at which people use groundwater compare with the rate at which it is replenished by rainfall? The answer is that we often use water in an unsustainable way: we pump it from *aquifers* (underground bodies of water, sometimes bounded by impermeable layers of rock) much more rapidly than it is renewed.

In Texas, for example, for 100 years water has been pumped from the vast Ogallala aquifer faster than it has been replenished,

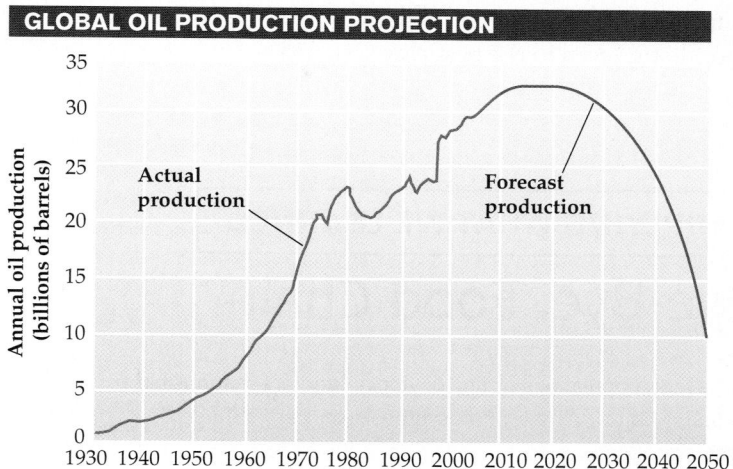

**GLOBAL OIL PRODUCTION PROJECTION**

Actual production

Forecast production

**FIGURE 1  Running Out of Oil**
Many experts predict that the annual global production of oil will peak, and then decline, sometime before 2020.

causing the Texan portion of the aquifer to lose half its original volume. If that rate of use were to continue, in another 100 years the water would be gone, and many of the farms and industries that depend on it would collapse. Texas is not alone. Rapid drops in groundwater levels (about 1 meter per year) in China pose a severe threat to its recent agricultural and economic gains; and at current rates of use, large agricultural regions in India will completely run out of water in 5–10 years. In Mexico City, pumping has caused land within the city to sink by an average of 7.5 meters (more than 24 feet) since 1900, damaging buildings, destroying sewers, and causing floods.

Sustainability is one aspect of ecology where each of us has a role. We can build a more sustainable society by supporting legislation that fosters less destructive and more efficient use of natural resources; by patronizing businesses that take measures to lessen their negative impact on the planet; by supporting sustainable agriculture; and by modifying our own lifestyle to reduce our Ecological Footprint (see the box in Chapter 22, p. 496). For example, we can increase our use of renewable energy and energy-efficient appliances; reduce all unnecessary use of fossil fuels—for instance, by biking to work or using public transportation; support organic farming; buy seafood from sustainable fisheries; use "green" building materials; and reduce, reuse, and recycle waste. Experts estimate that more than 200 million women around the world wish to limit their family size but have no access to family planning. Those of us who live in wealthy countries can support aid efforts that provide education, health care, and family-planning services in poor countries.

**FIGURE 2  The Most Inconvenient Truth: Climate Change Is Caused by Overpopulation and Overconsumption**

**Concept Check Answers**

1. Humans contribute through production and use of fertilizer, which is manufactured by conversion of $N_2$ into $NH_4^+$ and $NO_3^-$; and through combustion, which converts $N_2$ into gases such as NO and $NO_2$. Human inputs substantially exceed natural inputs.

2. Increase in atmospheric levels of greenhouse gases, including $CO_2$, increases global temperatures by intensifying the greenhouse effect, which is caused by the trapping of infrared radiation by greenhouse gases in Earth's atmosphere.

3. Shrinking of polar ice sheets and glaciers; sea level increases; change in rainfall patterns; increased frequency of severe storms; poleward migration of many warm-weather species; earlier blooming of many northern plants; extinction of some species.

warming will have social and economic costs, but any delay will most likely lead to even greater costs in the future. Because some of the effects of climate change are inevitable, we must also develop plans to cope with the floods, storm surges, violent weather, increased fire risk, water shortage, reduced crop yields, and harm to human health that are the expected outcomes of climate change even in the best-case scenario.

## Concept Check

1. How do human activities contribute nitrogen to the environment? How do human inputs of nitrogen compare with natural inputs?

2. How do greenhouse gases contribute to global warming?

3. Which predicted effects of global warming are already apparent?

## APPLYING WHAT WE LEARNED

# Bye-Bye, Food Chain?

Vast ocean populations of tiny phytoplankton are responsible for about half of all the photosynthesis on Earth. These tiny organisms supply all of the energy for marine food chains and about half of all newly generated oxygen. Every organism on Earth depends on the work of phytoplankton—not just polar bears and fish, but humans too. Every time we take a breath, some of the oxygen comes from phytoplankton. Earth's atmosphere of 20 percent oxygen comes largely from the work of photosynthetic ocean bacteria that lived 2.5 billion years ago.

And much of the oil that fuels our vehicles and industry comes from phytoplankton that lived millions of years ago.

So, when researchers at Dalhousie University in Nova Scotia reported in 2010 that marine phytoplankton were in sharp decline, ecologists were shocked. Without phytoplankton, life in the oceans would die and atmospheric oxygen levels would begin to decline. Were things that bad?

Oceanographers have known for a long time that rising temperatures hurt phytoplankton. As ocean waters warm up, they become more "stratified," with warmer water sitting on top of colder water. These static layers of water don't turn over, so there's no upwelling of nutrients from the seafloor. And without nutrients, phytoplankton can't grow. Data collected from satellite images suggest that warming oceans have caused a 6 percent drop in phytoplankton numbers since the early 1980s.

How does global warming hurt phytoplankton? We know that burning fossil fuels adds carbon dioxide to the atmosphere and that a lot of the extra carbon dioxide is absorbed by the oceans. Since carbon dioxide is a building block used for making sugars during photosynthesis, you might think that more carbon dioxide would be good for phytoplankton. And in some parts of the ocean, populations of phytoplankton *are* increasing. Unfortunately, carbon dioxide dissolved in water also makes water acidic, and acidic water contains less iron, an essential nutrient for phytoplankton.

Different parts of the ocean are warming up faster or slower, and some parts are becoming more acidic while other parts are not. Oceanographers aren't sure where all this is going. To find out whether the changes they had observed were short-term (and perhaps not serious) or long-term (and serious), the Nova Scotia researchers looked at half a million measurements of ocean clarity and color collected since the late 1800s. In general, cloudier water and greener water contain more phytoplankton. After analyzing the data, these researchers concluded that phytoplankton had declined by 40 percent since 1950.

Not so fast! said researcher Mark Ohman, of the Scripps Institution of Oceanography. Ohman didn't object to the data, but he claimed that the other researchers' analysis contained errors. For example, those researchers had reported that 59 percent of local ocean measurements had shown declines in phytoplankton. But Ohman pointed out that only 38 percent of those declines were statistically significant. The researchers had also used measurements from only the top 20 meters of the ocean, but in some parts of the ocean, phytoplankton live much deeper than that.

Although ocean scientists agree that phytoplankton have declined in number, and that this decline is most likely related to global warming and ocean acidification, just how extensively phytoplankton populations have declined is, for now, unsettled. Marine fisheries *are* collapsing, but mostly because of overfishing. That said, we need to halt ongoing declines in phytoplankton. Otherwise, it will be bye-bye, marine food chain and bye-bye, polar bears.

# Goose Eggs Open Scientific Debate over Polar Bears' Future

BY BRIAN MAFFLY, *Salt Lake Tribune*

The effects of global warming can be seen everywhere in the Arctic, even in scat. Researchers are finding more egg shells and goose feathers in polar bear feces as ground-nesting birds increasingly replace seals on the polar bear menu, thanks to loss of pack ice, which forces bears ashore earlier in the spring.

This change in bears' foraging patterns has raised concerns that the now-abundant snow geese that use Canada's Hudson Bay as a breeding ground could face elimination. This is because their eggs easily double as phone-sized energy boosts for bears during a season when their caloric needs are highest.

But new research by a Utah State University wildlife biologist, with colleagues at the American Museum of Natural History, suggests the situation for either animal is not as dire as it may appear.

Goose numbers won't plummet because fluctuations in weather will guarantee nesting geese a periodic reprieve from bear predation, asserts David Koons, an assistant professor of wildland resources at USU's Ecology Center.

However, other scientists dismiss the new findings as "a fairy tale," contending they gloss over the peril polar bears face in a thawing world.

Geese and their eggs might be an "interesting snack," but they cannot replace blubbery seals as polar bears' main source of calories, said bear researcher Andrew Derocher, a professor of biology at the University of Alberta. Disappearing sea ice, which bears need as a platform for hunting seal, is the primary justification for polar bears' 2008 listing as a threatened species.

Koons' research, published last month in the journal Oikos, suggests geese could sustain Hudson Bay bears not every year, but in years when early ice breakup deprives them access to their preferred marine food sources.

---

Covering the northern end of our planet is the Arctic Ocean. Bigger than all of Canada, this shallow sea has been covered by floating sea ice every winter for a million years or more. The thick sheet of ice—called the Arctic ice cap—provides habitat for a diversity of marine life, including seals and polar bears. In the summer the ice cap thins and shrinks around the edges. How much depends on climate. Between 1870 and 1950, the summer ice cap averaged about 11 million square kilometers ($km^2$), but today it hovers at just 5 or 6 million $km^2$—half its former area. Unless something is done to slow climate change, the summer Arctic ice cap will probably melt completely within the lifetime of the average person reading this book.

Polar bears hunt around the edge of the Arctic ice cap, sniffing out the breathing holes of seals and waiting for a seal to pop up for a breath. But the thinning ice of recent years can collapse under the bears' enormous weight, and the bears have been known to drown trying to swim the increasingly long distances to land.

Koons and colleagues hope the bears can eke out a living in the summer by eating snow geese and their eggs. One of Koons's two coauthors on the paper, Linda Gormezano, trained her dog to find polar bear scat and help her collect hundreds of samples of scat over a 3-year period. By teasing apart the scat and looking for berry skins, bones, feathers, and eggshells, Gormezano was able to estimate how much of each food the bears had eaten. It's possible that if the bears eat geese during summers when the ice cap melts away, enough polar bears can make it through the summers to keep the species from becoming extinct.

Most bears are omnivorous, meaning they can eat almost anything. But polar bears have evolved to specialize on seals. Until recently, researchers believed that polar bears could not make the switch to being omnivores. But recent genetic studies suggest that polar bears are closely related to grizzly bears and the two species can interbreed and produce fertile offspring. Will polar bears go extinct, or will they adopt more omnivorous habits, interbreed with grizzly bears, and migrate south?

## Evaluating the News

**1.** From your reading of this chapter, what kinds of things do you think could be done to reverse the melting of the Arctic ice cap to save polar bears?

**2.** If the bears ate all the nesting snow geese each summer, the goose population would go extinct. Explain why Koons thinks the geese would get a "reprieve" in colder years. In what way would that be good for the polar bears?

**3.** The first sentence of this article says that the effects of global warming can be seen in scat. Explain what the author means. Think of another animal besides polar bears whose scat would reflect changes caused by global warming. What would be different, and why?

**SOURCE:** *Salt Lake Tribune*, December 25, 2010.

# Summary

## 25.1 Land and Water Transformation

- Many lines of evidence show that human activities are changing land and water worldwide.
- Land and water transformation has caused extinctions of species and has the potential to alter local and global climate.

## 25.2 Changes in the Chemistry of Earth

- Human activities are changing the way many chemicals, both natural and synthetic, are cycled through ecosystems.
- Bioaccumulation is the tendency of some chemicals to be deposited in the tissues of an organism at concentrations higher than those found in the surrounding environment. Methylmercury and PCBs are among the many persistent organic pollutants (POPs) that are bioaccumulated.
- In biomagnification, tissue concentrations of pollutants increase as biomass is transferred from one trophic level to another. Fishes, birds, and mammals that feed at the highest trophic levels in a food chain tend to accumulate the highest tissue concentrations of biomagnified chemicals.
- The release of chlorofluorocarbons (CFCs) into the atmosphere has thinned the ozone layer over Earth, posing a serious threat to all life.

## 25.3 Changes in Global Nutrient Cycles

- The extra nitrogen fixed by human activities has altered the global nitrogen cycle, leading to increases in productivity that can cause losses of species from ecosystems.

- Concentrations of atmospheric $CO_2$ have increased greatly in the past 200 years and are higher now than in the past 420,000 years. These $CO_2$ increases are caused by the burning of fossil fuels and the destruction of forests.
- Increased $CO_2$ concentrations can alter the growth of plants in ways that will probably cause changes in many ecological communities.

## 25.4 Climate Change

- Carbon dioxide and other greenhouse gases in the atmosphere trap heat that radiates from Earth's surface. As the concentration of greenhouse gases increases, average temperatures on Earth are expected to rise.
- Human activities have contributed to the global warming that has occurred in the past 100 years.
- Some predicted effects of global warming are already being witnessed, including the melting of polar ice sheets, sea level rise, ocean acidification, and migration of some species toward the poles.
- Climate models predict changes in rainfall patterns, increased frequency of severe weather, and species extinctions.
- Timely action may help avert the worst case scenarios. This will require reduced use of fossil fuels, increased energy efficiency, and increased reliance on renewable energy such as solar power.

# Key Terms

bioaccumulation (p. 550)
biomagnification (p. 550)
chlorofluorocarbon (CFC) (p. 551)

climate change (p. 548)
endocrine disrupter (p. 551)
global change (p. 548)
global warming (p. 555)

greenhouse effect (p. 554)
greenhouse gas (p. 554)
land transformation (p. 548)

persistent organic pollutant (POP) (p. 550)
water transformation (p. 548)

# Self-Quiz

1. Which of the following is most directly responsible for global warming?
   a. increased $CO_2$ concentration in the atmosphere
   b. sunspot activity
   c. climate change
   d. biomagnification

2. $CO_2$ absorbs some of the _____ that radiates from the surface of Earth to space.
   a. ozone
   b. infrared energy
   c. ultraviolet light
   d. smog

3. The conversion of $N_2$ gas to ammonia by bacteria is known as
   a. biological nitrogen fixation.
   b. fertilizer production.
   c. nitrogen cycling.
   d. denitrification.

4. Human activities can alter the natural cycling of nitrogen
   a. through biomagnifications of methylmercury.
   b. through pollution controls that reduce acid rain.
   c. by preventing fertilizer runoff from farm fields.
   d. by burning fossil fuels in automobile engines.

5. Most scientists think that three of the following four statements related to global warming are correct. Which one is *not* correct?
   a. The concentration of greenhouse gases in the atmosphere is not increasing.
   b. Dozens of species have shifted their geographic ranges to the north.
   c. Plant growing seasons are longer now than they were before 1980.
   d. Human actions, such as the burning of fossil fuels, contribute to global warming.

6. Substances that bioaccumulate
   a. are found in organisms at lower concentrations than in their abiotic surroundings.
   b. are readily eliminated by excretion, for example in urine.
   c. are invariably inorganic, rather than organic, substances.
   d. tend to be chemically stable, and are not degraded in the body or in the environment.

7. Mercury that enters waterways
   a. can undergo biomagnification, but not bioaccumulation.
   b. will always be more abundant in a larger animal than in a smaller one.
   c. occurs in increasingly higher concentrations at successively higher trophic levels in the food chain.
   d. is found in secondary consumers but not in primary consumers.

# Analysis and Application

1. Summarize the major types of global change caused by human activities. What consequences do such types of global change have for species other than humans?

2. Compare examples of human-caused global change with examples of global change not caused by people. What is different or unusual about human-caused global change?

3. How does the current atmospheric $CO_2$ concentration compare with concentrations over the past 420,000 years? How do scientists know what Earth's $CO_2$ concentrations were hundreds of thousands of years ago?

4. The future magnitude and effects of global warming remain uncertain. Do you think we should take action now to address global warming, despite those uncertainties? Or do you think we should wait until we are more certain what the ultimate effects of global warming will be? Support your answer with facts already known about global warming.

5. Would you be willing to pay a gasoline tax to help fund aggressive actions to reduce global warming? If so, how much tax per gallon would you be willing to pay—50 cents, one dollar, two dollars? If not, why not?

6. What changes to human societies would have to be made for people to have a sustainable impact on Earth?

7. The graph shows the change in ocean pH since the industrial revolution, computed on the basis of carbon dioxide emissions over this time span. Average ocean pH is projected to decline by about 0.5 pH unit below the preindustrial levels (pH 8.2) by the end of this century. Imagine you are taking tourists on an underwater dive to see

Australia's Great Barrier Reef in the year 2100. What kinds of changes are the tourists likely to see, compared to the state of this underwater wonderland at the turn of the twentieth century? What biological mechanisms would explain the observed changes? What can we do now to ensure that these tourists-of-the-future will have a more rewarding experience of nature under the waves?

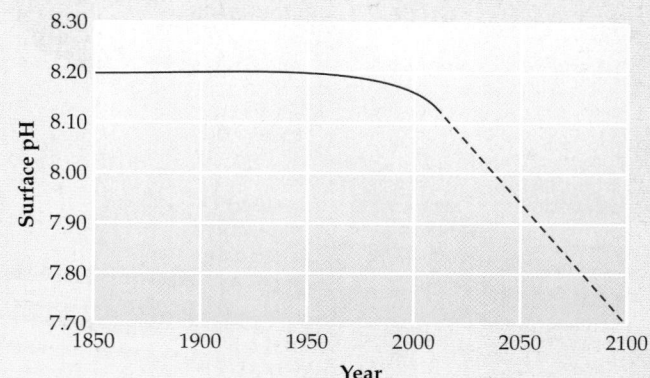

8. What are endocrine disrupters? Name two endocrine disrupters that are found in the tissues of most Americans. What industrial products or consumer items do these pollutants come from and what harmful effects are known to be associated with exposure to these chemicals?

9. List two sources of anthropogenic nitrogen, and explain how ecosystems can be disrupted when humans increase nitrogen inputs. How can anthropogenic nitrogen be harmful to an ecosystem, when nitrogen is an important nutrient for producers?

# Internal Organization and Homeostasis

**CAMELS ADAPTED TO THE DESERT.** A Bedouin leads a camel train through the desert in Algeria. Camels have extensive adaptations for thriving in rugged deserts. Traditional dress for Bedouin men is a loose white tunic, which allows enough air to circulate to cool the body but also protects the skin and limits the amount of water lost to evaporation. Humans survive in the desert only with the help of camels and technologies such as clothing and stored food and water.

# Beating the Heat

In the 1850s, the U.S. Congress appropriated $30,000 to buy and test a new kind of desert military equipment—Arabian camels. As soon as the 75 camels arrived in 1857, the army loaded them with supplies and personnel and deployed them into the blazing Arizona desert. The camels proved their mettle almost immediately; carrying up to a thousand pounds, they climbed craggy mountains too steep for horses or mules and withstood days without water.

Yet many soldiers hated the camels, which bit and spit; horses panicked when the aggressive camels were nearby; and the soldiers mistreated the camels, beating them and purposely releasing them into the desert. As late as the 1940s, those camels' wild descendants still survived in parts of Arizona and southern California.

Arabian camels are tough animals highly adapted to desert environments, where temperatures can reach 54°C (130°F) or more. Camels can withstand temperatures that would kill most other mammals, and they tolerate the loss of a third of the water in their bodies. Domesticated 5,000 years ago, trains of camels helped open early trade routes such as the Silk Road that connected China, India, Iran, Arabia, Europe, and Africa. An Arabian camel can carry a rider 150 kilometers through the desert with neither rest nor water. No other domesticated animal comes close to the toughness of a camel. Not even humans.

Where camels can survive up to 2 weeks without water or shade in the deadly heat of a Saharan summer, humans usually die after just 2 days. Most of the time, humans maintain a fairly constant body temperature. In contrast, camels survive desert heat by allowing their body temperature to rise. Camels can also lose 30 percent of their total body water, an amount that would kill a human.

How do camels survive such extreme conditions? How do animals survive first by maintaining internal conditions such as temperature (homeostasis) and second by doing the opposite—allowing internal conditions to change (tolerance)?

Before we address these questions, we will explore how animals use energy to maintain a stable internal environment when the outside world is changing—the process called homeostasis.

**MAIN MESSAGE** The animal body is organized into specialized tissues, organs, and organ systems; animals adjust the internal state to keep it relatively constant and therefore suitable for physiological processes.

## KEY CONCEPTS

- The animal body is made up of specialized cells and tissues. In most animals, different tissue types are organized into discrete organs, each with a unique set of functions.

- An organ system contains two or more organs. The human body has many different organ systems, ranging from the integumentary system to the immune system.

- Homeostasis is the process of maintaining a stable internal state in the organism despite fluctuations in the environment.

- Homeostasis requires that organisms be able to monitor and adjust the chemical and physical characteristics of their internal environment. Many homeostatic pathways have negative feedback loops.

- Homeostasis requires energy; the greater the difference between internal and external conditions, the higher the energy expenditure.

- Thermoregulation—the maintenance of body temperature—is an example of a homeostatic mechanism. Animals exchange heat with their external environment to maintain a suitable body temperature.

- Osmoregulation is another example of a homeostatic mechanism. Most animals regulate the levels of nutrients and wastes in their internal fluids. The kidneys play a central role in regulating water content and solute concentrations in the bodies of vertebrates.

**ANIMALS ARE MULTICELLULAR HETEROTROPHS** that ingest their food. As noted in Chapters 3 and 6, multicellularity enabled fungi, plants, and animals to evolve large bodies. Within limits, a larger body often means an enhanced ability to recruit resources from the environment—by enhancing the ability to absorb mineral nutrients from the soil or to capture and subdue prey, for example. But no matter how large the body is, the cells within it are generally microscopic. The plasma membranes of the many microscopic cells collectively present a large surface area for taking up nutrients and getting rid of wastes. Another adaptive benefit of multicellularity is efficient functioning through cell specialization: most cells in the animal body conduct a narrow set of functions, and they perform these functions far better than any jack-of-all-trades nonspecialist cell could. The multicellular body is therefore a highly efficient and well-coordinated community of specialized cell types.

Multicellularity mandates a certain level of organizational complexity that we can observe in even the simplest animals, such as sponges. In this chapter we present a broad overview of the structural and functional organization of the animal body, with a focus on the vertebrate body plan. The study of the structures that make up a complex multicellular body is known as **anatomy**. The science that focuses on the functions of anatomical structures is **physiology**. The chapters that follow in this unit explore the anatomy and physiology of the major organ systems in animals like us.

Most biological processes take place only within a certain temperature range, with the right amount of water, at the appropriate pH, and at a particular concentration of inorganic and organic chemicals. But rarely does the external environment continually match the conditions that are most favorable for processes inside a cell or for the coordinated function of organ systems in a multicellular organism. **Homeostasis** [HOH-mee-oh-STAY-sus] (*homeo*, "same"; *stasis*, "standing") is the process of maintaining a relatively constant internal state despite fluctuations in the external environment.

Homeostasis is vital for the proper function of most organisms. For example, cell-level homeostasis maintains cytosolic pH at about 7 in nearly all organisms, from bacteria to oak trees. Homeostasis at the whole-body level maintains our blood pH between 7.35 and 7.45, and a failure to stay within this narrow range can result in death. The greater the anatomical and physiological complexity of an organism, the greater the complexity of the homeostatic mechanisms that must be deployed to keep the internal environment stable. We close this chapter with a closer look at two examples of whole-body homeostasis in animals: the regulation of body temperature, and the balance of body fluids by the urinary system.

## 26.1 Internal Organization: Cells and Tissues

The multicellular body is a community of specialized cell types. The human body has at least 220 different specialized cell types. In humans, as in other animals, a majority of the different cell types are physically attached to surrounding structures. Recall that animal cells lack cell walls, but most have plasma membrane proteins that attach them to neighboring cells or pin them to the mat of proteins and carbohydrates that makes up the **extracellular matrix** (see Figure 6.7).

Cells that function in an integrated manner to perform a common set of functions constitute a **tissue**. A tissue may be composed of just one cell type, or it may contain multiple cell types; in either case, the cells that compose a tissue must cooperate to perform the distinctive functions of that tissue. The different types of tissues found in the animal body can be placed into four broad categories:

- Epithelial tissue
- Connective tissue
- Muscle tissue
- Nervous tissue

### Epithelial tissue covers organs and lines body cavities

As we noted in Chapter 1, an **organ** has a defined shape and location in the body and is composed of two or more tissues that function in a coordinated fashion to perform a unique set of functions. Some of our organs have internal spaces, such as the space inside the stomach or the four hollow chambers inside the human heart. *Blood vessels* and *lymphatic vessels* are tubelike structures with a hollow space (the lumen), through which fluids (blood and lymph, respectively) are transported. Many of our internal organs lie inside one of several large body cavities, such as the chest cavity (thoracic cavity) and the abdominal cavity. The brain lies inside yet another cavity, the braincase (cranium), formed by the flat bones of the skull.

**Epithelial tissues** are those that cover the surface of the body and the surfaces of organs, and that line the internal spaces inside organs, the lumen of fluid-transporting vessels, and all body cavities. Because it constitutes a boundary—demarcating the outside of the body or the surface of a particular organ, for example—epithelial tissue forms a closely packed sheet of cells held together by special cell attachment structures such as tight junctions (see Section 7.5 and Figure 7.11). Epithelial tissues form the outermost living layer of the skin, the lining of the digestive system, the gas exchange surfaces of the lungs, and fluid-filtering structures in the kidneys (**FIGURE 26.1**).

Most epithelial cells divide on a regular basis, in part because this protective layer often encounters physical stress (on the skin surface, for example) or wear and tear from the movement of fluids or exposure to various chemicals (in the stomach, for example). Frequent cell division makes this tissue more prone to becoming cancerous (see Chapter 11). About 80 percent of the cancers diagnosed in humans are *carcinomas*, which are cancers of epithelial tissue.

## Connective tissue forms a matrix that binds and supports tissues

Connective tissue consists of cells embedded in a matrix composed of a variety of biomolecules, especially proteins and carbohydrates. The main function of connective tissue is to bind cells and tissues together and to provide structural support for the body, although some types of connective tissue have other narrowly specialized functions. Connective tissue is classified and named in various ways, usually reflecting the chemical and physical properties or the distinctive functions of the tissue. **FIGURE 26.2** shows a traditional classification scheme.

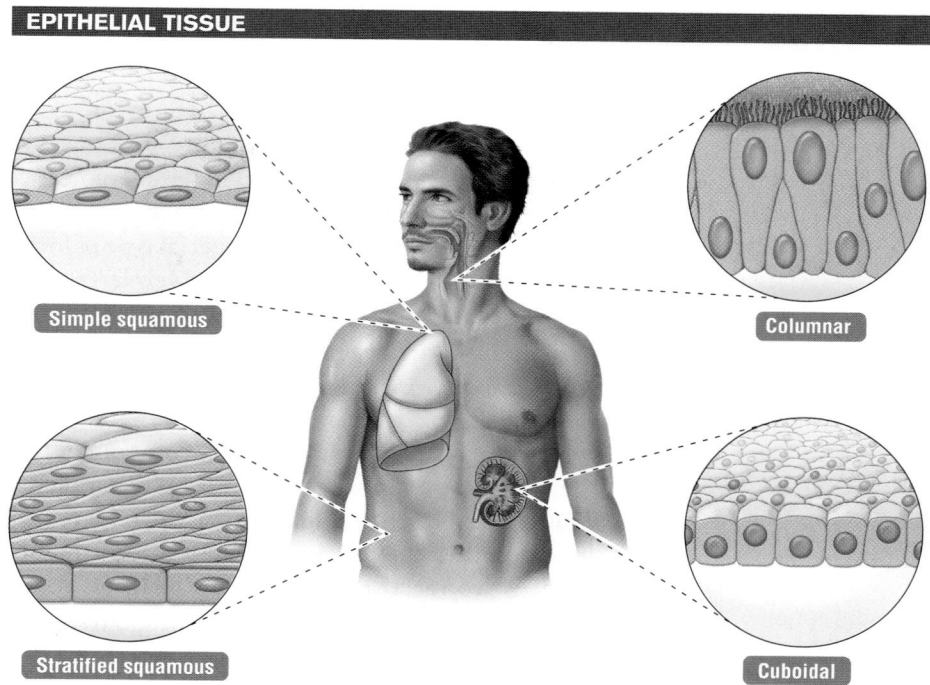

**FIGURE 26.1 Epithelial Tissue Covers and Protects**

Epithelial tissue are of three main types based on shape: squamous, columnar, and cuboidal. There are variations on these basic types, such as stratified (layered), squamous epithelium, and ciliated columnar epithelium.

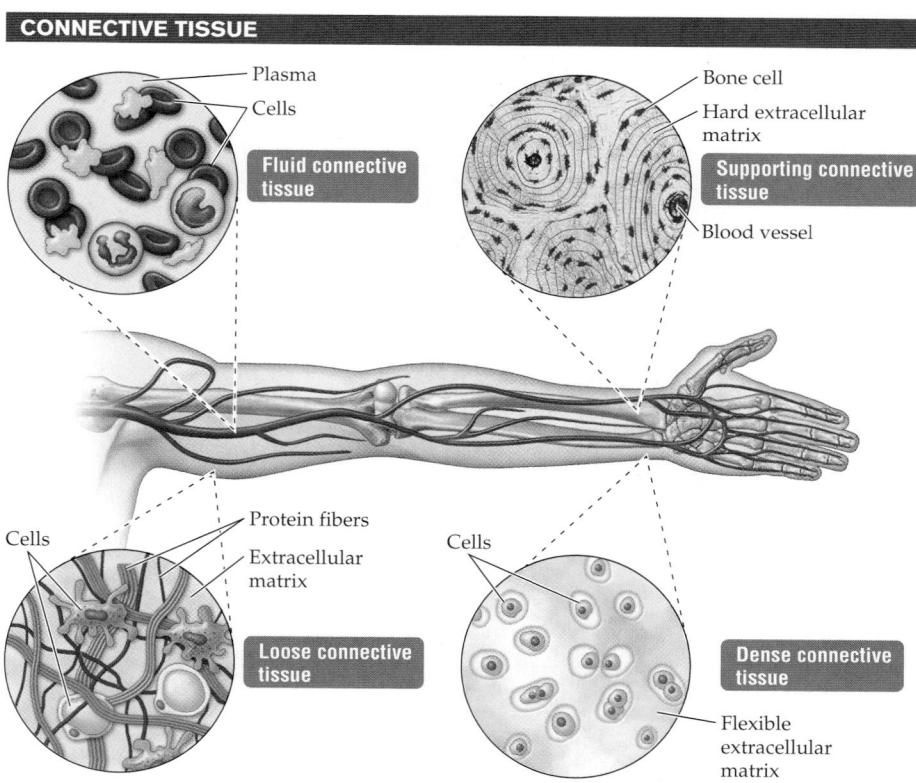

**FIGURE 26.2 Connective Tissue Binds and Cushions the Organs**

The four main categories of connective tissue are described below.

- *Loose connective tissue* is characterized by a soft matrix, usually with a meshwork of loosely arranged protein fibers embedded within it. Loose connective tissue holds organs in place and provides padding between them. *Adipose tissue* is a special type of loose connective tissue that contains relatively large cells filled with fat deposits (droplets of triglycerides).

- *Dense connective tissue* has a closely packed matrix containing strong cables of protein arranged in parallel arrays. *Collagen*, the most abundant protein in our body, is a common component of dense connective tissue. **Tendons** (which attach muscles to bone) and **ligaments** (which tie bones to each other at joints in the body) consist mainly of dense connective tissue rich in collagen.

- *Supporting connective tissue* has a semirigid to rigid matrix that creates strong structures such as *cartilage* and *bone*. **Cartilage**, which strengthens the earlobe and forms a cushion that keeps bones from rubbing against each other in many of our joints, has a highly hydrated matrix that is strong but flexible. In bone, the matrix is extensive compared to the volume occupied by the cells them-

selves. **Bone** matrix is exceptionally rigid because it is strengthened by deposits of calcium salts.

- *Fluid connective tissue* consists of cells suspended in a liquid matrix. Blood and *lymph* are examples of fluid connective tissue. The fluid portion of blood is called *plasma*. Suspended in it are several different kinds of cells, including *red blood cells*, which ferry oxygen; *white blood cells*, which fight invaders; and *platelets*, which are cell fragments that facilitate the clotting of blood.

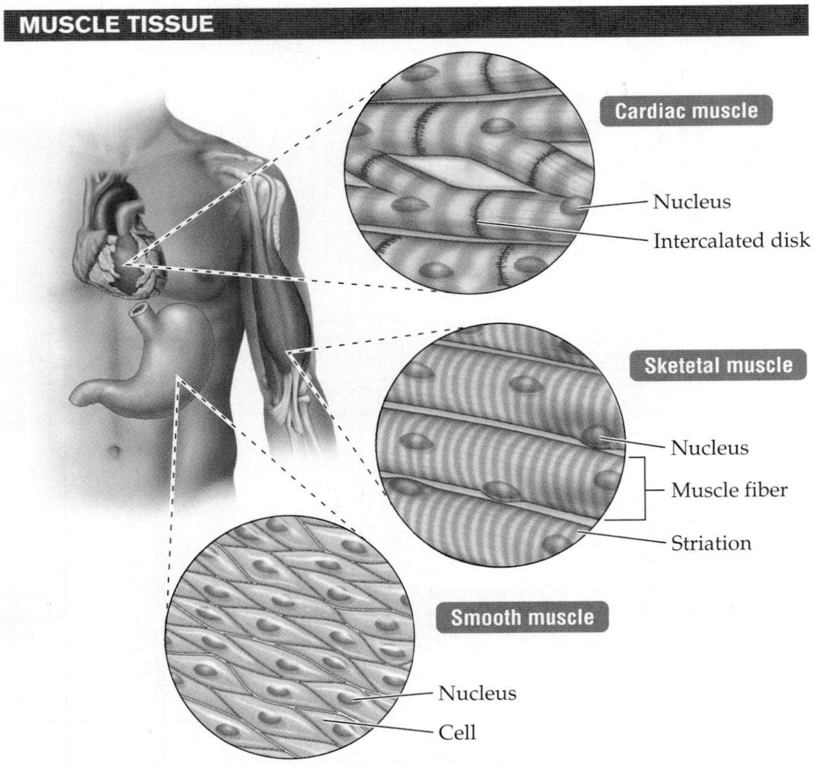

## Muscle tissue generates force by contracting

Muscle tissue consists of bundles of elongated cells known as muscle fibers. Each muscle fiber contains stacks of contractile proteins that can shorten (contract) the fiber by ratcheting against each other. Muscle contraction generates force, which can do a variety of work in the body, from squeezing blood out of the heart to pulling on bones that the muscle is attached to. Muscle contractions are most commonly triggered by a signal delivered by a nerve cell. There are three main types of muscle tissue:

- Skeletal muscle
- Cardiac muscle
- Smooth muscle

**Skeletal muscle** has a striated (striped) appearance caused by the regular pattern in which the contractile proteins are arranged within each muscle fiber (**FIGURE 26.3**). Skeletal muscle can be contracted voluntarily, meaning it is under our conscious control. **Cardiac muscle** is also striated, but it is arranged as branched bundles of muscle fibers and is not under our voluntary control. Cardiac muscle fibers are linked end to end and contract in unison upon receiving an electrical signal. Cardiac muscle is responsible for the blood-pumping action we call the heartbeat. **Smooth muscle** is so named because it lacks the striped pattern typical of skeletal muscle and cardiac muscle. Smooth muscle contracts involuntarily and is found in the walls of our intestines, blood vessels, and urinary bladder. Although slow to contract in response to a nerve or hormonal signal, smooth muscle can maintain a contraction longer than skeletal or cardiac muscle can.

## Nervous tissue communicates and processes information

Like all other organisms, animals have the ability to sense and respond to external and internal stimuli.

**MUSCLE TISSUE**

Cardiac muscle
- Nucleus
- Intercalated disk

Skeletal muscle
- Nucleus
- Muscle fiber
- Striation

Smooth muscle
- Nucleus
- Cell

**FIGURE 26.3 Muscle Tissue Is Contractile and Generates Force**

Nervous tissue enables animals to detect a signal and respond to it within a fraction of a second. **Neurons** are key players in nervous tissue because they can receive, integrate, and transmit information. Neurons have cytoplasmic projections, called *dendrites*, that receive signals (**FIGURE 26.4**). The signal is often propagated as a series of electrical pulses along the length of an exceptionally long extension of the nerve cell known as the *axon*. Because an axon can be very long, more than a meter in some cases, neurons can transmit signals over long distances through the body at lightning speed.

Nervous tissue contains other cell types, besides neurons, including support cells called *glia*. Nervous tissue is found in sensory organs such as the eyes, nose, ear, and mouth. It is the main tissue in nerves, which carry signals to and from the spinal cord and the *brain*. The **brain** is a central information-processing unit with a dense concentration of nervous tissue.

## NERVOUS TISSUE

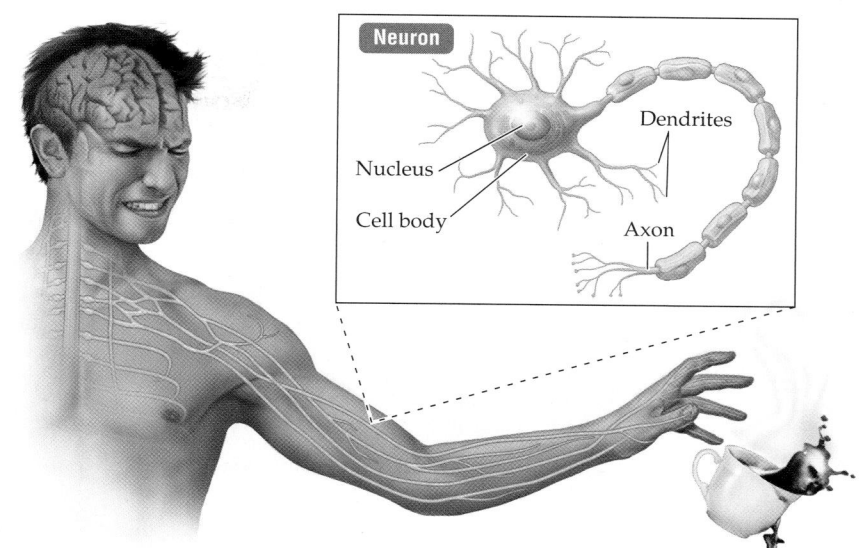

**FIGURE 26.4  Nervous Tissue Is Specialized for Communicating and Processing Information**

## 26.2 Internal Organization: Organs and Organ Systems

An organ has more than one tissue type and forms a functional unit with a distinctive shape and location in the body (**FIGURE 26.5**). The brain, stomach, liver, spleen, and kidneys are examples of organs in the vertebrate body.

An **organ system** is composed of two or more organs that work in a closely coordinated manner to perform a distinct set of functions in the body. The 11 major organ systems in the human body are illustrated in **FIGURE 26.6**.

The **integumentary system** is the largest organ system in the human body, covering and protecting the surface of the body. It consists of skin (see Figure 26.5), including structures embedded in the skin, such as hair and nails.

The **skeletal system** provides an internal framework to support the body of vertebrates. It consists of bones, cartilage, and ligaments. We would collapse into a formless mass of trillions cells without our bony internal skeleton to shore up the body from the inside.

The **muscular system**, consisting of the three types of muscle tissue, produces the force that moves structures within the body—the beating of the heart, the bellowslike movement of the chest cavity, the

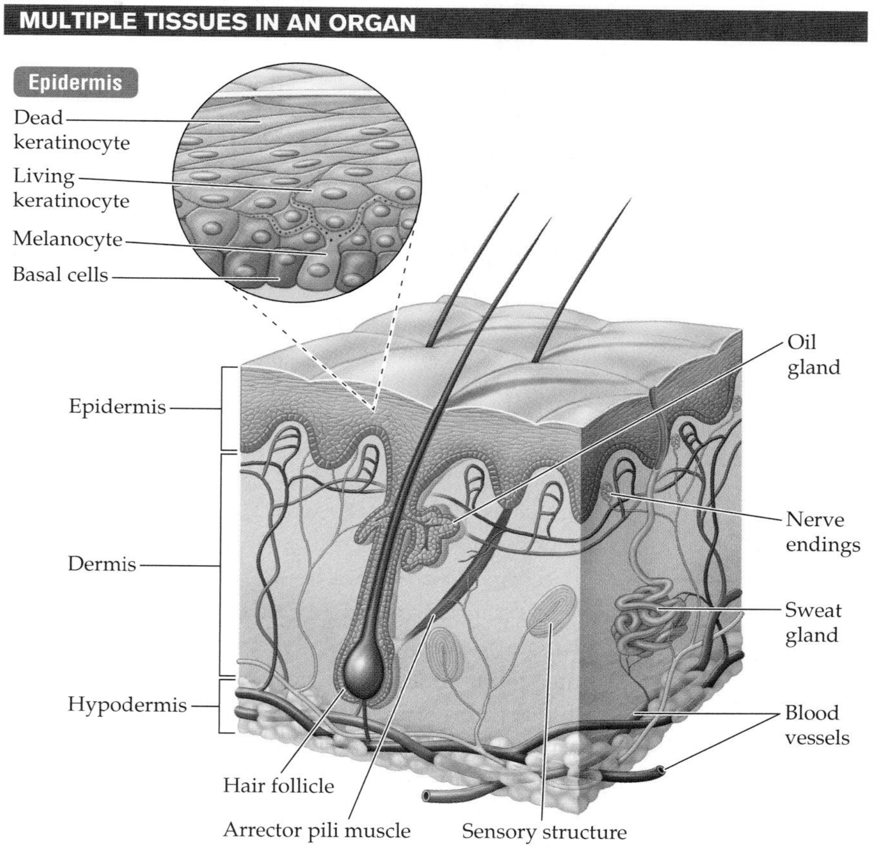

**FIGURE 26.5  Each Organ Is Composed of Multiple Tissue Types**
Many different tissue types are seen in skin, the largest organ in the human body. Note the multiple cell types in the epidermis, an example of epithelial tissue. The nerve endings are examples of nervous tissue. The erector pili muscle is composed of smooth muscle tissue. Much of the dermis is made up of connective tissue. Adipose tissue dominates in the hypodermis, which forms a thick insulating sheet under the skin.

FIGURE 26.6 The
Eleven Major Organ
Systems of the
Human Body Work
in an Integrated
Manner

## THE ORGAN SYSTEMS

Integumentary system
(Ch. 26)

Urinary system
(Ch. 26)

Digestive system
(Ch. 27)

Circulatory system
(Ch. 28)

Respiratory system
(Ch. 28)

Endocrine system
(Ch. 29)

Nervous system
(Ch. 30)

Skeletal system
(Ch. 31)

Muscular system
(Ch. 31)

Immune system
(Ch. 32)

Reproductive system
(Ch. 33)

squeezing action of the intestines, for example. Skeletal muscles work closely with the skeletal system to move body parts and to move the whole body in space.

The **digestive system** is essentially a food processor that extends from the mouth to the anus. Various large macromolecules are broken down in the mouth, stomach, and small intestines, and the smaller building blocks and minerals are absorbed in the small intestine. The *liver* and *pancreas* are accessory organs that assist in the digestion of food but also have many other critical roles in the body. The liver, for example, detoxifies potentially dangerous chemicals that enter the blood, and the pancreas plays a vital role in maintaining glucose homeostasis throughout the body.

The **urinary system** (also known as the excretory system) removes excess fluid from the body, along with waste products, toxins, and other water-soluble substances that are not needed. Kidneys and associated fluid-conducting tubes are key components of the urinary system. We examine the workings of the urinary system in Section 26.5.

The **respiratory system** brings in oxygen and expels carbon dioxide. The **lungs** are lined with a highly folded epithelial tissue that presents a large surface area for gas exchange.

The **circulatory system** whisks oxygen from the lungs to the heart, which then pumps oxygen-rich blood to the rest of the body through a closed network of vessels that includes *arteries*, *veins*, and *capillaries*. **Arteries** pump oxygen-rich blood from the heart to the rest of the body. **Veins** return oxygen-depleted, carbon dioxide–enriched blood to the heart.

The smallest branches of arteries and veins merge with each other in a fine meshwork of tiny tubes that make up a **capillary** bed. Carbon dioxide collected by the circulatory system is delivered to the respiratory system for exchange with the air outside. The circulatory system brings the many substances dissolved in blood to the urinary system, which discharges excess water, ions, water-soluble toxins, and organic wastes such as urea.

The **immune system** defends the body from invaders such as viruses, bacteria, fungi, and parasitic protists and worms. It is the most diffuse of the organ systems: it includes an army of lone white blood cells spread out all over the body on patrol for invaders. The *spleen* and *thymus* are important organs of the immune system because many immune cells arise through cell division in these organs. Immune cells also arise in the *bone marrow*. Although they are found in every tissue and organ in the body, immune cells are often concentrated in small pockets of tissue such as the *lymph nodes* and the tonsils.

The **reproductive system** generates gametes—eggs or sperm—and in many vertebrates it also facilitates the merger of the gametes inside the body of the female. In placental mammals, the young develop within a saclike uterus (the womb), with a substantial input of maternal resources.

The *endocrine system* and *nervous system* form a closely coordinated communication network that integrates and regulates all other organ systems. The **endocrine system** consists of a number of glands and secretory tissues in various locations in the main body cavities. A **gland** is a group of cells that manufactures and releases biomolecules into the blood, into body cavities, or on the body surface. The glands of the endocrine system produce signaling molecules, called hormones, that they discharge directly into the bloodstream. **Hormones** are signaling molecules that are distributed widely through the body and are active at very low concentrations.

The **nervous system** consists of sensory structures containing nervous tissue, and in vertebrates it includes organs such as the eyes, all the nerves, the spinal cord, and the brain. Neurons communicate with each other, and with other cell types, by receiving and sending electrical impulses and by picking up and releasing chemical messengers called *neurotransmitters*. Because the nervous system is a key player in sensing the external world and the internal state, and because it communicates with all the organ systems, it is the main agent of homeostasis in all but the simplest animals.

## 26.3 Maintaining the Internal Environment: Homeostasis

Whether an organism can survive in a particular habitat is determined by the *adaptations* it possesses. Recall that an **adaptation** is a genetic characteristic that matches an organism to its natural environment and that adaptations are a product of evolution by natural selection. Note that an adaptation is an inherent or inborn characteristic of an individual, not an adjustment triggered by the environment; individuals who lack the appropriate genes will fail to cope with adverse conditions and will most likely perish.

Some adaptive responses are expressed in the individual only when an environmental challenge presents itself—a process known as **acclimation** (or acclimatization). For example, most of us can *acclimate* to the "mile-high" altitude of Denver in a few days because most humans have the *adaptation* that enables us to function well at sea level and, given a long enough adjustment period, at moderate altitude as well. Most people acclimate to higher altitudes by making more hemoglobin, the oxygen-carrying protein that is abundant in our red blood cells. About 3,000 years ago the inhabitants of Tibet evolved additional adaptations that no other human group is known to possess. Tibetans have a form of hemoglobin that binds to oxygen more efficiently; because they possess this form of hemoglobin at all times in adulthood, the oxygen use efficiency of Tibetans is an adaptation, not an acclimation (**FIGURE 26.7**).

Homeostasis is the most widespread adaptive strategy for dealing with changing environmental conditions; it is seen in all types of organisms, including single-celled organisms. Metabolic activity cannot take place in the cytosol unless there is enough water, a near-neutral pH, and an optimum concentration of salts. Traffic across the plasma membrane is regulated, especially in terms of when, how much, and in what direction water, oxygen, solutes, and organic molecules are allowed to cross the membrane. Every organism therefore must manage the exchange that takes place over its surface to ensure that its internal state is compatible with life.

What is true of a single-celled organism is also true of a single cell buried deep within the body of a multicellular organism. Every cell in the animal body is bathed in a fluid of some sort, such as the *interstitial fluid* that seeps out of blood to fill the space between

**FIGURE 26.7**
**Adaptation versus Acclimation**
(*a*) Tibetans are adapted to high altitude at all times. (*b*) The skier has adaptations that enable him to acclimate to higher elevations.

cells in our tissues. In vertebrates, the interstitial fluid exchanges substances with the circulatory system, which in turn interacts with the respiratory system, the urinary system, and the integumentary system to distribute heat and to absorb vital gases and discharge wastes from and into the external world. Managing the exchange with the outside world is therefore a matter of life and death for all organisms.

## Body shape and size affect exchange rates

The cells of sponges and sea anemones are bathed in seawater (see Figure 4.3) and can therefore trade directly for vital materials such as oxygen and nutrients. Other animals, such as flatworms, have a body so thin and flat that all the cells are within close

reach of the body surface, which conducts the exchange with the outside world (see Figure 4.7). The body surface of many aquatic animals has elaborations that increase the trading potential. The feathery tufts on the back of a sea slug, for example, are external gills: the extensive branching increases the surface area available for absorbing oxygen from the water (see Figure 4.1*c*). Terrestrial animals need a large surface area for absorbing oxygen gas and expelling waste carbon dioxide, but they must prevent too much water from escaping the body across this surface. Lungs, with their moist exchange surfaces sheltered within the chest cavity, are one solution to these conflicting needs (**FIGURE 26.8**).

Large animals tend to exchange water, solutes, and heat with their environment more slowly than small animals do. That is because a large animal has a larger volume relative to its surface area than a smaller

**FIGURE 26.8 All Animals Exchange Gases, Nutrients, and Heat with Their External Environment**

A large surface area for exchange of materials can be beneficial, but it also carries the risk of excessive heat loss and, in terrestrial organisms, excessive water loss.

### GAS EXCHANGE SURFACES

(*a*)

Aquatic organisms have specialized gas exchange surfaces exposed to the outside of their bodies…

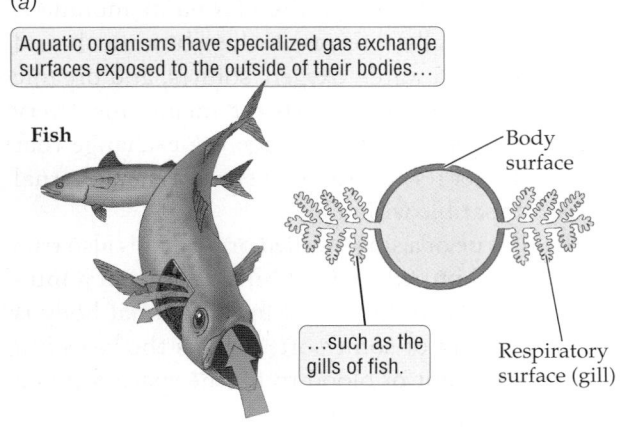

Fish

…such as the gills of fish.

Body surface

Respiratory surface (gill)

(*b*)

Land-dwelling organisms have specialized gas exchange surfaces inside the body…

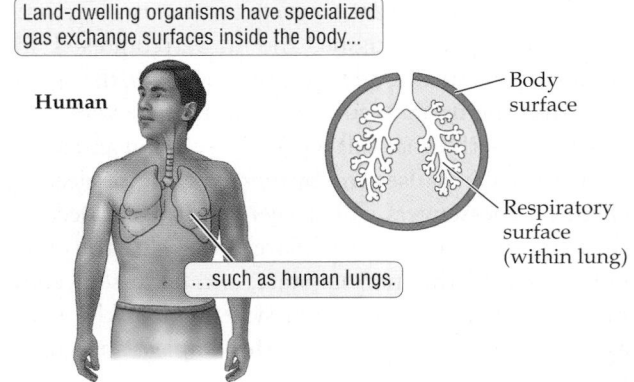

Human

…such as human lungs.

Body surface

Respiratory surface (within lung)

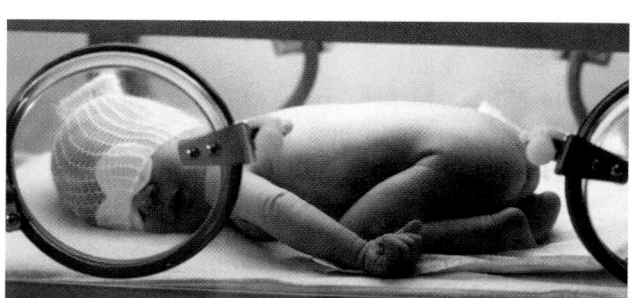

**FIGURE 26.9 Size Affects Homeostasis**
Newborn babies have a larger ratio of surface area to volume than adults do, so they lose heat and moisture to the environment faster. When babies are born prematurely, they spend time in an incubator until they gain enough weight (volume) to lower their disproportionately high ratio of surface area to volume.

animal has (see Figure 6.5 for an explanation of the relationship between surface area and volume). The ratio between these two quantities—surface area and volume—determines how quickly or slowly an animal can gain or lose water, solutes, or heat. When the ratio of surface area to volume is relatively high, gains and losses are rapid; when the ratio is relatively low, gains and losses happen more slowly.

The surface area–volume relationship has far-reaching implications for animals of any size. It explains, for example, why newborn babies are more susceptible to changes in temperature and to dehydration than adults are. A newborn has a large surface area compared to its body volume, so it loses heat and water more rapidly than an older child or an adult does. That is why we are careful to protect babies from temperature extremes (**FIGURE 26.9**). More generally, smaller animals exchange water and solutes rapidly but have a harder time maintaining temperature homeostasis than larger animals do.

## Homeostasis maintains the internal state at a level that is compatible with life

Homeostasis enables an organism to sense its internal environment continually and readjust it rapidly to maintain conditions best suited for life processes. Homeostasis buffers the internal environment, which shows minor changes, despite large fluctuation in the outside world.

On a wintry day we may not be able to produce heat as rapidly as we are losing heat to the environment. In that case, heat losses start to exceed heat gains, and the

body temperature begins to drop. Cells in the brain sense this drop and initiate processes that produce more heat. Our homeostatic processes now go into action, and we start shivering and get "goose bumps." Shivering and goose bumps are caused by muscle contractions, and those contractions use energy to produce heat. Blood flow to the skin decreases, so that less heat from the blood escapes at the body surfaces.

How does the body "know" when it is losing excessive amounts of heat, or moisture, or if the salt balance is off-kilter? Once it detects trouble, how does it correct the problem? Once it starts the correction process, how does it sense that it has gone far enough and that going further would be counterproductive? Certain principles of homeostasis that apply broadly to most animals help us understand how the body pulls off this remarkable balancing act.

## Homeostatic pathways have negative feedback loops

A homeostatic pathway is the sequence of steps that produces homeostasis. Homeostatic pathways have two basic features: (1) The physical and chemical characteristics of the internal environment are continually monitored. (2) Regulatory processes are triggered within the body if the monitoring system detects any departure from the normal state. In most vertebrates, characteristics such as body temperature, hydration level, pH of body fluids, and the concentration of salts, vital gases, and organic molecules in body fluids are constantly monitored. The overall effect, or output, of regulatory processes is an increase or decrease in the value of the characteristic being monitored. For example, the pH of our blood is continually monitored, and regulatory processes are set in motion if the blood becomes too acidic or too alkaline. Successful homeostasis returns the internal state to a genetically determined state of normality called a **set point**. The set point for blood pH ranges from 7.35 to 7.45 in humans.

The regulatory processes in homeostatic pathways commonly have **feedback loops**. Processes with feedback loops have multiple steps, and one step can "reach back" to control an earlier step in such a way that the output of the process as a whole is changed. The feedback loop can turn off the whole process, or reduce or enhance its output. A **negative feedback** loop turns off or reduces the output of a process; a **positive feedback** loop increases the output. Negative feedback loops are especially common in the regulatory mechanisms that restore homeostasis in animals.

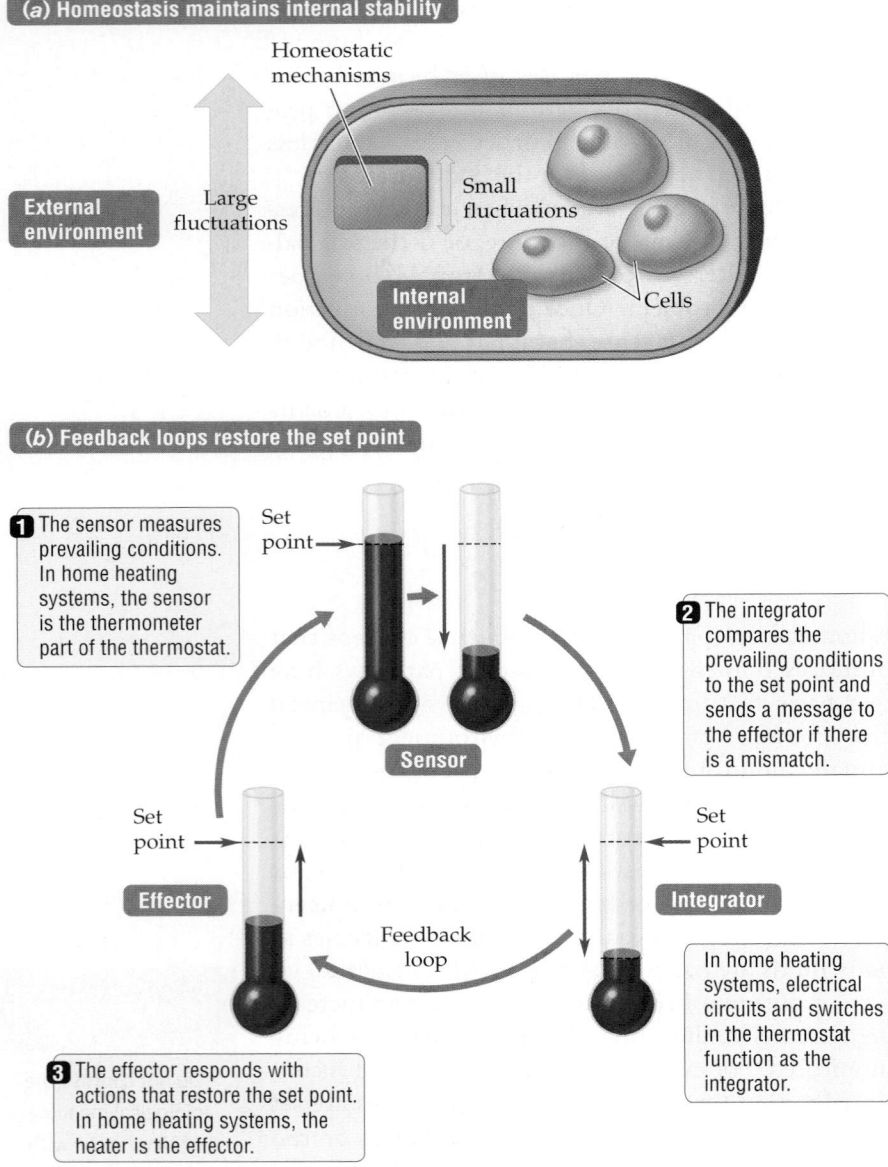

(a) Homeostasis maintains internal stability

Homeostatic mechanisms

External environment

Large fluctuations

Small fluctuations

Internal environment

Cells

(b) Feedback loops restore the set point

**1** The sensor measures prevailing conditions. In home heating systems, the sensor is the thermometer part of the thermostat.

Set point

**2** The integrator compares the prevailing conditions to the set point and sends a message to the effector if there is a mismatch.

Sensor

Set point

Effector

Set point

Feedback loop

Integrator

In home heating systems, electrical circuits and switches in the thermostat function as the integrator.

**3** The effector responds with actions that restore the set point. In home heating systems, the heater is the effector.

**FIGURE 26.10 Homeostatic Mechanisms Maintain Stable Internal Conditions despite Fluctuations in External Conditions**

(a) Because of homeostatic mechanisms, large fluctuations in external conditions produce little or no change in the overall state within the animal body. (b) A home heating system is analogous to the homeostatic mechanisms that maintain our body temperature. Temperature receptors in the skin function as sensors in the human body. A brain structure called the hypothalamus acts as the integrator, launching corrective action if core body temperature deviates from the set point (about 37°C). If body temperature declines, the hypothalamus activates a number of different effectors, including muscles that constrict surface blood vessels to reduce heat loss and those that produce shivering to generate extra metabolic heat.

Homeostatic mechanisms with negative feedback loops have three key components: (1) a *sensor*, (2) an *integrator*, (3) an *effector*. A nonliving example—a home heating system—shows how these components work (**FIGURE 26.10**). The thermostat of a home heating system includes a thermometer, which functions as the sensor. The electrical circuits and switches (often programmable in modern units) function as an integrator, the part of the system that compares incoming information with the set point and launches a response when conditions deviate from this reference state. The heating unit is the effector, the mechanism that restores the set point. We can program the thermostat to hold a certain temperature, which constitutes the set point for the system. The thermometer function of the thermostat monitors the temperature inside the house. If the inside temperature drops below the set point, the thermostat exerts its integrator function and turns on the heater. If the heater warms the house so much that the temperature rises above the set point, that higher temperature feeds back into the thermostat, which then turns off the heater.

As with the example of the thermostat and the home heating system, negative feedback provides a "steadying hand" in the homeostasis of many animals. Negative feedback tends to keep the internal environment in a near-constant state despite a broad range of external conditions. Such closely controlled homeostasis is dynamic, toning down the extent of internal change so that large shifts in the external environment produce only small, acceptable shifts internally. For example, when we get too hot, we sweat. When the body temperature drops to the set point, negative feedback stops the cooling process (sweating).

Positive feedback is the opposite of negative feedback in that the effects of a process increase the intensity or speed of that process. Positive feedback is not involved in homeostasis, because positive feedback pushes a process progressively in one direction, rather than maintaining a set point. The contractions of the uterus during childbirth are an example of positive feedback in a biological process. The first weak contractions, themselves launched by a complex set of hormonal signals, cause the baby's body to push against the wall of the uterus. That pressure activates membrane receptors in the wall of the uterus that send out nerve signals calling for yet more forceful contractions of the uterine muscles. Positive feedback loops must have an upper limit to counteract the runaway reinforcement of a process, since that could be dangerous. In our example, the baby's birth ends the escalation of the uterine contractions by positive feedback.

## Homeostasis requires energy

The more effectively an animal can regulate its internal environment, the closer it can keep its internal conditions to the ideal for its cellular activities—in particular, for its enzymes. Because we humans are nearly constantly at 37°C, we maintain the kind of internal environment in which our enzymes can function at their best. Animals such as butterflies, whose body temperatures change dramatically over the course of a day as environmental temperatures fluctuate, have less efficient enzyme function.

If maintaining near-constant internal conditions is best for cellular activities, why do only some animals closely regulate those conditions, while many others allow their internal conditions, including temperature, to change as their surroundings change? The answer is that homeostasis requires energy. The narrower the range of conditions the body tries to maintain, the more energy the body must expend. Put another way, the greater the difference between external conditions and ideal internal conditions, the greater the energy cost for the animal.

For some animals, especially small ones, the energy cost of regulating internal body temperature outweighs the benefits. A small butterfly, with its relatively large surface area, loses heat to the environment much more quickly than we do. Therefore, the butterfly must expend much more energy relative to its size to replace lost heat. The high energy cost cannot be supported without access to enough food calories, which are especially difficult to come by for nectar-sipping butterflies. Insects, amphibians, and reptiles that inhabit northern latitudes have no choice but to migrate to warmer regions or become dormant during the cold months.

Birds and mammals have evolved mechanisms that enable them to balance the pluses and minuses of regulating their internal temperature, and this ability has allowed them to be active in a greater range of habitats, often through quite large changes in the seasons. Mammals native to northern zones tend to be larger than the same or similar species found farther south, because their lower surface area relative to body volume is more efficient in conserving heat. However, despite evolving heavy insulation in the form of blubber and thick fur coats, some mammals must turn down their metabolism and hibernate for much of the winter because they cannot secure enough food calories during this part of the year to sustain their "metabolic furnace." Northern

bats, marmots, and ground squirrels are examples of mammals that go into true hibernation: the set point for body temperature plummets to near freezing and, in the case of the Arctic ground squirrel, even a few degrees below 0°C. The winter sleep of black bears is not regarded as "true hibernation" by many physiologists because the body temperature of a bear in its den drops only five or six degrees below the summertime norm of 37°C.

> ### Concept Check
>
> 1. Why are newborns more susceptible than older individuals to losing body heat and suffering from dehydration?
>
> 2. What is the difference between regulatory processes that have negative feedback loops and those that have positive feedback loops? Which type of feedback loop is found in homeostatic pathways?

## 26.4 Homeostasis in Action: Regulating Temperature

Many animals, especially terrestrial vertebrates, must gain or lose heat to maintain an internal temperature that is suitable for life processes. This process of controlling heat gain and loss is known as **thermoregulation**. Animals gain heat from two major sources: their internal metabolic processes and heat energy obtained from their environment. **Metabolic heat** is a by-product of an animal's metabolism, as we saw in Chapter 8. Skin, and the insulating layers of fat just underneath the skin, is effective at holding in much of this metabolic heat, and it helps maintain human body temperature at an average of 37°C. The sun is the other major source of heat. As we shall see shortly, energy from the sun can reach us directly, when we are exposed to sunlight; or indirectly, when we are exposed to air and the surfaces of objects that have been heated by the sun.

The surface area that an animal presents to a heat source affects the rate at which the animal exchanges heat with the external environment. If we want to warm ourselves in the heat of the sun, we face the sun fully to place more surface area of skin in full sunlight. But if we have to sleep in a cold room, we tend to curl up into a ball, minimizing our surface area and slowing down the rate at which we lose metabolic heat to our surroundings.

## Animals have two basic ways of regulating their internal temperature

Only a few groups of animals generate enough metabolic heat to contribute significantly to their internal body temperature. Animals that use internally generated heat to warm themselves are called **endotherms** (*endo*, "inside"; *therm*, "heat"). Mammals and birds are not only endotherms, but *homeotherms* as well: homeotherms maintain a nearly constant body temperature even in an environment with variable temperatures. Maintaining a high body temperature regardless of environmental temperatures is not a practical strategy for most animals. Instead, **ectotherms** (*ecto*, "outside") depend on environmental heat sources to regulate their body temperature. Ectotherms include most fishes, invertebrates, amphibians, and reptiles.

As endotherms, mammals and birds can remain active even when the lack of environmental heat sources forces ectotherms into inactivity. This advantage comes at a huge energetic cost, however. Endotherms use energy about 10 times as quickly as ectotherms of a similar size. As a result, endothermic animals must eat much more food in a given period of time than ectotherms of the same size require. For example, if your house were guarded by a 50-kilogram (kg) endothermic dog, you would have to feed it roughly 500 kg of dog chow each year (**FIGURE 26.11a**), whereas if you relied on a 50-kg ectotherm, such as an alligator, for the same service, you would have to provide it with a mere 50 kg of alligator chow annually (**FIGURE 26.11b**).

## Animals can gain or lose heat by conduction

Regardless of whether an animal is an endotherm or an ectotherm, heat can enter and leave its body by three different routes: *conduction*, *radiation*, and *evaporation* (**FIGURE 26.12**). Heat **conduction** can occur by direct contact between objects or substances. For example, if you put a heavy metal frying pan on a lit stove burner, the handle soon becomes too hot to hold. The heat moves by conduction from the pan along the handle and to your hand, if you touch the metal handle.

An animal often gains heat conductively by lying in contact with a rock or other surface that is warmer than its body (**FIGURE 26.13a**), or it gets rid of heat conductively by digging down into cooler desert sands during the hot desert daytime. Conduction can also happen within an object, as heat is transferred among atoms or molecules in one part of the object to atoms or molecules elsewhere in the object (as when it moves to the handle of the frying pan). Metabolic heat reaches an animal's surface partly by conduction of this kind as heat is passed from cell to cell and transported in body fluids such as blood.

Whether an animal gains or loses heat conductively depends on which has the higher temperature—the animal or its surroundings. That is because all heat exchanges in nature follow a universal rule: heat always flows from warmer areas to cooler areas. How fast heat is gained or lost conductively depends on several factors. The materials involved are one factor: solids and liquids conduct heat much more quickly than air does. We can remain comfortable indefinitely in 20°C (68°F) air, but we start to shiver with cold if we stay in a 20°C swimming pool for too long, because water conducts heat

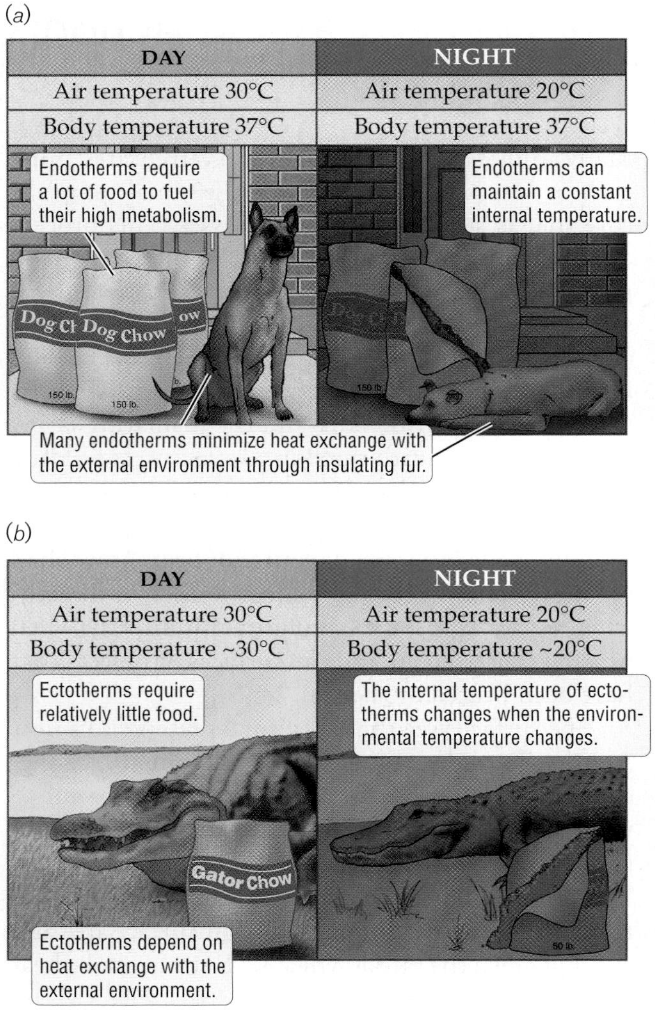

**FIGURE 26.11**
**Endotherms Generate Body Heat; Ectotherms Depend on Heat from the Environment**
(*a*) Endothermic dogs gain independence from environmental heat sources by using a lot of energy to generate metabolic heat. (*b*) By relying almost completely on environmental heat sources, ectothermic alligators use much less energy to regulate their internal temperature.

**ENDOTHERMS AND ECTOTHERMS**

(a)

| DAY | NIGHT |
|---|---|
| Air temperature 30°C | Air temperature 20°C |
| Body temperature 37°C | Body temperature 37°C |

Endotherms require a lot of food to fuel their high metabolism.

Endotherms can maintain a constant internal temperature.

Many endotherms minimize heat exchange with the external environment through insulating fur.

(b)

| DAY | NIGHT |
|---|---|
| Air temperature 30°C | Air temperature 20°C |
| Body temperature ~30°C | Body temperature ~20°C |

Ectotherms require relatively little food.

The internal temperature of ectotherms changes when the environmental temperature changes.

Ectotherms depend on heat exchange with the external environment.

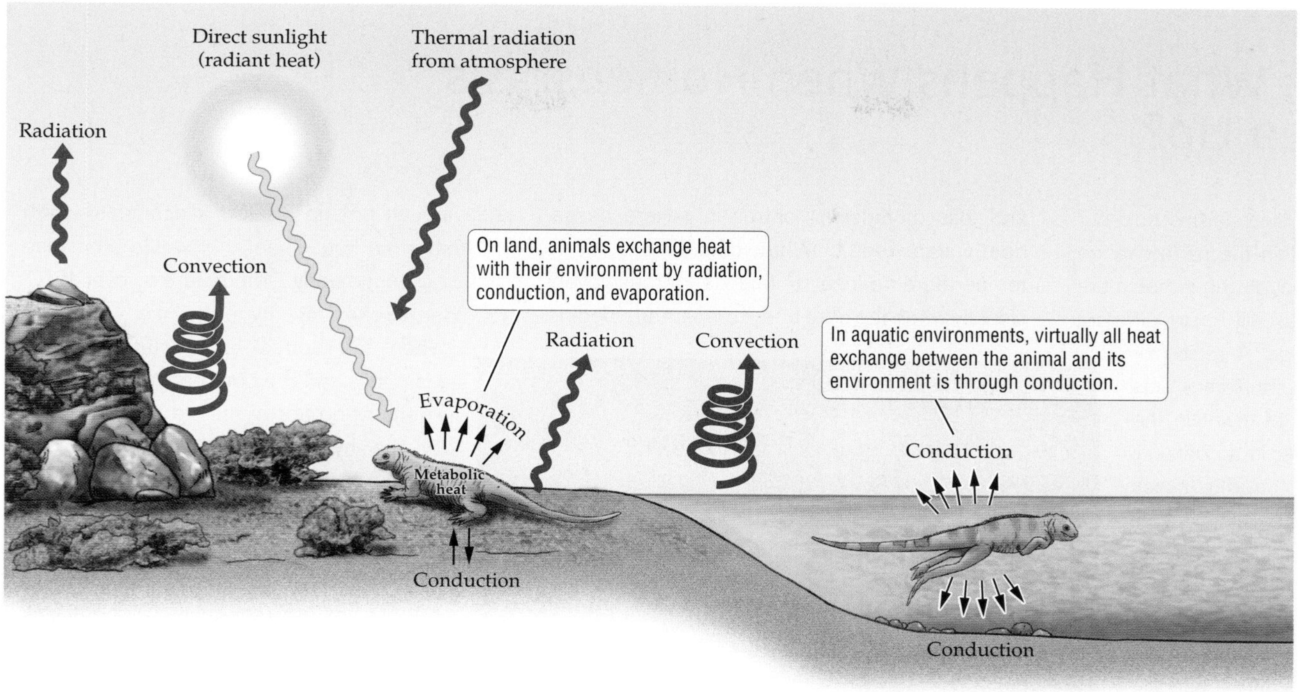

FIGURE 26.12

**Animals Gain or Lose Heat through Conduction, Radiation, and Evaporation**

As depicted in this scene from the Galápagos Islands, animals can gain heat in two different ways: externally from sunlight (directly or indirectly), and internally from metabolic heat. Regardless of the source of heat, animals regulate their body temperature by exchanging heat with their external environment by three different routes: conduction, radiation, and evaporation. In a mass of air, heat can also be transferred through a fourth mechanism, known as convection. Convection is the flow of particles, such as molecules in air, and is driven in many cases by differences in temperature between two regions.

from the body 25 times faster than air does. This difference in conduction rates means that aquatic animals lose or gain heat conductively much more quickly than land animals do. As a result, terrestrial animals can maintain a constant body temperature that differs from that of their environment relatively easily, whereas aquatic animals can do so only with great effort. For all but the aquatic mammals, such as seals and whales, the energy cost of raising and maintaining body temperature above the temperature of the surrounding water is too high. The bodies of most aquatic animals, therefore, remain at about the same temperature as their environment.

## Animals can exchange heat with their environment through radiation

**Radiation** is a means of exchanging heat in the form of light or infrared (heat) waves. The heat we feel from sunlight is radiant heat. All objects, including animals, can both absorb and emit radiant heat. During the day, radiant heat from the sun provides the major heat source for animals that rely heavily on an external source of heat. At night, without the sun's light, the animals that depend on environmental heat

sources slowly cool because they lose heat as infrared radiation.

As with conduction, the rate of radiant heat exchange depends on the materials involved. Objects with dark surfaces absorb sunlight better than do objects with light-colored surfaces, which reflect sunlight. A light-colored butterfly basking in the sun will take longer than a dark-colored butterfly to absorb heat (**FIGURE 26.13b**). Similarly, on hot summer days we wear light-colored clothing to stay cool because it absorbs less radiant heat than dark clothing does.

## Animals can cool themselves by evaporation

**Evaporation** enables animals to lose heat to their environment from a liquid (mostly water) present on a body surface that is exposed to air. The transition from a liquid (water) to a gas (water vapor) requires a lot of energy; water evaporating from an animal draws most of this energy from body heat. In this way the animal loses internal heat to the external environment.

Many animals living in hot terrestrial environments have ways of exposing wet surfaces to the external environment. When we sweat, we feel cooler

# Heatstroke: What Happens When Homeostasis Cannot Keep Up?

Normally, our homeostatic mechanisms hold our core body temperature within the narrow range of 36.5°C–37.5°C, even though the temperature of the air around us is usually quite different from this temperature range. A region of the brain known as the hypothalamus acts like the body's thermostat to maintain the temperature set point. Fever is an unusual situation in which the hypothalamus raises the set point in response to an infection; this situation is analogous to setting the thermostat for your home heating-cooling system to a higher temperature, so that it keeps the house warmer. A higher core temperature is often successful in limiting the replication of many viruses and some bacteria, which is why fever has evolved as a battle strategy against infection in mammals.

Heatstroke in humans illustrates the limits of homeostasis. Heatstroke occurs when the body temperature rises above 41°C. Symptoms include dizziness, nausea, confusion,

and loss of muscle control; in severe cases, death can result. What causes the body temperature to rise to this dangerous level? Usually when the core body temperature begins

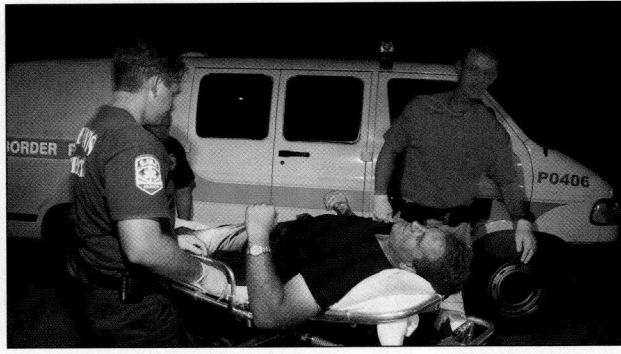

A US Border Patrol agent is evacuated to a hospital after suffering from heatstroke while on patrol in the Sonoran Desert near Tucson, Arizona.

to rise above normal, we start to sweat. As the water in the sweat evaporates, it takes with it a surprising amount of heat. A person sweating

heavily can get rid of 900 kilocalories' worth of heat in an hour—heat that would otherwise have raised the body temperature by over 12°C.

Heatstroke, or hyperthermia, happens when the body's homeostatic mechanism fails and the core body temperature keeps rising above the set point dictated by the hypothalamus. Heatstroke can happen if the external temperatures are so high that homeostatic mechanisms are unable to compensate for them. Homeostatic mechanisms can also fail if the sweating response does not kick in, which can happen in two situations: If the body has already lost a large amount of water, the sweating mechanism shuts down to prevent further water loss; this is the main cause of heatstroke in young people. Or the sweating response may be impaired by age or certain prescription drugs—the main cause of heatstroke in the elderly.

**FIGURE 26.13 Some Strategies for Dealing with Different Kinds of Heat Exchange**

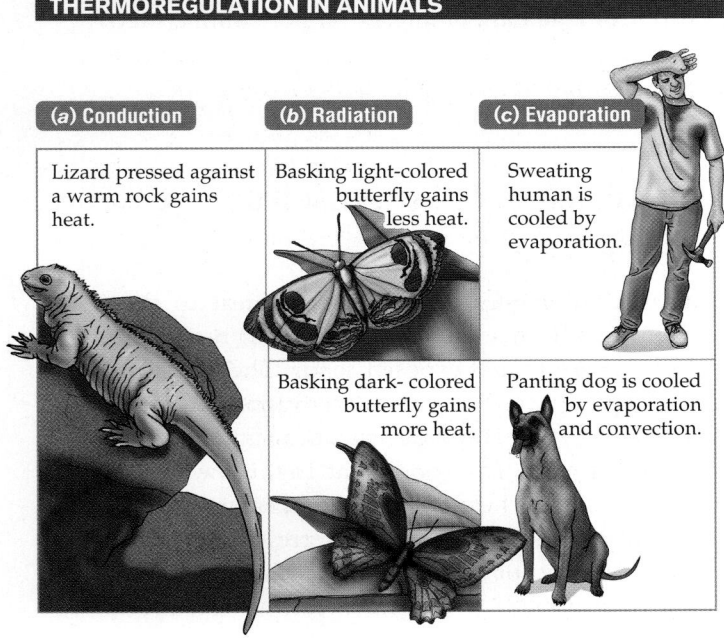

THERMOREGULATION IN ANIMALS

(a) Conduction — Lizard pressed against a warm rock gains heat.

(b) Radiation — Basking light-colored butterfly gains less heat. Basking dark-colored butterfly gains more heat.

(c) Evaporation — Sweating human is cooled by evaporation. Panting dog is cooled by evaporation and convection.

because the sweat removes heat as it evaporates from our skin (**FIGURE 26.13c**). Human skin makes a good evaporative surface because humans have relatively little hair. Most other mammals have evolved other forms of evaporation because their outer skin is covered by fur. Dogs, for example, rely on the tongue, whose wet surface loses moisture to the air when it is exposed by panting. Evaporation is enhanced by convective cooling as air moves back and forth over a panting dog's tongue.

The most important feature of evaporation is that it provides a way of losing heat that does not depend on the temperature of the animal relative to its environment. An animal can lose heat by conduction or radiation only if the temperature of the environment is lower than its body temperature. However, the same animal can lose heat evaporatively even in extreme heat. In deserts, where environmental temperatures often exceed internal temperatures, the value of cooling through evaporation often outweighs the accompanying loss of precious water. On the other hand, in cold terrestrial environments, animals tend to minimize evaporation to preserve heat; even so, the unavoidable evaporation of water during gas exchange (breathing) often represents an important source of heat loss.

## Convection increases heat exchange

**Convection** is the physical movement of molecules, and the heat contained in them, in a gas or liquid. Air currents and water currents are created by convection (see Figure 26.12). Convection in the environment can indirectly affect heat exchange in an animal by speeding up conduction, radiant heat loss, and evaporation. For example, a breeze cools us by whisking away the conductively heated layer of air next to our skin, replacing it with cooler air (see Figure 26.13). In this way, convection maintains a large temperature difference between the skin surface and the air immediately next to it. This difference speeds the conduction of heat from the skin to the air it touches. The animal is still losing heat by conduction and radiation, but those processes are helped along by convective currents in the air swirling around the body. In the same way, convection also speeds up the evaporative loss of water, which is why sweating cools us more under breezy conditions than in still air.

The nature of an animal's outer surface can affect convection. Anything that prevents the physical movement of air or water near the surface reduces convection. For example, fur, feathers, and the fill in sleeping bags trap a layer of conductively heated air next to the body surface (**FIGURE 26.14**), reducing the temperature difference between skin and adjacent air to slow heat loss in a cold environment.

## 26.5 Homeostasis in Action: Regulating Water and Solute Levels

Liquid water is the medium for the chemistry of life. Even so, organisms that live in aquatic habitats must constantly protect themselves from excessive water uptake. Too much water can cause animal cells to

**(a) Increased radiation**

**(b) Decreased convection**

burst because the delicate plasma membranes cannot hold up under the increased internal pressure (see Figure 7.5). Too little water causes the fluid inside cells to thicken and flow less easily, interfering with the movement of solutes within cells and the activity of many proteins and other biomolecules.

An excess of ions, salts, and organic molecules is disruptive in part because these compounds can poison vital cellular functions, particularly the working of enzymes. Enzymes are especially sensitive to the hydrogen ion concentration, or pH, of the cytosol. The control of internal water content and solute concentration by an organism is known as **osmoregulation**.

## The environment affects the regulation of body fluid composition

The challenges an animal faces in regulating its water and solute concentrations depend on its environment (**FIGURE 26.15**). The less complex sea animals, such as corals and jellyfish, have body walls that are no more than a few cells thick. Because most of their cells are in contact with seawater, they can exchange nutrients and wastes with the sea and they have no need to circulate fluids through the body. Larger marine animals, such as lobsters, do circulate special fluid through the body, but this fluid matches seawater in its saltiness. Such species, known as **osmoconformers**, do not have to spend much energy maintaining water and salt homeostasis, because their internal fluids resemble seawater quite closely (see Table 26.1 on p. 581). Osmoconformers do spend some energy fine-tuning the levels of certain other solutes, including nutrients and waste products.

**FIGURE 26.14**
**Releasing or Trapping Metabolic Heat**

(*a*) This athlete's flush comes from the widening of blood vessels close to the skin surface. The increased blood flow brings more of his body heat to the skin surface, where it is dissipated through radiation and evaporation. (*b*) In contrast, this wolf's fur reduces convective heat transfer from the skin surface to the cold air, and thereby helps keep heat inside its body.

Many other marine animals, including all marine vertebrates, are **osmoregulators**, which means their water and salt levels are not the same as those of their surroundings. The body fluids and cell contents of osmoregulating marine animals are not as salty as the sea. As a result, these species must spend a considerable amount of energy on water and salt homeostasis. All freshwater animals are osmoregulators because they live in an environment that is much less salty than their body fluids and the cytoplasm within their cells.

Marine fishes spend about 5 percent of their metabolic energy maintaining the composition of their body fluids. These saltwater fishes have body fluids that contain lower concentrations of solutes than seawater has. As a consequence, marine fishes tend to gain solutes from, and lose water to, the ocean (**FIGURE 26.15a**). They must expend energy to hold on to water and get rid of excess solutes. Freshwater fishes have the opposite problem: their body fluids contain solutes in much higher concentrations than are found in their external environment. As a result, they tend to lose solutes to, while constantly gaining water from, the environment (**FIGURE 26.15b**). Freshwater animals must expend energy to retain the solutes they

**FIGURE 26.15** How an Animal Regulates Its Water Content and Solute Concentrations Depends on Its Environment

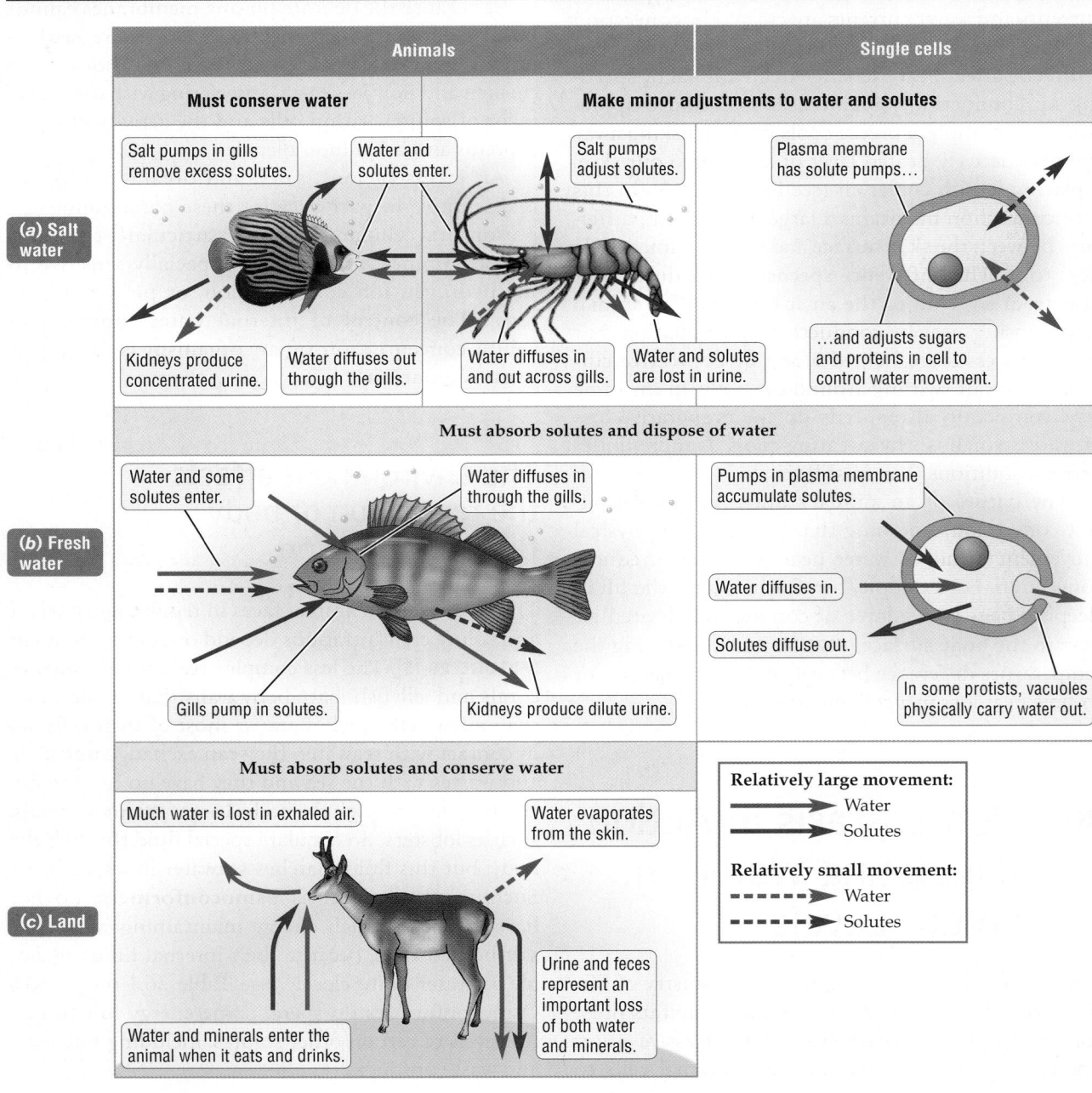

REGULATING WATER AND SOLUTES

| Animals | | Single cells |
|---|---|---|
| **Must conserve water** | **Make minor adjustments to water and solutes** | |

**(a) Salt water**

Salt pumps in gills remove excess solutes.

Water and solutes enter.

Salt pumps adjust solutes.

Plasma membrane has solute pumps…

Kidneys produce concentrated urine.

Water diffuses out through the gills.

Water diffuses in and out across gills.

Water and solutes are lost in urine.

…and adjusts sugars and proteins in cell to control water movement.

**Must absorb solutes and dispose of water**

**(b) Fresh water**

Water and some solutes enter.

Water diffuses in through the gills.

Pumps in plasma membrane accumulate solutes.

Water diffuses in.

Solutes diffuse out.

Gills pump in solutes.

Kidneys produce dilute urine.

In some protists, vacuoles physically carry water out.

**Must absorb solutes and conserve water**

**(c) Land**

Much water is lost in exhaled air.

Water evaporates from the skin.

Urine and feces represent an important loss of both water and minerals.

Water and minerals enter the animal when it eats and drinks.

Relatively large movement:
→ Water
→ Solutes

Relatively small movement:
- - - → Water
- - - → Solutes

| | CONCENTRATION (mM)[a] | | | | | |
|---|---|---|---|---|---|---|
| | BODY FLUID[b] | | | | | |
| SOLUTE | LOBSTER | FLOUNDER | GOLDFISH | HUMAN | SEAWATER | FRESH WATER |
| Sodium | 541 | 180 | 142 | 142 | 470 | 0.17 |
| Potassium | 8 | 4 | 2 | 4 | 10 | Not detectable |
| Chloride | 552 | 160 | 107 | 104 | 548 | 0.03 |

[a]The abbreviation mM stands for "millimolar," a unit that describes the concentration of a solute based on its mass (weight) in one liter of solvent.

[b]Lobsters, which are marine invertebrates, are osmoconformers. Flounder, which are marine vertebrates, are osmoregulators. Goldfish live in fresh water and are osmoregulators. Humans, like most other terrestrial vertebrates, are osmoregulators.

need and to keep too much water from accumulating in their bodies.

In terms of maintaining the composition of their body fluids, land animals live in perhaps the most challenging environment of all: they tend to lose both water and solutes (for example, salts lost in sweat) to their environment (**FIGURE 26.15c**). Water constantly escapes from a terrestrial animal's body. Land animals must offset these constant losses by drinking water and getting solutes from their food. Because they must constantly adjust the water and salt balance of their internal fluids, all land animals are osmoregulators.

## Metabolism changes the composition of body fluids

The solute composition of an animal is inevitably altered by metabolic activities within the body. As cells metabolize macromolecules, they use up some of the chemicals dissolved in body fluids and produce some new ones. Often these changes require only minor adjustments, but some waste products can pose serious problems. The metabolism of proteins, for example, produces *ammonia*, the same chemical found in many household cleaners. The buildup of ammonia can poison cells. To protect against this buildup, cells dump ammonia into the surrounding interstitial fluid, from which it diffuses into the blood plasma. Ammonia in the blood continues to pose a potential danger, however, so it must be removed from the body altogether.

Animals have evolved a number of different ways of eliminating ammonia from the body. Some flush out the ammonia directly, usually with large amounts

of water; others use energy to convert ammonia into less dangerous molecules. Animals living in fresh water do not have to spend energy to process ammonia; they simply expel the waste ammonia from the urinary system as a large volume of dilute urine. These animals can afford the large water loss that accompanies ammonia excretion because they need to get rid of water anyway (see Figure 26.15b).

Some terrestrial animals, including humans, that can get plenty of water to drink convert ammonia into a much less toxic compound called *urea*. Urea is highly soluble in water, and it can be allowed to linger in the body in small amounts because it is not toxic at low concentrations. The downside is that it takes a large volume of water to remove urea from the body to maintain low levels in blood plasma. Other terrestrial animals, such as birds, reptiles, and insects, conserve precious water by converting ammonia into *uric acid*, which they then excrete using relatively little water (**FIGURE 26.16**). The guano we might see at places

**FIGURE 26.16**
**Birds Convert Toxic Ammonia into Uric Acid**

The white color associated with bird droppings, such as these deposits made by seagulls, comes from uric acid. Birds and other animals, including reptiles and insects, use energy to convert the toxic ammonia produced by protein metabolism into white crystals of uric acid.

**Helpful to Know**

The endings "ule," "ula," "ulum," and "ulus" (Latin for "small one") are added to a number of anatomy terms to indicate a small version of something. For example, a tubule is a small tube, a capsule is a small container (from *capsa*, "box"), and a glomerulus is a small globe (from *glomus*, "ball").

where birds roost in large numbers is a deposit of the white crystals of uric acid. With its high nitrogen content, guano is an excellent fertilizer for plants.

## The urinary system regulates water and solutes

All animals must remove from their cells and body fluids certain metabolic products, such as ammonia, that, if left in the body too long, would become poisonous to their cells. All animals must also regulate the concentrations of solutes, such as sodium and calcium, in their body fluids. Terrestrial animals face the additional challenge of conserving water and retaining vital solutes. In vertebrates, water and solute homeostasis is maintained by a pair of organs known as the **kidneys**, key components of the urinary system.

Kidneys filter and regulate the composition of the blood as it moves through them, and these organs can process surprisingly large volumes of fluid. Though our paired, fist-sized kidneys make up only a tiny fraction (0.5 percent) of our total body weight, they process over 20 percent of the blood each time it

circulates through the body. That means that about a liter of blood enters and leaves our kidneys each minute. When blood leaves the kidneys and returns to the circulatory system, it is largely cleansed of metabolic wastes and is carrying water and solutes in normal amounts. The volume of blood leaving the kidneys is slightly smaller than the entering volume because some water is lost to make **urine**, the waste-carrying watery solution that is expelled from our bodies. On average, we lose about 1.5 liters (about a half gallon) of water each day in urine, and we replace it when we drink water and other fluids. That is one reason we get thirsty: thirst is a signal that we need to replenish lost water.

The blood-cleansing work of the kidneys—**filtration**—is performed by the kidney's basic functional unit, the **nephron** [*NEH*-frahn] (**FIGURE 26.17**). Each human kidney has about a million of these tiny filtration units. Blood enters and leaves each nephron through a bed of capillaries called a **glomerulus** [gluh-*MER*-yuh-lus] (plural "glomeruli") (see Figure 26.17). The glomerulus lies in a cup-shaped capsule that opens at its far end into a U-shaped **tubule**. Plasma from the blood flowing through the

**OSMOREGULATION IN THE KIDNEY**

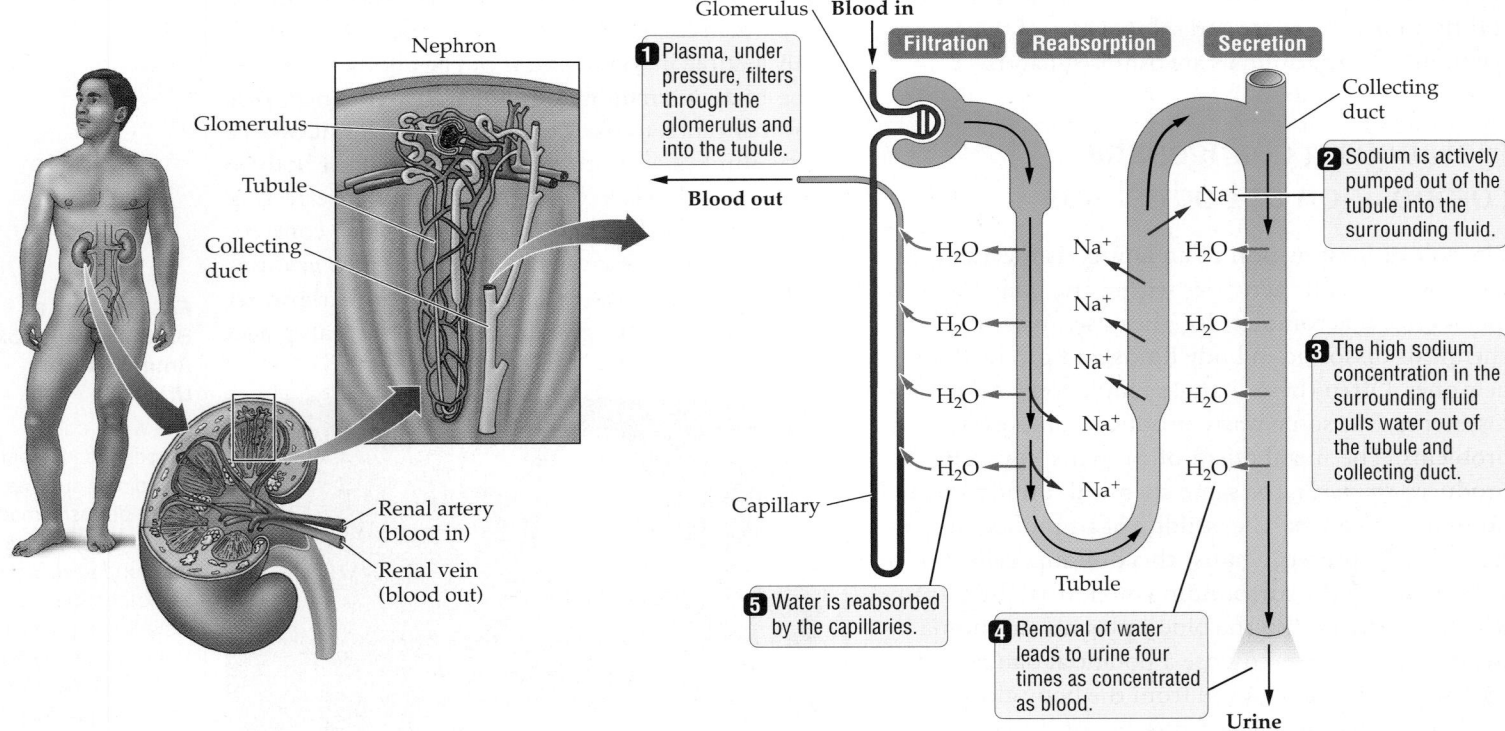

**FIGURE 26.17 The Regulating Functions of the Human Kidney**
The kidney regulates internal water content, balances solute concentrations, and removes toxic wastes.

glomerulus is filtered—forced by blood pressure to pass through tiny pores in the capillary walls into the tubule. All ions and molecules small enough to pass through these pores travel with the plasma into the tubule, and substances too large to pass through (such as large proteins, lipids, and blood cells) remain in the blood.

The **filtrate** (the filtered plasma in the tubule) includes many of the small solutes, such as urea, that the body needs to get rid of. As the filtrate travels down the tubule, urine gradually forms. Each tubule connects to a larger tube called a collecting duct, which leads to a series of ever-larger tubes. All these tubes eventually join the main tube, which drains urine from each kidney and carries it to the bladder. Here urine is stored until it can be excreted into the external environment.

Removing wastes from the blood plasma is only part of the kidney's job. In addition to urea and other waste substances, the filtrate just starting its journey down the tubule contains materials (solutes) that the body needs to take back; these materials include water, ions such as sodium and chloride, dissolved nutrients such as sugars, and small proteins. A second important function of the nephron is **reabsorption** of these valuable solutes and water from inside the tubule before they can drain out of the kidney with the rest of the urine. All along the length of the tubule within the nephron are sites where water and needed solutes are reabsorbed. These substances move out of the tubule and enter the extensive network of capillaries that wraps around each tubule.

The reabsorption of water causes urine to become increasingly concentrated in waste solutes as it moves down the tubule. The magnitude of this concentration is truly remarkable. Each day about 180 liters of water—enough to fill a large bathtub—is filtered out through the capillaries in the glomeruli. Of course, we cannot afford to lose that amount of water each day. And even if we could, it would not be easy to replace that much lost water by drinking it. So we are indebted to our kidneys for selectively reabsorbing almost all of the valuable nutrients and 99 percent of the water: only about 1.5 liters of urine actually leaves the average-sized person each day.

The tubules manage this impressive efficiency almost entirely through active transport: cells in the tubule walls pump sodium ions out of the tubules into the interstitial fluid outside their walls, causing water to diffuse from inside the tubules to these zones of high sodium concentration in the interstitial fluid. As Table 26.2 shows, water and valuable nutrients

| TABLE 26.2 | Substances Processed by the Human Kidney | |
|---|---|---|
| **SUBSTANCE** | | **PERCENTAGE REABSORBED** |
| **Nutrients** | | |
| Glucose | | 100 |
| Sodium | | 99.4 |
| Chloride | | 99.1 |
| **Wastes** | | |
| Urea (a waste product of the breakdown of proteins) | | 50 |
| Creatinine (a waste product produced by muscle tissue) | | 0 |

are very effectively reabsorbed from the tubule. Most wastes (such as creatinine) are reabsorbed inefficiently, or not at all; wastes remain in the tubules, become more concentrated as reabsorption progresses, and are finally eliminated in urine. It may be surprising to see from Table 26.2 that 50 percent of the urea is reabsorbed. However, urea is not toxic except at high concentrations, and moderate blood and tissue concentrations of urea might actually be beneficial. Urea is a potent antioxidant, a substance that protects tissues from damaging chemicals (called free radicals) that are generated as a by-product of certain metabolic reactions. We humans, and our closest primate relatives, have relatively high blood concentrations of urea compared to other animals. Biologists suggest that we may owe our longer life spans, at least in part, to the tissue-protective action of the urea that enters the bloodstream after being recovered from the fluid in the kidney tubules.

A third function of the kidney is **secretion**: certain substances, such as potassium ions, hydrogen ions, and some medications and toxins, are actively transported from the capillaries into the fluid inside the tubule. Secretion is not part of the initial filtration process in the glomerulus, but occurs in the last segment of the tubule, before the tubule joins the collecting duct. The collecting duct is the last segment of the excretory system that participates in fluid homeostasis. Typically, a small amount of water, sodium, and other ions is removed from the contents of the ducts and absorbed into blood. The concentrated fluid that remains is urine. Urine from the many collecting ducts in a kidney drains into a long tube, the *ureter*, which delivers the fluid to the urinary bladder for storage. The bladder is a stretchable, muscular sac that holds about 300 milliliters (10 ounces) of urine in most people. Nerve endings sense distension of the bladder and generate the urge

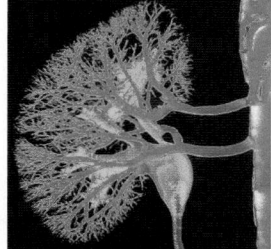

Network of blood vessels and capillaries in the human kidney

to urinate after the bladder is filled to about 25 percent of its capacity, with the urge getting stronger and harder to ignore as the bladder fills to capacity. In urination, the bladder empties through a tube called the *urethra*. The urethra is about 8 inches long in men and extends to the end of the penis. In women it is much shorter (1.5–2 inches) and opens just above the vagina.

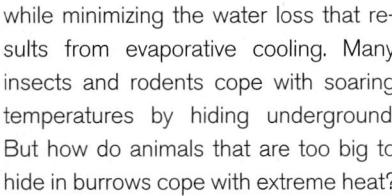

## APPLYING WHAT WE LEARNED

# How Camels Beat the Desert Heat

For any animal in a hot environment, the major challenge is to keep the body temperature from climbing to dangerous levels while minimizing the water loss that results from evaporative cooling. Many insects and rodents cope with soaring temperatures by hiding underground. But how do animals that are too big to hide in burrows cope with extreme heat?

When it's hot, humans begin to sweat almost immediately, using 3–15 liters of water per day to stay cool. Without water, we can quickly become dehydrated. As the amount of water in the blood falls, the total blood volume and blood pressure decline. Reduced blood flow inhibits the flow of nutrients to cells throughout the body. Dehydration leads to a cessation of sweating, a dry mouth, and concentrated urine—all changes that help conserve water. With continued high temperatures, the inability to sweat causes the body to overheat, well past the normal 37°C, leading to *heatstroke*—a core body temperature of 40°C or more. With heatstroke comes dizziness, nausea, rapid heartbeat and breathing, confusion, fainting, and, eventually, death.

Camels avoid heatstroke by means of adaptations that help them withstand much greater fluctuations in both core body temperature and dehydration than many other mammals can endure. When a camel is well hydrated, its body temperature varies by only about 2°C. This is the same temperature range that human bodies normally experience. But once a camel begins to be a little dehydrated, it stops sweating and its temperature can rise by as much as 7°C, to about 42°C (108°F), before it begins to sweat. Camels engage in thermoregulation (a form of homeostasis) just as all mammals do, but when air temperature soars, they can switch over to a *tolerance* approach and allow their body temperature to increase. Tolerance is the opposite of homeostasis.

The tolerance approach has several advantages over homeostasis. First, it reduces water loss. And conserving water is, of course, a form of homeostasis too. As a camel's temperature rises from 34°C in the cool desert morning to 42°C in the afternoon heat, its body stores about 2,900 kilocalories of excess heat. Dissipating that much heat through evaporation would require the camel to sweat about 5 liters of water. By allowing itself to heat up, the camel saves that water. When night falls and air temperatures drop below the camel's body temperature, the camel loses the excess heat through conduction and radiation. Depending on nighttime temperatures, the camel's temperature may be back down to a healthy 34°C by morning.

Heat tolerance has another benefit. When the difference between a camel's body temperature and air temperature is close to zero, the camel's body absorbs less heat than if it were cooler.

Once the camel's body temperature reaches 42°C, the overheated camel begins to sweat and lose valuable water at the rate of 5 liters a day. The animal switches to an approach that maintains body temperature (homeostasis), but at the expense of water—losing up to 30 percent of all the water in its body. A camel tolerates such severe dehydration by keeping what water it has in the blood and drawing water away from other parts of its body. Just like dehydrated humans, dehydrated camels have reduced blood volume, dry skin, dry mouth, and highly concentrated urine.

Because of camels' economical use of water, they can travel for more than 2 weeks through waterless tracts of desert. Only because humans have developed clothing to slow water loss from the skin and ways to store and carry water have we been able to follow camels into the desert.

# Relentless Heat Wave Roasts Russia

BY ANDREW FREEDMAN, *Washington Post*

While the mid-Atlantic has suffered through a sultry summer so far, conditions have not been nearly as extreme as in parts of Europe, particularly Russia. The Russian heat wave of July and August, along with related drought conditions and wildfires, have garnered international headlines in recent weeks, and will surely be studied by climate scientists and public health experts for years to come, both for its intensity and duration.

The statistics are staggering. According to meteorologist Jeff Masters of Weather Underground, at least 26 days in a row have had temperatures exceeding 86 degrees Fahrenheit in Moscow, and those conditions are expected to continue for at least the next several days. Moscovites are simply not used to such scorching temperatures. The city's typical July daily high temperature is just 74 F, with an average August daily high of 68 F.

Authorities have already estimated that about 5,000 people may have died from the heat in Moscow alone, but that figure is expected to rise considerably, especially considering that the city has met or exceeded what had been its all-time record high temperature of 99 F five times, and has been blanketed by acrid smoke from nearby wildfires.

Masters and his colleagues at Weather Underground have been providing some of the most in-depth reporting on the meteorological dynamics behind the heat wave, as well as the climate change context in which it is taking place. Here is what Masters wrote on Aug. 6.

> One of the most remarkable weather events of my lifetime is unfolding this summer in Russia, where an unprecedented heat wave has brought another day of 102°F heat to the nation's capital . . . Prior to this year, the hottest temperature in Moscow's history was 37.2°C (99°F), set in August 1920.

The Moscow Observatory has now matched or exceeded this 1920 all-time record five times in the past eleven days, including today. The 2010 average July temperature in Moscow was 7.8°C (14°F) above normal, smashing the previous record for hottest July, set in 1938 (5.3°C above normal) . . . It is stunning to me that the country whose famous winters stopped the armies of Napoleon and Hitler is experiencing day after day of heat near 100°F, with no end in sight.

Masters stated that the heat wave has been more intense than the infamous European heat wave of 2003, which killed an estimated 40,000 people. The death toll from the Russian event may wind up being even higher than the 2003 disaster, Masters noted. However, other elements, such as the response of the public health system, will also influence the death toll and could keep the numbers down.

According to the United Nations, the Russian heat wave of 2010 killed nearly 56,000 people. Why did a summer temperature spike kill so many people? Analysis of the European heat wave of 2003 provides some answers. Like the Russian heat wave, the 2003 heat wave in western Europe killed tens of thousands of people. Exact numbers are still hard to come by, but we know that most of them were sick or elderly. France was particularly hard hit, with a week of temperatures over 40°C (104°F) and nearly 15,000 deaths. People with dehydration and heatstroke (a body temperature greater than 40°C) swamped hospitals, and bodies piled up in morgues. Europeans have access to medical care as good as any in the world, yet a few days of hot weather took an awful toll.

Some of the causes were cultural. In countries with cooler summer temperatures, air-conditioning is rare, and houses and apartments became unbearably hot. Many people had never experienced such temperatures and simply did not know how to stay cool. And because the heat wave peaked during August, a time when Europeans traditionally take summer holidays, many families had left elderly relatives home alone.

Older people have a harder time regulating their body temperature, and many also take medications that interfere with their ability to regulate body temperature. A recent analysis of heat-related deaths worldwide concluded that the following groups are most at risk of fatal heatstroke during a severe heat wave: (1) elderly people, especially those confined to bed; (2) those on certain medications, especially long-term users of blood pressure–lowering drugs; (3) those with heart and respiratory diseases and mental illness.

## Evaluating the News

**1.** What is heatstroke, and why does it represent a failure of normal homeostasis?

**2.** Many of us have elderly relatives. What would you do to ensure their well-being during periods of hot weather?

**3.** A disproportionate number of heat-related deaths occur in cities rather than in the countryside. What are some of the physical and social aspects of cities that make their residents vulnerable to extreme summer heat?

**SOURCE:** *Washington Post*, Capital Weather Gang blog, August 9, 2010.

# CHAPTER REVIEW

## Summary

### 26.1 Internal Organization: Cells and Tissues

- Four main types of tissues are found in vertebrates: epithelial, connective, muscle, and nervous tissue.
- Epithelial tissue covers and protects surfaces and cavities. Connective tissue binds and cushions organs and tissues. Muscle tissue is contractile, and skeletal muscles enable movement. Nervous tissue communicates and processes information.

### 26.2 Internal Organization: Organs and Organ Systems

- An organ has more than one tissue type and forms a functional unit with a distinctive shape and location in the body. The brain, stomach, liver, spleen, and kidneys are examples of organs in the vertebrate body.
- An organ system is composed of two or more organs that work in a closely coordinated manner to perform a distinct set of functions in the body. The 11 major organ systems in the human body work together to accomplish all the necessary functions of life.

### 26.3 Maintaining the Internal Environment: Homeostasis

- Homeostasis is the process of monitoring the internal environment of an organism and keeping that environment stable so that it will be suitable for critical life processes. Cellular homeostasis is a universal property of all life-forms. Multicellular animals have complex homeostatic mechanisms.
- Organisms must keep their cells within a certain temperature range. Most cells cannot survive at temperatures below 0°C, at which point water freezes, destroying cell membranes; or at temperatures above 40°C, which deform proteins, destroying their function.
- Organisms must have the right amount of water in their cells to survive and remain active. Too much water can interfere with chemical reactions, alter protein function, and burst cell membranes. Too little water can also alter protein function.
- Homeostatic pathways have two basic features: sensors that monitor the internal environment, and regulatory processes that attempt to restore the normal internal state when deviations from optimal conditions are detected.
- Many homeostatic pathways are controlled by negative feedback, in which the results of a process cause that process to slow down or stop.

- Body size affects homeostasis. Large animals, which have a smaller surface area relative to their volume than small animals have, do not lose or gain heat as rapidly as small animals do. Large animals can maintain internal conditions different from those of their environment more easily than small animals can.
- Homeostasis requires energy. The greater the difference between external conditions and the desired internal conditions, the more energy it takes for the animal to maintain homeostasis.

### 26.4 Homeostasis in Action: Regulating Temperature

- Animals gain heat in two ways: externally from the sun (directly or indirectly) and internally from metabolic heat.
- Endotherms (including birds and mammals) depend on metabolic heat to maintain a nearly constant body temperature. This close regulation requires a lot of energy, but it enables animals to be active in a wide range of external temperatures.
- Ectotherms (fishes, invertebrates, amphibians, and reptiles) depend more on environmental heat than on metabolic heat, and their internal temperatures vary more than those of endotherms. This form of regulation requires less energy but forces animals to reduce their activity in cold environments.
- Animals can exchange heat with their environment by conduction, radiation, and evaporation, sometimes facilitated by convection.

### 26.5 Homeostasis in Action: Regulating Water and Solute Levels

- The content of interstitial fluid must be regulated to facilitate exchanges of water and solutes across the cell's selectively permeable plasma membrane.
- Marine fishes, freshwater fishes, and terrestrial animals each face different challenges in maintaining the composition of their body fluids.
- Metabolism changes the composition of body fluids. Some waste products, including ammonia, are toxic to cells, requiring continual removal.
- In many animals, including humans, kidneys regulate body water and solute concentrations. The kidney's basic unit, the nephron, performs three functions: filtration, reabsorption, and secretion. The resulting concentrated solution (urine) is carried by ducts to the bladder and excreted from the body.

## Key Terms

acclimation (p. 571)
adaptation (p. 571)
anatomy (p. 566)
artery (p. 570)
bone (p. 568)

brain (p. 569)
capillary (p. 571)
cardiac muscle (p. 568)
cartilage (p. 568)
circulatory system (p. 570)

conduction (p. 576)
convection (p. 579)
digestive system (p. 570)
ectotherm (p. 576)
endocrine system (p. 571)

endotherm (p. 576)
epithelial tissue (p. 567)
evaporation (p. 577)
extracellular matrix (p. 566)
feedback loop (p. 573)

filtrate (p. 583)
filtration (p. 582)
gland (p. 571)
glomerulus (p. 582)
homeostasis (p. 566)
hormone (p. 571)
immune system (p. 571)
integumentary system (p. 569)
kidney (p. 582)
ligament (p. 568)

lung (p. 570)
metabolic heat (p. 575)
muscular system (p. 569)
negative feedback (p. 573)
nephron (p. 582)
nervous system (p. 571)
neuron (p. 569)
organ (p. 566)
organ system (p. 569)
osmoconformer (p. 579)

osmoregulation (p. 579)
osmoregulator (p. 580)
physiology (p. 566)
positive feedback (p. 573)
radiation (of heat) (p. 577)
reabsorption (p. 583)
reproductive system (p. 571)
respiratory system (p. 570)
secretion (p. 583)
set point (p. 573)

skeletal muscle (p. 568)
skeletal system (p. 569)
smooth muscle (p. 568)
tendon (p. 568)
thermoregulation (p. 575)
tissue (p. 566)
tubule (p. 582)
urinary system (p. 570)
urine (p. 582)
vein (p. 570)

# Self-Quiz

1. Which organ system functions as in integrator and regulator of other organ systems?
   a. reproductive system.
   b. integumentary system.
   c. endocrine system.
   d. immune system.

2. Tendons and ligaments are examples of
   a. epithelial tissue.
   b. adipose tissue.
   c. loose connective tissue.
   d. dense connective tissue.

3. Endothermic animals
   a. are usually cooler than their surroundings.
   b. cannot gain heat by conduction.
   c. can generate body heat to warm themselves.
   d. generate relatively small amounts of metabolic heat.

4. Ectothermic animals
   a. produce large amounts of metabolic heat.
   b. rely primarily on external sources of heat.
   c. cannot radiate heat.
   d. cannot lose heat by convection.

5. The primary heat source for endothermic animals is
   a. metabolic heat.
   b. conduction.
   c. radiation.
   d. evaporation.

6. Fur and feathers
   a. reduce heat loss in endothermic animals.
   b. reduce convective heat transfer.
   c. reduce heat loss by trapping air, which conducts heat poorly.
   d. all of the above

7. Interstitial fluid is
   a. found only in kidney tubules.
   b. the fluid portion of blood.
   c. the fluid surrounding cells.
   d. found only in kidney capillaries.

8. Blood plasma transports
   a. waste products.
   b. water.
   c. solutes.
   d. all of the above

9. In the nephron, water is reabsorbed from the
   a. glomerulus.
   b. bladder.
   c. tubule.
   d. plasma membrane.

10. An important function of the human kidney is to
    a. produce a concentrated urine, after reabsorbing nearly all of the water.
    b. produce a dilute urine, after reabsorbing nearly all of the urea.
    c. convert toxic wastes into harmless ones.
    d. all of the above

# Analysis and Application

1. Describe the range of conditions under which life is known to exist, and explain why extreme conditions are a challenge to living organisms.

2. What is homeostasis? List three general characteristics of homeostasis in animals.

3. Imagine you are preparing for an expedition to the North Pole. What tactics might you use to maintain homeostasis in this bitterly cold environment? Discuss some of the ways humans use technology to assist homeostasis.

4. Describe negative feedback in terms of eating your favorite meal.

5. What are the three routes by which animals exchange heat with the external environment? Describe each route in terms of an animal's body and its surroundings.

6. Describe the processes of filtration and reabsorption in a nephron.

7. Describe heatstroke from the point of view of an internal cell and its immediate surroundings. How might heatstroke affect the cell's ability to function?

# 27 Nutrition and Digestion

**EAT YOUR GREENS.** A healthy diet begins with plenty of vegetables, whole grains, and fruit.

# The Cost of Eating in America

In the spring of 2011, congressional Democrats and Republicans were arguing about how to cut 38 billion dollars in spending to help reduce the country's 14-trillion-dollar debt. The $38 billion was pocket change, not only compared to the debt, but also compared to mounting health care costs. Indeed, staring Congress right in the face was a rapidly accumulating and unnecessary trillion-dollar-a-year health care expense.

Within 20 years, cardiovascular disease is expected to cost the United States a trillion dollars a year—mainly in health care costs but also in indirect costs such as lost workdays. Heart disease is a "lifestyle disease" that is about 90 percent avoidable. For most people, it's not caused by an unlucky set of genes or an infection. It's the result of smoking, poor diet, and sitting nearly all the time—sitting in a car, sitting in an office for 8 hours a day, and finally collapsing at home to watch television or chat online.

The story is even worse for type 2 diabetes, another lifestyle disease, whose incidence more than doubled between 1960 and 1980 and then tripled again between 1980 and 2010. All told, more than nine times as many people have type 2 diabetes today as did in 1960. By 2020, half of all Americans will have either type 2 diabetes (or prediabetes), and treating us will cost half a trillion dollars a year.

> Why have Americans become so much more likely to suffer from type 2 diabetes and heart disease? Can we turn back the clock; live longer, healthier lives; and reduce health care costs and the federal deficit all at the same time?

Type 2 diabetes is a chronic and dangerous increase in blood sugar that triples the risk of fatal heart attack and stroke. It can lead to kidney failure, blindness, and severe infections of the hands, feet, and limbs that can require amputation. It's also nearly completely preventable; when our grandparents were young, diabetes was a rare disease. Today, 11 percent of adults have diabetes and more than a quarter of those 65 and older have it. Our risk of these chronic diseases is related in part to how we eat and digest food—a topic we'll explore in this chapter.

## MAIN MESSAGE

Animals must break down ingested food to extract and absorb nutrients they need as an energy source and as chemical building blocks for the body.

## KEY CONCEPTS

- Nutrients provide animals with energy and with the chemical building blocks needed for growing and maintaining their bodies. Digestion is the chemical breakdown of food.

- The function of the digestive system is to break down large molecules, such as proteins, absorb useful nutrients and much of the water, and expel unusable material from the body. In most animals, the digestive system consists of a digestive tract and a number of accessory organs. The digestive tract is a long, hollow passageway in which food is processed.

- In humans, the processing of food begins in the mouth, where food is moistened and chewed into smaller pieces that are easier to swallow.

- The digestion of ingested protein begins in the stomach, which secretes an acidic fluid and protein-degrading enzymes.

- In the small intestine, lipids are broken down and the digestion of carbohydrates and proteins continues. Cells lining the cavity of the small intestine absorb the small molecules released by digestion and transfer them to the bloodstream.

- Despite large differences in diets, the digestive systems of most animals are similar in overall organization and general features. The digestive system of herbivores is adapted for processing large quantities of plant matter.

SOME PEOPLE MAY LIVE TO EAT, but all people must eat to live. The nutrients we get from food play two critical roles in our survival (**FIGURE 27.1**). Nutrients provide us with the energy we need to live; they also provide our bodies with the raw materials for constructing the molecules on which our survival depends. Like all other animals, we are consumers. This means we cannot get either the energy or the chemical building blocks we need directly from our nonliving physical surroundings, as plants do. Instead, we must get both by eating other organisms. The **nutrients** required by animals include organic molecules, such as carbohydrates, proteins, lipids ("fats"), and vitamins.

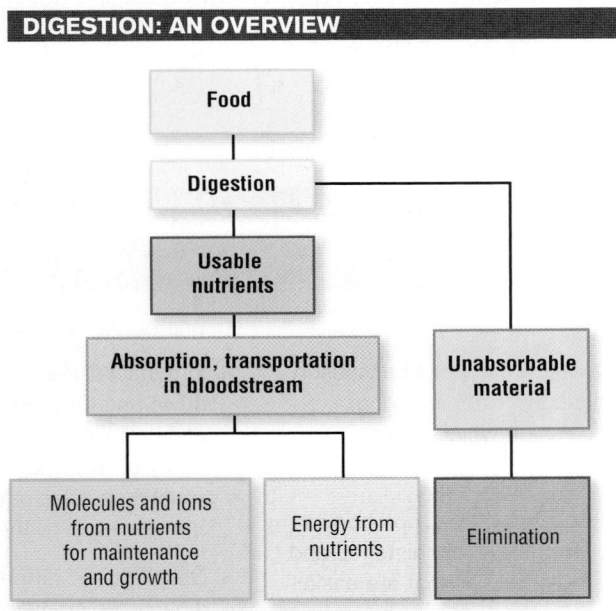

**FIGURE 27.1 The Fate of Food: A Brief Overview**
Food contains materials from which we obtain energy and the chemical building blocks needed for maintaining and growing the body. Digestion converts food into usable nutrients that can be absorbed into the body. Nutrients supply energy or become incorporated into the molecules and ions that make up the body. Material that is not absorbed is eliminated as feces.

We also need certain minerals, which are inorganic chemicals such as sodium, calcium, and iron.

The function of the digestive system is to process food, absorb useful nutrients, and eliminate unusable waste. The **digestive system** of most animals consists of a long, hollow passageway, known as the digestive tract, and a number of *accessory organs* such as the pancreas and liver. In most animals, the digestive tract takes the form of a muscular tube, usually interrupted by one or more sacs. It has two openings: the mouth, through which food enters; and the anus, through which unusable remains are expelled. The squeezing action of muscles that lie within the wall of the digestive tract pushes the food through the cavity in the center of the digestive tract. The cavity is lined by specialized cells, broadly classified as epithelial cells, that perform specific functions, such as secreting enzymes into the cavity or absorbing nutrients from the cavity.

Eating, or **ingestion**, is the first stage in the processing of food by the digestive system. **Digestion**, the chemical breakdown of food, begins almost immediately after ingestion in many species. Absorption is the uptake of small molecules and mineral ions by cells lining the cavity of the digestive tract. Most of the nutrients absorbed by the digestive tract are sent to the bloodstream, which eventually delivers them to every cell in the body. The final segment of the digestive tract specializes in absorbing water. Elimination, the last step in digestive functions, involves the expulsion of feces, which consist of the undigested remains of food, along with some of the bacteria that inhabit the digestive tracts of most animals.

In this chapter we look at how animals break up and digest the food they ingest, and how they use the nutrients released by digestion for their own life processes. We begin with an overview of the nutrients that animals need and the forms in which animals can use them. Next we take a tour of the human digestive system and compare it with those of other animals. We conclude by considering some of the special adaptations displayed by the digestive systems of different animal groups, including carnivores and herbivores.

## 27.1 Nutrients That Animals Need

Animals must eat other organisms, whether those organisms are microscopic algae that shrimp feed on, or the shrimplike krill that blue whales live on, or the prairie grasses that supported vast buffalo herds in the past, or the snowshoe hares that sustain the Canadian lynx through bitter winters. Animals that eat plants and fungi are called **herbivores**. **Carnivores** eat animals. Some animals, including humans and grizzly bears, eat both plants and animals and are known as **omnivores**. **Detritivores** eat the organic matter that is left over as dead organisms decompose into small particles. Earthworms and millipedes are examples of detritivores.

## All animals need carbohydrates, lipids, and proteins from food

Despite the variety in what they eat, all animals need three main categories of large organic molecules (macromolecules) from food: carbohydrates, lipids, and proteins. These macromolecules can serve as sources of energy, and they also furnish chemical building blocks such as sugars, fatty acids, and amino acids. In Chapter 5 we introduced nucleic acids as one of the macromolecules crucial for the function of every living cell. Although ingested nucleic acids are broken down in the digestive tract and their chemical components are absorbed and utilized by the animal's body, they are not usually singled out as a major category of food molecule. That is partly because nucleic acids are not a rich source of energy or a unique source of any nutrient, and also because they are present in relatively small amounts in most foods.

Most macromolecules are too large to cross the plasma membranes of the epithelial cells that form the absorbing surface of an animal's digestive system. These macromolecules must be broken into the simpler subunits of which they are built. Large carbohydrates, including all polysaccharides and disaccharides (see Chapter 5), are broken into monosaccharides, which are simple sugars such as glucose and fructose. Starch, a polysaccharide that is abundant in plant foods such as bread and potatoes, is broken down to glucose monomers.

Certain plant polysaccharides, including cellulose, cannot be digested by most animals. The term **dietary fiber** refers to indigestible plant polysaccharides. Although it does not contribute energy or chemical building blocks, dietary fiber is important for human health. Through its sheer bulk, and also because it is often interlinked with the more digestible polysaccharides, such as starch, fiber reduces the speed with which carbohydrates are digested. When we eat high-fiber foods, sugars are released over a

period of several hours, providing a sustained source of energy and reducing hunger cravings at the same time. In contrast, when we eat carbohydrates with little fiber (white rice or refined sugar, for example), the bloodstream must cope with a large surge of sugars within as little as 30 minutes. Such spikes in blood sugar level tax the body's sugar uptake system (controlled by a hormone called insulin) and, over time, can make us more susceptible to developing diabetes and high blood pressure.

The milling of wheat to make white flour strips away wholesome germ (the seed embryo from which a plant would grow) and bran (layers of protective tissue). Whole-grain foods, such as oatmeal, brown rice, and popcorn, are minimally processed, so they are more intact and retain more of the fiber that is naturally present in these foods (**FIGURE 27.2**). Whole-grain flour consists of ground whole kernels of wheat, barley, and other grains, and much of the fiber is retained

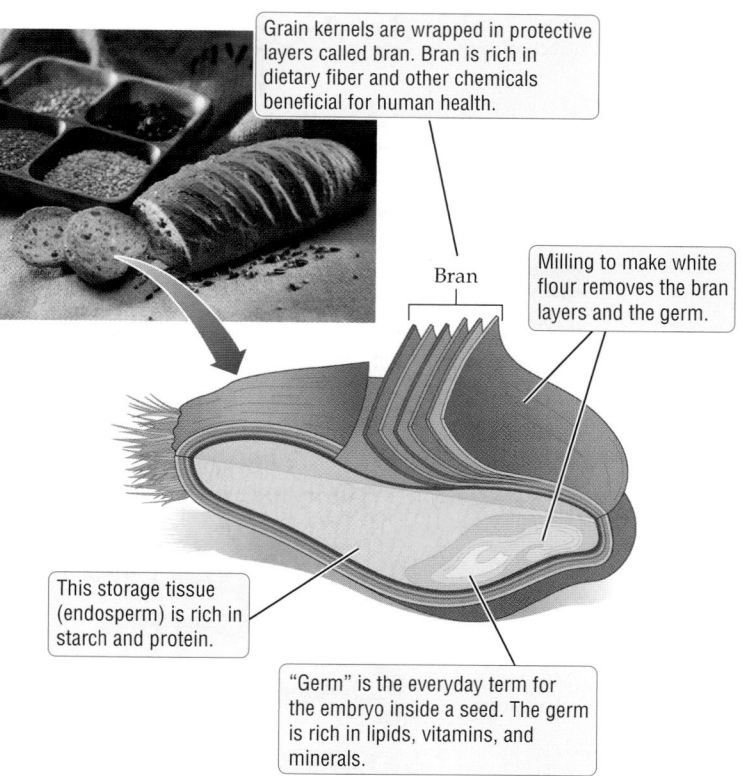

**BENEFITS OF WHOLE GRAIN**

Grain kernels are wrapped in protective layers called bran. Bran is rich in dietary fiber and other chemicals beneficial for human health.

Bran

Milling to make white flour removes the bran layers and the germ.

This storage tissue (endosperm) is rich in starch and protein.

"Germ" is the everyday term for the embryo inside a seed. The germ is rich in lipids, vitamins, and minerals.

**FIGURE 27.2 Energy and Building Blocks from Bread**
Whole-wheat flour is made from ground wheat kernels, and it includes the fiber-rich bran layers and the nutrient-rich "germ." Dietary fiber consists of cellulose and other indigestible plant polysaccharides that do not provide energy or building blocks to the human body but are vital for the health of the digestive system.

in the milled flour. Fiber is almost completely eliminated in the process of making refined or white flour.

Most animal fat consists of triglycerides, lipids composed of three fatty acids and a glycerol unit (see Figure 5.19). Each triglyceride is taken apart to produce two fatty acids and a fatty acid–glycerol unit called a monoglyceride [MAH-noh-GLIS-uh-ride]. Proteins are taken apart in the cavity of the digestive tract to release the different types of amino acids that make up a protein. The simple sugars, fatty acids, monoglycerides, and amino acids are absorbed by specialized epithelial cells lining the cavity of the digestive tract (FIGURE 27.3).

**FIGURE 27.3 Large Biomolecules Must Be Broken Down Before They Can Be Absorbed**
Enzymes degrade proteins, lipids, and large carbohydrates to their smaller building blocks, which are then absorbed by the cells that line the small intestines. Minerals cross membranes in their ionic form.

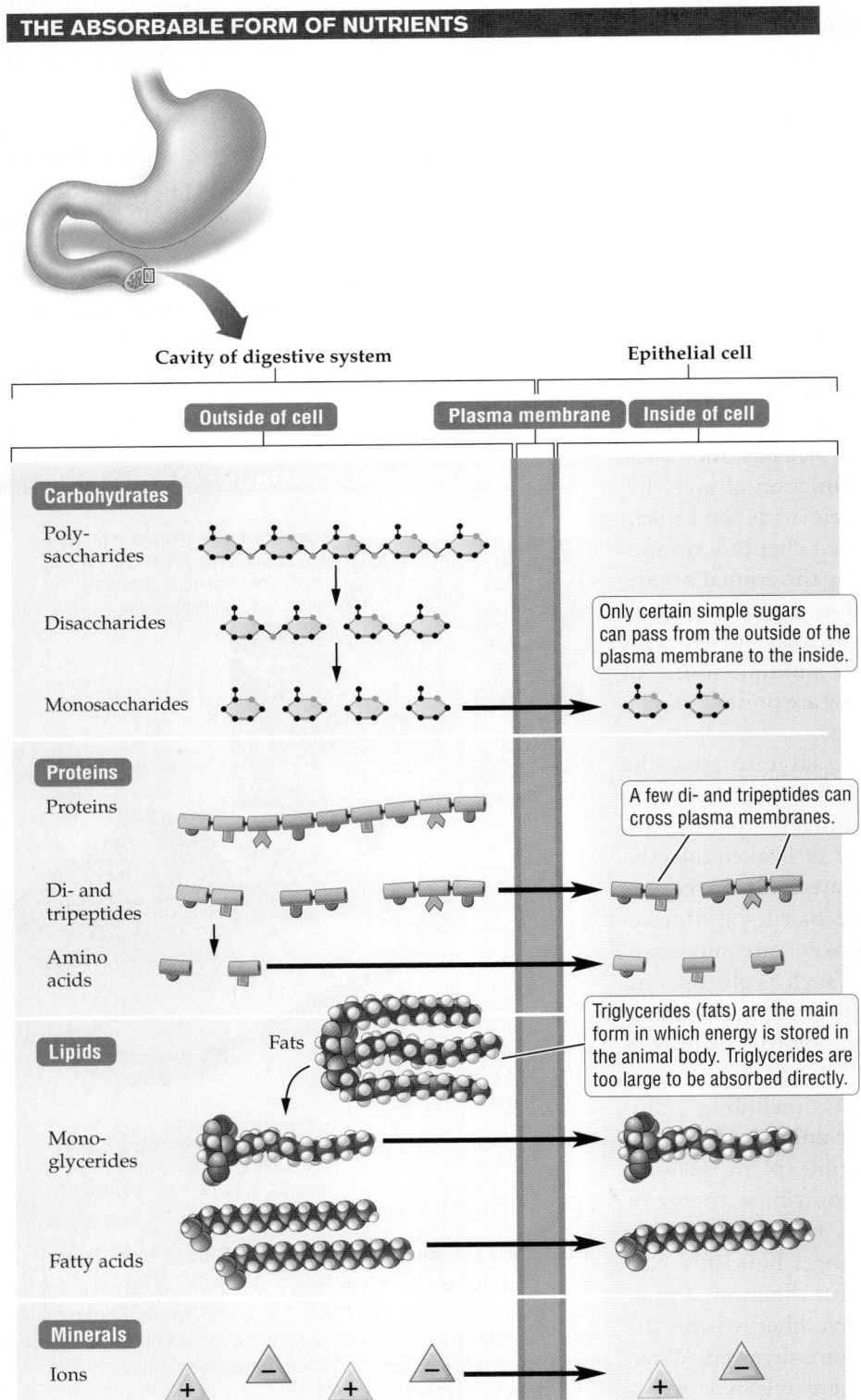

**THE ABSORBABLE FORM OF NUTRIENTS**

Cavity of digestive system

Epithelial cell

Outside of cell

Plasma membrane

Inside of cell

**Carbohydrates**

Poly-saccharides

Disaccharides

Monosaccharides

Only certain simple sugars can pass from the outside of the plasma membrane to the inside.

**Proteins**

Proteins

A few di- and tripeptides can cross plasma membranes.

Di- and tripeptides

Amino acids

**Lipids**

Fats

Triglycerides (fats) are the main form in which energy is stored in the animal body. Triglycerides are too large to be absorbed directly.

Mono-glycerides

Fatty acids

**Minerals**

Ions

| TABLE 27.1 | Energy and Building Blocks That Nutrients Provide to Animals | | |
|---|---|---|---|
| **NUTRIENT** | **ABSORBABLE UNITS** | **ENERGY CONTENT OF 1 GRAM (KCAL)** | **MAJOR USE** |
| Carbohydrates | Monosaccharides | 4 | Energy; building other macromolecules and cell structures |
| Fats | Fatty acids, monoglycerides | 9 | Energy storage; building other macromolecules and cell structures, especially cell membranes |
| Proteins | Amino acids, small peptides | 4 | Building other proteins and other organic molecules, such as signaling substances |

Once absorbed, monosaccharides, fatty acids, monoglycerides, and amino acids enter the bloodstream. Every cell in the body receives regular shipments of these small organic molecules through the blood. What does the cell do with them? These molecules are important building blocks for the manufacture of carbohydrates, lipids, and proteins. Monosaccharides are put together to create the disaccharides or polysaccharides needed by the cell. Liver and muscle cells store small amounts of glycogen, the storage polysaccharide of animal cells; if the stores are depleted—for example, by strenuous exercise—more glycogen can be manufactured as soon as new shipments of monosaccharides arrive. Fatty acids and monoglycerides delivered by the bloodstream can be converted into storage lipids (mainly triglycerides) or other types of lipids. Amino acids are linked to create the thousands of different types of proteins found in most animal cells.

The carbon-containing backbones of these small organic molecules can also be shaped into a great variety of other vital substances, including nucleic acids and signaling molecules, such as hormones and neurotransmitters. In this way, the organic molecules that were originally part of the body of the organism that was eaten become part of the body of the eater, illustrating that we are indeed, to a large extent, what we eat.

Monosaccharides, fatty acids, and monoglycerides serve another crucial function within the cell: energy is released when these molecules are broken down through glycolysis or cellular respiration. Chemists use kilocalories (kcal) as a unit of measure that gives the total energy in any substance. Carbohydrates and lipids provide us with most of the energy we need each day, which for unathletic people ranges from about 1,200 kilocalories for women to 2,200 kilocalories for men. Animals use carbohydrates (glycogen) primarily as a convenient source of quick energy. Lipids pack

more calories into each gram than carbohydrates do (Table 27.1), which is why nearly all animals store surplus energy in the form of lipids (triglycerides, in most cases).

Most animals do not use amino acids as a direct source of energy, except under conditions of starvation, when the internal stores of glycogen and triglycerides have been severely depleted. In other words, amino acids are not normally used as raw material for glycolysis or cellular respiration, the way sugars and fatty acids are (see Section 9.4.). But in most animals, including us, metabolic pathways will readily convert amino acids into storage lipids (usually triglycerides). Therefore, if we eat more protein than our bodies need, the bloodstream will deliver more amino acids than can be used by cells, thereby converting the surplus amino acids into fat.

Although an adult human can synthesize some of the 20 amino acids needed to make proteins, we must get 8 of them, called **essential amino acids**, from food (see Figure 5.15b for a refresher on the amino acids found in proteins). Infants need 9 essential amino acids because they cannot make histidine. The body can obtain all of the essential amino acids by breaking down proteins found in eggs, milk, meat, fish, and other animal food. However, some plant foods are deficient in certain amino acids and are therefore an incomplete source of the amino acids essential for balanced nutrition in humans. Most legume seeds (beans, lentils, peas) have low levels of two amino acids, and the cereal grains (wheat, barley, corn, rice) are deficient in a different pair of amino acids (**FIGURE 27.4**). The culinary traditions of many cultures routinely combine cereals and legumes (corn and beans, for example) to yield a complete source of amino acids. Some plant foods naturally contain a balanced complement of the essential amino acids; these foods include soybeans and tree nuts (walnuts and hazelnuts, for example).

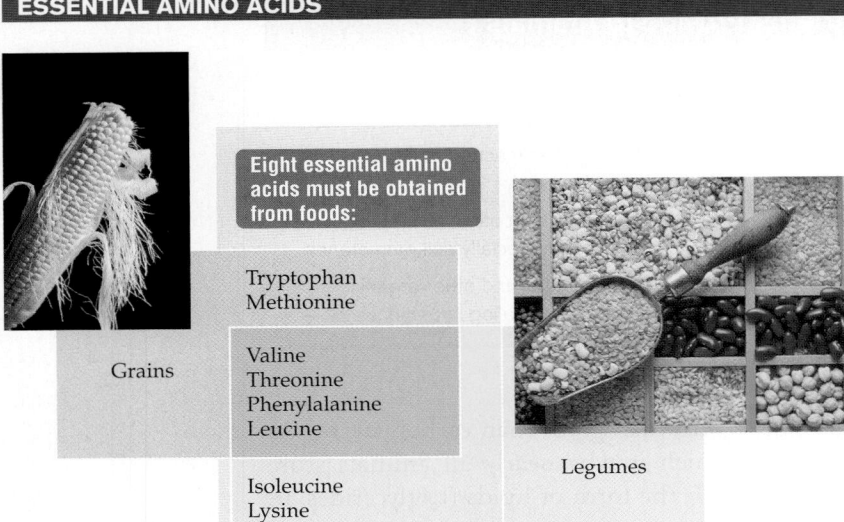

**Eight essential amino acids must be obtained from foods:**

Grains

Tryptophan
Methionine

Valine
Threonine
Phenylalanine
Leucine

Isoleucine
Lysine

Legumes

**FIGURE 27.4 Adult Humans Cannot Make 8 of the 20 Amino Acids That We Need to Build Proteins**
Meat contains all the essential amino acids, but vegetarians must combine different grains and legumes to get all of them. The pink box lists amino acids that both grains and legumes provide in adequate quantities. Legumes are plants in the pea family, which includes beans, garbanzos (chickpeas), and lentils.

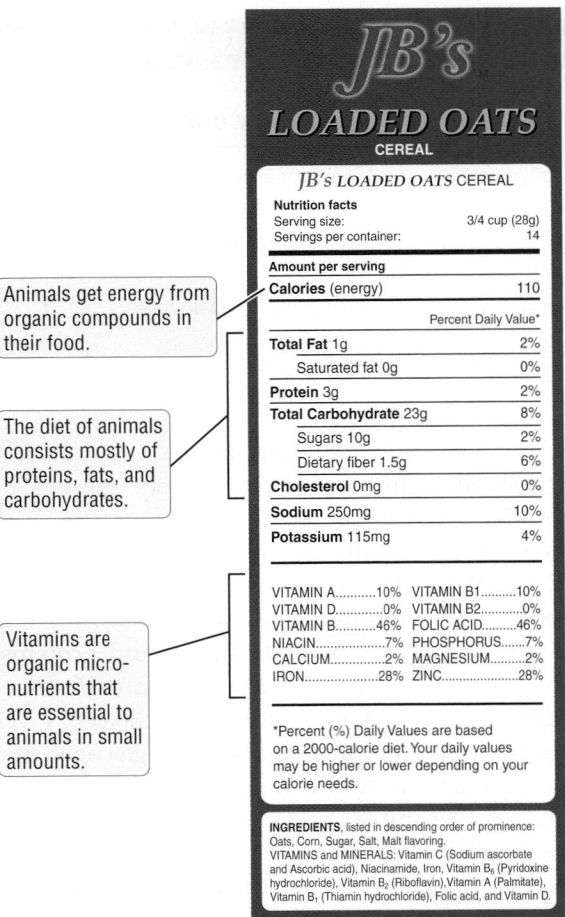

Animals get energy from organic compounds in their food.

The diet of animals consists mostly of proteins, fats, and carbohydrates.

Vitamins are organic micro-nutrients that are essential to animals in small amounts.

### JB's LOADED OATS CEREAL

**Nutrition facts**

| | |
|---|---|
| Serving size: | 3/4 cup (28g) |
| Servings per container: | 14 |

**Amount per serving**

| | | |
|---|---|---|
| **Calories** (energy) | | 110 |

| | | Percent Daily Value* |
|---|---|---|
| **Total Fat** 1g | | 2% |
| Saturated fat 0g | | 0% |
| **Protein** 3g | | 2% |
| **Total Carbohydrate** 23g | | 8% |
| Sugars 10g | | 2% |
| Dietary fiber 1.5g | | 6% |
| **Cholesterol** 0mg | | 0% |
| **Sodium** 250mg | | 10% |
| **Potassium** 115mg | | 4% |

| | | | |
|---|---|---|---|
| VITAMIN A............10% | | VITAMIN B1..........10% | |
| VITAMIN D............0% | | VITAMIN B2............0% | |
| VITAMIN B...........46% | | FOLIC ACID.........46% | |
| NIACIN................7% | | PHOSPHORUS......7% | |
| CALCIUM...............2% | | MAGNESIUM.........2% | |
| IRON.................28% | | ZINC...................28% | |

*Percent (%) Daily Values are based on a 2000-calorie diet. Your daily values may be higher or lower depending on your calorie needs.

INGREDIENTS, listed in descending order of prominence: Oats, Corn, Sugar, Salt, Malt flavoring.
VITAMINS and MINERALS: Vitamin C (Sodium ascorbate and Ascorbic acid), Niacinamide, Iron, Vitamin B₆ (Pyridoxine hydrochloride), Vitamin B₂ (Riboflavin),Vitamin A (Palmitate), Vitamin B₁ (Thiamin hydrochloride), Folic acid, and Vitamin D.

**FIGURE 27.5 Nutrition Labels Can Tell Us a Great Deal**
The label from a cereal box indicates the important sources of energy and nutrients for humans. We obtain energy primarily from carbohydrates, fats, and proteins.

## Most vitamins must be obtained from food

The nutrition labels on our foods give us a good overview of the kinds of nutrients we need. For example, the cereal box label shown in **FIGURE 27.5** first lists the nutrients we need in relatively large amounts—lipids ("fat"), proteins, and carbohydrates. Farther down, the Nutrition Facts label shows nutrients needed in much smaller quantities; these include *minerals* (such as sodium and potassium) and a number of *vitamins*. Let's take a closer look at vitamins first.

**Vitamins** are small organic nutrients that are needed by our bodies, but only in tiny amounts. Vitamins have two other characteristics that distinguish them from the groups of nutrients discussed so far: unlike proteins, vitamins are not used to generate the physical building blocks of the body. And unlike carbohydrates and fats, vitamins do not supply energy. Instead, they participate in a great variety of essential metabolic processes.

Some vitamins bind to enzymes and assist them in speeding up chemical reactions within the cell. Some vitamins act as a delivery service, supplying chemical groups needed in important metabolic reactions. Some vitamins act as signaling molecules. Others are believed to act as antioxidants, substances that protect body tissues from destructive chemicals, known as *free radicals*, that are generated as a by-product of metabolism. Most vitamins have multiple functions in the animal body (Table 27.2).

There are two main classes of vitamins: water-soluble and fat-soluble. Humans need nine water-soluble vitamins from the diet: vitamin C and eight different types of B vitamins. And we need four fat-soluble vitamins: vitamins A, D, E, and K. Water-soluble vitamins dissolve readily in water. Because they are easily excreted in urine, water-soluble vitamins tend not to accumulate in body tissues in appreciable amounts, which means the animal must obtain these vitamins from food on a regular basis. Fat-soluble vitamins are not excreted as readily and tend to accumulate in body fat. That is why excessive consumption of fat-soluble vitamins can lead to overdosing.

TABLE 27.2 Vitamins Needed in the Human Diet

| CLASS | VITAMIN | MAIN FUNCTIONS | POSSIBLE SYMPTOMS OF DEFICIENCY AND EXCESS | DIETARY SOURCES |
|---|---|---|---|---|
| **Water-soluble** | B vitamins: thiamine ($B_1$), riboflavin ($B_2$), niacin (nicotinamide), pyridoxine ($B_6$), pantothenic acid, folate (folic acid), cyanocobalamin ($B_{12}$), biotin | Act with enzymes to speed metabolic reactions, or act as raw materials for chemicals that do so. Work with enzymes to promote necessary biochemical reactions. | Deficiency: B vitamins act in concert; deficiency in one can cause symptoms related to deficiency in others. Deficiency diseases include pellagra and beriberi (damage to heart and muscles). *Excess: $B_6$ in excess can cause neurological damage.* | Folic acid, a B vitamin, is abundant in green vegetables, legumes, and whole grains. $B_{12}$, a B vitamin, is scarce in plant foods but abundant in milk, meat, fish, and poultry. |
| | Vitamin C (ascorbic acid)[a] | Assists in the maintenance of teeth, bones, and other tissues. | Deficiency: scurvy (teeth and bones degenerate), increased susceptibility to infection. *Excess: diarrhea, kidney stones with chronic overuse.* | Vitamin C is abundant in many fruits (e.g., kiwi, strawberry, citrus) and in many vegetables (e.g., bell peppers, broccoli, spinach). |
| **Fat-soluble** | Vitamin A (carotene) | Produces the visual pigment needed for good eyesight; also used in making bone. | Deficiency: poor night vision; dry skin and hair. *Excess: nausea, vomiting, fragile bones.* | Carotene is responsible for the color of yellow and orange fruits and vegetables. It is converted into vitamin A within our bodies. |
| | Vitamin D | Promotes calcium absorption and bone formation. | Deficiency: poor formation of bones and teeth, irritability. *Excess: diarrhea and fatigue.* | Fish is the richest source of vitamin D; shellfish and egg yolks provide smaller quantities; fortified foods (such as milk, soy milk, and breakfast cereals) are important sources for most people. |
| | Vitamin E | Protects lipids in cell membranes and other cell components. | Deficiency: neuromuscular problems (deficiency is very rare). *Excess: heart problems.* | Vitamin E is abundant in nuts, vegetable oils, whole grains, and egg yolk. |
| | Vitamin K | Produces clotting agent in the blood. | Deficiency: prolonged bleeding, slow wound healing. *Excess: liver damage.* | Leafy green vegetables and some fruits (avocado, kiwi) are rich in vitamin K, which is also manufactured by intestinal bacteria. |

**NOTE:** The human body cannot make these essential vitamins, or else makes them in insufficient amounts, so it must get what it needs from food.

[a]Vitamin C is a vitamin only for primates (including humans) and a few other animals (such as guinea pigs, bats, some birds, and some fishes). Most other animals can make vitamin C as needed.

Most animals cannot manufacture all of the vitamins needed by the body. Guinea pigs, bats, the great apes, and humans are among the few animals that need vitamin C but cannot make it within the body. Folic acid (vitamin $B_9$) is critical in the development of bone, muscle, and brain tissue; and inadequate intake of this B vitamin by pregnant women has been linked to increased risk for neural tube disorders and other birth defects. The neural tube starts developing in the first 4 weeks of pregnancy, when many women are unaware that they are pregnant. Several countries, including the United States and Canada, require food processors to add folic acid to flour, bread, breakfast cereals, and other grain products. The mandatory fortification program was launched in the United States in 1992, and in the decade that followed, neural tube birth defects declined by more than 50 percent.

Vitamin D is the only vitamin that we can manufacture entirely within our tissues, and ironically, it is the one vitamin many Americans get inadequate amounts of. Ultraviolet (UV) rays in sunshine prompt our skin cells to convert a cholesterol-like molecule into vitamin $D_3$. Vitamin $D_3$ is modified further in the liver and kidneys before it becomes the biologically active form of vitamin D. Vitamin D is crucial for the development of bone, teeth, and muscles. Fish and seafood are the only foods rich in vitamin D.

Scientists suggest that pale skin, which is more UV transparent than dark skin, may have evolved to enable humans to make enough vitamin D in the lower light of northern latitudes. People native to tropical regions need dark skin to protect their DNA from UV damage (FIGURE 27.6), but they are able to make enough vitamin D because tropical sunshine is intense through much of the year. About 50,000 years ago, ancestral groups that were almost certainly dark skinned began moving into northern parts of Eurasia. Vitamin D deficiency because of weaker sunlight, especially during the long winter, would have been a serious problem for these populations. Individuals with reduced skin pigmentation presumably survived and reproduced better as a result of natural selection, and their descendants went

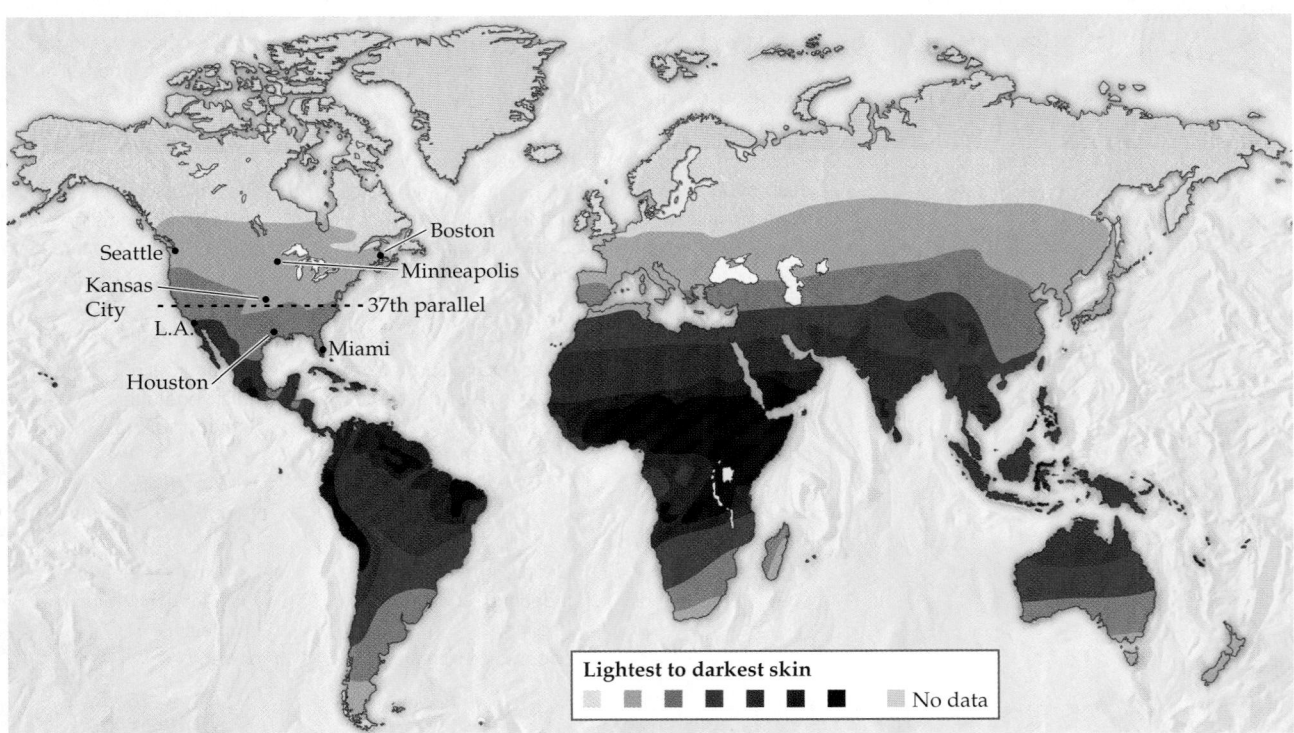

**GLOBAL DISTRIBUTION OF SKIN COLOR IN INDIGENOUS POPULATIONS**

Lightest to darkest skin

No data

**FIGURE 27.6 Human Skin Color Is an Adaptation to Prevailing Sunshine**

Humans adapted to tropical and subtropical environments have darker skin to protect against DNA damage from ultraviolet radiation in the intense sunshine of the lower latitudes. People who migrated north evolved lighter skin, enabling them to manufacture vitamin D more effectively in the weaker light of higher latitudes. However, if you live north of the thirty-seventh parallel, from November through February you may not make enough of the vitamin from sunshine alone.

on to populate Europe. The Inuit of northern Canada and other aboriginal peoples of the Arctic went north in more recent times and have traditionally depended on seafood for an abundant supply of vitamin D.

Vitamin deficiencies (**FIGURE 27.7**) producing symptoms such as those listed in Table 27.2 are rare in the United States and other wealthy countries. However, those who smoke, consume alcohol in excess, or adhere to certain fad diets could develop deficiencies. The efficiency with which we absorb certain vitamins declines with age, and the elderly are at greater risk for deficiency in vitamins such as $B_{12}$. Furthermore, nutritionists believe that many Americans have suboptimal amounts of certain vitamins; their vitamin intake is not so low that it would produce the classic deficiency symptoms, but it is not high enough for the best possible health outcomes. A large body of information now indicates that many Americans, perhaps most Americans, do not get enough vitamin D, the "sunshine vitamin."

If you live in America north of the latitude where Los Angeles is situated, then from November through March you do not receive enough sun to meet your daily vitamin D requirements. In addition, if you avoid the sun much of the year, or use sunscreen as recommended by most dermatologists, or if you are dark complexioned and live in a northern state, you need vitamin D from dietary sources to meet the recommended daily requirement. In the early 1940s, the government mandated the addition of vitamin D to all milk sold in the United States. The fortification program practically wiped out rickets, the most serious symptom of vitamin D deficiency in children. Although rickets may be a disease of the past, studies show that suboptimal levels of vitamin D may be contributing to bone fractures in older people and to some types of cancer, including breast cancer and colon cancer. Inadequate vitamin D intake may also increase our odds of developing diabetes, as well as immune system diseases such as multiple sclerosis and rheumatoid arthritis.

One cup of vitamin D–enriched milk has 300 units of vitamin D; a 3-ounce serving of catfish has 570 units. A light-skinned person in a bathing suit can make more than 6,000 units of vitamin D with just 10 minutes of sunshine on a summer day at a latitude comparable to that of New York. The more the body is covered, the darker a person is, the farther away it is from summer, and the farther north you go, the less vitamin D is made. The recommended dietary allowance for vitamin D is 600 international units (IU) daily for those under age 70, and 800 units daily for older people because the ability to make vitamin D from sunshine declines as we age. Some experts are calling for these levels to be raised to at least 1,000 units a day for everyone, and for the upper limit to be set at 10,000 units.

Nutritionists agree that we can obtain all necessary nutrients in optimal amounts from natural sources,

JAMES LIND

**FIGURE 27.7 Insufficient Vitamin Intake Can Produce a Deficiency Disease**

The painting shows James Lind treating scurvy on HMS *Salisbury* in 1747. Scurvy is a severe form of vitamin C deficiency, marked by bleeding and weakened muscles, bones, and teeth. It was common on sailing ships before Lind, an officer in the British Royal Navy, used controlled experiments to demonstrate that consuming citrus fruits on long sea voyages could prevent the nutritional deficiency.

provided we eat a well-balanced diet that includes a variety of foods. Recognizing that relatively few people meet that ideal in reality, some physicians advise their patients to take a multivitamin mineral supplement as "added insurance." The use of vitamin supplements raises the specter of toxicity from excessive intake, particularly of the fat-soluble vitamins. Most, but by no means all, manufacturers attempt to avoid high levels of the fat-soluble vitamins in their vitamin preparations, but the burden is largely on us to use supplements wisely, since this industry is not regulated by the government. It is important, for example, not to combine a standard multivitamin, which may have 100 IU of vitamin E, with a vitamin E capsule that has 400 IU. The safe limit of this fat-soluble vitamin is only about 1,000 IU, and if you also get vitamin E from foods such as vegetable oils, nuts, and whole eggs, you could find yourself in the danger zone.

Vitamin toxicity from natural sources is unknown, because the body appears to have built-in protective mechanisms. We cannot overdose on vitamin D from sunshine, because the body will turn off activation of vitamin $D_3$ in the liver once sufficient amounts of the active form are available. Carotenoids, the yellow and orange pigments in vegetables such as carrots, squash, and tomatoes, are turned into vitamin A in the body, but only in the amounts needed.

## Food supplies animal cells with essential minerals

Nutritionists have traditionally used the term **minerals** to refer to certain inorganic chemicals that have critical biological functions. Living organisms are built from a relatively small number of chemical elements (see Chapter 5). Of these, carbon, hydrogen, oxygen, and nitrogen alone make up about 93 percent of the animal body. By convention, these four elements are excluded from the category of dietary minerals.

More than 20 other elements are essential for the normal function of most animals, and these are classified as dietary minerals. Some dietary minerals are needed in larger amounts than others, and these are known as macrominerals. The 11 macrominerals together make up 6 percent of our body weight. Sodium, calcium, phosphorus, magnesium, chlorine, potassium, and sulfur are examples of macrominerals. Sodium ions ($Na^+$) and chloride ions ($Cl^-$) are crucial for the survival of every animal, and they are most commonly obtained from table salt (NaCl). Sodium ions in particular are used by animals for everything from maintaining water balance in cells to transmitting electrical signals along a nerve. Phosphorus and calcium are key components of the material that bones and teeth are made of, and the two are also important for signaling processes within cells. Sulfur atoms are found in some proteins and in many other organic molecules, such as vitamin $B_7$ (also known as biotin). A dietary mineral could be biologically active as an ion (such as $Na^+$), or it could be covalently bonded into a molecule (the way sulfur atoms are in proteins).

At least a dozen other elements, known as *trace minerals*, are essential for animals but are needed in very small amounts. Trace minerals make up less than 1 percent of an animal's body weight. The average human body, for example, contains an amount of iodine that would come to less than a quarter teaspoon. Without this minute quantity, however, the body cannot make the thyroid hormones that regulate how we burn food to obtain energy. This trace mineral enters the food chain when it is absorbed from soil by plants or from water by aquatic organisms. People who live in areas with extremely low levels of environmental iodine tend to develop goiters (see Figure 5.3b). Marked by enlargement of the thyroid gland, this disease has disappeared from parts of the world where table salt, NaCl, is fortified with iodine.

Many trace minerals bind to proteins and enable them to function normally. Hemoglobin, the protein that helps red blood cells carry oxygen throughout the body, cannot capture the gas without the help of iron atoms. Women of reproductive age lose iron in menstrual blood, and anemia induced by iron deficiency is common among women in the poor parts of the world. The recommended daily amount is 18 milligrams (mg) for women in their reproductive years and 8 mg for men, and the safe level is just 45 mg. Men who eat a lot of red meat, as well as people with a genetic susceptibility for iron overload, may get more iron from their diet than is good for them.

As vital as sodium and chloride ions are for the normal function of the human body, most Americans consume too much table salt (NaCl), almost all of it from processed and packaged foods such as salty snacks and canned soup. Health experts say that consuming more than 1,500 mg of salt per day predisposes us to high blood pressure, especially if we have a genetic susceptibility to "salt-induced hypertension." High blood pressure increases our risk for heart disease, stroke, and kidney failure. The average American takes in about 5,000 mg of added salt every day (Table 27.3).

Compared to vitamins, it is even easier to ingest harmful levels of some minerals from supplements, because the upper limits for some of them are quite

TABLE 27.3    Sodium Content of Some Common Foods

| | MEASURE | SODIUM CONTENT (MG) PER MEASURE |
|---|---|---|
| Onion soup, dry | 1 packet | 3,132 |
| Table salt | 1 teaspoon | 2,325 |
| Baking soda | 1 teaspoon | 1,259 |
| Baked beans, canned | 1 cup | 1,106 |
| Cheeseburger with condiments | 1 burger | 1,051 |
| Beef noodle soup, canned | 1 cup | 930 |
| Pasta with meatballs, canned | 1 cup | 733 |
| Sliced ham, extra lean | 2 slices | 601 |
| Swiss cheese | 1 ounce | 388 |
| Milk, 2% fat | 1 cup | 115 |
| Spinach, raw | 1 cup | 24 |
| Olive oil | 1 tablespoon | 0 |

**SOURCE:** Data from the USDA National Nutrient Database.

low. Zinc is necessary for many metabolic processes, and it supports the immune system if you receive the recommended daily dose, which is 8 mg for women and 11 mg for men. But getting more than 40 mg a day can reduce the availability of copper, another trace mineral; 300 mg a day will actually depress the immune system. Selenium has been touted as an anticancer mineral, but a recent study showed that subjects who took 200 micrograms ($\mu$g) of selenium in a pill every day for 8 years had a 50 percent higher risk of diabetes compared to subjects receiving a placebo ("dummy pill"). Calcium, in combination with vitamin D, strengthens our bones; and women need 1,000–1,500 mg a day, depending on their age (higher levels if they are older than 50). Men, however, face an increased risk of prostate cancer if they exceed 2,000 mg of calcium per day.

### Concept Check

1. Which of the following is an organic molecule that can be directly absorbed by epithelial cells lining the cavity of the digestive tract: triglyceride, fatty acid, protein, sodium ion?

2. Which vitamin can be manufactured entirely within the human body? What environmental conditions are necessary for the manufacture of this vitamin in human tissues? Which tissues are involved in producing a biologically active form of this vitamin?

3. Which mineral is consumed in excessive amounts by most Americans? What potential health risks are posed by this overconsumption?

## 27.2 The Human Digestive System

Although nutrients are contained in the food ingested by animals, many of those nutrients are not in a form that can be used directly by cells in the body. The process of digestion breaks down large macromolecules into smaller organic molecules that are more easily absorbed by the digestive tract and then transferred to body fluids.

The digestive tract is the main component of the human digestive system (**FIGURE 27.8**). It consists of a passageway that begins at the mouth and ends at the other opening, the anus. Associated with the digestive system are a number of accessory glands and organs. (A gland is an organ that secretes one or more substances into a body cavity, the bloodstream, or the skin surface.) The digestive tract consists of several regions, each of which plays a specific role in processing food. Epithelial tissues lining the tract are specialized to carry out the functions associated with the particular region of the tract. As food passes through the tract, it is broken down both physically and chemically, until it is converted into small organic molecules such as simple sugars, amino acids, and fatty acids. These smaller molecules cross the lining of the digestive system and eventually move into the bloodstream to be distributed to all the cells within the body, which are crucially dependent on this supply route.

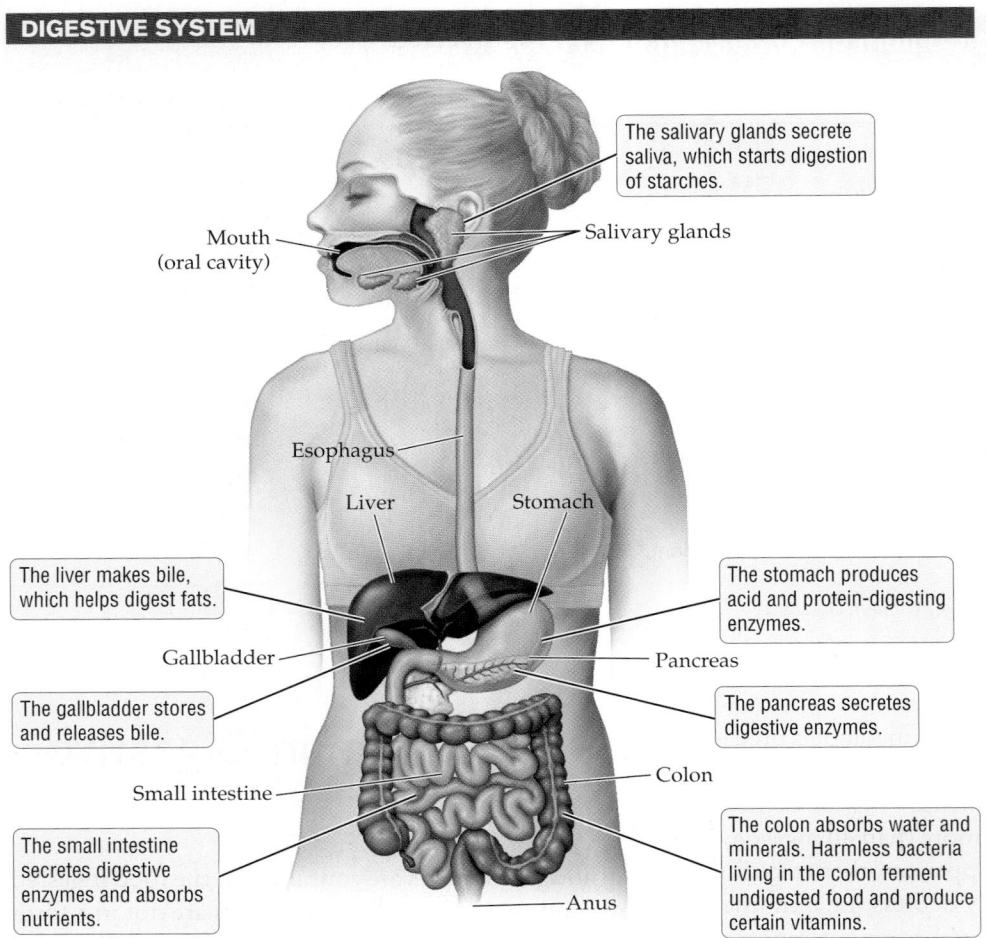

Mouth
(oral cavity)

Salivary glands

The salivary glands secrete saliva, which starts digestion of starches.

Esophagus

Liver

Stomach

The liver makes bile, which helps digest fats.

The stomach produces acid and protein-digesting enzymes.

Gallbladder

Pancreas

The gallbladder stores and releases bile.

The pancreas secretes digestive enzymes.

Small intestine

Colon

The small intestine secretes digestive enzymes and absorbs nutrients.

Anus

The colon absorbs water and minerals. Harmless bacteria living in the colon ferment undigested food and produce certain vitamins.

**FIGURE 27.8 The Human Digestive System Converts Food into Absorbable Nutrients**

As it moves through the digestive system, food is broken down into small molecules that can be absorbed by the lining of the intestine. The liver, gallbladder, and pancreas are accessory organs of the digestive system. They play a key role in aiding the digestion and absorption of food molecules.

## Digestion begins in the mouth

During ingestion (eating), food enters the **oral cavity** (mouth). In the mouth food is broken down mechanically, and chemical digestion of carbohydrates is initiated. The different types of teeth are shaped to cut, crush, or grind food into smaller pieces. Many small pieces provide a greater surface area for digestive enzymes to work on, compared to a few large pieces. The muscular tongue mixes these food particles with **saliva**, a fluid made and delivered by six **salivary glands** associated with the mouth. Saliva contains enzymes that start to break down starches in the food; it also turns the food into a moist mass that can slip easily down the throat.

The tongue assists in pushing food into the throat, or **pharynx** (plural "pharynges"), where the back of the mouth and the nasal cavity come together. The

pharynx is a common entryway leading to the air tube (trachea), as well as the food tube (esophagus). When the food mass makes contact with the wall of the pharynx, it stimulates nerves that launch the **swallowing reflex**. As part of this involuntary action, entry into the air passages is sealed off by a flap of tissue (the epiglottis), and the food is pushed into the tubular **esophagus [uh-SAHF-uh-gus]** (plural "esophagi"). Waves of muscular contractions then carry the food to the **stomach**. Protein digestion begins in the stomach, which secretes acid and pepsin, an enzyme that begins to break down proteins into shorter chains of amino acids. The acidity promotes the breakdown of food and stimulates the activity of pepsin. Muscles in the wall of the stomach alternately contract and relax to mix the food particles with the acid and pepsin. The mixture is stored in the stomach until it can move into the small intestine.

## The pancreas and liver assist digestion in the small intestine

The **small intestine** is a thin tube (about 1 inch, or 2.5 centimeters, in diameter) that is highly coiled; if straightened, it would extend about 20 feet (6 meters). The upper and lower regions of the small intestine serve different functions. The upper region, which lies nearest the stomach, digests large molecules into simpler forms that the body can absorb. Here a number of enzymes that specifically digest proteins, carbohydrates, and lipids complete the digestion of food molecules. The small intestine itself produces some of these enzymes, but many are produced in accessory organs closely associated with the small intestine. The **pancreas** produces a broad range of enzymes that reach the upper small intestine through a duct.

The digestion of fats poses a particular problem because fats are not soluble in water. Fats form large globules that must be broken down and made to mix with the watery contents of the digestive tract. **Bile** is a fluid that helps dissolve fat globules. It acts in much the same way that dishwashing detergent does on greasy dishes: it creates a coating on fat globules that enables them to interact with water molecules and that partially dissolves them. Large globules break into tiny droplets, which offer a larger work surface for lipid-degrading enzymes. Bile is produced by the **liver**, an organ that serves a multitude of functions (including homeostasis, as described in Chapter 26). Some of the bile made by the liver is stored in the **gall-bladder**, which dispenses the bile into the small intestine as needed.

## Nutrients and water are absorbed in the small and large intestines

The lower region of the small intestine is specialized for absorption, and its innermost lining presents a large surface area to the food mass in the cavity of the digestive tract. Nutrients are absorbed by large numbers of fingerlike projections called **villi** (singular "villus"). Each villus (**FIGURE 27.9**) is about 1 millimeter long, with a surface consisting of cells specialized for nutrient absorption. The plasma membrane of each of these cells has many tiny projections (called microvilli), each less than one-thousandth of a millimeter long. The membrane projections collectively give the outer surfaces of the villi a fuzzy, brushlike appearance under the microscope. The "fuzzy" cell layer that makes up the surface of the villi is called the **brush border**. This complexity in the folding of the intestinal lining produces an impressive surface area for absorption—almost 300 square meters, roughly equivalent to the area of a tennis court!

For absorption to occur, the digested nutrients must move across the epithelial lining of the small intestine and into the body. The inside of each villus has two ways of carrying the absorbed nutrients off to other cells: blood capillaries (part of the circulatory system) whisk away the amino acids and sugars; a thin-walled lymph vessel (part of the lymphatic system) picks up the monoglycerides and fatty acids first, but the lymphatic system eventually releases them into the bloodstream.

By the time the material remaining in the digestive system arrives in the large intestine, it contains

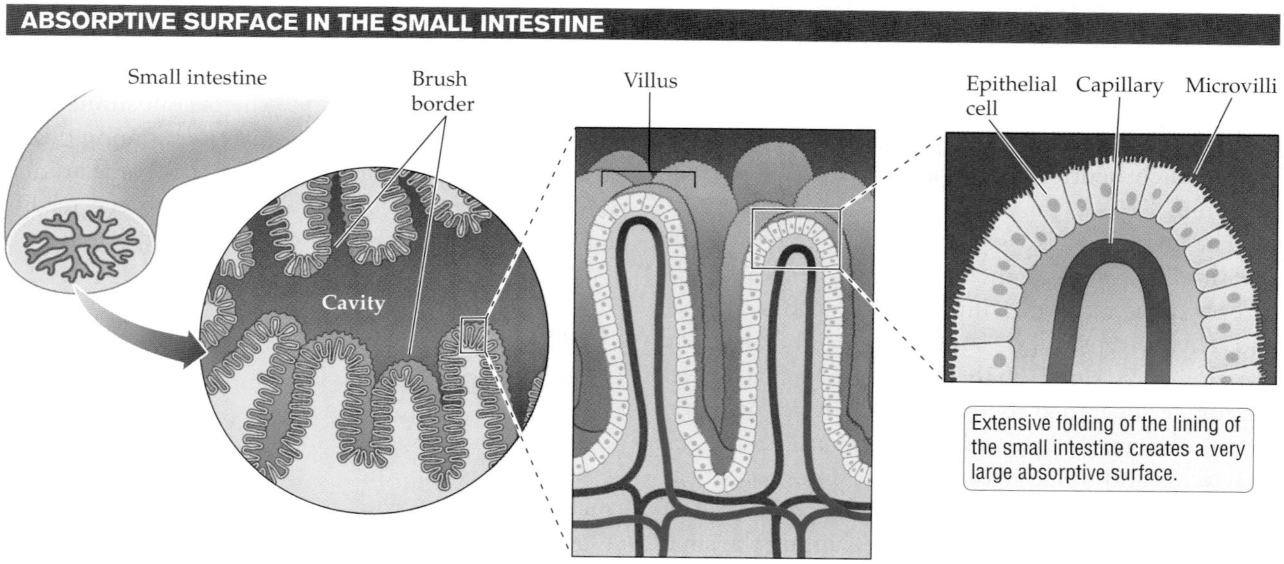

**ABSORPTIVE SURFACE IN THE SMALL INTESTINE**

Small intestine · Brush border · Villus · Epithelial cell · Capillary · Microvilli

Cavity

Extensive folding of the lining of the small intestine creates a very large absorptive surface.

**FIGURE 27.9 The Lining of the Small Intestine Presents a Large Surface Area**

The lining of the small intestine has ridges and furrows. The surface of this lining is densely covered with fingerlike projections, called villi, which are like the pile of a carpet. The cells of the villi have microscopic extensions of their plasma membranes (microvilli). Collectively, these structures create a very large absorptive surface.

relatively few nutrients. In the large intestine, also called the **colon**, this residual matter is prepared for removal from the body as waste, a process that involves three different functions: First, the colon absorbs almost all remaining minerals from the waste as ions. Second, the colon absorbs almost all of the remaining water. Finally, large numbers of bacteria living in the colon break down some of the remaining waste to release additional nutrients that the body can absorb. These bacteria also produce certain organic molecules (such as vitamin K and biotin) that are absorbed into the body from the colon. The end result is a relatively solid waste product, politely known as **feces** [*FEE*-seez], that consists mostly of indigestible material and bacteria. The feces leave the body through a muscle-lined opening called the **anus**.

## Digestive enzymes speed up the chemical breakdown of food

The specialized regions of the digestive tract secrete a variety of enzymes to digest carbohydrates, lipids, and proteins. (For a review of how enzymes work, see Chapter 8.) Each digestive enzyme acts on a specific type of chemical bond. For example, the saliva contains amylase, which breaks starches into sugars.

The stomach produces acid and the enzyme pepsin, which together start the process of breaking proteins into shorter strings of amino acids. Pepsin is specialized to work well in the acidic (low-pH) environment of the stomach. The partially digested food leaving the stomach is also very acidic, creating a potential problem as it enters the small intestine: the enzymes that carry out most of the digestion of our food in the small intestine do not work well in a highly acidic environment. To adjust the pH, the pancreas secretes a solution containing sodium bicarbonate—perhaps more familiar in its powdery form as baking soda—which neutralizes the acid in the upper small intestine.

The pancreas also secretes several enzymes into the small intestine. Pancreatic amylase breaks long carbohydrate molecules into simpler sugars. Trypsin and chymotrypsin [*KYE*-moh-*TRIP*-sin] break down proteins into small chains of amino acids, which are further degraded to amino acids by yet other enzymes. Pancreatic lipases break fats into absorbable fatty acids and monoglycerides. The digestion of carbohydrates and proteins started by the pancreatic enzymes is completed by epithelial cells lining the small intestine. These cells secrete a variety of enzymes that split short chains of sugars into absorbable monosaccharides and short chains of amino acids into single amino acids. The brush border also secretes nucleases, enzymes that digest nucleic acids into the nucleotides from which these macromolecules are built.

> ### Concept Check
>
> 1. How does chewing help in the digestion of food?
> 2. Which of the following macromolecules are digested in the stomach: carbohydrates, lipids, proteins?
> 3. What is the function of bile?
> 4. What feature of the small intestine is responsible for its large absorptive capacity?

## 27.3 Special Adaptations of Animal Digestive Systems

The first animals to evolve tissues specialized for digestion probably resembled present-day cnidarians, such as jellyfish and coral polyps: digestion was achieved in a simple internal cavity lined with cells specialized for secretion and absorption. A true digestive tract, a tubelike passageway with two openings at opposite ends, did not evolve until about 500 million years ago, when protostomes such as roundworms and mollusks first appeared (see Section 4.4). In these groups, and all the other invertebrates and vertebrates that follow them in evolutionary history, the digestive tract is organized like a food-processing assembly line, with a one-way flow of material and different regions of the digestive system conducting specialized functions (**FIGURE 27.10**). Despite the overall similarity in the organization of the digestive tract among diverse animal groups, the digestive systems of some animals have special features, usually related to the feeding habits and dietary preferences of these species.

## The crop stores food, and the gizzard grinds it

In many animals, from insects to birds, part of the esophagus is modified into a thin-walled compartment, called the crop (see Figure 27.10), that is used like a shopping basket. The animal fills the muscular, expanded crop with food as it forages, delaying the processing of that food until a suitable time and safe roost can be found. Some birds will regurgitate their crop contents if startled by a predator—a behavior that lightens the load when a quick getaway is needed. Pigeons, penguins, and flamingos produce a fluid from their crops, known as crop milk, that they feed to their nestlings.

The less complex animals, such as roundworms, lack specialized mechanisms for breaking up the food they ingest. The digestive tracts of nematodes, a type of roundworm, even lack muscles to push the food along, relying instead on the squeezing action of the body wall and squirming body movements. Some species, including animals that lack teeth or even jaws, depend completely on chemical digestion to break down food. Their food must be ingested in small particles or as a liquid. For example, orb-weaving spiders cannot mechanically break up their food, because they lack chewing mouthparts; however, their piercing jaws can inject venom into prey. After the venom has liquefied the tissues of its prey, the spider sucks out the fluid and discards all of the hard parts.

Most animals, both invertebrate and vertebrate, have special mechanisms for breaking down ingested food into smaller particles. Relative to their volume, smaller food particles expose more surface area to acid and enzymes, greatly speeding their digestion. Humans and many other animals use hard teeth set in strong jawbones to crush, rip, tear, or grind large food particles into smaller ones. Other animals use muscles in combination with hard surfaces, such as stones stored in a special sac, to break up food particles mechanically. In earthworms, insects, many reptiles, and birds, part of the digestive system is modified into a muscular **gizzard**, which grinds food with small pebbles or sand grains that the animal ingests for this very purpose (see Figure 27.10).

## Herbivore digestive systems are modified for digesting plants

Herbivores have digestive systems that are specialized to extract nutrients from plant tissues. These animals face a different set of digestive challenges from carnivorous animals. Plant tissues contain much less protein and much more indigestible cellulose than animal tissues do. In terms of bulk, herbivores must eat more food than carnivores eat, and they must invest more time in and have more tactics for breaking down their

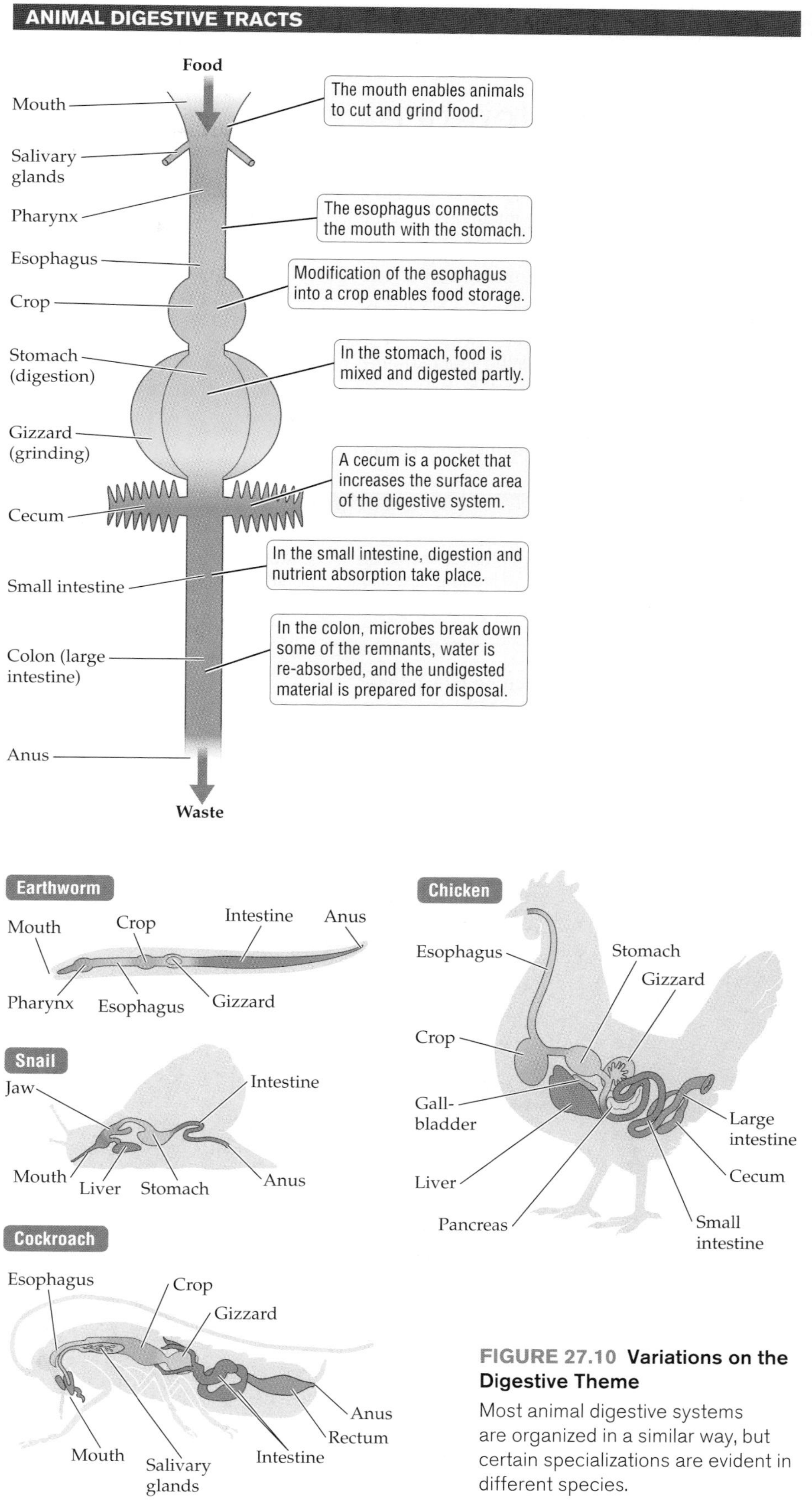

**ANIMAL DIGESTIVE TRACTS**

The mouth enables animals to cut and grind food.

The esophagus connects the mouth with the stomach.

Modification of the esophagus into a crop enables food storage.

In the stomach, food is mixed and digested partly.

A cecum is a pocket that increases the surface area of the digestive system.

In the small intestine, digestion and nutrient absorption take place.

In the colon, microbes break down some of the remnants, water is re-absorbed, and the undigested material is prepared for disposal.

**FIGURE 27.10 Variations on the Digestive Theme**

Most animal digestive systems are organized in a similar way, but certain specializations are evident in different species.

# Lactose Intolerance

Among the many enzymes secreted in the upper portion of the small intestines is one called *lactase*. Lactase digests the disaccharide lactose, breaking it into the monosaccharides glucose and galactose (Figure 1). Lactose is the most abundant sugar in milk produced by mammals. Human infants, like the young of other mammals, produce lactase when nursing so that they can digest the milk sugar and derive nutritional benefit from it. Until humans began dairy farming and consuming cow's milk, however, adults rarely encountered lactose in their diets. Accordingly, most humans stop producing lactase as they mature.

However, individuals of European, northern Indian, Arabian, and central African descent frequently retain the ability to produce lactase as adults (Figure 2). That is because these groups domesticated cattle and began consuming dairy products about 10,000 years ago—a development that must have given lactase-producing adults a fitness advantage. With the particular range of food options available to them in these habitats, individuals who retained lactase production into adulthood may have gained a large nutritional edge over those who did not. A chance mutation probably gave rise to an allele of the lactase gene that caused the gene to remain active in adults.

Adults who do not produce lactase may experience digestive problems, known as lactose intolerance, if they drink milk. Because the lactose escapes digestion in the small intestine, it arrives in the colon intact. The unusually large sugar supply triggers a population explosion among bacteria inhabiting the colon. The by-products of bacterial fermentation lead to painful gas and diarrhea. Many lactose-intolerant people can eat dairy products such as cheese and yogurt because the bacteria used in the production of these foods convert much of the lactose to a digestible form.

## LACTASE ACTIVITY

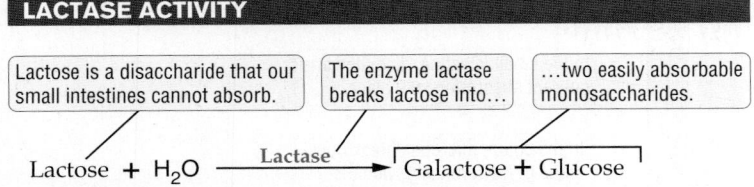

Lactose is a disaccharide that our small intestines cannot absorb. | The enzyme lactase breaks lactose into... | ...two easily absorbable monosaccharides.

$$\text{Lactose} + \text{H}_2\text{O} \xrightarrow{\text{Lactase}} \text{Galactose} + \text{Glucose}$$

**FIGURE 1 Digestion of Lactose by Lactase**
Lactase is the enzyme that digests lactose, the major sugar in milk. Accumulation of undigested lactose in individuals who do not produce enough lactase gives rise to the intestinal problems known as lactose intolerance.

## GLOBAL DISTRIBUTION OF LACTOSE INTOLERANCE IN INDIGENOUS POPULATIONS

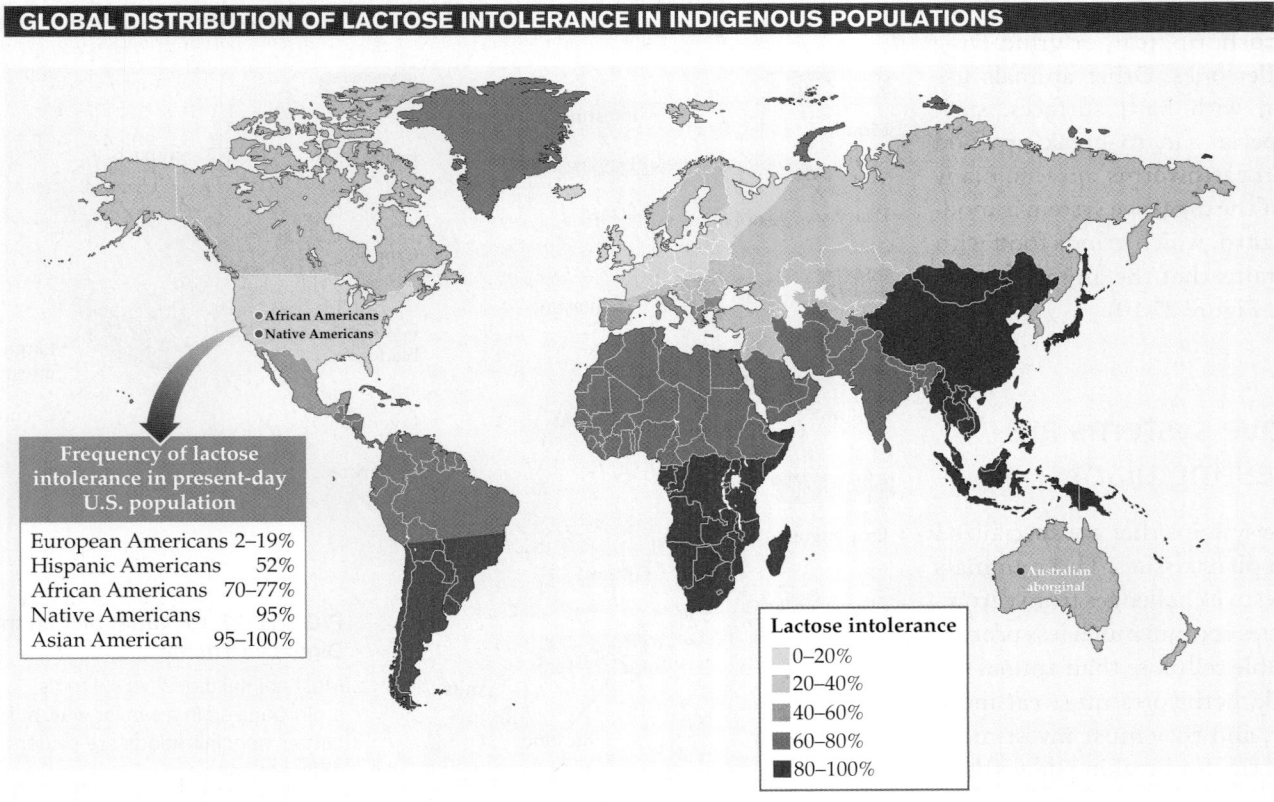

Frequency of lactose intolerance in present-day U.S. population

European Americans 2–19%
Hispanic Americans 52%
African Americans 70–77%
Native Americans 95%
Asian American 95–100%

Lactose intolerance
0–20%
20–40%
40–60%
60–80%
80–100%

**FIGURE 2 The Frequency of Lactose Intolerance**
About 70 percent of the world's population is unable to produce lactase in adulthood. This map shows the frequency of the allele encoding a nonfunctional adult lactase in populations native to various regions of the world. A very small proportion of adults in East Asia and sub-Saharan Africa produce lactase. This is also true of the original natives of the Americas and the Pacific islands.

food. To highlight the evolutionary adaptations for herbivory, let's compare two similarly sized mammals: a sheep (an herbivore) and a wolf (a carnivore).

Several parts of the digestive systems of wolves and sheep reflect the differences in their diets (FIGURE 27.11). These differences begin in the mouth. Wolves have spiked and bladelike front teeth suited for slicing and tearing meat. Their flattened back teeth come into contact with their food only briefly, just long enough to grind the torn pieces of meat into chunks small enough to swallow. In contrast, sheep have broad, flattened back teeth that can grind tough plant tissues into small pieces.

There are also noticeable differences in the stomachs of herbivores and carnivores. Pound for pound, a carnivore's diet has a much higher protein content than an herbivore's diet. The single-chambered stomach of a wolf produces enzymes that partially break down meat.

Animals secrete no enzymes that can degrade cellulose. This complex carbohydrate is a major component of plant cell walls. It is composed of long chains of glucose molecules bound together by chemical bonds (see Figure 5.14a). Certain prokaryotes, including both bacteria and archaeans, and many species of fungi, can make cellulase, an enzyme that digests cellulose into glucose units. Although animals cannot themselves make cellulase, the digestive tracts of some herbivores, from termites to sheep, are colonized by symbiotic prokaryotes that can do so and that share the bounty of energy-rich glucose with their hosts.

The sheep's stomach, for example, has evolved into a complex, four-chambered structure that harbors thriving populations of bacteria, fungi, and single-celled eukaryotes. These microorganisms break down the cellulose in plant matter, making the nutrients and energy contained in the cellulose available to the sheep. The sheep provides the microbes with a well-maintained home and abundant food, so both parties benefit.

Wolves have a small intestine that, relative to the sheep's, provides a small surface area for absorption. The difference is due primarily to length: sheep have a small intestine that is six times as long as that of a similarly sized canine. This design provides a huge surface area over which sheep can absorb the nutrients released by the relatively scarce protein in their diet. The elaborate digestive system of sheep is what enables them to survive on a diet that could not support a wolf.

**HERBIVORES VERSUS CARNIVORES**

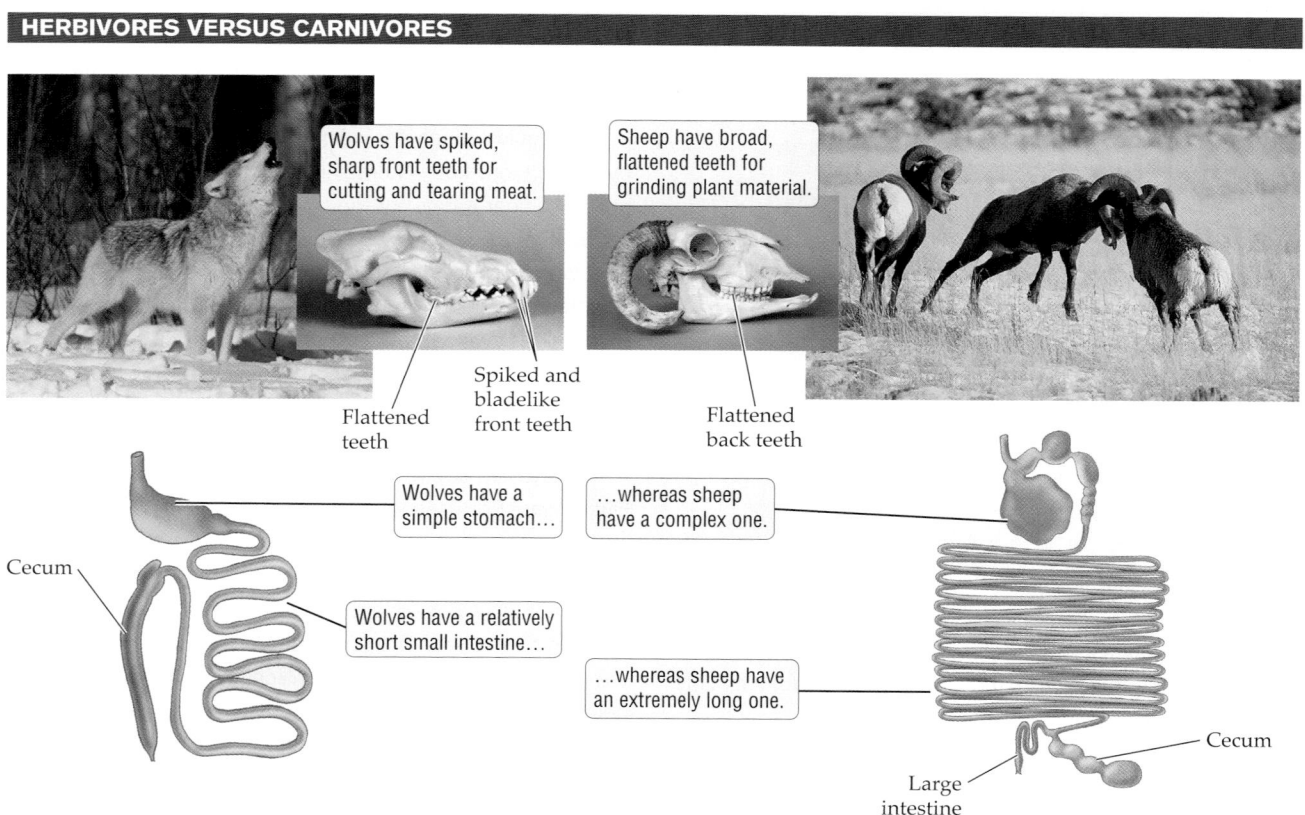

**FIGURE 27.11**

**Differences in the Digestive Systems of Carnivores and Herbivores Reflect Their Different Diets**

# How Can We Eat Better, Live Longer, and Save Money?

In 1960, about 1 percent of the U.S. population had type 2 diabetes; today the disease is spreading so fast that by 2020, nearly half the population is expected to be ill with diabetes or pre-diabetes. Millions of adults who would not have been considered at risk for diabetes as children in 1975 now have the disease. It's the result of a dramatic change in the way Americans live. What happened to transform a mostly healthy population into a chronically ill one?

Type 2 diabetes—also called "adult-onset diabetes"—results from poor regulation of blood sugar. Whenever we eat, sugars move from food in the digestive tract into the blood. As we'll see in Chapter 29, the body responds by secreting the hormone insulin, which stimulates body cells to take up blood sugar (glucose), which in turn triggers body responses that lower blood sugar. People with type 2 diabetes develop "insulin resistance," in which body cells ignore insulin and fail to take up blood sugar. Instead, blood sugar spikes whenever people eat, and the body secretes more and more insulin.

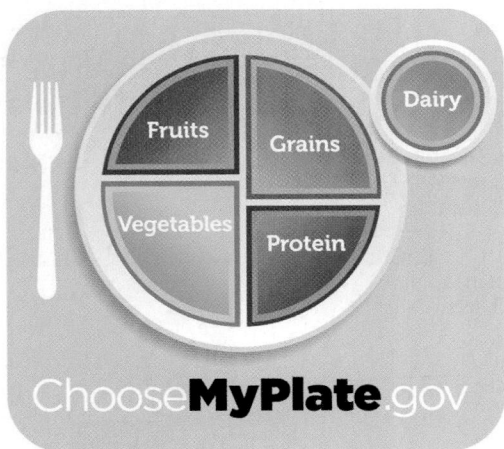

The new dietary guidelines issued by the USDA (United States Department of Agriculture) emphasize whole grains, vegetables, and fruits, which most Americans do not consume in sufficient quantities.

Four major risk factors for developing diabetes are age, obesity, lack of exercise, and a collection of abnormalities called "metabolic syndrome," which include insulin resistance, high blood pressure, abnormal cholesterol, and abdominal obesity—all major risk factors for cardiovascular disease. Metabolic syndrome increases the risk of developing type 2 diabetes by as much as 30 times, and the single greatest risk factor for metabolic syndrome is obesity. Thirty percent of Americans are now obese (not just overweight) and rates have doubled or nearly doubled in 39 states in just 15 years.

Exercise helps. People who exercise 30 minutes four to six times a week have half the rate of diabetes as those who don't exercise. Yet, most of us find ourselves in situations where we drive to work and then sit all day at work or school. In families where both parents work and commute, there's little time or energy for shopping and cooking meals from fresh ingredients. Rushed and tired, families naturally gravitate to convenience foods. Thirty percent of American children eat fast food, which gives them an extra 187 calories *every day*. In 2003, Americans ate 523 calories a day more than they did in 1970. Meanwhile, the foods that are best for us—fresh fruits and vegetables, whole grains, beans, nuts, and polyunsaturated fats—are inconvenient and usually more expensive.

We could reduce our weight and control our blood sugar by eating more fiber. Nutritionists recommend we eat 20–35 grams of fiber a day, but the typical American manages only about 12–18 grams. Instead of eating fresh fruits and vegetables, we grab a candy bar or a container of sweetened yogurt, both high in sugar and usually fiber-free. Instead of oatmeal for breakfast, we eat "Sugar-Os"; and instead of a bean burrito for lunch, we have a turkey-and-cheese wrap. Low-fiber foods cause a quick spike in blood sugar and a rapid drop-off that leaves us hungry soon after.

It's easy to say we *should* eat better and exercise more. But for most of us, such changes are difficult. It's not only a matter of "good" or "bad" habits, but also a matter of budgets, jobs, and time constraints. The United States must find ways to make it more convenient for people to eat better and exercise more—not only to improve the health and happiness of the people, but also to save $4 trillion in health care costs and help balance the federal budget.

# Want a Long Life? No Worries

BY ALEX BEAM, *Boston Globe*

I am sick to death of the purported science of longevity. Just this month, one team of researchers proclaimed that driving slowly lengthens life spans, while others said watching too much TV will drive you to an early grave. Scientists at Canada's McMaster University just announced a "dietary formula that maintains youthful function into old age." A few days later, at Great Britain's Newcastle University, "Scientists Solve Ageing Puzzle," according to their own press release.

Meanwhile, genius at large Ray Kurzweil promotes his formula for immortality, while Big Pharma greed-heads in Cambridge and elsewhere are still hoping to commercialize compounds like resveratrol, the ingredient in red wine that may—or may not—prolong life. Then there is the 1,000-calorie-a-day longevity gang, heartened by research showing that half-starved mice live longer than their better-fed cagemates.

Loma Linda, Calif., is only one place where a large population enjoys startling longevity, with so many healthy 100-year-olds that it qualifies as what demographers call a statistical "blue zone" . . . The reason for the longevity: not miracle pills, and not starvation diets. Loma Linda is a center of the Seventh Day Adventist religion, which counsels its members to live pretty much the way your doctor would want you to. "Their approach to food is pretty simple and extremely sound," says Dr. Kenneth Minaker, chief of geriatrics at Massachusetts General Hospital. "And in their community, they reinforce themselves in terms of positive health behaviors."

What behaviors are those? Adventism forbids smoking, frowns on alcohol, and observes the kosher laws in Leviticus 11, urging its adherents to shun pork, shellfish, and "unclean" foods. As with the Mormon religion, caffeine is out. What should you eat? "Grains, fruits, nuts, and vegetables constitute the diet chosen for us by our Creator," according to Ellen White, the Mary Baker Eddy of the Adventist Church.

The Adventist diet is very close to veganism or vegetarianism. The church runs a supermarket in Loma Linda, and there is no meat or poultry department, no ice cream, and vast arrays of bins filled with dried legumes, the favored staple of strict Adventists. "As you move from veganism over to conventional meat consumption, there is always a step increase in body mass index," Dr. Serena Tonstad, a professor of public health at the Adventist Loma Linda University, explained to me.

Nutritionists often study populations with unusually good health and longevity in an attempt to identify dietary practices that could improve human health. The Seventh-Day Adventist religion emphasizes good nutrition and abstinence from alcohol, tobacco, and caffeinated drinks. Followers eat legumes, nuts, whole grains, vegetables, and fruits; moderate amounts of low-fat dairy products; and minimal amounts of saturated fat, cholesterol, sugar, and salt. Except for its emphasis on vegetarianism, and even veganism, the Seventh-Day Adventist diet closely mirrors nutritional guidelines set by the U.S. Department of Agriculture.

Seventh-Day Adventists and others who eat a similar diet have lower rates of heart disease, cancers, stroke, and diabetes, compared to the overall population. The protective effect of a plant-based diet is well established. Abundant evidence shows that the omega-3 fatty acids in oily fish (such as salmon, herring, and sardines), flaxseed, walnuts, seaweed, and algae promote heart health and may even protect against depression, ADHD, and Alzheimer's disease.

Studies of the Amish, another American religious group, show that it's not only diet that promotes good health, but also exercise. On average, the Amish eat 3,600 calories per day, half again as much as the average American. The calorie-rich Amish diet is laced with saturated fat from bacon, eggs, and sausages. Yet the great majority of Amish stay slim and have low rates of heart disease and diabetes compared to other Americans. The secret? Hard work. The Amish engage in about six times as much physical activity as other Americans. American adults walk, on average, only about 2,000–3,000 steps a day, while their Amish counterparts walk 14,000–18,000 steps.

## Evaluating the News

**1.** Whole-grain foods, such as brown rice, are often called "slow carbs" because the sugars from their carbohydrates are released into the bloodstream much more slowly than are those from "fast carbs" such as mashed potatoes or white bread. Given what you learned in this chapter about the human digestive system, explain why a 100-calorie serving of mashed potatoes produces a sharp spike in blood glucose half an hour after it is eaten, while a 100-calorie serving of brown rice leads to a much smaller rise in blood glucose that is sustained over two hours.

**2.** Compare your diet with that of Seventh-Day Adventists. How is your diet more or less healthy?

**3.** How have work-saving devices affected our lifestyle and health? List some specific examples.

**SOURCE:** *Boston Globe*, February 23, 2010.

# CHAPTER REVIEW

## Summary

### 27.1 Nutrients That Animals Need

- Life depends on only a few chemical elements. Some mineral elements are needed in relatively large amounts; others are needed in only tiny quantities, although they are still essential to survival.
- Plant-eating animals are herbivores, animals that eat other animals are carnivores, and animals (including humans) that eat both plant and animal foods are omnivores.
- Animals must break down carbohydrates into monosaccharides, fats into fatty acids and monoglycerides, and proteins into amino acids before these nutrients can be absorbed into the body.
- Animals rely on nutrients for both chemical building blocks and energy. Respiration releases the energy in carbohydrates and fats. Proteins are used mostly as a source of amino acids.
- Human bodies cannot make certain amino acids. These essential amino acids must be obtained from our food.
- Vitamins are organic compounds obtained from food that help regulate metabolic processes in the animal body. Minerals are small inorganic molecules needed by the body in small amounts. For optimal function, vitamin and mineral intake should be neither too low nor too high.

### 27.2 The Human Digestive System

- The digestive system is a tubular passageway that, in conjunction with associated organs, processes ingested food.
- Ingested food is broken down into smaller pieces in the oral cavity through the grinding action of teeth. Salivary glands secrete saliva, which moistens the food and begins the chemical breakdown of starch.

- Food passes from the oral cavity to the pharynx, and then down the esophagus and into the acid environment of the stomach, where protein digestion begins.
- Partially digested food moves from the stomach into the upper region of the small intestine. Here, enzymes secreted by the pancreas and the intestine itself complete digestion of the food. The liver produces bile (stored and delivered by the gallbladder), which helps digest fats.
- In the lower region of the small intestine, digested nutrients are absorbed, via capillaries and lymph vessels, into the body. The lining of the small intestine is highly folded and bears fingerlike projections (villi) that present a large surface area for absorbing nutrients.
- From the small intestine, any unabsorbed material moves into the colon, where much of the water and minerals remaining in the material are absorbed. Here, bacteria break down some of the waste and release nutrients that the body can absorb.

### 27.3 Special Adaptations of Animal Digestive Systems

- The digestive systems of most animals share a similar design, based on similar function.
- Most animals use muscles and hard surfaces to break food into increasingly smaller pieces to speed digestion. In many animals, a gizzard helps grind food.
- Herbivores extract nutrients from plants through special adaptations of their digestive systems. The teeth of herbivores are shaped for thorough grinding of tough plant tissues, and they have a relatively long small intestine, which facilitates the absorption of nutrients released by the breakdown of proteins. Some herbivores have multichambered stomachs that house cellulose-degrading microorganisms.

## Key Terms

anus (p. 602)
bile (p. 601)
brush border (p. 601)
carnivore (p. 591)
colon (p. 602)
detritivore (p. 591)
dietary fiber (p. 591)
digestion (p. 590)

digestive system (p. 590)
esophagus (p. 600)
essential amino acid (p. 593)
feces (p. 602)
gallbladder (p. 601)
gizzard (p. 603)
herbivore (p. 591)
ingestion (p. 590)

liver (p. 601)
mineral (p. 598)
nutrient (p. 590)
omnivore (p. 591)
oral cavity (p. 600)
pancreas (p. 601)
pharynx (p. 600)

saliva (p. 600)
salivary gland (p. 600)
small intestine (p. 601)
stomach (p. 600)
swallowing reflex (p. 600)
villus (p. 601)
vitamin (p. 594)

# Self-Quiz

1. Which of the following is an essential mineral for humans?
   a. hydrogen
   b. triglyceride
   c. cellulose
   d. iodine

2. During digestion, proteins are broken down into
   a. monosaccharides.
   b. fats.
   c. sugars.
   d. amino acids.

3. Which of the following would contain all eight essential amino acids?
   a. all-wheat cereal
   b. corn chips
   c. beans and rice
   d. roasted peanuts

4. Digestion
   a. reassembles nutrients into molecules for use by the body.
   b. breaks down nutrients into a form that can be absorbed.
   c. forms proteins from amino acids.
   d. forms fats from carbohydrates.

5. Which of the following contains a strongly acidic environment?
   a. mouth
   b. small intestine
   c. stomach
   d. large intestine

6. Digestive enzymes are secreted by the
   a. salivary glands.
   b. pancreas.
   c. small intestine.
   d. all of the above

7. Bacteria that assist in human digestion are found in the
   a. stomach.
   b. pancreas.
   c. gallbladder.
   d. colon.

8. Which of the following *cannot* be digested by animals?
   a. proteins
   b. cellulose
   c. fats
   d. carbohydrates

9. In the digestive systems of some animals, gizzards
   a. secrete digestive enzymes.
   b. absorb nutrients.
   c. grind food.
   d. store bile.

10. Which of the following is an adaptation that enables herbivores to digest cellulose?
    a. short intestines
    b. intestinal villi
    c. the pancreas
    d. resident bacteria

# Analysis and Application

1. What is meant by "dietary fiber," and why is it considered essential to human health? Name four foods that are a rich source of dietary fiber.

2. Why are large doses of fat-soluble vitamins more likely to be toxic than large amounts of water-soluble vitamins? Give one example of each type of vitamin.

3. A typical apple, including its core, contains less than 1 gram of protein and fat, 32 grams of carbohydrates, and 6 grams of fiber. Describe the journey of an apple through the digestive system as it is eaten, digested, and absorbed. Identify the different regions of the digestive system and what happens in each region. What do you think an apple contains that would be indigestible to humans?

4. Describe the role of accessory organs in the digestion of food in the human gastrointestinal tract.

5. Which organ manufactures bile? Which organ stores bile? What role does bile play in the digestion of food?

6. In celiac disease, an autoimmune disorder, an overreaction to gluten, a protein found in wheat and barley, damages intestinal villi. What symptoms would you predict in a child with celiac disease and why?

7. Describe the general features common to the digestive systems of most animals. Why do you think animals share these features?

8. Describe the features of an herbivore's digestive system that are special adaptations for a plant-based diet.

# Circulation and Gas Exchange

**RISING TO THE TOP.** A giraffe's heart must generate enough pressure in the blood to pump it all the way up to the giraffe's head, 2 meters or more above the heart.

# Pumping It Up: High Blood Pressure on the African Savanna

On the African savanna, giraffes move gracefully across sweeping grasslands, picking leaves from high, flat-topped acacia trees. To pump blood up their long necks to heads more than 2 meters above their hearts, giraffes' hearts must generate blood pressure that is about 260/160, double the pressure generated by a healthy human heart.

Like a giraffe's heart, a human heart generates enough pressure to drive blood through a vast network of blood vessels. Doctors routinely measure blood pressure to assess the health of our heart and blood vessels. A resting blood pressure of less than 120/80 generally indicates that our heart and blood vessels are in good working order. The high and low numbers are the maximum and minimum pressures that our blood vessels experience. As the heart contracts, it generates the high number; as the heart relaxes, it generates the low number. Blood pressure is expressed in units of millimeters of mercury (mm Hg), a standard measure of pressure referring to how high a tube of mercury is pushed upward against gravity.

In humans, blood pressures above 120/80 are considered too high. About half of North American adults suffer from mild high blood pressure (greater than 120/80 but less than 140/90), and about a fifth of the population suffers from severe high blood pressure, with blood pressure readings greater than 160/95. If blood pressures remain above normal levels for an extended time, the risk of serious health problems such as heart attack, stroke, and kidney damage increases. For most people, high blood pressure can be reduced through weight reduction and exercise, although drugs may be needed to manage moderate to severe high blood pressure. Yet giraffes withstand far higher blood pressures without any medical assistance.

How do giraffes manage blood pressures that would kill a human? And how do they prevent excessive blood flow to the brain when they bend over to take a drink?

Before we try to answer these questions, we'll take a tour of the human circulatory system and examine the workings of the human heart, before we turn our attention to how humans, and other animals, bring life-giving oxygen into the body.

---

**MAIN MESSAGE**

The circulatory system delivers nutrients and removes wastes through a network of fluid-transporting vessels. The respiratory system absorbs oxygen from and expels carbon dioxide into the environment; the circulatory system exchanges these gases with respiring cells throughout the body.

---

## KEY CONCEPTS

- The human circulatory system, like that of many other animals, depends on a muscular heart that contracts and relaxes as it pumps blood through a vast network of vessels.

- Blood enriched with oxygen is delivered to the cells in the body by arteries. All respiring cells obtain oxygen and nutrients from the blood and release into it carbon dioxide and waste. Veins return oxygen-depleted blood to the heart.

- The heart pumps oxygen-poor blood to the lungs, where excess carbon dioxide is released and oxygen absorbed. The oxygen-laden blood then returns to the heart so that it can be pumped to respiring cells throughout the body.

- The movement of the rib cage and diaphragm pulls and pushes air into and out of our lungs.

- Gas exchange occurs over the large surface area presented by the alveolar sacs in our lungs.

- Oxygen from the inhaled air diffuses into the capillaries that surround the alveoli. Carbon dioxide diffuses out of the blood plasma and into the air space in the lungs, and much of it is expelled with each exhalation.

- Hemoglobin, packed inside red blood cells, binds oxygen strongly but releases the gas near respiring cells.

## CELLULAR RESPIRATION

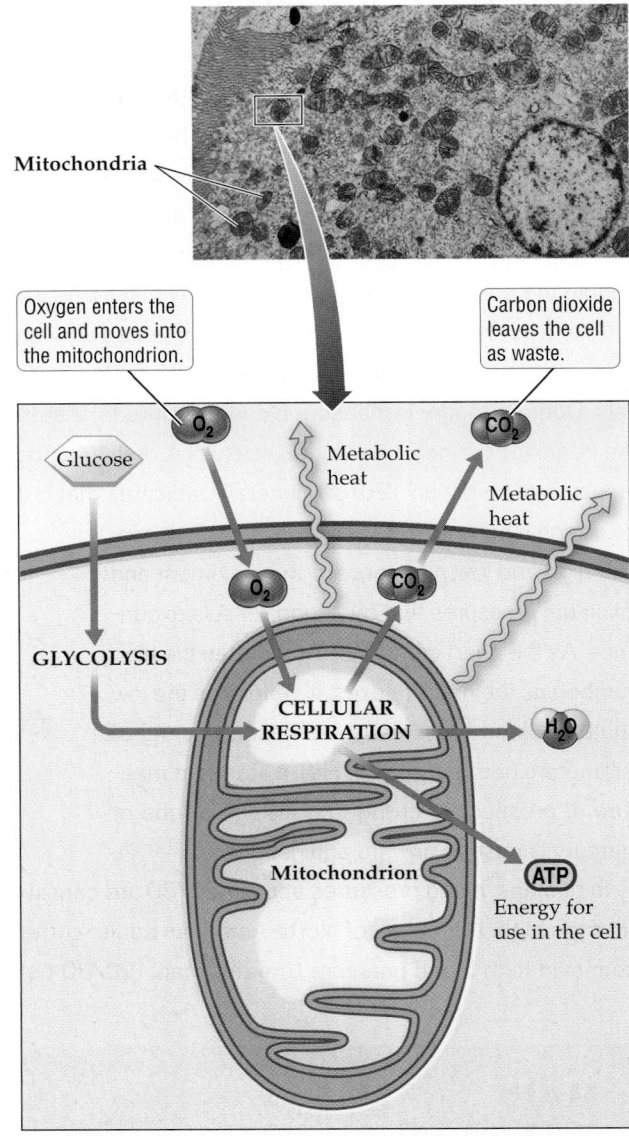

**FIGURE 28.1 Cellular Respiration Occurs in Mitochondria**

During cellular respiration, oxygen is consumed as energy is extracted from glucose for use by the cell. Heat, water, and carbon dioxide are also produced. The inset shows mitochondria in a liver cell.

**MOVING MATERIALS WITHIN THE BODY** is an important function that all complex multicellular organisms, including animals, must accomplish. Coordinated activities such as growth and homeostasis depend on the transfer of ions, waste products, signaling molecules, transport proteins, and other substances within the body. Cellular respiration requires the delivery of oxygen to mitochondria and the removal of carbon dioxide generated as waste (**FIGURE 28.1**).

In previous chapters we discussed the importance of diffusion as a means of transporting materials in animal bodies (see Section 7.1). Diffusion over short distances—such as across plasma membranes, or within a single cell, or across the small distance between a blood vessel and interstitial fluid—takes no more than a fraction of a second. However, the distances between sources and destinations in most animals are far too great for diffusion to be an effective means of distributing gases, nutrients, signaling substances, and other materials. As multicellular animals evolved from their single-celled ancestors, they eventually developed an internal transport system, known as the **circulatory system**, consisting of tubelike **blood vessels**, muscular **hearts**, and fluids that carry nutrients and wastes over long distances.

To support cellular respiration, animals need a way of taking in oxygen and getting rid of carbon dioxide. The organ system that enables an animal to bring oxygen into the body while expelling carbon dioxide from it is the **respiratory system**. The respiratory system provides a large, specialized surface that gases can diffuse across. The circulatory system picks up oxygen that diffuses across this surface and delivers it to all the cells in the body. It also absorbs carbon dioxide released by these cells and brings it the respiratory surfaces for discharge into the environment.

In this chapter we consider how animals transport vital materials through the circulatory system. We start by introducing the structure and function of the human circulatory system. We examine how the

structure and size of the blood vessels affect blood flow, and consider some of the health problems that commonly beset the human circulatory system.

Next we look at how the respiratory system enables animals to exchange oxygen and carbon dioxide between the outside and inside of their bodies. We discuss breathing and gas exchange in humans, and the general principles that apply to gas exchange in all animals. We will see that vertebrates use special oxygen-binding pigments, packaged in specialized blood cells, to move

### Helpful to Know

In everyday use, "respiration" refers to the process of breathing in (inhalation) and out (exhalation). The term "cellular respiration" refers to the energy-extracting reactions in cells that require oxygen and release carbon dioxide. The respiratory system links both these processes in the human body because it exchanges the gases needed and released by cellular respiration with the environment.

oxygen from the gas exchange surfaces to the cells in which cellular respiration takes place.

## 28.1 The Human Cardiovascular System

In vertebrates, the circulatory system is called the **cardiovascular system** (from *kardia*, "heart"; *vasculum*, "small vessel"). It consists of a muscular heart, a complex network of blood vessels that collectively form a closed loop, and a fluid called **blood** that circulates through the heart and blood vessels. To explore how animals move materials within the body, we begin with a look at the structure and function of the human cardiovascular system.

### Blood is a watery fluid containing many different cell types

About 45 percent of blood is composed of cells and cell fragments that float in a watery fluid known as **plasma** (**FIGURE 28.2a**). **Red blood cells**, or erythrocytes, are specialized for the transport of oxygen throughout the body. They make up about 95 percent of the cellular component of blood and are the most numerous cell type in the human body. A mature red blood cell lacks a nucleus, and its cytoplasm is packed with millions of molecules of an oxygen-binding protein called *hemoglobin*.

The cellular fraction of blood also contains a variety of immune cells, including several different types of white blood cells that help defend the body from invading organisms (see Figure 28.2a). **Platelets**, which are small cell fragments produced by a larger cell, are also part of the cellular fraction of blood. They can clump together to help stanch the loss of blood if a blood vessel is damaged. Platelets also release substances that stimulate plasma proteins, called *clotting factors*, to create a meshwork of protein strands, platelets, and trapped blood cells that collectively constitute a blood clot (**FIGURE 28.2b**). Most blood cells and cell fragments have a short life, and replacements are made through mitotic divisions of stem cells in the bone marrow.

Blood plasma is 92 percent water and contains dissolved gases, ions, and molecules that are critical for homeostasis, as nutrients, or as signaling molecules. Much of the carbon dioxide carried in the blood is dissolved in plasma in the form of bicarbonate ions ($HCO_3^-$). Other critical ions include sodium ($Na^+$), potassium ($K^+$), chloride ($Cl^-$), and calcium ($Ca^{2+}$). Vital nutrients found in the plasma include glucose, amino acids, various lipids, and vitamins. Blood plasma is a rapid means

---

**COMPOSITION OF HUMAN BLOOD**

**(a) Components of blood**

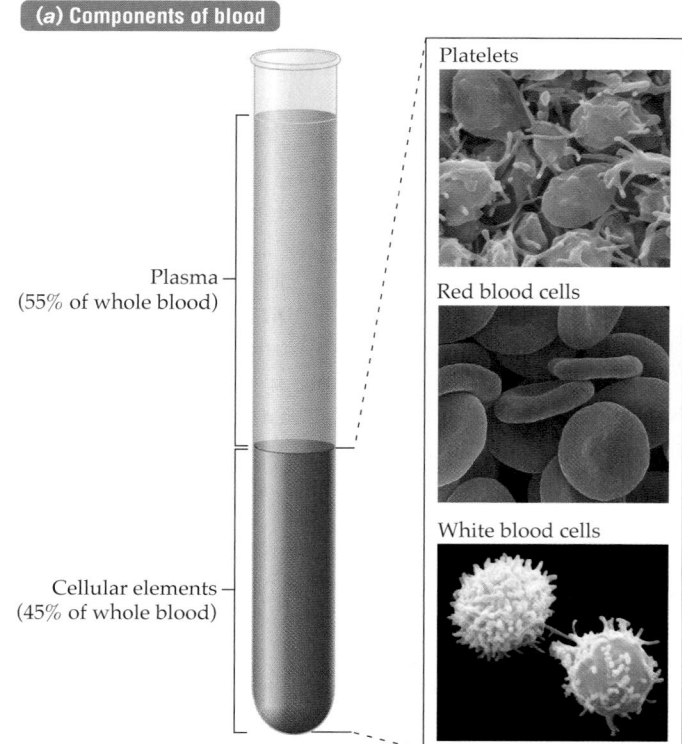

Platelets

Red blood cells

White blood cells

Plasma
(55% of whole blood)

Cellular elements
(45% of whole blood)

**(b) Blood clot**

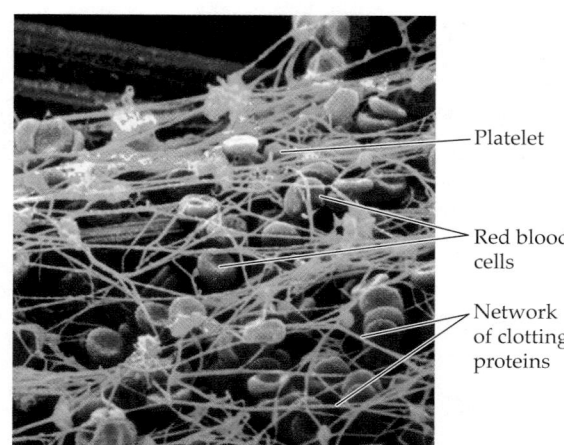

Platelet

Red blood cells

Network of clotting proteins

**FIGURE 28.2 Human Blood Consists of a Fluid and Several Different Cell Types**

(*a*) The cellular fraction of whole blood consists of many different types of cells and cell fragments, three of which are shown here. Red blood cells account for about 95 percent of the cellular fraction by volume. (*b*) Chemicals produced by platelets help in the formation of blood clots, which are clumps created by a meshwork of clotting factors (plasma proteins) and trapped blood cells. After a wound is sealed, blood clots are usually cleared from the site by clot-dissolving enzymes.

of broadcasting signaling molecules, such as the many hormones produced in the human body, to all tissues. Proteins are the most abundant organic molecules in the plasma. Some of these proteins are signaling molecules, including certain protein hormones; others include antibodies (involved in immune defense), clotting factors, and proteins (such as the albumins and lipoproteins) that assist in the transport of other substances.

## The human heart pumps blood to the body through two circuits

In the human cardiovascular system, the heart pushes blood through two main pathways, or circuits, both of which begin and end with the heart. One pathway, the **pulmonary circuit** (from *pulmon*, "lung") consists of

blood streaming from the heart to the lungs and then back from the lungs to the heart (**FIGURE 28.3**). To the lungs the heart pumps blood that it has received from all regions of the body. This blood is depleted in oxygen and rich in carbon dioxide ($CO_2$) released by all the respiring tissues in the body. In the lungs this oxygen-poor blood flows through the fine capillaries that surround the epithelial tissues lining the lung cavities. Oxygen diffuses from the air space in the lungs to the epithelial cells and then across the thin capillary walls and into the blood. At the same time, the $CO_2$ in the blood diffuses across the capillaries, through the epithelial cells, and into the air space in the lungs. The blood, now rich in oxygen and low in carbon dioxide, flows back to the heart, completing the pulmonary circuit. The heart then pumps this blood into the **systemic circuit**, which delivers oxygen to cells throughout the body and picks up carbon dioxide from them (upper and lower part of Figure 28.3). The gases are exchanged across an extensive network of fine blood vessels in the systemic circuit. The return of oxygen-poor, $CO_2$-enriched blood to the heart completes the systemic circuit.

In a typical adult, 5–6 liters of blood circulates continually through the blood vessels. Just about all of our cells lie within 0.03 millimeter (less than one-third the thickness of this page) of a blood vessel with which they can exchange materials. Over these very short distances, diffusion works well. Carrying blood so close to all the trillions of cells in our bodies requires an extensive system of blood vessels. If laid end to end, the approximately 17 billion blood vessels in the human body would stretch an amazing 20,000 kilometers, or approximately from Seattle to London and back again.

## Blood vessels are the highways for fluid transport

The human body has three major kinds of blood vessels: *arteries*, *veins*, and *capillaries* (**FIGURE 28.4**). **Arteries** carry blood from the heart for distribution to the body. These vessels branch many times into smaller and smaller vessels. The finer branchings (called **arterioles**) are only 0.3 millimeter in diameter, about the same as the diameter of a fine thread used for sewing. Arterioles control the flow of blood to the finest vessels, the capillaries.

**Capillaries**, less than 0.01 millimeter in diameter, allow the exchange of materials between the blood and the surrounding interstitial fluid and cells. A piece of muscle tissue the size of a pencil tip may contain more than 1,000 capillaries. Taken together, the capillaries in our bodies provide a combined surface area equivalent to the size of nearly three tennis courts.

**HUMAN CARDIOVASCULAR SYSTEM**

Respiring tissues in upper body

Exchange of gases between the blood and the surrounding cells occurs in tiny capillaries.

In the pulmonary circuit, blood that has delivered oxygen to cells and picked up carbon dioxide from them is pumped to the lungs, where gas exchange takes place.

Capillaries

Right lung

Left lung

Gas exchange in lung capillaries

Gas exchange in lung capillaries

The muscular heart contracts to generate the pressure that pushes blood through the circulatory system.

Right

Left

Heart

Artery

Vein

Respiring tissues in lower body

In the systemic circuit, blood that has picked up oxygen from the lungs is pumped into body cells, where gases are exchanged.

Arterial circulation

Venous circulation

Capillaries

Gases are exchanged across capillary walls.

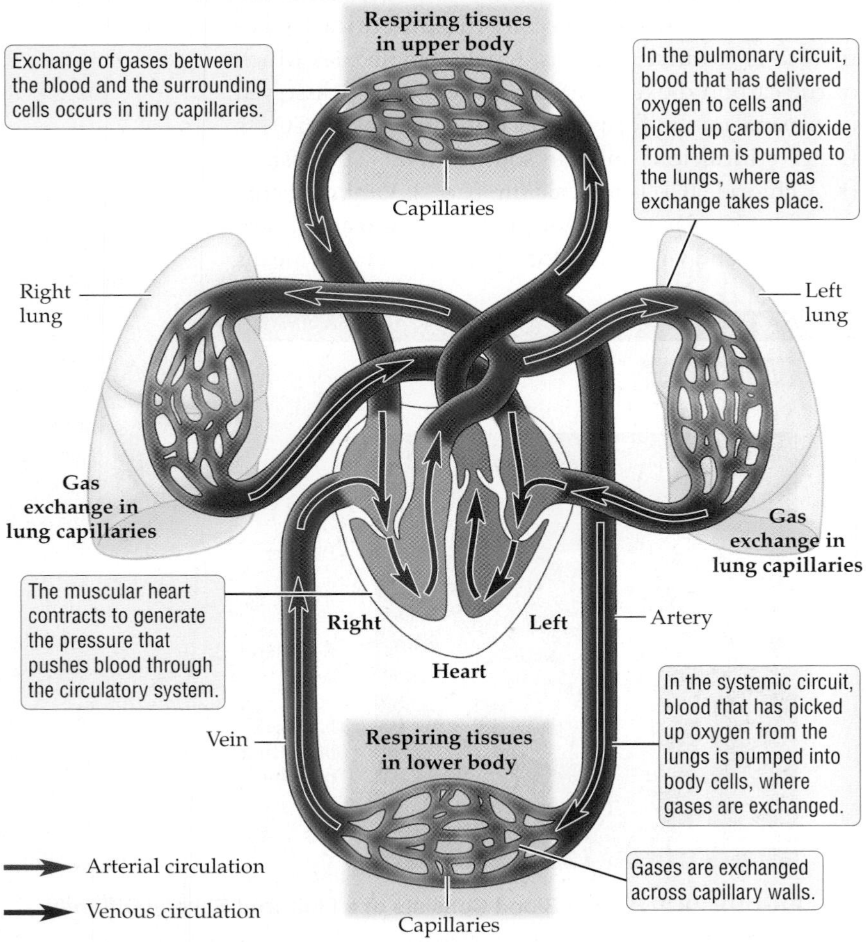

**FIGURE 28.3  The Heart Pushes Blood through the Systemic and Pulmonary Circuits**
The systemic circuit carries oxygen-rich blood away from the heart, circulates it throughout the body, and returns oxygen-depleted blood back to the heart. The pulmonary circuit brings oxygen-poor blood from the heart to the lungs and returns oxygen-rich blood to the heart.

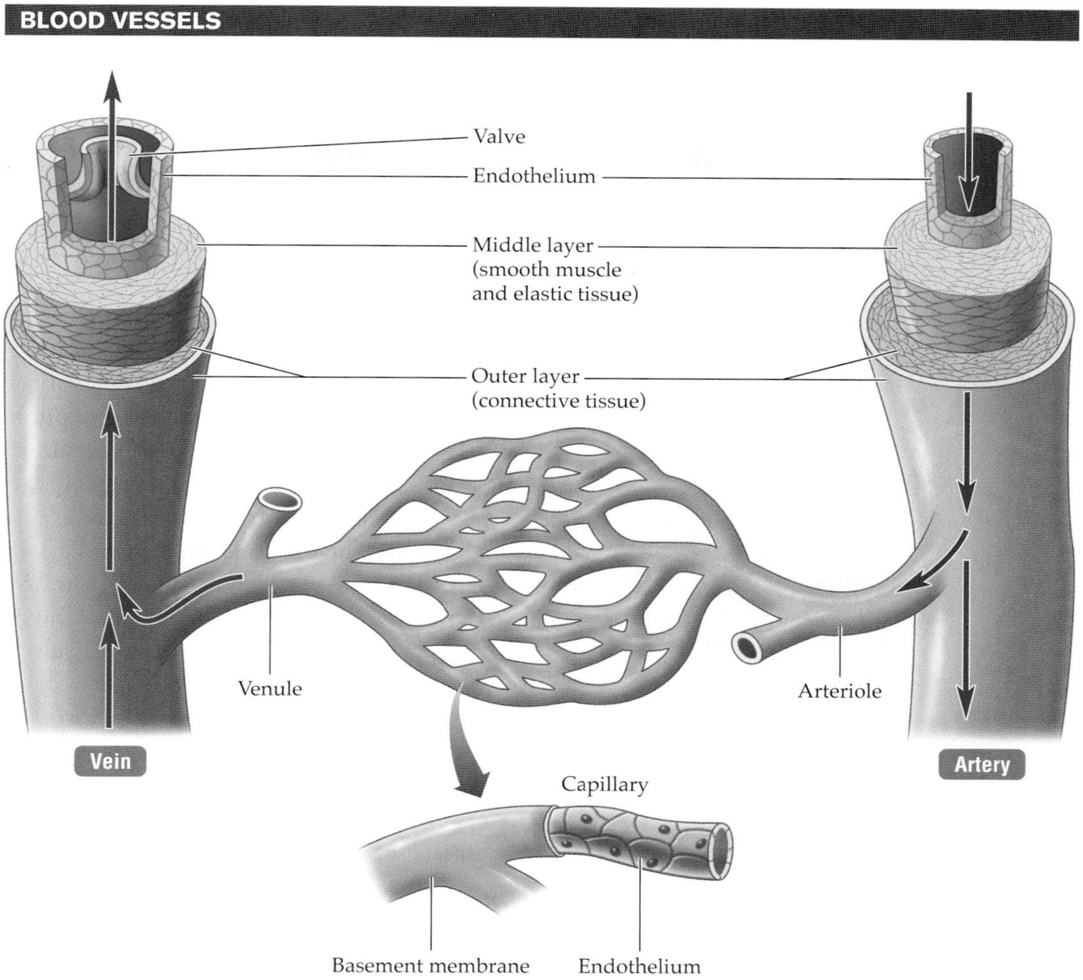

FIGURE 28.4

**Comparison of Arteries, Veins, and Capillaries**

Arteries carry blood away from the heart. In veins, blood flows in the direction of the heart. Arterioles and venules are smaller arteries and veins, respectively. The narrowest arteries and veins connect with each other in a fine network known as a capillary bed.

Valve
Endothelium
Middle layer (smooth muscle and elastic tissue)
Outer layer (connective tissue)

Vein
Venule
Arteriole
Artery

Capillary

Basement membrane    Endothelium

**Veins** carry blood from the body back to the heart. The return journey begins as blood leaves the capillaries in tiny veins called **venules**, which join together into ever-larger veins, eventually collecting all the blood from the body for return to the heart.

The human cardiovascular system is said to be a **closed circulatory system** for two reasons: blood is enclosed in the vessels and the heart, and together the two circuits form a closed loop: from heart to arteries to capillaries to veins and back to the heart again.

## The human heart has two pumps and four chambers

About the size of a clenched fist, the human heart lies at an angle under the chest bone, with about two-thirds of it to the left of the center of the body and the remaining third to the right of an imaginary center line. The conical lower end of the heart points toward the left hip, and the blunt upper end toward the right shoulder. By convention, what is referred to as the "right

side" of the heart is that which is toward the right side of the body being observed (not the observer's right).

Like the hearts of all other mammals, and birds as well, the human heart is divided into four chambers that create two distinct, physically independent pumping units. The two chambers on the right side of the heart form one pumping unit. The two chambers on the left form the other. The chambers on the right and left sides are separated by a thick wall of tissue called the septum (**FIGURE 28.5a**), and they pump blood in independent but coordinated ways (**FIGURE 28.5b**). The chambers on the left side of your heart receive oxygenated blood returning from the lungs and pump it through the systemic circuit to respiring cells. The two chambers on the right side receive blood returning from the systemic circuit that is low in oxygen and laden with $CO_2$, and pump it through the pulmonary circuit for gas exchange in the lungs.

As we have seen, each side of the heart is divided into two chambers. The upper chamber is the smaller of the two and is called the **atrium** (plural "atria"); the lower chamber is called the **ventricle** (see Figure 28.5a).

**Helpful to Know**

In the *systemic circuit*, arteries carry oxygenated blood, and veins carry oxygen-poor blood. But because arteries and veins are defined in terms of direction of flow from and to the heart (arteries away from the heart, veins toward the heart), the arteries and veins in the *pulmonary circuit* have the opposite pattern: the pulmonary artery carries oxygen-poor blood from the heart to the lungs, and the pulmonary vein carries oxygen-rich blood from the lungs back to the heart.

**(a) Four chambers of the heart**

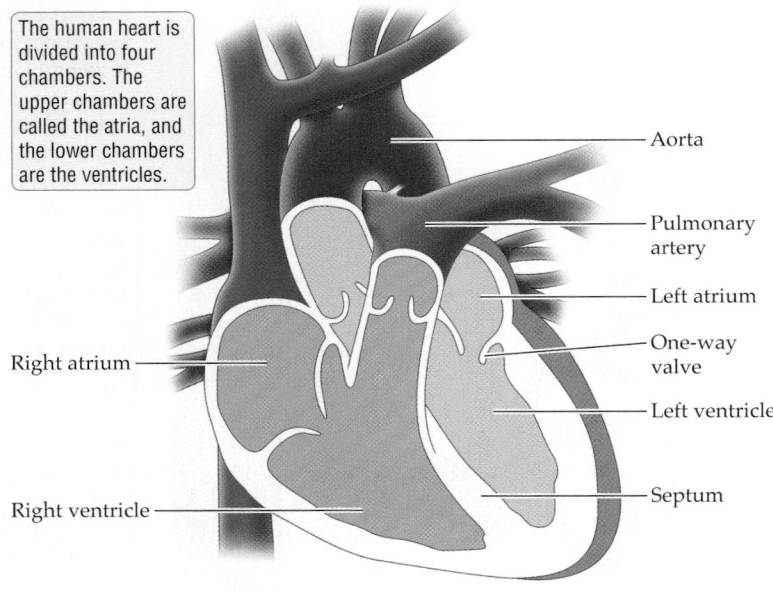

The human heart is divided into four chambers. The upper chambers are called the atria, and the lower chambers are the ventricles.

Aorta

Pulmonary artery

Left atrium

One-way valve

Left ventricle

Right atrium

Septum

Right ventricle

**(b) Two pumps of the heart**

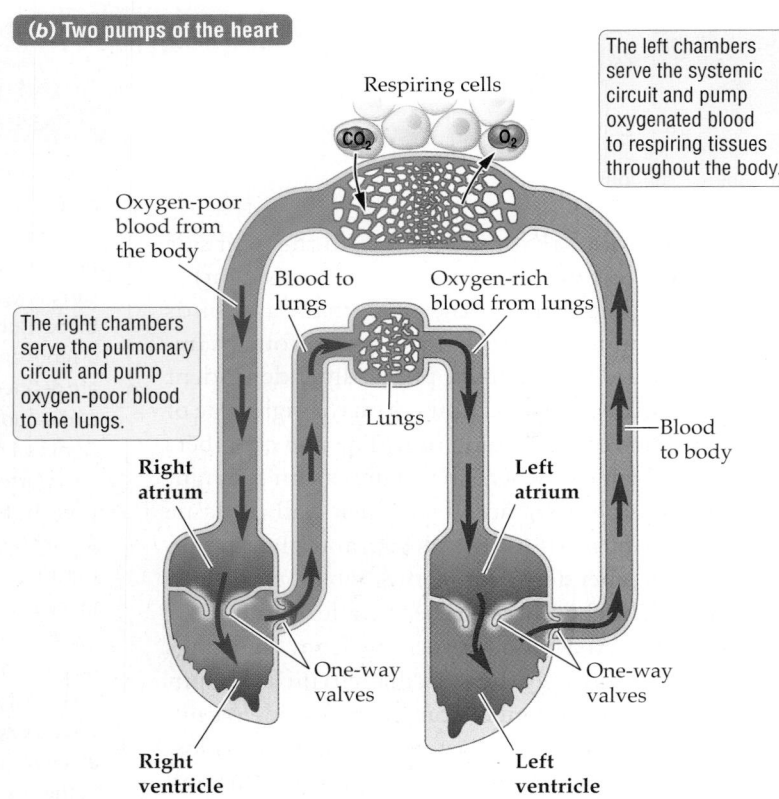

The left chambers serve the systemic circuit and pump oxygenated blood to respiring tissues throughout the body.

Respiring cells

$CO_2$    $O_2$

Oxygen-poor blood from the body

Blood to lungs

Oxygen-rich blood from lungs

The right chambers serve the pulmonary circuit and pump oxygen-poor blood to the lungs.

Lungs

Blood to body

**Right atrium**

**Left atrium**

One-way valves

One-way valves

**Right ventricle**

**Left ventricle**

**FIGURE 28.5  Blood Flows from Atria to Ventricles**
The right and left sides of the heart function as two separate pumps, although the two atria contract in unison, as do the two ventricles.

Contraction of the smaller, thinner-walled atrium pushes blood into the ventricle below it. Ventricles have thicker, more muscular walls because they must pump blood out of the heart and into the blood vessels heading for either the lungs or the rest of the body. The left ventricle is slightly larger than the right ventricle, and its muscular walls are especially thick—about three times thicker than the walls of the right ventricle. The left ventricle must generate higher pressures to distribute blood across the long and complex network of the systemic circuit. In contrast, the right ventricle pumps blood to the nearby lungs and the relatively simple system of blood vessels that make up the pulmonary circuit. Even though the two sides of the heart send blood to different destinations, they pump together at the same time. So the left and right atria contract at the same time, and then the left and right ventricles contract together.

One-way valves separate the atria from the ventricles, and the ventricles from the arteries leading out (**FIGURE 28.6**). These one-way valves ensure that blood flows through each side of the heart in one direction only—from the atria to the ventricles to the lungs or the rest of the body, and not the reverse. Heard through a stethoscope, the *lub-dub* sound of the heartbeat is actually the sound of these valves snapping shut. The *lub* sound is made when the valve between the atrium and the ventricle closes; the pressure that develops in the ventricle as it contracts snaps this valve closed. The *dub* sound occurs when the valve separating the ventricle from the artery closes; this valve closes when the back pressure in the artery becomes greater than the pressure in the relaxing ventricle.

## Rhythmic contractions of the heart produce the heartbeat

Our hearts beat, without fail, about 60–75 times each minute, amounting to about 3 billion beats in a 70-year lifetime. The number of beats per minute is referred to as the **heart rate**. It can be felt easily and counted as the pulse rate if the tips of the fingers are placed on the artery located on the inside of the wrist on the side of the thumb. In a human body at rest, the heart circulates *all* of the blood (5–6 liters) through the heart and throughout the body in just 1 minute. In other words, the equivalent of 7,000 liters of blood passes through the cardiovascular system each day. Each contraction of the heart generates a sharp increase (spike) in pressure that propels the blood (**FIGURE 28.7**).

The blood pressure measured in a doctor's office reflects the pressure in the arteries leading to the body

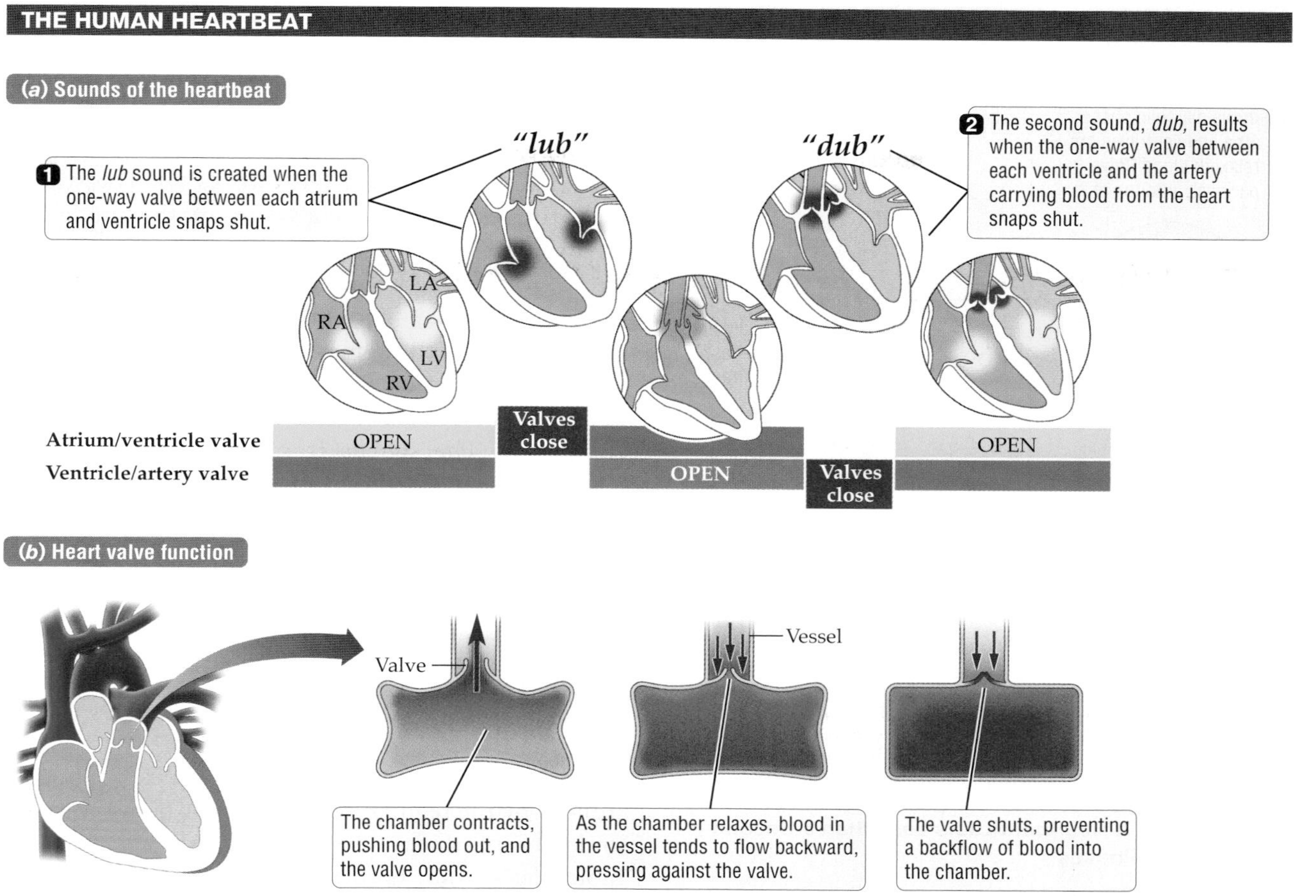

**(a) Sounds of the heartbeat**

"lub"

"dub"

**1** The *lub* sound is created when the one-way valve between each atrium and ventricle snaps shut.

**2** The second sound, *dub*, results when the one-way valve between each ventricle and the artery carrying blood from the heart snaps shut.

LA
RA
LV
RV

| Atrium/ventricle valve | OPEN | Valves close | | | OPEN |
| Ventricle/artery valve | | | OPEN | Valves close | |

**(b) Heart valve function**

Valve

Vessel

The chamber contracts, pushing blood out, and the valve opens.

As the chamber relaxes, blood in the vessel tends to flow backward, pressing against the valve.

The valve shuts, preventing a backflow of blood into the chamber.

**FIGURE 28.6 The Opening and Closing of One-Way Valves Produce an Audible Heartbeat**

(a) One-way valves in the heart direct the flow of blood in response to contraction of the heart muscle, resulting in the familiar *lub-dub* sound of the human heartbeat. Arrows pointing to the red regions show the direction of blood flow that closes the one-way valves. LA, left atrium; LV, left ventricle; RA, right atrium; RV, right ventricle. (b) A one-way valve allows blood flow out, but not back in.

from the left ventricle. A **blood pressure** reading of 120/80, for example, means that contraction of the left ventricle generates 120 millimeters of mercury (mm Hg) of pressure in the arteries, followed by a drop to 80 mm Hg when the ventricles relax and refill. Similar measurements made in the arteries leading from the smaller right ventricle into the shorter pulmonary circuit would reveal lower pressures, changing between only 8 and 25 mm Hg.

By the time the blood enters the capillaries, the pressure has dropped to 35 mm Hg. When the blood leaves the capillaries, the pressure is only 10 mm Hg. The drop in pressure as the blood progresses through the circulatory system reflects the cumulative effect of friction between the blood and the blood vessel walls.

Each **heartbeat** we experience lasts (for most people) a little less than 1 second and consists of a series of events called the **cardiac cycle** (**FIGURE 28.8**). This cycle has two phases: relaxation and contraction. The relaxation phase is called **diastole [dye-AS-toh-lee]** from the Greek word for "expansion." The lower number in a blood pressure reading represents the pressure at diastole. Diastolic pressure is the number that doctors worry about more if it is high, because it represents

**PRESSURE CHANGES IN THE CIRCULATORY SYSTEM**

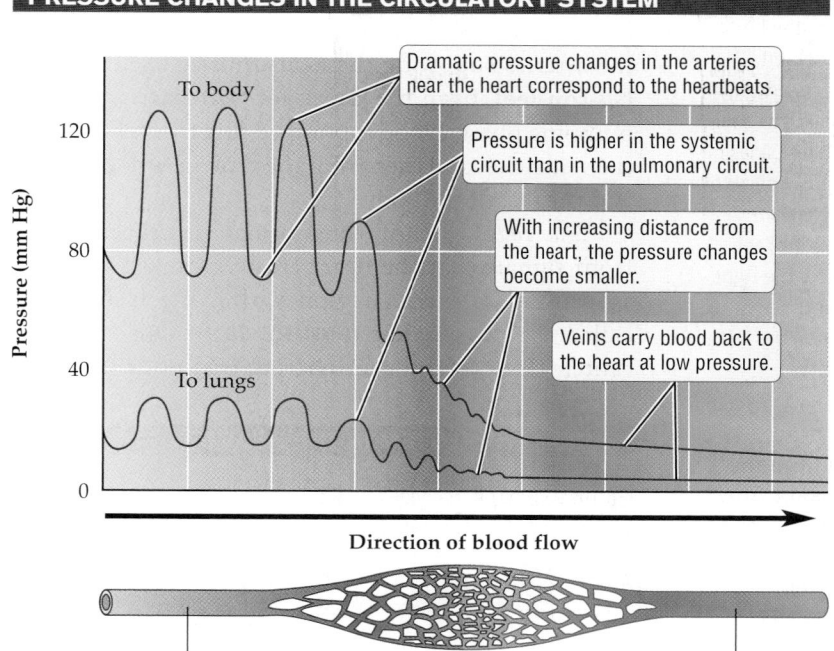

To body

To lungs

Dramatic pressure changes in the arteries near the heart correspond to the heartbeats.

Pressure is higher in the systemic circuit than in the pulmonary circuit.

With increasing distance from the heart, the pressure changes become smaller.

Veins carry blood back to the heart at low pressure.

Pressure (mm Hg)

120

80

40

0

Direction of blood flow

Arteries    Arterioles    Capillaries    Venules    Veins

**FIGURE 28.7 Blood Pressure Drops with Increasing Distance from the Heart**

Both the pressures and the pressure differences in human blood vessels decrease as the blood moves through the circulatory system.

## THE CARDIAC CYCLE

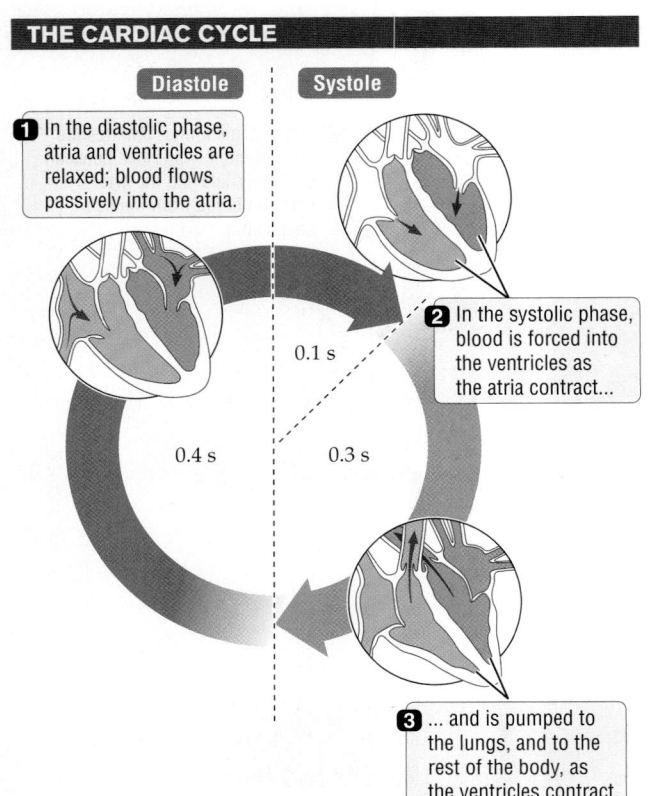

| Diastole | Systole |

**1** In the diastolic phase, atria and ventricles are relaxed; blood flows passively into the atria.

**2** In the systolic phase, blood is forced into the ventricles as the atria contract...

0.1 s

0.4 s    0.3 s

**3** ... and is pumped to the lungs, and to the rest of the body, as the ventricles contract.

**FIGURE 28.8  A Heartbeat Lasts Less than a Second and Has Two Phases**

**FIGURE 28.9  A Signal from the Pacemaker Spreads over the Heart in an Orderly Way**

AV, atrioventricular; SA, sinoatrial.

the "background" pressure that is constantly pushing on the arterial walls. During diastole, blood fills the relaxed heart chambers and also allows blood to flow into the arteries supplying the heart muscle itself. **Systole** [SIS-toh-lee], from the Greek for "contraction," is the pumping phase: blood is pumped first from the atria to the ventricles, and then from the ventricles to the lungs and the rest of the body.

The regularity of the normal heartbeat comes from signals that the heart receives from a group of specialized cells in a region within the heart itself, called the **pacemaker** (**FIGURE 28.9**). This signaling center, also called the **sinoatrial** (**SA**) **node** [SYE-noh-

AY-tree-ul...], is responsible for initiating contraction of the whole heart. Signals from the SA node travel to both atria, causing them to contract simultaneously and pump blood into the ventricles below. If the SA node is not working correctly, an artificial pacemaker is sometimes implanted (**FIGURE 28.10**). The signals from the SA node are relayed to another node: the **atrioventricular** (**AV**) **node** [AY-tree-oh-ven-TRIK-yoo-ler...], located between the atria and the ventricles. After a short delay—about one-tenth of a second—the AV node passes the signal on to the ventricles, causing them to contract and send blood to the lungs and body. The short delay allows the atria to empty more completely. The signals from these two nodes are what an **electrocardiogram** (**ECG**, also abbreviated as EKG) measures. In a disease condition, such as a heart attack, the passage of this signal is changed, and these changes can be detected in an ECG.

## The cardiovascular system responds to the body's changing needs

Although the rhythm of the heartbeat is locally controlled—by the regular firing of the SA node—it is also under the influence of the nervous system and signaling molecules such as hormones. Input from the nervous system underlies the racing heart you might experience in response to a stressful stimulus, such as a sudden loud noise or a 100-point pop quiz. Stressful or exciting stimuli can activate nerves that deliver a neurotransmitter called *norepinephrine*. The action of norepinephrine on the cells of the SA node causes the heart to beat faster and more forcefully. The resulting increase in blood flow is part of an ancient fight-or-flight response that we share with many vertebrate species. The heart also responds to hormones such as *epinephrine* (more widely known by its old name, adrenaline), which is produced by the adrenal glands in response to stressful situations. An increase in body temperature (as in a fever) tends to

## SIGNALING SYSTEM IN THE HEART

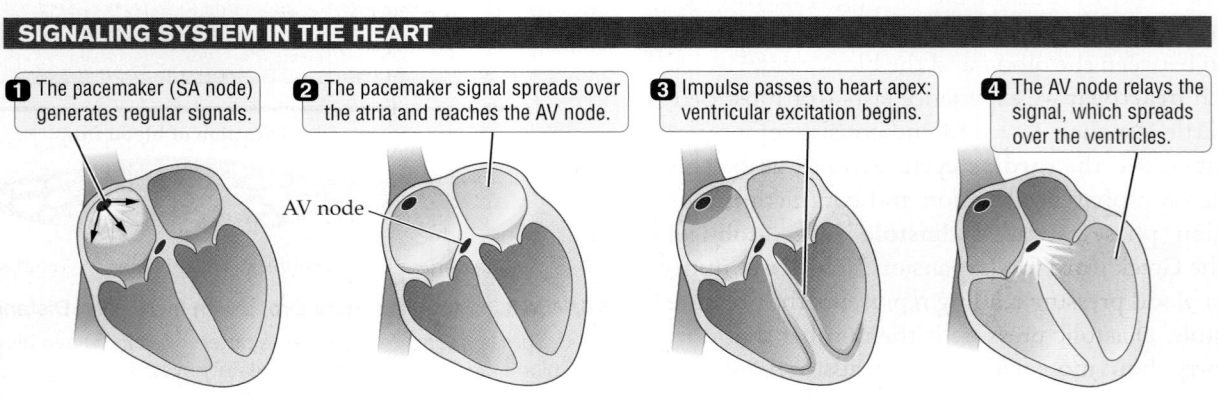

**1** The pacemaker (SA node) generates regular signals.

**2** The pacemaker signal spreads over the atria and reaches the AV node.

AV node

**3** Impulse passes to heart apex: ventricular excitation begins.

**4** The AV node relays the signal, which spreads over the ventricles.

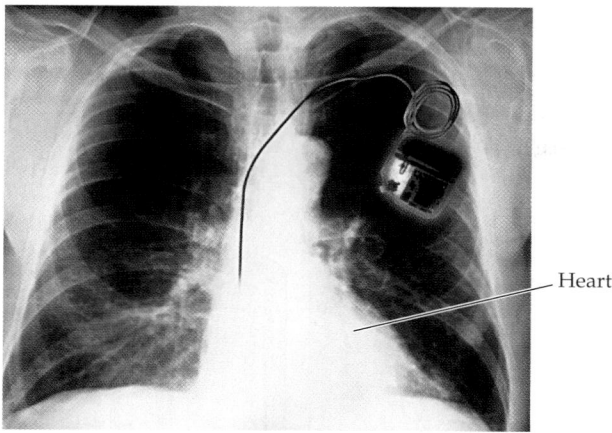

Heart

**FIGURE 28.10 Artificial Pacemakers Can Save Lives**
When the heart's natural pacemaker fails to function, an artificial pacemaker can be implanted surgically in a patient's chest, where it generates regular electrical signals that have the same effect as those produced by the natural pacemaker.

increase the heart rate, while a lowering of body temperature generally slows the heart.

During strenuous exercise, the number of times the heart beats per minute can double. Doubling the heart rate can increase the blood pressure in the arteries from 120 to about 180 mm Hg. The increased blood pressure stretches the blood vessels, enabling them to carry more blood. The distribution of blood also changes: more blood is sent to blood vessels supplying the muscles involved in exercise, and less blood is sent to tissues not involved in exercise, such as those of the digestive system. As a result of these changes, the supply of blood to the active muscles may increase nearly ten-fold.

## 28.2 Blood Vessels and Blood Flow

Blood vessels, as we have seen, serve different purposes, depending on whether they are arteries, veins, or capillaries, large or small. In both size and structure, blood vessels match their function.

### Arteries and veins are specialized for transporting blood in quantity

Large vessels—both arteries and veins—carry blood in large volumes. Their purpose is not to exchange significant amounts of materials directly with the interstitial fluid around the cells, but rather to transport blood rapidly and in large amounts throughout the body. Vessels specialized for the bulk transport of blood have relatively large diameters and elastic walls (see Figure 28.4).

The complex walls of the largest arteries equip them to regulate blood distribution. Layers of muscle tissue in the walls contract and relax to alter the amount of blood flowing through the arteries by changing the diameter of these vessels. Blood flow to different parts of the body can be altered during rest or exercise through a decrease or increase in vessel diameter. The elasticity of the artery walls also enables them to stretch when a heart contraction increases the blood pressure. When we feel the pulse in an artery, we are actually feeling the artery walls bulge and relax in response to the surge of pressure generated by contraction of the heart. The ability of the artery walls to stretch and relax reduces these pressure changes as distance from the heart increases, helping to protect the most delicate blood vessels, the capillaries, from damaging pressure changes. As we grow older, this elasticity decreases and the resulting hardening of the vessel walls contributes to higher blood pressure, which can increase the risk of heart disease, stroke, and kidney malfunction (see the box on page 621).

Veins are not as muscular as arteries, because they do not have to sustain pressures as large as those in most arteries. Since venous blood pressure is relatively low (see Figure 28.7) and blood returning to the heart, in species like us, often must rise against the force of gravity, veins have an added feature that arteries do not need: one-way valves to keep blood flowing toward the heart (**FIGURE 28.11a**). These valves have flaps that work like the ones in heart valves. Blood actively flowing toward the valve forces the flaps open. When the pressure declines below a certain level, some blood starts to flow in reverse because of gravity, forcing the flaps closed and arresting any appreciable backflow. When these valves fail to close properly, as can happen in older or overweight people, some backflow does occur, causing *varicose veins*. These abnormally swollen and twisted-looking veins, caused by the pooling of venous blood, are especially common in the legs.

The cycle of inhaling and exhaling during breathing also helps in transporting venous blood back to the heart. Each inhalation produces a drop in pressure in the chest cavity but increases abdominal pressure because as the lungs expand they compress the contents of the abdominal cavity. Blood in the abdominal veins is therefore pushed upward toward the chest, where the larger veins entering the heart are under less pressure.

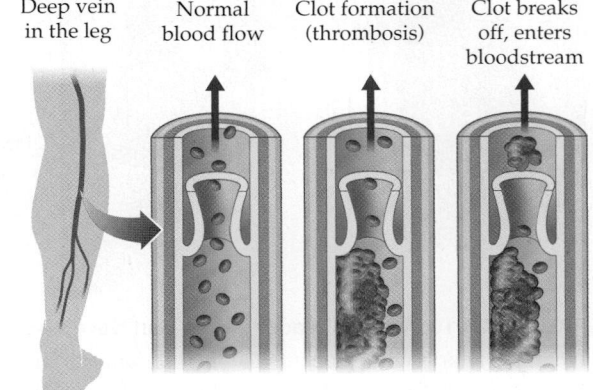

**(a) Valve function in veins**

Blood flows forward...

Valve open

Muscles are contracted

Valve closed

...but cannot flow backward.

Valve closed

Muscles are relaxed

Valve open

**(b) Venous thrombosis**

Deep vein in the leg

Normal blood flow

Clot formation (thrombosis)

Clot breaks off, enters bloodstream

**FIGURE 28.11 One-Way Valves and Muscle Contractions Help Push Venous Blood toward the Heart**

The veins of the systemic circuit, which return blood to the heart, are under much lower pressure than arteries are. Many veins have one-way valves that prevent the backflow of blood. (*a*) Leg veins are surrounded by skeletal muscles, and the squeezing action created by the contraction and relaxation of these muscles helps push blood toward the heart. (*b*) Prolonged inactivity—from being bedridden or during transcontinental flights, for example—raises the risk of clot formation in the leg veins, a potentially life-threatening condition known as venous thrombosis.

Contractions of the muscles in the lower body further assist in returning venous blood to the heart. When we walk, the calf muscles squeeze the venous blood rhythmically and drive the blood upward against the pull of gravity. If we are immobile for an extended period—while recovering from an injury or illness, for example—the return of venous blood to the heart slows down. The increased tendency of blood to pool in the veins of the leg raises the risk that blood clots will form, precipitating a potentially dangerous condition called *venous thrombosis* (**FIGURE 28.11*b***). If a clot breaks off and travels to the lungs, it can block the pulmonary circuit, producing the medical emergency known as a pulmonary embolism. Being immobile for many hours on long flights also increases the risk of clot formation in the leg veins, the so-called economy class syndrome. Smokers, women who are on birth control pills or are pregnant, people with varicose veins, and those with a genetic predisposition for rapid blood clotting are at the highest risk. Passengers on flights that take 4 hours or longer are now advised to get up and walk around when they have the chance or to do seated leg exercises that stimulate the flow of venous blood from the legs to the heart.

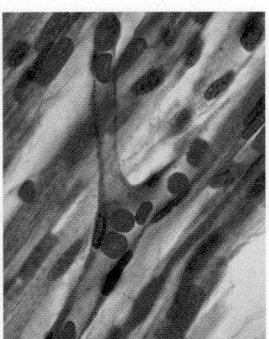

Red blood cells in a narrow capillary

## Capillaries are specialized for exchanging materials

Human circulatory systems, like those of vertebrates generally, exchange materials across the thin walls of tiny capillaries, which complete the circulatory loop by connecting arteries to veins. Closed circulatory systems provide a distribution system that can attend to the finest details at the cellular level, such as the complex exchange of water and solutes that occurs in the nephrons of the kidneys (see Figure 26.17 for a review of this process).

Compared with arteries and veins, capillaries have extremely thin, porous walls, formed exclusively of epithelial cells, across which materials diffuse easily (see Figure 28.4). In some places, the entire circumference of the capillary wall is composed of just one epithelial cell! In addition, some capillaries are so small that red blood cells must pass through them single file.

Their tiny diameters give capillaries a large surface area relative to the volume of blood passing through them. We can understand the relationship between large vessels and capillaries by picturing a telephone cable containing hundreds of small wires.

# Hypertension and Cardiovascular Disease

Hearts need to generate pressure to keep blood circulating, but sometimes that pressure can be excessive. The technical term for abnormally high blood pressure is "hypertension." Hypertension in humans can have several causes. It can result from an upset of the water and solute balance in the body. If we ingest more salt than our kidneys can readily excrete, the higher solute concentration causes more water to be drawn in and held in blood plasma and interstitial fluid. This osmotic buildup of water increases the volume of the blood plasma beyond its normal 5–6 liters. Increasing the volume of any

closed system means increasing the pressure within that system.

The buildup of fatty plaques in the arteries (see Chapter 7) can lead to atherosclerosis [*ATH-uh-roh-skluh-ROH-sus*], or "hardening of the arteries." Plaque buildup decreases the diameter of a blood vessel, which increases the pressure the heart must generate to maintain an adequate flow of blood. Over time, a plaque can rupture, triggering the formation of blood clots that in turn can completely block the flow of blood (Figure 1).

The heart responds to sustained high blood pressure as any muscle would to increased work—by bulking up. Rather than helping the situation, however, this thickening eventually interferes with the ventricle's ability to pump blood. Over many years, the heart's ability to pump blood is reduced, predisposing us to cardiovascular disease, stroke, and kidney damage.

One in three adult Americans has high blood pressure. Genes do affect our risk of developing high blood pressure, and some people have to take medication to keep their blood pressure in the healthy range (Figure 2). How-

ever, nearly everyone with high blood pressure can benefit from lifestyle changes that include regular aerobic exercise, losing excess weight, and healthy eating. Exercise makes the heart more efficient so that it takes less pressure, and fewer contractions per minute, to push enough blood through the body. Athletes commonly have blood pressures lower than 100/60, and endurance athletes often have a heart rate below 50 beats per minute. Eating a lot of salty food can increase fluid retention, raising blood pressure, especially in people with a genetic predisposition to hoard salt. People who eat more than four daily servings of nonstarchy vegetables have lower rates of heart disease than those who have one or fewer. Eating two servings of cold-water oily fish, such as salmon or sardines, per week also protects the heart. Experts say that, for better heart health, we should limit our intake of saturated fat, keeping it no more than 7 percent of the total calories we consume. As we discussed in Chapter 5, trans fats are particularly damaging to the cardiovascular system.

(*a*)

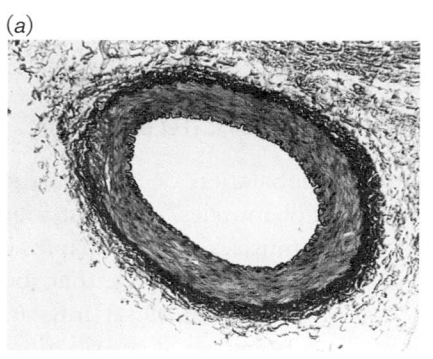

(*b*)

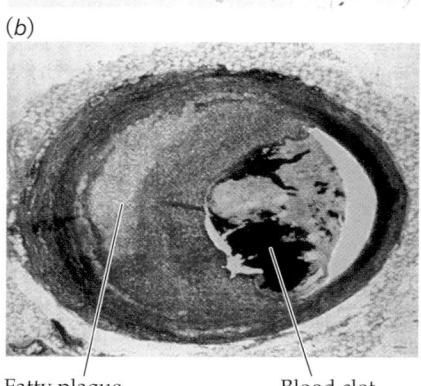

Fatty plaque          Blood clot

**FIGURE 1  Healthy and Clogged Arteries**
(*a*) Cross section of a healthy artery. (*b*) Artery dangerously clogged by plaque and a blood clot. A heart attack is likely if the arteries supplying blood to the heart muscle become clogged. Similar clogging of arteries in the brain results in a stroke, a condition in which brain tissue is killed when blocked arteries fail to deliver enough blood.

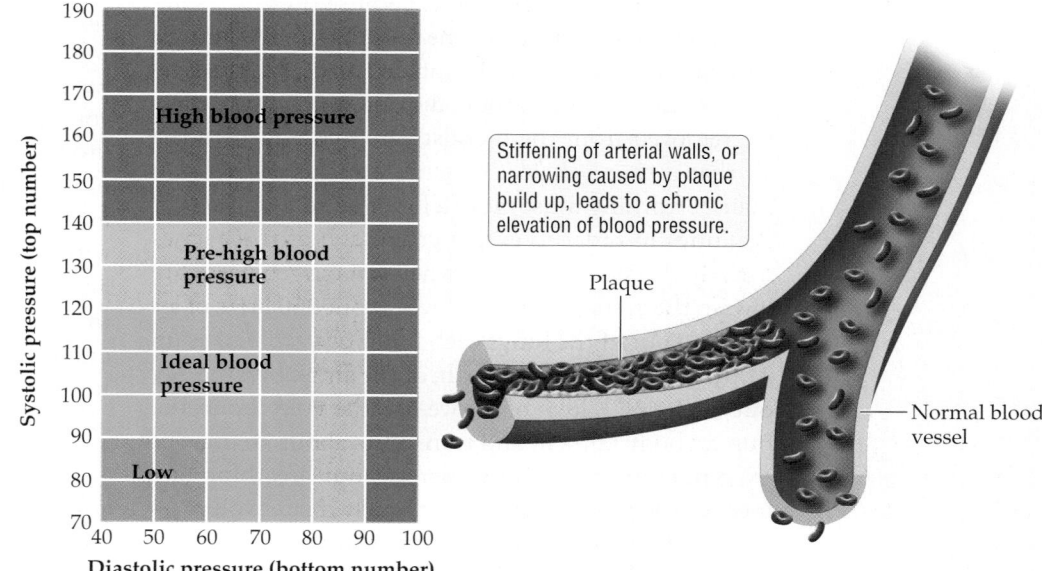

Stiffening of arterial walls, or narrowing caused by plaque build up, leads to a chronic elevation of blood pressure.

Plaque

Normal blood vessel

**FIGURE 2  Adult Blood Pressure Ranges**

If we removed all the wires from the cable and compared their volumes and surfaces, we would find that the volume of the many small wires taken together is essentially equal to the volume of the empty cable sheath, but that the combined surface area of the wires is far greater than the surface area of the sheath.

The large surface area of capillaries is critical to their function: the exchange of gases, nutrients, and other materials with the interstitial fluid surrounding the capillaries, and in turn with respiring cells in contact with the interstitial fluid. Blood flows through capillaries slowly enough to ensure adequate time for the exchange of substances with nearby cells.

> **Concept Check**
>
> 1. What structural features of a red blood cell reflect the special function of oxygen transport that this cell performs?
>
> 2. Why is the left ventricle larger than the right ventricle, and why are its walls thicker and more muscular?
>
> 3. A nurse records a patient's blood pressure as 110/70. What, in terms of heart function, do these numbers represent?
>
> 4. How are arteries and veins different?

## 28.3 Breathing in Humans

The process of taking air into the lungs (inhaling) and expelling air from them (exhaling) is called **breathing**. The air we breathe in is rich in oxygen, but the air we breathe out is rich in carbon dioxide and water vapor (water in the gas state). That is because these gases are exchanged at the surface of the cells that line our lungs: oxygen is removed from the inhaled air and sent to the bloodstream, while carbon dioxide and water vapor are removed from the bloodstream and added to the air that is then exhaled. In active animals such as humans, breathing must move a lot of air into and out of the lungs in order to exchange enough of these gases with the outside air. At rest, when our oxygen needs are lowest, the average human moves about 360 liters of air into and out of the lungs each hour. Oxygen makes up 21 percent, or about one-fifth, of the air we breathe. So, out of the 360 liters of air we breathe every hour, the human body takes in and transports about 76 liters of oxygen. During strenuous exercise, our breathing rate increases so that we can move a much greater volume—up to 6,000 liters per hour in trained athletes.

Although we could live for more than a week without food and a few days without water, a mere 4 minutes without oxygen would very likely produce irreversible brain damage, and death would follow in the next few minutes. A key part of giving first aid to someone who has stopped breathing is to "check their ABCs": Airway, Breathing, and Circulation. To check the ABCs, emergency personnel examine the pathway that gases take through the respiratory system. This pathway begins with a system of entryways, chambers, and tubes that are broadly known as the "airways" of the respiratory system.

The airways allow air to move between the external environment and the gas exchange surfaces in the lungs that lie inside the body. When we breathe in, air passes through a series of passageways in the head, neck, and chest before moving into the lungs, where the actual exchange of gases takes place. The circulation of blood then carries the gases to and from cells throughout the body. Much of the time, breathing is controlled automatically by sensory systems located in the heart and brain. If we choose, we can also control our breathing—for example, by holding the breath, or by taking rapid, shallow breaths or long, deep breaths.

## Rib muscles and the diaphragm help us breathe

How do we control an inhalation or an exhalation? We do so with a system of muscles, the most important of which are the rib muscles and the diaphragm. The diaphragm is a thick sheet of muscle that forms the floor of the chest cavity (**FIGURE 28.12**). Inhalation is made possible by the contraction of the rib muscles and the diaphragm: as these muscles contract, the rib cage moves outward and the diaphragm moves downward, increasing the volume of the chest cavity. The lungs, which lie flush against the chest wall, expand in volume as well, and as a result the pressure in the lungs drops below atmospheric pressure. Since gases will move from a region of higher pressure to a region of lower pressure, outside air rushes in and the lungs fill with oxygen-rich air. To exhale, we relax the rib muscles and the diaphragm: that action compresses the space in the chest cavity, raising the pressure inside the lungs and forcing the air out of them. With each breath, the lungs alternately draw in and force out 0.4–0.5 liter of air, depending on whether we are female or male.

## The nose and mouth are part of the upper respiratory system

The upper segment of the airways, known as the **upper respiratory system**, includes airways in the nose,

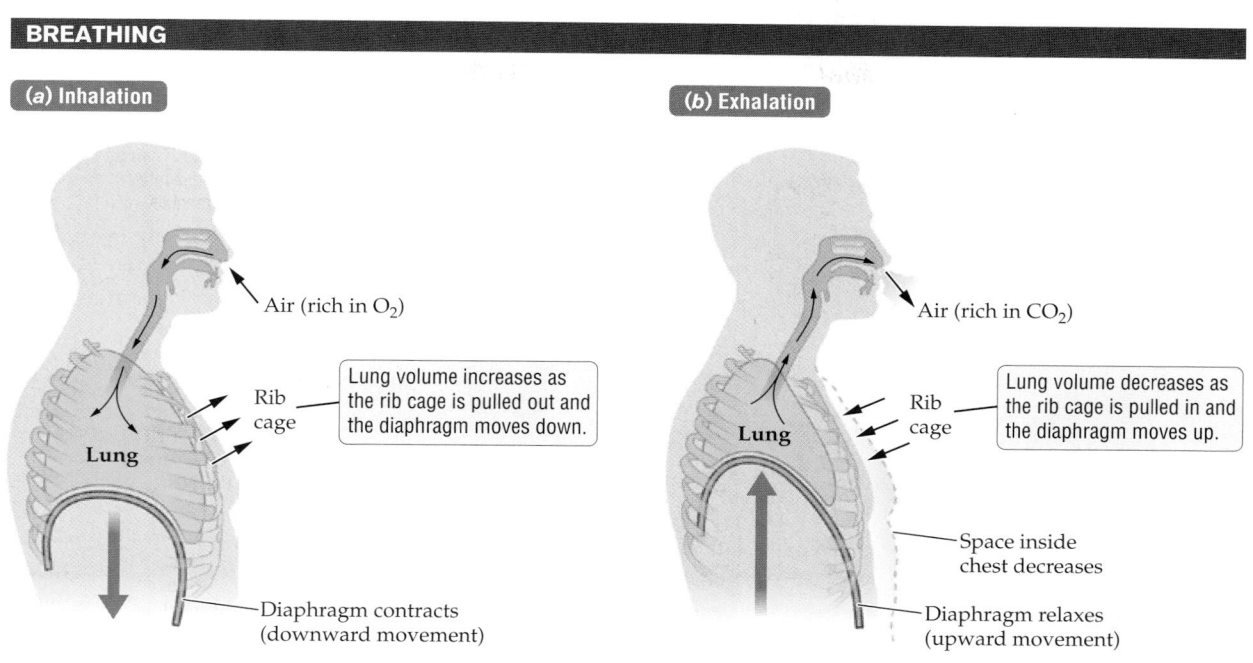

Breathing involves two main steps: (*a*) inhalation, when air is pulled into the lungs; and (*b*) exhalation, when air is pushed out of the lungs.

mouth, and pharynx (**FIGURE 28.13a**). When we inhale, air enters through the two openings of the nose and moves into each nasal cavity. The two nasal cavities are separated by a partition that is made of cartilage toward the tip of the nose and is bony toward the base. (Cartilage is a dense but flexible tissue containing large amounts of extracellular proteins such as collagen.) The walls of each nasal cavity are lined by a moist layer of surface (epithelial) cells that make up the mucous membranes. Entering air is warmed and moistened by heat and water vapor released by the mucous membranes, which are in turn richly supplied by a network of tiny blood vessels. Next the air enters the throat, or **pharynx [*FAIR*-inx]**, an area where the back of the mouth and the two nasal cavities join together into a single passageway. (When we breathe through the mouth, however, the air passes directly from the mouth into the pharynx, bypassing the nasal cavities.)

Some of the bones around the nose and in the forehead and cheeks have spacious cavities, called **sinuses**, that open into the pharynx. The hollow spaces of the sinuses act as resonating chambers when we make sounds, and the tone qualities they add are part of the reason why your friends and family members can pick out your voice from that of other people the same age and gender as you. The mucous membranes of the nasal cavities are continuous with those that line the sinuses. We experience the pain and stuffiness of *sinusitis* when these membranes become inflamed from a viral infection, allergy, or other irritation.

## The windpipe and lungs are part of the lower respiratory system

From the pharynx, air moves into the **larynx** (plural "larynges"), or voice box, which forms the entryway to the windpipe, or **trachea** (plural "tracheae"). The sound-producing structures are two shelflike extensions of cartilage, known as the vocal cords, that project from the wall of the larynx. The trachea is the largest breathing tube in our respiratory system. If you touch the front of your neck, you will feel stiff ridges under the skin. These C-shaped bands of cartilage reinforce the trachea, giving it the strength to maintain its shape during breathing. The mucous membranes lining the trachea have many microscopic hairs, or *cilia*, that trap small particles such as dust and smoke and reduce the chance that these potential irritants will pass into the delicate tissues of the lungs.

Within the chest, the trachea branches into two smaller tubes, called **bronchi [*BRAHNG*-kye]** (singular "bronchus"). Each bronchus leads to one of the paired **lungs**, the organs where gases are exchanged. Together, the trachea, bronchi, and lungs make up the **lower respiratory system** (see Figure 28.13a).

Inside the lungs, the bronchi divide into **bronchioles**, a series of branching, ever-smaller tubes (**FIGURE 28.13b**). The tiniest bronchiole extensions open into the **alveoli [al-*VEE*-oh-lye]** (singular "alveolus"), small clusters of minute sacs that resemble a bunch of grapes. Each alveolus is only about 0.05 millimeter wide but inflates to twice that volume when the lungs fill with air as we inhale. Gases are

**(a) The respiratory system**

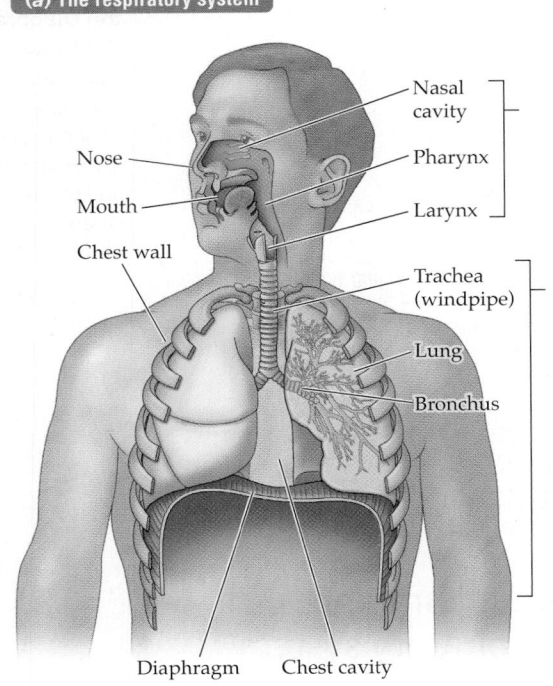

- Nasal cavity
- Nose
- Mouth
- Chest wall
- Pharynx
- Larynx
- Trachea (windpipe)
- Lung
- Bronchus
- Diaphragm
- Chest cavity

**(b) Gas exchange in the alveoli**

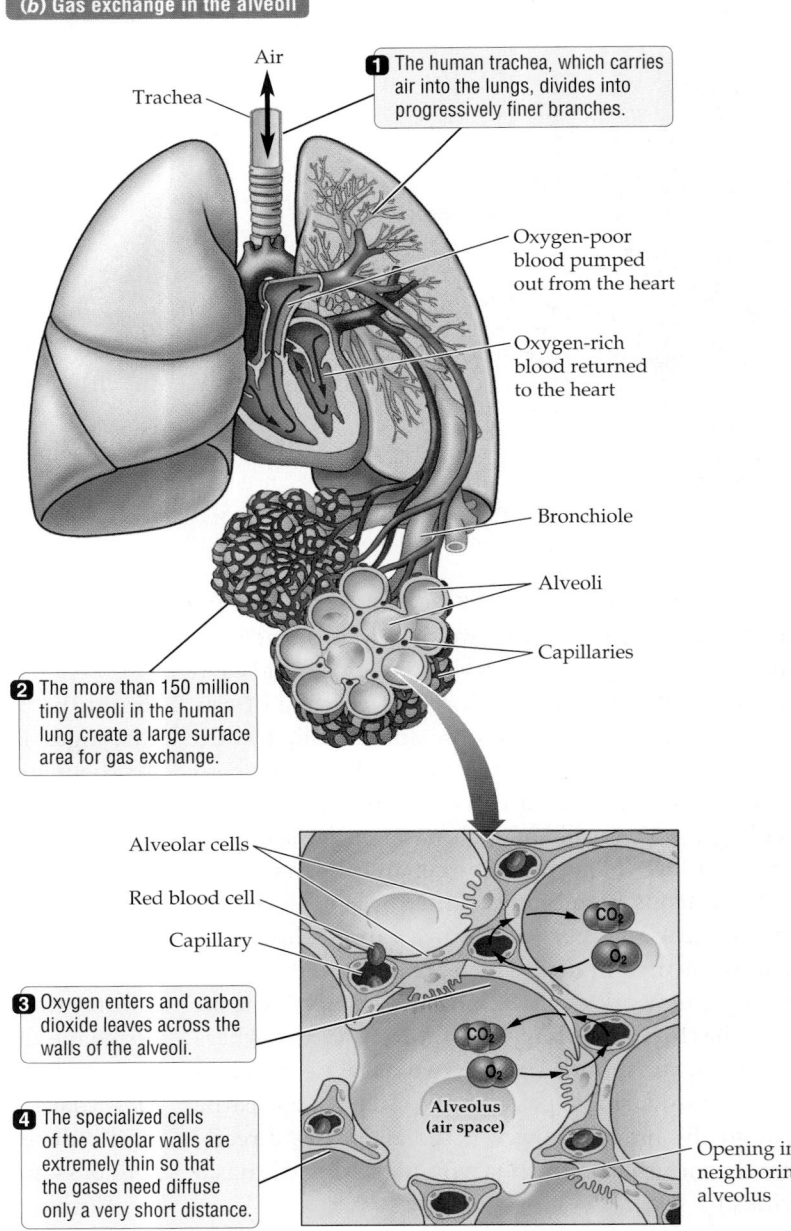

Air

Trachea

**1** The human trachea, which carries air into the lungs, divides into progressively finer branches.

Oxygen-poor blood pumped out from the heart

Oxygen-rich blood returned to the heart

Bronchiole

Alveoli

Capillaries

**2** The more than 150 million tiny alveoli in the human lung create a large surface area for gas exchange.

Alveolar cells

Red blood cell

Capillary

**3** Oxygen enters and carbon dioxide leaves across the walls of the alveoli.

CO₂

O₂

CO₂

O₂

Alveolus (air space)

Opening into neighboring alveolus

**4** The specialized cells of the alveolar walls are extremely thin so that the gases need diffuse only a very short distance.

**FIGURE 28.13  A Large Surface Area Enables Rapid Diffusion of Gases in the Lungs**

The structure of the lungs speeds the diffusion of oxygen and carbon dioxide into and out of the body by providing a large surface area for gas exchange.

exchanged across the moist surface of the thin layer of epithelial cells that line each alveolar sac. Each alveolus is wrapped in a dense network of capillaries. Oxygen and carbon dioxide are exchanged between the capillaries and the epithelial cells and then again between the epithelial cells and the air inside each alveolar sac. During an inhalation, there is an overall movement of carbon dioxide from the capillaries into the alveolar air space, and a net movement of oxygen from the air space into the epithelium and the capillaries.

After crossing the thin cells that form the walls of the alveoli, oxygen moves into the tiny blood capillaries that surround the alveoli, and from there it is carried by the blood to the heart and then distributed to respiring cells throughout the body. With each breath, a new supply of oxygen arrives at the surface of each alveolus to replace the oxygen of the previous breath that has already moved across the alveolar cells and into the blood. Damage to these delicate gas exchange surfaces—by cigarette smoking, for example—can have serious health consequences.

## 28.4 Principles of Gas Exchange

In many animals, including humans, breathing involves the muscle-driven pumping of air into and out of the lungs. However, at the surface of an alveolus or at the plasma membrane of any cell within the body, oxygen and carbon dioxide are exchanged solely through the passive process of diffusion. The three basic principles that govern diffusion help us understand the mechanisms that animals have evolved to facilitate gas exchange in diverse and sometimes quite challenging environments.

### Gases diffuse from areas of high concentration to areas of low concentration

Two important points follow from the fact that oxygen and carbon dioxide diffuse from areas of high concentration to areas of low concentration. First, the concentration of a gas at its source must be higher than that at its destination, in order for it to diffuse toward its destination. For example, oxygen diffuses into a cell if there is more of it (a higher concentration) outside the cytosol than there is inside. Second, greater differences in concentration mean larger and more rapid flow of gases from source to destination.

Let's consider the concentration gradients in the alveoli. The concentration of oxygen is higher in newly inhaled air in the alveoli than in the oxygen-poor blood arriving at the alveoli, so oxygen moves from the alveolar air (the source) into the blood (the destination). This blood, now rich in oxygen (highly oxygenated), is next carried to respiring cells by the circulatory system. The concentration of oxygen is higher in this highly oxygenated blood (now the source) than in the respiring cells (the destination), where oxygen is being consumed. So oxygen moves out of the oxygenated blood and into the cells. As oxygen moves into a cell, its cytosol becomes richer in oxygen. Mitochondria in the cell are constantly using oxygen. When the concentration of oxygen in a mitochondrion becomes lower than the concentration in the surrounding cytosol, oxygen diffuses into the mitochondrion, and cellular respiration can proceed.

The same rule holds for the journey of carbon dioxide ($CO_2$), but the direction is reversed. While using oxygen, the mitochondria are producing $CO_2$. When the concentration of $CO_2$ in a mitochondrion becomes greater than that in the cytosol, $CO_2$ diffuses out of the mitochondrion into the cytosol and then out of the cell. The concentration of $CO_2$ near respiring cells is higher than that in the blood, so $CO_2$ moves into the blood.

### A large surface area facilitates diffusion

The surface area that an organism has available for gas exchange is one factor that determines the capacity of that organism to absorb oxygen from its environment. Twice as much gas will diffuse across an area of 4 square feet than an area of 2 square feet, over the same span of time. Most animals cannot supply enough oxygen to their cells if they simply rely on diffusion across the whole body surface. In humans, for example, even if the skin were specialized for gas exchange, it would not have enough surface area to supply the oxygen needs of all our 100 trillion cells.

Although each alveolus in our lungs is extremely small, together the 150 million alveoli in a single lung provide a combined surface area for gas exchange that is 90 times the surface area of the skin. If the surfaces of the alveoli in just one lung could be spread out flat, they would cover about the same area as a tennis court. Without this specialized and very large gas exchange surface, we could never acquire enough oxygen to meet the very high demands of our large, active bodies. Like that of the alveoli in human lungs, the architecture of the specialized gas exchange structures in other animals maximizes surface area; most such gas exchange surfaces consist of folded sheets of specialized epithelial tissue that pack an immense surface area into a small space.

The oxygen demand of an animal depends on its size and its level of activity. The larger an animal is, the greater the number of cells it possesses, and therefore the greater its needs for oxygen to fuel energy-delivering cellular respiration in these many cells. A more active animal needs more energy, generated through oxygen-consuming cellular respiration, to support those high levels of activity. Consequently, an animal that is larger and more active needs a greater surface area for gas exchange than does an animal that is smaller and/or less active. For example, the surface area of the gills in fish and other aquatic animals matches the level of activity of those animals very

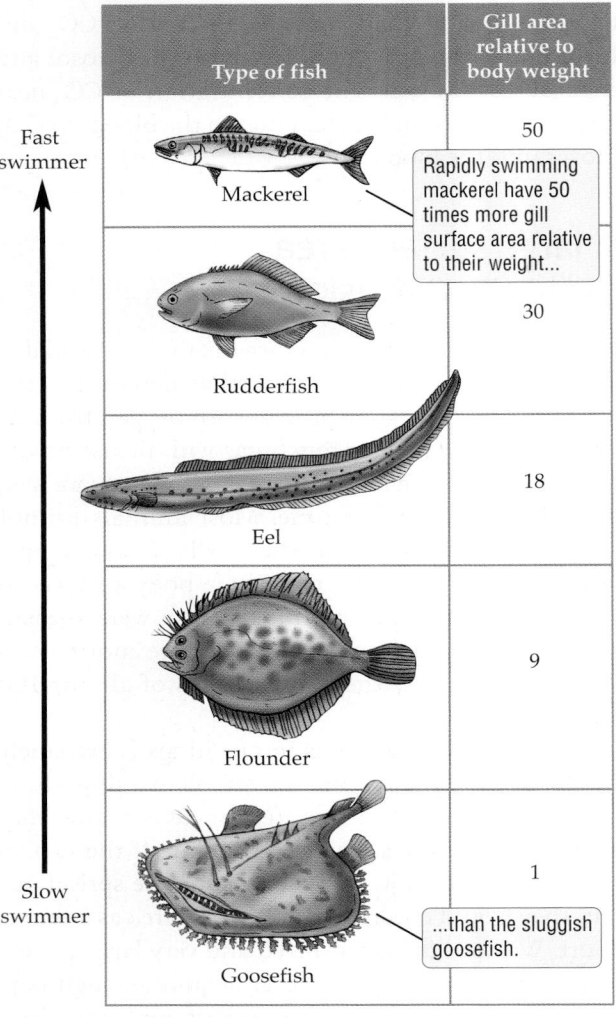

**GILL SURFACE AREA AND LEVEL OF ACTIVITY**

| Type of fish | Gill area relative to body weight |
|---|---|
| Mackerel | 50 |
| Rudderfish | 30 |
| Eel | 18 |
| Flounder | 9 |
| Goosefish | 1 |

Fast swimmer → Slow swimmer

Rapidly swimming mackerel have 50 times more gill surface area relative to their weight...

...than the sluggish goosefish.

such as those across plasma membranes or within a cell, gases and other dissolved materials, such as oxygen, can diffuse within a fraction of a second (**FIGURE 28.15**). However, even for animals that we would consider small—such as an insect a few millimeters in diameter—it can take more than a few seconds for gases to diffuse from one end of the body to the other. In biological terms, that is a long time. On the scale of a human, it would take months at that rate for oxygen to diffuse from our lungs to the tips of our toes.

Because rates of diffusion decrease with distance, most animals must have either a shape that brings all their cells within a few millimeters of their surroundings or a way of transporting gases quickly within their bodies. Very small animals such as coral polyps, or sheet-shaped animals such as flatworms, can survive by allowing $O_2$ and $CO_2$ to diffuse across the body wall and into their respiring cells, since the route is quite short. Because of their larger size and more complex organization of internal body tissues, other animals must have special mechanisms for bringing gases from the external environment to the plasma membranes of the many respiring cells deep within the body—which, as we have seen, is the purpose of the circulatory system. Many animals transfer gases from their gas exchange surfaces to blood, which is then transported under pressure to respiring cells throughout the body.

closely (**FIGURE 28.14**). As we saw in Chapter 26, endothermic animals (such as humans and dogs), which use metabolic heat to maintain high body temperatures, respire 10 times as quickly as ectothermic animals (such as alligators, frogs, and insects), which rely primarily on environmental heat. As a result, endothermic animals have a much higher oxygen demand, so they must have a larger surface area for gas exchange than ectothermic animals of similar size require.

## The shorter the diffusion distance, the faster a gas can reach its destination

Oxygen and carbon dioxide cannot diffuse over long distances fast enough to meet the metabolic needs of large animals. Over short distances (about 0.001 millimeter),

## 28.5 How Animals Transport Gases to Respiring Cells

After oxygen reaches the gas exchange surfaces (whether they are external tissues like gills or lie inside internal organs such as lungs), it must find its way to the many respiring cells in the animal body.

As we have seen, in most groups of invertebrates and vertebrates, the circulatory system distributes gases throughout the body. Blood is the fluid component in the circulatory system of all vertebrates and most invertebrates. In many invertebrates, oxygen dissolves directly in blood plasma—the liquid component of blood. However, plasma is poorly suited for transporting oxygen because water, the main component of plasma, cannot hold much dissolved oxygen; this is especially true at the high body temperatures of endotherms. Larger animals, including all vertebrates, have solved the problem of blood plasma's poor oxygen-carrying capacity by two means: oxygen-binding pigment molecules, and specialized blood cells for transporting great quantities of these molecules.

## Pigment molecules increase the efficiency of oxygen transport

One way that vertebrates have overcome the poor oxygen-carrying capacity of plasma is by employing **oxygen-binding pigments** to transport oxygen gas. What makes these pigments useful is not just their ability to bind oxygen, but also their ability to let go of it. In addition to having this capacity for **reversible binding** with oxygen, the pigments are very good at picking up oxygen in regions of high oxygen concentration and at letting oxygen go in regions of relatively low oxygen concentration. This ability to pick up or let go of oxygen, depending on the conditions—is necessary for the successful transport of oxygen from outside the body to respiring cells.

The oxygen-binding pigment used by most vertebrates, including humans, is **hemoglobin**. Each hemoglobin molecule can carry up to four oxygen molecules (**FIGURE 28.16**). Because each human red blood cell contains about 250,000 hemoglobin molecules, a single one of these cells can bind up to a million molecules of oxygen! Hemoglobin molecules contain iron atoms, and it is to these iron atoms that oxygen atoms bind. When we do not get enough iron in our diet, the body cannot produce enough hemo-

globin. In the condition known as iron deficiency anemia, the oxygen-carrying capacity of the blood is reduced, often accompanied by constant fatigue.

## Hemoglobin is carried in the blood by red blood cells

The second way that some animals, including humans, have overcome the oxygen-carrying limitations of their blood plasma is by packaging their oxygen-

### BODY SIZE AND DIFFUSION TIME

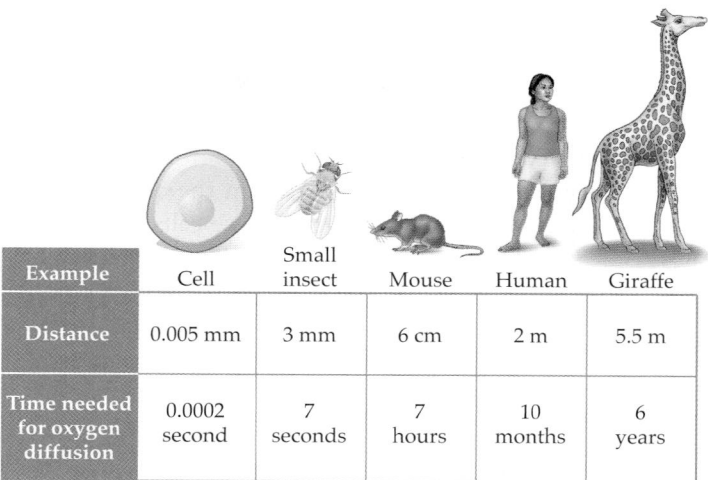

| Example | Cell | Small insect | Mouse | Human | Giraffe |
|---|---|---|---|---|---|
| Distance | 0.005 mm | 3 mm | 6 cm | 2 m | 5.5 m |
| Time needed for oxygen diffusion | 0.0002 second | 7 seconds | 7 hours | 10 months | 6 years |

**FIGURE 28.15 Diffusion Time Increases Greatly with Distance**
The times reported here are based on the speed of oxygen diffusion through water, the main component of living things.

### OXYGEN BINDING BY HEMOGLOBIN

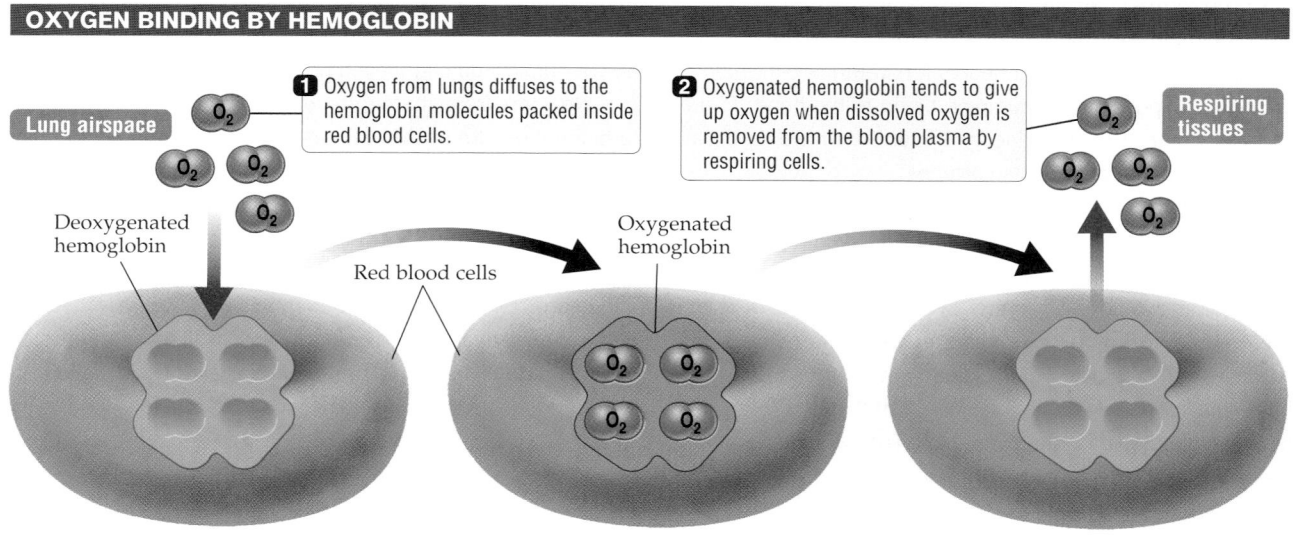

**Lung airspace**

1 Oxygen from lungs diffuses to the hemoglobin molecules packed inside red blood cells.

2 Oxygenated hemoglobin tends to give up oxygen when dissolved oxygen is removed from the blood plasma by respiring cells.

**Respiring tissues**

Deoxygenated hemoglobin

Oxygenated hemoglobin

Red blood cells

**FIGURE 28.16**
**Hemoglobin Can Carry More Blood Than Plasma Can**
Hemoglobin picks up or releases oxygen molecules depending on the amount of oxygen dissolved in the blood plasma around it.

binding pigment into cells specialized for oxygen delivery. In animals that use hemoglobin, these cells are the red blood cells. Each human red blood cell contains about 250,000 hemoglobin molecules. That adds up to a million molecules of oxygen that can bind to the hemoglobin molecules in one red blood cell! Red blood cells lack nuclei and are shaped like flattened discs that are thinner in the center than at the edges—a shape that gives these cells a larger surface area than a spherical cell of similar volume has. The thinness of the cell facilitates the rapid exchange of oxygen.

## APPLYING WHAT WE LEARNED

# Why Giraffes Have Such High Blood Pressure

A patient who came into a doctor's office with a blood pressure reading of 260/160 would shock some doctors into sending the patient to the emergency room. Yet, such readings are surprisingly common in Americans. And without other symptoms, such high blood pressure is not actually a medical emergency. Still, superhigh blood pressure can damage delicate capillary beds throughout the body and constitutes a dangerous risk factor for stroke, heart attack, arterial aneurysm, or kidney failure. Any good doctor would immediately help the patient address this severe health risk.

Yet for a giraffe, a blood pressure reading of 260/160 is normal and healthy. The animal's powerful heart easily pumps blood up to the head, which towers 4 or more meters above the ground. In fact, the longer a giraffe's neck is, the thicker is the muscular wall of the heart's left ventricle—the part of the heart responsible for pumping blood to the head. To protect the arteries from bursting, giraffe arteries have far thicker walls than comparable human arteries.

If the blood pressure at the giraffe's heart is 260 mm, think how high the pressure must be in the animal's feet and legs. A standing person with a blood pressure of 100 mm at the heart would have higher blood pressure down in the feet—about 183 mm. In a standing giraffe, the numbers are much higher. In addition to the blood pressure exerted by the heart muscle, all the blood in the giraffe's head, neck, and body weighs down on the blood in its legs, resulting in blood pressures that would rupture human vessels. How do giraffes keep the blood vessels in their legs from bursting?

Surprisingly, the solution to this problem lies in the giraffe's skin. A tight, thick sheath of skin around the legs counters the high pressures inside the blood vessels of the legs. A giraffe's tight skin acts like the pressurized flight suit a pilot wears. In pilots who lack such pressure suits, the acceleration forces generated during aerobatic rolls and dives can pull blood away from pilots' brains, causing blackouts.

In the same way, giraffes also have to limit blood pressure in their heads. If each time a giraffe bent over to take a drink at a watering hole the blood pressure in its head rose as high as the pressure in its feet, the pressure would burst the capillaries in its brain.

When you are lying down, the blood pressure in your head is nearly the same as at your heart—let's say 100 mm. If you stand up, the blood pressure in your head drops to about 50 because the heart has to work against gravity to push blood up to your head. When you bend over to touch your toes, however, gravity is working with the heart to pressurize the blood in your brain. You might even be able to feel the difference.

Now let's look at a giraffe with its head in the trees. With a blood pressure of 260 mm at its heart, the blood pressure in its head is only about 100 mm. When the giraffe leans over to drink, however, it places its head down by its feet. Without some way to reduce the pressure, the blood pressure in its head would skyrocket and burst the capillaries in its brain. Once again, giraffes have an adaptation that solves the problem: one-way valves in the long veins of the neck that stop blood from flowing back into the brain. Giraffe circulation offers a host of interesting evolutionary adaptations to a high-pressure life on the savanna.

# WADA Reports Breakthrough in Gene Doping Tests

BY STEPHEN WILSON, *Sports News*

Two groups of scientists have developed tests for gene doping in what the World Anti-Doping Agency hailed Friday as a major breakthrough in fighting the next frontier in cheating in sports.

Scientists in Germany said they have come up with a blood test that can provide "conclusive proof" of gene doping, even going back as far as 56 days from when the doping took place.

And a U.S.-French research team has devised its own method for detecting genetic doping in muscles.

The discoveries raise the possibility that a valid gene-doping test can be implemented by the 2012 London Olympics.

"This is a really significant and major breakthrough," WADA director general David Howman told The Associated on Friday in a telephone interview. "This

is a project we've been engaged in since 2002. Now we've reached the situation where we're pretty certain that it can be detected."

Gene doping is the practice of using genetic engineering to artificially enhance athletic performance. It is a spinoff of gene therapy, which alters a person's DNA to fight disease. The method is banned by WADA and the International Olympic Committee . . .

The [German] study said the test provides clear "yes or no answers" on whether DNA in blood samples has been transferred into the body to create performance-enhancing substances such as the endurance-boosting hormone EPO.

"The body of a gene-doped athlete produces the performance-enhancing hormones itself without having to intro-

duce any foreign substances to the body," Prof. Perikles Simon of Johannes Gutenberg University in Mainz said. "Over time, the body becomes its own doping supplier."

Foreign genetic material was inserted into the muscles of laboratory mice, triggering excess production of a hormone that creates new blood vessels. Two months later, researchers could still differentiate between the mice that underwent gene doping and those that didn't.

. . . [Meanwhile, the U.S.-French] researchers showed that monkeys genetically doped with EPO have an altered form of the substance in their blood. The EPO gene was injected into the monkeys' muscle, believed to be the most likely target for gene dopers.

In the highly competitive world of endurance sports, any means of improving performance offers a significant advantage. Of the 24 Tour de France winners since 1961, 13 have either tested positive for performance-enhancing drugs or admitted using them. Cyclists Bjarne Riis, Floyd Landis, Tyler Hamilton, and David Millar have all said they injected the hormone erythropoietin (EPO) [eh-*RITH*-roh-*POY*-uh-tin]. EPO, which is naturally made in the kidneys, increases the blood's oxygen-carrying capacity, making it valuable in endurance sports such as marathon races, cross-country skiing, and bicycle racing.

Human red blood cells last only about 17 weeks before they need to be replaced with new ones, which develop from stem cells in our bone marrow. EPO increases the oxygen-carrying capacity of blood by stimulating red blood cell production. In addition to increasing the production of red blood cells, EPO also stimulates the growth of capillaries that carry oxygen to tissues.

A synthetic form of EPO developed by drug companies is used to treat patients with anemia, kidney damage, and malaria. When athletes inject synthetic EPO, they can increase the concentration of red blood cells from about 45 percent to 60 percent or more. Such "blood doping" comes with health risks. Too many red blood cells can make the blood so thick that it clots or fails to flow easily through the heart. Nearly two dozen endurance athletes are thought to have died of heart attacks caused by doping with EPO.

Because EPO is a naturally occurring hormone, identifying it in blood or urine samples of athletes has been difficult. Many tests focus on whether an athlete has an exceptionally high red blood cell count. The "gene doping" described in this article presents one way for athletes to induce their own bodies to make extra EPO. But the genetically modified cells express a subtly different form of EPO that tests can detect. As the technology for doping has become more sophisticated, so has the technology for combating it.

## Evaluating the News

**1.** What is EPO, and why does it offer a performance advantage in sports, especially endurance events such as cycling and rowing? How could taking EPO kill a person?

**2.** Some cyclists increase their red blood cell counts by training at high altitudes. The low oxygen content of mountain air triggers the natural release of EPO. Other athletes have accomplished the same thing by spending time in special low-oxygen tents. Do you think either of these approaches is more acceptable than injecting either EPO or cells engineered to express EPO? Where would you draw the line, and why?

**SOURCE:** *Sports News*, Associated Press Online, September 3, 2010.

# Summary

## 28.1 The Human Cardiovascular System

- Humans and other vertebrates have a cardiovascular system, a closed circulatory system with a chambered heart that pumps blood through a complex network of blood vessels.
- The human cardiovascular system has two main circuits. In the pulmonary circuit, oxygen-deficient blood is pumped to the lungs. In the systemic circuit, oxygenated blood returning from the lungs is pumped out to body tissues.
- Arteries are large vessels that transport blood away from the heart. Veins are large vessels that carry blood back to the heart. Capillaries, the smallest vessels, exchange materials by diffusion with nearby cells.
- The vertebrate heart is composed of four chambers that make up two separate muscular pumps, each composed of an atrium and a ventricle. The left atrium and ventricle pump blood to the body; the right atrium and ventricle pump blood to the lungs.
- The heart rate is the number of times a heart beats per minute. A heartbeat, regulated by the heart's natural pacemaker, consists of one cardiac cycle, which has two phases. During diastole, the heart rests briefly; during systole, the heart muscle contracts and pumps blood.
- The human circulatory system adjusts the heart rate and patterns of blood distribution according to the body's needs.

## 28.2 Blood Vessels and Blood Flow

- Large vessels (arteries and veins) are built for mass transport of blood. Capillaries are built for slower movement of blood, and their large surface area facilitates the exchange of materials with surrounding cells.
- The walls of large vessels are thick and strong. Muscle tissue in artery walls enables arteries to contract and relax to control the flow of blood to different parts of the body. Veins have valves to keep blood flowing in one direction: back to the heart.
- In capillaries, thin, porous walls enable easy diffusion. The small diameters of capillaries produce high ratios of surface area to volume, providing a large area across which diffusion can occur.

## 28.3 Breathing in Humans

- Inhalation and exhalation are controlled by the contraction of muscles, especially muscles of the diaphragm and the rib cage.
- The human respiratory system carries air from the nose (or mouth) to the lungs through a series of tubular passageways, eventually reaching clusters of tiny sacs in the lungs called alveoli.
- Actual gas exchange takes place in the alveoli, where oxygen diffuses into the blood and carbon dioxide diffuses out of it.

## 28.4 Principles of Gas Exchange

- Oxygen and carbon dioxide enter and leave a cell, or a body fluid, exclusively by diffusion.
- Gases diffuse from areas of high concentration to areas of low concentration. This principle has two biological consequences: First, the concentration of a gas at its source must be lower than that at its destination. Second, the higher the concentration of a gas in the environment compared with that inside an organism, the faster the gas will diffuse into the organism.
- The greater the surface area available for diffusion, the more gas can diffuse across it per unit of time.
- The shorter the distance over which a gas must diffuse, the faster it can reach its destination.

## 28.5 How Animals Transport Gases to Respiring Cells

- Many groups of animals rely on a circulatory system that pumps liquid blood through the body to carry gases between gas exchange surfaces and respiring cells more quickly than is possible by diffusion alone.
- Blood plasma has a low capacity for transporting dissolved oxygen. To compensate, animals use oxygen-binding pigments, such as hemoglobin, which greatly increase the oxygen-carrying capacity of blood by means of reversible binding.
- In some animals, including humans, hemoglobin molecules are housed in red blood cells, further increasing the oxygen-carrying capacity of the blood.

# Key Terms

alveolus (p. 623)
arteriole (p. 614)
artery (p. 614)
atrioventricular (AV) node (p. 618)
atrium (p. 615)
blood (p. 613)
blood pressure (p. 617)
blood vessel (p. 612)
breathing (p. 622)
bronchiole (p. 623)
bronchus (p. 623)

capillary (p. 614)
cardiac cycle (p. 617)
cardiovascular system (p. 613)
circulatory system (p. 612)
closed circulatory system (p. 615)
diastole (p. 617)
electrocardiogram (ECG) (p. 618)
heart (p. 612)
heart rate (p. 616)
heartbeat (p. 617)
hemoglobin (p. 627)

larynx (p. 623)
lower respiratory system (p. 623)
lung (p. 623)
oxygen-binding pigment (p. 627)
pacemaker (p. 618)
pharynx (p. 623)
plasma (p. 613)
platelet (p. 613)
pulmonary circuit (p. 614)
red blood cell (p. 613)
respiratory system (p. 612)

reversible binding (p. 627)
sinoatrial (SA) node (p. 618)
sinus (p. 623)
systemic circuit (p. 614)
systole (p. 618)
trachea (p. 623)
upper respiratory system (p. 622)
vein (p. 615)
ventricle (p. 615)
venule (p. 615)

# Self-Quiz

1. Which blood vessels carry blood back toward the heart?
   a. ventricles
   b. arteries
   c. veins
   d. capillaries

2. The pulmonary circuit of the human cardiovascular system moves blood to and from the
   a. kidneys.
   b. digestive system.
   c. systemic circuit.
   d. lungs.

3. In the human cardiovascular system, blood flows from
   a. artery to atrium to ventricle to vein.
   b. vein to atrium to ventricle to artery.
   c. vein to ventricle to atrium to artery.
   d. artery to ventricle to atrium to vein.

4. During diastole,
   a. only the atria are pumping blood.
   b. only the ventricles are pumping blood.
   c. neither the atria nor the ventricles are pumping blood.
   d. both the atria and the ventricles are pumping blood.

5. The signals that establish a regular heartbeat come from the
   a. ventricles.
   b. atrium.
   c. SA node.
   d. AV node.

6. Which of the following statements is correct?
   a. The circulation to the lungs works at a lower pressure than the circulation to the body.
   b. Both atria pump blood to the lungs, and both ventricles pump blood to the body.
   c. Blood returning to the heart from the lungs carries little oxygen, while blood returning to the heart from the body carries a lot of oxygen.
   d. Both atria pump blood to the body, and both ventricles pump blood from the lungs.

7. The largest tube in our respiratory system is the
   a. bronchiole.
   b. alveolus.
   c. trachea.
   d. bronchus.

8. In humans, gas exchange takes place in the
   a. bronchioles.
   b. trachea.
   c. alveoli.
   d. bronchi.

9. The rate at which oxygen moves from the alveoli of our lungs into our blood depends on
   a. the difference in oxygen concentration between the alveoli and the blood.
   b. the distance between the alveoli and the respiring cells.
   c. the availability of energy to transport gases across the alveoli.
   d. none of the above

10. In the human respiratory system, there is net diffusion of carbon dioxide from
    a. hemoglobin to red blood cells.
    b. blood to mitochondria.
    c. alveoli to blood.
    d. blood to air in the alveoli.

11. Oxygen is transported throughout the human body primarily
    a. as a dissolved gas in the blood plasma.
    b. bound to hemoglobin molecules in red blood cells.
    c. through tracheoles.
    d. by diffusion.

12. Hemoglobin
    a. binds strongly to nitrogen gas ($N_2$).
    b. binds strongly to oxygen gas ($O_2$).
    c. is the enzyme that generates ATP.
    d. contains red blood cells.

# Analysis and Application

1. Describe the voyage of a drop of blood from the left atrium of the heart and back again.

2. Describe the cardiac cycle, and explain how regularity of the heartbeat is controlled.

3. Compare blood flow through an artery and a vein and relate the flow characteristics to the structure of these blood vessels.

4. We learned about kidney nephrons in Chapter 26 (see p. 582). Describe how high blood pressure could damage a nephron and its function.

5. Describe the human respiratory system from nose and mouth to alveoli.

6. Describe the transport of oxygen from an alveolus to a respiring cell and the movement of carbon dioxide from a cell to the air in an alveolus. Include the directions of diffusion for oxygen and carbon dioxide in an alveolus and in the blood plasma near a respiring cell.

7. Hemoglobin greatly increases the amount of oxygen our blood can carry. Describe why this is so, as well as the conditions under which oxygen is loaded and unloaded from the hemoglobin molecule.

# Animal Hormones

**INSECTS AND HORMONES.**
The beautiful swallowtail caterpillar (*Papilio machaon*) eats a diet of milk parsley. When the caterpillar has grown large enough, it forms a pupa, which then transforms into an adult butterfly under the influence of a specific set of hormones.

# Come to Pupa: Caterpillars Grow Up Fast

Butterflies have four lives. Their mothers lay a clutch of eggs, often on the underside of a leaf, where the embryonic butterflies develop into tiny caterpillars. These caterpillars hatch out with tiny bodies and powerful cutting jaws and immediately begin to eat the leaf. For days or weeks, the caterpillars devote themselves entirely to eating, shedding a skin every few days to allow their growing bodies more room to eat. When they have eaten and grown all they can, they close themselves up inside a saclike pupa. Inside the pupa, the caterpillar's stubby legs and chewing mouthparts vanish and the animal completely transforms itself into an adult butterfly—often a spectacularly beautiful one.

As adults, these same butterflies devote themselves to reproduction. Their broad, colorful wings send them fluttering great distances to find mates, nectar from flowers, and healthy plants on which to lay their eggs. The sugary flower nectar they drink is a rich source of energy for powering flight, but the nutrients that help form a butterfly's eggs come not from the nectar but from the plant diet the individual ate when it was a caterpillar.

Most people know the two active stages of a butterfly's life—caterpillar and adult—but butterflies (and many other insects) actually experience four distinct stages of development: embryo, caterpillar, pupa, and adult. The transition from one stage to another is coordinated by the release of hormones, the subject of this chapter. For example, during the pupal stage the structure of the caterpillar breaks down completely. If you open a pupa, you might find what looks like fluid. But new tissues and body parts are developing to form the adult butterfly. This dramatic change, known as metamorphosis [MEH-tuh-MOHR-fuh-sus], is coordinated by hormones.

What kind of internal changes does it take to transform a caterpillar into an adult? How do hormones facilitate this transformation?

In this chapter we will see that hormones are a group of signaling molecules that coordinate body functions at all stages in the life of an animal. Among their many functions, hormones control metabolism, water balance, and homeostasis of key ions such as calcium. In humans, and in butterflies, specific hormones orchestrate growth and all the developmental transitions in the life cycle, from birth to death.

**MAIN MESSAGE** Animals use signaling molecules called hormones to coordinate many of the functions necessary for life.

## KEY CONCEPTS

- Hormones affect almost all aspects of body function, including metabolism, nutrient homeostasis, embryonic development, growth through the juvenile stages, and sexual reproduction.

- Hormones are chemical signals that circulate in body fluids and bind to receptors on target cells to regulate the activities of those cells.

- Hormones can act locally or at a distance. They enable widely separated cells within an organism to communicate and coordinate their activities. The circulatory system carries hormones from sites of production (endocrine glands or cells) to sites of action (target cells).

- The pituitary gland produces growth hormone, which lengthens limb bones in children, among many other effects.

- Insulin increases the storage of glucose in liver cells, while glucagon causes the liver to release glucose into the bloodstream.

- Under stressful conditions, epinephrine from adrenal glands cause the liver to release glucose, providing quick energy for escape from danger.

- Calcitonin and parathyroid hormone are antagonistic hormones that regulate blood levels of calcium.

- Androgens, produced mainly in the pair of testes, control male sexual characteristics. Estrogens, made mainly in a pair of ovaries, control female characteristics.

- The menstrual cycle is launched by gonadotropins, which also trigger ovulation. Estrogens and progestogens play key roles in preparing for and maintaining a pregnancy.

# 29.1 How Hormones Work

Animal hormones are produced by specialized cells that are often organized into discrete organs called **endocrine glands**. **FIGURE 29.1** shows the location of the major endocrine glands in the human body and lists some of the hormones they produce. Unlike many other glands, endocrine glands do not have ducts—special tubes that deliver secretions from the gland directly to the site of action. Instead, endocrine glands release hormones into body fluids, such

ALL MULTICELLULAR ORGANISMS MUST COORDINATE the functions of their many specialized cell types; animals are no exception. Just as a committee of humans cannot effectively complete a task without exchanging information, the many cells, tissues, and organs in the animal body need to communicate so that the whole organism can develop and function effectively. As in most multicellular organisms, the animal body contains specialized cells whose main job is to produce signaling molecules that tell other cells what to do under specific situations or at certain times in the life cycle of the individual.

In most animals, it is the endocrine system and the nervous system, often working in close cooperation, that function as the "great communicators" at the whole-body level. The **endocrine system** consists of different types of secretory cells that release a variety of *hormones*. **Hormones** are signaling molecules produced by specialized cells and distributed through body fluids. Hormones may act locally, or they may circulate within the animal body to control the behavior of distant tissues. A distinctive feature of hormones as signaling molecules is that they are effective at very low concentrations. Hormones are also highly specific: although all the cells in the body may be awash in a particular hormone, only those cell types that have the right "communication gear" actually sense and respond to the signal.

In this chapter we explore the role that hormones play in coordinating the lives of animals, especially humans. We begin with an overview of how hormones work. Then we introduce examples of how animals use hormones to coordinate essential functions over a short time span or on a long-term basis. We close with a look at what we know about how hormones control the life cycles of butterflies and how we humans can exploit that knowledge for the biological control of insect pests.

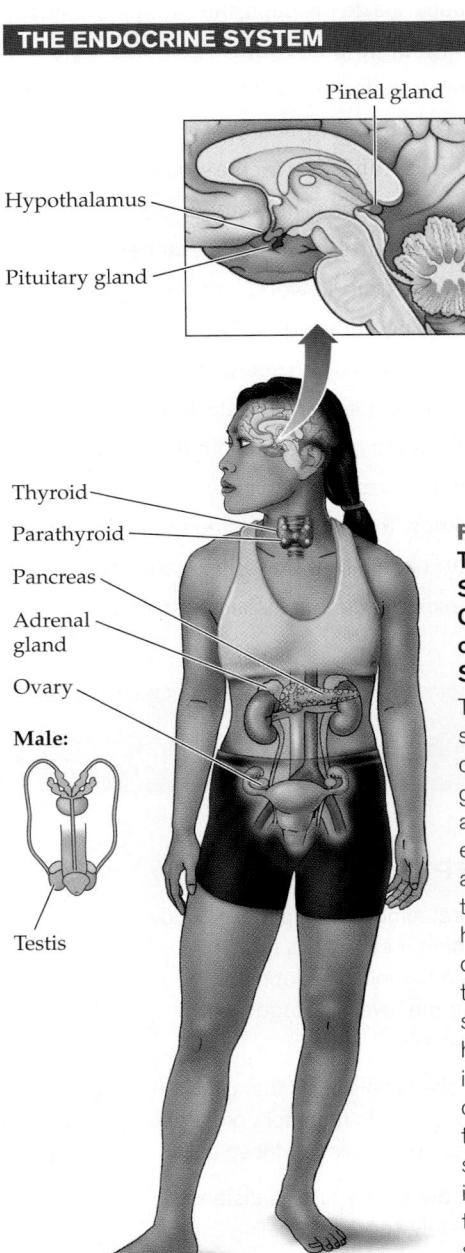

**THE ENDOCRINE SYSTEM**

**FIGURE 29.1
The Endocrine System Is Composed of Hormone-Secreting Cells**

The endocrine system consists of ductless glands, as well as scattered endocrine cells and tissues, that release hormones directly into the circulatory system. The hypothalamus is the main coordinator of the endocrine system; it also integrates the endocrine system with the nervous system.

as the bloodstream. The body fluids then carry these chemical messengers throughout the body.

Some hormone-secreting cells are not organized into distinct glands but are instead embedded as single cells or clusters of cells within other specialized tissues and organs. Endocrine cells are scattered throughout the lining of the stomach and intestine, for example. These endocrine cells secrete a variety of hormones that stimulate the release of digestive juices or regulate appetite. Because they are products of endocrine cells, these hormones enter the bloodstream instead of being secreted into the gut cavity. Some organs, such as the pancreas, function as endocrine glands and also as ducted (*exocrine*) glands. Clusters of endocrine cells inside the pancreas release hormones such as insulin directly into the bloodstream. The pancreas also has a system of tubelike ducts that discharge digestive juices into the intestine—an exocrine function.

Endocrine glands and the endocrine cells embedded in other organs together make up the endocrine system (see Figure 29.1). The *hypothalamus*, a small organ at the base of the vertebrate brain, coordinates the endocrine system and also integrates it with the other major communication highway, the nervous system (discussed in Chapter 30). The hypothalamus contains both neurons that interact with the brain and also endocrine cells that produce hormones. For example, if the water content of your blood declines, neurons in your hypothalamus will detect the change and launch both an endocrine response and a neural response. Endocrine cells in the hypothalamus will release *antidiuretic hormone*, which prods the kidney to conserve water. If the water conservation efforts are not enough to restore normal water balance, neurons in the hypothalamus will trigger pathways in the brain that will make you feel thirsty. The hypothalamus therefore deploys both neural and endocrine communication pathways to maintain water homeostasis. Many other endocrine glands are integrated with the nervous system in a similar fashion.

## Most hormones travel through the circulatory system to act on distant cells

In most animals, hormones are distributed through the body by the circulatory system (**FIGURE 29.2**). In the human body, most hormones can travel only as fast as the blood moves, which means that they take several seconds or more to arrive at their target cells.

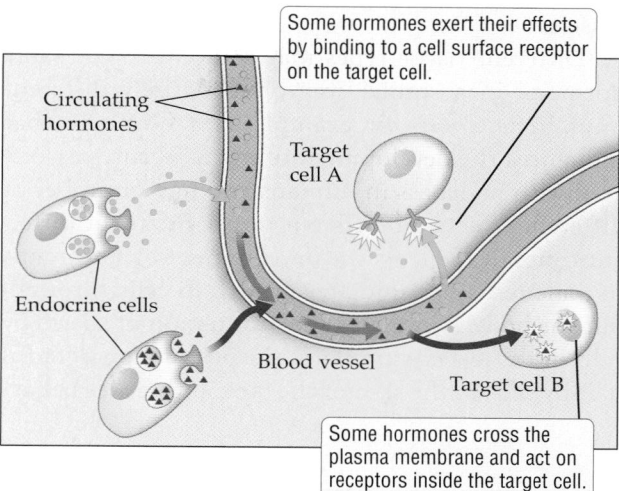

**HORMONE PRODUCTION AND TARGETING**

Some hormones exert their effects by binding to a cell surface receptor on the target cell.

Circulating hormones

Target cell A

Endocrine cells

Blood vessel

Target cell B

Some hormones cross the plasma membrane and act on receptors inside the target cell.

**FIGURE 29.2 Hormones Enable Cells to Communicate with One Another**
Hormones released by endocrine cells travel through the circulatory system to produce a response in target cells often located at a distance in the body.

A few seconds may not seem like a long time, but it means that hormones cannot coordinate activities that require very quick responses. Only the nervous system can coordinate activities, such as muscle contraction and movement, that require virtually instantaneous reactions. Hormones coordinate functions that take place over timescales of seconds to months.

Typically, hormones become greatly diluted after they are released into the circulatory system. They must therefore be able to exercise their effects at very low concentrations. Hormones are effective in small amounts because they bind to their targets with great specificity and tenacity. Despite being broadly distributed, hormones must act on specific cells, leaving nontarget cells alone.

## Each hormone triggers one or more specific responses in each target cell

A hormone released by one cell causes one or more specific responses in another cell, the target cell (see Figure 29.2). The target cells for a particular hormone may lie in distinctly different tissue types. For example, target cells for the hormone *epinephrine* are found in a wide variety of tissues, including those of the brain, heart, liver, skeletal muscles, and blood vessels.

Different tissue types may respond to the same hormone in a similar way or in entirely different ways. *Testosterone*, for example, is a hormone that influences the development of many features associated with maleness in humans and has a number of different effects on different target tissues at different times. It controls the development of sex organs in male fetuses, stimulates growth in cells throughout the body, stimulates the production of sperm by cells in the testes, stimulates the production of facial hair by cells in the skin of the face, and causes behavioral changes through interactions with cells in the brain. Testosterone enhances libido and the buildup of bone and muscle in both men and women, although the average woman produces about 50 times less of this hormone than does the average man. The multiple effects of testosterone underscore the multitasking nature of most hormones: animals need fewer distinct types of hormones because a single hormone often produces a diversity of effects in a variety of potential target cells.

Hormones can act on target cells in one of two different ways:

1. *By binding to plasma membrane receptors.* Hydrophilic ("water-loving") hormones cannot pass through the hydrophobic environment presented by the lipid bilayer of the plasma membrane (see Figure 7.12). They can exert their activity only by binding to cell surface receptor proteins embedded in the target cell's plasma membrane (**FIGURE 29.3**). Epinephrine, for example, binds to specific plasma membrane proteins (known as adrenergic receptors) found in all its target cells.

2. *By binding to intracellular receptors.* Some hormones, particularly hydrophobic molecules such as steroid hormones, can pass through the plasma membrane of a target cell to act inside that cell. Such hormones usually bind to a receptor protein in the cytoplasm and are then moved into the nucleus (see Figure 29.3), where they increase or decrease the expression of one or more genes. Testosterone, for example, exerts its effects by binding to an intracellular receptor protein (known as the androgen receptor) that then activates various cellular processes, including changes in gene expression.

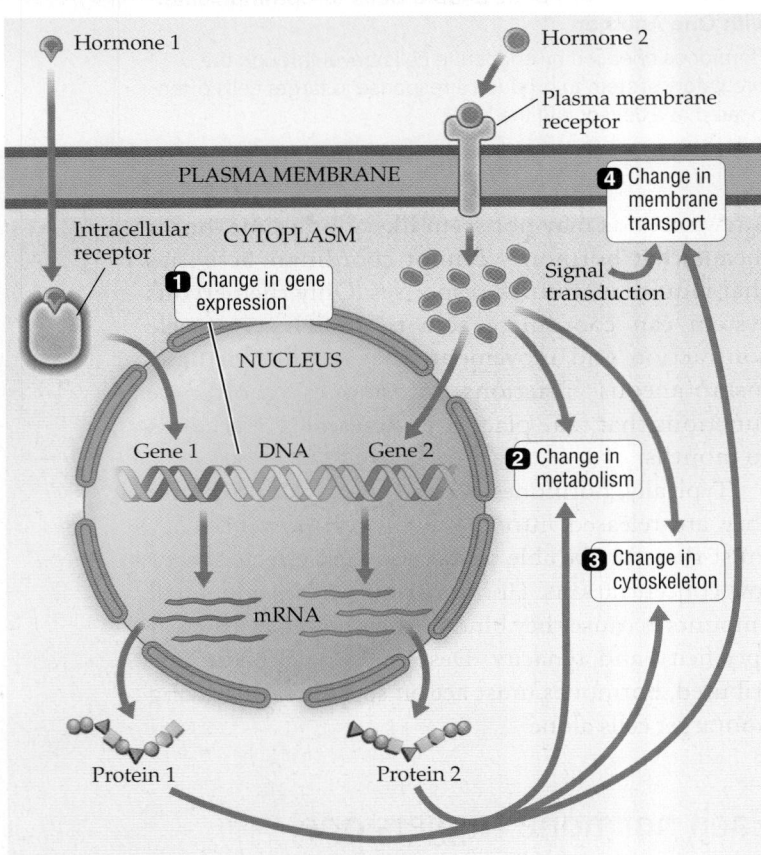

**CELLULAR RESPONSE TO HORMONES**

**FIGURE 29.3 Hormonal Signals Are Amplified within the Cell and Elicit Specific Responses in the Target Cell**
Hormones are active at very low concentrations because tiny amounts of a hormone can generate a large internal signal within a target cell. A hormone that docks to a plasma membrane receptor triggers an internal signal relay, known as a signal transduction pathway, in the course of which the original signal becomes magnified. In addition to changing gene expression (1), hormones can bring about a change in cell activity by altering metabolism (2), cytoskeletal organization (3), or membrane transport (4).

## Hormonal signals are amplified inside the target cell and alter key cellular processes

Animals usually produce hormones in tiny amounts measured in micrograms. (The smallest *single* grain of table salt we can see weighs about 10 micrograms.) Over the course of a day, the average adult human female produces no more than 200 micrograms of estrogens, a class of steroidal hormones that have many effects throughout the body. How can such minuscule doses of a hormone so dramatically regulate the function and appearance of an animal? When a single hormone molecule binds to its receptor in the target cell, it sets in motion a chain of events that may ultimately activate thousands of protein molecules in the target cell. This signal amplification means that just a few hormone molecules can have a substantial impact on a target cell. Through its effects on many such cells, a hormone can exert a profound influence on the body as a whole.

The signal received by a membrane receptor must be relayed within the cytoplasm to produce one or more responses from the target cell. This internal relay, conducted mostly by specialized molecules called second messengers and their associated proteins, is known as **signal transduction** (see Figure 29.3). Many of the steps in the signal transduction pathway are set into motion by a few ions or second-messenger molecules, but they can result in the activation of a large number of molecules, such as a metabolic enzyme. If we imagine the initial signal, the hormone, as a whisper, the signal transduction pathway turns that whisper into a roar within the cell. The receipt of just a few hormone molecules triggers a strong internal signal and launches a robust response in the target cell.

Some signal transduction pathways produce a relatively rapid response by affecting metabolic enzymes or cytoskeletal proteins (see Figure 29.3). But many hormones that bind to membrane receptors, and nearly all that bind to intracellular receptors, exert their influence by altering gene expression in the nucleus. Both types of hormones can cause a gene to be switched on or turned on more strongly; they can also turn off or turn down the expression of a gene. A single hormone can turn on a number of genes simultaneously, even while turning down the expression of other genes. Activation of a single gene results in the production of many messenger RNA (mRNA) molecules, each of which in turn codes for many molecules of protein (as discussed in Chapter 15). Therefore, the hormonal signal is further amplified at the level of gene expression.

Whether received by an extracellular or intracellular receptor, a hormonal signal can produce any one or more of these four main categories of cellular responses:

1. Changes in the expression of one, few, or many genes
2. Changes in metabolism (altered enzyme activity, for example)
3. Changes in the activity of the cytoskeleton (delivery of transport vesicles to the plasma membrane, for example)
4. Changes in membrane transport

The effects of a hormone can be rapid—occurring within seconds—if the cellular response is induced by signal transduction pathways acting directly on cellular processes such as membrane transport or vesicle fusion with the plasma membrane. However, hormone effects may take an hour or more if they are brought about through changes in gene expression.

> **Concept Check**
>
> 1. What are the characteristics of an animal hormone?
> 2. Melatonin is a water-soluble hormone, derived from the amino acid tryptophan, that controls our sleep-wake cycle. Is this hormone likely to bind an intracellular receptor? Why or why not?

## 29.2 Regulating Short-Term Processes: Glucose and Calcium Homeostasis

Hormones that help regulate homeostasis must have the capacity to act relatively quickly to maintain the constancy of the internal environment. In this section we examine how two hormones—insulin and glucagon—participate in energy homeostasis; and how other endocrine hormones regulate the fluid and tissue concentrations of a crucial mineral nutrient: calcium. We illustrate how one hormone can affect different target tissues differently, and how multiple hormones can interact to coordinate a biological process.

### Blood glucose levels are regulated by pancreatic hormones

Glucose is one of the most important sugars in animal nutrition. It is the starting point for glycolysis and cellular respiration, which generate ATP, the energy-rich molecule that is crucial for cell survival (see Chapter 9). Most animals, including humans, obtain glucose from the food they ingest. Glucose is absorbed from digested food in the small intestine, from which it travels through the circulatory system to respiring cells. Animals, including humans, can store excess glucose in the form of glycogen, a storage polysaccharide (see Figure 5.14c). Surplus glucose can also be stored as triglycerides. Commonly known as "fats," triglycerides contain three fatty acid chains held together by a glycerol molecule (see Figure 5.19). All animals must have the ability to quickly break down

## Helpful to Know

The prefixes "glyco" (meaning "sweet") and "gluco" show up in many terms relating to sugars in general or glucose in particular. Examples in this chapter include "glucagon," a hormone involved in regulating glucose; "glycogen," the storage form of glucose in the liver; and "glycolysis," the first stage in the chemical breakdown of glucose in the cytosol.

stored glycogen and triglycerides (even proteins, in times of starvation) and ship the energy-rich organic molecules that are released to where they are needed in the body. Four hormones—*insulin, glucagon, epinephrine,* and *norepinephrine*—coordinate the storage and release of energy-rich molecules in the human body.

The pancreas, in addition to producing digestive enzymes (see Chapter 27), contains clusters of endocrine cells called **islet cells** [*EYE*-let...] that produce and release insulin and glucagon. These two hormones act in opposite ways to maintain homeostasis in blood glucose levels.

Digestible carbohydrates in the food we eat are degraded when they arrive in the small intestines, and blood glucose levels start rising as the sugars released from the food are absorbed by the bloodstream. The spike in blood glucose levels is sensed by certain islet cells, which begin secreting insulin. **Insulin** is a hormone that acts on target cells throughout the body, but especially in the liver and skeletal muscles, and prompts these cells to increase their uptake of glucose from the blood. The target cells use the glucose they absorb for their energy needs and store any surplus as **glycogen** [*GLYE*-kuh-jin], a polymer made up of long chains of glucose. Insulin can also induce the target cells (especially liver cells and fat cells) to convert some of the absorbed sugars (and even amino acids) into stored triglycerides (fat).

By prodding cells into taking up glucose, insulin keeps blood glucose levels from rising excessively. Insulin helps ensure that we store extra glucose and other carbon-based nutrients rather than losing them by eliminating them from the body. When levels of glucose in the blood drop, the pancreas produces less insulin. The blood glucose level has a homeostatic set point that shuts off the production of insulin when a threshold low concentration is reached—an example of negative feedback at work (see Chapter 26 for a review of feedback control).

Other islet cells produce and release **glucagon**, a hormone that induces cells to release glucose from storage depots within the cell. When blood glucose falls to low levels, in addition to secreting less insulin the pancreas releases more glucagon, thereby stimulating cells in the liver to convert stored glycogen into glucose and release it into the bloodstream. **FIGURE 29.4** illustrates the insulin-glucose-glucagon relationship. As glycogen stores are depleted, which

**FIGURE 29.4**
**Balancing Levels of Glucose in the Blood**

Two hormones—insulin and glucagon—work in opposite ways to regulate blood levels of glucose. The width of the red circuit indicates the relative amounts of glucose in the bloodstream at different times.

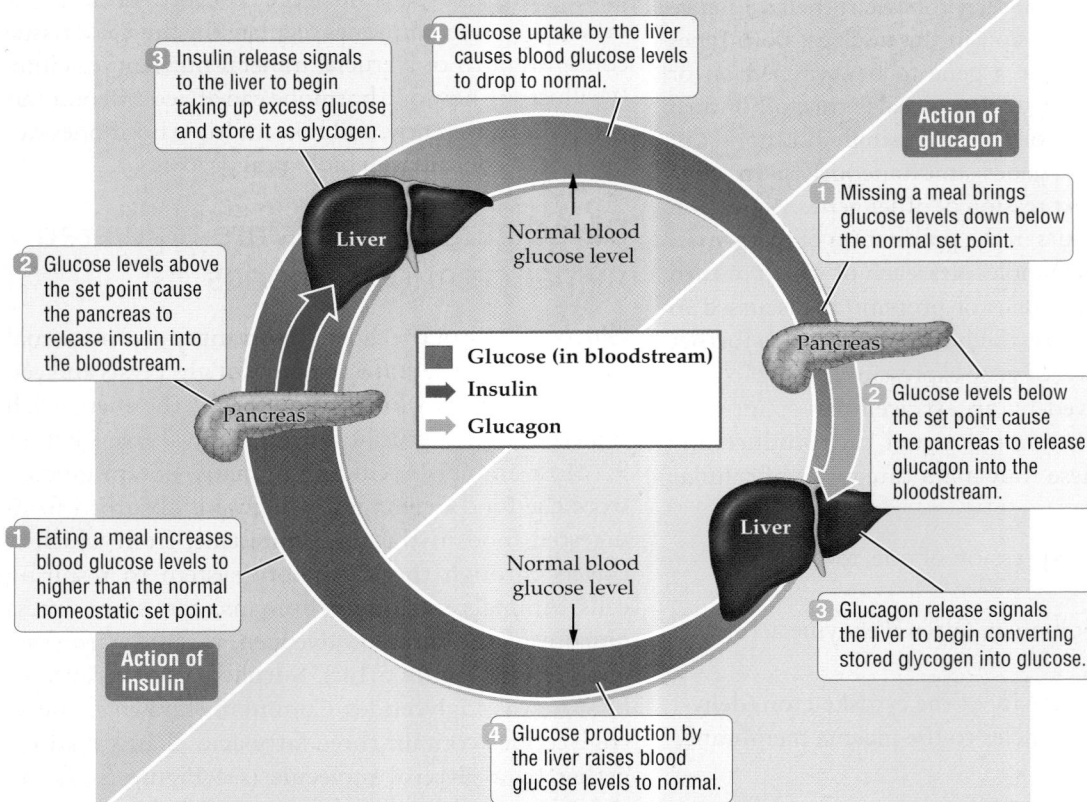

**GLUCOSE HOMEOSTASIS**

3 Insulin release signals to the liver to begin taking up excess glucose and store it as glycogen.

4 Glucose uptake by the liver causes blood glucose levels to drop to normal.

**Action of glucagon**

1 Missing a meal brings glucose levels down below the normal set point.

2 Glucose levels above the set point cause the pancreas to release insulin into the bloodstream.

Liver

Normal blood glucose level

Pancreas

Pancreas

■ Glucose (in bloodstream)
➡ Insulin
⇨ Glucagon

2 Glucose levels below the set point cause the pancreas to release glucagon into the bloodstream.

Liver

1 Eating a meal increases blood glucose levels to higher than the normal homeostatic set point.

Normal blood glucose level

**Action of insulin**

3 Glucagon release signals the liver to begin converting stored glycogen into glucose.

4 Glucose production by the liver raises blood glucose levels to normal.

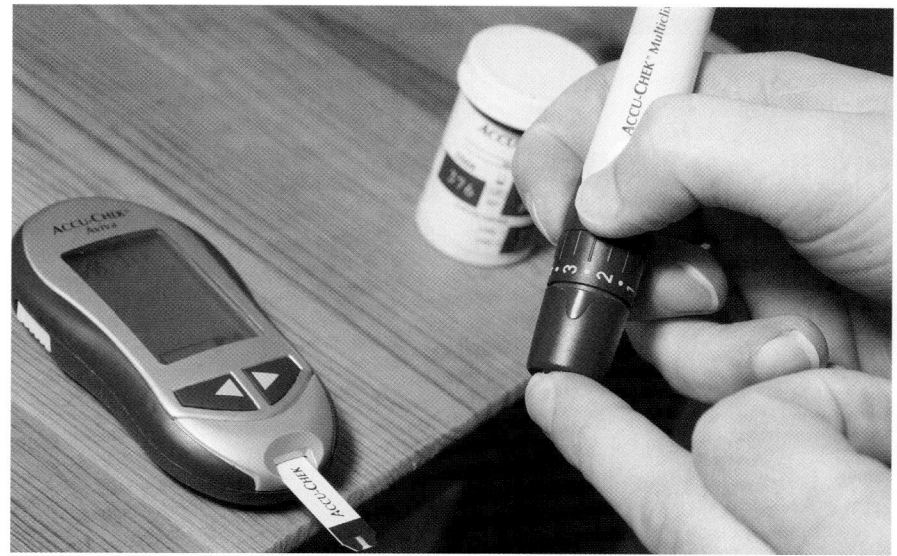

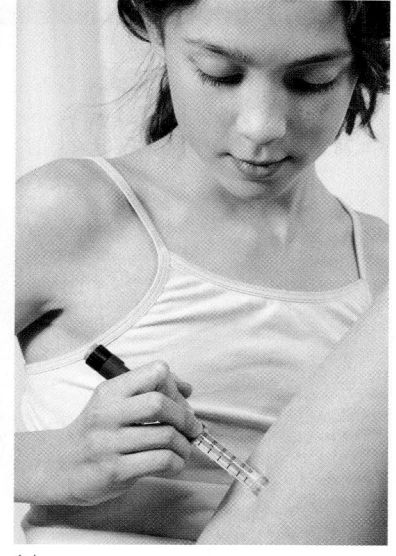

(a)

(b)

FIGURE 29.5
**Compensating for a Defective Pancreas**
(*a*) People with diabetes must monitor their blood glucose levels, such as with this glucometer. (*b*) Patients with insulin-dependent diabetes must inject measured quantities of insulin to prevent abnormally high levels of blood glucose.

can happen in about 20 minutes of vigorous exercise, glucagon also stimulates the breakdown of triglycerides to release fatty acids that can be "burned" through cellular respiration to meet the body's energy needs.

Blood always needs to carry some glucose, but too much glucose in the blood can be harmful. In a condition called **diabetes [dye-uh-BEE-deez]**, not enough glucose is moved from the blood into the cells. In **type 1 diabetes** (also known as juvenile-onset diabetes), either damage to the islet cells prevents insulin production altogether or the pancreas produces a defective form of insulin that cannot bind to receptor proteins on the surfaces of target cells. Type 1 diabetes is usually an inherited condition, and those diagnosed with it must receive insulin injections for the rest of their lives. In **type 2 diabetes**, either the pancreas produces too little insulin, or the receptors on target cells respond poorly to insulin. Over time, the resulting high levels of glucose in the blood damage capillaries, causing poor circulation and slowing the healing of injured tissues. Damage to blood vessels in the eyes can cause vision problems (in severe cases, blindness).

Thanks to our understanding of the biology of insulin and our ability to produce the hormone in commercial quantities through genetic engineering techniques (see Section 16.4), most people with diabetes can now monitor and manage their own glucose levels by modifying their diet and injecting additional insulin as needed (**FIGURE 29.5**). In addition, by eating well and exercising, people can reduce their chances of developing type 2 diabetes, which is closely linked with accumulating excess body fat.

## The adrenal glands trigger fight-or-flight responses

The **adrenal glands** are a pair of endocrine glands that sit on top of the kidneys. They release **epinephrine [EH-puh-NEH-frun]** (also known as adrenaline) and **norepinephrine [NOHR-eh-puh-NEH-frun]**, a pair of similar hormones that coordinate our response to sudden stress. The hormones are released in response to nerve signals from the brain that report on the body's stress levels or warn of danger (**FIGURE 29.6**). The release of the hormones launches a number of rapid physiological responses, which include boosting blood glucose levels.

Upon release from the adrenal glands, the hormones travel through the bloodstream until they reach their target cells. While broadly similar in their structure and function, the two hormones vary in terms of the precise tissues they target and the precise responses they elicit. Epinephrine stimulates glycogen breakdown in liver and skeletal muscle cells, causing glucose to be released into the bloodstream. The hormone also speeds up the heartbeat, and the force with which the heart contracts, so that glucose is delivered throughout the body more rapidly. In this way, glucose becomes available to fuel a rapid response to a stressful situation.

If we see a rattlesnake in front of us, ready to strike, we are likely to jump back or at least freeze in place, with the heart racing. We can thank the nervous system and the adrenal hormones for the quick response. The nervous system processes the visual information and transmits an alarm signal to the adrenal glands

### Helpful to Know

Some hormones go by more than one name. In Britain, epinephrine and norepinephrine are called adrenaline and noradrenaline, respectively. "Adrenaline" is not the official name of the hormone in the United States, because that label is very similar to a registered trademark owned by a pharmaceutical company. Nevertheless, "adrenaline" is popular in ordinary English, even in the United States, as in "adrenaline rush," the sensation evoked by an exciting or potentially dangerous situation.

**FIGURE 29.6**

**Adrenal Hormones Produce a Rapid Response to Stress**

The adrenal glands produce epinephrine (adrenaline) and norepinephrine (noradrenaline), which trigger rapid release and delivery of stored energy.

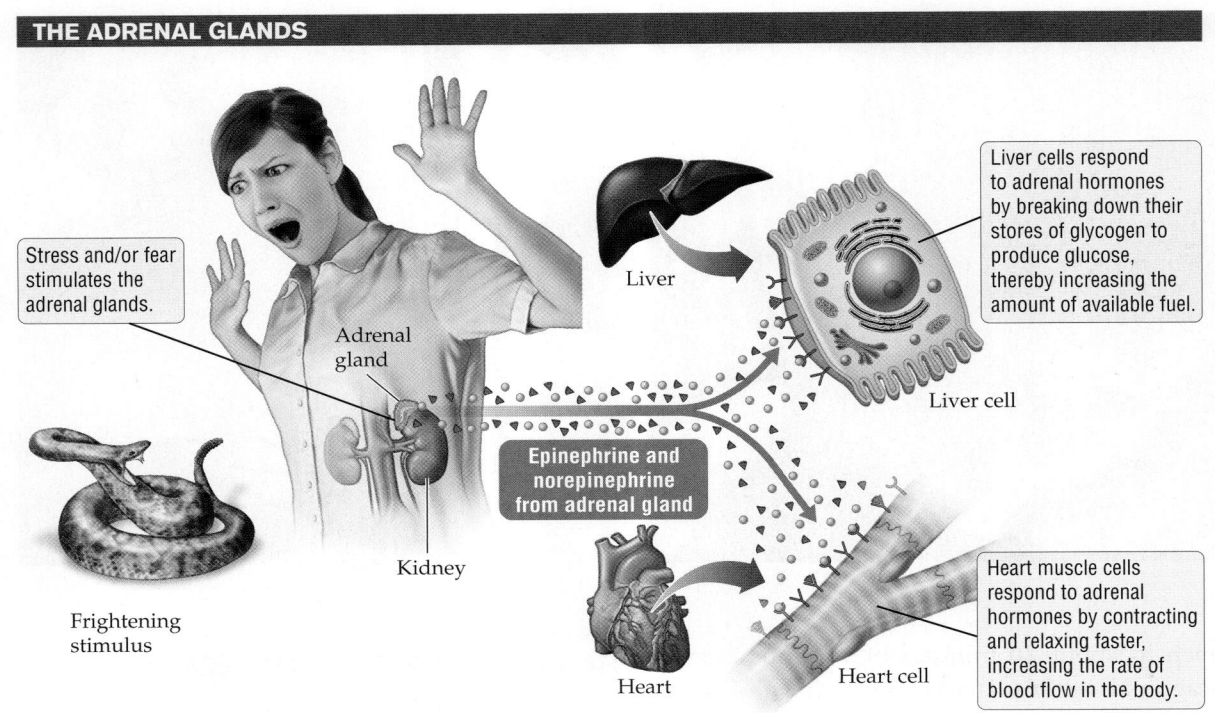

**THE ADRENAL GLANDS**

Stress and/or fear stimulates the adrenal glands.

Adrenal gland

Kidney

Frightening stimulus

Liver

Liver cell

Epinephrine and norepinephrine from adrenal gland

Liver cells respond to adrenal hormones by breaking down their stores of glycogen to produce glucose, thereby increasing the amount of available fuel.

Heart

Heart cell

Heart muscle cells respond to adrenal hormones by contracting and relaxing faster, increasing the rate of blood flow in the body.

within a fraction of a second. The adrenal glands kick in right away, pouring epinephrine and norepinephrine into the blood. Within just a few seconds, these hormones increase the pumping of blood and trigger the release of glucose, all the better to support the next move: arming ourselves with a stout stick or running away quickly.

## Calcium is regulated by hormones from the thyroid and parathyroid glands

In addition to regulating energy-supplying nutrients such as glucose, hormones coordinate the uptake and release of mineral nutrients, such as calcium. Calcium in a mineral form is a major component of human bones. In blood and other body fluids, and within the cell, it exists as calcium ions, $Ca^{2+}$. Calcium ions are important in the function of muscles and nerves. Two antagonistic hormones—*calcitonin* and *parathyroid hormone*—regulate the amount of calcium circulating in the human bloodstream. (Antagonistic hormones work in opposing ways. Insulin and glucagon are also antagonistic hormones.)

An excess of calcium in the blood triggers the **thyroid gland**, an endocrine gland in the neck, to release *calcitonin*. **Calcitonin** removes calcium from the blood by promoting its storage in bones and by stimulating the kidneys to reabsorb less calcium, allowing more calcium to be lost in urine. When there is too little calcium in the blood, the **parathyroid glands**, which consist of at least four patches of tissues lying behind the thyroid gland, release **parathyroid hormone** (**PTH**). PTH stimulates the release of calcium from bones and increases the reabsorption of calcium in the kidneys, so that less is excreted from the body in the urine. The hormone also activates vitamin D, which increases the uptake of calcium from food in the intestines. **FIGURE 29.7** illustrates how calcitonin and PTH maintain proper levels of calcium in the blood while storing the excess and strengthening bones. Negative feedback comes into play with each hormone: when either hormone causes blood calcium to reach a normal level, the gland producing that hormone slows down or stops.

## 29.3 Regulating Long-Term Processes: Growth

Hormones regulate such long-term processes as the overall growth of animals: when and how much an animal grows. The pituitary gland produces a hormone

FIGURE 29.7
**Balancing Calcium Levels in the Blood**
The thyroid and parathyroid glands secrete hormones that control calcium levels in the blood. The width of the red circuit indicates the relative levels of calcium in the bloodstream at different times.

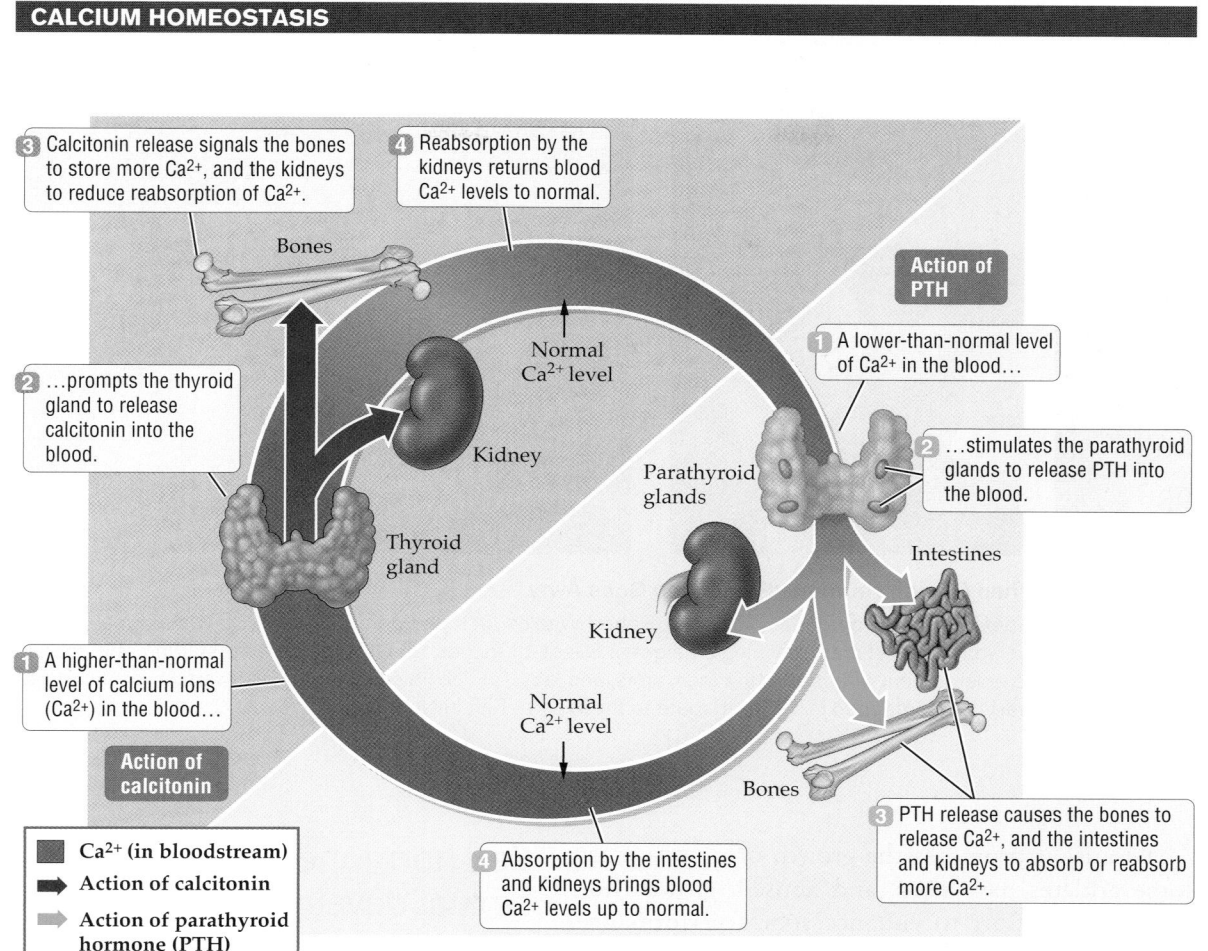

3 Calcitonin release signals the bones to store more Ca²⁺, and the kidneys to reduce reabsorption of Ca²⁺.

4 Reabsorption by the kidneys returns blood Ca²⁺ levels to normal.

**Action of PTH**

Bones

2 ...prompts the thyroid gland to release calcitonin into the blood.

Normal Ca²⁺ level

Kidney

1 A lower-than-normal level of Ca²⁺ in the blood...

2 ...stimulates the parathyroid glands to release PTH into the blood.

Parathyroid glands

Thyroid gland

Kidney

Intestines

1 A higher-than-normal level of calcium ions (Ca²⁺) in the blood...

Normal Ca²⁺ level

**Action of calcitonin**

Bones

3 PTH release causes the bones to release Ca²⁺, and the intestines and kidneys to absorb or reabsorb more Ca²⁺.

■ Ca²⁺ (in bloodstream)
➡ Action of calcitonin
➡ Action of parathyroid hormone (PTH)

4 Absorption by the intestines and kidneys brings blood Ca²⁺ levels up to normal.

appropriately called **growth hormone** (GH). Its main target cells lie in bones and in muscles associated with bones. GH promotes the growth of bones and stimulates increases in muscle mass. Normally, the human pituitary gland produces GH in amounts that promote adequate growth during childhood and puberty, but the output drops when we become adults and continues to decline with age.

Too much or too little GH can have lifelong consequences. If too much growth hormone is present before a person reaches adulthood, the result is gigantism (**FIGURE 29.8a**). People with this condition can grow to be more than 8 feet tall. On the other hand, if the pituitary begins producing excess growth hormone after a person has reached adulthood, the person will not become a giant, because the growth centers in the bones that contribute the most to giving us our height shut down during puberty. Instead, bones in other parts of the body (in particular, the hands, feet, and parts of the face and

head) will begin to grow again, becoming enlarged. This condition is called acromegaly [ak-roh-*MEH*-guh-lee] (**FIGURE 29.8b**).

Other conditions can develop if the body receives too little GH. Dwarfism results when too little GH is available as growth progresses during childhood and adolescence (**FIGURE 29.8c**). Adult dwarfs are usually under 4 feet tall. Growth hormone supplements are now being manufactured by processes like those described in Chapter 16, using genetically engineered bacteria. Under careful medical guidance, young people producing too little GH—whether simply shorter than normal, or potential dwarfs—can be helped to increase their height somewhat.

Although the use of growth hormone in the production of beef, pork, and lamb is controversial, no connections have been found linking these uses with human health problems. But there is an increasing demand for organic meats from livestock that have been raised without the use of added hormones.

**(a) A pituitary giant**

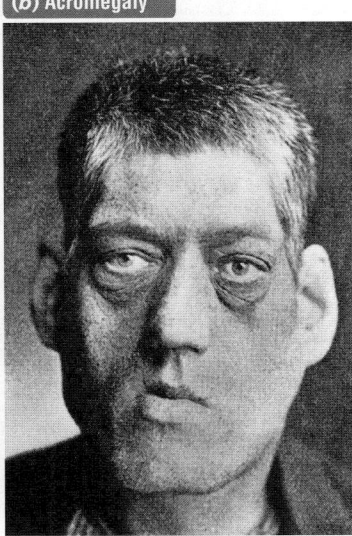

**(b) Acromegaly**

**(c) Pituitary dwarfs**

**FIGURE 29.8 When Growth Hormone Production Goes Awry**
(a) Too much GH before puberty produces gigantism. Robert Wadlow, in a photograph from 1938 with normal-sized women, was reportedly the world's tallest man, reaching a height of 8 feet 11 inches. (b) GH production restarting after maturity results in acromegaly, the enlargement of bones in the face, arms, and legs. (c) Too little GH produces dwarfism, as is the case with these pituitary dwarfs photographed in 1934 en route to the World's Fair in New York.

Because it promotes the growth of muscle mass, some athletes have used, and abused, growth hormone in an effort to enhance sports performance. Used in this way, growth hormone can have serious side effects, including increased risk of cancer and heart failure.

## 29.4 Regulating Long-Term Processes: Reproduction

Animals rely on hormones to regulate nearly all aspects of reproduction, from mating behaviors to the development and birth of offspring. In humans, hormones influence nearly all aspects of sexual development and reproduction. The emergence of sex-specific characteristics in the fetus, and the maturation of reproductive organs during puberty, are examples of long-term aspects of reproduction that are controlled by hormones. The regular stimulation of sperm production in males, and the monthly cycle of hormonal changes that control menstruation in females, are other physiological processes that are regulated by hormones.

## Sex hormones play a role in sexual development before birth

Hormones influence human sexual development even before birth. The genetic program for developing into one sex or the other in humans is determined by the presence or absence of a Y chromosome. In mammals, females are XX and males are XY (as discussed in Chapter 13). It is through the action of hormones circulating in its blood, however, that a fetus develops into either a male or a female. If a human fetus has a Y chromosome, testes (the male reproductive organs) begin to form just 4–6 weeks after fertilization of the egg. In the absence of a Y chromosome, ovaries (the female reproductive organs) develop instead. These sex glands—testes and ovaries—are more generally called **gonads**. By the seventh week of the fetus's development, its gonads are producing steroid hormones specific to its sex. These **sex hormones** signal genes in their target cells to begin the process of sexual development.

The gonads produce three major types of hormones: estrogens, progestogens, and androgens. Both males and females produce all three, but in different ratios: males have more androgens than estrogens, and females have more estrogens than androgens, for example. **Estrogens** play a role in determining female characteristics such as wide hips, a voice that is pitched higher than that of males, and the development of breast tissues. **Progestogens** have a number of functions in the female body, including thickening the lining of the uterus and increasing the blood supply to it to create a suitable environment for a developing fetus. **Progesterone** is the most important of the progestogens. **Androgens** (*andro*, "male") stimulate cells to develop the characteristics of maleness, such as beard growth and the production of sperm. The primary androgen is **testosterone**.

The testes secrete three types of hormones that together coordinate the development of male reproductive structures. Testosterone, together with another closely related androgen, directs the development of internal reproductive structures such as the sperm ducts and prostate gland; the third androgen directs the development of external structures such as the penis. Similarly, the development of all female reproductive structures, both internal and external, falls under the control of certain estrogens.

## Sex hormones coordinate sexual maturation at puberty

At an age of about 10–13 years, young humans begin making the transition to sexually mature adults capable of reproduction. During this transition, called **puberty**, the levels of sex hormones produced by the gonads rise markedly. Moreover, the hypothalamus activates the production of two other hormones by the **pituitary gland**, a double-lobed gland that lies at the base of the brain. These two hormones—**luteinizing hormone (LH)** and **follicle-stimulating hormone (FSH)**—are referred to as the **gonadotropins** [goh-NAD-uh-TROH-pinz]. Together they coordinate the development of sperm in males and play a role in regulating the menstrual cycle in females. Acting in concert with growth hormone, the sex hormones also control the development of other characteristics associated with sexual maturity: a deepening voice, the development of facial and pubic hair, and growth of the penis and scrotum in males; and the development of breasts and pubic hair, as well as changes in body shape, in females.

Gonadotropins help maintain the functioning of the reproductive organs and glands throughout an individual's lifetime. In males, gonadotropins further stimulate the production of sperm in the presence of testosterone. At birth, the ovaries of a female already contain her entire lifetime supply of immature egg cells (primary oocytes). These cells remain in a state of suspended development until the production of gonadotropins at puberty stimulates one, or occasionally two, of them to mature fully each month in preparation for ovulation (described next).

## Sex hormones coordinate the menstrual cycle of human females

In human males, sperm are produced continually form puberty to old age. Human females, however, do not produce mature eggs continually. Instead, individual eggs mature and are released in a hormone-driven sequence of events known as the **menstrual cycle**. The menstrual cycle averages about 28 days, but cycle lengths from 21 to 35 days are considered normal. **FIGURE 29.9** summarizes the fluctuations in hormone levels that drive this cycle.

A menstrual cycle begins with the first day of bleeding, which marks the end of the previous cycle. Over the next few weeks, a succession of hormones stimulates the release of an egg and prods the uterine lining to grow and thicken in preparation for a potential pregnancy. If the pregnancy fails to materialize, hormone levels plummet and the lining is sloughed off as menstrual flow, ending that menstrual cycle. We examine this sequence of events in greater detail next.

A new menstrual cycle begins as gonadotropins released by the pituitary trigger follicle development in the ovaries. An *ovarian follicle* is a cluster of cells with a large immature egg in the center. Under the influence of FSH, the follicle enlarges and the primary oocyte inside begins to mature into an egg. As the follicle develops, it secretes increasing amounts of estrogens. Estrogen levels peak in about 12 days, causing a massive and sudden release of LH by the pituitary gland. The LH surge triggers **ovulation**, the release of the egg from the follicle. The cells of the ruptured follicle, which remain behind in the ovary after ovulation, now develop into a **corpus luteum** (Latin for "yellow body"). The same spike in gonadotropin levels that triggered ovulation also stimulates the corpus luteum to begin secreting large amounts of a progestogen called progesterone and smaller amounts of estrogen.

The lining of the uterus thickens in response to rising levels of estrogen, and now progesterone as well. If fertilization does occur, the embryonic tissues will produce a hormone (*human chorionic gonadotropin, hCG*) that maintains the corpus luteum. Progesterone is made, first by the corpus luteum and later by the

**FIGURE 29.9 The Human Menstrual Cycle Depends on the Sequential Release of Several Hormones**

Follicle-stimulating hormone (FSH) launches a new menstrual cycle by stimulating the growth of ovarian follicles. Developing follicles produce estrogen. When estrogen levels reach a certain threshold, they trigger release of luteinizing hormone (LH) from the pituitary, which triggers ovulation. An estrogen, especially the large amounts of progesterone secreted by the corpus luteum, stimulates a buildup of the uterine lining in preparation for a possible pregnancy. If fertilization fails to occur, the corpus luteum dies after about 14 days, hormone concentrations crash, and the uterine lining is sloughed off.

## HORMONAL CONTROL OF THE MENSTUAL CYCLE

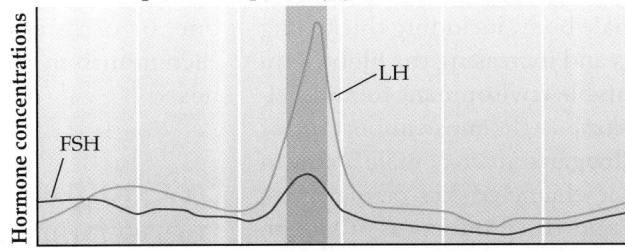

Gonadotropins (from pituitary gland)

FSH released by the pituitary triggers follicle development.

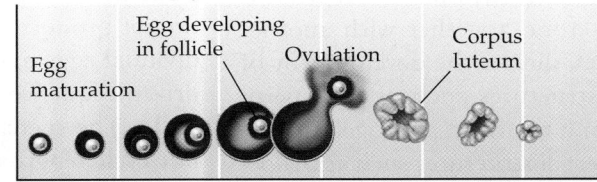

Events in the ovary

The developing follicle releases estrogen.

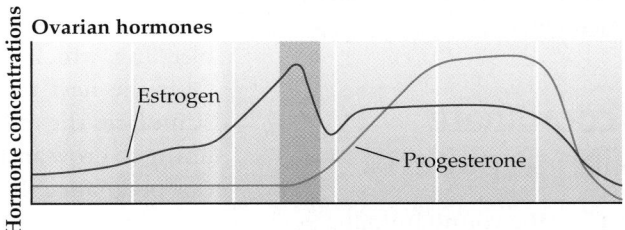

Ovarian hormones

High concentrations of estrogen lead to a surge in LH, which induces ovulation.

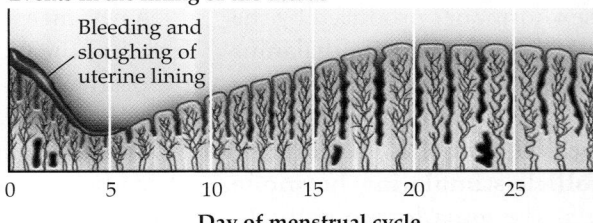

Events in the lining of the uterus

Estrogen and progesterone produced by the corpus luteum thicken the lining of the uterus. If fertilization does not take place, the corpus luteum breaks down, shutting down the supply of hormones. As a consequence, the uterine lining is sloughed off.

placenta, throughout pregnancy and is necessary for maintaining the lining of the uterus. One reason birth control pills usually contain synthetic hormones that mimic progesterone is that these mimics trick the body into "thinking" it is already pregnant and therefore blocking the monthly cycle of egg maturation and ovulation (see Section 33.4). A successful pregnancy is marked by high levels of both estrogens and proges-

terone, which not only maintain the pregnancy but also turn off the now unnecessary rounds of monthly egg maturation and release by inhibiting the release of gonadotropins from the pituitary.

If the egg is not fertilized, no embryo develops; and without the embryonic hormones to maintain it, the corpus luteum starts to break down. As a result, progesterone levels drop. Without high levels of progesterone to support them, the new blood vessels and thickened lining of the uterus do not last long. They separate from the uterus and are expelled from the body during menstruation.

High levels of estrogen and progesterone inhibit FSH production through a negative feedback loop, but the concentrations of these hormones crash as the corpus luteum breaks down. The pituitary is therefore no longer inhibited by these hormones; it begins releasing FSH again and a new cycle begins.

After about 500 cycles of menstruation, a change in hormonal events occurs that is, in some ways, the reverse of puberty. Typically when a woman reaches between 40 and 50 years of age, her ovaries begin to produce less of the estrogen hormones and progesterone. Eventually the drop in the levels of these hormones results in **menopause**, in which the menstrual cycle ceases permanently.

> ### Concept Check

1. Shortly after you eat a large slice of pizza, your blood insulin levels rise. Why?

2. Explain whether the following statement is true or false and why: Some adult athletes use human growth hormone to gain a height advantage in their sport.

---

**APPLYING WHAT WE LEARNED**

# Metamorphosis: A Hormonal Symphony

The development of a human from a fertilized egg to adulthood requires that thousands of genes turn on and off in sequence from conception to adulthood. The genetic orchestration needed to guide the development of a caterpillar into a butterfly is equally complex and, just as in other animals, much of the development is orchestrated by hormones.

Adult butterflies develop from clusters of cells, called imaginal discs, that reside within the growing caterpillar and, later, the pupa. During the pupal stage, the structure of the caterpillar body is literally disassembled. The cells of the imaginal discs then begin to divide and grow to form adult structures, scavenging materials from the broken-down tissues and cells of the former caterpillar. Without hormonal coordination, the disassembly of the caterpillar and subsequent assembly of the adult during metamorphosis would lead to a hopeless mess instead of a functioning butterfly.

Amazingly, just three hormones coordinate all this activity (**FIGURE 29.10**). Molting hormone (ecdysone), which is released by the prothoracic glands, plays different roles at different stages. For example, during the caterpillar stage, molting hormone stimulates molting, the repeated shedding of the outer layers protecting the caterpillar's body (the exoskeleton). During the pupal stage, molting hormone promotes the development of the imaginal discs.

A second hormone, juvenile hormone, which is released by glands just behind the caterpillar's brain, controls the outcome of the molt. If the concentration of juvenile hormone is high, the next stage will be a larger caterpillar. If the concentration of juvenile hormone is low, the next stage will be a pupa. When the concentration of juvenile hormone drops to zero, the pupa develops into an adult.

The timing of events during metamorphosis depends both on hormones and on signals from the nervous system. When the nervous system detects that a caterpillar has

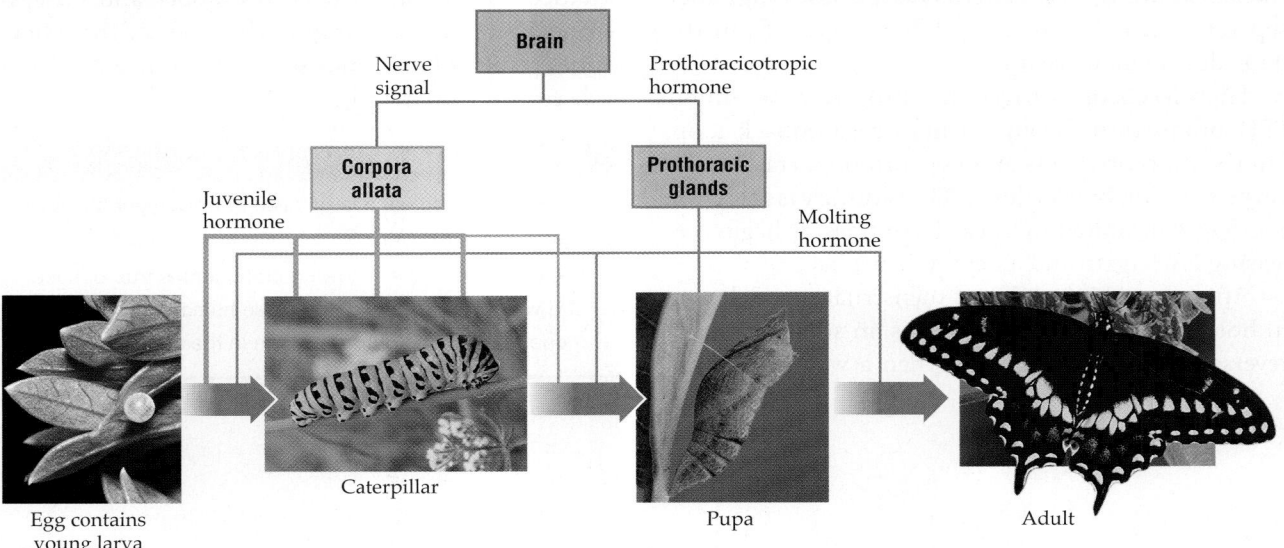

**FIGURE 29.10  Three Hormones Interact to Coordinate Butterfly Metamorphosis**
Hormone-producing cells link the nervous and endocrine systems to control the timing of events in metamorphosis. The release of molting hormone in response to the release of prothoracicotropic hormone determines when the insect sheds its exoskeleton. The amount of juvenile hormone released by the corpora allata (singular "corpus allatum") determines whether the individual takes the form of a caterpillar, pupa, or adult following the molt.

reached the proper stage of development, the brain stops the release of juvenile hormone, triggering metamorphosis. The timing of the release of molting hormone depends on a more complex series of events. The brain initiates the release of a third hormone—called prothoracicotropic [**proh-thuh-***RAS*-**ih-koh-***TROH*-**pik**] hormone—that stimulates the prothoracic glands to release molting hormone—but only if two conditions are met: (1) The level of juvenile hormone indicates to target cells in the brain that the caterpillar has reached the right stage of development to begin metamorphosis. (2) Information reaching the brain indicates that the release of prothoracicotropic hormone will cause the molt to the pupal stage to take place at a safe time of day.

Three factors enable butterflies to coordinate metamorphosis: a combination of interacting chemical messages; target tissues that respond differently to each hormone; and hormones whose function is integrated with the nervous system.

Applied researchers have used knowledge of insect hormones to develop ways to kill fleas, cockroaches, and agricultural pests. These researchers have found molecules that mimic or block insect hormones and so keep the insects from completing their life cycle. For example, *S*-methoprene, a product used to kill fleas and ticks, acts like juvenile hormone. When the pupae of fleas and ticks are exposed to *S*-methoprene, they become so overloaded with synthetic juvenile hormone that they are unable to molt into adults. Hydroprene, another synthetic juvenile hormone, keeps immature cockroaches from growing and shedding their exoskeleton, preventing them from developing into adults.

Still another insecticide, azadirachtin, has a chemical structure like that of molting hormone; this chemical seems to prevent the production and release of the insect's own molting hormone. In doing so, azadirachtin blocks the metamorphosis of the caterpillar into a butterfly. It works against cabbage worm caterpillars, aphids, whiteflies, mealybugs, gypsy moths, and many beetles.

# Earlier Hormone Therapy Elevates Breast Cancer Risk, Study Says

BY DENISE GRADY, *New York Times*

Growing evidence about the risks of breast cancer and other serious illnesses posed by hormone therapy for menopause has led many women to give up the drugs, and many doctors to stop recommending them.

But there has been a lingering belief that for younger women in the early stages of menopause, hormone risks may be negligible, at least for a while . . . Some researchers are even testing an idea, called the timing hypothesis, that starting hormone treatment early in menopause may help protect women from heart disease.

Now, information from a huge study in Britain suggests that the women thought to be at the lowest risk from hormones may actually be at the highest risk, at least when it comes to breast cancer. The study found that women with the greatest risk of breast cancer from hormones were those who took

them earliest—before or soon after menopause began.

The new findings, published on Friday in The Journal of the National Cancer Institute, are not the strongest type of evidence: They do not come from a randomized trial, an experiment in which people are picked at random to take either a drug or a placebo and are then studied and compared over time. Instead, the British study was observational, meaning that the women being studied had made their own decisions about whether and when to take hormones. There is always a chance, in observational studies, that there is some underlying difference between people who choose to take drugs and those who do not, and that the difference—not just the drugs they took—may help account for different health outcomes. Observational findings are sometimes disproved by randomized trials.

But this particular observational study also has a unique strength—it included more than a million postmenopausal women, one in four British women who were aged 50 to 64 during the enrollment period, from May 1996 to December 2001.

The research, called the Million Women Study, found that in women ages 50 to 59 who had never taken hormones, 0.3 percent a year developed breast cancer. The rate was higher, 0.46 percent a year, in women who started taking the most commonly used hormones—estrogens combined with progestin—five or more years after menopause began. But it was highest of all—0.61 percent a year—in women who started taking the drugs before or less than five years after menopause began. And the risk was increased even in women who took the drugs for less than five years.

Until recently, hormone replacement therapy (HRT) was a widely prescribed treatment for the symptoms of menopause. Menopause, as discussed in this chapter, is a normal stage in human development and aging. Nonetheless, the physiological changes of menopause include an increased risk of heart disease and osteoporosis [AHS-tee-oh-puh-ROH-sus] and a loss of calcium from the bones, which can lead to fractures in the hips and backbone. Some women also experience unpleasant symptoms such as headaches, hot flashes, and poor sleep that decrease the quality of life.

Beginning in the 1960s, doctors began prescribing hormone pills to millions of menopausal women in an attempt to reduce hot flashes and other symptoms. Most of the pills contained either estrogens or a combination of estrogen and progestin. In some cases, the hormones were prescribed for only a few months, while in others women continued to take them long past menopause, into their sixties.

In recent decades, studies have shown that long-term exposure to these hormones significantly increases the risk of heart attack by 29 percent, the risk of stroke by 41 percent, and the risk of breast cancer by 26 percent. In 2002, the U.S. National Heart, Lung, and Blood Institute halted a large clinical study of women taking estrogen-plus-progestogen HRT when re-

searchers discovered these increased risks. The treatment also had benefits, including a 37 percent decrease in cases of colon cancer and a 33 percent reduction in hip fractures. Overall, there was no difference in the death rate between women on the estrogen-plus-progestogen treatment and those taking a placebo pill. The greater risk of heart attack, stroke, and breast cancer was countered by a reduced risk of other causes of death.

## Evaluating the News

**1.** Name one long-term function regulated by estrogen and one medium-term function regulated by progestin described in the main text of this chapter.

**2.** HRT seeks to replace compounds that the body no longer produces. Such treatment would seem relatively innocuous. Why might the long-term use of HRT cause health problems?

**3.** Can you think of other treatments that involve significant benefits as well as significant risks? What would you want to know from your doctor before following such a treatment?

**SOURCE:** *New York Times*, January 28, 2011, http://www.nytimes.com.

# Summary

## 29.1 How Hormones Work

- A hormone is a signaling molecule that is distributed through the body by the circulatory system. Hormones are active at low concentrations. A single hormone may affect many different kinds of target cells, potentially triggering a different response in each.
- Hormones act on target cells either by moving through the plasma membrane to the cell's interior or by acting on receptors embedded in the plasma membrane.
- Hormones typically move through the circulatory system to reach distant target cells. Because they can move only as quickly as the blood moves, they tend to coordinate slower, longer-term functions than the nervous system.
- The hormonal signal is usually amplified in target cells; the binding of a single hormone molecule to its receptor may activate thousands of protein molecules in the target cell.
- Animal hormones are produced by groups of specialized hormone-secreting cells within other kinds of tissues, as well as in endocrine glands, which release their products directly into the bloodstream. The glands and specialized cells together make up the endocrine system.
- The hypothalamus coordinates the endocrine system and integrates it with the nervous system.

## 29.2 Regulating Short-Term Processes: Glucose and Calcium Homeostasis

- Hormones often act together to coordinate biological processes. A pair of hormones with opposite effects is said to be antagonistic.
- A pair of antagonistic hormones produced by islet cells in the pancreas—insulin and glucagon—regulate glucose levels in human blood. Insulin increases the storage of glucose in liver cells, while glucagon causes the liver to release glucose into the bloodstream.
- Under stressful conditions, the adrenal glands produce epinephrine and norepinephrine, which cause the liver to release glucose, providing quick energy for escape from danger.
- In the disease diabetes, the pancreas produces either too little insulin or insulin that does not bind effectively to its target cells. The high levels of blood glucose that result can damage tissues.
- Calcitonin and parathyroid hormone (PTH) are antagonistic hormones that regulate blood levels of calcium. Calcitonin (produced by the thyroid gland) lowers calcium levels by promoting its storage in bones and by stimulating the kidneys to reabsorb less calcium. PTH, released by the parathyroid gland, raises calcium levels by stimulating bones to release calcium, by prompting the kidneys to reabsorb more calcium, and by causing the intestines to take up more calcium during digestion.

## 29.3 Regulating Long-Term Processes: Growth

- The pituitary gland produces growth hormone (GH), which promotes the growth of bones and stimulates increases in muscle mass.
- Too much GH present in the body during childhood and puberty results in gigantism. Exposure to GH after puberty results in acromegaly, a condition in which bones in the face, hands, and feet begin to grow again. Too little GH during childhood and puberty results in dwarfism.
- Young people who naturally produce too little GH can now receive GH supplements derived from genetically engineered bacteria.

## 29.4 Regulating Long-Term Processes: Reproduction

- In humans, hormones influence both long-term and medium-term aspects of sexual development, pregnancy, and birth.
- Sex hormones guide the development of a fetus's sexual characteristics according to its genotype (XX for a female; XY for a male). By the seventh week of embryonic development, the sex organs start producing steroid hormones specific to the sexual genotype.
- The gonads (testes in males, ovaries in females) produce three major types of hormones: estrogens stimulate the development of female characteristics, progestogens prepare the uterine wall to support a developing fetus during pregnancy, and androgens stimulate the development of male characteristics. Both men and women produce all three types of sex hormones, but in different proportions.
- At puberty, humans make the transition to reproductive maturity. Puberty is caused by rising levels of sex hormones and by the production of gonadotropins, two hormones that coordinate the development of sperm in males and eggs in females.
- Produced by the pituitary gland, gonadotropins also help maintain the functioning of the reproductive organs and glands throughout an individual's lifetime.
- Hormones coordinate the menstrual cycle of human females. Individual eggs mature and are released in a 28-day menstrual cycle. During the first 2 weeks of the menstrual cycle, the immature egg completes its development in a follicle in the ovary, prodded by follicle-stimulating hormone (FSH) released from the pituitary. Developing follicles secrete estrogen, whose levels increase steadily.
- High levels of estrogen stimulate a massive, sudden release of luteinizing hormone (LH), which triggers ovulation, the release of the egg from the ovary. The remains of the "empty" follicle become the corpus luteum, which begins releasing large amounts of progesterone.
- If the egg is fertilized, progesterone levels remain high throughout the pregnancy, keeping the uterine lining well developed to nourish the developing fetus. If the egg is not fertilized, no embryo develops, and about 12 days after the egg's release, progesterone levels drop, initiating menstruation, in which the additional blood vessels and thickened lining of the uterus are expelled from the body.
- High levels of estrogen and progesterone inhibit FSH production through negative feedback. When the levels of these hormones decline, the pituitary starts secreting FSH, and the cycle begins again.

# Key Terms

adrenal gland (p. 639)
androgen (p. 643)
calcitonin (p. 640)
corpus luteum (p. 643)
diabetes (p. 639)
endocrine gland (p. 634)
endocrine system (p. 634)
epinephrine (p. 639)
estrogen (p. 642)

follicle-stimulating hormone (FSH) (p. 643)
glucagon (p. 638)
glycogen (p. 638)
gonad (p. 642)
gonadotropin (p. 643)
growth hormone (GH) (p. 641)
hormone (p. 634)
insulin (p. 638)

islet cell (p. 638)
luteinizing hormone (LH) (p. 643)
menopause (p. 645)
menstrual cycle (p. 643)
norepinephrine (p. 639)
ovulation (p. 643)
parathyroid gland (p. 640)
parathyroid hormone (PTH) (p. 640)
pituitary gland (p. 643)

progesterone (p. 643)
progestogen (p. 643)
puberty (p. 643)
sex hormone (p. 642)
signal transduction (p. 637)
testosterone (p. 643)
thyroid gland (p. 640)
type 1 diabetes (p. 639)
type 2 diabetes (p. 639)

# Self-Quiz

1. Which of the following is true of target cells?
   a. They are endocrine cells that release hormones.
   b. They can amplify the chemical message carried by hormones.
   c. Some have cell surface receptor proteins embedded in the plasma membrane that bind a specific hormone.
   d. both b and c

2. Endocrine glands in humans secrete hormones directly into the
   a. kidneys.
   b. pancreas.
   c. bloodstream.
   d. hypothalamus.

3. Which of the following is an endocrine gland?
   a. thyroid gland
   b. adrenal gland
   c. pituitary gland
   d. all of the above

4. Which one of the following increases the uptake of glucose by the liver?
   a. glucagon
   b. insulin
   c. calcitonin
   d. parathyroid hormone

5. Islet cells in the pancreas produce
   a. epinephrine.
   b. norepinephrine.
   c. insulin.
   d. growth hormone.

6. Adrenal glands secrete
   a. epinephrine.
   b. calcitonin.
   c. glucagon.
   d. insulin.

7. Estrogens play a role in
   a. developing female characteristics.
   b. promoting beard growth.
   c. regulating butterfly metamorphosis.
   d. stimulating sperm production.

8. The pituitary gland produces
   a. glucagon.
   b. insulin.
   c. gonadotropins.
   d. calcitonin.

9. Progesterone prepares the uterus for
   a. producing gonadotropins.
   b. receiving a fertilized egg.
   c. initiating puberty.
   d. producing more testosterone.

10. Acromegaly results from
    a. not enough growth hormone after puberty is complete.
    b. not enough growth hormone during puberty.
    c. too much growth hormone before and during puberty.
    d. too much growth hormone after puberty is complete.

# Analysis and Application

1. Describe the journey that insulin takes from where it is produced in the body to its target cells, and explain what it does when it reaches its destination.

2. Testosterone has many effects. Describe how it affects human males.

3. Hormones can work quickly, but not as quickly as the nervous system. Describe how your adrenal glands and nervous system would work together if you suddenly came face-to-face with an angry, snarling dog.

4. During the menstrual cycle in human females, fluctuating levels of several hormones work together to create a series of events that repeats about every 28 days. If you were "inventing" an animal like a female white-tailed deer, which breeds just once a year, how would you design a hormonal program for it? Describe the hormones involved and the timing of the shifts in their levels.

5. Growth hormone has lifelong effects. What are three possible consequences of abnormalities in the production of growth hormone in humans?

# 30 Nervous and Sensory Systems

**EYE ON THE FUTURE.** Biomedical engineer Shawn Kelly holds a prosthetic eye—or "retinal implant"—that hooks up to the nervous system and may enable a blind person to see.

# A Vision of the Future

When Terry Byland first began having trouble driving at night, he figured he had a minor problem the eye doctor would fix. He was shocked when his doctor told him he was going blind and there was no cure. Seven years later, he was completely blind; at 45, his whole world was dark, a result of a rare genetic disease that destroys the light-sensing cells in the retina of the eye. Angry and disappointed with his life, it took him years to accept his fate.

Then, a decade after he went blind, he heard about an experimental device that might enable him to see again. A small camera mounted on a pair of dark glasses scans the world and transmits images to the few light-sensing nerve cells that remain in the retinas of the eyes. Nerve cells in the eye can, in turn, stimulate the optic nerve, which leads directly to the brain.

Researcher Mark Humayun of the University of Southern California wasn't promising Byland anything. For one thing, the quality of the signal coming into Byland's brain was limited both by the number of pixels transmitted by the prosthetic device's sensor and by the number of nerves left in the retinas of Byland's eyes. In addition, if the brain doesn't receive information from, say, the eyes or a hand for a long time, it can lose its ability to interpret signals from that part of the body. After 10 sightless years, it was possible that Byland's brain had lost the knack for interpreting visual information. Even if the implant worked perfectly, his brain might have to learn how to see again.

The implant couldn't guarantee that he would see again, but Byland was hungry for the chance to see anything at all and to lead the way for others who might benefit from the research.

Is it possible to create an artificial eye that can help people see? How does the eye detect and transmit images to the brain and how does the brain interpret those signals?

Before we address these questions, we will explore the workings of the sensory system and the rest of the nervous system, which, along with the endocrine system, coordinates the functioning of the body.

---

**MAIN MESSAGE** The animal nervous system rapidly transmits and integrates information and initiates responses to that information.

---

## KEY CONCEPTS

- The two main components of the vertebrate nervous system are the peripheral nervous system (PNS) and the central nervous system (CNS).

- The PNS gathers information from the external and internal environment and sends it on to the CNS. The CNS integrates and processes the information and often generates a signal in response. The PNS relays the signal to the body parts that will execute an action dictated by the CNS.

- Neurons transmit information in the form of electrical signals called action potentials. Most neurons pass that information on to other cells by means of chemical signals.

- Sensory receptors detect stimuli in the environment, enabling us to hear, see, taste, smell, and feel. Sensory receptors are clusters of cells, single cells, or parts of a single cell that respond to stimuli by generating a nerve impulse that carries sensory information back to the central nervous system, including the brain.

- Photoreceptors respond to light and are responsible for vision. The lens of the eye focuses the incoming light on the retina, where light-sensitive cells send nerve impulses to the brain, which interprets the information to construct an image.

- Mechanoreceptors respond to physical changes in the environment. They are responsible for touch and hearing, as well as other kinds of perception, such as balance, body position, temperature, and pain.

**MULTICELLULAR ORGANISMS MUST COORDINATE** the functions of their cells and tissues. As we saw in Chapter 29, animals use hormones to send information between different parts of the body through the bloodstream. In addition to this slower endocrine system, animals have a unique high-speed internal communication system, the **nervous system**. The nervous system consists of a variety of nerve cells, including those that provide nutrition, support, or defense against foreign invaders such as viruses. Most distinctively, the nervous system contains cells called **neurons** (from *neuro*, "nerve, sinew") that can transmit signals from one part of the body to another in a fraction of a second. Animals need this almost instantaneous form of communication to coordinate the rapid contractions of their muscles, which play such an essential role in their ability to run, swim, or fly. The nervous system also enables an animal to detect food, find a mate, size up rivals, avoid predators, and respond to extremes of heat and cold. The elaborate nervous system that humans possess makes it possible for us to think, remember, empathize, write poetry, make music, and enjoy the aroma of fresh-brewed coffee or the rich smoothness of a chocolate cheesecake.

In this chapter we explore the basic organization of the nervous system and the structure and function of neurons, before examining the mechanisms that enable the nervous system to receive and transmit information and to elicit responses with lightning speed. We consider the structure and function of the human brain, so critical in processing the enormous quantity of information transmitted through the nervous system. We also explore what happens at the "receiving ends" of the nervous system: how sensory input is generated in response to different kinds of environmental stimuli. We will explore the three most widespread types of sensory receptors found in animals—chemical, visual, and mechanical—with an emphasis on how these receptors work in humans to produce the familiar "five senses."

## 30.1 An Overview of the Nervous System

All animals, except sponges, have at least a rudimentary nervous system that enables them to receive information from the external and internal world, to process that information, and to respond to it in ways that promote survival and reproduction. Invertebrates with a simple body plan, such as jellyfish and sea anemones, have a network of interlinked neurons that enables them to sense and respond to a stimulus such as the presence of food (see Section 4.3). Other invertebrates, such as worms, mollusks, and insects, have a more sophisticated nervous system that includes at least one centralized cable of tissue, the *nerve cord*, from which *nerves* (nerve cell extensions bundled with other tissues) branch out to different body parts.

The nervous system of many invertebrates, and all vertebrates, can be divided into two main structural and functional units: the **peripheral nervous system** (**PNS**) and the **central nervous system** (**CNS**). The PNS gathers information from the external and internal environment and sends it on to the CNS. The CNS integrates and processes the information and often generates a signal in response. The PNS relays the signal to the body parts that will actually execute the action dictated by the CNS (**FIGURE 30.1**). The CNS is rather like the central processing unit (CPU) in a computer: it processes information received from "peripherals" the way a CPU processes input from a keyboard; and it sends out directions to body parts the way a CPU controls the display that appears on the monitor screen.

## The neuron is specialized for processing information

A neuron is a specialized cell that can receive information, transmit that information rapidly over a distance, and then pass that information on to another cell. Acting collectively, neurons can interpret information (pressure waves felt on the eardrum and perceived as sound, for example), integrate information from multiple origins (from the eyes, as well as the fingertips, when we write, for example), and store processed information (in the still mysterious process of memory formation, for example). Neurons interact

## CENTRAL AND PERIPHERAL NERVOUS SYSTEMS

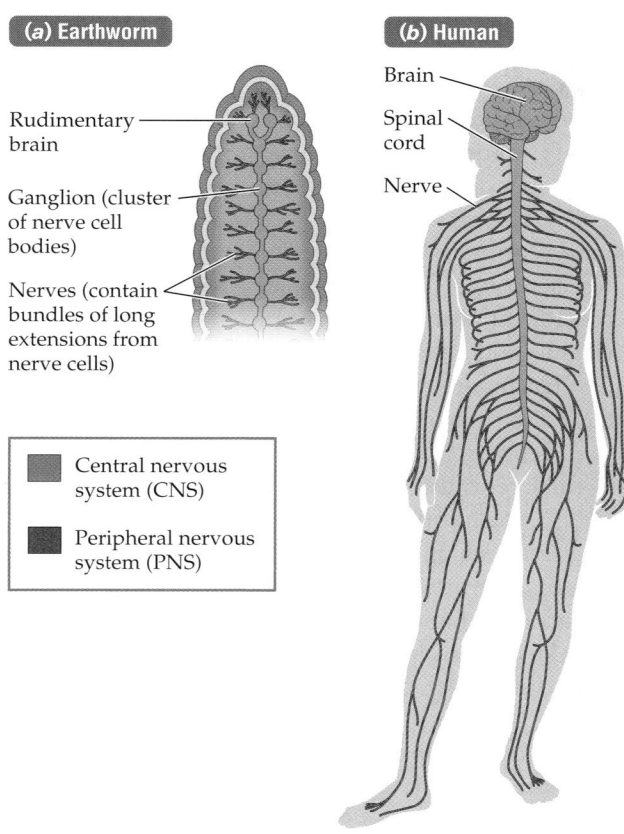

**(a) Earthworm**

Rudimentary brain

Ganglion (cluster of nerve cell bodies)

Nerves (contain bundles of long extensions from nerve cells)

**(b) Human**

Brain

Spinal cord

Nerve

Central nervous system (CNS)

Peripheral nervous system (PNS)

**FIGURE 30.1 The Nervous System Is a Communication Network**

(a) The earthworm's central nervous system consists of a central bundle of nerve tissues, the nerve cord, and a cluster of large ganglia near the head that function as a rudimentary brain. The peripheral nervous system consists of nerves emanating from the nerve cord, with one pair of nerves branching into each of the body segments. (b) The central nervous system in a human consists of the brain and the spinal cord (blue). Some of the paired nerves that make up the peripheral nervous system (red) are shown. Twelve pairs of nerves branch out from the brain; the remainder emanate from the spinal column.

with other neurons, with muscle cells, and with endocrine cells. A single neuron may receive information from thousands of other cells and send out information to thousands of recipient cells.

The structure of a neuron reflects its unique function (**FIGURE 30.2**). The neuron's cell body contains a nucleus and the same organelles found in any other animal cell. In addition, a neuron has multiple extensions from the cell body, which are identified by different terms depending on whether they receive signals from or send signals to other cells. A neuron's many, highly branched **dendrites** (from *dendron*, "tree") re-

ceive signals from adjacent cells. Signals from the dendrites travel to the cell body, where they are collected and transmitted down an **axon**, a long, thin extension from the cell body that specializes in transmitting signals to other cells, sometimes over long distances. Although we think of cells as small structures, the axons of some neurons are extremely long. Some neurons at the base of the spine have axons that extend all the way into the toes, as much as 1 meter (a little over 3 feet) away in the average person.

Nervous system tissues also contain a variety of support cells, known as *glial cells*. In fact, glial cells outnumber neurons by 10–50 times in various parts of the vertebrate nervous system. One type of glial cell that is closely associated with neurons creates an insulating sheath made of a fatty material called

## NEURON STRUCTURE AND FUNCTION

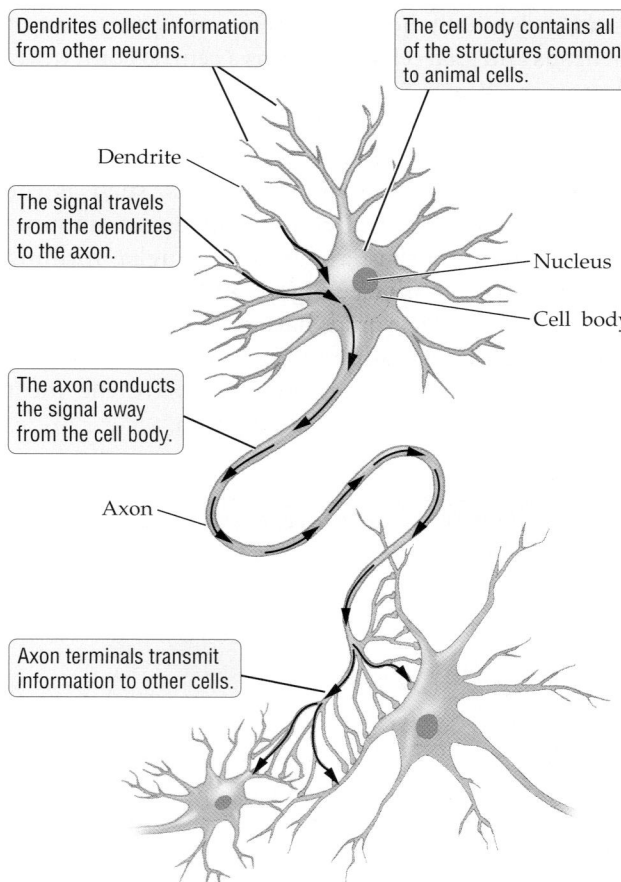

Dendrites collect information from other neurons.

The cell body contains all of the structures common to animal cells.

Dendrite

The signal travels from the dendrites to the axon.

Nucleus

Cell body

The axon conducts the signal away from the cell body.

Axon

Axon terminals transmit information to other cells.

**FIGURE 30.2 Neurons Transmit Signals from One Cell to Another**

A neuron receives information from other cells, including other neurons, through one or more dendrites. The neuron pictured here has a single long axon with branched endings (multiple termini). Axons carry signals away from the cell body and to another cell (two target cells are shown here).

**myelin**. Myelin encases the axons of most neurons in vertebrates, including humans. Neurons with a myelin sheath are said to be *myelinated*; those without one are *unmyelinated*. As we shall see, myelinated axons can carry signals more rapidly than unmyelinated ones.

At its end, a single axon usually divides into many axon terminals, each of which transmits the signal to other cells (see Figure 30.2). The axons of many individual neurons may be bundled together (along with supporting cells, blood vessels, and connective tissue) to form a major communication pathway called a **nerve** (FIGURE 30.3). Just as axons are bundled together with certain tissues to form nerves, neuronal cell bodies combine in various ways to create larger structures. A **ganglion** (plural "ganglia") is a cluster of neuronal cell bodies that serves as a local or regional integrating center. The neural interconnections between the PNS and CNS often form prominent ganglia (see Figure 30.1a).

In animals that have a distinct head and tail, the head end usually houses a dominant aggregation of neurons and glial cells that forms a central computing center, the **brain**. The brain has a large capacity for processing diverse types of sensory information, and it controls and coordinates nerve signals throughout the body. In relatively simple animals, such as the earthworm, this central computing center amounts to nothing more than a cluster of ganglia, forming a rudimentary brain (see Figure 30.1a). Mollusks, especially octopi and squid, have the largest and most highly elaborated brains among the invertebrates (see Figure 4.13c). Vertebrates, however, have the most complex brains in the animal kingdom.

## The central nervous system of vertebrates consists of a brain and a spinal cord

The vertebrate brain serves as the major clearinghouse for information. You may have noticed that most of the sensory organs—eyes, ears, and nose, for example—are located near the "brainy end" in vertebrates like us. The evolutionary advantage to having the brain near these sensory structures is that it reduces the time it takes for the brain to receive sensory information. For humans and many other vertebrates, this arrangement has increased our ability to avoid danger and to detect and capture food more quickly.

Vertebrates, including humans, have an exceptionally thick central nerve cord, known as the **spinal cord**, that is continuous with the brain (see Figure 30.1). The vertebrate spinal cord contains large concentrations of dendrites and axon terminals that enable information exchange among huge numbers of neurons along largely myelinated pathways. Both components of the vertebrate central nervous system—the brain and the spinal cord—are protected within bony enclosures that are part of the skeletal system. In most vertebrates, the brain is encased in bony plates that together make up the skull, or *cranium*. The spinal cord is surrounded by stacks of ring-shaped bones, each known as a *vertebra* (plural "vertebrae"), that collectively make up the *spinal column*, or backbone.

The peripheral nervous system of vertebrates consists of branching nerves that carry information into and out of the spinal cord. In the human body, the PNS consists of 31 pairs of nerves, one member of each pair supplying information from and to one side of the body. Some types of nerves specialize in bringing information (input) to the CNS. Others specialize in carrying signals from the CNS to other parts of the body, including muscles, skin, or one of the internal organs.

## The peripheral and central nervous systems exchange information

As we saw earlier, the central nervous system (CNS) consists of all the neurons (and their supporting cells) in the brain and the spinal cord, and its neural pathways are devoted to the integration and processing of information (FIGURE 30.4). The peripheral nervous system (PNS) acts as a kind of messenger service, ferrying signals to and from the CNS. The PNS consists

**NERVE STRUCTURE**

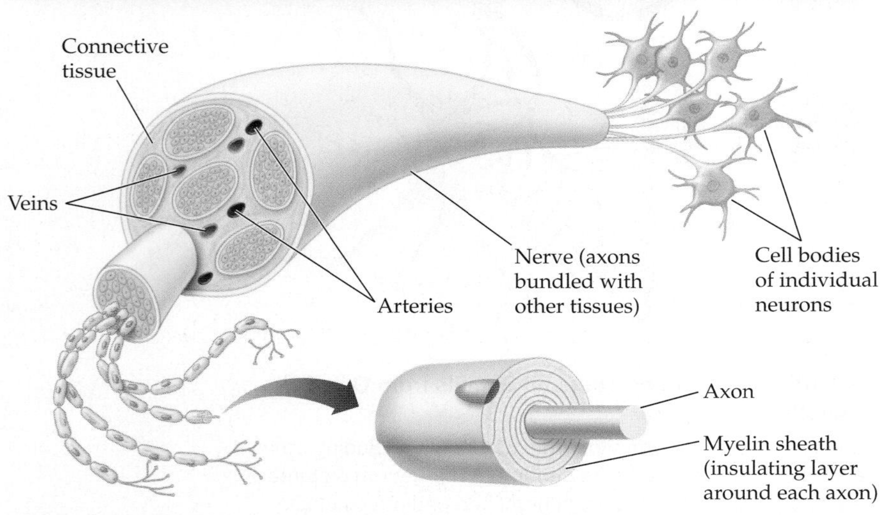

**FIGURE 30.3 Neurons Are Bundled Together with Supporting Tissues and Blood Vessels into Cablelike Nerves**

of the sensory structures (organs such as the eyes and ears), plus all the nerves that are not part of the CNS.

Different types of neurons are involved in transmitting sensory information. **Sensory neurons** in the PNS convey sensory input to the CNS. Most signals first pass to **interneurons**, which lie wholly within the CNS, primarily in the spinal cord. Interneurons process the sensory input from sensory neurons, and then pass the signals on to other neurons in one of several ways. First, they may send signals directly to **motor neurons**. These neurons in the PNS carry signals back out to the body, thereby causing changes in muscles, organs, or endocrine glands in response to the initial stimuli. Second, interneurons may send the signals up the spinal cord to the brain for further processing; the brain may (or may not) respond by sending new signals down the spinal cord for relay out through motor neurons to the body (see Figure 30.4). Finally, a single interneuron in the spinal cord may send its output up to the brain and, *at the same time*, out to the PNS. This simultaneous flow of interneuronal output to both the brain and the PNS is possible because a neuron can be part of a divergent circuit. Its many dendrites can simultaneously receive inputs from many different neurons, and its branched axon terminals usually make contact with many different target cells. Because most neurons have branched axon endings (see Figure 30.2), they can "talk" to several different targets simultaneously.

## Nervous system responses may be voluntary or involuntary

The peripheral nervous system converts stimuli into **sensory input**—electrical and chemical signals that are received, transmitted, and processed by neurons. The CNS integrates and processes the sensory input, and then sends out commands that are carried out by the PNS (PNS output). You can easily demonstrate this for yourself by taking a moment to perform a simple experiment: bring the tips of your two index fingers together in front of you until they touch. You have just converted visual stimuli (printed words) gathered through your eyes (sensory input) into a motor output (bringing your index fingers together). How did your nervous system act as a unified whole to convert billions of signals—each propagated within a fraction of a second, received from multiple sources, and relayed in diverse directions—into a coordinated response?

**THE CENTRAL NERVOUS SYSTEM AS INTEGRATOR**

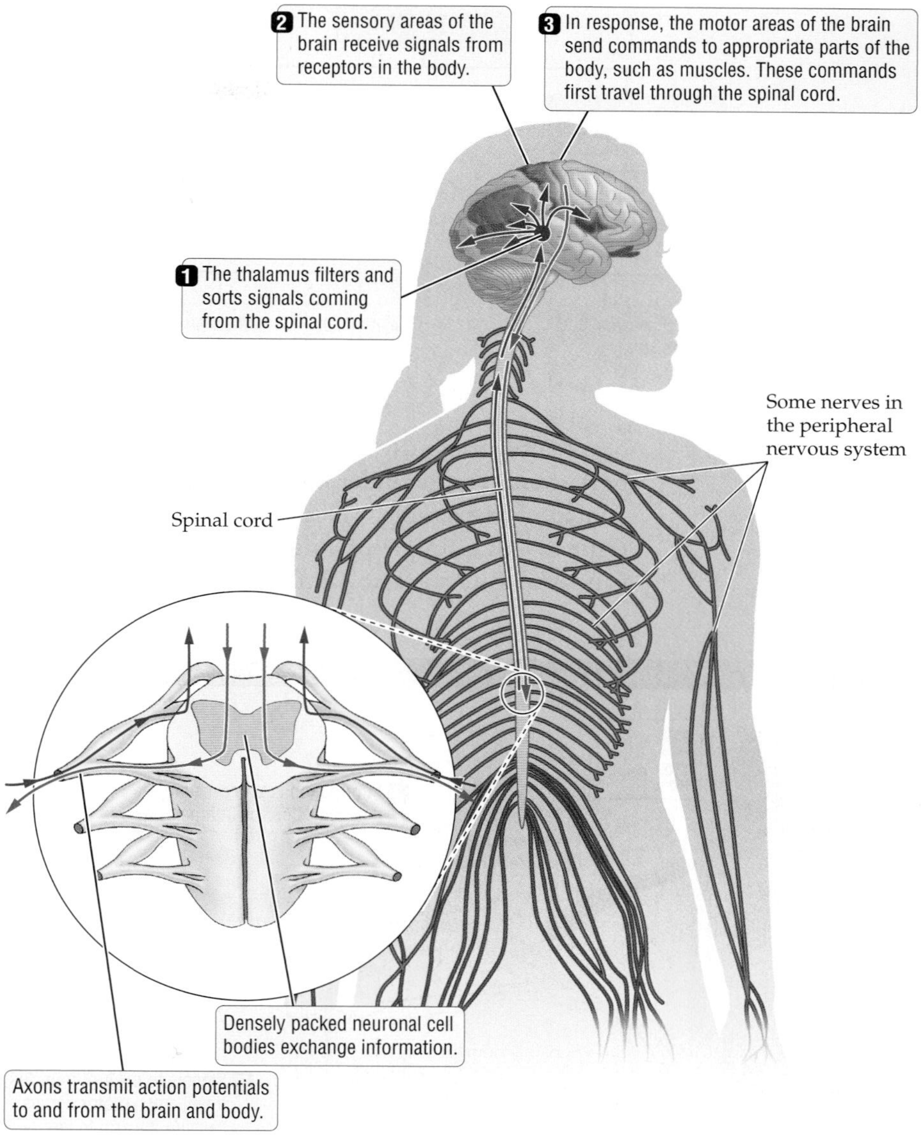

**2** The sensory areas of the brain receive signals from receptors in the body.

**3** In response, the motor areas of the brain send commands to appropriate parts of the body, such as muscles. These commands first travel through the spinal cord.

**1** The thalamus filters and sorts signals coming from the spinal cord.

Some nerves in the peripheral nervous system

Spinal cord

Densely packed neuronal cell bodies exchange information.

Axons transmit action potentials to and from the brain and body.

**FIGURE 30.4**
**The CNS Receives, Processes, and Sends Out Information**

Sensory input (red arrows) travels along sensory neurons in the PNS to the spinal cord. There the information may be processed and acted on or, more often, sent on to the brain for additional integration and interpretation. Different parts of the brain are devoted to processing different kinds of information. Command signals from the brain (blue arrows) travel back down the spinal cord, where they are passed on to motor neurons in the PNS, leaving the spinal cord by a slightly different route. In the body, these motor signals may stimulate or inhibit muscles, organs, or endocrine cells; they may also override the actions triggered by spinal cord processing. Only some of the PNS nerves from the neck down are shown here.

Some types of PNS output are under our own control, and some are not. Bringing the tips of our fingers together is an example of voluntary, or **somatic**, control over outgoing PNS messages. Many PNS signals

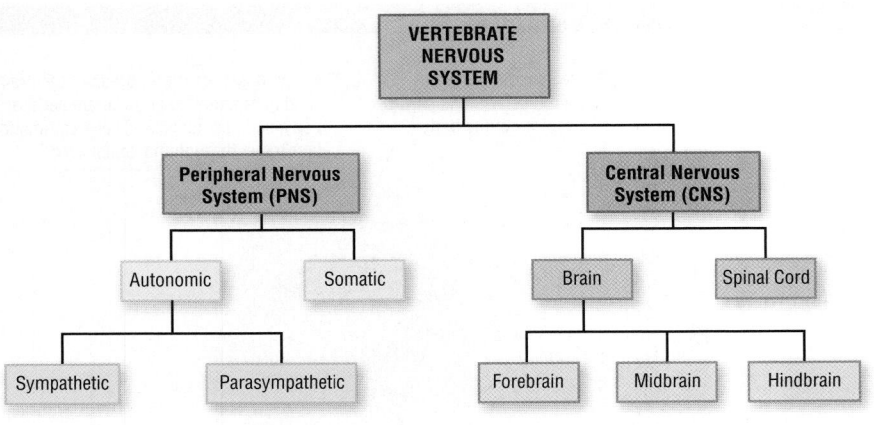

Organization of the
nervous system

minute that we are awake, usually without being aware that we are doing so—an example of involuntary control of PNS output. Heart rate and blood pressure are other examples of body functions that are under autonomic control. The autonomic nervous system has two broad divisions: sympathetic and parasympathetic. The *sympathetic division* orchestrates the body's fight-or-flight response to stressful situations, among other functions. The nerves of the sympathetic division make you break out in a sweat or make your heart race when you are nervous or scared. The *parasympathetic division* opposes many of the actions of the sympathetic division—reducing heart rate, stimulating digestion, and promoting relaxation, for example.

occur subconsciously, however; such "hardwired" signals are said to be under involuntary, or **autonomic**, control. For example, we can close our eyes when we want to—an example of voluntary control over body function. But we also blink about 15 times every

## Sensory information can be processed with or without the brain

The human spinal cord—a collection of nerve tissue about as thick as a finger—is an organized bundle of neurons and associated cells. Dense collections of neuronal cell bodies, dendrites, and axon terminals exchange information there, and large bundles of axons carry this information to and from the brain (see Figure 30.4). The vertebrate spinal cord forms an intermediate information-processing site that acts as a kind of filter between the brain and the sensory neurons bringing input from the far reaches of the body. Although much information needs to be passed on to the more sophisticated processing centers in the brain, some simpler processing can be taken care of by the spinal cord neurons themselves.

You know from experience that if you touch something hot, such as a candle flame, you jerk back your hand without having to think about it. A pain message is sent by sensory neurons to your spinal cord, and a "get the hand out of there" message reaches the muscles in your arm via motor neurons (**FIGURE 30.5**). You have just experienced the action of a **reflex arc** (also called a *spinal reflex*). This reflex arc consists of just three neurons: a sensory neuron that sends the message to the spinal cord, an interneuron, and a motor neuron that creates a response in the body by sending a message to, in this case, a muscle. All of the information processing takes place in the spinal cord; the brain is not required.

Even though a reflexive motor response to pain does not require the brain's involvement, some of the sensory input is sent to the brain. We are conscious of the pain from the candle flame, and the brain may process that information into additional responses,

### THE REFLEX ARC

CNS
PNS

**2** Information is processed in the spinal cord by interneurons.

Cross section of spinal cord

Interneuron

Sensory neuron

Motor neuron

**3** Signals to the arm muscles travel down motor nerves, causing the arm to jerk away from the stimulus.

Muscles

Painful stimulus

**1** A pain message goes up a sensory nerve to the spinal cord.

**FIGURE 30.5 Reflex Arcs Do Not Require Processing in the Brain**

In situations that call for a quick response, sometimes a simple circuit of neurons can generate quick action without us having to "think" about it.

including a stream of words, perhaps some unfit to print here. However, because the brain requires more time to process information than a reflex arc does, these brain-directed responses come later—perhaps only a matter of milliseconds later, but that time lag is significant. If our bodies relied on the brain to process the motor part of the reflex arc—jerking the hand away from a painful stimulus—we might be burned more severely.

The human body has many reflex arcs. One of the best known is the "knee-jerk response," which you may have encountered at the doctor's office: the physician taps just below the kneecap, which automatically forces the lower leg to twitch forward. We also have a number of reflex arcs that respond to internal stimuli, such as those involved in fine-tuning the operation of some of our organs.

# 30.2 Signal Transmission by Neurons

How does a neuron receive and send signals? The different types of neurons respond to different types of stimuli, which range from chemicals and rays of light to the stretching of the plasma membrane by applied pressure. A stimulus can alter the flow of ions across the plasma membrane. Because ions carry electrical charge, either positive or negative, a change in ion flow amounts to an electrical disturbance. In neurons, the electrical disturbance can travel down the length of an axon as a pulse of electrical activity known as an *action potential*. Upon reaching the end of the axon, the action potential triggers the release of chemical messengers, called *neurotransmitters*, that signal to the next cell in this line of communication. We will first examine the mechanism that generates action potentials in a neuron and then see how such a change in electrical activity launches the outflow of neurotransmitters at the axon terminal.

## Action potentials are propagated along the length of an axon

A neuron can transmit information to other cells using a self-sustaining electrical signal called an **action potential** that travels rapidly in one direction—away from the cell body—along the length of an axon (**FIGURE 30.6**). Action potentials depend on the rap-id movement of positively charged ions across the plasma membrane of the neuron. To understand how this works, first imagine an unstimulated neuron—one that has not received a stimulus and has therefore not produced an action potential. Such a cell has a higher concentration of positively charged ions outside, in the extracellular fluid, compared to the concentration of this ion within the cytoplasm. The concentration difference is maintained through the continual pumping of ions by a membrane transport protein complex called the sodium-potassium pump (see Section 7.3). In part because there are many more positively charged ions on the outside relative to the cytoplasmic side, the plasma membrane is in a **polarized** state; that is, there is an imbalance in electrical charge across it so that the inside surface of the membrane is relatively more negative.

A difference in electrical charge across a biological membrane is similar to the difference in the electrical charge between the positive and negative terminals of a battery, and it can be measured as a voltage difference expressed in millivolts (mV). The electrical charge that exists across the plasma membrane of an unstimulated neuron is known as the **resting potential**.

A stimulus activates a neuron by changing its resting potential momentarily, which it does by altering the activity of one or more ion channels or pumps. A stimulus **depolarizes** a neuron if ion flow is changed in such a way that many more positively charged ions enter that cell than leave it. The consequence of this altered ion flow is that the charge inside is not as negative as before, so the voltage difference across the plasma membrane has lessened. Stimuli that generate action potentials produce a strong depolarization that flings open previously closed sodium channels, causing $Na^+$ to rush in to the cytoplasm from the surrounding fluid.

The voltage-sensitive sodium channels stay open for only a millisecond and then close, but in that time so many sodium ions enter so quickly that the local charge difference across the plasma membrane actually reverses; for a short time, the inside of the affected portion of the membrane takes on a relatively positive charge. This reversal of charge depolarizes the adjoining portion of the plasma membrane, which in turn depolarizes the next portion, and so on down the length of the axon away from the cell body. The signal cannot reverse, heading back toward the cell body, because the channels that just opened in each depolarized segment of the axon close and become briefly unresponsive (**FIGURE 30.6a**). The only

FIGURE 30.6 How Axons Transmit Signals

Action potentials (red) move down axons differently, depending on whether the axon is myelinated. (*a*) Unmyelinated axons carry self-sustaining action potentials that rely on changes in the electrical charge across their plasma membranes. (*b*) Myelinated axons transmit action potentials very rapidly because the signal leaps from node to node. The arrow from A to B shows the time taken for the action potential to travel one meter (1m).

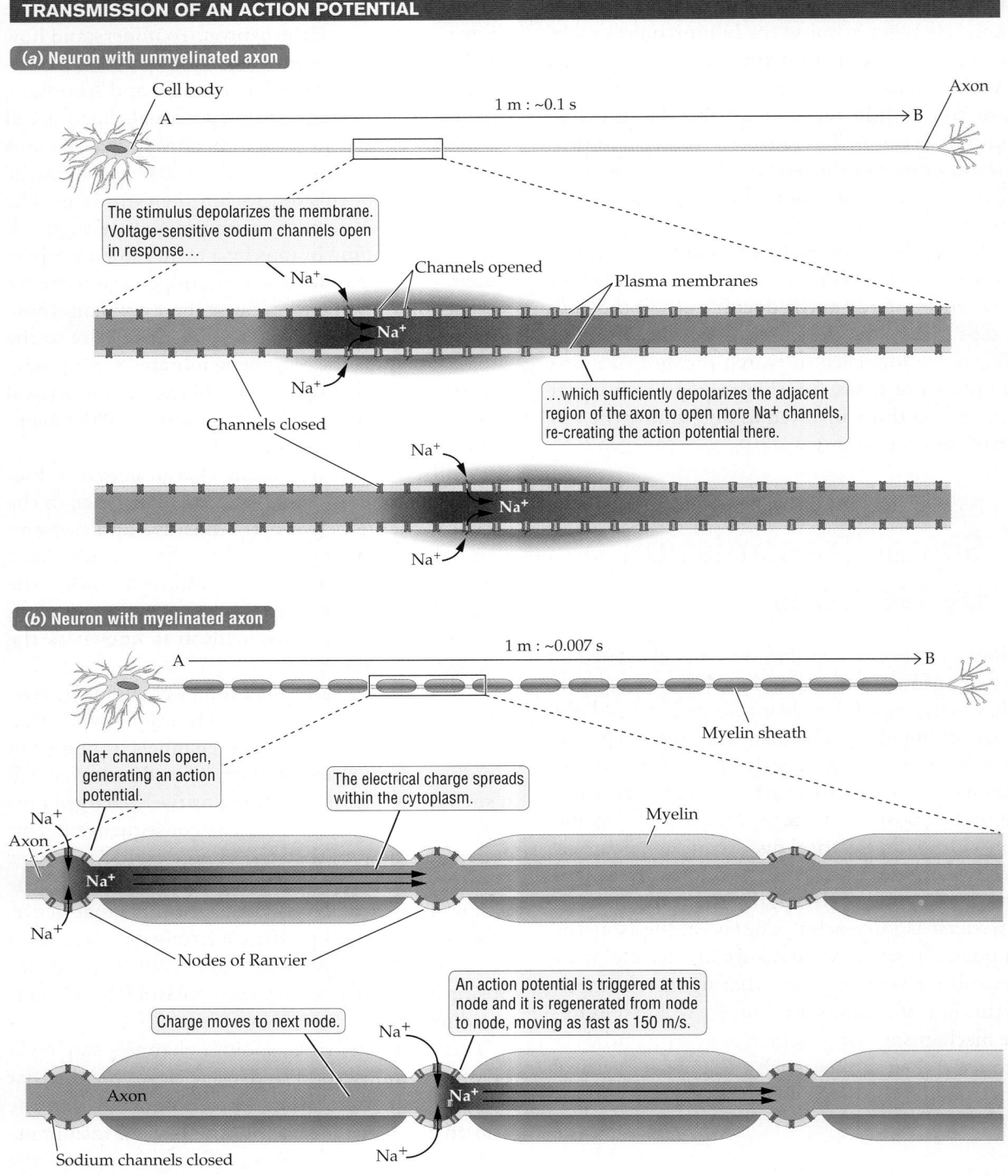

**TRANSMISSION OF AN ACTION POTENTIAL**

**(*a*) Neuron with unmyelinated axon**

Cell body

A     1 m : ~0.1 s     B

Axon

The stimulus depolarizes the membrane. Voltage-sensitive sodium channels open in response…

Channels opened

Plasma membranes

Na⁺

Na⁺

Na⁺

Channels closed

…which sufficiently depolarizes the adjacent region of the axon to open more Na⁺ channels, re-creating the action potential there.

Na⁺

Na⁺

Na⁺

**(*b*) Neuron with myelinated axon**

A     1 m : ~0.007 s     B

Myelin sheath

Na⁺ channels open, generating an action potential.

The electrical charge spreads within the cytoplasm.

Myelin

Na⁺

Axon

Na⁺

Na⁺

Nodes of Ranvier

Charge moves to next node.

An action potential is triggered at this node and it is regenerated from node to node, moving as fast as 150 m/s.

Na⁺

Axon

Na⁺

Na⁺

Sodium channels closed

voltage-sensitive sodium channels that are available to respond to the local depolarization are the ones located "downstream," away from the cell body. Once the action potential has moved on, other channels and pumps restore the resting potential in that portion of the membrane.

## Myelinated axons transmit action potentials especially rapidly

In vertebrates and a few invertebrates, a fatty layer called the *myelin sheath* surrounds the axons of most neurons and greatly speeds signal transmission (see Figure 30.3). The myelin sheath covers segments of the axon, leaving unmyelinated gaps between the myelinated zones, giving the axon the appearance of a chain of sausage links (**FIGURE 30.6b**). The gaps between myelinated segments are called **nodes of Ranvier**. Action potentials can occur only at these nodes because there are no sodium channels along the portions of an axon's plasma membrane that are surrounded by myelin. Action potentials move faster in myelinated fibers because they leap from node to node, instead of traveling from one depolarized patch of the membrane to the next all along the length of the axon.

Membrane depolarization at the node opens voltage-sensitive sodium channels, and an action potential is launched as sodium ions flood in. However, instead of depolarizing an adjacent patch of membrane, the electrical charge flows rapidly *inside* the axon, from the depolarized node to the next node along the axon (see Figure 30.6b). As a result, the axon depolarizes at the next node, and a new action potential begins there. Because the action potential is regenerated at each node, the signal does not decay as it jumps rapidly from one node of Ranvier to the next node down the axon.

The action potential can move down the interior of a myelinated axon much more quickly than can a wave of charge reversals along the plasma membrane of an unmyelinated neuron. Unmyelinated neurons transmit action potentials down axons at a maximum speed of about 30 meters per second (m/s) and an average speed of about 5 m/s. In myelinated axons, action potentials average about 120 m/s and can jump along at up to 150 m/s. Disruption of the myelin sheath can lead to serious problems in humans. Diseases that affect the myelin sheath, such as multiple sclerosis and Guillain-Barré [ghee-*LAN*-bah-*RAY*] syndrome, can disrupt vision, speech, balance, and muscular coordination, and may eventually kill their victims.

## Action potentials have several important features

Action potentials—whether in myelinated or unmyelinated axons—have several important characteristics that make them well suited for sending information rapidly throughout the body. First, as we just saw, action potentials move along the axon in only one direction, which tracks the original stimulus along a set pathway. This precise targeting of the action potential keeps the signal from "going astray" and producing garbled information in the process.

Second, an action potential remains consistently strong as it moves from one end of the axon to the other; it does not weaken with distance. If these signals weakened as they moved along, they could lose their capacity to initiate events in other neurons or cells, or disappear altogether before reaching their target cells. Maintaining signal strength is a critical feature for any long-distance communication system.

Third, action potentials are "all-or-none" events: if the initial depolarization is large enough, an action potential is triggered; otherwise, it is not. Any action potential in a particular cell type has the same strength. A stronger stimulus will initiate action potentials more often, but any individual action potential will be no stronger. This all-or-none type of trigger allows information to be "digitized" in much the same way that computers digitize information as series of zeros and ones. A zero is an absence of an action potential, and a one is an action potential. There is no potentially confusing in-between. The sound of a whisper generates action potentials that have the same strength as those generated by a shout, but there are fewer of them. A whisper could be described as 100010001, a shout as 101010101, and an explosion as 111111111.

## Neurotransmitters transmit signals between adjacent cells

To be part of a whole-body communication network, neurons must be able to pass their signals along to other cells. Once an action potential reaches the end of an axon, the signal is usually converted into a chemical message and relayed to the next cell at a junction called a **synapse** (**FIGURE 30.7**). Neurons form synapses with other neurons, with muscle cells (**FIGURE 30.8**), and with endocrine cells. In a synapse between two neurons, the axon terminal of the sending cell may interact with a dendrite or with the cell body of the receiving cell.

The transmission of information at a synapse takes place across a tiny fluid-filled gap between the two cells, called the **synaptic cleft**. In the axon terminal, the electrical signal is transformed into a chemical signal in the form of molecules called **neurotransmitters**. The arrival of an action potential at the axon terminal causes it to release neurotransmitters into the synaptic cleft by exocytosis. The neurotransmitters diffuse virtually instantaneously across the narrow cleft and bind to receptor proteins in the plasma membrane of the target cell.

FIGURE 30.7
## A Neuron Communicates with Other Cells across a Synapse

Most neurons transmit information to another cell, including other neurons, at junctions called synapses. At a synapse, electrical signals in the axon terminals are converted into chemical signals through the release of a specific neurotransmitter. Neurotransmitters are chemical messengers that diffuse across the small gap (the synaptic cleft) between two neurons, or a neuron and its target cell such as a muscle cell.

**NEURONAL SYNAPSE**

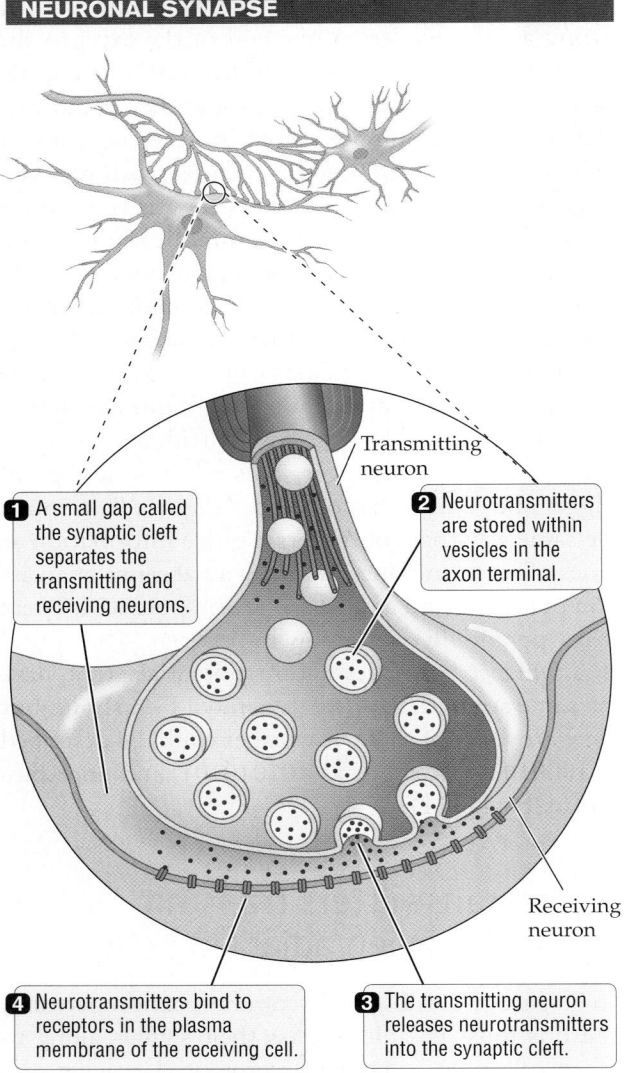

Transmitting neuron

Receiving neuron

**1** A small gap called the synaptic cleft separates the transmitting and receiving neurons.

**2** Neurotransmitters are stored within vesicles in the axon terminal.

**4** Neurotransmitters bind to receptors in the plasma membrane of the receiving cell.

**3** The transmitting neuron releases neurotransmitters into the synaptic cleft.

If enough molecules of the appropriate neurotransmitter bind to enough target cell receptors, they will affect the electrical charge of the target cell. In some cases the plasma membrane of the target cell depolarizes, triggering an action potential that begins its rapid journey along the dendrite or across the cell body. Such a neurotransmitter is said to produce excitation of the target cell. In other cases the neurotransmitter inhibits the target cell, making it less likely that it will produce an action potential. Such a neurotransmitter is said to have an inhibitory effect on the target cell. Neurotransmitters excite or inhibit non-neuronal target cells, such as muscle cells, by binding to specific receptor proteins on the plasma membrane. Neurotransmitter binding alters ion movement across the target cell's plasma membrane, which in turn can trigger a specific cell behavior, such as the contraction of a muscle cell or the release of a specific hormone by an endocrine cell.

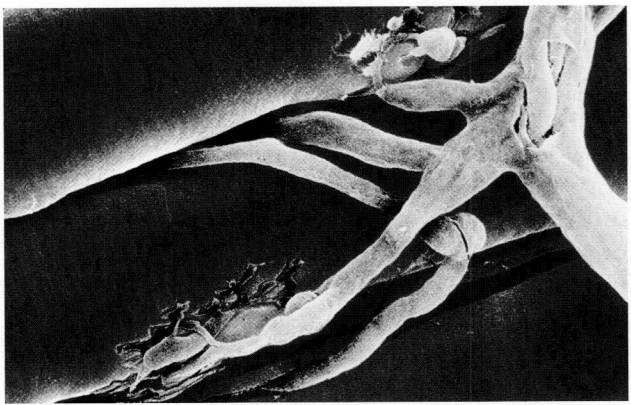

**FIGURE 30.8 Synapse at a Muscle Junction**
This color-enhanced photomicrograph shows neurons (green) associated with muscle fibers (red) via synapses. A neurotransmitter called acetylcholine is released into the synaptic cleft by the neuron. Acetylcholine triggers muscle contractions when it binds to receptors in the plasma membrane of the muscle fibers.

More than 70 different types of neurotransmitters are known to be naturally active in the human body; Table 30.1 lists a small number of them. Some neurons both produce and respond to more than one neurotransmitter. The types of receptors on the target cell plasma membrane determine which neurotransmitters it will respond to and in what way. One neurotransmitter may bind to several different receptor types and trigger different responses in different target cells. Acetylcholine, for example, is the major neurotransmitter responsible for signaling skeletal muscles to contract, while in the heart the same neurotransmitter signals muscles to slow down—to contract less often and with less strength.

Neurotransmitters released into a synaptic cleft must be quickly removed once the message from the original stimulus has been conveyed to the target cell. Otherwise, the target cell would continually respond to the neurotransmitter and the signaling pathway would be lost, the way a phone left off the hook becomes useless as a communication device. Neurotransmitters are cleared from the synaptic cleft through uptake by the neuron that released them or through uptake by special glial cells found in the vicinity. They can also be destroyed or inactivated by specific enzymes present in the synaptic cleft. Many drugs used to treat depression and anxiety, or attention deficit hyperactivity disorder (ADHD), work by blocking the removal of certain neurotransmitters (serotonin or dopamine, for example) from the synaptic cleft, which prolongs the biological action of these chemical messengers.

Knowing that the nervous system transmits information much more quickly than does the endocrine system, it may surprise you that neuronal communi-

TABLE 30.1    **Functions of Common Human Neurotransmitters**

| NEUROTRANSMITTER | MAJOR FUNCTIONS |
| --- | --- |
| Acetylcholine [**uh**-SEH-**tul**-KOH-**leen**] | Controls muscle contractions |
| Epinephrine [EH-**puh**-NEH-**frun**] (adrenaline) | Increases heart rate, elevates blood pressure, as part of fight-or-flight response |
| Norepinephrine (noradrenaline) | Stimulates fight-or-flight response; increases alertness |
| Dopamine | Affects muscle activity; stimulates neurons in the pleasure center of the brain |
| GABA (gamma-aminobutyric acid) [...-**uh**-MEE-**noh-byoo**-TEER-**ik** . . .] | Assists muscle coordination by suppressing the activity of certain neurons |
| Serotonin [SIHR-**uh**-TOH-**nun**] | Regulates temperature, sleep, mood |
| Melatonin [MEH-**luh**-TOH-**nun**] | Helps regulate day-night cycles, including sleep |
| Enkephalins and endorphins | Inhibit transmission of signals from pain sensors, pain perception |
| Substance P | Regulates transmission of signals from pain sensors |

cation depends on the chemical diffusion at synapses between cells, which is slower than the propagation of electrical disturbances. However, the journey that neurotransmitters make across the synaptic cleft is extremely short—less than a millionth of a millimeter—so they can reach their destinations in tiny fractions of a second. Moreover, the very long axons of some neurons ensure that action potentials need to cross only a few synapses to reach their destinations.

The subtlety and specificity of communication through the nervous system comes from the capacity of each neuron to generate large numbers of action potentials in a fraction of a second and to aim those signals narrowly at specific target cells. Every action potential is a small unit of information, just as a byte is in computer talk. A single neuron can generate up to 4,800 action potentials per minute. That means 4,800 "bytes" of signal information from a single neuron! Like the bodies of most vertebrates, the human body has many billions of neurons. Some neurons in humans may interact with as many as 10,000 other neurons in a dense tangle of dendrites and axon terminals. The scale of these numbers brings the operation of nervous systems into clearer perspective.

### Concept Check

1. What are the two main branches of the human nervous system, and what are their roles in routing and processing information?
2. What is an action potential?
3. List three distinctive characteristics of signaling mediated by action potentials.

## 30.3 Organization of the Human Brain

In contrast to the exchange of information among a few neurons in a reflex arc, the transmission of information to the brain is quite complex. The brain receives information in the form of action potentials from millions of sensory neurons, and it can direct its response to millions of motor neurons. The human brain contains billions of neurons organized and interconnected to enable the efficient exchange of information. Its vast numbers of neurons can sort and interpret a bewildering array of incoming information.

Far from being a homogeneous mass of neurons, the brain has three distinct regions: the *forebrain*, the *midbrain*, and the *hindbrain* (**FIGURE 30.9**). Each region has a different set of functions.

The **forebrain** might be described as the supercomputer region. It has three important components: the *cerebrum*, the *thalamus*, and the *hypothalamus*. The **cerebrum** forms the most conspicuous portion of the brain in humans and many mammals (though not in many other vertebrates). The cerebrum has four major lobes, as shown in **FIGURE 30.10**. The *frontal lobe* does most of the thinking. The *parietal lobe* handles aspects of speech, as well as taste, reading, and sensation. The *occipital lobe* is the region that deals with vision. And the *temporal lobe* processes sound and smell. Bundles of axons carry information between the left and right halves of the cerebrum.

**Concept Check Answers**

1. The central nervous system (brain and spinal cord) and the peripheral nervous system (sensory organs and all remaining nerves). The CNS integrates and processes information; the PNS transmits information between the CNS and other body parts.
2. A self-sustaining wave of electrical signals propagated along an axon in response to a localized depolarization initiated by a stimulus.
3. They are all-or-none events, travel in one direction only, and do not weaken as they travel along an axon.

## MAIN REGIONS OF THE BRAIN

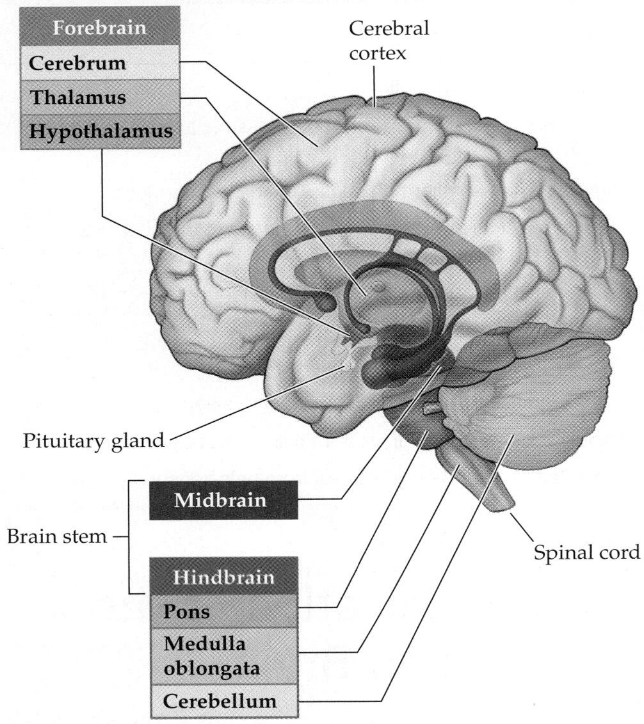

Forebrain
- Cerebrum
- Thalamus
- Hypothalamus

Cerebral cortex

Pituitary gland

Brain stem

Midbrain

Hindbrain
- Pons
- Medulla oblongata
- Cerebellum

Spinal cord

**FIGURE 30.9 The Human Brain Has Three Specialized Regions**

Shown here are the major parts of the brain viewed from the left side (with the back of the brain to the right). Certain portions of the forebrain (shown in purple and orange) and midbrain (red) together make up the limbic system. The limbic system regulates emotions, motivations, and behavioral drives, and also participates in memory formation.

## INFORMATION PROCESSING BY CEREBRAL CORTEX

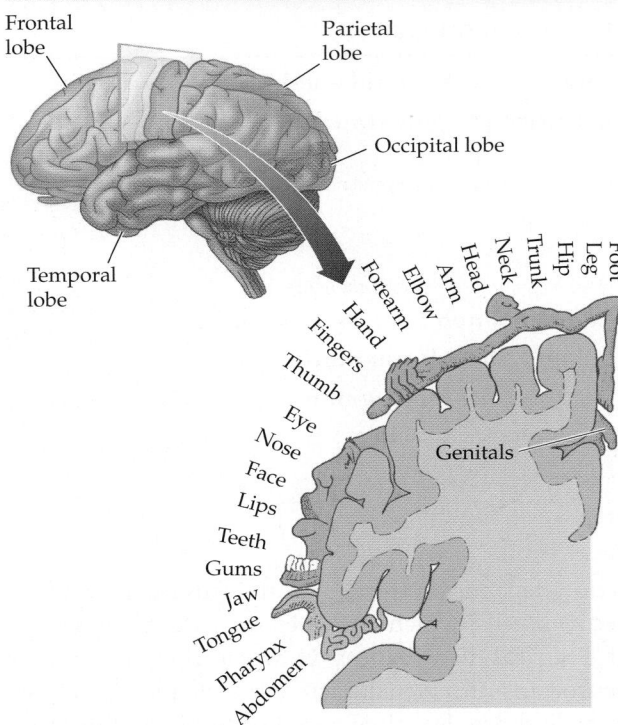

Frontal lobe

Parietal lobe

Occipital lobe

Temporal lobe

Foot, Leg, Hip, Trunk, Neck, Head, Arm, Elbow, Forearm, Hand, Fingers, Thumb, Eye, Nose, Face, Lips, Teeth, Gums, Jaw, Tongue, Pharynx, Abdomen

Genitals

**FIGURE 30.10 How the Brain "Sees" the Body**

Different parts of the cerebral cortex process sensory information received from different parts of the body. The four lobes of the cortex represent physically distinct areas of the brain. The cross section of the brain shown here "maps" body parts to different parts of the brain. The size distortion of the body parts reflects the relative numbers of neurons in the brain devoted to sensory information received from those parts. For example, more neurons process sensory information from the foot, a relatively small body part, than from the leg, a much larger body part.

The outer layer of the cerebrum, called the **cerebral cortex**, is only about 5 millimeters thick, but because it is highly and intricately folded, it accounts for approximately 80 percent of our brain mass. In its billions of neurons and even more billions of connections rest our personalities—our capacity to talk, calculate, create art, and consciously sense the world around us. When the thalamus sends signals to the cerebrum, it usually directs them to the region of the cerebral cortex that specializes in processing information from the relevant area of the body (the part of the body where the stimulus originated). These sensory signals then become part of our conscious awareness. Coordinated responses to these signals arise in the motor areas of the cerebrum.

The **thalamus** is the central switching area of the brain. It acts as a critical filter, saving the cerebrum from processing all the stimuli our bodies are constantly receiving from both within and without.

The thalamus determines which of the sensory signals coming from the spinal cord should go to the conscious perception centers in the cerebrum and which should go to the portions of the brain that act on sensory inputs without our being aware. For example, when we touch a candle flame, the thalamus passes the pain message along to one part of the brain, and we are very much aware of that. But it also passes information along to other parts of the brain that we probably will not be aware of until the next time we see a lit candle. Then we will remember what *not* to do.

The **hypothalamus** functions as an integrator of the nervous system and the endocrine system. It regulates homeostasis in blood pressure, body temperature, sex drive, hunger, and thirst. Along with

# The Neurobiology of Addiction

All of us find pleasure in *something*, whether it is listening to music, sipping a favorite drink, hiking in the mountains, playing with a pet, or hanging out with someone we care about. Researchers have identified regions distributed in the forebrain and midbrain that respond to pleasurable experiences, and these brain regions are collectively known as the *pleasure centers* of the brain.

Two of the best-studied regions—the ventral tegmental area and the nucleus accumbens—generate feelings of pleasure in response to stimuli that include food, sex, music, exercise, and even charitable actions such as donating to the needy. These pleasure centers probably evolved to reinforce behaviors with survival value, such as seeking food and mates and building social networks. Pleasurable stimuli trigger the release of dopamine in these pleasure centers. The neurotransmitter activates complex *reward circuits* that act on the hypothalamus to generate memories of a pleasant experience and, often, the desire for more of the same.

Certain chemical stimulants flood the pleasure centers with neurotransmitters, evoking a 2- to 10-times stronger response in the reward circuits than is generated by everyday stimuli such as fresh-brewed coffee in the morning. Not only is the response stronger, but often the effect lasts much longer as well. Cocaine and amphetamines trigger a large release of dopamine and also prevent the reuptake of dopamine from the synaptic cleft, prolonging its availability among pleasure-sensing neurons. Nicotine acts on a number of brain areas where it hyperactivates acetylcholine receptors, which elevate dopamine levels indirectly. The strong sense of well-being it generates makes nicotine one of the most addictive substances known. Alcohol, opioids such as heroin, and the psychoactive compounds in cannabis, also raise dopamine levels through indirect pathways.

The pleasure centers of some individuals seem to be especially susceptible to seeking the high levels of stimulation generated by certain plant chemicals, such as heroin and cocaine, and many synthetic ones, such as painkillers and antianxiety medications. A person who develops a deep-seated psychological and physical dependency on a chemical is said to be addicted to that substance. *Addiction* produces cravings so strong and obsessive that most affected persons are unable to control their drug-seeking behavior, even if they recognize its destructiveness. Addictive substance abuse is usually marked by a tendency to seek higher and higher doses of the substance, as the brain centers habituate to the lower or more moderate dose. The same reward circuits can also produce behavioral addictions, such as compulsive gambling or shopping.

Addiction has a strong genetic component, but environmental factors play an important role. Though substance use is initially influenced mainly by availability of the drug, continued use and progression to addiction is determined more by genetic vulnerability and psychosocial factors such as chronic stress. Drug abuse alters brain chemistry further, worsening chemical imbalances and setting up a vicious cycle that increases craving and makes quitting harder. Being able to identify those at risk and preventing them from entering this self-destructive loop is therefore critical. Most neurobiologists today believe that addiction should be treated as medical condition, instead of being seen as a moral failing. However, there is no social consensus on this issue.

the rest of the **limbic system** (see Figure 30.9), parts of the hypothalamus give rise to emotions, such as pleasure, anger, or fear. Most of us have regular patterns of sleep and wakefulness; the timing of these patterns rests in the biological clocks of the hypothalamus.

The **midbrain** helps maintain muscle tone and sends some sensory data to higher brain centers in the forebrain. Higher brain centers are called "higher" because they process information in much more sophisticated ways than "lower" brain centers do; they are also literally higher up when we are standing.

The **hindbrain**—which includes the medulla oblongata, the pons, and the cerebellum—coordinates information dealing with breathing rhythms, blood pressure, heart rate, and balance, among many other functions. The midbrain and hindbrain together make up the **brain stem**.

> ### Concept Check
>
> 1. Name the three main types of neurons and the role played by each in neural communication pathways.
> 2. What is the thalamus, and what is its function?

## 30.4 Sensory Structures: Making Sense of the Environment

We, like all other animals, are constantly bombarded with information in the form of stimuli that can be converted into perceptions. We can see images, hear sounds, perceive smells and flavors, and feel when we are touched. And, like all other animals, we need to sense our "inner environment" as well, to check on the status of various functions in the body.

Humans have five basic types of **sensory receptors** (not to be confused with the membrane-localized protein receptors described in Chapter 7). Sensory receptors may be part of a cell, a whole cell, or a group of cells. For example, the pain receptors in our skin consist of the highly branched processes (dendrites) that extend from special neurons. The two types of light-sensitive cells in our retina are examples of whole cells that function as sensory receptors. One type of pressure-sensing receptor, especially abundant in our fingertips, is composed of dendrites intermixed with certain highly specialized cell types.

While sensory receptors detect the stimulus and transmit information about its intensity and persistence, it is usually the brain that interprets the information as a conscious sensation (what we call perception) or registers the information at a subconscious level. There are five main classes of sensory receptors in humans; some additional categories are found in other animals but are not known to be active in humans.

1. **Chemoreceptors** [*KEE*-moh-rih-*SEP*-terz] are found on cells that respond to chemicals. Chemoreceptors are involved in two types of sensory perception: receptors on the tongue give us our sense of taste (**FIGURE 30.11**); receptors in the nasal passages give us our sense of smell (**FIGURE 30.12**). Taste involves sensing chemicals in direct contact with our bodies (food in the mouth). Smell involves sensing molecules that are vaporized (are in gas form), like the scent molecules released by a rose, for example.

2. **Photoreceptors** are light-sensitive cells that contain light-absorbing pigments located in stacks of membranes. In humans, only one sense (vision) uses photoreceptors.

3. **Mechanoreceptors** [*MEH*-kuh-noh-rih-*SEP*-terz] detect various kinds of physical stimuli. Mechanoreceptors are responsible not only for sensing physical changes directly affecting the surface

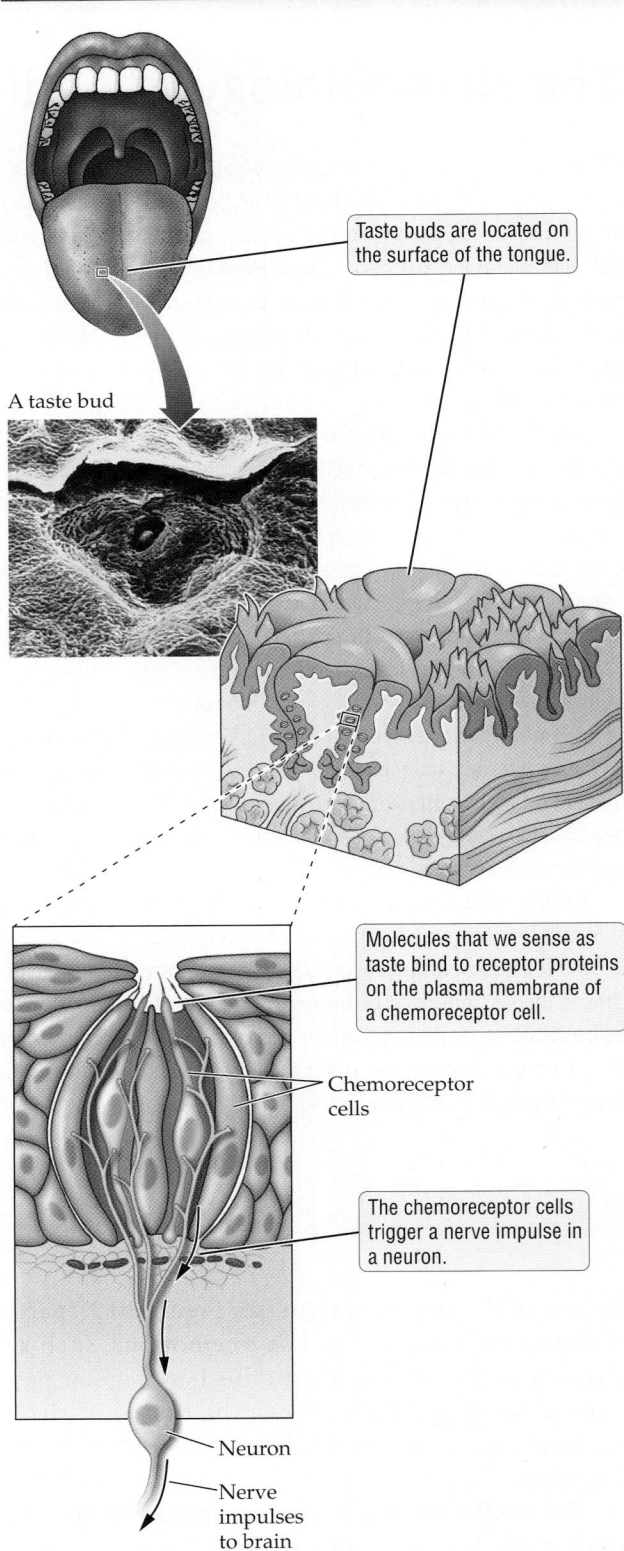

**CHEMORECEPTORS AND TASTE**

Taste buds are located on the surface of the tongue.

A taste bud

Molecules that we sense as taste bind to receptor proteins on the plasma membrane of a chemoreceptor cell.

Chemoreceptor cells

The chemoreceptor cells trigger a nerve impulse in a neuron.

Neuron

Nerve impulses to brain

**FIGURE 30.11 Taste in Humans**

Taste buds consist of groups of chemoreceptor cells and supporting cells. The receptor cells respond to chemicals that lodge in the tongue during eating.

**FIGURE 30.12** The Human Sense of Smell

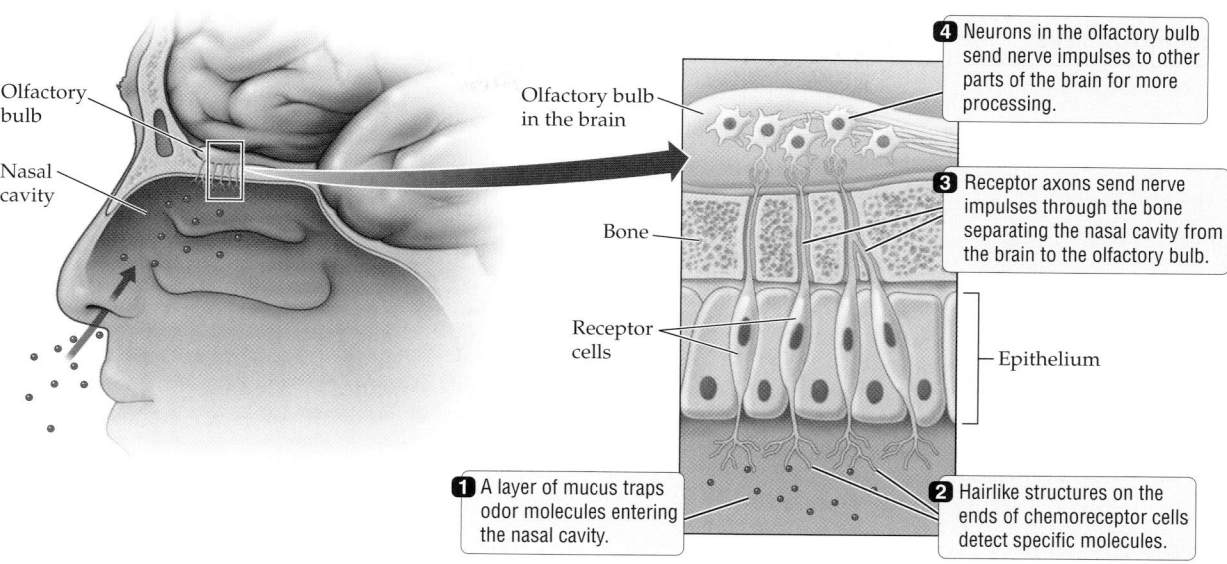

**4** Neurons in the olfactory bulb send nerve impulses to other parts of the brain for more processing.

Olfactory bulb in the brain

**3** Receptor axons send nerve impulses through the bone separating the nasal cavity from the brain to the olfactory bulb.

Bone

Receptor cells

Epithelium

Olfactory bulb

Nasal cavity

**1** A layer of mucus traps odor molecules entering the nasal cavity.

**2** Hairlike structures on the ends of chemoreceptor cells detect specific molecules.

and interior of the body (through our senses of touch, body position, and balance), but also for detecting physical changes happening outside of our bodies, as in hearing. The multicellular pressure-sensitive receptors in our fingertips are one type of a mechanoreceptor (**FIGURE 30.13**).

4. **Thermoreceptors** are found in a variety of tissues, including the skin, mouth, and some internal organs. They consist of special areas on sensory neurons that contain membrane channels activated by temperatures in the moderate range (about 10–30°C). As we shall see next, extreme temperatures are sensed by a subclass of pain receptors, rather than by thermoreceptors.

5. **Pain receptors**, technically known as nociceptors [*NOH*-sih-*SEP*-terz], are found in just about every tissue type inside and on the surface of the body. Different types of nociceptors detect different types of noxious stimuli (stimuli that are potentially damaging to the body). They can sense intense heat or cold, poison gases such as mustard gas, and even chemicals in hot peppers and wintergreen! Our perception of "itch" also comes from receptors that belong in the family of pain receptors.

In addition to perceiving stimuli sensed by humans, many animals respond to stimuli that we cannot sense. Some animals, for example, are able to detect the tiny electrical currents generated by the muscular contractions of other animals. Sharks, rays, and salmon are among the organisms that have the ability to sense electrical fields. Sharks may use this sense to detect prey

that are not visible, such as flounder buried just under the sand on the ocean bottom. Table 30.2 summarizes the various senses commonly found in the animal kingdom and the receptors that mediate them.

**MECHANORECEPTORS IN SKIN**

Hair

Light touch receptor

Bare nerve endings sense pain, itch, and temperature.

Touch and pressure receptor

Strong pressure and touch receptor

Sweat gland

Dendrites that wrap around the base of the hair sense its movement.

Neurons

**FIGURE 30.13 Some Mechanoreceptors in Human Skin**
Human skin contains many different types of neurons and sensory cells that provide detailed information about mechanical stimuli.

| TABLE 30.2 | Different Ways to Sense the World | |
| --- | --- | --- |
| **RECEPTOR TYPE** | **STIMULUS** | **SENSE(S)** |
| Chemoreceptors | Chemicals | Taste, smell |
| Photoreceptors | Light | Vision |
| Mechanoreceptors | Physical changes | Touch, hearing, proprioception (body position), balance |
| Thermoreceptors | Moderate heat and cold | Thermoreception (gradations of heat and cold) |
| Pain receptors | Injury, noxious chemicals, chemical and physical irritants | Pain, itch |
| Electroreceptors* | Electrical fields (especially those generated by muscle contractions of other animals) | Electrical sense |
| Magnetoreceptors* | Magnetic fields | Magnetic sense |

*The sensory receptors listed here are found in many animals, including most vertebrates. Those marked with an asterisk, however, are not known to be active in humans.

**Helpful to Know**

Note that the second half of "proprioceptor," unlike the names of other receptor types, is "ceptor," not "receptor." This variation in spelling has no significance, though it may make the word a little easier to pronounce.

## 30.5 Photoreceptors: Vision

At the simplest level, light detection requires a light-sensitive pigment. If the pigment molecule changes shape in response to light and that change in turn triggers either a signaling molecule or a nerve impulse, then the organism can respond to light or dark. For certain organisms, including some protists and simple animals, this rudimentary form of vision is adequate.

### Forming images requires a way to focus light

Most animals go well beyond merely responding to light and dark: their eyes can form images. The great variety of image-forming eyes in the animal world have two features in common:

1. The eye gathers light and focuses it onto an array of photoreceptor cells in such a way that all of the light from a single point in the animal's field of view goes to a single point in the array of photoreceptors.

2. The photoreceptor cells convert light energy into nerve impulses that are sent by sensory neurons to the brain for processing.

Animals have developed these two features in several different ways, one of which is the **single-lens eye**. In a typical single-lens eye, a **lens** focuses incoming light point for point, and all the photoreceptors are clustered at the back of the eye in a sheet called the **retina**, which converts the light stimuli into nerve impulses. All vertebrates, including humans, have single-lens eyes (**FIGURE 30.14**). Let's take a look at vision in a single-lens eye, using the human eye as an example.

The formation of a sharp image on the retina requires a lens. The shape of the lens allows it to bend light so that the light coming from each particular point on an object is focused on a corresponding particular point on the retina, stimulating the photoreceptors there (**FIGURE 30.15a**). The result is a pattern of stimulated photoreceptors that re-creates the object. The image of the object that forms on the retina is upside down; further processing in the brain turns the image "right side up" again.

In order to keep an image focused on the retina, the lens must bend light from nearby objects more strongly than light from distant objects. In many vertebrates, including humans, the focus is adjusted by

**STRUCTURE OF THE HUMAN EYE**

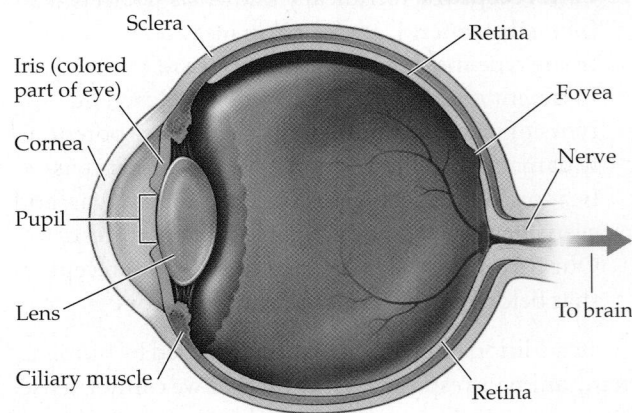

**FIGURE 30.14 Major Features of the Human Eye**

A pupil adjusts the amount of light entering through the lens. The lens focuses images on the retina, where photoreceptor cells generate nerve impulses in sensory neurons that are closely associated with them.

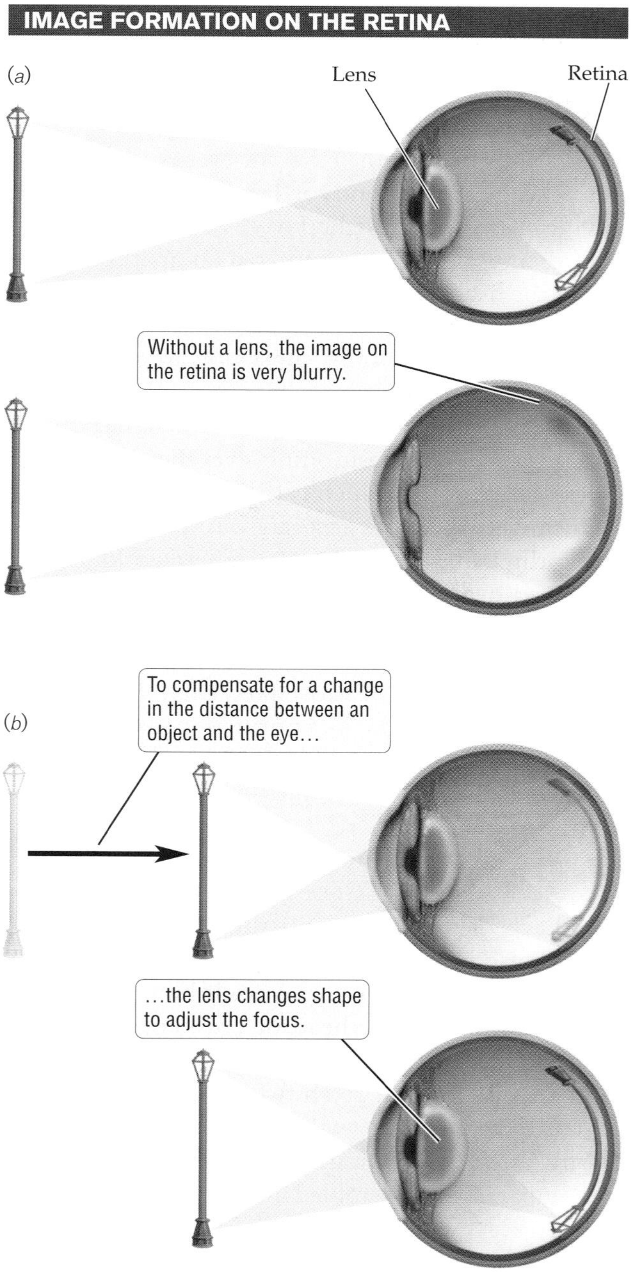

(a)

Lens

Retina

Without a lens, the image on the retina is very blurry.

(b)

To compensate for a change in the distance between an object and the eye…

…the lens changes shape to adjust the focus.

**FIGURE 30.15  How the Human Eye Forms Sharp Images**

(a) The lens brings the image into focus on a layer of photoreceptor cells in the retina. (b) As the distance to an object changes, the eye must adjust to keep the image focused on the retina. It does this primarily by changing the shape of the lens.

changing the shape of the lens (**FIGURE 30.15b**). Such adjustments require intricate neuronal connections between the retina and the brain that enable the nervous system to continually monitor the quality of the image and send instructions to muscles that control the shape of the lens.

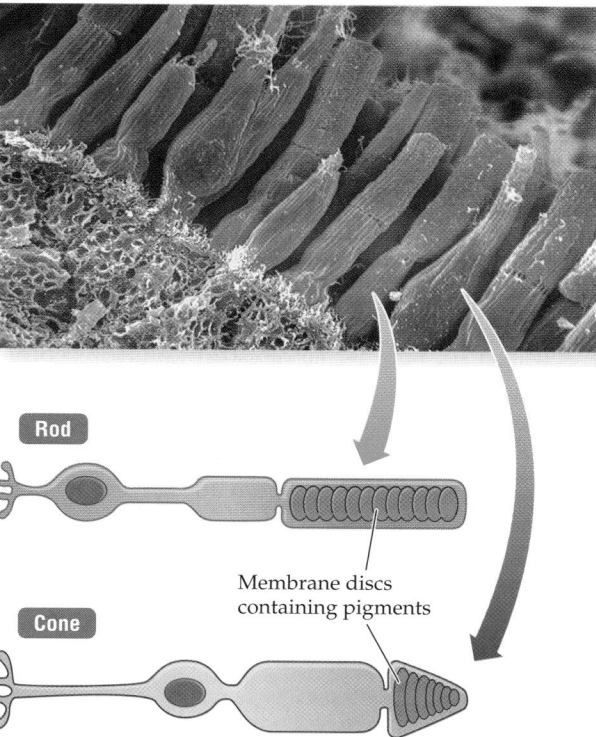

Rod

Membrane discs containing pigments

Cone

**FIGURE 30.16**
**Human Photoreceptors**
These diagrams and the scanning electron micrograph show photoreceptors found in the human retina. Rods (green in micrograph) detect dim light; cones (blue) detect color in relatively bright light.

Vertebrates, especially those living on land, encounter a wide range of light intensities. The amount of light reaching the retina needs to be regulated. Too much light hitting the retina can overload the photoreceptors, but too little light makes for poor images. An adjustable opening called the **pupil** controls the amount of light entering the eye (see Figure 30.14). In bright external light, the pupil becomes smaller, while in dim light it opens wider to let more light in.

The human retina has two classes of photoreceptor cells, *rods* and *cones*, that get their names from their shapes (**FIGURE 30.16**). **Rods**, which are more numerous than cones, cannot detect color, but they do enable us to see images in shades of gray in dim light. **Cones**, on the other hand, transmit information in color, but they function best in bright light. Both rods and cones contain light-sensitive pigments. At night, we can detect shapes only in shades of black and gray because only the rods receive enough light to function.

In addition to our lenses, several other features contribute to the ability of our eyes to produce sharp, crisp images. One is having about 100 million photoreceptors in each retina. Just as the number of pixels affects how sharp a computer image or digital camera image will be, the number of photoreceptors in an eye helps to determine how detailed the image can be. Furthermore, due to efficient processing by other cells in the retina, the information gathered by this huge number of photoreceptors is condensed so that "only" about a million axons are needed to carry signals from rods and cones back to the brain for further processing and integration (**FIGURE 30.17**).

Our vision depends not only on how many photoreceptors we have, but also on where they are located on the retina. The more densely packed the photoreceptors are, the crisper the image. The center of the human retina has a region, called the **fovea** (see Figure 30.14), where rods and cones are the most densely packed. This higher density of photoreceptors enables us to form especially sharp, colorful images of whatever is in the center of our view. For example, as you read this page, you are automatically moving your eyes so that each word you read is projected onto the fovea. To see the importance of the fovea, look at a particular word and then, without moving your eyes, try to make out surrounding words. You can read the word in the center of your view—the one projected onto your fovea—but not the words off to the sides.

## Having two eyes facilitates three-dimensional perception

Most animals capable of forming images tend to have two eyes on their heads. How the eyes are positioned on the head often depends on how the animal "makes a living." Because most herbivores are also prey animals, their eyes tend to be situated far apart on either side of the head. This positioning gives them the widest possible panoramic view—the better to watch out for predators. In contrast, most predators have eyes situated close enough together on the front of the head to enable both eyes to focus on the same point at the same time. This positioning gives them exceptional depth perception, which is the ability to determine the distances of objects accurately in space—that is, to have a three-dimensional view of their world. Highly developed depth perception is a great advantage for chasing and successfully catching prey.

Primates (including humans) have exceptionally good depth perception. Most likely this is because our primate ancestors, like most primates today, lived in trees, where the need for depth perception was (and still is) essential to avoid falling while moving—often leaping—from branch to branch. This evolutionary perspective helps explain why even primates that are not predators, such as gorillas, have depth perception. Each of our eyes sees a slightly different view of the world because our eyes are positioned slightly apart. You can observe for yourself how this works: Without moving your head or eyes, cover first one eye and then the other with your hand. Objects in your view will appear to shift from side to side, and objects that are nearby will seem to shift more than distant ones. In the brain, these two different views are compared and assembled into a three-dimensional image.

## 30.6 Mechanoreceptors: Hearing

The sounds we hear are the result of collisions among the molecules in our external environment. We call these collisions sound waves. Sound waves can travel through any state of matter—solid (such as the soil under your feet or the bones in your head), liquid (water), or gas (air). Our discussion here will focus primarily on the perception of sounds traveling

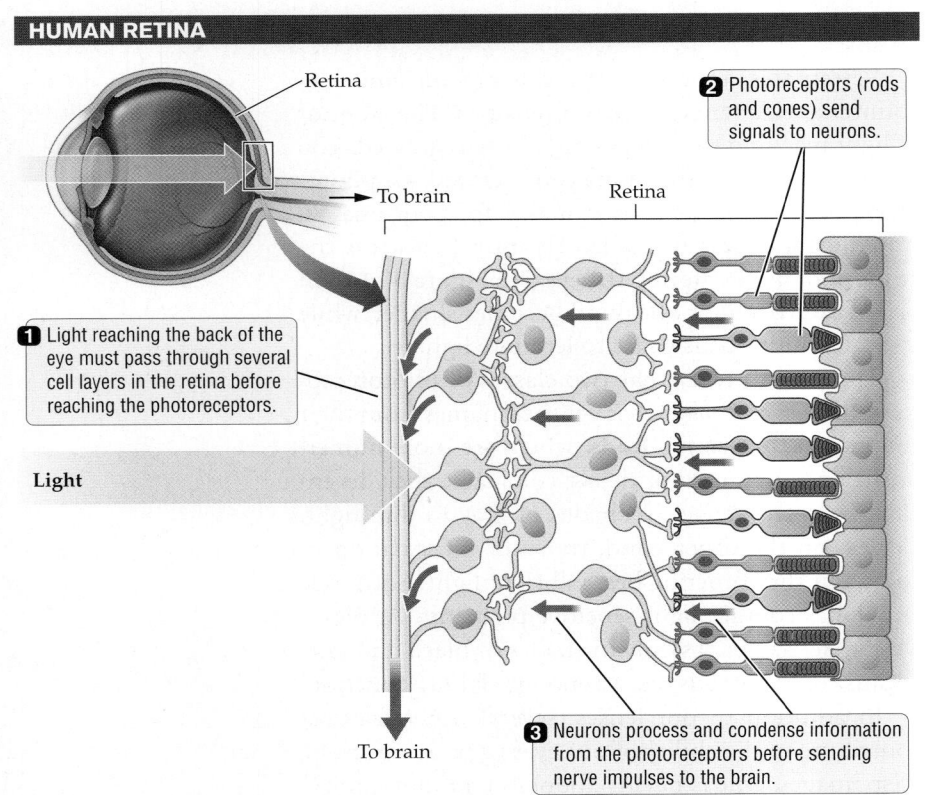

**HUMAN RETINA**

Retina

To brain

**1** Light reaching the back of the eye must pass through several cell layers in the retina before reaching the photoreceptors.

Light

**2** Photoreceptors (rods and cones) send signals to neurons.

Retina

**3** Neurons process and condense information from the photoreceptors before sending nerve impulses to the brain.

To brain

**FIGURE 30.17 Sensory Neurons in the Retina Send Visual Information to the Brain**

Layers of sensory neurons stacked above the rods and cones process visual information to some degree before sending it to the brain via the optic nerve.

through air. Sound waves spread out in all directions from the sound source, much as ripples on the surface of a still pond spread out from a fallen raindrop (**FIGURE 30.18a**). We can hear a drum whether we are over it, under it, behind it, or in front of it.

Hearing involves detecting waves of pressure change in air. For example, when we clap our hands one time, we hear a "smacking" sound. Before we can perceive the sound, however, several events occur. First, the air near the collision is compacted, producing a wave of compressed air that travels outward in all directions at the speed of sound (about 756 miles an hour, or 1,209 kilometers an hour). The pressure is greater within the wave of compressed air generated by the clap than it is in the air just in front of it and just behind it. When the wave of higher-pressure air enters our ears, it is transmitted to mechanoreceptors that transform it into a pattern of electrical impulses conveyed by neurons to the brain; then, after some processing in the brain, we "hear" the clap. But how about a continuous stream of sound? When we listen to a guitar being played, we are receiving a series of pressure waves, thousands

per second, which are created as the vibrating strings compress air (**FIGURE 30.18b**). Sound, then, consists of pressure changes that can occur singly (as with a clap) or many times a second (as with the strings of a guitar). Sound waves diminish in intensity as they travel, which explains why distant sounds are softer than the same sounds originating nearby.

Humans can distinguish two aspects of sound: pitch and loudness. In the case of a sustained sound, the number of pressure changes per second is known as the sound's frequency, which we perceive as pitch. The higher the frequency, the higher the pitch we hear. The highest pitch the human voice can generate is a frequency of about 1,000 pressure changes per second (1,000 hertz [Hz]). It would take a soprano with exceptional gifts to sing a note that high. Male voices in general have a lower pitch than female voices, but the deepest bass voice is unlikely to reach lower than 70 Hz. The lowest sounds we can hear are about 20 Hz; the highest, about 20,000 Hz. Some animals, such as elephants, can hear sounds at lower frequencies than we can. They can pick up pressure waves traveling many miles through the ground, enabling a stray animal to use these low-frequency sounds to locate its herd. Dolphins can produce and perceive sounds as high as about 180,000 Hz. Like most whales and most bats, dolphins use sound not only to communicate but also to find prey and to navigate through their watery world. They produce high-frequency sounds and listen to the echoes to generate a "sound image" of their surroundings through the process known as echolocation.

The loudness of a sound reflects the intensity of the pressure changes in the sound waves. Clapping loudly creates a greater pressure change than clapping softly. The ability to detect sound depends on the fact that the mechanoreceptors in the ear are arranged so that they can respond to rapid and regular pressure changes as these changes arrive in waves.

## Hearing is a highly refined form of mechanoreception

Animals use various ear designs to concentrate and convert waves of changing air pressure into meaningful patterns of nerve impulses. The human ear illustrates the essential features of many vertebrate hearing systems, in which sound waves are concentrated by an outer ear and funneled to thin, delicate membranes and mechanoreceptors (**FIGURE 30.19a**). Let's look more closely at how this works.

**SOUND WAVES**

(a)

(b)

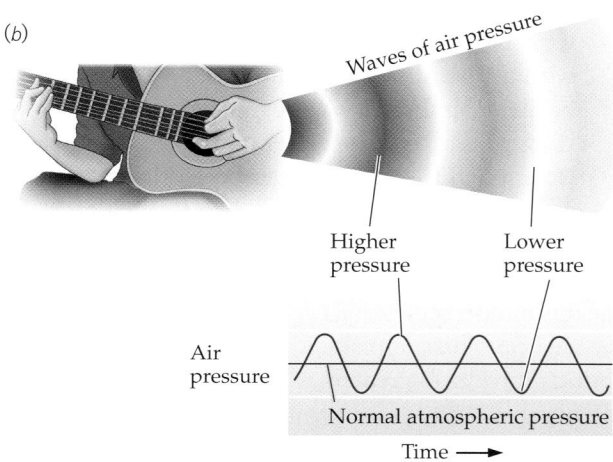

**FIGURE 30.18 Sound Is Composed of Pressure Waves**
(a) Waves on a pond's surface from a raindrop radiate outward much like sound waves. (b) Sound caused by a vibrating object consists of rapidly changing waves of higher and lower air pressure.

(a)

The outer ear funnels sound to the eardrum.

Tiny bones in the middle ear transmit vibrations of the eardrum to the cochlea.

Mechanoreceptor cells in the inner ear detect pressure changes, which the brain interprets as sounds of different frequencies.

Eardrum

Nerve going to brain

Cochlea

Pinna

Eustachian tube

Auditory canal

Outer ear

Middle ear

Inner ear

(b)

Overlying membrane

Mechanoreceptor cells with hairlike projections

Basilar membrane

Dendrites of sensory neurons

To brain

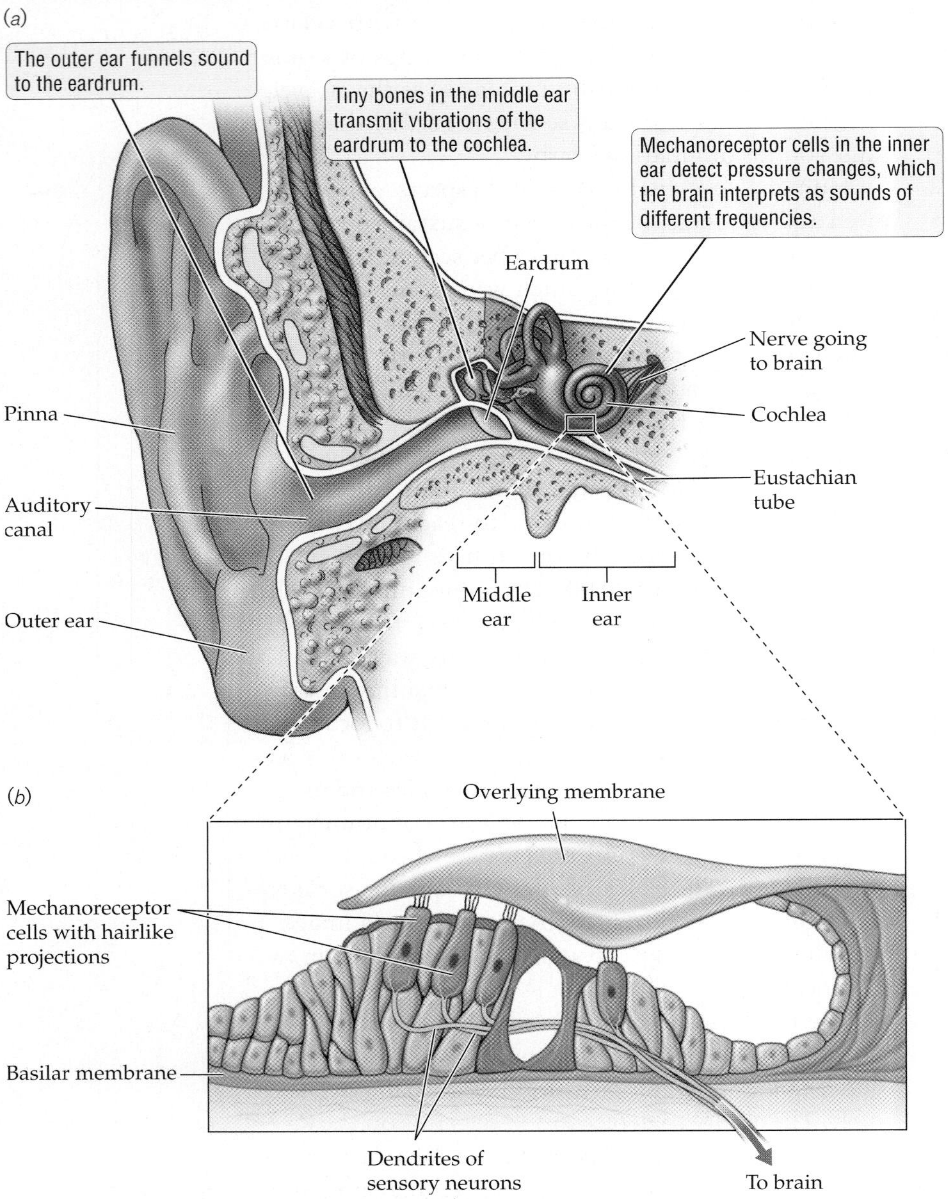

**FIGURE 30.19 The Human Ear Detects Minute Pressure Changes**

(a) The anatomy of the human ear enables it to convert tiny, rapid changes in air pressure into movements that mechanoreceptors can detect. The outer ear acts as a funnel to collect sound over a large area and concentrate it onto the small area of the eardrum. In the middle ear, the eardrum converts changes in air pressure into physical movements, which three tiny bones transmit to the cochlea, in the inner ear. (b) In the cochlea, patches of mechanoreceptor cells in the organ of Corti convert these movements into nerve impulses, which are sent to the brain for processing.

The **outer ear** consists of the **pinna** (plural "pinnae")—the curled, cartilaginous part that we identify as "the ear"—and the **auditory canal**. The characteristic funnel shape of our outer ear collects sound over an area about 20 times that of the **eardrum**, the delicate membrane at the end of the auditory canal. The pinna concentrates sound energy on the eardrum, enabling us to detect fainter sounds than would otherwise be possible. By vibrating in response to rapid changes in air pressure, the eardrum converts sound energy into physical movements. This mechanical energy passes across the eardrum and into the middle ear.

For our hearing apparatus to work properly, the air pressures between the middle and outer ears must be equal. The **eustachian tube [yoo-STAY-shun...]**, which runs from the middle ear to the throat, enables us to equalize these pressures. The need to do this is particularly noticeable during airplane flights, which can cause us to experience painful or unpleasant pressure imbalances when the plane is ascending or descending. These are usually relieved if our ears "pop."

In the **middle ear**, the three tiniest bones in the body receive the vibrations from the eardrum and transmit them to a second membrane, which channels them to the **cochlea [KOH-klee-uh]** (Latin for "snail shell"). The cochlea is a coiled, fluid-filled tube in the **inner ear**. Within the cochlea lies the **organ of Corti [KOHR-tee]**, which converts mechanical sound (sensory input) into a pattern of nerve impulses.

In the organ of Corti, mechanoreceptor cells with hairlike projections lie in close association with the **basilar membrane** (**FIGURE 30.19b**). The basilar membrane vibrates in response to vibrations passed to it from the middle ear. These vibrations push the mechanoreceptor cells up against an overlying membrane, which bends their hairlike projections. This bending triggers nerve impulses in individual sensory neurons that synapse directly with the mechanoreceptor cells.

We can distinguish loudness because louder sounds produce greater movement of the basilar membrane. And we can perceive pitch because different regions of the basilar membrane vibrate in response to different sound frequencies. The internal ears of elephants and dolphins are structurally similar to ours, as are those of most vertebrates. Elephants can position their pinnae, the largest in the animal kingdom, to gather airborne sounds. They seem to have the ability to pick up sound vibrations through their large feet as well. Although dolphins do not have pinnae, they have a short auditory canal, situated on the head in line with the eyes, that conveys sounds to an eardrum and the middle ear. In addition, sound waves are conducted to the middle ear through the dolphin's fat-filled lower jawbone.

## Specialized mechanoreceptors tell us about body position and balance

An often overlooked aspect of mechanical stimuli is the role they play in telling us what our own bodies are doing at any given moment. We are constantly informed of the position of our bodies in space by vast numbers of **proprioceptors** [PROH-**pree-oh**-SEP-**terz**], mechanoreceptors that respond to stretching of or pressure on muscles, tendons, and joints. Proprioceptors tell you that your arm is extended or bent at the elbow, that your left foot has just hit the ground, or that you are smiling.

In addition to proprioceptors, a structure in our ears, called the **vestibule**, helps us maintain our sense of balance (**FIGURE 30.20a**). The mechanoreceptors in the vestibule fall into two classes: those involved in telling up from down, and those involved in detecting head movements. The mechanoreceptor cells have hairlike projections extending into a gelatinous layer topped by a layer of small, dense crystals of calcium carbonate called **otoliths** (from *oto*, "ear"; *lith*, "stone"). The otoliths tend to settle in the direction of the pull of gravity. As they settle, they push on and bend down the hairs in the direction gravity is pulling them (**FIGURE 30.20b**). The pattern of stimulation of the mechanoreceptor cells enables us to distinguish up from down even when our eyes are closed.

The vestibular apparatus also contains three fluid-filled **semicircular canals** (see Figure 30.20). The canals are oriented at right angles to one another, forming a three-dimensional grid. When you rotate or nod your head, the fluid inside the semicircular canals moves relative to the walls of the canals. Mechanoreceptors with hairlike projections inside these canals detect the direction of flow of the fluid. For example, when you shake your head from side to side, the fluid in the canal most nearly parallel to the

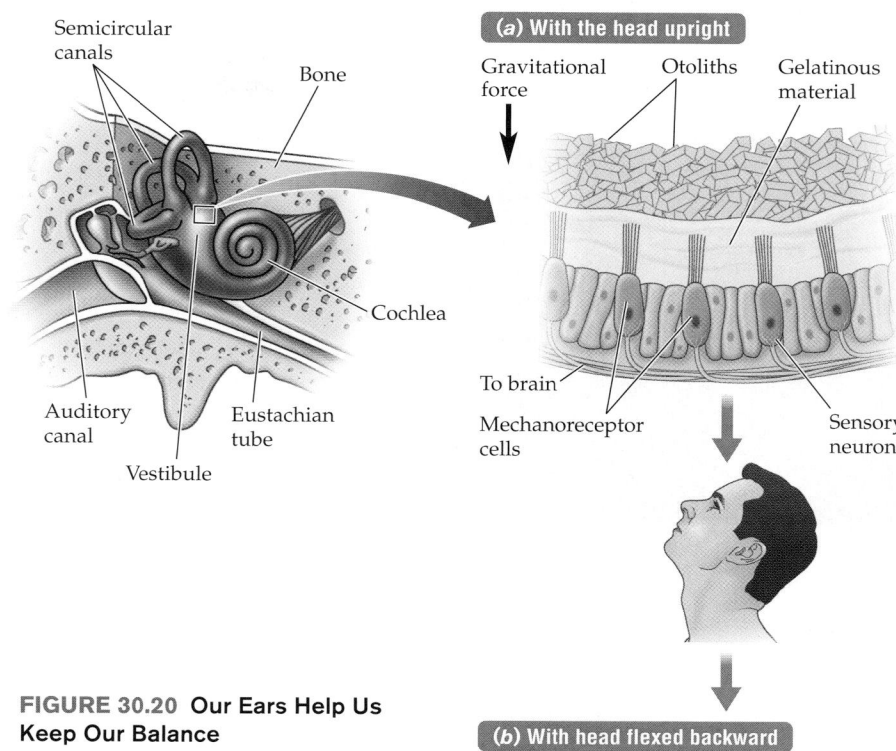

Semicircular canals

Bone

Cochlea

Auditory canal

Eustachian tube

Vestibule

(a) With the head upright

Gravitational force

Otoliths

Gelatinous material

To brain

Mechanoreceptor cells

Sensory neuron

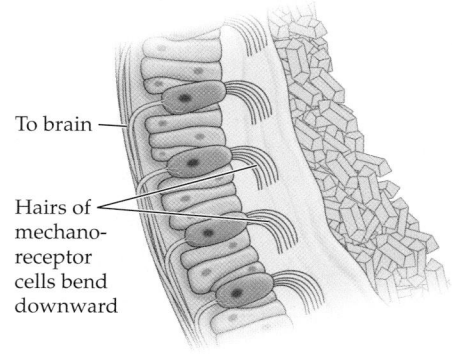

(b) With head flexed backward

To brain

Hairs of mechano-receptor cells bend downward

**FIGURE 30.20 Our Ears Help Us Keep Our Balance**
The vestibule of the inner ear contains otoliths, or "ear stones." (a) When the head is upright, even distribution of the otoliths over the gelatinous material produces no deformation of the hairs of the mechanoreceptor cells. (b) When the head is tilted, the otoliths tumble in response to gravity, pulling some of the gelatinous material downward. This pull, in turn, bends the hairs of the mechanoreceptor cells, triggering nerve impulses in sensory neurons associated with them.

plane of movement flows to the left and right relative to the hairs lining the wall of that canal. The characteristic pattern of nerve impulses from the mechanoreceptor cells lets you know that you are shaking your head.

### Concept Check

1. What is the cause underlying the "popping" of the ears we might experience in an airplane ascending quickly into the skies?

2. Compare and contrast rods and cones.

**Concept Check Answers**

1. The rapid ascent creates an imbalance in air pressure between the outer and middle ear; the pressure equalizes when the ears "pop."

2. Both are pigment-containing, light-sensitive photoreceptor cells. Rods are more numerous and enable vision in dim light; cones are responsible for color vision and work best in bright light.

# Making Eyes, Evolving Eyes

Vision provides our quickest access to the world. When someone brings a rose into a room, we see it long before its fragrance reaches us. When Terry Byland lost his vision, he lost a major source of information about the world. Eager to see again, he volunteered for a surgical implant he hoped would restore his vision. To make a visual prosthesis work, surgeons connect tiny electrodes to a person's retina; a camera mounted on eyeglasses transmits a signal to the electrodes; and the retina transmits a signal to the brain by way of the optic nerve. What would Byland's brain make of such signals? Would he see anything?

At first, Byland saw only blotches of light. He couldn't tell left from right or vertical bars from horizontal ones. It was exciting to see anything at all, but not very useful. It took his brain a long time to learn to make sense of the signals the optic nerve was transmitting from the camera. But after 3 years, he could see a tree branch, an overhead light coming on, and even his teenage son's shadow. He was utterly delighted.

The human eye does such an amazing job of gathering images and transmitting a complex signal that many people find it hard to imagine how it ever evolved. But, step by step, it did. The simplest kind of a light-sensing structure does not focus an image. In flatworms, crabs, and sea snakes, the whole surface of the skin is sensitive to light—a trait called "dermal light sense." Dermal light sense allows animals to tell sun from shade and day from night. Some aquatic animals rely on dermal light sense to detect sunlight, so that they know when to descend into deeper water to avoid predators.

Jellyfish and some worms have eyespots (**FIGURE 30.21**), patches of photoreceptors in pairs or groups that allow animals to compare light on different sides of the body. Eyespots don't generate an image, but they can tell a jellyfish if there's more light to the left or to the right or if a predator is casting a shadow. A step up from eyespots are pit eyes, eyespots at the bottom of a depression in the skin. Pit eyes, which had already evolved in ancient snails half a billion years ago, allow animals to pinpoint the *direction* of a light source.

Two more evolutionary innovations were the lens and the retina. Recall that the retina is a thin layer of photoreceptors that lies at the back of the eye and transmits light signals to the optic nerve (see Figure 30.17 ). In addition, many lineages of animals have evolved lenses that focus light onto the retina. Some lenses concentrate light on relatively insensitive receptors without generating an image. Others focus a high-resolution image onto the retina.

Living organisms have evolved innumerable different eyes over millions of years. Some lineages have stuck with simple light-detecting eyespots or pit eyes. Others have evolved complex and sophisticated eyes—combining deep pits, retinas, and lenses into camera eyes, like those of octopi and vertebrates.

The artificial eye that Terry Byland received was capable of transmitting only 16 pixels of information at a time—a tiny fraction of the information transmitted by even a cheap video camera today. Researchers are working on wireless retinal implants that will enable recipients to read facial expressions, and one day Terry Byland may be able to see his adult son's face for the first time.

**FIGURE 30.21  Eyespots on a Jellyfish**
Jellyfish have clusters of photoreceptor cells known as eyespots (black dots near the rim of the "bell"). Eyespots were an evolutionary stage in the development of the camera eye of vertebrates and of octopi.

# Crank That iPod: Hearing Loss Rates Lower Than Thought

BY JACQUI CHENG, *wired.com*

We all surely remember what our parents drilled into our brains about listening to loud music: Turn that . . . down or you'll go deaf! As it turns out, the prevalence of young people suffering from hearing loss thanks to loud music may be much lower than previously believed, according to a new report published in the *Journal of Speech, Language, and Hearing Research*. Although the latest findings go against recent research, the researchers warned that we should still be cautious of our exposure to loud noises over time.

The paper's authors, from the University of Minnesota, believe that conventional hearing tests are producing false positives when measuring low levels of hearing loss in children and teenagers. According to U of M Department of Speech-Language-Hearing Sciences professor Bert Schlauch, who headed the study, 10 percent or more of children are falsely identified as having noise-related hearing loss this way . . .

Concerns about childhood hearing loss have been amplified in recent years thanks to the proliferation of personal music players. In 2006, Apple was sued for selling a device—the iPod—that could result in hearing loss, even though the plaintiff in that case did not claim to have suffered any kind of hearing loss of his own. That case was eventually dismissed because an iPod can be used in a manner that wouldn't cause hearing loss, but debates about whether music players need lower default music settings have stayed strong.

Even though the real problem may be far lower than what [a recent *Journal of the American Medical Association* (*JAMA*)] study claimed, the U of M researchers warn that we shouldn't just start cranking our iPods back up again . . .

"Our findings do not mean that people should not be concerned about exposure to loud sounds, such as those from personal stereo devices, live music concerts or gun fire," Schlauch said. "The damage may build up over time and not appear until a person is older. For all sounds, the risk increases the more intense the sound and the longer the exposure, particularly from sustained or continuous sounds."

In many areas, schoolchildren are routinely tested for hearing loss. While it's usually clear if a child has moderate to severe hearing loss, this article suggests that tests have overestimated *low-level* hearing damage because of "false positives." Of the 28 million people in the United States who suffer from hearing loss, about one-third have noise-related hearing damage. Noise can cause mild to profound hearing loss, as well as tinnitus [**tin**-*EYE*-**tus**], the perception of constant noise—whether ringing, buzzing, or roaring—that only the sufferer can hear. Explosions and gunfire can permanently damage the sensitive mechanoreceptor hair cells in the inner ear, as well as the auditory nerve, which carries signals to the brain. But long exposure to everyday sounds such as loud traffic noise can also permanently damage hearing.

How much noise is too much? Scientists measure sound intensity in decibels (dB). Every 10-dB increase is 10 times louder. In other words, 60 dB is 10 times louder than 50 dB. We can usually ignore background noise such as the humming of a refrigerator at 45 dB or less. It's harder to ignore conversation (50–60 dB), loud snoring (75 dB or more), or city traffic (85 dB). Extremely loud noise, such as a poorly muffled motorcycle or a jet taking off, can measure 120–150 dB and can immediately damage hearing.

Finally, long or repeated exposure to noise greater than 85 dB also damages hearing. Working long hours in a noisy environment or regularly playing portable music devices at 90 dB or more can permanently damage the delicate hair cells. To protect hearing, earplugs and earmuffs should be worn around constant loud noise or extremely loud impulse noises such as gunfire. Music should not be played longer than an hour or louder than 85 dB—a volume that is much softer than the maximum volume of most devices.

## Evaluating the News

**1.** How are sensory structures in the ear affected by sound waves?

**2.** Legislation often requires hearing protection for workers who spend an 8-hour workday exposed to sound louder than 90 dB (as loud as a lawn mower). Make an educated guess about your sound exposure in your work or school environment and your nonwork environment (including listening to music through earphones). Does your exposure meet these guidelines? Are there any activities that pose a particular risk to your hearing? If so, what could you do to protect your hearing?

**3.** Many communities in recent years have enacted restrictions on noise. Does your city or town have such restrictions? What do they allow and restrict? Are the restrictions enforced? Do you think they're fair?

**SOURCE:** *Wired.com*, September 20, 2010.

# CHAPTER REVIEW

## Summary

### 30.1 An Overview of the Nervous System

- Neurons carry signals along dendrites and axons.
- Axons are bundled together with supporting tissues into nerves.
- Ganglia (clusters of neuronal cell bodies) are integrating centers.
- The vertebrate nervous system is divided into the central nervous system (CNS), consisting of the brain and spinal cord, and the peripheral nervous system (PNS), consisting of the sensory organs and all the remaining nerves.
- Sensory neurons carry sensory input from the PNS to the CNS, interneurons in the CNS process sensory input, and motor neurons carry command signals from the CNS back to the PNS.
- Reflex arcs process sensory input from the PNS without input from the brain. Reflex arcs give the body a much quicker response to danger than would be possible if processing by the brain were required.

### 30.2 Signal Transmission by Neurons

- The electrical signals carried by neurons are called action potentials.
- The difference in electrical charge that exists across the membrane of an unstimulated axon is called its resting potential.
- An action potential is a brief reversal of the charge difference across the membrane; it is propagated in a single direction down an axon, and is an "all-or-nothing" event.
- In myelinated axons, action potentials travel quickly, leaping from one node of Ranvier (a gap in the myelin sheath) to the next.

### 30.3 Organization of the Human Brain

- The human brain has three major regions: forebrain, midbrain, and hindbrain. The major components of the forebrain are the cerebrum, the thalamus, and the hypothalamus. The cerebrum is responsible for higher order brain functions, including learning and memory, and voluntary actions. The midbrain helps maintain muscle tone and relays some sensory input to higher brain centers. The hindbrain coordinates breathing rhythms and balance.
- The thalamus filters and sorts incoming signals and sends those that need further processing to the appropriate part of the brain.

- The hypothalamus regulates homeostasis, sleep and wakefulness, and many endocrine functions.

### 30.4 Sensory Structures: Making Sense of the Environment

- All human senses rely on input from just five types of sensory receptors. Chemoreceptors respond to chemical stimuli and are used in smell and taste. Mechanoreceptors respond to physical changes in the environment and are found in our senses of touch, pain, temperature, hearing, body position, and balance. Photoreceptors respond to light and are used only in vision.
- All receptors convert environmental stimuli into nerve impulses that are carried by sensory neurons to the CNS.

### 30.5 Photoreceptors: Vision

- Most image-forming eyes can (1) focus light onto photoreceptors so that the pattern of stimulation of the photoreceptors re-creates the object or scene being viewed, and (2) convert light energy into nerve impulses, which are conveyed by sensory neurons to the brain for processing.
- There are two types of photoreceptors: rods and cones. Rods produce images in dim light, but not color. Cones respond to color but require brighter light to function.

### 30.6 Mechanoreceptors: Hearing

- Mechanoreceptors tell us about our immediate internal and external environments, and help us perceive sound.
- Hearing involves detecting waves of rapid pressure changes. Vibrations in the eardrum are passed along to tiny bones in the middle ear.
- From the middle ear, vibrations are passed to the cochlea in the inner ear. The organ of Corti in the cochlea contains mechanoreceptor cells with hairlike projections that respond to vibrations of the basilar membrane.
- The vestibular apparatus, part of the inner ear, helps with our sense of balance.

## Key Terms

action potential (p. 657)
auditory canal (p. 670)
autonomic (p. 656)
axon (p. 653)
basilar membrane (p. 670)
brain (p. 654)
brain stem (p. 663)
central nervous system (CNS) (p. 652)
cerebral cortex (p. 662)

cerebrum (p. 661)
chemoreceptor (p. 664)
cochlea (p. 670)
cone (p. 667)
dendrite (p. 653)
depolarized (p. 657)
eardrum (p. 670)
eustachian tube (p. 670)
forebrain (p. 661)
fovea (p. 668)

ganglion (p. 654)
hindbrain (p. 663)
hypothalamus (p. 662)
inner ear (p. 670)
interneuron (p. 655)
lens (p. 666)
limbic system (p. 663)
mechanoreceptor (p. 664)
midbrain (p. 663)
middle ear (p. 670)

motor neuron (p. 655)
myelin (p. 654)
nerve (p. 654)
nervous system (p. 652)
neuron (p. 652)
neurotransmitter (p. 659)
node of Ranvier (p. 659)
organ of Corti (p. 670)
otolith (p. 671)
outer ear (p. 670)

# Self-Quiz

1. In neurons, dendrites
   a. send signals to other neurons.
   b. secrete neurotransmitters.
   c. receive signals from other neurons.
   d. secrete hormones.

2. Action potentials are converted from an electrical to a chemical signal at
   a. a dendrite.　　　c. a node of Ranvier.
   b. an axon.　　　　d. a synapse.

3. Neurotransmitters
   a. move across synaptic clefts by diffusion.
   b. cannot diffuse out from axons that are unmyelinated.
   c. are found only in interneurons.
   d. move across synaptic clefts only by active transport.

4. The part of the human brain that is proportionally much larger than in other primates is the
   a. thalamus.　　　c. neocortex.
   b. pituitary gland.　d. midbrain.

5. The brain stem contains the
   a. cerebral cortex.　c. cerebrum.
   b. pons.　　　　　　d. thalamus.

6. The function of the hypothalamus
   a. is to release hormones that control the function of the pituitary gland.
   b. is to release hormones that directly control the thyroid gland.
   c. was largely responsible for the increase in human brain size over evolutionary time.
   d. is to serve as a neurotransmitter.

7. In our eyes, light is focused by the
   a. rods.　　　　　　b. cones.
   c. retina.　　　　　d. lens.

8. In our eyes, rods and cones are found in the
   a. lens.　　　　　　c. cochlea.
   b. retina.　　　　　d. pupil.

9. We see images in three dimensions because
   a. our lenses can change shape.
   b. rods are in one eye and cones are in the other.
   c. our eyes are positioned slightly apart.
   d. our retinas are curved.

10. Hearing is a sophisticated form of
    a. mechanoreception.
    b. chemoreception.
    c. photoreception.
    d. none of the above

11. Which of the following would detect the position of your head?
    a. cochlea
    b. vestibular apparatus
    c. organ of Corti
    d. pinna

12. The mechanoreceptors that convert sound into nerve impulses are located in the
    a. eardrum.
    b. semicircular canals.
    c. cochlea.
    d. vestibular apparatus.

# Analysis and Application

1. Describe the movement of action potentials down the axon of one neuron and across a synapse to the next neuron.

2. A simple reflex arc can coordinate an appropriate response to a stimulus such as heat with just two or three neurons. If the incoming information says "too hot," then the instruction "move hand" is sent out. If no "too hot" message is sent, then a "move hand" instruction does not get sent. Using four neurons, including one interneuron, design a nervous system that could decide to send the message "move" or "stay still" in response to two different stimuli that do not necessarily occur at the same time.

3. The pigments in the rods and cones of our eyes briefly change chemically when light hits them. The chemical change creates a nerve impulse that travels to the brain. Do you think these receptors should be called "chemoreceptors" instead of "photoreceptors"? Why or why not?

4. Suppose you are listening to a guitar being played. Describe how the sense of touch in the player's fingers and your sense of hearing are related to the performance. What mechanoreceptors are involved in the player and in the listener?

# 31 Skeletons, Muscles, and Movement

**HOW HIGH?** Cats sometimes survive spectacularly high falls. How do they do it?

# Falling Cats

Urban life can be tough. Humans living in cities must withstand polluted air, violent crime, and speeding taxis. As our companions, domestic cats face their own hazards in a landscape of tall buildings and hard pavement. Cats love to sit in high places and watch the world go by—a fact that lures them to open windows and balconies several stories above street level. Unfortunately for many young cats, a high-altitude lunge at a passing fly, a catnap on a balcony railing, or a moment of simple inattention can sometimes result in a quick vertical trip that ends abruptly on a sidewalk several stories below.

Amazingly, cats can survive falls that would kill most people. Falls are a major cause of death in children and teens, and the death rate increases rapidly as people fall from greater heights—up to the seventh story, after which the chance of dying levels off at almost 100 percent. Just as in humans, most cats that fall from buildings are young. Yet, for a variety of reasons, they survive better than people.

In a study of 132 cats that showed up at a New York City animal hospital for injuries from falls, many of the cats did surprisingly well. The cats fell an average distance of five and a half stories. Of these 132 cats, 104 (79 percent) survived their injuries. One cat even fell an astounding 32 stories and lived. Incredibly, according to the study's conclusions, cats seemed to stand a better chance of surviving a fall from above the seventh story than from below the seventh story. Only 5 percent of cats that fell from between 7 stories and 32 stories died. By comparison, 10 percent of cats that fell from between 2 stories and 6 stories died.

How could so many cats survive falls that would kill most people? Why did falling seven floors appear to be less risky for a cat than falling three floors?

Before we can consider these questions, we must examine the structure and function of animal skeletons and muscles, explore the constraints on movement in certain environments, and consider the energy-saving adaptations that some organisms have evolved to meet those challenges.

## MAIN MESSAGE

The skeletons and muscles of animals work together to make movement possible.

### KEY CONCEPTS

- In humans, the axial skeleton supports and protects the long axis of the body. It consists of the skull, the ribs, and a long, bony spinal column. The bones of the arms, legs, and pelvis make up the appendicular skeleton, which facilitates movement.

- Bone and cartilage are living tissues—light, strong, flexible, and well adapted to the functions they serve. Our bones are constantly changing in response to how we live.

- Joints are combinations of stiff and flexible materials that enable motion in animals.

- Tendons connect muscle to bone. Ligaments hold bones together at joints.

- Skeletons and muscles work together to control the strength and speed of movement.

- In muscles, two types of protein filaments pull against each other, with the help of ATP, to power muscle contractions.

- Internal movements are created by specialized muscle types. Cardiac muscle is found in the heart, where its contractions pump blood. Smooth muscle, found in the digestive tract and blood vessels, can sustain contractions longer than skeletal muscles can.

- Skeletons can be internal or external. External skeletons, called exoskeletons, provide a protective armor, and in terrestrial invertebrates they prevent excessive moisture loss. Animals that outgrow their exoskeletons must shed them periodically, in the process known as molting.

UNLIKE PLANTS, ANIMALS CAN MOVE through their environment at least at some stage of their life cycle. The ability of animals to walk, fly, or swim provides an almost endless list of potential advantages. Think how different our own lives would be if we could not move from place to place—that is, if we lacked the capacity for **locomotion**. We could not explore our surroundings, we could not dance or play sports, we could not seek mates, or even find and consume food and water.

Animals can move so well because they have a unique combination of structural support tissues, or **skeletons**, connected to **muscle** tissues that can contract and relax. We begin this chapter with a brief introduction to the skeletons of humans and other animals, concentrating on the features that give an animal a strong framework and enable it to move. We consider how muscles work at the microscopic level and how the concerted action of many bundles of muscle cells moves the skeletal framework. Finally, with this basic understanding in hand, we examine some of the physical forces that affect how animals move through different environments by walking, swimming, or flying.

## 31.1 Basic Features of the Human Skeleton

We will use the human skeleton as a starting point in our overview of animal skeletons because it is most familiar to us. Understanding some basic features of the human skeleton will give us a point of reference for exploring some general patterns in the way skeletons work in other animals.

## The human skeleton protects organs and enables complex movements

Like all vertebrates, we humans have a bony internal skeleton that supports the body, gives it shape, and protects soft tissues and organs. Working in conjunction with muscles, this strong internal framework also enables movement, from the precise strokes of an artist's brush to a dancer's gravity-defying leap into the air. The skeleton has two major components: the *axial skeleton* and the *appendicular skeleton* (**FIGURE 31.1a**). The **axial skeleton** supports and protects the long axis of the body. As its name indicates, it extends from head to "tail" end, and consists of the skull, the ribs, and a long, bony spinal column. Although the axial skeleton plays a role in movement—we can, for instance, nod and turn our heads because of the way the spinal column joins up with the skull—its primary purpose is to protect vital organs. Ribs form a strong cage around the heart, liver, pancreas, and lungs. The skull is a concrete-hard assembly of bones "welded" together to protect the delicate brain from injury. The hard bones of the spinal column surround the nerves in the spinal cord that send and receive vital information.

The bones of the arms and legs (our appendages) and the pelvis make up the **appendicular skeleton** [AP-un-DIK-yoo-ler . . .]. These bones have more to do with motion than with protection. In sports, arms and legs, with their attached muscles, do most of the throwing and running, while the axial skeleton does most of the protecting. But both components of the skeleton work together to create the graceful, fluid motions that we see in a running back or a ballet dancer. A wide receiver's head fake is mostly axial. A dancer's long leap is mostly appendicular.

## The human skeleton is composed of bone tissue

Although much of bone is made up of nonliving material, bone is a living tissue that has a blood and nerve supply. Specialized bone cells, called **osteocytes**, surround themselves with a hard, nonliving mineral matrix composed largely of calcium and phosphate compounds that we accumulate from our diet. Bones are built for strength. The long bones in our body (such as the upper arm bone, the humerus, shown

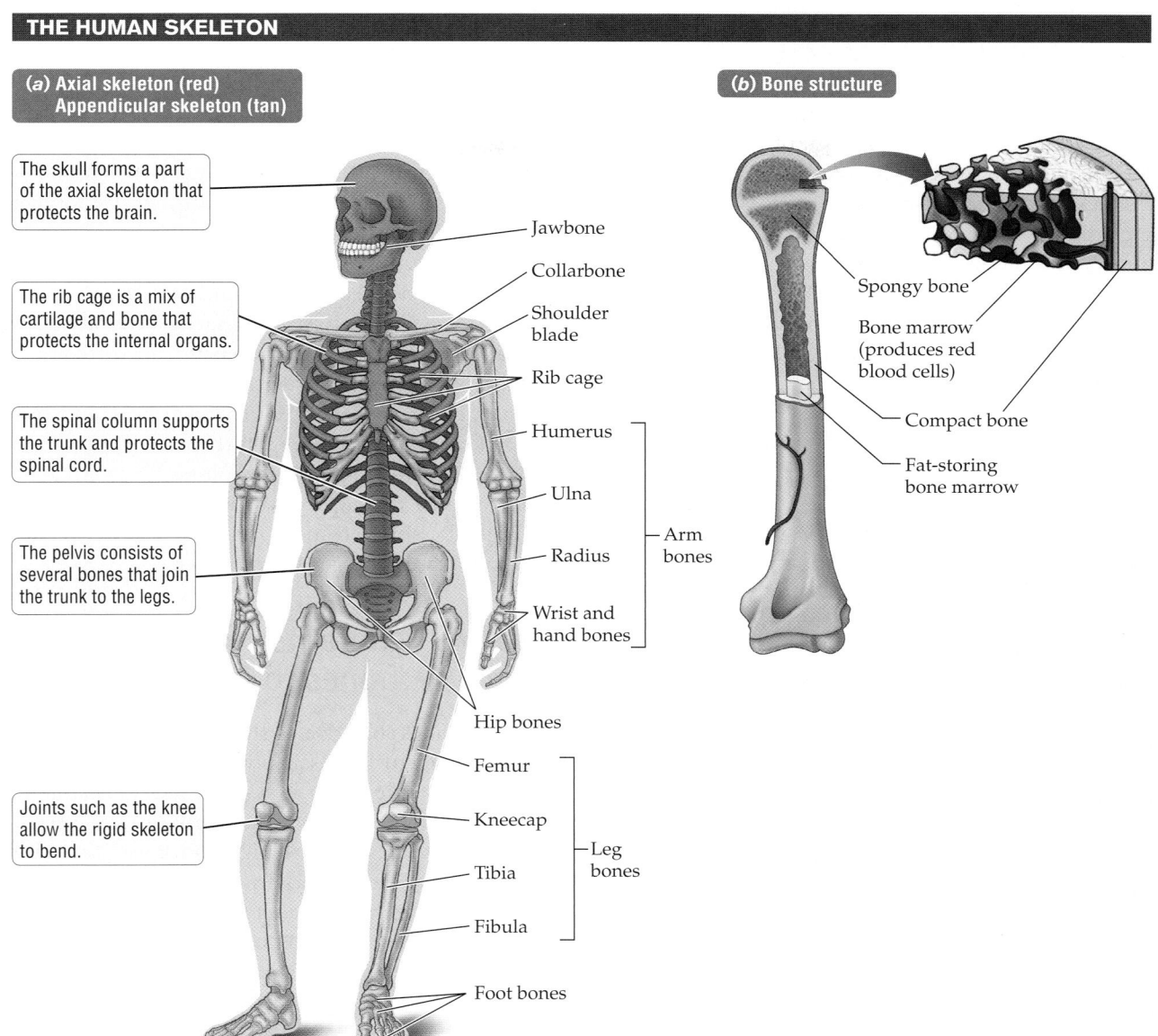

**(a) Axial skeleton (red)**
**Appendicular skeleton (tan)**

The skull forms a part of the axial skeleton that protects the brain.

The rib cage is a mix of cartilage and bone that protects the internal organs.

The spinal column supports the trunk and protects the spinal cord.

The pelvis consists of several bones that join the trunk to the legs.

Joints such as the knee allow the rigid skeleton to bend.

Jawbone
Collarbone
Shoulder blade
Rib cage
Humerus
Ulna
Radius
Wrist and hand bones
} Arm bones

Hip bones
Femur
Kneecap
Tibia
Fibula
} Leg bones

Foot bones

**(b) Bone structure**

Spongy bone
Bone marrow (produces red blood cells)
Compact bone
Fat-storing bone marrow

**FIGURE 31.1 The Human Skeleton Has Two Major Components Made Up of Bones**

(*a*) The axial human skeleton (red) protects vital organs; the appendicular skeleton (tan) facilitates movement. Cartilage is shown in gray. (*b*) Most of our bones, like the humerus detailed here, have an intricate inner structure.

### Helpful to Know

Many medical and anatomical terms relating to bone contain the prefix "osteo" (meaning "bone"). For example, an osteocyte (from *cyte,* "cell") is a bone cell. Osteoarthritis (from *arthr,* "joint") is a form of arthritis marked by degeneration of the cartilage and bone in joints. And osteoporosis (literally "porous bone") is a condition in which the bones become increasingly brittle because of calcium loss.

in **FIGURE 31.1***b*), and some others, such as the ribs and breastbone, have a hollow interior. The hollow construction makes these bones light but strong—an engineering principle we see replicated in the hollow tubes used for bicycle frames. The cavities inside hollow bones contain **marrow**, a tissue that, depending on the type of bone, produces blood cells, such as red blood cells, or stores fat.

Most bones are made of two major types of bone tissue. **Compact bone** forms the hard, white outer region. **Spongy bone** lies inside the compact bone and is most extensive at the knobby ends of our long bones (see Figure 31.1*b*). The osteocytes in spongy bone are more loosely arranged, forming a meshwork that encloses many small cavities. While the large

cavity in the center of a long bone resists bending and twisting forces exceptionally well, the honeycomb construction of the spongy bone is especially well suited to absorbing crushing forces, such as those received at the elbow end of your humerus if you do a handstand.

Our bones are constantly changing in response to how we live. Physical activity builds stronger bones. The bones in a pitcher's throwing arm, for example, are stronger and have larger ridges to which muscles can attach than do the same bones in the other arm. For this reason doctors encourage older women to exercise to help offset the effects of osteoporosis, a condition common in postmenopausal women that leads to a gradual weakening of the bones. Inactivity leads to weaker bones because special osteocytes (known as osteoclasts) step up the rate at which they remove tissue from the bone when the skeleton is not under physical stress; such weakening is seen, for example, in a person confined to bed by an injury or illness.

Bone is stiff yet somewhat flexible, meaning that it does not snap under moderate stress and it acts like a shock absorber to some extent, so it is well suited to providing a framework for the body. If you snap your fingers, you can feel the force you are exerting, yet your finger and thumb maintain their overall shapes. Without bones for support, your thumb and forefinger would simply collapse against one another as if made of jelly.

## Cartilage provides additional support and cushioning

**Cartilage** is a dense tissue that combines strength with flexibility. As in bone, extracellular material predominates in this type of tissue: only about 5 percent of it consists of cells, which secrete the protein-polysaccharide material that surrounds them and which, together with bound water, make up 95 percent of the tissue's bulk. The extracellular material contains bundles of **collagen [KAH-luh-jun]**, a tough but pliable protein that makes up 25 percent of all the protein in the human body. Collagen is found in a great variety of tissues, including skin, blood vessels, bones, teeth, and the lens of the eye. The material surrounding cartilage cells also contains large amounts of *proteoglycans* ("sugary proteins"); the proteins have so many branched polysaccharides (carbohydrate chains) covalently attached to the amino acid backbone that they look like bottle brushes when examined with powerful microscopes. The sugar chains form many hydrogen bonds with water molecules (see Chapter 5), which is why healthy cartilage is a highly hydrated tissue.

In the human skeleton, cartilage gives form to the nose, to the pinnae of our ears, and to part of the rib cage (see Figure 31.1a). Compared with bone, cartilage is more pliable, as you can observe by gently bending your ear or the tip of your nose. Cartilage is found at nearly every point in the body where two bones come into contact; it creates a smooth surface that prevents the two bony surfaces from grinding against each other. Cartilage covering the upper end of the humerus enables this bone to rotate freely within the shoulder joint, for example. The saclike discs that are sandwiched between each vertebra in the spiny column are also made of cartilage (shown in gray in Figure 31.1a).

**FIGURE 31.2**
**Cartilage Piercing**

The central nose ring and the two rings at the top of the ear pass through holes made in cartilage. The other metal adornments are located in soft tissue lacking both bone and cartilage. While perforations in soft tissue can be repaired through surgery, those in cartilage are not easily repaired.

Although true bones are absent in invertebrates, some invertebrates have cartilage (examples include squid and horseshoe crabs). The first fishes to appear in the fossil record and their present-day relatives—sharks, skates, and rays—have internal skeletons made out of cartilage, not bone. Early in our fetal development, the skeletal system consists almost entirely of cartilage. As development progresses, the cartilage is replaced by bone; at birth, the skeleton is mostly bony, but some *ossification* (conversion of cartilage to bone) continues until puberty.

Cartilage is thin tissue and lacks a blood supply. The living cells in cartilage depend on diffusion to acquire oxygen and nutrients and remove wastes. As a result, cartilage is slow to grow and heal, and some types of cartilage cannot repair themselves after injury (**FIGURE 31.2**). Cartilage in the joints thins with aging; it can also be damaged by mechanical stresses, such as might be suffered in sports, and it is defective in certain diseases and genetic disorders.

## Flexible ligaments connect inflexible bones to form joints

The stiff bones that give us shape and strength would make movement impossible were it not for our *joints*. **Joints** are junctions in the skeletal system that let the skeleton move in specific ways. Walking, for example, requires movement at the hips and knees, as well as many other joints. Joints in our shoulders, elbows, wrists, and fingers enable us to throw a ball, swing a bat, or hold a pen (**FIGURE 31.3**). The lower jaw, which connects to the skull at a joint, can move relative to

**FIGURE 31.3 Our Jointed Skeletons Allow Many Different Movements**

The joints in our body enable the movements required of a gymnast.

the rest of the skull so that we can chew and talk. Specialized, flexible bands of tissue called **ligaments** join bone to bone to help hold a joint together.

## 31.2 How Joints Work

Although our knees differ in detail from other joints in the body and from the joints in many other animals, they make a good general model for how flexible and stiff materials are combined in a joint to allow controlled motion (**FIGURE 31.4**).

The bones that meet in the human knee joint support body weight and define how the leg bends. The lower end of the femur (the thighbone) rides in a pair of grooves at the end of the tibia, which is the larger of the two lower leg bones. This arrangement allows the lower leg to swing forward and back like a hinge, but not side to side, thereby providing both motion and stability. The knee, however, must allow for the slight twisting of the tibia relative to the femur that takes place when we walk.

Ligaments connect bone to bone in the knee, as they do in other joints. Two pairs of ligaments, one on the front and back of the knee and a second on the sides of the knee, connect the femur to the lower leg bones. These ligaments prevent sliding along the joint surface when the knee bends, keeping the femur and the lower leg bones in their proper place. The collagen of the ligaments can stretch slightly, allowing some bending and twisting motion in the knee. **Tendons**, which are also rich in collagen, connect muscle to bone; tendons help hold the knee together by connecting the upper leg muscles to the bones of the lower leg.

Wherever two moving parts rub against each other, as in a joint, wear can erode bone, and friction can waste energy. Layers of cartilage in the knee cushion the points where the femur meets the tibia. The joint is lined with a sheet of tissue, called the synovial membrane, that creates the joint cavity, or **synovial sac**. The space inside the synovial sac is filled with a lubricating fluid that reduces friction between the two bony surfaces, allowing femur and tibia to glide over each other more easily than a skate slides on ice.

When all these components—bone, cartilage, ligaments, tendons, and synovial sacs—work together, the result is a joint that can move with precise control and withstand motion over many decades. When one of

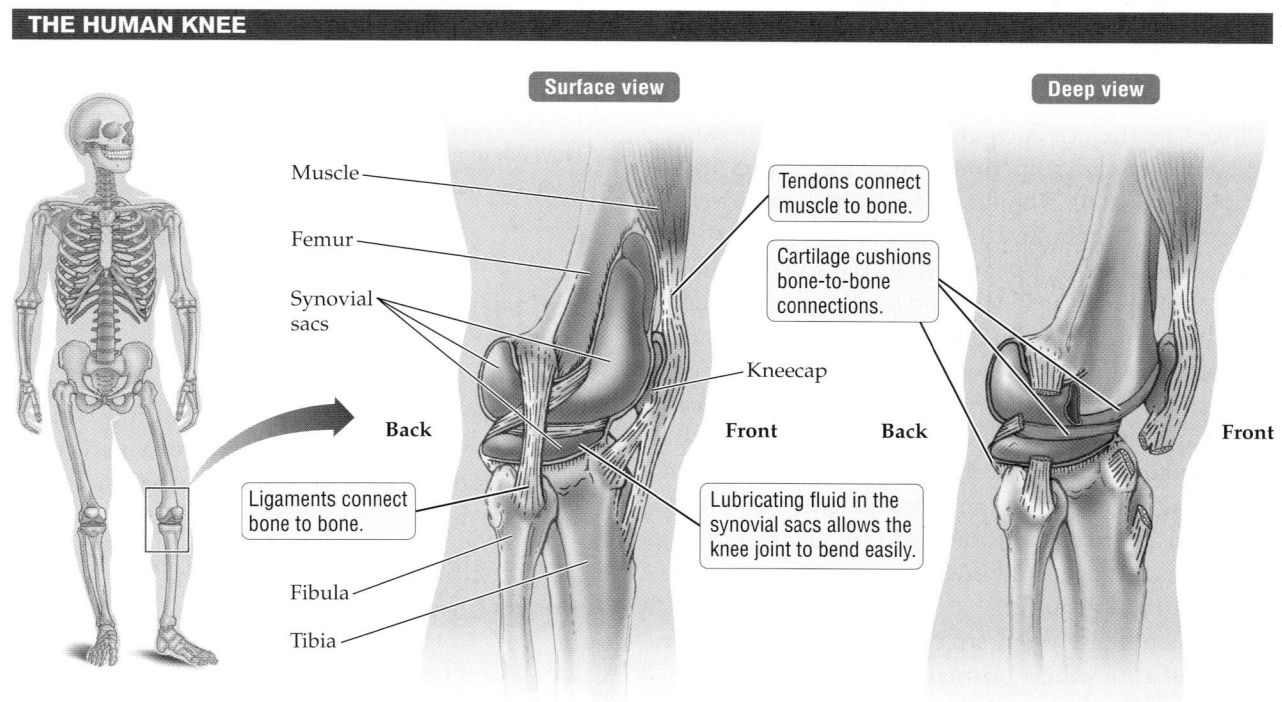

**THE HUMAN KNEE**

Surface view

Deep view

Muscle

Femur

Synovial sacs

Tendons connect muscle to bone.

Cartilage cushions bone-to-bone connections.

Kneecap

Back            Front        Back            Front

Ligaments connect bone to bone.

Lubricating fluid in the synovial sacs allows the knee joint to bend easily.

Fibula

Tibia

**FIGURE 31.4  Joints Combine Rigidity and Flexibility to Enable Movement**
The human knee illustrates how rigidity and flexibility together enable movement. Fluid-containing synovial sacs are shown in purple; cartilage, in gray.

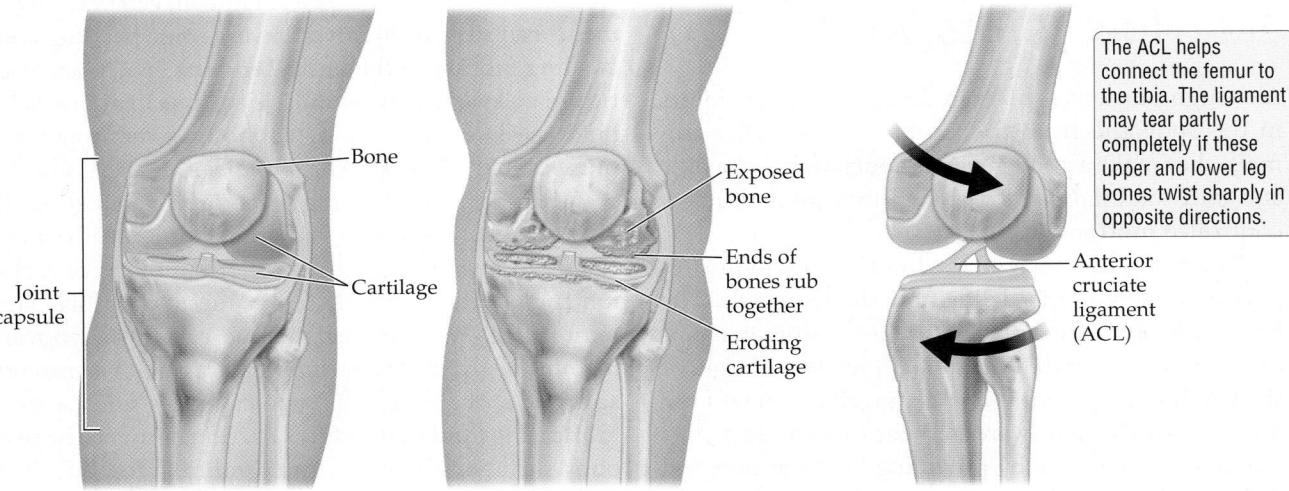

**(a) Healthy joint (left) Osteoarthritic joint (right)**

**(b) Ligament injury**

Bone

Joint capsule

Cartilage

Exposed bone

Ends of bones rub together

Eroding cartilage

The ACL helps connect the femur to the tibia. The ligament may tear partly or completely if these upper and lower leg bones twist sharply in opposite directions.

Anterior cruciate ligament (ACL)

**FIGURE 31.5 The Knee Joint Is Susceptible to Disabling Injury**

(a) Osteoarthritis is a progressive, degenerative condition that affects 12 percent of the general population and 80 percent of those over 75. It is marked by a wearing down of cartilage and the underlying bone because of injury, age, or genetic disorders affecting the skeletal system. (b) The cruciate ligaments, one in the front and the other in the back of the knee, help stabilize the knee. Anterior cruciate ligament (ACL) injuries are most common in sports that involve jumping and pivoting, such as basketball, soccer, tennis, and volleyball. Smaller tears can heal on their own, but if the ligament is completely severed, a new one must be reconstructed using grafts from other ligaments.

these components fails, however, potentially serious medical problems can result.

Tears in the cartilage and ligaments of the knee are common and sometimes disabling sports injuries (**FIGURE 31.5**). The knee joint can handle vertical forces up to about 10 times body weight, which is quite remarkable, but it is less able to resist horizontal forces, such as blocks and tackles in football or being hit in the knee by a hockey puck. Injuries to the knee can also happen when a person rapidly speeds up or slows down while turning. Torn cartilage can lead to increased wear on the knee joint. Unless the tear is repaired surgically, the remaining cartilage and underlying bone may become permanently damaged, leading to arthritis. In various kinds of arthritis, damage to the cartilage and bone in the joint make bending difficult and painful.

Ligaments tear when stretched to more than one and a half times their normal length, either by a severe blow to the front or sides of the knee or by severe twisting of the lower leg relative to the femur. Torn ligaments cannot hold the femur in place relative to the tibia and fibula, leading to movement in the knee joint that may make walking impossible. Like torn cartilage, torn ligaments require surgery to repair.

**Helpful to Know**

Many terms related to muscles contain the prefix "myo" (meaning "muscle"), including "myofibril," "myosin," and "myoglobin," all found in this chapter.

**Concept Check**

1. Compare cartilage and bone.

2. How are tendons and ligaments similar to, and different from, each other?

## 31.3 How Muscles Work

Muscle tissue is unique to animals. The skeleton and its joints are the framework for motion, but muscles provide the power necessary for movement. Muscle tissue possesses a crucial property: it can contract and relax.

### Muscle tissue can generate force by contracting

We are all familiar with muscles in a general way. Most of the features that relate to the role of muscles as a source of power for movement, however, remain hidden to the unaided eye. Skeletal muscle—for example, the biceps muscle of the upper arm (**FIGURE 31.6a**)—consists of many bundles of **muscle fibers**. A muscle fiber is made up of several muscle cells that fused together during the development of the muscle fiber. Each muscle fiber is packed with cylindrical structures containing proteins that can contract by bracing against each other. Each such cylinder is known as a **myofibril**. Myofibrils are organized as a series of contractile units called **sarcomeres** (*sarco*, "flesh"; *mere*, "part").

Sarcomeres are visible as bands when seen through a microscope (**FIGURE 31.6b**). At extremely high mag-nification, additional details of their structure become evident. Two kinds of protein filaments—*actin* and *myosin* filaments—are arranged in a very specific manner inside each sarcomere. Each **actin filament** consists of the protein actin, and each **myosin filament** is made up of many molecules of the protein myosin. In muscle tissues prepared for microscopy, the ends of each sarcomere appear as a dark line, called the **Z disc**. The two

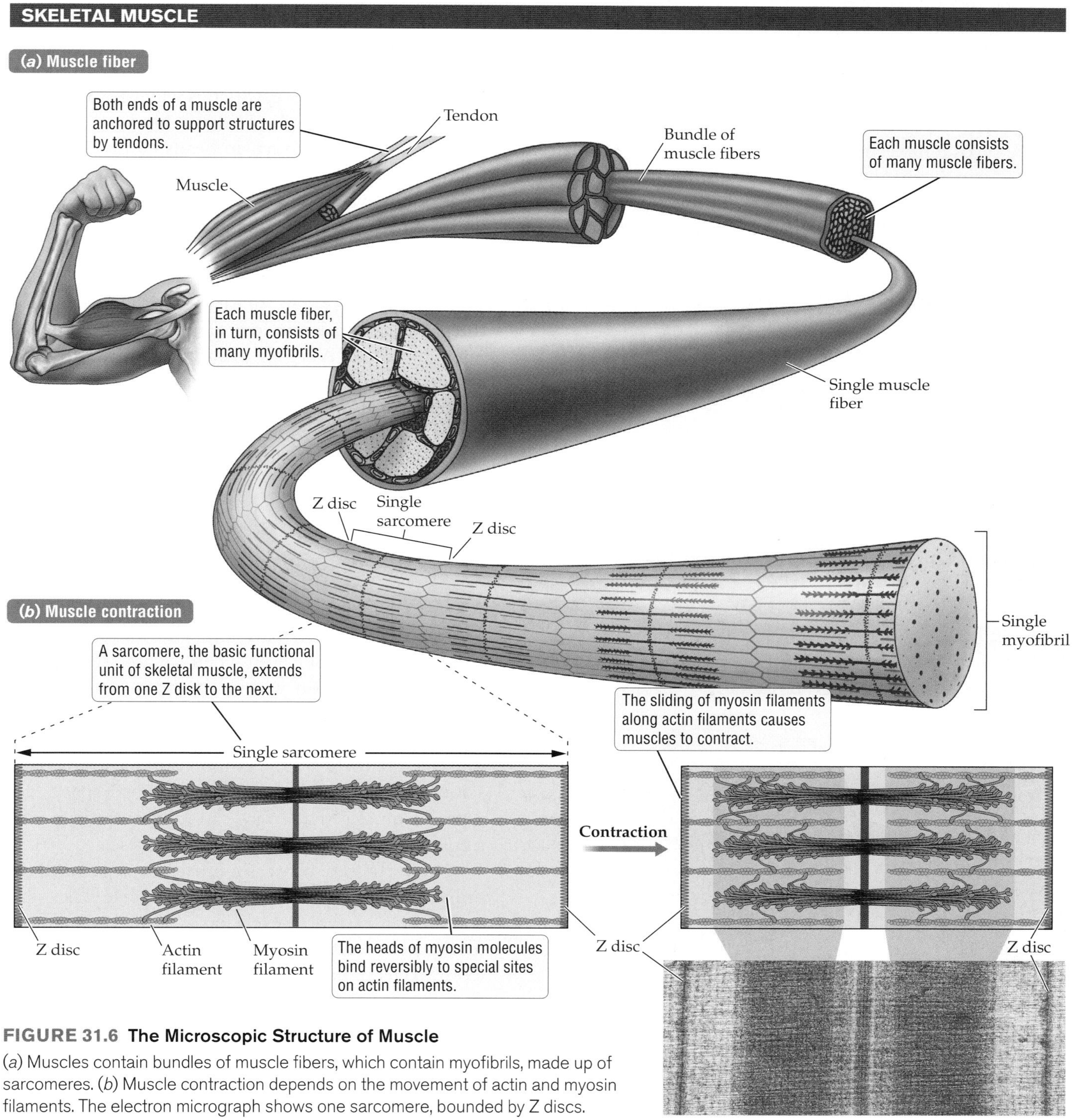

**SKELETAL MUSCLE**

**(a) Muscle fiber**

Both ends of a muscle are anchored to support structures by tendons.

Tendon

Muscle

Bundle of muscle fibers

Each muscle consists of many muscle fibers.

Each muscle fiber, in turn, consists of many myofibrils.

Single muscle fiber

Z disc  Single sarcomere  Z disc

Single myofibril

**(b) Muscle contraction**

A sarcomere, the basic functional unit of skeletal muscle, extends from one Z disk to the next.

The sliding of myosin filaments along actin filaments causes muscles to contract.

Single sarcomere

Contraction

Z disc   Actin filament   Myosin filament

The heads of myosin molecules bind reversibly to special sites on actin filaments.

Z disc

Z disc

Z disc

**FIGURE 31.6 The Microscopic Structure of Muscle**

(a) Muscles contain bundles of muscle fibers, which contain myofibrils, made up of sarcomeres. (b) Muscle contraction depends on the movement of actin and myosin filaments. The electron micrograph shows one sarcomere, bounded by Z discs.

Z discs that mark the ends of each sarcomere contain a large protein that provides anchor points for actin filaments. Between the free ends of the actin filaments, attached to the middle of the sarcomere, lie the thicker myosin filaments. The sliding of the myosin filaments against the actin filaments enables a sarcomere to contract. The simultaneous contraction of all the sarcomeres, usually taking no more than a tenth of a second, produces the contraction of a whole muscle.

If you could watch your biceps muscle at high magnification while lifting a weight, you would see that the two Z discs of every sarcomere are pulled closer together (see Figure 31.6b). An even higher magnification would reveal that the protruding "head" on each myosin molecule can bind to specific sites on adjacent actin filaments. In slow motion, muscle contraction works like this: Some of the many myosin heads at each end of the myosin filaments attach to binding sites on the surrounding actin filaments. As it engages an actin filament, each myosin head changes shape in a series of steps that move it a small distance (about 5 nanometers) along the surface of the actin filament in the direction of the Z disc. The myosin head then detaches from its binding site, before attaching to a new binding site that is now closer to the Z disc (see Figure 31.6b).

At any moment during the contraction of a sarcomere, some myosin heads are attaching and others are detaching from the actin filament, so the two types of filaments remain linked in a contracting muscle. The actin filaments, and the Z discs to which they are attached, are free to move, which is why they are pulled inward when thousands of myosin heads exert a force against them. As the myosin heads "walk" their way from binding site to binding site along the actin filaments, the overall effect is that the two Z discs are pulled closer together. When contracted to the maximum extent, each sarcomere is typically shortened by about a third of its length.

Every "walking step" of each myosin head requires the presence of calcium ions and the energy stored in one molecule of ATP. The result of countless myosin and actin filaments interacting in this way is a muscle contraction. Contracting muscles bulge because all the actin and myosin filaments must squeeze into a shorter length of muscle.

## Involuntary muscle contractions keep us alive and well

When we walk, run, or jump we are using *voluntary* contractions of **skeletal muscle**—movements we choose to make. But even when we are still, there is movement going on within the body. The heart is pumping blood, and food is being moved along a digestive tract by muscular contractions. *Involuntary* muscles do their work without our having to consciously think about moving them. Two specialized types of muscle that engage in involuntary contraction are cardiac muscle and smooth muscle.

The vertebrate heart is a muscular organ and the only organ that contains **cardiac muscle**. Like skeletal muscle, cardiac muscle has a banded appearance (compare **FIGURE 31.7a** and *b*). It is differentiated from skeletal muscle, however, in that its muscle fibers are joined by interconnecting branches that help them produce the coordinated contractions we know as heartbeats.

**Smooth muscle** lacks the bands associated with other muscle types (**FIGURE 31.7c**), and its contractions

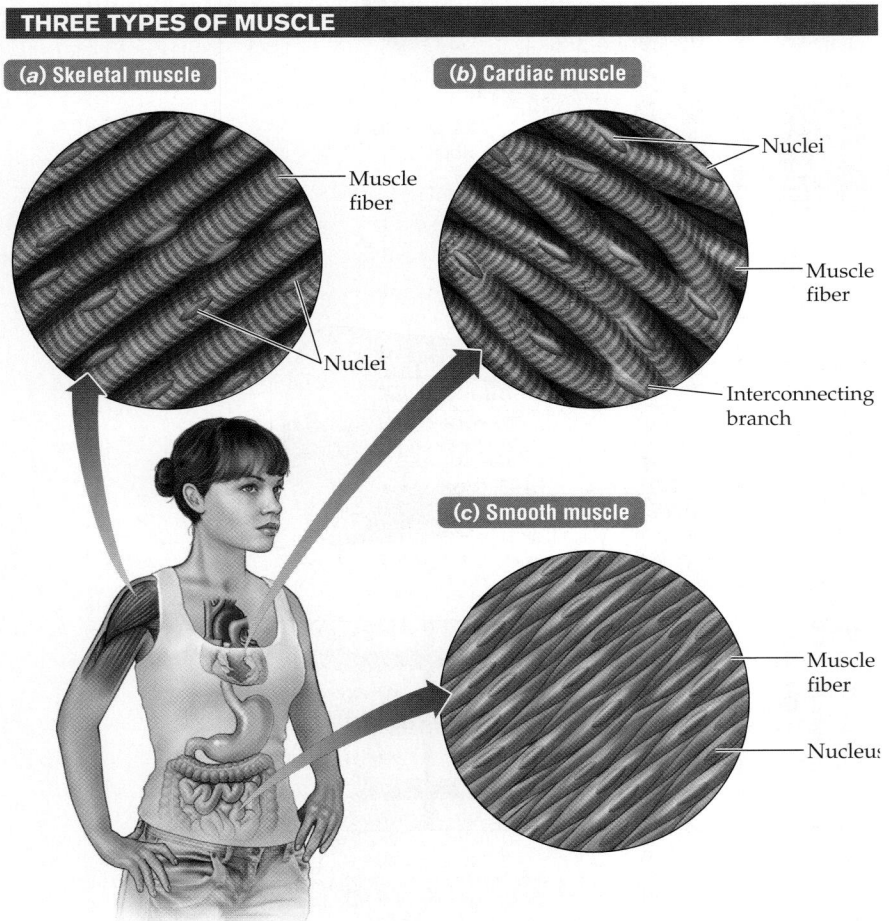

**THREE TYPES OF MUSCLE**

**(a) Skeletal muscle**

Muscle fiber

Nuclei

**(b) Cardiac muscle**

Nuclei

Muscle fiber

Interconnecting branch

**(c) Smooth muscle**

Muscle fiber

Nucleus

**FIGURE 31.7 Specialized Types of Muscles for Different Types of Movement**
(*a*) Muscles associated with the skeleton have a banded, or striated, appearance. (*b*) Cardiac muscle is also banded, and its muscle fibers are branched. (*c*) Smooth muscle does not have bands.

are entirely involuntary. The contractile proteins of smooth muscle cells interlink the plasma membrane on opposite sides of these spindle-shaped cells. When the contractile proteins slide against each other, they pull in the plasma membrane on both sides, shrinking the diameter of the cell. When many such cells shrink in unison, they can exert a squeezing action in the tissues where they occur. Some smooth muscles can contract as rapidly as skeletal muscle fibers do. Some types of smooth muscle can sustain a contraction for hours.

For digestion to proceed, food is moved along the digestive tract in coordinated waves by the contractions of smooth muscle. Smooth muscle is also found in the walls of blood vessels, the respiratory tract, and the urinary bladder. Tiny strips of smooth muscle are responsible for narrowing the iris, the pigmented circular ring that surrounds the lens of the eye. The rapid contraction of tiny smooth muscles in the skin is responsible for "goose bumps" and the "hair-raising" aspects of a scary experience or an encounter with cold.

## Muscles that move body parts often work in opposing pairs

Because the myosin heads can only pull the Z discs together, a contracted muscle cannot relax itself. For this reason, most muscles that move limbs and other body parts are arranged in opposing pairs so that the contraction of one causes the other to "relax," or stretch back to its condition before contraction. For example, the triceps muscle (on the lower side of your upper arm) contracts to stretch the biceps muscle (on your upper arm), and vice versa (**FIGURE 31.8**). This pattern of paired muscles is repeated in other parts of the body. To kick your lower leg forward, you contract your quadriceps, the muscle in the front of the thigh, while relaxing the hamstring, the muscle in the back of your thigh. The reverse set of contractions and relaxations—contracting the hamstring and relaxing the quadriceps—draws your lower leg back behind you, if you begin from a standing position.

## 31.4 Converting Muscle Contraction into Motion

Within the animal kingdom, mechanisms have evolved that convert simple muscle contractions into the wide range of motions that most animals are capable of. Muscle contractions enable a leaf-cutting

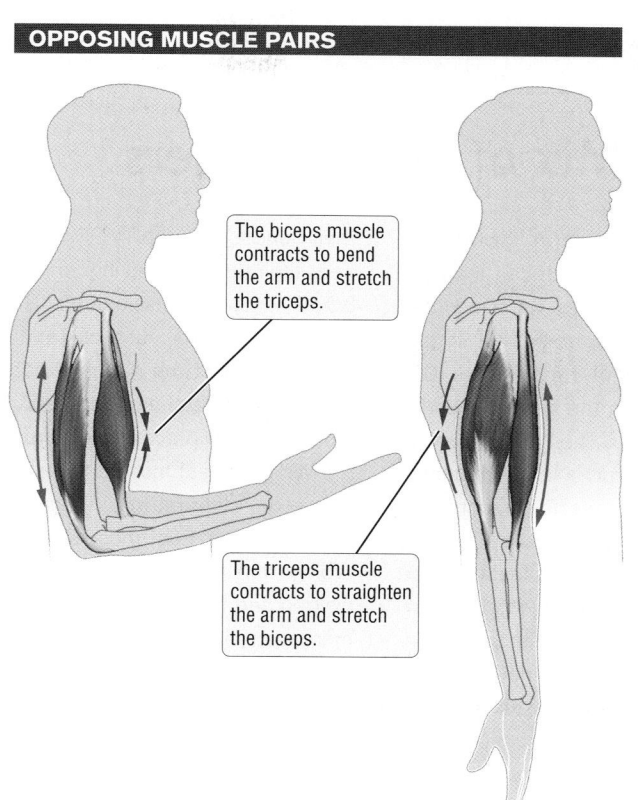

**OPPOSING MUSCLE PAIRS**

The biceps muscle contracts to bend the arm and stretch the triceps.

The triceps muscle contracts to straighten the arm and stretch the biceps.

**FIGURE 31.8 Most Muscles That Move Body Parts Work in Pairs**

When one member of a pair contracts, the other is relaxed.

ant to hoist leaf fragments that weigh 50 times more than the diminutive insect, and they enable a hummingbird to beat its wings 70 times a second so that it can hover and sip nectar from a flower.

All animals have essentially identical sarcomeres that contain the same actin and myosin molecules arranged in much the same pattern. Because of this similarity, a muscle fiber from an ant can lift about the same weight as a muscle fiber from an elephant, if the fibers are of the same thickness.

Differences in strength between species and between muscles within an individual depend almost entirely on the cross-sectional area of the muscles. (Cross-sectional area is the area of the exposed surface in a three-dimensional object that has been sliced through. Assuming a muscle is approximately cylindrical at its widest part, the cross-sectional area of that muscle is given by the area of a circle, $\pi r^2$, where $r$ is the radius, or half the width of the muscle.) The greater the cross-sectional area (or, more simply, the thickness) of the muscle, the more muscle fibers there are in it. These muscle fibers, lying side by side, contract almost simultaneously, and the more of them there are in a muscle, the greater the contractile force the muscle can generate. In other words, thicker muscles are stronger than thinner muscles.

# A Winning Mix of Muscle Fibers

There are some 650 muscles in the human body. The gluteus maximus, located in the buttocks, is the largest. The smallest muscle in the human body, the stapedius, is located in the middle ear and is only about 5 millimeters (0.2 inch) long. Muscles vary in the type of muscle fiber they are composed of. The type of muscle fibers found in a muscle has a large influence on the speed with which a muscle can contract.

Two main types of fibers are found in human skeletal muscles: slow (type I) and fast (type II). Different types of muscles have a different mix of slow and fast muscle fibers depending on the main function of that muscle. The two types of fibers differ in how fast they respond to the nerve impulses that trigger contractions and how long they can sustain a muscle contraction. Slow muscle fibers (also known as slow-twitch muscle fibers) respond to a nerve impulse by contracting in a series of small steps, taking longer to achieve a full contraction than the rapid "firing" of a fast muscle fiber. Slow muscle fibers do not produce as large an output of power, but their contractions can be sustained over a much longer time, even hours. Muscle fibers that contract quickly, like those that generate the explosive power of a runner sprinting from the starting blocks, are fast muscle fibers (also known as fast-twitch muscle fibers). Fast muscle fibers contract completely in a quick, but short-lived, all-or-nothing response to nerve impulses. The table summarizes the characteristics of these two types of muscle fibers.

Long-distance runners have a very high proportion of slow muscle fibers in their leg muscles, partly because of training but largely because of their genetic makeup. In Olympic marathon runners, the muscles used in running contain about 80 percent slow muscle fibers. Those who excel in strength sports, such as sprinting or weight lifting, have just the opposite pattern: a majority of their skeletal muscle fibers are of the fast type (about 80 percent in elite athletes).

Whereas fast muscle fibers are specialized for power, slow muscle fibers sacrifice power for endurance. Fast muscle fibers contract so rapidly that they quickly use up the available oxygen and have to rely on a less efficient fermentation pathway for supplemental energy (see Section 9.4). Fermentation yields less than a tenth of the energy that is delivered by oxygen-dependent (aerobic) respiration. Slow muscle fibers, in contrast, rely on cellular respiration, an aerobic process that yields much more energy. To ensure a steady oxygen supply, slow muscle fibers contain large amounts of a pigment called *myoglobin*. This reddish pigment, similar to hemoglobin, binds and holds oxygen until it is needed by the muscle.

We can usually distinguish slow muscle tissue by its darker color, which reflects its myoglobin content. The white meat of a turkey or grouse consists mostly of fast muscle fibers used to power the wings; the dark meat of the legs consists mostly of slow muscle fibers that must contract for long periods as the birds forage on the ground. On the other hand, migrating ducks have dark breast meat, because they need a high proportion of slow fibers to sustain the many wing beats it will take to get them from their summer breeding grounds to their winter homes.

Muscles respond to training. If you want to develop slow-twitch fibers for endurance sports, you will want to get a lot of aerobic exercise. If strength, power, and bulk are your goal, weight training can put you on a fast track to developing your fast muscle fibers.

## A Comparison of Slow and Fast Muscle Fibers

| CHARACTERISTIC | TYPE OF MUSCLE FIBER | |
| --- | --- | --- |
| | SLOW (TYPE I) | FAST (TYPE II) |
| Speed of contraction | Slow | Fast |
| Force of contraction | Less power | More power |
| Length of contraction | Sustained | Brief |
| Response of sarcomeres | Partial contraction possible | Either no contraction or complete contraction |
| Source of ATP | Cellular respiration | Fermentation |
| A human muscle especially rich in these fibers, in the average person | Gluteus maximus | Quadriceps (in the thigh) |

Weight training increases the thickness of individual muscle fibers, which increases muscle strength because it increases the overall cross-sectional area of the muscle. With prolonged weight training, the *number* of these muscle fibers also increases, leading to the large gains in muscle strength and the muscle bulk for which bodybuilders are known. The extent to which muscle fibers increase in width and number in response to weight training depends on gender, genetic makeup, age, and nutritional status.

## 31.5 Comparing Skeletons

Other animals have skeletons that may differ from ours in appearance but share two basic functional features with ours: stiff structures that maintain shape, and joints that let the stiff parts of the skeleton move relative to one another.

### Most skeletons depend on tissues that produce a stiff matrix

Although animals have evolved a diversity of skeletons, all animals, including humans, rely on specific tissue types to support the rest of the body. The support tissues consist of specialized cells that produce a strong, stiff matrix around themselves. This extracellular matrix can be hardened with minerals (as in human bones) or tough proteins (as in cartilage) or carbohydrates (as in the exoskeletons of arthropods such as spiders and beetles). Among animals as a whole, the most widespread nonmineral substances in the matrices of support tissues are *collagen* and *chitin*. **Chitin** [KYE-tin] is a tough polysaccharide, often interlinked by proteins, that forms the exoskeletons of arthropods.

Terrestrial animals need strong skeletons to counter the pull of gravity and to support their weight. However, the support tissues themselves add weight to the body. As a general principle, the skeletons of larger animals make up a greater proportion of the animal's body weight than do those of smaller animals. For example, a shrew, a tiny animal that weighs about 6 grams (about as much as a couple of pennies), has a skeleton that makes up only 4.6 percent of its body weight, whereas the skeleton of a 6,600-kilogram elephant makes up 24 percent of its body weight. Elephants need heavier skeletons in part to support their heavier skeletons (**FIGURE 31.9**).

### Many invertebrates depend on hydrostatic skeletons for structure and movement

A large number of invertebrates, including familiar animals such as earthworms and octopi, have soft bodies that appear to lack an obvious means of support or movement, since they have neither an endoskeleton nor an exoskeleton. These animals rely on a hydrostatic skeleton, or **hydrostat**, consisting of a fluid-filled compartment surrounded by elastic and muscular tissues. Fluid pressure within the body chamber gives shape to the bodies of soft-bodied animals, such as sea anemones, the

**SKELETAL WEIGHT RELATIVE TO BODY WEIGHT**

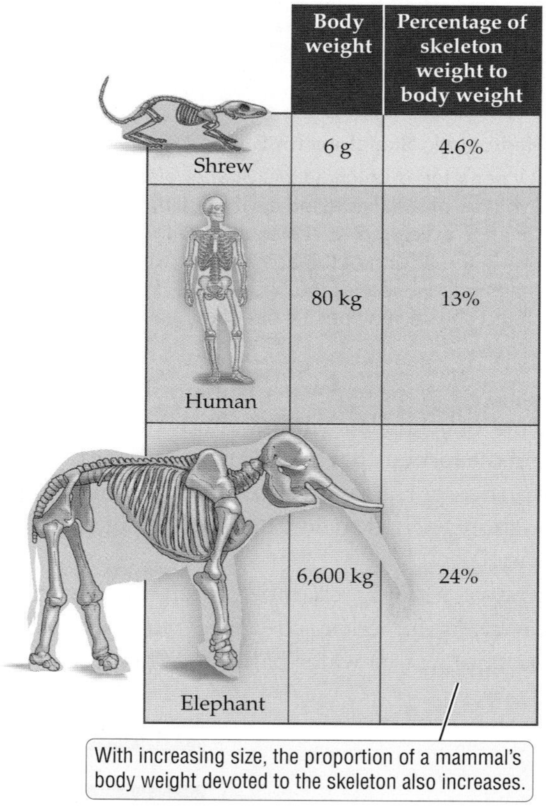

| | Body weight | Percentage of skeleton weight to body weight |
|---|---|---|
| Shrew | 6 g | 4.6% |
| Human | 80 kg | 13% |
| Elephant | 6,600 kg | 24% |

> With increasing size, the proportion of a mammal's body weight devoted to the skeleton also increases.

**FIGURE 31.9 Size Matters**
A shrew's skeleton is stronger than an elephant's in the sense that the ratio of the skeletal weight to total body weight is smaller in the shrew (4.6 percent) than in the elephant (24 percent); this means that a lighter skeleton supports a relatively large total body weight in the small rodent. For an elephant's skeleton to support as large a fraction of the total body weight as the shrew's skeleton does, it would have to be almost five times larger than it is. A large skeleton makes larger size possible, but geometry and the laws of physics dictate that the weight of the skeleton becomes an even larger proportion of the total body weight as the size of the mammal body increases.

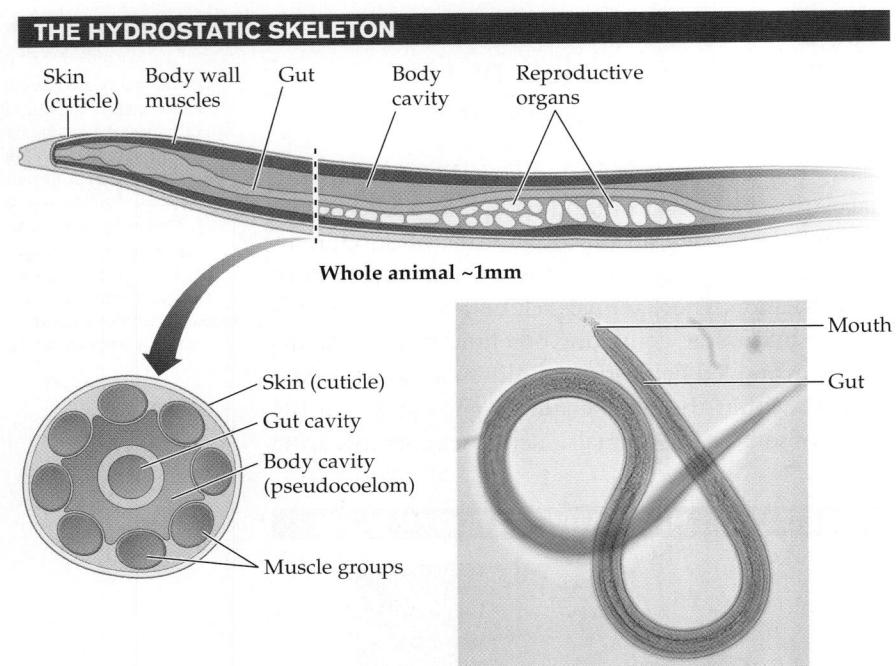

Skin (cuticle)    Body wall muscles    Gut    Body cavity    Reproductive organs

**Whole animal ~1mm**

Skin (cuticle)
Gut cavity
Body cavity (pseudocoelom)
Muscle groups

Mouth
Gut

**FIGURE 31.10 Nematodes Use Hydrostatic Skeletons for Support and Movement**
When muscles in the body wall of tiny nematodes (such as this *Strongyloides filariform*) contract, they decrease the internal volume of the fluid-filled cavity within. The fluid inside the nematode's body cannot change in volume, so the squeezing force is resolved by a bending of the cavity, and through it, the flexing of the body wall as well. Waves of muscle contractions and relaxations along the length of the body cause it to bend back and forth in motions that slowly track the animal in a specific direction.

way water gives shape to a water balloon. The wriggling movements of animals such as roundworms (**FIGURE 31.10**), and the movements of body parts such as the tentacles of an octopus, are powered by muscle contractions and relaxations exerted against pressurized fluid inside a body cavity.

A fluid-containing chamber, such as a worm's body cavity, is an enclosed space with a flexible envelope, similar to a water balloon. When the mus-

cles of the body wall contract around this chamber in any one spot, they squeeze the fluid and push it outward in another region of the chamber, the way pinching a balloon at one end makes it bulge out at the other end. The shifts in fluid pressure change the shape of the body cavity, which in turn flexes the body wall. Two or more muscle layers surround the body cavities of animals that use a hydrostatic skeleton to move; the coordinated contractions and relaxations of these muscle layers bend the body wall or shorten or lengthen it.

## Skeletons can be internal or external

In the course of evolution, two distinct ways of arranging strong support tissues have arisen in the animal kingdom. While humans and other vertebrates have an **endoskeleton** (*endo*, "inner, inside"), in which the support tissues lie inside the body, many other animals, such as lobsters, insects, and many other invertebrate groups—have an **exoskeleton** (*exo*, "outer, outside"), an external skeleton that surrounds and encloses the soft tissues it supports (**FIGURE 31.11**).

Exoskeletons provide a protective armor for many animals and also protect terrestrial invertebrates from excessive moisture loss. The rigidity of an exoskeleton means that immature animals that outgrow their exoskeletons must shed them periodically, in the process known as molting. Until it regrows a new, rigid exoskeleton, the juvenile is vulnerable to predators and is at risk of drying out. A significant advantage of the endoskeleton is that it can grow continually throughout the juvenile phase of development.

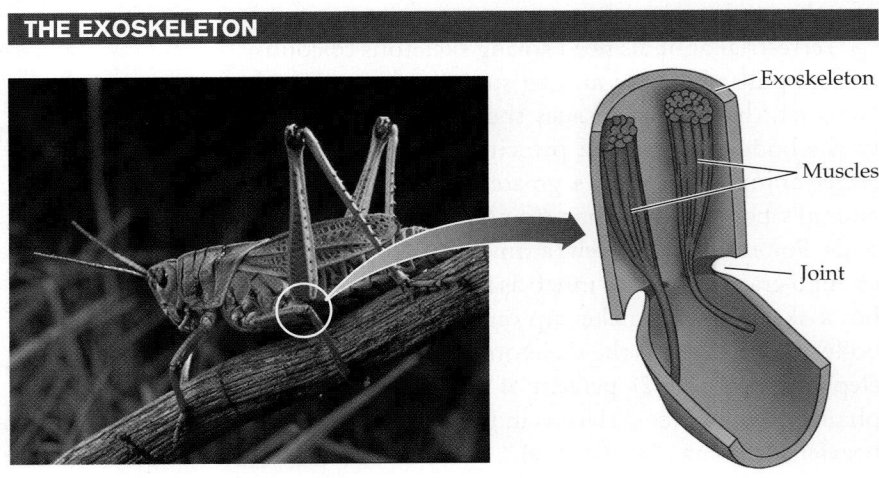

Exoskeleton
Muscles
Joint

**FIGURE 31.11 Grasshoppers and Most Other Arthropods Have External Skeletons with Joints**

## Natural selection favors streamlining in swimmers and fliers

As animals move through their environment, they convert the contractions of their muscles into a force that propels them forward. However, an animal's motion is always resisted by drag, the force that resists the movement of an object through a fluid medium such as air or water. The greater the surface area that is exposed to the medium, the greater the drag. The thicker, or more viscous, a medium is, the greater is the resistance that it offers.

Increasing the total surface area of an organism increases drag, as does increasing the viscosity (stickiness) of the fluid it is moving through. As a result, drag has its greatest effect on animals with a large surface area and on those that live in water. Drag can become a serious burden in water because water is a much denser fluid than air. If you ride your bicycle through a deep puddle, you can feel the much greater drag offered by water. Small swimming animals must expend relatively more energy overcoming drag because they have a larger surface area, relative to their volume, than do larger aquatic animals.

Drag is costly in a biological sense because organisms expend energy to generate forward movement. Evolution favors adaptations that reduce drag because they free up energy for other uses. The most effective strategy to minimize drag seems to be a streamlined shape (**FIGURE 31.12**). In spite of their very different evolutionary histories, fast-moving members of bird, aquatic mammal, and fish groups have all converged on a similar body shape. This shape makes it easy for water or air to flow smoothly around the animal. Animals can also reduce drag by having "nonstick" surfaces. The smooth, slimy surface that makes many aquatic animals so unappealing to touch effectively reduces the "clinginess" of water molecules, lowering the friction between the medium and the body surface, which is one aspect of drag.

## Different modes of locomotion have different energetic costs

How much energy do animals expend to move from one place to another? Scientists have calculated energy expenditures for animals of different sizes and using different means of locomotion. Not surprisingly, heavy animals use more energy to move a given distance than do lighter animals. Pound for pound, however, heavy animals expend *less* energy in relation to their body weight. A cat, for example, may weigh 10 times as much as a squirrel, but it uses less than 10 times as much energy to run a given distance. Why do heavier animals move more efficiently than lighter animals? In part, larger animals have less surface area per unit of body weight than smaller ones have and therefore are subjected to relatively less drag.

How bodies move in different fluid environments determines how much energy they use. Runners expend more energy moving a given distance than do fliers of the same size. Swimmers use the least energy. Half the energy spent on running is "wasted" moving the body up and down, and on changing the momentum of the legs as they alternate between moving forward and moving backward. Bicycling, which involves fewer changes in momentum, is more efficient than running. Animals that fly use less energy to change momentum than do runners, but fliers must expend energy to keep themselves aloft. Despite the movement of their fins and tails, swimmers lose little energy to changes in momentum. Although they must invest energy in propelling themselves through a more viscous medium, their special adaptations (streamlining, low-friction body surfaces) help reduce the drag offered by this medium. Furthermore, while runners and fliers must devote energy toward fighting the pull of gravity, the body weight of aquatic animals is substantially supported by the buoyancy of water, aided by such additional mechanisms as air bladders and stored oils that some species possess.

> **Helpful to Know**
>
> Even though the term "hydrostat" (from *hydro*, Latin for "water") implies that water is involved, the fluid in a hydrostat can also be a gas, as in hydrostats that depend on air.

**FIGURE 31.12**
**Whales, Fishes, and Birds Have Streamlined Bodies**

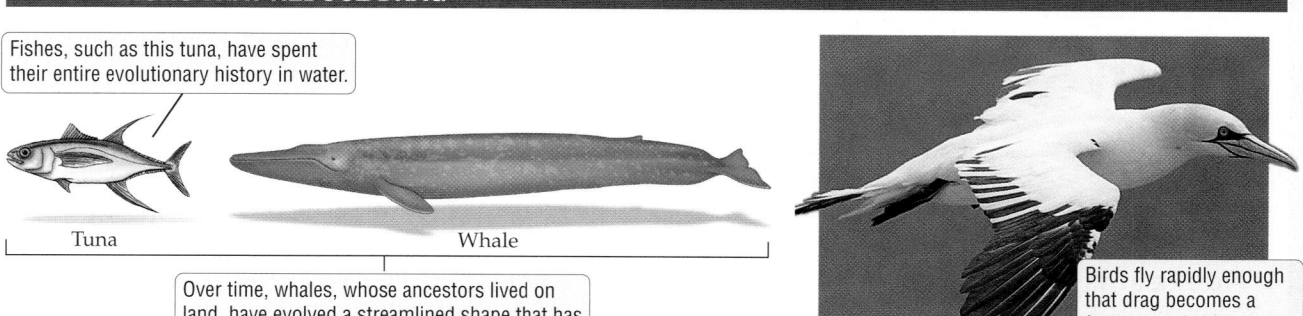

**ADAPTATIONS THAT REDUCE DRAG**

Fishes, such as this tuna, have spent their entire evolutionary history in water.

Tuna

Whale

Over time, whales, whose ancestors lived on land, have evolved a streamlined shape that has converged on that of fast-swimming fishes.

Birds fly rapidly enough that drag becomes a factor even in air—hence their streamlined shape.

# The Ninth Life of Cats

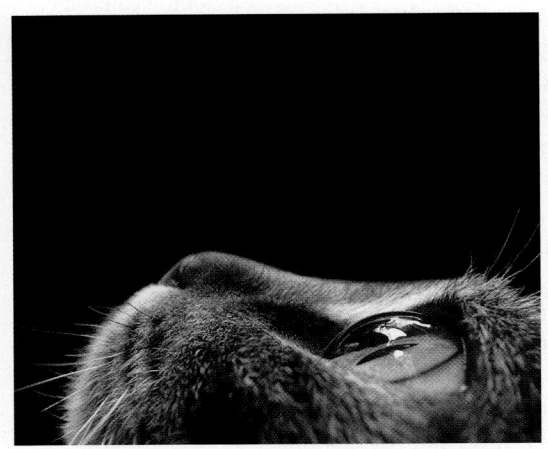

Skydiving domestic cats sometimes survive falls from great heights. The secret to their survival lies in their size and the physics of their movement through the air. One reason cats may do so well is their small size. When an object falls, whether a stone or an elephant, it accelerates. Gravity makes things fall about 10 meters per second faster each second. If you fell from a tall building, you'd fall 10 meters in the first second, 20 meters in the second second, and 40 meters in the third second. To fall 7 stories (25 meters) takes a little over 2 seconds.

How badly a cat or human is hurt by a fall depends partly on the force at impact. And you may remember from physics that force is proportional to mass and acceleration ($F = ma$). With gravity, acceleration is a constant: 10 meters per second per second. So force at impact depends heavily on mass. An elephant that weighed, say, 2,500 kilograms (kg) would hit the ground with a force 50 times greater than a human that weighed 50 kg. By the same reasoning, a 75-kg human falling from a tall building would hit the ground with a force that was 15 times greater than that experienced by a 5-kg cat. And greater force translates into greater injury.

Cats have other advantages besides merely being lightweight. Quick reflexes and unusually flexible backbones help cats twist around in midair to land on their feet, which protects them from landing on (and breaking) backs or necks and also spreads the force of impact over four paws (**FIGURE 31.13**). Cats also prepare for hard landings in other ways. They relax some muscles a bit during the fall and bend their legs as they land so that the ligaments and tendons associated with the joints can act as springs. The spread-eagle form of a falling cat slows the cat's rate of descent the way an open parachute would.

It's easy to see why cats would survive long falls better than people. But the study cited in the opener claimed that cats survived longer falls over 7 stories (about 25 meters) better than they survived shorter falls. This claim led some people to conclude that cats are adapted to falling long distances, even suggesting that longer falls give cats more time to turn over and prepare for impact. But cats can turn over to land on their feet in about one meter of falling distance, less than even one story.

Media reports have focused on cats' ability to survive longer falls. But even when cats survive, they are usually badly hurt, taking the brunt of the impact on their legs, chests, and faces. Of the cats taken to the vet in the New York study, more than 60 percent had collapsed lungs, 57 percent had broken bones in their faces, and 57 percent had broken or dislocated limbs. More than a third had life-threatening injuries that required immediate medical care. And another third had less-serious injuries, such as broken legs.

Finally, the cats in the New York study were not a random sample of cats that fell. For example, cats that fell 8 or 10 stories and died immediately would not have been taken to the vet. Cats that fell only one story were excluded. Several owners decided to have their cats put to sleep, and those cats were not included in the data. Another similar study concluded that cats that fell from greater than seven stories had more serious injuries than cats that fell shorter distances.

**A FLEXIBLE SKELETON**

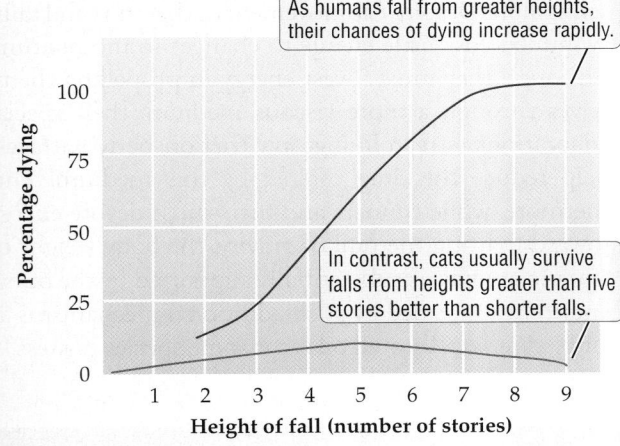

**FIGURE 31.13  A Cat Finds Its Feet**
Cats' legendary ability to land on their feet may help them survive long falls.

# How Olympics Loom, but Nobody Is Pushing Usain Bolt

BY EDDIE PELLS, *Associated Press*

Even when he's not running, Usain Bolt is there.

His name and his world records hover over every track meet in every corner of the globe.

The shadow he casts is even more pronounced at big meets like this—the U.S. nationals, where Bolt's key challenger, Tyson Gay, pulled out with an injury and the other would-be contenders are a 29-year-old on a comeback from a doping conviction and an Olympic bronze medalist who still needs to find a faster gear.

There are 13 months to go until the London Olympics. And in America—supposedly a breeding ground for some of the world's fastest track stars—there are few signs that anyone is ready to give Bolt a go.

"They won't do it running 9.9," said Ato Boldon, the former Olympic medal winner from Trinidad and Tobago.

Bolt's world record is 9.58. In the 100-meter finals Friday, on one of the fastest tracks in the country and with a light breeze at their back, the best time was the 9.94 by Walter Dix. Don't know much about Walter Dix? Not your fault. He was part of the "Footnotes To History" collection in Beijing, winning bronze medals in the 100 and 200 on the same nights Bolt was rewriting the record book.

In Eugene this week, Dix barely beat Justin Gatlin, who is returning to the sport after serving a four-year doping ban that deprived him of his prime years.

And Gay?

He wasn't in the final—victim of a nagging hip injury that will deprive him, and the world, of a showdown with Bolt at world championships, nine weeks from now.

Every year, athletes set new world records. The overall trend in times recorded in the men's 100-meter sprint over the past 40 years would suggest that we can expect more world records in the years to come. In 1968, Jim Hines became the first sprinter to break the 10-second barrier by running 100 meters in 9.95 seconds. Before that run, no one had ever covered 100 meters in less than 10 seconds. Since then, various sprinters have done it well over 250 times. Nowadays, a time of less than 10 seconds is still good, but not exceptional, as described in this news article.

Does this mean that humans will continue to improve their times indefinitely in the 100-meter sprint and other races? Probably not. Some of the improvement in times over the past several decades has come from improvements in technology. For example, the relatively basic running shoes of years past have been replaced by more sophisticated shoes that combine light weight with stiffness and suppleness in just the right places. Similarly, the simple shorts and tops that served sprinters well in the 1960s have morphed into aerodynamic bodysuits.

Other advances have come from an improved understanding of the physics and psychology of running, fast and slow muscle types, and nutrition. Sprinters today have their stride analyzed by computer so that they can adjust their stride to maximize efficiency and convert as much muscle force into forward motion as possible. And sports psychologists help sprinters overcome mental obstacles that might take only a thousandth of a second from their time. In some sports, competition is so tight that performance-enhancing drugs have become an accepted part of a training regimen (see the "Biology in the News" feature in Chapter 28, page 629).

Even though the record time for the 100-meter sprint continues to fall, overall performance may have peaked. In 1998, the average of the 10 fastest times was 9.86 seconds. In 2003, this average increased to 9.97 seconds, over a tenth of a second longer (slower) than 5 years earlier. Perhaps the technology of running has begun to approach the limits of what it can do, and runners may now be facing biological limits.

## Evaluating the News

**1.** Which muscles and joints play an especially critical part in running? What might be the biological limit that will keep humans from running a 100-meter dash in less than 9.50 seconds?

**2.** When a world record is broken, it is always big news. Do you think that when the limit of human performance has been reached in a particular sport, and world records in that sport are being broken very rarely, the sport will become less exciting to watch? If world records were no longer being broken, would track-and-field fans lose interest in the sport?

**3.** If improved training methods and improved technologies have played an important part in the ever-faster times recorded by runners, how is this different from the use of ever more effective drugs to enhance performance? How should we view the roles of different ways of enhancing performance in athletics?

**SOURCE:** *Associated Press*, June 25, 2011. Copyright © 2011 The Associated Press. All rights reserved.

691

# CHAPTER REVIEW

## Summary

### 31.1 Basic Features of the Human Skeleton

- The axial skeleton supports and protects vital organs along the long axis of the body. The appendicular skeleton is composed of the bones in the arms, legs, and pelvis, and is primarily involved with movement.
- Compact bone consists of dense deposits of hard calcium and phosphate compounds. Spongy bone lies inside compact bone and contains bone marrow, which makes red and white blood cells, in spaces between the hard materials.
- Cartilage, which is more pliable than bone, supplements the support provided by bone.
- Ligaments hold together the bones at a joint.

### 31.2 How Joints Work

- In our knees, as well as in other joints, ligaments connect bone to bone. Tendons help hold the knee together by connecting upper leg muscles to the bones of the lower leg. Tendons and ligaments are made of collagen.
- Cartilage in the knee cushions the points where femur and tibia meet. Fluid-filled synovial sacs lubricate the cartilage, reducing friction.
- Tears in cartilage and ligaments, which are common sports injuries, often require surgery to repair.

### 31.3 How Muscles Work

- Muscles provide the power necessary for movement.
- Muscles consist of muscle fibers, each of which is packed with myofibrils. Myofibrils contain repeating units called sarcomeres that do the work of contraction.
- Sarcomeres are visible as a pattern of light and dark bands when seen through a microscope. A sarcomere extends between two Z discs, which are the attachment points of actin filaments. The thicker myosin filaments lie in the center of the sarcomere.

- During contraction of a muscle, the many myosin heads on each end of the myosin filaments make contact with the nearby actin filaments, pulling them in toward the center of the sarcomere in the process. The sliding of the myosin filaments against the actin filaments pulls the Z discs closer together, shortening the sarcomere. The energy stored in ATP powers the interaction of myosin and actin.
- Internal movements are created by specialized muscle types. Cardiac muscle is found in the heart, where its contractions pump blood. Smooth muscle, which is found in the digestive tract and blood vessels, contracts in waves.
- Most muscles that move body parts work in pairs. When one muscle in a pair contracts, the other is relaxed.

### 31.4 Converting Muscle Contraction into Motion

- All skeletal muscles have essentially identical sarcomeres with the same type of contractile proteins arranged in much the same way.
- Muscle strength increases in proportion to the cross-sectional area of the muscle.
- Nerve impulses arriving at neuromuscular junctions cause muscle fibers to contract.

### 31.5 Comparing Skeletons

- An exterior skeleton provides an animal with support and armor. In arthropods, the support tissue is made up of chitin, a complex polysaccharide.
- In general terms, the larger the animal, the greater its ratio of skeletal weight to total body weight.
- Many soft-bodied animals depend on hydrostats to provide support. In a hydrostat, a fluid (usually water) exerts pressure on elastic or muscular tissues to support structures or produce movement.

## Key Terms

actin filament (p. 683)
appendicular skeleton (p. 678)
axial skeleton (p. 678)
cardiac muscle (p. 684)
cartilage (p. 680)
chitin (p. 687)
collagen (p. 680)

compact bone (p. 679)
endoskeleton (p. 688)
exoskeleton (p. 688)
hydrostat (p. 687)
joint (p. 680)
ligament (p. 681)
locomotion (p. 678)

marrow (p. 679)
muscle (p. 678)
muscle fiber (p. 682)
myofibril (p. 683)
myosin filament (p. 683)
osteocyte (p. 678)
sarcomere (p. 683)

skeletal muscle (p. 684)
skeleton (p. 678)
smooth muscle (p. 684)
spongy bone (p. 679)
synovial sac (p. 681)
tendon (p. 681)
Z disc (p. 683)

# Self-Quiz

1. Which of the following would you find in the appendicular skeleton?
   a. legs
   b. arms
   c. toes
   d. all of the above

2. A hydrostat
   a. is found only in animals with an endoskeleton.
   b. contains a fluid-filled cavity.
   c. is absent in soft-bodied animals.
   d. enables movement in animals that lack muscles.

3. Ligaments connect
   a. muscles to other muscles.
   b. muscles to bones.
   c. bones to other bones.
   d. tendons to muscles.

4. Which of the following would have the strongest skeleton relative to its body weight?
   a. cat
   c. squirrel
   b. human
   d. buffalo

5. Which of the following would *not* be found in the human knee joint?
   a. synovial sac
   b. chitin
   c. cartilage
   d. ligaments

6. Sarcomeres contain
   a. muscle fibers.
   b. myofibrils.
   c. actin.
   d. cartilage.

7. Which of the following is *not* part of a sarcomere?
   a. actin          c. Z disc
   b. myosin         d. collagen

8. Which of the following athletes would have the most muscles containing large amounts of myoglobin?
   a. marathon runner
   b. 100-meter sprinter
   c. high jumper
   d. gymnast

9. Which type of muscle lacks a pattern of light and dark bands (striations) when viewed at very high magnification?
   a. cardiac muscle
   b. skeletal muscle
   c. smooth muscle
   d. none of the above

10. Cartilage
    a. is composed of large cells surrounded by a mineralized extracellular matrix.
    b. is more flexible than bone is.
    c. has a rich supply of blood vessels.
    d. is dead tissue and therefore lacks blood vessels.

# Analysis and Application

1. Describe how bone and cartilage provide a framework of support for the human body.

2. Knee joints are designed for one type of "back and forth" motion. Design a joint that would allow circular motion, such as swinging an arm to throw an underhand softball pitch. Where would you place cartilage?

3. Turn your hand palm up and bring it to your shoulder. Draw the arrangement of bones, joints, and muscles that you used to carry out this action. Include tendons and ligaments.

4. The sarcomere is the basic unit of muscle contraction. Describe how a sarcomere works.

5. Compare a hydrostatic skeleton with an exoskeleton.

6. Describe how drag would work as a bird flaps its wings and moves forward. How could drag actually assist a bird's motion as a wing flaps?

7. What function is performed by the anterior cruciate ligament (ACL)? What type of injury is the ACL prone to and why?

8. The table shows the skeletal location of the main groupings of bones in the human body. Mark each group with an (a) for axial or (app) for appendicular to show which part of the skeleton it belongs to. Choose one large bone in the axial skeleton and one in the appendicular skeleton and describe their main functions.

| SKELETAL LOCATION OF BONES | NUMBER OF BONES |
|---|---|
| Skull | 22 |
| Ears (two) | 6 |
| Vertebrae | 26 |
| Sternum (breastbone) | 3 |
| Ribs | 24 |
| Throat | 1 |
| Shoulder (pectoral) girdle | 4 |
| Arms (two) | 60 |
| Hips (pelvis) | 2 |
| Legs (two) | 58 |
| Total | 206 |

# 32 Defenses against Disease

# The Spread of HIV

Virtually every nation in the world has been touched by AIDS. In the United States, more than half a million Americans have died of the disease and a million more are infected, often without knowing it. But the region most devastated by AIDS is sub-Saharan Africa, where the disease probably originated in the mid-1950s. In nine countries in the southern tip of Africa, more than 10 percent of each population is infected with the AIDS virus—human immunodeficiency virus, or HIV. In some countries, one-quarter of people aged 15–49 test positive for HIV. In Zimbabwe, 15 percent of children have lost one or both parents to AIDS. In the hardest-hit nations, increasing numbers of orphaned children and a workforce decimated by deaths are destroying the social fabric.

The AIDS pandemic in Africa has been attributed to two main causes. First, since as far back as the 1950s, a lack of funds has forced hospitals and health care workers to reuse improperly sterilized needles for blood transfusions and vaccinations. As a result, the AIDS virus has been transmitted directly from one person to another by means of thousands of dirty needles. In Africa, nonsterile medical practice may be responsible for over half of all adult AIDS cases.

Second, the withdrawal of European colonial powers from Africa during the 1960s led to unstable governments and severe social upheaval. One result of such turmoil was large groups of men in isolated labor camps. In such places, the men's main access to women was women offered for sale, which led to skyrocketing rates of sexually transmitted diseases among both men and women. The AIDS virus normally passes through mucous membranes during sexual intercourse, but sexually transmitted diseases such as herpes and syphilis cause open sores that greatly accelerate the transmission of HIV from one person to another by allowing HIV-infected blood and semen to come into contact. Compounding the problem, famines and wars have forced populations to migrate from one region to another, spreading AIDS from one region to another.

Is the relentless spread of AIDS inevitable? What makes AIDS such a deadly disease? Can the disease be cured?

Before we address these questions, we will examine the three lines of defense that protect the body from potentially harmful infectious agents.

**MAIN MESSAGE** We have physical and chemical barriers and powerful internal defenses that limit pathogen entry and attack pathogens if they do gain entry.

## KEY CONCEPTS

- A pathogen is an infectious agent that invades an organism and causes harm that can range from mild illness to death.

- Our first line of defense against pathogens consists of external mechanisms, mainly physical and chemical barriers that keep pathogens from entering the body.

- Defense cells, in both the innate and the adaptive immune systems, have the ability to distinguish self (the body's own chemical composition) from nonself (any foreign substance).

- If the first line of defense fails, the body mounts internal defenses against nonself cells and foreign substances. The responses launched by the innate immune system are immediate and nonspecific in that they target broad classes of invaders such as bacteria or viruses.

- Inflammation is a response to tissue damage that often manifests with redness, warmth, and swelling at the site of injury.

- The defensive actions of the adaptive immune system are slower and specific to particular pathogens, such as a particular strain of virus or bacterium.

- Adaptive immunity often provides long-term protection: immune cells that preserve a memory of a specific pathogen are retained, and these memory cells help launch a faster and stronger response to a repeat infection by the same pathogen.

**THE AIR WE BREATHE IS FULL OF GERMS**, and almost any surface we touch is likely to be covered in germs. Even hospitals are not exempt. Germs that can cause serious infections have been found on doctors' stethoscopes and on hospital uniforms. Scientists began using the term "germs" long ago for agents that cause disease. The more technical term for a disease-causing agent is **pathogen**. The individual that becomes infected by a pathogen is the **host**. Pathogens of humans include viruses, bacteria, and protists, as well as some fungi and multicellular animals, such as parasitic worms (**FIGURE 32.1**).

Different genetic versions (genotypes) of a pathogen are known as strains. For example, *Escherichia coli* (*E. coli*) is a species of bacterium that is a normal resident in the large intestine of humans, cows, and many other animals. Hundreds of strains of *E. coli* have been identified. One of them, *E. coli* O157:H7,

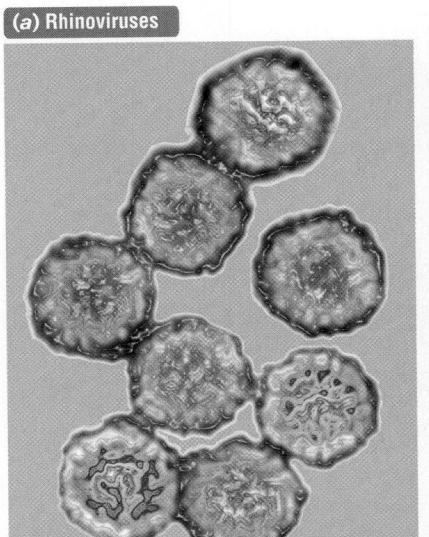
*(a)* **Rhinoviruses**

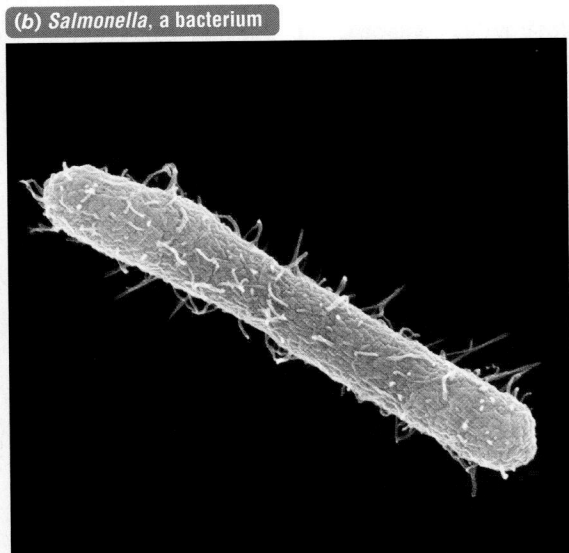
*(b)* ***Salmonella*, a bacterium**

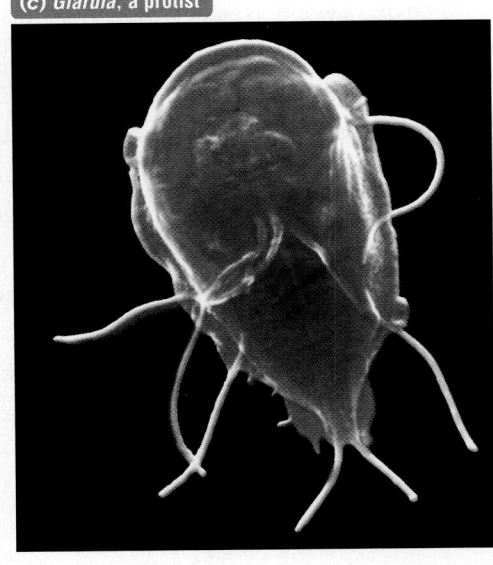
*(c)* ***Giardia*, a protist**

*(d)* ***Trichophyton*, a foot fungus**

*(e)* ***Enterobius*, human pinworms**

**FIGURE 32.1 Pathogens Are Viruses or Small Organisms That Harm Their Host**

*(a)* Approximately 200 different viruses produce the symptoms that we lump together as the "common cold." Shown here is a color-enhanced transmission electron micrograph of rhinoviruses, which are responsible for about half of all cases of the common cold. The pink material in the core of the virus particle is single-stranded RNA, which serves as the genetic material of the virus. It is surrounded by a shell of proteins and lipids (green and blue). *(b) Salmonella typhimurium*, a bacterium infamous for its involvement in large outbreaks of food poisoning. *(c) Giardia lamblia*, a flagellated protist that causes diarrhea and fatigue in campers and hikers who drink contaminated water. *(d) Trichophyton mentagrophytes*, the fungus that causes athlete's foot and scalp ringworm. The threadlike hyphae (green) and yellow spores of the fungus are seen in this color-enhanced scanning electron micrograph. *(e)* Human pinworm, *Enterobius vermicularis*, is the cause of the most common worm infection in North America, affecting mainly children in crowded conditions such as day care facilities.

is harmless to cows but can cause bloody diarrhea, kidney failure, and death in people. As a pathogen accumulates mutations in its genetic material, it may evolve into a new, genetically different, strain. Just about every flu season witnesses at least one or two freshly mutated flu virus strains, for example. Occasionally the mutations generate an exceptionally dangerous strain that devastates populations as it sweeps across the globe.

Though pathogens abound in nature, animals possess a remarkable defense system—the **immune system**—that protects them against most infectious agents. *External defenses*, usually consisting of physical and chemical barriers on the surface of the body, are the first line of defense in animals. The external defenses reduce the likelihood that a harmful organism or virus will gain access to internal tissues. Most animals also have an *internal defense* mechanism—known as the **innate immune system** or innate immunity—that is immediately set in motion when the external barriers fail to stop a pathogen. The external defenses and innate immunity are both nonspecific in their response to pathogens; they deploy a preset repertoire of defensive actions against infectious agents, rather than "made-to-order" attacks on specific species or strains of pathogens.

The immune system of vertebrates includes another layer of internal defense—known as the **adaptive immune system** or adaptive immunity—which acts against pathogens in a highly specific manner, with specialized defense cells mobilized against particular strains of pathogens. Another distinctive feature of the adaptive immune system is **immune memory**, the capacity of this defense system to remember a first encounter with a specific strain of pathogen and to mobilize an especially speedy and precisely targeted response to a repeat infection by the same strain. Immune memory is the reason you are unlikely to get chicken pox again if you contracted this viral infection as a child. The adaptive immune system relies on two main weapon systems: *antibody-mediated immunity*, which uses powerful and versatile antipathogen protein complexes called *antibodies*; and *cell-mediated immunity*, whose role is to destroy cells harboring pathogens and other substances that are sensed as foreign by the body.

We examine the external defenses and both types of internal defenses in greater detail in this chapter, focusing mostly on our own highly sophisticated immune system. The chart in **FIGURE 32.2** summarizes the multilayered organization of the human immune system and provides a road map for our discussion.

We close the chapter with a look at the subversion of cell-mediated immunity by HIV, the particularly devious virus that is responsible for AIDS.

## 32.1 Defense Systems: Distinguishing Self from Nonself

The external defense system, as we shall see in greater detail in Section 32.2, consists of protective layers of tissue on those surfaces of the body that come in direct contact with the environment. The skin, which acts as both a physical and a chemical barrier, is the most important of our external defenses. If a pathogen succeeds in entering the body by overwhelming these external barriers, defense becomes a much more complicated matter. Internal defense systems must meet two distinct challenges: first, they must detect the presence of pathogens inside the body; second, they must attack the pathogens from within without harming the healthy tissues of the body.

Before we can mount an internal defense that kills, disables, or isolates invading pathogens, our body must have a way of recognizing that an invader is present. Although we are not consciously aware of it, the body can distinguish foreign invaders (nonself) from our own cells (self). If our internal defenders could not tell self from nonself, they would attack our own cells—and this sometimes happens, as we describe in the box on page 702.

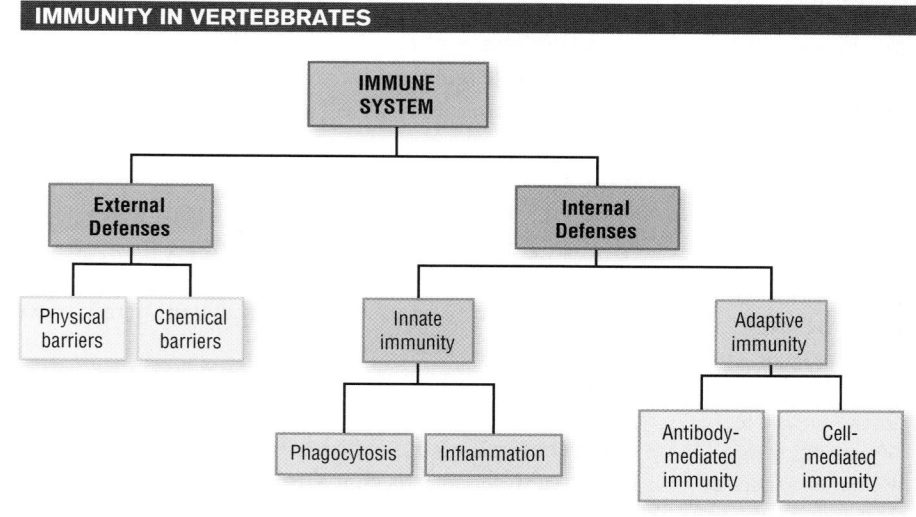

FIGURE 32.2 An Overview of the Immune System in Animals

We make the same sort of distinction in team sports. We know which team is which because each team wears a different uniform. We know which players are on our home team and which are on the opposing team by the uniforms they wear. Similarly, the cells in our body wear our "home team uniform," and each kind of invading pathogen wears the uniform of its own specific team. The "uniforms" for both the home team and the opposing teams usually consist of proteins or carbohydrates associated with the plasma membrane of cells or the outer coat of viruses.

Innate immunity, our second line of defense, is immediately deployed when the external defenses are breached. Like the external defense system, innate immunity is a **nonspecific response**. However, as an internal defense mechanism, the ability to distinguish friend from foe—self from nonself—is a key characteristic of innate immunity. As soon as the body detects cells or molecules that do not belong, certain defense cells and defensive proteins take measures to eliminate the pathogens and nonself substances. Using the team sport analogy again, the innate immune system functions like a defending team that is prepared to play against any team not wearing the home team uniform. But because it is a nonspecific response mechanism, innate immunity uses much the same strategy ("same play") against any opponent.

Vertebrates like us have a third line of defense, adaptive immunity, which consists of **specific responses**: the body goes beyond simply recognizing something as nonself; it identifies particular invaders, such as the H10407 strain of *E. coli* or the OKA strain of the chicken pox virus. To do this, the body has separate teams of defensive cells, each specialized to recognize and play against only one particular opposing team (for example, the H10407 strain, but not any other strain of *E. coli*). In addition to being highly specific, these teams of cells have long memories: each team can remember being attacked by a particular kind of pathogen in the past, enabling the body to mount an exceptionally effective defense if the same opposing team shows up again. The immune system can "hold a grudge" for decades. This memory of our own specific response to a pathogen is what enables us to become immune to attack by the same strain after we suffer the disease a first time.

The players for the home team include several types of *white blood cells* and a battery of defensive proteins and other molecules. **White blood cells** are a vital family of defense cells that are found in the interstitial fluid (the fluid between cells) and in the blood, where they constantly intercept invading pathogens. As we

shall see, some types of white blood cells participate in only innate immunity, others participate in only the specific responses of adaptive immunity, and a few participate in both of these internal defense systems. Most of these players have several roles; although they specialize, they also back up other players.

The innate and adaptive immune systems also cooperate in recognizing and destroying cancer cells—a strategy known as **cancer immunosurveillance**. Cancer cells not only behave abnormally (see Chapter 11), but they typically display cell surface features (such as unusual proteins or carbohydrates) that set them apart from healthy, noncancerous cells. These abnormalities in the structure and behavior of cancer cells are interpreted as a nonself signal, and such cells are therefore subject to attack by both the innate and adaptive defenses. A diagnosed cancer represents a failure of cancer immunosurveillance.

## 32.2 First Line of Defense: Physical and Chemical Barriers

Our first line of defense against invading pathogens consists of barriers that work to keep them from entering the tissues of the body. Some of these barriers are physical, just like the walls, closed windows, and locked doors of a home. Some are sticky traps that bind up invaders so that they can be disposed of. External defense mechanisms also include chemical agents (such as enzymes) or chemical environments (such as acidic conditions) that keep the invaders from attaching to or growing on body surfaces.

## Linings in the lungs, digestive system, and skin keep most pathogens at bay

Pathogens can infect us only if they can find a way into the body. Linings that separate the "outside" from the "inside" of the body make up the first line of defense against invaders. These linings—skin and the linings of the lungs and digestive system, for example—act as a barrier to keep out most potentially dangerous pathogens. Our skin may not seem like a strong defense, but unless it is torn, its outer layer of dead cells and the interlocking live cells just beneath it make the skin virtually impenetrable to bacteria and other pathogens. The acidic pH of the skin and the salty secretions from skin glands create an environment that is inhospitable to many potential parasites. Like the cells that line our lungs and digestive system, skin cells also secrete **defensins**, which are small chains of amino acids that can destroy many types of bacteria, and even certain types of viruses (those that have a lipid membrane), by creating holes in their membranes.

Our eyes are bathed by the same fluid that forms tears. Bacteria that land on the outer surfaces of our eyes are often washed away in this fluid. Bacteria that remain are usually destroyed by antibacterial enzymes (such as lysozyme) secreted by the tear gland above each eyelid. Similarly, many bacteria entering our bodies in food or drink are killed by chemicals in our saliva. Hairs in the nose filter out large airborne particles, such as dust, that may carry pathogens. The tubes of the respiratory system, as we saw in Chapter 28, trap invading pathogens in mucus, which is then removed, with its contents, from the respiratory system. In women of reproductive age, the vagina—the part of the female reproductive system that is exposed to the outside—maintains an acidic pH (about pH 4), which discourages the growth of some bacteria and yeasts while supporting beneficial bacteria such as lactobacilli.

The digestive system is protected by a chemical barrier (including defensins) that prevents most potentially harmful organisms from colonizing the stomach and intestines. The acids and digestive enzymes in the stomach destroy most of the pathogens we swallow. Even so, a variety of bacteria, viruses, protists, and intestinal worms can infect the digestive tract. The CDC (Centers for Disease Control and Prevention) estimates that 76 million people in the United States get food poisoning every year, mostly caused by bacteria such as *Campylobacter* and *Salmonella* in contaminated food or drink. Most people recover from the diarrhea and vomiting typical of food-borne bacterial infections. However, about 5,000 people die every year from such infections, either because they are unusually vulnerable (the very young and the elderly, for example) or they become infected with a particularly aggressive strain (such as *E. coli* strain O157:H7).

## Some pathogens can overcome our external defenses

Although our external defenses do a good job of keeping out most pathogens, the body is still vulnerable. Consider the skin. Wounds, in the form of cuts, abrasions, and punctures, are common. Many pathogens take advantage of breaks in the skin to gain entry to their hosts. For example, the tetanus-causing bacterium, *Clostridium tetani*, occurs commonly in soil and on other surfaces worldwide. Although we constantly come in contact with this bacterium, it becomes a threat only if it enters the body through a wound that is deep and that tends to shut out air, such as a deep puncture wound. In low oxygen environments, *C. tetani* can multiply rapidly, producing a **toxin** (a poisonous chemical) that causes severe muscle spasms. In most cases our body's defenses alone cannot act quickly enough to prevent fatal amounts of this toxin from being produced; for those who do not receive regular tetanus shots and boosters, death is the usual outcome. In parts of the world where infants are born under poor sanitary conditions, *C. tetani* kills more than 200,000 babies annually by entering through the cut made at birth to remove the umbilical cord.

Many important infectious diseases and parasites of humans and other animals rely on blood-feeding insects to penetrate the host's skin. For example, the saliva of certain mosquito species carries *Plasmodium*, the protist that causes malaria, an infectious disease that kills 1–2 million people annually. When biting through skin to take a blood meal, these mosquitoes inject *Plasmodium* into the bloodstream. Flea bites were responsible for spreading bubonic plague from person to person in Europe in past centuries, and lice have done the same during typhus epidemics. Other well-known insect-borne diseases include West Nile virus, yellow fever, and sleeping sickness.

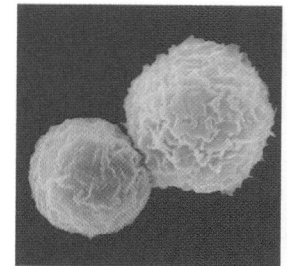

T cells (pink) and B cells (green) are part of the adaptive immune system. Both cells originate in the bone marrow. T cells mature in the thymus and attack abnormal or foreign cells. B cells mature in the bone marrow and secrete antibodies against specific foreign substances.

## 32.3 Second Line of Defense: The Innate Immune System

When a pathogen succeeds in getting past our external defenses, the nonspecific responses of the innate immune system are deployed as the next line of defense against the invader. Such responses are said to be innate (inherent) because the necessary components—defense cells, proteins, and other molecules—are held constantly at the ready for deployment against an invading organism, virus, or other substance perceived as nonself. The defense responses may be local, occurring at the point of entry; or they may involve the whole body. Unlike the adaptive immune system, the innate immune system does not provide long-lasting protection against pathogens encountered on a previous occasion. Innate immunity is an ancient defense mechanism found in both invertebrates and vertebrates.

## Phagocytes engulf or encapsulate stray pathogens

Backing up the mechanical and chemical protection that our external barriers provide (our first line of defense), the cells and defensive proteins of the innate immune system patrol these borders from within, forming a second line of defense. In the layers of our living skin tissue, the alveoli of our lungs, our intestinal walls, and many other places in the body, white blood cells called **phagocytes** (*phago*, "to devour"; *cyte*, "cell") are constantly on the prowl. There are several different types of phagocytes, including the relatively large *macrophages* (*macro*, "big"; *phage*, "eater") and the smaller but more numerous *neutrophils*, so named because they can be readily stained with a colored dye that has a neutral pH (**FIGURE 32.3**).

Phagocytes destroy invading cellular organisms the same way many single-celled eukaryotes, such as amoebas, eat their food: they engulf their target by **phagocytosis** (see Figure 7.10*c* and *e*). Once inside the cytoplasm of the phagocyte, the engulfed pathogen is confined to a membrane-enclosed compartment (*phagolysosome*), where it is destroyed with a toxic brew of chemicals. Any pathogens that the phagocytes find indigestible can be walled off from other cells by *encapsulation*—that is, by the formation of a protective capsule around the invaders.

Individual phagocytes die when they are full of pathogens. This cellular debris is removed from the body by various means. Dead phagocytes from the lining of the alveoli, for example, are swept out of the lungs, along with whatever they have ingested, in the mucus that leaves the respiratory system.

## Inflammation shields an injury site from potential pathogens

Tissue damage often results when cells in the body have been irritated, damaged, or killed, either because of pathogen invasion or because of wounding by a physical or chemical agent. The innate immune system responds to tissue damage by mounting an immediate and coordinated sequence of events known as **inflammation**. The nonspecific events of inflammation occur regardless of whether any pathogens are present: the immune system swings into action to clean up the damaged tissues and to prevent the entry or spread of a pathogen whenever cellular damage is detected. Although inflammation can occur anywhere inside the body, let's consider what happens following a deep cut or puncture wound to the skin. (Note that even though we lay out these events sequentially, many of them actually happen at the same time.)

The inflammatory response begins as damaged tissue or pathogens in the wound release chemical signals (**FIGURE 32.4**). Direct injury or activation by chemical signals stimulates **mast cells**, which develop from a type of phagocyte that is especially abundant at boundaries between body tissues and the outside world. Activated mast cells release **histamine** and a variety of other alarm signals, including small protein molecules broadly classified as **cytokines** (*cyto*, "cell"; *kinein*, "to move"). These chemical signals trigger the three classic signs of inflam-

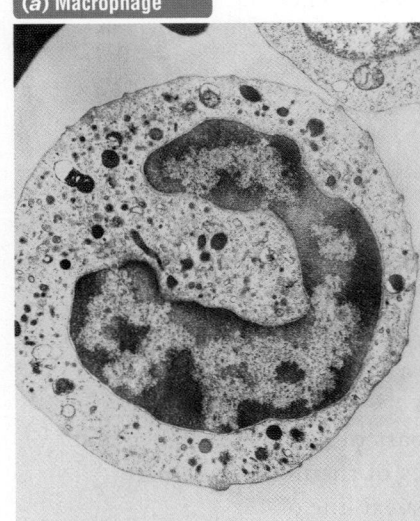

**(a) Macrophage**

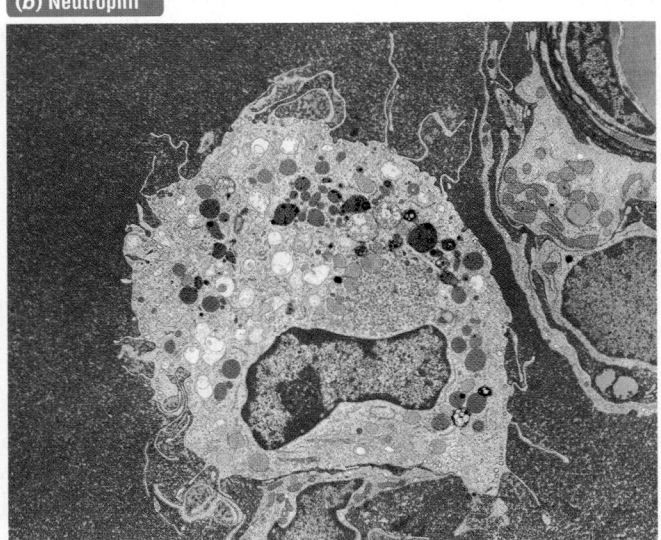

**(b) Neutrophil**

**FIGURE 32.3 Phagocytes Engulf Pathogens and Foreign Substances**

Two types of phagocytes are seen in these light micrographs. (*a*) Macrophages are relatively large and may take an hour or more to engulf a pathogen. (*b*) Neutrophils—the most abundant type of white blood cell—attack pathogens with antimicrobial chemicals and destroy bacteria by engulfment. Phagocytes are critical for innate immunity, and some of them also assist in adaptive immune responses.

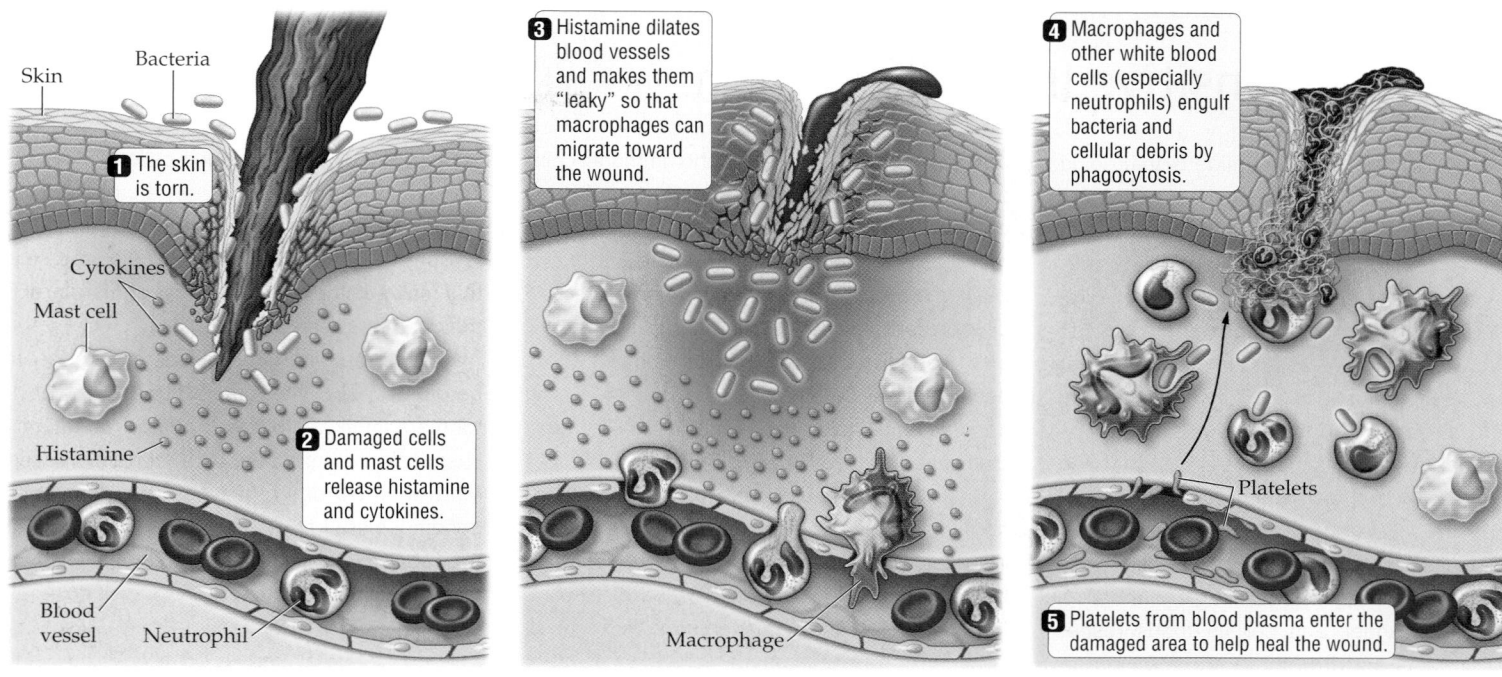

**FIGURE 32.4** The Inflammatory Response Activates Several Kinds of Cells and Proteins That Act against Invading Pathogens

mation: *redness*, *warmth*, and *swelling*. Histamine causes the dilation of smaller blood vessels, bringing more blood to the area around the wound. We experience the increased supply of blood warmed by metabolic heat as reddening and warmth in the injured area. The stepped-up blood flow serves an important purpose: oxygen and nutrients, which are needed for repair, reach the area more quickly and in larger quantity. Increased blood flow also means that immune cells, such as phagocytes, are delivered more quickly to the site of inflammation.

In addition to increasing blood flow by dilating blood vessels, histamines make the capillary walls more porous. Immune cells are able to squeeze between the cells that make up the walls of the blood vessels to enter the inflamed tissues. Blood plasma also leaks out of these capillaries to mingle with the interstitial fluid in the area. We experience this pooling of fluid as swelling at the site of injury or infection. The swelling characteristic of inflammation has important functions: it can dilute any toxins that invaders may be producing, and it delivers a variety of chemicals that promote healing and repair of damaged tissues. Histamines can also irritate nerve endings, leading to itching or pain. The itchy red bumps you may have experienced from mosquito bites are a

manifestation of inflammation launched by the innate immune system.

## Innate immunity is involved in blood clots and fever

The innate immune system is also responsible for the crucial task of clotting the blood to close a wound. Sealing an open wound serves the purpose of reducing blood loss, as well as restoring the integrity of external defense barriers. Tissue damage triggers a cascade of events that cause proteins normally dissolved in the

Platelets (blue) and fibrin (yellow) form a blood clot to seal broken skin.

# The Hypersensitive Immune System: Allergies and Autoimmunity

With the billions of foreign antigens that could potentially invade the body, our immune system has the herculean task of discriminating between self and nonself and selectively attacking dangerous agents only. Sometimes our immune system overreacts. Allergies are caused when our immune system launches an all-out attack against harmless antigens. Autoimmune diseases result when the immune system mistakenly attacks normal body cells.

An *allergy* is a rapid and predictable inflammatory response to common environmental substances, or *allergens*, that generally pose no significant threat to nonallergic individuals. Pollen, cat dander, and certain foods (especially peanuts and tree nuts) are common allergens. Mast cells respond to an allergen by rapidly releasing large quantities of histamines and cytokines that precipitate an inflammatory response. The runny nose that accompanies hay fever is caused by the histamine-stimulated loss of fluid from leaky blood vessels. Histamines released by mast cells also irritate certain nerves, which allergy sufferers may experience as itchiness or pain. Many over-the-counter allergy medications act by blocking the action of histamines (as in the case of antihistamines like loratadine) or certain cytokines (as in the case of drugs like montelukast).

Sometimes the immune system mistakenly responds to some of the body's own cells as if they were foreign. The result is an *autoimmune disease*, in which the immune system attacks tissues that should otherwise have been rec-

ognized as "self" and left alone. Immune cells that are specialized to recognize self antigens do arise, but normally such self-directed immune cells are suppressed or deleted from the body in a process known as *immunological tolerance*. For reasons that are poorly understood, in some individuals significant numbers of antiself immune cells remain, and this failure in immunological tolerance results in autoimmune diseases.

The list of autoimmune diseases is long, and it includes some familiar afflictions. Type 1 diabetes, also known as insulin-dependent diabetes, is caused by the destruction of insulin-producing cells in the pancreas. It is the less common form of diabetes; more than 90 percent of the 26 million Americans who have diabetes have the type 2 form (see Chapter 29). Type 1 diabetes usually develops in childhood or young adulthood, and it is treated with injections of genetically engineered human insulin (see Section 16.4). In multiple sclerosis (MS), the immune system attacks neurons in the brain, causing them to lose some of the myelin sheath that is essential to their proper function. Many sufferers of MS experience a loss of muscular coordination that eventually confines them to a wheelchair. In rheumatoid arthritis, another autoimmune disease, the immune system attacks the cartilage that lines the joints, leading to crippling pain and loss of joint function. The organs and tissues of people with lupus are chronically inflamed because they are targeted by the person's own immune system.

In celiac disease, immune cells in the small intestine react to dietary gluten, a protein found in wheat and related grains; then, for unknown reasons, the immune system turns on the cells that make up the intestinal lining, compromising their nutrient-absorbing function.

Some autoimmune diseases appear to be genetic, but others are influenced by environmental risk factors that range from dietary chemicals to infection with certain viruses. More than 80 different types of autoimmune conditions are known, and they remain among the most mysterious of the diseases that afflict humans.

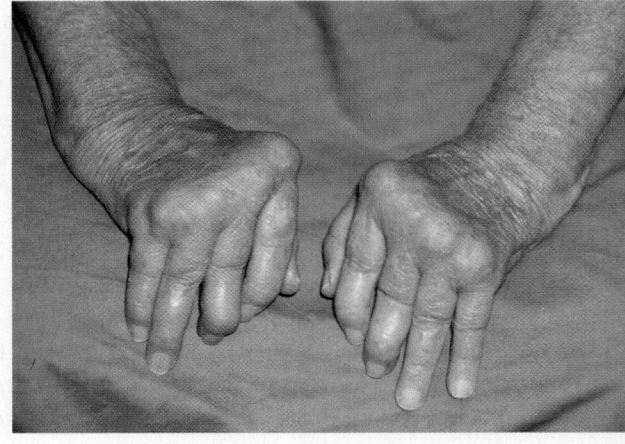

**Rheumatoid Arthritis Is a Common Autoimmune Disease**
The disfigurement and pain caused by rheumatoid arthritis can be severe.

---

blood to combine with sticky cell fragments, called **platelets**. The platelets, which also circulate in the blood, interlink with the clotting proteins to form a gel-like mesh that traps blood cells. The result is a blood clot, which helps prevent any pathogens that may be present in the wound from spreading to surrounding tissues. Clotting can begin as quickly as 15 seconds after tissue damage occurs. The subsequent

growth of new tissues then repairs the wound more permanently.

Phagocytes activated by the presence of pathogens may also launch the fever and malaise that is familiar to everyone who has had the flu. Activated phagocytes release chemicals such as cytokines, which in turn trigger the production of *prostaglandins*. **Prostaglandins** are a family of lipid molecules manufactured by many different types of tissues in the body, including certain phagocytes and the cells that line our blood vessels. Prostaglandins stimulate the hypothalamus, the brain region that functions as the body's thermostat (see Chapter 26). The hypothalamus raises the set point for core body temperature above the average 37°C (98.6°F), and the body responds by both generating and conserving more heat. The resulting fever may be uncomfortable, but as long as the temperature increase is moderate, fevers are beneficial: they limit the growth of many pathogens, enhance phagocytosis, and speed the repair of damaged tissues. The drowsiness, pain, and fever initiated by prostaglandins prompt us to slow our pace of activity, allowing the body the time and metabolic resources it needs to heal. Many of the medications commonly used to alleviate a high fever or pain act by inhibiting the synthesis of prostaglandins. They include over-the-counter drugs such as aspirin, ibuprofen, and acetaminophen (which may be more familiar to you by its various brand names, such as Tylenol).

## Complement proteins and direct attack are additional strategies

The innate immune system also has a "mark and destroy" game plan that is carried out by a suite of pathogen-recognizing proteins. Circulating in the blood plasma are about 25 different **complement proteins**, so called because they work together with antibodies that are produced by certain white blood cells. When they arrive at the wound, complement proteins bind to the invading cells and destroy them by creating holes in their plasma membranes. Some pathogens broadly resemble normal body cells, and complement proteins have the additional task of labeling invaders to differentiate them more clearly from normal cells. This would be like putting an "X" on the backs of every member of an opposing team whose uniform looked a bit too much like the home team's. Some of the complement proteins may remain bound to the invaders to help phagocytes and white blood cells identify and destroy foreign cells that have survived the initial attack by complement proteins. For example, complement proteins marking the surfaces of invading *Clostridium tetani* cells help phagocytes bind to these cells in preparation for phagocytosis.

When the pathogen is a virus rather than a bacterium or protist, the innate immune system responds somewhat differently. Viruses can replicate themselves only if they enter a host cell and take over its metabolic machinery. The cells that mediate innate immunity respond not to the viral particles directly, but to cells that have become infected by viruses. Bits of the viral protein coat, abandoned on the plasma membrane when the virus fuses to a cell, mark infected human cells. Specialized phagocytes identify these marked cells and then proceed to devour them. In addition, cells infected by viruses release **interferon**. This chemical attaches to the plasma membranes of nearby cells, interfering (hence its name) with the ability of the virus to enter and infect those cells. Interferon also summons another type of white blood cell—**natural killer (NK) cells**—to the site. NK cells destroy any cell whose plasma membrane is marked with foreign proteins, including viral proteins. NK cells kill either by causing so many holes to form in the other cell's plasma membrane that it disintegrates, or by inducing the infected cell to destroy itself.

> ## Concept Check
>
> 1. What role do phagocytes play in innate immunity?
> 2. What are the three main signs of the inflammatory response, and how does this response combat invading pathogens?

## 32.4 Third Line of Defense: The Adaptive Immune System

In a puncture wound like the one described earlier, the chances are high that some live *Clostridium tetani* are thrust deep into the wound, in this way bypassing the first line of defense. The bacteria then find themselves in a favorable environment for proliferating and producing their deadly toxin. In addition to our second line of defense—the various nonspecific responses of the innate immune system—the body automatically puts into play a third line of defense: the adaptive immune system. The adaptive immune system is a coordinated network of interacting cells and defensive molecules that give us the ability to resist specific pathogens on a long-term basis. Compared to innate immunity, adaptive immunity is slower to mobilize,

Concept Check
Answers

1. Phagocytes engulf and destroy pathogens or isolate them by encapsulation, release fever-inducing chemical signals, and remove dead cells and debris at injury sites via phagocytosis.
2. Redness, swelling, and warmth at the site of injury or infection are the effect of blood vessel dilation, which delivers pathogen-fighting white blood cells and repair proteins.

often taking more than 2 weeks to successfully engage and destroy a first-time invader. However, the selectivity with which the adaptive immune system attacks a particular invader, and the memory it carries of that encounter, make it the most sophisticated and effective of animal defense systems.

## The lymphatic system supports immunity

The human adaptive immune system consists of about 7 trillion white blood cells of several different types and an assortment of defensive proteins. Many of these cells and proteins circulate in the body in an inactive form. When activated, they move out of the blood into the interstitial fluid to fight an infec-

tion. From there, these defensive proteins and white blood cells are collected, along with the interstitial fluid, in the **lymphatic ducts**, a network of tubes that returns the interstitial fluid to the circulatory system. At various points along the lymphatic ducts lie **lymph nodes**, pockets of tissue containing large numbers of white blood cells that trap bacteria, viruses, and foreign proteins. Lymph nodes (popularly known as "glands") often swell in response to an infection, as the white blood cell population in these tissues increases rapidly to fight the invaders. With our fingers we can feel lymph nodes located near the skin surface—under the jaw and chin and behind the ears, for example—when they become enlarged. Lymph nodes, the lymph ducts that connect them, and several other organs, such as the spleen, collectively make up the **lymphatic system** (**FIGURE 32.5a**). The lymphatic system assists both

## THE LYMPHATIC SYSTEM

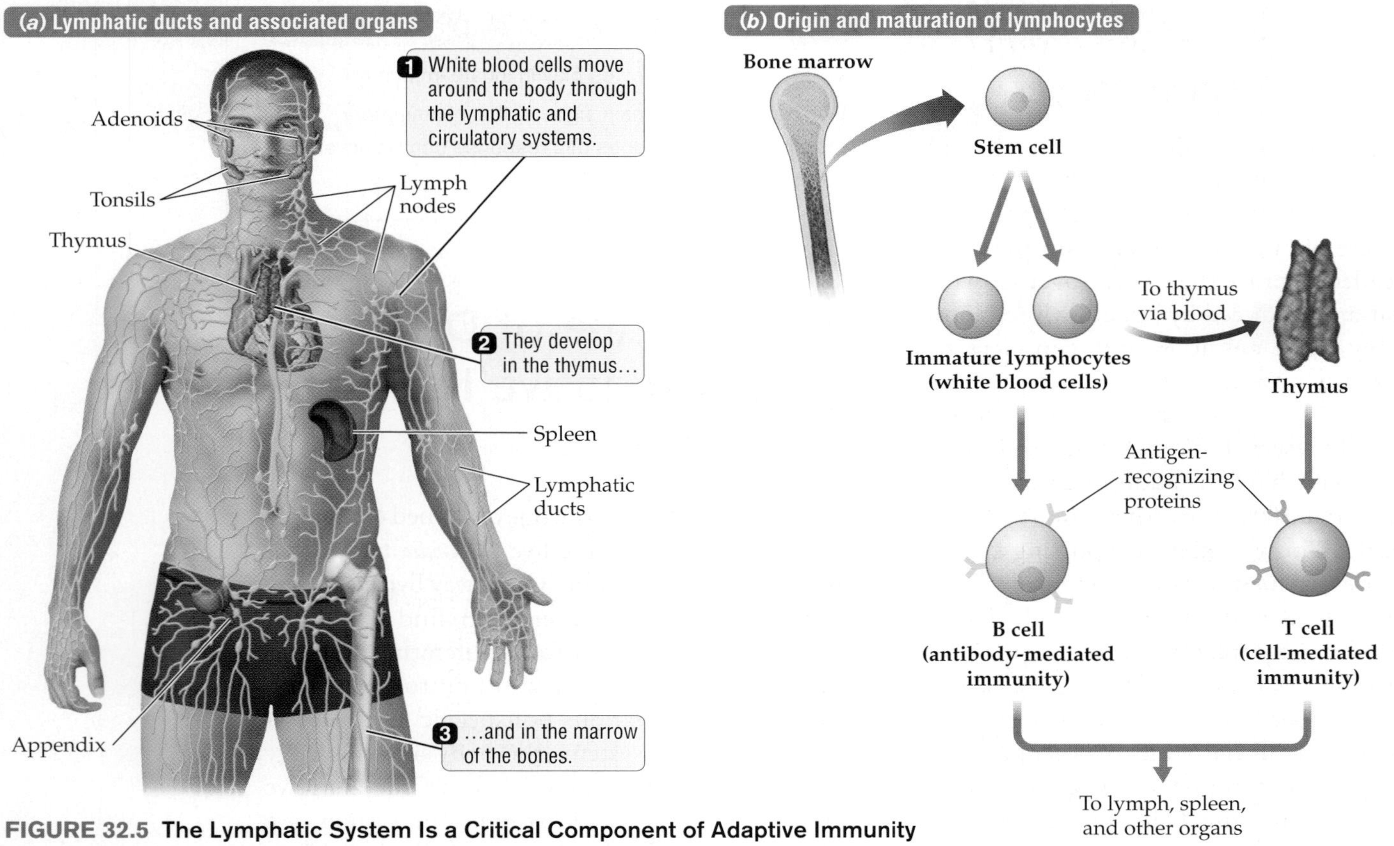

(a) Lymphatic ducts and associated organs

Adenoids
Tonsils
Thymus
Appendix

1 White blood cells move around the body through the lymphatic and circulatory systems.

Lymph nodes

2 They develop in the thymus…

Spleen

Lymphatic ducts

3 …and in the marrow of the bones.

(b) Origin and maturation of lymphocytes

Bone marrow

Stem cell

Immature lymphocytes (white blood cells)

To thymus via blood

Thymus

Antigen-recognizing proteins

B cell (antibody-mediated immunity)

T cell (cell-mediated immunity)

To lymph, spleen, and other organs

**FIGURE 32.5  The Lymphatic System Is a Critical Component of Adaptive Immunity**
(a) The lymphatic system consists of lymphatic ducts, lymph nodes, and associated organs.
(b) Lymphocytes originate from stem cells in bone marrow. B cells mature in the bone marrow; T cells mature in the thymus. Lymphocytes circulate in the lymphatic and circulatory systems and accumulate in lymph nodes and other organs, such as the spleen and the tonsils.

innate and adaptive immunity, and many of the cells that mediate these types of immunity are found in lymphatic tissue.

**Lymphocytes** (**FIGURE 32.5b**) are a class of white blood cells that are found primarily in the lymphatic system and play a critical role in adaptive immunity. They enable us to mount specific responses to particular species and strains of invading pathogens. Lymphocytes that bind to *C. tetani*, for example, do not bind to other pathogens, so these lymphocytes distinguish *C. tetani* from all other invaders. Similarly, lymphocytes that specialize in recognizing *E. coli* strain O157:H7 do not recognize other strains of *E. coli*.

## B and T lymphocytes recognize pathogens with great specificity

Lymphocytes develop from stem cells in bone marrow (see Section 11.1). Immature lymphocytes differentiate into two basic types of mature lymphocytes: **B cells** ("B" for "bone") mature in the bone marrow; **T cells** ("T" for "thymus") migrate from the bone marrow to mature in the thymus, a gland found in the chest (see Figure 32.5b). Once mature, both types of lymphocytes move into the lymphatic system, congregating at places where foreign cells and foreign proteins tend to accumulate. Large numbers of mature B cells and T cells are found in the lymph nodes and other organs associated with the lymphatic system.

As it matures, a B or T cell becomes specialized to recognize a very specific chemical signature, or *antigen*. An **antigen** consists of one or more molecules displayed on the surface of a particular pathogen or foreign substance that can be recognized by lymphocytes. Antigens characterize pathogens in much the same way that unique combinations of letters and colors on their uniforms distinguish different opposing teams in our team sport analogy. Different lymphocytes have unique receptor proteins embedded in their plasma membranes that can bind with different antigen molecules. Like a key that fits a particular lock, each lymphocyte is individually specialized to recognize and bind with only its corresponding antigen. There are a huge number of antigen-binding receptor proteins, but each "team" of lymphocytes carries only one specific kind in its plasma membrane. For example, only one group of B cells and one group of T cells bear the membrane receptor protein that recognizes and binds with antigen molecules on the surface of

*Clostridium tetani*. Other B cells and T cells may recognize a specific cold virus, or the bacterium that causes strep throat, or even proteins on the surface of a particular kind of pollen grain.

Lymphocytes, each programmed to recognize a particular antigen, circulate through the body in an inactive state. A lymphocyte becomes activated when it encounters and binds its specific antigen in the spleen or lymph nodes. In a process known as **clonal selection**, each activated lymphocyte divides rapidly to make many identical copies, all of them targeted against the same specific antigen (**FIGURE 32.6**).

**CLONAL SELECTION**

Vertebrates produce a great diversity of lymphocytes that differ in their unique membrane proteins.

Invading pathogens produce characteristic antigens.

Of the millions of lymphocytes produced, only a few bear membrane proteins that can bind to the invader's antigen.

The lymphocyte that successfully binds to an invader's antigen proliferates to form a large number of clones.

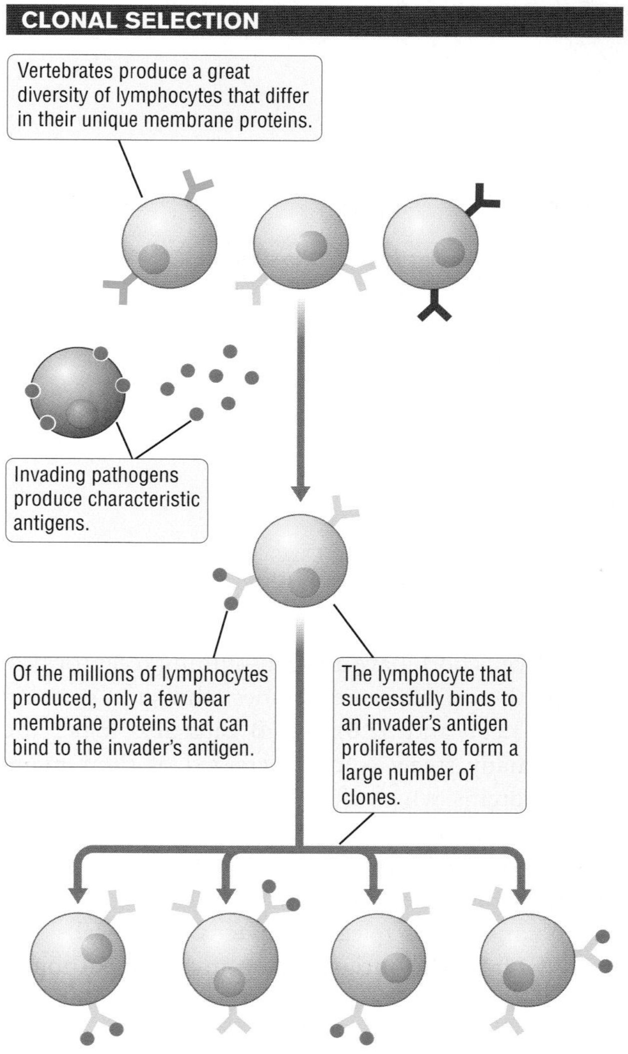

**FIGURE 32.6 Lymphocytes Are Activated by Antigen Recognition**
Lymphocytes that bind to an antigen on an invading pathogen multiply rapidly, giving rise to identical lymphocytes, all of which can bind specifically to the same antigen.

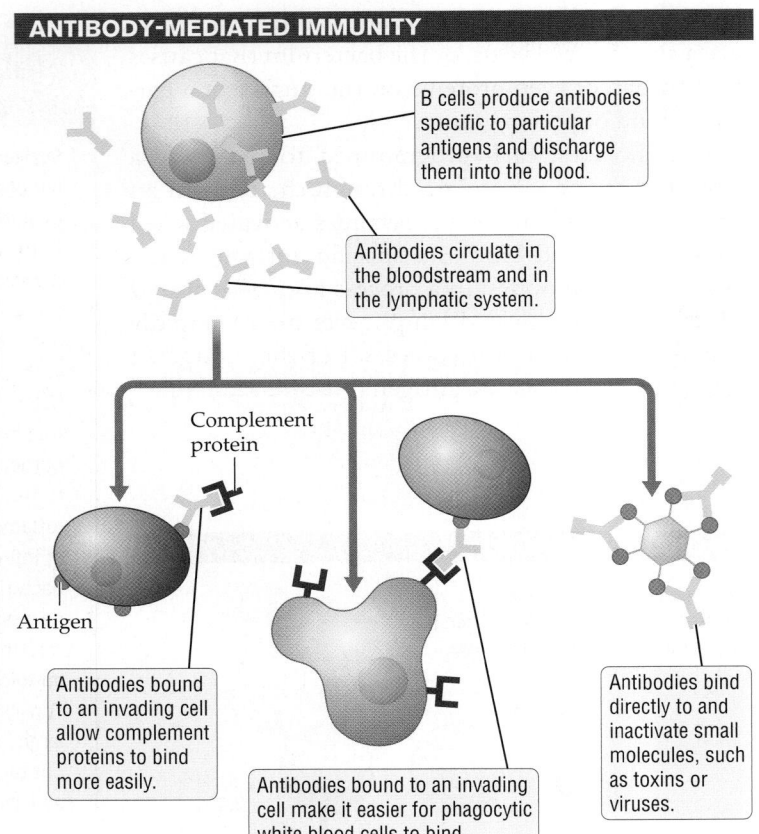

B cells produce antibodies specific to particular antigens and discharge them into the blood.

Antibodies circulate in the bloodstream and in the lymphatic system.

Complement protein

Antigen

Antibodies bound to an invading cell allow complement proteins to bind more easily.

Antibodies bound to an invading cell make it easier for phagocytic white blood cells to bind.

Antibodies bind directly to and inactivate small molecules, such as toxins or viruses.

**FIGURE 32.7 How Antibodies Work**

The binding of antibodies (green) to invading cells (red) enables phagocytic white blood cells (macrophages and neutrophils) and complement proteins to bind to the invaders. Antibodies can also bind to and neutralize toxins and viruses.

## B cells produce antibodies

B and T cells both have antigen-binding receptor proteins in their plasma membranes (shown in green in Figures 32.5*b* and 32.6), but B cells also make and release many freely soluble versions of their membrane proteins, which circulate in the blood and lymphatic system as **antibodies**. Because B cells can attack invaders with antibodies, they are said to deploy **antibody-mediated immunity**.

The antibodies released by activated B cells specifically target the antigen on an invading pathogen, such as *Clostridium tetani*. When antibody molecules bind to an invading cell, they target the invader for destruction by other members of the immune system defense team. Pathogens tagged with antibodies are targeted by complement proteins, for example. Antibodies also increase the ease with which macrophages and neutrophils can bind to and destroy invaders. In

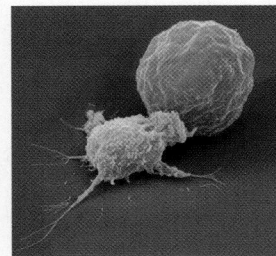

**FIGURE 32.8 T Cells Engage in Cell-to-Cell Combat**

A cytotoxic T cell (yellow) is seen attacking a cancer cell (pink) in this color-enhanced scanning electron micrograph.

addition, antibodies can bind to and neutralize small foreign molecules, such as toxins (like those produced by *C. tetani*) and viruses (**FIGURE 32.7**), which then cannot penetrate a host cell.

## T cells target infected or abnormal cells

T cells are specialized for taking down infected or abnormal cells in a one-on-one play. In contrast to B cells, T cells keep all their antigen-binding receptor proteins in their plasma membranes. Each T cell's receptor protein is specialized for recognizing a specific antigen displayed on the surface of an infected cell or an abnormal cell such as a tumor cell. When the T cell receptor proteins bind to their target antigens, the whole T cell becomes attached to the target cell (**FIGURE 32.8**). Because the result is a cell-to-cell bond, immunity involving the action of T cells is called **cell-mediated immunity**.

A T cell becomes activated when it encounters its specific target antigen displayed on the surface of a whole cell. Immune cells that display antigens are called **antigen-presenting cells** (**APCs**), and they include certain phagocytes that specialize in displaying ingested antigens on their cell surface. A virus engulfed by a phagocyte, for example, is broken up inside the phagolysosome, and small pieces of the invader are then inserted in the cell's plasma membrane. The T cell is activated as its antigen-recognizing receptors dock with the displayed antigen. Activated T cells proliferate through clonal selection, producing a population boom in T cells directed specifically against the antigen presented by the phagocyte or other APC (**FIGURE 32.9**).

T cells come in two main categories of effector cells: *helper T cells* and *cytotoxic T cells*. **Helper T cells** are so called because they act as guides for other cells in the immune system. When helper T cells become activated by binding to their specific antigen, they not only proliferate rapidly by clonal selection, but also stimulate the activation and proliferation of B cells and cytotoxic T cells that bind to the same antigen. Activated helper T cells also help macrophages bind more rapidly to invaders. **Cytotoxic T cells** are so named because they destroy the body's own cells that are damaged, that exhibit the cell surface signature of a cancer, or that have been infected by viruses. Because viruses need working host cells to reproduce, every infected host cell that is destroyed lessens the chance that the viral particles inside it will be able to complete their reproductive cycle. **FIGURE 32.10** illustrates one mechanism by which cytotoxic T cells destroy their target cells.

## The first infection produces a slower, milder immune response

Our first exposure to a particular antigen sets into motion the **primary immune response** of adaptive immunity. It takes time—more than 2 weeks sometimes—to get this primary response up to full steam. Because the primary immune response takes place relatively slowly, and because pathogens like *Clostridium tetani* multiply so rapidly, people infected with an aggressive pathogen can lose the race, become ill, and die. For this same reason, any pathogen that is new to humans is particularly dangerous. However, the combined action of innate immunity and the primary response of adaptive immunity usually prevails against most pathogens.

As noted earlier, once a particular B cell or T cell has bound to its specific antigen, it reproduces rapidly through clonal selection. The process yields large numbers of identical lymphocytes that carry exactly the same antigen-recognizing proteins as their parent cells. The great majority of these cloned lymphocytes become *effector cells*, those that are ready to engage the antigen-bearing pathogen, tumor cell, or foreign substance. A small number of the cloned lymphocytes become *memory cells*, which, as we shall see next, are held in reserve for a rapid response to a repeat invasion by the same foreign particle.

**CELL-MEDIATED IMMUNITY**

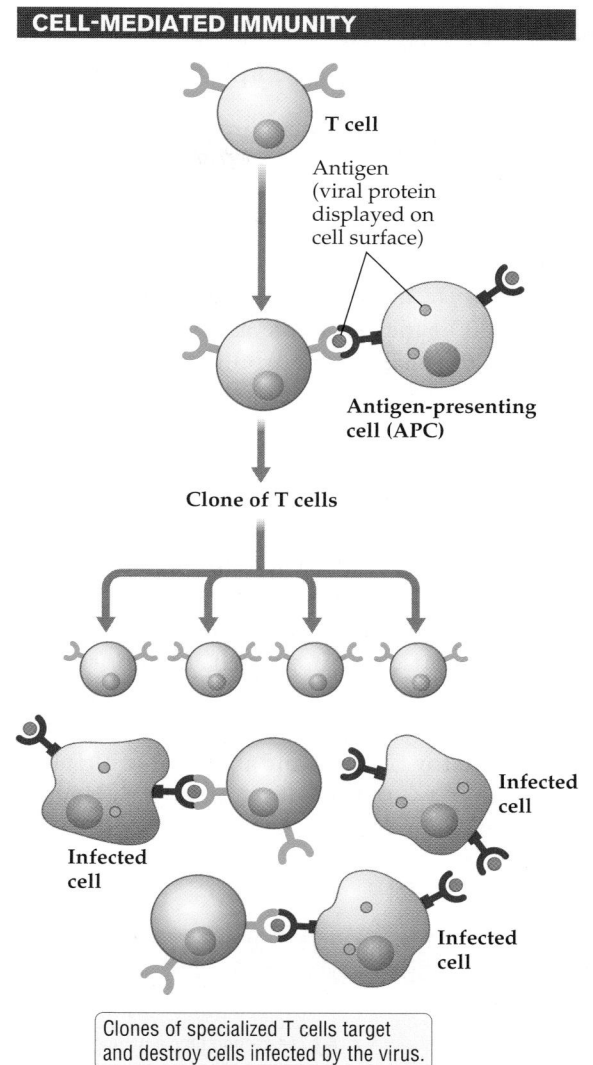

**FIGURE 32.9**
**Antigen Binding Stimulates Clonal Selection of T Cells**
Cell-mediated immunity involves antigen-recognizing proteins that have the same specificity as the antibodies in the antibody-mediated system have, but these proteins remain bound to the surface of the T cells that produce them.

**FUNCTION OF CYTOTOXIC T CELLS**

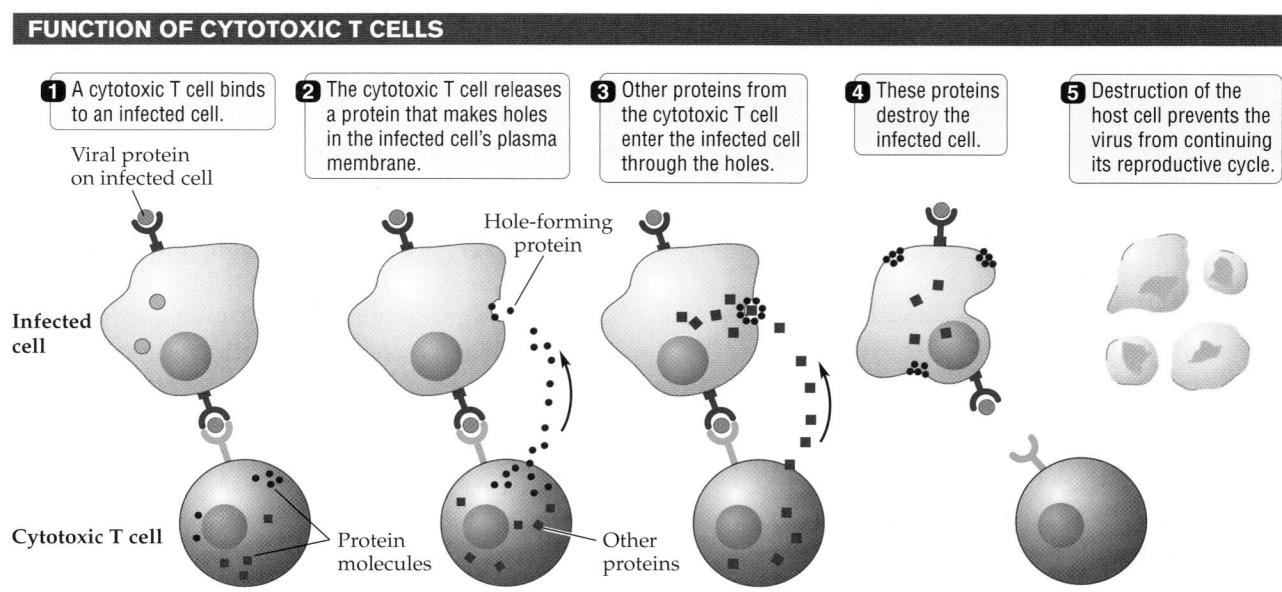

**FIGURE 32.10**
**Cytotoxic T Cells Destroy Infected or Damaged Host Cells**
One type of protein released by cytotoxic T cells forms a hole in the plasma membrane of a host cell that has been infected by a virus, enabling other proteins from the cytotoxic T cell to enter the infected cell and destroy it.

## Subsequent exposures provoke a faster, stronger response

The adaptive immune system produces a faster, more dramatic response to pathogens when it encounters them a second time (**FIGURE 32.11**). This subsequent response is called a **secondary immune response**. For example, after a first exposure to the toxin produced by *Clostridium tetani*, we can survive an exposure more than 100,000 times as severe as the one that would have killed us the first time around.

The key to the body's ability to learn from its experience with a particular invader lies with the lymphocytes that recognize particular antigens. Some of the T cells and B cells that recognized *C. tetani* and underwent clonal selection during the first infection turn into memory cells. **Memory cells** are clonally selected lymphocytes that remain in the body in small numbers as a long-term record of the primary immune response to an antigen. Memory cells are the members of the home team that hold a grudge. During a second exposure to *C. tetani*, the memory cells specific for *C. tetani* and its toxin multiply rapidly, forming a vast fleet of lymphocytes. Having so many *C. tetani*–specific lymphocytes this time around dramatically speeds up all steps of the immune

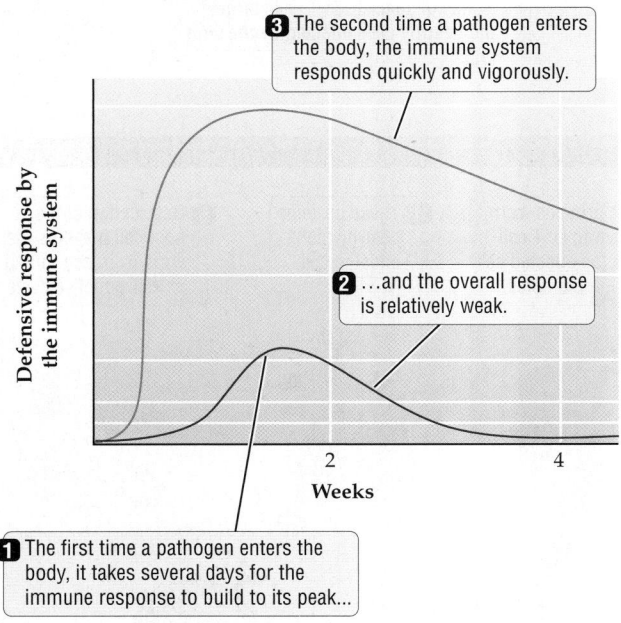

**PRIMARY VERSUS SECONDARY IMMUNE RESPONSE**

**3** The second time a pathogen enters the body, the immune system responds quickly and vigorously.

**2** ...and the overall response is relatively weak.

Defensive response by the immune system

Weeks

2      4

**1** The first time a pathogen enters the body, it takes several days for the immune response to build to its peak...

**FIGURE 32.11 Lymphocytes Defend the Body More Effectively after the First Time**

response and increases the strength of the response as well. Later exposures to this pathogen produce yet more memory cells.

## Immunity may be active or passive

We can acquire immunity in two ways: either actively or passively. We acquire **active immunity** to a particular pathogen when the antibodies to its antigens are produced by our own bodies, not received from an outside source. We can acquire active immunity naturally, as when we are exposed to certain diseases, such as measles. We can also acquire active immunity to certain diseases through vaccination. A **vaccine** consists of antigen-containing preparations that are administered to a person to prevent future infections with the pathogen from which the antigens are derived.

Vaccines are made using intact pathogenic cells or viral particles, but the pathogen is first killed or weakened by treatment with heat or chemicals so that it cannot make us ill. Even though the pathogen has been disarmed, it nevertheless displays its characteristic antigens; so when a vaccine is administered, the body launches a primary immune response to it. The primary response to a vaccine leads to the production of antibodies and memory cells directed against those specific antigens. If you have had a flu shot, you probably received a vaccine containing a killed influenza virus. The viral particles contained in nasal-spray versions of the flu vaccine are live, but they are too weak to make us ill and instead they confer immunity against the virus strain used in the vaccine. Highly antigenic parts of a pathogen, such as surface proteins or distinctive chemicals secreted by a particular pathogen, are also used to make vaccines. Tetanus shots, for example, contain an inactive form of the *Clostridium tetani* toxin, but not the live bacterium or the active form of the toxin.

Our bodies respond to the antigens in a vaccine as if they were the real thing, signaling the presence of an invader. Within about 2 weeks of a vaccination, the adaptive immune system generates antibodies and memory cells directed against the antigen. This primary immune response enables the body to launch a speedy secondary immune response if the same antigens are encountered later. Put another way, vaccination causes a primary immune response so that any subsequent natural exposure to a pathogen will produce a vigorous secondary response.

Preventive medicine relies heavily on vaccines. Most vaccines are given to children, when the adaptive immune system is strong enough to mount a response but before we are likely to encounter the pathogen naturally. The primary immune response is strong enough to give us lifelong immunity in the case of some vaccines, but in other cases the levels of circulating antibodies and memory cells decline over the years. **Booster shots**, which are repeat vaccinations, restore immunity by raising antibody concentrations and memory B cell numbers through fresh exposure to the antigens. Health professionals recommend booster shots against tetanus and diphtheria every 10 years, but many people skip the boosters, thereby diminishing their protection against these infections.

We acquire **passive immunity** by receiving antibodies that were not made by our own bodies. A human fetus receives all the nutrients and oxygen it requires from exchanges between its blood and its mother's blood. In the same way, the fetus acquires antibodies from its mother's blood. Mother's milk, especially the first milk (colostrum) produced by a woman who has just given birth, is rich in antibodies. Because the mother's immune system has encountered many antigens and made many antibodies in her lifetime, a nursing baby receives passive immunity to a broad range of potential pathogens. Passive immunity produces no memory cells, so it wears off as the received antibodies degrade, usually within a few weeks or months.

Passive immunity can also be delivered artificially, by injection of a concentrated dose of premade antibodies (immunoglobulins) into the bloodstream. This is sometimes done in an effort to combat a severe infection, when any time lost puts the patient at much higher risk of death. It is used, for example, in treating a person without immunity to *C. tetani* who shows signs of tetanus. It is also used to treat life-threatening infections that people are not routinely immunized against, such as rabies, and to combat certain types of poisonings. Snakebite antivenom, for example, contains immunoglobulins (usually obtained from immunized horses) that are specific to the antigens found in a particular snake's venom.

## Concept Check

1. How would you explain the observation that many infections, from strep throat to mononucleosis, are marked by the swelling of lymph nodes?

2. List at least three differences between innate and adaptive immunity.

## RECOMMENDED ADULT IMMUNIZATION SCHEDULE — UNITED STATES, 2010

| VACCINE | AGE GROUP | 19–26 years | 27–49 years | 50–59 years | 60–64 years | ≥ 65 years |
|---|---|---|---|---|---|---|
| Tetanus, diphtheria, pertussis (Td/Tdap)* | | Substitute one-time dose of Tdap for Td booster; then boost with Td every 10 years | | | | Td booster every 10 yrs |
| Human papillomavirus* | | 3 doses (females) | | | | |
| Varicella* | | 2 doses | | | | |
| Zoster | | | | | 1 dose | |
| Measles, mumps, rubella* | | 1 or 2 doses | | 1 dose | | |
| Influenza* | | 1 dose annually | | | | |
| Pneumococcal (polysaccharide) | | 1 or 2 doses | | | | 1 dose |
| Hepatitis A* | | 2 doses | | | | |
| Hepatitis B* | | 3 doses | | | | |
| Meningococcal* | | 1 or more doses | | | | |

*Covered by the Vaccine Injury Compensation Program

[  ] For all persons in the category who meet the age requirements and who lack evidence of immunity (e.g., lack documentation of vaccination or have no evidence of prior infection)

[  ] Recommended if some other risk factor is present (e.g., based on medical, occupational, lifestyle, or other indications)

[  ] No recommendation

# Why Is AIDS So Deadly, and How Can It Be Treated?

In the early 1980s, doctors in the United States began to notice that gay men were dying of a variety of rare diseases, including a skin cancer called Kaposi sarcoma, an unusual kind of pneumonia, and other infections that most people ordinarily shake off. By the mid-1980s it was clear that patients with the syndrome—named acquired immunodeficiency syndrome, or AIDS—had broken immune systems, the result of infection by a virus called the human immunodeficiency virus, or HIV.

In North America and Europe, the number of new cases rapidly increased, claiming the lives of tens of thousands of people each year. Initially, most new cases were limited to gay men, intravenous drug users, and people who had received blood transfusions. The common denominator was contact with the blood or body fluids of others: gay men during sex, drug users when sharing used needles, and surgical and hemophilia patients who received blood transfusions contaminated with HIV.

In time, safe-sex education and clean-needle programs reduced the rates of infections among gay men and blood transfusion patients, but the virus spread to other populations. Globally, 25 million people have died of AIDS, leaving 16.6 million orphaned children. In just one year, 2009, 1.8 million people died of AIDS and 2.6 million more became infected.

Inside the bloodstream, HIV enters two kinds of immune system cells: macrophages, which devour cellular debris and pathogens; and helper T cells, which stimulate cytotoxic T cells (also known as "killer T cells"), macrophages, and the production of antibodies by B cells. HIV reproduces inside these cells, eventually killing so many immune cells that the body's defenses are crippled.

Early in the infection, HIV preferentially infects the large population of helper T cells in the gut and destroys them. In the blood, meanwhile, the body's killer T cells track down HIV-infected helper T cells and destroy them. Because the killer T cells do such a good job of killing HIV-infected cells in the blood, most people with HIV have about a decade of normal health before they become ill.

Over time, however, the AIDS viruses in the body evolve. First, antigens on the viral protein coat mutate. Individual viruses with mutations that help them evade killer T cells survive longer and reproduce more, while killer T cells continually remove viruses that lack the mutations. Before long, the viruses adapted to evading killer T cells outnumber the original ones. As HIV evolves in the body, the T cells no longer recognize and kill the virus. The population of HIV viruses increases and begins destroying helper T cells faster than they can multiply. And with the helper T cells go the antibodies and killer T cells that the body needs to fight off infections by bacteria, yeasts, and other viruses. Once the immune system collapses, a person is vulnerable to any "opportunistic" infection.

So far, there is no effective vaccine or complete cure for HIV. But a variety of new drugs enable people with AIDS to live years longer with fewer symptoms. The "HIV cocktails," as the standard mixture of therapeutic drugs are called, prevent the viral genetic material from replicating or else prevent the virus from merging with plasma membranes and entering cells through that mechanism. These drugs limit transmission of the virus from a mother to her newborn child or nursing infant, and some drugs reduce the rate of new infections. But the drugs can cost hundreds or thousands of dollars a month. Because treatment is so costly, only one in five AIDS patients in Africa and Asia receives effective treatment. For now, the best way to slow the spread of the disease remains safe-sex education, the free availability of condoms, and clean-needle programs.

# Lack of Success Terminates Study in Africa of AIDS Prevention in Women

BY DONALD G. MCNEIL, JR., *New York Times*

In an unexpected setback for a new form of AIDS prevention, scientists on Monday halted a study in Africa intended to find out whether a daily antiretroviral pill can prevent women from becoming infected with the AIDS virus.

Early data showed no evidence that the pill was working.

Women taking the medication, Truvada, were just as likely as those taking a placebo to become infected, according to an independent panel that analyzed the results after the study had enrolled about half the 4,000 women researchers had hoped to enlist.

Of the 1,900 women taking Truvada or a placebo, 28 in each group [28 receiving Truvada; 28 receiving a placebo] had become infected as of last week, according to FHI, formerly Family Health International, the nonprofit group that was conducting the study in South Africa, Kenya and Tanzania...

The inconclusive results of the latest trial are a surprise to many experts. A study published last November found that Truvada protected gay men against infection. Men who took their pills faithfully were shown to have better than 90 percent protection, a result hailed as a breakthrough for AIDS prevention.

And results published last summer from a trial in South Africa showed that a vaginal gel containing tenofovir, one of the two antiretroviral drugs in Truvada, reduced the chances of infection by 54 percent in the women who used it faithfully before and after sex.

Because the current study... was stopped early and abruptly, many questions about it remain.

One is whether there was any difference in how often the women given Truvada and those given a placebo actually took their pills. More women taking Truvada complained of unpleasant side effects, and therefore a greater number might have failed to adhere to the program...

Another question is how much Truvada infused the walls of the vagina, where the initial infection takes place.

According to Dr. Robert M. Grant, a California AIDS researcher who led the study of Truvada in gay men, other research has shown that vaginal gels are as much as 100 times more effective than pills at getting protective antiretroviral drugs into the vaginal walls. Pills and gels are about equally effective in getting them into rectal tissue, he said.

---

The treatment described in this article was intended to prevent new HIV infections in women. But the researchers halted the study early because the treatment appeared not to work. The same treatment had already been shown to prevent HIV infection in gay men who took the drugs in pill form and in women who used a vaginal gel containing one of the ingredients in the pill. Yet, when 2,000 women took the pill, the women were just as likely to be infected with HIV as if they had received no treatment at all.

There are several reasons that this treatment, known to be effective, might not have worked. For example, unpleasant side effects can cause people in the treatment group to avoid the treatment, but that argument should apply to gay men as well. Another possibility is that contraceptive drugs the women were taking to prevent pregnancy interfered with the anti-HIV drugs.

Another possible complication is the way in which the drugs were delivered to the body. The women in this study took pills, which don't deliver as much of the drug to the wall of the vagina as the gel used in the other study does. When we take any pill, the drug goes into the stomach and into the digestive tract. Depending on how easily it's absorbed into the bloodstream, the drug may reach all the tissues in the body or pass through the gastrointestinal tract and into the rectum. For gay men who engage in anal sex, a drug that ends up in the rectum may work quite well. But for women who engage in vaginal sex, the same pill might not work at all. It's possible that there wasn't enough drug getting to the vagina to prevent infection.

## Evaluating the News

**1.** Explain why the study described in the article was halted.

**2.** How likely do you think it is that the drug simply doesn't work? How would you test the other alternative explanations offered in the analysis above?

**3.** In the United States, AIDS originally surfaced in major cities such as New York City and San Francisco. Because the virus is well established in urban areas, New York, Florida, California, and Texas still have the highest rates of new infections. Yet, in the last few years, 16 southern states have accounted for a larger and larger share of new AIDS cases. Discuss factors you think might be contributing to the rise of AIDS in the South.

**SOURCE:** *New York Times*, April 18, 2011, http://www.nytimes.com.

# CHAPTER REVIEW

# Summary

## 32.1 Defense Systems: Distinguishing Self from Nonself

- The vertebrate immune system possesses three layers of defenses. The first layer consists of external physical and chemical barriers. The innate and adaptive immune systems are the second and third lines of defense, respectively.
- Cells involved in innate immunity and adaptive immunity have the ability to distinguish self (the body's own cells) from nonself (cells and molecules that are not part of the body).
- The innate immune system is a series of immediate, short-term, non-specific responses.
- The adaptive immune system deploys various slower, longer-term, but specific responses. Several types of white blood cells and many defensive proteins are involved in the precisely targeted recognition of specific pathogens.

## 32.2 First Line of Defense: Physical and Chemical Barriers

- Our skin and the linings of our respiratory and digestive systems form barriers against pathogens.
- Many pathogens take advantage of breaks in the skin to gain entry to their hosts. Some pathogens rely on other organisms, such as blood-feeding insects, to penetrate the skin.

## 32.3 Second Line of Defense: The Innate Immune System

- Several types of blood cells and molecules produce the nonspecific responses of the innate immune system. Phagocytes such as macrophages engulf or encapsulate stray pathogens that get past the physical barriers of the skin and lungs. Complement proteins tag pathogens for engulfment by phagocytes or destruction by white blood cells.
- Inflammation occurs in response to tissue damage caused by pathogens or wounding. It is commonly characterized by redness, warmth, and swelling, caused by increased blood flow to the wound.
- Histamine is released during inflammation, and this signaling molecule causes blood vessels to dilate, bringing blood, nutrients, and oxygen to the injured area. Redness and warmth follow when blood warmed by metabolic heat arrives at a faster rate. The leakiness of capillaries allows blood plasma to move into the area, causing swelling.
- Tissue damage stimulates platelets and proteins in the blood to form a blood clot, which helps prevent any pathogens present in the wound from spreading.
- Prostaglandins released by macrophages cause fever and drowsiness, and they initiate pain signals. These effects encourage us to slow our ordinary activities so that the body can heal.

- Complement proteins circulate to the wound, where they kill invading cells by destroying their plasma membranes and also tag pathogens for destruction by white blood cells.
- Virus-infected body cells release interferon, which interferes with the infection of nearby cells by viruses. Interferon also summons NK cells, which destroy any cells (including body cells infected by viruses) that display foreign proteins on their surfaces.

## 32.4 Third Line of Defense: The Adaptive Immune System

- The adaptive immune system provides vertebrates with long-term defenses against specific pathogens and parasites.
- The lymphatic system provides the primary sites for adaptive immunity. White blood cells called lymphocytes confer specific immunity. Immature lymphocytes differentiate into B cells in the bone marrow and T cells in the thymus. Each lymphocyte has special membrane proteins that bind only to a specific antigen of a specific pathogen.
- B cells provide antibody-mediated immunity: they make and release antibodies, which are soluble versions of their antigen-binding membrane proteins. Antibodies move about in the circulatory and lymphatic systems, where they bind to antigens on invading pathogens, "tagging" them for destruction by other immune system cells.
- T cells confer cell-mediated immunity: the whole T cell binds to the pathogen or infected body cell that displays a pathogen-specific antigen. T cells come in two types: helper T cells, which help other defensive cells in various ways, and cytotoxic T cells, which destroy body cells damaged or infected by viruses.
- First exposure to a pathogen brings on a primary immune response to that pathogen, which is followed by clonal selection in which many copies of B cells or T cells are made. Some of these clones become effector cells, which then combat the new pathogen; others become memory cells.
- A primary immune response is relatively slow and mild. A secondary immune response is a faster, stronger response to a pathogen that has been encountered one or more times.
- Active immunity can be acquired through natural exposure to a pathogen or through a vaccine. Vaccines "prime" the immune system by provoking a mild primary immune response so that during subsequent natural exposures to the pathogen, the immune system can respond immediately and forcefully. Passive immunity comes from receiving antibodies that were not made by our own bodies, as when a fetus acquires antibodies from its mother.

# Key Terms

active immunity (p. 708)
adaptive immune system (p. 697)
antibody (p. 706)
antibody-mediated immunity (p. 706)
antigen (p. 705)
antigen-presenting cell (APC) (p. 706)
B cell (p. 705)
booster shot (p. 709)
cancer immunosurveillance (p. 698)
cell-mediated immunity (p. 706)
clonal selection (p. 705)

complement protein (p. 703)
cytokine (p. 700)
cytotoxic T cell (p. 706)
defensin (p. 699)
helper T cell (p. 706)
histamine (p. 700)
host (p. 696)
immune memory (p. 697)
immune system (p. 697)
inflammation (p. 700)
innate immune system (p. 697)

interferon (p. 703)
lymph node (p. 704)
lymphatic duct (p. 704)
lymphatic system (p. 704)
lymphocyte (p. 705)
mast cell (p. 700)
memory cell (p. 708)
natural killer (NK) cell (p. 703)
nonspecific response (p. 698)
passive immunity (p. 709)
pathogen (p. 696)

phagocyte (p. 700)
phagocytosis (p. 700)
platelet (p. 702)
primary immune response (p. 707)
prostaglandin (p. 702)
secondary immune response (p. 708)
specific response (p. 698)
T cell (p. 705)
toxin (p. 699)
vaccine (p. 708)
white blood cell (p. 698)

# Self-Quiz

1. Which of the following contains a tissue layer that constitutes a first line of defense?
   a. liver
   b. lung
   c. spleen
   d. thymus gland

2. Which of the following would engulf and digest bacteria in a wound?
   a. cytotoxic T cell
   b. B cell
   c. mast cell
   d. macrophage

3. The inflammatory response
   a. occurs only in lymph nodes.
   b. includes the release of histamine.
   c. requires helper T cells.
   d. involves the constriction of small blood vessels.

4. Macrophages
   a. are specialized for the secretion of histamines.
   b. produce antibodies.
   c. engulf and digest microscopic invaders.
   d. produce antigens.

5. Which of the following do *not* participate in adaptive immunity?
   a. B cells
   b. mast cells
   c. macrophages
   d. all of the above

6. Which statement about antigens is true?
   a. They are produced by helper T cells.
   b. They are present on the surfaces of pathogens.
   c. They are part of a nonspecific response.
   d. They are produced by neutrophils.

7. Which statement about antibodies is true?
   a. They are produced by helper T cells.
   b. They are produced by cytotoxic T cells.
   c. They are produced by B cells.
   d. all of the above

8. Which of the following would be able to recognize a specific kind of invading bacterium, such as *E. coli* strain H10407?
   a. B cells
   b. helper T cells
   c. cytotoxic T cells
   d. all of the above

9. Cell-mediated immunity involves
   a. neutrophils.
   b. B cells.
   c. T cells.
   d. histamine.

10. Memory in B cells
    a. prevents autoimmune diseases.
    b. increases the speed and strength of the immune response.
    c. produces specific interferons.
    d. is stored in stem cells.

# Analysis and Application

1. Describe the first line of defense in humans that blocks the entry of most pathogens. Does it constitute a specific or a nonspecific response? Explain your answer.

2. How do disease-fighting cells in the body distinguish between foreign invaders and our own cells?

3. Suppose you were an intelligent bacterium that had penetrated a breach in a person's skin. How would you go about defeating the forces arrayed against you in the second and third lines of defense (responses launched by innate and adaptive immunity)?

4. Compare and contrast B and T lymphocytes.

5. Explain how and why a secondary immune response is more effective than a primary immune response.

# Reproduction and Development

**DRINKING AND DEVELOPING.** The U.S. surgeon general recommends that women who are pregnant or could become pregnant refrain from consuming any alcohol.

# Bad Genes or Bad Habits?

In 1912, a New Jersey psychologist described a family he called the "Kallikaks" whose long pedigree of "feeble-mindedness" and low moral character demonstrated, he claimed, the shocking effects of heredity. According to Henry Goddard, such mentally inferior families should be prevented from having children and passing on their weak genes. Goddard's arguments about the Kallikaks helped strengthen the American eugenics movement, which flourished in the first five decades of the twentieth century and resulted in the forced sterilization of 60,000 Americans, one-third in California alone.

Modern reevaluation of the Kallikaks suggests that the family's mental retardation and odd faces were the result not of "weak genes" but of malnutrition and alcohol use during pregnancy—both related to poverty. Alcohol is a teratogen, a chemical that can cause birth defects, also called malformations, in developing embryos or fetuses. Alcohol use during pregnancy can cause a broad range of malformations, including small size at birth; hip dislocation; microcephaly (a small head); defects of the heart, eyes, and genitals; hearing loss; a distinctive flat face and thin upper lip; as well as poor coordination, mental retardation, poor impulse control, hyperactivity, and palsy. The range of birth defects, known as fetal alcohol spectrum disorder (FASD), is the most common preventable group of birth defects.

In early 2005, the U.S. surgeon general issued an advisory recommending that any woman who was pregnant or might become pregnant should abstain from drinking alcohol—a policy position far stronger than one issued 24 years earlier. Since about half of all pregnancies are unplanned and women typically don't know they're pregnant in the first weeks of pregnancy, the government was recommending, in effect, that most women between the ages of about 15 and 44 should abstain from alcohol.

What made the government take a stronger position on alcohol use during pregnancy? How does alcohol cause birth defects, and how much alcohol is required to have an effect?

Before we address these questions, we'll examine how humans and other animals reproduce, trace the developmental journey that begins when egg and sperm fuse, examine some of the processes that govern development, and consider how relatively small changes in development can forge dramatic evolutionary changes.

> **MAIN MESSAGE** Sexual reproduction and the development of specialized cells, tissues, and organ systems are of fundamental importance in the life cycles of animals.

## KEY CONCEPTS

- In sexual reproduction, gametes from two parents of different sexes join to create a genetically unique individual offspring.

- In human males, sperm formation begins at puberty, as each diploid precursor divides by meiosis to produce four haploid sperm cells.

- In females, meiosis I begins in the ovaries even before birth. At puberty, fewer than half the primary oocytes remain, and usually one of these resumes meiosis each month during a woman's reproductive years. Meiosis is completed, producing a large ovum, only after a sperm cell fuses with the ovum.

- During fertilization, one sperm, aided by enzymes in its acrosome, enters the ovum. The sperm nucleus then fuses with the egg nucleus to create a diploid zygote. From zygote to birth, development is marked by rapid cell division, cell differentiation, and organ formation. Cell division and differentiation in the early embryo is more rapid than at any other stage of life. As development continues, cell types become more defined and organ systems emerge.

- Pregnancy can be prevented by contraceptives. Some contraceptives can help prevent the transfer of sexually transmitted diseases, but most do not.

- Development is controlled by master genes and inductive signals, and is often influenced by environmental factors as well.

**REPRODUCTION IS A BASIC FEATURE OF ALL LIFE.** Generating new individuals from existing ones is a driving force for all life-forms that inhabit Earth. The new individuals are called offspring or progeny. In most animals, including humans, an offspring is a unique blend of genetic information received from two different individuals, the parents. The genes of each parent are reshuffled in the process of generating gametes—eggs in female animals and sperm in male animals. The genetic reshuffling, and the random inheritance of the parental chromosomes, produces offspring with unique combinations of the parental genes. Natural populations tend to be genetically variable because they are composed of individuals with diverse genetic makeup and therefore diverse traits. This genetic diversity is the raw material for evolutionary change through natural selection.

Because an animal's evolutionary success hinges on its reproductive success, the animal devotes much of its time and energy to producing offspring and ensuring that those offspring will themselves survive to reproduce. In a sense, the function of all the structures and processes described so far in Unit 6 is to enable animals to accumulate and manage the resources they need to reproduce.

In this chapter we explore how animals—particularly humans—reproduce and develop. We begin with a basic overview of general strategies used by animals for sexual and asexual reproduction and look at the advantages and disadvantages of each. We consider how the **gametes**—eggs and sperm—originate and how animals bring eggs and sperm together. Then, in the context of human development, we follow the steps by which the fertilized egg (the zygote) develops into a complex individual consisting of many different cell types. We take a brief look at human development following birth. And we conclude with a description of the regulatory mechanisms that control developmental processes in animals.

# 33.1 Sexual and Asexual Reproduction in Animals

Animals have two basic means of reproducing. In **sexual reproduction**, the mode of reproduction used by humans and most other animals, haploid gametes from a male and a female combine to form a diploid **zygote**, which develops into a multicellular individual that is genetically unique and different from either parent. In **asexual reproduction**, cells from only one parent are involved in producing the offspring, and all of the offspring's genes come from that parent.

When male and female gametes combine to create a zygote, we say that **fertilization** has occurred. Fertilization can occur internally or externally (**FIGURE 33.1**). With internal fertilization, common among animals that live on land, sperm are released inside a female

**Internal fertilization**

On land, males typically transfer sperm directly into the female's body to avoid exposing them to the environment.

**External fertilization**

Aquatic animals such as these sockeye salmon release vast quantities of sperm and eggs into the water, where fertilization takes place.

**FIGURE 33.1 Fertilization Can Be Internal or External**
Internal fertilization, usually involving copulation, is common among land animals. External fertilization, which involves releasing sperm near eggs laid separately by a female, is common among aquatic animals.

reproductive tract. External fertilization occurs outside the body and is common in aquatic animals. Male and female salmon, for example, after traveling treacherous miles upstream to the spawning grounds of their birth, pair up and together release sperm and eggs in depressions in streambed gravel.

The two modes of reproduction have different advantages. Sexual reproduction produces offspring with a new mix of genes—their genotypes differ not only from those of their parents, but also from those of their siblings. The resulting genetic variation increases the chances that some of the parents' offspring will survive in a changing environment. The offspring of asexually reproducing animals do not have this genetic variation. They all arise from the same set of genes, and if their environment changes enough, all, not just some, of the offspring are threatened.

The benefits of asexual reproduction are that it occurs more rapidly and at a lower energy cost. For example, the parent does not have to expend energy to find a mate. Asexually reproducing animals tend to be found in environments stable and steady enough that genetic variation is not essential for survival.

## Asexual reproduction involves only one parent

In asexual reproduction there is no fertilization, and the offspring have the same DNA as their single parent. Animals reproduce asexually in several different ways. Some, like the hydra, reproduce asexually by *budding*, in which a fully formed offspring grows on the parent animal (**FIGURE 33.2a**). Other animals, such as sea anemones (**FIGURE 33.2b**), can reproduce asexually by *fission*: the anemone grows and a piece of the body simply breaks off and develops into a whole new anemone. In *parthenogenesis*, an embryo forms in the female in the absence of fertilization and develops into a new offspring. This form of asexual reproduction has been reported in diverse animal groups, including vertebrates such as fishes and lizards.

Although some animals rely exclusively on asexual reproduction, most asexually reproducing species switch between sexual and asexual reproduction, depending on environmental conditions. Aphids, the small insects that sometimes infest our greenhouses or garden plants, commonly reproduce through parthenogenesis. The crowds of aphids that we see sucking sap from a plant in the summer are all genetically identical copies of a female that originally colonized

**(a) Budding in *Hydra***

Budding offspring

**(b) Fission in sea anemone**

**FIGURE 33.2**
**Fertilization Is Not Needed for Asexual Reproduction**
(a) The common freshwater cnidarian *Hydra* reproduces by budding. Here the parent is the larger individual on the left, and the offspring is emerging on the right. (b) Some animals, such as this aggregating sea anemone on the Pacific coast of North America, reproduce by fission.

that plant. Aphids switch to sexual reproduction, however, when the temperature drops and the days grow shorter as winter approaches, driving them from their summer food plant.

## Sexual reproduction requires meiosis

Early in development, most animals set aside a small group of cells, known as **germ line cells**, that will develop into gamete precursor cells. These unique cells remain separate from all other cells of the embryo and remain unspecialized until the reproductive organs develop, at which time they migrate to take up residence inside the egg-producing ovaries or sperm-producing testes. The germ line cells resident in these organs eventually undergo meiosis to produce gametes.

Gametes differ in chromosome number from other cells in the animal body: they are **haploid** (denoted $n$), containing just half the chromosome set found in other cells, which are **diploid** ($2n$). Meiosis, a special type of cell division (see Figure 10.14 for a review), is required to generate a haploid egg or sperm from a diploid precursor. The fusion of a haploid egg and a haploid sperm during fertilization produces a diploid zygote—the first step in generating a new individual.

## Some animals can switch gender

Sex in animals is more variable than our human perspective might lead us to expect. As already noted, humans are generally sperm-producing males or egg-producing females. Many animal groups, however, include species in which some individuals produce both functional testes and functional ovaries and are therefore both male and female. We call such individuals **hermaphrodites [her-MAF-roh-dytes]**. Hermaphrodites live in a wide variety of habitats. They are most common among invertebrates—for example, the common earthworm and most flatworm species are hermaphrodites—but they also exist among vertebrates.

A common misconception about hermaphrodites is that they fertilize their own eggs. Most hermaphrodites—even those, like the earthworm, that have functional testes and ovaries in the same individual at the same time—must still mate with another individual. Nonetheless, unlike animals that are only male or only female throughout their lives, hermaphrodites can mate with an individual of either gender. Among vertebrates, the gender of an individual is changeable in some reptiles and fishes; such animals are called *sequential hermaphrodites*. Many fish species can change their gender (developing female reproductive organs after hatching as males, for example) depending on their size, their stage of life, or environmental conditions. Some fishes can also be *simultaneous hermaphrodites*, bearing male and female reproductive organs at the same time.

## 33.2 Human Reproduction: Gamete Formation and Fertilization

Reproduction in humans, as in most other animals, begins with a sequence of activities that increase the chance that a sperm will encounter an egg and that their fusion will result in a zygote capable of developing into a new individual, who in turn can grow up to reproduce.

In human males, meiosis occurs inside tubelike structures (*seminiferous tubules*) that permeate the testes. In response to male hormones that surge at the onset of puberty, diploid germ line cells in the tubules start dividing to form sperm in a sequence of steps known as **spermatogenesis** (**FIGURE 33.3a**). After males enter puberty and for the rest of their life span, germ line cells in the tubules will regularly divide through mitosis to produce **primary spermatocytes** ($2n$). Each primary spermatocyte divides to complete the first division of meiosis (meiosis I), resulting in two haploid ($n$) **secondary spermatocytes**. The two secondary spermatocytes divide again during the second division of meiosis (meiosis II) to produce a total of four haploid cells that mature into sperm. The average man produces about 300 million sperm each day. Surplus sperm that accumulate over the days are degraded and reabsorbed by the cells that line the tubules.

Eggs also develop through a series of cell divisions. **Oogenesis [OH-uh-JEN-uh-sus]**, the production of mature eggs capable of being fertilized (**FIGURE 33.3b**), begins before birth, when germ line cells multiply and develop into diploid cells called **primary oocytes** ($2n$). Even before a human female is born, these primary oocytes begin, but do not complete, the first meiotic division on the way to producing haploid eggs. At birth, a human female has in her ovaries between 1 million and 2 million primary oocytes whose development has been suspended at an early stage of meiosis. By the time she reaches puberty, at about 10–12 years of age, approximately 400,000 viable primary oocytes remain—still more than she will use in her lifetime.

At puberty, the primary oocytes resume their development into mature eggs. On average, during a woman's reproductive years, one primary oocyte a month successfully develops into a mature egg. The release of follicle-stimulating hormone (see Section 29.4) stimulates several of the small primary oocytes to resume meiosis. Of these, only the most rapidly growing oocyte completes meiosis; the others degenerate. Meiosis in the fastest-growing primary oocyte produces two haploid cells of very different sizes: a large **secondary oocyte** and a smaller **polar body** (see Figure 33.3b), which plays no further role. The haploid secondary oocyte then begins a second meiotic division that will ultimately generate a small, nonfunctional polar body and one large haploid egg cell, or **ovum** (plural "ova"). But with meiosis only halfway finished, the secondary oocyte is released into the oviduct. The second meiotic division is completed—and the ovum is formed—only if the secondary oocyte encounters and binds a sperm.

The production of ova and of sperm differ in several important ways. The supply of a female's primary oocytes is limited, and once a primary oocyte develops into a mature ovum, it is deleted from the supply. In

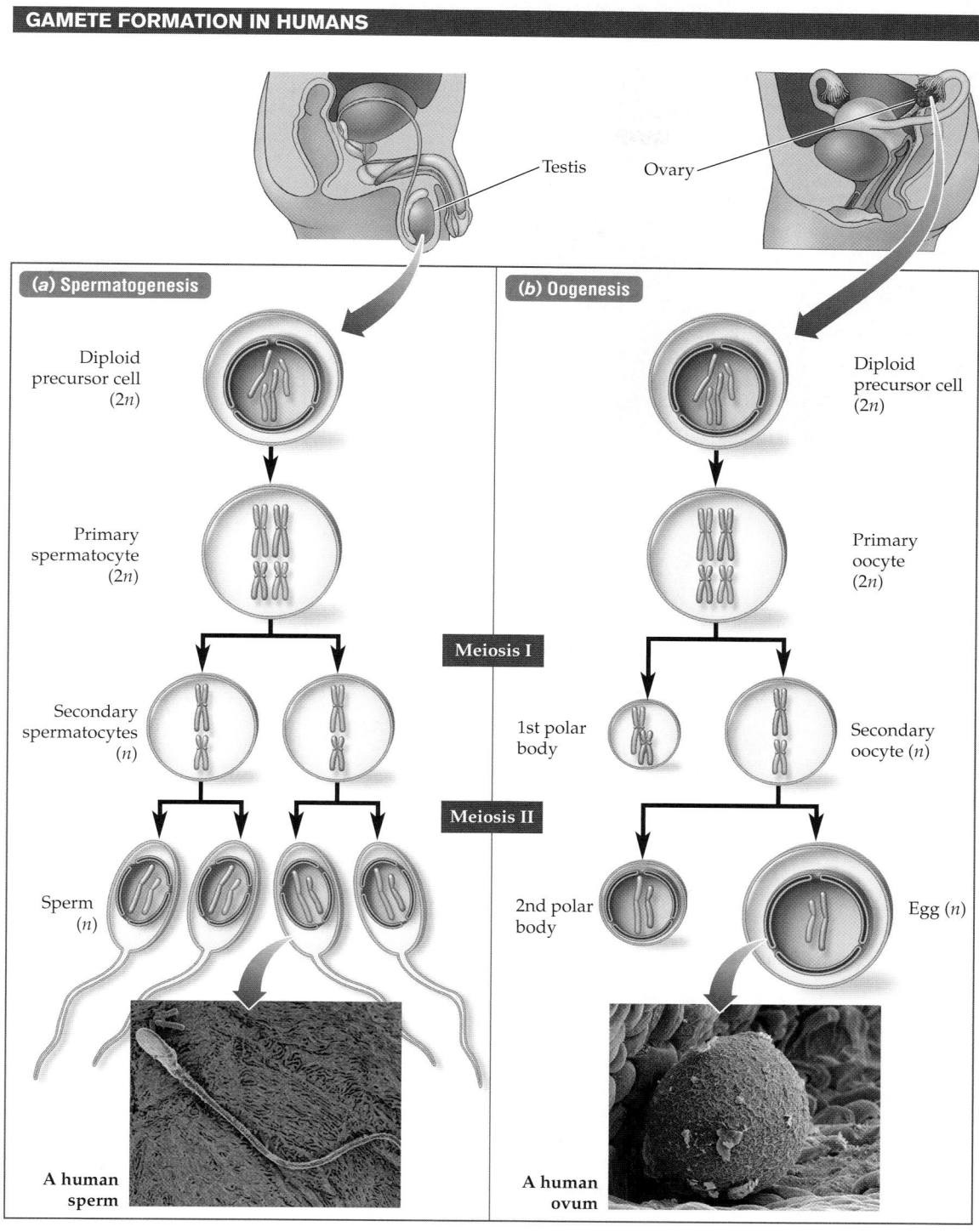

**FIGURE 33.3**

**Sexual Reproduction Requires the Production of Haploid Gametes through Meiosis**

(a) Spermatogenesis produces haploid sperm. (b) Oogenesis produces haploid eggs (ova).

Testis

Ovary

**(a) Spermatogenesis**

Diploid precursor cell (2n)

Primary spermatocyte (2n)

Secondary spermatocytes (n)

Sperm (n)

**A human sperm**

**(b) Oogenesis**

Diploid precursor cell (2n)

Primary oocyte (2n)

Meiosis I

1st polar body

Secondary oocyte (n)

Meiosis II

2nd polar body

Egg (n)

**A human ovum**

contrast, the supply of sperm precursor cells in males constantly replenishes itself because when these cells divide, only some of the resulting cells are primary spermatocytes; the others are new precursor cells. It is also noteworthy that in a normal 28-day menstrual cycle (see Figure 29.9), a human female produces only one mature egg cell, while a male produces about 300 million sperm every day. These two observations tell us that, in evolutionary terms, a single egg is far more precious than a single sperm.

In addition to these differences, ova are typically much larger than sperm (**FIGURE 33.4a**). The human egg is visible (just barely) to the naked eye, but seeing individual sperm requires a microscope. Sperm contain little substance beyond chromosomes and the cellular machinery needed to move, attach to

## FIGURE 33.4
### Fertilization Occurs When a Single Sperm Enters the Egg

(*a*) Sperm are many times smaller than eggs, as can be seen in this scanning electron micrograph of a swarm of sperm surrounding a single human egg. (*b*) The tail of a sperm consists of a whiplike flagellum, which helps it wriggle into the uterus and through the oviduct to meet the egg. The acrosome contains enzymes that help the sperm penetrate the egg's outer covering. (*c*) Although many sperm are present on the surface of the egg, only one can penetrate to fertilize the egg.

(*a*)

(*b*)

(*c*)

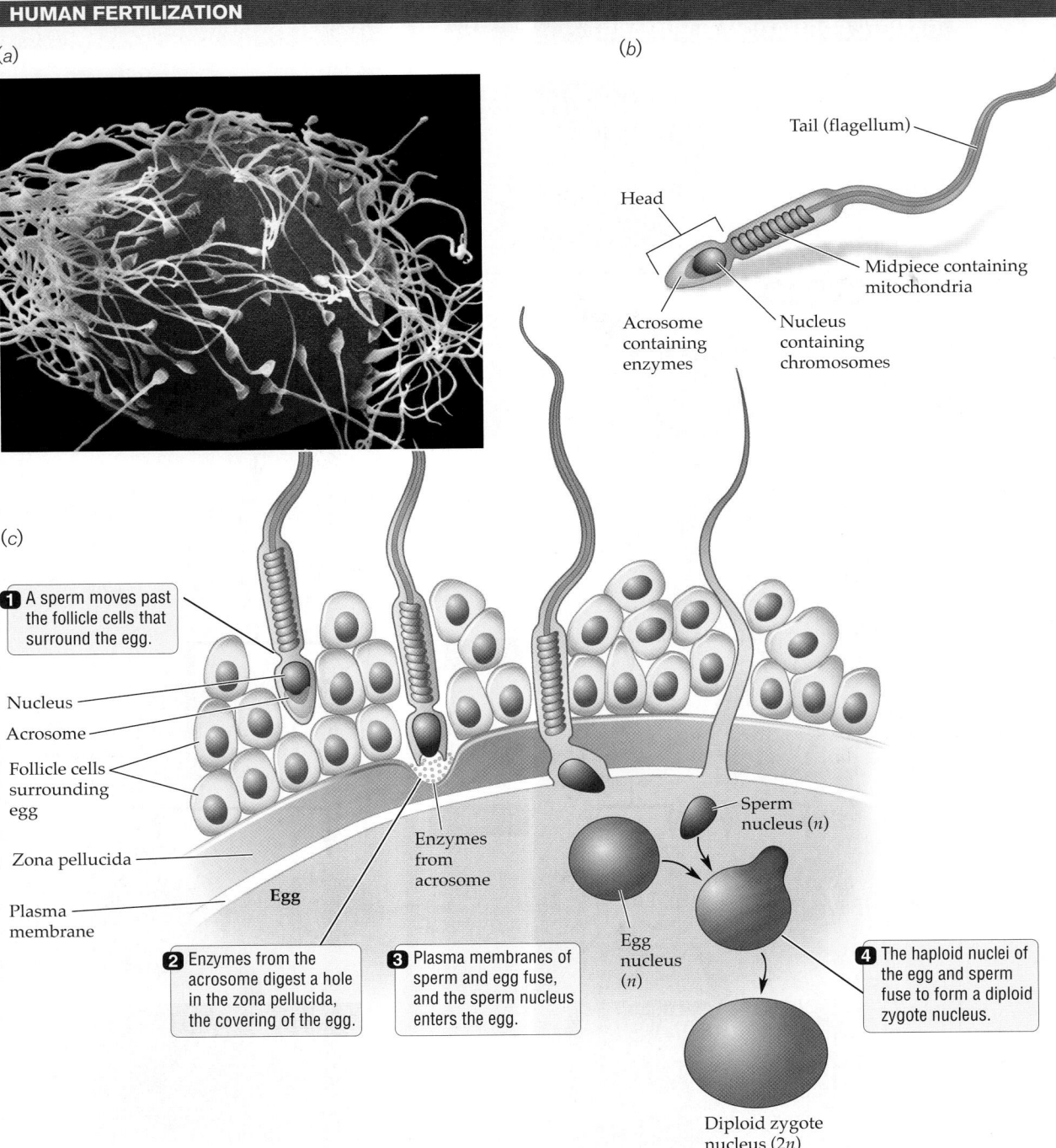

Tail (flagellum)

Head

Midpiece containing mitochondria

Acrosome containing enzymes

Nucleus containing chromosomes

**1** A sperm moves past the follicle cells that surround the egg.

Nucleus

Acrosome

Follicle cells surrounding egg

Zona pellucida

Plasma membrane

**Egg**

Enzymes from acrosome

**2** Enzymes from the acrosome digest a hole in the zona pellucida, the covering of the egg.

**3** Plasma membranes of sperm and egg fuse, and the sperm nucleus enters the egg.

Sperm nucleus (*n*)

Egg nucleus (*n*)

**4** The haploid nuclei of the egg and sperm fuse to form a diploid zygote nucleus.

Diploid zygote nucleus (2*n*)

an egg, and mobilize the sperm chromosomes into the egg's cytoplasm (**FIGURE 33.4b**).

For fertilization to take place (**FIGURE 33.4c**), a sperm must first work its way through the follicle cells surrounding the outer surface of the egg and then penetrate the egg's outer, gel-like covering, called the *zona pellucida*. The zona pellucida has carbohydrate-containing proteins (glycoproteins) that trap sperm cells in a species-specific manner: the glycoproteins recognize and bind to proteins found only in the sperm head of the same species. This species specificity

in sperm binding is especially important for animals with external fertilization (from sponges to fishes), whose eggs could encounter sperm of other species.

The **acrosome**—the front tip of a sperm—contains enzymes that start digesting the zona pellucida, enabling the sperm to burrow through to the jellylike coating on the surface of the egg. The interlocking of special proteins present in the plasma membranes of the sperm and egg facilitates the fusion of the two membranes. The sperm nucleus now separates from the rest of the sperm (which degenerates) and fuses with the egg's nucleus to form a genetically unique diploid zygote. Although the chromosomes of the zygote are contributed equally by the two parents, the organelles and other cellular machinery of the zygote come almost entirely from the female. For example, all of your mitochondria, which generate life-sustaining energy for your body in the form of ATP, came from your mother.

> ### Concept Check

1. What are the adaptive advantages of sexual and asexual reproduction, respectively?

2. What are some key differences between human egg production (oogenesis) and sperm production (spermatogenesis)?

## 33.3 Human Reproduction: From Fertilization to Birth

In this section we consider the human reproductive system in greater detail, focusing on the developmental processes that unfold after fertilization and culminate in the birth of a child. **FIGURE 33.5** shows the general anatomy of the human male and female reproductive systems and the series of events that lead to a fertilized ovum.

A woman releases an egg from her ovaries, or ovulates, about once every 28 days. The egg moves down a tube called the **oviduct** (or fallopian tube). If the woman has sexual intercourse, the man's **penis**

Concept Check
Answers

1. Sexual reproduction produces genetic variation, which is adaptive in a changing environment. Asexual reproduction is possible with just one parent and creates large numbers of offspring at a lower energy cost; however, offspring are genetically identical, so the strategy is likely to be successful only in a stable environment.

2. Meiosis to produce eggs begins in the female fetus; oogenesis resumes at puberty, with typically one oocyte released each month until menopause. Sperm are produced in large amounts starting at puberty and continuing through much of the life span of the adult male.

## HUMAN SEXUAL REPRODUCTION

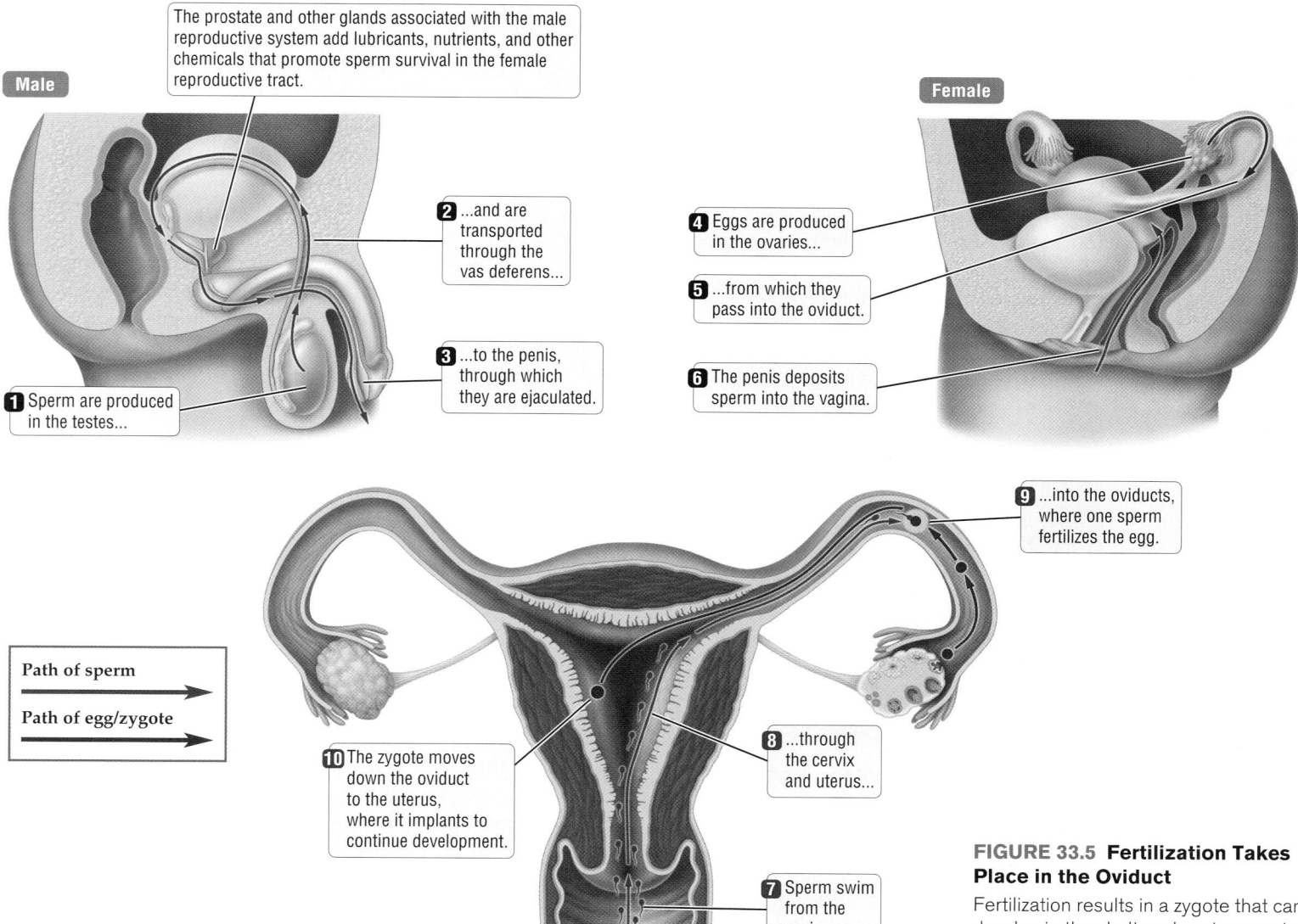

**Male**

The prostate and other glands associated with the male reproductive system add lubricants, nutrients, and other chemicals that promote sperm survival in the female reproductive tract.

**2** ...and are transported through the vas deferens...

**3** ...to the penis, through which they are ejaculated.

**1** Sperm are produced in the testes...

**Female**

**4** Eggs are produced in the ovaries...

**5** ...from which they pass into the oviduct.

**6** The penis deposits sperm into the vagina.

**9** ...into the oviducts, where one sperm fertilizes the egg.

Path of sperm

Path of egg/zygote

**10** The zygote moves down the oviduct to the uterus, where it implants to continue development.

**8** ...through the cervix and uterus...

**7** Sperm swim from the vagina...

**FIGURE 33.5 Fertilization Takes Place in the Oviduct**

Fertilization results in a zygote that can develop in the sheltered environment of the uterus.

ejaculates almost 300 million sperm into her **vagina**. Huge numbers of sperm swim from the vagina through the opening of the uterus, called the cervix, into the uterus, and then up the oviduct in response to a chemical signal released by the ovary. Human sperm can swim through fluid, but human eggs cannot actively propel themselves. The woman aids the progress of the swimming sperm by a combination of muscular contractions of the reproductive tract and the release of chemical signals that stimulate sperm to swim. Of the many millions of sperm ejaculated into the vagina, only a few hundred manage to reach the egg in the oviduct. Only one of those sperm fertilizes the egg to create a zygote. Fertilization is often called "the moment of conception."

## Rapid cell division characterizes early development

Although it is a critical step in human reproduction, fertilization marks only the beginning of a 9-month-long period of development within the mother's uterus. The single-celled zygote divides by mitotic divisions to form a ball of cells called a *morula* (Latin for "mulberry"). About 2 days after fertilization, the morula consists of eight identical, totipotent [toh-TIP-uh-tent] cells (**FIGURE 33.6**). *Totipotent* cells have the broadest developmental potential in that they can give rise to *all* the cell types found in the organism.

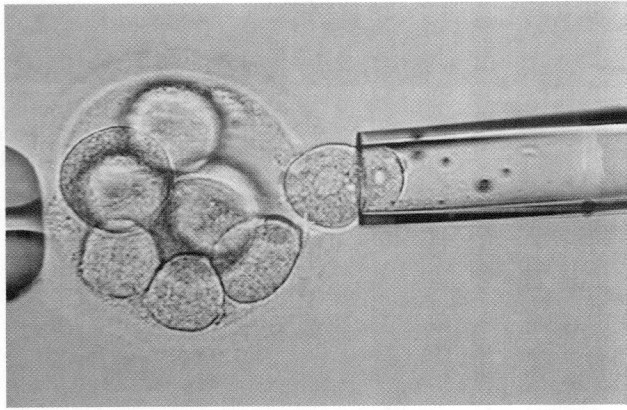

**FIGURE 33.6 From Zygote to Morula**
Immediately after fertilization, the zygote divides to form a 2-celled morula. These cells divide rapidly without increasing in size, so the size of the morula does not increase appreciably as it grows to about 50 cells. A human morula consisting of 8 cells is shown in this micrograph. The cells are loosely held at this stage, and one is being removed for genetic testing. The remaining cells will develop normally.

As cell divisions continue, the morula is transformed into a **blastocyst**, a hollow ball with a fluid-filled cavity in the center. Approximately 7 days after fertilization, this ball of about 100 cells attaches to the inner lining of the **uterus** (womb) in the process known as **implantation**. The mammalian blastocyst is not a simple, hollow sphere, but one that contains two distinct tissue types: there is an outer layer of cells, the **trophoblast**; and inside it, clustered at one end, are cells of the **inner cell mass** (**FIGURE 33.7**). The trophoblast contributes to nutritive tissue that envelops the developing young, and the inner cell mass grows and differentiates into the embryo proper.

When the blastocyst encounters the **endometrium**, the layer of cells lining the wall of the uterus, cells in the trophoblast secrete enzymes that enable the blastocyst to attach to the endometrium (see Figure 33.7) and gradually burrow into those tissues (implantation). The cells of the trophoblast expand into the endometrium as implantation proceeds and eventually contribute to formation of the *placenta* in cooperation with maternal tissues. The **placenta** consists of tissues derived from both the mother's uterus and the blastocyst. The function of this uniquely mammalian structure is to facilitate the transfer of nutrients and wastes between the circulatory system of the mother and that of the developing offspring.

The inner cell mass develops into the **embryo**, an immature state of the offspring that lacks most of the major organ systems. As development proceeds, the cells in the embryo begin to acquire specific identities through **cell differentiation**, the process by which unspecialized cells acquire distinctive properties and special functions. Nerve cells and muscle cells are examples of differentiated cells. In contrast, the cells in the morula and inner cell mass are undifferentiated.

## Cell differentiation and cell migration produce the three main tissue layers

As embryonic development continues, two critical events take place. First, depending on their location, the cells in the embryo differentiate to give rise to one of the three main tissue layers of animal embryos. These layers follow three distinctly different developmental paths, each giving rise to different sets of tissues in the mature animal.

- The **endoderm** gives rise to the epithelial tissues of the gut and lungs, the liver, and the glands of the endocrine system, among other structures.

Oviduct

Uterus

Endometrium

**Blastocyst**

Trophoblast

Mother's blood vessels

**Day 6**

Fertilization of ovum

Inner cell mass

Trophoblast cells, together with endometrial cells, give rise to the placenta...

Endometrium

Developing placenta

**Day 7**

Ovary

Implantation takes place in the endometrium about 7 days after fertilization.

...while the inner cell mass starts developing into the embryo.

**Day 8**

**Helpful to Know**

Knowing how the names of the three tissue layers are derived reminds us where each layer is found. Endoderm (from *endo*, "inner") is the innermost layer, ectoderm (*ecto*, "outer") mostly forms the outer boundary of the gastrula, and mesoderm (*meso*, "middle") is generally found sandwiched between the other two layers.

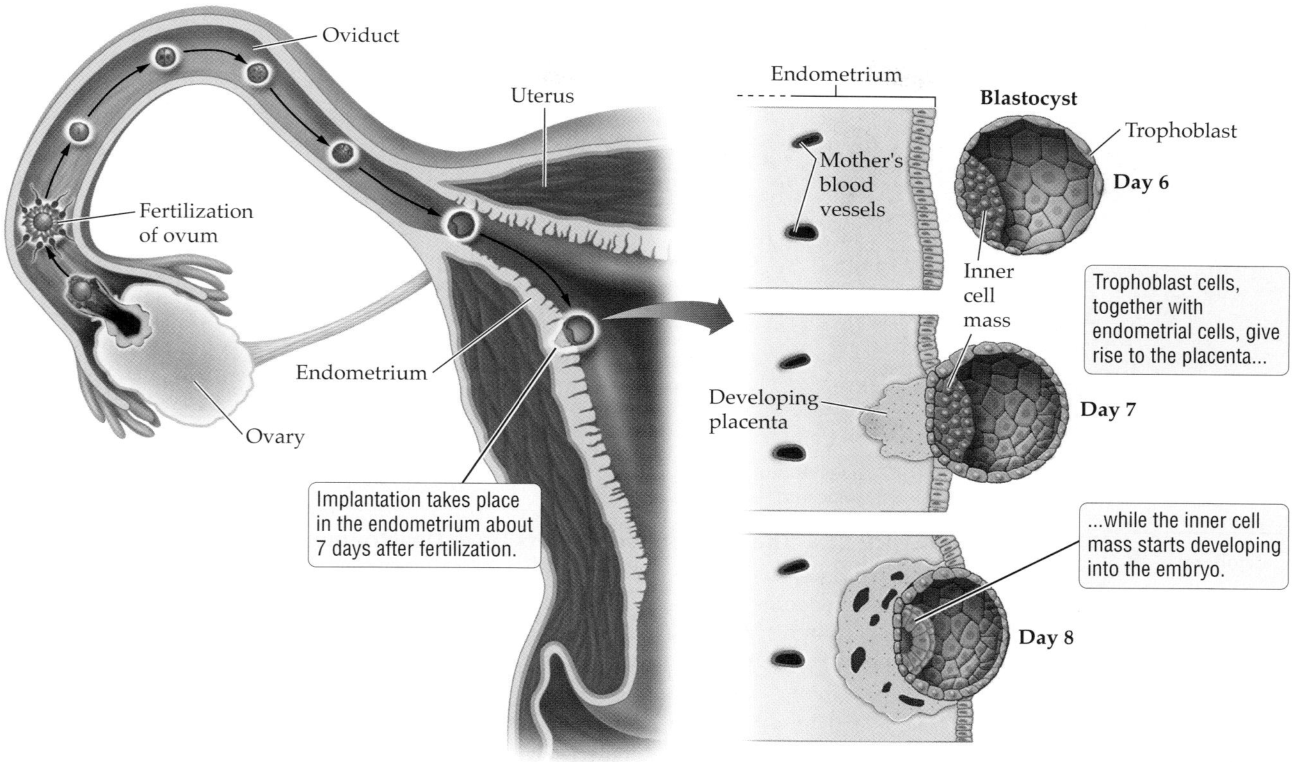

**FIGURE 33.7 Cell Differentiation in the Blastocyst Produces the Embryo and Portions of the Placenta**
The mammalian blastocyst has two types of tissue, whose fates become evident after the blastocyst is implanted in the endometrium. The inner cell mass begins to differentiate into the actual embryo. The trophoblast expands into the endometrium to begin forming a placenta in conjunction with endometrial tissues.

- The **mesoderm** gives rise to the muscles, heart, kidneys, reproductive system, and skeletal structures, along with other tissues and organs.
- The **ectoderm** gives rise to the nervous system and the outer layer of the skin, among other tissues located near the body surface (Table 33.1).

These three tissue types are found in most animals and represent the first visible evidence of a trend that continues throughout the embryo's development: that the identity, or role, of any particular cell becomes more narrowly defined as development progresses.

The second important change results from a remarkable sequence of cell movements within and among these layers. Initially, all three tissue layers of the human embryo are arranged in a flat, disclike structure, the *embryonic disc*, with no infoldings or outpouches. However, if the endoderm is to give rise to the lining of the gut and the mesoderm to internal organs, these tissue layers must be rearranged so that

they can occupy new positions inside the embryo. The elaborate migrations of these cells take place during **gastrulation**, a sequence of events in which the three layers move relative to one another. The resulting embryonic stage is called the **gastrula**. As gastrulation proceeds, the endoderm comes to lie inside the embryo and becomes surrounded, at least partially, by mesoderm; the ectoderm forms the entire outer layer of the embryo. In humans, differentiation of the three

| TABLE 33.1 | The Fates of the Three Tissue Layers in the Developing Embryo |
|---|---|
| **TISSUE LAYER** | **CORRESPONDING ADULT STRUCTURES** |
| Endoderm | Liver, pancreas, thyroid, and linings of gut and lungs |
| Mesoderm | Skeleton, muscles, reproductive structures, kidneys, circulatory and lymphatic systems, blood, and inner layer of skin |
| Ectoderm | Skull, nerves and brain, outer layer of skin, and teeth |

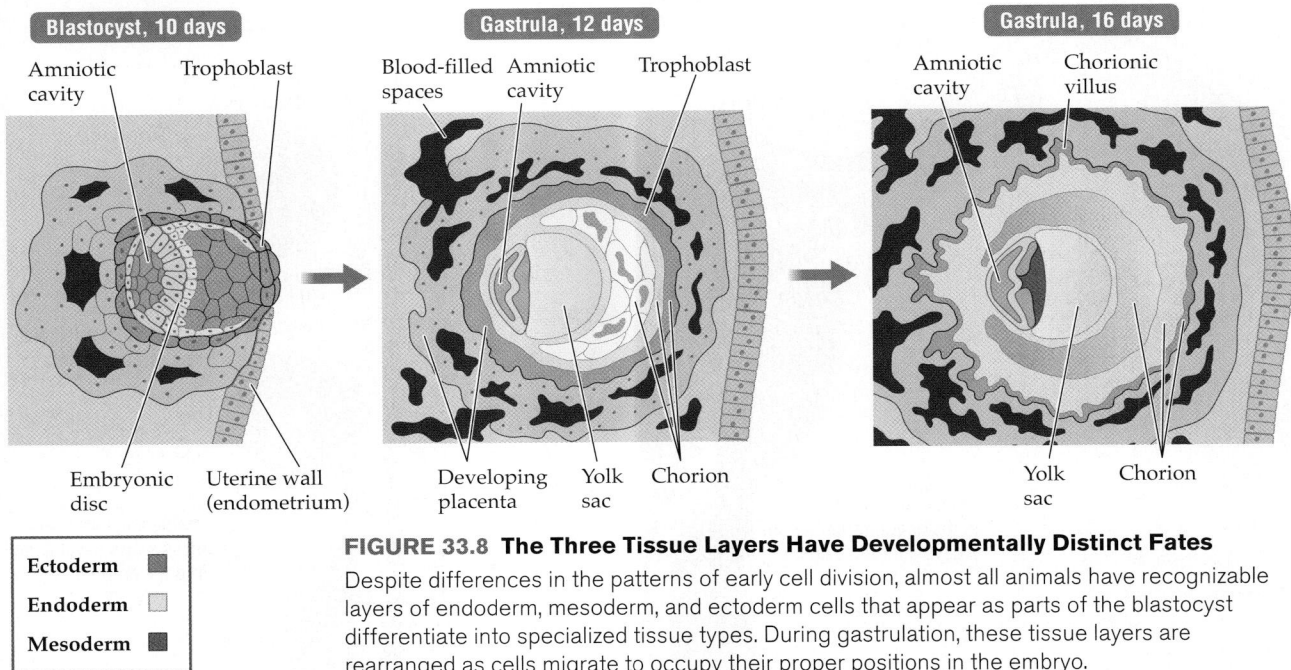

**FIGURE 33.8 The Three Tissue Layers Have Developmentally Distinct Fates**
Despite differences in the patterns of early cell division, almost all animals have recognizable layers of endoderm, mesoderm, and ectoderm cells that appear as parts of the blastocyst differentiate into specialized tissue types. During gastrulation, these tissue layers are rearranged as cells migrate to occupy their proper positions in the embryo.

Ectoderm
Endoderm
Mesoderm

tissue layers and their rearrangement during gastrulation (**FIGURE 33.8**) take place during the 2 weeks following implantation.

## Development in the uterus is divided into trimesters

Human development in the uterus averages about 38 weeks and is divided into three stages known as **trimesters**, each about 3 months long. (Human pregnancy is commonly timed from the start of the woman's last menstrual period and is therefore said to last about 40 weeks.) During the first trimester, the zygote develops from a single cell into an embryo possessing all the main tissue types. All of the organ systems are established by the third month, and the developing individual, now known as a **fetus**, has recognizably human features. The organs grow and the fetus increases in size during the next 3 months, which constitute the second trimester. The largest weight gain occurs in the final trimester, as the fetus grows and also accumulates fat. By the third trimester, fetal development has progressed to the point that, with the help of modern technology, the fetus has reasonably good odds of surviving outside the mother's body.

We have seen that the human blastocyst attaches to the endometrium about 7 days after fertilization, as the trophoblast cells adjacent to the endometrium

start forming the embryonic side of the placenta (see Figure 33.7). About 10 days after fertilization, the embryonic disc starts differentiating into the three main tissue layers of the embryo (see Figure 33.8). The embryonic disc pulls away from the outer layer of blastocyst cells to create the **amniotic cavity**. Some cells from the embryonic disc move out to line the cavity, creating a membrane, the **amnion**, that will later (in the second month) expand over the entire embryonic disc to encase the developing embryo.

Certain endodermal cells on the other side of the embryonic disc migrate to line the original large cavity of the blastocyst, thereby creating the **yolk sac**. Next, starting about 12 days after fertilization, the trophoblast cells pull away to create yet another cavity, the chorionic cavity, which almost completely envelops the amnion, embryonic disc, and yolk sac. The tissues that form the wall of this cavity are known as the **chorion**. They produce fingerlike growths, called *chorionic villi*, that extend into the blood-filled spaces between the maternal and embryonic tissues that make up the placenta (**FIGURE 33.9**). During the third month of development, the circulatory system differentiates in the developing embryo. Two arteries and a vein extend through a long tube, called the **umbilical cord**, to connect the fetus's blood supply to the placenta.

If you look closely at the placental structure shown in Figure 33.9, you will notice that there is no direct intermingling of the maternal and embryonic circulations. On the maternal side, the placenta consists of endometrial tissue richly supplied with small arteries and veins. The chorion, along with other cells from the trophoblast, makes up the embryonic side of the placenta. The maternal and embryonic sides of the placenta are separated by a blood-filled gap called the *intervillous space*. Oxygen and nutrients from the endometrial arteries diffuse into this space. They are then absorbed by the capillary network in the chorionic villi and delivered to the embryo by the vein in the umbilical cord. The two umbilical arteries carry carbon dioxide and wastes from the embryo to the chorionic capillaries; from there, these substances diffuse into the intervillous space and then into endometrial veins, which whisk them away for disposal by the mother's lungs and kidneys.

The human embryo, and later the fetus, floats inside the fluid-filled amniotic cavity, lined by the amnion (see Figure 33.9). The watery fluid cushions the developing tissues and organs from mechanical shock and sudden temperature changes, while allowing the developing fetus room to grow and move about. Just before birth the sac usually ruptures and the amniotic fluid flows out of the mother's vagina. This is what has just happened when a pregnant woman announces, "My water broke!"

The yolk sac is a prominent part of the 16-day mammalian embryo (see Figure 33.8), but it pulls away from the embryonic disc and starts to shrink as the amnion expands around the embryo in the seventh week of development. The mammalian yolk sac contains no yolk, the stored food that nourishes the developing embryos of nonplacental mammals (such as the platypus) and of birds, reptiles, amphibians, and fishes (see Section 4.6). Instead, the mammalian yolk sac produces blood cells and gives rise to germ line cells, the gamete precursors that migrate into the reproductive organs of the embryo, which start to form about 7 weeks after fertilization.

After 8 weeks of development, the developing human is 3.5 centimeters long—little more than an inch. The head is clearly identifiable, the endoderm has differentiated into an identifiable liver, the mesoderm has differentiated into red blood cells and kidneys, and the beginnings of all the organs found in an adult are present. From this point on, the developing human is called a fetus and shows some of the distinctive characteristics of human form (morphology). Near the end of the first trimester, the fetus is 12 centimeters long, the fingers and toes have nails, the external genitalia are recognizably male or female, and the gut has developed from the endoderm to the point that it can absorb sugars (**FIGURE 33.10**).

## THE PLACENTA

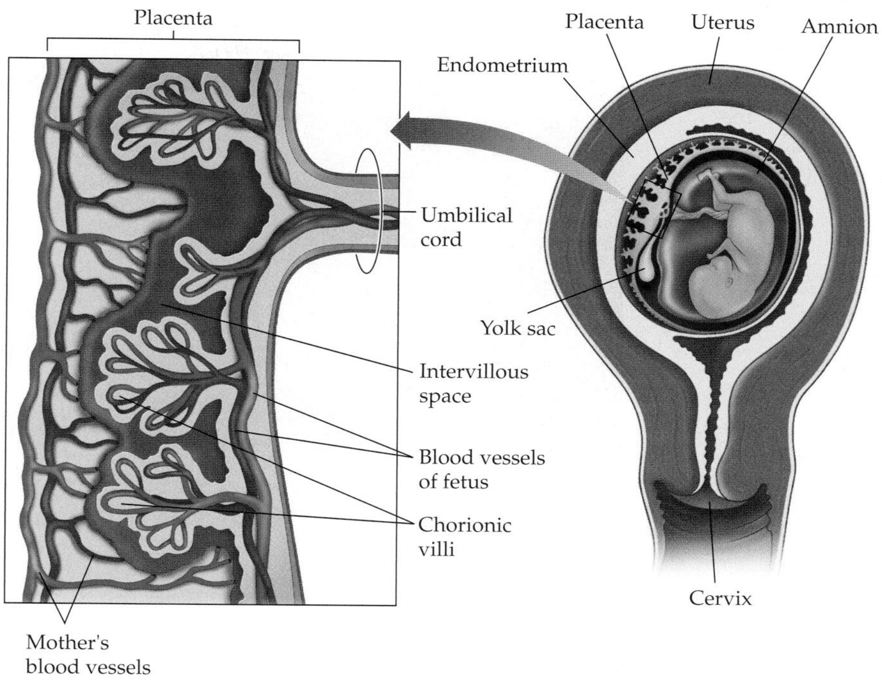

**FIGURE 33.9  The Placenta Is the Life-Sustaining Connection between Mother and Developing Fetus**
Through the placenta, the fetus receives oxygen and nutrients from the mother's bloodstream and disposes of wastes. Note that there is no intermingling of the maternal and fetal blood vessels. Gases, nutrients, and wastes diffuse into and out the intervillous space, the pools of blood that bathe the chorionic villi.

Human development is particularly vulnerable to disruption during the first trimester. Anything that alters the normal course of events during this critical time of rapid cell division and differentiation can lead to serious problems. In humans, most miscarriages take place during the first trimester of pregnancy because any genetic problems in the developing embryo are most likely to be expressed at this time. Nutritional deficiencies, inhaled or ingested toxins, drug abuse, and other environmental factors can disrupt development. The embryo is especially vulnerable because crucial events in tissue differentiation, such as brain development, take place in the first trimester.

In the second and third trimesters, the emphasis switches from rapid differentiation and the emergence of discrete structures to growth and further development of already existing organs. During this time the fetus grows from the size of a large mouse to the typical birth weight of about 3.4 kilograms (about 7.5 pounds). During the second trimester, weight increases faster than length. By the end of that trimester, the developing fetus has gained about half of the

FIGURE 33.10
**Nine Months in the Womb**

For convenience, this period is traditionally divided into three trimesters, each 3 months long.

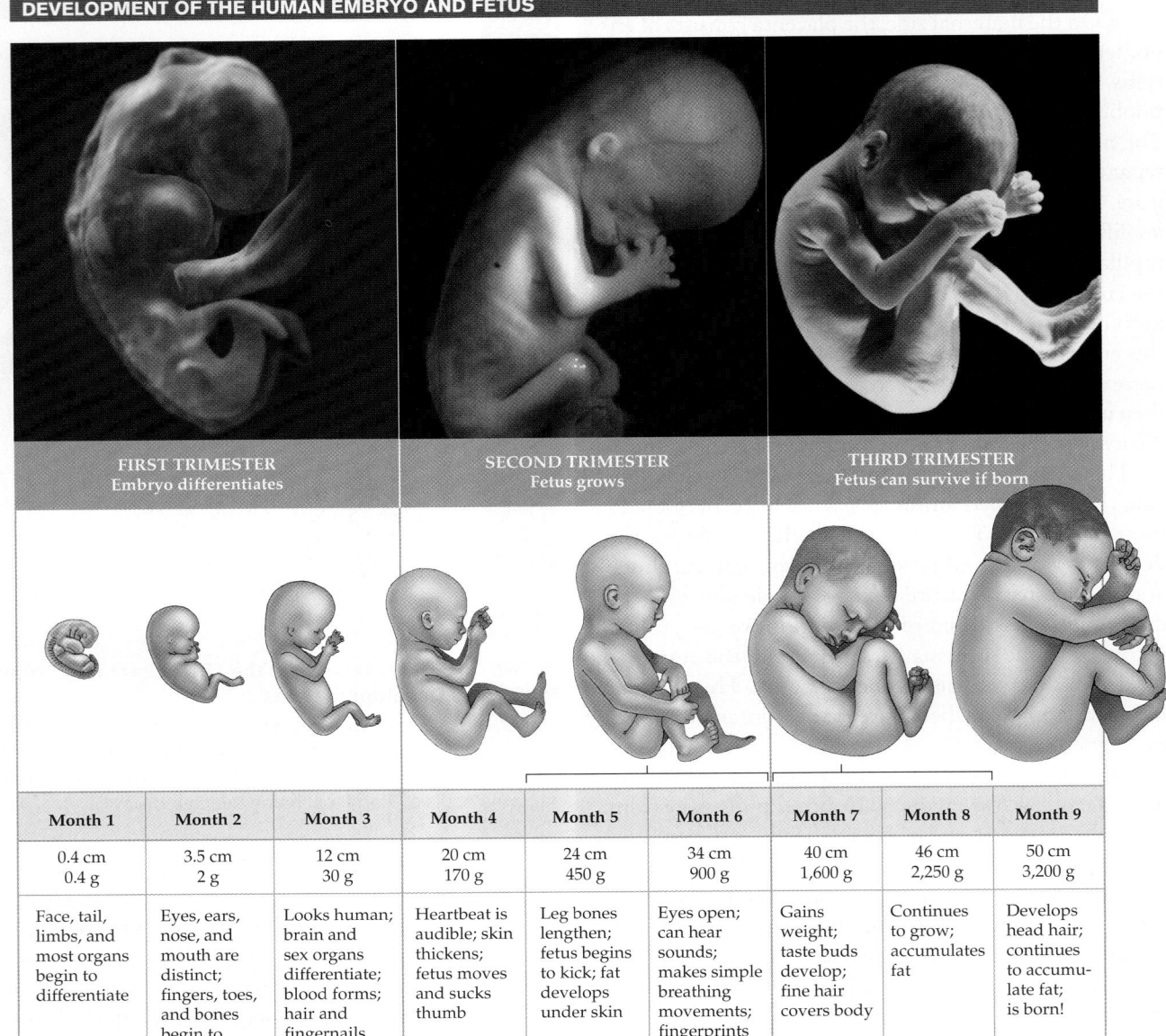

### DEVELOPMENT OF THE HUMAN EMBRYO AND FETUS

| FIRST TRIMESTER Embryo differentiates | | | SECOND TRIMESTER Fetus grows | | | THIRD TRIMESTER Fetus can survive if born | | |
|---|---|---|---|---|---|---|---|---|
| **Month 1** | **Month 2** | **Month 3** | **Month 4** | **Month 5** | **Month 6** | **Month 7** | **Month 8** | **Month 9** |
| 0.4 cm 0.4 g | 3.5 cm 2 g | 12 cm 30 g | 20 cm 170 g | 24 cm 450 g | 34 cm 900 g | 40 cm 1,600 g | 46 cm 2,250 g | 50 cm 3,200 g |
| Face, tail, limbs, and most organs begin to differentiate | Eyes, ears, nose, and mouth are distinct; fingers, toes, and bones begin to develop; heart beats | Looks human; brain and sex organs differentiate; blood forms; hair and fingernails develop | Heartbeat is audible; skin thickens; fetus moves and sucks thumb | Leg bones lengthen; fetus begins to kick; fat develops under skin | Eyes open; can hear sounds; makes simple breathing movements; fingerprints develop | Gains weight; taste buds develop; fine hair covers body | Continues to grow; accumulates fat | Develops head hair; continues to accumulate fat; is born! |

length and about 25 percent of the weight it will have at birth. It becomes active in the uterus and can kick. At the end of the third trimester, the fetus is ready for its sudden transition from the uterus to the outside world. It gains a good deal of weight during the third trimester, and its circulatory and respiratory systems prepare for living in a gaseous atmosphere rather than the watery world of the amniotic fluid.

## Childbirth occurs in stages

The last few weeks of pregnancy are marked by a sharp rise in the levels of certain hormones, including estrogen. The higher estrogen levels in the mother's blood make the muscles of the uterus more sensitive to **oxytocin**, a hormone secreted by the fetus and, later in the birth process, by the mother's pituitary gland. Oxytocin stimulates the uterine muscles and causes the placenta to secrete prostaglandins, which reinforce the contractions. Labor begins when the muscles of the uterus begin to contract in response to these hormones (**FIGURE 33.11**). In a positive feedback loop (see Section 26.3), more contractions cause the production of more oxytocin, and the strength of the contractions increases as more oxytocin is produced. The cervix begins to open, and the increasingly strong contractions eventually push the fetus out of the mother's body. At this point the positive feedback

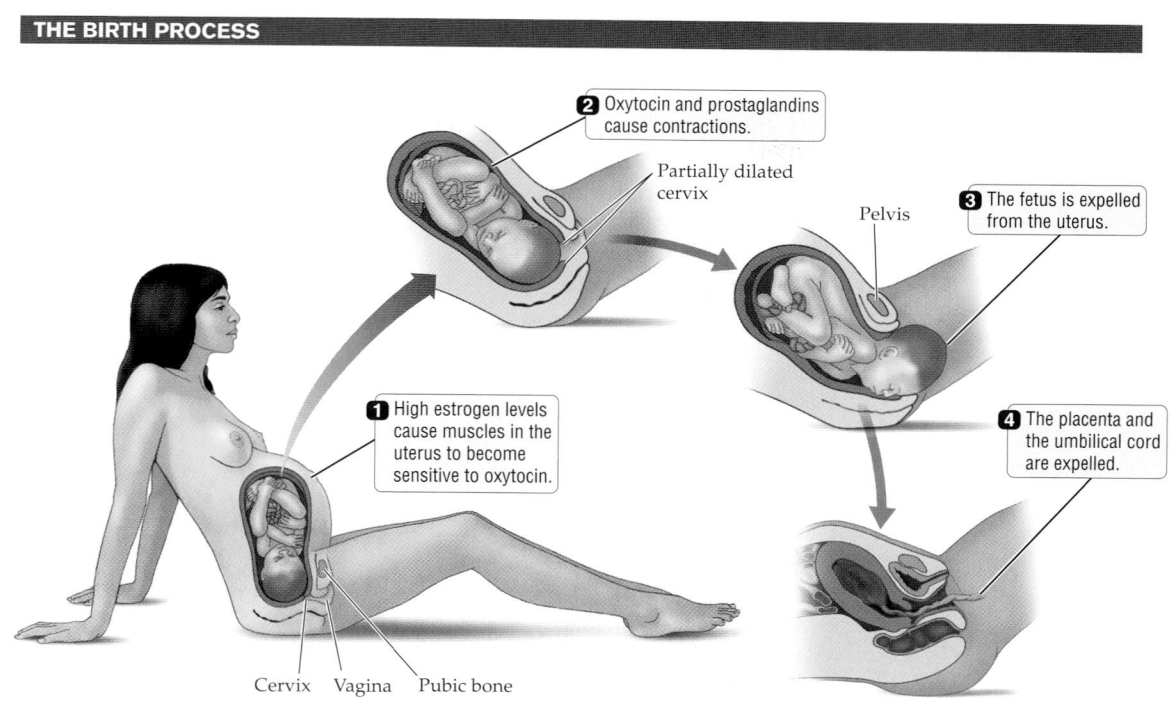

**2** Oxytocin and prostaglandins cause contractions.

Partially dilated cervix

Pelvis

**3** The fetus is expelled from the uterus.

**1** High estrogen levels cause muscles in the uterus to become sensitive to oxytocin.

**4** The placenta and the umbilical cord are expelled.

Cervix  Vagina  Pubic bone

**FIGURE 33.11**
**Ready or Not, Here I Come!**
Hormones initiate contractions of the uterus that eventually push the baby out through the cervix and vagina and into the outside world.

ends, and the contractions subside as oxytocin levels decrease. The placenta, often referred to as the "afterbirth," is expelled during the last stage of childbirth.

At birth, a baby becomes physically independent of its mother. It no longer obtains its oxygen and nutrients directly from her blood, and it must eat, breathe, and maintain homeostasis on its own.

## 33.4 Fertility and Contraception

Pregnancy can be prevented. Abstinence—refraining from sexual intercourse—is one way to avoid pregnancy, as well as sexually transmitted disease. Another option for preventing pregnancy is to use a **contraceptive**. A wide variety of contraception methods are available (see Table 33.2).

Oral contraceptives are among the most popular and effective methods of contraception. Oral contraception, or "the Pill," was approved by the FDA (Food and Drug Administration) in 1960. Although it immediately secured a place in history as one of the most widely used prescription drugs, and is believed by some to have worked a social transformation by giving women control over their reproductive life, the oral contraceptive was controversial from the start (it could not, for example, be prescribed to unmarried women until a Supreme Court ruling struck down the ban in 1972).

When used correctly, the Pill has proven highly reliable in preventing unwanted pregnancies. The most commonly used oral contraceptives contain a combination of synthetic versions of two types of hormones: progestogens and estrogens. These "combination" pills prevent the release of mature eggs each month. But if an egg does happen to become fertilized, they provide a "backup" action by making the lining of the uterus inhospitable to a fertilized egg. Without a receptive uterine wall, a pregnancy cannot become established.

Among women who take the Pill exactly according to directions, its failure rate is much less than 1 percent. But "exactly according to directions" is an important caveat. Success or failure depends more on the person than on the Pill. Only about one in 1,000 women who take the Pill as directed become pregnant. But among a population of less careful "typical" women, the failure rate is 50 times greater.

The Pill reliably prevents pregnancy, but it is not the best option for some. For example, smoking while taking an oral contraceptive greatly increases the risk of heart attack, stroke, and blood clots, especially in women over 35 who smoke 15 or more cigarettes a day. Women who have certain preexisting conditions, such as high blood pressure, heart disease, liver disease, or a personal history of breast cancer, are usually advised to use alternative methods of birth control. Certain drugs and medications, including some antibiotics and painkillers, can decrease the effectiveness of the Pill. Contrary to popular belief, modern oral

## TABLE 33.2 Birth Control Methods

| METHOD | MECHANISM | EFFECTIVENESS (IF USED AS DIRECTED)[a] |
|---|---|---|
| Oral contraceptives: can be ingested in pill form (the Pill); applied on the skin as a patch; implanted under the skin; or inserted in the vagina (vaginal ring) | Hormones (often a combination of estrogens and progestogens) suppress ovulation by mimicking a pregnancy. | 99% |
| Birth control sponge | Sponge containing spermicide is inserted deep in the vagina before intercourse. | 90% |
| Female condom | Plastic pouch, inserted before intercourse, lines the vagina. | 95% |
| Male condom | Plastic or latex pouch covers the penis to keep sperm from entering the vagina. | 98% |
| Diaphragm | Dome-shaped latex cup filled with spermicide, inserted before intercourse, covers the cervix and keeps sperm out of the uterus. | 92% |
| Intrauterine device (IUD) | T-shaped plastic device must be positioned inside the uterus by a health care professional. | 99% |
| Female sterilization (tubal ligation) | The oviducts are sealed with clamps or by other surgical means. | 100% |
| Male sterilization (vasectomy) | The tubes that carry sperm are sealed surgically by a health care professional. | 100% |

[a]The actual effectiveness of some methods may be lower, especially if correct use depends on individual users. For example, some studies have found that the "real-world" effectiveness of birth control pills that women must take through the month is about 95 percent.

contraceptives, which have a much lower dose of hormones than the formulations used in the 1960s had, do not lead to weight gain.

Birth control pills are known to reduce acne, anemia, and premenstrual syndrome (PMS), and they often reduce the symptoms of conditions such as polycystic ovary syndrome. Regular users have a 40–80 percent lower risk of ovarian and uterine cancers, which are relatively rare cancers. Some studies have shown a slight increase in breast cancer risk among long-term users, but other studies fail to find such an effect. A recent large study revealed slightly lower all-cause mortality (death from any cause) among women who took birth control pills for 5 years or longer compared to women who had never used them.

Oral contraceptives can have a "hidden" side effect: feeling "safe" can mean an increase in sexual activity, and that can mean increased exposure to sexu-

ally transmitted diseases (STDs). By the age of 24, one in three sexually active Americans will have contracted an STD; moreover, at least one in four—perhaps as many as one in two—will contract an STD at some point in their lives (see the box on page 729). Oral contraceptives offer no protection against these diseases.

## 33.5 Development after Birth

Development does not end when a human child—or a juvenile of most other animal species—is born or hatched. We humans, in particular, spend about a quarter of our lives reaching our full adult size. Most of this growth occurs during childhood, before we become sexually mature.

# Sexually Transmitted Disease

According to the Centers for Disease Control and Prevention (CDC), 19 million Americans have a sexually transmitted disease. That statistic includes one in four teenage girls (approximately 3 million). More than 30 different types of sexually transmitted diseases (STDs) affect females; some of these also produce serious illness in males. Surveys reveal that several STDs, such as gonorrhea and syphilis, have been on the rise in the general population, especially among younger people, in recent years.

Trichomoniasis, a disease caused by a protist, *Trichomonas vaginalis*, is the most common STD in North America. However, it is less common among young people than among older adults (about 2 percent of teenagers were infected, in a recent CDC survey). Both sexes can be infected; men usually have no significant symptoms, but they can transmit the parasite to a sexual partner.

Infection with HPV is the second most common STD in the United States (see the table). Females in the 15–26 age group account for 75 percent of all new cases. Both males and females can be infected without experiencing any symptoms. Over time, however, the virus can produce genital warts and lead to various cancers, including cervical, penile, and anal cancers.

In a recent CDC study, 18 percent of the teenagers tested had chlamydia, a bacterium that causes no symptoms in most men and women when they first become infected. However, an untreated infection progresses to serious health problems in about 40 percent of women who harbor this "silent disease." The bacterium can enter the uterus and spread to the fallopian tubes, causing scarring and tissue damage. Damage to the reproductive organs can result in sterility or complications in pregnancy that can be fatal for the mother and her fetus.

About 2 percent of the teenagers in the CDC study were infected with the HSV-2 strain of the herpes simplex virus, which can cause genital herpes. More than 45 million Americans harbor this strain of the virus, but most have no symptoms. Some infected people experience outbreaks of skin sores, usually in the genital area, and the infection can be deadly to the fetus of an infected mother or to a person whose immune system is compromised. Infection with this virus is also known to make people more susceptible to the HIV-AIDS virus.

Sexual abstinence is one way to keep from contracting an STD. People in long-term monogamous relationships are also at lower risk than individuals who have multiple sexual partners. Latex condoms protect against some STDs, such as chlamydia and trichomoniasis. Most STDs can be treated, and most of them can be completely cured if detected early, underscoring the importance of testing and diagnosis before long-term damage occurs or more people catch the disease from an untreated infected partner.

## Incidence of Some Common Sexually Transmitted Diseases

| SEXUALLY TRANSMITTED DISEASE | ESTIMATED INCIDENCE (NEW CASES PER YEAR)[a] |
| --- | --- |
| Trichomoniasis | 7.4 million |
| Human papillomavirus (HPV) | 6 million |
| Chlamydia | 1.2 million |
| Herpes (HSV) | 1 million |
| Gonorrhea | 301,174 |
| Hepatitis B (HBV) | 38,000 |
| Hepatitis B (HCV) | 18,000 |
| Syphilis | 13,997 |
| HIV-AIDS | 1,092 |

SOURCE: CDC, "Transmitted Disease Surveillance 2009" (http://www.cdc.gov).
[a]Prevalence, which is the total number affected at any time, is usually much higher. For example, 20 million people were estimated to be infected with HPV in 2009.

In many ways, the development of the human baby is a continuation of its fetal development. The developmental changes that take place after birth follow a sequence as predictable as the sequence of events before birth. During the first 2 years of life, a child's body continues to increase in size at a rate comparable to that of a fetus, rather than at the slower rate typical of infant development in other mammals. Our brains also continue to increase in weight at the fetal rate for the first 2 years of life and do not reach their mature weight for another 8 years.

Many cells in the human body have a limited life span and must be replaced as they wear out or die. Human skin, for example, must regenerate itself as layer after layer of skin cells is worn away. Similarly, the food-absorbing cells lining the small intestine live less than a week and so must be continually replaced with new cells. Depending on the tissue involved, cells are regenerated in one of two ways. Some tissues, such as the cells that form the inner walls of blood vessels, regenerate by simple cell division (mitosis) of preexisting specialized cell types. Other tissues, such as the

## Fertility declines with age

In humans, both males and females reproduce less effectively as they age. When they pass 40 years of age, females show a clear drop in their ability to produce normal eggs and bear children (Table 33.3). If eggs from younger women are implanted into women over 40, the pregnancy rate equals that for the age group of the women who donated the eggs. This finding suggests that as women age, it is their eggs that decline in quality—a conclusion supported by the increased risk of birth defects in children born to older mothers. For example, Down syndrome (trisomy 21) occurs much more commonly in fetuses conceived by women over the age of 40 than in younger women. Human females reach *menopause*, the end of their reproductive lives, around the age of 50 (see Section 29.4).

Similar patterns have been found for sperm produced by older and younger men. Children fathered by older men tend to live shorter lives than children fathered by the same men at a younger age, suggesting that the genetic quality of sperm declines during the father's lifetime. In addition, men produce fewer sperm as they age, decreasing their chances of fertilizing an egg. Males do not undergo the clearly identifiable menopause that is characteristic of females. Instead, their ability to produce sperm and their sex drive slowly decrease as they age.

Why does fertility decline with age? Egg and sperm precursor cells are present in our bodies from birth, so they accumulate mutations over time from exposure to various environmental factors. The older these cells are, the more likely they are to have accumulated mutations harmful or even lethal to a developing fetus. This observation raises a question that has intrigued biologists, anthropologists, demographers, and the general public for more than 50 years: If the buildup of gametic mutations curtails fertility, then why do so many humans, almost uniquely in the animal world, continue to live a few decades past their reproductive years? Could we as a species benefit from the presence of older, nonfertile individuals in our society? Could menopause be an adaptive trait?

## Humans can live past their reproductive years

Other than elephants, whales, and humans, few species live significantly beyond their fertile years. Even among the other apes, which are otherwise so similar to us, no species has a significant period of life after reproduction ends. When chimpanzees reach the end of their reproductive years, which can be as long as five decades in captivity, they are frail with age, and death follows in short order. Anthropologists note that even in preindustrial societies, such as hunter-gatherer tribes, at least one-third of the adults live past 55 years in good health. The life expectancy for men and women in developed countries is 75–85 years. Women in particular, because they typically live longer than men, may have half their life ahead of them after reaching 40, and only the last few of those years are likely to be marked by frailty.

A number of evolutionary models have been proposed to explain postreproductive longevity in humans, the best known of which is the **grandmother hypothesis**. According to this hypothesis, menopause in older women is an adaptive trait because it contributes to the survival of grandchildren. For a woman in middle age who has raised a number of children of her own, investing in the reproductive success of her existing daughters is more adaptive than giving birth to yet more children of her own. Before modern times, childbirth was a substantial metabolic burden on a mother and left her vulnerable to life-threatening infections and complications of pregnancy. In that environment, an older woman's genetic interest was furthered more by nurturing her grandchildren, and enhancing their survival, than by risking childbirth at an older age.

Proponents of the grandmother hypothesis maintain that the adaptive value of grandmothering is greatest in social animals, like humans and whales, that have a long juvenile period and significant reliance on learning certain behaviors from their elders (**FIGURE 33.12**). Menopause evolved, with its abrupt cessation of fertility, because it freed older women from mothering and enabled them to invest in the success of their genetic line through grandmothering instead. Advocates of the hypothesis maintain that, in

| TABLE 33.3 | The Effect of Age on the Fertility of Human Females | |
|---|---|---|
| AGE (YEARS) | MONTHLY CYCLES THAT RESULT IN A PREGNANCY (%) | MONTHLY CYCLES THAT RESULT IN AN EMBRYO THAT FAILS TO COMPLETE DEVELOPMENT (%) |
| Under 30 | 29.0 | 14.9 |
| 30–35 | 19.8 | 16.5 |
| 35–40 | 17.1 | 22.4 |
| 40–45 | 12.8 | 33.2 |

favoring postmenopausal grandmothering, natural selection also favored an increase in the average adult life span. That is because a long-lived population of humans had more long-lived grandmothers in it; and because these grandmothers helped improve the survival odds of many grandchildren, the adaptive fitness of such a population increased compared to a population with a shorter postreproductive life span.

## 33.6 How Development Is Controlled

The development of animals like us is complex, orderly, and exquisitely precise. There must be precise control, for example, over the fates of the billions of cells that are produced as a single-celled zygote gives rise to an embryo, which in turn develops into a fetus. How does one end of the embryo "know" to become the head, and not the tail, of the animal? How does a particular ectoderm cell "know" to become a nerve cell, not a skin cell? How does a single fertilized egg become a human adult with trillions of cells that can be categorized into at least 220 different types according to their structure and function? These are some of the deepest, most complex questions in biology. In this section we offer a brief glimpse of what has been learned recently about these mysteries.

### Cell fate is controlled by differential gene expression

Cell types become arranged into different tissues in an organism not because they contain different DNA (just about all cells in an individual's body have the same DNA sequence), but because a unique subset of the total DNA information is "read out," or expressed, in each of the different cell types. A major advance in biology in the past few decades has been the rapid growth in our understanding of how expression of the genetic information in cells is controlled during development. Some invertebrates, such as nematode worms and fruit flies, that have a simpler body plan and fewer types of tissues and organs have played key roles in these discoveries.

In the embryo, the potential fates of a cell depend on the particular genes in its DNA that can be activated. We noted in Chapter 15 that most genes code for proteins, and that transcription and translation are the two processes that turn the DNA code of a particular gene into a specific protein. Transcription converts the nucleotide sequence information in DNA to

**FIGURE 33.12 The Grandmother Effect**
According to the grandmother hypothesis, natural selection has favored menopause, and as a consequence a longer average life span in humans because the species benefits when older women invest in the rearing and education of their grandchildren instead of bearing children themselves in their later years. Here, 86-year-old Ameia Tepano demonstrates an ancient form of storytelling, called Kai Kai, to her granddaughter. The two live on Easter Island.

the nucleotide sequence of an RNA molecule. Protein-coding RNA, known as messenger RNA (mRNA), is extensively processed before it is exported from the nucleus to the cytoplasm. In the cytoplasm, mRNA serves as a template for the manufacture of protein in the process known as translation. Preventing or promoting the transcription of certain genes, or altering their processing into mRNA, can cause identical genetic information in each cell's DNA to be expressed differently in different cells or at different times in the same cells.

Let's look at an example of how the timing of gene expression can affect development. Humans and other mammals need different kinds of oxygen-binding hemoglobin pigments at different stages of their development: embryonic hemoglobin at 0–8 weeks, fetal hemoglobin at 8–12 weeks, and then progressively greater proportions of adult hemoglobin at 12 weeks and beyond. The embryonic and fetal hemoglobins have a greater affinity for oxygen than adult hemoglobin does, enabling the developing baby to efficiently pull oxygen from its mother's blood by way of the placenta. The genes that encode the embryonic, fetal, and adult hemoglobin proteins are associated with other DNA sequences that, when bound by regulatory proteins, enable transcription of those genes (see Section 14.8). At certain stages during development, transcription of the appropriate hemoglobin genes is activated in red blood cells, while the other hemoglobin genes remain inactive.

Another way that animals can regulate gene expression during development is to cut up and re-join (splice) the transcribed RNA in different ways to make different mRNA products (see Section 15.3). For example, male and female fruit flies form different mRNAs from the transcription of a single gene that they both express. The resulting proteins, which determine sex in the developing flies, play an important role in controlling the sexual development of these animals.

## Specific signaling molecules activate genes during development

How are the various regions of the body specified: front and back, head and tail, the chest cavity and the abdominal cavity? How do two identical daughter cells become different from one another? How might one daughter cell become a neuron, while the other daughter cell becomes a skin cell? Tissues in certain parts of the body progress along a particular pathway, and a particular daughter cell differentiates into a specific cell type, guided by the *positional information* gathered from the surroundings. **Positional**

**information** is like a GPS for the cell: it consists of physical and chemical cues that inform a cell as to its location in the body. A cell develops along a particular path in response to the type, and even the strength, of the positional cues it receives. *Inductive signals* are one example of positional cueing that guides embryonic development. **Inductive signals** are short-range chemical cues that are passed to embryonic cells from neighboring cells or from the maternal environment and that "tell" the embryonic cell what to do next. For example, an inductive signal may communicate which cells in the developing gastrula are to develop into the head end of the organism and which into the tail.

Inductive signals commonly work by activating or inactivating the transcription of genes that play a critical role in development. One method by which inductive signals can work is to spread outward from a cell that produces them, decreasing in concentration with increasing distance from the source. Cells with the proper receptor proteins in their plasma membranes have the capacity to respond to the signals. A specific concentration of the inductive signal elicits a specific pattern of gene expression in a responding cell, through the activation or inactivation of specific sets of genes. In this way, cells located at different points along the concentration gradient can express very different sets of genes, which would nudge them along different pathways of differentiation in the developing animal.

Inductive signals of a specific class known as **morphogens** play a crucial role in controlling embryonic development in many animals, including fruit flies of the genus *Drosophila*. In the embryonic fruit fly, patterns in the concentrations of several morphogens provide developing cells with information about their position relative to the head (**FIGURE 33.13a**). If the signal that identifies the head end is injected into the opposite end of the embryo, for example, a two-headed embryo will result (**FIGURE 33.13b**).

The bodies of insects are divided into segments, and studies of the development of segments in fruit flies have revealed how morphogens work in those insects: they act on a family of *homeotic genes* that distinguish the different segments in the body of the fly (**FIGURE 33.14**). These master genes regulate a suite other genes that in turn control the form of a particular segment. Homeotic genes and morphogens that resemble those in fruit flies play a major role in directing "head-to-tail" differences in the development of tissues in all animals, including humans.

In some cases inductive signals simply act on adjacent tissues to provide their cells with information on the identity of the surrounding cells. In birds, for ex-

**FIGURE 33.13**
**Morphogens Create Developmental Patterns in the Early Embryo**

(a) Patterns of increasing or decreasing morphogen concentrations provide cells with information about their position relative to the head and tail ends of a developing fruit fly embryo. (b) Experimental manipulation of morphogen concentration gradients can dramatically disrupt development. If a high concentration of morphogen A (which identifies the head) is injected into the tail end of an embryo, cells near the tail end can be "fooled" into developing into head tissues.

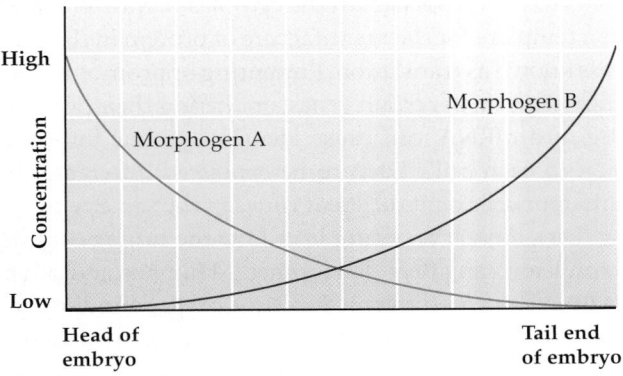

**HOW MORPHOGENS INFLUENCE DEVELOPMENT**

(a) Morphogen concentrations from head to tail

Morphogen A

Morphogen B

High / Low — Concentration

Head of embryo

Tail end of embryo

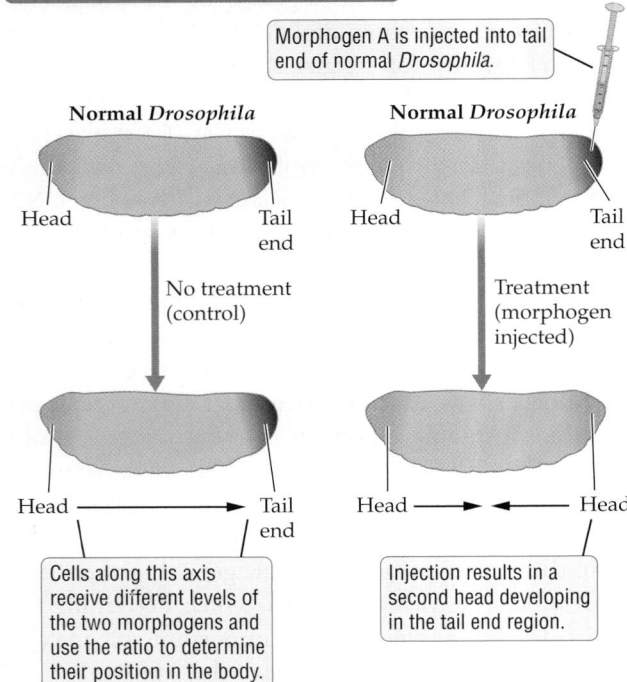

(b) Adding head morphogen to tail end

Morphogen A is injected into tail end of normal *Drosophila*.

Normal *Drosophila*

Head — Tail end

No treatment (control)

Head → Tail end

Cells along this axis receive different levels of the two morphogens and use the ratio to determine their position in the body.

Normal *Drosophila*

Head — Tail end

Treatment (morphogen injected)

Head → ← Head

Injection results in a second head developing in the tail end region.

ample, the interaction between mesoderm-derived and ectoderm-derived tissues in the skin determines what ectodermal cells produce, as scientists have shown by transplant experiments (**FIGURE 33.15**). Chicken ectoderm from the wing develops into wing feathers when placed next to mesoderm from the wing, but it develops into scales and claws when placed next to mesoderm from the foot. Chemical interactions between the mesoderm and ectoderm tissues determine which genes in the ectoderm tissue are activated.

Hormones also help control animal development. Sex hormones, for example, influence the development of reproductive structures and determine whether individuals will become males or females, as described in Chapter 29.

## The environment can influence development

Genes and their protein products do not work in isolation during development. The developing animal also responds to various environmental influences. These influences include cues from the internal environment—for example, the amounts of stored resources available—as well as external cues such as temperature, day length, or even the chemicals produced by potential predators and competitors. We offer just a few examples here.

Gender in many turtles and in alligators depends on the temperature of the nest in which the eggs develop: turtle eggs from cool nests tend to develop into males; alligator eggs from cool nests tend to develop into females. Many aquatic invertebrates develop spines or bristles when exposed to chemicals released by nearby predators in their environment. The spines and bristles make it harder for the predators to eat their prey. Aphids show several environmental influences on their development: scarcity of food and high densities of other aphids nearby stimulate the development of winged forms, and temperature and day length determine whether the newly formed individuals will reproduce sexually or asexually.

## Alterations in developmental patterns contribute to the evolution of novel structures

Turning a single gene on or off at a given point in development can lead to complex changes in an animal. Changes in developmental genes have therefore played an important role in the evolution of life. Consider the following two examples.

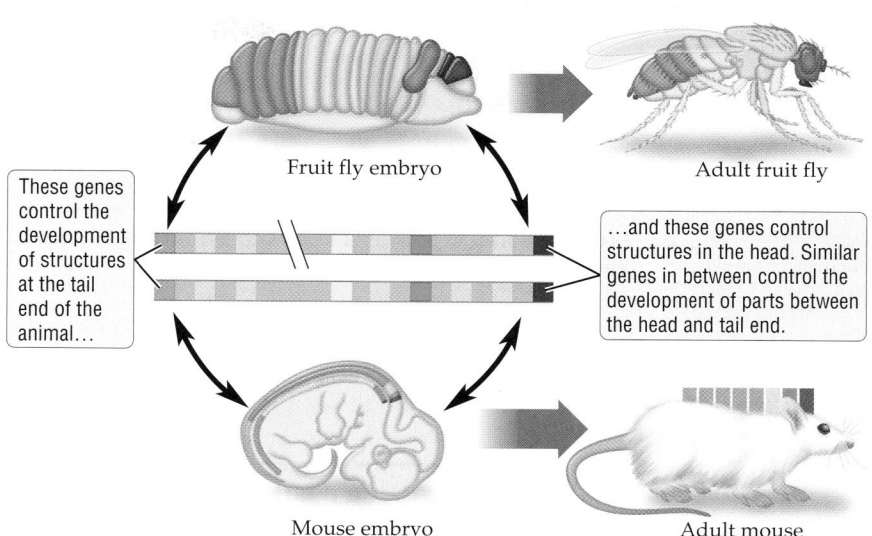

FIGURE 33.14 **Homeotic Genes Control Patterns of Development**

Chicken feet have four distinct toes. Chick embryos, however, have webbed feet more like those of an adult duck than those of an adult chicken. The disappearance of the webbing between the chick embryo's toes is believed to depend on a single gene. This particular gene, when active in a cell, causes the cell to die in what is called programmed cell death, or **apoptosis** [*AP*-uh-*TOH*-sus]. During chick development, this gene is activated specifically in the cells of the webbing, but not in the toes. A chick embryo with a mutant form of

**INDUCTIVE SIGNALS DIRECT DEVELOPMENT**

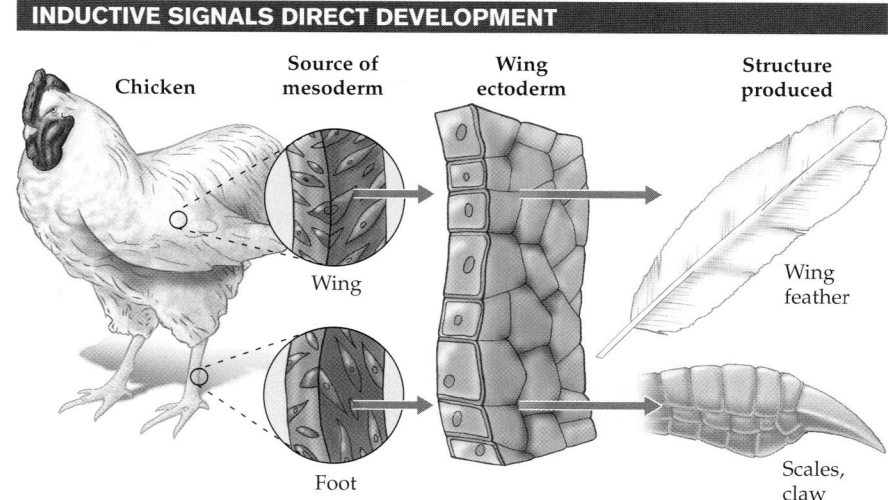

FIGURE 33.15 **Surrounding Cells Influence Development**
Cells in the ectoderm-derived layers of chicken skin develop differently depending on the source (wing or foot) of the adjacent mesoderm-derived layers of chicken skin. When placed next to ectoderm from the wing, mesoderm from the wing stimulates the production of feathers typical of wings. When placed next to the same wing ectoderm, mesoderm from the foot stimulates the production of scales and claws typical of feet.

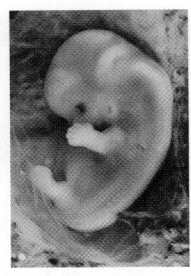

**FIGURE 33.16 Apoptosis Leads to Separate, Rather Than Webbed, Fingers and Toes**

this gene—one that failed to be activated in cells in the webbed areas of the embryonic feet—might well grow up to have feet resembling those of a duck rather than

normal, chickenlike feet. Apoptosis plays a key role in the development of almost all animals. Human fingers and toes are webbed in the human embryo, just as they are in developing chicken feet (**FIGURE 33.16**). Because the gene activating apoptosis is turned on in the cells of the webbed hand and foot before birth, human babies are born with beautifully sculpted fingers and toes.

> **Concept Check**

1. How is a human embryo different from a human fetus?

2. What role do inductive signals play in controlling development?

> **APPLYING WHAT WE LEARNED**

# How Much of a Threat Is Alcohol to Normal Development?

In 2005, the surgeon general recommended total abstinence from alcohol for women who might become pregnant—a far stricter position than that taken two decades earlier. One reason for the government's extreme position was that warning labels on containers of alcohol and a campaign to raise awareness of the dangers of drinking while pregnant had failed to stop many women from heavy drinking during pregnancy.

According to a 2005 study, 20–30 percent of pregnant women drink at some time during their pregnancy, and 2 percent engage in binge drinking—drinking five or more drinks at one sitting. Binge drinking or drinking every day or every weekend greatly increases the risk of severe alcohol-related birth defects. Of the 4 million babies born in the United States each year, about 40,000 are born with mild to severe birth defects that result from the mother's abuse of alcohol during pregnancy. Of these, 4,000 have fetal alcohol syndrome (FAS), a specific set of alcohol-related birth defects.

Scientifically, the topic is controversial. There's no question that heavy drinking, whether binge or otherwise, is extremely bad for a developing fetus. The surgeon general rightly concluded that women were not getting the message that they should not drink if there is any possibility they're pregnant. At the same time, light drinking—for example, half a glass of wine several times a month—has not been shown to cause harm (or, for that matter, *shown* to be harmless). Studies that report fetal effects from "light" drinking sometimes turn out to have averaged binge drinking over a long period of time. For example, one subject in an alcohol study reported drinking a gallon of wine and half a case of beer every Friday and Saturday evening during the first 3 months of her pregnancy and no alcohol at all during the second

3 months; researchers rated her drinking as averaging only one drink per day over the course of her pregnancy.

Women who might become pregnant are advised not to drink, because many women do not know they're pregnant in the first few weeks of pregnancy, during "critical periods" of fetal development. The heart, for example, is most sensitive to disruption by teratogens between the third and sixth weeks after conception; and the genitals are most sensitive between the seventh and ninth weeks. The first 2 months of pregnancy are rich with critical events, as the embryo forms a rough outline of each of the major parts of the body—including the heart, kidneys, arms, and legs. An embryo exposed to a teratogen during this time is much more likely to end up with a major birth defect than if exposed later. But during this time, may women do not know they're pregnant.

Alcohol can have negative effects throughout development. A fetus exposed to alcohol late in development might not have facial or heart malformations, but it could be smaller than normal—a result of the alcohol interfering with growth in the last month or two. The central nervous system is sensitive to alcohol throughout development, so children can seem normal when they are babies and toddlers, but go on to have behavioral problems as they grow older.

Alcohol does its dirty work first by interfering with the mother's blood flow, slowing the delivery of oxygen and nutrients to a developing embryo. Alcohol also interferes with the migration of cells as they form the embryo's nervous system, interferes with cells' ability to stick together, kills cells critical to development, and interferes with a vast array of signaling molecules that regulate development. For most women, the decision to not drink during pregnancy is a no-brainer.

# CDC: One-Third of Sex Ed Omits Birth Control

BY MIKE STOBBE, *Washington Post*

Almost all U.S. teens have had formal sex education, but only about two-thirds have been taught about birth control methods, according to a new government report released Wednesday.

Many teens apparently are not absorbing those lessons—other recent data shows that after years of steady decline, the teen birth rate rose from 2005 to 2007. It dipped again in 2008, to about 10 percent of all births.

The report from the Centers for Disease Control and Prevention is based on face-to-face interviews with nearly 2,800 teenagers in their homes from 2006 through 2008. Female interviewers from the University of Michigan asked the questions for the CDC.

About 97 percent of teens said they received formal sex education by the time they were 18. Formal sex education was defined in the report as instruction at a school, church, community center or other setting teaching them how to say no to sex or about birth control and sexually transmitted diseases.

Lessons about saying no and STDs were more common than instruction on how to use a condom or other birth control, the study found.

Overall, about two-thirds of teens got birth control instruction by the end of high school—about 62 percent of boys and 70 percent of girls.

In contrast, about 92 percent of boys and girls reported being taught about sexually transmitted diseases, and almost that many learned about preventing infection with the AIDS virus. And about 87 percent of girls and 81 percent of boys were taught how to say no to sex.

In 2009, 750,000 U.S. teens of ages 15–19 became pregnant—82 percent of the pregnancies unintended. The good news is that, overall, the teen birth rate in the United States had declined by 37 percent since 1991. Even so, the U.S. teen birth rate—at about 39 births per 1,000 teens—is the highest teen birth rate in the developed world, about three times the rate in most other developed countries. Teen parents are less likely to go to college, more likely to have children with health problems, and more likely to live in poverty—all predictors for later health and social problems for their children.

What makes the United States different from other developed countries in this respect? An important cause of the high rate of teen pregnancy in the United States is the quality of sex education. U.S. teens have similar rates of sexual activity as teens in places such as Canada, England, France, and Sweden. But U.S. teens are more likely to have short-term relationships and less likely to use contraceptives. Only about 20 percent of sexually active U.S. teens use the most reliable forms of birth control, such as birth control pills or injectable contraceptives.

U.S. parents are notoriously poor sources of information about pregnancy prevention. Among teens who talked to their parents about sex-related topics in one survey, 80 percent of boys and 62 percent of girls were not told where to get birth control. In part because parents generally fail to fill the information gap, most communities offer formal sex education. But even so, among sexually experienced teens, nearly half of sexually active teenage boys and one-third of sexually active teenage girls do not learn about birth control.

One reason is that many classes emphasize sexual abstinence without discussing contraception. In 2006–2008, for example, about a quarter of teens received abstinence education without receiving any instruction about birth control. Such "abstinence-only" education delays teens' first sexual experiences, but once they begin having sex, they are less likely to know how to prevent pregnancy than are teens who learn about contraception.

Regional differences play a part. The states with the lowest teen birth rates are northeastern states such as New Hampshire, Massachusetts, and Vermont, averaging about 23 births per 1,000 teens—a rate comparable to that of Iceland or Canada. The states with the highest teen birth rates are southern states that emphasize abstinence, including Mississippi, Texas, Arkansas, Tennessee, and Kentucky. Such states have nearly three times the teen birth rate that northeastern states have.

## Evaluating the News

**1.** Discuss why the teen birth rate in the United States is so much higher than in other developed countries.

**2.** Should sex education be primarily the responsibility of the educational system (schools and colleges), rather than parents and family members? Can peers make a useful contribution? What are some pros and cons of students learning about sexual health from these sources?

**3.** How would you propose lowering America's high teen birth rate? Use the information that you've learned about reproduction and contraception, as well as some of the ideas presented in this article and analysis. Explain your rationale.

**SOURCE:** *Washington Post*, October 22, 2009, http://washingtonpost .com.

# CHAPTER REVIEW

# Summary

## 33.1 Sexual and Asexual Reproduction in Animals

- Most animals produce offspring through sexual reproduction, although some can produce genetically identical offspring through asexual means.
- Sexual reproduction tends to be found in variable environments, where the genetic variation of the offspring provides a hedge against unpredictable changes. Asexual reproduction requires less time and energy than sexual reproduction and tends to be found in stable environments where genetic variation is not required to ensure reproductive success.
- Fertilization can occur either externally or internally. Most land-dwelling animals have internal fertilization. Most aquatic animals have external fertilization, releasing their gametes directly into the water near potential mates.
- Some animals are hermaphrodites, producing both eggs and sperm. They can mate with others of their species, but not with themselves. Some hermaphrodites can switch from male to female, or vice versa.

## 33.2 Human Reproduction: Gamete Formation and Fertilization

- Sexual reproduction involves the union of male and female gametes (fertilization). Spermatogenesis and oogenesis—the production of male and female gametes, respectively—involves meiosis. Meiosis reduces diploid ($2n$) precursor cells to mature haploid ($n$) gametes (sperm and eggs). Fertilization fuses the haploid sperm and egg to produce a diploid ($2n$) zygote.
- In humans, male and female gametes are called sperm and eggs (ova), respectively. Males produce sperm in testes, and females produce eggs in ovaries.
- In human males, sperm formation begins at puberty, as each diploid precursor divides by meiosis to produce four haploid sperm cells.
- Meiosis I begins in the ovaries of a human female even before she is born. At birth, a baby girl has between 1 million and 2 million primary oocytes arrested at an early stage of meiosis. At puberty, only about 400,000 viable primary oocytes remain, and usually one of these resumes meiosis each month during a woman's reproductive years. Meiosis is completed, producing a large ovum, only after a sperm cell fuses with the ovum.
- During fertilization, one sperm, aided by enzymes in its acrosome, enters the ovum. The sperm nucleus then fuses with the egg nucleus to create a diploid zygote.

## 33.3 Human Reproduction: From Fertilization to Birth

- An egg released from a woman's ovary moves into the oviduct, where it can be fertilized. During sexual intercourse, the man's penis releases into the woman's vagina nearly 300 million sperm, only one of which can fertilize the egg.
- In the first stage of animal development, the single-celled zygote divides rapidly to form a multicellular morula. Cell divisions in the morula produce the blastocyst, which has two tissue types: the inner cell mass, which will develop into the actual embryo; and the trophoblast, which will become part of the placenta.

- Upon implantation in the endometrium, trophoblast cells in the blastocyst contribute to formation of the placenta. Found only in mammals, the placenta is a two-way exchange surface for receiving nutrients and oxygen and expelling wastes via the mother's blood.
- The inner cell mass produces an embryo with three tissue layers: endoderm, mesoderm, and ectoderm. The three layers move relative to one another to produce the complex tissue arrangements of the gastrula. After gastrulation, the three tissue layers differentiate into specialized organs.
- As development progresses, most cells become increasingly specialized. Cell fates are determined in part by inductive signals released by neighboring cells and the external environment.
- During the first trimester of human development in the uterus, embryonic cells differentiate rapidly into the various organs and structures present at birth. From the ninth week of development on, the developing human is called a fetus. During the second and third trimesters the fetus grows rapidly.
- Childbirth occurs in stages. The hormone oxytocin, secreted by both mother and fetus, signals uterine muscles to contract. The contractions become stronger as positive feedback increases the amounts of oxytocin produced. The mother's cervix opens, and the fetus is eventually expelled from the uterus. The placenta soon follows.

## 33.4 Fertility and Contraception

- Pregnancy can be prevented by contraceptives. Some contraceptives, such as condoms, can help prevent the transfer of sexually transmitted diseases, but most do not.
- Oral contraceptives—which contain progestogens, with or without estrogens—are among the most popular and effective methods of contraception.

## 33.5 Development after Birth

- Development for most animals continues after birth. Human babies grow at rates similar to that of a mammalian fetus.
- Unlike most other animals, modern humans live part of their lives past their reproductive years. According to the grandmother hypothesis, menopause is adaptive because postreproductive women further their genetic interests by trading their midlife reproductive potential for fostering the survival of their grandchildren.

## 33.6 How Development Is Controlled

- Development of a zygote into a mature animal requires precise control over cell fates. Gene expression is regulated by which genes are transcribed and when, and by how mRNA, transcripts are processed.
- Genes are switched on or off by inductive signals. Patterns of inductive signal concentrations in tissues inform cells of their position within the developing embryo and influence the expression of homeotic (master) genes.
- Changes in the pattern of apoptosis (programmed cell death) during development can influence structure and function, as can changes in the rate of growth of one structure in relation to another.
- Small changes in how genes influencing development are expressed can have profound effects on form, which in turn can result in evolutionary change.

# Key Terms

acrosome (p. 720)
amnion (p. 724)
amniotic cavity (p. 724)
apoptosis (p. 733)
asexual reproduction (p. 716)
blastocyst (p. 722)
cell differentiation (p. 722)
chorion (p. 724)
contraceptive (p. 727)
diploid (p. 717)
ectoderm (p. 723)
embryo (p. 722)
endoderm (p. 722)

endometrium (p. 722)
fertilization (p. 716)
fetus (p. 724)
gamete (p. 716)
gastrula (p. 723)
gastrulation (p. 723)
germ line cell (p. 717)
grandmother hypothesis (p. 730)
haploid (p. 717)
hermaphrodite (p. 718)
implantation (p. 722)
inductive signal (p. 732)

inner cell mass (p. 722)
mesoderm (p. 723)
morphogen (p. 732)
oogenesis (p. 718)
oviduct (p. 721)
ovum (p. 718)
oxytocin (p. 726)
penis (p. 721)
placenta (p. 722)
polar body (p. 718)
positional information (p. 732)
primary oocyte (p. 718)

primary spermatocyte (p. 718)
secondary oocyte (p. 718)
secondary spermatocyte (p. 718)
sexual reproduction (p. 716)
spermatogenesis (p. 718)
trimester (p. 724)
trophoblast (p. 722)
umbilical cord (p. 724)
uterus (p. 722)
vagina (p. 722)
yolk sac (p. 724)
zygote (p. 716)

# Self-Quiz

1. Which of the following is a gamete?
   a. ovary
   b. testis
   c. sperm
   d. stem cell

2. A zygote is formed
   a. just prior to fertilization.
   b. when more than one sperm fertilizes the same egg.
   c. within the ectoderm of a blastula.
   d. when sperm and egg nuclei fuse.

3. Hermaphrodites
   a. always reproduce asexually.
   b. have both testes and ovaries.
   c. cannot produce zygotes.
   d. result from unfertilized eggs.

4. Human fertilization normally takes place in the
   a. oviduct.
   b. vagina.
   c. ovary.
   d. uterus.

5. Which of the following stages comes first in the development of a human?
   a. gastrula
   b. zygote
   c. fetus
   d. blastocyst

6. During implantation, the blastocyst implants itself in
   a. the endometrium.
   b. the oviduct.
   c. an ovum.
   d. the trophoblast.

7. A human baby's skin develops from
   a. ectoderm.
   b. mesoderm.
   c. endoderm.
   d. trophoderm.

8. As development of an embryo continues,
   a. pluripotent stem cells develop from the endoderm.
   b. gastrulation produces a blastula.
   c. mesoderm gives rise to heart muscle.
   d. endoderm gives rise to ectoderm.

9. Inductive signals play a role in directing head-to-tail differences in
   a. fruit flies only.
   b. humans only.
   c. most animals.
   d. birds only.

10. Apoptosis describes a process of
    a. cell differentiation.
    b. cell proliferation.
    c. cell specialization.
    d. cell death.

# Analysis and Application

1. Describe the gametes produced by animals in terms of their structure and function.

2. Compare sexual and asexual reproduction. What are some advantages and disadvantages of each?

3. Describe the fertilization of a human egg by a human sperm.

4. The development of a human fetus follows a pattern. Describe that pattern from implantation of the blastocyst to birth. Which trimester is most susceptible to disruption? Why?

5. During development, a cell's fate becomes more defined. Explain what this means and how it happens.

6. Terrestrial vertebrates are thought to have evolved from fish. Describe a way in which apoptosis could have led to an adaptation that enabled better mobility on land.

7. If intelligent species inhabited other planets, do you think they would live beyond their reproductive years? What would some of the advantages be?

# Animal Behavior

**DOMESTICATED SIBERIAN SILVER FOXES.** Not just hand-raised descendants of wild foxes, these foxes have been bred since 1959 *only* to respond positively to humans. In the process of genetic domestication, the foxes also evolved multicolored coats, curly tails, broader heads, and other traits typical of domesticated animals. Domesticated foxes, unlike their wild cousins, make great pets.

# The Evolution of Niceness

Dogs are famous for their affectionate and loyal natures. Playful and relaxed around people, "man's best friend" will approach total strangers, lick faces, and welcome impromptu belly rubs. In contrast, wild wolves and coyotes avoid humans and even at their least shy are wary, unfriendly, and often aggressive. It's easy to assume that dogs are friendly just because they've been raised by humans from puppies. But coyotes, wolves, and foxes that have been raised by humans are not like dogs. The submissive and friendly behavior of most domesticated dog breeds is fundamentally different from the wary alertness of their wild counterparts, and the difference is genetically based.

One of the most dramatic demonstrations of the genetic basis of domestic behavior came from an unlikely source—a Soviet research program that had been virtually stripped of genetics. Beginning in the 1930s, the powerful Soviet scientist Trofim Lysenko launched a campaign to wipe out the science of genetics—at least in the USSR. Lysenko rejected the very idea of genes and, with the support of the ruthless Soviet premier Joseph Stalin, eliminated departments of genetics throughout the Soviet Union. An entire generation of Soviet geneticists died in Soviet prisons, were executed, or else fled to the West. One canny survivor was geneticist Dmitry Belyaev. His own brother, also a geneticist, died in a concentration camp. After Dmitry lost his job as head of an animal breeding lab in Moscow in 1948, he accepted a transfer to Siberia and reported that he was switching to research in physiology.

Inside, however, Belyaev remained a geneticist, hanging on until Stalin's death in 1953, when Lysenko's hold on Soviet science began to weaken. In 1959, Belyaev began a multidecade effort to domesticate silver foxes, valued for their beautiful fur. Belyaev was convinced that behavior had a genetic basis, and he planned to select for just one behavior—tameness—and no other traits.

Did Belyaev succeed in taming the foxes? What did his experiment tell us about the domestication of dogs and other animals and about the genetics of behavior? Is behavior controlled by genes? Is behavior learned?

Before we address these questions, we will examine the main types of behaviors displayed by animals, including humans.

> **MAIN MESSAGE** Animal behavior is the way an animal responds to information obtained from its environment; patterns of behavior can be fixed or learned.

## KEY CONCEPTS

- Behavior is a predictable response to external stimuli.

- Behavior can be fixed (inborn) or learned. Most animals have some fixed behaviors, which are usually triggered by simple external stimuli.

- Learned behaviors enable individual animals to respond to their environment flexibly.

- All behaviors have a genetic basis. Differences in the genetic makeup of individuals can be observed as differences in their behaviors or abilities. Natural selection can work on both behaviors and physical traits.

- Communication is an important type of behavior. Communication ranges from simple chemical signals to elaborate displays and songs to the intricate languages of humans.

- Group living and social behavior offer many advantages, including extra protection from predators and greater access to resources. Social behavior involves cooperation among members of a group, usually of the same species.

- Altruistic individuals increase the adaptive fitness of the recipient at an actual or potential cost to themselves. Altruism is usually extended toward close relatives and therefore increases the fitness of the social group as a whole.

- "Eusociality" refers to the cooperative behaviors of close-knit social groups in which there is division of labor and some individuals give up reproduction to further the welfare of the group. Eusociality is uncommon because it takes many gene mutations and complex routes of natural selection to evolve such elaborate interactions.

ORGANISMS REACT IN PREDICTABLE WAYS to one another and to the nonliving environment. Animals acquire information from the environment and respond to it in predictable ways, and such responses to external stimuli constitute **behavior**. The capacity to react to external signals is at least partly inherited, although it may be influenced by nongenetic factors as well. This definition of animal behavior encompasses responses that are familiar—for example, courtship in humans and song sparrows, and scent marking by territorial animals. But as we shall see in this chapter, animals can also respond in ways that are more subtle and sometimes nonintuitive to us.

We begin this chapter with a brief look at human behaviors compared to those of nonhuman animals. We then introduce two different kinds of behaviors: *fixed behaviors*, which are expressed from birth; and *learned behaviors*, which animals acquire as they gain experience with their surroundings. Next we delve into the genetic bases of both fixed and learned behaviors. Then we consider communication, which enables two or more animals to coordinate their behavior. We end with a look at group living and the adaptive value of social behavior.

## 34.1 Sensing and Responding: Behavior in Humans and Other Animals

Behavioral scientists are interested in learning how a specific behavior increases or decreases an animal's *adaptive fitness*—its ability to survive and reproduce in its particular habitat. Another objective of behavioral science is to dissect the genetic basis of behavior and understand how nongenetic influences, such as the

stress experienced by an individual, might affect behavior. Biologists also want to know the functional basis of behavior: how the stimulus is perceived, what information is processed by the nervous system, and how various other organ systems execute a coordinated response. Finally, reconstructing the evolutionary history of specific behaviors, such as tool use, and the role of natural selection in shaping those behaviors, is an important goal in the study of animal behavior.

Behavior enables animals to respond quickly to changes in the environment, such as the presence of a predator or the changing of the seasons. Moreover, it enables animals to coordinate their own activities with those of other individuals of the same or different species. Through appropriate behavioral responses, honed by natural selection, animals can find food, defend territory and other resources, avoid predators, choose mates, and care for their young. Single-celled organisms, such as amoebas, also exhibit behaviors, since they can sense and respond to their external environment in a manner that is controlled mainly by genes. However, behavioral responses are most obvious in complex animals because their nervous systems, and their muscles and bones, assist them in processing sensory information rapidly and in generating behaviors involving quick movements.

Behavior is a crucial part of human life, which is why it is studied not just by biologists, but also by scholars who range from behavioral psychologists to political analysts. Direct analysis of human behavior is challenging, in part because ethical concerns keep us from carrying out certain types of controlled experiments on humans. In addition, humans have a complex, highly developed nervous system that can receive and process diverse external stimuli and launch diverse and complex responses to them. As a general principle, the most complex animal behaviors are found among species that have the most complex nervous systems. Compounding the challenge is the fact that many, perhaps most, human behaviors are influenced by human culture. *Culture* consists of those aspects of society, such as a specific language or tribal traditions, that are passed from one generation to the next without being encoded in the DNA.

Biologists observe and analyze the behaviors of a broad variety of other animals, in both the lab and the natural habitats in which those species live. Analyzing animal behavior in its ecological and evolutionary context helps us understand how life is organized and how it has evolved. What we learn from observing the behavior of nonhuman animals, especially mammals and primates, can enhance our understanding of human society as well. Do our closest relatives display empathy or vengefulness or a sense of self? Could what we learn about aggression in nonhuman animals help us understand and avert psychopathic behavior in humans, resolve everyday conflicts, or avoid wars? As you might imagine, these questions have relevance for a broad cross section of scholarship, which is why animal behavior is of interest to many ecologists, evolutionary biologists, psychologists, anthropologists, policy makers, political scientists, and even market researchers and economists.

## 34.2 Fixed and Learned Behaviors in Animals

Behavioral patterns can be either fixed (innate) features of an animal's behavior or developed through the animal's experiences. For instance, males of most bird species sing a distinctive song to attract females. Males of some species know their courtship songs from birth, but others must learn them. Male brown-headed cowbird chicks that are raised in captivity and never hear an adult male cowbird sing still give a flawless rendition of their species' courtship song when they mature. In contrast, male white-crowned sparrows raised in captivity fail miserably in their attempts to sing a song attractive to females. To learn their courtship song, male white-crowned sparrow chicks must hear other males of their species sing.

### Fixed behaviors are elicited at the first encounter with a stimulus

A **fixed behavior** is an innate (that is, inborn) behavior in which the first encounter with the appropriate stimulus leads predictably to a well-defined response. Prior experience of the stimulus is not a prerequisite for an animal to display a fixed behavior. Such behaviors are strongly rooted in genetics and are commonly described as instinctive or "hardwired." In humans, fixed behaviors are most readily seen in very young babies and include some that are displayed only in the first few weeks or months after birth. Pediatricians commonly test newborns for some of these behaviors, known as neonatal reflexes, to determine whether a baby's nervous system is healthy. Healthy newborns display the **grasp reflex**, for example, which enables them to hold on tightly to anything that will fit in their tiny fists (**FIGURE 34.1**).

Fixed behaviors are common in the young of other animals as well. Kittens reflexively cover their droppings when first introduced to a litter box—even kittens removed from their mothers too young to have ever seen their mothers demonstrate this behavior. Pipevine swallowtail caterpillars can distinguish between food plants that are suitable for them and those that are not.

Fixed behaviors enable animals to behave appropriately when they have no chance to learn from experience and the risks associated with the wrong behavior are great. Consider female cowbirds, which lay their eggs in the nests of other bird species. Because of the nest-parasitizing behavior of their mothers, male cowbird chicks grow up hearing only the courtship songs of their foster parents, and they have no opportunity to learn their own species' song from their fathers. Male cowbirds have to be able to sing the cowbird-specific courtship song to have any chance of attracting female cowbirds as mates, so they must ignore any singing lessons they might receive while being reared in another species' nest and instead depend on their fixed, genetic programming to produce their song. The ability to sing the "cowbird song" is therefore a fixed behavior that is automatically displayed when male cowbirds become sexually mature and receive the appropriate stimulus, such as the lengthening days of spring.

The stimuli that trigger fixed behaviors can be quite simple. Herring gull chicks, for example, beg eagerly

## Helpful to Know

Behavior researchers must be careful to avoid *anthropomorphism*— that is, ascribing humanlike attitudes or thought processes to nonhuman animals when there is no evidence to support such ascriptions. Modern behavior researchers use standardized protocols to minimize subjectivity and individual or group bias. For example, *ethologists*, who study animal behavior in the field, avoid interacting with their study subjects and try to keep an emotional distance from them.

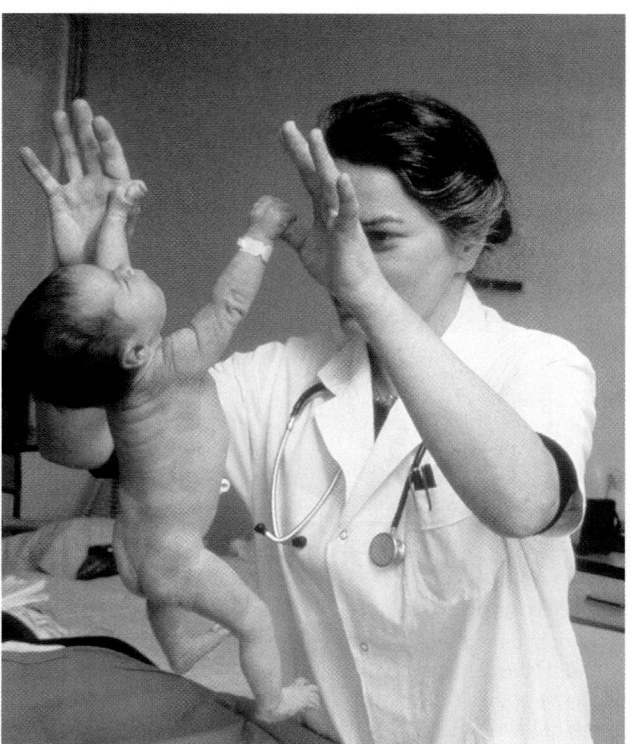

**FIGURE 34.1 The Grasp Reflex in a Human Newborn**
Healthy newborns will cling tightly to any object placed in the palm, an innate behavior that disappears 5–6 months after birth.

Brown-headed cowbird

for food whenever their parents return to the nest. In a famous series of experiments, Nobel prize–winning behavioral scientist Niko Tinbergen showed that the chicks "aim" their begging behavior at a conspicuous red dot on their parents' bills. Oddly colored or shaped models of herring gull heads trigger begging behavior just as effectively as lifelike models do, as long as they feature the red dot (**FIGURE 34.2**). A simple stimulus that causes a fixed behavior, such as the red dot in herring gulls, is called a **releaser**. Similarly, the sight of the red patches on the wings of a male red-winged blackbird stimulates neighboring male red-winged blackbirds to defend their nesting territories vigorously. Research has shown that it is the color of the wing patch (**FIGURE 34.3**), not the overall appearance of the bird, that stimulates this behavior.

The stimuli that trigger fixed behaviors often do so only under certain circumstances. An individual's physical condition may affect whether a stimulus triggers a fixed behavior. For example, the aggressive response of male red-winged blackbirds to the color red occurs only during the breeding season, when levels of the male sex hormone testosterone are at their highest.

## Learned behaviors add flexibility to animal responses

In **learned behaviors**, the response to a stimulus depends on an animal's past experience. For instance, we can train a dog to sit in response to a whistle or to the

The red patches on the wings of male red-winged blackbirds play a key role in territorial behavior.

**FIGURE 34.3 Seeing Red**
The color red is a releaser for male red-winged blackbirds.

command "Sit." A dog trained to the oral command sits in response to "Sit"; a dog trained to the whistle also sits, but in response to a different stimulus. The two dogs learn to respond appropriately to their different learning experiences.

Much of our own behavior is learned rather than fixed. We learn appropriate behaviors for many different social situations. We learn rules of games and how to play musical instruments. We refine learning by practice—that is, more learning. Students learn to answer questions correctly on biology exams in response to information gained from lectures and textbooks. Many other animals also modify their behavior in light of their previous experiences.

The feeding behavior of rats illustrates some advantages of learned behavior. Rats in laboratory cages can learn to press a lever that releases food pellets. When first placed in the cage, a rat will sometimes accidentally bump into the lever and discover the food pellet. The rat soon learns to press the lever deliberately to get more pellets. This capacity to learn from experience may be related to one aspect of the biology and behavior of rats that has made them so successful: they thrive on the ever-changing variety of garbage that humans discard. However, that garbage is a mix of edible and inedible substances. Because of the diversity of their diet, rats cannot have a set of fixed rules about what to eat and what to avoid. Instead,

**A FIXED BEHAVIOR**

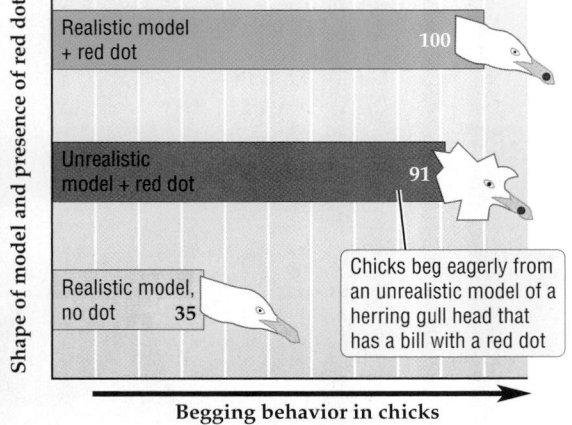

Shape of model and presence of red dot

Realistic model + red dot — 100

Unrealistic model + red dot — 91

Realistic model, no dot — 35

Chicks beg eagerly from an unrealistic model of a herring gull head that has a bill with a red dot

Begging behavior in chicks (percent begging)

**FIGURE 34.2 Simple Stimuli Trigger Fixed Behaviors**
Herring gull chicks will beg almost as eagerly from an unrealistic model of an adult gull head with the critical red dot (the releaser for begging behavior) as they will from a realistic one. The red dot on the adult's bill (photo), not the appearance of the whole head, is what triggers begging behavior.

they combine a specific behavior with the capacity to learn. This learn-as-you-go approach to eating is especially important in rodents because they lack the vomiting reflex. Their anatomy and brain stem circuitry is such that they cannot forcefully expel the contents of the stomach if nerve receptors in the gut detect that a food item is poisonous.

When rats encounter a food for the first time, they sample only an amount small enough to avoid serious consequences should the material prove inedible. If the food sample sickens them, they avoid that food in the future. Rats will sniff the mouths of littermates returning to the nest and stay away from any food that has an odor they smelled on the breath of a sick rat. If a food causes no harm, they add it to their learned list of good things to eat. By learning what to avoid from their experiences, rodents can cope with new and unexpected kinds of food—something they could not do with inflexible fixed behaviors.

## Through imprinting, offspring learn who their parents are

An interesting type of learning, called **imprinting**, occurs in animal species in which parental involvement in rearing the young is important. You may have seen a mother goose with her goslings trailing along behind her. Within a short time after hatching, the goslings develop a "vision" of who their mother is and what a goose is supposed to look like. This sense of "mother goose" is critical to the survival of young animals that must spend time with parents in order to learn how to survive on their own. Imprinting usually takes place only during a specific period of time in the early life of offspring. In ducklings, for example, imprinting occurs during (but not before or after) the period about 7–23 hours after hatching, with a peak sensitivity to imprinting at about 15 hours (**FIGURE 34.4**).

Generally this form of learning works well, because after the young hatch the mother is close by and is the animal the young are most likely to encounter. In a series of experiments conducted in the mid-twentieth century, however, behavioral scientist Konrad Lorenz found that when goslings saw him and heard "goose" sounds only from him—and not from their actual mother goose—in their early hours after hatching, they imprinted on him. These goslings treated him as "mother goose" and followed him, not their mother.

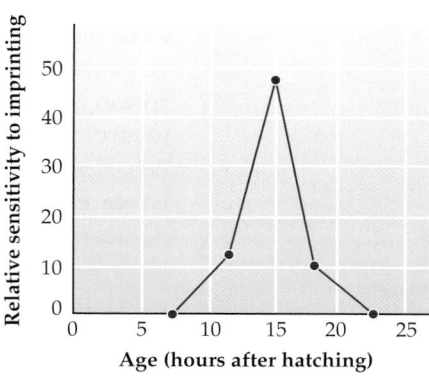

**FIGURE 34.4 Some Animals Learn Who Their Parents Are by Imprinting**

Imprinting usually takes place only during a brief "window" of sensitivity. A study of ducklings found that their peak sensitivity to imprinting occurred at about 15 hours after hatching. Young geese that imprinted on the famous behaviorist Konrad Lorenz (photo) considered him their "mother" and followed him everywhere.

### Concept Check

1. Why are human behaviors particularly difficult to analyze?
2. Compare fixed (innate) behaviors and learned behaviors.

## 34.3 The Genetic Basis of Animal Behavior

Although an environmental stimulus is necessary to elicit a behavior, the capacity to display a particular behavior is at least partly governed by genetics. The role of genes in controlling a behavioral display is most prominent in fixed behaviors and is weaker in learned behaviors such as imprinting, where environmental experiences exert a strong influence. Because genes are involved to a lesser or greater degree, natural selection acts on behaviors just as it does on physical or biochemical traits—by selecting behaviors that contribute to survival and reproduction.

## Genes control fixed behaviors

We can easily appreciate the genetic control of fixed behaviors. One of the clearest examples of the direct genetic control of a fixed behavior is demonstrated by the nest-cleaning behavior of honeybees. Hives of honeybees sometimes fall victim to a contagious bacterial disease that kills larvae (immature bees) in the hive. The hives of some genotypes of honeybees, however, rarely suffer from the disease. Hives of these "resistant" genotypes do become infected, but the bees reduce the spread of the bacteria by quickly removing infected larvae from the hive. This nest cleaning involves two fixed behaviors: cutting open cells of the hive that contain larvae killed by the bacteria, and removing the dead bodies from the hive. Genetic crosses have revealed that the two behaviors are under the control of different genes. In some "susceptible" genotypes, the honeybees neither cut open cells nor dispose of dead larvae. In other susceptible genotypes, the bees either fail to open cells containing infected larvae or fail to remove dead larvae from opened cells (**FIGURE 34.5**).

Genes influence behavior, but more typically, interactions between genes and the environment ultimately determine the behavioral patterns shown by an individual. Careful studies of identical and fraternal twins in humans, for example, have revealed that schizophrenia—a mental illness in which a person hallucinates, feels persecuted, and behaves strangely—has a strong genetic component. However, unknown environmental factors control whether a genetically susceptible person will actually display the condition.

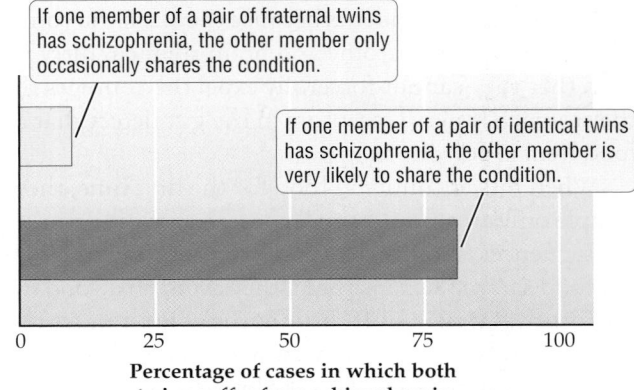

If one member of a pair of fraternal twins has schizophrenia, the other member only occasionally shares the condition.

If one member of a pair of identical twins has schizophrenia, the other member is very likely to share the condition.

Percentage of cases in which both twins suffer from schizophrenia

**FIGURE 34.6 Behavioral Studies of Twins Can Reveal a Genetic Basis of Behavior**

Schizophrenia is a genetically based mental disorder in humans. However, the environment may affect the expression of the genes causing schizophrenia, because even among genetically identical twins, in many cases only one twin develops the condition.

Identical twins arise from a single fertilized egg. The resulting embryo splits into two genetically identical individuals early in embryonic development. Fraternal twins result when the ovaries release two eggs simultaneously, and these eggs are then fertilized by two different sperm. As a result, fraternal twins are no more similar genetically than are typical brothers and sisters. If genes controlled schizophrenia, we would predict that either both members of a pair of identical twins should have schizophrenia or both should be healthy. In addition, we would expect it to be more rare for both members of a pair of fraternal twins to have schizophrenia, since they differ genetically.

Researchers have found that it is indeed much more common for both members of a pair of identical twins to have schizophrenia than for both members of a pair of fraternal twins to have the disorder. However, if one member of a pair of identical twins has schizophrenia, the other may still be healthy, even though he or she carries all the same genes. This observation indicates that environmental influences play a role in determining whether a person with the genes responsible for schizophrenia actually develops the disease (**FIGURE 34.6**).

## Genes also influence learned behaviors

Learned behaviors, as we have seen, are strongly influenced by the environment: what animals learn depends on what they experience and pay attention

A *UuRr* × *ur* cross yields the following phenotypes:

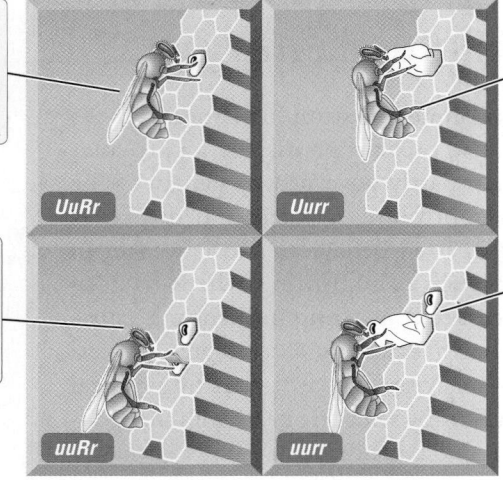

These bees are susceptible because they carry both the *U* and the *R* alleles. They will neither uncap cells nor remove dead larvae.

*UuRr*

These bees are susceptible because they carry the *U* allele, but they will remove dead larvae if cells are uncapped.

*Uurr*

These bees are susceptible because they carry the *R* allele. They will uncap cells of dead larvae, but won't remove them.

*uuRr*

These bees are resistant because they carry neither the *U* nor the *R* allele.

*uurr*

**FIGURE 34.5 Nest-Cleaning Behavior Has a Genetic Basis**

A cross between a *UuRr* female bee and a *ur* male bee reveal that nest-cleaning behavior in honeybees is based in genetics. Male honeybees are haploid and hence have just one allele for each gene instead of two alleles, as diploid females have. Female bees that carry either the *U* allele or the *R* allele are susceptible to a bacterial disease because they do not perform behaviors that keep their hive disease-free. Female bees with the *U* allele do not uncap cells containing larvae killed by the disease; female bees with the *R* allele do not remove the dead larvae from uncapped cells.

to during their lifetime. Nonetheless, differences in what, how much, and when animals can learn reflect genetic differences in their learning patterns and capacities. Most laboratory rats, for example, can learn to negotiate mazes, but some rats learn faster than others. By mating fast learners with other fast learners, researchers can produce offspring that also learn to negotiate mazes quickly. Mating slow learners with other slow learners produces rats that have a difficult time learning to find their way through a maze.

Some species of birds, such as Clark's nutcrackers, excel at remembering where they have stored food. These birds have brains that differ structurally, and therefore presumably genetically, from the brains of closely related species with only an average ability to find stored food. Researchers also have identified genes that seem to control the ability of birds to learn the songs that are characteristic of their species.

## 34.4 Facilitating Behavioral Interactions through Communication

**Communication** is a type of behavior that enables one individual to exchange information with another, thereby making it possible for an animal to coordinate its activities with those of other individuals. By communicating with one another, animals in groups can do things that no individual could do on its own, such as fend off large predators, capture large prey, or deliver emergency assistance (**FIGURE 34.7**).

## One individual produces signals that stimulate responses in others

The communication behavior of animals varies widely in its complexity and includes just about any type of signal that other animals can sense: sounds, visual signals, odors, electrical pulses, touch, and tastes.

At the simple end of the spectrum, the release of a chemical signal, called a **pheromone**, by one individual informs others of the same species about its identity, its location, its physical condition, or a situation in the environment. For example, female silkworm moths release bombykol, a sex pheromone, to attract males. The pheromone not only reveals to a male that the individual releasing the chemical is a female of his species and that she is interested in finding a mate; it also tells him where she is. The male can locate the female by moving toward an area with a higher level of the pheromone, since the concentration of the chemical increases as he gets closer to the female. Evidence suggests that we humans also produce airborne signals that may influence sexual behavior.

Pheromones communicate other conditions besides sexual readiness. For example, alarm pheromones released by ants, honeybees, and other social insects when a colony or hive is disturbed inform other individuals in the community about the situation. Those individuals interpret the disturbance as an attack and respond accordingly.

An example of a more complex signal is the intricate dance used by honeybees to communicate the distance and location of food to other members of

(a)

(b)

(c)

**FIGURE 34.7**
**Communication Is Vital to Animals**
Howling is important for pack cohesion among wolves and may warn outsiders to stay clear (a). The "jump-yip" display of black-tailed prairie dogs maintains group-cohesion in these social ground squirrels (b). Verbal and nonverbal communication is vital to these traders in Brazil's Bovespa exchange (c).

Sun

Flowers

The "waggle" conveys direction and distance to the flowers.

45°

Up

Hive

**1** Honeybee workers dance, on the vertically oriented hive, in a figure eight pattern...

45°

The angle of the "waggle" relative to straight up the honeycomb is the angle of the flowers relative to the direction of the sun.

**2** ...centered on a "waggle" portion in which the worker vibrates her body from side to side.

The greater the number of "waggles," the greater the distance to the food.

**3** Other workers watch closely to learn where to find food.

Pattern of waggle dance

**FIGURE 34.8 Honeybees Dance to Communicate**
Bees use a dance language to communicate complex information about the location of distant food sources to other worker bees in their hive.

their hive (**FIGURE 34.8**). And at the extreme end of the complexity spectrum of communication lies human **language**. Most human languages consist of thousands of words representing everything from objects to actions to abstract ideas.

**FIGURE 34.9 Showing Off to Attract Mates Is a Form of Communication**

(a) A singing male frog advertises to listening females its species, its location, and its quality as a mate.
(b) Male bighorn sheep communicate their genetic "quality" to potential mates by engaging in head-butting contests. Their backward-curved horns enable them to participate in these contests with a relatively low risk of injury.

## Identification is a key function of animal communication

Animals most often communicate to identify themselves to other animals, to avoid conflict, or to coordinate their activities. Signaling an individual's identity is probably the most common function of communication.

An animal uses a variety of signs and signals to inform other members of its species that it is a member of the same species, as well as to indicate its sex, its physical condition, and its location. When a dog sprays a fire hydrant with its urine, the scent communicates the dog's sex, breeding condition, health, and status to other dogs. The mating songs and behaviors of male birds, frogs, and crickets tell females their species, location, and potential quality as mates (**FIGURE 34.9a**).

A second important function of communication in animals is to avoid potentially harmful conflicts. Physical conflict over food or mates can lead to serious injury for both the winner and the loser. To reduce the risk of injury in such encounters, many species communicate their fighting ability through ritual displays. When male red-winged blackbirds flash red wing patches at one another, for example, their brightness and size signal a male's quality as a fighter to other males. This ritual allows the less able fighter to back down without engaging in a potentially dangerous fight. The males of many mammalian species mark their territory by depositing their scent on objects throughout their "beat." The scent marking serves as a "keep off my territory" sign to potential competitors.

(a)

(b)

# Drinking and the Dark Sides of Human Behavior

Human behavior changes dramatically under the influence of alcohol, as does our ability to regulate that behavior. Among college students, the consequences of abusing alcohol go far beyond the fatalities and injuries caused by drinking and driving. Each year in the United States, approximately 600,000 physical assaults and 70,000 cases of sexual assault involving college students are alcohol related. In addition, 400,000 students have unprotected sex, and 100,000 students report that they were too intoxicated to know whether they had consented to sexual intercourse. About one in 100 students attempt suicide because of alcohol or other drugs. On the academic side, students who drink to excess miss more classes, get lower grades, and are more likely to fall behind in their work.

Furthermore, consequences of drinking spill over into the lives of those who are not even doing the drinking. Among students who live on campus or in sororities or fraternities and do not drink, or drink moderately, 60 percent have had

sleep interrupted because of drinking behavior by others, almost half have had to take care of a drunken fellow student, and 15 percent have had property damaged as a result of drunken behavior. One in three has been insulted or humiliated by or has had a serious quarrel with a drunken student, and almost one in 10 has been pushed or hit in an alcohol-related incident. One study found that alcohol is one of the most significant contributing factors to sexual aggression by male college students. One in 100 college-age women have been victims of alcohol-related sexual assault or acquaintance rape. The damage to the psychological well-being and academic performance of these victims is difficult to measure.

Alcohol, when consumed in excess, very often causes drastic and dangerous changes in human behavior. An awareness of the costs and personal consequences associated with those changes can build a better appreciation for the advice we often see in the media: "If you drink, drink responsibly."

**DUI Arrest**

An officer from the Miami Beach police department arrests a motorist who failed a Breathalyzer test. Two readings with the device measured her blood alcohol concentration (BAC) at 0.19 and 0.183, about twice the legal limit in Florida.

---

Sheep, antelope, and their relatives bear horns that vary greatly in shape from species to species—ranging from stout, straight weapons capable of doing great damage to impressive but ineffective weapons. The species with the most lethal horns usually engage in displays that communicate their suitability as mates rather than in actual fights, whereas species with less lethal horns more often engage in physical combat (**FIGURE 34.9b**).

Animals that live in groups communicate with one another to coordinate their behaviors to accomplish a shared goal. Humans do this when they work as a team. Wolf packs and lion prides communicate when they hunt for food. And, as we saw in Figure 34.8, the dances performed by honeybees enable the members of a hive to coordinate their foraging activities.

Animals can even communicate across species lines, as some birds do when they mob a predator. The loud mobbing calls of a chickadee will attract other chickadees, but they may also draw other tiny birds, such as warblers, all of whom will join in harassing and dive-bombing a hawk that has come too close to

their nesting sites. Mobbing behavior deters or distracts predators, but it may also be a teaching tool: mobbing by adults may help young nestlings learn to identify the species they need to be wary of.

## Language may be a uniquely human trait

Humans rely heavily on spoken, written, and grammatically based languages for communication. Much of our human identity depends on our ability to express complex and abstract ideas through language. One of our fixed behavioral characteristics is that we are problem solvers. Language enables us to exchange ideas and solve problems, even in very stressful or dangerous situations.

There is no strong evidence of language among other animals. Many birds and mammals can produce a variety of sounds, each conveying a particular message, but do not assemble the sounds into ideas. Research on the ability of chimpanzees to communicate indicates that they can string together symbols provided by humans to express abstract ideas, but there is no indication that chimpanzees use such sophisticated communication in nature.

Recent research on humans suggests that language goes beyond communicating simple information such as "I'm hungry" or "I'm angry." Language can affect our ability to form abstract concepts, such as beauty in art and the meaning of numbers. For example, one study indicated that the concept of counting beyond 2 may require words for those numbers. The language of the Pirahã, a group of indigenous people living in the Amazon basin of Brazil, has only a few words to represent numbers of things. In particular, the language can express only quantities of "one," "two," or "many." Experiments involving the Pirahã suggest that their limited counting vocabulary affects their ability to work with abstractions based on counting. For example, some Pirahã speakers were asked to line up a number of objects that matched the number of objects set out by the researchers. When that number exceeded two (the limit of the ability of the language to communicate a number precisely), they started to have problems with the task. In the world of the Pirahã, quantities of three or more objects are all included in the category "many." These results suggest that the lack of appropriate words for numbers higher than 2 limits the ability to form a concept of numbers greater than 2.

## 34.5 Social Behavior in Animals

Many animals besides humans live in closely interacting groups. **Social behavior**—behavioral interactions among members of a group, usually of the same species—offers its members advantages not available to solitary animals, but its success requires communication.

## Group living is an effective survival strategy

The many benefits of group living can more than compensate for the increased competition for resources that results from living in close proximity to other animals of the same species. Groups can find some foods better than individuals can. Groups can stir up and capture prey more effectively than individuals can, and unsuccessful hunters or hunters-to-be can watch and learn while others in their group demonstrate successful hunting behaviors. By working together, members of a group can acquire foods that are not available to individuals. The hunting of large mammals by groups of cooperating wolves or humans illustrates such collaborative behavior well (**FIGURE 34.10a**).

Living in groups also gives animals two effective defenses against predators that are not available to single individuals. First, the larger the group, the more eyes, ears, and noses it contains, making early detection of approaching predators more likely. Second, the individuals on the edge of a group are more likely to be attacked than those on the inside, offering members of the group in the middle additional warning and the advantage of being in less vulnerable positions. Gazelles in the middle of a herd, for example, benefit because a cheetah is most likely to kill an individual at the edge of the herd. Moreover, all gazelles in the herd benefit because at least one is likely to notice and warn of approaching cheetahs before the cheetahs are close enough to attack.

In some cases, living in a group enables animals to use scarce resources in a more efficient manner. Florida scrub

**FIGURE 34.10**
**The Benefits of Group Life**

(a) As a group, wolves can hunt larger prey than they can as individuals. (b) Family groups of scrub jays work together to feed young, defend nests, and hold on to valuable breeding territories.

(a)

(b)

jays live in shrubby oak thickets that are dry for much of the year and that have only a limited supply of both food and nest sites. They form lifelong monogamous pairs, and DNA tests show that there is little "cheating" among mated scrub jays (**FIGURE 34.10***b*). After fledging, both the male and female offspring of a breeding pair may remain with the parents as nonbreeding members of a cooperative group. The nonbreeding members help care for the newly hatched young and join in defending the breeding territory against intruders. When they cannot find territories of their own, it is in the genetic interest of the nonbreeding birds to promote the survival of their closest relatives. As with gazelles that live in a herd, belonging to a group makes the nonbreeding birds less vulnerable to predators than a solitary bird would be. Further, belonging to a cooperative group puts the helpers on a "waiting list" to inherit the territory when the breeding pair becomes too old to reproduce or if one or both birds are lost to predators. The fledglings produced by a former helper are themselves likely to display the helping behavior, boosting the reproductive success of the bird that inherits the parental territory.

## Individuals in groups may act to benefit other members more than themselves

Animals that live in groups often display **altruism**; that is, they do things that help other members of their group survive or reproduce while decreasing their own chances of doing so. The scrub jays mentioned earlier provide a good example of altruism: nonbreeding members of a group help the breeding pair raise their young while possibly forgoing reproduction themselves. At first glance, altruism seems to contradict Darwin's idea that only traits that improve the *individual's* reproductive success can spread through a population. However, a deeper examination reveals that altruistic individuals further their genetic interests by improving the reproductive success of individuals that are genetically related to them.

In general, social groups consist of closely related individuals. A pride of lions, for example, centers around a group of closely related females. Similarly, among scrub jays the helpers are most often older offspring of the breeding pair, so the young birds they are helping to raise are their younger siblings. In these cases, the individuals benefiting from altruistic behavior carry many of the same genes that the altruistic individuals carry, so altruistic behavior often helps to spread many of the same genes that the altruist carries.

## Group living is highly developed in social insects

Social insects such as ants, bees, and termites include some of the most successful species on Earth. Although ants, bees, and termites evolved social behavior independently of one another, these groups share some remarkable traits. Each of these groups lives in large colonies containing individuals belonging to distinct classes that serve distinct functions within the colony. The workers collaborate to build complex nests (**FIGURE 34.11**), forage widely for food, and defend the colony against predators in ways that individuals acting by themselves never could. The queen spends her life producing massive numbers of eggs. The workers in a hive are the female offspring of the queen and are therefore closely related to her and to one another. The workers are sterile; they do not reproduce, but they contribute to reproduction by carefully tending to the queen and her larvae. Their altruistic behaviors, including defending to the death the queen and the colony or hive, do not reduce the reproductive capacity of the colony but rather increase its chances of survival.

The term **eusociality** [yoo-soh-shee-AL-uh-tee] (*eu*, "real") is used to describe the cooperative behaviors that organize life in complex social groups. A hallmark of eusociality is the division of labor among different classes of individuals in the group, including altruism by nonreproductive members. Eusociality is best known in hymenopteran insects (bees, wasps, and ants) and in termites. But it is also reported in a few other invertebrates, such as sponge-dwelling pistol shrimp, whose colonies are defended by sterile males. And depending on how broadly one defines the term, eusociality is seen in at least a few vertebrates: naked mole rats and Damaraland mole rats both live in extensive underground colonies inhabited by a large fertile female whose needs

**FIGURE 34.11**
**Cooperation Creates Termite Mounds**
This giant termite mound was built through the cooperative effort of millions of closely related, sterile termite workers. Such workers make up most of the population of each colony. Some individual termite mounds have existed for hundreds of years.

are catered by many smaller, nonreproductive workers. A common nest, or shared living space, is the one feature that unites the known examples of eusocial animals, both invertebrate and vertebrate.

Resting as it does on cooperative survival and reproduction, eusociality is clearly a highly effective strategy. However, eusociality is relatively rare in the animal world. The behavior is rare most likely because it takes many random mutations in the many genes controlling reproductive development and social interactions, followed by natural selection through complex routes, to evolve something as elaborate as a functional eusocial system. In other words, many low-probability events must take place in a specific sequence for eusociality to evolve in a particular group; and such a confluence is bound to be uncommon, just as it is uncommon for one person to draw the winning number in two different state lotteries.

### APPLYING WHAT WE LEARNED

# The Genetics of Domestication

In 1959, a Russian geneticist committed to the idea that genes influence behavior began an experiment to breed a strain of friendly foxes from 130 farm-raised silver foxes (*Vulpes vulpes*). In simple terms, the selection for friendly foxes was ruthless. Each litter of fox puppies was tested several times for friendliness. The 10 percent of fox puppies that didn't growl, bare their teeth, or show too much fear were selected for breeding; for the rest, it was off to the fur coat factory. So that the animals didn't become tame—that is, unafraid of humans because of associating with humans—all the foxes had minimal contact with humans.

Over the years, Dmitry Belyaev and his colleagues selected the friendliest foxes from 45,000 foxes, breeding those that showed the least aggression and fear. In just 10 generations, 18 percent of foxes were extremely friendly, whining for attention and licking experimenters just like dogs. By 1985, all the foxes were as tame as dogs. Although less than 100 years removed from wild foxes, today Belyaev's domesticated Siberian foxes are as friendly as dogs; they play together, whine for human attention, and lick faces. And their behavior is completely unlike that of wild foxes, who—even when hand-raised by humans—are skittish, liable to bite, and not eager to please.

In less than 40 years, Soviet researchers succeeded in fully domesticating a wild animal—a process that has generally been assumed to have taken thousands of years in dogs, horses, and other domestic animals.

To check whether the change in behavior really was genetic, Belyaev's successor to the breeding program (Lyudmila N. Trut) and her colleagues transplanted embryos from genetically domesticated foxes into genetically wild fox mothers. They also performed the reverse experiment—transplanting wild fox embryos into genetically domesticated mothers. In both cases, the embryos developed into foxes like their original mothers, not their surrogate mothers, showing that it was genes that caused the difference in behavior. The researchers calculated that about 35 percent of a typical domestic fox's friendliness is genetic, meaning that the other 65 percent is environmental.

Although the researchers selected only for friendliness, the domesticated foxes differ from wild foxes in other ways as well. Many have white stars on their heads, white feet and chests, spots, floppy ears, broad heads that give them more puppyish faces, and curly tails. What's remarkable about this suite of traits is how many other domesticated animals share similar color changes. Dogs, cats, horses, goats, and pigs all share this suite of traits, including spotted or piebald coats, white stars on their heads, floppy ears, and broader heads than their wild cousins. It is as if all domesticated animals experienced a similar selection program.

Belyaev expected this result. Charles Darwin had noticed that all domestic animals show these traits to some degree and wild animals almost never do. Belyaev and Trut believed that selecting for genetically based tameness was the same as selecting for a major difference in animal development.

When they looked more closely, the researchers found that domesticated foxes show a lower activity of the adrenal glands, which produce the "stress" hormone adrenaline. As puppies, domesticated foxes develop a fear response later than regular foxes. Their brains also produce more serotonin, a neurotransmitter connected to happiness and appetite. Many of the traits that domesticated animals display are likely related to delays in development, a phenomenon called *pedomorphosis* (keeping juvenile traits into adulthood). Some researchers hypothesize that humans are themselves "domesticated" in similar ways, with a suite of traits that enable large groups of strangers to assemble without aggression.

# Take Play Seriously

BY ROBIN MARANTZ HENIG, *New York Times*

On a drizzly Tuesday night in late January, 2008, 200 people came out to hear a psychiatrist talk rhapsodically about play—not just the intense, joyous play of children, but play for all people, at all ages, at all times . . . Stuart Brown, president of the National Institute for Play, . . . called play part of the "developmental sequencing of becoming a human primate. If you look at what produces learning and memory and well-being, play is as fundamental as any other aspect of life, including sleep and dreams" . . .

This is part of a larger conversation Americans are having about play. Parents bobble between a nostalgia-infused yearning for their children to play and fear that time spent playing is time lost to more practical pursuits . . . Armed with research grounded in evolutionary biology and experimental neuroscience, some scientists . . . have spent the past few decades learning how and why play evolved in animals, generating insights that can inform our understanding of its evolution in humans too. They are studying, from an evolutionary perspective, to what extent play is a luxury that can be dispensed with when there are too many other competing claims on the growing brain, and to what extent it is central to how that brain grows in the first place . . .

But the growing science of play does have much to add to the conversation.

Armed with research grounded in evolutionary biology and experimental neuroscience, some scientists have shown themselves eager—at times perhaps a little too eager—to promote a scientific argument for play. They have spent the past few decades learning how and why play evolved in animals, generating insights that can inform our understanding of its evolution in humans too. They are studying, from an evolutionary perspective, to what extent play is a luxury that can be dispensed with when there are too many other competing claims on the growing brain, and to what extent it is central to how that brain grows in the first place.

"Play behaviors" are activities that lack an immediate purpose such as finding food, shelter, or mates. In most wild species, only juvenile animals play, although sometimes adults join in. Young mammals and birds play most when they are comfortable and well fed, least when times are hard. Young rats injected with the stress hormone cortisone lose interest in play.

Play can include frolicking, playing with objects, splashing water, sliding down mud banks, or wrestling with other juveniles or friendly adults. In humans, children's play may also include pretending and role-playing, as well as activities that engage higher-order mental functions such as empathy (bandaging a teddy bear's "boo-boo," for example).

Play is risky, however, and comes with a fat energy price. Biologists estimate that play may use 2–15 percent of all the calories a young mammal consumes. Play also exposes young animals to falls and other accidents. Seal pups absorbed in play are more likely to be killed by predators than are pups doing other things. Despite the downsides, most juvenile mammals play, suggesting that the benefits of play outweigh its many costs.

Why do animals play? According to one hypothesis, play may function as rehearsal for adult life, helping young animals develop the physical coordination needed to escape predators or catch prey. Play also seems to foster cooperation and social problem solving, skills that are vital to social animals. And play contributes to growth and development of the nervous system. Researchers suggest that the experience of diverse kinds of play may mold a more flexible and well-trained brain—one that is better able to meet the challenges of life.

## Evaluating the News

**1.** Cougar kittens raised in captivity enjoy pouncing on Ping-Pong balls (showing a pronounced preference for balls that roll easily). Do you think this is a fixed behavior or a learned behavior? Explain why.

**2.** Some commentators have expressed concerns that children who spend a lot of time playing computer or video games do not receive the benefits of traditional play (making mud pies or building a fort, for example). From your own experience, do you think playing video games is an acceptable substitute for at least some kinds of more physical play? If so, which kinds? What might be some advantages and disadvantages of playing video games?

**3.** Play often has a "dark side": the teasing, hurt feelings, and bullying that can accompany it. Some experts consider such negative behaviors part of how children learn about social dynamics and develop appropriate coping skills. If you were in the presence of one child verbally abusing another, how do you think you might react and why?

**SOURCE:** *New York Times*, February 17, 2008, http://www.nytimes.com.

# CHAPTER REVIEW

## Summary

### 34.1 Sensing and Responding: Behavior in Humans and Other Animals

- A behavior is a coordinated response made by one animal in reaction to another organism or to the physical environment. The capacity to display a specific behavior is at least partly inherited.
- Human behaviors are difficult to study because our highly developed nervous system generates complex behaviors, those behaviors are influenced in complicated ways by human culture, and for ethical reasons scientists cannot conduct certain types of experiments on humans.

### 34.2 Fixed and Learned Behaviors in Animals

- Behaviors can be fixed (inborn or "hardwired," requiring no learning) or learned (requiring response to changes in the environment based on a memory of previous experiences).
- Fixed behaviors have a simple pattern: a single (usually simple) stimulus (called a releaser) leads to a single response. Fixed behaviors enable animals to behave appropriately when they have no chance to learn by experience or when the risks associated with the wrong behavior are great. Fixed behaviors may be stimulated only under certain conditions.
- Learned behaviors are responses based on the past experience of the individual animal. The same response can be brought about by different stimuli, depending on previous experiences. Much of human behavior is learned.
- Imprinting is a type of behavior in which offspring learn to identify their parents and other members of their species according to who is nearby during a certain period in their development.

### 34.3 The Genetic Basis of Animal Behavior

- Both fixed and learned behaviors have a genetic basis, so all behaviors can evolve.
- Behaviors involve interactions between genes and the environment. Fixed behaviors are more strictly governed by inheritance than are learned behaviors, but the capacity to learn may also be under genetic control.

### 34.4 Facilitating Behavioral Interactions through Communication

- Communication enables individuals to coordinate their activities with those of other individuals. Information is exchanged through the production of signals by one animal that stimulate a response in another.
- Animals communicate to identify themselves (as individuals and as members of a species), to avoid conflict, to coordinate their activities in order to perform shared tasks, to indicate their sex and sexual readiness, and to convey information about their physical condition and location.
- Communication can range from simple to highly complex. Pheromones are a simple form of communication that can be used to communicate identity, location, physical condition, or a situation in the environment. Language is the most complex form of animal communication.
- Language may be a uniquely human trait. It enables humans to form abstract concepts such as numbers.

### 34.5 Social Behavior in Animals

- Social behavior—the interactions among members of a group of the same species—provides several different advantages. For prey animals, it provides added protection against predation. Social behavior also enables animals to gain access to resources, including food and breeding territories, more effectively—in some cases enabling animals to obtain foods not available to individuals working alone.
- Individuals living in groups often act in ways that benefit other group members more than themselves. Such altruistic behavior can evolve within groups of closely related animals. Some individuals sacrifice their lives or reproductive success to increase the overall survival and success of the group.
- "Eusociality" refers to the cooperative behaviors of close-knit social groups in which there is division of labor and some individuals give up reproduction to further the welfare of the group. Eusociality is uncommon because it takes many gene mutations and complex routes of natural selection to evolve such elaborate interactions.
- Eusocial insects, including ants, bees, and termites, are among the most successful animals. Sterile female workers forgo reproduction to promote the success of the colony or hive as a whole. They carry out specific tasks, such as foraging for food, protecting the colony or hive, and tending to the needs of the queen, which lays large numbers of eggs.

## Key Terms

altruism (p. 749)
behavior (p. 740)
communication (p. 745)
eusociality (p. 749)
fixed behavior (p. 741)
grasp reflex (p. 741)
imprinting (p. 743)
language (p. 746)
learned behavior (p. 742)
pheromone (p. 745)
releaser (p. 742)
social behavior (p. 748)

# Self-Quiz

1. The grasp reflex
   a. is seen in both children and healthy adult humans.
   b. is an example of a fixed behavior.
   c. occurs in response to a visual stimulus.
   d. is an example of a learned behavior.

2. Behavior
   a. is seen in vertebrates, but not in invertebrates.
   b. always involves communication.
   c. enables animals to evolve.
   d. enables animals to respond quickly to changes in their environment.

3. Fixed behaviors
   a. derive from experience.
   b. are always imprinted.
   c. are genetically inherited.
   d. require language.

4. Learning
   a. occurs only in humans.
   b. overrides all genetic control of behavior.
   c. depends on an animal's past experience.
   d. requires pheromones.

5. Genes
   a. have no influence on behavior.
   b. influence only social behavior.
   c. affect only fixed behaviors.
   d. can affect fixed and learned behaviors.

6. The dance of a honeybee worker
   a. communicates the location of a food source.
   b. communicates the location of a mate.
   c. communicates the sex of the dancer.
   d. has no influence on fixed behaviors.

7. Which of the following animals would rely most on learned behaviors?
   a. termite
   b. frog
   c. rat
   d. spider

8. Which of the following is the most complex form of communication?
   a. mating displays
   b. the Pirahã language
   c. pheromones
   d. birdsong

9. Which of the following is *not* an example of altruistic behavior?
   a. A butterfly lays an egg.
   b. A scrub jay chases a snake away from its parents' nest.
   c. A sterile worker ant defends the nursery of its anthill from invading ants.
   d. A lioness regurgitates food for her sister's cubs.

10. Animals that live in groups
    a. have no fixed behaviors.
    b. usually rely on language for communication.
    c. can reap benefits that compensate for increased competition among group members.
    d. have no genetically controlled behaviors.

# Analysis and Application

1. Give one example of a fixed behavior, and one example of a learned behavior, in humans.

2. Do you think humans are born with an innate fear of spiders? Explain and defend your answer.

3. Some people are born with athletic or musical talent. But even the most talented must learn and practice in order to excel. Pick a sport or a musical instrument, and describe in detail the dimensions of learning that must take place in order to fully develop a talent. Keep in mind that learning includes more than just practice.

4. Language is a major component of the human ability to communicate. But we also communicate without words. What nonverbal forms of human communication have you responded to or used?

5. What kinds of fixed behaviors might be advantageous to animals that live in a group, and why?

# 35 Plant Structure, Nutrition, and Transport

**FOREST IN WINTER.** Long, harsh winters make this boreal forest in Finland a challenging environment for plants.

# On Being Green in a Frozen Landscape

Plants solve tough environmental challenges in ways that might seem strange and even surprising to us animals. Many tundra plants survive temperatures as low as −50°C to −100°C (−58°F to −148°F), and they do so without any fur or blubber or the internal metabolic furnace that endotherms like us depend on. To appreciate the magnitude of the challenge plants face in Arctic or alpine winters, imagine being stranded outside in your pajamas for a few minutes during a winter night in central Canada. If there is no wind and the air temperature is −40°C (−40°F), how long do you think it will take for frostbite to set in? The answer, according to U.S. Army data, is not long: just 11 minutes in most people, and less than an hour in even the toughest individuals.

Plants have several ways to cope with cold. First, as winter approaches, days begin to get shorter and temperatures drop close to 0°C (32°F). Cold-tolerant plants respond by preparing to enter dormancy, a state of depressed metabolism and growth. Braodleaved plants prepare for dormancy with specific physical and chemical changes: they pull valuable proteins and sugars out of their leaves for storage in the root or stem, and then drop their leaves, which are now little more than a skeleton of cellulose, lignin, and some other unrecyclable polymers. All overwintering plants cover their buds with thick, leathery scales that protect the apical meristem inside from drying winds and provide some insulation from the cold.

How do plants cells keep from freezing solid when temperatures plunge below 0°C? What do these cells have in common with the radiator of a winter-ready automobile? How can plants tolerate temperatures below −40°C?

Before we consider these questions, we will introduce the two main groups of flowering plants and see how they differ in form. We will describe how plants are built, how they acquire the nutrients they need, and why some of them eat (small) animals. We will study how tall trees such as redwoods, eucalyptus, and Douglas fir can raise water to enormous heights, including the world record for height set by a 116-meter (379-foot) coast redwood in northern California.

## MAIN MESSAGE

The root and shoot systems of plants are built from three main tissue types, including vascular tissues that transport food, water, and dissolved minerals throughout the plant body.

## KEY CONCEPTS

- The plant body has a belowground root system and an aboveground shoot system. It has three basic organs: roots in the root system, and stems and leaves in the shoot system. Roots absorb water and nutrients, anchor the plant, and store food. Stems support the plant, and leaves are the primary site of photosynthesis. Roots, stems, and leaves can be modified to perform additional specialized tasks.

- Each organ consists of three main types of tissue: Dermal tissues protect the plant and control the flow of materials in and out of the plant. Ground tissues support the plant, heal wounds, and perform specialized function such as photosynthesis. Vascular tissues transport food and water throughout the plant.

- Plants get essential mineral nutrients from the soil solution. Most of the dry weight of a plant comes from carbon dioxide absorbed from the air and converted to carbohydrates by photosynthesis.

- Nitrogen, phosphorus, and potassium are the three mineral nutrients that plants need in largest quantity.

- The transport of food in the phloem requires energy and relies on water pressure. The transport of water and mineral nutrients in the xylem does not depend on metabolic energy and is driven instead by the evaporation of water from the surface of the shoot (transpiration).

EARTH HAS A RICH DIVERSITY OF PLANTS—over 280,000 species—ranging from low-lying mosses to ferns to giant redwood trees to spectacular flowering plants. The main characteristics of the major groups of plants—bryophytes, seedless vascular plants, gymnosperms, and angiosperms—are described in Chapter 3. In terms of sheer numbers, flowering plants dominate the planet (see Figure 20.3 for the time periods when other plant groups dominated life on Earth). Today there are over 250,000 species of flowering plants, compared with 18,000 bryophytes, 13,000 seedless vascular plants, and 720 gymnosperms. Flowering plants literally cover the land; they include the most common plant species found in tundra, tropical forests, temperate forests (many of which are dominated by trees that lose their leaves each year), grasslands, deserts, and other plant communities. Flowering plants are also of overwhelming importance in terms of what we eat: worldwide, people get over 80 percent of their calories from flowering plants such as grasses (wheat, rice,

corn), legumes (peas, beans, peanuts), potatoes, manioc (a starchy root vegetable), and sweet potatoes.

Flowering plants have traditionally been classified into two main groups—*dicots* and *monocots*—on the basis of their external form and internal structure (**FIGURE 35.1**). The **dicots** are the larger of these two informal categories and include about 175,000 species. Magnolias, dandelions, roses, maples, and oak trees are familiar examples of dicots. Leaf veins, which contain vascular tissues, are usually arranged in a netlike pattern in dicots (see Figure 35.1). If you dig up a dicot plant, you are likely to see a main root (called a **taproot**) with many side branches. If you slice through a dicot stem, you will find the vascular tissues arranged in a discrete ring.

The typical dicot flower has four or five of each of the main flower parts. (For a generalized diagram of a flower, see Figure 3.13.) A geranium flower, for example, has five petals, five sepals (the small leafy structures that cover a flower bud), and five stamens (the pollen-producing male parts of a flower). The name "dicot" comes from the presence of two *cotyledons* (*di*, "two"), in dicot seeds. **Cotyledons**, or "seed leaves," are food-storing organs that are part of the tiny, embryonic seedling that lies inside a seed. In some species, such as beans and squash, the cotyledons emerge aboveground as small green flaps attached to the stem. These aboveground cotyledons nourish the seedling for a short time, and then wither and fall off as the plant starts photosynthesizing and becomes nutritionally independent.

**Monocots** include all the grasses (bamboo being the tallest), members of the lily family, palm trees, and banana plants. The leaf veins of monocots are not in a netlike pattern; in most monocots, the veins are arranged roughly parallel to each other—a pattern that

**FIGURE 35.1**
**Comparing Dicots and Monocots**

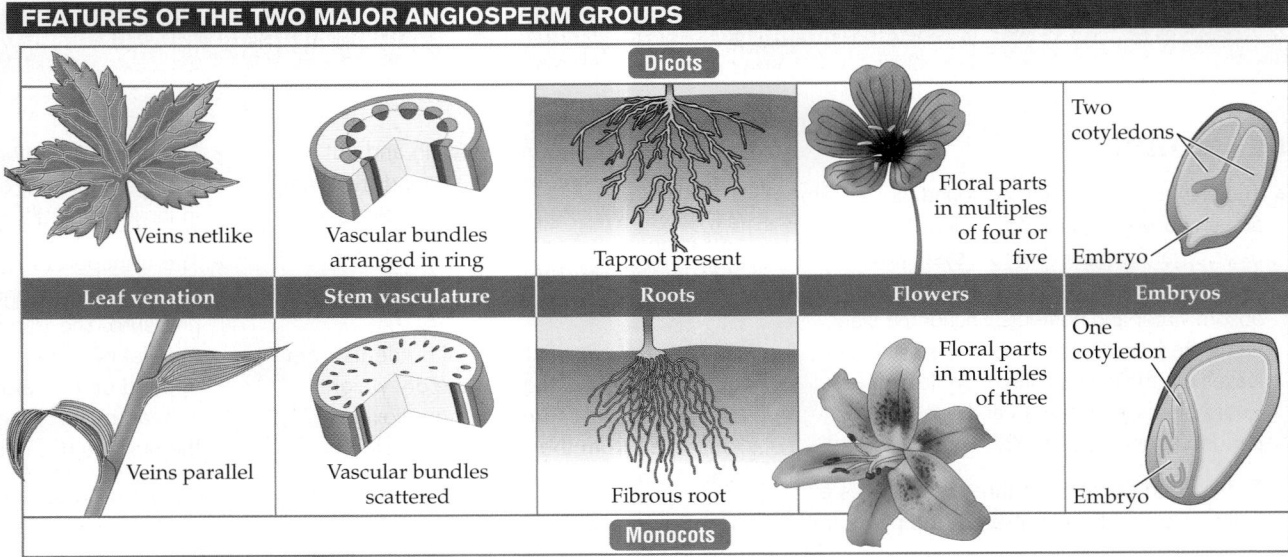

FEATURES OF THE TWO MAJOR ANGIOSPERM GROUPS

Dicots

| Veins netlike | Vascular bundles arranged in ring | Taproot present | Floral parts in multiples of four or five | Two cotyledons Embryo |
|---|---|---|---|---|
| **Leaf venation** | **Stem vasculature** | **Roots** | **Flowers** | **Embryos** |
| Veins parallel | Vascular bundles scattered | Fibrous root | Floral parts in multiples of three | One cotyledon Embryo |

Monocots

is readily apparent in a blade of grass. If you dig up a monocot, you are likely to see a tuftlike mass of roots without a main, dominant root among them. This type of root, known as a **fibrous root**, is characteristic of monocots (see Figure 35.1). The vascular tissues are distributed in a scattered pattern in the stem of most monocots, instead of being arranged in a discrete ring. Usually there are three, or multiples of three, of each of the main parts in a monocot flower. Most ornamental lilies, for example, have three petals, three sepals that are unusually showy and petal-like, and six stamens. A monocot seed has a single, small cotyledon (*mono*, "one") that remains belowground in the seed husk as the seedling sends a shoot aboveground.

In this chapter we take a closer look at how plants are built, how they obtain nutrients, and how they transport materials throughout their bodies. We focus mostly on flowering plants, both dicots and monocots, because these are the plants that are most familiar to us and that sustain us on a daily basis.

## 35.1 An Overview of the Plant Body

The plant body is relatively simple in its organization, compared to the body of vertebrate animals. The body of a flowering plant can be divided into two basic organ systems, the belowground **root system** and the aboveground **shoot system** (FIGURE 35.2). These two systems are specialized for life in two very different environments: roots in soil, shoots in air.

Compared with most animals, plants also have fewer types of cells, tissues, and organs. Three types of organs are found in virtually all flowering plants: roots, stems, and leaves. Each of these organs is made of three main types of complex tissues, each consisting of multiple different cell types that collectively carry out a set of related functions. The **dermal tissue** forms the outermost layer, or "skin" of the plant. **Vascular tissues** contain tubelike cells that conduct water or food, as well as other cell types that assist by strengthening the organ or by storing food or performing other essential functions. There are two main categories of vascular tissue: *xylem*, which transports water; and *phloem*, which conducts food. Any tissue that is not part of the dermal or vascular tissue is classified as a component of the **ground tissue** (shown in yellow in Figure 35.2).

Working from the bottom up, the root system contains a single type of plant organ: the roots. Roots anchor the plant, absorb water and nutrients from the soil, transport food and water, and (often) store food. The shoot system contains two main types of plant organs: stems and leaves. Stems provide the plant with structural support and also transport food and water. Leaves are the site of photosynthesis, so their main function is to produce food. The growing parts of the shoot system have buds, which are growing points located at the tip of each shoot and also at the base of many leaf stalks (see Figure 35.2).

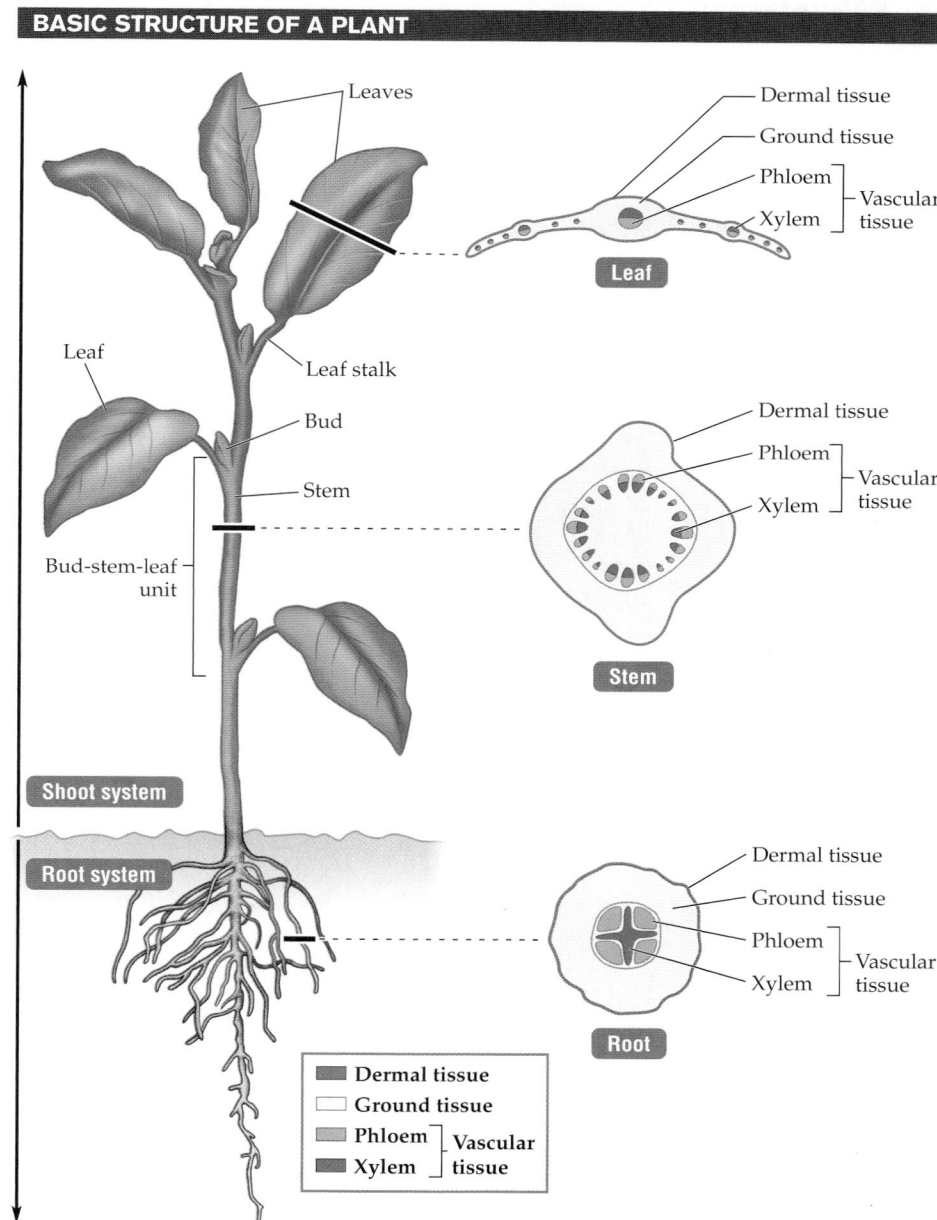

**BASIC STRUCTURE OF A PLANT**

Leaves

Dermal tissue
Ground tissue
Phloem ⎤ Vascular
Xylem ⎦ tissue

Leaf

Leaf

Leaf stalk

Bud

Stem

Bud-stem-leaf unit

Dermal tissue
Phloem ⎤ Vascular
Xylem ⎦ tissue

Stem

Shoot system

Root system

Dermal tissue
Ground tissue
Phloem ⎤ Vascular
Xylem ⎦ tissue

Root

■ Dermal tissue
☐ Ground tissue
■ Phloem ⎤ Vascular
■ Xylem ⎦ tissue

**FIGURE 35.2 How Plants Are Built**

Plants have three main types of organs (roots, stems, and leaves), each of which is composed of the same three basic tissue types: dermal, ground, and vascular. Belowground, plants grow by extending old roots and producing new lateral roots. Aboveground, plants grow by adding new sets of the bud-stem-leaf unit shown in the diagram.

Each bud contains dormant tissue that, under the right conditions, can produce either a shoot or flowers.

Flowers and fruits, which carry out the all-important function of reproduction in flowering plants, are other examples of plant organ systems. Although they look very different, flower petals are actually modified leaves. Other unique-looking plant structures can also develop from one of the three basic plant organs.

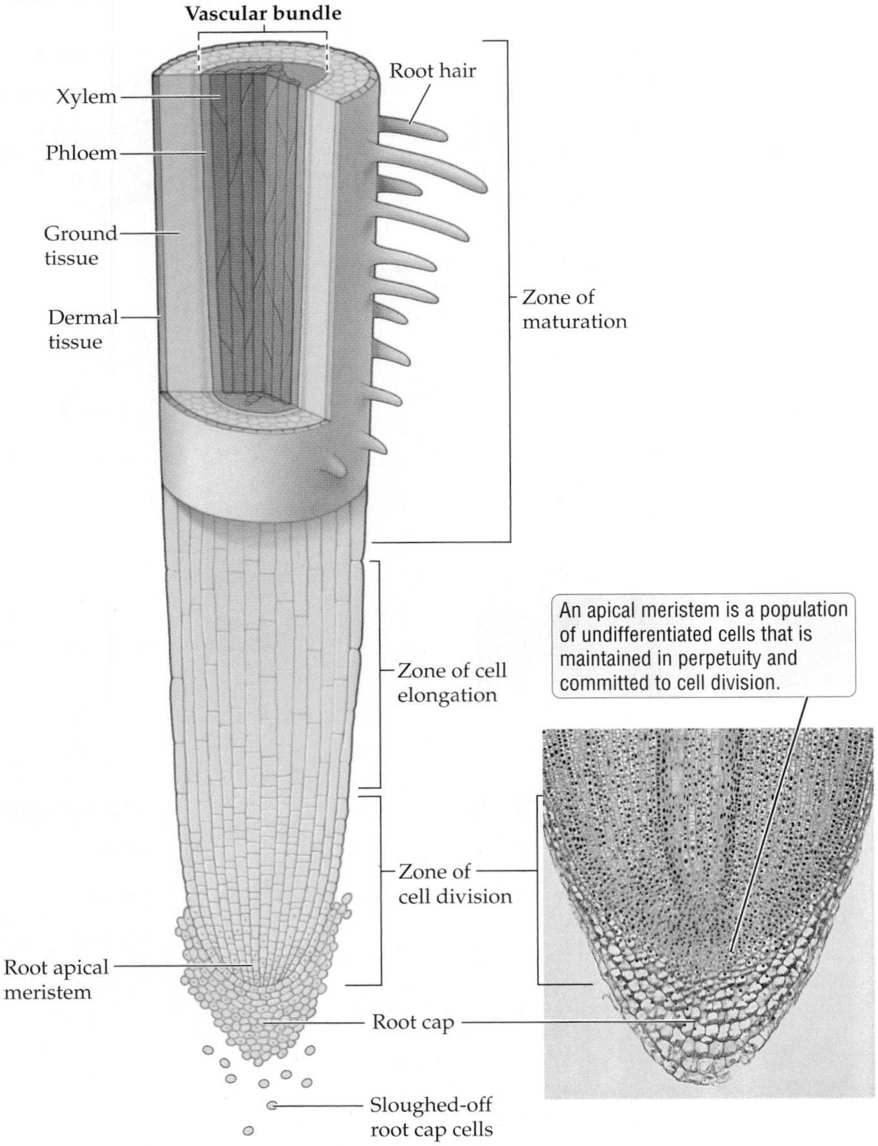

**FIGURE 35.3 Anchoring the Plant and Absorbing Nutrients**

Roots produce numerous outgrowths from dermal cells called root hairs, which aid in nutrient absorption. Plant roots have a region of active cell division, protected by the root cap, and a region of cell elongation in which cells increase in size and complete their development. The dermal, vascular, and ground tissues of the root are shown in this diagram.

For example, the edible part of the carrot plant is a modified root, and thorns are modified branch stems. The bulb of the onion consists of the enlarged, tightly packed bases of onion leaves; the inner ones store starch and sugars, and the dry, scaly layers on the outside protect the inner, food-storing layers of the leaf bases. We will see other examples of modified stems and leaves throughout this chapter.

Plants differ from animals not only in the way their bodies are constructed, but also in how they grow and develop through their life span. Unlike most animals, plants grow throughout their lives. Furthermore, plants grow in a unique way: aboveground, the same basic unit, the bud-stem-leaf unit shown in Figure 35.2, is added over and over again. As we will see in Chapter 36, this modular approach to building an organism gives plants great flexibility in responding to changing environmental conditions.

## 35.2 Plant Organs

Roots, stems, and leaves are three organs you will find in almost all flowering plants. In this section we take a closer look at the structure of these vital plant organs.

### Roots absorb water and nutrients, anchor the plant, and store food

Plants invest a great deal of resources in making roots, testifying to the importance of their various functions: absorption of water and nutrients, physical anchoring of the plant, and food storage. In terms of structure, roots have a zone of active cell division protected by a conical patch of tissue known as the **root cap** (**FIGURE 35.3**). The root hairs, which greatly increase the surface area over which water and nutrients can be absorbed from the soil, are produced in the zone of maturation, which lies above a zone of rapid cell expansion. Mature roots can also produce many lateral (side) roots, forming an extensive network that anchors the plant. Roots are made up of an outer layer of dermal cells, then the cortex (which consists of ground tissue cells that often store starch), and toward the center, bundles of vascular tissues composed of phloem and xylem.

The dominant taproot of dicots grows directly downward, producing lateral roots that extend to the sides before also descending downward (**FIGURE 35.4a**). In the fibrous root system of monocots, no single root predominates (**FIGURE 35.4b**). Fibrous roots

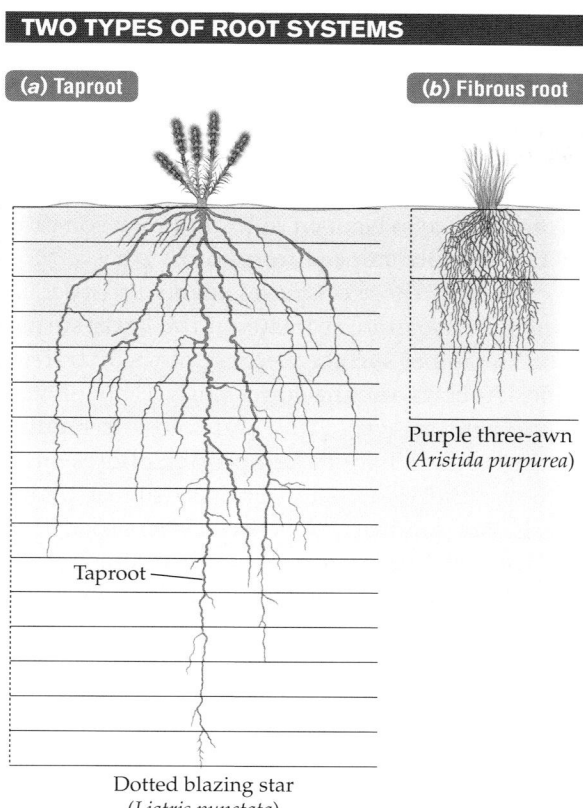

**(a) Taproot**

**(b) Fibrous root**

Purple three-awn
(*Aristida purpurea*)

Taproot

Dotted blazing star
(*Liatris punctata*)

**FIGURE 35.4 Dicots Have Taproots; Monocots Have Fibrous Roots**

(*a*) The dotted blazing star, a wildflower found on the prairie, has a taproot system. (*b*) Purple three-awn, a prairie grass, has a fibrous root system.

can form dense mats that hold soil firmly in place. Taproots, on the other hand, usually hold the soil less firmly but penetrate more deeply than fibrous roots. Some taproots descend great distances, reaching a record depth of 53.3 meters (nearly 175 feet) in mesquite, a desert shrub. Perhaps even more astonishing is the vast amount of root material that even a small plant can produce. For example, one 4-month-old rye plant grown in 52 liters (about 14 gallons) of soil made enough roots to cover 639 square meters; about 400 square meters of this area resulted from its 14 billion root hairs, which, if placed end to end, would have extended over 10,000 kilometers (a little over 6,000 miles—approximately twice the distance between the East and West coasts of the United States).

## Stems support the plant

Stems provide support to the plant, enabling it to grow vertically against the force of gravity. Stems also hold leaves up to intercept light. Viewed in cross section, stems show many similarities to roots in their structure. As in other parts of the plant body, the arrangement of tissues in the stem consists of dermal tissue at the external surface and ground tissue in the interior, with bundles of vascular tissue embedded in the ground tissue. In dicots, the vascular bundles in the stem are arranged in a single ring (**FIGURE 35.5**). In most monocots, the stem vascular bundles are scattered throughout the ground tissue (as seen in Figure 35.1). Stems can be highly modified to perform specialized tasks, including protection (the thorns of a hawthorn bush are actually modified stems), climbing (grape tendrils are stems), and belowground food storage (potatoes, believe it or not, are the swollen tips of underground stems that are specialized for food storage).

Hawthorn branch with thorns

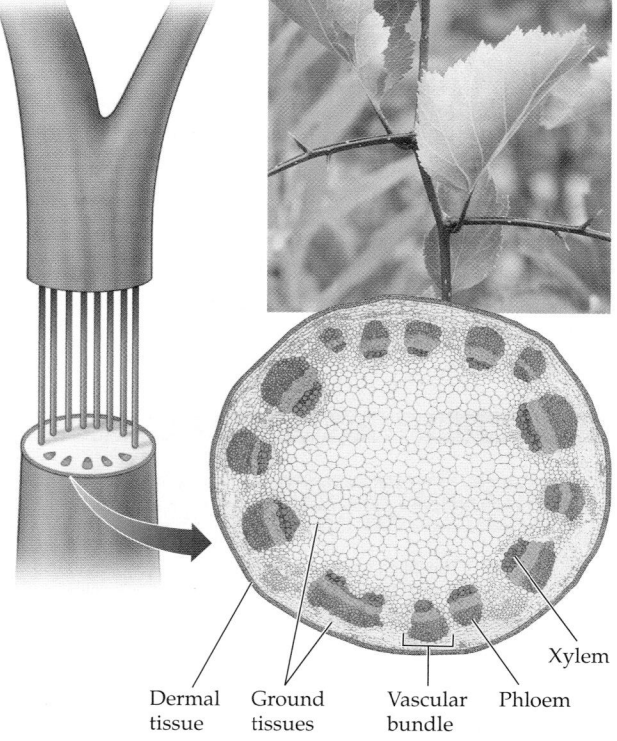

Dermal tissue

Ground tissues

Vascular bundle

Phloem

Xylem

**FIGURE 35.5 Stems Support the Plant**

Ground tissues in stems support the plant against gravity. The vascular bundles resemble a set of pipes running through the stem. In most dicots, the vascular bundles are arranged in a roughly circular shape, as shown here; in monocots, they are scattered throughout the ground tissue. Some plants use modified stems for functions such as protection (thorns) and storage (the potatoes we eat).

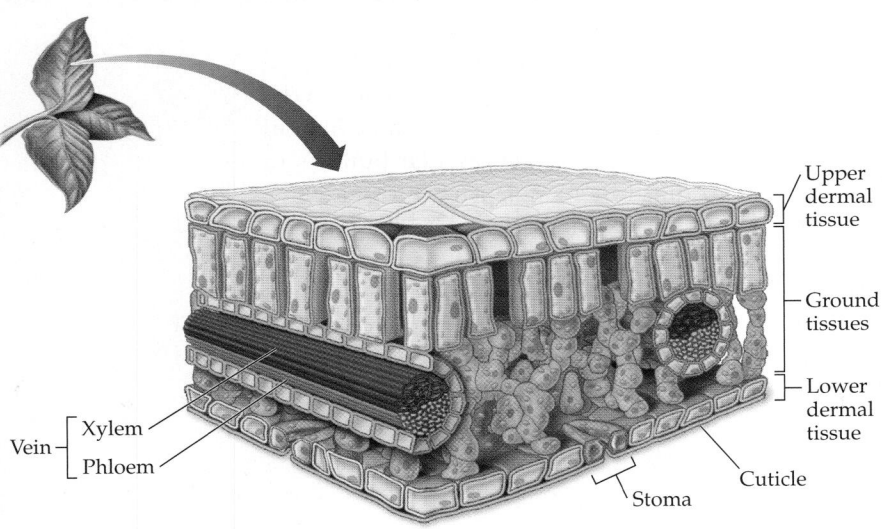

Vein — [ Xylem
Phloem ]

Upper dermal tissue

Ground tissues

Lower dermal tissue

Cuticle

Stoma

**FIGURE 35.6  Leaves Feed the Plant**

Plants feed themselves through their leaves, the primary site of photosynthesis. $CO_2$ enters the leaf through small air pores (stomata) that are typically more numerous on the lower side of the leaf.

## Leaves feed the plant

Although cells in the stems of many plants can perform photosynthesis, in a typical plant most plant sugars are produced by photosynthesis in the leaves. In cross section, leaves look different from roots or stems, but the basic structural arrangement of tissues is the same (**FIGURE 35.6**). The outer layer—the top and bottom of the leaf—is made up of dermal tissues.

The dermal layer of leaves includes the regulated air pores, known as **stomata [stoh-***MAH***-tuh]** (singular, "stoma"), that control gas exchange. Vascular plants control the size of the stomatal opening by regulating a pair of cells, called guard cells, that form the border each stoma (see Figure 3.11).

Stomata have to be open to let in the carbon dioxide ($CO_2$) that plants need for photosynthesis. Most leaves, especially those of species adapted to drier environments, have more stomata on the lower surface than on the upper surface. Because of their strategic location, stomata on the lower surface are shaded from the sun, so evaporative water loss is reduced when the stomata have to be open for photosynthesis. The ground tissues—the primary site of photosynthesis—are sandwiched between the two dermal tissue layers. The vascular tissues, containing phloem and xylem, are located roughly in the middle of the ground tissue. Many of the vascular bundles are bulky enough that we can see them from the outside, especially on the lower side of the leaf, and these bundles are commonly known as the veins of a leaf.

Leaves are often covered with fuzzlike dermal hairs that can be deadly to insects. Leaf cells also contain chemicals (such as anthocyanins) that act as sunscreens, protecting the plant from damage by UV light (see the box in Chapter 9, page 216). The structure of some leaves is modified greatly to perform functions other than photosynthesis (**FIGURE 35.7**). Examples of modified leaves include the debris-collecting "pots" of the flower pot plant, which obtain mineral nutrients from the debris; the tendrils of peas (used for climbing); and the protective spines of cacti and other plants.

**FIGURE 35.7
Modified Leaves**

Some plants use modified leaves for functions such as nutrient capture. For example, the pouches of the flower pot plant contain its roots and collect debris (a), spines provide cacti with protection (b), and tendrils enable pea plants to climb (c).

**DIVERSITY IN LEAF FUNCTIONS**

*(a)* The "pot" of the flower pot plant (*Dischidia rafflesiana*)

This plant has aerial roots that grow into these pots (the pots collect debris).

A modified leaf or "pot" of the flower pot plant, a species that grows on other plants, not in soil.

*(b)* Spines of a saguaro cactus (*Carnegiea gigantea*)

Spines are modified leaves that defend the plant from herbivores.

*(c)* Tendril of a pea plant (*Pisum sativum*)

The pea tendril is a modified leaf. It wraps around nearby objects, including other plants, enabling the plant to stay upright.

## 35.3 Plant Tissues

As we have seen, flowering plants have three main tissue systems: dermal, ground, and vascular. Each of these is composed of different types of cells that work together to perform particular functions. Dermal tissues, which form the outermost layer of the plant, protect the plant from the outside environment and control the flow of materials into and out of the plant. Ground tissues, which form the intermediate layer, make up the bulk of the plant body and perform a wide range of functions, including support, wound repair, and photosynthesis. Finally, in or near the center of the plant body are the vascular tissues, which transport food and water throughout the plant.

### Dermal tissues interact with the environment

Dermal tissues play a key role in the ways plants meet their many tough environmental challenges. They help protect the plant from enemies such as herbivores (plant-eating organisms) or pathogens (disease-causing organisms). They also increase the uptake of water and nutrients, control gas exchange, and limit water loss. Most dermal tissue consists of a single layer of rectangular protective cells that form the *epidermis* (**FIGURE 35.8**). In aboveground plant parts, these cells are covered with a waxy **cuticle** that helps prevent water loss and keep enemies such as fungi from invading the plant.

Certain dermal tissue cells have special shapes for performing particular functions. For example, the leaves, stems, and even fruits of many plants are covered with **dermal hairs** (more technically known as trichomes). The "fuzz" on a peach fruit is a mat of dermal hairs (in fact, a nectarine is really a peach variety that lacks dermal hairs on the fruit). Dermal hairs have a variety of functions, including shield-

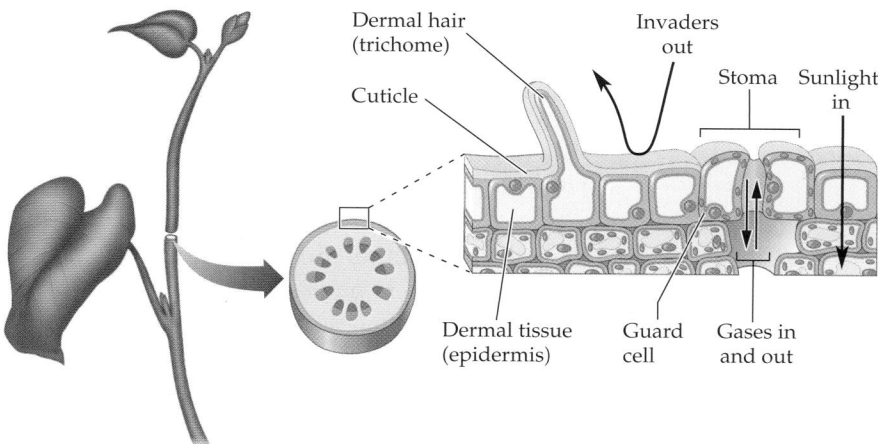

**DERMAL TISSUE STRUCTURE**

**FIGURE 35.8 Dermal Tissue Has Contact with the Environment**
Dermal tissues protect the plant from attack and control the flow of materials into and out of the plant.

ing the plant from excessive ultraviolet radiation, as in such woolly looking alpine plants as edelweiss (**FIGURE 35.9a**). Discouraging herbivory by insects is the most important and widespread function of dermal hairs on aboveground parts of the plant. Small insects can become entangled in the hair. Some types of dermal hairs secrete chemicals that are toxic to insects and other herbivores (**FIGURE 35.9b**). **Root hairs** are single-celled extensions of root dermal cells. They play an essential role in the uptake of water and nutrients.

As mentioned earlier, stomata are regulated air pores on leaves, and on green stems, that control the rates at which $CO_2$ is brought into the plant and $O_2$ and water are lost from the plant (**FIGURE 35.9c**). Each stoma is bordered by a pair of cells, called **guard cells**, that can inflate and deflate almost like water balloons. The guard cells swell as they fill up with water (which they store inside membrane-enclosed sacs called vacuoles). Because the guard cells are firmly attached at the ends, the force generated by the

(a)

(b)

(c)

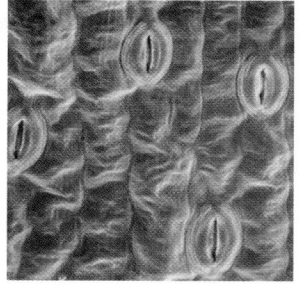

**FIGURE 35.9 Dermal Tissues Provide Protection and Gas Exchange**
(a) Dermal hairs serve a wide range of functions, including helping alpine plants such as the edelweiss to deflect the intense ultraviolet radiation at high altitudes. (b) Dermal hairs protect many plants from insects (like the ant seen here in a tangle of dermal hairs). (c) Stomata open and close to promote the exchange of gases such as $CO_2$ and $O_2$ and to limit the loss of water.

entering water cannot be resolved by an increase in the length of each guard cell. Instead, the two guard cells buckle out, revealing an opening (the stomatal pore) that lies between the cells. When these cells lose water, they lose their "plumpness," which causes them to collapse against each other and close off the air pore between them.

Because excessive water loss can doom a plant, stomata are carefully regulated by water availability. As we have seen, plants must open their stomata to admit $CO_2$, which is necessary for photosynthesis. Most plants open their stomata in the daytime and close them at night, when photosynthesis is not an option. However, a plant experiencing water stress (inadequate water supply) will immediately close its stomata to conserve water, no matter what time of day it is.

## Ground tissues have many essential roles

Have you ever wondered why nonwoody plants—which do not have an obvious support system like the skeleton of an animal or the woody tissue of a tree—do not just flop over? Pressure from water within the cells is needed to support any living plant cell, much as air within a balloon keeps it from collapsing. When this pressure is lost, as when a plant has gone too long without receiving water, the leaves and tender tips of the stem have the droopy look of a wilted plant. The older parts of the stem may remain stiff, however, because the ground tissue in such stems contains cells with thick walls that provide mechanical support (**FIGURE 35.10**).

Plants have three main cell types, and all three can be found in the ground tissue of many species.

**GROUND TISSUE**

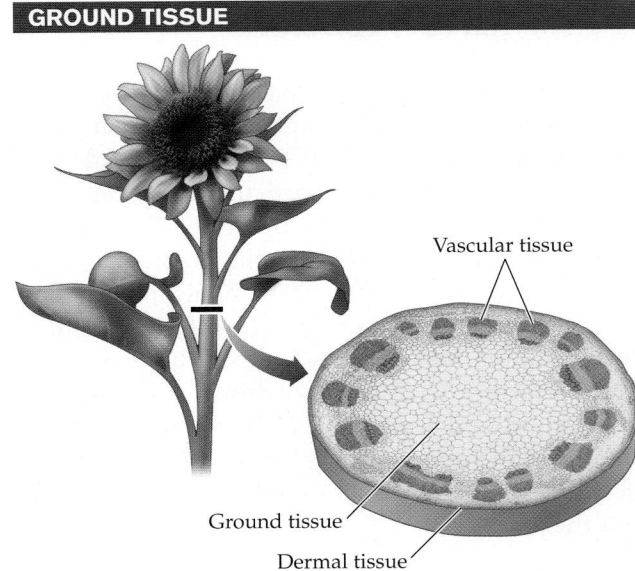

FIGURE 35.10 **Ground Tissues Play Essential Roles in Plants**
The roles of ground tissue in plants include support, wound repair, and photosynthesis.

**Parenchyma [puh-*RENG*-kuh-muh]** (*para*, "beside"; *enchyma*, "tissue") is made up of relatively large cells with thin cell walls (**FIGURE 35.11a**). Parenchyma is the most abundant cell type in the ground tissue of most plants. Parenchyma cells may be photosynthetic, as in leaves and in the outer layers of the ground tissue in a green stem (the inner layers are invariably nongreen, because they do not receive enough light for photosynthesis). They play an important role in wound repair and in enabling plants to regrow new organs or damaged parts. Parenchyma cells often store surplus food and other substances (such as defensive compounds). Much of the fresh produce you consume is composed of parenchyma cells.

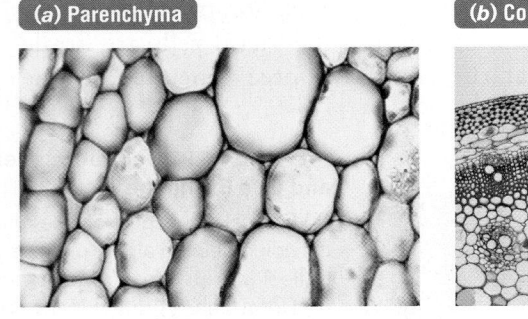

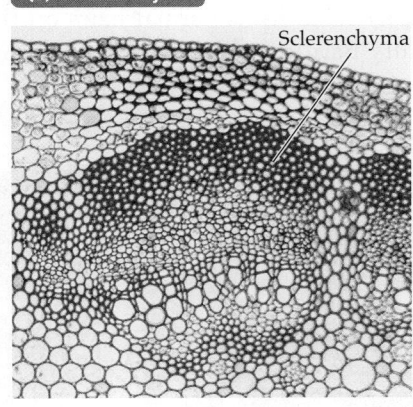

FIGURE 35.11
**The Three Cell Types Found in Ground Tissue**
Parenchyma and sclerenchyma are also common in dermal and vascular tissue.

**Collenchyma** [kuh-*LENG*-kuh-muh] (kolla, "glue") is made up of living cells that provide mechanical reinforcement in young, growing parts of the plant, such as tender stems and leaf stalks. These cells are best identified by their unusual cell walls, which are thick in some places and thin in others (**FIGURE 35.11b**). In a celery stick, collenchyma occurs in patches directly underneath the ridges that are such prominent features of these leaf stalks. When you pull the "string" off a celery stick, you are pulling off nearly pure bundles of collenchyma cells.

**Sclerenchyma** [skluh-*RENG*-kuh-muh] (*scleros*, "hard") consists of thick-walled cells whose function is to provide mechanical strength to plant parts (**FIGURE 35.11c**). Sclerenchyma cells are often dead when they are fully functional, because their function is best served by tough, thickened cell walls but thick cell walls are not compatible with maintaining a living cytoplasm; thick walls would hinder the uptake of essential nutrients. Sclerenchyma may be found in both ground and vascular tissues of stems, roots, and leaves. Sclerenchyma cells of a type known as **fibers** are among the longest cells found in plants. Linen fabric is made from stem fibers of the flax plant. Stem fibers from the hemp plant are used to make ropes. Occasionally, sclerenchyma is found in the ground tissue of fruit. The gritty texture that gives the crunch to European and Asian pears is created by patches of sclerenchyma cells (of a type called "stone cells") that are distributed throughout the flesh of these fruit. Sclerenchyma is even found in the dermal tissue sometimes; the red, papery "skin" you might peel off a peanut is almost entirely the sclerenchyma layer that makes up the dermal tissue in the seeds of most legumes (members of the bean family).

## Vascular tissues transport food and water

Like any other multicellular organism, plants must transport fluids and food throughout their bodies. We will discuss how they do this later in the chapter; here we describe how the vascular tissues that make such transport possible are put together.

Plants have two types of vascular tissues: **phloem** [*FLOH*-em], which transports food (mainly sugars made through photosynthesis); and **xylem** [*ZYE*-lum], which transports water and mineral nutrients. Phloem and xylem both contain stacks of long cells that form continuous tubes specialized for carry-

ing either food or water (**FIGURE 35.12**). Like an interstate system of highways and expressways, these conducting tubes run throughout the plant body, linking all organs of the root and shoot systems. Food must be shipped from the leaves, where it is produced, to living cells in every part of the plant. This task is performed by food-conducting tubes found in the phloem. Water and minerals absorbed from the soil must move upward from the roots and outward from the central stem to the leaves. This essential function is carried out by water-conducting tubes found in xylem.

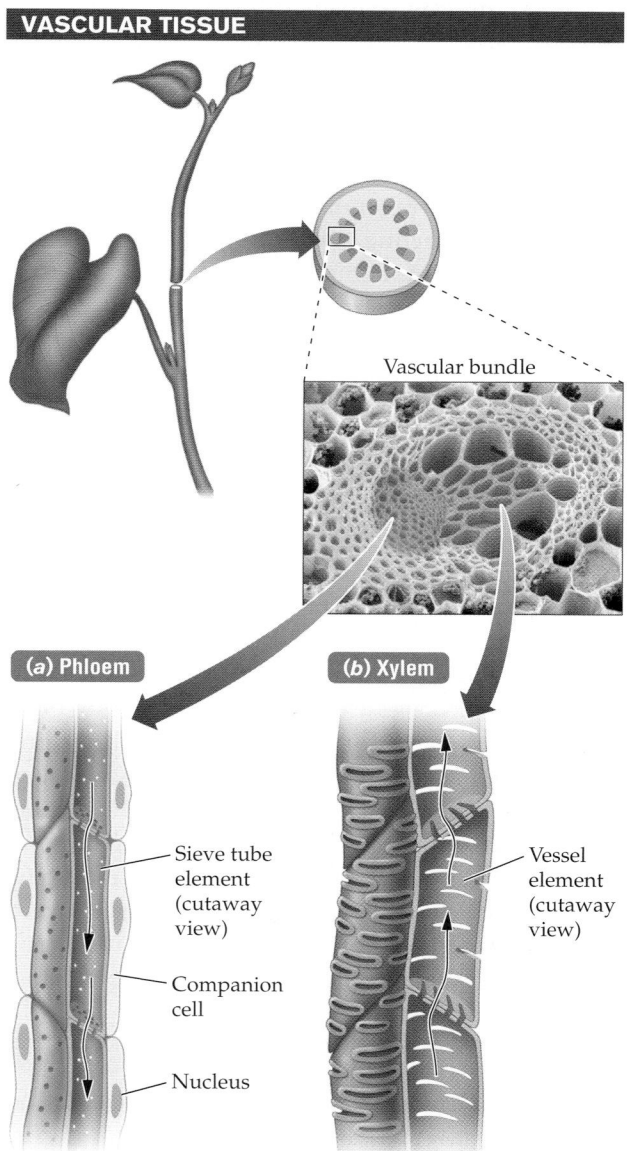

**VASCULAR TISSUE**

Vascular bundle

**(a) Phloem**

**(b) Xylem**

Sieve tube element (cutaway view)

Companion cell

Nucleus

Vessel element (cutaway view)

**FIGURE 35.12**
**Vascular Tissues Transport Food and Water**

Vascular tissues form a continuous set of pipes that run through the plant. (*a*) Food is transported in the phloem, which contains sieve tube elements and their associated companion cells. Note the presence of a nucleus in companion cells and the absence of this organelle in sieve tube elements. (*b*) Water and mineral nutrients are transported in the xylem. The xylem contains two types of water-conducting tubes: tracheids and vessels. The cells that make up tracheids and vessels are dead by the time they become functional in water transport. Vessels, illustrated here, are wider than tracheids. Water and mineral nutrients move easily from one vessel element to the next because there are large holes, or perforations, in their end walls (four perforations are seen in the end walls of the vessel elements shown in the cutaway view).

The food-conducting tubes in phloem are called **sieve tubes**. They are made up of living cells (sieve tube elements, seen in **FIGURE 35.12a**) joined end to end to create a continuous tube. The cytoplasm of one sieve tube element is continuous with that of its neighbor because these cells are linked through relatively large openings in their end walls (the walls at the top and bottom of each long cell). The sieve pores, as the openings are known, form open channels lined by the plasma membrane of the interconnected cells. Sieve tube elements are specialized for moving sugars and other organic molecules; they lack a nucleus and many other organelles that could "get in the way" of rapid transport. Each sieve tube element is extensively connected to one or more companion cells through fine cytoplasmic tunnels called **plasmodesmata [plaz-moh-*DEZ*-muh-tuh]** (singular "plasmodesma") (see Figure 7.11*b*). Companion cells do have a nucleus and are active in the manufacture of proteins and other macromolecules. The role of companion cells is to supply sieve tube elements with macromolecules that the nucleus-deficient cells cannot make for themselves; in many plants, companion cells also help by gathering sugars from the surrounding ground tissue cells and shipping them to the sieve tubes.

Unlike sieve tube elements, the cells that make up the water-conducting tubes of xylem are dead by the time they become fully functional. Xylem conducting elements, as these cells are known, have thick walls and hollow interiors when they are fully mature. They are stacked one on top of another to form long hollow cylinders, rather like the sections of pipe used in household plumbing. There are two types of xylem conducting tubes: **tracheids**, which are narrow; and **vessels**, which are much wider and therefore carry a larger volume of water in a given amount of time. **FIGURE 35.12b** illustrates the structure of vessel elements. The end walls of vessel elements have one or more large openings, or perforations; four end wall perforations are seen in the cutaway view of the vessel in Figure 35.12*b*. Because the individual vessel elements are interconnected through the perforations, water moves unimpeded in a vessel, the way it moves in plumbing pipes. The thick walls of tracheids and vessels keep these vital structures from collapsing under the strong lifting forces that are exerted on them (Section 35.6).

> ### Concept Check
>
> 1. What is the function of root hairs, and under which tissue system are they classified?
>
> 2. Compare the conducting elements of xylem and phloem.

## 35.4 How Plants Obtain Nutrients

Because of the importance of plants in our diets, people have long been interested in how plants function, including how they get nutrients. Aristotle thought plants got everything they needed for growth from the soil. Almost 2,000 years later, however, the Belgian physician Jan Baptista van Helmont showed that soil alone is not enough. He did this by planting a small willow tree in a pot and adding only water to the pot. After 5 years of growth, the willow had increased in weight by 74.4 kilograms, but the soil had decreased in weight by only 0.06 kilogram. Van Helmont concluded that soil had nothing to do with plant growth, and that plants needed only water to grow!

Roughly 100 years later, many scientists in the 1700s suggested that plants get most of what they need to grow from air, not soil or water. It turns out that Aristotle, van Helmont, and these scientists were all partly right (**FIGURE 35.13**). Plants get

**RAW MATERIALS FOR PLANT GROWTH**

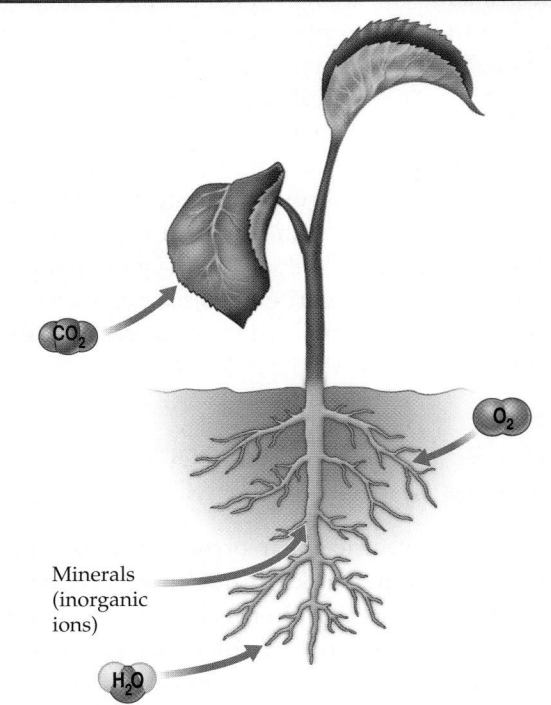

**FIGURE 35.13 Plants Need Water, Minerals, and CO$_2$**

To grow, plants need CO$_2$, mineral nutrients, and water. CO$_2$ enters the plant through the leaves and is converted to sugars through photosynthesis. Mineral nutrients and water enter the plant through the roots. As a by-product of photosynthesis, plants release oxygen (O$_2$) to the air, some of which diffuses from the air into the soil, where it can be used by root cells for cellular respiration.

essential mineral nutrients (inorganic ions) from the soil, as Aristotle thought. But nutrients from the soil contribute only a small amount to the weight of the plant, as van Helmont showed in his experiment. Instead, much of the weight of a living plant comes from water, most of which is held in the central vacuoles of its cells (see Figure 6.14). Finally, if we weigh a plant after removing its water (using a drying oven), we are left with mostly carbon-based material known as dry biomass. Scientists in the 1700s realized that this dry biomass is built from something plants get from the air, since it cannot be accounted for by the water added to the plant or the minuscule loss in soil weight. We now know that most of the plant's dry weight comes from $CO_2$ absorbed from the air by the leaves and used during photosynthesis to produce sugars.

As we noted in Chapters 6 and 9, photosynthesis is a series of chemical reactions that use sunlight, $CO_2$ from the atmosphere, and water from the soil to produce sugars, releasing $O_2$ as a by-product. We described the process of photosynthesis in detail in Chapter 9; for the remainder of this section, we will focus on the mineral nutrients needed by plants.

## Plants need mineral nutrients to grow

The label on a typical package of plant fertilizer shows that it provides plants with three mineral macronutrients—nitrogen (N), phosphorus (P), and potassium (K)—and an assortment of mineral micronutrients (**FIGURE 35.14**). Plants require **macronutrients** in relatively large amounts (at least 1,000 milligrams of the nutrient per kilogram of plant dry weight), but they require much smaller amounts of **micronutrients** (less than 100 milligrams per kilogram of plant dry weight). In total, plants need nine macronutrients (carbon, oxygen, hydrogen, nitrogen, phosphorus, potassium, calcium, sulfur, and magnesium) and at least eight micronutrients (including iron, zinc, and copper). Carbon, oxygen, and hydrogen are obtained from air or water; the rest of the macronutrients and micronutrients must be absorbed from soil.

If we grow garden plants or crop plants in the same soil for several seasons, the mineral nutrients in that soil are likely to become depleted as they enter the plant biomass and are then removed with the harvest. Macronutrients, especially nitrogen, phosphorus, and potassium, are the first to become deficient in culti-

vated soils. Plants draw larger quantities of these minerals from the soil because they need them in greater amounts to sustain their growth and development. That is why plant fertilizers contain more nitrogen (N), phosphorus (P), and potassium (K) than any other mineral.

The NPK label that is marked on all fertilizer containers gives the relative amounts of the "big three" minerals in that fertilizer. For example, an NPK ratio of 10-15-10 means 10 percent of the fertilizer (by weight) is nitrogen, 15 percent is phosphorus, and another 10 percent is potassium (see Figure 35.14). The remaining 65 percent of the fertilizer may consist of other macronutrients, micronutrients, and ingredients such as inert material ("fillers") and chemicals that adjust the acidity of the soil.

Plants can absorb these minerals only if the minerals are dissolved in water. To "feed" plants with fertilizer, we mix it into the soil, allowing the nutrients in the fertilizer to dissolve in the water found in soil. We can also add the fertilizer to water first and then apply the solution to the soil. As long as the right

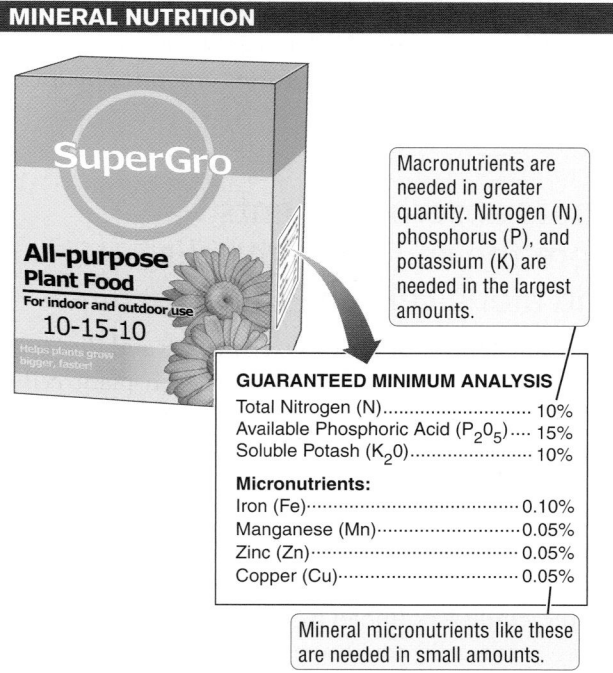

**FIGURE 35.14 Mineral Nutrients Are Essential for Plant Function**

The label on this box of fertilizer ("plant food") shows that it contains three essential plant macronutrients—nitrogen, phosphorus (as phosphoric acid), and potassium (as potash)—and four essential micronutrients (iron, manganese, zinc, and copper).

Growing plants hydroponically

nutrients are present, it is also possible to grow plants without soil—a technique known as *hydroponics*.

The ingredient label in Figure 35.14 comes from a synthetic fertilizer that contains no organic (carbon) compounds. Other commonly used fertilizers, such as compost and manure, contain many organic compounds and provide relatively small amounts of the inorganic nutrients that plants need. Nevertheless, such organic fertilizers improve plant nutrition in two ways. First, the spongelike texture of rotting organic material soaks up and holds water into which mineral nutrients can dissolve. These minerals tend to be released gradually over the course of a growing season instead of entering the soil in one large, potentially wasteful flood. Second, particles in organic fertilizers act as "magnets" that hold mineral nutrients in the soil, keeping them from being washed out of reach of the plant roots. In contrast, inorganic fertilizer is often applied in excessive amounts, and the excess is more likely to run off and contaminate surface water and groundwater.

Note that fertilizers are not a source of energy for plants. Plants do not need energy from chemical sources, because photosynthesis enables them to transform light energy into chemical energy as they convert carbon dioxide into sugars. Humans and other animals, in contrast, must eat foods that contain energy stored in chemical forms—in carbohydrates, fats, and proteins.

## Roots absorb nutrients from the soil through cells and along cell walls

To understand how roots absorb water and dissolved nutrients from the soil, recall that each plant cell is surrounded by a cell wall (see Chapters 6 and 7). As shown in **FIGURE 35.15a**, the plasma membrane lies just inside the cell wall, enclosing the gel-like cytosol. Within the cytosol of plant cells are one or more saclike structures known as vacuoles, which are filled with water and various solutes (see also Figure 6.14). Each vacuole is bounded by a single membrane. Many living plant cells are connected by tiny tunnels (the plasmodesmata) that enable transportation and communication between adjacent plant cells.

There are two pathways by which water and dissolved nutrients absorbed by roots reach other cells of the plants: the *cell-interior route* and the *cell-wall route* (**FIGURE 35.15b**). The cell-interior route is

served by the plasmodesmata, through which materials that enter the cytosol of one root cell can pass to other root cells without having to repeatedly cross cell walls and plasma membranes. Of the two pathways by which roots absorb nutrients, this cell-interior route is more common: mineral nutrients enter the root through root hairs and then travel from one cell's cytosol to another's via the plasmodesmata. However, roots can also absorb nutrients by a pathway that bypasses cell interiors; in this cell-wall route, nutrients dissolved in water travel from one cell wall to the next.

Nutrients traveling by either pathway must cross at least one plasma membrane before they enter the vascular tissues. An encounter with this barrier allows the plant to block harmful materials from reaching aboveground parts. Nutrients traveling by the cell-interior route must pass through a plasma membrane (that of the root hair) right away. Any nutrients that have followed the cell-wall route exclusively will have to cross a plasma membrane when they reach the *endodermis*; otherwise they will be excluded from the vascular tissues. The **endodermis** is a layer of ground tissue cells that surrounds the vascular tissue in the root and functions as a gatekeeper for nutrients entering the vascular tissues. Endodermal cells have a waxy deposit in their cell walls that acts like the weather stripping in a window or door. The waxy layer, known as the *Casparian strip*, is laid down as a continuous band, or sometimes as a continuous sheet, on the four sides of an endodermal cell that touch other endodermal cells. The overall effect is that the water-repelling Casparian strip prevents water from creeping *between* endodermal cells. As a result, water, and any substance dissolved in it, cannot enter the vascular system without having crossed at least one plasma membrane. In effect, the endodermis ensures that all material entering the vascular system is "screened" by passage through at least one plasma membrane.

Nutrient concentrations in roots can be 10–10,000 times greater than those in soil. As we saw in Chapter 7, molecules can cross a plasma membrane up a concentration gradient (to a region of higher concentration) only by **active transport**, which requires energy. Therefore, plants must expend energy to absorb nutrients into their roots. If no energy is available, nutrient uptake stops. For example, when scientists reduced the energy stores of corn plants (by placing them in the dark for 4 days), the amount of phosphate ions absorbed by their roots dropped to just 5 percent of nor-

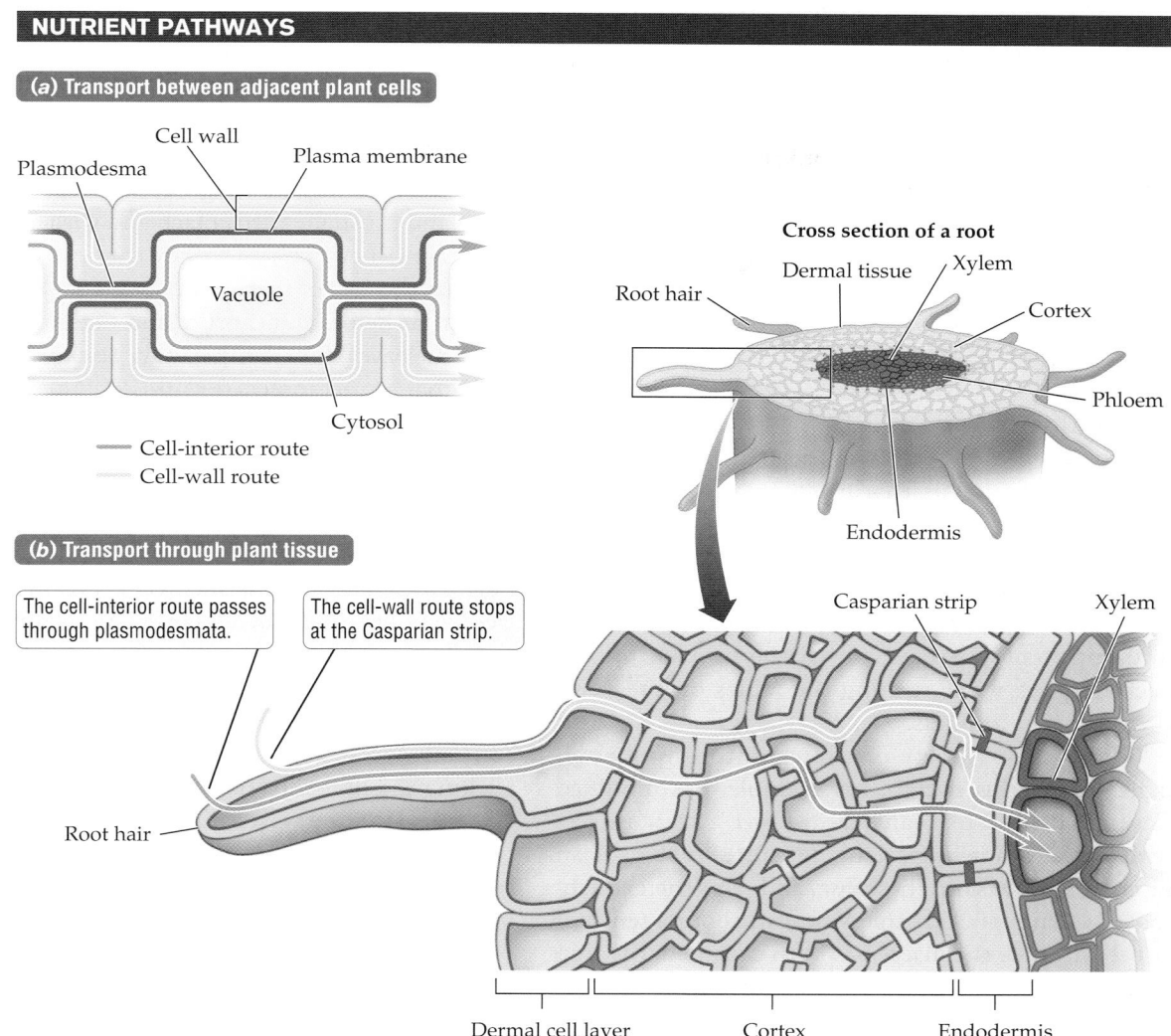

## NUTRIENT PATHWAYS

### (a) Transport between adjacent plant cells

Cell wall

Plasmodesma

Plasma membrane

Vacuole

Cytosol

— Cell-interior route
— Cell-wall route

### (b) Transport through plant tissue

The cell-interior route passes through plasmodesmata.

The cell-wall route stops at the Casparian strip.

Root hair

Dermal cell layer

Cortex

Endodermis

**Cross section of a root**

Dermal tissue

Root hair

Xylem

Cortex

Phloem

Endodermis

Casparian strip

Xylem

**FIGURE 35.15**

**How Plants Absorb Nutrients from Soil**

Mineral nutrients enter plant roots in two ways: by the cell-interior route (shown in orange) and by the cell-wall route (shown in aqua).

mal levels; absorption levels returned to normal after 2 days of sunlight. If plants did not expend energy to absorb nutrients, eventually the nutrients would leak (by diffusion) from the plant back into the soil—just the opposite of what the plant needs.

Adding complexity to this story of nutrient absorption from the soil is the fact that most plants receive help from a mutually beneficial relationship with fungi. Plant roots unite with fungal hyphae to form structures called **mycorrhizae** [*MYE*-koh-*RYE*-zee] (singular "mycorrhiza"), which were discussed in Chapter 3. The fungus benefits from this arrangement because the plant provides it with carbohydrates produced by photosynthesis, and the plant benefits because mycorrhizae increase the surface area over which the plant can absorb nutrients. Benefits for the plant can be substantial—up to 3 *meters* of hyphae can extend from only 1 centimeter of plant roots!

Of all the mineral nutrients, plants need nitrogen in the largest amounts. Although nitrogen is extremely abundant in the atmosphere in the form of nitrogen gas ($N_2$), this form of nitrogen is not directly available to plants. Plants can absorb nitrogen only in the form of dissolved nitrate ($NO_3^-$) or ammonium ($NH_4^+$) ions. Some plants, most notably the legumes (members of the bean family), have a mutually beneficial relationship with bacteria that have the extraordinary ability to convert atmospheric nitrogen gas into ammonium—a process known as **nitrogen fixation**. The host plant harbors the nitrogen-fixing bacteria within its tissues, usually in lumpy structures called root nodules. The bacteria receive metabolic energy in the form of carbohydrates that the plant manufactures through photosynthesis. The bacteria use some of that energy for nitrogen fixation, which is a complex and energy-intensive reaction driven by a remarkable bacterial enzyme called *nitrogenase*. As part of this mutually beneficial trade, the plant receives ammonium that the bacterium makes, with the help of nitrogenase, from atmospheric $N_2$. Nitrogen fixation, and its role in the cycling of nitrogen through ecosystems, is discussed in greater detail in Chapter 24.

## 35.5 Plants That Digest Animals

We can think of plants as living "between the devil and the deep blue sea." The "devil" they face is the vast array of animals that eat them, including over 300,000 species of herbivorous insects. Plants must also cope with a "deep blue sea" of soil in which essential nutrients (such as nitrogen and phosphorus) are scarce. About 600 plant species that live in very nutrient-poor soils have solved the problem of the deep blue sea while turning the tables on the devil: they consume animals.

Plants use a variety of strategies to capture animals. The Venus flytrap, for example, has modified leaves that look like fanged jaws, yet attract insects with a sweet-smelling nectar (**FIGURE 35.16a**). If an insect is lured within these jaws and trips their touch-sensitive hairs, the leaf snaps shut in about a tenth of a second. Once the insect has been captured, the trap tightens further, forming an airtight seal around its prey. The leaf then secretes enzymes that digest the plant's prey over the course of 5–12 days. The dermal

system in these traps is unusual in that it can absorb the released nutrients, especially amino acids (rich in nitrogen) and nucleotides (rich in phosphorus). Once the insect is digested and useful nutrients absorbed, the trap opens to wait for another insect. After three to five meals, the trap no longer functions to capture insects, and it usually turns black and withers away. The plant grows new traps to replace the dead ones.

Like the Venus flytrap, other plants have moving parts that help them capture and digest animal prey. Sundew leaves, for example, have sticky, club-shaped hairs that attract insects (**FIGURE 35.16b**). When an insect becomes stuck on these hairs, they bend inward, causing the leaf to curl around its hapless prey, which is then slowly killed and digested. A group of aquatic flowering plants, members of the bladderwort genus *Utricularia* [yoo-trik-yuh-*LAIR*-ee-uh], are even more aggressive: they literally suck their prey into a digestive chamber (**FIGURE 35.16c**). This chamber has a trapdoor entry with trigger hairs on the outside. If a small animal, such as a mosquito larva, bumps into these hairs, the door swings open and then—in just a thirtieth of a second—water rushes into the chamber, sucking the prey into the trap. The door then closes, and the plant digests its animal prey.

Other plants lack moving parts yet still manage to trap animals. Pitcher plants use nectar to lure insects into a pitcher-shaped trap (**FIGURE 35.16d**). The inside of the pitcher may have downward-facing hairs, which make it easy to crawl in but hard to crawl out. And in many pitcher plants, the surface of the pitcher about halfway down is covered with flaky wax. Any insect

**FIGURE 35.16**
**Death Awaits**

Most carnivorous plants capture and digest small animals such as insects (*a–c*), but tropical pitcher plants (*d*) have been known to digest vertebrates, including frogs, lizards, birds, and mice.

(a)

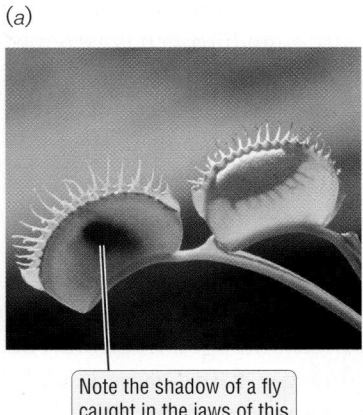

Note the shadow of a fly caught in the jaws of this Venus flytrap (*Dionaea*).

(b)

The long, sticky dermal hairs on the leaves of the sundew (*Drosera*) are touch-sensitive. They wrap around small insects and digest them with the help of secreted enzymes.

(c)

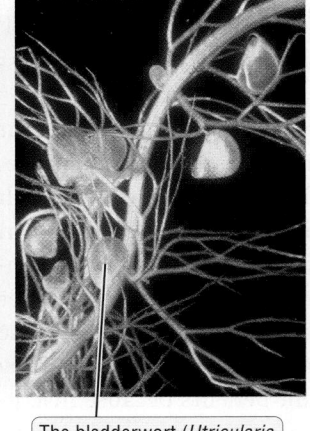

The bladderwort (*Utricularia neglecta*) sucks underwater animals into its bladders.

(d)

Perfect mouse skeletons have been discovered inside tropical pitcher plants, such as this *Nepenthes rafflesiana*.

# Maple Syrup and the Rites of Spring

Commercial maple syrup is made mainly from the sugar maple (*Acer saccharum*) and the black maple (*Acer nigrum*). In early spring, a sugary solution (maple sap) drips out from cuts or holes made in the tree trunk. To produce syrup, the sap is boiled down. Maple sap is about 2.5 percent sugar, so typically it takes about 40 gallons of sap to make 1 gallon of maple syrup, which must have 66.6 percent sugar to be legally sold as maple syrup. The hole for tapping maple sap is drilled into the xylem of the tree trunk, *not* the phloem. It would be very difficult for us to tap into the sugary solution that flows through the phloem because phloem sieve tubes are thin-walled, flimsy cells that clog up quickly if punctured. In addition, tree trunks contain much less phloem than xylem.

Maple sap flows in the trees for only a few weeks in early spring, when the trees are still leafless and there may be snow on the ground. Sap flow is the result of stem pressure. As the days start to warm up in spring, starch stored inside parenchyma cells in the stem and root is converted to sugar, mainly sucrose (the same type of carbohydrate as in cane sugar). A large amount of sugar is released into the surroundings, and much of it makes its way into tracheids and vessels simply because these hollow conducting tubes make up much of the tree trunk. The sugar release is greatest on the sun-warmed parts of the tree trunk, which is why sugar tappers place their holes on southern exposures.

The high concentration of sugar draws water from the surroundings and ultimately from the roots, which in turn absorb the water from the thawing soil. Water is drawn to the sugars by osmosis—the same reason that sieve tubes start soaking up water after sugars have been loaded into them. Just as uptake of water increases pressure inside a sieve tube, a large pressure, known as *stem pressure*, builds up inside the xylem conducting elements. Stem pressure pushes the sugary solution up the tree trunk and along the branches to the very tips, where the buds are located. These growing points of the plant (more about buds in the next chapter) lie dormant through the cold winter, and metabolism starts to revive within their cells with the return of daytime warmth. This "reawakening" of metabolism, and the subsequent growth, is fueled by the sugars delivered by stem pressure. As soon as the buds start consuming the sugar, the concentration of carbohydrate in the xylem conducting elements declines sharply and stem pressure disappears as quickly as it appeared. As the buds burst out in the first flush of growth, sugaring season draws to a close.

Tapping xylem sap from a sugar maple.

that steps onto this wax is doomed: the wax sticks to its feet, causing it to lose its grip and tumble into a vat of deadly digestive juices. Although pitcher plants usually digest small animals such as insects, they hold the record for the biggest animals captured and consumed by plants: some species occasionally trap and digest vertebrate animals, including frogs, lizards, birds, and even members of the class of animals to which we belong, the mammals.

Notice that all the carnivorous plants in Figure 35.16 are green. Carnivorous plants do not need carbohydrates or lipids from their animal prey; they can manufacture these macromolecules readily through photosynthesis. Enzyme analysis shows that "flesh-eating" plants, such as the pitcher plants, secrete primarily protein-degrading and nucleic acid–degrading enzymes into their traps. The protein-degrading enzymes release amino acids, a rich nitrogen source, and the nucleic acid–degrading enzymes release phosphate from the backbones of DNA and RNA. Carnivory appears to have evolved in plants exclusively for acquiring minerals in habitats that are deficient in these essential nutrients.

## 35.6 How Plants Transport Food and Water

The human circulatory system is impressive in terms of the pressure it can generate, but it pales in comparison to what trees can do. Against the pull of gravity, trees must lift water and dissolved nutrients from their roots to their leaves, which often are 20 meters (65 feet) or more above the ground. To manage this feat, a 20-meter-tall tree must generate a pressure of 4,500 millimeters of mercury (mm Hg), which is more than 25 times the 175 mm Hg of pressure generated by the human heart during exercise. Now consider the world's tallest living tree, a 116-meter-tall coast redwood (*Sequoia sempervirens* [sih-*KWOY*-uh *SEM*-per-*VEER*-unz]) growing in California. To supply its uppermost leaves with water and nutrients, this tree must generate an astonishing 25,500 mm Hg of pressure—more than enough to burst a human heart.

Plants do not have a muscular pump like a heart to transport water from their roots to their leaves; indeed, such a pump would have to be unrealistically large to produce the pressures needed in trees. Nor do plants use pumps to transport food from the leaves to the rest of the body. Instead of relying on pumps to move food and water, plants use clever alternative approaches that develop huge pressures at the cost of relatively little metabolic energy, or with no investment of metabolic energy at all.

## Phloem transport is driven by water pressure

Plants transport food—that is, the sugars that are produced by photosynthesis—in the sieve tubes of phloem tissue (FIGURE 35.17). Active transport is required to pump the sugars into the sieve tubes from the surrounding space. Active transport is necessary because the concentrations of sugar in the sieve tubes are much higher than in the surrounding leaf tissue. In fact, the contents of sieve tubes may be 10–30 percent sugar. As sugar is loaded into a sieve tube element, the sugar concentration rises. As a consequence, water molecules diffuse from the surrounding leaf tissue and enter the sugar-rich sieve tube by osmosis (see Section 7.2 for a review of osmosis). The phloem sieve

**FIGURE 35.17**
**Sugars Are Transported in Phloem Sieve Tubes**

Sugars are moved from tissues where they are produced (leaves in this example) to parts of the plant that consume sugar, such as buds. Fluid flow within a sieve tube is driven by a difference in water pressure, which is highest where sugars are loaded into the sieve tube (leaves in this example) and lowest where sugars are unloaded from the same sieve tube (buds in this example).

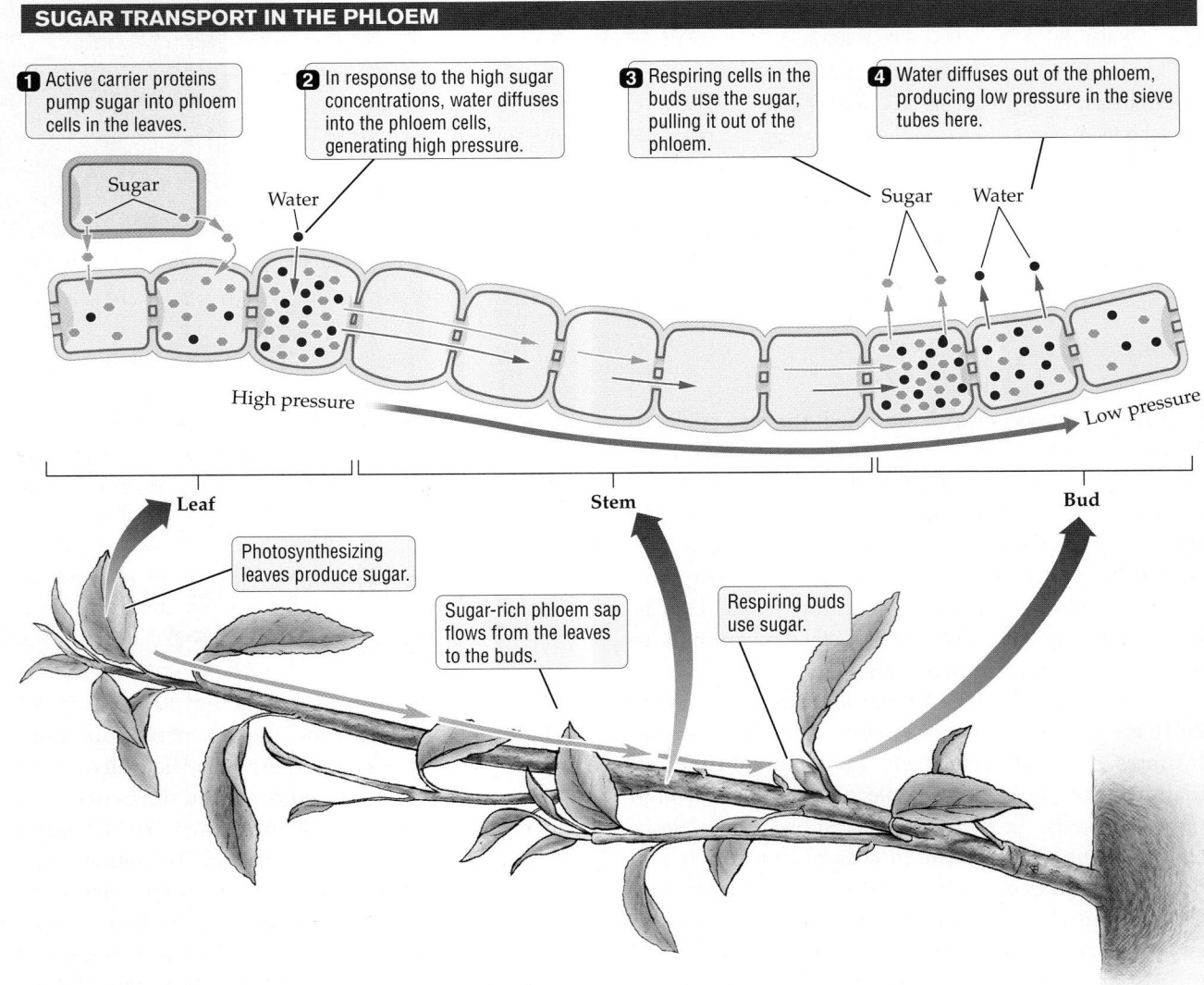

**SUGAR TRANSPORT IN THE PHLOEM**

1. Active carrier proteins pump sugar into phloem cells in the leaves.
2. In response to the high sugar concentrations, water diffuses into the phloem cells, generating high pressure.
3. Respiring cells in the buds use the sugar, pulling it out of the phloem.
4. Water diffuses out of the phloem, producing low pressure in the sieve tubes here.

Sugar

Water

Sugar      Water

High pressure

Low pressure

Leaf          Stem          Bud

Photosynthesizing leaves produce sugar.

Sugar-rich phloem sap flows from the leaves to the buds.

Respiring buds use sugar.

tubes fill with water and hence develop high pressures at the points where sugar is actively loaded into them. The region where sugar is being loaded into a sieve tube is like the faucet end of a garden hose: water is being poured into it from the faucet, so pressure is highest in this part of the hose.

In nonphotosynthesizing plant parts, which may be distant from the leaves, cells take up sugars from nearby phloem tissue. Once the sugars have been removed from the sieve tubes, water diffuses from the now sugar-poor tubes into the respiring cells. This osmotic loss of water from the sieve tube elements leads to a region of relatively low pressure in this part of the sieve tube. This region of a sieve tube, where sugars and water are being lost from the tubes, is like the far end of a garden hose where water gushes out, perhaps through a sprinkler head.

The difference in pressure between sugar-rich and sugar-poor regions of sieve tubes can easily reach 700 mm Hg—much greater than the pressure a human heart generates during exercise. This difference in pressure pushes food in the phloem from the leaves (where the pressure is high) to all plant tissues, such as buds, that use sugars for energy (where the pressure is low). Water moves in a garden hose for much the same reason: there is high pressure where water enters the hose from the faucet and low pressure where it exits the hose at the sprinkler end. The difference in water pressure is what pushes water from the faucet end to the sprinkler end and not in the opposite direction. In the same way, the difference in pressure between a sugar-loading zone and a sugar-unloading zone is what drives the long-distance transport of sugars through a sieve tube.

## Xylem transport is driven by transpiration

The process that lifts water and minerals in the xylem is remarkable: it generates enormous pressure, yet at no energy cost to the plant. No metabolic energy is needed because the process is driven by physical forces that result when water evaporates from the surface of the plant—it is driven ultimately by energy from the sun.

The evaporation of water from the shoot surface of a plant is known as **transpiration**. Transpiration takes place mostly through open stomata, since the aboveground parts of most plants are covered in a layer of wax (the cuticle) that is largely impermeable to water molecules. Transpiration rates are highest when stomata are open, which is in daytime, and also

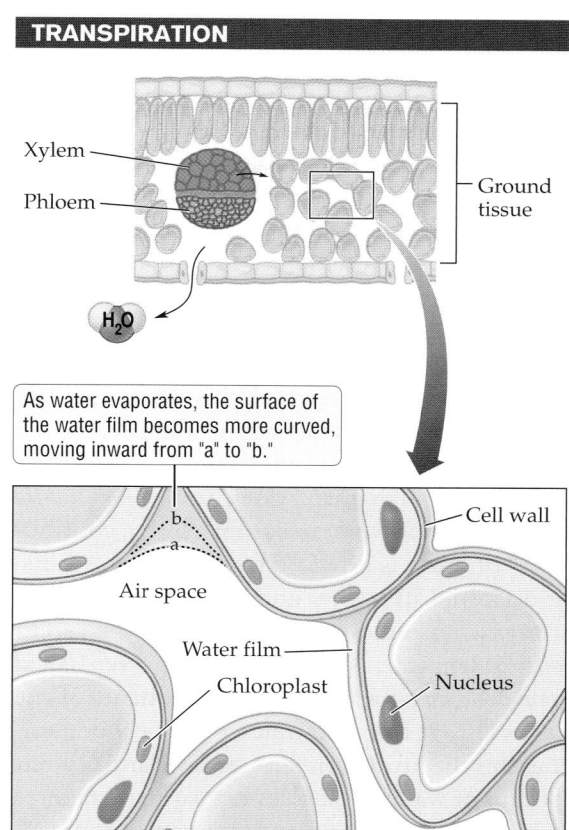

TRANSPIRATION

Xylem

Phloem

H₂O

Ground tissue

As water evaporates, the surface of the water film becomes more curved, moving inward from "a" to "b."

Cell wall

Air space

Water film

Chloroplast

Nucleus

Cytosol

**FIGURE 35.18**

**Transpiration Generates a Force That Lifts Water in Xylem Conducting Elements**

Water coats the surfaces of cells within a leaf, which also contains many air spaces. A pulling force, called surface tension, exists at all air-water boundaries. Surface tension generates a curved surface in the water film at an air-water boundary. Transpiration, caused by the loss of water through open stomata, causes the water film to recede into microscopic cracks in plant cell walls. As water is lost from the water film, it becomes more curved and its surface tension becomes stronger. The surface tension from numerous curved surfaces is transmitted to the water column in the xylem as a strong pulling force, or tension. Enormous pressures can be generated in this way, requiring no input of energy from the plant. The cohesion of water molecules maintains a continuous water column in the xylem.

on drier and windier days because these conditions increase the rate of evaporation.

Transpiration from leaf surfaces generates a lifting force called **tension** that tugs on the water column in the xylem. Xylem tension is the outcome of the *surface tension* that exists at all the countless air-water boundaries in a transpiring leaf (**FIGURE 35.18**). In a well-hydrated plant, water molecules stick to the hydrophilic cell walls of leaf cells because they display a property known as **adhesion** (the tendency of a substance to cling to other substances). Because leaf tissue encloses many air spaces, the film of water that adheres to cell walls is in direct contact with air. **Surface tension** is a pulling force that exists at all air-water boundaries. As water evaporates from cell surfaces into the leaf air spaces and then to the outside air, the remaining water film is drawn into microscopic cracks between cell walls. The more curved the air-water boundary becomes, the stronger the surface tension it generates. Because the water clinging to leaf cells is continuous with the water in xylem, the surface tension generated at leaf air-water boundaries is transmitted to the water column in the xylem. The strength of this pull, or tension, can be measured in the same units as pressure. It turns out that the tensions generated by millions and millions

## Concept Check Answers

2. **True.** The long-distance transport of water in xylem conducting elements is driven by tension, which in turn is generated by transpirational water loss. The evaporation of water, and the resulting pulling force (tension), is likely to be greater when stomata are open.

1. The uptake of inorganic minerals, such as potassium ions, is an active process; energy is needed for these ions to cross the plasma membrane and enter the cell-interior pathway. Light is the only source of energy for plants. Mineral uptake is hampered when carbohydrate supplies dwindle in the absence of photosynthesis.

of curved air-water boundaries can be very large: more than −67,000 mm Hg in some species. (When a force pulls rather than pushes, the pressure is reported as a negative number, hence the minus sign.)

The tension generated in the leaf pulls a column of water up the xylem, all the way from the roots, against the force of gravity. This water column is strong enough to withstand very high tensions, such as the −25,500 mm Hg tension in a coast redwood. **Cohesion**, the tendency of water molecules to cling to each other because of hydrogen bonding (see Section 5.3) is responsible for the strength of this water column. The molecules in a column of water hold together so tightly that they can withstand greater tensions than a steel cable of equal diameter does.

Overall, as water evaporates from a leaf, a continuous "cable" of water is lifted up from the roots, transporting dissolved nutrients and replacing the water lost through transpiration. This model of the ascent of water in xylem tracheids and vessels is often called the *tension-cohesion theory*. The key components of the theory are (1) that transpiration from the shoot surface generates the lifting force, or tension; and (2) that the cohesion of water molecules transmits that force all the way down to the root, so that a continuous column of water is lifted all the way from the roots to the evaporative surfaces in the leaf air spaces.

Anything that breaks the continuous column of water, such as an air bubble, breaks the cable and disrupts the lifting of water in that particular water column. Tall plants such as trees have billions to trillions of water columns and hence have multiple backup transport routes available when a particular column is blocked. Both drought and freezing temperatures promote air bubble formation, thereby blocking columns; if enough columns are blocked, xylem transport may become impossible.

If you snip a flower stalk, it is likely that the tracheids and vessels at the cut end will become clogged with air bubbles. Such flowers will wilt if you arrange them in a vase because the air-clogged xylem cannot draw any water. The standard advice from florists is to recut the stalk in water. Doing so removes the clogged portions of tracheids and vessels, while ensuring that the newly exposed xylem pulls in water instead of air.

### Concept Check

1. Potassium levels were found to decline in the leaves of a corn plant after it was kept in the dark for 4 days. Explain this observation.

2. True or false: Water in xylem vessels moves more rapidly when stomata are open than when they are closed. Explain your answer.

---

### APPLYING WHAT WE LEARNED

# How Plants Survive Extreme Cold

For many plants, the hazard of winter is ice. Warm-climate plants cannot tolerate temperatures at or below the freezing point of water (0°C). In such species, temperatures of 0°C (32°F) or lower that last for more than a few minutes can cause ice to form inside cells. Ice crystals shred cellular structures the way a chef's knife chops vegetables for a stir fry. Cold-climate plants that manage to cope with temperatures below 0°C do so by packing their cells with chemicals that, like car antifreeze, limit ice formation. A mixture of sugars and amino acids lowers the temperature at which the cell contents freeze. Oak, maple, and fir trees all use this approach, which is so effective that water does not freeze inside their cells until outside temperatures reach −40°C.

Once temperatures dive below −40°C, however, ice forms spontaneously, no matter how much antifreeze a plant has inside its cells. So how do plants survive extreme cold—temperatures of −50°C or even −100°C? The main strategy is to prevent ice from forming *inside* the cells, which would burst the cells. In the hardiest cold-climate plants, once temperatures drop below −40°C, ice crystals begin to form, but the crystallization begins *outside* the cells where there is little of the antifreeze chemicals. The ice that forms in the extracellular spaces draws water vapor out of the cells (the same phenomenon will dry laundry that's hung out on a sub-zero day). As water moves out of the cells and crystallizes outside the cell wall, the inside of the cell becomes more and more dehydrated, essentially becoming freeze-dried. Plants that tolerate extreme cold can tolerate having the cytoplasm dried. In fact, trees that are highly resistant to cold are also highly resistant to drought.

# The Claim: Exposure to Plants and Parks Can Boost Immunity

BY ANAHAD O'CONNOR, *New York Times*

### The Facts

This time of year [spring and summer], allergies and the promise of air-conditioning tend to drive people indoors.

But for those who can take the heat and cope with the pollen, spending more time in nature might have some surprising health benefits. In a series of studies, scientists found that when people swap their concrete confines for a few hours in more natural surroundings—forests, parks and other places with plenty of trees—they experience increased immune function.

Stress reduction is one factor. But scientists also chalk it up to phytoncides, the airborne chemicals that plants emit to protect them from rotting and insects and which also seem to benefit humans.

One study published in January included data on 280 healthy people in Japan, where visiting nature parks for therapeutic effect has become a popular practice called "Shinrin-yoku," or "forest bathing." On one day, some people were instructed to walk through a forest or wooded area for a few hours, while others walked through a city area. On the second day, they traded places. The scientists found that being among plants produced "lower concentrations of cortisol, lower pulse rate, and lower blood pressure," among other things.

A number of other studies have shown that visiting parks and forests seems to raise levels of white blood cells, including one in 2007 in which men who took two-hour walks in a forest over two days had a 50-percent spike in levels of [immune cells called] natural killer cells. And another found an increase in white blood cells that lasted a week in women exposed to phytoncides in forest air.

### The Bottom Line

According to studies, exposure to plants and trees seems to benefit health.

---

Every animal that lives on land ultimately depends on plants for food. We humans eat all kinds of plants—from the grains made by grasses to the leaves of lettuces, the roots of carrots and potatoes, and the flower buds of broccoli plants. And even when we eat animals, we are usually eating creatures that have themselves derived all their energy and nutrients from plants. Plants withstand constant predation by means of thorns and other physical defenses. But the number one way plants defend themselves from plant-eating organisms is chemicals.

Plants contain thousands of substances that help them defend themselves from attack by bacteria, fungi, and plant-eating animals. These substances, called secondary plant compounds, include, for example, dangerous toxins like strychnine and nicotine that make animals sick; the pungent flavors of onions and garlic; the resins secreted by pine and fir trees; and essential oils such as peppermint, sage, and lemon. Plants also use secondary compounds as fragrances or brightly colored pigments to attract pollinators.

The secondary plant compounds known as *phytoncides* are essential oils found in woods that have attracted the interest of practitioners of alternative medicine and aromatherapy. Like other secondary plant compounds, many phytoncides discourage bacteria, fungi, and insects from feeding on wood. This *New York Times* article reports that phytoncides released into the air by trees seem to stimulate immune function in humans.

The evidence described in the article supports the idea that people who go for a walk in the woods have reduced stress levels and elevated immune function. But do these effects really result from exposure to phytoncides? Maybe the enhanced immune function results from the exercise of walking in the woods, or from being relaxed by the walk in the woods away from everyday cares. Maybe immune function is enhanced by simply not being in a stressful office work environment. Perhaps the reduced stress and elevated immune function result from multiple factors.

It's possible that lack of normal stresses, rather than the presence of phytoncides, caused the elevated immune response. To find out, the same researchers studied 12 men who stayed for three nights in hotel rooms where oils from cypress tree roots were wafted around the room. The men's immune function was elevated—even though they weren't in a forest. But this study had holes in it, too. The control for the experimental "treatment" was their immune function on a normal workday. It's still possible that the hotel room was simply more relaxing than a regular day at work and the phytoncides did nothing.

### Evaluating the News

**1.** What are the two main effects reported for "forest bathing"?

**2.** Propose two different experiments that would help reveal whether plant compounds in the air are actually causing the elevated immune function or if it's just looking at the trees, being away from home, or being relaxed that brings about the change. *Hint:* Think about ways you could take a relaxing nature walk without smelling trees and other plants; now think about ways you could smell living trees without changing anything else about your life.

**SOURCE:** *New York Times*, July 5, 2010, http://www.nytimes.com.

# CHAPTER REVIEW

## Summary

- There are two main categories of flowering plants: dicots and monocots. Dicot plants usually have a netlike pattern of veins in the leaf, a taproot, a ring of vascular bundles in the stem, flower parts in fours or fives, and two cotyledons in the embryo inside each seed. Monocots typically have parallel leaf veins, fibrous roots, scattered vascular bundles in the stem, flower parts in threes, and one cotyledon.

### 35.1 An Overview of the Plant Body

- The plant body contains two basic systems: the belowground root system and the aboveground shoot system.
- Plant bodies are made of three basic tissue types: dermal, ground, and vascular tissues.
- Root, stem, and leaf are three of the basic plant organs. Flower and fruit are other examples of plant organs.
- Plants are built differently from animals: aboveground, plants grow in size by the repeatedly adding the same basic bud-stem-leaf unit. Unlike most animals, plants grow throughout their lives.

### 35.2 Plant Organs

- Roots absorb water and mineral nutrients, anchor the plant, and store food. Root hairs greatly increase the surface area through which plants can absorb water and mineral nutrients.
- Stems support the plant against gravity. In many species, stems also contain ground tissue cells that perform a limited amount of photosynthesis. Stems can be highly modified to perform specialized tasks such as protection, climbing, and food storage.
- Leaves feed the plant: ground tissue cells in the leaf perform photosynthesis and produce most of the plant's food. Leaves can be highly modified to perform specialized tasks such as protection, climbing, food storage, and collection of mineral nutrients.
- Guard cells in leaf dermal tissues control the rate of gas exchange by regulating the width of the stomata (air pores). $CO_2$ is absorbed from the air, and $O_2$ and water are lost to the air when stomata are open.

### 35.3 Plant Tissues

- Dermal tissues interact with the environment, controlling the flow of materials into and out of the plant and protecting the plant from attack. Dermal cells include dermal hairs, root hairs, and guard cells.
- Ground tissues make up the bulk of the plant body and play many essential roles, including support, wound repair, and photosynthesis.
- Phloem and xylem are the two types of vascular tissue, which transport sugars and water, respectively, through the entire plant body.
- Phloem conducting tubes (known as sieve tubes) are made up of living cells connected to one another via pores through which the food moves.
- Xylem conducting tubes are of two types: tracheids, which are long and narrow; and vessels, which are shorter and wider. Both tracheids and vessels are made up of hollow, dead cells arranged end to end.

### 35.4 How Plants Obtain Nutrients

- To grow, plants need $CO_2$, water, and mineral nutrients. Most of the dry weight of a plant comes from $CO_2$ converted into organic material.
- Plants need nine macronutrients (especially N, P, and K) in large quantities, and eight micronutrients in small quantities.
- Plant roots absorb water and mineral nutrients from the soil solution by two routes: a cell-interior pathway dependent upon plasmodesmata that connect the cytosol of one cell to that of the next; and a cell-wall pathway in which nutrients move along plant cell walls.
- Because mineral nutrient concentrations are much higher within the cytosol of root cells than in the soil, plants must expend energy (use active transport) to absorb essential minerals.
- Most plants form structures, called mycorrhizae, in which plant roots are united with fungal hyphae. Mycorrhizae greatly increase the ability of plants to absorb nutrients from soil.
- Some plants, most notably those in the bean family, have a mutually beneficial relationship with nitrogen-fixing bacteria that can turn $N_2$ into ammonium ($NH_4^+$), which is a usable form of nitrogen.

### 35.5 Plants That Digest Animals

- Some plants that live in environments with nutrient-poor soils obtain mineral nutrients by capturing and digesting animals.
- Amino acids (rich in nitrogen) and nucleotides (rich in phosphorus) are the main substances carnivorous plants obtain from their animal prey.

### 35.6 How Plants Transport Food and Water

- Tall trees require much higher pressures to transport food and water than animals do.
- Plants use water pressure to transport food through the phloem. Cells in the leaves use active transport to pump sugar into phloem cells, lowering the concentration of water in those cells. Water then flows into the sugar-rich phloem cells, creating a zone of high pressure that pushes food from the leaf to parts of the plant that absorb the sugar (where the pressure is low).
- The transport of water and mineral nutrients through the xylem relies on tension created by transpiration (evaporative water loss from the leaf surface) and two of the distinctive properties of water molecules: adhesion and cohesion. Surface tension, which is a pulling force, exists at all air-water boundaries. As water is lost via transpiration, the water film becomes more curved and surface tension grows stronger (more negative). The cohesion of water molecules maintains the continuity of the water column in xylem conducting elements.

# Key Terms

active transport (p. 766)
adhesion (p. 771)
cohesion (p. 772)
collenchyma (p. 763)
cotyledon (p. 756)
cuticle (p. 761)
dermal hair (p. 761)
dermal tissue (p. 757)
dicots (p. 756)
endodermis (p. 766)

fiber (p. 763)
fibrous root (p. 757)
ground tissue (p. 757)
guard cell (p. 761)
macronutrient (p. 765)
micronutrient (p. 765)
monocots (p. 756)
mycorrhiza (p. 767)
nitrogen fixation (p. 767)
parenchyma (p. 762)

phloem (p. 763)
plasmodesma (p. 764)
root cap (p. 758)
root hair (p. 761)
root system (p. 757)
sclerenchyma (p. 763)
shoot system (p. 757)
sieve tube (p. 764)
stoma (p. 760)
surface tension (p. 771)

taproot (p. 756)
tension (p. 771)
tracheid (p. 764)
transpiration (p. 771)
vascular tissue (p. 757)
vessel (p. 764)
xylem (p. 763)

# Self-Quiz

1. The waxy covering on leaves and stems is called the
   a. dermal tissue.
   b. cuticle.
   c. endodermis.
   d. Casparian strip.

2. Water and mineral nutrients are transported throughout the plant body mainly in
   a. phloem tissue.
   b. ground tissue.
   c. dermal tissue.
   d. xylem tissue.

3. The most common route by which plants absorb mineral nutrients into their roots is
   a. directly into the phloem.
   b. directly into the xylem.
   c. through root hairs.
   d. passively, by diffusion from the soil.

4. Compared with taproot systems, fibrous root systems usually
   a. hold soil more firmly.
   b. penetrate deeper into the soil.
   c. have no ground tissues.
   d. transport only water, not mineral nutrients.

5. The small openings that connect adjacent cells throughout the plant body are called
   a. dermal hairs.
   b. guard cells.
   c. plasmodesmata.
   d. xylem.

6. Which property of water molecules is responsible for maintaining the continuity of the water column in xylem conducting tubes?
   a. dehydration
   b. cohesion
   c. surface tension
   d. embolism

7. The vast majority of plants alive today are
   a. nonvascular plants.
   b. gymnosperms.
   c. conifers.
   d. angiosperms.

8. Outgrowths of dermal cells that have special functions in protection and nutrient absorption are called
   a. cell walls.
   b. dermal hairs.
   c. guard cells.
   d. dermal meristemoids.

# Analysis and Application

1. Compare monocots and dicots. Name three members of each of these plant groups.

2. List the three organs and three tissue types found in almost all flowering plants. For each of the tissue types, describe at least two functions. Similarly, for each organ, list typical and specialized tasks performed by that organ.

3. Describe the two routes by which mineral nutrients are absorbed by plant roots.

4. By placing a plant in a drying oven to remove its water, we can determine what accounts for most of the (dry) weight of the plant. Do mineral nutrients absorbed by the roots account for this weight? Explain your answer.

5. What is distinctive about the habitats where carnivorous plants are typically found? What types of nutrients do they obtain from the animals they prey on?

6. How do tall trees generate enough pressure to move water from their roots to their uppermost leaves? Similarly, how do plants generate enough pressure to move food produced by their leaves to other parts of their bodies?

# 36 Plant Growth and Reproduction

**FALL COLOR IN THE SAN JUAN MOUNTAINS, COLORADO.** Each clump of aspens in this photograph is a different color. Aspen trees grow in groups of genetically identical individuals, or clones. Each genetic clone responds to its environment differently and turns color at its own particular pace.

# A Forest Walks

Villains in literature have a problem with forests that walk. Consider the lead character in *Macbeth*, Shakespeare's tale of ambition, murder, and madness. Obsessed with power after he kills the king and usurps the throne, Macbeth counts on the prophecy that he will never be defeated until a forest called Birnam Wood "shall come against him." But when an army carrying branches from Birnam Wood marches on his castle, Macbeth's complacency turns to horror.

Similarly, in Tolkien's *Lord of the Rings*, the evil plans of the wizard Saruman are doomed when a forest of walking trees surrounds his fortress and rips down its walls. Of course, forests do not walk around, let alone wage war, in the sense of *Macbeth* and *The Lord of the Rings*. A tree cannot pull up its roots and walk from one place to another. But if we look into the biology of trees more deeply, we indeed find many forests "on the move."

For example, individual aspens colonize the space around them by producing genetically identical extensions of themselves—called clones—that eventually separate from the original tree. Trees do this by producing rhizomes, underground stems that grow from the base of the trunk. After growing underground for a few meters, the rhizomes turn upward and sprout, giving rise to a new tree that is genetically identical to the original tree from which it came. The original tree has reproduced asexually by making a genetic copy, or clone, of itself, and the tree has also advanced or "walked" a short distance.

How far can clones move, and how long can they live? How can forests travel the long distances that will be necessary as climate change forces them to live farther north? How can aspen clones help us think about the longevity of trees and other plants, as well as how they migrate?

Many other plants besides trees travel about, by sending out rhizomes or in other ways. Before we tackle these issues, let's explore the varied and flexible ways in which plants grow, develop, and reproduce.

## MAIN MESSAGE

Flowering plants compensate for their inability to move by having a flexible pattern of growth and a variety of ways for reproducing, interacting with the environment, and defending themselves.

## KEY CONCEPTS

- Meristems, which are groups of undifferentiated cells, enable plants to grow throughout their lives and to add new body parts as needed. Plants increase in length when apical meristems, located at the tips of the shoots and roots, divide. Many plants can increase in thickness through cell divisions in a lateral meristem.

- The plant life cycle is characterized by an alternation of haploid and diploid generations.

- Flowers are a means for bringing sperm and egg together. The male part of the flower produces pollen, which transports sperm cells to the ovary, which houses egg cells.

- Fertilization of an egg cell by a sperm cell produces the zygote; another type of fertilization simultaneously gives rise to the endosperm, a tissue that stores food for the developing embryo.

- The zygote undergoes many mitotic divisions to create the embryo, which, together with the surrounding tissues, matures into a seed. When conditions are favorable, the seed germinates and the embryo grows out as a seedling.

- Plant hormones control how plants grow, develop, and respond to their environment.

- Plants defend themselves against herbivores and pathogens using physical and chemical weapons. These weapons include a tough outer surface and nonspecific chemical defenses.

THERE ARE NEARLY 300,000 SPECIES OF PLANTS, and they can be grouped into three categories on the basis of their life history: *annuals, biennials,* and *perennials.* **Annuals** (from *annus,* "year") are plants that complete their entire life cycle in one year. In angiosperms (flowering plants), an annual has 1 year to grow from a seed into a mature plant, produce flowers, and make the seeds that will start the next generation. Some plants are **biennials** (*bi,* "two"; *ennial,* "year"), which means they need 2 years to complete their life cycle. In its first year, a biennial grows and matures in all aspects but one: it does not reproduce. Reproduction takes place in the second year of growth. Finally, many angiosperms are **perennials** (*per,* "throughout"), which means that they live 3 years or more—sometimes hundreds, even thousands, of years.

In this chapter we use these three terms to help us understand how flowering plants grow and reproduce. We begin with a look at the unique ways in which plants grow. Then we examine how plants reproduce, how new individuals begin their development, and what roles hormones play in the lives of plants. Finally, we discuss the powerful mechanisms that plants use to defend themselves from attack.

## 36.1 How Plants Grow: Indeterminate Growth

To appreciate how plants grow, first consider your own body. Like all other humans, you grew rapidly from a fertilized egg according to a fixed developmental plan (one head, two arms, two legs, and so on). Once you reach adulthood (if you haven't already), you stop growing. You may gain weight, but you will not get taller, and you will not add new arms or heads or other major body parts.

Plants are completely different because they do not grow according to a rigid developmental plan. Instead, the plant body exhibits **indeterminate growth**, which means it can grow throughout its life, adding new body parts as needed. The plant body is *modular,* formed by the repeated addition of the same basic units. Aboveground, plants get bigger through the repeated addition of the bud-stem-leaf unit shown in **FIGURE 36.1**. Belowground, plants get bigger through the repeated formation of a root axis with lateral roots branching off from it.

### PLANT MERISTEMS

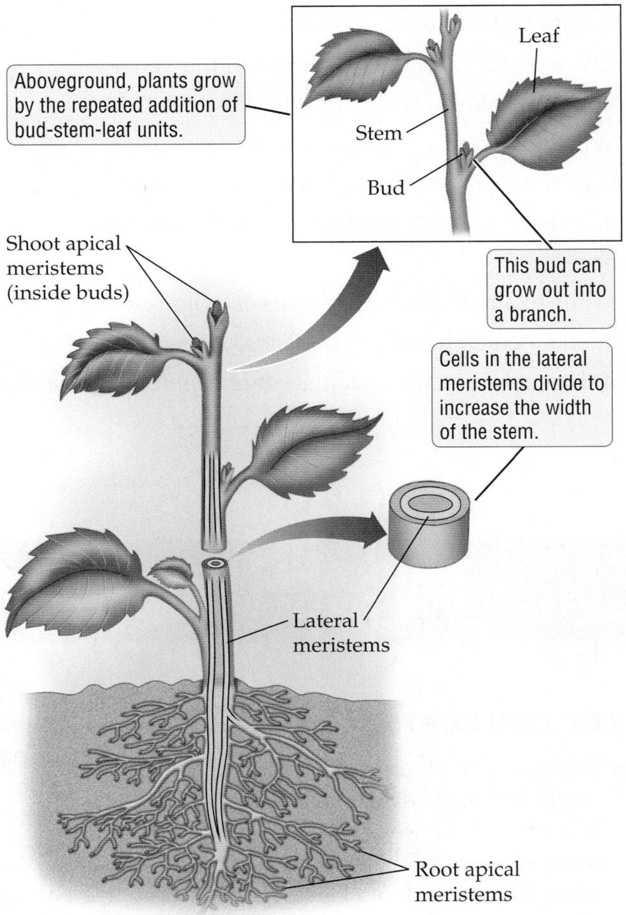

FIGURE 36.1 **How Plants Grow**
Plants grow throughout their lives, increasing in size and adding new parts as needed. They can do this because they have two types of undifferentiated (perpetually young) meristem tissue: apical meristem tissue (shown in red) and lateral meristem tissue (shown in yellow). Apical meristems enable the plant to increase in length both aboveground and belowground. Lateral meristems enable many plants to increase the thickness of their stems and roots as they grow.

The indeterminate approach to growth helps plants adjust to varying environmental conditions, such as high or low levels of sunlight, water, or nutrients. If conditions are favorable, a plant may add many new body parts and take on one shape, but if conditions are unfavorable, the plant may add few new body parts and take on another shape. Chicory, for example, is an herbaceous (nonwoody) plant that grows 4 feet tall in good conditions, producing dozens of flowers. However, when the top of a chicory plant is repeatedly cut off (as by a cow or a lawn mower), the plant eventually "gives up" its attempts to grow to its usual height. Instead, it reaches a maximum height of only a few inches, grows bushier, and produces fewer flowers. Though stunted, the plant survives and reproduces. Similarly, when two trees grow very close together, each may resemble half of a typical tree, and from a distance the two trees look like one tree. The two trees may not take the shape that is normal for the species and best suited for collecting sunlight, but the flexibility of their developmental plan enables these plants to cope with a less-than-ideal situation. Developmental flexibility also enables plants to replace damaged tissues and organs.

How are plants able to grow and add new body parts throughout their lives? The reason is that they have several different types of **meristems**, which are groups of perpetually young, undifferentiated cells that divide to give rise to new cells (see Figure 36.1). Vertebrates, including humans, have similar cells, called stem cells. We have a number of different types of stem cells, but the ones that can differentiate into nearly any type of human cell are found only in embryos, not in adults (see Chapter 11). In contrast, meristem cells remain active in adult plants, enabling the plant to add new body parts as needed. Like stem cells, meristem cells renew themselves through mitosis and also produce daughter cells that give rise to the specialized cell types. In fact, plants are so flexible in their development that most living cells in the adult plant body can generate whole new plants.

## 36.2 How Plants Grow: Primary and Secondary Growth

Populations of meristem cells are present at specific locations throughout the shoot and root system. Meristems located at the tips of shoots and roots increase the length of the shoot and root system, respectively. Most perennial plants also have a hoop-like band of meristem cells that divide to increase the thickness of the stem or root.

## Plants increase in length by primary growth

If you were to watch a tree grow, you would find that the tree increases in length (upward and outward) only at the tip of its central stem (trunk) and the tips of its branches (**FIGURE 36.2**). A group of meristem cells, called an **apical meristem**, is located at the tip (apex) of every stem and every root. A shoot apical meristem lies at the tip of every main stem and every side branch in the shoot system. Similarly, every root, whether it is a taproot or a side (lateral) root, has a root apical meristem at its end (see Figure 36.1). Cells in these apical meristems divide to produce new cells, and the increase in length that results is known as **primary growth**.

When an apical meristem cell divides, some of the daughter cells remain as undifferentiated meristem tissue. Other daughter cells divide repeatedly to produce descendant cells that begin to differentiate into new

### PRIMARY AND SECONDARY GROWTH

If you drive a nail into a tree…

…and return some years later, you'll notice:
**1** The nail remains at the same height even though the tree is taller.
**2** The nail is partially buried in the trunk.

**FIGURE 36.2  Growing Up and Out**
This simple experiment illustrates that as a tree grows taller, not all of its parts increase in length; otherwise the nail would move upward with time. Plants get taller through primary growth, by growing at their tips. Many plants also exhibit secondary growth, increasing in thickness over time, as the fact that the nail becomes partially buried demonstrates.

Secondary growth is generated by a vascular cambium. Secondary xylem, produced by a vascular cambium, is called wood. One annual ring consists of all the wood made in spring and early summer (early wood) plus all the wood made in late summer and early fall (late wood).

**SECONDARY GROWTH**

**(a) Primary growth**

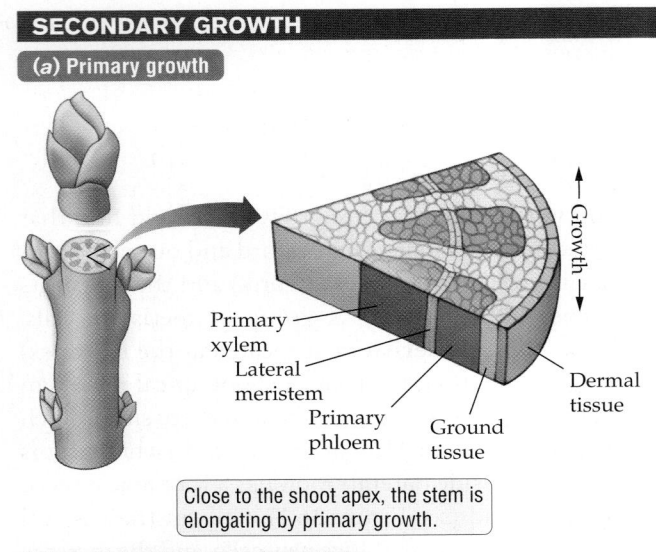

Primary xylem
Lateral meristem
Primary phloem
Ground tissue
Dermal tissue
Growth

Close to the shoot apex, the stem is elongating by primary growth.

**(b) First year of secondary growth**

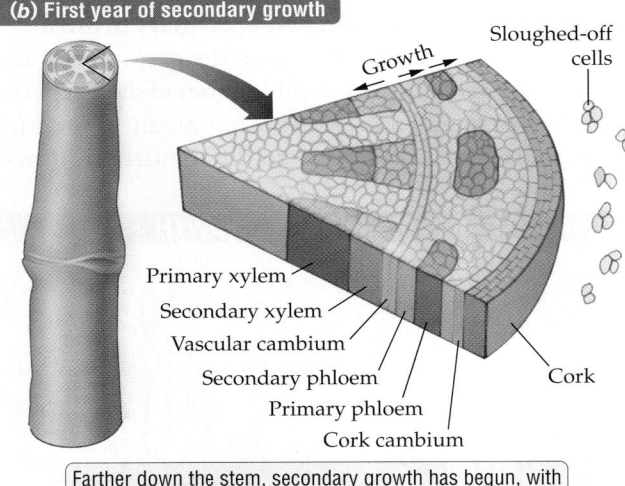

Growth
Sloughed-off cells

Primary xylem
Secondary xylem
Vascular cambium
Secondary phloem
Primary phloem
Cork cambium
Cork

Farther down the stem, secondary growth has begun, with the vascular cambium producing secondary xylem toward the inside and secondary phloem toward the outside.

**(c) Second year of secondary growth**

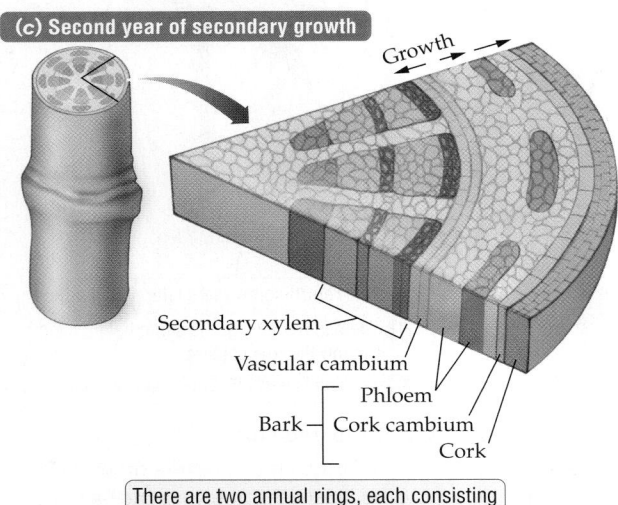

Growth

Secondary xylem
Vascular cambium
Phloem
Bark — Cork cambium
Cork

There are two annual rings, each consisting of early wood (pink) and late wood (red).

body parts. Aboveground, for example, some descendant cells form a new bud, others form a new leaf, and still others add to the main stem. As long as growth conditions remain favorable, a plant can increase in length by adding new bud-stem-leaf units to the tips of existing ones. It can also get bushier by branching: each bud contains a shoot apical meristem that can grow out as a branch. Belowground, a parallel process adds greater length and new branches to the root system.

## Woody perennials increase in thickness by secondary growth

As they grow, many plants increase in thickness as well as in height—that is, in girth as well as length. The increase in thickness is most obvious in woody perennials such as trees, so our discussion will focus on trees. And although we describe this process of growing thicker for stems, bear in mind that a similar process thickens the root system.

Trees and other woody plants grow outward when a second type of meristem tissue—the **lateral meristem**—begins to divide. A lateral meristem called the **vascular cambium** is responsible for most of the thickening in woody stems and roots. As the cells of the vascular cambium divide, their daughter cells take on one of three identities: they remain undifferentiated meristem cells, develop into new xylem tissue, or develop into new phloem tissue (**FIGURE 36.3**). Whenever the vascular cambium divides to increase the thickness of a plant, **secondary growth** is said to have occurred. Xylem and phloem produced by the vascular cambium are called *secondary xylem* and *secondary phloem* to distinguish them from the original, or primary, xylem and phloem that formed during primary growth (when the plant was increasing in length but not thickness).

As it grows each year, a tree adds new layers of secondary xylem inside the vascular cambium and new layers of secondary phloem outside the vascular cambium. Most trees produce more secondary xylem (for transporting water) than secondary phloem (for transporting food). What we call **wood** is actually many layers of secondary xylem.

In woody plants that grow in regions with a cold winter, there is a drastic change in the appearance of the secondary xylem from early to late in the growing season. **Early wood**, produced in spring and early summer, is lighter in color because it contains tracheids and vessels with a wider diameter (and therefore larger air spaces in their hollow interiors) and relatively thin cell walls. **Late wood**, made from midsummer to early fall, contains narrower conducting elements that have relatively thick walls, which makes the wood appear darker. Late wood is the "safer" option toward the close of the growing season, when soil water is depleted by the vigorous growth of vegetation and plants often experience water stress. The narrower conducting elements are less liable to become clogged with air bubbles (see Section 35.6). One ring of early and late wood constitutes one **annual ring**, which is all the wood produced in a single growing season. In **FIGURE 36.4a**, each light band of secondary xylem, and the dark ring immediately outside it, together make up one annual ring.

In many trees, the older secondary xylem in the center of the trunk becomes clogged and its color darkens as chemicals such as tannins are deposited in it. This darkly stained xylem at the center of the tree is called **heartwood** (see Figure 36.4a); although heartwood is too clogged to transport water, it provides support to the tree. Because of all the defensive plant chemicals deposited in it, heartwood tends to be resistant to decay. Newer secondary xylem that can transport water ("sap") is called **sapwood**.

All the tissues outside the vascular cambium are collectively known as the **bark**, which has two broad regions: an inner and an outer bark (see Figure 36.4a). The outer bark contains a lateral meristem called the **cork cambium**. As the cork cambium divides, some of the daughter cells differentiate into a tissue called cork. Cork cells have thick, water-repellent cell walls. They replace the original (primary) dermal layer to create a tough barrier that deters most herbivores and keeps out most pathogens (disease-causing organisms such as fungi and bacteria). As secondary tissues accumulate, the cork layers often develop cracks, creating the furrows we commonly see on the surfaces of tree trunks.

The inner bark contains the functioning, most recently made, secondary phloem. A tree can be killed if the bark on its main trunk is cut all the way around and deep enough to slice entirely through the functioning secondary phloem (**FIGURE 36.4b**). A tree killed in this manner is said to have been "girdled."

## Concept Check

1. What are meristems?

2. Compare primary growth and secondary growth.

---

## TREE RINGS

### (a) Structure of a tree trunk

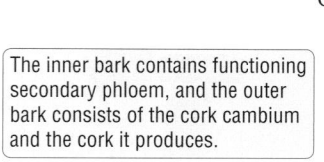

The inner bark contains functioning secondary phloem, and the outer bark consists of the cork cambium and the cork it produces.

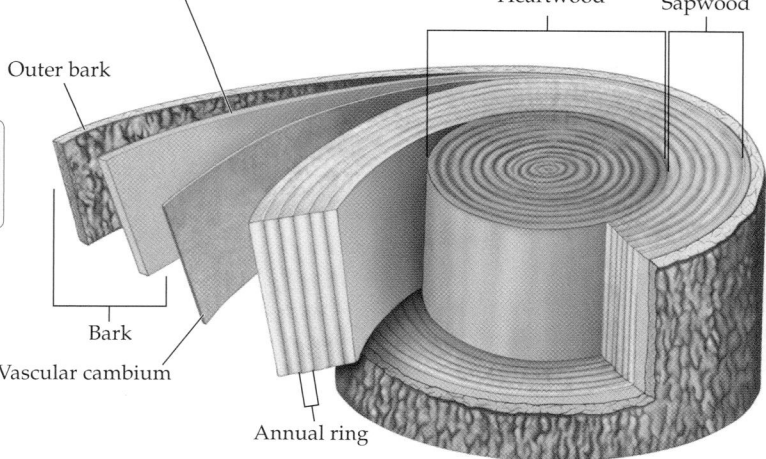

Inner bark

Outer bark

Heartwood

Sapwood

Bark

Vascular cambium

Annual ring

### (b) Girdled tree trunk

**FIGURE 36.4 The Insides of a Tree**

Cutting through a tree trunk or branch reveals several distinct layers. (a) Most of a tree's thickness is due to added layers of secondary xylem. The "tree rings" we use to age a tree are caused by the abrupt change in the appearance of successive annual rings, as one year's early wood is viewed adjacent to the previous year's late wood. Sapwood can transport water, whereas heartwood is xylem that has become clogged and can no longer transport water, although it retains its strengthening function. The inner bark contains functioning secondary phloem, and the outer bark consists of the cork cambium and the cork it produces. (b) Girdling a tree—cutting through the layer of inner bark all the way around the main trunk—will kill it.

## 36.3 Producing the Next Generation: Flower Form and Function

Many plants can reproduce asexually, as when a parent plant forms a genetically identical clone. Most plants can also reproduce sexually. The overall principle of sexual reproduction in plants is similar to that in animals: A haploid male gamete (**sperm**) fuses with a haploid female gamete (**egg**) to give rise to a diploid cell, the **zygote**, which undergoes mitosis to create a multicellular diploid **embryo**. In time, the embryo develops into an individual offspring, which represents the next generation. As noted in Section 10.5, haploid cells have half the chromosome set of diploid cells, and the haploid set is represented by *n*.

## The life cycle of plants is characterized by an alternation of generations

Plant and animal life cycles differ in one key respect: in animals, meiosis creates gametes, and nothing but gametes. In plants, meiosis generates haploid cells called **spores**. Each spore undergoes mitotic divisions to create a haploid, multicellular tissue known as a **gametophyte** (literally "gamete-bearing plant"). Specialized cells in the gametophyte differentiate to produce sperm or egg cells.

Gametes are the only haploid cells in an animal, and animals have no such thing as a gametophyte. However, gametophytes—the haploid, multicellular products of meiosis—are part of the life cycle of every plant. In nonseed plants, such as mosses and ferns, the spores generated by meiosis are released from the mother plant and give rise to free-living gametophytes (**FIGURE 36.5**). Because these gametophytes live

**FIGURE 36.5 Plants Display Alternation of Generations**

The life cycle of plants is characterized by the alternation of two phases: a haploid phase (shown in purple) and a diploid phase (orange). The fern plant, with its green fronds, is a diploid, multicellular individual (a sporophyte). If you turn over the fronds (leaves), you might see raised, brown patches. Meiosis occurs in these structures, called sporangia (singular "sporangium"), giving rise to haploid spores that are shed into the air. When they encounter a suitable environment, the spores produce a multicellular gametophyte. Certain cells in the gametophyte differentiate into sperm or egg cells. In some species, both types of gametes are formed on the same gametophyte; other species produce separate male and female gametophytes. Fern gametophytes live independently of the sporophyte and are often photosynthetic. Fern species that are commonly grown as houseplants produce gametophytes that are about 8 millimeters (0.3 inch) wide. If you examine the soil around an old potted fern, you might find the flat, green, heart-shaped gametophytes on the soil surface.

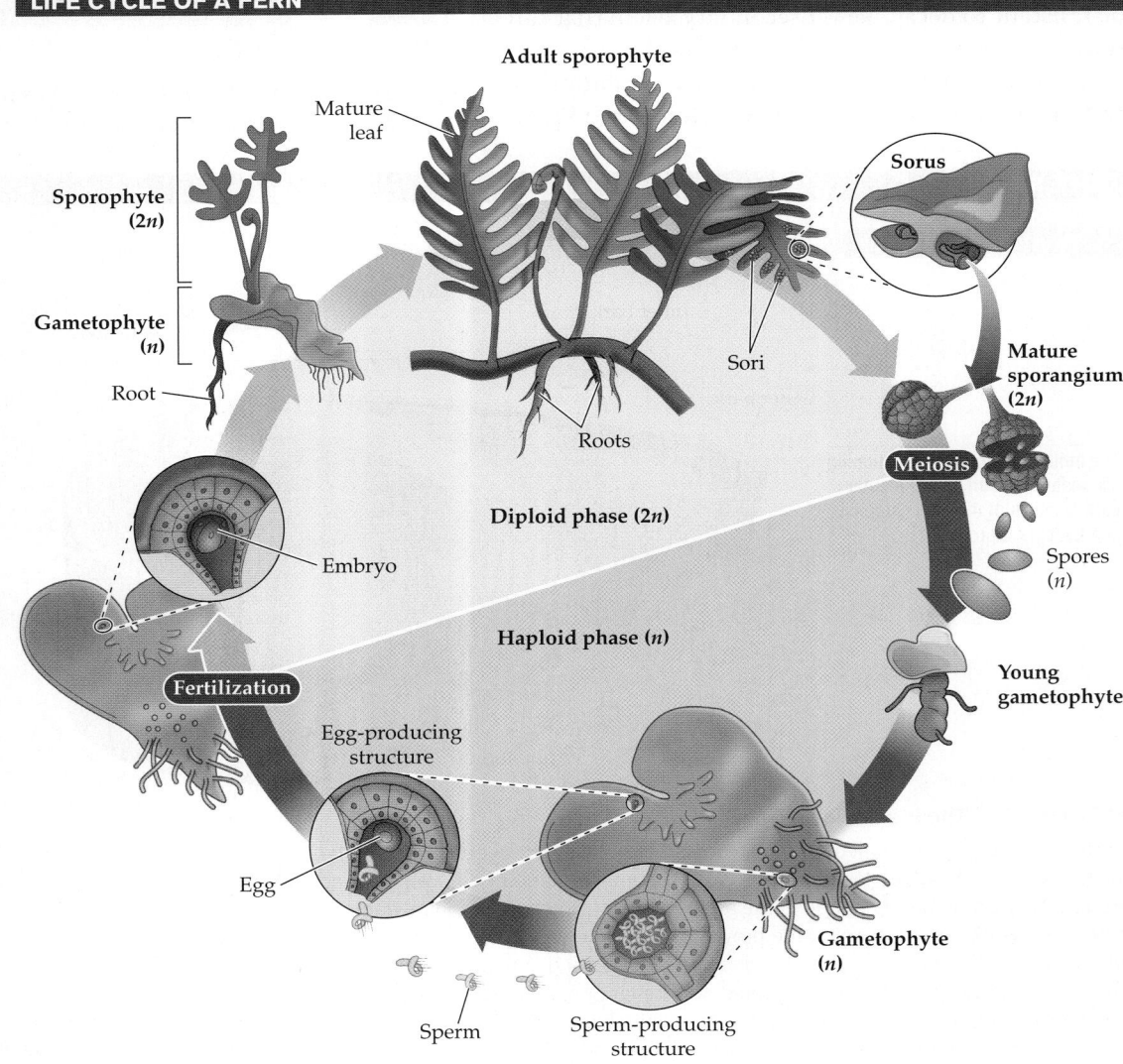

**LIFE CYCLE OF A FERN**

independently, as separate "gamete-bearing plants," they are regarded as a new generation, the gametophytic generation. Some seedless plants, and all seed-producing plants, produce separate male and female gametophytes that produce sperm and egg, respectively. When an egg is fertilized by sperm to give rise to a zygote, the first diploid (2n) cell of the next generation is created.

The zygote divides mitotically to create the embryo, which then grows into a diploid, multicellular individual that, in plants, is known as a **sporophyte**. The sporophyte of a plant is analogous to an individual animal in that both are multicellular, diploid organisms. We need a special term for such an individual in plants because the plant life cycle includes another multicellular structure, the haploid one known as the gametophyte. The life cycle of a plant is said to exhibit an **alternation of generations** because to complete its life cycle, every plant must produce two distinctly different multicellular "generations": haploid, multicellular gametophytes, which then give rise to the diploid, multicellular sporophytes of the next generation (see Figure 36.5). When you look at a flowering plant—dandelion, rosebush, or oak tree—you are looking at a sporophyte. The gametophytes of seed plants are microscopic, and the female gametophytes are held on the mother plant instead of living independently, the way moss gametophytes do. In contrast, the male gametophyte of seed plants is released, sometimes in huge numbers, by the mother plant. The male gametophyte of seed plants is familiar to us as pollen.

## Pollen delivers sperm cells to egg cells that reside in the ovary

Flowers and fruits, the two evolutionary innovations that are unique to flowering plants, are credited with the spectacular success of the angiosperms (see Chapter 3). Mosses and ferns rely on water to splash the sperm toward the eggs but, as you might imagine, this is not the most efficient way to get the job done for any plant that does not form a ground-hugging mat in moist places. In angiosperms, sperm are delivered to the eggs in a more efficient manner, and the flower is the structure that makes such efficiency possible. The flower houses the structures that produce male and female gametes and, in many species, also facilitates the delivery of the sperm-bearing gametophytes (pollen) to the female reproductive organs (carpels). Portions of the carpel develop into the fruit and, as we will see shortly, fruits help disperse the seed, which represents the next generation, in highly effective ways.

The typical flower has four types of structures, each arranged in a circular pattern known as a whorl (**FIGURE 36.6**). Moving from the outermost whorl inward, the four whorls are *sepals*, *petals*, *stamens*, and *carpels*. **Sepals** [SEE-pulz] enclose and protect the flower before it opens; **petals**, if they are colorful or scented or both, attract animals to the flower. **Stamens** [STAY-munz]—the male reproductive part of the flower—consist of an anther on top of a slender stalk, the filament. The anther produces pollen, which contains sperm. **Carpels**—the female reproductive part of the flower—consist of a stigma, style, and ovary. The egg is contained within the ovule, which is located inside the ovary. The female gametophyte of seed plants is called the **embryo sac**.

**FLOWER STRUCTURE**

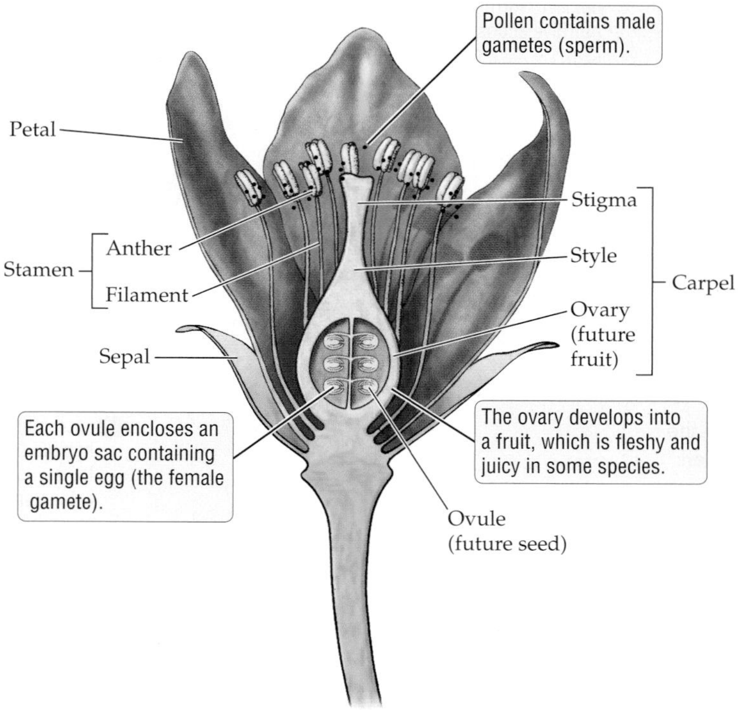

Pollen contains male gametes (sperm).

Petal

Stigma

Stamen — Anther

Style

Filament

Carpel

Sepal

Ovary (future fruit)

Each ovule encloses an embryo sac containing a single egg (the female gamete).

The ovary develops into a fruit, which is fleshy and juicy in some species.

Ovule (future seed)

**FIGURE 36.6 Four Whorls Make a Flower**
The various parts of a flower are arranged in concentric rings, or whorls. From the outermost whorl inward, a typical flower consists of four whorls: sepals, petals, stamens, and carpels. All the petals in a flower are collectively known as the "corolla." The collective term for all the sepals is the "calyx."

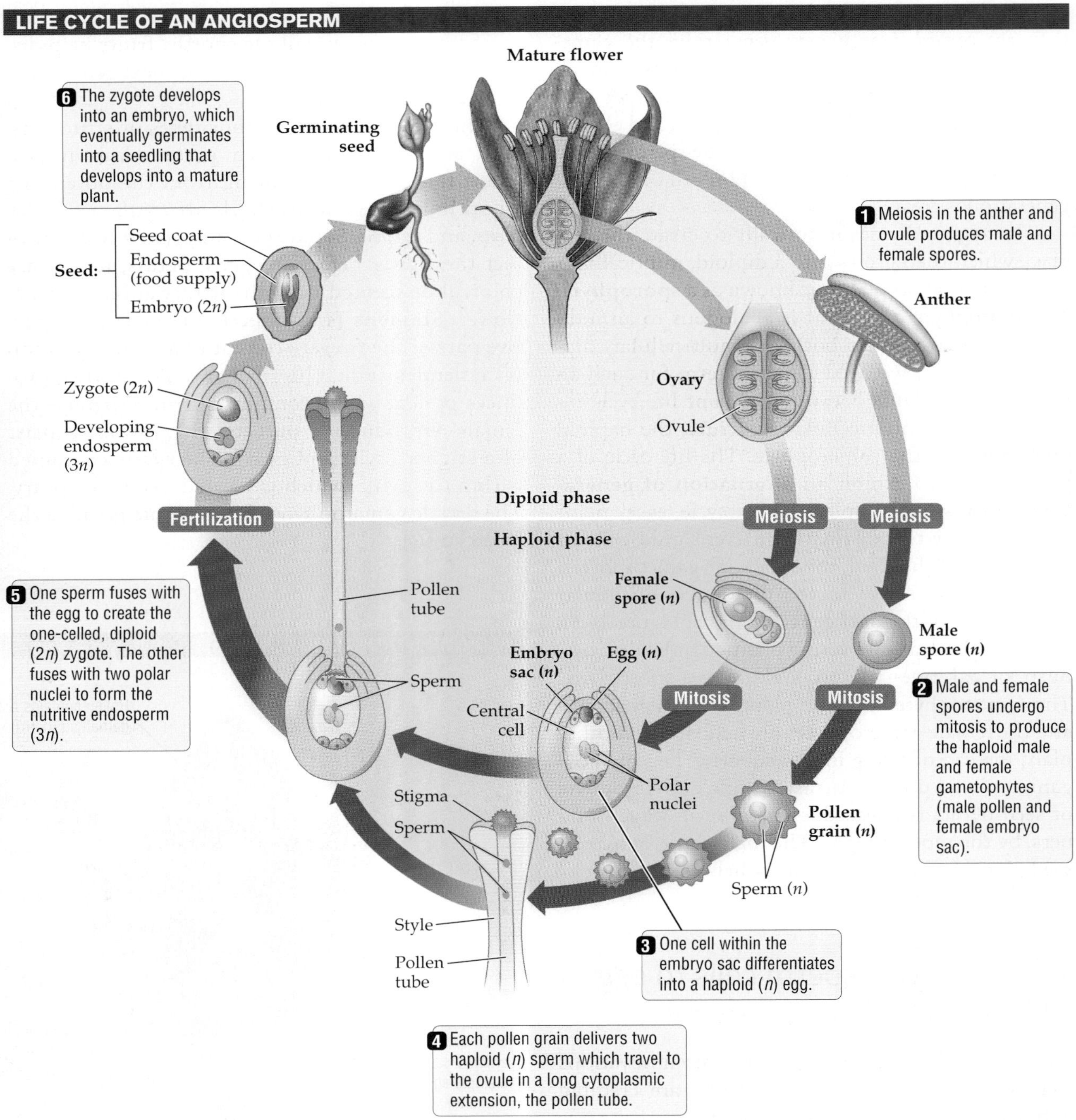

**FIGURE 36.7** From **Generation to Generation**

The life cycle of flowering plants is marked by alternation of generations. Haploid (*n*) stages of the life cycle are shown in purple; diploid (2*n*) stages, in orange.

**Coevolution of Plant and Pollinator**

Charles Darwin observed an orchid in Madagascar having an extremely long spur in the back of the flower, and predicted that its pollinator would have a very long tongue. His prediction was substantiated years later with the discovery of a long-tongued hawk moth, *Xanthopan morganii*, whose proboscis is a foot and a half long! The interaction between plant and pollinator is the result of coevolution.

## LIFE CYCLE OF AN ANGIOSPERM

**Mature flower**

**6** The zygote develops into an embryo, which eventually germinates into a seedling that develops into a mature plant.

**Germinating seed**

**1** Meiosis in the anther and ovule produces male and female spores.

**Anther**

Seed coat
**Seed:** Endosperm (food supply)
Embryo (2*n*)

**Ovary**

Ovule

Zygote (2*n*)

Developing endosperm (3*n*)

**Diploid phase**

Fertilization

Meiosis    Meiosis

**Haploid phase**

**Female spore (*n*)**

Pollen tube

**Male spore (*n*)**

**5** One sperm fuses with the egg to create the one-celled, diploid (2*n*) zygote. The other fuses with two polar nuclei to form the nutritive endosperm (3*n*).

Sperm

**Embryo sac (*n*)    Egg (*n*)**

Central cell

Mitosis    Mitosis

**2** Male and female spores undergo mitosis to produce the haploid male and female gametophytes (male pollen and female embryo sac).

Polar nuclei

**Pollen grain (*n*)**

Stigma

Sperm

Sperm (*n*)

Style

Pollen tube

**3** One cell within the embryo sac differentiates into a haploid (*n*) egg.

**4** Each pollen grain delivers two haploid (*n*) sperm which travel to the ovule in a long cytoplasmic extension, the pollen tube.

The major events in the life cycle of a flowering plant are diagrammed in **FIGURE 36.7**:

1. Male and female meiosis in the anther and ovary, respectively, lead to the production of male and female spores.

2. Male meiosis gives rise to a male spore that divides through mitosis to make a haploid, multicellular gametophyte, which, in seed plants, is known as

**pollen**. (In Latin, *pollen* means "flourlike dust"; it is a collective noun, like "sheep" or "species.") The function of pollen is to deliver the two sperm it contains to egg cells located deep within the carpel. Where did those egg cells come from? A diploid cell inside each ovule undergoes meiosis to give rise to female spores. When a female spore divides through mitosis, the haploid, multicellular embryo sac (the female gametophyte) is formed.

3. One cell in each embryo sac differentiates into an egg cell. The embryo sac of flowering plants contains a pair of nuclei lying free in the cytoplasm of a larger cell (the *central cell*). The two nuclei are called *polar nuclei* and, as we will see shortly, they participate in creating nutritive tissue for the embryo.

4. When pollen lands on a stigma, the receptive surface of the carpel, it produces a long extension called the pollen tube. The pollen tube grows through the style, with the two sperm positioned near the tube tip. When the pollen tube reaches the ovary, it enters the embryo sac inside an ovule, and as it does so, the tip bursts open, releasing the two sperm.

5. One sperm fertilizes the egg, creating the zygote. The other fertilizes the two polar nuclei to create the **endosperm**, a triploid (3*n*) tissue that provides food for the developing embryo. Because there are *two* types of fertilization events when flowering plants reproduce sexually, they are said to have **double fertilization**.

6. When conditions are right, the seed germinates and the embryo within emerges as a seedling, the beginning of a new plant.

## Many plants use animals to bring pollen to flowers

Given that plants cannot travel to find mates, how does pollen from one flower reach the carpels of another flower? In some species, like grasses and pine trees, pollen is transported by wind. The air is filled with pollen through much of the growing season, causing hay fever in millions of people who have an allergic reaction to the proteins that cover the outside of pollen grains. Wind can carry pollen long distances, but most pollen blown by wind lands in inhospitable places (such as parking lots or lakes), not on the stigma of a flower. Many flowering plants sidestep this problem by not relying on wind: instead, their pollen is transported by animals such as insects, birds, and even mammals (bats and opossums, for example).

How do angiosperms get animals to transport their sperm? The answer is bribery: they use their flowers to attract **pollinators**, which are animals that carry pollen grains from the stamens of one flower to the stigmas of other flowers of the same species

(**FIGURE 36.8**). The relationship between pollinator and plant is *mutualistic* because both species benefit (see Section 23.1). Plants bribe pollinators to visit their flowers by producing sugary nectar or protein-rich pollen. Because pollinators are often attracted to just one or a few types of flowers, the transport of pollen by animal pollinators is much more efficient than by wind. For example, if you follow a honeybee through a field, you will find that it visits only a few of the available flower types. As it collects nectar and pollen for food, the honeybee picks up pollen that gets caught in the hairs on its body. That pollen is likely to be deposited on the stigma of the next flower the insect visits. Some plant-pollinator interactions are exquisitely fine-tuned by **coevolution**, in which an adaptation in one species leads to a reciprocal adaptation in its "partner." For example, some moth-pollinated plants have nectar-containing tubes or "spurs" exactly the same length as their pollinator's "tongue."

Honeybees carry the pollen that dusts their bodies from flower to flower as they search for food.

Birds have good color vision and favor red flowers with long floral tubes.

The distinctive colors, shapes, and smells of flowers often attract highly specific pollinators that are most efficient as pollen dispersal agents.

**FIGURE 36.8**
**Bribing Animals to Do the Work**
Pollinators provide stationary plants with a way of transporting sperm to eggs. The spectacular colors, shapes, and odors of flowers, in combination with food rewards such as nectar, lure pollinating animals into visiting several flowers of the same species, accidentally transferring pollen in the process.

## 36.4 Producing the Next Generation: From Zygote to Seedling

When fertilization occurs, a single-celled zygote is formed, and the diploid chromosome number ($2n$) is restored by the union of a haploid sperm ($n$) and a haploid egg ($n$). Next the zygote begins to divide, forming a heart-shaped embryo (**FIGURE 36.9**). Cell division and development continue, eventually forming a mature embryo that can grow into a seedling and then into a whole plant. The mature embryo consists of the bare essentials: the shoot apical meristem, the root apical meristem, and the cotyledons ("seed leaves"), which serve as stored food for the embryo.

Let's look at these stages of development in a little more detail. To help you keep your bearings, remember that the zygote and developing embryo are located within the ovule, which is contained within the ovary. As the embryo develops inside the ovule, the parent plant transfers nutrients to the embryo and to the dividing endosperm cells, the food source for the developing embryo. As the embryo and endosperm are maturing, the outer tissue layers of the ovule differentiate into the seed coat, the usually tough outermost covering of a seed. A **seed**, then, is a mature ovule, consisting of an embryo, a food supply, and a protective seed coat. The endosperm is prominent in monocot seeds, and it is the main food source for the embryo when the embryo begins to grow out as a seedling. The endosperm is less prominent in most fully mature dicot seeds (see Figure 36.9) because food reserves are transferred from the endosperm to the two large cotyledons of the embryo during the later stages of seed maturation.

The ovary surrounding the developing seed becomes a **fruit**. The role of the fruit is to facilitate dispersal of the seeds so that the next generation of plants can colonize new areas in the landscape without competing with the parent plant. Fruit structure can vary widely (**FIGURE 36.10**). Some fruits are dry, rather than fleshy and juicy; they may be dispersed by wind or by attaching to the fur of animals. Some plants produce fleshy fruits that attract animals and are eaten by them. Seeds typically pass unharmed through the animal's digestive system and then are deposited on the ground along with a ready sup-

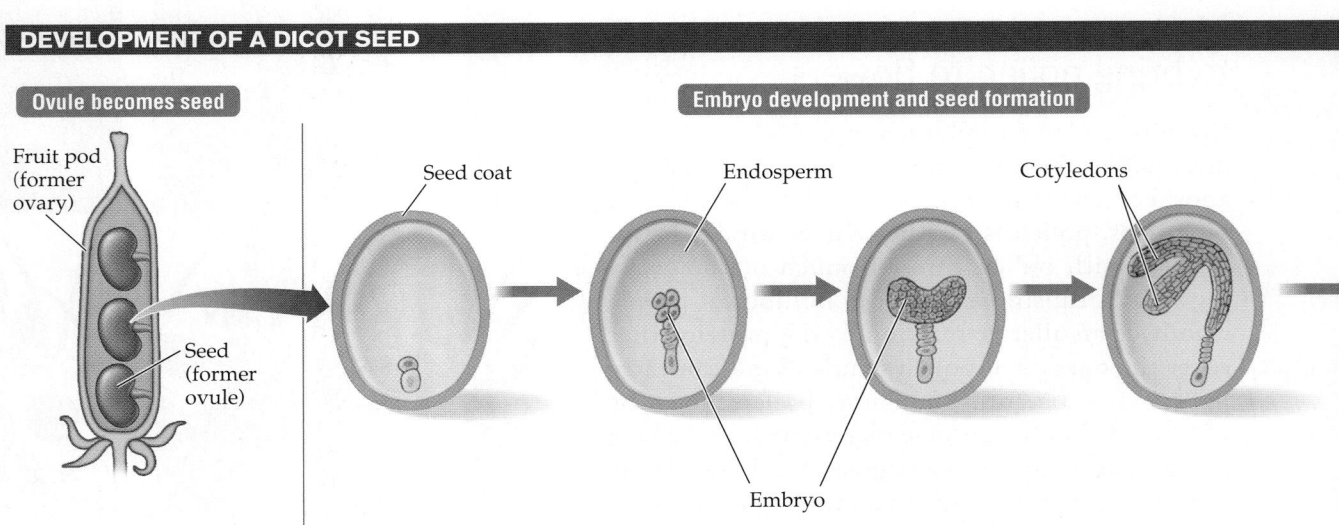

### DEVELOPMENT OF A DICOT SEED

**Ovule becomes seed**

**Embryo development and seed formation**

Fruit pod (former ovary)

Seed (former ovule)

Seed coat

Endosperm

Cotyledons

Embryo

**FIGURE 36.9  From Zygote to Seedling**

After fertilization, the single-celled zygote divides to produce two cells, one of which goes on to produce the embryo (shown in dark green); the other produces a structure (yellow) that anchors the embryo to the ovule. (For simplicity, only one ovule is shown developing within this flower's ovary, as in Figure 36.6; however, an ovary may contain many ovules, each with its own developing embryo.) The ovule develops into a seed, which consists of the embryo, its food source (the endosperm), and a protective outer coat (the seed coat). The development of a dicot seed is represented here. Much of the food stored in the endosperm is transferred to the cotyledons during the development of a typical

(a)

Dandelion fruits have a parachute-like structure and are spread by wind.

(b)

Fruits of the African plant *Harpagophytum* have "grappling hooks" that attach to the fur of large mammals.

(c)

Seed

Fruit

This seed developed from a fertilized ovule; it is surrounded by fleshy layers and a leathery covering, both of which come from the ovary wall.

**FIGURE 36.10**
**Plants Produce a Vast Variety of Fruits**

ply of fertilizer: animal feces. After the embryo inside is fully matured and well protected by the seed coat, the fruit wall (former ovary wall) may become brightly colored as a cue to animals that it is ready to be eaten.

Peaches, cherries, grapes, mangoes, and papayas are examples of fleshy fruits that are sweet because they accumulate sugars as they ripen. Not all fleshy fruits are sweet. Cucumbers, tomatoes, and bell peppers are com-monly thought to be vegetables because they are not sweet, but each is as much a fruit as an apricot or an orange, having been derived from a fertilized ovary. The word "vegetable" holds no scientific meaning, although in everyday usage it is applied to any edible plant part that is not sweet, including starchy roots, leafy greens, and nonsweet, fleshy fruits. If it has seeds inside, you can be sure you are looking at a fruit. Keep in mind, though, that horticulturists have created seedless fruits

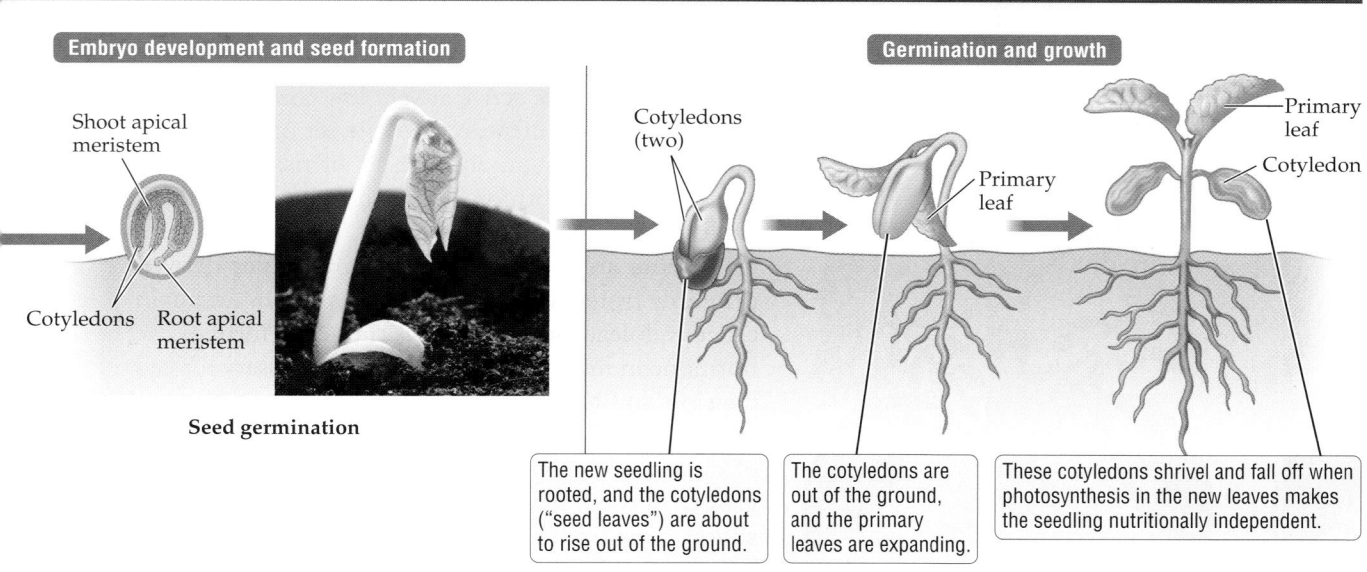

**Embryo development and seed formation**

Shoot apical meristem

Cotyledons    Root apical meristem

**Seed germination**

**Germination and growth**

Cotyledons (two)

Primary leaf

Primary leaf

Cotyledon

Primary leaf

Cotyledon

The new seedling is rooted, and the cotyledons ("seed leaves") are about to rise out of the ground.

The cotyledons are out of the ground, and the primary leaves are expanding.

These cotyledons shrivel and fall off when photosynthesis in the new leaves makes the seedling nutritionally independent.

dicot seed, which is why most dicot seeds have little endosperm but large cotyledons. Once the seed is fully formed, the embryo enters dormancy until it is stimulated by environmental conditions to begin growing again. During seed germination the seed coat softens and breaks open, and the growing embryo emerges as a young seedling. In many dicot seeds, the portion of the stem that is below the cotyledons grows rapidly to push the shoot out of and above the soil. In seeds of this type, the cotyledons are also pushed aboveground. In the light, they become green and photosynthetic. As the seedling produces new leaves, it becomes photosynthetically independent, and the cotyledonary "leaves" shrivel and fall off.

through artificial selection and special hybridization techniques. Seedless oranges, watermelons, and bananas are quite literally the fruits of such experimentation. Seedless varieties must be propagated through asexual reproduction, such as cuttings.

Once the embryo within a seed is mature, it enters **dormancy**, halting its growth for a period of time. Seeds may remain dormant from a few days in some species to one to many years in other species. Dormancy ends once the embryo is stimulated by favorable conditions, such as rainfall (in a desert) or warm weather after months of cold (in a region with cold winters). The seed is said to germinate when the embryo absorbs water and starts growing until it breaks out of the seed coat and emerges as a seedling (see Figure 36.9). Seedlings often die because the seed from which they grew happened to land in a place without enough soil, water, or sunlight. Other seeds land in a favorable environment that enables the seedling to grow into a plant that matures and produces flowers and fruits, the beginning of yet another generation.

| TABLE 36.1 | Some Key Plant Hormones and Their Primary Functions |
|---|---|
| **HORMONE(S)** | **PRIMARY FUNCTION(S)** |
| Auxins | Cell division; root formation; apical dominance; phototropism and gravitropism; stem elongation; inhibition of aging |
| Cytokinins | Cell division; shoot formation; greening and inhibition of aging |
| Gibberellins | Stem growth (through both cell elongation and cell division); stimulation of seed germination |
| Abscisic acid | Mediation of adaptive responses to stress, including stomatal closure under drought conditions; induction of dormancy in seeds and winter buds |
| Ethylene | Mediation of adaptive responses to stress; fruit ripening; aging (senescence) of plant parts and promotion of leaf fall |

## 36.5 Plant Hormones

All multicellular organisms must coordinate their body activities. Animals do this in two ways: through chemical signaling with hormones and through electrical signaling with nerve impulses. Plants lack a nervous system, but like animals and other organisms, they have hormones that control how they grow, reproduce, and respond to their environment. Like animal hormones, plant hormones are organic substances that are active at very low concentrations. They may be manufactured by one set of tissues but exert their activity on another set of tissues. Five of the main plant hormones and their chief functions are listed in Table 36.1.

**Auxins** are a family of small molecules derived primarily from the amino acid tryptophan. An auxin called indoleacetic acid (IAA) is the active form of the hormone in most plants. Auxins are necessary for cell division and the formation of organs such as roots. They are involved in *phototropism*, which is the growth of shoots toward light, and in *gravitropism*, which is the growth of roots toward the ground and the growth of shoots away from the ground. The shoot apical meristem, young leaves, and developing seeds are the major centers of auxin production.

Auxins made by the shoot apex suppress the outgrowth of buds, preventing them from growing out as side branches. This inhibition of buds by auxins is known as **apical dominance** (**FIGURE 36.11**). The

**APICAL DOMINANCE**

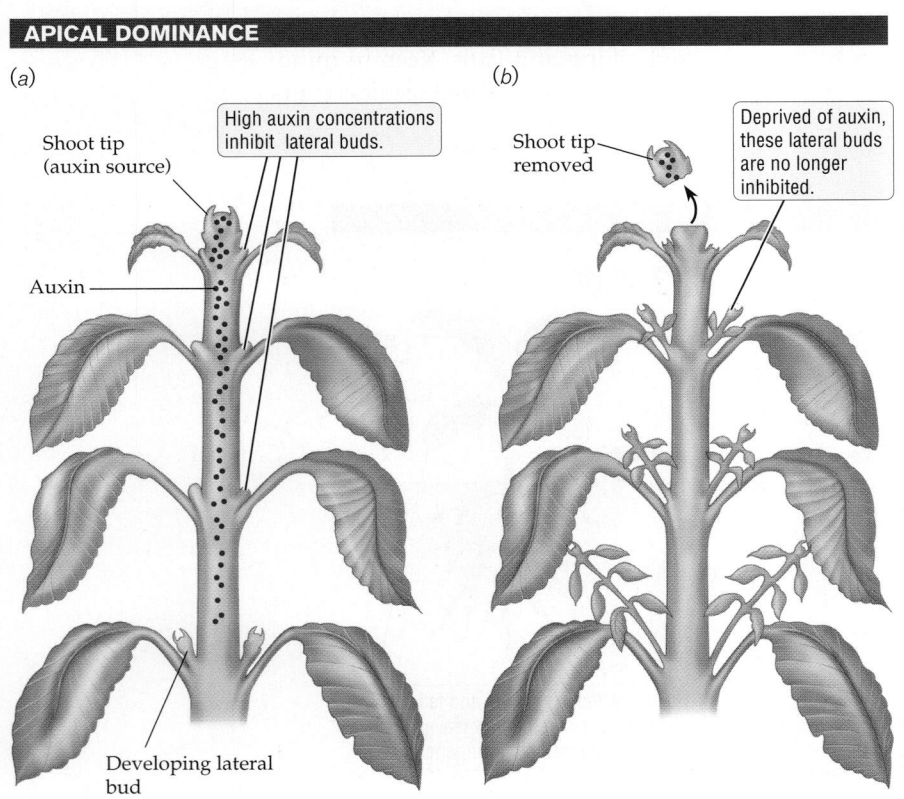

(a) High auxin concentrations inhibit lateral buds.

Shoot tip (auxin source)

Auxin

Developing lateral bud

(b) Shoot tip removed

Deprived of auxin, these lateral buds are no longer inhibited.

**FIGURE 36.11 Removing the Tip Lets the Branches Grow**

(a) Auxin hormones (shown in red) are produced in apical meristem cells and ordinarily act to inhibit the growth of lateral buds. (b) When the auxin concentration drops for any reason (here, because the shoot tip has been removed), the lateral buds begin to grow, producing branches more rapidly.

# Fall Colors Unmasked

The riot of colors we associate with fall foliage is part of the winter survival strategy of temperate, broad-leaved plants. Retaining its broad leaves through a subfreezing winter, when it is much too cold for photosynthesis, can be a risky venture for a plant. Moisture loss from the large surface area of the leaves, at a time when liquid water is scarce, could severely dehydrate a plant. Evergreens like pine and fir can afford to keep their foliage through bitterly cold winters because their needlelike leaves present a much smaller surface area for water loss. The needles are covered in a thick layer of wax, which further reduces the risk of excessive water loss.

Broad-leaved plants detect the approach of autumn by sensing the colder temperatures, the ever-shorter days, or both. Maple and sumac, for example, change color and drop their leaves in response to the shorter days and longer nights of early fall. But it takes a sharp frost to prod most oaks into changing.

These external stimuli trigger a decline in auxin and cytokinin levels in the leaf, which in turn activates the aging (*senescence*) of the leaf. As cytokinins disappear, enzymes that destroy chlorophyll are activated. The loss of chlorophyll unmasks the color of other pigments, especially the yellow and orange hues of *carotenoids* found in chloroplasts. These pigments are responsible for the golden leaves of birch and aspen in fall.

The red foliage of maple and sumac comes from water-soluble pigments called *anthocyanins* that are newly made in these species as leaf senescence gets under way.

Experiments with genetically modified plants suggest that anthocyanins protect senescing leaves from sun damage. Before the leaves fall off, most of the recyclable macromolecules, such as starch and many proteins, are broken down and their building blocks (sugars and amino acids, for example) shipped to the root and tree trunk for storage. Anthocyanins absorb ultraviolet (UV) radiation and are believed to act like sunscreens that protect the cell's DNA from damage. The shielding effect of anthocyanins may protect cellular processes in senescing leaf cells long enough to facilitate the salvage operation. Anthocyanins are also known to be powerful antioxidants, substances that detoxify highly reactive chemicals known as free radicals. Drought, nutrient stress, and sun damage are believed to increase the output of free radicals. Plants experiencing stress, such as nutrient deficiency, often show increased anthocyanin accumulation. Increased anthocyanin production might be a mechanism for coping with destructive free radicals, while also acting as a UV shield, in leaves undergoing genetically controlled self-destruction.

A shape of firs, spruces, and other "Christmas trees" is produced by the progressively greater outgrowth of lower branches as apical dominance weakens farther away from the shoot apex. To counteract apical dominance and make a plant bushier, a gardener snips off the shoot tip, reducing the auxin supply and thereby promoting the growth of side branches.

Because auxins promote root formation, they are used in rooting powders that gardeners and plant growers use to induce roots on stem or leaf cuttings. High auxin concentrations are toxic to dicot plants, but not to monocots such as grasses. This is why auxins are found in selective weed killers—those that target broad-leaved weeds like dandelions but do not harm our lawns.

**Cytokinins** are a family of small molecules that are similar in structure to a nitrogenous base (adenine) found in DNA and RNA. Cytokinins are necessary for cell division, and they promote shoot formation. Cytokinins are the "greening hormones" because, acting in concert with auxin, they help maintain organs such as leaves. The levels of both hormones decline sharply just before plants drop their leaves in the fall (see the box on page 789). **Gibberellins** are a family of small hydrophobic molecules that stimulate the elongation of stems. Many of the dwarf varieties of garden plants—dwarf peas, for example—are mutants that cannot grow tall because they are deficient in gibberellin production.

**Abscisic acid** (**ABA**) is a small molecule that can be made by nearly all cells in a plant, although roots are the richest source in most species. The hormone mediates adaptive responses to drought, cold, heat, and other stresses. Roots sense water scarcity in drying soils and increase their output of ABA in response. The hormone travels from the root to the shoot, where it acts on guard cells to close the stomata. Stomatal closure reduces transpiration (see Chapter 35) and thereby conserves the moisture within the plant.

**Ethylene** is a gaseous hormone, made up of just two carbon atoms and four hydrogen atoms ($C_2H_4$). All parts of the plant are capable of making ethylene. Ethylene stimulates the ripening of some types of fruit, including peaches, pears, cantaloupes, bananas, avocadoes, and tomatoes. In response to the hormone, enzymes are activated that convert starches into sugars, resulting in a sweeter fruit. You can speed up the ripening of bananas, cantaloupes, and other ethylene-responsive fruit by putting them in a bag. The bag traps ethylene, increasing the concentration of the hormone and thereby accelerating ripening. Remember, this trick does not work with all types of fruit; cherries, strawberries, watermelon, and citrus are examples of fruits that are not responsive to ethylene.

Spring ephemerals

**FIGURE 36.12 Plants Can Detect Day Length**

(*a*) Like many other plants that bloom in summer, irises flower when nights become *shorter* than a critical length. (*b*) If nights are longer than the critical length, the plant does not flower. (*c*) In the laboratory, a flash of light during an otherwise "too long" night can cause flowering in such plants. In plants that flower when nights become *longer* than a critical length, a flash of light during the night *prevents* flowering.

**PHOTOPERIODISM**

(*a*)    (*b*)    (*c*)

Darkness

Light

Flash of light

Critical night length

The iris needs long days and short nights to flower.

## 36.6 How Flowering Is Regulated

The timing of flowering is a critical event in the life cycle of a plant. In some species, flowering is controlled exclusively by internal signals. Some plants flower as soon as they achieve a certain level of developmental maturity. For example, some species flower after they reach a certain size or produce a certain number of leaves or have lived a certain number of years.

Many species, however, use external cues to synchronize their bloom period with a time of year that is best suited for their growth and reproduction. In some regions, for example, small herbaceous plants known as spring ephemerals grow rapidly and flower in the spring (before new tree leaves block most of the sunlight). Other species flower in the fall, when their specific pollinators are most abundant. Plants native to areas with pronounced wet and dry seasons usually time their blooms for the arrival of the rains. Many temperate species, such as apples, must be exposed to a few months of cold temperatures before they will bloom in spring. *Vernalization*, as this mechanism is known, enables the plant to sense the passage of winter and ensures that a brief, unseasonal warm spell in January will not be "mistaken" for spring.

Many plants, especially species native to temperate regions, perceive the seasons by sensing the length of the day. This is possible because day length varies with the season—shorter in winter and longer in summer. This sensing of the duration of light and dark in a 24-hour cycle is known as **photoperiodism** (from *photo*, "light"). Plants also use day length to sense when conditions are favorable for seed germination. The dormancy of buds through fall and winter, as well as their regrowth in spring, is also influenced by photoperiod. The consequences could be disastrous if responses like these were initiated at the wrong time of year.

Leaf cells detect the photoperiod with the help of a pigment-protein complex called *phytochrome*. Some plants (for example, irises, spinach, and some cereal grains) flower if the period of nighttime darkness is *shorter* than a certain number of hours (**FIGURE 36.12**). Long-night plants (such as poinsettias and chrysanthemums) start making flowers if nighttime darkness is *longer* than a particular number of hours.

Once a leaf detects that it is the appropriate time of year for flowering, it generates a chemical signal that travels to the shoot apical meristem. Scientists have long known that the signal that induces flowering must be mobile, since it is readily transmitted across a graft. After nearly eight decades of searching, the identity of the flowering stimulus was finally established in 2005. The signal for flowering is a protein (called FLT in *Arabidopsis*, the plant in which it was first discovered). It is transported in phloem conducting tubes from the leaf to the shoot apical meristem. The protein combines with other regulatory proteins in the cells of the shoot apical meristem to change the activities of genes. The new pattern of gene activity prods the meristem into producing flowers instead of making leafy shoots only.

> ## Concept Check
>
> 1. Compare seeds and fruits.
> 2. Why does removal of the shoot tip lead to increased branching in most plants?

## 36.7 How Plants Defend Themselves

For hundreds of millions of years, life on Earth has been a struggle between species trying to eat and species trying to avoid being eaten. Plants are no exception. Over time, plants have developed a rich variety of mechanisms to deter attacks by herbivores and by pathogens such as fungi, bacteria, and viruses.

### Plants use physical and chemical defenses against herbivores

If you walk through a thicket of thorny plants, you will probably experience one of the many physical defenses that plants have against herbivores: your foot may be punctured (right through the shoe), clothes ripped, or body covered with scratches. Plants are also well protected by defenses that seem harmless to humans, such as dermal hairs that feel like "fuzz" to us but can be fatal to insect herbivores, as we saw in Chapter 35.

In addition to their physical defenses, many plants contain chemical substances that are toxic to herbivores. Nicotine, the addictive chemical in cigarette smoke, protects tobacco plants from most insects. In fact, nicotine is sold as a potent insecticide, and a few drops of pure nicotine will kill a human. Similarly, plants in the mustard family (such as cabbage and horseradish) are protected by chemicals so pow-

erful that most insects, even if they are starving, will not eat their leaves. However, the same chemicals that drive most herbivores away may serve as attractants for the few species that are adapted to tolerate the plant's toxic compounds. For example, caterpillars of the cabbage white butterfly, when presented with the chemicals from cabbage that repel most insect herbivores, extend their mouthparts and act as they would when feeding.

### Plants have three lines of defense against pathogens

Humans are protected against pathogens by a sophisticated defense system, including mechanisms that give us increased protection against previous attackers (see Chapter 32). Like animals, plants have both physical and chemical defenses to protect themselves against pathogens. First, as we have seen, plants have tough outer surfaces that keep pathogens out. As in animals, however, that line of defense can be broken, as when a plant is wounded. Once a pathogen enters the plant body, it encounters two other lines of defense: specific and nonspecific chemical defenses.

The specific chemical defense is based on single genes in what is called **gene-for-gene recognition**: plants have a number of genes (known as *R* genes, for "resistance") that respond to complementary genes in a pathogen. Each *R* gene codes for a protein that recognizes specific chemicals, such as small carbohydrates, produced by a particular pathogen. If a specific pathogen is detected, a range of protective defenses, including toxic chemicals, are unleashed. This system does not work if the plant lacks the appropriate *R* gene and therefore cannot recognize the pathogen in question and "sound the alarm" when it is present. In addition, a plant that was resistant to a specific pathogen may lose that resistance if the pathogen mutates in such a way that its distinctive chemistry can no longer be recognized by the plant's *R* gene system.

Plants also deploy general, or nonspecific, chemical defenses, which involve a range of broadly targeted antipathogen and antiherbivore chemicals. The antipathogen chemicals include those that attack the cell walls of bacteria. Others are hormones that signal other parts of the plant to manufacture their chemical weapons and have them "battle ready." Still other nonspecific chemicals cause cells around the point of attack to be sealed off, in effect setting up a barricade to limit the invader's spread.

Concept Check Answers

1. Seeds are mature, fertilized ovules; they contain the embryo. Fruits develop from the ovary wall and usually contain seeds.

2. It releases buds from apical dominance, enabling them to grow out as branches.

# Of Ancient Plants and Firmly Rooted Wanderings

Plants, unlike most animals, cannot run from their enemies, seek shelter from heat or cold, or travel to find food or mates. Although plants are challenged by their environment in as many ways as animals, there is one sense in which plants have a clear advantage over animals: longevity.

Even the oldest animals have life spans of only 250 years (a tube worm, pictured in **FIGURE 36.13a**), 190 years (the Galápagos land tortoise), 220 years (a species of clam), or 122 years (humans; **FIGURE 36.13b**). Although a life span of 100–200 years seems like a long time, it pales in comparison with the life spans of many plants. One of the world's oldest living trees is a bristlecone pine—named "Methuselah" after the biblical patriarch said to have lived 969 years old. But the bristlecone Methuselah is far older; it's been growing in the White Mountains of California for more than 4,800 years (**FIGURE 36.13c**). This ancient tree was already 183 years old when Egyptians began construction on the oldest known Egyptian pyramid, in 2648 BC.

Other plants are far older. Contenders for the world's oldest living plant include "King Clone," a creosote bush in California's Mojave Desert that is nearly 12,000 years old (**FIGURE 36.13d**); a 13,000-year-old box huckleberry in Pennsylvania; and a grove of quaking aspens in Utah—called Pando (Latin for "I spread")—that is at least 80,000 years old and possibly as old as a million years.

How can plants grow so old? Each of the plants just described germinated from a seed and then grew into a mature plant. Once grown, each plant reproduced asexually by forming clones. The asexual offspring slowly "moved" from the point where the original seedling had sprouted, forming a circular clone about 50 feet in diameter in the case of King Clone and covering 107 acres in the case of Pando, the army of quaking aspens. Pando consists of about 47,000 individual trees, with an average age of only about 130 years. Individually, the trees are not that old, but new trees keep sprouting, so the genetic clone taken as a whole is far older.

To put it another way, in plants that reach ages of 10,000 years or more, it is the asexual offspring that are still alive, not the original stem (which is long dead). So in one sense, the bristlecone pine Methuselah is the oldest individual plant. But if you think of an individual as beginning life when sperm and egg fuse to form a new genetic individual, and ending life when the last cell that is part of that genetic individual dies, then the cloned shrubs and trees described here are the oldest plants. These shrubs and trees can keep forming new, genetically identical copies of themselves indefinitely and slowly move across the landscape like Macbeth's walking forest.

Finally, forests move in a different way. Over long periods of time, tree species can migrate north or south, following changes in temperature and precipitation as the climate changes. Of course, it is not individual rooted trees that move, but their seeds. As you might expect, most seeds fall close to the parent tree. But others may sail aloft in the wind, float down a river, or stick to the fur or feathers of animals—traveling 25 miles or more. Squirrels and birds that store food for the winter bury seeds in food caches, carrying even large seeds such as acorns uphill. And birds and mammals that eat fruit can deposit seeds in a rich mound of fertilizer, sometimes hundreds or thousands of miles away.

**(a)** Tube worm (*Lamellibrachia luymesi*): up to 250 years old

**(b)** Jeanne Calment: lived to be 122 years old (120 in this photo)

**(c)** Bristlecone pine (*Pinus aristata*): up to 4,800 years old

**(d)** Creosote bush clonal ring, "King Clone" (*Larrea divaricata*): 12,000 years old

**FIGURE 36.13 How Long Can Organisms Live?**

# As Earth Warms, Move Species to Save Them?

BY BALLIATEWARI, "Hum Kisise Kam Nahin" blog

On naked patches of land in western Canada and [the western] United States, scientists are planting trees that don't belong there. It's a bold experiment to move trees threatened by global warming into places where they may thrive amid a changing climate.

Take the Western larch with its thick grooved bark and green needles. It grows in the valleys and lower mountain slopes in British Columbia's southern interior. Canadian foresters are testing how its seeds will fare when planted farther north—just below the Arctic Circle . . .

About 20 to 30 percent of species worldwide face a high risk of becoming extinct possibly by 2100 as global temperatures rise, estimated a 2007 report by the Nobel-winning international climate change panel . . .

"A tree that we plant today better damn well be adapted to the climate for 80 years, not just the climate today," said Greg O'Neill, a geneticist with the British Columbia Ministry of Forests and Range. "We really have to think long-term."

O'Neill is heading the government-funded experiment that will transform certain North American forests into climate change laboratories. The large-scale, first-of-its-kind test involves purposely planting seeds from more [than] a dozen timber species outside their normal comfort zone to see how well they survive decades from now . . .

Outsiders are also keenly watching the experiment as a test case for what is professionally known as "assisted migration." "We'd all prefer species to move naturally," said Duke conservation biologist Stuart Pimm. But "sometimes you just can't get there from here. Some species are going to be isolated and they're going to get stuck". . .

This spring, crews fanned across rugged mountains and began the first dozen plantings on cleared forest land in British Columbia's southern interior and on a private plot near Mount St. Helens in Washington state.

Each test site contains some 3,000 seedlings, on average a foot tall, planted side-by-side on five acres. Fluorescent pin-flags and aluminum stakes dot the corners so that scientists can come back every five years to document their health.

The project will eventually include 48 plots around British Columbia, Washington state, Oregon, Montana and Idaho. It will test the ability of 15 tree species to survive in environments colder and hotter than they're used to.

Although the individual trees don't walk like the Ents in *Lord of the Rings*, forests really do move. The fossil record shows that over the last 18,000 years, the geographic range of spruce trees has shifted northward more than 1,000 kilometers. In fact, in the United States alone, dozens of tree species are heading north at an average speed of 62 miles per year as climate change makes northern climes warmer.

Since the global climate is heating up faster than many plants can evolve and some tree species will not survive the coming drought and heat to the south, some researchers have decided to help these trees make the trek north by moving their seeds north. For people who grow and harvest timber, and for anyone who uses wood products, a reliable source of trees is important. We don't know how moving tree species into new ecosystems will affect those ecosystems, but timber researchers believe that moving populations of trees is better than losing them altogether.

A forest ecosystem is not just a group of species of trees and shrubs that can be carted to a new home; ecosystems typically comprise hundreds of coevolved species—including, for example, fungi that grow on plant roots, and pollinating insects. So even if the experimenters are successful in moving a few tree species northward, they will not have moved a forest ecosystem. And moving forests to the far north could have a downside. Recent climate models suggest that as trees migrate into the Arctic tundra—a land of grasses and knee-high shrubs—the advancing forest will accelerate global warming. Forests accumulate humid air, which holds more heat than the dry air of the tundra, and the darker color of forests absorbs and retains more solar heat, which also contributes to global warming.

## Evaluating the News

**1.** This article mentions that researchers planted 3,000 tree seedlings. What kind of experimental control would the researchers need to be able to tell whether the relocated forests are doing as well in their new location as they would at home?

**2.** Some trees are wind-pollinated, while others are insect-pollinated. What could happen to insect-pollinated trees if they were transplanted to a place where their pollinator did not live?

**3.** Discuss the value of the tundra and how the encroachment of trees will affect it.

**SOURCE:** Balliatewari's "Hum Kisise Kam Nahin" blog, July 20, 2009, http://pramodtrivedi.blogspot.com.

# CHAPTER REVIEW

# Summary

## 36.1 How Plants Grow: Indeterminate Growth

- Plants have indeterminate growth; that is, they can grow throughout their lives. The plant body is modular, formed by the repeated addition of the same basic units.
- Indeterminate and modular growth habits enable plants to respond flexibly to their environment. Plants tend to add many new parts when conditions are favorable and few new parts when conditions are not.

## 36.2 How Plants Grow: Primary and Secondary Growth

- Plants increase in length by primary growth, which results from the division of apical meristem cells located at the tip of each stem and each root.
- During primary growth, the plant increases in length by repeated addition of the same basic units: the bud-stem-leaf unit aboveground and new lateral roots belowground.
- Plants increase in thickness by secondary growth. In woody perennials, this growth results mainly from cell divisions in a lateral meristem called the vascular cambium.
- Dividing vascular cambium cells produce new vascular tissues: secondary xylem (wood) and secondary phloem. Cell divisions in another lateral meristem, the cork cambium, produce a tough new set of dermal tissues that form the outer bark.
- Secondary growth in trees produces wood. Wood consists of many layers of secondary xylem tissue. New layers of secondary xylem produce the tree rings we use to age a tree.

## 36.3 Producing the Next Generation: Flower Form and Function

- Meiosis in plants produces a single-celled spore, which divides through mitosis to create a haploid, multicellular structure, the gametophyte. Cells in the gametophyte differentiate into egg or sperm.
- The fertilization of the egg by sperm generates a diploid zygote, which gives rise to the diploid, multicellular individual known as a sporophyte. This alternation of gametophytic and sporophytic generations is a hallmark of plant life cycles.
- In angiosperms, male and female reproductive parts are contained in flowers, which consist of four whorls: sepals, petals, stamens, and carpels.
- Meiosis in plants produces a spore, which undergoes mitosis and grows into a small, multicellular haploid stage—a pollen grain in male reproductive organs (anthers) and an embryo sac in female reproductive organs (ovaries). Pollen grains contain sperm, and embryo sacs produce eggs.
- Flowering plants have double fertilization. Angiosperm pollen contains two sperm: one that fertilizes the egg to produce a zygote, and another that fertilizes two polar nuclei in the embryo sac to produce endosperm (stored food for the developing embryo).
- Animal pollinators provide immobile plants with a way of transporting sperm-containing pollen to eggs.

## 36.4 Producing the Next Generation: From Zygote to Seedling

- Fertilization produces a single-celled, diploid ($2n$) zygote, which begins to divide and develop into an embryo. The embryo is located within the ovule, which is contained within the ovary.
- The parent plant provides nutrients to the developing embryo and the developing endosperm cells. The outer layers of the ovule harden into a protective seed coat. Each seed (mature ovule) contains the ingredients for growing a young plant of the next generation: a mature embryo, a food source (which may be cotyledons primarily, or the endosperm primarily), and the seed coat. The ovary surrounding the seed forms a fruit.
- Once a seed is formed, the embryo enters dormancy, and the seed is dispersed from its parent. Seeds are often dispersed by wind, or by attaching to or being eaten (and excreted) by animals.
- Dormancy ends when the embryo is stimulated by favorable conditions to start growing again. Then the seed germinates, and a seedling begins to grow.

## 36.5 Plant Hormones

- Hormones control how plants grow, reproduce, and respond to their environment.
- Auxins influence apical dominance, in which an apical meristem inhibits the growth of buds. Apical dominance enables the plant to control how much primary growth occurs at any one time.
- Cytokinins promote cell division. Gibberellins enhance stem elongation.
- Abscisic acid (ABA) mediates several types of stress responses, including stomatal closure in response to drought.
- Ethylene is a gaseous plant hormone that stimulates the ripening of some types of fruit.

## 36.6 How Flowering Is Regulated

- In many plants, a light-sensitive pigment-protein complex (phytochrome) in the leaves detects changes in day length. Once the appropriate day length is detected, a chemical signal (a protein called FLT) initiates flowering.
- Day length also influences other important plant responses, such as when seeds germinate and when buds go dormant for the winter. Such responses to day length are called photoperiodism.

## 36.7 How Plants Defend Themselves

- Plants defend themselves against herbivores with physical weapons (such as thorns and dermal hairs) and a rich variety of toxic chemicals.
- Plants defend themselves against pathogens with a tough outer covering and well-developed chemical defenses.
- The specific chemical defense system of plants is based on gene-for-gene recognition, by which plants with specific resistance (R) genes recognize pathogens with complementary genes and trigger defensive responses when the specific pathogen is detected.
- Nonspecific chemical defenses include producing antipathogen and antiherbivore chemicals, signaling nearby cells of an impending attack, and sealing off the point of invasion.

# Key Terms

abscisic acid (ABA) (p. 790)
alternation of generations (p. 783)
annual (p. 778)
annual ring (p. 787)
apical dominance (p. 788)
apical meristem (p. 779)
auxin (p. 788)
bark (p. 781)
biennial (p. 778)
carpel (p. 783)
coevolution (p. 785)
cork cambium (p. 781)

cytokinin (p. 789)
dormancy (p. 788)
double fertilization (p. 785)
early wood (p. 780)
egg (p. 782)
embryo (p. 782)
embryo sac (p. 783)
endosperm (p. 785)
ethylene (p. 790)
fruit (p. 786)
gametophyte (p. 782)
gene-for-gene recognition (p. 791)

gibberellin (p. 790)
heartwood (p. 781)
indeterminate growth (p. 778)
late wood (p. 780)
lateral meristem (p. 780)
meristem (p. 779)
perennial (p. 778)
petal (p. 783)
photoperiodism (p. 790)
pollen (p. 784)
pollinator (p. 785)
primary growth (p. 779)

sapwood (p. 781)
secondary growth (p. 780)
seed (p. 786)
sepal (p. 783)
sperm (p. 782)
spore (p. 782)
sporophyte (p. 783)
stamen (p. 783)
vascular cambium (p. 780)
wood (p. 780)
zygote (p. 782)

# Self-Quiz

1. The increase in length that results from the division of apical meristem cells is called
   a. primary growth.
   b. secondary growth.
   c. apical dominance.
   d. secondary xylem.

2. In flowering plants, each pollen tube delivers two sperm cells to the embryo sac. What is the fate of these sperm cells?
   a. One fertilizes an egg to produce a zygote; the other fertilizes a cell to produce fruit.
   b. Each one fertilizes a different egg, producing two genetically distinct zygotes.
   c. Both fertilize a single egg, producing two genetically identical zygotes.
   d. One fertilizes an egg to produce a zygote; the other fertilizes two nuclei to produce the endosperm.

3. In terms of plant structure, what we call the wood of a tree consists of
   a. layers of secondary xylem.
   b. layers of secondary phloem.
   c. endosperm.
   d. cork cambium.

4. A mature ovule is called
   a. an ovary.
   b. an embryo.
   c. a seed.
   d. the endosperm.

5. Plants have a specific chemical defense that depends on
   a. cytokinins.
   b. gene-for-gene recognition.
   c. apical dominance.
   d. antibodies.

6. Apical dominance
   a. occurs when apical meristems inhibit the growth of buds.
   b. is caused by gibberellins.
   c. enables the plant to detect day length.
   d. is inhibition of root growth.

7. Most of the increase in the thickness of tree trunks and branches comes from the division of cells in the
   a. apical meristems.
   b. phloem sieve tube elements.
   c. cork.
   d. lateral meristems.

8. In flowering plants, which of the following is produced *directly* by meiosis?
   a. haploid spores
   b. pollen grains
   c. embryo sacs
   d. endosperm

# Analysis and Application

1. What is meant by "indeterminate growth"? Explain why this growth strategy is crucial for the survival of plants but is not seen in most animals.

2. Identify the flower parts in the diagram: anther, ovary, style, and sepal.

3. Describe the main events in the life cycle of a flowering plant. Is the pollen grain haploid or diploid? Explain your answer.

4. What is wood, and how does it form? What is the difference between sapwood and heartwood?

5. In plants with fleshy fruits that attract animals, the fruits often taste bitter until the seeds they contain are fully mature.

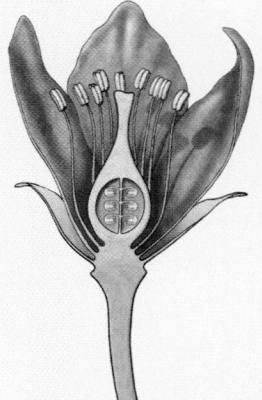

Explain why this characteristic might increase the plant's chance of reproducing successfully.

6. People argue endlessly about whether the tomato is fruit or a vegetable. Having studied sexual reproduction in plants, how would you settle this argument?

7. Consider a cultivated plant species that has two genetic mutations: one that results in a feature that people value (such as a hypothetical genetic mutation in clover that causes the plant to produce four leaves consistently, instead of the usual three), and another that impairs the ability of the plant to detect day length. If the plant could be mass-produced (say, by a method of cloning), and if people throughout the world planted it in their gardens, in what parts of the world would the mutant be likely to thrive? In what parts of the world would it be likely to grow and reproduce poorly? Explain your answer.

8. Describe how plants defend themselves against herbivores and pathogens.

# Appendix
## The Hardy–Weinberg Equilibrium

In this appendix we describe the conditions under which populations do not evolve. Specifically, we discuss the conditions for the Hardy–Weinberg equation, a formula that allows us to predict genotype frequencies in a hypothetical nonevolving population. As described in Chapter 18 (see the box on page 412), this equation provides a baseline with which real populations can be compared in order to figure out whether evolution is occurring.

A population can evolve as a result of mutation, gene flow, genetic drift, or natural selection. Put another way, a population does *not* evolve when the following four conditions are met:

1. There is no net change in allele frequencies due to mutation.
2. There is no gene flow. This condition is met when new alleles do not enter the population via immigrating individuals, seeds, or gametes.
3. Genetic drift does not change allele frequencies. This condition is met when the population is very large.
4. Natural selection does not occur.

The Hardy–Weinberg equation is derived from the assumption that all four of these conditions are met. In reality, these four conditions are rarely met completely in natural populations. However, many populations meet these conditions well enough that the Hardy–Weinberg equation is approximately correct, at least for some of the genes within the population.

To derive the Hardy–Weinberg equation, consider a hypothetical population of 1,000 moths. The dominant allele for orange wing color ($W$) has a frequency of 0.4, and the recessive allele for white wing color ($w$) has a frequency of 0.6. What we seek to do now is predict the frequencies of the $WW$, $Ww$, and $ww$ genotypes in the next generation for a population that is not evolving.

If mating among the individuals in the population is random (that is, if all individuals have an equal chance of mating with any member of the opposite sex), and if the four conditions just described are also met, we can use the approach described in the accompanying figure to predict the genetic makeup of the next generation. This approach is similar to mixing all the possible

### The Hardy–Weinberg Equation

When mating is random and certain other conditions are met, allele and genotype frequencies in a population do not change. $p$ = frequency of the $W$ allele; $q$ = frequency of the $w$ allele.

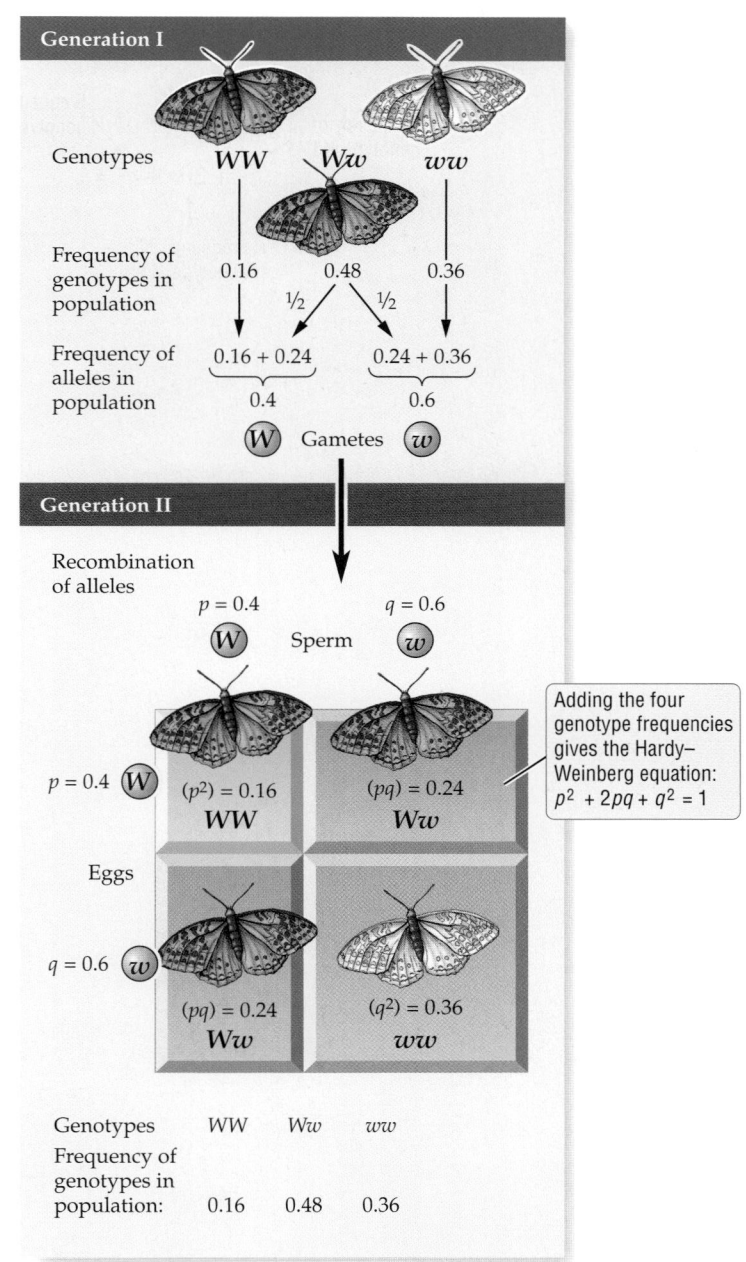

gametes in a bag and then randomly drawing one egg and one sperm to determine the genotype of each offspring. With such random drawing, the allele and genotype frequencies in our moth population do not change from one generation to the next, as the figure shows.

Because the *WW*, *Ww*, and *ww* genotypes are the only three types of zygotes that can be formed, the sum of their frequencies must equal 1. As the figure shows, when we sum the frequencies of the three genotypes, we get the Hardy–Weinberg equation:

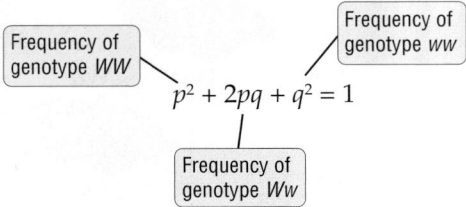

In this equation, the frequency of the *W* allele is labeled $p$ and the frequency of the *w* allele is labeled $q$.

In general, once the genotype frequencies of a population equal the Hardy–Weinberg frequencies of $p^2$, $2pq$, and $q^2$, they remain constant over time if the four specified conditions continue to be met. A population in which the observed genotype frequencies match the Hardy–Weinberg predicted frequencies is said to be in Hardy–Weinberg equilibrium

# Table of Metric–English Conversion

## Common conversions

| Length | | To convert | Multiply by | To yield |
|---|---|---|---|---|
| nanometer (nm) | $0.000000001\ (10^{-9})$ m | inches | 2.54 | centimeters |
| micrometer (μm) | $0.000001\ (10^{-6})$ m | yards | 0.91 | meters |
| millimeter (mm) | $0.001\ (10^{-3})$ m | miles | 1.61 | kilometers |
| centimeter (cm) | $0.01\ (10^{-2})$ m | | | |
| meter (m) | — | centimeters | 0.39 | inches |
| kilometer (km) | $1{,}000\ (10^{3})$ m | meters | 1.09 | yards |
| | | kilometers | 0.62 | miles |

| Weight (mass) | | | | |
|---|---|---|---|---|
| nanogram (ng) | $0.000000001\ (10^{-9})$ g | ounces | 28.35 | grams |
| microgram (μg) | $0.000001\ (10^{-6})$ g | pounds | 0.45 | kilograms |
| milligram (mg) | $0.001\ (10^{-3})$ g | | | |
| gram (g) | — | grams | 0.035 | ounces |
| kilogram (kg) | $1{,}000\ (10^{3})$ g | kilograms | 2.20 | pounds |
| metric ton (t) | $1{,}000{,}000\ (10^{6})$ g $(=10^{3}$ kg) | | | |

| Volume | | | | |
|---|---|---|---|---|
| microliter (μl) | $0.000001\ (10^{-6})$ l | fluid ounces | 29.57 | milliliters |
| milliliter (ml) | $0.001\ (10^{-3})$ l | quarts | 0.95 | liters |
| liter (l) | — | | | |
| kiloliter (kl) | $1{,}000\ (10^{3})$ l | milliliters | 0.034 | fluid ounces |
| | | liters | 1.06 | quarts |

| Temperature | | |
|---|---|---|
| degree Celsius (°C) | — | To convert Fahrenheit (°F) to Celsius (°C): $°C = \frac{5}{9}°F - 32°$ |
| | | To convert Celsius (°C) to Fahrenheit (°F): $°F = \frac{9}{5}°C + 32°$ |

# PERIODIC TABLE OF THE ELEMENTS

**Legend:**

- 1 — Atomic number
- H — Symbol
- Hydrogen — Name
- 1.00794 — Average atomic mass

Metals
Metalloids
Nonmetals

| Group | 1 / 1A | 2 / 2A | 3 / 3B | 4 / 4B | 5 / 5B | 6 / 6B | 7 / 7B | 8 / 8B | 9 / 8B | 10 / 8B | 11 / 1B | 12 / 2B | 13 / 3A | 14 / 4A | 15 / 5A | 16 / 6A | 17 / 7A | 18 / 8A |
|---|---|---|---|---|---|---|---|---|---|---|---|---|---|---|---|---|---|---|
| **1** | 1 H Hydrogen 1.00794 | | | | | | | | | | | | | | | | | 2 He Helium 4.002602 |
| **2** | 3 Li Lithium 6.941 | 4 Be Beryllium 9.012182 | | | | | | | | | | | 5 B Boron 10.811 | 6 C Carbon 12.0107 | 7 N Nitrogen 14.0067 | 8 O Oxygen 15.9994 | 9 F Fluorine 18.9984032 | 10 Ne Neon 20.1797 |
| **3** | 11 Na Sodium 22.98976928 | 12 Mg Magnesium 24.3050 | | | | | | | | | | | 13 Al Aluminum 26.981586 | 14 Si Silicon 28.0855 | 15 P Phosphorus 30.973762 | 16 S Sulfur 32.065 | 17 Cl Chlorine 35.453 | 18 Ar Argon 39.948 |
| **4** | 19 K Potassium 39.0983 | 20 Ca Calcium 40.078 | 21 Sc Scandium 44.955912 | 22 Ti Titanium 47.867 | 23 V Vanadium 50.9415 | 24 Cr Chromium 51.9961 | 25 Mn Manganese 54.938045 | 26 Fe Iron 55.845 | 27 Co Cobalt 58.933195 | 28 Ni Nickel 58.6934 | 29 Cu Copper 63.546 | 30 Zn Zinc 65.38 | 31 Ga Gallium 69.723 | 32 Ge Germanium 72.64 | 33 As Arsenic 74.92160 | 34 Se Selenium 78.96 | 35 Br Bromine 79.904 | 36 Kr Krypton 83.798 |
| **5** | 37 Rb Rubidium 85.4678 | 38 Sr Strontium 87.62 | 39 Y Yttrium 88.90585 | 40 Zr Zirconium 91.224 | 41 Nb Niobium 92.90638 | 42 Mo Molybdenum 95.96 | 43 Tc Technetium [98] | 44 Ru Ruthenium 101.07 | 45 Rh Rhodium 102.90550 | 46 Pd Palladium 106.42 | 47 Ag Silver 107.8682 | 48 Cd Cadmium 112.411 | 49 In Indium 114.818 | 50 Sn Tin 118.710 | 51 Sb Antimony 121.760 | 52 Te Tellurium 127.60 | 53 I Iodine 126.90447 | 54 Xe Xenon 131.293 |
| **6** | 55 Cs Cesium 132.9054519 | 56 Ba Barium 137.327 | 57 La Lanthanum 138.90547 | 72 Hf Hafnium 178.49 | 73 Ta Tantalum 180.94788 | 74 W Tungsten 183.84 | 75 Re Rhenium 186.207 | 76 Os Osmium 190.23 | 77 Ir Iridium 192.217 | 78 Pt Platinum 195.084 | 79 Au Gold 196.966569 | 80 Hg Mercury 200.59 | 81 Tl Thallium 204.3833 | 82 Pb Lead 207.2 | 83 Bi Bismuth 208.98040 | 84 Po Polonium [209] | 85 At Astatine [210] | 86 Rn Radon [222] |
| **7** | 87 Fr Francium [223] | 88 Ra Radium [226] | 89 Ac Actinium [227] | 104 Rf Rutherfordium [261] | 105 Db Dubnium [262] | 106 Sg Seaborgium [266] | 107 Bh Bohrium [264] | 108 Hs Hassium [277] | 109 Mt Meitnerium [268] | 110 Ds Darmstadtium [271] | 111 Rg Roentgenium [272] | 112 Cn Copernicium [285] | | 114 Uuq Ununquadium [289] | | 116 Uuh Ununhexium [292] | | |

**6 Lanthanides**

| 58 Ce Cerium 140.116 | 59 Pr Praseodymium 140.90765 | 60 Nd Neodymium 144.242 | 61 Pm Promethium [145] | 62 Sm Samarium 150.36 | 63 Eu Europium 151.964 | 64 Gd Gadolinium 157.25 | 65 Tb Terbium 158.92535 | 66 Dy Dysprosium 162.500 | 67 Ho Holmium 164.93032 | 68 Er Erbium 167.259 | 69 Tm Thulium 168.93421 | 70 Yb Ytterbium 173.05 | 71 Lu Lutetium 174.967 |
|---|---|---|---|---|---|---|---|---|---|---|---|---|---|

**7 Actinides**

| 90 Th Thorium 232.03588 | 91 Pa Protactinium 231.03588 | 92 U Uranium 238.02891 | 93 Np Neptunium [237] | 94 Pu Plutonium [244] | 95 Am Americium [243] | 96 Cm Curium [247] | 97 Bk Berkelium [247] | 98 Cf Californium [251] | 99 Es Einsteinium [252] | 100 Fm Fermium [257] | 101 Md Mendelevium [258] | 102 No Nobelium [259] | 103 Lr Lawrencium [262] |
|---|---|---|---|---|---|---|---|---|---|---|---|---|---|

We have used the U.S. system as well as the system recommended by the International Union of Pure and Applied Chemistry (IUPAC) to label the groups in this periodic table. The system used in the United States includes a letter and a number (1A, 2A, 3B, 4B, etc.), which is close to the system developed by Mendeleev. The IUPAC system uses numbers 1–18 and has been recommended by the American Chemical Society (ACS). While we show both numbering systems here, we use the IUPAC system exclusively in the book. Elements with atomic numbers higher than 112 have been reported but not yet fully authenticated.

# Answers to Self-Quiz Questions

## Chapter 1
1. *a*
2. *b*
3. *d*
4. *c*
5. *d*
6. *a*
7. *c*
8. *b*
9. *c*

## Chapter 2
1. *c*
2. *c*
3. *c*
4. *c*
5. *d*
6. *b*
7. *d*
8. *c*

## Chapter 3
1. *c*
2. *d*
3. *c*
4. *c*
5. *c*
6. *a*
7. *a*
8. *c*

## Chapter 4
1. *a*
2. *b*
3. *d*
4. *b*
5. *b*
6. *b*
7. *a*
8. *d*

## Chapter 5
1. *a*
2. *c*
3. *d*
4. *a*
5. *c*
6. *b*
7. *c*
8. *b*
9. *d*
10. *c*
11. *c*
12. *d*
13. *a*
14. *b*

## Chapter 6
1. *b*
2. *a*
3. *c*
4. *d*
5. *b*
6. *a*
7. *b*
8. *a*
9. *b*
10. *c*

## Chapter 7
1. *d*
2. *b*
3. *a*
4. *c*
5. *c*
6. *d*
7. *b*
8. *a*
9. *d*
10. *b*

## Chapter 8
1. *c*
2. *a*
3. *b*
4. *c*
5. *d*
6. *a*
7. *c*
8. *d*
9. *d*
10. *c*

## Chapter 9
1. *d*
2. *b*
3. *d*
4. *b*
5. *b*
6. *c*
7. *a*
8. *d*
9. *b*
10. *c*

## Chapter 10
1. *b*
2. *a*
3. *b*
4. *d*
5. *c*
6. *c*
7. *d*
8. *a*

## Chapter 11
1. *d*
2. *b*
3. *c*
4. *d*
5. *c*
6. *c*
7. *d*
8. *c*
9. *c*

## Chapter 12
1. *a*
2. *b*
3. *d*
4. *d*
5. *a*
6. *d*
7. *d*

## Chapter 13
1. *c*
2. *b*
3. *c*
4. *c*
5. *a*
6. *c*
7. *d*

## Chapter 14
1. *c*
2. *c*
3. *b*
4. *a*
5. *c*
6. *d*
7. *c*
8. *c*

## Chapter 15
1. *b*
2. *c*
3. *b*
4. *d*
5. *a*
6. *d*
7. *b*
8. *c*

## Chapter 16
1. *c*
2. *d*
3. *d*
4. *a*
5. *c*
6. *a*
7. *d*

## Chapter 17
1. *d*
2. *d*
3. *c*
4. *a*
5. *b*
6. *c*
7. *d*
8. *a*
9. *b*
10. *c*

## Chapter 18
1. *b*
2. *a*
3. *b*
4. *d*
5. *c*
6. *a*
7. *c*
8. *d*

## Chapter 19
1. *b*
2. *b*
3. *c*
4. *a*
5. *d*
6. *d*
7. *b*
8. *c*
9. *d*
10. *c*

## Chapter 20

1. d
2. d
3. b
4. a
5. c
6. c
7. a
8. d
9. b
10. b

## Chapter 21

1. c
2. d
3. c
4. b
5. c
6. d
7. a
8. b
9. c
10. b

## Chapter 22

1. d
2. b
3. b
4. c
5. d
6. a
7. d
8. b

## Chapter 23

1. d
2. b
3. d
4. c
5. c
6. b
7. d
8. a

## Chapter 24

1. c
2. a
3. d
4. a
5. b
6. b
7. c
8. d
9. d

## Chapter 25

1. a
2. b
3. a
4. d
5. a
6. d
7. c

## Chapter 26

1. c
2. d
3. c
4. b
5. a
6. d
7. c
8. d
9. c
10. a

## Chapter 27

1. d
2. d
3. c
4. b
5. c
6. d
7. d
8. b
9. c
10. d

## Chapter 28

1. c
2. d
3. b
4. c
5. c
6. a
7. c
8. c
9. a
10. d
11. b
12. b

## Chapter 29

1. d
2. c
3. d
4. b
5. c
6. a
7. a
8. c
9. b
10. d

## Chapter 30

1. c
2. d
3. a
4. c
5. b
6. a
7. d
8. b
9. c
10. a
11. b
12. c

## Chapter 31

1. b
2. d
3. c
4. a
5. c
6. b
7. d
8. b
9. c
10. b

## Chapter 32

1. b
2. d
3. b
4. c
5. b
6. b
7. c
8. d
9. c
10. b

## Chapter 33

1. c
2. d
3. b
4. a
5. b
6. a
7. a
8. c
9. c
10. d

## Chapter 34

1. b
2. d
3. c
4. c
5. d
6. a
7. c
8. b
9. a
10. c

## Chapter 35

1. b
2. d
3. c
4. a
5. c
6. b
7. d
8. b

## Chapter 36

1. a
2. d
3. a
4. c
5. b
6. a
7. d
8. a

# Answers to Review Questions

## Chapter 1

1. Science is a body of knowledge about the natural world and an evidence-based process for generating that knowledge. The characteristics of science include these: (1) Science deals with the natural world, which can be detected, observed, and measured. (2) Scientific knowledge is based on evidence that can be demonstrated through observations and/or experiments. (3) Scientific knowledge is subject to independent validation and peer review. (4) Science is open to challenge based on evidence by anyone at any time. (5) Science is a self-correcting enterprise.

   Science cannot answer all types of questions that humans might raise. Science is restricted to seeking natural causes to explain the workings of our world. Science cannot tell us what is morally right or wrong, or speak to the existence of God or any supernatural being. Science can only attempt to find the objective truth; it cannot address subjective issues such as what is beautiful or ugly.

2. "Correlation" means that two or more aspects of the natural world behave in an interrelated manner: if one variable shows a particular value, we can predict a particular value for the other. However, correlation between two variables does not necessarily imply that one is the cause of the other.

   As one example, studies show that the incidence of skin cancer has gone up sharply since the 1950s, after sunscreens were introduced and their use climbed in the decades that followed. Does that mean that the use of sunscreen causes skin cancer? Further analysis revealed that post–World War II generations spend much more time in the sun, and that people who sunburn easily and would have spent less time in the sun in the past are more sun-exposed now because of a false sense of security. Further, dermatologists say that most people use sunscreen inappropriately, usually applying too little of it and too infrequently for maximum protection.

   As another example, a Hungarian study reported that men who carry their cell phones in their pants pockets have a 30 percent lower sperm count than do men who keep their cell phones in a jacket pocket. The media ran with the assumption that cell phones reduce sperm count. Shortly thereafter, a physician pointed out that men who use their pants pockets for cell phones are disproportionately smokers, who commonly save their jacket pockets for their cigarette packs because there they are less likely to get crushed. That smoking reduces sperm count is already well known.

3. **Observation:** Huge numbers of fish were being found dead; their bodies, covered with bleeding sores, were found floating by the millions in the estuaries of North Carolina.

   **Hypothesis:** On the basis of a previous experience in which laboratory fish had died suddenly after exposure to local river water, Dr. Burkholder hypothesized that *Pfiesteria*, a protist found in high numbers in the tanks containing the dead lab fish, was also responsible for the fish die-offs in local estuaries.

   **Experiment:** Dr. Burkholder isolated samples of *Pfiesteria* and exposed healthy fish to the samples. The fish were quickly killed, upholding the prediction that Dr. Burkholder had formulated on the basis of her hypothesis.

4. A scientific fact is a direct and repeatable observation of any aspect of the physical world. A scientific theory is a component of scientific knowledge that has been repeatedly confirmed in diverse ways and is provisionally accepted by the experts of the discipline because it has stood the test of time.

5. Energy flows from the sun to photosynthesizing organisms such as grasses. The grasses, which are producers, use the sun's energy to produce chemical energy in the form of sugars and starches. Antelope, which are consumers, feed on the grasses to produce energy for their own use. Lions, also consumers, then eat the antelope. Ticks, consumers as well, feed on both the antelope and the lions. The grasses are producers because they capture sunlight and convert it to energy. The antelope, lions, and ticks are consumers because they eat either plants or other organisms that derive energy from plants. The food chain looks like this:

6. Levels of the biological hierarchy, from smallest to largest (with examples), are: atom (carbon); molecule (DNA); cell (bacteria); tissue (muscle tissue); organ (heart); organ system (stomach, liver, and intestines in the digestive system); individual (human); population (field mice in one field); community (different species of insects living in a forest); ecosystem (a river and the communities of organisms living in and around it); biome (the Arctic tundra or a coral reef); biosphere (Earth).

7. The American Council on Exercise (ACE) and *Consumer Reports* have both found fault with the claims made by makers of toning shoes. Check ACE's website to see how it designed and conducted experiments to test the claims.

# Chapter 2

1. At the website maintained by the Integrated Taxonomic Information System (http://www.itis.gov), you can find taxonomic information about more than half a million species belonging to the three domains. You could also try the website of the International Union for Conservation of Nature (http://www.iucn.org) to learn more about the status of your favorite organism, especially if it is endangered.

   From smallest to largest, the groupings of the Linnaean hierarchy are species, genus, family, order, class, phylum, kingdom. Any two species that belong to the same family must also belong to the same order because order is a more inclusive category than family. The order encompasses not only all members of a family classified within it but also all members of other closely related families.

2. Prokaryotes perform key tasks in ecosystems, including carrying out photosynthesis, providing nitrate to plants, and decomposing dead organisms. Prokaryotes are at the bottom of many food chains that sustain a great diversity of life. Prokaryotes are useful to humanity in many ways (for example, in cleaning up oil spills and helping with our digestion), but they also cause deadly diseases.

3. In designing your experiment, keep in mind the following characteristics of living organisms: (1) They are built of cells. (2) They reproduce themselves using DNA. (3) They develop. (4) They capture energy from their environment. (5) They sense their environment and respond to it. (6) They show a high level of organization. (7) They evolve. Also keep the characteristics of viruses in mind, since viruses lack many of the qualities present in living organisms. Your experiment could choose to investigate any one of these characteristics.

4. No, cell shape is not a suitable criterion for classifying these archaeans, because not all the genera with the same shape (coccus, for example) form a distinct branch (clade) on this DNA-based evolutionary tree. In other words, a particular shape (coccus) is not a shared derived trait for any of these groups. (Unique features that have evolved in a group's most recent common ancestor and are then shared by the descendant species of that ancestor are known as shared derived traits.) For the same reason, habitat preference is not a suitable criterion for classifying these groups. Sister groups that share a most recent common ancestor can have very different shapes (coccus and rod, for example) and different habit preferences (halophile and thermophile, for example).

   No, we cannot assume that a taxon shown to the right of a fork is more complex or recently evolved than a taxon depicted to the left of the node. Any node can be swiveled 180 degrees, without changing the meaning of the evolutionary tree.

5.

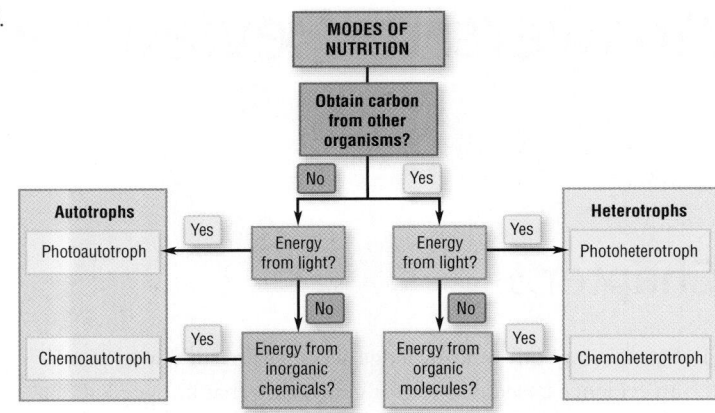

# Chapter 3

1. Although prokaryotes are known to acquire genetic information through a process called lateral (or horizontal) gene transfer, they do not exhibit sexual reproduction. Sexual reproduction is common in each of the four kingdoms of the Eukarya. Sexual reproduction combines genetic information from two parents to produce offspring that are genetically different from each other and from both parents. Because it promotes genetic diversity in the offspring, sexual reproduction increases the odds that the population will be varied enough to adapt to environmental changes such as an attack by a new strain of bacterium or virus.

2. Examples of multicellular protists include *Volvox carteri* and other green algae, as well as the red and brown algae that are large enough to be seen with the naked eye and are popularly called seaweed. Multicellularity enabled organisms to grow larger, and large size can be helpful in gathering resources (such as sunshine) and storing them (as surplus food or mineral nutrients) and in escaping potential predators or surviving predation. Multicellularity also enabled more complex and efficient functioning by enabling the division of labor among cell types dedicated to specific functions.

3. A red tide is a population explosion, or bloom, of photosynthetic plankton, usually toxic dinoflagellates. Blooms of these protists sometimes cause the water to turn red because of a reddish pigment concentrated inside their cells. Dinoflagellates and other photosynthetic plankton can produce a variety of toxins, including some that cause nerve and muscle paralysis in humans and other mammals. Some of these toxins accumulate in shellfish and cause paralytic shellfish poisoning in humans and wild animals who eat the shellfish. Scientists do not have a full understanding of what causes the sudden blooms, but the frequency of such blooms has been rising around the world, and pollution from fertilizer runoff and sewage are thought to be among the culprits.

4. Fungi are absorptive heterotrophs, whereas animals (including humans) are ingestive heterotrophs. Fungi are more closely related to animals than they are to plants. Fungi and animals share certain similarities, including these: In the cell walls fungi have chitin, the chemical that also strengthens the exoskeleton of most insects and crustaceans. Fungi and animals both store surplus carbohydrate as a molecule called glycogen. Most telling of all, DNA comparisons show that fungi and animals are closer than fungi and plants.

5. Great piles of unrotted plant material will have accumulated everywhere—in backyards, city parks, and forests. The landfills will be overflowing with undecomposed woody material, and disposing of manure from farms will be even more of a problem. The loss of their mycorrhizal partners will cause nutritional deficiencies in many plants, and some of these plants may become extinct. On the other hand, farmers will raise bumper crops with synthetic fertilizer, since they will lose much less of their crop to the scourge of fungal diseases. There will be no bread leavened with yeast, no beer and wine as we know them, and sliced mushrooms will not be an option for pizza topping. On the other side of the ledger, we will not have to confront moldy fruit and cheese in the neglected corners of the refrigerator.

6. When plants colonized land, they had to evolve in ways that enabled them to persist in an environment where they were no longer surrounded and supported by water. To deal with the problem of obtaining and retaining water, plants evolved root systems (which enable them to absorb water and nutrients from soil) and the waxy covering over their stems and leaves known as the cuticle (which prevents their tissues from drying out when exposed to sun and air). To combat gravity and grow taller, plants evolved a tough chemical polymer called lignin, which strengthened cell walls, especially those of the water-conducting tubes of xylem.

7. A lichen is a mutualistic association between a photosynthetic microbe and a fungus. The lichen-forming microbe can be a unicellular green alga, a cyanobacterium, or both together. The lichen body is thin and has no protective sheath like the cuticle of a plant or the "armor plating" of some protists. Nor does it have any mechanism for ridding the body of wastes or toxic substances. Therefore, lichens readily absorb and accumulate air- or water-borne pollutants. Lichens are destroyed by acid rain, heavy-metal pollutants, and organic toxins, which is why lichens tend to disappear in heavily industrialized areas with poor pollution controls.

# Chapter 4

1. The risk of desiccation (drying out) and the need for mechanical reinforcement of the body were some of the environmental challenges faced by the first insects. The evolution of a chitin-reinforced exoskeleton helped solve both problems. The exoskeleton prevents the soft body tissues from drying out and also strengthens the body. In contrast to the external gills of crustaceans, the gas exchange surfaces for insects are internal, which protects the moist surfaces from desiccation.

2. Sponges lack true tissues and distinct body symmetry. Snails, in contrast, have true tissues, organs and organ systems, bilateral symmetry, a true coelom, segmentation, and cephalization. Created by former marine biologist Stephen Hillenburg, *SpongeBob SquarePants* features several protostome characters: Gary is a snail; Squidward Tentacles is a squid, although he fancies himself an octopus; Mr. Krabs is a red crab. As a sea star, Patrick is an echinoderm, and therefore a member of the deuterostomes. The only chordate character is Sandy Cheeks, a squirrel. Plankton, the arch villain, is a protist, not a member of the kingdom Animalia.

3. A segmented body plan, in which the body consists of repeated units known as segments, was a significant evolutionary innovation because it paved the way for the specialization of body segments, and the appendages that arise from them, for diverse, specialized functions; these in turn enabled animals to adapt to new habitats or to acquire new modes of life. The varied uses of segments and their appendages are exemplified by the body plan of a lobster or crayfish.

4. Muscle and nerve tissues were critical to the development of efficient modes of locomotion, and they are found in all animals except sponges. A fluid-filled body pseudocoelom enabled locomotion through a hydrostatic skeleton. Another evolutionary advance was the evolution of the exoskeleton in many protostomes and the endoskeleton in all chordates; the skeleton provides anchoring for muscles, and collectively the musculoskeletal system enables efficient movement. Bilaterality combined with paired appendages, and cephalization, were adaptations for faster and more efficient modes of locomotion.

   Fish propel themselves through water by moving paired pectoral and pelvic fins and making powerful side-to-side movements with their caudal fins. Tetrapod limbs have to support the animal's weight and move that weight. Accordingly, their limbs are jointed for greater range of motion and strengthened internally by strong bones. Locomotion helps an animal capture prey, eat prey, avoid being captured, attract mates, care for young, and migrate to new habitats.

5. Birds have hollow bones that are light but strong for their weight. They have toothless beaks, and many of their internal organs are reduced (for example, female birds have a single ovary), which lessens body weight. The feathers attached to their forelimbs not only act as insulation to hold body heat, but also enable strong sustained flight, with the tail feathers acting as stabilizers.

   Like dinosaurs and other reptiles, birds lay amniotic eggs and have scaly skin on parts of the body. Unlike reptiles, however, birds have a four-chambered heart that resembles ours: oxygenated blood and nonoxygenated blood are kept in separate chambers, and the blood that is pumped to the rest of the body therefore contains high levels of oxygen, supporting a high metabolic rate. The respiratory system of birds is considered even more efficient than ours. Birds have a one-way flow of air through their respiratory passages, so that incoming air never mixes with outgoing air. In contrast, the air we inhale and exhale takes the same route (although in opposite directions), so there is some mixing of the incoming and outgoing airstreams. Like most mammals, most birds are endotherms and homeotherms.

6. The embryo developing inside a bird's egg is surrounded and protected by layers of extraembryonic membranes, including one called the amnion, that promote gas exchange and waste removal. The calcium-rich protective shell retards moisture loss but is porous enough to allow the entry of life-giving oxygen and the release of waste carbon dioxide. The amniotic egg hoards food in the form of a large yolk mass, which enables the young to achieve a relatively advanced level of development before emerging (hatching) from the shell.

7. Feathers evolved among the theropod dinosaurs as a layer of insulation that kept the animals warm. In time, the feathers attached to the forelimbs became modified for sustained flight. Feathered fossil theropods with limbs too weak for flight are a clue that feathers evolved as insulation initially and were later adapted for gliding and flying.

8. The multistep developmental changes through which immature forms of animals are transformed into adults is called metamorphosis. When the developmental changes are gradual, the species are said to have incomplete metamorphosis. When the transition from one developmental stage to the next is dramatic, with the immature forms bearing little resemblance to adults, the species are said to have complete metamorphosis.

   By having two very different body forms in their life cycle, species with complete metamorphosis can pack very different but highly successful and highly specialized modes of living into the life cycle of one animal. Among lepidopterans, for example, the body plan of the larval stage (the caterpillar) is well suited for voracious eating in a limited area, whereas the winged adult (the butterfly) is best suited for seeking mates and finding the optimal locations for depositing eggs. Together, these very different modes of living acquire a greater variety, and therefore quantity, of available resources than the one body form could on its own.

9. Most mammals are endotherms and homeotherms: they use metabolic energy to generate heat, and they maintain a near-constant body temperature. They can trap body heat with hair, keratin-containing strands on the skin that are effective insulators. Muscles in the skin can raise the hair to trap a thicker layer of air next to the skin, which increases the insulating properties of body hair. Only mammals have sweat glands. Evaporation of sweat cools terrestrial

mammals and enables them to maintain a moderate body temperature even in extremely hot and dry environments such as the desert.

10. Mammalian mothers nurse their young with milk, a liquid from their mammary glands that is rich in fat, proteins, salts, and other nutritive substances. Monotremes lay eggs, and the newborns that hatch from them are relatively undeveloped. The mothers secrete milk from their mammary glands directly on the fur, where it is lapped up by the newborns after they hatch from their shells. Marsupial females give birth to somewhat more developed young (joeys) that are further nurtured in a ventral pouch. Marsupial mothers have a nipple in each pouch that the newborn attaches to and feeds from. Eutherians have a longer gestation and give birth to relatively well-developed young. They nurse their newborns from two or more nipples on the ventral side of the body.

# Chapter 5

1. Monomers are small molecules that serve as repeating units in a larger molecule (macromolecule). Macromolecules that contain monomers as building blocks are known as polymers. Lipids, such as triglycerides and sterols, are macromolecules, but since they are not built from discrete units that are repeated multiple times, they are not usually regarded as polymers.

2. The pH of pure water should be 7. Units on the pH scale represent the concentration of free hydrogen ions in water. In the presence of a base, the pH of a solution will be above 7, indicating that there are more hydroxide ions than hydrogen ions, so the solution is basic. In the presence of an acid, the pH will be below 7, indicating more free hydrogen ions than hydroxide ions, so the solution is acidic. Pure water has equal amounts of hydrogen and hydroxide ions and is therefore neutral.

3. A hydrogen bond is a noncovalent bond created by the electrical attraction between a hydrogen atom with a partial positive charge and any other atom that has a partial negative charge. Hydrogen bonds are weaker than ionic bonds, which are weaker than covalent bonds. Water molecules are polar: the region around the oxygen atom is slightly negative, and the regions around the two hydrogen atoms are slightly positive. This property provides for the formation of hydrogen bonds between water molecules, since each partially charged hydrogen atom in a water molecule is attracted to any atom with a partial negative charge, including the oxygen atom of a nearby water molecule.

4. Each carbon atom can form strong covalent bonds with up to four other atoms, including other carbons, creating large molecules that contain hundreds, even thousands, of atoms. These molecules play many different roles critical to life.

5. Cells use carbohydrates as a readily available energy source. Some carbohydrates, such as the cellulose found in plant cell walls, have structural functions. Nucleic acids such as DNA and RNA, which carry genetic information, are polymers of nucleotides. Some nucleotides act as energy carriers. Proteins make up the physical structures of organisms, as well as the enzymes that catalyze biochemical reactions. Lipids such as triglycerides are common means of long-term energy storage, and lipids such as phospholipids are important components of cell membranes.

6.

| FOOD INGREDIENT | MACRO-MOLECULE | BUILDING BLOCK(S) | FUNCTION |
|---|---|---|---|
| Hamburger bun | Starch (polysaccharide) | Glucose (sugar) | Energy source; building block for many other biomolecules |
| Lettuce | Cellulose (polysaccharide) | Glucose (sugar) | Non-nutritive because undigested; important for intestinal health |
| Meat patty | Protein | Amino acids | Building block for proteins, as well as other biomolecules, such as neurotransmitters |
| Cheese | Triglyceride (fat) | Fatty acids; glycerol | Energy source; building block for other biomolecules, including membrane phospholipids |

7. Elements that are abundant in the human body but scarce in rocks are carbon and nitrogen. Amino acids contain both of these elements. Proteins are polymers of amino acids.

8. The solvent in black coffee is water. In a cup of coffee that is sweetened and topped with whipped cream, the hydrophilic solutes are the hundreds of organic molecules found in coffee (including caffeine) and sugar (sucrose, a disaccharide). The hydrophobic substances, perhaps partially mixed with the coffee but mainly floating on the top, are the whipped-cream constituents, which are mostly lipids (triglycerides and fatty acids).

9. Ice is 9 percent less dense than liquid water, and that is why ice floats on water. If ice were denser than liquid water, it would sink in this lake in winter, and the lake would then freeze from the bottom up. Instead, the ice that forms on the lake surface acts like an insulating blanket, enabling aquatic organisms to survive the winter in the liquid water below.

   Water molecules in liquid water move about vigorously, constantly forming and breaking hydrogen bonds with neighboring water molecules. The water molecules in ice have less energy and cannot move about as vigorously; they form a more stable network of hydrogen bonds. A given mass of liquid water occupies more space when it turns into ice because the molecules are spaced farther apart in ice, locked into an orderly pattern known as a crystal lattice.

# Chapter 6

1. One feature of all cells, the plasma membrane, provides a necessary boundary between a cell and its surrounding environment. The plasma membrane is selectively permeable, controlling what gets in and what flows out. Both prokaryotic and eukaryotic cells also contain DNA, cytosol, and ribosomes. DNA contains the information for producing the proteins needed by each cell. Cytosol is the watery medium in which biochemical reactions take place. Ribosomes are the workbenches for producing proteins.

2. The major components of a plasma membrane are a phospholipid bilayer and an assortment of proteins. The phospholipid molecules are oriented so that their hydrophilic heads are exposed to the watery environments both inside and outside the cell. Their hydrophobic fatty acid tails are grouped together inside the membrane away from the watery surroundings. Some membrane proteins extend all the way through the phospholipid bilayer and act as gateways for the passage of selected ions and molecules into and out of the cell. Other membrane proteins are used by the cell to detect changes in and signals from the environment outside the cell. Proteins that are not anchored to structures within the cell are free to move sideways within the phospholipid bilayer. This freedom of movement supports what is known as the fluid mosaic model, which describes the plasma membrane as a highly mobile mixture of phospholipids and proteins. This mobility is essential for many cellular functions, including movement of the cell as a whole and the ability to detect external signals.

3. Chloroplasts are found in photosynthetic cells of algae and plants. Each chloroplast has two concentric membranes and an internal network of membranes (thylakoids) that contain the light-absorbing green pigment chlorophyll. Chloroplasts harness light energy to convert carbon dioxide ($CO_2$) to sugars, splitting water molecules ($H_2O$) and releasing oxygen gas ($O_2$) in the process of photosynthesis. Mitochondria are found in nearly all cell types in all eukaryotes, both producers and consumers. Each mitochondrion has two membranes, the inner of which is thrown into many folds (cristae) and contains proteins and other components that enable the organelle to generate the energy carrier ATP, through the oxygen-dependent process of cellular respiration. In this process, the chemical energy of organic molecules (such as sugars) is converted into energy stored within ATP molecules, and $CO_2$ and $H_2O$ are released as $O_2$ is consumed.

4. Most cells are small because the ratio of surface area to volume limits cell size. As a cell's width increases, its volume increases vastly more than its surface area, so a larger cell has proportionately less plasma membrane area to import and export substances but must support a much larger cytoplasmic volume.

5. Multicellularity enables organisms to attain a larger size while still being composed of small cells that collectively present a large surface area for the exchange of materials. Multicellularity also

offers the advantage of greater efficiency by enabling the division of labor among the highly specialized cell types.

6. The partitioning of the cytoplasm into a variety of highly specialized membrane-enclosed compartments confers speed and efficiency through an intracellular division of labor. The different types of membrane-enclosed organelles serve very specific and unique functions. Unique chemical environments can be maintained within a membrane-enclosed compartment. For example, the reactions that break apart polymers work best under exceptionally acidic conditions, so the organelles that specialize in that task maintain a very low pH, even though the pH of the cytosol outside the organelles is close to neutral. Some chemical reactions produce by-products that could interfere with other vital reactions or even poison the cell. Locking interfering or toxic substances into special compartments avoids such "collateral damage."

7.

| CELLULAR STRUCTURE | MAIN FUNCTIONS | FOUND IN: PROKARYOTES? EUKARYOTES? PLANTS? ANIMALS? |
|---|---|---|
| Plasma membrane | Boundary of cell; controlling movement of substances in and out of cell | Yes, in all |
| Cytoplasm | Water-based medium; site of thousands of critical chemical reactions | Yes, in all |
| Nucleus | Repository of DNA, the genetic material; site for replication of DNA and expression of coded information in the form of RNA | Not in prokaryotes, but in all others |
| Ribosome | Protein-synthesizing unit in the cytoplasm | Yes, in all |
| Endoplasmic reticulum | Site for synthesis of most lipids and many proteins exported to organelles or the cell exterior | In all eukaryotes, including plants and animals |
| Golgi apparatus | Site for tagging of lipids and proteins, and for shipping to their final destinations | In all eukaryotes, including plants and animals |
| Lysosome | Degradation and recycling of molecules and even whole organelles | In animals and some other eukaryotes, such as fungi and some protists |
| Mitochondrion | Energy production in the form of the energy-rich molecule ATP, which is needed by all cells | In all eukaryotes, including plants and animals |
| Chloroplast | Synthesis of carbohydrate from carbon dioxide and water using light energy | In plants and some protists (algae); not in any other eukaryote |
| Cytoskeleton | Cell shape; internal organization; intracellular movement of organelles; whole-cell movement | In all eukaryotes, including plants and animals |

# Chapter 7

1. Yes, the net movement of scent molecules throughout the room is an example of diffusion because it is a passive process driven by the difference in the concentration of molecules from one part of the room to another. Diffusion ceases at equilibrium because net movement ceases. However, the molecules continue to move about because of their kinetic energy. Despite the continuation of this molecular motion in the room, the averages of these movements are equal in all directions and therefore cancel themselves out.

2. The pond water surrounding the *Paramecium* is hypotonic relative to the cell. The cell has a tendency to gain water through osmosis. It has special organelles, known as contractile vacuoles, that regularly collect excess water and discharge it to the outside of the cell.

3. Phagocytic white blood cells will engulf the invading bacteria. They use a special form of endocytosis, called phagocytosis, to internalize large particles and whole cells. Membranes play a crucial part in the process, which begins when receptor proteins on the outer surface of the plasma membrane recognize surface characteristics of the material to be brought into the cell. Extensions of the membranes called pseudopodia then encircle the invading bacterium to enclose it completely in a vesicle. The vesicle eventually fuses with a lysosome, where the engulfed bacterium is destroyed.

4. Epithelial cells in animals can form leakproof barriers because of tight junctions, which bind cells together with belts of protein embedded in their plasma membranes. The tight junctions prevent molecules from slipping between the cells of the intestinal epithelium to enter the bloodstream on the other side of the cell layer. Molecules and ions can pass to the other side only if they are selectively taken up by the epithelial plasma membrane that faces the space inside the gut.

5. The role of LDL (low-density lipoprotein) particles is to deliver lipids (including cholesterol) from the liver, where these lipids are manufactured, to other cells in the body. A region of the apolipoprotein in the LDL particle is recognized by the LDL receptor. The docking of an LDL particle with an LDL receptor triggers endocytosis of the entire complex. Endocytotic compartments eventually deliver the LDL particle–LDL receptor complex to the lysosome, an organelle that specializes in taking apart biomolecules and releasing the building blocks into the cytosol for reuse.

# Chapter 8

1. Hopefully you are not among the 4 in 10 Americans who skip breakfast most days. According to an ABC News poll, 31 percent of Americans who eat breakfast regularly go for cold breakfast cereal, the top choice. The macromolecular carbohydrates, proteins, fats, and nucleic acids in a typical breakfast cereal are broken down

to small organic molecules (sugars, amino acids, fatty acids and monoglycerides, nucleotides, and phosphate groups) that are then absorbed into the bloodstream. Some of the chemical energy in these macromolecules is released as heat when they are digested, but beyond warming us slightly, this heat energy is not available for cellular work. In cellular respiration, some of the chemical energy in sugars and fats is turned into the chemical energy of ATP molecules, but a large amount is also released as metabolic heat in accordance with the second law of thermodynamics.

ATP is used in myriad ways: to contract the heart muscle, thereby pushing blood through the body (chemical energy to kinetic energy); to fire brain cells that help us think (chemical energy to electrical energy); to supply the sodium-potassium pumps that maintain the osmotic balance of every cell (chemical energy being transformed into the potential energy of an electrical gradient); and to synthesize every major biomolecule somewhere in the body (chemical energy being stored as the potential energy of larger molecules). The average cell in the human body uses up about 1–2 billion molecules of ATP per minute, and while some of the chemical energy released in the breaking of a phosphate bond is used to fuel cellular activities, a large proportion is released as metabolic heat. These are just some of the many different kinds of energy transformations that take place in your body every second of your life.

2. The second law of thermodynamics holds that systems tend toward disorder. In a living system such as a cell, the order maintained by chemical reactions is counterbalanced by the release of heat energy (disorder) into the surroundings.

3. "Anabolism" refers to metabolic processes that manufacture larger molecules from smaller units. Anabolic pathways need an input of energy. "Catabolism" refers to metabolic processes that break down macromolecules, releasing energy and small organic molecules. Photosynthesis is an anabolic process because it creates large molecules such as sugars from smaller units such as carbon dioxide and water. It is driven by an input of light energy.

4. Enzymes are biological catalysts that speed up chemical reactions by positioning bound reactant molecules in such a way that they collide more often and more accurately, thereby increasing the reaction rate. Like any other catalyst, an enzyme lowers the activation energy barrier of a reaction to make a reaction proceed faster, but it does not provide energy to make that reaction happen. In other words, it cannot make a chemical reaction happen that would not proceed without an input of external energy.

5. According to the induced fit model of enzyme action, the binding of the substrate to the active site of an enzyme further molds the active site to create a more stable interaction between an enzyme and its substrate. The bound substrate reshapes the binding site of the enzyme slightly, the way your hand shapes a properly sized glove when you put it on.

6. Ephedrine, the active ingredient in the herbal medicine, is a stimulant that activates the nervous system in ways that increase the heart rate and metabolic rate. The increased output of metabolic heat leads to flushing and sweating in most users. The drug was quite effective in weight loss because it increased the metabolism of food stores while also suppressing appetite. However, it also produced serious side effects, and after a number of deaths from heatstroke or fast and irregular heartbeats were linked to the use of ephedrine, sale of the drug was banned by the FDA (Food and Drug Administration).

# Chapter 9

1. No, the Calvin cycle cannot be sustained in a plant that is kept in the dark for more than a very short period (a few hours), because these enzymatic reactions depend on the chemical energy delivered by NADPH and ATP that are generated by the light reactions.

2. The transfer of electrons down an electron transport chain (ETC) produces a proton gradient in both chloroplasts and mitochondria. The protons move down that gradient through a membrane channel protein known as ATP synthase. The movement of the protons releases energy, which is used by ATP synthase to phosphorylate ADP to form ATP.

3. The postglycolytic fermentation reactions do not generate any energy. Fermentation swings into action in some cells when the oxygen supply is low, and its main function is to support increased ATP production through increased rates of glycolysis. The only role of the postglycolytic fermentation reactions is to regenerate $NAD^+$, which is essential for the continued operation of glycolysis.

4. ATP production through oxidative phosphorylation is critically dependent on a proton ($H^+$) gradient generated by the energy released by electrons as they travel down the electron transport chain (ETC). Electron transport pumps protons from the matrix into the intermembrane space of the mitochondrion, and the potential energy of this gradient powers the phosphorylation of ADP to generate ATP by ATP synthase.

If the proton gradient collapses before ATP can be generated, *all* of its potential energy is released as heat, instead of some being transformed into the chemical energy of ATP through oxidative phosphorylation. By collapsing the proton gradient, DNP causes all of its energy to be released as heat. The overall effect is that at low concentrations of DNP, some of the chemical energy in food molecules (the ultimate source of the electrons that travel down the ETC) is released as heat. That means some of the energy in food is "burnt off" by the body, so less of it is available as surplus energy to be stashed away as stored fat. This is a dangerous way of tweaking mitochondrial energy output because if the DNP levels become too high in critical tissues, the molecule can kill cells by destroying their ability to generate the life-giving molecule ATP.

5. Mitochondrial respiration is more productive than glycolysis alone. For each molecule of glucose consumed during glycolysis, there is a net yield of 2 ATP molecules and 2 NADH molecules. Much energy remains in the two pyruvate molecules furnished by the glycolytic degradation of one glucose molecule.

   Mitochondria break up pyruvate through a series of reactions (glycolysis, the Krebs cycle, and oxidative phosphorylation) to yield about 30 ATP molecules per molecule of glucose.

6. Mitochondrial membranes play a crucial role in oxidative phosphorylation, which takes place in the many folds (cristae) of the inner mitochondrial membrane. The folds in the inner mitochondrial membrane create a large surface area on which are embedded many electron transport chains (ETCs) and many units of ATP synthase. In contrast, the outer membrane is unfolded because it does not bear any of the components that are responsible for ATP generation. The inner mitochondrial membrane also serves the important function of forming a barrier across which a proton gradient can develop during electron transport. As described in the answer to question 4, the potential energy of this gradient is absolutely necessary for ATP synthesis and hence the energy output from cellular respiration. The thylakoids serve the same function in chloroplasts as the space created by the inner mitochondrial membrane in mitochondria.

7. Photosynthesis occurs in chloroplasts, uses light energy, synthesizes sugars from carbon dioxide and water, and releases oxygen as a by-product. Cellular respiration occurs in mitochondria, releases energy from organic molecules such as sugars in an oxygen-dependent process, and generates carbon dioxide and water as by-products. Photosynthesis is an anabolic process, while cellular respiration is a catabolic process. Photosynthesis occurs only in producers (algae and plants), while cellular respiration takes place in both producers and consumers. The energy carriers (ATP and NADPH) generated by the light reactions provide the energy and the protons and electrons necessary for converting carbon dioxide into sugar. All three stages of cellular respiration (glycolysis, Krebs cycle, and oxidative phosphorylation) generate ATP and NADH (closely related to NADPH).

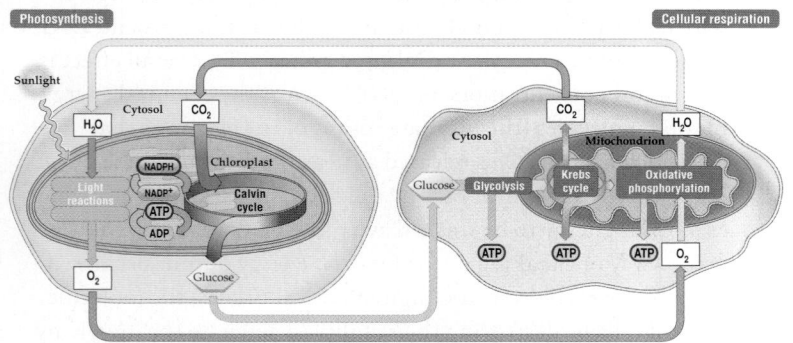

# Chapter 10

1. Regular checkpoints ensure that the right cell is dividing at the right time and that cell division occurs with high fidelity. This system of checks and balances is important because cell division is metabolically expensive. Cells with an abnormal chromosome number are also generally nonfunctional and would reduce the fitness of the organism. Finally, in vertebrates runaway divisions can result in cancer. The cell cycle can progress in only one direction, and because there is no "reverse button," checks are made to ensure that conditions are optimal *before* a cell commits to cell division by moving from $G_1$ to S phase. Cell cycle regulatory proteins can halt the cell cycle if the cell is too small, the nutrient supply is inadequate, or DNA is damaged. $G_2$ arrests in the same circumstances, as well as when the DNA duplication that begins in S phase is incomplete for any reason.

2. A horse cell has a total of 128 DNA molecules in the $G_2$ phase prior to mitosis. At the end of meiosis I, each daughter cell would contain 64 DNA molecules in the form of 32 pairs of linked sister chromatids.

3. Anaphase in mitosis involves (1) the separation of the two chromatids that make up every replicated chromosome and (2) their symmetrical segregation to opposite poles of the cell. The chromatids cannot be segregated evenly unless they are properly aligned at the cell center, which is the main outcome of metaphase. To reduce the risk of mis-segregation of the chromatids, a special anaphase checkpoint called the spindle checkpoint (not described in the chapter) ensures that anaphase does not begin until all replicated chromosomes are properly aligned at the metaphase plate.

4. 

   |   | S | s |
   |---|---|---|
   | S | SS | Ss |
   | s | Ss | ss |

5. If gametes of sexually reproducing organisms were produced by mitosis, the offspring of each succeeding generation would have double the number of chromosomes of the parent generation.

6. Homologous chromosome pairs are segregated during meiosis I, whereas sister chromatids are segregated during meiosis II. Meiosis I is therefore the *reduction division*, resulting in haploid daughter cells (cells with only one chromosome set, instead of two). Tetrads are formed and crossing-over occurs during prophase of meiosis I, but there is no equivalent event in prophase of meiosis II. Meiosis II is similar to mitosis in that sister chromatids are segregated into separate daughter cells after cytokinesis is complete. When meiosis begins in a single diploid cell, two cells result at the end of meiosis I, but four daughter cells are seen at the end of meiosis II.

7. Crossing-over occurs during prophase of meiosis I (prophase I). It is the physical exchange of chromosomal segments between non–sister chromatids in paired-off paternal and maternal homologues (tetrads). Crossing-over exchanges alleles between the paternal and maternal homologues. Therefore, chromatids produced by crossing-over are genetic mosaics, bearing new combinations of alleles compared to those originally carried by the paternal and maternal homologues in the diploid parent cell. The mosaic chromatid is said to be recombined, and the creation of new groupings of alleles through the exchange of DNA segments is known as genetic recombination. The biological significance of crossing-over is that it generates genetically diverse daughter cells every time meiosis occurs in the diploid cells of a given individual.

# Chapter 11

1. Embryonic stem cells are derived from the inner cell mass in the blastocyst and are pluripotent in that they can give rise to all cell types in the body. Adult stem cells arise in the fetus and persist in small numbers in various tissues and organs in children and adults; they are multipotent or unipotent. Induced pluripotent stem cells (iPSCs) are generated in the lab from ordinary differentiated cells, such as living skin cells from an adult. These iPSCs have the developmental flexibility of embryonic stem cells because they are treated in a way that reprograms their developmental potential, "resetting" it to a stage that resembles the embryonic stage.

2. Currently, technical issues related to health, the environment, agriculture, and the safety of food and consumer products are evaluated by committees of experts that advise the various arms of the federal and state governments. You can explore decision making on biology-related issues in the United States by researching federal agencies such as the FDA (Food and Drug Administration), EPA (Environmental Protection Agency), USDA (U.S. Department of Agriculture), HHS (U.S. Department of Health and Human Services), NIH (National Institutes of Health), and BLM (Bureau of Land Management). There are similar agencies in Canada (Environment Canada and Health Canada, for example) and most other developed countries. These agencies, and the lawmakers they report to, are known to be influenced by special interests, which may include commercial or public interest groups. Finally, some issues (such as public support for stem cell research in California) are put to the ballot in some states so that citizens can vote on them directly. For examples of such ballot measures and citizen initiatives, see the table in the box on page 13. You will have your own opinion about which of these avenues is best suited for decision making on a particular issue that affects all of us.

3. Preimplantation genetic testing is a widely used technology that makes use of just such methods. Figure 33.6 shows how a single cell can be removed from an eight-cell-stage morula without damaging the remaining embryo. Whether it is permissible to remove such a cell for genetic testing (a routine and legal procedure currently), while not using it to generate embryonic stem cell lines, is for you to decide.

4. Colon cancer begins with a benign growth of cells called a polyp, which is commonly caused by mutations that inactivate both copies of a tumor suppressor gene and/or transform a proto-oncogene into an oncogene. In many patients, loss of a part of chromosome 18 deletes two other tumor suppressor genes, allowing for more aggressive and rapid cell division in the tumor. In a majority of cases, the protective activity of the tumor suppressor gene *p53* is lost, which removes all remaining cell division controls and enables the cancer cells of the now malignant tumor to metastasize (spread to other parts of the body).

5. You will have your own opinion on this issue, but all 50 states in the United States place some restrictions on the sale and advertising of tobacco products (for example, a ban on the sale of cigarettes to minors, or a ban on the use of cartoon characters in tobacco advertising). Most states restrict smoking in public places and require warning labels about the harmful effects of tobacco. Public health experts have proposed additional regulations, which include a ban on candy-flavored tobacco products that might be attractive to children; a ban on the labeling of cigarettes as "light" and "mild" because of concerns that such products might be incorrectly perceived as safer; limits on the tar and nicotine content in cigarettes; and larger and more informative warning labels. Beginning in September 2012, the FDA (Food and Drug Administration) will require new health warnings on all cigarette packaging and advertisements. These new warnings include very graphic (some say too horrifying) pictures of the damage caused by tobacco use. Are these justified, in your opinion? Will they have any useful impact? Here's a fact to consider: more than 50 percent of American males smoked in the 1950s; slightly less than 20 percent do so now.

6. It is important to provide sufficient information about cancer risks so that consumers can make informed decisions about their product purchases. Food labels could be simplified to provide more concise information about possible cancer risks, and public health programs could raise public awareness of the cancer risks present in certain foods and other products.

# Chapter 12

1. Genes are the basic units of inheritance; as such, they carry genetic information for specific traits. Genes are composed of DNA and are located on chromosomes. Most genes govern a genetic trait by influencing the production of specific proteins. There are two copies of every gene in all the diploid cells in a woman's body because she has two of each type of chromosome (a paternal

homologue and a maternal homologue), each with one copy of the gene. Like women, men have two copies of all the genes that are located on autosomal chromosomes. However, they have only one copy of the genes that are unique to the X chromosome, since they have only one X chromosome.

2. New alleles arise when genes mutate. A mutation is any change in the DNA that makes up a gene. When a mutation occurs, the new allele that results may contain instructions for a protein with a form different from that of the protein specified by the original allele. By specifying different versions of proteins, the different alleles of a gene cause hereditary differences among organisms.

3. To determine the genotype of the purple-flowered plant (*PP* or *Pp*), you could cross it with a white-flowered plant (*pp*). When either genotype is crossed with the homozygous recessive (the white-flowered plant), the resulting phenotypes of the offspring will indicate whether the purple-flowered parent plant is heterozygous or homozygous. A plant with the genotype *Pp*, when crossed with a plant possessing the *pp* phenotype, will produce offspring that have an equal chance of being either white-flowered or purple-flowered (diagram 1). A dominant homozygous plant (*PP*), when crossed with a recessive homozygous plant (*pp*), will always produce purple-flowered offspring (diagram 2).

(1)

|   | *P* | *p* |
|---|-----|-----|
| *p* | *Pp* | *pp* |
| *p* | *Pp* | *pp* |

(2)

|   | *P* | *P* |
|---|-----|-----|
| *p* | *Pp* | *PP* |
| *p* | *Pp* | *PP* |

4. Although identical twins are genetically identical, their phenotypes can differ because environmental factors can alter the effects of genes. One twin who is well nourished in childhood, for example, may grow up to be tall; while the other of the pair, if malnourished in childhood, may grow up to be short. Similarly, exposure to different amounts of sunlight will cause twins' skin color to differ. The phenotypes of twins who are predisposed to a particular genetic disorder can differ radically if one of them is exposed to environmental factors that trigger the onset of that disorder but the other is not.

5. According to the CDC (Centers for Disease Control and Prevention), the four diseases that kill most people in the United States are heart disease, stroke, cancer, and diabetes, with the first three accounting for more than 50 percent of all deaths. Our risk for these diseases is influenced by genes, but lifestyle factors also have a large impact. All four diseases show complex patterns of inheritance because the genetic risk is controlled by numerous, mostly unidentified, genetic loci, and the disease risk is strongly affected by the environment.

Complex traits are influenced by multiple genes that interact with one another and with the environment. The inheritance of complex traits cannot be predicted by Mendel's laws, because

Mendelian traits are controlled by a single gene that is little affected by other genes or by environmental conditions. One way to explain complex traits in simple terms is to say that complex traits are genetic characteristics that are so complicated that we really can't predict how they will be inherited from one generation to the next. The practical value of understanding that many human traits are complex traits is that it can help us avoid the oversimplifications of "pop genetics": claiming to be able to predict a child's height from that of her parents, or setting the odds that someone will get skin cancer because both parents had skin cancer, for example. It also helps us debunk the claims that someone has found, for instance, "a gene for diabetes" or "a gay gene." Finally, complex traits remind us that genes are not destiny; our environment, including our lifestyle, has a large impact on some of the phenotypes that are critical for our health and well-being.

# Chapter 13

1. A gene is a small region of the DNA molecule in a chromosome. Genes are located on chromosomes.

2. Human females have two X chromosomes, while human males have one X and one Y chromosome; therefore, human males have only one copy of each gene that is unique to either the X or the Y chromosome. As a result, patterns of inheritance for genes located on the X chromosome may differ between males and females. A mother can pass an X-linked allele, such as one for a genetic disorder, to her male or female offspring. A male can pass an X-linked allele only to his female offspring (because his male offspring receive his Y chromosome, not his X chromosome).

3. Yes, a woman can have red-green color blindness if she inherits the allele from both her color-blind father ($X^cY$) and her carrier or color-blind mother ($XX^c$ or $X^cX^c$), which would make her a homozygote for this recessive allele ($X^cX^c$). All the daughters of this woman and a man with normal vision will be carriers, and all their sons will be color-blind.

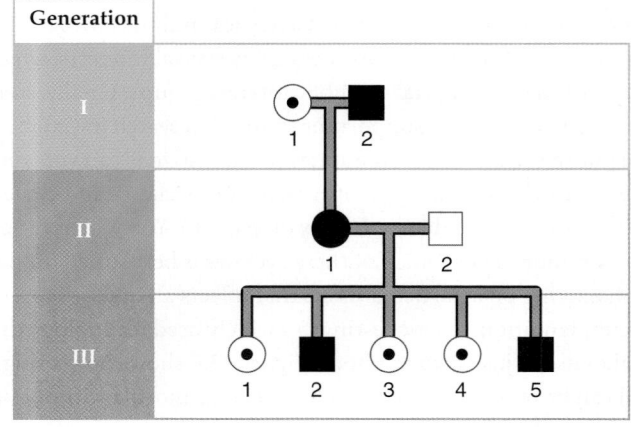

Note: In a pedigree, children are represented by birth order, with the oldest on the left and the youngest on the right. You may imagine any birth order for the three girls and two boys in this pedigree. In addition, carriers can be represented by filling in half of the relevant of circle or square, instead of placing a dot in the center of these shapes.

4. Genes located on different chromosomes separate into gametes independently of one another during meiosis; hence, such genes are not linked. If the genes for the traits shown in Figure 13.6 were inherited independently of each other, Morgan would have obtained approximately equal numbers of flies for each of the four genotypes shown in the figure. Since the numbers of the two parental genotypes outnumbered the other two genotypes by a wide margin, Morgan concluded that the genes must be located on the same chromosome. Because they are physically connected to each other, they are inherited together, or linked.

5. Crossing-over occurs when genes are physically exchanged between homologous chromosomes during meiosis. Part of the chromosome inherited from one parent is exchanged with the corresponding region from the other parent. Two genes that are far apart on a chromosome are more likely to be recombined by crossing-over than are two genes that are close to each other. Understanding this, one can assume that genes A and C are more likely to be separated into different gametes because of crossing-over than are genes A and B.

6. Nonparental genotypes may arise because of crossing-over. The exchange of genes that takes place during crossing-over makes possible the formation of gametes with combinations of alleles that differ from those found in either parent (see Figure 10.15). The independent assortment of homologous chromosome pairs into different gametes is another mechanism that shuffles the allelic combinations found in each parent, generating genetically diverse gametes that each have a random mix of the various paternal and maternal homologues.

7. Relatively few human genetic disorders are caused by inherited chromosomal abnormalities, probably because most large changes in chromosomes kill the developing embryo. Genetic disorders caused by single-gene mutations appear to be more common, because the survival rate of embryos with single-gene mutations is higher.

# Sample Genetics Problems, Chapter 12

1. a
2. d
3. a. *A* and *a*

b. *BC*, *Bc*, *bC*, and *bc*

c. *Ac*

d. *ABC*, *ABc*, *AbC*, *Abc*, *aBC*, *aBc*, *abC*, and *abc*

e. *aBC* and *aBc*

4. a. Genotype ratio: 1:1. Phenotype ratio: 1:1.

|   | A | a |
|---|---|---|
| a | Aa | aa |

b. Genotype ratio: 1:0. Phenotype ratio: 1:0.

|   | B |
|---|---|
| b | Bb |

c. Genotype ratio: 1:1. Phenotype ratio: 1:1.

|   | AB | Ab |
|---|---|---|
| ab | AaBb | Aabb |

d. Genotype ratio: 1 *BBCC*: 1 *BBCc*: 2 *BbCC*: 2 *BbCc*: 1 *bbCC*: 1 *bbCc*. Phenotype ratio: 6:2, reduced to 3:1.

|   | BC | Bc | bC | bc |
|---|---|---|---|---|
| BC | BBCC | BBCc | BbCC | BbCc |
| bC | BbCC | BbCc | bbCC | bbCc |

e. Genotype ratio: 1 *AABbCC*: 2 *AABbCc*: 1 *AABbcc*: 1 *AAbbCC*: 2 *AAbbCc*: 1 *AAbbcc*: 1 *AaBbCC*: 2 *AaBbCc*: 1 *AaBbcc*: 1 *AabbCC*: 2 *AabbCc*: 1 *Aabbcc*. Phenotype ratio: 6:2:6:2, reduced to 3:1:3:1.

|   | ABC | ABc | AbC | Abc | aBC | aBc | abC | abc |
|---|---|---|---|---|---|---|---|---|
| AbC | AABbCC | AABbCc | AAbbCC | AAbbCc | AaBbCC | AaBbCc | AabbCC | AabbCc |
| Abc | AABbCc | AABbcc | AAbbCc | AAbbcc | AaBbCc | AaBbcc | AabbCc | Aabbcc |

5.

|   | S | s |
|---|---|---|
| S | SS | Ss |
| s | Ss | ss |

Genotype ratio: 1 *SS*: 2 *Ss*: 1 *ss*. Phenotype ratio: 3 healthy: 1 sickle-cell anemia.

Each time two *Ss* individuals have a child, there is a 25 percent chance that the child will have sickle-cell anemia.

6. 100 percent of the offspring should be chocolate labs.

7. a. *NN* and *Nn* individuals are normal; *nn* individuals are diseased.

b.

|   | N | n |
|---|---|---|
| N | NN | Nn |
| n | Nn | nn |

Genotype ratio: 1 *NN*: 2 *Nn*: 1 *nn*. Phenotype ratio: 3 normal: 1 diseased.

c.

|   | N | n |
|---|---|---|
| N | NN | Nn |

Genotype ratio: 1:1. Phenotype ratio: 2 healthy: 0 diseased.

8. a. *DD* and *Dd* individuals are diseased; *dd* individuals are normal.

b.

|   | D | d |
|---|---|---|
| D | DD | Dd |
| d | Dd | dd |

   Genotype ratio: 1*DD*: 2*Dd*: 1*dd*. Phenotype ratio: 3 diseased: 1 normal.

c.

|   | D | d |
|---|---|---|
| D | DD | Dd |

   Genotype ratio: 1:1. Phenotype ratio: 2 diseased: 0 healthy.

9. The parents are most likely *BB* and *bb*. The white parent must be *bb*. The blue parent could potentially be *BB* or *Bb*, but if it were *Bb*, we would expect about half of the offspring to be white. Therefore, if the cross yields many offspring and all are blue, it is extremely likely that the blue parent's genotype is *BB*.

10. The allele for green fruit pods is dominant. Since each parent breeds true, that means each parent is homozygous. When a homozygous recessive parent is bred to a homozygous dominant parent, the F$_1$ generation will exhibit only the dominant phenotype. Therefore, the phenotype of the F$_1$ generation—green fruit pods—is produced by the dominant allele.

# Sample Genetics Problems, Chapter 13

11. a. A male inherits his X chromosome from his mother, since his Y chromosome must come from his father. His mother does not have a Y chromosome to give him, and he must have one in order to be male.

   b. No, she does not have the disorder. If she has only one copy of the recessive allele, her other X chromosome must then have a copy of the dominant allele. She is a carrier, but she does not have the disorder herself.

   c. Yes, he does have the disorder. The trait is X-linked, he has only one X chromosome, and that X chromosome carries the recessive, disorder-causing allele. His Y chromosome does not carry an allele for this gene, so it cannot contribute to the male's phenotype relative to this trait.

   d. If the female is a carrier of an X-linked recessive disorder, her genotype is X$^D$X$^d$, where *D* is the dominant allele and *d* is the recessive, disorder-causing allele. This means she can produce two types of gametes relative to this trait: X$^D$ and X$^d$. Only the X$^d$ gamete carries the disease-causing allele.

   e. None of their children will have the disorder, because the mother will always contribute a dominant, non-disorder-causing allele to each child. However, all of the female children will be carriers, because their second X chromosome comes from their father,

who has only one X chromosome to contribute, and it carries the disorder-causing allele.

12. a. 50 percent chance of the *aa* cystic fibrosis genotype:

|   | A | a |
|---|---|---|
| a | Aa | aa |
| a | Aa | aa |

   b. 0 percent chance of the *aa* cystic fibrosis genotype:

|   | A | A |
|---|---|---|
| A | AA | AA |
| a | Aa | Aa |

   c. 25 percent chance of the *aa* cystic fibrosis genotype:

|   | A | a |
|---|---|---|
| A | AA | Aa |
| a | Aa | aa |

   d. 0 percent chance of the *aa* cystic fibrosis genotype:

|   | A | A |
|---|---|---|
| a | Aa | Aa |
| a | Aa | Aa |

13. a. 50 percent chance of the Huntington's disease genotype *Aa*:

|   | A | a |
|---|---|---|
| a | Aa | aa |
| a | Aa | aa |

   b. 100 percent chance of the Huntington's disease genotype *AA* or *Aa*:

|   | A | A |
|---|---|---|
| A | AA | AA |
| a | Aa | Aa |

   c. 75 percent chance of the Huntington's disease genotype *AA* or *Aa*:

|   | A | a |
|---|---|---|
| A | AA | Aa |
| a | Aa | aa |

   d. 100 percent chance of the Huntington's disease genotype *Aa*:

|   | A | A |
|---|---|---|
| a | Aa | Aa |
| a | Aa | Aa |

14. a. 0 percent chance of a hemophilia genotype:

|   | X$^a$ | Y |
|---|---|---|
| X$^A$ | X$^A$X$^a$ | X$^A$Y |
| X$^A$ | X$^A$X$^a$ | X$^A$Y |

b. 50 percent chance of the hemophilia genotype $X^aX^a$ or $X^aY$:

|  | $X^a$ | Y |
|---|---|---|
| $X^A$ | $X^AX^a$ | $X^AY$ |
| $X^a$ | $X^aX^a$ | $X^aY$ |

c. 25 percent chance of the hemophilia genotype $X^aY$:

|  | $X^A$ | Y |
|---|---|---|
| $X^A$ | $X^AX^A$ | $X^AY$ |
| $X^a$ | $X^AX^a$ | $X^aY$ |

d. 50 percent chance of the hemophilia genotype $X^aY$:

|  | $X^A$ | Y |
|---|---|---|
| $X^a$ | $X^AX^a$ | $X^aY$ |
| $X^a$ | $X^AX^a$ | $X^aY$ |

e. No, male and female children do not have the same chance of getting the disease. Male children are more likely to have hemophilia, because they do not possess a second allele for this trait to mask a recessive allele that they may inherit.

15. The terms "homozygous" and "heterozygous" refer to pairs of alleles for a given gene. Since a male has only one copy of any X-linked gene, it does not make sense to use these pair-related terms when talking about X-linked traits in males.

16. Although neither the mother nor the father expresses the trait in question, some of their children do, so the disease-causing allele ($d$) is recessive and is carried by both parents. The disease-causing allele is located on an autosome. If it were on the X chromosome, the father would express the gene, since we have already determined that he must carry one recessive copy of the gene and he would not have another copy of the gene to mask this recessive allele. Both individuals 1 and 2 of generation I have the genotype $Dd$. Individual 2 of generation I has the disease, so she must have the genotype $dd$, not $Dd$.

17. Designate the dominant, X-linked allele $D$ and the recessive normal allele $d$. According to the Punnett squares shown in (a) and (b) below, males are not more likely than females to inherit a dominant, X-linked genetic disorder.

a. There are two possible Punnett squares, depending on whether the affected female has genotype $X^DX^d$ or genotype $X^DX^D$:

|  | $X^d$ | Y |
|---|---|---|
| $X^D$ | $X^DX^d$ | $X^DY$ |
| $X^d$ | $X^dX^d$ | $X^dY$ |

or

|  | $X^d$ | Y |
|---|---|---|
| $X^D$ | $X^DX^d$ | $X^DY$ |

b. This cross is $X^dX^d \times X^DY$, which gives the following Punnett square:s

|  | $X^D$ | Y |
|---|---|---|
| $X^d$ | $X^DX^d$ | $X^dY$ |
| $X^d$ | $X^DX^d$ | $X^dY$ |

18. The disorder allele is a recessive allele, located on the X chromosome. We know the allele is recessive because individual 2 in generation II carries the allele but does not have the condition. If the allele were located on an autosome, the parents in generation I would be of genotype $AA$ (the male) and $aa$ (the female). In this case, none of the individuals in generation II could have the condition—yet two of them do have the condition, implying that the allele is on a sex chromosome. Finally, we know that the allele is on the X chromosome, because otherwise only males could get the condition.

19. a. If the two genes are completely linked:

|  | $AB$ | $ab$ |
|---|---|---|
| $aB$ | $AaBB$ | $aaBb$ |

b. If the two genes are on different chromosomes:

|  | $AB$ | $Ab$ | $aB$ | $ab$ |
|---|---|---|---|---|
| $aB$ | $AaBB$ | $AaBb$ | $aaBB$ | $aaBb$ |

20. Eight genotypes of gametes are possible: $ADEg$, $AdEg$, $ADeg$, $Adeg$, $aDEg$, $adEg$, $aDeg$, $adeg$. If $D$ and $E$ are linked, we expect only four genotypes: $ADEg$, $Adeg$, $aDEg$, $adeg$.

# Chapter 14

1. The genetic information of the alleles is contained in the sequence of the nitrogenous bases—adenine (A), cytosine (C), guanine (G), and thymine (T)—found within the segment of DNA that constitutes each allele. At any genetic locus, different alleles differ in the sequence of bases they contain. Therefore, the DNA segments of the two codominant alleles $A^1$ and $A^2$ differ in the sequence of bases found in each allele.

2. The double helical structure of DNA and the base-pairing rules theorized by Watson and Crick suggested a simple way that genetic material could be copied. Because A pairs only with T and C pairs only with G, each strand of DNA contains the information needed to produce the complementary strand. In DNA replication, the two strands of the helix separate, and each strand then serves as a template for the construction of its complementary strand, resulting in two identical copies of the original DNA molecule.

3. The sequence of bases in DNA is the basis of inherited variation. A change in the sequence, whether because of an error during replication or because of exposure to a mutagen, is a mutation. Such a change could result in a new allele that would encode a new version of the protein encoded by the gene in which the mutation occurred. If the new allele produced a protein that did not function properly (or at all), serious damage could be done to the cell, and consequently to the organism; such an allele could cause a genetic disorder.

4. DNA is repaired by protein complexes that include enzymes. When DNA is being replicated, enzymes check for and immediately correct mistakes in pair bond formation. Mistakes that escape this process, called mismatch errors, are caught and corrected by repair proteins.

    DNA repair is essential for cells to function normally because DNA is constantly being damaged by chemical, physical, and biological agents. If none of this damage were repaired, genes that encoded proteins critical to life would eventually cease to function, thereby disabling the production of those proteins and killing the cell, and ultimately the organism.

5. Two features enable cells to pack an enormous amount of DNA into a very small space: the thinness of the DNA molecule and a highly organized, complex packing system. Each portion of a chromosome, which contains one DNA molecule, consists of many tightly packed loops. Each loop is composed of a chromatin fiber consisting of many nucleosome spools, which are made of proteins called histones. A segment of DNA winds around each spool, and if that DNA were unwound it would reveal its double helical structure.

6. Prokaryotes have less DNA than eukaryotes have. All DNA in a prokaryote is located on one chromosome; in eukaryotes, by contrast, the DNA is distributed among several chromosomes. Eukaryotes have more genes than prokaryotes do, and genes constitute only a small portion of the eukaryotic genome. Most prokaryotic DNA encodes proteins, and very little of it consists of noncoding DNA and transposons. Functionally related genes in prokaryotes are grouped together on the chromosome; eukaryotic genes with related functions often are not located near one another.

7. The bacterium would begin expressing the gene responsible for encoding the enzyme that breaks down arabinose. Organisms can turn genes on and off in response to short-term changes in food availability or other features of the environment.

8. In multicellular organisms, different cell types express different genes by controlling transcription, along with other methods. By switching specific genes on or off, cells can vary their structure and perform specialized metabolic tasks, even though each has exactly the same genes (and alleles).

9. From gene to protein, the steps at which gene expression can be controlled are as follows: (1) The expression of tightly packed DNA can be prevented, in part because the proteins necessary for transcription cannot reach them. (2) Transcription can be regulated by regulatory proteins that bind to regulatory DNA, effectively switching a gene on or off. (3) The breakdown of mRNA molecules can be regulated such that mRNA is destroyed hours or weeks after it is made. (4) Translation can be inhibited when proteins bind to mRNA molecules to prevent their translation. (5) Proteins can be regulated after translation, either when the cell modifies or transports them or when they are rendered inactive by repressor molecules. (6) Synthesized proteins can be destroyed.

# Chapter 15

1. A gene is a DNA sequence that contains information for the synthesis of one of several types of RNA molecules used to make proteins. A gene stores information in its sequence of nitrogenous bases.

2. Genes control the production of a variety of RNA products, including mRNA, rRNA, and tRNA. Messenger RNA (mRNA) encodes the amino acid sequence of proteins, ribosomal RNA (rRNA) is an essential component of ribosomes (the site of protein synthesis), and transfer RNA (tRNA) carries amino acids to the ribosomes during protein construction. Therefore, each of the RNA products specified by genes functions in the synthesis of proteins. Proteins are essential for many functions that support life. In cells and organisms, proteins provide structural support, transport materials through the body, and defend against disease-causing organisms. Enzymes are a class of proteins that speed up chemical reactions.

3. Genes commonly contain instructions for the synthesis of proteins. Each gene is composed of a segment of DNA on a chromosome and consists of a sequence of the four bases adenine (A), cytosine (C), guanine (G), and thymine (T). The sequence of bases specifies the amino acid sequence of the gene's protein product. Through transcription and translation, proteins are produced from the information stored in genes. In transcription, mRNA is synthesized directly from the sequence of bases in one DNA strand inside the nucleus of a cell. Translation occurs in the cytoplasm and converts the sequence of bases in an mRNA molecule into the sequence of amino acids in a protein. Proteins, by

their many and various functions, influence the phenotype of an individual.

4. For a protein to be made in eukaryotes, the information in a gene must be sent from the gene, which is located in the nucleus, to the site of protein synthesis, on a ribosome. This transfer of information requires an intermediary molecule because DNA does not leave the nucleus but ribosomes are located in the cytoplasm. In eukaryotes, a newly formed mRNA molecule usually must be modified before it can be used to make a protein. The reason is that most eukaryotic genes contain internal sequences of bases (introns) that do not specify part of the protein encoded by the gene. DNA sequences copied from introns must be removed from the initial mRNA product if the protein encoded by the gene is to function properly.

5. RNA splicing is a step in RNA processing in which introns are removed from a newly transcribed mRNA and the remaining exons are joined together to create the mature, export-ready form of the mRNA. RNA splicing is not known to occur in prokaryotes, but is common among eukaryotes. The great majority of our protein-coding genes produce mRNA that must undergo splicing before the RNA can exit the nucleus.

6. Messenger RNA is the product of transcription, and is a version of the genetic information stored in a gene. The mRNA moves from the nucleus to the cytoplasm, where it binds with a ribosome to guide the construction of a protein.

   Ribosomal RNA is a major component of ribosomes. Translation occurs at ribosomes, which are molecular machines that make the covalent bonds linking amino acids to form a particular protein.

   Transfer RNA molecules carry the amino acids specified by mRNA to the ribosome. At the ribosome, a three-base sequence (anticodon) on the tRNA binds by means of complementary base pairing with the appropriate codon on the mRNA. Each tRNA molecule carries the amino acid specified by the mRNA codon to which its tRNA anticodon can bind.

7. If a tRNA molecule does not function properly because of a mutation, each protein that it helps to build will be altered in some way. By failing to bind properly with the mRNA codons of many different genes, a mutant tRNA may significantly affect the structure of many different protein products. Because their structure is altered, the function of these protein products may be impaired. Since proteins are key components of many metabolic reactions, changing the function of many different proteins can result in a series of metabolic disorders.

8. A mutation is any alteration in the information coded within an individual's DNA. Sometimes the effects of that change can be detected as a change in the inherited characteristics of that individual (a phenotypic change). But in other cases, the mutation may be "silent," with no outward sign that it has occurred. Most mutations are neutral in their impact on the individual, neither benefiting it nor harming it. Some mutations have harmful effects, and rarely, a mutation might produce a change that enhances the individual's ability to survive and reproduce in a particular environment.

   A very large number of our genes carry information for the construction of specific proteins. That information is carried in the form of a sequence of chemical units, called nitrogenous bases, that in turn specify the sequence in which the amino acids in a protein are strung together (proteins are built from amino acids, and each unique protein has a unique sequence of amino acids). When the base sequence of a protein-coding gene changes, the amino acid sequence of its protein is altered as well. Every protein has special chemical and biological properties that are critical to its function, and most of those properties stem from the precise sequence of amino acids in it. If the amino acid sequence of a protein changes because of a mutation in the gene that codes for it, the biological function of that protein may change as well.

# Chapter 16

1. To produce domesticated species, humans have manipulated the reproduction of other organisms, selecting for desirable qualities that, over time, have become standard in domesticated species. Although such selection practices do lead to changes in the DNA of organisms (that is, they lead to an increase in the frequencies of alleles that control the inheritance of the traits we select for), genetic engineering enables us to make much greater changes in a much shorter span of time. Using such methods, we can manipulate the DNA of organisms directly, and we can transfer genes from one species to another. Transfers of DNA from, say, a human to a bacterium (as is done in the production of human insulin) far exceed the scope of DNA transfers that occur in nature or are possible through conventional breeding of crops and farm animals. We can also selectively change specific DNA sequences—something we could never do before. Overall, we can now manipulate DNA with greater power and precision than we could when we domesticated species such as dogs, corn, and cows.

2. By using restriction enzymes and gel electrophoresis together, geneticists can examine differences in DNA sequences. Judy and David could be tested for the sickle-cell allele by use of the restriction enzyme *Dde*I, which cuts the normal hemoglobin allele into two pieces but cannot cut the sickle-cell allele. Their doctor might also want to use a DNA probe to test for the sickle-cell allele in these would-be parents. A DNA probe is a short, single-stranded segment of DNA with a known sequence, usually tens to hundreds of bases long. A probe can pair with another single-stranded segment of DNA if the sequence of bases in the probe is complementary to the sequence of bases in the other segment.

3. DNA cloning is the introduction of a DNA fragment into a host cell that can generate many copies of the introduced DNA. The purpose of DNA cloning is to multiply a particular DNA fragment, such as a specific gene, so that a large amount of this DNA is made available for further analysis and manipulation. Bacteria are the most common host cells in DNA cloning. Two of the most common methods of cloning a gene are constructing a DNA library and using the polymerase chain reaction (PCR). To build a DNA library, a vector such as a plasmid is used to transfer DNA fragments from the organism whose gene is to be cloned to a host organism, such as a bacterium. To clone a gene by PCR, primers are synthesized, enabling DNA polymerase to produce billions of copies of the gene in a few hours.

4. The advantage of DNA cloning is that it is easier to study a gene and its function, and to manipulate that function for practical benefits, once you have many copies of it. Once a gene has been cloned, it can be sequenced, transferred to other organisms, or used in various experiments. Today, many lifesaving pharmaceuticals, such as human insulin, human growth hormone, human blood-clotting proteins, and anticancer drugs, are manufactured by bacteria that have been genetically modified by having cloned human genes inserted into them.

5. Genetic engineering is the permanent introduction of one or more genes into a cell, a certain tissue, or a whole organism, leading to a change in at least one genetic characteristic in the recipient. The organism receiving the DNA is said to be genetically modified (GM) or genetically engineered (GE). To create GM organisms, a DNA sequence (often a gene) is isolated, modified, and inserted back into the same species or into a different species.

    Fish such as salmon have been genetically engineered to grow much more rapidly than their unaltered counterparts. Farm-raised, genetically modified fish provide low-cost, high-quality protein with a lower carbon footprint than cattle raising. Furthermore, consumption of GM fish reduces fishing pressure on wild stocks. However, there are concerns that escaped GM fish could threaten wild fish stocks not only by interbreeding with them and thereby reducing their natural diversity, but also by outcompeting them for resources and thereby driving them toward extinction.

6.–8. These answers depend on your viewpoint, which we hope is at least in part guided by scientific facts.

# Chapter 17

1. Evolution is change in the genetic characteristics of a population over time, which can occur through mechanisms such as genetic mutation or natural selection. Since the genotypes of individuals do not change, a population can evolve but an individual cannot.

2. In the new habitat, larger lizards will have an advantage over smaller ones. The larger lizards of the species will therefore be more likely to pass the trait of large size on to their offspring, and the average size of lizards in the population will increase over time because of this selective advantage.

3. Each of these aspects of life on Earth can be explained by evolution.
   a. Adaptations, which improve the performance of an organism in its environment, result from natural selection.
   b. The diversity of life results from speciation, which occurs when one species splits to form two or more species.
   c. Organisms can share puzzling characteristics because of common descent. Consider the wing of a bird, the flipper of a whale, and the arm of a human. Even though these appendages are used for very different purposes—and hence we would not expect them to be structurally similar—they are composed of the same set of bones. The reason is that birds, whales, and humans share a common ancestor that had these bones. Organisms can also share less puzzling characteristics, because of convergent evolution resulting from similar selective pressures.

4. Overwhelming evidence indicates that evolution occurred and continues to occur. Support for evolution comes from five lines of evidence: (1) The fossil record provides clear evidence of the evolution of species over time and documents the evolution of major groups of organisms from previously existing organisms. (2) Organisms contain evidence of their evolutionary history. For example, scientists find that studies of proteins and DNA support the evolutionary relationships determined by anatomical data; that is, the proteins and DNA of closely related organisms are more similar than those of organisms that do not share a recent common ancestor. In this and many similar examples, the extent to which organisms share characteristics other than those used to determine evolutionary relationships is consistent with scientists' understanding of evolution. (3) Scientists' understanding of evolution and continental drift has enabled them to predict the geographic distributions of certain fossils, depending on whether the organisms evolved before or after the breakup of Pangaea. (4) Scientists have gathered direct evidence of small evolutionary changes in thousands of studies by documenting genetic changes in populations over time. (5) Scientists have observed the evolution of new species from previously existing species.

5. In any area of science, new pieces of information are continually being added to our knowledge. The debate among scientists as to which mechanisms of evolution are most important means only that evolution is not fully understood, not that it does not occur.

6. Genetic drift has a greater effect on smaller populations. If the plant population were larger, the likelihood that all plants of a

certain genotype would die in a windstorm would be smaller, and the dramatic shift in the frequencies of the A and a alleles would therefore be less likely.

# Chapter 18

1. *Mutation:* A nonlethal mutation of a particular allele can be inherited by offspring, thereby increasing the frequency of the mutant allele in a population over time.

    *Gene flow:* The exchange of alleles between populations can change the frequencies at which alleles are found in those populations by introducing new alleles. Populations affected by gene flow tend to become more genetically similar to one another.

    *Genetic drift:* Random events (such as chance events that influence the survival or reproduction of individuals) can cause one allele to become dominant in a small population. By chance alone, drift can lead to the fixation of alleles in small populations; if these alleles are harmful, the population may decrease in size, perhaps to the point of extinction.

    *Natural selection:* If an inherited trait provides a selective advantage for individuals in a certain population, individuals with that trait will be more likely to reproduce, and the frequency of the allele for that trait will increase in succeeding generations. Likewise, an allele for a disadvantageous trait will be selected against, and will be found with decreasing frequency over time.

2. *Gene flow* is the exchange of alleles between populations. Gene flow makes populations more similar to one another in their genetic makeup. *Genetic drift* is a process by which alleles are sampled at random over time. Genetic drift can have a variety of causes, such as chance events that cause some individuals to reproduce and prevent others from reproducing. *Natural selection* is a process in which individuals with particular inherited characteristics survive and reproduce at a higher rate than other individuals do. *Sexual selection* is a form of natural selection in which individuals with certain traits have an advantage in attracting mates, and consequently in passing those traits on to offspring.

3. The potential benefits of intentionally introducing new individuals into a population include making the population larger, and therefore less susceptible to genetic drift, and providing an input of new alleles on which natural selection can operate. The potential drawbacks include the introduction of individuals with genotypes that are not well matched to the local environmental conditions of the smaller population. Throughout time, some species have gone extinct locally or worldwide.

    Extinction is a natural process. However, humans have greatly increased the rate at which populations and species become extinct. If numerous other populations of a species exist, it may not be worth introducing new members to the smaller population. If the smaller population is one of the few populations of that species left, however, it may be important to introduce new individuals in an attempt to help the population recover and survive.

4. Genotype frequencies for the original population:

$$AA: \frac{280}{280 + 80 + 60} = 0.67$$

$$Aa: \frac{80}{280 + 80 + 60} = 0.19$$

$$aa: \frac{60}{280 + 80 + 60} = 0.14$$

Allele frequencies for this population:

Frequency of *A* allele =

$$p = \frac{2(280) + 80}{2(280 + 80 + 60)} = 0.76$$

Frequency of *a* allele =

$$q = \frac{2(60) + 80}{2(280 + 80 + 60)} = 0.24$$

The Hardy-Weinberg equation predicts that the frequency of genotype *AA* should be

$$p^2 = (0.76)(0.76) = 0.58,$$

that the frequency of genotype *Aa* should be

$$2pq = (2)(0.76)(0.24) = 0.36,$$

and that the frequency of genotype *aa* should be

$$q^2 = (0.24)(0.24) = 0.06.$$

Note the sum of the genotype frequencies,

$$p^2 + 2pq + q^2 = 1.0.$$

These calculated genotype frequencies do not match those of the original population. This difference could be due to mutation, nonrandom mating, gene flow, a small population size, and/or natural selection.

5. Using an antibiotic drug to kill large numbers of bacteria tends to give a considerable reproductive advantage to bacteria that possess resistance to the drug. Since bacteria reproduce extremely rapidly, the entire population of bacteria will soon be resistant. Reducing human exposure to the bacteria would not enable resistant strains to have as great a degree of reproductive advantage over normal bacterial strains. Slowing the growth of bacteria would likewise help limit the reproductive advantage of resistant strains.

6. The chance events that cause genetic drift are much more important in small populations than in large populations. In natural populations, the number of individuals in a population has an effect similar to the number of times a coin is tossed. A small population is analogous to a coin being tossed just a few times. By chance alone, some individuals in a small population may

leave offspring while others do not. One allele may be lost when certain individuals fail to reproduce, and another may become fixed when the individuals bearing it contribute disproportionately to the next generation. When a population has many individuals, the likelihood that each allele will be passed on to the next generation greatly increases. In a large population, it is unlikely that chance events could have caused a dramatic change in allele frequencies in a short time (just as a fair coin tossed hundreds of times is unlikely to give you any outcome other than approximately 50-50 heads and tails). Genetic drift can occur in large populations, but in these cases its effects are more easily overcome by natural selection and other evolutionary mechanisms. In large populations, genetic drift causes little change in allele frequencies over time.

# Chapter 19

1. For examples of adaptations in the pronghorn, take a look at Table 1.1 in Chapter 1 (p. 17).
2. Individuals with inherited traits that enable them to survive and reproduce better than other individuals replace those with less favorable traits. This process, by which natural selection improves the match between organisms and their environment over time, is called adaptive evolution. In our efforts to kill or control bacteria that cause infectious diseases, we are creating a new environment in which bacteria that cannot withstand antibiotics are eliminated. Often, some bacteria in the population are not killed by antibiotics; these bacteria reproduce, increasing the frequency of resistant bacteria in the population. These evolutionary changes in disease-causing bacteria are harmful to us because more and more of the diseases we encounter will be resistant to medical treatment.
3. Species that hybridize in nature may still be distinct species, because of a host of alleles that do not affect their ability to interbreed but may cause them to look different or to differ from each other ecologically. For this reason, many people would argue that the rare species is separate from the common species and should remain classified as rare and endangered.
4. Defining species by their inability to reproduce sexually with other species is convenient, but many alleles do not affect reproductive isolation yet could cause the two oaks to be different enough that they could be classified as separate species, even though they would be able to produce hybrids.
5. Because this storm-blown population is now geographically isolated from other populations of its species, there will be little or no gene flow to such populations. As a result, genetic changes due to mutation, genetic drift, and natural selection will accumulate over time. Natural selection is likely to cause genetic change

in the population because the new environment is different from the parent population's environment; genetic drift will probably also be important, because the island population is small (making drift more likely). As a by-product of genetic changes due to selection, drift, and mutation, the island population may become reproductively isolated from the parent population. If the island population remains isolated long enough, a sufficient number of genetic changes may accumulate for it to evolve into a new species.
6. Some of the cichlid populations of Lake Victoria may have had so little contact with one another that they can be said to have evolved into separate species in geographic isolation, even though they live in the same lake. Other populations may have evolved into new species in the absence of geographic isolation.
7. New plant species can form in the absence of geographic isolation as a result of polyploidy, a condition in which an individual has more than two sets of chromosomes. Strong evidence suggests that sympatric speciation occurred in the cichlids that live in Lake Victoria, and evidence is accumulating that apple and hawthorn populations of the apple maggot fly are diverging into two species, despite living in the same area. In apple maggot flies and cichlids, sympatric speciation is promoted by ecological factors (such as selection for specialization on different food items) and sexual selection.

   There is a greater potential for gene flow between populations whose geographic ranges overlap than between populations that are geographically isolated from one another. Gene flow tends to cause populations to remain (or become) similar. Therefore, in the absence of geographic isolation, it can be difficult for genetic differences great enough to cause reproductive isolation to accumulate over time. As a result, sympatric speciation occurs less readily than allopatric speciation.

# Chapter 20

1. One example of one group of organisms evolving from another is the emergence of mammals from reptiles. The emergence of the mammalian jaw and teeth illustrates the steps in this process. The first step was the development of an opening in the reptilian jaw behind the eye. Then, more powerful jaw muscles and specialized teeth appeared with the therapsids. Finally, a subgroup of these reptiles, the cynodonts, emerged with more specialized teeth and a more forward hinge of the jaws, completing this aspect of mammalian evolution.
2. The emergence of photosynthesis in ancient organisms gradually led to the buildup of $O_2$ in the atmosphere, which killed many organisms to which oxygen was toxic. However, the oxygen supplied by photosynthetic organisms made possible the evolution of eukaryotes and later multicellular life-forms.

3. The Cambrian explosion was a large increase in the diversity of life-forms over a relatively short time about 530 million years ago. Larger organisms of most phyla emerged during this period, setting the stage for the colonization of land.

4. The colonization of land led to another great increase in the diversity of life-forms. Life on land required different means of mobility and reproduction, adaptations to obtain and retain water, and ways to breathe in air rather than water. Early terrestrial organisms had the opportunity to expand into new types of largely unoccupied habitat, which provided ample resources for organisms that were able to survive the challenges of life on land.

5. A mass extinction event may be associated with rapid environmental changes that have no relation to the conditions that favor a particular adaptation. Organisms with wonderful adaptations can (and have) become extinct during mass extinction events.

6. Although speciation can happen within a single year, it often takes hundreds of thousands to millions of years to occur. It is not surprising, then, that it usually takes 10 million years for the number of species found in a region to rebound after a mass extinction event. The time required to recover from mass extinction events provides a powerful incentive for humans to halt the current, human-caused losses of species; otherwise, it will take millions of years for biodiversity to recover.

7. If other human species shared the world with us today, there would undoubtedly be instances of social or cultural friction between them and modern humans, just as there exist tensions between different ethnic groups today. In order for there to be peaceful cohabitation, societies would need to recognize the humanity of other human species and discuss ways of cooperation in sharing living space and resources.

# Chapter 21

1. The description of the biosphere as an "interconnected web" is apt because all organisms within it are connected by their interactions, as shown by examples ranging from various food chains to symbiotic relationships between species. The global spread of invasive species and their often detrimental effects on native ecosystems, and the change in red kangaroo populations in response to the "dingo fence," are two of the case histories discussed in this chapter that underscore the interrelatedness of organisms throughout the biosphere.

2. Giant convection cells in the atmosphere and ocean currents carry the results of local events (such as a volcanic eruption or an oil spill) to distant areas around the globe. For example, oil spilled into an ocean current next to one continent's shore may be carried by that current and end up coating the shores of other continents. If shorebirds on those continents are killed when they become coated with oil from the spill, they may no longer keep populations of their food organisms under control, and they will no longer be available as a food source for their predators.

3. These are some of Earth's terrestrial biomes: tropical forest, temperate deciduous forest, grassland, chaparral, desert, boreal forest, and tundra. Identify the terrestrial biome in this list that is closest to where you live. Reread Section 21.3 for a description of the climatic and ecological characteristics of this biome.

4. The defining feature of a desert is the scarcity of moisture (less than 25 centimeters of precipitation), not high temperatures. Antarctica, which receives less than 2 centimeters of precipitation per year, is the largest cold desert in the world. The Sahara desert in northern Africa is the largest hot desert. Because desert air lacks moisture, it cannot retain heat, and therefore it cannot moderate daily temperature fluctuations. As a result, temperatures may be above 45°C (113°F) in the daytime and then plunge to near freezing at night.

5. Desert plants have small leaves because such leaves present a smaller surface area for evaporative water loss than do large leaves. Plants native to tropical rainforests do not need special adaptations to reduce water loss, because they live in a habitat that is wet year-round. Some desert plants produce enormously long taproots that are able to reach subsurface water. Desert animals often have light-colored fur, kidneys that help them conserve water, and behavioral adaptations such as hiding in a cool burrow during the hottest parts of the day. Some, like jackrabbits, have large ears that act like a radiator to help dissipate heat.

6. In temperate regions, seasonal changes in temperature cause the oxygen-rich water near the top of a lake to sink in the fall and the spring, bringing oxygen to the bottom of the lake. This seasonal turnover also delivers nutrients from the bottom sediments to the surface layers, where they enhance the growth of photosynthetic organisms. By delivering oxygen to lake-bottom animals and nutrients to producers at the lake surface, lake turnover increases both primary and secondary productivity.

7. An estuary, where a river empties into the sea, is a shallow marine ecosystem. The plentiful light, the abundant supply of nutrients delivered by the river system, and the regular stirring of nutrient-rich sediments by water flow creates a highly productive community of photosynthesizers. The coastal region, which stretches from the shoreline to the continental shelf, is among the most productive marine ecosystems because of the ready availability of nutrients and oxygen. Nutrients delivered by rivers and washed off the surrounding land accumulate in the coastal region. Nutrients that settle to the bottom are stirred up by wave action, tidal movement, and the turbulence produced by storms. The nutrients support the growth of photosynthetic producers, which inhabit the well-lit upper layers of coastal waters to a depth of about 80 meters (260 feet). The vigorous mixing by wind and waves also adds atmospheric oxygen to the water.

# Chapter 22

1. A population may be difficult to define if the boundaries of its range are unclear, if its members move around frequently, or if its members are small and hard to count.

2. If a population is threatened with extinction, possible options for saving it might be to protect it from human disturbances, to treat diseases, to reduce the number of predators, to move the population to an area with greater food supply (limiting death or emigration), to introduce individuals from other populations of the species (increasing immigration), or to institute captive breeding programs (increasing the birth rate).

3.

4. a. Some factors that limit population growth include availability of habitat, availability of food and water, disease, weather, natural disturbances, and predators.

   b. Species new to an area often do not have established predators. In addition, they have not yet reached the carrying capacity of their habitat.

5. A density-dependent factor is one whose intensity increases as the density of the population increases. An example of a density-dependent factor is an infectious disease that spreads more rapidly in densely populated areas. Plants that become overcrowded in a field compete with each other for nutrients, and therefore nutrient availability is another example of a factor that limits population growth in a density-dependent manner.

   A density-independent factor is not affected by the density of the population. Temperature is a density-independent factor: if the temperature drops below what a certain plant species can tolerate, the plants will die no matter how sparse or dense that plant population is. Natural disturbances such as fires and floods also limit the growth of populations in a density-independent way.

6. If the pattern of population growth is understood, managers may be able to manipulate the factors that most directly affect the population's growth rate. If a population of organisms were more successful because, for example, it had adequate access to water, a manager would be sure that nearby rivers were not drained off for agricultural needs. Understanding population growth is especially valuable in pest management. Knowing the population cycles of pests—for example, whether their populations display boom-and-bust growth patterns—can inform growers about when to spray pesticide or release natural predators for biocontrol for maximum effectiveness.

7. Specific actions that could limit the negative impact of humans on our world include the following: (1) Reducing the consumption of unnecessary goods. (2) Reusing and recycling items to promote the sustainable use of resources. (3) Working to develop and follow environmentally friendly policies and activities (for instance, using energy-efficient cars and lightbulbs). (4) Purchasing goods that have a lower impact on the environment, including organically grown clothing and food items. Finally, the one factor that will affect all others is this: limiting reproduction to no more than one child per parent (zero population growth). Limiting human population growth will reduce all human impacts on the environment.

# Chapter 23

1. Mutualism is common because its costs are outweighed by the benefits it provides. Yucca plants, for example, may lose a few seeds to the offspring of their moth pollinators, but they still end up with more seeds than if the moths had not pollinated them to begin with.

2. Organisms eaten by consumers are under selection pressure to develop defenses against those consumers. Likewise, consumers experience selection pressure to overcome the defenses of their food organisms. Adaptations that improve the survival of individuals in either group will therefore be likely to spread throughout the population. An example would be the poison of the rough-skinned newt, which can kill nearly all predators; garter snakes, however, have evolved the capacity to tolerate the toxin and eat the newt.

3. The plant community would have fewer species (a). When the rabbits were removed, the grass they prefer would no longer be eaten, so that grass would assert its dominance as the superior competitor and would probably drive some of the other grass species in the area to extinction. The inferior competitor is still using resources that the superior competitor needs, thereby possibly limiting the superior competitor's distribution or abundance.

4. a. Food webs influence the movement of energy and nutrients through a community. Some species, called keystone species, have a disproportionately large effect, relative to their abundance, on the types and abundances of other species in the community.

   b. Disturbances such as fire occur so often in many ecological communities that the communities are constantly changing and hence may never establish climax communities (relatively stable end points of ecological succession). Depending on the type and severity of the disturbance, a given community may or may not be able to recover.

    c. Climate is a key factor in determining which organisms can live in a given area, so if the climate changes, the community changes.

    d. As continents move to different latitudes, their climates—and therefore their communities—change.

5. The introduction of beard grass to Hawaii has increased the frequency and size of fires on the island. This change is due to the large amount of dry matter that the grass produces, which burns more easily and hotter than does the native vegetation. In this way, the presence of one species in the community has profoundly altered its disturbance pattern, leading to other large changes in the community.

6. The disturbance described in (b) would probably require a longer recovery time, assuming that no other disturbances, such as fire, were to occur. In the disturbance described in (a), the soil and ground cover would be left intact, so new trees would be able to sprout according to natural succession, eventually growing to replace the trees that were removed. In (b), however, the pollutant would have damaged the soil, which would hinder the ability of the trees and ground vegetation to grow. The soil chemistry would need to return to normal before the forest vegetation would be able to grow back and thrive again.

7. Change is a part of all ecological communities. However, human-caused change is unique in that we can consider the impact of our actions, and we can decide whether or not to take actions that cause community change. Whether or not a particular change is viewed as ethically acceptable will depend on the type of change, the reason for it, and the perspective of the person evaluating the change. For example, a person might find it ethically acceptable to alter a region so as to produce a long-term source of food for the growing human population, yet not ethically acceptable to take actions that result in short-term economic benefit but cause long-term economic loss and ecological damage.

# Chapter 24

1. An ecosystem consists of a community of organisms together with the physical environment in which those organisms live. The organisms in an ecosystem interact with one another in various ways; organisms can also move from one ecosystem to another. For this reason, determining the boundaries of protection for a particular ecosystem would probably be difficult. Such a plan would require an understanding of the roles that certain organisms play in the overall function of an ecosystem.

2. Energy captured by producers from an external source, such as the sun, is stored in the bodies of producers in chemical forms, such as carbohydrates. At each step in a food chain, a portion of the energy captured by producers is lost from the ecosystem as metabolic heat. This steady loss prevents energy from being recycled.

3. The productivity of marine ecosystems tends to be high close to land but relatively low in the open ocean because nutrients needed by aquatic photosynthetic organisms are in short supply in the open ocean. Nutrients delivered by streams and rivers account for the high productivity of many coastal areas. Nutrients drained off the land stimulate the growth and reproduction of phytoplankton, the small photosynthetic producers that form the foundation of aquatic food webs. Estuaries are some of the most productive habitats because the rich nutrient supply supports large populations of producers, which in turn nourish large populations of consumers. Wetlands such as swamps and marshes can also match the productivity levels of tropical forests and farmland. They trap soil sediments rich in nutrients and organic matter, thereby promoting the growth of flooding-tolerant plants and phytoplankton, which in turn feed a complex community of consumers.

    The relatively high productivity of the Gulf of Alaska is attributed to upwelling. In upwelling, wind and ocean currents drive cold, nutrient-rich layers of water to the surface to replace warm, nutrient-depleted water. Regions that experience upwelling have high NPP because producers that live there are less nutrient-limited than are producers in similar ecosystems that lack upwelling.

4. Decomposers break down the tissues of dead organisms into simple chemical components, thereby returning nutrients to the physical environment so that they can be used again by other organisms.

5. Nutrients can cycle on a global level. When sulfur dioxide pollution, for example, enters the atmosphere in one area of the world, winds can move that pollution around the world, where it can affect other ecosystems.

6. Human economic activity is interwoven with several key ecosystem services. Pollination is essential for the productivity of both commercial crops and backyard gardens. Floodplains act as safety valves for major floods, provided we do not build on them or separate them from the bodies of water they help control. Forests act as water filtration systems. We rely on nutrient cycling to keep us alive. When ecosystem services such as these are damaged, human economic interests are damaged as well.

7. An acre of agricultural land can support many more vegetarians than it can support people who live mainly on animal products such as eggs, poultry, milk, or beef. The reason is the pyramid of energy transfer in food chains. Only about 10 percent of the energy at any trophic level is available to consumers at the next-higher trophic level in a food chain. More of the net primary productivity (NPP) generated by plants is available to primary consumers, less to secondary consumers, and even less to tertiary consumers. Therefore, it takes more land area to grow the same mass of an animal-based food than a plant-based food. According to David Pimentel at Cornell University, all the grain that is currently fed to livestock in the United States would feed nearly 800 million if these people obtained their calories directly from the grain. Raising beef is less efficient in converting plant biomass into human food than is raising chickens. Part of the reason for the reduced efficiency is that

less of the larger animal is available as edible food, but the methods used in intensive livestock production are a contributing factor. For example, grass-fed cattle are more efficient in converting plant biomass into meat than are grain-fed cattle.

# Chapter 25

1. Major types of global change caused by humans include global warming, land and water transformation, and changes in the chemistry of Earth (for example, changes to nutrient cycles). By altering the conditions under which species live, all of these changes could result in the increased dominance of certain types of species and the disappearance of others from various ecosystems.

2. Human-caused global changes often happen at a much more rapid rate than do changes due to natural causes. Continental drift and natural climate change happen much more slowly than do the measurable increases humans have caused in atmospheric carbon dioxide levels and nitrogen fixation. In addition, humans have a choice about the global changes we cause.

3. The present levels of atmospheric $CO_2$ are higher than any seen in the previous 420,000 years. Since the middle of the twentieth century, atmospheric $CO_2$ levels have been rising at about 2 parts per million (ppm) per year, and they stood at 385 ppm toward the end of 2008. Scientific instruments can directly measure the amount of carbon dioxide in the atmosphere. By measuring $CO_2$ levels in bubbles of air trapped in ancient ice, scientists can estimate the amount of $CO_2$ that was present in the atmosphere up to hundreds of thousands of years ago.

4. It would be prudent to take action on global warming sooner rather than later, despite present uncertainties as to its extent. There is already evidence of climate changes that are consistent with the predicted effects of global warming. In addition, the correlation of rising $CO_2$ levels and worldwide temperature increases suggests that these increases will continue in the future if carbon dioxide emissions are not reduced. If action is delayed too long, it may be too late to undo many of the effects of global warming.

5. According to a New York Times/CBS News poll in 2006, a majority of Americans (55 percent) will support an increase in the federal gasoline tax (currently 18.4 cents per gallon) as long as the money raised is used to reduce global warming or to reduce U.S. dependence on foreign oil. Advocates of the tax increase say it will lower carbon emissions by reducing gasoline consumption as individuals and companies reduce waste and inefficiency in transportation. Many economists argue for a gas tax of up to one dollar a gallon phased in over 5 years, and maintain that most low-income and middle-income citizens will actually stand to benefit, provided some of the tax revenue is invested in improving public transportation.

6. For people to have a sustainable impact on Earth, we must reduce the rate of growth of the human population, and we must reduce the rate of resource use per person. To achieve these goals, many aspects of human society would have to change. For example, we would need to alter our view of nature from looking at it as a limitless source of goods and materials that can be exploited for short-term economic gain to accepting nature's limits and seeking always to take only those actions that can be sustained for long periods of time. Many specific actions would follow from such a change in our view of the world, such as an increase in recycling, the development and use of renewable sources of energy, a decrease in urban sprawl, an increase in the use of technologies with low environmental impact (such as organic farming), and a concerted effort to halt the ongoing extinction of species.

   Examples of actions you could take to make your impact on Earth more sustainable include the following: reducing the quantity of nonfood items purchased; reusing items until they are no longer usable; buying used items rather than getting everything new; recycling paper, plastic, glass, and metal; taking along reusable cloth bags when shopping; rarely using paper cups, plates, or towels; planting trees and other native plants, especially those that help feed native wildlife; reducing water use by not leaving water running when brushing teeth, by adjusting the water level of washing machines to match the size of the load, and by using water-saving fixtures; reducing fossil fuel use by choosing a fuel-efficient car and by using household heating and air-conditioning only as needed; using compact fluorescent lightbulbs and turning off lights that are not in use; supporting organic farmers by purchasing organically grown food.

7. Although ocean acidification by half a pH unit may not seem like much, studies show that the ability of the coral animal (polyp) to calcify is substantially reduced and reef growth slows dramatically. Other reef-building organisms, such as coralline algae, are also debilitated by the acidity. Because coral reefs provide diverse habitats, biodiversity will decline as the reefs decline. A coral reef acts like a breakwater, and biodiversity in sheltered bays and lagoons is likely to be devastated, endangering sea grasses and mangroves, while eroding the shoreline. Tourism, fisheries, and other economic activities supported by coral reefs are valued at $400 billion globally, and all of them are likely to be affected if the current trend in ocean acidification continues.

8. An endocrine disrupter is a chemical that interferes with hormone function to produce negative effects, such as reduced fertility, developmental abnormalities, immune system dysfunction, and increased risk of cancer. DDT and bisphenol A are endocrine disrupters that can be detected in the tissues of most Americans. DDT is a pesticide that is also a POP (persistent organic pollutant) that is bioaccumulated and biomagnified along a food chain. Until its use was banned in 1972, DDT was extensively sprayed in the United States to control mosquitoes and protect crops from insect pests. DDT disrupts reproduction in a

variety of animals, but predatory birds were especially hard-hit. The chemical interferes with calcium deposition in the developing egg, resulting in thin, fragile eggshells that break easily. Bisphenol A is found in many types of plastics (including those marked with the recycling number 7). In laboratory animals, bisphenol A increases the risk of diabetes, obesity, reproductive problems, and various cancers.

9. Industrial nitrogen fixation to make agricultural fertilizer is one source of anthropogenic nitrogen in ecosystems. Another major source of nitrogen fixation is car engines, in which heat from combustion converts some of the $N_2$ found in the air to nitrogen monoxide (NO) and nitrogen dioxide ($NO_2$). These gases enter the atmosphere as engine exhaust, combine with oxygen and water in the air, and then fall to the ground as nitrate ($NO_3^-$) dissolved in rain. In recent years, the amount of nitrogen fixed by human activities has exceeded the amount fixed by all natural processes combined. When nitrogen is added to terrestrial communities, NPP usually increases, but the number of species often decreases. Species richness is affected because species best able to use the extra nitrogen outcompete other species. When nitrogen is added to nitrogen-poor aquatic ecosystems, such as many ocean communities, productivity increases but many species are lost if eutrophication follows, resulting in oxygen-poor "dead zones" where most animal species cannot survive.

# Chapter 26

1. Living organisms are known to survive, even thrive, in some extreme environments, such as the −70°C cold of Siberian winters or the +70°C heat of desert sands. Some species live in extremely dry areas—those that receive less than 10 centimeters of rainfall per year, for example—and others thrive in forests that are drenched in more than 1,000 centimeters of rain through the year. Most organisms, however, live in more moderate conditions, and the cells within their bodies can tolerate only a narrow range of internal conditions. Most animals must have the right amount of water in their cells (typically 40–65 percent of their body weight) to survive and remain active, although some can remain dormant with almost no water. Too little water impedes the movement of solutes and the action of enzymes in cells; too much water can interfere with biochemical reactions, alter protein structure, and cause cells to burst. The freezing of water at 0°C can kill cells by bursting plasma membranes. Above about 40°C, many enzymes become denatured and unable to catalyze biochemical reactions. Extreme environments are a threat to survival because maintaining an optimal internal state through homeostasis is difficult under those conditions.

2. "Homeostasis" refers to the processes that stabilize conditions within an organism so that they will be appropriate for vital biochemical reactions and critical interactions among cells. An organism must maintain this favorable internal state while surrounded by a very different, often hostile, and potentially change-able, external environment. Here are the general characteristics of homeostasis in animals: (1) Homeostatic pathways generally operate through negative feedback loops that attempt to restore the normal internal state if the monitoring system detects a deviation from optimal conditions. (2) Body size affects homeostasis, such that large animals, whose surface area relative to their volume is smaller than that of small animals, do not lose or gain heat as rapidly as small animals do. (3) Homeostasis requires energy. The greater the difference between external conditions and the desired internal conditions, the greater the energy cost to the animal.

3. In a cold environment such as that of the North Pole, you might maintain homeostasis by wearing several layers of insulating clothing to minimize heat loss, drinking sufficient fluids, and increasing caloric intake. Technology, such as indoor heating and air-conditioning, can assist humans in their maintenance of homeostasis by minimizing differences between body conditions and the external environment. Special fabrics used in clothing can help dissipate excess heat or prevent heat loss.

4. When your body senses it needs food, certain signals (including hormones released by the gut) are sent to the hunger control centers in the brain (the hypothalamus, specifically) that increase your appetite. When enough food is eaten, the brain turns off the hunger signals and generates a sense of satiety instead.

5. The three routes are conduction, radiation, and evaporation. In conduction, heat flows between an animal's body and a substance with which it is in direct contact. In radiation, an animal's body absorbs radiant (infrared) energy from, or emits radiant energy to, its surroundings. In evaporation, the body loses heat by the conversion of water to water vapor on a wet body surface.

6. In a nephron, plasma from blood that enters the capillaries of the glomerulus is filtered through pores in the capillary walls. The filtrate enters and flows down a tubule. Along the way, valuable solutes, water, and nutrients are reabsorbed into the bloodstream, and the filtrate becomes more concentrated. The tubules join to form larger tubes, and the end product of the filtration and reabsorption process, called urine, flows down these tubes and out of the body.

7. The loss of water and the high body temperature that result from heatstroke would disrupt metabolic reactions in a cell and its surroundings as enzymes became denatured and solute movement was impeded.

# Chapter 27

1. "Dietary fiber" refers to indigestible plant polysaccharides. Although it does not contribute energy or chemical building blocks, dietary fiber is important for human health. Through its sheer bulk, and also because it is often interlinked with the more digestible polysaccharides, such as starch, fiber reduces the speed with which carbohydrates are digested. Good sources of fiber include oatmeal, brown rice, popcorn, and whole-grain bread.

2. Water-soluble vitamins dissolve readily in water. Fat-soluble vitamins are not excreted as readily and tend to accumulate in body fat. That is why excessive consumption of fat-soluble vitamins can lead to overdosing. In contrast, water-soluble vitamins are easily excreted in urine and therefore tend not to accumulate in body tissues in appreciable amounts. We obtain nine water-soluble vitamins from the diet: vitamin C and eight different types of B vitamins. We need four fat-soluble vitamins: A, D, E, and K.

3. In the mouth, teeth bite and chew pieces of the apple. The apple pieces are mixed with saliva by the tongue and then swallowed. They travel down the esophagus to the stomach, where acid and pepsin begin breaking them down as the stomach muscles churn the mixture. From here the mixture is released into the small intestine, where digestive enzymes released by the pancreas, gallbladder, and wall of the small intestine break down the apple's carbohydrates, proteins, and fats. Cells in the lower small intestine absorb the nutrients and water into the bloodstream. The remaining fiber from the apple is consolidated in the colon and passed out of the body as feces. Indigestible material in the apple would probably include the skin and fiber in the fruit.

4. Accessory organs secrete digestive enzymes and other fluids that aid digestion. The pancreas produces a variety of enzymes, including those that degrade carbohydrates and proteins. The liver serves a multitude of functions in the human body, and assists digestion by producing bile. Some of the bile made by the liver is stored in the gallbladder, which dispenses the bile into the small intestine as needed.

5. The liver produces bile. The gallbladder stores bile. Bile is a fluid that helps dissolve fat globules. It acts in much the same way that dishwashing detergent acts on greasy dishes: it creates a coating on fat globules that allows them to interact with water molecules and that partially dissolves them. Large globules break into tiny droplets, which offer a larger work surface for lipid-degrading enzymes.

6. Celiac disease is marked by malabsorption, the inability to absorb nutrients from food, and results in malnutrition if untreated. The symptoms stem from damage to the intestinal villi, which are critical for absorption of the sugars, amino acids, and fatty acids released by digestion. The accumulation of undigested lipids leads to fatty stools. The presence of undigested food, including lactose, triggers a population increase among colonic bacteria, leading to abdominal bloating and diarrhea, symptoms also experienced by people with lactose intolerance when they consume dairy products.

7. The digestive systems of most animals have hard surfaces that mechanically grind food into smaller pieces; enzymes that break down complex carbohydrates, proteins, and fats; and mechanisms for the absorption of nutrients. These features are widespread because all animals must convert the large, complex molecules present in food into small molecules that can be absorbed by the cells of the body.

8. Herbivores have broad, flat teeth suitable for extensive grinding of plant material. Microorganisms in the multichambered stomach of an herbivore break down the cellulose from plant material. Finally, the relatively large surface area of an herbivore's small intestine maximizes the absorption of the scarce protein in its plant diet.

# Chapter 28

1. From the left atrium, the drop of blood enters the left ventricle and is then pumped through the pulmonary artery to the lungs. The blood releases $CO_2$ and picks up $O_2$ in the alveoli of the lung, and then flows back to the heart through the pulmonary vein, completing the pulmonary circuit. It enters the right atrium and then the right ventricle. Contraction of the right ventricle pumps the oxygenated blood into the aorta, beginning the systemic circuit. From there the blood travels through arteries and successively smaller arterioles, eventually reaching the capillaries surrounding respiring cells. Here the red blood cells deliver $O_2$ and pick up $CO_2$, and then they flow through successively larger venules and veins as the blood is carried away from the tissues. The blood then flows through the major veins back to the left atrium of the heart.

2. A heartbeat consists of one cardiac cycle, which has two phases: diastole and systole. During diastole, the heart rests briefly; during systole, the heart muscle contracts and pumps blood. The heartbeat's regularity is maintained by electrical signals from the heart's natural pacemaker, the sinoatrial node. The electrical signal from this node stimulates the contraction of the two atria. The signal is then relayed to the atrioventricular node, which introduces a delay of about 0.1 second before triggering the simultaneous contraction of the two ventricles.

3. Arteries carry oxygenated blood away from the heart; veins carry oxygen-poor blood toward the heart (the pulmonary artery and pulmonary vein are exceptions to this rule). In comparison to veins, arteries have thicker, more flexible walls that can withstand higher pressures. Most veins have valves to keep blood flowing in one direction: back to the heart.

4. In nephrons, large volumes of blood are filtered through the tiny, thin capillaries of the glomerulus. Blood passing through these capillaries under high pressure could cause tears in the capillaries, resulting in leakage and less capacity for filtration. This failure of kidney function could cause toxic wastes to build up in the blood.

5. Air arrives at the pharynx after passing through either the mouth or the nasal cavity. From the pharynx, air passes down the trachea and into the bronchi leading to the lungs. In the lungs, each bronchus branches into many smaller bronchioles, which end at the balloonlike alveoli, where gas exchange occurs.

6. Oxygen from the air diffuses through the thin wall of the alveolus and into the capillaries surrounding it. In the oxygen-deficient blood, it is picked up by hemoglobin molecules in the red blood cells. The red blood cells are carried through the bloodstream to respiring cells, where oxygen diffuses to the cells and carbon dioxide from the cells diffuses into the blood plasma. When the blood returns to the lungs, carbon dioxide diffuses out of the blood plasma into the air in the alveoli.

7. A molecule of hemoglobin can bind up to four molecules of oxygen, so hemoglobin has a much higher oxygen-carrying capacity than does blood plasma. This binding is reversible:

hemoglobin not only picks up oxygen easily from oxygen-rich surroundings (as in the alveoli of the lungs), but releases it easily to surroundings with relatively low oxygen concentrations (near respiring cells).

# Chapter 29

1. Islet cells in the pancreas produce insulin and release it directly into the bloodstream. The insulin molecules circulate in the bloodstream until they reach their target cells, such as those in the liver. In response to the insulin signal, liver cells absorb higher amounts of glucose from the blood and assemble glucose molecules into chains of glycogen for storage.
2. In human males, testosterone controls the development of the male sex organs before birth; interacts with brain cells to bring about behavioral changes; and stimulates sperm production in the testes, growth of facial hair, and growth in cells throughout the body.
3. When you are confronted with a threatening dog, nerve signals from your brain travel to the muscles of your arms and legs, causing you to begin running away. At the same time, the brain signals the adrenal glands to secrete epinephrine, which stimulates glycogen breakdown in your liver and speeds up your heartbeat, providing the quick energy boost you need for running.
4. In an animal that breeds only once a year, a rise in estrogen levels could be triggered by an annual environmental signal, such as warm temperatures or long days in the spring. The high estrogen levels in the female animal would trigger an increase in the levels of gonadotropins, hormones that would induce ovulation and prepare the uterus for a pregnancy. The production of these hormones would be regulated at much lower levels during the rest of the year.
5. If too much growth hormone is produced in the years before adulthood, gigantism results; if too little is produced, dwarfism results. An excess of growth hormone after a person reaches adulthood results in acromegaly, a condition characterized by enlargement of the hands, feet, and bones of the head and face.

# Chapter 30

1. When a stimulus of sufficient strength is received, channels in the neuron's plasma membrane open, causing an influx of $Na^+$ ions that results in a localized depolarization in the axon (a reduction in the electrical charge difference across the axon's plasma membrane at a point close to the cell body). Depolarization causes sodium channels to open in adjacent portions of the axon's cell membrane, leading to depolarization at "downstream" locations in the axon's plasma membrane. In this way the action potential travels down the length of the neuron. When the action potential reaches an axon terminal, neurotransmitters are released from vesicles by exocytosis into the synaptic cleft between the neuron and its target cell. The neurotransmitter molecules diffuse across the cleft and bind to receptor proteins on the plasma membrane of the target cell, where they can either trigger or inhibit action potentials.
2. In this nervous system, two sensory neurons—one responding to each of the two different stimuli—would send their signals to an interneuron that would relay those signals to a motor neuron, which would convey the "move" message to the appropriate muscles. The interneuron could also process the signals so that no signal (the "stay still" message) was sent to the motor neuron.
3. The pigments in the eye are more properly called photoreceptors because even though a chemical change occurs in response to a stimulus, the stimuli *received* are actually photons of light, not chemicals.
4. The guitar player relies in part on the feel of the fingers on the strings to create the music in the performance. The listener perceives the sounds produced by the vibrating guitar strings as music. The player uses mechanoreceptors near the surface of the fingertips that sense slight changes in pressure. The listener depends on hair cells in the basilar membrane of the organ of Corti to convert the pressure waves from the vibrating strings into nerve impulses that the brain interprets as sounds of different frequencies (pitch).

# Chapter 31

1. The scaffoldlike arrangement of bones, and their rigidity and strength, enables body parts to maintain their shape and withstand stress. Bones also enable movement in vertebrates because they provide attachment sites for muscles and ligaments and support for muscle contraction. Cartilage supports structures that need to be stiff but flexible, such as the nose and ear pinnae, and provides cushioning in joints and other places where one bone meets another.
2. A joint allowing circular motion would be similar to a ball-and-socket joint, such as the one found in the hips. In such a joint, cartilage would line the entire socket to facilitate rotation of the connected bone.
3.

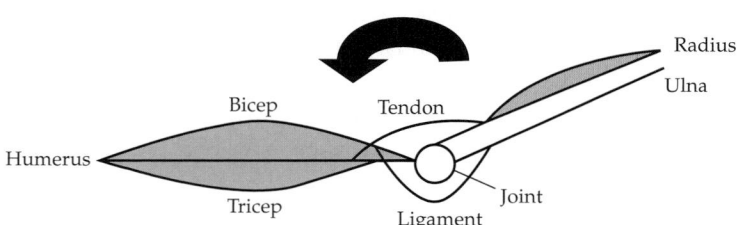

4. A sarcomere contains stacks of actin and myosin filaments. The actin filaments are anchored to Z discs at the ends of the sarcomere. During contraction, the heads of the myosin molecules "walk" down the actin filaments by attaching to and detaching from binding sites. This action pulls the Z discs closer together, and the sarcomere contracts.

5. Muscles are necessary for generating force in both an exoskeleton and a hydrostatic skeleton. In an exoskeleton, however, body parts move when muscles attached to different parts of the skeleton contract. In contrast, in a hydrostatic skeleton muscle contraction squeezes fluid in a body cavity and the fluid pressure causes body parts to move.

6. As a bird flaps its wings, it experiences drag when the pressure from air buildup in front of the wings pushes backward as the bird moves its wings forward. Alternately, the drag from air pushing against the wings can help move the bird forward as it pulls its wings backward.

7. The cruciate ligaments, one in the front and the other in the back of the knee, help stabilize the knee. These ligaments tear when stretched to more than one and a half times their normal length, either by a severe blow to the front or sides of the knee or by severe twisting of the lower leg relative to the femur. Anterior cruciate ligament (ACL) injuries are most common in sports that involve jumping and pivoting, such as basketball, soccer, tennis, and volleyball. Smaller tears can heal on their own, but if the ligament is completely severed, a new one must be reconstructed using grafts from other ligaments.

8.

| SKELETAL LOCATION OF BONES | NUMBER OF BONES | |
|---|---|---|
| Skull | 22 | (a) |
| Ears (two) | 6 | (a) |
| Vertebrae | 26 | (a) |
| Sternum (breastbone) | 3 | (a) |
| Ribs | 24 | (a) |
| Throat | 1 | (a) |
| Shoulder (pectoral) girdle | 4 | (app) |
| Arms (two) | 60 | (app) |
| Hips (pelvis) | 2 | (app) |
| Legs (two) | 58 | (app) |
| Total | 206 | |

The vertebrae are the hard bones of the spinal column that surround and protect the spinal cord while enabling supple movements of the back. The tibia is the larger of the two long bones in the lower leg and the second-largest bone in the human body (the femur is the largest). The tibia supports the weight of the body, forms a hinge joint with the femur and helps stabilize the knee, and in conjunction with the muscles attached to it and other bones, it enables us to walk, run, and jump.

# Chapter 32

1. The first line of defense against pathogens consists of physical and chemical barriers. Physical barriers include the skin and the linings of the lungs and intestines, which most pathogens are unable to penetrate. Chemicals in tears and saliva kill pathogens that come into contact with them, and mucus in the respiratory system traps invading pathogens. These physical and chemical barriers constitute a nonspecific defense response because their action is not directed against particular species or strains of pathogens and is instead exerted against nearly all potential pathogens in much the same way.

2. Disease-fighting cells can recognize distinctive proteins and carbohydrates on the plasma membranes of other cells in the body; and because such normal constituents of the body are perceived as "self," the immune system refrains from attacking them. Immune defense cells also recognize the unique types of proteins and carbohydrates on the surface of foreign substances, such as pathogens and viruses. Because such foreign substances are perceived as "nonself," the immune system mounts an attack against them.

3. An intelligent bacterium might produce large amounts of toxins that would overwhelm the nonspecific responses that constitute the second line of defense. The bacterium could also produce proteins that would mimic the antigens on the host body's cells ("self" antigens), thereby avoiding detection and subsequent destruction by the specific immune responses of adaptive immunity.

4. Both are lymphocytes, a class of white blood cells that play a critical role in adaptive immunity. Both are produced in the bone marrow. T cells migrate from the bone marrow to mature in the thymus. Activated B cells secrete antibodies and orchestrate humoral (antibody-mediated) immunity, while T cells orchestrate cell-mediated immunity by attacking cells in the body that have been infected by a pathogen. B cells recognize the free (soluble) form of an antigen; antigens recognized by T cells are fragments of the pathogen that are displayed on the surface of other cells in the body. Activated B cells and T cells can turn into long-lived memory cells, which enable a rapid secondary immune response to a repeat infection by a pathogen encountered earlier.

5. Our first exposure to a particular antigen sets into motion the primary immune response of adaptive immunity. It takes time—more than 2 weeks sometimes—to activate this response. However, the primary immune response leads to clonal selection, in which many copies of B cells or T cells are made, with some of the effector cells becoming memory cells that persist in the body. Because the memory cells can recognize the pathogen in a repeat infection, they can launch a faster and stronger response that is known as a secondary immune response.

# Chapter 33

1. Sperm, the male gametes, are small and mobile. They deliver the male parent's genetic material. The female gamete—the ovum, or egg—is large and does not move on its own. It contributes the female parent's genetic material and nearly all of the cellular organelles to the offspring.

2. Sexual reproduction requires meiosis and results in genetically unique individuals. In sexual reproduction, genetic material from two individuals is combined; in asexual reproduction, offspring are genetically identical to their parent. The genetic variation of sexually produced offspring makes it more likely that some individuals in a population will be able to survive environmental changes, while a population of offspring produced asexually is more vulnerable to such changes. Sexual reproduction requires time and energy; asexual reproduction takes less energy and can produce offspring more rapidly.

3. Sperm swim into the oviduct to reach the egg, following a chemical signal released by the ovary. A sperm makes its way through the follicle cells surrounding the egg to the outer covering of the egg, where it releases enzymes from its acrosome. The enzymes digest the egg's outer covering and allow the sperm to enter the egg, where its nucleus detaches and fuses with the nucleus of the egg, forming a diploid zygote.

4. The blastocyst first implants itself in the wall of the uterus. The inner cell mass grows and divides, forming the three main tissue layers that give rise to the embryo itself and some of the membranes (extraembryonic membranes) that will surround the developing embryo and fetus. The heart is identifiable in the third week, and all the major organs have begun to form by the end of the eighth week. From this point on, the developing individual is considered a fetus, rather than an embryo. By the end of the first trimester, external structures such as nails and genitalia have formed. In the second and third trimesters, the existing organs grow and develop further, and the fetus gains weight. Before birth, the circulatory and respiratory systems prepare for the transition to a gaseous atmosphere. The developing fetus is most susceptible to disruption in the first trimester, when cells are differentiating and organs are developing rapidly.

5. At the beginning of development, cells are undifferentiated and have the ability to become cells of any type. As development progresses, the range of cell types that a cell can give rise to becomes increasingly limited. This restriction of a cell's fate is achieved by the activation and expression of specific genes. Gene expression can be controlled by morphogens, hormones, and environmental factors.

6. The apoptosis of skin cells between the rays in a fin (especially the paired pectoral fins and pelvic fins, located near the front and back, respectively, of the lower body) would create structures resembling fingers or toes. Over time, such structures could have evolved into the forelimbs and hind limbs of terrestrial vertebrates. Limbs would have enhanced mobility on land, enabling the animals to find food and mates and to escape predators more effectively.

7. Intelligent species on other planets would live beyond their reproductive years if natural selection acted on their populations to favor longer average life spans, enabling many individuals to survive past the optimal reproductive age. Advantages of living longer could include societal benefits from the ability of older generations to pass down skills and knowledge to younger generations.

# Chapter 34

1. Fixed behaviors are less obvious in adults, although we are all familiar with some—for example, our tendency to smile when we see friends, or to frown in response to annoying stimuli. "Infectious yawning," the tendency to yawn in response to yawning by another individual, is a fixed behavior seen in humans, as well as in many other mammals. In humans, fixed behaviors are most apparent in newborns and include the grasping reflex, which is the tendency of babies under 6 months of age to hold on tightly to any object placed in their fists.

   A majority of the behaviors that are common in humans are learned behaviors, and many of them are strongly influenced by the time and society we happen to live in. Familiar learned behaviors in Western societies include these common stimulus-response patterns: our mouth waters at the sight of a raspberry cheesecake; we stop at a red light, and go when it turns green; and, we say "thank you" to a person who does something nice for us.

2. Fear of spiders (arachnophobia) is regularly listed by people as among the things they are most fearful of, on a par with fear of heights and fear of enclosed spaces. Behavioral psychologists suggest that a fear of spiders is innate in many people, although there is broad variation in humans and across different cultures. Certainly a person who has suffered a spider bite is likely to develop an aversion, if not a fear, of spiders. Fearful behaviors are difficult to analyze in humans because of the complexity and diversity of human experiences and because we cannot do the necessary controlled experiments in people. Animal studies show that some insects who are preyed on by spiders also have an innate fear of spiders, while in other species the fear is learned by young observing the reactions of their mothers.

3. The details of your answer will depend on which sport or musical instrument you have chosen to discuss. In general terms, learning to play an instrument or a sport requires learning the techniques and strategies needed to be effective (how to hold the instrument or how to kick a soccer ball) and learning which forms or positions are most effective (which sounds made by the instrument sound nice to the ears or which poses in dance look the most graceful). However, the most talented among us very likely have genetically

determined advantages as well. For example, having perfect pitch is a genetic trait, but it becomes evident only in a person with musical training. As an example from sports, the proportion of fast-twitch and slow-twitch muscles we have can determine our prowess in strength versus endurance sports (see the box on page 686), but once again training plays a large role too.

4. There are many forms of nonverbal communication, some of which tend to be culture-specific. Examples include smiling, frowning, winking, waving goodbye, giving someone an inquiring look, making a "peace" sign, and wearing a uniform (such as a police officer's uniform).

5. Fixed behaviors that might be important to group members could involve responses to stimuli that an organism might never have a chance to learn about. For instance, group-living animals might have fixed behaviors that enable them to identify members of their own group (for example, by recognizing an odor unique to the group). Or a group-living animal might have a fixed behavior that leads it to avoid snakelike objects or to hide at the sight of a silhouette resembling that of a predatory bird. There might not be a chance to learn the appropriate behaviors from others, because one incorrect response could easily be fatal. Behaviors associated with surviving during the first few days following birth or hatching, before the creature has a chance to learn, are likely to be fixed behaviors.

# Chapter 35

1. The dicots are the larger of the two groups. They usually have the following characteristics: a netlike pattern of leaf veins; stem vascular bundles arranged in a discrete ring; a main root (taproot) with many side branches; flower parts, such as petals, in multiples of four or five; and embryos with two cotyledons. Monocots usually have these characteristics: parallel leaf veins; stem vascular bundles distributed in a scattered pattern; a fibrous root; flower parts in three, or multiples of three; and embryos with a single, small cotyledon. Examples of dicots include pea plants, apple trees, and dandelions. Examples of monocots include lawn grasses, tulips, and palm trees.

2. The three plant organs are (1) roots, which anchor the plant, absorb and transport water and nutrients, and can be specialized to store food; (2) stems, which give plants support, transport water and nutrients, and can be specialized to form structures such as thorns; and (3) leaves, which provide food by means of photosynthesis and can be specialized to form reproductive organs (flowers) or spines. The three plant tissue types are (1) dermal tissues, which protect the plant and control the exchange of gases and water; (2) ground tissues, which support the plant, heal wounds, and perform photosynthesis; and (3) vascular tissues, which transport food and water throughout the plant.

3. Mineral nutrients dissolved in water can be absorbed by plant roots through cells (the "cell-interior route") or along cell walls (the "cell-wall route").

4. Most of the dry weight of a plant comes not from nutrients absorbed by the roots, but from sugars synthesized by photosynthesis using $CO_2$ obtained from the air and water from the soil. The atoms found in $CO_2$ and $H_2O$ are assimilated into carbohydrates, which account for most of the dry weight of any plant. Two of the mineral elements—nitrogen and phosphorus—are assimilated into organic compounds such as proteins and nucleic acids, and they also contribute a significant amount to a plant's dry weight. Other mineral nutrients are used in relatively small amounts, and they add very little to the overall dry weight of plant tissues (less than 5 percent in most plants).

5. Most carnivorous plants are found in nutrient-poor habitats, such as acid bogs, which tend to be deficient in nitrogen and phosphorus. As they digest their prey, carnivorous plants absorb nitrogen from the amino acids released by the breakdown of proteins, and phosphorus from the breakdown of nucleic acids.

6. Transpiration, the evaporative loss of water from the surface of leaves, creates uncountable curved surfaces at every air-water boundary within the leaf tissue. The surface tension at each of these curved air-water boundaries collectively exerts a very large pulling force, or tension, on the column of water in the xylem that extends to the roots of the tree. The cohesion of water molecules creates a continuous and strong "cable" of water in the xylem conducting tissues. As water molecules leave the upper part of the plant through transpiration, more water is drawn up from the roots to replace the lost water. Plants generate the pressure required to transport food from the leaves to other areas by pumping sugars into the phloem. The movement of sugar into the phloem cells decreases the concentration of water in those cells, causing water to diffuse into them through osmosis. At points where sugars are unloaded from the phloem for consumption or storage by tissues, there is osmotic loss of water, which produces a drop in phloem pressure in these regions. The result is a pressure difference between regions where sugars are loaded into the phloem and regions where sugars are unloaded from the phloem. This pressure difference drives the long-distance transport of sugars in phloem conducting tubes.

# Chapter 36

1. "Indeterminate growth" refers to a pattern of highly flexible development in which the organism continues to grow throughout its life, adding new body parts as needed. Indeterminate growth enables plants to adjust to varying environmental conditions, such as high or low levels of sunlight, water, or nutrients. If conditions

are favorable, a plant may add many new body parts and take on one shape, but if conditions are unfavorable, the plant may add few new body parts and take on another shape. Most animals can move away from adverse environmental conditions and therefore do not need the developmental flexibility that indeterminate growth provides.

2.

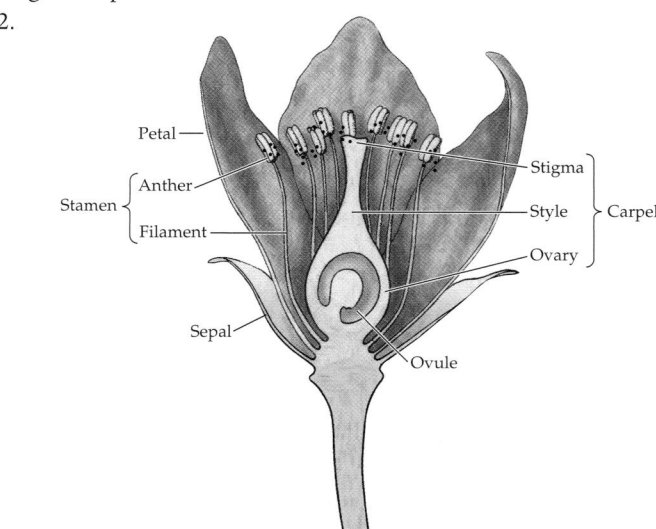

3. A flowering plant germinates from a seed when conditions are suitable for growth. The plant grows and produces flowers, which contain the reproductive structures of the plant. Small, haploid pollen grains (in which sperm are produced) are formed in the anther of a flower; small, haploid embryo sacs (in which eggs are produced) are formed inside the ovules, which lie within the ovary. When a pollen grain from a male anther lands on the stamen of a flower of the same species, a pollen tube containing two sperm cells grows downward to the ovule. Double fertilization occurs in the ovule: one sperm fertilizes the egg, producing an embryo; and the other fuses with the two polar nuclei within the large central cell in the embryo sac, producing endosperm (a tissue that will provide food for the embryo). The embryo becomes encased in a protective seed coat, and the surrounding ovary forms a fruit. The fruit and its enclosed seeds are dispersed by means of wind, water, or animals. When a seed germinates, the cycle begins again. Each pollen grain is a haploid structure (male gametophyte), created through mitoses in a haploid spore. Spores are haploid cells because they are direct products of meiosis.

4. Wood consists of layers of secondary xylem, which are formed as new rings of cells are put down each year; these cells are produced by the vascular cambium during secondary growth. Sapwood is younger secondary xylem that provides strength and also conducts water. Heartwood is older secondary xylem that lies at the center of a tree trunk; it provides strength but does not conduct water, because its water-conducting tubes are clogged. Heartwood is darker in color than sapwood because of chemical deposits (tannins and resins, for example); the chemicals in heartwood make it more resistant to decay by wood rot fungi and other decomposers.

5. Young fruits, containing still-developing seeds, are often bitter or otherwise unpalatable to animals. If an animal were to feed on an immature fruit, it could irreversibly damage the developing embryo in that fruit, harming the plant's reproductive potential. When the embryo is fully developed and the seed better protected (for instance, with a tough seed coat), the fruit starts to ripen and the bitter compounds disappear. The other changes that accompany fruit ripening (changes in color and aroma, for example) advertise its palatability to animal seed dispersal agents. Animals attracted to the ripened fruit are likely to disperse mature, well-protected seeds, increasing the chance that the plant will reproduce successfully.

6. A fruit is a plant organ that develops from an ovary containing fertilized ovules (seeds). From the scientific standpoint, the edible part of the tomato, bearing seeds inside, is undeniably a fruit, because it develops from an ovary. The juicy red part—the fruit wall—develops from the ovary wall, and the many small, yellow seeds are mature, fertilized ovules. The word "vegetable" holds no scientific meaning, although in everyday usage it is applied to any edible plant part that is not sweet, including starchy roots, leafy greens, and nonsweet, fleshy fruits. To combine both scientific accuracy and popular usage, one might say that the tomato is both a fruit and a vegetable.

7. Plants with an impaired ability to detect day length would probably survive best in tropical areas of the world, where the climate is favorable for plant growth year-round. The plants would probably do poorly at more northerly or southerly latitudes or in other places where there is more seasonal variation in environmental conditions. In these areas, the plant's inability to respond to seasonal changes, such as the onset of cold winters, would interfere with its reproducing at the proper time.

8. Plants may protect themselves from herbivores with thick bark, spines or thorns, or toxic compounds in their leaves. Plants also possess specific and nonspecific immune responses to pathogens. The R alleles of plant defense genes can recognize specific pathogens and trigger the production of toxic defensive chemicals. Other chemical signals produced by the pathogen, or by the damage created by the pathogen, cause the sealing off of potential entry points or activate chemicals that attack bacterial cell walls.

# Glossary

**ABA** See *abscisic acid*.

**abiotic** Of or referring to the nonliving environment that surrounds the community of living organisms in an ecosystem; the atmosphere, water, and Earth's crust. Compare *biotic*.

**abscisic acid (ABA)** A plant hormone that mediates adaptive responses to drought, cold, heat, and other stresses.

**abundance** The number of individuals of a species in a defined habitat.

**abyssal zone** The deeper waters of the open ocean, beyond the continental shelf, at depths greater than 6,000 meters (nearly 20,000 feet).

**acclimation** Also called *acclimatization*. The process by which an adaptive response is expressed in an individual in response to an environmental challenge. Compare *adaptation* (definition 1).

**acid** A chemical compound that can give up a hydrogen ion. Compare *base* (definition 1) and *buffer*.

**acid rain** Precipitation with an unusually low pH, compared to that of unpolluted rain (which has a pH of about 5.2). Acid rain is a consequence of the release of sulfur dioxide and other pollutants into the atmosphere, where they are converted to acids that then fall back to Earth in rain or snow.

**acrosome** The front tip of a sperm, which contains enzymes that help digest the outer covering (zona pellucida) of the unfertilized ovum.

**actin** The protein component of actin filaments.

**actin filament** The thinnest of the three main types of protein fibers that make up the cytoskeleton in eukaryotic cells (see also *intermediate filament* and *microtubule*). Actin filaments confer cell shape, enable whole-cell movement in crawling cells, and assist in muscle contraction (see also *myosin filament*).

**action potential** An electrical signal generated by the flow of ions across the plasma membrane of a neuron. Action potentials are self-sustaining and can travel down a neuron in only one direction.

**activation energy** The small input of energy required for a chemical reaction to occur at a noticeable rate.

**active carrier protein** Also called *membrane pump*. A cell membrane protein that, using energy from an energy-rich molecule such as ATP, changes its shape to transfer an ion or a molecule across the membrane. Compare *passive carrier protein*.

**active immunity** Immunity that depends on the production of antibodies specific to a particular pathogen by the organism's own body. Compare *passive immunity*.

**active site** The specific region on the surface of an enzyme where substrate molecules bind.

**active transport** Movement of ions or molecules across a biological membrane that requires an input of energy because it is "uphill" (against an electrochemical gradient). Compare *passive transport*.

**adaptation** 1. The evolutionary process by which a population as a whole becomes better matched to its habitat. Compare *acclimation*. 2. An inherited characteristic—structural, biochemical, or behavioral—that enables an organism to function well and therefore to survive and reproduce better than competitors lacking the characteristic; also called *adaptive trait*.

**adaptive evolution** See *natural selection*.

**adaptive immune system** Also called *adaptive immunity*. An internal defense system in vertebrates that acts against pathogens in a highly specific manner, with specialized defense cells mobilized against particular strains of pathogens. Its immune memory allows it to remember a first encounter with a specific strain of pathogen and to mobilize an especially speedy and precisely targeted response to a repeat infection by the same strain. Compare *innate immune system*.

**adaptive radiation** An evolutionary expansion in which a group of organisms takes on new ecological roles and forms new species and higher taxonomic groups.

**adaptive trait** See *adaptation* (definition 2).

**adenosine diphosphate** See *ADP*.

**adenosine triphosphate** See *ATP*.

**adhesion** The tendency of a substance to cling to other substances. Compare *cohesion*.

**ADP** Adenosine diphosphate, the low-energy form of the energy carrier *ATP*, which constitutes the "energy currency" in all living cells. When ADP becomes phosphorylated (acquires an additional phosphate group), it is transformed into ATP.

**adrenal gland** One of a pair of endocrine glands that sit atop the mammalian kidney and release the hormones epinephrine and norepinephrine.

**adrenaline** See *epinephrine*.

**adult-onset diabetes** See *type 2 diabetes*.

**adult stem cell** Also called *somatic stem cell*. A cell that retains the capacity for self-renewal and persists into adulthood. Compare *embryonic stem cell*.

**aerobe** An organism that requires oxygen to survive. Compare *anaerobe*.

**aerobic** Of or referring to a metabolic process or organism that requires oxygen gas. Compare *anaerobic*.

**afterbirth** See *placenta*.

**alga (pl. algae)** Any photosynthetic protist. Algae may or may not be motile. Compare *protozoan*.

**allele** One of several alternative versions of a gene. Each allele has a DNA sequence that is somewhat different from that of all other alleles of the same gene.

**allele frequency** The proportion (percentage) of a particular allele in a population.

**allopatric speciation** The formation of new species from populations that are geographically isolated from one another. Compare *sympatric speciation*.

**alternation of generations** A type of life cycle, found in all plants, in which two distinctly different multicellular "generations" must be produced: the haploid, multicellular gametophytes of one generation give rise to the diploid, multicellular sporophytes of the next generation.

**altruism** A behavior that benefits another individual but has a cost to the individual performing the behavior.

**alveolus (pl. alveoli)** Any of the small sacs in the mammalian lung where gas exchange takes place.

**amino acid** A nitrogen-containing, small organic molecule that has an amino group, a carboxyl group, and a variable R group, all covalently attached to a single carbon atom. Proteins are polymers of amino acids.

**amnion** The membrane enclosing the amniotic cavity.

**amniotic cavity** The cavity that is created when the embryonic disc pulls away from the outer layer of blastocyst cells. The human embryo, and later the fetus, floats inside the fluid-filled amniotic cavity.

**amniotic egg** An egg in which the developing embryo is surrounded and protected by layers of extraembryonic membranes (including the amnion) that promote gas exchange and removal of waste.

**anabolism** Also called *biosynthesis*. Metabolic pathways that build macromolecules. Compare *catabolism*.

**anaerobe** An organism that can survive without oxygen. Compare *aerobe*.

**anaerobic** Of or referring to a metabolic process or organism that does not require oxygen gas. Compare *aerobic*.

**analogous** Of or referring to a characteristic shared by two groups of organisms because of convergent evolution, not common descent. Compare *homologous*.

**anaphase** The stage of mitosis during which sister chromatids separate and move to opposite poles of the cell.

**anatomy** The study of the structures that make up a complex multicellular body. Compare *physiology*.

**anchorage dependence** The inability of a cell to divide if it is detached from its surroundings. Most human cells exhibit this trait.

**anchoring junction** Also called *desmosome*. A protein complex that acts as a "hook" between two animal cells or between a cell and the extracellular matrix, linking them to brace against rupturing forces. Compare *gap junction* and *tight junction*.

**androgen** Any of a class of hormones that is one of three major types produced by the gonads in males and, in much smaller amounts, in females. Androgens stimulate cells to develop the characteristics of maleness, such as beard growth or the production of sperm. The primary androgen is testosterone. Compare *estrogen* and *progestogen*.

**angiogenesis** The formation of new blood vessels.

**angiosperms** Also called *flowering plants*. One of four major groups of plants. Angiosperms have vascular tissues, seeds, flowers, and fruits. They include most plants on Earth today. Compare *bryophytes* and *gymnosperms*.

**Animalia** The kingdom of Eukarya that is made up of animals—multicellular heterotrophs that have evolved specialized tissues, organs and organ systems, body plans, and behaviors.

**Annelida** Annelids. The eukaryotic phylum consisting of all worms with true coeloms and segmented bodies. Earthworms are the most familiar example.

**annual** A plant that completes its entire life cycle in one year. Compare *biennial* and *perennial*.

**annual ring** All the wood produced in a single growing season of a tree.

**antenna complex** A disclike grouping of pigment molecules, including chlorophyll, that harvests energy from sunlight in the thylakoid membrane of a chloroplast.

**antibiotic** A molecule that is secreted by one microorganism to kill or slow the growth of another microorganism.

**antibody** A protein that is produced by a B cell and binds specifically to a particular *antigen*.

**antibody-mediated immunity** The part of the adaptive immune system that relies on antibodies to fight off pathogens. By binding with the antigens of invading pathogens, antibodies tag the alien substances or cells for destruction by other members of the immune system defense team. B cells play an integral role in antibody-mediated immunity. Compare *cell-mediated immunity*.

**anticodon** A sequence of three nitrogenous bases on a transfer RNA molecule that enables it to form complementary base pairs with a corresponding *codon* on an mRNA molecule.

**antigen** A characteristic protein or other molecule produced by an invading pathogen that is recognized as alien by particular lymphocytes and that incites B cells to produce *antibodies* directed against it.

**antigen-presenting cell (APC)** An immune cell that displays an antigen.

**anus** The opening of the digestive system through which undigested food and other solid waste leaves the body.

**APC** See *antigen-presenting cell*.

**apical dominance** Inhibition of the growth of lateral buds by the apical meristem.

**apical meristem** A population of perpetually undifferentiated cells in the tips of plant branches and roots. These cells divide to renew the population, and also produce descendants that give rise to new stem and root tissues. Compare *lateral meristem*.

**apoptosis (pl. apoptoses)** Programmed cell death—that is, cell death that generally benefits the organism.

**aposematic coloration** See *warning coloration*.

**appendicular skeleton** The bones of the arms, legs, and pelvis, collectively. Compare *axial skeleton*.

**Archaea** One of the three domains of life, encompassing the microscopic, single-celled prokaryotes that arose after the Bacteria. The domain Archaea is equivalent to the kingdom Archaea. Compare *Bacteria* and *Eukarya*.

**arteriole** A fine branch of an artery (only 0.3 millimeter in diameter). Arterioles control the flow of blood into the finest vessels, the capillaries. Compare *venule*.

**artery** A blood vessel that carries blood from the heart. Compare *vein*.

**Arthropoda** Arthropods. The largest eukaryotic phylum, consisting of animals characterized by a hard exoskeleton and jointed body parts. Arthropods include millipedes, crustaceans, insects, and spiders.

**artificial selection** A process in which only individuals that possess certain characteristics are allowed to breed. Artificial selection is used to guide the evolution of crop plants and domestic animals in ways that are advantageous for people. Compare *natural selection*.

**ascomycetes** Also called *sac fungi*. One of three major groups of fungi. Compare *basidiomycetes* and *zygomycetes*.

**asexual reproduction** The production of genetically identical offspring without the exchange of genetic material with another individual. Compare *sexual reproduction*.

**atmospheric cycle** A type of nutrient cycle in which the nutrient enters the atmosphere easily. Compare *sedimentary cycle*.

**atom** The smallest unit of a chemical element that still has the properties of that element.

**atomic mass number** The sum of the number of protons and neutrons found in the nucleus of an atom of a particular chemical element.

**atomic number** The number of protons found in the nucleus of an atom of a particular chemical element.

**ATP** Adenosine triphosphate, a molecule that is commonly used by cells to store energy and to transfer energy from one chemical reaction to another. ATP fuels a wide variety of cellular activities in every organism on Earth.

**ATP synthase** A large channel-containing protein complex that spans the thylakoid membrane and catalyzes the conversion of ADP to ATP.

**atrioventricular (AV) node** A signaling center in the heart that relays the signal from the *sinoatrial node* to the ventricles, triggering muscle contraction in both ventricles.

**atrium (pl. atria)** In reference to the heart, a chamber that receives blood from the body and pumps it into a ventricle. Compare *ventricle*.

**auditory canal** The tubular opening leading from the ear pinna to the eardrum.

**autonomic** Involuntary. Compare *somatic*.

**autosome** Any chromosome that is not a *sex chromosome*.

**autotroph** See *producer*. Compare *heterotroph*.

**auxin** Any of a group of plant hormones that promote growth, cause apical dominance, and have other important effects, such as controlling the way plants bend toward light.

**AV node** See *atrioventricular node*.

**axial skeleton** The skull, spinal column, and ribs, collectively. Compare *appendicular skeleton*.

**axon** An extension of a neuron that carries action potentials away from the cell body of a neuron and toward another neuron or other target, such as muscle or endocrine tissue.

**B cell** One of two basic types of mature lymphocytes. B cells mature in the bone marrow, and they are involved in antibody-mediated immunity. Compare *T cell*.

**Bacteria** One of the three domains of life, encompassing the microscopic, single-celled prokaryotes that were the first organisms to arise. The domain Bacteria is equivalent to the kingdom Bacteria. Compare *Archaea* and *Eukarya*.

**bacterial conjugation** The process in which a bacterium actively trades DNA with another bacterium, usually of the same species.

**bacterial culture** A lab dish that contains bacteria living off a specially formulated food supply.

**bark** The protective outer covering of stems in woody plants that contains cork cells, the cork cambium, and the secondary phloem.

**base** 1. A chemical compound that can accept a hydrogen ion. Compare *acid* and *buffer*. 2. See *nitrogenous base*.

**base pair** Also called *nucleotide pair*. A pair of complementary nitrogenous bases connected by hydrogen bonds. Base pairs form the "rungs" of the DNA double helical "ladder." In DNA, adenine pairs with thymine (A-T) and cytosine pairs with guanine (C-G); in RNA, uracil replaces thymine.

**basidiomycetes** Also called *club fungi*. One of three major groups of fungi. Compare *ascomycetes* and *zygomycetes*.

**basilar membrane** A thin membrane lying in the inner ear that vibrates in response to vibrations passed to it from the middle ear.

**behavior** A coordinated response to a stimulus; particularly, a response that involves movement.

**benign tumor** A relatively harmless cancerous growth that is confined to a single tumor and does not spread to other tissues in the body.

**benthic zone** The bottom surface of any body of water.

**biennial** A plant that completes its life cycle in two years. Compare *annual* and *perennial*.

**bilateral symmetry** A body arrangement in which only one plane passes vertically from the top to the bottom of the animal, dividing the body into two halves that mirror each other. Bilateral animals have distinct right and left sides, with near-identical body parts on each side. Compare *radial symmetry*.

**bile** A lipid-based substance produced by the liver that helps break up large fat globules in the small intestine.

**binary fission** A form of asexual reproduction in which a cell divides to form two genetically identical daughter cells that replace the original parent cell.

**bioaccumulation** The deposition of a substance within an organism at concentrations higher than in the surrounding abiotic environment. Compare *biomagnification*.

**biodiversity** The variety of organisms on Earth or in a particular location, ranging from the genetic variation and behavioral diversity of individual organisms or species through the diversity of ecosystems.

**biogeochemical cycle** See *nutrient cycle*.

**biological evolution** A change in the overall genetic characteristics of a group of organisms over multiple generations of parents and offspring.

**biological hierarchy** The nested series in which living things, their building blocks, and their living and nonliving surroundings can be arranged—from atoms at the lowest level to the entire biosphere at the highest level.

**biological species concept** The idea that a species is defined as a group of populations that can interbreed but are reproductively isolated from other such groups. Compare *morphological species concept*.

**biology** The study of life.

**biomagnification** An increase in the tissue concentrations of a bioaccumulated chemical at successively higher trophic levels in a food chain. Compare *bioaccumulation*.

**biomass** The mass of organisms per unit of area.

**biome** A large area of the biosphere that is characterized according to its unique climatic and ecological features. Terrestrial biomes are usually classified according to their dominant vegetation; aquatic biomes, on the basis of physical and chemical features.

**biomolecule** Any molecule found within a living cell.

**bioremediation** The use of organisms to clean up environmental pollution.

**biosphere** All living organisms on Earth, together with the environments in which they live. Compare *ecosystem*.

**biosynthesis** See *anabolism*.

**biotic** Of or referring to the assemblage of interacting organisms—prokaryotes, protists, animals, fungi, and plants—in an ecosystem. Compare *abiotic*.

**bipedal** Of or referring to an organism that walks upright on two legs.

**bivalent** See *tetrad*.

**blastocyst** A hollow ball of cells that forms the stage following a morula in mammalian development.

**blood** A fluid tissue consisting of liquid plasma and the blood cells that it carries. Blood plays a central role in carrying nutrients, oxygen, and wastes through the circulatory system.

**blood pressure** The pressure that pushes the blood from the heart to the respiring tissues. In humans, it is typically measured as the pressure in the arteries leading from the left ventricle.

**blood vessel** In the circulatory system, any of the tubelike structures that transport blood throughout the body. The major blood vessels in the human body are arteries, veins, and capillaries.

**bone** A strong and rigid supportive structure in animals that is strengthened by deposits of calcium salts. Compare *cartilage*.

**booster shot** A repeat vaccination administered periodically to maintain immunity to specific antigens by raising antibody concentrations and memory B cell numbers through fresh exposure to those antigens.

**boreal forest** Also known as *taiga*. A terrestrial biome dominated by coniferous trees that grow in northern or high-altitude regions with cold, dry winters and mild summers.

**brain** A complex collection of neurons located at the head end of an animal that serves as the major clearinghouse for information in the nervous system.

**brain stem** The midbrain and hindbrain collectively, which together control many of the functions of the body that are not under conscious control.

**branch** See *clade*.

**breathing** The process of drawing air into and out of the lungs to extract oxygen and expel carbon dioxide.

**bronchiole** Any of the small, highly branched tubes inside each lung that carry air from the bronchus to the many alveoli in that lung.

**bronchus (pl. bronchi)** One of two tubes that branch off from the trachea and carry air to each of the two lungs in a vertebrate.

**brush border** A series of microscopic, fingerlike extensions from the cells that line the inner surface of the small intestine. These extensions of the plasma membrane greatly increase the surface area available for nutrient absorption.

**bryophytes** One of four major groups of plants. Bryophytes include mosses, liverworts, and hornworts. They lack true vascular tissues. Compare *angiosperms* and *gymnosperms*.

**buffer** A chemical compound that can both give up and accept hydrogen ions. Buffers can maintain the pH of water within specific limits. Compare *acid* and *base*.

**calcitonin** A hormone produced by the mammalian thyroid gland that promotes the storage of calcium in bones and stimulates the kidneys to reabsorb less calcium. Compare *parathyroid hormone*.

**Calvin cycle** The series of chemical reactions in photosynthesis that manufacture sugar. The Calvin cycle takes place in the stroma of the chloroplast and synthesizes sugars from carbon dioxide and water. Compare *light reactions*.

**Cambrian explosion** A major increase in the diversity of life on Earth that occurred about 530 million years ago, during the Cambrian period. The Cambrian explosion lasted 5–10 million years, during which time large and complex forms of most living animal phyla appeared suddenly in the fossil record.

**cancer cell** Also called *malignant cell*. A tumor cell that breaks loose from the tumor, invades neighboring tissues, and disrupts the normal function of tissues and organs.

**cancer immunosurveillance** The cooperative effort of the innate and adaptive immune systems to recognize and destroy cancer cells.

**capillary** A tiny blood vessel in a closed circulatory system, across the walls of which all exchange with the surrounding tissues takes place.

**carbohydrate** Any of a class of organic compounds that includes sugars and their polymers, in which each carbon atom is usually linked to two hydrogen atoms and an oxygen atom. See also *sugar*.

**carbon cycle** The transfer of carbon within biotic communities, between living organisms and their physical surroundings, and within the abiotic world.

**carbon fixation** The process by which carbon dioxide is incorporated into organic molecules. Carbon fixation occurs in the chloroplasts of plants and results in the synthesis of sugars.

**carcinogen** A physical, chemical, or biological agent that causes cancer.

**cardiac cycle** The alternate contraction and relaxation of the heart muscle, which is necessary for the pumping of blood by the heart. The cycle consists of two phases: systole and diastole.

**cardiac muscle** The specialized muscle of the vertebrate heart. Cardiac muscle has bands like the muscles associated with the skeleton, and its muscle fibers are joined by interconnecting branches that help them produce the coordinated contractions known as heartbeats. Compare *skeletal muscle* and *smooth muscle*.

**cardiovascular system** The closed circulatory system of vertebrates, which consists of a muscular, chambered heart and loops of blood vessels.

**carnivore** A consumer that eats animals and digests them internally to extract nutrients. Compare *herbivore*, *omnivore*, and *detritivore*.

**carpel** The female reproductive part of a flower, consisting of a stigma, style, and ovary. Carpels form the innermost of the four whorls of modified leaves that make up a flower. Compare *petal*, *sepal*, and *stamen*.

**carrier protein** A cell membrane protein that recognizes, binds, and transports a specific cargo molecule across the membrane. See also *active carrier protein* and *passive carrier protein*.

**carrying capacity** The maximum population size that can be supported indefinitely by the environment in which the population is found.

**cartilage** A strong, but relatively flexible supportive structure in animals. Cartilage is made up of cells that secrete an extracellular matrix rich in the protein collagen. Compare *bone*.

**catabolism** Metabolic pathways that take macromolecules apart. Compare *anabolism*.

**catalyst** A substance that speeds up a specific chemical reaction without being permanently altered in the process. Enzymes, which are usually proteins, are an example of biological catalysts.

**cell** The smallest self-contained unit of life, enclosed by a membrane.

**cell culture** A laboratory procedure in which stem cells are spread in a special nutrient broth in a plastic petri dish, where they undergo mitotic divisions to make more stem cells, which can then be induced to differentiate into specific cell types.

**cell cycle** A series of distinct stages in the life cycle of a cell that is capable of dividing. Cell division is the last stage in the cycle.

**cell differentiation** The process through which a daughter cell becomes different from its parent cell and acquires specialized functions.

**cell division** The final stage in the life of an individual cell, in which a parent cell is split up to generate two (mitotic cell division) or four (meiotic cell division) daughter cells.

**cell junction** A structure that anchors a cell, connects it with a neighbor, or creates communication passageways between two cells.

**cell-mediated immunity** The part of the adaptive immune system whose role is to destroy cells harboring pathogens and other substances that are sensed as foreign by the body. T cells play an integral role in cell-mediated immunity. Compare *antibody-mediated immunity*.

**cell plate** A partition, consisting of membrane and cell wall components, that appears during cytokinesis in dividing plant cells. The cell plate matures into a polysaccharide-based cell wall flanked on either side by the plasma membranes of the two daughter cells.

**cell theory** The theory that every living organism is composed of one or more cells, and that all cells living today came from a preexisting cell.

**cellular respiration** A metabolic process that extracts chemical energy from organic molecules, such as sugars, to generate the universal energy carrier ATP, consuming oxygen and releasing carbon dioxide and water in the process. Cellular respiration has three phases: glycolysis, the Krebs cycle, and oxidative phosphorylation. Compare *photosynthesis*.

**cellulose** An extracellular polysaccharide (a type of carbohydrate), produced by plants and some other organisms, that strengthens cell walls.

**central nervous system (CNS)** The portion of the nervous system devoted to the exchange and integration of information among neurons. The central nervous system consists of the brain and spinal cord. Compare *peripheral nervous system*.

**centromere** A physical constriction that holds sister chromatids together.

**centrosome** A cytoskeletal structure in the cytosol that helps organize the mitotic spindle and defines the two poles of a dividing cell.

**cephalization** In bilateral animals, the evolutionary trend in which structures used for eating are concentrated at the anterior end, thereby allowing the animal to look where it is going.

**cerebral cortex** The highly folded outer portion of the human brain that contains neurons specialized for higher-order functions such as thought, memory, learning, and language.

**cerebrum** The largest part of the human forebrain; it includes the cerebral cortex.

**CFC** See *chlorofluorocarbon*.

**channel protein** Also called *membrane channel*. A cell membrane protein that forms an opening through which specific ions or molecules can pass.

**chaparral** A terrestrial biome characterized by shrubs and small nonwoody plants that grow in regions with rainy winters and hot, dry summers.

**character displacement** A process by which intense competition between species causes the forms of the competing species to evolve to become more different over time.

**chemical bond** The attractive interaction that causes two atoms to associate with each other.

**chemical compound** An association of atoms of different chemical elements linked by ionic or covalent bonds.

**chemical energy** Energy that is stored in atoms because of their position in relation to other atoms in the system under consideration.

**chemical formula** The simple shorthand that is used to represent the atomic composition of salts and molecules. A subscript number to the right of the symbol of each element shows how many atoms of that element are contained in the salt or molecule.

**chemical reaction** A process that creates or rearranges chemical bonds between atoms.

**chemoautotroph** An organism that obtains its energy from chemicals and derives its carbon from carbon dioxide in the air. All chemoautotrophs are prokaryotes. Compare *chemoheterotroph* and *photoautotroph*.

**chemoheterotroph** An organism that obtains its energy from chemicals and derives its carbon from carbon-containing compounds found mainly in other organisms. All fungi and animals, as well as many protists and prokaryotes, are chemoheterotrophs. Compare *chemoautotroph* and *photoheterotroph*.

**chemoreceptor** A sensory cell that initiates an action potential in response to a chemical stimulus.

**chemotherapy** The administration of high doses of chemical poisons to destroy cancer cells by killing all rapidly dividing cells.

**chitin** A carbohydrate that serves as an important support material in the cell walls of fungi and in animal exoskeletons.

**chlorofluorocarbon (CFC)** Any of a class of synthetic chemical compounds whose release into the atmosphere can damage the ozone layer.

**chlorophyll** A green pigment that is used to capture light energy for photosynthesis.

**chloroplast** An organelle, found in plants and algae, that is the primary site of photosynthesis.

**chordate** Any of a group of animals that have a notochord, pharyngeal pouches, and a post-anal tail.

**chorion** The tissues that form the wall of the chorionic cavity, which almost completely envelops the developing embryo, and later the fetus. Fingerlike extensions from the chorion, the chorionic villi, help in the transfer of nutrients from the mother's circulation to the fetus.

**chromatin** DNA complexed with packaging proteins. Chromosomes are made up of compacted chromatin.

**chromosome** A threadlike structure composed of a single DNA molecule packaged with proteins. Chromosomes achieve the highest level of compaction when prophase begins during mitosis or meiosis.

**chromosome theory of inheritance** A theory, supported by much experimental evidence, stating that genes are located on chromosomes.

**cilium (pl. cilia)** A hairlike structure, found in some eukaryotes, that uses a rowing motion to propel the organism or to move fluid over cells.

**circulatory system** The organ system in animals that consists of interconnected tubular vessels through which blood is pumped by the heart; it is the main internal transport system.

**citric acid cycle** See *Krebs cycle*.

**clade** Also called *branch*. The part of an evolutionary tree that consists of a given ancestor and all its descendants.

**class** In reference to biological classification systems, the level in the Linnaean hierarchy above order and below phylum.

**climate** The prevailing weather conditions experienced in an area over relatively long periods of time (30 years or more). Compare *weather*.

**climate change** Significant and long-term change in the average climatic conditions in our biosphere, such as global warming.

**clonal selection** The process by which a lymphocyte that has bound to its specific antigen reproduces rapidly, yielding large numbers of genetically identical lymphocytes with exactly the same antigen-recognizing proteins that the parent cell has.

**clone** An offspring that is genetically identical to its parent.

**cloning** Making a copy (of a gene, a cell, or a whole organism) that is genetically identical to the original.

**closed circulatory system** An internal transport system of animals in which blood vessels carry blood throughout the body and back to the heart.

**club fungi** See *basidiomycetes*.

**CNS** See *central nervous system*.

**coastal region** An ecosystem of the marine biome defined as the area of the ocean stretching from the shoreline to the continental shelf, which is the undersea extension of a continent's edge.

**cochlea** A coiled, fluid-filled tube in the inner ear.

**codominance** A type of allele interaction in which the effects of both alleles at a given genetic locus are equally visible in the phenotype of a heterozygote. In other words, the influence of each codominant allele is fully displayed in the heterozygote, without being diminished or diluted by the presence of the other allele (as in *incomplete dominance*).

**codon** A sequence of three nitrogenous bases in an mRNA molecule. Each codon specifies either a particular amino acid or a signal to start or stop the translation of a protein. Compare *anticodon*.

**coelom** A body cavity that lies within the mesoderm.

**coevolution** The process by which interactions among species drive evolutionary change in those species.

**cohesion** The tendency of substances (such as water molecules) to cling to each other (because of hydrogen bonding in the case of water molecules). Compare *adhesion*.

**collagen** A protein produced by the connective tissues of animals.

**collenchyma** A plant tissue made up of living cells that provide mechanical reinforcement in young, growing parts of the plant, such as tender stems and leaf stalks. Compare *parenchyma* and *sclerenchyma*.

**colon** Also called *large intestine*. The portion of the digestive system in which undigested material is prepared for release from the body.

**commensalism** An interaction between two species in which one partner benefits while the other is neither helped nor harmed. Compare *competition*, *exploitation*, and *mutualism*.

**communication** A type of behavior that allows one individual to exchange information with another, thereby making it possible for an animal to coordinate its activities with those of other individuals.

**community** An association of populations of different species that live in the same area.

**compact bone** The hard, white outer region of vertebrate bone. Compare *spongy bone*.

**competition** An interaction between two species in which each has a negative effect on the other. Compare *commensalism*, *exploitation*, and *mutualism*.

**competitive exclusion** Competition in which one species is so superior as a competitor that it ultimately drives the other species extinct.

**complement protein** A type of protein that circulates in the blood plasma and concentrates at sites of tissue damage, stimulating macrophage and neutrophil production and activity.

**complex trait** A genetic trait whose pattern of inheritance cannot be predicted by Mendel's laws of inheritance.

**compound eye** An eye, common in insects, that consists of many individual light-receiving units, each with its own lens and small cluster of photoreceptors. Honeybees use their compounds eyes to see patterns of color on flowers. Compare *simple eye*.

**conduction** The transfer of heat by direct contact between two materials. Compare *convection*, *evaporation*, and *radiation* (definition 2).

**cone** A type of photoreceptor cell in the vertebrate eye that detects color and functions best in bright light. Compare *rod*.

**consumer** Also called *heterotroph*. An organism that obtains its energy by eating other organisms or their remains. Consumers include herbivores, carnivores, and decomposers. Compare *producer*.

**continental drift** Also called *plate tectonics*. The movement of Earth's continents over time.

**contraceptive** Any means of preventing pregnancy.

**control group** A group of participants in an experiment that are subjected to the same environmental conditions as the *treatment group(s)*, except that the factor or factors being tested in the experiment are omitted.

**controlled experiment** An experiment in which the researcher measures the value of the dependent variable for two groups of study subjects that are comparable in all respects except that one group is exposed to a systematic change in the independent variable and the other group is not.

**convection** The physical movement of heat in air or water. Compare *conduction*, *evaporation*, and *radiation* (definition 2).

**convection cell** A large and consistent atmospheric circulation pattern in which warm, moist air rises and cool, dry air sinks. Earth has four stable giant convection cells (two in tropical regions and two in polar regions) and two less stable cells (located in temperate regions).

**convergent evolution** Evolutionary change that occurs when natural selection causes distantly related organisms to evolve similar structures in response to similar environmental challenges.

**cork cambium** A type of lateral-meristem tissue that contributes to secondary growth by dividing to produce tough dermal tissues (including the outer surface of bark). Compare *vascular cambium*.

**corpus luteum (pl. corpora lutea)** The cells of the ruptured follicle that remain behind in the ovary after ovulation to produce the hormone progesterone.

**correlation** A statistical relation indicating that two or more phenomena behave in an interrelated manner. Correlation does not establish causation.

**cotyledon** A food-storing organ that is part of the embryo that lies inside a seed.

**covalent bond** A strong chemical linkage between two atoms that is based on the sharing of electrons. Compare *ionic bond*.

**Cretaceous extinction** A mass extinction that occurred 65 million years ago, wiping out many marine invertebrates and terrestrial plants and animals, including the last of the dinosaurs.

**crista (pl. cristae)** A fold in the inner mitochondrial membrane.

**crossing-over** A physical exchange of chromosomal segments between paired paternal and maternal members of homologous chromosomes. Crossing-over takes place during prophase of meiosis I.

**crustacean** Any of a group of aquatic arthropods that are especially diversified in the marine environment. Shrimp, lobster, and crabs are among the most familiar crustaceans.

**cryosurgery** Surgical procedures that make use of extremely cold temperatures to kill abnormal cells, usually precancerous cells or cancers that are confined to a small region, such as the cervix.

**cuticle** A waxy layer that covers aboveground plant parts, helping to prevent water loss and to keep enemies, such as fungi, from invading the plant.

**cynodont** A member of a group of mammal-like reptiles from which the earliest mammals arose, roughly 220 million years ago.

**cytokine** Any of a class of small protein cells that participate in inflammation.

**cytokinesis** The stage of the cell cycle following mitosis, during which the cell physically divides into two daughter cells.

**cytokinin** Any of a group of plant hormones that are necessary for cell division and promote shoot formation.

**cytoplasm** The contents of a cell enclosed by the plasma membrane but, in eukaryotes, excluding the nucleus. Compare *cytosol*.

**cytoskeleton** A complex network of protein filaments found in the cytosol of eukaryotic cells. The cytoskeleton maintains cell shape and is necessary for cell division and movement.

**cytosol** The water-based fluid component of the cytoplasm. In eukaryotes, the cytosol consists of all the contents of a cell enclosed by the plasma membrane, but excluding all organelles. Compare *cytoplasm*.

**cytotoxic T cell** A lymphocyte that binds to cells that have been infected by a virus and helps destroy them.

**data (s. datum)** Information. Typical data answer questions such as where, when, or how much.

**decomposer** An organism that breaks down dead tissues into simple chemical components, thereby returning nutrients to the physical environment.

**defensin** Any of a variety of small chains of amino acids, secreted by skin cells, that can destroy many types of bacteria, and even certain types of viruses, by creating holes in their lipid layers.

**deletion** In genetics, a mutation in which one or more nucleotides are removed from the DNA sequence of a gene, or a piece breaks off from a chromosome and is lost. Compare *insertion*.

**denaturation** The destruction of a protein's three-dimensional structure, resulting in loss of protein activity.

**dendrite** An extension of a neuron that receives signals from adjacent cells.

**density-dependent** Of or referring to a factor, such as food shortage, that limits the growth of a population more strongly as the density of the population increases. Compare *density-independent*.

**density-independent** Of or referring to a factor, such as weather, that can limit the size of a population but does not act more strongly as the density of the population increases. Compare *density-dependent*.

**deoxyribonucleic acid** See *DNA*.

**dependent variable** Also called *responding variable*. Any variable that responds, or could potentially respond, to changes in the *independent variable*.

**depolarization** A loss or reduction in the electrical charge difference that exists across the plasma membrane in many cell types. In neurons, depolarization can trigger an action potential if the charge difference is sufficiently reduced. Compare *polarization*.

**dermal hair** Also called *trichome*. Hair that is found on the outer covering of leaves, stems, and fruits of many plants. The fuzz on a peach is an example.

**dermal tissue** The tissue type in plants that forms the outer covering of the body. Compare *ground tissue* and *vascular tissue*.

**desert** A terrestrial biome dominated by plants that grow in regions with low precipitation, usually 25 centimeters per year or less.

**desmosome** See *anchoring junction*.

**detritivore** A consumer that eats the organic matter left over as dead organisms decompose into small particles. Examples include earthworms and millipedes. Compare *carnivore*, *herbivore*, and *omnivore*.

**deuterostome** Any of a group of animals, including sea stars and vertebrates, in which the second opening to develop in the early embryo becomes the mouth. Compare *protostome*.

**development** The sequence of predictable changes that occur over the life cycle of an organism as it grows and matures to the reproductive stage.

**diabetes** A disease in which not enough glucose is moved from the blood into the cells. See also *type 1 diabetes* and *type 2 diabetes*.

**diastole** The relaxation phase of the cardiac cycle. The diastolic pressure corresponds to the second and lower of the two values reported in a blood pressure reading. Compare *systole*.

**dicots** One of two main groups of flowering plants, characterized by the presence of two cotyledons in the seed, netlike leaf veins, vascular tissues arranged in a ring, and (typically) a main root with many side branches. Familiar examples of dicots include magnolias, dandelions, roses, maples, and oak trees. Compare *monocots*.

**dietary fiber** Indigestible plant polysaccharides, such as cellulose. Dietary fiber is important for human health because it slows the digestion of carbohydrates, providing a sustained source of energy and reducing hunger cravings. Whole grains are a good source of dietary fiber.

**differentiation** See *cell differentiation*.

**diffusion** The passive movement of a substance from an area of high concentration of that substance to an area of low concentration. See also *facilitated diffusion* and *simple diffusion*.

**digestion** The chemical breakdown of food.

**digestive system** The organ system in animals that consists of a tubular passageway through which food generally moves in a one-way flow and is converted into a form that the animal's body can absorb and use.

**dihybrid** An individual that is heterozygous for two traits. Compare *monohybrid*.

**diploid** Of or referring to a cell or organism that has two complete sets of homologous chromosomes (2*n*). Compare *haploid* and *polyploid*.

**directional selection** A type of natural selection in which individuals with one extreme of an inherited characteristic have an advantage over other individuals in the population, as when large individuals produce more offspring than do small and medium-sized individuals. Compare *disruptive selection* and *stabilizing selection*.

**disaccharide** A molecule made up of two *monosaccharides*. Some common examples are sucrose, lactose, and maltose. Compare *polysaccharide*.

**disruptive selection** A type of natural selection in which individuals with either extreme of an inherited characteristic have an advantage over individuals with an intermediate phenotype, as when both small and large individuals produce more offspring than do medium-sized individuals. Compare *directional selection* and *stabilizing selection*.

**distribution** The geographic area over which a species is found.

**diversity** The composition of an ecological community, which has two components: the number of different species that live in the community and the relative abundances of those species.

**DNA** Deoxyribonucleic acid, a double-stranded molecule consisting of two spirally wound polymers of nucleotides that store genetic information, including the information needed to synthesize proteins. Each nucleotide in DNA is composed of the sugar deoxyribose, a phosphate group, and one of four nitrogenous bases: adenine, cytosine, guanine, or thymine. Compare *RNA*.

**DNA cloning** Also called *gene cloning*. The introduction of recombinant DNA into a host cell (usually a bacterium), followed by the copying and propagation of the introduced DNA in the host cells and all its offspring.

**DNA fingerprinting** The use of DNA analysis to identify individuals and determine the relatedness of individuals.

**DNA hybridization** An experimental procedure in which DNA from two different sources bind to each other through complementary base pairing.

**DNA polymerase** The key enzyme that cells use to replicate their DNA. In DNA technology, DNA polymerase is used in the polymerase chain reaction to make many copies of a gene or other DNA sequence.

**DNA primer** A short segment of DNA used in PCR amplification that is designed to pair with one of the two ends of the gene being amplified.

**DNA probe** A short sequence of DNA (usually tens to hundreds of bases long) that is labeled with a radioactive or fluorescent tag and then allowed to hybridize with a target DNA sample, with the objective of determining whether the two share sequence similarity.

**DNA repair** A three-step process in which damage to DNA is repaired. Damaged DNA is first recognized, then removed, and then replaced with newly synthesized DNA.

**DNA replication** The duplication, or copying, of a DNA molecule. DNA replication begins when the hydrogen bonds connecting the two strands of DNA are broken, causing the strands to unwind and separate. Each strand is then used as a template for the construction of a new strand of DNA.

**DNA technology** The set of techniques that scientists use to manipulate DNA.

**domain** A level of biological classification above the kingdom. The three domains are Bacteria, Archaea, and Eukarya.

**dominant** Of or referring to an allele that determines the phenotype (dominant phenotype) when it is paired with a *recessive* allele in a heterozygous individual.

**dormancy** In plants, a state of arrested growth. Seed dormancy occurs when a mature plant embryo (inside of a seed) halts its growth, and it ends when the embryo is stimulated by favorable conditions to begin growing again. In some plant species, bud dormancy occurs during winter and ends when the plant resumes growth in the spring.

**double bond** The covalent sharing of two pairs of electrons between two atoms. Compare *triple bond*.

**double fertilization** The two fertilization events that occur when a flowering plant reproduces sexually. In one of these events, a sperm cell fertilizes the egg, producing a diploid cell (zygote) that will develop into the embryo that lies within each seed; in the other, a second sperm cell fertilizes a pair of polar nuclei within the embryo sac, producing the endosperm that will nourish the developing embryo.

**doubling time** The time it takes a population to double in size. Doubling time can be used as a measure of how fast a population is growing.

**duplication** In genetics, a type of mutation in which an extra copy of a gene or DNA fragment appears alongside the original, increasing the length of the chromosome.

**eardrum** A delicate membrane at the end of the auditory canal that converts pressure changes in the environment into physical changes in the structures found within the middle ear.

**early wood** Wood of a tree that is produced in spring and early summer. Early wood is light in color because it contains tracheids and vessels with a wide diameter (and therefore large air spaces in their hollow interiors) and relatively thin cell walls. Compare *late wood*.

**ECG** See *electrocardiogram*.

**ecdysis** See *molting*.

**ecology** The scientific study of interactions between organisms and their environment.

**ecosystem** A community of organisms, together with the physical environment in which the organisms live. Global patterns of air and water circulation link all the world's organisms into one giant ecosystem, the *biosphere*.

**ectoderm** The cell layer in animal development that forms the exterior of the gastrula and gives rise to the epidermis and nervous system. Compare *endoderm* and *mesoderm*.

**ectotherm** An organism that relies on environmental heat for most of its heat input. Compare *endotherm*.

**egg** The large, immobile, haploid gamete that is produced by sexually reproducing female eukaryotes. Compare *sperm*. See also *ovum*.

**EKG** See *electrocardiogram*.

**electrocardiogram (ECG, EKG)** A recording of the signals from the sinoatrial and atrioventricular nodes of the heart.

**electron** A negatively charged particle found in atoms. Each atom of a particular element contains a characteristic number of electrons. Compare *proton* and *neutron*.

**electron shell** One of several defined volumes of space in which electrons move around the nucleus of an atom.

**electron transport chain (ETC)** A group of membrane-associated proteins and other molecules that can both accept and donate electrons. The transfer of electrons from one ETC component to another releases energy that can be used to drive protons across a membrane, and ultimately, in both chloroplasts and mitochondria, to manufacture ATP.

**element** In reference to chemicals, a pure substance made up of only one type of atom, with a characteristic number of protons. The physical world is made up of 92 naturally occurring elements.

**EM** See *extracellular matrix*.

**embryo** The early stage of development in plants and animals, extending from the zygote to the early development of organs and organ systems. Compare *fetus*.

**embryo sac** The haploid, multicellular, female gametophyte of seed plants, in which eggs are produced. Compare *pollen*.

**embryonic stem cell** A pluripotent stem cell in or derived from the inner cell mass, which exists only at the blastocyst stage in mammalian embryos. Compare *adult stem cell*.

**endocrine disrupter** A chemical that interferes with hormone function to produce negative effects, such as reduced fertility, developmental abnormalities, immune system dysfunction, and increased risk of cancer.

**endocrine gland** An organ or discrete tissue that produces hormones and releases them directly into the circulatory system.

**endocrine system** The organ system in animals that produces signaling molecules called hormones and releases them directly into the circulatory system without using ducts.

**endocytosis** A process by which a section of a cell's plasma membrane bulges inward as it envelops a substance outside of the cell, eventually breaking free to become a closed vesicle within the cytoplasm. Compare *exocytosis*.

**endoderm** The cell layer in animal development that forms the interior of the gastrula and gives rise to the gut and associated organs. Compare *ectoderm* and *mesoderm*.

**endodermis** A layer of ground tissue in plants that surrounds the vascular tissue in the root and controls the pathway by which water and dissolved substances enter the vascular tissue system.

**endometrium (pl. endometria)** The layer of cells lining the wall of the mammalian uterus.

**endoplasmic reticulum (ER)** An organelle composed of many interconnected membrane sacs and tubes. The ER is the main site for the synthesis of lipids, and certain types of proteins, in eukaryotic cells.

**endoskeleton** An internal framework of hard material (such as bone) that is surrounded by the soft tissues it supports. Compare *exoskeleton*.

**endosperm** The nutritive tissue of plants that supports the growth of an embryo within a seed, or the growth of a seedling following seed germination.

**endosymbiont theory** The theory that eukaryotic cells are descended from a predatory ancestral eukaryote that engulfed prokaryotic cells that survived, instead of being digested for food, and evolved into endosymbionts (intracellular symbiotic partners).

**endotherm** An organism that generates metabolic heat in order to warm itself, instead of depending mainly on heat gain from the environment. Compare *ectotherm*. See also *homeotherm*.

**endotoxin** A component of bacterial cell walls that triggers illness. Endotoxins can produce fever, blood clots, and toxic shock. Compare *exotoxin*.

**energy** The capacity of any object to do work, which is the capacity to bring about a change in a defined system.

**energy carrier** A molecule that can store energy and donate it to another molecule or to a chemical reaction. ATP is the most commonly used energy carrier in living organisms.

**energy pyramid** A hierarchical representation of the amounts of energy available to organisms in an ecosystem in which each level corresponds to a step in a food chain (a trophic level).

**enzyme** A macromolecule, usually a protein, that acts as a catalyst, speeding the progress of chemical reactions. Nearly all chemical reactions in living organisms are catalyzed by enzymes.

**epinephrine** Also called *adrenaline*. A hormone produced by the adrenal glands that regulates the fight-or-flight response to stressful situations in conjunction with *norepinephrine*.

**epistasis (pl. epistases)** A gene interaction in which the phenotypic effect of the alleles of one gene depends on which alleles are present for another, independently inherited gene.

**epithelial tissue** Also called *epithelium*. A type of animal tissue that is specialized to regulate how the animal's internal environment exchanges water, solutes, and heat directly with the external environment. Epithelial tissue lines body cavities and covers the surfaces of many organs, including the skin.

**ER** See *endoplasmic reticulum*.

**erythrocyte** See *red blood cell*.

**esophagus (pl. esophagi)** The portion of the digestive system that conveys food from the mouth to the stomach.

**essential amino acid** An amino acid that a consumer cannot synthesize and must therefore obtain from its food.

**estrogen** Any of a class of hormones that is one of three major types produced by the gonads in both males and females. Among many functions, estrogens play a role in determining female characteristics, such as the development of breast tissues, and in the release of mature "eggs" (ova) from the ovaries. Compare *androgen* and *progestogen*.

**estuary** An ecosystem of the marine biome defined as a region where a river empties into the sea.

**ETC** See *electron transport chain*.

**ethylene** A gaseous plant hormone that stimulates the ripening of some types of fruit.

**Eukarya** One of the three domains of life, encompassing the eukaryotes. Four kingdoms are included: Animalia, Plantae, Fungi, and Protista. Compare *Archaea* and *Bacteria*.

**eukaryote** A single-celled or multicellular organism in which each cell has a distinct nucleus and cytoplasm. All organisms other than the Bacteria and the Archaea are eukaryotes. Compare *prokaryote*.

**eukaryotic flagellum** A hairlike structure, found in eukaryotes, that propels the cell or organism through a whiplike action, with waves passing from the base to the tip of the flagellum. Compare *prokaryotic flagellum*.

**eusociality** The cooperative behaviors that organize life in complex social groups.

**eustachian tube** A thin tube that connects the middle ear to the throat, allowing mammals to equalize the pressure in these two structures.

**eutherian** Any of a group of mammals whose young are nourished inside the mother's body through the placenta and are therefore born in a relatively well-developed state. Most mammals, including humans, are eutherians. Compare *marsupial* and *monotreme*.

**eutrophication** A process in which enrichment of water by nutrients (often from sewage or runoff from fertilized agricultural fields) causes bacterial populations to increase and oxygen concentrations to decrease.

**evaporation** The conversion of a substance from a liquid to a gas. Because this process requires a lot of energy, evaporation is an effective cooling device for animals, transferring heat from water on the body surface to the air in the surrounding environment. Compare *conduction*, *convection*, and *radiation* (definition 2).

**evolution** See *biological evolution*.

**evolutionary tree** A diagrammatic representation showing the order in which different lineages arose, with the lowest branches having arisen first.

**exchange pool** In reference to nutrients, a source such as the soil, water, or air where nutrients are available to producers.

**excretory system** See *urinary system*.

**exocytosis** A process by which a vesicle approaches and fuses with the plasma membrane of a cell, thereby releasing its contents into the cell's surroundings. Compare *endocytosis*.

**exon** A DNA sequence within a gene that encodes part of a protein. Each exon codes for a stretch of amino acids. Compare *intron*.

**exoskeleton** An external framework of stiff or hard material that surrounds the soft tissues of an animal. Compare *endoskeleton*.

**exotoxin** A poison that a bacterium releases into its surroundings. Exotoxins kill tissues, and because the bacteria can spread rapidly, the tissue death can be extensive enough to kill a person in a day or two after the symptoms appear. Compare *endotoxin*.

**experiment** A controlled manipulation of nature designed to test a hypothesis.

**experimental group** See *treatment group*.

**exploitation** An interaction between two species in which one species benefits (the consumer) and the other species is harmed (the food organism). Exploitation includes the killing of prey by predators, the eating of plants by herbivores, and the harming or killing of a host by a parasite or pathogen. Compare *commensalism*, *competition*, and *mutualism*.

**exploitative competition** A type of competition in which species compete indirectly for shared resources, with each reducing the amount of a resource available to the other. Compare *interference competition*.

**exponential growth** A type of rapid population growth in which a population increases by a constant proportion from one generation to the next. Exponential growth is represented graphically by a J-shaped curve. Compare *logistic growth*.

**extracellular matrix (EM)** A coating of nonliving material, released by the cells of multicellular organisms, that often helps hold those cells together.

**extremophile** An organism, such as many archaeans, that lives in an extreme environment, such as in a boiling hot geyser or on salted meat.

**F₁ generation** The first generation of offspring in a genetic cross. Compare *P generation*.

**F₂ generation** The second generation of offspring in a genetic cross. Compare *P generation*.

**facilitated diffusion** The passive transmembrane movement of a substance with the assistance of membrane transport proteins. Compare *simple diffusion*.

**fallopian tube** See *oviduct*.

**family** In reference to biological classification systems, the level in the Linnaean hierarchy above genus and below order.

**fatty acid** An organic molecule with a long hydrocarbon chain and a hydrophilic head group. Fatty acids are found in phospholipids, glycerides such as triglycerides, and waxes.

**feces** The solid waste product of animal digestion, which consists mostly of indigestible food and bacteria.

**feedback loop** A loop within a process in which one step can "reach back" to control an earlier step in such a way that the output of the process as a whole is changed. Feedback loops can turn off an entire process, or reduce or enhance its output.

**fermentation** A series of catabolic reactions that produce small amounts of ATP through glycolysis and can function without oxygen. In most fermentation pathways, pyruvate from glycolysis is converted to other organic molecules, such as ethanol and carbon dioxide, or lactic acid.

**fertilization** The fusion of two different haploid gametes (egg and sperm) to produce a diploid zygote (the fertilized egg).

**fetus** The stage of animal development extending from the time when most of the major organs and tissues are identifiable to birth. Compare *embryo*.

**fiber** In plant tissues, a type of sclerenchyma cell that is one of the longest cell types found in plants.

**fibrous root** A tuftlike mass of roots without a main, dominant root among them. Compare *taproot*.

**filtrate** The liquid that passes through a filtration system, such as the nephron in a vertebrate kidney, to appear on the other side of the screen (in the tubule, in the case of the nephron).

**filtration** The process of removing certain substances from a liquid by passing it through a filter, such as a semipermeable membrane.

**first law of thermodynamics** Also called *law of conservation of energy*. The law stating that energy can be neither created nor destroyed, but only transformed or transferred from one molecule to another. Compare *second law of thermodynamics*.

**fixation** In genetics, the removal of all alleles within a population at a particular genetic locus except one. The allele that remains has a frequency of 100 percent.

**fixed behavior** A predictable response to a particular, often simple stimulus that does not involve learning. Compare *learned behavior*.

**flagellum (pl. flagella)** A long extension from the cell that is lashed or rotated to enable that cell to move.

**flower** A specialized reproductive structure that is characteristic of the plant group known as the angiosperms, or flowering plants.

**flowering plants** See *angiosperms*.

**fluid mosaic model** The concept of the plasma membrane as a phospholipid bilayer containing a variety of other lipids and embedded proteins, some of which can move laterally in the plane of the membrane.

**follicle-stimulating hormone (FSH)** A hormone produced by the pituitary gland that works with *luteinizing hormone* to regulate sperm development in males and the menstrual cycle in females.

**food chain** A single sequence of feeding relationships describing who eats whom in a community. Compare *food web*.

**food web** A summary of the movement of energy through a community. A food web is formed by connecting all of the *food chains* in the community.

**forebrain** The part of the brain that consists of the thalamus, the cerebrum, and the hypothalamus and is involved in many of the higher-order functions (such as thought and memory) of the brain. Compare *hindbrain* and *midbrain*.

**fossil** Preserved remains or an impression of a formerly living organism. Fossils document the history of life on Earth, showing that many organisms from the past were unlike living forms, that many organisms have gone extinct, and that life has evolved through time.

**founder effect** A genetic bottleneck that results when a small group of individuals from a larger source population establishes a new population far from the original population.

**fovea** A region of especially densely packed photoreceptors in the retina that allows for the formation of sharp images.

**frameshift** In genetics, the large change in coding information that results when the deletion or insertion in a gene sequence is not a multiple of three base pairs (a codon). The amino acid sequence of the protein that is translated from such a gene is severely altered in most cases, and protein function is typically lost.

**fruit** A mature ovary surrounding one or more seeds. The ovary wall becomes the fruit wall, which is often juicy and sweet and therefore attractive to animals, which then function as seed dispersal agents.

**FSH** See *follicle-stimulating hormone*.

**functional group** A specific cluster of covalently bonded atoms, with distinctive chemical properties, that forms a discrete subgroup in a variety of larger molecules.

**Fungi** The kingdom of Eukarya that is made up of absorptive heterotrophs (consumers that absorb their food after digesting it externally). Fungi include mushroom-producing species, yeasts, and molds, most of which make their living as decomposers.

**G₀ phase** A resting state during which the cell withdraws from the cell cycle before S phase. $G_0$ cells are not competent to divide, unless they receive and respond to signals that direct them to reenter the cell cycle and proceed through S phase.

**G₁ phase** The stage of the cell cycle that follows mitosis and precedes S phase. The cell grows in size during $G_1$ phase and makes a commitment to enter S phase if it receives and responds to cell division signals.

**G₂ phase** The stage of the cell cycle that follows S phase and precedes mitosis. $G_2$ phase serves as a checkpoint ensuring that mitosis will not be launched under inappropriate conditions (such as inadequate nutrient supply, DNA damage, or incomplete DNA replication).

**gallbladder** A small sac where bile secreted by the liver is stored before being released into the small intestine.

**gamete** A haploid sex cell that fuses with another sex cell during fertilization. Eggs and sperm are gametes. Compare *somatic cell*.

**gametophyte** The haploid, multicellular tissue in plants that is created by the mitotic divisions of a spore (itself, a product of meiosis). Specialized cells in the gametophyte differentiate to produce sperm or egg cells. Compare *sporophyte*.

**ganglion (pl. ganglia)** A structure consisting of densely packed neuronal cell bodies that allows the integration of signals.

**gap junction** A cytoplasmic channel, created by a narrow cylinder of proteins, that directly connects two animal cells and allows the passage of ions and small molecules between them. Compare *anchoring junction* and *tight junction*.

**gastrula** The stage of vertebrate development that results from the rearrangement of the three tissue layers (endoderm, mesoderm, and ectoderm) during gastrulation.

**gastrulation** The movement of the three tissue layers (endoderm, mesoderm, and ectoderm) during animal development to the positions appropriate for the tissues to which these layers give rise.

**gel electrophoresis** A procedure in which DNA fragments are placed in a gelatin-like substance (a gel) and subjected to an electrical charge, which causes the fragments to move through the gel. Small DNA fragments move farther than large DNA fragments, causing the fragments to separate by size.

**gene** The smallest unit of DNA that governs a genetic characteristic and contains the code for the synthesis of a protein or an RNA molecule. Genes are located on chromosomes.

**gene expression** The creation of a functional product, such as a specific protein or RNA, using the coding information stored in a gene. Gene expression is the means by which a gene influences the cell or organism in which it is found.

**gene cloning** See *DNA cloning*.

**gene flow** The exchange of alleles between different populations.

**gene pool** The sum of all the genetic information carried by all the individuals in a population.

**gene therapy** A treatment approach that seeks to correct genetic disorders by repairing the genes responsible for the disorder.

**gene-for-gene recognition** In plants, a specific immune response in which many genes, which collectively have hundreds of alleles (known as *R* alleles), provide resistance against specific attackers.

**genetic bottleneck** A drop in the size of a population that results in low genetic variation or causes harmful alleles to reach a frequency of 100 percent in the population.

**genetic carrier** A heterozygous individual (*Aa*) that carries the allele for a recessive genetic disorder but, because the allele is recessive, does not get the disorder.

**genetic code** The code that specifies how information in mRNA is translated to create the specific sequence of amino acids found in the protein encoded by that mRNA. The genetic code consists of all possible three-base combinations (codons) of each of the four nitrogenous bases found in RNA. Of the 64 possible codons, 60 specify a particular amino acid, 3 serve as a "stop translation" signal, and 1 (AUG) acts as a "start translation" signal.

**genetic cross** A controlled mating experiment, usually performed to examine how a particular trait may be inherited.

**genetic drift** The natural process in which chance events cause certain alleles to increase or decrease in a population. The genetic makeup of a population undergoing genetic drift changes at random over time, rather than being shaped in a nonrandom way by natural selection.

**genetic engineering** The process in which a DNA sequence (often a gene) is isolated, modified, and inserted back into an individual of the same or a different species. Genetic engineering is commonly used to change the performance of the genetically modified organism, as when a crop plant is engineered to resist attack from an insect pest.

**genetic linkage** The situation in which different genes that are located close to one another on the same chromosome are inherited together; that is, they do not follow Mendel's law of independent assortment.

**genetic recombination** The creation of new groupings of alleles through the breaking and re-joining of different DNA segments, as in the crossing-over that takes place between paired homologues during meiosis I.

**genetic trait** An inherited characteristic of an organism, such as its size, color, or behavior.

**genetic variation** The allelic differences among the individuals of a population.

**genetically modified organism (GMO)** Also called *genetically engineered organism (GEO)* or *transgenic organism*. An individual into which a modified gene or other DNA sequence has been inserted, typically with the intent of improving a particular aspect of the recipient organism's performance.

**genetics** The scientific study of the inheritance of characteristics encoded by DNA.

**genome** All the DNA of an organism, including all its genes; in eukaryotes, the term refers to the DNA in a haploid set of chromosomes, such as that found in a sperm or egg.

**genotype** The allelic makeup that is responsible for a particular *phenotype* displayed by an individual.

**genotype frequency** The proportion (percentage) of a particular genotype in a population.

**genus (pl. genera)** In reference to biological classification systems, the level in the Linnaean hierarchy above species and below family.

**GEO** See *genetically modified organism*.

**geographic isolation** The physical separation of populations from one another by a barrier such as a mountain chain or a river. Geographic isolation often causes the formation of new species, as when populations of a single species become physically separated from one another and then accumulate so many genetic differences that they become reproductively isolated from one another. Compare *reproductive isolation*.

**germ line cell** A cell that will develop into a gamete precursor cell. Germ cells remain separate from all other cells of the embryo, and they stay unspecialized until the reproductive organs develop, at which time they migrate to and take up residence inside the egg-producing ovaries or sperm-producing testes.

**GH** See *growth hormone*.

**gibberellin** Any of a group of plant hormones that stimulate the elongation of stems.

**gizzard** The portion of the digestive system that breaks down food by grinding it against rocks or sand collected from the environment.

**gland** A group of cells that manufactures and releases biomolecules into the blood, into body cavities, or on the body surface.

**global change** Worldwide change in the environment. There are many causes of global change, including climate change caused by the movement of continents and changes in land and water use by humans.

**global warming** A worldwide increase in temperature. Earth appears to be entering a period of global warming caused by human activities—specifically, by the release of large quantities of greenhouse gases such as carbon dioxide into the atmosphere.

**glomerulus (pl. glomeruli)** The tiny capillary bed through which blood is filtered from the circulatory system into the tubules of the vertebrate kidney.

**glucagon** A hormone produced by the pancreas that induces cells to release glucose from storage depots within the cell. Compare *insulin*.

**glucose** A monosaccharide that is the primary metabolic fuel in most cells.

**glycogen** The storage carbohydrate in animals. Found in humans primarily in the liver and skeletal muscles, glycogen is a polysaccharide and is structurally similar to starch, the storage carbohydrate of plants.

**glycolysis** A series of catabolic reactions that splits glucose to produce pyruvate, which is then used in fermentation in the cytosol or degraded further in the mitochondrion. The glycolytic breakdown of one molecule of glucose yields two molecules of the energy carrier ATP. Glycolysis is the first of the three major phases of cellular respiration, preceding the Krebs cycle and oxidative phosphorylation.

**GMO** See *genetically modified organism*.

**Golgi apparatus** An organelle composed of flattened membrane sacs that routes proteins and lipids to various parts of the eukaryotic cell.

**gonad** An animal sex gland that produces gametes and sex hormones. See also *ovary* and *testis*.

**gonadotropin** Either of two hormones—luteinizing hormone and follicle-stimulating hormone—that are produced by the pituitary gland and regulate the development and function of the reproductive organs.

**grandmother hypothesis** A hypothesis stating that menopause in older women is an adaptive trait because it contributes to the survival of grandchildren by eliminating the possibility of childbirth and enabling older women to further their genetic interests by nurturing their grandchildren.

**grasp reflex** A reflex displayed by newborns that enables them to hold on tightly to anything that will fit in their tiny fists.

**grassland** A terrestrial biome dominated by grasses. Grasslands occur in relatively dry regions, often with cold winters and hot summers.

**greenhouse effect** The increase in Earth's temperature when absorbed heat that is reemitted by greenhouse gases becomes trapped because it lacks the energy to escape into outer space.

**greenhouse gas** Any of several gases in Earth's atmosphere that let in sunlight but trap heat. Examples include carbon dioxide, water vapor, methane, and nitrous oxide.

**ground tissue** The tissue in plants that is part of neither the *dermal tissue* nor the *vascular tissue*. Ground tissue constitutes the bulk of the tissue in nonwoody organs and includes all of the photosynthetic tissue in leaves.

**growth factor** Any of a class of signaling molecules that play an especially important role in initiating and maintaining cell proliferation in the human body.

**growth hormone (GH)** A hormone produced by the pituitary gland that plays a key role in regulating the growth of bones and muscles.

**guard cell** A specialized cell type found on leaves and stems of plants. Guard cells can inflate and deflate in response to their water content, and through this action they regulate the opening and closing of stomata, thereby controlling the rates at which $CO_2$ is brought into the plant and $O_2$ and water are lost from the plant. See also *stoma*.

**gymnosperms** One of four major groups of plants. Gymnosperms are represented by conifers such as pine or spruce trees. They have vascular tissues and seeds but lack flowers and fruits. Compare *angiosperms* and *bryophytes*.

**habitat** A characteristic place or type of environment in which an organism lives.

**haploid** Of or referring to a cell or organism that has only one set ($n$) of homologous chromosomes, such that only one member of each homologous pair (either the paternal or maternal member) is represented in that set. Compare *diploid* and *polyploid*.

**heart** The muscular pump that pushes blood through the animal circulatory system.

**heart rate** The number of heartbeats per minute.

**heartbeat** The series of events that constitute one cardiac cycle.

**heartwood** Older, centrally located secondary xylem tissue that provides support to a tree but has become clogged over time and hence no longer functions in the transport of materials. Compare *sapwood*.

**heat capacity** The heat energy required to raise the temperature of a specific volume of water by a fixed amount.

**heat energy** Also called *thermal energy*. Energy inherent in the random motion of particles in a system that can be transferred to other particles in the system; the portion of the total energy of a particle of matter that can flow from one particle of matter to another.

**helper T cell** A type of lymphocyte that stimulates cytotoxic T cells and B cells.

**hemoglobin** An oxygen-binding pigment, found in most vertebrates, that is used to carry oxygen from the gas exchange surfaces to the tissues.

**herbivore** A consumer that relies on living plant tissues for nutrients. Compare *carnivore*, *omnivore*, and *detritivore*.

**hermaphrodite** An individual that can produce both eggs and sperm.

**heterotroph** See *consumer*. Compare *autotroph*.

**heterozygous** Of or referring to an individual that carries one copy of each of two different alleles (for example, an *Aa* individual). Compare *homozygous*.

**hindbrain** The part of the brain that includes the medulla oblongata, pons, and cerebellum and coordinates breathing rhythms and balance. Compare *forebrain* and *midbrain*.

**histamine** A protein released by mast cells that causes blood vessels to dilate and capillaries to become more porous.

**homeostasis** The process of maintaining appropriate and constant conditions inside cells.

**homeotherm** An endotherm that maintains a near-constant internal temperature.

**hominid** Any member of the ape family. Compare *hominin*.

**hominin** Any member of the branch of the ape family (the *hominids*) that includes humans and our now extinct relatives.

**homologous** Of or referring to a characteristic shared by two groups of organisms because of their descent from a common ancestor. Compare *analogous*.

**homologous chromosomes** or **homologous pair** Also called *homologues*. A pair of matched chromosomes in diploid cells, one of which is inherited from the individual's female parent and the other from its male parent.

**homozygous** Of or referring to an individual that carries two copies of the same allele (for example, an *AA* or an *aa* individual). Compare *heterozygous*.

**horizontal gene transfer** See *lateral gene transfer*.

**hormone** A signaling molecule released in very small amounts into the circulatory system of an animal, or into a variety of tissues in a plant, that affects the functioning of target tissues.

**hormone therapy** Manipulation of the hormone environment in the body to stop or slow cancer cells. This treatment is used in some hormone-responsive cancers, which include some types of breast and prostate cancer.

**host** The individual, or organism, in which a particular parasite or pathogen lives.

**housekeeping gene** A gene that has an essential role in the maintenance of cellular activities and is expressed by most cells in the body.

**hybrid** An offspring that results when two different species, or two different varieties or genotypic lines, are mated.

**hybridize** To cause hybrid offspring to be produced.

**hydrogen bond** A weak electrical attraction between a hydrogen atom that has a slight positive charge and another atom with a slight negative charge.

**hydrophilic** Of or referring to substances, both salts and molecules, that interact freely with water. Hydrophilic molecules dissolve easily in water but not in fats or oils. Compare *hydrophobic*.

**hydrophobic** Of or referring to molecules or parts of molecules that do not interact freely with water. Hydrophobic molecules dissolve easily in fats and oils but not in water. Compare *hydrophilic*.

**hydrostat** Also called *hydrostatic skeleton*. A support structure that enables movement through the squeezing action of muscle layers surrounding a fluid-filled body cavity.

**hypertonic solution** A solution that has a higher solute concentration than the cytosol of a cell, causing more water to flow out of the cell than into it. Compare *hypotonic solution* and *isotonic solution*.

**hypha (pl. hyphae)** In fungi, a threadlike absorptive structure that grows through a food source. Mats of hyphae form mycelia, the main bodies of fungi.

**hypothalamus (pl. hypothalami)** A structure at the base of the vertebrate brain that controls the release of hormones by the pituitary gland. Along with the pituitary gland, the hypothalamus helps regulate interactions between the nervous and endocrine systems.

**hypothesis** See *scientific hypothesis*.

**hypotonic solution** A solution that has a lower solute concentration than the cytosol of a cell, causing more water to flow into the cell than out of it. Compare *hypertonic solution* and *isotonic solution*.

**immune memory** The capacity of the adaptive immune system to remember a first encounter with a specific strain of pathogen and to mobilize an especially speedy and precisely targeted response to a repeat infection by the same strain.

**immune system** The organ system in animals, consisting of defensive proteins and specialized immune cells (such as white blood cells in vertebrate animals), that destroys invading pathogens.

**implantation** The process by which the blastocyst attaches to the inner lining of the uterus.

**imprinting** A type of learning in which an offspring forms an association or bond with its parent early in its development.

**incomplete dominance** A type of allelic interaction in which heterozygotes (*Aa* individuals) are intermediate in phenotype between the two homozygotes (*AA* and *aa* individuals) for a particular gene. Compare *codominance*.

**independent assortment of chromosomes** The random distribution of maternal and paternal chromosomes into gametes during meiosis.

**independent variable** Also called *manipulated variable*. The single variable that is manipulated in a typical scientific experiment. Compare *dependent variable*.

**indeterminate growth** The ability of a plant to grow throughout its life, adding new body parts as needed.

**individual** A single organism, usually physically separate and genetically distinct from other individuals.

**induced defense** A defensive response in plants that is directly stimulated by an attack from herbivores.

**induced fit model** A model of substrate-enzyme interaction stating that as a substrate enters the active site, the parts of the enzyme shift about slightly to allow the active site to mold itself around the substrate.

**induced pluripotent stem cell (iPSC)** Any cell, even a highly differentiated cell in the adult body, that has been genetically reprogrammed to mimic the pluripotent behavior of embryonic stem cells.

**inductive signal** A short-range chemical cue that is passed to an embryonic cell from a neighboring cell or from the maternal environment and that "tells" the embryonic cell what to do next.

**inflammation** Part of the nonspecific response to invading pathogens, allergens, or tissue damage. Inflammation is characterized by redness, swelling (due to fluid accumulation), and warmth (local increase in temperature).

**ingestion** The act of taking food into the mouth, or eating.

**innate immune system** Also called *innate immunity*. An internal defense system that is immediately set in motion when the external barriers fail to stop a pathogen. Its response to pathogens is nonspecific, deploying a preset repertoire of defensive actions. Compare *adaptive immune system*.

**inner cell mass** The cluster of cells inside the blastocyst that eventually develops into the embryo and some of the membranes that surround a mammalian embryo and fetus. Compare *trophoblast*.

**inner ear** The portion of the ear where physical movements in the middle ear are converted into the action potentials that vertebrates detect as sound. Compare *middle ear* and *outer ear*.

**inorganic molecule** A molecule that lacks carbon atoms or has no more than one carbon atom. Examples include iron ore, sand, and carbon dioxide. Compare *organic molecule*.

**insertion** In genetics, a mutation in which one or more nucleotides are inserted into the DNA sequence of a gene. Compare *deletion*.

**insulin** A hormone produced by the pancreas that induces cells, especially those of the liver and the skeletal muscles, to increase their uptake of glucose from the blood. Compare *glucagon*.

**insulin-dependent diabetes** See *type 1 diabetes*.

**integumentary system** Skin; the organ system in animals that covers and protects the surface of the body.

**interference competition** A type of competition in which one organism directly excludes another from the use of resources. Compare *exploitative competition*.

**interferon** A chemical released by virus-infected cells that attaches to the plasma membranes of nearby cells, interfering with the ability of the virus to enter and infect those cells.

**intermediate filament** One of a diverse class of ropelike protein filaments that serve as structural reinforcements in the cytoskeleton. Compare *actin filament* and *microtubule*.

**intermembrane space** The space between the inner and outer membranes of a chloroplast or a mitochondrion.

**interneuron** A neuron in the central nervous system that processes signals from sensory neurons and then passes those signals on to other neurons. Compare *motor neuron* and *sensory neuron*.

**interphase** The period of time between two successive mitotic divisions, during which the cell increases in size and prepares for cell division.

**intertidal zone** The part of the coastal region that is closest to the shore, where the ocean meets the land. It extends from the highest tide mark to the lowest tide mark.

**intron** A sequence of nitrogenous bases within a gene that does not specify part of the gene's final protein or RNA product. Enzymes in the nucleus must remove introns from mRNA, tRNA, and rRNA molecules for these molecules to function properly. Compare *exon*.

**inversion** In genetics, a mutation in which a fragment of a chromosome breaks off and returns to the correct place on the original chromosome, but with the genetic loci in reverse order.

**invertebrate** Any animal that is not a chordate possessing a backbone. Compare *vertebrate*.

**ion** An atom or group of atoms that has either gained or lost electrons and therefore has a negative or positive charge.

**ionic bond** A chemical linkage between two atoms that is based on the electrical attraction between positive and negative charges. Compare *covalent bond*.

**iPSC** See *induced pluripotent stem cell*.

**irregular fluctuation** In reference to natural populations, a pattern of population growth in which the number of individuals in the population changes over time in an irregular manner.

**islet cell** Any of the specialized endocrine cells within the pancreas that produce and release the hormones insulin and glucagon.

**isotonic solution** A solution that has the same solute concentration as the cytosol of a cell, resulting in an equal amount of water flowing into the cell and out of it. Compare *hypertonic solution* and *hypotonic solution*.

**isotope** A variant form of a chemical element that differs in its number of neutrons, and therefore in its atomic mass number, from the most common form of that element.

**J-shaped curve** The graphical plot that represents exponential growth of a population. Compare *S-shaped curve*.

**joint** In reference to the skeletal system, a flexible connection between the rigid elements that make up an animal skeleton.

**jumping gene** See *transposon*.

**juvenile-onset diabetes** See *type 1 diabetes*.

**karyotype** A display of the specific number and shapes of chromosomes found in the diploid cells of a particular individual, or of a species in general.

**keystone species** A species that, relative to its own abundance, has a large effect on the presence and abundance of other species in a community.

**kidney** An organ that regulates the body's supply of water and solutes and helps dispose of metabolic wastes.

**kinetic energy** The energy of motion.

**kingdom** In reference to biological classification systems, the highest taxonomic category in the Linnaean hierarchy. Generally six kingdoms are recognized: Bacteria, Archaea, Protista, Plantae, Fungi, and Animalia. A higher level of classification, the domain, is now also recognized.

**Krebs cycle** Also called *citric acid cycle*. The second of three major phases of cellular respiration, following glycolysis and preceding oxidative phosphorylation. This series of enzyme-driven oxidation reactions takes place in the mitochondrial matrix and yields many molecules of NADH (and a few of ATP and $FADH_2$).

**lake** An ecosystem of the freshwater biome defined as a standing body water of variable size ranging up to thousands of square kilometers.

**land transformation** Changes made by humans to the land surface of Earth that alter the physical or biological characteristics of the affected regions. Compare *water transformation*.

**language** The complex system of human communication, consisting of thousands of words representing everything from objects to actions to abstract ideas.

**large intestine** See *colon*.

**larynx (pl. larynges)** Also called *voice box*. The voice-producing structure that contains the vocal cords and forms the entryway to the trachea.

**late wood** Wood of a tree that is produced from midsummer to early fall. Late wood is dark because it contains narrow conducting elements that have relatively thick walls. Compare *early wood*.

**lateral gene transfer** Also called *horizontal gene transfer*. The transfer of genetic material between different species under natural conditions.

**lateral meristem** A population of perpetually undifferentiated cells that cause a plant to increase in thickness when they divide. There are two types of lateral meristem tissue: vascular cambium and cork cambium. Compare *apical meristem*.

**law of conservation of energy** See *first law of thermodynamics*.

**law of independent assortment** Mendel's second law, which states that when gametes form, the separation of alleles of one gene is independent of the separation of alleles of other genes. We now know that this law does not apply to genes that are linked.

**law of segregation** Mendel's first law, which states that the two copies of a gene separate during meiosis and end up in different gametes.

**learned behavior** A predictable response acquired by trial and error or by watching others. Compare *fixed behavior*.

**lens** A structure of the eye that concentrates light on photoreceptors.

**LH** See *luteinizing hormone*.

**lichen** A mutualistic association between a photosynthetic microbe (alga and/or cyanobacterium, kingdom Protista) and a fungus (kingdom Fungi).

**ligament** A collagen-rich connective structure that attaches bone to bone in vertebrate skeletons. Compare *tendon*.

**ligase** An enzyme that can connect two DNA fragments. Ligases are used in DNA technology when a gene from one species is inserted into the DNA of another species.

**light reactions** The series of chemical reactions in photosynthesis that harvest energy from sunlight and use it to produce energy-rich compounds such as ATP and NADPH. The light reactions occur at the thylakoid membranes of chloroplasts and produce $O_2$ as a by-product. Compare *Calvin cycle*.

**lignin** A substance that links cellulose fibers in plant cell walls to create a rigid strengthening network. Lignin is one of the strongest materials in nature.

**limbic system** A functional region of the forebrain and midbrain that includes portions of the hypothalamus. The limbic system generates emotional states such as pleasure or anger, and behavioral drives such as thirst and hunger. It also participates in memory formation.

**lineage** A group of closely related individuals, species, genera, or the like, depicted as a branch on an evolutionary tree.

**linked genes** See *genetic linkage*.

**Linnaean hierarchy** The classification scheme used by biologists to organize and name organisms. Its seven levels—from the most inclusive to the least—are kingdom, phylum, class, order, family, genus, and species.

**lipid** A hydrophobic molecule made by living cells and built from chains or rings of hydrocarbons. Lipids are a key component of cell membranes. See also *phospholipid*.

**liver** An organ associated with the upper small intestine in the digestive system and with the circulatory system. The liver produces bile and plays an important role in carbohydrate and lipid metabolism.

**locomotion** The ability to move.

**locus (pl. loci)** The physical location of a gene on a chromosome.

**logistic growth** A type of population growth in which a population increases nearly exponentially at first but then stabilizes at the maximum population size that can be supported indefinitely by the environment. Logistic growth is represented graphically by an S-shaped curve. Compare *exponential growth*.

**lower respiratory system** Collectively, the trachea, bronchi, and lungs. Compare *upper respiratory system*.

**lumen** The space enclosed by the membrane of an organelle, or the cavity inside an organ.

**lung** The saclike gas exchange organ of a terrestrial animal, formed by an infolding of epidermal tissue.

**luteinizing hormone (LH)** A hormone produced by the pituitary gland that works with the *follicle-stimulating hormone* to regulate sperm development in males and to regulate the menstrual cycle in females.

**lymph node** A pocket of tissue lying along a lymphatic duct that contains large numbers of white blood cells and traps pathogens.

**lymphatic duct** One of a network of tubes that return interstitial fluid to the circulatory system.

**lymphatic system** The network of vessels that returns interstitial fluid from the body to the circulatory system.

**lymphocyte** Any of several types of white blood cells that bind to specific antigens and then contribute in various ways to the destruction of the pathogens that bear those antigens.

**lysosome** A specialized vesicle with an acidic lumen containing enzymes that break down macromolecules.

**macroevolution** The rise and fall of major taxonomic groups due to evolutionary radiations that bring new groups to prominence and mass extinctions in which groups are lost; the history of large-scale evolutionary changes over time. Compare *microevolution*.

**macromolecule** A large organic molecule formed by the bonding together of small organic molecules.

**macronutrient** A chemical element required by an organism in relatively large amounts. Compare *micronutrient*.

**malignant cell** See *cancer cell*.

**mammary gland** A modified sweat gland that is the most distinctive feature of mammals. Mammary glands produce a liquid rich in fat, proteins, salts, and other nutritive substances that nourish the newborn.

**manipulated variable** See *independent variable*.

**marrow** A tissue within vertebrate bone that, depending on the type of bone, produces blood cells or stores fat.

**marsupial** Any of a group of mammals that protect and feed their newborns with milk in an external pocket or pouch (marsupium). Marsupials include kangaroos and opossums. Compare *eutherian* and *monotreme*.

**mass extinction** An event during which large numbers of species become extinct throughout most of Earth.

**mast cell** A type of white blood cell that releases histamine and is found in a number of tissues, including blood, as well as epithelial and connective tissues.

**maternal homologue** In a homologous pair of chromosomes, the one that comes from the mother. Compare *paternal homologue*.

**matrix (pl. matrices)** The space interior to the cristae of the inner mitochondrial membrane.

**matter** Anything that has mass and occupies space.

**mechanoreceptor** A sensory cell that initiates an action potential in response to a mechanical stimulus.

**meiosis** A specialized process of cell division in eukaryotes during which diploid cells divide to produce haploid cells. Meiosis has two division cycles, and in animals it occurs exclusively in cells that produce gametes. Compare *mitosis*.

**meiosis I** The first cycle of cell division in meiosis, in which the members of each homologous chromosome pair are separated into different daughter cells. Meiosis I produces haploid daughter cells, each with half of the chromosome set found in the diploid parent cell.

**meiosis II** The second cycle of cell division in meiosis, in which the sister chromatids of each duplicated chromosome are separated into different daughter cells. Meiosis II is essentially mitosis, but in a haploid cell.

**membrane channel** See *channel protein*.

**membrane pump** See *active carrier protein*.

**memory cell** A type of B cell that remains in the body as a long-term record of the primary immune response to a particular pathogen. During a second exposure to that pathogen, the memory cells specific to the pathogen multiply rapidly to produce large numbers of lymphocytes that can attack it.

**menopause** The long-term cessation of menstrual cycles, marking the end of a human female's reproductive life.

**menstrual cycle** A series of hormonally controlled, cyclical changes that take place in the reproductive system of human females, which include the shedding of the lining of the uterus (menstruation) about every 28 days.

**meristem** A group of perpetually young, undifferentiated plant cells that can renew themselves and that can give rise to new cell types.

**mesoderm** The cell layer in animal development that lies between the *endoderm* and *ectoderm* in the gastrula and gives rise to muscle tissue, connective tissue, and the kidney.

**messenger RNA (mRNA)** A type of RNA that specifies the order of amino acids in a protein.

**metabolic heat** Heat generated as a by-product of metabolic processes, including cellular respiration.

**metabolic pathway** A series of enzyme-controlled chemical reactions in a cell in which the product of one reaction becomes the substrate for the next.

**metabolism** All the chemical reactions in a cell that involve the acquisition, storage, or use of energy.

**metamorphosis (pl. metamorphoses)** A dramatic developmental transformation from a reproductively immature to a reproductively mature form, involving great change in the form and function of the animal.

**metaphase** The stage of cell division during which chromosomes become aligned at the equator of the cell.

**metastasis** The spread of a disease from one organ to another.

**methanogen** An archaean that produces methane gas as a by-product of its metabolism.

**microevolution** Changes in allele or genotype frequencies in a population over time; the smallest scale at which evolution occurs. Compare *macroevolution*.

**microfilament** A protein fiber composed of actin monomers. Microfilaments are part of a cell's cytoskeleton and are important in cell movements.

**micronutrient** A chemical element required by an organism in tiny amounts. Compare *macronutrient*.

**microtubule** A protein cylinder composed of tubulin monomers. Microtubules are part of the cell's cytoskeleton. Compare *actin filament* and *intermediate filament*.

**midbrain** The part of the brain that helps maintain muscle tone and sends some sensory data to higher brain centers. Compare *forebrain* and *hindbrain*.

**middle ear** The portion of the vertebrate ear that contains three tiny bones that receive vibrations from the eardrum and transmit them to the inner ear. Compare *inner ear* and *outer ear*.

**mimicry** A type of adaptation arising from predator-prey interactions in which a species evolves to imitate the appearance of something unappealing to its would-be predator.

**mineral** An element (such as copper) or a small inorganic compound (such as a phosphate). Many minerals are needed by organisms in small amounts for proper nutrition.

**mitochondrion (pl. mitochondria)** An organelle with a double membrane that is the site of cellular respiration in eukaryotes. Mitochondria break down simple sugars to produce ATP in an oxygen-dependent (aerobic) process.

**mitosis** The process of cell division in eukaryotes that produces two daughter nuclei, each with the same chromosome number as the parent nucleus. Compare *meiosis*.

**mitotic division** The process that generates two genetically identical daughter cells from a single parent cell in eukaryotes. A mitotic division consists of mitosis followed by cytokinesis.

**mitotic spindle** An football-shaped array of microtubules that guides the movement of chromosomes during mitosis.

**mixotroph** A nutritional opportunist, an organism that can use energy and carbon from a variety of sources (functioning as both autotroph and heterotroph) to fuel its growth and reproduction.

**molecule** An association of atoms in which two or more of the atoms are linked through covalent bonds.

**Mollusca** Mollusks. After Arthropoda, the most diverse phylum of animals, characterized by a body that has a muscular foot, a visceral mass, and a mantle. Mollusks include clams and other bivalves (familiar "shellfish"), squid and octopi (cephalopods), and snails (gastropods).

**molting** Also called *ecdysis*. The process by which juvenile ecdysozoans (a group of protostomes) shed the cuticle that encases them.

**monocots** One of two main groups of flowering plants, characterized by the presence of only one cotyledon in the seed, roughly parallel leaf veins, vascular tissues distributed in a scattered pattern, and (typically) a fibrous root system. Monocots include all the grasses, members of the lily family, palm trees, and banana plants. Compare *dicots*.

**monohybrid** An individual that is heterozygous for only one trait. Compare *dihybrid*.

**monomer** A molecule that can be linked with other related molecules to form a larger *polymer*.

**monosaccharide** A simple sugar that can be linked to other sugars, forming a *disaccharide* or *polysaccharide*. Glucose is the most common monosaccharide in living organisms.

**monotreme** Any of a group of mammals that have no placenta and lay eggs. The platypus and echidnas are monotremes. Compare *eutherian* and *marsupial*.

**morphogen** A chemical signal that influences the developmental fate of a cell by cueing a cell as to its position in the body.

**morphological species concept** The idea that most species can be identified as a separate and distinct group of organisms by the unique set of morphological characteristics they possess. Compare *biological species concept*.

**most recent common ancestor** The ancestral organism from which a group of descendants arose.

**motor neuron** A neuron that carries signals from the central nervous system to the body, thereby causing changes in muscles, organs, or endocrine glands. Compare *sensory neuron* and *interneuron*.

**mRNA** See *messenger RNA*.

**multicellular organism** An organism made up of more than one cell.

**multipotent** Of or referring to a cell that can differentiate into only a relatively narrow range of cell types. Compare *pluripotent*, *totipotent*, and *unipotent*.

**multiregional hypothesis** A hypothesis stating that anatomically modern humans evolved from *Homo erectus* populations scattered throughout the world. According to this idea, worldwide gene flow caused different human populations to evolve modern characteristics simultaneously and to remain a single species. Compare *out-of-Africa hypothesis*.

**muscle** A tissue unique to animals that can contract, or shorten, to produce movement.

**muscle fiber** The basic unit of muscle tissue, consisting, in vertebrates, of many fused cells that form a long cylindrical structure bounded by a plasma membrane and packed in the center with bundles of contractile proteins (myofibrils).

**muscular system** The organ system in animals, consisting of three types of muscle tissue (cardiac, skeletal, and smooth), that produces the force that moves structures within the body.

**mutagen** A substance or energy source that alters DNA.

**mutation** A change in the sequence of an organism's DNA. Because new alleles arise only by mutation, mutations are the original source of all genetic variation.

**mutualism** An interaction between two species in which both species benefit. Compare *commensalism*, *competition*, and *exploitation*.

**mycelium (pl. mycelia)** The main body of a fungus, composed of threadlike hyphae.

**mycorrhiza (pl. mycorrhizae)** A mutualism between a fungus and a plant, in which the fungus provides the plant with mineral nutrients while receiving organic nutrients from the plant.

**myelin** The fatty material that forms an insulating sheath around the axon of a neuron and greatly speeds the rate at which action potentials move along the axon.

**myofibril** One of many bundles of contractile proteins inside a muscle fiber. Each myofibril is made up of thousands of contractile units, called sarcomeres, arranged end to end.

**myosin filament** One of the two main types of protein filaments involved in the contraction of muscles. They consist of many molecules of the protein myosin, and assist in muscle contraction by interacting with *actin filaments*.

**NADH** The reduced form of nicotinamide adenine dinucleotide (NAD$^+$), an energy carrier that delivers electrons and protons (H$^+$) to catabolic reactions (such as cellular respiration, which produces ATP from the breakdown of sugars into water and carbon dioxide). Compare *NADPH*.

**NADPH** The reduced form of nicotinamide adenine dinucleotide phosphate (NADP$^+$), an energy carrier molecule that delivers electrons and protons (H$^+$) to anabolic reactions, such as the Calvin cycle reactions of photosynthesis. Compare *NADH*.

**natural killer (NK) cell** A type of white blood cell that destroys any cell whose plasma membrane is marked with foreign proteins, including viral proteins.

**natural selection** Also called *adaptive evolution*. An evolutionary mechanism in which the individuals in a population that possess particular inherited characteristics survive and reproduce at a higher rate than other individuals in the population because those characteristics enable the individuals to function optimally in their particular habitat. Natural selection is the only evolutionary mechanism that consistently improves the survival and reproduction of the organism in its environment. Compare *artificial selection*.

**negative feedback** A mechanism for slowing down or damping the fluctuations in a process. In response to a change in conditions, negative feedback acts to bring conditions back to a set point. Compare *positive feedback*.

**negative growth regulator** Any of a variety of external and internal signals and regulatory proteins that control the cell cycle by halting cell division. Compare *positive growth regulator*.

**nephron** The basic unit of the vertebrate kidney involved in urine production, consisting of a glomerulus, across which blood is filtered under pressure into a U-shaped tubule, which facilitates the reabsorption of water and valuable solutes from the filtrate.

**nerve** A major communication pathway in the animal body made up of the axons of many individual neurons bundled with other support cells.

**nerve cell** See *neuron*.

**nervous system** The organ system in animals, consisting of sensory structures containing nervous tissue, that facilitates high-speed internal communication through specialized cells called neurons.

**net primary productivity (NPP)** The amount of energy that producers capture by photosynthesis, minus the amount lost as metabolic heat. NPP is usually measured as the amount of new biomass produced by photosynthetic organisms per unit of area during a specified period of time. Compare *secondary productivity*.

**neuron** Also called *nerve cell*. A type of animal cell that is highly specialized for transmitting action potentials from one part of the body to another.

**neurotransmitter** Any of various signaling molecules that transmit signals across the gaps, or synapses, that separate neurons from their target cells.

**neutron** A particle, found in the nucleus of an atom, that has no electrical charge. Compare *electron* and *proton*.

**niche** The sum total of the conditions and resources a species or population needs in order to survive and reproduce successfully in its particular habitat.
**niche partitioning** The differential use of space or resources in a common niche by competing species that enables species to coexist despite their potential for competition.

**nicotinamide adenine dinucleotide** See *NADH* and *NADPH*.

**nitrogen fixation** The process by which nitrogen gas (N$_2$), which is readily available in the atmosphere but cannot be used by plants, is converted to ammonium (NH$_4^+$), a form of nitrogen that can be used by plants. Nitrogen fixation is accomplished naturally by bacteria and by lightning, as well as by humans in industrial processes such as the production of fertilizer.

**nitrogenous base** Any of the five nitrogen-rich compounds found in nucleotides. The four nitrogenous bases found in DNA are adenine (A), cytosine (C), guanine (G), and thymine (T); in RNA, uracil (U) replaces thymine.

**NK cell** See *natural killer cell*.

**node** The moment in evolution, depicted as a point on an evolutionary tree, at which one lineage split, or diverged, into two separate lineages (such as Archaea and Eukarya).

**node of Ranvier** An unmyelinated portion of the axon of a vertebrate neuron. Action potentials jump from node to node, greatly speeding signal transmission.

**noncoding DNA** A segment of DNA that does not encode proteins or RNA. Introns and spacer DNA are two common types of noncoding DNA.

**nonpolar molecule** A molecule that has an equal distribution of electrical charge across all its constituent atoms. Nonpolar molecules do not form hydrogen bonds and tend not to dissolve in water. Compare *polar molecule*.

**nonspecific response** An immune response found in most animals that leads to the destruction of cells not recognized as belonging to the organism. Innate immunity is a nonspecific response. Compare *specific response*.

**norepinephrine** Also called *noradrenaline*. A hormone produced by the adrenal glands that regulates the fight-or-flight response to stressful situations in conjunction with *epinephrine*.

**notochord** A structure composed of large cells that collectively form a strong but flexible bar running dorsally along the length of an animal, providing support for the rest of the body.

**NPP**  See *net primary productivity*.

**nuclear envelope**  The double membrane that forms the outer boundary of the nucleus, an organelle found only in eukaryotic cells.

**nuclear pore**  One of many openings in the nuclear envelope that allow selected molecules, including specific proteins and RNA, to move into and out of the nucleus.

**nucleic acid**  A polymer made up of nucleotides. There are two kinds of nucleic acids: DNA and RNA.

**nucleotide**  Any of a class of organic molecules that serve as energy carriers and as the chemical building blocks of nucleic acids (DNA and RNA). A nucleotide is made up of a phosphate group, a five-carbon sugar, and one of four nitrogenous bases (see *nitrogenous base*). Nucleotides are linked to form a single strand of DNA or RNA.

**nucleotide pair**  See *base pair*.

**nucleus (pl. nuclei)**  The organelle in a eukaryotic cell that contains the genetic blueprint in the form of DNA.

**nutrient**  In an ecosystem context, an essential element required by a producer. See also *macronutrient* and *micronutrient*.

**nutrient cycle**  Also called *biogeochemical cycle*. The cyclical movement of a nutrient between organisms and the physical environment. There are two main types of nutrient cycles: *atmospheric cycles* and *sedimentary cycles*.

**nutrient recycling**  The breakdown by decomposers of dead organisms or their waste products that releases the chemical elements locked in the biological material and returns them to the environment, where these elements (for example, carbon dioxide, nitrogen, or phosphorus) are used by autotrophs and eventually heterotrophs.

**observation**  A description, measurement, or record of any object or phenomenon. Facts learned in this manner are subsequently used to formulate hypotheses.

**oceanic region**  Also called *open ocean*. An ecosystem of the marine biome defined as the part of the ocean beginning about 40 miles offshore, where the continental shelf, and therefore the coastal region, ends.

**offspring**  A new individual that was generated through reproduction and therefore received DNA from its parents.

**omnivore**  An organism that uses both plant and animal tissues as a source of nutrition. Compare *carnivore*, *herbivore*, and *detritivore*.

**oncogene**  A mutated gene that promotes excessive cell division, leading to cancer.

**oogenesis**  The production of a mature ovum capable of being fertilized in animals. Compare *spermatogenesis*.

**open ocean**  See *oceanic region*.

**operator**  In prokaryotes, a regulatory DNA sequence that controls the transcription of a gene or group of genes.

**opposable**  In primates, of or referring to a thumb (or big toe) that moves freely and can be placed opposite other fingers (or toes).

**oral cavity**  The mouth, where food first enters an animal's digestive system.

**order**  In reference to biological classification systems, the level in the Linnaean hierarchy above family and below class.

**organ**  A self-contained collection of different types of tissues, usually of a characteristic size and shape, that is organized for a particular set of functions.

**organ of Corti**  A structure in the cochlea of the inner ear that converts mechanical energy into a pattern of action potentials that vertebrates detect as sound.

**organ system**  A group of organs of different types that work together to carry out a common set of functions.

**organelle**  A discrete cytoplasmic structure with a specific function. Some cell biologists use the term only for membrane-enclosed cytoplasmic compartments; others include other cytoplasmic structures, such as ribosomes, in the definition.

**organic molecule**  A molecule that contains at least one carbon covalently bonded to one or more hydrogen atoms. Before modern chemistry, organic molecules on Earth were exclusively of biological origin, but now chemists can create many organic molecules artificially (synthetically). Compare *inorganic molecule*.

**osmoconformer**  A marine animal whose internal fluids resemble seawater quite closely in terms of salt concentration and that therefore does not have to spend energy directly for maintaining water and salt homeostasis. Examples include coral and jellyfish. Compare *osmoregulator*.

**osmoregulation**  The process of maintaining an internal water and salt balance that supports biological processes.

**osmoregulator**  An aquatic animal whose water and salt levels are not the same as those of its surroundings and therefore must spend a considerable amount of energy maintaining water and salt homeostasis. Most marine vertebrates and all freshwater animals are osmoregulators. Compare *osmoconformer*.

**osmosis**  The passive movement of water across a selectively permeable membrane.

**osteocyte**  A cell that either produces or reabsorbs the matrix of calcium and phosphorus compounds that makes up the greater part of a vertebrate bone.

**otolith**  One of many dense crystals of calcium carbonate in the vestibular apparatus that signal the direction of gravitational pull.

**outer ear**  The outer portion of the mammalian ear, which consists of the pinna and the auditory canal and which funnels pressure changes from the environment to the eardrum. Compare *inner ear* and *middle ear*.

**out-of-Africa hypothesis**  A hypothesis stating that anatomically modern humans evolved in Africa within the past 200,000 years and then spread throughout the rest of the world. According to this idea, as they spread from Africa, modern humans completely replaced older forms of *Homo sapiens*, including advanced forms such as the Neandertals. Compare *multiregional hypothesis*.

**oviduct**  Also called *fallopian tube*. The tube down which ova produced by female animals pass after leaving the ovary.

**ovulation**  The release of an ovum from the ovary.

**ovum (pl. ova)**  The female gamete, the *egg*.

**oxidation**  The loss of electrons by one atom or molecule to another. Compare *reduction*.

**oxidation-reduction reaction**  See *redox reaction*.

**oxidative phosphorylation**  The third of three major phases of cellular respiration, following glycolysis and the Krebs cycle. The shuttling of electrons down an electron transport chain in mitochondria that results in the production of ATP.

**oxygen-binding pigment**  Any of various complex molecules, often colored proteins, that increase the oxygen capacity of the tissues in which they occur. One example is hemoglobin.

**oxytocin**  A hormone released by the pituitary gland that triggers, among other things, milk production in mammals.

**P generation** The parent generation in a genetic cross. Compare $F_1$ generation and $F_2$ generation.

**pacemaker** See *sinoatrial node*.

**pain receptor** A sensory cell that initiates an action potential in response to extreme heat or cold or to injury.

**pancreas** An organ associated with the digestive system that produces a number of important digestive enzymes, as well as hormones.

**Pangaea** An ancient supercontinent that contained all of the world's landmasses. Pangaea formed 250 million years ago and began to break apart 200 million years ago, ultimately yielding the continents we know today.

**parasite** An organism that lives in or on another organism (its host) and obtains nutrients from that organism. Parasites harm and may eventually kill their hosts but do not kill them immediately.

**parathyroid gland** A gland lying next to the thyroid gland that produces parathyroid hormone.

**parathyroid hormone (PTH)** A hormone produced by the mammalian parathyroid gland that stimulates the release of calcium from bones and increases the reabsorption of calcium in the kidneys. Compare *calcitonin*.

**parenchyma** A plant tissue made up of relatively large cells with thin cell walls. Parenchyma cells are the most abundant cell type in the ground tissue of most plants. Compare *collenchyma* and *sclerenchyma*.

**passive carrier protein** A cell membrane protein that, without the input of energy, changes its shape to transport a molecule across the membrane from the side of higher concentration to the side of lower concentration. Compare *active carrier protein*.

**passive immunity** Immunity gained by an individual through antibodies that it receives from another individual (a nursing infant from its mother, for example) or another organism (for example, horses, which are used in the production of snakebite antivenom). Compare *active immunity*.

**passive transport** Movement of ions or molecules across a biological membrane that requires no input of energy because it is "downhill" (in the same direction as an electrochemical gradient). Compare *active transport*.

**paternal homologue** In a pair of homologous chromosomes, the one that comes from the father. Compare *maternal homologue*.

**pathogen** An organism or virus that infects a host and causes disease, harming and in some cases killing the host.

**PCR** See *polymerase chain reaction*.

**pedigree** A chart that shows genetic relationships among family members over two or more generations of a family's history.

**penis** A reproductive structure used by male animals to introduce sperm directly into a female's reproductive tract.

**peptide bond** A covalent bond between the amino group of one amino acid and the carboxyl group of another. Peptide bonds link amino acids together.

**perennial** A plant that lives for 3 or more years. Compare *annual* and *biennial*.

**peripheral nervous system (PNS)** The portion of the nervous system that carries signals to and from the central nervous system. The peripheral nervous system consists of the sensory organs and all the nerves that are not part of the central nervous system. Compare *central nervous system*.

**permafrost** Permanently frozen soil that is found below the surface layers and may be a quarter of a mile deep.

**Permian extinction** The largest mass extinction in the history of life on Earth; it occurred 250 million years ago, driving up to 95 percent of the species in some groups to extinction.

**persistent organic pollutant (POP)** Any long-lived organic molecule of synthetic origin that bioaccumulates in organisms, and that can have harmful effects. Some of the most damaging and widespread POPs are PCBs (polychlorinated biphenyls) and dioxins.

**petal** A part of a flower that often serves to attract pollinators to the flower. Petals form the second (from the outside) of the four whorls that make up a flower. Compare *carpel*, *sepal*, and *stamen*.

**pH scale** A scale that indicates the concentration of hydrogen ions in a solution. The pH scale runs from 1 to 14. A pH of 7 is neutral; values below 7 indicate acids, and values above 7 indicate bases.

**phagocyte** A type of white blood cell that destroys invading cellular organisms by engulfing them (phagocytosis). Macrophages and neutrophils are two examples of phagocytes.

**phagocytosis** A form of endocytosis by which a cell engulfs a large particle, such as another cell; "cell eating." Compare *pinocytosis*.

**pharyngeal pouch** A structure, found in the early embryo of all chordates, that first appears as a pocket of tissue on either side of the embryonic pharynx and later develops into structures such as gill slits in fish and larval amphibians, or the larynx and trachea in mammals.

**pharynx (pl. pharynges)** The portion of the respiratory system where the mouth and the nasal cavity join to form a single passageway.

**phenotype** The specific version of a genetic trait that is displayed by a given individual. For example, black, brown, red, and blond are phenotypes of the hair color trait in humans. Compare *genotype*.

**pheromone** A chemical signal produced by one individual to communicate its identity and location to another individual.

**phloem** A tissue composed of living cells through which a plant transports sugars and other organic and inorganic substances. Compare *xylem*.

**phosphate group** A functional group consisting of a phosphate atom and four oxygen atoms.

**phospholipid** A lipid consisting of two fatty acids, a glycerol, and a phosphate as part of the hydrophilic head group. Phospholipids are the main component in all biological membranes.

**phospholipid bilayer** A double layer of phospholipid molecules arranged so that their hydrophobic "tails" lie sandwiched between their hydrophilic "heads." A phospholipid bilayer forms the basic structure of all biological membranes.

**photoautotroph** An organism that obtains its energy from sunlight and derives its carbon from carbon dioxide in the air. Examples include cyanobacteria, green algae, and plants. Compare *photoheterotroph* and *chemoautotroph*.

**photoheterotroph** An organism that obtains its energy from sunlight and derives its carbon from carbon-containing compounds found mainly in other organisms. All photoheterotrophs are prokaryotes. Compare *photoautotroph* and *chemoheterotroph*.

**photoperiodism** A physiological response by a plant to day length; examples include the timing of flowering, seed germination, and bud dormancy (during winter).

**photoreceptor** A sensory cell that initiates an action potential in response to light.

**photosynthesis** A metabolic process by which organisms capture energy from sunlight and use it to synthesize sugars from carbon dioxide and water. Compare *cellular respiration*.

**photosystem** A large complex of proteins and chlorophyll that captures energy from sunlight. Two distinct photosystems (I and II) are present in the thylakoid membranes of chloroplasts.

**photosystem I** The photosystem that is primarily responsible for the production of NADPH. Compare *photosystem II*.

**photosystem II** The photosystem in which light energy is used to initiate an electron flow along the electron transport chain, resulting in ATP synthesis and the release of oxygen gas ($O_2$) as a by-product. Compare *photosystem I*.

**phylum (pl. phyla)** In reference to biological classification systems, the level in the Linnaean hierarchy above class and below kingdom.

**physiology** The study of the functions of anatomical structures. Compare *anatomy*.

**phytoplankton** Free-floating, single-celled algae that drift at or near the surface of water bodies. Compare *zooplankton*.

**pilus (pl. pili)** A short hairlike projection found on the surface of some bacteria.

**pinna (pl. pinnae)** The curled, cartilaginous part of the ear that is visible externally. The pinna funnels pressure changes from the environment into the auditory canal.

**pinocytosis** A form of nonspecific endocytosis by which cells take in fluid; "cell drinking." Compare *phagocytosis*.

**pituitary gland** A gland associated with the vertebrate brain that releases hormones that control the release of hormones by other glands. Along with the hypothalamus, the pituitary helps regulate interactions between the nervous and endocrine systems.

**placenta** Also called *afterbirth*. A structure found in mammals that transfers nutrients and gases from the blood of the mother to the blood of the fetus developing in her uterus.

**plankton** Microbes that drift at or near the surface of water bodies. See also *phytoplankton* and *zooplankton*.

**Plantae** The kingdom of Eukarya that is made up of plants—multicellular autotrophs that live mainly on land and photosynthesize.

**plasma** The watery, fluid portion of the blood, in which cells and cell fragments float.

**plasma membrane** The phospholipid bilayer that forms the outer boundary of any cell.

**plasmid** A small circular segment of DNA found naturally in bacteria. Plasmids are involved in natural gene transfers among bacteria and can be used as vectors in genetic engineering.

**plasmodesma (pl. plasmodesmata)** A tunnel-like channel between two plant cells that provides a cytoplasmic connection allowing the flow of small molecules and water between the cells.

**plate tectonics** See *continental drift*.

**platelet** A type of sticky cell fragment that circulates in the blood and helps form a blood clot by combining with proteins that also circulate in the blood to form a gel-like mesh that traps blood cells.

**pleiotropy** A type of genetic control in which a single gene influences a variety of different traits.

**pluripotent** Of or referring to a cell that can differentiate into any of the cell types in the adult body. Compare *multipotent*, *totipotent*, and *unipotent*.

**PNS** See *peripheral nervous system*.

**point mutation** A mutation in which only a single base is altered.

**polar body** The smaller of the two haploid cells that result from the first meiotic division of the primary oocyte. The polar body plays no role in development. Compare *secondary oocyte*.

**polar molecule** A molecule that has an uneven distribution of electrical charge. Polar molecules can easily interact with water molecules and are therefore soluble. Compare *nonpolar molecule*.

**polarization** The existence of an electrical charge difference across the plasma membrane, such that there are many more positive ions on the outside relative to the cytosolic side of the membrane. Many cells, including neurons, are polarized in their resting state. Compare *depolarization*.

**pollen** The haploid, multicellular, mobile male gametophyte of seed plants, in which sperm are produced. Compare *embryo sac*.

**pollinator** An animal that carries pollen grains from the stamens of one flower to the stigmas (see *carpel*) of other flowers of the same species.

**polygenic** Of or referring to inherited traits that are determined by the action of more than one gene.

**polymer** A large organic molecule composed of many *monomers* linked together.

**polymerase chain reaction (PCR)** A method of DNA technology that uses the DNA polymerase enzyme to make multiple copies of a targeted sequence of DNA.

**polypeptide** A polymer consisting of covalently linked linear chains of amino acids.

**polyploid** Of or referring to a cell or organism that has three or more complete sets of chromosomes (rather than the usual two complete sets). Populations of polyploid individuals can rapidly form new species without geographic isolation. Compare *diploid* and *haploid*.

**polysaccharide** A polymer composed of many linked *monosaccharides*. Examples include starch and cellulose. Compare *disaccharide*.

**POP** See *persistent organic pollutant*.

**population** A group of interacting individuals of a single species located within a particular area.

**population cycle** A pattern in which the population sizes of two species increase and decrease together in a tightly linked cycle; this pattern can occur when at least one of the two species involved is very strongly influenced by the other.

**population density** The number of individuals in a population, divided by the area covered by the population.

**population ecology** A branch of science concerned with questions that relate to how many organisms live in a particular environment, and why.

**population size** The total number of individuals in a population.

**positional information** Physical and chemical cues that inform a cell as to its location in the body.

**positive feedback** A regulatory mechanism that amplifies a particular process. In response to a change in conditions, positive feedback pushes a process progressively in one direction, rather than maintaining a set point. Compare *negative feedback*.

**positive growth regulator** Any of a variety of internal signals, including growth factors, hormones, and regulatory proteins, that control the cell cycle by stimulating cell division. Compare *negative growth regulator*.

**postzygotic barrier** A barrier that prevents zygotes from developing into healthy offspring. Compare *prezygotic barrier*.

**potential energy** Stored energy.

**precancerous cell** An abnormal cell, such as a cell in a benign tumor, whose descendants can, with time, become increasingly abnormal, changing shape, increasing in size, and quitting their normal job.

**predator** An organism that kills other organisms (called *prey*) for food.

**prey** Animals that *predators* kill and eat.

**prezygotic barrier** A barrier that prevents a male gamete (like a human sperm cell) and a female gamete (like a human egg cell) from fusing to form a zygote. Compare *postzygotic barrier*.

**primary consumer** An organism that eats a producer and is eaten by *secondary consumers*. Compare also *tertiary consumer* and *quaternary consumer*.

**primary growth** An increase in the length (or height) of a plant that results from cell divisions in the apical meristem. Compare *secondary growth*.

**primary immune response** The relatively slow mobilization of B cells and T cells following the first exposure to a pathogen. Compare *secondary immune response*.

**primary oocyte** A diploid animal cell that can undergo meiosis to form mature eggs. Compare *primary spermatocyte* and *secondary oocyte*.

**primary spermatocyte** A diploid animal cell that can undergo meiosis to form mature sperm. Compare *primary oocyte* and *secondary spermatocyte*.

**primary structure** In reference to proteins, the sequence of amino acids in a protein. Compare *secondary structure*, *tertiary structure*, and *quaternary structure*.

**primary succession** Ecological succession that occurs in newly created habitat, as when an island rises from the sea or a glacier retreats, exposing newly available bare ground. Compare *secondary succession*.

**primary tumor** The tumor that is located at the initial site of a cancer. Compare *secondary tumor*.

**primate** Any of an order of mammals whose living members include lemurs, tarsiers, monkeys, humans, and other apes. Primates share characteristics such as flexible shoulder and elbow joints, opposable thumbs or big toes, forward-facing eyes, and brains that are large relative to body size.

**producer** Also called *autotroph*. An organism that uses energy from an external source, such as sunlight, to produce its own food without having to eat other organisms or their remains. Compare *consumer*.

**product** A substance that is formed by a chemical reaction. Compare *reactant*.

**progesterone** The main hormone, among the progestogens, that helps maintain pregnancy in female mammals.

**progestogen** Any of a class of hormones that is one of three major types produced by the ovary in female mammals. Progestogens have multiple functions, including thickening the lining of the uterus and increasing blood supply to it to create a suitable environment for a developing fetus, and suppressing ovulation. Compare *androgen* and *estrogen*.

**programmed cell death** See *apoptosis*.

**prokaryote** A single-celled organism that does not have a nucleus. All prokaryotes are members of the domains Bacteria or Archaea. Compare *eukaryote*.

**prokaryotic flagellum** A hairlike structure, found in prokaryotes, that propels the cell or organism through a whiplike action. Prokaryotic flagella lack a membrane covering and have a very different internal structure from that of *eukaryotic flagella*. They are believed to have evolved separately from eukaryotic flagella.

**promoter** In genetics, the DNA sequence in a gene that RNA polymerase binds to in order to begin transcription, and that therefore controls gene expression at the transcriptional level.

**prophase** The stage of mitosis or meiosis during which chromosomes first become visible under the microscope.

**proprioceptor** A pressure-sensitive sensory receptor that provides information on the position of an animal's body relative to its surroundings and of the animal's body parts relative to one another.

**prostaglandin** Any of a class of lipid-based signaling molecules with varied functions, which include raising body temperature and dilating blood vessels to increase blood flow during inflammation.

**protein** A polymer of amino acids that are linked in a specific sequence. Most proteins are folded into complex three-dimensional shapes.

**proteomics** The study of the full set of proteins encoded by genes.

**Protista** The oldest kingdom of Eukarya, made up of a diverse collection of mostly single-celled but some multicellular organisms. Protista is an artificial grouping, defined only by what members of this group are not: protists are not plants, animals, fungi, bacteria, or archaeans.

**proton** A positively charged particle found in atoms. Each atom of a particular element contains a characteristic number of protons. Compare *electron* and *neutron*.

**proton gradient** An imbalance in the concentration of protons across a membrane.

**proto-oncogene** A gene that promotes cell division in response to growth signals as part of its normal cellular function. Compare *tumor suppressor gene*.

**protostome** Any of a group of animals, including insects, worms, and snails, in which the first opening to develop in the early embryo becomes the mouth. Compare *deuterostome*.

**protozoan** Any nonphotosynthetic protist. All protozoans are motile. Compare *alga*.

**PTH** See *parathyroid hormone*.

**puberty** In humans, the transition from childhood to reproductive maturity.

**pulmonary circuit** The loop of the circulatory system through which blood travels from the heart to the lungs and back. Compare *systemic circuit*.

**Punnett square** A diagram in which all possible genotypes of male and female gametes are listed on two sides of a square, providing a graphical way to predict the genotypes of the offspring produced in a genetic cross.

**pupil** The opening through which light enters the eye.

**pyruvate** A three-carbon molecule produced by glycolysis that is processed in the mitochondria to generate ATP.

**quaternary consumer** An organism that eats a *tertiary consumer*. Compare also *primary consumer* and *secondary consumer*.

**quaternary structure** In reference to proteins, the three-dimensional arrangement of two or more separate chains of amino acids into a functional protein complex. Compare *primary structure*, *secondary structure*, and *tertiary structure*.

**quorum sensing** A system of cell-to-cell communication that enables some prokaryotes to sense and respond to bacterial density.

**radial symmetry** A body arrangement in which an animal could be sliced symmetrically along any number of vertical planes passing through the center of the animal to produce body parts that are nearly identical. Compare *bilateral symmetry*.

**radiation** 1. The waves of energy, such as light or infrared (heat), that are released by an object and that can be absorbed by another object. 2. A means of exchanging heat in the form of light or infrared waves. Compare *conduction*, *convection*, and *evaporation*.

**radiation therapy** The administration of high-energy radiation to destroy cancer cells by killing all rapidly dividing cells.

**radioisotope** An unstable, radioactive form of an element that releases energy as it decays to more stable forms at a constant rate over time.

**rain shadow** An area on the side of a mountain facing away from moist prevailing winds where little rain or snow falls.

**reabsorption** In kidney function, the removal of valuable solutes and water from the filtrate in the tubules of the kidney, before the filtrate leaves the kidney in the form of urine.

**reactant** A substance that undergoes a chemical reaction. Compare *product*.

**reaction center** A cluster of chlorophyll molecules within an antenna complex whose electrons become excited and are passed to an electron transport chain when the pigment molecules absorb light energy.

**receptor** See *receptor protein*.

**receptor protein** Also called simply *receptor*. A protein in the plasma membrane or cytoplasm of a target cell that binds signaling molecules, allowing those molecules to indirectly affect processes inside the cell.

**receptor-mediated endocytosis** A form of endocytosis in which receptor proteins embedded in the plasma membrane of a cell recognize certain surface characteristics of materials to be brought into the cell by endocytosis.

**recessive** Of or referring to an allele that does not have a phenotypic effect when paired with a *dominant* allele in a heterozygote.

**recombinant DNA** An artificial assembly of genetic material created by the enzyme-mediated linking of DNA fragments.

**red blood cell** Also called *erythrocyte*. A type of cell that circulates freely with the blood and that contains an oxygen-binding pigment, hemoglobin, that greatly increases the amount of oxygen that the blood can carry.

**redox reaction** An oxidation-reduction reaction, a chemical reaction in which electrons are transferred from one molecule or atom to another.

**reduction** The gain of electrons by one atom or molecule from another. Compare *oxidation*.

**reflex arc** Also called *spinal reflex*. A neuronal connection in the spinal cord between a sensory cell and a muscle cell that allows a simple stimulus to be translated rapidly into a movement.

**regenerative medicine** A field of medicine that uses stem cells to repair or replace tissues and organs damaged by injury or disease.

**regulatory DNA** A DNA sequence that can increase, decrease, turn on, or turn off the expression of a gene or a group of genes. Regulatory DNA sequences interact with regulatory proteins to control gene expression.

**regulatory protein** Also called *transcription factor*. A protein that signals whether or not a particular gene or group of genes should be expressed. Regulatory proteins interact with regulatory DNA to control gene expression.

**relative species abundance** The number of individuals of one species in a community, compared to the number of individuals of other species.

**releaser** In animal behavior, a simple stimulus that triggers a fixed behavior.

**repressor protein** A protein that prevents the expression of a particular gene or group of genes.

**reproduction** The generation of a new individual like oneself.

**reproductive cloning** A technology used to produce an offspring that is an exact genetic copy (a clone) of another individual. Rather than stem cells being removed from the embryo as they are in therapeutic cloning, the embryo is transferred to the uterus of a surrogate mother, where, if all goes well, the birth of a healthy offspring ultimately results; this offspring is genetically identical to the individual that provided the donor nucleus.

**reproductive isolation** A condition in which barriers to reproduction prevent or strongly limit two or more populations from reproducing with one another. Many different kinds of reproductive barriers can result in reproductive isolation, but it always has the same effect: no or few genes are exchanged between the reproductively isolated populations. Compare *geographic isolation*.

**reproductive system** The organ system in animals that generates gametes (eggs or sperm) and, in many vertebrates, facilitates the merger of the gametes inside the body of the female.

**respiration** See *cellular respiration* or *respiratory system*.

**respiratory system** The organ system in animals that delivers oxygen to the cells and removes carbon dioxide.

**responding variable** See *dependent variable*.

**resting potential** The electrical state of an axon when it is not transmitting an action potential.

**restriction enzyme** Any of a class of enzymes that cut DNA molecules at a specific target sequence; a key tool of DNA technology.

**retina** A field of photoreceptors in the eye that enables organisms to form an image.

**reversible binding** The capacity of a molecule to bind another molecule and, under different conditions, to release that molecule.

**ribonucleic acid** See *RNA*.

**ribosomal RNA (rRNA)** A type of RNA that is an important component of ribosomes.

**ribosome** A particle composed of proteins and RNA at which new proteins are synthesized. Ribosomes can be either attached to the endoplasmic reticulum or free in the cytosol.

**ring species** A species whose populations loop around a geographic barrier (such as a mountain chain) and in which the populations at the two ends of the loop are in contact with one another, yet cannot interbreed.

**river** An ecosystem of the freshwater biome defined as a body of water that moves continuously in a single direction.

**RNA** Ribonucleic acid, a single-stranded polymer of nucleotides that is necessary for the synthesis of proteins in living organisms. Each nucleotide in RNA is composed of the sugar ribose, a phosphate group, and one of four nitrogenous bases: adenine, cytosine, guanine, or uracil. Compare *DNA*.

**RNA interference (RNAi)** A mechanism for selectively blocking the expression of a given gene in which small chunks of RNA silence genes that share nucleotide sequence similarity with them.

**RNA polymerase** The key enzyme in DNA transcription. RNA polymerase links together the nucleotides of the RNA molecule specified by a gene.

**RNA splicing** The process by which mRNA introns are snipped out and the remaining pieces of mRNA are re-joined.

**RNAi** See *RNA interference*.

**rod** A type of photoreceptor cell in the vertebrate eye that detects light and dark and functions well in dim light. Compare *cone*.

**root cap** A cone-shaped tissue at the tip of the root that protects the root apical meristem.

**root hair** A long, thin extension of a root dermal cell that increases the surface area for absorption of water and dissolved nutrients.

**root system** The branched, nongreen underground organ system of a plant, which is specialized for absorbing water and mineral nutrients and for anchoring the plant in soil. Compare *shoot system*.

**rough ER** A region of the endoplasmic reticulum that has attached ribosomes and that specializes in protein synthesis. Compare *smooth ER*.

**rRNA** See *ribosomal RNA*.

**rubisco** The enzyme that catalyzes the first reaction of carbon fixation in photosynthesis.

**S phase** The stage of the cell cycle during which the cell's DNA is replicated.

**SA node** See *sinoatrial node*.

**sac fungi** See *ascomycetes*.

**saliva** A fluid that contains enzymes that help digest carbohydrates and that is released into the mouth by the salivary glands.

**salivary gland** A gland in the mouth that produces saliva.

**salt** A compound consisting of ions held together by the mutual attraction between their opposite electrical charge.

**sapwood** Relatively new secondary xylem that functions to transport water and also provides strength to the tree trunk. Compare *heartwood*.

**sarcomere** The contractile unit of a myofibril, consisting of many bundles of actin and myosin filaments that extend between the two Z discs, which mark the boundaries of a sarcomere.

**saturated fatty acid** A fatty acid that has no double bonds between the carbon atoms in its hydrocarbon backbone. Compare *unsaturated fatty acid*.

**science** A method of inquiry that provides a rational way to discover truths about the natural world.

**scientific fact** A direct and repeatable observation of any aspect of the physical world.

**scientific hypothesis (pl. hypotheses)** An informed, logical, and plausible explanation for observations of the natural world. An "educated guess." Compare *scientific theory*.

**scientific method** A series of steps in which the investigator develops a hypothesis, tests its predictions by performing experiments, and then changes or discards the hypothesis if its predictions are not supported by the results of the experiments.

**scientific name** The unique two-part name given to each species that consists of, first, a Latin name designating the genus and, second, a Latin name designating that species. Scientific names are traditionally italicized.

**scientific theory** A component of scientific knowledge that has been repeatedly con-firmed in diverse ways and is provisionally accepted by those who have knowledge of the discipline as the best available description of the truth about the phenomenon in question. Compare *scientific hypothesis*.

**sclerenchyma** A plant tissue made up of thick-walled cells whose function is to provide mechanical strength to plant parts. Sclerenchyma cells are often dead when fully functional. Compare *collenchyma* and *parenchyma*.

**second law of thermodynamics** The law stating that all systems, such as a cell or the universe, tend to become more disordered, and that the creation and maintenance of order in a system requires the transfer of disorder to the environment. Compare *first law of thermodynamics*.

**secondary consumer** An organism that eats a *primary consumer* and is eaten by *tertiary consumers*. Compare also *quaternary consumer*.

**secondary growth** An increase in the thickness of a plant that results from cell divisions in the lateral meristem. Compare *primary growth*.

**secondary immune response** The rapid defensive response, following a second exposure to a pathogen, that is mediated by memory B cells produced during the first exposure to that pathogen. Compare *primary immune response*.

**secondary oocyte** The larger of the two cells resulting from the first meiotic division of the *primary oocyte*. The secondary oocyte gives rise to the mature ovum through the second meiotic division. Compare *polar body* and *secondary spermatocyte*.

**secondary productivity** The rate of new biomass production by consumers per unit of area. Compare *net primary productivity*.

**secondary spermatocyte** One of the two haploid cells that result from the first meiotic division of the *primary spermatocyte*, and that give rise to mature sperm through the second meiotic division. Compare *secondary oocyte*.

**secondary structure** In reference to proteins, the patterns of local three-dimensional forms in segments of a protein. Spiral shapes and pleated sheets are common forms in the secondary structure of many proteins. Compare *primary structure*, *tertiary structure*, and *quaternary structure*.

**secondary succession** Ecological succession that occurs as communities recover from a disturbance, as when a forest grows back after a field ceases to be used for agriculture. Compare *primary succession*.

**secondary tumor** A tumor that is spawned by a *primary tumor* at a site distant from the primary tumor's location.

**secretion** Active transport of materials out of an organ or an organism.

**sedimentary cycle** A type of nutrient cycle in which the nutrient does not enter the atmosphere easily. Compare *atmospheric cycle*.

**seed** The embryo of a plant, encased in a protective covering.

**selectively permeable membrane** A membrane that controls which materials can pass through it. An example is the plasma membrane.

**semicircular canal** One of the three fluid-filled tubes in the vestibular apparatus of the ear that are oriented at right angles to one another, forming a three-dimensional grid. Movement of the fluid within the tubes informs the animal about head movement and orientation.

**semiconservative replication** DNA replication in which one "old" (template) strand is retained, or conserved, in each new helix.

**sensory input** An action potential that conveys information about stimuli.

**sensory neuron** A neuron that carries sensory information to the central nervous system. Compare *motor neuron* and *interneuron*.

**sensory receptor** Part of a cell, a whole cell, or a group of cells that responds to environmental stimuli by generating a nerve impulse that carries sensory information back to the central nervous system, including the brain. Sensory receptors allow us to hear, see, taste, smell, and feel. Examples include the pain receptors in skin and the light-sensitive cells of the retina.

**sepal** A part of a flower that encloses and protects the flower before it opens. Sepals form the outermost of the four whorls that make up a flower. Compare *carpel*, *petal*, and *stamen*.

**serotype** See *viral strain*.

**set point** A genetically determined state of normality that is regulated by homeostasis. The set point for blood pH, for example, ranges from 7.35 to 7.45 in humans.

**sex chromosome** Either of a pair of chromosomes that determines the sex of an individual. Compare *autosome*.

**sex hormone** Any of a class of steroid hormones that regulate the development of sex and sexual behavior in animals.

**sex-linked** Of or referring to genes located on a sex chromosome. See also *X-linked* and *Y-linked*.

**sexual dimorphism** A distinct difference in appearance between the males and females of a species.

**sexual recombination** The genetic processes (crossing-over, independent assortment of chromosomes, and fertilization) that lead to offspring having novel groupings of alleles that differ from the allelic combinations in either parent.

**sexual reproduction** The combining of genes from two individuals to give rise to a new individual, known as the offspring. Compare *asexual reproduction*.

**sexual selection** A type of natural selection in which individuals that differ in inherited characteristics differ, as a result of those characteristics, in their ability to get mates.

**shared derived trait** An evolutionary novelty shared by an ancestor and its descendants but not seen in groups that are not direct descendants of that ancestor.

**shoot system** The aboveground system of a plant, which is specialized for life in the air and, in vascular plants, consists of the stems and leaves. Compare *root system*.

**sieve tube** A food-conducting tube, created by stacks of living cells, in the phloem of plants.

**signal transduction** The relay, within the cytoplasm of a cell, of a signal received by a membrane receptor to produce one or more responses from a target cell. Signal transduction is conducted mostly by specialized molecules called second messengers and their associated proteins.

**signal transduction pathway** A series of cellular events that relay receipt of a signal from protein receptors on the plasma membrane to the cytoplasm.

**signaling molecule** A molecule produced and released by one cell that affects the activities of another cell (referred to as a *target cell*). Signaling molecules enable the cells of a multicellular organism to communicate with one another and coordinate their activities.

**simple diffusion** The passive transmembrane movement of a substance without the assistance of membrane transport proteins. Compare *facilitated diffusion*.

**simple eye** A non-image-forming eye that can distinguish light from dark. Compare *compound eye*.

**single-lens eye** An eye in which a lens focuses light on an image-forming surface called a retina.

**sinoatrial (SA) node** Also called *pacemaker*. A signaling center in the heart that is responsible for initiating the contraction of the whole heart by relaying a signal to the *atrioventricular node*.

**sinus** Any of several spacious cavities in the bones around the nose and in the forehead and cheeks that open into the pharynx.

**sister chromatids** A pair of identical double helices that are produced by the replication of DNA during the cell cycle.

**skeletal muscle** The specialized muscle associated with the skeleton. It has a striated (striped) appearance caused by the regular pattern in which the contractile proteins are arranged within each muscle fiber. Compare *cardiac muscle* and *smooth muscle*.

**skeletal system** The organ system in animals, consisting of bones, cartilage, and ligaments, that provides an internal framework to support the body of vertebrates.

**skeleton** The structural support tissue of animals.

**small intestine** The portion of the digestive system where digestion of food and most nutrient absorption takes place.

**smooth ER** A region of the endoplasmic reticulum that is specialized for lipid synthesis. It is "smooth" because it does not have attached ribosomes. Compare *rough ER*.

**smooth muscle** The specialized muscle found in the walls of the digestive system and blood vessels, so called because it lacks the banding pattern associated with other muscle types when viewed under a microscope. Compare *cardiac muscle* and *skeletal muscle*.

**social behavior** Behavior that involves cooperation among members of a group of animals, usually of the same species.

**soluble** Of or referring to a chemical that will dissolve (mix) in water.

**solute** A dissolved substance. Compare *solvent*.

**solution** Any combination of a solute and a solvent.

**solvent** A liquid (in biological systems, usually water) into which a *solute* has dissolved.

**somatic** Voluntary. Compare *autonomic*.

**somatic cell** Any cell in a multicellular organism that is not a *gamete* or part of a gamete-making tissue.

**somatic mutation** A mutation that occurs in a cell other than a sex cell and hence is not passed down to offspring.

**somatic stem cell** See *adult stem cell*.

**spacer DNA** A region of noncoding DNA that separates two genes. Spacer DNA is abundant in eukaryotes, but not in prokaryotes.

**speciation** The splitting of one species to form two or more species that are reproductively isolated from one another.

**species (pl. species)** A group of interbreeding natural populations that is reproductively isolated from other such groups.

**species richness** The total number of different species that live in a community.

**specific response** An immune response in which an organism recognizes and responds much more rapidly to a particular parasite or pathogen to which it has had previous exposure. Adaptive immunity consists of specific responses. Compare *nonspecific response*.

**sperm** The small, mobile, haploid gamete that is produced by sexually reproducing male eukaryotes. Compare *egg*.

**spermatogenesis** The production of sperm in animals. Compare *oogenesis*.

**spinal cord** A dense collection of neuronal cell bodies and axons that carries information between the brain and the tail end of a vertebrate. The spinal cord of vertebrates forms an intermediate site of information processing that acts as a filter between the brain and the muscles and sensory neurons.

**spinal reflex** See *reflex arc*.

**spongy bone** The porous bone that lies within the compact bone of vertebrates. Compare *compact bone*.

**spore** A reproductive structure of fungi and plants, usually thick walled, that can survive for long periods of time in a dormant state and will sprout under favorable conditions to produce the body of the organism.

**sporophyte** The diploid, multicellular individual in plants. It develops from the embryo, which in turn develops from the single-celled, diploid zygote created by fertilization. Compare *gametophyte*.

***SRY* gene** A gene, located on the Y chromosome, that functions as a master switch, committing the sex of a developing embryo to "male." *SRY* is short for "<u>s</u>ex-determining <u>r</u>egion of <u>Y</u>."

**S-shaped curve** The graphical plot that represents logistic growth of a population. Compare *J-shaped curve*.

**stabilizing selection** A type of natural selection in which individuals with intermediate values of an inherited characteristic have an advantage over other individuals in the population, as when medium-sized individuals produce offspring at a higher rate than do small or large individuals. Compare *directional selection* and *disruptive selection*.

**stamen** The male reproductive part of a flower, consisting of an anther on top of a slender filament. Stamens form the third (from the outside) of the four whorls that make up a flower. Compare *carpel*, *petal*, and *sepal*.

**start codon** A three-nucleotide sequence on an mRNA molecule (usually the codon AUG) that signals where translation should begin. Compare *stop codon*.

**statistics** A mathematical science that uses probability theory to estimate the reliability of data.

**stem cell** An undifferentiated cell that can renew itself through cell division, theoretically indefinitely. Some of the daughter cells generated by stem cells may differentiate into specialized cell types.

**sterol** Also called *steroid*. Any of a class of lipids whose fundamental structure consists of four hydrocarbon rings fused to each other.

**stoma (pl. stomata)** An air pore, found on the leaf or green stem of a plant, that can open and close, thereby controlling the rates at which $CO_2$ is brought into the plant and $O_2$ and water are lost from the plant. The flexing of the pair of guard cells that border the stoma determines whether it is open or closed.

**stomach** The portion of the digestive system that is specialized for mixing and storing food.

**stop codon** A three-nucleotide sequence on an mRNA molecule that signals where translation should end. Compare *start codon*.

**stroma** The space enclosed by the inner membrane of the chloroplast, in which the thylakoid membranes are situated.

**substitution** In genetics, a mutation in which one nitrogenous base is replaced by another at a single position in the DNA sequence of a gene.

**substrate** In reference to enzymes, the particular substance on which an enzyme acts. Only the substrate will bind to the active site of the enzyme.

**succession** A process by which species in a community are replaced over time. For a given location, the order in which species are replaced is fairly predictable. See also *primary succession* and *secondary succession*.

**sugar** A simple carbohydrate, generally a monosaccharide or a disaccharide, that has the general chemical formula $(CH_2O)_n$, where $n$ is less than 7 for monosaccharides and about 12 for most disaccharides.

**surface tension** A force that tends to minimize the surface area of water at an air-water boundary. Surface tension holds the water surface taut, resisting any stretching or breaking of that surface, and it is strong enough to support very light objects.

**swallowing reflex** An involuntary action, stimulated when a food mass makes contact with the wall of the pharynx, in which the epiglottis seals off the air passages and the food is pushed into the esophagus.

**symbiosis** A relationship in which two or more organisms of different species live together in close association.

**sympatric speciation** The formation of new species from populations that are not geographically isolated from one another. Compare *allopatric speciation*.

**synapse** The narrow gap that separates a neuron from its target cell.

**synaptic cleft** The minute gap between a neuron and its target cell across which neurotransmitters, rather than action potentials, carry the signal.

**synovial sac** A fluid-filled sac that forms a cushion in vertebrate joints.

**systemic circuit** The loop of the circulatory system through which blood travels from the heart to respiring tissues and back to the heart. Compare *pulmonary circuit*.

**systole** The pumping phase of the cardiac cycle. The systolic pressure corresponds to the first and higher of the two values reported in a blood pressure reading. Compare *diastole*.

**T cell** One of two basic types of mature lymphocytes. T cells migrate from the bone marrow to mature in the thymus, and they are involved in cell-mediated immunity. Compare *B cell*. See also *helper T cell* and *cytotoxic T cell*.

**taiga** See *boreal forest*.

**taproot** A main root that grows directly downward and has many side branches. Compare *fibrous root*.

**target cell** A cell that receives and responds to a *signaling molecule*.

**taxon (pl. taxa)** Also called *taxonomic group*. A group defined within the Linnaean hierarchy—for example, a species or a kingdom.

**taxonomy** The branch of biology that deals with the naming of organisms and with their classification in the Linnaean hierarchy.

**technology** The practical application of scientific techniques and principles.

**telomere** A unique DNA sequence, associated with special proteins, that caps each end of a chromosome and protects it.

**telophase** The stage of mitosis or meiosis during which chromosomes arrive at the opposite poles of the cell and new nuclear envelopes begin to form around each set of chromosomes.

**temperate deciduous forest** A terrestrial biome dominated by trees and shrubs that grow in regions with cold winters and moist, warm summers.

**template strand** In gene transcription, the strand of DNA (of the two strands in a DNA molecule) that is copied into RNA and is therefore complementary to the RNA synthesized from it.

**tendon** A collagen-rich connective structure that attaches muscle to bone in vertebrate skeletons. Compare *ligament*.

**tension** In reference to water transport in vascular plants, the collective pulling force exerted on the water column in the xylem by the surface tension that exists at each of the countless air-water boundaries within a transpiring leaf.

**terminator** In bacterial gene transcription, a DNA sequence that, when reached by RNA polymerase, causes transcription to end and the newly formed mRNA molecule to separate from its DNA template.

**tertiary consumer** An organism that eats a *secondary consumer* and is eaten by *quaternary consumers*. Compare also *primary consumer*.

**tertiary structure** In reference to proteins, the overall three-dimensional form of a protein, created and stabilized by chemical interactions between distantly placed segments of the protein. Compare *primary structure*, *secondary structure*, and *quaternary structure*.

**testosterone** One of a group of steroid hormones, called androgens, that generate and maintain male sexual and behavioral characteristics.

**tetrad** Also called *bivalent*. The paternal and maternal members of a pair of homologous chromosomes that are aligned parallel to each other during meiosis I.

**tetrapod** Any terrestrial vertebrate that has four limbs. Amphibians, birds, and mammals are all tetrapods.

**thalamus (pl. thalami)** The portion of the vertebrate brain that relays sensory information to the cerebral cortex.

**thermal energy** See *heat energy*.

**thermoreceptor** A sensory cell that initiates an action potential in response to heat.

**thermoregulation** The process by which animals control the gain and loss of heat in order to maintain an internal temperature that is suitable for life processes.

**thylakoid** One of a series of flattened, interconnected membrane sacs that lie one on top of another within a chloroplast in stacks called grana.

**thyroid gland** An endocrine gland in the neck that releases calcitonin.

**tight junction** A structure made up of rows of proteins associated with the plasma membranes of adjacent cells. Neighboring cells are held together tightly by the interlocking of the membrane proteins of adjacent cells. Cells connected by tight junctions form impermeable sheets that do not permit ions and molecules to pass from one side of the sheet to the other by slipping between the cells. Compare *anchoring junction* and *gap junction*.

**tissue** A collection of coordinated and specialized cells that together fulfill a particular function for the organism.

**totipotent** Of or referring to a cell that can differentiate into any cell type. Compare *multipotent*, *pluripotent*, and *unipotent*.

**toxin** A poison produced by a living organism. Toxins are small to large molecules and include poisonous proteins.

**trachea (pl. tracheae)** In vertebrates, the wide, rigid tube, also called the windpipe, that connects the pharynx with the bronchi.

**tracheid** One of two types of tubes that conduct water in the xylem of a plant. Tracheids are narrow, with tapered ends. Compare *vessel* (definition 1).

**transcription** In genetics, the synthesis of an RNA molecule from a DNA template. Transcription is the first of the two major steps in the process by which genes specify proteins; it produces mRNA, tRNA, and rRNA, all of which are essential in the production of proteins. Compare *translation*.

**transcription factor** See *regulatory protein*.

**transfer RNA (tRNA)** A type of RNA that transfers the amino acid specified by mRNA to the ribosome during protein synthesis.

**transgenic organism** See *genetically modified organism*.

**translation** In genetics, the conversion of a sequence of nitrogenous bases in an mRNA molecule to a sequence of amino acids in a protein. Translation is the second of the two major steps in the process by which genes specify proteins; it occurs at the ribosomes. Compare *transcription*.

**translocation** In genetics, a mutation in which a segment of a chromosome breaks off and is then attached to a different, nonhomologous chromosome.

**transpiration** The evaporation of water from the surface of a plant shoot.

**transport protein** A membrane-spanning protein that provides a pathway by which materials can enter or leave cells.

**transport vesicle** A vesicle that specializes in moving substances from one location to another within the cytoplasm and to and from the exterior of the cell.

**transposon** Also called *jumping gene*. A DNA sequence that can move from one position on a chromosome to another, or from one chromosome to another.

**treatment group** Also called *experimental group*. A group of participants in an experiment that are subjected to the same environmental conditions as the *control group* but are exposed to a specific treatment in the form of a change to the independent variable.

**tree of life** A branched, treelike diagram that depicts the evolutionary history of life from its origin to the present time.

**trichome** See *dermal hair*.

**triglyceride** A lipid in which all three hydroxyl groups in glycerol are bonded to a fatty acid. Animals store most of their surplus energy in the form of triglycerides.

**trimester** One of the three stages of human pregnancy, each lasting about 3 months.

**triple bond** The covalent sharing of three pairs of electrons between two atoms. Compare *double bond*.

**trisomy** In diploid organisms, the condition of having three copies of a chromosome (instead of the usual two).

**tRNA** See *transfer RNA*.

**trophic level** A level or step in a food chain. Trophic levels begin with producers and end with predators that eat other organisms but are not fed on by other predators.

**trophoblast** The outer layer of the mammalian blastocyst, which contributes to the placenta and extraembryonic membranes. Compare *inner cell mass*.

**tropical forest** A terrestrial biome dominated by a rich diversity of trees, vines, and shrubs that grow in warm, rainy regions.

**tubule** The portion of the nephron in the vertebrate kidney where valuable water and solutes are reabsorbed before the filtrate entering the nephron is released as urine.

**tubulin** The protein monomer that makes up microtubules.

**tumor** A solid cell mass formed by the inappropriate proliferation of cells.

**tumor suppressor gene** A gene that inhibits cell division under normal conditions. Compare *proto-oncogene*.

**tundra** A terrestrial biome dominated by low-growing shrubs and nonwoody plants that can tolerate extreme cold.

**type 1 diabetes** Also called *juvenile-onset diabetes* or *insulin-dependent diabetes*. Diabetes in which either damage to the islet cells prevents insulin production altogether, or the pancreas produces a defective form of insulin that cannot bind to receptor proteins on the surfaces of target cells. Type 1 diabetes is usually an inherited condition, and those diagnosed with it must receive insulin injections for their whole lives. Compare *type 2 diabetes*.

**type 2 diabetes** Also called *adult-onset diabetes*. Diabetes in which either the pancreas produces too little insulin, or the receptors on target cells respond poorly to insulin. Over time, the resulting high levels of glucose in the blood damage capillaries, causing poor circulation and slowing the healing of injured tissues. Damage to blood vessels in the eyes can cause vision problems. Type 2 diabetes is closely linked with accumulating excess body fat during adulthood. Compare *type 1 diabetes*.

**umbilical cord** The structure that connects a developing mammalian fetus to the placenta, and that contains the veins and arteries that convey materials to and from the placenta.

**unipotent** Of or referring to a stem cell that can differentiate into only one cell type. Compare *multipotent*, *pluripotent*, and *totipotent*.

**unsaturated fatty acid** A fatty acid that has one or more double bonds between the carbon atoms in its hydrocarbon backbone. Compare *saturated fatty acid*.

**upper respiratory system** Collectively, the nose, mouth, and pharynx. Compare *lower respiratory system*.

**upwelling** The stirring of bottom sediments in a body of water by air or water currents, or by shifts in water temperature (as in northern lakes during fall and spring).

**urinary system** Also called *excretory system*. The organ system in animals that adjusts the volume and composition of body fluids, discharging excess water, ions, water-soluble toxins, and organic wastes such as urea. In vertebrates, the system consists of a pair of kidneys and an elaborate system of tubes and sacs.

**urine** The filtrate from the kidneys that animals expel from their bodies.

**uterus** The organ in female mammals in which the blastocyst implants and embryonic and fetal stages of development take place.

**vaccine** A preparation of killed or weakened pathogens that is used to stimulate the vertebrate immune system as protection against future attack by that pathogen.

**vacuole** A large, water-filled vesicle found in plant cells. Vacuoles help maintain the shape of plant cells and can also be used to store various molecules, including nutrients and antiherbivory chemicals.

**vagina** The reproductive structure in female animals into which the penis of a male deposits sperm.

**variable** A characteristic of any object or individual organism that can change.

**vascular cambium** A type of lateral-meristem tissue that contributes to secondary growth by dividing to form new vascular tissues. Compare *cork cambium*.

**vascular system** The tissue system in plants that is devoted to internal transport.

**vascular tissue** The tissue in plants that contains tubelike cells that conduct water or food, as well as other cell types that assist by strengthening the organ or by storing food or performing other essential functions. The two main categories of vascular tissue are xylem and phloem. Compare *dermal tissue* and *ground tissue*.

**vein** A blood vessel that carries blood to the heart. Compare *artery*.

**ventricle** In reference to the heart, a chamber that receives blood from an atrium and contracts to propel blood away from the heart. Compare *atrium*.

**venule** A tiny vein. Venules control the flow of blood out of the finest vessels, the capillaries. Compare *arteriole*.

**vertebrate** Any chordate that has a backbone. Vertebrates include fishes, amphibians, mammals, birds, and reptiles. Compare *invertebrate*.

**vessel** 1. In plants, one of two types of tubes that conducts water in the xylem. Vessels are wide, with one or more open perforations in their end walls. Compare *tracheid*. 2. See *blood vessel*.

**vestibule** In reference to the inner ear, the portion of the vestibular apparatus that contains the otoliths and mechanoreceptors that make it possible to detect gravitational pull.

**vestigial organ** A structure or body part that served a purpose in an ancestral species but is currently of little or no use to the organism that has it.

**villus (pl. villi)** One of numerous tiny projections that increase the absorptive surface of the small intestine.

**viral strain** Also called *serotype*. Any of the variant forms of a particular type of virus.

**virus** An infectious particle consisting of nucleic acids and proteins. A virus cannot reproduce on its own, and must instead use the cellular machinery of its host to reproduce.

**vitamin** Any of a class of small organic molecules that play important roles in diverse metabolic processes but are required by consumers in small amounts.

**voice box** See *larynx*.

**warning coloration** Also called *aposematic coloration*. Bright colors or striking patterns evolved by many prey organisms to warn potential predators that they are heavily defended.

**water transformation** Changes made by humans to the waters of Earth that alter their physical or biological characteristics. Compare *land transformation*.

**weather** Temperature, precipitation, wind speed, humidity, cloud cover, and other physical conditions of the lower atmosphere at a specific place over a short period of time. Compare *climate*.

**wetland** An ecosystem of the freshwater biome defined as standing water shallow enough that rooted plants emerge above the water surface. Bogs (stagnant, acidic, and oxygen-poor), marshes (grassy), and swamps (dominated by trees and shrubs) are all wetlands.

**white blood cell** Any of several types of cells that are part of an animal's nonspecific and specific internal defenses against pathogens.

**wood** Layers of secondary xylem. Each new layer forms a ring around the previous year's layer, producing the "tree rings" that can be used to age a tree.

**X-linked** Of or referring to sex-linked genes located on an X chromosome. Compare *Y-linked*.

**xylem** A plant tissue composed of a number of cell types, including tubelike conducting cells that are dead at maturity and that transport water and dissolved minerals from the soil to the leaves. Compare *phloem*.

**Y-linked** Of or referring to sex-linked genes located on the Y chromosome. Compare *X-linked*.

**yolk sac** The structure in most developing animals that becomes the yolk of the egg, but that in the developing mammal gives rise to blood cells and the cells that produce gametes.

**Z disc** A complex of large proteins at each of the two boundaries of a sarcomere. Z discs appear as dark lines in electron micrographs of muscle tissue. They serve as an anchor for many of the proteins that make up the contractile machinery, including actin and myosin filaments.

**zooplankton** Heterotrophic prokaryotes and protists, together with microscopic animals, that drift at or near the surface of water bodies. Compare *phytoplankton*.

**zygomycetes** One of three major groups of fungi. Zygomycetes were the first fungal group on land. Compare *ascomycetes* and *basidiomycetes*.

**zygote** The diploid ($2n$) cell formed by the fusion of two haploid ($n$) gametes; a fertilized egg.

# Credits

## Photography Credits

right: Thinkstock; **p. 785**: Hans Pfletschinger/Peter Arnold/Getty Images; **p. 785**: Anthony Marceica/National Audubon Society Collection/Photo Researchers; **p. 787 a**: Lynwood M. Chace/Photo Researchers; **b**: Martin Harvey/Peter Arnold/Getty Images; **c**: Lois Ellen Frank/Corbis; bottom: Sheila Terry/SPL/Photo Researchers; **p. 789** Jorgeantonio/Dreamstime; **p. 790**: fotolia; **p. 792** trees: Jorgeantonio/Dreamstime; tubeworm: Ian R. MacDonald, Texas A&M University, Corpus Christi; woman: Pascal Parrot/Corbis Sygma; pine tree: Tony Craddock/SPL/Photo Researchers; creosote bush: Jerry L. Ferrara/Photo Researchers.

## Text Credits

# Index

Page numbers in *italics* refer to illustrations and tables.

eukaryote cell evolution and, 40, 54–55
evolution of, 106
gene expression in, 336–40, *337*
multicellularity advantages and, 147–49
mutations in, 282
sponges as, 107
multigenic disorders, 312
multiple sclerosis (MS), 659, 702
multipotent cells, 257–58, *258, 259*
multiregional hypothesis, 462
muscle fibers, 682, 684–85, 686
muscle junctions, *660*
muscles, 659, 678, 740
breathing and, 622
contraction and motion of, 660, *660*, 685–87
exercise and, 686–87
function of, 685–87
internal movement and, 684, *684*
involuntary, 684
lever systems and, *684*
paired, 685
specialized types of, 684–85
structure of, *683*, 684–85
muscle tissues, 568, *568*
types of, 568
muscular dystrophy, 313
muscular system, 85, 569–70
mushrooms, 68
mussels, 517, *517*
mutagens, 330
mutations, 330, *330*, 435
age of cells and, 730
of alleles, 245, 246, 249, 281–82
of autosomes, 309–11
cancer and, 260, 264–65, *265*
definition of, 281, 389
deletion, 354–55
of DNA, 245, 332, 354–57, *355*
DNA and, 264, 265
evolution and, 245, 393, 410–11, 412, 413, 418
frameshift, 355, *355*
gene expression and, 264
genetic disorders and, 308–11
insertion, 354–55, *355*
in protein synthesis, 355, *355*, 357
sex-linked, 313–14
single-gene, 309–11
somatic, 309
speciation and, 435–37
mutualisms, 68, 72, 73, *73*, 162, 507, *509*
photosynthesis and, 509
types of, 507–9
mycelium, 68–69, *69, 70*, 71, 72
mycobacteria, *408*
mycorrhizae, 767
mycorrhizal fungi, 72, 73–74, *73*
myelin, 654, 659
myelinated axon, 654, 657–59, *658*
myofibrils, 683
myosin filament, 683–84

myriapods, *92, 95*
mytonic dystrophy, 312

Nabisco, 111
NADH, 208, 217, 218, 219, 220, *220*, 221, *222*
NADPH, 208, 209–10, 211–13, *214, 215*, 217
naked mole-rats, 219–20, *220*
nasal cavity (nose), 623
National Aeronautics and Space Administration (NASA), 3, 20, 21
National Cancer Institute, 260
National Center for Science Education (NCSE), 402
National Heart, Lung, and Blood Institute, 647
National Institutes of Health (NIH), 13
National Restaurant Association, 111
National Science Foundation (NSF), 13
natural killer (NK) cells, 703
natural selection, 12, *17*, 387–88, 389, 390–93, *392, 394*, 395, 398, 402, 403, *403, 408*, 410, 411, 417–20, 421–22
as adaptive evolution, 428
behavior and, 740
competition and, 515, 516
definition of, 16, 387–88, 417
environment and, *392*
evolution and, 390–91, *392, 394*, 398, 402, 410, 411, 417–20
mammalian leg arrangement and, *684*
medium ground finch example of, 403, *404*
speciation and, 435–37
types of, 418–20
*see also* adaptations; evolution
Neandertals, 367, 462–63
humans and, 441
Neaves, Bill, 249
nectar, 64
neem tree, 646
negative feedback, 573–74, 638, 640
Nematodes, 92, *92*
*see also* roundworms
nematode worms, 333, *688*, 731
*Nemoria arizonaria, 429*
*Neoceratodus fosteri* (lungfish), 401, *401*
neonatal reflexes, 741
*Nepenthes rafflesiana* (pitcher plant), 768–69, *768*
nephron, 582, 620
nerve cords, 89
nerves, 651, 652, 654, *654*
nervous system, 85, 571, 634, 635, 652–53, 740
autonomic, 656
basic features of, 652–57, *653*
brain and, *see* brain
central (CNS), 652, *653*, 654–55
endocrine system and, 651, 652, 660
hormones and, 660
involuntary responses, 655–56

parasympathetic autonomic control, 656
peripheral (PNS), 652, *653*, 654–55, *655*
somatic responses, 655
sympathetic autonomic control, 656
transport of electrical signals in, *see* action potentials
voluntary responses, 655–56
nervous tissues, 568–69, *569*
Netherlands, 552
net primary productivity (NPP), 530–32, *531*, 537–39, 549–50, 553
human activity and, 537–39
secondary productivity and, 533
neurodegenerative diseases, 157
neurons, 652–61
brain and, *166*, 182
definition of, 652
diffusion process in, 659–61
function of, 652, *653*
motor, 655, *655*, 656, 664
sensory, 146, 655, *655*, 656
synapses and, 659, *660*
transport of electrical signals by, *see* action potentials
neurotransmitters, 181, 571, 657–61, *660*, 663
neutrons, 112, 113, *113*
neutrophils, 236, 700, *700*, 706
New Guinea, 437
New Mexico, 522
Newton, Isaac, 210
New York, N.Y., 111, 541, *541*
New York City Board of Health, 111
niche:
ecological, 513
partitioning, 515
Nicholas II, Tsar of Russia, *276*, 277, 296, *296*
nicotine, 791
addiction to, 663
Nile perch, 440
nitrate, 533, 539, *540*
nitrogen, 116, 122, 123, 480, 533, 534, 535–36, *538*, 539, *539*, 550, *552*, 582, 598, 765, 767
nitrogen cycle, *536*
nitrogen fixation, 41, 535–36, 552–53, 767
nitrogenous base, 135, 327
nitrosamines, 270
nitrous oxide, 554, 558
Nixon, Richard, 268
nodes, 27
nodes of Ranvier, 659
noggin gene, *337*
noncoding DNA, 334, *334*, 368
noncoding genes, 346, 357
nonpolar molecules, 119
nonspecific responses, 698
norepinephrine, 638, 640
North America, 452, 540
changing climate of, 519, *520*

northern elephant seal, 416
Norway, 474
nose, 623
*see also* nasal cavity; smell, sense of
Nostoc forms, 35
notochords, 96, *96*
Nova Scotia, 560
nuclear envelope, 152, 160
nuclear pores, 153
nucleases, 602
nucleic acids, 123, 135–36
nucleoid region, 153
nucleolus, 153
nucleotides, 124, 135–36, *137*, 198, 254, 325, 346, 373, 602, 768
as building blocks of DNA, 324–25, 327, *327*
components of, 135, *136*
functions of, 135–36
nucleus, of atom, 113, *113*
nucleus, of cell, 14, 30, 144, 150, *152*, 348
nucleus accumbens, 663
nutrient cycle, 529, 530, 533–38, *533*, *540*, 550, 552–54
atmospheric, 534, 536–37
exchange pools and, 533
human activity and, 539–42
phosphorus, 537–38, *537*
sedimentary, 534, 537
sulphur, 536–37, *537*
nutrient recycling, 40
nutrients, nutrition, 38, 39, 40, 612, 613
animals' need for, 312, 590–99, *592, 593*
energy and, 590–99, *593*
healthy diets and, *588*
nutritionists and, 607
overview of, 590, *590*
in plants, 764–68, *765, 766, 767*
*see also* digestive systems
Nutrition Facts label, 594, *594*

oak trees, 772
obesity, 271, 551
blood pressure and, 611
metabolism and, 187
objectivity, 10
observation, 6, *6*, 7–8, 10
reproducibility of, 6
occipital lobe, 661, *662*
ocean acidification, 556, 557, 560
ocean currents, 474, *474*
detritus and, 485
oceanic region, 481, *482*, 483
ocean species, conservation of, 100
octopi, 91, *91*, 687–88
eyes and, 91
locomotion and, 91
Odysseus, 323
offspring, 14
Ogallala aquifer, 559
Ohman, Mark, 560

vertebrates, 81, 626, 627, 666–68
   brain of, 654
   deuterostomes, 97–106
   genome of, 333
   nervous system of, 656
very long chain fatty acids (VLFAs), 157
vesicles, 52, 154, 154, 160, 177–78, 241
   transport, 158, 159, 177–78
vessel elements, 763
vessels, plant, 764
vestibular apparatus, 671
vestibule, of ear, 671
vestigial organs, 394, 394, 398
Vibrio cholerae, 42
Victoria, Lake, 427, 439, 440
villi, 601
viruses, 312, 370, 373, 413, 430, 710
   as anti-cancer weapon, 269
   avian, 43
   body's response to, 703
   definition of, 42
   DNA of, 324
   Ebola, 43
   emergence of, 231
   evolution of, 44, 44
   features of, 42–44, 42
   in gene therapy, 372, 380
   immune system and, 697, 699
   infection and, 43–44
   influenza, 42
   life cycle of, 44
   plant, 791
   proteins and, 42
   reproduction of, 44
   as retroviruses, 43
   as rhinoviruses, 43
   shape classification and, 43
   strains of, 43
   structure of, 42–44, 42
   treatment for, 44
visceral mass, 89
vision, sense of, 666–68
   see also eye

vitamin A, 594, 595, 598
vitamin C, 312, 594–95, 595, 597
vitamin D, 134–35, 594–98, 595, 640
   cancer and, 597
   skin color and, 596, 596
   sun and, 597
   ultraviolet (UV) rays and, 596, 596
vitamin E, 594, 595, 598
Vitamin K, 602
vitamins, 138, 590, 594–98, 595, 597, 602
   deficiency in, 597, 597
   fat-soluble, 594
   supplemental, 598
   urine and, 594
   water-soluble, 594
volume, surface area and, 146, 147, 572–73
Volvox carteri, 147–48, 158

Wadlow, Robert, 642
Wallace, Alfred, 17, 18, 386, 387–88, 390
Wang, Jian, 224
warning coloration, 510, 511
wasps, 429, 514, 515
water, 194, 450, 531, 628
   acidification of ocean, 556, 557, 560
   in cells, 168, 171–74, 172
   in cellular respiration, 41, 206, 209, 210, 221
   chemistry of life and, 112
   climate change and, 559
   digestion and, 590
   drinking, 541
   evaporative cooling of, 120–21
   heat capacity of, 120
   in homeostasis, 579–84, 580
   hydrogen bonds and, 118, 120
   as ice, 119–20, 120
   molecule of, 112, 116, 116, 118
   pH of, 122
   in photosynthesis, 206, 212, 213, 221
   plant routes for absorption of, 766
   plants and, 761–62, 763–64, 789

   as polar molecule, 118
   properties of, 118–21
   as solvent, 119
   surface tension and, 121, 121
   temperatures and, 120–21
water transformation, 548–49
Watson, James, 326–29, 327
wavelengths, color and, 210–11, 211
waves, 669
weather, definition of, 471
weaver ants, 428, 428
webs, spider, 95
Weismann, August, 302
western meadowlark, 434, 434
West Nile virus, 557, 699
wetlands, 480, 481–82, 481, 531, 531, 548–49
whales, 223, 398, 445, 510, 510, 517, 730
   evolution of, 447–49
wheat, 756
whiptail lizard, 249
White, Ellen, 607
white blood cells, 179, 371, 372, 380, 568, 613, 698, 700, 700, 704, 773
whole-grain foods, 591, 591, 607
whorls, 783, 783
Wiens, John, 465
Wilkins, Maurice, 327
Wilson, Edward O., 100
Winberg, Wilhelm, 412
windpipe (trachea), 623
winds, 472–74, 473
wings, 94
Wolfe-Simon, Felisa, 2, 3, 20, 20, 21
Wolf Island, 404
wolves, 364, 512, 605, 739, 745, 747
   domestication of, 399–400
women, AIDS and, 711
wood, 780–81
   early, 780
   late, 780–81
wood pigeons, 512, 512
work, as exercise, 607

World Anti-Doping Agency (WADA), 629
World Health Organization (WHO), 163, 396

X chromosome, 243, 303, 304, 313–14, 313, 315–16
xeroderma pigmentosum (XP), 333
X-SCID, 372–73
xylem, 63, 757, 758, 760, 763, 763, 769
   secondary, 780–81, 780, 781
   transport in, 771–72, 771, 791

Y chromosome, 243, 303, 304, 313–14, 315–16, 642
yeasts, 25, 68, 69, 144, 710
   fermentation and, 218, 219
   gene interaction in, 292
   genome of, 333
yellow cinchona, 65
yellow fever, 699
Yellowstone National Park, 519
yeti crab, 80
Yi, Xin, 224
yolk sac, 724, 725
Yosemite National Park, 100
Yssichromis pyrrhocephalus, 440
yucca moth, 508–9, 508
yucca plant, 508–9, 508

Z discs, 683–84
zebra mussel, 468, 469, 470, 484
Zimbabwe, 695
zinc, 599, 765
zooplankton, 60, 551
   see also plankton
zoospores, 54
zygomycetes, 68, 68, 70
zygote, 83, 233, 242–43, 243, 248, 257, 285, 716, 717, 717, 721
   of plants, 61, 782–83, 785, 786, 786
   see also development